# 11th International Conference on Experimental Fluid Mechanics (EFM 2016)



EPJ Web of Conferences Volume 143 (2017)

Marienbad, Czech Republic
15-18 November 2016

**Editor:**

**P. Dancova**

ISBN: 978-1-5108-4044-7

**Printed from e-media with permission by:**

Curran Associates, Inc.
57 Morehouse Lane
Red Hook, NY 12571

**Some format issues inherent in the e-media version may also appear in this print version.**



Printed by Curran Associates, Inc. (2017)

For additional information, please contact EDP Sciences – Web of Conferences at the address below.

EDP Sciences – Web of Conferences
17, Avenue du Hoggar
Parc d'Activité de Courtabœuf
BP 112
F-91944 Les Ulis Cedex A
France

Phone: +33 (0) 1 69 18 75 75
Fax: +33 (0) 1 69 28 84 91

contact-edps@webofconferences.org

**Additional copies of this publication are available from:**

Curran Associates, Inc.
57 Morehouse Lane
Red Hook, NY 12571 USA
Phone: 845-758-0400
Fax: 845-758-2633
Email: curran@proceedings.com
Web: www.proceedings.com

# TABLE OF CONTENTS

EPJ Web of Conferences **143**, 00001 (2017) DOI: 10.1051/epjconf/201714300001
*EFM 2016*

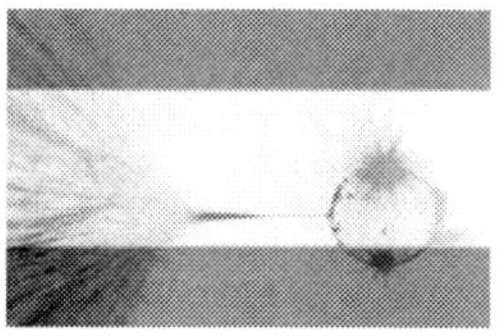

# Experimental Fluid Mechanics 2016

Marienbad, Czech Republic, 15th–18th November 2016

efm.kez.tul.cz

11th International Conference „Experimental Fluid Mechanics 2016" was held at 15th–18th November 2016 in The Casino Cultural and Conference Centre in Marienbad, Czech Republic.

The EFM conference is focused on obtaining overviews of projects under investigation and experimental methods used in the field of fluid mechanic and thermodynamic. Its aim is to organize a network of research workers who are interested in experimental, and also theoretical, work in the field of fluid mechanic and thermodynamic. The conference should contribute the creation of closer contacts and sharing experience in application of various experimental methods, preparation and implementation of experiments, processing of results and numerical simulations of experiments.

*The main topics of the conference are:*

- Experimental methods used in fluid mechanics and thermodynamics.
- Preparation and implementation of experiments.
- Application of numerical simulation for preparation of experiments, comparison of the results obtained by numerical or analytical methods with experiments.
- Processing the results.
- Experiments in the field of compressible and uncompressible fluid flow, stability of flow, heat and mass transfer, cavitation etc.
- Experiments and measuring carried out on power and regulating equipment.
- Cryogenic fluid dynamics.

*Members of Sciences Committee are:*

- Assoc. Prof. Tomáš Vít – Head of the committee (TU Liberec, Czech Republic)
- Simone Ferrari (University of Cagliari, Italy)
- Jan Hrubý (Academy of Science of Czech Republic)
- Hendric de Lange (Eindhoven Univ. of Technology, Netherlands)
- Václav Kopecký (TU Liberec, Czech Republic)
- Robert Pastuszko (Kielce University of Technology, Poland)
- Kazimierz Peszynski (UTP University of Science and Technology, Poland)
- Ludvík Prášil (TU Liberec, Czech Republic)
- Jaromír Příhoda (Academy of Science of Czech Republic)
- Rudi Santbergen (Delft University of Technology, Netherlands)
- Ladislav Skrbek (Charles University, Czech Republic)
- Zdeněk Trávníček (Academy of Science of Czech Republic)
- Václav Uruba (Academy of Science of Czech Republic)
- Magda Vestfálová (TU Liberec, Czech Republic)
- An-Bang Wang (National Taiwan University, Taiwan)
- Yoshifumi Yokoi (National Defense Academy of Japan, Japan)

*Steering Committee:*

- Petra Dančová – Head of the committee (TU Liberec, Czech Republic)
- Jan Novosád (TU Liberec, Czech Republic)
- Martin Veselý (TU Liberec, Czech Republic)

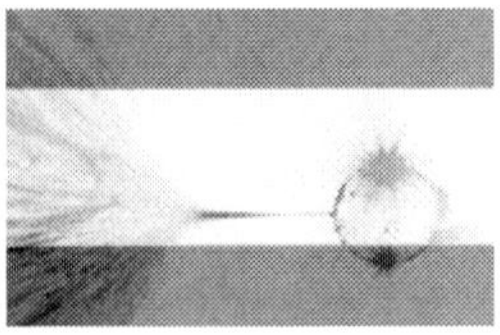

*Keynote lectures for 2016:*

- Prof. Simone Ferrari (University of Cagliari, Italy)
  *Image Analysis techniques for the study of turbulent flows*
- Prof. Muk Chen Ong (University of Stavanger, Norway)

*CFD Applications in Offshore Engineering*

*List of Sponsors (in alphabetical order):*

- DANTEC DYNAMICS GmbH (www.dantecdynamics.com)
- LAVISION (http://www.lavision.de/en/)
- MIT s.r.o. (www.mit-laser.cz)
- TSI GmbH (www.tsi.com)

*Conference venue:*

Every year, the organizers want to prepare the conference at nice and interesting place of the Czech Republic. In 2016, the conference venue was Marienbad.

Marienbad means beautiful spa parks, romantic colonnades, charming pavilions, pleasant cafes and cozy hotels. The town is endowed with many healing mineral springs and revels in a unique atmosphere that attracts thousands of visitors every year who come to soak up this atmosphere. Within the town and its surroundings rise around one hundred mineral springs, containing carbon dioxide and mineral salts. They are cold ferrous mineral waters. Drinking treatments mostly use the main 6 mineral springs: Cross (Křížový), Rudolf, Karolina, Forest (Lesní), Ambrose and Ferdinand.

**The organizers would like to invite you for the next year of the EFM conference, which will be held in Mikulov, November 21$^{st}$ – 24$^{th}$, 2017 (see our web pages efm.kez.tul.cz).**

Marienbad

View into Room A

Keynote lecture – prof. Simone Ferrari

Poster section

Informal Gathering

# Image analysis techniques for the study of turbulent flows

*Simone* Ferrari[1,*]

[1] DICAAR - Dipartimento di Ingegneria Civile, Ambientale e Architettura, University of Cagliari, via Marengo 2, 09123 Cagliari, Italy

**Abstract.** In this paper, a brief review of Digital Image Analysis techniques employed in Fluid Mechanics for the study of turbulent flows is given. Particularly the focus is on the techniques developed by the research teams the Author worked in, that can be considered relatively "low cost" techniques. Digital Image Analysis techniques have the advantage, when compared to the traditional techniques employing physical point probes, to be non-intrusive and quasi-continuous in space, as every pixel on the camera sensor works as a single probe: consequently, they allow to obtain two-dimensional or three-dimensional fields of the measured quantity in less time. Traditionally, the disadvantages are related to the frequency of acquisition, but modern high-speed cameras are typically able to acquire at frequencies from the order of 1 KHz to the order of 1 MHz. Digital Image Analysis techniques can be employed to measure concentration, temperature, position, displacement, velocity, acceleration and pressure fields with similar equipment and setups, and can be consequently considered as a flexible and powerful tool for measurements on turbulent flows.

## 1 Introduction: Image Analysis

Turbulence is present in almost every field of the human life and, consequently, it has many practical applications in Engineering, Medicine, Geophysics, etc. For instance, turbulent flows play a huge role in the motion of air around and behind cars, airplanes, boats, and other moving objects, because of drag and wake; in the reduction of the concentration of pollutants released in the atmosphere or in the sea (Ferrari and Querzoli, 2010 [1]) by human activities; in the weather forecasts, in the study of the motion of hurricanes, water and air masses in the ocean and in the atmosphere; in the optimization of mixing and combustion and in flow control (Rossi et al 2009 [2]) in engines and chemical reactors; in biological flows such as the air flowing through the lungs (Ramuzat and Riethmuller, 2001 [3]) or the blood motion in the aorta (Elkins et al., 2005 [4]) or downstream a stenosis (Zhu et al., 2011 [5]). The scalar quantity transport by turbulent flows plays a relevant role in many environmental phenomena, such as particle transport in geophysical flows (Boffetta et al., 2000 [6]). Following Meng et al (2016 [7]), turbulence is also one of the key phenomena in the study of the process of flotation for the separation of minerals. Moreover, the turbulent flow conditions caused by the motion of a swimmer in water are studied to predict the hydrodynamic forces and the swimmer's potential speeds (Takagi et al 2016 [8]). Turbulence plays also a fundamental role in the air quality and human comfort in built environment (Blocken et al, 2016 [9], Fu et al 2015 [10]).

Even if the computational capabilities of modern computers are always improving, the direct numerical simulation of high Reynolds number turbulent flows is still very difficult. Consequently, the role of experimental measures in the laboratory are at this stage still essential.

Traditional systems to perform measures in the laboratory are the single point probes, such as resistive probes to measure concentrations or Laser-Doppler Velocimetry (LDV, also known as Laser Doppler Anemometry LDA), Pitot tubes, Hot Wire Anemometer (HWA), Hot-film Anemometer (HFA) and Pulsed Wire Anemometer (PWA) to get the time history of velocity in a single point of space. They have the advantage of a high acquisition frequency (order of $10^5$ Hz) but in order to obtain a field of the measured quantity, a number of probes high enough must be used or, when the experiment is repeatable, at the end of each run they can be moved to another position and the experiment repeated many times. On one hand, as the single point probes are intrusive (i.e. they alter the flow they should measure), it is nearly impossible to sample simultaneously many points in the flow, particularly if they are close to each other. On the other hand, the repetition of an experiment for as many times as the number of points needed to well define in space the investigation area can be very time-consuming. Moreover, a hot probe induces thermal convection on the flow, causing another local perturbation. A non-intrusive technique is the Laser Doppler Anemometry (LDA) that is still a point-wise technology. All these techniques perform measures in a fixed point in space, so they allow to obtain only Eulerian information. As stated by Cao et al (2014 [11]), to obtain accurate and detailed measures of the global turbulent airflow field by means of point-wise velocimetry is still excessively difficult and time-consuming.

* Corresponding author: ferraris@unica.it

For this reason, optical techniques, such as Digital Image Analysis (DIA) one, have gradually become alternative tools for measuring turbulent flow quantities. In this techniques, a camera with one or more sensors (CCD - Charge-Coupled Device or CMOS - (Complementary Metal-Oxide-Semiconductor) shoots images of the flow, previously seeded with non-buoyant particles or dyes and lighted by a light source, at a given frequency. The first advantages of DIA techniques are that they are non-intrusive, quasi-continuous in space (as every pixel on the camera sensor works as a single probe): consequently, they allow to obtain two-dimensional or three-dimensional fields of the measured quantity in less time, if compared with single point probes. Moreover, even if the directly measured quantity is the light intensity, many quantities can be measured indirectly, such as the position or the size of objects in the investigation area, concentration of substances, temperature, or, when comparing two consecutive images, the displacement, velocity and acceleration of moving objects. Moreover, the pressure on solid interface or in the flow filed can be measured as well.

Another advantage of DIA techniques is that some of them give access to Lagrangian information. Moreover, they can be performed employing relatively low cost equipment. Traditionally, the disadvantages of DIA techniques are related to the frequency of acquisition, but modern high-speed cameras are typically able to acquire at frequencies from the order of 1 KHz to the order of 1 MHz. A disadvantage that, on the contrary, cannot be overtaken is that an optical access to the region of interest is needed, so transparent materials must be employed. Moreover, a certain care must be given to the index of refraction, in particular with curve surfaces.

In the following chapters, a brief review of Digital Image Analysis techniques employed to measure various quantities relevant for the study of turbulent flows is given, with a particular focus on the techniques developed by the research teams the Author worked in, that can be considered relatively "low cost" techniques.

## 2 Concentration

DIA techniques to measure the concentration of a certain substance in a flow is based on the characteristic of some dyes, in given quantities, to absorb or radiate light. In the first case, we have Light attenuation (LA), in the second case Light/Laser Induced Fluorescence (LIF).

### *2.1.1 Light Attenuation*

The Light Attenuation (LA) technique has been used since the 1990s and it is based on the fact that the light from the back-lighting is attenuated (absorbed) by the dye as it passes through the flow, so the light source is typically opposite to the camera. Allgayer and Hunt (2012 [12]) has described the application of LA, focusing on the selection of chemical dye, dye concentration, illumination and optics. A typical dye for LA is the methylene blue. Anyway, a suitable dyeing agent for LA has to be a strongly light-absorbing substance (even at low concentrations) which does not react with the fluid in which it is dissolved. As the light is passing throw the flow, the measured concentration is usually the integral of the concentration in the direction of the light.

Hacker et al (1996 [13]) employed LA to determine the mixing of fluid properties transported by gravity currents.

Choi et a. (2016 [14]) has studied with this technique the layer-averaged concentration fields in of an inclined buoyant jet discharged into a co-flowing current

### *2.1.2 Laser Induced Fluorescence*

The Light/Laser Induced Fluorescence (LIF or PLIF, i.e. Planar LIF) has been employed since the 1970s to measure concentration of certain substances in fluid flows (e.g., Stewart and Judeikis, 1974 [15]). The principle is the flow is seeded with a dye (such as Fluorescein or Rhodamine, see Melton and Lipp -2003 [16]- for a review of the techniques, dyes and criteria for quantitative PLIF measures) that, when meet an incident light at a certain frequency, radiates light at a different frequency. Provided the density of the substance is in the correct range, a linear relation between the intensity of the emitted light and the concentration of the dye holds. The dye concentration has usually to be very low, in order to avoid nonlinearity between light intensity and dye concentration and to minimize the light attenuation (Sutton et al. 2008 [17]). So in LIF a light/laser sheet illuminates the region of interest (typically, a plane section of the flow) perpendicularly to the camera, in order that the flow itself results on the image as a light flow on a dark background. When a laser light is not available, the light sheet can be generated employing one or more slide projectors, each one equipped with a black slide with narrow, vertical slit at its centre and focused in the observational area (see Ferrari and Querzoli -2010 [1]- for more details): in this way, a very inexpensive set-up to perform quantitative measurements of flow density can be obtained. A reference image of the lighted background has to be acquired to measure the distribution of the light intensity in the observational area: in this way, the intensity of the fluorescent light varies only because of the incident light. Filters on the camera lens can be employed to filter out all the frequencies but the one coming from the dye, even if this reduces the amount of light recorded by the camera sensors.

Ferrari and Querzoli (2010 [1]) employed this set-up to measure the dilution of the salt (directly proportional to the fractional volume of the fluorescein) in a negatively buoyant jet released in a stagnant environment.

A similar set-up was then employed to measure the same phenomenon released in a wavy environment (Ferrari and Querzoli 2015 [18]).

Bouche et al (2012 [19]) studied the mixing in a swarm of bubbles rising in a confined cell using a PLIF employing two different dyes.

Techniques that couple LIF and Particle Image Velocimetry (PIV, see hereafter), allowing to obtain simultaneous measures of concentration and velocity, have been developed (e.g., Charogiannis et al 2015 [20]) using a two-cameras set-up.

## 3 Temperature

A typical DIA technique to measure the temperature in a flow is the use of temperature sensitive particles (TSPs), able to change colour with temperature, to put them into the flow and to record the flow with a camera. This allows the simultaneous and non-intrusive measurement of temperature and velocity (e.g. with Particle image velocimetry PIV or Particle Tracking velocimetry, see beyond) fields. For instance, Massing et al., 2016 [21], obtained the temperature fields from the intensity (thermographic phosphors) or frequency of the light emitted or reflected by the tracer particles (polymer particles doped with the temperature sensitive dye europium III thenoyltrifluoroacetonate –EuTTA- and the nearly temperature insensitive reference dye perylene).

Someya et al., 2010 [22], employed phosphor thermometry to visualize temperature fields (luminophors, including phosphors, are typically rare earth-doped ceramics that emit light when suitably excited).

Also Temperature Sensitive Paint (TSP) techniques, providing spatial temperature distributions on surfaces, can be employed. For instance, Ozawa (2016 [23]), employed TSP to experimentally study the unsteady aerothermodynamic phenomena occurring on a shock-tube wall.

Moreover, thermal camera con be used, as done by Sagaidachnyi et al (2016 [24]), who employed cooled thermal imaging cameras to estimate the blood flow oscillations at the surface of living body

## 4 Position

If the flow is seeded with an emitting light dye, in addition to concentration, the position of a lighted fluid or particle on a dark flow or background can be detected with a DIA technique. For instance, Ferrari et al (2016 [25]), evaluated the wave motion transformations above a breakwater, both submerged and emerged, with a DIA technique. The fluid was seeded with a fluorescent dye, and the free surface was identified with a robust algorithm, that showed to properly work also in prohibitive situations for traditional resistive probes (e.g., very shallow waters and/or breaking waves) and to be able to measure the free surface all over the investigation field in a non-intrusive way.

Mendez et al (2016 [26]) measured in a similar way the thickness of a vertical falling liquid film, as well as the wave celerity and the wave profile.

Brady et al (2009 [27]) gave a review of different methods for particle image sizing, with a particular regard to the improvement of accuracy and robustness. This measure is quite relevant in multi-phase flows, that can be encountered in various industrial applications (e.g., the atomization of droplets in sprays, particulate flows, powders, fluidization, bubbly flows, flotation, etc.), where the features of the flow can be determined by the particle distribution and by the interaction of them with the fluid.

## 5 Displacement

If the position of an object, such a particle, can be spotted in an image, its displacement between two images taken at different times can be measured and, consequently, the trajectory of the object in time can be reconstructed. Moreover, dividing the displacement by the interval time, an indirect measure of the velocity can be attained. If the measure is good enough, the object acceleration can be measured as well.

### 5.1. Velocity

DIA techniques suitable for velocity measurement can be basically classified as Particle Streak Velocimetry (PSV), Particle Tracking velocimetry (PTV), Particle Image Velocimetry (PIV) and Feature Tracking (FT). All these techniques evaluate the velocity measuring the distance covered by a feature (a particle or a marker for PSV, PIV and PTV, a point of high gradient of the intensity for FT) on the images in a given time.

These DIA techniques for velocimetry give measures all over the investigation field (or at least where particles/features are found), do not need a probe (so they are essentially not invasive, at least if the particles are non-buoyant) and do not need any particular calibration (as the relation between the measured quantity, a displacement, and the required one, a velocity, is linear).

In all these techniques, the set-up consists of a fluid seeded with neutrally buoyant particles, illuminated with a light source and of one or more cameras shooting pictures at known temporal frequency.

The relevant scientific and technical progress attained in the last years in lasers, cameras, post-processing, velocimetry codes and computer capabilities has led to the widespread diffusion of DIA velocimetry techniques.

#### *5.1.1 Particle Streaks Velocimetry*

In Particle Streak Velocimetry (PSV) the velocity is measure along the streaks left by moving particles on a long exposure image (e.g., Bergthorson and Dimotakis, 2006 [28]).

In this technique, the ambiguity between particles (typical of PIV or PTV) can be reduced, but strong issues can arise when particles enter or exit the investigation area or when particle trajectories have a strong curvature.

Moreover, ambiguity on the particle direction can arise.

### *5.1.2 Particle Tracking Velocimetry*

In Particle Tracking Velocimetry (PTV), the particle centroid is identified and tracked from one image to the following one (e.g. Dalziel, 1992 [29], Querzoli, 1996 [30], Virant and Dracos, 1997 [31]). In this way, PTV allows to reconstruct trajectories and to attain a Lagrangian measure and it differs from Particle Image Velocimetry (PIV) because this last relies on pattern matching in an essentially Eulerian way.

PTV information achieve a high local accuracy in every seeded point of the flow, with the possibility of solving small velocity variations and of obtaining a higher spatial resolution compared to PIV; on the other side, PTV is much more sensitive to noise. Moreover, a Eulerian description can be extracted from the Lagrangian information if there are enough samples to cover the investigation field. On the opposite, it is not possible to get reliable Lagrangian information starting from PIV data, because the transformation from the Eulerian description to the Lagrangian one implies an integration in time of the velocity field, not very reliable especially when considering turbulent flows with high sensitivity to initial conditions.

In PTV, the density of particles has to be low enough to allow the recognition of the single particles.

The main drawbacks of PTV are linked to the particle appearing/disappearing, to the difficulties in distinguishing overlapping particles and o the amount of data needed to cover the whole investigation field. So, PTV is complementary to PIV, with its ability to cope with low and medium particle concentration, to reach a higher spatial resolution and to cope with specific cases like regions with strong velocity gradients.

PTV shows clear advantages to PIV in near-wall regions with high velocity gradients as they can provide measurements nearest to the wall (Stanislas et al., 2005 [32]).

After a decade where PIV was often preferred, as it was easier to obtain results than with PTV, this last is becoming again popular, in particular with the development of three-dimensional measurements.

As stated by Fu et al (2015 [10]) in their review of PTV and PSV techniques for indoor airflow field investigation, the PTV technique has gradually become again a powerful tool, as the quantitative and detailed turbulent flow information obtained by PTV measurement is critical for analysing turbulent properties and developing numerical simulations.

As a matter of fact, one of the conclusions of the last PIV Challenge has been that sophisticated PTV algorithms can outperform state-of-the-art PIV algorithms (Kahler et al 2016 [33]).

### *5.1.3 Particle Image Velocimetry*

The term Particle Image Velocimetry (PIV) first appeared in the literature in 1980s: reviews can be found by Adrian, 1991 [34], Buchave, 1992 [35], Adrian, 2005 [36], Adrian and Westerweel, 2011 [37]. An extensive review of the PIV techniques employed for the experimental wind comfort assessment at pedestrian level has been done in Blocken et al (2016 [9]).

In PIV, a spectral analysis or an auto/cross correlation is performed on a small interrogation window on two consecutive images in order to get a single velocity for each interrogation window. PIV algorithms obtain velocity fields comparing windows of successive frame on a regular grid and maximizing the correlation function to obtain their displacement. PIV is consequently fairly robust to noise and has a spatial resolution that is inversely proportional to the interrogation window size.

PIV allows higher seeding levels when compared with PTV.

As stated by Kahler et al. (2016 [33]), in the last decade, worldwide PIV development efforts have resulted in significant improvements in terms of accuracy, resolution, dynamic range and extension to higher dimensions.

As stated by Cao et al. (2014 [11]), PIV has gradually become the most popular and promising technique for airflow field measurement in indoor environment during the last decade. The main reasons for this are the well-developed technologies, a lot of experimental experiences and a large availability of commercially codes and systems.

For resolving the directional ambiguity, Stucky et al (1994 [38]) employed a two-colour PIV system: two differently coloured sheets of light illuminate the flow field at a different time, in order to distinguish the first image from the second image. Filters are employed to separate the two colours during interrogation, thus creating two images that are then cross-correlated with each other.

Falchi et al. (2006 [39]) has proposed a Robust Image Velocimetry (RIV), where the measure of the dissimilarity between interrogation windows is employed (instead of the auto/cross correlation typical of traditional PIV) which is based on a more robust estimator than cross-correlation.

### *5.1.4 Feature Tracking*

In Feature Tracking (Lucas and Kanade, 1981 [40], Shi and Tomasi, 1994 [41], Miozzi et al, 2008 [42]) the object or region to be tracked is not chosen a priori, like in PTV and PIV, but by the method itself, as a feature that can be well tracked: a certain number of points in the picture with a high spatial gradients of light intensity are usually selected.

As stated by Miozzi et al. (2008 [42]), the displacements are obtained by searching the features between the first and second image, in a way which minimize a correlation distance between them (the minimum of the sum of squared differences of interrogation windows intensities). The linearized equation governing the minimization problem is solved in an iterative way only where the solution is guaranteed to exist. FTV provides the velocity in a Lagrangian framework.

#### *5.1.5 Hybrid methods*

Some hybrid schemes (e.g. the one by Cowen et Monismith, 1997 [43]) combining PIV and PTV have been proposed.

Besalduch et al. (2013 [44] and 2014 [45]) has developed and tested a Feature Tracking Velocimetry (FTV), attempting to combine in a single scheme the advantages of the previous presented techniques. FTV compares windows only where the motion detection may be successful (as in FT, where there are high luminosity gradients in two directions, i.e. corners) FTV procedure consist of:

- identification of the features using the Harris corner detection (Harris & Stephens, 1988 [46]);
- ordering of the features according to their cornerness;
- selection of the first N features and computation of the velocity by the comparison a window centered with the $i^{th}$ feature ($W_t$) with windows ($W_{t+1}$) within a range of displacements, ($d_i$, $d_j$) in the next frame;
- for each displacement, a measure of the dissimilarity, $D(d_i, d_j)$, between $W_t$ and $W_{t+1}(d_i, d_j)$ is computed using the Lorentzian estimator and the displacement is consequently minimizing the dissimilarity;
- displacements are validated by means of algorithm based on Gaussian filtering of first neighbors (defined by the Delaunay triangulations), and simulated annealing.

FTV has shown to be suitable for measurements on a wide range of seeding density (from the high levels of PIV to the low ones of PTV), for the analysis of images with intermediate seeding density or non-homogeneous seeding and in presence of different seeding density.

### 5.2 Acceleration

As the material acceleration in a fluid flow is an intrinsically Lagrangian quantity, being the sum of a local (Eulerian) acceleration and of a convective acceleration, it cannot be reliably derived from traditional PIV but has to be derived from PTV Lagrangian measurements. Unfortunately, as before stated, PTV is very sensitive to measurement noise, which tends to grow at every derivation in time (from displacement to velocity to acceleration).

To reduce this problem, low-pass filter on the tracked particle positions have been proposed (e.g., La Porta et al, 2001 [47], Voth et al., 2002 [48], Mordant et al., 2004 [49]). The main parameter is the temporal window size of the filter: on one side, when it is too large, small fluctuations of the measured values can be deleted; on the other, when it is too small, the noise is not removed.

Ferrari and Rossi (2008 [50]) has developed a self-adaptive iterative method (Particle Tracking Velocimetry and Accelerometry, PTVA) that relies on an adaptive number of tracked positions, for the trajectories approximation, depending only on the local flow properties. The criterion proposed holds the potential to withdraw some convection/sweeping information and thus to extract more accurately the acceleration.

In a first step, the position extracted by PTV are approximated into the trajectory components via a least square approximation, making use of a fourth order moving polynomial, with a variable number of positions depending only on a minimum and maximum: from the temporal derivatives of this approximated trajectory components, a first guess velocity components are computed via a temporal derivative.

In the second step, the same approximation is performed starting again from the positions extracted by PTV, again using the least square approximation of a fourth order moving polynomial, but with a criterion that adapts the number of positions for the approximation in order to have a parameter (called the convected displacement), able to take into account the local flow properties, as close as possible to a target value, equal to the perimeter of the smallest eddy to be measured. The components of the velocity and acceleration are then extracted via temporal derivatives.

From the comparison with methods using a constant number of positions, PTVA errors are about six times smaller for velocity and about four times smaller for acceleration. The high quality of this acceleration measurement allowed to even measure the divergence of the acceleration, strictly related to the stirring in the flow.

PTVA has been applied to extract the single terms of the Navier-Stokes equations (Lardeau et al., 2009 [51]), for flow control purposes (Rossi et al. 2010 [2]) and to investigate time dependent quasi-two-dimensional flows (Ferrari et al., 2007 [52] and 2008 [53]).

Ultrasonic techniques employing fast silicon strip detectors have been used to optically image short particle (tiny transparent beads) tracks and measure particle acceleration (La Porta et al., 2001 [47], Voth et al, 2002 [48]), reaching a very high time resolution (up to 70,000 frames per second) but in a small measurement volume (particles can be followed for only few Kolmogorov time scales). The technique is based on the capability to time-resolve the Doppler shift of the sound scattered by continuously insonified particles.

## 6 Pressure

Traditional techniques for pressure measurement away from solid boundaries are usually based on Pitot-tube type of probes (as previously stated, intrusive and point-wise).

Another technique used in the past to measure pressure away from the boundaries was based on microscopic bubbles as pressure sensors (Ran and Katz, 1994 [54]), but they provided only a very limited number of data points for each run of measure.

DIA techniques have reached a level of development that allows to even measure the pressure fields in a flow.

For instance, the PTVA technique (Ferrari and Rossi 2008 [50]), allowing to measure all the terms in the Navier-Stokes equation, gives also access to the pressure gradient term all over the investigation field.

Another non-intrusive technique for measuring the instantaneous spatial pressure distribution in a flow field is the four-exposure PIV system developed by Liu and Katz (2006 [55]): they measure the distribution of the material acceleration by comparing the velocity of the same group of particles at different times and then integrating it to obtain the pressure distribution. Exposing both cameras to the same particle field at the same time and cross-correlating the images enables precision matching of the two fields of view.

A last technique is the Pressure Sensitive Paint (PSP), a non-intrusive optical pressure measurement technique that provides spatial pressure distributions but only on surfaces of solid objects. Gregory et al. (2014 [56] have reviewed fast-responding PSP techniques for applications to aerodynamic and acoustic investigations. PSP determines surface pressure distributions by measuring changes in the intensity of emitted light, while fast-responding PSP is an extension applicable to unsteady flows and acoustics. Most fast-responding PSP paints are based on the development of porous binders allowing a fast oxygen diffusion and interaction with the chemical sensor.

## 7 Conclusions

Digital Image Analysis techniques for the measure of different quantities in turbulent flows have been briefly reviewed. The relevant scientific and technical progress achieved in recent years in lasers, cameras, post-processing, velocimetry codes and computer capabilities has taken to the widespread diffusion of DIA techniques. They have the advantage, when compared to the traditional techniques employing physical point probes, to be non-intrusive and quasi-continuous in space. DIA techniques can be employed to measure concentration, temperature, position, displacement, velocity, acceleration and pressure fields with similar equipment and setups, and can be consequently considered as a flexible and powerful tool for measurements on turbulent flows.

The Author would like to acknowledge Prof. Maria Grazia Badas, Dr. Luigi Antonio Besalduch, Mr. Antonio Mascia, Prof. Giorgio Querzoli, Dr. Lionel Rossi and Prof. John Christos Vassilicos.

## References

1. Ferrari, S., Querzoli, G., Mixing and re-entrainment in a negatively buoyant jet, Journal of Hydraulic Research, 48 (5), pp. 632-640. (2010)
2. Rossi, L., Bocquet, S., Ferrari, S., Garcia de la Cruz, J.M., Lardeau, S., Control of flow geometry using electromagnetic body forcing, International Journal of Heat and Fluid Flow, 30 (3), pp. 505-513. (2009)
3. Ramuzat, A., Riethmuller, M.L. Steady and unsteady PIV investigations of flows within a 3D lung bifurcation model, DLR-Mitteilung, (3), pp. 109-119 (2001)
4. Elkins, C.J., Iyengar, A.S., Draney, M.T., Dake, M.D., Medina, F., Wicker, R.B. Rapid in-vitro MRV and PIV measurements in anatomically accurate human thoracic aorta phantoms, Proceedings of the 2005 Summer Bioengineering Conference, 2005, pp. 235-236 (2005)
5. Zhu, H., Qian, M., Zou, Y., Song, R., Niu, L., Jiang, B., Guang, Q., Zheng, H. The validation of Echo-PIV technique used in a stenosis model, 5th International Conference on Bioinformatics and Biomedical Engineering, iCBBE 2011, art. no. 5780292 (2011).
6. Boffetta G., Celani A. and Vergassola M., Inverse energy cascade in two-dimensional turbulence: deviation from Gaussian behaviour, Physical Review E, 61 (1), R29-R32, (2000).
7. Meng, J., Tabosa, E., Xie, W., Runge, K., Bradshaw, D., Manlapig, E. A review of turbulence measurement techniques for flotation, Minerals Engineering, 95, pp. 79-95 (2016)
8. Takagi, H., Nakashima, M., Sato, Y., Matsuuchi, K., Sanders, R.H. Numerical and experimental investigations of human swimming motions, Journal of Sports Sciences, 34 (16), pp. 1564-1580, (2016).

Blocken, B., Stathopoulos, T., van Beeck, J.P.A.J. Pedestrian-level wind conditions around buildings: Review of wind-tunnel and CFD techniques and their accuracy for wind comfort assessment, Building and Environment, 100, pp. 50-81 (2016).

9. Fu, S., Biwole, P.H., Mathis, C. Particle tracking velocimetry for indoor airflow field: A review, Building and Environment, 87, pp. 34-44 (2015).
10. Cao, X., Liu, J., Jiang, N., Chen, Q. Particle image velocimetry measurement of indoor airflow field: A review of the technologies and applications, Energy and Buildings, 69, pp. 367-380 (2014).
11. Allgayer D, Hunt GR. – On the application of the light attenuation technique as a tool for non intrusive buoyancy measurements, Exp. Thermal Fluid Science 38, 257-261 (2012).
12. Hacker J, Linden PF, Dalziel SB. Mixing in lock-release gravity currents, Dynamics of Atmospheres and Oceans 24, 183-195 (1996).
13. Choi, K.W., Lai, C.C.K., Lee, J.H.W. Mixing in the intermediate field of dense jets in cross currents, Journal of Hydraulic Engineering, 142 (1), art. no. 04015041 (2016).
14. Stewart, T.B., Judeikis, H.S. Measurements of spatial reactant and product concentrations in a flow reactor using laser induced fluorescence, Review of Scientific Instruments, 45 (12), pp. 1542-1545 (1974)
15. Melton LA, Lipp C.W., Criteria for quantitative PLIF experiments using high-power lasers, Experiments in Fluids 35, 310–316 (2003)
16. Sutton, J.A., Fisher, B.T., Fleming, J.W. A laser induced fluorescence measurement for aqueous fluid flows with improved temperature sensitivity. Exp. Fluids 45(5), 869–881 (2008).
17. Ferrari, S., Querzoli, G., Laboratory experiments on the interaction between inclined negatively buoyant

jets and regular waves, EPJ Web of Conferences, 92, art. no. 02018 (2015)
18. Bouche E., Cazin S., Roig V., Risso F and Billet A.M., Mixing in a swarm of bubbles rising in a confined cell measured by mean of PLIF with two different dyes, 16th Int Symp on Applications of Laser Techniques to Fluid Mechanics Lisbon, Portugal, 09-12 July, 2012
19. Charogiannis, A., An, J.S., Markides, C.N. A simultaneous planar laser-induced fluorescence, particle image velocimetry and particle tracking velocimetry technique for the investigation of thin liquid-film flows. Experimental Thermal and Fluid Science, 68, pp. 516-536 (2015).
20. Massing, J., Kaden, D., Kahler, C.J., Cierpka, C. Luminescent two-color tracer particles for simultaneous velocity and temperature measurements in microfluidics, Measurement Science and Technology, 27 (11) (2016).
21. Someya S., Tominaga K, Li Y, Okamoto K, Combined velocity and temperature measurements of natural convection using temperature sensitive particles, 15th Int Symp on Applications of Laser Techniques to Fluid Mechanics Lisbon, Portugal, 05-08 July, 2010
22. Ozawa, H. Experimental study of unsteady aerothermodynamic phenomena on shock-tube wall using fast-response temperature-sensitive paints, Physics of Fluids, 28 (4) (2016)
23. Sagaidachnyi, A.A., Volkov, I.U., Fomin, A.V. Influence of temporal noise on the skin blood flow measurements performed by cooled thermal imaging camera: Limit possibilities within each physiological frequency range, Progress in Biomedical Optics and Imaging - Proceedings of SPIE, 9917, art. no. 99170N (2016).
24. Ferrari, S., Badas, M.G., Querzoli, G., A non-intrusive and continuous-in-space technique to investigate the wave transformation and breaking over a breakwater, EPJ Web of Conferences, 114, art. no. 02022 (2016).
25. Mendez, M.A., Németh, L., Buchlin, J.-M. Measurement of Liquid Film Thickness via Light Absorption and Laser Tomography, EPJ Web of Conferences, 114, art. no. 02072 (2016)
26. Brady, M.R., Raben, S.G., Vlachos, P.P. Methods for Digital Particle Image Sizing (DPIS): Comparisons and improvements, Flow Measurement and Instrumentation, 20 (6), pp. 207-219 (2009).
27. Bergthorson J. M. and Dimotakis P. E., Particle velocimetry in high-gradient/high-curvature flows, Experiments in Fluids, 41 (2), 255-263, (2006).
28. Dalziel S.B., Decay of rotating turbulence: some particle tracking experiments, Applied Scientific Research, 49, 217-244, (1992).
29. Querzoli G, A Lagrangian study of particle dispersion in the unstable boundary layer. Atmos Environ 30(16):2821–2829 (1996)
30. Virant M. and Dracos T., 3D PTV and its application on Lagrangian motion, Measurement Science and Technology, 8, 1539-1552, (1997).
31. Stanislas M., Okamoto K., Kähler C.J. and Westerweel J., Main results of the Second International PIV Challenge, Experiments in Fluids, 39, 170-191, 2005.
32. Kähler, C.J., Astarita, T., Vlachos, P.P., Sakakibara, J., Hain, R., Discetti, S., La Foy, R., Cierpka, C. Main results of the 4th International PIV Challenge, Experiments in Fluids, 57 (6) (2016)
33. Adrian R.J., Particle-imaging techniques for experimental fluid mechanics, Annual Review of Fluid Mechanics, 261-303, (1991).
34. Buchhave, P. Particle image velocimetry-status and trends, Experimental Thermal and Fluid Science, 5 (5), pp. 586-604 (1992).
35. Adrian R.J., Twenty years of particle image velocimetry, Experiments in Fluids, 39 (2) 159–169 (2005)
36. Adrian R.J., Westerweel J., Particle image velocimetry. Cambridge (2011)
37. Stucky, M.J., Nino, E., Gajdeczko, B., Felton, P.G. Two-color particle image velocimetry technique for an internal combustion engine, Experimental Thermal and Fluid Science, 8 (4), pp. 305-314 (1994)
38. Falchi, M., Querzoli, G., Romano, G.P. Robust evaluation of the dissimilarity between interrogation windows in image velocimetry, Experiments in Fluids, 41 (2), pp. 279-293 (2006)
39. Lucas B.D. and Kanade T., An Iterative Image Registration Technique with an Application to Stereo Vision, Proceedings of Imaging Understanding Workshop, Washington, D.C. (USA), 121-130, 1981.
40. Shi J. and Tomasi C., Good features to track, Proceedings of the IEEE Conference on Computer Vision and Pattern Recognition (CVPR94), Seattle, USA, 1994.
41. Miozzi, M., Jacob, B., Olivieri, A. Performances of feature tracking in turbulent boundary layer investigation, Experiments in Fluids, 45 (4), pp. 765-780 (2008).
42. Cowen EA, Monismith SG. A hybrid digital particle tracking velocimetry technique. Experiments in Fluids;22(3):199-211 (1997)
43. Besalduch, L.A., Badas, M.G., Ferrari, S., Querzoli, G., Experimental Studies for the characterization of the mixing processes in negative buoyant jets, EPJ Web of Conferences, 45, art. no. 01012 (2013)
44. Besalduch, L.A., Badas, M.G., Ferrari, S., Querzoli, G., On the near field behavior of inclined negatively buoyant jets, EPJ Web of Conferences, 67, art. no. 02007 (2014)
45. Harris, C. & Stephens, M. A combined corner and edge detector. In: Proceedings of the 4th Aivey Vision Conference, Manchester, 147-151, 1988
46. La Porta A, Voth GA, Crawford AM, Alexander J, Bodenschatz E, Fluid particle accelerations in fully developed turbulence. Nature 409:1017–1019 (2001)
47. Voth GA, La Porta A, Crawford AM, Alexander J, Bodenschatz E, Measurement of particle

accelerations in fully developed turbulence. J Fluid Mech 469:121–160 (2002)
48. Mordant N., Pinton J.-F. and Michel O., Time-resolved tracking of a sound scatterer in a complex flow: nonstationary signal analysis and applications, Journal of the Acoustic Society of America, 112(1), 108-118, (2002).
49. Ferrari, S., Rossi, L., Particle tracking velocimetry and accelerometry (PTVA) measurements applied to quasi-two-dimensional multi-scale flows, Experiments in Fluids, 44 (6), pp. 873-886 (2008).
50. Lardeau, S., Ferrari, S., Rossi, L., Three-dimensional direct numerical simulation of electromagnetically driven multiscale shallow layer flows: Numerical modeling and physical properties, Physics of Fluids, 20 (12), art. no. 127101, (2008)
51. Ferrari, S., Rossi, L., Vassilicos, J.C., Acceleration measurements in turbulent-like flows, Advances in Turbulence XI - Proceedings of the 11th EUROMECH European Turbulence Conference, pp. 485-487 (2007).
52. Ferrari, S., Kewcharoenwong, P., Rossi, L., Vassilicos, J.C., Multi-scale flow control for efficient mixing: Laboratory generation of unsteady multi-scale flows controlled by multi-scale electromagnetic forces, Solid Mechanics and its Applications, 7, pp. 267-272 (2008).
53. Ran B. and Katz J., Pressure fluctuation and their effect on cavitation inception within water jets, Journal of Fluid Mechanics, 262, 223-263, (1994).
54. Liu X. and Katz J., Instantaneous pressure and material acceleration measurements using a fourexposure PIV system, Experiments in Fluids, 41, 227-240, (2006).
55. Gregory J.W., Sakaue H., Liu T, and Sullivan J.P., Fast Pressure-Sensitive Paint for Flow and Acoustic Diagnostics, Annual Review of Fluid Mechanics, Vol. 46: 303-330 (2014)

# CFD applications in offshore engineering

*Muk Chen* Ong[1,*]

[1]Department of Mechanical and Structural Engineering and Materials Science, University of Stavanger, 4036 Stavanger, Norway

**Abstract.** This paper gives a discussion on the Computational Fluid Dynamics (CFD) applications in offshore engineering. Accurate hydrodynamic quantities are essential for engineering design. Offshore structures are generally subject to high Reynolds number flows. These high Reynolds number flow conditions (Re > $10^6$) are difficult and expensive to achieve in an experimental setup. Therefore, it is attractive to use CFD to provide the essential hydrodynamic quantities for practical design. Verification and validation studies are important for determining the validity of the CFD prediction. A procedure of performing CFD simulation is shown. Different types of turbulence modelling are discussed. Three examples of high Reynolds number CFD simulations, covering flow around offshore structural components and waves past partially-submerged horizontal cylinders, are shown and discussed.

## 1 Introduction

The rapid developments of offshore oil and gas industry for drilling, well intervention, production and storage at sea, aquaculture industry and the exploitation of renewable energy from wind, waves and current have increased the importance of flow-structure interactions and sediment transport around offshore structures such as platform legs/columns, offshore foundations, offshore floating structures, pipelines and risers. These offshore structures are often subject to very high Reynolds number flows, especially in the open sea environment. The flow mechanisms around these structures are complex. Hence, more research of the high Reynolds number flows in the ocean is necessary in order to gain a better understanding of the flow characteristics and dynamic responses of these structures. Furthermore, high Reynolds number flows related to offshore structures are hard and expensive to measure. Therefore, an attractive alternative is to use Computational Fluid Dynamics (CFD) to obtain the essential hydrodynamic quantities needed for engineering design.

The definition of Reynolds number is $Re = UL/v$, where $U$ is the flow velocity, $L$ is the characteristic length and $v$ is the kinematic viscosity of the fluid. Generally the characteristic length of the offshore structures is large. Thus, the order of magnitude of $Re$ can range from $10^3$ to $10^8$. For cases of $Re > 10^6$, the cost of carrying out the experiments in water basin/tank or wind tunnels is high. Moreover, scaling effect is usually an important concern for model testing of large scale structures, since it is very difficult to achieve the same flow condition as in the real life engineering application.

It is always possible to carry out field measurement. However, the data analysis for field measurement is demanding and full of uncertainties. Interpretation of the field measured data can be subjective if the amount of data is insufficient. However, when the amount of data is large, judgement of the measurement accuracy may be subjected to a vast variation.

Since there are plenty of difficulties to carry out high Reynolds number model testing and field measurement, numerical simulations should be used as an alternative to reduce the burden of experiments and field measurements. Generally detailed numerical simulations are computationally costly. It is essential to understand the flow physics before carrying out the numerical simulations, so that the complex engineering problem can be simplified without losing the essence of engineering. Verification and validation studies are important to ensure the validity of the CFD model. For example, in order to capture the boundary layer flow accurately, the near-wall mesh should be sufficiently fine. The demand of the computational resource will then be increased tremendously. Hence, engineers should have understood the flow physics well before they start performing the numerical modelling.

In this paper, applicability of turbulence models for high Reynolds will be discussed. An engineering procedure to perform CFD simulations will be discussed. Three examples of flows related to offshore engineering applications are mentioned in this paper to demonstrate the CFD applicability at high Reynolds number flow, i.e., a) flow around rectangular cylinders with different aspect ratios, (b) flow around offshore wind turbine foundations, and (c) waves past partially-submerged cylinders.

* Corresponding author: muk.c.ong@uis.no

## 2 Numerical Models

The understanding of the flow physics around the offshore structures is important for achieving accurate numerical prediction. For example, the flow condition of the boundary layer around the structure is important to determine the location of separation point, pressure distribution and skin friction distribution around the structures. This will subsequently influence the hydrodynamic quantities such drag forces, lift forces and vortex shedding frequencies. First we should identify out several important non-dimensional parameters, such as Reynolds number and Keulegan-Carpenter number, which will be the governing parameters for the variation of hydrodynamics quantities in different flow regimes. For example, for flow around circular cylinder, the time-averaged drag coefficient is dependent on the change of Reynolds number, see Sumer and Fredsøe [1] and Ong et al. [2] for details.

Effects of turbulence on the hydrodynamic coefficients should be investigated. The geometry of the offshore structures should be also taken in consideration. The boundary layer around cylindrical-like structure will change from laminar to turbulent dependent on the corresponding Reynolds number, see Sumer and Fredsøe [1] for details. Therefore, appropriate turbulence modelling becomes essential when the boundary layers need to be resolved and computed correctly. Three dimensionality of the flow should also be considered. When the 3D effect of the flow is prominent, 3D simulations should be carried out in order to capture the correct flow physics, see Section 3 for further discussion.

Figure 1 shows a flow chart describing a procedure to perform CFD simulations including both verification and validation studies. Generally, there are four major stages: 1) Problem identification, 2) Pre-processing, 3) Solving, and 4) Post-processing. Grid and time-step convergence studies need to be carried out; and then the numerical results can be validated against available published experimental or field data.

Based on Figure 1 "Item 4: Identify flow physics", it includes the selection of turbulence modelling. For turbulent flows, complexity of flow phenomena and different range of length scales make most of the modelling approaches computationally expensive. The computational cost for resolving all the length scales in turbulence is tremendously high. Hence, a feasible approach is to create turbulence models to approximate unresolved parts in the flow. Generally, turbulence models can be categorized based on the corresponding computational cost. The more turbulence length scales are resolved, the finer the simulation resolution is, and consequently the higher the computational cost will be. However, if most of the turbulent length scales are not resolved (i.e. low computational cost), the poor numerical accuracy is expected to be obtained. Moreover, the Navier-Stokes equations contain non-linear convection and pressure gradient terms; and these nonlinear terms have to be solved numerically with appropriate boundary conditions.

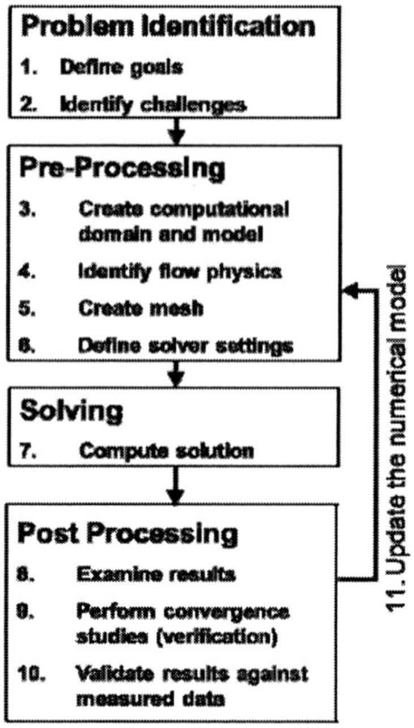


**Fig. 1.** A procedure to perform CFD simulations including both verification and validation studies.

In operating conditions, offshore structures are often subject to high Reynolds number flow conditions. At high Reynolds numbers, flows are turbulent. Turbulent flows are highly irregular; however, they are governed by the exact Navier-Stokes equations. Efficient numerical algorithms solving the Navier-Stokes equations and the turbulence models for high Reynolds number flows are still considered as a major challenge although many numerical methods (e.g. SIMPLE-like class of algorithms and projection methods) have been developed. Furthermore, the appropriateness of turbulence models for the targeted engineering applications should be considered carefully. The computer and cost efficiency of the computational method is also important, especially for engineering applications.

Direct Numerical Simulation (DNS) resolves the whole range of spatial and temporal scales of the turbulence in the flow. A DNS simulation requires a number of mesh points more than $Re^{2.25}$ [3]. It makes the computational cost extremely high for flow around complex geometries at high Reynolds number. Hence, hydrodynamic analyses based DNS simulations are currently not feasible for offshore industry.

Reynolds-averaged Navier-Stokes (RANS) model is the most commonly-used numerical approach for offshore engineering applications. All of the unsteadiness in the Navier-Stokes equations is taken as part of turbulence and is averaged out in the Reynolds-averaged approach. In the averaging process, the non-linearity of the Navier-Stokes equations gives the fluctuation terms which needs to be modelled. Turbulence models, such as k-ε model, k-ω model, k-ω SST model and Spalart-Allmaras model, can be used as a turbulence closure. More details can be referred to Ferziger and Perić [4].

Large-eddy simulation (LES) is a numerical approach, in which large-scale motions are explicitly computed and small-scale eddies are modeled with subgrid scale models. LES is three-dimensional, time dependent and less expensive than DNS for the same flow. LES is the preferred method when the Reynolds number is too high or the geometry is too complex to allow the application of DNS [4].

Detached Eddy Simulation (DES) is a hybrid RANS-LES model, in which the model switches from RANS to LES (subsgrid scale formulation) in regions where the mesh resolution is sufficiently fine for LES. Therefore, the grid resolution for DES is not as high as pure LES, and subsequently the computational cost is reduced.

Three examples related to offshore engineering applications at high Reynolds number flow will be shown in Section 3. RANS, LES and DES are employed in these simulations to demonstrate the applicability of CFD approach.

# 3 Examples

## 3.1 Flow around rectangular cylinders with different aspect ratios

Many marine and offshore structures have structural components with square or rectangular cross-sections, such as columns of semi-submersibles and tension-leg platforms, as well as heave damping plates of truss spar platforms. Two-dimensional flow normal to the longer edge of rectangles with different aspect ratios ($R = B/H$), i.e. from a square to a thin plate, has been investigated numerically in Tian et al. [5] at $Re$ = 21400. Here $B$ is the breadth along the direction of flow and $H$ is the height of the cross-section. The objective of the study is to evaluate the validity of 2D unsteady RANS (URANS) simulation with k-ω SST model for flow around rectangular cylinders. Figure 2 shows the comparison of the mean drag coefficient $\overline{C_D}$ between 2D URANS results reported by Tian et al. [5] and the published 2D and 3D numerical results as well as experimental results [6-12]. The drag foces acting on the cylinders with high aspect ratios (i.e. $R$ = 1, 0.8 and 0.6) are well predicted by the numerical simulations; however, the drag forces are over-predicted for low aspect ratios (i.e. R=0.4, 0.2, 0.1 and 0.05).

Tian et al. [13] performed both 2D URANS and 3D LES simulations to investigate the 3D effects of the flow normal to a flat plate with an infinite spanwise length at $Re = 1.5\times10^5$. The ratio of the plate thickness to the plate height is 0.02. The hydrodynamic results (such as time-averaged drag coefficient, Strouhal number and mean recirculation length) were compared with the published experimental data reported by Fage and Johansen [8] and Leder [14]. The 3D LES results are in good agreement with the published experimental data; however, the 2D URANS shows a poor comparison with experimental data. This feature is similar to the conclusion obtained by Tian et al [5] for rectangular cylinders with small aspect ratio. Figure 3 shows the instantaneous vortical structures in the wake reported by Tian et al. [13]. A highly 3D and chaotic wake flow is observed (especially the secondary flow in the spanwise direction) although the plate with infinite spanwise length is nominally 2D. Figure 4 shows the streamlines of the mean velocity field obtained in the 2D and 3D simulations (taken from Tian et al. [13]). It shows that the recirculation length in the 2D simulation is approximately two times larger than that in the 3D simulation. The 3D flow effects are prominent for high Reynolds number flow normal to a flat plate. Tian et al. [13] reported that the reliability of 2D simulations for the nominally 2D problem is questionable and more attention should be made to investigate the 3D effects.

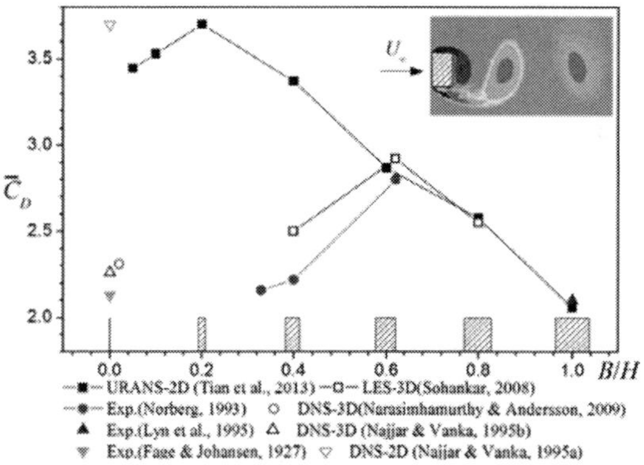


**Fig. 2.** Comparison of mean drag coefficients between 2D URANS results reported by Tian et al. [5] and published experimental and numerical results [6-12] for flow around rectangular cylinders with different aspect ratios.

**Fig. 3.** Instantaneous vortical structures in the wake of flow normal to a thin flat plate (taken from Tian et al. [13]).

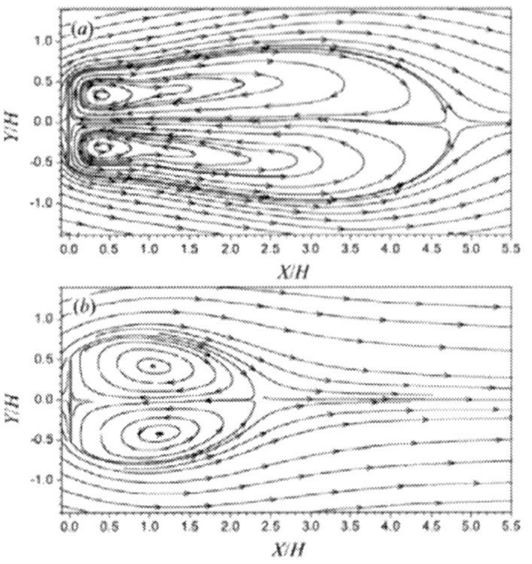


**Fig. 4.** Streamlines of the mean velocity field obtained in the 2D (top) and 3D (bottom) simulations (taken from Tian et al. [13]).

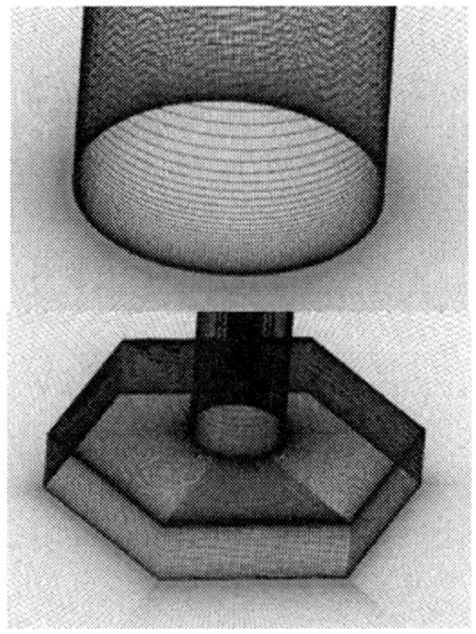

**Fig. 5.** 3D views of the mesh on the surface of the monopile and hexagonal gravity-based foundations (taken from Ong et al. [15]).

## 3.2 Seabed boundary layer flow around offshore wind turbine foundations

Ong et al. [15] investigated the formation of horseshoe vortex and flow structures around monopile and gravity-based offshore wind turbine foundations in order to provide an assessment of potential scour for engineering design. Three-dimensional CFD simulations were performed using Spalart-Allmaras Delayed Detached Eddy Simulation (SADDES) at $Re = 4\times10^6$ based on the free stream velocity and the diameter of the monopile foundation. Grid and time-step convergence studies (verification study) were performed. Figure 5 shows 3D views of the mesh on the surface of the monopile and hexagonal gravity-based foundations (Ong et al. [15]).

Figures 6(a) and (b) show 3D views of instantaneous vortical structures around the monopile and hexagonal gravity-based foundation, respectively (Ong et al. [15]). A distinct horseshoe vortex is found in front of the monopile foundation. Two small horseshoe vortices are found in front of the hexagonal gravity-based foundation, i.e. one is on the top of the bottom slab ("1" marked in Figure 6(b)) and one is near the seabed in front of the bottom slab ("2" marked in Figure 6(b)). This flow feature reduces the strength of the horseshoe vortex on the seabed for the hexagonal gravity-based foundation.

The size of the horseshoe vortex can be determined from the time-averaged bed shear stress $\tau$ along the centreline from the inlet to the location of the foundation toe. Figures 7 (a) and (b) shows the normalized time-averaged bed shear stress $\tau/\tau_\infty$ along the centreline for the monopile and hexagonal gravity-based foundations, respectively, where $\tau_\infty$ is the time-averaged undisturbed bed shear stress. Ong et al. [15] reported that the hexagonal gravity-based foundation shows a reduction of the horseshoe vortex size as compared with that for the monopile foundation under the same flow condition.

(a) Monopile foundation

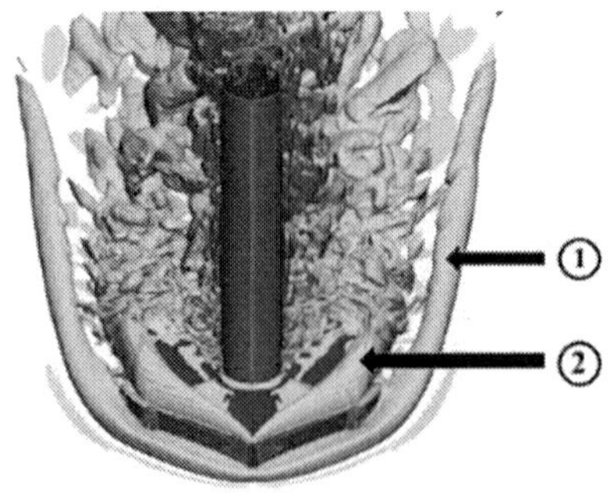


(b) Hexagonal gravity-based foundation

**Fig. 6.** 3D views of instantaneous vortical structures around the monopile (top) and hexagonal gravity-based (bottom) foundations (taken from Ong et al. [15]).

 DOI: 10.1051/epjconf/201714301002

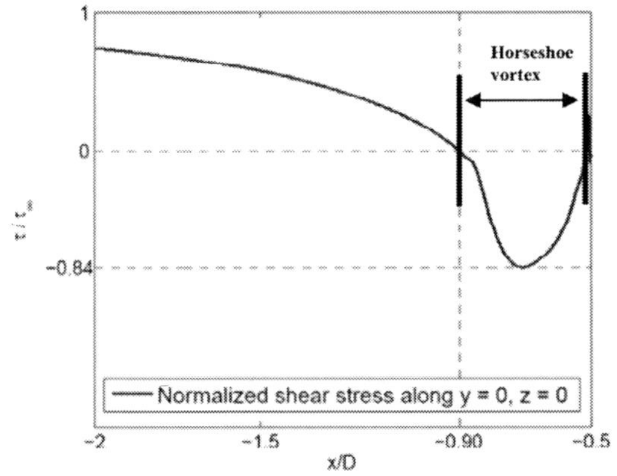


(a) Monopile foundation

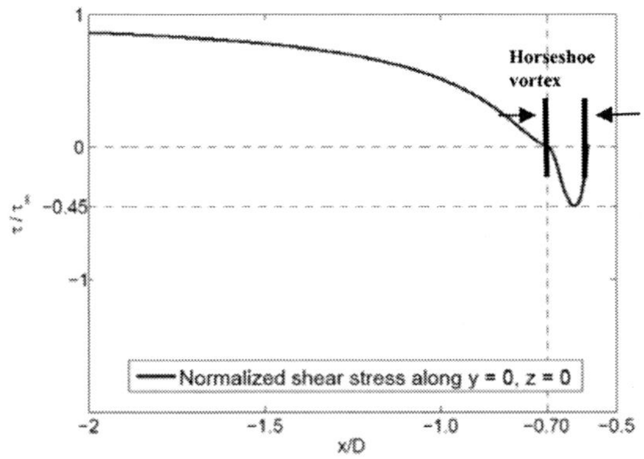


(b) Hexagonal gravity-based foundation

**Fig. 7.** The normalized time-averaged bed shear stress along the centreline for the monopile (top) and hexagonal gravity-based (bottom) foundations (taken from Ong et al. [15]).

### 3.3 Free Surface Waves past Partially Submerged Horizontal Cylinders

Partially-submerged bluff bodies are often found in offshore structures, e.g. wave energy converter and fish cages. Circular cylinders are usually one of the important components in these structures. Bihs and Ong [16] performed 2D CFD simulations to investigate the flows past partially-submerged circular cylinders in free surface waves. The simulations were carried out by solving the URANS equations with the k-ω model. The level set method was employed to model the free-surface waves. Vertical wave forces on the partially-submerged cylinders have been calculated numerically and compared with experimental data reported by Dixon et al. [17] under regular-wave conditions. A Grid convergence study was performed to obtain sufficient mesh resolution. The investigated Keulegan-Carpenter number ranged from 1.26 to 3.14. Figure 8 shows the normalized vertical hydrodynamic force (*Fv'*) versus the non-dimensional time (*t'*) over one wave period. The wave amplitude is half of the cylinder diameter (*D*) and the wave length is 15.62*D*. Bihs and Ong [16] reported that the numerical prediction of the vertical hydrodynamic forces is in good agreement with the experimental data reported by Dixon et al. [17]. Figure 9 shows the free-surface elevations around the cylinders. Both wave run-up and overtopping actions are observed clearly in the simulation.

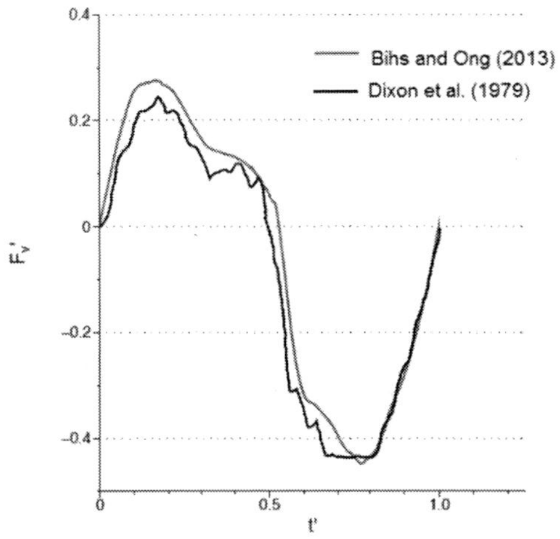


**Fig. 8.** The normalized vertical hydrodynamic force versus the non-dimensional time over one wave period (taken from Bihs and Ong [16]).

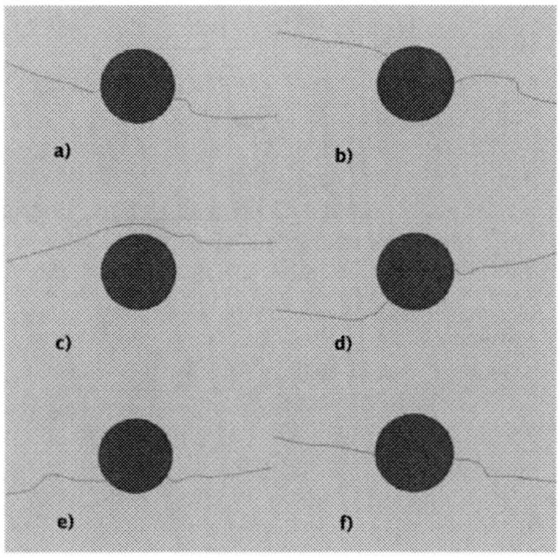


**Fig. 9.** Time history of free surface elevation over a wave period for (a) t'=0.0, (b) t'=0.12, (c) t'=0.36, (d) t'=0.6, (e) t'=0.73, (f) t'=1.0

## 4 Conclusions

Offshore structures are generally subject to high Reynolds number flows. These high Reynolds number flow conditions ($Re > 10^6$) are difficult and expensive to achieve in an experimental setup. An alternative method is to employ CFD modelling to obtain the essential hydrodynamic coefficients for practical engineering design.

A procedure of performing CFD simulations has been shown with the emphasis on verification and validation studies. Grid and time convergence studies (verification) and comparisons with available experimental data (validation) are important to ensure the validity of the CFD prediction. Due to the demand of high mesh resolution, DNS simulations are not feasible for applications in offshore industry. RANS and URANS simulations are commonly used to predict high Reynolds number engineering flows due to lower computational cost and reasonable accuracy of hydrodynamic prediction. DES and LES simulations are employed when detailed flow structures are essential for practical design.

Three examples, i.e. i) flow around rectangular cylinders with different aspect ratios, ii) seabed boundary layer flow around offshore wind turbine foundations, iii) free surface waves past partially-submerged horizontal circular cylinders, have been shown and discussed to demonstrate the applicability of CFD simulations for offshore engineering design.

## References

1. B.M. Sumer, J. Fredsøe, Hydrodynamics around Cylindrical Structures: Advanced series on ocean engineering, **12** (1997).
2. M.C. Ong, T. Utnes, L.E. Holmedal, D. Myrhaug, B. Pettersen, Marine Structures, **22**, 142-152 (2009).
3. S.B. Pope, Turbulence Flows (2000).
4. J.H. Ferziger, M. Perić, Computational Methods for Fluid Dynamics (2001).
5. X. Tian, M.C. Ong, J. Yang, D. Myrhaug, Ocean Engineering, **58**, 208-216 (2013).
6. C. Norberg, Journal of Wind Engineering and Industrial Aerodynamics, **49** (1-3), 187-196 (1993).
7. D.A. Lyn, S. Einav, W. Rodi, J.H Park, Journal of Fluid Mechanics, **304**(1), 285-319 (1995).
8. A. Fage, F.C. Johansen, Proceedings of the Royal Society of London. Serie A, **116**, 170-197 (1927).
9. A. Sohankar, Journal of Wind Engineering and Industrial Aerodynamics, **96**, 640-655 (2008).
10. V.D. Narasimhamurthy, H.I. Andersson, B. Pettersen, International Journal of Heat and Fluid Flow, **30**, 1037-1043 (2009).
11. F.M. Najjar, S.P. Vanka, Physics of Fluids, **7**, 2516-2518 (1995).
12. F.M. Najjar, S.P. Vanka, International Journal of Numerical Methods in Fluids, **21**, 525-547 (1995).
13. X. Tian, M.C. Ong, J. Yang, D. Myrhaug, G. Chen, OMAE Conference, OMAE2012-83730 (2012).
14. A. Leder, Physics of Fluids A: Fluid Dynamics, **3**, 1445 (1991).
15. M.C. Ong, E. Trygsland, D. Myrhaug, OMAE Conference, OMAE2016-54643 (2016).
16. H. Bihs, M.C. Ong. OMAE Conference, OMAE2013-10529 (2013).
17. A.G. Dixon, C.A. Greated, S.H. Salter, Journal of the Waterway, Port, Coastal and Ocean Division, **105** (4), 421-438 (1979).

# Experiments of tuned liquid damper (TLD) on the reduced shear frame model under harmonic loads

*Ersin* Aydın[1,*] , *Baki* Öztürk[2], *Maciej* Dutkiewicz[3] , *Hüseyin* Çetin[1], *Ozan* Okkay[1], *Uğur* Ohancan[1], *Yunus Emre* Şirin[1]

[1] Engineering Faculty, Civil Engineering Department, Nigde University, 51240 Nigde, Turkey
[2] Engineering Faculty, Civil Engineering Department, Hacettepe University, 06800 Ankara, Turkey
[3] Faculty of Civil, Environmental Engineering and Architecture,Department of Building Construction, University of Science and Technology in Bydgoszcz, ul. Prof. S. Kaliskiego 7, 85-796 Bydgoszcz, Poland.

**Abstract:** In this study shaking table tests which are applied on 3-Storey reduced shear frame models with TLD's subjected to harmonic loadings are presented. Firstly, free vibration experiments are conducted on the structure and 1st free vibration frequency of the structure is determined. The structure is shaken under harmonic loading at a frequency equal to 1st frequency of the structure which provides the resonance condition. Displacements and accelerations are measured at storey levels of the structure. A container in a rectangular prism shape is manufactured as a TLD model. Liquid is poured in the container and the same experiments are repeated at different liquid heights. The effect of TLD application on the structural models considering displacement and acceleration of the structure are investigated. In addition the effect of TLD application and its allocation at different storey levels are calculated experimentally. As a result of the conducted experiments, most convenient TLD models considering both displacement and acceleration behavior are determined. It is observed that all the damping models cause significant levels of reduction in seismic behavior of the structure under harmonic loading.

## 1 Introduction

In recent days, a number of methods have been developed in order to reduce dynamic excitation such as wind and earthquake excitations. In civil engineering the use of tuned mass dampers (TMD's) and tuned liquid dampers (TLD's) are comprehensively approved by many researchers, so use of such control devices have been more and more preferred instead of conventional design [1,3,5,7]. As shown for a single degree of freedom system with TLD model in Figure 1, Tuned Liquid Dampers (TLD's) are passive energy dissipation systems which consist of water tank and sloshing viscous fluid. These devices absorb dynamic inputs; therefore, structures are protected against damage or collapse. For the condition of lateral accelerations of water tank, while part of water level rises, another part decreases simultaneously. This sway movement supplies energy absorption arising from dynamic forces. In these devices, wave movements and natural vibration periods depend on water tank geometry, length of dimensions, water depth and acceleration of gravity. When periods of TLD's are tuned to a particular value, they resonate out of phase with structural movement. Two types of Tuned liquid dampers are commonly used which are called (i) tuned liquid dampers which consist of rectangular section tank and (ii) tuned liquid column dampers which have two limbs of tubes filled with same kind of liquids as seen schematically in Figures 2 (a) and 2 (b).

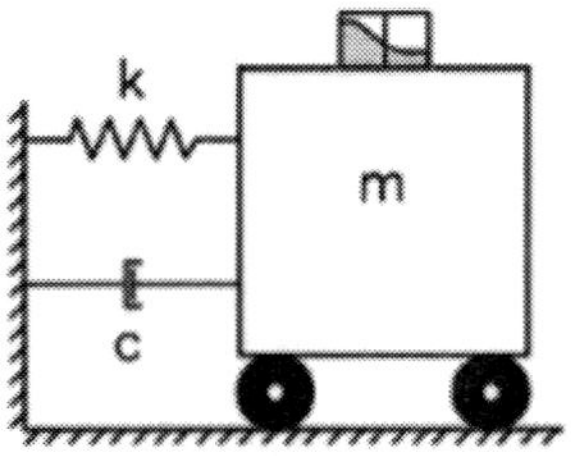


**Fig. 1.** Single degree of freedom system with TLD model

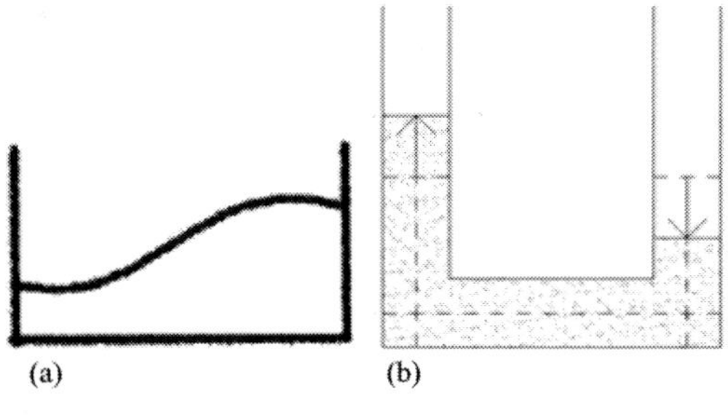

(a) (b)

**Fig. 2** (a) Schematic demonstration of TLD and (b) Schematic demonstration of TLCD

* Corresponding author: eaydin@nigde.edu.tr

While TLD's use wave breaking in order to absorb input energy effect the system, TLCD's absorb this energy by adjusting atmosphere pressure at limbs of TLCD. In resonance condition, because of undesirable fluid behaviors of TLD's, TLCD's are superior to TLD devices.

Bauer [2] used very different two kinds of liquids by fully filled rectangular tanks in order to supply damping response through the movement of interfaces. Welt and Modi [8,9] proposed to use tuned liquid dampers (TLD's) enhancing ability of structure against wind and earthquake induced excitations. Owing to resemblance of these dampers, tuned liquid column dampers (TLCD's) system having partially filled tube including some internal orifices is suggested by Xu et al. [8]. Characteristics of multiple tuned liquid column dampers (MTLCD) were searched by Gao et al. [12] for suppressing structural vibration. They have discussed advantages and disadvantages of multiple tuned liquid column dampers in their research. They indicate that a multiple tuned liquid column damper (MTLCD) is less sensitive than a single tuned liquid column damper (TLCD). They also found out that increasing number of TLCD enhances efficiency of MTLCD. However, more than five TLCD's do not give contribution to augment performance behavior of structure against vibratory excitation forces. When MTLD's are taken into account, Fujino and Sun [4] found out that weakly excited small amount of liquid movement has similar feature covering efficiency and robustness to frequency ratio of device. For the condition of strongly forced much more liquid motion appearance, MTLD can't be more effective than TLD. On the other hand an MTLD has almost equal influence, if breaking waves has already existed.

## 2 General Formulation of TLD

As seen in Figure 3, rectangular section of liquid filled tank with width b has 2a length along x direction, liquid level h which is at rest condition. Sun et al. [6] proposed some formulation to define and design TLD's. These formulations are determined as:

$$\frac{\partial \eta}{\partial t} + h\sigma \frac{\partial(\phi u)}{\partial x} = 0 \qquad (1)$$

$$\frac{\partial u}{\partial t} + (1-T_H^2)u\frac{\partial u}{\partial x} + C_{fr}^2\, g\frac{\partial \eta}{\partial x} + gh\sigma\emptyset\frac{\partial^2 \eta}{\partial x^2}\frac{\partial \eta}{\partial x} = -C_{da}\,\lambda u - \ddot{x}_s \qquad (2)$$

Where $\eta(x,t)$ and $u(x,\eta,t)$ are independent variables which are elevation of undistributed water free surface and particle velocity of free water surface, respectively. TLD's base acceleration is equal to base shear force acceleration, $\ddot{x}_s$ and $g$ is acceleration of gravity.

By using Equations (1) and (2), $\sigma$, $\phi$ and $T_H$ can be calculated as:

$$\sigma = \frac{tanh(kh)}{kh}, \qquad (3a)$$

$$\phi = \frac{tanh[k(h+\eta)]}{tanh(kh)}, \qquad (3b)$$

$$T_H = tanh(k+\eta) \qquad (3c)$$

Where $k$ is number of wave, $\lambda$ is damping parameter depending on boundary layer through bottom of tank, side walls and water free surface contamination. $\lambda$ can be given as:

$$\lambda = \frac{1}{\eta+}\frac{1}{\sqrt{2}}\sqrt{\omega_1\, v}\left[1 + \left(\frac{2\Box}{b}\right) + s\right] \quad (4)$$

Where $\omega_1$is linear fundamental frequency of sloshing water in tank which defined equation, $\vartheta$ is kinematic viscosity of water and s is unity surface contamination factor, respectively. Fundamental frequency of TLD can be written as:

$$\omega_1 = \sqrt{\frac{\pi g}{2a}}\tanh(\pi\Delta) \qquad (5)$$

in which, $\Delta = h/2a$ is called water depth ratio.

In Equation (2), when considering breaking wave coefficients $C_{fr}$ and $C_{da}$, η is taken bigger than h. If breaking wave doesn't exist, $C_{fr}$ is empirically calculated as 1,05 and $C_{da}$ is calculated from Equation (6) which is calculated according to top of the building maximum displacement $(x_s)_{max}$ which is considered without TLD. This equation is given as:

$$C_{da} = 0,57\sqrt{\frac{h^2\omega_1}{a\, v}(x_s)max} \qquad (6)$$

To calculate approximate F, shear force at base, Equations (1) and (2) are used to obtain $\eta(x,t)$ which is called free surface elevation. Neglecting some higher order terms, F can written as:

$$F = \frac{\rho g b}{2}\left[(\eta_n + \Box)^2 - (\eta_0 + \Box)^2\right] \qquad (7)$$

in which b is width of tank, $\rho$ is mass density of water, $\eta$ and $\eta_0$ are free surface elevations of right and left walls of water tank, respectively.

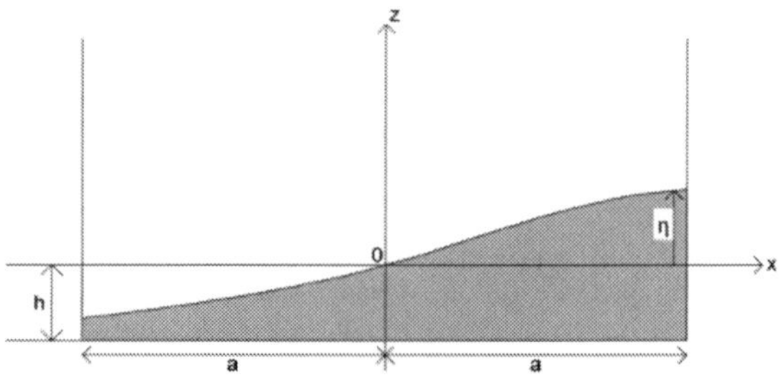


**Fig. 3.** Rectangular section of liquid filled tank

## 3 Experimental study

As seen in Figure 4, 3 storey Building frame shear model with TLD, which is used for experiment setup, is made of S235JR steel. Physical features of this model is summarized in Table 1 as shown below.

**Table 1** Physical features of structural model

| | |
|---|---|
| Shear frame weight | 6500 gr |
| Weight located at 1st storey of model | 10812 gr |
| Weight located at 2nd storey of model | 8346 gr |
| Weight located at 3rd storey of model | 6718 gr |
| Each storey height | 35 cm |
| Shear frame column dimensions | 1.0 cm x 0.3cm, thickness=0.1 cm |
| Shear frame beam dimensions | 1.5 cm x 1.5 cm , thickness=0.1 cm |
| Shear frame floor dimensions | 24 cm x 24 cm |
| 1st floor mass of Building frame shear model with connection element and added masses to floors | 13 kg |
| 2st floor mass of Building frame shear model with connection element and added masses to floors | 10.5 kg |
| 3st floor mass of Building frame shear model with connection element and added masses to floors | 8.9 kg |
| Total mass of Building frame shear model with connection element and added masses to floors | 35 kg |
| Dimensions of TLD cup | 22.5 cm x 22.5 cm x 27.5 cm |

Shaking table used in this experiment has 0-400 Hz frequency range. This liner-shaking table has a compact shape and its dimensions are 50 cm x 50 cm. It also has sensitive position taking capacity. Shaking table can produce one dimensional horizontal vibration that can produce desired number and range (1-10 Hz) of frequency and its amplitude changes in between -200 mm to + 200 mm. Gathering given data, general dynamic device is used. Displacement and acceleration measurements are connected to this device and a software data recorded from this device can be saved and used.

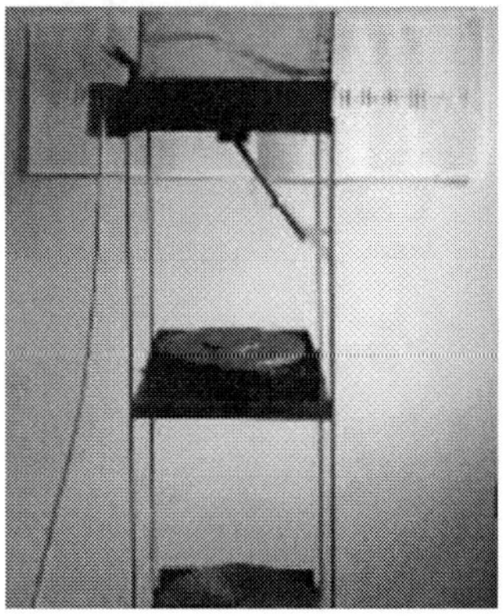

**Fig. 4.** 3-storey Building shear frame model with TLD

This experiment is designed for 3 mm amplitude and 1.5 Hz frequency which is also fundamental natural frequency of building model and 10 cycle harmonic load is applied to 3-storey model with and without TLD by means of shaking table. After this application, 3 storey building model displacements are measured without TLD and with TLD which are separable for three different TLD liquid height levels in cups. At each story of TLD location for 3 story model, experiments are repeatedly conducted. The reason for choosing 1.5 Hz shaking table frequency is to create an unfavorable condition which resonates 3 storey models. After measuring displacements of each cup, these displacements are plotted with respect to time in order to find most appropriate model. Rectangular sections of liquid filled tanks are shown in Figure 5 while TLD's models of experiment are provided in Table 2.

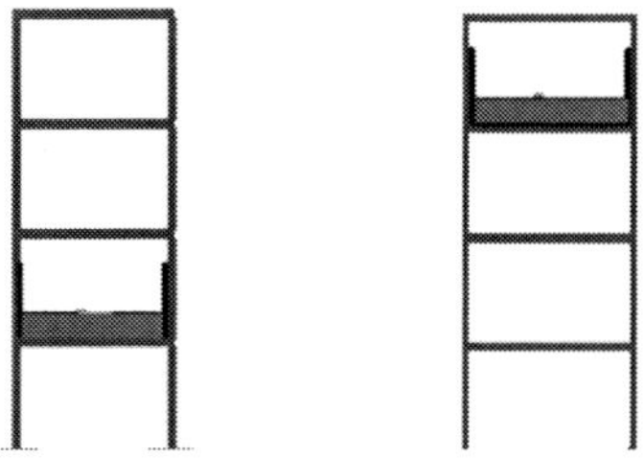

**Fig. 5.** Rectangular sections of liquid filled tanks

**Table 2.** TLD's models of experiment

| Type of TLD model | Figure of model | Total mass of cup and water (kg) |
|---|---|---|
| MODEL 1 | 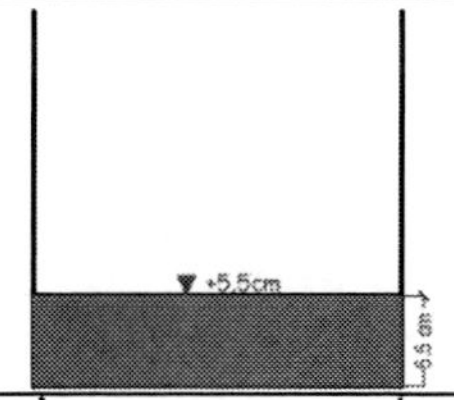 | 5.692 kg |
| MODEL 2 | 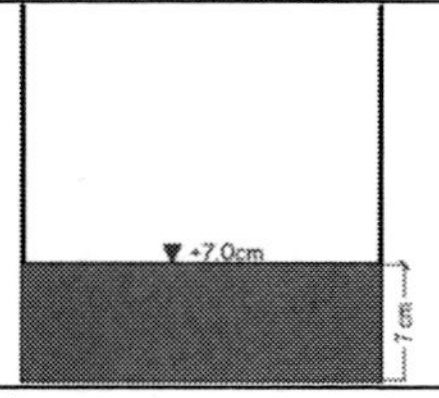 | 6.223 kg |
| MODEL 3 | 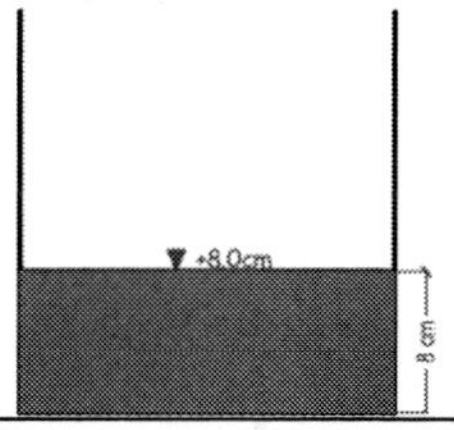 | 6.717kg |

As seen in Figure 6 and Table 3, when TLD cup is located at first floor of the model, model 2-1 gives better performance than other cups in terms of percent decrease in performance which are 75.6838 % at third floor,

74.5454 % at second floor, and 72.9729 % at first floor, respectively.

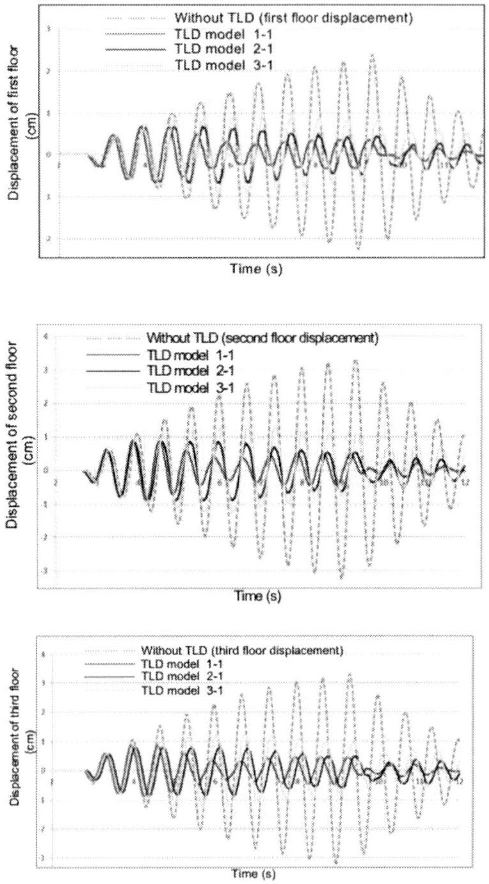


**Fig. 6**. Displacement of first, second and third storey of structural model without TLD and with TLD models which are located at first storey

**Table 3.** Comparison of first storey displacement of structural model considering different TLD models

| Type of TLD model | Third floor Peak displacement of structural model | Second floor Peak displacement of structural model | First floor Peak displacement of structural model | Decrease in floor displacement due to TLD (%) | | |
|---|---|---|---|---|---|---|
| Without TLD | 3,29 cm | 3,3 cm | 2,368 cm | 3rd Storey | 2nd Storey | 1st Storey |
| TLD model 3-1 | 1.21 | 1.28 | 1.004 | 63.22188 | 61.2121 | 57.6013 |
| TLD model 2-1 | 0.8 | 0.84 | 0.64 | 75.68389 | 74.5454 | 72.9729 |
| TLD model 1-1 | 0.87 | 0.89 | 0.66 | 73.55623 | 73.0303 | 72.1283 |

As can be understood from Figure 7 and Table 4, when TLD cup is located at third floor of the model, model 2-3 gives better performance than other cups in terms of percent decrease in performance which are 82,0668 % at third floor, 81,5151 % at second floor, and 82,68581 % at first floor, respectively.

For this experiment, model 2-3 gives best performance among all TLD's cups with respect to decrease in displacements of all storeys of structural models. These displacements are about 82% less.

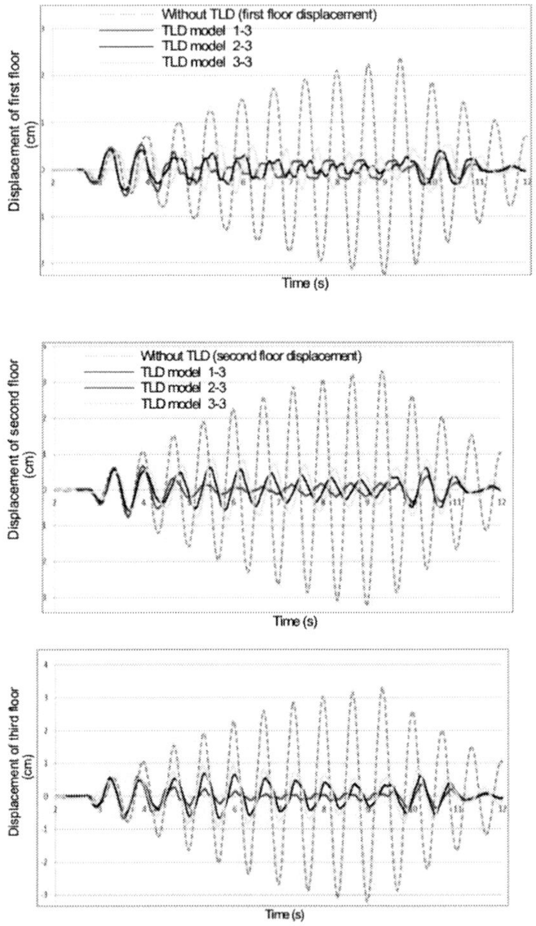


**Fig. 7.** Displacement of first, second and third storey of structural model without TLD and with TLD models which are located at third storey

## 4 Conclusions

Structural control systems are modern systems that develop ability to strengthen the structure against dynamic effects. TLD systems are also kinds of structural control systems for which deriving formulation and solution is not easy due to complex behavior of liquid sloshing in cup. However, these systems are very useful for lessening vibration effects on structures, even for their complex applications. TLD formulation can be simplified in order to explain their behavior associated to tuned mass damper (TMD) system. In this study, experiments for such complex devices are conducted in which harmonic effectiveness at resonance condition for fundamental frequency of structural system is evaluated. Considering the results of the presented study, the following conclusions can be expressed.

1) For this study, when considering achievement of the best displacement decrease, TLD model 2-3 which is model 2 located at 3rd storey has the best performance compared to other TLD models. As shown in results for this model, decrease in quantities as percentages are 79,02 % at 3rd floor, 81,51% at 2nd floor and 82,68 % at 1st floor, respectively.

2) If TLD models are evaluated relatively,

a) Location of TLD model-1 at 3rd floor has the best performance considering displacement reduction for 3th floor,

b) Location of TLD model-2 at 3rd floor has the best performance considering displacement reduction for 2nd floor,

c) Location of TLD model-3 at 2nd floor has the best performance considering displacement reduction for 2nd and 3rd floors.

3) As can be understood from the conducted experiments, TLD is very effective in reducing harmonic force effects at resonance conditions. The ranges of these reductions are approximately 79% -80 %. The shear forces lessen such that elements of structure are exposed to less strain. Accordingly, element sections can be chosen smaller than conventional design such that TLD's derogate the cost function of structure.

**Table 4.** Comparison of first storey displacement of structural model considering different TLD models

| Type of TLD model | Third floor Peak displacement of structural model | Second floor Peak displacement of structural model | First floor Peak displacement of structural model | Decrease at floor displacement due to TLD (%) | | |
|---|---|---|---|---|---|---|
| Without TLD | 3,29 cm | 3,3 cm | 2,368 cm | **3.storey** | **2. storey** | **1. storey** |
| TLD model 3-3 | 0.94 cm | 0.86 cm | 0.54 cm | 71.42857 | 73.93939 | 77.19594 |
| TLD model 2-3 | 0,69 cm | 0.61 cm | 0.41 cm | 79.02735 | 81.51515 | 82.68581 |
| TLD model 1-3 | 0.59 cm | 0.66 cm | 0.52 cm | 82.06686 | 80 | 78.04054 |

## References

1.Almazan, J.L., G. Espinoza, and J.J. Aguirre, *Torsional balance of asymmetric structures by means of tuned mass dampers*, Engineering Structures, **42**, 308–328, 2012.

2.Bauer, H.F., *Oscillations of Immiscible Liquids in Rectangular Container: A New Damper for Vibration Control of Towers*, Engineering Structures, **17**(9): 668-675, 1984.

3.Daniel, Y., and O. Lavan, *Gradient based optimal seismic retrofitting of 3D irregular buildings using multiple tuned mass dampers*, Computers and Structures, **139**, 84-97, 2014.

4.Fujino Y., and Sun LM, *Vibration control by multiple tuned liquid dampers (MTLDS),* Journal of Structural Engineering, ASCE, **119**(12): 3482-3502, 1993.

5.Sakai F, Takaeda S, Tamaki T., *Tuned liquid column damper –new type device for suppression of building vibration*, Proceedings of the International Conference on High-rise Buildings, Nanjing, China, 926–31, 1989.

6.Sun LM, Fujino Y., Koga K., *A model of tuned liquid damper for suppressing pitching motions of structures*, Earthquake Engineering & Structural Dynamics, **24**, 5, 625–636, 1995.

7.Tamura Y, Fujii K, Ohtusi T, Wakahara T., Kohsaka R, *Effectiveness of tuned liquid dampers under wind excitations*, Engineering Structures, **17(9)**, 609-621, 1995.

8.Xu, Y.L., Samali, B. and Kwok, K.C.S., *Control of Along Wind Response of Structure by Mass and Liquid Dampers*, Journal of Engineering Mechanics, **17(9)**: 668-675, 1992.

9.Welt,.F and Modi, V.J., *Vibration damping Through Liquid Sloshing: Part I-A Nonlinear Analysis*, Proc Diagnostic, Vehicle Dyn and Special Topics, ASME, Design Engineering Division(DE),**18-5**,148-156, 1989.

10.Welt,.F and Modi, V.J., *Vibration damping Through Liquid Sloshing: Part II-Experimental Results*, Proc Diagnostic, Vehicle Dyn and Special Topics, ASME, Design Engineering Division (DE),**18-5**,157-165, 1989.

11.Yu J, Wakahara T, Reed DA, *A non-linear numerical model of the tuned liquid damper*, Earthquake Engineering and Structural Dynamics, **28**, 671-686, 1999.

12. Gao H., Kwok K.C.S., Samali B., *Optimization of tuned liquid column dampers,* Engineering Structures, **19**, 6, 476-486, 1997

DOI: 10.1051/epjconf/201714302002

# Statistics for comparison of simulations and experiments of flow of blood cells *

K. Bachratá, H. Bachratý, M. Slavík

Cell-in-fluid research group, Department of Software Technologies, Faculty of Management Science and Informatics, University of Žilina, Slovakia, katarina.bachrata@fri.uniza.sk

**Abstract.** In this article we propose statistical method for comparison of simulation and real biological experiments of elastic objects moving in fluid. Our work is focused on future optimization of microfluidic devices used for capture of circulating tumor cells from blood samples. Since the design optimization using biological experiments is both time consuming and expensive, in silico experiments with a broad spectrum of complex and computationally simulations are intensely performed. Necessary verification if simulation models, hitherto mainly realised by comparision of individual cells properties must be extended to more complex simulations. We present our first results with characteristics designed for this purpose.

## 1 Introduction

Investigation of properties of elastic particles in the blood flow is a very broad field of research. Our research group focuses on the area of optimization of microfluidic devices intended to capture circulating tumor cells (CTCs) from blood samples. In this article, we use results of simulation experiments of blood flow through devices with system of internal obstacles and dimensions at the order of tens to hundreds of micrometers. For this purpose we have used a model implemented in an open source scientific package ESPResSo. The model consists of two parts: the fluid representing blood plasma and the immersed objects representing blood cells. The fluid is modeled using the lattice-Boltzmann method and the objects are modeled by the immersed boundary methods implemented by our group. Objects are represented by triangular meshes of their surface and its elastic behavior is governed by five elastic moduli applied between nodes of the mesh. In this study we use only models of red blood cells (RBCs), which are the main component of the solid part of the blood suspen-

* All authors were supported by the Slovak Research and Development Agency [contract number APVV-15-0751].

sion and determine its behavior. More detailed description of our simulation model can be found in [1], [2], [7], [5]. Despite these specific conditions we believe that here presented results can be useful also for other types of experiments.

The main objective of simulations is to replace the biological experiments and bring results and knowledge about devices in cases where the biological experiments would be technologically too complex, too time-consuming or expensive. It is necessary to ensure and continuously maintain the accuracy and consistency of the simulation models and algorithms with the reality. This is accomplished mainly by comparing the simulation behavior to the biological experiments. Also in our research several (successful) comparisons were done, but almost all of them are related to the behavior and characteristics of the individual elastic objects in certain specific situations [9], [8], [4], [6]. Thanks to the research progress and to the increasing computational capability, it is also possible and necessary to verify the quality and precision of the simulation models in experiments with the flow of tens or hundreds of objects in larger devices. In addition to the comparison with biological experiments, there is also the possibility of comparing the results of identical in-silico experiments realised by different simulation tools and methods.

In both cases, our fundamental objective is to find methods, which gives a qualified answer to the question whether two experiments are consistent or not. According our main goals, we focus on results describing behavior and movement of microsized elastic particles in a steady, not turbulent flow. The main challenge is the big amount and rich variety of data needed to be collected from the simulation and then subsequently processed into a form, which allows to judge and measure the match of experiments. We have identified a relatively simple set of data for comparisons that can be obtained as output of simulation algorithms and also as results of image processing of visual records of biological experiments.

An example of the snapshot from real experiment with microfluidic devices is shown in the Figure 1.

Despite their simplicity, the data sufficiently describe the dynamics of the movement of tens or hundreds of objects in the experiment. They incorporate the stochastic features of experiment, the interactions between objects and device. It is clear, that direct comparison with a similar data set from another experiment is not possible. To enable the comparison, the data require a careful and sophisticated statistical processing including significant reduction into a suitable format. The aim of our paper is to present our first experience with designing, realization and implementation of such methods.

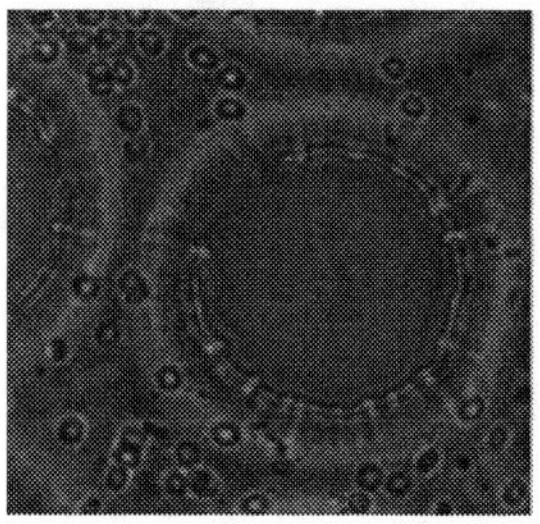

**Fig. 1.** Cells in microfluidic device, snapshot from the real experiment (personal communication with Dr. Giulia Mazza, Danube University Krems, July 2016).

## 2 Description of experiments and data

Simulation experiments described in this work were performed in the microfluidic chamber in form of a box with a height of $30\mu m$ ($z-$axis), width of $50\mu m$ ($y$-axis) and a length of $100\mu m$ ($x-$axis). The simulations included a cell-wall and fluid-wall interaction with top, bottom and sidewalls. To reduce the computational complexity, the simulation was periodic in the $x$ direction and its length could theoretically be unlimited. The cells in performed simulations used the lenght of channel up to $800\mu m$ in the $x$ direction. The assumption that the periodicity does not affect the results was also experimentally verified. Inside the microfluidic device, was placed a simple regular grid of cylindrical obstacles with diameter $10\mu m$ (Figure 2).

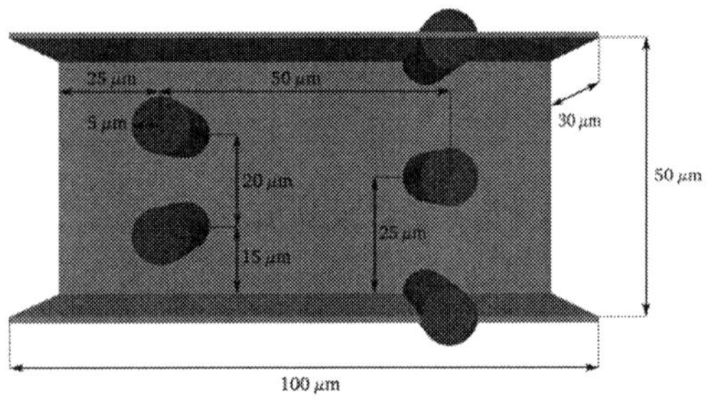


**Fig. 2.** Channel with five obstacles and its positions

Several different initial seedings were used to place 50 RBCs into the 'first period' of device (i.e. at the $x-$positions ranging from $0\mu m$ to $100\mu m$). Each cell was modeled using a triangulation of its surface with 141 nodes and identical elastic properties. In experiments $A$ and $B$ (Figure 3), the initial positions and skew of RBCs were generated randomly. To check the expected differences, in experiment $C$ the cells were placed in regular grid positions and for experiment $D$ we have randomly generated 50 cells (as in $A$, $B$), but with a modified channel topology where the central cylinder in the second row of obstacles was removed.

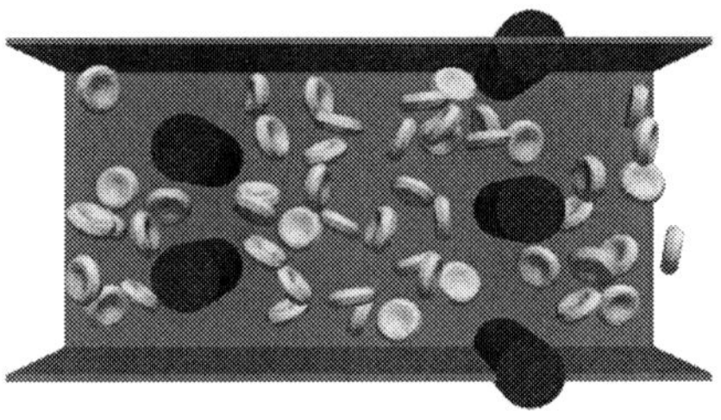

**Fig. 3.** Channel with five obstacles and randomly seeded cells for experiment $A$

The time step of simulation corresponds to $0.2\mu s$ and according to 381 000 simulation steps of each experiment the total simulated time was 0.0762 s. The flow of blood plasma was initialized with fluid density 1, fluid viscosity 1.5 and spatial step $1\mu m$. External volume force 0.0005 was used to induce the flow in the $x$ direction.

### 2.1 Experiment snapshots

We have obtained two sets of data from the simulations. The first set are 'simulation experiment snapshots'. Their meaning and content is motivated by the actual experiments, which are indeed often captured by one or more static images. For each snapshot at a fixed time we acquire the following information about each of the 50 RBCs:

– minimum and maximum value of the $x$, $y$, $z$- coordinates of cell surface (thus a cuboid containing the entire RBC and oriented in the direction of main the axes is defined)
– the position ($x$, $y$, $z$ coordinates) of RBC center, defined as the center of circumscribed cuboid

The data are deliberately very simple. It should be troublefree get their counterparts from digital image processing of biological experiments. In this case, oftentimes $z-$coordinate will be missing. In the following, we will work with all three data dimensions, reducing them to only $x-$ and $y-$ dimension should not bring major problems.

### 2.2 RBC movement tracking

The second set of data is created by tracking the movement of all individual cells throughout the whole simulation. The aim was to continuously monitor particular characteristics of each RBC velocities in simulation. These data for in-silico experiment can be obtained with arbitrary precision with respect to the simulation algorithm step. For our needs it was sufficient to sample data every 200 simulation steps, which corresponds to time of 40 microseconds and RBC movement less than 0.5 micrometers. As the total length of the simulation was 381,000 steps, the movement data of each cell were tabulated at 1,909 points. For each RBC, we have obtained a sequence of the following data:

– The RBC center position ($x$, $y$, $z$ coordinates) calculated as an average of coordinates of all surface triangulation nodes
– $x$, $y$, $z$ component of the RBC center velocity calculated as an average of the corresponding velocity components of all surface triangulation nodes
– absolute speed of the center calculated from $x$, $y$, $z$ speed components
– extremal points of RBC surface in terms of $x-$axis (the 'first' and 'last' surface point as seen the $x$ direction of the channel), $y-$axis ('most left' and 'most right' point) and $z-$axis ('top' and 'bottom' point)
– $x$, $y$, $z$ component of the velocity of the six extremal points of the RBC surface

We expect that these data should be available as outputs from other simulation tools as well. To obtain comparable data from biological experiments (with the possible loss of information about the $z-$coordinate), digitall processing of video experiment or a sufficiently detailed sequence of images must be performed. In any case, it is necessary to ensure the identification and tracking for each individual RBC. In this case, it will be much easier to obtain the necessary data for comparison of in-silico experiments than for comparison of two in-silico to biological experiments.

## 3 Cell-tracking data analysis

### 3.1 Absolute cell velocity

As mentioned above, the values of velocity components along the $x$, $y$ and $z$ axes tabulated every 200 simulation steps were used for calculation of the absolute velocity for each RBC. This way we obtained a sufficiently precise vector of velocities for each of the 50 simulated cells.

The analysis of the individual velocity components indicates that the $x$ and $y$ directions are crucial, and neglecting the $z-$component would not cause a significant change of the following conclusions. This fact is important for comparing in-silico and biological experiments, where often only the top view is recorded and the $z-$axis coordinates information is lost.

The velocity plots (Figure 4) show a large diversity of the data and markedly different behavior of individual cells. To obtain the overall RBC movement characteristics, usage of statistical methods with further reduction of amount and relation of available data is needed.

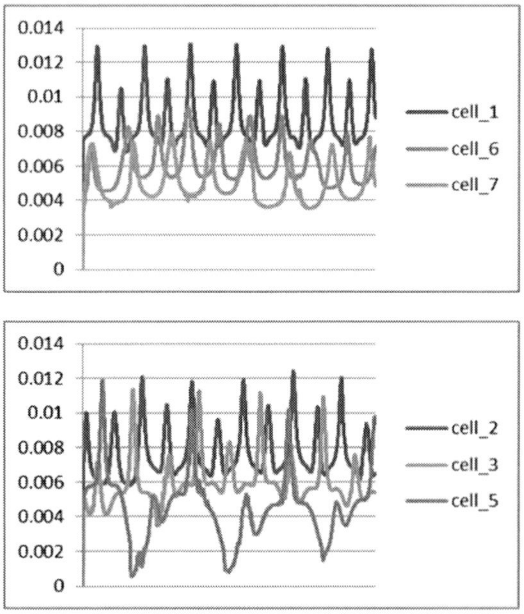


**Fig. 4.** Plot of speed graph for several RBCs

### 3.2 Basic statistical analysis of the velocity data

In the first step, we simply process the basic quantitative attributes of velocity vectors for each individual RBCs. For each velocity record we determined its minimum, maximum, range (i.e. the difference of maximum and minimum) and average of its values. (This average is not identical to the average value of the cell velocity, since the cell trajectory can be just extrapolate by linear curve dependent on the sampling frequency.)

Despite these simplification, the minimal value for each RBC captures the deceleration during a collision with an obstacle or when the cell is moving at the widest point of the channel. The maximum captures high velocity when the cells pass unconflictly the narrow parts of the channel. In range value the variability in cell movement and the overall ratio of 'slow' and 'fast' parts of its path is assessed.

Each of the monitored characteristics was stored in a separate vector and these were independently sorted by the magnitude of their components. These four vectors ($v\_min, v\_max, v\_rng, v\_avg$) represent the first overall characteristics of the movement of elastic objects. The comparison of their charts for simulations *A*-*D* (Figure 5) shows the expected similarity and differences of the individual experiments.

DOI: 10.1051/epjconf/201714302002

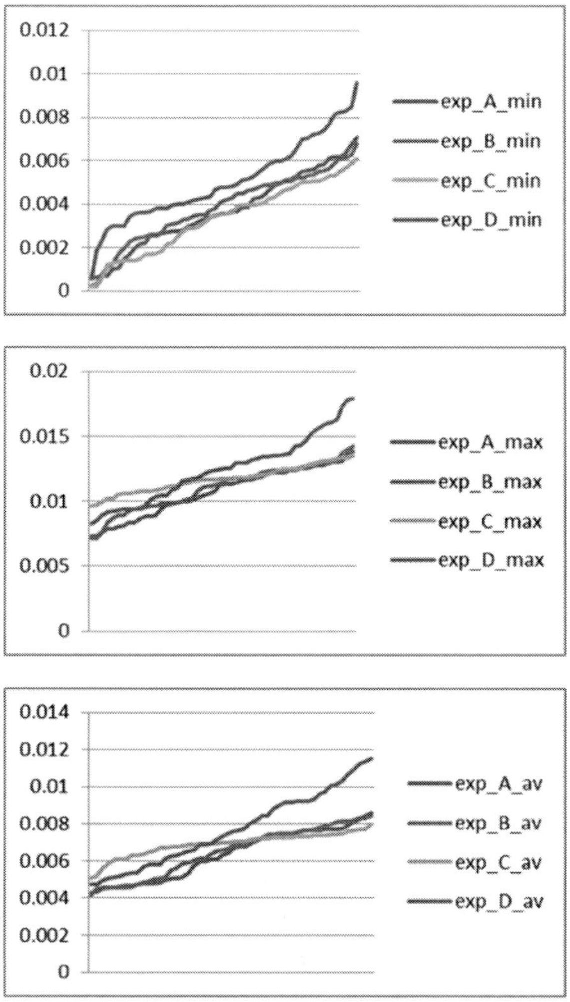


**Fig. 5.** Plot of vectors v_min, v_max and v_aver for simulations *A* to *D*).

We have quantitatively characterized this similarity using the standard Kolmogorov-Smirnov test (KS-test). Results for two of this characteristics are listed in the Table 1 and Table 2. We have denoted as $h_0$ the statistic results, for which we do not reject the hypothesis $H_0$ that the measured data come from the same distribution, using the $\alpha = 0.05$ significance level against alternative $H_1$. If we reject $H_0$, we denote it $h_1$.

**Table 1.** Kolmogorov Smirnov test for $v\_max$ characteristic

| $v\_max$ | A | B | C | D |
|---|---|---|---|---|
| **A** | - | 0.6779 | 0.0044 | 0.0021 |
| **B** | $h_0$ | - | 0.0089 | 0.0089 |
| **C** | $h_1$ | $h_1$ | - | 0.0044 |
| **D** | $h_1$ | $h_1$ | $h_1$ | - |

**Table 2.** Kolmogorov Smirnov test for $v\_aver$ characteristic

| $v\_aver$ | A | B | C | D |
|---|---|---|---|---|
| **A** | - | 0.9541 | 0.0021 | 0.0010 |
| **B** | $h_0$ | - | 0.0021 | 0.0004 |
| **C** | $h_1$ | $h_1$ | - | 0.0001 |
| **D** | $h_1$ | $h_1$ | $h_1$ | - |

### 3.3 Analysis of the periodic characteristics of RBCs movement

The velocity graphs clearly indicate that in addition to the extreme and average values, the periodicity of their behavior is another key feature. It can better characterize the topology of the channel, including the pattern of the obstacle array. We see an accelerated movement of cells in the narrow places and slower movement in wider parts of the channel. The topology is reflected in the trajectory of individual RBCs (how they slow down during collision with an obstacle, how they accelerate when passing obstacles without contact etc.). The irregularities in velocity

DOI: 10.1051/epjconf/201714302002

periodicity correspond to collisions with other objects and their consequences. The number of periods also characterizes the overall velocity of the cell and its total covered distance.

The possibility to design our own criterions for periodicity of RBCs movement was very titilating, but we decided to use the Discrete Fourier Transform (DFT) as a standard method for periodic processes analysis. For each of the 50 RBCs we have determined the three most important frequencies. For most of them, the first one was highly dominant.

This way we obtained a set of 50 dominant and 150 important frequencies. The frecvencies of their values were sorted and saved into vectors $v\_freq\_dom$ and $v\_freq\_imp$. Thus we obtained a second much finer characteristic of the objects in flow velocities. The figures 6 and 7 show $v\_freq\_dom$ and $v\_freq\_imp$ and illustrate the degree of (dis)similarity of this characteristic for separate experiments. Even more significant results may be obtained when the frequencies are aggregated into groups of five (Figure 8). Comparison of experiments $A$ and $C$ shows that initial regular seeding of RBCs simplifies the frequency distribution. Finally, we have performed the KS test to find out whether the data come from the same distribution. This test has confirmed the results observed from graphs.

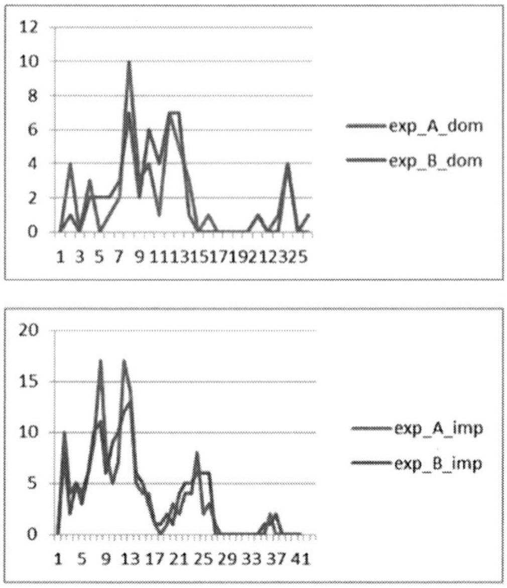


**Fig. 6.** Match in number of dominant and important frequencies for experiments $A$ and $B$

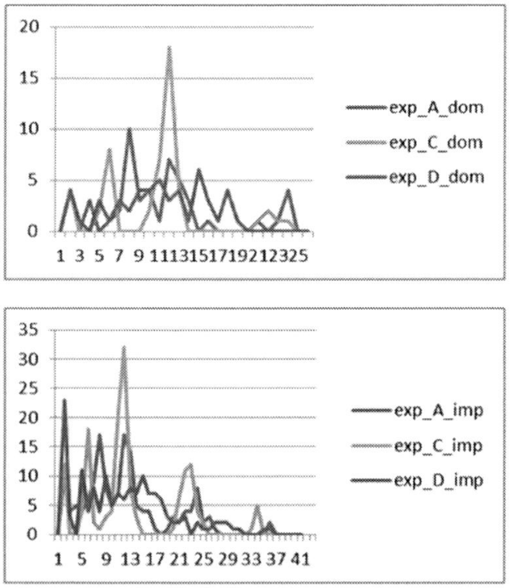


**Fig. 7.** Different course in number of dominant and important frequencies for experiments $A$, $C$ and $D$

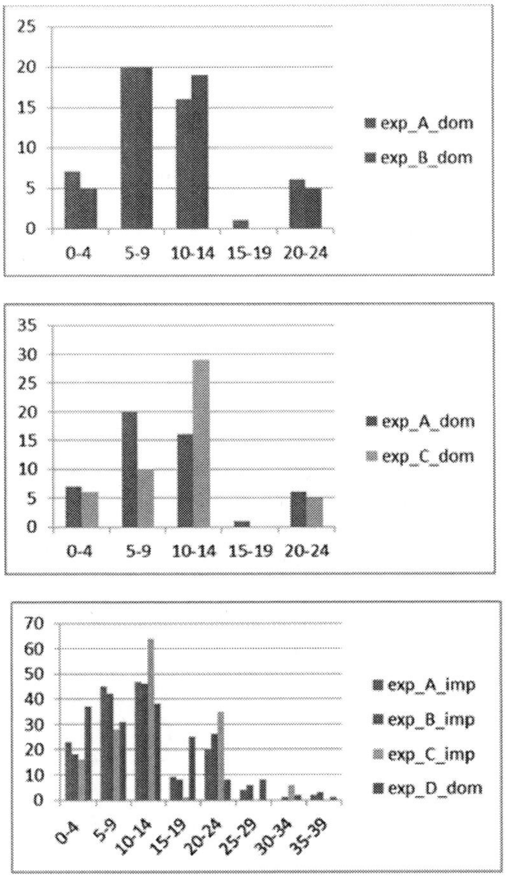


**Fig. 8.** Match of cumulated dominant frequencies for experiments $A$ and $B$, difference in cumulated dominant frequencies for experiments $A$ and $C$, numbers of cumulated important frequencies for all experiments

### 3.4 Analysis of the cell rotation

RBC rotation in the flow is another important characteristic of the experiment. It allows us to capture mainly the quality and quantity of particle interactions with walls and obstacles, their response to a non-constant velocity profile of the fluid flow in the device and in case of dense suspension also the interaction among the particles. Proper modeling of cells rotation is crucial for long-term research of CTC adhesion to the surface of microfluidic devices.

To measure the rotation, we use the velocity component data in at the extremal points of the RBC surface. In this article we focus on one possible combination, which is obviously crucial and allows the analysis even in case of missing data in $z$ direction. If necessary, we prove the possibility to process other combinations of parameters in the same way.

Here, the extremal points were the minimum and maximum point of the cell surface in the $y$ direction ('right' and 'left' point) and we considered the $x-$component of velocity at these points. If these two values are equal, the interpretation is that the RBC does not rotate in the $xy-$plane. In the experimental data these velocities mostly differ. Their difference corresponds to the instantaneous rotation of the cell and the sign of the difference to the spin of rotation. Therefore, we have chosen the vector of differences of the $x-$components of velocities in $y-$extremal points as the basis for further statistical analysis. This vector is available for each of the 50 RBCs and its length corresponds to the number of recorded values of cell motion.

The graphs of these differences (Figure 9) are very informative. One can observe significantly periodic behavior, surprisingly frequent changes in spin of rotation, as well as a large variation of this characteristic. This situation is very similar to the processing of vectors of absolute velocities of the RBCs in part 3.3, so we have also used similar methods for processing of gained data.

DOI: 10.1051/epjconf/201714302002

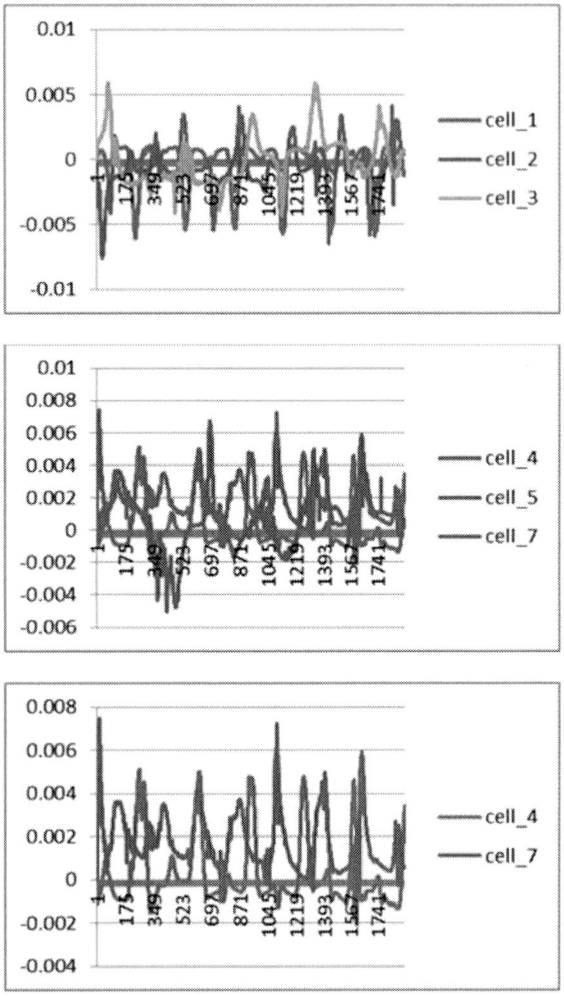


**Fig. 9.** Plots of rotation speed for several RBC's

First we have found the minimum and maximum value for each RBC (i.e. the maximum rotation in both spin directions), the average of the recorded values, the number of changes of spin and an integral (area between the curves of the $x-$component of the velocity at the $y-$extremal points). The latter value should correspond to the overall rate of rotation. (In fact, the integral was calculated as an integral sum of the absolute values of the tabulated velocity differences with width of 200 simulation steps and numericaly estimates this integral.)

The obtained vectors *rot_min*, *rot_max*, *rot_avg*, *rot_switch* and *rot_integral* with length 50 were independently sorted by the magnitude of their elements. These five vectors are the first rotation characteristic of the whole experiment. Their plots (Figure 10) confirm the expected (dis)similarities of experiments supported also by the KS-test.

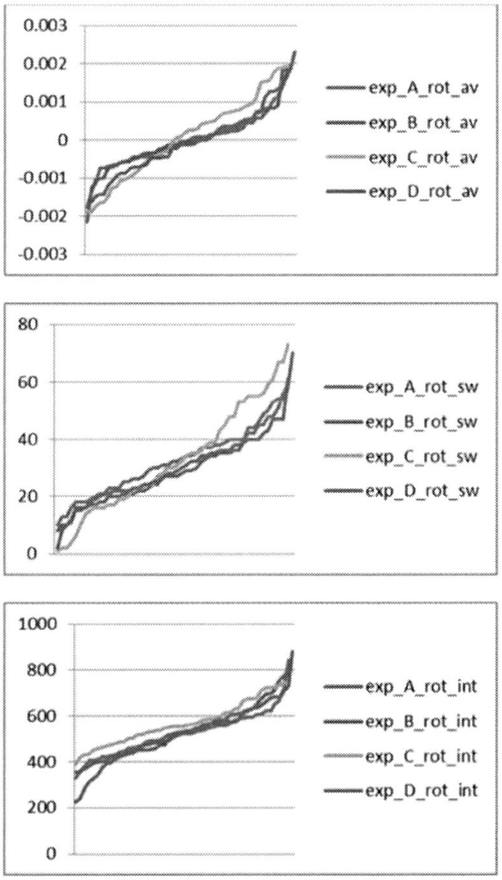


**Fig. 10.** Plots of rotational speeds, numbers of rotational shifts and integral for experiments $A$, $B$, $C$ and $D$

DOI: 10.1051/epjconf/201714302002

Another characteristic focused on periodic behavior of rotation velocity was investigated using DFT similarly as in the part 3.3. The results are shown in the following graphs (Figures 11 and 12) as frecvency vectors $rot\_freq\_dom$ and $rot\_freq\_imp$. Surprisingly, there is a great similarity between the frequencies of velocity differences and absolute velocities, and as expected, the number of rotations changes $rot\_switch$ corresponds to double of frequencies $rot\_freq\_dom$ values.

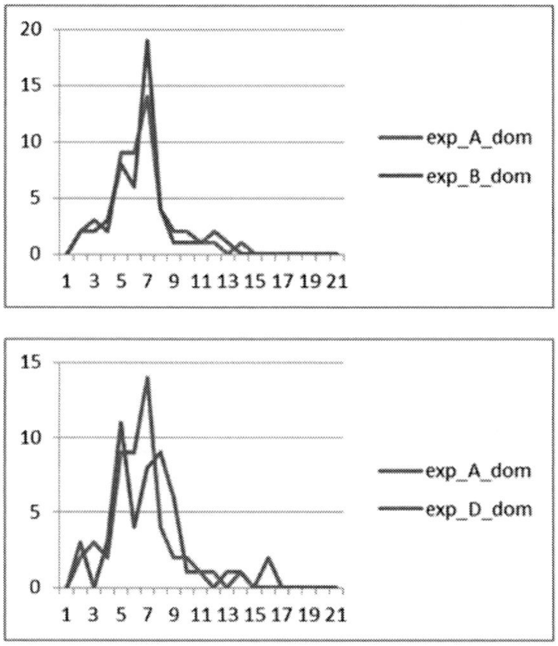


**Fig. 11.** Match of number of dominant frequencies for experiments $A$ and $B$ and their differencies for experiments $A$ and $D$

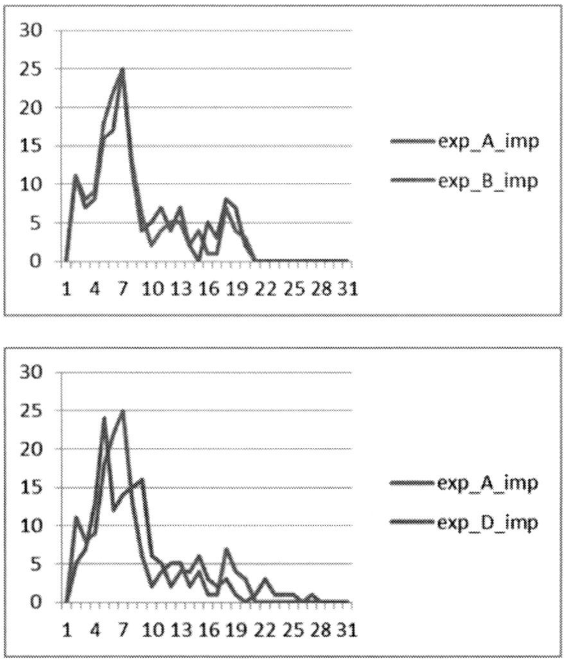


**Fig. 12.** Match of number of important frequencies for experiments $A$ and $B$ and their differences for experiments $A$ and $D$

The last graphs (Figure 13) show the results of a simple comparative experiment for some selected cells. If the idea of measuring the RBC rotation is correct, the differences in $x-$components of velocity in the $y-$extremal points should correspond to the differences in $y-$components of velocity in the $x-$extremal points ('first' and 'last' point of the cell surface). Pictured characteristics confirm the assumption.

Some differences arise just when part of the RBC is fixed or slowed down as a result of contact with an obstacle or with another object. These results are illustrated characteristics for three randomly selected RBCs.

DOI: 10.1051/epjconf/201714302002

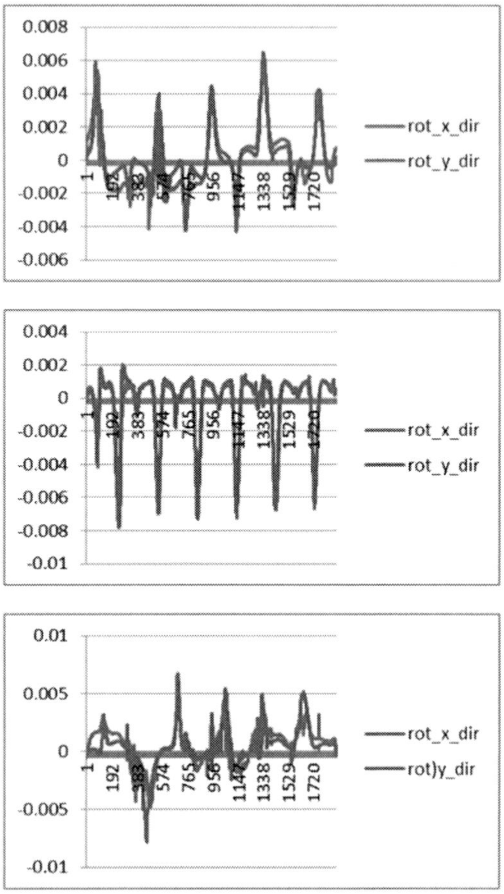


**Fig. 13.** Difference of rotational speed in $x-$axis and $y-$axis direction

## 4 Snapshot Data Analysis

As mentioned above, the data obtained from the snapshots are much simpler, static and provide fewer options for the characterization of cell motion during the simulation experiment. On the other hand, this kind of data is much easier to obtain using digital image processing of biological experiments. During our simulation experiments we acquire the snapshots after every 1000 simulation steps, but we have used only a small partion of them. For monitoring the development and stabilization of the individual characteristics, we have used snapshots after every 100,000 steps, and after 300,000 steps, we have recorded the snapshot every 10,000 steps.

### 4.1 Location of the cells in the device

The first characteristic of the cell flow comes from data of the center position of individual RBCs (center of the circumscribed box with sides parallel to the axes). In this section we limit ourselves to analysis of their $x-$coordinates. If necessary, the method can be extended to the other coordinates. At the beginning of the simulation, the $x-$coordinates of RBCs range from $0\mu$m to $100\mu$m. At the end of the simulation, after 381,000 simulation steps, depending on the starting position, trajectory and number of collision, the cells $x-$coordinates range between $352\mu$m and $718\mu$m. The following figure (Figure 14) illustrates the developement of $x-$coordinates values for the whole RBCs 'bunch' during the simulation run.

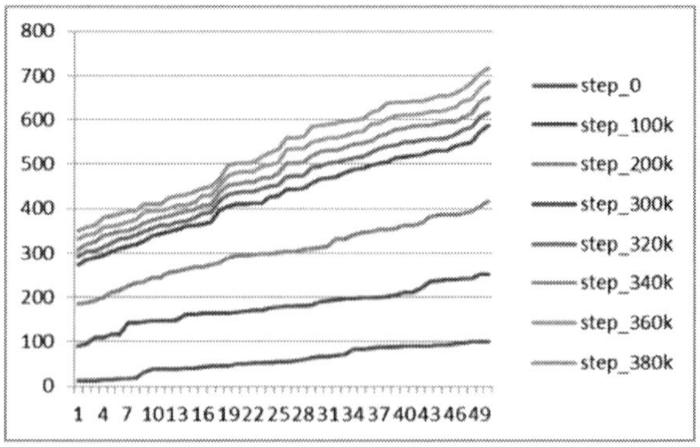


**Fig. 14.** Plot of $x-$coordinates of RBC's center for different simulation steps (in thousands)

However, this characteristic is not our goal. In the biological experiments, we expect a much larger microfluidic device (with a periodic obstacle array) and the snapshot containing a large number of elastic objects continuosly filling interior of the device. Therefore, not the global RBC positions, but their positions in typical regions of obstacle array will by the important characteristic. Those would have to be defined specifically depending on the array type.

For our experiment, we have processed the RBCs $x$ coordinates in two steps. First, we replaced them by their remainder after devision by 100, which 'shifts' all coordinates to a common rectangle representing the basic periodic block of the channel. Then we divided them into three typical zones: entry zone which consists of two $10\mu m$ strips (15 to $25\mu m$ and 65 to $75\mu m$), through which the RBCs enter the narrow part in front of the obstacle belt. The exit zone consists of two $10\mu m$ strips (25 to $35\mu m$ and 75 to $85\mu m$), through which the cells leave these narrow parts. Third transit zone corresponds to the remaining 60 micrometers of the channel length without obstacles with a maximum flow space. The entry zone corresponds to the area with fast acceleration of the fluid flow, where also cell-cell and cell-wall collisions are the most frequent. Output zone is the area of cell flow deceleration, in which the collision situations are much less frequent. The transit zone correspond to the relatively steady state with a smooth flow of cells.

Our next proposed characteristic of the cell flow is the distribution of cells count in these zones. The first graph (Figure 15) shows the time-evolution of this characteristic for the experiment $A$ using the snapshots after 300k, 320k, 340k, 360k, 380k and 381k simulation steps. The distribution look quite stable, which confirms the suitability of this characteristic. The second graph compares these distributions for experiments $A$, $B$, $C$, $D$ after 381k simulation steps with high degree of similarity.

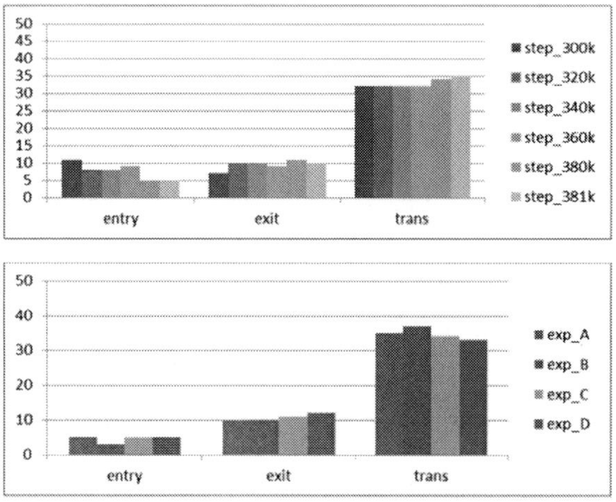


**Fig. 15.** Number of RBCs in zones for experiment $A$ during different time snapshots, comparison of numbers of cells in zones in simulation step 381k for experiments $A$, $B$, $C$ and $D$

A similar statistics can be created with lower respect to the specific shape of the experimental device. The following figures (Figure 16) show similar results for the distribution of cells count in sketchy created $10\mu m$ zones, covering whole length of basic chanel box. The values are not as stable as in the previous distribution, however, this may be due to the small number of RBCs in the experiment. We have performed a control experiment identical to the previous one but including only 20 RBCs. It turns

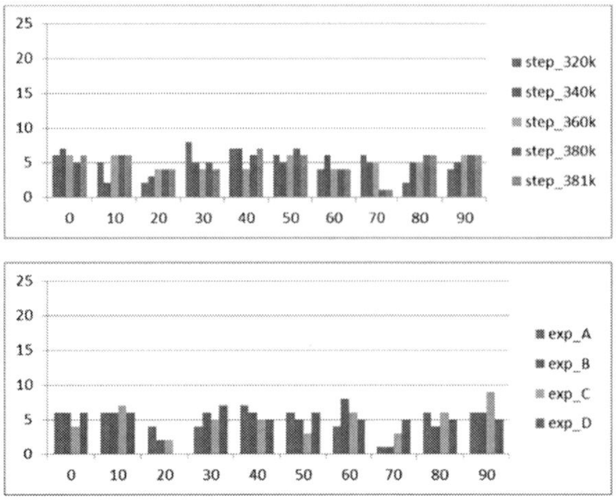


**Fig. 16.** Number of RBCs in $10\mu m$ bands for experiment $A$ during different time snapshots, comparison of numbers of cells in bands in simulation step 381k for experiments $A$, $B$, $C$ and $D$

out that the spatial distribution of cells has unsteady but periodic nature. This effect disappears when the number of cells in the model is increased to 50.

For our type of channel topology and flow within the channel, the $x-$coordinates of the RBCs are crucial for description of the main directions of the cells movement. Another control experiment has shown that similar results may be obtained in an analogous analysis of the $y-$coordinates of cell positions.

### 4.2 Cell skew in the device

In section 3.4, we have described how important the investigation of the cell rotation characteristics is. The snapshots give us only limited options in this field, because they do not capture the dynamics of the simulations. Therefore, we have proposed a method for describing the RBC skew using the dimensions of a bounding box, defined as a cuboid containing the entire RBC and oriented in main axes direction, as mentioned in section 2.1. If the RBC is oriented by one way in the directions of the $x$, $y$, $z$ axes, the dimensions of the box are the same as the dimensions of the RBC, i.e. $7.82\mu m \times 7.82\mu m \times 2.56\mu m$. When the cell is skewed, the bounding box has different dimensions, which represent the rotational characteristics of the cell.

The obtained bounding box sizes were processed in two ways. For the simpler one, the values of the $x$, $y$ and $z$ dimensions (after 380k simulation steps) for all cells were independently sorted and stored in arrays *cub_x*, *cub_y* and *cub_z*. These arrays form the first proposed characteristic. The graph (Figure 17) show a significant match of *cub_x* vectors for experiments $A$, $B$, $C$ and $D$, but for the values of *cub_z* vector some difference between the couple $A$, $B$ and couple $C$, $D$ of experiments arise.

It relates to interesting observation, that the distributions of $x$, $y$ and $z$ dimension of cuboids in a single simulation are not the same, i.e. the RBC skew is different for different axes. Comparing the graphs for simulation $A$ and $D$ (Figure 18) shows, then the differences between *cub_x*, *cub_y* and *cub_z* distribution (mind the switch of *cub_y* and *cub_z* graphs position) can characterize the experiment too.

The second skew characteristic measures nearly standardly the normalized square differences of the current bounding box dimensions and dimensions of the bounding box of RBC aligned with the coordinate axes. The small difference consist in fact, that the values of the box

 DOI: 10.1051/epjconf/201714302002

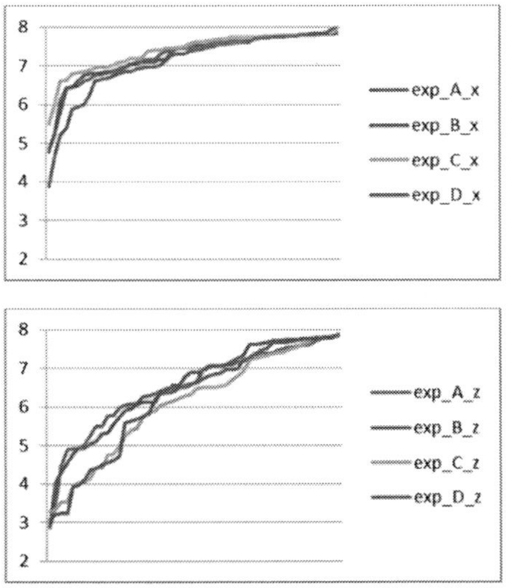


**Fig. 17.** Plots of $x-$sizes and $z-$sizes of cuboids for experiments $A$, $B$, $C$ and $D$

dimensions are first sorted by size. Than the two largest values are compared to the diameter of RBC ($7.82\mu$m) and the smallest one is compared to the thickness of the RBC ($2.56\mu$m). In this way, in order to simplify, we intentionally decrease mentioned difference of the dimension in the direction of the individual axes. Nevertheless, the acquired characteristic still has very good properties. The measured values are stored in a vector *cub_sq_err*. After being sorted by size, this vector represents another skew characteristic of the RBC in the snapshots of simulation experiments.

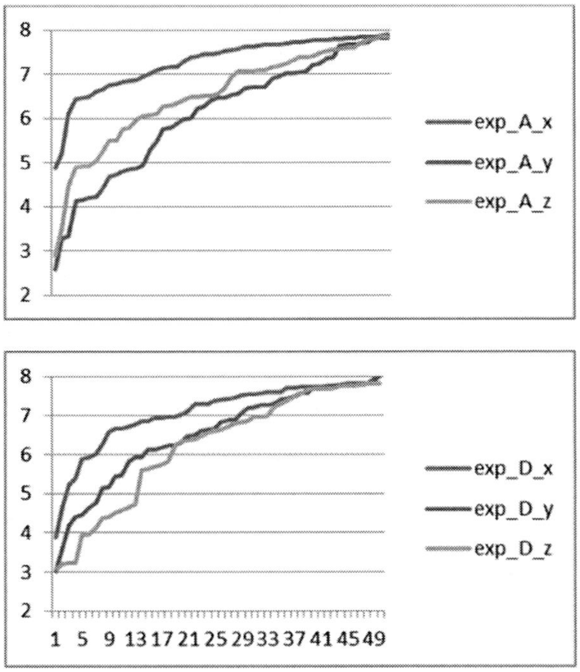


**Fig. 18.** Comparison of $x$, $y$ and $z-$sizes of cuboids for experiments $A$ and $D$

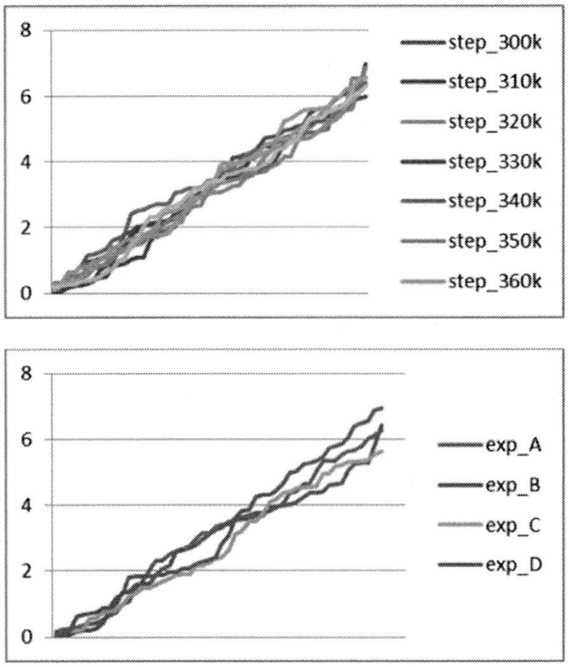


**Fig. 19.** Stability of characteristic for all snapshots from experiment $A$, comparison of square errors of cuboid rotation for experiments $A$, $B$, $C$ and $D$

The first graph (Figure 19) illustrates the stability of the vector in snapshots between 300,000 and 380,000 simulation steps for simulation $A$. The second graph shows a comparison for the snapshots of experiments $A$, $B$, $C$ and $D$ after the 380,000 simulation steps.

## 5 Used technologies

For the simulation experiments, we have used the version 3.2.0 of computational package *Esspresso*, especially its $OIFframework$. The generated data files were subsequently processed by algorithms developed in $C\#inVisualStudio2010$ into the data files suitable for presentation of results and further statistical processing. The statistical processing was done in *Matlab*, version 7.11.0.

The following simulation parameters were used: $fluidfriction\_coeff = 0.025865531914893616$, $fluid\_density = 1.0$, $fluid\_viscosity = 1.5$, $fluid\_force = 0.0005$, $timestep = 0.2microseconds$, $lb\_agrid = 1.0$, $number\_simulation\_steps = 381,000$.

The RBC was modeled with standard triangulation with 141 nodes and with its external dimensions of $7.82\mu m \times 7.82\mu m \times 2.56\mu m$.

For the elastic modulus of RBC, the used coefficients were $ks = 0.0044$, $kb = 0.0715$, $kal = 0.005$, $kag = 1$, $kv = 1.25$.

## 6 Conclusions

The proposal of statistical methods for mutual comparison of conformity or similarity within the simulations and real experiments, with the flow of a large number of objects, are in our opinion a relatively new area of research. The characteristics suggested in the article represent our first attempt at defining and processing methodology. The presented results indicate their stability and satisfactory resolution of the proposed methods. They become a good starting point for further development of the topic. It should focus on the following areas:

- More detailed verification of stability of characteristics and verification of distinguishing capabilities in a case of a fine changes of the parameters of the experiment,
- Collection of comparable data from real experiments and simulations of other simulation tools, and comparing the characteristics.
- Further development and widespread use of characteristics.
- Extending the use of the characteristics to measure the quality of simulations in terms of optimizing the properties of microfluidic devices.

## Acknowledgements

The authors acknowledge the experimental data from Dr. Giulia Mazza, Center for Integrated Sensor Systems, Danube University Krems, Austria. Her work is supported by the NÖ Forschungs- und Bildungsges.m.b.H. (NFB) through the Life Science Calls (Project ID: LSC13-024).

## References


1. K. Bachratá and H. Bachratý, On modeling blood flow in microfluidic devices, ELEKTRO 2014: 10th International Conference, IEEE, ISBN 978-4799-3720-2, 2014, pp. 518-521
2. I. Cimrák, K. Bachratá, H. Bachratý, I. Jančigová, R. Tóthová, M. Bušík, M. Slavík and M. Gusenbauer, Object-in-fluid framework in modeling of blood flow in microfluidic channels, Comunications, Scientific Letters of the University of Zilina, vol. 18/1a, 2016, pp. 13-20
3. M. Bušík, I. Jančigová, R. Tóthová and I. Cimrák, Simulation study of rare cell trajectories and capture rate in periodic obstacle arrays, Journal of Computational Science, 2016
4. X. Li, Z. Peng, H. Lei, M. Dao and G. E. Karniadakis, Probing red blood cell mechanics, rheology and dynamics with a two-component multi-scale model, Philosophical Transactions Series A, Mathematical, Physical and Engineering Sciences, 2014
5. D. A. Fedosov, J. Fornleitner, J. L. McWhirter, K. Müller, H. Noguchi, M. Peltomäki and G. Gompper, Blood Flow in silico: From Single Cells to Blood Rheology, 4th Micro and Nano Flows Conference UCL, 2014
6. R. Tóthová, I. Jančigová and M. Bušík, Calibration of elastic coefficients for spring-network model of red blood cell, International Conference on Information and Digital Technologies (IDT), 2015, pp. 376-380
7. I. Cimrák, M. Gusenbauer and I. Jančigová, An ESPResSo implementation of elastic objects immersed in a fluid, Computer Physics Communications, vol. 185, 2014, pp. 900-907
8. H. Basu, A. K. Dharmadhikari, J. A. Dharmadhikari, S. Sharma and D. Mathur, Tank treading of optically trapped red blood cells in shear flow, Biophysical Journal, vol. 101, 2011, pp. 1604-1612
9. K. Tsubota and S. Wada, Elastic force of red blood cell membrane during tank-treading motion: Consideration of the membrane's natural state, International Journal of Mechanical Sciences, vol. 52, 2010, pp. 356-364

# Simulation of the acoustic wave propagation using a meshless method

*J.* Bajko[1a], *P.* Niedoba[1], L. Čermák[1], and *M.* Jícha[1]

[1]Energy Institute, Department of Thermodynamics and Environmental Engineering, Brno University of Technology, Technická 2896/2, 616 69 Brno, Czech Republic

**Abstract.** This paper presents numerical simulations of the acoustic wave propagation phenomenon modelled via Linearized Euler equations. A meshless method based on collocation of the strong form of the equation system is adopted. Moreover, the Weighted least squares method is used for local approximation of derivatives as well as stabilization technique in a form of spatial filtering. The accuracy and robustness of the method is examined on several benchmark problems.

## 1 Introduction

Numerical study of sound generation and propagation phenomenon plays a major role in Computational Aeroacoustics (CAA), especially in predictions of sound generated by turbulent flows and its environmental consequences. This paper is aiming to contribute to the development of suitable methods for sound propagation simulations governed by the Linearized Euler equations (LEE), cf. [1,2].

Various numerical mesh-based methods, cf. [3,4], have been investigated in this context, but there is also growing interest in meshless methods, cf. [5–11], due to their potential advantages. Depending on applications, these methods can be superior in accuracy, stability, robustness and can deal with complex geometries, multiphase flow, problems with free boundaries and moving objects. In this paper, the development of a meshless method adjusted for linear systems is outlined.

Firstly, LEE as a linear hyperbolic system is introduced, followed by the Weighted Least SQuares (WLSQ) method which is employed in two ways - for local interpolation of acoustic variables and for spatial filtering, cf. [12,13]. The WLSQ filtration technique serves primarily for the stabilization of the numerical scheme and can be utilized under the assumption that the governing equations are linear and the solution is smooth enough.

Finally, abilities of the method are demonstrated on various benchmark problems for the acoustic wave propagation, cf. [14–16]. The method is tested on unstructured point distributions and several types of boundary conditions (BC) are implemented, namely the non-reflecting BC realized by the Perfectly Matched Layer (PML), cf. [17–19].

## 2 Linearized Euler equations

The hyperbolic non-homogeneous LEE represent one of the suitable models for the acoustic wave propagation phenomenon. We consider the 2D matrix form ($\mathbf{x} = (x, y) \in \Omega$, $t > 0$) written as

$$\frac{\partial \mathbf{w}'}{\partial t} + \mathbb{A}_1(\mathbf{w}_0)\frac{\partial \mathbf{w}'}{\partial x} + \mathbb{A}_2(\mathbf{w}_0)\frac{\partial \mathbf{w}'}{\partial y} = \mathbf{S}, \qquad (1)$$

where

$$\mathbf{w}'(\mathbf{x},t) = \begin{pmatrix} \rho'(\mathbf{x},t) \\ u'(\mathbf{x},t) \\ v'(\mathbf{x},t) \\ p'(\mathbf{x},t) \end{pmatrix}, \quad \mathbf{w}_0(\mathbf{x}) = \begin{pmatrix} \rho_0(\mathbf{x}) \\ u_0(\mathbf{x}) \\ v_0(\mathbf{x}) \\ p_0(\mathbf{x}) \end{pmatrix}, \qquad (2)$$

Vector function $\mathbf{w}'(\mathbf{x}, t)$ denotes the time dependent acoustic variables (density, velocity components and pressure) and $\mathbf{w}_0(\mathbf{x})$ steady mean flow variables corresponding to the underlying flow field.

Jacobian matrices of the system (1) are given as

$$\mathbb{A}_1 = \begin{pmatrix} u_0 & \rho_0 & 0 & 0 \\ 0 & u_0 & 0 & \frac{1}{\rho_0} \\ 0 & 0 & u_0 & 0 \\ 0 & \gamma p_0 & 0 & u_0 \end{pmatrix}, \mathbb{A}_2 = \begin{pmatrix} v_0 & 0 & \rho_0 & 0 \\ 0 & v_0 & 0 & 0 \\ 0 & 0 & v_0 & \frac{1}{\rho_0} \\ 0 & 0 & \gamma p_0 & v_0 \end{pmatrix}, \qquad (3)$$

where $\gamma = 1.4$ is the ratio of specific heats and $\mathbf{S} = \mathbf{S}(\mathbf{x}, t)$ represents the acoustic source term, cf. [1,2].

The initial–value boundary problem for LEE then consists of the equation system (1) with the initial and boundary condition

$$\text{(IC)} \quad \mathbf{w}'(\mathbf{x},0) = \mathbf{w}'_{\text{in}}(\mathbf{x}), \quad \mathbf{x} \in \Omega, \qquad (4)$$

$$\text{(BC)} \quad \mathbf{w}'(\mathbf{x},t) = \mathbf{w}'_{\Gamma}(\mathbf{x},t), \ \mathbf{x} \in \Gamma = \partial\Omega, \ t > 0. \qquad (5)$$

[a] *Present address:* bajko@fme.vutbr.cz

## 3 Meshless method with WLSQ filtering

In the context of meshless methods we distiguish between the global cloud $\hat{\Omega}$ defined as a set of $n$ points discretizing a domain of interest $\overline{\Omega}$ and local clouds $\hat{\Omega}_i, i = 1, \ldots, n$ that are calculated in the pre-processing step of the simulation. An $i$-th local cloud usually consists of a set of $n_i$ neighbours chosen by a certain rule. Practically, efficient algorithms for nearest neighbor search such as the k-d tree are used. Therefore, the number of points $n_i$ is controlled by the radius of an open ball or it can be prescribed manually.

### 3.1 Weighted Least Squares Method

For every local cloud $\hat{\Omega}_i$ we wish to find a local approximation $\hat{w} : \Omega_i \to \mathbb{R}$ in the form

$$\hat{w}(\mathbf{x}) = \sum_{l=1}^{m} \alpha_l p_l(\mathbf{x}) = \mathbf{p}^T(\mathbf{x})\boldsymbol{\alpha} \tag{6}$$

where

$$\mathbf{p}^T(\mathbf{x}) = (p_1(\mathbf{x}), p_2(\mathbf{x}), \ldots, p_m(\mathbf{x})) \tag{7}$$

and

$$\boldsymbol{\alpha} = (\alpha_1, \alpha_2, \ldots, \alpha_m)^T \in \mathbb{R}^m. \tag{8}$$

Functions $p_l : \mathbb{R}^d \to \mathbb{R}$, $l = 1, \ldots, m$, form a basis $\mathcal{B}$ of an approximation space $\mathcal{F} = \text{span}(\mathcal{B})$. The complete (multivariate) polynomial basis of degree $\nu$, cf. [5,10] is adopted in this work. For $d = 2$ and $\nu = 3$, the basis

$$\mathcal{B} = \left\{1, x, y, x^2, xy, y^2, x^3, x^2y, xy^2, y^3\right\}. \tag{9}$$

### 3.2 WLSQ approximation

It is well known that WLSQ method finds the coefficients $\boldsymbol{\alpha}$ in (6) by minimizing the objective function $J(\boldsymbol{\alpha})$ with weights

$$J(\boldsymbol{\alpha}) = (\mathbf{P}\boldsymbol{\alpha} - \mathbf{w})^T \boldsymbol{\Phi} (\mathbf{P}\boldsymbol{\alpha} - \mathbf{w}), \tag{10}$$

where $\mathbf{P}$ denotes the moment matrix of type $(n_i \times m)$, $\boldsymbol{\Phi}$ the diagonal weight matrix of order $n_i$ and $\mathbf{w}$ the vector of given function values. The weighting function $\phi : \mathbb{R}^d \to \mathbb{R}$ prescribes the contribution of every point to $J(\boldsymbol{\alpha})$ in the local cloud according the distance to the star point $\mathbf{x}_1^i$, cf. [10].

Solution to the system of *normal equations*

$$\mathbf{P}^T\boldsymbol{\Phi}\mathbf{P}\boldsymbol{\alpha} = \mathbf{P}^T\boldsymbol{\Phi}\mathbf{w}. \tag{11}$$

then yields

$$\boldsymbol{\alpha} = \left(\mathbf{P}^T\boldsymbol{\Phi}\mathbf{P}\right)^{-1}\mathbf{P}^T\boldsymbol{\Phi}\mathbf{w} \tag{12}$$

assumming the invertibility of the above mentioned matrix. We suppose that the number of points $n_i$ in $\hat{\Omega}_i$ is greater than the number of basis functions $m$, i.e. $n_i > m$ holds.

### 3.3 WLSQ interpolation

By adding the *interpolation constraint*

$$\hat{w}(\mathbf{x}_1^i) = w_1, \tag{13}$$

to the Normal equations (11), the value $w_1$ is reproduced at the *star point* $\mathbf{x}_i$ exactly. This modification can be gently implemented and allows to use the WLSQ method for multiple purposes, namely the spatial discretization of the governing equations (interpolation) as well as numerical filtering (approximation).

### 3.4 Spatial discretization of LEE

In order to obtain the semi–discrete form of governing equations (1), the collocation method is adopted, i.e. the expression

$$\left[\frac{\partial \hat{\mathbf{w}}'}{\partial t} + \mathbb{A}_1\frac{\partial \hat{\mathbf{w}}'}{\partial x} + \mathbb{A}_2\frac{\partial \hat{\mathbf{w}}'}{\partial y} - \mathbf{S}\right]_{\mathbf{x}=\mathbf{x}_i} = 0 \tag{14}$$

has to be satisfied. Acoustic variables $\hat{\mathbf{w}}_i$ are locally approximated using WLSQ method with interpolation condition described in section (3.1).

The collocation then leads to a system of ordinary differential equations (ODE) in form

$$\frac{\mathrm{d}\mathbf{w}_i(t)}{\mathrm{d}t} = \mathbf{RHS}_i(t) + \mathbf{S}(\mathbf{x}_i, t) \tag{15}$$

where $\mathbf{RHS}_i(t)$ contains contributions to the right-hand side from all neighbours of the star point $\mathbf{x}_i$, cf. [7,10,11]. The interpolation condition in the WLSQ approach allows to simplify the left side of the equation system (15) avoiding the necessity to solve the system of $n$ linear equations. Therefore, using WLSQ with interpolation condition greatly reduces the computation costs and improves stability, cf. [10,11]. The system of ODE (15) is then solved with high order low-dissipation and low-dispersion runge-kutta scheme optimized for wave propagation problems, cf. [20].

### 3.5 Numerical filtering

The numerical scheme (15) can be viewed as a generalized finite differences scheme based on central differences which is not stable in general, cf. [8,9]. For linear problems, e.g. the acoustic wave propagation described by LEE, the stabilization can be gently achieved in the context of WLSQ method used as a filter. In general, filters based on least squares minimization are known as *Savitzky–Golay filters*, cf. [12,13]. It is not only the stability, but also robustness and accuracy of the meshless method is then affected by the WLSQ filter.

The $i$-th local cloud $\hat{\Omega}_i$ is utilized for filtering after every time step in a way that $\mathbf{w}_i(t) \approx \mathbf{w}'(\mathbf{x}_i, t)$ is replaced by the filtered value

$$\mathbf{w}_i(t) = \mathbf{p}^T(\mathbf{x}_i)\boldsymbol{\alpha} \tag{16}$$

at the star point $\mathbf{x}_i$.

### 3.6 Boundary conditions

Firstly, solid–wall slip BC is prescribed at points where the acoustic waves should be reflected. The acoustic velocity vector $\mathbf{v}' = (u', v')$ at the boundary point $\mathbf{x} \in \Gamma$ is forced to stay orthogonal to the normal vector $\mathbf{n} = (n_x, n_y)$, i.e.

$$\mathbf{v}' \cdot \mathbf{n} = 0. \tag{17}$$

Secondly, non–reflecting BC is prescribed at domain boundaries, where waves should leave the computational domain without any reflection. It can be achieved artificially by designing a layer of points around the domain of interest that will affect the solution in such a way that the major spectrum of incoming waves will dissipate. The PML is implemented in our numerical experiments, cf. [17–19].

## 4 Numerical experiments

### 4.1 2D Wall-reflected acoustic pulse problem

This benchmark problem is designed to verify the implementation of the wall boundary condition and the accuracy and stability of a numerical method when a reflection of an acoustic wave from the solid wall occur. Let us consider a 2D domain $\Omega = (-100, 100) \times (0, 150)$ that is depicted in Fig. 1 together with an initial pulse located at $(x_p, y_p) = (0, 25)$. The wall boundary condition is prescribed at line $y = 0$. We suppose the underlying medium at rest, which implies that the mean flow velocity equals zero and therefore the Mach number $M = 0$ and for $\mathbf{w}_0$ it holds

$$\rho_0 = 1, \quad u_0 = v_0 = 0, \quad p_0 = \frac{\rho_0}{\gamma} = \frac{5}{7}. \tag{18}$$

The Initial-value boundary problem consists of the equation system (1) with the source term $\mathbf{S} = \mathbf{0}$, above mentioned boundary conditions and initialized by the initial condition $\mathbf{w}'(x, y, 0)$ given as follows

$$\mathbf{w}'_{\text{in}}(x, y) = \varepsilon \exp(-\kappa r_p^2)\,(1, 0, 0, 1)^T, \tag{19}$$

where the radius $r_p = \sqrt{(x - x_p)^2 + (y - y_p)^2}$ and $\kappa = (\ln 2)/b^2$. The amplitude and the half-width of the acoustic initial pulse are determined by $\varepsilon = 1$ and $b = 5$, respectively.

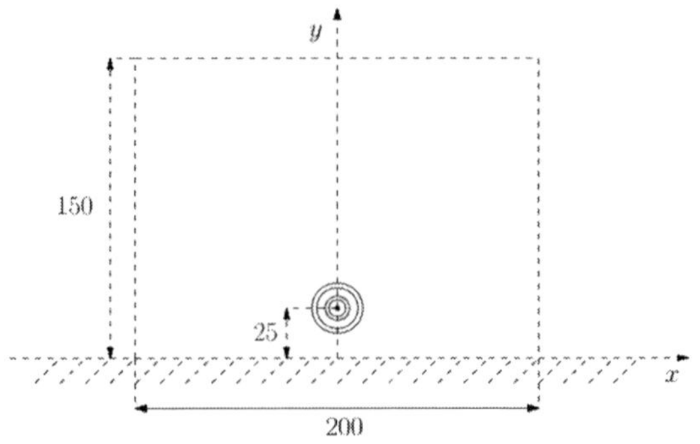


**Fig. 1.** Rectangular domain $\Omega$, wall (solid line) and external boundary (dashed line).

#### 4.1.1 Numerical solution

Results of the acoustic wave simulation in terms of the acoustic pressure contours are plotted ase a series of Fig. (2–5) at particular times $T = 10, 25, 50, 75$. The time integration was performed with time step $dt = 0.025$ that corresponds to the unstructured point distribution with $n = 36006$ number of points.

**Fig. 2.** Ac. pressure $p'(x, y, T)$ contours at time $T = 10$.

The comparison of the acoustic pressure $p'(x, y, T)$ along the line $x = 0$ and analytical solution at time $T = 75$ is shown in Fig. 6.

### 4.2 2D Convected acoustic monopole

The domain of interest for the following benchmark problem is simple square $\Omega = (-100, 100)^2$ extended by the PML $\Omega_e$, cf. Fig. 7. Furthermore, stretching of points towards outlet boundaries and artificial dissipation is applied in order to suppress all incoming waves and eliminate spurious reflections.

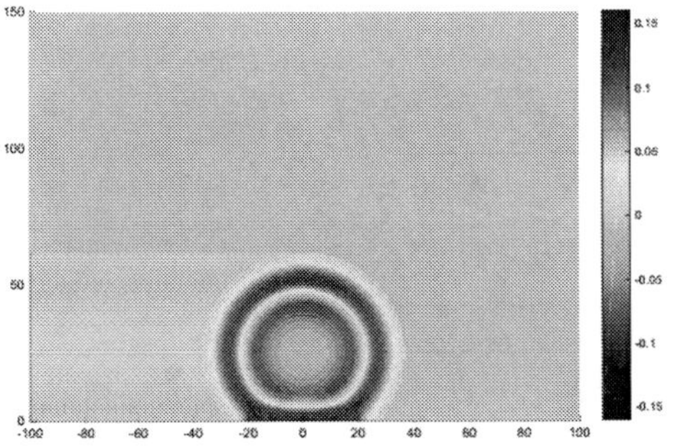

**Fig. 3.** Ac. pressure $p'(x, y, T)$ contours at time $T = 25$.

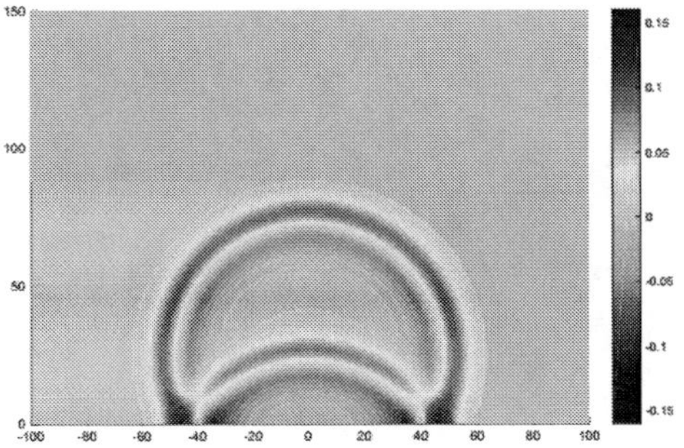

**Fig. 4.** Ac. pressure $p'(x, y, T)$ contours at time $T = 50$.

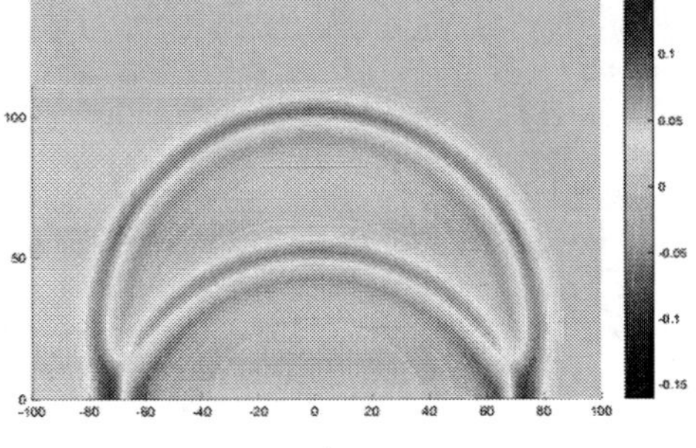

**Fig. 5.** Ac. pressure $p'(x, y, T)$ contours at time $T = 75$.

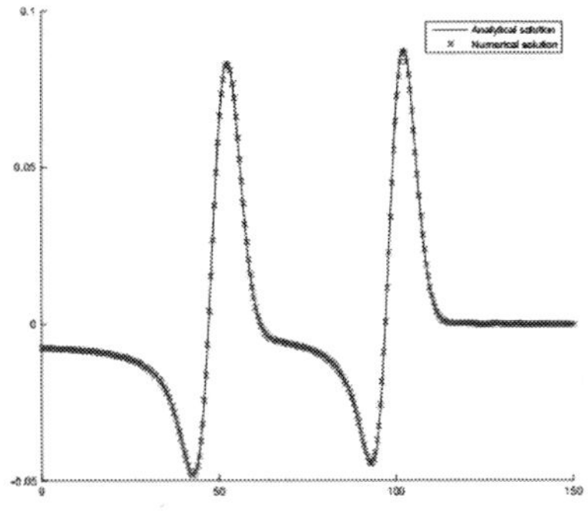

**Fig. 6.** Ac. pressure $p'(x, y, T)$ along the line $x = 0$.

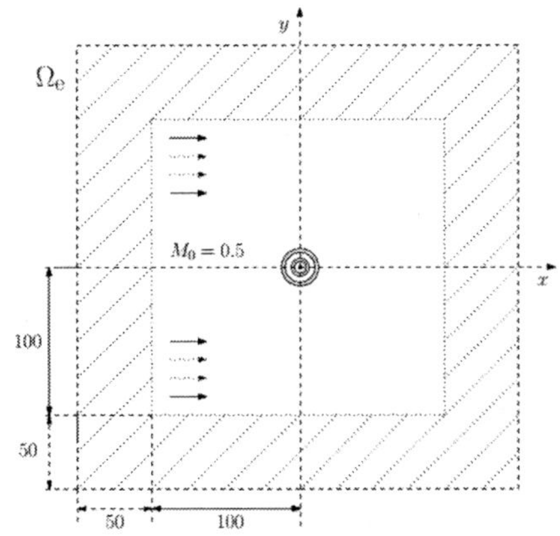


**Fig. 7.** Square domain $\Omega$, PML layer $\Omega_e$ (shaded region) and external boundary (dashed line).

We will consider subsonic underlying mean flow in $x$-direction with Mach number $M_0 = 0.5$ as well as supersonic flow with $M_0 = 1.5$. In this example, the acoustic source term on the right-hand side of the system (1) is the periodic perturbation of acoustic density and pressure at the origin $x_s = y_s = 0$ which is given as

$$\mathbf{S}(\mathbf{x}, t) = e^{-\alpha\left((x-x_s)^2+(y-y_s)^2\right)} \sin(\omega t)(1, 0, 0, 1)^T \tag{20}$$

where $\alpha = \ln(2)/2$, amplitude $\varepsilon = 0.5$ and angular frequency $\omega = 2\pi/30$.

The temporal computation was performed with time step $dt = 0.1$, number of points $n = 50181$ irregularly distributed and stretched in the PML. Contours of the acoustic pressure $p'(x, y, T)$ at time $T = 270$ for the subsonic case are shown in Fig. 8 and for the supersonic case in Fig. 9. Waves moving in the PML layer clearly dissipate and there are no spurious waves returning to the domain of interest.

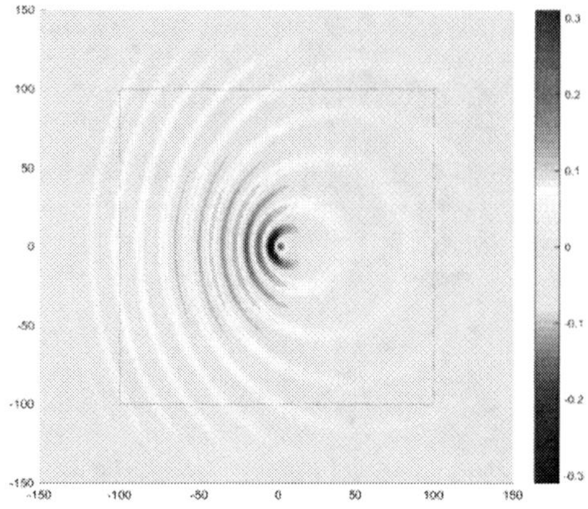

**Fig. 8.** Subsonic flow - acoustic pressure $p'(x, y, T)$ contours at time $T = 270$.

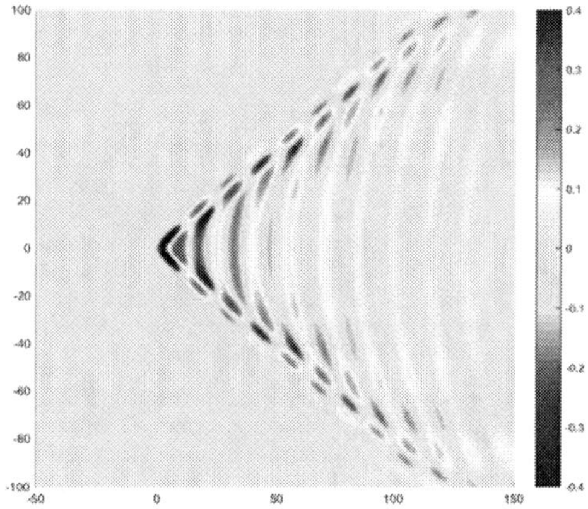

**Fig. 9.** Supersonic flow - detail of the acoustic pressure $p'(x, y, T)$ contours at time $T = 270$.

### 4.3 2D Acoustic monopole in a sheared mean flow

Prescribing the velocity components in (18) in form

$$u_0(x, y) = u_{max} \tanh(2y/\delta), \quad v_0 = 0 \qquad (21)$$

we get the non-constant underlying flow known as sheared mean flow, cf. [1, 15]. In this example, the peak velocity is taken as $u_{max} = 0.5$ and the shear layer thickness $\delta = 150$. The convection velocity profile is depicted in Fig. 10.

Instantaneous acoustic pressure contours at time $T = 270$ are plotted in Fig. 11.

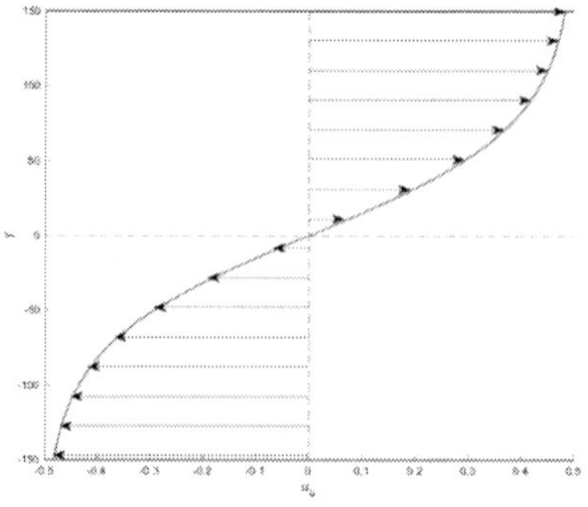

**Fig. 10.** Velocity profile $u_0(x, y)$

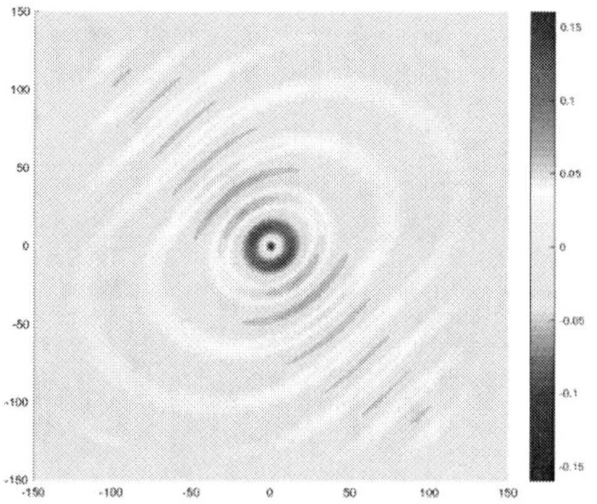

**Fig. 11.** Acoustic pressure $p'(x, y, T)$ at time $T = 270$.

## 5 Discussion

Linearity of governing equations (LEE) and therefore the absense of shock waves and strong disturbances in the solution allowed to develop a new stabilization technique in a meshless framework. Instead of the upwind approximation of derivatives widely used, e.g. in Finite volume methods, the stability of the meshless method was achieved by the WLSQ filtering of the numerical solution after each time step.

The method was tested on acoustic benchmark problems, such as the 2D wall bounded acoustic pulse problem and the acoustic monopole in a free stream and sheared mean flow. In order to suppress waves leaving the domain, the non-reflecting boundary condition in form of the PML was sucessfully utilized, cf. Fig. 8. To conclude, the method proved its ability to solve the wave propagation problems governed by LEE in various settings and is prepared for further applications.

## Acknowledgement

Authors greatly acknowledge support from the project No. GA 13-27505S funded by the Czech Grant Agency.

## References

1. Ch. Bailly, D. Juvé, AIAA **38** (2000).
2. W.D. Roeck, *Hybrid Methodologies for the computational Aeroacoustic Analysis of Confined, Subsonic Flows*, Katholieke Universiteit Leuven (2007).
3. Ch.K.W. Tam, J.C. Webb, J. Comput. Phys. **107** 262–281 (1993).
4. X. Nogueira, S. Khelladi, I. Colominas, L. Cueto-Felgueroso, J. París, H. Gómez, Arch. Comput. Meth. Eng. **18** 315–340 (2011).
5. G.R. Liu, Y.T. Gu, *An Introduction to Meshfree Methods and Their Programming* (2005).
6. G.E. Fasshauer, *Meshfree Approximation Methods with Matlab*, **6**, ISBN: 978-981-270-634-8, (2007).
7. J. Bajko, L. Čermák, M. Jícha, Comput. Meth. Appl. Mech. Eng. **280** 157–175 (2014).
8. A.J. Katz, A. Jameson, AIAA **699** 2008.
9. A.J. Katz: *Meshless methods for computational fluid dynamics*, Stanford University (2009).
10. E. Ortega, E. Oñate, S. Idelsohn, Int. J. Numer. Meth. Fluids **60** 937–971 (2009).
11. R. Löhner, C. Sacco, E. Oñate, S. Idelsohn, Int. J. Numer. Meth. Eng. **53**, 1765–1779 (2002).
12. A. Savitzky, M. J. E. Golay, Anal. Chem. **36(8)** 1627–1639 (1964).
13. P.A. Gorry, Anal. Chem. **62(6)** 570–573 (1990).
14. Ch. K. W. Tam, *Benchmark Problems and Solutions*, 1-13 (1995).
15. R. Ewert, W. Schröder, J. Comput. Phys. **188(2)** 365–398 (2003).
16. K. Li, Q.B. Huang, J.L. Wanga, L.G. Lin, Eng. Anal. Boundary Elem. **35** 47–55 (2011).
17. T. Colonius, Annu. Rev. Fluid Mech. **36** 315–45 (2004).
18. F. Q. Hu, J. Comput. Phys. **175** 455–480 (2001).
19. F. Q. Hu, J. Comput. Phys. **208**, 469–492 (2005).
20. F.Q. Hu, M.Y. Hussainy, J.L. Manthey, J. Comput. Phys. **124** 177-191 (1996).

DOI: 10.1051/epjconf/201714302004

# Study of flow around model of cooling tower by means of 2D Particle Image Velocimetry measurement

*Veronika* Barraclough [1], *Jan* Novotný [2], and *Pavel* Šafařík[2]

[1] Aerospace Research and Test Establishment, PLC. (ARTE), Aerodynamics Department, Prague, Czech Republic
[2] Czech Technical University in Prague, Faculty of Mechanical Engineering, Department of Fluid Dynamics and Thermodynamics, Prague, Czech Republic

**Abstract.** This paper deals with flow around a bluff body of hyperboloid shape. It combines results gathered in the course of research by means of Particle Image Velocimetry (PIV). The experiments were carried out by means of low-frequency 2D PIV and the Reynolds number was 43 000.

## 1 Introduction

In this contribution, research in mechanics of fluids field, that developed from experimental testing for civil engineering, will be presented. Originally one of the well-known, and often investigated, problems of civil engineering - wind loading on cooling towers - had been investigated. There are plenty of articles dealing with this problem, whose enumeration can be found, for instance, in a summary article with a symptomatic name - A Survey on Hyperbolic Cooling Towers [1]. This paper is not going to deal with the cooling towers, it only originated there.

The object of measurement consisted of four models of hyperboloid cooling towers, whose centres were located in the corners of the square, and were tested in a wind tunnel designed for civil engineering experiments. According to the original measurement plans and requests from the industrial partner, the pressures inside and outside the model were measured and also calculated by means of mathematical modelling and numerical calculation. These two attitudes were compared and evaluated.

The experimental part described above was later extended to the pure mechanics of fluids discipline. The original test plan had determinated pressures on the models by classical means of fluid mechanics testing (pressure taps) which was completed by research of wakes with the help of a modern tool of non-intrusive measuring, Particle Image Velocimetry (PIV). This paper hopes to deal with some of the results gathered on this journey.

There are various paths the researcher can go by on the way to the results for the hyperboloid object. To get a comprehensive picture of flow it is possible to come out of the mean flow topology of the cylinder, as depicted for instance in [2], page 14, Picture 2. Therefore, in flow around the hyperboloid object we could expect horseshoe vortex, tip vortex, arch vortex and trailing vortex and some clues of these patterns were indicated in the presented measurement. The aim of this work is to test possibilities of low frequency 2D PIV for the measurement of 3D structures of flow field and to identify areas of later research by means of hot wire anemometry and numerical calculation.

## 2 Experimental Facility

The experimental part accomplished so far was provided by means of 2D Particle Image Velocimetry (PIV), being the quickest tool for instant viewing of the problem. The measurement was provided in a range of Reynolds numbers varying from 43000 up to 130000; this contribution contains results for Re=43000. The Reynolds numbers varied by the change of free stream velocity $U_\infty$ in the wind tunnel. The tunnel used for the measurement was the Boundary Layer Wind Tunnel (BLWT). It is equipped with a special surface helping the development of the boundary layer, so that the intensity in the main measured plane reaches 13% [3]. The ratio of the height of the model to the thickness of the modelled boundary layer in the BLWT is $\frac{1}{3}$. The BLWT dimensions of the rectangular measurement area are 1,8 x 1,5 metres.

The scale of the measured model of the cooling tower was 1:400. The model was printed using a 3D printer as 0,33 m tall plastic object.

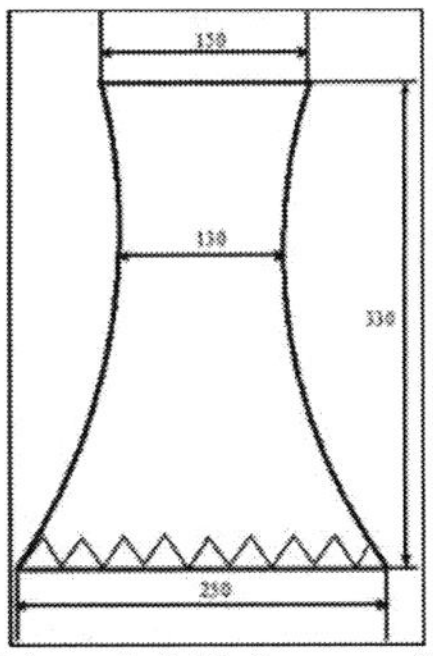


**Fig. 1**. Model of hyperboloid-shape cooling tower and its dimensions.

The PIV experiment was provided in a classical way - the measurement area was seeded with olive oil droplets produced by Laskin nozzles and a green laser light was used to illuminate the area. The main 2D experimental part at the beginning was provided with a laser with Nd:YAG crystals and low frequency (15 Hz) of the pulses.

The main characteristics of PIV measurements:

| | |
|---|---|
| PIV method used: | 2D |
| Max. in-plane velocity: | 5 m/s |
| Max. cross-plane vel.: | 5 m/s |
| Field of view: | See Results |
| Interrogation area: | 32 x 32 pixels |
| Observation distance: | Maximum 1000 mm |
| Recording method: | double-frame, single expos. |
| Recording medium: | CMOS sensor, 1280 x 1024 pxl, pixel size 12 x 12 μm |
| Recording lens: | FL = 50 mm, 35 mm for details, f♯ = 1,2 - 2 |
| Illumination: | YAG laser, 2x 60mJ/pulse |
| Pulse delay: | 100 μs |
| Recording frequency: | 15 Hz |

For data evaluation, Dynamic Studio software was used. All image processing, and later calculations of particle movements in two consequentive images were provided with the help of this post-processing tool. For this work, adaptive correlation was used with these parameters: final interrogation area was 32 x 32 pixels, which represented 2.3 to 6.5 times 1.9 to 6.3 square millimetres of area. Horizontal and vertical overlap were 50%. Evaluated vectors were filtered by range validation and universal outlier detection. Usually, the vector statistics representing averaged velocity field was calculated.

## 3 Measurements and results

### 3.1 Incoming flow

Measurements were carried out in front of the object, on its side and in its wake. In the case of the front part, the measurement plane was set to the axis of the model and parallel to the flow, perpendicular to the ground. Note the small icons in the velocity fields for positioning of the illuminated area. There were vortices observed in instant velocity fields nearby the bottom of the object (Fig. 2) and the velocity component extracts show some movement against the flow direction (Fig. 3). The averaged velocity field (Fig. 4) doesn´t indicate the presence of the structure, which would encircle the bottom part of the object, though. The cause of it could apparently consist of the airway of the bottom part of the model, or could be the result of an inappropriate scale of the measurement. This area will definitely be a subject of later, more detailed research.

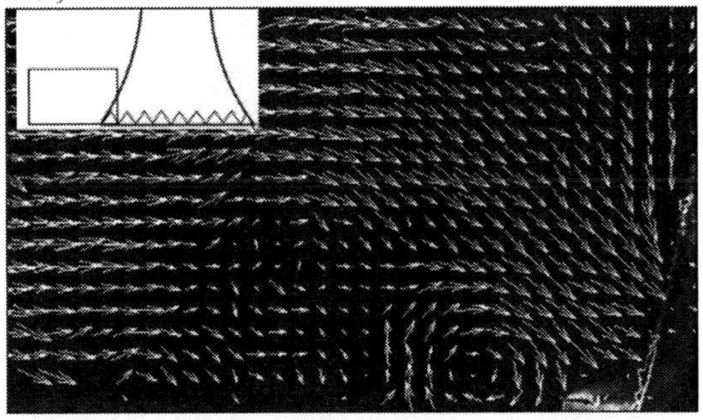

**Fig. 2.** Instant velocity field of the front bottom part of the model. Parallel to the flow, perpendicular to the ground. The field of view 94 x 62 mm.

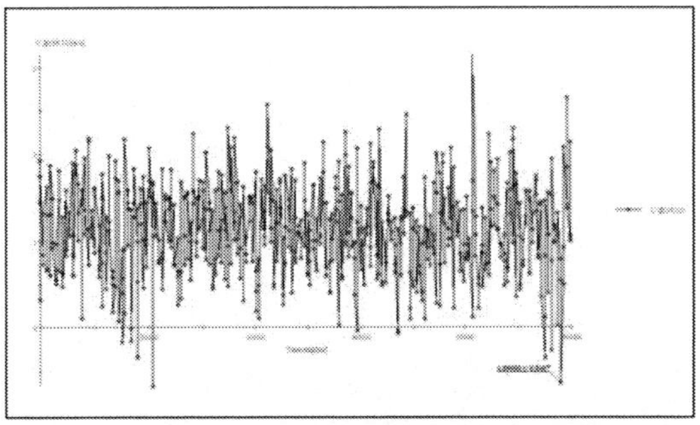

**Fig. 3.** Velocity x-component development in time. The extract was made in the front of the model and close to the ground (in the centre of the vortex depicted on Fig.2).

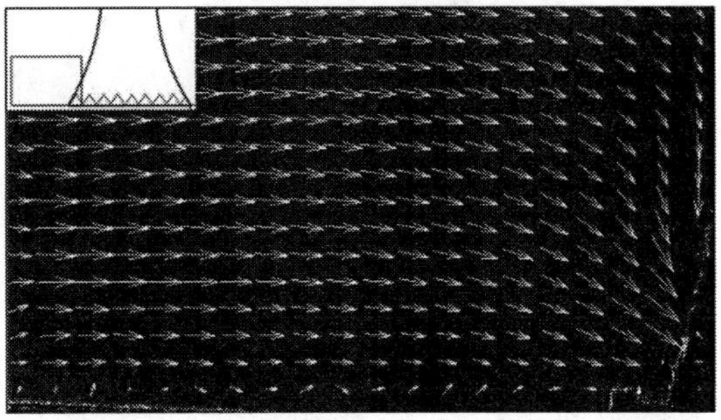

**Fig. 4.** Average velocity field of the front bottom part of the model. Parallel to the flow, perpendicular to the ground. The field of view 94 x 62 mm.

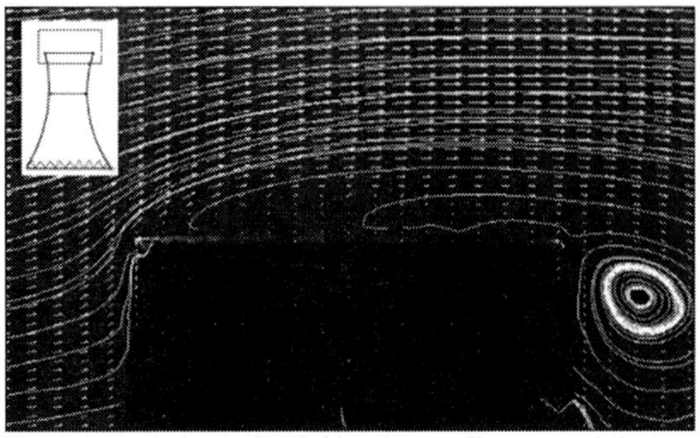

**Fig. 5.** Averaged velocity field and streamlines around the top of the object. Parallel to the flow, perpendicular to the ground. The field of view 260 x 208 mm.

### 3.2 The wake

There was special attention paid to the measurement in the wake. The aim of the first part of the measurement was to compare the influence of the turbulence intensity on a wake character, so two types of wind tunnels with different rates of turbulence were used. The evaluation of this measurement is not part of the paper.

Measurement planes for the wake were parallel to the flow direction in the object axis and perpendicular to it. For the case of measurement in the object axis, the whole picture was put together from two positions of the camera. This solution was chosen due to lack of space in the measurement area of BLWT for all equipment needed.

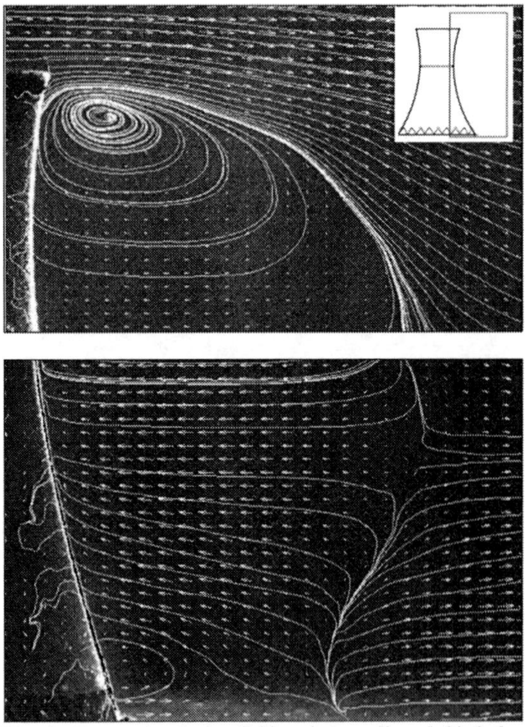

**Fig. 6.** Averaged velocity field and streamlines behind the model. The plane parallel to the flow, perpendicular to ground. The field of view 286 x 229 mm.

The view into the wake, the perpendicular plane was supposed to offer some proof of existence of the typical 3D flow structures. It was later realized that the PIV measurement near the ground of BLWT, covered with special material to induce natural-like conditions, is very difficult due to strong light reflection from this rough, uneven surface. This problem leads to some captured data being worthless and needs to be solved more carefully in future with other tools of experimental fluid mechanics.

There were measurements in two planes in different distances from the bottom of the model provided and several characteristics evaluated (Fig. 7,8, in this paper only more interesting data from the plane closer to the model are introduced).

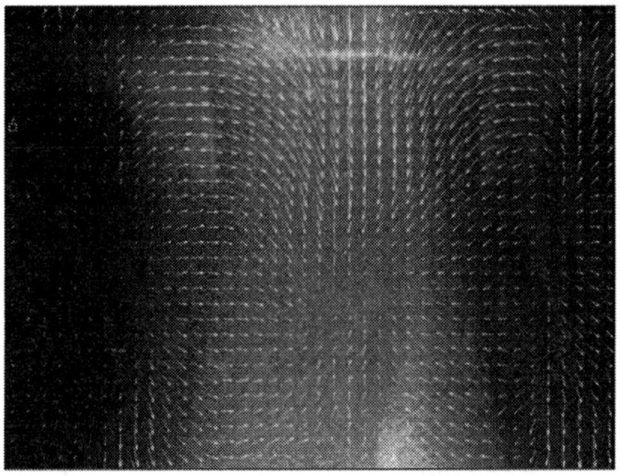

**Fig. 7.** Averaged velocity field, distance 56 mm from the bottom of the model. The plane perpendicular to the flow and the ground. The field of view 286 x 229 mm.

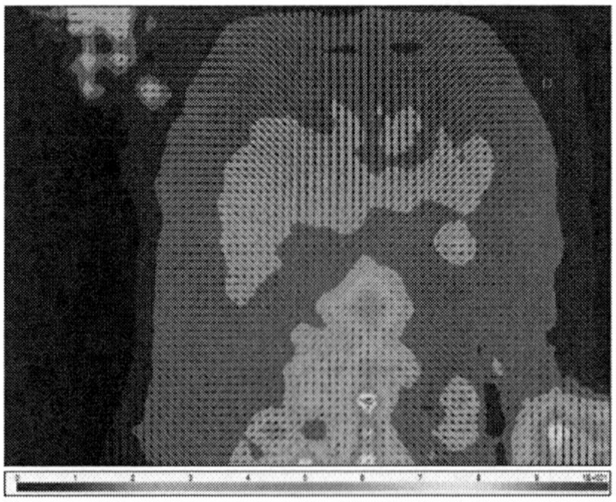

**Fig. 8.** Sum of variances, which is proportional to turbulent kinetic energy in units $\left(\left(\frac{m}{s}\right)^2\right)$. The plane perpendicular to the flow direction and to the ground. The distance 56 mm from the bottom of the model.

It has to be emphasized here, that the software evaluates the velocities from a spatial average over the interrogation area, while the classical attitude would be infinitely small elements. Therefore, the Fig. 8 shall serve for later comparison to numerical calculation.

And in the end, the measurement in the plane parallel to the ground where the object has the narrowest diameter shouldn´t be omitted. The measuring plane was set at an altitude of 0.2 m from the ground desk.

As expected, the averaged velocity field and the streamlines show two symmetric eddies. Dimensionless recirculation length in this case is less than 1; such limited recirculation area is caused by turbulence in the free stream supressing the wake. The shape of the wake shows turbulent separation with the separation point behind the widest area of the model.

For later research, this part could be even more interesting: according to [4], shedding of vortices changes from Kármán asymmetric type to symmetric arch type at a geometrical aspect ratio of an object (cylinder) of 2.5; as the aspect ratio of our measured object varies from 1.3 to 2.54, the narrowest area of it could be measured with respect to this fact.

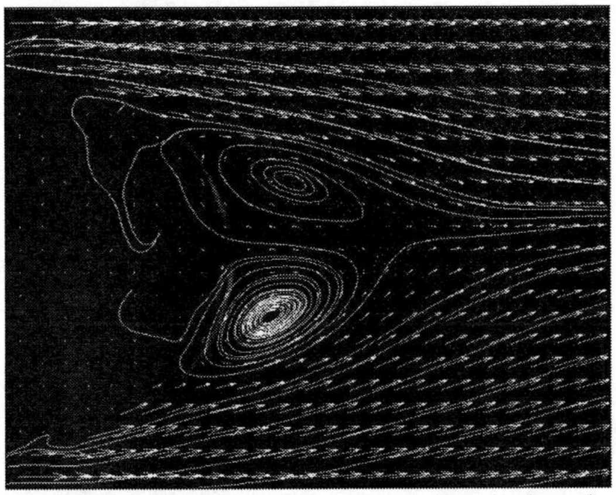

**Fig. 9.** Averaged velocity field and streamlines in the wake, plane parallel to the ground. The field of view 250 x 200 mm.

## 4 Conclusions

The model of hyperboloid shaped object was measured by means of 2D Particle Image Velocimetry in the Wind Tunnel with simulated atmospheric boundary layer. The model was tested in subcritical range of Reynolds numbers and results from it were presented. Various planes around the model were captured: in the front of the model, the plane parallel to the flow and perpendicular to the ground was measured and a more detailed view to the flow was obtained as well, with the extraction of velocity components in small distances from the object bottom. The bottom part of the model consists of pillars and is penetrable for the flow, which might influence the formation of horseshoe vortex. The measurement didn´t show this structure, which might be caused by the inappropriate scale of the view or bad placing of the capturing camera. The model is hollow, so the suction effect of this part will somehow influence an outfall and the flow around it.

There were several planes measured on the side of the model as well. All planes were parallel to the flow, perpendicular to the ground. The strong 3D character of the flow requests the information of third velocity component or at least other measurements of the same kind, but in the plane perpendicular to the flow direction.

Special care was dedicated to the wake. There were three planes chosen to be measured. One was set in the model axis and perpendicular to the ground. The averaged velocity field of this measurement revealed a possible arch vortex behind the model or highlighted point of later research. Next, the plane perpendicular to the ground was measured. As expected, the averaged velocity field showed two symmetrical formations. And eventually, the plane perpendicular to both flow direction and the ground completed the measurement in the wake. Instant velocity fields were used for testing of vorticity identification tools, but all these tools used in the software calculated with strong uncertainties and didn´t show us any useful results. The average velocity field served for calculation of sum of variances proportional to turbulent kinetic energy in units.

In conclusion, it could be said that from all the 2D PIV measurements considering spatial formations around such a bluff body raised more questions than answers.

The research was performed with financial institutional support from government budget through the Ministry of Industry and Trade of the Czech Republic.

The support from the Technology Agency of the Czech Republic under Project No. TE 01020036 is gratefully acknowledged.

## References

[1] E. Asadzadeh, M. Alam, A Survey on Hyperbolic Cooling Towers, Intern. Journal of Civil, Envir., Struct., Construct. and Arch. Engineering **8**, (2014)

[2] R. J. Pattenden, S. R. Turnock, X. Zhang, Measurements of the flow over a low-aspect-ratio cylinder mounted on a ground plane, Experiments in Fluids **39**,10–21(2005)

[3] P. Michálek, D. Zacho, R-6123 Report, ARTE, Plc, Prague, Czech Republic (2014)

[4] H. Sakamoto, M. Arie Vortex shedding from a rectangular prism and a circular cylinder placed vertically in a turbulent boundary layer. J Fluid Mech **126**, 147–165 (1983)

[5] N. P. van Hinsberg, J.-M. Lueckhof, J. Nuhn, The Reynolds number effect from subcritical to high transcritical on steady and unsteady loading on a rough circular cylinder, Proceedings of the 9th Intern. Confer. on Struct. Dyn., EURODYN (2014)

[6] X. P. Liu, L. Zhao, Y. J. Ge, Extreme value distribution of surface aerodynamic pressure of hyperbolic cooling tower, The 7th Inter. Colloq. on Bluff Body Aerodynamics and Applications (BBAA7) Shanghai, China; (2012)

[7] A. Roshko, On the development of turbulent wakes from vortex streets, NACA Report 1191 (1954)
[8] P. Chakraborty, S. Balachandar, R. J. Adrian, Kinematics of Local Vortex Identification Criteria, Journal of Visualization, Vol. 10, No. 2 **137-140**, (2007)
[9] V. Holmén, Methods for Vortex Identification, Lund University Libraries, lup.lub.lu.se/student-papers/record/3241710 (2012)
[10] M. Jirsák, Physical Principles of Work with Boundary Layer Wind Tunnel, ARTE Plc., Prague, Czech Republic (2012)

DOI: 10.1051/epjconf/201714302005

# Experimental and numerical studies of rotating drum grate furnace

*Grzegorz* Basista[1,*], *Mateusz* Szubel[1], *Mariusz* Filipowicz[1] , *Bartosz* Tomczyk[1], and *Joanna* Krakowiak[1]

[1] *AGH University of Science and Technology, Faculty of Energy and Fuels, al. Mickiewicza 30, 30-059 Krakow, Poland*

**Abstract.** Waste material from the meat industry can be taken into account as a biofuel. Studies confirm, that calorific value is higher and ash content is lower comparing to some conventional fuels. EU directives regulate details of thermal disposal of the waste material from the meat industry - especially in range of the process temperature and time of the particle presence in area of the combustion zone. The paper describes design of the rotating drum grate stove, dedicated to thermal disposal of the meat wastes as well as solid biomass (pellet, small bricket, wood chips) combustion. Device has been developed in frames of cooperation between AGH University of Science and Technology (Krakow, Poland) and producer focused on technologies of energy utilization of biomass in distributed generation. Results of measurements of selected operational parameters performed during startup of the furnace have been presented and discussed. Furthermore, numerical model of the combustion process has been developed to complement experimental results in range of the temperature and oxygen distribution in the area of the combustion chamber. ANSYS CFX solver has been applied to perform simulations including rotational domain related with specifics of operation of the device. Results of numerical modelling and experimental studies have been summarized and compared.

## 1 Introduction

Waste material from the meat industry is characterized by a high content of organic matter, from 50%, to almost 80%. This is extremely important for the natural environment, due to bacterial and epidemiological risk [1]. Due to this fact, mentioned wastes should be disposed of as soon as possible, to avoid problems resulting from their storage [2].

There are many ways to exploit various types of waste from the meat industry. They are used in the production of fertilizers, animal feed and the pharmacological and cosmetic industry [3]. However, some components of wastes cannot be processed in any ways, so they have to be utilized [4].

From the energy industry point of view, waste material from the meat production can be used as a biofuel [5]. Due to the necessity of increasing the contribution of renewable energy sources (RES) in the national energy balance, this fact could be important for many countries of European Union (EU). An additional advantage of using meat wastes as a fuel are savings, which result of using a heat of incineration process [6].

Conditions of thermal utilization of this type of wastes are strictly regulated by EU directives 89/369/EEC, 94/67/EC and 2010/75/EU. They define some legal regulations, as well as the specific details of combustion process. Gas temperature behind the last point of air supplying should be at least 850°C for at least 2 seconds, with a minimum oxygen content of 6% [7]. Also a content of organic compounds (defined as a total content of carbon) and carbon monoxide (CO) are regulated [8]. Thermal devices for meat waste utilization should therefore be designed to meet mentioned regulations.

The required temperature of 850°C has to be kept on the same level during whole process of combustion. Furthermore, dedicated fuel feeder, which will provide material only in case of sufficient enough temperature in the combustion chamber is also necessary. Each decrease of temperature increases risk of incomplete combustion. All above mentioned facts are reason of necessity to use stoves designed and constructed in specific way. Usually combustion is performed in devices equipped with rotating drum grate. During rotation material is combusted, but also crumbled, warmed, dried and pyrolised. In case of fats combustion, temperature of the process in regions near to the stove walls can be higher than 1000°C [9]. Due to the high intensity of pyrolysis it is necessary to apply appropriate air staging system. Dynamic control of the operation of this system provides high efficiency of oxidation of components of the gas in the combustion chamber.

Virmond et al. [10] performed studies of possibility of using biosolids originating from the wastewater treatment process of two meat processing plants, as an alternative fuels. The authors performed an analysis of fuels properties, and the emissions during

* Corresponding author: basistag@agh.edu.pl

combustion in the pilot scale cyclone combustor. Measured lower heating value of some of the investigated fuels was about almost 26 [MJ kg-1, daf]. Temperature of combustion during the test in cyclone combustion was between approximately 750 and 1200°C. The concentration of CO, $CO_2$, CxHy and total organic carbon during the test remained below the limits of strict German regulations for waste incineration. As a main result, the authors have found, that the tested fuels have a high potential for energetic use.

Numerical modelling with CFD methods is a useful tool for advanced design process. It allows to predict some characteristics of combustion process inside the combustion chamber, that cannot be investigated with any experimental methods [11-13].

Ariyaratne et all. [14] have investigated process of meat and bone meal (MBM) combustion in a cement rotary kiln, using a three-dimensional, steady state CFD model. The authors have carried out set of simulations, to find out an effect of various fuel feeding positions and fuel particle size for combustion characteristics. Coal was used in a study as a reference fuel. As a result of the simulations, it was found, that MBM combustion gives a lower gas temperature than coal combustion. This is due to the some parameters of the waste-derived fuel, such as higher moisture and ash content, and lower oxygen content compared to the coal. Lower particle size and temperature in MBM combustion process results also in poor char burnout, but due to the low content of fixed carbon, the impact on gas temperature is insignificant.

## 2 Test stand

The main element of test stand was patented ceramic rotational furnace, designed specifically to thermal utilization of organic wastes from meat industry. It is designed to keep required conditions during the combustion process. The system consists also: heat exchanger at the chimney, two screw feeders driven by electric engines, twelve independent fans of air supply system, electric igniter, three thermocouples, power supply and a control system equipped with PLC controllers.

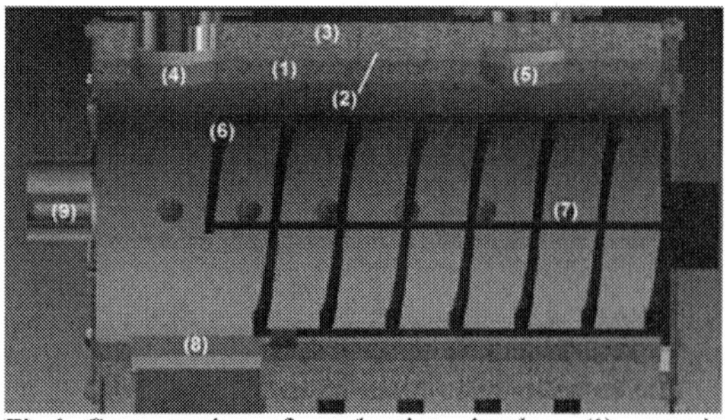


**Fig.1** Cross section of combustion chamber: (1) ceramic material, (2) insulation, (3) metal cover, (4) gas exhaust, (5) fuel inlet, (6) rotational grate, (7) electric igniter, (8) ash outlet, (9) place of thermocouple installation.

Furnace is constructed of three layers (Fig. 1). Internal layer, which is exposed to high temperatures, is made of ceramic material. External layer is a metal cover, and there is also insulation between cover and ceramic material. Rotational grate is placed inside the combustion chamber. It transports the fuel and ash along the furnace, to ash settling. To provide accurate conditions of combustion process, and continues operation in required temperature range, air is supplied by twelve independent ducts, located at both sides of furnace. Each of these pipes has individually controlled fan.

**Fig.2** General view of test stand: (1) heat exchanger, (2) tank of fuel, (3) flue pipe, (4) furnace, (5) chimney, (6,7) electric engines, (8) fans of air supply system.

Gas exhaust is located under the ash settling. Heat exchanger with water jacket is located at the flue pipe, due to high temperature of exhaust gas. Fuel is supplied by the feeder located at the top of furnace, at its beginning part (Fig. 2). Combustion process is initialized next to first air duct.

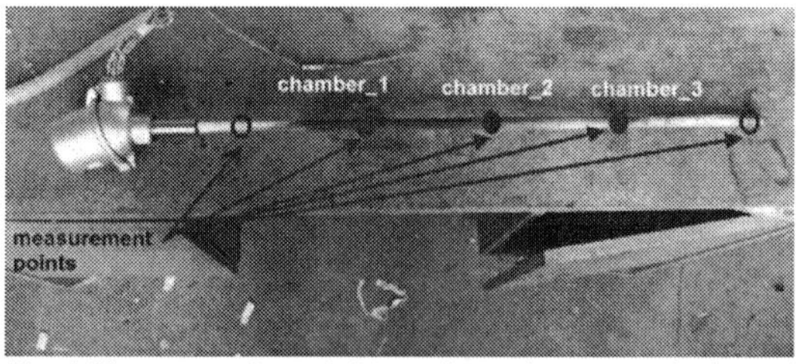


**Fig. 3** Thermocouple no 2. Measuring points are marked at red color.

Set of instruments was installed in test stand, to make measurements of temperature in important points of furnace. Two of them was located at the flue pipe, behind gas exhaust (flue_1) and behind heat exchanger (flue_2). The third one (Fig 3.) was located in the central area of combustion chamber. This one had five measuring points, and due to possibility of installation along the furnace, it provided detailed information about temperature inside the combustion chamber.

The supporting electric system includes inverters controlling engines of rotational grate and screw fuel feeder, igniter switch, power supply, and control modules for each of air fans.

**Fig. 4** Switch cabinet: (1) power supply, (2) control modules, (3) igniter switch, (4) fuses, (5) inverters, (6) fan controllers.

Operation of devices and data acquisition system are controlled using programmable logic controllers, which transmit control signals to inverters and suppliers of air fans. This system is placed in switch cabinet (Fig 4.).

# 3 Experimental and numerical research

## 3.1 Experimental tests

The aim of presented study was to analyze the combustion process in a ceramic rotational furnace. The main subject was an examination of temperature distribution in combustion chamber of the device, to achieve the conditions required for thermal utilization of wastes of meat industry.

Biomass pellets with calorific value about 18 $MJ \cdot kg^{-1}$ and low ash content was used as a fuel. The granulated form of fuel was an additional advantage for combustion process in the device, due to an easy application using a screw feeder. Studies with some conventional type of fuel was necessary to get operational characteristics of device before using organic wastes.

At the first stage of research, the ignition process and the way of controlling of the air supply system were investigated and analyzed. Suitable amount of the fuel at the beginning of the combustion process as well as the rate of the feeding during the process were selected. Firstly, the series of measurements aimed to achieve the highest temperature in the combustion chamber were made. However, it was necessary to take into account a resistance to high temperature of materials used to construction of some elements of prototype. The following measurement series were focused at keeping a proper characteristics of combustion process in required conditions, with 850 Celsius degrees.

Fig 5. shows the chart of temperature distribution in combustion chamber, and flue gas temperature, during the first measurement series. It was investigated to analyze an influence of rotational grate operation on combustion process. Rotating of the grate was periodic, as well as fuel feeding. Full turn of the grid on both sides have lasted for 25 to 30 seconds, and was performed every 3 minutes. Fuel in amount of 480 grams was delivered in a two-minute intervals. Rotation of the grid resulted in a rapid rise of temperature in the combustion chamber, as can be seen in the graph as a group of peaks at all measurement points. After reaching the maximum local temperature levels due to rotation of the grid, there were significant decreases of temperature (even almost 200°C). Movement of the grid results in a better mixing of the fuel with the air, and prevents the deposition of ash on the bottom of the combustion chamber. It causes better access of an oxygen to the unburned fuel particles and results in a greater level of post-combustion of these particles. Due to this fact, also temperature value is higher during the rotation of the grid, what can be seen on the graph at Fig. 5.

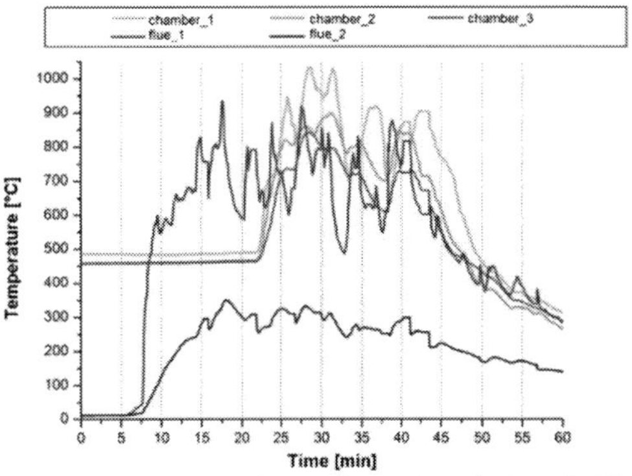


**Fig. 5.** Temperature value at five points during the first measurement series

The maximum temperature value in the first measurement series in the point no. 2 was 900°C, while in the point no. 1 slightly exceeded 1000°C. The temperature of the exhaust gases before the heat exchanger was oscillating from 600°C to 900°C. Due to the limitations of the materials (not insulated regions), after reaching mentioned thermal conditions, between 30 and 35 minute of the process, it was necessary to temporarily suspend delivering of the fuel, to slightly decrease temperature at flue gas exhaust. Due to the fact, that the furnace is still under the research process, the design of both the furnace and the flue is not adapted to operation at temperatures exceeding 1000°C. Operation of heat exchanger have allowed to decrease flue gas temperature to the level about 300°C.

During the first measurement series, the expected temperature of 850°C in the combustion chamber have occurred for 10 minutes, so the first objective has been achieved. Moreover, the impact of the rotation of the grate to the combustion process has been investigated, and some data has been collected.

The aim of the second measurement series was also to obtain maximum possible temperature in the combustion chamber. The mass flow of the air, as well as the amount of the fuel, was the same as in the first series. However, the time between the turns of the grate was shortened to two minutes. The turn of the grate was performed immediately after the delivering of the fuel.

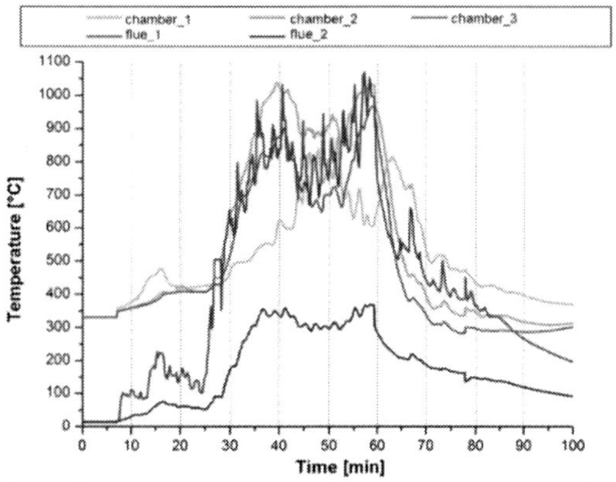


**Fig. 6**. Temperature value at five points during the first measurement series

Figure 6 shows temperature values at five points during the second measurement series. In this case, the value of temperature at measurement point no. 2, located nearby gas exhaust, next to the last stage of air supply system, was significantly higher than in first series.

In the second measurement series, the required conditions (temperature exceeding 850°C) have persisted much longer than in first series – from 30 to 60 minute of the process, which is the entire main phase of combustion process. Similarly to the first series, firing up the furnace has lasted about 25 minutes, while blanking started in 60 minute. It was caused by the increase of the flue gas temperature above 1000°C. Due to the mentioned before material limitations, after reaching temperature of 1000°C in combustion chamber at 40 minute of the process, the amount of the fuel was reduced by half, to 240 grams every two minutes.

Just as in the first measurement series, periodic rotation of the grate directly affected the large temperature fluctuations (most of all the flue gas temperature). However, the effect of shortening the time between turns of the grid is noticeable. This procedure has allowed achieving significantly higher temperature.

The increase in number of rotations of the grate resulted in the shift of reaction zone to the further part of combustion chamber, what can be identified with the temperature curve of measurement point no. 1. The values of these temperature during the entire process was lower comparing to first measurement series. As a result of movement of the grate, the fuel was constantly transferred to the end part of the furnace. Due to this fact, higher temperature has occured at the required region, nearby the gas exhaust, after the last stage of air supplying system.

The first part of the research has shown, that it is possible to achieve temperature conditions required to safe incineration wastes of meat industry. By appropriate controlling, it was possible to maintain required conditions during the main part of the combustion process at the second measurement series. On the base of obtained results, it can be concluded, that after the adaption of the furnace to the higher temperature (using appropriate insulation and materials), it will be possible to exceed the temperature of 1000°C, and maintain these conditions for a longer time. In the existing device, excessive temperature growth has forced to interrupt the process, due to the risk of damaging some parts of the device. However, temperature curves at this points have shown, that the temperature should continue to grow.

To provide efficient and effective operation it is also necessary to develop appropriate algorithms of combustion controlling – for both rotating grate and the fuel feeder. The results show, that more stable conditions of combustion process can be achieved by shortening the length of periods between rotations of the grate, so it can be concluded, that continuous operation would be the most efficient. Due to limited ability of controlling of a grate operation, it was not possible at this stage of research, so it was not examined.

The next stage of research was focused on maintaining the operating conditions for a longer period of time.

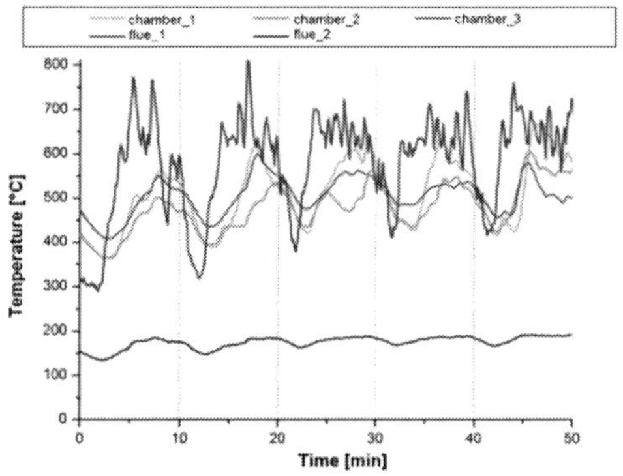


**Fig. 7.** Temperature value at five points during the first measurement series

During the third measurement series, the temperature on combustion chamber did not exceed 500°C, to avoid excessive growth over 1000°C. It would have adversely affect the measurement series, due to the necessity of reduction of temperature. Figure 7 shows the temperature in combustion chamber and flue pipe during the third measurement series. Due to the delivering of the fuel in 10 minutes intervals, there are huge fluctuations in temperature. On this stage, the rotation grate operation was continuous. The results clearly shows, that it is also necessary to provide and automatic, continuous feeding of the fuel to achieve constant conditions in combustion chamber. For thermal utilization of wastes of meat industry, which requires the temperature above 850°C during the entire process, such large falls of temperature would require an operation of an additional burner.

### 3.1 Numerical simulation

To identify in more detailed way the characteristic of combustion process on investigated combustion chamber, the three-dimensional CFD model has been developed. A steady-state simulation has been prepared and solved using ANSYS CFX solver. The main purpose of the simulation was investigation of influence of location of air feeding system nozzles and grate rotation, on local temperature and turbulence level. As a first stage of modeling, some simplifications has been applied. The numerical model includes only homogenous gas-phase reactions.

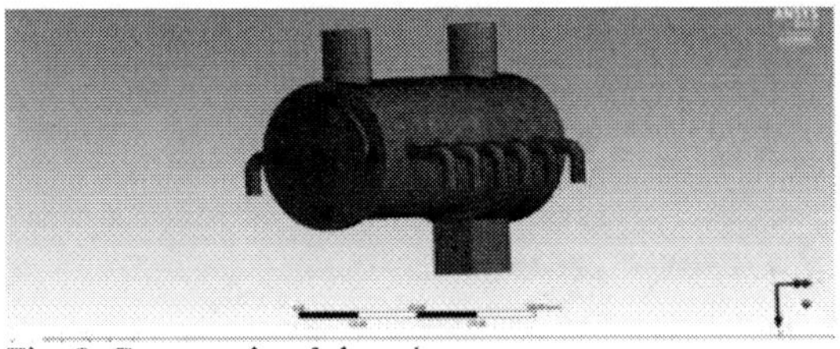

Fig. 8. Computational domain

The computational domain has been defined as a continuous fluid filling the combustion chamber of tested device. The domain was divided into two parts (Fig. 8.). The central region was defined as a rotational domain.

Numerical model is based on CFD governing equations, which include continuity, momentum and energy equations. The "k-ε" turbulence model has been applied. Combustion process was simulated using „Eddy Dissipation" with "Finite Rate Chemistry" model. It is based on two different mechanisms of combustion, including the rate constant of chemical reactions and level of components mixing. The rate constant is defined by Arrhenius equation:

$$k = A \cdot \exp[-E_a \cdot (RT)^{-1}] \tag{1}$$

k - rate constant,
A - pre-exponential factor,
$E_a$ - activation Energy [7],,
R - universal gas constant [$J \cdot mol^{-1} \cdot K^{-1}$],
$T$ - temperature [8]

The numerical model have considered only homogenous combustion in gas phase, excluding gasification process of solid fuel. The set of chemical reactions was implemented based on [15]:

$$CO + 0{,}5\ O_2 \rightarrow CO_2 \tag{2}$$
$$CO_2 \rightarrow CO + 0{,}5\ O_2 \tag{3}$$
$$H_2 + 0{,}5\ O_2 \rightarrow H_2O \tag{4}$$
$$CH_4 + 1{,}5\ O_2 \rightarrow CO + 2\ H_2O \tag{5}$$
$$NH_3 + O_2 \rightarrow 0{,}5\ H_2 + H_2O + NO \tag{6}$$
$$NH_3 + NO \rightarrow 0{,}5\ H_2 + H_2O + N_2 \tag{7}$$
$$HCN + O_2 \rightarrow CO + 0{,}5\ H_2 + NO \tag{8}$$
$$HCN + NO \rightarrow CO + 0{,}5\ H_2 + N_2 \tag{9}$$

The material of the computational domain was reacting mixture of all chemical compounds involved in reactions described in (2-9), and inert gas - nitrogen Heat transfer has been described with convection mechanisms. On this early stage of modeling, radiation model and heat transfer through the external walls was not included.

Set of boundary conditions has been defined in computational domain (Fig. 9). The first one was inlet of reacting mixture to combustion chamber. It was placed on the front wall of furnace (INLET_FUEL). This surface has been selected as an inlet to describe combustible gases flow into computational domain in the most accurate way. Mass flow of combustible gases was set to 0,025 kg/s, and it's temperature was 800°C. The compounds of reacting mixture was: methane ($CH_4$), CO, carbon dioxide ($CO_2$), hydrogen ($H_2$), water vapor ($H_2O$), hydrogen cyanide (HCN), and ammonia vapor ($NH_3$).

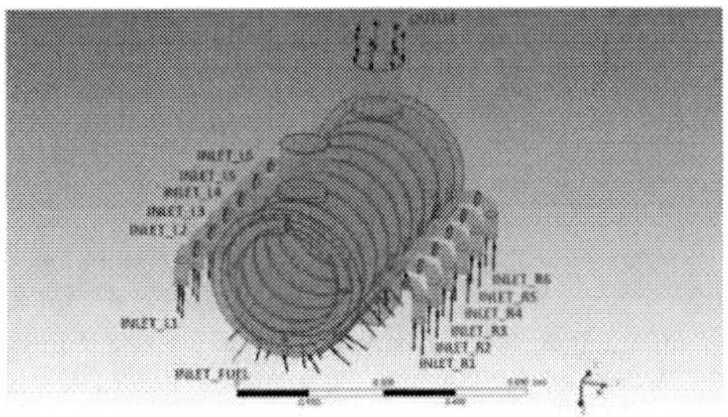

Fig. 9. Locations of boundary conditions

The next boundaries were air velocity on the all of 12 inlets to combustion chamber (INLET_R1 – INLET_R6, INLET_L1-INLET_L6). The value was 10 m/s in each case. It has been defined according to the conditions obtained at experimental stand, described at the first part.

As it was mentioned before, the radiation model as well as heat transfer by conduction through the walls has not been considered. The external wall has been defined as an adiabatic wall. Due to this fact, temperature obtained in computational process are slightly higher than real values.

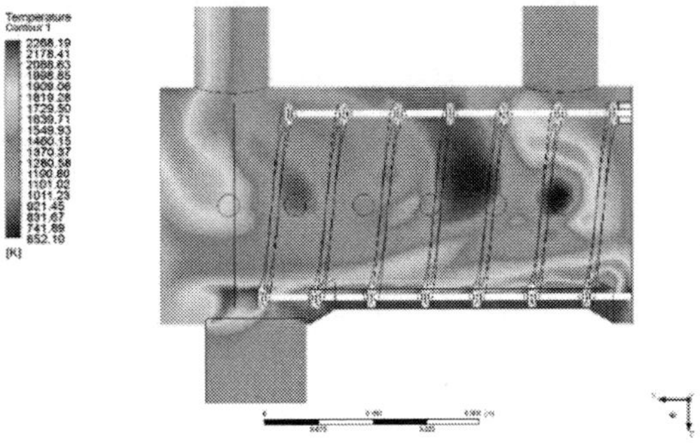

Fig. 10. Temperature at cross-section of combustion chamber

As a result of the simulation, a series of data relating to operation of examined device has been analyzed. Figure 10 shows the temperature distribution at cross section of combustion chamber. An influence of turbulence coming from both, side air feeding, and rotation of the grid, on the temperature in combustion chamber can be observed. At the central region of combustion chamber, in points of temperature measured during the experimental tests, the temperature values obtained on numerical model ranges between about 600°C and nearly 1500°C. There are some zones in initial part of combustion chamber with lower temperature, not exceeding 600°C. However, appropriate conditions nearby flue gas outlet have been achieved, what is the most important due to legal regulations. Average temperature on this region was about 1200°C. This demonstrates a high potential of the device in case of meat waste incineration, even taking into account some inaccuracy of numerical model.

## Conclusions

As a subject of the research, a pilot small-scale installation for meat waste thermal utilization has been tested, to obtain required temperature level. The studies were carried out at the test stand with rotational drum grate furnace. Simultaneously, the numerical model has been prepared. The main purpose of the numerical simulations was analysis of processes occurring inside the combustion chamber, which is not possible using experimental methods.

Experimental test were carried out using biomass pellets, characterized by similar to meat wastes calorific value. It allowed to obtain similar conditions of combustion process without using harmful fuel in not tested device.

As a result of experimental tests it was found, that the design of tested device allows to carrying out the combustion process on required conditions, with temperature exceeding 850°C.

Based on the collected data it was also found, that appropriate conditions can be maintained in the case of continuous operation of the device. Adverse influence of both periodic delivering the fuel and periodic rotation of the grate on temperature values has been noticed. They have resulted in huge fluctuations of temperature, and some decreases below required values.

The numerical model describing processes occurring inside the combustion chamber has confirmed the results of the experimental studies. The model has also indicated the possibility of obtaining higher temperature values, which was not possible during the experimental tests at the current stage of research, due to material limitations in some regions of the device.

As the first stage of research has indicated a high potential of the device for meat waste incineration, it is planned to carry out the next, more advanced stage of research. Providing a possibility of continuous operation of the device by modifying the fuel feeding system, the use of sensors adapted to operation in high temperature, and carrying out the measurement of emissions of particular chemical compounds (primarily CO) would be included. However, the most significant step would be use of meat waste as an addition for biomass fuel under appropriate conditions. It will require also installing an additional burner.

The numerical model will also be developed in the next stage of research. All heat exchange mechanisms should be included, with radiation model inside the computational domain as well as on external walls. More suitable combustion model with developed chemistry kinetics will be applied. Boundary conditions would also be adapted. Both mass flow and composition of the fuel delivered to the combustion chamber would be calculated using experimental data. Important stage of the numerical model would be a calculation of residence time of the fuel particles in the combustion chamber on required temperature.

## Acknowledgment

This study was carried out under the "BioORC: Construction of cogeneration system with small to medium size biomass boilers" project.

## References

1. F. Toldra, M. C. Aristoy, L. Mora, M. Reig, T, Meat Sci., **92**, 290-296, (2012)
2. W. Russ, R.M. Pittroff, Crit. Rev. Food Sci., **44**, 57-62, (2004)
3. U. Rahman, A. Sahar, M. A. Khan, Food Res. Int., 65, 322-328, (2014)
4. E. Cascarosa, A. Boldrin, T. Astrup, Waste. Manage. **33**, 2501-2508, (2013)
5. L. Fryda, K. Panapoulos, P. Vourliotis, E. Pavlidou, E. Kakaras, Fuel, **85**, 1685-1699, (2006)
6. K. Jayathilakan, K. Sultana, K. Radhakrishna, A.S. Bawa, J Food Sci., 49, 278-293, (2012)
7. Council Directive 94/67/EC of 16 December 1994 on the incineration of hazardous waste
8. Directive 2010/75/EU of the European Parliament and of the Council of 24 November 2010 on industrial emissions
9. C.G. Soni, Z. Wang, A.K. Dalai, T. Pugsley, T. Fonstad, Fuel, 88, 920-925, (2009)
10. E. Virmond, R. L. Schacker, W. Albrecht, C. A. Althoff, M. de Souza, R. F. P. M. Moreira, H. J. Jose, Energy, **36**, 3897-3906 (2011)
11. E. Mastorakos, A. Massias, C.D. Tsakiroglou, D.A. Goussis, V.N. Burganos, A.C. Payatakes, Appl. Math. Model., **23**, 55-76, (1999)
12. J. Chaney, H. Liu, J. Li, Energ. Convers. Manage., **63**, 149-156, (2012)
13. J. Collazo, J. Porteiro, J.L. Miguez, E. Granada, M.A. Gomez, Energ. Convers. Manage., **64**, 87-96, (2012)
14. W.K. Hiromi Ariyaratne, A. Malagalage, M. C. Melaaen, L.A. Tokheim, Chem. Eng. Sci., **123**, 596-608, (2015)
15. M. Miltner, A. Makaruk, M. Harasek, A. Friedl, Clean Technol. Envir., **10**, 165–174 (2008)

# Measurement of an electronic cigarette aerosol size distribution during a puff

*Miloslav* Belka[1,*], *Frantisek* Lizal[1], *Jan* Jedelsky[1], *Miroslav* Jicha[1], and *Jiri* Pospisil[2]

[1]Dept. of Thermodynamics and Environmental Sciences, Energy Institute, Faculty of Mechanical Engineering, Brno University of Technology, 616 69 Brno, Czech Republic
[2]Dept. of Power Engineering, Energy Institute, Faculty of Mechanical Engineering, Brno University of Technology, 616 69 Brno, the Czech Republic

**Abstract.** Electronic cigarettes (e-cigarettes) have become very popular recently because they are marketed as a healthier alternative to tobacco smoking and as a useful tool to smoking cessation. E-cigarettes use a heating element to create an aerosol from a solution usually consisting of propylene glycol, glycerol, and nicotine. Despite the wide spread of e-cigarettes, information about aerosol size distributions is rather sparse. This can be caused by the relative newness of e-cigarettes and by the difficulty of the measurements, in which one has to deal with high concentration aerosol containing volatile compounds. Therefore, we assembled an experimental setup for size measurements of e-cigarette aerosol in conjunction with a piston based machine in order to simulate a typical puff. A TSI scanning mobility particle sizer 3936 was employed to provide information about particle concentrations and sizes. An e-cigarette commercially available on the Czech Republic market was tested and the results were compared with a conventional tobacco cigarette. The particles emitted from the e-cigarette were smaller than those of the conventional cigarette having a CMD of 150 and 200 nm. However, the total concentration of particles from e-cigarette was higher.

## 1 Introduction

Roughly 6 million deaths around the world are attributable to tobacco smoking annually. More than 5 million of those deaths are outcomes of direct tobacco smoking while more than 0.6 million are caused by exposure to secondhand smoke [1]. Secondhand smoke or sometimes referred as "environmental tobacco smoke" is a combination of exhaled mainstream smoke (released from the cigarette end during a puff) and sidestream smoke (released from the cigarette between puffs). A chemical composition of both mainstream and secondhand smoke is almost identical. However, the total particle concentrations vary greatly [2]. Cigarette smoke contains around 4000 constituents including many carcinogens, such as polynuclear aromatic hydrocarbons (PAHs) or aldehydes [2,3].

Because of many adverse effects of smoking tobacco on human health, there are various tobacco control policies being implemented to reduce the tobacco use. The most effective policy is raising tobacco taxes, other can be the introduction of mandatory smoke-free bars and restaurants [4]. However, this is a long-time process and for example, smoke-free bars are not mandatory in the Czech Republic yet. Another mentioned way to avoid using tobacco products can be a use of alternatives, such as electronic cigarettes.

E-cigarettes are advertised as a healthier alternative to conventional cigarettes and they are categorized as nicotine delivery devices. They usually consist of three parts: a battery, an atomizer and a cartridge. The battery provides power to the heating element/atomizer, which atomizes a liquid solution inside the cartridge and thus produces vapors that are inhaled by a user [5]. Major compounds of the liquid solution are propylene glycol, glycerol, various amounts of nicotine and flavors.

An e-cigarette smoke is an aerosol comprised of fine particles (smaller than 1 μm). Knowledge of the particle size distribution and chemical composition is essential because it mainly determines deposition hot-spots and the subsequent impact on human health. A chemical analysis of an e-cigarette smoke has been carried out in several studies [3,6-8] and the results were compared with conventional cigarettes concerning the presence of carcinogens and other harmful constituents. The results showed that although the levels of harmful constituents varied considerably among different brands of e-cigarettes, they were around 1 to 2 orders lower than in conventional cigarette aerosol. On the other hand, a study from Wieslander et al. [9] concludes, that inhalation of highly concentrated propylene glycol can cause irritation of upper respiratory airways in non-asthmatic subjects and some subjects can react with lower airway obstruction or cough. Moreover, since e-cigarette popularity spread only recently, there are no

* Corresponding author: Belka@fme.vutbr.cz

direct studies engaged in the long-term use of e-cigarettes and its biological effects on human health [5].

Particle concentration and size distribution measurements encompass many obstacles and complications because one has to deal with high concentration aerosol containing volatile compounds. Ingebrethsen et al. [10] employed spectral transmission procedure to analyze particle concentration and sizes of undiluted e-cigarette aerosol. Fuoco et al. [11] compared several e-cigarettes and conventional tobacco products using Fast Mobility Particle Sizer. Results from both studies showed that e-cigarette aerosol is highly concentrated (order of $10^9$ particles/cm$^3$) with particle diameters around 200 nm. Manigrasso et al. [12] used similar experimental setup and the results implemented in a model for prediction of particle deposition in human respiratory airways. Schripp et al. [13] studied indoor air quality in conjunction with e-cigarette smoking by placing a volunteering smoker into 8 m$^3$ chamber.

E-cigarettes are becoming more and more popular nowadays. Therefore, it is very important to study their properties and evaluate their potential health threats. In this paper we carried out concentration and size measurements of particles emitted by e-cigarette commonly available on the Czech Republic market. Moreover, the results were compared with conventional tobacco cigarettes.

## 2 Methods

The aim of the experiments was to measure total particle concentration and sizes of mainstream cigarette and e-cigarette smoke. An experimental setup is depicted in Fig. 1. The smoke was analyzed for a 4 s square puff of total volume 60 cm$^3$, which was similar to that reported by Ingebrethsen et al [10]. The puff was performed by a "puffing machine", which was a piston placed in a cylinder and driven by an electromotor. The motions of the motor were set via laptop. After the puff was drawn into a sampling chamber, a control valve at the e-cigarette was closed and the one connecting the sampling chamber to a Scanning Mobility Particle Sizer 3936 (TSI, Shoreview, MN, USA) was opened.

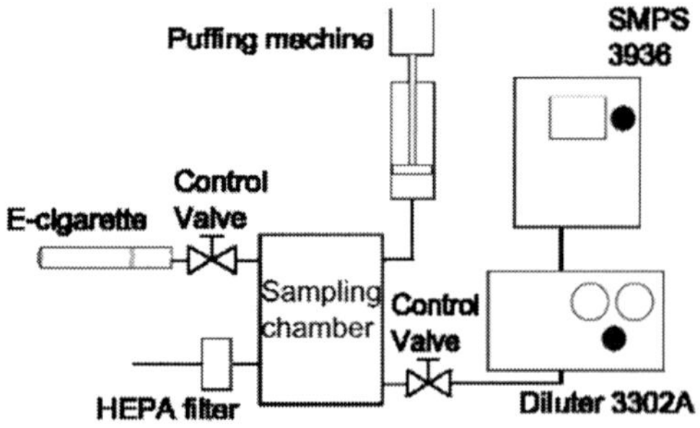


**Fig. 1.** Experimental setup for particle concentration and size measurement during a cigarette puff

In the SMPS system, the particles were neutralized using X-Ray neutralizer 3087 and then classified according to their electrical mobility inside a Differential Mobility Analyzer 3081. The classification is executed by changing voltage of an electrode. The classified particles continued to a Condensational Particle Counter 3775, where they grew via condensation of butyl alcohol and were counted afterwards. The SMPS requires some time to "scan" a chosen size distribution and the more time, the better accuracy. On the other hand, e-cigarette smoke is high concentration aerosol with volatile compounds. Therefore, fast measurement is required. The interval of 90 s was set as a compromise between sufficient sampling accuracy and requirement of the shortest possible duration of measurement. The flow rate of the SMPS was set to 0.3 LPM.

Because the particle concentration was expected to exceed the limit of the SMPS system, two-step dilution was applied. The first step took place in a 2.5 liters big sampling chamber and the second step was Aerosol Diluter 3302A (TSI, Shoreview, MN, USA). A dilution factor of the aerosol diluter was set to 100 and together with the sampling chamber, the total dilution factor was roughly around 4000. In order to preserve constant pressure inside the sampling chamber during the SMPS measurement, ambient air filtered through HEPA filter was supplied.

An e-cigarette by Joyetech was tested during the measurements and it was filled with two different liquids with nicotine content of 0 and 16 mg/ml. The e-cigarette was fully charged and its reservoir was filled up to three–quarters before the measurements. The e-cigarette has variable wattage output, which changes power output from the battery. The wattage output was adjusted to a maximum value, 9.6 W, and thus delivering the largest amount of vapor. The results were compared with Light Marlboro cigarette, which has a nicotine content of 0.6 mg. Four puffs were measured for every sample while the first puff was not included into averaged results as being considered a "warm up" puff. After every puff, a filtered air was drawn into the sampling chamber by a vacuum pump in order to remove any remaining aerosol.

## 3 Results

The particle number concentrations and size distributions of e-cigarettes can be seen in Fig. 2. and Fig. 3. Particle number concentrations were in the order of $10^9$ part/cm$^3$, which is comparable to results by Fuoco et al [11]. Size

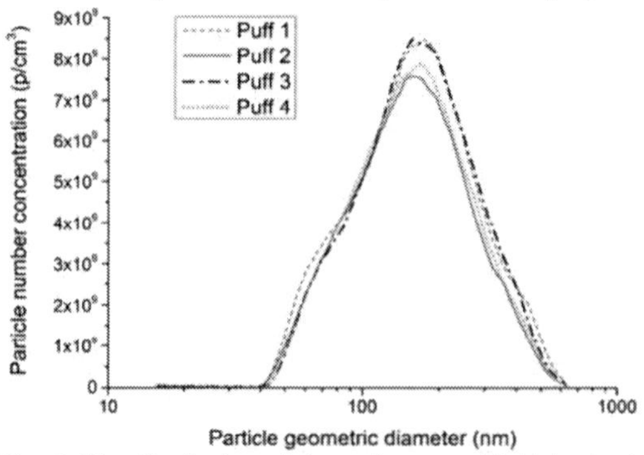


**Fig. 2.** Size distribution of the e-cigarette with high nicotine content

distributions of single puffs are rather identical and there is no noticeable difference for the first "warm up" puff, which was expected considering the e-cigarette working principle. The main modes for the e-cigarette with high nicotine content were 156–174 nm, for the e-cigarette with zero nicotine content they were in the range of 145–168 nm, which indicated that there was no significant dependence of particle sizes on nicotine content.

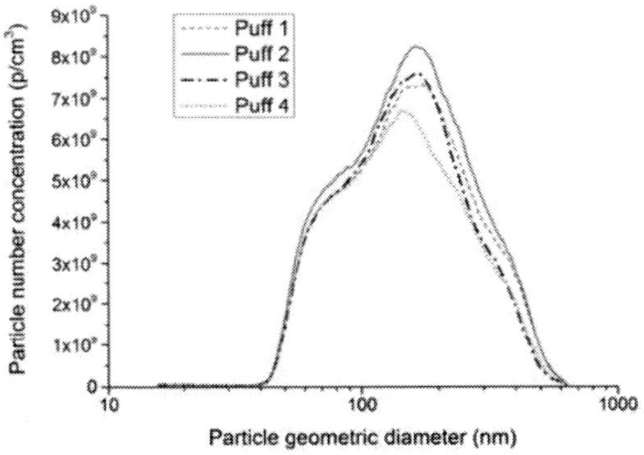


**Fig. 3.** Size distribution of the e-cigarette with zero nicotine content

The particle number concentration and size distribution of conventional cigarette is depicted in Fig. 4. The particle number concentrations varied greatly between individual puffs. The particle number concentration increased with increasing puff number. This phenomenon was observed in other studies as well [14,15] and it is probably caused by the actual length of the remaining smoldering cigarette. On the other hand, the size distributions were comparable between the puffs except for the first puff. The main mode for the first puff was 156 nm while for the rest of the puffs the main modes covered the range of 187–201 nm.

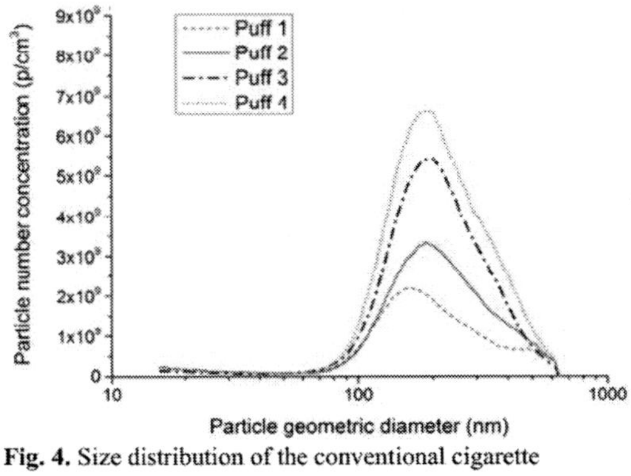


**Fig. 4.** Size distribution of the conventional cigarette

A comparison of size distributions of the e-cigarette and the conventional cigarette can be seen in Fig. 5. Size distribution for every type was calculated as the mean average of puffs 2 to 4. The averaged total number concentrations and count median diameters (CMDs) are listed in Table 1. All the distributions were log-normal with positive skewness. Comparing two liquids for the e-cigarette, there is a minor change in particle number concentration, in particular higher nicotine content resulted in higher particle number concentration for the main mode. However, as can be seen in Table 1, the total number concentration is higher for the zero nicotine content liquid. It is worth noting that there is a small deviation at 70 nm diameter for zero nicotine content e-cigarette, which increases the total number concentration and decreases a CMD.

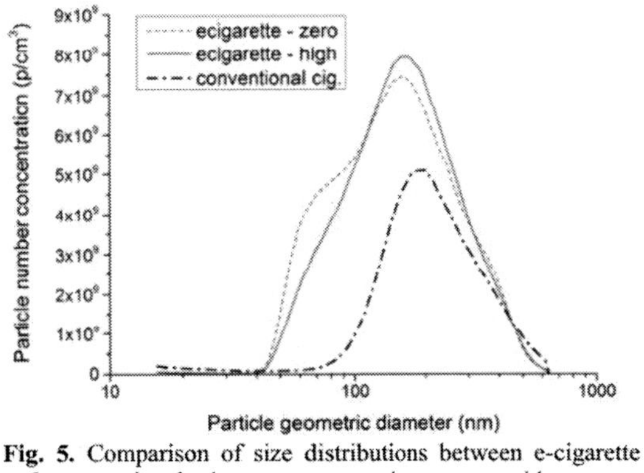


**Fig. 5.** Comparison of size distributions between e-cigarette and conventional cigarette; two e-cigarette cartridges were tested (high and zero nicotine content)

The total particle concentrations of e-cigarette were roughly two times higher than that of a conventional cigarette. This indicates that although e-cigarette smoke contains less harmful constituents than conventional cigarette smoke, the particle numbers for e-cigarettes are higher and human respiratory airways can be exposed to high doses of compounds such as propylene glycol. The median size of the particles was a little bit smaller for the e-cigarettes and ranged around 150 nm. This means that as well as for conventional cigarettes, e-cigarette smoke particles are in fine or ultrafine range and thus can easily penetrate deep into human lungs and deposit there mostly by diffusion and sedimentation.

**Table 1.** Comparison of total particle concentrations and CMDs (count median diameters); Value±Geometric Standard Deviation

| | Averaged total number concentration (p/cm³)·10⁹ | Averaged CMD (nm) |
|---|---|---|
| E-cigarette zero nicotine | 4.81±0.1 | 147.1±4,5 |
| E-cigarette high nicotine | 4.63±0.05 | 157.6±3,8 |
| Conventional Cigarette | 2.41±0.15 | 201.6±0.3 |

## Discussion

In this paper we carried out measurements of particle number concentrations and size distributions of e-cigarettes using the SMPS 3936 system. The measured cigarette mainstream smoke was sampled into the chamber and analyzed within 90 s. The total concentration and size distributions were found to be in acceptable agreement with other studies [11,14].

DOI: 10.1051/epjconf/201714302006

However, there are some uncertainties in the measurement. Since the SMPS system divides a size distribution into channels and then scans the concentration in every channel step by step, it requires constant size distribution in analyzed sample during scan time. However, an e-cigarette smoke is high concentration aerosol with volatile compounds. That means high probability of coagulation, which is a collision between particles and subsequent adherence. The coagulation rate is time-dependent, therefore it is convenient to measure the particle sizes very fast, but measurement of the size distribution within few seconds is not possible with the SMPS 3936 system.

Coagulation can be prevented by proper dilution, such as sampling of particles into a chamber. On the other hand, Ingebrethsen et al. [10] stated that high dilution can cause evaporation of large particles and biases in the results. Moreover, dilution of e-cigarette smoke can result in condensation of volatile gaseous compounds and occurrence of small diameter (less than 50 nm) mode in size distribution. This bimodal distribution was observed for example by Schripp et al. [13]. Small diameter mode was not found in our measurements, but there was a slight deviation around 70 nm for zero nicotine content e-cigarette.

Another issue is the sedimentation and deposition of particles on the chamber walls, which can alter the size distribution during measurement.

To summarize the facts above, the concentration and size distribution of e-cigarette was measured with sufficient accuracy. However, more measurements will be needed to quantify the influence of various aerosol phenomena.

## Conclusions

The total concentrations and size distributions of e-cigarette mainstream smoke were measured by the SMPS system. The e-cigarettes emitted smoke with concentrations in the order of $10^9$ particles/cm$^3$. The main mode was approximately 170 nm and the count median diameter was approximately 150 nm. There was only a minor change in concentrations and sizes between different liquids used in the e-cigarette.

Comparing the e-cigarette smoke with that of a conventional cigarette, the number concentration was two times higher for the e-cigarette and the count median diameter was smaller than that of the conventional cigarette, 150 against 200 nm. This indicates that some e-cigarettes can deliver considerably more particles into a human respiratory system than conventional cigarettes, even though the e-cigarette smoke contains less harmful constituents. All the particles of the e-cigarette smoke are smaller than 1 μm and some of them are in ultrafine particle range (smaller than 0.1 μm), thus able to penetrate deep into a human respiratory system, deposit there and potentially get into the bloodstream. However, these measurements are only a first step in evaluation of possible health outcomes. It is necessary to evaluate a long-term health effects of e-cigarettes

## Acknowledgement

This work was supported by the Czech Science Foundation under the grant GA16-23675S and by the project FSI-S-14-2355.

## References

1. World Health Organization., *WHO report on the global tobacco epidemic, 2011: warning about the dangers of tobacco.* (World Health Organization, Geneva, 2011).
2. IARC Working Group on the Evaluation of Carcinogenic Risks to Humans : Tobacco Smoke and Involuntary Smoking and International Agency for Research on Cancer., *Tobacco smoke and involuntary smoking.* (International Agency for Research on Cancer, Lyon, 2004).
3. R. R. Baker, E. D. Massey, and G. Smith, Food and Chemical Toxicology **42**, S53 (2004).
4. World Health Organization., *WHO Report on the Global Tobacco Epidemic, 2009 : implementing smoke-free environments.* (World Health Organization, Geneva, 2009).
5. R. Grana, N. Benowitz, and S. A. Glantz, Circulation **129** (19), 1972 (2014).
6. C. Hutzler, M. Paschke, S. Kruschinski, F. Henkler, J. Hahn, and A. Luch, Arch Toxicol **88** (7), 1295 (2014).
7. R. Tayyarah and G. A. Long, Regulatory Toxicology and Pharmacology **70** (3), 704 (2014).
8. T. R. Cheng, Tobacco Control **23**, 11 (2014).
9. G. Wieslander, D. Norback, and T. Lindgren, Occup Environ Med **58** (10), 649 (2001).
10. B. J. Ingebrethsen, S. K. Cole, and S. L. Alderman, Inhalation Toxicology **24** (14), 976 (2012).
11. F. C. Fuoco, G. Buonanno, L. Stabile, and P. Vigo, Environ Pollut **184**, 523 (2014).
12. M. Manigrasso, G. Buonanno, F. C. Fuoco, L. Stabile, and P. Avino, Environ Pollut **196**, 257 (2015).
13. T. Schripp, D. Markewitz, E. Uhde, and T. Salthammer, Indoor Air **23** (1), 25 (2013).
14. S. K. Sahu, M. Tiwari, R. C. Bhangare, and G. G. Pandit, Aerosol Air Qual Res **13** (1), 324 (2013).
15. S. L. Alderman and B. J. Ingebrethsen, Aerosol Science and Technology **45** (12), 1409 (2011).

# New methodology for the walls design in buildings by numerical simulation of the thermal convection

*Elhadj* Benachour[1,2,a], Belkacem Draoui[1], *Bachir* Imine[2], *Khadidja* Asnoune[1], *Elmir* Mohamed[1]

[1]Laboratory of Energy in Arid Regions, University of Tahri Mohamed - Bechar, P.O. P 417. 08000 , Road Kenedza, Bechar, Algeria
[2]Department of Mechanical Engineering, University of Sciences and Technology of Oran-Mohamed Boudiaf, P.O.1505 El-Mnaouar, Oran, Algeria

**Abstract.** Buildings are complex systems composed of several elements, which are assembled to respond to a number of needs functional and symbolic according to set of legal and environmental requirements and potentially accommodate users with different levels of demand. Predicting the conception of the external wall is beneficial in the design of house and building structures.in this study, an analogy was used for the functions which are discretized by the finite difference method and integrated in the CFD code which is based on the finite volume method. The CFD software is used as a technique to modelling the behaviour of fluid and the thermal convection in the external wall of the house with different Rayleigh numbers [$10^3 \leq$ Ra $\leq 10^5$]. In the second phase, we change the thickness of the wall several times and calculate the Nusselt number and exchange coefficient of heat transfer aims to find a cloud point respectively for the thicknesses e = 0, L /40, L /20 and L /10. After, we developed a relationship that helps us to know the exchange ratio for each thickness ( e ) belongs to the interval [0, L /10] by the Lagrange polynomial interpolation method for Rayleigh number equal $10^4$ , and then we developed a FORTRAN program to control the nonlinear equation of order three. This method for predicting exchange coefficient of convection for to optimize the design of walls in buildings.

## 1 Introduction

The thermal behavior of buildings is a current problem Actually, In Algeria, The construction industry must start to take into account the changing needs of the population in its projects. It's not just homes that need to be adapted to ensure a safe and inclusive environment, but also the wider surroundings. Local shops and town centres have to be accessible and to accommodate the needs of people. This survey indicates that the biggest improvements could be made to the building sector, which must be properly adapted to ensure a safe environment for a workforce .The building sector today is known to be consuming 40% of the world energy [1,2], and in turn, supports 23–40% of the world's greenhouse gas emission, particularly CO2 [3]. It has become common in recent years to gather data on human attitudes and behaviour in building energy research through interviews with building occupants and other relevant actors. Examples of this kind of research are found in the main journals which deal with technical aspects of energy in buildings, including Building and Environment [4–12], Building and Environment [13-15], Energy Storage [16-20], Energy and Buildings [21–33]. In most situations, the mechanical cooling devices offer solutions that are neither environment friendly nor energy sustainable. The mechanical devices are non-functional and cannot offer thermal comfort without energy input. Hence utilization of advanced building materials and passive technologies in buildings may offer the solution for thermal comfort demands, substantially reduce the energy demand, impact on the environment and carbon footprint of building stock worldwide [34].Thermal comfort is dependent and influenced by a range of environmental factors viz. air temperature, radiant temperature, humidity, air movement, metabolic rate or human activity [35,36]. Also, numerous studies across the world have shown the impacts of hot working environments on the working population [37–43]. The choice of which material or combination of materials is used depends on a wide variety of factors. Some insulation materials have health risks, some so significant the materials are no longer allowed to be used but remain in use in some older buildings such as asbestos fibers and urea. Also, the cost can be high compared to the traditional insulation. The interior or external insulation does not often become the only possible solution in particular with the old buildings in the Sahara of Algeria. Hygrothermal parameters of the existing wall material should be reported and well known for designing powerful and durable walls in the time

[a] Corresponding author: benachour_elhadj@yahoo.fr

study. First, in this paper, CFD software is used as a technique to modelling the behaviour of fluid and the thermal convection in the external wall of the house with different Rayleigh numbers $[10^3 \leq Ra \leq 10^5]$, In this context, an analogy was used for the functions which are discretized by the finite difference method and integrated in the CFD code which is based on the finite volume method. Secondly, The most important part in this work is to vary the thickness of the building material of the outer wall four times and calculate the Nusselt number and exchange coefficient of heat transfer aims to find a cloud point respectively for the thicknesses e = 0, L / 40, L / 20 and L / 10. After, we developed a relationship that helps us to know the exchange ratio for each thickness e belongs to the interval [0, L / 10] by the Lagrange polynomial interpolation method, and then we developed a FORTRAN program to control the nonlinear equation of order three. This method for predicting exchange coefficient of convection to optimize the design of walls in buildings before starting the wall construction for Rayleigh number equal $10^4$.

## 2 Geometric configuration

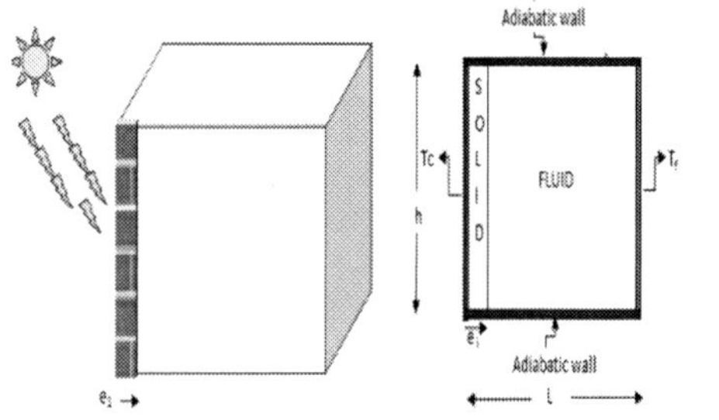


**Fig. 1.** Schematic of the studied configuration

To simplify the problem, assume that:
- The fluid is Newtonian and incompressible.
- The heat dissipation by viscous friction is neglected.
- The Boussinesq approximation is considered

## 3 Objectives

We can express our aim for this study in the following points:

- Prediction of the wall design through numerical evaluation of the convection in buildings by the Lagrange polynomial interpolation method.
- Modelling Study of the convection – conduction coupling.
- Study of the effect of the distribution of the heat inside buildings on the convection.
- Comparative study between the construction of buildings with different thicknesses and without insulation.
- Study the effect of the Rayleigh number on the convection in building.

## 4 Mathematical model

The fluid is assumed incompressible and obeys the Boussinesq approximation. In these cases, continuity in two dimensions and the equations governing the flow and energy is given by:

**Continuity.**

$$\frac{\partial U}{\partial X} + \frac{\partial V}{\partial Y} = 0 \qquad (1)$$

**X-momentum.**

$$\frac{\partial U}{\partial t} + U\frac{\partial U}{\partial X} + V\frac{\partial U}{\partial Y} = -\frac{1}{\rho}\frac{\partial P}{\partial X} + \upsilon \nabla^2 U \qquad (2)$$

**Y-momentum.**

$$\frac{\partial V}{\partial t} + U\frac{\partial V}{\partial X} + V\frac{\partial V}{\partial Y} = -\frac{1}{\rho}\frac{\partial P}{\partial X} + \upsilon \nabla^2 V + g\beta\frac{\partial T}{\partial X} \qquad (3)$$

**Energy.**

$$\frac{\partial T}{\partial t} + U\frac{\partial T}{\partial X} + V\frac{\partial T}{\partial Y} = \frac{\lambda}{\rho c_p}\nabla^2 T + \frac{1}{\rho c_p}\varphi \qquad (4)$$

The derived equation of motion Eq. (2) over Y and the equation of motion Eq. (3) by contributing to x, then, after subtracting the two equations obtained, we obtain the equations dimensionless variables in writing Helmotz in terms of vorticity and stream function formulation are as follows:

$$\frac{\partial \omega}{\partial t} + U\frac{\partial \omega}{\partial X} + V\frac{\partial \omega}{\partial Y} = \Pr \nabla^2 \omega + Ra \Pr \frac{\partial T}{\partial X} \qquad (5)$$

$$\frac{\partial \theta}{\partial t} + U\frac{\partial T}{\partial X} + V\frac{\partial T}{\partial Y} = \frac{\partial^2 T}{\partial X^2} + \frac{\partial^2 T}{\partial Y^2} + \frac{\varphi}{\rho Cp}\frac{L^2}{a\Delta T} \qquad (6)$$

$$\frac{\partial^2 \psi}{\partial X^2} + \frac{\partial^2 \psi}{\partial Y^2} = -\omega \qquad (7)$$

The stream function and vorticity are related to the velocity components by the following expressions:

$$U = \frac{\partial \psi}{\partial Y},\ V = -\frac{\partial \psi}{\partial X} \ \ and \ \ \omega = \frac{\partial V}{\partial X} - \frac{\partial U}{\partial Y} \qquad (8)$$

The dimensionless parameters in the equations above are defined as follows:

$$\begin{cases} X = \dfrac{x}{L}, \quad Y = \dfrac{y}{L}, \quad U = \dfrac{u}{u_i} \\ V = \dfrac{v}{u_i}, P = \dfrac{P}{\rho u_i^2}, T^* = \dfrac{T - T_i}{T_c - T_i}, t^* = \dfrac{t}{L^2 / a} \end{cases} \quad (9)$$

For solid (Concrete), we are interested only in the following heat equation:

$$\frac{\partial T}{\partial t} = \frac{1}{a}\left(\frac{\partial^2 T}{\partial X^2} + \frac{\partial^2 T}{\partial Y^2}\right) \quad (10)$$

## 5 Procedure of simulation

First, The numerical calculation was conducted using the GAMBIT and FLUENT software. The numerical procedure used in this work is that of finite volume. It involves the integration of differential equations of mathematical model on finite control volumes for the corresponding algebraic equations. In this context, an analogy was used for the functions which are discretized by the finite difference method and integrated in the CFD code which is based on the finite volume method. The SIMPLE algorithm [44] was chosen for the coupling speed pressure in the Navier-Stokes equations on a staggered grid. The convective terms in all equations are evaluated using the schema apwind first order. The discretization of the time term is made in a totally implicit scheme. The convergence of the solution is considered reached when the maximum relative change of all variables (u, v, w, p, t) between two successive time is not less than 104. With an aim of following well any variation of the fields thermal and hydrodynamic, we used a uniform grid of 14241 nodes and 14480 Elements in non-stationary mode. Secondly, we developed a FORTRAN program to control the nonlinear equation of order three. This method of interpolation was used for predicting exchange coefficient of convection to optimize the design of walls in buildings

## 6 Validation of the Code of Calculation

### 6.1 Validation of the model

The computer code was validated on a natural convection problem of stale air in a square cavity with vertical walls differentially heated and adiabatic horizontal walls. Our results were compared with those obtained by De Vahl Davis (1983) [45] .The latter dealt with the same problem by adopting the finite difference method with the vorticity-stream function formulation (see Table 1).

**Table 1 Comparisons of the results of validation**

| | $Ra = 10^3$ | $Ra = 10^4$ | $Ra = 10^5$ |
|---|---|---|---|
| Nu (average number) (hot wall) | present study: 1.2 | present study : 2.257 | Present study : 4.64 |
| | | | |
| Nu (average number) (hot wall) | De Vahl Davis [45] 1.12 | De Vahl Davis [45]: 2.243 | De Vahl Davis [45] : 4.52 |

## 7 Mathematical model for the lagrange polynomial interpolation method

The Lagrange interpolating polynomial is the polynomial of degree ≤ (n-1) that passes through the n points (x1, y1=f(x1)) , (x2, y2=f(x2)) , ..., (xn, yn=f(xn)), the interpolation polynomial in the Lagrange form is a linear combination and is given by :

$$P_n(x) = \sum_{i=0}^{n} y_i \; L_i(x) \quad (11)$$

For Lagrange basis polynomials

$$L_i(x) = \frac{\prod_{\substack{j=0 \\ j\neq i}}^{n}(x - x_j)}{\prod_{\substack{j=0 \\ j\neq i}}^{n}(x_i - x_j)} = \prod_{\substack{j=0 \\ j\neq i}}^{n} \frac{(x - x_j)}{(x_i - x_j)} \quad (12)$$

Where

$$\begin{cases} L_i(x_j) = 0 \;\; if \;\; i \neq j \\ L_i(x_j) = 1 \;\; if \;\; i = j \end{cases} \quad (13)$$

### 7.1 The polynomial interpolation points

We wish to find the polynomial interpolating the points

**Table2 The sets of polynomial interpolation points.**

| B=e/L | 0 | 1/40 | 1/20 | 1/10 |
|---|---|---|---|---|
| $Nu_a$ | 2.24 | 1.59 | 1.35 | 0.98 |

Where, the points of Nu (e) are obtained by numerical simulation of the FLUENT software, for the Rayleigh number equal $10^4$.

## 8 Results and discussion

The boundary conditions have been established to simulate a geometric configuration used frequently in two-dimensional approximation .The structure of the flow, the temperature field and heat transfer through the hot wall are discussed in this section.In this study, to target the most important goal, we will show and studied the dynamic and thermal behavior of the fluid in the cavity.

### 8.1 Isotherms

The isotherms are shown in Figures [2]. The heat distribution in the cavity is in accordance with the fluid circulation revealed by isotherms and iso-currents. Indeed we find a heating fluid from the interface, if it causes the change of the heat distribution in the cavity (see figure) for different numbers of Rayleigh For a fixed value of the number of Prandtl equal 0.71 (air fluid). Gradually, as the Rayleigh number has increased, the isotherms become increasingly wavy and heat transfer increases, so the flow intensifies and natural convection is expanding and predominates (natural convection is predominant).

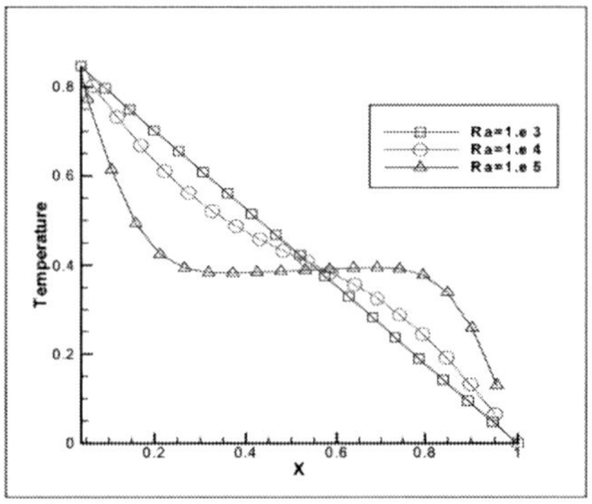


**Fig. 2. The Isotherms for wall thickness e = L/20, and for different Rayleigh Numbers, Pr=0.71.**

### 8.2 Lagrange polynomials

The points of Nu (e) are obtained by numerical simulation of the FLUENT software, for the Rayleigh number equal $10^4$. (See table 3)

**Table3 Nusselt number for different values of thickness. and for Ra=$10^4$**

| Ra | i | Xi=e | Yi=Nu (e) Average Nusselt number |
|---|---|---|---|
| $10^4$ | 0 | 0 | **2.24** |
| $10^4$ | 1 | e=L/40 | **1.59** |
| $10^4$ | 2 | e=L/20 | **1.35** |
| $10^4$ | 3 | e=L/10 | **0.98** |

If (0, 2.24),(L/40, 1.59), (L/20, 1.35), (L/10, 0.98), are given data points, then the cubic polynomial passing through these points can be expressed as,

$$P_3(x) = y_0 l_0 + y_1\ l_1 + y_2\ l_2 + y_3\ l_3$$

We would have the four basis polynomials

$$l_0 = \frac{(x - \frac{L}{40})\ (x - \frac{L}{20})\ (x - \frac{L}{10})}{(0 - \frac{L}{40})\ (0 - \frac{L}{20})\ (0 - \frac{L}{10})} \quad (14)$$

$$l_1 = \frac{(x - 0)\ (x - \frac{L}{20})\ (x - \frac{L}{10})}{(\frac{L}{40} - 0)\ (\frac{L}{40} - \frac{L}{20})\ (\frac{L}{40} - \frac{L}{10})} \quad (15)$$

$$l_2 = \frac{(x - 0)\ (x - \frac{L}{40})\ (x - \frac{L}{10})}{(\frac{L}{20} - 0)\ (\frac{L}{20} - \frac{L}{40})\ (\frac{L}{20} - \frac{L}{10})} \quad (16)$$

$$l_3 = \frac{(x - 0)\ (x - \frac{L}{40})\ (x - \frac{L}{20})}{(\frac{L}{10} - 0)\ (\frac{L}{10} - \frac{L}{40})\ (\frac{L}{10} - \frac{L}{20})} \quad (17)$$

We can see that each term is of degree three, so the entire sum has degree at most three.
The polynomial P(x) given by the above formula is called Lagrange's interpolating polynomial and the functions Eq. (l4, Eq). (l5), Eq). (l6), Eq.( l7) are called Lagrange's interpolating basis functions.

### 8.3 Nusselt number correlations

It should be noted that the numerical result given by the equation Eq.(20). Remarkably similar to the estimate of the average Nusselt number by the Lagrange interpolation method for each wall thickness between [0, L / 10] but only for Ra=$10^4$.

$$P_3(x) = y_0 l_0 + y_1\ l_1 + y_2\ l_2 + y_3\ l_3 \quad (18)$$

$$P_3(x) = 2.24 l_0 + 1.59\ l_1 + 1.35\ l_2 + 0.98\ l_3 \quad (19)$$

$$P_3(x) = -\frac{8960}{3L^3}x^3 + \frac{1656}{3L^2}x^2 - \frac{113.8}{3L}x + 2.24 \quad (20)$$

Therefore, we can write,

$$\overline{Nu}(e) = -\frac{29866}{L^3}e^3 + \frac{552}{L^2}e^2 - \frac{37.93}{L}e + 2.24 \quad (21)$$

Where:

$\overline{Nu}$ : Average Nusselt number.
e : Wall thickness.
Also, we can write:

$$\overline{Nu} = -2986.6\ B^3 + 552\ B^2 - 37.93\ B + 2.24 \quad (22)$$

Where: B: e/L

In heat transfer within a fluid, the Nusselt number (Nu) is the ratio of convective to conductive heat transfer. The average value of the heat transfer coefficient is often needed in process engineering and design applications. In this context, the average Nusselt number is given by:

$$\overline{Nu} = \frac{\bar{h} H}{\lambda} \Rightarrow \bar{h} = \frac{\lambda \overline{Nu}}{H} \quad (23)$$

Where h is the convective heat transfer coefficient of the flow, H is the characteristic length, λ is the thermal conductivity of the fluid.

### 8.4 Average Nusselt numbers

The influence of the Rayleigh number and Wall thickness on the average Nusselt number is shown in Fig. 5 and fig 6. The Nusselt number decreases with increasing of the wall thickness. Therefore, the heat transfer also decreases because the inertia and the thermal resistance of the wall increases. The average value of the Nusselt number

 DOI: 10.1051/epjconf/201714302007

increases as the value of the Rayleigh number (Ra) increases. Broadly, advection becomes stronger and thus heat transfer increases. This trend is qualitatively consistent with that seen in the convection for important values of Ra.

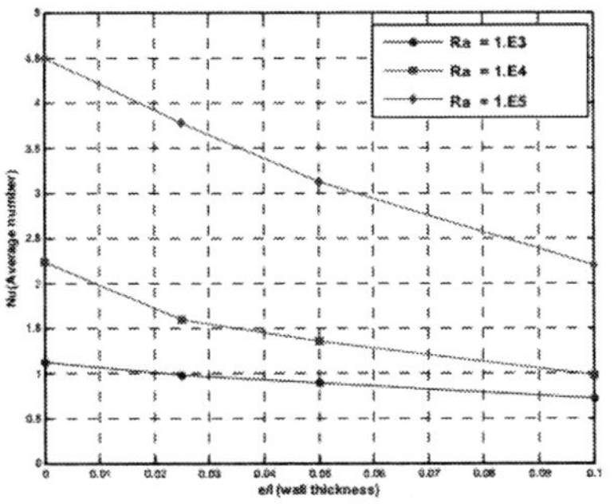


**Fig. 5.** Profiles of average Nusselt number for different wall thickness and for different Rayleigh number, $Ra=10^3, 10^4, 10^5$, Pr = 0.71.

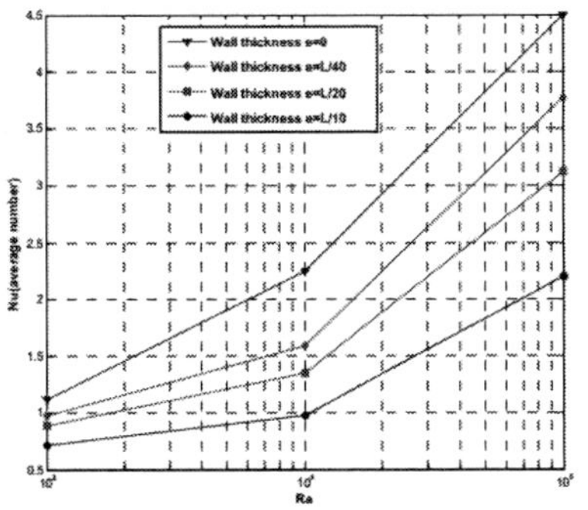


**Fig. 6.** Profiles of average Nusselt number for different Rayleigh number, $Ra=10^3, 10^4, 10^5$, Pr = 0.71.

## 9 Conclusion

Extensive numerical results for the laminar natural convection heat transfer from a heated wall are obtained for the range of conditions as: Rayleigh number, $10^3 \leq Ra \leq 10^5$ and Prandtl number Pr = 0.71, Concrete wall for different thicknesses is viewed with $0 \leq e \leq L/10$. We are interested in convection conduction coupling. First, CFD software is used as a technique to modelling the behaviour of fluid and the thermal convection in the external wall of the house, in this context, an analogy was used for the functions which are discretized by the finite difference method and integrated in the CFD code which is based on the finite volume method. Secondly, The most important part in this work is to vary the thickness of the building material of the outer wall four times and calculate the Nusselt number and exchange coefficient of heat transfer aims to find a cloud point respectively for the thicknesses e = 0, L / 40, L / 20 and L / 10. After, we developed a relationship that helps us to know the Nusselt number and exchange ratio for each thickness ( e ) belongs to the interval [0, L / 10] by the Lagrange polynomial interpolation method, and then we developed a FORTRAN program to control the nonlinear equation of order three (equation Eq.(21).or. Eq.(22). This method for predicting exchange coefficient of convection for to optimize the design of walls in buildings. Since, Interpolation is the process of defining a function that takes on specified values at specified points.

Predictive simple correlation was developed to estimate the value of the average Nusselt number and the coefficient of heat transfer for exchange any thickness ranging from 0 to L/10 .it just enough to replace the thickness value in equation Eq.(21), to calculate the Nusselt number planned before the design and construction of walls.

## Acknowledgment

This work supported by the Research program sponsored by the Faculty of Science and Technology, Department of technology, ENERGARID laboratory, Tahri Mohamed University, Bechar, Algeria.
I would like to thank all reviewers for taking the time and energy to review our work.

## References

1. Zhang, F., Cooke, P., (2012) Green Building and Energy Efficiency. Center for Advanced Studies, Cardiff University, United Kingdom.,
2. A.P. Olukoya Obafemi, Sevinç Kurtb, (2016) Environmental impacts of adobe as a building material: The north cyprus traditional building case, Case Studies in Construction Materials,4, 32-41.
3. Gunnell, K., Chrisna, D.P., Gibberd, J.,( 2009.) Green Building In South Africa: Emerging Trends. Department of Environmental Affairs and Tourism, The Republic of South Africa,
4. Ray Galvin, (2015) How many interviews are enough? Do qualitative interviews in building energy consumption research produce reliable knowledge?, Journal of Building Engineering. 1 , 2–12
5. P. Davies, M. Osmani, (2011) Low carbon housing refurbishment challenges and incentives: architects' perspectives, Build. Environ. 46, 1691–1698.
6. C. Isakssona, F. Karlsson, (2006) Indoor climate in low-energy houses—an interdisciplinary investigation, Build. Environ. 41, 1678–1690.
7. R. Kyrö, J. Heinonen, S. Junnila, (2012) Housing managers key to reducing the greenhouse gas emissions of multi-family housing companies? A mixed method approach, Build. Environ. 56, 203–210.
8. B. Meerbeek, M. te Kulve, T. Gritti, M. Aarts, E. van Loenen, E. Aarts, (2014) Building automation and perceived control: a field study on motorized exterior blinds in Dutch offices, Build. Environ. 79, 66–77.

9. Meier, C. Aragon, T. Peffer, D. Perry, M. Pritoni, (2011) Usability of residential thermostats: preliminary investigations, Build. Environ. 46 1891–1898.
10. W. Miller, L. Buys, J. Bell, (2012) Performance evaluation of eight contemporary passive solar homes in subtropical Australia, Build. Environ. 56, 57–68.
11. M. Osmani, A. O'Reilly, (2009),Feasibility of zero carbon homes in England by 2016: a house builder's perspective, Build. Environ. 44 , 1917–1924.
12. M. Pelenur, H. Cruickshank, (2013) Investigating the link between well-being and energy use; an explorative case study between passive and active domestic energy management systems, Build. Environ. 65 26–34
13. Hyo Seon Park, Changyoon Ji, Taehoon Hong,( January 2016), Methodology for assessing human health impacts due to pollutants emitted from building materials, Building and Environment, Vol 95, PP. 133-144
14. Andrey Tsapalov, Konstantin Kovler, (January 2016), Revisiting the concept for evaluation of radon protective properties of building insulation materials, Building and environment , 95, PP 182-188
15. Nadja Lynge Lyng, Lars Gunnarsen, Helle Vibeke Andersen, Vivi Kofoed-Sørensen, Per Axel Clausen, ( May 2016), Measurement of PCB emissions from building surfaces using a novel portable emission test cell, Building and Environment, 101, PP 77-84
16. Muthuvelan Thambidurai, Karthik Panchabikesan, krishna Mohan N, Velraj Ramalingam,( December 2015) Review on phase change material based free cooling of buildings—The way toward sustainability, Journal of Energy Storage, Vol 4, , PP 74-88
17. N Heinrich Badenhorst, Carl Sandrock, ( 2016) Novel method for thermal conductivity measurement through flux signal deconvolution, ournal of Energy Storage, 6,PP. 32-39
18. Najoua Bader, Abdelmottaleb Ouederni, (2016), Optimization of biomass-based carbon materials for hydrogen storage, Journal of Energy Storage, 5, PP 77-84
19. Amr Owes Elsayed,( December 2015), Numerical study on performance enhancement of solid–solid phase change materials by using multi-nanoparticles mixtures, Journal of Energy Storage, Volume 4, PP 106-112
20. Satyender Singh, Prashant Dhiman, ( 2016), Exergoeconomic analysis of recyclic packed bed solar air heater- sustained air heating system for buildings, Journal of Energy Storage, 5, PP. 33-47
21. T. Chiang, S. Natarajan, I. Walker, (2012) , A laboratory test of the efficacy of energy display interface design, Energy Build. 55 ,471–480.
22. M. Coleman, N. Brown, A. Wright, S. Firth, (2012), Information, communication and entertainment appliance use—insights from a UK household study, Energy Build. 54 , 61–72.
23. R. Galvin, (2013) , Impediments to energy-efficient ventilation of German dwellings: a case study in Aachen, Energy Build. 56 ,32–40.
24. K. Gram-Hanssen, T. Christensen, P. Petersen, (2012), Air-to-air heat pumps in real-life use: are potential savings achieved or are they transformed into increased comfort? Energy Build. 53, 64–73.
25. Jovanovíc, P. Pejíc, S. Djoríc-Veljkovíc, J. Karamarkovíc, M. Djelíc, (2014), Importance of building orientation in determining daylighting quality in student dorm rooms: physical and simulated daylighting parameters' values compared to subjective survey results, Energy Build. 77 , 158–170.
26. J. Kanters, M. Horvat, M.-C. (2014) Dubois, Tools and methods used by architects for solar design, Energy Build. 68, 721–731.
27. M. Santamouris, S. Alevizos, L. Aslanoglou, D. Mantzios, P. Milonas, I. Sarelli, S. Karatasou, K. Cartalis, J. Paravantisca, (2014) Freezing the poor—indoor environmental quality in low and very low income households during the winter period in Athens, Energy Build. 70 ,61–70.
28. S. Karjalainen, (2011), Consumer preferences for feedback on household electricity consumption, Energy Build. 43, 458–467.
29. S. Karjalainen, (2013), Should it be automatic or manual—the occupant's perspective on the design of domestic control systems, Energy Build. 65 119–126.
30. G. Menzies, J. (2005) Wherrett, Windows in the workplace: examining issues of environmental sustainability and occupant comfort in the selection of multiglazed windows, Energy Build. 37, 623–630.
31. Ridley, J. Bere, A. Clark, Y. Schwartz, A. Farrd, (2014), The side by side in use monitored performance of two passive and low carbon Welsh houses, Energy Build. 82 ,13–26.
32. B. Risholt, B. Time, A. Hestnes, (2013) Sustainability assessment of nearly zero energy renovation of dwellings based on energy, economy and home quality indicators, Energy Build. 60 ,217–224.
33. T. Sookchaiy, V. Monyakul, S. Thep, (2010) , Assessment of the thermal environment effects on human comfort and health for the development of novel air conditioning system in tropical regions, Energy Build. 42 ,1692–1702.
34. P.K. Latha , Y. Darshana , Vidhya Venugopal , (2015), Role of building material in thermal comfort in tropical climates – A review , Journal of Building Engineering, 3, 104–113.
35. K. Parsons, Human, (2003), Thermal Environment: The Effects of Hot, Cold and Moderate Environment on Human Health, Comfort and Performance, second edition,. Taylor and Francis, London.

36. Y. Epstein, Daniel S. Moran, (2006) Thermal comfort and the heat stress indices, J. Ind. Health 44, 388–398.
37. S.A. Arngrïmsson, D.S. Petitt, M.G. Stueck, D.K. Jorgensen, K.J. Cureton, (2004), Cooling vest worn during active warm-up improves 5-km run performance in the heat, J. Appl. Physiol. 96 1867–1874.
38. González, G. Craig, M.J. John, (2008), The cardiovascular challenge of exercising in the heat, J. Physiol. 586 45–53.
39. Wesseling, J. Crowe, C. Hogstedt, K. Jakobsson, R.A. Lucas, D.H. Wegman, (2014), Resolving the enigma of the Mesoamerican nephropathy-men – a research workshop summary, Am. J. Kidney Dis. 63 (3) ,396–404.
40. R.S. Bridger, (2003), Introduction to Ergonomics, second ed., Taylor and Francis, London, p. 780.
41. G.P. Bates, J. Schneider, (2008), Hydration status and physiological workload of UAE construction workers – a prospective longitudinal observational study, J. Occup. Med. Toxicol. 3 21–26.
42. P. Dutta, V. Chorsiya, (2013) Scenario of climate change and human health in India, Int. J. Innov. Res. Dev. 2 (8) ,157–160.
43. Benachour E. and al,(2015)., Study of the Influence of the Mural Heating on the Convection in the Building in Arid Area, international review of mechanical engineering (IREME), vol.9N.1,
44. S. V. Patankar, ( 1980)., Numerical Heat Transfer and Fluid Flow, McGraw-Hill New York,
45. De Vahl Davis G, (1983), Natural convection of air in a square cavity. A bench mark numerical solution.

# Experimental and simulation flow rate analysis of the 3/2 directional pneumatic valve

*Slawomir* Blasiak [1,*] *Jakub E.* Takosoglu [1], *Pawel A.* Laski [1], *Dawid S.* Pietrala [1], *Jaroslaw* Zwierzchowski [1], *Gabriel* Bracha [1], *Lukasz* Nowakowski [1], *Malgorzata* Blasiak [1]

[1] Kielce University of Technology, Faculty of Mechatronics and Mechanical Engineering,
Aleja Tysiaclecia Panstwa Polskiego 7, 25-314 Kielce, Poland

**Abstract.** The work includes a study on the comparative analysis of two test methods. The first method - numerical method, consists in determining the flow characteristics with the use of ANSYS CFX. A modeled poppet directional valve 3/2 3D CAD software – SolidWorks was used for this purpose. Based on the solid model that was developed, simulation studies of the air flow through the way valve in the software for computational fluid dynamics Ansys CFX were conducted. The second method - experimental, entailed conducting tests on a specially constructed test stand. The comparison of the test results obtained on the basis of both methods made it possible to determine the cross-correlation. High compatibility of the results confirms the usefulness of the numerical procedures. Thus, they might serve to determine the flow characteristics of directional valves as an alternative to a costly and time-consuming test stand.

## 1 Introduction

Pneumatics along with hydraulics and electrics is the most commonly used technique for automation of industrial processes. Practically, each production machine or production line consists of drive and pneumatic control elements. The popularity of pneumatic drives results from many advantages which include i.a. availability of pressurized air almost in every industrial facility. Pneumatic drives are characterized by high dynamics, reliability and long life, ease of use, non-complicated control systems, simple automation of processes and the fact that they can withstand strong forces. There is no need to use drive overload safety systems as in the case of the electric drives.

Pneumatic drives are controlled with the use of pneumatic divide valves. The main parameters of the divide valves include: the number of channels, number of positions, maximum pressure of working medium, nominal flow rate, switching time and idle time, as well as the size of their channel openings. With respect to the technical solutions applied therein, valves are classified into: spool, poppet and ball valves. In terms of their manner of operation, they are divided into: electrically and mechanically operated, hand- or foot-operated, pneumatically operated via pressure increase/decrease, and spring operated valves. Taking the applied materials into consideration [1,2], valves are the most commonly classified as well into: aluminum valves, plastic valves, valves made of brass or other similar materials allowed in hazardous areas with risk of explosion, e.g. mines. Valves are manufactured using various technologies. The oldest technologies include casting and machining, whereas the most modern ones include production with the use of injection plastics-moulding. The variety of the divide valve parameters, environments they can operate in, production processes or manufacturing techniques and technologies poses many problems [3–20]. Classic approach of designing divide valves used currently is time consuming and expensive. It requires developers to make projects, technical drawings, technical documentation, develop technologies, create valve prototypes, test them under laboratory conditions in order to determine the technical and operational parameters, and conduct durability tests. If the valve meets the assumed requirements, the process is completed and the prototype can be implemented into the manufacturing process, and if not – the whole process starts over again.

Recently there has been many research works concerning the flow analysis [8,16,18,19,21–33], optimization of working medium flow and different types of valves. Programs based on Computational Fluid Dynamics allowing to conduct computer simulations, which reflect the nature of the physical phenomena quite well, are becoming increasingly important. The authors used that type of calculation tools in order to determine the local changes of flow, pressure etc. Studies which intended to compare the results obtained on the basis of numerical tests with the results achieved on a test stand were also conducted [34].

## 2 Valve solid model

The assembly drawing and the solid model of the 3/2 valve is presented in Fig. 1. In the initial position, the

* Corresponding author: sblasiak@tu.kielce.pl

medium flows between channels 1-2, whereas channel 3 is cut-off, and after setting the piston in the opposite extreme position - the medium flows through channels 2-3, while channel 1 is cut-off.

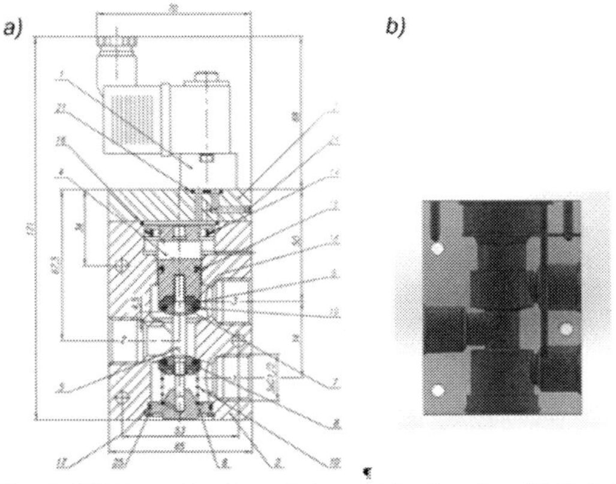

**Fig. 1.** 3/2 Poppet valve, a) Assembly drawing, b) Solid model.

3/2 divide valves currently belong to the group of the most frequently used valves in pneumatic drive and control systems.

**Table 1.** Poppet valve operating parameters.

| | |
|---|---|
| **Working medium** | **Pressurized air or gas with similar properties, filtered with max. size of solids of 40 μm, lubricated or not lubricated with oil mist** |
| Operating pressure range | 0.25 ÷ 1.0 MPa |
| Operating temperature range | -10 ÷ +60 °C |
| Type of design | Mushroom |
| Position of operation | Any |

## 3 Mathematical model of fluid flow

In the calculation process, two principles of fluid movement were used, building a mathematical model of the divide valve in the form of equations describing the physics of the flow. Turbulent movement of viscous fluid is described with the use of Reynolds average Navier-Stokes equations (2)-(4), which together with continuity equation (1) constitute a complete system of relations which allows to determine pressure as well as velocity field of the fluid flow.

$$\frac{\partial \rho}{\partial t} + div(\rho \mathbf{U}) = 0 \tag{1}$$

Reynolds average Average Navier-Stokes equations.

$$\frac{\partial(\rho U)}{\partial t} + div(\rho U \mathbf{U}) = -\frac{\partial P}{\partial x} + div(\mu\, grad\, U) + \left[-\frac{\partial(\rho \overline{u'^2})}{\partial x} - \frac{\partial(\rho \overline{u'v'})}{\partial y} - \frac{\partial(\rho \overline{u'w'})}{\partial z}\right] \tag{2}$$

$$\frac{\partial(\rho V)}{\partial t} + div(\rho V \mathbf{U}) = -\frac{\partial P}{\partial y} + div(\mu\, grad\, V) + \left[-\frac{\partial(\rho \overline{u'v'})}{\partial x} - \frac{\partial(\rho \overline{v'^2})}{\partial y} - \frac{\partial(\rho \overline{v'w'})}{\partial z}\right] \tag{3}$$

$$\frac{\partial(\rho W)}{\partial t} + div(\rho W \mathbf{U}) = -\frac{\partial P}{\partial z} + div(\mu\, grad\, W) + \left[-\frac{\partial(\rho \overline{u'w'})}{\partial x} - \frac{\partial(\rho \overline{v'w'})}{\partial y} - \frac{\partial(\rho \overline{w'^2})}{\partial z}\right] \tag{4}$$

k-ε model is one of the most popular and commonly used flow turbulence models. Two variables *(k-ε)* require additional two transport equations, which might be written down as:

$$\frac{\partial(\rho k)}{\partial t} + div(\rho k \mathbf{U}) = div\left(\frac{\mu_t}{\sigma_t} grad\, k\right) + \mu_t \phi - \rho\varepsilon \tag{5}$$

$$\frac{\partial(\rho \varepsilon)}{\partial t} + div(\rho \varepsilon \mathbf{U}) = div\left(\frac{\mu_t}{\sigma_t} grad\, \varepsilon\right) + C_1 \mu_t \frac{\varepsilon}{k}\phi - C_2 \rho \frac{\varepsilon^2}{k} \tag{6}$$

Variable $k$ means kinetic energy of the turbulence, while $\varepsilon$ - dissipation of that kinetic energy. The partial differential equations shown above are implemented in the calculation module of Ansys CFX programme. In order to provide a final solution to the system of equations describing turbulence flow of the fluid, it is necessary to determine boundary conditions which ensure unambiguity of the solution and determine the course of the calculation process inside the area in question.

*a)*

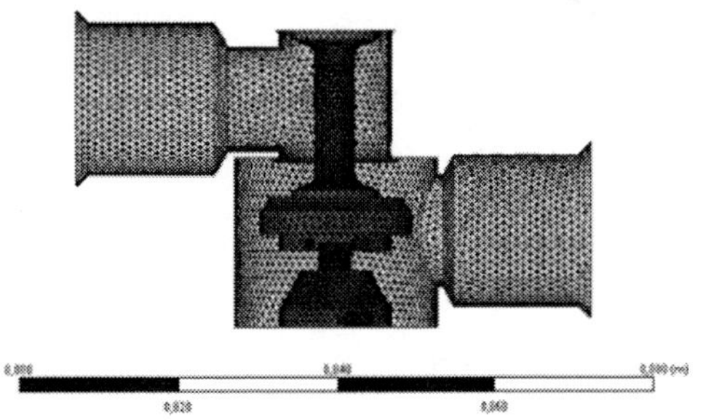

*b)*

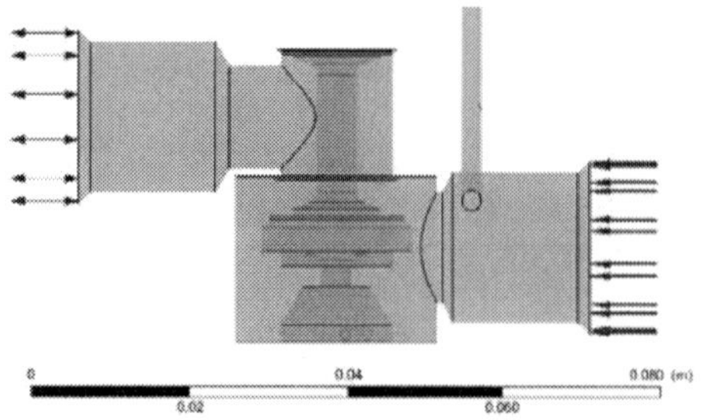

**Fig. 2.** a) Calculation Grid, b) Boundary Conditions.

A non-structural calculation grid with 170236 elements and 36640 nodes was used.

For further calculations, the following individual physical quantities were adopted: ambient temperature T=25 °C, density $-\rho$ = 1.185 (kg / m$^3$), reference pressure – p = 101325 Pa.

## 4 Test stand

The measuring system scheme, where a part of the elements responsible for safety was omitted, is presented below. A filter of sufficient effectiveness must be installed at the test stand. No condensate should accumulate in the connection outlets. The pipes should have a radial, smooth cross-section on the external surface. Those need to be straight and have a fixed diameter along their full length. A simplified scheme of the test stand is shown below. During measuring at the test stand, it was necessary to maintain a constant pressure at the inlet, i.e. not less than 4 bar. Subsequently, the pressure at the outlet was decreased through the throttle valve, until the mass flow rate was not increased any more. It meant that the critical flow had been achieved. At that moment, readings of the temperature and pressure at the inlet, and the flow rate and pressure at the outlet were taken.

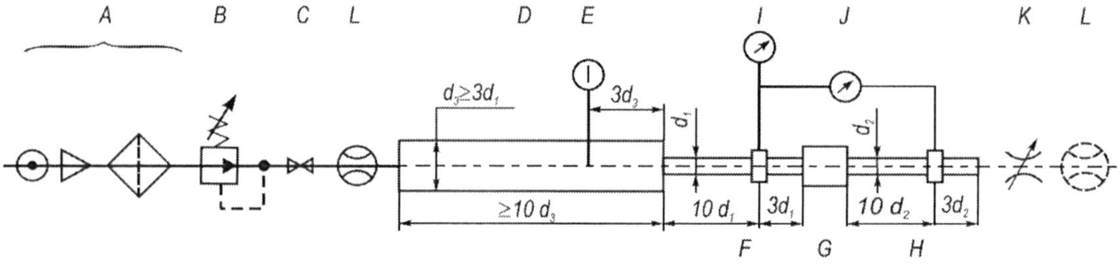


**Fig. 3.** Scheme of the Flow Element Measuring Test Stand.

**Table. 2.** Key to the test stand symbols.

| Symbol | Name |
|---|---|
| A | Pressurized gas source and filter |
| B | Reduction valve |
| C | Shut-off valve |
| D | Temperature measuring tube |
| E | Temperature measuring tool |
| F | Inlet Pressure measuring tube |
| G | Studied element |
| H | Outlet pressure measuring tube |
| I | Inlet pressure measuring tool |
| J | Pressure difference measuring tool |
| K | Throttle Valve |
| L | Flowmeter |

In the next phase of the study, the throttle valve was gradually being closed in order to decrease the flow rate. The readings of all studied parameters were taken. The steps were repeated for a different flow rate. By using the reduction valve, the inlet pressure was maintained constant throughout the entire study.

## 5 Comparison of the results

This section presents a comparison of the results obtained from the numerical tests with the results achieved from the stand tests for the poppet valve.
Fig. 4 presents the characteristics obtained on the basis of the model tests and the experimental stand tests. The diagram also shows inconformity between the model test results and the experimental test results.

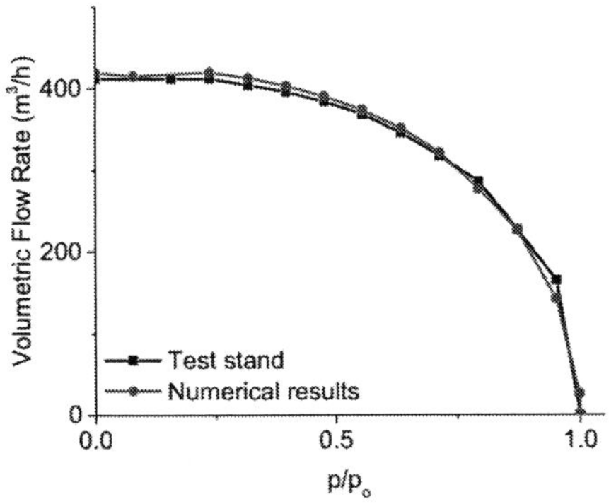


**Fig. 4.** Volumetric flow rate.

The differences in volumetric flow rates are small and there is a large convergence of the received results.

A relative error was calculated on the basis of the following formula:

$$\sigma_{er} = \frac{|x_n - x_{ts}|}{x_{ts}} \cdot 100\% \qquad (7)$$

where:
$x_n$ – value from the numerical calculation,
$x_{ts}$ – value from the test stand.

and amounts to approx. 1.8 %.

## 6 Conclusions

The use of numerical fluid mechanics allowed to determine the pressurized air flow that goes through the valve as well as the primary flow characteristics already in the initial phase of designing the pneumatic valve. The comparison of the simulation and stand tests presented herein determined a large convergence of the results acquired. Hence, the method based on CFD might constitute an alternative to time-consuming and costly stand tests.

***This research was supported in part by PLGrid Infrastructure.***

## References

1. J. Bochnia, *Procedia Eng.* **39** (0) (2012), 98–110.
2. S. Adamczak, J. Bochnia, B. Kaczmarska, *Metrol. Meas. Syst.* **22** (1) (2015), 127–138.
3. S. Blasiak, C. Kundera, J. Bochnia, *Procedia Eng.* **39** (2012), 366–378.
4. S. Blasiak, A. Pawinska, *Int. J. Heat Mass Transf.* **90** (2015), 710–718.
5. D. Gapinski, I. Krzysztofik, (IEEE, 345 E 47TH ST, NEW YORK, NY 10017 USA, 2014).
6. D. Gapinski, I. Krzysztofik, Z. Koruba, *J. Theor. Appl. Mech.* **52** (3) (2014), 629–639.
7. D. Janecki, J. Zwierzchowski, L. Cedro, *Bull. POLISH Acad. Sci. Sci.* **63** (3) (2015), 771–779.
8. P.A. Laski, (ACAD SCI CZECH REPUBLIC, INST THERMOMECHANICS, DOLEJSKOVA 5, PRAGUE 8, 182 00, CZECH REPUBLIC, 2016).
9. E. Miko, L. Nowakowski, *Procedia Eng.* **39** (2012), 395–404.
10. E. Miko, L. Nowakowski, *Procedia Eng.* **39** (2012), 405–413.
11. L. Nowakowski, M. Miesikowska, M. Blasiak, (ACAD SCI CZECH REPUBLIC, INST THERMOMECHANICS, DOLEJSKOVA 5, PRAGUE 8, 182 00, CZECH REPUBLIC, 2016).
12. L. Nowakowski, E. Miko, M. Skrzyniarz, (ACAD SCI CZECH REPUBLIC, INST THERMOMECHANICS, DOLEJSKOVA 5, PRAGUE 8, 182 00, CZECH REPUBLIC, 2016).
13. L. Nowakowski, M. Wijas, (ACAD SCI CZECH REPUBLIC, INST THERMOMECHANICS, DOLEJSKOVA 5, PRAGUE 8, 182 00, CZECH REPUBLIC, 2016).
14. M. Piasecka, *Int. J. Refrig.* **56** (2015), 198–212.
15. M. Piasecka, *Int. J. Heat Mass Transf.* **81** (2015), 114–121.
16. D.S. Pietrala, (ACAD SCI CZECH REPUBLIC, INST THERMOMECHANICS, DOLEJSKOVA 5, PRAGUE 8, 182 00, CZECH REPUBLIC, 2016).
17. J.E. Takosoglu, R.F. Dindorf, P.A. Laski, *Int. J. Adv. Manuf. Technol.* **40** (3–4) (2009), 349–361.
18. J.E. Takosoglu, P.A. Laski, S. Blasiak, *Proc. Inst. Mech. Eng. Part I J. Syst. Control Eng.* **226** (10) (2012), 1335–1343.
19. J.E. Takosoglu, P.A. Laski, S. Blasiak, G. Bracha, D. Pietrala, *Meas. Control.* **49** (2) (2016), 62–71.
20. J. Zwierzchowski, (ACAD SCI CZECH REPUBLIC, INST THERMOMECHANICS, DOLEJSKOVA 5, PRAGUE 8, 182 00, CZECH REPUBLIC, 2016).
21. W. Inman, K. Domansky, J. Serdy, B. Owens, D. Trumper, L.G. Griffith, *J. Micromechanics Microengineering.* **17** (5) (2007), 891–899.
22. B. Li, L. Gao, G. Yang, *Energy Convers. Manag.* **67** (2013), 92–102.
23. E. Lisowski, W. Czyżycki, J. Rajda, *Energy Convers. Manag.* **70** (2013), 220–229.
24. M.R. Mokhtarzadeh-Dehghan, N. Ladommatos, T.J. Brennan, *Appl. Math. Model.* **21** (7) (1997), 437–445.
25. X. Pan, G. Wang, Z. Lu, *Energy Convers. Manag.* **52** (10) (2011), 3249–3256.
26. J. Qian, L. Wei, Z. Jin, J. Wang, H. Zhang, A. Lu, *Energy Convers. Manag.* **87** (2014), 220–226.
27. C. Srikanth, C. Bhasker, *Adv. Eng. Softw.* **40** (3) (2009), 193–201.
28. L. WANG, J. CUI, K. YAO, *Chinese J. Chem. Eng.* **16** (4) (2008), 541–546.
29. Q. Yang, Z. Zhang, M. Liu, J. Hu, *Procedia Eng.* **23** (2011), 543–550.
30. G.F. Bracha, (ACAD SCI CZECH REPUBLIC, INST THERMOMECHANICS, DOLEJSKOVA 5, PRAGUE 8, 182 00, CZECH REPUBLIC, 2016).
31. M. Blasiak, (ACAD SCI CZECH REPUBLIC, INST THERMOMECHANICS, DOLEJSKOVA 5, PRAGUE 8, 182 00, CZECH REPUBLIC, 2016).
32. P.A. Laski, J.E. Takosoglu, S. Blasiak, (ACAD SCI CZECH REPUBLIC, INST THERMOMECHANICS, DOLEJSKOVA 5, PRAGUE 8, 182 00, CZECH REPUBLIC, 2014).
33. P.A. Laski, J.E. Takosoglu, S. Blasiak, *Rob. Auton. Syst.* **72** (2015), 59–70.
34. S. Blasiak, J.E. Takosoglu, P.A. Laski, (ACAD SCI CZECH REPUBLIC, INST THERMOMECHANICS, DOLEJSKOVA 5, PRAGUE 8, 182 00, CZECH REPUBLIC, Svratka, CZECH REPUBLIC, 2014).

# The influence of physical properties of materials used for slide rings on the process of heat transfer in the non-contacting face seals

*Slawomir* Blasiak [1,*]

[1] Kielce University of Technology, Faculty of Mechatronics and Mechanical Engineering, Department of Mechanical Technology and Metrology, Aleja Tysiaclecia Panstwa Polskiego 7, 25-314 Kielce, Poland

**Abstract.** The paper presents the results of analytical solution of the model of heat transfer for non-contacting face seals. Comparative analyses were performed for various physical properties of materials used for slide rings. A mathematical model includes a series of differential equations of partial derivatives with generally used boundary conditions, i.e. the Reynold's equation, energy equation and heat transfer equations, which describe the heat transfer in sealing rings with surrounding medium. Heat transfer equation is written in the Cartesian coordinate system and solved using the Green's functions method. Theoretical studies made it possible to draw a number of practical conclusions on the phenomena of heat transfer in the node seal. The presented model will allow more accurate identification of the heat transfer mechanism in the node seal. The results will help to select appropriate materials for sealing rings, depending on operating conditions of non-contacting face seals.

## 1 Introduction

Sealing technique used in flow machines is one of the key and steadily developing branches of applied research. A wide spectrum of the inter-connected physical phenomena occurring in sealing nodes poses serious problems. Over the past years, many research works considered the results of the bench testing and of the numerical studies obtained via solutions based on complex mathematical models. The scope of the conducted research included, among others, the sealing rings vibration model [1–6], thermal and fluid flow dynamics [7–26], material properties [27–30], measurements and control systems [31–43], as well as the heat transfer in the sealing nodes, the works of: [44–46] but also in experimental studies covering measurements of geometrical quantities and surface structures, materials used [47–52] or studies aimed at measuring temperatures on test rigs [53–55].

The studies published in the above mentioned works are especially important due to the wide range of temperatures for contactless face seals. Manufacturers of the contactless face seals declare working temperature ranges of those components to be between -20 and 220 °C for gas dynamics seals, and between -54 and 650 °C for contactless face seals co-functioning with other media [20] and under extremely different operating parameters [14]. Such a wide range of working temperatures creates a series of problems that researchers have to deal with. Because the assumed height of the radial gap varies by several micrometres, there might occur a rapid evaporation of the working medium in the fluid film. This determines how research is undertaken and conducted. It also determines how to develop improved mathematical models describing the phenomenon of two-phase radial gap flow [29], and how to conduct research on those phenomena under laboratory conditions [41]. By carrying out more specific literature analysis in terms of heat transfer of the contactless face seals, for both the fluid film and co-functioning rings, it is appropriate to mention the works that present complex thermodynamic (THD) and thermo-elastohydrodynamic (TEHD) models, such as the works of: [8] and [13] as well as the works characterized by numerical solutions of the proposed models of heat transfer and thermoelastic deformations of the ring seals [12].

The development of new technologies allows for preparation of materials with physical properties that enable dissipation of huge amounts of heat flux into the environment. It is especially important in cases, where contactless face seals are used in devices that pump hazardous media, such as chemically aggressive or explosive substances whose leakage could lead to contamination of the natural environment or other accidents.

## 2 Formulation of the problem

Depending on the type of its design, contactless face seal must consist of at least one pair of main rings, i.e.: one stationary (1) and one vibrating (2) ring, with radial gap filled with working medium, the so-called fluid film. The

* Corresponding author: sblasiak@tu.kielce.pl

diagram of the FMR-type (Flexible Mounted Stator) contactless seal is shown in Fig. 1.

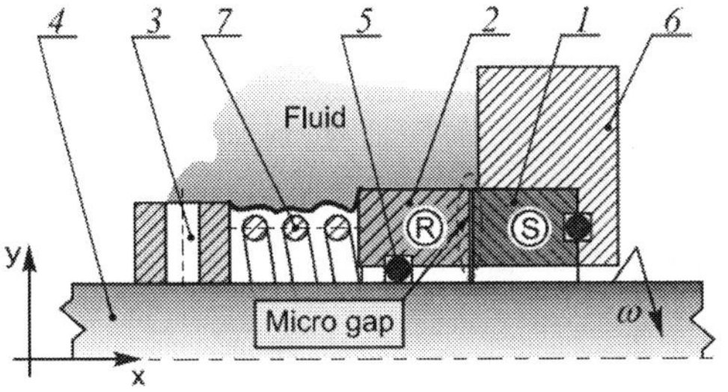


**Fig. 1.** Diagram of the FMR-type contactless face seal. 1 – Stator, 2 – Rotor, 3 – Connecting element, 4 – Shaft, 5 – O-ring, 6 – Casing, 7 – Spring.

The primary condition that determines proper functioning of the contactless face seals is the maintenance of gap that keeps the co-functioning rings apart by few micrometres. That condition can be met only when balance of forces acting on the rings system is maintained. At this point one should distinguish two main types of forces acting on the susceptibly positioned rotor, i.e. the closing force, which comes mainly from the compression spring, and the opening force generated in the fluid film. The opening force is dependent on the shape of the gap and on the pressure in the fluid film that results from it. The mathematical model of the contactless face seals is commonly known and has been presented in the previous works of the author. At this point, the only cited elements are the primary formulas that govern the processes taking place in the fluid film and ring seals.

Reynolds' single-dimensional (in the radial direction) equation describes pressure layout in the fluid film:

$$\frac{d}{dy}\left(\frac{\rho h(y)^3}{\mu}\frac{dp}{dy}\right)=0 \tag{1}$$

Height of the radial gap for the smooth gliding surface – without modifications:

$$h=h(y)=h_o \tag{2}$$

Temperature layout in the fluid film is determined based on a simplified energy equation:

$$\mu\left(\frac{\partial v_\varphi}{\partial x}\right)^2+\lambda^f\frac{\partial^2 T^f}{\partial x^2}=0 \tag{3}$$

A linear change in the fluid flow $v_\varphi$ along the height of the fluid film (Couette's Flow) was adopted. Change in the fluid speed by the coordinate x is noted as:

$$\frac{\partial v_\varphi}{\partial x}=\frac{\omega(r_i+y)}{h} \tag{4}$$

Boundary conditions used in the solution of mathematical model are illustrated in Fig. 2:

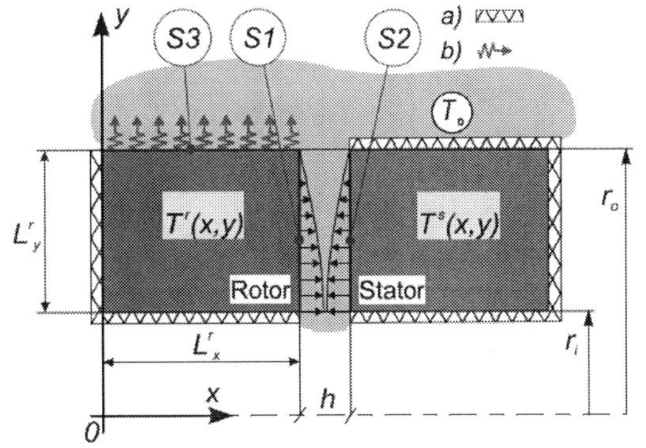


**Fig. 2.** Heat transfer conditions in the contactless face seal, a) Isolated surfaces, b) Heat transfer through natural convection.

The equality of temperatures– $T^r$ (y) = $T^f$ (y) as well as heat flux: $\lambda^r\left(\frac{\partial T^r}{\partial x}\right)=\lambda^f\left(\frac{\partial T^f}{\partial x}\right)\Big|_{x=0}$ between the main ring and the fluid film was adopted for the *S1* surface. On the *S2* surface – it is assumed similarly that the temperatures of the fluid film and the stationary ring are equal – $T^f$ (y) = $T^s$ (y) for $x = h$, and that the temperature on contact between the fluid film and the stator reaches the maximum value: $\left(\frac{\partial T^f}{\partial x}\right)_{x=h}=0$. A heat transfer through natural convection on the S3 surface was assumed according to the following dependency: $-\lambda^r\frac{\partial T^r}{\partial y}\Big|_{y=L^r_y}=\alpha^r\left(T^r-T_o\right)\Big|_{y=L^r_y}$.

The mathematical model formulated in this way has been solved analytically using the Green's Function Method.

## 3 Analytical Solution

The first stage of the analytical solution consisted in determining the temperature layouts in the ring seal by outlining the general functions satisfying the Laplace's Equation (5) for both main rings. The solution included the above boundary conditions and employed a coordinate system to simplify analytical calculations.

$$\frac{\partial^2\theta}{\partial x^2}+\frac{\partial^2\theta}{\partial y^2}=0 \tag{5}$$

Further part of the study will include analytical solution for the stator. Calculations for the rotor are characterized with full analogy, and hence final formulas will be quoted. The separation of variables method was used in the course of solving the Laplace's Equation. Two harmonic functions *X(x)* and *Y(y)* were introduced, being added that each of these had only one variable, while the sought temperature layout $\theta(x,y)$ was presented as their product:

$$\theta(x,y)=X(x)\cdot Y(y) \tag{6}$$

After placing the adopted functions in the differential equation (5), the outcome was: $\frac{X''(x)}{X(x)} = \beta^2$ and $\frac{Y''(y)}{Y(y)} = \gamma^2$. After the transformation, two independent equations were determined:

$$X''(x) + \beta^2 X(x) = 0 \tag{7}$$

$$Y''(y) + \gamma^2 Y(y) = 0 \tag{8}$$

General solutions of the (7) and (8) equations take the overall shape of:

$$X(x) = A_m \cos(\beta_m x) + B_m \sin(\beta_m x) \tag{9}$$

$$Y(y) = C_n \cos(\gamma_n y) + D_n \sin(\gamma_n y) \tag{10}$$

By implementing the $\theta|_{x=0} = 0$ for $0 < y < L_y$ condition in the equation (9), it has been determined that the coefficient $A_m$ amounts to 0. Subsequently, by taking into account the general condition: $\left.\frac{\partial\theta}{\partial x}\right|_{x=L_x} = 0$ for $0 < y < L_y$ it can be assumed that the solutions for the equation $\beta_m B_m \cos(\beta_m L_x) = 0$ are the following values: $\beta_m = \frac{\left(m-\frac{1}{2}\right)\pi}{L_x}$ for $m$=1,2,3,….
Therefore, the equation (9) takes the following form:

$$X(x) = \sum_{m=1}^{\infty} B_m \sin(\beta_m x) \tag{11}$$

By implementing the boundary condition: $\left.\frac{\partial\theta}{\partial y}\right|_{y=0} = 0$ for $0 < x < L_x$ in the dependency (10), the outcome was that $D_n$ = 0. By further inserting the boundary condition: $\left.\frac{\partial\theta}{\partial y}\right|_{y=L_y} = 0$ for $0 < x < L_x$ into the equation (10), the coefficient $\gamma_n$ took the value of $\gamma_n = \frac{n\pi}{L_y}$ for $n$=1,2,3,…. Whereas the equation (10) adopted the form of:

$$Y(y) = \sum_{m=1}^{\infty} C_n \cos(\gamma_n y) \tag{12}$$

By using the equations (11) and (12), the dependency (6) describing the temperature layout in the stator has been determined as: $c_{n,m} = B_m C_n$:

$$\theta(x,y) = \sum_{m=1}^{\infty}\sum_{n=1}^{\infty} c_{n,m} \sin(\beta_m x)\cos(\gamma_n y) \tag{13}$$

The $c_{n,m}$ coefficient has been determined similarly, as in the work of [56].

$$c_{n,m} = \frac{\int_0^{L_x}\int_0^{L_y} \theta(x',y')\sin(\beta_m x')\cos(\gamma_n y')dy'dx'}{\int_0^{L_x}\sin^2(\beta_m x)dx \int_0^{L_y}\cos^2(\gamma_n y)dy} \tag{14}$$

The temperature layout in the cross-section of the stator has been noted as:

$$\theta^s(x,y) = \int_0^{L_x}\int_0^{L_y} \frac{\partial G^s(x,y\mid x',y')}{\partial x'}\theta(x',y')dy'dx' \tag{15}$$

The Green's Function for the $G^s(x, y \mid x', y')$ is presented by the following dependency:

$$G^s(x,y\mid x',y') = \frac{4}{L_x L_y} \sum_{m=1}^{\infty}\sum_{n=1}^{\infty} \frac{\sin(\beta_m^s x)\sin(\beta_m^s x')\cos(\gamma_n^s y)\cos(\gamma_n^s y')}{(\beta_m^{s\,2} + \gamma_n^{s\,2})} + \frac{2}{L_x L_y}\sum_{m=1}^{\infty}\frac{\sin(\beta_m^s x)\sin(\beta_m^s x')}{\beta_m^{s\,2}} \tag{16}$$

In the following part of the work, the derivative of the Green's Function in relation to the $x'$ coordinate has been determined:

$$\left.\frac{\partial G^s(x,y\mid x',y')}{\partial x'}\right|_{x'=0} = \frac{\partial}{\partial x'}\left(\frac{4}{L_x L_y}\sum_{m=1}^{\infty}\sum_{n=1}^{\infty}\frac{\sin(\beta_m^s x)\sin(\beta_m^s x')\cos(\gamma_n^s y)\cos(\gamma_n^s y')}{(\beta_m^{s\,2}+\gamma_n^{s\,2})} + \frac{2}{L_x L_y}\sum_{m=1}^{\infty}\frac{\sin(\beta_m^s x)\sin(\beta_m^s x')}{\beta_m^{s\,2}}\right)$$

In the $x'$ = 0 point, it takes the value of:

$$\left.\frac{\partial G}{\partial x'}\right|_{x'=0} = \left(\frac{4}{L_y\pi}\sum_{n=1}^{\infty}\sum_{m=1}^{\infty}\frac{\cos\left(\frac{n\pi}{L_y}y\right)\cos\left(\frac{n\pi}{L_y}y'\right)\cdot\left(m-\frac{1}{2}\right)\sin\left(\frac{\left(m-\frac{1}{2}\right)\pi}{L_x}x\right)}{\left(\left(m-\frac{1}{2}\right)^2+\left(\frac{L_x n}{L_y}\right)^2\right)} + \frac{2}{L_y\pi}\sum_{m=1}^{\infty}\frac{\left(m-\frac{1}{2}\right)\sin\left(\frac{\left(m-\frac{1}{2}\right)\pi}{L_x}x\right)}{\left(m-\frac{1}{2}\right)^2}\right) \tag{18}$$

Using the substitution:

$$\sum_{m=1}^{\infty} \frac{\left(m-\frac{1}{2}\right)\sin\left(\frac{\left(m-\frac{1}{2}\right)\pi}{L_x}x\right)}{\left(\left(m-\frac{1}{2}\right)^2+\left(\frac{L_x n}{L_y}\right)^2\right)} = \frac{\pi}{2}\,\frac{\sinh\left(\frac{L_x n\pi}{L_y}\left(1-\frac{x}{L_x}\right)\right)}{\sinh\left(\frac{L_x n\pi}{L_y}\right)} \tag{19}$$

In the end, the dependency describing the temperature layout in the stator takes the following form:

$$\begin{aligned}\theta^s(x,y) &= \frac{2}{L_y}\sum_{n=1}^{\infty}\cos\left(\frac{n\pi}{L_y}y\right)\frac{\sinh\left(\frac{L_x n\pi}{L_y}\left(1-\frac{x}{L_x}\right)\right)}{\sinh\left(\frac{L_x n\pi}{L_r}\right)}\cdot \\ &\cdot\int_0^{L_y}\left(\theta(x',y')\big|_{x'=0}\cos\left(\frac{n\pi}{L_y}y'\right)\right)dy' + \\ &+\frac{2}{L_y\pi}\sum_{m=1}^{\infty}\frac{\sin\left(\frac{\left(m-\frac{1}{2}\right)\pi}{L_x}x\right)}{\left(m-\frac{1}{2}\right)}\int_0^{L_y}\left(\theta(x',y')\big|_{x'=0}\right)dy'\end{aligned} \tag{20}$$

By using the dependencies set out in the work of [56], the Green's Function for the rotor takes the following form:

$$\begin{aligned}G^r(x,y\,|\,x',y') &= \frac{2}{L_x}\sum_{m=1}^{\infty}\sum_{n=1}^{\infty}\frac{2\left(\gamma_n^{r2}+\left(\frac{\alpha^r}{\lambda^r}\right)^2\right)}{L_y\left(\gamma_n^{r2}+\left(\frac{\alpha^r}{\lambda^r}\right)^2\right)+\left(\frac{\alpha^r}{\lambda^r}\right)}\cdot \\ &\cdot\frac{\cos\left(\beta_m^r x\right)\cos\left(\beta_m^r x'\right)\cos\left(\gamma_n^r y\right)\cos\left(\gamma_n^r y'\right)}{\left(\gamma_n^{r2}+\beta_m^{r2}\right)} \\ &+\frac{1}{L_x}\sum_{n=1}^{\infty}\frac{2\left(\gamma_n^{r2}+\left(\frac{\alpha^r}{\lambda^r}\right)^2\right)}{L_y\left(\gamma_n^{r2}+\left(\frac{\alpha^r}{\lambda^r}\right)^2\right)+\left(\frac{\alpha^r}{\lambda^r}\right)}\frac{\cos\left(\gamma_n^r y\right)\cos\left(\gamma_n^r y'\right)}{\left(\gamma_n^{r2}\right)}\end{aligned} \tag{21}$$

The values of the coefficient $\beta_m^r$ are $\beta_m^r = \frac{m\cdot\pi}{L_x}$ for $m$=1,2,3… The values of the coefficient $\gamma_n^r$ were determined by solving the equation $\gamma_n^r\tan\left(\gamma_n^r L_y\right)=\frac{\alpha^r}{\lambda^r}$. This is a transcendental equation, and it can be solved by iteration, where $\lambda^r$ is the coefficient of rotor heat conductivity. The temperature layout in the rotor has been formulated in way convenient for use in mathematical calculations:

$$\begin{aligned}\theta^r(x,y) &= \frac{2}{L_x}\sum_{m=1}^{\infty}\sum_{n=1}^{\infty}\frac{2\left(\gamma_n^{r2}+\left(\frac{\alpha^r}{\lambda^r}\right)^2\right)}{L_y\left(\gamma_n^{r2}+\left(\frac{\alpha^r}{\lambda^r}\right)^2\right)+\left(\frac{\alpha^r}{\lambda^r}\right)}\cdot \\ &\cdot\frac{\cos\left(\beta_m^r x\right)\cos\left(\beta_m^r L_x\right)\cos\left(\gamma_n^r y\right)}{\left(\gamma_n^{r2}+\beta_m^{r2}\right)}\cdot\int_0^{L_y}\left(\frac{d\theta(x',y')}{dx}\bigg|_{x'=Lx}\cos\left(\gamma_n^r y'\right)\right)dy' + \\ &+\frac{1}{L_x}\sum_{n=1}^{\infty}\frac{2\left(\gamma_n^{r2}+\left(\frac{\alpha^r}{\lambda^r}\right)^2\right)}{L_y\left(\gamma_n^{r2}+\left(\frac{\alpha^r}{\lambda^r}\right)^2\right)+\left(\frac{\alpha^r}{\lambda^r}\right)}\frac{\cos\left(\gamma_n^r y\right)}{\left(\gamma_n^{r2}\right)}\cdot \\ &\cdot\int_0^{L_y}\left(\frac{d\theta(x',y')}{dx}\bigg|_{x'=Lx}\cos\left(\gamma_n^r y'\right)\right)dy'\end{aligned}$$

(22)

The further part of the study involved an analysis regarding how physical properties of materials used for manufacturing of the ring seals influence the heat transfer in the fluid film-main rings system.

## 4 Results and Discussion

Numerical calculations of the heat transfer in mechanical seals have been conducted with the assumption that the stator is made of resin-impregnated carbon, whereas the rotor, whose properties are compiled in Table 1, is made of the most commonly used materials in manufacturing ring seals.

**Table 1.** Properties of the materials used for ring seals manufacturing.

| **Material** | **Thermal expansion coefficient α [$10^{-6}$ $K^{-1}$] for 20 °C** | **Thermal conductivity [W $m^{-1}$ $K^{-1}$]** | **Thermal shock resistance [°C]** |
|---|---|---|---|
| Resin-impregnated carbon | 5.0 | 14 | 200 |
| Chemically bound silicon carbide (SiC-Si) | 4.4 | 120 | 250 |
| Sintered silicon carbide SiC | 4.1 | 100 | 250 |
| Tungsten carbide (WC) | 4.8 | 70 | 300 |
| Ceramics (AL203) | 7.4-7.6 | 40 | 150 |
| Metallized carbon graphite (SCG-M) | 3.7 | 25 | 250 |

The parameters from Table 2 were adopted for the purpose of calculations.

**Table 2.** Geometry- and performance-related parameters.

| **Geometry and performance-related parameters** | |
|---|---|
| Inner radius $r_i$ | 0.040 *(m)* |
| Outer radius $r_o$ | 0.045 *(m)* |
| Thickness of the rings | 0.005 *(m)* |
| Angular velocity ω | 650 *(rad/s)* |
| Fluid temperature $T_o$ | 20 (°C) |
| Fluid thermal conductivity λf | 0.65 $(W\,m^{-1}\,K^{-1})$ |
| Convection coefficient (water) α | 18 000 $(W\,m^{-2}\,K^{-1})$ |
| Nominal clearance height $h_o$ | $1\cdot10^{-6}$ *(m)* |

In Table 2, the working medium is water with temperature of 20 °C on the process side – $T_0$. It has been assumed that the external radius and the internal radius have the length of 45 mm and 40 mm respectively, for both co-functioning ring seals. A radial gap with the height of 1 μm was used in the calculations.

Fig. 3 presents the temperature layouts in the cross-sections of ring seals and fluid film on the example of silicon carbide (SiC-Si) and metallized carbon graphite, in other words, materials with the highest and the lowest coefficient of heat conductivity $\lambda$ (according to the parameters of Table 1).

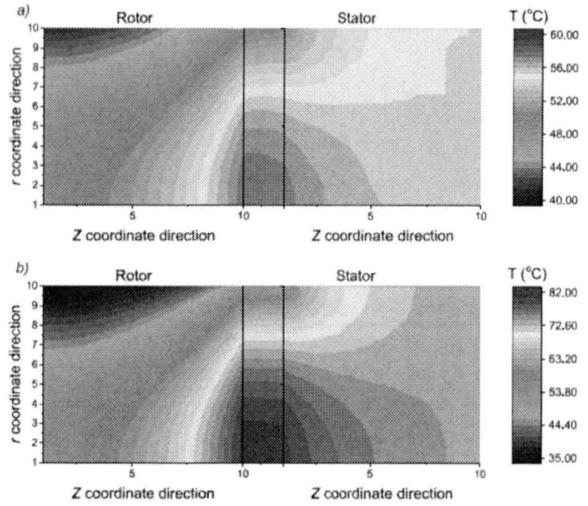


**Fig. 3.** Temperature distribution in the fluid film and sealing rings for, a) Chemically bound silicon carbide (SiC-Si), b) Metallized carbon graphite.

It can be observed (Fig. 3a) that the rotor, made of a material with high heat conductivity $\lambda$, has the maximum temperature lower by 22 °C in the fluid film-ring seals system.

The next figure presents the temperature changes in the radial direction on the verge of the rotating ring and the fluid film.

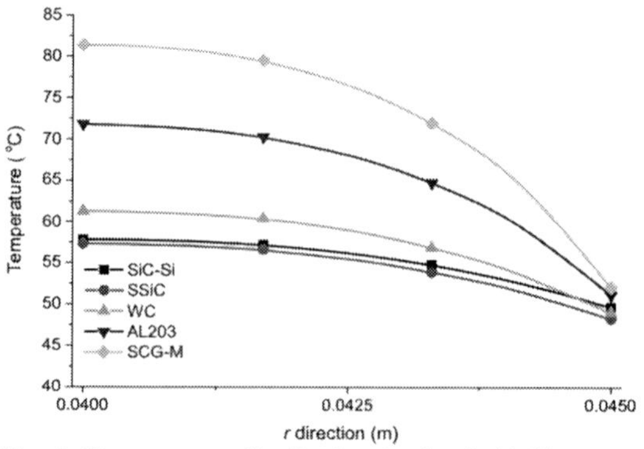


**Fig. 4.** Temperature distribution in the fluid film-rotor interface

The analysis of the results shown in Fig. 4 was followed by a determination of the smallest differences in temperature layouts of materials with similar physical properties. This concerns chemically bound silicon carbide and sintered silicon carbide. The thermal resistance of the whole studied system increases when the heat transfer coefficient decreases.

## 5 Conclusion

The main task of the contactless face seals is to keep tightness, regardless of external conditions. There is a close relation between the geometry of the radial gap and the occurrence of a leakage. The major factors connecting the above mentioned parameters are any deformations caused by uneven temperature layout in the ring seals. A development of more precise mathematical models and calculation apparatuses, in the course of the layout design process, will allow to develop the temperature of the rings-fluid film system. That will allow researchers and constructors to estimate the temperature range and select proper materials for the manufacture of ring seals. This knowledge, in turn, will lead to the achievement of high reliability and a long operational life of the developed structures.

### Nomenclature

$h(y)$ – clearance geometry in the $y$ direction,
$h_o$ – nominal clearance height,
$T^i$ –temperature,
$T_o$ – temperature of the surrounding fluid,
$i = f, r, s$ – fluid, rotor and stator respectively.

**Greeks**
$\alpha$ – convection coefficient,
$\lambda^r$ – thermal conductivity for the rotor,
$\mu$ – dynamic viscosity,
$\omega$ – angular velocity,
$\rho$ – density.

## References


1. D. Gapinski, I. Krzysztofik, (IEEE, 345 E 47TH ST, NEW YORK, NY 10017 USA, 2014).
2. D. Gapinski, I. Krzysztofik, Z. Koruba, *J. Theor. Appl. Mech.* **52** (3) (2014), 629–639.
3. D. Janecki, L. Cedro, J. Zwierzchowski, *Metrol. Meas. Syst.* **22** (2) (2015), 289–302.
4. Z. Koruba, I. Krzysztofik, *Proc. Inst. Mech. Eng. PART K-JOURNAL MULTI-BODY Dyn.* **227** (K1) (2013), 12–16.
5. I. Krzysztofik, (ACAD SCI CZECH REPUBLIC, INST THERMOMECHANICS, DOLEJSKOVA 5, PRAGUE 8, 182 00, CZECH REPUBLIC, 2016).
6. I. Krzysztofik, Z. Koruba, *J. Appl. Math.* (2014),.
7. S. Blasiak, J.E. Takosoglu, P.A. Laski, (ACAD SCI CZECH REPUBLIC, INST THERMOMECHANICS, DOLEJSKOVA 5, PRAGUE 8, 182 00, CZECH REPUBLIC, Svratka, CZECH REPUBLIC, 2014).
8. S. Blasiak, *J. Therm. Sci. Technol.* **10** (1) (2015), JTST0016-JTST0016.
9. S. Blasiak, C. Kundera, *Procedia Eng.* **39** (2012), 315–326.
10. S. Blasiak, C. Kundera, J. Bochnia, *Procedia Eng.* **39** (2012), 366–378.
11. S. Blasiak, A. Pawinska, *Int. J. Heat Mass Transf.* **90** (2015), 710–718.
12. S. Blasiak, J.E. Takosoglu, P.A. Laski, *J. Therm. Sci. Technol.* **9** (2) (2014), JTST0011-JTST0011.
13. N. Brunetiere, *Proc. Inst. Mech. Eng. PART J-JOURNAL Eng. Tribol.* **224** (J12) (2010), 1221–1233.
14. S. Kavinprasad, S. Shankar, M. Karthic, *Ind. Lubr. Tribol.* **67** (2) (2015), 124–132.
15. P.A. Laski, (ACAD SCI CZECH REPUBLIC, INST THERMOMECHANICS, DOLEJSKOVA 5, PRAGUE 8, 182 00, CZECH REPUBLIC, 2016).
16. P.A. Laski, J.E. Takosoglu, S. Blasiak, (ACAD SCI CZECH REPUBLIC, INST THERMOMECHANICS, DOLEJSKOVA 5, PRAGUE 8, 182 00, CZECH REPUBLIC, 2014).
17. P.A. Laski, J.E. Takosoglu, S. Blasiak, *Rob. Auton. Syst.* **72** (2015), 59–70.
18. S.C. Lee, X.L. Zheng, *Comput. Fluids.* **88** (2013), 326–333.
19. Y. Liu, W. Liu, Y. Li, X. Liu, Y. Wang, *Tribol. Int.* **90** (2015), 43–54.
20. X. Meng, S. Bai, X. Peng, *J. ZHEJIANG Univ. A.* **15** (3) (2014), 172–184.
21. A.P. Nyemeck, N. Brunetiere, B. Tournerie, *Tribol. Trans.* **58** (5) (2015), 836–848.
22. M.D. Pascovici, I. Etsion, *ASME J. Tribol.* **1992** (vol.114) (1992), 639–645.
23. D.S. Pietrala, (ACAD SCI CZECH REPUBLIC, INST THERMOMECHANICS, DOLEJSKOVA 5, PRAGUE 8, 182 00, CZECH REPUBLIC, 2016).
24. T. Wang, W. Huang, X. Liu, Y. Li, Y. Wang, *Tribol. Int.* **72** (2014), 90–97.
25. G. Zhang, W. Zhao, *Tribol. Lett.* **53** (2) (2014), 497–509.
26. H. Zhang, R.G. Landers, B.A. Miller, *J. Dyn. Syst. Meas. Control. ASME.* **132** (4) (2010),.
27. M. Blasiak, (ACAD SCI CZECH REPUBLIC, INST THERMOMECHANICS, DOLEJSKOVA 5, PRAGUE 8, 182 00, CZECH REPUBLIC, 2016).
28. J. Bochnia, *Procedia Eng.* **39** (0) (2012), 98–110.
29. J. Bochnia, T. Kozior, *Solid State Phenom.* **223** (2014), 199–208.
30. M. Blasiak, R. Kotowski, *Prz. Elektrotechniczny.* **85** (12) (2009), 40–43.
31. S. Adamczak, J. Bochnia, B. Kaczmarska, *Metrol. Meas. Syst.* **22** (1) (2015), 127–138.
32. S. Adamczak, J. Bochnia, B. Kaczmarska, *Metrol. Meas. Syst.* **21** (3) (2014), 553–560.
33. G.F. Bracha, (ACAD SCI CZECH REPUBLIC, INST THERMOMECHANICS, DOLEJSKOVA 5, PRAGUE 8, 182 00, CZECH REPUBLIC, 2016).
34. D. Janecki, J. Zwierzchowski, L. Cedro, *Bull. POLISH Acad. Sci. Sci.* **63** (3) (2015), 771–779.
35. J.E. Takosoglu, R.F. Dindorf, P.A. Laski, *Int. J. Adv. Manuf. Technol.* **40** (3–4) (2009), 349–361.
36. J.E. Takosoglu, P.A. Laski, S. Blasiak, *Proc. Inst. Mech. Eng. Part I J. Syst. Control Eng.* **226** (10) (2012), 1335–1343.
37. J.E. Takosoglu, P.A. Laski, S. Blasiak, (ACAD SCI CZECH REPUBLIC, INST THERMOMECHANICS, DOLEJSKOVA 5, PRAGUE 8, 182 00, CZECH REPUBLIC, 2014).
38. J.E. Takosoglu, P.A. Laski, S. Blasiak, G. Bracha, D. Pietrala, *Meas. Control.* **49** (2) (2016), 62–71.
39. R. Trochimczuk, T. Kuźmierowski, *Int. J. Appl. Mech. Eng.* **19** (2014), 841.
40. R. Trochimczuk, *Int. J. Appl. Mech. Eng.* **19** (2014), 771.
41. R. Trochimczuk, in: Mechatron. Syst. Mater. IV, 2013: pp. 3–8.
42. R. Trochimczuk, M. Gawrysiak, in: Gosiewski, Z and Kulesza, Z (Ed.), Mechatron. Syst. Mater. III, 2009: pp. 107–112.
43. J.E. Takosoglu, (ACAD SCI CZECH REPUBLIC, INST THERMOMECHANICS, DOLEJSKOVA 5, PRAGUE 8, 182 00, CZECH REPUBLIC, 2016).
44. Huang Weifeng, Liao Chuanjun, Liu Xiangfeng, Suo Shuangfu, Liu Ying, Wang Yuming, *CHINESE J. Mech. Eng.* **27** (5) (2014), 949–957.
45. S. Bai, C. Ma, X. Peng, S. Wen, *J. Tribol. ASME.* **137** (2) (2015),.
46. C. Ma, S. Bai, X. Peng, *Tribol. Int.* **95** (2016), 44–54.
47. E. Miko, L. Nowakowski, *Procedia Eng.* **39** (2012), 405–413.
48. E. Miko, L. Nowakowski, *Procedia Eng.* **39** (2012), 395–404.
49. L. Nowakowski, M. Miesikowska, M. Blasiak, (ACAD SCI CZECH REPUBLIC, INST THERMOMECHANICS, DOLEJSKOVA 5, PRAGUE 8, 182 00, CZECH REPUBLIC, 2016).
50. L. Nowakowski, E. Miko, M. Skrzyniarz, (ACAD SCI CZECH REPUBLIC, INST THERMOMECHANICS, DOLEJSKOVA 5, PRAGUE 8, 182 00, CZECH REPUBLIC, 2016).
51. L. Nowakowski, M. Wijas, (ACAD SCI CZECH REPUBLIC, INST THERMOMECHANICS, DOLEJSKOVA 5, PRAGUE 8, 182 00, CZECH REPUBLIC, 2016).
52. J. Zwierzchowski, (ACAD SCI CZECH REPUBLIC, INST THERMOMECHANICS, DOLEJSKOVA 5, PRAGUE 8, 182 00, CZECH REPUBLIC, 2016).
53. M. Piasecka, *Int. J. Heat Mass Transf.* **81** (2015), 114–121.
54. M. Piasecka, *Int. J. Refrig.* **56** (2015), 198–212.
55. M. Piasecka, B. Maciejewska, *Exp. Therm. Fluid Sci.* **68** (2015), 459–467.
56. M.N. Özişik, (2nd ed., Wiley, New York, 1993).

# The characteristics of pneumatic piston drive

*Gabriel F.* Bracha[1,*]

[1]Faculty of Mechatronics And Machine Design, Kielce University of Technology, al. 1000-PP 7; 25-314 Kielce; PL

**Abstract.** The article presents a study on the rodless pneumatic piston actuator. Mathematical model of the pneumatic actuator, static and dynamic characteristics are obtained and compared with the real drive characteristics.

## 1 Introduction

One of the primary actuators used in industry is a pneumatic drive. Due to its construction and used working medium, the drive is characterized by high dynamics of movements, resistance to overload and not very high positioning accuracy. Easy to use and high speed drive systems, especially in performing simple two state movements made pneumatic drive essential in every mass production factory. More advanced use of pneumatic drives is construction of pneumatic manipulator, which requires a much more detailed study of drive parameters, and control systems that will take into account many factors such as compressibility of working medium, friction, various cross-sections of the drive and connectors. As shown in the work of P. A. Laski[1,2], developing appropriate trajectory of pneumatic manipulator requires more advanced control system such as Fuzzy-logic controller[3].

## 2 Mathematical model

A common mathematical model of pneumatic piston drive is a model based on energy balance, which can be represented as follows:

$$Q_i = Q_o + dL + dU + Q_T \qquad (1)$$

Where: : $Q_i$ – flowing energy, $Q_o$ – outflowing energy, $dL$ – work done by the actuator, $U$ – changes of internal energy, $Q_T$ – energy flow to the environment.

Assuming that the temperature in the chambers is uniform, and that the working medium is a gas which meets the ideal gas law, it can be derived differential equation determining the changes of pressure in a single chamber:

$$\dot{p} = \frac{1}{V}\left(kRT_i\dot{m}_i - kRT\dot{m}_o - kp\dot{V} - \frac{pV}{T}\dot{T} - (k-1)\alpha\pi\left(Dx + 0,25D^2\right)\Delta\mathrm{T}\right) \qquad (2)$$

Where heat capacity ratio $k = c_p / c_v$, the gas constant $R = c_p - c_v$, $D$ – diameter of the piston, $x$ – the current position of the piston, $\Delta T$ – the temperature difference between the actuator and the environment.

Changes in temperature inside the chamber is obtained from the ideal gas equation differentiation:

$$\frac{dT}{dt} = T\left(\frac{1}{p}\frac{dp}{dt} + \frac{1}{x}\frac{dx}{dt} - \frac{\dot{m}RT}{pV}\right) \qquad (3)$$

On the piston acts forces resulting from the pressure difference between the chambers as well as the friction:

$$\frac{d^2x}{dt^2} = \frac{1}{m}\left[A\left(p_1 - p_2\right) - F_T\right] \qquad (4)$$

$m$ – mass of piston, $A$ – surface on which pressure is acting, $p_1$ – pressure in chamber, $F_t$ – friction forces.

### 2.1 Friction

Friction is one of the most inlfuential force occurring in the piston pneumatic drives, we can distinguish the static friction - important at the beginning of the movement, and dynamic friction that occurs during the entire drive operation. Moreover compressibility of the working medium causes a phenomenon known as stick-slip, which makes piston moves in a stepped manner when at low velocity (Figure 1).

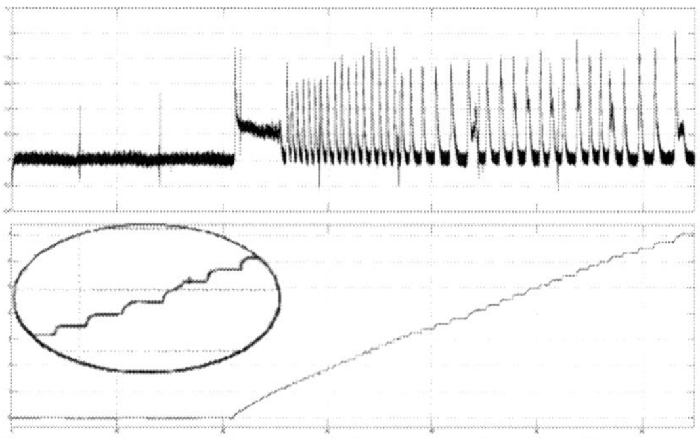

**Fig. 1.** Real pneumatic piston drive with stick – slip effect. Top picture – velocity, bottom – position, approximate velocity 9.34 mm/s.

There are several basic models of friction which are used to simulate behaviour of pneumatic actuators, one

* Gabriel Bracha: gbracha@tu.kielce.pl

of them is a Lund – Grenoble model[4]. Equations describing LG model:

$$\frac{dz}{dt} = v - \sigma_0 \cdot \frac{|v|}{g(v)} \cdot z \qquad g(v) = F_c + (F_s - F_c) \cdot e^{-(v/v_s)^2}$$

$$F_f = \sigma_0 z + \sigma_1 \dot{z} + F_v v \qquad (5)$$

$z$ – average deviation bristles, $v$ – velocity of the piston, $\sigma_0$ – stiffness of bristles, $\sigma_1$ – damping, $F_v$ – viscous friction, $F_c$ – Coulomb force, $F_s$ – stiction force.

This model reflects the behaviour of friction in the pneumatic drives very well, and also allows to simulate the stick – slip phenomenon, which for a very low speed of the piston makes positioning accurate very difficult.

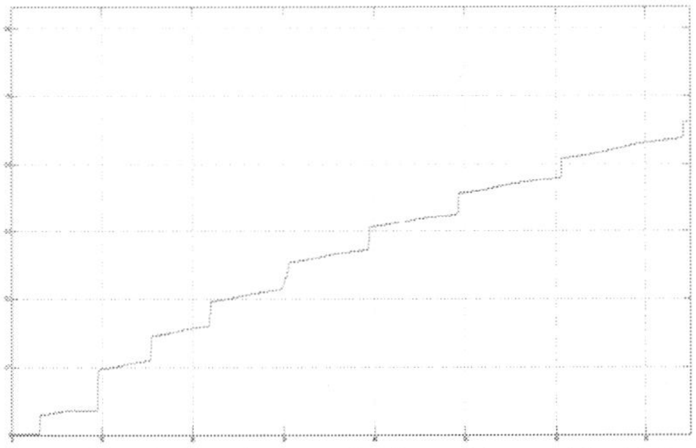

**Fig. 2.** Simulation of pneumatic piston drive model with Lund – Grenoble friction model (stick-slip effect).

Another much simpler approach that allows to simulate friction in pneumatic drives is a combination of Coulomb model and Stribeck curve. Usage of such a simulation model is shown on the Figure 3.

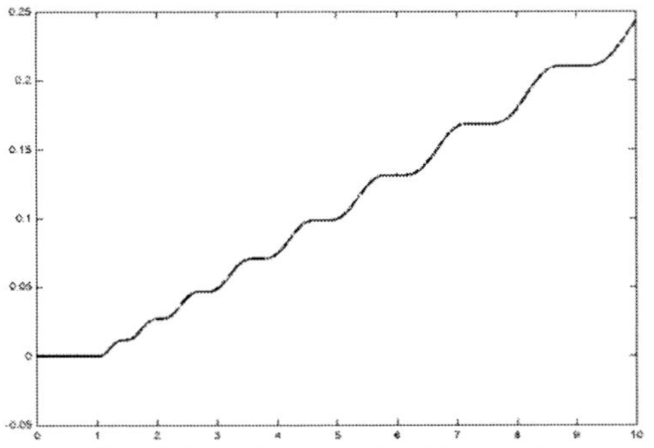


**Fig. 3.** Simulation of pneumatic drive model for slow movement with combined Stribeck and Coulomb friction.

## 2.1 Measurements

In order to present certain characteristics of the pneumatic actuator, the test rig consisting of a linear servo and pneumatic drive combined with a force sensor was developed.

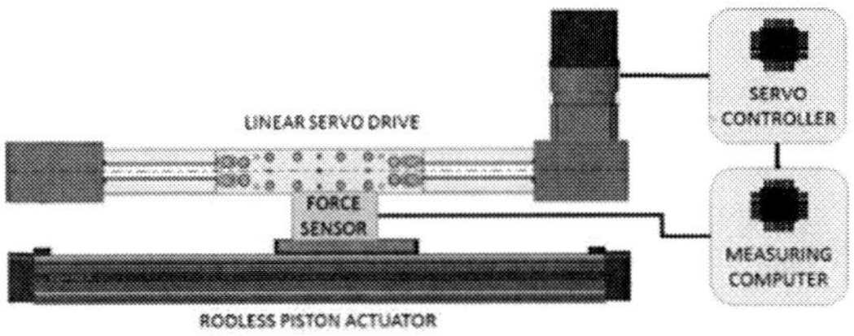


**Fig. 4.** Test rig for friction measurements.

The measurement process takes place automatically by increasing the speed from 1 to 450 mm / s and continuous readings measurements from the force sensor. Every single cycle generates about 5 000 measurements of friction force. In order to reduce the impact of movement changes on the measurements, 5% of the initial values are ignored and the others are subjected to moving average filter with a period equal to 5. Figure 4 shows a part of measurement cycle on which we can see changes in friction force shortly after the start of the movement.

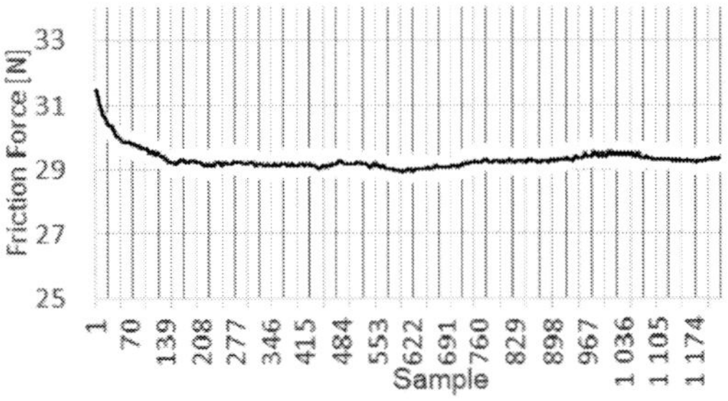


**Fig. 5.** Friction measurement cycle for the 2 mm/s speed.

As we can notice, the friction value is the largest right after the change of movement direction of the piston, is above all due to the construction of a seal, and a static friction value - which is highly dependent on the stopping time.

From one measurement cycle we obtain a mean value of the force. All cycles form a graph showing the relationship of the friction and velocity (Figure 5).

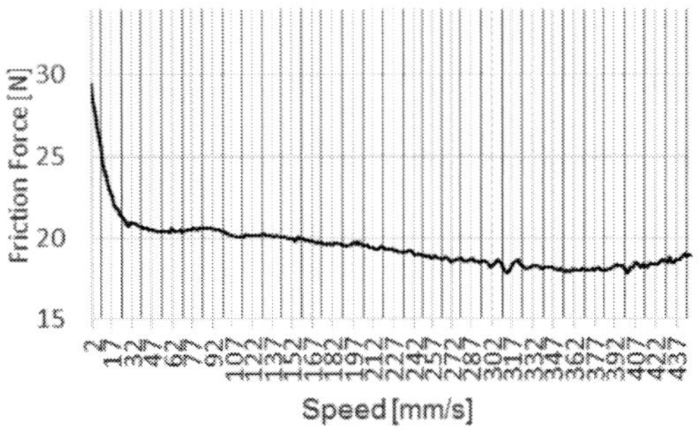


**Fig. 6.** The results of measuring the friction force as a function of speed.

DOI: 10.1051/epjconf/201714302010

As can be seen from the measurements, for small values of speed the resistance of the drive is the biggest. As well, it is difficult to see on the figure a point of Stribeck speed directly. This may be caused by: the surface quality of the cylinder [5], impurities, worn seals or even corrosion.

To check the impact of pneumatic actuator seals on motion resistance, the friction force measurements for the different speed of the piston subjected to varying pressure value was made.

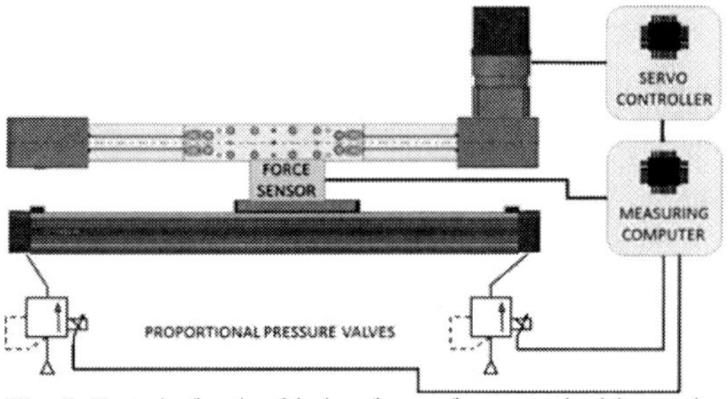


**Fig. 7.** Test rig for the friction force of pneumatic drive under pressure measurements.

The piston velocity was changed in the range of 1 - 500 mm / sec, and the pressure from 0.1 to 1.5 Bar which was fed to both chambers. Unfiltered measurement results shown in Fig 6.

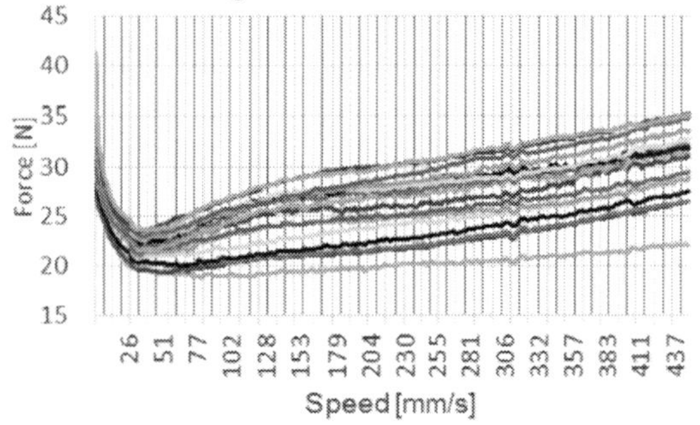


**Fig. 8.** The results of measuring the friction force as a function of speed for a pressure from 0.1 to 1.5 Bar. Lower line is lower pressure.

It can be seen that as the pressure increases the characteristics of the friction force are more like Stribeck curve. As a result of drive seal deformation, for low velocities, the friction value increases with increasing pressure.

## 3 Conclusions

The study showed that the pneumatic actuator is a very complex and sensitive to many physical elements. The performance of this drive is affected by many factors such as changes in temperature, working fluid properties, and especially the velocity of piston and how long the drive is stopped. While simple applications are not so demanding, so far attempts to precise positioning can be challenging. However, if the control of manipulator consisting of a pneumatic drive would take place along a predefined trajectory it is possible to use the predictive control based on complex mathematical model which takes into account the impact of known physical factors.

The current study was conducted in one cycle without stopping, we can say for sure that due to such action drive characteristics can maintain a specific trend directly resulting from continuous operation. Even short-term stop of the drive during the test causes a complete change of drive characteristics, an example of such middle-stop presents Figure 9.

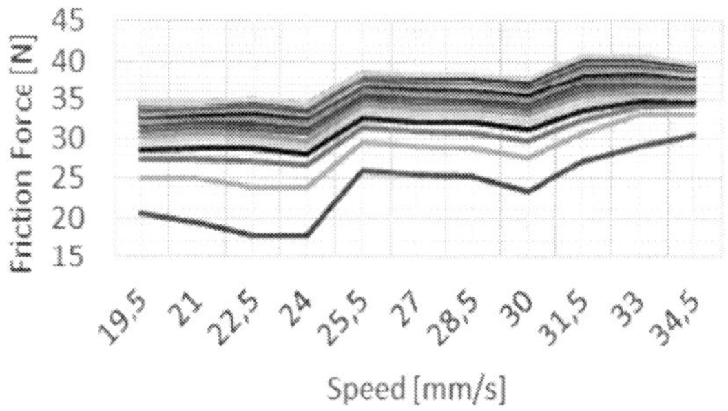


**Fig. 9.** Changes of friction forces during 15-seconds stops.

A particular impact on results (Figure 9) has work of the seals, which are often intentionally designed in two directions: reducing friction or reduction of leakage. Currently, technology of various kinds of seals is widely developed, which may result in the use of more complex sealing systems with reduced influence on the drive work[7-12]. Also, the use of new methods to improve the surface properties cooperating with the seal, could significantly improve the operating parameters [13]. The complexity of pneumatic drive cylinder quality, can be connected not only to the manufacturing method, but also to verification methods of desired cylinder parameters [13,14]. Another element affecting the operation of the drive is the size of connectors [15] and length of pneumatic lines whereby improper selection significantly reduces the maximum speed or dynamics of the piston.

After analysing most of factors affecting action of pneumatic piston drive, it can be concluded that for a complete description of phenomena occurring in drive, it is required to carry out a more detailed study which will take into account the measurements of friction for various durations of stopping the piston, in the function of operating pressure. In addition, measurements must be taken for a wider range of pressure connected asymmetrically, which much more corresponds to a typical drive operation. This measurement due to occurring moments that influence the non-uniform deformation of the sealing should be based also on the method of affixing the drive to the measuring system.

## References

1. P. A. Laski, J. E. Takosoglu, S. Blasiak. *Design of a 3-DOF tripod electro-pneumatic parallel manipulator*. Robotics and Autonomous Systems Vol. 72, pp. 59-70, (2015)
2. P. A. Laski, J. E. Takosoglu, S. Blasiak. *Delta type closed kinematics chain with pneumatic muscle actuator manipulator*. In Fuis, V (Ed.), *ENGINEERING MECHANICS 2014* (pp. 360–363). (2014).
3. J. E. Takosoglu, P. A. Laski, S. Blasiak. *A fuzzy logic controller for the positioning control of an electro-pneumatic servo-drive*. Proceedings of the Institution of Mechanical Engineers, Part I: Journal of Systems and Control Engineering Vol. 226, no. 10: pp. 1335-1343, (2012)
4. Olsson, H. *Control Systems with Friction*. Department of Automatic Control, Lund Institute of Technology, (1996).
5. L. Nowakowski, E. Miko, M. Skrzyniarz, *The analysis of the zone for initiating the cutting process ofx37crmov51 steel*, Engineering Mechanics (2016).
6. L. Nowakowski, M. Wijas, *The Evaluation of the process of surface regeneration after laser cladding and face milling*, Engineering Mechanics (2016).
7. S. Blasiak, J. E. Takosoglu, P. A. Laski. *Heat transfer and thermal deformations in non-contacting face seals*. Journal of Thermal Science and Technology Vol. 9, no. 2, (2014)
8. S. Blasiak, P. A. Laski, J. E. Takosoglu. *Parametric analysis of heat transfer in non-contacting face seals*. International Journal Of Heat And Mass Transfer, Vol.57, Issue 1, pp. 22-31, (2013)
9. S. Blasiak. *The two dimensional thermohydrodynamic analysis of a lubrication in non-contacting face seals*. Journal of Thermal Science and Technology, *10*(1), JTST0016 (2015).
10. S. Blasiak. *An analytical approach to heat transfer and thermal distortions in non-contacting face seals*. International Journal of Heat and Mass Transfer, *81*, 90–102. (2015).
11. S. Blasiak., A. Pawinska, *Direct and inverse heat transfer in non-contacting face seals*. International Journal of Heat and Mass Transfer, *90*, 710–718. (2015).
12. S. Blasiak., A. V. Zahorulko, *A parametric and dynamic analysis of non-contacting gas face seals with modified surfaces*. Tribology International, 94, 126–137. (2016).
13. D. Janecki, J. Zwierzchowski, L. Cedro, *A problem of optimal cylindricity profile matching,* Bulletin Of The Polish Academy Of Sciences-Technical Sciences Vol: 63, Pages: 771-779 (2015)
14. D. Janecki, J. Zwierzchowski, *The bird-cage method used for measuring cylindricity. A problem of optimal profile matching,* XIX IMEKO World Congress: Fundamental And Applied Metrology, Proceedings Pages: 1784-1789 (2009).
15. S. Blasiak, J. E. Takosoglu, P. A. Laski. *Optimizing the flow rate in a pneumatic directional control valve*. Proceedings of 20th International Conference on Engineering Mechanics 2014, Brno University of Technology, Czech Republic, pp. 96-99, (2014)

# Active control of the wake behind the cylinder

*Zuzana* Broučková [1], *Erik* Flídr[1,3], *Pavel* Šafařík[2] and *Zdeněk* Trávníček[1]

[1]Institute of Thermomechanics, The Czech Academy of Sciences, Prague, Czech Republic
[2]Czech Technical University, Faculty of Mechanical Engineering, 166 07 Prague, Czech Republic
[3]VZLÚ, Beranových 130, 199 05 Prague – Letňany

**Abstract.** Active control of the flow around a circular cylinder using a synthetic jet (SJ) array was experimentally investigated in a small wind tunnel. The prevailing method was flow visualization. Modification of the wake was investigated at two values of a dimensionless momentum coefficient ($C_\mu = 0.21$ and 1.02). Various $\alpha$ and $C_f$ parameters were tested (where $\alpha$ is the angle between SJs and the main flow direction and $C_f$ is the frequency ratio). It was found that the control moves a separation point more downstream, i.e. the width of the wake can be reduced. Moreover, increasing frequency ratio $C_f$ causes refining of the vortex structure. When the SJs are placed at the front or rear stagnation points, the wake tends towards the symmetrical pattern.

## 1 Introduction

Based on a need of energy input flow control methods can be categorized into passive and active [1]. Synthetic jet (SJ) is one of the active methods used for flow control. Formation and evaluation of the SJ was investigated in [2, 3]. An influence of the active flow control using SJs on the flow around a circular cylinder was extensively studied by many researches. Some important findings concerning these cases were summarized in [4], more detailed view on the topic can be found in [5].

In [6] was shown if the angle between the SJ and oncoming flow $\alpha$ (see Figure 1) and the dimensionless momentum coefficient $C_\mu$ (see Eq.(4) in the text below) were optimal, then the drag coefficient was reduced about 30%. If the SJ was placed in the rear stagnation point, small close envelope was formed during the extrusion part of the period [7]. During the intrusion part of the period the vortex pair is convected downstream from the both sides of the cylinder and interacts with the wake. In [8] it was found that vortex shedding under the flow control can be categorized into six regimes.

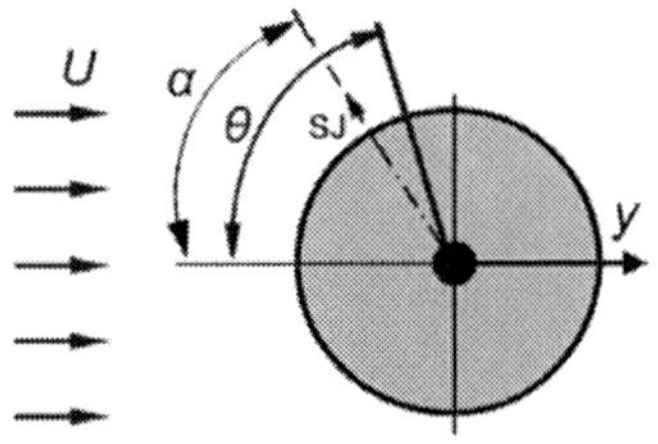


**Fig. 1.** Coordinate system of the cylinder in cross flow.

All the above mentioned investigations used harmonic driving signal (sinusoidal wave). Another approach was tested in [9, 10]. If the extrusion and intrusion part of period is changed (typically, the duty cycle is changed), the $C_\mu$ coefficient is changed too without frequency changing. With increasing frequency of synthetic jet, the lock-on effect was observed in [11, 18, 19]. In [12] two roughness strips (i.e. passive control) were used in combination with synthetic jet (i.e. active control).

In this paper, an influence of the angle $\alpha$, SJ frequency and coefficient $C_\mu$ was studied by means of flow visualization.

## 2 Parameters

For description of the flow around a cylinder two dimensionless numbers were introduced in [13]. Based on the oncoming flow velocity $U$ and the cylinder diameter $d$ the Reynolds number is defined as $Re = U{\cdot}d/\nu$, where $\nu$ is the kinematic viscosity The Strouhal number is defined as $St = f{\cdot}d/U$, where $f$ is the natural vortex shedding frequency.

For SJ the Reynolds number is defined as $Re_{U_0} = U_0 D_o / \nu$, where $U_0$ is time-mean velocity based on the extrusion part of the period and $D_o$ is the diameter of the actuator output,

$$U_0 = \frac{1}{T}\int_0^{T/2} u_0(t)dt, \qquad (1)$$

where $u_0(t)$ is the SJ exit velocity in time ($t$), and $T$ is the period.

Strouhal number of the SJ is defined as $St_{U_0} = f_{sj} D_o / U_0$, where $f_{sj} = 1/T$ is the SJ frequency. The jet momentum is generated during the extrusion part of the actuating period. Assuming the slug flow model

* Corresponding author: brouckova@it.cas.cz

and the sinusoidal waveform of the actuation signal, in general, the jet momentum can be calculated as

$$M_{sj} = \frac{1}{T}\int_0^{T/2} \rho A u^2(t)dt \text{,} \quad (2)$$

where $\rho$ is the fluid density, $A$ is the area of the actuator output.

For describing of the flow control it is appropriate to defined some dimensionless ratios. The frequency ratio is defined as

$$C_f = \frac{f_{sj}}{f} \text{.} \quad (3)$$

The momentum coefficient is defined in [6] as

$$C_\mu = \frac{M_{sj}}{M} \text{,} \quad (4)$$

where $M$ is the oncoming flow momentum.

## 3 Experimental setup and methods

### 3.1 Synthetic jet actuator

Figure 2 shows a schematic view of the SJ actuator including a coordinate system. The actuator consisted of a cavity with a height of 10 mm, which was equipped with an electrodynamically driven diaphragm. Diaphragm originated from loudspeaker Monacor SP-7/4S, powered by function generator FG7002C. The diaphragm was characterized by its diameter $D_D = 53$ mm and rigidity $K_p = 4.8{\cdot}10^5$ N·m$^{-3}$ [14]. The cavity was connected with the surroundings by the pipe with the inner diameter $D = 3$ mm, the outer diameter $d = 4$ mm and the length $L = 54$ mm. End of the pipe was placed into the flow. The pipe was placed perpendicularly to the main flow, its end part touched the front wall of the tunnel. In the part of the pipe which was submerged in the main flow ($l = 29$mm) 14 actuator outputs with diameter $D_o = 1.05$ mm were placed in one row with distance of 2 mm (see Figure 2).

For the particular case of the present actuator, the momentum coefficient, defined by Eq. (4), was evaluated in [20] as

$$C_\mu = \frac{7\pi^3}{8}\frac{D_o^2}{dl}\left(\frac{U_0}{U}\right)^2 \text{.} \quad (5)$$

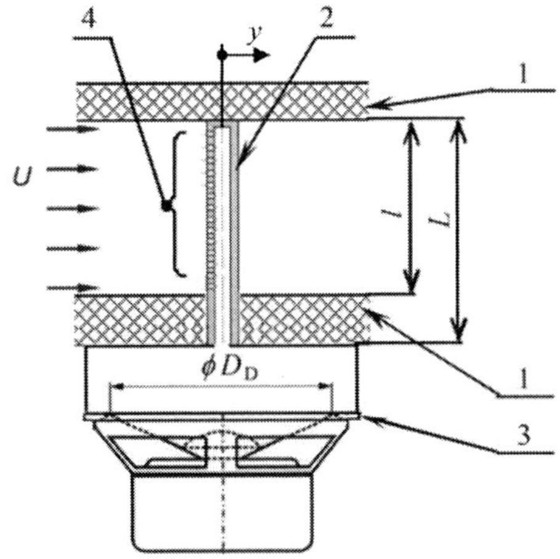


**Fig. 2**. Top-view scheme of the cylinder in the wind tunnel: 1: tunnel side walls, 2: cylinder (outer diameter of $d = 4$ mm), 3: SJ actuator, 4: row of 14 SJs (diameters of $D_o = 1.05$ mm, equidistantly distributed with spacing of 2 mm).

### 3.2 Flow visualization equipment

Flow visualization was performed in a small wind tunnel with a test section (460 × 350 × 29) mm = (length × height × depth). The free stream velocity was $U = 0.77$ m·s$^{-1}$, i.e. the corresponding Reynolds number was $Re = 193$. Visualization medium was water fog created by the ultrasonic generator MiniNebler. During the experiments continual or stroboscopic light was used for illumination of the test section.

The light system for the continual or stroboscopic regime is an inhouse-built system based on the battery of 10 LEDs (Hi-Power HPA8B-49k5WHC, 5W of each) designed for the operation frequency of 0-500Hz and the light pulse duration of 75-3500μs. The system enables to synchronize the stroboscopic light with the driven frequency at an arbitrary phase of the driven cycle.

Pictures were taken by digital camera Canon PowerShot G7 connected to a PC via USB cable, using exposures of 1s. The photos were multi-exposures of a large number of frames (in range from 27 to 162 frames).

### 3.3 Pitot tube

The time mean velocity was measured by the Pitot tube (blunt-nosed impact tube, with the outer and inner diameter of 2 mm and 1.5 mm, respectively), connected to an electronic manometer GMH 3156 with GMSD 2.5 transducer with resolution of 0.1 Pa. From Bernoulli's equation, the flow velocity can be calculated as $u = c_p[2(p\text{-}p_s)/\rho]^{0.5}$ where $p$ is the total pressure, $p_s$ is the static pressure (the static pressure was considered equal to the barometric pressure) and $c_p$ is the pressure coefficient (approximately 1.0 in agreement with the Bernoulli's equation).

In the present study, the Pitot tube was used for measurement of frequency and power characteristic. In [14] was shown that the Pitot tube placed far enough from the actuator output can be used for pulsatile flow measurement. Therefore, the SJ time-mean velocity, $U_0$, was estimated from the Pitot tube measurement (similarly as in [14]).

## 4 Results and discussion

### 4.1 Frequency and power characteristics

In order to keep the momentum coefficient $C_\mu$ constant, the power of the SJ actuator was varied to compensate velocity changes caused by changes of the frequency. For that reason, the frequency and power characteristics were measured.

The Pitot tube was placed at $y/D_o = 6$ and $z = 0$ mm. The measurement points were distributed equidistantly along the $x$-axis, with the spacing 2 mm. The first point was placed at $x = 14$ mm and the last one at $x = 40$ mm. The frequency characteristic was evaluated over the range of $f = 30$–$150$ Hz. During the measurement, an input power of the SJ actuator was set as the constant of $P = 2$ W. It vas evaluated as the product of the measured supply current and voltage (both effective values; the phase angle between them was neglected similarly as in [14]).

Figure 3 shows the frequency characteristic evaluated experimentally. Moreover, the theoretical resonance frequency is shown too (for derivation details see [20]):

$$f_{teor} = \frac{7}{\pi}\left[\frac{D_o^2}{D_D}\right]\sqrt{\frac{K_p}{\rho\left(D^2L + 14D_o^2L_e\right)}}, \tag{6}$$

where $L_e$ is the effective length calculated from [15]:

$$L_e = H + \frac{8}{3\pi}D_o, \tag{7}$$

where $H$ is the thickness of the pipe wall.

Experimentally evaluated resonance frequency is approximately 45 Hz, this is in good agreement with the theoretical value ($f_{theor} = 41.9$ Hz).

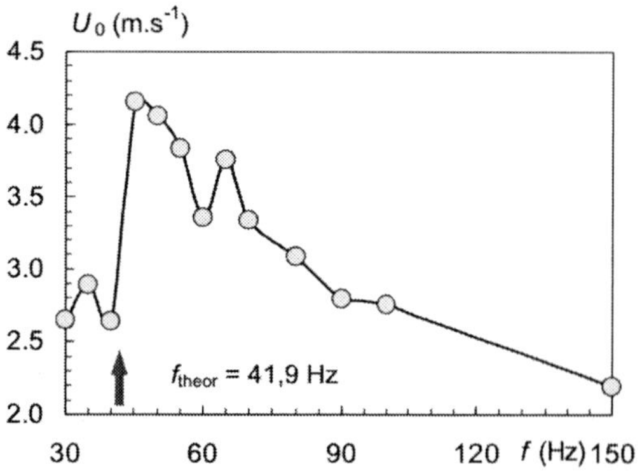


**Fig. 3.** Frequency characteristic.

The power characteristics were measured for five different frequencies $f = 30$, 45, 60, 110 and 160 Hz. The Pitot tube was again placed at $y/D_o = 6$ and $z = 0$ mm. Characteristics were measured at $x = 18$, 28 and 38 mm.

Figure 4 shows the power characteristics. Figure 4b shows in detail the range of 0 – 0.2 W. It can be seen that with decreasing power the velocity of SJ decreases too. For very low power input, the SJ was weak and probably near the formation criterion [16, 17].

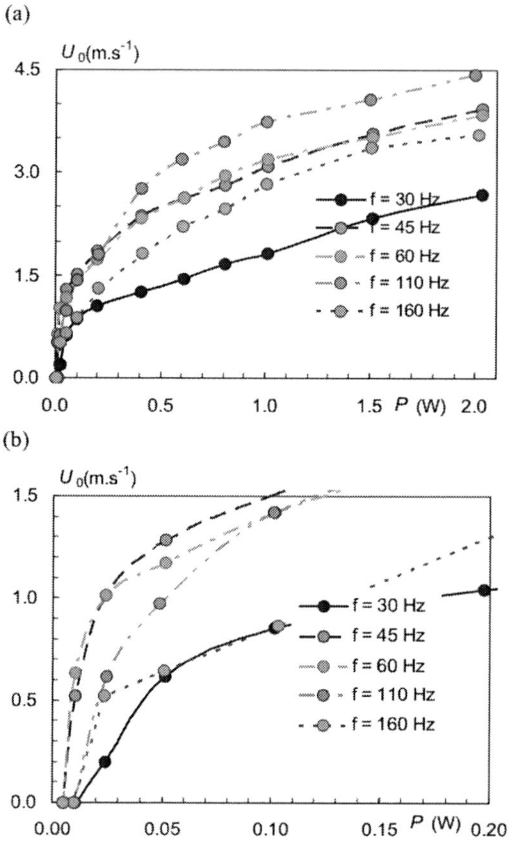


**Fig. 4.** Power characteristics a) whole characteristic, b) detail of the area with low values of the power.

### 4.2 Flow visualization

Figure 5a shows a visualization of the flow around the cylinder at the regime of the von Kármán vortex street; Fig. 5a and 5b were made under continual and stroboscopic light, respectively. The natural vortex shedding frequency was $f = 27.12$ Hz and $St = 0.141$. In these cases, the SJs were off.

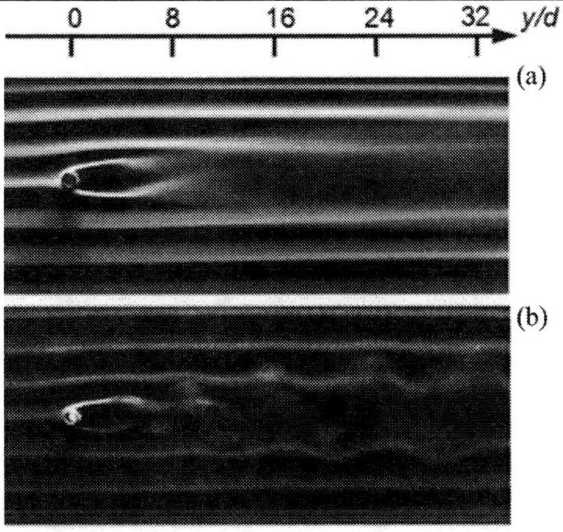


**Fig. 5.** Flow visualization of cylinder in cross-flow under (a) continual light, (b) stroboscopic light.

Visualization of the flow under the SJ control was made for five different $C_f$, four different angles $\alpha = 0$, 75, 120 and 180° and two values of $C_\mu = 0.21$ and 1.02.

Parameters of the flow visualization experiments for $C_\mu = 0.21$ are summarized in Table 1.

**Table 1.** Parameters of the visualization experiments for $C_\mu = 0.21$.

| Fig. | $\alpha$ (°) | $P$ (W) | $f_{sj}$ (Hz) | $St_{U0}$ (1) | $C_f$ (1) |
|---|---|---|---|---|---|
| 6a | 0 | 0.080 | 27.2 | 0.041 | 1.00 |
| 6b | 75 | | | | |
| 6c | 120 | | | | |
| 6d | 180 | | | | |
| 7a | 0 | 0.015 | 41.2 | 0.062 | 1.52 |
| 7b | 75 | | | | |
| 7c | 120 | | | | |
| 7d | 180 | | | | |
| 8a | 0 | 0.015 | 53.9 | 0.081 | 1.99 |
| 8b | 75 | | | | |
| 8c | 120 | | | | |
| 8d | 180 | | | | |
| 9a | 0 | 0.030 | 108.4 | 0.163 | 4.00 |
| 9b | 75 | | | | |
| 9c | 120 | | | | |
| 9d | 180 | | | | |
| 10a | 0 | 0.080 | 162.4 | 0.244 | 5.99 |
| 10b | 75 | | | | |
| 10c | 120 | | | | |
| 10d | 180 | | | | |

Figures 6(a-d) show flow visualizations for $C_f = 1.00$. In all cases the separation point moved downstream under the flow control condition which caused reduction of the width of the wake. If the SJs were placed at the front stagnation point ($\alpha = 0°$), small closed region was formed as a result of interaction between the SJs and the oncoming main flow as shown in Figure 6a.

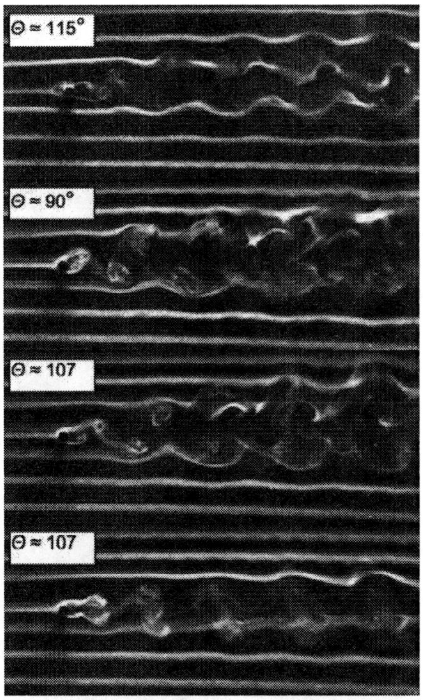


**Fig. 6.** Flow visualization for $C_\mu = 0.21$ and $C_f = 1.00$ at (a) $\alpha = 0°$, (b) $\alpha = 75°$, (c) $\alpha = 120°$, (d) $\alpha = 180°$;(parameters are summarized in Table 1).

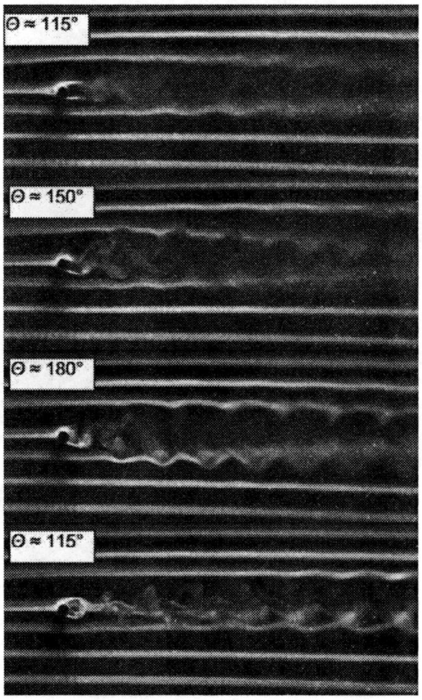


**Fig. 7.** Visualization for $C_\mu = 0.21$ and $C_f = 1.52$ at (a) $\alpha = 0°$, (b) $\alpha = 75°$, (c) $\alpha = 120°$, (d) $\alpha = 180°$; (parameters – see Table 1).

DOI: 10.1051/epjconf/201714302011

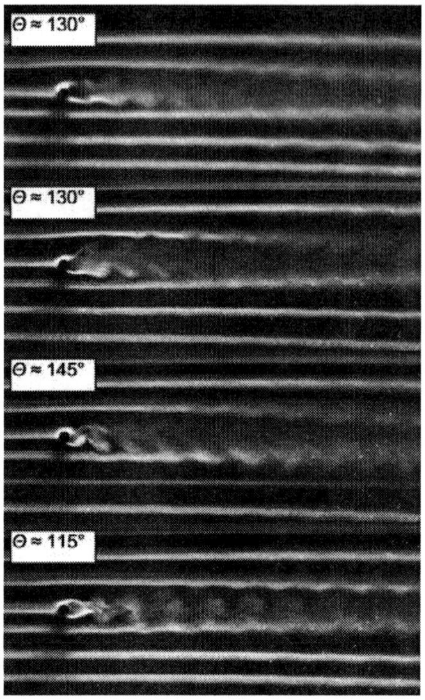


**Fig. 8.** Visualization for $C_\mu = 0.21$ and $C_f = 1.99$ at (a) $\alpha = 0°$, (b) $\alpha = 75°$, (c) $\alpha = 120°$, (d) $\alpha = 180°$; (parameters – see Table 1).

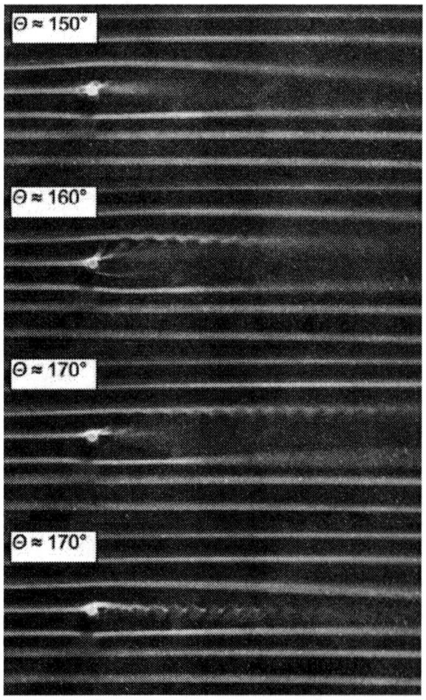


**Fig. 9.** Visualization for $C_\mu = 0.21$ and $C_f = 4.00$ at (a) $\alpha = 0°$, (b) $\alpha = 75°$, (c) $\alpha = 120°$, (d) $\alpha = 180°$; (parameters – see Table 1).

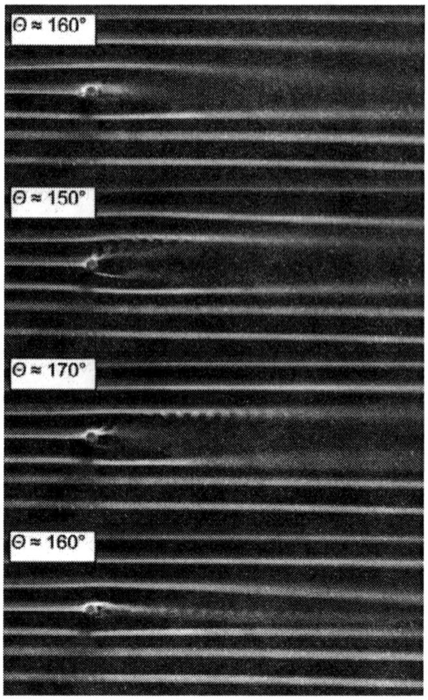


**Fig. 10.** Visualization for $C_\mu = 0.21$ and $C_f = 5.99$ at (a) $\alpha = 0°$, (b) $\alpha = 75°$, (c) $\alpha = 120°$, (d) $\alpha = 180°$; (parameters – see Table 1).

This finding is in agreement with [7]. If the SJs were placed at the rear stagnation point ($\alpha = 180°$) the wake behind the cylinder was symmetric.

Figures 7(a-d) show flow visualization for $C_f = 1.52$. With increasing frequency of the SJs, structures in the wake were refined and the separation point were moved more downstream.

With further increasing SJ frequency, a further refinement of the wake structure is visible as shown in Figures 8(a-d) for $C_f = 1.99$. Apparently, this flow field pattern indicates the lock-on effect [11, 18, 19]. Closed region at the front stagnation point ($\alpha = 0°$) is not so obvious, because the SJs were captured in the suction part of period. similarly as e.g. in [7].

Figures 9(a-d) show flow visualization for $C_f = 4.00$. In these cases, the von Kármán vortex street was completely suppressed. Structures in the wake were created by the SJs during the extrusion part of the period. If the SJs were placed at the stagnation points ($\alpha = 0$ and 180°), width of the wake was approximately equal to the diameter of the cylinder.

Figures 10(a-d) show flow visualization of the flows with $C_f = 5.99$. This increase of frequency yielded no significant changes compared to the visualization for $C_f = 4.00$. Only structure in the wake was finer as a manifestation of the lock-on effect.

Another visualization experiments were made for stronger SJ at $C_\mu = 1.02$. The parameters for this visualization experiments are sumarized in Table 2.

**Table 2.** Parameters of the visualization experiments for $C_\mu = 1.02$.

| Fig. | $\alpha$ (°) | $P$ (W) | $f_{si}$ (Hz) | $St_{U0}$ (1) | $C_f$ (1) |
|---|---|---|---|---|---|
| 11a | 0 | | | | |
| 11b | 75 | 0.75 | 27.3 | 0.041 | 1.01 |
| 11c | 120 | | | | |
| 11d | 180 | | | | |
| 12a | 0 | | | | |
| 12b | 75 | 0.12 | 41.0 | 0.062 | 1.51 |
| 12c | 120 | | | | |
| 12d | 180 | | | | |
| 13a | 0 | | | | |
| 13b | 75 | 0.15 | 54.2 | 0.081 | 2.00 |
| 13c | 120 | | | | |
| 13d | 180 | | | | |
| 14a | 0 | | | | |
| 14b | 75 | 0.16 | 108.2 | 0.163 | 3.99 |
| 14c | 120 | | | | |
| 14d | 180 | | | | |
| 15a | 0 | | | | |
| 15b | 75 | 0.32 | 162.3 | 0.244 | 5.98 |
| 15c | 120 | | | | |
| 15d | 180 | | | | |

If the momentum coefficient was increased from $C_\mu = 0.21$ to 1.02, the interaction between the SJs and oncoming flow can no longer be considered a flow control, it is rather an interaction between two comparable flows.

Figures 11(a-d) show flow fields for $C_f = 1.01$. If the SJs were placed at the front stagnation point ($\alpha = 0°$) the entire flow field in the observed area was disturbed. Because of higher $C_\mu$, the separation angle $\theta$ was almost same as the angle $\alpha$. If the SJs were placed at the rear stagnation point ($\alpha = 180°$), flow separation was not observed either on the upper, or on the bottom sides of the cylinder.

Figures 12(a-d) show flow visualization for $C_f = 1.51$. The increase of the SJ frequency reduced the disturbed region in front of the cylinder.

Figures 13(a-d) show interactions between the SJs and the oncoming flow for $C_f = 2.00$. Open envelope was observed if the SJs were placed at the front stagnation point ($\alpha = 0°$). This is in good agreement with [7]. Figures 14 and 15 show visualization for $C_f = 3.99$ and $C_f = 5.98$, respectively. Structures in the wake were more refined because of the lock-on effect. As shown in Figures 12-15, the separation point on the surface of the cylinder was moved downstream.

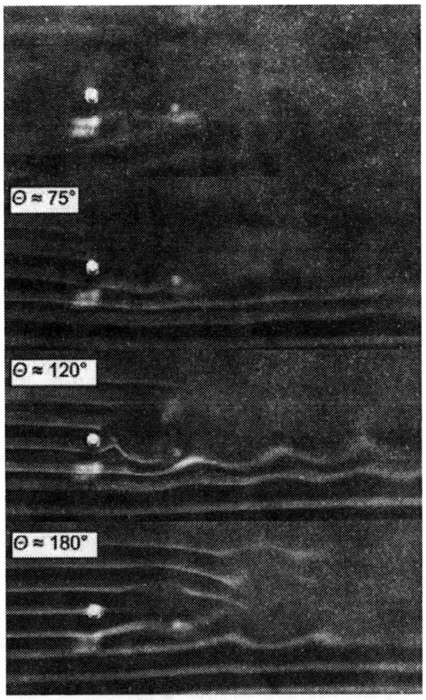


**Fig. 11.** Visualization for $C_\mu = 1.02$ and $C_f = 1.01$ at (a) $\alpha = 0°$, (b) $\alpha = 75°$, (c) $\alpha = 120°$, (d) $\alpha = 180°$; (parameters – see Table 2).

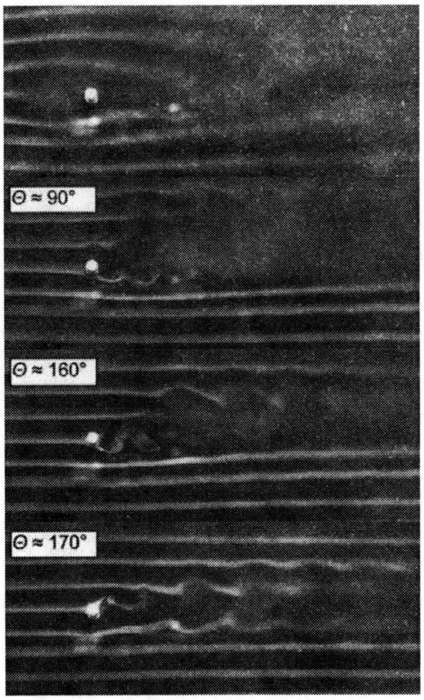


**Fig. 12.** Visualization for $C_\mu = 1.02$ and $C_f = 1.51$ at (a) $\alpha = 0°$, (b) $\alpha = 75°$, (c) $\alpha = 120°$, (d) $\alpha = 180°$; (parameters – see Table 2).

DOI: 10.1051/epjconf/201714302011

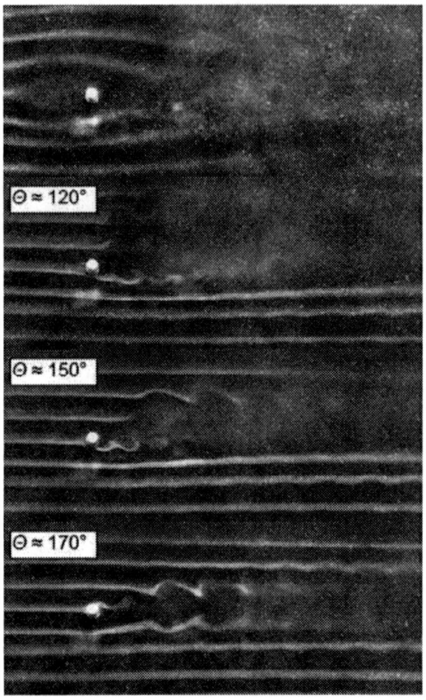


**Fig. 13.** Visualization for $C_\mu = 1.02$ and $C_f = 2.00$ at (a) α = 0°, (b) α = 75°, (c) α = 120°, (d) α = 180°; (parameters – see Table 2).

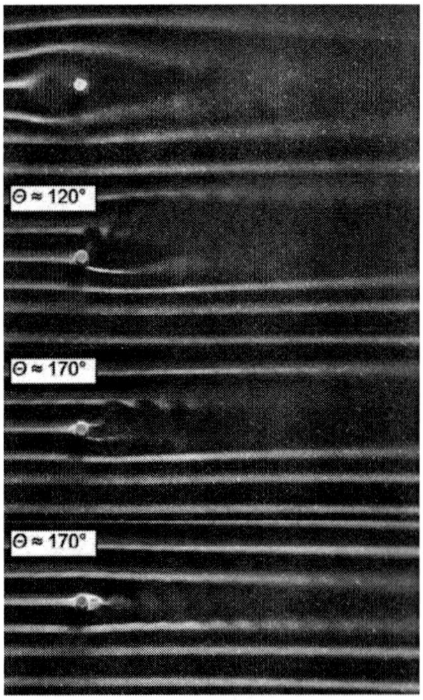


**Fig. 14.** Visualization for $C_\mu = 1.02$ and $C_f = 3.99$ at (a) α = 0°, (b) α = 75°, (c) α = 120°, (d) α = 180°; (parameters – see Table 2).

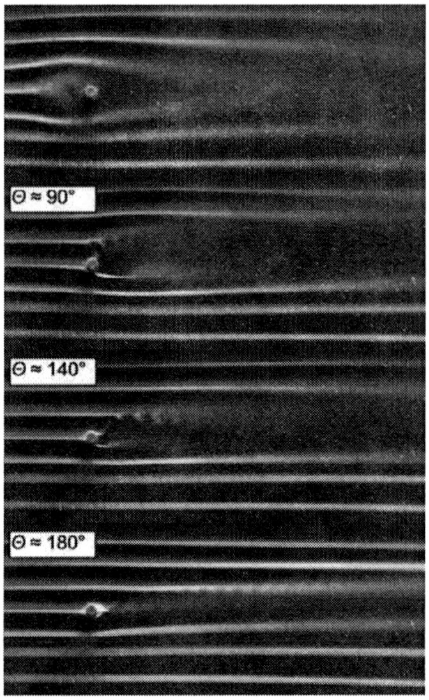


**Fig. 15.** Visualization for $C_\mu = 1.02$ and $C_f = 5.98$ at (a) α = 0°, (b) α = 75°, (c) α = 120°, (d) α = 180°; (parameters – see Table 2).

## 5 Conclusions

The aim of this work was an experimental investigation of the flow field around the circular cylinder under the active flow control. The row of the 14 synthetic jets issuing from the cylinder surface was used for this purpose.

Firstly, the frequency and power characteristics were experimentally evaluated. Then a flow visualization study in the wind tunnel was performed. The most significant effects of the SJ control on the flow field are briefly summarized in the following points:

- An increase of the frequency ratio $C_f$ influenced vortex structures in the wake. More specifically, the large vortex structures were refined with increasing $C_f$.
- The lock-on effect was indicated for majority of experiments at $C_f \geq 1.5$.
- The separation point was moved downstream. When the momentum coefficient was $C_\mu = 0.21$, the separation angles were $\theta \approx 130 - 160°$. For $C_\mu = 1.02$ the separation angles were $\theta \approx 90 - 160°$. If the SJs were placed at the rear stagnation point ($\alpha = 180°$), no separation was observed in some cases, see Figures 11d and 15d.
- For $C_\mu = 0.21$, the flow control always reduces the wake width. The same observation was made for $C_\mu = 1.02$ except the cases of the SJs at the front stagnation point ($\alpha = 0°$).

- If the SJs were placed at the front or rear stagnation points ($\alpha = 0$ or 180°), the wake was symmetric in majority cases as shown in Figures 6d, 7(a, d), 9(a, d) 10d, 11d, 14d and 15d.

## Acknowledgments

We gratefully acknowledge the support of the GACR – Czech Science Foundation (projects no. 14-08888S and 16-16596S) and the institutional support RVO: 61388998.

## References

1. M. Gad-El-Hak, *Flow Control* (Cambridge, University Press, New York, 2000)
2. B.L. Smith, A. Glezer, Phys. Fluids **10**, pp. 2281-2297 (1998)
3. J. E. Cater, J. Soria, J. Fluid Mech. **472**, pp. 167 – 200 (2002)
4. L-H. Feng, J-J. Wang, *Proceedings of 16th International Symposium on Flow Vizualization, June 24-28*, Okinawa, Japan (2014)
5. P-F. Zhang, J-J. Wang, L-H. Feng, Sci. China-Technol. Sci. **51**, pp. 1315 -1344 (2008)
6. M. Amitay, A. Honohan, M. Trautman, A. Glezer, AIAA Paper 97-2004 (1997)
7. J-J. Wang, L-H. Feng, Ch-J. Xu, Sci. China-Technol. Sci. **50**, pp. 550 –559 (2007)
8. L-H. Feng, J-J. Wang, Eur. J. Mech. B-Fluids **43**, pp. 14 – 32 (2014)
9. L-H. Feng, J-J. Wang, Ch. Pan, J. Fluids Struct. **26**, pp. 900 – 917 (2010)
10. L-H. Feng, J-J. Wang, Exp. Fluids **53**, pp. 467 – 480 (2012)
11. L-Q. Ma, L-H. Feng, Sci. China-Technol. Sci. **56**, pp. 158 – 170 (2013)
12. J. Tensi, I. Boué, F. Paillé, G. Dury, J. Vis. **5**, pp. 37 – 44 (2002)
13. M. M. Zdravkovich, *Flow around a Circular Cylinders* (Oxford University Press, 1997)
14. Z. Broučková, Z. Trávníček, J. Vis. **18**, pp. 581-593 (2015)
15. Z. Škvor, *Akustika a elektroakustika* (Acoustic and Electro-acoustic, Praha: Academia 2001, in Czech)
16. R. Holman, Y. Utturkar, R. Mittal, B.L. Smith, L. Cattafesta, AIAA J. **43**, pp. 2110 – 2116 (2005)
17. Z. Trávníček, Z. Broučková, J. Kordík, AIAA J. **50**, pp. 2012 – 2017 (2012)
18. L. H. Feng, J. J. Wang, J. Fluid Mech. **662**, pp. 232 – 259 (2010)
19. W. Kim, J. Y. Yoo, J. Sung, Phys. Fluids **18**, pp. 1 – 20 (2006)
20. E. Flídr, *Kruhový válec v homogenním proudovém poli s vlivem aktivního řízení* (Circular Cylinder in Homogeneous Flow Field under an Active Control Influence, Master Thesis, Faculty of Mechanical Engineering, CTU in Prague, 2016, in Czech).

# PIV and LIF study of flow and thermal fields of twine plumes in water

*Zuzana* Broučková [1] and *Zdeněk* Trávníček[1]
[1]Institute of Thermomechanics, The Czech Academy of Sciences, Prague, Czech Republic

**Abstract.** Flow and thermal fields of a pair of plane plumes in water are investigated by means of PIV and LIF experiments. The plumes are generated from thermal line sources, which are made out of electrically heated cylinders with a diameter of $D$ = 1.21 mm. A cylinder-to-cylinder distance was 17.9 $D$. Either continuous or pulsating heating were used with the same heating input power. Because the cylinder-to-cylinder distance is moderately small, deflections of plumes from a vertical direction occur and the plumes are inclined together. This behavior is caused by a confined entrainment from a space between the both plumes. For a continuous heating, low frequency oscillations were identified and the natural frequency was evaluated as 0.5 Hz. Based on this finding, pulsating heating was used at the subharmonic frequency of 0.25 Hz. The maximum time-mean velocity magnitude at the continuous and pulsating heating were commensurable, approximately 0.007 m/s. On the other hand, pulsating heating achieves by 36 % higher velocity peaks. A very strong velocity oscillations were generated by pulsating heating at the distance approximately 8.3 $D$ above the cylinders, where the velocity maxima oscillate along the time-mean value of 0.0057 m/s from -30% to +70 %. Temperature fields reasonably agree with this findings, despite a relatively fast equalization of the temperature field was concluded. The results demonstrate enhancement effects of pulsations in flow/thermal fields.

## 1 Introduction

A buoyant jet is a type of fluid motion and one of basic examples of free shear flows – see Schlichting [1], Blevins [2]. It is generated by differences in fluid density. In other words, the driven force (buoyancy) results from density differences. A buoyant jet whose initial momentum is approximately zero is called a plume. In the case of a thermal source, the plume is developed because a local temperature increase causes a decrease of density. For a line thermal source, a plane plume is developed. In the present study, a twin of plane plumes has been investigated by means of PIV and LIF experiments in water.

## 2 Experiments

### 2.1 Experimental setup and methods

Figure 1 shows a schematic view of the present experimental setup. The experiments are made in a glass water tank with the following dimensions: height × length × depth = 150 × 245 × 140 mm. The working fluid is distilled water.

The device frame is made out of PMMA (polymethyl methacrylate). A pair of line thermal sources (pos. 1 in Fig. 1) are used to generate a twine of the plane plumes. These thermal sources are made from a smooth stainless steel tubes, its outside and inside diameters are $D$ = 1.21 mm and 0.91 mm, respectively. The cylinders are mounted horizontally at the center-to-center distance of 21.7 mm. Four metal brackets (2) fix the cylinders at the ends. The total length of the cylinders between the brackets is $L$ = 60.7 mm, i.e. the aspect ratio is $L/D$ = 50.2. The immersion depth of cylinders is approximately 40 mm. The coordinate system $x$, $y$, $z$ is also shown in Fig. 1.

The cylinders are heated by the Joule effect of direct or pulsating current $I$ from the brackets and Fig. 1 schematically shows the electrical connections. The PC-controlled power supply (Voltcraft CPPS, maximum DC parameters: 42 V, 20A, and 320W) can operate in two regimes to produce either direct or pulsating current. The maximum current was 12.9 A in this study.

For DC-supply heating (continuous heating, CH), the current $I$ is measured by a multimeter (Metex M 4600M). Voltage differences along the cylinders ($V_L$ and $V_R$) are measured by two multimeters (Metex M 3800) via the 4-terminal sensing (considering that the 4-point probes method allows more accurate measurements). The total voltage difference is $V = V_L + V_R$. The maximum uncertainties of the current and voltage measurements were less than ±0.3%, and ±0.5%, respectively.

For pulsating-supply heating (pulsating heating, PH), the current and voltage are measured by means of

* Corresponding author: brouckova@it.cas.cz

 DOI: 10.1051/epjconf/201714302012

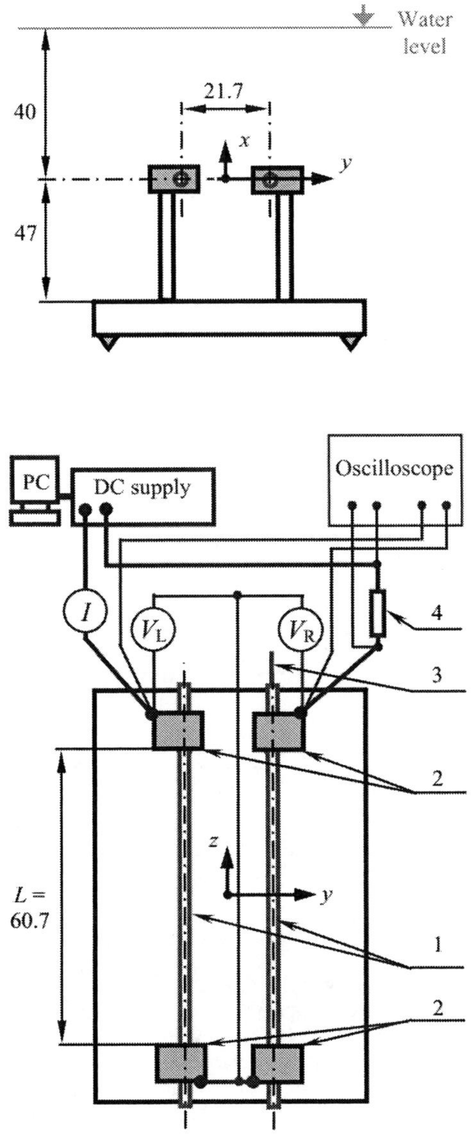


**Fig. 1.** Experimental device, sizes are in millimetres; 1: heated cylinders, 2: brackets, 3: thermocouples for $T_w$ measurement, 4: shunt 0.100 Ω.

the two channel oscilloscope (Agilent DSO3102A) with a 0.100 Ω shunt (4) – see Fig. 1.

The LIF technique is based on the fluorescence effect. A solution of Rhodamine B is inserted into the water. The dye-stained flow is lit by the laser sheet (Nd:Yag pulsed laser, Litron, NANO S 65-15, with a wavelength of 532 nm and 65 mJ output energy of a 6-8 ns pulse, with cylindrical optics). The laser is placed approximately 1.8 m from the water tank. The repetition rate is 15 Hz (i.e. the maximum of the laser possibility).

The flow field pictures are recorded with a digital camera (HiSense Neo, 2560×2160 pixels, 16 bit) equipped with the lens (Tokina 100 mm F2.8 Macro D) and an optical filter (high pass in wavelength with cut-off 570 nm) which allows transmission of the fluorescent light only while filters out the laser-scattered light. The camera is placed approximately 350 mm from the water tank, producing approximately 62×52 mm field of view. The system is controlled by DynamicStudio v3.40 Software (Dantec Dynamics). More details were described previously by Broučková et al. [3].

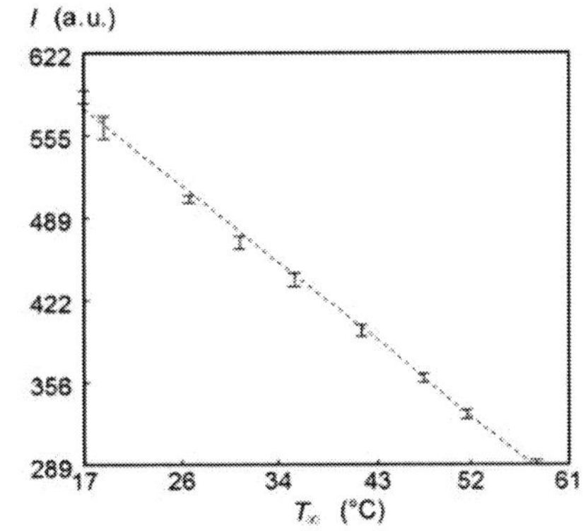


**Fig. 2**. The relation between the fluorescence intensity and the temperature.

To quantify temperature field by means of the LIF technique, an in-situ calibration was performed in the range of 17–58°C. A typical calibration curve is shown in Fig. 2 as a relationship between the fluorescence intensity (in arbitrary units) and temperature at the given laser conditions and given Rhodamine B concentration. To enhance reliability of the calibration experiment, each calibration point were measured repeatedly three times and the error bars in Fig. 2 indicate reasonably small data scattering (causing mainly by a slight variation of the laser intensity with time). The relationship is linear in character which well agrees with expectations based on literature results; see e.g. Seuntiëns et al. [4].

The PIV experiments are made with the same techniques (laser, camera with the color filter, and software) as described in the text above. Fluorescent seeding particles (PS-FluoRed-FRAK-Fi204, 1–20 μm in diameter) are used. The wavelength peaks of the absorption and emission lights are 530 nm and 607 nm, respectively. The adaptive PIV technique is used with the minimum and maximum interrogation areas 32×32 pixels and 64×64 pixels, respectively. The resulting velocity maps are averaged from a sequence of 50 or 90 images.

The temperature of the right cylinder ($T_W$) is measured by a J type (iron–constantan) thermocouple connected with a thermometer Omega DP41-B. The thermocouple was inserted through the cylinder tube on the coordinate $z$ = 0, see pos. 3 in Fig. 1. The temperature difference between inside and outside of cylinder surface has been neglected since the cylinder wall had only 0.15 mm thickness. Therefore, the measured temperature $T_w$ is considered as the average surface temperature of the cylinder. A similar approach was used in the recent forced convection study by Broučková and Trávníček [5].

The bulk temperature of quiescent water in the tank $T_\infty$ is evaluated as the average of two temperatures, measured by two NTC thermistor sensors connecting width a thermometer (Ahlborn AMR, Therm 2280-3). The sensors are placed below the tested cylinders, approximately in the places of $x = -30$ mm, $y = \pm 11$mm and $z = -10$mm. During the present experiments, the maximum differences between the both bulk temperatures were 0.3°C. The temperatures of the cylinders and water ranged $T_w = 31–64$°C and $T_\infty = 23–26$ °C, respectively. All the thermometers with probes were calibrated and the maximum uncertainties of the $T_w$ and $T_\infty$ are less than ±0.2°C and ±0.1°C, respectively.

### 2.2 Parameters and data reduction

The heating input power $P$ is evaluated from the measured electric current $I$ and voltage $V$. For DC-supply heating (continuous heating, CH), the input power is $P = I \cdot V$. For pulsating-supply heating (PH), the input power can be evaluated by integration

$$P = \int_0^T i(t)\, v(t)\, \mathrm{d}t\,, \qquad (1)$$

where $i(t)$ and $v(t)$ are the instantaneous current and voltage, $t$ is time and $T$ is the cycle period. In the present case, the rectangular signal is used with the electric current and voltage alternating between and the values $I$ an $V$, respectively, and zeros. The duty cycle was chosen 50%, thus the heating is off/on during the cycle parts of $t/T = 0–0.5/\ 0.5–1.0$, respectively. Therefore the input power can be evaluated as $P = I \cdot V/2$. The experiments were made in a range of $P = 6–41$ W, and the present study deals with the case of $P = 14.7$ W.

Based on an experimental evaluation of the natural frequency of the present plumes, the thermal actuation was chosen as a subharmonic action with the driven frequency of $f = 1/T = 0.25$ Hz; more details of this approach and findings are explained in part 3.1 bellow.

The measured temperature $T_w$ is the average surface temperature of the cylinder, as written in the text above. The thermal boundary condition at the cylinder surface is approximately uniform heat flux, similarly as in [5]. Small differences from the heat flux uniformity which result from the local temperature differences are negligible because of small sensitivity of the stainless steel electrical resistivity to temperature. Namely, the temperature coefficient of the resistivity of the present stainless steel was evaluated as 0.0009 °C$^{-1}$, hence the temperature differences of order 10°C can cause the resistivity changes less than 1 %. Thus the Joule's first law gives negligible differences (less than 1 %) in the local heat generation rates and local heat fluxes.

The governing parameter for natural convection is the Grashof number, which is defined as $Gr = g\beta(T_w - T_\infty)D^3/\nu^2$, where $g$ is the acceleration of gravity and $\beta$ is the volumetric thermal expansion coefficient. The both material parameters $\beta$ and $\nu$ are evaluated at the film temperature $T_f$.

Another parameter important in natural convection is the Rayleigh number, which is defined as $Ra = Gr\ Pr$, where $Pr$ is the Prandtl number. Again, an evaluation of the $Pr$ is made at the film temperature $T_f$.

## 3 Results and discussion

The following experiments were performed after well stabilized flow and temperature fields, i.e. at least 1 hour after filling the water tank with water of the laboratory temperature, and at least 5 minutes after heating was turned on.

### 3.1 Velocity flow field: PIV experiment

The measurements were performed for the investigated twin plumes, which are generated by either continuous or pulsating heating (CH or PH, respectively). The operating parameters are summarized in Table 1 including the temperature differences $T_w - T_\infty$ and the Grashof and Rayleigh numbers. Note that the both cases of heating, which are presented in this study, have identical heating input power of $P = 17.4$ W.

**Table 1.** Operating parameters of the present experiments

| Case | $f$ (Hz) | $T_w - T_\infty$ (°C) | $Gr$ | $Ra$ |
|---|---|---|---|---|
| **CH** (continuous heating) | 0 | 19.8 | 203 | 958 |
| **PH** (pulsating heating) | 0.25 | 17.7–20.1 | 203–249 | 915–1086 |

Figure 3 shows a typical PIV result of the twin plumes at CH case as the time-mean velocity vector map colored by the velocity magnitude. Figure 3 was made by averaging over 50 instantaneous flow fields. Two small white crosses indicate the positions of the cylinders (they are not visible here because they are hidden behind their brackets, as can be clear from Fig. 1). A small white ring in Fig. 3 indicates the position in the shear layer in which the velocity magnitude data were analyzed to obtain the time behavior and power spectral density (PSD), as will be discussed in the following text (Figs. 5 and 6).

As is presented in Fig. 3, the plumes issuing from the cylinders are laminar in character. They are inclined together. The main reason is a fluid entrainment from surroundings to the plumes. The entrainments generate a very weak under pressure between the plumes, however the plumes restrict each other to entrain fluid from there. Note that another mechanism causing a deformation of plumes is linked with the 3-dimensional end effects. The fluid entrainments are not confined at the plume ends, therefore the plume volume flux is enhanced at the ends.

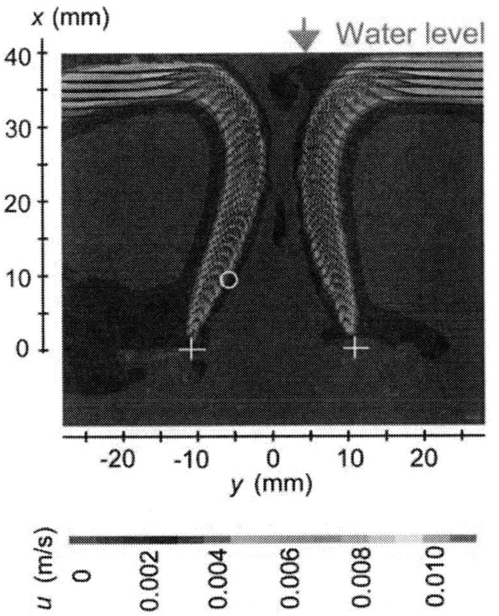


**Fig. 3**. Time-mean velocities of CH case, presented as the vector map colored by the velocity magnitude; parameters are summarized in Table 1.

It is worthy of notice here that this mechanism is well know from rectangular fluid jets, where this process basically changes a two-dimensional (plane) jet flow causing a three-dimensional flow character with a saddle-back velocity profile – see Vouros et al. [6] and Broučková et al. [3].

It can be imagined that the both plumes could merge in the far field. However, the present geometry and given moderate immersion depth of cylinders (40 mm i.e. 33 *D*, see Fig. 1) causes impingements of the plumes to the water surface and plume turning apart from each other. Finally, a pair of the wall jets is generated to the opposite directions along the water surface.

Figures 4(a–c) show the twin plumes generated by pulsating heating (PH). Meaning of the small white crosses and the small white ring are the same as in Fig. 3, i.e. the cylinders positions and a point for time and frequency analyzes of velocities, respectively. A comparison of the both time-mean cases CH and PH (Figs. 3 and 4(a), respectively) indicate more stronger character of the time-mean PH plumes including wider wall jets outgoing from the test section. To demonstrate more deeply this behavior, Figs. 4(b and c) show the flow fields at heating and non-heating parts of the pulsating-supply cycle, respectively: the time-averaging was made for 10 instantaneous flow fields at $t/T = 0.32$–$0.47$ and $0.72$–$0.87$, respectively. As is shown, ascending flows from the cylinders are generated during the heating moment – see Fig.4(b). Obviously, warmed fluid volumes continue the advection motion during the non-heating moment – see Fig.4(c). The maximum time-mean velocity magnitude of CH and PH cases are commensurable – they are approximately 0.007 m/s, see

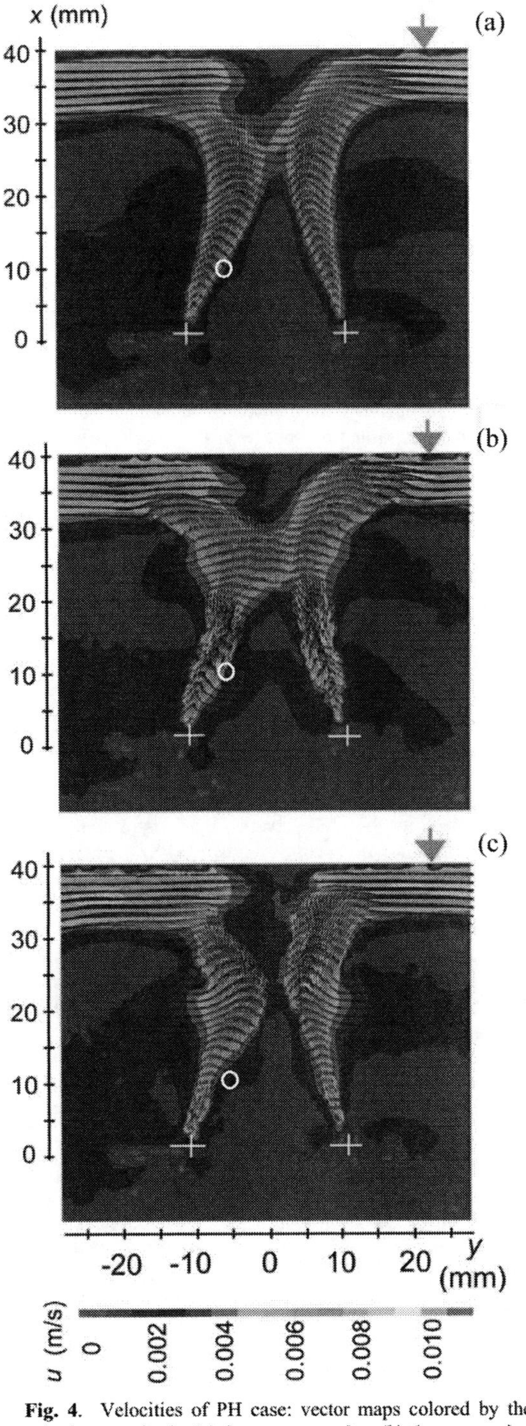


**Fig. 4**. Velocities of PH case: vector maps colored by the velocity magnitude; (a) time-mean results, (b) time-averaging for heating part of the cycle, at $t/T = 0.72$–$0.87$, (b) time-averaging for non-heating part of the cycle at $t/T = 0.32$–$0.47$; (parameters – see Table 1).

Figs. 3 and 4(a). On the other hand, the both heating and non-heating moments always generate higher velocities, namely 0.010 and 0.008 m/s, see Figs. 4(b and c), respectively. In other words, the PH generates by 36 % higher velocity magnitude peaks in comparison with CH case – see a comparison of Figs. 4(a) and Fig. 3, respectively. An obvious reason is a concentration of the heating input power into one half of the cycle.

Unsteady components of velocities result from the natural frequency of plume flows and from the thermal periodical actuation. The former source can be an exclusive reason of oscillations (if any exists) for CH case and the latter one can be a prevailing source of oscillations for PH case. To demonstrate this opinion, a temporal evolution of the velocity magnitude was extracted from the PIV data in a given representative point, located in the shear layer of plumes, where an intense receptivity to disturbances can be expected. Namely, the point was chosen at the coordinates of $x$ = 9.4 mm and $y$ = – 5.3 mm. This point is marked by a small white ring in Figs. 3 and 4(a–c). The temporal evolutions of the velocity magnitude for the both cases CH, PH are shown in Fig. 5. The velocity magnitude at CH oscillates between 0.0015 and 0.006 m/s. On the other hand, velocities at PH clearly demonstrates the predetermined actuation cycle oscillating between the non-heating and heating parts of the cycle, i.e. from approximately 0.003 m/s at $t/T$ = 0–0.5 to 0.006 m/s at $t/T$ = 0.5–1.0, respectively. The time-mean velocity magnitudes in the given point are plotted in Fig. 5 too: they are 0.0042 m/s and 0.0045 m/s for CH and PH, respectively – i.e. the time-mean velocity generated by PH is by 8 % higher in the given point.

The time behaviors from Fig. 5 were analyzed to obtain the power spectral density (PSD) and Fig. 6 shows the results. For CH case, the spectrum low frequency part indicates that the natural frequency is approximately 0.5 Hz. Based on this finding, the thermal actuation was chosen as a subharmonic action with the driven frequency of 0.25 Hz. Evidently, the peak in the low frequency part of the spectrum for PH well confirms this value.

Figure 7 shows the cross-stream velocity profiles at $x$ = 10 mm, i.e. 8.3 $D$. A comparison of the time-mean velocity magnitude of CH and PH show that the CH case

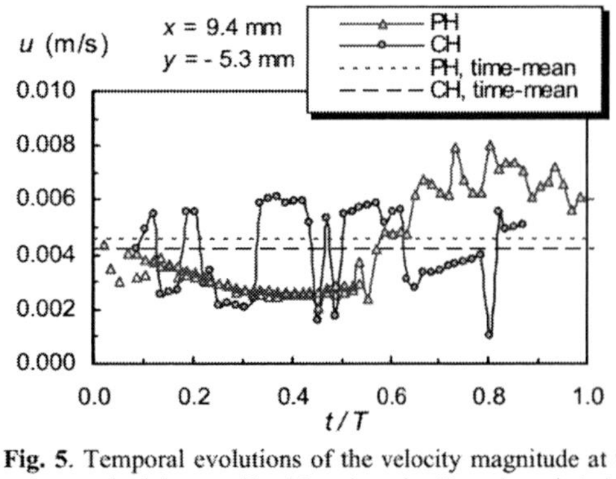


**Fig. 5**. Temporal evolutions of the velocity magnitude at the point marked by small white rings in Figs. 3 and 4 ($x$ = 9.4 mm, $y$ = – 5.3 mm).

exhibits by 15 % higher velocity peaks. To better illustrate PH velocity cycle, the velocity magnitude profiles for heating and non-heating parts of the pulsating-supply cycle were evaluated from PIV data presented in Figs. 4(b and c). Figure 7(a) shows that

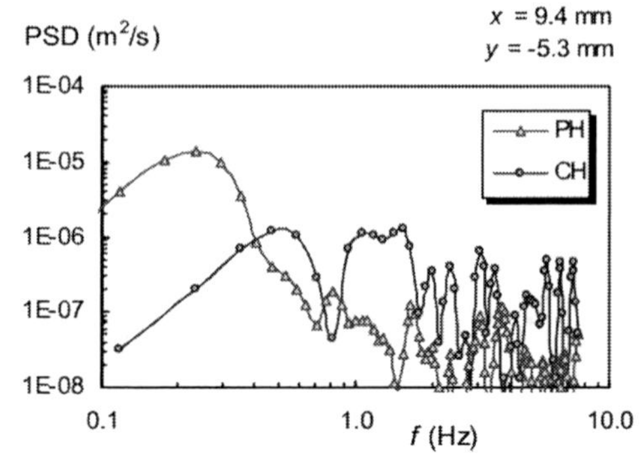


**Fig. 6**. PSD evaluated from the velocity temporal evolutions presented in Fig. 5.

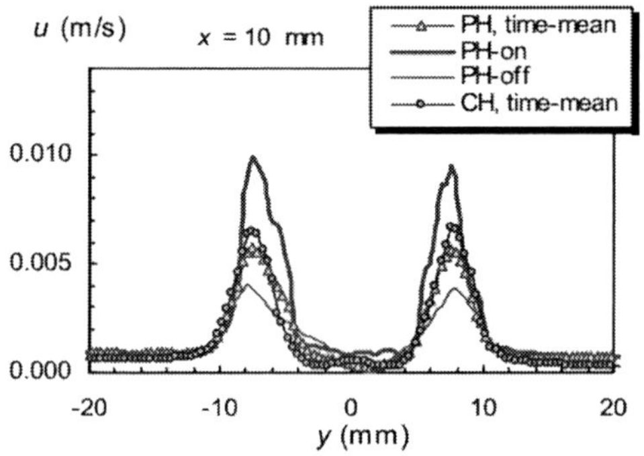


**Fig. 7**. Cross-stream velocity profiles at $x$ = 10 mm.

velocity magnitude of PH significantly oscillates along the time-mean value (0.0057 m/s) - namely, the peak velocities oscillate from -30% to +70 % .

Figure 8 shows another presentation of PIV data from Figs. 3 and 4: the streamwise velocity profiles are plotted along the lines, where the peaks of the time-mean velocity are achieved. Namely the velocity magnitude

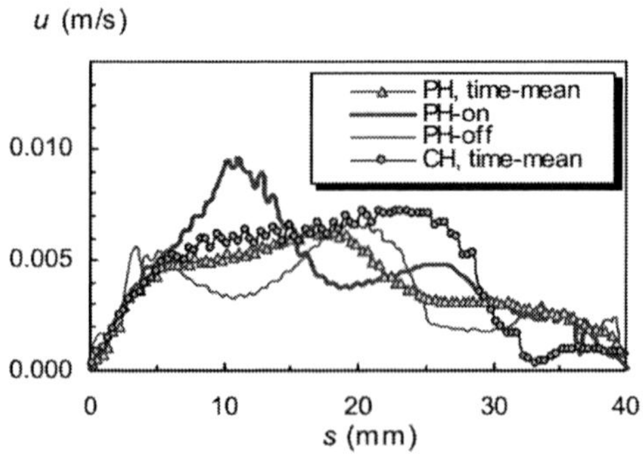


**Fig. 8**. Streamwise velocity profiles plotted along the line of the time-mean velocity magnitude peaks.

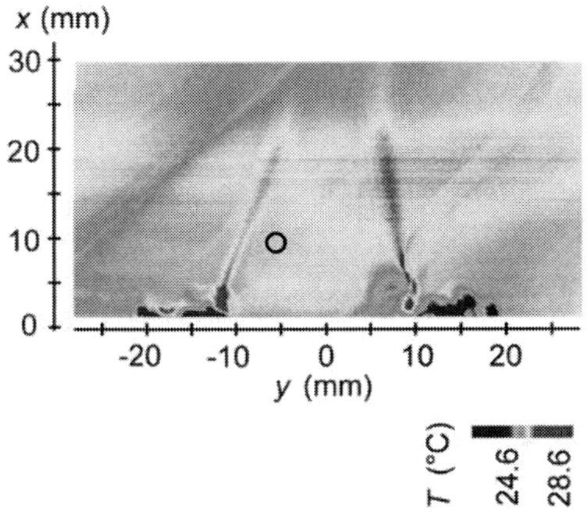


**Fig. 9**. Time-mean temperature distribution of CH case; parameters – see Table 1.

maxima of the left plumes from Figs. 3 and 4 were extracted along the line, which originated at the cylinder vicinity and which is inclined by approximately 20° from the vertical direction to the right. The length coordinate along this line is named $s$. Figure 8 shows that the velocity of the CH case gradually increase to the maximum of approximately 0.0072 m/s at $s$ = 20–25 mm. Then a rapid velocity decrease occurs. This decrease is rather a manifestation of the plume turning to the water surface, than a plume flow decay during its streamwise development (it is obvious from Figs. 3 and 4).

Similarly, the PH plume velocity gradually increases to the maximum at $s$ = 17–20 mm, then a rapid velocity decrease occurs as a manifestation of the plume turning. The largest oscillations in Fig. 8 are indicated at $s$ = 11 mm, where the peak velocities oscillate from -37% to +78 %. It reasonably agrees with Fig. 7: very strong velocity oscillations of PH case occur in the near field at approximately $x = 8.3\ D$.

### 3.2 Temperature field: PLIF experiment

Figure 9 shows a typical PLIF result of the twin plumes generated by continuous heating (CH) as the time-mean temperature distribution. Figure 9 was made by averaging over 50 instantaneous temperature field records.

Unlike the velocity flow filed shown in Fig. 3, the temperature field is equalized much faster. Despite the original temperature difference is rather big ($T_w - T_\infty$= 19.8 °C, see Table 1), the maximum differences between the plume and its surroundings ale less than 3–4 °C. Moreover, the temperature differences diminish during a distance of $x$ = 0–25 mm, hence a thermal trace of the plume is not identifiable for larger distances $x$ > 25 mm. Therefore Figs. 9 and 10 show a detail of the test section up to $x$ = 0–30 mm only.

A comparison of the both time-mean cases CH and PH (Figs. 9 and 10(a), respectively) indicate a basically different character of the both thermal fields. While the CH case manifests a monotonically decreasing temperatures in both streamwise and cross-stream directions, the PH case shows warmed fluid volumes in ascending motion - see the mushroom-like shaped volumes in Fig. 10(a). To indicate this fact more clearly, Fig. 10(b) shows an instantaneous temperature field at the heating moment of the cycle (approximately at $t/T \sim$ 0.8). The result is consistent with the above discussed findings of the PIV experiments - cf. Figs. 3 and 4(b).

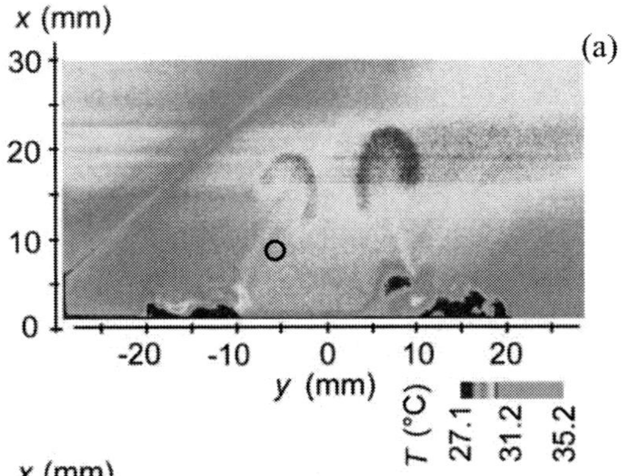


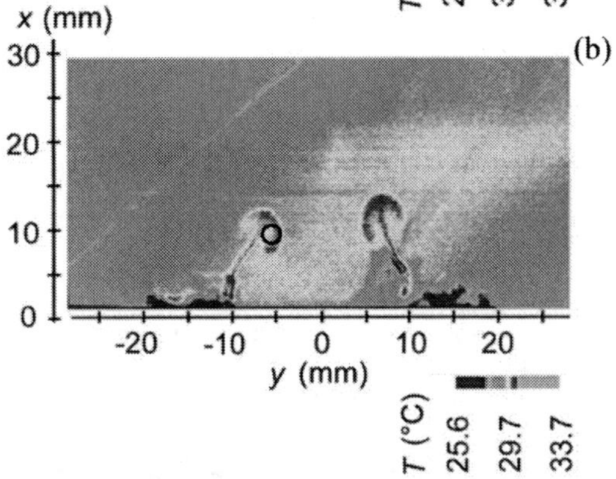


**Fig. 10**. Temperature distribution of PH case; (a) time-mean results, (b) instantaneous temperature field at the heating part of the cycle.

The PH case concentrates the heating input power into one half of the cycle. It implies an enhancement of heat and mass transfer processes. Therefore higher velocities and higher local temperature gradients can be achieved at PH, in comparison with CH.

Note that the oblique lines in Figs. 9 and 10 are just a consequence of an optical disturbances (mainly reflections from the water surface at $x \sim 40$ mm), which penetrate across the whole test section as unwanted defects. Certainly, they are not linked with the investigated task.

In Figs. 9 and 10, a small black ring indicates the position of $x$ = 9.4 mm, $y$ = – 5.3 mm in which the temperature data will be analyzed in the following text (the same point was used for the PIV data analysis, as was discussed in the text above – see Figs. 3 and 4). A time behavior of the time-mean temperature in this point is shown in Fig. 11. The temperature of the PH case is a bit higher than that of the CH (30.9 °C and 27.3 °C, respectively). A reason is a bit higher bulk temperature in the water tank during the PH experiment. Obviously, the both temperatures oscillate along their time-mean value. Surprisingly, a low frequency thermal actuation cannot be found in Fig. 11 – contrary to the well identifiable period of PH case in the velocity time record in Fig. 5.

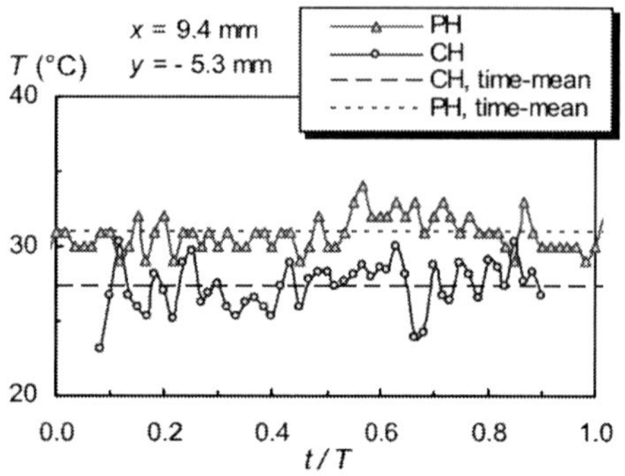


**Figure 11.** Temporal evolutions of temperatures at the point marked in Figs. 9 and 10 ($x$ = 9.4 mm, $y$ = - 5.3 mm).

Figures 12 (a and b) show the spatial distribution of the time-mean temperature of the both cases CH and PH, similarly as above discussed distributions of velocities in Figs. 7 and 8. Similarly as in Fig. 11, a bit higher time-mean temperature of PH in Fig. 12(a) is a consequence of a bit higher bulk temperature in the water tank during the PH experiment. More important are the local maxima on the temperature profiles in Fig. 12(a), because they are linked with the investigated task: For the CH case, two local maxima occur at $y$ = –8.1 mm and +8.2mm. These maxima are clearly distinguishable – a local preheating is about 1.5 °C for both of them. They indicate crossings of the plumes through the line of x = 10 mm – consistently with Fig. 9.

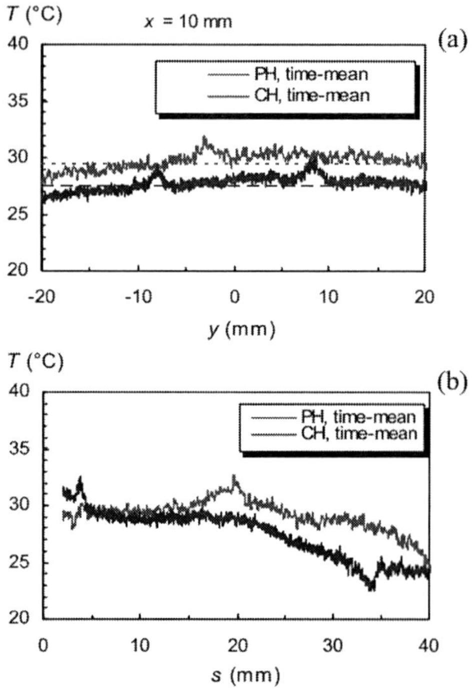


**Fig. 12.** Temperature profiles, (a) cross-stream profiles at $x$ = 10 mm. (b) streamwise profiles plotted along the line of the temperature peaks.

Note that no similar pair of the local maxima exists for the PH case in Fig. 12(a), because of different character of warmed fluid propagation. Namely, an unsteady ascending motion of mushroom-like shaped warmed volumes was shown in Figs. 10(a and b).

Figure 12 (b) shows the streamwise time-mean temperature profiles plotted along the lines, where the peaks of the time-mean temperature are achieved (similarly, the streamwise velocity profiles were shown in Fig. 8). For a small distance $x$ < 15 mm, the time-mean temperature of the CH a PH cases are nearly the same. For $x$ > 15 mm, the CH temperatures go to decline. A similar decline of PH temperatures occurs 10 mm later. This behavior is another manifestation of an advection unsteady mechanism of ascending motion of warmed fluid volumes. The local maximum on the PH temperature curve at $x$ ~ 20 mm well indicates this warmed fluid volume - consistently with the mushroom-like shaped volume in the left plume in Fig. 10(a).

## 4 Conclusions

A pair of line thermal sources (electrically heated cylinders with a diameter of $D$ = 1.21 mm) were used to generate a twine of the plane plumes in water. The distance between the cylinders was 17.9 $D$. Either continuous or pulsating heating were used with the same heating input power 17.4 W. Velocity and temperature fields were investigated by means of PIV and PLIF experiments, respectively.

For the investigated geometry with a moderately small cylinder-to-cylinder distance, both plumes influence each other. Namely, deflections of plumes from a vertical direction occur and the plumes are inclined together. This behavior is caused by effects of fluid entrainments from surroundings versus a confined entrainment from a space between the both plumes.

For a continuous heating, low frequency oscillations were identified and the natural frequency of the plume was evaluated as 0.5 Hz. To promote the unsteady flow character, pulsating heating was used at the subharmonic frequency of 0.25 Hz.

While the maximum time-mean velocity magnitude at the continuous and pulsating heating were commensurable (approximately 0.007 m/s), the pulsating heating causes significant oscillations of the flow field. Namely, the velocity peaks at pulsating heating were by 36 % higher than the velocity peaks at continuous heating. A very strong velocity oscillations were generated by the pulsating heating in the near field at the distance approximately 8.3 $D$ above the cylinders, where the velocity maxima oscillate along the time-mean value 0.0057 m/s from -30% to +70 %.

Behavior of temperature fields consistently agrees with this findings, despite a relatively fast equalization of the temperature field was concluded. The results demonstrate possibilities of a pulsating control on the flow and thermal fields. The pulsations implied an enhancement of heat and mass transfer processes, therefore higher velocities and higher local temperature gradients were achieved

## Acknowledgments

The authors gratefully acknowledge the support from the Grant Agency CR–the Czech Science Foundation (Projects No. 14-08888S and 16-16596S) and the institutional support RVO:61388998.

## References

1. M. H. Schlichting, K. Gersten, Boundary-Layer Theory (8th ed., Springer-Verlag, Berlin, 2000, 1st ed in German: H. Schlichting, Grenzschicht-theorie, Braun, 1951).
2. R. D. Blevins, Applied Fluid Dynamics Handbook (Florida: Krieger Publishing Company 2003)
3. Z. Broučková, S-S. Hsu, A.-B.Wang, and Z.Trávníček, EPJ Web of Conferences 114 02007 (2016)
4. H.J. Seuntiëns, R.N. Kieft, C.C.M. Rindt, and A.A. van Steenhoven, Exp. Fluids 31 pp. 588–595 (2001)
5. Z. Broučková and Z. Trávníček, Proceedings of the 12th International Conference on Heat Transfer, Fluid Mechanics and Thermodynamics HEFAT2016 (Malaga, Spain, July 11-13, 2016, pp. 1791–1796)
6. A. P. Vouros, T. Panidis, A. Pollard, R. Schwab, Int. J. Heat Fluid Flow 51 pp. 383–394 (2015)

# The calibration of fluid-object interaction in immersed boundary method

*Martin* Bušík, *Ivan* Cimrák

Cell-in-fluid research group, University of Žilina, Slovakia, martin.busik@fri.uniza.sk

**Abstract.** In modelling of elastic objects immersed in a fluid, very important part of the model is the interaction between objects and fluid. In models based on immersed boundary method and lattice-Boltzmann method, this interaction is mediated by friction coefficient between fluid and particles. The friction coefficient is phenomenological and needs to be properly calibrated. In this work we design an experiment, the results of which can be predicted by theoretical computations. Based on the theory, we calibrate friction coefficient to properly describe the interaction between fluid and objects. The simulation experiments are performed in Object-in-fluid framework. We calibrate it for objects immersed in flow with low Reynolds number to model flow of cells inside microfluidic devices. We analyze dependence of interaction method on number of nodes and size of spherical object.

## 1 Introduction

We consider model of elastic objects immersed in a fluid based on the lattice-Boltzmann (LB) method coupled with the immersed boundary (IB) method. Both methods have its own mesh, one for the fluid and one for the object.

In the LB method, fluid is modeled by fictive particles that travel and collide over a fixed cubic lattice mesh [?]. The fluid dynamics are computed using the LB equation. The unknown variable in this equation is the density function for the fictive particles. Macroscopic properties of the fluid can be afterwards recovered from the density function.

In the IB method, the objects are characterized by triangulation of their surface. The discretization mesh contains boundary points linked together by elastic forces along the edges and faces of the mesh. These forces move the IB points in the space according to Newton equation of motion. This equations include the mass of individual IB points. This way, the object's boundary may change the shape according to the elastic forces. The IB mesh is not fixed, the positions of the IB points are determined freely in the space.

To maintain mechanical properties of the cell's membrane, five elastic moduli are considered, each defining different forces. These moduli govern the stretching and bending of the membrane, as well as local and global preservation of the surface and volume of the cell.

The interaction between the fluid and the object is provided by coupling of LB and IB method. From the fluid velocity field given over the LB fixed mesh, the corresponding fluid forces are applied to IB points, adding up with elastic forces. Reversely, the change of object's shape generates forces that are applied back to the fluid by entering the external forces in the governing LB equation for the LB method. Since the fixed LB mesh and free IB mesh do not overlap, we need to interpolate the values of respective forces and velocities between the IB points and the lattice points as depicted in Figure ??.

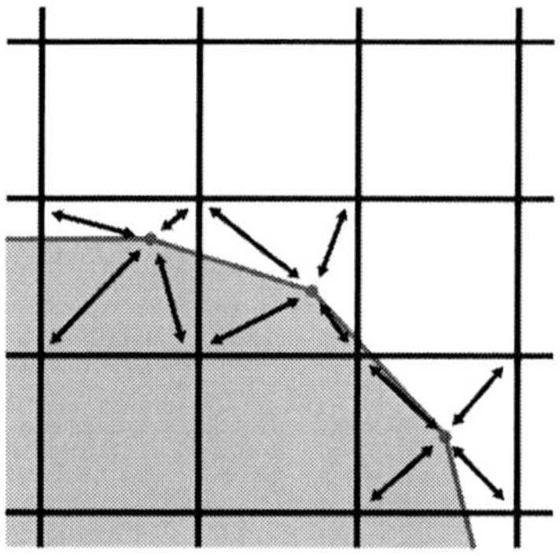

**Fig. 1.** Two dimensional version of the LB-IB method coupling. The respective forces and velocities are transferred from and to the vertices of LB lattice squares where the IB point is currently located.

Basic principle for objects immersed in a fluid is the so called no-slip boundary condition. The aim of this method is to keep the zero difference between the fluid velocity and the velocity of IB points on the surface of the objects. Several approaches have been established to achieve this e.g. bounce back scheme, implicit treatment of restoring force or the direct forcing method [?].

In our case, we use transfer force between the objects and the fluid similarly as in [?] by following equation

$$F_i = \xi(v_i - u_i), \tag{1}$$

where $v_i$ is the velocity of the IB point $i$, $u_i$ is the interpolated velocity of the fluid at the particle position $i$ and $\xi$ is phenomenological friction coefficient. Then, total force exerted on each IB mesh point consists of elastic forces and force of fluid-object interaction [?].

Current implementation of the Object-in-fluid framework distributed as a part of ESPResSo [?] requires to set the friction coefficient. This phenomenological constant was previously calibrated by the following experiment: A rigid ball with the same density as surrounding fluid was put into a static fluid. The ball was pushed with given nonzero initial velocity. The resulting drag force of the fluid on the sphere may be computed explicitely. This drag force slows the sphere down and the velocity evolution of the sphere may be evaluated by theoretical computations. The computational simulations of this experiment were performed using the Object-in-fluid framework and the corresponding friction coefficient was determined to fit the theoretical data [?].

The computational experiment however was performed according to the authors not correctly. The initial speed of the sphere was set to the IB points only and the inner fluid of the sphere was initiated with zero speed. This way, the whole movement had to be mediated by the IB points only, while we argue that the inner fluid should have been initiated with the same initial speed as the IB points.

The remedy for this situation is however not trivial. In the LB method, setting the nonzero initial speed to inner LB points of the sphere while preserving zero speed to the neighbouring outer LB lattice points, is unphysical and leads to shocks. Therefore we present new experiment with physically correct assumption in Section **??**.

### 1.1 Theoretical considerations

In general, the motion of any object under the application of given force, is described by the Newton's second law of motion

$$F = mv', \tag{2}$$

where $m$ is the mass of the object and $v'$ is the acceleration of the object. We consider small solid rigid sphere relatively slow speed. The drag force of the fluid exerted on moving objects with very low Reynolds number ($R_e < 1$) is described by Stoke's law as

$$F_d = 6\pi\nu r K v, \tag{3}$$

where $r$ is the radius of the object, $v$ is the relative velocity of the object to the fluid, $K$ is the shape dependent constant, for spherical object $K = 1$, and $\nu$ is the dynamic viscosity of the fluid. This equation assumes the object to be immersed in an unbounded fluid. Otherwise, complex boundary effects will come into play.

## 2 Simulation setup

Based on the previous theoretical considerations, we designed the following simulation experiment.

### 2.1 Terminal velocity experiment

We put a spherical object into a static flow. We exert a horizontal constant force $F_0$ on the sphere that causes the sphere to move horizontally. The static fluid will reversely exert a drag force on the sphere. As mentioned above, the drag force is directly proportional to the velocity of the sphere. Therefore the drag force is increasing and we can compute the velocity profile.

The Newton equation of motion for such sphere is

$$mv' = F_0 - 6\pi\nu r v. \tag{4}$$

This differential equation has the following solution

$$v(t) = \frac{F_0}{6\pi\nu r} - \frac{F_0}{6\pi\nu r} exp\left(\frac{-6\pi\nu r t}{m}\right). \tag{5}$$

We can see that the velocity depends on the mass of the sphere. In the coupled LB-IB method, there is fluid contributing to the object's mass as well as the mass of IB points. There is however no direct way how to interpret the physical object's mass in terms of model components (LB fluid mass and IB points mass).

We can however get rid of the object's mass in the velocity expression by considering the terminal velocity as limit $t \to \infty$

$$v_\infty = \lim_{t\to\infty} v(t) = \frac{F_0}{6\pi\nu r}, \tag{6}$$

where $v_\infty$ is terminal velocity. This way we do not have the problem with overall mass of the object, since the terminal velocity do not depend on the object's mass.

When simulating this experiment, we obtain terminal velocity of the simulated sphere that should be the same as in (**??**). This will however hold true only if the interaction between fluid and object is calibrated properly. In other ways, the simulated terminal velocity will equals to theoretical $v_\infty$ only if the friction coefficient is correct.

### 2.2 Restriction on practical implementation

The theoretical expressions for forces and velocities derived from (**??**) are only valid under the assumptions that no boundary effects are involved. This means that in theory the unbounded domain must be used. This is of course problematic in simulations and we must be careful when choosing the size of computational domain.

### 2.3 Implementation details

We simulate experiments by software ESPResSo with framework Object-in-fluid. It allows to simulate elastic object described by triangulation its surface filled with the same fluid as outside. The simulation box has a cubical shape with uniform discretization. The discretization grid is chosen such that we do not have significant boundary effects.

The boundary conditions for the fluid on the all sides of the simulation box are set to be zero.

Every time a particular force must be exerted on a spherical object, this force is divided by the number of IB mesh points and only afterwards it is applied to each IB point.

We use different meshes with different number of nodes, lengths of edges and sizes of the object. All meshes however are quasi-uniform in a sense that they are composed of almost equilateral triangles.

Since the model we are working with is primarily used for elastic objects, we set quite high elastic coefficients to ensure that simulated object behaves like a rigid one. To control this effect we record the differences in the volume, surface area and maximal diameter of the object at the beginning and at the end of the simulation. We keep all these differences under the threshold 5%.

## 3 Computational analysis and results

### 3.1 Size of the simulation box

Theoretical considerations in Section ?? assume the immersed sphere in a unbounded fluid to eliminate boundary effects. In practice, we need to set up the simulation box large enough so that boundary effects are negligible. We follow the numerical simulation reported in [?] where the ratio between the size of the sphere and the simulation box was approximately 1:30. With the size of our sphere having diameter $8\mu m$ we set the size of the box to $200\mu m$. Moreover we perform an analysis of the influence of the simulation box on the resulting drag force.

Setting the friction coefficient to 1.0 we choose four different sizes of cubic box: $50\mu m$, $100\mu m$, $200\mu m$ and $400\mu m$. We perform the simulation for the terminal velocity. The results are presented in Table ??.

**Table 1.** Terminal velocities for different box sizes for fixed friction coefficient $\xi = 1.0$

| $s$ (size of box) | $50\mu m$ | $100\mu m$ | $200\mu m$ | $400\mu m$ |
|---|---|---|---|---|
| $v_s$ (term. vel.) | 0.00304 | 0.00356 | 0.00381 | 0.00395 |
| $\delta_{s\to 2s}$ (rel. diff.) | +14.61% | +6.56% | +3.54% | |

The differences reported in Table ?? are defined as relative difference when doubling the size of the box by expression

$$\delta_{s\to 2s} = \frac{v_{2s} - v_s}{v_{2s}}$$

We can see that the differences between the terminal velocities when changing the size of the box from $200\mu m$ to $400\mu m$ is less then 5%. Therefore we will perform all subsequent simulation using the size of simulation box $200\mu m$.

### 3.2 Results for fixed mesh and fixed size of the sphere

Although we assume that the friction coefficient will change for different triangulations, first we determined the friction coefficient for mesh with 393 points and with radius $4\mu m$. We set density of fluid to $1025 kg.m^{-3}$, dynamic viscosity to $1.5375 \times 10^{-3} Pa.s$ and force exerted on the sphere to $0.393 \times 10^{-9} N$. We performed a series of simulations of terminal velocity experiment described in Section ??. We varied the value of friction coefficient and for each value we obtained different terminal velocity. In Figure ?? we can see the results with value of friction coefficient computed from theoretical consideration indicated by red horizontal line. The intersection of the line and the curve determines the calibrated friction coefficient with value 1.76. In the

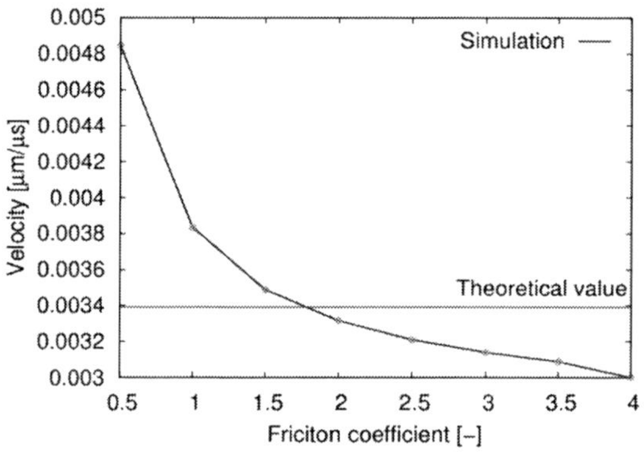


**Fig. 2.** The dependence of friction coefficient on the terminal velocity experiment.

figure we can see that the terminal velocity is more sensitive for lower values of friction coefficients and increasing the friction coefficient above certain level does not influence the terminal velocity much.

### 3.3 The dependence of friction coefficient on the number of nodes and radius of the sphere

In Section ?? we determined friction coefficient for one specific mesh and for one specific radius of the sphere. If we use different mesh or different size of the sphere, the friction coefficient changes. We would like to derive an expression for dependence of the friction coefficient on the number of mesh points and radius of the sphere. Knowing this dependence, we do not need to calibrate the friction coefficient each time the mesh or size changes.

### Heuristics

First we analyze situation for two different meshes with number of mesh points $n_1$ and $n_2$ with the radius of the sphere fixed. We want to find out what is the relation between corresponding friction coefficients $\xi_1$ and $\xi_2$ in order to achieve the same simulated behaviour.

As soon as the sphere reaches terminal velocity, all forces exerted on the mesh points come to equilibrium. There are three types of forces exerted on the mesh points: elastic forces, dragging force $F_0$, and fluid forces that act in the opposite direction to $F_0$. The elastic forces just keep the sphere almost solid and do not contribute to the balance between the dragging force and the fluid force. So basically, in each mesh point we have the equilibrium of the dragging force and the fluid force. We can use (??) to calculate the fluid force and thus we have for both meshes

$$F_0 = \sum_{i=1}^{n_1} \xi_1(u_i - v_i) = \sum_{i=1}^{n_2} \xi_2(u_i - v_i). \quad (7)$$

Our aim is to get the same behaviour of simulated sphere in the computational experiment and thus the velocities $u_i$ of mesh points for both meshes will be the same. The same holds for the fluid interpolated velocities $v_i$ and thus we can conclude that

$$\sum_{i=1}^{n_1} \xi_1 = \sum_{i=1}^{n_1} \xi_2, \qquad \text{or} \qquad n_1.\xi_1 = n_2.\xi_2. \quad (8)$$

Second, we analyze the situation when we keep the number of mesh points fixed equal to $n$ and we have two different radii $r_1$ and $r_2$. Assume that we want to use two different dragging forces $F_1$ and $F_2$ to get the same terminal velocity for both spheres. In the equilibrium, the dragging forces $F_1, F_2$ are compensated by the drag force computed from (??) and thus we get

$$F_1 = 6\pi\nu r_1 v_\infty K = \sum_{i=1}^{n} \xi_1(u_i - v_i)$$

$$F_2 = 6\pi\nu r_2 v_\infty K = \sum_{i=1}^{n} \xi_2(u_i - v_i)$$

With similar argumentation as before we note that both spheres are moving with the same terminal velocity and thus $u_i$ and $v_i$ are the same. Using this result after dividing the previous two equations we end up with

$$\frac{r_1}{r_2} = \frac{n.\xi_1}{n.\xi_2} = \frac{\xi_1}{\xi_2} \quad (9)$$

### Hypotheses

From relations (??) and (??) that we have obtained using heuristic considerations, we can formulate the following two hypotheses:

(a) the friction coefficient increases linearly with the increasing radius of the sphere

(b) the friction coefficient decreases linearly with the increasing the number of nodes

Both hypotheses may be merged into one expression. Starting from the already calibrated value of friction coefficient $\xi_{393,4}$ for mesh with 393 nodes and radius $4\mu m$, we can write down the following expression for mesh with $n$ nodes and radius of sphere $r$

$$\xi_{n,r} = \frac{393}{n}\frac{r}{4}\xi_{393,4}. \quad (10)$$

This expression reflects both hypotheses and satisfies both (??) and (??).

### Verification

To confirm these hypotheses, we performed number of simulations. We picked 4 different meshes with number of nodes 126, 393, 500, 1182. With each mesh we considered three different radii 2, 4 and 8 $\mu m$. For each of these 12 cases we computed the friction coefficient using (??). Then we performed our computational experiment with fixed $F_0$ and we recorded the obtained terminal velocities.

In first two columns of Table **??** we can see the parameters of the simulations. In column 3 we see values of friction coefficient computed from (??). In next column we see the obtained terminal velocity that can be compared to the theoretical value of terminal velocity computed from (??) presented in column 5. In the last column we can see relative error in $v_\infty$.

The relative error in all simulations is low (under 5%) which confirms the validity of (??).

**Table 2.** The comparative result for different radiui and number of mesh points. The calibrated case is emphasized with boldface.

| *mesh* | $r$ | $\xi$ | $v_{sim}$ | $v_{exact}$ | $\delta_{sim\to exact}$ |
|---|---|---|---|---|---|
| 126 | 2 | 2.74 | 0.006532 | 0.006780 | 3.66 % |
| 126 | 4 | 5.49 | 0.003428 | 0.003390 | 1.10 % |
| 126 | 8 | 10.89 | 0.001668 | 0.001695 | 1.60 % |
| 393 | 2 | 0.88 | 0.006532 | 0.006780 | 0.72 % |
| **393** | **4** | **1.76** | **0.003404** | **0.003390** | **0.41 %** |
| 393 | 8 | 3.52 | 0.001646 | 0.001695 | 2.89 % |
| 500 | 2 | 0.69 | 0.006515 | 0.006780 | 3.92 % |
| 500 | 4 | 1.38 | 0.003399 | 0.003390 | 0.26 % |
| 500 | 8 | 2.77 | 0.001645 | 0.001695 | 2.98 % |
| 1182 | 2 | 0.29 | 0.006513 | 0.006780 | 4.18 % |
| 1182 | 4 | 0.59 | 0.003400 | 0.003390 | 0.30 % |
| 1182 | 8 | 1.17 | 0.001644 | 0.001695 | 3.01 % |

## 4 Summary and discussion

In this work we designed simulation experiment for calibration of interaction between object and fluid in the lattice-Boltzmann – immersed boundary method. We calibrated friction coefficient and we derived an explicit formula describing its dependence on size and number of

DOI: 10.1051/epjconf/201714302013

mesh points. This formula was confirmed by series of computational simulations.

The results presented in Table ?? indicate two phenomena. First, when keeping the radius fixed to value $4\mu m$ for which the friction was calibrated, the relative error between simulated and exact terminal velocity is very low, under 1%. This suggests, that the change of the mesh density with size of the object fixed is a safe operation.

Second, when keeping the number of mesh points fixed and changing the radius, relative errors are bit higher, but still acceptable. This indicate that the behaviour of the sphere is more sensitive to the changes in mesh density.

## References

1. P. Ahlrichs, B. Dunweg, Lattice-Boltzmann simulation of polymer-solvent systems, Internat. J. Modern Phys. C vol. 8 (1998) 1429 - 1438
2. I. Cimrák, M. Gusenbauer, T. Schrefl, Modelling and simulation of processes in microfluidic devices for biomedical applications, Computers and Mathematics with Applications vol. 64 (2012) 278 - 288
3. I. Jančigová, Modeling elastic objects in fluid flow with biomedical applications, Dissertation Thesis, Faculty of Management Science and Informatics, University of Žilina, 2015
4. B. Dunweg and A. J. C. Ladd, Lattice-Boltzmann simulations of soft matter systems, Advances in Polymer Science, vol. 221 (2009) 89–166
5. D. A. Jones and D. B. Clarke, Simulation of Flow Past a Sphere using the Fluent Code, Maritime Platforms Division, Defence Science and Technology Organisation, DSTO-TR-2232, AR-014-365, 2008
6. A. Arnold, O. Lenz, S. Kesselheim, R. Weeber, F. Fahrenberger, D. Roehm, P. Koovan, and C. Holm, ESPResSo 3.1 - molecular dynamics software for coarsegrained models, in Meshfree Methods for Partial Differential Equations VI, Lecture Notes in Computational Science and Engineering, M. Griebel and M. Schweitzer, Eds., vol. 89, 2013, pp. 123.
7. I. Jančigová, On mass distribution in ESPResSo simulations of elastic objects, MiST 2014: proceedings of Mathematics in Science and Technologies 2014 conference, 2014

# Finite volume - space-time discontinuous Galerkin method for the numerical simulation of compressible turbulent flow in time dependent domains

Jan Česenek[1,a]

[1]Aerospace Research and Test Establishment, Beranových 130, 199 05 Praha - Letňany, Czech Republic

**Abstract.** The article is concerned with the numerical simulation of the compressible turbulent flow in time dependent domains. The mathematical model of flow is represented by the system of non-stationary Reynolds-Averaged Navier-Stokes (RANS) equations. The motion of the domain occupied by the fluid is taken into account with the aid of the ALE (Arbitrary Lagrangian-Eulerian) formulation of the RANS equations. This RANS system is equipped with two-equation $k - \omega$ turbulence model. These two systems of equations are solved separately. Discretization of the RANS system is carried out by the space-time discontinuous Galerkin method which is based on piecewise polynomial discontinuous approximation of the sought solution in space and in time. Discretization of the two-equation $k - \omega$ turbulence model is carried out by the implicit finite volume method, which is based on piecewise constant approximation of the sought solution. We present some numerical experiments to demonstrate the applicability of the method using own-developed code.

## 1 Introduction

During the last decade the space-time discontinuous Galerkin finite element method (ST-DG), which is based on piecewise polynomial discontinuous approximations of the sought solution, became very popular in the field of numerical simulation of the fluid flow. This method of higher order was successfully used for simulation of the Navier-Stokes equations ([2],[4],[5]). Unfortunately this method became unstable, when it was used for the whole system of the RANS equations with $k - \omega$ equations. Thus we solve this two systems separately. We use the ST-DG for the discretization of the RANS system of equations and the finite volume method for the equations of the $k - \omega$ turbulence model. Due to the time dependent domain the RANS equations are transformed in the ALE (Arbitrary Lagrangian-Eulerian) formulation. In this article Wilcox $k - \omega$ turbulence model was choosen ([10]). For numerical experiments will be used flow around moving airfoil. A solid airfoil with two degrees of freedom performs rotation around elastic axis and oscillations in the vertical direction. The discrete flow problem is coupled with the system of ordinary differential equations describing airfoil vibrations. Results of numerical simulation are presented.

## 2 Formulation of the $k - \omega$ turbulence model

We consider compressible turbulent flow in a bounded domain $\Omega(t) \subset R^2$, depending on time $t \in [0, T]$. We assume that the boundary $\Omega(t)$ consists of three disjoint parts $\partial\Omega(t) = \Gamma_I \cup \Gamma_O \cup \Gamma_W(t)$, where $\Gamma_I$ is the inlet, $\Gamma_O$ is the outlet and $\Gamma_W(t)$ is impermeable wall, whose parts may move.

The time dependence of the domain is taken into account with the aid of a regular one–to–one ALE mapping

$$\mathcal{A}_t : \Omega(0) \rightarrow \Omega(t).$$

We define the ALE velocity $z$ by the relations

$$\tilde{z}(X, t) = \frac{\partial}{\partial t}\mathcal{A}_t(X), \quad t \in [0, T], X \in \Omega(0),$$
$$z(x, t) = \tilde{z}(\mathcal{A}_t^{-1}(x), t), \quad t \in [0, T], x \in \Omega(t)$$

and the ALE derivative of a vector function $\mathbf{w} = \mathbf{w}(x, t)$ defined for $x \in \Omega(t)$ and $t \in [0, T]$:

$$\frac{D^{\mathcal{A}}}{Dt}\mathbf{w}(x, t) = \frac{\partial \tilde{\mathbf{w}}}{\partial t}(X, t),$$

where

$$\tilde{\mathbf{w}}(X, t) = \mathbf{w}(\mathcal{A}_t(X), t), \; X \in \Omega(0), \; x = \mathcal{A}_t(X).$$

Then the system of the RANS equations describing the compressible turbulent flow can be written in the ALE form

$$\frac{D^{\mathcal{A}}\mathbf{w}}{Dt} + \sum_{s=1}^{2} \frac{\partial \mathbf{g}_s(\mathbf{w})}{\partial x_s} + \mathbf{w}\mathrm{div}\, z + \sum_{s=1}^{2} \frac{\partial \mathbf{f}_s^P(\mathbf{w})}{\partial x_s} = \sum_{s=1}^{2} \frac{\partial \mathbf{k}_s(\mathbf{w})}{\partial x_s} + \sum_{s=1}^{2} \frac{\partial \mathbf{R}_s(\mathbf{w}, \nabla\mathbf{w})}{\partial x_s}, \tag{1}$$

where for $s = 1, 2$ we have

$$\mathbf{w} = (w_1, ..., w_4)^T = (\rho, \rho v_1, \rho v_2, E)^T \in R^4,$$
$$\mathbf{f}_s(\mathbf{w}) = (f_{s,1}, ..., f_{s,4})^T$$

[a] e-mail: jan.cessa@seznam.cz, cesenek@vzlu.cz

$$= (\rho v_s, \rho v_1 v_s + \delta_{1s} p, \rho v_2 v_s + \delta_{2s} p, (E+p) v_s)^T,$$

$$\mathbf{f}_s^p(\mathbf{w}) = (f_{s,1}^p, ..., f_{s,4}^p)^T = (0, \frac{2}{3}\rho k \delta_{1s}, \frac{2}{3}\rho k \delta_{2s}, \frac{2}{3}\rho k v_s)^T,$$

$$\mathbf{g}_s(\mathbf{w}) = \mathbf{f}_s(\mathbf{w}) - z_s \mathbf{w},$$

$$\mathbf{R}_s(\mathbf{w}, \nabla\mathbf{w}) = (R_{s,1}, ..., R_{s,4})^T = \left(0, \tau_{s1}^V, \tau_{s2}^V, \sum_{r=1}^{2} \tau_{sr}^V v_r + \left(\frac{c_p \mu_L}{Pr} + \frac{c_p \mu_T}{Pr_T}\right)\frac{\partial \theta}{\partial x_s}\right)^T,$$

$$\mathbf{k}_s(\mathbf{w}) = (k_{s,1}, ..., k_{s,4})^T = \left(0, ..., 0, \left(\mu_L + \sigma_k \frac{\rho k}{\omega}\right)\frac{\partial k}{\partial x_s}\right)^T,$$

where

$$\tau_{sr}^V = -\frac{2}{3}(\mu_L + \mu_T)\, \mathrm{div} \boldsymbol{v}\, \delta_{sr} + 2(\mu_L + \mu_T)\, d_{sr}(\boldsymbol{v}),$$

$$d_{sr}(\boldsymbol{v}) = \frac{1}{2}\left(\frac{\partial v_s}{\partial x_r} + \frac{\partial v_r}{\partial x_s}\right).$$

We use the following notation: $\boldsymbol{v} = (v_1, v_2)$ - velocity, $\rho$ - density, $p$ - pressure, $\theta$ - absolute temperature, $E$ - total energy, $\gamma$ - Poisson adiabatic constant, $\kappa$ - heat conduction coefficient, $c_v$ - specific heat at constant volume, $c_p$ - specific heat at constant pressure, where $c_p = \gamma c_v$, $Pr$ is the laminar Prandtl number, which can be express in the form $Pr = \frac{c_p \mu_L}{\kappa}$, $Pr_T$ is the turbulent Prandtl constant number, $\mu_T$ is the eddy-viscosity coefficient, $\mu_L$ is the dynamic viscosity coefficient dependent on temperature via Sutherland's formula. The above system is completed by the thermodynamical relations

$$p = (\gamma - 1)\left(E - \frac{1}{2}\rho |\boldsymbol{v}|^2 - \rho k\right), \quad \theta = \frac{1}{c_v}\left(\frac{E}{\rho} - \frac{1}{2}|\boldsymbol{v}|^2 - k\right)$$

and is equipped with the initial condition

$$\mathbf{w}(x, 0) = \mathbf{w}^0(x), \quad x \in \Omega$$

and the following boundary conditions

$$\text{a)}\ \rho|_{\Gamma_I} = \rho_D,$$

$$\text{b)}\ \boldsymbol{v}|_{\Gamma_I} = \boldsymbol{v}_D = (v_{D1}, v_{D2})^T, \qquad (2)$$

$$\text{c)}\ \sum_{s,r=1}^{2} \tau_{sr}^V n_s v_r + \left(\frac{c_p \mu_L}{Pr} + \frac{c_p \mu_T}{Pr_T}\right)\frac{\partial \theta}{\partial n} = 0 \quad \text{on } \Gamma_I,$$

$$\text{a)}\ \boldsymbol{v}|_{\Gamma_{W(t)}} = z_D = \text{velocity of a moving wall},$$

$$\text{b)}\ \left.\frac{\partial \theta}{\partial n}\right|_{\Gamma_{W(t)}} = 0,$$

$$\text{a)}\ \sum_{s=1}^{2} \tau_{sr}^V n_s = 0,\ r = 1,2 \quad \text{on } \Gamma_O, \qquad \text{b)}\ \left.\frac{\partial \theta}{\partial n}\right|_{\Gamma_O} = 0,$$

with given data $\mathbf{w}^0, \rho_D, \boldsymbol{v}_D, z_D$. It is possible to show that $\mathbf{f}_s(\alpha \mathbf{w}) = \alpha\, \mathbf{f}_s(\mathbf{w})$ for $\alpha > 0$. This property implies that

$$\mathbf{f}_s(\mathbf{w}) = \mathbf{A}_s(\mathbf{w})\mathbf{w}, \quad s = 1, 2, \qquad (3)$$

where

$$\mathbf{A}_s(\mathbf{w}) = \frac{D\mathbf{f}_s(\mathbf{w})}{D\mathbf{w}}, \quad s = 1, 2,$$

are the Jacobi matrices of the mappings $\mathbf{f}_s$. Similary we can express

$$\mathbf{f}_s^p(\mathbf{w}) = \mathbf{A}_s^p(\mathbf{w})\mathbf{w}, \quad s = 1, 2.$$

The viscous terms $\mathbf{R}_s(\mathbf{w}, \nabla\mathbf{w})$ can be expressed in the form

$$\mathbf{R}_s(\mathbf{w}, \nabla\mathbf{w}) = \sum_{k=1}^{2} \mathbf{K}_{s,k}(\mathbf{w})\frac{\partial \mathbf{w}}{\partial x_k}, \quad s = 1, 2, \qquad (4)$$

where $\mathbf{K}_{s,k}(\mathbf{w}) \in R^{4\times4}$ are matrices depending on $\mathbf{w}$.

The aforesaid system of the RANS equations is completed by the Wilcox's turbulence model (see [10]) for $\mu_T$:

$$\frac{\partial \tilde{\mathbf{w}}}{\partial t} + \sum_{s=1}^{2} \frac{\partial \tilde{\mathbf{f}}_s(\tilde{\mathbf{w}})}{\partial x_s} = \sum_{s=1}^{2} \frac{\partial \tilde{\mathbf{R}}_s(\tilde{\mathbf{w}}, \nabla\tilde{\mathbf{w}})}{\partial x_s} + \tilde{\mathbf{s}}(\tilde{\mathbf{w}}), \qquad (5)$$

where for $s = 1, 2$ we have

$$\tilde{\mathbf{w}} = (\tilde{w}_1, \tilde{w}_2)^T = (\rho\omega, \rho k)^T \in \boldsymbol{R}^2,$$

$$\tilde{\mathbf{f}}_s(\tilde{\mathbf{w}}) = (\tilde{f}_{s,1}, \tilde{f}_{s,2})^T = (\rho\omega v_s, \rho k v_s)^T,$$

$$\tilde{\mathbf{R}}_s(\tilde{\mathbf{w}}, \nabla\tilde{\mathbf{w}}) = (R_{s,1}, R_{s,2})^T = \left((\mu_L + \sigma_\omega \frac{\rho k}{\omega})\frac{\partial \omega}{\partial x_s}, (\mu_L + \sigma_k \frac{\rho k}{\omega})\frac{\partial k}{\partial x_s}\right)^T,$$

$$\tilde{\mathbf{s}}(\tilde{\mathbf{w}}) = (P_\omega - \beta\rho\omega^2 + C_D, P_k - \beta^* \rho\omega k)^T.$$

Here $\omega$ is the turbulence dissipation, $k$ is the turbulence kinetic energy and $\mu_T$ is the eddy viscosity. We can write the production terms as

$$P_k = \sum_{s=1}^{2}\sum_{r=1}^{2} \tau_{sr}^T \frac{\partial v_s}{\partial x_r},$$

$$P_\omega = \alpha_\omega \frac{\omega P_k}{k},$$

where

$$\tau_{sr}^T = -\frac{2}{3}\mu_T\, \mathrm{div}\boldsymbol{v}\, \delta_{sr} - \frac{2}{3}\rho k\, \delta_{sr} + 2\mu_T\, d_{sr}(\boldsymbol{v}),$$

$$d_{sr}(\boldsymbol{v}) = \frac{1}{2}\left(\frac{\partial v_s}{\partial x_r} + \frac{\partial v_r}{\partial x_s}\right),$$

$$\mu_T = \frac{\rho k}{\tilde{\omega}}.$$

Limited eddy viscosity $\tilde{\omega}$ is given by the term

$$\tilde{\omega} = \max\left\{\omega, C_{lim}\sqrt{\frac{1}{2\beta^*}\sum_{r,s=1}^{2} \tilde{\tau}_{rs}\tilde{\tau}_{rs}}\right\},$$

$$\tilde{\tau}_{rs} = -\frac{2}{3}\mathrm{div}\boldsymbol{v}\, \delta_{rs} + 2d_{rs}(\boldsymbol{v}).$$

The cross-diffusion term $C_D$ is defined as

$$C_D = \sigma_D \frac{\rho}{\omega}\max\left\{\sum_{s=1}^{2} \frac{\partial k}{\partial x_s}\frac{\partial \omega}{\partial x_s}, 0\right\}.$$

The coefficients $\beta, \beta^*, \sigma_k, \sigma_\omega, \sigma_D, \alpha_\omega, C_{lim}, Pr_T$ are chosen by [10]:

$$\alpha_\omega = \frac{13}{25}, \quad \beta = 0.0708, \quad \beta^* = 0.09, \quad \sigma_k = 0.6, \quad \sigma_\omega = 0.5,$$

$$C_{lim} = \frac{7}{8}, \quad \sigma_D = \frac{1}{8}, \quad Pr_T = \frac{8}{9}.$$

This system is also equipped with the initial condition

$$\tilde{\mathbf{w}}(x,0) = \tilde{\mathbf{w}}^0(x), \quad x \in \Omega$$

and the following boundary conditions

$$\text{a)}\ \omega|_{\Gamma_I} = \omega_D, \quad \text{b)}\ k|_{\Gamma_I} = k_D, \tag{6}$$

$$\text{a)}\ \omega|_{\Gamma_W} = \omega_{wall}, \quad \text{b)}\ k|_{\Gamma_W} = 0, \tag{7}$$

$$\text{a)}\ \left.\frac{\partial \omega}{\partial n}\right|_{\Gamma_O} = 0, \quad \text{b)}\ \left.\frac{\partial k}{\partial n}\right|_{\Gamma_O} = 0, \tag{8}$$

with given data $\tilde{\mathbf{w}}^0, \omega_D, k_D, \omega_{wall}$.

## 3 Discretization

### 3.1 Space discretization of the flow problem

By $\Omega_h(t)$ we denote polygonal approximation of the domain $\Omega(t)$. Let $\mathcal{T}_h(t)$ be a partition of the domain $\Omega_h(t)$ into finite number of closed elements with mutually disjoint interiors such that $\overline{\Omega_h(t)} = \bigcup_{K\in\mathcal{T}_h(t)} K$. In 2D problems, we usually choose $K \in \mathcal{T}_h(t)$ as triangles or quadrilaterals. By $\mathcal{F}_h(t)$ we denote the system of all faces of all elements $K \in \mathcal{T}_h(t)$. Further, we introduce the set of boundary faces $\mathcal{F}_h^B(t) = \{\Gamma \in \mathcal{F}_h(t); \Gamma \subset \partial\Omega_h(t)\}$ and the set of inner faces $\mathcal{F}_h^I(t) = \mathcal{F}_h(t)\backslash\mathcal{F}_h^B(t)$. Each $\Gamma \in \mathcal{F}_h(t)$ is associated with a unit normal vector $\boldsymbol{n}_\Gamma$. For $\Gamma \in \mathcal{F}_h^B(t)$ the normal $\boldsymbol{n}_\Gamma$ has the same orientation as the outer normal to $\partial\Omega_h(t)$. For each $\Gamma \in \mathcal{F}_h^I(t)$ there exist two neighbouring elements $K_\Gamma^L, K_\Gamma^R \in \mathcal{T}_h(t)$ such that $\Gamma \subset \partial K_\Gamma^L \cap \partial K_\Gamma^R$. We use the convention that $K_\Gamma^R$ lies in the direction of $\boldsymbol{n}_\Gamma$ and $K_\Gamma^L$ lies in the opposite direction to $\boldsymbol{n}_\Gamma$. If $\Gamma \in \mathcal{F}_h^B(t)$, then the element adjacent to $\Gamma$ will be denoted by $K_\Gamma^L$.

We shall look for an approximate solution of the problem in the space of piecewise polynomial functions

$$\mathbf{S}_h^p(t) = (S_h^p(t))^4, \quad S_h^p(t) = \{v;\ v|_K \in P^p(K), \forall K \in \mathcal{T}_h(t)\},$$

where $p > 0$ is an integer and $P^p(K)$ denotes the space of all polynomials on $K$ of degree $\le p$. A function $\Phi \in \mathbf{S}_h^p(t)$ is, in general, discontinuous on interfaces $\Gamma \in \mathcal{F}_h^I(t)$. By $\Phi_\Gamma^L$ and $\Phi_\Gamma^R$ we denote the values of $\Phi$ on $\Gamma$ considered from the interior and the exterior of $K_\Gamma^L$, respectively, and set

$$\langle\Phi\rangle_\Gamma = \frac{1}{2}\left(\Phi_\Gamma^L + \Phi_\Gamma^R\right),$$

$$[\Phi]_\Gamma = \Phi_\Gamma^L - \Phi_\Gamma^R,$$

which denotes the average and jump of $\Phi$ on $\Gamma$.

The discrete problem is derived in the following way: For arbitrary $t \in [0, T]$ we can multiply the system by a test function $\mathbf{S}_h^p(t)$, integrate over $K \in \mathcal{T}_h(t)$, apply Green's theorem, sum over all elements $K \in \mathcal{T}_h(t)$, use the concept of the numerical flux and introduce suitable terms mutually vanishing for a regular exact solution. Moreover, we carry out a suitable partial linearization of nonlinear terms.

In order to evaluate the integrals over $\Gamma \in \mathcal{F}_h(t)$ in inviscid term we use the approximation

$$\mathbf{H}(\mathbf{w}_\Gamma^L, \mathbf{w}_\Gamma^R, \boldsymbol{n}_\Gamma) \approx \sum_{s=1}^{2} \mathbf{g}_s(\mathbf{w})(\boldsymbol{n}_\Gamma)_s,$$

where $\mathbf{H}$ is a numerical flux. For the construction of the numerical flux we use the properties (3) of $\mathbf{f}_s$. Let us define the matrix

$$\mathbf{P}(\mathbf{w}, \boldsymbol{n}) := \sum_{s=1}^{2}(\mathbf{A}_s(\mathbf{w}) - z_s\mathbf{I})n_s,$$

where $\boldsymbol{n} = (n_1, n_2)$, $n_1^2 + n_2^2 = 1$. Then we have

$$\mathbf{P}(\mathbf{w}, \boldsymbol{n})\mathbf{w} := \sum_{s=1}^{2}\mathbf{g}_s(\mathbf{w})n_s.$$

It is possible to show that the matrix $\mathbf{P}$ is diagonalizable. It means that there exists a nonsingular matrix $\mathbf{T} = \mathbf{T}(\mathbf{w}, \boldsymbol{n})$ and a diagonal matrix $\Lambda = \Lambda(\mathbf{w}, \boldsymbol{n})$ such that

$$\mathbf{P} = \mathbf{T}\Lambda\mathbf{T}^{-1}, \quad \Lambda = \text{diag}(\lambda_1, ..., \lambda_4),$$

where $\lambda_i = \lambda_i(\mathbf{w}, \boldsymbol{n})$ are eigenvalues of the matrix $\mathbf{P}$. Then we can define the "positive" and "negative" parts of the matrix $\mathbf{P}$ by

$$\mathbf{P}^\pm = \mathbf{T}\Lambda^\pm\mathbf{T}^{-1}, \quad \Lambda^\pm = \text{diag}(\lambda_1^\pm, ..., \lambda_4^\pm),$$

where $\lambda^+ = \max(\lambda, 0)$, $\lambda^- = \min(\lambda, 0)$. Using this concept, we introduce the so-called Vijayasundaram numerical flux

$$\mathbf{H}(\mathbf{w}^L, \mathbf{w}^R, \boldsymbol{n}) = \mathbf{P}^+\left(\frac{\mathbf{w}^L + \mathbf{w}^R}{2}, \boldsymbol{n}\right)\mathbf{w}^L + \mathbf{P}^-\left(\frac{\mathbf{w}^L + \mathbf{w}^R}{2}, \boldsymbol{n}\right)\mathbf{w}^R.$$

This numerical flux has suitable form for a linearization. Now we can define inviscid form in the following way:

$$\begin{aligned}
&b_h(\bar{\mathbf{w}}_h, \mathbf{w}_h, \Phi_h, t) := \\
&- \sum_{K\in\mathcal{T}_h(t)} \int_K \sum_{s=1}^{2} (\mathbf{A}_s(\bar{\mathbf{w}}_h) - z_s(t)\mathbf{I})\,\mathbf{w}_h \cdot \frac{\partial\Phi_h}{\partial x_s}\,\mathrm{d}\boldsymbol{x} \\
&+ \sum_{\Gamma\in\mathcal{F}_h^I(t)} \int_\Gamma \left(\mathbf{P}^+\left(\langle\bar{\mathbf{w}}_h\rangle, \boldsymbol{n}_\Gamma\right)\mathbf{w}_h^L + \mathbf{P}^-\left(\langle\bar{\mathbf{w}}_h\rangle, \boldsymbol{n}_\Gamma\right)\mathbf{w}_h^R\right)\cdot[\Phi_h]\,\mathrm{d}S \\
&+ \sum_{\Gamma\in\mathcal{F}_h^B(t)} \int_\Gamma \left(\mathbf{P}^+\left(\langle\bar{\mathbf{w}}_h\rangle, \boldsymbol{n}_\Gamma\right)\mathbf{w}_h^L + \mathbf{P}^-\left(\langle\bar{\mathbf{w}}_h\rangle, \boldsymbol{n}_\Gamma\right)\bar{\mathbf{w}}_h^R\right)\cdot\Phi_h\,\mathrm{d}S
\end{aligned}$$

The boundary state $\bar{\mathbf{w}}_h^R$ is evaluated with the aid of the local linearized Riemann problem described in [9].

For the discretization of the viscous terms we use the property (4) and get the viscous form

$$\begin{aligned}
&a_h(\bar{\mathbf{w}}_h, \mathbf{w}_h, \Phi_h, t) := \\
&+ \sum_{K\in\mathcal{T}_h(t)} \int_K \sum_{s=1}^{2}\sum_{k=1}^{2} \mathbf{K}_{s,k}(\bar{\mathbf{w}}_h)\,\frac{\partial\mathbf{w}_h}{\partial x_k}\cdot\frac{\partial\Phi_h}{\partial x_s}\,\mathrm{d}\boldsymbol{x} \\
&- \sum_{\Gamma\in\mathcal{F}_h^I(t)} \int_\Gamma \sum_{s=1}^{2}\left\langle\sum_{k=1}^{2}\mathbf{K}_{s,k}(\bar{\mathbf{w}}_h)\frac{\partial\mathbf{w}_h}{\partial x_k}\right\rangle(\boldsymbol{n}_\Gamma)_s\cdot[\Phi_h]\,\mathrm{d}S \\
&- \sum_{\Gamma\in\mathcal{F}_h^B(t)} \int_\Gamma \sum_{s=1}^{2}\sum_{k=1}^{2}\mathbf{K}_{s,k}(\bar{\mathbf{w}}_h)\frac{\partial\mathbf{w}_h}{\partial x_k}\,(\boldsymbol{n}_\Gamma)_s\cdot\Phi_h\,\mathrm{d}S \\
&- \Theta \sum_{\Gamma\in\mathcal{F}_h^I(t)} \int_\Gamma \sum_{s=1}^{2}\left\langle\sum_{k=1}^{2}\mathbf{K}_{k,s}^T(\bar{\mathbf{w}}_h)\frac{\partial\Phi_h}{\partial x_k}\right\rangle(\boldsymbol{n}_\Gamma)_s\cdot[\mathbf{w}_h]\,\mathrm{d}S \\
&- \Theta \sum_{\Gamma\in\mathcal{F}_h^B(t)} \int_\Gamma \sum_{s=1}^{2}\sum_{k=1}^{2}\mathbf{K}_{k,s}^T(\bar{\mathbf{w}}_h)\frac{\partial\Phi_h}{\partial x_k}\,(\boldsymbol{n}_\Gamma)_s\cdot\mathbf{w}_h\,\mathrm{d}S
\end{aligned}$$

We set $\Theta = 1$ or $\Theta = 0$ or $\Theta = -1$ and get the so-called symmetric version (SIPG) or incomplete version (IIPG) or nonsymetric version (NIPG), respectively, of the discretization of the viscous terms.

Further, we define the turbulent forms $p_h, k_h$, the interior and boundary penalty form $J_h^\sigma$ and the right-hand side form $l_h$ in the following way:

$$
\begin{aligned}
p_h(\bar{\mathbf{w}}_h, \mathbf{w}_h, \Phi_h, t) := & \sum_{\Gamma\in\mathcal{F}_h^I(t)} \int_\Gamma \sum_{s=1}^{2} \mathbf{A}_s^P(\bar{\mathbf{w}}_h^L)\mathbf{w}_h^L \cdot \Phi_h^L\,(\boldsymbol{n}_\Gamma)_s\,\mathrm{d}S \\
& - \sum_{\Gamma\in\mathcal{F}_h^I(t)} \int_\Gamma \sum_{s=1}^{2} \mathbf{A}_s^P(\bar{\mathbf{w}}_h^R)\mathbf{w}_h^R \cdot \Phi_h^R\,(\boldsymbol{n}_\Gamma)_s\,\mathrm{d}S \\
& + \sum_{\Gamma\in\mathcal{F}_h^B(t)} \int_\Gamma \sum_{s=1}^{2} \mathbf{A}_s^P(\bar{\mathbf{w}}_h)\mathbf{w}_h \cdot \Phi_h\,(\boldsymbol{n}_\Gamma)_s\,\mathrm{d}S \\
& - \sum_{K\in\mathcal{T}_h(t)} \int_K \sum_{s=1}^{2} \mathbf{A}_s^P(\bar{\mathbf{w}}_h)\mathbf{w}_h \cdot \frac{\partial \Phi_h}{\partial x_s}\,\mathrm{d}\boldsymbol{x},
\end{aligned}
$$

$$
\begin{aligned}
k_h(\bar{\mathbf{w}}_h, \Phi_h, t) := & \sum_{\Gamma\in\mathcal{F}_h^I(t)} \int_\Gamma \sum_{s=1}^{2} \mathbf{k}_s(\bar{\mathbf{w}}_h^L) \cdot \Phi_h^L\,(\boldsymbol{n}_\Gamma)_s\,\mathrm{d}S \\
& - \sum_{\Gamma\in\mathcal{F}_h^I(t)} \int_\Gamma \sum_{s=1}^{2} \mathbf{k}_s(\bar{\mathbf{w}}_h^R) \cdot \Phi_h^R\,(\boldsymbol{n}_\Gamma)_s\,\mathrm{d}S \\
& + \sum_{\Gamma\in\mathcal{F}_h^B(t)} \int_\Gamma \sum_{s=1}^{2} \mathbf{k}_s(\bar{\mathbf{w}}_h) \cdot \Phi_h\,(\boldsymbol{n}_\Gamma)_s\,\mathrm{d}S \\
& - \sum_{K\in\mathcal{T}_h(t)} \int_K \sum_{s=1}^{2} \mathbf{k}_s(\bar{\mathbf{w}}_h) \cdot \frac{\partial \Phi_h}{\partial x_s}\,\mathrm{d}\boldsymbol{x},
\end{aligned}
$$

$$
\begin{aligned}
J_h^\sigma(\mathbf{w}_h, \Phi_h, t) := & \sum_{\Gamma\in\mathcal{F}_h^I(t)} \int_\Gamma \sigma[\mathbf{w}_h]\cdot[\Phi_h]\,\mathrm{d}S \\
& + \sum_{\Gamma\in\mathcal{F}_h^B(t)} \int_\Gamma \sigma\mathbf{w}_h\cdot\Phi_h\,\mathrm{d}S, \\
l_h(\bar{\mathbf{w}}_h, \Phi_h, t) := & \sum_{\Gamma\in\mathcal{F}_h^B(t)} \int_\Gamma \sigma\mathbf{w}_B\cdot\Phi_h\,\mathrm{d}S \\
& - \Theta \sum_{\Gamma\in\mathcal{F}_h^B(t)} \int_\Gamma \sum_{s=1}^{2}\sum_{k=1}^{2} \mathbf{K}_{k,s}^T(\bar{\mathbf{w}}_h)\frac{\partial \Phi_h}{\partial x_k}(\boldsymbol{n}_\Gamma)_s\cdot\mathbf{w}_B\,\mathrm{d}S,
\end{aligned}
$$

where $\sigma|_\Gamma = \frac{C_W}{Re\, d(\Gamma)}$, where $d(\Gamma)$ is the diameter of $\Gamma \in \mathcal{F}_h(t)$, $Re$ is the Reynolds number and $C_W > 0$ is a suitable sufficiently large constant. The boundary state $\mathbf{w}_B$ is defined on the basis of the Dirichlet boundary conditions and extrapolation:

$$
\begin{aligned}
&\mathbf{w}_B = (\rho_D, \rho_D v_{D1}, \rho_D v_{D2}, c_v\rho_D\theta_\Gamma^L + \frac{1}{2}\rho_D\,|\boldsymbol{v}_D|^2) \quad \text{on } \Gamma_I, \\
&\mathbf{w}_B = \mathbf{w}_\Gamma^L \quad \text{on } \Gamma_O, \\
&\mathbf{w}_B = (\rho_\Gamma^L, \rho_\Gamma^L z_{D1}, \rho_\Gamma^L z_{D2}, c_v\rho_\Gamma^L\theta_\Gamma^L + \frac{1}{2}\rho_\Gamma^L\,|z_D|^2) \quad \text{on } \Gamma_W(t).
\end{aligned}
$$

In the vicinity of discontinuities or steep gradients non-physical oscillations can appear in the approximate solution. In order to overcome this difficulty we employ artificial viscosity forms, see [8]. They are based on the discontinuity indicator

$$
g(t)(K) := \frac{1}{d(K)\,|K|^{3/4}} \int_{\partial K} [\bar\rho_h]^2 dS, \quad K \in \mathcal{T}_h(t),
$$

where $[\bar\rho_h]$ is the jump of the function $\bar\rho_h$ (= the first component of the vector function $\bar{\mathbf{w}}_h$) on the boundary $\partial K$, $d(K)$ denotes the diameter of $K$ and $|K|$ denotes the area of the element $K$. Then we define the discrete discontinuity indicator

$$
\begin{aligned}
G(t)(K) &:= 0 \quad \text{if } g(t)(K) < 1, \\
G(t)(K) &:= 1 \quad \text{if } g(t)(K) \geq 1,\ K \in \mathcal{T}_h(t),
\end{aligned}
$$

and the artificial viscosity forms

$$
\begin{aligned}
&\beta_h(\bar{\mathbf{w}}_h, \mathbf{w}_h, \Phi_h, t) := \\
&\nu_1 \sum_{K\in\mathcal{T}_h(t)} d(K)G(t)(K) \int_K \nabla\mathbf{w}_h\cdot\nabla\Phi_h\,\mathrm{d}\boldsymbol{x}, \\
&\tilde J_h(\bar{\mathbf{w}}_h, \mathbf{w}_h, \Phi_h, t) := \\
&\nu_2 \sum_{\Gamma\in\mathcal{F}_h^I(t)} \frac{1}{2}\left(G(t)(K_\Gamma^L) + G(t)(K_\Gamma^R)\right) \int_\Gamma [\mathbf{w}_h]\cdot[\Phi_h]\,\mathrm{d}S,
\end{aligned}
$$

with constants $\nu_1$ and $\nu_2$.

All these forms are linear with respect to $\mathbf{w}_h$ and nonlinear with respect to $\bar{\mathbf{w}}_h$.

Finally, we set

$$
(\varphi, \psi)_t = \int_{\Omega_h(t)} \varphi\,\psi\,\mathrm{d}x.
$$

### 3.2 Full space-time DGM discretization

Let $0 = t_0 < t_1 < \ldots < t_M = T$ be a partition of the interval $[0, T]$ and let us denote $I_m = (t_{m-1}, t_m)$, $\tau_m = t_m - t_{m-1}$ for $m = 1, \ldots, M$. We define the space $\mathbf{S}_{h,\tau}^{p,q} = (S_{h,\tau}^{p,q})^4$, where

$$
S_{h,\tau}^{p,q} := \left\{\phi\,;\,\phi|_{I_m} = \sum_{i=0}^{q} \zeta_i\phi_i, \text{ where } \phi_i \in S_h^p(t),\ \zeta_i \in P^q(I_m)\right\}.
$$

with integers $p, q \geq 1$. $P^q(I_m)$ denotes the space of all polynomials in $t$ on $I_m$ of degree $\leq q$. Moreover for $\Phi \in \mathbf{S}_{h,\tau}^{p,q}$ we introduce the following notation:

$$
\begin{aligned}
\Phi_m^\pm &= \Phi(t_m^\pm) = \lim_{t\to t_{m\pm}} \Phi(t), \\
\{\Phi\}_m &= \Phi_m^+ - \Phi_m^-.
\end{aligned}
$$

Approximate solution $\mathbf{w}_{h\tau}$ of the problem will be sought in the space $\mathbf{S}_{h,\tau}^{p,q}$. Since the functions of this space are in general discontinuous in time, we ensure the connection between $I_{m-1}$ and $I_m$ by the penalty term in time

$$
\left(\{\mathbf{w}_{h\tau}\}_{m-1}, \Phi_{h\tau}(t_{m-1}^+)\right)_{t_{m-1}}.
$$

The initial state $\mathbf{w}_{h\tau}$ is included by the $L^2(\Omega_h(t_0))$-projection of $\mathbf{w}^0$ on $\mathbf{S}_h^p(t_0)$:

$$
\left(\mathbf{w}_{h\tau}(t_0^+), \Phi_{h\tau}(t_0^+)\right)_{t_0} = \left(\mathbf{w}^0, \Phi_{h\tau}(t_0^+)\right)_{t_0} \quad \forall \Phi_{h\tau} \in \mathbf{S}_{h,\tau}^{p,q}.
$$

Now we introduce a suitable linearization. We can use two possibilities.

1) We put $\bar{\mathbf{w}}_{h\tau}(t) := \mathbf{w}_h(t_{m-1}^-)$ for $t \in I_m$.

2) We prolong the solution from the time interval $I_{m-1}$ to the time interval $I_m$.

We say that a function $\mathbf{w}_{h\tau} \in \mathbf{S}_{h,\tau}^{p,q}$ is the approximate solution of the problem (1) obtained by the ST-DG method, if it satisfies the following conditions

$$
\begin{aligned}
&\sum_{m=1}^{M}\int_{I_m}\left(\left(\frac{D^{\mathcal{A}}\mathbf{w}_{h\tau}}{Dt},\Phi_{h\tau}\right)_t + ((\mathbf{z}(t)\cdot\nabla)\mathbf{w}_{h\tau},\Phi_{h\tau})_t\right)\,dt\\
&+\sum_{m=1}^{M}\int_{I_m}(a_h(\bar{\mathbf{w}}_{h\tau},\mathbf{w}_{h\tau},\Phi_{h\tau},t)+b_h(\bar{\mathbf{w}}_{h\tau},\mathbf{w}_{h\tau},\Phi_{h\tau},t))\,dt\\
&+\sum_{m=1}^{M}\int_{I_m}\left(\beta_h(\bar{\mathbf{w}}_{h\tau},\mathbf{w}_{h\tau},\Phi_{h\tau},t)+\tilde{J}_h(\bar{\mathbf{w}}_{h\tau},\mathbf{w}_{h\tau},\Phi_{h\tau},t)\right)\,dt\\
&+\sum_{m=1}^{M}\int_{I_m}(J_h^{\sigma}(\mathbf{w}_{h\tau},\Phi_{h\tau},t)+p_h(\bar{\mathbf{w}}_{h\tau},\mathbf{w}_{h\tau},\Phi_{h\tau},t))\,dt\\
&+\sum_{m=2}^{M}\left(\{\mathbf{w}_{h\tau}\}_{m-1},\Phi_{h\tau}(t_{m-1}^+)\right)_{t_{m-1}}+\left(\mathbf{w}_{h\tau}(t_0^+),\Phi_{h\tau}(t_0^+)\right)_{t_0}\\
&=\sum_{m=1}^{M}\int_{I_m} l_h(\bar{\mathbf{w}}_{h\tau},\Phi_{h\tau},t)\,dt+\left(\mathbf{w}^0,\Phi_{h\tau}(t_0^+)\right)_{t_0}\\
&+\sum_{m=1}^{M}\int_{I_m} k_h(\bar{\mathbf{w}}_{h\tau},\Phi_{h\tau},t)\,dt\quad\forall\Phi_{h\tau}\in\mathbf{S}_{h,\tau}^{p,q}.
\end{aligned}
$$

## 4 Finite volume discretization of the $k-\omega$ turbulence model

The finite volume method is applied on the same mesh $\mathcal{T}_h(t)$ as ST-DG method. So, similarly as in Section 3.1, let $\mathcal{T}_h(t) = \{K_i(t)\}_{i\in I}$ be a partition of the domain $\Omega_h(t)$ into finite number of closed elements with mutually disjoint interiors such that $\overline{\Omega}_h(t) = \bigcup_{i\in I} K_i(t)$, where $I \subset Z^+ = \{0,1,2,3,...\}$ is a suitable index set. If two elements have a common face, then we call them neighbours and put $\Gamma_{ij}(t) = \Gamma_{ji}(t) = \partial K_i(t) \cap \partial K_j(t)$. For each $i \in I$ we define the index set $s(i) = \{j \in I; K_j(t) \text{ is a neighbour of } K_i(t)\}$. The boundary $\partial\Omega_h(t)$ is formed by a finite number of sides of elements $K_i(t)$ adjacent to $\partial\Omega_h(t)$. We denote all these boundary sides by $S_j(t)$, where $j \in I_b \subset Z^- = \{-1,-2,-3,...\}$ and set $\gamma(i) = \{j \in I_b; S_j(t) \text{ is a side of } K_i(t)\}, \Gamma_{ij}(t) = S_j(t)$ for $K_i(t) \in \mathcal{T}_h(t)$ such that $S_j \subset \partial K_i, j \in I_b$. For an element $K_i(t)$, not containing any boundary side $S_j(t)$, we set $\gamma(i) = \emptyset$. Moreover we define $S(i) = s(i) \cup \gamma(i)$ and $\boldsymbol{n}_{ij}(t) = ((n_{ij}(t))_1, (n_{ij}(t))_2)$ as the unit outer normal to $\partial K_i(t)$ on the side $\Gamma_{ij}(t)$. We again consider a partition $0 = t_0 < t_1 < ... < t_M = T$ of the interval $[0,T]$ and denote $I_m = (t_{m-1}, t_m)$, $\tau_m = t_m - t_{m-1}$ for $m = 1,...,M$.

Now we can integrate the system (5), apply Green's theorem and approximate the integrals with the one-point rule and we get the following implicit finite volume scheme

$$
\begin{aligned}
&|K_i(t_m)|\,\tilde{\mathbf{w}}_i^m - |K_i(t_{m-1})|\,\tilde{\mathbf{w}}_i^{m-1} - \int_{t_{m-1}}^{t_m}\int_{K_i(t)}\tilde{\mathbf{s}}(\tilde{\mathbf{w}}_i^m)\,dS\,dt\\
&+\sum_{j\in S(i)}\int_{t_{m-1}}^{t_m}\int_{\Gamma_{ij}(t)}\tilde{\mathbf{H}}(\tilde{\mathbf{w}}_i^m,\tilde{\mathbf{w}}_j^m,\boldsymbol{n}_{ij}(t))\\
&-\sum_{j\in S(i)}\int_{t_{m-1}}^{t_m}\int_{\Gamma_{ij}(t)}\sum_{s=1}^{2}\tilde{\mathbf{R}}_s(\tilde{\mathbf{w}}_{\Gamma_{ij}}^m,\nabla\tilde{\mathbf{w}}_{\Gamma_{ij}}^m)(n_{ij}(t))_s\\
&=0\qquad\forall i\in I.
\end{aligned}
$$

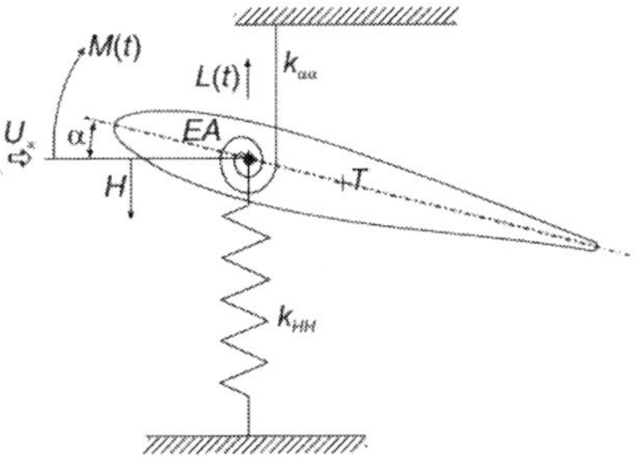


**Figure 1.** The scheme of the vibrating airfoil.

Here $\tilde{\mathbf{w}}_{\Gamma_{ij}}^m$ resp. $\nabla\tilde{\mathbf{w}}_{\Gamma_{ij}}^m$ denote $\tilde{\mathbf{w}}^m$ resp. $\nabla\tilde{\mathbf{w}}^m$ at the center of the edge $\Gamma_{ij}$ at time instant $t_m$. By $\tilde{\mathbf{w}}_i^m$ we shall denote approximate solution of $\tilde{\mathbf{w}}$ on the element $K_i$ at time instant $t_m$

$$
\tilde{\mathbf{w}}_i^m \approx \frac{1}{|K_i(t_m)|}\int_{K_i(t_m)}\tilde{\mathbf{w}}(x,t_m)\,dx.
$$

For the numerical flux $\tilde{\mathbf{H}}$ we again use the concept of the Vijayasundaram numerical flux in the following way:

$$
\tilde{\mathbf{H}}(\tilde{\mathbf{w}}_i^m,\tilde{\mathbf{w}}_j^m,\boldsymbol{n}_{ij}(t)) = (\langle\mathbf{v}-\mathbf{z}\rangle\cdot\boldsymbol{n}_{ij})^+\tilde{\mathbf{w}}_i^m + (\langle\mathbf{v}-\mathbf{z}\rangle\cdot\boldsymbol{n}_{ij})^-\tilde{\mathbf{w}}_j^m
$$

for $j \in s(i)$ and

$$
\tilde{\mathbf{H}}(\tilde{\mathbf{w}}_i^m,\tilde{\mathbf{w}}_j^m,\boldsymbol{n}_{ij}(t)) = ((\mathbf{v}-\mathbf{z})\cdot\boldsymbol{n}_{ij})^+\tilde{\mathbf{w}}_i^m + ((\mathbf{v}-\mathbf{z})\cdot\boldsymbol{n}_{ij})^-\tilde{\mathbf{w}}_B
$$

for $j \in \gamma(i)$, where $\tilde{\mathbf{w}}_B$ denotes boundary conditions. One possibility of linearization of $\tilde{\mathbf{R}}_s$ and $\tilde{\mathbf{s}}$ is via Taylor expansion which is used in our case. For more details see [6].

## 5 Equations for the moving airfoil

We shall simulate the motion of a profile in 2D with two degrees of freedom: $H$ - displacement of the profile in the vertical direction (positively oriented downwards) and $\alpha$ - the rotation of the profile around the so-called elastic axis (positively oriented clockwise), see Figure 1. The motion of the profile is described by the system of ordinary differential equations

$$
\begin{aligned}
m\ddot{H} + S_\alpha\ddot{\alpha} + k_{HH}H &= -L(t),\\
S_\alpha\ddot{H} + I_\alpha\ddot{\alpha} + k_{\alpha\alpha}\alpha &= M(t),
\end{aligned}
\qquad (9)
$$

where we use the following notation: $m$ - mass of the airfoil, $L(t)$ - aerodynamic lift force, $M(t)$ - aerodynamic torsional moment, $S_\alpha$ - static moment of the airfoil around the elastic axis (EA), $k_{HH}$ - bending stiffness, $k_{\alpha\alpha}$ - torsional stiffness. We define $L(t)$ and $M(t)$ by the terms

$$
L(t) := -l\int_{\Gamma_{W(t)}}\sum_{r=1}^{2}\tau_{2r}n_r\,dS,
$$

$$
M(t) := l\int_{\Gamma_{W(t)}}\sum_{s,r=1}^{2}\tau_{sr}n_r(-1)^s\left(x_{1+\delta_{1s}} - x_{1+\delta_{1s}}^{EA}\right)\,dS,
$$

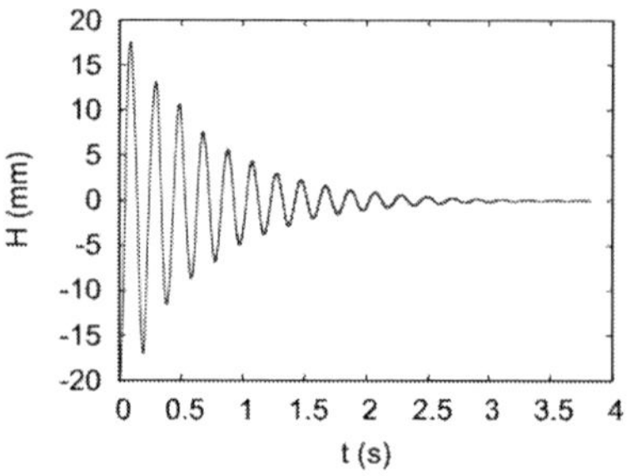


**Figure 2.** Displacement for far-field velocity 10 m s$^{-1}$.

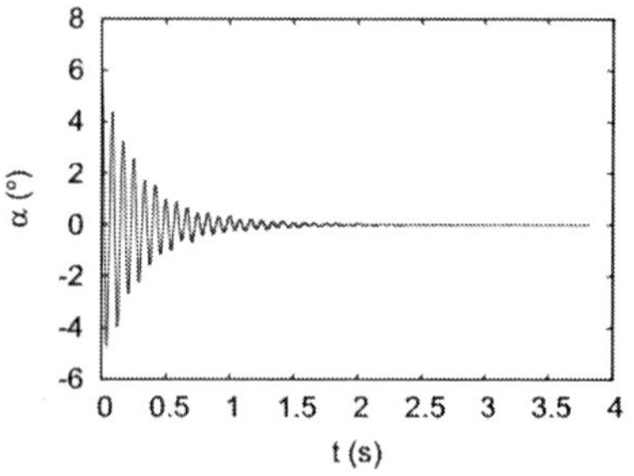


**Figure 3.** Rotation angel for far-field velocity 10 m s$^{-1}$.

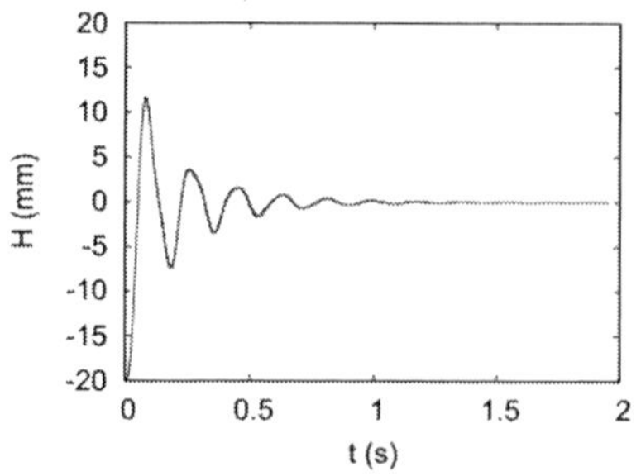


**Figure 4.** Displacement for far-field velocity 20 m s$^{-1}$.

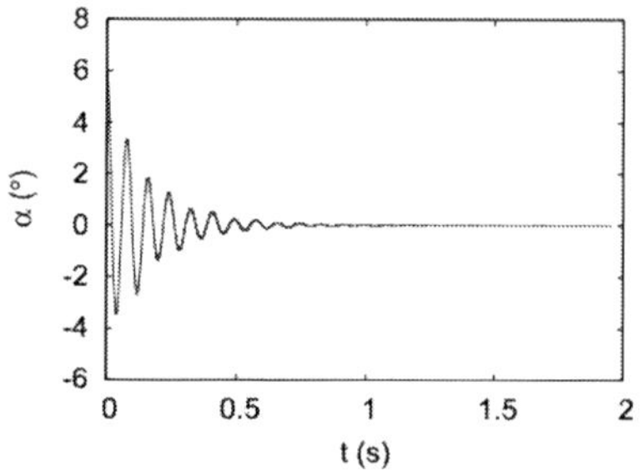


**Figure 5.** Rotation angel for far-field velocity 20 m s$^{-1}$.

where $l$ is the depth of the airfoil, $x_s^{EA}$ are the coordinates of the elastic axis and $\tau_{sr} := -p\delta_{sr} + \tau_{sr}^V$ are the components of the stress tensor. For the derivation of the system (9), see e.g. [11].

The system (9) is transformed to a first-order system and solved by the fourth-order Runge-Kutta method together with the discrete flow problem.

## 6 Numerical experiments

We performed numerical simulations in 2D for the profile NACA0012 with the following data and initial conditions: $m = 0.086622$ kg, $S_\alpha = -0.000779673$ kg m, $I_\alpha = 0.000487291$ kg m$^{-2}$, $k_{HH} = 105.109$ N m$^{-1}$, $k_{\alpha\alpha} = 3.696682$ N m rad$^{-1}$, far-field pressure $p = 101325$ Pa, airfoil depth $l = 0.05$ m, airfoil length $c = 0.3$ m, far-field density $\rho = 1.225$ kg m$^{-3}$, Poisson adiabatic constant $\gamma = 1.4$, specific heat $c_v = 721.428$ m$^2$ s$^{-2}$ K$^{-1}$, heat conduction coefficient $\kappa = 2.428 \cdot 10^{-2}$ kg m s$^{-2}$ K$^{-1}$, $H(0) = -20$ mm, $\alpha(0) = 6°$, $\dot{H}(0) = \dot{\alpha}(0) = 0$. In the case of space-time discontinuous Galerkin method we employ linear polynomials in space and in time. Far field-velocity was chosen in the range 10 m s$^{-1}$ to 40 m s$^{-1}$. Figures 2-13 show the computed displacement and rotation of the profile. Figures 14 - 17 show different quantities for far-field velocity 37.5 m s$^{-1}$ at the time instant 0.34 s. Figure 18 shows detail of the mesh which consist of 69766 elements in total.

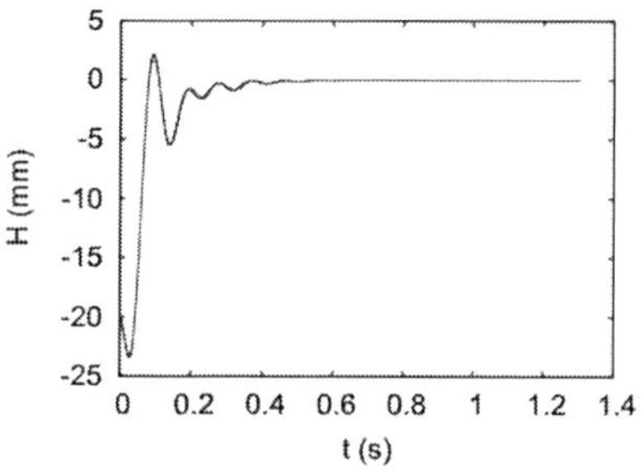


**Figure 6.** Displacement for far-field velocity 30 m s$^{-1}$.

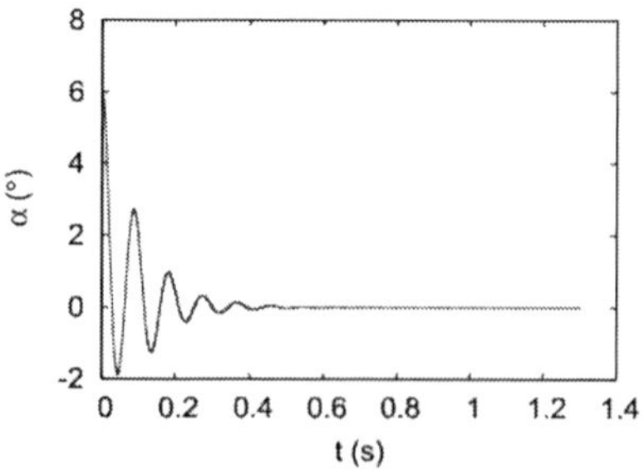


**Figure 7.** Rotation angel for far-field velocity 30 m s$^{-1}$.

## 7 Conclusion

In this paper we dealt with the finite-volume - space-time discontinuous Galerkin method for the numerical solution

DOI: 10.1051/epjconf/201714302014

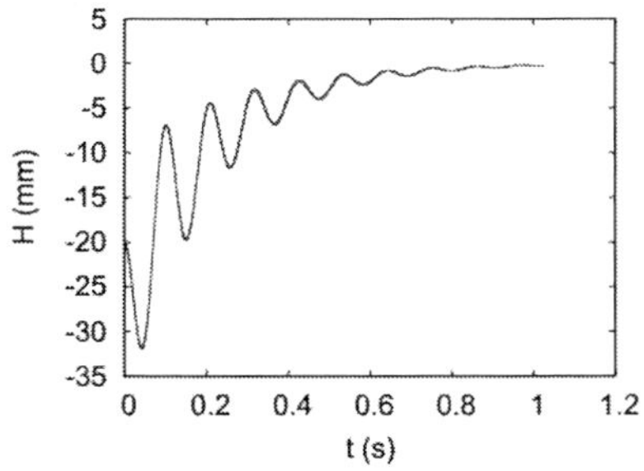


**Figure 8.** Displacement for far-field velocity 35 m s$^{-1}$.

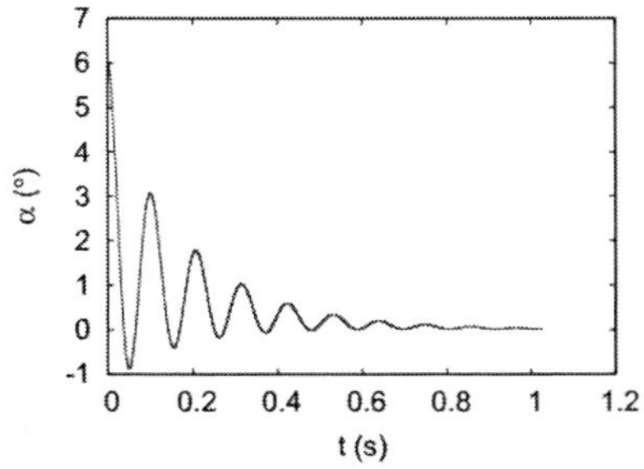


**Figure 9.** Rotation angel for far-field velocity 35 m s$^{-1}$.

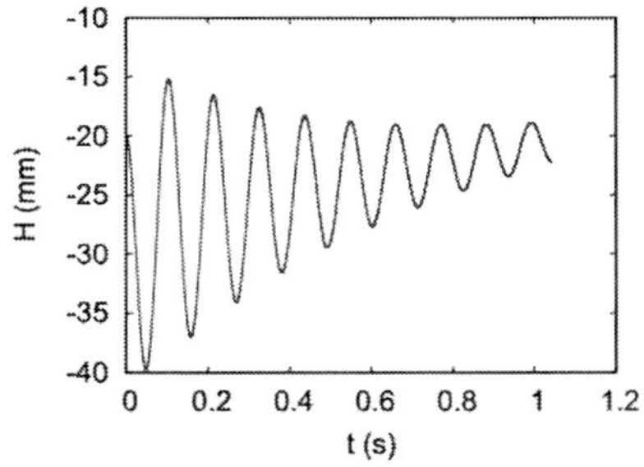


**Figure 10.** Displacement for far-field velocity 37.5 m s$^{-1}$.

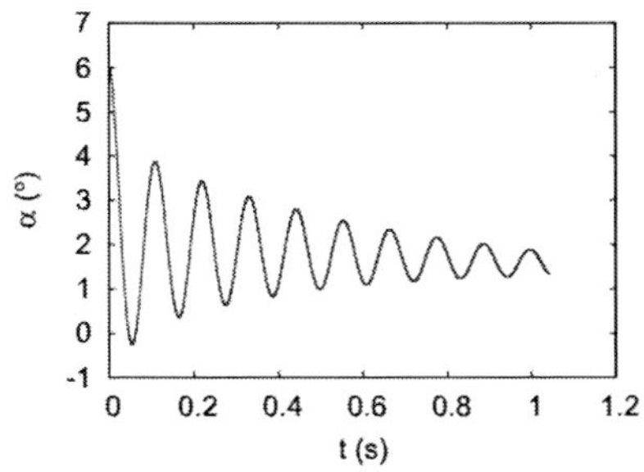


**Figure 11.** Rotation angel for far-field velocity 37.5 m s$^{-1}$.

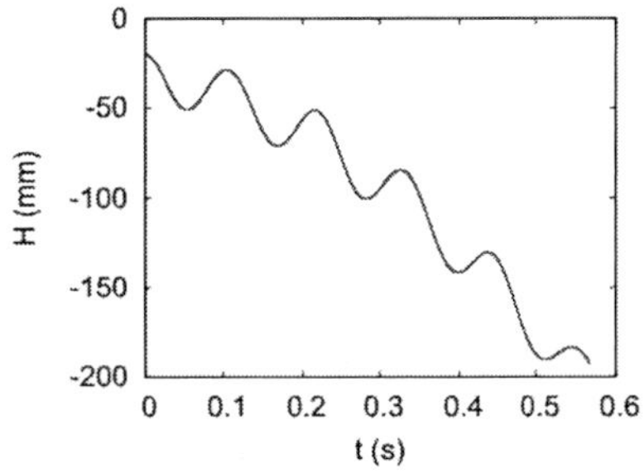


**Figure 12.** Displacement for far-field velocity 40 m s$^{-1}$.

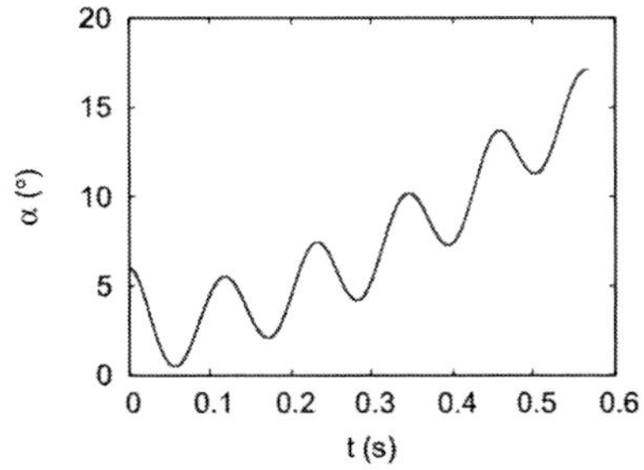


**Figure 13.** Rotation angel for far-field velocity 40 m s$^{-1}$.

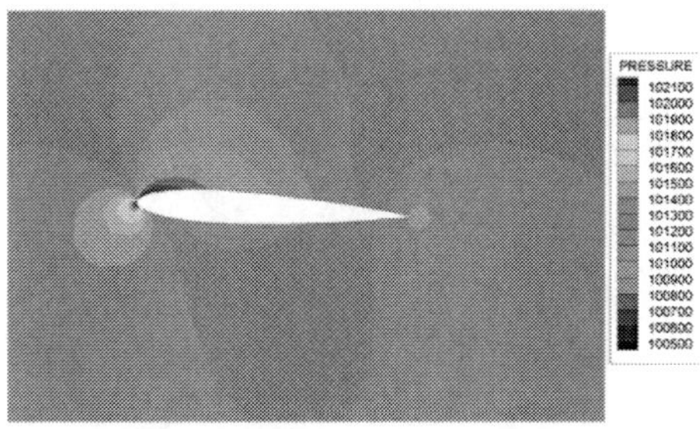


**Figure 14.** The distribution of the pressure.

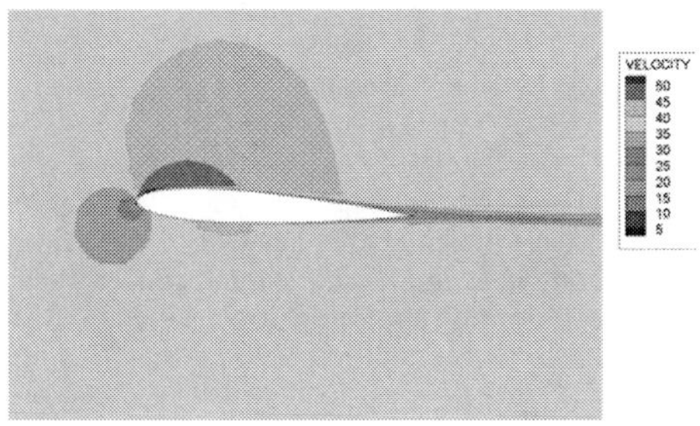


**Figure 15.** The distribution of the velocity.

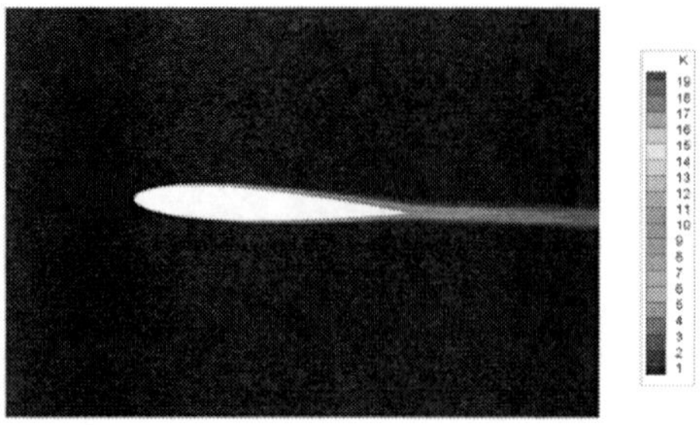


**Figure 16.** The distribution of the turbulent kinetic energy.

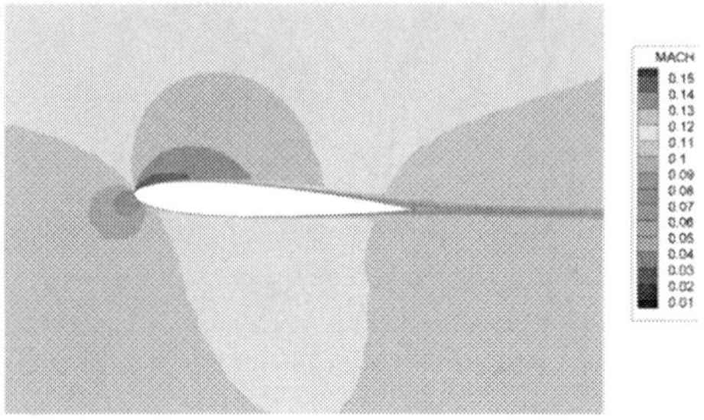


**Figure 17.** The distribution of the Mach number.

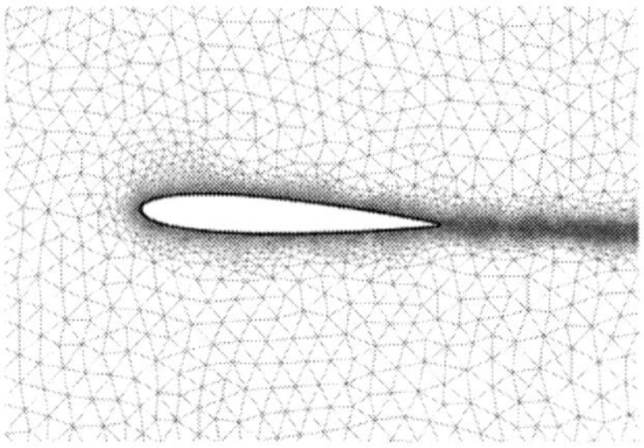

**Figure 18.** Detail of the mesh.

of compressible turbulent flow in a time dependent domain. The applicability of the proposed method was demonstrated on the examples of the interaction of compressible turbulent flow and a moving airfoil.

## Acknowledgment

This result originated with the support of Ministry of Industry and Trade of the Czech Republic for the long-term strategic development of the research organization. The authors acknowledge this support.

## References

1. Cook P. H., McDonald M. A. and Firmin M. C. P. AGARD Advisory Report 138, (1979)
2. Česenek J. *Discontinuous Galerkin method for solving compressible viscous flow*. Doctoral Thesis, (2011)
3. Česenek J. et al., Appl. Math. Comput., doi:10.1016/1.amc.2011.08.077 (2011)
4. Česenek J., Feistauer M., Kosik A. ZAMM – Z. Angew. Math. Mech. **93**, 6 – 7, 387 – 402 (2013)
5. Česenek J. *Experimental fluid mechanics*,142-147 (2013)
6. Česenek J. *Combined finite volume - space-time discontinuous Galerkin method for the 2D compressible turbulent flow*. Technical report R-6154, VZLÚ, Beranových 130, Prague (2014)
7. Česenek J. *Experimental fluid mechanics*,112-116 (2015)
8. Dolejší V., Feistauer M. *Discontinuous Galerkin Method*. Springer (2015)
9. Feistauer M., Felcman J., and Straškraba I. *Mathematical and Computational Methods for Compressible Flow*. Clarendon Press, Oxford (2003)
10. Wilcox, D.C.: AIAA Journal, Vol. **46**. No. 11., (2008)
11. Sváček P., Feistauer M., Horáček J. J. Fluids Struc., **23**, 391-411 (2007)

 DOI: 10.1051/epjconf/201714302015

# Cooling tower plume - model and experiment

*Jan* Cizek[1], *Jiri* Gemperle[1], *Miroslav* Strob[1] and *Jiri* Nozicka[1]

[1]CTU in Prague, Faculty of Mechanical Engineering, Technicka 4, 166 07 Prague, Czech Republic

**Abstract.** The paper discusses the description of the simple model of the, so-called, steam plume, which in many cases forms during the operation of the evaporative cooling systems of the power plants, or large technological units. The model is based on semi-empirical equations that describe the behaviour of a mixture of two gases in case of the free jet stream. In the conclusion of the paper, a simple experiment is presented through which the results of the designed model shall be validated in the subsequent period.

## 1 Introduction

The cooling towers are used in many industrial plants as a system for efficient and reliable liquidation of low-potential waste heat. Their history dates back far into the past and first mention of the basic principles of evaporative cooling (respectively, the fact that it is possible to achieve a certain level of cooling by evaporation) can be found already more than 2,500 years BC. Application of this type of cooling in industrial practice started in the second half of the 19th century. At present, cooling towers working on the evaporative cooling principle are relatively standard technological units, which can be supplied on contractor basis without any major problems in a practically arbitrary range of required outputs (the smallest fan cooling towers are currently made from app. dimensions of 0.5 x 0.5 m).

Cooling tower issues are discussed in many professional publications. The basic publications may be considered as the books [1] and [2], and from the ranks of professional articles, these are, for instance, [3], [4], [5], or [6]. This contribution is then from the viewpoint of the field of evaporative cooling tied to the relatively narrow issue of the creation of the steam plume. Few global professional publications exist about their characteristics and behaviour, however, worth mentioning are publications [7] and [8] which although a bit old, nevertheless, present an idea in the given issue.

## 2 Velocity field

The mathematical model used to get the basic idea of the behaviour of the steam plume is based on the solution presented in the paper [9]. It is a semi-empirical model of the free stream jet flow from the circular or rectangular nozzle. Schematically, the geometrical characteristics of this flow are given in Fig. 1, where the detail in the vicinity of its outlet is shown on the right side. The model in itself practically respects on the retention of the overall momentum of the flow without considering any other influences. Although it is a relatively trivial model, its resultant values (e.g. in the form of the shape of the velocity profile, or progress of the decline of the mean velocity on the flow axis) do not differ much from the more sophisticated models.

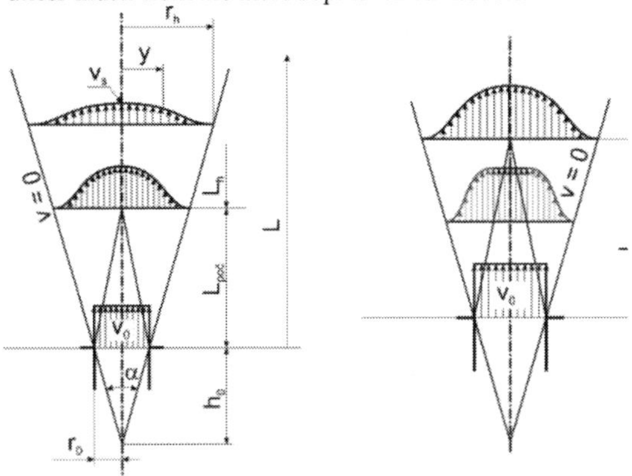

**Fig. 1:** Geometrical characteristics of the considered model

The velocity field is subsequently described in the system of linear algebraic equations 1 to 5

$$L_{po\check{c}} = (0.67 \cdot r_0) / a \qquad (1)$$

$$L_{po\check{c}} = (0.29 \cdot r_0) / a \qquad (2)$$

$$r_h = 3.3 \cdot r_0 \cdot v_0 / v_s \qquad (3)$$

$$v_s = (0.96 \cdot v_0) / [(a \cdot L) / r_0 + 0.29] \qquad (4)$$

$$v = v_s \cdot [1 - (y / r_h)^{(3/2)}]^2 \qquad (5)$$

Symbol ***a*** shown in equations 1 to 5 is the, so-called, coefficient of the structure of the flow and an adequately large Reynolds Number (Re > 2.104) it is not dependent on the Reynolds Number. In the publication [9], it is recommended for the circular nozzle section to use the value ***a = 0.07 ÷ 0.08***. The values given for the velocity at an arbitrary flow point (5), respectively, for the velocity in the flow axis (4), it is then comprehensively valid only for the area where $L \geq L_{po\check{c}}$.

From the given equations (1) and (2), it is also good to notice the facts that the values $L_{po\check{c}}$ and $h_0$ are dependent only on the parameters $r_0$ and $a$. Due to the potential recommended distribution of the values of parameter $a$ it is possible to state that the size of the area, which is affected by the exit flow depends only on its diameter.

The advantage of the solution in [9] is, among other things, that it is possible to expand it further with the velocity of the ambient air in a case where the spreading of the steam plume is influenced by cross wind. In this situation, it is therein recommended for calculation of the centre line of the derived velocity profile to use the equation (6) in the shape

$$a \cdot y / (2 \cdot r_0) = 195 \cdot (\rho_{ok} \cdot v_{ok}^2 / (\rho_0 \cdot v_0^2))^{1.3} \cdot (a \cdot x / (2 \cdot r_0))^3 + a \cdot x / (2 \cdot r_0) \cdot cotg\alpha. \quad (6)$$

The given equation was approximated from the experimental data and its area of validity is therefore very limited, specifically up to an ambient air velocity of about 0.3-fold the velocity of the main flow. In real cases, a horse-shoe shaped vortex is also created after the outlet from the cooling tower and the whole issue is therefore generally more complicated. For the conclusions derived here, however, this statement suffices. In order to allow, also in this case, for design of the shape of the velocity profile at the individual points in the evaluated area, the velocity profiles obtained in the equation (5) are transformed into the curvilinear coordinate system defined by equation (6) (Fig. 3).

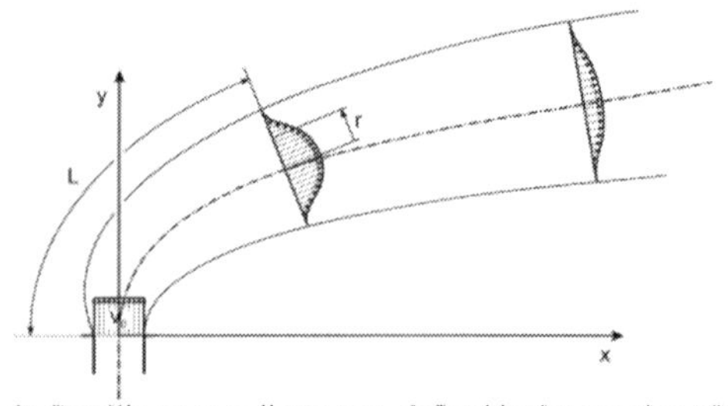


**Fig. 2:** Curvilinear coordinate syst. defined in the equation (6)

The results of the given model are schematically shown in Fig. 3.

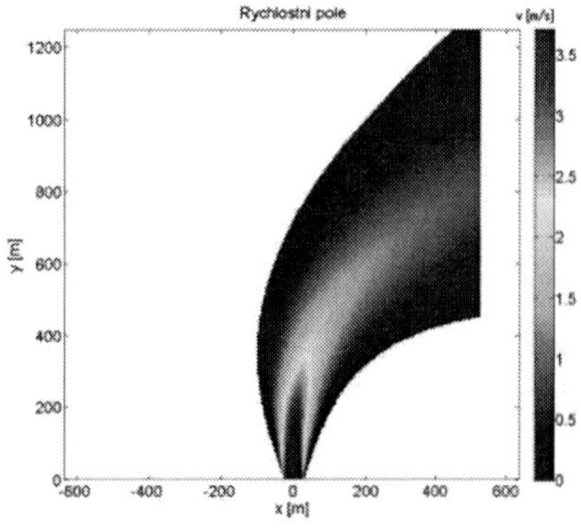


**Fig. 3:** Evaluated velocity fields

## 3 Concentration field

For the concentration field of two gases of same density and known velocity field shown in the preceding chapter, [9] recommends use of the relationships on the analogy between the dimensionless velocity and concentration field in the form:

$$c_{i,s} = 0.96 / [(a \cdot L) / r_0 + 0.26] \quad (7)$$

If the velocity of the ambient air shall be zero and the individual specific humidities of the concerned media shall be equal to $x_1$ for the ambient air, respectively, $x_2$ for the effluent air from the cooling tower, we can, due to the fact that equation (7) is written for molar concentration, directly derive the equation for the specific air humidity in the area influenced by the flow stream that is defined by the equations (1) to (6). The resultant specific humidity shall be given by the equation

$$x = x_1 + (x_s - x_1) \cdot (v / v_s)^{0.5} \quad (8)$$

where we determine the value $x_s$ from the equation

$$x_s = 0.96 \cdot (x_2 - x_1) / [(a \cdot L) / r_0 + 0.26] + x_1 \quad (9)$$

The specific humidity fields calculated on the basis of the velocity field and equations (8) and (9) is given in the graph in Fig. 4 for the case with side wide wind.

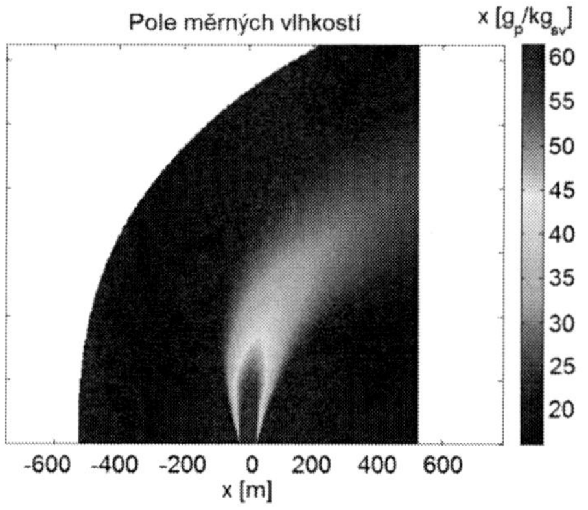


**Fig. 4:** Specific humidity field

The acquired velocity and in relation to it, also the humidity field in the area above the cooling tower makes it possible in the following paragraphs to make an elementary determination of the assumed shape of the plume, which is directly tied to the specific humidity values. It is clear that if the specific humidity at a given point is higher than the one that would correspond to the supersaturated vapour pressure at a given pressure and temperature, water vapour condenses in the space above the cooling tower in the form of droplets and thus creates a visible plume.

## 4 Plume

In order to be able to use the specific humidity field to determine the condensed water volume, it is first necessary to know either the temperature or enthalpy at the given point. In publication [9], practically the same equation is given for the temperature as in the equation (9) for concentration. Even here, it is however necessary to realise that unlike the concentrations, which are expressed as molar and equation (9), it applies for them that the temperature variable cannot be valid. This is due to the fact that in the entire process, a change occurs in state of matter and part of the heat energy, which is necessary for such change, is saturated just by the local drop in temperature. Due to the above-stated, for description of the local humid air condition, it is thus necessary to determine the enthalpy value.

Just line the corresponding specific humidity, we can also determine the enthalpy as such from the Mollier $h_{1+x}$-x diagram using the lever rule; clearly, the analogous equations in equations (8) and (9) must apply to it. By using enthalpy in place of the specific humidity, we thus get

$$h = h_1 + (h_s - h_1) \cdot (v / v_s)^{0.5} \quad (8)$$

where we determine the value $\boldsymbol{h_s}$ from the equation

$$h_s = 0.96 \cdot (h_2 - h_1) / [(a \cdot L) / r_0 + 0.26] + h_1 \quad (9)$$

If it is now possible on the basis of equations (6) to (9) in the monitored space for all its points to determine the image in the Mollier diagram $h_{1+x}$-x, it is possible to finally calculate the local temperature value, but simultaneously also the specific humidity corresponding to the satration curve. This concerns points with the same specific enthalpy and different specific humidity. In areas where the difference of the two stated humidity values is positive, it is thus possible to assume the presence of the steam plume (see Fig. 5).

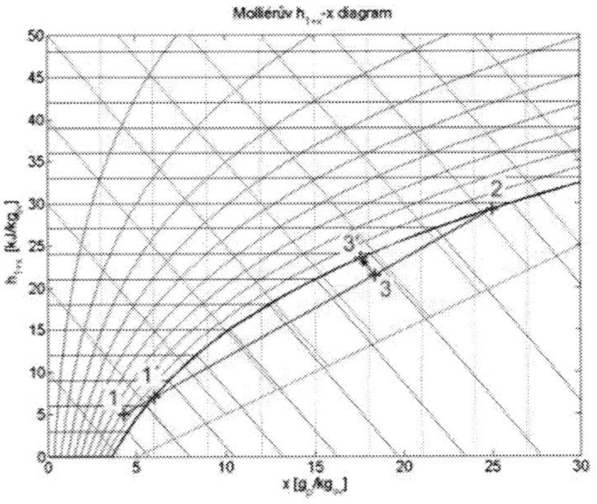


**Fig. 5:** The plume shall appear at the given point if such point shall lie on the connecting lime of points 1´and 2.

The results of the aforementioned procedure are given in Fig. 6 for the normally achieved values of the individual variables.

Here, it is necessary to mention the fact that this case concerns a maximalist estimate. For evaluation of the shape and size of the plume, its visibility is relatively important. It is always the result of a combination of a number microscopic droplets in a unit volume, their average size and light conditions. It generally applies that the plume shall be visible in case of an adequately large number of particles of adequate size in a unit volume. In this context, it is necessary to realise that it is not true to state that condensation occurs for the points of the connecting line 1´and 2. In fact, it is true that between points 1´and 2 the air is in state of supersaturation even in a situation in which the water evaporates from the surface of the droplets present. For this reason, the overall volume of plume evaluated from Fig. 6 shall be the maximum possible volume.

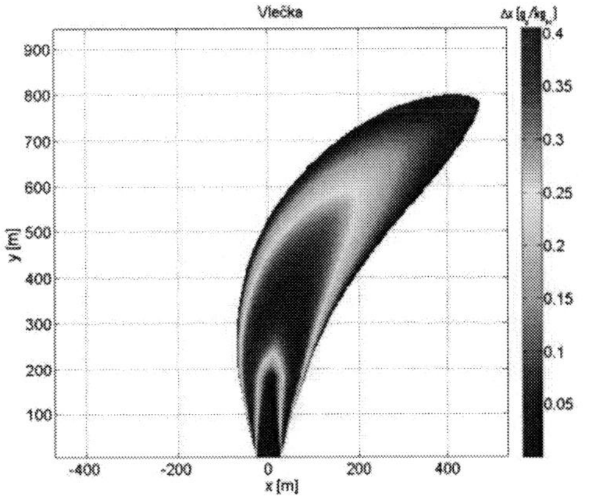


**Fig. 6:** Plume presence area.

Due to the above-stated, the given plume model can be supplemented, for instance, with the visibility condition stated above and analysed further. It is then possible to deduce many interesting facts, for instance, the fact that the overall volume, which the plume covers in the space depends only on the output diameter of the cooling tower and the states of the two mixed air masses (effluent air from the cooling tower and ambient air). In detail, the entire deduction is given, for instance, in [10]; for this reason, we limit ourselves in this case only to statement of the resultant relationship (10)

$$V_{plume} = C \cdot [(x_2 - x_1') / (x_1' - x_1)]^3 \cdot D^3 \quad (10)$$

It is just the constant $\boldsymbol{C}$ in the given equation (10) that is influenced by the above-stated facts relating to the visibility of the plume. It is possible to demonstrate that in an idealised case, its value should be equal to $\boldsymbol{C = 2.5}$.

## 5 Verification experiment

In order to be able to assess whether and eventually to what extent the results of the above-stated model correspond to reality, a simple experiment was designed. Currently, relevant experimental data are not available in professional literature. The only available publication, which discusses the given issue (8) refers to the results obtained from experiments in the USA in the seventies of the last century. The second possibility would thus be to compare the results to some other computational procedure, e.g. (11); however, a risk exists here consisting in the unknown precision of such a model.

The experiment is based on use of simple visualisation techniques together with the ambient conditions data (temperature, pressure, humidity and wind direction and velocity) and the data on the effluent air flow from the cooling tower. These values cannot be measured directly in a simple way. For this reason, an experiment is currently being designed in which the parameters describing the thermodynamics of the condition of the cooling tower shall be monitored (flow rate and input and output temperature of the water), from which by means of known tower characteristics, the condition of the air at its outlet shall be calculated additionally.

**Fig. 7:** The device for experimental determination of the size of the plume in varying atmospheric conditions

The device itself (Fig. 7) is based on a microprocessor with ARM core by ST Microelectronics and is used to record the wind speed and direction data from the cup anemometer (Fig. 8); it is further possible to take a photograph or make a short video recording using the external camera. After installation of an external sensor, we additionally get atmospheric press, temperature and humidity data.

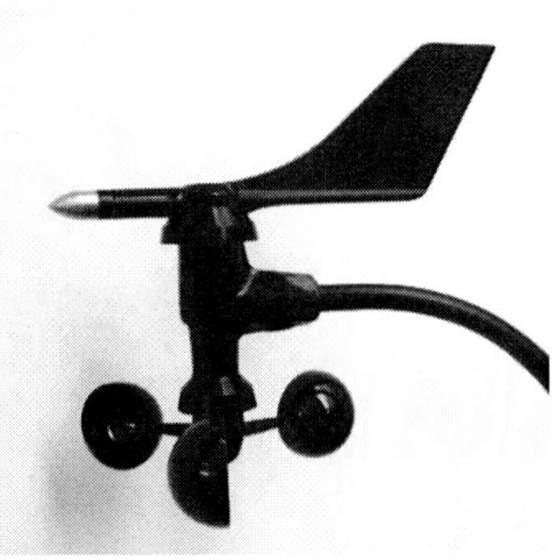

**Fig. 8:** Cup anemometer

The device was designed for easy operation and reliability – the camera is selected as a generally available product for amateur photography with a high resolution, the camera power supply is solved by means of an in-built Li-Po accumulator and the data is stored on a standard microSD card. A similar solution is also applied to the design device that is powered by a single-cell Li-Po accumulator; the record is stored on a microSD card with FAT16 file system that can be processed easily on a PC. A record is made within the required time intervals. To save power, the processor is kept in standby mode during idle time and is activated only upon measurement query. The microprocessor contains a real-time circuit with external battery back-up that keeps the clock running even when the accumulator is disconnected. The time can be set on the real time circuit via standard serial line RS-232 using a set AT commands; in future, it will be possible to equip the device with a DCF-77 precision time receiver for automatic time setting and adjustment.

## References


1. D. G. Kröger: *Air-cooled heat exchangers and cooling towers: thermal-flow performance evaluation and design, Volume I*, Penn Well Corporation, Oklahoma (2004)
2. D. G. Kröger: *Air-cooled heat exchanger and cooling towers: thermal-flow performance evaluation and design, Volume II*, Penn Well Corporation, Oklahoma (2004)
3. F. Merkel: *Verdunstungskühlung*, VDI Zeitchrift Deustscher Ingenieure, Vol. 70, pp. 123-128, (1925)
4. J. C. Kloppers, D. G. Kröger: *Cooling tower performnnce: A critical evaluation of the Merkel assumptions*, International Journal of Thermal Science, Vol. 44, (2005)
5. Hyhlík, T., *Quasi One-Dimensional Model of Natural Draft Wet-Cooling Tower Flow*, Heat and Mass Transfer (2015) EPJ Web of Conferences, 92, art. no. 02027,
6. Hyhlík, T. *Concept of CFD model of natural draft wet-cooling tower flow* (2014) EPJ Web of Conferences, 67, art. no. 02044,
7. W. E. Dunn, L. Coke, A. J. Policastro: *Cooling-tower-plume prediction code,* Engineering and Environmental Science (1987), Illinois, US
8. A. Kouchi, R. Ohba, K. Okabayashi, T. Nonaka: *Prediction technique for visible plume from cooling tower*, Mitsubishi Heavy Industries Ltd., Technical Review, Vol. 36, No. 3, (1999)
9. M. Rédr, M. Příhoda: *Základy tepelné techniky*, SNTL, Praha (1991)
10. Čížek, J. - Nožička, J.: *Cooling tower plume*, 35. Stretnutie kateder mechaniky tekutín a termomechaniky . AIP Conference Proceedings, (to be published)
11. Hyhlík, T.: *Mathematical model of unsaturated and supersaturated moist air flow based on two pasive scalar transport equation*, 35. Stretnutie kateder mechaniky tekutín a termomechaniky . AIP Conference Proceedings, (to be published)

# Active control of continuous air jet with bifurcated synthetic jets

*Petra* Dančová[1], *Tomáš* Vít[1], *Darina* Jašíková[2], *Jan* Novosád[1]

[1]Department of Power Engineering Equipment, Technical University of Liberec, Studentská 2, 46117 Liberec 1, Czech Republic
[2]Institute for Nanomaterials, Advanced Technology and Innovation, Technical University of Liberec, Studentská 2, 46117 Liberec 1, Czech Republic

**Abstract:** The synthetic jets (SJs) have many significant applications and the number of applications is increasing all the time. In this research the main focus is on the primary flow control which can be used effectively for the heat transfer increasing. This paper deals with the experimental research of the effect of two SJs worked in the bifurcated mode used for control of an axisymmetric air jet. First, the control synthetic jets were measured alone. After an adjustment, the primary axisymmetric jet was added in to the system. For comparison, the primary flow without synthetic jets control was also measured. All experiments were performed using PIV method whereby the synchronization between synthetic jets and PIV system was necessary to do.
Keywords: Synthetic jet, primary flow, unforced flow, flow control, pulse modulation, PIV, synchronization.

## 1 Introduction

Flow field control is a part of fluid mechanics which is intensively studied in recent time. Based on the need to supply the energy, control can be active or passive, [1]. In case of active control, it is necessary to supply energy to the system. As an example can be the utilization of synthetic jet, see [2] or [3]. Passive control can be based e.g. on a change in geometry of a body or its surface roughness, [4].

Flow field control finds the applications in external and internal aerodynamics - with the appropriate option of the parameters, the forces acting on the bluff body can be significantly affected.

Active control is applicable e.g. in aviation, where another change of wing geometry does not achieve significant improvement of aerodynamic properties. In work [5], the synthetic jet was used for controlling of boundary layer separation on a wing of Cessna 182 and therefore the lift force increased. Further applications of active controlling with synthetic jet in can be found on airfoils [6] or propellers of helicopters; some of these examples are inscribed as a "virtual shaping effect" of an airfoil [7]. In automotive industry, active control with synthetic jet used to reduce drag force and thereby decrease fuel consumption was firstly presented concept incorporated in the Renault Altica sports car in 2006 [8]. Reduction of drag force acting on the car model with use of synthetic jet array is described in [9]. Another example of active control with use of synthetic jet array is in work [10].

Control of a primary flow with the synthetic jets used for the heat/mass transfer distribution is studied in [11]. Works [12, 13 and 14] dealt with active control of the primary laminal channel flow with synthetic jet array.

Works [15 and 16] describe control of a flow in an ejector with multiple synthetic jets.

### 1.1 Synthetic jet

The synthetic jets (SJs) are generated by the periodic motion of an actuator oscillating membrane. The SJs are synthesized by the interactions within a train of vortex rings or counter-rotating vortex pairs in axis-symmetric or two-dimensional geometry; see Smith and Glezer [17]. Vortex rings are formed at the lip of the orifice (see Fig. 1). These rings move in $x$ direction with a velocity, which must be high enough to prevent interaction with suction in the orifice. It was observed that an SJ far enough from the orifice has a character of a conventional steady jet. This is caused by the development and dissipation of vortexes.

The equipment for the SJs can have various designs, but the main mechanism and principle is primarily the same: sealed cavity equipped with an orifice at one end and the periodically moving membrane at the second end.

Figure 2 shows details of the working cycle of a SJ. The working cycle starts with a membrane motion from its zero position (position 0a) in $-x$ direction. This motion causes fluid movement into an actuator cavity. If the diaphragm deviation is maximum (pos. 1), the fluid is extruded from the orifice of an actuator. The highest velocity of extrusion is when the diaphragm experiences zero deviation (pos. 0b). Then the diaphragm moves in $+x$ direction to pos. 2. When the diaphragm reaches position 2, the fluid is sucked in again, and the cycle is repeated.

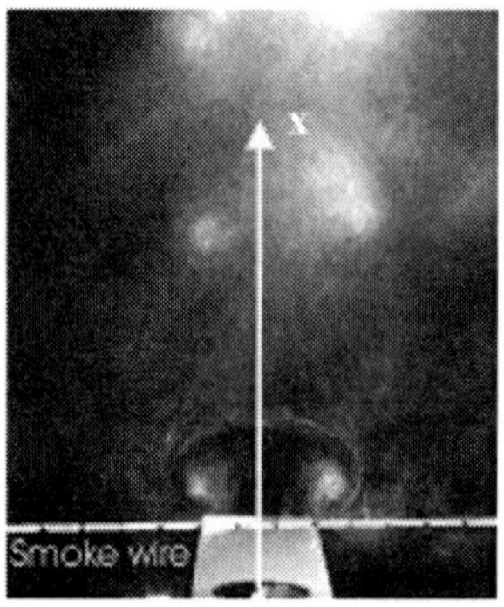


**Fig. 1.** Creation of vortex rings by fluid extrusion from the cavity of an actuator, results of "smoke wire" visualization [20].

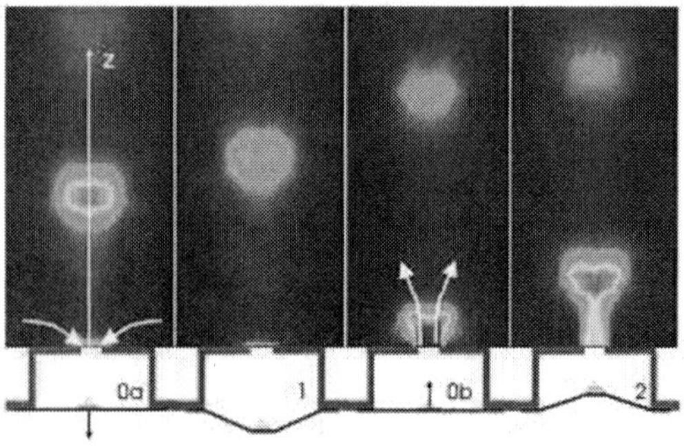


**Fig. 2.** One of the working cycles of a SJ.
Experimental results measured by CTA with an X-wire probe, colorized according to the velocity magnitude.
2→0a→1 suction; 1→0b→2 extrusion, [19].

# 2 Experiments

## 2.1 Experimental setup and technique

This paper continues research in [11]. In that work the used experimental setup consisted of the primary, continuous axisymmetric nozzle equipped with four synthetic jet (SJ) actuators placed around its exit orifice, visibly in Fig. 3(a). The synthetic jet actuators worked in two modes: the bifurcated mode and helical one. SJ actuators were fed with sinusoidal signal with frequency of 41, 82 and 164 Hz.

In this paper, the experimental setup described in [11] is used. In our research, the primary flow is controlled with two SJ actuators working also in a bifurcated mode, but the actuators were fed with modulated sinusoidal signal (signal parameters are described in chapter 3).

The bifurcated mode means that only two SJ actuators situated in a plane of investigation (right and left) worked in antiphase, another two actuators (front and rear ones) were turned off, see Fig. 3(b).

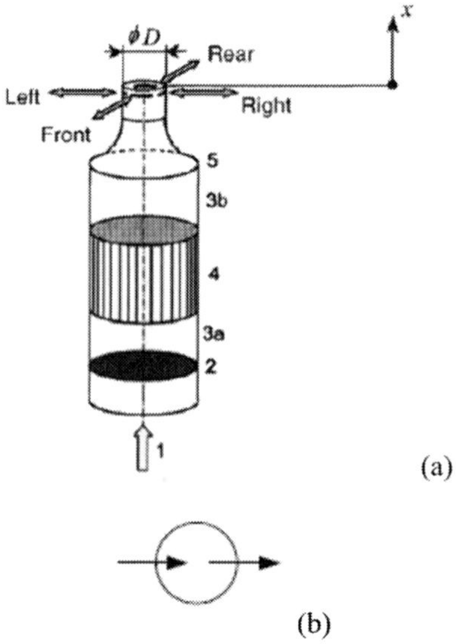


**Fig. 3.** Schematic view of the tested configuration.
(a) The main, quadrant-contoured nozzle generating the main flow, with the four SJ actuators acting at its exit;
1: primary air flow supply, 2: grid, 3a, 3b: settling chambers, 4: honeycomb, 5: contoured contraction, [11].
(b) The bifurcated mode.

The experiments were performed in air as the working fluid. The continuous nozzle was supplied with air from the compressor and requested volume flow set with system of valves. The time-mean orifice velocity $U$ was evaluated from the volume flow measured by the flow meter in the supply pipe. As visible in Fig. 3(a), from the supply pipe air entered the nozzle then passes the grid (position 2 in this figure, and setting chambers (pos. 3a, 3b), among which is inserted a honeycomb (4). The exit orifice has a shape of short cylinder with diameter $D = 10$ mm and length of 13 mm. As is written above, this exit orifice is equipped with four SJ actuators located around its circumference. Loudspeakers Monacor SP-7/4S were used as a moving membrane for a SJ actuator and were driven with the modulated sinusoidal signal. The SJ exhausting orifices are slots with dimensions 1.1×6 mm each in perpendicular direction to the axis of the continual nozzle.

The experiments were performed with the particle image velocimetry method (PIV). Experiments were carried out in a Plexiglas box with dimensions of (0.5×0.5×1) m. The olive oil particles with diameters of (2-4) μm generated from the Scitec aerosol generator were used for flow saturation. These particles were illuminated by a double pulse laser (New Wave Gemini Nd:Yag laser) with a maximum 100 mJ per 10 ns pulse at a repetition rate of 2 × 3 Hz. The typical delay time between two pulses was 70 μs or 400 μs (for SJs in operation, or unforced flow respectively).

The laser beam was expanded by a cylindrical lens into a light sheet of approx. 1 mm thick. The image pairs were acquired using the HiSense NEO camera (DANTEC) with 5 Mpx resolutions.

The resulting vector maps were averaged over 100 PIV double images. The velocity vectors were determined by cross-correlation using interrogation windows of (32 x 32) pixels at a 50 % overlap. Data processing was done using DynamicStudio v.4.10 commercial software (DANTEC).

*Laser and system synchronization*: In our experiments, a unit named Timer box and DynamicStudio v.4.10 software were used. All the control boards of the lasers, cameras, and timers of the input and output signals were placed directly into the main computer. DynamicStudio is a control and processing program for the timer box unit.

In this case, the PIV system was set as a master and the synthetic jet actuators as a slave. The signal generator received a TTL pulse from the timer box, from the PIV system respectively. This TTL pulse started the modulated sinusoidal signal (prescribed in the pulse generator) which went then via amplifier to the SJs actuators.

The delay inside the electronic synchronization devices does not exceed 3μs.

Figure 4 shows the basic principle of synchronization. Recorded phase $\varphi(°)$ of the SJs (visible in Fig. 4) can be calculated as

$$\varphi = \left(\frac{T_L - d}{T_{mod}} - n\right)360 \qquad (1)$$

where $n$ is an integer indicating the number of periods $T_{mod}$ during the time $T_L$.

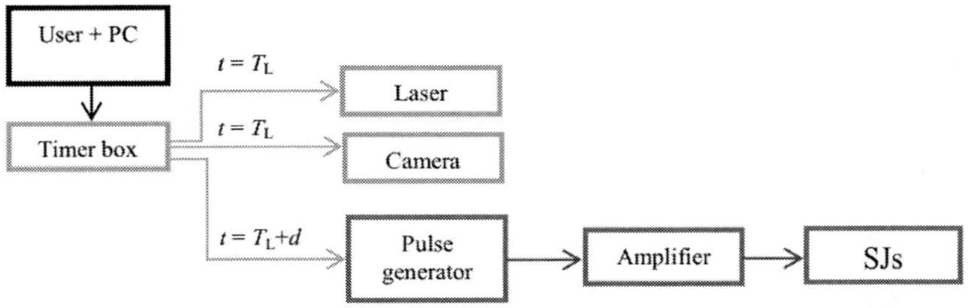


**Fig. 4.** Schematic view of the basic synchronization for PIV measurements.
$f_L$ = 3Hz is a working frequency of a laser ($T_L = 1/f_L$), $d$ is a signal delay from the laser pulse (prescribed by the experimenter to shift throughout the SJs (modulated) period).

## 2.2 Problem parameterization

The primary, continuous axisymmetric jet is characterized by the Reynolds number $\mathrm{Re} = \frac{U_N D}{\nu}$, where $U_N = (2/3) \cdot U_{N\_max}$ (for simplification calculated as two-dimensional case).

The synthetic jet is defined with the time-mean orifice velocity $U_0 = \frac{1}{T}\int_0^{T_E} u_0(t)dt$, where $T$ is the time period, i.e. $T = 1/f$ and $f$ is the frequency, $T_E$ means the extrusion time, $u(t)$ is the periodical axial orifice velocity. The characteristic length scale of SJ is the "stroke length" $L_0 = U_0 \cdot T$.

The Reynolds number of the SJ is defined as $Re_{SJ} = U_0(2h)/\upsilon$, where $h$ = 1.1 mm and it is the width of the SJ outlet slot.

The strength of the control SJs can be quantified relative to the primary axisymmetric flow in terms of the ratios of velocities $c_U$, flow rates $c_Q$, and momentum $c_M$ as $c_U = \frac{1}{n}\sum_{i=1}^{n}\left(\frac{U_{0i}}{U_N}\right)$, $c_Q = \frac{A_{SJ}}{A}\sum_{i=1}^{n}\left(\frac{U_{0i}}{U_N}\right)$, and $c_M = nM_{SJ}/(\rho A U_N^2)$, where $A_{SJ}$ is the cross-section area of the SJ actuator slot (1.1×6.0 mm), $n$ is the number of control SJs ($n$ = 2 in current study), $A = \pi D^2/4$ is the cross-section area of the main axisymmetric nozzle, $\rho$ is the working fluid density (air in our case), $M_{SJ}$ is the time-averaged SJ momentum flux during the extrusion stroke averaged over the entire period. Assuming the "plug flow model", i.e. the uniform profile of the instantaneous velocity $u_0(r,t) = u_0(t)$ and the sinusoidal waveform in time $u_0(t) = U_{max}\sin(2\pi f t)$, where $U_{max}$ is the velocity amplitude, the time-averaged SJ momentum flux can be expressed as $M_{SJ} = \frac{1}{T}\rho A_{SJ}\int_0^{T_E} u^2(t)dt$. Then the form of $c_M$ can be derived as:

$$c_M = n\left(\frac{\pi}{2}\right)^2\left(\frac{A_{SJ}}{A}\right)\left(\frac{U_0}{U_N}\right)^2. \qquad (2)$$

### *2.2.1 Velocity decomposition*

The phase averaging was used for the velocity analysis. The definition of the velocity components in the stream-wise direction is the following:

$$u(t) = U + U_P\,(t/T) + u'(t/T), \qquad (3)$$

where $u(t)$ is the instantaneous velocity, $U$, $U_P$, and $u'$ are the time-mean, periodic (coherent), and fluctuation (incoherent, random) components, respectively, and $t/T$ indicates the phase during the cycle.

A principle of the conditional sampling was used for the velocity decomposition of the PIV data, i.e. sampling has to always occur at the same phase in a process. The sampling starts with the beginning of the period given by a pulse signal from the PIV system (Timer box). In accordance with the pulse signal, the synchronization is performed. The image pairs were acquired at ten equally spaced intervals throughout the actuation cycle, $T_{mod}$ respectively.

## 3 Results and discussion

### 3.1 Unforced flow

The primary flow from the axisymmetric nozzle without SJs control was set at Re = 1600; the time-mean velocity through the nozzle cross-section is calculated as approx. $U_N = 2.46$ m/s.

Fig. 5 shows the fog visualization of the unforced jet.

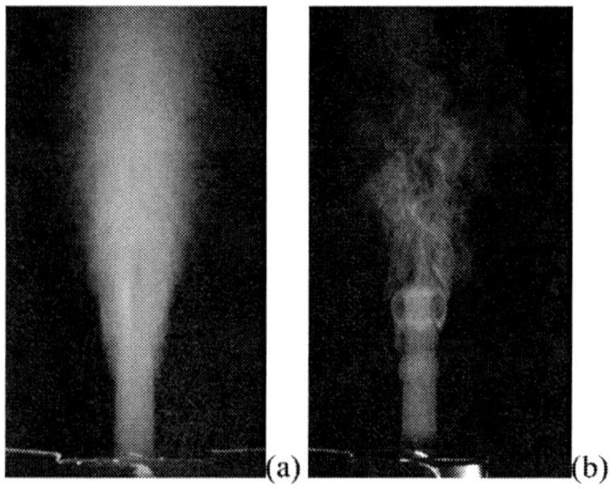


**Fig. 5.** Fog visualization of the unforced jet, (a) averaged, (b) instantaneous (pictures taken by Z. Trávníček).

For the PIV experiments, the delay time between two laser pulses was 400 μs. Decreasing of the primary flow time-mean velocity stream-vise is visible in Fig. 6(a). Exponent $n = -1.08$ confirms the continual (steady) jet; see e.g. Schlichting and Gersten [18]. Fig. 6(b) shows the velocity profiles of the primary flow in distance $x/D = 1$, 2, 5, 7, and 10. As is visible, in the distance nearest to the orifice, the profile has a character of a "top hat profile". Axial velocity decreasing and mixing of the air from the outer part of a stream with the ambient air start in distance $x/D = (3\text{-}7)$. Fully developed flow starts from the distance approx. $x/D > 7$ and velocity profiles have similar shape.

### 3.2 Synthetic jets development

The primary flow was controlled with two SJ actuators worked in the bifurcated mode (Fig. 3(b)). The supply signal of these actuators from the signal generator was set with following parameters: carrier sinusoidal signal with frequency of $f = 850$ Hz, modulation frequency $f_{mod} = 21.25$ Hz, number of waves $N = 6$, voltage peak-to-peak of the supplying signal, which was leaving the amplifier, was set as 9.60 V. Fig. 7 represents the SJs driving signal modulation. The position of $t/T_{mod}$ (in range from 0 to 0.9) is also marked there.

In case of working SJs, the delay time between two pulses was set as 70 μs. The SJs without the primary flow in $t/T_{mod} = 0$, 0.1, 0.4, and 0.9 is shown in the form of phase-averaged velocity magnitude contours in discrete drawing style in Fig. 8 and in the form of phase-averaged velocity magnitude profiles in Fig. 9.

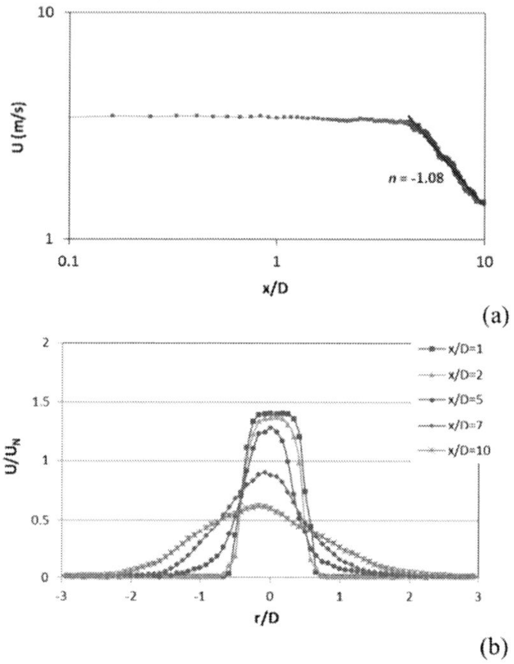


**Fig. 6.** (a) Dependence of the primary flow time-mean velocity on the distance from the orifice (logarithmic scale), (b) velocity profiles, Re = 1600.

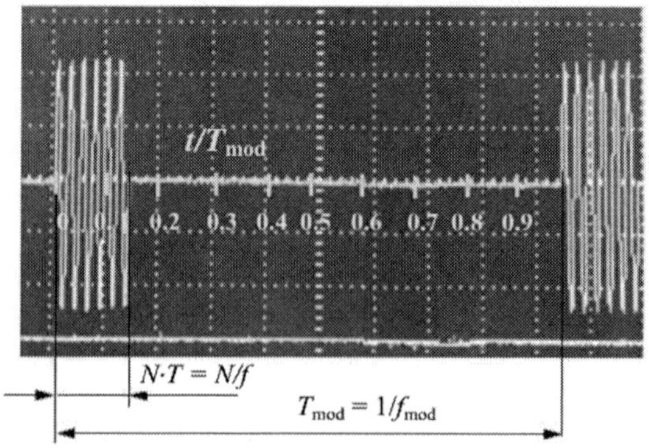


**Fig. 7.** Modulated driving signal for SJs.

Fig. 8 shows the bending of the jet in range of 13-18 degrees. In $t/T_{mod}$ = (0 and 0.1), this bending is caused because the images were taken at the instant of extrusion from the right SJ actuator and suction to the left one. In the interval $N \cdot T$ (visible in Fig. 7) each SJ actuator works as with non-modulated sine wave supply. In $N \cdot T$, the time of extrusion as $T_E = T/2$ can be supposed. Outside the interval $N \cdot T$, the behavior the SJ actuators moving membranes is different from typical operation of actuator membranes, described e.g. in [19]. The movement of the membrane is stopped here and there is no suction to the orifice. Vortex rings formed at the lip of the orifice during the actuation cycle are moving in $x$ direction without the interaction with suction into the orifice. This vortex movement, looks like a phase of extrusion, takes in interval $t/T_{mod} = (0.3 - 0.9)$.

On these figures is also visible the maximum extrusion – vortex velocity in $t/T_{mod} = 0.4$. Suction can be considered in $t/T_{mod} = 0.1$.

Fig. 10 shows the SJ phase averaged velocity magnitude profiles at $t/T_{mod} = 0.1$ and 0.4 at different distances from the orifice. As the distance from the orifice increased, the velocity decreased and velocity profiles widened. The bending of SJs is demonstrated with profile's peak shift to the negative values of $r/D$.

### 3.3 Axisymmetric jet under bifurcated mode excitation

Fig. 11 – 14 show the results of the experiments of the primary flow control with activated SJs worked in the bifurcated mode. These experiments were set at the same conditions as the both previous cases: Re of the primary flow was set as 1600, and SJs were fed with the sinusoidal modulated signal described in the previously text.

Fig. 11 demonstrates the development of controlled axisymmetric jet during the SJs actuation period in form of velocity magnitude. The vortex rings causes in the primary flow the effect called "vena contra". The action of vortex ring leads to the local narrowing of the primary flow and increasing of the velocity inside the ring. Compared with the unforced jet, the velocity inside the ring increased by approx. 20% (in case of $t/T_{mod} = 0.4$). In comparison with [11], where the SJs actuation caused periodic zig-zag vectoring of the primary jet towards left and right direction, in our case due to the actuation frequency and its modulation no or only small bifurcation is manifested. The change of the jet shape and formation and development the vortexes are visible.

Fig. 12 shows phase-averaged velocity magnitude profiles during the actuation cycle in different distances from the orifice. The same width of the profiles is visible in distance $x/D = 1$. The significant jet narrowing occurs in distance $x/D = (2.5\text{-}6)$ for $t/T_{mod} = 0.4$.

Phase averaged velocity magnitude profiles in $t/T_{mod} = 0.1$ and 0.4 at different distances from the orifice are shown in Fig. 13. The profile in the distance $x/D = 5$ is slightly shifted to the left because of the vortex which leans the jet to the negative values of $r/D$.

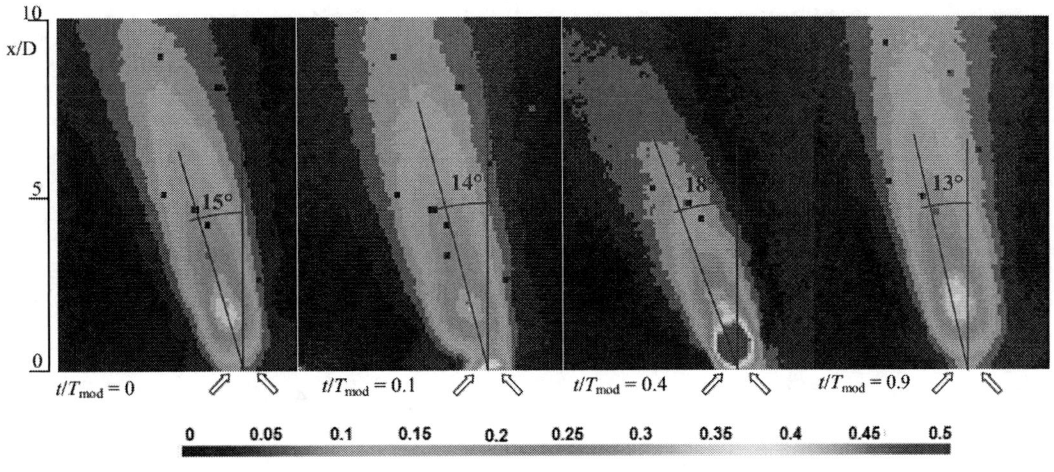


**Fig. 8.** SJs without the primary flow: phase-averaged velocity magnitude, PIV experiments, discrete drawing style.

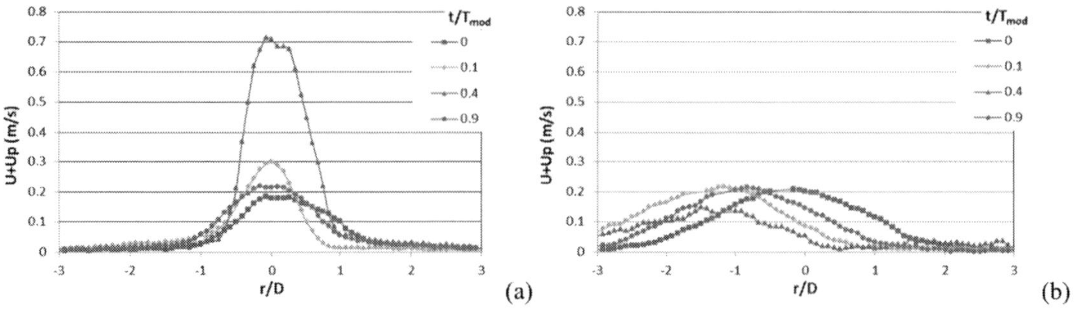


**Fig. 9.** SJs without the primary flow: phase-averaged velocity magnitude profiles during the actuation cycle, (a) $x/D = 1$, (b) $x/D = 5$.

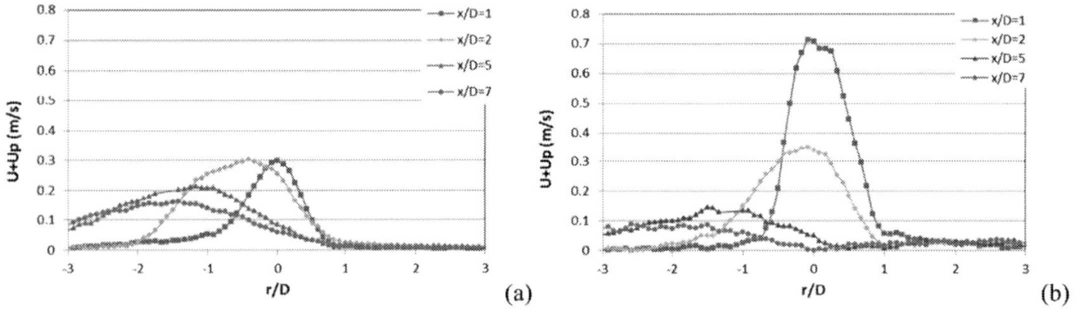


**Fig. 10.** SJs without the primary flow: phase-averaged velocity magnitude profiles at the instant of (a) maximum suction velocity ($t/T_{mod}$ =0.1) and (b) maximum extrusion velocity ($t/T_{mod}$ = 0.4).

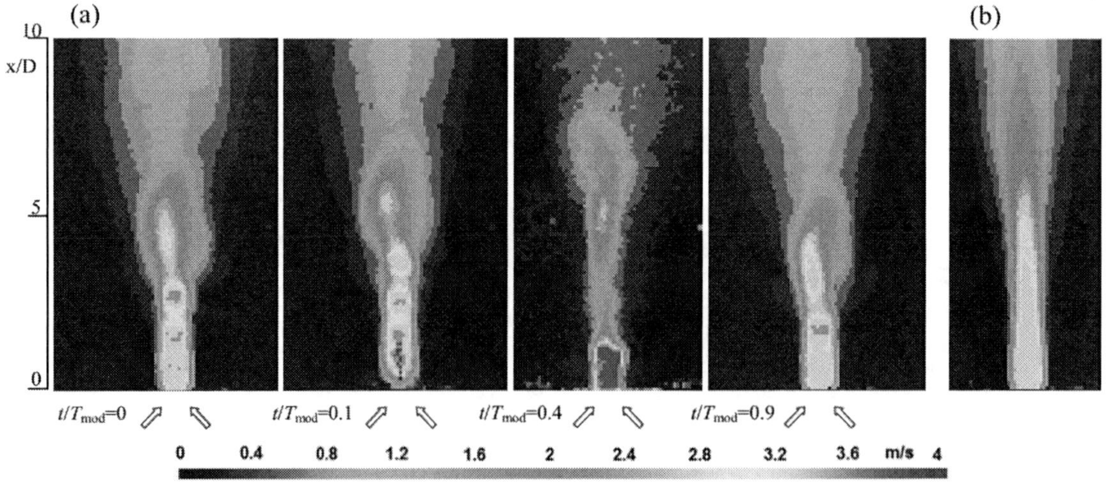


**Fig. 11.** (a) SJs with the primary flow: phase-averaged velocity magnitude, (b) unforced primary axisymmetric flow. PIV experiments, discrete drawing style.

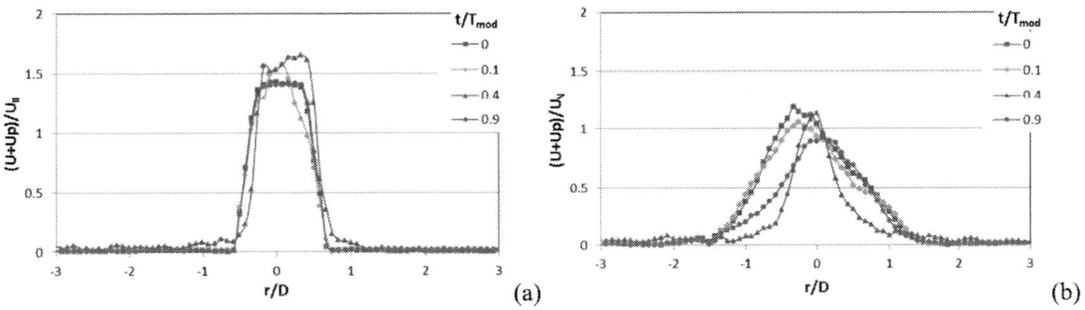

**Fig. 12.** SJs with the primary flow: phase-averaged velocity magnitude profiles during the actuation cycle, (a) $x/D = 1$, (b) $x/D = 5$.

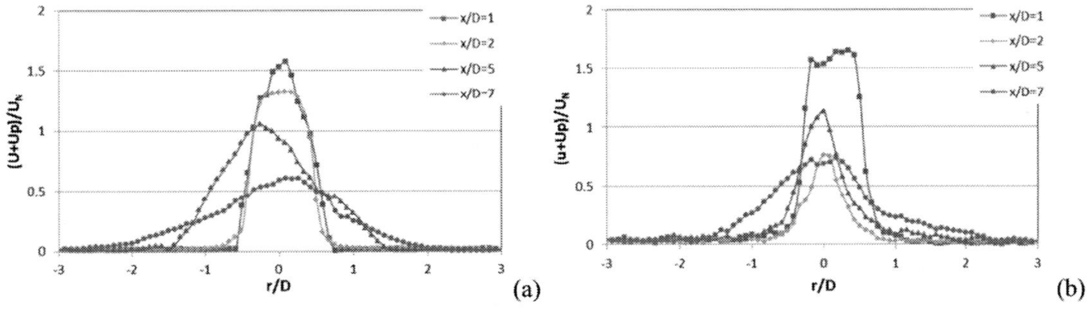

**Fig. 13.** SJs with the primary flow: phase-averaged velocity magnitude profiles at the instant of (a) maximum suction velocity ($t/T_{mod}$=0.1) and (b) maximum extrusion velocity ($t/T_{mod}$ = 0.4).

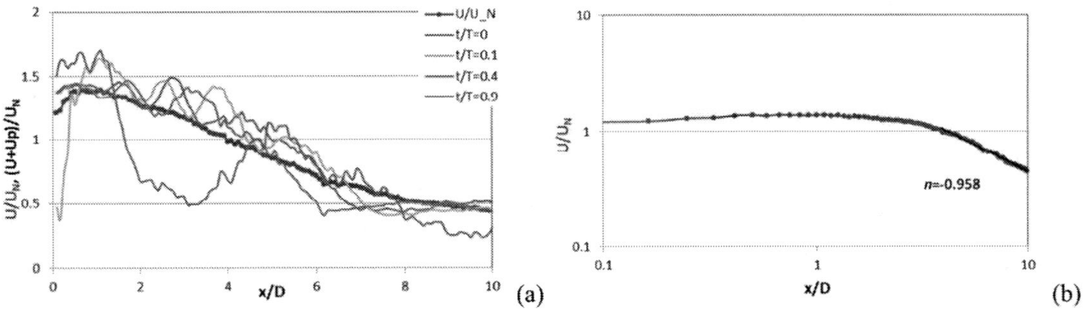


**Fig. 14.** SJs with the primary flow: (a) dependence of the phase-averaged velocity magnitude at different instances in the period and the time-mean velocity on the distance from the nozzle orifice, (b) dependence of the time-mean velocity on the distance from the nozzle orifice on a logarithmic scale.

Fig. 14(a) demonstrates the dependence of the phase-averaged velocity $U_P$ in a different instant of the period and the time-mean velocity $U$ on the distance from the nozzle orifice (the time-mean velocity of the controlled flow was calculated by averaging of all phases during the SJs actuation period $T_{mod}$). There was a visible decrease in velocity as the distance increased along $x$ direction. At a distance of $x/D = 8$, the flow oscillation practically disappeared except $t/T_{mod} = 0.4$ (maximum extrusion), and the flow can be considered as a steady jet flow. Further downstream from the orifice, the vortex structure disappeared owing to dissipative processes, thereby decreasing the velocity. The dependence of the time-mean velocity on the $x$ in the logarithmic scale (in dimensionless form in Figure 14(b)) shows the slope of the line of controlled flow. The exponent of the velocity decay $U{\sim}x^n$ was $n = -0.958$, which confirms that controlled flow in the distant field approaches with its character to the continual fluid jet. This is caused with part of the period outside interval $N \cdot T$, where the new vortex rings are not generated.

## 4 Conclusions

A round air jet was controlled using two synthetic jets set in the bifurcated mode and located on the circumference of the nozzle exit. The setup was taken from authors of work [11]. The control synthetic jets, the primary axisymmetric jet, and the complex case of the primary jet under the bifurcated mode were studied experimentally by means of PIV method. During the experiments the primary flow was set at Re = 1600, and the synthetic jets actuators were driven with modulated sinusoidal signal. During the experiments it was necessary to synchronize PIV system with SJs actuators.

For the primary flow without affection of SJs, the results correspond with the literature. The affection of SJs enhances the fluid mixing and increases the fluid velocity in the vortex rings, especially at the instant of extrusion. Further downstream the velocity decreases.

## Acknowledgements

We gratefully acknowledge the support of the Grant Agency of the Czech Republic (Project No. 16-16596S).

## References

[1] M. Gad-El-Hak, *Flow control*, Cambridge University Press, New Yourk, 2000.
[2] D.A. Tamburello, M. Amitay, Three-dimensional interactions of a freejet with a perpendicular synthetic jet, *Journal of Turbulence*, Vol. 8, No. 38, 2007.
[3] D.A. Tamburello, M. Amitay, Interaction of a free jet with a perpendicular control jet, *Journal of Turbulence*, Vol. 8, No. 21, pp. 1-27, 2007.
[4] E.L. Houghton, P.W. Carpenter, *Aerodynamics for engineering students*, Butterworth-Heinemann, Great Britain, 2006.
[5] M. Ciuryla et al., Flight control using synthetic jet on a Cessna 182 model, *Journal of Aircraft*, Vol. 44, pp. 642-653, 2007.
[6] A. Nishizava et al.: Toward smart control of separation around a wing – Active separation control system part 2. Proc. 5th Symp. Smart Control of Turbulence, Tokyo, pp. 7-14, 2004.
[7] R. Mittal, P. Rampunggoon, On the virtual aeroshaping effect of synthetic jets. *Phys. Of Fluids*, Vol. 14, pp. 1533-1536, 2002.
[8] Renault Altica: 44MPG Diesel Concept with Active Airflow Management. http://www.greencarcongress.com/2006/02/renault_altica_html.
[9] H. Park et al, Experimental study on synthetic jet array for aerodynamic drag reduction of a simplified car, *Journal of Mechanical Science and Technology*, Vol. 27, pp. 3721-3731, 2013.
[10] M. Amitay et al., Separation control in duct flows, *Journal of Aircraft*, Vol. 39, pp. 616-620, 2002.
[11] Z. Trávníček et al., Axisymmetric impinging jet excited by a synthetic jet system, *International Journal of Heat and Mass Transfer*, Vol. 55, 2012, pp. 1279-1290.
[12] Z. Trávníček, P.Dančová, J. Kordík, T. Vít, M. Pavelka, Heat and mass transfer caused by a laminar channel flow equipped with a synthetic jet array, *Transaction of the ASME*, Vol. 2, 2010
[13] P. Dančová, Z. Trávníček, T. Vít, Experimental investigation of a synthetic jet array in a laminar channel flow, *EPJ Web of Conferences*, Vol. 45, Article Number: 01002, 2013.
[14] Z. Trávníček, P.Dančová, J.H. Lam, V. Timchenko, J. Reizes, Numerical and experimental studies of a channel flow with multiple circular synthetic jets, *EPJ Web of Conferences*, Vol. 25, Article Number: 01094, 2012.
[15] V. Dvořák, P. Dančová, P. Švarc, Experimental investigation into flow in an ejector with four synthetic jets, *EPJ Web of Conferences*, Vol. 25, Article Number: 02003, 2012.
[16] V. Dvořák, P. Novotný, P. Dančová, D. Jašíková, PIV and CTA measurement of constant area mixing in subsonic air ejector, *EPJ Web of Conferences*, Vol. 45, Article Number: 01003, 2013.
[17] B.L. Smith, A. Glezer, The formation and evolution of synthetic jets, *Phys. Fluids*, Vol. 10, pp. 2281-2297, 1998.
[18] H. Schlichting, K. Gersten, *Boundary-Layer Theory*, Springer-Verlag, Berlin, 2000.
[19] T. Vít, P. Dančová, Z. Trávníček, Syntetizovaný proud (Synthetic jet), *MM Průmyslové spectrum*, Vol. 3, pp. 104-106, 2007 (in Czech).
[20] P.Dančová, Analysis of the "Synthetic Jet", Master thesis, TUL, 2006 (in Czech).

EPJ Web of Conferences **143**, 02017 (2017) DOI: 10.1051/epjconf/201714302017
*EFM 2016*

# Vibrations of hydraulic pump and their solution

Lenka Dobšáková[1,a], Naděžda Nováková[1], Vladimír Habán[1], Martin Hudec[1] and Pavel Jandourek[1]

[1] *Victor Kaplan Dept. of Fluid Engineering, Faculty of Mechanical Engineering, Brno University of Technology, Czech Republic*

**Abstract.** The vibrations of hydraulic pump and connected pipeline system are very problematic and often hardly soluble. The high pressure pulsations of hydraulic pump with the double suction inlet are investigated. For that reason the static pressure and accelerations are measured. The numerical simulations are carried out in order to correlate computed data with experimental ones and assess the main source of vibrations. Consequently the design optimization of the inner hydraulic part of pump is done for further experimental measurements with the variable pump discharge and speed. Several optimizations were considered – guide ribs in pump suction, an optimized diffuser and modified shaft bearings. The influence of optimized designs on the pump vibrations is discussed.

## 1 Introduction

The solution of vibrations generated in the pump and piping system connected to it is often an exacting problem. But to find a source of vibrations takes a lot of time and a lot of strength. There could be many theoretical sources of vibrations. So, the discussed pump should be put to the demanding tests. [1]

An example of a solution of excess vibrations may be the case of pumps in the pumping station of a contract owner. In the pumping station high-pressure pumps with double-side inlet were used. In order to design an appropriate adjustment of the pump, which would reduce the size of vibrations; it was necessary to determine the vibration source.

## 2 First measurement

One of the pumps of the pumping station (called pump B) was subjected to a dynamic test to determine the resources of pressure pulsations. Measurements were performed on the part. Pressure pulsation and absolute vibration were measured on the bearings of the pump and the suction and the discharge piping. Transducer of pressure and transducer of acceleration were used for measuring of the required values.

The entire measuring process was continuously recorded and stored for 1 minute. The record contains four values of pressures and eight values of acceleration. The rotor and an impeller changed by the diameter were removed from the pump after the measuring and an impeller, not changed by diameter, was fitted. After balancing of the impeller and refitting further measurement was taken. When values were evaluated, options were designed that would help improve the behaviour of the whole system and not produce undesirable vibrations. [2]

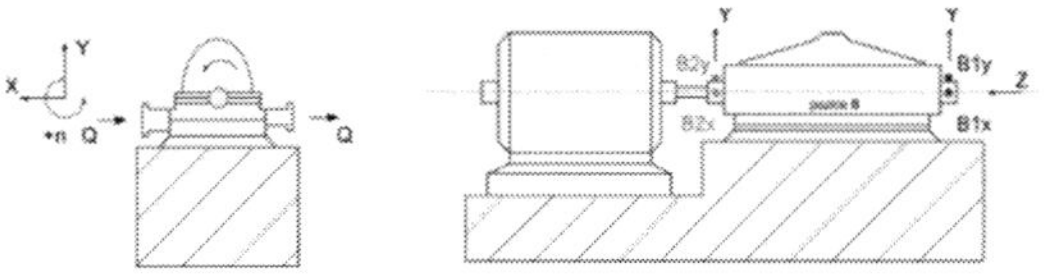


**Figure 1.** Representation of each measured acceleration on the pump B

### 2.1 The pump parameters

As was said before, the measured pump was high-pressure, with double-side inlet. The pump includes the double-side impeller with 7 blades and diameter D2 = 460 mm, one diffuser and one spiral case. Revolutions of the pump are $n = 2980\ \text{min}^{-1}$. The optimum of this pump is $Q_{opt} = 1500\ \text{m}^3\cdot\text{h}$ with head $H_{opt} = 270$ m. Power needed for running in the optimum is $P_{opt} = 1320$ kW.

### 2.2 Evaluation of the first measurement

Effective rate was evaluated from measured values by using accelerometers. First there was performed DFT (discrete Fourier transform). Amplitude of velocity was obtained by integrating from the calculated amplitudes of acceleration. In the frequency space it is dealt with the division of the amplitude of acceleration by angular frequency (*1*).

[a] Corresponding author: lenkarehak@seznam.cz

DOI: 10.1051/epjconf/201714302017

$$v(f) = \frac{a(f)}{2 \cdot \pi \cdot f}. \tag{1}$$

Effective values can be computed from equation (2):

$$v_{ef}^{\,2} = \frac{1}{T} \cdot \int_0^T v^2(t)dt. \tag{2}$$

Components of velocity at a frequency less than 10 Hz and greater than 1 kHz are necessary to filter out to the effective velocity values that were obtained. There was performed DFT, and effective velocity values were calculated from the amplitudes of velocity in selected frequency range.

$$v_{ef}^{\,2} = \sum_{f=f_1}^{f_2} \frac{v^2(f)}{\sqrt{2}}. \tag{3}$$

The equation (3) was also used for calculation of the other frequency bands than according to the standard; it is from 10 Hz to 1 kHz. Especially the equation (3) was used for separation of non-synchronous and synchronous vibration and for monitoring the effective contributions of the speed values in the individual frequency bands. [3]

The effective velocity values on the bearings of the pump B are shown in Figures 2 and 3. It is obvious that velocity values in the horizontal direction reach 4 mm/s during the low flow in the impeller changed by diameter. When the impeller not changed by the diameter was used, the effective velocity values were just below 4 mm/s.

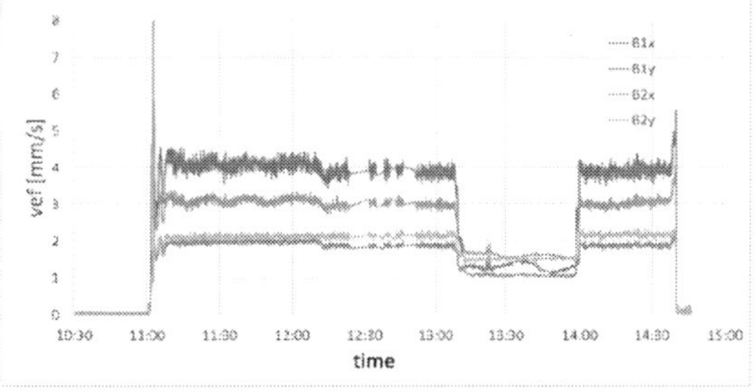


**Figure 2.** Effective velocity values on the bearings of the pump B. The time step is 5 s. The impeller changed by diameter.

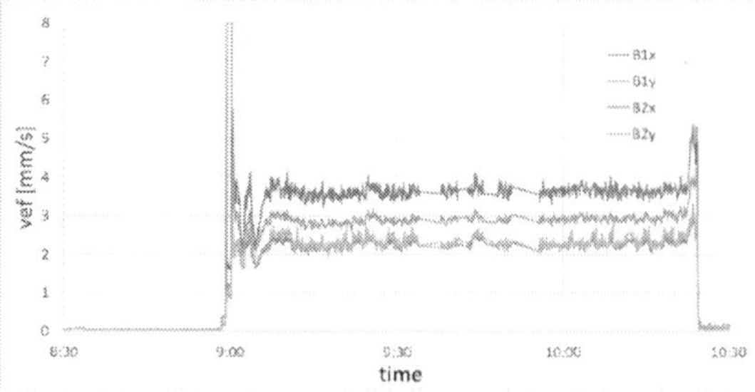


**Figure 3.** Effective velocity values on the bearings of the pump B. The time step is 5s. The impeller not changed by diameter.

Effective velocity values in different nodes of the bearing are shown on graphs in Figures 4 - 7. Values are plotted separately for x-direction and y-direction. They are differentiated by origin as synchronous and asynchronous. All multiples of the speed frequency ± 1.5 Hz are considered to be synchronous. The component of non-synchronous vibration is evident when measuring the impeller changed by the diameter; it is in the half of the nominal flow, which is twice greater than the component of synchronous vibration. This also applies vice versa. Influence of the flow on the component of synchronous vibration is negligible, but it is very important on the component of non-synchronous vibrations. Its value varies up about 8 times by changing the flow rate from 1300 on 2350 $m^3 \cdot h$.

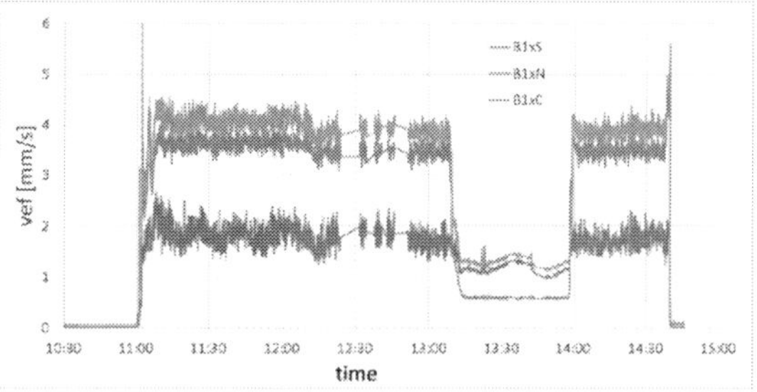


**Figure 4.** Effective velocity values in different nodes. Bearing B1 in the x direction for the impeller changed by diameter.

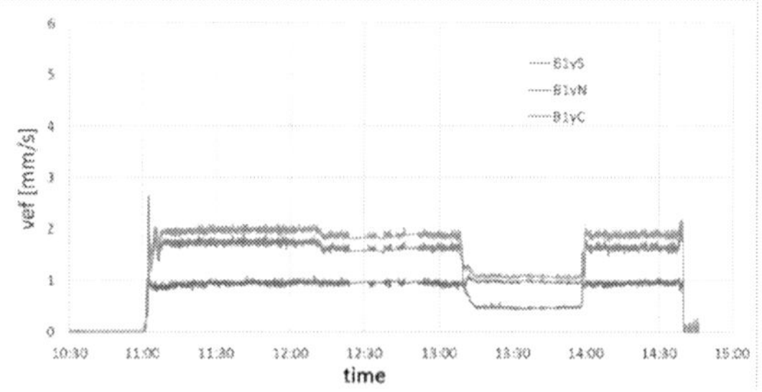


**Figure 5.** Effective velocity values in different nodes. Bearing B1 in the y direction for the impeller changed by diameter.

DOI: 10.1051/epjconf/201714302017

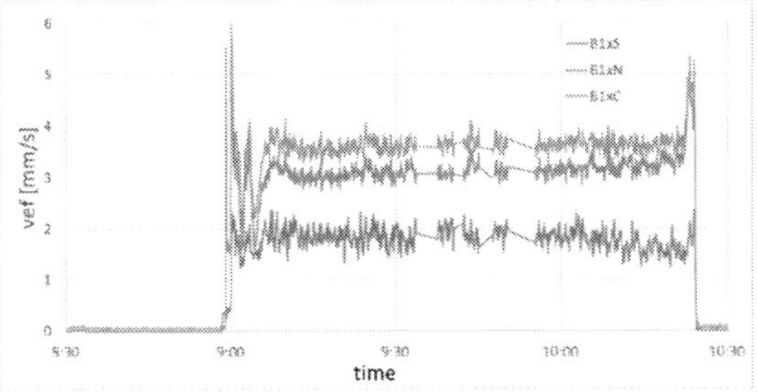


**Figure 6.** Effective velocity values in different nodes. Bearing B1 in the x direction for the impeller not changed by diameter.

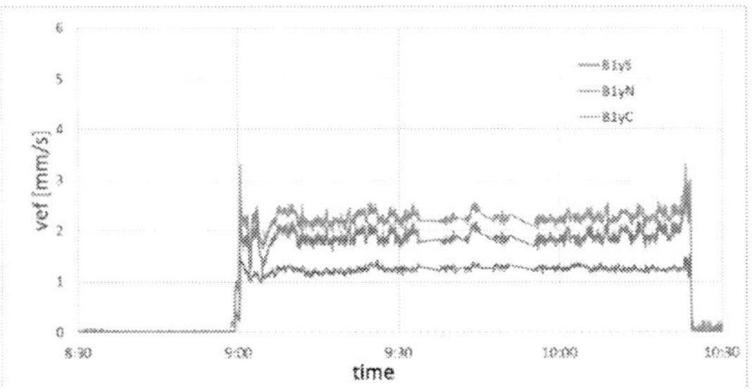


**Figure 7.** Effective velocity values in different nodes. Bearing B1 in the y direction for the impeller not changed by diameter.

On previous graphs the components of velocity were differentiated as synchronous and non-synchronous. Based on the results, we can say that the synchronous components of velocity have no effect on the total value of the effective rate. Therefore they are not considered in further processing and analysis will be performed only by contributions of non-synchronous components of velocity in the frequency domain.

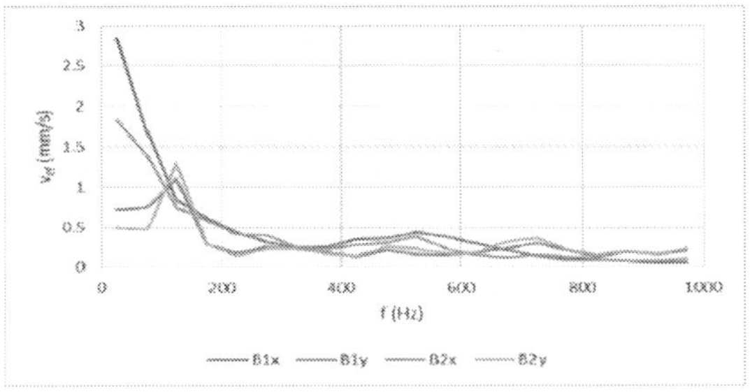


**Figure 8.** Effective values of velocity on the bearings of the pump B for each frequency band between multiple speeds frequencies. Synchronous components are neglected.

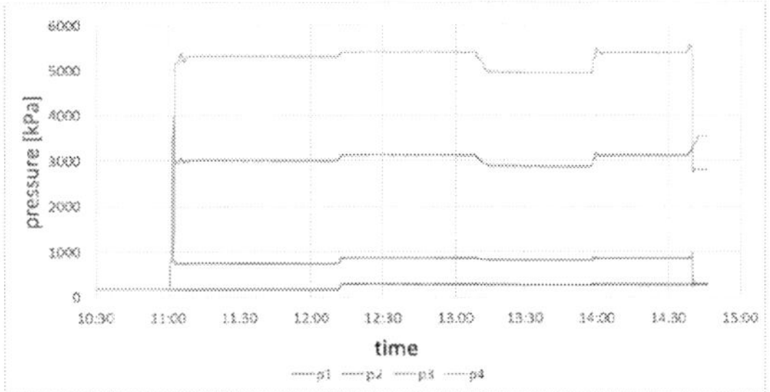


**Figure 9.** The time sequence of pressure measurement

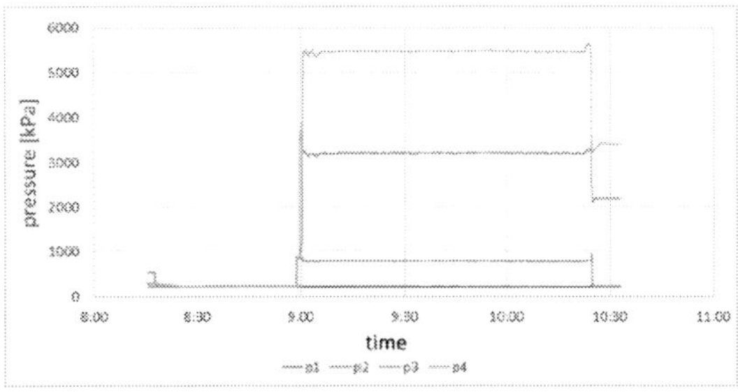


**Figure 10.** The time sequence of pressure measurement

Several variants have been proposed on the basis of these measured values. All these variants were measured and evaluated.

## 3 Numerical simulations

In all probability, the pump diffuser was the source of pressure pulsations. The diffuser was numerically simulated, to find out unwanted effects which were inside the diffuser. The diffuser was simulated as a 3D problem, because the canal width had a non-linear shape (see diagram in Figure 11, where is a diffuser with the linear change of width).

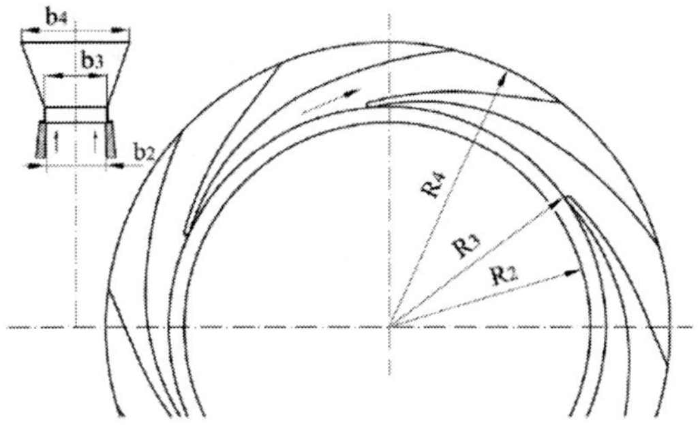


**Figure 11.** The schema of diffuser with the linear change of width [4].

The numerical simulation was made with the use of Ansys Fluent. Only the diffuser was simulated, without impeller and spiral case. (This could lead to a little mistake - there was set a uniform flow from an impeller.) This simplification was done for reasons of time – the pump was working in the pumping station and there was exerted pressure on quick resolution.

The velocity was set as a boundary condition. The velocity was determined by calculation from already measured characteristics. The impeller before the diffuser causes the fluid to enter the diffuser under specific angle. Velocity components (radial and tangential) must be set as the boundary condition on inlet. The pressure was set as the boundary condition on outlet.

The diffuser was computed for three flow rates - for the flow rate corresponding with the maximal flow rate $Q_3$ = 2250 $m^3 \cdot h$ used on a locality, with the minimum flow rate $Q_2$ = 1300 $m^3 \cdot h$ used on a locality and with the chosen flow rate $Q_1$ = 800 $m^3 \cdot h$. Only three flow rates were simulated, there was no effort to do the characteristics; there was a need to find out the shape of the fluid flow inside canals. Small flow rates caused problems with vibrations, so the flow rate $Q_1$ was chosen as a smaller one than the flow rate $Q_2$.

Results of numerical simulation showed, that a bad fluid flow doesn't cause hydraulic shocks on leading edges of diffuser blades. Nevertheless, the diffuser contributed to vibrations of the pump. In all probability, the flow separation inside the canal could cause vibrations. Values of total velocity on the diffuser sectional view are shown in Figure 12. Apparently, the fluid flow is stabilized at the maximal used flow rate $Q_3$. By reducing the flow rate, the fluid flow inside the canal begins tearing off. The canal at minimum used flow rate $Q_1$ is in Figure 13, the swirling is considerable. Vortexes from the flow separation could cross in the spiral case and they could cause pressure pulsations of the pump and the connected piping system. [5] The reason for the flow separation is obvious - a rapid expansion of the canal.

**Figure 12.** Values of total velocity on the diffusers sectional view – original diffuser, flow rate $Q_3$ = 2250 $m^3 \cdot h$.

**Figure 13.** Values of total velocity on the diffusers sectional view – original diffuser, flow rate $Q_1$ = 800 $m^3 \cdot h$.

Based on the simulation of the original diffuser, several simulations of adjusted diffuser were done. One of those simulations was the diffuser with half the number of diffuser blades. Another adjustment changed the width b of the diffuser - the diffuser with constant width and the diffuser with the linear change of width (see Figure 11). The most suitable modification was the change of a shape of the diffuser blade (see Figure 14). The blade was adjusted to reduce the flow area of the canal, to obviate the flow separation. In comparison with the original shape of the blade (Figure 12), a significant difference is observed.

Values of total velocity in the adjusted diffuser are shown in Figure 14. Velocities in the canal are slightly increased, compared with the original diffuser, but power values needed for the working point attainment changed inconsiderable.

The adjusted diffuser was made and new measurements of pump were done.

**Figure 14.** Values of total velocity on the diffusers sectional view – adjusted diffuser, flow rate $Q_2$ = 1300 $m^3 \cdot h$.

## 4 Solving of problems and other measurements

The diffuser was designed for preliminary source of the vibration and it was suitably modified (see previous section). There may be even more potential sources of vibration. All these theoretical resources were addressed during the last measurements that were made to the test of the company. Measurements were carried out with several pumps of modified variants. [6]

Measured variables:

1st variant - the original pump (this variant was followed by a change of the electric motor bearings)

2nd variant - adjusting the suction inlet impeller, ribs

3rd option - adjusting the suction inlet impeller, ribs and a split diffuser used

4th variant - a new diffuser used as proposed by the University and the original suction

5th variant - the same as 4th variant, but with the new bearings on the pump

Vibrations on the pump bearings, vibrations on base of pump and vibration at the end of the tube from 2nd variant were measured for all these measurement variations. The measurements were carried out at variable engine speeds for 9 flows for each variant. The measuring of steady state longer than 20 seconds followed after a flow setting at maximum speed of approximately 24.8 Hz (1488 $min^{-1}$) and settling transient analysis. Subsequently the speeds were reduced and increased again back to the maximum value.

### 4.1 Evaluation of modified versions

Evaluation of the measured values was performed according to the same relationship as for the first measurement. The difference was that in the evaluation only measured vibrations were considered on the bearings of the pump from a steady mode.

Furthermore there applies that disturbing frequencies are filtered off before performing the calculation by equation (3). The filtration consisted of the removal at all multiples of the amplitudes of 50 Hz ± 1 Hz – there were filtered frequency 49-51 Hz, 99-101 Hz and so on. These disturbing frequencies were particularly noticeable on the bearing 1 in the x-direction, and at the individual measurements varied greatly. There were also determined a rotation frequency, which was used to calculate the effective velocity of the synchronous frequency (Fig. 15) and the subsynchronous frequency (Fig. 17). Rotation frequency and all multiples of the rotation frequency plus or minus one step on a frequency, which is about 0.5 Hz, were regarded as synchronous frequency. [7]

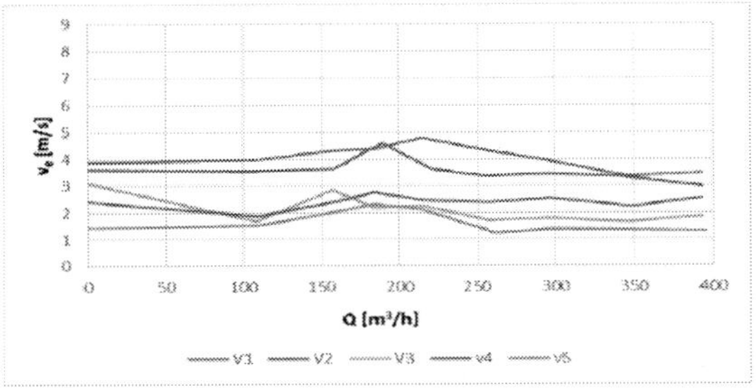


**Figure 15.** The bearing 1 in x-direction, the synchronous frequency.

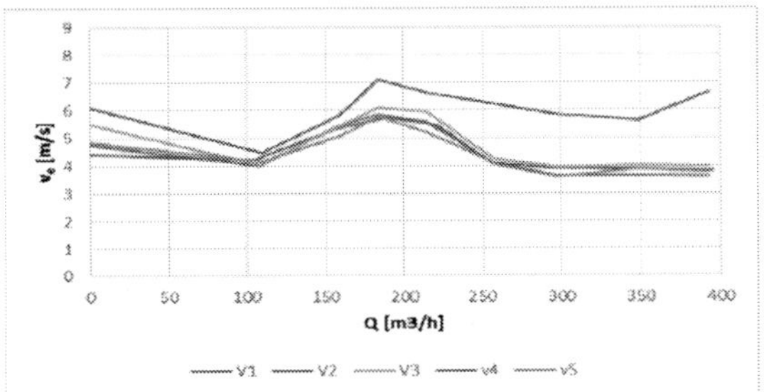


**Figure 16.** The bearing 1 in x-direction, the non-synchronous frequency.

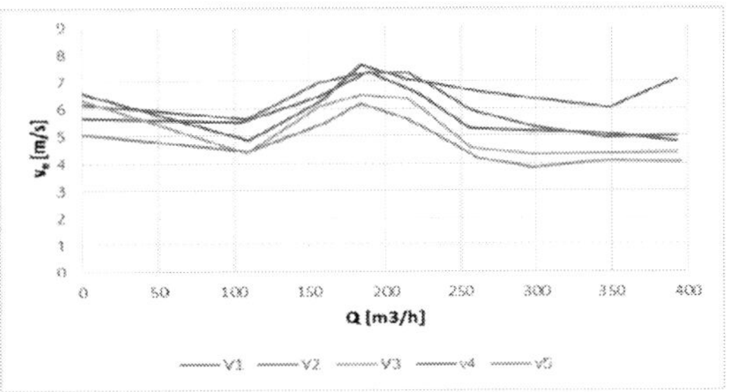


**Figure 17.** The bearing 1 in x-direction, the subsynchronous frequency.

## 5 Conclusions

The variant of the new adjusted diffuser was recommended based on the results of CFD flow modelling. Measurement on bearing 1 in the x direction were in most measurements burdened by significant noise at a frequency of 50 Hz and its multiples, even if those frequencies are filtered, probably distorting the results.

Dependence of the effective velocity on the flow can be ascertained by the following:

A comparison of 1st and 2nd variants shows that the impact on subsynchronous effective velocity is negligible and the impact on the synchronous effective velocity is distinct. 2nd variant came out better.

A comparison of 2nd and 4th variants shows that we do not resolve bearing 1 in the x direction because of the high noise it can be considered that the new diffuser is a slightly reduced vibration.

2nd and 3rd variants show that the divided diffusor reduced synchronous components; subsynchronous components increased slightly, overall vibration used by divided diffuser decreased. There can be traced the uniformity of vibrations for the divided diffuser at both bearings in both directions. It means that the effective velocities are the same for the one flow everywhere.

4th and 5th variants show that the vibration of pump slightly decreased after replacing the bearings. A significant decline was on bearing 1 in x direction, but there we are concerned that greater importance was the noise from the outside than bearings. The effect of bearings was on other measurements of vibration at the level of measurement accuracy.

## References

1. Taneja, S.: *Effect of unbalance on performance of centrigugal pump.* International Journal of Scientific & Technology Research, vol.2, issue 8, August 2013, pp.56-60, ISSN 2277-8616
2. Matoušek, R.; Dobrovský, L.; Minář, P.; Mouralová, K.: *A Note about Robust Stabilization of Chaotic Hénon System Using Grammatical Evolution.* Advances in Intelligent Systems and Computing, 2014, roč. 2014, č. 289, s. 219-228. ISSN: 2194- 5357.
3. Dvořák, J.; Dvořáková, J.; Mouralová, K.: *Application of fuzzy logic in analysis of input parameters in AWJ technology.* In Mendel 2015 - 21st International

Conference of Soft Computing - Mendel Journal Series. Mendel Journal series. 2015. s. 97-104. ISBN: 978-3-319-19823- 1. ISSN: 1803- 3814.
4. Dobšáková, L.: *Hydraulický návrh difuzoru čerpadla pro dva provozní body*. Master's thesis. Brno university of technology. Brno 2012.
5. S.-S. Yang, H.-L. Liu, F.-Y. Kong, B. Xia, and L.-W. Tan,: *Effects of the radial gap between impeller tips and volute tongue influencing the performance and pressure pulsations of pump as turbine*. Journal of Fluids Engineering, Transactions of the ASME, vol. 136, no. 5, Article ID 054501, 2014.
6. Sedlák, J.; Drábek, T.; MouralováK.; Chladil, J.; Kouřil, K.: *Machining Issues of Titanium Alloys*. International Journal of Metalcasting, 2015, roč. 9, č. 2, s. 41-50. ISSN: 1939- 5981.
7. Mouralová, K.; Bednář, J. *Determination of significant process parameters of WEDM*. Preliminary Study. In MENDEL 2013. Mendel Journal series. ISCC. Brno: BUT, 2013. s. 451-454. ISBN: 978-80-214-4755-4. ISSN: 1803- 3814.
8. Habán, V.: *Experimentální výzkum tlakových pulsací na čerpací stanici*. Technical report, Brno university of technology, Brno 2014.
9. Habán, V.: *Výzkum tlakových pulsací na čerpací stanici – CFD modelování rozvaděče a měření efektivních rychlostí na ložiskách na zkušebně*. Technical report, Brno university of technology, Brno 2014.

## Acknowledgement

Presented research was supported by centre of competence of Technology Agency of the Czech Republic TE02000232 "Rotary machines".

DOI: 10.1051/epjconf/201714302018

# Calculations of parameters and mathematical model of rotary air motor

*Lukáš* Dvořák[1,*], *Kamil* Fojtášek[2], *Vojtěch* Řeháček[3]

[1,2,3] VŠB-Technical University of Ostrava, Faculty of Mechanical Engineering, Department of Hydrodynamics and Hydraulic Equipment, 17. listopadu 15/2172, Ostrava 70833, Czech Republic

**Abstract.** The article describes calculations of the parameters of rotary air motors, focuses on the mathematical model of air motor using Matlab Simulink-Simscape and concludes with the comparison of the results of the experiment and a the mathematical model.

## 1 Introduction

The rotary pneumatic motors are not used to drive machines so often. It is primarily due to the relatively low efficiency of the pneumatic actuator. Other disadvantages are the speed change when the load changes and the noise. Even so, in some applications they are irreplaceable. In these cases on the other hand the benefits are manifested as: simple operation and maintenance, operational reliability, ability to work in potentially explosive atmospheres, the overload does not demage the motor etc. From the perspective of design the pneumatic motors can be divided into vane, piston and gear types. All types may be included in the group of volumetric motors. Turbine motors are in the separate category.

## 2 Calculation of basic parameters of rotary motors

According to the utilization of air in the working space of motor, the motors can be divided into full-pressure motors and motors with partial expansion types, Fig. 1.

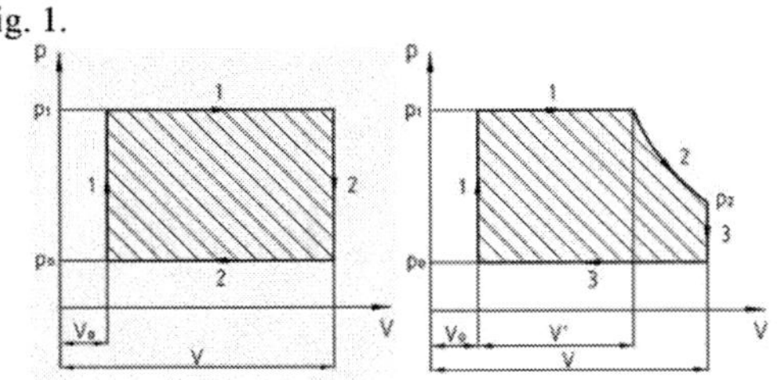

**Fig. 1.** p-V (indicator) diagram of full-pressure motors and motors with partial expansion

For motors with partial expansion the ratio of the volumes $(V - V_0)\ /\ V)$ is called coefficient of filling ($\delta$) and it ranges from 0.45 to 0.7.

Main value for the calculation of the motor parameters is called the indicated pressure and its theoretical value can be calculated for full-pressure motors from equation 1

$$p_{it} = (p_1 - p_0) \tag{1}$$

and for motors with partial expansion from equation 2.

$$p_{it} = p_1 \cdot \delta \cdot \frac{n - \delta^{n-1}}{n-1} - p_0 \tag{2}$$

A polytropic exponent is in the range of $n = 1.25$ to 1.32 depending on the motor type. The problem is that the real value of the indicated pressure will be lower, because at high speeds the working space will not be filled enough, as it is shown in Fig. 2.

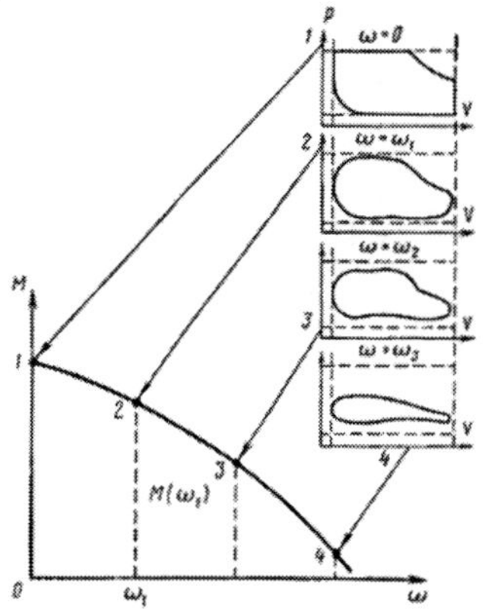

**Fig. 2**. Influence of speed to filling of work space [3]

The final shape of the indicator diagram and thus the value of mean indicated pressure is also affected by the size and shape of distribution channels and their resistance to airflow.

* Corresponding author: lukas.dvorak@vsb.cz

The true value of indicated pressure we can obtained from equation 3,

$$p_i = p_{it} \cdot \eta_p \tag{3}$$

where $\eta_p$ is filling efficiency of workspace. The efficiency curves for some designs of rotary motors are shown in Fig. 3.

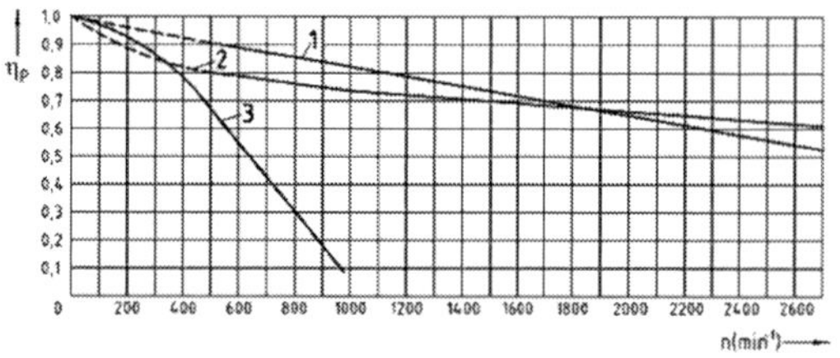


**Fig. 3.** Filling efficiency 1 – gear motor with spiral gearing, 2 – gear motor with herring-bone gearing 3 – piston motor [3]

The filling efficiency affects the calculation of all the basic parameters throughout the operating speed range. An examples of the torque and power characteristics together with the air consumption are in Fig. 4.

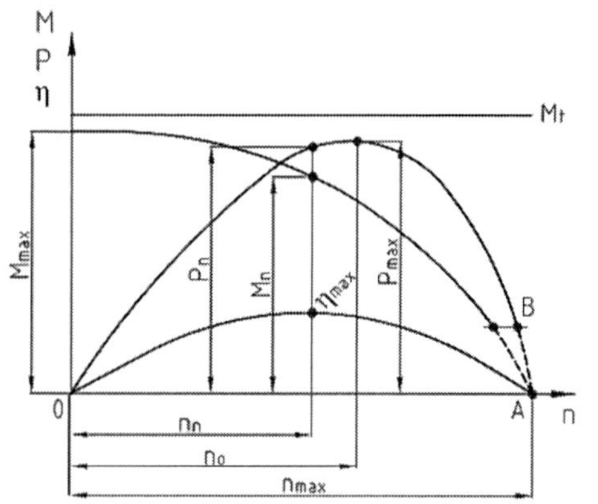


**Fig. 4.** Characteristics of rotary motors depending on speed

For the calculation of individual parameters we can use the following equations [1]. The theoretical torque is not dependent on the speed. But the calculation of the real torque is influenced by the filling efficiency therefore the characteristic has a downward trend.

$$M = \frac{1}{2\pi} \cdot V_g \cdot p_{it} \cdot \eta_p \cdot \eta_m \tag{4}$$

In the equation $V_g$ is displacement of motor, $p_{it}$ is theoretical indicated pressure, $\eta_p$ is filling efficiency and $\eta_m$ is mechanical efficiency, which has value for piston motors with connecting rod mechanism $\eta_m = 0.85 \div 0.9$ and for motors with a sliding mechanism $\eta_m = 0.9 \div 0.92$.

Power can be calculate from equation 5.

$$P_s = V_g \cdot n \cdot p_{it} \cdot \eta_p \cdot \eta_m \tag{5}$$

From the analysis of experimental measurements of all types of air motors it is evident that the filling efficiency $\eta_p$ depending on the speed varies approximately linearly.

This fact can be expressed as follows $\eta_p = 1 - a \cdot n$ where $a$ expresses the inclination of line. After substitution it is therefore possible to calculate the power from the equation 6.

$$P_s = V_g \cdot n \cdot P_{it} \cdot \eta_m - V_g \cdot n^2 \cdot p_{it} \cdot a \cdot \eta_m \tag{6}$$

The air consumption recalculated to atmospheric normal conditions can be approximately calculated from equation

$$Q_n = V_g \cdot n \cdot \frac{p_1}{p_N} \cdot \delta \cdot \frac{1}{\eta_Q} \tag{7}$$

where $\eta_Q$ is flow efficiency, which has a value from about 0.85 to 0.95, depending on the design and motor wear.

## 3 Comparison of measured and calculated characteristics

The object of our research were mainly small radial piston motors with power up to 500 $W$. For characteristics measuring test equipment has been compiled, Fig. 5.

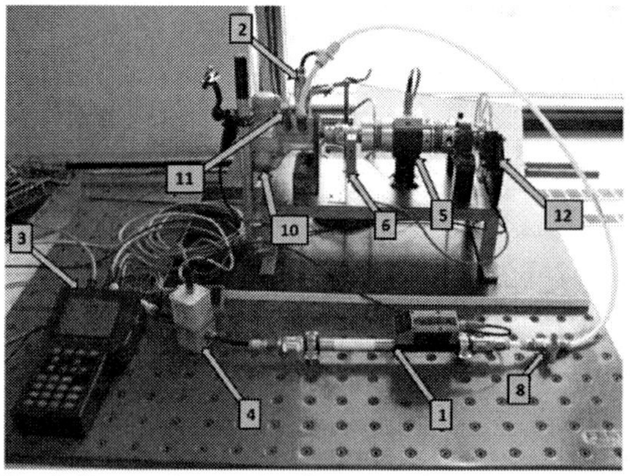


**Fig. 5.** Test equipment, 1 - flowmeter, 2 – pressure sensor, 3 - recording device, 4 - reduction valve, 5 - torque sensor, 6 - speed sensor, 7 - FRL unit, 8,9 - ball valve, 10 - pneumatic rotary motor, 11 - silencer, 12 - disc brake

The measurement results are the characteristics of the torque, power, air consumption and overall efficiency.

From the measurement results the filling efficiency of radial piston motor was found, Fig. 6. Parameter $a$ expressing the inclination of line in this case has a value $a = 0.028$. Further, based on the measurement the mechanical efficiency of the motor was determined. Its value is $\eta_m = 0.86$.

DOI: 10.1051/epjconf/201714302018

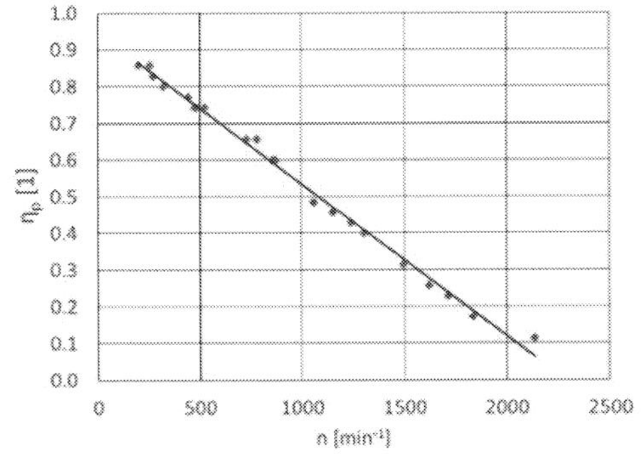


**Fig. 6.** Filling efficiency of measured motor

When we compare power calculated using equation 6 it well corresponds with the experimental results, Fig. 7. The difference is noticeable at higher speeds, but this area is already outside of the optimum operating range.

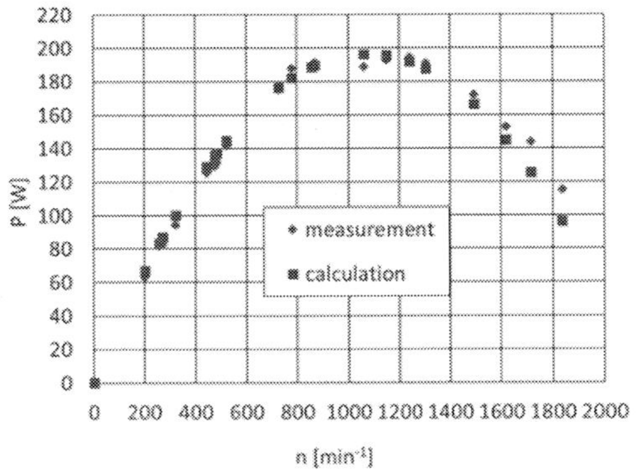


**Fig. 7.** Power of measured motor

## 4 Modelling of a pneumatic system with a rotary motor

In addition to the measurements we carried out verification of the possibility of mathematical modeling of pneumatic systems with the rotary motor. The Matlab-Simulink Simscape was used for modeling. Simscape contains models of the basic pneumatic components including the rotary air motor model "Rotary Pneumatic Piston Chamber". Into this model, it is necessary to specify Displacement (volume per unit angle), Initial angle, Dead volume and Chamber orientation (direction of rotation). Mechanical neither flow efficiency is not included into the mathematical model of the motor. The efficiencies must be added in the calculation externally. But even after these adjustments we have not achieved good results. The torque at zero speed corresponds to the real motor but the torque characteristics depending on the speed do not decrease linearly. Larger differences are then in power characteristics. Maximum calculated power is several times greater than the real power and it is also achieved at higher speeds. With the above model we have not achieved real results.

The manual for Simulink (Help) contains another model, "Pneumatic motor". There is chosen a completely different approach of motor parameters definition. Into the model it is necessary to enter vector of rotational speeds, vector of torque values, vector of volumetric flow rates and pressure differential at which the torque and flow data were measured. By this way the characteristics of the motor can be very precisely defined. The disadvantage is in obtaining these values by measurement. On the other hand, the simulation results are quite accurate. Table 1 shows the values that were entered into the model of rotary motor.

It was also necessary to add the source of constant pressure, which was set at 5 bar, and the models of directional valve and compressed air lines in the model of system, Fig.8.

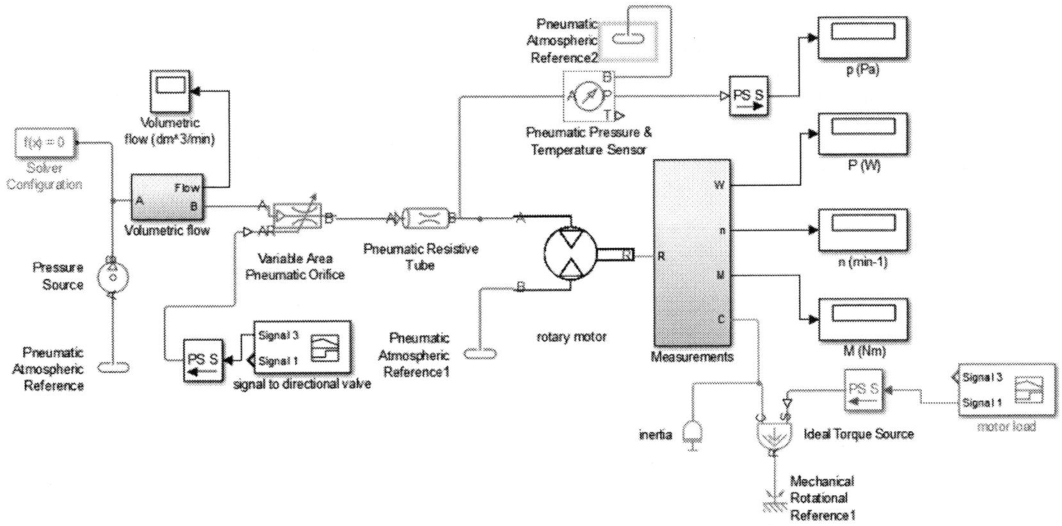


**Fig.8.** Block diagram of the mathematical model [2]

The elements for measuring the flow, the pressure, the torque, the speed and the power were also included in the model. The load was simulated in two ways. Firstly, the block "Inertia" was set to 0.001 $kg \cdot m^2$ . This value corresponds to the inertia of the motor and other rotating parts of the test equipment (torque sensor, clutch, shaft and brake disc). It was also necessary to define the load torque. This was realized by the block "Ideal Torque Source". The signal which corresponded with the motor load during the measurement was fed on the input of this block.

**Table 1.** Values entered into the model

| *n* [$min^{-1}$] | *M* [*Nm*] | *Q* [$dm^3.min^{-1}$] (ANR) |
|---|---|---|
| 0 | 3.329 | 55 |
| 200 | 3.029 | 119 |
| 400 | 2.729 | 177 |
| 600 | 2.429 | 228 |
| 800 | 2.129 | 273 |
| 1000 | 1.829 | 311 |
| 1200 | 1.529 | 343 |
| 1400 | 1.229 | 369 |
| 1600 | 0.9287 | 388 |
| 1800 | 0.6287 | 401 |
| 2000 | 0.3287 | 407 |
| 2136 | 0.1247 | 435 |

The measurement and the modeling was performed for several different ways of the load. The first of these was the case when the motor was loaded by a constant torque and the motor accelerated from the zero speed. An example of measured and calculated values is in Fig. 9. In this case, at the time of 1.4 *s* the valve has opened. At the time of approximately 2 *s* the steady speeds was achieved. During the acceleration the torque has risen to 1.9 *Nm* due to the inertia of the rotating parts, after the acceleration the torque reached the level of 1.1 *Nm*.

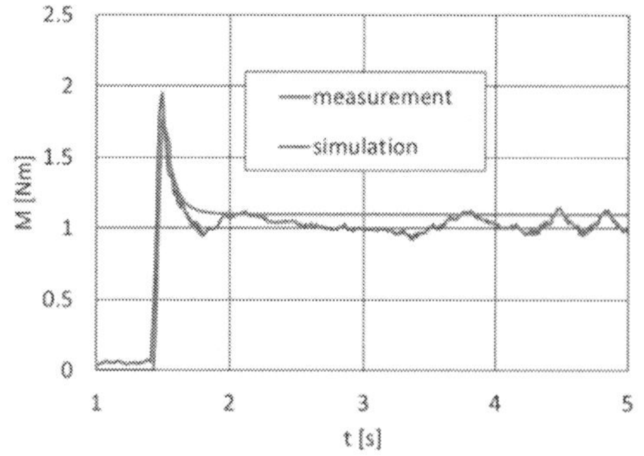


**Fig. 9.** Torque during motor acceleration

Subsequently, the accuracy of the simulation was validated in the case where during the rotation the load was increased, for example from 0.7 *Nm* to 0.95 *Nm*. This change in torque has resulted in the decrease of speed from 1245 $min^{-1}$ to 980 $min^{-1}$, see Fig. 10. This behavior is typical for the rotary pneumatic motors. Similar results were achieved in all other investigated cases.

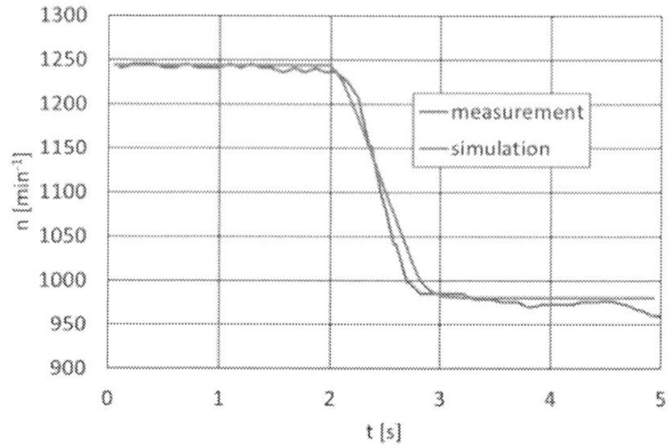


**Fig. 10.** Speed during load increase

## 5 Conclusion

At the beginning of the article the procedure for calculating the main parameters and characteristics of the rotary air motors was stated. This problem is quite difficult, because the results are influenced by the filling efficiency. The efficiency varies according to the type of engine. For this reason, it is very difficult to predict the characteristics of the motor during its design. Another part is devoted to the possibilities of mathematical modeling of pneumatic system with rotary pneumatic motor. From the calculations it follows that model results, when the basic data are entered, are not very accurate. Relatively accurate results can be achieved with a model in which speed vector and corresponding vectors of torque and air consumption are inserted. The disadvantage of this model is that these values must be detected experimentally.

## Acknowledgements

This work was supported by the project SP2016/70 of VŠB-Technical University of Ostrava.

## References

1. J. Kopáček: *Pneumatické mechanismy* (1998)
2. V. Řeháček: *Design of Test Equipment of Pneumatic Rotary Motors: Master Thesis* (2016)
3. K.S. Borisenko: *Pnevmatičeskije dvigateli gornych mašin* (1958)

DOI: 10.1051/epjconf/201714302019

# Experimental equipment for measuring of rotary air motors parameters

*Lukáš* Dvořák[1,*], *Kamil* Fojtášek[2], *Vojtěch* Řeháček[3]

[1,2,3] VŠB-Technical University of Ostrava, Faculty of Mechanical Engineering, Department of Hydrodynamics and Hydraulic Equipment, 17. listopadu 15/2172, Ostrava 70833, Czech Republic

**Abstract.** In the article the construction of an experimental device for measuring the parameters of small rotary air motors is described. Further a measurement methodology and measured data processing are described. At the end of the article characteristics of the chosen air motor are presented.

## 1 Introduction

Rotary pneumatic motors can be divided into several categories by motor design. In this case we can divide the rotary pneumatic motors in vane, axial and radial piston, gear and turbine designs, Fig. 1.

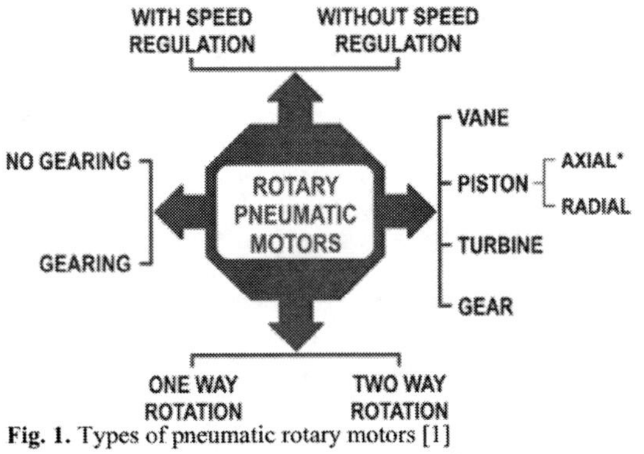


**Fig. 1.** Types of pneumatic rotary motors [1]

The objects of our research were mainly small radial piston motors with power up to 500 W and nominal speed up to 1000 $min^{-1}$. An example of design of these motors is shown in Fig.2.

During the rotary air motors characteristics measuring it is necessary to determine the following parameters: an inlet pressure and eventually an outlet pressure, a motor torque, a rotation speed and an air consumption. The processing the measured data is described below. The result of measurements is the rotary motor characteristic, which is mainly a torque and power dependence on a rotational speed, Fig. 3. This characteristic may be complemented by an air consumption and an efficiency, this is then called a complex characteristic of the rotary motor.

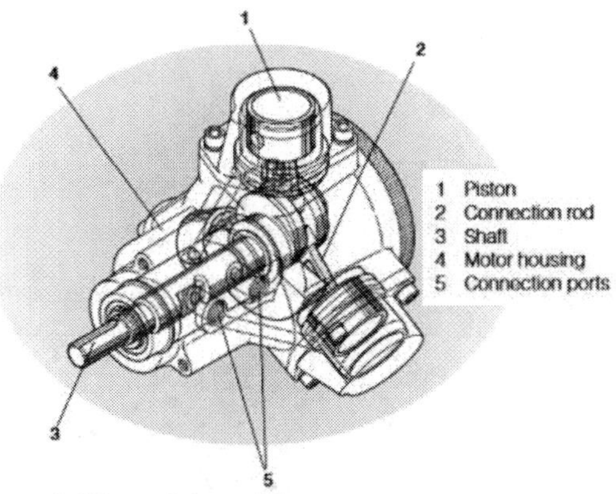


**Fig. 2.** Radial piston air motor [2]

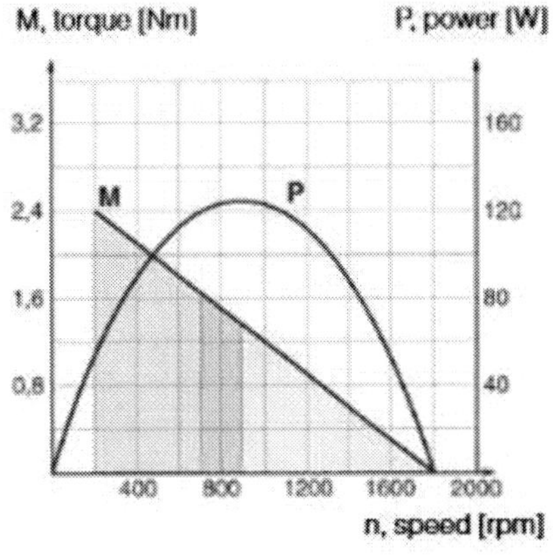


**Fig. 3.** Torque and power characteristics of motor [2]

* Corresponding author: lukas.dvorak@vsb.cz

## 2 Test equipment

During the measurement of rotary motors it is necessary to change the load of the output shaft continuously. This can be achieved by some types of brakes. Table 1 provides an overview of possible ways of brake implementation. The values in the table show that suitable break for the motors with power up to 500 *W* is a friction brake. Due to its low power and simplicity of design a bicycle break has been chosen. Specifically type-ZEE BR-M640 from Shimano in combination with disc SM-RT 66 with a diameter of 160 *mm* was chosen. According to the producer the brake was tested for speeds up to 1350 $min^{-1}$ and guaranteed braking power is 800 *W*. This brake therefore fully complies with our requirements.

It was also necessary to choose sensors. To measure a torque the sensor T22 / 50 from HBM with a measuring range of +/- 50 *Nm* was chosen. Furthermore, the pressure sensor PR 15 with a measuring range of -1 to 6 *bar*, the speed sensor DS 03 and the recording device M5050 ware chosen, all made by company Hydrotechnik. The flow measurement was realized by the sensor SD 6000 from IFM elekrtotechnik with measuring range $4 - 1250\, dm^3 \cdot min^{-1}$(ANR). A frame to which were mounted all the necessary parts was also designed and built. The device is shown in Fig. 4, a circuit diagram is in Fig. 5.

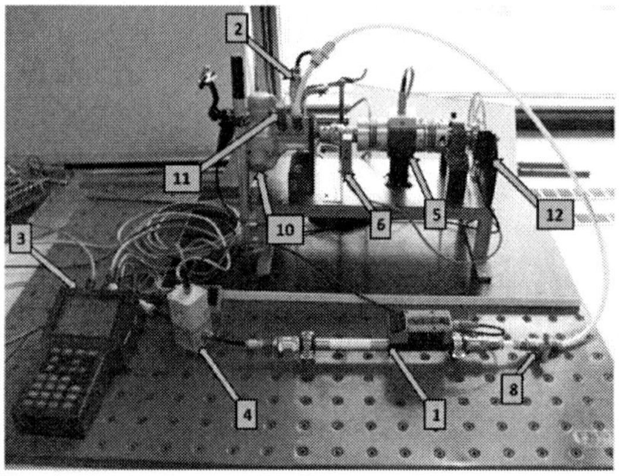


**Fig. 4.** Test equipment, 1 - flowmeter, 2 – pressure sensor, 3 - recording device, 4 - reduction valve, 5 - torque sensor, 6 - speed sensor, 7 - FRL unit, 8.9 - ball valve, 10 - pneumatic rotary motor, 11 - silencer, 12 - disc brake

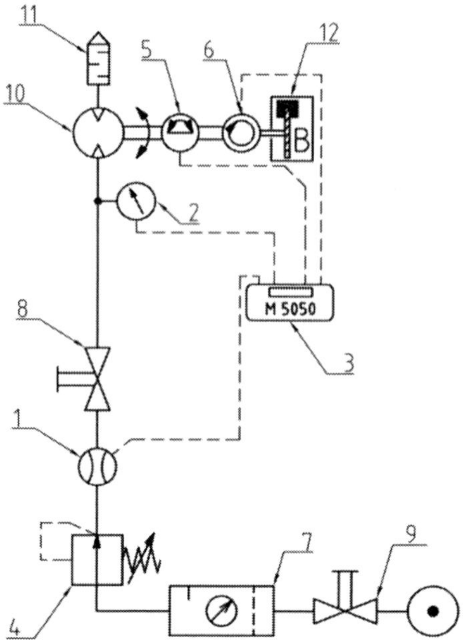


**Fig. 5.** Diagram of the test equipment (the description is the same as in Fig. 4)

## 3 Measurement methodology and measured data processing

The basic requirement for the measurement is to keep a constant pressure at the inlet of the motor. After the first few tests it was proved that the mechanically controlled pressure reducing valves are not suitable for this application. Therefore, in the subsequent measurements the pressure regulation was realized by proportional pressure valve VPPE-3-1-1 / 8-10-010-E1 made by Festo. By selecting the control voltage the measurements ware then carried out at a pressure level of 3, 4, 5 and 6 *bar*.

At each pressure level the torque was increased using the friction brake and after each change a record of all measured variables (pressure - $p$, torque - $M$, speed -$n$, flow rate $Q_N$) was made for 3 seconds with sampling period of 0.1 second. From the measured values the mean values were calculated. These were directly plotted into the graph, or were used to calculate the power $P$, the overall efficiency $\eta$ and specific consumption $\overline{m}$. An equation for the calculation of individual parameters is shown in equations 1 to 3.

**Table 1.** Types of brakes

| Brake type | Range of parameters | | |
|---|---|---|---|
| | Power [*kW*] | Torque [*Nm*] | Rotation speed [$min^{-1}$] |
| Elektrodynamometr | 3 - 800 | 20 - 8 000 | 300 - 3 000 |
| Hydrodynamic brake | 2 - 1 000 | 10 - 6 000 | 200 - 10 000 |
| Hydrostatic brake | 0.5 - 50 | 1 - 100 | 100 - 8 000 |
| Friction brake | 0.1 - 5 | 0.5 - 50 | 10 - 2000 |

The motor mechanical power on the output shaft can be calculated from the torque and speed

$$P = \frac{2 \cdot \pi \cdot n \cdot M}{60} \quad [W] \tag{1}$$

where *M* is torque [*Nm*] and *n* is speed [$min^{-1}$].

The motor overall efficiency is the ratio of mechanical output power to pneumatic input power. Due to the volume flow rate dependence on the pressure, it is necessary to recalculate the flow rate to the level of working pressure.

$$\eta = \frac{P}{p \cdot Q_p} \tag{2}$$

In the equation *P* is power [*W*], *p* working pressure [*Pa*] and $Q_p$ is volume flow rate of compressed air [$m^3.s^{-1}$]

Specific air consumption is air consumption related to output power. In this case, it is necessary to use flow rate recalculated to normal atmospheric conditions (ANR – atmospheric normal reference).

$$\bar{m} = \frac{Q_N}{P} \; \left[dm^3 \cdot min^{-1}(ANR) \cdot W^{-1}\right] \tag{3}$$

In the equation *P* is power [*W*] and $Q_N$ is air consumption $[dm^3 \cdot min^{-1}(ANR)]$

## 4 Measurement results

Fig. 6 shows the torque characteristic of the measured motor. The torque significantly decreases from a maximum value at zero speed (starting torque). The value of the starting torque is depending on the size of the pressure. At the inlet pressure of 3 to 6 *bar* the torque varies from 2 to 3.8 *Nm*. From the course of the power in Fig. 7 it is evident a parabolic shape of characteristic, which is typical for rotary pneumatic motors.

Table 2 shows the maximum power values at different working pressures. From the values it is obvious that the maximum power is achieved in the range of revolutions from 1100 to 1200 $min^{-1}$.

**Table 2.** The maximum measured power

| *p [bar]* | *n* [$min^{-1}$] | *P [W]* |
|---|---|---|
| 3 | 1180 | 128 |
| 4 | 1210 | 170 |
| 5 | 1100 | 192 |
| 6 | 1120 | 229 |

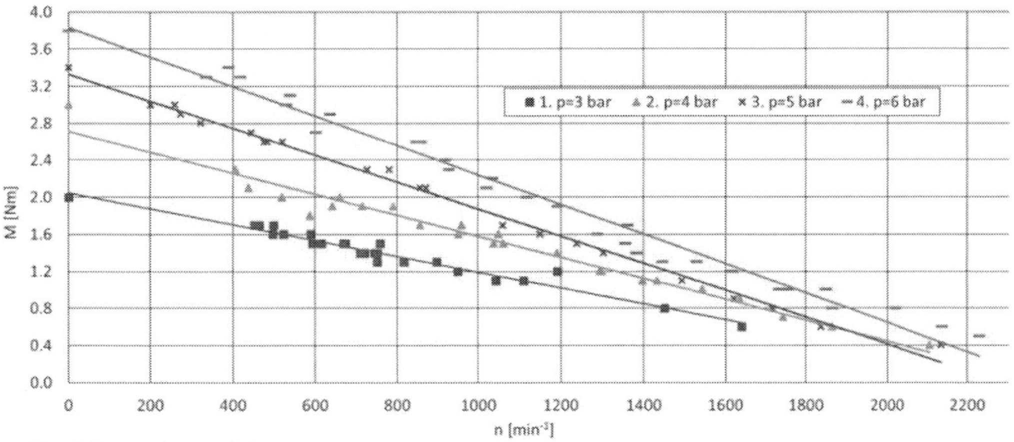


**Fig. 6.** Torque characteristic

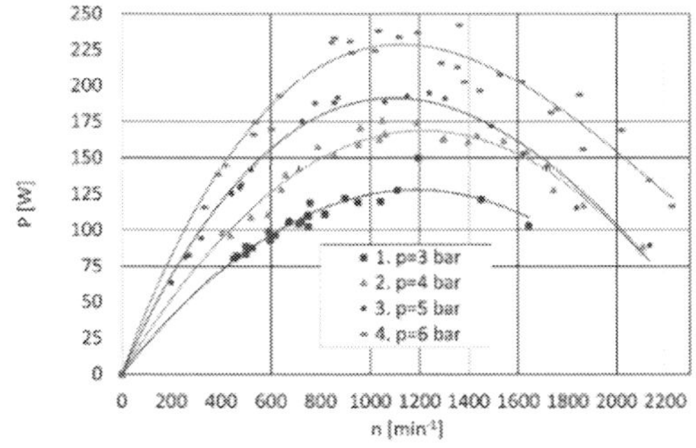


**Fig. 7.** Power characteristic

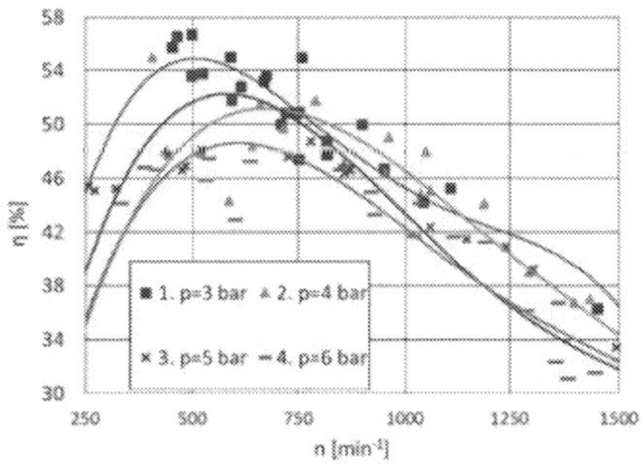


**Fig. 8.** Overall motor efficiency

Fig. 8 shows the course of overall efficiency of the measured motor. The maximum efficiency for different pressure values is in the range of 48-55%. These values are achieved in the speed range from 500 to 750 *min* $^{-1}$.

From the above it is clear that the area of maximum efficiency is shifted in comparison with the area of maximum power to the lower speed. Generally, the optimum working range of the motor is located between maximum efficiency and maximum power. The maker states that the optimum range is from 700 to 900 *min* $^{-1}$, which corresponds with the measured values.

## 5 Conclusion

The article described the measuring of basic characteristics of small rotary air motors. The device can be used for motors with power up to 500 *W*. The equipment can be used further for measuring of the impact of speed regulation. The motor speed control can be easily realized with a throttle valve, Fig. 9. The valve can be connected into the circuit so that it regulate air flow on inlet to the motor (supply throttling) or some motor designs allows regulation on the outlet (outlet throttling). Throttle valve and its location affects the torque and power characteristics. Fig. 10 and Fig. 11 shows the measured characteristics for the different settings of throttle valve.

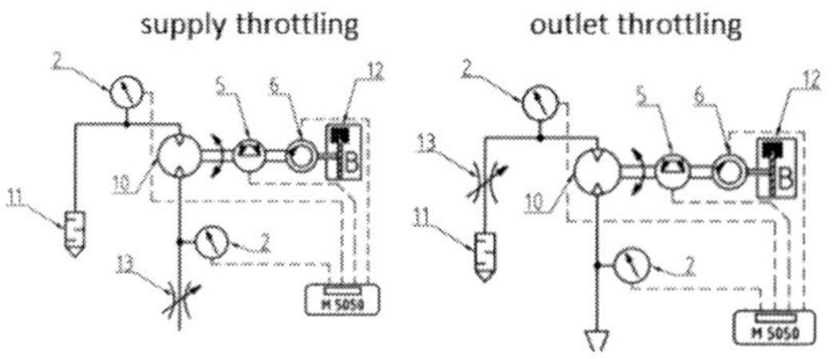

**Fig. 9.** Location of the throttle valve

## Acknowledgements

This work was supported by the project SP2015/95 of VŠB-Technical University of Ostrava.

## References

1. P.Tuleja. Application Possibilities of Rotary Pneumatic Drives Using. *American Journal of Mechanical Engineering*. Vol.3 (2015). online: http://pubs.sciepub.com/ajme/3/6/25/index.html
2. Rotary Actuators Catalogue. Online: http://www.parker.com/literature/Pneumatics%20Division%20Europe/PDE-Documents/Rotary_Actuators_Platform_Catalogue_PDE2613TCUK.pdf
3. J. Kopáček: *Pneumatické mechanismy* (1998)

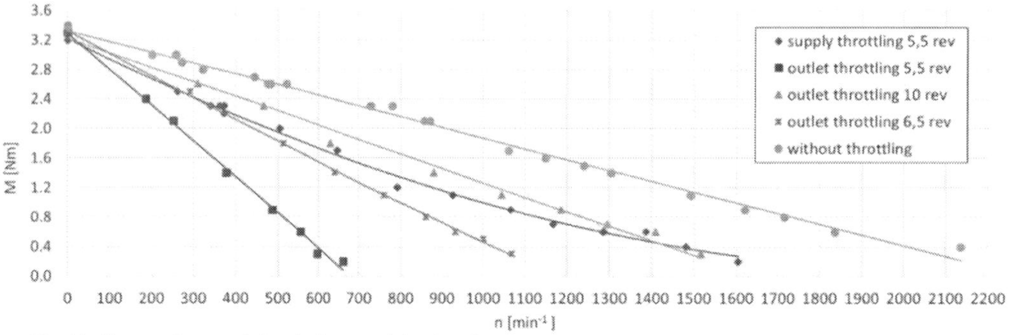

**Fig. 10.** Torque characteristic – influence of the throttling

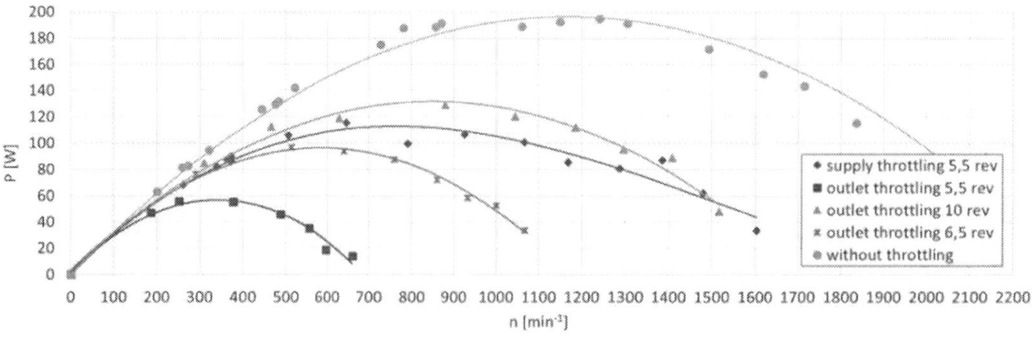

**Fig. 11.** Power characteristic – influence of the throttling

# Comparison of turbulent models in the case of a constricted tube

*Jakub* Elcner[1,*], *Frantisek* Lizal[1] and *Miroslav* Jicha[1]

[1]Brno University of Technology, Faculty of Mechanical Engineering, Energy Institute, Technicka 2896/2, Brno 616 69, Czech Republic

**Abstract.** The validation of a proper solution is an indispensable phase of every numerical simulation. Nowadays, many turbulent models are available, whose application leads to slightly different solution of flow behaviour depending on the boundary conditions of a specific problem. It is essential to select the proper turbulence model appropriate for the given situation. The aim of this study is to select the most suitable two-equation eddy-viscosity model, which can be further used during calculations of airflow in human airways. For this purpose, geometry of a constricted tube with well-documented experimental measurements was chosen. The flow in the constricted tube was calculated using Spallart-Almaras, k-omega, k-epsilon and SST model approach using commercial software. The outcome of the comparison is a choice of the suitable model which is capable of simulating the transition of the boundary layer from laminar to turbulent flow. This transition typically arises in the upper part of the respiratory system, where the airways are constricted, specifically in the area, where the oral cavity continues through the glottis to trachea. The simulations were performed in a commercial solver Star-CCM+.

## 1 Introduction

The human airways comprise extremely complicated system of branches with many variable parameters. The route for the oral inhalation of pharmaceutical aerosols consists of the oral cavity, larynx, pharynx, glottis, trachea and bronchi, where the flow divides in the system of bifurcations with branches of decreasing dimensions.

A part of airways that serves for delivery of aerosol into the lungs can be modelled as a tube with a constriction in the area of glottis. This constriction significantly influences the downstream flow [1]. The most common flow rates that are being used for the breathing simulation are 15 and 30 l/min. The flow in the upper airways is laminar under these conditions, however, in glottis, where the vocal cords are located, is the flow accelerated above the critical Reynolds number due to the constriction, and therefore the transition to turbulent flow occurs. This condition must be reflected in the process of selection of the turbulent model during the numerical simulation.

It is therefore necessary to select such a model, which accurately simulates flow in a circular tube with low Reynolds numbers, but at the same time is capable of precise modelling of the boundary layer transition from laminar to turbulent flow. Previous studies aiming at similar topics stated, that it is necessary to use the models of turbulence, which do not utilize wall functions, because they are not suitable for our type of flow [2].

The aim of this study is to select the most suitable two-equation eddy-viscosity model integrated in the commercial solver Star-CCM+ (CD Adapco), which can be further used during calculations of airflow in human airways. Accordingly, the models supporting the low wall y+ approach were selected based on the recommendations given by [3] for human airway flow simulations.

The results of this study will bring valuable information for the numerical simulations of human airways. According to Longest [4] is the accurate solution of the flow field necessary for calculations of local deposition characteristics and therefore the correct turbulent model is a necessary condition.

## 2 Methods

The geometry of a circular tube with 50 % and 75 % constriction was created in a software Rhinoceros (McNeel, Seattle, USA) for the numerical simulation. Several models of turbulence were used and compared on the geometry.

### 2.1 Model geometry

The test geometry comes from the paper of Ahmed and Giddens [5]. A smooth pipe with diameter of 50.8 mm is constricted to 50% of its diameter. The length of the constriction is 101.6 mm. The upstream length of the pipe is 4877 mm to allow the velocity profile to fully develop. The downstream length of the pipe is 1829 mm (see Fig. 1).

* Corresponding author: elcner@fme.vutbr.cz

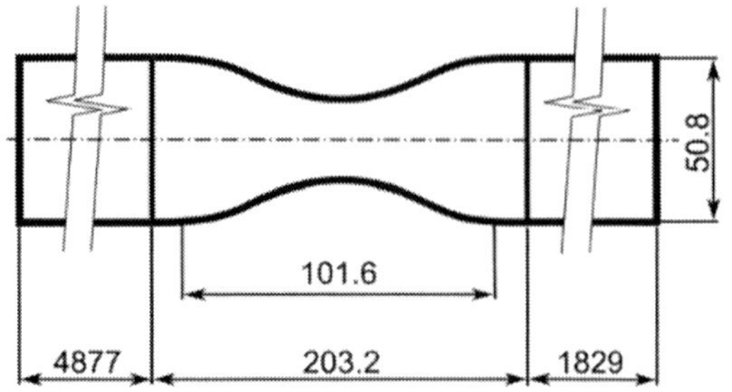


**Fig. 1.** A scheme.

Owing to the use of turbulent models based on low wall y+ approach, close attention was payed to the boundary layer treatment. A near wall layer consisting of 10 layers of prism cells was created. The value of wall y+ was approximately 1 on whole length of used geometry. The computational mesh was made of polyhedral elements and the prism layer contained 13.2 million cells. (see Fig. 2).

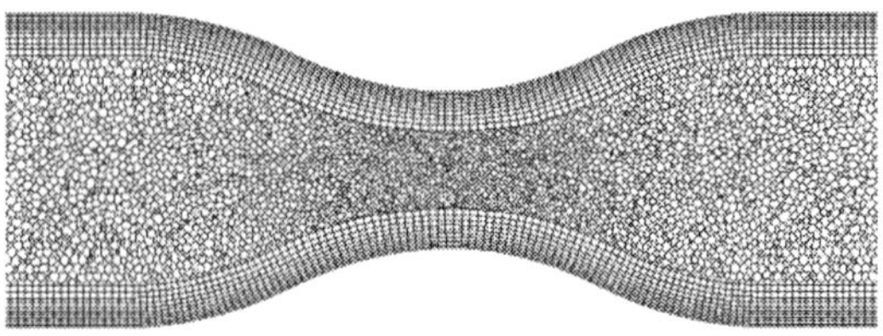

**Fig. 2.** Mesh.

## 2.2 Numerical method

The Star-CCM+ solver contains many turbulence models, whose application depends on the specific requirements of a given problem. Only the models supporting the low wall y+ approach were eligible from the range of available models in our case. Furthermore, just the models suitable for the internal pipe flow with a constriction could have been selected on the basis of their description given in the Star-CCM+ manual [6].

**The following turbulence models were chosen:**

**Standard Spalart-Allmaras model** is a good choice for applications in which the boundary layers are largely attached and separation is mild if it occurs. The Spalart-Allmaras turbulence models solve a single transport equation that determines the turbulent viscosity. In its standard form, the model is applied without wall functions. According to the formulation of the model, the entire turbulent boundary layer, including the viscous sublayer, can be accurately resolved and the model can be applied on fine meshes.

**Standard k-ε model.** Generally, k-ε models provide a good compromise between robustness, computational cost and accuracy. The k-ε turbulence model is a two-equation model in which transport equations are solved for the turbulent kinetic energy and its dissipation rate. In its original form, the k-ε turbulence model was applied with wall functions, but was later modified to use the Low Re approach for resolving the viscous sublayer.

**V2F Low-Reynolds Number k-ε model** is known to capture the near-wall turbulence effects more accurately, which is crucial for the accurate prediction of flow separation. This model solves two more turbulence quantities, namely the normal stress function and the elliptic function, in addition to turbulent kinetic energy and dissipation rate. This model is designed to handle wall effects in turbulent boundary layers and to accommodate non-local effects.

**Standard k-ω model** is similar to k-ε models in that two transport equations are solved, but differ in the choice of the second transported turbulence variable. The transport equations that are solved are for the turbulent kinetic energy and a specific dissipation rate. One reported advantage of the k-ω model over the k-ε model is its improved performance for boundary layers under adverse pressure gradients. Perhaps the most significant advantage, however, is that it may be applied throughout the boundary layer, including the viscous-dominated region, without further modification.

**SST k-ω model.** The problem of sensitivity to free-stream/inlet conditions was addressed by Menter [7], who recognized that the ε transport equation from the standard k-ε model could be transformed into an ω transport equation by variable substitution. Menter suggested using a blending function that would include the cross-diffusion term far from walls, but not near the wall. This approach effectively blends a k-ε model in the far-field with a k-ω model near the wall. In addition, Menter also introduced a modification to the linear constitutive equation and dubbed the model containing this modification the SST (shear-stress transport) k-ω model. The SST model has fairly wide application in situations, where viscous flows are typically resolved and turbulence models are applied throughout the boundary layer.

The above mentioned models were applied in default settings without changes in their specific constants. Axial velocities (i.e. the velocity component in the direction parallel with the axis of the tube with the stenosis) were observed. The flowing media in the tube was 63 % solution of glycerol in water at 33 °C, while the Reynolds number was Re = 2000.

# 3 Results and discussion

The observed values were compared in the nondimensional coordinate Z = x/D, where x is the axial distance from the throat of the stenosis and D is the diameter of the unconstricted tube (see Fig. 3). The calculated values of the axial velocities for each turbulence model were normalized according to [5] and their values were compared.

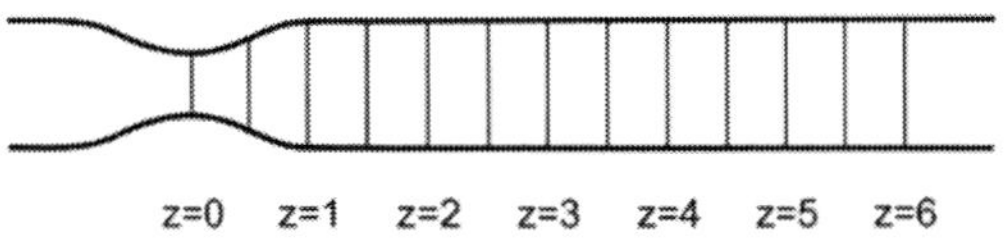


**Fig. 3.** Cross-sections used for comparison

The results of the comparison of velocity profiles in the axis of the stenosis can be found in Fig. 4. It is obvious from the graph, that different turbulent models estimate the flow behavior with various accuracy. The most reliable models are k-ω SST model (SST) and V2F Low-Reynolds Number k-ε model (V2F), which give the best agreement with experiment in terms of the axial velocity, as documented in the Fig. 4. The worst agreement was found for Spallart-Allmaras (SA) and Standard k-ε model, which predict faster decrease of velocity downstream of the stenosis in the distance 2 – 3 diameters from the stenosis throat. The wrong prediction was probably caused by an inaccurate estimate of the shear layer separation.

The figures 5 and 6 show the comparison of velocity profiles from the tube axis (0) to the wall (1) in the distances 0, 1 and 2 diameters (Fig. 5) and in the distances 3, 4 and 5 diameters from the stenosis throat (Fig. 6). Both the graphs illustrate gradual transition of the velocity profile from almost flat profile (Z = 0) to less turbulent profile. Fairly good agreement of all the models on the velocity profile was found in the location Z = 0, i.e. exactly in the stenosis throat. However, despite the fact, that all the models predicted the same velocity profile, the actual measured velocity profile was

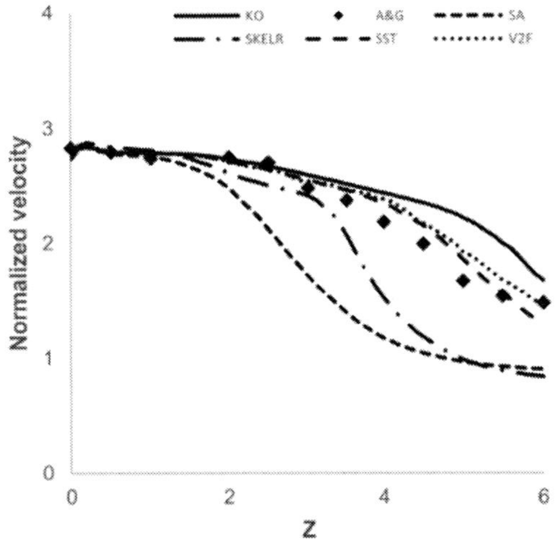


**Fig. 4.** Comparison of centreline velocities

slightly different (Fig. 5, Z = 0). This small discrepancy was reported also by Tan [3].

The velocity profile changes close to the wall due to the recirculation zone in the location Z = 1. The normalized velocity turns into negative values there. This phenomenon was captured by Spalart-Allmaras turbulence model, whose prediction corresponds most with the experiments. All other models also predict recirculation zone, however they significantly underpredict axial velocity values.

The Spalart-Allmaras model agrees well with the experiment also in the following location in the distance of 2 diameters from the throat, nevertheless, all the other models approximate the experimentally measured velocity profile as well. The main difference is that both the Spalart-Allmaras model and experiment show positive values of the normalized velocity. It means that

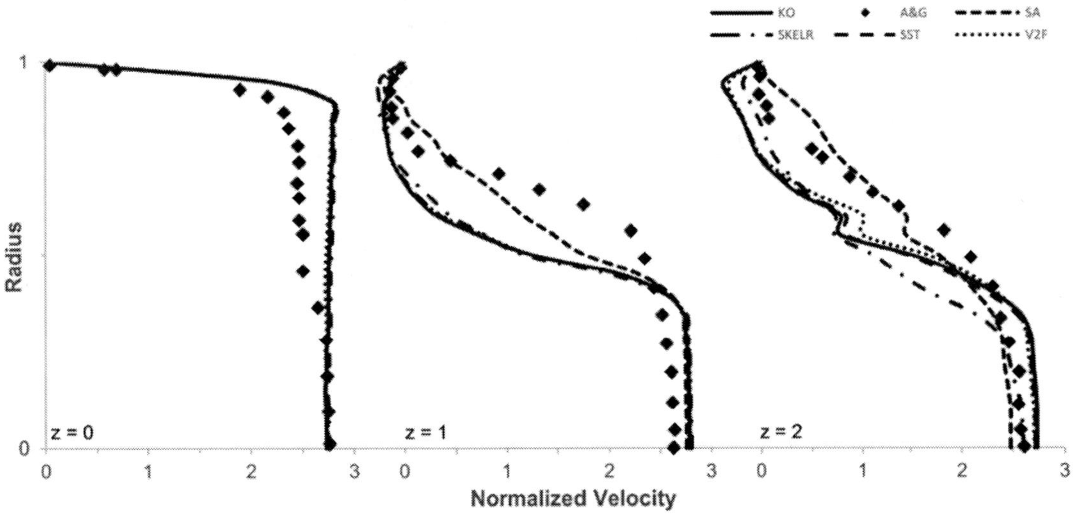


**Fig. 5.** Comparison of normalized velocities in z = 0 to z = 2

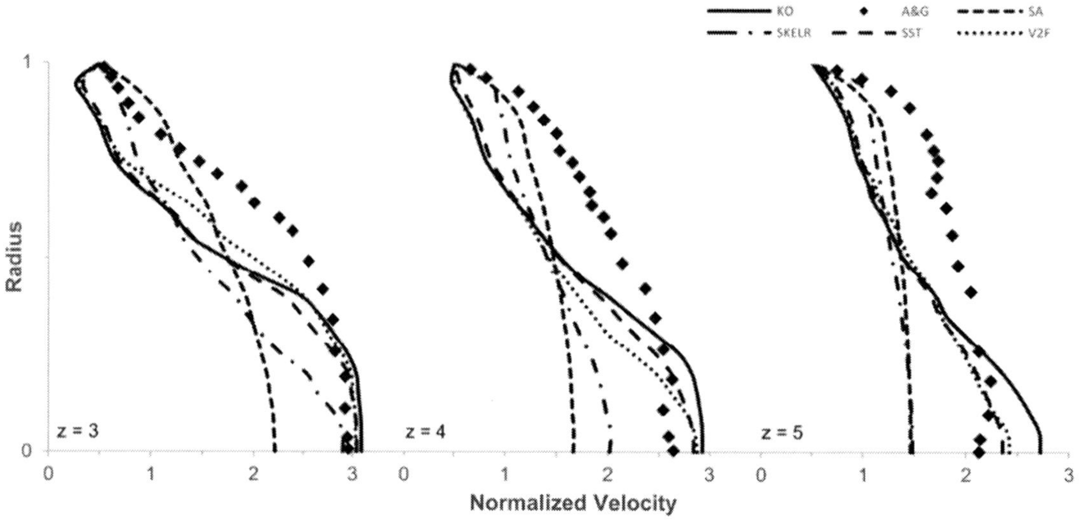


**Fig. 6.** Comparison of normalized velocities in z = 3 to z = 5

the flow separation and the recirculation region occur immediately downstream of the throat and continues to the distance of 2 diameters downstream of the throat. All other models predict larger recirculation region length.

Velocity profiles in distances 3 to 5 diameters downstream of the stenosis throat are plotted in Fig. 6. The graph confirms that k-ω (KO), V2F, SKELR and SST models indeed predict larger recirculation region length compared to the actually measured values. They all predict the end of the recirculation zone in the Z = 5 distance. After that the transition to the ideal turbulent profile can be seen. Spalart-Allmaras model in the locations Z = 4 and 5 shows gradual transition of the velocity profile to the shape qualitatively similar to the experimental values, however, the axial velocity values are significantly underpredicted.

## 4 Conclusions

The results show that the Spalart-Allmaras model can well predicts the behaviour in near-wall region, but it is unable to predict flow in axis of pipe. This model was able to capture the separation of boundary layer with subsequent recirculation downstream of the constriction. Best models for predicting of flow in axis of pipe were models based on k-ω approach. The least suitable models for this case can regarded a models based on k-ε approach, which are mostly applied to cases focused on free. Since the bronchial tree, which is part of the respiratory tract consists of a series of consecutive similar cases it is advisable to choose the model that best predicts this behaviour, thus preventing a significant increase of computational errors.

This work was supported by the Czech Science Foundation under the grant GA16-23675S, by the project FSI-S-14-2355 and by the project LO1202 NETME CENTRE PLUS with the financial support from the Ministry of Education, Youth and Sports of the Czech Republic under the "National Sustainability Programme I".

## References

1. J. Elcner, F. Lizal, J. Jedelsky, M. Jicha M. Chovancova, Biomechanics and Modeling in Mechanobiology, *15*(2), 447–469, (2016).
2. C. Kleinstreuer, Z. Zhang, International Journal of Multiphase Flow, 29, 271–289, (2003).
3. F. P. P. Tan, G. Soloperto, S. Bashford, N. B. Wood, S. Thom, A. Hughes, X. Y. Xu, Journal of Biomechanical Engineering, *130*(6), 61008. (2008).
4. P. W. Longest, L. T. Holbrook, Advanced Drug Delivery Reviews, 64(4), 296–311, (2012).
5. S. A. Ahmed, D. P. Giddens, Journal of Biomechanics, 16(7), 505–516, (1983).
6. Manual Star-CCM+ ver. 9.
7. F. R. Menter, AIAA Journal, 32(8), 1598–1605, (1994).

# Experimental investigation of liquid-liquid system drop size distribution in Taylor-Couette flow and its application in the CFD simulation

*Reza* Farzad[1, 2*], *Stefan* Puttinger[2], *Stefan* Pirker[2] and *Simon* Schneiderbauer[1, 2]

[1]Christian Doppler Laboratory for Multi-Scale Modeling of Multiphase Processes, Johannes Kepler University, 4040 Linz, Austria
[2]Department of Particulate Flow Modelling, Johannes Kepler University, 4040 Linz, Austria

**Abstract.** Liquid-liquid systems are widely used in the several industries such as food, pharmaceutical, cosmetic, chemical and petroleum. Drop size distribution (DSD) plays a key role as it strongly affects the overall mass and heat transfer in the liquid-liquid systems. To understand the underlying mechanisms single drop breakup experiments have been done by several researchers in the Taylor-Couette flow; however, most of those studies concentrate on the laminar flow regime and therefore, there is no sufficient amount of data in the case of in turbulent flows. The well-defined pattern of the Taylor-Couette flow enables the possibility to investigate DSD as a function of the local fluid dynamic properties, such as shear rate, which is in contrast to more complex devices such as stirred tank reactors. This paper deals with the experimental investigation of liquid-liquid DSD in Taylor-Couette flow. From high speed camera images we found a simple correlation for the Sauter mean diameter as a function of the local shear employing image processing. It is shown that this correlation holds for different oil-in-water emulsions. Finally, this empirical correlation for the DSD is used as an input data for a CFD simulation to compute the local breakup of individual droplets in a stirred tank reactor.

## 1 Introduction

The mixture of two or more immiscible liquids, is called emulsion. Emulsions are important in several processes such as food, pharmaceutical, polymer and chemical industries. Drop size distribution (DSD) plays an essential role as it controls the overall heat and mass transfer inside the reactor [1]. Several studies are done to measure the DSD in the system. On one hand, some of these works are just focused on the single drop experiment [2–4] to define the break up kernel for the Population balance equation [5] . On the other hand, there are studies focused on the drop size distribution in the stirred tank reactor regardless of the event happens to each droplet [6,7]. The well-defined pattern of the Taylor-Couette flow enables the possibility to investigate DSD as a function of the local fluid dynamic properties, such as shear rate, which is in contrast to more complex devices such as stirred tank reactors.

Equation (1) is one of the first expressions to evaluate the Sauter mean diameter in a stirred tank reactor that was developed by the work of Shinnar and Church [8] and Chen and Middleman [6].

$$\frac{D_{32}}{L} = A\, We^{-3/5} \quad (1)$$

Where A and L denote a constant value, which can be found experimentally and diameter of the impeller (characteristics length), respectively.

There are several models which follow the same idea as equation (1) to predict the global Sauter mean diameter[9]; however, they are not able to predict the local drop size distribution as they use global flow values such as the average dissipation rate and the average velocity in the stirred tank reactor. In this study we present an overview of the experiment of emulsion formation in a Taylor-Couette flow device to find a correlation as function of local shear rate, then implement it beside a new idea for the droplet break up in order to simulate the liquid-liquid system in a stirred tank reactor.

## 2 Theory and experiment setup

Fig.1 shows the schematic view of the experiment. The $R_{in} = 112.5mm$ and $R_{out} = 120mm, h = R_{out} - R_{in} = 7.5mm$ is the gap between the cylinders Characteristic length.

* Corresponding author: reza.farzad@jku.at

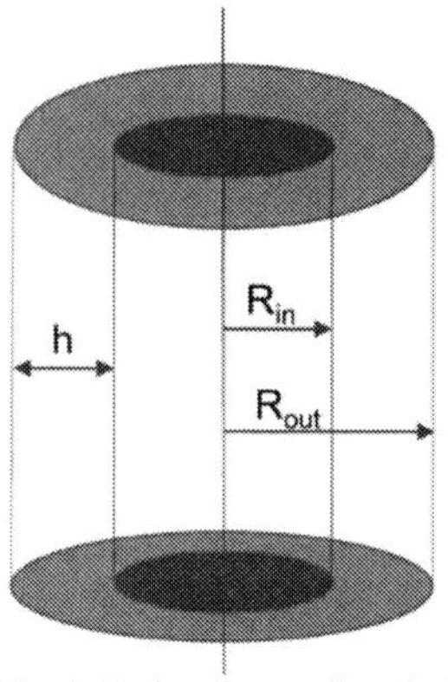


**Fig. 1.** Taylor-Couette flow device

shear rate in Couette flow is defined, $a = \frac{U}{h}$ where $U$ is the linear velocity of the inner cylinder. Thus, the dimensionless numbers can be defined as below:

$$Re = \frac{\rho_c a\, h^2}{\mu_c} \qquad (2)$$

$$Ca = \frac{\mu_c a h}{\gamma} \qquad (3)$$

$$We = \frac{\rho_c a^2 h^3}{\gamma} \qquad (4)$$

Where $\mu_c$,$\rho_c$ and $\gamma$ denote continuous phase viscosity, continuous phase density and interfacial tension, respectively. The liquid-liquid system in this work consists of 1000 ml of continuous phase and 10 ml of dispersed phase. Volume fraction of the dispersed phase was kept under 1% in order to neglect the coalescence [1]. Therefore, pure break up process was considered in the current work. Table 1 shows the oils at temperature of 21 °C which were used as the dispersed phase and water was the continuous phase, respectively.

**Table 1.** Liquid-Liquid system properties at 21°C

| | Density ($Kg/m^3$) | Viscosity (cP) | Interfacial tension (mN/m) |
|---|---|---|---|
| Raps oil/w | 907 | 70 | 23.71 |
| Soybean oil/w | 916 | 47 | 20.81 |
| Silicon oil/w | 760 | 0.49 | 15.9 |

A high speed camera (Photron FASTCAM SA2) was employed along with a 1200 W photography lamp to capture the movie from the droplets. ImageJ (Non-commercial scientific program) was used for post processing. Diameter was calculated based on the area of the droplets ($D = \sqrt{\frac{4A}{\pi}}$).

## 3 Experimental investigation

Experiment was started by the lowest possible rotational speed and it was increased gradually till maximum speed can be reach by the electromotor. Movie was captured on the specific rotational speeds for each oil.

### 3.1. Results and discussion

Fig.2 shows the log-normal cumulative drop volume distribution of raps oil.

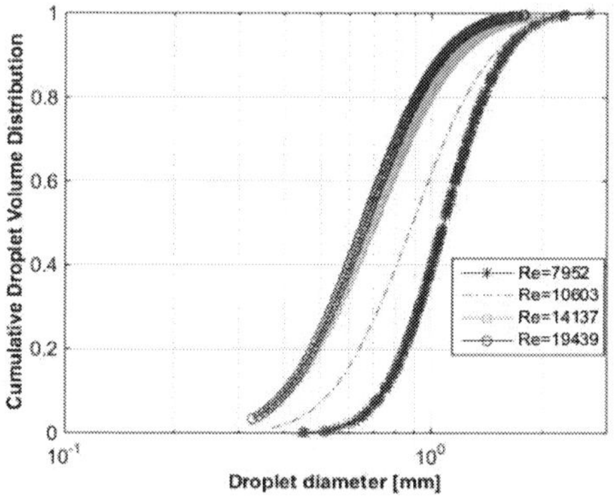


**Fig. 2.** Raps oil- lognormal cumulative drop volume distribution

As it can be seen in Fig.2, drop size distribution moves towards to the smaller sizes as the rotational speed (Re) increases. It is the same trend for the other oils. We observed that the fluid properties such as interfacial tension and viscosity of the dispersed phase are important to keep the interface stable against the disruptive forces which mainly comes from dynamic pressure of the continuous phase.

Fig.3 illustrates the dimensionless Sauter mean diameter of the raps oil against shear rate for all cases. Sauter mean diameter reduces as the shear rate increases, however the trend is not linear. In particular, as the shear rate increases to very high values, it appears that the Sauter mean diameter approaches a constant value.

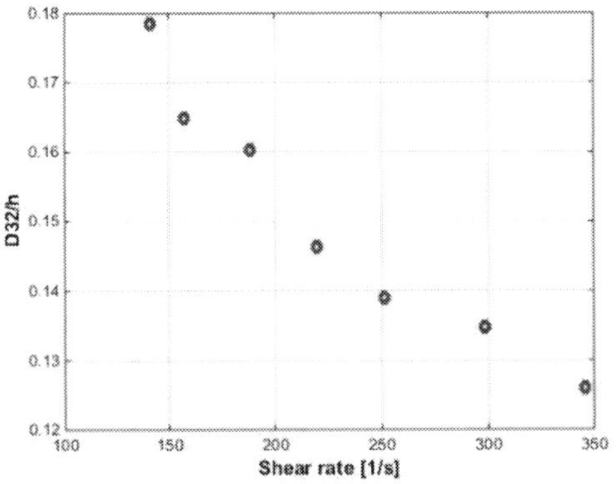


**Fig. 3.** Raps oil dimensionless Sauter mean diameter vs. shear rate

From data fitting by employing a least-squares method, the equation below was obtained,

$$\frac{D_{32}}{h} = We^{-0.0045}\, Re^{-0.4059} \left(\frac{\mu_d}{\mu_c}\right)^{0.3217} \left(\frac{\rho_d}{\rho_c}\right)^{-6.2485} \quad (5)$$

Finally, we found that the standard deviation of the droplet size distribution increases linear with the Sauter mean diameter, which yields $\sigma = 0.3\, D_{32}$.

## 4 CFD simulation

The main goal is to predict the drop size distribution inside a stirred tank reactor. Thus, efficient and accurate breakup and coalescence models are required in order to simulate the drop size distribution of the liquid-liquid system.

### 4.1. Break up model

The breakup model follows the experimental configuration, where the droplet size distribution can be obtained for a specific shear rate. Thus, from the local shear rate computed from the CFD, the corresponding local equilibrium droplet size distribution due to breakup can be obtained. In particular, fitted correlation for the Sauter mean diameter (equation (5)) is used in each cell (computational domain) inside the stirred tank to produce the local drop size distribution. Finally, actual size of a droplet (which is computed by employing a discrete phase model) is compared to the local equilibrium drop size distribution, which comes from the correlation. If the actual drop is larger than $D_{90}$, then it is very likely that the drop breaks. It has to be noted that this is actually realized by a random process, where on the one hand the size of the daughter droplets is sampled from the equilibrium distribution and on the other hand a random number is sampled to initialize the breakup itself.

### 4.2. CFD Simulation

To verify the presented approach we compare our numerical simulations with Roudsari et al [10] and Boxall et al [7], who provide detailed simulations and measurements of droplet size distributions in liquid-liquid stirred tank reactor. It has to be noted that their system is water-in-oil; however, it is a good verification system, which we can be used as a case study. Roudsari et al [10] used PBE with multiple reference frame method to model the impeller's rotation which was run in the steady state model.

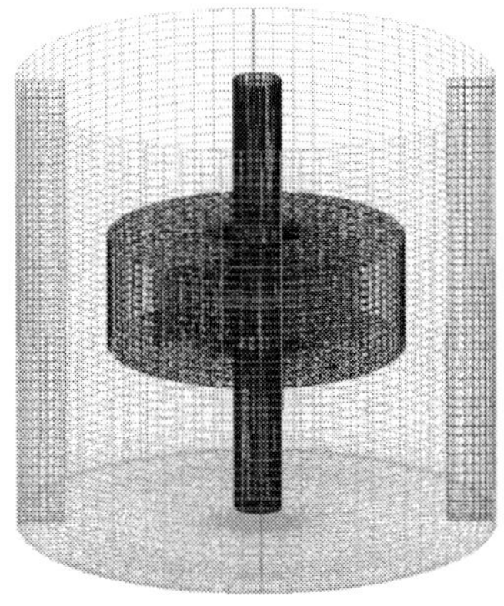

**Fig. 4.** Stirred tank reactor mesh

Hexahedral mesh of the stirred tank reactor was generated by ANSYS ICEM (Fig.4) and simulations were performed by ANSYS FLUENT 16. A specific oil so-called 'Conrone oil' was used as the continuous phase From Boxall et al [7] and water was selected as the dispersed phase. The Conrone oil density, viscosity and interfacial tensions are 842 kg/m$^3$, 3.1 cP and 20 mN/m, respectively. The volume fraction of water is 15% in the Boxall et al [7] paper, however in order to avoid the implementation of a model for the coalescence, a linear correction factor (a function of dispersed phase volume fraction) was applied to the Sauter mean diameter (equation (5)), since the correlation was founded for the dispersed phase which has the volume fraction of 1%. Break up model was written as a user-defined function (UDF) to be implemented in ANSYS FLUENT. Eulerian-Lagrangian scheme along with the K-ε turbulence model was used to simulate the emulsion in the stirred tank reactor. Rotation of the impeller was modelled by sliding mesh method.

### 4.3. Results and discussion

Initial droplets with diameter of 0.3 mm were injected at t=0. Simulation were performed for 300 rpm and 600 rpm. Simulation stopped when the final drop size distribution did not change any more. Fig.4, 5 show the cumulative drop size distribution for 300 rpm and 600 rpm, respectively.

DOI: 10.1051/epjconf/201714302021

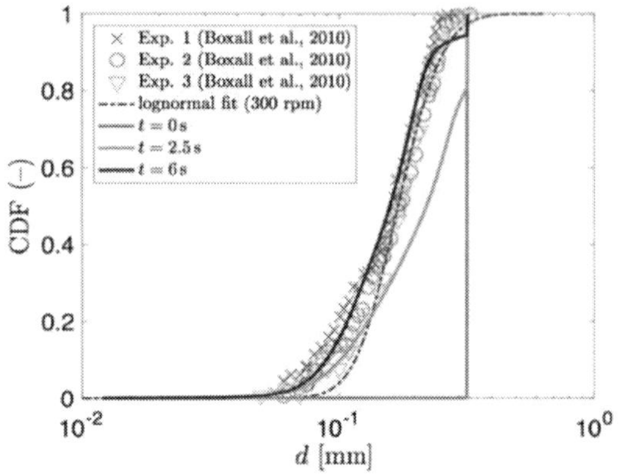


**Fig. 5.** Cumulative drop size distribution- 15 vol% water in oil-300 rpm

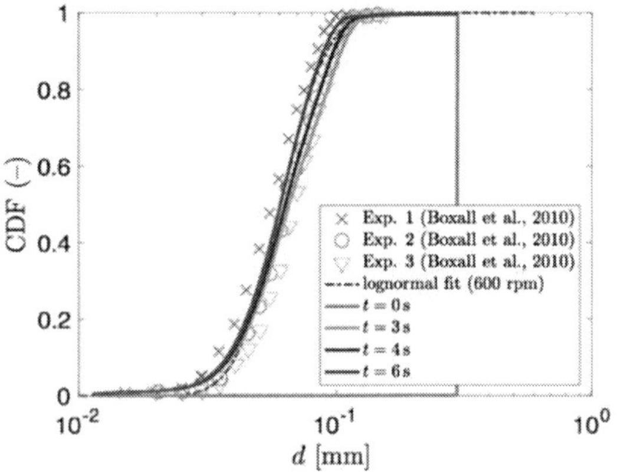


**Fig. 6.** Cumulative drop size distribution-15 vol% water in oil-600 rpm

Green line indicates the drop size distribution at t=0. After performing the simulation for 6 seconds in both cases, final results were obtained. Results are in a good agreement with the experimental data from Boxall et al [7]. This model works if the rotational speed goes from the lower speeds to higher one, since there is no coalescence model then simulation is not able to find the correct drop size distribution if the rotational speed is switched from higher speeds to a lower speed.

## 5 Conclusion

Experimental investigation of drop size distribution in a Taylor-Couette device was performed in the current work. Results show that by increasing the Reynolds number, drop size distribution shifts towards the smaller droplets. Also, Sauter mean diameter becomes smaller as shear rate increases. Fitted correlation was used to predict the local drop size distribution in the stirred tank reactor in cooperation with the break up idea. CFD simulation of the liquid-liquid system was done for a specific case and the final results look promising. Current strategy works only if the rotational speed is going upward and the other way around requires coalescence model.

## References

1. D. E. Leng and R. V. Calabrese, in *Handb. Ind. Mix.*, edited by E. L. Paul, V. A. Atiemo-Obeng, and S. M. Kresta (John Wiley & Sons, Inc., 2004), pp. 639–753.
2. J. Solsvik and H. A. Jakobsen, Chem. Eng. Sci. **131**, 219 (2015).
3. S. Maaß, N. Paul, and M. Kraume, Chem. Eng. Sci. **76**, 140 (2012).
4. J. Solsvik, P. J. Becker, N. Sheibat-Othman, I. Mohallick, R. Farzad, and H. A. Jakobsen, J. Dispers. Sci. Technol. **36**, 577 (2014).
5. D. Ramkrishna, *Population Balances* (2000).
6. H. T. Chen and S. Middleman, AIChE J. **13**, 989 (1967).
7. J. A. Boxall, C. A. Koh, E. D. Sloan, A. K. Sum, and D. T. Wu, Ind. Eng. Chem. Res. **49**, 1412 (2010).
8. R. Shinnar and J. M. Church, Ind. Eng. Chem. **52**, 253 (1960).
9. G. Zhou and S. M. Kresta, Chem. Eng. Sci. **53**, 2063 (1998).
10. S. F. Roudsari, G. Turcotte, R. Dhib, and F. Ein-Mozaffari, Comput. Chem. Eng. **45**, 124 (2012).

# Energy dissipation of the liquid in systems with a hydrophobic surface

*Simona* Fialová, *František* Pochylý, *Martin* Hudec

Faculty of mechanical engineering, Brno University of Technology, Technicka 2896/0, Kralovo Pole, 61669, Brno, Czech Republic

**Abstract:** The work is divided into theoretical and experimental parts. In the theoretical part the viscous forces performance for both laminar as well as turbulent flow of compressible fluids are defined. It is split into the viscous forces useful power and the power corresponding to mechanical losses. This part is expressed by the dissipation function. It is shown that the dissipation function considering hydrophobic surfaces cannot be determined from the pressure difference. It is necessary to take into account the slip of the liquid over the hydrophobic surface which depends on the value of surface energy. The values of surface energy for different types of hydrophobic surfaces, obtained by experiment, are shown within this work. The new definition of the dissipation function for hydrophobic surfaces is presented depending on the adhesion coefficient, also determined by an experiment. The theoretical part of the work is complemented by experiment. The experiment is aimed at the differential pressure measurement in a circular cross section pipe. The results are processed for both laminar and turbulent stationary flow in relation to the flow. Both hydrophilic and hydrophobic surfaces and their mutual correlation are taken into consideration. On the basis of the new mathematical model for hydrophobic surfaces the value of dissipation function is determined depending on the slip velocity of the liquid along the hydrophobic surface.

## 1 Introduction

Hydrophobicity 8 is related to the effect of adhesion forces between the surface and the liquid. The degree of hydrophobicity is assessed on the basis of the adhesion of the droplet to the surface, represented by the contact angle $\theta$ 1, 2. Angles $\theta > 80^\circ$ characterize hydrophobic surfaces, angles $\theta > 120^\circ$ characterize so called superhydrophobic surfaces and surfaces associated with angles $\theta > 150^\circ$ are ultrahydrophobic. This characteristic of the surface to repel liquid was assumed by Navier 7, 200 years ago. For liquid on hydrophobic surface, he predicted the shear stress to be proportional to the slipping velocity.
Within the experimental part of the woldwide work the description of the behavior of liquids in the vicinity of hydrophobic walls affected only 13, who processes the analytical solutions of the motion of a group of continuous variable systems - everything is simplified to the plane; and 13 who shows two types of determination of the motion tensor that characterizes straight hydrophobic surface for any slip length. It is clear that the experimental determination of the influence of hydrophobicity is clearly easier than its analytical description, which would lead to the possibility of using for simulations in commercial softwares. Experiments describing the flow behavior after hydrophobic and superhydrophobic surfaces are covered by series of authors. E.g. 10 who studied the wettability effect on the stability of flow and transition to turbulent flow. The slip boundary condition on the wall is given as a velocity proportional to the velocity gradient and slip length. Calculations on the wall used linearized Navier-Stokes equations for 2D flow (Orr-Sommerfeld equations). Their conclusion is that the transition to turbulence is slowed in the direction of the streamline and accelerated in the direction perpendicular to it.
Expected reduction in losses due to friction decrease on the wall by hydrophobic and ultrahydrophobic surfaces are experimentally verified eg. in 13 where the authors studied the ability of superhydrophobic

materials to reduce friction losses in laminar and turbulent flow regimes. They predict the emergence of the air layer on the superhydrophobic surface. They used PIV and pressure measurement in the droplets. They also use the Navier model [7] of the slippage (again velocity gradient and length of slip).

The fact that the superhydrophobic materials have the potential to reduce friction by forming an air layer (plastron) on the surface refers e.g. 12. Numerical simulation is performed using the two-phase laminar flow. Using Navier [7] slip model (velocity gradient and length of slip) they point out that the Navier model simulates only the mean value but does not simulate the interaction between the surface and the air and between the surface portions and the outer liquid stream. The results show losses reduction in Stokes flow by 19% in the water-air system. When increasing the Reynolds number to 100, the losses will reduce by 50% (assuming a viscous friction and suppressing the plastron separation). Constant velocity slippage and losses reduction due to plastron cause a complete suppression of the turbulent flow when flow is torn off, but is still stable (24 <Re <130). Important is the recognition that helps to the losses reduction is more the detailed surface geometry than averaged influence, which is expected to reduce overall losses.

Reduction of losses of 40% and the length of the slip to 20 microns for ultrahydrophobic surface states as a result of a series of experiments 9. The losses reduction by using a hydrophobic surface has not been proven.

The importance of hydrophobic surfaces for the flow in tube of 16 mm diameter, where occurred under laminar flow the losses reduction of 14% is demonstrated in 5. Mathematically, again using the Navier-Stokes equations, where is introduced the slip coefficient. The same coefficient is implemented in the loss coefficient λ. They conducted experiments to verify the theory in small diameter tubes (6 and 12 mm) and the coating to about 10 microns.

In fact all the known research of the viscous liquid flow over hydrophobic surfaces was made in small diameters tubes and low Reynolds numbers (including the uncerntainty of the Reynolds number value for the flow with assumed air plastron) The degree of hydrophobicity determines the magnitude of slippage velocity of the liquid over the surface 3. Slip conditions vary for laminar and turbulent flow 3. In case of laminar flow a thin layer of air between the surface and the liquid is formed as an effect of adhesion and intermolecular forces. The thickness of the layer depends on the Reynolds number. At higher Reynolds numbers, the layer gets thinner until it dissappears. This corresponds to the size of the slip velocity on the wall. When the air layer disappeares, the liquid is slipping, provided that the shear stress $\tau$ exceeds a certain critical value $\tau_k$. The slippage exists, if it holds:

$$|\tau| > |\tau_k|. \tag{1.1}$$

The slip of the liquid is in this case governed by the new boundary condition 1. For the flow in the pipe of constant cross-section the following applies:

$$\boldsymbol{\tau} = -k v. \tag{1.2}$$

Another situation may occur when a turbulent flow develops. Given that the velocity on the hydrophobic surface is not zero, there will be nonzero values of turbulent viscosity in contrast to hydrophilic surfaces whereupon the liquid adheres to. And this fact can significantly affect energy dissipation during the flow of a liquid over a hydrophilic/hydrophobic surface.

## 2 Energy dissipation

The power consumption $P$ required to cover the hydraulic losses is determined by the dissipation function $\mathcal{D}$ in the form described in the 5 as Rayleigh friction:

$$\mathcal{D} = \frac{1}{2}P = \frac{1}{2}\int_V \Pi_{ij}\frac{\partial v_i}{\partial x_j}dV. \tag{2.1}$$

A turbulent flow is characterized by significant unsteady pulsating velocity components $v_i'(\bar{x}, t)$, that can have both positive and negative values. It is assumed that the average value for a sufficiently long period of time $T$ is equal to zero. Thus:

$$\frac{1}{T}\int_{t-\frac{T}{2}}^{t+\frac{T}{2}} v_i'(\boldsymbol{x},\tau)d\tau = 0. \tag{2.2}$$

When we decompose the resulting velocity $\boldsymbol{v}(\boldsymbol{x},t)$ to its components:

$$\boldsymbol{v} = \widehat{\boldsymbol{v}}(\boldsymbol{x},t) + \boldsymbol{v}'(\boldsymbol{x},t), \tag{2.3}$$

Than for $\widehat{\boldsymbol{v}}$ , in regards to (2.2), holds the relation:

$$\widehat{\boldsymbol{v}} = \frac{1}{T}\int_{t-\frac{T}{2}}^{t+\frac{T}{2}} \boldsymbol{v}(\boldsymbol{x},\tau)d\tau. \tag{2.4}$$

For the average value of the dissipation function can be analogously written:

$$2\hat{\mathcal{D}} = \hat{\boldsymbol{P}} = \\ = \frac{1}{T}\int_{t-\frac{T}{2}}^{t+\frac{T}{2}}\int_V \hat{\Pi}_{ij}(\boldsymbol{x},\tau)\frac{\partial \hat{v}_i(\boldsymbol{x},\tau)}{\partial x_j}dVd\tau. \quad (2.5)$$

The term (2.5) can be in the case of steady flow corrected to:

$$2\hat{\mathcal{D}} = \hat{\boldsymbol{P}} = \int_V \hat{\Pi}_{ij}(\boldsymbol{x})\frac{\partial v_i(\boldsymbol{x})}{\partial x_j}dV. \quad (2.6)$$

$$\hat{\Pi}_{ij} = 2(\eta + \eta_T)\hat{\mathrm{v}}_{ij} - \frac{2}{3}\rho k\delta_{ij} \quad (2.7)$$

where $\eta$ – molekular viscosity, $\eta_T = \eta_T(\boldsymbol{x})$ is the turbulent viscosity, $k$ is the kinetic energy density of dissipation for the turbulence model $k - \varepsilon$. [4]
It holds:

$$k = \frac{1}{T}\int_{t-\frac{T}{2}}^{t+\frac{T}{2}} v_i' v_i' d\tau. \quad (2.8)$$

$$\eta_T = \rho C_\mu \frac{k^2}{\varepsilon}; \quad C_\mu = konst. \quad (2.9)$$

$$\varepsilon = \nu\frac{1}{T}\int_{t-\frac{T}{2}}^{t+\frac{T}{2}} \frac{\partial v_i'}{\partial x_j}\frac{\partial v_i'}{\partial x_j}d\tau\,, \qquad \nu = \frac{\eta}{\rho}. \quad (2.10)$$

From the definition of (2.8) ÷ (2.10) result the following conclusions.

**Hydrophillic surface**
The condition of liquid adhesion holds for the hydrophillic surface. On the immobile wall then holds:

$$\boldsymbol{v} = 0 \Longrightarrow \widehat{\boldsymbol{w}} = 0\,;\ \boldsymbol{v}' = 0\,; \\ k = 0\,;\ \eta_T = 0\,;\ \varepsilon \neq 0. \quad (2.11)$$

**Hydrophobic surface**
Nonwettable wall is characterized by so called hydrophobic surface that repels the liquid which slides over the surface and thus the velocity is considered as non-zero. Contrary to the wettable wall, the following conditions can be written:

$$\boldsymbol{v} \neq 0\,;\ \hat{\boldsymbol{v}} \neq 0\,;\ \boldsymbol{v}' \neq 0\,; \\ k \neq 0\,;\ \eta_T \neq 0\,;\ \varepsilon \neq 0. \quad (2.12)$$

These conditions will significantly affect the value of the dissipation function.
The dissipation function can be set from the Navier-Stokes average equations, considering for example the $k = \varepsilon$ turbulence model. For the incompressible fluid they are expressed in the form:

$$\rho\frac{\partial}{\partial x_j}(\hat{v}_i, \hat{v}_j) - \frac{\partial \hat{\Pi}_{ij}}{\partial x_j} + \frac{\partial \hat{p}}{\partial x_i} = 0 \quad (2.13)$$

$$\hat{p} = \frac{1}{T}\int_{t-\frac{T}{2}}^{t+\frac{T}{2}} p(\boldsymbol{x},\tau)\,d\tau = \hat{p}(\boldsymbol{x}). \quad (2.14)$$

From the expression (2.13), the dissipation function can be derived by integration over the multiple contiguous area $V$ as:

$$2\hat{\mathcal{D}} = \hat{P} = \\ = -\left(\frac{\rho}{2}\int_S \hat{v}_m\,\hat{v}^2 dS + \int_S \hat{p}\,\hat{v}_m dS\right) + \\ + \int_S \hat{\Pi}_{ij} m_j\,\hat{v}_i dS + \int_\Gamma \hat{\Pi}_{ij}\,\hat{v}_i m_j d\Gamma \quad (2.15)$$

$$\int_S p m_i dS = \int_S \hat{\Pi}_{ij} m_j\,dS + \\ + \int_\Gamma \Pi_{ij} m_j\,d\Gamma - \rho\int_S \hat{v}_i\,\hat{v}_m dS \quad (2.16)$$

Where $\hat{v}_m = \hat{v}\cdot\mathbf{m}$, $\mathbf{m}$ – the nominal vector of the outer normal to the liquid, $S$ – the input/output surface, $\Gamma$ – rigid surface (wall).

If the term (2.15) is simplified for the simply contiguous area formed by the pipe of constant crossection, it holds:

$$2\hat{\mathcal{D}} = \hat{P} = \int_\Gamma \hat{\Pi}_{ij}\,\hat{v}_i m_j d\Gamma - \int_S \hat{p}\,\hat{v}_m dS \quad (2.17)$$

$$\int_S p m_i dS = \int_\Gamma \Pi_{ij} m_j\,d\Gamma \quad (2.18)$$

For a cirrcular cross section of the pipe, (2.18) can be written in the form:

$$\Delta\hat{p} = \frac{\Gamma}{S}(\eta + \eta_T)|tg\alpha|, \quad (2.19)$$

where $S = \pi R^2$ ; $\Gamma = 2\pi RL$, $R$ – pipe radius, $\alpha$ – velocity profile angle on surface $\Gamma$, $|tg\alpha| = \frac{dv_T}{dr}$, $L$ – pipe length.

Considering that in the given case it holds $v_T = (v_1, 0,0)$ , the term (2.17) can be modified to the form:

$$2\hat{\mathcal{D}} = \hat{\mathrm{P}} - v_T\Gamma(\eta + \eta_T)|tg\alpha| + \Delta\hat{p}v_sS \quad (2.20)$$

After substitution in equation (2.19) we obtain a very important result:

$$2\hat{\mathcal{D}} = \hat{\mathrm{P}} = (\eta + \eta_T)\Gamma(v_S - v_\Gamma)|tg\alpha| \quad (2.21)$$

It is obvious from terms (2.19), 2.20) that the turbulent viscosity value and $|tg\alpha|$ will significantly affect the dissipation function.

## 3 The dependence of the dissipation function on the velocity profile shape

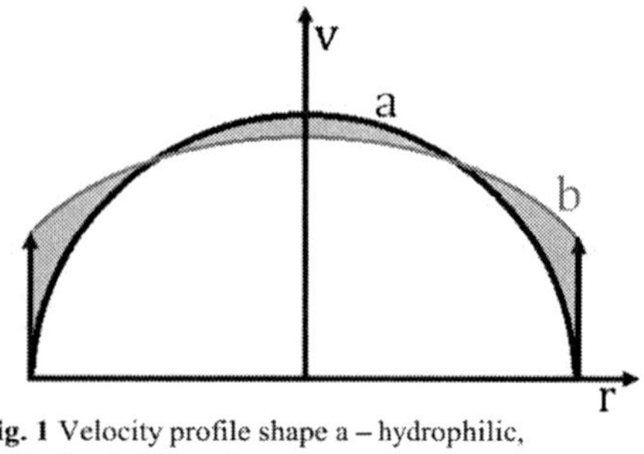


**Fig. 1** Velocity profile shape a – hydrophilic, b – hydrophobic surface

When laminar flow is considered, terms (2.18), (2.20) can be written in the form:

$$\Delta p = \frac{\Gamma}{S}\eta|tg\alpha| \quad (3.1)$$

$$2\mathcal{D} = \eta\Gamma(v_S - v_\Gamma)|tg\alpha|. \quad (3.2)$$

From Fig. 1, it is obvious that:

$$|tg\alpha_a| > tg\alpha_b \quad (3.3)$$

Given that for laminar flow $\eta = const.$ it is clear that slippage of the liquid will reduce the pressure difference $\Delta p$ as well as the hydraulic losses and the value of the dissipation function. This is valid for the case of slippage due to an air layer, as well as when it disappears.

Another situation occurs for the turbulent flow, when both the pressure difference and the dissipation function will be affected by nonzero turbulent viscosity.

In that case, it holds:

$$\Delta\hat{p} = \frac{\Gamma}{S}(\eta + \eta_T)|tg\alpha| \quad (3.4)$$

$$2\hat{\mathcal{D}} = \hat{\mathrm{P}} = (\eta + \eta_T)|tg\alpha|(v_S - v_\Gamma) \quad (3.5)$$

From here it is visible that if $|tg\alpha_b| < |tg\alpha_a|$ then probably $\eta_T > \eta$ . Therefore despite of slippage on the wall, it can not occur that both the differential pressure and dissipation function will grow compared to the case of laminar flow.

For example:

$$\Delta p_b - \Delta p_a = \frac{\Gamma}{S}[\eta(|tg\alpha_b| - |tg\alpha_a|) + \eta_T tg\alpha_b] \quad (3.6)$$

$$\Delta p_b - \Delta p_a = [(\eta + \eta_T)tg\alpha_b - \eta tg\alpha_a] \quad (3.7)$$

## 4 Boundary condition on the hydrophobic surface

In the above mentioned hypotheses we have focused only on the fluid slip along the hydrophobic surface.

For these considerations it was not necessary to pay attention to the shape of the boundary condition.

Lets now suppose that the fluid flows in an area with a hydrophobic surface. Further, it is assumed that in a certain sub-region the shear stress vector $\sigma_A$ exceeds a critical size, at which the liquid slips.

Then the following inequality is obtained:

$$|\sigma_A| > |\sigma_K|. \quad (4.1)$$

On the boundary, we assume the validity of the following boundary condition 1, when we change $\bar{v} = 0$ for hydrophilic surface and $\kappa$ represents the adhesion coefficient.

$$\boldsymbol{\sigma}_A - (\sigma \times m) \times \boldsymbol{m} = \kappa \boldsymbol{v}_\Gamma. \quad (4.2)$$

Because it holds:

$$\sigma_i = \Pi_{ij}m_j, \quad (4.3)$$

The expression (4.2) can be written in a form:

DOI: 10.1051/epjconf/201714302022

$$\hat{\Pi}_{jk} m_k m_j m_i - \hat{\Pi}_{ij} m_j = \kappa v_{\Gamma i} \quad (4.4)$$

where $\hat{\Pi}_{jk} = 2\eta_T \hat{v}_{ij} - \frac{2}{3}\rho k \delta_{ij}$. (4.5)

We note that for the numerical implementation of this boundary condition it is necessary to extend the model of turbulence with an additional boundary condition for $k$. Since $k$ is important for the pressure function, we propose the boundary condition for $k$ in the form:

$$k = \xi \rho \hat{v}_{\Gamma i} \hat{v}_{\Gamma i}. \quad (4.6)$$

The value of $\xi$ will depend on the surface quality and it must be obtained from an experiment.

## 5 Experiment

Verification of the effect of hydrophobic surface on hydraulic losses was performed on a tube of circular cross section with a diameter of 42.6 mm and a length of 6 m with water as as a viscous liquid representative. The differential pressure $\Delta p$ was measured by pressure sensors and the value of the specific energy was evaluated as:

$$Y_z = \frac{\Delta p}{\rho}, \quad (5.1)$$

For 4 types of surfaces - one hydrophilic - stainless steel $\theta = 80° \pm 3°$
and three hydrophobic surfaces:
Steel UED - treated by the product Ultra Ever Dry developed by UltraTech international, Inc.: $\theta = 158° \pm 1{,}5°$
Ceramics: $\theta = 95° \pm 3°$
Ceramics + UED: $\theta = 157° \pm 3°$
The results are shown in the following Figures 5.1, 5.2, 5.3.

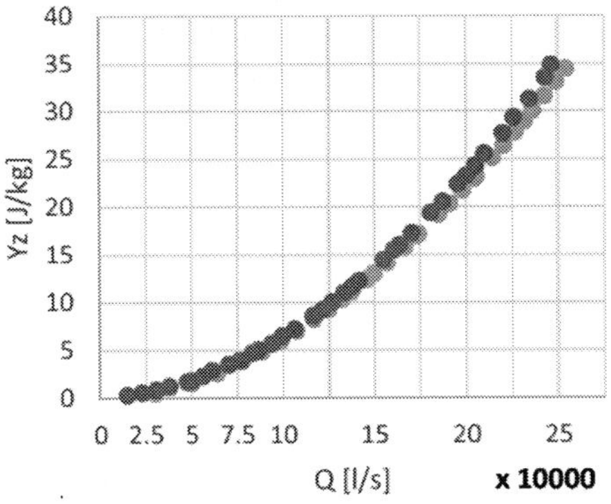


**Fig. 5.1** Comparison of experimental results of the dissipation energy for two types of surfaces – green stailess steel, blue treated with UED

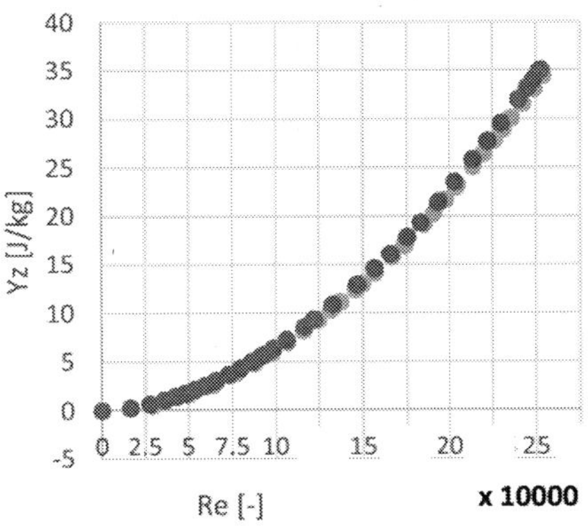


**Fig. 5.2** Comparison of experimental results of the dissipation energy for three types of surfaces – green stailess steel, blue treated with UED, red ceramics

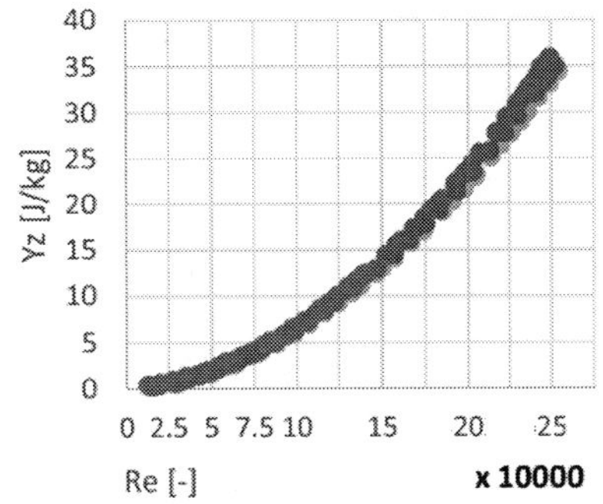


**Fig. 5.3** Comparison of experimental results of the dissipation energy for four types of surfaces – green stailess steel (lowest in the graph), blue steel treated with UED, red ceramics, violet ceramics treated with UED (most upper in the qraph)

From Fig 5.1 it is clear that along a hydrophobic surface, the differential pressure increases with increasing flow rate and thus with increasing Reynolds number as compared to the hydrophilic surface. This result is very surprising since it is usually assumed that the differential pressure will be reduced when the slip of the liquid occurs close to the hydrophobic surface, compared with the hydrophilic surface.
However, considering the expression (2.19) for the pressure difference, it is clear that its value can be significantly affected by the size of turbulent viscosity. Turbulent viscosity on the wall is precisely the result of turbulent fluctuations that are caused by the liquid slip. That is the reason, that the pressure losses can be higher for higher Reynolds numbers. The value of the dissipation function may be in this case determined from relations (2.20) or (2.21).

## 6 Conclusion

Within this work two very important relations were derived for the differential pressure values and dissipation function determination. Dissipation function characterizes the conversion of part of mechanical energy into heat:

$$\Delta \hat{p} = \frac{\Gamma}{S}(\eta + \eta_T)|tg\alpha| \tag{6.1}$$

$$2\hat{\mathfrak{D}} = \hat{P} = (\eta + \eta_T)\Gamma(v_S - v_\Gamma)|tg\alpha|. \tag{6.2}$$

From here it is clear that the values of $\Delta\hat{p}, \hat{P}$ stongly depend on the value of the angle $\alpha$ of the velocity profile on the wall of the tube, slip velocity and the value of the viscosity at the wall.

In the case of laminar flow $\eta_T = 0$ and as a consequence of the liquid slip, angle $\alpha$ decreases, and thus also the value of $\Delta\hat{p}$ and $\hat{P}$ decreases in comparison with a hydrophilic surface.

Another situation occurs when the flow becomes turbulent. The angle α, even though due to the slippage at the hydrophobic surface will decrease, the viscosity $(\eta + \eta_T)$ on the wall will rise up, so that $\Delta\hat{p}$ and $2\hat{\mathfrak{D}}$ might increase.

However, situations may occur where in turbulent flow the value of $\Delta\hat{p}$ and $\hat{P}$ can drop down. This relates to both the shape of the area and the surface quality as well as the method of surface loading by shear stress; for example, a rotating disc near a hydrophobic surface.

In connection with the effects of a hydrophobic surface it is, however, important to realize that due to slippage of the liquid on the surface the vortex structures inside the area may significantly change. This explaines how the dissipation energy can differ in comparison with a hydrophilic surface.

Grant Agency of the Czech Republic, within the project GA101/16-18316S, is gratefully acknowledged for support of this work. Special thanks to Masaryk University, Faculty of Science, Dpt. of Physical Electronics and company ASIO, spol. s r.o..

## References

1. Fialová S., Pochylý F., Identification and Experimental verification of the adhesive coefficient of hydrophobic materials, *Wasserwirtschaft Extra,* 1/2015, ISSN 0043 0978, pp. 125-129 (2015)
2. Jašíková, D. a Kotek, M., The Estimation of Dynamic Contact Angle of Ultra-hydrophobic Surfaces Using Inclined Surface and Impinging Droplet Methods. *EPJ Web of Conferences. Vol. **67***, n. 0. pp. 1 – 6. ISSN 2100-014X. (2014)
3. Pochylý F., Fialová S., Kotek M., Zavadil L., Habán V., Volkov A.V., Parygin A.G., Utilization of hydrophobic layers in the design of hydraulic machines, *EkopumpRus'2015 - sbornik dokladov*, ISBN 978-5-9903138-5-9, printed, pp 77-83 (2015)
4. Versteeg H.K., Malalasekera W., An introduction to computational Fluid Mechanics. *Longman*, ISBN 0-582-21884-5, 257pages, (1995)
5. De Groot S.R., Mazur P.: Nonequilibrium thermodynamics, *North-Holland Publishing*, ISBN 0486647412, (1962)
6. Watanabe K., Yanuar,H. Udagawa, Drag reduction of Newtonian fluid in a circular pipe with a highly water-repellent wall, *J. Fluid Mech. vol. **381***, pp 225-238, (1999)
7. C. L. M. H. Navier, *Mem Acad. Sci. Inst. France, vol. **1***, pp. 414-416, (1823).
8. Olga I. Vinogradova, Slippage of water over hydrophobic surfaces, *Int. J. Mineral. Process.* 56, pp 31-60, (1999)
9. J.Ou, B.Perot, J.P.Rothstein, Laminar drag reduction in microchannels using ultrahydrophobic surfaces, *Phys. Fluids 16*, 4635-4643, (2004)
10. T.Min and J.Kim: Effects of hydrophobic surface on stability and transition. Physics of Fluids 17, 108106 (2005)
11. K.Kamrin, M.Z.Bazant,H.A.Stone, Effective slip boundary conditions for arbitrary periodic surfaces: the surface mobility tensor, J. Fluid Mech. vol. 658, pp 409-437, (2010)
12. B.R.K.Gruncelli, N.D.Sandham, G.McHale, Simulations of laminar flow past a superhydrophobic sphere with drag reduction and separation delay, *Phys. Fluids 25*, 043601, (2013)
13. P.Six, K.Kamrin, Some exact properties of the effective slipover surfaces with hydrophobic patterning, *Phys. Fluids 25*, 021703, (2013)
14. Daniello RJ, Waterhouse NE and Rothstein JP, Drag reduction in turbulent flows over superhydrophobic surfaces. *Phys Fluids 21*: 085103, (2009)

# Velocity profiles of fluid flow close to a hydrophobic surface

*Simona* Fialová[1], *František* Pochylý[1], *Michal* Kotek[2], *Darina* Jašíková[2]

[1] Faculty of mechanical engineering, Brno University of Technology, Technicka 2896/0, Kralovo Pole, 61669, Brno, Czech Republic
[2]Department of Physical Measurement, Technical University of Liberec, The Institute for Nanomaterials, Advanced Technology and Innovation, Studentska 1402/2 Liberec 1 461 17, Czech Republic

**Abstract:** The results of research on viscous liquid flow upon a superhydrophobic surface are presented in the paper. In the introduction, the degrees of surface hydrophobicity in correlation with an adhesion coefficient are defined. The usage of the adhesion coefficient for the definition of a new boundary condition is employed for expressing the slip of the liquid over the superhydrophobic surface. The slip of the liquid was identified on a special experimental device. The essence of the device consists of a tunnel of rectangular cross section whose one wall is treated with a superhydrophobic layer. The other walls are made of transparent organic glass whose surface is hydrophilic. Velocity profiles are measured by PIV. The methodology is drawn so that it allows the speed determination at the closest point to the wall. The measurements were performed for different Reynolds numbers for both laminar and turbulent flow. Based on the measured velocity profiles, marginal terms of use have been verified, expressing slippage of the liquid on the wall. New forms of velocity profiles considering superhydrophobic surfaces are shown within the work.

## 1 Introduction

It is nowadays possible to generalize the solution of hydrodynamic problems and assume hydrophobic character of the surface in the solid/liquid interaction studied by [5], [7], [8]. This characteristic of the surface to repel liquid was assumed by Navier [9], 200 years ago. For liquid on hydrophobic surface, he predicted the shear stress to be proportional to the slipping velocity.

The findings from experimental research on superhydrophobic surfaces is presented in the paper. Experiments describing the flow behavior over the hydrophobic and superhydrophobic surfaces were presented by numerous authors. For example [5], studies the wettability effect on the flow stabilty and the transient process. The slip boundary condition is written that a velocity is proportional to the velocity gradient and the slip lenght.

Calculations on the wall were made by linearized Navier-Stokes equations for 2D flow (Orr-Sommerfeld equation). Their conclusion states that the transition to turbulent flow is slowed down in the streamline direction and quickend perpandiculary on it. Friction losses were stuied within the work [6], where the PIV (Particle Image Velocimetry) method was used and studied the water flow over superhydrophobic surface in the channel of small crossection. Based on our experiment, a contact angle of water droplets on a flat surface [1], [2], [10] is determined. A methodology for the determination of an adhesion coefficient is outlined [1], [2], [10]. For such case applies the boundary condition $\boldsymbol{\sigma}_A = (\boldsymbol{\sigma} \times \boldsymbol{n}) \times \boldsymbol{n} = -k\boldsymbol{v}$ with the rationale and the results of velocity profiles measurements using PIV method being introduced. Using PIV methods, the curves of the velocity profiles in the channel of large rectangular cross section for the various types of superhydrophobic surfaces are measured.

## 2 Experimental setup – Contact angles

The contact angles and adhesion coefficient on superhydrophobic (SH) surfaces was measured using optical shadowgraphy methods. The steel sample was situated between the tabular LED panel and the digital camera. The LED light pointed over the sample straight to the camera. The droplet on the SH surface was represented by a dark shadow object on the white background, see Fig. 2.1.

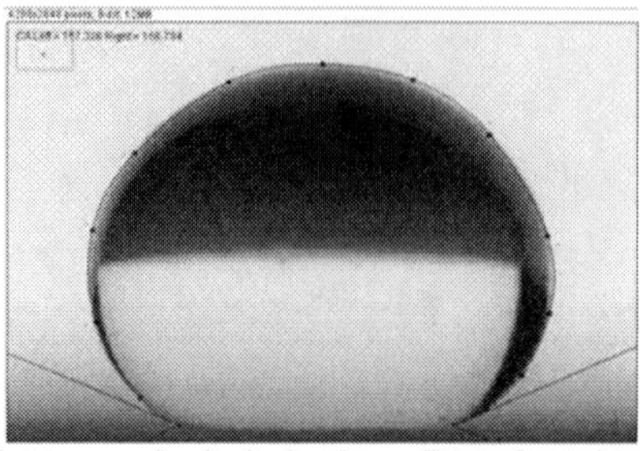

**Fig.2.1** Image of a single droplet on SH surface with determined contact angles CA (left and right)

Because of the high background-object contrast and significant object border identification, this method is suitable for object geometry analysis and curve measurement.

The measurement of the contact angle was a static process applied on one image of a single droplet of 1μl to avoid the gravity effect of the droplet wight on the result. The image records were analyzed by a unique processing algorithm. The contact angles were measured on both sides of each droplet (left and right CA). Ten separated droplets were recorded to ensure a reliable result of the contact angle on the surface under investigation. The contact angle measurements were first made on a dry sample, see Tab. 2.1 and then on a wet sample after measurement of the flow in the water channel, see Tab. 2.2. The dry and wet surfaces provided significantly different results for most samples.

**Tab.2.1** Measurements of CA on air before measurements in the channel

| sample | Left CA | Right CA |
|---|---|---|
| 1 | 159,07 | 161,01 |
| 2 | 150,88 | 157,09 |
| 3 | 157,45 | 154,65 |
| 4 | 157,25 | 152,37 |
| 5 | 159,63 | 142,95 |
| 6 | 151,06 | 147,13 |
| 7 | 165,98 | 149,74 |
| 8 | 155,90 | 156,33 |
| 9 | 158,97 | 152,88 |
| 10 | 151,31 | 161,23 |
| **average** | **156,75** | **153,54** |
| ± | 4,49 | 5,53 |

**Tab.2.2** Measurements of CA after measurements in the channel filled with water

| sample | Left CA | Right CA |
|---|---|---|
| 1 | 125,58 | 123,16 |
| 2 | 125,15 | 134,83 |
| 3 | 123,73 | 119,62 |
| 4 | 133,77 | 128,40 |
| 5 | 145,54 | 140,01 |
| 6 | 141,64 | 131,66 |
| 7 | 148,12 | 129,79 |
| 8 | 124,99 | 115,98 |
| 9 | 125,75 | 132,77 |
| 10 | 121,19 | 125,37 |
| **average** | **131,55** | **128,16** |
| ± | 9,48 | 6,89 |

## 3 Slip coefficient measurement

Measurement of the slip coefficient was also provided by the optical shadowgraphy method. The metal sample was fixed to a base with an adjustable angle of inclination. Liquid injector was used to drop a droplet to the sample surface; upper Fig. 3.1. When the surface was inclined enough, the droplet started to move on the sample; bottom Fig. 3.1. The acceleration of the droplet was analyzed from the images recorded on a high speed camera. The next series of images shows the process of image preprocessing and analysis of the droplet movement.

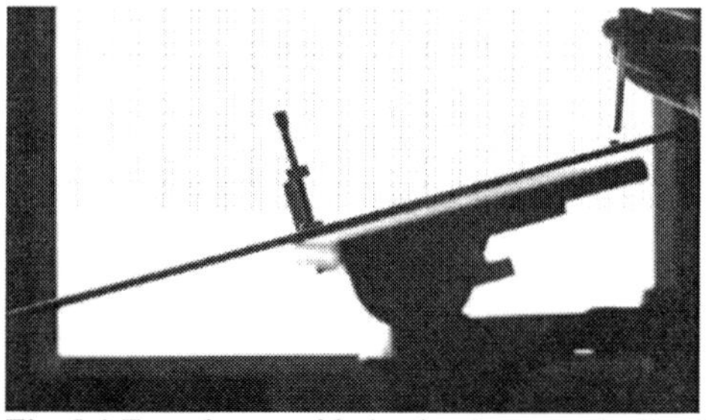

**Fig. 3.1** Experiment with the droplet on an inclined plane – original image of the process

The segmentation algorithm was applied on every recorded image . to find the center of each droplet. Images of the droplet movement were recorded with 500 frames per second. The movement of the center of the droplet between every 5 images is calculated.

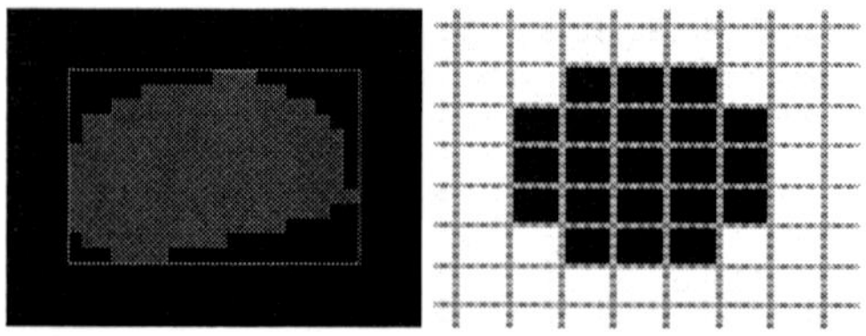

**Fig. 3.2** Image preprocessing – obstructive objects elimination

Using this movement, the instantaneous speed of the droplet is determined. The increase in speed of the droplet gives an information about the acceleration and provides important data to calculate the slip coefficient.

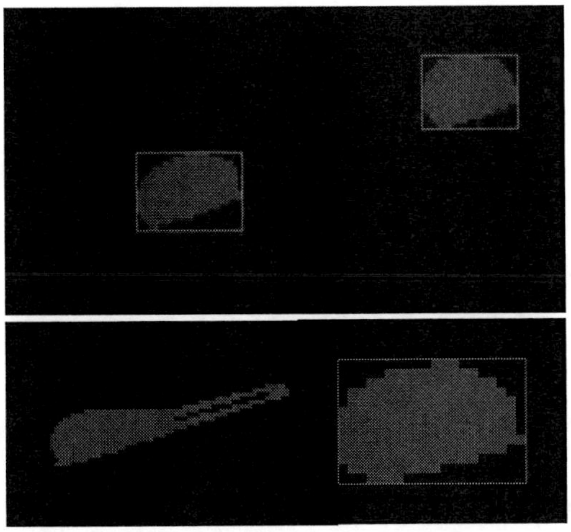

**Fig 3.3** Droplet image recognition and tracking process

The adhesion coefficient is finally determined from the mathematical model according to the equation (3.1), based on regression analysis [1], [2]:

$$\frac{dv_1}{dt} = \left(v_1 - \frac{|\sigma_A|_\kappa}{\kappa}\right) e^{-\frac{S\kappa}{m}t} \qquad (3.1)$$

where $v_1$ – velocity of the droplet, $\sigma_A$ – shear stress, $\kappa$ – adhesion coefficient, $m$ – mass of the droplet, $t$ – time, $S$ - solid/liquid contact surface

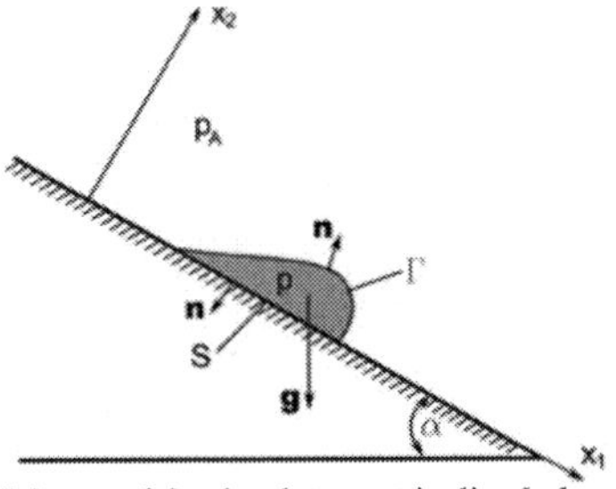


**Fig 3.4** Scheme of the droplet on an inclined plane
p – pressure in the droplet, $p_A$ – atmospheric pressure,
n – nominal vector, S – solid/liquid contact surface,
Γ – air/liquid contact surface, g – gravity acceleration,
α – angle of plane inclination

## 4 Flow profiles measurement

Flow profiles above the SH surfaces were measured using Particle Image Velocimetry (PIV) method. The PIV system was built by Dantec Dynamics, consisting of the high sensitive HighSense NEO camera, NewWave Gemini laser (100mJ on 15Hz) and Dynamic Studio software for data acquisition and analysis [4]. Images were recorder with the frequency of 4Hz. The HighSense NEO camera captured the images with the resolution of 2500x2000 pixels. For PIV correlation analysis, the interrogation areas of 32x32 pixels were used, ino rder to provide a sufficient spatial resolution of the resulting flow field. The standard PIV analysis was applied on the recorded data. In the first step, the light scattered from walls was masked. After that the double images were correlated using the PIV cross-correlation method. The vector maps were validated with the Peak and Range validation method. For each flow condition 250 records were captured and statistically analyzed. Amap of mean velocity vectors was used to obtain the profile plot in the selected position.

The setup of the horizontal laser shield was chosen to eliminate the minor temperate influence on the flow in the channel. The scheme of the PIV system setup is shown in Fig. 4.1.

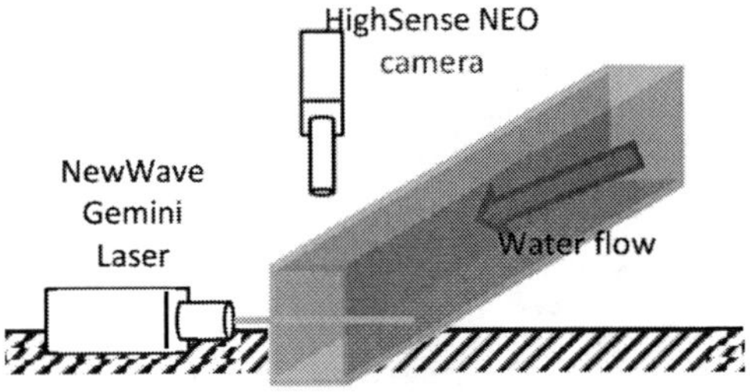


**Fig. 4.1** Scheme of the PIV setup for flow profiles measurement

**Fig. 4.2** Experimental device for the flow velocity profiles measurements

Emphasis was placed on flow settling in the water channel. The channel has the cross-section dimensions 75x80mm, so the settling length was established to 4 meters, as is visible from Fig. 4.2. To eliminate pulsations of the flow caused by a water pump, the flow was driven by gravity. A pump was used to deliver the water to the top reservoir with the level 2 meters above the experiment and the lower reservoir. The flow was controlled by a valve and measured on an induction flowmeter. The flow profiles were measured in the selected range of Reynolds numbers.

## 5 Velocity profiles

As previously stated, the velocity profiles were measured in the rectangular area, made out of plexiglas (perspex) with a contact angle of approximately 70°, for two tasks. In the first task, all inner walls were hydrophilic (blue line) whereas in the second case, one out of four sides of the channel is provided with a superhydrophobic coating (red line) while the three other walls were wettable. From here the asymmetry of the velocity profile is apparent. A comparison of the velocity profiles was performed at a constant flow rate, always for both variants. The following figures show the velocity profiles in the selected plane, but also the details of velocity profiles near the wall. From the graphs it is apparent that indeed in the vicinity of the wall a slippage of the liquid on the superhydrophobic surface occurs. The velocity of the liquid increases with increasing flow rate. See Figures 5.1 – 5.6; where for all holds that the x-axys is distance (canal width in mm) and y-axis is velocity (m/s).

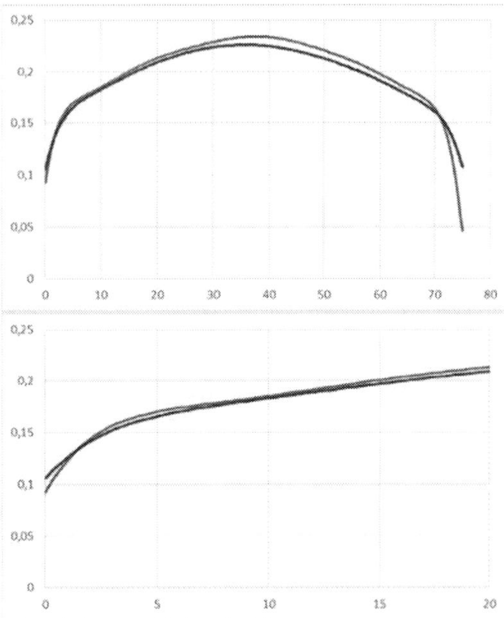


**Fig.5.1** Velocity profile and the detail close to the wall for Re = 13780 - red superhydrophobic, blue hydrophilic

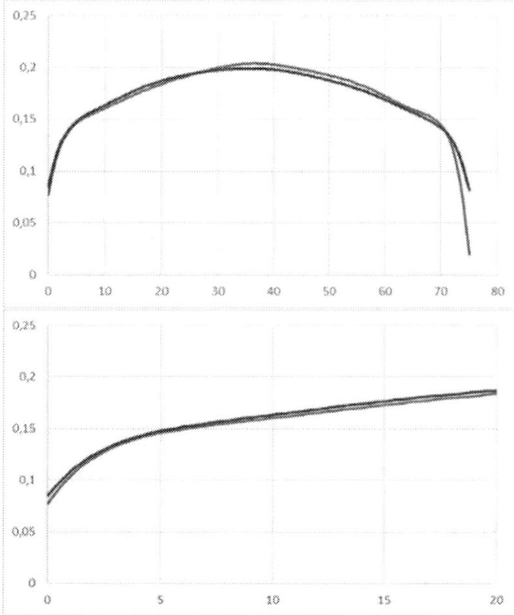


**Fig.5.2** Velocity profile and the detail close to the wall for Re = 12000 - red superhydrophobic, blue hydrophilic

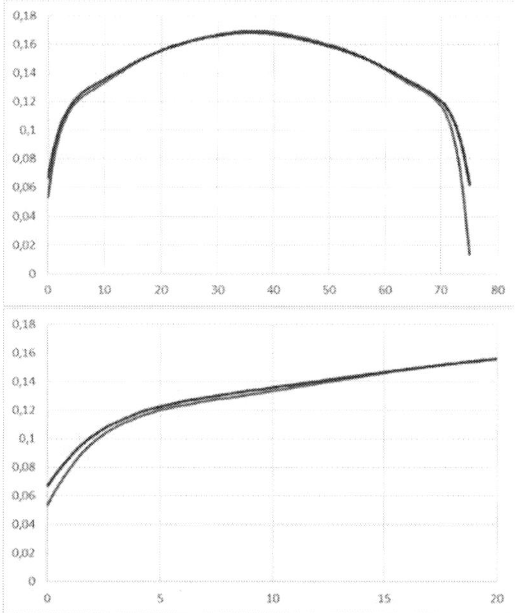


**Fig.5.3** Velocity profile and the detail close to the wall for Re = 10000 - red superhydrophobic, blue hydrophilic

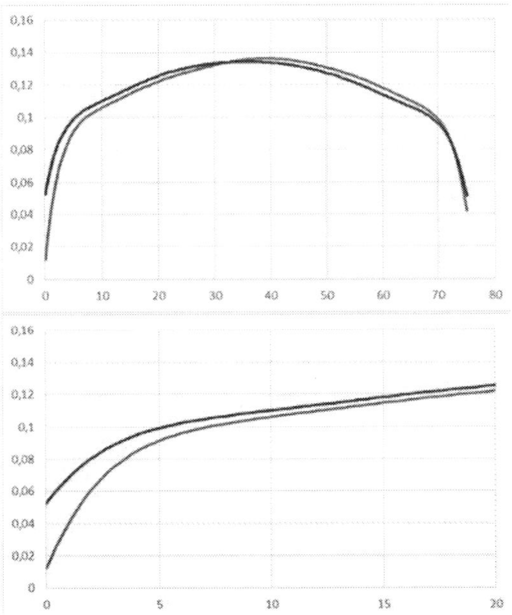


**Fig.5.4** Velocity profile and the detail close to the wall for Re = 8000 - red superhydrophobic, blue hydrophilic

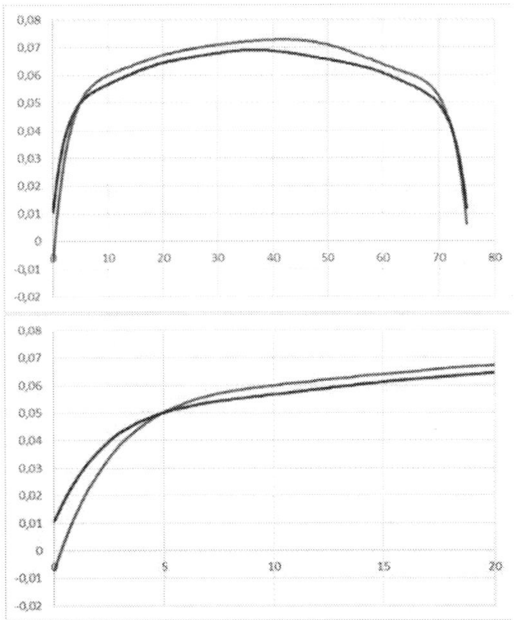


**Fig.5.6** Velocity profile and the detail close to the wall for Re = 4000 - red superhydrophobic, blue hydrophilic

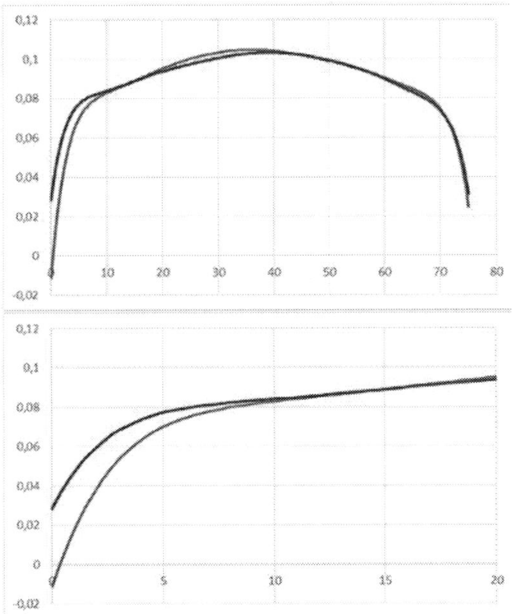


**Fig.5.5** Velocity profile and the detail close to the wall for Re = 6000 - red superhydrophobic, blue hydrophilic

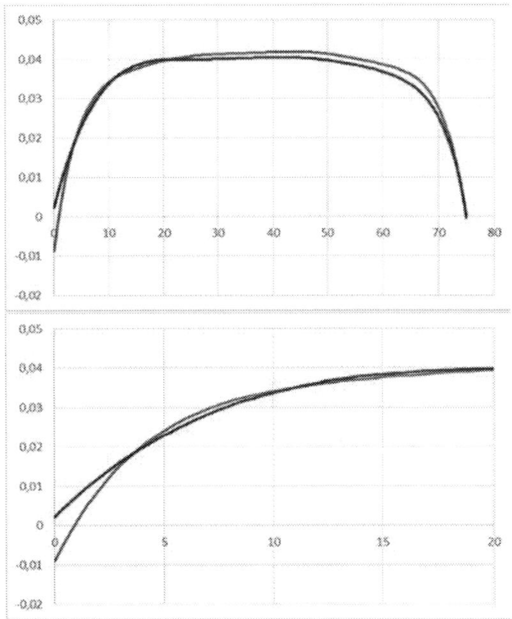


**Fig.5.7** Velocity profile and the detail close to the wall for Re = 2000 - red superhydrophobic, blue hydrophilic

## 6 Conclusion

The methodology for the measurement of the contact angle of water droplets on a superhydrophobic surface was presented within this

DOI: 10.1051/epjconf/201714302023

work. The adhesion coefficient and velocity profiles with the implementation of superhydrophobic surfaces were shown [3]. Measurements demonstrate the slip of the liquid along the superhydrophobic surface with a contact angle of approximately 140°. The contact angle did not change significantly during the experiments. The curves of the velocity profiles are shown in Fig 5.1 – 5.7., where the velocity profile near the wall is highlighted. From the figures the slippage of the liquid is apparent close to the superhydrophobic surface. From this we can obtain the velocity slip value that is needed for the computations of the flow affected by hydrophobic surfaces. The measured curves of the velocity profiles were evaluated based on regression analysis. Fig. 6.1 shows the originally measured velocity profile curves near the wall. Obtained data presented within the graphs were used for the aim of the dissipative energy setting and from the approximations we can obtain the value of the liquid slippage on the wall.

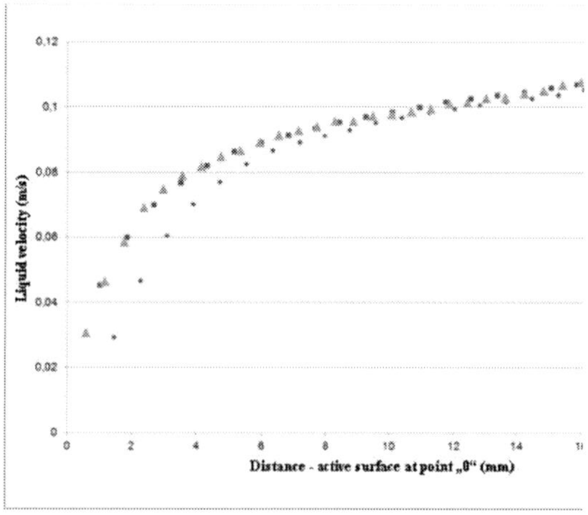


**Fig. 6.1** Water velocity profiles close to the superhydrophobic surface (blue dots– plexiglas, CA $\theta = 80°$, red squares and green triangles– plates with different technologies of superhydrophobic coating, CA $\theta = 145°$ - show similar effect on the profile

Grant Agency of the Czech Republic, within the project GA101/16-18316S, is gratefully acknowledged for support of this work. Special thanks to Masaryk University, Faculty of Science, Dpt. of Physical Electronics and company ASIO, spol. s r.o.

## References

1. Fialová S., Pochylý F., Identification and Experimental verification of the adhesive coefficient of hydrophobic materials, *Wasserwirtschaft Extra, 1/2015*, ISSN 0043 0978, pp. 125-129 (2015)
2. Jašíková, D. a Kotek, M.: The Estimation of Dynamic Contact Angle of Ultra-hydrophobic Surfaces Using Inclined Surface and Impinging Droplet Methods. ed. FR: *EPJ Web of Conferences, Vol. **67***, n. 0. pp. 1 – 6. ISSN 2100-014X (2014)
3. Pochylý F., Fialová, S., Kotek M., Zavadil L., Habán V., Volkov A.V., Parygin A.G., Utilization of hydrophobic layers in the design of hydraulic machines, *EkopumpRus'2015 - sbornik dokladov*, ISBN 978-5-9903138-5-9, printed, pp 77-83 (2015)
4. Jašíková, D., Němcová, L., Kopecký V., The Methodic for Study of Smart Surfaces Using PIV Technique. *AIP Conf. Proc. vol. **80***, nr. 8. pp. 80 – 87 (2014)
5. Min T. and Kim J.: Effects of hydrophobic surface on stability and transition. *Physics of Fluids **17***, 108106 (2005)
6. Daniello RJ, Waterhouse NE and Rothstein JP. Drag reduction in turbulent flows over superhydrophobic surfaces. *Phys Fluids **21***: 085103 (2009)
7. Vinogradova O. I., Slippage of water over hydrophobic surfaces, *Int. J. Mineral. Process.* ***56***, pp 31-60 (1999)
8. B.R.K.Gruncelli, N.D.Sandham, G.McHale, Simulations of laminar flow past a superhydrophobic sphere with drag reduction and separation delay, *Phys. Fluids **25***, 043601, (2013)
9. C. L. M. H. Navier, *Mem Acad. Sci. Inst. France*, vol. ***1***, pp. 414-416, (1823).
10. Fialova S., Identification of the properties of hydrophobic layers and its usage in technical practice. *Habilitation thesis, VUTIUM.* (2016)

# Experimental device for measuring the dynamic properties of diaphragm motors

*Kamil* Fojtášek[1,a], *Lukáš* Dvořák [1] and *Jan* Mejzlík [1]

[1]VŠB-Technical University of Ostrava, Faculty of Mechanical Engineering, Department of Hydrodynamics and Hydraulic Equipment, 17. listopadu 15/2172, Ostrava 70833, Czech Republic

**Abstract.** The subject of this paper is to design and description of the experimental device for the determination dynamic properties of diaphragm pneumatic motors. These motors are structurally quite different from conventional pneumatic linear cylinders. The working fluid is typically compressed air, the piston of motor is replaced by an elastic part and during the working cycle there is a contact of two elastic environments. In the manufacturers catalogs of these motors are not given any working characteristics. Description of the dynamic behavior of diaphragm motor will be used for verification of mathematical models.

## 1 Introduction

The object of this article is the design of the experimental device to determine dynamic properties of pneumatic motors. Design of this equipment with a verification of its functionality was a part of the solution in the magister thesis. Pneumatic systems are used in many industrial applications. They are part of automated production lines, plays an irreplaceable role in the handling of materials or products. Bearer of energy and information (control signal) is compressed air. This fact brings considerable advantages in using of these systems. Internal overpressure prevents to get dirt (contaminants) into system, itself clean of working medium offers the use of these systems in the food industry, or pharmaceutical industry. Compressibility of air allows overloading of pneumatic motors without consequences for their damage, absence of electrical wiring allows work in explosive environments.

Pneumatic mechanisms usually work with a maximum pressure $1\ MPa$. Common working pressure is $0{,}6\ MPa$, to this pressure is also adapted all elements of the pneumatic circuit. This relatively low pressure, together with the previously mentioned compressibility of the working medium, limits the maximum force generated by a pneumatic motors. Common mass produced linear motor generates a maximum force $30\ 000\ N$ (with pressure $0{,}6\ MPa$) and the cylinder diameter $250\ mm$. Pneumatic motors with larger diameters are generally not produced. As a possible alternative we can use a pneumatic linear motors with special construction. One of the special construction are pneumatic motors with elastic working elements. In these motors is the working element (piston), or another part of motor, replacing a flexible (usually rubber) part. When working, therefore to comes into contact of two flexible environment. Compressed air on one side and a flexible rubber part on the other. A group of these motors include air bellows and diaphragm motors. Generally these engines are classified as low-stroke. Stroke of motors is limited by extend and deformation of the flexible part. Air bellows motors work practically without friction. They are made in the basic design up to diameter $400\ mm$, and can be implemented force to $70\ 000\ N$. In this design, allow the use of up to 20 ° tilting [1]. Pneumatic diaphragm motors Fig.1 are used as a linear single acting or double-acting low-stroke motors, especially in the regulatory systems. The elastic (rubber) membrane is fixed to the edge of the motor housing and in the middle on both sides by metal rings, to which is attached the piston rod. When working causes to movement of the diaphragm (and its deformation) in the motor housing. Using diaphragm motors can be realized force to $30\ 000\ N$.

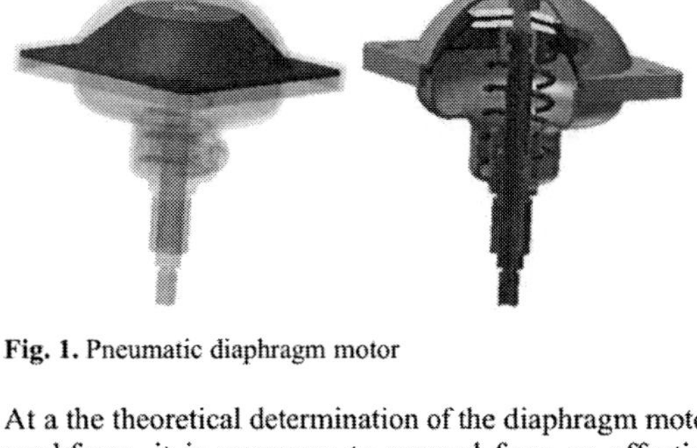

**Fig. 1.** Pneumatic diaphragm motor

At a the theoretical determination of the diaphragm motor workforce, it is necessary to proceed from an effective membrane area (which is changed during the stroke). A detailed description of the calculation of the effective area was made in the post [2]. A group of of pneumatic motors with an elastic working element includes fluid muscles. But they have the theoretical force only about

[a] Corresponding author: kamil.fojtasek@vsb.cz

10 000 $N$. However, looking at the above types of motors, it is evident that among themselves greatly differ.

## 2 Experimental device

When designing experimental equipment must be taken into account constructional and functional differences between different pneumatic motors. For a description of dynamic properties is considered to conventional linear motors, air bellows, diaphragm motors and fluid muscles. Measuring device must be universal. In the past, was in the laboratory of pneumatic mechanisms assembled device for the measurement of static characteristics of these engines [3]. On this measurement device it was possible to measure only static characteristics, at higher load occurred to weightlifting of guide rods, and small overall stiffness of the device not to guarantee accurate results of the measured values. Load was realized by combining mass load and pneumatic motor. Pneumatic part of the load was very difficult to regulate and device at higher pressures has not been able to work with constant load. In view of the fact that the air bellows and diaphragm motors have a higher force effects than conventional pneumatic motors. It is necessary for the measurement of static and primarily dynamic characteristics to ensure adequate load with sufficient stiffness and fast accurate regulation.

At the final design were compared two options of experimental device. Screwed vertical variation Fig.2 (left), and welded horizontal variant Fig.2 (right).

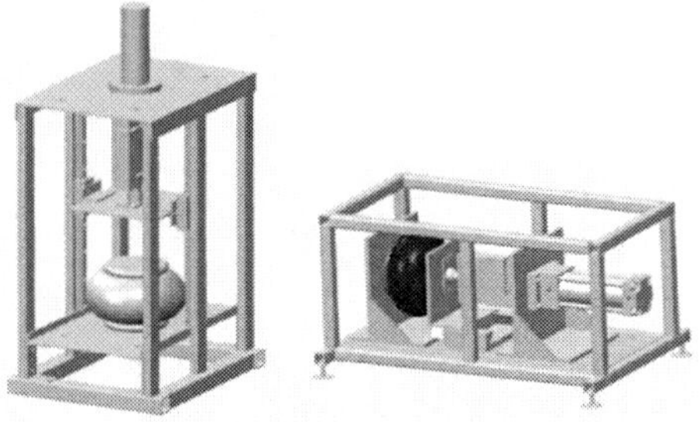

**Fig. 2.** Variants of solution experimental device, vertical variation (left), horizontal variation (right)

The first construction design solutions of test device is considered as screwed or possibly welded construction, the main force effects on testing pneumatic motor will acting in the vertical axis. The main frame is made of steel tubes of rectangular profile. On the upper frame part is mounting a plate, on which is mounted by screws hydraulic cylinder as the main generator of force effects on the test motors. On the bottom plate is mounted tested pneumatic motor, which is from the upper side fixed to the central plate. The central plate is fixed to the two slideways, which ensure smooth movement and reduce the deflection caused by the load. To the central plate is mounting from above the load cell, for measuring the force effect and its mounting to piston rod of hydraulic cylinder. The lower plate is variably adjustable to a several heights according to the type of tested motor. For larger air bellows motors is more advantageous location in the lower position and saved space serves for motor for installation and measurement equipment. In the case if is tested diaphragm motor which has most of the construction parts under mounting plate, this plate is fixed in the upper position. Fluid muscles can be suspended on the upper part of the frame. The second variant has a welded frame construction which has in a bottom plate opposite T shaped channel. This allows quick and easy mounting of working parts. In this case, there are mounted three parts on which there are mounted pneumatic motors and a center plate. The elements are attached so that the main force effects acting in the horizontal plane. This may result in deflection of the outer element which are mounted only on one side (by the influence of gravitational forces). The whole assembly is welded which not allows the variability of test motors. The whole device is undetachable except the parts placed on the grooved plate. This fact for any change of the workplace generates a few problems from dimensional to weight. Comparison of the above was carried out using the value analysis where, as evaluation criteria were e.g. weight, method of production, distribution of mass or price. Based on this comparison, for the realization was selected a vertical variation of the experimental device in screwed design. After the dimensional the proposal of the chosen solution was the construction checked on strength-static loads in software ANSYS. The control of tension and deformation was focused on the top and bottom plate on which are mounted the test and load motors.

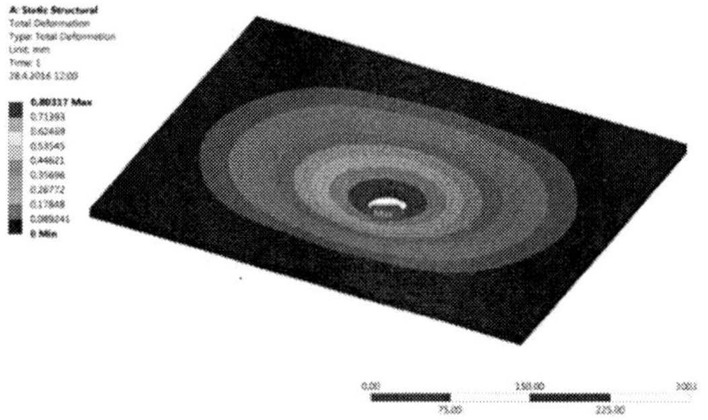

**Fig. 3.** Deformation of top plate

The analysis was set to for the load 20000 $N$ acting in the axis of device where also acting the hydraulic motor. Deformation at the most critical point not exceed the value 1 $mm$ Fig.3. which can be considered as sufficient and for the measurement these minimum values will have no relevant influence [4].

That the equipment can be considered fully operational, it was necessary to ensure that during measurement avoid friction in places where a central plate is fitted with a motion sensor. This is achieved by using a linear ball guide. As already mentioned, the original measuring device not ensure sufficient stiffness of the load (pneumatic) motor which in measurement replaces the mass load. As a source of static and dynamic load will be used the hydraulic motor with a small stroke. For the conditions in the laboratory and the elements that are available, it was selected maximum working load

$20000\,N$ and working pressure of approximately $10\,MPa$. At this load was dimensioned hydraulic motor with a piston diameter $50\,mm$. With regard to the measured pneumatic motors with elastic working elements which can be described as low-stroke, it was selected stroke of hydraulic motor $80\,mm$. Stroke and loading force of the hydraulic motor can be changed at any time by selecting a different type of motor. The speed of movement of the piston has been pre-selected to $0{,}1\,m\cdot s^{-1}$ and for this parameters were designed compact hydraulic power unit with small space requirements.

## 3 Experimental measurements

Functionality of designed experimental device was tested for measuring static characteristics of the diaphragm motor. When measuring static characteristics, we based on the fact that the tested motor is loaded with mass load. This load have a constant value and measured variables are force *F*, which affected by mass load on the measured motor, depending on the change of stroke motor *x*. Pressure *p* in measured motor which must be also constant. These measurements are repeated for the other values of mass load and a series of test pneumatic motors.

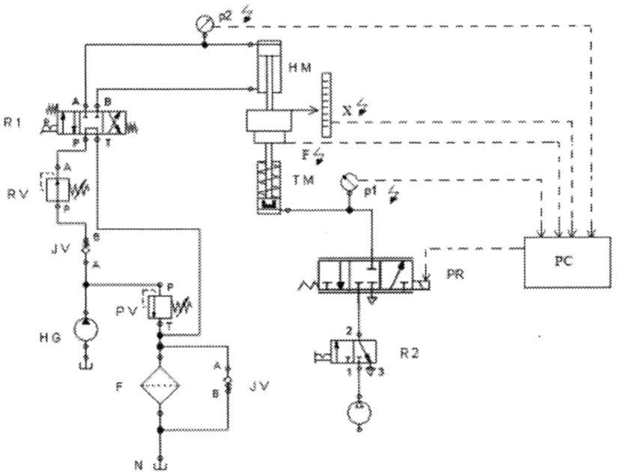


**Fig. 4.** Diagram of hydraulic and pneumatic circuit

On diagram Fig.4. the test circuit is divided into two parts – pneumatic and hydraulic. The pneumatic part is connected to the central supply of compressed air with pressure $0{,}6\,MPa$ and through a valve *R2* is controlled air bellows or diaphragm test motor *TM*. To ensure a constant pressure in the pneumatic motor is in circuit proportional pressure regulator *PR* with bidirectional flow. Because the both pneumatic actuators are single-acting is necessary to ensure discharge of air during reverse movement. The hydraulic circuit is ensured by a hydraulic unit *HG* which supplies pressure fluid on the maximum pressure $12\,MPa$. This value is limiting by safety valve *PV*. For correct measurements it must be load force from the hydraulic motor for fractional measurement constant. This condition ensures a pressure reducing valve *RV*. Hydraulic motor *HM* is double acting and its control is ensured by valve 4/3 actuated by button *R1*. To measure static and dynamic characteristics of motors with elastics working elements was created a control program in Matlab Simulik. With this program you can acquire and evaluate the measured variables, concretely force $F$, piston rod stroke of tested motor $x$ and pressure in tested motor $p_1$. As described above, when measuring the static characteristics is kept a constant pressure in the measured motor. This pressure is controlled by a proportional pressure regulator and its value is set to as a constant progression in the program. For measuring the dynamic characteristics is possible in this program enter any course of the signal and regulate the pressure in the measured motor.

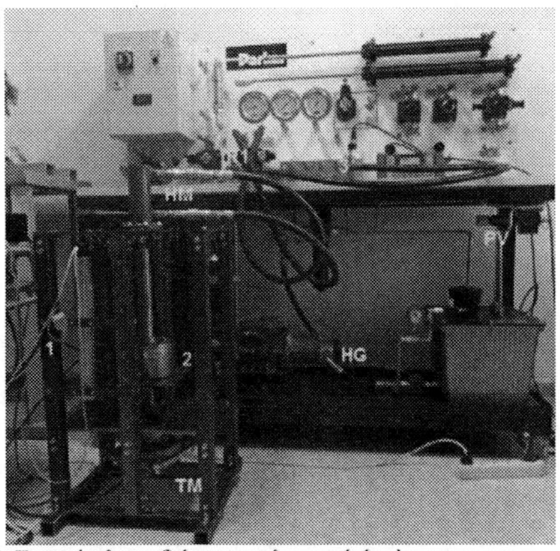


**Fig. 5.** Description of the experimental device
*R1 – directional valve, HM – hydraulic motor, HG – hydraulic pump, PV – safety valve, TM –tested motor, 1 – pressure $p_1$ sensor, 2 – force sensor, 3- pressure $p_2$ sensor*

The measurement process is described by Fig.5. and Fig.6. In this measurement was used in the hydraulic part an educational stand Parker. For the measurement of static characteristics is constant pressure $p_1$ in tested motor *TM*. This is achieved by setting the desired pressure in the control program which sends through a measurement card *5* signal to proportional control valve *PR*. In hydraulic circuit is on the safety valve *PV* set up required pressure value $p_2$ and in interaction of hydraulic motor *HM* and tested motor *TM* we find force effects using by sensor *2* and the displacement of piston rod by sensor *4*. At the beginning of measurement is the pressure $p_2$ on safety valve *PV* set to the minimal value and progressively this value increases. This leads to ejection of piston rod *HM*. This way you can achieve a range of values with constant load for which you can save an individual measurements at different stroke of measuring motor.

The duration of one measurement is 30 seconds, the values used for the creation of characteristics are recorded only the last 10 seconds of measurement due to stabilization at a constant position. When measuring the dynamic characteristics the circuit is a same. In the control program sets the pressure change in the measured or hydraulic motor. Duration of the measuring is selected according to the character of load.

Sensor values of air pressure, force and stroke are through the measuring card sent to *PC*, where they are processed in the form of static and dynamic characteristics.

**Fig. 6.** Description of the experimental device
*PC – control computer, K – compressor, PR – proportional pressure regulator, 4 – position sensor, 5 – measuring card*

In view of the fact that in manufacturer catalog of diaphragm motors does not list any output characteristics, the results of the test measurements were compared with the results of previous measurements (on the previous experimental stand) [5]. From the comparison Fig.7. is obvious that the measured values are lower about *500 N*. At the maximum stroke of the motor there is no significant decrease in force. It is also evident that even at higher pressures it was possible to measure the force in full stroke of motor.

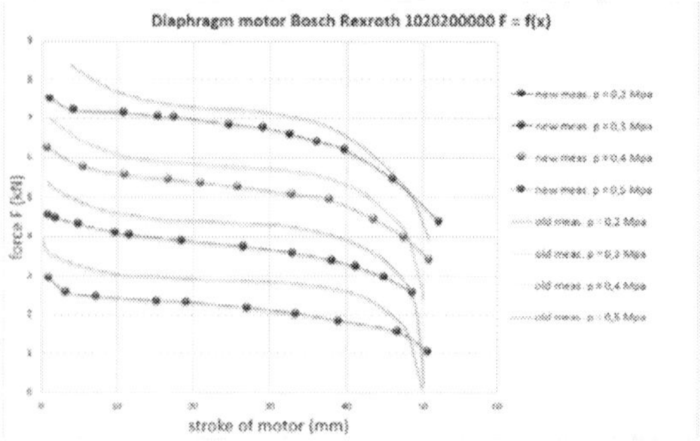


**Fig. 7.** Static characteristics – the force depending on the stroke

In the case of dynamic characteristics are variables changing with a time and the it is necessary to regulate (control) the pressure in the measured motor or resize the mass load, depending on the type of measurement. The involvement would be similar to Fig.4. In the case of pressure control in the measured motor *TM* would be used proportional valve *PR*. In the case of regulation on a mass load would be reducing valve *RV* replaced by proportionally controlled reducing valve.

For test measurement of dynamic characteristics has been created a control signal which controlled the pressure in the measured motor. The duration of measurement was 14 seconds, when in third second was abruptly increased pressure $p_1$ in measured motor. Time of pressure increase from the minimum value to the maximum was *0,1 s*. At the time of 10 *s* was again abruptly pressure drop to a minimum value. Progression of pressure $p_1$, force *F* and stroke *x* depending on time *t* from software Matalb Simulink are shown in Fig.8. From progression of force is evident relatively significant oscillation about a mean value, this problem could be solved by using force sensor with a lower working range. To stabilizing of force at the end of stroke occurs after less than 1 *s*, which I consider as sufficient and selected mass load can be used to measure the dynamic characteristics of pneumatic motors. Speed of extension and retraction of both motors can be controlled by throttle valves.

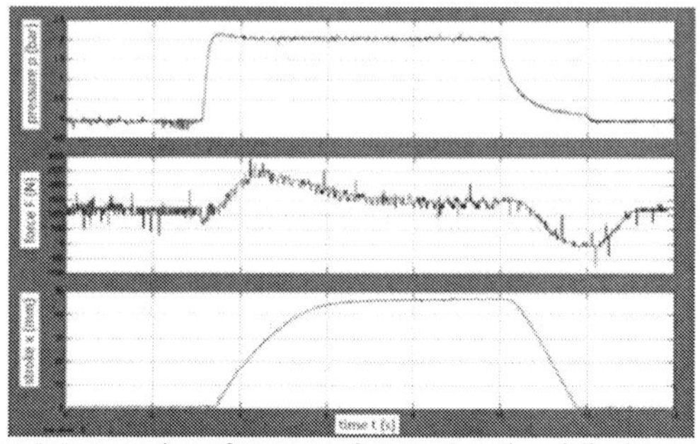


**Fig. 8.** Progression of pressure, force and stroke of diaphragm motor in dynamic measurement

## 4 Conclusion

In this paper was described proposal of experimental device for measuring the dynamic properties of pneumatic motors. It was chosen universal vertical design experimental device with the possibility to measure a different types of pneumatic motors. As a mass load was used hydraulic motor which is sufficiently tough and its load can be continuously controlled. The function of the proposed device was tested at measuring of diaphragm motor. In the future the equipment will be used to determine the dynamic behavior diaphragm motors. The measured data will serve for verification and verification of mathematical models.

## Acknowledgements

This work was supported by the project SP2015/95 of VŠB Technical University of Ostrava.

## References

1- K. Fojtášek, J. Kopáček, L. Dvořák, Pneumatické motory s elastickými pracovními členy, *Sborník příspěvků 30. Setkání kateder mechaniky tekutin a termomechaniky*, p.37 (2011)
2- K. Fojtášek, J. Kopáček, Pneumatic motors with elastic working elements and their mathematical modeling, *Journal of the* **21**. *International conference on hydraulic and pneumatic*, p.171 (2011)
3- K. Fojtášek, L. Dvořák, J. Kopáček, Měření statických charakteristik pneumatických motorů s elastickými pracovními členy, *Hydraulika a pneumatika*, **1-2**, p.9, (2012)
4- J. Mejzlík, Zkušební zařízení pneumatických motorů, Magister Thesis, VŠB-TU Ostrava, (2016)
5- K. Fojtášek, L. Dvořák, Mathematical Modeling of Diaphragm Pneumatic Motors, in *EPJ Web of Conferences* **67**, 02028 (2014)

DOI: 10.1051/epjconf/201714302025

# The proposal of the closure element pipe

*Kamil* Fojtášek[1,a], *Jaroslav* Krutil[1] and *Lukáš* Dvořák [1]

[1]VŠB-Technical University of Ostrava, Faculty of Mechanical Engineering, Department of Hydrodynamics and Hydraulic Equipment, 17. listopadu 15/2172, Ostrava 70833, Czech Republic

**Abstract.** The present paper summarizes the basic equations describing the behavior of butterfly valves during the fluid flow. Butterfly valves are used as a safety element in the energy industry, petrochemical industry, in steam lines etc. The paper describes the theoretical determination of the control torque on the closure element, the torque causes the opening and closing of butterfly valve. Determination of this torque depends on many factors, geometry of the closure element, misalignment to the axis of fluid flow, control method of valve, method of imposition of the moving member and of course on the operating parameters. Correct determination of the torque affects the proper design butterfly valves and ensure its functionality, which can prevent damage to other elements of the monitoring system.

## 1 Introduction

The armatures are a part of many devices in various industry sectors. Very often they ensure the safety and reliability of a much larger and more complex unit. The group of closing valves include various constructions of gate valves, shut-off valves, ball valves and butterfly valves. Their primary function is to ensure reliable opening and closing of the flow with a desired tightness, whereby is basically controlled the movement of the transported medium. Besides the already mentioned tightness, is in the choice of the closure element must be also take account an additional attributes such as: pressure loss in the flow of the transported media (usually they are characterized by loss coefficient), the inner diameter of the pipe and nominal working pressure, the size of actuating force, the possibility of control, constructional length of closure element, its durability and repairability.

Next we will devote the butterfly valves, the use of which has recently been significantly expanded. Butterfly valve has a small constructional length, weight and relatively small loss coefficient (hydraulic resistance). It consists of a valve body, rotatable circular plate, drive pin and elastic cuff seal. The functional principle is quite simple. The flow of transported medium is caused by the rotating of circular plate around its axis (up to 90 degrees). This is achieved by a drive pin, which is located in the middle of plate Fig.1 (left) or in off-axis of plate Fig.1 (right) (eccentric butterfly valves), depending on the construction of the valve. Tightness of valve is ensured by the elastic cuff seal. Cuff material is usually elastomer depending on the type of working medium. Operating pressures for simple valves are in the range $1 \div 2{,}5\ MPa$, temperatures to $130°C$ and the inner diameters $1500\ mm$. Recommended velocity of transported media depends on the inner diameter of the valve. The permissible velocity for liquids is $3\ m \cdot s^{-1}$ and for gases $50\ m \cdot s^{-1}$, at valves with inner diameter up to $450\ mm$. With increasing inner diameters of valves are reduced the permissible velocities. For maximal inner diameter of valve $1500\ mm$ is recommended flow velocity for liquids only $1\ m \cdot s^{-1}$. The flow velocity of the working medium affects the desired control torque. Loss coefficient is featured for maximum inner diameter in range $\zeta(0{,}12 \div 0{,}25)$, for smaller diameters its value increases [1].

**Fig. 1.** Simple butterfly valve CEREX-L (left), Double eccentric butterfly valve EBRO (right)

At a simple valve and valve with eccentric fitting of plate occurs when they are closing to the deformation of the elastic sealing material. This fact must be included in the size of the control torque a negatively affects the size of working pressure. For this reason, there is used constructional variant of valves with double or triple eccentricity. In this construction there are inserted conical sealing surface of the plates in the conical sealing

[a] Corresponding author: kamil.fojtasek@vsb.cz

surfaces in the valve body. This modification allows the use of a metal spring elements and achieve higher operating pressures and temperatures. However, increases loss coefficient.

Constructionally very similar to butterfly valves are check valves. With check valves the main function is to ensure the flow of working medium in one direction only. In the other direction valve must prevent from flow. The function of these valves must be automatic and reliable. Usually they work as a fully open, which places emphasis on a low loss coefficient values. The valve disc is suspended on a lever which is pivotable about a pivot. For closing of valves acting torque, which is determined by the weight of plate and its length centre of gravity and the lever from the axis of rotation. This torque is increases with the opening angle of the plate. The opposite moment is deduced as the hydraulic force of the flowing medium. If the valve is fully open, the valve disc is located almost tangentially to the flow direction of working medium. For closing the valves is used primarily mechanical torque from the weight of plate. Hydraulic torque in reverse flow at closing mainly involved in the final phase of closing and causes rapid closing movement of the valve. An important factor is the time, for which the valve is closed. If the valve is closed fast enough, than it is possible to prevent a back-flow of the working medium. However, the rapid closure of valves may result in pressure surges. To prevent reverse flow and accelerate the closing of the plate can be used an external lever with mass weight. This weight may be sliding on the lever. Thereby can be regulate the closing moment and the resulting closing speed of valves, with regard to working conditions [5].

**Fig. 2.** Double eccentric check valve with mass weight and hydraulic motor APCO-CVS

Check valve can be complemented by external control Fig. 2, due to the character of the flow of the working medium and the function of valves in the circuit [6]. This control may be in the form of electric, pneumatic or hydraulic motor.

## 2 Theoretical analysis of moments

In this section will be described the theoretical determination of the control torque on closure element (butterfly valve), this torque causes the opening and closing of valve. In analysis is considered butterfly valve with double eccentricity, with external positionable weights on lever. Pivot is supported in bearings which are equipped with seals. Seals prevent leaks of working medium. Along the perimeter of the plate is placed gasket. For opening the valve (depending on the character of flow) is used a hydraulic motor Fig.3.

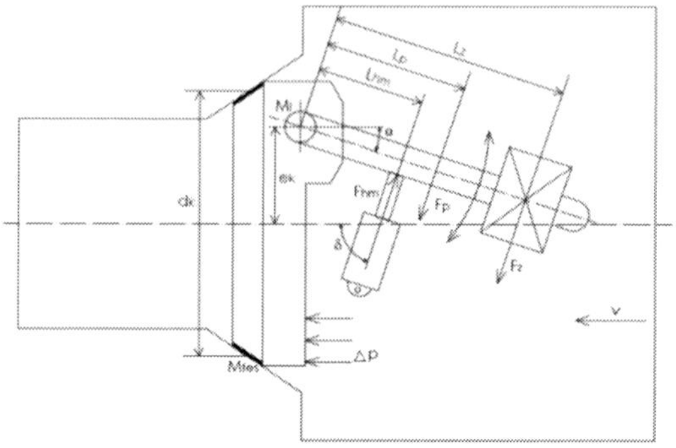

**Fig. 3.** Scheme for calculating butterfly valve

When opening valves is required to overcome the frictional moment on gasket [2]

$$M_{tes} = \frac{(2{,}4 \cdot \Delta p + 1{,}6) \cdot d_k^2}{1000}, \qquad (1)$$

where $\Delta p$ is pressure differnece and $d_k$ is diameter of valve plate.
Moment generated by friction in bearings

$$M_{tl} = F_l \cdot \frac{d_l}{2}, \qquad (2)$$

where $d_l$ is a diameter of bearings and $F_l$ is the force acting on bearings. The bearings are placed on both sides of the control pin. The force from the weight of plate is equally divided on both bearings. On the one bearing is still a force from the weight of control device (lever and mass weight)

$$F_l = F_{l1} + F_{l2}$$

$$F_l = \left(\frac{1}{2} \cdot m_t \cdot g + \frac{1}{2} \cdot m_t \cdot g + m_{ov} \cdot g,\right) \cdot f_l$$

$$F_l = (m_t + m_{ov}) \cdot g \cdot f_l, \qquad (3)$$

where $m_t$ is the weight of plate, $m_{ov}$ is the weight of control device, $g$ is gravity acceleration and $f_l$ is coefficient of friction.
Moment of friction in seals

$$M_{tu} = F_u \cdot \frac{d_u}{2}. \qquad (4)$$

where $d_u$ is diameter of seals and $F_u$ is the force acting on seals

$$F_u = \frac{\pi \cdot d_u \cdot l_u \cdot p \cdot f_u}{2}, \qquad (5)$$

where $l_u$ is length of seals, $p$ is pressure of working medium and $f_u$ coefficient of friction of seals (depends on the material of seal). In calculation is for simplification considered an inner cylindrical surface of seals. Manufacturers of closure elements usually to calculate the moment on seals used experimentally observed empirical equation, which is under their production know-how.
Moment by the weight of lever

$$M_p = F_p \cdot L_p, \tag{6}$$

where $L_p$ is length of centre of gravity lever and $F_p$ is the force acting in this centre of gravity

$$F_p = m_p \cdot g \cdot \cos\alpha, \tag{7}$$

where $m_p$ is weight of lever and $\alpha$ is the angle of rotation lever (and mass weight).
Moment by the mass weight

$$M_z = F_z \cdot L_z, \tag{8}$$

where $L_z$ is length of centre of gravity mass weight (this value can be changed) and $F_z$ is the force acting in centre of gravity mass weight

$$F_z = m_z \cdot g \cdot \cos\alpha, \tag{9}$$

where $m_z$ is the amount of mass weight.
The total moment on the lever and mass weight

$$M_{zc} = M_p + M_z, \tag{10}$$

Finally, hydraulic moment from the flowing medium. In the case that the valve is completely closed ($\beta = 0$), this is a static moment

$$M_{ts} = \pi \cdot \frac{d_k^2}{4} \cdot \Delta p \cdot e_k, \tag{11}$$

where $e_k$ is eccentricity, distance from the axis of pin. When the valve is opened, depending on the opening acting hydrodynamic moment [3]

$$M_{td} = k \cdot \frac{d_k^3}{12} \cdot \frac{\gamma \cdot v^2}{20 \cdot g} \cdot \frac{9{,}81}{100}, \tag{12}$$

where $k$ is resistance coefficient, $\gamma$ $[kg \cdot m^{-4} \cdot s^2]$ includes the density of the flowing medium and $\rho = \gamma / g$ [4], $v$ is velocity of flowing medium. If we are substituting in basic units and substitute in formula the density $\rho$ of the flowing medium, we can modify the equation (12) to

$$M_{td} = k \cdot \frac{d_k^3}{12} \cdot \frac{\rho \cdot v^2}{20} \cdot 9{,}81 \tag{13}$$

Resistance coefficient includes the effect of opening the valve. Its value can be determined experimentally, or by using mathematical models. In the model calculation were considered the value of the resistance coefficient $k$ from the Fig. 4 [4].

Summing up all these moments we get the minimum required moment of hydraulic motor

$$M_{hm} = M_{tes} + M_{tl} + M_{tu} + M_{zc} + M_t. \tag{14}$$

Minimum required force of hydraulic motor (not tilted into the axis of the hydraulic motor)

$$F_{hm} = \frac{M_{hm}}{L_{hm}}, \tag{15}$$

where $L_{hm}$ is the attaching distance of hydraulic motor to the lever. After tilting the forces to the axis of the hydraulic motor

$$F_{hms} = \frac{\cos\delta \cdot F_{hm}}{\cos\alpha}. \tag{16}$$

where $\delta$ is the tilt angle of the hydraulic motor to the axis of valve (axis of flow) and $\alpha$ is the tilt angle of lever and mass weight to the axis of valve. Based on the resulting force $F_{hms}$, it is necessary to select a higher working force of hydraulic motor.

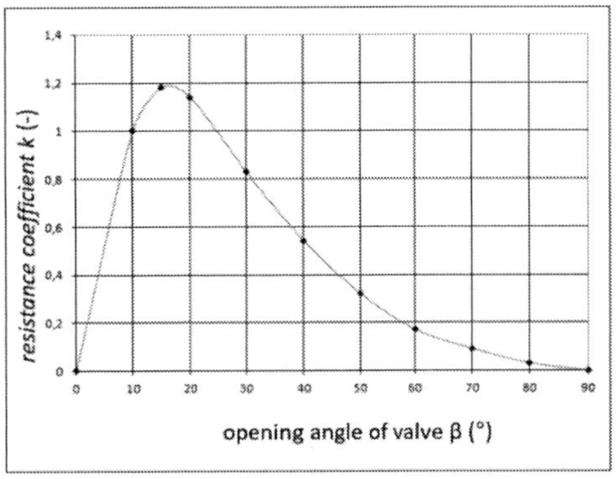


**Fig. 4.** Graph of the resistance coefficient depending on the opening of valve

## 3 Examples of calculation

As an example of the design of butterfly valve and control hydraulic motor, we can mention a model case. The parameters required for the calculation are shown in Tab.1.

**Tab. 1.** Parameters required for the example calculation

| Given parameters | Value of parameter |
|---|---|
| Diameter of valve | $d_k = 1000\ mm$ |
| Eccentricity of valve | $e_k = 200\ mm$ |
| Pressure drop | $\Delta p = 0{,}2\ MPa$ |
| Opening angle of valve | $\beta = 0 \div 90°$ |
| Length of mass weight | $L_z = 800\ mm$ |
| Mass of the weights | $m_z = 210\ kg$ |
| Length of lever | $L_p = 450\ mm$ |
| Mass of lever | $m_p = 70\ kg$ |
| Mass of plate | $m_t = 800\ kg$ |
| Mass of control | $m_{ov} = 300\ kg$ |
| Length of hydraulic motor | $L_{hm} = 300\ mm$ |

| | |
|---|---|
| Diameter of seals | $d_u = 70\ mm$ |
| Length of seals | $l_u = 80\ mm$ |
| Coefficient of friction of seals | $f_u = 0{,}2$ |
| Diameter of bearings | $d_l = 70\ mm$ |
| Friction coefficient in bearings | $f_l = 0{,}16$ |

Working medium is water, the flow velocity is considered $v = 1\ m \cdot s^{-1}$. Orientation of flow direction corresponding to Fig.3. Realistically, there may be two basic states. The first case when the valve is closed and to its opening occurs due to moment from control hydraulic motor. This moment was determined based on the equation (14) and the required force of hydraulic motor from equation (16).

This force was for our model case $F_{hms} = 117545{,}2\ N$, on the basis of that was selected a greater force of the hydraulic motor $125000\ N$. For this force would be designed hydraulic motor, when at the working pressure in the hydraulic circuit $12\ MPa$, and the piston diameter of the hydraulic motor would be $D = 115\ mm$. The sizes of moments are shown in Fig.5. Where $M_n$ is size of moment required to opening valve and $M_{hm}$ is the size of moment from our designed hydraulic motor.

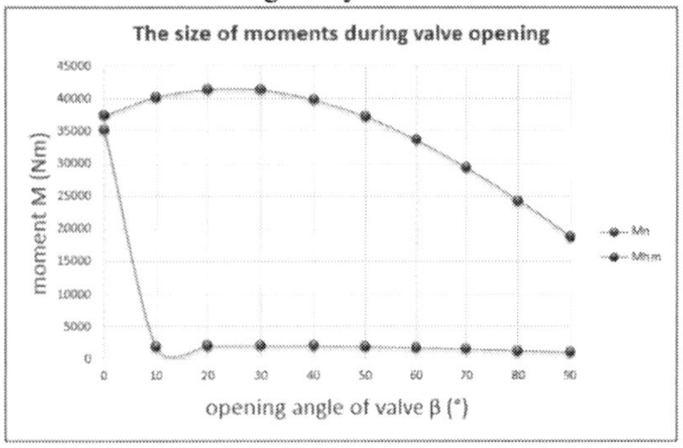


**Fig. 5.** The size of moments during the opening of valve

Tab. 2 shows the maximum values of the individual components of moments, for our model case.

**Tab. 2.** The maximum values of the individual components of moments

| Individual components of moments | Size of moments |
|---|---|
| Moment on gasket | $M_{tes} = 2080\ N \cdot m$ |
| Moment from friction in bearings | $M_{tl} = 60{,}42\ N \cdot m$ |
| Moment of friction in seals | $M_{tu} = 12{,}31\ N \cdot m$ |
| Moment on the lever and mass weight | $M_{zc} = 1694{,}89\ N \cdot m$ |
| Static moment from the flowing medium | $M_{ts} = 31415{,}9\ N \cdot m$ |
| Dynamic moment from the flowing medium | $M_{td} = 46{,}59\ N \cdot m$ |

In the case of closing valve, it is necessary to obtain a closing moment $M_z$, which must be higher than the moment of passive resistances $M_n$. In this case, the flowing medium helps with closing of valve and the equation (14) is changing to

$$M_{zc} + M_t \geq M_{tes} + M_{tl} + M_{tu} + M_{hmp}, \qquad (17)$$

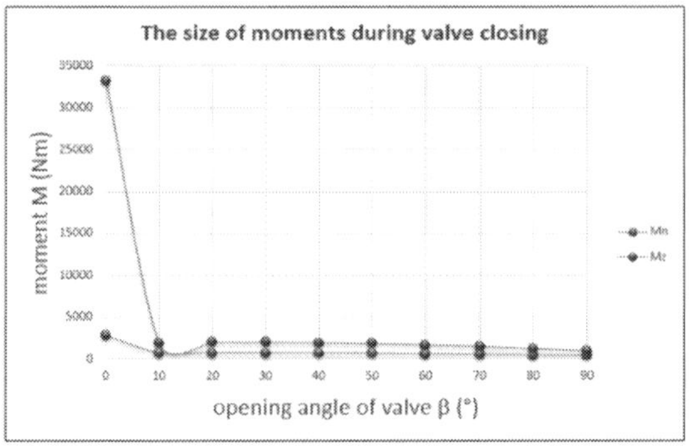


**Fig. 6.** The size of moments during the closing of valve

Where $M_{hmp}$ is moment from passive resistances of hydraulic motor, its size depends on the type of hydraulic motor and control of hydraulic circuit. The courses of moments when the valve is closing are shown for model case in Fig.6.

## 4 Conclusion

In this paper was presented a possible procedure for the design of butterfly valves and their controls. The manufacturers of closure elements in their design often use empirical equations which determined experimentally. Realization of experimental measurements primarily for large diameter of valves is relatively complicated and expensive. For a description of the behavior of valve is possible based from the theory of hydraulic similarity. From this perspective, it seems like the ideal solution to use the numerical simulations. In the future we plan to deal with the issue of using CFD modeling.

## Acknowledgements

This work was supported by the project SP2016/70 of VŠB Technical University of Ostrava.

## References

1- J. Rončеk, *Průmyslové armatury*, (2002)
2- J. Melecký, *Návrh pohonů pro uzavírací bezpečnostní klapku*, (2013)
3- D.F. Gurevič, *Raschet i konstruirovanie truboprovodnoy armatury*, (1966)
4- C. Keller, F. Salzmann, *Luft-Modellversuche an Drosselklappen für Druckleitungen von Wasserkraftanlagen*, (1936)
5- M. Nechleba, *Vodní turbíny, jejich konstrukce a příslušenství*, (1962)
6- J. Jablonská, M. Kozubková, Flow characteristics of control valve for different strokes, *EPJ Web of Conferences*. Vol. **114**. EDP Sciences (2016)

DOI: 10.1051/epjconf/201714302026

# Impact of measurable physical phenomena on contact thermal comfort

*Miloš* Fojtlín[1,*], *Jan* Pokorný[1], *Jan* Fišer[1], *Róbert* Toma[1], and *Ján* Tuhovčák[1]

[1] *Department of Thermodynamics and Environmental Engineering, Energy Institute, Faculty of mechanical engineering, Brno University of Technology, The Czech Republic.*

**Abstract.** Cabin HVAC (Heating Ventilation and Air-conditioning) systems have become an essential part of personal vehicles as demands for comfortable transport are still rising. In fact, 85 % of the car trips in Europe are shorter than 18 km and last only up to 30 minutes. Under such conditions, the HVAC unit cannot often ensure desired cabin environment and passengers are prone to experience thermal stress. For this reason, additional comfort systems, such as heated or ventilated seats, are available on the market. However, there is no straightforward method to evaluate thermal comfort at the contact surfaces nowadays. The aim of this work is to summarise information about heated and ventilated seats. These technologies use electrical heating and fan driven air to contact area in order to achieve enhanced comfort. It is also expected, that such measures may contribute to lower energy consumption. Yet, in real conditions it is almost impossible to measure the airflow through the ventilated seat directly. Therefore, there is a need for an approach that would correlate measurable physical phenomena with thermal comfort. For this reason, a method that exploits a measurement of temperatures and humidity at the contact area is proposed. Preliminary results that correlate comfort with measurable physical phenomena are demonstrated.

## 1 Introduction

Development of personal transport is greatly influenced by efforts to reduce energy consumption, emissions and increase safety on roads. However, these demands are contradictory when thinking in conventional technologies. Therefore, new concepts of efficient energy utilisation are investigated. One of the indispensable systems that contribute to enhanced comfort in a cabin, as well as to considerable energy consumption, is the system of HVAC (Heating Ventilation and Air-conditioning). HVAC systems drain at least 7 % of total energy consumption of a vehicle, whereas, in peak loads this could be quadruped [1, 2]. Particularly a driving range of hybrid or electric vehicles is sensitive to climate, as it is demonstrated in the work of Kambly and Bradley [1]. The more energy needed to manage cabin microclimate, the lower the driving range. A current concept of the HVAC systems continually treats the incoming air and distribute it according to passengers' demands into the cabin. On the other hand, there are more energy efficient ways to satisfy passengers' needs, such as contact heating or cooling. In fact, such technologies are routinely installed into modern vehicles. Most common are heated or ventilated seats, a heated steering wheel and armrests. However, the greatest potential to enhance cabin comfort has the seat, as it creates a considerable contact area with a passenger's body. To these days, complex knowledge to assess contribution of seats to thermal comfort is missing.

Available standards, ISO 14505:2006, ISO 7730, and thermo-physiological models, such as Fiala model, are considered to be the most widely accepted approaches to evaluate thermal comfort in car cabins [3]. Yet, the first two methodologies are not ultimately applicable to contact parts and the Fiala model requires detailed inputs that are not easily accessible in field conditions. Cengiz and Babalik [4, 5] carried out an on-the-road experiment with the aim of examining impact of seat cover materials on the thermal comfort. Even thought, they did not investigate heated or ventilated seats, the properties of cover materials influence heat and moisture transport from a human body. The results of their studies clearly shows two important findings: (a) there are significantly different thermal sensations perceived on the back and the seat; (b) even when the overall perceived thermal sensation was between *neutral to* slightly warm, contact comfort on the back and the seat was compromised due to skin wettedness. This is, however, expectable, as the seat acts like an additional thermal insulation and a barrier to water vapour. As a result, heat and moisture accumulates in the clothing, what contributes to discomfort [6]. Thermal and moisture equilibrium is, therefore, needed. This is also the first condition to achieve thermal comfort [7].

A human body generates heat of around 60 $W.m^{-2}$ in a calm sitting posture. Similarly, insensible perspiration occurs at all times because of water vapour diffusion.

[*] Corresponding author: milos.fojtlin@vutbr.cz

Under thermal neutral conditions a person perspires at least 30 g of water vapour per hour [8]. Heavier rate of sweating occurs in situations, when a body cooling is insufficient. Therefore, seat ventilation is meaningful if excessive heat and moisture need to be led away from the contact area. Generally, two types of seat ventilation systems are used: (a) push systems, blowing the air into the contact area; (b) pull systems, extracting the air from the contact area. However, differences in efficiency have not been found in the literature.

In the Nilsson's thesis [7], effects of ventilated seats on seat thermal insulation properties were tested. It was found that the additional thermal insulation of a ventilated seat running at the highest power level is negligible. However, quantification of efficiency of ventilation is in this case problematic. The air flows through the seat cover material and layers of clothing. Further, the effective ventilation area depends on occupant's body proportions. Additional tests with added moisture (a piece of wet cloth placed on the seat) showed that ventilated seat helps reducing humidity at the contact area, but with undesirably cold sensation for the seat occupant.

Another system of enhancing comfort on the seat is a system of surface heating. A heated seat should ensure quick heat up in cold conditions to desired surface temperature. Oi et al. [9] recommend temperatures on the heated seat of around 34 °C depending on the ambient conditions. On the other hand, long exposures to temperatures above 43 °C may cause skin burns.

Distribution of heat on the surface of the seat is accomplished by resistance wires, which are embedded under the seat cover. Typical temperature fields on the surface of the heated seat are depicted in the Fig. 1. On the left side, the heating is on without an occupant, and a temperature field of the same heated seat after 30 min of occupation by a passenger is on the right hand side.

Next, findings in literature suggest that not only optimal temperature on the seat is necessary, but also moisture management is crucial to achieve thermal comfort. For example, Cengiz and Babalik [4] used skin wttedness sensor attached to the skin to measure the amount of moisture on the back. Skin wittedness $w$ (-) is defined as follows:

$$w = \frac{p_{sk} - p_a}{p_{ssk} - p_a} \quad (1)$$

where $p_{sk}$ is water vapour pressure at skin; $p_{ssk}$ is saturated vapour pressure at skin; $p_a$ is ambient vapour pressure; all units are in $Pa$. Capability to relate thermal comfort with skin wittedness is limited, because of its relativity to temperature. Moreover, the results from the study of Cengiz and Babalik [4] are presented as average values. Therefore, it is not possible to identify skin wettedness in time and its correlation with comfort.

Pasut [10] demonstrated an enhancement of comfort on heated and ventilated office chairs. In addition, power saving potential was stated. However, the participants involved in the experiment were asked to adjust heating, ventilation and clothing to match their needs. On the other hand, automatic or semi-automatic control of the seat heating and ventilation should be available in the cabins to avoid distraction of the driver.

Based on the literature findings, heated and ventilated seats can clearly enhance human thermal comfort. However, the intensity of heating or ventilation must be applied with respect to possible health risks. The aim of the experimental work of this paper is to gather pilot data about comfort and thermal sensation on a heated seat. Two major contact parts will be examined: the seat and the back surface temperatures. Subjective votes will be obtained from a pool of participants.

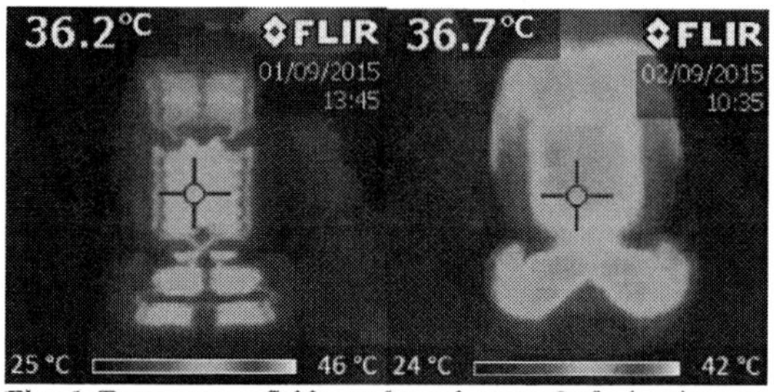


**Fig. 1** Temperature fields on heated seats. Left: heating on; Right: heating on after 30 min of occupation.

## 2 Methods

The experiment was designed to verify correlation of a subjective perception with objective (measurable) phenomena at the contact surfaces. To do so, objective parameters – temperatures, were measured directly on the surface of a textile covered heated seat. Distribution of the temperature sensors is depicted in the Fig. 2. Accuracy of temperature measurements was ± 0.4 °C, such precision should be sufficient for the given purpose.

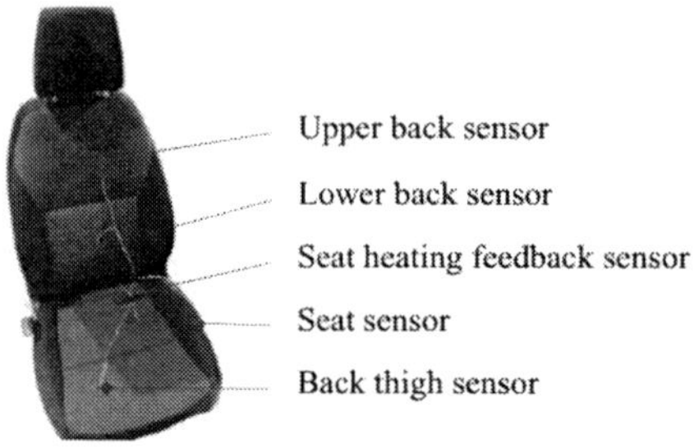


**Fig. 2** Distribution of temperature and humidity sensors on the seat.

Next, the subjective data, thermal comfort and thermal sensation, were harvested from a pool of volunteers. The participants involved in the study were 14 males, and 6 females, all of them active drivers. Age of the subjects covered a span of 20 to 60 years old. Average body mass index (BMI) was 25 for men and 23 for women. Clothing of participants consisted of, underwear, socks, shoes, trousers, a cotton T-shirt and a light pullover or a cotton shirt. This was found to be the standard light winter clothing in Central Europe suitable for driving. Before the actual measurements, it was important to ensure similar metabolic rate of the volunteers at the beginning of the tests. Otherwise, there might be numerous unknown personal factors involved in the results of the study. Therefore, no beverages, food or smoking was allowed two hours before the measurements (lukewarm water only). It was also requested to come to the session rested; no minor or major sickness was acceptable.

Further, the experimental procedure consisted of preconditioning phase and measurement phase. Both activities took place at Brno University of Technology, in a laboratory equipped with a climate chamber.

The preconditioning was mandatory for all participants, who were asked to stay in a defined environment (23 °C, Rh 50 %) at least one hour in a calm sitting position. After this, a group of four volunteers was seated on the seats, which were arranged in the vertices of the 3 × 3 m square in the climate chamber. Preconditioning of the chamber took at least 90 minutes to reach uniform ambient temperature of 16 °C ± 1 °C. Relative humidity was set to 50 % and the airspeed in the chamber was less than 0.1 $m.s^{-1}$.

Volunteers took seats according to a random seating plan to avoid systematic errors. The seat heating was switched on right after the participants were seated. The seat heating was regulated by a feedback sensor attached to the seat cushion (Fig. 2). The prescribed seat surface temperature was 38 °C and the participants were not able to change this temperature pre-set. In case of emergency, they were allowed to switch off the heating themselves. The same temperature pre-set on back and seat cushion was used.

One measurement sitting lasted 30 minutes. The first subjective evaluation took place right before entering the chamber (time -3 min). The following evaluation continued right after sitting on the seat (time 0). Then the seat heating was switched on. Next, the evaluation continued in 6 minutes intervals. One extra 3 minute interval was added to allow detailed evaluation of the seat heat up period.

The according subjective scales are shown in the Table 1; please note that comfort scale has not a zero value to distinguish between comfortable and uncomfortable states.

**Table 1.** Scales of thermal sensation and comfort.

| Thermal sensation | | Thermal comfort | |
|---|---|---|---|
| +3 | Hot | +3 | Very comfortable |
| +2 | Warm | +2 | Comfortable |
| +1 | Slightly warm | +1 | Just comfortable |
| 0 | Neutral | - | - |
| -1 | Slightly cool | -1 | Just uncomfortable |
| -2 | Cool | -2 | Uncomfortable |
| -3 | Cold | -3 | Very uncomfortable |

## 3 Results and discussion

Results are presented separately for back and seat cushion. Average values from twenty subjective votes are presented together with the standard deviation values (SD) in Fig. 3 and 4.

Time course, in the graphs, begins with value -3 min, which represents value when the test subjects were in the preconditioning room just before entering the climate chamber. At this point, most of the participants were feeling comfortable, average thermal comfort vote was around 1.5. The preconditioning phase was successful. The reported thermal sensation was between neutral and slightly warm. At this point, the seat and the back surface temperatures were matching with the ambient temperature inside the chamber, 16 °C.

In the next time step, time 0 – right after the seating in the chamber, the participants reported drop in thermal the comfort and the sensation. The back part was evaluated as slightly colder because of its greater sensitivity to the cold stimulus. Average value of thermal comfort fell below 1. However, the SD of the comfort votes was approximately ± 1.5 at both body parts. This means that there was large diversity in the comfort perception, from very uncomfortable to comfortable. Such findings were stated also in other papers [4, 9, 10].

Between the third and the sixth minute, the seat surface temperature reached its stable average value of around 41.2 °C on the back part and 39.9 °C on the seat cushion. Fluctuations in the surface temperatures, expressed by SD, were observed and were typically less than 2 °C. As a cause of this, variability in a body constitution and a seating style was identified.

During the seat heat up period, the comfort votes reached the highest rating. However, the further development of the votes on the seat cushion shows gradual decrease of comfort, whereas, the thermal sensation remains constant between slightly warm and warm. Moreover, the comfort votes were becoming less uniform towards the end of the experiment.

Slightly different situation was found on the seat. The mean value of the thermal sensation was lower than that of the back, but still within the borders of slightly warm and warm. Thermal comfort votes were also gradually decreasing, however, at the end of the experiment, the mean value was still above 1, just comfortable. Possible answer to these discrepancies between the seat and the back might be a pressure distribution on the seat that influences heat transfer to the body. The greater the pressure on the seat cushion, the better the contact with the seat heating. Opposite situation is on the back, where the contact pressure is lower. However, even the participants were instructed to sit straight on the seats, they tended to bend their backs. This also causes imperfect contact with the seat surface, what was also reflected by higher average temperature on the back.

Next, it was important to find an optimal placement of the feedback sensor. Even the surface temperature was set to 38 °C, the temperature peaks reached more than 44 °C in some cases. Variability in the seat surface temperature might be avoided, in the future, if at least two feedback sensors were used, one for the back and

 DOI: 10.1051/epjconf/201714302026

one for the seat cushion. However, this test simulated real conditions with one sensor only.

Comfortable skin temperature on the heated seat is approximately 34 °C [9]. In this experiment, the comfort was scarified by higher skin temperatures to reveal a comfort vote development under warmer conditions. Preliminary results show that there is a good correlation between the contact temperature and thermal sensation. On the other hand, thermal comfort cannot be correlated only with temperature. Some additional parameter is needed, such as skin wittedness or other non-temperature related index. To prove this, further tests are needed.

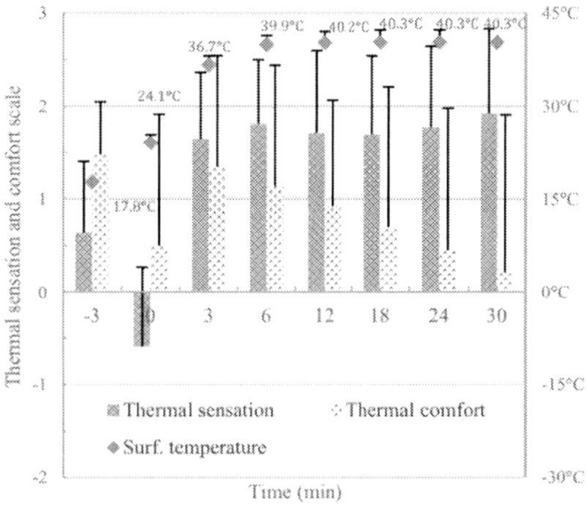


**Fig. 3** Ambient 16°C - **Seat** - thermal sensation , thermal comfort, and surface temperature; average values with standard deviation.

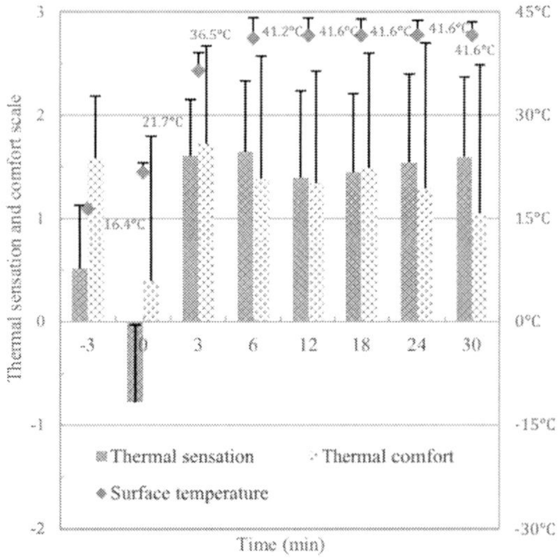


**Fig. 4** Ambient 16°C - **Back** - thermal sensation , thermal comfort, and temperature; average values with standard deviation.

## 4 Conclusion

Heated and ventilated seats were proved to be useful in enhancing thermal experience. The findings from the experiment and the literature suggest that there is a good correlation between the surface temperatures and thermal sensation. Yet, thermal sensitivity of the back differs from that of the seat. Further, the standard deviation of the thermal comfort votes was gradually rising throughout the experiment. This means that there is a rising diversity in the perception of thermal comfort. The reason for this should be explained in further studies with additional objective parameters, such as skin wettedness.

Next, one temperature feedback sensor was found to be insufficient to control the seat surface temperature effectively. The surface temperature was set to 38 °C, whereas, the resulting contact temperatures at the end of the experiment were: 40.3 °C ±1.3 °C on the seat cushion, 41.6 °C ± 2.3 °C on the back rest.

## Acknowledgement

The research was support from the project RP-906-230-0713/1120, Brno University of Technology, The Czech Republic.

## References

1. K. R. Kambly, T. H. Bradley, J. Power Sources **259**, 117-124 (2014)
2. C. De Cauwer, J. Van Mierlo, T. Coosemans, Energies **8**,8573–8593 ( 2015)
3. C. Croitoru, I. Nastase, F. Bode, A. Meslem, A. Dogeanu, Renew. Sustain. Energy Rev. **44**, 304–318 (2015)
4. T. G. Cengiz, F. C. Babalik, Appl. Ergon. **38**, 337–347 (2007)
5. T. G. Cengiz, F. C. Babalik, Int. J. Ind. Ergon. **39**, 287–294 (2009)
6. J. Toftum, A. S. Jørgensen, P. O. Fanger, Energy Build. **28**, 1–13 (1998)
7. H. O. Nilsson, *Comfort climate evaluation with thermal manikin methods and computer simulation models* (Thesis, University of Gavle, 2004)
8. V. T. Bartels, Appl. Ergon. **34**, 393–399 (2003)
9. H. Oi, K. Tabata, Y. Naka, A. Takeda, Y. Tochihara, Appl. Ergon. **43**, 360–367 (2012)
10. W. Pasut, H. Zhang, E. Arens, Y. Zhai, Build. Environ. **84**, 10–21 (2015)

EPJ Web of Conferences **143**, 02027 (2017) DOI: 10.1051/epjconf/201714302027
*EFM 2016*

# On the implicit density based OpenFOAM solver for turbulent compressible flows

Jiří Fürst[a]

Czech Technical University in Prague, Karlovo nám. 13, 121 35 Praha 2, Czech Republic

**Abstract.** The contribution deals with the development of coupled implicit density based solver for compressible flows in the framework of open source package OpenFOAM. However the standard distribution of OpenFOAM contains several ready-made segregated solvers for compressible flows, the performance of those solvers is rather week in the case of transonic flows. Therefore we extend the work of Shen [15] and we develop an implicit semi-coupled solver. The main flow field variables are updated using lower-upper symmetric Gauss-Seidel method (LU-SGS) whereas the turbulence model variables are updated using implicit Euler method.

## 1 Introduction

The OpenFOAM package [20] is widely used open source toolbox for computational fluid dynamics (CFD). The keystone of the package is a library written in object oriented C++ alowing simple manipulation with fields related to cells, vertices, or nodes of three-dimensional polyhedral meshes. The toolbox contains also several ready-made solvers for incompressible as well as compressible flows based on the finite volume method. The OpenFOAM originated from the so-called pressure based segregated velocity-pressure correction methods devised originally for the incompressible Navier-Stokes equations, see eg. SIMPLE [4] or PISO [7]. However these segregated methods have been later extended to compressible flow problems including transonic or supersonic flows [12], the efficiency of the segregated methods is in the case of high speed flows usually worse when comparing with the so-called density based or coupled methods available in some commercial packages and in some academic in-house codes.

So far several attempts have been done in order to develop density based solvers for OpenFOAM. First of all, there is already a solver based on central-upwind scheme of Kurganov and Tadmor [6] in the standard distribution. Next, the OpenFOAM-extend version of the package contains some coupled solvers based on AUSM+up [9], HLLC [17], Roe or Rusanov fluxes combined with an explicit Runge-Kutta method in time. Similar capabilities has also the AeroFOAM package developed at the Politecnico di Milano. The main deficiency of the above mentioned solvers is the explicit time integration which puts a severe stability bound on the maximum time step and the efficiency of these explicit methods is much lower than the efficiency of implicit methods. This is especially true for high speed flows at high Reynolds number where the correct resolution of thin boundary layer requires very fine meshes. Let's note also the hybrid pressure-density based method [21], which is based on the AUSM+up fluxes, but still uses a pressure correction loop similar to PISO algorithm and the maximal time step is therefore also limited.

The fully implicit coupled methods (see eg. [5], [3]) provides much better efficiency usually at the price of much higher complexity of the code and higher memory requirements. Besides the fully implicit method the matrix-free LU-SGS (lower-upper symmetric Gauss-Seidel, [3]) method shares the appealing properties of fully implicit methods (i.e. the unconditional stability) with low code complexity and low memory overhead. In their recent paper [15] Shen at al. give detailed description of the implementation of LU-SGS method in the framework of OpenFOAM for the case of inviscid compressible flows. We follow their work and extend the LU-SGS based solver to turbulent flows described by the set of Reynolds-averaged Navier–Stokes equations combined with an additional turbulence model provided by the OpenFOAM package.

## 2 Mathematical model

### 2.1 Governing equations

The governing equations expressing the conservation of mass, momentum, and energy are given as follows

$$\frac{\partial \rho}{\partial t} + \nabla(\rho \boldsymbol{U}) = 0, \quad (1)$$

$$\frac{\partial(\rho \boldsymbol{U})}{\partial t} + \nabla(\rho \boldsymbol{U} \otimes \boldsymbol{U}) + \nabla p = \nabla \boldsymbol{\tau}, \quad (2)$$

$$\frac{\partial(\rho E)}{\partial t} + \nabla\left[(\rho E + p)\boldsymbol{U}\right] = \nabla(\boldsymbol{\tau}\boldsymbol{U}) + \nabla(\lambda \nabla T), \quad (3)$$

[a] Corresponding author: Jiri.Furst@fs.cvut.cz

where $\rho$ is the density, $\boldsymbol{U}$ is the velocity vector, $p$ is the pressure, $\boldsymbol{\tau}$ is the (effective) stress tensor, $E$ is the specific total energy, $\lambda$ is the (effective) thermal conductivity, and $T$ is the temperature. The system is closed by the equation of state for ideal gas $p/\rho = rT$ and constant specific heat capacity $c_p$.

Spatial derivatives at the left hand side of the previous equations represent inviscid terms whereas the right hand side represents viscous terms. The above equations can be expressed in the integral form suitable for finite volume approximation as follows (assuming fixed control volume $\Omega$)

$$\frac{d}{dt}\int_{\Omega} \boldsymbol{W}\, d\Omega + \oint_{\partial\Omega} (\boldsymbol{F}^c - \boldsymbol{F}^v)\, dS = 0, \tag{4}$$

where $\boldsymbol{W} = [\rho, \rho\boldsymbol{U}, \rho E]$ is the vector of conservative variables and $\boldsymbol{F}$ and $\boldsymbol{F}^v$ represent inviscid and viscous fluxes.

In the case of turbulent flow the above mentioned system of equations is coupled to an additional turbulence model (e.g. a two-equation SST model [10]) which provides components of the Reynolds stress tensor and the turbulent heat flux.

## 2.2 Spatial discretization

The numerical solution is obtained with the finite volume method in semi-coupled fashion. It means that the code updates in each step the main flow variables $\boldsymbol{W}$ and then it updates the turbulence model variables. Since we are using turbulence models provided by the OpenFOAM package, we do not give detailed description of its discretization and we focus the attention to the discretization of the equations for main flow variables (4).

The integral form of the equations (4) discretized in space using collocated finite volume method (FVM) by taking

$$\boldsymbol{W}_i(t) = \frac{1}{|\Omega_i|}\iiint_{\Omega_i} \boldsymbol{W}(\boldsymbol{x}, t)\, d\Omega, \tag{5}$$

where $\Omega_i$ is the mesh cell $i$. Hence the equation (4) can be approximated as

$$|\Omega_i|\frac{d\boldsymbol{W}_i}{dt} = -R(\boldsymbol{W})_i = -\sum_{j\in N_i}(\boldsymbol{F}_{ij} + \boldsymbol{F}^v_{ij}), \tag{6}$$

where $N_i$ is the set of indices of neighbor cells and $\boldsymbol{F}_{ij}$ and $\boldsymbol{F}^v_{ij}$ are the numerical fluxes approximating the inviscid and viscous fluxes. Viscous fluxes are discretized directly with OpenFOAM built-in schemes (e.g. central scheme). On the other hand the discretization of inviscid fluxes is made using well established approximate Riemann solvers. The first order Rusanov convective flux is

$$\boldsymbol{F}^{Rus}_{ij} = \frac{\boldsymbol{F}(\boldsymbol{W}_i) + \boldsymbol{F}(\boldsymbol{W}_j)}{2}\cdot \boldsymbol{S}_{ij} - \frac{\lambda_{ij}}{2}(\boldsymbol{W}_i - \boldsymbol{W}_j), \tag{7}$$

where the inviscid flux matrix $\boldsymbol{F}$ is

$$\boldsymbol{F}(\boldsymbol{W}) = \begin{bmatrix} \rho\boldsymbol{U} \\ \rho\boldsymbol{U}\otimes\boldsymbol{U} + p\boldsymbol{I} \\ (\rho E + p)\boldsymbol{U} \end{bmatrix}, \tag{8}$$

and the stabilization parameter $\lambda_{ij}$ is

$$\lambda_{ij} = |\boldsymbol{U}_{ij}\cdot \boldsymbol{S}_{ij}| + |\boldsymbol{S}_{ij}|a_{ij}. \tag{9}$$

Here $\boldsymbol{U}_{ij}$ is the velocity at the face shared by cell $i$ and $j$ (it can be taken as an arithmetic average of $\boldsymbol{U}_i$ and $\boldsymbol{U}_j$) and $a_{ij}$ is the sound speed at the face. The Rusanov flux together with the viscous terms forms a low order rezidual

$$\boldsymbol{R}^{lo}(\boldsymbol{W})_i = \sum_{j\in N_i}\left(\boldsymbol{F}^{Rus}(\boldsymbol{W}^L_{ij}, \boldsymbol{W}^R_{ij}, \boldsymbol{S}_{ij}) + \boldsymbol{F}^v_{ij}\right). \tag{10}$$

The residual can be further simplified using the fact that $\sum_j \boldsymbol{S}_{ij} = 0$, so

$$\boldsymbol{R}^{lo}(\boldsymbol{W})_i = \frac{1}{2}\sum_{j\in N_i}(\boldsymbol{F}(\boldsymbol{W}_j)\cdot\boldsymbol{S}_{ij} + \lambda_{ij}\boldsymbol{W}_j) - \\ -\frac{1}{2}\left(\sum_{j\in N_i}\lambda_{ij}\right)\boldsymbol{W}_i + \sum_{j\in N_i}\boldsymbol{F}^v_{ij}. \tag{11}$$

It is well known that the Rusanov flux is overly diffusive and it is not a good choice for high speed flows. On the other hand thanks to its simplicity one can easily calculate the Jacobian matrix $\partial\boldsymbol{R}/\partial\boldsymbol{W}$, see next sections. In order to improve the accuracy a better Riemann solver should be used in the evaluation of the residual. In our the AUSM+up flux [9] or HLLC [17] flux is employed together with a piece-wise linear reconstructions with Barth [2] or Venkatakrishnan [18], thus the high order rezidual is

$$\boldsymbol{R}^{ho}(\boldsymbol{W})_i = \sum_{j\in N_i}\left(\boldsymbol{F}^{ho}(\boldsymbol{W}^L_{ij}, \boldsymbol{W}^R_{ij}, \boldsymbol{S}_{ij}) + \boldsymbol{F}^v_{ij}\right). \tag{12}$$

Here $\boldsymbol{F}^{ho}$ is the AUSM+up or HLLC flux evaluated from the reconstructed states at the left ($\boldsymbol{W}^L$) and right ($\boldsymbol{W}^R$) hand side of the face.

## 2.3 Temporal discretization

The temporal discretization is achieved by using the implicit Euler method, thus

$$\boldsymbol{W}^{n+1}_i = \boldsymbol{W}^n_i - \frac{\Delta t}{|\Omega_i|}\boldsymbol{R}^{ho}(\boldsymbol{W}^{n+1})_i. \tag{13}$$

The large system of non-linear equations is then linearized

$$\boldsymbol{W}^{n+1}_i \approx \boldsymbol{W}^n_i - \frac{\Delta t}{|\Omega_i|}\left(\boldsymbol{R}^{ho}(\boldsymbol{W}^n)_i + \sum_{j\in N_i\cup\{i\}}\frac{\partial\boldsymbol{R}^{ho}_i}{\partial\boldsymbol{W}_j}\Delta\boldsymbol{W}_j\right), \tag{14}$$

with $\Delta\boldsymbol{W} = \boldsymbol{W}^{n+1} - \boldsymbol{W}^n$, thus

$$\sum_{j\in N_i\cup\{i\}}\left[\frac{|\Omega_i|}{\Delta t}\delta_{ij} + \frac{\partial\boldsymbol{R}^{ho}_i}{\partial\boldsymbol{W}_j}\right]\Delta\boldsymbol{W}_j = -\boldsymbol{R}^{ho}(\boldsymbol{W}^n)_i. \tag{15}$$

Finally the Jacobian matrix of high order residual at the left hand side is replaced by its low order approximation, so

$$\sum_{j\in N_i\cup\{i\}}\left[\frac{|\Omega_i|}{\Delta t}\delta_{ij}+\frac{\partial \boldsymbol{R}_i^{lo}}{\partial \boldsymbol{W}_j}\right]\Delta \boldsymbol{W}_j=-\boldsymbol{R}^{ho}(\boldsymbol{W}^n)_i. \quad (16)$$

The matrix of the linear system (16) can be divided into block diagonal part $D$, the lower triangular matrix $L$ and the upper triangular matrix. Replacing the Jacobian matrices of viscous fluxes by their spectral radii [3] and assuming local time step one finally arrives to diagonal operator $D$ in the form

$$\boldsymbol{D}_i=\left(\frac{\Omega_i}{\Delta t_i}+\frac{1}{2}\sum_{j\in N_i}\lambda_{ij}^*\right)\boldsymbol{I}. \quad (17)$$

The spectral radius $\lambda_{ij}^*$ is composed of the inviscid contribution $\lambda_{ij}$ given by (9), an over-relaxation parameter $\omega$ in the range $1\le\omega\le 2$, and the viscous spectral radius $\lambda_{ij}^v$ [3]

$$\lambda_{ij}^v=\frac{|\boldsymbol{S}_{ij}|}{||\boldsymbol{x}_i-\boldsymbol{x}_j||}\max\left(\frac{4}{3\rho_{ij}},\frac{\gamma}{\rho_{ij}}\right)\left(\frac{\mu}{Pr}+\frac{\mu_T}{Pr_T}\right), \quad (18)$$

thus

$$\lambda_{ij}^*=\omega\lambda_{ij}+\lambda_{ij}^v. \quad (19)$$

Let us define the sets of cells belonging to lower and upper part of the matrix:

$$L_i=\{j\in N_i: j<i\}, \quad (20)$$
$$U_i=\{j\in N_i: j>i\}. \quad (21)$$

Then the matrix-free LU-SGS scheme can be written using the following two step procedure:

$$\boldsymbol{D}_i\Delta\boldsymbol{W}_i^{(1)}=-\boldsymbol{R}_i^{ho}-\frac{1}{2}\sum_{j\in L_i}\left[\Delta\boldsymbol{F}_j^{(1)}\cdot\boldsymbol{S}_{ij}+\lambda_{ij}^*\Delta\boldsymbol{W}_j^{(1)}\right],$$
$$\boldsymbol{D}_i\Delta\boldsymbol{W}_i=\boldsymbol{D}_i\Delta\boldsymbol{W}_i^{(1)}-\frac{1}{2}\sum_{j\in U_i}\left[\Delta\boldsymbol{F}_j\cdot\boldsymbol{S}_{ij}+\lambda_{ij}^*\Delta\boldsymbol{W}_j\right].$$

Here

$$\Delta\boldsymbol{W}_i^{(1)}=\boldsymbol{W}_i^{(1)}-\boldsymbol{W}_i^n, \quad (22)$$
$$\Delta\boldsymbol{F}_i^{(1)}=\boldsymbol{F}(\boldsymbol{W}_i^{(1)})-\boldsymbol{F}(\boldsymbol{W}_i^n), \quad (23)$$
$$\Delta\boldsymbol{F}_i=\boldsymbol{F}(\boldsymbol{W}_i^{n+1})-\boldsymbol{F}(\boldsymbol{W}_i^n). \quad (24)$$

### 2.4 Coupling with a turbulence model

The LU-SGS solver is combined with an additional turbulence model in a weakly-coupled manner. It means that the turbulence quantities (e.g. $k$ and $\omega$ in the case of two-equation SST model) are kept fixed during one iteration of LU-SGS model. Then these variables are updated using appropriate model keeping the vector of conservative variable constant. Hence (for the case of SST model:

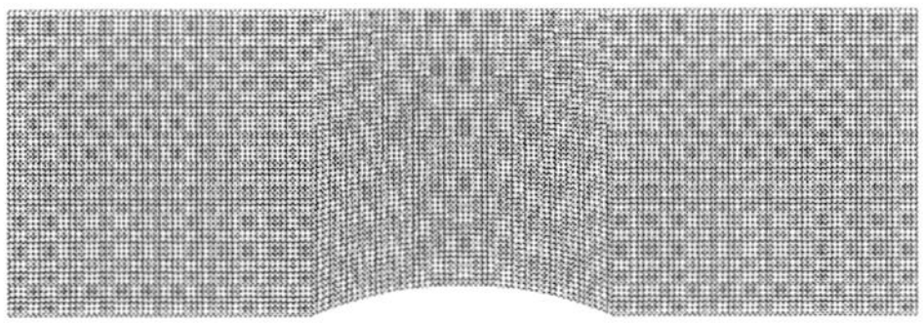

**Fig. 1.** Structured mesh for the flow over a bump.

1. calculate $\boldsymbol{W}^{n+1}$ from $\boldsymbol{W}^n$ using both steps of LU-SGS method (22) and assuming fixed $k^n$ and $\omega^n$,
2. calculate $k^{n+1}$ and $\omega^{n+1}$ using fixed $\boldsymbol{W}^{n+1}$.

The advantage of the semi-coupled approach is that it allows inclusion of built-in OpenFOAM turbulence models. Moreover, the implementation is much simpler than it would be in the fully-coupled scheme.

## 3 Results

### 3.1 Subsonic flow over a bump

The two-dimensional subsonic flow over a 10% circular bump has been chosen as a first benchmark. The inviscid flow passes through a channel of height $H = 1\,\mathrm{m}$ and length $L = 3\,\mathrm{m}$. The total pressure at the inlet was $p_{tot,in} = 100\,\mathrm{kPa}$, the total temperature $T_{tot,in} = 293.15\,\mathrm{K}$ and the inlet flow direction was parallel to $x$-axis. The average outlet pressure corresponds to isentropic Mach number $M_{is,out} = 0.1$. The solution was obtained using a structured mesh with $150\times 50$ quadrilateral cells, see fig. 1.

The figure 2 shows contours of Mach number obtained with LU-SGS scheme combined with AUSM+up flux and piece-wise linear reconstruction with Venkatakrishnan limiter. The figure 3 shows the comparison of Mach number along the lower wall obtained with the LU-SGS scheme using AUSM+up flux and HLLC flux with the result of segregated solver based on SIMPLE loop. One can see that the HLLC scheme slightly underestimates the maximal Mach number with respect to other mentioned methods. The figure 4 compares the convergence history of LU-SGS scheme with AUSM+up and HLLC fluxes using either Barth or Venkatakrishnan limiter with the convergence history of segregated solver. One can see that the segregated solver is in this case more efficient than the LU-SGS scheme due to low Mach number character of the flow. Note that the efficiency of the LU-SGS scheme can be still improved using a proper preconditioning technique, see e.g. [14].

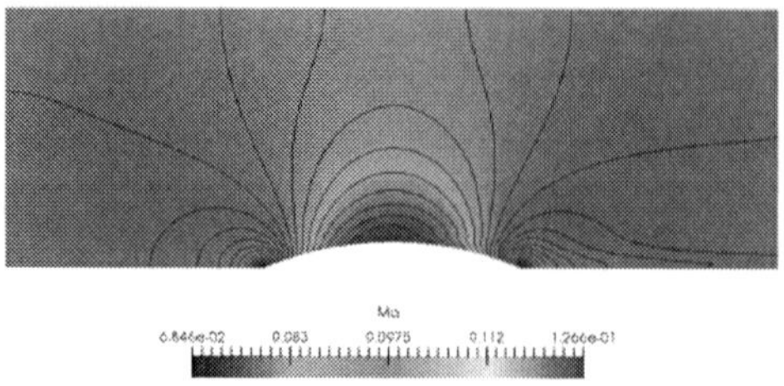


**Fig. 2.** Mach number distribution for subsonic flow over a bump.

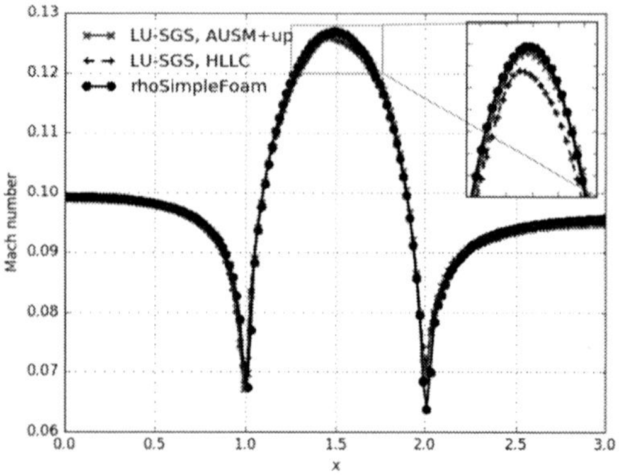


**Fig. 3.** Mach number distributen over the lower wall for subsonic flows over a bump.

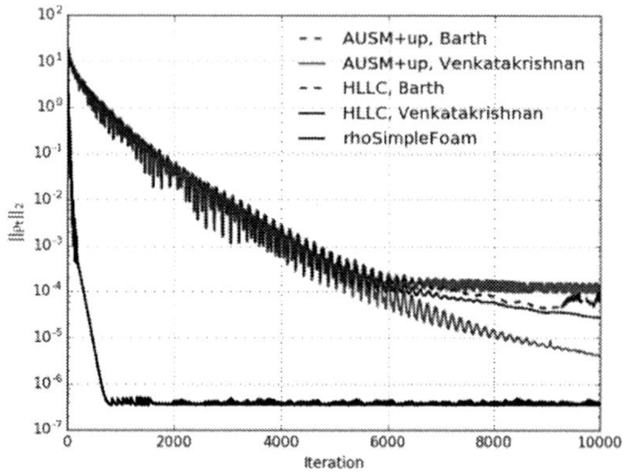


**Fig. 4.** Convergence history for subsonic flows over a bump.

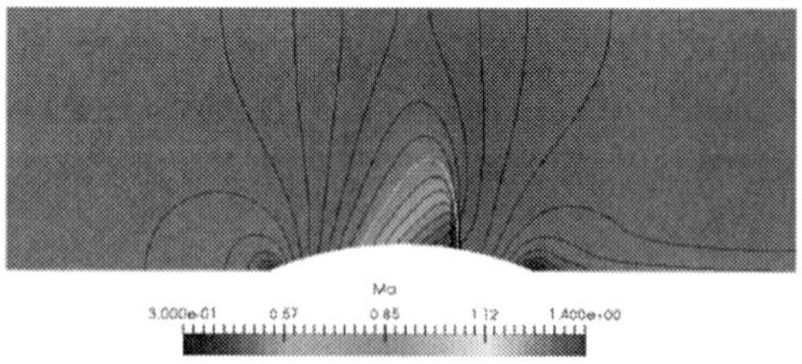


**Fig. 5.** Mach number distribution for transonic flow over a bump.

### 3.2 Transonic flow over a bump

The second benchmark is a transonic flow over the bump using the same geometry as in the previous case. The same boundary conditions were used at the inlet and the average outlet pressure was set to a value corresponding to isentropic Mach number $M_{is,out} = 0.675$. In this case the flow accelerates over the bump and reaches supersonic speed. Then it passes through a shock wave and continues towards the outlet.

The figure 5 shows the contours of Mach number obtained with LU-SGS scheme with AUSM+up flux and piecewise linear reconstruction with Venkatakrishnan limiter. The sonic line (i.e. the line where the magnitude of the velocity equals to local sound speed is plotted in other color in order to depict the shape of the supersonic region. The comparison of the Mach number over the lower wall is shown in the figure 6. One can see that all methods including the transonic variant of SIMPLE solver give similar results. Nevertheless the resolution of the shock wave is much better with the AUSM or HLLC based LU-SGS scheme than with the SIMPLE solver which spreads the shock over 4 cell due to higher amount of numerical viscosity. The figure 7 compares the convergence history for all solvers. One can see that the LU-SGS scheme is more efficient than the segregated approach in this case. Note that the CPU time for 10000 iterations was approximately the same using all methods including segregated SIMPLE solver.

### 3.3 Transonic flow through a turbine cascade

The next benchmark is two-dimensional compressible turbulent flow through test turbine cascade SE-1050. The geometry and boundary conditions correspond to the test from ERCOFTAC QNET database [13]. The inlet total presuure and total temperature were $p_{tot,in} = 100\,\text{kPa}$ and $T_{tot,in} = 293.15\,\text{K}$ and the inlet angle was $\alpha_{in} = 19.3°$. The average outlet pressure corresponds to isentropic outlet Mach number $M_{is,out} = 1.2$, the Reynolds number related to outlet flow conditions and blade pitch was $Re = 1.2 \times 10^6$ and the inlet turbulence intensity was $Tu_{in} = 2\,\%$.

DOI: 10.1051/epjconf/201714302027

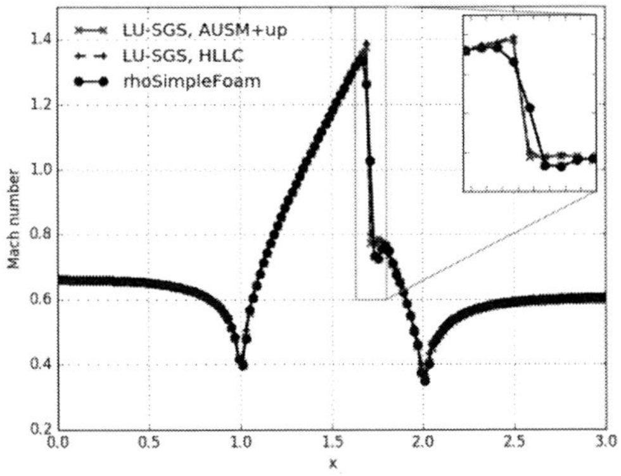


**Fig. 6.** Mach number distributen over the lower wall for transonic flows over a bump.

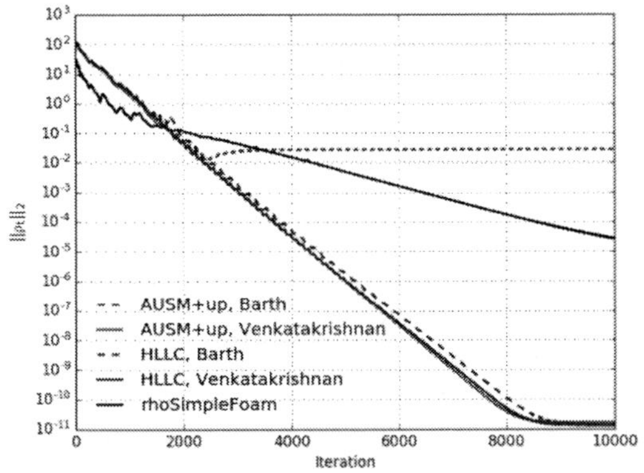


**Fig. 7.** Convergence history for transonic flows over a bump.

The hybrid mesh was composed of structured quadrilateral layer around the blade and unstructured triangular part in the rest of the domain. The mesh contained approximately $66 \times 10^3$ cells and the near-wall refinement corresponded to $y_1^+ \approx 0.2$. Details of the mesh around leading and trailing edges is given at the figure 8. The figure 8 shows contours of the Mach number in two periods of the mesh obtained using LU-SGS scheme with HLLC flux. One can see here both branches of the shock emanating from the trailing edge as well as the interaction of the inner branch of the shock wave with the blade. The figure 10 shows the comparison of the pressure along the blade with the results obtained with segregated SIMPLE solver as well as with the experimental data [13]. One can see that both numerical methods shows very good agreement with experimental data.

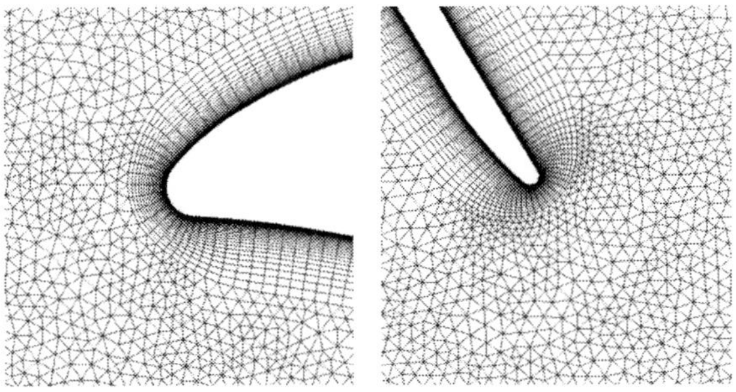

(a) Leading edge (b) Trailing edge

**Fig. 8.** Hybrid mesh for test turbine cascade.

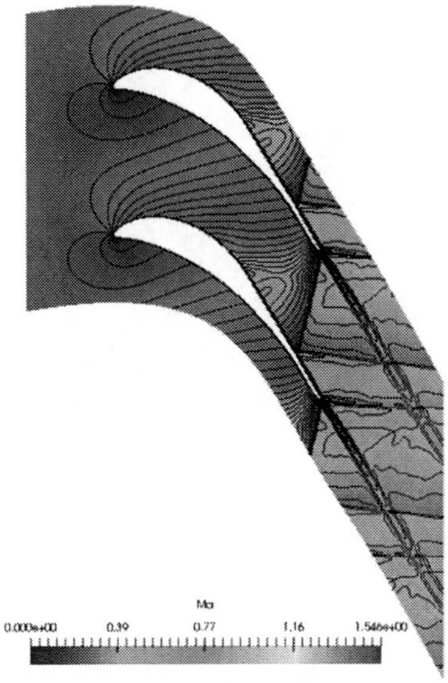


**Fig. 9.** Calculated Mach number distribution in the test turbine cascade.

### 3.4 Transonic flow through 3D test turbine cascade

The next case is the 3D flow through a test turbine cascade with prismatic blades. The geometry of blades correspond to the previous case and the blade has finite span, see e.g. [16]. The flow regime corresponds to outlet isentropic Mach number $M_{is,out} = 1.2$ and Reynolds number related to chord length $Re = 1.34 \times 10^6$. Inlet conditions include a simple model of boundary layer corresponding to conditions in the experiment, see [16] for details. The numerical solution was obtained on an unstructured mesh with $3.3 \times 10^6$ cells with near wall refinement to $y_1^+ \approx 0.2$. The figure 11 shows the distribution of the pressure at the blades and the side wall and the distribution of entropy in the test plane. The regions with higher level of entropy show re-distribution of the energy losses

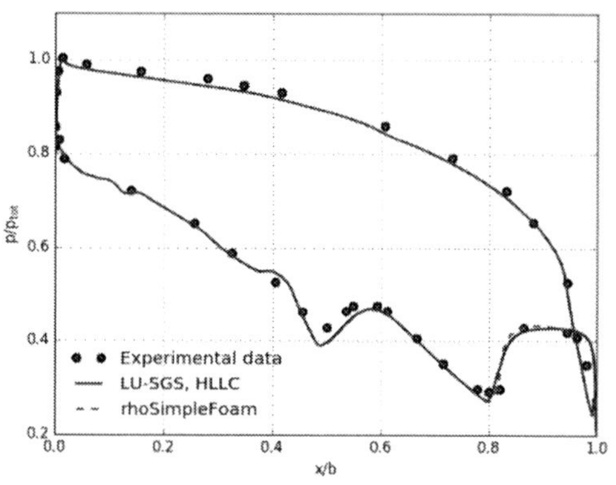


**Fig. 10.** Pressure distribution over the test turbine blade, comparison with experimental data [13].

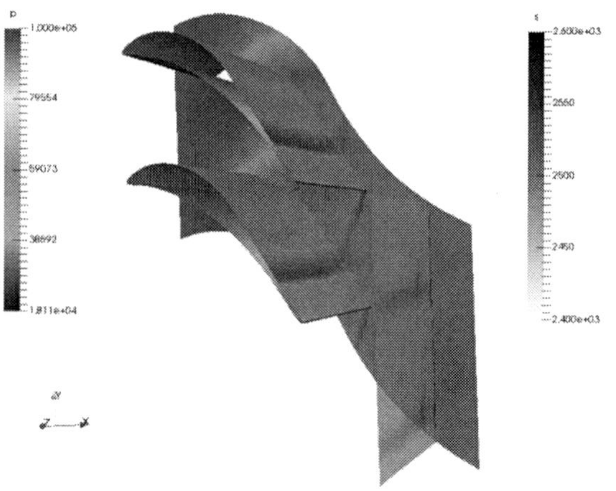

**Fig. 11.** Pressure distribution at the blades and side wall and the entropy distribution in test plane for the 3D flows over test turbine cascade.

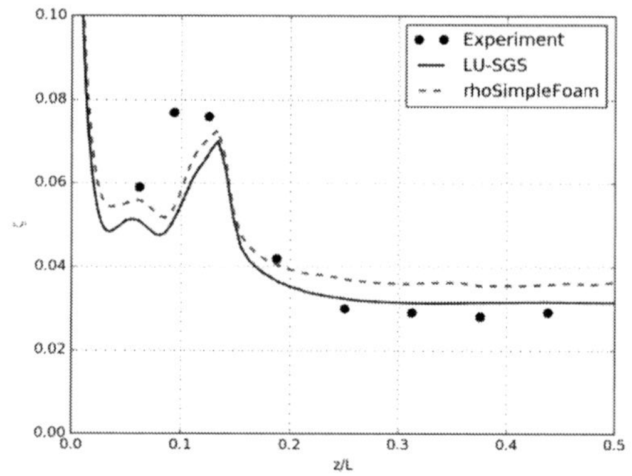


**Fig. 12.** Spanwise distribution of pitch averaged loss coefficient $\zeta$, comparison with experiment.

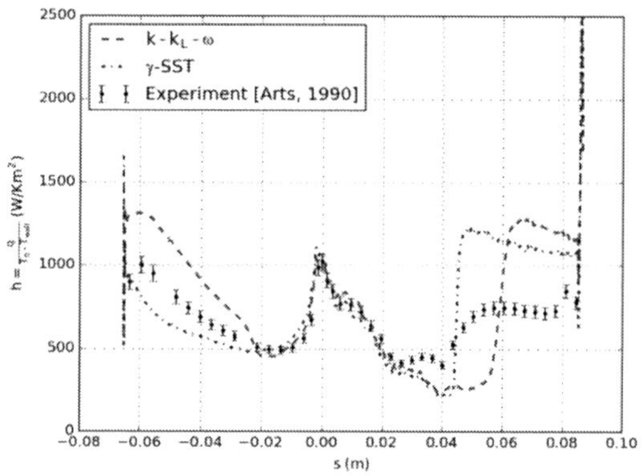


**Fig. 13.** Surface heat transfer coefficient for VKI MUR 241 Cascade.

in the vicinity of side walls. The figure 12 shows the comparison of the spanwise distribution loss coefficient $\zeta$ (see [16]) obtained using LU-SGS scheme with HLLC flux and the transonic version of segregated SIMPLE solver with experimental data. One can see that the loss coefficient is better predicted by LU-SGS method especially in the middle of the channel.

### 3.5 Transitional flow through VKI turbine cascade

The last case demostrates the combination of the LU-SGS scheme for main flow variables with transitional turbulence models. The test case corresponds to the experimental study of Arts et al. [1] performed at the von Karman Institute for Fluid Dynamics. The 2D flow through a turbine cascade was measured in terms of surface heat flux. The numerical solution was obtained with 62 000 cells for the regime corresponding to MUR 241 case from [1], i.e. the outlet Mach number was $M_2 = 1.089$, the Reynolds number related to outlet parameter and chord length was $Re = 2.1 \times 10^6$ and the free-stream turbulence intensity was $Tu = 6\,\%$. The figure 13 shows the comparison of experimental data with numerical results obtained with a three-equation $k - k_L - \omega$ model developed by Walters and Cokljat, see [19] or [8], and with a three-equation $\gamma - SST$ model by Menter et al. [11]. One can see that both models capture very well the heat flux in the laminar part of the flow, the transition onset at the suction side is better predicted by the $\gamma - SST$ model, and the heat flux is over-predicted by both models in the rear part of the suction side.

## 4 Conclusion

The matrix-free steady-state LU-SGS solver for compressible turbulent flow is described. The solvers shows very good accuracy for several test cases ranging from inviscid low Mach number flow over a bump to transonic turbulent flow through turbine cascade. The convergence to steady state is superior to segregated SIMPLE solver in the case of transonic flows. The convergence is a bit worse for low Mach number flows but it should be possible to improve it using a preconditioning. The LU-SGS solver provides very robust method which doesn't need any case dependent tuning (like e.g. relaxation factors in SIMPLE method for transonic flows). The main advantage of OpenFOAM platform is that the LU-SGS solver can be combined with a lot of built-in turbulence models as well as with the post-processing or parallel computing provided by the platform.

**Acknowledgements.** This research has been realized using the support of the Czech Science Foundation under grant P101/12/1271.

## References

[1] ARTS, T., LAMBERT DE ROUVROIT, M., and RUTHERFORD, A.W. *Aero-thermal investigation of a highly loaded transonic linear turbine guide vane cascade: A test case for inviscid and viscous flow computations.* Technical Note 174. Von Karman Institute for Fluid Dynamics, 1990, p. 97.

[2] BARTH, Timothy and JESPERSEN, Dennis. "The design and application of upwind schemes on unstructured meshes". In: *27th Aerospace Sciences Meeting.* Reston, Virigina: American Institute of Aeronautics and Astronautics, 1989.

[3] BLAZEK, Jiri. *Computational fluid dynamics : principles and applications.* Butterworth-Heinemann, 2015, p. 466.

[4] FERZIGER, Joel H. and PERIC, Milovan. *Computational Methods for Fluid Dynamics.* Springer Science & Business Media, 2012, p. 426.

[5] FÜRST, Jiří. "A weighted least square scheme for compressible flows". In: *Flow, Turbulence and Combustion* 76.4 (2006), pp. 331–342.

[6] GREENSHIELDS, Christopher J. et al. "Implementation of semi-discrete, non-staggered central schemes in a colocated, polyhedral, finite volume framework, for high-speed viscous flows". In: *International Journal for Numerical Methods in Fluids* 63.1 (2010), pp. 1–21.

[7] ISSA, R.I. "Solution of the implicitly discretised fluid flow equations by operator-splitting". In: *Journal of Computational Physics* 62.1 (1986), pp. 40–65.

[8] KOŽÍŠEK, Martin et al. "Implementation of k-kL-$\omega$ Turbulence Model for Compressible Flow into OpenFOAM". In: *Applied Mechanics and Materials* 821 (2016), pp. 63–69.

[9] LIOU, Meng-Sing. "A sequel to AUSM, Part II: AUSM+-up for all speeds". In: *Journal of Computational Physics* 214.1 (2006), pp. 137–170.

[10] MENTER, F. R. "Two-equation eddy-viscosity turbulence models for engineering applications". en. In: *AIAA Journal* 32.8 (1994), pp. 1598–1605.

[11] MENTER, Florian R. et al. "A one-equation local correlation-based transition model". In: *Flow, Turbulence and Combustion* 95.4 (2015), pp. 583–619.

[12] MOUKALLED, F. and DARWISH, M. "A High-Resolution Pressure-Based Algorithm for Fluid Flow at All Speeds". In: *Journal of Computational Physics* 168.1 (2001), pp. 101–130.

[13] ŠAFAŘÍK, Pavel. "Experimental data from optical measurement tests on a transonic turbine blade cascade". In: *13th Symposium on Measuring Techniques for Transonic and Supersonic Flow in Cascades and Turbomachines.* Ed. by G. Gyarmathy) C. GROSSWEILER. ETH Zurich, 1996, pp. 0–14.

[14] SHEN, Chun, SUN, Fengxian, and XIA, Xinlin. "Implementation of density-based solver for all speeds in the framework of OpenFOAM". In: *Computer Physics Communications* 185.10 (2014), pp. 2730–2741.

[15] SHEN, Chun et al. "Implementation of density-based implicit LU-SGS solver in the framework of OpenFOAM". In: *Advances in Engineering Software* 91 (2016), pp. 80–88.

[16] ŠIMURDA, David, FÜRST, Jiří, and LUXA, Martin. "3D flow past transonic turbine cascade SE 1050 - Experiment and numerical simulations". In: *Journal of Thermal Science* 22.4 (2013), pp. 311–319.

[17] TORO, Eleuterio F. "The HLL and HLLC Riemann Solvers". In: *Riemann Solvers and Numerical Methods for Fluid Dynamics.* Berlin, Heidelberg: Springer Berlin Heidelberg, 1997, pp. 293–311.

[18] VENKATAKRISHNAN, V. "On the accuracy of limiters and convergence to steady state solutions". In: *31st Aerospace Sciences Meeting & Exhibit* (1993), pp. 1–10.

[19] WALTERS, D. Keith and COKLJAT, Davor. "A three-equation eddy-viscosity model for Reynolds-averaged Navier-Stokes simulations of transitional flow". In: *Journal of Fluids Engineering* 130.12 (2008), pp. 121401–121401–14.

[20] WELLER, H. G. et al. "A tensorial approach to computational continuum mechanics using object-oriented techniques". In: *Computers in Physics* 12.6 (1998), p. 620.

[21] XISTO, Carlos M., PÁSCOA, José C., and OLIVEIRA, Paulo J. "A pressure-based high resolution numerical method for resistive MHD". In: *Journal of Computational Physics* 275.151 (2014), pp. 323–345.

EPJ Web of Conferences **143**, 02028 (2017) DOI: 10.1051/epjconf/201714302028
*EFM 2016*

# Air flow around suspended cables

*Irena* Gołębiowska[1], *Maciej* Dutkiewicz[1,*]

[1] Faculty of Civil, Environmental Engineering and Architecture, Department of Building Construction, University of Science and Technology in Bydgoszcz, ul. Prof. S. Kaliskiego 7, 85-796 Bydgoszcz, Poland.

**Abstract.** The impact of wind on construction structures is essential issue in design and operation. In particular, the wind can cause the dengerous vibrations of slender structures with low rigidity, eg. vibrations of cables of suspension and cable-stayed bridges or high voltage transmision lines, thus understanding of wind flow around such constructions is significant. In the paper the results of the analysis of wind flow around the cables for different Reynolds number is presented. The analysed flow meets the Navier-Stokes and continuity equations. The circle and elipse section of the cable is analysed. The discusion of vorticity, drag and lift coefficients and cases due to different angle of wind flow action is presented. The boundary layer and its infuence on total flow is analysed.

## 1 Introduction

High-rise buildings, chimneys, suspended and stayed bridge, voltage transsmision lines are the examples of slender structures. As vibration due to wind flow can significantly reduce the time of operating of the objects, hence this impact must be taken into account at the design stage and later during operation.

From the point of aerodynamics, the bodies are divided on: streamlined - elongated in the direction of the fluid flow such as cross sections used for wings of aircrafts, and bluff bodies. Above listed objects belong to the bluff bodies group. The fluid flow around these objects is characterized by numerous physical complex phenomena. This flow is determined, when three parameters are known: pressure, velocity, and density that describe the fluid. In the general case, these parameters are functions of time and space. When the parameters depend on time, the flow is unsteady (non-stationary), if depend on space, the flow is un-uniform. The nature of the fluid flow is characterized by Reynolds number which presents the relations between inertial and viscous forces.

Fluid flow analysis around the smooth cylinder were the subject of many authors' studies [8-13]. In the literature flow regimes are devided in several states [1-7]. According to [3] for lower values of the Reynolds number such as Re<< 1, the flow is steady and balanced. For Re $\in$ (5; 40) - the flow is steady, laminar, some vortices form themselves symmetrically in wake. For Re $\in$ (40; 100 000) flow is in subcritical range, flow is unsteady, turbulent, vortical and unsymmetrical. The Karman street is formed, the frequency of vortices is described by Strouhal number. For Re $\in$ ($10^5$; $3{,}5*10^6$) flow is in critical and supercritical range, unsteady, turbulent. For Re > 3,5 * $10^6$ the flow is supercritical.
Location of points of separation of the boundary layer on the walls of the body is critical from the point of aerodynamical influence on the body.

* Corresponding author: macdut@utp.edu.pl

The phenomena of interaction between solid bodies and the flow of air is significant from the point of induced vibrations.

In this paper the air flow past the cable of circle and ellipse section for Reynolds numbers in the range of Re $\in$ (10; $10^5$) is analysed. The ellipse section can present the variation in cross shape due to e.g. icing of the construction of the cable located in the open space.

## 2 Wind flow around the cable's cross-section

Analysed model with the mesh of flow domain and general assumption of inflow direction and section of cable is shown in Fig.1. The analysis is performed for different fixed Reynolds numbers for circle and ellipse sections. The assumption that air is incompressible is made.

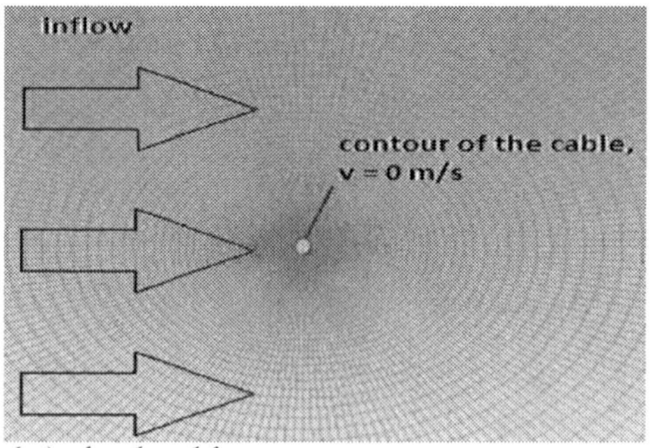


**Fig.1.** Analysed model

Navier-Stokes equations of motion for description of incompressible and viscous air is used. In the analyzed state, the interial force, pressure and viscosity forces are balanced.

The basic equation of motion follows from the momentum conservation principle. This principle implies that the derivative of the velocity of air inside the volume V with respect to time is equal to the sum of the external forces acting on the area. According to d'Alembert, formulation of this principle can be written as follows [1]:

$$\iiint\limits_V \rho \frac{d\boldsymbol{v}}{dt} dV = \iiint\limits_V \rho \boldsymbol{F}_m \, dV + \iint\limits_A \boldsymbol{p}_A \, dA \quad (1)$$

where

$\iiint_V \rho \frac{dv}{dt} dV$ - means inertia forces,

$\iiint_V \rho \boldsymbol{F}_m \, dV$ - means the sum of the mass forces acting on the volume 'V',

$\iint_A \boldsymbol{p}_A \, dA$ - means the sum of the pressure acting on the entire surface 'A',

The folowing equation was obtained using the Gauss-Ostrogradski theorem for equation and $\mathbf{p}_A = - \mathbf{n}\, p$ (1):

$$\iiint\limits_V \rho \frac{d\boldsymbol{v}}{dt} dV - \iiint\limits_V \rho \boldsymbol{F}_m \, dV + \iiint\limits_V grad\, p \; dV = 0 \quad (2)$$

The sum of integrands functions must be equal to zero for any volume 'V':

$$\frac{d\boldsymbol{v}}{dt} = \boldsymbol{F}_m - \frac{1}{\rho} grad\, p \quad (3)$$

Using Newton's hypothesis concerning the relationship between stress and deformation of air, the Navier-Stokes equation is obtained in the form:

$$\frac{d\boldsymbol{v}}{dt} = \boldsymbol{F}_m - \frac{1}{\rho} grad\, p + \mu \nabla^2 v \quad (4)$$

where $\rho$ is air density, $v$ - velocity of flow, t time, p is air pressure and $\mu$ is the coefficient of viscosity.

The purpose of the analysis of the above equation is to determine the relationship between pressure and velocity of flow, in particular the analysis of phenomena occurring at the air / cable contact. The above equation for each direction has the following form:

$$\frac{\partial v_x}{\partial t} + v_x \frac{\partial v_x}{\partial x} + v_y \frac{\partial v_x}{\partial y} + v_z \frac{\partial v_x}{\partial z} = X - \frac{1}{\rho}\frac{\partial p}{\partial x} + \mu \left( \frac{\partial^2 v_x}{\partial x^2} + \frac{\partial^2 v_x}{\partial y^2} + \frac{\partial^2 v_x}{\partial z^2} \right) (5a)$$

$$\frac{\partial v_y}{\partial t} + v_x \frac{\partial v_y}{\partial x} + v_y \frac{\partial v_y}{\partial y} + v_z \frac{\partial v_y}{\partial z} = Y - \frac{1}{\rho}\frac{\partial p}{\partial y} + \mu \left( \frac{\partial^2 v_y}{\partial x^2} + \frac{\partial^2 v_y}{\partial y^2} + \frac{\partial^2 v_y}{\partial z^2} \right) (5b)$$

$$\frac{\partial v_z}{\partial t} + v_x \frac{\partial v_z}{\partial x} + v_y \frac{\partial v_z}{\partial y} + v_z \frac{\partial v_z}{\partial z} = Z - \frac{1}{\rho}\frac{\partial p}{\partial z} + \mu \left( \frac{\partial^2 v_z}{\partial x^2} + \frac{\partial^2 v_z}{\partial y^2} + \frac{\partial^2 v_z}{\partial z^2} \right) (5c)$$

For the constant density the continuity equation has the following form:

$$\frac{\partial v_x}{\partial x} + \frac{\partial v_y}{\partial y} + \frac{\partial v_z}{\partial z} = 0 \quad (6)$$

Among parameters that describe the fluid flow is vorticity.

Vorticity is a measure of the rotation of the air element as it moves in the flow field, and is defined as the curl of the velocity vector:

$$\boldsymbol{\xi} = \nabla \, x \, \boldsymbol{v} \quad (7)$$

where: $\nabla$ is the operator, $\boldsymbol{v}$ is flow velocity.

In a two-dimensional flow where the velocity is independent of the $'z'$ coordinate and has no $z$ component, the following equation has the form:

$$\boldsymbol{\xi} = \left( \frac{\partial v_y}{\partial x} - \frac{\partial v_x}{\partial y} \right) \boldsymbol{z} \quad (8)$$

This parameter is analysed in the next clause of the paper in the context of circle and ellipse section of the cylinder for the fix Reynolds number.

## 3. Results and discusion

The wind around the cable, that is modeled as smooth cylinder, of infinite length of circle and ellipse section was analysed. Diameter of the circle is 120mm, for ellipse: major diameter is 180mm and minor diameter is 120 mm. The 'x' and 'y' direction of the wind is analysed and 'z' component is zero, so two dimensional character of the phenomen is considered. The model presented in the paper was created with use of Ansys Fluent. The unsteady flow for stream simualtion around cylinders with fixed Reynolds numbers was used. For improving accuracy of results, 22650 elements were applied. Navier Stokes equations were solved numerically with use of Semi Implicit Method for Pressure Linked Equations with sequenced calculations of velocity and pressure's components. The coefficients of subrelaxations were additionally used for stabilizing the calculations process. Additionaly, second order upwind method for momentum equations was adopted. At the inlet the velocity is equal to 1m/s and at the outlet the gauge pressure is equal to zero.

When Reynolds number is progressively increased from Re = 0, the flow experiences the transition and diverse flows are observed. At lower Reynolds number the viscous forces are significant and they keep the air in line and thus flow is laminar. The viscous forces are

DOI: 10.1051/epjconf/201714302028

much more bigger then inertia forces. At high Reynolds number the inertia forces will dominate the viscous forces. Depending on Reynolds numbers of the flow, different flow regimes will occur.

For fixed Reynolds number Re=0,01 and both circle section as well as ellipse one, the distribution of pressure and velocity, is symmetrical and boundary layer do not separate from cylinder wall at any point. Flow field is steady (Fig. 2 and 3).

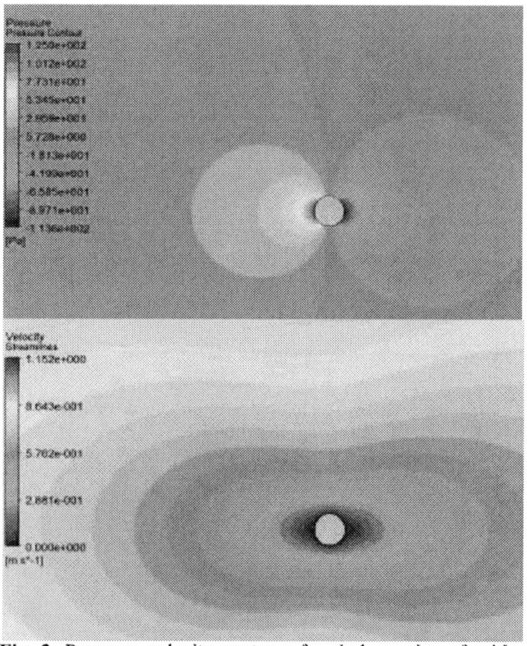


**Fig. 2.** Pressure, velocity contours for circle section of cable for Re=0,01

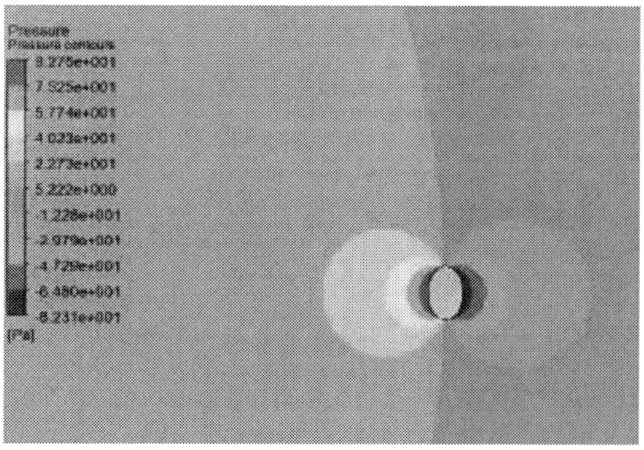


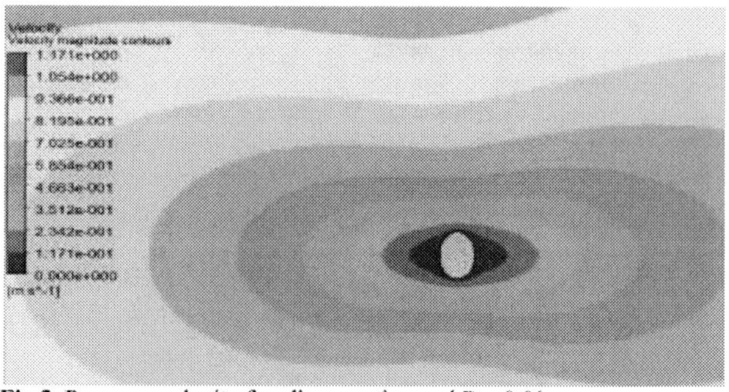


**Fig.3.** Pressure, velocity for elipse section and Re=0,01.

For Re = 200 wake becomes unsteady and vortices appear (Fig. 4 i 5, Re=200).

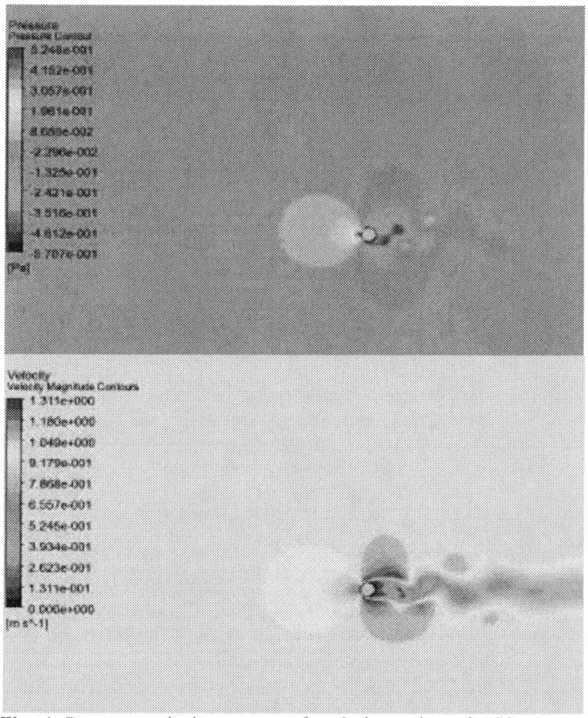


**Fig. 4.** Pressure, velocity contours for circle section of cable for Re=200

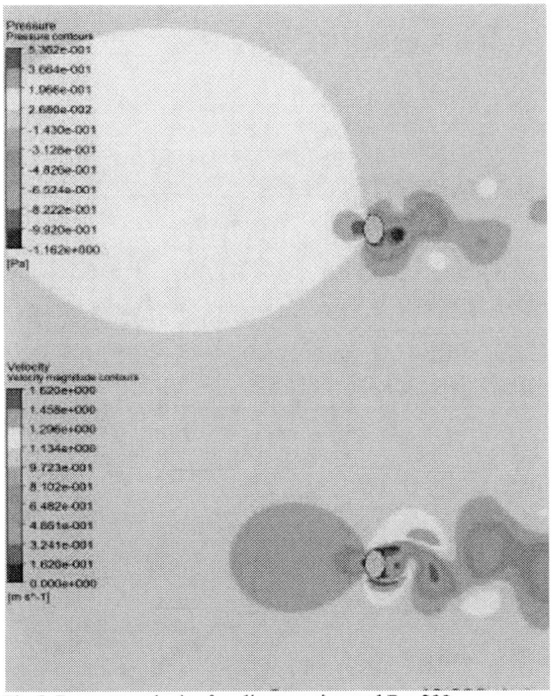

**Fig.5.** Pressure, velocity for elipse section and Re=200

Further increase of Reynolds number results in a decrease of the distance between the vortices, vortices tear-off time is shorter and wake increases for the circle and ellipse sections (Fig. 6 and 7).

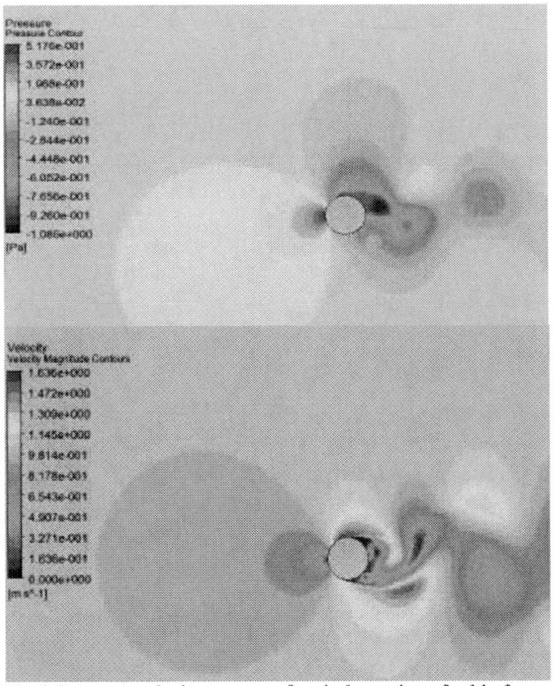

**Fig.6.** Pressure, velocity contours for circle section of cable for Re=25000

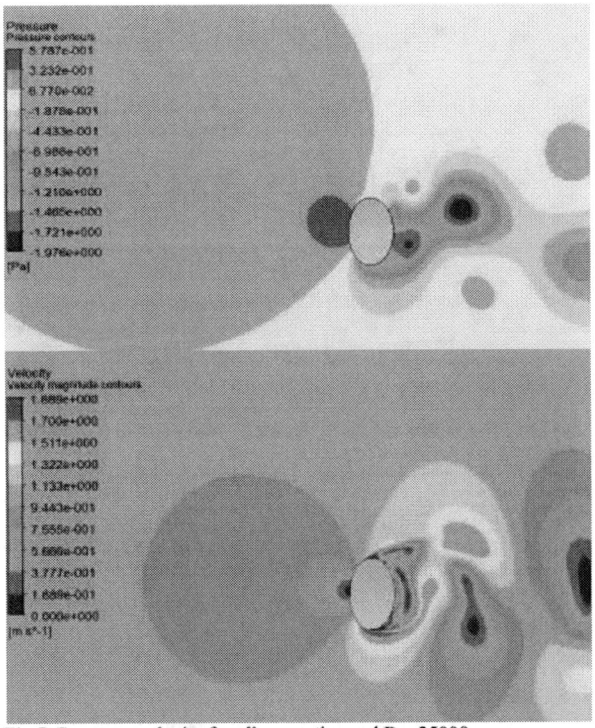

**Fig.7.** Pressure, velocity for elipse section and Re=25000

Figure 8 shows the results of calculations of drag and lift coefficients for various values of the Reynolds number and circle and ellipse cross-sections.

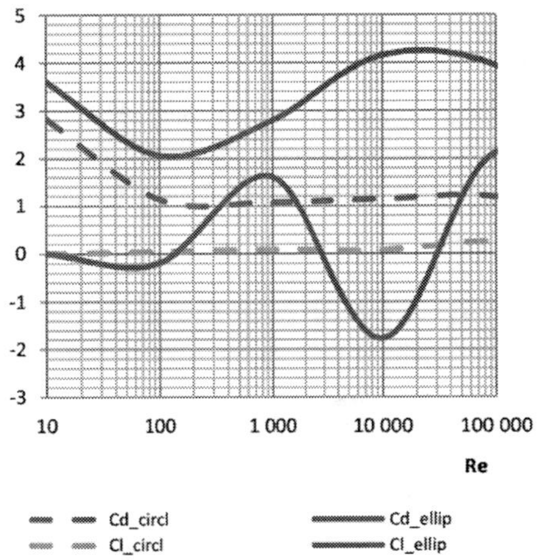


**Fig.8.** Drag coefficient Cd and lift coefficient Cl against Reynolds number, for ellipse and circle cross-sections.

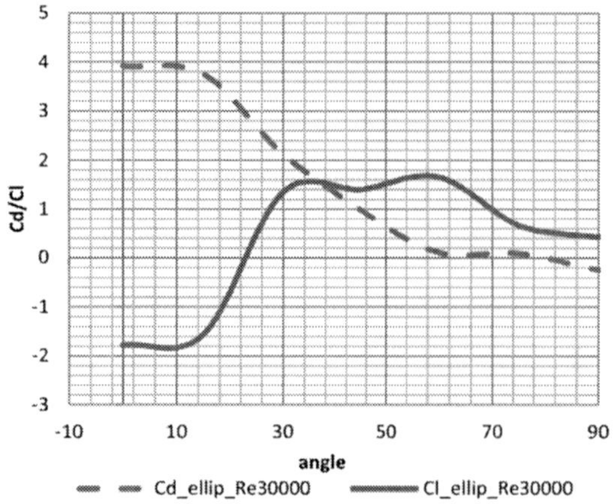


**Fig. 9.** Drag coefficient Cd and lift coefficient Cl against angle of wind action, for ellipse cross-section and Re=30000

For ellipse the Cd coefficient has much higher value than for the circle in the analysed range of Reynolds number. Coefficient Cl takes positive and negative values for ellipse section.

Figure 9 shows the drag and lift coefficients for different angle of air inflow for ellipse section and Re = 30000.

Due to contact of the air with the cylinder, the boundary layer is formed. At a small distance from the surface, a strong velocity gradient is formed. For Re = 25 000 the maximal width of the boundary layer is 8mm for and ellipse section and 5mm for circle section.

Dependence of drag, lift coefficients and flow time are shown on figures 10 and 11. Both for circle and elipse sections drag and lift coefficients show the periodic oscillations with higher frequency for circle than ellipse section. Some disturbances come from vortex shedding process from both side of the cylinder.

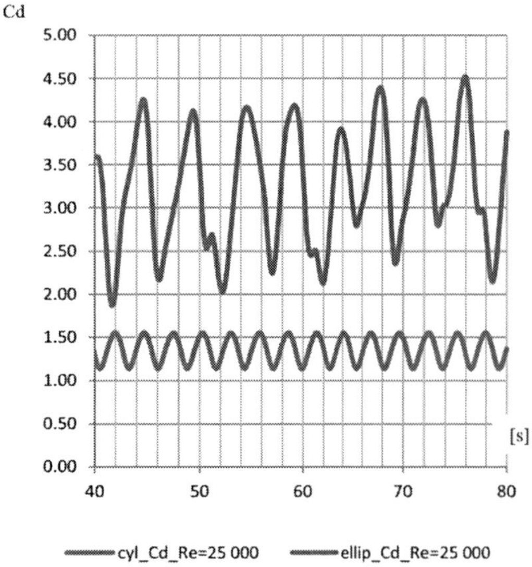


**Fig.10** Flow time of drag coefficient Cd, for circle and ellipse section, Re=25 000

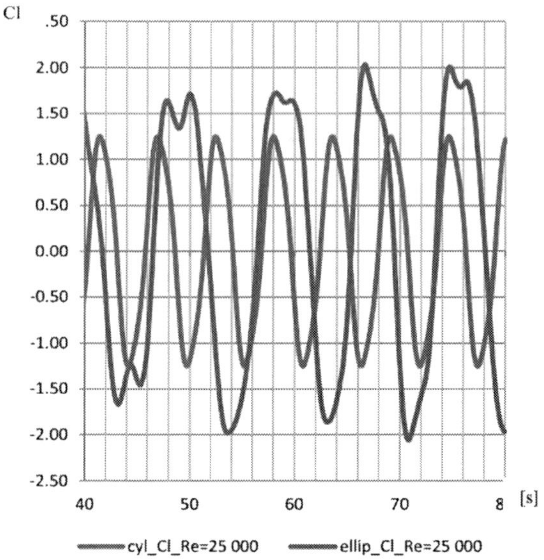


**Fig.11** Flow time of lift coefficient Cl, for circle and ellipse sections, Re=25 000

Figure 12 present the magnitude of the vorticity vector for circle and ellipse section and Re = 25 000.

Fluid around the ellipse section show the higher vorticity magnitude compared to circle section.

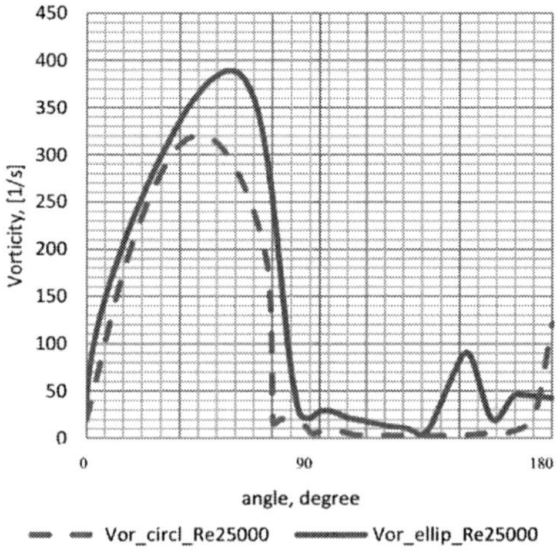


**Fig.12** Dependence of vorticity magnitude for circle and ellipse sections, Re=25 000 on angle describing the analysed point on cylinder

## 5. Summary and Conclusions

In the paper the flow past cable with circle and ellipse cross-section was discussed. Drag and lift coefficients for different Reynolds numbers and angle of wind action were analyzed. These coefficients were analyzed in the time domain. For ellipse, the Cd coefficient has much higher values than for the circle one, in the analysed range of Reynolds number.

Both in the case of circle and ellipse cross-section, oscillation frequency of drag coefficient is higher than oscillation frequency of lift coefficient Besides that, oscillation frequency of drag and lift coefficients for circle section is higher than oscillation frequency of drag and lift coefficients for ellipse section.

Also the vorticity magnitude was analysed. Fluid around the ellipse section show the higher vorticity compared to circle section.

## References

1. Zdravkovich, M. M. (1997). *Flow around cylindrical structures*. **Vol 1**: Fundamentals. Oxford University Press.
2. Williamson, C. H. K. (1996). *Vortex dynamics in the cylinder wake.* Annu.Rev. Fluid Mech., **28:**477–539.
3. A. Flaga, *Inżynieria Wiatrowa,* Arkady, Warszawa 2008
4. J.A.Żurański, *Obciążenia wiatrem budowli i konstrukcji,* Arkady Warszawa 1978
5. Z.Orzechowski, J.Prywer, R.Zarzycki, *Mechanika płynów w inżynierii środowiska*, WNT, Warszawa 1997
6.J.Bukowski *Mechanika Płynów*, PWN Warszawa 1975
7. H.Rouse *Elementary Mechanics of Fluids,* John Wiley and Sons, New York 1948
8. B. Hunt *Fluid Mechanics for Civil Engineers,* Christchurch, New Zealand, 1995
9. B.R.Munson, *Fundamentals of fluid mechanics,* 6th ed., New York, John Wiley, 2009
10. S.C.R. Dennis, G.Z. Chang, *Numerical solutions for steady flow past a circular cylinder at Reynolds numbers up to 100*, J. Fluid Mech. **42** (1970) 471.
11. S.Y. Tuann, M.D. Olson, *Numerical studies of the flow around a circular cylinder by a finite element method*, Comput. Fluids **6** (1978) 219.
12. M.M. Alam, M. Moriya, K. Takai, H. Sakamoto, *Fluctuating fluid forces acting on two circular cylinders in a tandem arrangement at a subcritical Reynolds number*, Journal of Wind Engineering Industrial Aerodynamics, **91**, 139–154, 2003
13.M.Matsumoto, *Aerodynamic damping of prisms*, Journal of Wind Eng. and Industrial Aerodynamics, **59**, 159-175, 1996

DOI: 10.1051/epjconf/201714302029

# The effectiveness of vibration damper attached to the cable due to wind action

*Irena* Gołębiowska[1], *Maciej* Dutkiewicz[1,*]

[1] Department of Building Construction, University of Science and Technology in Bydgoszcz, ul. Prof. S. Kaliskiego 7, 85-796 Bydgoszcz, Poland.

**Abstract.** In the paper the analysis of vibration of cable with attached damper is performed. Source of vibrations comes from the wind. The level of reduction of vibrations is also analysed. Damper attached to the cable is characterized by mass, stiffness and damping ratio. The paper also presents the equations of motion for the cable with damper, taking into account the initial parameters: cable cross-section and the location of the damper. The analysis is performed in time domain. It is discussed that in real situations the different types of cable vibration due to their amplitudes and frequencies is met: short-term oscillations of high amplitude, caused by a single impulse force, called bouncing cable, low amplitude and low wave length and the high frequency, called aeolian vibration, large oscillation amplitude, long wave length and low frequency, called galloping or dancing cables.

## 1 Introduction

### 1.1 Bridge cable vibrations

Due to construction benefits and architectural view in recent years a significant number of suspension and cable-stayed bridges were built. From the point of load and work of bridge, the wind load can be reason of vortex induced vibrations that can adversely effect on strength, durability of cables and connection failures [1,2]. In some cases local fatigue cracks are formed and further destruction of the structure is performed. Methods of reduction vibrations are applied to avoid such situations as cables have a very low level of damping an low bending stiffness [3]. Connection of cables to each other is another possible solution but this solution is used very rarely because of the adverse visual impact. Application of passive vibration damper, usually mounted at a distance not exceeding 5% of the length from the anchorage cable construction platform is another solution that is often used. Aerodynamic cable surface modification is another method of vibration reduction [1].

The selection of appropriate parameters of the vibration damper is associated with the characteristics of the damped structure working under the dynamic load. Cables perform uninterrupted vibrations both at minimum parameters of the damping system as well as when damping parameter is large, so it is essential to perform the tuning of the damping system [4].

### 1.2 Transmission lines vibrations

Vibrations of transmission lines due to wind actions are characterized by high frequency and small amplitudes of vibrations. They are called aeolian vibrations. Apart from them, oscillations of high amplitude and low frequency called galloping vibrations and secondary induced vibrations of average amplitude but lower than aeolian frequency are also observed. Aeolian vibrations are in the range of 3 to 50 Hz at wind speed from 0,5 to 7 m/s acting perpendicular to the cable, which usually happens in open plots, in the area without buildings and infrastructure. The amplitude of oscillation does not exceed 2.5 cm [5,6]. These vibrations are difficult to observe due to the fact that transission lines diameters are from 6 to 50 mm, which exceeds the amplitudes of vibrations.

Galloping of cables appears at winds speed from 7 to 25 m/s. Cables oscillate according to first or second form as one or two half-waves. The frequency of oscillation is small, less than 1 Hz [5].

The maximum amplitude of vibrations occurs at resonance, and increases when the natural frequency is close to the forced frequency called Strouhal frequency. Vibrations can be controlled by the use of different dampers. The use of Stockbridge'a damper is a very popular solution.

Stockbridge damper is constructed from a flexible cable or rod and the masses at its ends [2]. The vibrations of the cable are passed through the clamp mounted the damper's rod to the cable and consequently the masses on the ends of the damper begin to vibrate. By careful selection of parameters of the damper such as mass at the ends of the rod and stiffness of the damper's rod, one can tune the optimal damping of the transmission energy line. First Stockbridge damper held at the ends of the rod masses made of concrete blocks. Nowadays steel masses at end of the damper's rod are used and galvanized steel is applied for damper's rod. In the modelling of Stockbridge damper the massless rod element is

* Corresponding author: macdut@utp.edu.pl

analysed with masses at the ends. From the point of reduction of vibrations, the location of damper on the cable is significant [5].

The aeolian vibrations were analysed in the paper [8]. Analysis confirms that use of damper do not cause a dissipation of energy and reduction of vibrations in any case.

In the paper [6] the aplication of Stockbridge damper caused the feedback of the damper with the cable and disturbance of aeolian vibrations. Deterministic or probabilistic models are often applied among computational models. In the paper [5] a probabilistic model of the aeolian vibration with regard to fatigue strength was performed.

The purpose of the dampers is reduction of vibration by absorbing energy from the wind and stabilization of the movement of the cable to avoid deformation that can cause the damage. According to analysis performed in the paper [5] the value of safe limit deformation of the cable is 200 microns/m. The effectiveness of the damper depends on its response to the wind force at a certain frequency.

## 2 Modelling and analysis of cable's vibrations

### 2.1 Assumptions of the analysed system

The issue of inextensible cable's free vibration was analyzed by Hajduk [7]. Vibration equations shown below fulfill the law of change of the amount of motion, law of conservation of mass and the law of change of the body deformation under load.

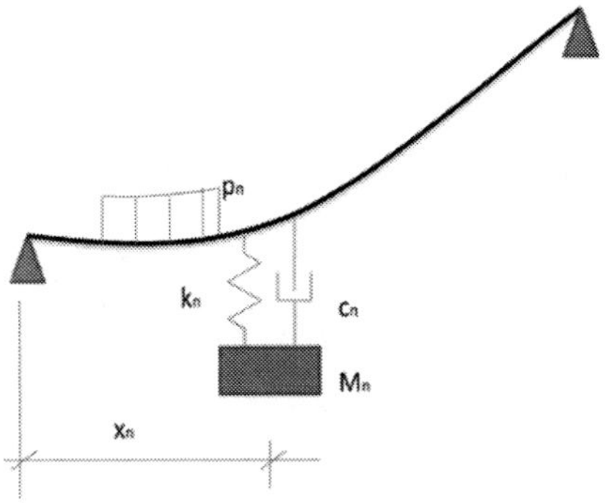


**Fig.1.** System of the cable with the damper

The general idea of the system equipped with the damper, that is the subject of the paper, is shown on figure 1. It is shown the cable of length 'l' under the load $p_n$ with the system of dampers represented by 'n-th' damper with parameters: stiffness $k_n$, damping $c_n$, mass $M_n$, m - mass of the cable distributed uniformly on the cable.

The purpose of the paper is to demonstrate the need of protection against vibrations of cables acting as bearing structures and proper selection of vibration dampers' parameters.

Numerical simulation is carried out for 20 seconds of vibration excitation, for span of the cable of 20 m. The cable is modelled as a strand, non-displaceably fixed to the structure. It is assumed that path of the cable is in-plane and the chord is horizontal. It is analysed the continous distribution of the mass along the length of the cable.

The analysis is limited to vertical vibrations occurring in the plane of the cable sag. According to Kirchhoff's hypothesis, because the sag of the cable is small, the effects of horizontal displacements is omitted. It is assumed that the material of the cable is linear-elastic and fulfill Hooke's law, i.e. the relationship between displacements and deformations is linear.

Taking into consideration the above assumptions the solution of dynamic displacement is taken in the form:

$$w(x,t) = \sum_{j=1}^{k} \Phi(t) X_j(x) \qquad (1)$$

where:

$\Phi$ (t) is unknown function of time, treated as the coefficient of series:

$$\Phi(t) = \phi_0 e^{i\omega t} \qquad (2)$$

and $\phi_0$ is the amplitude coefficient of the function $\Phi(t)$, 'i' is the imaginary part of the complex number, $\omega$ is the forced frequency, t - means time of the analysed process. $X_j(x)$ shown in the equation no.1 is function that meets all boundary conditions of the analysed problem:

$$X_j(x) = \sum_{j=1}^{k} sin \frac{j\pi x}{l} \qquad (3)$$

This assumption is made according to Galerkin's method for cable's vibration of an infinite number of degrees of freedom [7]. It is assumed that functions $X_j(x)$, j = 1,2,...,n, are linearly independent.

Vibration of the single damper is presented in the following equation:

$$w_1(t) = \phi_1 e^{i\omega t} \qquad (4)$$

where $\phi_1$ is the amplitude coefficient of the function $w_1(t)$.

### 2.2 Solution and discussion

Taking into account the approximation (1) and equations from (2) to (4), the equation of motion of the system can be written in Langrange'a coordinates:

$$M\,\ddot{\Phi}(t) + C\,\dot{\Phi}(t) + K\,\Phi(t) = P(t) \qquad (5)$$

where M, C, K are respectively matrices: mass, damping and stiffness, P (t) is the external load, i.e:

$$M = \begin{bmatrix} m_{red} & \cdots & 0 \\ \vdots & \ddots & \vdots \\ 0 & \cdots & M_n \end{bmatrix}, C = \begin{bmatrix} C_{11} & \cdots & C_{1n} \\ \vdots & \ddots & \vdots \\ C_{n1} & \cdots & C_{nn} \end{bmatrix},$$

$$K = \begin{bmatrix} K_{11} & \cdots & K_{1n} \\ \vdots & \ddots & \vdots \\ K_{n1} & \cdots & K_{nn} \end{bmatrix}, P = \begin{bmatrix} p \\ \dots \\ 0 \end{bmatrix} \quad (6)$$

Orthogonality condition for the equation describing the vibrations of the cable takes the following form:

$$\int_0^l \sum_{p=1}^{s} \mathrm{L}(w) X_p(x) dx = 0 \quad (7)$$

where:

$$L(w) = m\frac{\partial^2 w(x,t)}{\partial t^2} - H_{st}\frac{\partial^2 w(x,t)}{\partial x^2} - \frac{EF}{l}y''(x)\int_0^l \frac{\partial w(x,t)}{\partial x}y'(x)dx + \gamma_1 D\frac{\partial w(x,t)}{\partial t} - \frac{\eta F}{l}y''(x)\int_0^l \frac{\partial^2 w(x,t)}{\partial x \partial t}y'(x)dx + \sum_{n=1}^{m} k_n\left(w(x,t) - w_n(t)\right)\delta(x - x_n) + \sum_{n=1}^{m} c_n\left(\frac{\partial w(x,t)}{\partial t} - \frac{\partial w_n(t)}{\partial t}\right)\delta(x - x_n) + p(x,t) \quad (8)$$

E - modulus of elasticity of the cable, $H_s$ - tension force in the cable, F - cross-sectional area of the cable, l - the length of the cable, $\gamma_1$- drag coefficient of the air, D - cable diameter, $\eta$ - internal damping coefficient of the cable, $k_n$- stiffness coefficient of the "n-th" damper, $c_n$ - damping coefficient of the "n-th" damper, $\delta$ - Dirac delta function.

It was assumed the following load:

$$p(x,t) = p_0 e^{i\omega t} \quad (9)$$

The equation for "n-th" dynamic damper takes the following form:

$$\sum_{n=1}^{m} \frac{\partial^2 w_n(t)}{\partial t^2} M_n + \sum_{n=1}^{m} k_n\left(w_n(t) - w(x,t)\right)\delta(x - x_n) + \sum_{n=1}^{m} c_n\left(\frac{\partial w_n(t)}{\partial t} - \frac{\partial w(x,t)}{\partial t}\right)\delta(x - x_n) = 0 \quad (10)$$

where $M_n$ is the mass of "n-th" dynamic damper.

After calculations, the solution of the system of equations (5), taking into consideration equation (1), the deflection function takes the form:

$$w(x,t) = Re(w) + Im(w) \quad (11)$$

where:

$$Re(w) = p_0 \sum_{j=1}^{k}\sum_{n=1}^{m} sin\frac{j\pi x}{l}\frac{s_{1n}cos\omega t + s_{2n}sin\omega t}{s_{1n}^2 + s_{2n}^2} \quad (12)$$

$$Im(w) = p_0 \sum_{j=1}^{k}\sum_{n=1}^{m} sin\frac{j\pi x}{l}\frac{s_{1n}sin\omega t - s_{2n}cos\omega t}{s_{1n}^2 + s_{2n}^2} \quad (13)$$

In equations (12) and (13) $s_1$ and $s_2$ denote functions dependent on the known parameters of stiffness, damping, frequency domains, the mass of the cable and the masses of dampers.

## 3 Numerical analysis

Numerical analyses were performed for various parameters and the location of dampers. It was assumed the length of the steel cable l=20m, cross section of the cable F = 1,13 cm$^2$, force of tension $H_{st}$=20 kN.

Figure 2 shows that the use of the dynamic vibration damper reduces the dynamic displacement from the value of 62.12 mm to the value of 10,53mm. After using the damper placed in the middle of the cable, the displacement have been reduced by almost 6 times. The above reduction of vibration was achieved by choosing the optimal parameters as follows: $M_1 = 48{,}3\ kg$, $k_1 = 13{,}03\ kN/m$, $c_1 = 10{,}097\ N * s/m$.

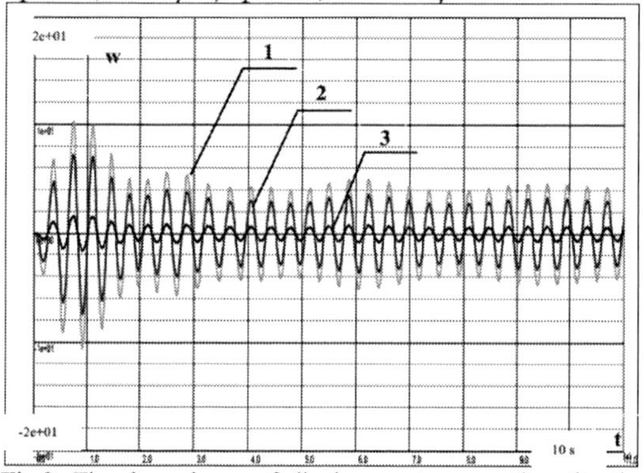


**Fig.2.** The dependence of displacement versus time for the vibration damper in the middle of the span. Curve 1 - vibration in the middle of the span of the cable, curve 2 - the vibration in the 1/4 of span of the cable, curve 3 - vibrations close to the support.

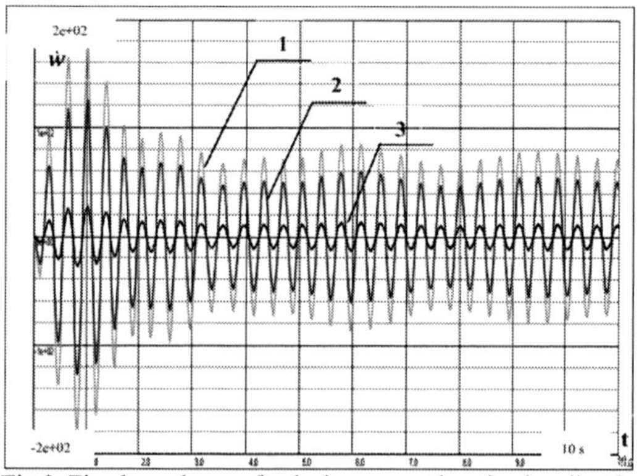


**Fig.3.** The dependence of velocity versus time for the vibration damper in the middle of the span. Curve 1 - vibration in the middle of the span of the cable, curve 2 - the vibration in the 1/4 of span of the cable, curve 3 - vibrations close to the support.

Figure 3 shows the velocity versus time for the parameters of the damper as in the case shown in Figure 2.

DOI: 10.1051/epjconf/201714302029

Figure 4 shows the acceleration versus time for the parameters of the damper as in the case shown in Figure 2.

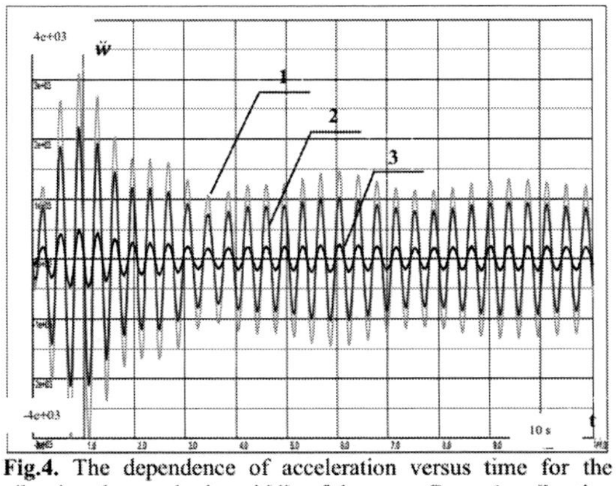

**Fig.4.** The dependence of acceleration versus time for the vibration damper in the middle of the span. Curve 1 - vibration in the middle of the span of the cable, curve 2 - the vibration in the 1/4 of span of the cable, curve 3 - vibrations close to the support.

Figure 5 shows the displacement depending on the frequency of vibration of cable with damper and without damper.

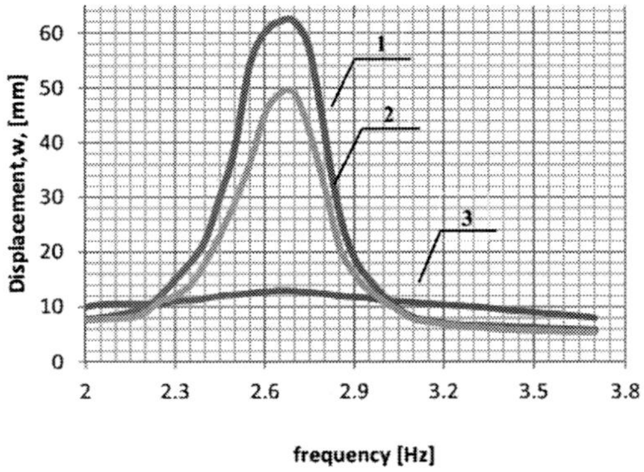


**Fig. 5.** The dependence of the displacement versus frequency, for a cable with damper and without damper. Curve 1 - vibrations of the cable without damper, curve 2 - cable vibration for optimal parameters of the damper $M_1 = 48{,}3\ kg$, $k_1 = 13{,}03\ kN/m$, $c_1 = 5{,}00\ N*s/m$, curve 3 - cable vibration for parameters of the damper $M_1 = 48{,}3\ kg$, $k_1 = 13{,}03\ kN/m$, $c_1 = 10{,}097\ N*s/m$.

Figure 6 shows the function of displacement versus the location of the vibration damper on the cable.

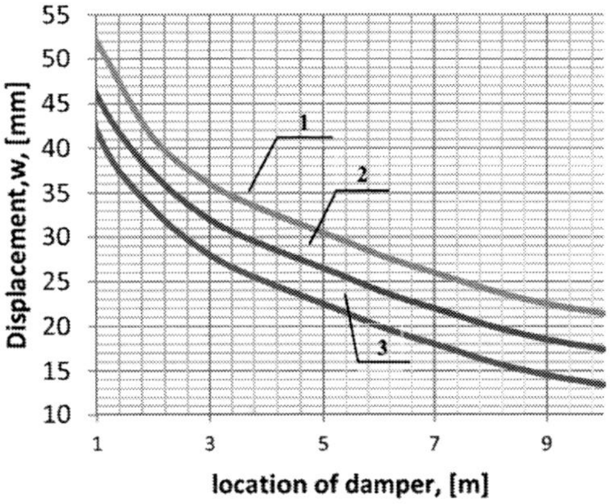


**Fig.6.** Dependence of displacement as a function of the location of the damper, Curve 1 - vibration of the cable for the parameters of the damper $M_1 = 48{,}3\ kg$, $k_1 = 13{,}03\ kN/m$, $c_1 = 2{,}50\ N*s/m$ , curve 2 - vibration of the cable for the parameters of the damper $M_1 = 48{,}3\ kg$, $k_1 = 13{,}03\ kN/m$, $c_1 = 5{,}00\ N*s/m$, curve 3 - vibration of the cable for the parameters of the damper $M_1 = 48{,}3\ kg$, $k_1 = 13{,}03\ kN/m$, $c_1 = 10{,}097\ N*s/m$.

## 4 Summary and Conclusion

In the paper the importance of cable structures' analysis is emphasized. Cables in structures play an important supporting function. They are part of the construction of suspension bridges and the power transmission lines. From the point of view of the structural analysis, cables are important element in the design process to keep the sufficient ultimate and serviceability limit states. These constructions have large spans and low rigidity susceptible to deformation what was discussed in the paper. Because of low amplitudes often less than the diameter of the cable, these vibrations are slightly perceptible, but may lead to material fatigue and then to failure.

In the paper the influence of the parameters of the damper and the location on cable is shown and analysed. The proper location and optimization of damper's parameters is shown as requirement to reduce the vibration of the cable.

## References

1. I.Gołębiowska, K.Peszyński, *Reduction of Cable Vibrations Caused by Wind*, Engineering Mechanics, 22th International Conference Engineering Mechanics, Svratka 2016
2. I.Gołębiowska, M.Dutkiewicz, B.Usewicz, *Sposoby tłumienia drgań napowietrznych linii elektroenergetycznych*, TTS Technika Transportu Szynowego - 2015, 12, 2544-2548
3. B.M. Pacheco,Y. Fujino, A. Sulekh, *Estimation Curve for Modal Damping in Stay Cables with*

*Viscous Damper,* J. Struct. Engrg., ASCE, **119(6),** (1993).

4. K. Uno, S. Kitagawa, H. Tsutsumi, A.Inoue, S.Nakaya, *A simple method of designing cable vibration dampers of cable-stayed bridges*, J. of Struct.l Eng., JSCE, **Vol. 37A**, pp.789-798, (1991)
5. H. Q. Wang, J. C. Miao, J. H. Luo, F. Huang, and L. G. Wang, *The free vibration of long-span transmission line conductors with dampers,* Journal of Sound and Vibr., **208(4):501** (1997).
6. H. Meynen, P.Verma, P. Hagedorn, M. Schafer, *On the numerical simulation of vortex-induced vibrations of oscillating conductors*, J.of Fluids and Struct., **21(1)** (2005).
7. J.Hajduk, J. Osiecki, *Ustroje cięgnowe, teoria i obliczanie*, WNT Warszawa (1970)
8. J.Vecchiarelli, I.G. Curie, D.G. Havard, *Computational analysis of aeolian conductor vibration with a stockbridge-type damper*, J. of fluids and struct. **14**, 489-509 (2000)

# Estimating the uncertainty of the liquid mass flow using the orifice plate

*Anna* Golijanek-Jędrzejczyk[1,*], *Dariusz* Świsulski[1], *Robert* Hanus[2], *Marcin* Zych[3], and *Leszek* Petryka[4]

[1]Gdańsk University of Technology, Faculty of Electrical and Control Engineering, 80-233 Gdańsk, Poland
[2]Rzeszów University of Technology, Faculty of Electrical and Computer Engineering, 35-959 Rzeszów, Poland
[3]AGH University of Science and Technology, Faculty of Geology, Geophysics and Environmental Protection, 30-059 Kraków, Poland
[4]AGH University of Science and Technology, Faculty of Physics and Applied Computer Science, 30-059 Kraków, Poland

**Abstract.** The article presents estimation of measurement uncertainty of a liquid mass flow using the orifice plate. This subject is essential because of the widespread use of this type of flow meters, which renders not only the quantitative estimation but also qualitative results of this type so those measurements are important. To achieve this goal, the authors of the paper propose to use the theory of uncertainty. The article shows the analysis of the measurement uncertainty using two methods: one based on the "Guide to the expression of uncertainty in measurement" (GUM) of the International Organization for Standardization with the use of the law of propagation of uncertainty, and the second one using the Monte Carlo numerical method. The paper presents a comparative analysis of the results obtained with both of these methods.

## 1 Introduction

The knowledge of the accuracy of the mass flow of liquids is significant due to the fact that it occur in most processes and is a significant research topic. Especially experimental or particularly accredited laboratories should focus on full and correct estimation of measurement uncertainty of mass flow.

Modern metrology [1,2] requires that in addition to the quantitative measure of the particular value, its quality should be given as well – preferably the uncertainty of measurement. For this reason, in each thematic area, both in the measurement of power [3,4], biomedical [5], and electric values [6,7], an estimate of the uncertainty of the measured value is provided.

This publication presents the methodology for estimating uncertainty of measurement of the mass flow of the liquid measured with the reducing flow meter. There are various methods and measuring instruments engaged to measure the mass flow [8-12] such as: orifice, classical Venturi reducer, thermoanemometer, Prandtl Pitot tube or the radiation methods [13]. Due to the nature of the analyzed issue these methods continue to be developed. The reducing flow meter was selected for analysis because of its most common application thanks to its simple structure, reliable operation and the usability in a wide range of pressures and temperatures of transported substances.

There is an available study addressing this significant issue of selecting an appropriate method [14]. However, this analysis is incomplete due to the lack of estimation of uncertainty of Type A.

The following paper presents the complete methodology for calculating the uncertainty of measurement of the mass flow of the liquid reducing flow meter using both: the analytical method [1] and Monte Carlo simulation – the numerical one [15].

The analytical method is based on a convolution of the input distribution values, using a mathematical model. In this case, the designated measure of uncertainty is the expanded uncertainty, calculated as the product of the expansion coefficient $k$ and the standard uncertainty value.

The numerical method is based on determining the uncertainty of measurement based on the extension range, which is determined by probability distribution of the measured value quantiles. The parameters of this distribution are: the expected value - as the estimate of the output value, the standard deviation and the confidence interval for a given probability level.

## 2 The study

The most common reducer of a stream is the orifice. It is a thin disc mounted in a pipe so that the axis of the bore coincides with the axis of the tube (Fig. 1).

The liquid stream flowing through the pipe becomes reduced ahead of the orifice. In contrast, the liquid stream insight the reducer reaches a minimum cross-section, and behind it the flow gradually expands to fill the entire volume of the pipe. The orifice fluid pressure increases slightly and is reduced to a minimum in the orifice at the narrowest flow cross-section.

* Corresponding author: anna.golijanek-jedrzejczyk@pg.gda.pl

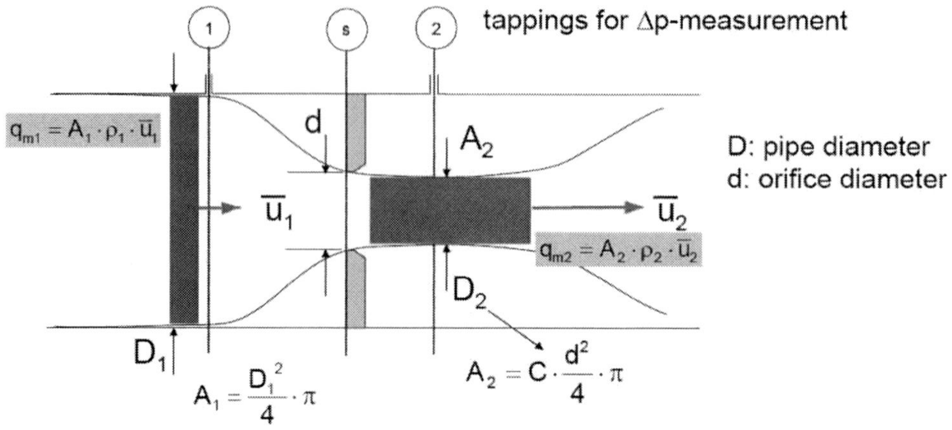


**Fig. 1.** Fluid flow through a measurement orifice [11].

The relation between the mass flow $q_m$ [kg/s] and the pressure difference $\Delta p$ can be determined by the transformation of pressure energy into kinetic energy (speed) at a flow through the reducer. During this process, it is assumed that the continuity of the fluid stream is constant and its density does not change during the passage through different cross sections.

Those conditions allow the determination of the mass flow $q_m$ as follows [2]:

$$q_m = \frac{C}{\sqrt{1-\beta^4}} \varepsilon \frac{\pi}{4} d^2 \sqrt{2\Delta p \rho_1} \quad (1)$$

where:
$C$ – a discharge coefficient [-],
$d$ – orifice diameter [m],
$\beta$ – relation between orifice and pipe diameter:

$$\beta = \frac{d}{D} \quad (2)$$

$D$ – pipe diameter [m],
$\varepsilon$ - compressibility of different fluids. For water and other liquids $\varepsilon$=1.
$\Delta p$ – pressure difference [Pa],
$\rho_1$ – density of medium [kg/m$^3$].

When substituting equation (2) into (1):

$$q_m = \frac{C}{\sqrt{1-\left(\frac{d}{D}\right)^4}} \varepsilon \frac{\pi}{4} d^2 \sqrt{2\Delta p \rho_1} \quad (3)$$

The mass flow $q_m$ is determined indirectly, and the function of the measurement is dependent on the following parameters $q_m = f(C, d, D, \Delta p, \rho_1)$.

The analysis of the uncertainty of measurement of the mass flow $q_m$ presented further in this publication was carried out under the following assumptions:

- the medium is a fluid, that is a ratio $\varepsilon$=1,
- at any point in any cross section perpendicular to the axis of the flow, a continuous stream of the fluid flow is maintained and equal to the speed,
- incompressible fluid without internal friction and a constant density $\rho_1 = \rho_2$ = const.,
- a minimum of 30 measurements were made on the basis of which the $n$ = 30 observations of mass flow $q_m$ were achieved.

### 2.1 Analysis of the measurement uncertainty according to GUM

Complex uncertainty $u_c(q_m)$ determining the mass flow $q_m$ is defined as follows [1]:

$$u_c(q_m) = \sqrt{u_A^2(q_m) + u_B^2(q_m)} \quad (4)$$

where: $u_A(q_m)$ - uncertainty Type A, $u_B(q_m)$ - uncertainty Type B.

To determine Type A uncertainty, the probability distribution of values of the observations was examined. Most frequently normal distribution is assumed 'a priori' (especially when the number of measurements is greater than 30).

The estimate of the mass flow was determined as [1]:

$$\overline{q_m} = \frac{1}{n} \sum_{i=1}^{n} q_{mi} \quad (5)$$

The standard deviation $s$ of an observation was determined from the dependency:

$$s = \sqrt{\frac{1}{(n-1)} \sum_{i=1}^{n} (q_{mi} - \overline{q_m})^2} \quad (6)$$

The standard uncertainty $u_A(q_m)$ of the measurement is evaluated as [1]:

$$u_A(q_m)=\frac{s}{\sqrt{n}} \tag{7}$$

Assuming no correlation between the uncertainties of measured quantities, according to the law of the uncertainty propagation [1], the one of Type B determining the mass flow $q_m$ is defined as follows:

$$u_B(q_m)=\sqrt{\begin{matrix} c_1{}^2u^2(C)+c_2{}^2u^2(d)+ \\ c_3{}^2u^2(D)+c_4{}^2u^2(\Delta p)+ \\ c_5{}^2u^2(\rho_1)\end{matrix}}= \\ =\sqrt{\begin{matrix}\left(\frac{\partial q_m}{\partial C}\right)^2u^2(C)+\left(\frac{\partial q_m}{\partial d}\right)^2u^2(d)+ \\ \left(\frac{\partial q_m}{\partial D}\right)^2u^2(D)+\left(\frac{\partial q_m}{\partial \Delta p}\right)^2u^2(\Delta p)+ \\ \left(\frac{\partial q_m}{\partial \rho_1}\right)^2u^2(\rho_1)\end{matrix}} \tag{8}$$

where:
$c_1$ to $c_5$ – sensitivity coefficients,
$u_B(q_m)$ – the estimated uncertainty of measurement of the mass flow $q_m$ with the Type B method,
$u(C)$ – the uncertainty in the discharge coefficient $C$,
$u(d)$ – the uncertainty of orifice diameter measurement,
$u(D)$ – the uncertainty of pipe diameter measurement,
$u(\Delta p)$ – the uncertainty of the differential pressure measurement,
$u(\rho_1)$ – the uncertainty of the density of medium measurement.

In order to estimate the uncertainty of Type B, the equations describing the sensitivity coefficients appearing in equation (8) were established first. Following that the values of sensitivity coefficients are summarized in Table 1 along with their units.

The weight coefficients in equation (8) (partial derivatives) were determined and are also included in Table 1.

The next step of the uncertainty analysis was to estimate the following variances:

a) discharge coefficient $C - u^2(C)$,
b) orifice diameter $u^2(d)$,
c) pipe diameter $u^2(D)$,
d) differential pressure $u^2(\Delta p)$,
e) density of a medium $u^2(\rho_1)$.

The variance of the discharge coefficient $u^2(C)$ was estimated on the assumption of the normal probability distribution for estimation of the discharge coefficient $C$, as:

$$u^2(C)=\left(\frac{\Delta C}{2}\right)^2 \tag{9}$$

where $\Delta C$ is the estimation error of the coefficient $C$.

**Table 1**. The weight coefficients arranged to equation (8) (partial derivatives).

| | Partial derivatives | Formula |
|---|---|---|
| 1 | $c_1\left[\frac{kg}{s}\right]$ | $\frac{1}{\sqrt{1-\left(\frac{d}{D}\right)^4}}\frac{\pi}{4}d^2\sqrt{2\Delta p\rho_1}$ (10) |
| 2 | $c_2\left[\frac{kg}{m\cdot s}\right]$ | $\pi C\sqrt{2\Delta p\rho_1}\frac{2d^3}{D^4\left(1-\left(\frac{d}{D}\right)^4\right)^{\frac{3}{2}}}$ (11) |
| 3 | $c_3\left[\frac{kg}{m\cdot s}\right]$ | $\pi Cd^2\sqrt{2\Delta p\rho_1}\frac{-d^4}{2D^5\left(1-\left(\frac{d}{D}\right)^4\right)^{\frac{3}{2}}}$ (12) |
| 4 | $c_4\left[m\cdot s\right]$ | $\frac{C}{\sqrt{(2\Delta p\rho_1)\left(1-\left(\frac{d}{D}\right)^4\right)}}\frac{\pi}{4}d^2\rho_1$ (13) |
| 5 | $c_5\left[\frac{m^3}{s}\right]$ | $\frac{C}{\sqrt{(2\Delta p\rho_1)\left(1-\left(\frac{d}{D}\right)^4\right)}}\frac{\pi}{4}d^2\Delta p$ (14) |

Consequently the variances of the orifice diameter $u^2(d)$ and pipe diameter $u^2(D)$ were estimated on the assumption of the rectangular probability distribution of the orifice $d$ and pipe $D$ diameters tolerances, as:

$$u^2(d)=\left(\frac{\Delta d}{\sqrt{3}}\right)^2 \tag{15}$$

$$u^2(D)=\left(\frac{\Delta D}{\sqrt{3}}\right)^2 \tag{16}$$

where $\Delta d$ and $\Delta D$ are tolerances of the apropriate diameters.

The differential pressure $\Delta p$ is usually measured by differential pressure transducer, which is made with the accuracy $\Delta(\Delta p)$. The differential pressure variance $u^2(\Delta p)$, resulting from the processing error at the pressure transducer was determined on the assumption of the rectangular probability distribution, as:

$$u^2(\Delta p)=\left(\frac{\Delta(\Delta p)}{\sqrt{3}}\right)^2 \tag{17}$$

The following discussion assumed that density of the medium $\rho_1$ is constant. The variance of this density $u^2(\rho_1)$ was determined also on the assumption of the rectangular probability distribution, as:

$$u^2(\rho_1)=\left(\frac{\Delta\rho_1}{\sqrt{3}}\right)^2 \quad (18)$$

After determining the combined uncertainty $u_c(q_m)$ according to formula (4), one should calculate the expanded uncertainty $U_p$, which is a measure of the required quality of mass flow $q_m$. The expanded uncertainty $U_p$ is determined following the formula:

$$U_p(q_m)=k_p\cdot u_c(q_m) \quad (19)$$

The most commonly accepted value of the coverage factor $k_p$ is equal to 2 for confidence level $p = 95\%$.

However, it is more accurate to determine the coverage factor $k_p=t_p(\nu_{eff})$ on the adopted probability distribution and the basis of the combined uncertainty values $u_c(q_m)$, as well as the knowledge concerning the number of degrees of freedom for the uncertainty Type A - $\nu_A$ and Type B - $\nu_B$.

The $\nu_{eff}$ signifies effective degrees of freedom and can be obtained from the expanded Welch-Satterthwaite formula [1]:

$$\nu_{eff}=\frac{u_c^4(q_m)}{\frac{u_A^4(q_m)}{\nu_A}+\frac{u_B^4(q_m)}{\nu_B}} \quad (20)$$

The number of degrees of freedom $\nu_A$ was designated as [1]:

$$\nu_A=n-1 \quad (21)$$

Consequently the number of degrees of freedom $\nu_B$ can be estimated from the squared relative uncertainty $\delta^2_{rel,B}$ according to the following formula [1]:

$$\nu_B=\frac{1}{2\cdot\delta^2_{rel,B}} \quad (22)$$

The values of $\nu_{eff}$ were read out from t-Student's table as corresponding to coverage factors $k_p$.

Those results should be presented as the mass flow estimate $q_m$ with the expanded uncertainty $U_p(q_m)$:

$$\left(q_m\pm U_p(q_m)\right)\ \frac{kg}{s} \quad (23)$$

and with information about the confidence level $p$, the effective degrees of freedom $\nu_{eff}$, coverage factor $k_p$ and probability distribution.

### 2.2 Analysis of the measurement uncertainty according to Monte Carlo method

The Monte Carlo numerical method [15] is used for simulations of different kinds of phenomena, it is especially effective in case where the analytical description of the phenomenon is complicated.

This method is a numerical tool which simulates generally an unlimited number of unique measurements by random sampling from the known probability density function of all input quantities and propagates their distributions for the measurement model as the output.

The Monte Carlo procedure is conducted as follows:

1) the number $M$ of the trials should be selected,
2) generate $M$ vectors by random sampling from the probability density function for the (set of $N$) input quantities,
3) for each vector, evaluate the model to give the corresponding output quantity,
4) estimate of the output of the model,
5) sort the model values into non-decreasing order,
6) use the sorted values to estimate the uncertainty for the output.

In metrology it is used to verify the estimates of analytical uncertainty, especially in cases when dealing with an indirect measurement of the measured value [16-17] and when the measuring function is non-linear.

## 3 The example

In order to better illustrate the methodology presented above for estimating the uncertainty of measurement, it was decided to attach an example of the mass flow $q_m$ measurement. This measurement is carried out as shown in Fig. 1. The following data was adopted - Table 2 (measurement conditions are compliant with the ISO Standards [18-19]).

**Table 2.** Data.

| | Parameter | Unit | Value |
|---|---|---|---|
| 1 | $C$ | [-] | 0.605070 |
| 2 | $d$ | [m] | 0.073648 |
| 3 | $D$ | [m] | 0.100051 |
| 4 | $\varepsilon$ | [-] | 1.000000 |
| 5 | $\Delta p$ | [Pa] | 2753.400 |
| 6 | $\rho_1$ | [kg/m$^3$] | 1.109800 |
| 7 | $\beta$ | [-] | 0.736100 |

Assumptions raised in paragraph 2 are still applied. A series of 40 mass flow rate $q_m$ results were obtained and are summarized in Table 3. An estimated value of the mean flow rate $q_m$ was calculated according to the relation (5), on the basis of which the value of $239.57\cdot10^{-3}$ kg/s was reached.

**Table 3**. Data of the mass flow $q_m$.

| $i$ | $q_{mi}$ [kg/s] | $i$ | $q_{mi}$ [kg/s] | $i$ | $q_{mi}$ [kg/s] |
|---|---|---|---|---|---|
| 1 | 0.237787 | 14 | 0.238402 | 27 | 0.241761 |
| 2 | 0.238122 | 15 | 0.240914 | 28 | 0.238990 |
| 3 | 0.240419 | 16 | 0.238900 | 29 | 0.239754 |
| 4 | 0.240966 | 17 | 0.240062 | 30 | 0.238643 |
| 5 | 0.241060 | 18 | 0.243598 | 31 | 0.239063 |
| 6 | 0.237631 | 19 | 0.240143 | 32 | 0.239783 |
| 7 | 0.239521 | 20 | 0.238266 | 33 | 0.237751 |
| 8 | 0.239253 | 21 | 0.237290 | 34 | 0.242054 |
| 9 | 0.239305 | 22 | 0.237893 | 35 | 0.238818 |
| 10 | 0.236826 | 23 | 0.241714 | 36 | 0.241229 |
| 11 | 0.238927 | 24 | 0.236430 | 37 | 0.239436 |
| 12 | 0,.238068 | 25 | 0.240820 | 38 | 0.238690 |
| 13 | 0,.240917 | 26 | 0.241837 | 39 | 0.241107 |
| | | | | 40 | 0.240567 |

## 3.1 Results of the analysis of the measurement uncertainty according to GUM

The first step in order to determine the mass flow $q_m$ measurement uncertainty was the choosing of uncertainty Type A. According to the model equation (7) the value $u_A(q_m)$ equals $0.25 \cdot 10^{-3}$ kg/s.

The next step was to estimate the uncertainty of Type B, basing on the sensitivity coefficients appearing in equation (8). Their values are summarized in Table 4.

**Table 4**. The values of the partial derivatives.

| | Partial derivatives | Value |
|---|---|---|
| 1 | $\left(\frac{\partial q_m}{\partial C}\right)\left[\frac{kg}{s}\right]$ | 3.96E-01 |
| 2 | $\left(\frac{\partial q_m}{\partial d}\right)\left[\frac{kg}{m \cdot s}\right]$ | 9.22E+00 |
| 3 | $\left(\frac{\partial q_m}{\partial D}\right)\left[\frac{kg}{m \cdot s}\right]$ | -1.99E+00 |
| 4 | $\left(\frac{\partial f}{\partial \Delta p}\right)[m \cdot s]$ | 4.35E-05 |
| 5 | $\left(\frac{\partial f}{\partial \rho_1}\right)\left[\frac{m^3}{s}\right]$ | 1.08E-01 |

The next step in the estimation of measurement uncertainty was to estimate the following variances; discharge coefficient $C$ – $u^2(C)$, orifice diameter $u^2(d)$, pipe diameter $u^2(D)$, differential pressure $u^2(\Delta p)$, and density of the medium $u^2(\rho_1)$.

Table 5 shows the values of the variance for the assumed probability distributions and the relative errors $\delta$ of the various parameters measurements.

**Table 5**. The values of the partial variances.

| | Variance | Error δ [%] | Distribution | Value |
|---|---|---|---|---|
| 1 | $u^2(C)$ [–] | 0.73 | normal | 4.88E-06 |
| 2 | $u^2(d)$ $[m^2]$ | 1.36 | rectangular | 4.11E-05 |
| 3 | $u^2(D)$ $[m^2]$ | 0.50 | | 8.34E-08 |
| 4 | $u^2(\Delta p)$ $[Pa^2]$ | 0.40 | | 3.30E-09 |
| 5 | $u^2(\rho_1)$ $\left[\frac{kg^2}{m^6}\right]$ | 1.00 | | 4.04E+01 |

Finally, Type B uncertainty $u_B(q_m)$ was $1.39 \cdot 10^{-3}$ kg/s, which is more than 5 times greater than the Type A uncertainty $u_A(q_m)$.

Figure 2 shows the share of particular elements of Type B mass flow uncertainty $u_B(q_m)$, according to formula (8).

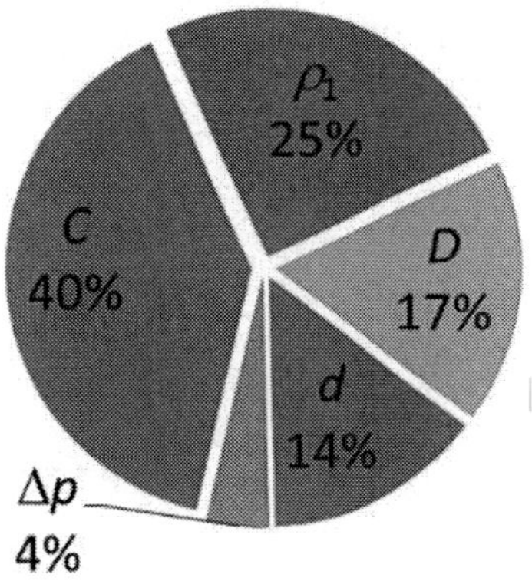


**Fig. 2.** Share of particular components of the mass flow uncertainty $u_B(q_m)$.

Summarizing above the most essential uncertainty factor is the discharge coefficient $C$. It brings the greatest information about the measurement precision. Contribution of the rest parameters in the $q_m$ uncertainty in descending order is as follows: density of the medium $\rho_1$, pipe diameter $D$ and orifice diameter $d$.

Using the formula (4), the complex uncertainty $u_c(q_m)$ was $1.41 \cdot 10^{-3}$ kg/s.

The last step in estimating the uncertainty of measurement of the mass flow $q_m$ is the calculation of the expanded uncertainty.

The expanded uncertainty of the mass flow $U_p(q_m)$ measurement for the coverage factor $k_p$ = 2 (which corresponds to approximately 95% probability of expansion) is:

$$U_p(q_m) = 2 \cdot u_c(q_m) = 2.83 \cdot 10^{-3}\ \frac{kg}{s} \quad (24)$$

On the other hand, assuming coverage factor $k_p = t_p(\nu_{eff}) = 2.01$, and that the relative uncertainty $\delta_{rel,B}$ of estimated uncertainty Type B is 10%, the estimated at the same probability the expanded uncertainty of the mass flow $U_p(q_m)$ measurement is $2.88 \cdot 10^{-3}$ kg/s.

Table 6 shows the uncertainty values collected during the mass flow estimation on base of the attached data.

**Table 6.** Uncertainty budget of mass flow $q_m$ estimate.

| Source of uncertainty | Value | Standard uncertainty | Probability distribution | Sensitivity coefficient $c_n$ | Variance |
|---|---|---|---|---|---|
| $q_m$ | 0.2396 | 2.54E-04 | t-Student | 1 | 6.45E-08 |
| $\Delta C$ | 0 | 2.21E-03 | normal | 3.96E-01 | 7.66E-07 |
| $\Delta d$ | 0 | 5.74E-05 | rectangular | 9.22E+00 | 2.80E-07 |
| $\Delta D$ | 0 | 2.89E-04 | | -1.99E+00 | 3.31E-07 |
| $\Delta(\Delta p)$ | 0 | 6.36E+00 | | 4.35E-05 | 7.67E-08 |
| $\Delta \rho_1$ | 0 | 6.41E-03 | | 1.08E-01 | 4.79E-07 |
| | | | Standard uncertainty $u_c(q_m)$ | | 1.41E-03 |
| | | | Expanded uncertainty $U_p(q_m)$ | | 2.83E-03 |

In summary, by using the analytical method in accordance with the guidelines contained in [1] for the case under consideration, the following results were obtained:

$$(239.57 \pm 2.83) \cdot 10^{-3}\ \frac{kg}{s},$$

for $p = 95\%$, expansion coefficient, respectively $k_p = 2$ and

$$(239.57 \pm 2.88) \cdot 10^{-3}\ \frac{kg}{s},$$

for $k_p = t_p(\nu_{eff})$ and t-Student distribution.

A greater value of the uncertainty of measurement of the mass flow $q_m$ was achieved when the coverage factor $k_p$ was calculated according to the formula (20) – that is, the range in which 95% of the measurement results are wider.

The relative uncertainty of the mass flow $q_m$ measurement in the analyzed example, did not exceed 0.5%, however this value is highly dependent on the relative uncertainty $\delta_{rel,B}$.

## 3.2 Results of the analysis of the measurement uncertainty according to Monte Carlo method

In order to verify results of the estimated uncertainty obtained in the previous section, the Monte Carlo simulaton was performed [9]. This approach is based on the definition of the range expansion, and limits of them were defined by quantile of the probability distribution associated with the measured values.

For this purpose, the random number generator from Microsoft Excel was used. It is assumend that the function of measuring mass flow, according to equation (3) is:

$$q_m = \overline{q_m} + c_0 u_A(q_m) + c_1 u(C) + c_2 u(d) + + c_3 u(D) + c_4 u(\Delta p) + c_5 u(\rho_1) \quad (25)$$

where $c_1$ to $c_5$ marks sensitivity coefficients, and the coefficient $c_0$ is equal to 1. The rest of the coefficients are defined in Table 1, so their values for the considered example are reported in Table 4.

Following above the probability distributions for the input quantities were assumed : for discharge coefficient $C$ the normal distribution, while for other parameters: $d$, $D$, $\Delta p$, $\rho_1$ the rectangular distributions.

Estimation of uncertainty using Monte Carlo was performed in Microsoft Excel for the number of samples $M$ equal to $10^4$. The value of the expected mass flow $q_m$ and its density function for a confidence level $p = 95\%$ and $k_p = t_p(\nu_{eff}) = 2.01$ was determined and applied to the following presentation.

Figure 3 shows the probability density function of the simulated numerical values of the mass flow $q_m$. Based on these results a histogram established with channel $0.4 \cdot 10^{-3}$ kg/s, was plotted in Fig. 4.

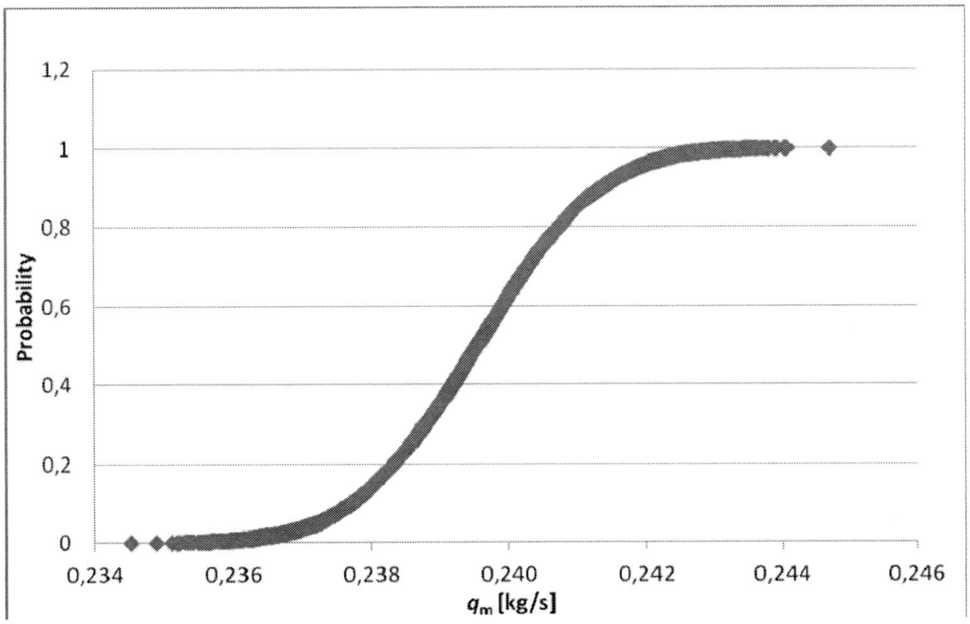


**Fig. 3**. Simulated the outflow distribution of the $q_m$.

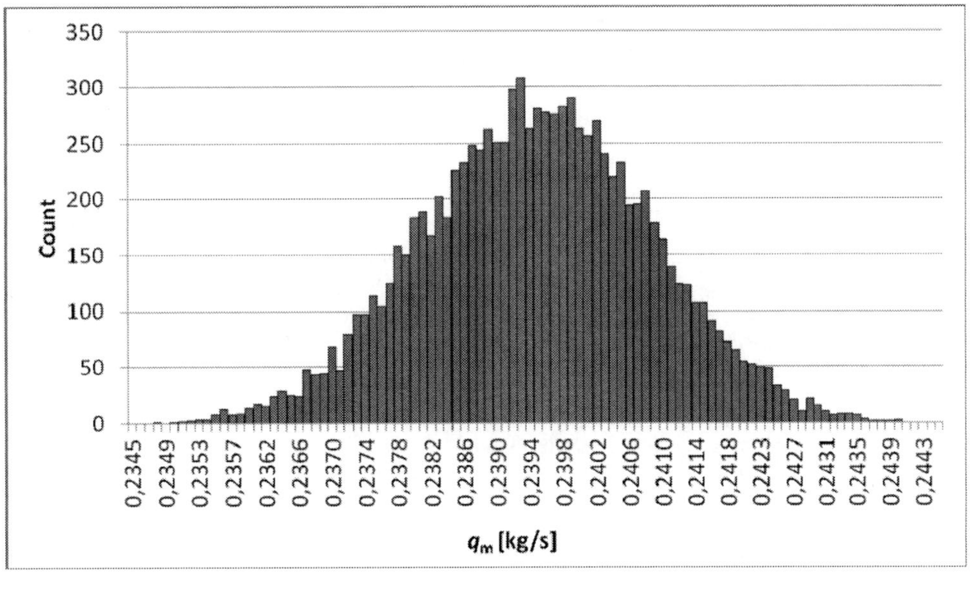


**Fig. 4**. Histogram for $q_m$.

Table 7 summarizes the results obtained for estimation of value of mass flow $q_m$ and the expanded uncertainty $U_p(q_m)$ estimated with both; the method of propagation of uncertainties (traditional one), and the Monte Carlo one.

It can be observed that the results of measurement uncertainty of the mass flow by both methods are similar. The relative uncertainty is equal to, respectively: 1.20% - by the traditional method, and 1.15% by the Monte Carlo one.

**Table 7**. The expanded standard uncertainty of the mass flow.

| **Mass flow estimation $q_m$** | | | |
|---|---|---|---|
| $p$ [%] | $k_p$ | **Traditional method** $10^{-3}$[kg/s] | **Monte Carlo** $10^{-3}$ [kg/s] |
| 95 | 2.01 | (239.57±2.88) | (239.54±2.77) |

## 4 Summary

The article presents the problem of the mass flow uncertainty estimation using two methods: the method based on the GUM Guide using the law of propagation of uncertainty and the Monte Carlo numerical method.

The authors presented a comprehensive methodology for estimating uncertainty of measurement of the fluid flow through the orifice measurement and calculations carried out for the sample data. The results obtained using the traditional method have been verified by numerical Monte Carlo, and obtained by the one convergence of the estimated measurement uncertainty; those confirmed the correct determination of the analytical method.

The uncertainty of estimating methodology described in the paper can be adapted for use in the metering stations with free selected parameters not only those presented in the article.

The article shows a procedure for estimating uncertainty of the liquid mass flow measurement using the orifice plate. All results refer to a specific case, but the used methodology can be easy extended to other methods.

Standard estimation of the measurements uncertainty is essential in scientific research. Fulfill these conditions allow the assessment of measurements made in different institutions and allow the experimental verification of announced hypotheses in any place in the world.

## References

1. *Guide to the Expression of Uncertainty in Measurement* (JCGM 100 2008).
2. *Handbook of uncertainty calculations fiscal orifice gas and turbine oil metering stations* http://nfogm.no/wp-content/uploads/2014/04/Handbook-USM-fiscal-gas-metering-stations.pdf
3. A. Golijanek-Jędrzejczyk, Przegl. Elektrotech. **85** (2), 25-28 (2009) (in Polish)
4. A. Golijanek-Jędrzejczyk, MAM **58** (11), 987-990 (2012)
5. A. Dzwonkowski, A. Golijanek-Jędrzejczyk, L. Rafiński, Zeszyty Naukowe Wydziału Elektrotechniki i Automatyki Politechniki Gdańskiej **34**, 21-24 (2013) (in Polish)
6. A. Dzwonkowski, A. Golijanek-Jędrzejczyk, Przegl. Elektrotech. **91** (10), 166-169 (2015)
7. A. Dzwonkowski, MAM **59** (5), 394-397 (2013)
8. R.C. Baker, *An introductory guide to flow measurement* (Alden Press, Oxford, 1989)
9. JP. De Carlo, *Fundamentals of flow measurement Instrument Society of America* (Research Triangle Park, NC, 1984)
10. E.L. Upp, P.J. LaNasa, *Fluid Flow Measurement; a Practical Guide to Accurate Flow Measurement*, (Gulf Professional Publishing, MA, 2002)
11. M. Reader-Harris, *Orifice Plates and Venturi Tubes*, (Springer International Publishing, Switzerland, 2015), DOI 10.1007/978-3-319-16880-7_2
12. A. Harrouz, O. Harrouz, A. Benatiallah, *IJPEDS* **3**(4), 450-468 (2013)
13. L. Petryka, M. Zych, R. Hanus, J. Sobota, P.Vlasak, M. Sleziak, D. Świsulski, *EPJ WoC* **114**, 02093 (2016)
14. http://www.ematem.org/Dokumente/2008_lau_calculat.pdf
15. *Supplement 1 to the Guide to the expression of uncertainty in measurement – Propagation of distributions using a Monte Carlo method* (JCGM 101 :2008)
16. P. Fotowicz, Pomiary Automatyka Robotyka 1/2005, 5-9 (2005)
17. A. Dzwonkowski, A. Golijanek-Jędrzejczyk, L. Rafiński, Przegl. Elektrotech. **91** (8), 11-14 (2015)
18. *ISO Measurement of fluid flow by means of pressure differential devices inserted in circular cross-section conduits running full—Part 1: general principles and requirements.* (International Organization for Standardization, Geneva, ISO 5167-1:2003)
19. *ISO Measurement of fluid flow by means of pressure differential devices inserted in circular cross-section conduits running full—Part 2: Orifice plates.* (International Organization for Standardization, Geneva, ISO 5167-2:2003)

# Experimental and numerical analysis of cylindrical straw drying

*Wojciech Goryl*[1,*], *Mariusz Filipowicz*[1]

[1]AGH University of Science and Technology, Faculty of Energy and Fuels, A. Mickiewicza Av. 30, 30-059 Krakow, Poland

**Abstract.** The paper presents experimental and numerical results of the heat and mass transfer in a cylindrical bale of straw. The experimental measurements were made in a specialized stand of straw driers. Flue gasses, comes from straw combustion in the biomass boiler, are used as a drying medium. There were made measurements of humidity and temperature inside the cylindrical straw bale during the drying process. The results were used to prepare the drying rate curve. Moreover, data were used to validate the numerical model of straw drying. The numerical model was performed to depict the heat and mass transfer inside the straw bale. Furthermore, the model was used to optimize the drying process. The paper presents result of experimental and numerical drying rates of cylindrical straw bale and heat and mass transfer in its interior. As a result of the work numerical model was obtained. It satisfactorily describes the mechanisms inside the drying straw bale.

## 1 Introduction

Poland, as an agricultural country, and having a large area of forestation has a pretty large biomass resources, which can be used for generating electricity or heat production. One of the basic problems of the energy use of biomass is the bulk density as well as the calorific value. In the given biomass volume there is approx. 4-8 times less energy than in the same coal volume. This causes a number of problems, and the main one is the transportation problem – the costs related to biomass transport over long distances and emissions associated with the biomass transportation from the place of harvesting to the place of utilization. One of the most important factors influencing on the biomass calorific value is humidity. The higher humidity of the biomass the calorific value of biomass is lower. It should be mentioned that there is a lower and upper limit of humidity, which should not be exceeded in energy use of biomass. This has negative effect on biomass combustion process and construction elements of the biomass boiler.

Because of the energy crisis in the 70s, straw has been used on an increasing scale, mainly in Denmark. This country is a precursor of the energy use of straw. The first straw-fired boilers showed very poor working conditions, low efficiency (30-40%) and high emission of carbon monoxide (above 10 000 ppm in the exhaust gas) [1]. Then, in the 80s of the XX century the construction of boilers was improved and the combustion efficiency was achieved within 75%. In the 90s, the Danish government introduced subsidies for the purchase of straw-fired boilers, which resulted in a further development of this fuel. At the beginning of the XXI century the combustion efficiency of 87% was achieved and the CO content in the exhaust gases less than 1 000 ppm [1].

The easiest way is to use shredded straw in large power boilers, where the fuel moisture is not so relevant [2,3,4]. In addition, in the case of shredded straw, drying of the fuel is relatively simple, because the solutions for drying loose materials have been known for many years and they are still used, for example: rotary dryers [5,6,7], pneumatic dryers [8, 9], fluidized bed dryers [10,11,12], belt dryers [12,13,14].

The bigger problem is the use of the bales of straw in batch boilers. In this case, drying is not a simple issue [15]. Most studies on drying bales talks about the modification of the structure of the dryer itself, and not the drying process. This study aims to provide an effective and economical process of drying cylindrical bales of straw. Developing an optimised drying process can be performed by conducting long-lasting studies on the experimental stand. However, currently we have got a possibility to apply the advanced computer tools, which assist in conducting numerical calculations. The use of computational fluid dynamics (CFD) [16,17], in this case allows to optimise the drying process without the need to perform multiple series of experimental measurements.

## 2 Measurement stand

The main elements of the dryer used in the study include the drying chamber, the transport table, the discharge fan, the fire prevention system, the regulation assembly and the system for the control of operational parameters, the system of pre-extraction and separation of sparks and the exhaust gasses system.

* Corresponding author: wgoryl@agh.edu.pl

 DOI: 10.1051/epjconf/201714302031

The heat used in the drying process comes from the process of biomass combustion in the boiler. Such utilised waste heat increases the efficiency of the entire system and helps to obtain a much more efficient fuel. The drying process involves the injection of waste gas characterized by a properly controlled temperature by the automatic system using a specially designed nozzle with a diameter of 80 cm into the bale of straw.

For safety reasons, the exhaust gases coming from the boiler are pre-dusted in the sedimentation chamber (cyclone) due to the presence of large, sometimes incandescent stalks of unburned straw. Subsequently, the gases are transported to a special mixer (Fig. 1), where, if necessary, the process of mixing of the exhaust gas with ambient air takes place in order to obtain the desired temperature of the drying medium. After the mixing process, the fan introduces the previously prepared mixture using the specially designed nozzle into the cylindrical bale of straw. The schematic diagram of the system has been presented in EFM2015 [18].

The effectiveness of drying cylindrical straw bales (diameter: 160 cm, length: 140 cm) with exhaust gases was measured using a specially prepared recorder equipped with eight measuring probes. The probes were calibrated by the manufacturer using a sample of straw characterized by varying humidity. Capacitive relative humidity sensors and temperature sensors with the SHT11 device were installed.

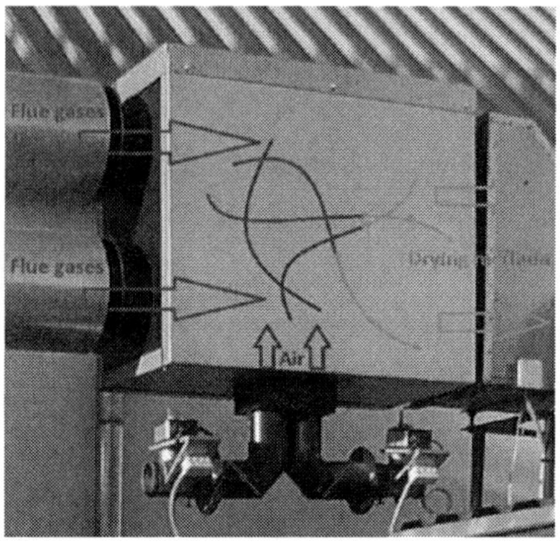


**Fig. 1.** General view of the medium mixer

The measuring probes varied in length (80 cm, 60 cm, 40 cm and 20 cm denoted respectively by lowercase "a", "b", "c" and "d"), so as to make it possible to measure the humidity and temperature in locations at a preset distance from the axis of the straw bale. Figure 2 presents the visualisation of the placement of the right probe in the bale during the measurement.

The measurements were always carried out in two planes. The measurement points were located at a distance (x=0 – the place where the drying mixture was pumped to the bale) of 40 cm (plane A, x=40) and 100 cm (plane B, x=100) from one end of the bale (Fig. 2).

The results of the experiment have been presented below. The curves show the change of temperature and humidity during the drying process. The prefixes "A" and "B" designating the probes refer to the measurement plane according to Figure 2b, while the prefixes "T" and "H" refer to the measurement of temperature and humidity, respectively.

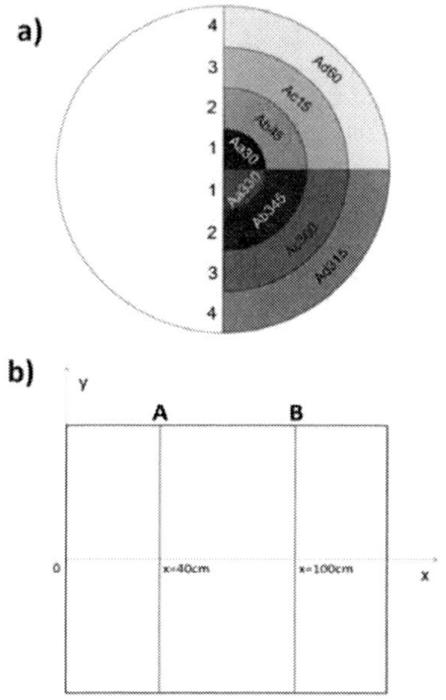


**Fig 2.** Visualisation of the locations of measurement probes: a) locations of individual probes, b) coordinates of the A and B planes.

This article presents the possibilities of drying straw using a modified drying process. After gaining some experience in the process of drying straw bales [18,19,20], some modifications of this process have been implemented, which involved preparing a duct along the bale axis with the diameter of 63mm and applying a metal plate with the diameter of 1,000 mm at the side opposite to the fitted drying medium. This modification was designed to allow for the highest radial propagation of the medium and to obtain a uniform temperature and humidity inside the bale.

## 3 Results and discussion

A study of changes in the humidity and temperature inside the bale of straw were conducted in the A plane (Fig. 2b), which is presented below. During the drying process, the temperature of the drying medium injected into the bale was recorded. The exhaust gas temperature was presented in Fig. 3-4 to illustrate its dynamics due to operation of the control system. It is presented in the form of a dotted curve with the values given on the right axis. The process have been carried out by 140 min.

The graph below (Fig. 3) presents the change in humidity inside the bale of straw with one hole in the centre of the bale and with the applied metal plate.

The starting humidity of the bale, on which the measurements were performed for the modified process of drying with one hole inside the bale of straw, varied

DOI: 10.1051/epjconf/201714302031

between 25-45%. The internal layers of the bale were drier, and the further from the centre of the bale, the humidity started to rise. This was due to the storage of straw uncovered and absorbing moisture by the outer layers.

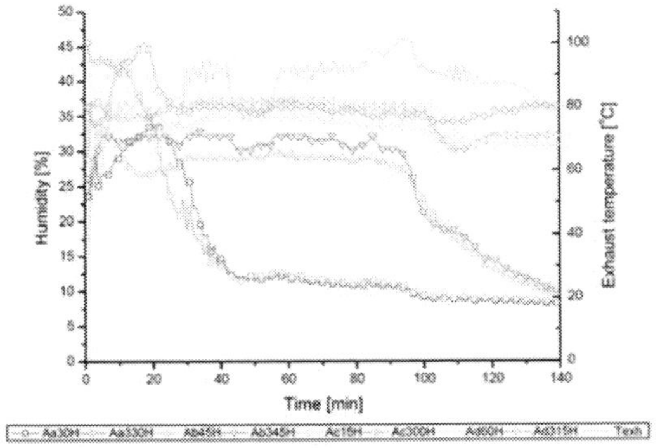


**Fig. 3**. Humidity measurement.

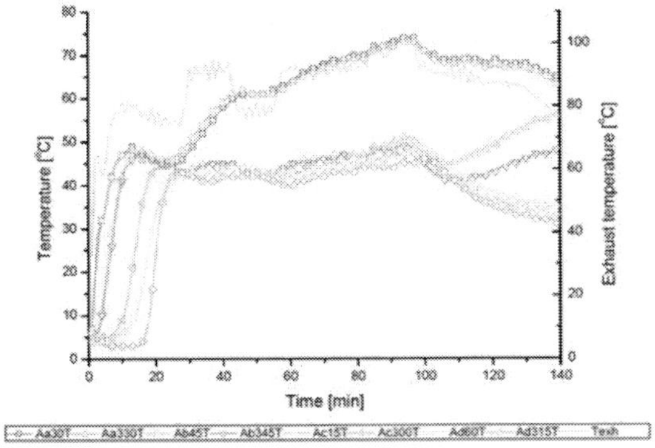


**Fig. 4**. Temperature measurement.

In the initial phase from approx. 10 min we can see an increase in moisture in all planes of the bale. This is caused by the rapid rise in temperature in the bale, which contributed to the temporary increase in humidity due to reaching the dew point temperature. After 20 minutes from starting the measurement, the moisture in the entire bale started decreasing. The most rapid drop is visible in Plane 1 and it lasts up to 40 minutes of measurement. Then, moisture in this plane stabilises and at the end of the drying process reaches the value of approx. 8% of moisture. For Plane 2 to 90 minutes of the process, the moisture remains constant in the range of 28-33%. After 90 minutes, the humidity starts dropping and reaches the values like Plane 1. Plane 3 and 4 behaves similarly like during the drying in the basic drying process, that is, it remains constant in the range of 30-35%.

The above chart (Fig. 4) presents the temperature change inside the bale of straw. After 2 minutes from the start of the drying process we can observe a rapid increase in temperature in plane 1. Then, after another two minutes plane 2 is quickly heated up. Plan 3 and 4 because of a significant distance from the bale axis are heated up much longer, and the beginning of the temperature growth in these planes is observed, respectively, from 10 and 15 minutes. The maximum temperature for Plan 1 is achieved in 90 minutes of the measurement, with the continuous increase during the process of combustion, and its values was 74°C. For subsequent planes, the temperature remains at a similar level and fluctuates around 40-50°C, an its maximum is achieved similarly like in the case of Plane 1, in the 90 minute of the measurement. The flue gas temperature during whole drying process varied between 80-95°C.

## 4 Numerical model

A preliminary mathematical model was created in the ANSYS FLUENT 15.0, describing the movement of the mass and heat transfer inside the bale of straw in question. Compared to the results presented during EFM2016 the model was modified by adding evaporation-condensation model and Eulerian model. The previous model took into account only the change of the temperature field and the velocity of the drying medium flow. Currently, also the existence of 3 phases was taken into account: air, water and water vapour.

The model contains 239 274 nodes and 276 844 elements, mostly hexagonal. The computing grid was an average quality. This was dictated by some initial simplification for the acceleration of numerical calculations. Moreover, the transient model, SST turbulence model and Eulerian model were used. Three phases were entered in the model, which are: the air phase, the water vapour phase and the water in the liquid state phase. In addition, the bi-directional model was applied: evaporation-condensation model describing the phase transition.

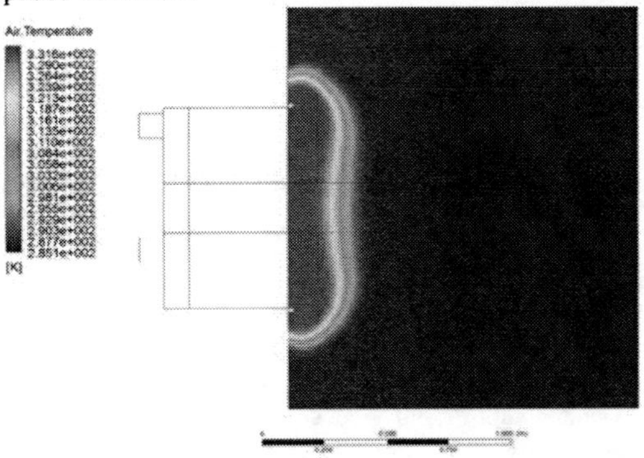


Figure 5. Temperature of the drying medium in straw bale.

The model provided two domains: 1 – intake of drying air, 2 – a bale of straw, the porous domain [18]. The time step was 1s, the total number of iterations 100. The bale domain has the ability of mass and heat transfer with the surroundings. Initial conditions were: 1) temperature – 285K, 2) temperature of dry air – 333K, 3) pressure – 101325Pa, 4) the volume content of water in the bale – 2%, 5) the amount of vapour in the bale – 0%.

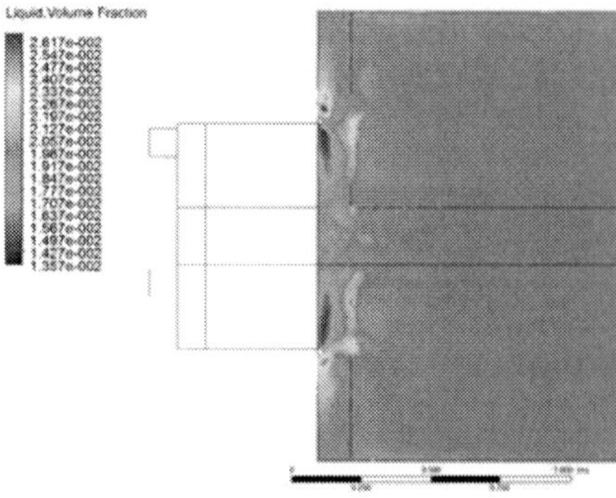

**Fig. 6**. Liquide volume fraction in straw bale.

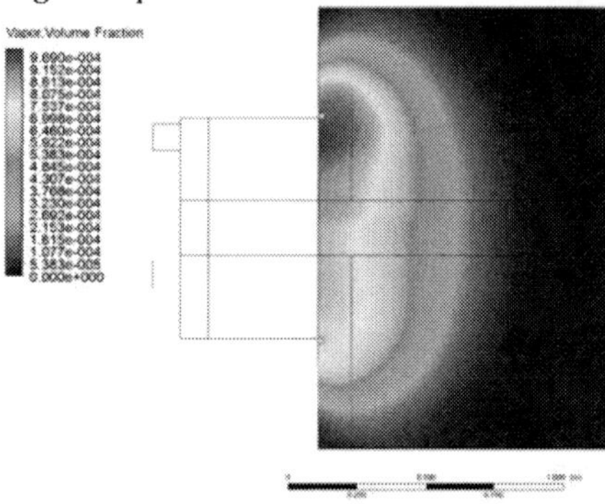

**Fig. 7**. Vapour volume fraction in straw bale.

Figures 5-7 present the preliminary results of the developed model. It can be observed that in the inlet portion of the drying medium there is the temperature increase of the air domain. In the area of high temperature we can observe the decrease of the water content in the bale and the increase of the water vapour content, what allows to assume that the evaporation process takes place in this place. Additionally, we can note an increased content of water in the area of high thermal gradient. In this case it can be concluded that there is the condensation of water vapour. Similar observations were made during the performance of experimental measurements. It can be assumed that this model describes the process of phase change from water into water vapour and vice versa. Currently, the model should be modified so that the injected drying medium flow through the whole model of straw, what will cause the heating of the whole bale of straw and evaporation of water from its interior. To improve the model the viscous and inertial resistance coefficients should be determined.

## 5 Conclusions

The preliminary results of the mathematical model show that the assumed objective was achieved, which was to develop the model of water evaporation and condensation of water vapour in the bale of straw. The experimental results confirm that in the place of a large temperature gradient there is a rapid increase of the moisture contents, that is, the condensation of water vapour. Currently, the works are conducted, which involve the introduction of modifications in the model, so that the process of drying takes place in the entire volume of the bale.

## Acknowledgment

This work was financially supported by the statutory founding of Faculty of Energy and Fuels at AGH University of Science and Technology and it was carried out by BioORC Project (KIC InnoEnergy Programme).

## References

1. E. F. Kristensen, J. K. Kristensen, Biomass Bioenerg, **26**, 561 – 569 (2004)
2. J. Skvaril, A. Avelin, J. Sandberg, E. Dahlquist, Energy Procedia, **61**, 643 – 647 ( 2014 )
3. L.A.C. Tarelho, E.R. Teixeira, D.F.R. Silva, R.C.E. Modolo, Energy, **90**, 387-402 (2015)
4. H. Kuuluvainen, P. Karjalainen, C.J.E. Bajamundi, J. Maunula, Fuel, **139**, 144–153 (2015),
5. F. J. G.Cruz, P. J. Casanova-Pelaez, J. M. Palomar-Carnicero, Appl. Therm. Eng., **80**, 362-373 (2015)
6. C. Gu, X. Zhang, B. Li, Z. Yuan, Powder Technol., **267**, 234–239 (2014)
7. F. Geng, Z. Yuan, Y. Yan, D. Luo, H. Wang, B. Li, D. Xu, Powder Technol., **193**, 50–58 (2009)
8. Yefri Chan, T. M. Nining Dyah, A. Kamaruddin, Energy Procedia, **65**, 378 – 385 (2015)
9. Thermal Drying of Wet Fuels: Opportunities and Technology, Vancouver, Canada, December 1996
10. I. Ceylan, A. E. Gürel, Appl. Therm. Eng., **106**, 899–905 (2016)
11. F. Geng, G. Luo, Y. Li, L. Yuan, Z. Yuan, X. Wu, Powder Technol., **267**, 322–332 (2014)
12. A Review of Drying Technologies, ed. Dr Hanning Li, Dr K Finney, Sheffield, February 2010,
13. A. Alamia, H. Strom, H. Thunman, Biomass Bioenerg, **77**, 92-109 (2015)
14. E.F. Zanoelo, A. Abitante, L.A.C. Meleiro, J. Food Eng., **84**, 458–468 (2008),
15. F. Roman, V. Strahl-Schafer, O. Hensel, Biosyst. Eng., **112**, 359-369 (2012)
16. Y. Amanlou, A. Zomorodian,, J. Food Eng., **101**, 8–15 (2010)
17. F. D. Roman, O. Hensel, Biosyst. Eng., **122**, 1-15 (2014)
18. W. Goryl, M. Filipowicz, EPJ Web of Conferences, **114**, 02033 (2016), DOI: http://dx.doi.org/10.1051/epjconf/201611402033
19. W Goryl, M. Filipowicz, E3S Web of Conferences (2016), In press
20. W. Goryl, *Proceedings of 2nd International Conference RENEWABLE ENERGY SOURCES engineering, technology, innovations - scientific conference,* Krynica, Poland (2015)

# The effect of the Saharan climate on the indoor environment of a station of electric energy transformation MV / LV

*Amina* Hammadou[1], *Abdelkrim* Missoum[1], *Elmir* Mohamed[1] Belkacem Draoui[1]

[1] Laboratory of Energetic in Arid Zone ENERGARID, Mohamed Tahri University Bechar, Algeria

**Abstract.** The proposed work studies numerically the mixed convection heat transfer in a station electric transformer MV / LV for the public distribution of a neighborhood located in the town of Bechar (Sahara region) in the south-west Algeria. Taking into account the climatic conditions measured during the summer period [1], introduced as boundary conditions using the Fluent CFD based on the finite volume method. In order to get results presented in form the isothermal and the stream function and the profile of the evolution of the temperature and velocity components of the air inside the station.The objective of this study is to analyze the results of the distribution of temperature, aeration of the position and the flow of air to the interior of the station.The influence of the number and location of openings in the station is studied.

## 1 Introduction

Algeria is a vast country it is between 18 ° and 38 ° north latitude and between 9 ° west longitude and 12 ° east longitude (the International meridian Greenwich (0 °) passing near the town of Mostaganem).

Bechar is located in southwest from ALGERIA: latitude : 31 ° 37 'N 31.6167 and longitude 2 ° 14' W - 2.23333; it is characterized by a continental climate desert type, with rainfall of about 40 mm / year. Sandstorms are very common, and the average temperature is 35 ° C (+ 45 ° C in summer and - 5 ° C in winter).

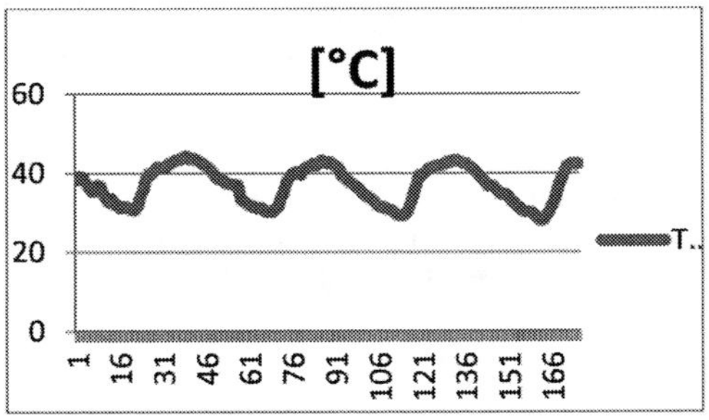


**Fig. 1.** Outside temperature

The Bechar region therefore has a hot and arid climate. It is then placed in the zone D1 (see figure I.2)

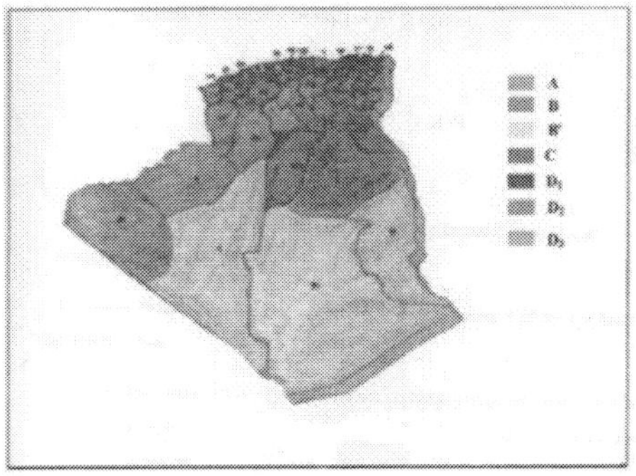


**Fig. 2.** Climate zones in Algeria

The study is a numerical modeling of heat transfer by mixed convection of transformer station of the electric energy ; taking into account the climatic conditions of the region of Bechar.

Mixed convection in rectangular spaces is a very important subject of investigation for its presence in various industrial applications such as : the cooling of electronic components, heat losses in solar collectors and ventilation of rooms and buildings. The laminar flow of forced or natural convection has limited value because the number of Reynolds or Grashof / square Rayleigh must not exceed a certain threshold [2-3].

The mixed convection heat transfer, is of considerable interest for technological applications such as chemical deposition of thin layers, cooling of electronic components [4-10]

A large number of numerical and experimental studies [11-16] relate to heated or isothermal tubes mixed convection. Most of these studies considers the bi-dimensional case with vertical tubes. They studied the effect of dimensionless numbers on the flow behavior in mixed convection.

The effect of the tilt was considered by [17] and [18] in 3D which assume that the flow has a predominant direction in order to simplify the Navier Stokes equation .

The ventilation of rooms habitat types for buildings is studied by [19] due to the significant increase in temperature in summer in Saharan zones and because of the bad mastery of ventilation in the cavity where there is the electrical transformers there will have nefarious consequences of the quality of indoor air, built conservation and energy consumption. In order to treat this problem, we examine in this work numerically the various positions of the openings in the station and their influence on the inside heat transfer .

## 2 Configuration studied

The configuration studied is shown in Figure 1. This is an electric power processing station for the public distribution of length L = 3.4 m and height H = 2.8 m, equipped with two openings same heights h1 = h2 = 0.44 m. One is located at the bottom for the admission of air and the other at the top for the hot air outlet. The wall near the transformer is at a warm temperature Tc and the other wall kept at a cold temperature Tf, too, the floor and the ceiling are supposed Job thermally insulated. The station contains a transformer dimension e1 = 1.7m, e2 = 1.85m. The boundary conditions of the station studied are taken from the climatic data of the region of Bechar.

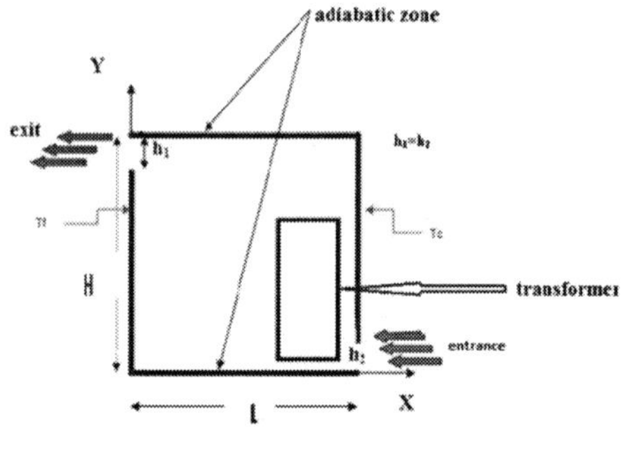


**Fig. 3.** Geometric Configuration

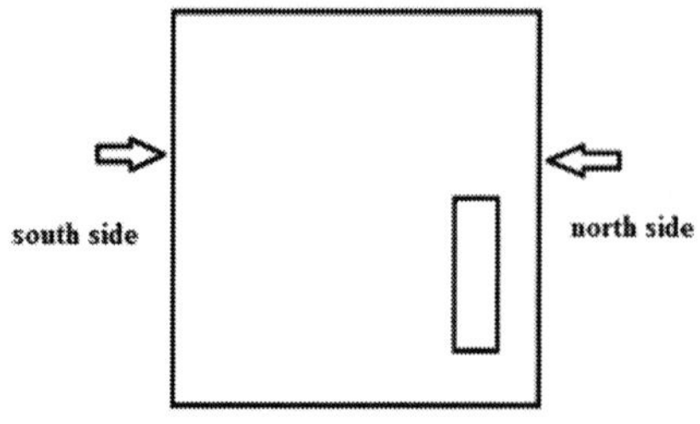


**Fig. 4**. Sides of the station

To simplify the problem, assume that:
• The fluid is Newtonian and incompressible.
• Heat transfer by radiation is negligible.
• The heat dissipation by viscous friction is neglected.
• The Boussinesq approximation is considered.
• The flow is bidimensional.

## 3 Mathematical model

The system of equations governing the model is written in the following dimensionless form:

**a.mass conservation equation (continuity) :**

$$\frac{\partial u}{\partial x}+\frac{\partial v}{\partial y}=0 \qquad (1)$$

**a. Equations of conservation of momentum :**

$$\left(\frac{\partial u}{\partial t}+u\frac{\partial u}{\partial x}+v\frac{\partial u}{\partial y}\right)=-\frac{\partial P}{\partial x}+\frac{1}{\mathrm{Re}}\left(\frac{\partial^2 u}{\partial x^2}+\frac{\partial^2 u}{\partial y^2}\right) \qquad (2)$$

$$\left(\frac{\partial v}{\partial t}+u\frac{\partial v}{\partial x}+v\frac{\partial v}{\partial y}\right)=-\frac{\partial P}{\partial y}+\frac{1}{\mathrm{Re}}\left(\frac{\partial^2 v}{\partial x^2}+\frac{\partial^2 v}{\partial y^2}\right)+\frac{Gr}{\mathrm{Re}^2}T \qquad (3)$$

**b. Energy conservation equation:**

$$\frac{\partial T}{\partial t}+u\frac{\partial T}{\partial x}+v\frac{\partial T}{\partial y}=\frac{1}{\mathrm{Re\,Pr}}\left(\frac{\partial^2 T}{\partial x^2}+\frac{\partial^2 T}{\partial y^2}\right) \qquad (4)$$

**The initial and boundary conditions:**

The previous system of equations must be solved under certain specified conditions

**At the entrance :**

$U = V_{max}$ ; $V = 0$ ; $T = T_f$ (5)

**At the exit:**

Outflow (all gradients are taxed except zero pressure)

**Hot walls:**

$U = V = 0$ ; $T = T_C$ (6)

**Vertical walls:**

$U = V = 0$ ; $T = T_f$ (7)

**Horizontal walls (adiabatic) :**

$$u=v=\frac{\partial T}{\partial x}=0 \qquad (8)$$

The numerical procedure used in this work is that of finite volume. It involves the integration of differential equations of mathematical model on finite control volumes for the corresponding algebraic equations. The SIMPLE algorithm was chosen for the coupling speed pressure in the Navier Stokes equations on a staggered grid. Convective terms in all equations are evaluated using the scheme apwind1er order. In order to follow any changes in hydrodynamic and thermal fields, we used a uniform mesh.

## 4 Results and discussion

**A) Without insulation**

**1) Isotherms and streamlines:**

The isotherms are shown in Figure (5) by varying the number and location of the openings of the station for the fixed climatic conditions; which shows that the temperature exceeds 65 ° C on the transformer and 55 ° C on the sunny sides with an external temperature equal to 45 ° C. As it's remarked, the addition of openings reduces slightly the maximum temperature.

The figure (6) shows the streamlines for the same cases; we see that the flow is well distributed around the transformer. We observe the appearance of a vortex which is on the left of the heating element caused by natural convection in the station, therefore its size is considerably weakened by adding openings.

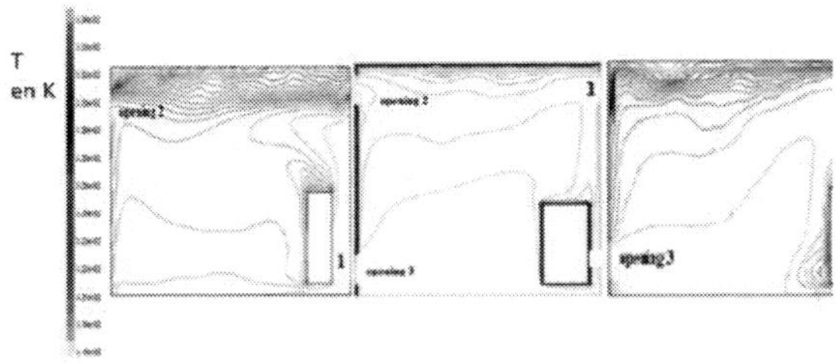


**Fig. 5 : Isotherms in the station**

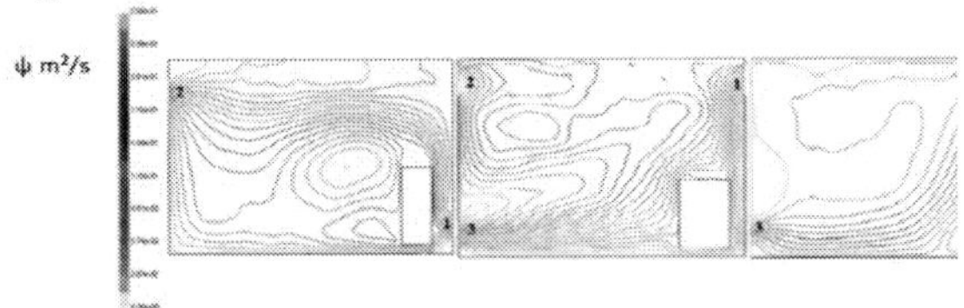


**Fig. 6** . Streamlines in the station

The figure (7) show the air circulation in the station.Fresh air enters through the openings at the bottomit it flow over the wall of the transformer, it is heated and by natural convection it heads upwards to come out. In all cases, the top of the transformer is a recirculation zone which is not well refreshed.

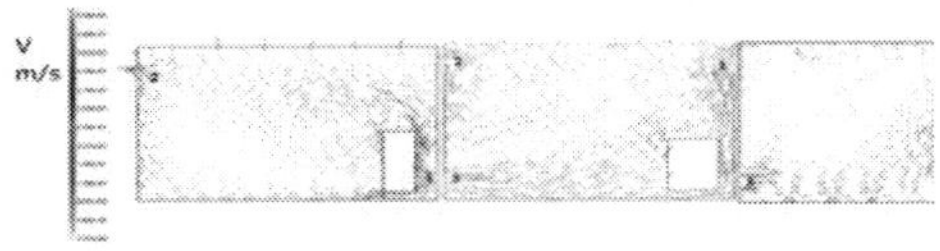


**Fig.** 7 : Air circulation in the station

**B)with insulation**

We tried to insulate the system by a screen in cases of 2 and 4 openings.

**1)Isotherms and streamlines :**

By comparing the two cases, it is found that the removal of the solar gain through the insulation lowers the temperature.

The simulation for the previous existing stations fitted with double screen (insulation) on the sunny sides reduces the flow absorbed dramatically. It can be estimated with a dual screen, the temperature applied on protected sides is the same as the temperature of the atmosphere of the time (45 ° C). The heating of the transformer is caused by its own power generation.

The greatest value in the station is located on the main of the heat generator transformer. The average temperature of the air inside the station is of the order of 45 ° C to an external temperature of 40 ° C and generally of the order of 5 ° C higher than the external temperature.

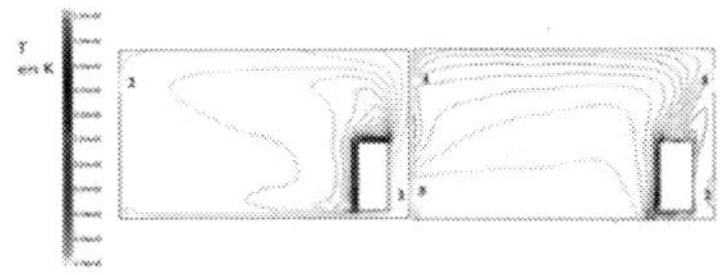

**Fig. 8** : Isotherms in the station (with insulation)

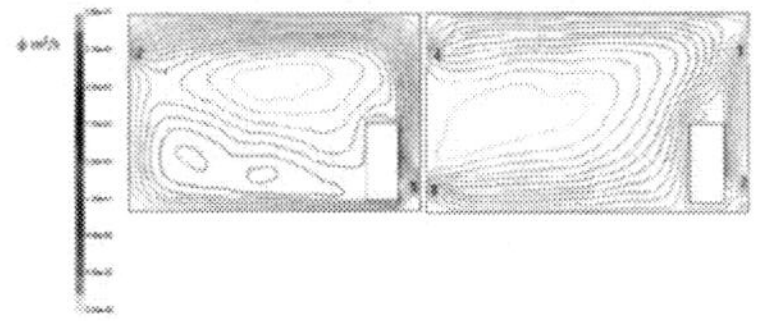

**Fig. 9** : Streamlines in the station (with insulation)

The figure (10) shows that the case where there are four opening the air circulation around the transformer is more regular and causes more uniform cooling than in cases where there are only two or three.

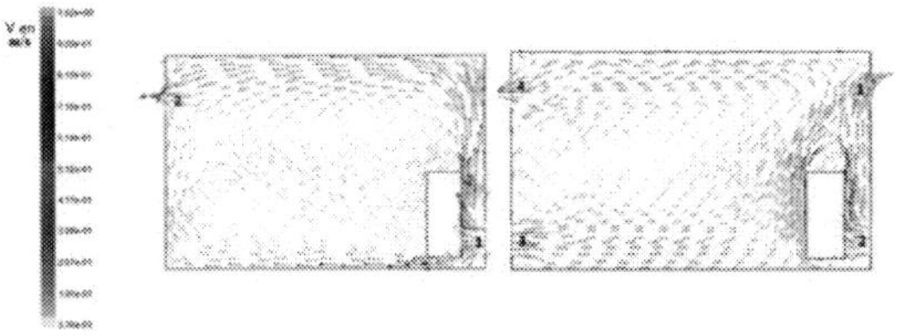

**Fig. 10** : Air circulation in the station (with insulation)

**C)Temperatures and velocity Profiles:**

Figure (11) illustrates the evolution of the temperature on the line Y = 2 m in case of two and three openings; the temperature distribution according to the flow of fluid which it is noted that the temperature does not exceed the ambient (Ta = 45 ° C) throughout the area of the transformer in parallel. In the case of low temperatures, we see an increase in temperature up to Tc = 55 ° C just after the position X = 2.3 m to stabilize until almost hot wall characterizes the heat transfer is dominated by conduction. As against where was added an additional opening the temperature area of renewal increases rapidly from the position X = 3.3 m to reach the hottest value which is why the transfer by convection is predominant.throughout the area of the transformer in parallel. In the case of low temperatures, we see an increase in temperature up to Tc = 55 ° C just after the position X = 2.3 m to stabilize until almost hot wall characterizes the heat transfer is dominated by conduction. As against where was added an additional opening the temperature area of renewal increases rapidly from the position X = 3.3 m to reach the hottest value which is why the transfer by convection is predominant.

The profile of the average vertical velocity at a height of Y = 2 m in both cases is shown in Figure (12). A sharp decrease of the vertical velocity is noted

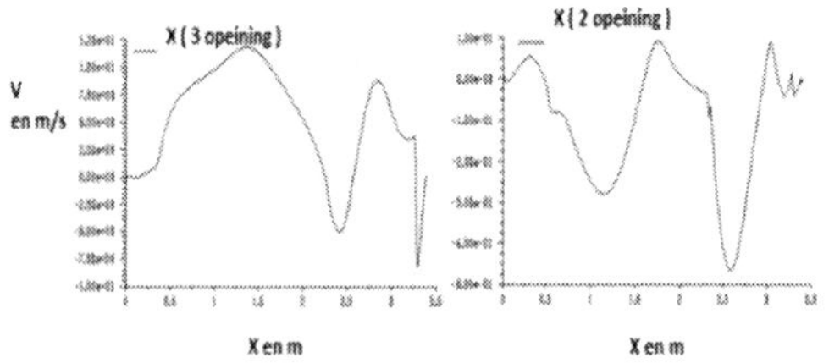


**Fig. 12 :** the temperature profile at 2m height along the station

## Conclusion

The maximum temperature reached 55 ° C greater than ambient temperature that is to say that the heat flow from the transformer increases the heating of indoor air.

The multiplication of openings and regular cleaning are the first inexpensive solution to reduce the heat load on the stations.

A more interesting solution is to have on the sunny faces, in addition to openings ; Double lightweight material screen that has the effect of removing the solar heat flux incident on these surfaces and is the main cause of the warming of the stations.

## Nomenclatures

g Gravitational acceleration [m / s2]

H Height of the cavity [m]

L The cavity length [m]

h1 evacuation length [m]

h2 Length of entered [m]

e1 The thickness of the heating element [m]

e2 The height of the heating element [m]

to achieve minimum values about X = 0.5. However the highest rate values are met in the range 0.8 < X < 2.5 (i.e. of near heating corpd). These values indicate where the fluid particles follow the streamlines in ascending and descending movement.

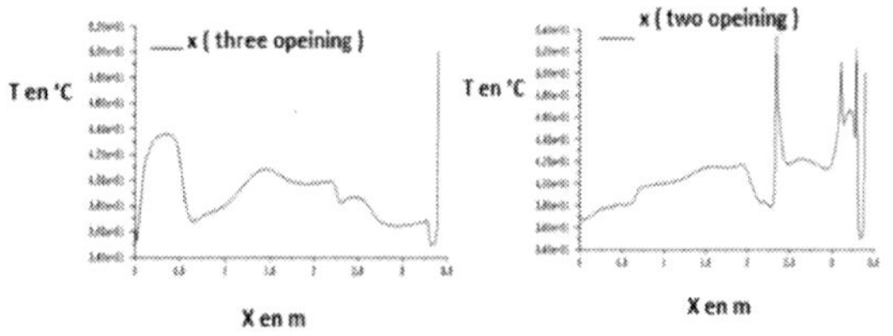


**Fig.** 11 : the temperature profile at 2m height along the station

U, V velocity vector Components [m / s]

**Greek symbols**

**λ** Thermal conductivity [W / mK]

**μ** Dynamic viscosity [Kg M-1 / s-1]

**υ** Kinematic viscosity [m2 / s]

**ρ** Density [kg / m3]

**Δ** Temperature difference [K]

## References

[1] climatic data of the region of Bechar. weather station of Bechar university

[2] Sefcik, D. M., Webb, B. W., Heaton, H. S., " Analysis of Natural
Convection in Vertically-Vented Enclosures, " Int. J. Heat Mass
Transfer 34: 3037-3046,1991.

[3] Yu, E., Joshi, Y., " A Numerical Study of Three-Dimensi-onal Laminar
Natural Convection in a Vented,Enclosure, "Int. J. Heat and Fluid Flow
18: 600-612. 1997.

[4] A. Zebib, Y.K. Wo, "A two-dimensional conjugate heat transfer model
for forced air cooling of an electronic device, "Transactions of ASME J.
Electronic Packaging 111 (1989) 41–45,1989.

[5] K. Hijikata, W. Nakayama, T. Nagasaki, R. Kurazume, K. Fushinobu, "
A study on heat transfer from small heating element in an integrated
circuit chip, " Proc. of the 3rd ASME/JSME Thermal Engineering Joint
Conf. 4, , pp. 93–98,1991.

[6] T. Nagasaki, K. Hijikata, K. Fishinobu, P.E. Phelan, "Numerical
simulation of the conjugate direct cooling of a micro heat generating

element, " Proc. ASME/JSME Conf. on Electronic Packaging 1, pp.
217–223, 1992.
[7] K. Fushinobu, P.E. Phelan, T. Nagasaki, K. Hijikata, M.I.Flik, " Periodic
steady-state thermal analysis of a high-Tc superconducting
microbolometer, " Trans. ASME J. Heat Transfer. 116 (1994) 278–
275,1994.
[8] K. Fushinobu, K. Hijikata, Y. Kurosaki, "Heat transfer regime map for
electronic devices cooling, Int. J. Heat Mass Transfer. 39 (1996) 3139–
3145,1996.
[9] B.A. Jubran, S.A. Swiety, M.A. Hamdan., Convective heat transfer and
pressure drop characteristics of various array configurations to simulate
the cooling of electronic modules, "Int. J. Heat Mass Transfer. 39 (1996)
3519–3529,1996.
[10] M. Hasnaoui, E. Bilgen, P. Vasseur, "Natural convection above an array
open cavities heated from below, "Numerical Heat Transfer. Part A 18
(1991) 463–482, 1991.
[11] M. Zghal, N. Galanis et C.T. Nguyen, "Developing mixed convection
with aiding buoyancy in vertical tubes: a numerical investigation of
different flow regions, "Int. J. Therm. Sci., 40, pp 816-824, 2001.
[12] T. Maré , I. Voicu et P. Hue, "Visualisation d'un écoulement en
convection mixte pour tube isotherme incliné, " Cit 26-28 Mai , Québec,
40-07, 2003.
[13] A. Behzadmehr, N. Galanis et A. Laneville, " low Reynods Nguyen,
Developing mixed convection with aiding buoyancy in vertical tubes: a
numerical investigation of different flow regions, "Int. J. Heat. Mass.
Transfer, 46, pp 4823-4833, 2003.
[14] J. Orfi, N. Galanis , " Developing Laminar Mixed Convection with Heat
and Mass Transfer in horizontal and vertical tubes, " Int. J. Thermt.Sci.,
41, pp 319-331, 2002.
[15] D. D. Joye, " Pressure drop correlation for laminar, mixed convection,
aiding flow heat transfer in a vertical tube, " Int. J. Heat Fluid Flow, 24,
pp 260-266, 2003.
[16] C. T. Nguyen, S. B. Maiyer, M. Landry, N. Galanis et G. " Roy
Numerical investigation of flow reversal and instability in mixed laminar
vertical tube flow, " Int. J. Therm. Sci., 43, pp 797-808, 2004.
[17] A. Barletta et E. Zanchini, "Time -periodic laminar mixed convection in
a inclined channel, " Int. J. Heat. Mass. Transfer, 46, pp 551-563, 2003.
[18] J. Orfi, N. Galanis et C. T. Nguyen , "Développement simultané
hydrodynamique et thermique d'un écoulement laminaire dans un tube
incliné en régime de convection mixte, " Rev. Gén. Therm., 36, pp 83-
92, 1997.
[19] E. Benachour "simulation numérique de la convection naturelle et mixte
dans une cavité carée (type habitat) avec la présence d'un élément de
climatisation, " mémoire de magister, Université de Béchar, Algerie,07
mars 2010.
[20] Mr. Modest "Radiative heat transfer", The McGraw-Hill Companies ed. (2002)
[21] R. Siegel and J. R. Howell, "Thermal radiation heat transfer", Taylor and Francis ed. (1992).
[22] E. Benachour, Numerical Simulation of the Mixed convection of a room (habitat type) with the presence of a heating element, J S R **No. 0 Vol. 1** (2010)

EPJ Web of Conferences **143**, 02033 (2017) DOI: 10.1051/epjconf/201714302033
*EFM 2016*

# Application of ANN and PCA to two-phase flow evaluation using radioisotopes

*Robert* Hanus[1,*], *Marcin* Zych[2], *Leszek* Petryka[3], *Dariusz* Świsulski[4] and *Anna* Strzępowicz[2]

[1]Rzeszów University of Technology, Faculty of Electrical and Computer Engineering, 35-959 Rzeszów, Poland
[2]AGH University of Science and Technology, Faculty of Geology, Geophysics and Environmental Protection, 30-059 Kraków, Poland
[3]AGH University of Science and Technology, Faculty of Physics and Applied Computer Science, 30-059 Kraków, Poland
[4]Gdańsk University of Technology, Faculty of Electrical and Control Engineering, 80-233 Gdańsk, Poland

**Abstract.** In the two-phase flow measurements a method involving the absorption of gamma radiation can be applied among others. Analysis of the signals from the scintillation probes can be used to determine the number of flow parameters and to recognize flow structure. Three types of flow regimes as plug, bubble, and transitional plug – bubble flows were considered in this work. The article shows how features of the signals in the time and frequency domain can be used to build the artificial neural network (ANN) to recognize the structure of the gas-liquid flow in a horizontal pipeline. In order to reduce the number of signal features the principal component analysis (PCA) was used. It was found that the reduction of signals features allows for building a network with better performance.

## 1 Introduction

Two-phase liquid-gas flow commonly occurs in nature and in industry, especially in: mining, nuclear, chemical, thermal and petrochemical engineering.

This type of flow may be studied using several methods, such as computer tomography, Coriolis flowmeters, optical equipment, PIV, LDA and nuclear techniques [1-8].

Knowledge of a two-phase flow structure is significant to properly conduct a number of industrial processes, so flow regime identification inspires many studies. Recent works show use of machine learning methods as e.g. artificial neural networks (ANN) [9-18] for this purpose. The flow structures classification with ANN is comprised of three steps: data acquisition and pre-processing, feature extraction, and structures classification. Generally, computational intelligence methods exploit various features of signals in the time, frequency and state-space domain.

The article shows how features of the signals from scintillation probe, analyzed in the time and frequency domain, can be used to build the ANN to recognize the structure of the gas-liquid flow in a horizontal pipeline. In order to reduce the number of signal features the principal component analysis (PCA) was used.

## 2 Gamma-absorption method

The gamma absorption technique is based on exponential decreasing of a gamma beam in function of composition and geometry of the absorbent [8]. The changes of the intensity of radiation are registered by the scintillation probes and converted into output electrical impulses [6].

The typical gamma-absorption set for two-phase gas-liquid flow measurement is presented in Fig. 1.

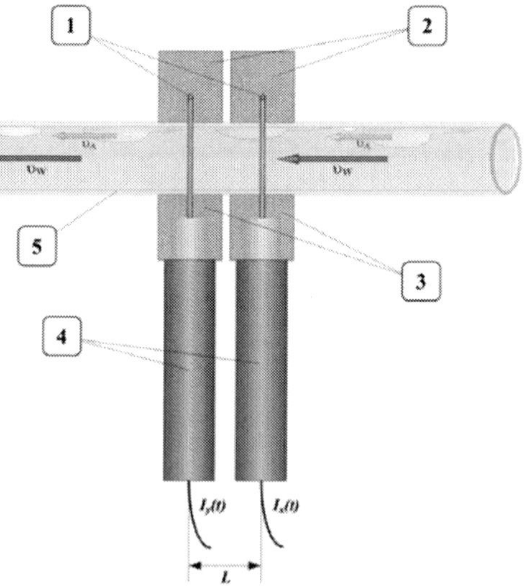


**Fig. 1.** The gamma-absorption measuring set: 1 – gamma radiation source, 2 – collimator of source, 3 – collimator of detector, 4 – scintillation probe, 5 – pipeline, $\upsilon_A$, $\upsilon_W$ – velocities of air and water respectively. [20]

Two sealed radioactive sources (1) emit gamma radiation beams with an energy of 59.5 keV shaped by collimators (2). Photons pass through the pipeline with

---
[*] Corresponding author: rohan@prz.edu.pl

flowing compound (5) and detector's collimators (3) before achieving probes (4). Two sets consist of linear $^{241}Am$ sources with an activity of 100 mCi and probes with 2'' NaI(Tl) scintillation detectors [19]. For flow velocity measurement, the sets are placed at a distance of $L = 97$ mm between them [20]. Count signals $I_x(t)$ and $I_y(t)$ are recorded at the outputs of scintillation probes.

Measuring equipment described above was applied in the experimental set-up, built in Industrial Radiometry Laboratory of the AGH University of Science and Technology in Krakow, Poland (Fig. 2). Measurement data was collected in experiments carried out on a laboratory hydraulic installation with a horizontal pipe of 4.5 m length and inner diameter of 30 mm. Detailed specification of experimental set-up presented in Fig. 2 is described in publications [19,20].

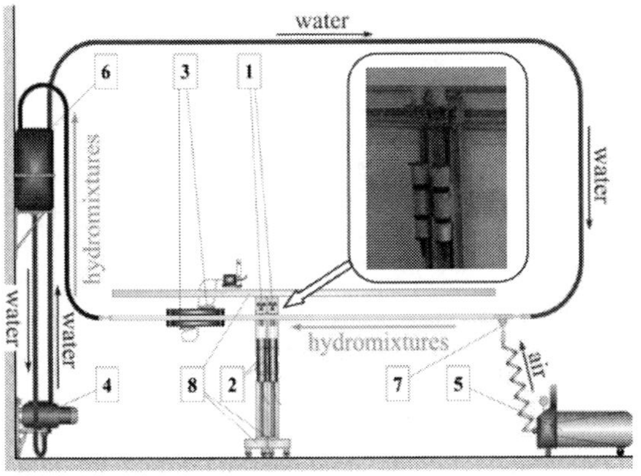


**Fig. 2.** Scheme of experimental set-up: 1 - gamma ray sealed source, 2 – scintillation probe, 3 – ultrasonic flow meter, 4 – pump, 5 – compressor, 6 – air-removing container, 7 – air nozzle, 8 – shifting system of gamma-absorption set. [20]

## 3 Recorded signals

In this work, three structures of air-water flow in horizontal pipeline are being considered: a plug flow, bubble flow, and transitional plug-bubble flow. For these structures of flow, the 300,000 sample signals were recorded in experiments denoted as: BUB6, BUB4, and BUB1 with a sampling rate of 1 kHz [19]:

- BUB6 - plug flow: $\upsilon_W = 0.90$ m / s, $\upsilon_A = 0.710$ m / s,
- BUB4 - transitional plug – bubble flow: $\upsilon_W = 1.36$ m / s, $\upsilon_A = 1.066$ m / s,
- BUB1 - bubble flow: $\upsilon_W = 1.92$ m / s, $\upsilon_A = 1.446$ m / s,

where $\upsilon_W$ – velocity of water was measured by ultrasonic flowmeter Uniflow 990. Velocity of gas phase $\upsilon_A$ was appointed using cross-correlation technique [21-24].

Fig. 3 presents photos of particular types of flow structures and time records of the signals obtained in experiments BUB6, BUB4, and BUB1 [19].

a)

b)

c)

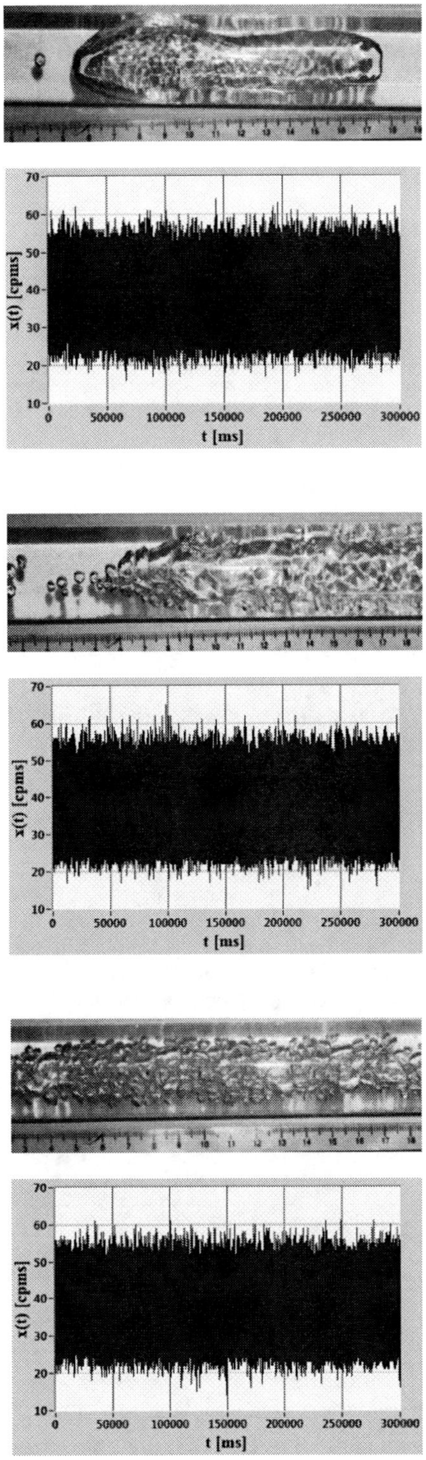


**Fig. 3.** Flow structures, and time records of signals obtained from scintillation probe in experiments (a) BUB6, (b) BUB 4 and (c) BUB1. [19]

## 4 Signals features extraction

Based on the analysis of the data records registered in the experiments the following parameters (features) were calculated for the three types of the flow regimes (classes): "plug flow", "transitional plug – bubble flow", and "bubble flow":

a) in the time domain: mean value, variance, standard deviation, 4$^{th}$ order moment, root mean square (RMS), skewness, kurtosis [19];

b) in the frequency domain: maximum value of the autospectral density function (ADF), maximum value of the cross-spectral density function (CSDF) magnitude, sum and relative sum of the ADF in the selected frequency range, sum and relative sum of the CSDF in the selected frequency range, the maximum value of the sum of selected amplitudes of Short-Time Fourier Transform (STFT) spectrogram, and area and relative area under the graph of STFT spectrogram sum in the selected frequency range [25].

A total of 16 features were extracted for data segments with a length of 10,000 samples and 50% overlapping. The data analysis and computation of the features (predictors) were performed using LabVIEW software. Applied signal processing procedures are described in publications [19, 25].

## 5 ANN and PCA application to recognition of the flow regime

In this work STATISTICA 12 Automated Neural Networks software was used for the construction of ANN [26]. Different artificial neural network structures were tested for obtaining the best one.

The best ANN configuration for total of 16 predictors proved to be Multi-Layer Perceptron (MLP) 16-8-3 (digits indicate the number of neurons in the input, hidden and output layer respectively). The number of samples for learning, testing and validation sets were 235 (about 70%), 50 (about 15%), 50 (about 15%) respectively. Table 1 shows the specification and results of test for the obtained ANN. Tables 2-4 present results for learning, testing and validation respectively.

**Table 1.** Specification of MLP 16-8-3 ANN

| Parameter | Specification |
|---|---|
| Quality (training) | 99.57% |
| Quality (test) | 100% |
| Quality (validation) | 100% |
| Learning algorithm | BFGS 16 |
| Activation function (hidden) | Tanh |
| Activation function (output) | Linear |

In order to reduce the number of predictors, the principal component analysis (PCA) was used. PCA explains the correlation structure of a set of predictors using a smaller set of linear combinations of them [27]. In general, the first principal component is the best summary of correlations among the predictors. This particular linear combination of the variables accounts for more variability than any other conceivable linear combination.

**Table 2.** Results for MLP 16-8-3 ANN (learning)

| | BUB1 | BUB4 | BUB6 |
|---|---|---|---|
| **Total** | 76 | 80 | 79 |
| **Correct** | 76 | 79 | 79 |
| **Incorrect** | 0 | 1 | 0 |
| **Correct, %** | 100 | 98.75 | 100 |
| **Incorrect, %** | 0 | 1.25 | 0 |

**Table 3.** Results for MLP 16-8-3 ANN (testing)

| | BUB1 | BUB4 | BUB6 |
|---|---|---|---|
| **Total** | 16 | 18 | 16 |
| **Correct** | 16 | 18 | 16 |
| **Incorrect** | 0 | 0 | 0 |
| **Correct, %** | 100 | 100 | 100 |
| **Incorrect, %** | 0 | 0 | 0 |

**Table 4.** Results for MLP 16-8-3 ANN (validation)

| | BUB1 | BUB4 | BUB6 |
|---|---|---|---|
| **Total** | 20 | 13 | 17 |
| **Correct** | 20 | 13 | 17 |
| **Incorrect** | 0 | 0 | 0 |
| **Correct, %** | 100 | 100 | 100 |
| **Incorrect, %** | 0 | 0 | 0 |

The second principal component is the second-best linear combination of the predictors, on the condition that it is orthogonal to the first principal component. The third component is the third-best linear combination of the predictors, on the condition that it is orthogonal to the first two components. The third component is derived from the variance remaining after the first two components have been extracted, and so on [27].

STATISTICA software was used for PCA after standardization of predictors [28]. The number of principal components was limited to four based on the scree plot (Fig. 4).

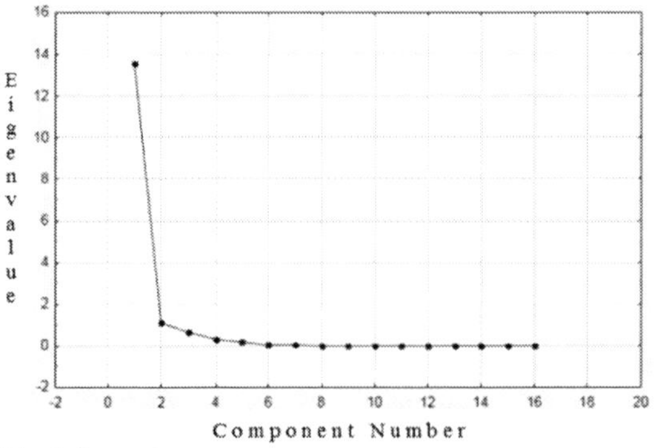

**Fig. 4.** Scree plot.

The selected principal components were used as predictors for ANN constructed using STATISTICA

Automated Neural Networks software. The obtained specification and results for selected ANN MLP 4-4-3 with the highest performance are shown in Tables 5-8. The type of flow structure was recognized with 100% accuracy using this ANN.

**Table 5.** Specification of MLP 4-4-3 ANN

| Parameter | Specification |
|---|---|
| Quality (training) | 100% |
| Quality (test) | 100% |
| Quality (validation) | 100% |
| Learning algorithm | BFGS 23 |
| Activation function (hidden) | Exponential |
| Activation function (output) | Softmax |

**Table 6.** Results for MLP 4-4-3 ANN (learning)

| | BUB1 | BUB4 | BUB6 |
|---|---|---|---|
| **Total** | 76 | 80 | 79 |
| **Correct** | 76 | 80 | 79 |
| **Incorrect** | 0 | 0 | 0 |
| **Correct, %** | 100 | 100 | 100 |
| **Incorrect, %** | 0 | 0 | 0 |

**Table 7.** Results for MLP 4-4-3 ANN (testing)

| | BUB1 | BUB4 | BUB6 |
|---|---|---|---|
| **Total** | 16 | 18 | 16 |
| **Correct** | 16 | 18 | 16 |
| **Incorrect** | 0 | 0 | 0 |
| **Correct, %** | 100 | 100 | 100 |
| **Incorrect, %** | 0 | 0 | 0 |

**Table 8.** Results for MLP 4-4-3 ANN (validation)

| | BUB1 | BUB4 | BUB6 |
|---|---|---|---|
| **Total** | 20 | 13 | 17 |
| **Correct** | 20 | 13 | 17 |
| **Incorrect** | 0 | 0 | 0 |
| **Correct, %** | 100 | 100 | 100 |
| **Incorrect, %** | 0 | 0 | 0 |

## 6 Conclusions

In this paper an application of the gamma-absorption method in combination with ANN and PCA to recognize the gas-liquid flow regime in a horizontal pipeline has been presented.

A total of 16 features was extracted for data obtained from the gamma-absorption set based on $^{241}$Am radioactive sources and NaI(Tl) scintillation probes. In order to reduce the number of predictors the principal component analysis was conducted. Next, four selected components were used as predictors for the construction of ANN.

It was found, that the reduction of predictors using PCA allows to build ANN with better performance (simpler structure and better accuracy) than without such a reduction.

## References

1. G. Falcone, G.F. Hewitt, C. Alimonti, *Multiphase flow metering: principles and applications* (Elsevier, Amsterdam 2009)
2. M. Bottin, et al, Int. J. Multiphas. Flow **60,** 161-179 (2014)
3. T. Xue, et al, Flow Meas. Instrum. **27,** 29–36 (2012)
4. P. Vlasak, et al, J. Hydrol. Hydrom. **62,** 241-247 (2014)
5. L. Petryka, M. Zych, R. Murzyn, Nukleonika **50,** 43–46 (2005)
6. Z. Yu, B. Qincheng, H. Richa, Appl. Therm. Eng. **60,** 398-410 (2013)
7. S.L. Soo (ed.), *Instrumentation for fluid-particle flow* (Noyes Publications, New Jersey, 1999)
8. W.A.S. Kumara, B.M. Halvorsen, M.C. Melaaen, Int. J. Multiphas. Flow **36,** 467-480 (2010)
9. S. Blaney, H. Yeung, Flow Meas. Instrum. **19,** 57–66 (2008)
10. O. Arjmandi-Tash, A. Zarghami, R. Zarghami, Particul. Sci. Technol. **32,** 498–505 (2014)
11. Y. Chen, et al, Chem. Eng. Sci. **129,** 156–167 (2015)
12. F. Johnsson, Int. J. Multiphas. Flow **26,** 663–715 (2000)
13. C.M. Salgado, Prog. Nucl. Energy **52,** 555–562 (2010)
14. Z. Sun, H. Zhang, Meas. Sci. Technol. **19,** 015401 (2008)
15. G.H. Roshani, et al, Measurement **62,** 25–32 (2015)
16. C.M. Salgado, Appl. Radiat. Isot. **67,** 1812–1818 (2009)
17. B.M. Abbagoni, H. Yeung, Meas. Sci. Technol. **27,** 084002 (2016)
18. M. Khorsandi, et al, Radiat. Meas. **59,** 183–187 (2013)
19. R. Hanus, et al, *EPJ WoC* **114,** 02035 (2016)
20. R. Hanus, et al, *EPJ WoC* **92,** 02021 (2015)
21. M.S. Beck, A. Plaskowski, *Cross correlation flowmeters - their design and application* (Adam Hilger, Bristol, 1987)
22. W.Q. Yang, M.S. Beck, Flow Meas. Instrum. **8,** 77-84 (1998)
23. L. Petryka, J. Oszajec, Nucl. Geoph. **7,** 323-333 (1993)
24. J.S. Bendat, A.G. Piersol, *Random data - analysis and measurement procedures* (John Wiley, New York, 2010)
25. R. Hanus, et al, *EPJ WoC* **114,** 02036 (2016)
26. http://www.statsoft.com/Products/STATISTICA/Automated-Neural-Networks
27. D.T. Larose, *Data mining - methods and models* (Wiley - Interscience, New Jersey, 2006)
28. http://www.statsoft.com/Portals/0/Products/PCA_Analysis_with_STATISTICA.pdf

EPJ Web of Conferences **143**, 02034 (2017) DOI: 10.1051/epjconf/201714302034
*EFM 2016*

# Correction coefficient for see-through labyrinth seal

*Dan* Hasnedl[1,*], *Premysl* Epikaridis[1], and *Vaclav* Slama[1]

[1]Doosan Skoda Power, R&D Department, Tylova 1/57, Pilsen, Czech republic

**Abstract.** In a steam turbine design, the flow-part design and blade shapes are influenced by the design mass-flow through each turbine stage. If it would be possible to predict this mass-flow more precisely, it will result in optimized design and therefore an efficiency benefit. This article is concerned with improving the prediction of losses caused by the seal leakage. In the common simulation of the thermodynamic cycle of a steam turbine, analytical formulas are used in order to simulate the seal leakage. Therefore, this article describes an improvement of analytical formulas used in a turbine heat balance calculation. The results are verified by numerical simulations and experimental data from the steam test rig.

## Nomenclature

| | | |
|---|---|---|
| m | mass flow | [kg.s$^{-1}$] |
| S | flow through area | [m$^2$] |
| $p_0$ | inlet absolute pressure | [bar$_a$] |
| $p_2$ | outlet absolute pressure | [bar$_a$] |
| $(p_2/p_0)_{cr}$ | critical pressure ratio | [-] |
| $p_i$ | absolute pressure after the tooth "i" | [bar$_a$] |
| $p_n$ | absolute pressure after the last tooth | [bar$_a$] |
| $v_0$ | inlet specific volume | [m$^3$.kg$^{-1}$] |
| κ | Poisson's constant | [-] |
| δ | radial clearance | [mm] |
| s | span of the teeth | [mm] |
| z | number of the teeth | [-] |
| k | see-through seal correction coefficient | [-] |
| μ | flow contraction correction coefficient | [-] |

## 1 Introduction

Classical labyrinth seals are still widely used due to their cost, simple and robust design despite of new types of seals, such as brush seals. In the turbine heat balance or efficiency calculations, an estimation of the leakage through these seals is represented by analytical formulas , which can be used to carry out calculations fast enough in order to save time especially during the bidding procedure. For these purposes, the time consuming numerical calculations are not feasible.

Several authors managed to estimate the labyrinth seal leakage by various formulas, such as Zalf [5], Kearton [6], Samoylovich [4] and others [1], [2], [7], [8]. These formulas are different in an approach and results. Over a number of years, Zalf formula was used due to its simplicity and conservative results in Doosan Skoda leakage predictions. In a modern turbine design, where a high efficiency is demanded, it is crucial to use more enhanced and precise way to estimate the leakage.

Therefore, a large set of experiments and numerical calculations were performed to compare the results. Stepped labyrinth and see-through labyrinth seals with different clearances were tested in steam environment to get as close to the real operating conditions as possible.

All geometric combinations were also simulated by CFD software (FLUENT, SEAL) as the alternative sources of leakage values.

## 2 Analytical formulas

As a representative of the common practice, the Zalf formula was brought for comparison. Zalf formula takes into consideration only pressure ratio and number of the teeth. The leakage is estimated as:

$$m = 1.5 \cdot S \cdot \sqrt{\frac{p_0}{v_0}} \cdot \sqrt{\kappa \cdot \left(\frac{2}{\kappa+1}\right)^{\frac{\kappa+1}{\kappa-1}}} \cdot \sqrt{\frac{1-(p_2/p_0)^2}{z-\ln(p_2/p_0)}} \quad (1)$$

The see-through labyrinth is simulated by using half number of the teeth (rounded) as long as the teeth span is lower than 16 mm. For larger teeth span, the full number of the teeth is used. This rule comes from the technical practice where either much smaller span is used in combination with a high number of teeth or a large span is used in combination with a low number of the teeth.

A better correlation with experimental data is expected from Samoylovich formula, which keeps a reasonable level of simplicity despite consideration of pressure ratio, number, thickness, span and shape of the teeth.

$$m = k \cdot \mu \cdot S \cdot \sqrt{\frac{p_0}{v_0}} \cdot \sqrt{\kappa \cdot \left(\frac{2}{\kappa+1}\right)^{\frac{\kappa+1}{\kappa-1}}} \cdot \sqrt{\frac{1-(p_2/p_0)^2}{z\cdot\left(1-(p_2/p_0)_{cr}\right)} - \frac{(p_2/p_0)_{cr}\cdot(1-p_2/p_0)^2}{\left(1-(p_2/p_0)_{cr}\right)^2\cdot z^2}} \quad (2)$$

* Corresponding author: dan.hasnedl@doosan.com

The see-through labyrinth is represented by a correction coefficient $k$, which is a function of the number of teeth, radial clearance and teeth span.

Added to this, the Kearton formula was brought for comparison despite its complexity, because it gives results in each inter-teeth chamber that can be later utilized for the seal mechanical design. The formula is:

$$m = k \cdot \mu \cdot S \cdot \sqrt{\frac{p_0}{v_0}} \cdot \sqrt{F \cdot \frac{1-(p_z/p_0)^2}{z}} \quad (3)$$

where

$$F = \frac{\sum_{i=1}^{z} \varepsilon_i}{z} \quad (4)$$

where

$$\varepsilon_i = \frac{p_i}{p_{i-1}} \quad (5)$$

$$p_i = \sqrt{p_0^2 - \frac{i}{z} \cdot (p_0^2 - p_n^2)} \quad (6)$$

### 2.1 Correction coefficients

Coefficients $k$, $\mu$ are used to correct the leakage due to flow behaviour.

Flow contraction is represented by the correction coefficient $\mu$ that comes from the interpolation of the Sceglajev figure.

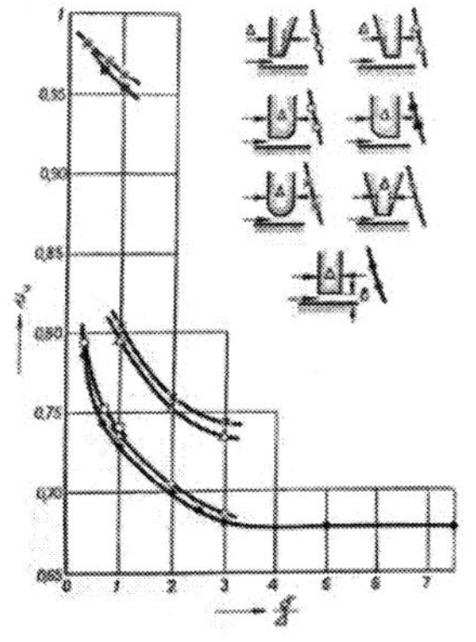

Fig. 1. Sceglajev figure for the evaluation of flow contraction correction coefficient

For calculation of the correction coefficient $k$, which takes in consideration flow in the see-through labyrinth seal, it was used the analytical formula in the first step:

$$k = 1 + 16 \cdot \left(1 - e^{-0,1234 \cdot z}\right) \cdot \frac{\delta}{s} \quad (7)$$

However, the experimental results didn't correlate with analytical approach (especially in cases where ratio δ/s is more than 0.1). Therefore, a new analytical formula for correction coefficient $k$ has been designed:

$$k = 1 + 3,27 \cdot \left(1 - e^{-0,4 \cdot z}\right) \cdot \left(\frac{\delta}{s}\right)^{\frac{2}{3}} \quad (8)$$

## 3 Experiments

Experiments were carried out on the steam test rig which is situated in the R&D laboratory of Doosan Skoda Power.

### 3.1 Measured seal geometry

Examples of typical stepped and see-through labyrinth seals are shown below. Figure 2 shows a typical stepped labyrinth while Figure 3 shows a see-through design. The seals were tested with radial clearances varying from 0.2mm to 0.9mm and number of teeth from 2 to 6.

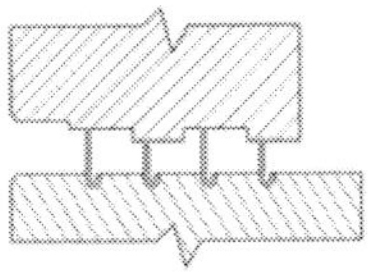

Fig. 2. Typical stepped labyrinth seal

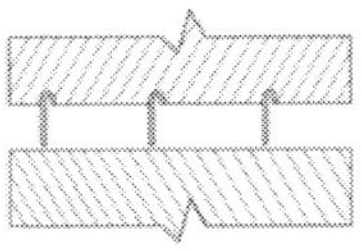

Fig. 3. Typical see-through labyrinth seal

### 3.2 Steam test rig

The steam test rig (shown on Figure 4) is used for a mass flow evaluation by measuring of the condensate volume (representing steam leakage through a seal). All steam, which passed through a seal, goes into an atmospheric condenser. From there, the steam condensate is collected in the tank, where the water level is measured. Calibrated volume between two water levels in combination with the time from the bottom to upper level gives the mass flow. The equipment for measuring the amount of the condensate is shown on Figure 5.

The test rig medium is a medium pressure steam supplied by the next-door power-plant at maximum parameters of 18bar$_g$ and 350°C. Nevertheless, typical operating conditions are usually lower.

The test rig itself consists of a stationary casing and a rotating overhung shaft. The test section consists of an exchangeable stationary part as well as a rotating part. Therefore, various geometries can be tested. The seal diameter is (depending on the seal type) around 290mm.

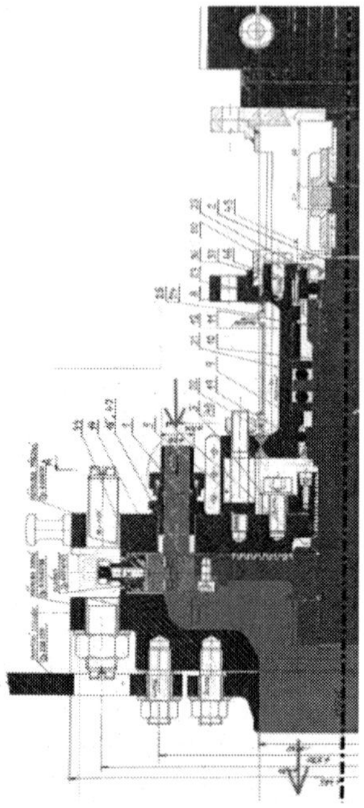

**Fig. 4.** Test rig section: motor(violet), shaft(green), stator(brown), test section(orange), steam flow-path(blue)

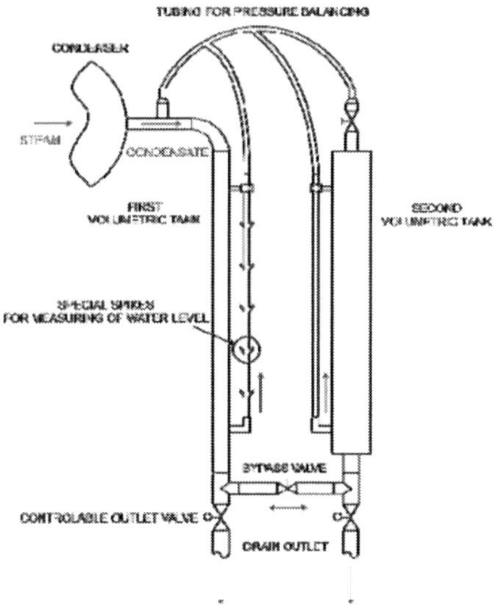


**Fig. 5.** Equipment for measuring the condensate volume (the leakage through the seal)

## 4 Numerical calculations

2D (SEAL) and 3D (FLUENT) calculation models were created. The models corresponded with real seals geometries which were tested by using the steam test rig described in previous chapter. An example of the 2D calculation model of a typical stepped labyrinth seal with four rotor teeth is shown on Figure 6.

**Fig. 6.** 2D calculation model of a typical stepped labyrinth

The 3D calculation model was created as one quarter of the test rig included tested seal and measuring planes in the same location as real measuring pressure and temperature taps. An example of the model is shown on Figure 7.

For calculations carried out by FLUENT, the k-ω turbulence model for the steady compressible turbulent flow was used.

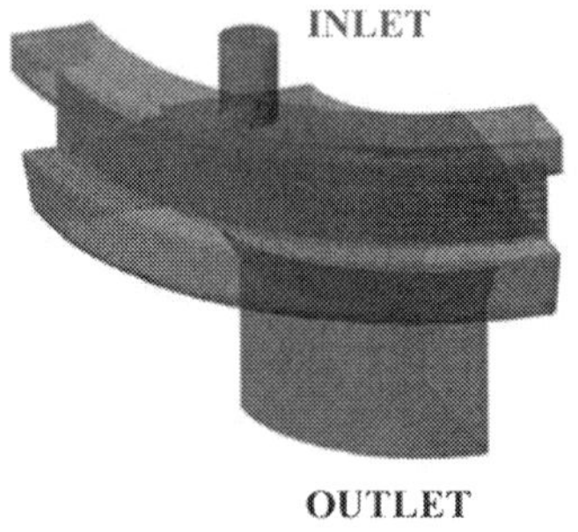


**Fig. 7.** 3D calculation model of one quarter of the test rig

## 5 Results

12 different seal geometries and clearances were tested, calculated and compared in a wide range of pressure ratios. For data summary, there were calculated average deviation values between experiment, analytical and numerical results.

The mass flow predicted by the Zalf formula is higher in case of the stepped as well as see-through labyrinth type. It is considerably more than the measured values. The experimental results are almost in agreement with the numerical calculations, which confirm that the Zalf formula really overestimates the seal leakage.

The results predicted by Samoylovich formula are significantly closer to the tested value with consideration the measurement uncertainty ±5%.

**Table 1.** Average deviations of the leakage from experiments

| Seal Type | Zalf [%] | Samoylovich [%] | FLUENT [%] | SEAL [%] |
|---|---|---|---|---|
| Stepped | +46 | +5 | -12 | -6 |
| See-through | +24 | +2 | +5 | +5 |

The experimental results (Figure 8. and 9.) show some practical findings. For example, for leakage reduction, it is more effective to lower the radial clearance than to increase the number of the teeth.

Another finding is connected with the see-through seal. Using a higher number of the teeth, while reducing span, has a very limited effect. This often happens when trying to keep the overall size of the seal.

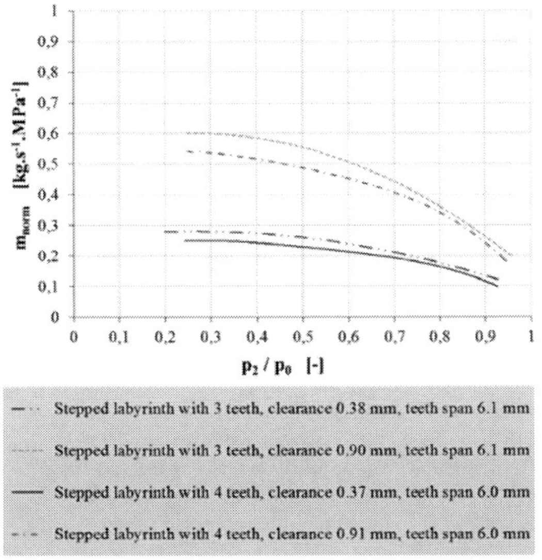


**Fig. 8.** Experimental results for the stepped labyrinth seal

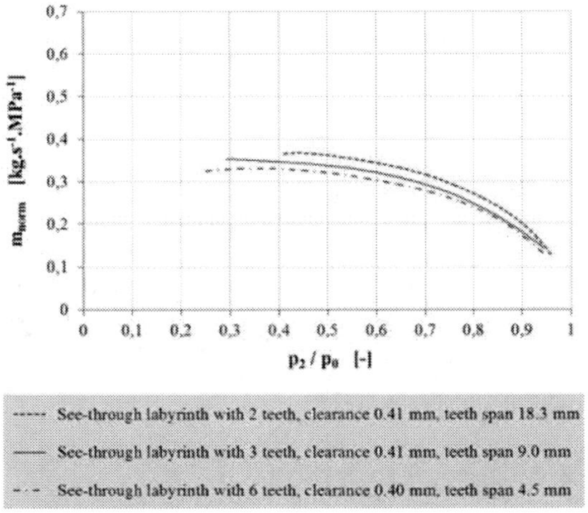


**Fig. 9.** Experimental results for the see-through labyrinth seal

## 6 Conclusions

The state of the art method (Zalf) for predicting seal leakage is limiting the achievable steam turbine efficiency. Therefore it was carried out the comparison between the experiments, analytical and numerical calculations. The experiments confirmed that the Zalf formula is too conservative and an alternative analytical formula has to be introduced to allow an improved turbine design.

The mass flow predicted by the Samoylovich empirical formula is very close to the experimental results. The approximate deviation ±10 % of the numerical data from the experimental and analytical data is sufficient in case of CFD calculations.

The new empirical formula for the see-through labyrinth seal correction coefficient was derived according to the results. Now, it is possible to better describe the flow behaviour in the see-through type of labyrinth seal by the new coefficient which ensures that the leakage prediction is more precise.

These conclusions have changed Doosan Skoda former analytical calculations of the seal losses. As a result, the leakage through the stepped and the see-through labyrinth seals should be calculated by the Samoylovich empirical formula instead of the Zalf approach.

## References

1. Aungier, R., H.: Turbine Aerodynamic: *Axial-Flow and Radial-Inflow Turbine Design and Analysis*. New York: ASME Press, 2006.
2. Stodola, A.: *Dampf und Gas-Turbinen. Berlin*, 1922.
3. Bednar, L.; Hoznedl, M.; Tajc, L.: *Ucinnost turbinovych stupnu, III. upravene vydani*. Vyzkumna zprava VZTP 1047, Plzen: Doosan Skoda Power, s.r.o., 2010.
4. Deych, M., E., Samoylovich, G., S.: *Bases of Aerodynamics of Axial Turbines*. Moscow, 1959.
5. Zalf, G., A., Zvyagintsev V., V.: *Thermal calculation of Steam Turbines*. Moscow, 1961.
6. Kearton, W., J.; Keh, T., H.: *Leakage of Air through Labyrinth Glands of Staggered Type*. University of Liverpool: 2013.
7. Shcheglyaev, A.V.: Steam Turbines. Moscow, 1967.
8. Egli, A.: *The Leakage of Steam through Labyrinth Seals*. Transactions ASME Trans, vol. 57, pp. 115-122.
9. Kearton W. J.: *The Flow of Air through Radial Labyrinth Glands*. Proceedings of the Institution of Mechanical Engineers, Vol. 169, 1955.
10. Trutnovsky, K.: Contactless Seals, Foundations and Applications of Flows through Slots and Labyrinths. NASA TT F 17, 352, April 1977.
11. Yoon, S.: *The Effect of the Degree of Reaction on the Leakage Loss in Steam Turbines*. ASME Turbo Expo, GT2012-69602, Copenhagen, June 2012.

# Experimental determination of the heat transfer coefficient in shell-and-tube condensers using the Wilson plot method

*Jan* Havlik[1,*], *Tomas* Dlouhy[1]

[1]Czech Technical University in Prague, Faculty of Mechanical Engineering, Department of Energy Engineering,, Technicka 4, 16607 Prague 6, Czech Republic

**Abstract.** This article deals with the experimental determination of heat transfer coefficients. The calculation of heat transfer coefficients constitutes a crucial issue in design and sizing of heat exchangers. The Wilson plot method and its modifications based on measured experimental data utilization provide an appropriate tool for the analysis of convection heat transfer processes and the determination of convection coefficients in complex cases. A modification of the Wilson plot method for shell-and-tube condensers is proposed. The original Wilson plot method considers a constant value of thermal resistance on the condensation side. The heat transfer coefficient on the cooling side is determined based on the change in thermal resistance for different conditions (fluid velocity and temperature). The modification is based on the validation of the Nusselt theory for calculating the heat transfer coefficient on the condensation side. A change of thermal resistance on the condensation side is expected and the value is part of the calculation. It is possible to improve the determination accuracy of the criterion equation for calculation of the heat transfer coefficient using the proposed modification. The criterion equation proposed by this modification for the tested shell-and-tube condenser achieves good agreement with the experimental results and also with commonly used theoretical methods.

## 1 Introduction

The heat transfer mechanism by convection is an energy transfer between a solid surface and a moving fluid induced by a temperature difference between the solid surface and the fluid. Heat transfer by convection is indeed a combination of conduction and fluid motion. The analytical treatment of convection problems requires the solution of a system of mass, momentum and energy conservation equations. Although these procedures are elaborate, analytic solutions are available for simple geometries under restrictive assumptions. Most of the convective heat transfer processes inherent to heat exchangers usually involve complex geometries and complicated flows so analytical solutions are not possible. Therefore, an approach based on Newton's law of cooling (see Eq. (1)), provides a simple alternate technique. Newton's law of cooling established an algebraic relation between the heat flow by convection $Q$, the surface area $A$, the heat transfer coefficient $h$ and the temperature difference between the solid surface $T_s$ and the fluid $T_f$ .

$$Q = h \cdot (T_f - T_s) \cdot A \tag{1}$$

For a given flow and surface geometry, the experimental data is usually obtained by measuring the surface and fluid temperatures. Then the heat transfer coefficient (HTC) may be calculated from Eq. (1). However, the main difficulty of this methodology lies in the measurement of the surface temperature, because the surface temperature varies from point to point, and the flow profile could be altered by the presence of the temperature sensors. Moreover, in many cases of heat exchangers the heat transfer surface is not accessible. Therefore, any alternative method conducive to the HTC calculation is attractive because of widespread use and practical applications [1].

## 2 Wilson plot method

The Wilson plot method is used for the experimental determination of the HTC or coefficients of a criterion equation, which describes the HTC calculation. This method avoids the direct measurement of the surface temperature and consequently any disturbance of the fluid flow and heat transfer introduced while attempting to measure those temperatures. Therefore, the method constitutes a suitable technique for estimating HTCs in a variety of convective heat transfer processes [1].

The original method has been derived for the processes of transferring heat by convection, when the thermal resistance on one side of the heat transfer surface remains constant (in the case of condensation[2]) while varying the mass flow on the other side changes the total thermal resistance in the heat exchanger. This relies on the fact that the overall thermal resistance can be taken from experimental measurements in a reliable

* Corresponding author: jan.havlik@fs.cvut.cz

manner [1]. This method was originally derived for determining the coolant HTC in steam condensers where steam condenses outside the tubes and the coolant flows inside. The sensitivity of the coolant HTC value to the overall HTC is more significant than the condensation HTC [2]. The mean value of the HTC on the coolant side can be described as a function of the speed of the coolant [1].

The aim of the method is to calculate the convection coefficients (or thermal resistance) of the fluid in the criterion equation for the defined type of convection heat transfer. Modifications of the Wilson plot method were proposed to enhance its accuracy. The principle of the original method is still maintained. The application of the Wilson plot method is based on the measurement of the experimental mass flow rate and temperatures.

## 3 Tested heat exchanger

The tested heat exchanger is a vertical shell-and-tube model in which the condensing water vapour flows downwards inside vertical tubes and the cooling water flows in a counter current in the outer shell (see Fig. 1). The shell side has a complex geometry with complicated flow patterns so using the Wilson plot method is suitable for determining the shell-side HTC.

The tube bundle is formed by 49 tubes 865 mm in length, 28 mm in external diameter and 24 mm in internal diameter. The tubes are arranged in staggered arrays with a triangular tube pitch of 35 mm. The cross-section of the shell is rectangular in shape, 223 mm by 270 mm in size. Seven segmental baffles (223 x 230 mm) are used in the shell section. The tubes are made from stainless steel 1.4301 (AISI 304).

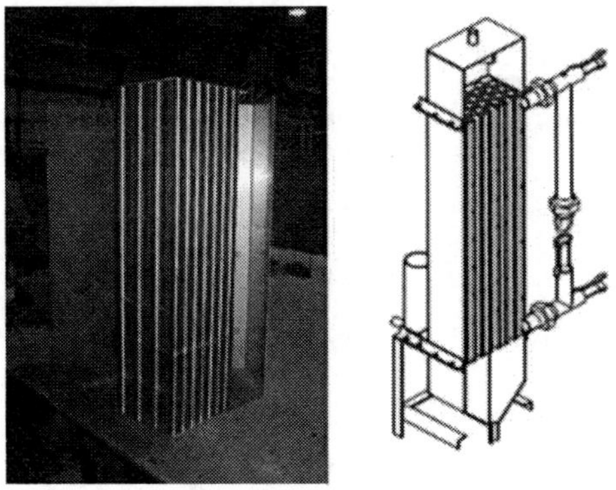

**Fig. 1.** Vertical tube condenser

The testing loop is shown in Fig. 2. Steam is produced in a steam generator. Before the steam enters the vertical tube condenser, its parameters are reduced to the required values. Hot water recirculation enables the cooling water temperature and the cooling water flow rate in the condenser to be regulated. The measured parameters are the inlet cooling water temperature $T_{w1}$, the temperature of the water after recirculation $T_{w2}$, the outlet cooling water temperature $T_{w3}$, the cooling water flow rate $Q_w$, the inlet steam pressure $p_s$, the inlet steam temperature $T_s$ and the amount of steam condensate $M_c$.

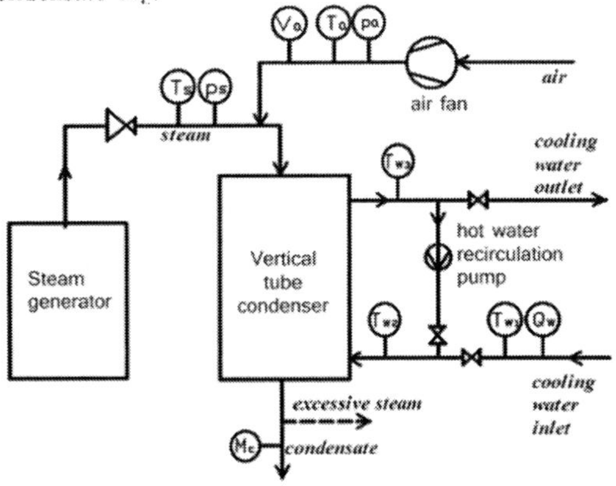


**Fig. 2.** Scheme of the testing loop

## 4 Original Wilson Plot method

Derivation of the Wilson-plot is well described in [1]. The overall thermal resistance of the condensation process in a shell-and-tube condenser $R_{ov}$ can be expressed as the sum of three thermal resistances corresponding to the internal convection $R_i$, the tube wall $R_s$ and the external convection $R_o$ as

$$R_{ov} = R_i + R_s + R_o \tag{2}$$

For the sake of simplicity, the thermal resistances of the fluid fouling are neglected. Wilson theorized that if the mass flow of the cooling liquid was modified, then the change in the overall thermal resistance would be mainly due to the variation of the coolant HTC, while the remaining thermal resistances remained nearly constant. The HTC varies with the fluid mass flow for many cases of the convection heat transfer process excluding the phase changes. It is valid also for complex heat exchanger geometries. An illustrative example of the experimental determined dependence of the HTC on the fluid mass flow is well described in [3].

Therefore, the thermal resistances on the condensation side and the tube wall could be considered constant. In the case of a condenser where steam condenses inside the tubes is considered

$$R_i + R_s = C_1 \tag{3}$$

where $C_1$ is a constant. For the thermal resistance on the cooling water side $R_o$ is considered

$$R_o = \frac{1}{hA} \tag{4}$$

where $A$ is the heat surface area and the HTC on the cooling water side $h$ is proportional to the power of the fluid velocity $v$

$$h = C_2 \cdot v^n \tag{5}$$

where $C_2$ is a constant and $n$ is the velocity exponent.

Upon combining Eq. (2), Eq. (3), Eq. (4) and Eq. (5), the overall thermal resistance turns out to be a linear function of $v^n$

$$R_{ov} = \frac{1}{C_2 A} \cdot \frac{1}{v^n} + C_1 \qquad (6)$$

This is presented graphically in Fig. 3.

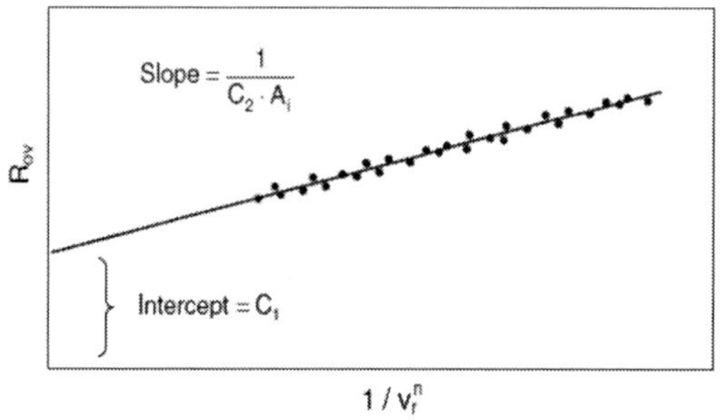


**Fig. 3**. A linear function of the thermal resistance [1]

From here, the straight line equation that fits the experimental data can be deduced by applying simple linear regression. The overall thermal resistance may be expressed as a function of the overall heat transfer coefficient $U$ and surface area $A$

$$R_{ov} = \frac{1}{U \cdot A} \qquad (7)$$

Then, for each set of experimental data corresponding to each mass flow rate, the overall thermal resistance is provided by substituting Eq. (7) to Eq. (7). This equation describes the calculation of the overall heat transfer coefficient from the performance of the heat exchanger and the logarithmic mean temperature difference of the fluids $\Delta T_{log}$

$$U = \frac{Q}{A \cdot \Delta T_{log}} \qquad (8)$$

The heat exchanger performance $Q$ is provided by the equation

$$Q = M_w \cdot c_w \cdot (T_{w,out} - T_{w,in}) \qquad (9)$$

where $M_w$ is the cooling water flow, $c_w$ is the specific heat capacity of water, $T_{w,out}$ is the outlet cooling water temperature and $T_{w,in}$ is the inlet cooling water temperature [1].

## 5 Proposed modification of the Wilson plot method

The determination of the HTC on the cooling side is based on the change in thermal resistance for different conditions (fluid velocity and temperature). The modification considers the presumption of validity of the Nusselt theory for calculation of the HTC on the condensation side. The thermal resistance on the condensation side varies and it is part of the calculation.

### 5.1 Non-dimensional parameters of the heat exchangers

*Reynolds Number*

The Reynolds number $Re$ is a function of the fluid flow rate. The mathematical formula for the Reynolds number is provided in Eq. (10). The ratio represents momentum to viscous forces, where $v$ is the fluid velocity, $D$ is characteristic diameter and $\nu$ is the kinematic viscosity [4].

$$Re = \frac{v \cdot D}{\nu} \qquad (10)$$

*Prandtl Number*

The Prandtl number $Pr$ is a function of two important physical properties (thermal and momentum). The Prandtl number may be seen as the ratio of the rate that viscous forces penetrate the material to the rate that thermal energy penetrates the material [4]. In Eq. (11), $\mu$ is the dynamic viscosity, $c_p$ the material heat capacity and $k$ is thermal conductivity.

$$Pr = \frac{\mu \cdot c_p}{k} \qquad (11)$$

*Nusselt Number*

The Nusselt number $Nu$ is equal to the dimensionless temperature gradient at the surface and it essentially provides a measure of the convective heat transfer. The Nusselt number may be viewed as the ratio of the conduction resistance of a material to the convection resistance of the same material [4].

$$Nu = \frac{h \cdot D}{k} \qquad (12)$$

*Nusselt Number Correlations*

In single phase fluid flow heat transfer, the Nusselt number is generally represented by a pragmatic expression in the form as seen in (13). The term $(\mu/\mu_s)$ is accountable for the variable viscosity effect for the case that the viscosity varies from the centre of the tube towards the tube wall [4].

$$Nu = C \cdot Re^n \cdot Pr^m \cdot \left(\frac{\mu}{\mu_s}\right)^{0.14} \qquad (13)$$

### 5.2 Derivation of the proposed modification

The overall thermal resistance $R_{ov}$ (Eq. (2)) for the described shell-and-tube condenser can be rewritten as

DOI: 10.1051/epjconf/201714302035

$$R_{ov} = \frac{1}{h_c \cdot A_i} + \frac{ln(d_o/d_i)}{2 \cdot \pi \cdot k_t \cdot L_t} + \frac{1}{h_w \cdot A_o} \quad (14)$$

where $h_c$ and $h_w$ are the condensation and cooling water HTCs, $d_i$ and $d_o$ are the inner and outer tube diameters, $k_t$ is the tube thermal conductivity, $L_t$ is the tube length and $A_i$ and $A_o$ are the inner and outer tube surface areas, respectively. By combining Eq. (8) and (14) is given

$$\frac{1}{U \cdot A} = \frac{1}{h_c \cdot A_i} + \frac{ln(d_o/d_i)}{2 \cdot \pi \cdot k_t \cdot L_t} + \frac{1}{h_w \cdot A_o} \quad (15)$$

where $h_w$, can be expressed combining Eq. (12) and (13)

$$h_w = \frac{k_w}{D} \cdot C \cdot Re^n \cdot Pr^m \cdot \left(\frac{\mu}{\mu_s}\right)^{0.14} \quad (16)$$

and $h_c$ is defined by the Nusselt condensation theory [5], [6] as

$$h_c = 0.943 \left[\frac{\rho_L g(\rho_L - \rho_p) h'_{fg} k_L^3}{\mu_L \Delta T_{sat} L}\right]^{1/4} \quad (17)$$

where $\rho_L$ is the density of the condensate, $\rho_p$ is the density of the water vapour, $h'_{fg}$ is the latent heat of condensation, $k_L$ is the thermal conductivity of the condensate, $\mu_L$ is the dynamic viscosity of the condensate, $\Delta T_{sat}$ is the saturation and wall temperature difference, $L$ is the wall length. By a substitution of terms in (8) and (16) to equation (15) is given

$$\frac{1}{\left(\frac{Q}{A \cdot \Delta T_{log}}\right) \cdot A} = \frac{1}{h_c \cdot A_i} + \frac{ln(d_e/d_i)}{2 \cdot \pi \cdot \lambda_s \cdot L_s} + \frac{1}{\frac{k_w}{D} \cdot C \cdot Re^n \cdot Pr^m \cdot \left(\frac{\mu}{\mu_s}\right)^{0.14} \cdot A_o} \quad (18)$$

It is possible to write Eq. (17) in linear form $Y = a \cdot X$

$$\left(\frac{\Delta T_{log}}{Q} - \frac{1}{h_c \cdot A_i} - \frac{ln(d_o/d_i)}{2 \cdot \pi \cdot \lambda_s \cdot L_s}\right) = \frac{1}{C} \cdot \frac{1}{\frac{k_w}{D} \cdot Re^n \cdot Pr^m \cdot \left(\frac{\mu}{\mu_s}\right)^{0.14} \cdot A_o} \quad (19)$$

Where the terms $Y, a$ and $X$ have the following forms

$$Y = \left(\frac{\Delta T_{log}}{Q} - \frac{1}{h_c \cdot A_i} - \frac{ln(d_o/d_i)}{2 \cdot \pi \cdot \lambda_s \cdot L_s}\right); \quad a = \frac{1}{C}; \; X = \frac{1}{\frac{k_w}{D} \cdot Re^n \cdot Pr^m \cdot \left(\frac{\mu}{\mu_s}\right)^{0.14} \cdot A_o} \quad (20)$$

The setting of $Y$ and $X$ values given for the set of experimental data can be fitted by the linear regression for an estimated value of the coefficient $n$. The value for the coefficient $m$ is set at 0.33 in accordance with standard practice in many transfer processes. The $X$ and $Y$ regression starts with an initial $n$ value. The $C$ coefficient is found based on a linear regression with the highest coefficient of determination for various values of the parameter $n$.

## 6 Measurements

Measurements of the operating parameters required for the calculation were carried out for two sizes of the exchanger heat transfer area, for the original number of 49 tubes and for a changed heat transfer area of 21 tubes with 28 tubes sealed off. The inlet steam velocity in both cases was maintained between 2 and 3.5 m/s. The steam is saturated at the atmospheric pressure. There were measured in 59 different operating states with changes in the cooling water flow and the cooling water temperature. The inlet cooling water temperature ranged between 60 and 85 ° C, the outlet temperature between 80 and 95 ° C. The cooling water flow rate was regulated by a circulating pump set to a Reynolds number value ranging from 1000 to 8000. The heat exchanger performance was from 14 to 56 kW, which corresponds with the amount of condensed steam between 21 and 90 kg/h, where the smaller values are related to the measurement with 21 tubes and the higher values to the measurement with 49 tubes. The condensate film thickness was calculated and it ranged from 0.063 to 0.11 mm, which corresponds to a Reynolds number from 38 to 183 confirming the wavy-laminar condensate film flow.

## 7 Results

The result of the dependence $Y = a \cdot X$ introduced in Eq. (19) is shown in Fig. 4. The values were fitted by a linear regression with the highest coefficient of determination.

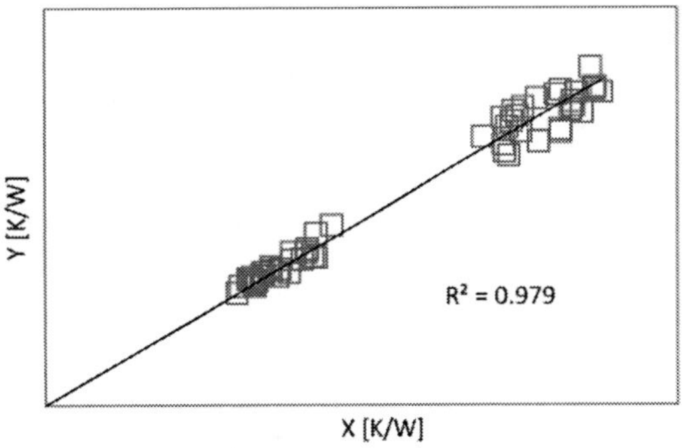


**Fig. 4.** Linear regression of thermal resistance according to the Wilson Plot method

In Fig. 4, there are two groups of points which differ in the number of tubes utilized (49 resp. 21). This is due to the significant difference in heat exchanger performance caused using different heat transfer surfaces. Based on the obtained linear regression, it is evident that the number of cooled tubes in the heat exchanger has no apparent effect on the dependency derived from the balance of heat resistance respective on the heat transfer in the shell-side of the heat exchanger. The coefficients evaluated from the linear regression are as follows

$$n = 0.56; m = 0.33; C = 0.2702 \quad (21)$$

 DOI: 10.1051/epjconf/201714302035

The resulting equation for the cooling water side determined by the Wilson plot method has the form

$$Nu = 0.270Re^{0.56}Pr^{0.33}\left(\frac{\mu}{\mu_s}\right)^{0.14} \quad (22)$$

Fig. 5 shows a comparison of the measured values of the HTC with the calculated values by the proposed equation. A correspondence between the calculated and experimentally determined values fits within the range of ± 8%.

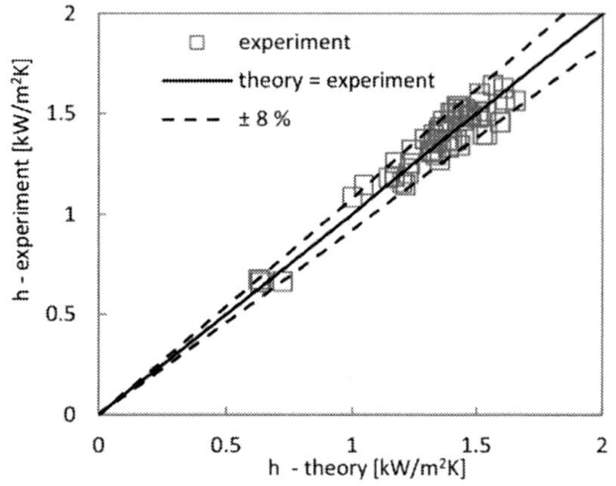


**Fig. 5.** Comparison of HTC experimental values with the results calculated by the proposed equation

## 8 Discussion

The proposed criterion equation is compared with the most commonly used theoretical methods – the methods by Kern, Bell-Delaware and Stream flow analysis (see Table 1). These methods are described in detail in [6]. For simplicity, the term $\left(\frac{\mu}{\mu_s}\right)^{0.14}$ is not shown in the table.

The proposed coefficients $C$ and $n$ reach similar values as the coefficients used in the theoretical methods. Fig. 6 shows that the proposed equation achieves good agreement with the experimental results and also with the Bell-Delaware method and Stream flow analysis. The Kern method does not take into account the heat exchanger geometry, e.g. manufacturing clearances and resulting leakages, and it gives higher values.

## 9 Conclusion

A modification of the Wilson plot method for shell-and-tube condensers was proposed. The modification is based on the validation of the Nusselt theory for calculation of the HTC on the condensation side. The thermal resistance on the condensation side varies depending on different conditions of the total heat transfer process and is part of the calculation. It is possible to improve the determination accuracy of the convection coefficients using the proposed modification.

For the described heat exchanger, the criterion

**Table 1** Comparison of criterion equation for calculation of the Nusselt number

| **Equation** | $Nu = C \cdot Re^n \cdot Pr^m$ | **note** |
|---|---|---|
| Proposed | $Nu = 0.270 \cdot Re^{0.56} \cdot Pr^{0.33}$ | valid for the described exchanger |
| Kern | $Nu = 0.36 \cdot Re^{0.55} \cdot Pr^{0.33}$ | it does not take into account the specific geometry |
| Bell-Delaware | $Nu = 0.273 \cdot Re^{0.635} \cdot Pr^{0.34} \cdot J$ | J is a correction coefficient for baffle geometry and manufacturing accuracy |
| Stream flow analysis | $Nu = 0.273 \cdot Re_{cr}^{0.635} \cdot Pr^{0.34}$ | $Re_{cr} = F_{cr} \cdot Re$, where $F_{cr}$ is a correction coefficient for idealized fluid flow |

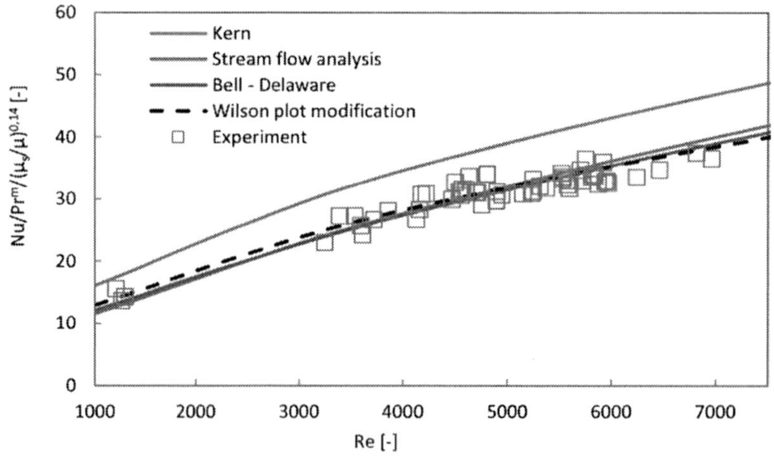


**Fig. 6.** Comparison of methods for the calculation of the HTC

equation for the calculation of the Nusselt number was proposed. The theoretical results of the proposed equation were compared with the measured values. Correspondence between the theoretically and the experimentally determined values is located in the range of ± 8%. The proposed coefficients C and n reach similar values as the coefficients found in the most commonly used theoretical methods. The proposed equation achieves better agreement with the experimental results as well as good compliance with the Bell-Delaware method and Stream flow analysis. The Kern method does not take into account the heat exchanger geometry and its results have higher values take into account the heat exchanger geometry and Stream flow analysis.

The modified method for the determination of the shell-side HTC was used in the subsequent research of the conditions of steam-air mixture condensation in the tubes, where the effect of noncondesable air on the condensation process was analysed.

This work has been supported by Norway Grants 2009-2014, project ID: NF-CZ08-OV-1-003-2015 and the Grant Agency of the Czech Technical University in Prague, grant no. SGS 14/183.

## Nomenclature

A heat transfer area ($m^2$)
C constant of geometry (-)
c specific heat capacity (J/kgK)
d diameter (m)
D characteristic diameter (m)
g gravitational acceleration ($m/s^2$)
h convection coefficient ($W/m^2K$)
$h_{fg}$ latent heat of condensation (kJ/kg)
k thermal conductivity (W/mK)
L tube length (m)
$\Delta T_{log}$ logarithmic mean temperature difference (K)
M mass flow rate (kg/s)
Nu Nusselt number (-)
Pr Prandtl number (-)
R thermal resistance (K/W)
Re Reynolds number (-)
Q rate of heat transfer (W)
S flow cross-section ($m^2$)
T temperature (K)
U overall heat transfer coefficient ($W/m^2K$)
v velocity (m/s)
μ dynamic viscosity (Pa·s)
ν kinematic viscosity ($m^2/s$)

Subscripts
f fluid
i inner
in inlet
l liquid
o outer
out outlet
ov overall
s wall
t tube
w water

## References

1. J. Fernández-Seara, F. J. Uhía, J. Sieres, A. Campo, Applied Thermal Engineering **27**, 2745 (2007).
2. J. Havlik, T. Dlouhy, Acta Polytechnica **55**, 306 (2015).
3. P. Kracik, J. Pospisil, Acta Polytechnica **55**, 329 (2015).
4. V. R. Naik, V. K. Matawala, IJEAT **2**, 362 (2013).
5. F. Hewitt, G. L. Shires, T. R. Bott, *Process Heat Transfer* (Begell House, New York, 2000).
6. F. P. Incropera et al., *Principles of Heat and Mass Transfer* (John Whiley, Singapore, 2013).

# Experimental investigation of the check valve behaviour when the flow is reversing

D. Himr[a], V. Habán, M. Hudec, and V. Pavlík

Brno University of Technology, Faculty of Mechanical Engineering, Energy Institute, Victor Kaplan Department of Fluid Engineering, Technická 2, 616 69, Brno, Czech Republic

**Abstract.** Check valve in a pipeline is supposed to prevent the reverse flow and to allow the flow in the positive direction. The construction of check valves follows these requirements, but the check valve must not cause pressure pulsations in transients. It means when the fluid is accelerating or decelerating. The article describes an experimental investigation of a swing check valve when the flow is changing its direction. The check valve was placed in an experimental circuit, where the pressure on the upstream and downstream side of the valve was measured and the current value of flow rate was determined. The goal was to simulate conditions in the real system, where the check valve slam had been observed.

## 1 Introduction

The check valve is an important part of hydraulic systems and allows flow in one direction and prevent the reverse flow through the pump. The check valve can be also installed in the system to limit the pressure surge induced by the pump failure [1]. The basic requirements on the check valve are:

- the low pressure loss in the positive direction,
- no flow in the opposite direction, it means good sealing when the valve is closed.

Producers, usually, give static characteristics of check valves such as a pressure drop dependence and opening dependence on the flow rate, the cracking pressure (the pressure when the check valve starts opening) and the minimal flow rate when the check valve is fully open.

The producer also can say whether the valve is appropriate for the pulsating flow, sludge water, horizontal or vertical pipe and so on. These parameters together with the reliability and maintenance difficulty allow choosing the right check valve to reduce operational costs.

But, there is a problem with predicting the dynamic behaviour of the check valve in the particular system. It is, maybe, more important than properties written above. When the fluid flow changes its orientation, the check valve is desired to close before the back flow velocity becomes too high otherwise the check valve disc slams and makes the high pressure surge, which is often connected with the column separation particularly on the upstream side of the check valve. The lower static pressure in the pipeline the greater column separation probability. The slam can lead to severe damages of the check valve and/or the whole system [2], [3], [4].

The check valve also can make problems with self-excited pressure pulsations: The oscillating control valve at the downstream end of pipe causes troubles with check valve slam at the upstream end [5]. An example of the pressure surge due to the check valve slam after the pump stoppage is shown in the figures 1 and 2. The slam also causes strong mechanical vibrations of pipeline with acceleration about 20 G in this case [4].

McElhaney made an extensive analysis of check valve failures in the nuclear industry, because these failures are well described, and they focused on correlation between the valve design and failure mode and failure distribution [6].

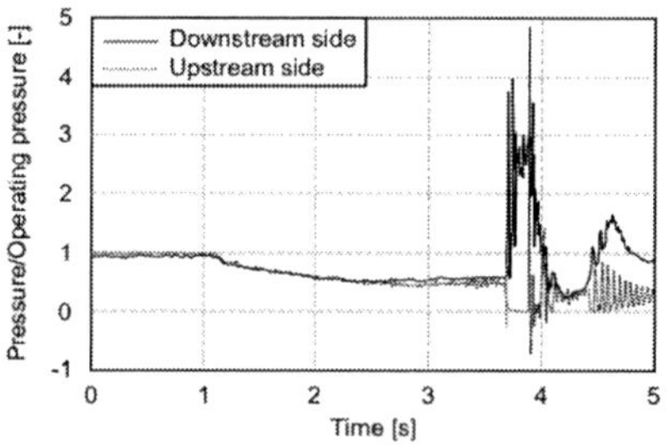


**Fig. 1.** Pressure surge at the check valve due to slam [4]

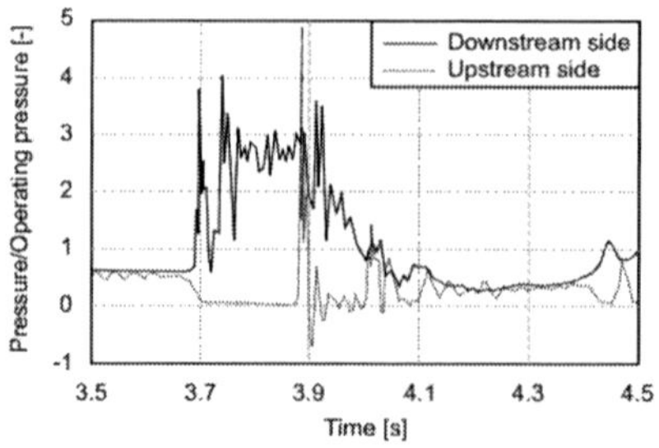


**Fig. 2.** Pressure surge at the check valve due to slam – detail [4]

[a] e-mail: `himr@fme.vutbr.cz`

*In severe pumping applications, almost all basic check valves will slam, and in extremely mild applications, hardly any check valves will slam* [7]. But it is difficult to predict probability of the check valve slam between these two extremes.

The calculation of the system deceleration is important to say whether the check valve will slam or not. The deceleration is calculated from the original forward flow velocity and time period when the velocity reaches zero. It depends on many factors: complexity of the system, friction, pump inertia, static head and so on. Thus, the deceleration can be hardly determined without direct measurement or at least a numerical simulation of the flow in the pipe line system.

The maximal velocity of the reverse flow depends on the check valve construction and on the system deceleration. The higher deceleration the higher reverse velocity can develop before the check valve closes. Then, the pressure surge follows the Joukowski equation.

The reverse velocity dependence on the flow deceleration according to [8] is shown in the figure 3 and a dependence published in [9] is plotted in the figure 4.

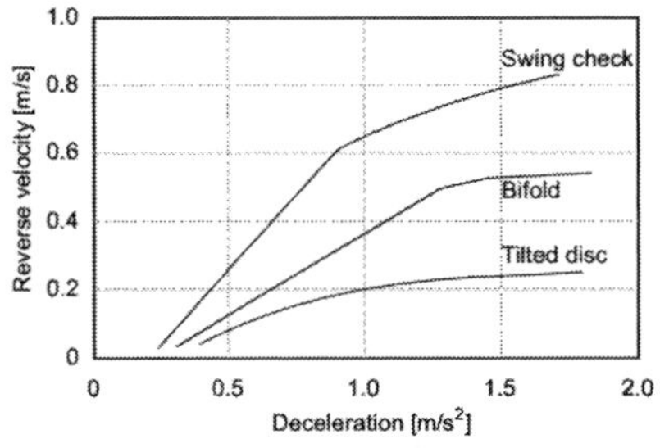


**Fig. 3.** Reversal velocity for different constructions of the check valve by [8]

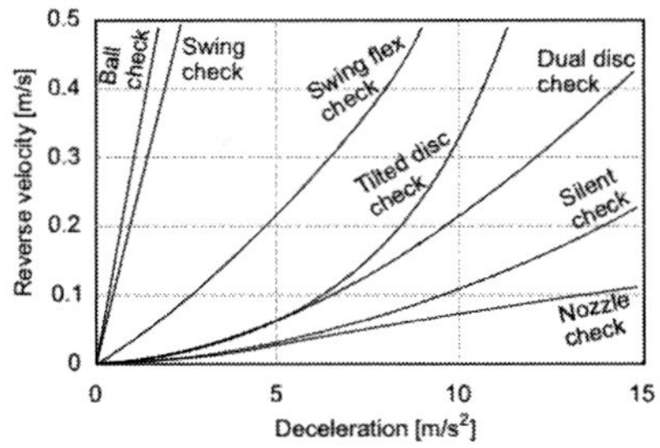


**Fig. 4.** Reversal velocity for different constructions of the check valve by [9]

One can see that results (e. g. of the tilted disc check valve) are quite different so it is obvious that the reversal velocity does not depends only on the deceleration and type of the check valve, but, probably, also on the specific construction, size, mass, placement,... The system deceleration gives just a basic hint whether the check valve tends to slam or not.

**Table 1.** Stand specifications

| Entry | Value | Unit |
|---|---|---|
| Pipe diameter | 0.1 | m |
| Maximal flow rate | 31 | $\mathrm{l\,s^{-1}}$ |
| Maximal flow velocity | 3.95 | $\mathrm{m\,s^{-1}}$ |
| Length of discharge pipe | 17.3 | m |
| Static head | 7.5 | m |
| Volume of bottom tank | 1 | $\mathrm{m^3}$ |
| Volume of top tank | 0.8 | $\mathrm{m^3}$ |
| Pipe material | steel | |
| Maximal pump speed | 1450 | rpm |

## 2 Experiment

An experimental stand has been built and serves for the examination of the check valve behaviour, when the flow orientation is changing. The experiment is a part of cooperation with MSA company, a producer of valves, and the goal is to design a check valve, which does not slam in extreme operating conditions. MSA company provided a swing check valve DN 100, which tends to slam and it is the first check valve of a row, which is going to be tested.

The stand consists of the pump, which collects water from the bottom tank. The tested check valve is placed in the horizontal part of the discharge pipe 2.3 m above the water level in the bottom tank. The discharge pipe ends in the top tank. The water from top tank returns back through the returning pipe and the overspill pipe (see figure 5 and table 1).

Pressure sensors (range 0 – 1 MPa, uncertainty 0.25% of the range, sampling frequency 1 kHz) are placed 0.3 m, 1.4 m and 2.5 m before the check valve and 0.41 m, 1.4 m and 2.38 m behind the check valve in the steady flow direction. The flow rate is measured with the electromagnetic flowmeter (0 – 80 l/s, uncertainty 0.5% of measured value). One accelerometer (-50 G – 50 G, uncertainty 1% of measured value, sampling frequency 1 kHz) is placed on the check valve body and another one is placed on the pump. Accelerometers should help to identify an exact time of some events after pump disconnection. The pump speed is measured with the laser and one mark on the motor shaft.

The pump speed is controlled with a frequency converter to control the flow rate and flow deceleration. Unfortunately, there was a problem with the converter during the experiments and it was not possible to control the deceleration rate.

The check valve design does not allow measuring position of the disc, but check valves, which are going to follow, have windows to see the exact position of the disc and allow PIV measurement of the flow field in the check valve body.

Three pressure sensors on both sides of the valve allow calculating the wave speed and flow rate during deceleration by time-pressure method. Using the flowmeter is not possible due to its great integration constant so the flowmeter can measure just the steady flow rate.

The experiment procedure is following:

1. Start of pump and run with the specific speed to have a desired flow rate (from 10 l/s to 30 l/s). It was not possible to keep the flow for time longer than 30 s, because the overspill aerated water. The air decreases the wave

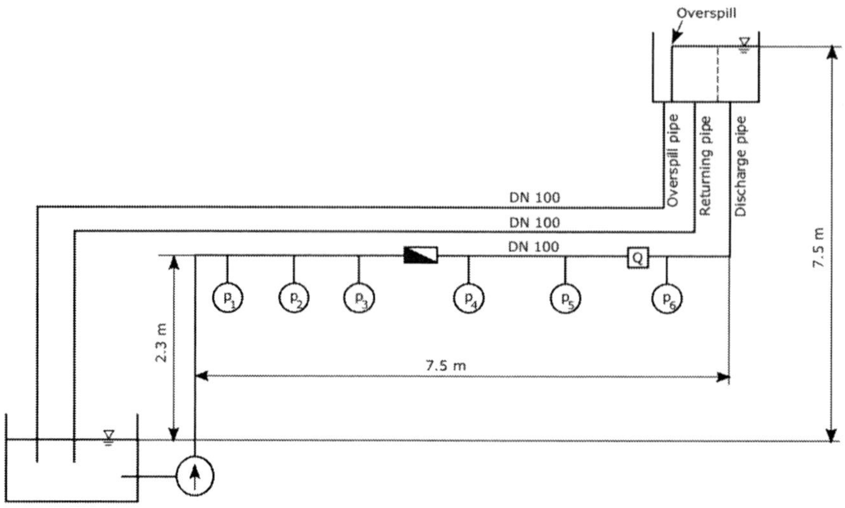


**Fig. 5.** Experimental stand

speed, which directly influences the height of pressure peaks according to Joukowski equation, and changes the system response [10], [11].

2. As soon as the steady flow was reached the pump was disconnected. Since the pipe line is quite short, it was possible to turn off the pump after five seconds after the start.
3. The measurement was recorded from the start of pump until all transients after the pump stop were finished. Then, it is possible to evaluate: cracking pressure, flow deceleration, maximal back flow velocity, pressure surge in front of and behind the check valve, wave speed and acceleration connected with mechanical vibration on the check valve body.

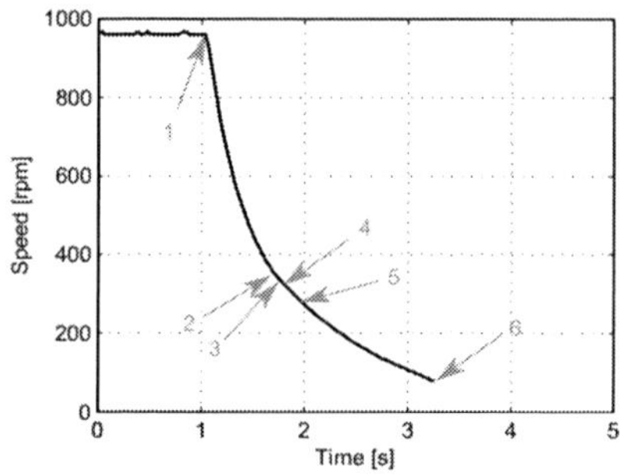


**Fig. 6.** Pump deceleration from steady flow rate 15 l/s

## 3 Results

The pump was stopped from the four different steady flow rates: 10 l/s, 15 l/s, 20 l/s, 27 l/s and 30 l/s. An example of results for stop from 15 l/s is shown in the figures 6 to 12.

The instant pump speed is evaluated from scanning the spot on the motor shaft. It allows identifying the time of pump disconnection from the grid and the time, when the shaft definitely stopped. These points are, respectively, labelled with numbers 1 and 6 in the figures.

Since there was only one mark on the rotor shaft, there is not problem to count number of revolutions and calculate the exact speed (there is not any demand on the exact position of the mark), but (because we are talking about an unsteady rotation) the slower shaft rotates the less accurate evaluated speed is. That is the reason why the last evaluated speed of the shaft is 85 rpm. The last mark was read at the time 3.2 s, the shaft stopped after that.

Pressure upstream of the check valve (figure 7) exhibits the pump disconnection (point 1), when the pressure started decreasing. Time, when the flow changes orientation and starts flowing backward, is marked with point 2. This very weak peak can be easily overlooked.

Point 3 labels time instant, when the check valve starts closing and point 4 is the moment, when the disc hits the seat. (The column separation occurred here in cases with greater initial flow rate).

Then, the highest peak emerges and pressure pulsations are damped, but the pump shaft still rotates. The shaft stop (point 6) caused amplification of the following pulsations about 100%, but the absolute value was still low, thus it did not cause any significant event. These amplified pulsations were damped within 1 second.

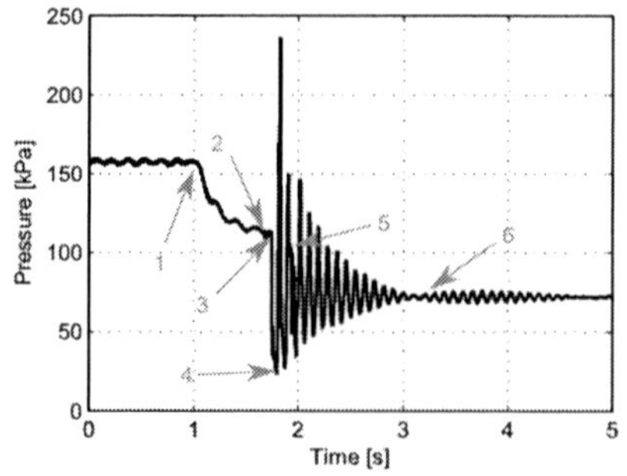


**Fig. 7.** Pressure in front of the check valve during the transient

The pressure downstream of the check valve (figure 8) allows identifying the time when the pump was disconnected (point 1). Points 3 and 4 mark moment when the disc started and finished closing.

The following pressure drop goes to the value, which is lower than one would expect (point 5). The same point in the figure 7 shows that the pressure peak is lower than it should be, when the previous and subsequent peaks are considered. This is the time instant when the disc reopens (pressure at the upstream side of the disc is greater than at the downstream side) and closes again. The figure 9 shows the pressure difference on the check valve. A positive value corresponds to the situation, when the pressure downstream of the valve is greater than upstream of the valve, so the check valve is closed. Negative value at the point 5 supports the statement that the disc reopens at that moment.

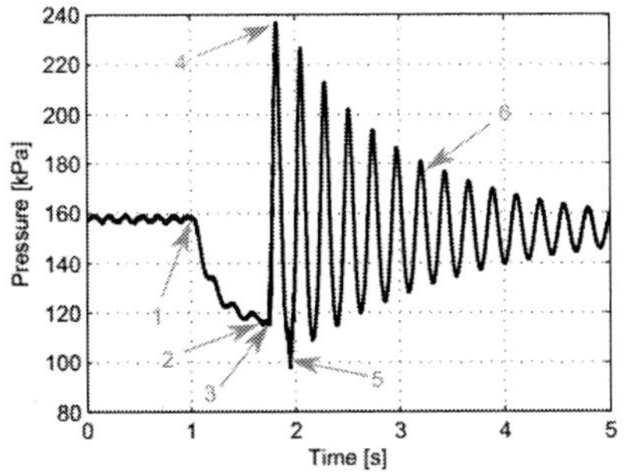


**Fig. 8.** Pressure behind the check valve during the transient

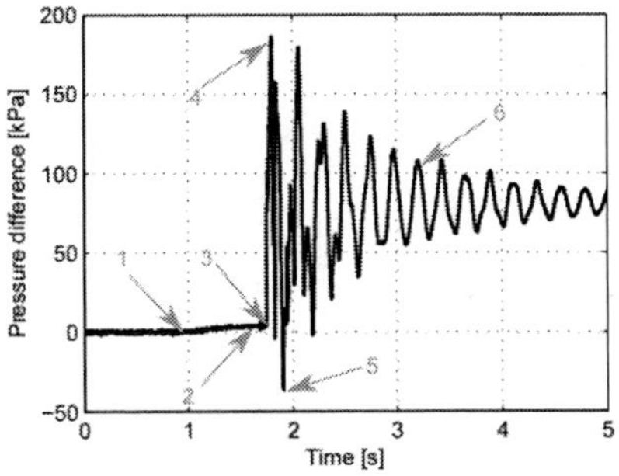


**Fig. 9.** Pressure difference on the check valve during the transient

The flow rate was evaluated from the pressure records $p_4$ and $p_6$ by the time-pressure method:

$$Q(t+\Delta t) = \frac{\Delta t S}{L\rho}\left[p_4 - p_5 - RQ^2(t)\right] + Q(t), \quad (1)$$

where the resistance $R$ is not a function of the time, but depends on the initial flow rate (2), which is subject to the numerical optimization.

$$R = \frac{p_4(0) - p_5(0)}{Q^2(0)}. \quad (2)$$

The dependence of the velocity on the time (figure 10) is computed from (1) by an appropriate numerical method, when the flow rate at the last time instant equals zero. The solution also includes the initial flow rate and resistance. Then, the start of deceleration (point 1) can be easily found as well as the time, when the flow changed its orientation (point 2). The maximal backward velocity (point 4) is the last information obtained from this graph.

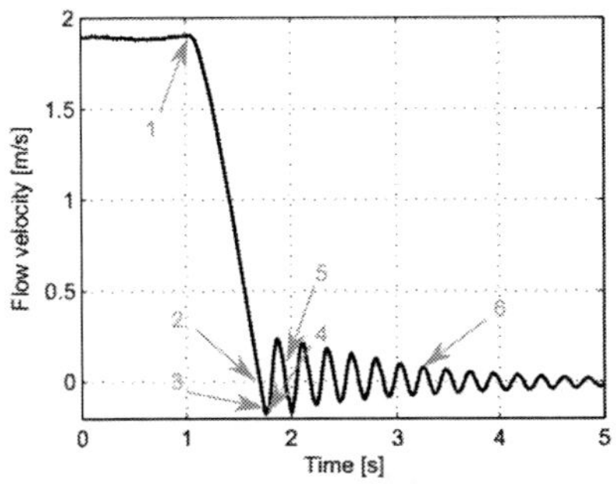


**Fig. 10.** Flow velocity during the transient

The measurement of the acceleration on the pump body also gives some information about events in the pipeline (see figure 11). The pump disconnection is not visible here (there is not any change of the signal at point 1), because all moving parts kept their moving. The first change is obvious when the speed decreases about 50%. The acceleration amplitude becomes noticeably lower 0.3 s after the pump disconnection.

Closure of the check valve disc is well visible (point 4) as well as its second closure after reopening (point 5). The pump shaft stop (point 6) is better recognizable than pump disconnection.

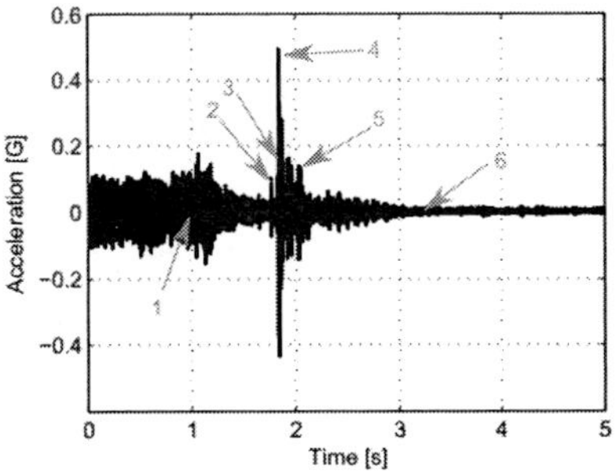


**Fig. 11.** Measurement of the acceleration on the pump body

The signal from the accelerometer on the check valve body (figure 12) contains strong noise so even the identification of the exact time of disc closure (point 4) is difficult. There is a question whether the noise came from mechanical reasons (e. g. pipeline vibrations) or from signal interference. The signal interference is the most probable, because mechanical vibrations would be visible also on the acceleration signal from pump body.

It is interesting that the strongest peak corresponds to the second pressure peak downstream of the check valve (see figure 8), but, at the same moment, the pressure upstream of the check valve is very low. It is the time, when the disc closes for the second time.

 DOI: 10.1051/epjconf/201714302036

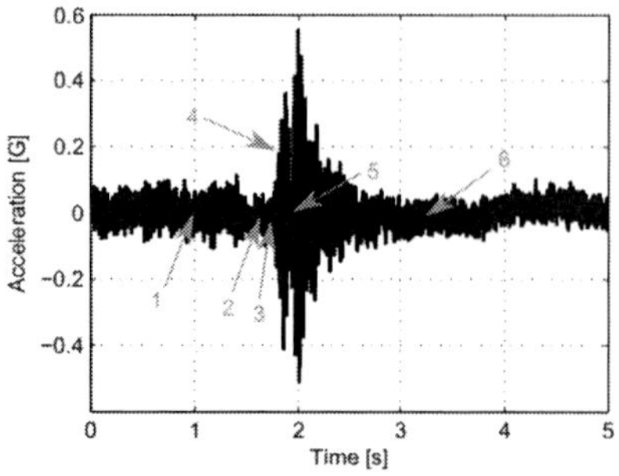


**Fig. 12.** Measurement of the acceleration on the check valve body

**Table 2.** Computed deceleration and the maximal reverse velocity

| Initial flow velocity ($m\,s^{-1}$) | Deceleration ($m\,s^{-2}$) | Reverse velocity ($m\,s^{-1}$) |
|---|---|---|
| 1.3 | 2.4 | 0.15 |
| 1.9 | 2.9 | 0.22 |
| 2.5 | 3.5 | 0.26 |
| 3.4 | 4.1 | 0.34 |
| 3.8 | 4.7 | 0.43 |

The system deceleration can be easily computed from the figure 10 using points 1 and 2. Maximal reverse flow velocity can be read at the same figure. These data are listed in the table 2 for different initial flow velocities.

Figure 13 shows measured data compared with data provided by [9]. The obtained characteristics is found between characteristics of swing check valve and swing flex check valve.

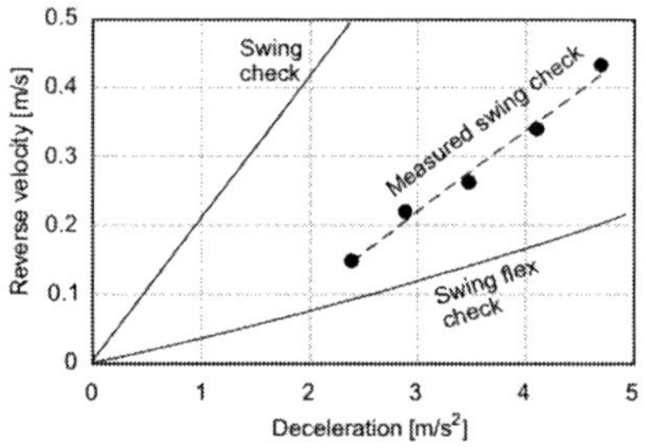


**Fig. 13.** Maximal reverse velocity dependence on the system deceleration

## 4 Discussion

Following list summarizes important events in the system (the numbering corresponds to the numbers in the figures):

1. **Pump disconnection** – the pump starts decelerating, which visible on the direct measurement of pump revolutions and pressure, because the pressure change is proportional to the second power of speed change. This point is also well recognizable on the computed flow velocity, but acceleration on the pump body does not give any clue.
2. **Change of the flow orientation** – can be identified from the computed flow velocity. A low peak can be found also in the signal from the pressure transducer upstream of the check valve.
3. **Start of the disc closure** – is easily visible on the pressure record at both sides (up- and downstream) of the check valve.
4. **End of the disc closure** – is easily visible on the pressure record and also on the computed flow rate. It corresponds to the maximal reverse flow velocity. This point could be also identified on the acceleration of the pump body, but, surprisingly, it was very difficult to find exact point also on the acceleration of the check valve body. This record is very noisy.
5. **Disc reopening and closing** – can be found on the pressure signals and it is visible also on the acceleration of the pump body.
6. **Pump shaft stop** – is definitely recognizable with scanning the shaft, but its speed before the stop is questionable, because only one mark on the shaft has been used. It is also possible to identify this point on the signal from the accelerometer on the pump body and on the pressure at downstream side of the check valve.

Results show that the higher initial flow the more intense is deceleration after pump disconnection. It can be caused by higher friction loss. It is also apparent that the higher deceleration the higher reverse velocity develops before the check valve closing, because the disc shuts with longer delay. The check valve might not be fully open for lower flow rates, so closes faster. Another explanation could be that higher deceleration causes swirls in the check valve space. These swirls might support the disc and cause delayed closing.

## 5 Conclusion

The paper is focused on the measurement of dynamic characteristic of the swing check valve. An experimental circuit has been built and the swing check valve behaviour was observed during transient events occurring after pump disconnection from the electric network.

Investigated check valve is first of a row of check valves with different constructions, which are going to be tested. This test was supposed to show whether we are able to measure dynamic characteristic of the check valve and if it is possible to describe transient event occurring in the system. The following experiments are going to include also recording of the check valve disc motion.

## Acknowledgement

This work has been supported by Technology Agency of the Czech Republic under the project Innovative research of check valves for extreme operating conditions in energetics TH01011352.

## Nomenclature

| | | |
|---|---|---|
| $L$ | (m) | pipe length between pressure transducers, |

DOI: 10.1051/epjconf/201714302036

| | | |
|---|---|---|
| $p$ | (Pa) | pressure, see figure 5 |
| $Q$ | ($m^3\ s^{-1}$) | flow rate, |
| $R$ | (Pa $s^2\ m^{-6}$) | resistance |
| $S$ | ($m^2$) | pipe cross-section, |
| $t$ | (s) | time, |
| $\Delta t$ | (s) | time step, |
| $\rho$ | (kg $m^{-3}$) | density. |

## References

1. B. W. Karney, A. R. Simpson, Journal of Hydraulic Research **45** 547 – 554 (2007)
2. C. Chiu, American Nuclear Society **54**, 289 – 291 (1987)
3. J. R. Travis, M. D. Torrey, *American Society of Mechanical Engineers winter annual meeting*, 1 – 9 (Miami 1985)
4. D. Himr, V. Habán, P. Dokoupil EPJ Web of Conferences **114** (2016) 02038
5. D. Himr, V. Habán, Applied Mechanics and Materials **630**, 375 – 382 (2014)
6. K. L. McElhaney, Nuclear Engineering and Design **197**, 169 – 182 (2000)
7. J. V. Ballun Journal of AWWA **99**, 3 (2007)
8. W. Rahmeyer *Nuclear Industry Check Valve Group, 1996 Winter Meeting*, 1 – 10 (St. Petersburg1996)
9. Val-Matic Valve and Manufacturing Corp. *Dynamic Characteristics of Check Valves* (2003)
10. J. Jablonská, M. Kozubková, EPJ Web of Conferences **114** (2016) 02049
11. M. Vašina, L. Hružík, A. Bureček, EPJ Web of Conferences **114** (2015) 02131

# Radial basis functions in mathematical modelling of flow boiling in minichannels

*Sylwia* Hożejowska[1,*], *Leszek* Hożejowski [1] and *Magdalena* Piasecka[2]
[1] Faculty of Management and Computer Modelling, Kielce University of Technology, Al. 1000-lecia P.P. 7, 25-314 Kielce, Poland
[2] Faculty of Mechatronics and Mechanical Engineering, Kielce University of Technology, Al. 1000-lecia P.P. 7, 25-314 Kielce, Poland

**Abstract.** The paper addresses heat transfer processes in flow boiling in a vertical minichannel of 1.7 mm depth with a smooth heated surface contacting fluid. The heated element for FC-72 flowing in a minichannel was a 0.45 mm thick plate made of Haynes-230 alloy. An infrared camera positioned opposite the central, axially symmetric part of the channel measured the plate temperature. K-type thermocouples and pressure converters were installed at the inlet and outlet of the minichannel. In the study radial basis functions were used to solve a problem concerning heat transfer in a heated plate supplied with the controlled direct current. According to the model assumptions, the problem is treated as two-dimensional and governed by the Poisson equation. The aim of the study lies in determining the temperature field and the heat transfer coefficient. The results were verified by comparing them with those obtained by the Trefftz method.

## 1 Introduction

Recently, research on heat transfer has focused on phenomena occurring in minichannels so that the solutions can be used in cooling systems for electronic devices. A review of literature relating to flow boiling heat transfer in minichannels was presented in [1-4].

The recent years have shown a dynamic development of meshless numerical methods which do not require troublesome and time-consuming mesh generation like e.g. Finite Element Method (FEM). Two typical examples could be the Trefftz method [5-10] and the Kansa method [11,12]. Analytical-numerical Trefftz method consists in approximating the solution of a partial differential equation with a linear combination of certain basis functions (named T-complete functions) satisfying the given equation. Similarly, in the Kansa method the solution is expressed by a linear combination of the radial basis functions, dependent on the distance of the argument from the chosen points called the centres. In order to specify the unknown coefficients of linear combinations we have to use (in both methods) the prescribed boundary conditions and also (in the Kansa method) the governing equation. The present study employed both these methods for determining two-dimensional distributions of the heating surface temperature. Thus the calculations (by both methods) based on empirical input data gave two alternative, but only slightly different, temperature distributions. Having determined the heating surface temperature distribution, one could estimate the local heat transfer coefficients at the interface between the heated plate and the working fluid.

## 2 Experimental background

Experimental data was collected in the measurement stand whose essential part is a tested module with a vertical minichannel 1.7 mm deep, 16 mm wide and 180 mm long, asymmetrically heated. The heated element for a boiling liquid flowing in a minichannel was a 0.45 mm thick plate made of Haynes-230 alloy.

The most important elements of the measurement stand are presented in figure 1a, the cross-section of the test module with a minichannel is shown in figure 1b. The experimental setup was discussed in [1-4, 13].

The working fluid circulating in a flow loop is FC-72 Fluorinert. The temperature of the outer side of the heated plate was measured by an infrared camera. The camera is positioned opposite the central, axially symmetric part of the minichannel. The plate surface was coated with a black paint with known emissivity [3]. Thermal accuracy of the infrared FLIR E60 camera is ± 1°C or ± 1% within the temperature range of 0÷120°C. K-type thermocouples and pressure converters were installed at the inlet and outlet of the minichannel. The two-phase flow patterns on the plate surface in contact with the fluid in the minichannel were observed through a glass pane, using digital SLR camera and illuminated by high power LEDs.

* Corresponding author: ztpsf@tu.kielce.pl

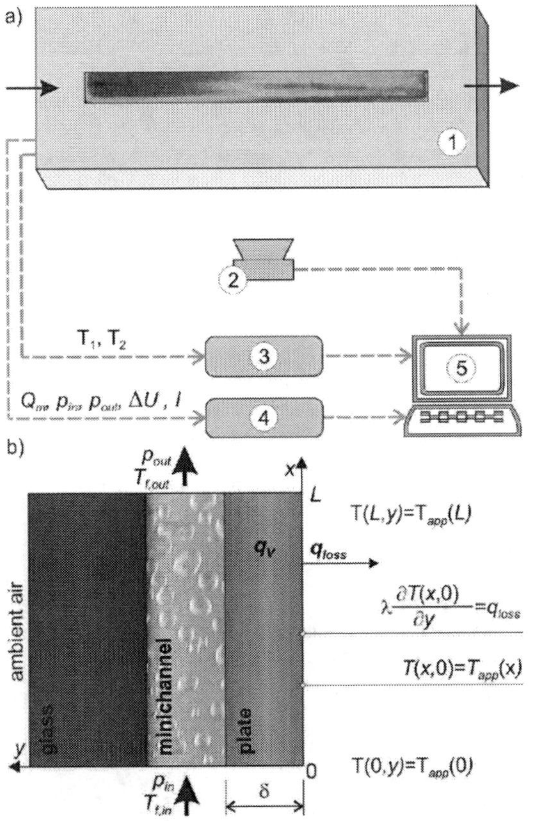


**Fig. 1.** The schematic diagrams of: a) the measurement stand, b) cross-section of the tested module with a minichannel;
1 - test module with a minichannel, 2 - infrared camera,
3, 4-data acquisition station, 5- computer with special software,
$L$ - minichannel length, $\lambda$- heated plate conductivity,
$q_V$- volumetric heat flux, $\delta$ - heated plate thickness,
experimental parameters: $Q_m$ - mass flow rate,
$p$ - pressure, $I$ - current, $\Delta U$- voltage drop.

## 3 Mathematical model and method

### 3.1. Model

The mathematical model refers the heat transfer problem in flow boiling of FC- 72 fluorinert flowing through the minichannel. The assumptions include steady state in the test module and slight temperature changes of the plate and fluid along the minichannel width. The proposed approach considered two dimensions: dimension $x$ in the direction of the flow and dimension $y$ perpendicular to the flow, related to the thickness of the heated plate ($\delta$), and to the depth of the minichannel. Further considerations refer only to the central section of the test module. The plate temperature was assumed to satisfy the Poisson equation with an adequate set of boundary conditions

$$\Delta T = -\frac{q_V}{\lambda}\ , \ (x,y)\in\Omega \tag{1}$$

where $\Omega = \{(x,y)\in R^2 : 0 < x < L,\ \ 0 < y < \delta\}$, figure 1b. The adequate boundary conditions can be written as

$$T(x,0) = T_{app}(x) \tag{2}$$

$$\lambda \frac{\partial T(x,0)}{\partial y} = q_{loss} \tag{3}$$

$$\text{a)}\ T(0,y) = T_{app}(0),\ \text{b)}\ T(L,y) = T_{app}(L) \tag{4}$$

where $T_{app}(x)$ denotes a polynomial approximating the temperature measurement of the heated plate, $q_{loss}$- heat flux density transferred from the plate into surrounding and determining as in [3]. The problem described by equation (1) with conditions (2) – (4) is an inverse heat conduction problem.

For the subcooled boiling region, when the void fraction is very low, the fluid temperature $T_f$ changes linearly from the minichannel inlet temperature $T_{f,in}$ to the outlet temperature $T_{f,out}$. For the saturated boiling region, when the bubbly or bubbly-slug boiling occurred in the considerable part of the minichannel domain, we assumed that the fluid temperature is equal the saturation temperature $T_{sat}$, [4]. Saturation temperature is determined from the assumption of linear distribution of the fluid pressure along the minichannel. The known temperature field of the fluid is used to determine the heat transfer coefficient $\alpha(x)$ at the plate - fluid interface from the following equations:

- for the subcooled boiling region

$$\alpha(x) = \frac{q_V\,\delta - q_{loss}}{T(x,\delta) - T_f(x)} \tag{5a}$$

- for the saturated boiling region

$$\alpha(x) = \frac{q_V\,\delta - q_{loss}}{T(x,\delta) - T_{sat}(x)} \tag{5b}$$

### 3.2 Numerical method

When considering some problems of mechanics, it often happens that the measurement of a physical quantity inside a region of interest is (for some reason) impossible. In such situations, one could measure the desired parameters on the boundary of a region and then try to search for their distributions (or histories) inside the region. A task like that is a typical inverse problem in which the reasons are to be inferred from the results. The present study addresses two kinds of the inverse problems considering identification of (i) the boundary temperature and (ii) the heat transfer coefficient. Generally, inverse problems are ill-posed [14] and they put great demands (concerning accuracy and stability) on the solution method.

#### *3.2.1 Radial basis function (RBF) method*

The concept of RBFs dates from the 1970s when R.L.Hardy [12] suggested using multiquadric functions in cartography for interpolating multidimensional scattered data. Since then, RBFs have found various useful applications including, among others, numerical solution of differential equations.

Consider a linear differential equation (1) subject to the boundary conditions (2) – (4). For the problem (1) – (4) we can use Kansa's collocation method [11], approximating the solution by a sum

$$T_1(x,y) \approx \sum_{j=1}^{M} a_j \phi(r_j(x,y)) \tag{6}$$

which we refer to a continuous function $\phi : [0,\infty) \to R$ as a radial basis function and where $r_j(x, y)$ denotes the Euclidean distance between $(x, y)$ and the $j^{th}$ node point $(x_j, y_j)$, while $a_j$ – the unknown coefficients.
Some of the nodes are selected to lie in the interior of $\Omega$ (for $j = 1, 2, \ldots, K$) while others on the boundary (for $j = K+1, K+2, \ldots, M$). It is worth noting that an extra term (constant or polynomial) can be added to expansion (6), this however will not be done in this study.

As to RBFs, some of the most commonly used in practical applications are the multiquadrics

$$\phi(r) = \left(r^2 + c^2\right)^{1/2} \tag{7}$$

where $c > 0$ is a shape parameter. These functions will be chosen for further processing. For determining the unknown coefficients $a_j$ we claim that the approximation (6) satisfies the governing equation (1) at interior nodes $(x_j, y_j)$ and the boundary conditions (2) – (4) at boundary nodes $(x_j, y_j)$. Thus we obtain the collocation system which we solve numerically.

### *3.2.2 Trefftz Method*

Two-dimensional temperature distribution $T_2$ of the heated surface was found by the Trefftz method with use of harmonic polynomials which are T-complete functions for Laplace's equation. More specifically, the solution of the problem (1) – (4) was being sought in the form of a linear combination of harmonic polynomials $h_i(x,y)$ with an added particular solution $\tilde{h}$ of equation (1), [7-9,15]:

$$T_2(x,y) \approx \tilde{h}(x,y) + \sum_{i=0}^{N} b_i h_i(x,y) \tag{8}$$

The unknown coefficients $b_i$ were determined on the basis of the known boundary conditions, as described in [8,15].

## 4 Results

Numerical calculations were performed for the data coming from the experiments where the volumetric heat flux $q_V$ varies from $4.51{\cdot}10^4$ kWm$^{-3}$ to $1.83{\cdot}10^5$ kWm$^{-3}$, average Reynolds number $Re$ = 2100, average inlet pressure $p_{in}$ = 110 kPa, average inlet liquid subcooling 40 K, flow velocity 0.24 m·s$^{-1}$. An infrared camera measured the temperature of the heated plate in contact with the ambient air. The temperature measurement was approximated with a 5$^{th}$ degree polynomial $T_{app}(x)$ for which the coefficient of determination $R^2$ had values no less than 0.9. Applying the Kansa method, we took $M$ = 67 and $K$ = 4. The profile of the function (7) depends on the shape parameter $c$. With the growth of $c$, the function becomes more and more flat and thus less sensitive to the changes of the distance between the point $(x, y)$ and the $j^{th}$ center $(x_j, y_j)$. Basically, the proper choice of $c$ is a problem itself, which has not yet been solved and is still being studied. The computation was performed for different values of $c$ ranging from 0.0001 to 0.1 and the results to a large extent agreed with one another. Below we present the numerical results for $c$ = 0.001. Applying the Trefftz method, there were used 10 harmonic polynomials to compute approximate temperature of the heated plate. Figure 2 shows an illustrative two-dimensional distribution of the heated plate temperature calculated with the Kansa method (figure 2a) and the Trefftz method (figure 2b).

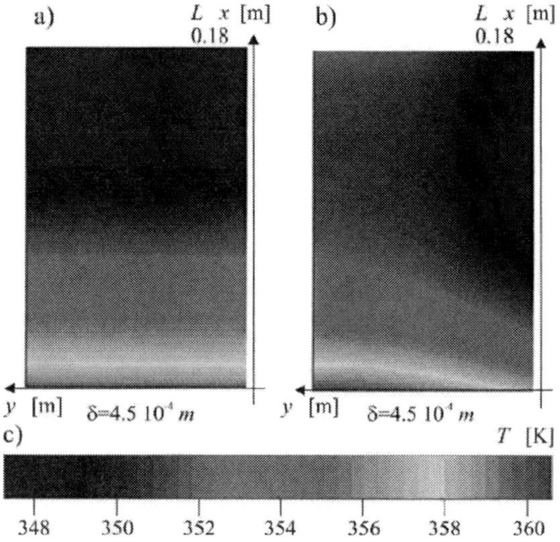


**Fig. 2.** Temperature of the heated plate obtained by:
a) Kansa method, b) Trefftz method, c) temperature scale, main experimental parameters: $q_V$=1.56·10$^5$ kWm$^{-3}$ , $p_{in}$ = 153 kPa.

In the calculation we considered two boiling regions: subcooled and saturated. It was assumed that in the subcooled boiling region the fluid temperature varied linearly from the measured value at the inlet to the value at the outlet of the minichannel, while in the saturated boiling region it was equal to local saturation temperature. Figures 3 and 4 show the heat transfer coefficient as a function of the distance from the minichannel inlet, calculated from (5), where the plate temperature was determined from equation (6) (figure 3a, figure 4a) or equation (8) (figure 3b, figure 4b).

For the subcooled boiling region, the values of the heat transfer coefficient obtained from both employed methods do not significantly differ from each other, however higher values were obtained using the Trefftz method, see figure 3. The differences between the heat transfer coefficient calculated from the Trefftz and

Kansa methods become more significant in the saturated boiling region, figure 4. The maximum relative error (MRE) of estimating the plate temperature, calculated from the formula

$$MRE = max\left\{\begin{array}{l}\left[\dfrac{\iint\limits_{\Omega}(T_1(x,y)-T_2(x,y))^2dxdy}{\iint\limits_{\Omega}(T_1(x,y))^2dxdy}\right]^{0.5}\\ \left[\dfrac{\iint\limits_{\Omega}(T_1(x,y)-T_2(x,y))^2dxdy}{\iint\limits_{\Omega}(T_2(x,y))^2dxdy}\right]^{0.5}\end{array}\right\} \quad (9)$$

ranges from 0.61% to 1.21% in the subcooled boiling region and from 0.91% to 1.24% in the saturated boiling region, for each setting of the volumetric heat flux.

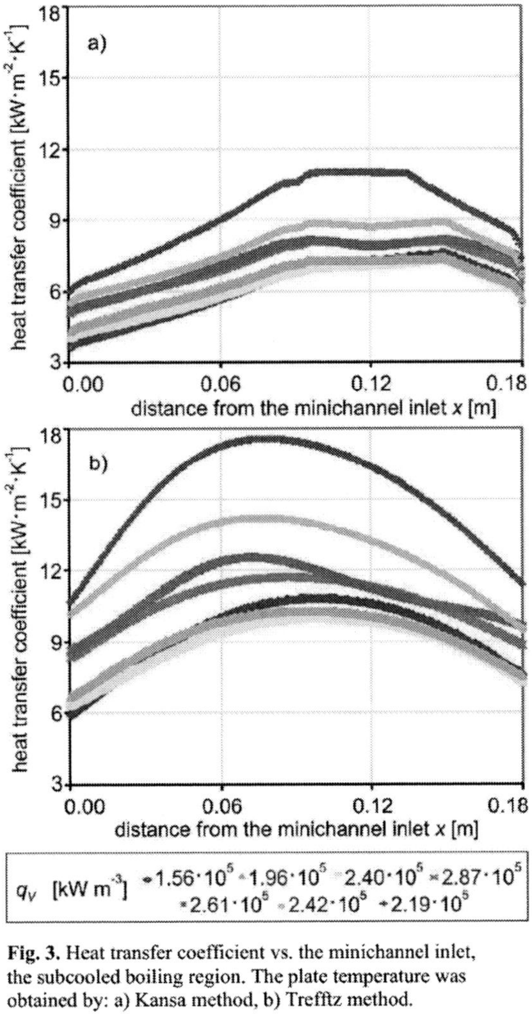


**Fig. 3.** Heat transfer coefficient vs. the minichannel inlet, the subcooled boiling region. The plate temperature was obtained by: a) Kansa method, b) Trefftz method.

The error of estimating the heat transfer coefficient turned to be less than 16.5%, no matter the plate temperature was obtained by the Kansa method or the Trefftz method. Similar estimates for the heat fluxes $q_{loss}$ and $q_V$ (in both boiling regions) have errors not exceeding 16.3% and 38%, respectively.

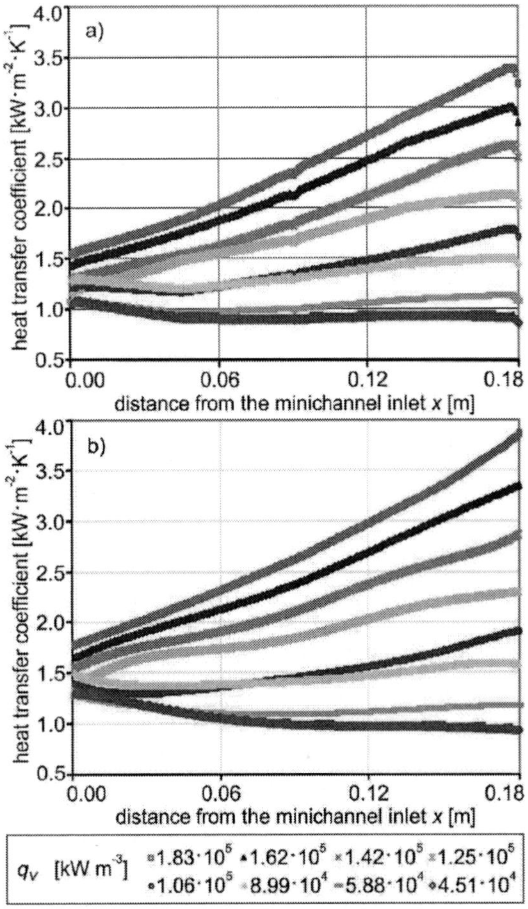


**Fig. 4.** Heat transfer coefficient vs. the minichannel inlet, the saturated boiling region. The plate temperature was obtained by: a) Kansa method, b) Trefftz method.

## 5 Conclusions

The paper focused on FC-72 flow boiling heat transfer in a vertical asymmetrically heated minichannel of 1.7 mm depth. Two numerical methods were applied to solve the considered inverse problem: the Kansa method and the Trefftz method. Both gave two-dimensional temperature distributions which agreed with each other, with MRE of the plate temperature no greater than 1.24%. Moreover, there was no significant disagreement between the heat transfer coefficients for the subcooled boiling region obtained by these methods. However greater differences between them could be observed in the saturated boiling region. Besides, in both considered boiling regions the heat transfer coefficient had higher values when its estimation was based on the Trefftz method.

**Acknowledgements**

The research reported herein was partly supported by a grant from the Polish National Science Centre (No. DEC-2013/09/B/ST8/02825).

## References

1. M. Piasecka, Int. J. Heat Mass Transf. **81**, 114-121 (2015)
2. M. Piasecka, Int. J. Refrig. **56**, 198-212 (2015)
3. M. Piasecka, K. Strąk, B. Maciejewska, Heat Transf. Eng. **38**, DOI: 10.1080/01457632.2016. 1189272 (2017)
4. M. Piasecka, B. Maciejewska, Exp. Therm. Fluid Sci. **68**, 459-467 (2015)
5. E. Trefftz, Proc. 2nd Int. Con. App. Mech. 131-137 (1926)
6. M.J. Ciałkowski, A. Frąckowiak, J. Therm. Sci. **11**, 148-162 (2002)
7. S. Hożejowska, R. Kaniowski, M.E. Poniewski, Exp. Therm. Fluid Sci. **78**, 18–29 (2016)
8. S. Hożejowska, R. Kaniowski, M.E. Poniewski, Int. J. Numer. Methods Heat & Fluid Flow **24**, 811-824 (2014)
9. S. Hożejowska, M. Piasecka, Heat Mass Transf. **50**, 1053-1063 (2014)
10. K.Grysa, A. Maciąg, A. Pawińska, Int. J. Heat Mass Transf. **55**, 7336-7340 (2012)
11. E. J. Kansa, Comput. Math. Apll. **19**, 147 – 161 (1990)
12. R. L. Hardy, J. Geop. Research. **76**, 1905–1915 (1971)
13. B. Maciejewska, K. Strąk, M. Piasecka, Procedia Eng. 157, 82-88 (2016)
14. J. Hadamard, *Lectures on the Cauchy problem in linear partial differential equations* (Yale University Press, New Haven 1923)
15. S. Hożejowska, L. Hożejowski, EPJ Web of Conf. **1**, 56-63 (2009)

# Trefftz method in solving Fourier-Kirchhoff equation for two-phase flow boiling in a vertical rectangular minichannel

*Sylwia* Hożejowska[1,*], *Magdalena* Piasecka[2], *Artur* Piasecki[3]

[1]Faculty of Management and Computer Modelling
[2]Faculty of Mechatronics and Mechanical Engineering
[3] contractor of the project from the Polish National Science Centre (No. DEC-2013/09/B/ST8/02825)
Kielce University of Technology, Al. 1000-lecia P.P. 7, 25-314 Kielce, Poland

**Abstract.** This paper presents the results of investigations into flow boiling heat transfer in an asymmetrically heated vertical minichannel of 1.7 mm depth. The heated element for FC-72 flowing in the minichannel was an alloy plate 0.45 mm thick, microstructured on one side, in direct contact with the flowing fluid. The computational part of the study contains approximate steady state solutions of the heat transfer problems described by Poisson's equation and the energy equation for the heated plate and the fluid, respectively. For both equations, the boundary conditions were specified on the basis of experimental data. Temperature of the outer plate surface, measured by infrared thermography, and heat losses to ambient air were included in the calculations. For the energy equation we assumed parabolic profile of fluid velocity and the equality of temperatures and heat fluxes at the interface between the heated surface and the fluid. The void fraction was taken from a single-phase flow model. Two-dimensional temperature distributions were obtained by the Trefftz method and, due to the Robin condition at the interface between them, it was possible to calculate the heat transfer coefficient. Its values were compared to those obtained by other correlations known from literature.

## 1 Introduction

The importance of flow boiling heat transfer applications to mini spaces of different geometries has recently increased considerably. New generations of cooling systems have been designed and their huge heat dissipation capacity has greatly increased heat removal rate per unit heat transfer area. Effective cooling schemes using flow boiling heat transfer in mini- and microchannels to maintain device temperature are being developed. The application of microstructured heated surfaces in minichannels attracts attention due to theoretical enhancement of their potential for transferring large heat fluxes. Consequently, this issue has become a promising aim of the study.

A review of literature relating to flow boiling heat transfer in minichannels with microstructured surfaces, smooth and enhanced, was presented in [1-3]. The literature fails to offer any generalized conclusions for predicting heat transfer in small channels. Focusing research on microstructured surface of heat exchangers with minichannels seems interesting because these surfaces provide a potential for further heat transfer enhancement.

In [2,3,5] the authors proposed one- and two-dimensional heat transfer model for a minichannel with an enhanced heated surface. The two-dimensional temperature distribution on the heated plate was determined by the Finite Element Method combined with Trefftz functions as shape functions (FEMT). In this paper, two sets of Trefftz functions were used to calculate 2D temperature distributions both for the heated plate and for the working fluid, FC-72. The numerical results obtained for the heat transfer coefficient were compared with the correlations reported in the literature.

## 2 Experimental test

The study was conducted for FC-72 boiling in a rectangular, vertical and asymmetrically heated minichannel with a depth (height) of 1.7 mm, a width of 18 mm and a length of 180 mm, figure 1. Perfluoro-compound FC-72 is 3M Fluorinert™ Electronic Liquid, thermally and chemically stable, compatible with sensitive materials, nonflammable, practically nontoxic and leaves essentially no residue upon evaporation. The experimental setup was presented in [1-6]. The heated element for FC-72 flowing in the minichannel (1) was the plate made of Haynes-230 alloy with thickness of 0.45 mm (2) which was microstructured on one side in direct contact with the fluid flowing in the channel. Cavities (8) in the plate surface were made by electromachining. An image of the section with cavities

* Corresponding author: ztpsf@tu.kielce.pl

is also shown in figure 1. The depth of the cavity craters is usually below 1÷2 μm [1,2,6]. It was possible to observe the plate surface through a glass pane (4) using a digital SLR camera (10). The camera allowed conducting two-phase flow visualizations on the microstructured plate side. K-type thermocouples (7) and pressure transducers were installed at the inlet and outlet of the minichannel. The temperature of the smooth side of the heated plate was measured by FLIR E60 infrared camera (9) in the central axially symmetric part of the minichannel. The plate was coated with a black paint (3) with a known emissivity [2,5]. Thermal accuracy of the infrared camera was ± 1°C or ± 1% ± 1 °C or ± 1% within the temperature range of 0÷120°C (temperature used in the experiment was within this range).

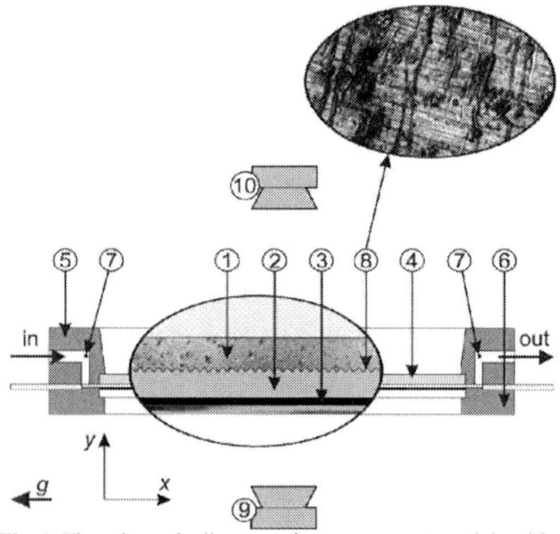


**Fig. 1.** The schematic diagram of a measurement module with a minichannel, 1-minichannel, 2-heated plate, 3-black paint, 4-glass panel, 5-channel body, 6-front cover, 7-thermocouple, 8-microstructured surface with cavities, image of the fragment with cavities, 9-infrared camera, 10-SLR camera.

## 3 Mathematical model and calculations

We assumed that physical phenomena on the side edges of the module did not affect the thermodynamic parameters in the segment of concern. From here our considerations focused on the central part of the measurement module for which only two dimensions were taken into account: dimension $x$ along the flow, and $y$ relating to the thickness of the module composed of the heated plate and the minichannel. The steady state heat transfer process in the heated surface was described by two-dimensional Poisson's equation with an adequate system of boundary conditions, as shown in [2,5]. It was assumed that the flow of FC-72 was laminar and stationary with a constant mass flux density and that its flow velocity in the minichannel has one component parallel to the heated plate (the other component takes the value of zero). The temperature of liquid at the interface with the heated surface depends on the temperature of the heated plate ($T_H$) and the temperature of saturation at a given pressure ($T_{sat}$) in the way presented in [7]. In addition, the liquid temperature at the minichannel inlet ($T_{l,in}$) and outlet ($T_{l,out}$) is known. The two-phase mixture that formed in the minichannel contained a gaseous phase and the liquid phase in proportions dependent on void fraction $\varphi(x)$. For the two-phase flow it is assumed that the entire heat flux from the heated plate is transferred to the liquid phase in proportions dependent on the void fraction

$$\lambda_l \frac{\partial T_l}{\partial y} = \lambda_H \left(1 - \varphi(x)\right) \frac{\partial T_H}{\partial y} \tag{1}$$

where $\lambda$ denotes thermal conductivity. The void fraction $\varphi(x)$ was determined from the Spedding correlation [8]

$$\varphi(x) = \left(1 + 2.22 \left(\frac{1 - X(x)}{X(x)}\right)^{0.65} \left(\frac{\rho_v}{\rho_l}\right)^{0.65}\right)^{-1} \tag{2}$$

Local thermodynamic vapour quality $X(x)$ was determined from the formula [9]

$$X(x) = \frac{(q_F - q_{loss})(x - x_{sat})}{W G h_{lv}} \tag{3}$$

where the heat flux generated by the heated plate $q_F$ was reduced by the heat flux transferred to the environment $q_{loss}$, [2]. In (3) $W$ is the heated plate width, $G$ – mass flux, $h_{lv}$ – heat of vaporization, $\rho$ - density, $x$ – the distance from the heated section starting point, subscripts: $l$ - is intended for liquid phase and $v$ - for vapour phase. The location where boiling starts $x_{sat}$ was given by the formula

$$x_{sat} = \frac{G A c_p \left(T_{sat} - T_{l,in}\right)}{\left(q_F - q_{loss}\right) L} \tag{4}$$

where $L$ is the minichannel length, $A$ – field, $G$ – mass flux, $c_p$– specific heat. Where the heated plate temperature and the liquid temperature distributions and gradient $\lambda_H \partial T_H / \partial y$ were known, the heat transfer coefficient $\alpha(x)$ at the plate-liquid interface could be determined using the condition

$$\alpha(x)_{exp} = \frac{-\lambda_H \frac{\partial T_H}{\partial y}(x, \delta_H)}{T_H(x, \delta_H) - T_{l,ave}(x)} \tag{5}$$

where $\delta_H$ is the plate thickness. The reference temperature $T_{l,ave}$ was taken as the mean liquid temperature in the interval $\langle 0, 0.5\delta_M \rangle$ where $\delta_M$ is the minichannel depth, [9]. The unknown approximations of the heated plate and liquid temperatures were determined using the Trefftz method [10]. First, the approximation of the plate partition temperature $T_H$ was found by solving the inverse heat conduction problem in the procedure described in [11]. These approximations were calculated using the Trefftz functions for Laplace's equation. Then, the approximation of the liquid temperature $T_l$ was determined by solving another inverse problem.

The Trefftz functions for energy equation with the parabolic liquid velocity were used to determine liquid temperature [9, 11]. The solutions obtained with Trefftz method fulfil exactly the governing differential equation are continuous functions and has a differentiable form.

## 4 Results

Figure 2 shows the heat transfer coefficient, calculated from (5), which depends on the distance along the minichannel length and on the increasing heat flux supplied to the heated plate. Moreover, figure 3 shows the heat transfer coefficient versus the vapour quality calculated according to equation (3), obtained for four distances from the channel inlet: 0.045 m, 0.09 m, 0.135 m and 0.153 m.

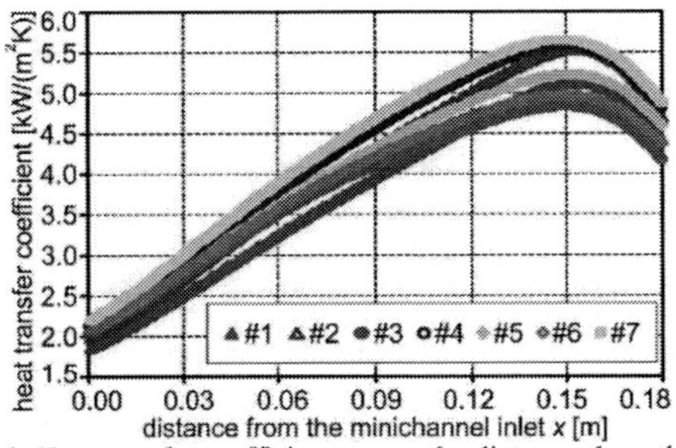


**Fig. 2.** Heat transfer coefficient versus the distance along the length of the minichannel, experimental parameters (average): mass flux 420 kg/(m$^2$ s), inlet pressure 110 kPa, inlet liquid subcooling 40 K, heat flux densities: #1) $q_w$ = 58 kW/m$^2$, #2) $q_w$ = 66 kW/m$^2$, #3) $q_w$ = 76 kW/m$^2$, #4) $q_w$ = 86 kW/m$^2$, #5) $q_w$ = 90 kW/m$^2$, #6) $q_w$ = 94 kW/m$^2$, #7) $q_w$ = 102 kW/m$^2$.

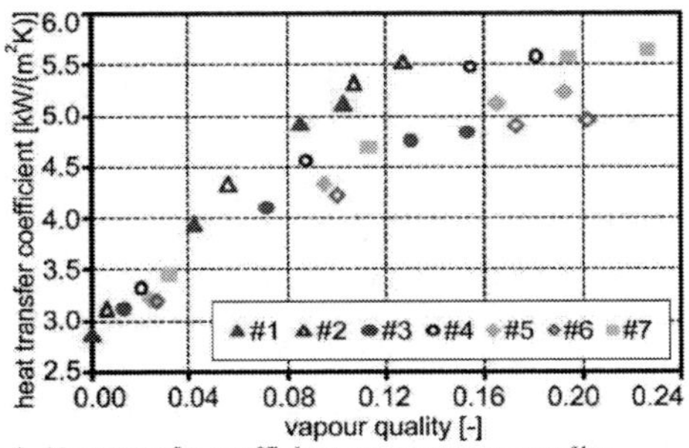


**Fig. 3.** Heat transfer coefficient versus vapour quality, experimental parameters as for figure 2, data for 0.045 m, 0.09 m, 0.135 m and 0.153 m distances from the minichannel inlet.

The experimental data show a monotonic increase in the values of local heat transfer coefficient with the distance from the minichannel inlet up to the distance of 0.15 m. At that point a drop in the coefficient value is observed.

## 5 Comparative analysis with other correlations

To compare the data obtained, five correlations known from literature were selected: the Lazarek and Black correlation [12], the Liu and Winterton correlation [13], the Warrier et al. correlation [14], the Sun and Mishima correlation [15] and the Bertsch et al. correlation [16], presented in [1]. All the selected correlations were dedicated to flow boiling heat transfer in channels.

In the correlations presented below, dimensionless numbers should be computed according to the equations known from literature; $d_h$ is a hydraulic diameter, the other quantities were defined earlier in this paper.

### 5.1 Lazarek and Black correlation

Lazarek and Black [12] proposed the following correlation:

$$\alpha = \left(30\ Re_l^{\ 0.857}\ Bo^{0.714}\right)\left(\lambda_l / d_h\right) \tag{6}$$

where *Re* - Reynolds number, *Bo* - Boiling number.

### 5.2 Liu and Winterton correlation

Liu and Winterton [13] presented the correlation in the following form:

$$\alpha_{th} = \sqrt{\left(F\ \alpha_l\right)^2 + \left(S\ \alpha_{nb}\right)^2} \tag{7}$$

where

$$F = \left[1 + X \cdot Pr_l \cdot \left(\rho_l / \rho_v - 1\right)\right]^{0.35} \tag{8}$$

$$S = \left(1 + 0.055 \cdot F^{0.1} \cdot Re_l^{0.16}\right)^{-1} \tag{9}$$

$\alpha_l$ - should be calculated from the Dittus-Boelter correlation

$$\alpha_l = \left(0.023\ Re_l^{\ 0.8}\ Pr_l^{\ 0.4}\right)\left(\lambda_l / d_h\right) \tag{10}$$

$\alpha_{nb}$ - should be determined according to the Cooper correlation [17].

### 5.3 Warrier et al. correlation

The Warrier et al. correlation has the following form [14]:

$$\alpha = \left[1 + 6Bo^{1/16} - 5.3\left(1 - 855Bo\right)X^{0.65}\right]\alpha_l \tag{11}$$

where *Bo* - Boiling number, *X* - vapour quality, $\alpha_l$ - should be calculated from the Dittus-Boelter correlation - eq. (10).

### 5.4 Sun and Mishima correlation

The Sun and Mishima correlation [15] was based on the Lazarek and Black correlation

$$\alpha_{th} = \frac{6Re_l^{1.05}\,Bo^{0.54}}{We_l^{0.191}\,(\rho_l/\rho_v)^{0.142}}\frac{\lambda_l}{d_h} \quad (12)$$

where $We$ - Weber number, other quantities such as density and similarity numbers as listed above.

### 5.5 Bertsch et al. correlation

Bertsch et al. [16] proposed the following dependence:

$$\alpha_{th} = S\alpha_{nb} + F\alpha_{SP} \quad (13)$$

where

$$S = 1 - X \quad (14)$$

$$\alpha_{SP} = \alpha_v + (1 - X)\alpha_l \quad (15)$$

$$F = 1 + 80(X^2 - X^6)\exp(-0.6Co) \quad (16)$$

- for turbulent flow $\alpha_l$ and $\alpha_v$ should be calculated from the Dittus-Boelter correlation, eq. (10),
- for laminar flow:

$$\alpha_{th\_v/l} = \frac{\lambda_{v/l}}{d_h}\left[3.66 + \frac{0.0668Re_{v/l}Pr_{v/l}\,d_h/L}{1 + 0.04(Re_{v/l}Pr_{v/l}\,d_h/L)^{2/3}}\right] \quad (17)$$

In eq. (17) dimensionless numbers should be computed according to the equations known from the literature.

Figure 4 compares the experimental heat transfer coefficient $\alpha_{exp}$ to the results of calculations based on theoretical correlations $\alpha_{th}$ (discussed above). The data shown in figure 3 were chosen as experimental heat transfer coefficients used for the comparison.

The best agreement was achieved by using the Liu and Winterton correlation within the margin of error of ±50 %. Good compliance was also found when the Bertsch et al. correlation was used.

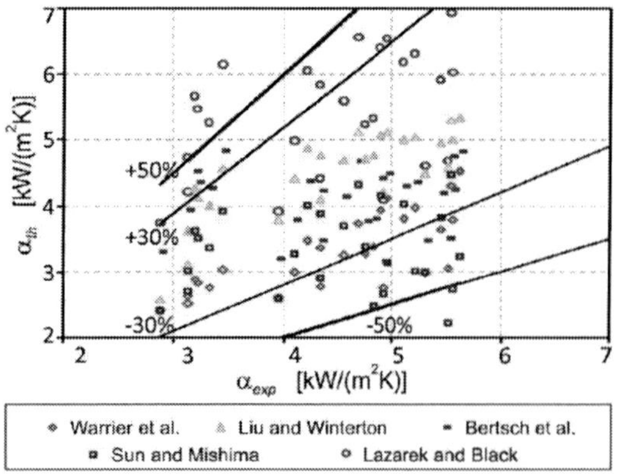


**Fig. 4.** Comparison of the experimental heat transfer coefficient with theoretically determined values according to selected correlations from literature.

## 6 Conclusions

The results for flow boiling heat transfer in an asymmetrically heated vertical minichannel were presented. The proposed mathematical model described well the heat transfer phenomenon in flow boiling in a vertical minichannel. The Trefftz method applied effectively computes two-dimensional temperature distributions in flow boiling in minichannels. The heat transfer coefficients determined based on the experimental data are consistent with the correlations known from the literature. The best agreement was achieved by using the Liu and Winterton correlation within the margin of error of ±50 %. Good compliance was also found when the Bertsch et al. correlation was used.

**Acknowledgments**

The research reported herein was partly supported by a grant from the Polish National Science Centre (No. DEC-2013/09/B/ST8/02825).

## References

1. M. Piasecka, Int. J. Refrig. **56**, 198-212 (2015)
2. M. Piasecka, K. Strąk, B. Maciejewska, Heat Transf. Eng. **38**, DOI: 10.1080/01457632.2016. 1189272 (2017)
3. M. Piasecka, B. Maciejewska, Exp. Therm. Fluid Sci. **68**, 459-467 (2015)
4. M. Piasecka, Int. J. Heat Mass Transf. **81**, 114-121 (2015)
5. B. Maciejewska, K. Strąk, M. Piasecka, Procedia Eng. **157**, 82-88 (2016)
6. R. Pastuszko, M. Piasecka, J. Physics Conf. Ser. **395**, paper No. 012137 (2012)
7. S. Hożejowska, M. Grabowski, EPJ Web of Conf. **114**, 02041 (2016)
8. P.L. Spedding, D.R. Spence, Int. J. Eng. Fluid Mech. **2**, 109–118 (1989)
9. S. Hożejowska, R. Kaniowski, M.E. Poniewski, Exp. Therm. Fluid Sci. **78**, 18–29 (2016)
10. E. Trefftz, Proc. 2nd Int. Con. App. Mech. 131-137 (1926)
11. S. Hożejowska, R. Kaniowski, M.E. Poniewski, Int. J. Numer. Methods Heat & Fluid Flow **24**, 811-824 (2014)
12. G.M. Lazarek, S.H. Black, Int. J. Heat Mass Transf. **25**, 945-960 (1982)
13. Z. Liu, R.H.S. Winterton, Int. J. Heat Mass Transf. **34** 2759-2766 (1991)
14. G.R. Warrier, V.K. Dhir, L.A. Momoda, Exp. Therm. Fluid Sc. **26**, 53-64 (2002)
15. L. Sun, K. Mishima, Int. J. Heat Mass Transf. **52**, 5323-5329 (2009)
16. S.S. Bertsch, E.A Groll, S.A. Garimella, Int. J. Heat Mass Transf. **51**, 4775-4787 (2009)
17. M.G. Cooper, Adv. Heat Transf. **16**, 157-239 (1984)

# Thermodynamic aspects of an LNG tank in fire and experimental validation

*Corina* Hulsbosch-Dam*, *Bilim* Atli-Veltin, *Jerry* Kamperveen, *Han* Velthuis, *Johan* Reinders, *Mark* Spruijt and *Lex* Vredeveldt

TNO, Van Mourik Broekmanweg 6, 2628 XE Delft, The Netherlands

**Abstract.** Mechanical behaviour of a Liquefied Natural Gas (LNG) tank and the thermodynamic behaviour of its containment under extreme heat load – for instance when subjected to external fire source as might occur during an accident - are extremely important when addressing safety concerns. In a scenario where external fire is present and consequent release of LNG from pressure relief valves (PRV) has occurred, escalation of the fire might occur causing difficulty for the fire response teams to approach the tank or to secure the perimeter. If the duration of the tank exposure to fire is known, the PRV opening time can be estimated based on the thermodynamic calculations. In this paper, such an accidental scenario is considered, relevant thermodynamic equations are derived and presented. Moreover, an experiment is performed with liquid nitrogen and the results are compared to the analytical ones. The analytical results match very well with the experimental observations. The resulting analytical models are suitable to be applied to other cryogenic liquids.

## 1 Introduction

Liquefied natural gas (LNG) is the cleanest burning fossil fuel and the demand for its use and transportation has been increasing in the last years due to the demanding reduced emission requirements on land and sea. For heavy duty truck fleets use of LNG as fuel is financially and environmentally very attractive, and nowadays many commercial options for vehicles working with LNG are available [1]. Similarly, in marine applications, there is significant number of examples of inland waterway vessels and short sea vehicles mainly in the Netherlands and USA [2]. Meanwhile, risks associated with the storage and transportation of large amounts of cryogenic liquid – not only LNG - have been the topic of discussions. Fire response of LNG tanks is one of the topics that harbour many questions, one of which is the opening time of the pressure relief valves (PRVs). The opening time of the PRV located on an LNG tank that is subjected to external fire is crucial information that fire response teams would like to have before approaching an accident. One of the first steps to reach that information is by investigating the thermodynamics of a cryogenic liquid tank that is subjected to external heat source.

The heat transfer calculation in a scenario where the outside temperatures reach more than 900°C and the inner liquid is at -163°C is a complex topic. LNG storage and transport tanks are usually double walled cylindrical structures made of stainless steel. The annular space between the walls is filled with insulation material, in combination with vacuum (low) pressure. The most common tank design consists of a perlite insulation at vacuum pressures. More recent designs, incorporate multilayer insulation (MLI) technology for insulation [3].

The operating pressures of cryogenic tanks are usually below 10 bar. Due to the fact that some heat transfer between the outside and the inside is always present, constant evaporation of the LNG occurs forming boil-off gas (BOG). Therefore, in an LNG tank, almost always vapour and liquid are present. Detailed design of tanks varies with manufacturer; however one of the common target is limiting the number of heat leak points between the inner and outer tanks. Stationary LNG tanks are generally supported with balsa wood located in the insulation space, and mobile tanks have skirts to limit the motion of the inner tank and to transfer the loads of the weight of the inner tank and the LNG to the outer tank.

In case of an external fire, heat transfer to the inner tank, and therefore the BOG, increases. Increasing gas in a limited space causes inner pressures to rise. PRVs are usually set to release this gas for safety reasons, circumventing a violent burst of the tank. In a scenario where external fire is present and consequent release of LNG from PRV has occurred, escalation of the fire might occur causing difficulty for the fire response teams to approach the tank or to secure the perimeter. If the duration of the tank exposure to fire is known, given the intensity of the fire, the PRV opening time can be estimated based on the thermodynamic calculations. In this paper, such a scenario is considered; relevant thermodynamic equations are derived and presented. Moreover, an experiment is performed with liquid

* Corresponding author: corina.hulsbosch@tno.nl

nitrogen and the results are compared to the analytical ones. The resulting analytical models are suitable to be applied to other cryogenic liquids.

## 2 Experiments

Any tank filled with refrigerated or liquefied gas is susceptible to heat influx, due to the storage conditions and physical properties of the substance inside the tank. Natural gas (methane) has to be refrigerated below its critical point of -82.6°C (at 45 bar) to be liquefied to LNG (approx. -163°C at 1 bar). Transport and storage tanks for cryogenic fluids (e.g. LNG) are provided with thermal insulation to prevent or at least minimize the heat influx from a source in the surroundings – ambient air, solar radiation, and fire exposure – into the liquid. A temperature increase of the liquid will result in a pressure increase in the tank. Pressures above the maximum allowed working pressure (MAWP) should be prevented by activation of installed PRV. In general, a PRV is a spring loaded safety valve. Furthermore, the yield and tensile strengths of the pressurized shell, carbon or stainless steel construction, will start to decrease at temperatures above 300°C [4]. Progressive weakening of the shell strength could lead to catastrophic tank rupture above its design and test pressures.

The resistance of a pressurized tank against fire exposure cannot be demonstrated by separately testing the elements of the fire protection system. The tank construction (wall thickness, insulation) and its safety accessories (heat resistant coating, double walls, PRV) need to be tested as a whole and subjected to heat radiation or direct flame impingement in order to bring the stress conditions on materials. Heat load tests where the tank is fully engulfed in flames are called bonfire tests. In 2015, a bonfire test has been performed by TNO at BAM facilities in Germany [5]. The goal of this test was to study the thermodynamic and mechanical behaviour of a partially filled tank in fire.

A double-walled tank with an inner volume of 3 $m^3$ (inner diameter: 1.2 m, outer diameter: 1.6 m; length: 2.384 m) was placed in fire. The tank consisted of an inner tank and an outer tank. The test tank was made of low temperature resistant stainless steel (type 304) to contain cryogenic fluids (LIN, LNG, etc.). The wall thickness for both tanks was 3 mm, which is thinner than for commercially available tanks (~6 mm). The thickness of the end caps of the outer tank were 5 mm. This also reduces the maximum allowed working pressure from 10 bar to 7 bar. A drawing of the tank can be seen in Fig. 1.

The inner and outer tanks were separated by 20 cm of perlite, which was the insulation material chosen for this experiment. The volume filled with perlite was vacuumed to 200 mbar. (For commercial tanks this annular space is usually ~ 10 mbar or below, in which case the thermal conductivity improves significantly.) The pressure inside the tank was monitored during the test. In addition, the temperature of the outer tank wall was monitored. Furthermore, thermocouples are placed to measure the temperature of the fluid (3 locations) and the inner wall (7 locations).

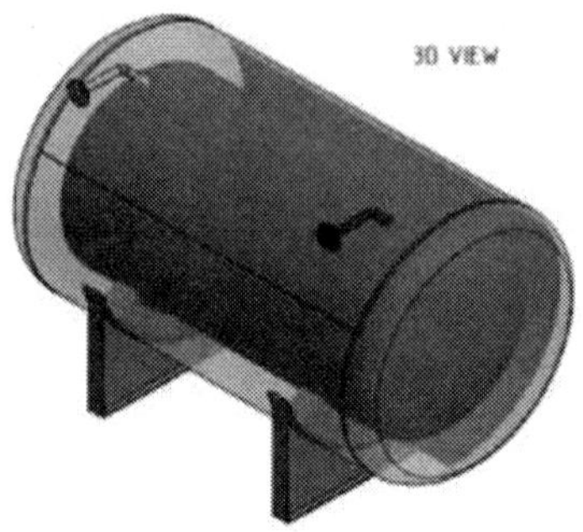


**Fig. 1.** 3D schematic view of the double wall tank used at the fire experiments

For safety, liquid nitrogen (LIN) was used in the experiments described in the current paper. The tank was filled with 1400 kg of LIN, which corresponds to a filling level of 66%. The filling level of the tank was controlled by the instrumentation on the filling truck as well as the load cells placed under the tank. After some settling time, the fire was initiated and the tank was completely engulfed in fire within seconds. The flame temperatures reached up to 800 - 900 °C at the bottom of the tank, and 450-600 °C at the top. An impression of the test is given in **Fig. 2**.

**Fig. 2.** Double walled tank containing cryogenic liquid engulfed in fire during the experiment

A double walled LNG tank that is engulfed in fire will have an outer tank temperature that is beyond 600°C within minutes from the start of the fire. Mechanical strength of the mild and stainless steel at these high temperatures is poor. Depending on the compressive performance of the insulation layer and the presence (or not) of the stiffening rings on the outer tanks interior, buckling of the outer tank is the most likely failure mode of the outer tank. The driving force to such an event is the vacuum (very low pressure) on the interior face and the atmospheric pressure on the exterior face of the external tank. The subsequent behaviour of the tank depends on the scenario; for instance a case where the PRV is not working properly, or there is damage on the inner tank which may result in different behaviour. In case of an accidental failure of the outer tank and a consequent loss of vacuum (or any other cause to an increase of the tank's overall thermal conductivity) the LNG in the inner tank will warm up faster than during normal use, and therefore its vapour pressure will increase faster. Considering such a scenario, the failure mode of the inner tank could be burst or rupture, at a weak spot due to increase of the inner pressure. Before performing the experiments, the burst pressure of the tank was calculated, and the tests were stopped when the inner pressures reached a predetermined value, which was below the calculated burst pressure.

## 3 Numerical models

Two different numerical models are described. The first is an equilibrium model which assumes that the vapour and liquid in the tank have the same temperature and pressure. The second model is a non-equilibrium model which allows temperature difference between the liquid and the vapour. The non-equilibrium model is the more realistic model of the two, however, it also requires more input parameters.

The developed models, in principle, can be applied to any (cryogenic) substance such as LIN or LNG. In order to be able to compare the results of the models to the experiments, the properties of LIN were used in the numerical models. The models use thermodynamic properties of LIN (and LNG) which were taken from the NIST database [6] for the equilibrium model and from the coolprop database [7] for the non-equilibrium model. The following section describes the difference of these two conditions.

### 3.1 Equilibrium model

The simplest model to describe the tank is an equilibrium model in which the liquid and vapor are considered as one node, i.e. the liquid and vapor have the same temperature and pressure. A schematic representation is given in **Fig. 3**. In this figure, $m$ is the mass inside the tank, $V$ is the tank volume, $T$ is the temperature of the tank contents, $p$ is the pressure inside the tank, $m_{out}$ and $h_{out}$ are the mass and the enthalpy leave the system through the PRV, respectively, and $Q_{in}$ is the heat input into the tank.

For this system both mass and energy conservation are applied (neglecting changes in tank volume and assuming no mass leaves the tank):

$$\mathrm{d}m/\mathrm{d}t = 0 \qquad (1)$$

$$\mathrm{d}(mu)/\mathrm{d}t = Q_{in}, \qquad (2)$$

in which $t$ is time and $u$ is the internal energy. These equations are only valid until the PRV opens and mass starts to leave the system.

The heat input depends on the fire temperature and the insulation properties. As a first step, a simple approximation can be taken as:

$$Q_{in} = (A\ k/d)\ (T_{fire} - T_{tank}) \qquad (3)$$

in which $A$ is the surface area of the tank, $k$ is the thermal conductivity, $d$ is the thickness of the insulation material, $T_{fire}$ is the flame temperature and $T_{tank}$ is the temperature inside the tank.

Having no mass leaving the tank and no volume changes of the tank results in a overall density in the tank which is constant in time. The heating process of the tank can then be described by following the corresponding constant density line in an pressure-enthalpy diagram.

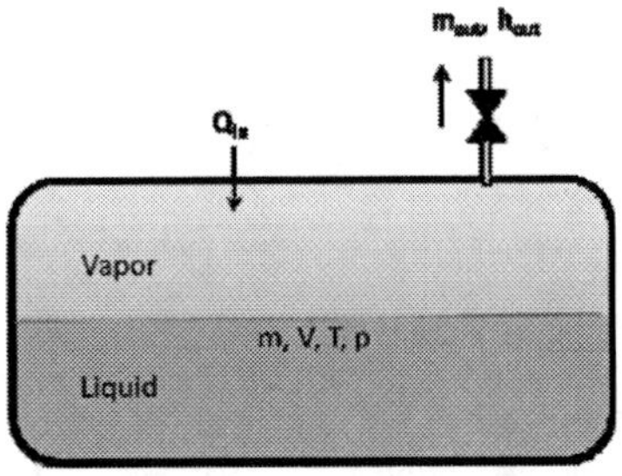


**Fig. 3.** Equilibrium tank model

The numerical relations for the equilibrium case that are described above were modelled at TNO in Matlab commercial code. This model takes less than 20 seconds to run. The main steps of this TNO model are summarized in the schematic below:

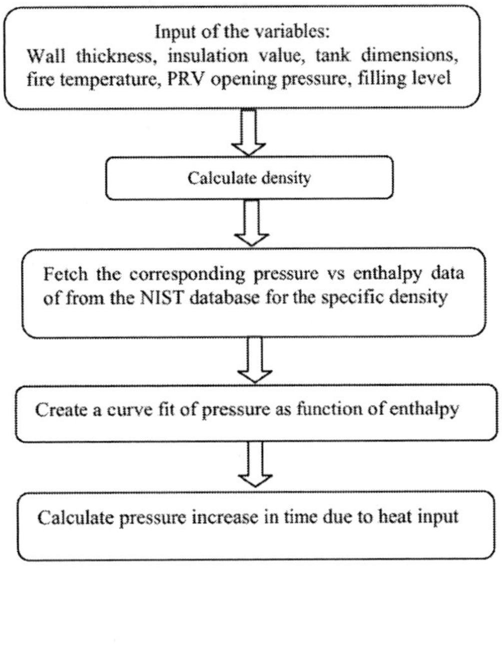

DOI: 10.1051/epjconf/201714302039

### 3.2 Non-equilibrium model

The non-equilibrium model consists of several nodes, similar to the one shown in Fig. **4**. The vapour and liquid are described as separate nodes, also the heat transfer through the tank inner and outer wall is taken into account.

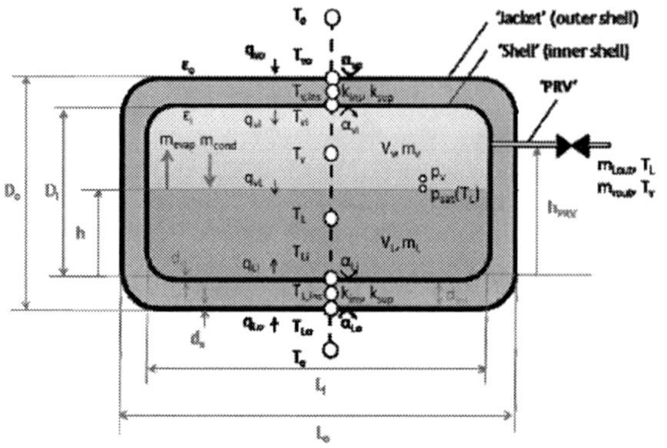


**Fig. 4.** Non-equilibrium tank model including tank walls and insulation.

For the vapour, liquid and total fluid mass the conservation of mass and energy are considered, including mass transfer between the vapour and liquid phase. This results in 4 equations for 5 unknowns: liquid and vapour mass: $m_L$, $m_v$; liquid and vapor temperature: $T_L$, $T_v$; and the pressure: $p_v$. To complete the set of equations the conservation of total volume is added:

$$V_L + V_v = \text{constant} \tag{4}$$

The set of equations has to be supplemented with expressions for the liquid and vapor discharge rates, the evaporation/condensation rates and the energy sources due to wall or interface interactions. For the evaporation/condensation rate ($\dot{m}_{net}$) the Hertz-Knudsen-Langmuir-equation is used:

$$\dot{m}_{net} = A \cdot \sqrt{\frac{M_v}{2\pi \cdot R}} \cdot \left( \beta_e \cdot \frac{p_{sat}(T_L)}{\sqrt{T_L}} - \beta_c \cdot \frac{p_v}{\sqrt{T_v}} \right) \tag{5}$$

in which the evaporation and condensation coefficients, $\beta_e$ and $\beta_e$, can be used to better fit the experimental data with the numerical model. The default value of 1 for both coefficients is used.

The energy sources in the conservation equations due to wall or interface reactions can further be described by several other processes. The first is the heat input of the fire, which can be described as radiation, or temperature or a combination of radiation and temperature. The walls of the tank will heat up and will release heat to the adjacent fluid.

As long as the fluid is quiescent, the heat is transferred, to the fluid and inside the fluid, through conduction. When the density differences in the liquid increase, a gravity driven flow pattern develops. This leads to a free convective heat transfer near the tank walls which depends on the Nusselt number.

The non-equilibrium model is programmed in Mathcad and takes 30 seconds to 1 minute to run.

## 4 Results and Discussions

During the first 118 minutes of the exposure to the fire the pressure inside the tank increased from 2.0 bar to 7.6 bar, at which pressure the pressure relieve valve (PRV) opened. The fuel supply for the fire was switched off 122 minutes after it started. An image of the tank after the fire has been switched off is shown in Figure 5. Some buckling of the outer wall can be seen.

**Fig. 5.** Cryogenic tank after the fire test

The pressure increase over time during the fire experiment is shown in Fig. **6** with the solid line. The dashed and the dotted line represent the pressure curves obtained by the non-equilibrium model and the equilibrium model respectively. In this plot, t = 0 min corresponds to the start of the fire. At this point the pressure starts to slowly increase and after 40 minutes it starts to rise faster and after 120 min the pressure reaches 7.6 bar. At this point PRV opens and subsequently the pressure remains constant until the fire is switched off and the pressure decreases again.

The initial flat portion of the curve indicates a stable period in the heating during which it is safe for firefighters to approach the tank and extinguish the external fire. The rapid increase in pressure in the later stage is where it becomes more dangerous to approach the LNG tank in during an external fire due to the risk of mechanical failure (and explosion) or the opening of the PRV.

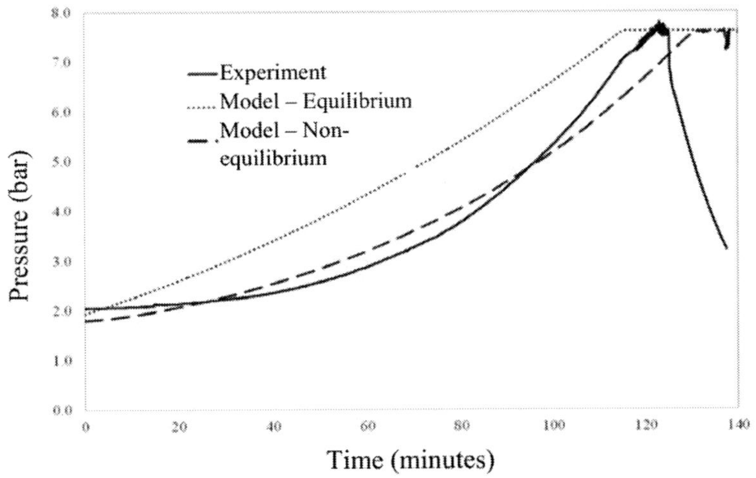


**Fig. 6.** Pressure vs time curves for the experiments, non-equilibrium model and equilibrium model.

Both the equilibrium and non-equilibrium models are able to capture the PRV opening time with acceptable error. The non-equilibrium model gives a better description of the slope of the pressure increase. The slow increase in pressure and temperature of the fluid directly at the start of the fire is caused by the heating of the wall and insulation before an effective heat transfer to the fluid starts. This delays the increase in temperature of the fluid, as is shown by the small slope at the start of the test. In the equilibrium model, the heating of the walls and the insulation are not included. For this reason the increase of temperature and pressure is nearly constant over time and doesn't show a speed up. The main parameter for this model is the heat input into the tank ($Q_{in}$). For the shown curve the heat input into the tank is taken constant from the beginning to the end. The model can be improved by using the time dependent tank temperature and temperature dependent heat conductivity. Adding these improvements allows a better match to the final slope of the experiments, however the initial slow heat up can't be properly captured by this model.The non-equilibrium model is the most complete model, and however, it also has most number of equations and needs several coefficients to describe the process. The actual solution is dependent on many coefficients and parameters. Some of which are: $\beta_e$ and $\beta_e$ for the evaporation/condensation model and thermal conductivity, $k$.

## 5 Conclusions

Fire response of cryogenic tanks was studied experimentally and numerically in this investigation. Experimental approach is based on subjecting a double walled, perlite insulated tank with 1.4 tons of liquid nitrogen to external fire until the pressure valve openings were observed. The numerical modelling are based on developing two models; one for assumption of equilibrium between the liquid and the vapour which was developed in Matlab and the second one assumed non-equilibrium and was developed in Mathcad.

The pressure evolution during the fire experiment was recorded and the compared to the numerical results. The increase in pressure starts slowly and at a later point starts to increase. The initial phase is used to heat up the tank itself and the insulation between the inner and outer wall. Both of the numerical models, capture the "Time to release" with small error. Two models have been described: equilibrium and non-equilibrium. The equilibrium model is able to describe the final, fast heat up of the fluid. However, as the thermal mass of the walls and insulation is not included in the model, the initial slow heat up is not captured well. The non-equilibrium model is able to describe both phases in the heat up and is able to follow the experimental results very well. In order to better validate the developed model more experimental data are needed. This can be data on LNG or LIN or any other cryogenic substance.

This work and the models developed by TNO can be used by the fire response teams to estimate the time to approach an accident area involving a cryogenic tank such as LNG fuel or transport tanks.

## Acknowledgements

The Dutch Ministry of Economic Affairs (Topsector Energy, TKI gas program) and Joint Industry Partners namely Rolande LNG, Engie, Shell, and Vopak are kindly acknowledged for their financial support.
TNO performed the experiments described in this paper at the test facility of BAM, the Bundesanstalt für Materialforschung und Prüfung (Berlin/Horstwalde, Germany). Special thanks go to their dedicated on site operational team.

## References

1. U.S. Department of Energy. [Online] https://www.portofrotterdam.com/en/cargo-industry/lng-liquefied-natural-gas/lng-as-a-fuel-for-vessels-and-trucks.
2. Port of Rotterdam. [Online] https://www.portofrotterdam.com/en/cargo-industry/lng-liquefied-natural-gas/lng-as-a-fuel-for-vessels-and-trucks.
3. R.G. Scurlock, *Stratification, Rollover and Handling of LNG, LPG and Other Cryogenic Liquid Mixtures.* London : Springer, (2016). ISSN 2191-5520.
4. J.F.M. Velthuis, and B. Atli-Veltin, *Thermal and Mechanical Model of a Vacuum Insulated Cryogenic Tank Subjected to External Tank.* Delft: TNO R10940, (2015).
5. J.P. Kamperveen, M.P.N. Spruijt, and J.E.A Reinders,. *Heat Load Resistance of Cryogenic Storage Tanks - Results of LNG Safety Program.* Utrecht : TNO, (2016).
6. NIST. National Institute of Standards and Technology. [Online] http://www.nist.gov/.
7. I.H. Bell, J. Wronski, S. Quoilin, V. Lemort,. **53**, (2014), I&ECR, pp. 2498-2508.

# Analysis of turbulent synthetic jet by dynamic mode decomposition

*Tomáš* Hyhlík[1,*], *Hana* Netřebská[1], *Jakub* Devera[1], and *Radomír* Kalinay[1]

[1] *CTU in Prague, FME, Department of Fluid Dynamics and Thermodynamics, Technická 4, 166 07 Prague*

**Abstract.** The article deals with the analysis of CFD results of the turbulent synthetic jet. The numerical simulation of Large Eddy Simulation (LES) using commercial solver ANSYS CFX has been performed. The unsteady flow field is studied from the point of view of identification of the moving vortex ring, which has been identified both on the snapshots of flow field using swirling-strength criterion and using the Dynamic Mode Decomposition (DMD) of five periods. It is shown that travelling vortex ring vanishes due to interaction with vortex structures in the synthesised turbulent jet. DMD modes with multiple of the basic frequency of synthetic jet, which are connected with travelling vortex structure, have largest DMD amplitudes.

## 1 Introduction

This work can be understood as a continuation of previous works dealt with modal decomposition and analysis of synthetic jets, see e.g. works [1, 2].

The existence of synthetic jet is related to vortex ring creation. The created vortex ring is induced velocity in the direction outside from the orifice. The most commonly used parameters of the synthetic jet are based on so-called slug model, see e.g. [3, 4]. Velocity scale is defined as time averaged blowing orifice centerline velocity over an entire cycle

$$U_0 = \frac{1}{T}\int_0^{T_E} u_0(r=0,t)\,\mathrm{d}t, \tag{1}$$

where $T_E$ is blowing time and $T = 1/f$ is oscillation period. Reynolds number can be determined using time averaged blowing orifice centerline velocity $U_0$ as

$$Re = \frac{2U_0D}{\nu}, \tag{2}$$

where $D$ is orifice diameter and $\nu$ is kinematic viscosity. Stokes number is defined as

$$S = D\sqrt{\frac{2\pi f}{\nu}}, \tag{3}$$

where $f$ is frequency.

The regime of synthetic jet with $Re = 898$ and $S = 19.7$ is choosen.

## 2 Numerical Simulation

Unsteady incompressible turbulent flow simulation has been performed using commercial solver ANSYS CFX. The turbulence is simulated using LES on the computational mesh of about 740000 cells. The subgrid-scale viscosity is solved with Smagorinsky model with constant of the model 0.1. Central difference scheme is chosen as advection scheme. The temporal scheme is set as second order backward Euler. Both the flow in the orifice and actuator cavity are included in the simulation to get more accurate results. The effect of oscillating diaphragm is replaced by velocity boundary condition which should guarantee achievement of required Reynolds number

$$u(t) = \frac{\pi}{2}\frac{Re\eta}{\rho D}\frac{D^2}{D_L^2}\sin(2\pi f t), \tag{4}$$

where $\eta$ is dynamic viscosity, $\rho$ is air density and $D_L$ is inlet diameter i.e. diameter of oscillating diaphragm. The time step is set to $1/1000$ of the period. The simulation is run for 65 periods to get the developed flow field.

Results of the numerical simulation are shown in the figure 1, where there are visualised vortex structures at five phases of the period after $1/5$ of the period. Vortex structures are visualised using swirling-strength criterion. It is possible to see the movement of the vortex ring which is going from left to right. The vortex ring created in the blowing part of the period is very stable structure because it is not possible to see its spontaneous brake up. There are vortex structures in the right part of the figure 1 connected with the turbulent jet which is created by travelling vortex ring. The tail like vortex structure is created behind travelling vortex ring. The travelling vortex ring seems to disappear due to interaction with vortex structures in the turbulent jet flow, as can be seen in the figure 1.

## 3 Dynamic Mode Decomposition

DMD is a kind of modal decomposition method. Let the velocity field be represented by ensemble of snapshots sampling at interval $\Delta t$ with a form of matrix $\mathbf{V}_1^N =$

* e-mail: tomas.hyhlik@fs.cvut.cz

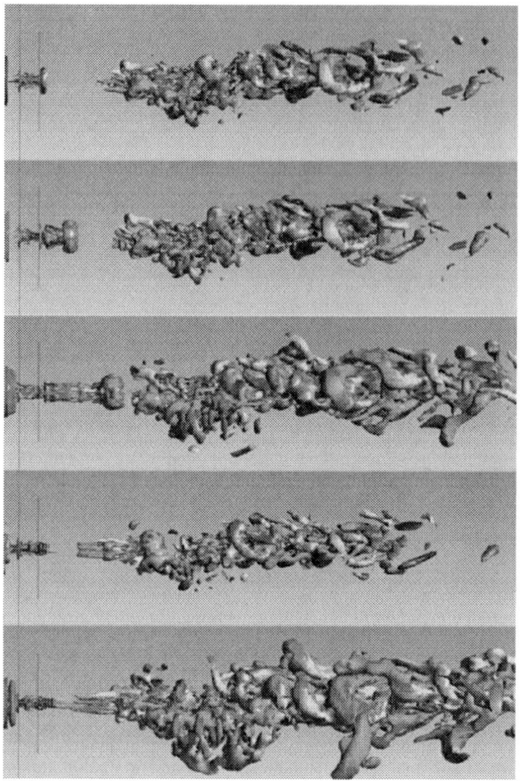

**Figure 1.** Vortex structures visualized using swirling-strength criterion; temporal step is 1/5 of the period

$\{\mathbf{v}_1, \mathbf{v}_2, \ldots, \mathbf{v}_N\}$, where the column vector $\mathbf{v}_j$ contains velocity field at $j$th time step. Let's consider linear mapping $\mathbf{A}$ that maps each snapshot to the next one in the following way

$$\mathbf{v}_{j+1} = \mathbf{A}\mathbf{v}_j. \tag{5}$$

This mapping is assumed to be same over the full sampling interval $[0, (N-1)\Delta t]$. The assumption of a constant mapping between the snapshots $\mathbf{v}_j$ allows us to formulate the sequence of flow fields as a Krylov sequence

$$\mathbf{V}_1^N = \left\{\mathbf{v}_1, \mathbf{A}\mathbf{v}_1, \mathbf{A}^2\mathbf{v}_1, \ldots, \mathbf{A}^{N-1}\mathbf{v}_1\right\}. \tag{6}$$

The goal of DMD is the extraction of dynamic characteristics such as eigenvalues and eigenvectors which are described by linear operator $\mathbf{A}$ based on sequence $\mathbf{V}_1^N$.

Robustness is achieved by preprocessing step using singular value decomposition of data sequence [5]. The algorithm is working with a similar matrix to matrix $\mathbf{A}$. Eigenvalues of similar matrix approximate eigenvalues of full mapping matrix $\mathbf{A}$ and provide temporal dynamics of the flow field.

## 4 Results of DMD

DMD decomposition of the flow field is based on unsteady data from last five periods. Only one hundred snapshots are taken from each period because of a large amount of the processed data. Figure 2 shows amplitudes of several DMD modes. Largest amplitude is connected with steady mode with zero frequency. There are a few DMD modes with low amplitude in figure 2, but largest amplitudes are connected with the basic frequency of synthetic jet $f = 96Hz$ and with its multiples. Figures 3 and 4 show the spatial structure of the DMD mode connected with synthetic jet frequency. Figures 5 and 6 show the spatial structure of the DMD mode connected with double of synthetic jet frequency. Figure 7 shows the spatial structure of the DMD mode connected with the triple of synthetic jet frequency. The real and imaginary part of the mentioned spatial DMD modes have similar scales. The spatial structures of real and imaginary parts are space shifted and are very similar, there are alternating local minimums, where the structures are finer for higher DMD mode frequency. The alternating local minimums like structure diminishes about fifteen diameters from the orifice. This kind of behaviour documents travelling vortex ring-like structure, where vortex ring diminishes about fifteen diameters from the orifice.

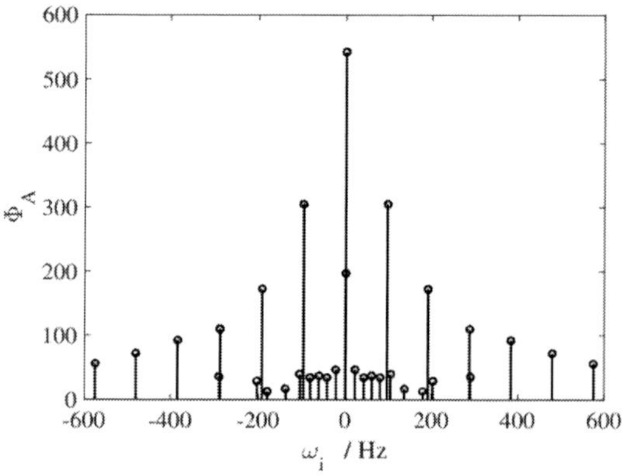


**Figure 2.** DMD amplitudes

## 5 Conclusions

The vortex structures in the turbulent synthetic jet are studied using swirling-strength criterion and using DMD decomposition method. DMD method is able to identify mainly the travelling vortex ring. The absence of DMD modes with a higher frequency which are connected with the turbulent jet is probably caused by low sampling frequency which is chosen with respect to the necessary amount of data. The results of vortex visualisation are based on the swirling-strength criterion, where the vortex structures calculated using LES are shown.

## Acknowledgement

The support from Czech Science Foundation under Grant No. 14-08888S is gratefully acknowledged.

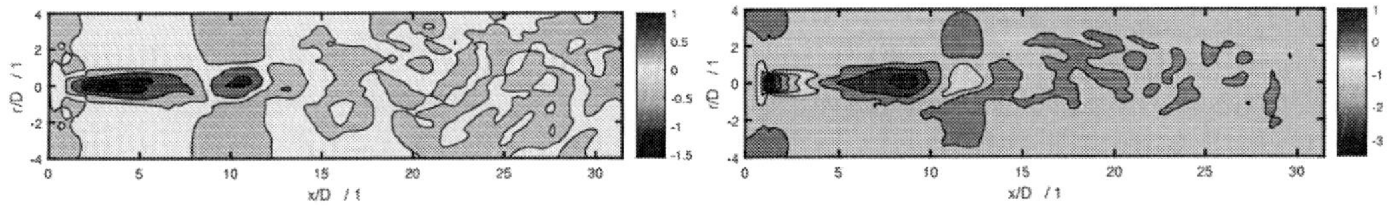


(a) xy plane, x velocity, real part of DMD mode

(b) xy plane, x velocity, imaginary part of DMD mode

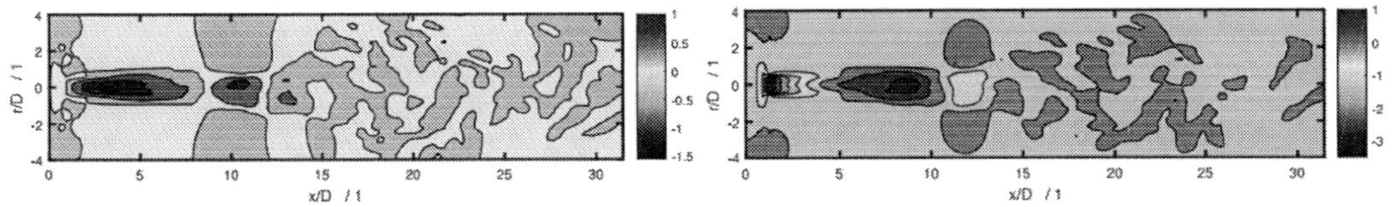


(c) xz plane, x velocity, real part of DMD mode

(d) xz plane, x velocity, imaginary part of DMD mode

**Figure 3.** Dynamic mode with $f = 96Hz$; axial velocity component x

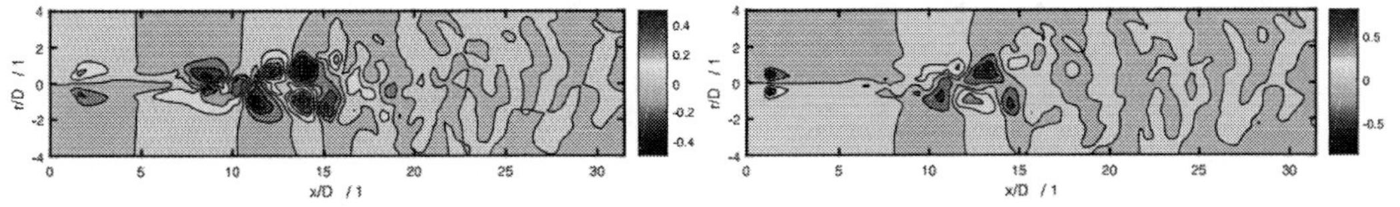


(a) xz plane, z velocity, real part of DMD mode

(b) xz plane, z velocity, imaginary part of DMD mode

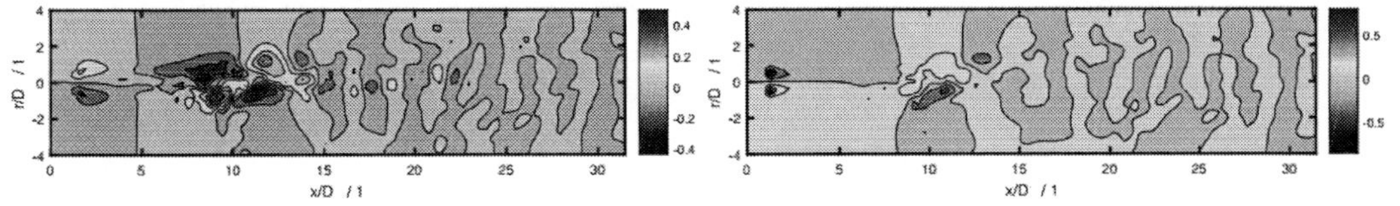


(c) xy plane, y velocity, real part of DMD mode

(d) xy plane, y velocity, imaginary part of DMD mode

**Figure 4.** Dynamic mode with $f = 96Hz$; y and z velocities

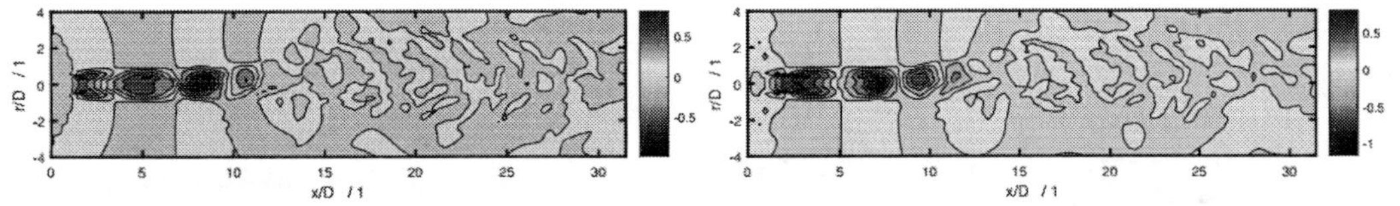


(a) xy plane, x velocity, real part of DMD mode

(b) xy plane, x velocity, imaginary part of DMD mode

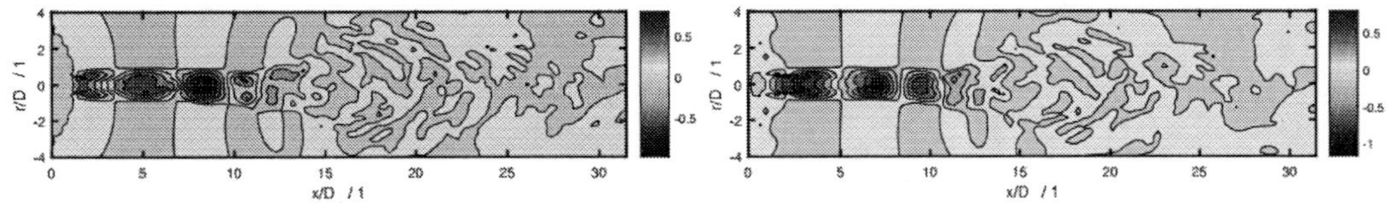


(c) xz plane, x velocity, real part of DMD mode

(d) xz plane, x velocity, imaginary part of DMD mode

**Figure 5.** Dynamic mode with $f = 192Hz$; axial velocity component x

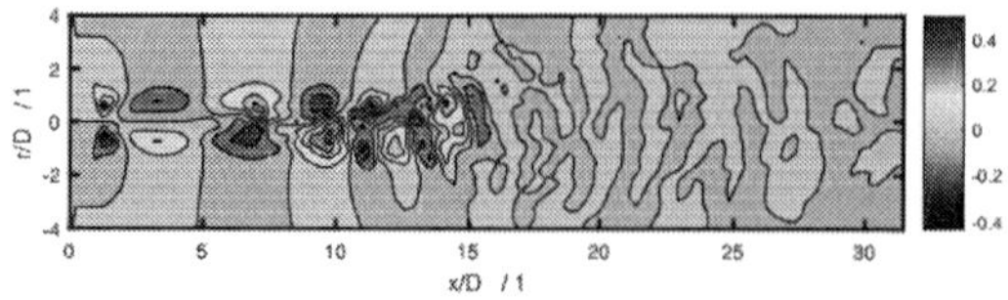

(a) xz plane, z velocity, real part of DMD mode

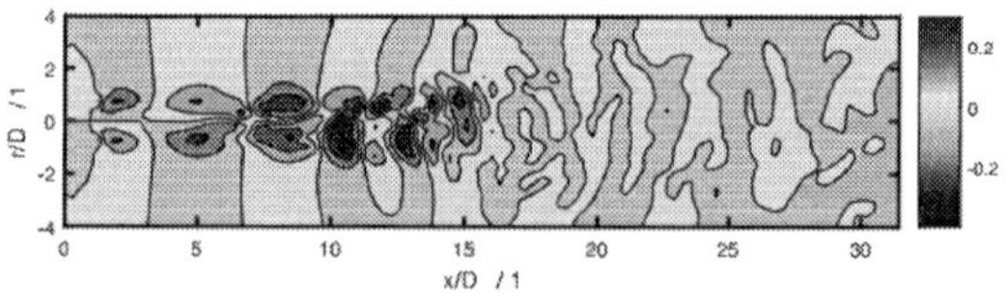

(b) xz plane, z velocity, imaginary part of DMD mode

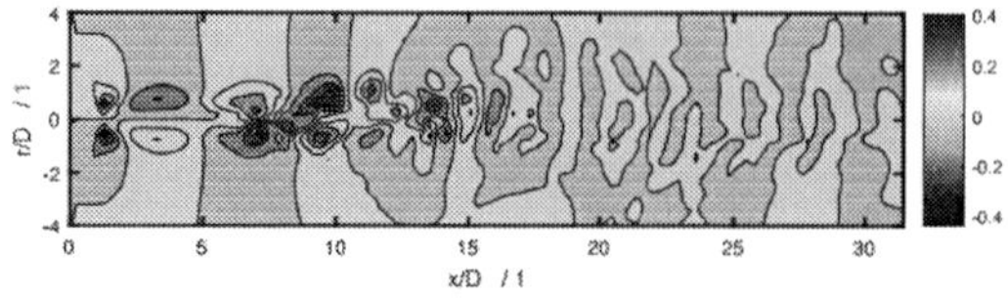

(c) xy plane, y velocity, real part of DMD mode

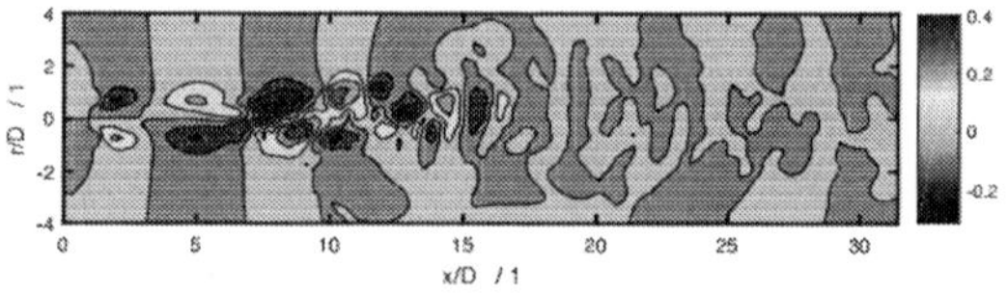

(d) xy plane, y velocity, imaginary part of DMD mode

**Figure 6.** Dynamic mode with $f = 192Hz$; y and z velocities

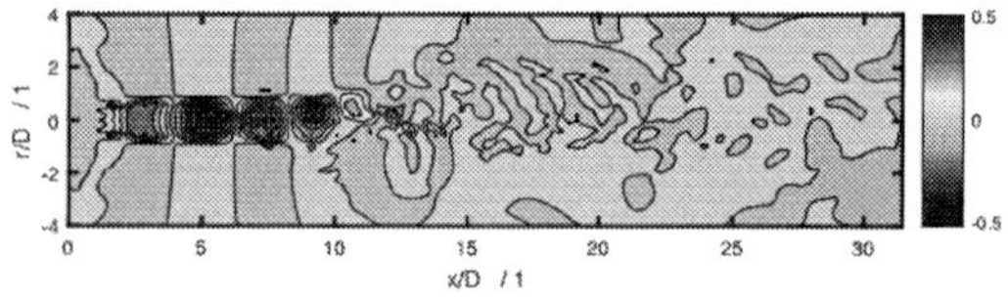

(a) xy plane, x velocity, real part of DMD mode

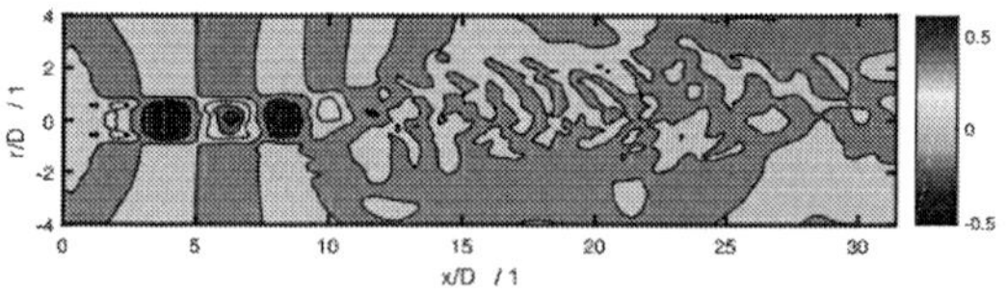

(b) xy plane, x velocity, imaginary part of DMD mode

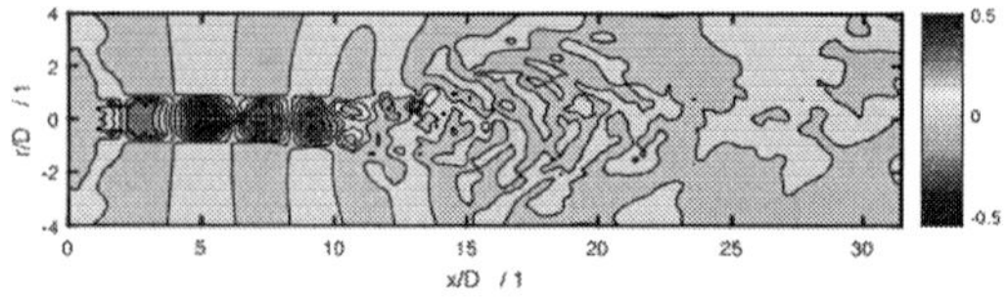

(c) xz plane, x velocity, real part of DMD mode

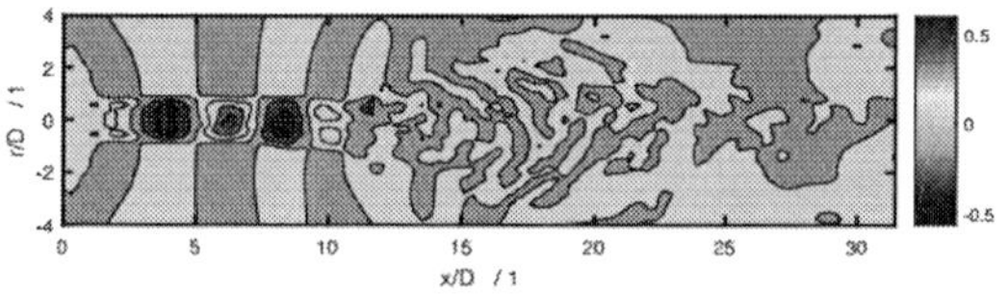

(d) xz plane, x velocity, imaginary part of DMD mode

**Figure 7.** Dynamic mode with $f = 288Hz$; axial velocity component x

## References

[1] T. Hyhlík, P. Železný, J. Čížek, EPJ Web of Conferences, **45**, (2013), 01043

[2] T. Hyhlík, H. Netřebská, EJP Web of Conferences, **114**, (2016), 02045

[3] Z. Trávníček, Z. Broučková, J. Kordík, AIAA Journal, **50(9)**, (2012), 2012-2017.

[4] R. Holman, Y. Utturkar, R. Mittal, B. L. Smith, L. Cattafesta, AIAA Journal, **43(10)**, (2005), 2110-2116.

[5] P. J. Schmid, J. Fluid Mech., **656**, (2010), 5-28.

DOI: 10.1051/epjconf/201714302041

# Engineering Model of High Pressure Moist Air

*Tomáš* Hyhlík[1,*]

[1] *CTU in Prague, FME, Department of Fluid Dynamics and Thermodynamics, Technická 4, 166 07 Prague*

**Abstract.** The article deals with the moist air equation of state. There are equations of state discussed in the article, i.e. the model of an ideal mixture of ideal gases, the model of an ideal mixture of real gases and the model based on the virial equation of state. The evaluation of sound speed based on the ideal mixture concept is mentioned. The sound speed calculated by the model of an ideal mixture of ideal gases is compared with the sound speed calculated by using the model based on the concept of an ideal mixture of real gases. The comparison of enthalpy end entropy based on the model of an ideal mixture of ideal gases and the model of an ideal mixture of real gases is performed. It is shown that the model of an ideal mixture of real gases deviates from the model of an ideal mixture of ideal gases only in the case of high pressure. An impossibility of the definition of partial pressure in the mixture of real gases is discussed, where the virial equation of state is used.

## 1 Basic parameters of moist air

This section covers only a few selected parameters of moist air. The detailed description can be found e.g. in [1]. Let us consider the moist as a binary mixture of dry air and water vapour. An amount of water vapour can change from zero (it is dry air) to some maximum, which depends on the temperature and pressure. Humidity ratio is defined as the ratio of mass of water vapour to the mass of dry air

$$x = \frac{V\rho_v}{V\rho_a} = \frac{\rho_v}{\rho_a}, \tag{1}$$

where $V$ is volume, $\rho_v$ is water vapour density (absolute humidity) and $\rho_a$ is dry air density. Let $x''$ is the humidity ratio of saturated moist air at the same temperature $t$ and pressure $p$. Degree of saturation is defined as the ratio of air humidity ratio $x$ to humidity ratio of saturated moist air $x''$

$$\mu = \left(\frac{x}{x''}\right)_{p,t}. \tag{2}$$

The value of saturated humidity ratio $x''$ is limited in the atmospheric moist air, e.g. that is in the case of the pressure $p = 98000$ Pa and temperature $t = 25°$C. There can by a different story in some technical application. Consider the case of the same pressure $p = 98000$ Pa and at the temperature $t = 150°$C, where is water vapour always superheated for arbitrary vapour pressure $p_v \leq p$, therefore moist air can not be saturated for arbitrary humidity ratio $x$. In such a situation is appropriate to define $x'' = \infty$ and degree of saturation will be zero.

Relative humidity $\varphi_R$ is frequently used parameter in psychrometrics. Relative humidity is defined as the ratio of absolute humidity $\rho_v$ to absolute humidity of saturated moist air $\rho_v''$ at the same temperature $t$ and pressure $p$

$$\varphi_R = \left(\frac{\rho_v}{\rho_v''}\right)_{p,t}. \tag{3}$$

Absolute humidity of moist air $\rho_v$ expresses weight of water vapour in cubic meter of moist air.

The partial pressure of water vapour in the saturated moist air is not in general equal to saturated water vapour pressure $p_{sat}$ at given temperature, which is related to pure water vapour [2]. Partial pressure of water vapour is

$$p_v = f_e p_{sat}, \tag{4}$$

where $f_e$ is so called enhancement factor, which has the value of 1.0041 at standard atmospheric pressure 101325 Pa and at temperature 0°C. In general, it is possible to say that the value of enhancement factor is going closer to one with increasing temperature and with decreasing temperature its value increasing. The value of enhancement factor is 1.46383 at 10 MPa a 0°C, and after temperature increases to 200°C is its value 1.21286 [2].

## 2 Model of ideal mixture of ideal gases

The most simplest moist air model is the model of ideal mixture of ideal gases. The pressure can be in ideal mixture expressed by using Dalton's law

$$p = p_a + p_v, \tag{5}$$

where $p_a$ is partial pressure of dry air and $p_v$ is partial pressure of water vapour. Partial pressure is hypothetical pressure of gas if it alone occupied the volume of the mixture at the same temperature. An individual components of ideal mixture exist independently without any interaction. The

*e-mail: tomas.hyhlik@fs.cvut.cz

moist air density is in the ideal mixture expressed as a sum of individual partial densities

$$\rho = \rho_a + \rho_v. \tag{6}$$

Partial densities can be expressed by using ideal gas equation of state as

$$\rho_a = \frac{p_a M_a}{RT}, \quad \rho_v = \frac{p_v M_v}{RT}. \tag{7}$$

Mass fractions of dry air and water vapour are

$$w_a = \frac{\rho_a}{\rho}, \quad w_v = \frac{\rho_v}{\rho}, \tag{8}$$

where

$$w_a + w_v = 1. \tag{9}$$

Mass fractions can be expressed using humidity ratio

$$w_a = \frac{1}{1+x}, \quad w_v = \frac{x}{1+x}. \tag{10}$$

The moist air equation of state of the mixture of dry air and water vapour can be derived using Dalton's law (5), the definition of humidity ratio (1) and the relation for moist air density (6) as

$$\rho = \frac{p}{RT}\frac{M_a(1+x)}{\left(1+\frac{M_a}{M_v}x\right)}, \tag{11}$$

where $M_a$ is molecular weight of dry air, $M_v$ is molecular weight of water vapour, $R$ is universal gas constant and $T$ is temperature.

To determine enthalpy and entropy it is necessary to choose reference state, which will determine zero value of that parameters. An appropriate choice is e.g. triple point of water, where $T_{ref} = 273.16$ K, $p_{ref} = 611.657$ Pa. Enthalpy per kilogramme of mixture is expressed as

$$h = w_a c_{p_a}(T - T_{ref}) + w_v\left[c_{p_v}(T - T_{ref}) + l_{ref}\right], \tag{12}$$

where $l_{ref}$ is specific latent heat at reference point, specific heat capacity of dry air $c_{p_a}$ and specific heat capacity of water vapour $c_{p_v}$ are expressed as

$$c_{p_a} = \frac{\kappa_a}{\kappa_a - 1} r_a, \quad c_{p_v} = \frac{\kappa_v}{\kappa_v - 1} r_v. \tag{13}$$

Specific gas constant of dry air $r_a$ and water vapour $r_v$ are

$$r_a = \frac{R}{M_a}, \quad r_v = \frac{R}{M_v}. \tag{14}$$

Poison's coefficient of dry air is $\kappa_a = 1.4$ and in the case of water vapour is $\kappa_v = 1.33$. Enthalpy is usually introduced in psychrometrics as enthalpy per kilogramme of dry air [1]

$$h_{1+x} = c_{p_a}(T - T_{ref}) + x\left[c_{p_v}(T - T_{ref}) + l_{ref}\right], \tag{15}$$

where the first part is related to dry air and the second part is related to water vapour, where there is $x$ kilograms of water vapour in one kilogram of dry air, and therefore is the moist air total enthalpy $H$ obtained by multiplying of $h_{1+x}$ with the weight of dry air in the mixture dry air and water vapour.

To determine specific entropy per kilogram of mixture, which is defined as

$$s = w_a s_a + w_v s_v, \tag{16}$$

is necessary to express partial entropies of individual components of mixture. Specific entropy of dry air is

$$s_a = c_{p_a} \ln\left(\frac{T}{T_{ref}}\right) - \frac{R}{M_a} \ln\left(\frac{p_a}{p_{ref}}\right) \tag{17}$$

and water vapour specific entropy is similarly

$$s_v = c_{p_v} \ln\left(\frac{T}{T_{ref}}\right) - \frac{R}{M_v} \ln\left(\frac{p_v}{p_{ref}}\right). \tag{18}$$

Speed of sound can be in the case of ideal mixture of ideal gases expressed by using its definition and finally by the relation for ideal gases

$$a = \sqrt{\left(\frac{\partial p}{\partial \rho}\right)_s} = \sqrt{\kappa r T}. \tag{19}$$

Specific gas constant of the mixture can be expressed using gas constants of individual components

$$r = w_a r_a + w_v r_v. \tag{20}$$

The Poisson exponent of the mixture is

$$\kappa = \frac{c_p}{c_v}, \tag{21}$$

where individual heat capacities are

$$c_p = w_a c_{p_a} + w_v c_{p_v}, \quad c_v = c_p - r. \tag{22}$$

## 3 Advanced engineering model

The advanced engineering model is the model of an ideal mixture of real gases, and therefore is based on the Dalton's law (5). To determine the moist air properties it is necessary to get partial pressures and partial densities $p_a$, $p_v$, $\rho_a$ a $\rho_v$ from the given parameters of moist air, and these are namely the moist air temperature $T$, pressure $p$ and humidity ratio $x$. The problem contains four unknowns, which can be obtained using the solution of the system of algebraic equations (5), (1),

$$p - p_v = p_{air}(T, \rho_a) \tag{23}$$

and

$$p - p_a = p_{vap}(T, \rho_v), \tag{24}$$

where $p_{air}(T, \rho_a)$ represents dry air equation of state e.g. according to [3] and $p_{vap}(T, \rho_v)$ represents water vapour equation of state e.g. according to [4]. The real gases equations of state are used to solve pressure depending on the temperature and the density. The moist air density is calculated using equation (6), where partial densities are used. The real gas equations of state [3, 4] allow calculating enthalpy and entropy of dry air

$$h_a = h_{air}(T, \rho_a), \quad s_a = s_{air}(T, \rho_a), \tag{25}$$

and enthalpy and entropy of water vapour

$$h_v = h_{vap}(T, \rho_v), \quad s_v = s_{vap}(T, \rho_v). \tag{26}$$

Moist air entropy is solved using equation (16). Enthalpy is

$$h = w_a h_a + w_v h_v. \tag{27}$$

The speed of sound of individual components of the mixture can be evaluated using real gas equations of state [3, 4]

$$a_a = a_{air}(T, \rho_a), \quad a_v = a_{vap}(T, \rho_v). \tag{28}$$

Reference [5] offers possibility to evaluate speed of sound using relation for ideal mixture

$$a = \frac{1}{\sqrt{M\left(\frac{\psi_v}{M_v a_v^2} + \frac{\psi_a}{M_a a_a^2}\right)}}, \tag{29}$$

where mole fractions are

$$\psi_a = \frac{n_a}{n}, \quad \psi_v = \frac{n_v}{n} \tag{30}$$

and molar mass of moist air is

$$M = \psi_a M_a + \psi_v M_v. \tag{31}$$

Molar fractions and mass fractions are related through the ratio of molecular weights

$$w_a = \frac{M_a}{M}\psi_a, \quad w_v = \frac{M_v}{M}\psi_v. \tag{32}$$

Equation of state of the real gas is possible to assume in the form

$$pV = nRTz(p, T), \tag{33}$$

where $z(p, T)$ is compressibility factor. It is possible to express equations of state for dry air and water vapour as

$$p_a V = n_a RT z_a(p_a, T), \quad p_v V = n_v RT z_v(p_v, T). \tag{34}$$

After addition of equations (34) and comparison with equation (33) is possible to express compressibility factor of the mixture as

$$z(p, T) = \psi_a z_a(p_a, T) + \psi_v z_v(p_v, T). \tag{35}$$

It means, that the consequence of utilising the Dalton's law for the mixture of real gases is, that compressibility factor of the mixture of real gases is the weighted average of individual compressibility factors [6].

## 4 Virial equation of state

Compressibility factor is possible to define in the real gas as

$$z = \frac{pV_m}{RT}, \tag{36}$$

where $V_m$ is molar volume

$$V_m = \frac{V}{n} = \frac{V}{n_a + n_v}. \tag{37}$$

Reference [7] shows, that it is possible to express compressibility factor by using power series depending on the molar concentration $n/V = 1/V_m$

$$z = 1 + B\left(\frac{1}{V_m}\right) + C\left(\frac{1}{V_m}\right)^2 + \ldots, \tag{38}$$

where $B(T)$ is the second and $C(T)$ is the third virial coefficient. The second virial coefficient of the mixture is possible to express according to [7] as

$$B = \psi_a^2 B_a + \psi_v^2 B_v + 2\psi_a \psi_v B_{av}, \tag{39}$$

where $B_a$ and $B_v$ are the second virial coefficients of the pure components and $B_{av}$ is the second virial coefficient which is connected with the interaction of dry air and water vapour. The relations for individual components are

$$p_a V = n_a RT + \frac{n_a^2 RT B_a}{V}, \quad p_v V = n_v RT + \frac{n_v^2 RT B_v}{V}. \tag{40}$$

If the Dalton's law is valid, then

$$pV = (n_a + n_v)RT + \frac{RT}{V}\left(n_a^2 B_a + n_v^2 B_v\right) \tag{41}$$

and after manipulation is

$$\frac{pV_m}{RT} = 1 + \left(\psi_a^2 B_a + \psi_v^2 B_v\right)\frac{1}{V_m}. \tag{42}$$

From the comparison of equation (42) with equations (39) and (38) is evident, that Dalton's law ignored molecular interaction [6]. If the virial equation of state is applied to the mixture is not Dalton's law taken into account and partial pressures are not clearly defined. Determination of partial pressures is possible by using molar fractions like when it is the mixture of ideal gases, where the molar fractions are determined using fractions of partial pressure and the mixture pressure

$$p_a^A = \psi_a p, \quad p_v^A = \psi_v p. \tag{43}$$

Determination of partial pressures is possible using classical definition, where it is the pressure of the mixture component at temperature $T$ occupying the overall volume $V$. From the equations of state for individual components (40) is

$$p_a^B = \frac{n_a RT}{V} + \frac{n_a^2 RT B_a}{V^2}, \quad p_v^B = \frac{n_v RT}{V} + \frac{n_v^2 RT B_v}{V^2}, \tag{44}$$

or by using of the mixture molar volume is

$$p_a^B = \psi_a\left[\frac{RT}{V_m}\left(1 + \psi_a \frac{B_a}{V_m}\right)\right], \quad p_v^B = \psi_v\left[\frac{RT}{V_m}\left(1 + \psi_v \frac{B_v}{V_m}\right)\right]. \tag{45}$$

It is clear, that square bracket in equations (45) does not correspond to the pressure of the mixture according to (38) and (39). Partial pressures based on both definitions are not in general equal

$$p_a^A \neq p_a^B, \quad p_v^A \neq p_v^B. \tag{46}$$

They agree only in the case of the ideal gas equation of state, where are virial coefficients zero.

Nowadays virial equation of state is standard in psychrometrics [2]. Virial coefficients are based on the general equation for water vapour according to [4] and on the general equation of state for dry air according to [3]. Equation of state according to [2] is based on the second and on the third virial coefficients

$$p(V_m, T, \psi_a) = \frac{RT}{V_m}\left(1 + \frac{B(\psi_a, T)}{V_m} + \frac{C(\psi_a, T)}{V_m^2}\right), \tag{47}$$

where for given pressure $p$, temperature $T$ and molar fraction of dry air $\psi_a$ is possible to solve iteratively molar volume $V_m$, and finally is the mixture density evaluated using molecular weight of moist air $M$ as

$$\rho = \frac{M}{V_m}. \tag{48}$$

Pressure is possible to solve directly in the case, where temperature $T$, density $\rho$ and molar fraction of dry air $\psi_a$ are given. The last possibility is an iterative solution of temperature $T$ as a function of the density $\rho$, pressure $p$ and dry air molar fraction. The equation of state according to [2] allows solving also other state variables like enthalpy $h$ and entropy $s$.

## 5 Results

The comparison of the model of an ideal mixture of ideal gases and the model of an ideal mixture of real gases is depicted in the figures. Figures 1 and 2 shows the dependence of enthalpy and entropy on the pressure and on the degree of saturation. It is visible, that even a minimal amount of water vapour influence at a high temperature significantly the values of enthalpy and entropy. It is clear, that both compared models give practically the same results at low pressures. There are significant discrepancies at a hight value of pressure such as the pressure 5MPa and 10MPa. Figure 3 documents deviations in the prediction of water vapour amount in the moist air, which is based on the both compared models. There are significant deviations at the pressure of 5MPa and 10MPa similarly like in the cases of enthalpy and entropy. Figure 4 shows the density comparison calculated by using tested models. It is possible to see, that there are significant discrepancies not only at a high pressure but also at a high value of temperatures. There is a high value of humidity ratio at high temperatures, and therefore it is possible to explain discrepancies using the inappropriate ideal gas model. There is a relation of the speed of sound on the pressure and the temperature in figure 4. It is clear from the figure, that the model of ideal gas mixture does not take into consideration the effect of pressure on the speed of sound. It is clear, that the model of an ideal mixture of ideal gases can be used at low pressures, but at the pressure of 1MPa are for the high value of humidity ratio evident discrepancies.

## 6 Conclusions

The model of an ideal mixture of ideal gases is applicable predominantly in the case of atmospheric and lower pressures such as in the case of evaporative cooling [8, 9]. The moist air model based on the virial equation of state is applicable in a wider range of pressures and temperatures, how is stated in [2]. The limit of applicability of this model can be maximal humidity ratio, which is according to [2] 10 $kg_p/kg_{sv}$. The model of an ideal mixture of real gases is in the range of validity of virial equation of state less accurate, but there is an advantage out of the range of its validity, where the real gas effect is taken into account. An advantage of this model can be seen e.g. in the case of a very high value of the water vapour mass fraction. There can be a problem with the speed of sound calculation because reference [2] does not contain any relevant equation, but the concept based on ideal mixture rule (29) is presented in this article.

## Acknowledgement

This work has been supported by *Technology Agency of the Czech Republic* under the project *Advanced Technologies for Heat and Electricity Production - TE01020036*.

## References

[1] *2013 ASHRAE Handbook-Fundamentals (SI)*, (ASHRAE, 2013)

[2] S. Herrmann, H. J. Kretzschmar, D. P. Gatley, *Thermodynamical Properties of Real Moist Air, Dry Air, Steam, Water, and Ice*, ASHRAE RP-1485, (ASHRAE, 2009)

[3] E. W. Lemmon, R. T. Jacobsen, S. G. Penoncello, D. G. Friend, J. Phys. Chem. Ref. Data, **29(3)**, (2000)

[4] *Revised Release on the IAPWS Formulation 1995 for the Thermodynamic Properties of Ordinary Water Substance for General and Scientific Use*, (IAPWS, 2014)

[5] K. Nerendra, B. Sudhamsa, B. M. Sarath, Research Journal of Chemical Sciences, **4(8)**, (2014), 42-45

[6] K. W. Woo, S. I. Yeo, The SNU Journal of Educational Research, **5**, (1995), 127-134

[7] B. E. Poling, J. M. Prausnitz, J. P. O'Connell, *The Properties of Gases and Liquids*, (The McGraw-Hill Companies, 2001)

[8] L. Dvořák, J. Čížek, J. Nožička, EJP Web of Conferences, **92**, (2015), 02014

[9] P. Vitkovič, EJP Web of Conferences, **114**, (2016), 02137

DOI: 10.1051/epjconf/201714302041

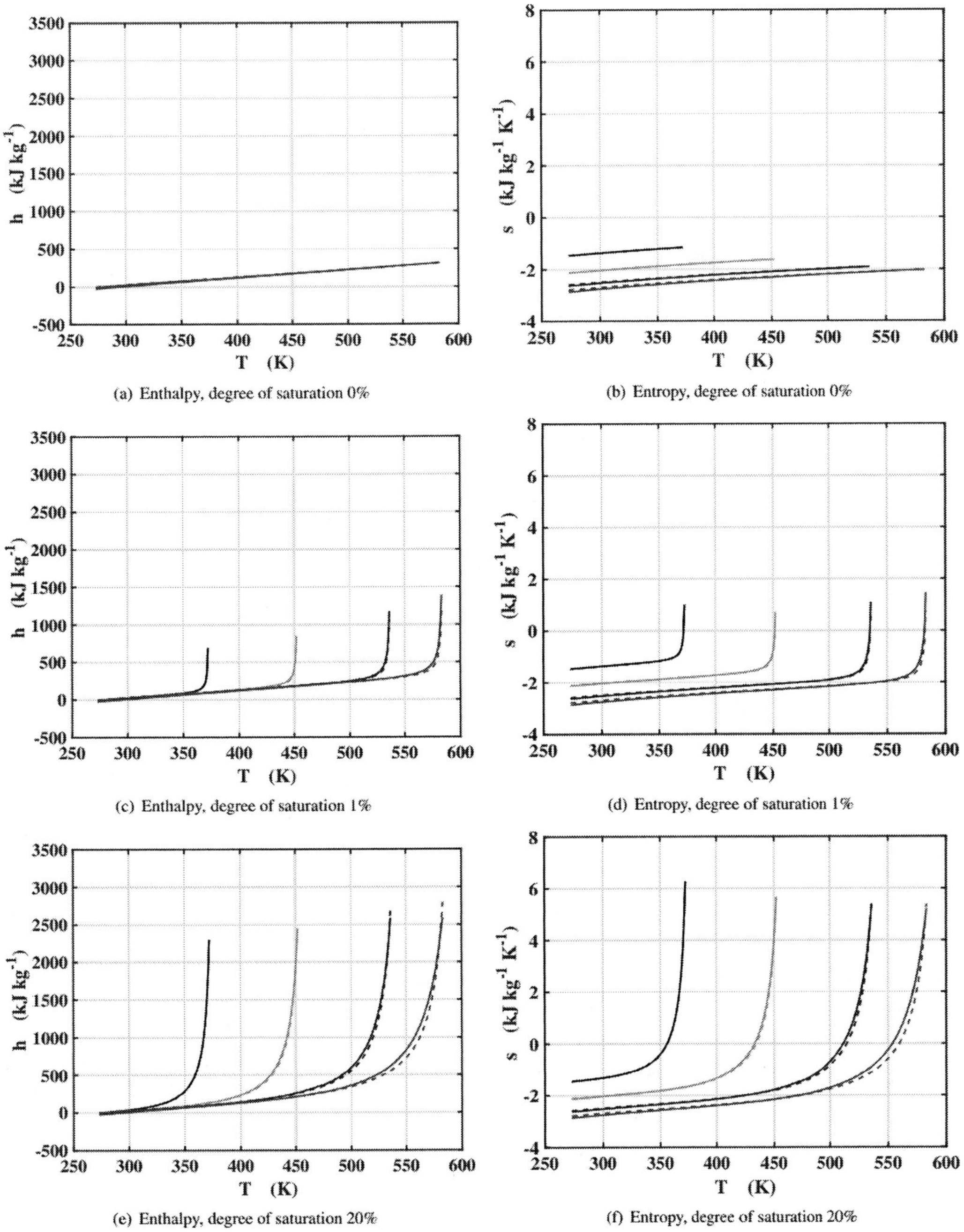


(a) Enthalpy, degree of saturation 0%

(b) Entropy, degree of saturation 0%

(c) Enthalpy, degree of saturation 1%

(d) Entropy, degree of saturation 1%

(e) Enthalpy, degree of saturation 20%

(f) Entropy, degree of saturation 20%

**Figure 1.** Enthalpy and entropy of moist air as a function of pressure, temperature and degree of saturation; for pressure 0.1MPa (black), 1MPa (green), 5MPa (blue) and 10MPa (black); solid line represents engineering model and dashed line represents ideal gas model

DOI: 10.1051/epjconf/201714302041

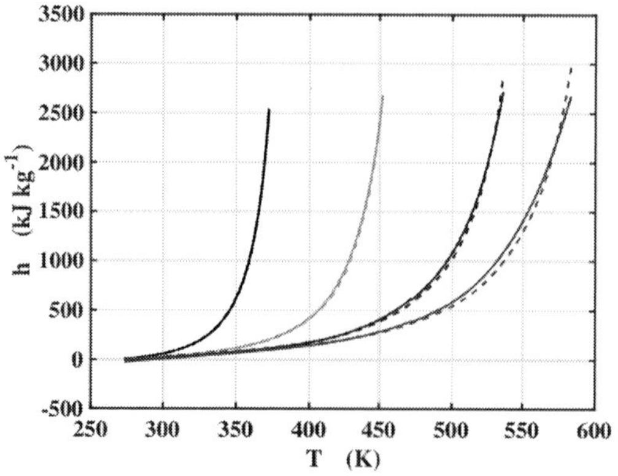


(a) Enthalpy, degree of saturation 60%

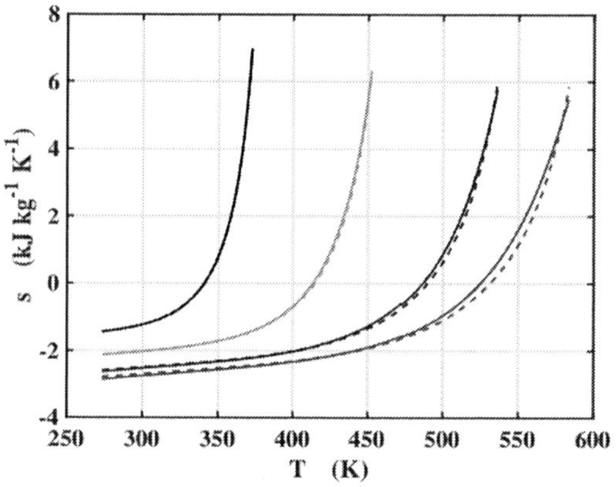


(b) Entropy, degree of saturation 60%

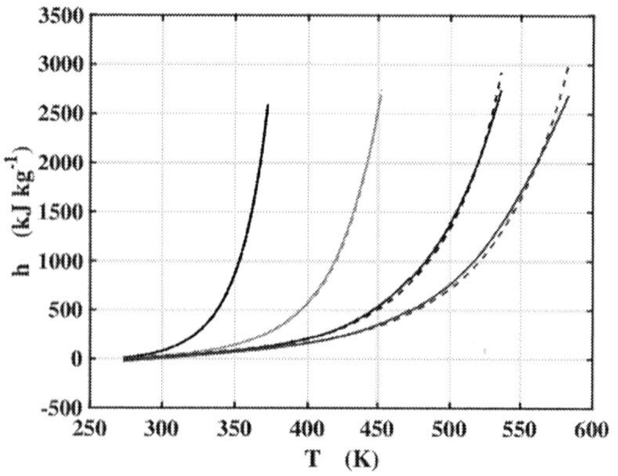


(c) Enthalpy, degree of saturation 100%

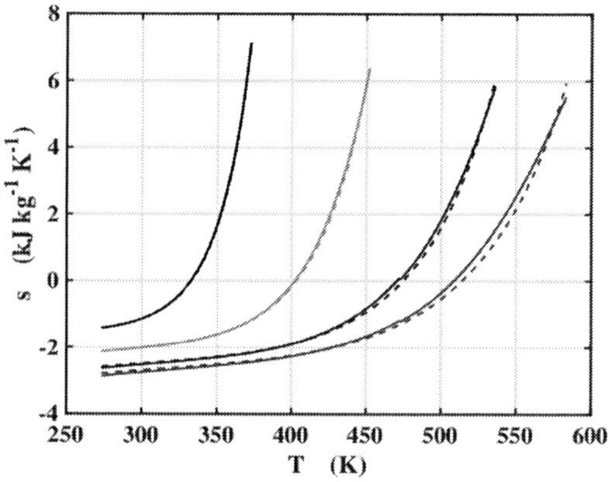


(d) Entropy, degree of saturation 100%

**Figure 2.** Enthalpy and entropy of moist air as a function of pressure, temperature and degree of saturation; for pressure 0.1MPa (black), 1MPa (green), 5MPa (blue) and 10MPa (black); solid line represents engineering model and dashed line represents ideal gas model

DOI: 10.1051/epjconf/201714302041

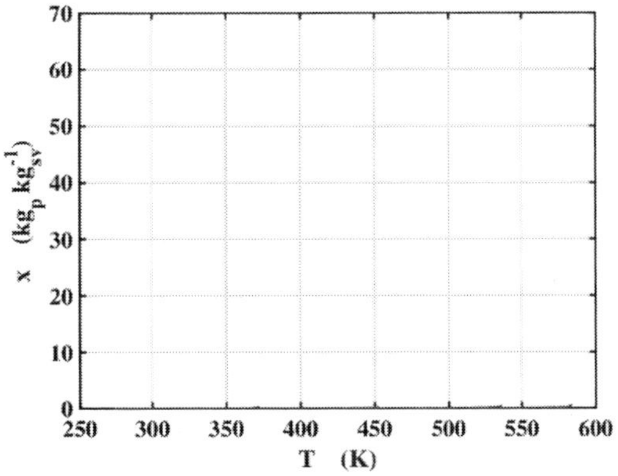

(a) Humidity ratio, degree of saturation 1%

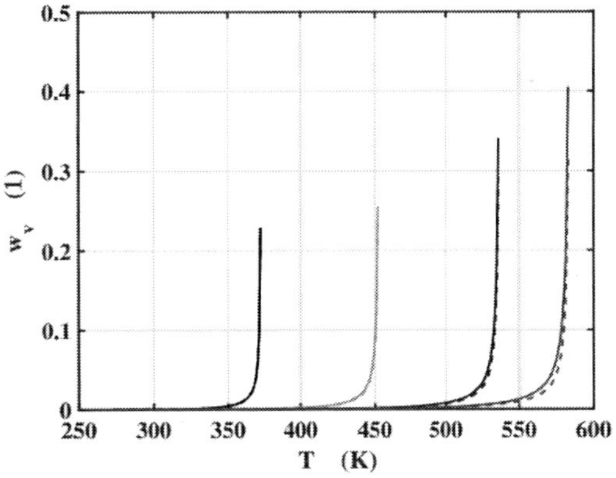

(b) Water vapour mass fraction, degree of saturation 1%

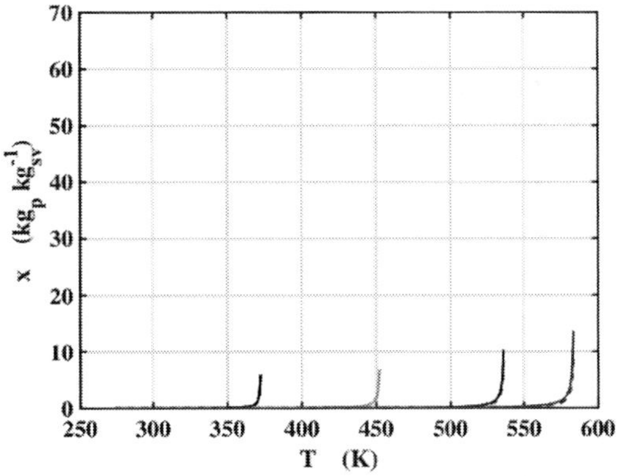

(c) Humidity ratio, degree of saturation 20%

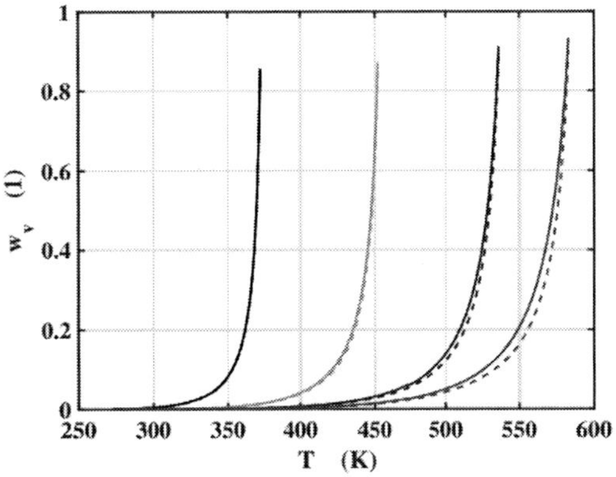

(d) Water vapour mass fraction, degree of saturation 20%

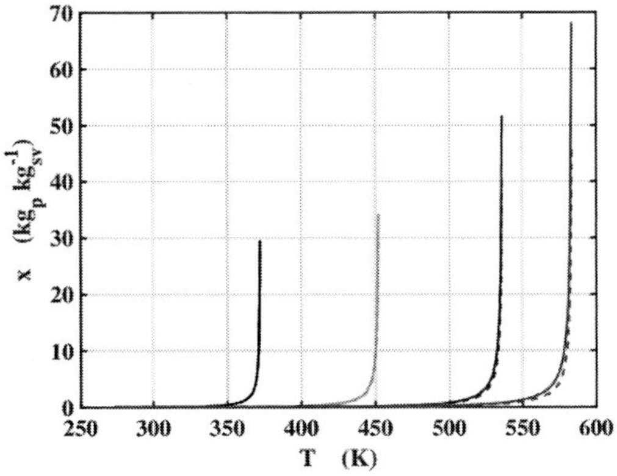

(e) Humidity ratio, degree of saturation 100%

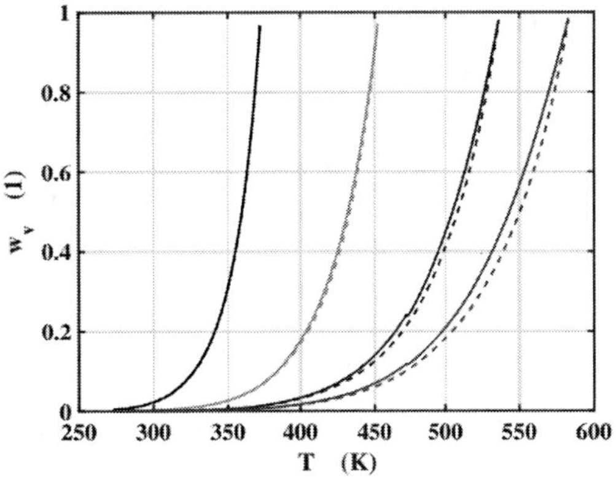

(f) Water vapour mass fraction, degree of saturation 100%

**Figure 3.** Humidity ratio and water vapour mass fraction as a function of pressure, temperature and degree of saturation; for pressure 0.1MPa (black), 1MPa (green), 5MPa (blue) and 10MPa (black); solid line represents engineering model and dashed line represents ideal gas model

DOI: 10.1051/epjconf/201714302041

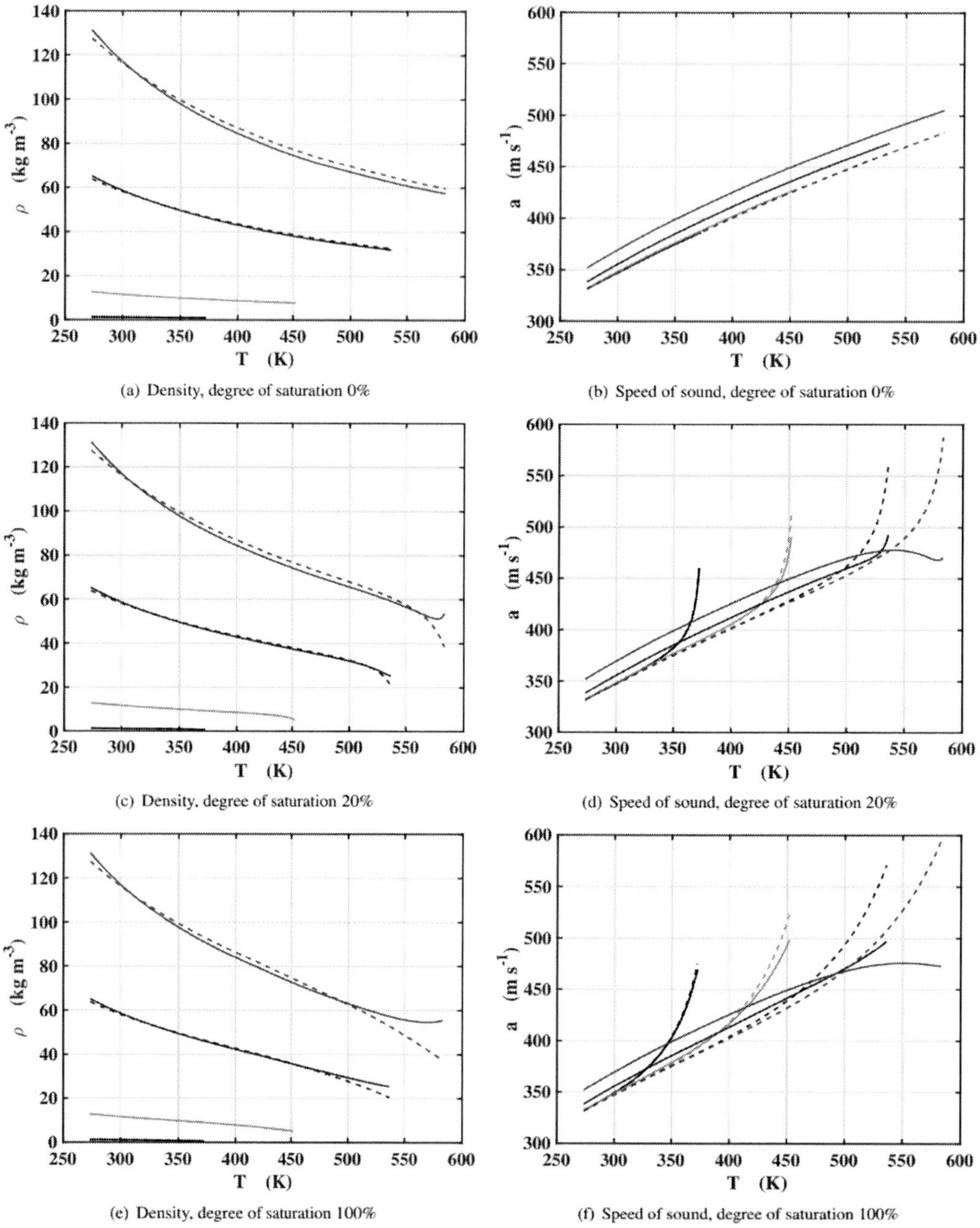


(a) Density, degree of saturation 0%

(b) Speed of sound, degree of saturation 0%

(c) Density, degree of saturation 20%

(d) Speed of sound, degree of saturation 20%

(e) Density, degree of saturation 100%

(f) Speed of sound, degree of saturation 100%

**Figure 4.** Density and speed of sound of moist air as a function of pressure, temperature and degree of saturation; for pressure 0.1MPa (black), 1MPa (green), 5MPa (blue) and 10MPa (black); solid line represents engineering model and dashed line represents ideal gas model

DOI: 10.1051/epjconf/201714302042

# Visual monitoring of the melting front propagation in a paraffin-based PCM

*Pavel* Charvát, *Josef* Štětina, *Tomáš* Mauder* and *Lubomír* Klimeš

Faculty of Mechanical Engineering, Brno University of Technology, 61669 Brno, Czech Republic

**Abstract.** Experiments were carried out in an environmental chamber with the aim to monitor the melting front propagation in a rectangular cavity filled with a paraffin-based Phase Change Material (PCM). The PCM was contained in transparent containers with the heat flux introduced by means of an electric heating element. The stabilized power source was used to maintain the constant heat output of the heating elements. The experiments were performed for the heat flux introduced at the side wall of the container and at the upper surface of the PCM. The paraffin-based PCM RT28HC with the phase change temperature of 28 °C was used in the experiments. The temperature in the environmental chamber was maintained at the melting temperature of the PCM. The propagation of the melting front was monitored with a digital camera and temperatures at several locations were monitored with RTDs and thermocouples. Significant natural convection was observed for the heat flux introduced at the side wall of the container. As a result the melting front propagated much faster at the top of the container than at its bottom. The heat flux introduced at the upper-surface of the PCM resulted in almost one-dimensional propagation of the melting front. The acquired data are to be used for validation of an in-house developed numerical model based on the front-tracking method.

## 1 Introduction

Thermal energy storage makes it possible to balance supply and demand of thermal energy over certain periods of time. Thermal energy storage (TES) is used in heating and cooling applications in order to increase the overall efficiency of these systems. At present, most thermal energy storage systems use sensible heat storage materials, such as water, to store thermal energy. Latent heat thermal energy storage (LHTES), which employs phase change of matter in order to store thermal energy, is much less commonly used. The phase change from solid to liquid state and vice versa is most commonly used in LHTES. The materials for LHTES employing solid-liquid phase change are called Phase Change Materials.

The advantages of LHTES in comparison to sensible heat storage are the reason for a widespread research in this area [1]. One of the advantages of LHTES over sensible heat storage is the high energy storage density in narrow temperature intervals. However, most PCMs have relatively low thermal conductivity and thus large heat exchange areas are needed to achieve high heat transfer rates between the heat transfer fluid and the PCM. Full scale LHTES units can contain large volumes of phase change materials. The melting and solidification of a PCM does not take place in the entire volume at the same time but rather the melting or solidification front starts from the heat exchange surface and then propagates from the surface into the volume of the PCM.

Melting of PCMs in the enclosures or containers of various shapes has been studied by many researchers in the last decade. There are two cases of melting in the enclosures – unconstrained melting and constrained melting. Unconstrained melting means that the solid PCM is not held in place and it can freely move within the enclosure or container due to gravity force or other forces. Since the PCMs have higher density in the solid state than in the liquid state the unconstrained solid PCM tends to sink to the bottom of the enclosure. In case of the constrained melting some measures are taken to prevent solid PCM from moving freely in the enclosure or container.

Kamkari et al. [2] reported experimental investigation of the effect of inclination angle on convection-driven melting of phase change material in a rectangular enclosure. Lauric acid was used as a PCM in the experiments. One of the walls of the enclosure was heated (at constant temperature) while other walls were thermally insulated. Three wall temperatures (55 °C, 60 °C and 70 °C) and three inclination angles (0°, 45°, and 90°) were studied. At 0° inclination the heated wall was horizontal (at the bottom of the cavity) and at 90° inclination the heated wall was vertical. The shortest melting time was achieved for 0° inclination for all three wall temperatures.

Shmueli et al. [3] carried out numerical investigation of melting in a vertical cylindrical tube. Previously obtained experimental results were used for verification of the simulation results. A commercially available PCM

* Corresponding author: mauder@fme.vutbr.cz

RT27 (Rubitherm GmbH), with the melting interval from 26 °C to 28 °C was used in the study. The authors report good quantitative agreement between the simulation and the experimental results in terms of the melt fraction as a function of time.

Fan et al. [4] reported experimental and numerical investigation of constrained melting heat transfer of a PCM in a circumferentially finned spherical capsule. The authors used a Plexiglas capsule with the inner diameter of 100 mm submerged in the Plexiglas water tank to allow for the visual observation. A transparent tube connected to the top of the sphere was used for indirect measurement of melt fraction based on the volumetric expansion of the PCM during melting. The PCM used in the study was n-octadecane ($C_{18}H_{38}$), with the nominal melting range from 28 °C to 30 °C. A thin straight circumferential fin was introduced along the equator of the capsule for augmentation of heat transfer.

The present paper deals with the visual monitoring of melting front propagation in paraffin-based PCM. The goal of the experiments was to obtain experimental data for validation of numerical models.

## 2 Experimental set-up

The goal of the experiments was to obtain experimental data for validation of in-house numerical models that are developed for simulation of melting front propagation in PCMs and other materials. Melting of a PCM in the enclosed cavity represents one of the simplest cases of melting front propagation. The real-life LHTES units often employ PCM-fluid heat exchangers with rather complex heat exchange surfaces where many melting fronts can propagate simultaneously in the volume of the LHTES unit.

### 2.1 Phase Change Material

Alkanes (paraffins) represent relatively non-expensive PCMs with the melting range suitable for thermal energy storage in building thermal systems. The alkanes are a group of hydrocarbons with the chemical formula $C_nH_{n+2}$. The melting temperature of an alkane depends on the number of carbon atoms. The melting temperature of alkanes generally increases with the increasing number of carbon atoms. Many producers of phase change materials use paraffin as the basic component in the so-called paraffin-based PCMs. Other components are usually added in order to modify the properties of the PCM.

The experiments presented in this paper were conducted in an environmental chamber. The environmental chamber allowed for the control of the ambient temperature, which was an important parameter in the experiments. The environmental chamber was a large one with the lengths of 8.85 m, width of 5 m and height of 3.8 m. The dimensions of the chamber allowed for the installation of all experimental equipment inside of the chamber. It was also possible for the investigators to be present in the chamber during the experiments or to periodically check on the progress of the experiments. For that reason several PCMs with the melting temperature between 15 °C and 35 °C were selected as suitable for the experiments. The properties of the paraffin-based PCM RT28HC that was used for the experiments reported in this paper are shown in Table 1.

**Table 1.** Properties of RT28HC

| | |
|---|---|
| Melting area | 27 - 29 °C<br>main peak 28 °C |
| Congealing area | 29 - 27 °C<br>main peak 27 °C |
| Heat storage capacity ±7.5%<br>(Combination of latent and sensible heat in a temperature range of 21 °C to 36 °C) | 245 kJ |
| Specific heat capacity | 2 $kJ \cdot kg^{-1} \cdot K^{-1}$ |
| Density solid (15 °C) | 880 $kg \cdot m^{-3}$ |
| Density liquid (40 °C) | 770 $kg \cdot m^{-3}$ |
| Heat conductivity | 0.2 $W \cdot m^{-1} \cdot K^{-1}$ |
| Volume expansion | 12.5% |
| Maximum operation temperature | 50 °C |

The data sheets of the PCMs mostly contain just the basic thermophysical properties such as the density in the solid and liquid state, thermal conductivity, specific heat, etc. Some characteristics, such as viscosity of a PCM in the liquid state as a function of temperature, which is an important input for CFD simulations, are very often difficult to obtain.

The phase change of most PCMs is not isothermal but it takes place in a certain temperature interval, as can be seen in Table 1. Latent heat, by definition, is the heat absorbed or released by matter at constant temperature (usually associated with the change of state or phase). To avoid confusion, some producers of PCMs provide heat storage capacity in a certain temperature interval around the phase change temperature instead of the latent heat.

The temperature-enthalpy curve is another way of providing information about the thermal capacity of a PCM in relation to temperature. The temperature-enthalpy curves are usually obtained by means of Differential Scanning Calorimetry where a small sample of PCM (usually a few milligrams) is heated up and cooled down at the given rate (e.g. 1 K per minute).

As the DSC uses very small samples of the PCM the method is suitable for microencapsulated PCMs were capsules hold very small amounts of PCM [5]. The microencapsulated phase change materials are often used in building applications as they can be mixed with construction materials such as concrete or plaster [6]. Other methods, such as T-history [7], were proposed for larger samples of PCMs.

### 2.2 Containers for the PCMs

In order to monitor the propagation of the melting front the PCM was contained in transparent containers during the experiments. The containers were made of 25 mm thick PMMA. An aluminium plate with the attached

 DOI: 10.1051/epjconf/201714302042

electric heating elements was used to introduce the heat flux to the block of PCM. The first set of experiments was performed with the heat flux introduced on the side wall of the container. The shape and dimensions of the container for these experiments can be seen in Fig. 1.

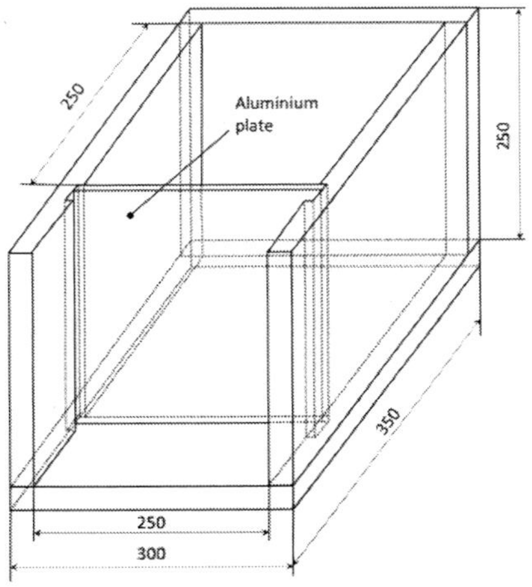


**Fig. 1.** Container with a heated side wall.

As the heat flux introduced on the side wall of the container led to a significant natural convection in the melted PCM, a container for the heating from the upper-surface was prepared. The aluminium plate in this case was 10 mm thick and it had the dimensions to fit into the container. It was important to make sure that the heating plate was in good contact with the PCM during the experiment.

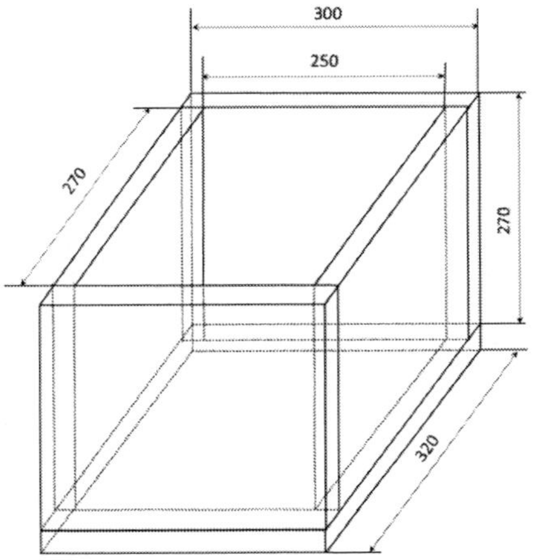


**Fig. 2.** Container for heating at the upper-surface.

### 2.3 Monitoring technique

The monitoring of the melting front propagation was done with the use of a digital camera. The progress of the melting front was relatively slow and it was sufficient to take pictures in 1 minute intervals in case of the side-wall heating and 5 minute intervals in case of the upper-surface heating. The pictures were taken with the Nikon D90 camera fitted with the Sigma macro lens The Aputure Pro Coworker II AP-WTR3N trigger was used to synchronize photo shooting with data acquisition. As the experiments took place in the environmental chamber an LED flash was used for the illumination of the containers.

The environmental chamber had stainless steel walls that created disturbing reflections of the photo flash. For that reason the container was placed in a photo shooting tent with the opaque black background (Fig. 3).

The data acquisition system for temperature measurements was based on the CompactDAQ Chassis connected to a laptop computer via an USB port. The NI 9219 module with four analogue channels was used for connecting the RTD (Pt100) temperature sensors. The Pt100 temperature sensors were used for the monitoring of the aluminium plate temperature. Thermocouples were used for monitoring of the PCM temperature at various positions. The positions of thermocouples differed in the side-wall and upper-surface heating case.

## 3 Results

An in-house image processing code was used for conversion of the photos into black and white (2 bit) images to make the propagation of the melting front better visible and also to assess the melt fraction.

It was assumed that the heat flux introduced on one side of the PCM block would lead to the quasi-2D propagation of the melting front. This assumption was confirmed by the experiments. However, it was important that the entire block of the PCM would have approximately the uniform temperature at the beginning of the experiment. The liquid PCM was poured into the container and then the container was placed in the environmental chamber where the ambient temperature was kept below the melting temperature. It could take two or more days for the PCM to fully solidify and to reach uniform temperature in the entire volume. The ambient temperature in the environmental chamber was set to 27 °C during the experiments.

### 3.1 Side-wall heating

The experiment with the heat flux introduced at the side-wall of the container can be seen in Fig. 3. As could be seen in Fig. 1 one wall of the container was made of an aluminium plate. The electric heating foil and thermal insulation were placed on the external side of the aluminium plate. The temperature of the PCM was measured with the Teflon coated T-type thermocouples at 5 locations at the midplane of the PCM block. Brass tubes were used to hold the thermocouples at the desired positions. The junctions of the thermocouples protruded a few millimetres from the tubes in order to be in the direct contact with the PCM. The thermocouples (sensing junctions) were positioned at the middle of the height of the container. The first thermocouple was 24 mm from the heated wall and the other four were 39 mm, 54 mm, 69 mm and 84 mm from the heated wall (the aluminium plate with the electric heating foil). Fig. 3 shows a situation where 4 thermocouples are in the melted PCM while the fifth thermocouple is still in the solid PCM.

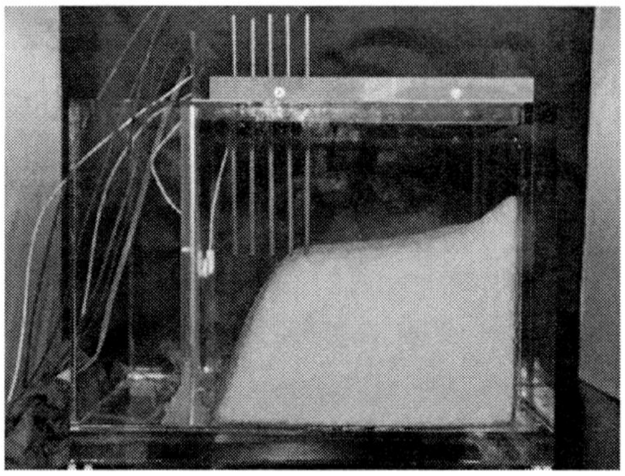

**Fig. 3.** Experiment with side-wall heating

The temperatures recorded in one of the experiments can be seen in Fig. 4. Though the power supplied to the heater was constant (30 W) the surface temperature of the heater $T_{surface}$ changed throughout the experiment. The temperature of the melted PCM increased during the experiment but was almost the same (thermocouples $T_1$ to $T_4$) due to natural convection. The temperature measured by the thermocouple $T_5$ does not change very much as the thermocouple was buried in the solid PCM. As can be seen in Fig. 4 the initial temperature of the PCM block was not uniform (temperatures at time 0 s).

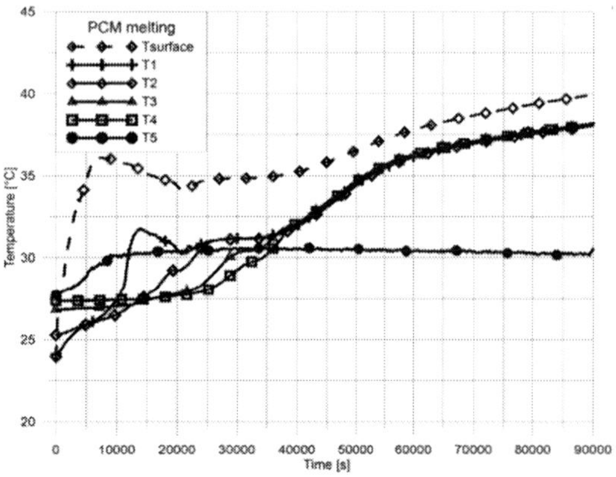


**Fig. 4.** Temperatures for side-wall heating

Propagation of the melting front can be seen in Fig. 5. Black regions represent the PCM in the liquid state and the white regions are the PCM in the solid state. The natural convection in the liquid state significantly contributed to the melting process. As could be seen in Fig. 5 the heat flux introduced at the side wall of the container induced significant natural convection in the liquid PCM and as a result the melting front propagated much faster at the top of the container than at its bottom.

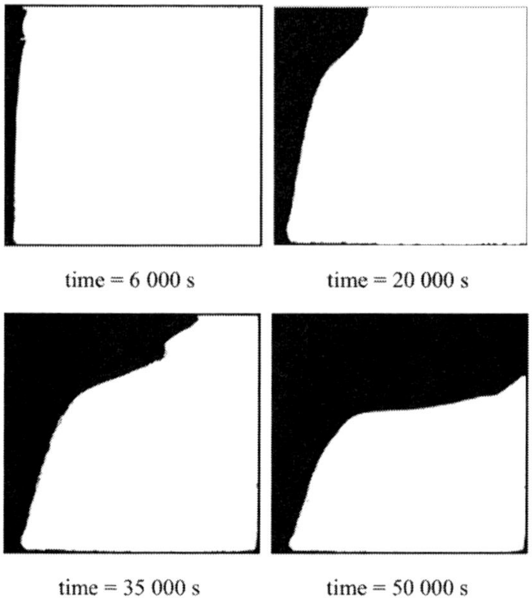


**Fig. 5.** Melting front propagation (side-wall heating)

### 3.2 Upper-surface heating

When the heater is positioned at the top of the container natural convection is significantly reduced and as a result the melting front propagates much slower than in case of the side-wall heating. The photo of the experiment with the heating plate at the upper surface of the PCM is shown in Fig. 6. The melting process in this case is almost one-dimensional and the melting front is a horizontal plane.

**Fig. 6.** Experiment with upper-surface heating

As in case of the side-wall heating the PCM temperature at various positions (distances from the aluminium plate) was monitored with thermocouples. The thermocouples were located in the vertical midplane of the container. The plot of the temperatures can be seen in Fig. 7. The surface temperature of the aluminium plate was higher than in case of the side-wall heating. Different heating elements were used in this case and the heating power was 60 W. In spite of that the melting front propagated much slower in case of the upper-surface heating as natural convection in melted PCM was significantly suppressed.

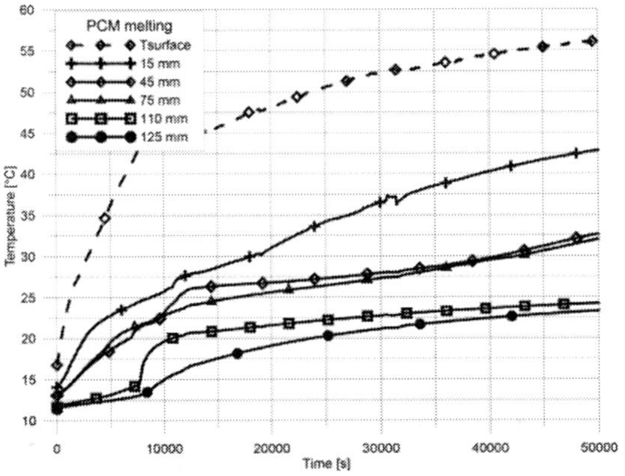


**Fig. 7.** Temperatures for upper-surface heating

The melting front propagation is shown in Fig. 8. The melt fraction at particular time intervals is much smaller than in case of the side-wall heating.

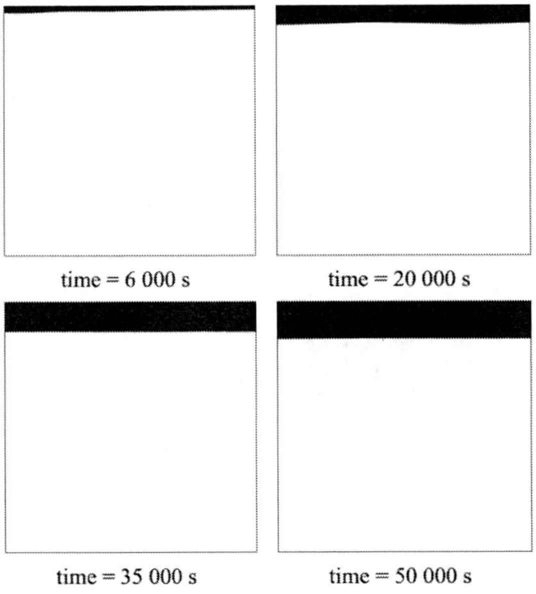


**Fig. 8.** Melting front propagation (upper-surface heating)

### 3.2 Melt fractions

The melt fractions for the two experiments are shown in Table 2. The melt fraction is the ratio of the PCM in the liquid state to the total volume of PCM. As can be seen in Table 2 the melt fraction at each time interval is much higher in case of the side-wall heating than in case of the upper-surface heating. Such result is not surprising as the heat flux direction from the top down is the worst case scenario. The experiment with the heat flux introduced at the bottom of the container is under preparation. As reported by Kamkari et al. [2] the heat flux introduced at the bottom of the enclosure should result in the largest melt fractions at the given time intervals.

**Table 2.** Melt fractions

| Time [s] | Melt fraction [-] | |
|---|---|---|
| | side-wall heating | upper-surface heating |
| 6 000 | 0.0435 | 0.0228 |
| 20 000 | 0.1636 | 0.0740 |
| 35 000 | 0.2949 | 0.1181 |
| 50 000 | 0.4763 | 0.1474 |

## 4 Conclusions

Monitoring of the melting front in the paraffin-based PCM was done with the use of a digital camera and image processing code. Two cases were studied: heat flux introduced at one side of the PCM block and the heat flux introduced at the upper surface of PCM block. Significant natural convection was observed in case of the heat flux introduced at the side wall of the container. As a result the melting front propagated much faster at the top of the container than at its bottom. The heat flux introduced at the upper-surface of the PCM resulted in almost one-dimensional propagation of the melting front. The acquired experimental data will be used in the development of in-house codes for simulation of melting front propagation.

This work was supported by the Czech Science Foundation under contract GA15-11977S.

## References

1. F. Agyenim, N. Hewitt, P. Eames, M. Smyth, *Renew. and Sustain. Energy Reviews* **14** (2010)
2. B. Kamkari, H. Shokouhmand, F. Bruno, *Int. J. of Heat and Mass Transfer* **72** pp. 186–200 (2014)
3. H. Shmueli, G. Ziskind, R. Letan, *International Journal of Heat and Mass Transfer* **53** pp. 4082–4091 (2010)
4. L-W. Fan, Z-Q. Zhu, S-L. Xiao, M-J. Liu, H. Lu, Y. Zeng, Z-T. Yu, K-F. Cen, *Applied Thermal Engineering* **100** pp. 1063–1075 (2016)
5. M. Ostry, R. Prikryl, P. Charvat, T. Mlcoch, B. Bakajova, *Materiali in Tehnologije*, **46**, pp. 813-816 (2012)
6. M. Ostry, P. Charvat, *Procedia Engineering*, **57**, 2013, pp. 837-843 (2013)
7. H. Hong, S. K. Kim, Y-S. Kim, *Int. J. of Refrigeration* **27** pp. 360–366 (2004)

EPJ Web of Conferences **143**, 02043 (2017) DOI: 10.1051/epjconf/201714302043
*EFM 2016*

# Valve exploiting the principle of a side channel turbine

*Pavel* Jandourek[1,a], *Vladimír* Habán[1], *František* Pochylý[1] and *Miloslav* Fic[1]

[1]Brno University of Technology, Faculty of Mechanical Engineering, Energy Institute, Victor Kaplan Department of Fluid Engineering, Technicka 2896/2, 616 69 Brno, Czech Republic

**Abstract.** The article deals with a side channel turbine, which can be used as a suitable substitute for a pressure reducing valve. Reducing valves are a source of hydraulic losses. The aim is to replace them by a side channel turbine. With that in mind, hydraulic losses can be replaced by a production of electrical energy at comparable characteristics of the valve and the turbine. The basis for the design is the loss characteristics of the valve. Thereby creating a kind of turbine valve with speed-controlled flow in dependence of runner revolution.

## 1 Introduction

In many technical solutions, it is necessary to decrease the pressure of liquids. Pressure reducing valves are normally used for this pressure reduction. However, it utilizes a considerable part of the pressure energy that could transform into e.g. electric energy. Part of the pressure energy which is not used, could be transformed e.g. into electric energy. The pumping technique uses side channel pumps for pumping small quantities of liquids to large transport heights. These pumps are characterized by low-specific speed $n_s = 4 \div 40$ min$^{-1}$ and after a small reconstruction they can be used in a turbine mode. It is possible to replace pressure reducing valves with side channel turbines. Thereby creating a new kind of turbine valve with speed-controlled flow dependent on the revolution of the runner.

At the pumped-storage power plant the water for cooling is taken directly from the penstock, see (Fig. 1.). The water taken for cooling is not used to produce electrical energy. During throttling in the valve, part of the energy is transferred in the form of heat to the cooling water and warms it up. The water pressure is throttled from 8.5 to 4 bars. With this comes the potential of energetic utilization.

**Fig. 1.** Pressure reducing valve in a cooling system

[a] Corresponding author: xm101051@ vutbr.cz

The flow is 0.145 m$^3$s$^{-1}$ and the possibility of regulation of the flow to 0.08 m$^3$s$^{-1}$ is required. Another utilization can be in the water mains, energetics or in the place where the conversion of energy into heat occurs due to throttling [1].

Pressure reducing valves are a source of high hydraulic losses and their replacement is possible thanks to the same resistance characteristics with the side channel turbine. Characteristics of the valve flow coefficient $K_v$ on the relative opening of the valve, see (Fig. 3.), is analogous to the turbine characteristic of unit flow $Q_{11}$ on unit speed $n_{11}$, see (Fig. 10.). A medium pressure drop across the valve is needed for energy recuperation.

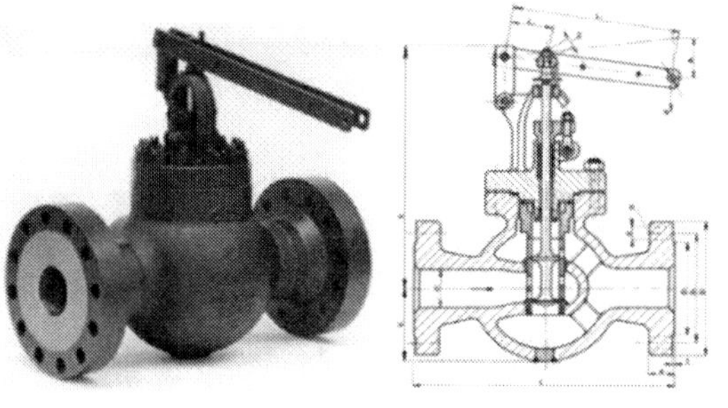

**Fig. 2.** Pressure reducing valve (http://www.ldmvalves.com)

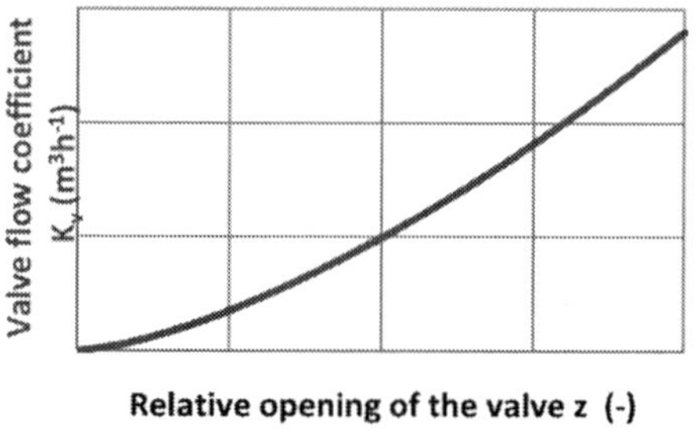


**Fig. 3.** Flow characteristic of a pressure reducing valve

A further use of side channel turbines may be in the replacement of multistage pumps in turbine mode, which are not suitable for the specific speed $n_s < 35$ min$^{-1}$, due to a largely problematic achievement of high efficiency. Under these parameters, a counter-rotating vortex will form in the channels between the impeller blades, disk and which will cause volume losses. This causes a significant drop in efficiency, even down to 30%. In such situations, it may be advantageous to use side channel turbines as they have a higher efficiency than the low-speed multistage pump and also a simpler construction.

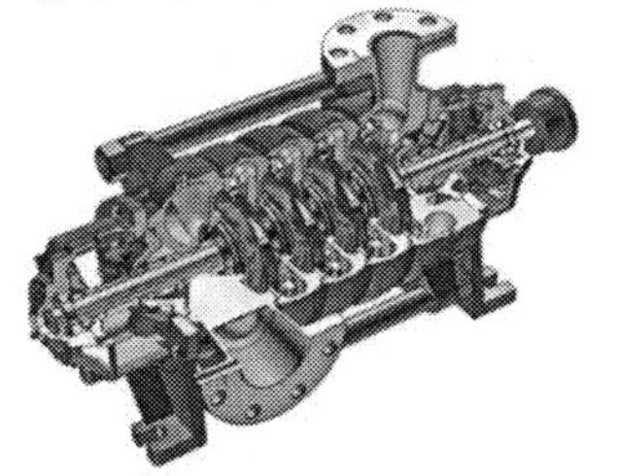

**Fig. 4.** Multistage pump (http://www.neptunopumps.com)

## 2 Side channel machines

Side channel machines are low-specific speed machines, which work with relatively high pressures and low flow rates and a power curve with very stable characteristics. They are smaller and simpler constructions than other comparable centrifugal pumps. The usability band with high efficiency for side channel machines is located on the border areas of hydrostatic and hydrodynamic hydraulic machines and strikes in both of these areas, see (Fig. 6.). Although side channel machines achieve relatively low efficiency, they are predominantly used for their reliability and simplicity.

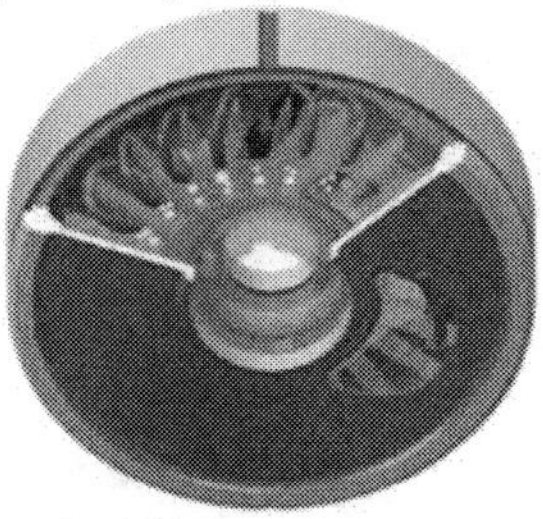

**Fig. 5.** Side channel machine

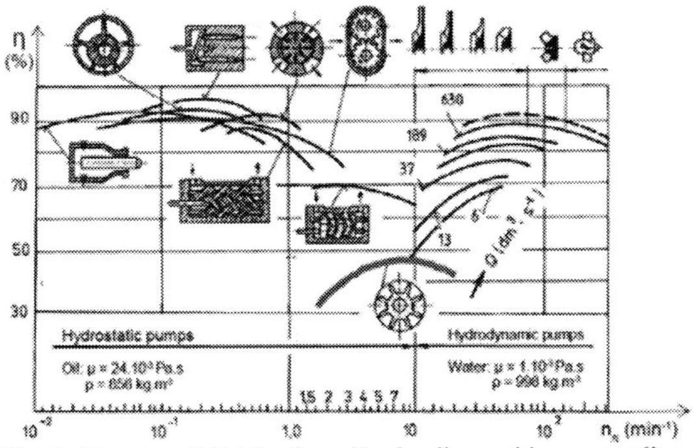


**Fig. 6.** Diagram of distribution of hydraulic machines according to their specific (red marked side channel machines) [2]

The design of the side channel machines is divided into two main types according to the impeller shape:

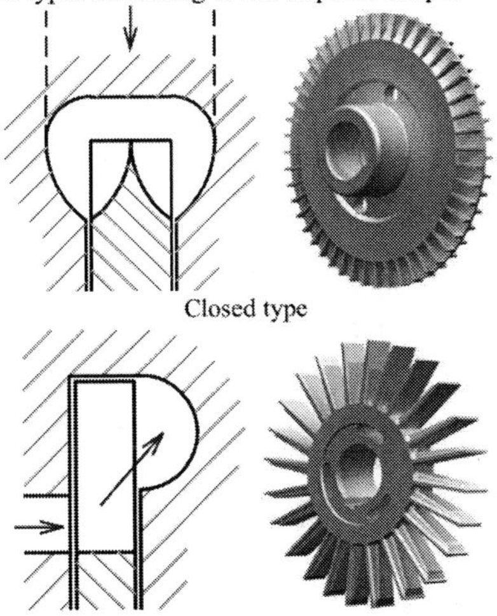


**Fig. 7.** The design of the impeller of side channel pumps

In the impellers of the closed type fluid is supplied directly to the channel. They are usually used for very low speed machines ($n_s = 5 \div 10$ min$^{-1}$), as turbines. Impellers of the open type inlet a liquid to the blades on a smaller radius, than is the radius of the channel. They are used rather in pumps because of their superior self-priming capability.

The principle of the side channel machines is based on a huge local swirl between the blades. (Fig. 8.) shows the swirl, as obtained by experiment in PIV. (Fig. 9.) shows the swirl obtained by using the calculation software Ansys - Fluent with the model of turbulence k-epsilon.

The low total efficiency achieved in side channel machines, usually from 20 to 50%, is caused by several factors. The biggest influence being from hydraulic efficiency. The very principle of operation is based on the swirl and causes high power dissipation. Therefore it is not possible for the side channel machines to reach high efficiencies. Another effect of the aspirated liquid is to impact on the impeller blades and exit from the impeller to the discharge side. Volumetric efficiency affects the width of the side gap between the impeller and the stator part. The mechanical efficiency is affected by large axial forces that may occur in circulation. Suitable type of symmetrical design can minimize them. Losses in storage and mechanical seals are about the same as for similar machines.

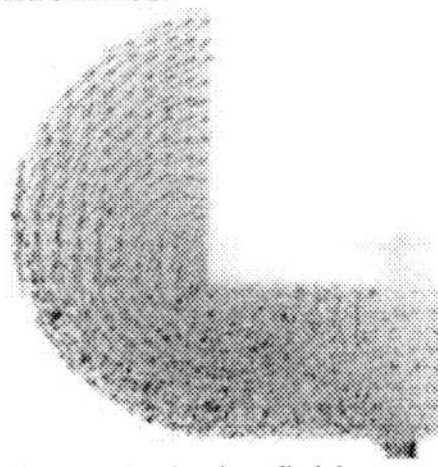

**Fig. 8.** Velocity field according to PIV [3]

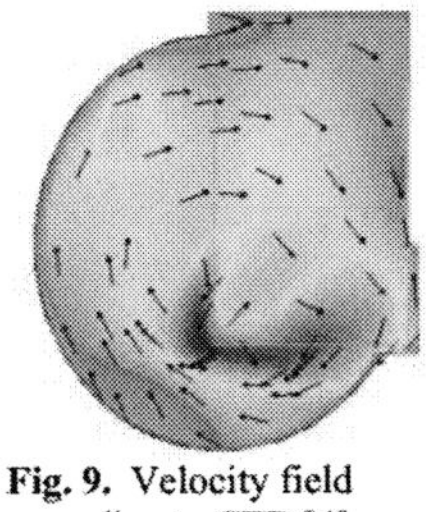

**Fig. 9.** Velocity field according to CFD [4]

### 2.1 Side channel turbines

Side channel turbine falls into the category of reaction turbines. They are intended for small discharges and medium heads. The working area is within the specific speed $n_{st} = 2 \div 30$ min$^{-1}$. The turbine can process only part of the head. The main disadvantage is low efficiency, which is usually within the range of 20 to 50%. Theoretical energy efficiency characteristics η, unit flow $Q_{11}$, power $P_{11}$ and torque $M_{11}$ are plotted in dependence on unit speed $n_{11}$ in (Fig. 10.). These machines find their use due to the fact that reaching specific speed $n_s < 35$ min$^{-1}$ with Francis' turbines at high efficiency is quite problematic. Under these circumstances, counteracting vortex starts to occur in the channels between the blades of the impeller. This causes a decrease in efficiency which thus can even get below 40%. In a situation like this, it might be worth using a side channel turbine due to their higher efficiency than the low specific speed Francis' turbines. Turbines are useful for regulation due to their flat curve of efficiency in the area of optimum point and their linear dependency of flow on revolutions.

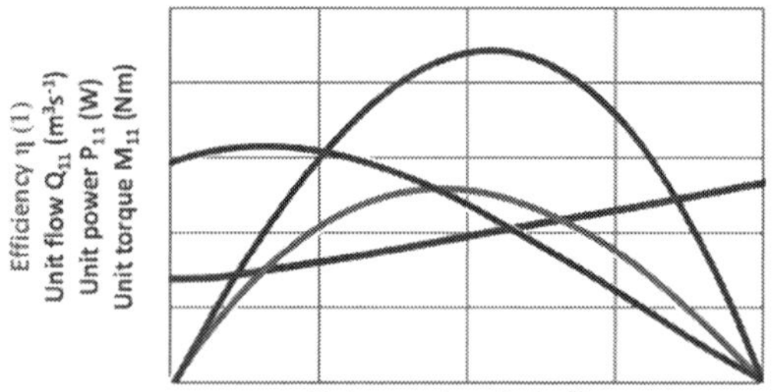


**Fig. 10.** The course of the energy characteristics of side channel turbines

Principle of operation of the turbine: when fluid flow occurs through the turbine runner, it is characterized by constant angular velocity. Whereas flow in the lateral channel depends on the principle of conservation of angular momentum. This causes a pressure differential between the side channels and the ends of the vanes, which causes a strong secondary liquid circulation. This movement is reinforced by circumferential movement of the blades formed by pushing on the liquid, which speeds up the entire radial width of the channel. Movement on the periphery of the runner results in a continuous shift of the difference in energy of the portion of the fluid leaving the impeller into the liquid portion in the side channels. This effect is increased due to the stroke in absence of liquid entering the side channels. Total increase in the energy of the fluid flow through the turbine is generated by pulses of exchange between the fluid in the space of the impeller and the lateral channels. The fluid passes through the runner during one revolution time [2].

Resistive characteristic of the side channel turbine - the dependence of pressure drop flow $p_T$ on turbine flow $Q_T$ (Fig. 11.). Pressure drop $p_T$ is measured at the inlet and outlet of the turbine.

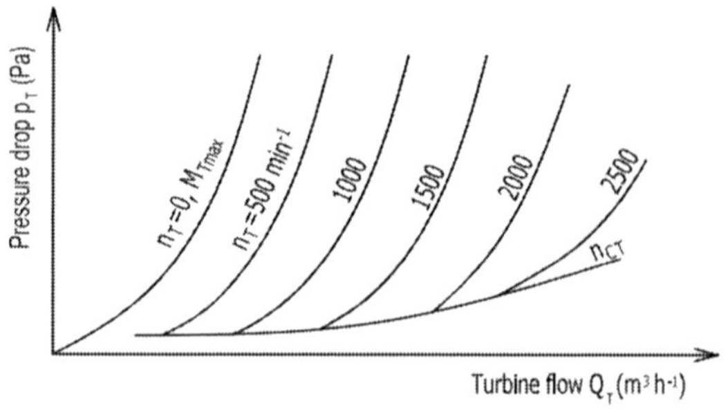


**Fig. 11.** Resistive characteristic of the side channel turbine [5]

## 3 Side channel turbine

### 3.1 The side channel pump in turbine mode

It is further examined, whether it is possible to use the side channel pump as a turbine for energy recovery in the piping system. For this experiment, a commercially produced side channel pump was purchased from the American company Delavan, labeled Turbo 90, see (Fig. 12.). The pump has a closed type impeller. This means that the water enters and leaks directly into the channel. The impeller is symmetrical and has 36 blades on each side. The opposing blades are rotated by half the angular pitch. The side channel is part of the iron casting cabinet, which according to the manufacturer should endure 7 bar overpressure. The size of the connecting pipe is 3/2'. The shaft is sealed with ceramic - carbon mechanical seals. The maximum speed of the wheel is 1200 min$^{-1}$. The width of the side gap δ is approximately 0.1 mm. The dimensions of the hydraulic parts are shown in (Fig. 13.). The impeller is made from nylon. It is connected with the hub straight-sided splines and can move in the axial direction. It is self-centered during operation.

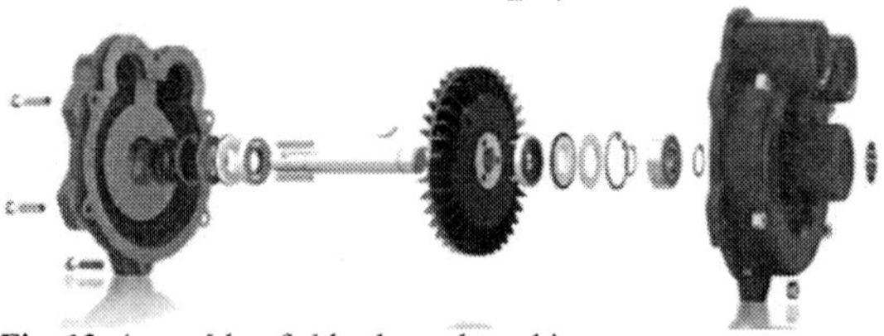

**Fig. 12.** Assembly of side channel machine (http://www.delavanagpumps.com)

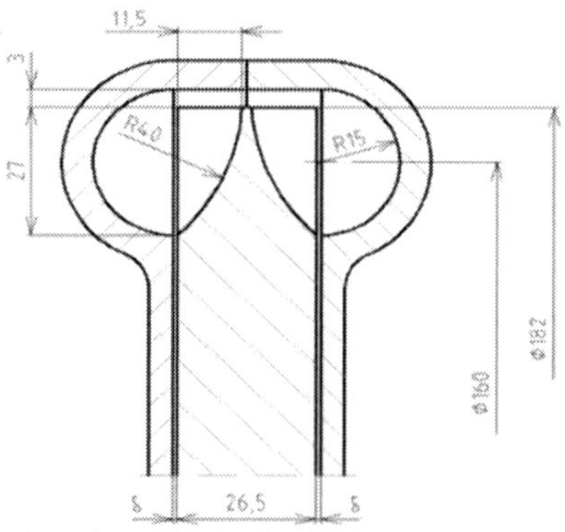


**Fig. 13.** Dimensions of the hydraulic shape

### 3.2 Measurement of the characteristics of the side channel turbine

The side channel pump is measured in turbine mode. A test rig for the measurement of the energy characteristics of the side channel turbine is located in the laboratory of Victor Kaplan dept. of fluid engineering (Fig. 14. and 15.). Frequency converters are used as a generator of the change of speed and a feed pump. Values of inlet and outlet pressure, torque, speed and flow rate are recorded by an industrial computer and our program, which is written in Labview.

**Fig. 14.** Test rig

**Fig. 15.** Turbine in the test rig

### 3.3 Characteristics of the turbine measured with a constant head

The turbine is measured at a constant head 10, 15, 25, 35 m with a changing speed of the runner. At head 10 m, the whole turbine characteristics as well as the stopped runner are re-measured. At head 35 m, it is not possible to develop sufficient pressure by the feed pump, and therefore the characteristics are measured only partially without the optimum point and the zone of higher efficiency. The measured values are recalculated to the unit values and plotted in (Fig. 16.).

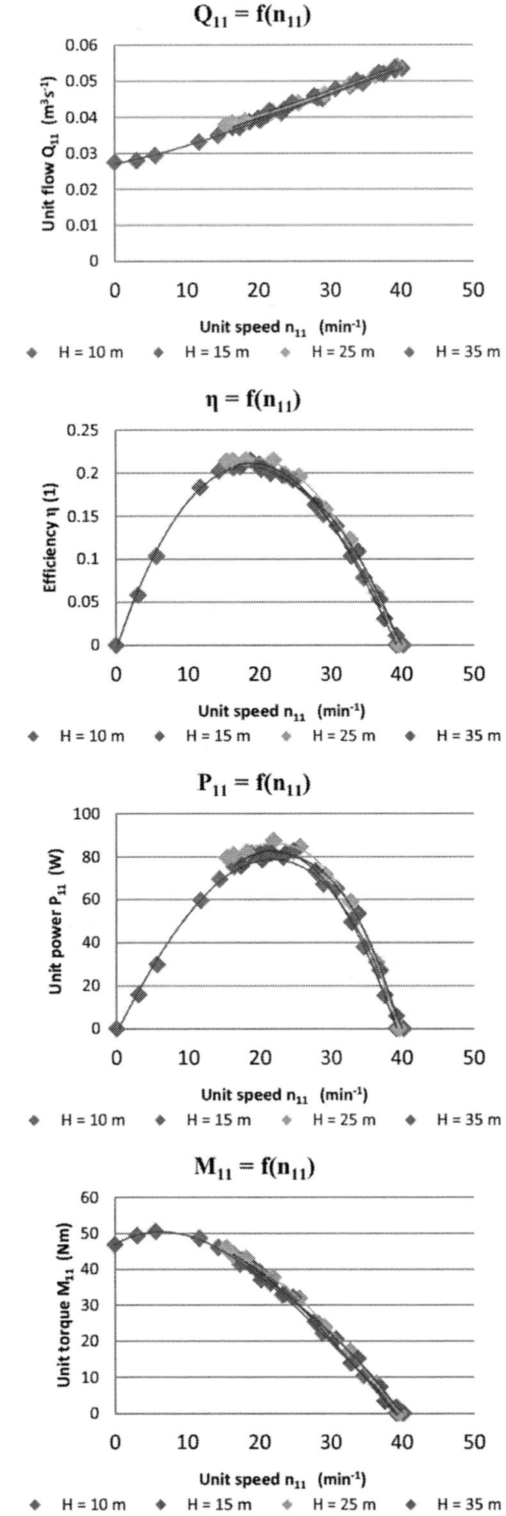


**Fig. 16.** Turbine graphs measured with constant heads

### 3.4 Characteristics of the turbine at constant heads with back-pressure in suction

This section investigates the change of the characteristics of the turbine interposed in the piping system as a replacement of a pressure reducing valve. Turbine works only with a part of the pressure energy from the system. The characteristics of the turbine are measured at a constant head 15 and a changing speed of the runner. The change in the absolute values of the back-pressure in the suction line during the measurement is shown in the graph (Fig. 17.) - $p_b = f(n_{11})$. The back pressure is not constant but rather depends on the flow rate. The values under backpressure are slightly lower than at atmospheric pressure on the suction.

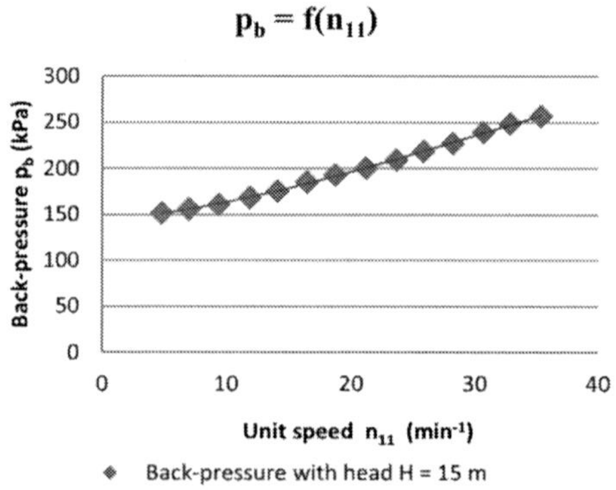


**Fig. 17.** Turbine graphs measured with constant heads and back-pressure in suction

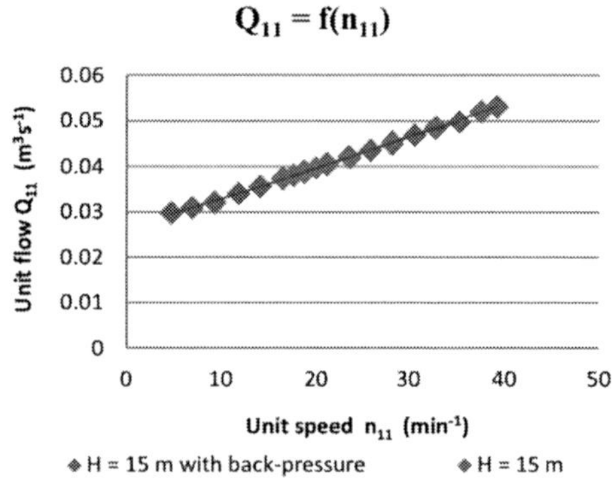


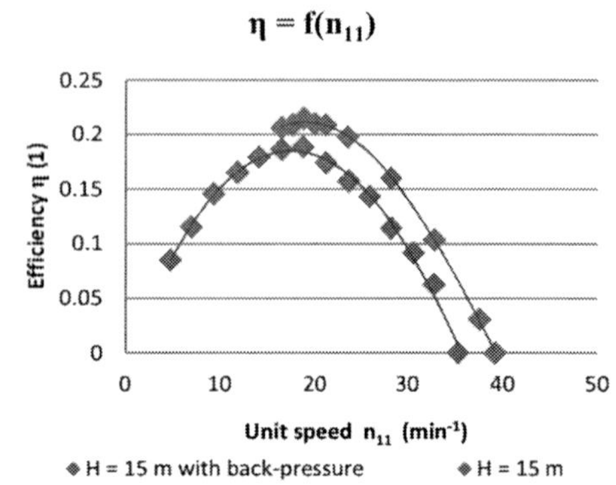


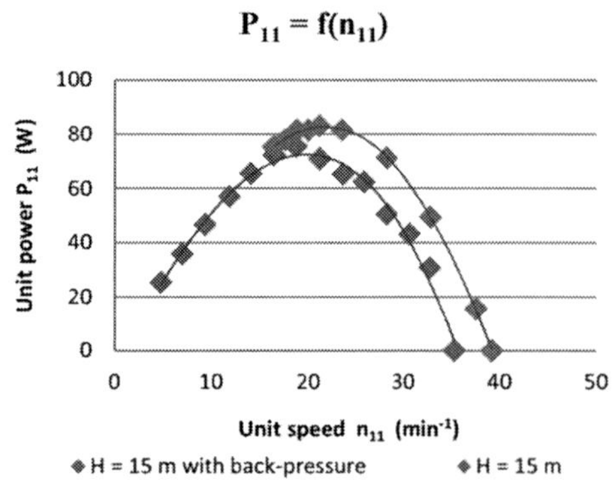


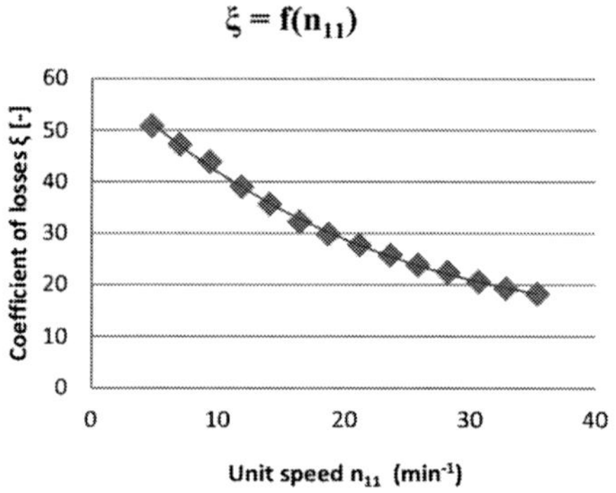


**Fig. 18.** Loss-making characteristic of the turbine

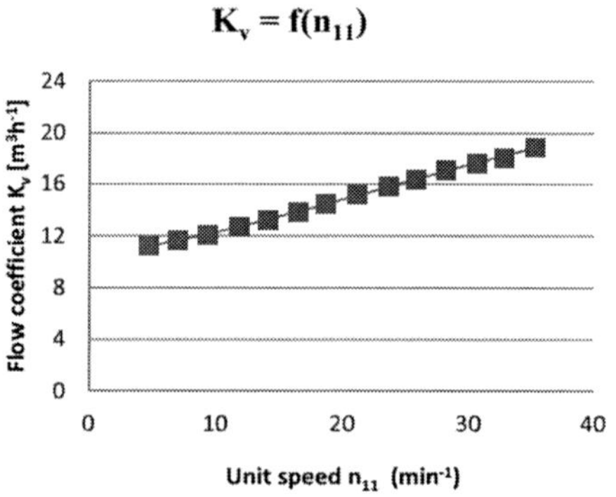


**Fig. 19.** Flow characteristic of the turbine

Characteristics of the turbine valve - during operation of the side channel turbine functioning as the pressure regulator, it is expected that only part of the pressure energy in the piping system will be used. Loss-making and flow characteristics of the side channel turbine Delavan Turbo 90 shown in (Fig. 12.). The characteristic curve of the turbine is the same as the curve of the pressure reducing valve.

## 4 Conclusion

In order to verify the theory, a commercially produced side channel pump was purchased which was used to measure turbine characteristics and verify the function of the partial recovery of energy. The efficiency of recovery is about 18.9% which opens a space for optimization of the hydraulic design of the turbine valve.

## Notation

| | | |
|---|---|---|
| $H$ | m | head of turbine |
| $K_v$ | $m^3 \cdot h^{-1}$ | valve flow coefficient |
| $M_{Tmax}$ | $N \cdot m$ | torque of turbine |
| $M_{11}$ | $N \cdot m$ | unit torque |
| $n_{CT}$ | $min^{-1}$ | continuous rotation of the runner |
| $n_T$ | $min^{-1}$ | revolutions of the runner |
| $n_s$ | $min^{-1}$ | specific speed |
| $n_{st}$ | $min^{-1}$ | specific speed of turbine |
| $n_{11}$ | $min^{-1}$ | unit speed |
| $p$ | Pa | pressure |
| $p_b$ | Pa | back-pressure in suction tube |
| $p_T$ | Pa | pressure drop |
| $P_{11}$ | W | unit power |
| $Q$ | $m^3 \cdot s^{-1}$ | flow |
| $Q_T$ | $m^3 \cdot s^{-1}$ | turbine flow |
| $Q_{11}$ | $m^3 \cdot s^{-1}$ | unit flow |
| $z$ | m | relative opening of the valve |
| $\delta$ | mm | width of the side gap |
| $\eta$ | 1; % | efficiently |
| $\eta_h$ | 1; % | hydraulics efficiently |
| $\mu$ | $Pa \cdot s$ | dynamic viscosity |
| $\xi$ | - | coefficient of hydraulic losses |
| $\rho$ | $kg \cdot m^{-3}$ | density |

| **Index** | **Significance** |
|---|---|
| PIV | particle image velocimetry |
| CFD | computational fluid dynamics |

## References

1. M. Haluza, F. Pochylý, V. Habán, M. Hudec, J. Kubálek: *Study of hydro-potential utilization of cooling water on the hydro-electric power plant Dalešice.* Brno: Brno University of Technology, Faculty of mechanical engineering, 2007. p. 1-16.
2. P. Jandourek: *Valve exploiting the principle of a side channel turbine.* Thesis. Brno: Brno University of Technology, Faculty of mechanical engineering, 2014. 31 p.
3. S. Fialová: *The extracorporeal blood pump.* Doctoral thesis. Brno: Brno University of Technology, Faculty of mechanical engineering, 2005. 68 p.
4. Z. Dančák: *Theoretical and experimental setting of the side channel pump specific energy.* Brno: Brno University of Technology, Faculty of mechanical engineering, 2011, p. 90.
5. J. Bláha, K. Brada: *Manual of pumping equipment.* ČVUT, Prag 1997, 289 s.

## Acknowledgement

This work is an output of research and scientific activities of NETME Centre, regional R&D centre built with the financial support from the Operational Programme Research and Development for Innovations within the project NETME Centre (New Technologies for Mechanical Engineering), Reg. No. CZ.1.05/2.1.00/01.0002 and, in the follow-up sustainability stage, supported through NETME CENTRE PLUS (LO1202) by financial means from the Ministry of Education, Youth and Sports under the „National Sustainability Programme I“ and by the project of science Fund of Faculty of Mechanical Engineering of Brno University of Technology FV 16-26 „Reynold's experiment“.

# Comparison of ultrasound and LIB generated cavitation bubble

*Darina* Jasikova*, *Petr* Schovanec, *Michal* Kotek, *Milos* Muller, *Vaclav* Kopecky

Department of Physical Measurement, Technical University of Liberec, The Institute for Nanomaterials, Advanced Technology and Innovation, Studentska 1402/2 Liberec 1 461 17, Czech Republic

**Abstract.** There has been tried many methods how to generate single cavitation bubble. The purpose of these experiments is to observe the surface impact of collapsing cavitating bubbles. The most of method are based on spark to induced superheat limit of liquid. Here we compare the one of these methods – laser induced breakdown (LIB) with more correct mechanical way using ultrasound as generator. There is described the setup for ultrasound cavitation effecting the sessile bubble as well as random generated and moving bubble. We visualize the cavitation bubble using high speed camera of speed 180 kHz. There are observed interface instabilities as the ultrasound response.

## 1 Introduction

The cavitation is in the scope of not only industrial application, but also biomedical such as medicine, biology, pharmacy, or tissue engineering. The very popular became ultrasonic cavitation in these days. It has a potential in drug delivering systems, cell walls disintegrations, or viruses and bacteria immobilization.

It may seem that the subject dealing with cavitation has reoriented, as applications are concerned. It is not, in engineering applications with more research it has shifted toward solving direct interactions of cavitating bubbles. This take shape in creating novel resistant materials, testing and development of new, more resistant structures or layers which better withstand the action of cavitating bubbles and prolong the service life of products. The research moved to the mathematical modelling and physical analysing.

Anyway, we dealing with great technique, there still exists a lack of information in experimental part of cavitating bubble investigation. There are many ways how to wake cavitating bubble, or the cloud of bubbles. Only few of them follow the postulate of cavitation bubble theory.

The most common method for bubble generation is spark in liquid, or a heated top of the wire. Once the bubble is stable, and of certain volume, it is either over heated, or expose to force impact. Convenient method for generating single cavitation bubble that can be very precisely geometrically placed in the volume of the liquid, and close to the sample, is Laser Induced Breakdown (LIB). [1-5]

To be consistent, cavitation bubble is wake through local decrease of pressure that is ordinary caused by local velocity increase of the liquid. The local velocity increase can be induced by acoustic wave or other mechanical acting. The generation of the single cavitation bubble with acoustic wave is challenging. There has to implemented conditions such as optimal basic wave frequency, the suitable amplitude, the acoustic wave dumping, and the dominating frequency taking in account the lifespan of cavitation bubble. We face many technical problems during implementation: the efficient power source working on variable set ultrasonic frequencies, the synchronizing elements, and the material resistivity. [6]

Cavitation can be defined as a collection of effects connected to the origin, activities, and collapse of macroscopic bubbles in liquid. Cavitation bubbles are usually not separated in real applications. The bubbles create structures, which acts collectively, however essential elements of these structures are individual bubbles.

The understanding of its dynamics and impact to the surroundings represents the key in the understanding of the cavitation phenomena in its complexity. Here we compare two methods of the impacting cavitation bubble. The first method is ultrasonic cavitation (SONO), the second LIB. The purpose of this study is to insight into the behaviour of bubbles.

## 2 Experimental

The bubbles can be generated by several physically different mechanisms. The most obvious in nature is the hydrodynamics cavitation, where bubbles are produced due to local pressure decrease caused by the flow acceleration in vicinity of obstacles. Acoustic cavitation is produced by imposing an intensive acoustic field into the bulk of liquid [1].

The acoustic field cause the local tension of the liquid and its rupture. Energy deposition represents another possibility for the bubble generation. Here, we used two ways for cavitation bubble generating in liquid. The spark based method LIB was compared to ultrasonic

* Corresponding author: darina.jasikova@tul.cz

wave impact to cavitating bubble. The working liquid was distillate water of 8 mg dissolved $O_2$/l at 20°C. Both experiments run in optical glass cuvette.

### 2.1 Laser Induced Breakdown

Optical breakdown in liquid is usually produced by focusing of the laser light trough suitably designed optics. Laser induced breakdown in aqueous media and its collateral effects are described in detail by Kennedy in [2]. The energy distribution during the growth and collapse of laser induced bubble was described e.g. by Vogel in [7] and [8]. Authors investigated the influence of laser pulse duration and input laser energy on the bubble dynamics and shock waves emission.

For the single bubble generation we used here the setup for LIB. The 10ns width laser pulse was generated using Q-switched Nd:YAG NewGemini pulse laser. This laser worked with one cavity for single shot generation on the wavelength 532 nm. The Q-switch signal synchronized the high speed camera running in triggering mode.

LIB was set as an optical direct way. The outlet diameter of laser beam was 5mm with Gaussian characteristics of the intensity. This setup was followed with concave lens $f_s$ = 200 mm, and convex lens $f_s$=25 mm of 1inch diameter. The focused laser beam created the laser point - probe (diameter < 0.1 mm). Due the losses in the optical path on each of optical elements, comparable to this, we had to increase the energy level that enters the whole system. The set output energy of the laser is taken in account in the relation to the bubble diameter. The energies that are required for the bubble generation are seen in Fig. 1.

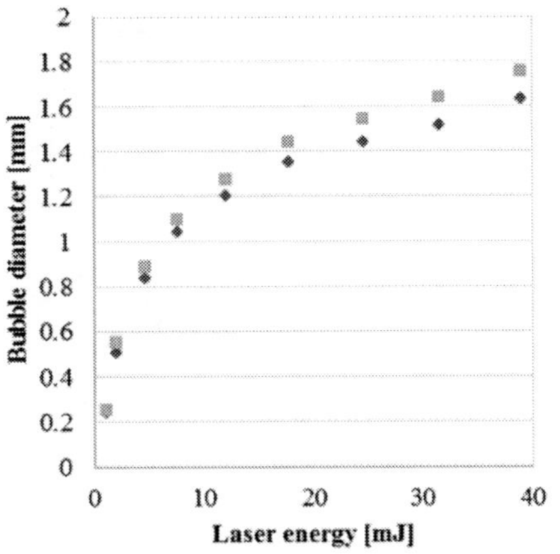


**Fig. 1.** The graph of maximal bubble diameter on laser energy.

For the relation between input laser energy and bubble size that was generated with our optical setup, we create dataset of size – measurement. The energy of laser beam was set and measured with Ophir pyroelectric energy sensor. The dataset of at least 20 cavitation processes for statistic evaluation was captured. This maximum bubble size was detected in each image and it was related to the energy of the laser beam. The bubble size was measured in vertical and horizontal axis and final value represents average value. The increase of laser energy causes the increase the horizontal value of cavitation bubble size. This effect corresponds with temporal and spatial plasma evolution at the very beginning of the process. The relation between the energy of laser beam and cavitation bubble size is asymptotic. This means that further increase of laser energy does not lead to significant increase of the bubble size.

According to Kennedy [9], plasma temperature shows asymptotic dependence on laser pulse energy as well. With higher laser energy we recognized negative influence of impurities and presence of segmentations on the bubble surface.

### 2.2 SONO setup

Ultrasound is an acoustic wave of frequency higher than 20 kHz. Its source can be mechanical, magnetostriction and piezoelectric generators.

Ultrasound passes through the material environment through the vibration spread over particles. Velocity of propagation is not frequency dependent, so the ultrasound propagates at same speed as sound. There are large differences in absorption in various environments. Ultrasound is strongly absorbed by gas, 100 kHz frequency reach 2.2 m in the air, and in water 4 km, whilst with lower wavelength absorption grows rapidly, so e.g. 1 MHz ultrasound reach only 22 mm in the air and 40 m in water. [10]

We used the ultrasonic setup for observing and visualization of the processes mentioned above. The cuvette was fulfilled with water, and there was place sonotrode above the steel sample. The ultrasound was generated using ultrasonic Processor UIP1000hd (1000W) working on 20 kHz frequency. The whole setup consisted of transducer and generator, and titanium horn as the sonotrode.

Cloud of bubbles was generated with te syringe pumping air into the liquid. Once the bubbles were spread close to the surface the ultrasonic generator was switched on.

### 2.3 Visualization setup

We used here shadowgraphy setup for the bubble visualization. This setup consists of continuous LED matrix daylight lamp Veritas, MiniConstellation 120 – 5000 K of illuminance 92 klux in 0.5 m, set with optical diffuse filter. Opposite to the light source was placed high speed CMOS camera SpeedSense. This camera is working on frequency 180 kHz with resolution of (128 x 128) px or lower frequency with higher resolution up to (1280 x 800) px, and the dynamic range 12 bit. The camera exposure time was 1 μs. The sub-pixel resolution was 20 μm. The camera was mounted with optical lens system INFINIPROBE™ TS-160 universal macro/micro imaging system that enables 4x, and 16x magnification. The camera was mounted with edge pass and long pass filter cut-on wavelength 550 nm low pass optical filter to reduce the backward laser flashes to the camera and also to eliminate the flash generated while plasmatic breakdown.

We also used a light reflecting setup for visualization of the outer bubble structures. We placed the light under 45° from the camera direction.

## 3 Results

The ultrasonic cavitation means active manifestations of oscillating gas bubbles in the liquid induced by ultrasound waves. Ultrasound causes periodic changes in the instantaneous pressure in the liquid, which are superimposed on the pressure. The small gas bubbles presented in the liquid begin to radically nonlinearly oscillate at certain pressured amplitude. The cavitation can be divided between collapse (transient) and resonant (stable). However, the resonating cavitation bubble can go to chaotic behaviour and collapse after a few cycles. Therefore the system is marked as cavitating or not. There have been discovered two main sources of physical damage. The first source is pressure pulse emitted into the liquid near the collapsing bubbles.

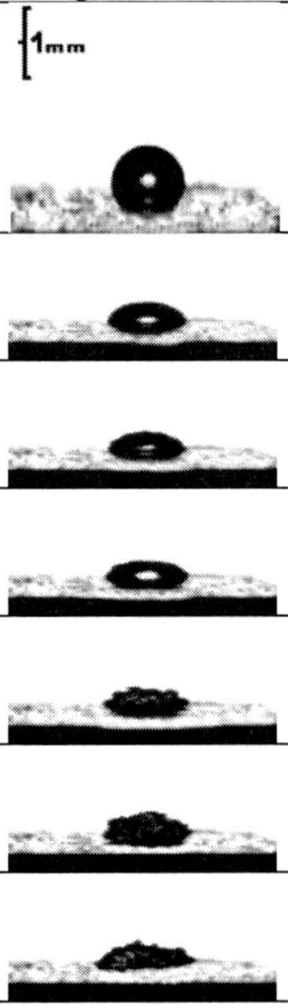


| Time | Image of the bubble |
|---|---|
| $t_0 = 0$ ms | |
| $t_1 = 6{,}7$ ms | |
| $t_2 = 8{,}3$ ms | |
| $t_3 = 10{,}0$ ms | |
| $t_4 = 11{,}7$ ms | |
| $t_5 = 15{,}0$ ms | |
| $t_6 = 35{,}0$ ms | |

**Fig. 2.** The effect of the ultrasound on the sessile bubble using shadowgraph method.

The spherical shock wave may have amplitude of up to 1 GPa. When the bubble collapses isolated, the shock wave is quickly suppressed. This bubble is damaging only touching surface structure in the immediate vicinity. [11, 12] On the other hand, in a concentrated burst of bubbles the collapse of one bubble may induce the collapse of other bubbles around. This "cavitation cloud" can cause mechanical damage at greater distances from the "epicentre". [13]

The ultrasound effecting on the bubble only in the peak of amplitude that is of 20 kHz, so one time period is 50 ms. Till the inner bubble pressure increase the certain value and collapse of bubble happens it takes a couple of periods.

The sessile bubble is directly acting at the surface. Finally, this leads to the bubble disintegration and breakup. For the better visualization of the bubble interface we used the reflective illumination method, the pictures taken with high speed camera are seen in Fig. 3. In this case, the bubble was directly generated with ultrasonic waves and moved towards the surface. The taken pictures are result of very stochastic process.

| Time | Image of the bubble |
|---|---|
| $t_0 = 0$ ms | |
| $t_1 = 15{,}0$ ms | |
| $t_2 = 17{,}5$ ms | |
| $t_3 = 25{,}0$ ms | |
| $t_4 = 27{,}0$ ms | |

**Fig. 3.** The effect of the ultrasound on the sessile bubble using light reflecting method.

The structures of the gas – liquid interface are se as the *orange-peel disturbance* that is said to be caused by evaporations. Frost, et.al used this term firstly. [6] The structure is seen in Fig. 2 at time $t_4$ = 11.7 ms from the Sono switch on. The instabilities are likewise Rayleigh-Taylors. We managed to capture the very beginning of the instability spreading from the top of the bubble. This more follows the behaviour of the ultrasound as it spreads over the liquid and refracts and reflects from the cuvette walls and the sonotrode itself. The force of waves is summed in small locals and it's projected on the bubble interface. The lifespan of the bubble toward the disintegration is strongly dependent on source frequency. There is seen the bubble response on the first half of the ultrasound wave. The main influence has positive part of the amplitude on emergence of failures.

The Sono bubble of comparable size has lifespan in one order more than LIB generated one. This can be change using 2.3 MHz ultrasound source, but of sufficient power.

Generate single cavitation bubble with LIB is limited by initial plasma probe. The geometry of the plasma predetermines the number of bubbles. If there is higher input of light energy the probe elongate and the energy diversify and there will be created more than one bubbles that finally conjunct into big one and collapse.

The LIB bubble lifespan is up to 120 μs. During the LIB the pressure increase is induced by the local temperature. This effect last 50 μs in average. The bubble capturing starts at the plasma point that is set as the zero time.

| Time | Image of the bubble |
|---|---|
| $t_0$ = 0 us | 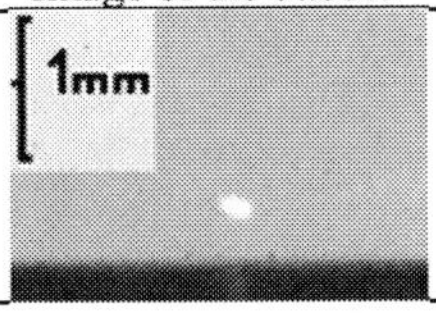  |
| $t_1$ = 5,6 us |  |
| $t_2$ = 27,8 us |  |
| $t_3$ = 55,6 us |  |
| $t_4$ = 66,7 us |  |
| $t_5$ = 77,8 us |  |
| $t_6$ = 100,0 us |  |
| $t_7$ = 122,2 us |  |

**Fig. 4.** LIB generated bubble and its implosion towards the surface.

The LIB bubble surface oscillations are developing faster than we were able to recognize. The maximum limit of the high speed camera was 180 kHz to keep sufficient imager resolution. The assumption of surface oscillations is supported by the observation that the bubble is exploded unsymmetrically. This theme should be of further examination at the edge of technical limits.

## 4 Conclusions

There has been done a series of experiments as the ultrasound cavitation is random process. The bubble response to the mechanical action is dependent of the upper half of amplitude and frequency. The lifespan of the cavitation bubble is also dependent on the working frequency. As the interface instabilities are seen right as the reaction on the force impact we suppose that the orange-peel disturbance is caused by ultrasound wave refractions and reflections.

The comparison with LIB bubbles is varying in range of lifespan. Also the range of bubble size is limited to the plasma probe geometry unlike the ultrasound is stochastic effect. The bubbles are created in the range of sizes and mostly occur as the cloud of bubbles. The comparable bubbles can be generated only with increased ultrasound frequency of with the pulse, amplitude modulation, and focused sonotrode.

Here, we managed to visualize single cavitation bubble that is affected by the ultrasonic waves. This first step is crucial for further investigation of the process itself in term of measuring cavitation impact force on the sample of material.

The results of this project LO1201 were obtained through the financial support of the Ministry of Education, Youth and Sports in the framework of the targeted support of the "National Programme for Sustainability I". Authors also acknowledge the institutional support of the Faculty of Mechatronics, Informatics and Interdisciplinary Studies of the Technical University of Liberec.

## References

1. Ch. E. Brennen, *Cavitation and bubble dynamics*, (Cambridge University Press, 2015)
2. P. K. Kennedy et.al, Prog.QuantElectr. **21**, 3, (1997)
3. E. A. Brujan, et.al, J.Fluid Mech. **433**, (2001)
4. W. Lauterborn, et.al, Archives of Acoustics **33**, (2008)
5. A. Shima, Y. Tomita, Ingenieur-Archiv **51**, (1981)
6. D. Frost, B. Sturtevant, ASME **108**, (1986)
7. A. Vogel, S. Busch, U. Parlitz, J. Acoust. Soc. Am. **100**, (1996)
8. A. Vogel, et. al, Applied Physics B: Lasers and Optics **68**, (1999)
9. P. K. Kennedy et al., Prog. Quant. Electr. **21**, 3, (1997)
10. I. Hrazdira, V. Mornstein, Medical biophysics and technology (Neptun, 2001)
11. R. Hickling, M. S. Plesset, Phys. Fluids, 7 (1964)
12. R. D. Ivany, Tech. Rep, 15, University of Michigan, Dept. Of nuclear engineering, 1965
13. B. Vias, C. M. Preece, J. Appl. Phys., **47** (1976)

# The interaction between fluid flow and ultra-hydrophobic surface in mini channel

*Darina* Jasikova[1,*], *Maja* Gašič [2], *Simona* Fialova [3], *Michal* Kotek[1] and *Vaclav* Kopecky[1]

[1]Department of Physical Measurement, Technical University of Liberec, The Institute for Nanomaterials, Advanced Technology and Innovation, Studentska 1402/2 Liberec 1 461 17, Czech Republic

[2]Department of Physical Electronics, Faculty of Science, Masaryk University, Kotlarska 267/2, 61137 Brno, Czech Republic

[3]Faculty of mechanical engineering, Brno University of Technology, Technicka 2896/0, Kralovo Pole, 61669, Brno, Czech Republic

**Abstract.** Interaction of liquid with ultra-hydrophobic surface is accompanied by creation of layer of air. The effect of the air film has a potential of use in industry in many applications. The quality of the surface is influenced by matrix roughness, the character of physical or chemical cover. There was developed a method for analysis of the liquid flow and the air film using the lighting in volume, visualization with CCD camera and long distance microscope, and optical filters. There were prepared four stainless steel samples of inner channel of dimensions (80 x 8 x 8) mm and initial surface roughness Ra 0.33, Ra 1.0, Ra 2.0, and Ra 2.2. The inner channel was treated with plasma and commercial hydrophobic coating Greblon (WEILBURGER Coatings GmbH). There was realized study focused on the liquid flow velocity profile close to the air film. There are present results for laminar, transient and turbulent flows. The study also estimated the air film thickness depending on the Re number. The knowledge of the air film behaviour helps applied suitable degree of processing and cover for the target application.

## 1 Introduction

The ultra-hydrophobic surfaces have the prospect of great importance in industry, both in applications demanding easy cleaning, and they are presumed to reduce loss when the active parts of hydraulic machines are treated. Interaction of fluids with ultra-hydrophobic surface is accompanied by creation of layer of air, which depends on the quality of the surface. The quality of the surface is influenced by the matrix roughness, the character of physical or chemical cover. This properties lead to monolithic air layer presented as air film, or lead to plurality of bubbles of various sizes seated upon the surface.

The air layer can be observed visually at sufficient magnification and the dynamic interaction between fluid flow and air film can be studied with Global Imaging methods, particularly PIV. There is the velocity profile in the vicinity of the air layer in the main interest of the research. There was already described in the previous publication, focusing on the initial identification of an air layer, that the dynamics of the air layer is dependent on the velocity of the flowing liquid. This speed is expressed as Reynolds number, which is taken as a hydraulic diameter of the duct without an air layer. [1]

The initial measurement and identification of the air layer was performed on the testing device of hydraulic diameter $D_h$ 0.00375 and extreme long entering distance – 80 times of the channel width to eliminate the effects of transient or underdeveloped flow pulsations. The test section was made of optically clear polycarbonate plastics with a defined roughness of interior walls. [1, 2]

There was developed the a method for analysis of the flow and the air film using the lighting in volume, visualization with CCD camera and microscope lens, optical filters based on defined depth of field measuring plane. [1, 2]

The next step was the experimental confirmation of theoretical assumptions about the nature of the air film on commercially used materials that have already been tested in hydraulic equipment – hydraulic pumps CKD Blansko. From industrial requirements resulted question, how is the air film depending on the roughness of the base material. There was a presumption of the smooth surface enabling more efficient and stable air film, but surface polishing increases production costs and it is not always consistently possible to all machined surfaces. There were fabricated four samples for testing purposes. These samples had defined inner duct's surface roughness corresponding to individual stages of material processing. The supposed result of measurement is to determine the nature of the air film depending on the Re number, the roughness of the material, and the character of the velocity field in the vicinity of the air film. [3]

* Corresponding author: darina.jasikova@tul.cz

## 2 Experimental setup

There were prepared four stainless steel samples of inner channel diameter (80 x 8 x 8) mm. The inner part of the channel was machined to corresponding surface roughness. The sample surface marking followed ISO CSN 1302:1992 norm. The Ra 0.33 (average surface roughness 0.33 μm) corresponds to very fine edging and brushing, Ra 1.0 represent the surface finished with fine machining, such as turning, milling work or drilling. The rough surface represented with Ra 2.0, and Ra 2.2 substitute materials machined as prepared for finishing. The roughness of each sample was measured using standard laboratory device Mitutoyo Surftest SV-2000 after the machinery was finished.

The inner duct was further surface modified using plasma treatment to increase the contact angle. The open channels were covered with PCB optically clear cover and set with connection hoses. The samples were joined to the water circuit.

The system was enhanced with a linear precision voltage source KORADO KA3005P controllable from a PC and an adjustable voltage with an accuracy of hundredths.

The liquid - water was draw from the reservoir. The entrance length of the inlet straight duct was 1m as well as the strait part on the outlet from the channel.

### 2.1 Materials and Sample preparation

Stainless steel samples (80 x 30 x 15) mm in size with commercial hydrophobic coating Greblon (WEILBURGER Coatings GmbH) were used as a substrate. The experiment was carried out in the channels with dimensions of (80 x 8 x 8) mm, which were placed in the middle of the sample. To obtain the super hydrophobic surface the samples were coated with super hydrophobic two-component (top and bottom) product Ultra Ever Dry developed by UltraTech international, Inc. Some samples (P-1L, P-2L) were pre-treated by plasma to improve durability of super hydrophobic layer. In order to assess the plasma modification effect, the samples without plasma treatment were also prepared (R-1L, R-2L). The samples R-1L and P-1L are coated only with one of the Ultra Ever Dry components (top) to increase the layer homogeneity.

#### *2.1.1 Plasma pre-treatment and super hydrophobic layer preparation*

The atmospheric plasma slit-jet [reference] was used for samples pre-treatment. Electrodes were connected via an impedance matching unit to a radio frequency generator (Dressler) which is operated at a frequency of 13.56 MHz. The applied power was 300 W. Argon (99.996 % Erproduct) was used as a working gas delivered at a flow rate of 50 ml/s into the plasma jet. The super hydrophobic product was subsequently applied according to the instructions using a commercial spray device. [4]

#### *2.1.2 Contact angle measurement*

Contact angle measurement was performed using SeeSystem 7.0 calculating the WCA on the basis of three-point interpolation of the drop height and width. Deionized water was used as a testing liquid and the volume of the drop was 2 μl.

**Table 1.** The sample marking, type of surface treatment, contact angle, and initial surface roughness.

| Sample | Treatment | Contact angle | Surface roughness |
|---|---|---|---|
| R-1L | Reference, 1 component | 155±1 | 0.33μm |
| R-2L | Reference, 2 components | 153±1 | 1μm |
| P-1L | Plasma pre-treatment, 1 component | 155±1 | 2.0μm |
| P-2L | Plasma pre-treatment, 2 components | 154±1 | 2.6μm |

Prepared Samples:
ČK-KAN-A = R-2L Reference, 2 components
ČK-KAN-B = P-2L Plasma pre-treatment, 2 components
ČK-KAN-C = R-1L Reference, 1 component
ČK-KAN-D = P-1L Plasma pre-treatment, 1 component

### 2.2 Visualization techniques

The visualization of the air film came out from the micro PIV measurement technique. This technique allows the visualization, tracking and evaluation processes within a very small scale of micrometres. The design of the measurement track and the samples were adapted to his method. The air film is shown as a light stripe that acts as lens. The air film refracts and reflects the illuminating light due its surface curvature. It also shows the mirroring effect. There is also perceptible surface structure in the picture. The effect of increasing flow rate is seen in Fig. 1.

The micro visualization technique provided us the information about the air film thickness. We used the long distance microscopy to cover the whole velocity profile across the channel – 5 mm, and 8 mm. The basis of micro PIV technique is a laser beam, which illuminates the entire volume of the examined space. For illumination of the evaluated area NewWave Gemini Nd:YAG pulsed laser is used, energy of 120mJ per pulse length of 10 ns, wavelength 532 nm.The images are captured with Neo CMOS chip 5.5 MPixel size of 6.5 microns camera. Laser and camera system is controlled from a computer and synchronized with the external TimerBox.

Unlike standard PIV method, where the depth of field depends on the distance to the measurement object from the camera lens, f-number, in the micro PIV method depth of field is dependent on the parameters of the microscope lens, index of refraction of the liquid, and the wavelength of used light. Other parameters are related to the system of lenses: NA - iris, M - total magnification microscope, and e - the distance between the pixels on the CCD camera.

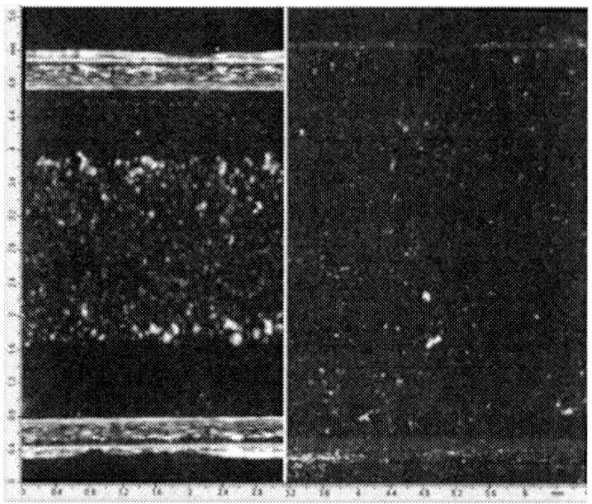

**Fig. 1.** View into the PCB channel: left side - air film upon the walls with the fluid flow Re 300, right side – the same channel with the Re 3000 - clear walls without air film.

The arrangement of micro PIV has a constant depth of field 4 μm at 10x magnification. This factor is important for choosing the size of seeding particles. Here in this investigating we used coated Rhodamine B 2 μm fluorescent particles emitting light on 570 nm. We used set of microscopic calibration targets for the calibration of the system to determine the magnitude coefficient.

The long distance microscopy method was used for the measurement of the velocity profile over the whole channel in one step. This optical setup was based on INFINIPROBE™ TS-160 universal macro/micro imaging system that enables 4x and 16x magnification. This probe was working in the set of compressor for the wide field of view, and the working distance was 100mm in the macro regime. The area of interest was illuminated with the laser light led with optical cable towards the sample.

The measurements run at 8 Hz frequency with 60 mW laser power at one flash. The light from the laser head was brought to investigated area with optical cable. For each measurement at least 290 records were always acquired and were further processed by the correlation function, validation and filtering before being evaluated statistics and estimated velocity profiles.

## 3 Results and discussion

The study of the velocity profile close to the air film, and the wall was based on the visualization using lightening in volume and the CCD camera with regulated shutter and defined depth of field. This setup allows us to visualize the thickness of the air film as well as the application of the cross-correlation analysing algorithms.

The entering distance was set according to the definition *L/Dh/Re = C*, for C > 0.03 in the turbulent flow and C > 0.08 for the laminar flow. [2] We elongated the entering distance to avoid the flow instabilities up to 1 m. The entering distance corresponds to 100times of channel widths.

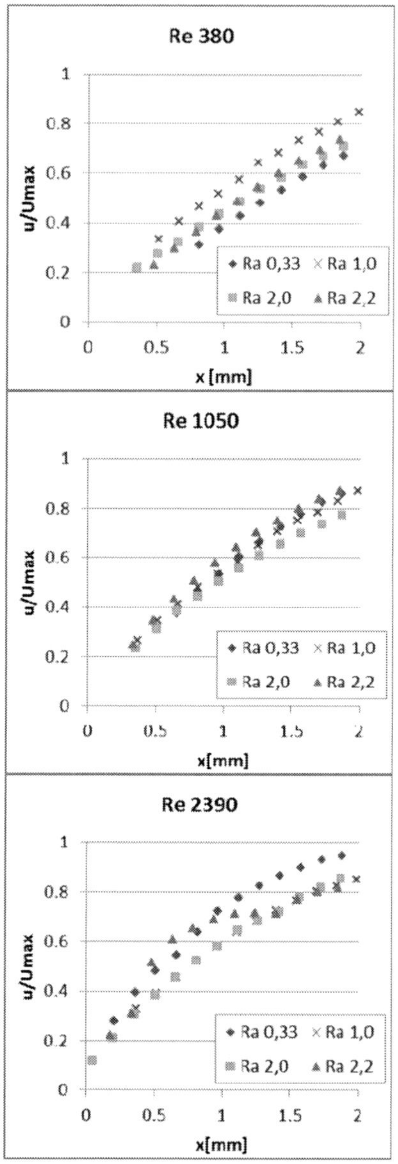


**Fig. 2.** The results of cross-correlation (32 x 32)px. The velocity profile normed to maximal velocity for the laminar flow Re 380, transient flow Re 1050 and the turbulent flow Re 2390. The x[mm] axis represents the distance from the surface/channel's wall. Umax represents the maximal velocity in the channel.

The water channel was fulfilled with water avoiding the bubbles. The Reynolds number was fluently increased on selected values. Each liquid flow was stabilized at least 15 minutes. There was observed and captured the air film. There were tracked significant changes in air film thickness in time using 1Hz sampling frequency. The air film was developed on each sample, but the quality of it varied depending on the surface roughness.

The thickness of air film on rough surface of Ra 2.0 and Ra 2.2 corresponds to the air film created on PCB with UED® nanoparticle. The average roughness of untreated surface was measured as 1.0 μm (Ra 1.0). The samples treated with single layer of UED® nanoparticles increased in the roughness to 1.6 μm (Ra 1.6), and those were double treated increased in surface roughness from 2.4 μm (Ra 2.4). The global increase of the surface roughness can be explained due the UED® nanoparticles coagulation during the process. The sample ČK-KAN-C and ČK-KAN-D with the surface roughness Ra 2.0 and Ra 2.2 are close those with particles coagulations.

The visualization was analysed using DantecStudio software. The images were processed to remove image noise, arithmetically analysed and masked. The processed images were cross-correlated using classical method at first. The resolution of the method corresponds to the camera resolution and magnifying lens system, i.e. after the calibration1px is 4.7 μm in the picture. The size of the interrogation area is (32 x 32) px. This means the size (150.4 x 150.4) μm in the picture. The selected velocity profiles close to the air film are shown in Fig. 2.

Therefore, change in the position of first valid vector indicates a change in the size of the air layer to 151.4 μm. This decrease corresponds to a change between Re of laminar flow, the critical flow, and the turbulent flow. If the initial thickness is smaller the shift in vector maps does not take effect. When selecting a smaller area we can no longer be sure of moving the particles within the evaluated areas, including appropriately chosen overlap.

The purpose of the analysis was the effect of air film on the velocity profile close to air layer. The other suitable method was the adaptive correlation. The setup of this analysing method was the initial interrogation area (32 x 32) px and the final interrogation area (16 x 16) px with 0% overlap calculated in 3 iterations. The real size of the interrogation area was (75.2 x 75.2) μm in this method so we got twice the resolution as in the first calculation. The changes in velocity profiles for selected Reynolds numbers are seen in figure 3.

There are changes in velocity profiles close to wall. We got higher resolution of results using adaptive correlation. This method uncovered tiny vortex structures that are developed in this stratified area of liquid flow. We also reached the calculation of first valid vector interrogation area closer to the wall, or air film. This effect is very well seen in Fig. 3 – C for Reynolds number 2390.

The initial velocity of the liquid flow was set repeatable for each sample and Reynolds number was calculation following the hydraulic diameter $D_h$ 0.008. Anyway, the velocity profile was influenced of the air film that is unique for each sample, so the maximal velocity was not constant. We normed the velocity profile related to the maximal velocity on the central line. The effect of the air film is already described. [1] The velocity of liquid flow close to the air film is round coefficient $u_x/U_{max}$ 0.3 for the laminar flow. This coefficient responds to each maximal velocity on the central line of each channel. The velocity profiles are changed in the transient area of liquid velocities. There are makeable shift in position that are caused by the different position of the stratified flows. The profile shift is also caused by different maximal velocities. As it is seen in Fig. 3 the sample with Ra 2.0 and Ra 2.2 exhibits the same trend.

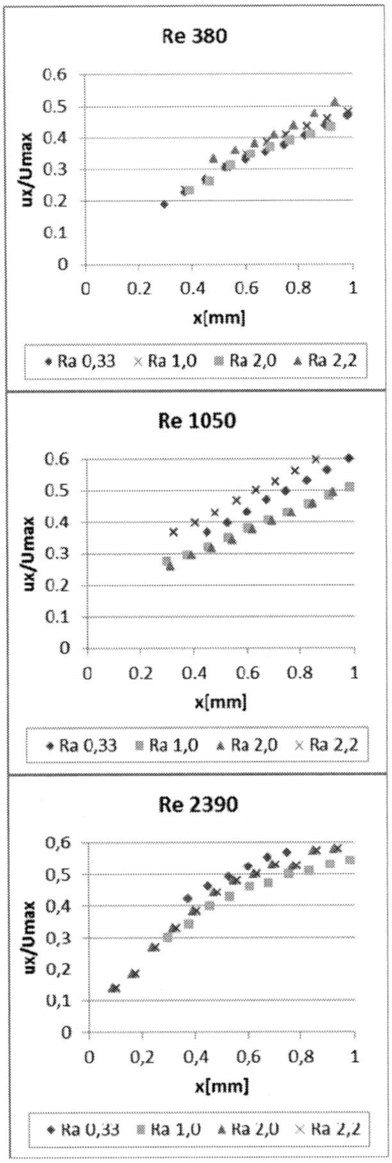


**Fig. 3.** The results of adaptive-correlation with initial interrogation area (16 x 16) px. The velocity profile normed to maximal velocity for the laminar flow Re 380, transient flow Re 1050 and the turbulent flow Re 2390. The x[mm] axis represents the distance from the surface/channel's wall. Umax represents the maximal velocity in the channel.

Markable difference is seen on turbulent flow. The border between the laminar flow and turbulent flow is in mini-channel shifted toward lower Reynolds number. Here we tried to reach the traditional edge of transient area that was limited by the pump. The velocity close to the disappearing air film is slowing down. Finally, the ratio between the local and maximal velocity decrease to 0.1. The change in the velocity profile is set more positional than due the interaction with the air film.

There is seen in Fig. 3 for Re 2390 that the velocity profile for Ra 0.33 starts at 350 μm far from the surface. The velocity of the initial point is only slightly higher comparing to the profiles without any air film. The maximal velocity on the central line is also higher as the channel cross-section is partly filled with air gap.

The second part of the research was dealing with the air film thickness estimation. The measurement of this property was based on image analysis of the intensity profiles over the selected line in the picture. Here we use the optical effect, that the air film reflect the light and can be visible as the light stripe. Due its geometrical character, it is rounded in the edges; it optically behaves as the lens. The air film is mirroring, and reflecting the emitted light from the seeding particles as well.

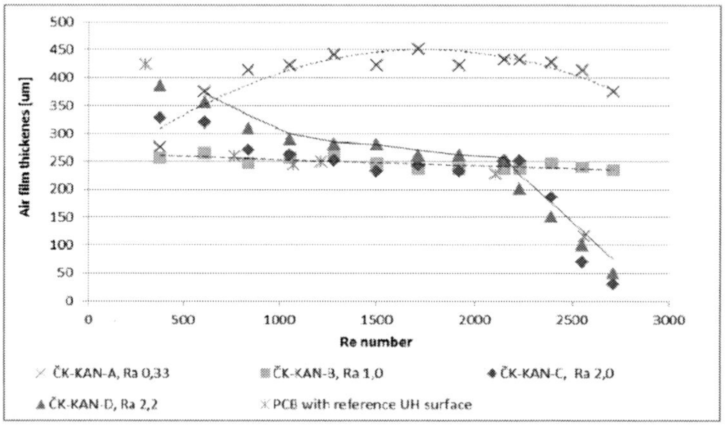


**Fig. 4.** The comparison of the air film thickness depending on the Reynolds number of the liquid flow for each sample. There is placed polycarbonate channel with UED® nanoparticles cover as a reference sample.

The difference in the intensities of pixels over the line is well determined. The thickness of the air film was calculated against the position of the surface. The values of air film was correlated to the Re number of liquid flow. The results are seen in the Fig. 4. The PCB sample covered with UED® nanoparticles is present as a reference sample. This sample has the surface roughness close to the stainless steel samples ČK-KAN-C with Ra 2.0, and ČK-KAN-D with Ra 2.2. This chart proves the dependence of the air film thickness on the surface roughness and it is not important, if the roughness is initial, ie. it is a material property, or it is developed due the covering process. This proof also corresponds with the theory of ultra-hydrophobicity on the rough defined surfaces. [5]

It is shown in Fig. 4 that the smooth surface of Ra 0.33 enables the air film existence over relatively high Re numbers. It was experimental prove the air film limitation due the liquid flow pulsation, flow instabilities, and Re numbers over 3000. There is step change in the air film thickness for the Re number over this limit.

## 4 Conclusions

Here we presented the testing of materials that has great potential of usage in industry. There was not significant border between the surface roughness that is influenced by machinery processing, and the effect of the physical or chemical treatment. We tested the machined samples to investigate the dominant role of the surface background. Presenting graph negates doubts about the advantages of coarse structure to increase the degree of hydrophobicity of the material. This knowledge is crucial in design of dynamic hydraulic part of the system.

The results of this project LO1201 were obtained with co-funding from the Ministry of Education, Youth and Sports as part of targeted support from the „National Programme for Sustainability I " program and the support of TA03010950 – "The use of hydrophobic and oleophobic surfaces during their interaction with liquids".

## References

1. D. Jasikova, M. Kotek, S. Fialova, V. Kopecky, EPJ Web of Conferences, **114** (2016)
2. D. Jasikova, M. Kotek, V. Kopecky, AIP Conf. Proc., **1745** (2016)
3. S. Fialova, F. Pochyly, Wasserwirtschaft Extra, **1** (2015)
4. M. Klíma, etal., Method of generating plasma at atmospheric pressure in a slot jet and device for performance the method, WO2014094695A1, WO2014094695A4 (2014)
5. J. Bicoa, U. Thieleb, D. Quérèc, Colloids and Surfaces A: Physicochemical and Engineering Aspects, **206** (2002)

# Direct numerical simulation of the passive heat transfer in a turbulent flow with particle

*Marek* Jaszczur[1,*], *Marcin* Zych[2], *Robert* Hanus[3]

[1]AGH University Science and Technology, Faculty of Energy and Fuels, 30-059 Kraków, Poland
[2]AGH University Science and Technology, Faculty of Geology, 30-059 Kraków, Poland
[3]Rzeszów University of Technology, Faculty of Electrical and Computer Engineering, 35-959 Rzeszów, Poland

**Abstract.** Turbulent non-isothermal fully developed channel flow with solid particles was investigated through Direct Numerical Simulation combined with the point-particle approach. The focus was on the interactions between discrete and continuous phase and their effect on the velocity and the temperature of the particles. It has been found that low momentum inertia particles have a mean temperature similar to the fluid temperature and this effect is almost independent of particle thermal inertia. For particles with larger momentum, the inertia thermal effect is more complex, particle temperature in the near-wall and buffer region is significantly lower than the fluid temperature. The difference between the fluid mean temperature and particle mean temperature increases along with the momentum response time. This may have important consequences on the chemical reactions, technological processes and on the accuracy of temperature measurement techniques based on seeding particle.

## 1 Introduction

Flow with small solid particles or droplets has recently received considerable attention for its relevance to numerous industrial processes, engineering applications and environmental flows. However, researchers have focused mainly on isothermal particle-laden flows. Experiments and numerical computations using Direct Numerical Simulations [1-3] demonstrate that isothermal shear flow has a very complex effect on particle behaviour. Due to particle-turbulence interaction, the distribution of particles and their properties can be highly non-uniform [4,5]. This non-uniformity can have very important consequences on the process—for example, affecting the efficiency of the chemical reaction or the properties of the mixtures.

A great deal of work in this field has been done towards understanding the influence of particle inertia on dispersion phenomena [6]. Even though we now have a better understanding of turbulence modification by small solid particles, the mechanisms are complex and not easy to predict [7].

In understanding particle-turbulence interactions, Direct Numerical Simulation (DNS) and Large Eddy Simulation (LES) plays an important role [8-10]. In the case of flows laden with a large number of small particles, Eulerian-Lagrangian point-particle DNS has proven quite successful in studies of particle-turbulence dynamics. Several studies have focused on the particle-turbulence interaction in homogeneous isotropic turbulence by two-way coupling [11]. They found that increasing the mass loading of small solid particles reduces turbulent kinetic energy. In this research, the preferential concentration of light particles led to an increase in the small-scale velocity fluctuation. The effect of the two-way coupling increases the rate of the kinetic energy of the turbulence transfer from the large scale towards the small scale. This may result in a larger dissipation rate [12].

An increasing amount of research has recently been reported for fully developed particle-laden turbulent channel flow. The application of DNS to the study of particle deposition in boundary layers by Wang and Squires [13] has clearly shown accumulation of heavy particles in the low-velocity streaks.

While there are a large number of applications from this including combustion, spray drying, and heat exchanger, very small attention has been paid to particle analysis in the non-isothermal flow [14-16]. This type of flow is known to be affected by the thermal layer through various mechanisms [17-19]. The modulation of the turbulent heat transfer due to particle–fluid interactions can be important at even low loadings [20,21]. The total heat capacity of the gas-particle mixture may be modified and because of that turbulent heat transfer modulation [22] may be the next important effect. The particle-wall and inter-particle collisions are the next effects expected to play a significant role [23]. The collision effect even in the flow with a very low volume fraction can seriously affect the dynamics of the two-phase flow and the heat transfer.But still today only few research has concentrated on the particle-laden flow in non-isothermal cases and the concentration is more on isolated effects than on overall particle-turbulence interaction [24].

* Corresponding author: marek.jaszczur@agh.edu.pl

Presented work thoroughly investigates the dynamics of small heavy inertial particles in non-isothermal turbulent fully developed channel flow. The cases considered are representative of a large family of phenomena known in the literature as multiphase flows, which contain both a continuous phase and a dispersed phase.

Fully-developed particle-laden non-isothermal turbulent channel flow [25] is investigated using Direct Numerical Simulation (DNS) combined with the Lagrangian particle tracking technique, which is used to track a large number of individual particles, using the point-particle approach.

In the configuration presented here, small solid dense particles carried by the fluid forces are influenced by the turbulent flow. The simulations are performed without gravity at shear Reynolds number $Re_\tau=150$ and for a molecular Prandtl number Pr=1.0. The effect of the particle diameter and density on the statistical quantities is considered. The key issue is to examine, on the basis of statistics, particle dispersion with reference to the particle hydrodynamic and thermal relaxation times.

## 2 Mathematical model and numerical methods

For computation of wall-bounded turbulent flow with particle driven by the uniform pressure gradient between the Eulerian-Lagrangian point-particle approach has been used. The continuous-phase is solved using DNS techniques for incompressible flow which is combined with the tracking of the individual particles. In the simulation, the transfer of momentum between the fluid and the point-particle is computed through a force located at the geometrical centre of the particle and is determined base on the velocities of the particle and of the surrounding fluid. The temperature is determined base on velocities of the particles and fluid and the temperatures of the particle and of the surrounding fluid. This approach is valid if the particles are significantly smaller than the smallest flow scales both for velocity and temperature. The flow in the channel (Fig.1.) is assumed to be fully developed both thermally and hydrodynamic.

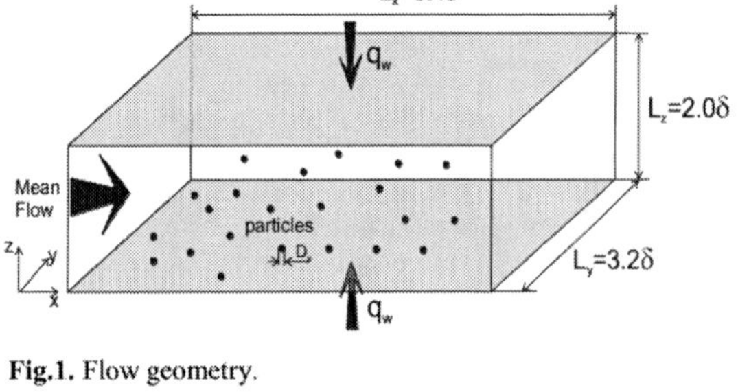


**Fig.1.** Flow geometry.

The streamwise wall-normal and spanwise coordinate are denoted x,y,z. The wall heat flux $q_w$ is assumed to be uniformly distributed on walls and the working fluid is assumed to be Newtonian fluid with constant properties. Temperature is treated as a passive scalar. Under this assumption, the continuity, Navier-Stokes and energy equations are described, as follows:

$$\nabla \cdot \vec{U} = 0 \tag{1}$$

$$\frac{\partial \vec{U}}{\partial t} + \left(\nabla \vec{U}\right) \cdot \vec{U} = -\frac{1}{\rho}\nabla P + \nu \nabla^2 \vec{U} \tag{2}$$

$$\frac{\partial T}{\partial t} = -\nabla\left(\vec{U} \cdot T\right) + \alpha \nabla^2 T \tag{3}$$

For small heavy particles with diameter smaller than Komogorov scale in the equation of particle motion the only significant force is the drag and the governing equations can be written as:

$$\frac{d\vec{v}}{dt} = C_d \frac{Re_p}{24} \frac{1}{\tau_V} (\vec{u} - \vec{v}) \tag{4}$$

$$\frac{dT_p}{dt} = \frac{Nu}{2} \frac{1}{\tau_T} \left(T - T_p\right) \tag{5}$$

Where $\vec{u}$ is the velocity of the surrounding fluid interpolated at the centre of the particle and $\vec{v}$ is the velocity of the particle. The particle Reynolds number, $Re_p$, and the hydrodynamic particle-relaxation-time $\tau_V$, are defined as:

$$Re_p = \frac{|(\vec{u} - \vec{v})| D_p}{\nu}, \quad \tau_V = \frac{\rho_p}{\rho} \frac{D_p^2}{18\nu}, \quad C_d = \frac{24}{Re_p} \tag{6}$$

where $\rho_p$ and $\rho$ are the particle and fluid densities, $D_p$ is the diameter of the particles and $C_d$ is the drag coefficient for Stokes flow. The volume fraction of particles is assumed to be small enough such that particle-particle interactions are negligible. The equation for the temperature of particles, assuming small Biot number and thermal particle-relaxation-time, $\tau_T$, can be written in the form:

$$\tau_T = \frac{\rho_p c_p D_p^2}{12k}, \quad Pr = \frac{\nu \rho c}{k} \tag{7}$$

where $T$ is the temperature of the surrounding fluid interpolated at the centre of the particle and $T_p$ is the temperature of the particle and $\tau_T$ is the thermal particle-relaxation-time. In order to take into account effect of particles relative velocity and fluid number dependence on the Nusselt number the value of the Nusselt is calculated from the Ranz-Marshall correlation:

$$Nu = 2 + 0.6 \mathrm{Re}_p^{1/2} \mathrm{Pr}^{1/3} \tag{8}$$

where the Pr is the Prandtl number.

The position, flow, and particle quantities are normalised by the channel half-width, $\delta$ the friction velocity, $u_\tau$ and the friction temperature, $T_\tau = q_w/(\rho \cdot c_p \cdot u_\tau)$. The governing equations (1)-(3) are solved numerically using the fractional step method on a staggered grid. Adams-Bashforth method was used for time advancement of all terms with the step determined by the Courant criterion. The Poisson equations for pressure is solved using Fast Fourier Solver. The equation of the particle motion (4) and the equation of the particle temperature (5) are solved use an Adams-Bashforth scheme for the time-advancement and a tri-linear interpolation for the velocity and temperature field at the centre of the particle. To force the fluid motion, a constant pressure gradient is imposed along the streamwise direction. Periodic boundary conditions are imposed in both streamwise and spanwise directions for the hydro and thermal fields. In order to apply periodicity in the streamwise direction, the following transformation of the pressure and temperature is made:

$$p_l = p - \left|\frac{d\langle\overline{p}\rangle}{dx}\right| x \quad , \quad T_l = T - \left|\frac{d\langle\overline{T}\rangle}{dx}\right| x \qquad (9)$$

where

$$\left|\frac{d\langle\overline{p}\rangle}{dx}\right| = -2\rho u_\tau^2 / \delta \quad , \quad \left|\frac{d\langle\overline{T}\rangle}{dx}\right| = \frac{q_w}{\rho c_p U_b \delta} \qquad (10)$$

## 3 Numerical Results

The numerical results of the most relevant statistics for non-isothermal particle-laden flow are presented and analysed. Fluid and particle statistics are presented at a statistically steady state at Reynolds number $Re_\tau$=150 and for Prandtl number Pr=1.0. Particle statistics for particle presented in Table 1 are computed assuming considered variable over particles in a certain test volume, constituted by wall parallel fluid slabs and by averaging the number of particles in the sampling volume.

**Table 1.** The particles and grid data.

| $d_p^+$ | St= $\tau_p^+$ | $St_T = \tau_T^+$ | $N_x \times N_y \times N_z$ |
|---|---|---|---|
| 0.15 | 1 | 0.1/1.0/10.0 | 128 × 128 × 64 |
| 0.15 | 1 | 0.1/1.0/10.0 | 256 × 256 × 256 |
| 0.342 | 5 | 0.5/5.0/50.0 | 128 × 128 × 64 |
| 0.765 | 25 | 2.5/25.0/250 | 128 × 128 × 64 |

It is much more time-consuming to get a statistically steady state for particle than for fluid. For the same Reynolds number and different Stokes number, the developed times are very different. For small particles (small St number), the time is shorter while for intermediate particles it is enormously long. In the Fig.2, the time evolution of the particle concentration in the near-wall region ($z^+$=1.0) has been presented up to $t^+$=30000. Particle field for St=1 was used as an initial condition to generate these profiles; for this value, particle immediately achieves a steady state. An increase or decrease in particle concentration in the near-wall region can be observed for other Stokes numbers.

Obviously, for the St=5 it takes significantly longer to reach a steady concentration at this location than for the St=1 where maximum concentration *C* is about 2.5. This is consistent with the results found by other researchers and shows that the maximum particle deposition rate occurs for the particle St=25. Then, as the relaxation time increases, particle concentration decreases. It was found by Elghobashi and Truesdell [26] that at $Re_\tau$=150 a peak of concentration occurs for St=25. The particle with this Stokes number is the most responsive to the turbulence structures in the near wall region.

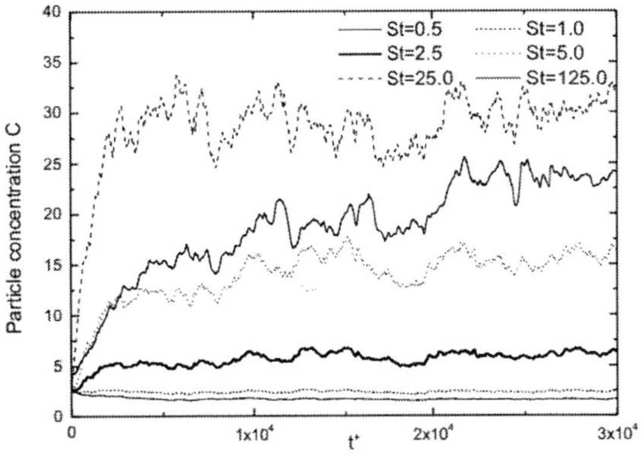


**Fig. 2.** Particle concentration evolution at the near-wall region ($z^+$=1.0) as a function of non-dimensional time $t^+$.

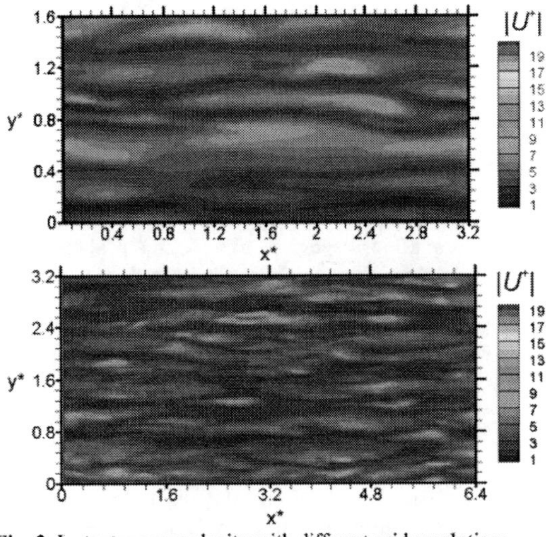


**Fig. 3.** Instantaneous velocity with different grid resolution: coarse(top), fine (bottom) in XY cross-section at $z^+$=3.6 for $Re_\tau$=150, Pr=1.

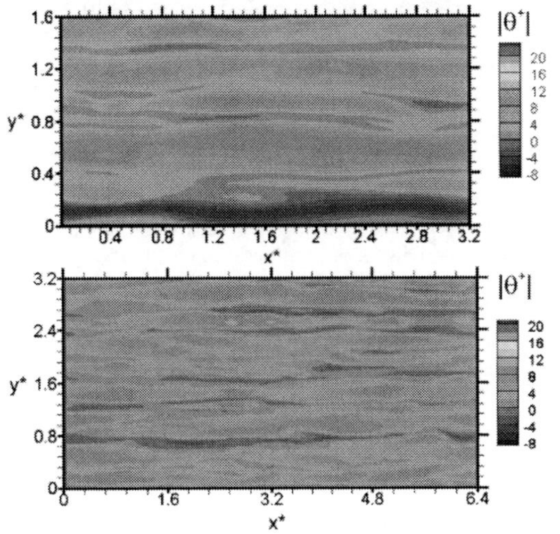


**Fig. 4.** Instantaneous temperature with different grid resolution: coarse(top), fine (bottom) in XY cross-section at $z^+$=3.6 for $Re_\tau$=150, Pr=1.

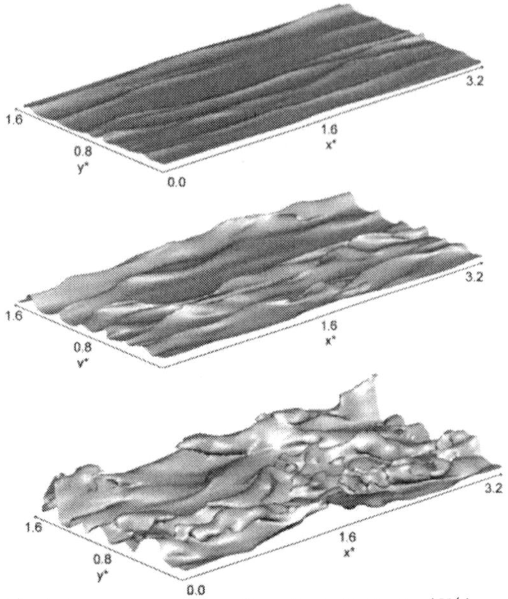


**Fig. 5.** Instantaneous iso-surface of velocity for the $|U^+| = \theta^+ = 3.6$; 10.0 and 15.0 (top, middle, bottom).

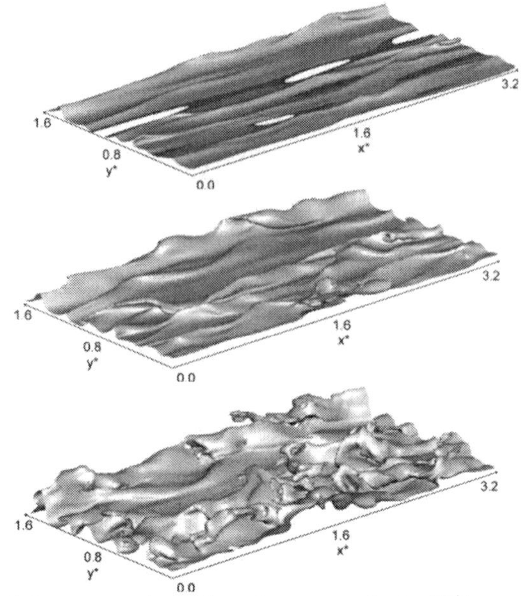


**Fig. 6.** Instantaneous iso-surface of temperature for the $|U^+| = \theta^+ = 3.6$; 10.0 and 15.0 (top, middle, bottom).

Fig. 3 and Fig. 4 show the effect of the grid resolution and the domain size, which was increased in streamwise and spanwise directions on the velocity magnitude. Two grid resolutions were tested at Reynolds number $Re_\tau$=150: (coarse) $N_x \times N_y \times N_z$ - 128x128x64 with the domain $L_x \times L_y \times L_z$ equal to $2\pi\delta\times1\pi\delta\times2\delta$; (fine) 256x256x256 and domain equal $4\pi\delta\times2\pi\delta\times 2\delta$. Each plot was obtained in different DNS computations and it is not possible to show simultaneously differently resolved fields. To show the flow and thermal field complexity, the iso-surface of the constant velocity magnitude $|U^+|$ and temperature $\theta^+$ are presented in Fig. 5 and Fig.6. for the bottom half of the channel. The iso-surface values for both properties are the same and are equal to 3.6, 10.0 and 15.0. Particularly the $|U^+|=\theta^+=3.6$ represents on average the mean temperature and mean velocity at position $z^+=3.6$. It can be seen that instantaneous fields have complex surfaces which reflect the influence of the large flow structures described previously. Flow and thermal fields look like valleys with peaks and troughs.

For the magnitude of velocity, the minimum instantaneous value is exactly zero (and obviously the mean value and its fluctuation at the wall are zero) which is a consequence of non-slip boundary conditions; however, the instantaneous temperature $\theta^+$ at the wall is not necessarily zero. In the introduced normalisation system, the mean value at the wall is zero but the instantaneous value can reach a much lower value. Temperature fluctuation at the wall is also non-zero (boundary condition – heat flux at the wall). This obvious difference causes "empty spaces" on the temperature isosurface.

Increasing the level of the isosurface up to $|U^+|=\theta^+=10.0$ does not cause the valley structures to become dissimilar, but the peaks are higher. The distance between the peaks here is similar to the distance between the peaks for iso-surface 3.6. The mean value for velocity magnitude and temperature equal to 10.0 is most representative for the location $z^+ \approx 11$. The last iso-surface presented here is plotted for 15.0. It can now be seen that the iso-surface for the temperature and velocity magnitude is very complex and in same areas the surface "starts to detach" from the main surface.

Obtaining particularly a statistically steady state for particle position is very difficult and required much more time than for any other development. From Fig.7 it can be seen that at the lowest Stokes number (St=1) the particles are nearly uniformly distributed in the near wall region and are not far from a random distribution in the center region. There are a few empty spots – places with a low number or no particles.

With increasing Stokes number up to St=5, more particles can be seen in the near-wall region. At the same time, a significantly smaller number of particles is observed in the centerline region, and with a distribution different from the other cases. St=5 particles in the center region are far from any uniformity and form structures at a very high concentration while at the same time leaving many areas entirely devoid of the 1.5 million particles. Further increases in the Stokes number up to St=25 show a more pronounced effect of the increasing concentration near the wall and many more particles are accumulated close to the wall. Additionally, in the near-wall region the main particle accumulations occur in nearly straight streaks elongated in the streamwise direction.

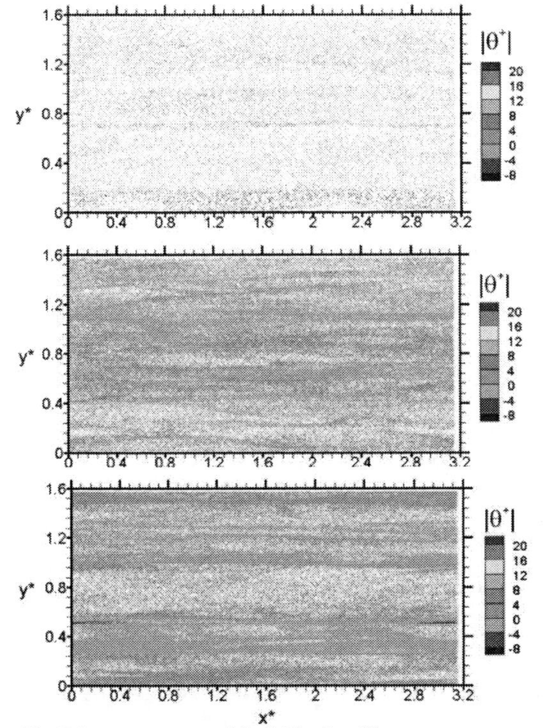


**Fig. 7.** Instantaneous particle field colored by temperature at $z^{+}$=3.6 for Stokes number $St=St_T$=1, 5, 25 (top, middle, bottom).

Fig. 8 presents profiles of the mean streamwise velocity and temperature for the fluid and particles non-dimensionalized by the wall shear velocity. The results show a modification of the mean particle dynamics with increasing particle.

For all of the momentum Stokes numbers considered here (St=1, 5, 25), the mean streamwise velocity lags the fluid velocity except for in the channel centre region. This lag is larger for particles with larger inertia. In fully developed channel flow independent of the particles relaxation time for long enough statistics to be determined, velocity (and temperature) could be expected to be identical for the fluid and particles.

For the St=1 particles with a response time equal to the fluid response time, the difference between fluid and particle velocity is very small. For 25 particles the difference became large and the particles' average velocity could become 20% lower than the fluid velocity. The difference between the fluid temperature and particle temperature for heavier particles is much larger than observed for the fluid.

In Fig.9 particle, velocity fluctuation for the streamwise, spanwise and wall-normal components are plotted as a function of the wall-normal coordinates in wall units. The results are presented for St=1,5,25. including the fluctuation of the fluid velocity components. For the streamwise component, the observed effect is the opposite of that for the particle mean streamwise velocity. The velocity fluctuation increases along with the Stokes number. Even for the lightest particles considered here the difference in fluid and particle fluctuation is large.

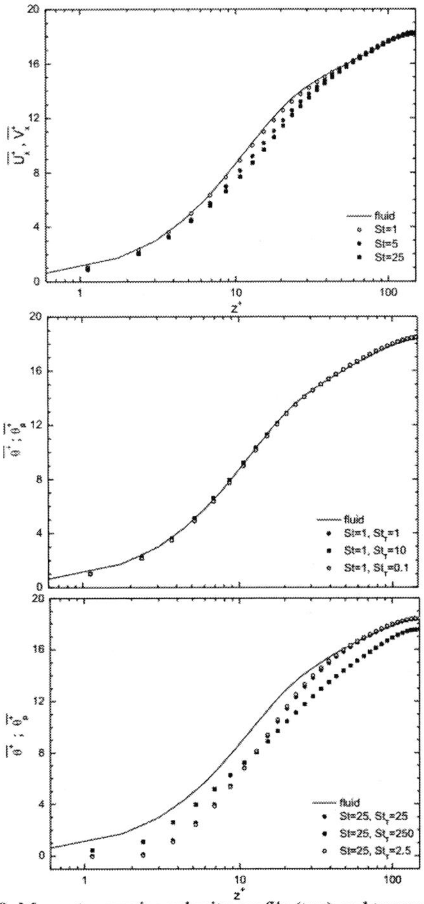


**Fig. 8.** Mean streamwise velocity profile (top) and temperature (middle, bottom) for the fluid (solid line) and for particles.

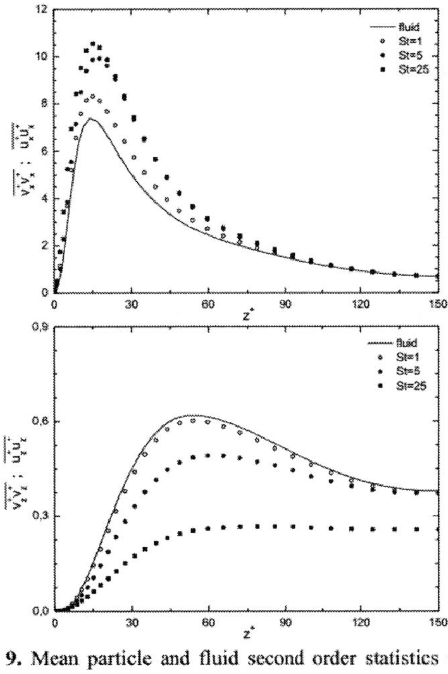


**Fig. 9.** Mean particle and fluid second order statistics for the St=1, 5, and 25 particles and for the fluid $Re_\tau$=150.

DOI: 10.1051/epjconf/201714302046

## 4 Conclusions

The DNS of solid particles without gravity was performed with the aim of extending knowledge about turbulence particle interaction in the non-isothermal flow.

The results presented in this paper for continuous phase show that the streaks are complicated but the spatial distance between them and the length is roughly constant and the thermal streaks are less elongated. The spatial distance in the near-wall region in the spanwise direction is about $\ell_y$=100, while the streamwise distance $\ell_x$ is about 10 times larger.

One and half million of particles introduced in the fully developed flow for each Reynolds number have been considered in this paper. Particles were characterised using particle relaxation time for momentum and for the thermal response, which is proportional to the Stokes numbers St and $St_T$. Particle inertia can be considered as a low-pass filter that strongly influences particle response to the turbulence, particularly to large-scale structures.

The results show that temperature and velocity, as well as concentration, are strongly influenced by the Stokes number with respect to the Reynolds number. In the near wall region, the particle is transported mainly by the large turbulent coherent structures and strongly depends on the ejections and sweeps. But because of particle inertia, the particle re-entrainment process is less promoted than deposition in the viscous region. Particle segregation occurs in particular places and has obtained what is call known in the literature as a preferential concentration. This mechanism influences not only particle position and, finally, particle distribution, but also influences particle temperature. Due to preferential concentration, a similar effect causes what can be called **preferential temperature**. Strong coherent structures segregate particle very effectively and by doing so also exert an influence on the particle temperature distribution. Preferential concentration occurs over a wide range of flows of many practical applications.

The mean temperature for the fluid in the boundary layer region is higher than the mean temperature for the particles. A similar effect can be found for the mean streamwise velocity, but the difference for the temperature is larger. In the centre region of the channel the particles, regardless of the thermal response time, are almost symmetrically distributed around the mean. Particles in this region form highly concentrated bands that influence the distribution, which became narrow and much higher. The difference between differently thermally responsive particles can also be very large. This may have important consequences on the chemical reactions, technological processes and on the accuracy of temperature measurement techniques based on seeding particle.

## Acknowledgements

The present work was supported by the Polish Ministry of Science (Grant AGH No. 11.11.210.312).

## References

1. R. D. Moser, J. Kim, N. N. Mansour, Phys. of Fluids. **11**, 943-945 (1999)
2. J. Kim, P. Moin, Turbulent Shear Flows **6**, 85–96 (1989)
3. N. Kasagi, Int. J. of Heat and Fluid Flow **19**, 125-134 (1998)
4. J.McLaughlin, Phys. Fluid A**1**, 1211-1224 (1998)
5. M. Rashidi, G. Hetsroni, S. Banerjee, Int. J. of Multiphase Flow **16**, 935-949 (1990)
6. J. G.M. Kuerten, C. W. M. van der Geld, B.J. Geurts, Phys. of Fluids **23**, 123301–123309 (2011)
7. N. Anbuchezhian, K. Srinivasan, K. Chandrasekaran, R. Kandasamy, Applied Mathematics and Mechanics **33**, 765–780 (2012)
8. L.M. Portela, R.V.A. Oliemans, Int. J. Numer. Fluids **43**, 1045-1065 (2003)
9. D. Rouson, J. Eaton, Proc. of 7th Workshop on Two-Phase flow Predictions, Erlangen (1994)
10. B. Lessani, M.H. Nakhaei, Int. J. of Heat and Mass Transfer **67**, 974–983 (2013)
11. L. Wang, M. Maxey, J. Fluid Mech. **256**, 27 (1993)
12. C.D. Dritselis, N.S.Vlachos, Int. J. of Multiphase Flow **37**, 706–721 (2011)
13. Q.Wang, D.Squires, J. of Multiphase flow **22**, 667-683 (2001)
14. N. Kasagi, Y. Tomita, A. Kuroda, Journal of Heat Transfer **114**, 598–606 (1992)
15. N. Kasagi, Y. Ohtsubo, Turbulent Shear Flows **8**, 97–119 (1993)
16. H. Kawamura, H. Abe, Y. Matsuo, Int. J. of Heat and Fluid Flow **20**, 196–207 (1999)
17. H. Kawamura, H. Abe, K. Shingai, Proceedings of the 3rd Int. Symp. on Turbulence, Heat and Mass Transfer, Japan (2000)
18. Y. Na, D. V. Papavassiliou, T. J. Hanratty, Int. J. of Heat and Fluid Flow **20**, 187–195 (1999)
19. L. Redjem-Saad, M. Ould-Rouiss, G. Lauriat, Int. J. of Heat and Fluid Flow **28**, 847–861 (2007)
20. M. Jaszczur, Archives of Mechanics **63**, 77–91 (2011)
21. M. Jaszczur, *Direct and Large-Eddy Simulation VIII*, Springer-Verlag, Berlin (2011)
22. F. Zonta, C. Marchioli, A. Soldati, Acta Mechanica **195**, 305–326 (2008)
23. Y.H. Dong, L.F. Chen, Int. J. of Heat and Mass Transfer **54**, 1168–1178 (2011)
24. I. Tiselj, E. Pogrebnyak, C. Li, A. Mosyak, G. Hetsroni, Physics of Fluids **13**, 1028–1039 (2001)
25. M. Ould-Rouiss, L. Redjem-Saad, G. Lauriat, A. Mazouz, Int. Comm. in Heat and Mass Transfer **37**, 958–963 (2010)
26. S. E. Elghobashi, G.C. Truesdell, Physics of Fluids, **A5**, 1790-1801 (1993)

# The study of the velocity field of the air flowing the swirl diffusers using PIV method.

*Marek* Jaszczur[1*], *Marek* Borowski[2], *Michal* Karch[2], *Marian* Branny[2]

[1] Faculty of Energy and Fuels, AGH University of Science and Technology, 30-059 Krakow, Poland,

[2] Faculty of Mining and Geoengineering, AGH University of Science and Technology, 30-059 Krakow, Poland

**Abstract.** The geometry of swirl diffusers plays a primary role in the effective airflow distribution. In this work, the airflow of swirl diffusers has been studied with the use of the Particle Image Velocimetry (PIV) method in order to analyse the mean swirling air flow and the geometry influence on the results. The presented laboratory model was a simplification of the real system and was made in the geometric scale 1:10. Based on the measurements, the velocity flow field was evaluated and the range of the effective operation of the diffuser in the axis of the flow was determined. This will enable a better definition of the conditions of ventilation in buildings and will help in the design of more efficient diffusers. Also, it will help in a better understanding of the swirl diffuser airflow and the physical effects of the key geometric features.

## 1 Introduction

Air swirl diffusers are popular for ceiling level air supply systems and they have been widely used for Indoor Air Quality (IAQ) and the occupant's comfort for more than three decades. They are nowadays among the most popular diffusers commonly used in air-conditioning systems. In the case of swirl diffusers, the fluid flow is usually significantly influenced by the characteristics of different diffuser designs. At the stage of the ventilation system design, a typical focus is on the ventilation effectiveness and it takes into account a large number of parameters. However, a discrepancy between the design and the practice is often experienced, mainly because the operating conditions do not correspond to the designing conditions.

The typical project validation methods are: the point-by-point method (anemometers methods), where detectors are moved randomly around the ventilated space, particle streak velocimetry (PSV), schlieren photography, particle tracking velocimetry (PTV) and particle image velocimetry (PIV) [9].

For the design of the air distribution in the room, it is necessary to know the distribution of the velocity field and the velocity profile which is generated by the swirl diffusers. This will enable a better definition of the conditions of ventilation in buildings. The existing anemometric methods make it possible only to determine the approximate range of operation of the diffuser. The information about the velocity field is required in order to determine the range of the diffuser's effective operation in the flow axis.

In the literature, several methods of fluid flow modelling in ventilation systems with diffusers have been proposed. Computational Fluid Dynamics is nowadays one of the most useful techniques used to predict the air distribution inside a ventilated space. However, it is known that the accuracy of CFD predictions is very sensitive to the boundary conditions assumed and supplied. On the other hand, due to the large scale differences between the diffuser size and the space dimension, it is common to simplify the geometric model of the diffuser. That, in combination with inadequate turbulence modelling, may generate a result which is far from the actual one. A CFD simulation needs to specify accurately the flow, the thermal and species conditions as well as the flow conditions, in order to ensure the correct heat momentum and the mass flow rate. In general, the approaches proposed in the literature [1,2,11,12] can be categorised into: direct description, simplified geometry, box, momentum and prescribed velocity methods.

The swirl method of air distribution causes an increased induction and a fast unification of the air temperature in the zones occupied by people. The presented diffusers are designed for applications in ventilation systems for low and medium pressures. This type of diffusers is especially recommended for large spaces with heights from 3 to 12 meters.

In the process of the installation design, an important issue is the proper selection of the type and details characteristic of the diffusers. The selection is based on determining the range of the air stream. The literature extensively presents the results of analyses of the flow of air through swirl diffusers with the use of CFD [3,4,5]. There are a few examples of experimental research related to this topic [6,7,13].

In the presented paper, Particle Image Velocimetry (PIV) has been used to investigate flow fields. The presented laboratory model was a simplification of the real system and was made in the geometric scale 1:10. The air flow conditions were adjusted to a reduced scale. Measurements were made for

* Corresponding author: marek.jaszczur@agh.edu.pl

various air flow speeds and for three different angles of the diffuser blades: 45°, 60° and 75°. Based on the measurements, the vector velocity field and the effective operation of the diffuser were determined.

## 2 Experimental setup

In Fig.1, a photograph of the test section (swirl diffuser, laser and two cameras) for swirl diffusers has been shown. The experimental measurement was performed with the use of the mono and stereo PIV technique [8,9,10]. The diffuser was installed at the end of a long circular channel. In order to apply different flow rates and to stabilise the flow rate at the main fan, a frequency inverter was used together with a high accuracy calibrated orifice and anemometry.

**Fig. 1.** Set-up of the measurement section: 1-swirl diffuser, 2-laser Nd:YAG, 3-camera CCD.

As the seeding for the particle image velocimetry method, DEHS oil was used, sprayed through a high accuracy atomiser, in order to obtain particles with the diameter of about 1 μm. The seeding particles were introduced to the main air stream. The test section plane was illuminated by two-impulse laser Nd:YAG and the particle images were registered by a monochromatic CCD (2014×2048 pixels) camera.

The test procedure was preceded by the calibration with the use of a 3D calibrating plate. In order to obtain accurate first order and second order fluid flow statistics, it was necessary to determine the number of the required non-correlated flow images. It was found that at least 1000 fields (double picture frames) are necessary for one flow field analysis.

Based on several tests and an analysis of the optimal interrogation window for the velocity vectors, evaluation by the cross-correlation method was performed. It was found that the best interrogation window is a dynamic window in the range between 32×32 and 64x64 pixels. The Davis 7 software was used to analyse the images and to find the velocity vector components.

It was decided to experimentally determine the velocity field of the air flowing through the swirl diffusers. In this measurement of the airflow in a laboratory model of swirl diffusers, the geometrical scale of the physical model was 1:10 and the measurements were performed for steady and isothermal flow conditions.

## 3 Measurement

In this research, three different models of swirl diffusers were used, i.e. with the blade angles of 45°, 60° and 75° (see Fig.2 for reference). The diameter of the diffuser's model at the diffuser's inlet was 50 mm, which corresponds to the actual 500mm diffuser. The diffusers were tested at four different volumetric flow rates of the air: 100, 185, 310 and 380 $m^3/h$, and these values were measured at the fan inlet section with the use of anemometry.

**Fig. 2.** Setting diffuser vanes: angle of 45 degrees (1), angle of 60 degrees (2) and angle of 75 degrees (3).

## 4 Results

The applied method enables the obtaining of images of the vector field velocity in the studied area at once. On this basis, the flow velocity can be analysed at any point of the test section. Fig. 3 shows an image of the measured velocity field and the velocity profiles created on the basis of this image. The presented results were obtained for the 45° diffuser and for the flow rate of 380 $m^3/h$.

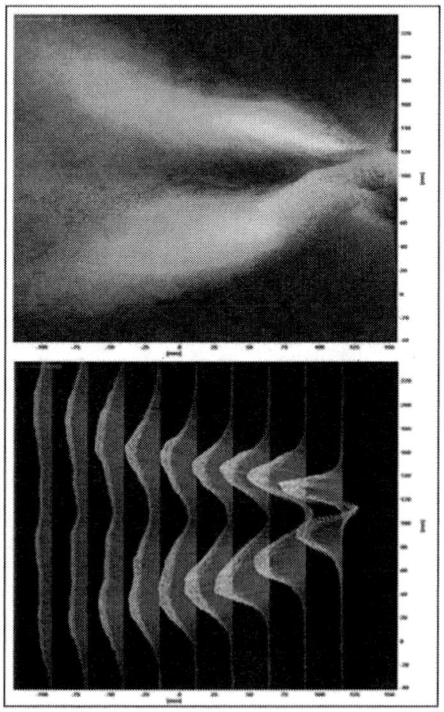

**Fig. 3.** The image of the measured velocity field (top) and velocity profiles (bottom). For the case: 45° diffuser and flow rate 380 $m^3/h$.

DOI: 10.1051/epjconf/201714302047

In Fig.4, the velocity vector component $V_x$ (axial) measured at the center-plane for the flow rate of 380 m$^3$/h and for three different diffuser angles is presented. As can be seen, the flow dispersion of the main jet flow in the direction of the outer region is the best pronounced for the angle of 45 degrees. The direction $x$ (axial) is the leading direction in accordance with the axis of the diffuser but decreasing blades angle radial component is increasing. The range of the effective operation of the diffuser in the axial direction for high angle diffusers is very narrow and does not guarantee a proper efficient hot/cold (or fresh/used) flow mixing.

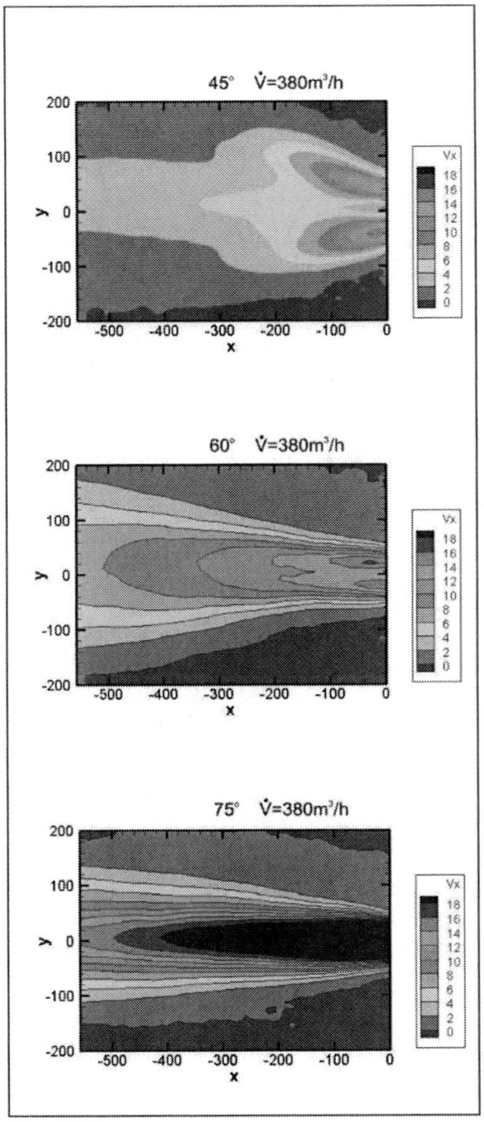


**Fig. 4.** The velocity vector axial component $V_x$ at the test section centre plane for air flow rate 385m$^3$/h and for different blade angles.

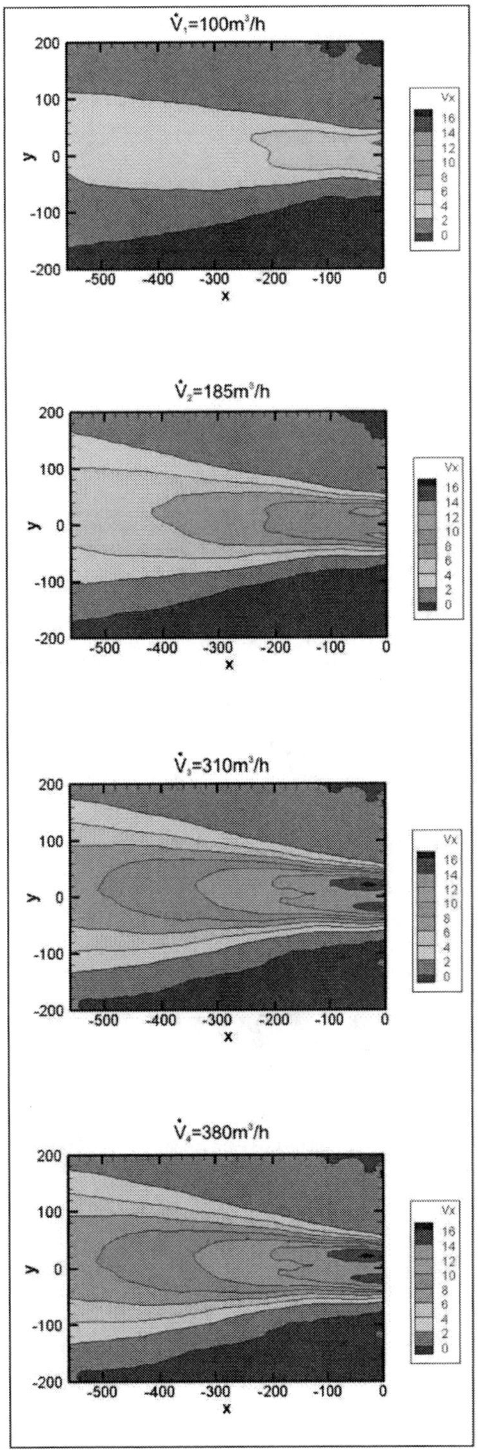


**Fig. 5.** The velocity vector axial component $V_x$ at the test section centre plane and for different air flow rates. All for the diffuser angle of 60 degrees.

Fig.5 shows the velocity vector axial component $V_x$ at the test section centre plane and for four volumetric flow rates: 100, 185, 310 and 380 $m^3/h$. The obtained velocity field for the diffuser angle 60° is similar for different flow rates, and after normalisation it becomes almost identical.

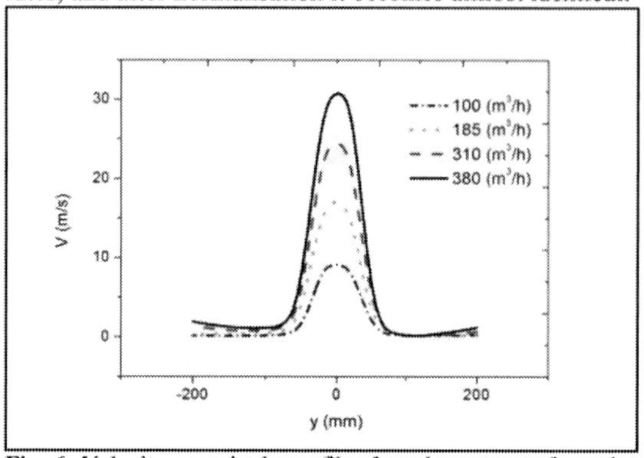


**Fig. 6.** Velocity magnitude profiles 2cm downstream from the diffuser outlet for a 75° diffuser and at different flow rates.

Fig. 6 shows the velocity for the 75° diffuser and at the distance of 2 cm downstream from the diffuser outlet. In Fig.7, a normalised flow pattern for this diffuser is presented. It can be seen that, for a specified diffuser type, the obtained flow pattern has a similar shape.

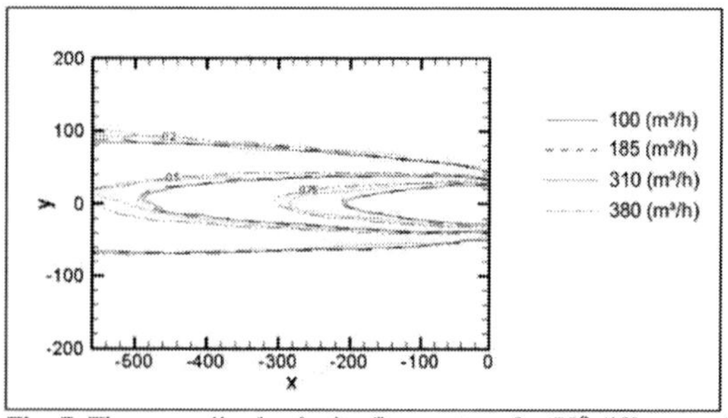


**Fig. 7.** The normalised velocity flow pattern for 75° diffuser and for different air flow rates.

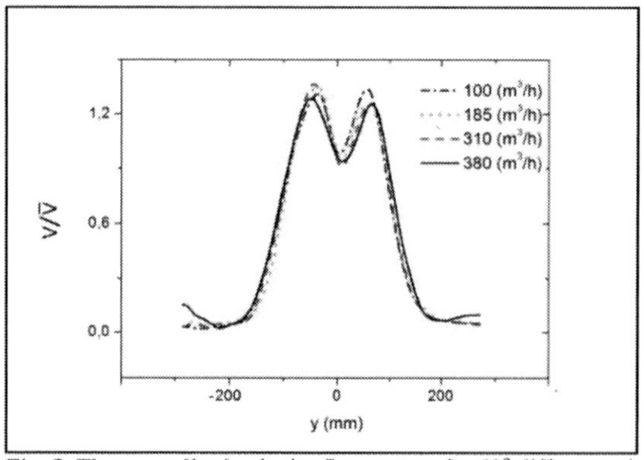


**Fig. 8.** The normalised velocity flow pattern for 60° diffuser and for different air flow rates.

Fig.8. shows the normalised velocity flow pattern for the 60° diffuser and for different air flow rates. These profiles were determined at the distance of 15 cm downstream from the diffuser outlet. As it is shown, the normalised profiles are quite similar.

## 5 Conclusions

The PIV method was used in order to determine the velocity field of the air flowing through a swirl diffuser. This method offers a good insight into the ventilation processes. The system is able to capture all the key features of the real flow with an accuracy higher than the one required for the needs of the ventilation practices. The knowledge of the shape of the air flow is important in order to provide adequate ventilation in people's workplaces.

An image of the velocity fields for different diffuser angles has shown that the widest air flow was obtained at the angle of 45 degrees and the narrowest air flow was observed at the angle of 75 degrees.

For a 45° diffuser, two symmetrical streams were observed in a 2D plane. For other angles, this phenomenon is not observed.

As shown in Fig. 7 and 8, the normalised velocity flow pattern is similar in the range of the analysed values of the flow rates. This is an important observation because it suggests that it will be possible to indicate a universal function describing the diffuser type velocity field. At present, the research continues aiming to analyse the fluid flow turbulence components.

## Acknowledgements

The research was supported by the Project No. 11.11.100.005.

## References

1. W. Songling, S. Zhe, W. Zhengren, Journal of North China Electric Power University **38**, 4, 81-85 (2011)
2. B.F. Yu, Z.B. Hu, M. Liu, H.L. Yang, Q.X. Kong, Y.H. Liu, Int. J. Refrig. **32**, 3-20 (2009)
3. A. Sami, AL-Sanea, M.F Zedan, M.B. Al Harbi, Int. J. Therm. Sci. **59**, 247-259 (2012)
4. R. Pandey, A Kumar Rao, V Yadav, IJSRD **3**, 671-677 (2015)
5. K.C. Ng, K. Kadirgama, E.Y.K. Ng, Energ. Buildings, **40**, 774-781 (2008)
6. R. Pandey, A. Kumar Rao, V. Yadav, IJESRT, **4**, 424-431 (2015)
7. M. Jaszczur, R. Nowak, J. Szmyd, M. Branny, M. Karch, W. Wodziak, J Phys Conf Ser. **318**, 012029,1-8 (2011)
8. M. Raffel et al. *Particle Image velocimetry: A practical guide*, Springer-Verlag, (2007)
9. J. Westerweel, *Digital particle image velocimetry: Theory and applications* Delft Uni. Press (1993)
10. LaVision Gmbh, *Flow Master Manual for Davis 7.0* (2007)
11. T. Zhang, K.S. Lee, Q. Chen, Indoor Air, **19**, 255-267 (2009)
12. A. Ch. Hviid, S. Petersen, CLIMA 2016, proc. of the 12th REHVA World Congress (2016)
13. B. Sajadi, A. Mohebbian, M.H. Saidi, Proc.of the 2nd Int. Conf. on Heating, Ventilation and Air-Conditioning, Tehran, Iran, June 1-3 (2010)

# Heat transfer research on enhanced heating surfaces in pool boiling

*Wojciech* Kalawa[1,*], *Tadeusz* M. Wójcik[1], *Magdalena* Piasecka[2]

[1]AGH University of Science and Technology, al A. Mickiewicza 30 , 30-059 Krakow, Poland
[2] Kielce University of Technology, Al. 1000-lecia P.P. 7, 25-314 Kielce, Poland

**Abstract.** The paper focuses on the analysis of the enhanced surfaces in such applications as boiling heat transfer. The testing measurement module with enhanced heating surfaces was used for pool boiling research. Pool boiling experiments were conducted with distilled water at atmospheric pressure in the vessel using an enhanced sample as the bottom heating surface. The samples are soldered to a copper heating block of the round cross-section .They were placed: in the fluid (saturation temperature measurement), under the sample for temperature determination. A vessel made of four flat glass panes was used for visualization. The heated surfaces in contact with the fluid differed in roughness were smooth or enhanced. This paper analyzes the effects of the microstructured heated surface on the heat transfer coefficient. The results are presented as relationships between the heat transfer coefficient and the heat flux and as boiling curves. The experimental data obtained for the two types of enhanced heated surfaces was compared with the results recorded for the smooth heated surface. The highest local values of the heat transfer coefficient were reported for the enhanced surfaces.

## 1 Introduction

With the ever-evolving state of the technology it is observed an increase in the intensity of heat flux in electronic systems and other modern technologies: military, space and medical. In addition, more and more often the focus is on minimizing the size of the heat exchangers, while maximizing leads heat flux. Therefore, there is a need to conduct research in the field of intensification of heat transfer.

One of the ways of intensification of heat transfer is the use of boiling phenomena, that is a phase change of the working fluid. Boiling is a process in which, due to bring heat to the liquid, the liquid inside the vapour phase formation of a liquid phase.

The boiling process is considered with regard to the classification of flow boiling, for which the geometry of space and hydrodynamics of the flow have a significant impact on the mechanisms of heat transfer and boiling in volume, for which these parameters have no effect on the course of the process. A special type of boiling is boiling in large volume - pool boiling. Such a process occurs, when due to the a much smaller size of the boundary layer in comparison with the dimension of the tank, there is no impact of the geometry of the tank on the boiling.

The development of technology, including electronics, requires providing huge heat fluxes, that boiling on the technically smooth surface becomes insufficient.

Among the methods of intensification of boiling are: active methods, requiring an additional supply of substances or energy into the liquid - heated surface, passive methods - without bringing the energy, involving the selection of the heat transfer surface or liquid, and complex methods, which are a combination of the listed above two methods. In recent years, mainly passive methods are being developed, both in the application of porous coatings (including thermal spraying methods, sintering of powders or fibers) and due to mechanical processing of the metal surface. The last of these methods involves changing of the shape of the heat transfer surface by producing e.g. holes, ribs or by increasing roughness of the surface. The extensive research on the passive methods shows that modifying the characteristics and structure of the heat transfer surface results in higher energy-efficiency and material saving.

In order to intensify boiling heat transfer tends to enlarge the heated surface, increasing the number of active nucleation centers, to ensure continuous operation and the activation at lower superheating of the heated surface in comparison to a smooth surface. Properly selected structure on the heated surface allows to significant increase in the heat transfer coefficient in comparison with obtained for the technically smooth surface.

Pool boiling process on the microstructured surface differs in comparison to the process taking place on the technically smooth surface. The values of the heat transfer coefficient differ strongly depends on the material characteristics of a porous layer.

* Corresponding author: *kalawa@agh.edu.pl*

In Figure 1 the results of experimental research on pool boiling heat transfer on microstructured surfaces laboratory-produced both in industry and commerce is shown. Results are presented as boiling curves (heat flux $q$ versus surface superheating $\Delta T$) for water with heat flux increasing, for selected surfaces.

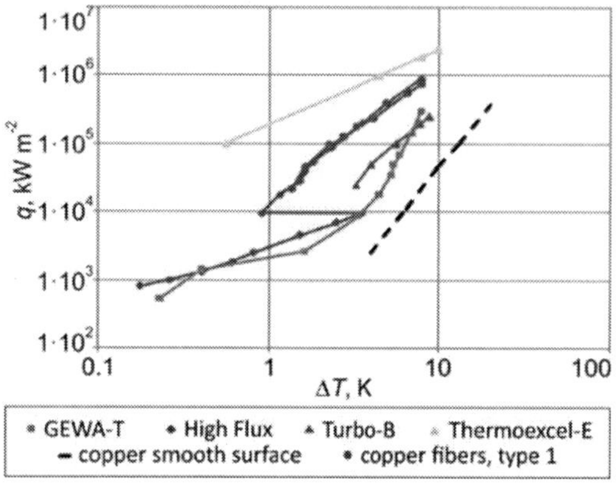


**Fig. 1.** Test results of experimental research on pool boiling heat transfer on microstructured surfaces [8].

Water has been chosen as a working fluid in the study, based on the literature review as an extensive database of experimental results can be found. Timeliness of the issue and the variety of available research results of own data and described in the literature allow to analyze the boiling process on different surfaces.

The analysis of experimental studies resulting from the author's experience in a flow boiling heat transfer in minichannels [1-3] indicates that lower heat flux was supplied using enhanced heated surfaces compared to the smooth surface. Values of the local heat transfer coefficient have proved higher when results from the enhanced heating surfaces were noticed [2-4]. Enhanced heating surfaces were produced using various techniques, e.g. laser texturing and modified spark erosion. Some research have been undertaken to produce porous coatings to be deposited on heating surfaces using an innovative technology which are mainly used in pool boiling research [1,5,11]. It involves sintering metallic powder in the dissociated ammonia atmosphere to generate diffusion-origin pores [6,7].

## 2 Research methodology

### 2.1 Object of study

In the current study, the tested porous surfaces were produced by sintering iron powder (Fe) of ASC100.29 with the stainless steel plate in the dissociated ammonia atmosphere and by soldering iron powder or stainless steel fibers to the stainless steel plate. Soldering was carried out in according to the patent application [8]: once the surface was covered with a thin film of solder and flux, the fragmented structures of metallic material (powders, fibres) were joined by soldering to the plate.

Figures 2 - 6 show 3D topographies and roughness profiles for five types of porous surfaces used in a pool boiling research. One of them (powder, type 1) was produced by sintering unsorted iron powder (size of the granulate 45÷250 μm) with the stainless steel plate in the dissociated ammonia atmosphere (Fig. 2). Next two types were produced by soldering sorted iron powders: of size of the granulate below 80 μm (powder, type 2, Fig. 3) and of size of the granulate above 95 μm (powder, type 3, Fig. 4), to the stainless steel plate.

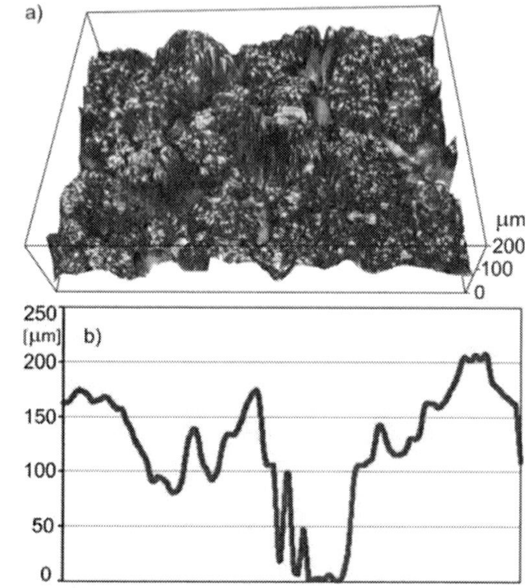


**Fig. 2.** Porous surface produced by sintering iron powder, type 1 (unsorted, size of the granulate 45÷250 μm): a) 3D topography, b) roughness profile.

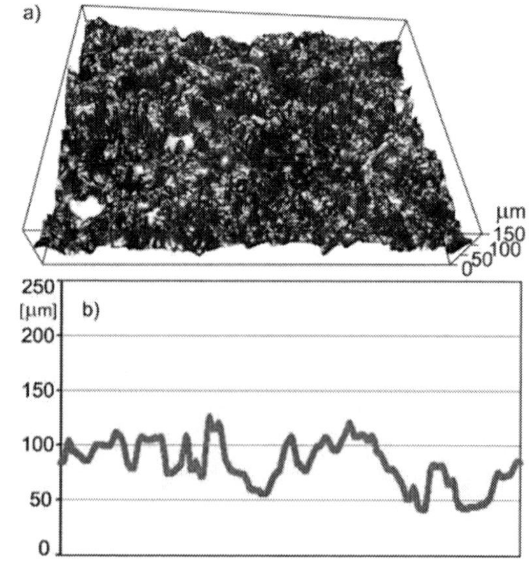


**Fig. 3.** Porous surface produced by soldering iron powder, type 2 (sorted, size of the granulate below 80 μm): a) 3D topography, b) roughness profile.

The last two types of porous surfaces were produced by soldering stainless steel fibres of different average fibre diameters: of 10 μm (fibre, type 1, Fig. 5) and of 25μm (fibre, type 2, Fig. 6) to the stainless steel plate.

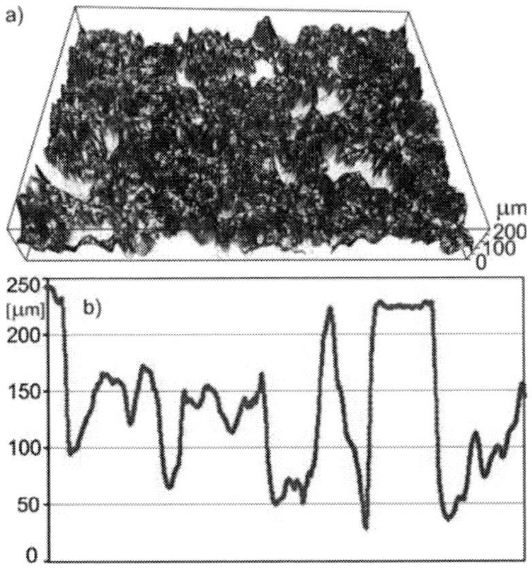


**Fig. 4.** Porous surface produced by soldering iron powder, type 3 (sorted, size of the granulate above 95 μm):
a) 3D topography, b) roughness profile.

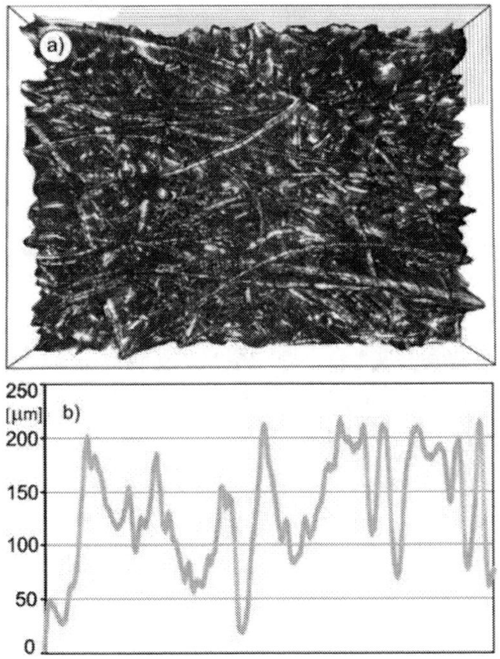


**Fig. 5.** Plate surface enhanced by soldering stainless steel fibres, type 1 (average fibre diameter - 10 μm):
a) 3D topography, b) roughness profile.

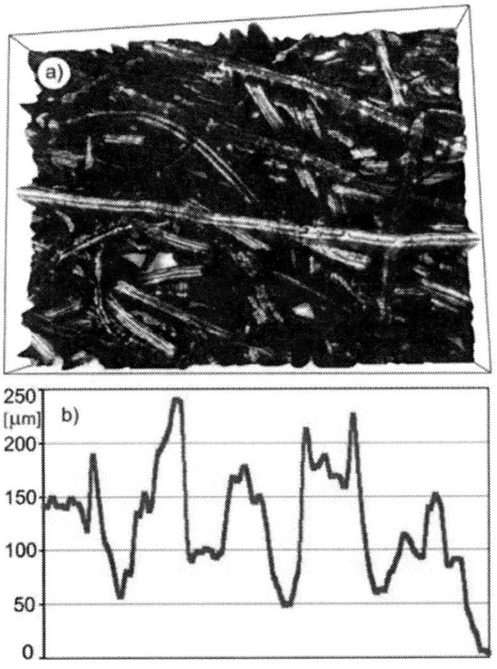


**Fig. 6.** Porous surface produced by soldering stainless steel fibres, type 2 (average fibre diameter - 25 μm):
a) 3D topography, b) roughness profile.

## 2.2 Experimental stand

The test stand facility was based on the literature review. In the process of the test stand constructing it was needed to eliminate its structural defects and to reduce material which were used in previous design.

Figure 2 shows the schema of the experimental stand. It consists of the following modules:

- the base module;
- the heating unit (along with the measurement of electrical power supply);
- the module of temperature measurement;
- the power measurement and control of supplied electric power module
- water supply, the cooling and condensate recovery system.

The base module consists of: an electric heater with a power of 1100 W, a copper core responsible for the accumulation and transport of heat between the heater and a sample and a copper disc (on which a sample of the porous structure is placed). In the core holes for thermocouples were drilled and grooves were incised under the spacer. In these grooves thermocouples were placed, located in the axis of the test sample. The base module are placed in thermal insulation made from carbon-fibre.

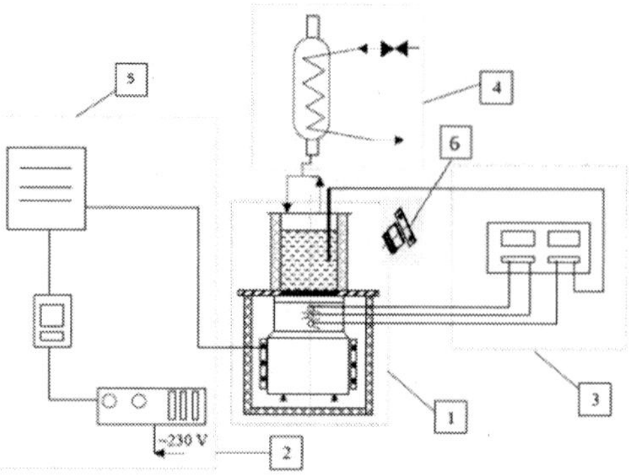


**Fig. 6.** Diagram of the experimental test of pool boiling heat transfer-[10] 1-base module; 2-heating unit; 3-the module of temperature measurement; 4-water supply, the cooling and condensate recovery system; 5-the power measurement and control of supplied electric power module; 6-the visualization of boiling.

### 2.3 Experimental procedure

Study on pool boiling on porous surfaces was carried out for water at atmospheric pressure. The test procedure has to precede the boiling heat transfer investigation in which the technically smooth surface made of stainless steel was used. On the basis of the test results boiling curves were constructed. Then, the experimental procedure of research on pool boiling heat transfer intensification is carried out and several microstructured surfaces of different structural parameters were tested.

After the test procedure, the intensive boiling is provided, once the fluid is degassed and fluid temperature achieved the saturation temperature. Then the process is carried out slowly, while the heat flux supplied to the sample are increased due to a power supply and a resistive heater placed on a copper cylinder. Gradually increasing of the heat flux in long intervals in the stationary state is conducted to stabilize the thermal condition in the sample. This is possible because a high thermal capacity of the heater (a heating roller), to which the sample is soldered. The heat flux is calculated from the electric power supplied to the heater and the comparative temperature distribution in the copper cylinder. Electric power during measurements is reduced by the heat losses to the environment, determined from the heat balance for the whole experimental stand. It allows to specify the amount of supplied heat flux. Figure 3 shows the position of the thermocouples in the heating roller. Such an arrangement of thermocouples allows to determine the temperature at the surface of the sample by extrapolation of the temperature distribution.

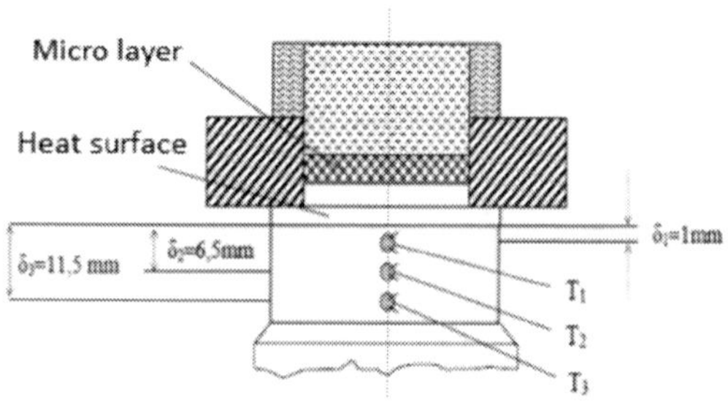


**Fig. 8.** The scheme of thermocouples arrangement in the basic module.

## 3 The results of experimental studies

Results are presented as boiling curves i.e. heat flux $q$ versus surface superheating $\Delta T$, with heat flux increasing, similarly as in Fig. 1. Results of experimental tests conducted using a technical smooth surface (Ra = 5) are shown in Fig. 9.

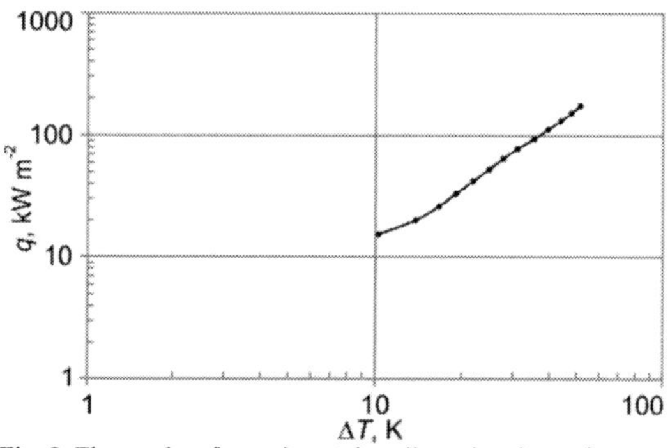


**Fig. 9.** The results of experimental studies using the surface made of stainless steel, technically smooth.

Test results of experiments carried out using the above-described modified surfaces were compared with the results for the sample of technically smooth surface, in order to illustrate the impact of the application of the developed surfaces, Figs. 10-14.

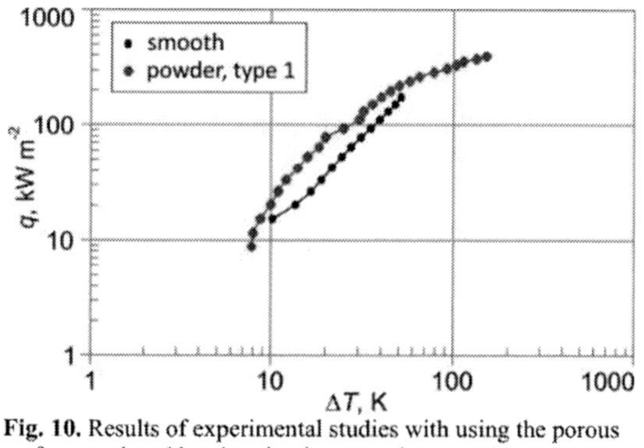


**Fig. 10.** Results of experimental studies with using the porous surface produced by sintering iron powder, type 1.

DOI: 10.1051/epjconf/201714302048

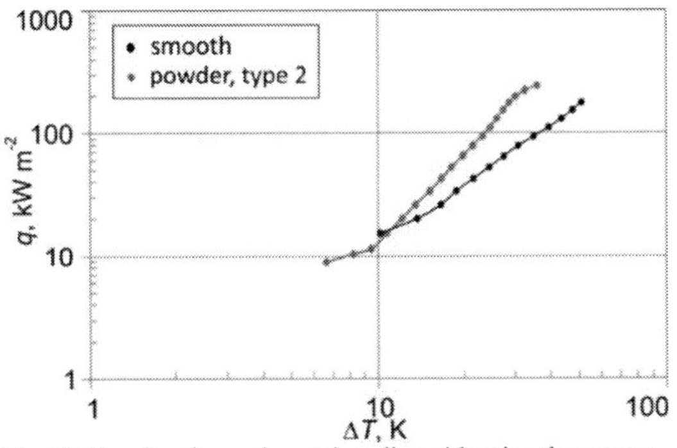


**Fig. 11.** Results of experimental studies with using the porous surface produced by soldering iron powder, type 2.

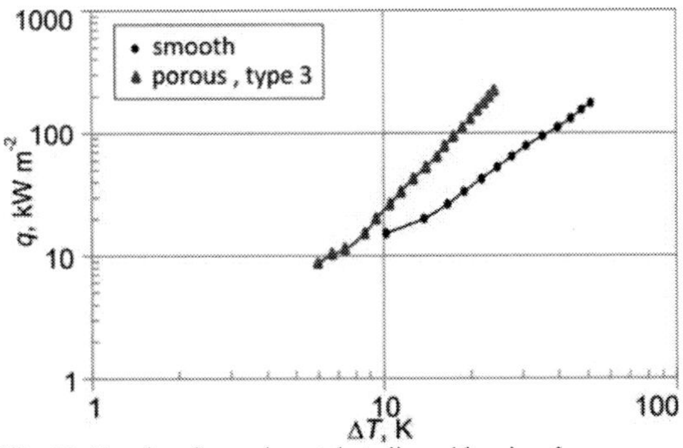


**Fig. 12.** Results of experimental studies with using the porous surface produced by soldering iron powder, type 3.

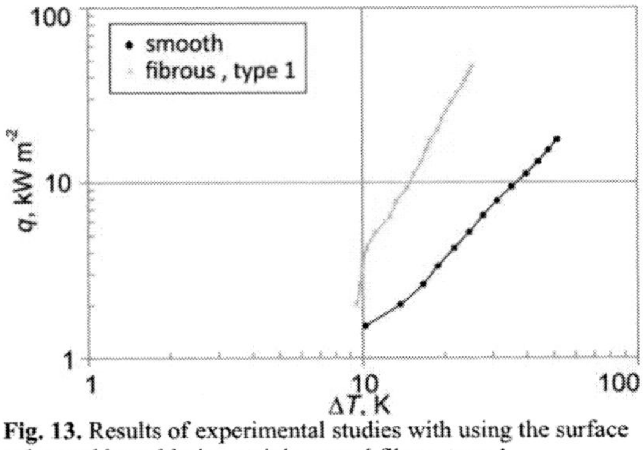


**Fig. 13.** Results of experimental studies with using the surface enhanced by soldering stainless steel fibres, type 1.

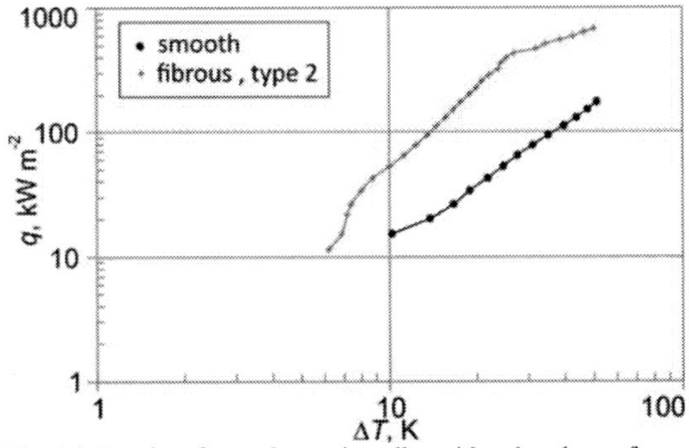


**Fig. 14.** Results of experimental studies with using the surface enhanced by soldering stainless steel fibres, type 2.

## 4 Analysis of the results

Table 1 shows a comparison of the values of the heat transfer coefficients obtained for modified surfaces-with the technically smooth one, for the constant surface superheating of 10 K.

**Table 1.** Heat transfer coefficient and heat flux for the constant surface superheating of 10 K, results from experiments with using different heated surfaces.

| Sample name | ΔT | *q* | α |
|---|---|---|---|
| - | K | W·m$^{-2}$ | W·m$^{-2}$·K$^{-1}$ |
| smooth surface | 10 | 15282 | 1498 |
| powder type 1 | 10 | 20163 | 2016 |
| powder type 2 | 10 | 12882 | 1288 |
| powder type3 | 10 | 23374 | 2337 |
| fibrous type 1 | 10 | 20197 | 2019 |
| fibrous type 2 | 10 | 52717 | 5271 |

Figure 15 presents the heat transfer coefficient versus the heat flux obtained for boiling of water at atmospheric pressure obtained from experiments with using different heated surfaces.

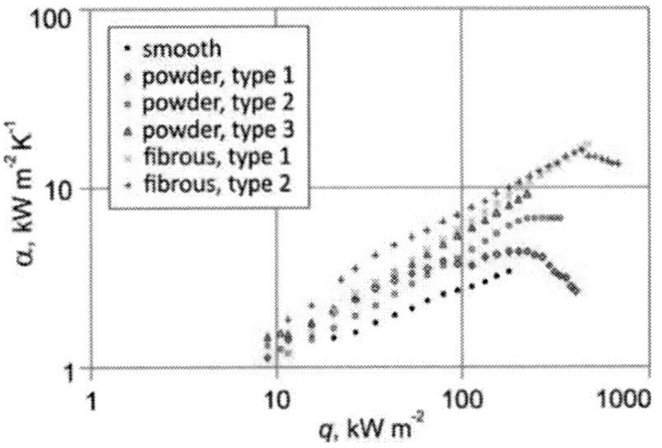


**Fig. 15.** Heat transfer coefficient versus the heat flux obtained for boiling of water at atmospheric pressure from experiments with using different heated surfaces.

Analysis of results has shown that all applied coatings on the heated surface causes heat transfer intensification in comparison to obtained for the smooth surface. In case of using the surfaces enhanced by soldering fibres in experiments, increasing diameters of fibres results in increasing values of the heat transfer coefficient at lower heat flux. At the same time, diminishing of the heat transfer coefficient is observed for this surface at higher heat flux, close to the critical heat flux.

It was also observed that when size of the powder granulate in the porous surface is larger, the higher values of the heat transfer coefficient are obtained. However, at higher heat flux, increasing the thickness of the porous powder layer significantly deteriorates the heat transfer.

## 5 Conclusion

With the rapid development of electronic, nuclear, space, military, medical and other technologies, accommodation of very high heat fluxes have become one of the most heavyweight problems to solve. A growing amount of heat emitted by large capacity computer processors creates a demand for more intensive cooling techniques. Heat transfer on smooth surfaces, even via boiling, is becoming insufficient and creates an urgent need for new solutions.

For more than twenty years, the authors have been conducting thorough theoretical and experimental studies of pool boiling heat transfer enhancement on extended surfaces, with the focus on developing new technologies, producing new structures on a lab scale, and application problems including the heat transfer regulation via boiling.

Results from the study have shown that the highest intensification of pool boiling heat transfer was achieved in experiments with using the porous surface enhanced by soldering stainless steel fibres, type 1. The heat transfer coefficient reached more than 500% of the value obtained for the smooth surface.

## Acknowledgments

The research has been financially supported by the National Scientific Center from funds granted by virtue of decision No. DEC-2013/09/B/ST8/02825.

## References

1. M. Piasecka, Ann. Nucl. Energy **73**, 282–293 (2014)
2. M. Piasecka, Int. J. Refrig. **56**, 198–212 (2015)
3. M. Piasecka, Int. J. Heat Mass Transf. **81**, 114–121 (2015)
4. M. Piasecka, K. Strąk, B. Maciejewska, Heat Transf. Eng. **38** (3) (2017), DOI: 10.1080/01457632.2016.118927, (in print)
5. R. Pastuszko, M. Piasecka, Journal of Physics: Conf. Ser. **395**, 012137 (2012)
6. W. Depczyński, R. Kazała, K. Ludwinek, K. Jedynak, Materials **9** (7), 567, 1–12 (2016)
7. W. Depczyński, *Proc. 23rd Int. Conf. Metallurgy Materials METAL 2014*, Poland (2014)
8. M. Piasecka, the Polish patent application *A method for porous structures producing on the surface of metal products*, No. A1 415204 (2015) (in Polish)
9. T.M. Wójcik, W. Kalawa, *M Balice, Energy and Fuels in XXI century*, 189-198 (2014) (in Polish)
10. T.M Wójcik, *Boiling incipience and hysteresis of heat transfer on porous coatings*, Series Hearing, Monographs **204** (the Publishing House of the AGH Univ. of Sc. and Techn. 2010) (in Polish)

# Effect of geometrical parameters of open microchannel surfaces on pool boiling heat transfer

*Robert* Kaniowski[1], *Robert* Pastuszko[1,*], and *Łukasz* Nowakowski[1]
[1]Kielce University of Technology, Faculty of Mechatronics and Mechanical Engineering, Al. 1000-lecia Państwa Polskiego 7, PL-25-314 Kielce, Poland

**Abstract.** This study focuses on the effect of channel depth on the heat transfer coefficient during nucleate pool boiling. Experimental studies were performed for saturated deionized water, and Novec-649 as working fluids at atmospheric pressure. Copper surfaces were modified to form microchannels with different geometrical properties. The microchannels were from 0.2 mm to 0.4 mm deep, 0.3 mm wide and spaced every 0.1 mm. The experiment was conducted for increasing heat flux up to the critical heat flux point. The surface modification provided an appreciably higher heat transfer coefficient compared to the smooth surface for all boiling liquids. The maximum heat transfer coefficient obtained exceeded 60 kW/m$^2$K.

## 1 Introduction

A high demand for effective heat removal from, in particular, microelectronic chips determines the development of small-size heat exchangers harnessing the liquid to gas phase transition. Thus far this has been one of the most effective forms of thermal energy transfer at the temperature of the heated surface maintained constant. The heated surface, however, needs to be modified and provided with an enhanced structure. A great many publications have reported the outcomes of boiling heat transfer research for various working fluids on different surface types and enhancement. Yet no simple structure has been proposed so far that would provide high heat transfer coefficients for different boiling liquids within a broad heat flux range. One of simpler but effective surface modifications involves cutting parallel grooves of 0.2 – 1 mm width and much the same depth range. As a result, a system of open microchannels is produced – the sites of vapour bubble intensive nucleation, growth and departure. Below, examples of the latest results from the literature-reported research on boiling heat transfer in open microchannels are discussed.

Jaikumar and Kandlikar [1] performed an experiment on surfaces that had three different inter-fin channel widths, 300 μm, 500 μm and 762 μm, with water at atmospheric pressure. They studied the effect of the inter-fin channel width on the surface performance and heat transfer mechanisms in open microchannel surfaces with three coating configurations: sintered-throughout, sintered-fin-tops, and sintered-channels. The heat transfer coefficients obtained, up to 2900 kW/m$^2$K, were large relative to the structured 10 x 10 mm$^2$ area, which gives about 1000 kW/m$^2$K relative to the total surface of heat transfer (17 x 17 mm$^2$)

In [2,3], the authors performed an experiment with water and ethanol at atmospheric pressure. The surfaces studied were Ω-shaped copper structures with circular cavities and narrow exit grooves upwards. These surfaces noticeably increase the number of active nucleation sites and are able to maintain a stable supply of liquid to the channels at large heat fluxes, which leads to high CHF.

The authors of [4] studied pool boiling of Novec-7300 fluid and self-rewetting water–heptanol mixtures on a bare copper surface and copper surface coated with copper-plated nanofibers. Heat transfer coefficients were found to be higher for the surface with the copper-plated nanofiber.

Research on pool boiling heat transfer with water, ethanol, FC-72 and Novec-649 from various structured surfaces [5,6] and boiling heat transfer with FC-72 flowing in narrow channels [7-9] has been conducted at the Kielce University of Technology for more a decade.

## 2 Experimental investigations

### 2.1 Test surfaces

The specimens with test surfaces were made of copper and had parallel grooves with a constant pitch and a cutter of 0.3 mm in diameter (CNC machining process), Fig.1. The test section consists of a 32.5 x 32.5 mm$^2$ square copper specimen with a 26.8 x 26.8 mm$^2$ boiling region as shown in Fig 1a. Table 1 compiles the surface codes and specifications according to Fig. 1b.

Talysurf CCI – Lite Non–contact 3D Profiler, a system providing optical, non-contact measurement of geometrical surface structures, was used to determine the roughness profile, shown in Fig. 2. Mean roughness was 0.127 μm.

* Corresponding author: tmprp@tu.kielce.pl

DOI: 10.1051/epjconf/201714302049

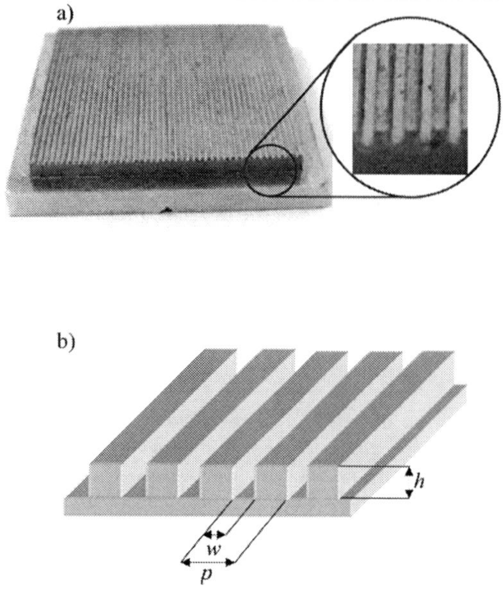


**Fig. 1.** a) View of MC-0.3-0.3-0.6 specimen, b) dimension symbols.

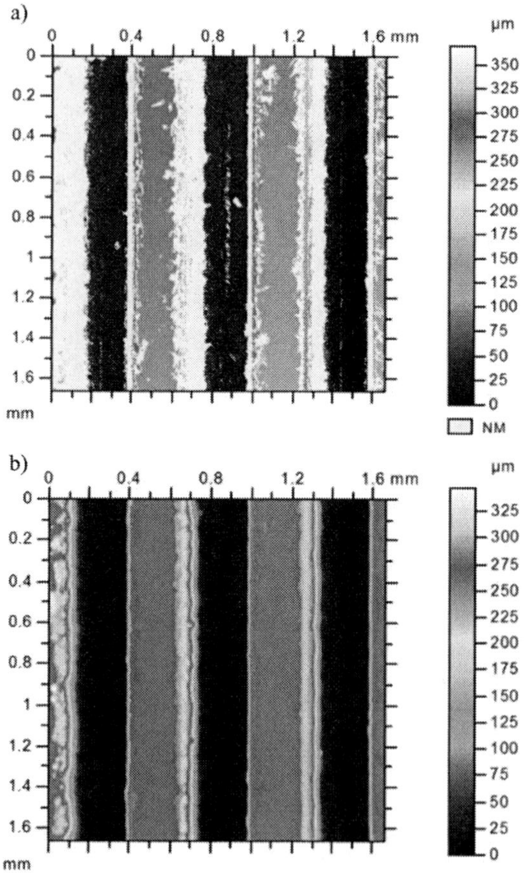


**Fig. 2.** Roughness profile of surface microchannel, a) microchannel top, b) microchannel bottom.

**Table 1.** Surface codes and specifications.

| Specimen code | *w* mm | *h* mm | *p* mm |
|---|---|---|---|
| MC-0.3-0.4-0.6 | 0.3 | 0.4 | 0.6 |
| MC-0.3-0.3-0.6 | 0.3 | 0.3 | 0.6 |
| MC-0.3-0.2-0.6 | 0.3 | 0.2 | 0.6 |

## 2.2 Experimental apparatus

The measurements were performed on the set-up shown in Figs 3 and 4. The set-up was designed to determine the temperature difference between the liquid and the heated surface and to find the heat flux. These parameters are necessary to be able to plot boiling curves. Heat transfer on the surfaces with microchannels was studied for nucleate pool boiling. The level of liquid above the specimen surface was more than 50 mm.

Prior to measurement, the fluid was degassed through boiling maintained for about 15 minutes. Measurement data were recorded after all temperature values stabilised. The heat flux was increased during the tests up to the boiling crisis level.

Diameter of the heating bar corresponded to the diagonal of the specimen base. A 1000 W electric cartridge heater 19 mm in diameter and 130 mm in length was installed into the bar.

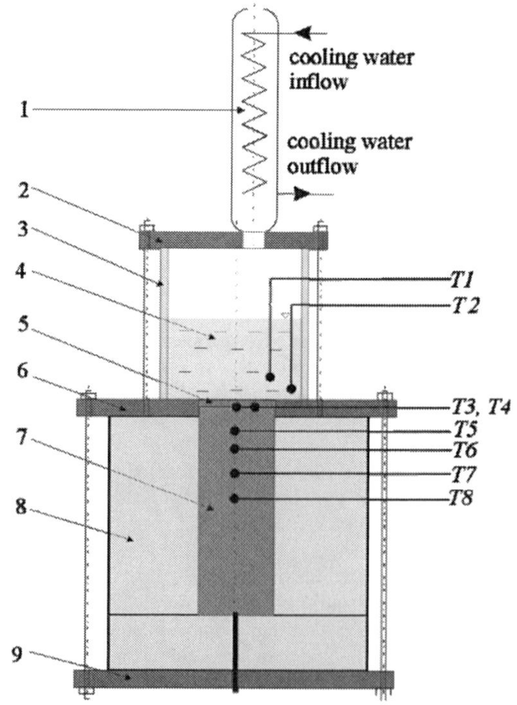


**Fig. 3.** Main module of pool boiling experimental stand: 1 – condenser, 2 – top flange 3 – glass vessel, 4 – test fluid, 5 – test surface, 6 – teflon lid, 7 – copper bar with cartridge heater, 8 – insulation, 9 – teflon base.

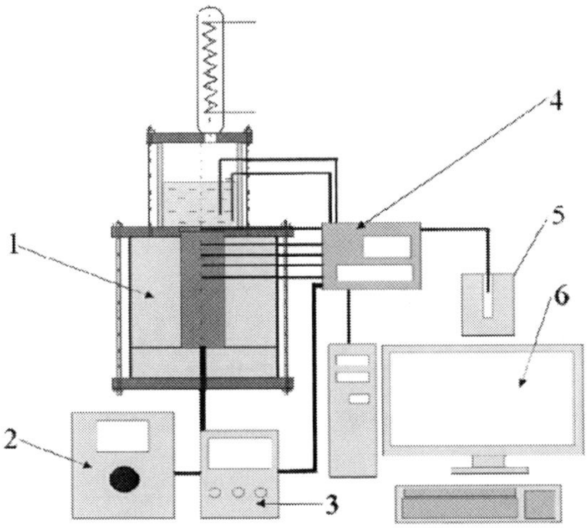


**Fig. 4.** Pool boiling measurement stand: 1 – main module, 2 – autotransformer, 3 – wattmeter, 4 – data logger, 5 – dry-well calibrator, 6 – PC.

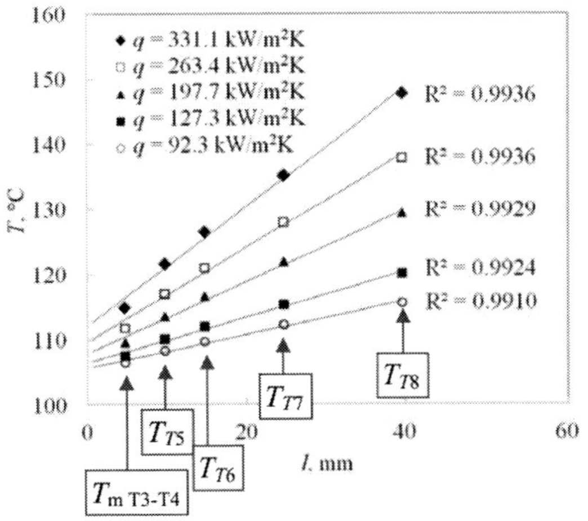


**Fig. 5.** Temperature vs. thermocouple location in the heating bar, water.

The estimated uncertainties were as follows:

- low heat flux (2 kW/m$^2$): heat flux ±35%, heat transfer coefficient ±40%,
- high heat flux (800 kW/m$^2$): heat flux ±2.1%, heat transfer coefficient ±2.4%.

## 3 Results

In order to mount the specimen on the heating bar, a layer of tin with thickness $\delta_{Sn}$ = 0.1 mm and thermal conductivity $\lambda_{Sn}$ = 66.5 W/(mK) was used. The linearity of temperature distribution between thermocouples T3 and T8 was checked, Fig. 5, and the coefficient of determination $R^2$ was between 0.9910 and 0.9936. This proves that one dimensional conduction heat transfer in the upper part of the heating bar is the reasonable dominant heat transfer mechanism.

With the assumption that the heat flow is one-dimensional, the heat flux was calculated on the basis of the temperature gradient in the copper bar from:

$$q = \lambda_{Cu} \frac{\Delta T_1}{l_{T5-T8}}, \tag{1}$$

where: $l_{T5\text{-}T8}$ = 30 mm is the distance between thermocouples T5 and T8, Fig. 3; the temperature difference is given by:

$$\Delta T_1 = T_{T8} - T_{T5}, \tag{2}$$

The heat transfer coefficient is expressed as:

$$\alpha = \frac{q}{\Delta T_{T2}}, \tag{3}$$

The difference between temperatures of the heated surface and liquid $\Delta T_2$ (superheat) is shown by the following equation, according to Fig. 3:

$$\Delta T_{T2} = \frac{T_{T3} + T_{T4}}{2} - \frac{T_{T1} + T_{T2}}{2} - q\left(\frac{\delta_s}{\lambda_{Cu}} + \frac{\delta_{Sn}}{\lambda_{Sn}}\right) \tag{4}$$

Figures 6 and 7 show the relationships between the heat transfer coefficient and heat flux and between the heat flux and temperature superheat at increasing heat flux on the MC surface for two boiling liquids. Compared with smooth surfaces, specimens with microchannels provided significant heat transfer enhancement, with heat transfer coefficient being dependent on the microchannel height. The best results were recorded for specimen MC-0.3-0.3-0.6, where the maximum heat transfer coefficient was $\alpha$ = 64.4 W/m$^2$K for $\Delta T \sim$ 10.5 K with water. With Novec-649, the highest coefficient was 7.7 W/m$^2$K for $\Delta T \sim$ 11.5 K.

For MC-0.3-0.3-0.6, the boiling incipience temperature was lower by about 5 K with water and 7 K with Novec-649.

Table 2 summarizes calculation results of constant C and the exponent n in the dependence for the value of the heat flux $q = C\Delta T^n$ for water and Novec-649 in relation to three MC surfaces and a smooth surface.

Heat transfer enhancement on the specimens with microchannels was confirmed by the ratio between heat transfer coefficient for the structured surface and that for the smooth surface. The coefficient values obtained were as follows: $\alpha_{enh}/\alpha_{smt}$ = 8 for MC-0.3-0.4-0.6, $\alpha_{enh}/\alpha_{smt}$ = 20.7 for MC-0.3-0.3-0.6, and $\alpha_{enh}/\alpha_{smt}$ = 5.2 for MC-0.3-0.2-0.6 at superheat $\Delta T \sim$ 8.5 K with water as the working fluid.

For Novec-649, the increase in coefficient values was similar. The $\alpha_{enh}/\alpha_{smt}$ ratio was 14 for specimen MC-0.3-0.4-0.6, 14.7 for MC-0.3-0.3-0.6, and 10 for MC-0.3-0.2-0.6 at superheat $\Delta T \sim$ 10.5 K.

**Table 2.** Constant C and exponent n for function $q(\Delta T)$

| Specimen code | fluid | $C$ | $n$ |
|---|---|---|---|
| MC-0.3-0.4-0.6 | water | 0.91 | 2.34 |
| MC-0.3-0.3-0.6 | water | 2.571 | 2.44 |
| MC-0.3-0.2-0.6 | water | 1.340 | 2.09 |
| Smooth surface | water | 0.038 | 3.03 |
| MC-0.3-0.4-0.6 | Novec-649 | 0.440 | 1.87 |
| MC-0.3-0.3-0.6 | Novec-649 | 0.095 | 2.67 |
| MC-0.3-0.2-0.6 | Novec-649 | 1.624 | 2.26 |
| Smooth surface | Novec-649 | 0.038 | 2.21 |

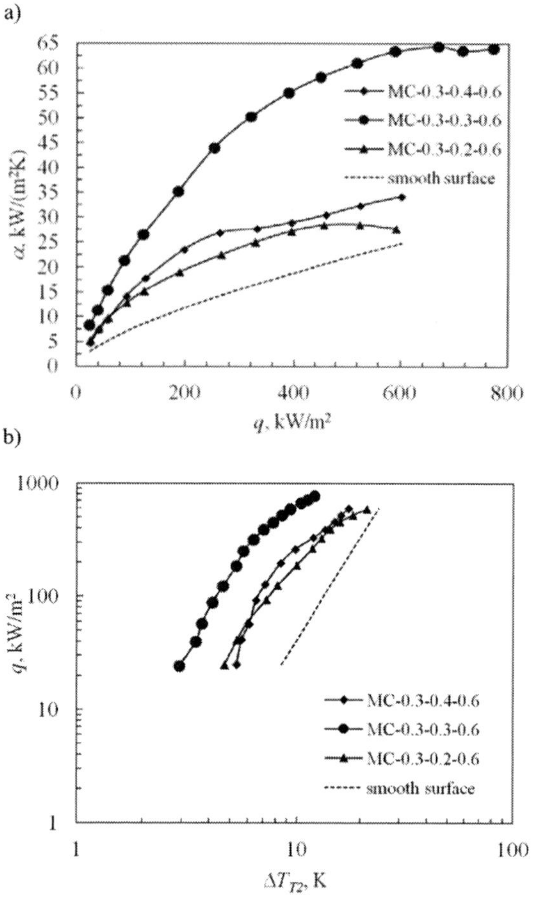


**Fig. 6.** Boiling curves for water, a) heat transfer coefficient vs. heat flux, b) heat flux vs. superheat.

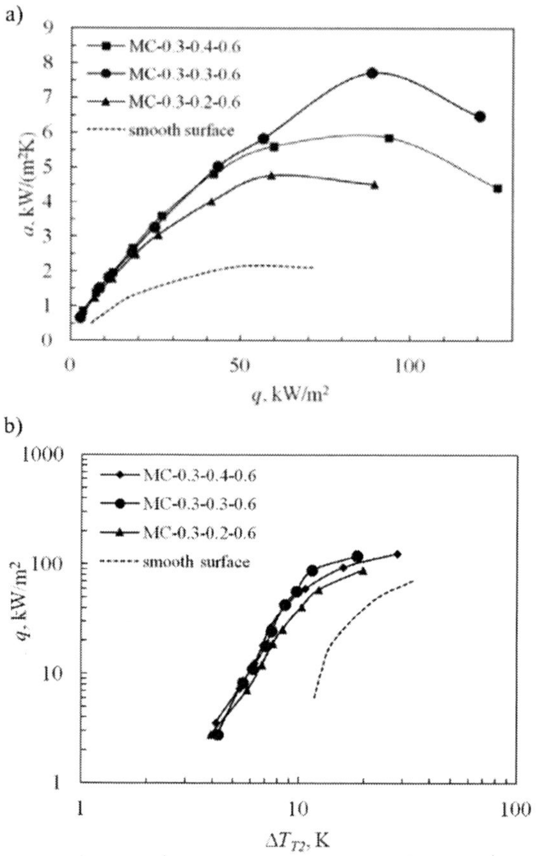


**Fig. 7.** Boiling curves for Novec-649, a) heat transfer coefficient vs. heat flux, b) heat flux vs. temperature difference.

The depth of the microchannels had a significant effect on heat transfer performance of the investigated surfaces. From among three surfaces with microchannels 0.2, 0.3 and 0.4 mm in depth, the best performance was provided by 0.3 mm deep microchannels. An increase in the depth of a microchannel leads to the rise in its bottom temperature, which is due to the reduced distance from the heater. As a result, nucleation is induced at lower heat fluxes and superheats and the heat transfer coefficient is higher. An increase in the microchannel depth implies an increase in the height of microfins forming the channel walls. Higher microfins cause a reduction in temperature at their tips, which reduces the intensity of vapour bubble generation from these surfaces.

## 4 Conclusions

The experimental studies discussed in this paper lead to the following conclusions:

- With water as the working fluid, heat transfer coefficients obtained for the surfaces with 0.3 mm deep microchannels were twice as high as those with microchannels 0.2 and 0.4 mm deep.

- With Novec-649 as the working fluid, a similar HTC increase for the surface with a 0.3 mm deep microchannels was observed on the surface whose channels were the least shallow.
- Compared with the smooth surface, the increase in the heat transfer coefficient value for the best surface was 3.5 times as high (at constant heat flux).
- At the stable microchannel pitch and width, the choice of channel depth is vital for the intensity of heating the inflowing liquid and for the higher vapour generation rate on the microchannel bottoms and microfin tips.
- Experimental studies for other liquids, channel widths and pitches are necessary to check the effect of at least three geometrical parameters on the heat transfer process.

## Nomenclature

$C$ – constant
CHF – critical heat flux
$h$ – microchannel height, m
$l$ – distance, m
$n$ – constant
$p$ – pitch, m
$q$ – heat flux, $kW/m^2$
$T$ – temperature, K
$w$ – width, m

**Greek symbols**

$\alpha$ – heat transfer coefficient, $W/(m^2K)$
$\Delta T$ – difference of temperature, K
$\delta$ – thickness, mm
$\lambda$ – thermal conductivity, W/(mK)

**Subscripts**

$Cu$ – copper
$enh$ – enhancement
$Sn$ – tin
$s$ – specimen
$smt$ – smooth

## References

1. A. Jaikumar, S. G. Kandlikar, Int. J. Heat Mass Transfer **95**, (2016)
2. D. Deng, J. Feng, Q. Huang, Y. Tang, Y. Lian, Int. J. Heat Mass Transfer **99**, (2016)
3. D. Deng, W. Wan, J. Feng, Q. Huang, Y. Qin, Y. Xie, App. Therm. Ing. **107**, (2016)
4. R. P. Sahu, S. Sinha-Ray, S. Sinha-Ray, A. L. Yarin, Int. J. Heat Mass Transfer **95**, (2016)
5. R. Pastuszko, Exp. Therm. Fluid Sci. **38**, (2012)
6. R. Pastuszko, T. Wójcik, Exp. Therm. Fluid Sci. **63**, (2015)
7. S. Hozejowska, R. Kaniowski, M. Poniewski, Exp. Therm. Fluid Sci.**78**, (2016)
8. M. Piasecka, Int. J. Heat Mass Transfer **81**, (2015)
9. M. Piasecka, B. Maciejewska, Exp. Therm. Fluid Sci.**78**, (2015)

# Pool boiling visualization on open microchannel surfaces

*Robert* Kaniowski[1], and *Robert* Pastuszko[1,*]
[1]Kielce University of Technology, Department of Mechanics, al. 1000-lecia Państwa Polskiego 7, PL-25-314 Kielce, Poland

**Abstract.** The paper presents visualization investigations into pool boiling heat transfer for open minichannel surfaces. The experiments were carried out wih saturated water at atmospheric pressure. Parallel microchannels fabricated by machining were about 0.3 mm wide and 0.2 to 0.4 mm deep. High-speed videos were used as an aid to understanding the heat transfer mechanism. The visualization study aimed at identifying nucleation sites of the departing bubbles and determining their diameters and frequency at various superheats.

## 1 Introduction

Knowledge about the conditions under which vapour bubbles form, the frequency of bubble detachment from a given nucleation site and their diameters at departure is the basis for the analysis of boiling mechanism. These quantities are determined, indirectly, with the use of high speed cameras.

Zhong *et al.* [1] studied pool boiling heat transfer with water as the working fluid. The inclination angles were 5°, 30°, 45°, 60° and 90° (vertical), two structured surfaces with interconnected-grooves and cavities were investigated. The boiling process was photographed at various heat fluxes. The authors found that the structured surface which consisted of cavities as stable cavity-centric vaporization cores and interconnected grooves as the cooling water supply pathways provided significant heat transfer enhancement and increase in CHF.

In [2], the authors experimented with water and ethanol at atmospheric pressure. The surface under test was a cooper specimen with microchannels with a Ω cross-section. Visualization studies revealed nucleation, growth and departure behaviours in the reentrant cavities and also that at higher heat fluxes, the bubbles tend to coalesce.

The authors of [3] used liquid nitrogen in their experiments at atmospheric pressure. At low heat fluxes, the bubbles were isolated and interactive effects were negligible. At high heat fluxes close to CHF, the bubbles interacted heavily and appeared in the form of a cluster with a neck-like structure. Periodical break-up of the neck led to shedding the upper part of the cluster. At intermediate heat fluxes, the bubbles interacted significantly and single bubbles were impossible to be seen – diameters and detachment frequencies of the bubbles could not be determined.

In [4], the authors used water at atmospheric pressure. The heated surface comprised two cooper specimens: a solid copper interconnected microchannel net and a porous interconnected microchannel net. Visualization of boiling was performed for both surface types focusing on the bubble growth rate and the influence of the heated surface in the vapour bubble diameter.

The Kielce University of technology has long been conducting visualization studies of boiling with water, ethanol, FC-72 and Novec -649 from various structured surfaces for both pool boiling and flow boiling [5-10].

## 2 Experimental set-up

The test setup for the visualization studies is shown in Fig. 1. The main stand module, Fig.2 consists of the specimen being investigated (5) soldered to the cylindrical copper bar (7) and placed under the glass vessel (3) filled with the working fluid (4). A 1200 W electric cartridge heater, 19 mm in diameter and 130 mm in length, was installed into the bar. The investigations were conducted with distilled water at atmospheric pressure. The setup was designed to allows determining the temperature difference between the liquid and the heated surface, the heat flux and the diameter and frequency of departing bubbles. Prior to the measurements, the fluid was degassed through preliminary boiling for about 15 minutes.

The images of the bubbles at the moment of their nucleation, growth and departure were obtained using a PHOT MV-D1024-160-CL (*Photonfocus*) digital monochromatic camera with a resolution of 1024x758 pixels. The camera recorded the photos at a speed of 300 fps at a resolution of 480x360 pixels with mean error 2 fps.

An EX-FH20 (*Casio*) camera was used to record the images of the entire surface of the specimen at a resolution of 480x360 and the recording speed of 210 fps, with the mean error of 2 fps. Lighting was provided by a halogen lamp with light pipes (front light and back light).

The study was carried out on copper specimens with parallel grooves cut in them. Figure 3 and Table 1 show geometrical dimensions of the grooves.

* Corresponding author: tmprp@tu.kielce.pl

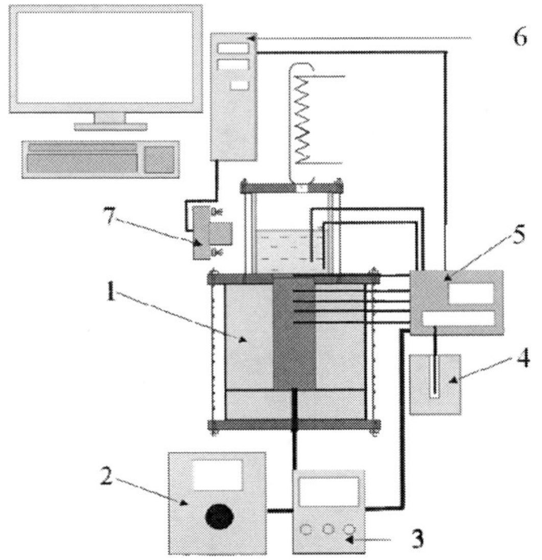


**Fig. 1.** Pool boiling visualization system: 1 – main module, 2 – autotransformer, 3 – wattmeter, 4 – dry-well calibrator, 5 – data logger, 6 – PC, 7 – high speed camera with lighting.

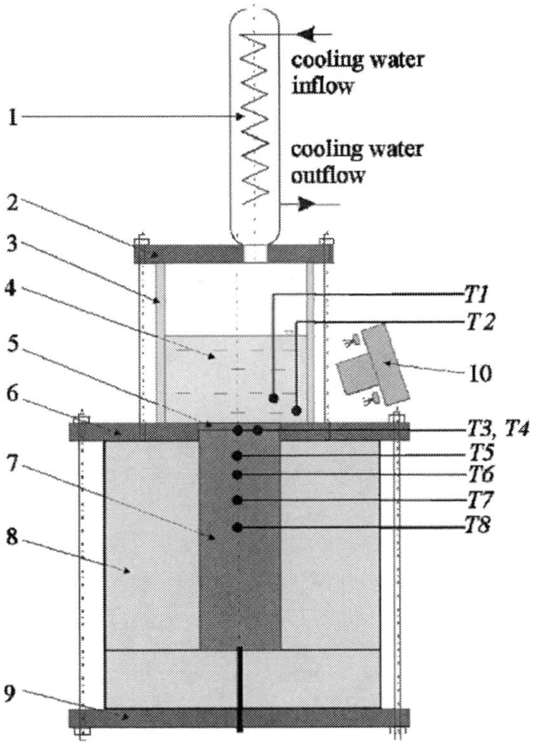


**Fig. 2.** Main module of pool boiling experimental stand: 1 – condenser, 2 – top flange 3 – glass vessel, 4 – test fluid, 5 – test surface, 6 – teflon lid, 7 – copper bar with cartridge heater, 8 – insulation, 9 – teflon base, 10 – high speed camera with lighting.

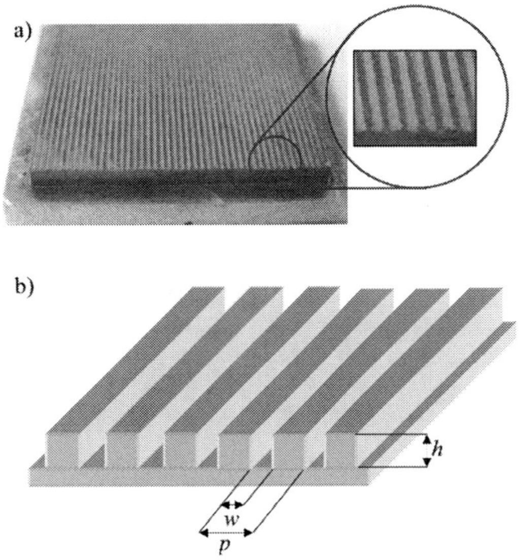


**Fig. 3.** Test surface: a) view of MC-0.3-0.2-0.6 specimen, b) dimension symbols.

**Table 1.** Surface codes and specifications.

| Specimen code | w mm | h mm | p mm |
|---|---|---|---|
| MC-0.3-0.4-0.6 | 0.3 | 0.4 | 0.6 |
| MC-0.3-0.3-0.6 | 0.3 | 0.3 | 0.6 |
| MC-0.3-0.2-0.6 | 0.3 | 0.2 | 0.6 |

## 3 Results

Figure 4 includes typical images of boiling of the working fluid at increasing heat flux. The photographs were taken after temperatures on the heating bar and under the specimen stabilised at constant heat flux. Bubble diameters and nucleation site number grew along with increasing heat flux. At small superheats, $\Delta T < 5.5$ K, on surfaces MC-0.3-0.2-0.6 and MC-0.3-0.4-0.6 and at $\Delta T < 3.5$ K on MC-0.3-0.3-0.6, the vapour bubbles had regular spherical shapes. When $\Delta T \approx 6.0$ K and $q \approx 57$ kW/m$^2$ for the microchannels 0.2 and 0.4 mm deep and at $\Delta T \approx 3.5$ K and $q < 57$ kW/m$^2$ for the surfaces with microchannels 0.3 mm deep, most bubbles had a form of a flattened sphere. Few bubbles coalesced because their diameters were larger than the distances between their formation sites. When the superheat was increased above 4.1 K and at $q > 88$ kW/m$^2$, most bubbles coalesced forming irregular, changing in time shapes, usually funnel-shaped from the lower part. At these vapour formations (bubble groups) the diameters and frequencies of departing bubbles became impossible to measure.

(a) $q = 24.3$ kW/m$^2$ (b) $q = 39.5$ kW/m$^2$ (c) $q = 57.3$ kW/m$^2$ (d) $q = 88.3$ kW/m$^2$

**Fig. 4.** Visualization observations of pool boiling of water on the microchannel surface MC-0.3-0.3-0.6.

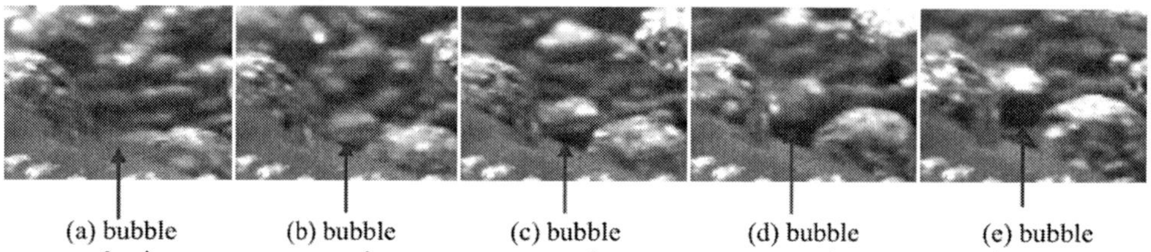

(a) bubble nucleation $t = 0$ ms (b) bubble growth $t = 3.3$ ms (c) bubble growth $t = 6.6$ ms (d) bubble growth $t = 9.9$ ms (e) bubble departure $t = 13.2$ ms

**Fig. 5.** Successive bubble formations on MC-0.3-0.3-0.6 specimen at $q = 24.3$ kW/m$^2$, water.

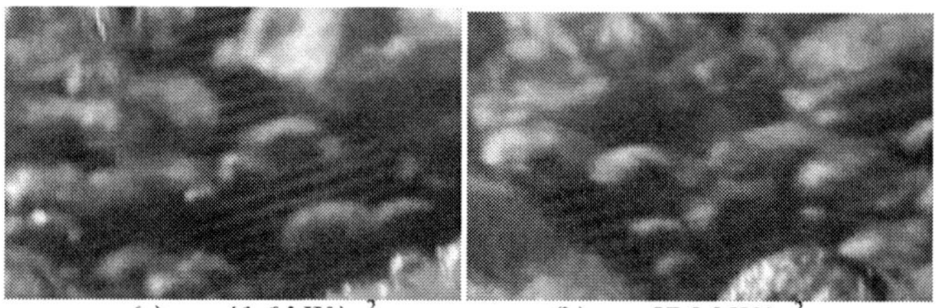

(a) $q = 41.6$ kW/m$^2$ (b) $q = 57.3$ kW/m$^2$

**Fig. 6.** Vapour bubble generation, MC-0.3-0.4-0.6, water.

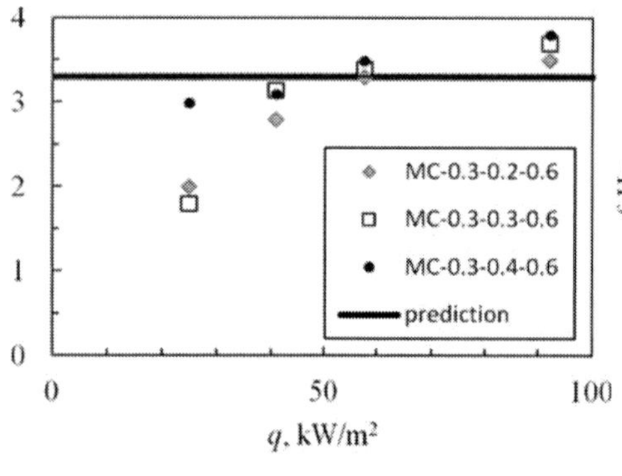


**Fig. 7.** Averaged departing bubble diameters for water.

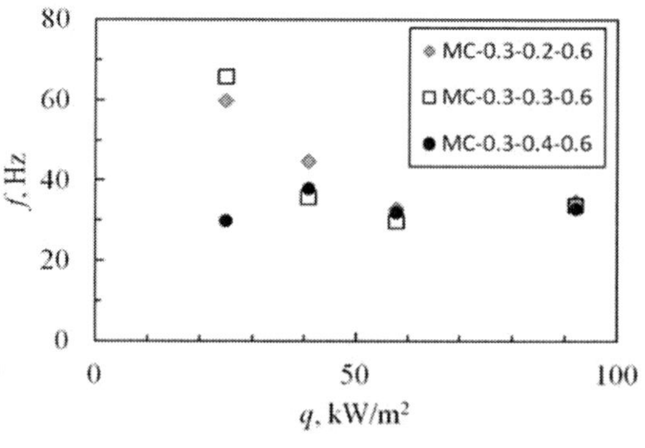


**Fig. 8.** Averaged departing bubble frequencies for water.

Figure 5 shows an example of bubble nucleation at small superheats, $\Delta T < 5.5$ K. After nucleation, the bubble grows up to the limiting value and detaches from the heated surface. Enlarged photographs of vapour forming on the specimen surface are shown in Fig. 6.

Bubble diameters and detachment frequencies were averaged from 8 – 10 nucleation sites for a minimum of 30 consecutive vapour bubbles. The diameter was measured along two directions. The quantities were determined at a constant heat flux. Absolute error of the diameter measurement was 2 pixels, which gives 0.2 mm [8].

Figure 7 shows the experimental results for the influence of heat flux on departing bubble diameter and Fig. 8 shows the frequency of bubble detachment from the heated surface during the boiling of water relative to the heat flux. Theoretical values of departing bubble diameters were calculated according to the dependence [5]:

$$d_b = \left( \frac{6\sigma d_h}{g(\rho_l - \rho_v)} \right)^{0.33} \tag{5}$$

DOI: 10.1051/epjconf/201714302050

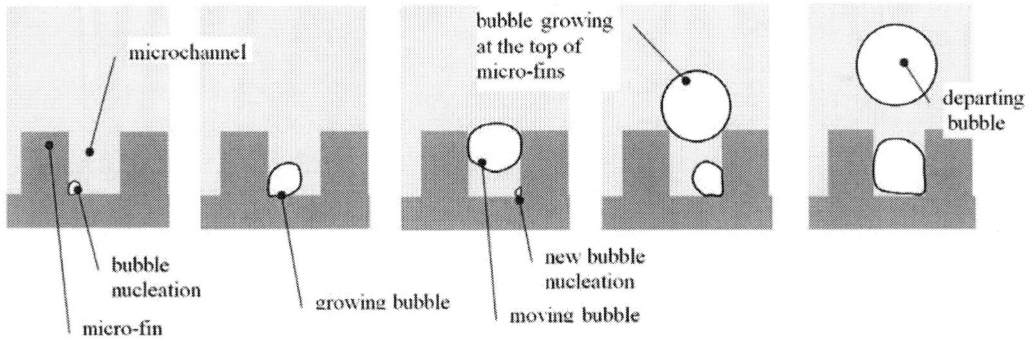


**Fig. 9**. Bubble growth in the microchannel space.

In addition, hydraulic diameter was assumed to be equal to the width of the microchannel, i.e., $d_h = w$.

At small heat fluxes ($q$ < 40 kW/m$^2$), bubble diameters are smaller than the calculated ones (Eq. 5) and detachment frequencies vary with surface type. At heat fluxes above 40 kW/m$^2$, averaged diameters are approximate in value to the predicted ones in equation 5 and the detachment frequencies stabilize at the level of 35 – 40 Hz.

On the basis of the visualization studies carried out by the authors of this paper and the visualization reported for a similar surface by Cooke and Kandlikar [4], a possible scenario of bubble growth phases in a microchannel is shown in Fig. 9. Nucleation starts in the corner between the channel bottom and the vertical side wall. The bubble grows to fill the microchannel space. At the same time, a new bubble at the bottom of the microchannel increases in volume and pushed larger bubble upwards toward the microfin tips. During this motion, intensive heat transfer from the microfin surfaces to the vapour leads to an increase in pressure and volume of the bubble. Further expansion of the bubble at the microfin tips contributes to its detachment.

## 4 Conclusions

Visualization of boiling mechanism allows formulating the following conclusions:

- At low superheats, the bubbles between the heated surface and liquid are spherical in shape and do not coalesce. At higher superheats, the bubbles interact with each other intensively, forming flattened spherical shapes funnel-shaped at the bottom.
- A minor increase in bubble diameter was observed at the moment of detachment at increasing heat flux, along with a substantial rise in the frequency of bubble generation at low $q$.
- In the preliminary model of vapour bubble formation and growth, nucleation was assumed to take place at the corners at the bottom of the microchannel, and the growing bubble forces its way through the channel space between the microfins in the direction of their tips, where it departs.

## Nomenclature

| | |
|---|---|
| CHF | – critical heat flux |
| $d$ | – diameter, m |
| $f$ | – frequency, Hz |
| $g$ | – gravity, m/s$^2$ |
| $h$ | – microchannel height, m |
| $p$ | – pitch, m |
| $q$ | – heat flux, kW/m$^2$ |
| $t$ | – time, s |
| $w$ | – width, m |

**Greek symbols**

| | |
|---|---|
| $\Delta T$ | – difference of temperature, K |
| $\rho$ | – density, kg/m$^3$ |
| $\sigma$ | – surface tension, N/m |

**Subscripts**

| | |
|---|---|
| $b$ | – bubble |
| $h$ | – hydraulic |
| $l$ | – liquid |
| $v$ | – vapour |

## References

1. D. Zhong, J. Meng, Z. Li, Int. J. Therm. Sci., **108** (2016)
2. D. Deng, W. Wan, J. Feng, Q. Huang, Y. Qin, Y. Xie, App. Therm. Ing. **107**, (2016)
3. X. Zhang, J. Chen, W. Xiong, T. Jin, Cryogenics, **72**, (2015)
4. D. Cooke, S. G. Kandlikar, Int. J. Heat Mass Transfer, **55**, (2012)
5. R. Pastuszko, Exp. Therm. Fluid Sci. **38**, (2012)
6. R. Pastuszko, T. Wójcik, Exp. Therm. Fluid Sci. **63**, (2015)
7. R. Pastuszko, R. Kaniowski, EPJ Web of Conferences, **25**, 02019 (2012)
8. S. Hozejowska, R. Kaniowski, M. Poniewski, Exp. Therm. Fluid Sci. **78**, (2016)
9. M. Piasecka, Int. J. Heat Mass Transfer **81**, (2015)
10. M. Piasecka, B. Maciejewska, Exp. Therm. Fluid Sci.**78**, (2015)

# Analysis of in-line check valve with respect to the pipeline dynamics

*Roman* Klas[1,*], *Vladimír* Habán[1], and *Pavel* Rudolf[1]

[1]Brno University of Technology, Faculty of Mechanical Engineering, Victor Kaplan Department of Fluids Engineering, Technická 2, 61669 Brno, Czech Republic

**Abstract.** The following study is focused on the flow analysis within the concentric in-line check valve. In-line check valve is part of the pipe and shut off the stream of fluid. In case that backflow occurs on the downstream side, the valve is fully or partially lifted and then abruptly closes. This effect brings undesirable pressure shocks in the entire system. The study therefore deals with the CFD analysis of steady and unsteady flow with respect to velocity and pressure field. Especially unsteady calculation mode reflects the influence of local inertial acceleration acting on the fluid and on the valve. Analysis is complemented by equation of motion of a valve which interacts with the fluid.

## 1 Introduction

It is well known that the field of industrial valves is very broad. The reason for this fact is that the valves are designed for different functions. In-line check valve is usually used as backflow damming element and is part of a complex piping system. Features of this system may be more important than static or dynamic characteristics of the in-line check valve itself. It is also possible to say that the static behavior of the valve is not usually problematic in terms of the entire facility. Nevertheless, it is of course necessary to know the static behavior of the valve, in order to assess e.g. size of hydraulic losses and other flow characteristics of the valve. Paradoxically, the simplicity of its design and the simplicity of its main function represent the biggest source of problems along with its automatic operation.

Usually, when dealing with the valves we get by with knowledge of opening pressures depending on the nominal size of the valve. Valve lift and shut off times are assumed to be very short. Any possible mixing of operating substances is small for this reason and the same should apply to the magnitudes of static pressure in the pipeline. However these assumptions are not always valid and automatic shutoff function of the valve is the source of adverse pressure shocks and backflow of the fluid reinforced by spring. Nevertheless, the pipeline operating properties, under which the valve is deployed, are decisive.

Therefore, the aim of the study is to examine the risk of the real operating conditions of the in-line valve including its basic features in the form of static characteristics.

## 2 Analytical and numerical methods and description of the valve

A one-dimensional approach in the form of differential equations of momentum and CFD simulations for two-dimensional and three-dimensional model were used to describe the behavior of the in-line valve. Parameters used in the differential equation are based on characteristics of the valve itself and with regard to the prediction of hydraulic losses also on CFD simulation. Therefore there is mutual effect of the approaches, but it is impossible to determine the hydraulic resistance of the valve otherwise than by experiments or CFD simulations [1, 2].

**Table 1.** Basic design parameters and description of numerical model.

| | 2D | 3D |
|---|---|---|
| Pipeline diameter | DN100 | |
| Number of computational cells | 750 000 | 9 millions |
| Turbulence model and near wall modeling | realizable $k-\varepsilon$ non equilibrium wall function | |
| Boundary conditions | Inlet: velocity inlet Outlet: pressure outlet | |
| Calculation mode | unsteady, sliding mesh incompressible flow | |

Examined in-line valve is a relatively new device, in which the entire experimental tests will be performed in the near future. Presently it is only known that operation

* Corresponding author: klas@fme.vutbr.cz

of the valve is problematic in terms of pressure pulses and the slight backflow.

Momentum equation (1), Navier-Stokes equation (2) and continuity equation (3) were employed to describe the movement of the valve poppet in a one-dimensional approach.

Fig. 1 captures a basic view of the valve, which is identical with a 3D CFD model. Simplified 2D model is shown in Fig. 2. The main parameters of the valve and CFD simulations are recorded in Tab. 1. Possibilities for choice of an appropriate turbulence model can be found in [3-6] and the influence of the number of computational cells on the results of CFD simulations in [7] as well.

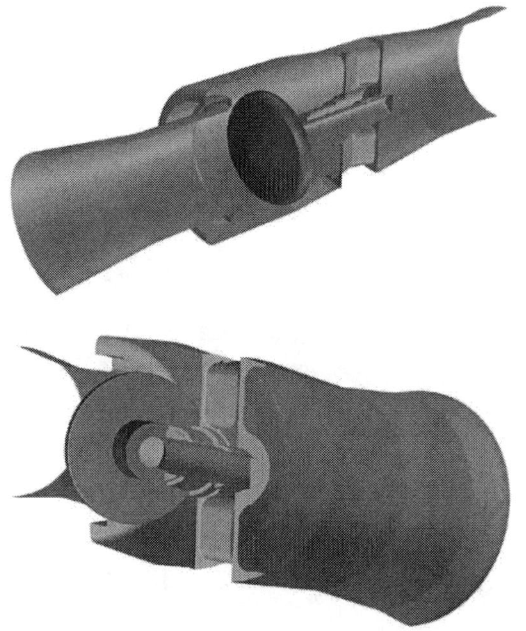

**Fig. 1.** The in-line valve, 3D simulation.

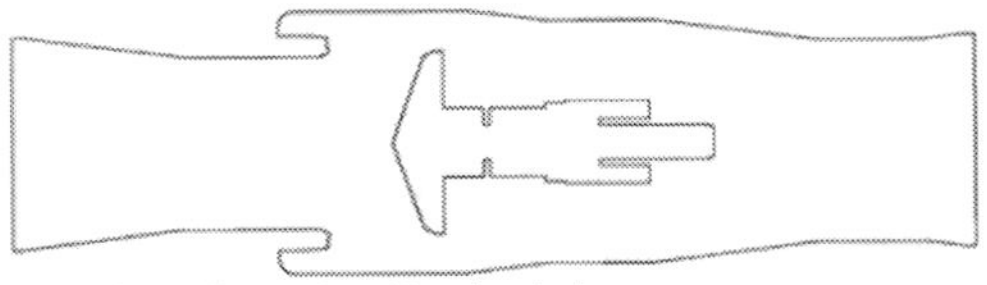

**Fig. 2.** The in-line valve, 2D simulation.

## 3 Equation of motion

One-dimensional solution of the basic momentum equation may not bring satisfactory results in many technical systems. However, the analytical solution of this equation provides the fundamental possibility of analyzing and understanding the phenomena and factors affecting the movement of poppet in liquid. Unfortunately, the complexity of determining of lift or valve poppet velocity as function of time is often so great that analytical solution is not always possible. Then it is necessary to rely on numerical solution of momentum equation.

Fig. 3 shows sketch of in -line valve in terms of the used symbols. The meaning all of the remaining used symbols is summarized in Tab.2.

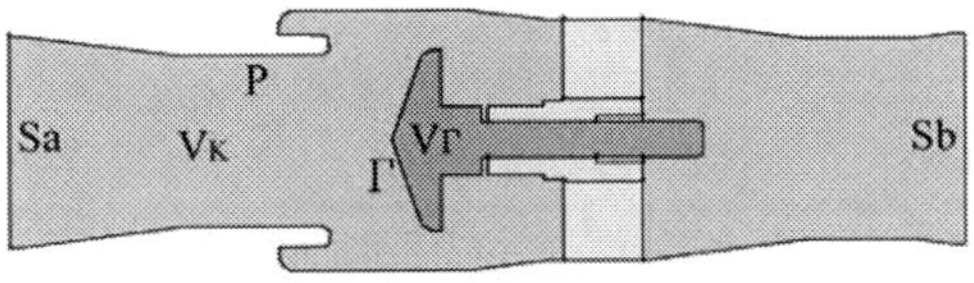


**Fig. 3.** Sketch of the in-line valve.

**Table 2.** List of symbols.

| | |
|---|---|
| $m_\Gamma$, $\rho_\Gamma$, $V_\Gamma$ | mass, density and volume of valve poppet |
| $m_K$, $\rho_K$, $V_K$ | mass, density and volume of liquid |
| $b_\Gamma$ | damping |
| k | stiffness of the spring |
| $x_0$ | minimum spring preload |
| Q, $Q_m$ | volumetric and mass flow rate |
| Sa, Sb, S | areas of the input and output surfaces of the valve |
| P | solid wall of the valve |
| Γ | moving wall of the poppet |
| $x_{aCG}$ | Sa surface center of gravity coordinates |
| $x_{bCG}$ | Sb surface center of gravity coordinates |
| $R_{ab}$ | Hydraulic resistance of the valve between the surfaces Sa and Sb |

Momentum equation:

$$m_\Gamma \ddot{x}_i + b_\Gamma \dot{x}_i + k\left(x_i + x_0\right) + F_K = G_{T_i} \;. \tag{1}$$

Integration of Navier-Stokes equation:

$$\iiint_{V_K} \rho_K \left( \frac{\partial v_i}{\partial t} + \frac{\partial v_i}{\partial x_j} v_j \right) dV_K = \iiint_{V_K} \rho_K g_i \, dV_K - \iiint_{V_K} \frac{\partial p}{\partial x_i} dV_K + \iiint_{V_K} \frac{\partial \tau_{ji}}{\partial x_j} dV_K \,. \tag{2}$$

Continuity equation:

$$\frac{\partial v_j}{\partial x_j} = 0 \,. \tag{3}$$

The final form of equation of motion:

$$\left(\rho_\Gamma + \rho_K\right) V_\Gamma \ddot{x}_3 + b_\Gamma \dot{x}_3 + k\left(x_3 + x_0\right) = \rho_K \left(x_{bCG3} - x_{aCG3}\right) \frac{\partial Q}{\partial t} - R_{ab}\, S\, Q \left|Q\right| . \tag{4}$$

The possibility how to determine $b_{\Gamma}$ or formulate the momentum equation is indicated here e.g. [8]. It is also necessary to consider the effects of possible deformations of the poppet rigid body and its seat in some cases [9, 10]. However deformation of the rigid body is not included in this study.

## 4 Static characteristics

The first of the characteristics describing the essential behavior of the valve in terms of its operational properties are those, which describe behavior of coefficients Kv for fully open valve. Data obtained from 2D and 3D CFD simulations are shown for comparison in Fig. 4 and Fig.5. Determination of the correct magnitude of the flow coefficient depends of course on the position of the considered boundary surfaces, which are assigned to the valve. They were chosen as surfaces defining inlet and outlet of the valve, i.e. Sa and Sb. See Tab. 2 and Fig. 3. Furthermore, the data corresponding to the surfaces on which the boundary conditions were located in CFD simulations, and which are situated at a distance of fifteen diameters of the pipe from the surfaces Sa and Sb are entered in the graphs.

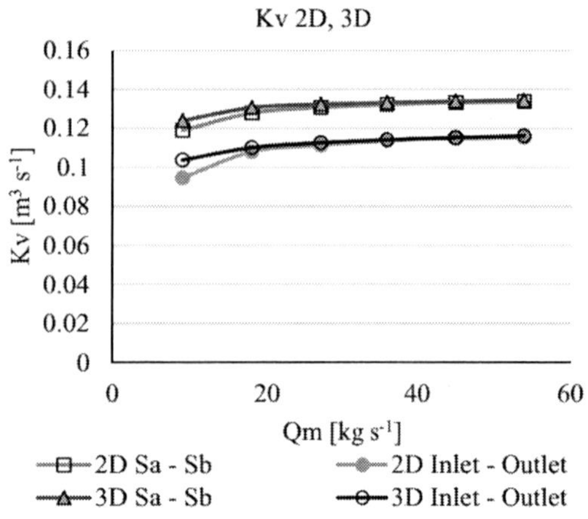


**Fig. 4.** Downstream flow coefficients.

It is apparent from Fig. 4 and Fig. 5 that the differences in 2D and 3D simulations in case of flow coefficients, especially at low mass flow rates, exist. . This effect is also evident in backflow, but obtained Kv values are almost constant.

The pressure drop in equation of motion is subsequently described by hydraulic resistance $R_{ab}$, see equations (4), and Tab. 2. The magnitude of the pressure drop is referenced to the same surfaces as in the case of flow coefficients Kv again, Fig. 6.

The axial forces depending on the static pressure difference are in both directions linear, as it was assumed, Fig. 7. Differences between 2D and 3D CFD simulations are greater due to the more complicated velocity and pressure fields around the valve poppet compared to the previous cases and other quantities.

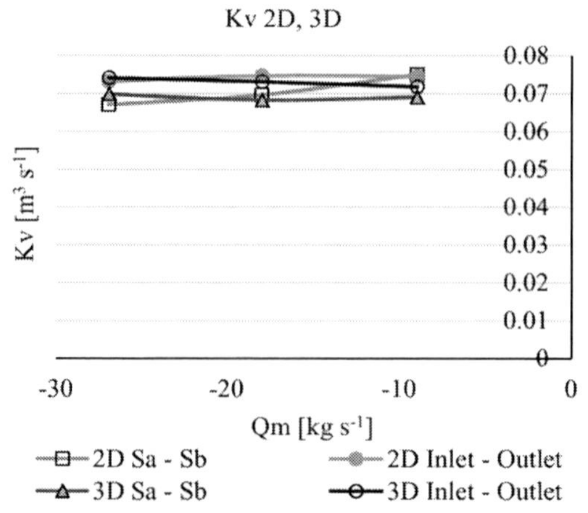


**Fig. 5.** Upstream flow coefficients.

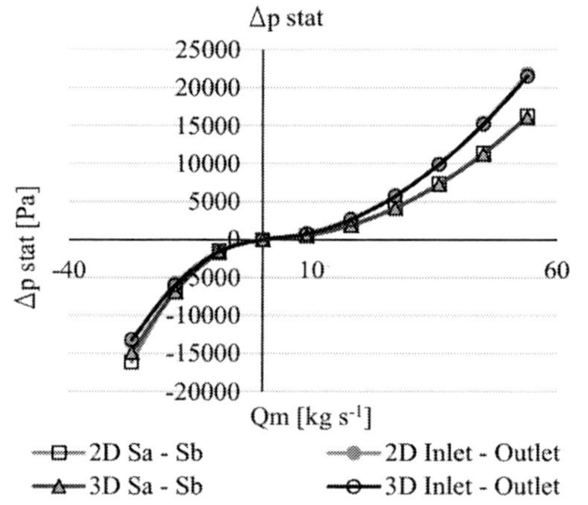


**Fig. 6.** Downstream and upstream static pressure differencies.

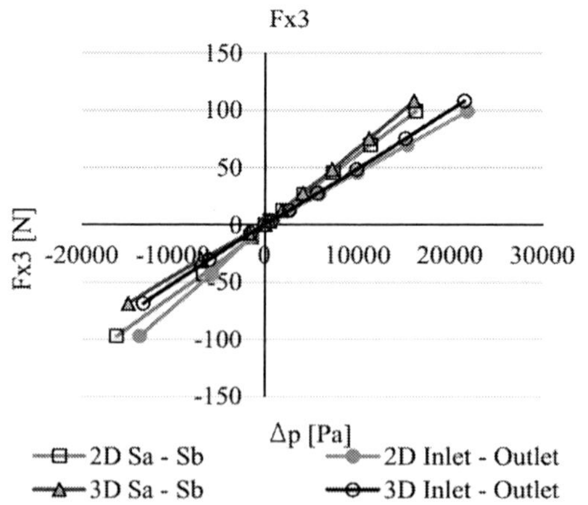


**Fig. 7.** The force magnitude along the axis of the valve acting on the poppet depending on the pressure drop.

The results are of course influenced by the static position of the poppet, which corresponds to a fully open valve, as already mentioned. It is also clear, with respect to the equation (4), that further study of valve properties requires a dynamic description of flow rate. Hitherto only a constant flow rate without the influence of time-dependent flow rate was considered.

The above characteristics, depending on the mass flow rate, will therefore be part of the static characteristics. The flow rate will be defined by the following equation (5).

It is apparent from the equation (5) that the course of flow rate is linear and its derivative with respect to time corresponds to constant value. Valve shutoff time Ts = 0.2 s is very short and for the current system is practically unrealistic. Shorter time changes were chosen primarily with respect to the CFD simulations, because a long time Ts would be very complicating as it will be explained later.

The following Fig. 8 represents an analogue of Fig. 6, which shows the pressure difference during the dynamic flow rate changes. The agreement of 2D and 3D CFD simulations is for the given scale almost absolute. It follows from the equation (4), that the offset of static pressure difference values should be constant with respect to the equation (5) compared to pressure difference in Fig. 6. Offset of curves Sa - Sb and Inlet - Outlet is given, of course, by their different distances from the valve poppet and different magnitudes of hydraulic losses in the piping outside the valve itself.

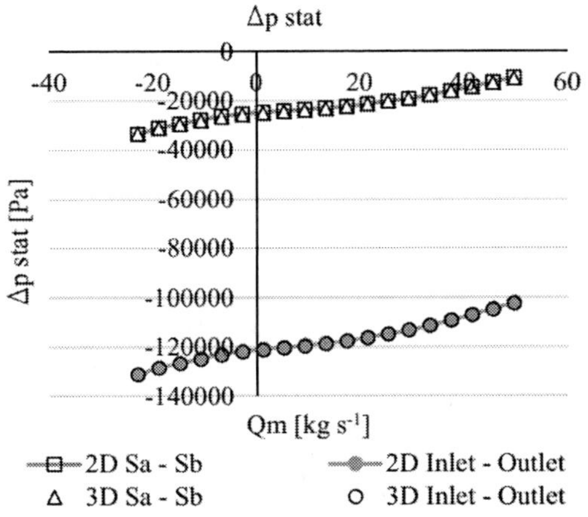


**Fig. 8.** Downstream and upstream static pressure difference.

It is apparent from a comparison of static pressure difference with and without considering time-dependent flow rate in Fig. 9, that there is a constant offset of the curves particularly in the area of positive mass flow rate values. This is in agreement with equations (4, 5). The differences are caused by different magnitudes of hydraulic losses, which result from different flow field around the poppet.

The magnitude of the force acting on the poppet depending on the static pressure difference is again approximately linear, Fig. 10.

It will be interesting to determine the magnitude of force acting on the valve poppet (Fig.11) in this context, if the poppet velocity and flow rate are prescribed according to equation (5). The magnitude of poppet force is obtained only from 2D CFD simulation. Velocity of poppet is equal to 0.2 $ms^{-1}$. Maximum distance of valve poppet motion corresponds to x3max in Tab.1.

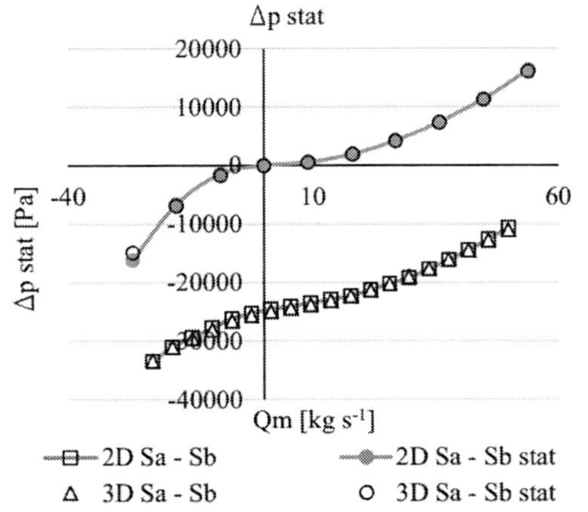


**Fig. 9.** Static pressure difference with and without time-dependent mass flow rate.

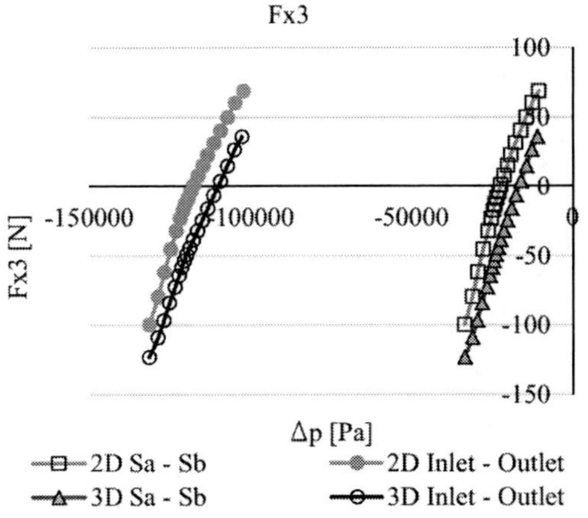


**Fig. 10.** The forces acting on the valve poppet.

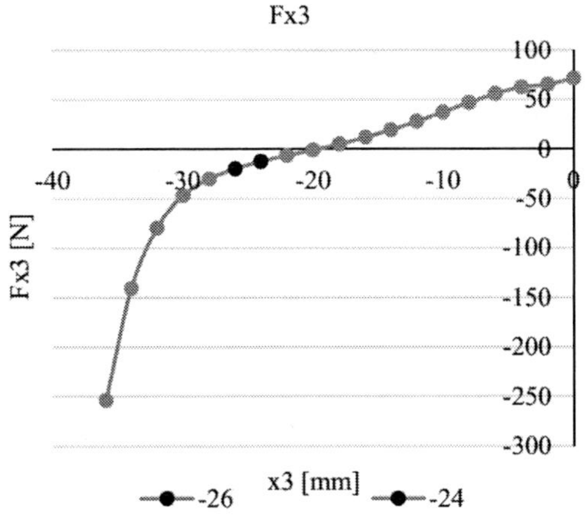


**Fig. 11.** The magnitude of forces acting on the valve poppet, depending on its position.

The marked values correspond to points in Fig. 11, between which a change occurs in the direction of flow. The positive direction of all parameters corresponds to the positive direction of flow. Changing the magnitudes of the force is up to these values approximately linear, then the hydrodynamic forces acting on poppet start to dominate.

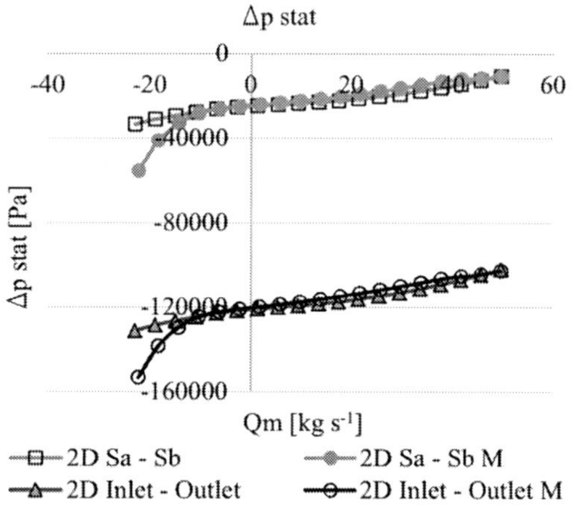


**Fig. 12.** Static pressure differences corresponding to the change of flow direction and movement of the valve poppet.

Fig. 12 depicts the change in static pressure between the surfaces Sa - Sb and Inlet - Outlet. The symbol "M" indicates data that were obtained in case of simultaneous changes of flow rate and poppet motion. It arises from Fig. 12 that the static pressure differences in 2D and 3D simulations correspond to each other up to the moment when the valve poppet is too close to the seat, and especially when a change of flow direction occurs. It follows that the poppet motion has for a certain operating range almost no influence on the pipeline system parameters. This was foreseen in introduction to the study and the data support this fact.

However, these findings also complicate the study of the poppet motion, because its behavior will largely depend on the dynamics of the whole pipeline.

## 5 Dynamic characteristics

Equation (4) was used to obtain the dynamic characteristics. Inclusion of following curves has a practical significance at this point, because to find solution of equation (4) it is necessary to know the resistance $R_{ab}$ or magnitude of the static pressure difference. Therefore, the results in the previous section will be used to determine the pressure difference. Complicated question is how to formulate a flow rate through the system.

Several test configurations have been chosen to describe the basic cases, which could occur during valve operation. The advantage is the possibility of a relatively quick and easy analysis of the poppet motion.

The simplest and well-known way of solving equation (4) is the case where the flow rate is constant, see Fig. 13.

$$Q_{0\,constant} = 0.04496011786\ m^3s^{-1}$$

Complications may arise in the formulation of resistance $R_{ab}$. It is possible to define the position of the poppet x3 in case of constant flow rate, and equation (4) also allows to check magnitude of damping $b_\Gamma$. We would obviously arrive to the same state, if we examined the case where the poppet is stationery and magnitude of flow rate corresponding to a specific position x3 would be unknown. The difference, of course, occurs with respect to the choice of initial conditions that define the transient state.

Another two cases defining the magnitude of the flow rate through the valve will now be closer to real conditions. The flow rate will be described by the equations (5) and (6). The curves in Fig. 13 labeled as cosine Qo1 and cosine Qo2 correspond to the equation (6). The only difference is in the initial flow rate Qo2. Graphical interpretation of the volume flow rate (5, 6) is entered in Fig. 13. The most controversial is the shutoff time of the valve, which was due to CFD simulations reduced ten times. However, there will be first presented results that correspond to the real time, Ts = 2s, and subsequently to a shorter shutoff time of the valve, Ts = 0.2s.

$$Q = Q_0 - \frac{Q_0 - Q_Z}{T_S}t \tag{5}$$

$$Q = \frac{1}{2}(Q_0 + Q_Z) + \frac{1}{2}(Q_0 - Q_Z)\cos\left(\frac{\pi}{T_Z}t\right) \tag{6}$$

Flow rate (6) corresponds best to the real system, in which the flow can be described by the following function (7) in the simplest cases.

$$Q = Q_0 e^{\alpha t}\cos(\beta t) \tag{7}$$

Function (7) will be replaced just by (6) for basic characterization of flow rate.

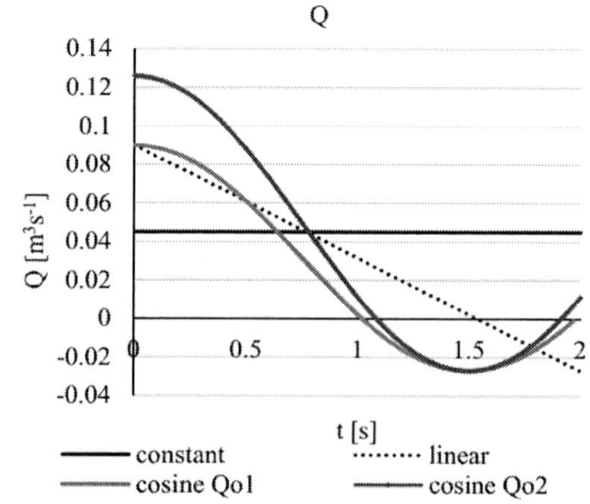


**Fig. 13.** Volume flow rates.

The constant flow rate is naturally chosen different from the one corresponding to fully open valve. Individual volume flow rates in Fig. 13 correspond to the time dependences of the poppet position x3 in Fig. 14 and the poppet velocity v3 in Fig. 15.

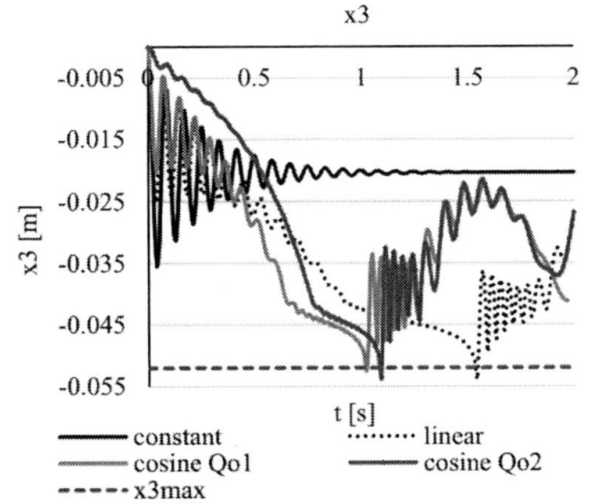


**Fig. 14.** Poppet position dependence on time and volume flow rate.

In Fig. 14, line x3max indicates the position of the poppet, which corresponds to the closed valve. If this value x3 is achieved and the direction of flow is positive, the backflow does not occur. Another part of the momentum equation has no importance in this case. The different maximum positions of cosine Qo1 and linear curves in Fig. 14, of course, correspond to the different points in which the flow rate reaches zero value, see Fig. 13.

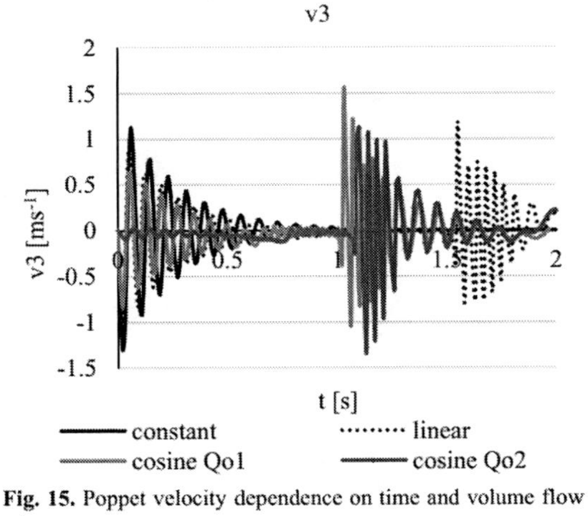


**Fig. 15.** Poppet velocity dependence on time and volume flow rate.

In Fig. 14 and 15 there are plotted data for the volume flow rates cosine Qo1 and cosine Qo2, which differ in magnitude of the initial flow rate. The influence of this fact on the x3 curve damping and a significant velocity v3 reduction is obvious. This is also related to the magnitude of attained overpressure [11]. Influences of other parameters, such as stiffness of the spring or the bellows can be found in [12, 13].

The shift of maximum value x3 on timeline is for these curves given by slightly different values of time when the flow rate Q reaches its zero values, see Fig. 13. Still, the results do not have a physical substantiation once x3 reaches x3max.

Equation (4) can be also modified by prescribing the poppet motion by poppet distance x3 and then time-dependent volume flow rate, Fig. 16, can be determined. Description of poppet position x3 obviously corresponds to a constant poppet velocity.

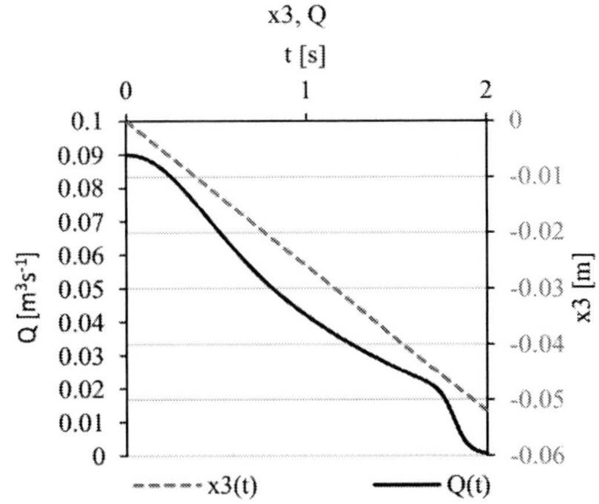


**Fig. 16.** The volume flow rate for a prescribed poppet position x3.

As it has already mentioned, CFD simulations were performed for the valve shutoff time Ts = 0.2s and for linearly changing volume flow rate Q in Fig. 13. Fig. 17 and Fig. 18 introduce position and velocity of the poppet for given shutoff time Ts of the valve.

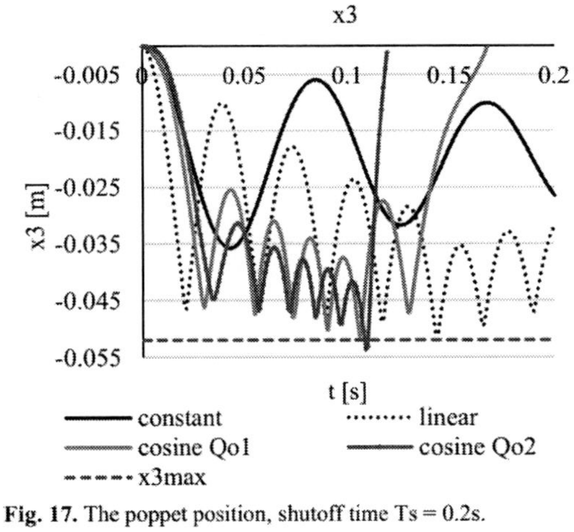


**Fig. 17.** The poppet position, shutoff time Ts = 0.2s.

Data are not available for the flow cosine Qo1 and cosine Qo2 (Fig. 17, 18) in entirety full range, because numerical solution of momentum equations does not exist. The reason is partly evident also from the following equation (8), which is obtained from the original equation (4).

The form of equation (4) can be described as a standard in terms of mechanical vibrations. Damping is now in less common form. Relationship of damping to vibration is also described in [14].

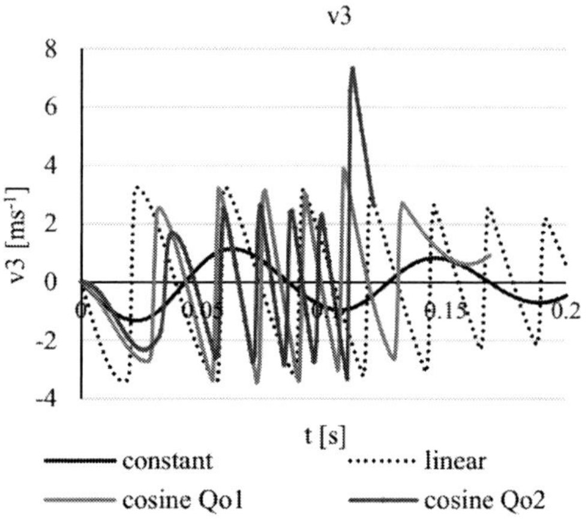


**Fig. 18.** The poppet velocity, shutoff time Ts = 0.2s.

$$x_3 + \frac{\left(b_\Gamma - \rho_K \left(x_{bCG3} - x_{aCG3}\right)\frac{\partial Q}{\partial x_3}\right)}{\left(\rho_\Gamma + \rho_K\right)V_\Gamma} x_3 + \frac{k}{\left(\rho_\Gamma + \rho_K\right)V_\Gamma} x_3 = \frac{-R_{ab}\, S\, Q\left|Q\right| - k\, x_0}{\left(\rho_\Gamma + \rho_K\right)V_\Gamma} \qquad (8)$$

Damping in equation (8) will, to large extent,depend on the nature of changes in the flow rate with respect to coordinate x3 because $\rho_K$ ($x_{bCGG3}$ - $x_{aCG3}$) is always positive. This change will be zero for a constant flow rate and control function will only be function of $b_\Gamma$, which depends on the viscosity of liquid and on the shape of valve inner surfaces. We can estimate from Fig. 13 and functions cosine Qo1 and cosine Qo2 that the derivative of volume flow rate with respect to coordinate x3 is positive in certain moments. This fact may indicate negative damping, which reveals itself in Fig. 14 and Fig. 17 for curves labeled cosine Qo1 and cosine Qo2 at times approaching Ts.

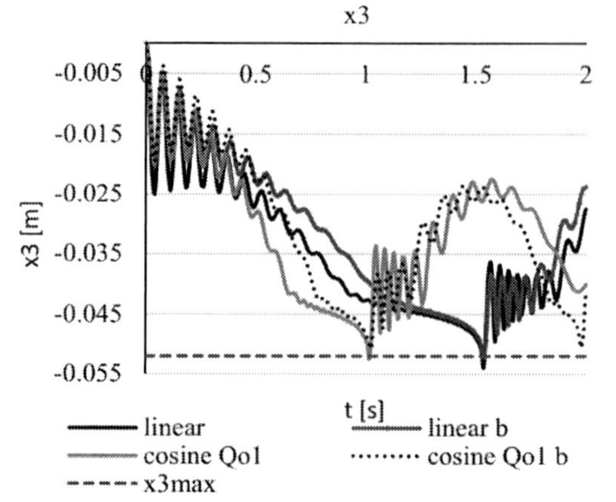


**Fig. 19.** The influence of damping on the poppet lift.

The general solution of equation (8) is well known and therefore will not be shown. Presence of flowing liquid, however, is the difference compared to the standard solution. Fig. 19 and Fig. 20 offer a comparison how the additional fluid damping effects the poppet motion. Variants in which only damping $b_\Gamma$ in equation (8) is considered are marked with the symbol "b".

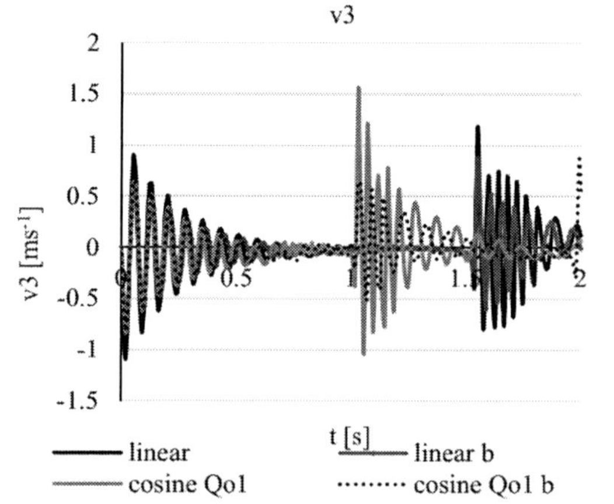


**Fig. 20.** The influence of damping on the poppet velocity.

It is apparent from Fig. 19 and Fig. 20 that the unsteady flow has rather a negative effect, because the vibration amplitudes are bigger.

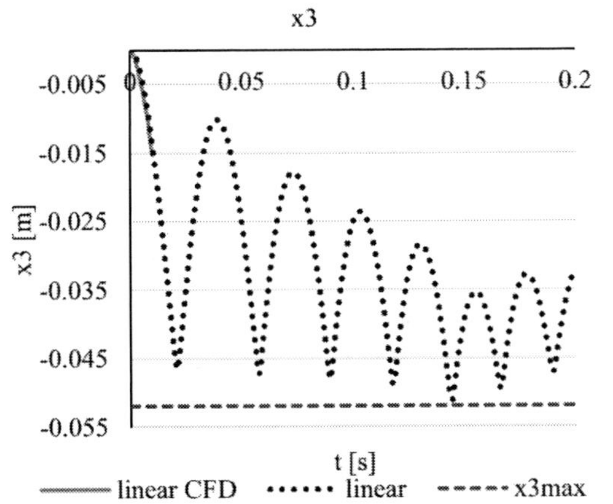


**Fig. 21.** The poppet lift comparison.

An attempt was also made to compare the results of time-dependent poppet lift with data from the 2D CFD simulation with one degree of freedom. The valve poppet moves by force resulting from the surrounding pressure field and the magnitude of its lift is recorded in time. One degree of freedom of the poppet is sufficient in this case, because the poppet moves only along the axis of the valve. However, the difficulty of this approach is with regard to the convergence of flow parameters and grid adaptation due to the great changes of poppet velocity and poppet lift. Therefore this type of simulation is unfeasible in full extent. Also, the influence of the spring is problematic because the spring complicates the code of UDF which defines the force acting on the poppet. The difficulty is not in the form of mathematical notation, but in the formulation of the spring force depending on the position itself. This spring force formulation causes a greater susceptibility of computational algorithm to instability and leads to disruption of the computing cycle. More can be found in [15].

From these reasons significantly shorter shutoff time of the valve was chosen, than are usual operating conditions. Of course, equation (4) quite simply allows this approach and for comparison is such approach beneficial. Fig. 21 therefore at least shows partial poppet lift comparison obtained from equation (4) and 2D CFD 1DOF simulation. The flow rate corresponds to the equation (5).

It seems there is only few CFD simulation data at first sight, but it is extremely time-consuming to obtain this data due to very short computational time step. However, the resulting consensus is at least in a small section of poppet lift very good. Relatively long total poppet lift x3max from fully open to closed valve is in this case problematic.

## 6 The velocity and pressure field

It will be important to find out how much the velocity and pressure fields are different due to differences between 2D and 3D CFD simulations. The fin is the most fundamental difference between both simulations, see Fig. 1. The housing of the valve poppet is attached just to the fin. The fin is obviously not placed in 2D simulations. The fin thickness is not too large, therefore assumption of flow axisymmetry in the rest of the valve domain might be plausible.

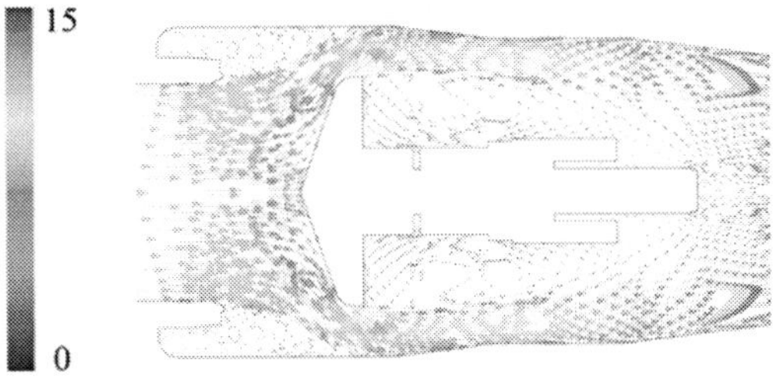


**Fig. 22.** Absolute velocity field obtained from 2D CFD simulation, flow rate Q1.

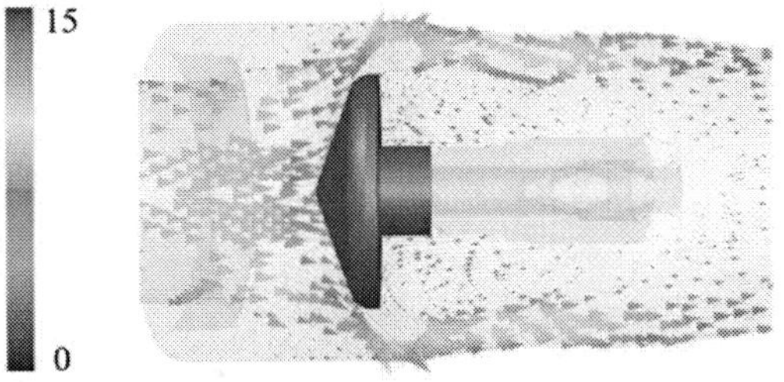


**Fig. 23.** Absolute velocity field obtained from 3D CFD simulation, flow rate Q1.

Cross-section outside of the fin of 3D valve body is depicted Fig. 22 - 25 to verify this fact. Figs. Fig. 22 - 25 show a comparison of velocity and pressure fields for the flow rates Q1 = 0.053913636 $m^3s^{-1}$ and Fig. 26 - 29 for flow Q2 = -0.026956818 $m^3s^{-1}$.

2D CFD simulation of the valve can be examined in more detail with respect to the discretization and solution methods here [16].

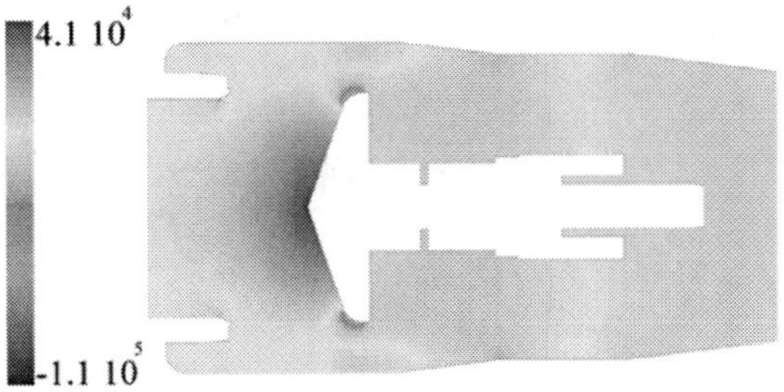


**Fig. 24.** Static pressure field obtained from 2D CFD simulation, flow rate Q1.

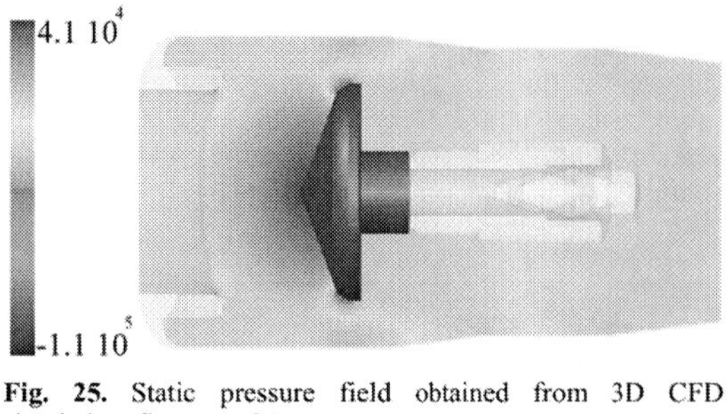


**Fig. 25.** Static pressure field obtained from 3D CFD simulation, flow rate Q1.

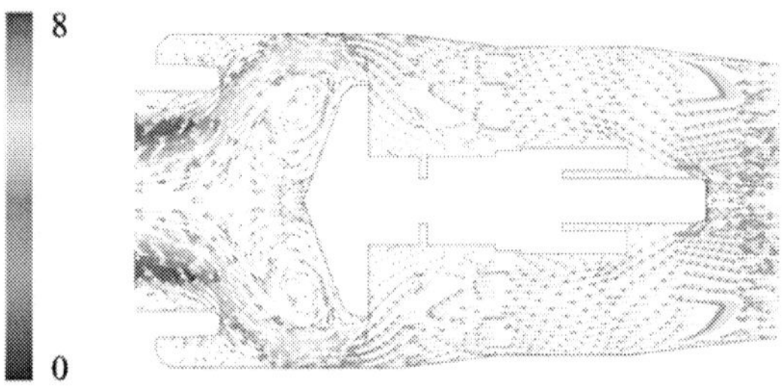


**Fig. 26.** Absolute velocity field obtained from 2D CFD simulation, flow rate Q2.

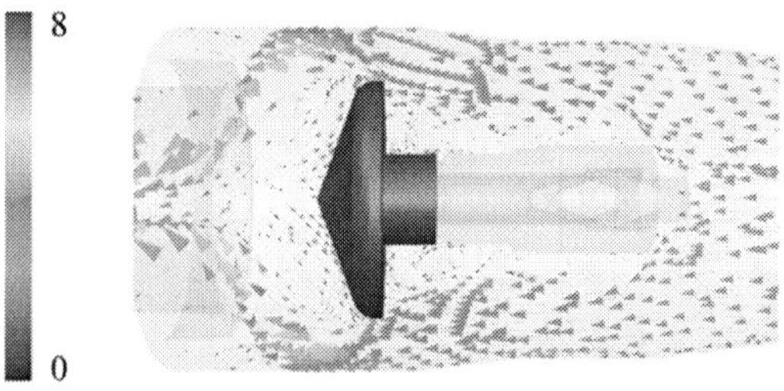


**Fig. 27.** Absolute velocity field obtained from 3D CFD simulation, flow rate Q2.

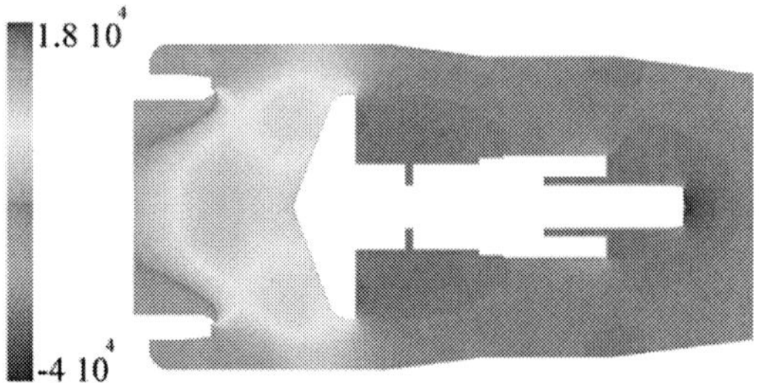


**Fig. 28.** Static pressure field obtained from 2D CFD simulation, flow rate Q2.

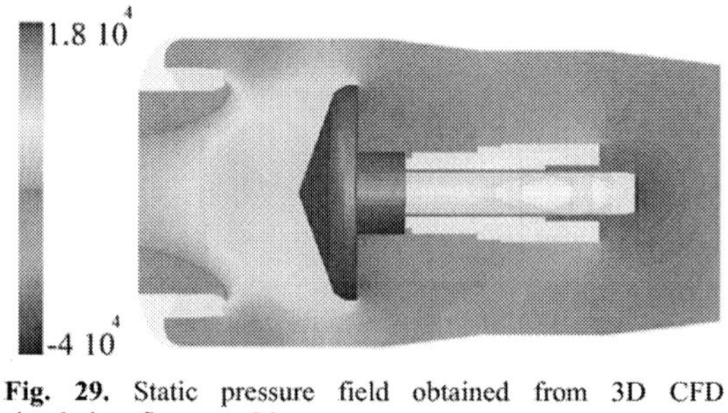


**Fig. 29.** Static pressure field obtained from 3D CFD simulation, flow rate Q2.

It is obvious that the agreement is very good. The same scales of static pressure and absolute velocity were chosen for this purpose. Similar results are also obtained for the other flow rate.

Note: It is necessary to monitor the flow around the poppet also in terms of water erosion and cavitation, [17].

## 7 Conclusions

Experiment, which has only partially been carried out so far, will decide about validity of presented data and characteristic curves. . It is important to assess at least the basic correspondence between equation of motion of valve poppet and CFD simulations of two-dimensional and three-dimensional models. Hydraulic resistance determination is also a problematic point, which can be defined only by experiment or simulation CFD.

From the analyzed alternatives 3D simulation should provide the most accurate results. It was therefore important to establish a comparison between 2D and 3D model in order to obtain some parameters from the simpler 2D model. Very good agreement was shown on the attached characteristics between 2D and 3D models. The same can be mentioned also for the comparison of the velocity and pressure fields.

The operating characteristics and force acting on valve poppet are essential in terms of valve properties. It can be stated about the force characteristic, that its dependence on the pressure difference remains linear as in the case of stationary flow.

From the case of prescribed poppet motion it can be concluded that poppet motion has no significant influence on the flow field except the region where hydrodynamic force considerably increases. The reason is that the poppet lift is relatively long.

Correct assessment of the unsteady flow influence is also essential in the formulation of the basic equations of motion in terms of its practical usability. Verification of influence of unsteady flow in CFD simulations revealed that the described simplified model is quite useful in the equation of motion and corresponds to the results of CFD simulations.

Effect of transient flow also causes a significant decrease in static pressure, which brings problems with cavitation. Cavitation mode was not simulated, but may obviously have an important influence on the dynamic properties of the valve. Cavitation can be simulated in later studies.

The dynamic lifts and velocities are the most important characteristics of the poppet, which of course greatly depend on the flow rate through the valve. The actual flow rate depends on the characteristics of the whole pipeline and the flow rate cannot be restricted to valve itself Therefore only few operating states corresponding to the basic cases be simulated. Especially those the which occur in real operation or which may be interesting for analyzing of the poppet motion.

The influence of unsteady flow on the damping of poppet motion is evident from the attached characteristics. Its magnitude for various positions of the poppet can cause oscillations. Also conditions in which the backflow should occur were simulated. Though, the situation of fully closed valve never occurs on the plotted characteristics. However, it is necessary to note the effect of cavitation, which occurs according to the rate of pressure drop, and which may affect the dynamics and function of the valve.

This work has been supported by Technology Agency of the Czech Republic under the project Innovative research of check valves for extreme operating conditions in energetics TH01011352.

## References

1. J. A. Davis, M. Stewart, J. Fluids Eng. **124**, 3, pp. 772-777 (2002)
2. J. A. Davis, M. Stewart, J. Fluids Eng. **124**, 3, pp. 778-783 (2002)
3. L. Davidson, *An Introduction to Turbulence Models* **97/2**, (2016)
4. A. Oza, S. Gosh, K. Chowdhury, *CFD Modeling of Globe Valves for Oxygen Application*, 1356-1363 (16th Australasian Fluid Mechanics Conference, 2007)
5. S. Sibilla, M. Gallati, J. Fluids Eng. **130**, 121101 (2008)
6. V. Yakhot, S. A. Orszag, S. Thangam, T. B. Gatski, C. G. Speziale, Phys. Fluids A 4_7_, pp. 1510–1520 (1992)
7. A. Budziszewski, L. Thorén, *CFD simulation of a safety relief valve for improvement of a one-dimensional valve model in RELAP5* (Department of Applied Physics, 2012)
8. M. Domagala, Czasopismo Techniczne **3-M**, pp. 24-29 (2008)
9. S. W. Lee, D. Y. Shin, C. W. Byun, H. J. Yang, I.C. Paek, J. of Achievements in Mat. and Manufact Eng. 23, pp. 67-70 (2007)
10. X. J. Yuan, K. H. Guo, SCIENCE CHINA Technological Sciences **58**, 6 (2015)
11. X. G. Song, J. H. Jung, H. S. Lee, D. K. Kim, Y. Ch. Park, *2-D Dynamic Analysis of a Pressure Relief Valve by CFD* (WSEAS Conferences, Hangzhou, China, 2010)
12. X. Song, L. Cui, M. Cao, W. Cao, Y. Park, W. M. Dempster, Energy Conversion and Management **81**, pp. 407-419 (2014)
13. S. Chabane, S. Plumejault, D. Pierrat, A. Couzinet, *Vibration and chattering of conventional safety relief valve under built up back pressure* (IAHR International Meeting, 2009)
14. A. Singh, Nuclear Eng. and Design **72**, pp.197-204 (1982)
15. X. G. Song, L. T. Wang, Y. C. Park, W. Sun, Procedia Engineering **130**, pp. 87-94 (2015)
16. A. Beune, *Analysis of high-pressure safety valves*, (Eindhoven University Press, 2009)
17. B. E. Dallstream, B. A. Fricke, B. R. Becker, *Swing Check Valve Design Criteria and CFD Validation* (ICONE14, 2006)

 DOI: 10.1051/epjconf/201714302052

# Control of the probe influence on the flow field in LP steam turbine

*Michal* Kolovratník[a], *Kukchol* Yun, *Ondřej* Bartoš

CTU in Prague, Department of Energy Engineering, Technická 4, Prague 6, 166 07, Czech Republic.

**Abstract.** For measuring the fine droplets properties in the wet steam expanding in the steam turbines the light extinction probes are usually used. The paper presents CFD modelling of the extinction probe influence on the wet steam flow field at the measurement position. The aim is to get a basic information about the influence of the flow field deviation on the measured data, in other words, of necessity to correct the measured data. The basic modelling procedure is described, as well as the supposed simplifications and the factor considering the change in the steam density in the measuring slot of the probe. The model is based on the experimental data that were achieved during the developmental measurements in the steam turbine 1090 MW in the power station Temelín. The experimental measurement was done in the cooperation with the Doosan Škoda Power s.r.o.

## 1 Introduction

Experimental research of the liquid phase of the wet steam expanding in a steam turbine is a challenging problem. Despite the fact that the research was started in the seventies of the last century [1], commercial instrumentation is not yet available. Several institutions focusing on this issue (e.g. EdF, RWTH Aachen, ITSM Stuttgart, USST, EPRI, ETH Zürich) are carrying out an individual and a partially published research of the probes for the measurement of the fine "submicron" droplets formed by the nucleation and condensation as well as of probes for measuring the coarse "submillimeter" droplets formed in the flow part during the breakup of water films.

In the Czech Republic has been developed such an optical measuring system at Czech Technical University in Prague (CTU) and used for research and development purposes in the low pressure parts of the steam turbines in the classical and nuclear power plants. These measurements have been mostly done in close cooperation with Doosan Škoda Power s.r.o. (DŠP). For the fine droplets size distribution measurement the extinction probes are commonly used. Their development in line with global trends can be described by gradual minimization of the dimensions and application of the modern optical and electronic elements.

For many reasons the probe head diameter was gradually lowered by 60 % and the measuring slot was shortened to 20 % of the original length. This shortening is logically associated with increasing of the proportion of the wall influenced areas in the probe head measuring slot. Thus research has been started at CTU to assess the probe influence on the flow field in the steam turbines. Here is briefly presented dependence between the probe influence on the flow field and the acquired/processed results of the wet steam wetness formed by the primary droplets. Research in further phases will be focused also on the issue of the influence of air flow for cleaning optical windows of the probe, the influence of probe's head shape and in the future also the probe head shape optimization.

## 2 The probe head shape design

The measurement of the wet steam in the newly accessible (since the end of 2015) position behind the last but one stage of the low pressure part (LP) of the steam turbine 1090 MW the power station Temelín (ETE) required shape adjustment of the measuring head of the CTU extinction probe. Regarding the higher pressure level the length of the measuring slot was shortened from 50 to 20 mm [2]. The shape solution and dimensions of the used measuring head are shown in Fig.1. The measuring slot 20 x 13 mm is cut in a cylindrical body with the diameter of 21 mm. The shape is symmetrical, i.e. the volume is on sides limited by carrying ribs of the same shape. Parallel to the head axis a light beam of 4 mm diameter goes through. It is limited in the inlet and outlet by the glass windows. To enable the purging air flow the windows are recessed 0.7 mm below measuring slot upper and lower wall plane. Thus the geometrically active measuring volume in the measuring slot through which the wet steam flows is a cylinder with a diameter of 4 and length of 21.4 mm.

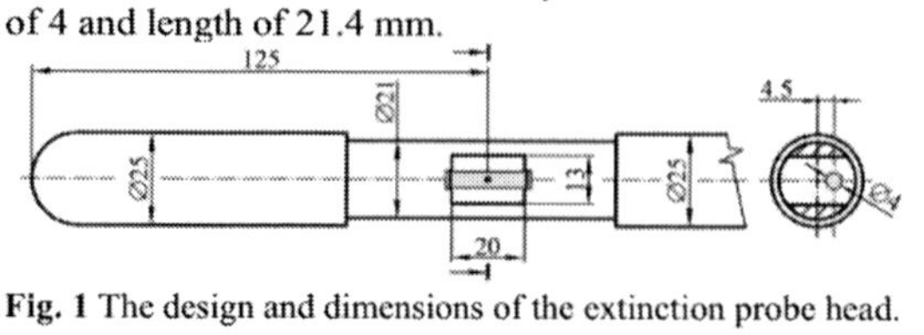


**Fig. 1** The design and dimensions of the extinction probe head.

[a] Corresponding author: michal.kolovratnik@fs.cvut.cz

In an ideal case without the flow field being influenced by the probe the wet steam parameters in this cylinder would be identical with parameters in the same place without the probe presence.

## 3 Placement and positioning of the probe in the steam turbine

The placement of the extinction probe behind the rotor blade of the penultimate (L-1) stage of the LP part in the measuring plane **A** with an angle ~10° from the radial is given in Fig. 2. Basic manipulation with the probe head in the steam turbine is enabled by a cylindrical carrier with the diameter 35 mm and the length about 4.5 m. It allows traversing the head from the outside in a number of positions along the blade length $z$ and turn it in the flow field (angle $\alpha_{probe}$) based on data from the pneumatic measurements (DŠP).

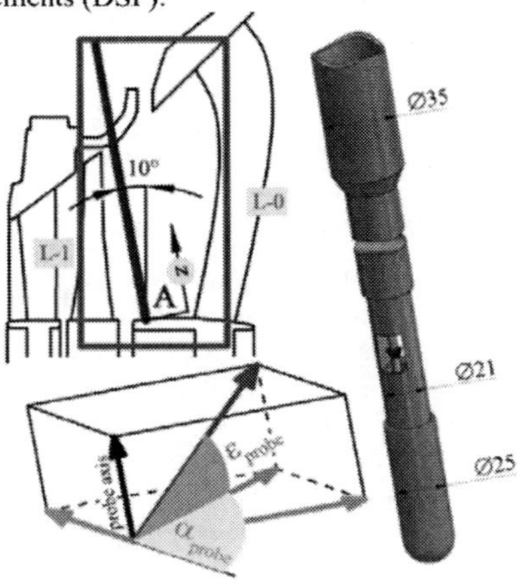


**Fig. 2** Measuring plane in the steam turbine and scheme of probe orientation

In an optimal case the probe is in each position situated according to the corresponding angle $\alpha_{probe}$, but adjustment of the measuring slot to the "radial" angle $\varepsilon_{probe}$ is not possible. The flow field is influenced by the probe head and also by the entire carrier.

## 4 Measuring method principle

For measuring, is used the dependence between a light attenuation and a dimensional structure of polydisperse system of droplets formed in the wet steam. It is expressed by known Lambert-Beer Law:

$$\frac{1}{\ell}\ln\left(\frac{I_0}{I}\right) = \frac{\pi}{4} N_v \int_0^{D_\infty} Q\left(\frac{\pi D}{\lambda}\right)\varphi(D) D^2\, dD \qquad (1)$$

where $\ell$ is the length of the measuring beam of the probe (or thickness of the wet steam lighted layer), $I_0$ intensity of light on the inlet of the measuring volume, $I$ intensity of light on the outlet of the measuring volume, $N_v$ is a number of droplets in the volume unit, $Q(\pi D/\lambda)$ coefficient of light extinction according to Mie theory [3], $\lambda$ is wavelength of light, $D$ is droplets diameter, $\varphi(D)$ normalized droplets size distribution.

Practically, in multispectral measuring of the light extinction, a beam of white light is used. The light intensity decreases as a result of light extinction on the droplets in the wet steam while passing through the probe measuring slot. The following spectral analysis of the light going out of the probe enables us to obtain the light intensity ratio decrease with dependence on wavelength $I/I_0 = f(\lambda_i)$. The data processing is done for the selected number of discrete wavelengths i=1, 2,...k and thus a following set of equations:

$$\frac{1}{\ell}\ln\left(\frac{I_0}{I}\right)_i = \frac{\pi}{4} N_v \int_0^{D_\infty} Q\left(\frac{\pi D}{\lambda_i}\right)\varphi(D) D^2\, dD \qquad (2)$$

For a numerical solution of this system of the ill conditioned integral equations it is necessary to apply special methods. In the Department of Energy Engineering on CTU in Prague a regularisation method RNL [4] has been developed for this purpose. This method seems to be suitable for solving integral characteristics of polydispersions , such as a mean Sauter diameter $D_{32}$ and a steam wetness $y$, from the measured light extinction.

For polydisperse droplet system the steam wetness is described in the following relation:

$$y = \frac{\pi}{6} N_v \frac{\rho'}{\rho_p} \int_{D_{min}}^{D_{max}} \varphi(D) D^3 dD = \frac{1}{1 + \left(\frac{\pi}{6} N_v \frac{\rho'}{\rho''} \int_{D_{min}}^{D_{max}} \varphi(D) D^3 dD\right)^{-1}} \qquad (3)$$

where $\rho'$ and. $\rho''$ is density of liquid and gaseous phase in the wet steam respectively.

## 5 Influencing parameters by the probe

In this phase of the research of probe influence on the flow field, which influences accuracy of the determined wetness $y$, two simplifying assumptions were chosen:

- Movement of the fine droplets and the carrying steam are identical.
- Droplets dimension do not change during the flow through the measuring slot of the probe, or they are identical with droplets dimension in the place of measurement without the probe influence.

When meeting these assumptions, wetness determined from the measurement can differ from the real wetness as a result of a change in wet steam density caused by the probe. Thus the CFD modelling described here focused on the change in steam density in passage through the slot of the probe, or the volume corresponding with the light beam.

## 6 CFD modelling of the probe influence

The probe presence disturbs the continual flow of the wet steam in the turbine. Aerodynamic parameters around the probe are locally changed. Logically changes of parameters occur in the slot of the probe head between the ribs as well. In practice, when using the extinction probe, the structure of the liquid phase of the wet steam is measured, whose aerodynamic parameters differ from parameters in the free flow. Thus it is necessary to evaluate the probe influence on the flow field. To get information about the intensity of this influence in the probe measuring position and about the possibilities of its compensation for the data processing the CFD simulations were used.

The computational model for defining the probe influence includes the area between the rotor blades of L-1 stage and the stator blades of L-0 stage. The modelled volume is in Fig. 2. marked in red rectangle.

The probe with a symmetrical type of head and two ribs was inserted into modelled volume. During the measurements the probe was, accordingly to measurement procedure, gradually shifted along the traversing axis in the A plane, which is inclined from the radial by 10 degrees (Fig. 2 and 3).

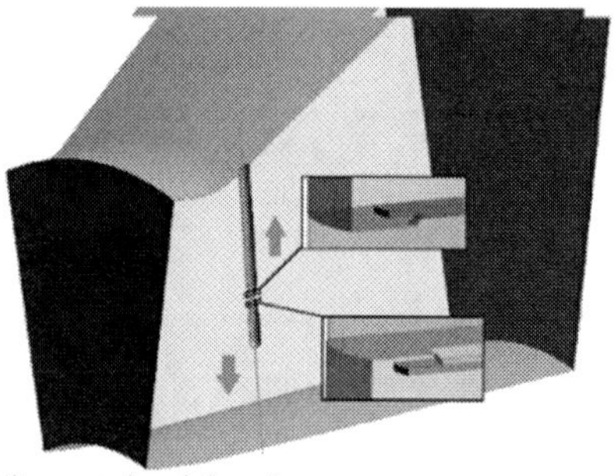

**Fig. 3** Computational domain

Computational mesh was prepared using ANSYS Meshing. Most of the computational model is meshed with hexahedral cells and only a part of the model with "complicated" geometry is meshed with tetrahedral or wedge cells.

At the inlet of the computational model the boundary condition of the „pressure–inlet" type is set with the profile of total pressures and corresponding flow directions. For this adjustment the values of total pressures and direction vectors were used. They were measured behind the L-1 stage of ETE steam turbine by the DŠP team using pneumatic probes. At the outlet of the computational model the boundary condition of the „pressure–outlet" type is adjusted with average static pressure before the L-0 stage. The set of boundary conditions is in Table 1.

Table 1: Survey of the boundary conditions

| | | | Inlet | Outlet |
|---|---|---|---|---|
| Boundary condition | | | Pressure inlet | Pressure outlet |
| Set values | Total pressure | [Pa] | Profile | - |
| | Static pressure | [Pa] | - | 28800 |
| | Total temperature | [K] | 315 | - |
| | Turb. intens. | [%] | 5 | - |

For each position $z$ along the blade, in which measurements were carried out using the extinction and pneumatic probe, an independent calculation was done. Altogether 18 calculations were carried out. In all calculations the two-equation k-ω turbulent model was used. As a working fluid in this phase of the research, water steam was chosen with the ideal gas properties.
All calculations were implemented on ANSYS CFX 17.0.

As an example of results obtained the distribution of density on the transection of the calculation domain with a probe in one of the solved positions is shown in Fig. 4. From this figure it is obvious that density distribution in the probe measuring position is strongly distorted in comparison with balanced conditions in the inlet of the computational volume.

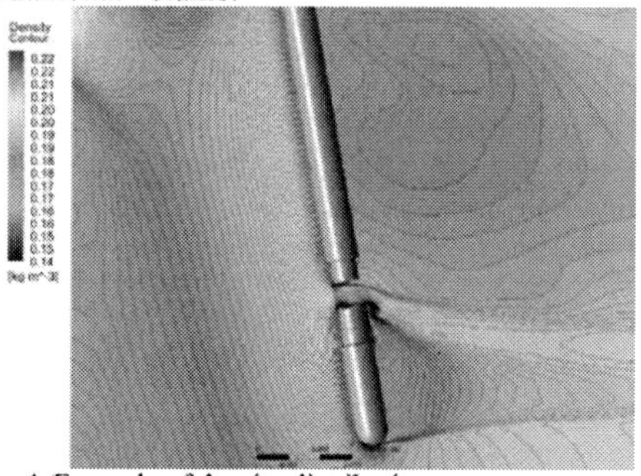

**Figure 4.** Example of density distribution

Results of implemented calculations were further processed in MATLAB.

## 7 Steam density in the probe slot

By processing the CFD calculation results of the wet steam density distribution was obtained in the probe measuring slot. As an example, a dimensionless steam density along the measuring beam $\rho=f(\ell)$ in one of the positions near the hub diameter of blading is shown in Fig. 5. This density profile is completed by mean value of density $\bar{\rho}$ (in blue).

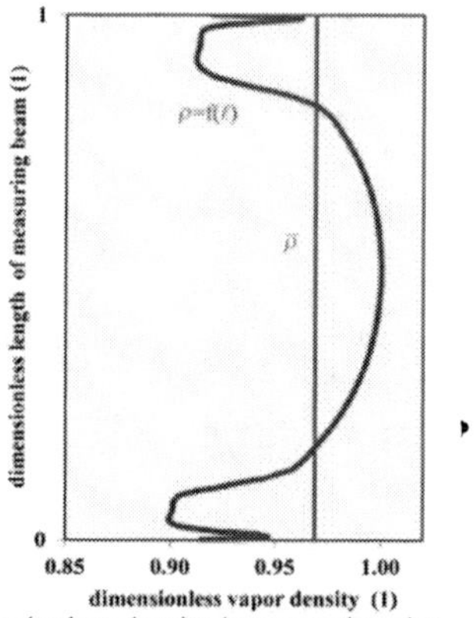


**Figure 5** Dimensionless density in measuring slot

In fact the steam density is changing along the light beam. Nevertheless for the extinction measurements is not a fundamental problem. Providing the steam wetness is constant in the measuring volume, the same value of the light intensity ratio $I/I_0$ occurs when the beam passes through an environment with variable density $\rho=f(\ell)$ as it does during the passage through an environment with constant density $\bar{\rho}$ .

$$\int_0^{\ell} \rho(\ell)d\ell = \bar{\rho}\ell \quad (4)$$

However, certain deviations of extinction against an ideal case with constant density can be caused by differences in density in cross sections of the beam. Thus in the modelled case in the cylindrical volume of the measuring beam with the diameter of 4 mm average density was defined in 16 control lines parallel to the probe axis. The differences in average density in these lines could be determined by the coefficient of variation $C_v < 0.15$ %. A detailed analysis of the issue goes beyond the paper's intention. But it is possible to state here that the non-homogeneity of average density identified is acceptable from the viewpoint of influence on results obtained from extinction measurements. At the same time, space opens here for further research focused on optimization of the measuring beam diameter in extinction probes, or looking for the dependence between the beam diameter and the influence of density non-homogeneity on measured extinction.

## 8 Probe influence

An important finding from the results of CFD calculations is confirmation of the assumption that average steam density in the probe measuring beam is not generally identical with the steam density in the measuring point uninfluenced by the probe [5]. This fact has been so far considered in the CTU procedure for extinction measurements by a simple correction that lowered the light intensity ratio $I/I_0$ [6]. However, CFD calculations provided more complex information.

In each measurement positions along the blade height $z$ the Probe Influence Factor (PIF) was defined as a ratio of average density in the place of measurement with the probe influence and without it. The dimensionless factor dependence on angle $\varepsilon_{probe}$ in individual positions is shown in Fig. 6.

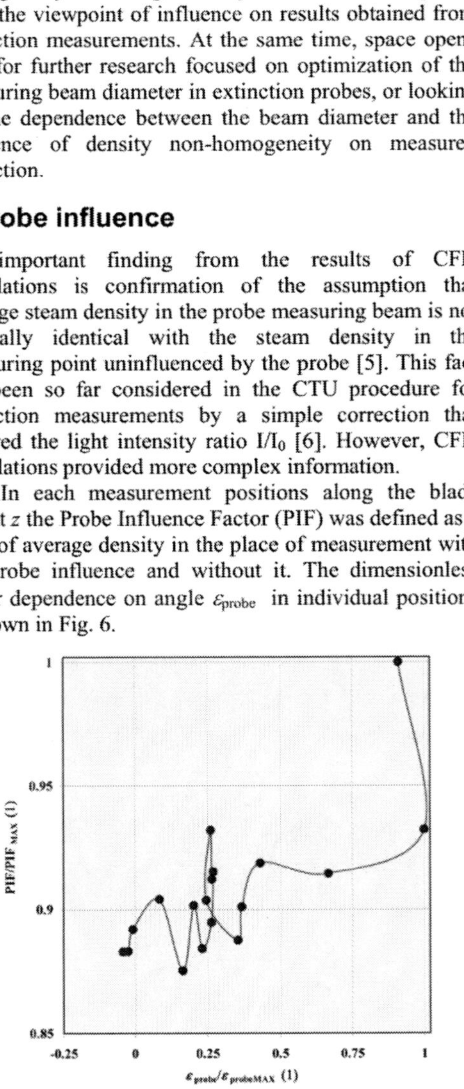


Figure 6. Dimensionless dependence of PIF on $\varepsilon_{probe}$ angle

From Fig. 6 it is obvious that for the same values of radial angle $\varepsilon_{probe}$ different PIF values were obtained in various measurement positions. This confirms the fact that the solved issue of probe influence cannot be reduced only to modelling the flow influence in the direction of angle $\varepsilon_{probe}$. On the contrary, it is necessary to include other influences in 3D flow field regarding interaction with the probe, with the carrier and limiting walls of the flow path.

Using the assumption that changes of density obtained by modelling of one-phase medium flow with ideal gas properties can be transmitted to conditions of real wet steam flow, a new calculation of steam wetness from extinction data corrected on defined change of density was carried out. In general it can be noted that, considering this correction, uncertainty of local wetness were in the range ±3.5 %. Considering these uncertainties are practically identical with the "exactness" of wetness measuring according to VDI 2043 [7], where it is said, that measurement uncertainty for the optical method measurements of wetness caused by fine droplets is in the steam turbines can be expected in the range 10-15 %. It is obvious that the examined probe influence on the flow field cannot be neglected. On the contrary, it is necessary to design and implement a procedure for processing and application of corrections on the influence of flow field parameters by optical probes.

## 9 Conclusions

The research of wet steam liquid phase structure in the flow part of steam turbines requires invasive measurements using an optical, usually extinction, method. An effort to increase accuracy of measurements leads to gradual improvement of instrumentation and procedure. One of the unresolved problems so far is the knowledge and possible correction of the probe influence on flow field parameters in the measurement area. Research was started based on numerical modelling with the aim to gain a basic idea of the importance of this influence. Using the knowledge of flow part geometry of LP part of steam turbine 1090 MW of ETE and experimental data obtained here from flow field research in the area between the L-1 and L-0 stages, the influence was modelled of CTU optical probe on the change of steam density in the measured volume. The changes of the yielded density in a number of positions along the blade show that the probe influence on flow field can, in the given modelled case, lead to relative uncertainty of measured local wetness up to ±3.5 %. Thus the probe influence on the flow field cannot be neglected.

The follow-up works will be focused on research of: this influence in the flow conditions behind the steam turbine last stage, influence of various shape modifications of optical probe heads (one or two rib probes), optimization of the probe shape and measuring volume shape, and development correction method to eliminate the probe's influence on the flow field.

## Acknowledgement

We gratefully acknowledge the support by Grant No. 16-20123S of the Grant Agency of the Czech Republic. Many thanks go to all the DŠP and ETE employees who enabled the developmental measurements on LP of the steam turbine and who participated in their implementation.

## References

1. Walters, P.T., Skingley, P.C.: *An Optical Instrument for Measuring the Wetness Fraction and Droplet Size of Wet Steam Flow in LP Turbines*, IMechE Proceedings: Steam Turbines for the 1980's, London, 1979, 337-348
2. Kolovratník, M. Bartoš O.: EPJ Web of Conferences 114, 02056, (2016)
3. Bohren, C.F., Huffman, D.R.: *Absorption and Scattering of Light by Small Particles*. John Wiley & Sons, Inc., USA, 1998
4. Kolovratník, M.: *Extinkční sondy pro měření v parních turbínách*. Habilitační práce, Praha, 2004
5. Walters, P. T.: *Wetness and Efficiency Measurements in L.P. Turbines with an Optical Probe as an Aid to Improving Performance*. Transactions of the ASME, Journal of Engineering for Gas Turbines and Power, 1987, January, Vol. 109, 85-91
6. Petr, V., Kolovratník, M.: *Heterogeneous Effects in the Droplet Nucleation Process in LP Steam Turbines* – Proceedings: 4th European Conference on Turbomachinery, Firenze-Italy, 2001, pp. 783-792
7. VDI Richtlinien - VDI 2043 Measurement of Steam Wetness Fraction.

 DOI: 10.1051/epjconf/201714302053

# Comparison of synthetic jet actuators based on sharp-edged and round-edged nozzles

J. Kordík[1,a] and Z. Trávníček[1]

[1]Institute of Thermomechanics, AS CR

**Abstract.** Axisymmetric synthetic jet actuators based on a loudspeaker and on two types of flanged nozzles were tested and compared experimentally. The first type of the nozzle was a sharp-edged circular hole. The second one had a special design with fillets at inner and outer nozzle exit and with a small step in the middle of the nozzle. The function of the step was to prevent the flow reattachment during the extrusion stroke. The actuators with the two types of nozzles were operated at resonance and were compared first qualitatively using a simple phase locked flow visualization. Then the hot-wire anemometer was used to measure velocity distributions along nozzle axis and velocity profiles at the nozzle exit. Comparison of the nozzles was based on evaluation of the characteristic velocity and integral quantities (volumetric, momentum, and kinetic energy fluxes). It was found out that these quantities, which were evaluated at the nozzle exit, differ substantially for both nozzles. On the other hand the velocity flow field in farther distances from the nozzle exit area did not exhibit such prominent differences.

## 1 Introduction

Synthetic jets (SJs) are fluid flows that are created by fluid pulsations and are formed in a free space behind a nozzle. The fluid is periodically pushed and pulled through the nozzle, whose one end exits into a cavity. The source of the pulsations is an oscillating diaphragm that bounds one side of the cavity. The nozzle is the only inlet/outlet in/from the cavity, therefore time mean mass flux through the nozzle is zero. It is the reason, why the SJs are sometimes called zero-net-mass-flux jets. The momentum flux along the nozzle axis is, however, non-zero. It causes the flow continuation ahead downstream from the nozzle exit and formation of a jet flow similar to the continuous jet.

Since the end of the last century, the topic of synthetic jets has been intensively investigated [1,2] and many potential applications of SJs have been developed. The potential applications have been found especially in boundary-layer separation control [3–10], jet vectoring [11–16], heat transfer enhancement [17–22], and mixing [23–25]. A variant of the SJ, namely a non-zero-net-mass-flux jet (or hybrid SJ), has been investigated more recently [26–28,21,29–31].

If the synthetic jet actuators based on elastic diaphragms (e.g. based on a loudspeaker or a piezoceramic membrane etc.) and input/output energy balance are considered, then the most advantageous operating conditions are found at a resonance. Therefore, many authors concentrated their efforts on location and prediction of the SJ actuator resonances (see e.g. [32–44]). Having a SJ actuator operating at a resonance frequency, there still exist other ways how to enhance its exploitable outlet quantities such as volumetric flux, momentum or kinetic energy fluxes. This will be the topic of this work. Particularly, this paper investigates the axisymmetric SJ actuators based on a loudspeaker and working at the first resonance. Under these conditions two types of flanged nozzles are tested within the actuator. The first type of the nozzle is a common sharp-edged circular hole and the second one possess fillets at inner and outer nozzle exit. The second nozzle was designed to have a lower hydraulic resistance and therefore, higher mentioned integral quantities (volumetric flux, momentum and kinetic energy flux) are anticipated. This paper is going to reveal whether this design is advantageous according to the expectations.

[a] e-mail: kordik@it.cas.cz

## 2 Observed Parameters

The first parameter, which is going to be observed and evaluated from the current experiments is the characteristic velocity of the SJ. This velocity is evaluated as follows:

$$U_0 = \frac{1}{TS_{\rm n}} \int_0^{T_{\rm E}} \int_{S_{\rm n}} u \mathrm{d}S\, \mathrm{d}t, \qquad (1)$$

where $S_{\rm n}$ is the nozzle exit cross-sectional area ($\mathrm{d}S$ is the differential element of $S_{\rm n}$), $u$ denotes the instantaneous velocity component aligned with the nozzle axis, $T_{\rm E}$ is the extrusion stroke duration time, $T$ is the actuation period, and the area-averaged velocity $\frac{1}{S_{\rm n}} \int_{S_{\rm n}} u \mathrm{d}S$ has the origin at $t = 0$ ($t$ is time).

A more simple evaluation of the SJ characteristic velocity can use a slug flow model approach, and the definition can be based on a point measurement of the centerline velocity waveforms only:

$$U_{0,\rm cl} = \frac{1}{T} \int_0^{T_{\rm E}} u(r = 0, x = 0) \mathrm{d}S\, \mathrm{d}t, \qquad (2)$$

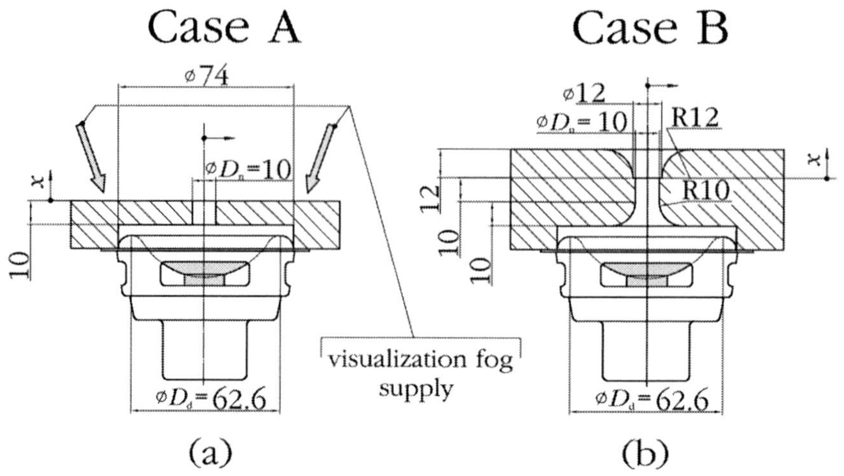


**Fig. 1.** Schema of the current actuators: (a) with a common, sharp-edged nozzle, (b) with a rounded nozzle.

Note that this relatively simple evaluation will be used for quantification of the velocity resonance curves – see paragraph 4.1 in the text below. The other parts of this study use more complex definition Eq. (1) which requires measurements of the velocity profiles across the jet.

The integral quantities, which are going to be analyzed, are: the time-average volumetric flux ($Q_0$), momentum flux ($M_0$), and kinetic energy flux ($E_0$):

$$Q_0 = \frac{1}{T}\int_0^{T_E}\int_{S_n} u \mathrm{d}S \, \mathrm{d}t = U_0 S_n, \quad (3)$$

$$M_0 = \frac{1}{T}\int_0^{T_E}\int_{S_n} \rho u |u| \, \mathrm{d}S \, \mathrm{d}t, \quad (4)$$

$$E_0 = \frac{1}{T}\int_0^{T_E}\int_{S_n} \frac{\rho u^3}{2} \mathrm{d}S \, \mathrm{d}t, \quad (5)$$

where $\rho$ is the working fluid density.

Note, that the interval $(0, T_E)$ can be a bit different for each quantity $Q_0$, $M_0$, and $E_0$. Generally, the interval represents time periods where the surface integrals from Eqs. (3,4,5) are greater or equal to zero.

## 3 Experimental Setup and Methods

### 3.1 The Synthetic Jet Actuators

Figure 1 shows the schematic view of the tested SJAs that were operated at standard room conditions (temperature 20 – 25 °C and barometric pressure 97 – 99 kPa) with air as the working fluid. The present actuators are based on a Aurasound NS3-193-8A loudspeaker, whose diaphragm generates the fluid oscillations in the nozzles. The main difference between the actuators dwells in their nozzle shapes. The first nozzle (case A) is a cylindrical hole – no special amendments lowering the hydraulic resistance are applied. The nozzle has a cylindrical shape and is sharp-edged at both flange surfaces. The second nozzle (case B) was designed with the aim to decrease the hydraulic resistance in both fluid flow directions. Therefore, the nozzle is rounded at its both surfaces. Moreover, as figure 1b shows, a small step in the middle of the nozzle that was designed to define flow separation from the step edge (i.e. to prevent the flow reattachment to the rounded wall) during the extrusion stroke. The narrowest cross-sectional area is the same for both tested nozzle shapes (the diameter is $D_n$ = 10mm).

During all experiments, the loudspeaker of the actuator was supplied by a harmonic electric current generated by Agilent 33220A function generator and gained in Pioneer A-209R amplifier. The experiments were performed at constant electrical power input into the actuator, $P_e$. Three input power levels were used $P_e$ = 0.11, 0.50, and 0.54W.

### 3.2 Data Acquisition and Hardware Setup

To get a qualitative view on flow field patterns issuing from both tested nozzles (case A, case B), a simple flow visualization was conducted before quantitative hot-wire measurements. The visualization method used cold water fog produced by an ultrasonic piezoelectric nebulizer (Mini Nebler). The fog was blown at speed of 0.2 m/s from the nebulizer using two supply pipes into the surroundings of the SJ orifice (fog supply spots are shown in figure 1a). The phase-locked SJ flow field patterns were observed under power-LED flashes synchronized with the actuator excitation frequency at an adjusted phase angle. Pictures were taken by a digital camera connected to a PC via an USB cable, using exposure time of 8.0s. The resulting photo is a multi-exposure of a large number of frames (typically 440 frames), which shows the phase-locked feature of the flow field and smoothing out of the deviation fluctuations in the individual cycles.

The MiniCTA 54T30 DANTEC hot-wire anemometer with a single wire probe 55P16 and the NI-PCI 6023E data acquisition device were used for this measurement. The hot-wire signal was sampled with a sampling frequency of

$f \cdot 128$ and the number of samples was 25600. The temperature of the working fluid (air) was measured using a fast-response Pt100 sensor (PT100-SMD0805) at the same time as the velocity samples. The measured temperature was used for the temperature correction of the hot-wire data. To evaluate and check the constant power input into the actuator, the supplied electric current and voltage were acquired concurrently with velocity and temperature samples.

After the identification of the velocity resonance frequency, the actuators were tuned to the resonance and velocity distributions along the $x$-axis were measured at power input of $P_e = 0.11$W. The same experimental equipment and hardware setup as during auxiliary experiments was used.

At the same power level and frequency the velocity profiles were measured across the jet at the distances of $x =$ 0.3 mm, $x = 13$ mm, and $x = 26$ mm. The velocity profiles measured at the distance $x = 0.3$ mm were acquired at 14 discrete positions along the nozzle diameter and the data were used for evaluation of $U_0$ and the integral quantities defined by Eqs. (3,4,5).

For all measurements the hot-wire probe was calibrated in the velocity range of 0.15 – 62 m/s. The maximum relative uncertainty of a single velocity sample was below 5% for velocities in range of 0.4 – 62 m/s. For very small velocities in the range of 0.15 – 0.4 m/s the uncertainty was 17.2%. A more detailed description of the anemometer settings, probe calibration method, and uncertainty evaluation is available in [42].

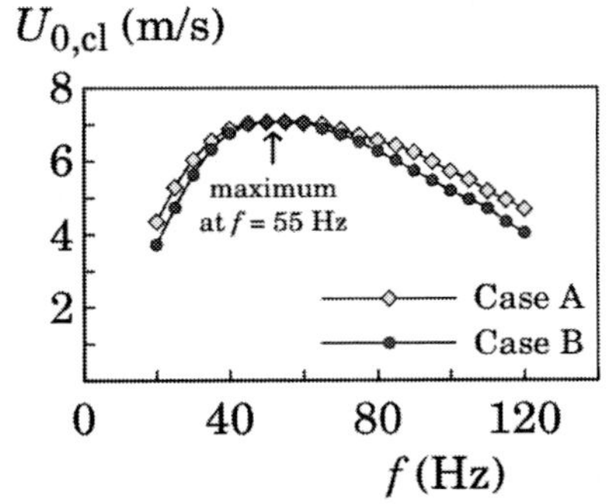


**Fig. 2.** Measured velocity resonance curves for both cases A and B.

## 4 Experimental Results

### 4.1 Auxiliary Experiments

The measured velocity resonance curves for identification of particular resonance frequencies are shown in figure 2. The two curves in figure 2 were acquired at constant power input into the actuator: $P_e = 0.5$W for both SJA cases A and B. It was found out that the resonance frequency was the same for both nozzle types: $f = 55$Hz. We can extend this result also for other tested power levels, because as was shown in [31] the resonance frequency is nearly independent of the supplied power-input.

### 4.2 Main Experiments and Discussion

After identification of the resonance frequencies, SJs from both actuators (case A and B) were visualized by a water fog. The visualizations were performed at one power input level: $P_e = 0.11$W and at $f = 55$Hz (the resonance frequency). The resultant pictures are shown in figure 3 for case A and in figure 4 for case B – see the photographs on left parts of the figures. Vortex formation process and its successive extinction can be observed in both figures. A phase shift of $\Delta\varphi \approx 30°$ between visualizations in figure 3 and 4 is remarkable. This is caused by the additional top plate in case B, which produces a spatial shift between the figures (the $x$-coordinate of the photographs in figure 4 starts at $x/D_n = 1$). Therefore, the initial vortex formation in the nozzle exit vicinity can not be observed. Apart from this phase shift there is no other remarkable qualitative difference between the visualized fluid flow patterns of both cases.

In the central column of Figs. 3 and 4 the distributions of instantaneous, $u$, and time-averaged velocities, $\bar{u}$, along the jet axis are shown. The traveling velocity maximum on these graphs roughly correspond with the vortex core position visualized by the water fog. Similarly as during the visualization, no distinct differences in the velocity distribution along the nozzle axis were found for both cases A and B.

The right parts of Figs. 3 and 4 show the velocity profiles measured at distances $x = 0.3$ mm, $x = 13$ mm, and $x = 26$ mm. The measurements of the velocity profiles revealed several differences in the flow fields formed downstream from the both nozzle types. The velocity profiles measured at $x = 0.3$ mm are more flat during the extrusion stroke for the case B (rounded nozzle type), this will be important in evaluation of the integral quantities $Q_0$, $M_0$, and $E_0$, as it is shown below.

The velocity profiles measured at $x = 13$ mm are also very different. While in case A (plain nozzle) the suction phase of the actuator does not affect the velocity profile at this distance, negative velocities for $\varphi = 180° - 360°$ can be observed in case B. This is caused by the proximity of the additional wall with the rounded edge (see figure 1).

The differences between velocity profiles nearly disappear at the distance of $x = 26$ mm. This can be also observed in Fig. 5 where the time-mean velocity profiles measured at $x = 26$ mm (i.e. $x/D_n = 2.6$) are displayed. The slight differences between the velocity profiles in figure 5 are within the measurement error, therefore, we can conclude that the profiles are practically identical.

The comparison of the integral quantities evaluated from the velocity profiles at the nozzle exit (more precisely at $x = 0.3$ mm) offers the most prominent differences. The results are listed in table 1. The nozzle associated with case B exhibited higher values in all observed parameters and quantities. The increase in $U_0$ and $Q_0$ was more than 18%, even higher increase was found for $M_0$

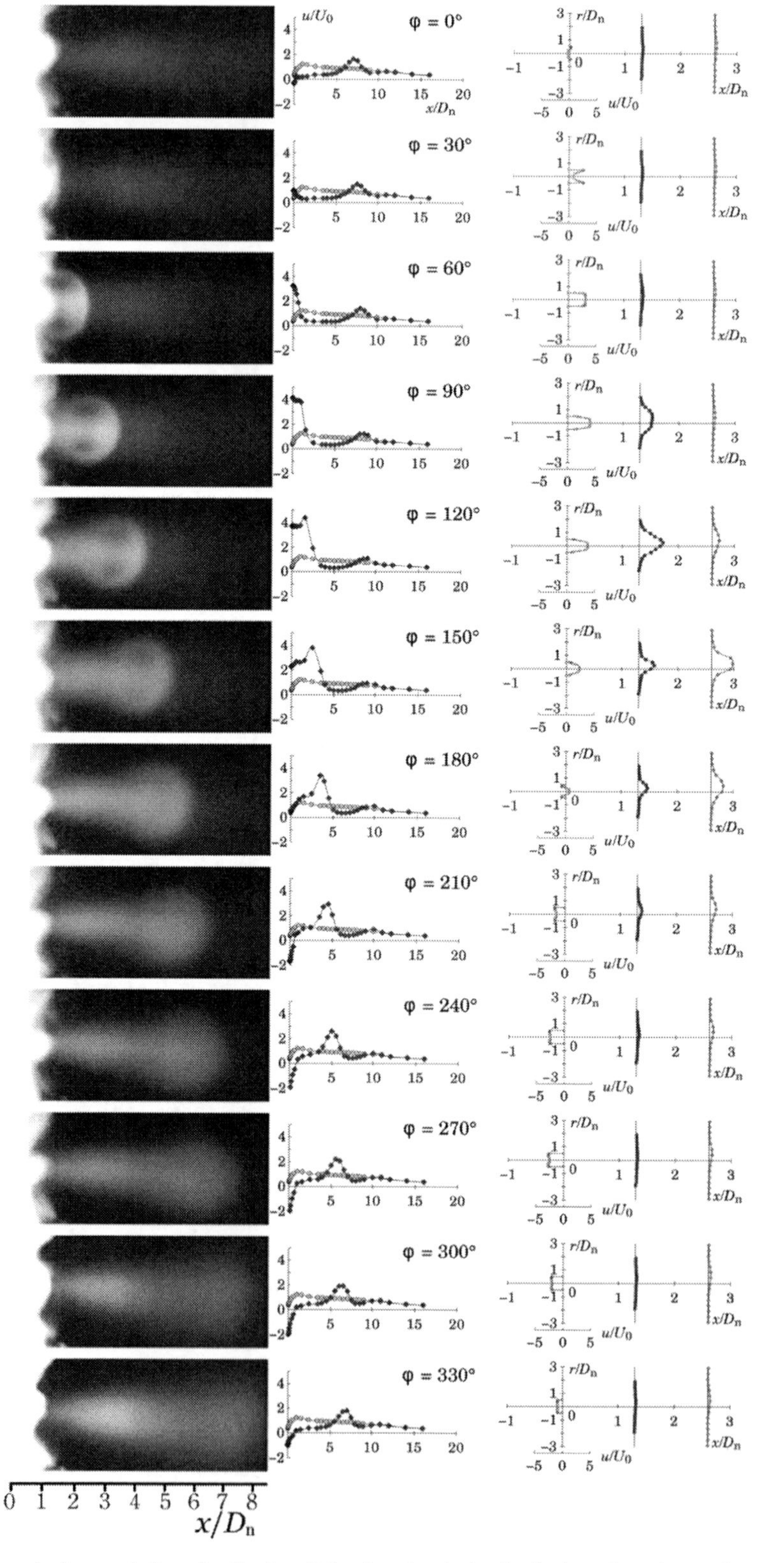


**Fig. 3.** Measurements results for case A, flow visualizations (left column), velocity distributions along the nozzle axis (central column), and velocity profiles (right column).

 DOI: 10.1051/epjconf/201714302053

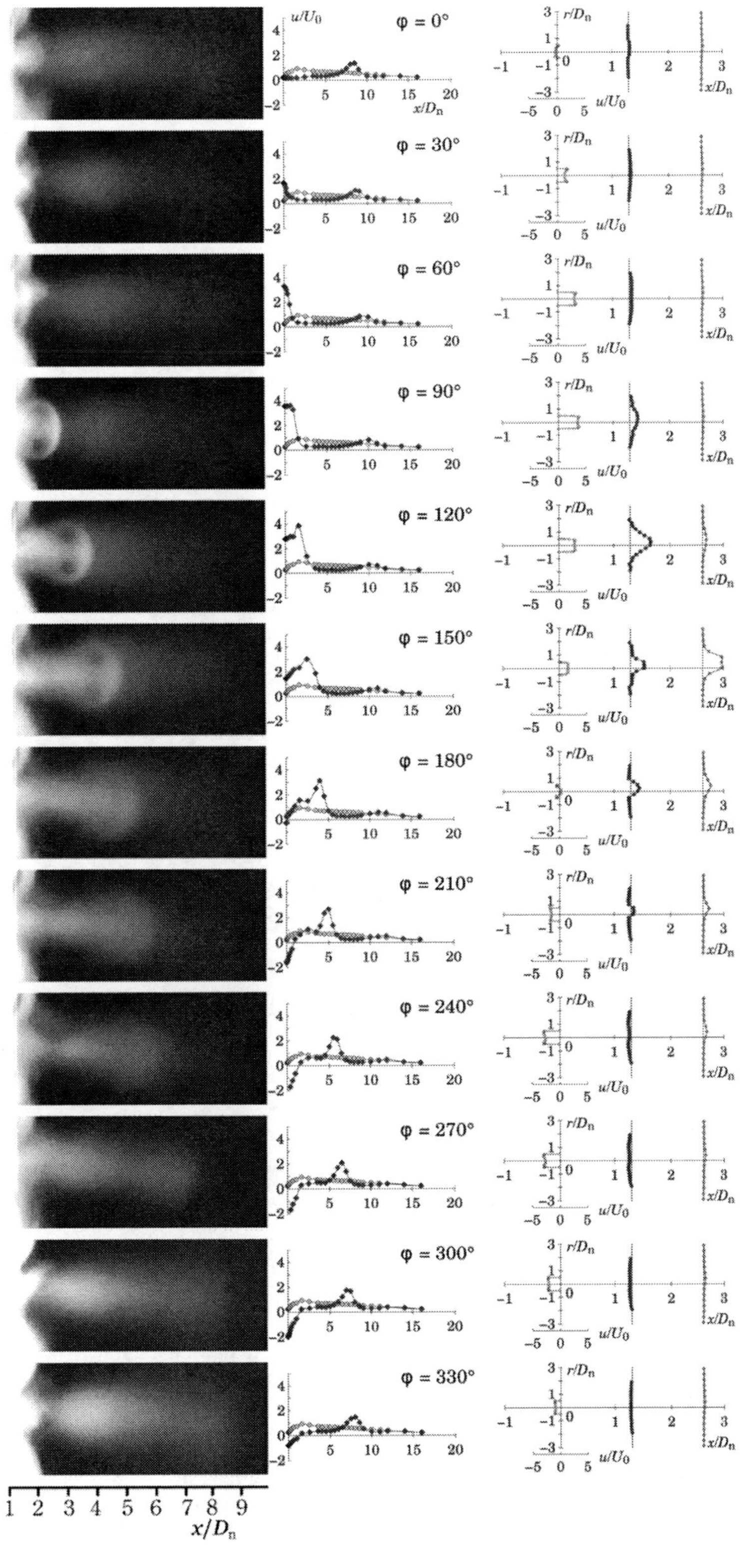

**Fig. 4.** Measurements results for case B, flow visualizations (left column), velocity distributions along the nozzle axis (central column), and velocity profiles (right column).

DOI: 10.1051/epjconf/201714302053

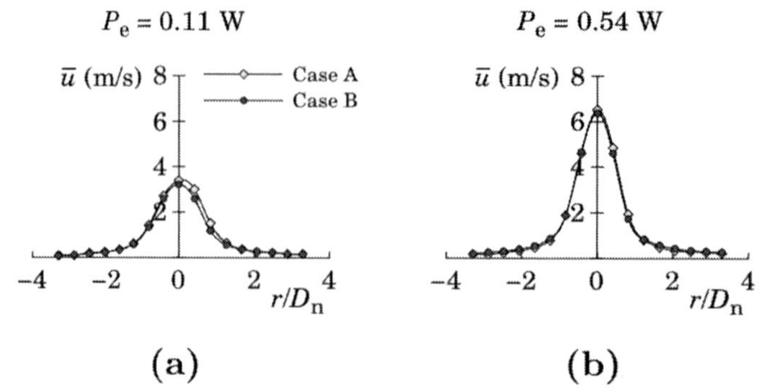


**Fig. 5.** Time-mean velocity profiles measured at the distance of $x/D_n = 2.6$ for input powers (a) $P_e = 0.11$W and (b) $P_e = 0.54$W.

| quantity | | case A | case B | increase |
|---|---|---|---|---|
| $U_0$ (m/s) | = | 3.4 | 4.1 | 18.5% |
| $Q_0$ (m3/s) | = | $2.7\times10^{-4}$ | $3.2\times10^{-4}$ | 18.5% |
| $M_0$ (N) | = | $2.8\times10^{-3}$ | $4.0\times10^{-3}$ | 43.2% |
| $E_0$ (W) | = | $1.4\times10^{-2}$ | $2.4\times10^{-2}$ | 70.8% |

**Table 1.** The characteristic velocities and the integral quantities evaluated from the velocity profiles measured at $x = 0.3$ mm and for $P_e = 0.11$W using Eqs. (1,3,4,5).

and $E_0$: 43.2% and 70.8%, respectively. This is a quite big difference, which shows the importance of evaluating these integral quantities from the velocity profile. For example, the velocities $U_0$ evaluated from the centerline waveforms at resonance – see e.g. figure 2 are practically identical, but the integral evaluation of $U_0$ shows the difference of 18.5%.

Although the differences in the quantities from table 1 are large, figure 5 shows that differences between velocity profiles measured at $x/D_n = 0.26$ are negligible. Obviously, these differences cannot be explained without additional and more complex measurements in the near future.

## 5 Conclusion

Two types of flanged nozzles for a SJA were tested. The first type of the nozzle was a sharp-edged circular hole (case A) and second was the rounded nozzle (case B), which had a special design with fillets at inner and outer nozzle exit and with a small step in the middle of the nozzle (case B). The investigated actuator that was based on a speaker and worked at resonance and at a constant power input. Three input power levels were used: $P_e = 0.11$W, 0.5W, and 0.54W.

A phase locked flow visualization and hot-wire measurements were used for the qualitative and quantitative testing. Velocity profiles for the rounded nozzle (case B) and acquired at $x/D_n = 0.03$ were more flat during the extrusion stroke. Moreover, in case B, fluid suction influenced the velocity profile measured at $x/D_n = 1.3$, nothing similar was observed with the plain nozzle (case A).

The evaluation of the characteristic velocity and the integral quantities (volumetric flux, momentum flux, and kinetic energy flux) showed the most prominent differences between the nozzles. These quantities were evaluated from the velocity profiles measured at $x/D_n = 0.03$ and were always higher for the rounded type of the nozzle. A relative increase in the integral quantities was following: 18% (volumetric flux), 43.2% (momentum flux), and 70.8% (kinetic energy flux). On the other hand, the velocity profiles of the both tested cases A and B are nearly identical in the far field. To explain these facts, more complex measurements seem to be needed.

## Acknowledgement

We gratefully acknowledge the support of the Grant Agency CR (project number 16-16596S) and the institutional support RVO:61388998.

## References

1. B.L. Smith, A. Glezer, Physics of Fluids **10**, 2281 (1998)
2. J.E. Cater, J. Soria, Journal of Fluid Mechanics **472**, 167 (2002)
3. G. Hong, Sensor Actuat. A-Phys. **132**, 607 (2006)
4. J. Dandois, E. Garnier, P. Sagaut, J. Fluid Mech. **574**, 25 (2007)
5. D. You, P. Moin, J. Fluid Struct. **24**, 1349 (2008)
6. S.G. Mallinson, J.A. Reizes, R. Hillier, Flow Turbulence and Combustion **66**, 1 (2001)
7. M. Amitay, A. Glezer, International Journal of Heat and Fluid Flow **23**, 690 (2002)
8. R. Rathnasingham, K.S. Breuer, Journal of Fluid Mechanics **495**, 209 (2003)
9. S. Lardeau, M.A. Leschziner, Journal of Fluid Mechanics **683**, 172 (2011)
10. C. Bernardini, M. Carnevale, M. Manna, F. Martelli, D. Simoni, P. Zunino, Journal of Thermal Science **21**, 404 (2012)
11. B.L. Smith, A. Glezer, Journal of Fluid Mechanics **458**, 1 (2002)

12. Z.B. Luo, Z.X. Xia, Modern Physics Letters B **19**, 1619 (2005)
13. Z.B. Luo, Z.X. Xia, Y.G. Xie, Chinese Journal of Aeronautics **20**, 193 (2007)
14. Z.X. Xia, Z.B. Luo, Applied Mathematics and Mechanics **28**, 907 (2007)
15. D.A. Tamburello, M. Amitay, International Journal of Heat and Fluid Flow **29**, 967 (2008)
16. M. Ben Chiekh, M. Ferchichi, J.C. Béra, Experiments in Fluids **53**, 1797 (2012)
17. M.B. Gillespie, M.S. thesis, Georgia Institute of Technology (1998)
18. Z. Trávníček, V. Tesař, International Journal of Heat and Mass Transfer **46**, 3291 (2003)
19. M.B. Gillespie, W.Z. Black, C. Rinehart, A. Glezer, Journal of Heat Transfer **128**, 990 (2006)
20. M. Arik, Applied Thermal Engineering **27**, 1483 (2007)
21. Z. Trávníček, V. Tesař, J. Kordík, Journal of Visualization **11**, 221 (2008)
22. M. Chaudhari, B. Puranik, A. Agrawal, International Journal of Heat and Mass Transfer **53**, 1057 (2010)
23. H. Wang, S. Menon, AIAA Journal **39**, 2308 (2001)
24. M. Al-Atabi, Journal of Fluids Engineering **133**, 094503 (2011)
25. Q. Xia, S. Zhong, International Journal of Heat and Fluid Flow **37**, 64 (2012)
26. Z. Trávníček, A.I. Fedorchenko, A.B. Wang, Sensors and Actuators A: Physical **120**, 232 (2005)
27. Z. Trávníček, T. Vít, V. Tesař, Physics of Fluids **18**, 081701 (2006)
28. V. Tesař, Sensors and Actuators A **138**, 394 (2007)
29. J. Kordík, Z. Trávníček, Journal of Fluids Engineering **135**, 101101 (2013)
30. Z. Trávníček, T. Vít, International Journal of Heat and Mass Transfer **85**, 473 (2015)
31. J. Kordík, Z. Trávníček, M. Pavelka, Experimental Thermal and Fluid Science **69**, 119 (2015)
32. Q. Gallas, M.S. thesis, University of Florida, Gainesville (2002)
33. Q. Gallas, J. Mathew, A. Kaysap, R. Holman, T. Nishida, B. Carroll, M. Sheplak, L. Cattafesta, *Lumped element modeling of piezoelectric-driven synthetic jet actuators*, in *40th AIAA Aerospace Sciences Meeting & Exhibit* (AIAA Paper 2002-0125, Reno, 2002)
34. Q. Gallas, R. Holman, T. Nishida, B. Carroll, M. Sheplak, L. Cattafesta, AIAA Journal **41**, 240 (2003)
35. Q. Gallas, M. Sheplak, L. Cattafesta, Final report for NAG-1-03031, University of Florida, Department of Mechanical and Aerospace Engineering, Gainesville (2005)
36. Q. Gallas, R. Holman, R. Raju, R. Mittal, M. Sheplak, L. Cattafesta, *Low dimensional modeling of zero-net mass flux actuators*, in *2nd AIAA Flow Control Conference* (AIAA Paper 2004-2413, Portland, 2004)
37. Q. Gallas, PhD. thesis, University of Florida, Gainesville (2005)
38. B.H. Kim, D.R. Williams, S. Emo, M. Acharya, AIAA Journal **43**, 314 (2005)
39. H. Tang, S. Zhong, *Lumped element modelling of synthetic jet actuators*, in *The 3rd AIAA Flow Control Conference* (AIAA Paper 2006-3696, San Francisco, 2006)
40. H. Tang, S. Zhong, M. Jabbal, L. Garcillan, F. Guo, N. Wood, C. Warsop, Flow, Turbulence and Combustion **78**, 309 (2007)
41. H. Tang, S. Zhong, Aerospace Science and Technology **13**, 331 (2009)
42. J. Kordík, *Synthetic and Hybrid Synthetic Jet Actutators – Theoretical and Experimental Analysis* (LAP LAMBERT Academic Publishing, 2013), ISBN 978-3-659-32910-4
43. J. Kordík, Z. Trávníček, AIAA Journal **51**, 2862 (2013)
44. J. Kordík, Z. Broučková, T. Vít, M. Pavelka, Z. Trávníček, Experiments in Fluids **55**, 1757 (2014)

DOI: 10.1051/epjconf/201714302054

# Experimental investigation of cavitation induced air release

*Karoline* Kowalski[1,*], *Stefan* Pollak[2], and *Jeanette* Hussong[3]

[1]Ruhr University Bochum, Chair of Process Technology, Universitätsstr. 150, 44801 Bochum, Germany
[2]Ruhr University Bochum, Chair of Particle Technology, Universitätsstr. 150, 44801 Bochum, Germany
[3]Ruhr University Bochum, Chair of Hydraulic Fluid Machinery, Universitätsstr. 150, 44801 Bochum, Germany

**Abstract.** Variations in cross-sectional areas may lead to pressure drops below a critical value, such that cavitation and air release are provoked in hydraulic systems. Due to a relatively slow dissolution of gas bubbles, the performance of hydraulic systems will be affected on long time scales by the gas phase. Therefore predictions of air production rates are desirable to describe the system characteristics. Existing investigations on generic geometries such as micro-orifice flows show an outgassing process due to hydrodynamic cavitation which takes place on time scales far shorter than diffusion processes. The aim of the present investigation is to find a correlation between global, hydrodynamic flow characteristics and cavitation induced undissolved gas fractions generated behind generic flow constrictions such as an orifice or venturi tube. Experimental investigations are realised in a cavitation channel that enables an independent adjustment of the pressure level upstream and downstream of the orifice. Released air fractions are determined by means of shadowgraphy imaging. First results indicate that an increased cavitation activity leads to a rapid increase in undissolved gas volume only in the choking regime. The frequency distribution of generated gas bubble size seems to depend only indirectly on the cavitation intensity driven by an increase of downstream coalescence events due to a more densely populated bubbly flow.

## 1 Introduction

Hydrodynamic cavitation in liquid flows can be produced by a pressure drop behind a flow constriction. Since most liquids used in technical applications contain dissolved gases both vaporous cavitation and degassing processes may take place simultaneously at different intensity depending on the pressure conditions and solubility of gases in liquids. The cavitation process is influenced by e.g. the liquid properties and often takes place as heterogeneous cavitation at nuclei such as impurities or the liquid-solid interface [1].

The two phase flow has an effect on the density, speed of sound and compressibility of the flowing liquid. This can lead to noise, vibrations and fluctuation of the mass flow rate in hydraulic systems and lower the efficiency of the system. Simulations of air release are mostly based on the assumption of a diffusion process which fails to make quantitative predictions on cavitation induced degassing [2]. Most flow cavitation experiments are realised in small scale and/or rectangular cross sectional channels, e.g. [3-6]. Optical measurements of air release provoked by cavitation in a rectangular microchannel of testing oil were performed by Freudigmann et al. [3]. As commonly known, cavitation phenomena cannot be simply scaled up [7]. A rectangular channel cross section enables a better optical access but leads to topology related corner flow and cavitation behaviour. Due to this fact, the experiments described in this paper are restricted to rotationally symmetric cavitation flows in generic geometries.

The general hydrodynamic behaviour of a non-cavitating orifice flow is well described by the Bernoulli equation. When choked flow conditions are reached no further increase in volume flow rate can be obtained and the mass flow rate stays at a constant value. The onset of a choked flow condition depends on the choice of upstream parameters only [8].

The dimensionless pressure drop can be described through the cavitation number $\sigma$ which is given by Nurick and others [9].

$$\sigma = \frac{P_{in} - P_v}{P_{in} - P_{out}} \tag{1}$$

$P_{in}$ denotes the inlet pressure and $P_v$ the vapour pressure. The pressure difference $P_{in}$ - $P_{out}$ is measured over the flow constriction. Decreasing cavitation numbers correspond to an increase in cavitation probability or intensity, respectively.

## 2 Experimental set-up

This section is divided into two parts. In section 2.1 the experimental set-up and its hardware components are described. In the following sections 2.2 and 2.3 details are given on the measurement procedure for flow characteristics and the optical measurements.

* Corresponding author: kowalski@vtp.rub.de

### 2.1 Cavitation channel

In the experimental set-up (Fig. 1) two stainless steel pressure vessels (585 l volume, 0 to 10 bar pressure range) are integrated in a closed fluid cycle. In the first vessel $V_1$ compressed air is used to set the pressure level $P_1$. The liquid can be circulated through a bypass BP to reach a faster gas saturation of the liquid using a centrifugal pump (MXH 803, Linn Pumpen). The second vessel $V_2$ can be either evacuated down to 0.2 bar using a vacuum pump (VACFOX VGD 10, Ehrler & Beck GmbH) or it can be set to higher pressures using compressed air. Different constrictions, e.g. orifices or venturi tubes can be placed in the test section. Due to the pressure difference between both vessels the overall pressure level $P_{in}$ and volume flow rate $\dot{V}$ can be set independently in the measurement section.

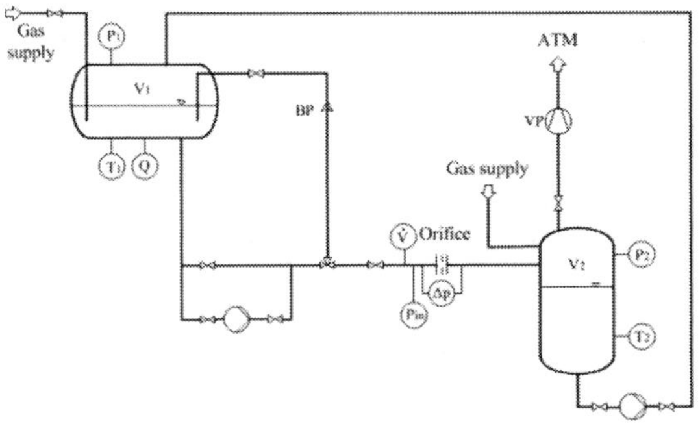


**Fig. 1:** Draft of the experimental set-up with pressure, flowrate, temperature and gas saturation sensors

The experiments are executed as a batch process to avoid influences of fluctuating gas content upstream of the test section. For every operation point the vessel pressures $P_1$ and $P_2$, as well as corresponding vessel temperatures $T_1$ and $T_2$, the inlet pressure $P_{in}$, the oxygen content Q in the first vessel, the volume flow rate $\dot{V}$ and the pressure difference $\Delta p$ at the orifice are recorded with 100 Hz for 2 minutes.

For examination of the orifice flow characteristics, measurements are performed for different inlet pressure levels $P_{in}$ at which the outlet pressure $P_2$ is lowered stepwise.

The orifice assumes a bore diameter of 3 mm with a 45° chamfer of approximately 0.63 mm at the inlet and a conical outlet of 60° (see Fig. 2). The flow direction is from left to right.

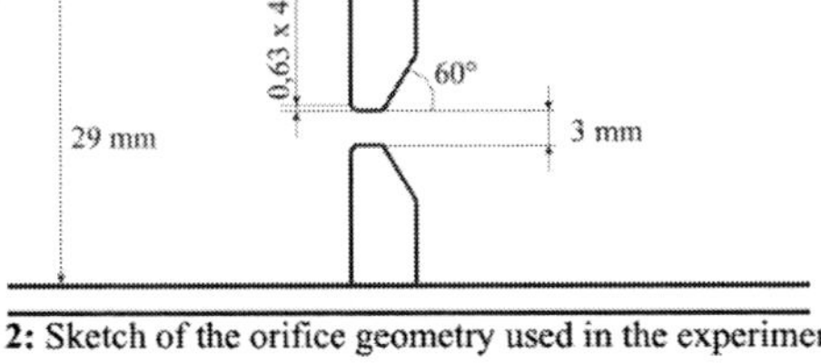


**Fig. 2:** Sketch of the orifice geometry used in the experiments

### 2.2 Flow characteristics

For each flow characteristic the inlet pressure is set to a fixed value. The measurement sequence is started with a maximum outlet pressure, which is decreased in 1 bar steps or less after every measurement point down to 0.2 bar. For every operation point data is recorded after the flow is fully developed and stationary flow conditions are reached. In the experiments the cavitation number is lowered in every measurement sequence by reducing the outlet pressure $P_2$.

### 2.3 Optical measurements

Flow characteristics and optical measurements were performed simultaneously. Shadowgraphy images were taken behind the orifice in an optically accessible test section using a CCD camera (ImagerProSX, LaVision) and a pulsed LED light source (Backlight blue, LaVision) as indicated in Fig. 3a.

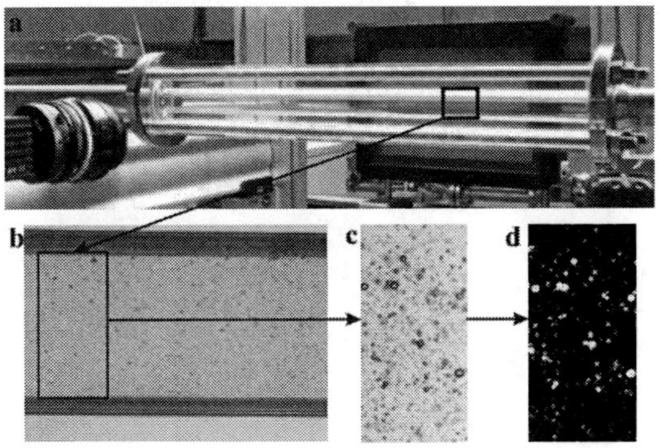


**Fig. 3:** a) Test section with LED light source and CCD camera, b) representative FOV of 40.64 x 34.03 mm² located 30 mm downstream of the orifice, c) region of interest selected for statistical evaluation, d) postprocessed image, detected bubble centers are marked red

Images are taken with a 50 mm focal distance lens (Nikon, Nikkor Lens AF 50 mm f/1.4D) and a 4x close-up lens. For every operation point, 200 images are recorded with a spatial resolution of 0.0166 mm per pixel covering a field of view (FOV) of 40.64 x 34.03 mm$^2$. Images are taken 30 cm behind the orifice exit in a cavitation free zone to record air bubbles only. Due to total reflection a section of approximately 4 mm of the upper and lower part cannot be accessed optically (dark areas in Fig. 3b).

The images are postprocessed by a self-written MATLAB algorithm. The bubble size distribution and bubble density for every operation point were determined. For preprocessing a sliding average filter of 2 x 2 pixel filter length was used to reduce the salt and pepper noise in the images. Hence, a constant image threshold was applied for a black-white transformation. Afterwards, all objects in the image plane were filled and labelled. Only objects with more than 4 pixels area size are used for a statistical analysis. For every measurement point shown in Fig. 4 the mean bubble size distribution was extracted from 200 images, corresponding to 40 s recording sequence. Furthermore, the ratio $\alpha$ between the area occupied by bubbles $A_{bubbles}$ and the total image area $A_{image}$ is depicted as measure for the released air volume fraction at different operation points.

$$\alpha = \frac{A_{bubbles}}{A_{image}} \quad (2)$$

## 3 Results

In this section flow characteristics and degassed volume fractions are measured for a cavitating orifice flow.

### 3.1 Flow characteristics

In Fig. 4 the hydraulic characteristics of the set-up with an orifice of 3 mm bore diameter are shown. Three different inlet pressure levels of 5.8, 7.8 and 9.8 bar are depicted. Each measurement was repeated three times.

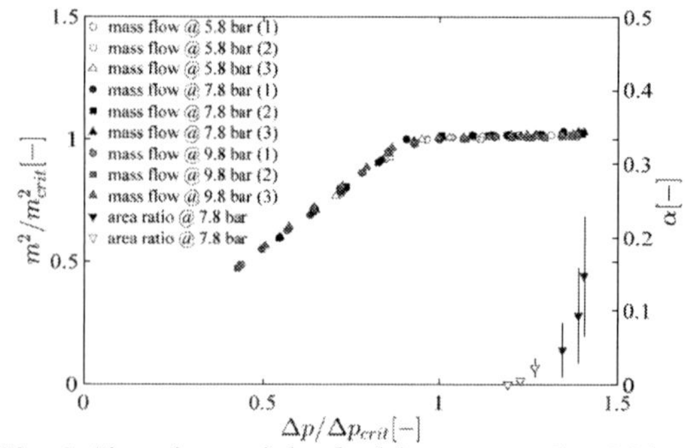


**Fig. 4**: Flow characteristics for inlet pressures $P_{in}$ = 9.8 bar, 7.8 bar, 5.8 bar (black labels) and results of air release fraction $\alpha$ for an inlet pressure of $P_{in}$ = 7.8 bar (blue labels)

Normalised squared volume flow rates $\dot{m}^2/\dot{m}^2_{crit}$ are plotted as a function of the normalised pressure differences $\Delta p/\Delta p_{crit}$ across the orifice. Critical values refer to the shift from the so called Bernoulli regime to the choked flow regime, apparent through a kink in the flow characteristics. Within the Bernoulli regime, the squared mass flow rate scales with the pressure drop across the flow constriction. The critical mass flow rate at which choking takes place depends on the inlet pressure. A standard deviation of mean mass flow rates $s(q)$ of less than 0.23 % was determined based on the arithmetic mean $q$ and standard deviation $s(q_k)$ of $n$ individual observations $q_k$ (see [10]).

$$q = \frac{1}{n} \cdot \Sigma q_k \quad (3a)$$

$$s(q_k) = \sqrt{\frac{1}{n-1}\,\Sigma(q_k - q)^2} \quad (3b)$$

$$s(q) = \frac{s(q_k)}{\sqrt{n}} \quad (3c)$$

The critical values for the mass flow rate $\dot{m}^2_{crit}$ and the pressure difference $\Delta p_{crit}$, respectively are given in Tab. 1.

**Table 1:** Critical values for choked flow conditions

| $P_{in}$ [bar] | $\Delta p_{crit}$ [bar] | $\dot{m}^2_{crit}$ [(g/s)²] | $P_{2crit}$ [bar] | $\sigma_{crit}$ [-] |
|---|---|---|---|---|
| 5.8 | 4.0 | 34000 | 2 | 1.44 |
| 7.8 | 5.4 | 45500 | 2.4 | 1.43 |
| 9.8 | 6.9 | 57750 | 3 | 1.42 |

For increasing inlet pressures $P_{in}$ the critical pressure drop $\Delta p_{crit}$, the critical squared mass flow rate $\dot{m}^2_{crit}$ and critical outlet pressure $P_{2crit}$ increase. The critical cavitation number $\sigma_{crit}$ shows a slight decrease but is nearly constant for the three examined inlet pressures. This is expected as only a small variation of inlet pressures $P_{in}$ was examined. The cavitation number is calculated as described in Eq. 1 with $\Delta p$ as the pressure difference at the flow constriction.

The experiments were carried out for cavitation numbers between $1.03 < \sigma < 3.34$. Due to a critical cavitation number of the system of approximately 1.4, non-cavitating flows for $\sigma > \sigma_{crit}$ and cavitating flows for $\sigma < \sigma_{crit}$ were realised.

### 3.2 Optical measurements

In this section, results obtained from optical measurements are presented. Representative shadowgraphy images for an inlet pressure $P_{in}$ = 5.8 bar and three different outlet pressures $P_{out}$ = 1.4, 0.7 and 0.2 bar are shown in Fig. 5a-c. The region of interest covers 500 pixels in width and 1250 pixels in height corresponding to 8.3 x 20.75 mm².

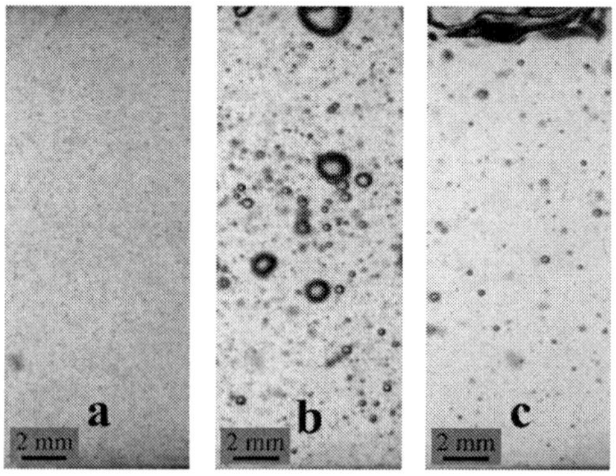


**Fig. 5:** Representative shadowgraphy images of a liquid flow 30 mm downstream of the orifice with an inlet pressure of $P_{in}$ = 5.8 bar and oxygen content of Q = 9.5 mg/l ± 0.3 mg/l: a) $P_2$ = 1.4 bar, b) $P_2$ = 0.7 bar, c) $P_2$ = 0.2 bar

A comparison of figures 5a and b show that the air bubble size increases with rising cavitation intensity. Due to buoyancy forces, bigger bubbles rise faster to the top of the pipe. Therefore the mean bubble size and density in the bulk tends to decrease downstream of the cavitation region, while larger air volumes accumulate at the upper wall as shown in Fig. 5c. The area ratio $\alpha$ of released air within the measurement volume is plotted for $P_{in} = 7.8$ bar in Fig. 4. Optical measurements show an increase of bubble filled areas with increasing pressure difference. The air release in the observed area begins later than the choked flow conditions are reached. The standard deviation of the area ratio $\alpha$ seems to grow with increasing pressure drop due to inhomogeneous bubble coalescence. Operation points at which such distinct air accumulation was observed are labelled with filled markers as systematic errors of $\alpha$ are expected due to a limited optical access within the chosen FOV. A first estimate of the buoyancy related error of $\alpha$ was performed. For this, we assume that bubbles move with the centreline velocity $u_c$ of a fully developed turbulent flow utilizing a drag coefficient of $c_D = 2F_d/\pi\rho u^2R^2$ [11]. Based on a force balance of buoyancy ($F_b$) and drag force ($F_d$) a critical bubble radius was determined to:

$$R = \sqrt{\frac{3 \cdot \mu \cdot u_c}{\Delta\rho \cdot g}} \quad (4)$$

Here, $\Delta\rho$ denotes the density difference, $g$ the gravity constant, $\mu$ the water viscosity and $u_c$ the centreline velocity. The Reynolds number is calculated as $Re = (\rho \cdot d \cdot u)/\mu$. Based on the centreline velocity of a fully developed turbulent flow ($u_c \approx 1.2 \cdot u_{bulk} = 0.3$ m/s) bubbles of $R \geq 0.3$ mm are expected to rise to the top of the pipe before reaching the downstream measurement section. As in the depicted FOV bubbles with a radius greater than the critical radius are detected in the bulk flow we assume that a delayed formation of these bubbles or a delayed coalescence further downstream the orifice takes place.

The image analysis is applied for operation points in the choked flow regime. As no released air bubbles could be observed close to the critical mass flow rate (see Fig. 4), we anticipate that this is either due to a limited spatial resolution or a diffusion process taking place such that small bubbles get into solution before entering the depicted FOV. Based on the spatial resolution limit a bubble has to exceed a minimum radius of $R_{min} = 0.0132$ mm, corresponding to a bubble image area of 4 pixel, to enter the statistical analysis. Based on the bulk velocity and time scales a conservative estimate can be made to determine the minimal radius of bubbles created behind the orifice that decrease their size during convection due to diffusion driven solution. For this, we assume that one bubble with radius $R_0$ at the orifice exit is filled with air only. This bubble is surrounded by degassed water (gas concentration $c_0 = 0$ mg/l). The time for the bubble to arrive in the measurement section amounts to approximately $t = 1$ s. The minimal bubble radius has to be $R_{min}$, for the given spatial resolution, to be analysed in the postprocessing. Using equation 5 the radius $R_0$ of the bubble can be calculated [12].

$$R_{min} = R_0^2 - \frac{2 \cdot \kappa}{\rho} \cdot (c_s - c_0) \cdot t \quad (5)$$

The diffusion coefficient is assumed to be $\kappa = 2.1 \cdot 10^9$ m²/s, the density $\rho = 1.189$ kg/m³ and the saturation concentration of oxygen in water $c_s = 9$ mg/l. Thus, a bubble with an initial radius of $R_0 \leq 15.8$ µm would dissolve to an extent that it is neglected in the evaluation procedure.

Fig. 6 displays the influence of the cavitation number on the volume faction of released air. Results are shown for an inlet pressure $P_{in} = 7.8$ bar whereas results for the buoyancy dominated regime are labelled with filled markers. The amount of released gas grows with decreasing cavitation number σ as the latter is associated with the increase in cavitation intensity. The present results show a similar trend to studies by Iben et al. [13], who measured an exponential increase in released gas inside a high pressure microfluidic device. However, for the present set-up studies at different downstream positions and under stereoscopic viewing angles are required in the future to gain statistically more reliable results for cavitation numbers below $\sigma = 1.1$ allowing to elaborate on the scaling behaviour in more detail.

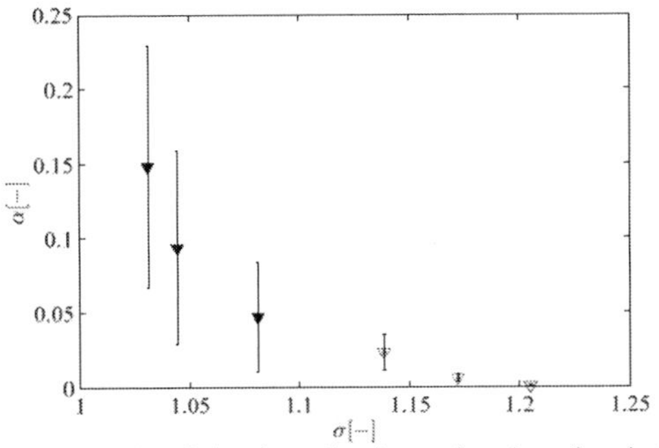


**Fig. 6:** Results of air release fraction as function of cavitation number for an inlet pressure of $P_{in} = 7.8$ bar

Released air fraction values were calculated from bubble frequency and size distributions. Corresponding bubble frequency histograms are plotted in Fig. 7 as function of bubble radius $R_b$ for different outlet pressures.

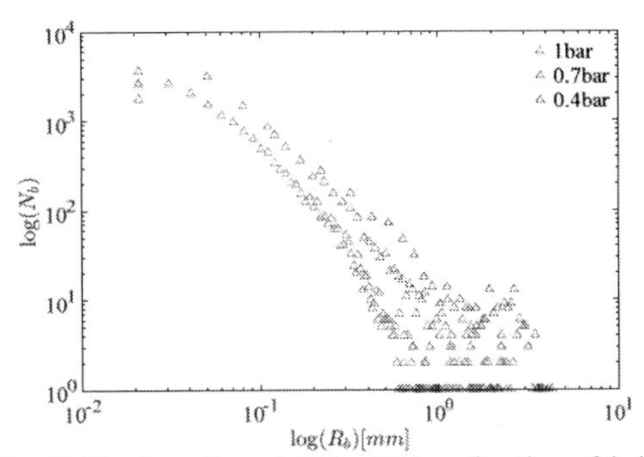


**Fig. 7:** Number of gas bubbles ($N_b$) as function of bubble radius ($R_b$) for 200 recordings, $P_{in} = 7.8$ bar and outlet pressures $P_2 = 1$, 0.7 and 0.4 bar

The bubble radius ranges from 0.02 to 4.23 mm. The frequency spectrum in the range between 0.03 mm < $R_b$ < 0.8 mm is very similar for all outlet pressures. Differences can be seen for the smallest radius

$R_b = 0.02$ mm, where the frequency of bubbles for an outlet pressure $P_{in} = 1$ bar (which corresponds to $\sigma = 1.14$) is 27% higher than for lower outlet pressures ($\sigma = 1.04$, 1.03). Main differences can be observed for larger bubble radii which have a strong influence on the results of the area ratio $\alpha$. For decreasing outlet pressures, and consequently also decreasing cavitation number, the frequency of bubbles with radii $R_b > 0.8$ mm increases. A general trend from smaller to higher bubble sizes with increasing cavitation intensity can be observed.

## 4 Conclusions

In the present study, a correlation between hydrodynamic behaviour and released air fraction of a cavitating orifice flow in the choking regime is investigated. For incipient choked flow conditions no air release could be observed up to approximately 1.2 times the critical pressure drop. A further decrease of the outlet pressure is associated with increasing air bubble volume fraction and a change in bubble size frequency. In alignment with results from Iben et al. [13] an increase in released air bubble fraction was observed for decreasing cavitation numbers.

In future, an extended, thorough experimental study will allow to gain statistical results of gas volume fractions also for lower cavitation numbers. For this, results have to be obtained in different measurement plains and viewing angles to cover the full cross sectional area of the flow. Detailed studies will provide further information on the dissolving process and representative time scales of the coalescence dynamics. In an attempt to find universal scaling relations of cavitation driven air release behind different generic flow constrictions studies will be repeated for different flow constriction geometries such as venturi tubes.

## Acknowledgment

This research is supported by the "Competence Center of Hydraulic Fluid Machinery".

## References

1. F. R. Young, *Cavitation*, World Scienific, (1989)
2. N. I. Kolev, *Multiphase Flow Dynamics 4*, Springer (2011)
3. H. A. Freudigmann, U. Iben, P. F. Pelz, Journal of Physics **656,** 012113 (2015)
4. P. F. Dunn, F. O. Thomas, M. P. Davis, I. E. Dorofeeva, Physics of Fluids **22**, 117102 (2010)
5. C. Mishra, Y. Peles, Journal of Microelectro-chemical Systems **14 (5)**, 987 - 999 (2005)
6. K. Ramamurthi, K. Nandakumar, Flow Measure-ment and Instrumentation **10**, 133–143 (1999)
7. C. E. Brennen, *Multiphase Fundamentals of Multiphase Flows*, Cambridge University Press, (2005)
8. I. H. Shames, *Mechanics of fluids*, 3 ed., (1992)
9. W. H. Nurick, Journal of Fluids Engineering **98 (4)**, 681-687 (1976)
10. *Guide to the expression of uncertainty in measurement*, JCGM 100, (2008)
11. R. Clift, *Bubbles, Drops, and Particles,* Academic Press, (1978)
12. L. Liebermann, Journal of Applied Physics **28**, 205 (1957)
13. U. Iben, F. Wolf, H.-A. Freudigmann, J. Fröhlich, W. Heller, Exp. Fluids **56**, 114 (2015)

# Transition of cavitating flow to supercavitation within Venturi nozzle – hysteresis investigation

*Kozák* Jiří[1,*], *Rudolf* Pavel[1], *Huzlík* Rostislav[2], *Hudec* Martin[1], *Chovanec* Radomír[1], *Urban* Ondřej[1], *Maršálek* Blahoslav[3], *Maršálková* Eliška[3], *Pochylý* František[1] and *Štefan* David[4]

[1]Brno University of Technology, FME, Victor Kaplan Dept. of Fluid Engineering, Brno, Czech Republic
[2]Brno University of Technology, FEEC, Dept. of Power Electrical and Electronic Engineering, Brno, Czech Republic
[3]The Czech Academy of Sciences, Centre for Cyanobacteria and Their Toxins, Brno, Czech Republic
[4]Laval University, Department of Mechanical Engineering, Québec, Canada

**Abstract.** Cavitation is usually considered as undesirable phenomena. On the other hand, it can be utilized in many applications. One of the technical applications is using cavitation in water treatment, where hydrodynamic cavitation seems to be effective way how to reduce cyanobacteria within large bulks of water. The main scope of this paper is investigation of the cavitation within Venturi nozzle during the transition from fully developed cavitation to supercavitation regime and vice versa. Dynamics of cavitation was investigated using experimental data of pressure pulsations and analysis of high speed videos, where FFT of the pixel intensity and Proper Orthogonal Decomposition (POD) of the records were done to identify dominant frequencies connected with the presence of cavitation. The methodology of the high speed (HS) records semi-automated analysis using the FFT was described. Obtained results were correlated and above that the possible presence of hysteresis was discussed.

## 1 Introduction

Cavitation plays a crucial role in design process as well as operation of many hydraulic machines such as turbines, pumps and valves. Despite the fact that occurrence of this phenomenon is usually considered as an undesirable condition, it is possible to utilize its concomitants in many positive ways.

Among the most significant side effects of cavitation belongs occurrence of pressure pulsations and erosion of the surfaces exposed to the cavitation bubbles implosions. These conditions lead to the reduction of the hydraulic machines efficiency, lifespan and operation range (Dörfler [1]).

On the other hand, the same effects of cavitation can be utilized for example in medical industry, chemical industry, cleaning and water treatment.

The main scope of this contribution is study of the transition between cavitation and supercavitation regime of flow within the axisymmetric Venturi tube, where the pressure drop in the constriction leads to inception of cavitation. This device can be exploited for the efficient reduction of the cyanobacteria bloom in large bulks of water (Jančura et al [2]). This contribution complements the previously presented papers, which were focused on the wider range of cavitation regimes. [3],[4]

Comprehensive study of cavitation and its detection within Venturi nozzle can be found in [5].

During the experimental research, the severe pressure pulsation as well as increase of the overall noise were observed prior the transition to the supercavitation regime of the flow. In addition, the correct evaluation of the experimental transition between fully developed cavitation and supercavitation has been considerably challenging compared to the other regimes with higher cavitation numbers.

These reasons led to decision to devote particular attention to the research of the transition using several methods of the experimental data evaluation. Especially methods of the high-speed video records postprocessing using the POD of the chosen region of interest and FFT of the pixel intensity fluctuations.

During the previous research utilizing the FFT of the pixel intensity, it became evident that the proper choice of the source pixel is the trickier, the lower is the cavitation number. Thus the evaluation of these regimes was usually excluded to prevent potentially misleading results. For these purpose, the automated choice of the source pixel based on root mean square of the pixel intensity within the chosen region was utilized. Results of all of the previously mentioned methods were compared and correlated. The software for the HS records analysis has been coded using the Python and OpenCV library.

Above the aforementioned verification of these methodologies of the cavitation dynamics evaluation, the hypothesis of the potential hysteresis during the transition caused by the residual cavitation nuclei has been done. Hysteresis during the investigation was examined from the hydraulic losses point of view as well as frequency of

* Corresponding author: jiri.kozak1@gmail.com

instabilities connected with the presence and motion of the cavitating structures.

## 2 Experimental set-up and Venturi nozzle geometry description

The experimental measurement has been carried out using the cavitation rig of the Victor Kaplan's Dept. of fluid engineering. The rig was equipped by the transducers recording the pressure upstream and downstream the Venturi nozzle, discharge within the system and temperature of the water in the reservoir.

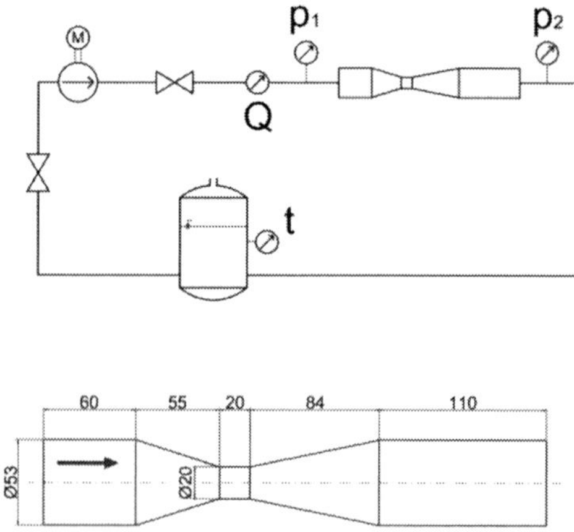


**Fig. 1**. Scheme of the cavitation rig and Venturi nozzle

Reservoir has been opened to the atmosphere, thus the cavitation number has been altered only by the change of the discharge. The experimental measurement has been carried out repeatedly. It should be noted that the presence of the cavitation nuclei has not been measured during the experiments. Water within the cavitation rig was deaerated by the operation in the supercavitation regime for 30 minutes prior the measurement. The list of the utilized transducers can be found in the Table 1

**Table 1.** List of transducers (Note: Acc – accuracy in % of the measured value* or % of the transducer measuring range**).

| Variable | Manufacturer | Type | Acc. |
|---|---|---|---|
| $p_1$ | BD SENSORS | DMP 331 | ± 0,25%* |
| $p_2$ | BD SENSORS | DMP 331 | ± 0,25%* |
| $Q$ | ELA Brno | MQI99-SN | ±0,5%* |
| $t$ | HIT Uherské Hradiště | HSO-502 1A2L | ±0,1%** |

HS Photron FastcamSA-X2 1000K has been utilized during the measurement A. Grayscale records with the sampling frequency of 20 000 fps and resolution of 1024x512 pixels captures 0.5 s of the cavitating flow have been captured.

Unfortunately, during this measurement the critical failure of the thermometer occurred, thus the saturated vapor pressure as well as water density was considered constant using the water temperature of 20°C.

Increase of the temperature has been captured during the measurement B, but without the connection with HS video records. This measurement has been carried out twice to partially prove reproducibility of the results ($B_1$ and $B_2$).

## 3 Computation of the water properties

The water properties necessary for the proper evaluation were computed using the recorded values of the temperature, discharge and pressures.

Density of the water $\rho$ was computed as a function of the water temperature using the following formula:

$$\rho = 1002 - 0.2716t + 0.01047t^2 - 0.00027t^3 \quad (1)$$

Value of saturated vapour pressure was computed using the exponential function of the thermodynamic temperature. [6]

$$p_{vap} = \frac{e^{77.3450 + 0.0057\,T - \frac{7235}{T}}}{T^{8.2}} \quad (2)$$

Using computed values of the water density, saturated vapor pressure and measured values of the pressure downstream the nozzle, the value of cavitation number can be expressed using the eq.3.

$$\sigma = \frac{p_2 - p_{vap}}{0.5\,\rho v_{th}^2} \quad (3)$$

Where $v_{th}$ represents the velocity of the flow within the constriction of the Venturi tube.

Hydraulic loss coefficient of the Venturi tube was computed and evaluated using the equation (4) in the whole range of the investigated cavitation regimes

$$\xi = \frac{p_1 - p_2}{0.5\,\rho v_{th}^2} \quad (4)$$

## 4 Methodology of the high speed records post processing

Various operating points have been investigated during the measurement A in order to capture transition between cavitation and supercavitation and vice versa. Two different methods of the HS records analysis have been introduced and their result have been correlated with the spectral analysis of the pressure fluctuations downstream the nozzle. The basic description of these methods is the content of the following sections.

It should be mentioned that although the sampling frequency of the camera and pressure transducers were different, the start of the data recording was synchronized using a voltage signal ramp.

### 4.1 FFT analysis of the pixel intensity fluctuations

Description of the theoretical and mathematical background of this method of the signal spectral analysis exceeds scope of this contribution, nevertheless it can be found in the literature. [7]
The source of the data for the spectral analysis is represented by the fluctuations of the pixel intensity which vary from 0 to 255 in case of grayscale HS video. The presence of interfacial interphase leads to a reflection of the light and therefore increases brightness of the region affected by the cavitation. Comparison of the HS video analysis with pressure fluctuations can be found in [8], where cavitation over NACA profile was investigated or in [4] where analysis of C-D nozzle was presented.
While in case of the initial phases of cavitation (e.g. higher values of the $\sigma$) the proper choice of the source pixel does not pose the problem, it became problematic with the fully developed stage of cavitation and its transition to the supercavitation. These regimes are distinguished by the significant amount of cavitation within the flow, which complicates proper illumination of the region of interest. It is usually not possible to supress occurrence of the over- and underexposed regions during recording of the cavitation. The improper choice of the pixel intensity can be source of the misleading results.
Manual choice of the proper source pixel can be time consuming and above that it can be misleading due to the introduction of human factor.
Thus the semi-automated method of the source pixel choice has been utilized. Larger region of interest was configured manually to prevent regions with significant presence of over exposed pixels and capturing characteristic fluctuations of the cavitation at the same time. The source pixel has been automatically selected from this region based on the largest intensity fluctuations root mean square value $RMS_{int}$ within the region. (eq. 5)

$$RMS_{int} = \sqrt[2]{\frac{1}{N}\sum_{i=1}^{N}\left(I_i - I_{avg}\right)^2} \quad (5)$$

Where: $N$ represents the number of samples of the record, $I_i$ is the greyscale intensity of the pixel in the sample and $I_{avg}$ represents the average intensity of the pixel computed using the whole length of record.

### 4.2 Proper Orthogonal Decomposition of the High speed records

POD represents an effective way how to decompose a complex physical phenomenon into the several POD modes while it can be assumed that the most of the information related to the dominant instability is contained in the first several POD modes. In other words, the most significant flow structures within the investigated flow can be relatively easily identified.
Although the fact that this method can be utilized for the analysis of the vector quantities and provide deeper insight in the physical nature of the phenomenon, the utilization of POD is limited to the processing of the scalar pixel intensity of the HS records in this contribution. Due to this fact, the brief description of this method will be limited to the analysis of scalar quantities.
Exhaustive description of this method can be found in [9].
Using the variant of POD called method of snapshot, the time dependent eigenfunctions $\boldsymbol{a}_i$ and spatio-dependent functions $\boldsymbol{\phi}_i$ are computed in order to find an approximation of a general spatio-temporal dependent variable $\boldsymbol{u}(\boldsymbol{x};\, \boldsymbol{t})$ (in this case pixel intensity $I$) within the domain $\Omega$. For this purpose, the correlation matrix $\boldsymbol{C}_{ij}$ is constructed using the samples of the record with known sampling frequency according to the next equation.

$$\boldsymbol{C}_{ij} = \frac{1}{vol\Omega}\int_{\Omega} \mathrm{u}(\mathbf{x};\, \mathrm{t}_i)\, \mathrm{u}(\mathbf{x};\, \mathrm{t}_j) \mathrm{d}\mathbf{x} \quad (6)$$

Solving the eigenvalue problem of the correlation matrix $\boldsymbol{C}_{ij}$, the eigenvalues and eigenvectors are obtained. Considering that the eigenvalue $\lambda$ is paired with corresponding eigenvector $\boldsymbol{e}$ and the fact that value of $\lambda$ is directly connected with contribution of the corresponding POD mode, the eigenvalues and eigenvectors are sorted by the magnitude of the eigenvalues. Temporal coefficients $a_i$ and spatial eigenmodes $\boldsymbol{\phi}_i$ are then computed using the following expressions:

$$\boldsymbol{a}_i = \boldsymbol{e}_i\sqrt{N\lambda_i} \quad (7)$$

$$\boldsymbol{\phi}_i = \boldsymbol{U}\frac{\boldsymbol{e}_i}{\sqrt{N\lambda_i}} \quad (8)$$

Where $N$ is the number of the samples and $\boldsymbol{U}$ is matrix consisting of all recorded samples transformed to the form of column vectors (matrix of size $M$ by $N$ size where $M$ is the number of pixels within one sample and $N$ is the number of the recorded samples).
Dominant frequency of the selected mode $i$ can be then evaluated by the spectral analysis of the correspondent temporal coefficient $a_i$.
Several applications of POD for the cavitating flow dynamics investigation can be found in the literature. Comprehensive study of the cavitating spiral vortex within the elbow draft tube was done by Štefan et al [10]. The POD analysis of the HS video capturing cavitation cloud over NACA profile was done by Rudolf et al [11]. Influence of the aeration on the cavitation patterns downstream within the Venturi tube was the scope of investigation which has been done by Tomov et al [12]. The operation regimes in this study corresponded to the partially developed stage of cavitation.
POD study of the cavitation regime detection in case of grooved C-D nozzle was done by Danlos et al [13].

## 5 Experimental measurement

As it was mentioned above, the presented contribution is focused on the cavitation regimes corresponding the transition from the fully developed cavitation to the supercavitation. Results of the two experimental sets focused on transition between fully developed cavitation and vice versa are described in the following chapters.

## 5.1 Experimental measurement A

During the first set of experimental measurements the HS camera has been utilized. Unfortunately, the temperature and its increase during the first part of experiments has not been captured. On the other hand, based on the measurement B, the growth of the water temperature should not exceed 1°C during the whole cycle of measurements (e.g from cavitation to supercavitation and back to the initial operating point). Thus the constant temperature 20°C has been considered. This approach neglects the change of water density as well as the change of the saturated vapor pressure and viscosity. Above that the potential nonlinearity of the temperature growth during the measured cycle has not been captured.
On the other hand, thanks to the known maxim magnitude of the temperature growth and the fact that the main scope of this part of measurement has been to correlate several methods of the cavitation dynamics investigation, the water temperature differences can be neglected.

### *5.1.1 Cavitation patterns during the transition*

Cavitation numbers $\sigma$ varied from the value of 0.27 corresponding to the supercavitation to 0.34 of the fully developed cavitation prior to the transition.
It should be noted that the behaviour of the cavitating flow was almost independent on the direction of the flowrate change. The only noticed difference was an amount of the small bubbles, which was slightly higher in case of increasing discharge (in the cavitation → supercavitation direction). It should be emphasized that this observation is only subjective.
The cycle of fully developed cavitation stage within Venturi tube starts with the inception near the throat edges (Figure 2).

**Fig. 2.** Start of the new cavitation cycle. (Q = 7.76 l/s, σ = 0.33)

The amount of observable cavitation grows in the diffuser downstream the throat and its partial separation is well observable at the certain moment (Figure 3).

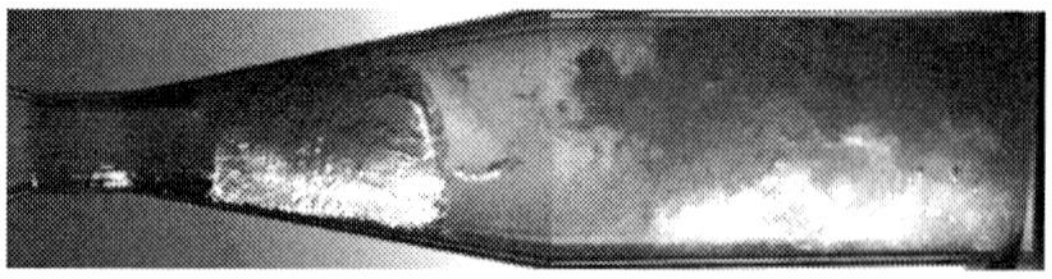

**Fig. 3.** Growth of the cavitation cloud within the diffuser (Q = 7.76 l/s, σ = 0.33)

The cavitation cloud is entrained by the flow downstream the nozzle, while the frontal part of cavitation collapses towards the throat (Figure 4).

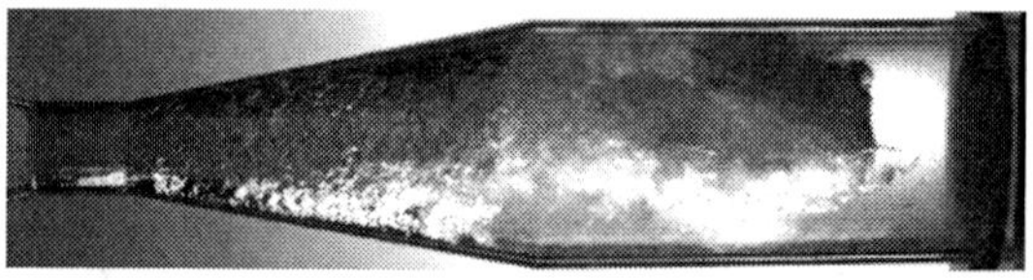

**Fig. 4.** Partial separation of the cavitation cloud (Q = 7.76 l/s, σ = 0.33)

Collapse of the cavitating cloud is accompanied by the creation of the significant amount of instabilities near the axis and cavitating vortices in the nozzle (Figure 5).

**Fig. 5.** Visible instabilities near the axis of the Venturi tube (Q = 7.76 l/s, σ = 0.33)

The cavitation near the throat disappears completely and new cycle starts prior the decay of the cavitating cloud downstream the nozzle (Figure 6).

**Fig. 6.** Complete decay of the cavitation within the constriction of the tube right prior the beginning of new cycle (Q = 7.76 l/s, σ = 0.33)

With the increasing discharge the instabilities near the axis became more coherent and occasionally start to form stabilized jet in the axis of the nozzle which is depicted in the Figure 7. Jet became the more dominant the lower is cavitation number. Above that cavitation cloud became more massive and its partial separation became less visible.

**Fig. 7.** Visible jet during the transition between cavitation regimes (Q = 8.34 l/s, σ = 0.29)

With the transition to the pure supercavitation regime (Figure 8), the noise as well as the pressure pulsations induced by the fluctuations of cavitation became significantly lower.
The jet near the axis of the nozzle is stable most of the time and collapses with very low frequency.

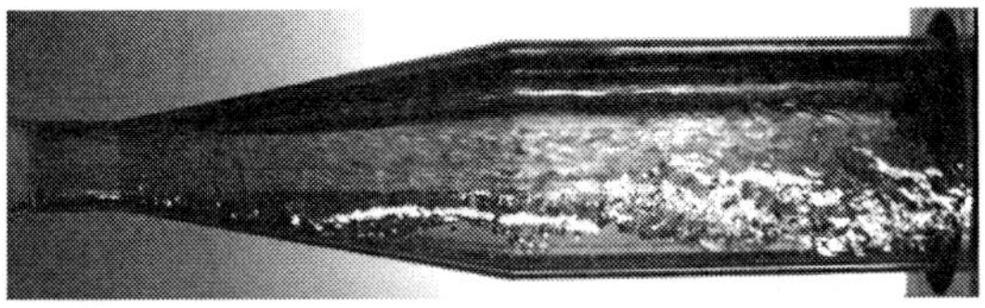

**Fig. 8.** Coherent jet visible during the supercavitation (Q = 8.56 l/s, σ = 0.27)

### 5.1.2 Spectral analysis of the pixel intensity fluctuations

The matrix of size 20 by 20 pixels has been chosen as the region, where the source pixel was selected based on the value of root mean square of the intensity fluctuations. Position of the source matrix in the diffuser downstream the constriction is in the Figure 9.

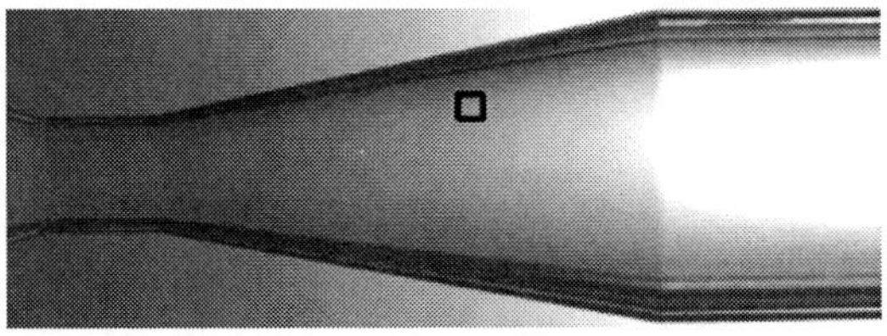

**Fig. 9.** Position of source matrix within the diffuser

It is clearly depicted that the RMS image of the pixel matrix is the more homogenous the higher cavitation number is (Figure 10).

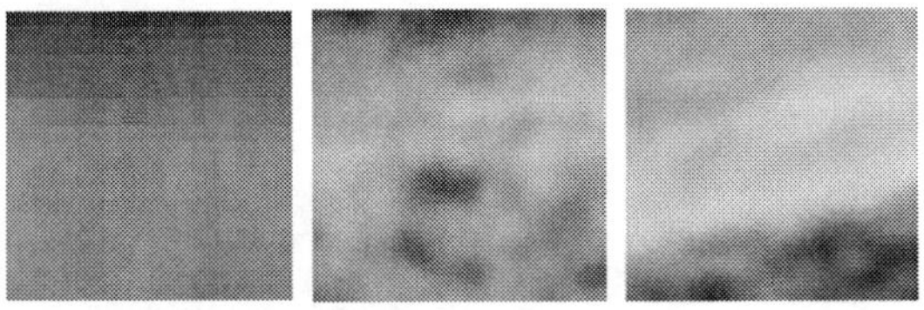

**Fig. 10.** RMS image of the source matrix in case of increasing Q. Fully developed cavitation: Q = 7.67 l/s, σ = 0.34; Transition Q = 8.34 l/s, σ = 0.29 and supercavitation Q = 8.55 l/s, σ = 0.27 (Note: Red – high intensity RMS, Blue – low intensity RMS)

While the value of intensity RMS within the matrix is nearly constant in case of fully developed cavitation, the proper choice of the pixel became more important in case of supercavitation, where the distribution of the intensity RMS value became rather uneven.
Results of pixel intensity fluctuations FFT analysis are depicted in the Figure 11. It should be emphasized that the results were not adjusted or filtered. Depicted dominant frequencies were obtained automatically based on the maximum value of the pulsations amplitudes without any further adjustment.
It is clearly depicted that the dominant frequencies are decreasing with decreasing cavitation numbers in whole spectrum of the investigated operating points. The only exception can be found in case of increasing discharge (in the chart marked by the red cross). Values of the pixel intensity fluctuations frequency are decreasing from approximately 50 Hz, which corresponds to the pulsations of the cavitation in case of fully developed cavitation to 2 Hz in case of supercavitation.
There is no significant difference between the results acquired for decreasing and increasing discharge. Therefore, a possible occurrence of the cavitation hysteresis cannot be confirmed using FFT analysis of the pixel intensity fluctuations. Results of the spectral analysis will be correlated with the results of other methods in one of the following chapters.

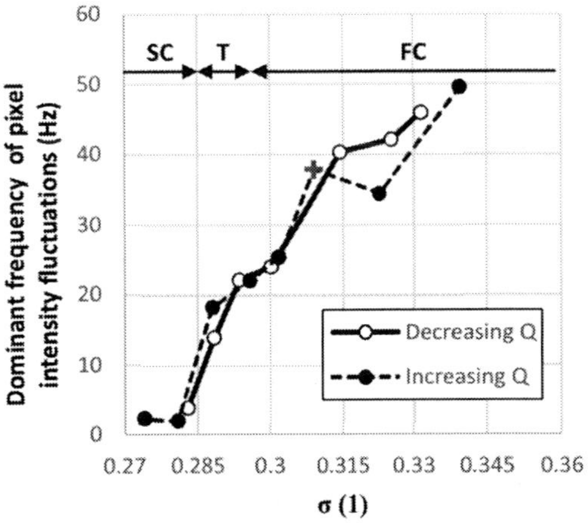


**Fig. 11.** Dominant frequencies of the pixel intensity fluctuations (S – supercavitation, T – transition, FC – fully dev. cavitation.)

### 5.1.3 POD analysis of the HS records

Region of interest (ROI) has been chosen for the purpose of the POD analysis as is depicted in the Figure 12. Position and size of the ROI was set-up so that typical patterns of cavitation were captured in whole spectrum of investigated cavitation numbers, while presence of the overexposed regions was minimized.

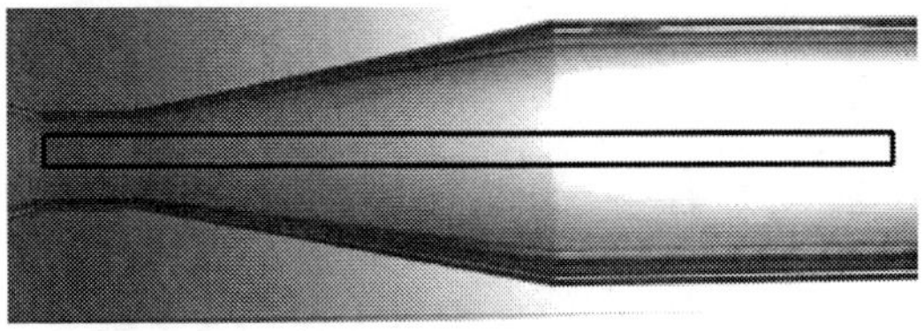

**Fig. 12.** Region of interest for the POD analysis

It should be noted that unlike the FFT of pixel intensity fluctuations, where all of the available samples were considered for the purpose of analysis, in case of POD only each tenth sample has been utilized in the decomposition. Skipping of the samples was motivated mainly by the reducing of the computational demands, considering fact, that lowered sampling frequency will be still more than sufficient for the purpose of the presented analysis. The POD analysis has been carried out using the values of pixel intensity fluctuations, e.g. the average image was obtained in the first step of analysis and then distracted from the record sample by sample.

 DOI: 10.1051/epjconf/201714302055

Results of the POD were analysed then. The comparison of the relative as well as cumulative contribution of the POD modes is depicted in the following charts.
It is clearly depicted that the overall tendencies are similar regardless the direction of discharge change. The lower is cavitation number, the lower significance of the most dominant POD modes is. In other words, with decreasing cavitation number more information about the captured flow is stored in less significant modes.
For the purpose of the results comparison the threshold value of 80% of cumulative significance has been chosen. While the threshold value is reached summing less than 50 POD modes in case of fully developed cavitation, more than 200 modes are needed for the same result in case of transition between cavitation and supercavitation. (Fig. 13 and 14)

Obtained results suggest, that the supercavitation regime of the flow has occurred earlier in case of the increasing discharge. Curve of the POD modes cumulative significance is considerably closer to the curve depicting supercavitation in case of the transition regime (Fig. 14) compared to the results of decreasing discharge (Fig. 13). Direct comparison of the results is shown in the Figure 15.
On the other hand, the correct evaluation of the results during the transition can be challenging and this conclusion will be verified in the following text using another experimental method. It should be also verified by the analysis of longer records. Larger amount of POD modes needed to reach 80% of cumulative significance also agree with the observation of slightly higher amount of small bubbles in case of records capturing cavitation during the cycle of increasing discharge.

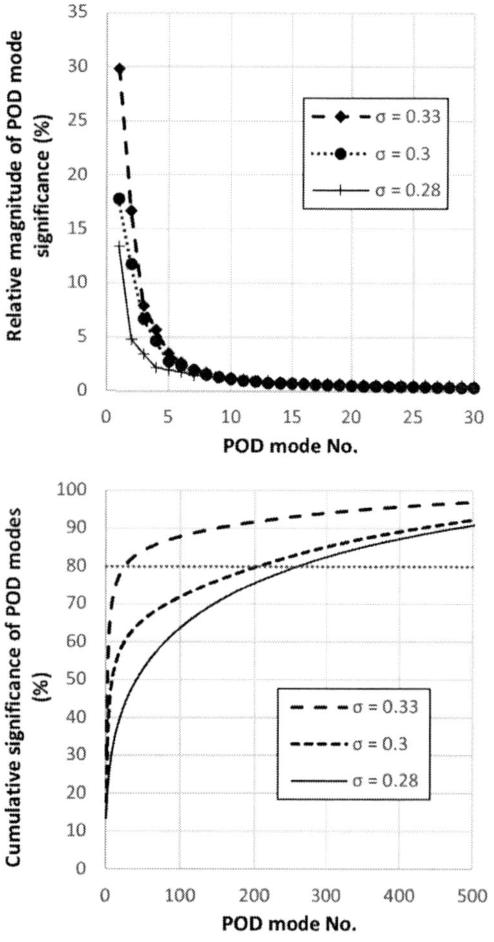


**Fig. 13.** Charts depicting relative and cumulative contributions of the POD modes significance in case of decreasing discharge.

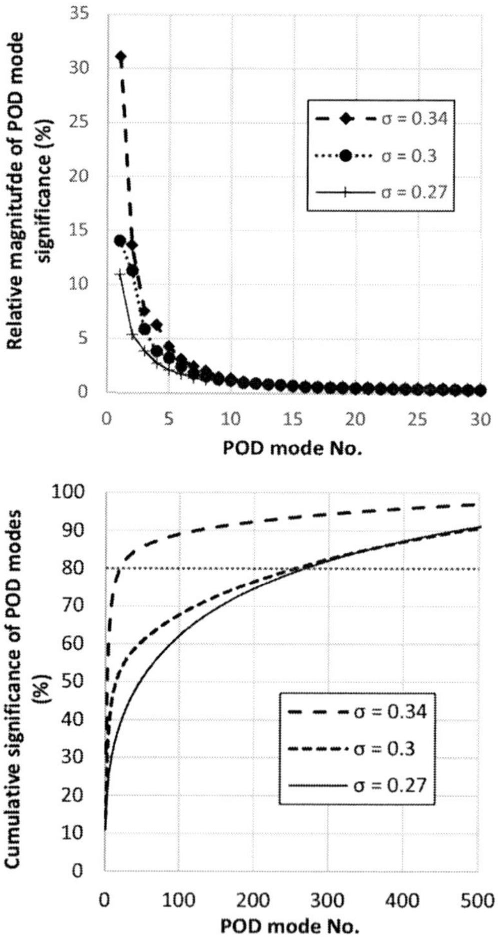


**Fig. 14.** Charts depicting relative and cumulative contributions of the POD modes significance in case of increasing discharge.

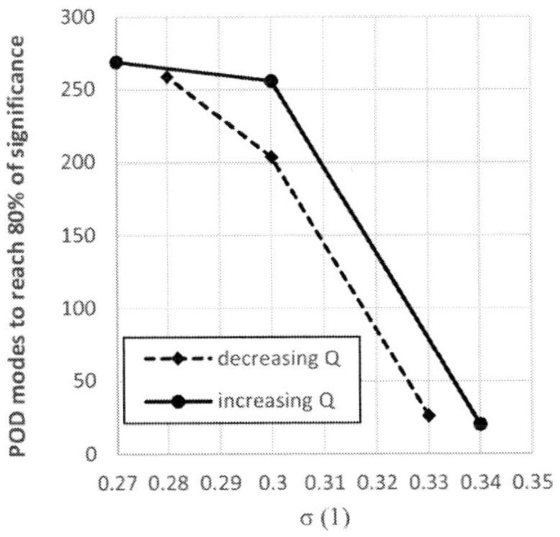


**Fig. 15.** Number of POD modes needed to reach 80% of cumulative significance.

Dynamics of the most significant POD modes has been investigated using FFT analysis of the corresponding time dependent eigenfunctions $a_i$. Ten most significant modes have been analyzed and the dominant frequencies were determined based on this analysis.

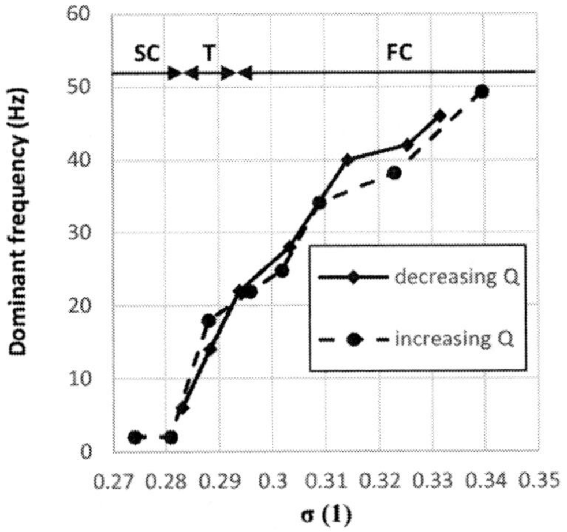


**Fig. 16.** Spectral analysis of the time-dependent eigenfunctions corresponding to the most significant POD modes.

Results of the spectral analysis depict decreasing frequency of dominant instabilities with the decreasing cavitation number similar to the results of pixel intensity fluctuations FFT analysis. Dominant frequencies corresponding to the increasing discharge are slightly lower compared to the decreasing Q in the region of fully developed cavitation. (Figure 16)

This result corresponds to the larger amount of bubbles which have been observed during the experiments. Small bubbles can act as an additional source of dampening, which decreases frequency of pulsations.

To complete the description of the POD, the most significant spatial eigenmode $\phi_0$ of selected operating point as well as the corresponding fluctuations (*a_fluct*) of time dependent eigenfunction $a_0$ and its spectral analysis are depicted in the Figure 17.

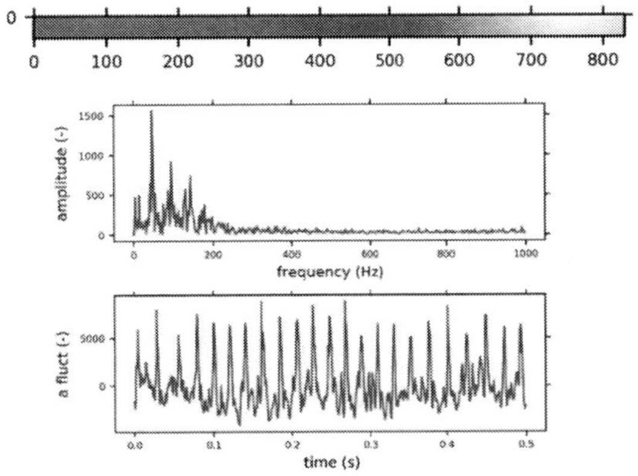


**Fig. 17.** The most dominant spatial eigenmode (scale in px) with corresponding time dependent eigenmode and its spectral analysis (Q = 7.76 l/s, σ = 0.33)

Compared to the most significant POD mode the spatial eigenmode corresponding to the 10th most significant POD mode is considerably less homogenous and frequency spectra of its time dependent eigenfunction is significantly noisier (Figure 18). This observation refers to the fact, that the less significant POD modes capture smaller instabilities within the recorded cavitating flow.

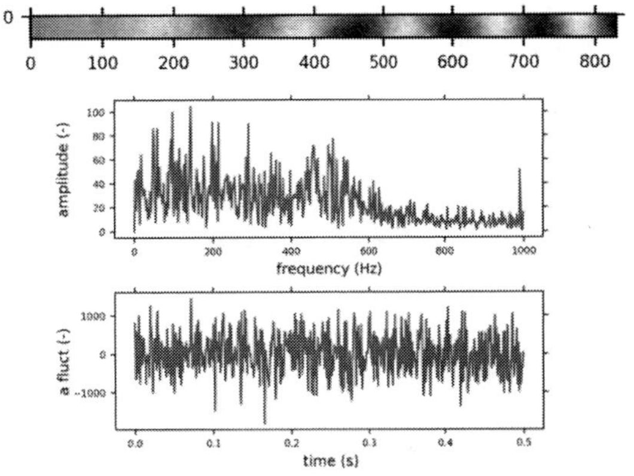


**Fig. 18.** 10th most significant spatial eigenmode (scale in px) with corresponding time dependent eigenmode and its spectral analysis (Q = 7.76 l/s, σ = 0.33)

### 5.1.4 Spectral analysis of pressure fluctuations and final correlation of results

Pressure fluctuations downstream the Venturi nozzle have been recorded and analysed to provide benchmark results. It should be emphasized, that the investigation of the cavitation dynamics using the pressure analysis can be more challenging compared to the analysis of the image data in case of lower cavitation number values. This became particularly evident in case of supercavitation when the pressure transducer is located in region of saturated vapour. The results of the pressure fluctuations analysis are depicted in the following Figure 19.

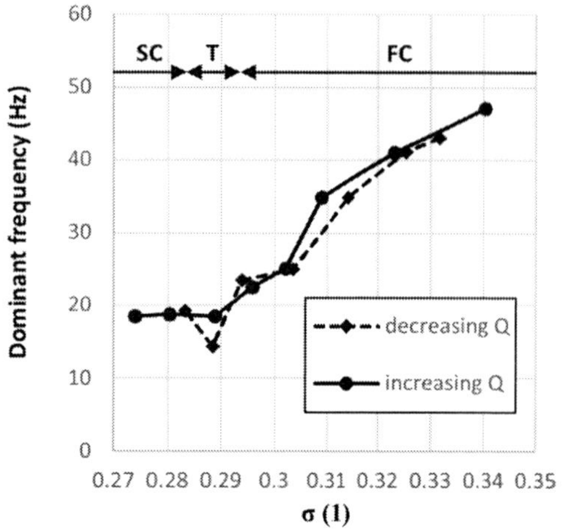


**Fig. 19.** Spectral analysis of pressure fluctuations downstream the nozzle

Decreasing values of dominant frequencies with the decreasing values of cavitation number have been captured again, similar to the analysis of the HS records. The values of dominant frequencies were slightly lower in case increasing discharge in the region of fully developed cavitation.

Correlation of the utilized methods is depicted in the following figures. Figure 20 is depicting results of increasing discharge. It is obvious that the dominant frequencies evaluated using the pressure fluctuations are considerably higher compared to the results of both methods of HS video analysis in case of supercavitation.

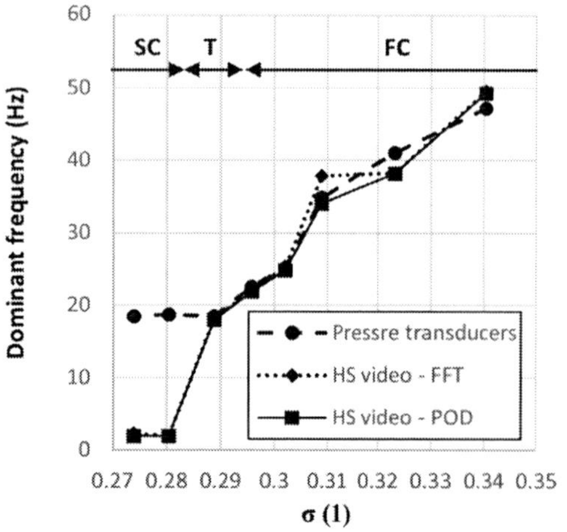


**Fig. 20.** Correlation of the methods of cavitation dynamics investigation – increasing discharge

Slightly different are the results of correlation in case of decreasing discharge (Figure 21). The more significant difference of image analysis methods compared each other can be found in the region of supercavitation.

Quite interesting is the fact that the results are in good agreement in case of transition between regimes.

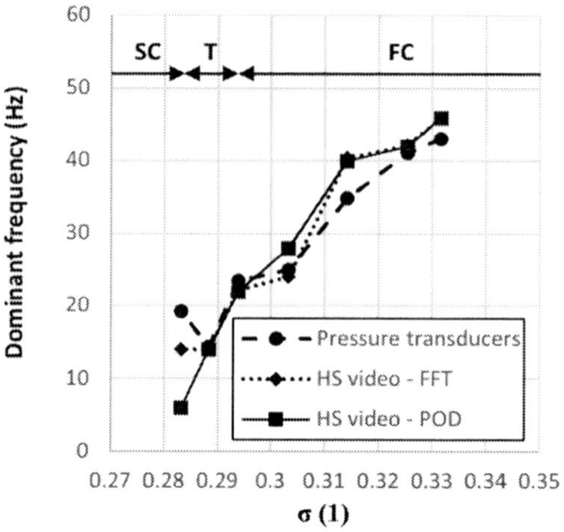


**Fig. 21.** Correlation of the methods of cavitation dynamics investigation – decreasing discharge

Based on the previous results it can be stated that the utilized methods of the cavitation dynamics analysis are fully comparable in case of higher cavitation numbers (fully developed cavitation and transition).

It should be noted that the frequencies captured by FFT of pressure fluctuations in case of supercavitation can be found as one of the significant frequencies in case of POD. Therefore, it is possible to assume that the most dominant source of pressure fluctuation is not identical to the most dominant source of image instability in case of supercavitation.

## 5.2 Experimental measurement B

Temperature of the water was not recorded during the measurement A. This led to the neglecting of the change of saturated vapor pressure as well as density of the water during the experiments as it was mentioned.

To capture the influence of the water properties change, the second measurement (B) has been carried out. Beside the measurement of temperature, the position of pressure transducers has been modified to provide undistorted data for the purpose of the hydraulic losses evaluation.

The measurement B has been done repeatedly to partially prove reproducibility of the results (B1 and B2).

The results of the pressure fluctuations spectral analyses of both of the measurement are depicted in the following figures (22 and 23).

Overall tendencies of the obtained dominant frequencies are similar to the measurement A. It should be noted that one of the operating point was excluded in case of measurement B1 considering increasing discharge (black cross in the Figure 22). Dominant frequency was significantly higher compared to the close neighborhood of the this operating point.

The transition between cavitation regimes started earlier in case of increasing Q. This behavior has been observed in both of the measurements B1 and B2 (Figure 22 and 23). On the other hand, the results suggest that the transition between regimes is more abrupt in case of the decreasing discharge and fully developed supercavitation can be observed prior to the increasing discharge.

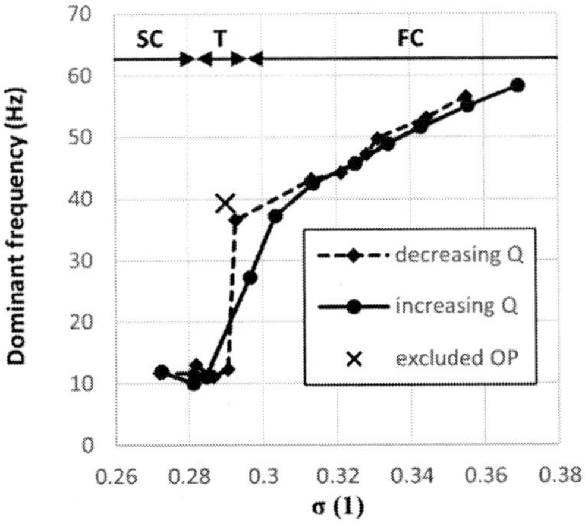


**Fig. 22.** Spectral analysis of the pressure pulsations recorded during the B1 measurement

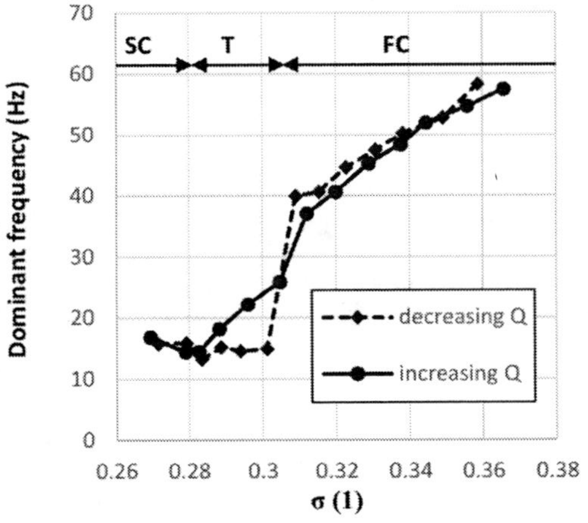


**Fig. 23.** Spectral analysis of the pressure pulsations recorded during the B2 measurement

As an additional information confirming the assumption that the neglecting of the temperature growth does not affect the obtained result significantly in case of measurement A, the change of the water temperature during the measurement B is depicted in the following figure. It can be seen that increase of the water temperature was approximately 1°C per experimental cycle (Figure 24).

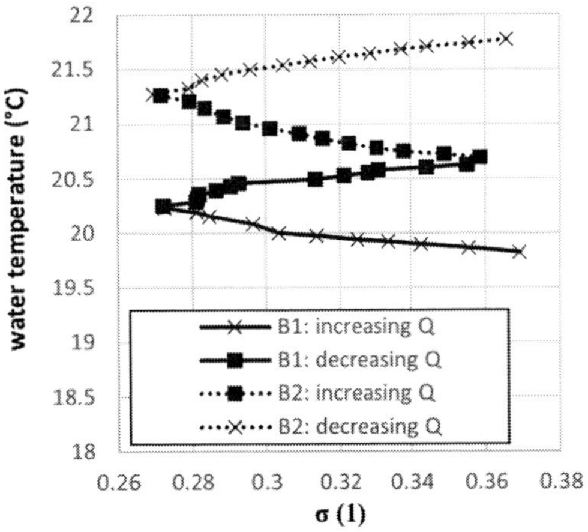


**Fig. 24.** Growth of the water temperature during the experimental cycles

Hydraulic losses have been evaluated in case of measurements B as it was mentioned above. The results are depicted in the following figures (25 and 26).
The results are fully comparable with each other from the direction of the discharge change as well as from the reproducibility point of view. It should be particularly emphasized that no significant difference between the hydraulic loss coefficient of corresponding cavitation number was observed depending on the direction of the discharge change.
Therefore, it can be stated that no significant signs of cavitation hysteresis have been observed during the evaluation of experiments. The only visible difference, which was observed and proved was the slightly earlier occurrence of the transition between cavitation regimes in case of increasing discharge. It has been possibly caused by larger amount of residual bubbles, attached to the walls of nozzle, which can play a role of cavitation nuclei.

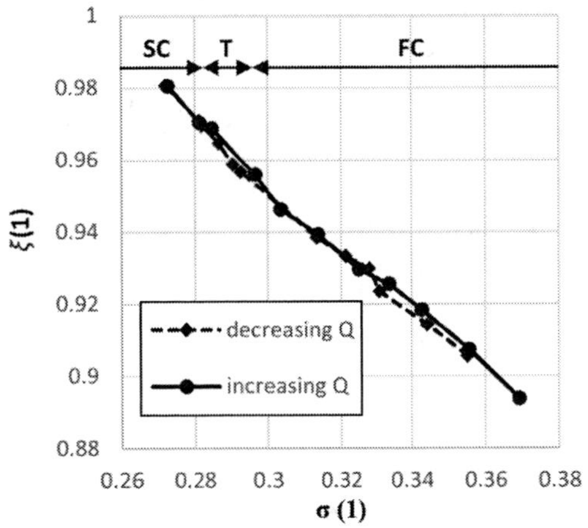


**Fig. 25.** Hydraulic losses evaluated using the results B1measurement

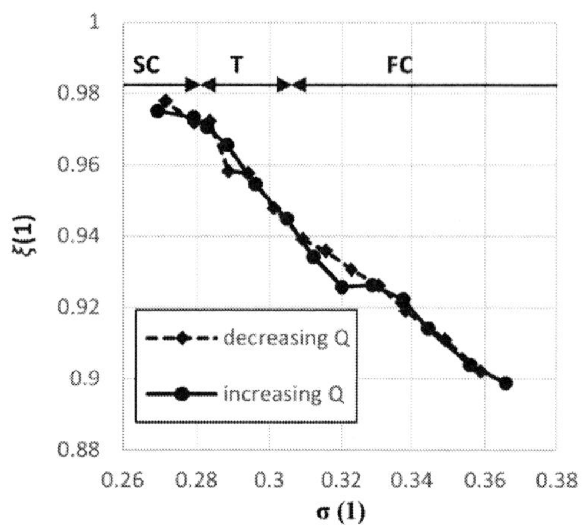


**Fig. 26.** Hydraulic losses evaluated using the results of B2 measurement

## 6 Future outlook

The presented methodology of the HS analysis will be applied for the investigation of the cavitation within the same Venturi tube, affected by the presence of the upstream swirl generator.

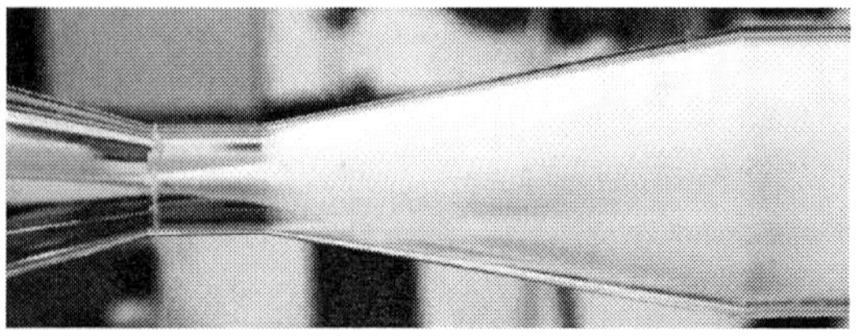

**Fig. 27.** Cavitation affected by the vortex within the Venturi tube (test measurement)

## 7 Conclusions

Dynamics of the cavitating flow within Venturi tube considering different gradient of the discharge change has been studied in the presented contribution. Above that, the hydraulic losses have been evaluated. Using these results, the possible occurrence of the hysteresis during the transition between fully developed cavitation and supercavitation has been discussed.

For the purpose of the investigation two experimental measurements have been carried out.

During the measurement A, the HS camera has been utilized to provide overall view of cavitation phenomenon as well as source data for the further image analysis.

Two different methods of the image analysis have been utilized: Spectral analysis of the pixel intensity fluctuations with semi-automated choice of the source pixel and proper orthogonal decomposition of the records. The choice of the source pixel for the purpose of FFT has been based on the maximum value of pixel intensity root mean square. This criterion provides quite robust as well as objective results.

The POD has been described in details. The provided results are in good agreement with the FFT analysis, but the additional information resulting from the decomposition are also useful. Distinguishing of the cavitation regimes using the analysis of the POD modes significance distribution was possible in this particular example of cavitating flow analysis.

Above that it was possible to state, that in case of the increasing discharge cycle the curve of the cumulative significance corresponding to the transition was significantly closer to the supercavitation curve compared to the results of decreasing discharge cycle analysis. This means that more significance (or information about the captured instabilities) is cumulated in the higher POD modes, e.g. the flow is more fluctuating.

It corresponds to the observation that the records of increasing discharge cycle contained larger amount of small bubbles. This results also suggests, that the transition to supercavitation occurred slightly earlier in case of increasing discharge cycle.

At the end of the measurement A, the results of the image analysis methods have been correlated together and with the results of the pressure fluctuations spectral analysis with a good degree of agreement.

To provide information about the hydraulic losses and to prove that the temperature growth during the measurement A can be neglected, the measurement B has been carried out. This measurement has been done twice to partially prove results reproducibility.

It is possible to state, that it is possible to reproduce the results of experiments, based on the analysis of the repeated measurements.

Approximate increase of temperature was 1°C per cycle of the measurement. Therefore, the temperature increase neglect should not significantly affect the results of the measurement A.

During the measurement B the hydraulic losses as well as spectral analysis of the pressure pulsation has been done. No significant difference has been observed from the hydraulic losses point of view. On the other hand, it has been clearly depicted that in case of the spectral analysis one can see that the transition regime began earlier in case of the increasing discharge while in case of the decreasing discharge cycle the cavitation regime change was more abrupt. This result is in good agreement with the conclusions of measurement A.

The final conclusion of the investigation is that no significant presence of the cavitation hysteresis has been observed or proved from the dominant frequencies related to the cavitation fluctuations point of view. Neither the analysis of the hydraulic losses has not proved any significant difference caused by the direction of the discharge change. The only captured and proved difference is the earlier occurrence of the transition between regimes in case of increasing discharge.

The presented methodology of the HS records analysis will be utilized in the further investigation of the cavitation within the Venturi tube.

## References


1. P. Dörfler, M. Sick M., A. Coutu, *Flow-induced pulsation and vibration in hydroelectric machinery: engineer's guidebook for planning, design and troubleshooting.* London: Springer, ISBN 978-1-4471-4251-5, (2013)
2. D. Jančura, P. Mikula, B. Maršálek, P. Rudolf, F. Pochýly, *Selective method for cyanobacterial bloom removal: hydraulic jet cavitation experience,* Aquaculture International, vol.: **22** Issue: **2** Pages: 509-521, DOI: 10.1007/s10499-013-9660-7, (2014)
3. P. Rudolf., M. Hudec, P. Zubík., D. Štefan, *Experimental measurement and numerical modeling of cavitating flow in converging-diverging nozzle.* EPJ Web of Conferences **25**, DOI: 10.1051/epjconf/20122501081, (2012)
4. J. Kozák, P. Rudolf, D. Štefan, M. Hudec, M. Gríger, *Analysis Of Pressure Pulsations Of Cavitating Flow In Converging-Diverging Nozzle.* 6th IAHR International Meeting of the Workgroup on Cavitation and Dynamic Problems in Hydraulic Machinery and Systems, (2015)
5. J. Jablonská, M. Kozubková, D. Himr, M Weisz, *Methods Of Experimental Investigation Of Cavitation In A Convergent - Divergent Nozzle Of Rectangular Cross Section.* Measurement Science Review, vol.: **16**, DOI: 10.1515/msr-2016-0024, (2016)
6. Engineering toolbox [online], cited 2016-07-20, <http://www.engineeringtoolbox.com/water-vapor-saturation-pressure-air -d_689.html>,( 2016)
7. Cooley, James W., and John W. Tukey. 1965. *An Algorithm For The Machine Calculation Of Complex Fourier Series.* In *Mathematics Of Computation*, vol. **19**, issue **90**, DOI:10.1090/S0025-5718-1965-0178586-1 (1965)
8. M. Sedlář, M. Komárek, P. Rudolf, J. Kozák, R. Huzlík, *Numerical and experimental research on unsteady cavitating flow around NACA 2412 hydrofoil.* IOP Conference Series: Materials Science and Engineering, vol. **72**, issue **2**, DOI: 10.1088/1757-899X/72/2/022014, (2015)
9. G. Berkooz, P. Holmes, J. Lumley, *The proper orthogonal decomposition in the analysis of turbulent flows,* Annual Rev. of Fluid Mechanics vol. **25**(1), Pages: 539-575, (1993)
10. D. Stefan,P. Rudolf, *Proper Orthogonal Decomposition of Pressure Fields in a Draft Tube Cone of the Francis (Tokke) Turbine Model,* Journal of Physics: Conference Series, vol. **579**, DOI : 10.1088/1742-6596/579/1/012002, (2015)
11. P. Rudolf, D. Štefan, M. Sedlář, J. Kozák, V. Habán, R. Huzlík, *Spatio-temporal description of the cavitating flow behavior around NACA 2412 hydrofoil,* Journal of Physics: Conference Series, vol. **656**, DOI: 10.1088/1742-6596/656/1/012168, (2015)
12. P. Tomov, A. Danlos, S. Khelladi, F. Ravelet, C. Sarraf, F. Bakir, *POD study of aerated cavitation in a venturi nozzle,* Journal of Physics: Conference Series, vol. **656**, DOI : 10.1088/1742-6596/656/1/012171, (2015)
13. A. Danlos, F. Ravelet, O. Coutier-Delgosha, F. Bakir, C. Sarraf, F. Bakir, *Cavitation regime detection through Proper Orthogonal Decomposition:Dynamics analysis of the sheet cavity on a grooved convergent–divergent nozzle,* International Journal of Heat and Fluid Flow, vol. **47**, DOI : 10.1016/j.ijheatfluidflow.2014.02.001, (2014)


## Acknowledgement

Czech Science Foundation is gratefully acknowledged for support of this research under project No 16-18316S „Principles and mechanisms causing microorganism elimination by hydrodynamic cavitation".

# Preliminary study of the primary nozzle position of a supersonic air ejector with a constant-area mixing chamber

*Jan* Kracik[1,*], and *Vaclav* Dvorak[1]

[1]Department of Power Engineering Equipment, Faculty of Mechanical Engineering, Technical University of Liberec, Studentska 2, 46117, Liberec, the Czech Republic

**Abstract.** This work aims at investigating the primary nozzle position in a proposed supersonic air ejector device. The ejector is primarily made up of a supersonic primary nozzle, which is located in the axis of the ejector, a suction chamber or secondary stream inlet, a mixing chamber and a diffuser. The ejector design allows to translate the primary nozzle in the axis direction and fix it in a chosen distance from the beginning of the mixing chamber and hence influence the secondary mass flow rate. In a limit case, it is possible to set the nozzle to such a position where no secondary flow occurs. If we ignore the case where no secondary flow occurs, five different nozzle distances have been investigated in this paper. Some cases seem to be alike and there are no significant dissimilarities between them. Courses of relative back-pressure ratio are carried out against the entrainment ratio and transition between on-design and off-design regimes is determined. Measurements of the mixed flow based on the standard ISO 5167 are performed by means of orifice plate method. In addition, a comparison between experiments and simulations performed by Ansys Fluent software is presented in order to indicate further improvements to the numerical model.

## 1 Introduction

These days, when environment-friendly technologies are highly desired, devices operating with low energy consumption should get more attention. Ejectors are not an exception. These devices, with their efficiency generally lower than 30 per cent, require more research effort and a detailed investigation. The low efficiency is the greatest drawback of these devices. On the other hand, there are some advantages which make the use of these devices more attractive. Ejectors are commonly used for their simplicity, reliability and lack of moving parts. In this regard, special attention should be devoted to the improvement of their design. Ejectors can be used for both incompressible and compressible fluids. But it should be mentioned that the flow and especially the mixing of compressible fluids is very difficult to analyse. Ejectors are widely used in applications where sufficient amount of the working fluid or some waste thermal energy is available. They have established a long history in refrigeration. In this regard, they can be employed in connection with solar-driven systems or absorption systems. Refrigeration is probably one of the most important application areas of ejectors. Other possible applications of ejectors are thrust augmentation of aircraft propulsion systems, improving turbine performance, transport of solid particles and more. To spread out the use of ejectors in other commercial sectors, more research in the field of design methods is needed.

Ejectors operating with incompressible fluids are often called as jet-pumps. Ejectors operating with compressible fluids are sometimes further distinguished as ejectors and injectors. The term ejector is used for a device that evacuates a vessel and an injector for a device that pumps mixed fluids against high pressure. One of the first works on ejectors was performed by Keenan [1].

Two major cases of mixing in ejector based on work [2] are distinguished: the constant-pressure mixing and the constant-area mixing. It should be noted that designing of constant-pressure ejectors is very complex and causes many problems for researchers. It is caused by the condition of constant static pressure which requires a precise determination of mixing chamber geometry to obtain constant static pressure. On the other hand, ejectors designed based on constant-pressure mixing seem to give superior performance in most cases. However, there has not been established a definite link between the performance of constant-area ejectors and constant-pressure ejectors yet. Another distinction of ejectors can be considered in terms of primary and secondary streams arrangement. The primary stream can either be supplied in the ejector axis and secondary stream is sucked on the periphery of the mixing chamber or primary stream enters the periphery of the mixing chamber and the secondary stream is sucked in the ejector axis. The latter design is commonly used for wind tunnel propulsions and can have two possible arrangements. The first arrangement uses a slot primary nozzle. The second one uses several primary nozzles and

* Corresponding author: jan.kracik@tul.cz

was investigated in works [3-6]. Many theoretical, experimental and numerical researches have been carried out to obtain the performance of ejector correctly so far.

In Fig. 1, the performance line of supersonic ejectors divided into three sections is shown. Generally, supersonic ejectors are designed for working within a range of back-pressures which corresponds to on-design section. This section is recognised as a part with a maximal entrainment ratio and it is also a sign that both primary and secondary flow are choked. There is no dependency of entrainment ratio on back-pressure in this section. The second section is often labelled as an off-design section. In this case, there is a significant dependency of entrainment ration on back-pressure. The last section is a section of reverse flow, i.e. malfunction working regime, and we always try to avoid working ejectors in this regime.

The ejector under investigation in this paper was proposed in work [7]. The geometry of the proposed ejector is shown in Fig. 2. The primary nozzle is designed to generate exit Mach number of 2.06. The ejector is aimed to be utilized for future investigations of optimal shape of the wall in the mixing chamber region, thus finding an optimum nozzle position (x) is needed. At present, the mixing chamber has constant-area shape, which will also be considered as an initial shape for possible future optimization. Air has been used as a working fluid. Optimization is intended to be done with a 3D model. Before optimization, a comprehensive study and research on this ejector is desired.

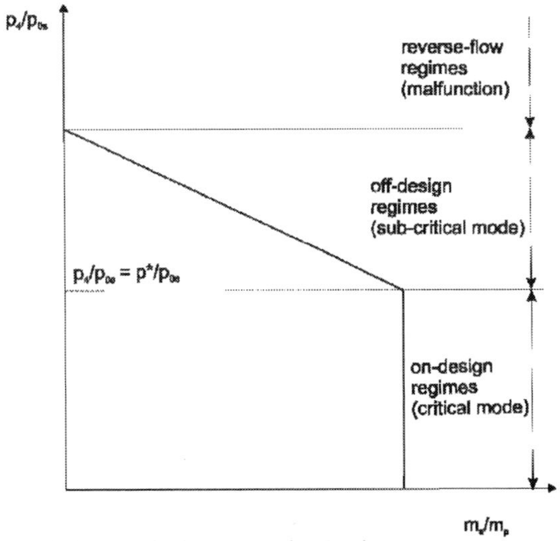


**Fig. 1.** Supersonic ejector operational regimes.

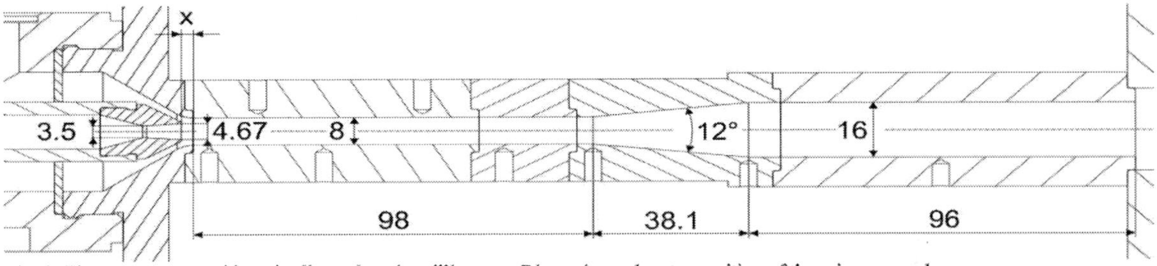


**Fig. 2.** Ejector geometry with main dimensions in millimetres. Dimension x denotes position of the primary nozzle.

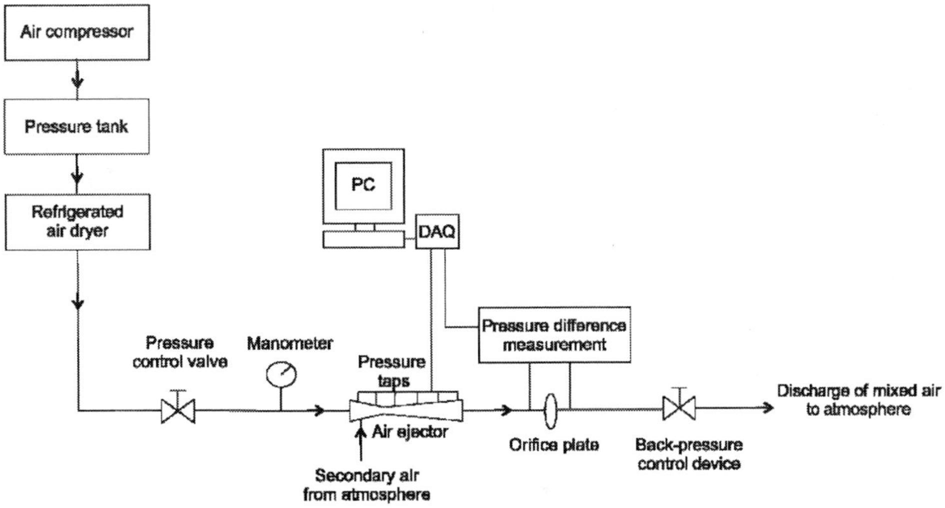


**Fig. 3.** Schematic diagram of experimental apparatus.

## 2 Experimental setup

In order to get performance lines of the ejector, the entrainment ratio must be calculated. The entrainment ratio is defined as a ratio of mass flow rates, i.e.

$$\omega = \frac{\dot{m}_s}{\dot{m}_p}, \tag{1}$$

where subscript $p$ denotes primary flow and $s$ stands for secondary flow, i.e. induced or sucked flow.

With regards to the experimental device arrangement shown in Fig. 3, the pressure difference of the mixed flow on the orifice plate is measured in case that the secondary flow is allow to be sucked into the mixing chamber. In the same manner and for no secondary flow, i.e. for closed secondary air inlet, the pressure difference of the primary flow can be measured. Secondary mass flow rate is then calculated as a difference of both and is expressed by

$$\dot{m}_s = \dot{m}_{total} - \dot{m}_p, \tag{2}$$

The primary and total mass flow rate are calculated according to the ISO 5167 standard as follows

$$\dot{m} = \frac{C}{\sqrt{1-\beta^4}}\,\varepsilon\,\frac{\pi}{4}\,d^2\sqrt{2\Delta p\,\rho_1}, \tag{3}$$

where $\Delta p = p_2 - p_1$ is pressure difference between the points located downstream and upstream the orifice plate respectively, $C$ is a discharge coefficient, $\varepsilon$ is an expansion coefficient, $d$ is the orifice diameter, $\rho_1$ is density upstream the orifice plate and $\beta$ is a relation between orifice and pipe diameter $D$ and is defined as

$$\beta = \frac{d}{D}. \tag{4}$$

Since a discharge coefficient is usually not known, the iterative procedure is needed for solving equation (3). The relative error between the calculated primary mass flow rate based on the introduced method and the primary mass flow rate obtained from numerical analysis is about 3 %. Even smaller relative error, approximately 0.1 %, is between the mass flow rate calculated according to relation (3) and the primary mass flow rate determined by theory.

Piezoelectric pressure sensors are used for acquisition of total pressure of primary air $p_{0p}$ and back-pressure downstream the diffuser $p_b$. Since the secondary air is sucked directly from atmosphere, total pressure of the secondary air is considered to be equal to the barometric pressure in the laboratory. Fig. 4 presents the ejector used for experiments. The total pressure of primary air is $p_{0p} = 397\ kPa$ with temperature $T_{0p} = 300\ K$. The total pressure of secondary air is equal to the ambient air pressure in laboratory, i.e. $p_{0s} = p_a = 97\ kPa$, temperatures are $T_{0s} = T_{0p} = 300\ K$. Relative back-pressure is changed within range (0 ÷ 65) kPa.

**Fig. 4.** Photo of the experimental ejector with its main parts: 1 – primary air inlet with the nozzle, 2 – secondary air inlet, 3 – mixing chamber, 4 - diffuser.

## 3 Numerical approach

The flow in the ejector is described by the compressible form of the conservation equations since they are more suitable for describing flows where density variations play a significant role. The influence of thermodynamics and transport properties of air was not proved to be significant, thus thermodynamics and transport properties are considered as constants during simulations. The air is modelled according to the perfect gas law. In order to limit complexity of the numerical model, two-dimensional axisymmetric models of the ejector is used to obtain numerical data, especially for calculating entrainment ratio. Although the steady state is assumed within the ejector, governing equations with an unsteady term are solved with a time marching technique. The governing equations can be written in their Cartesian form as follows:

$$\frac{\partial \rho}{\partial t} + \frac{\partial}{\partial x_i}(\rho u_i) = 0, \tag{5}$$

$$\frac{\partial}{\partial t}(\rho u_i) + \frac{\partial}{\partial x_j}(\rho u_i u_j) = -\frac{\partial}{\partial x_j}(p\delta_{ij}) + \frac{\partial \tau_{ij}}{\partial x_j}, \tag{6}$$

$$\frac{\partial}{\partial t}(\rho e_0) + \frac{\partial}{\partial x_j}(\rho u_j e_0 + u_j p + q_j - u_i \tau_{ij}) = 0, \tag{7}$$

$$p = \rho r T \tag{8}$$

and for a Newtonian fluid the viscous stress term is given by

$$\tau_{ij} = \mu_{eff}\left(\frac{\partial u_i}{\partial x_j} + \frac{\partial u_j}{\partial x_i}\right) - \frac{2}{3}\mu_{eff}\frac{\partial u_k}{\partial x_k}\delta_{ij}. \tag{9}$$

The governing equations are solved using the commercial CFD package Fluent provided by Ansys, Inc. Two-dimensional hybrid mesh with nearly 71,200 elements was created for calculations. However, most of the elements are quadrilaterals. Due to relatively small values of mass flow rates, double precision is used to get more precise solution. For flow variables a second-order upwind scheme is utilized. To take turbulence effects into account the k-ω SST turbulence model is used. This model is suitable for calculation of flows in supersonic ejectors as it was many times proved in previous works [8, 9 and 10]. For more information, see Ansys Fluent documentation [11].

## 4 Validation

In figures 5-9, comparison of experiment and numerical simulation in form of performance lines for different primary nozzle positions can be seen. In Fig. 5, performance lines for 1 mm are carried out to compare. Here can be seen relatively small discrepancies in on-design region, i.e. for the lowest values of back-pressure. In this case, the maximal entrainment ratio obtained from experiments is about 0.36. The performance lines for 2 mm show higher values of maximal entrainment ratio (Fig. 6). The maximal value for experiment is 0.74, which is approximately twice as high as in the case for 1 mm. However, the range of back-pressure to maintain this value is smaller. Fig. 7 to 9 show almost the same results for experiments and numerical analyses. For all three cases the maximal entrainment ratio is approximately 0.78 during measurements. The maximal entrainment ratio obtained from numerical analysis is a little higher (0.83). In Fig. 10, the performance lines for all five measured cases are carried out to compare. The same comparison for numerical analysis is shown in Fig. 11.

The relative error for all the investigated cases is depicted in Fig. 12. It is obvious for all the cases that for higher values of back-pressure the relative error is higher than 20 %. On the contrary, in the on-design region for lower back-pressures, the relative error lies within range ±10 %.

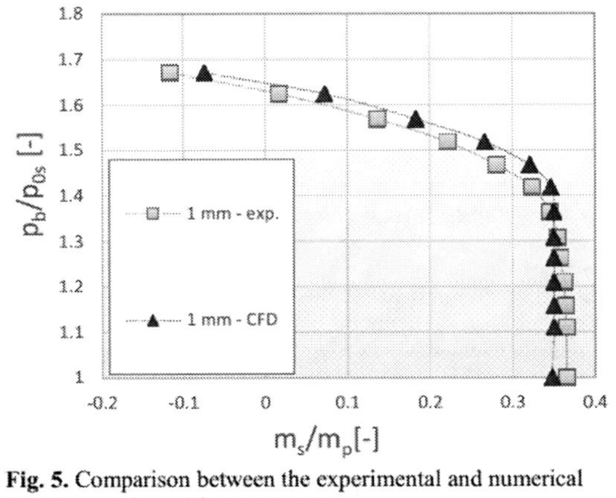


**Fig. 5.** Comparison between the experimental and numerical data for nozzle position x = 1 mm.

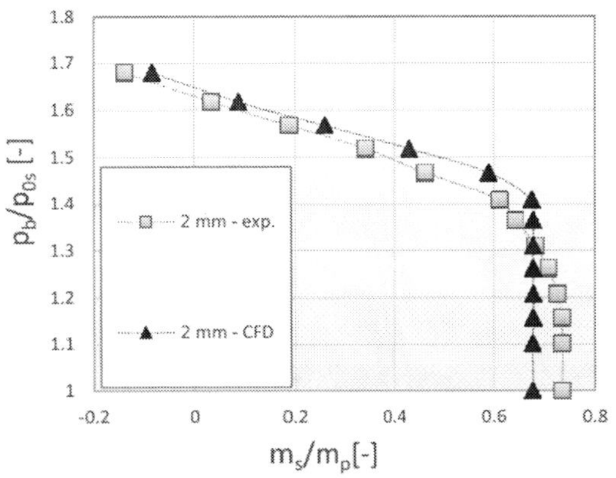


**Fig. 6.** Comparison between the experimental and numerical data for nozzle position x = 2 mm.

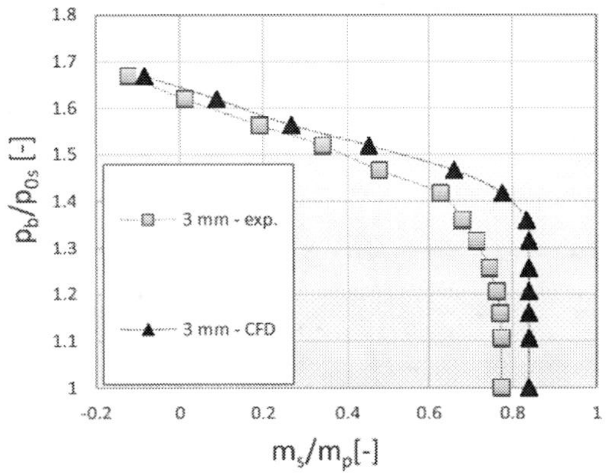


**Fig. 7.** Comparison between the experimental and numerical data for nozzle position x = 3 mm.

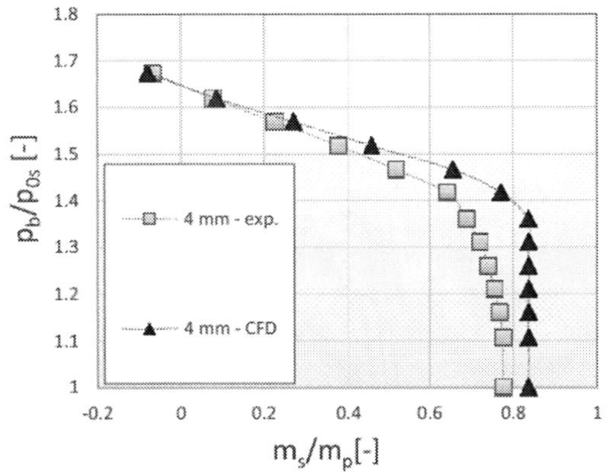


**Fig. 8.** Comparison between the experimental and numerical data for nozzle position x = 4 mm.

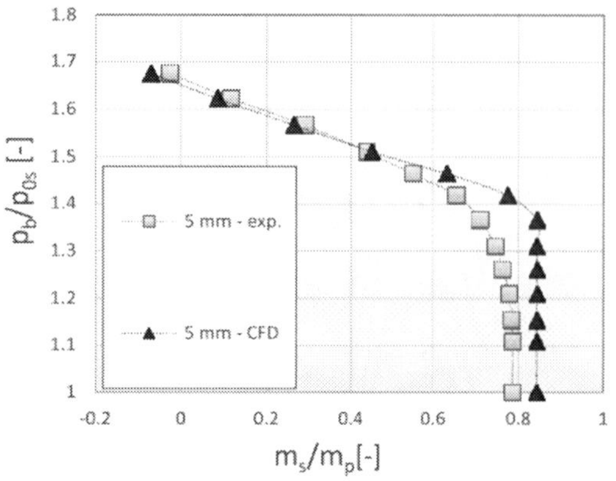


**Fig. 9.** Comparison between the experimental and numerical data for nozzle position x = 5 mm.

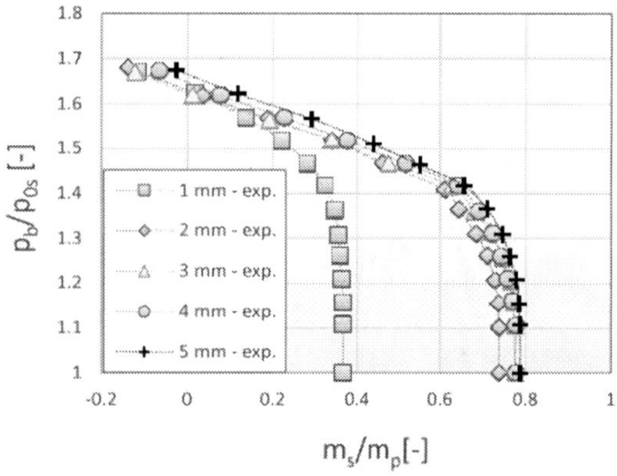


**Fig. 10.** Performance lines for all investigated cases obtained from experiment.

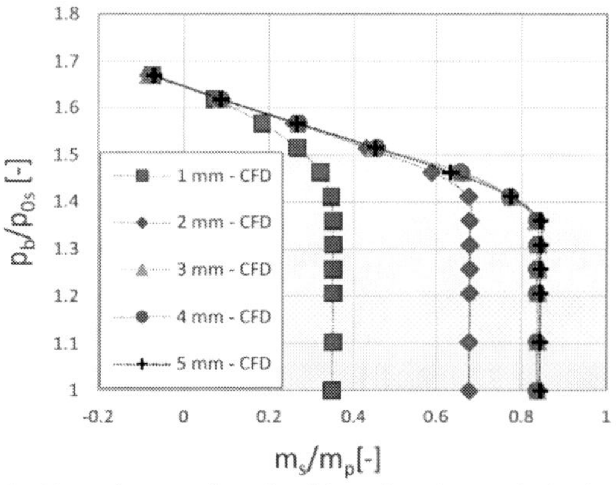


**Fig. 11.** Performance lines for all investigated cases obtained from numerical analysis.

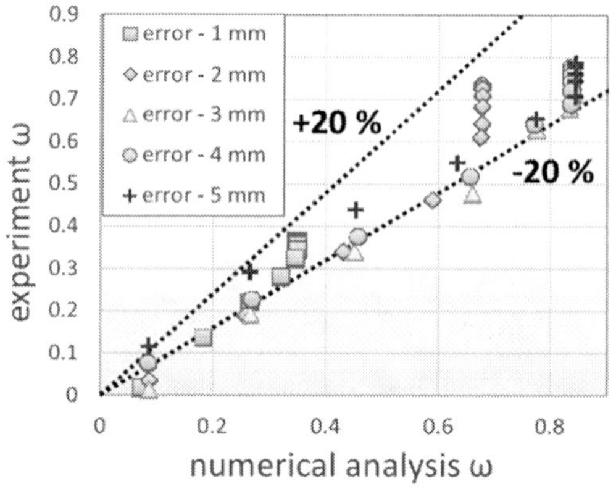


**Fig. 12.** Relative error between entrainment ratio ω obtained by the experiment and the numerical computation (CFD).

## 5 Summary and conclusion

The first experimentally obtained performance lines of a supersonic ejector with various primary nozzle positions have been presented in this paper. It turns out that from a certain distance of the primary nozzle from the beginning of the mixing chamber there is no significant increase in the maximal entrainment ratio (on-design regime) and the cases for distance of 3, 4 and 5 millimetres seem to be nearly the same. The measured maximal entrainment ratio is 0.78 for these cases.

The numerical results obtained on 2D axisymmetric models have been presented. The comparison with the experimental data shows some discrepancies, particularly in the off-design region. It could be caused by too much simplified numerical models or by other phenomena which occur for higher range of back-pressure. On the other hand, the largest error between experimental and numerical results in the on-design region is less than 20 % and most data points lies within ±10 % error. Anyway, there is a need for performing 3D calculations in the future. Furthermore, a comprehensive analysis of experimental and numerical data is necessary to understand flow processes that occur within the ejector, particularly the mixing chamber and diffuser sections.

This publication was written at the Technical University of Liberec as part of the project "Experimental and numerical investigation in applied fluid mechanics and energy devices, no. 21124" with the support of the Specific University Research Grant, as provided by the Ministry of Education, Youth and Sports of the Czech Republic in the year 2016.

## References

1. J.H. Keenan, E.P. Neumann, *A simple air Ejector,* Journal of Applied Mechanics, Trans. ASME **64**, pp A75 – A81 (1942)
2. J.H. Keenan, E.P. Neumann and F. Lustwerk, *An Investigation of Ejector Design by Analysis and*

*Experiment.* Journal of Applied Mechanics, Trans. ASME **17** (1950), pp 299 – 309.
3. J. Kracik, V. Dvorak, *Experimental and numerical investigation of an air to air supersonic ejector for propulsion of a small supersonic wind tunnel,* EPJ Web of Conferences **92** (2015) 02038
4. J. Kracik, V. Dvorak, J. Kolar, *Development of air to air ejector for supersonic wind tunnel,* Web of Conferences **67** (2014) 02059
5. J. Kracik, *Design of Supersonic Wind Tunnel* Diploma thesis, In Czech, (2014)
6. V. Dvorak, *Shape Optimization of Supersonic Ejector for Supersonic Wind Tunnel.* Applied and Computational Mechanics, Plzen, (2010)
7. A. M. S. Guanlao, *Design and verification of a test rig for research of supersonic ejectors,* Diploma thesis, (2016)
8. Y. Bartosiewicz, Z. Aidoun, P. Desevaux, Y. Mercadier, *Numerical and experimental investigations on supersonic ejectors,* International Journal of Heat and Fluid Flow **26**, pp 56 – 70 (2005)
9. J. Kolar, V. Dvorak, *Interaction of Shock Waves in Supersonic Ejector* (Proceedings of the 27th Meeting of Thermodynamics and Fluid Mechanics Departments, Plzen, 2008)
10. J. Simak, *Computation of the Flow and Interaction of Shock Waves in a 2D Supersonic Ejector* (Colloquium Fluid Dynamics 2009, Prague)
11. Ansys Fluent 15.0 *User's Guide*, Ansys Inc.

# Aerodynamics simulation studies on the observation head placed on a board of the unmanned aerial vehicle

*Izabela* Krzysztofik[1,*], *Slawomir* Blasiak[2]

[1,2] Faculty of Mechatronics and Mechanical Engineering, Kielce University of Technology, al. Tysiaclecia Panstwa Polskiego 7, 25-314 Kielce, Poland

**Abstract.** An important element of UAV equipment are observation heads. Observation head is multi-sensor optoelectronic system. It includes TV camera and infrared camera, safe for the eye rangefinder and laser pointer and a navigation system GPS. Image stability is provided by electro-mechanical gyro stability system. Optoelectronic head are designed for use in systems of detection, identification and observation of ground and air targets and in daylight and at night. In this paper simulation studies of the air masses flow around the observation head placed on the UAV deck, for different speeds of UAV movement were conducted. The results of numerical research were presented in a graphical form.

## 1 Introduction

Nowadays, modern equipment of a modern army includes unmanned aerial vehicles (UAV). The significant role is played by their independency during searching and tracking a target on the ground. One of the basic elements of UAV equipment is an observation head. The head is an optoelectronic system which integrates many sensors: an infrared camera, a daylight camera and an eye-safe laser rangefinder. It can be also equipped with a camera for low light performance, a laser pointer or an illuminator and GPS/IMU inner navigation system. The observation head is for automatic searching and tracking targets which have to be destroyed. It often performs its tasks in circumstances of vibration and interferences coming from UAV maneuvering board [1]. At the same time high accuracy and reliability are demanded. In this situation complex studies on the head are crucial. A wide range of research should include: material analysis, dynamics and control analysis, heat exchange, thermo-flexibility and analysis of developing the components and the elements of the described head, similarly to the studies [2-12]. In the paper [2] the general mathematic dynamics model and the algorithm for controlling the head placed on a moveable base are presented. The paper [3] also presents kinematic of mutual movement of UAV and a ground target and the research on dynamics of the controlled observation head placed on UAV during searching and tracking the target. PD controller was developed to control the head. The papers [4,5] present the analysis of head fuzzy logic controlling and controlling the head axis using the dynamics inverse problem. This paper presents simulation studies on air-masses flowing around the observation head placed on a board of UAV for different UAV speeds. A hypothetical set of observation head – UAV was taken under consideration.

## 2 Main aim

### 2.1 Modeling data

The object of the research is a volume of airspace inside which the aerial vehicle is located. The observation head is fixed to the vehicle. The paper describes simulation research for four cruise speeds of the aerial vehicle and determining their influence on the air stream behavior flowing around the head. Using the tool for solid modeling the researchers prepared the spatial model of the aerial object with the observation head (Fig. 1).

**Fig. 1.** The aerial object with the observation head.

Around the object the cuboid of dimensions (HxWxD) 35x40x50 m was generated. This space was so big to prevent the occurrence of uncontrolled turbulence. The whole aerial object-head system was implemented to Ansys CFX application which is presented in Fig. 2.

* Corresponding author: pssik@tu.kielce.pl

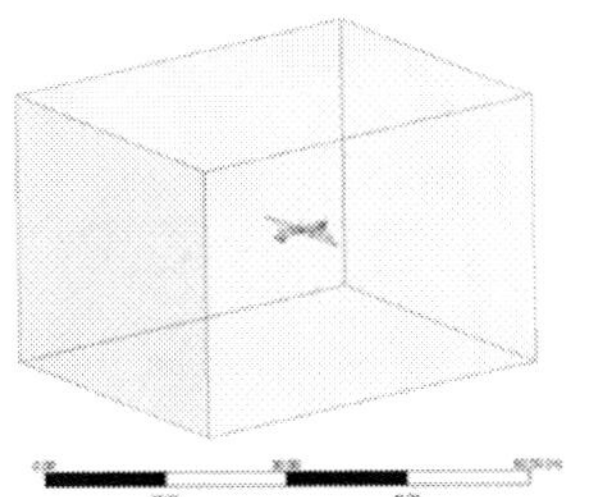

**Fig. 2.** Isometric view of the aerial object in the volume of air.

### 2.1 Numerical model

During the calculation process the fluid movement criterion was used to build mathematical model similarly as for pneumatic systems [13,14] in equations describing the physics of flow. In the analysis with using Computational Fluid Dynamics (CFD) the solving systems of transport differential equations are used to obtain detailed information on phenomena taking place in researched physical objects.

Mathematical description of researched phenomena is quite complicated therefore some assumptions and simplification conditions are used:

- Cartesian coordinate system (x, y, z),
- there are no chemical reactions and phase transition,
- gravity forces impact was ignored.

Fluid movement is described by Navier-Stokes equations (2)-(4) which together with the continuity equation (1) are the complete correlation system allowing to determine the pressure and the flow velocity field.

$$\frac{\partial \rho}{\partial t} + div(\rho \mathbf{U}) = 0 \tag{1}$$

Navier-Stokes equations.

$$\frac{\partial(\rho U)}{\partial t} + div(\rho U \mathbf{U}) = -\frac{\partial P}{\partial x} + div(\mu\, grad\, U) + \left[-\frac{\partial\left(\rho \overline{u'^2}\right)}{\partial x} - \frac{\partial\left(\rho \overline{u'v'}\right)}{\partial y} - \frac{\partial\left(\rho \overline{u'w'}\right)}{\partial z}\right] \tag{2}$$

$$\frac{\partial(\rho V)}{\partial t} + div(\rho V \mathbf{U}) = -\frac{\partial P}{\partial y} + div(\mu\, grad\, V) + \left[-\frac{\partial\left(\rho \overline{u'v'}\right)}{\partial x} - \frac{\partial\left(\rho \overline{v'^2}\right)}{\partial y} - \frac{\partial\left(\rho \overline{v'w'}\right)}{\partial z}\right] \tag{3}$$

$$\frac{\partial(\rho W)}{\partial t} + div(\rho W \mathbf{U}) = -\frac{\partial P}{\partial z} + div(\mu\, grad\, W) + \left[-\frac{\partial\left(\rho \overline{u'w'}\right)}{\partial x} - \frac{\partial\left(\rho \overline{v'w'}\right)}{\partial y} - \frac{\partial\left(\rho \overline{w'^2}\right)}{\partial z}\right] \tag{4}$$

where: $\rho$ – fluid density, kg/m$^3$, $\mu$ – dynamic viscosity, kg/ms.

Nowadays, $k$-$\varepsilon$ model is one of the most popular models and it is the most common model of the flow turbulence. Two mathematical units ($k$-$\varepsilon$) require two additional transport equations which can be described in the form:

$$\frac{\partial(\rho k)}{\partial t} + div(\rho k \mathbf{U}) = div\left(\frac{\mu_t}{\sigma_t} grad\, k\right) + \mu_t \phi - \rho\varepsilon \tag{5}$$

$$\frac{\partial(\rho\varepsilon)}{\partial t} + div(\rho\varepsilon \mathbf{U}) = div\left(\frac{\mu_t}{\sigma_t} grad\, \varepsilon\right) + C_1 \mu_t \frac{\varepsilon}{k}\phi - C_2 \rho \frac{\varepsilon^2}{k} \tag{6}$$

Quantity $k$ means kinetic energy of turbulence and quantity $\varepsilon$ means a dissipation of turbulence kinetic energy. The aforementioned partial differential equations are implemented in the calculation module of Ansys CFX application. For final solving of equations set describing turbulent fluid movements it is necessary to provide boundary conditions data which provides a clear solution and decides on a calculation process inside the researched area.

## 3 Ansys CFX analysis

In Ansys Workbench, Fluid Flow (CFX) -Meching module the non-structural computational grid was generated having 74523 junctions and 360140 elements. The computational grid structure is presented in Fig. 3.

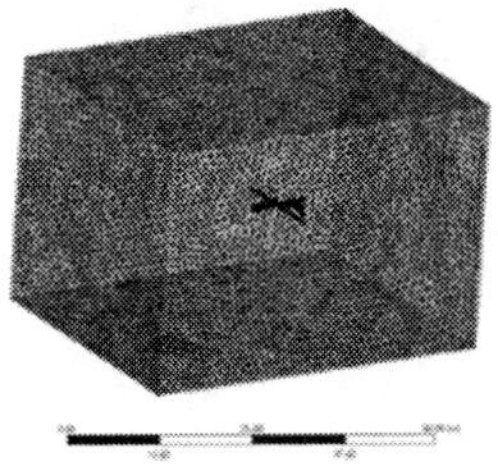

**Fig. 3.** MES model of the computational grid.

Due to the huge disproportion between the researched area and the aerial vehicle model the density of meshes was increased on the object surface as well as close to the object to – as accurately as possible – reflect its shapes.

Another preliminary stage to the numerical calculations was to determine boundary conditions.

1. At the entry point – Dirichlet criterion – there is the assumption that the air stream velocity (Fig. 4) is equal respectively for the four researched cases: a) 36.14 m/s, b) 40.31 m/s, c) 43.09 m/s and d) 45.87 m/s.
2. At the exit point Neuman Average Static Pressure criterion of 0.1 MPa was assumed.
3. For other surfaces on the cuboid circumference Free Slip Wall boundary conditions were used.

The values of physical quantities were also used for the calculations: ambient temperature T=25°C, density ρ=1.185 kg/m$^3$, reference pressure p=101325 Pa.

The aerial object cruise speed is much lower than the speed of sound and is within the limits of 130-165 km/h and therefore the air flow was assumed in the low turbulence level – $k$-$\varepsilon$ model.

DOI: 10.1051/epjconf/201714302057

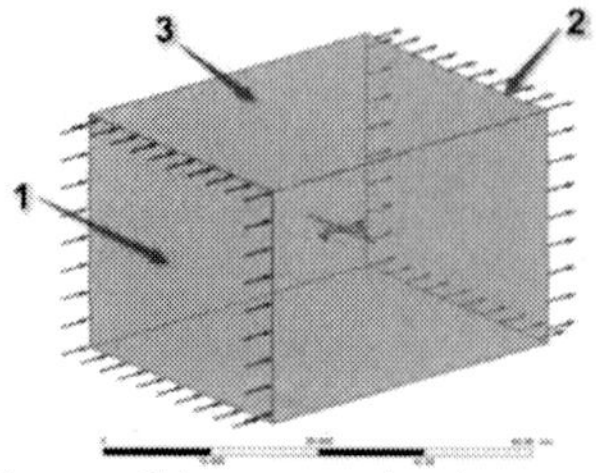


**Fig. 4.** Boundary conditions assumed to the numerical calculations.

## 4 Results and discussion

Because of conducting the numerical calculations considering boundary conditions the fields of velocity distribution in cross-section in its plane of the aerial object were determined as first.

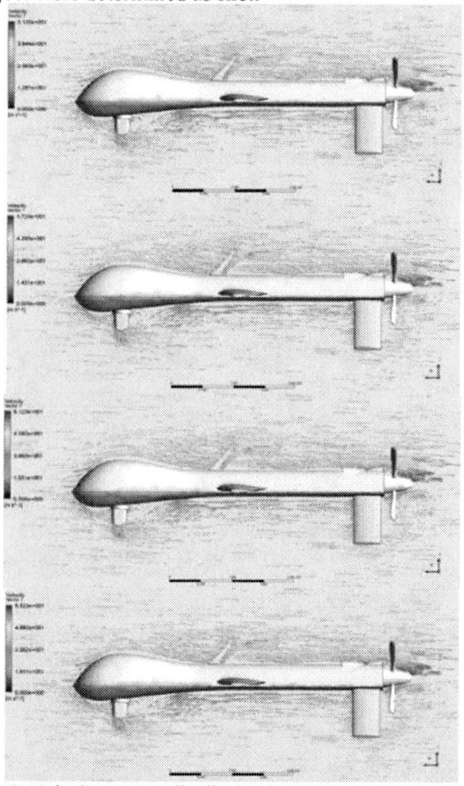

**Fig. 5.** Velocity vectors distribution in UAV cross-section.

The velocity vectors distribution in UAV cross-section is practically the same for all researched cruise speeds presented in Fig. 5. Only the maximal velocity of the air stream flow – from 51.25 to 65.23 m/s – is changed. The maximal air-masses turbulence can be observed behind the observation head and that may impact on the aerodynamic lift volume which have influence on the whole aerial object and resistances. In front of the head air movements are almost without turbulence.

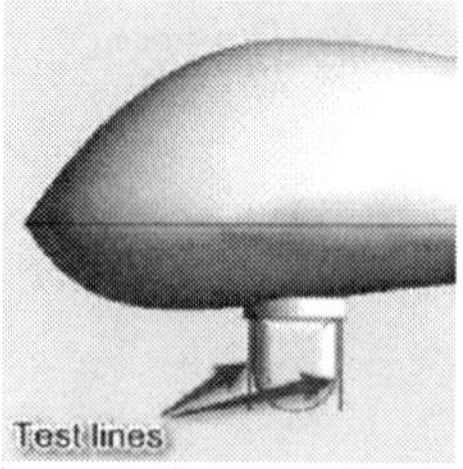


**Fig. 6.** Test lines for measuring the fluid velocity.

In Fig. 7 there are changes of the fluid velocity along the test lines for the chosen UAV cruise speeds.

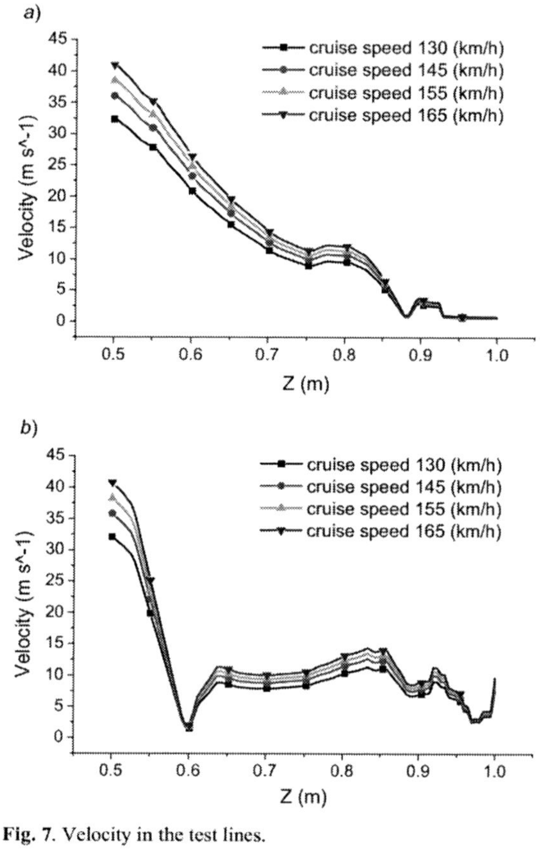


**Fig. 7.** Velocity in the test lines.

Maximal air-masses velocities in the front test line can be observed in the bottom part of the head. Depending on UAV cruise speed the values are from 32.5 to about 42.5 m/s. At distance of 0.8 m small flow velocity increasing can be observed. It is caused by the gap between the body and the rotating component of the observation head. Another situation of velocity distribution can be observed in the test line just behind the observation head. The velocity close to zero was experienced at distance of 0.6 m and at distance of 0.6-1.0 m the velocity did not exceed 15 m/s. The sharpest velocity change can be observed at distance of 0.5-0.6 m. In Fig. 8 there are the fields of pressure distribution impacting on the body of the aerial object and the head fixed on it for speeds of 36.14 m/s and 45.87 m/s.

Maximal pressure values (fields marked with red) can be observed in the vehicle part in front of the body forefront and in the part anterior to the observation head. For the chosen cruise speeds the maximal pressure values are almost the same (see Fig. 9).

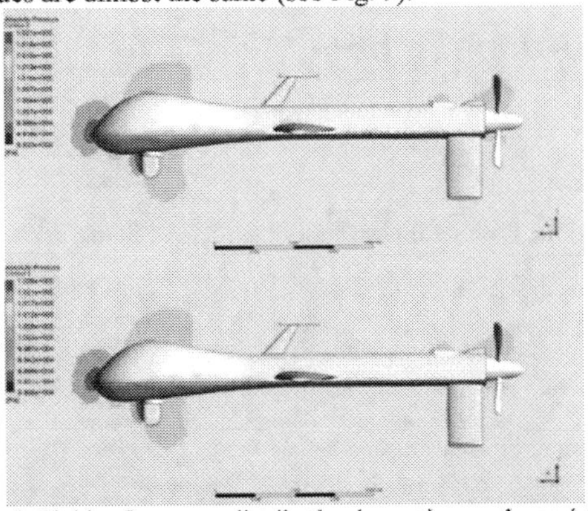

**Fig. 8.** Fields of pressure distribution impacting on the aerial object body and the head.

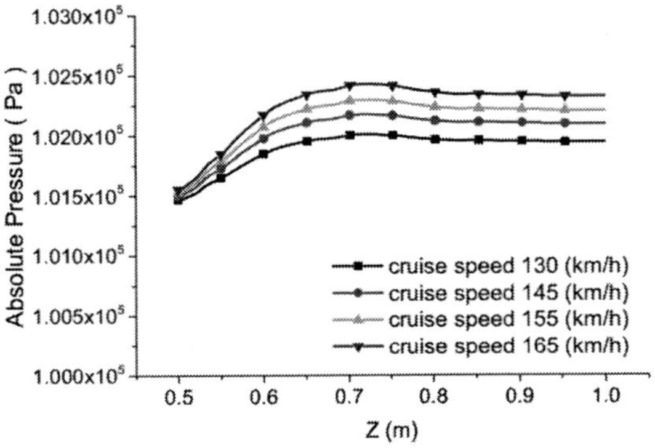


**Fig. 9**. Pressure in the front test line.

During the simulation studies it was assumed that the aerial object moved with low cruise speeds from 130 to 165 km/h. Choosing the range of cruise speeds was not without impact on Mach number distribution in the assessed area. In Fig. 10 it can be noticed that the maximal local Mach number for the speed of 130 km/h is 0.148 and for the speed of 165 km/h is 0.1884.

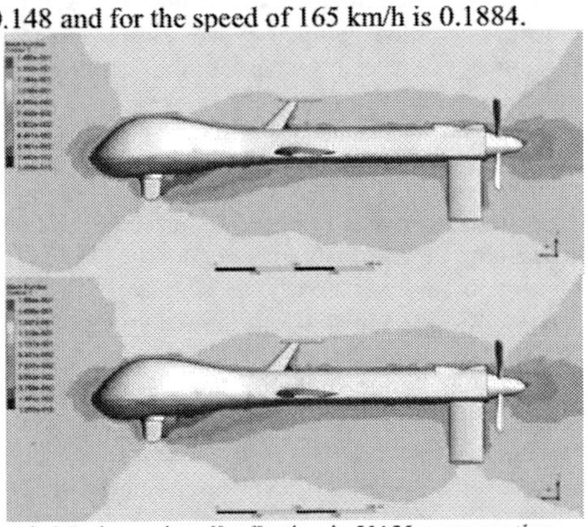

**Fig. 10**. Mach number distribution in UAV cross-section.

## 5 Conclusions

For the unmanned aerial vehicles the resistance values of the air-masses flowing around them are very important. As it is known, by minimising aerodynamic resistance the cruise range can be increased. Not without significance for the ground personnel controlling the object is the fact of possibility of speed increasing and making instantaneous acceleration. The prepared numerical model of air flowing around the whole aerial object and the observation head fixed to it which bases on CFD can be useful for shape designing and all UAV elements optimising. Using the presented model you can develop aerodynamic characteristics by changing their structures. In the further studies the shape and the construction of the observation head can be also considered. The prepared algorithm is a very useful calculation tool helping the designers of these types of devices. The obtained results should be experimentally verified in a wind tunnel.

## References

1. Z. Koruba, Ladyzynska-Kozdras E., J. Theor. Appl. Mech. **48**(3), 551-566 (2010)
2. I. Krzysztofik, Z. Koruba, J. Autom. Inf. Sci. **44**(5), 38-47 (2012)
3. I. Krzysztofik, Z. Koruba, J. App. Math. **2014**, Article ID 934250 (2014)
4. I. Krzysztofik, *Proceedings of 22nd International Conference on Engineering Mechanics 2016*, 326-329 (Academy of Sciences of the Czech Republic, 2016)
5. D. Gapinski, I. Krzysztofik, *Proceedings of the 15th International Carpathian Control Conference* (ICCC), 129-134 (2014)
6. J. E. Takosoglu, *Proceedings of 22nd International Conference on Engineering Mechanics 2016*, 546-549 (Academy of Sciences of the Czech Republic, 2016)
7. S. Blasiak J. Therm. Sci. Technol. **10**(1), JTST0016 (2015)
8. S. Blasiak, C. Kundera. Procedia Eng. **39**, 315-326 (2012)
9. S. Blasiak, J. E. Takosoglu, P.A. Laski, J. Therm. Sci. Technol. **9**(2), JTST0011 (2014)
10. M. Piasecka, Int. J. Heat Mass Transf. **81**, 114-121 (2015)
11. L. Nowakowski, M. Wijas, *Proceedings of 22nd International Conference on Engineering Mechanics 2016*, 430-433 (Academy of Sciences of the Czech Republic, 2016)
12. L. Nowakowski, E. Miko, M. Skrzyniarz, *Proceedings of 22nd International Conference on Engineering Mechanics 2016*, 426-429 (Academy of Sciences of the Czech Republic, 2016)
13. P. A. Laski, J. E. Takosoglu, S. Blasiak. Robot. Auton. Sys. **72**, 59-70 (2015)
14. J. E. Takosoglu, P. A. Laski, S. Blasiak, G. Bracha, D. Pietrala, Meas. Cont. **49**(2), 62-71 (2016)

***This research was supported in part by PLGrid Infrastructure.***

EPJ Web of Conferences **143**, 02058 (2017) DOI: 10.1051/epjconf/201714302058
*EFM 2016*

# Testing fireproof materials in a combustion chamber

Petr Kulhavy [1,a], Tomas Martinec[1], Ondrej Novak[1], Michal Petru[1] and Pavel Srb[1]

[1] Institute for Nanomaterials, Advanced Technologies and Innovation, Technical University of Liberec, Studentská 2, 461 17, Liberec 1, Czech Republic

**Abstract.** This article deals with a prototype concept, real experiment and numerical simulation of a combustion chamber, designed for testing fire resistance some new insulating composite materials. This concept of a device used for testing various materials, providing possibility of monitoring temperatures during controlled gas combustion. As a fuel for the combustion process propane butane mixture has been used and also several kinds of burners with various conditions of inlet air (forced, free) and fuel flows were tested. The tested samples were layered sandwich materials based on various materials or foams, used as fillers in fire shutters. The temperature distribution was measured by using thermocouples. A simulation of whole concept of experimental chamber has been carried out as the non-premixed combustion process in the commercial final volume sw Pyrosim. The result was to design chamber with a construction suitable, according to the international standards, achieve the required values (temperature in time). Model of the combustion based on a stoichiometric defined mixture of gas and the tested layered samples showed good conformity with experimental results – i.e. thermal distribution inside and heat release rate that has gone through the sample.

## 1 Introduction

Fire safety is an inseparable aspect during the buildings design. One of the basic requirements especially for industrial buildings is dividing the building into fire zones, each other protected by fire partition structures, to prevent spreading of a potential fire and minimization looses [1]. Also in this sector, reduction of weight used material plays an important role. It is necessary to minimize weight of parts without affecting their functionality and safety - often even on the contrary with their improvement. In our case some light fireproof materials that are used in fast fire barriers (e.g. movable gate, blinds) have been tested. Their function is to create a thermal shield between place which should stay safe and cool even if they are affected by high temperature radiation and convection from a near heat sources at the other side [2] for a by declared standard given time.

This article deals with numerical and experimental study of thermal transfer between flame of a propan butan jet inside a designed insulated chamber and measured layered samples. Solving of a similar problem requires simultaneously application of several disciplines. At first, it is necessary to consider the standardization perspective, experimental conditions and evaluating of results are described in the several standards [3]. Indicating the description, conditions (Eq. 1), implementation and evaluation of tests or process criteria for final certification. An overall description of fire, combustion and behavior of structures exposed extreme heat loading provide authors in books like e.g [1].

In studying numerical modeling of a combustion processes and spreading of fire we could meet with the classical heat transfer [2, 4, 5, 6] and also with the transient turbulent flow models. There are several possible models based on the Navier-Stokes equations that is possible to use [7, 8]. To their specific applications and description of using the so-called Eddy model that is most suitable for time unstable processes with amount of turbulence have written a lot of articles e.g. [8, 9, 10]. For proper adjustment of the model, especially in the definition of reaction media (air, gas), is very important come out of the fundamentals of chemistry, which defines the bonding forces [6, 11] and formed particles. The final model of gaseous components presented by particles participating in the reaction and their statistical distribution is described by [12, 13]. So the whole problem goes beyond simple mechanical definitions of material properties and boundary group members (geometry).

## 2 Materials and methods

Each manufacturer has to tested its products and make a certification for fire resistance in the state testing laboratory or another certifying authority [14]. During the verification is the sample exposed to the defined in time increasing temperature [3]. The recorded values are integrity, temperature, radiation and level of insulation [1, 14]. The test results are subsequently classified according to standard. The test results are subsequently classified, into categories according to their limit states that either prevent (EI) or reduce (EW) spreading of the fire at least for a declared time in minutes (15- 90min)

[a] Corresponding author: petr.kulhavy@tul.cz

[3]. Standardized test is conducted in a special chamber where the process of fire simulating two LPG burners while the mass flow of fuel and oxidant could be regulated.

A homogenous distribution of the temperature in testing chamber is required for corresponding data of tests that are used for construction optimization of fire doors. Temperature distribution in the testing chamber near of sample was tested during measurement cycle. The temperature curve is prescribed with equation (1).

$$T = 345 \log[8(t - 20) + 1] + 20 \quad (1)$$

where $t$ is time in minutes and $T$ is required temperature in the chamber in Celsius degree.

### 2.1 Fuels

Gaseous fuels are chemically mixture of natural gases. Its composition depends on the site, but their properties are not different from another. Even though their composition could differ, their main properties are generally quite similar.

In our case the liquefied petroleum gas, that consisting mainly propane and butane has been used. Propane and butane is a colorless, easily volatile liquid, belong to the group of hydrocarbon gases from three to four carbon atoms per molecule (Tab. 1), which are obtained during the processing of crude oil and natural gas. The combination of these gases falls within group D - ie. Very high calorific value (over 20.93 MJ / $m^3$). When for 0°C and 101.08 kPa is the measured calorific value of about 46.1 MJ / $m^3$ [1, 15].

**Table 1.** LPG composition

| Propan-butan | propane | butane |
|---|---|---|
| Chemical formula | C3H8 | C4H10 |
| Molecular weight | 44,094 | 58,12 |

Combustion meant a chemical physical process during which the combustible materials (gas) with an oxidant ($O_2$) are mixed. General stoichiometric equation of combustion hydrocarbon gas mixture (2, 3, 4), which can be for propane butane used is according to [16] given:

$$C_xH_y + zO_2 \rightarrow xCO_2 + \frac{y}{2}H_2O \quad (2)$$

When: $$z = x + {}^{1}\!/_{4}\,y \quad (3)$$

Which particularly for propane with oxygen is:

$$C_3H_8 + 5O_2 \rightarrow 3CO_2 + 4H_2O \quad (4)$$

### 2.2 Fireproof samples

For the first tests a layered sample, that provide us dependency between values and distribution of temperatures inside the chamber and at the outer side, was used. As the first prototypes combination of metal sheets and mineral foam were used (Fig. 1).

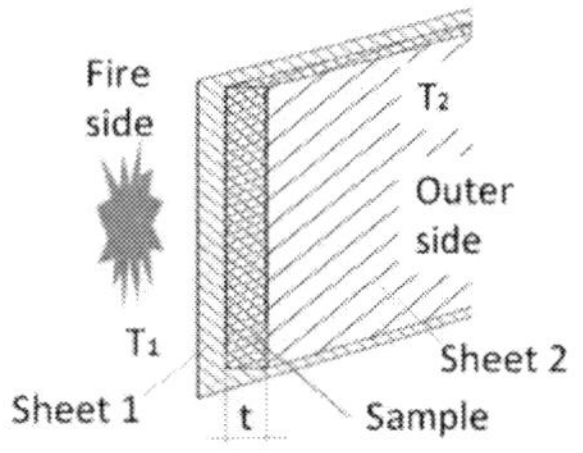


**Figure 1.** The measured layered sample

### 2.3 Experimental device

Set out the main geometry and dimensions was necessary to design the testing chamber (Fig. 2). Besides the isolated body these were primarily holes for the gas burners, adjustable air inlet (adjustable vents, fans, blowers) and simultaneously also the possibility of reduction the outgoing flue gases. Based on settings of all these conditions, we could distinguish between perfect and imperfect burning.

**Figure 2.** Scheme of the combustion chamber

When assuming the perfectly combustion it is possible according to [15] use the below approximated theoretical formula (5, 6).

$$V_{ssmin} = K_3 \frac{Q_n}{4186{,}7} + K_4 \;\; [m^3/kg],\; [m^3/m^3] \quad (5)$$

$$V_{vsmin} = K_1 \frac{Q_n}{4186{,}7} + K_2 \;\; [m^3/kg],\; [m^3/m^3] \quad (6)$$

Where the index **v** is volume of air, **s** is volume of dry soot, and $K_i$ coefficient could be based on heat ratio Q observed in Tab 2.

**Table 2.** LPG coefficients

| Gaseous fuels | K1 [m³/(m³kJ/m³)] | K2 [m³/m³] | K3 [m³/(m³kJ/m³)] | K4 [m³/m³] |
|---|---|---|---|---|
| Qn > 12,6 MJ /m³ | 1,07 | 0,0 | 1,09 | 0,47 |

DOI: 10.1051/epjconf/201714302058

## 3 Experiments

The combustion process in the chamber is accompanied by a flow of gas, air and combustion products. This process will affect the distribution of heat in the chamber and around the sample (Fig. 3). The temperature field on the surface of the sample facing towards the fire should not show significant variations with regard to position. Also the temperature in the chamber should meet a normalized temperature curve (Eq. 1).

**Figure 3.** Example of the burning chamber without a sample

For this purpose the verification measurement that will show the distribution of temperature field in observed area was carried out. The measurement procedure is described below. Temperature was measured by four wire thermocouples type K that were placed symmetrically 5 mm from sample inner surface. Three thermocouples marked ***a, b, c*** were placed directly on outer side of sample. Data collection was carried out with modules Advantech Adam. For online monitoring of data software in JAVA script was prepared. A distance of thermocouples from the sample surface was approximately 5 mm. After burner pre-heating outside of the chamber and the subsequent placement into the burner inlets the temperature inside the chamber was controlled by adjusting the gas pressure according to a prescribed curve. Recorded data from the individual thermocouples together with the control curve are shown in Fig. 4.

It could be seen that in the first minutes of measurement is not possible to ensure a desired course of the curve, which is due to the fact that thermocouples are very close to the flame or they lie on its edge. Maximal difference between 2 – 4 minutes is less than 50 °C. It should be noted that deviation in first 5 minutes of test is not in the standard determined; there is only recommendation to achieve temperature near to the required temperature curve. In the interval from 5 minutes to the end of the measurement is the difference among individual thermocouples less than 30 °C and about 15 °C among thermocouples and required temperature curve. Higher temperatures are measured by thermocouples placed in higher position.

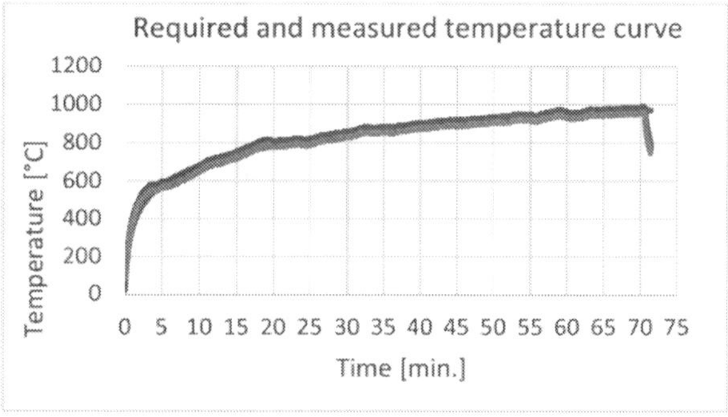


**Figure 4**. Temperature measured in the chamber

### 3.1 Temperature on a sample

Fig. 5 shows temperature on outer side of the sample. The thermocouple ***b*** is insert into a joint of door parts, where is not insulating layer. That is why a higher temperature is measured there.

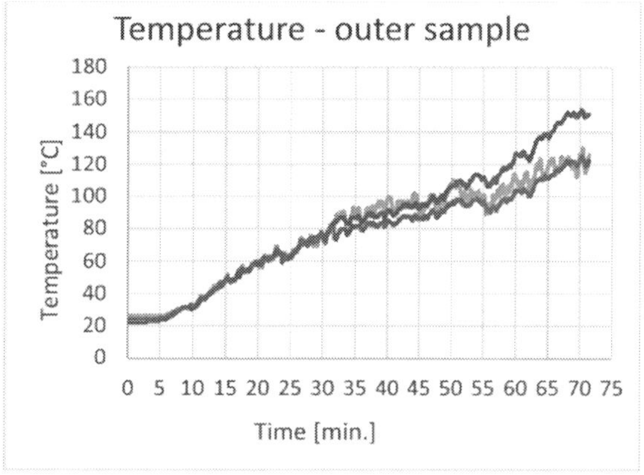


**Figure 5.** Temperature measured on outer surface of sample

## 4 Numerical simulation

Analytical or numerical solution of turbulent flow problems can be achieved using various levels of approximation, yielding more or less detailed descriptions of the flow. The easiest approach to modeling combustion is according to [8, 9] use semi-empirical correlations, Moody's diagram or a more sophisticated method involves the use of Reynolds' averaging (7):

$$\langle f \rangle = \frac{1}{T}\int_{t}^{t+T} f(\tau)d\tau \tag{7}$$

where $f$ is the long time average of quantity and T the time interval - longer than all the time scales of the turbulent flow.

If the averaging is applied to the equations of motion, we obtain the Reynolds-averaged Navier-Stokes equations (RANS) that describe the evolution of the mean quantities [7, 17].

The effect of turbulent fluctuations appears in a Reynolds stress term must be modeled to close the system [8, 9]. A very wide range of models for the Reynolds stresses is available - from simple algebraic models, K-ε models to full or algebraic Reynolds stress closures [9].

Direct numerical simulations of turbulence are the most relevant approach to the solution of turbulent flows. The governing equations are discretized directly, and solved numerically. If the mesh is fine enough to resolve the smallest scales of motion, one can obtain an accurate three dimensional, time-dependent solution [7, 10]. Used software PyroSim (Fire Dynamics Simulator developed at NIST) solves the Navier-Stokes equations using the Large Eddy Simulation method. This solution is transient and will vary with time at any point [9, 10]. In contrast, the Reynolds Averaged Navier-Stokes approach provides a single time averaged solution. Large-eddy simulations (LE) are a technique between the direct simulation of turbulent flows and the solution of the Reynolds-averaged equations. In LE the contribution of the large, energy carrying structures to momentum and energy transfer is computed exactly, and only the effect of the smallest scales of turbulence is modeled. LE simulations are based on the assumptions that the small scales are more isotropic, and less affected by the boundary conditions, than the large scales [7, 9, 10].

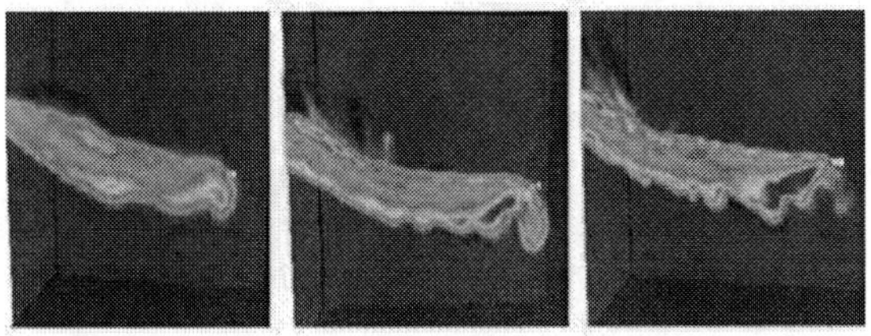

**Figure 6.** The combustion jet with different mesh size - 64, 512 a 1728 thousand cells (from left).

If a sufficiently refined mesh is used, FDS can accurately simulate a free jet. If the mesh refinement is too coarse, the solution will not observe turbulence in the shear layer that entertains the surrounding fluid and the predicted centerline velocity will not decay [9]. Generally, it is of course better to assess the quality of the mesh size in terms of nondimensional parameter, rather than an absolute size. Once mesh with size 5cm could be adequate for case of spreading fire through a building but absolutely inaccurately for small sprinkler or jet like in our case. In FDS is the cell ratio used to describe relative mesh size. Due to the accuracy the cell ratio should not been bigger than 1/5 of the characteristic fire diameter. According to [7] a mesh of even 1/16 the jet side is needed to obtain an reasonably accurate solution. This difference could be seen in the Fig. 6, where the same jet within different mesh cell ratio are shown. The cell number are 64k, 512k and 1728k elements. A twice smaller mesh mean decrease of the error 4x, at the other side cause increase the time for solving 16x [10].

## 4.1 Model definition

The non-premixed combustion model with a mixture of propane-butane (C2H6, CH4) has been created. The LE theorem and the solution with considered baroclinic torque and the non ideal energy approach, that consider arising CO, $H_2$ and soot, has been used. According to [6] until our knowledge of the disposition and motion of the electrons in atoms and molecules are not complete, we cannot make a direct calculation of the forces played an important role among molecules in a gas. One of methods is to assume a definite laq of force and then by the kinetic theory deduce the appropriate lay of dependence of the viscosity of a gas on temperature. For this reason it is necessary in the material definition set also the Lennard-Jones potential [6, 11]. The most common expressions of the L-J potential are:

$$V_{LJ} = 4\varepsilon\left[\left(\frac{\sigma}{r}\right)^{12} - \left(\frac{\sigma}{r}\right)^{6}\right] = \varepsilon\left[\left(\frac{r_m}{r}\right)^{12} - 2\left(\frac{r_m}{r}\right)^{6}\right] \quad (4)$$

where $\varepsilon$ is a depth of the potential well, $\sigma$ is the distance at which the inter-particle potential is zero, $r$ is the distance between particles and $r_m$ is the distance at which the potential reaches its minimum. At $r_m$, the potential function has the value $-\varepsilon$. The distances are related as $r_m = 2^{1/6}\sigma \approx 1.122\sigma$.

These parameters can be fitted to reproduce experimental data or accurate chemistry calculations. Due to its computational simplicity, the Lennard-Jones potential is used extensively in computer simulations even though more accurate potentials exist. This inter-molecular force may be attractive or repulsive, depending on the value of „r". From our point of view, using this parametr we could fit e.g. the „arising volume" of flame. The L-J potential has only two parameters (A and B) which determine the length and energy scales and may, without loss of generality, be set to unity. The potential is therefore unique, and cannot be fitted to properties of any real material [11]. For setting this parameter the studies of Heptane combustion [9, 10] has been used.

Three-dimensional flow was modelled by large Eddy model. Particle of defined gas, presented a moveable fragmented droplets (5000 per second), have been through nozzle sprayed into a model domain with defined volume flow rate (L/min), velocity (m/s) through a 12/10°conical jet. In specific processes such as atomization or spraying a wide variety of mathematical and empirical distribution functions, such as the normal, log-normal or Rosin-Rammler distribution of particles is used (Fig. 7) [18]. The particles with median diameter 50 μm via Rosin-Rammler log normal function have been distributed. This statistical model is suitable for description of materials composed of huge number of small particles like powder, emulsion and in our case small gas particles. More about using of this technique could be read there [12, 13, 18].

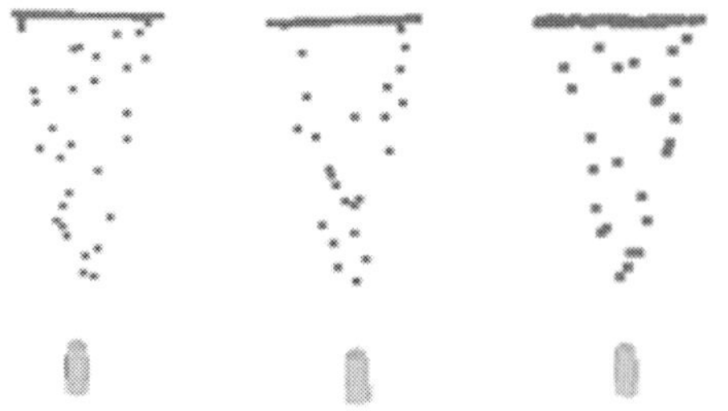

**Figure 7.** Constant, Rosin-Rammler and lognormal distribution of particles

At the other side, perhaps, due to the highly complicated model linked with a number of various boundary conditions and big concentration of the gas particles inside a relative small place (chamber), does not this parameter of used statistical approach seems to be as much important to the final results.

### 4.2 Results of numerical simulation

At the carried simulation (Fig. 8), we were primarily interested in the most accurate definition of the fire and subsequent in studying its effects on the tested layered sample (Fig. 1). The most important results for us is the inner heat (Fig. 9) is transferred through the sample and especially the temperature measured on the outside sheet (Fig. 10).

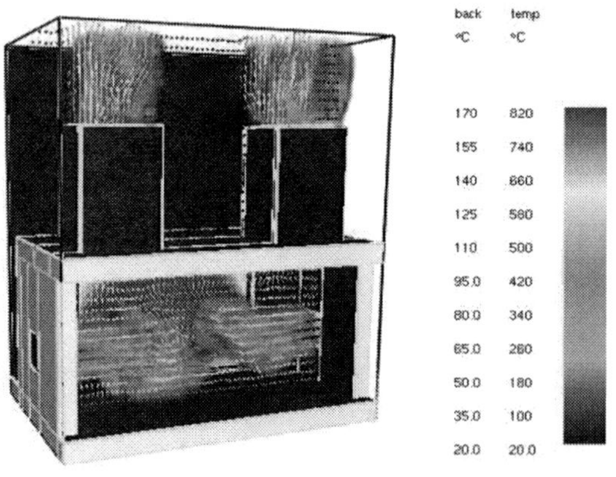


**Figure 8.** Temperature at the inner and outer sheets of the tested sample after 20 min

The temperature in the outer side of sample in dependency on temperature in the chamber could be seen in Fig. 10 for a clearer idea how the heat energy going through the sample and future material comparison.

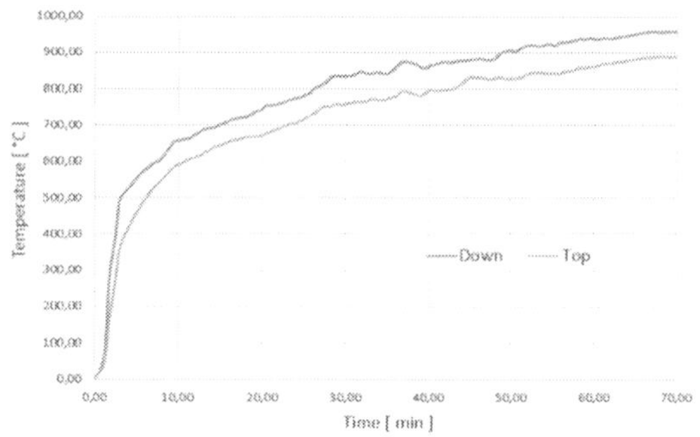


**Figure 9.** Graph of temperature in time – thermocouple near the top and the down side

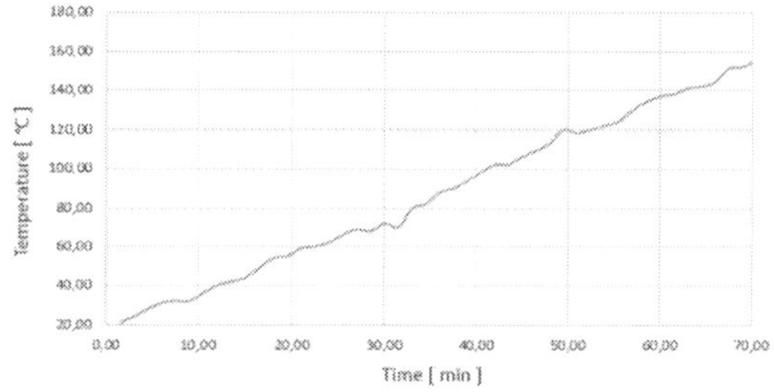


**Figure 10.** Temperature on the outer side of sample

## Results and discussion

The experimental and numerical determination of the temperatures arising on the sample are in quite a good conformity. Furthermore, we can not adequately define the place where interaction of tested materials and e.g. the sample holder are imperfectly overlapped. In a real situation just there arising the significant concentration and dissipation of heat. Due to the large time-consumption of the carried calculation, was also the whole body of the chamber considered as adiabatically inert only the front panel with the sample holder was adiabatic.

From the ordinary identifiable physical properties, which can be accurately determined experimentally it was in the fuel definition necessary to gone through the description of quantum chemical particles. At this moment, is due to the complexity of solved task necessary to rely on literature, which is based on empirically verified and tested data [6, 7, 11, 12]. After a thorough study of this issue is using and adjusting of these parameters advisable to tweak the simulation nearer the experiment.

## 5 Conclusion

In the case of a fire inside building it is necessary to limit the spreading by the safe isolation fire zones, ie. The closure of fire doors. Just so we can safely protect people's health and reduce the loss on property. In the presented work the method for testing some new heat-resistant materials was shown. Basis on the numerical model detailed idea about spreading of the gas mixture with air have been found. Using the software Pyrosim from the manufacturer Nist, the numerical model based on Large Eddy approach has been created. This software is one of the most appropriate tools for so complicated models, where various physical, chemical, thermal and hydrodynamic phenomenons are simultaneously solved. Work discussed mainly the possibilities of the materials (fuel, air, samples) definition. In comparison the simulation and experiment, we can talk about a relatively good agreement in the measured temperatures inside and outside the reference sample (inner and outer sheets). What remains, however problem for a similar simulation is modeling of the thermal degradation used materials over time. Use of simulation is therefore very suitable for direct determination of the moment and duration critical temperature values which must be coordinated with the knowledge obtained during the experiment. It can be also concluded that higher temperature on the beginning of the measurement has no influence of results, because higher temperature acts only short time.

In future work, efforts will be focused especially on definition of material models of tested insulating plates and improving their properties.

### Acknowledgements

The results of this project VI20152018046 were obtained with co-funding from the Ministry of the Interior of the Czech Republic as part of targeted support from the project "Využití moderních metod modelování při vývoji a zkoušení protipožárních uzávěrů".

## References

1. V. Kupilík, *Stavební konstrukce z požárního hlediska*, Grada Publishing , (2006)
2. J. Egert, K. Frana, S. Simon, S. Wichmann, *Heat Transfer Studies on Structured Metal Sheets.* KMUTNB: IJAST, **9**, (2016)
3. ISO 834-10:2014 Fire resistance tests - Elements of building construction, (2014)
4. P. Kulhavy, J. Kleckova, M. Petru and M. Havlicek. *Optimization of the heat output,* EPJ Web of Conferences, **114**, 02061 (2016)
5. R. L. Webb, N. KIM, *Principles of enhanced heat transfer*, Taylor & Francis, 2005.
6. *J. E. Lennard-Jones, On the Determination of Molecular Fields, Proc. R. Soc. Lond. A,* ***106****, (1924)*
7. B. J. M. Giesen, et. al., *Modelling and simulation of a jet fan*, Env. Modelling and Software, **26**, (2011)
8. U. Piomelli, *LE and direct simulation of turbulent flows*, University of Maryland, USA, (2001)
9. U. Piomelli, Large-eddy simulation: achievements and challenges, Progress in Aerospace Sciences, **35**, (1999)
10. K. McGrattan, et al. *Fire Dynamics Simulator Technical Reference Guide,* **6**, *(2016)*
11. D. Frenkel, B. Smit, *Understanding Molecular Simulation,* San Diego: Academic Press, **2**, (2002)
12. H. Sis, G. Kelbaliyev, S. Chandler, Kinetics of drop breakage under turbulent conditions, Sci. Technol., **26** (2005)
13. E. Jurado, et al, *Estimation of the distribution of droplet size*, Colloids and Surfaces A: Physicochem. Eng. Aspects, **295** (2007)
14. ČSN EN 1363-1 - Testing of fire resistance - Part 1, ÚNMZ, Praha, (2000)
15. P. Kól, Combustion dynamics, Combustion proceses, awailable at: http://Spalovaci-procesy.wz.cz/Statika horeni.html, (2011)
16. FDS-SMV Official Website. *Fire Dynamics Simulator.*, USA: National Ins. of Standards and Technology.
17. K. Akselvoll, P. Moin, *Large-eddy simulation of turbulent conned coannular jets*. J. Fluid Mech., (1996)
18. A. H. Lefebvre, *Atomization and Spray,* Hemisphere Publishing, Washington, (1989)

# Thermal properties of highly structured composite and aluminium sheets in an aerodynamic tunnel

Petr Kulhavy [1,a], Josef Egert[1]

[1] Faculty of Mechanical Engineering, Technical University of Liberec, Studentská 2, 461 17, Liberec 1, Czech Republic

**Abstract.** This article deals with the thermodynamic behaviour of heat shields - structured metal and composite plates. Experiments have been carried out in a wind tunnel with an additional heating, which simulates the heat source from engine or exhaust pipe and simultaneously the airflow generated during a car movement. The tested sheets with hexagonal structure were a standard commercial made of aluminium and a second manufactured by replication (lamination, diffusion) from glass fabric. The airflow in a parallel way along the sheets was analysed experimentally in order to determine the heat transfer efficiency between surfaces of sheets and surrounding airflow. The temperature on the sheets was chosen to observe the effects of different sheets material, various heat power and airflow velocity. During the experiment a thermal input below the sheets and airflow velocity through the tunnel have been changed. The thermal field distribution on the metal sheet is different than in case of composite sheet. For the composite material the thermal field distribution was more homogeneous. This article describe briefly also methods of obtaining real composite geometry based on scanned data and their reconstruction for using in some future numerical models.

## 1. Introduction

Due to the high energy demand and rising prices of materials plays the reduction of weight an important role through almost all industries. For vehicles it is necessary to minimize weight of parts without affecting their functionality and safety - often even on the contrary with their improvement. According to [1] an vehicle weight reduced by 100 kg saves 0.25 to 0.5 l/100 km of fuel that means reducing of $CO_2$ emissions from 3.5 to 8.4 g / km. Thin structured metal sheets (Fig. 1) become still more important for lot applications. One of their possible using could be as heat shields over the car exhaust and engine [2]. Their function is to create a thermal barrier between parts which should stay cool even if they are affected by high temperature radiation from a near heat sources [3]. The structured sheets offer us for particular applications the possibility of influencing on mechanical parameters, not only in terms of stress distribution, but also change in stiffness, shifting of natural frequencies or even customization of aerodynamic properties. They are used as a semi-finished product [4, 5]. From the point of the heat transfer mechanism, the structured surfaces intensified the convection heat transfer rate due to higher surface area and the suitable flow behaviour around the structured metal sheets as referred in [5, 6, 7] their papers focuses on heat shield application which is typically used in automotive.

For more efficient decreasing of the weight have been nowadays used also composite materials, which have considerably better specific strength than conventional. When compared with metal materials, composites are more appropriate especially in the terms of an efficient material using and saving weight. For this reason, composite parts whose properties could be modified directly with regard to the specific application, that significantly increases the utilization efficiency of materials are often used. Composite parts could be produced in many ways as winding, wrapping, pultruding or continuous laminating. Highly structured composite parts introduced in this work have been created by the method of steel templates replication by lamination and difusion method [8].

This article deals with experimental comparison of thermal transfer between steel and composite - glass fibre sheets, experimentaly measured in a wind tunnel with an additional heating, which simulates the heat source from engine or exhauster and simultaneously the airflow generated during driving a car.

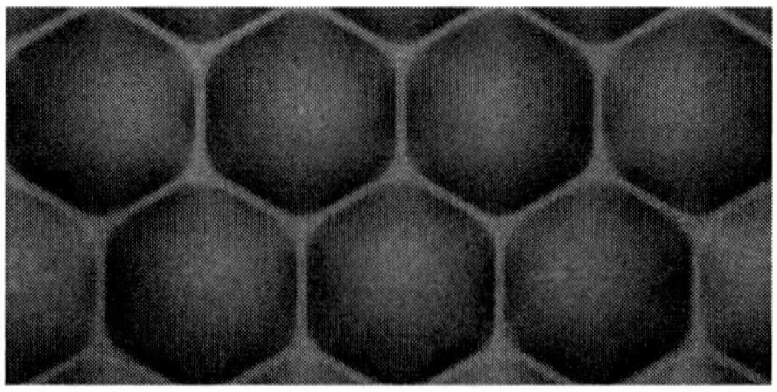

**Figure 1.** The hexagonal sheet structure

[a] Corresponding author: petr.kulhavy@tul.cz

## 2. Used materials

Using of the structured sheets (Fig. 1) could be as the heat shields over the car exhaust and engine, cover of movable parts, reflective layer of lamps or even a drum of wash machines. The replicated parts are structured glass or carbon composites created by replication of steel templates. Production of steel templates is a technological procedure mostly based on pressing or rolling [4, 5].

This way structured metal sheets were used as the shape template for replication on fiber fabrics, because it isn't possible produce such structure composite parts by a direct way. So, in this case instead of finding complicated methods for direct manufacturing composite parts, an sub step has been done with using the technologically simpler steel plate. Long fiber composites offer generally the greatest flexibility for design specification. However, the metal materials, have still lower price, also have great potential due to the possibility of they recycling and excellent technological processability [8].

**Figure 2.** The replication – manual saturation of plies

### 2.1 The tested prototype

An unilateral replication by stamping fabrics over the aluminium structure was carried out to create the FRP samples. Chemically treated surfaces of the aluminuim sheets were covered with four layers of glass or carbon fabrics (Fig. 2) and gradually saturated with an epoxy matrix. The created structure was uniformly without visible defects. Surely, according to [9, 10] the material properties determined from standard specimens tested in laboratory may often significantly deviate from components manufactured in factory.

**Figure 3.** The original and composite part

### 2.2 Digitization and meshing of samples

In contrast to a conventional metal parts, for the composite materials we are usually not able to accurate predict the final shape and dimensions of manufactured parts. This is caused by the fact, compared with e.g. precise machining, in this case we could speak moreover about copying shapes of used molds. Also the interleaving of individual plies including degree of resin saturation is not always the same and temperature or pressure could also differs. It means, we are not able to have the accurate and real CAD models, before we physically create the part. For the composite plate replicated by the above described method we need to know the exact dimensions and shape in order create a numerical model for simulation. For this reason the replicated plates have been scanned. At this moment, we got only a giant cloud of points that present just an undefined surface (Fig. 4 blue).

**Figure 4.** Part of the scanned and consequently styled geometry

To provide possibility import of a such data structure into a classic CAD their manufacturers have usually some own tools. In our case PTC Creo - Import data doctor has been used. These module contain algorithms for an additional editing, smoothing and cleaning of the imported meshes. It is also possible to cover the gaps, closed imperfect shapes or trim the surfaces. After the "coarse" repairing of the mesh we obtain a parametrically defined surface (Fig. 4 orange lines). Now, the CAD system is able to work with this way optimized data and use it as a design element for the projection, intersection, references or definition points, tangent properties etc.

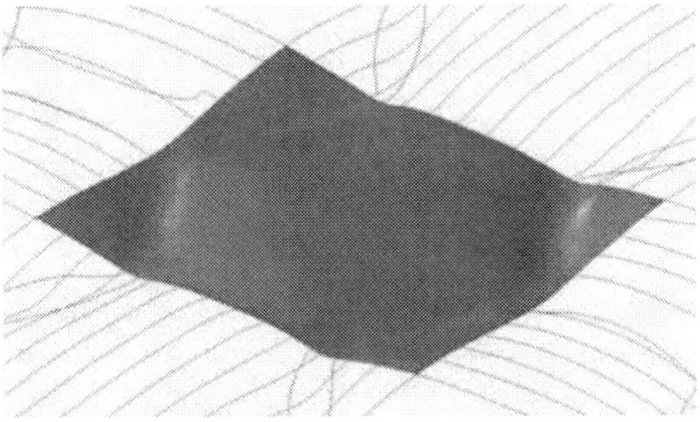

**Figure 5.** Boundary blend based on intersection surfaces

The most precisely defined fragments in a NURBS format, which are suitable for CAD data we could get by local cut outs and creation of a boundary blend as is seen in Fig. 5.

At this moment a real problem is that all the inaccuracies that arise during scanning and also geometric deviations of the measured parts are reflected and stacked up. Some unmeasurable errors in the intersection of supporting curves of created entities arise there almost always and this leading to a total or local collapse of the geometry. Sometimes, like in our case the inaccuracies are stacked and deform "only" the real shape locally - e.g. some spurs or holes could arise in the geometry as could be seen in Fig 6. About the reconstruction of similar geometries and their meshing it is possible to read more e.g. in [11, 12].

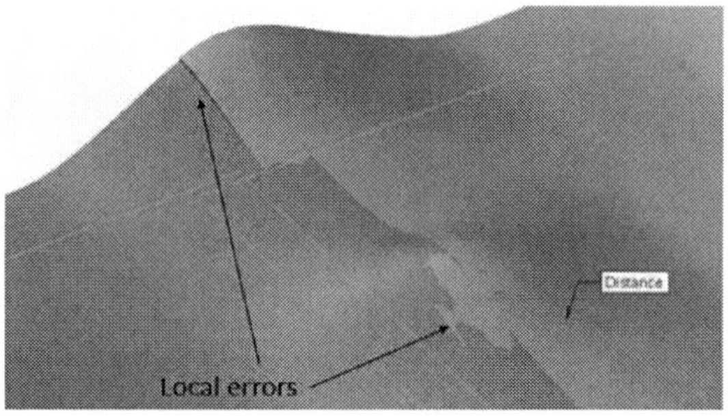


**Figure 6.** Local errors caused by stacking of inaccuracies

Now, probably the best variant how to repair the erroneous areas is to set them boundaries and fill with vector field consequently interleaved by surface. For the required vector definition was possible to simply use the adjustable parameters of normality, tangentiality or parametric settings of the transitions.

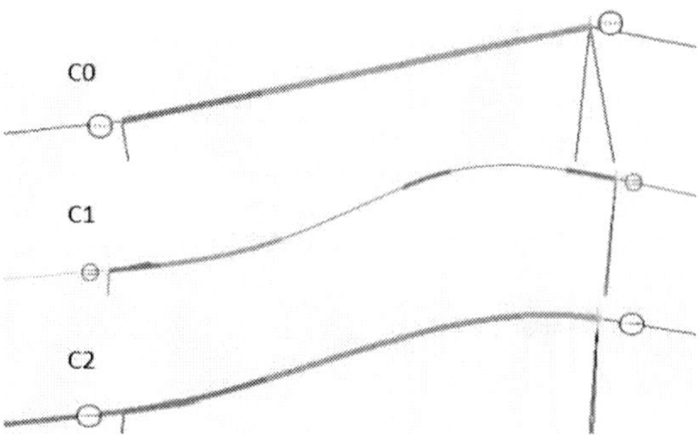


**Figure 7.** Classes *n* of the continuity transitions between individual segments

A very important parameter is settings of the parametric classes *n* of the continuity transitions between individual segments that defines us 3 classes of continuous. Connection $C^0$ means in common point two arbitrarily passing vectors and two curves with different radius of curvature. Continuity $C^1$ is characterized by the one common tangent vector $\overrightarrow{t_i}$ in the transition point, but different radii of curvature. Completely analogously then the last class $C^2$ provides a common tangent vector and value of the radius of curvature in the transition point as could be seen in the Fig. 7. Then, based on this surfaces, fully active CAD model could be created and exported to a required format. Six examples have been created (Fig. 8) in order to find out the dependency of geometric reconstruction (continuity vectors degree) and parametres of the computational mesh.

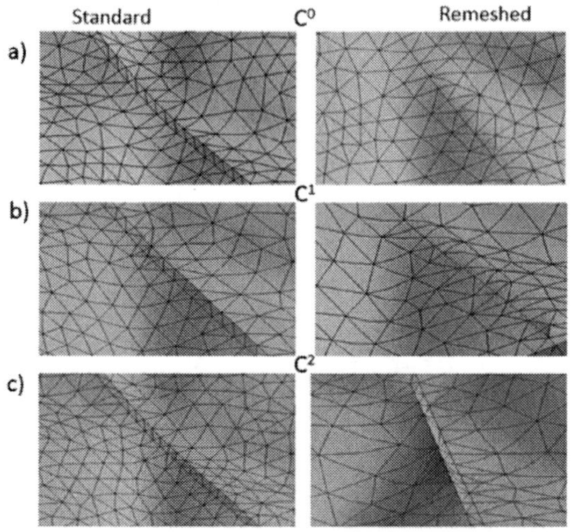


**Figure 8.** The mesh dependency on the vector continuity

The mesh dependency on the geometric vector continuity was compared in the commercial sw Ansys. First case was an automatically generated a tetrahedral mesh with +80% relevance of fine mesh (Fig. 9 Standard). In the second case the mesh was optimized with just several node and curve pinches and refined with virtual cells quite easily and fast. The results could be seen in Fig. 9 (column remeshed) or mesh the statistics in the Tab. 1. As could be recognized from the Table and Figure, the high degree of geometry optimization does not have the desired effect to the mesh quality and even just for the coarse transmission $\mathbf{C^0}$ we could obtain with using some remeshing tools almost equal quality with a considerable smaller effort.

**Table 1.** Mesh statistics

| | Method: | $C^0$ | $C^1$ | $C^2$ |
|---|---|---|---|---|
| Standard mesh | Nr. of elements | 9 576 | 10 268 | 10 017 |
| | Nr. Of nodes | 17 261 | 18 316 | 17 959 |
| Remeshed | Nr. of elements | 9 427 | 8 603 | 9 788 |
| | Nr. of nodes | 17 024 | 15 647 | 17 576 |

## 3. Experiment

The aim of the carried tests was to assess the thermodynamic behaviour for composite materials. In this research the results obtained with composite prototypes were compared with results obtained in previous research for steel prototype with the same structure shape [3]. The experimental device (Figure 9, 10) is combination of an open aerodynamic tunnel with a resistive heater located inside the box. Tested sample is placed on the bottom of test section of aerodynamic tunnel on the top of heater box.

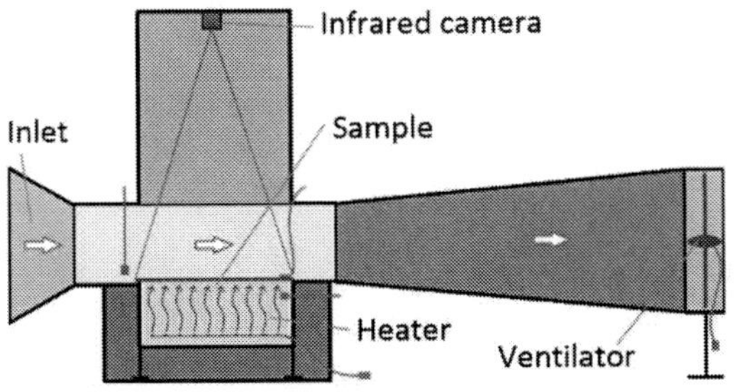


**Figure 9.** The experimental device

Detailed description of the device is available in previous study [3]. At the inlet of the model, there is a confusor with rectangular cross section which sucks the ambient air into the tunnel. The airflow speed was measured by a hot wire anemometer probe in front of the test section that is located near the bottom side of the tunnel. One side of the sheet was cooled by the airflow and the other side was heated by the convection from the hot air and by the heat radiation from heater. The global temperature field on the sheet surface was measured by the infrared camera placed on the top of the vertical part of the tunnel and locally with thermocouple. To minimize heat losses, the heating box was isolated by a polystyrene, mineral cotton, calcium-silicate layer and aluminium foil. The temperature inside the heat box was regulated varying voltages and measured with the thermocouple. The ventilator rotational speed was controlled by a potentiometer.

The tested sample was measured in the same way as in previous research to obtain comparable results. Sample was measured in the heating power range from 100 W to 500 W with step of 100 W and Reynolds numbers varied from 10000 to 145000. The Reynolds number range was given by the airflow speed which was limited by the power of used axial ventilator. The range of the heating was limited by range of infrared camera. For each of investigated state an infrared image was taken and the average and maximum temperature of the sheet's were determined. Also, the temperature inside the heating box and ambient temperature were recorded. The thermal camera emissivity for the lacquered aluminium sheet was based on the camera manufacturer tables and verified by measurement with thermocouples set to value 0.9 and for the epoxy composite to 0,85.

**Figure 10.** Assembly of the real experimental device

Velocity profile in the inlet of aerodynamics tunnel was measured with anemometric probe. Velocity was set to 13 m/s with probe in default probe position used for other regimes and then probe was moved across the cross section. Results are shown in Figure 11. Lover velocity near bottom wall is probably caused by the lead edge before the tested sample. Velocity profile in the rest of cross section is balanced.

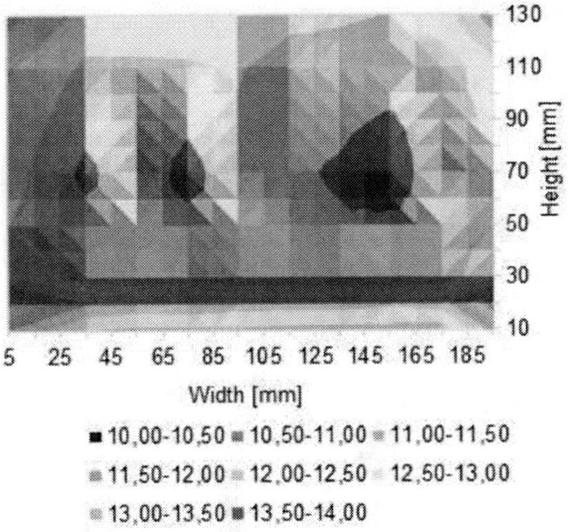


**Figure 11.** Inlet airflow velocity profile

### 3.1 Theoretical Equations and Parameters

For comparison with previous study, the theoretical equations and parameters were used in the same manner.

The Reynolds number was applied as a relevant number and it takes a form as

$$\mathrm{Re} = \frac{cL}{\upsilon} \tag{1}$$

where c [m/s] is the airflow velocity, L [m] is characteristic length and υ [$m^2$/s] is a kinematic viscosity of the surrounding air. The characteristic length is defined as equivalent hydraulic diameter

$$L = \frac{4A}{O} \tag{2}$$

where A [$m^2$] is a tunnel cross section and O [m] is circumference of the tunnel cross-section. To compare results with other studies every physical quantity was normalized. For temperature evaluations, the normalization to ambient temperature was used. The normalized temperature is defined as

$$T_N = \frac{T}{T^*} \tag{3}$$

where $T$ is measured temperature and $T^*$ is ambient temperature during the measurement. Assuming, that the lab space was large enough, the heat power could not influence the internal temperature in the surroundings. It means that this ambient temperature corresponded to the temperature of the surroundings. The heat transferred throughout the sheets can be expressed as follows

$$Q_S = Q_C - Q_Z \tag{4}$$

where $Q_C$ is the overall heat generated by the heating wire and $Q_Z$ is the heat loss through heating box walls. The heat loss is determined by heating box materials. The transferred heat from the sheet surfaces to the surrounding airflow is defined as

$$Q_S = \alpha(t - t_s)S \tag{5}$$

where $\alpha$ is the convective heat transfer coefficient, $t$ is average sheet temperature, $t_s$ is ambient temperature and $S$ is the sheet area. In case of convection, the heat transfer intensity is commonly express by the Nusselt number, which takes a form as

$$Nu = \frac{\alpha\, l}{\lambda} \tag{6}$$

Where l is the sheet length and λ is the thermal conductivity of the surrounding air.

## 4. RESULTS

Dependence of the sample´s surface average temperature to airflow velocity is shown in Figure 12. and in Figure 13. for surface maximal temperature. The results are combined with results of steel sample from previous study. Results shown that composite sample achieved lower both average and maximum temperatures than steel sample. For lower power inputs the temperature dependence on Reynolds number was similar to steel sample. For higher power inputs the dependence on composite sample was deformed in comparison with steel.

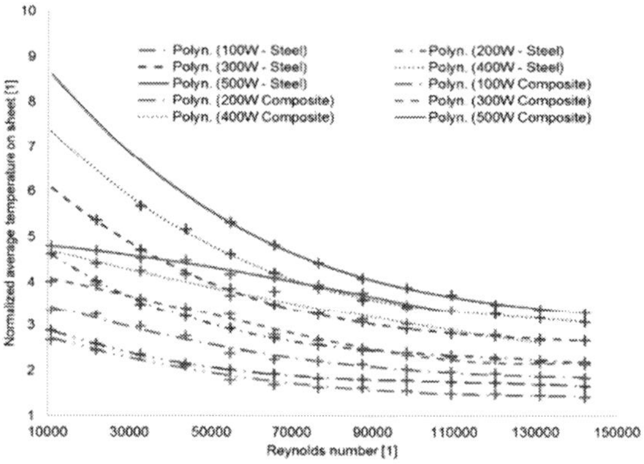


**Figure 12.** Comparison of average temperatures on sheets surfaces

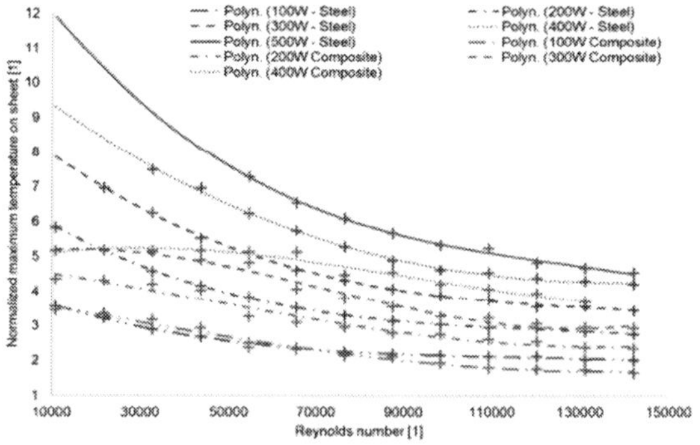


**Figure 13.** Comparison of maximum temperatures on sheets surfaces

Infrared image samples comparing the temperature distribution of composite and steel sample are shown in Figure 14. and Figure 15. Temperature distribution over composite sample is more homogenous. Distribution field is similar to previous results from numerical simulation described in [3]. For comparison, the image is shown in Figure 16. Unfortunately, the numerical simulation was realized for higher Reynolds number so it is not possible to compare in terms of temperature values, but only in terms of temperature distribution.

 DOI: 10.1051/epjconf/201714302059

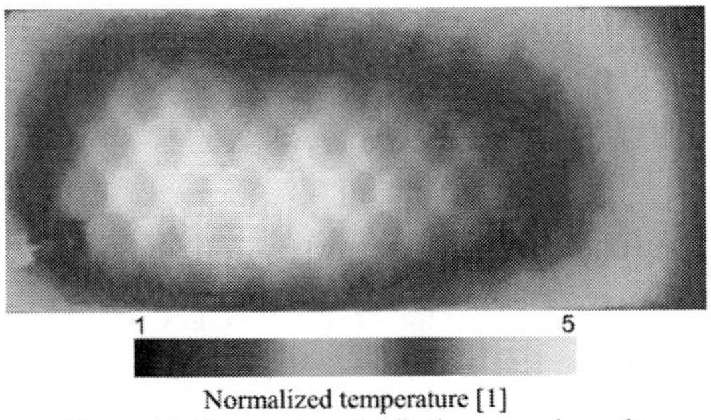


**Figure 14 :** Temperature distribution on steel sample 300 W, 5 m/s

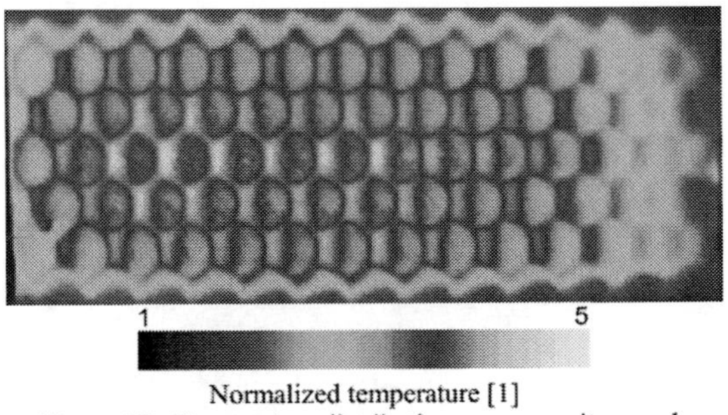


**Figure 15 :** Temperature distribution on composite sample 300 W, 5 m/s

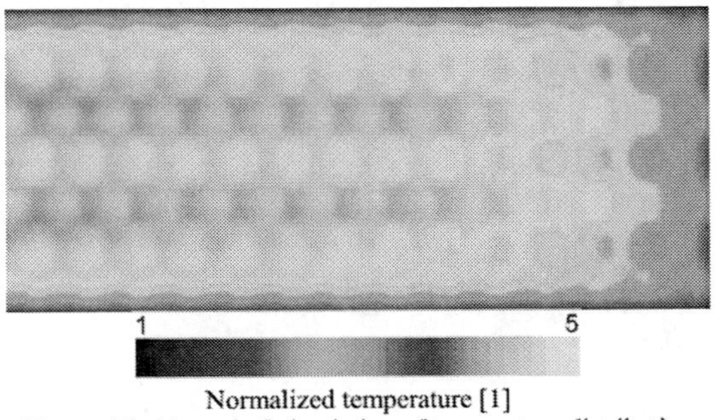


**Figure 16 :** Numerical simulation of temperature distribution on steel sample 300 W, 13 m/s

## 5. Conlusion

Based on the structural metal sheets, several samples from carbon and glass fabrics were created. To obtain an accurate 3D CAD models of this newly created parts, using of reverse engineering methods, namely the 3D scanning it was necessary. One of possible way, how to get the real geometry based on manual repairing of the scanned data by the vector blends definition is described. In respect to future numerical modeling the dependence of the geometry parameters to created mesh were compared. That could be even better and easier to use coarse geometry instead of highly optimized and smoothed one has been found.

Results of composite sample shown that temperatures on the sample was lower than in case of steel sample. This leads to conclusion that composite sample reached the better heat transfer than steel sample. This trend was not expected because with same surface structure the heat transfer of composite material should be lower than heat transfer of steel. It is possible that composite sample caused different airflow behaviour near the surface because the surface was not glossy as the steel sample, but it was covered by small roughness from the production. This could lead to higher turbulence intensity and higher heat transfer. For confirmation, it is planned to repeat the measurement. The fact that composite sample shown better homogeneity of temperature distribution than steel one will be studied in future research. Study also shown, that in conditions of higher ambient temperature the plastic made aerodynamics tunnel is not solid to stand under pressure in higher Reynolds number. This finding will be considered for future development of the experimental device. Also, numerical modeling of materials reinforced with fibers and dependency on the plies angles, numbers and alignment would be carried.

This publication was written at the Technical University of Liberec as part of the project "Innovation of technical systems structures with the use of composite materials" with the support of the Specific University Research Grant, as provided by the Ministry of Education, Youth and Sports of the Czech Republic in the year 2016.

## References

1. V. Malikov, *Umformung von strukturieiten Blechen.* Winter-Industries, **105** (2013)
2. Simon S., *Werkstoffgerechtes Konstruieren* BTU Cottbus, **208** (2005)
3. J. Egert, K. Frana, S. Simon, S. Wichmann, *Heat Transfer Studies on Structured Metal Sheets.* KMUTNB: IJAST, **9**, (2016)
4. Sterzing A., *Bewertung von Leichtbaupotenzial* Zwickau: Drecht aus dem IWU, (2005)
5. Thill C., Etches J.A., Bond I.P., *Trapezoidal corrugated laminates*, Appl. Science, **41**, (2010)
6. K. Frana, S. Simon, *Air Flows along Perforated Metal Plates*, Journal of Mech, **9**, (2015)
7. R. L. Webb, N. Kim, Principles of enhanced heat transfer, Taylor & Francis, (2005)
8. P. Kulhavy, J. Petrik, P. Srb, P. Lepsík: *Vibration response of composite structures*, EAN 2016, Uni. of West bohemia, (2016)
9. W. T. Wang, T. Y. Kam, Material characterization of laminated composite plates, Composite Structure, **50** (2000)
10. D. Briassoulis, *Equivalent orthotropic properties of corrugated sheets,* Comp. & Structures **23**, (1986)
11. ANSYS, Inc., Mechanical Structural Nonlinearities – *Meshing in mechanical*, (2010)
12. P. Kulhavy, P. Srb, P. Lepsík: *Digitization of structured composite, EVM*, Ústí n. L., **4**, (2016)

# CAA modeling of helicopter main rotor in hover

*Alexander N.* Kusyumov[1,a], *Sergey A.* Mikhailov[2], *Andrey S.* Batrakov[3], *Sergey A.* Kusyumov[4] and *George* Barakos[5]

[1-4]Kazan National Research Technical University n.a. A. N. Tupolev (KNRTU-KAI), Kazan, Russia
[5]University of Glasgow, Glasgow, UK

**Abstract.** In this work rotor aeroacoustics in hover is considered. Farfield observers are used and the nearfield flow parameters are obtained using the in house HMB and commercial Fluent CFD codes (identical hexa-grids are used for both solvers). Farfield noise at a remote observer position is calculated at post processing stage using FW–H solver implemented in Fluent and HMB. The main rotor of the UH-1H helicopter is considered as a test case for comparison to experimental data. The sound pressure level is estimated for different rotor blade collectives and observation angles.

## 1 Introduction

Main and tail rotors of conventional helicopters contribute to the generation of near-field and far-field noise [1]. Engine and fuselage noise are typically of secondary significance.

A helicopter main rotor generates primarily tonal and broadband noise. Tonal noise can be described as sum of two components: thickness and loading noise described by linear aerodynamic theory. The reason of thickness noise is the displacement of the fluid by the rotor blade and has a strong in-plane (rotor disk plane) contribution directed towards the flight direction. The term "loading noise" is usually a reference to the harmonic noise from non-impulsive loading sources. Loading noise is caused by the accelerating force generated by the moving blade surface. Basic loading noise during hover is generally dominant in a conical region directed 30 to 40 degrees downward from the rotor plane [2]. Thickness noise and loading noise are described by linear aerodynamic theory based on the Ffowcs Williams – Hawkings (FW–H) equation [2 - 4].

Additional sources, include Blade Vortex Interaction (BVI) noise and High Speed Impulsive (HSI) noise, and are dominant at specific flight regimes.

HSI noise is a particularly intense and is generated by helicopter rotors in high-speed forward flight. This HSI noise is closely associated with the appearance of shocks and transonic flow around the advancing rotor blades. Including the quadrupole sources in the FW–H equation [2] leads to account of nonlinearities in vicinity of the rotor blade. According to estimations (see, for example, [5]) omission of the quadrupole sources in the FW–H equation leads to inaccurate results for tip Mach numbers higher than 0.7.

Prediction of noise generation and propagation for aircraft design and certification falls in the area of Computational Aeroacoustics (CAA). CAA aims to combine flow data obtained by Computational Fluid Dynamics (CFD) for a nearfield area within a region near the aircraft surface. The sound perceived by a farfield observer is determined solving the wave propagation equation.

In the present work a task of design and validation of an in-house program on numerical solution of the FW–H equation is considered for a main rotor hover mode (unlike paper [6] that is based on a semi-analytical approach). The main rotor of UH–1H helicopter is considered as a test case for hover mode. Sound pressure level is estimated for different observation angles.

CFD data of the near field flow are obtained using the in-house HMB (University of Glasgow) code. In addition, results with the commercial ANSYS Fluent 13.0 code are obtained and the two data sets are compared. Farfield (thickness and load) noise obtained with an in-house FW–H solver (compatible with the HMB code) are compared to experimental data [7] for a remote observer position located at the rotor disc plane. For arbitrary observation angle the results of the in-house FW–H solver are compared to the data obtained with the FW–H method in Fluent.

[1a]Prof., department of aero-hydrodynamics, postbox7@mail.ru
[2]Prof., department of aero-hydrodynamics, sergey.mikhaylov@kai.ru
[3]PhD student, batrakov_a.c@mail.ru
[4]Teacher, department of aero-hydrodynamics, kusok88@yandex.ru
[5]Prof., School of Engineering, George.Barakos@glasgow.ac.uk

## 2 CFD Modelling

The geometry of the rotor is presented in table 1. Since temporal and spatial periodicity is assumed for the flow, the computational grid was constructed for a single rotor blade (with appropriate periodic boundary conditions). For the two-bladed rotor at hand, the computational domain is a half of a cylinder. The multi-block topology used in this paper, can be seen in Figure 1. C-type blocking is used around the blade. The structured and multi-block computational grid consisted of 88 blocks and 8.4 million cells. Along the aerofoil surface 218 points are located with concentration near the leading and trailing edges. Normal to the surface the first cell size is $10^{-5}$ of the blade's chord length and the cell ratio is less than 1.2.

**Table 1.** UH-1H operational characteristics

| Parameter | Value | |
|---|---|---|
| Number of blades $N$ | 2 | |
| Rotor radius $2R$, m | 14.63 | |
| Rotor solidity $\sigma$ | 0.0464 | |
| Blade chord $c$, m | 0.53 | |
| Blade airfoil | NACA 0012 | |
| Blade twist (root to tip), deg | -10.9 | |
| Max gross weight $W$, $N$ | 43000 | |

The multi-block topology used in this paper, can be seen in Figure 1.

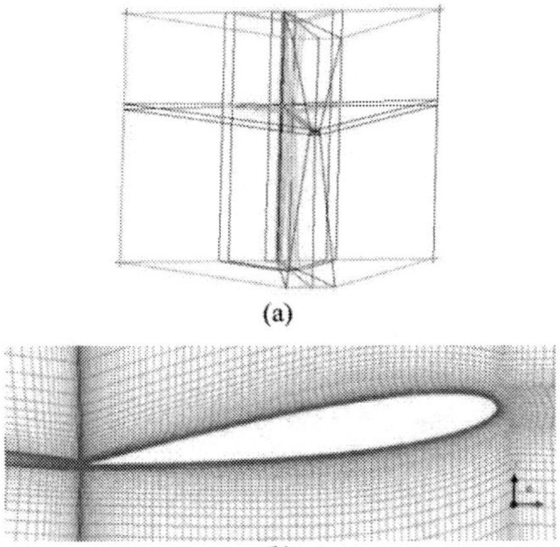

(a)

(b)

**Fig 1.** Multi-block topology (a) and mesh section (b)

The flow fields around the UH–1H helicopter rotor in hover were numerically simulated by solving 3D steady Reynolds-Averaged Navier-Stokes (RANS) equations.

Similar boundary conditions with some discrepancy at the bottom part of a modeling domain were used for the HMB and Fluent solvers.

For HMB the "source-sink" boundary conditions [8] were used at all surfaces of the computational domain apart from the symmetry plane. The "periodic" boundary condition provides the periodicity of the flow around the blade. Using the "inflow–outflow" boundary conditions allows for a reduction of the computational domain size. The inflow ($W_{in}$) and outflow ($W_{out}$) velocities were obtained from momentum theory according to the expressions:

$$W_{in} = -\frac{U}{8}\sqrt{C_T}\left(\frac{R}{d}\right)^2 ; \quad W_{out} = -U\sqrt{C_T}\ .$$

Here $U$ is the speed of the blade tip; $C_T$ is the thrust coefficient; $d$=2.5$R$ is the distance between the rotor centre and the top surface of the computational domain. The thrust coefficient ($C_T$) is determined by the expression:

$$C_T = \frac{T}{q_{tip}\pi R^2},$$

where $q_{tip}$ is the dynamic pressure at the blade tip and $T$ is the rotor thrust, obtained after computation in iteration process. The rotor thrust coefficient $C_{TW}$ (under the condition $T$=$W$) is first used as an initial guess and then the thrust coefficient value $C_T$ is recomputed using successive approximations.

For the bottom part of computational domain the «pressure-outlet» boundary condition was assigned.

Simulations were conducted with the $k$-ω SST turbulence model. The flow conditions, and the values of the trust coefficient obtained with HMB and Fluent are presented in Table 2. The collective pitch angle $\theta_{0.75}$ at the distance $r$=0.75$R$ was choosen according to an approximate trimming method [9]. The values $W_{in}$ used for computations in both HMB and Fluent are shown in table 2. Some discrepancy between HMB and Fluent results (including difference between $W_{in}$) can be explained, by the different type of boundary conditions at the bottom part of the computational domain.

**Table 2.** Conditions for computations

| Parameter | Value |
|---|---|
| Collective pitch, $\theta_{0.75}$ | 3.3° |
| Tip Mach number, $M_{tip}$ | 0.8 |
| Tip Reynolds number, Re | $9.5\times10^6$ |
| Thrust coefficient (for $T$=$W$), $C_{TW}$ | 0.0055 |
| Inflow velocity (for $T$=$W$), $W_{in}$ | 0.403 m/s |
| Inflow velocity (HMB), $W_{in}$ | 0.421 m/s |
| Inflow velocity (Fluent), $W_{in}$ | 0.2 m/s |
| Thrust coefficient (HMB), $C_T$ | 0.00577 |
| Thrust coefficient (Fluent), $C_T$ | 0.0054 |

Some distributed performances of flow around the rotor blades are presented in Figures 2 - 4. Figure 2 presents visualisation of the tip vortices structure using $Q$ criterion iso-surface based on the HMB code results.

DOI: 10.1051/epjconf/201714302060

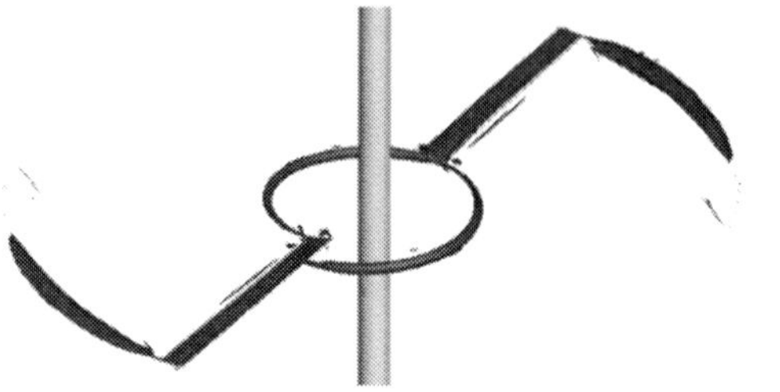

**Fig 2.** Tip vortex structure according to Q-criterion (Q=0.01)

Figure 3 presents results for the pressure coefficient distribution at different sections of the blade (normalized with the rotor radius, values of cross-sectional radius $\bar{r}$ ). Figure 3a shows the surface pressure coefficient distribution and location of the blade sections at $\bar{r}$ = 0.75, $\bar{r}$ = 0.85, $\bar{r}$ = 0.95.

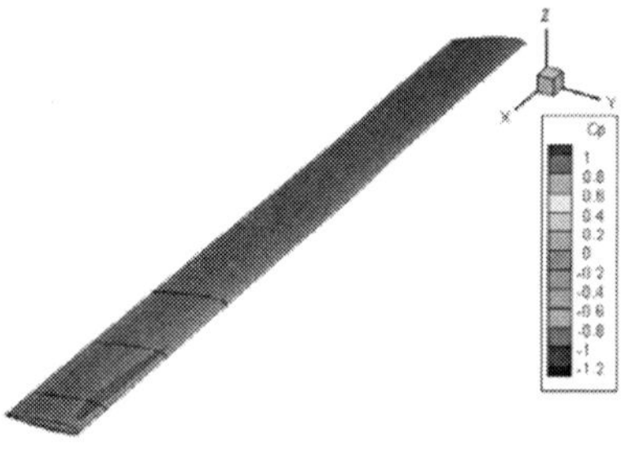

(a)

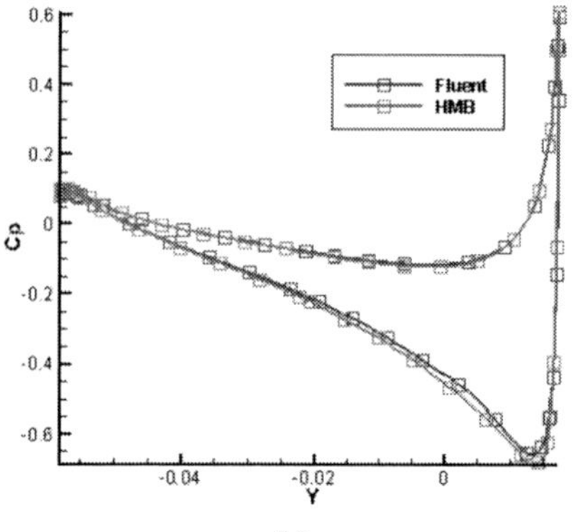


(b)

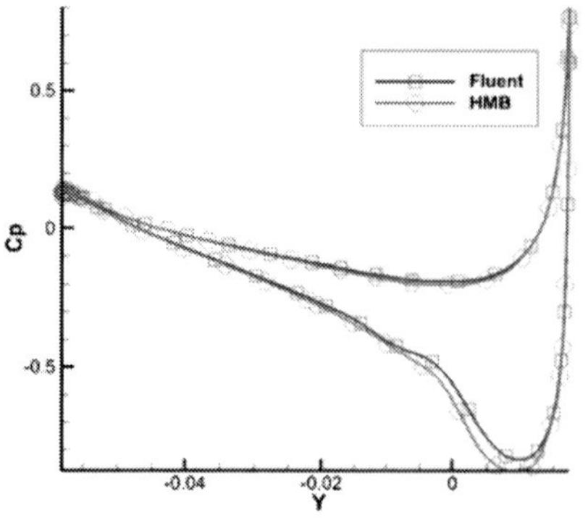


(c)

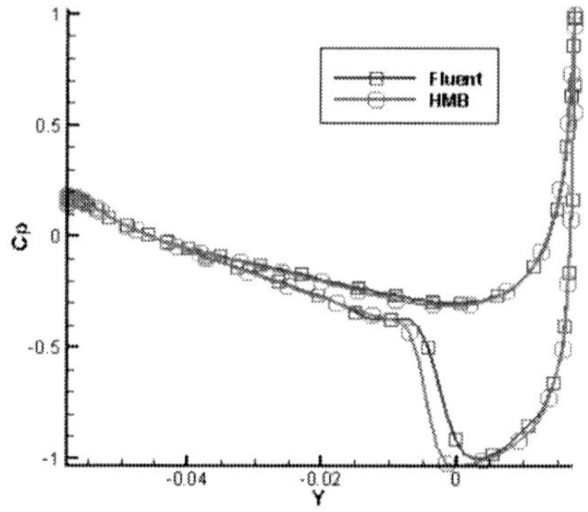


(d)

**Fig. 3.** Distribution of the pressure coefficient *Cp* on the UH-1H rotor blade surface (a) and at the blade sections: (b) $\bar{r}$ = 0.75; (c) $\bar{r}$ = 0.85; (d) $\bar{r}$ = 0.95.

Discrepancies between the pressure coefficient distributions obtained by HMB and Fluent lead to differences between the predicted values of the thrust coefficient (table 2).

The pressure coefficient *Cp* distribution (Figure 3) allows analysis of compressibility effects. From the comparison of *Cp* distribution on the top blade surface (*Cp*<0) a weak influence of compressibility effect is seen at sections $\bar{r}$ = 0.75, $\bar{r}$ = 0.85 (quite smooth *Cp* distribution). Stronger influence is seen at the section $\bar{r}$ = 0.95. This conclusion followed from the comparison of the local Mach number values distributions (HMB code results, Tecplot visualization) at the same sections in Figure 4: the local Mach number values $M \leq 0.4$ for $\bar{r}$= 0.75, $\bar{r}$ = 0.85 and $M > 0.4$ for $\bar{r}$ = 0.95.

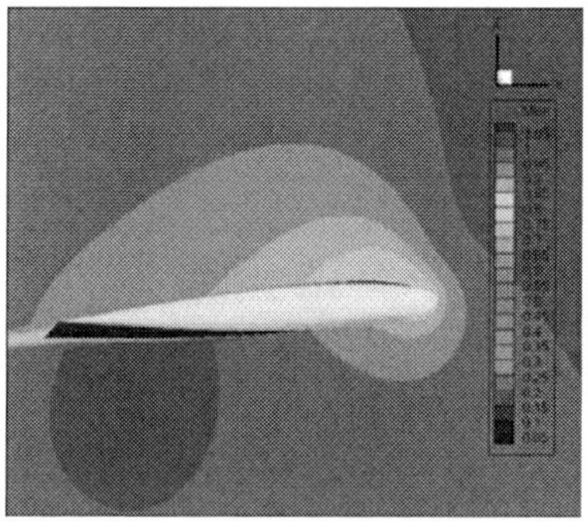

(a)

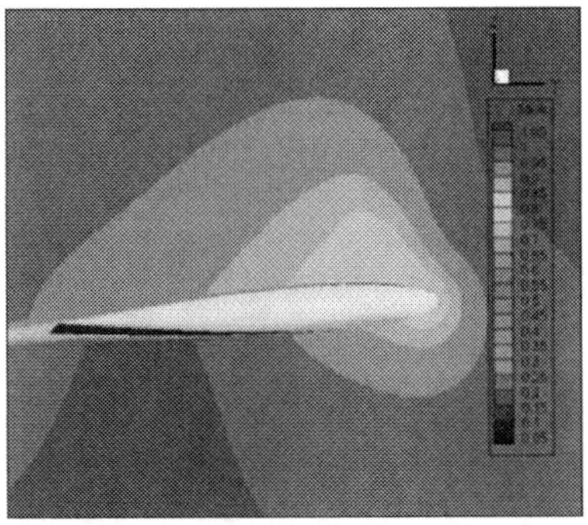

(b)

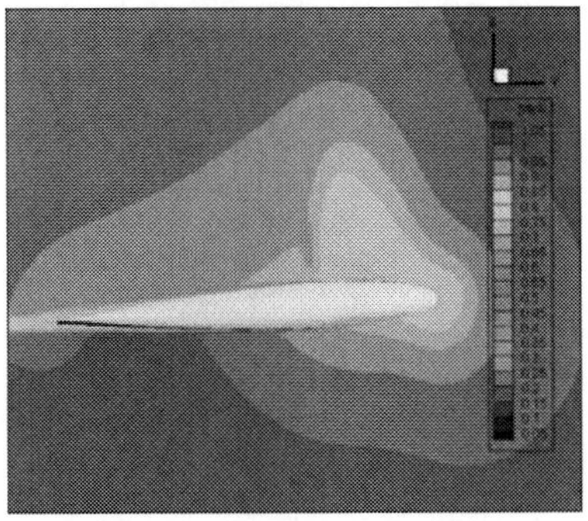

(c)

**Fig. 4.** Local Mach number distributions at the UH–1H blade sections: (a) $\bar{r}$ = 0.75; (b) $\bar{r}$ = 0.85; (c) $\bar{r}$ = 0.95.

From Figures 3 and 4 it follows that for the considered flow conditions around the UH-1H rotor blades the compressibility effect has significant influence on the flow parameters (and on the aeroacoustic emission power) close to the tip of blade ($\bar{r}$ > 0.85) only. The geometry of the rotor blade, the surface pressure distribution and the condition of the flow are used as initial data for integration of the sound propagation FW-H equation.

## 3 FW-H equations

A widely used method for predicting the aerodynamically generated rotor noise is based on the Ffowcs Williams – Hawkings equation [3]. Implementation of the FW–H equation to CFD methods is usually based on integral formulation of the FW–H equation. The FW–H equation integral formulation includes several terms, including the thickness, load and quadrupole terms. In the present work the linear formulation of task is considered (quadrupole term, corresponding to the HIS noise is neglected).

The far field retarded-time formulation by Farassat [2] for the sound pressure is commonly referred as:

$$4\pi p'(\mathbf{x},t) = \frac{1}{a_0}\frac{\partial}{\partial t}\int_{f=0}\left[\frac{\rho_0 a_0 v_n + l_r}{r|1-M_r|}\right]_{ret} dS. \quad (1)$$

Here $a_0$ is sound speed; $f$ = 0 is a function that describes the blade surface; $\rho_0$ is the density of medium; $v_n$ is the local normal velocity of source surface; $M_r = M_i\hat{r}_i$ is the source Mach number in radiation direction; $r$ is the distance between observer and source, $r = |\mathbf{x} - \mathbf{y}|$, $\mathbf{x}$ is the observer position vector, with components $x_i$, $\mathbf{y}$ is the source position vector, with components $y_i$; $p' = p - p_0$ is the acoustic pressure, $p_0$ is the pressure of the medium, $p$ is the surface pressure; $l_i$ are the components of the local pressure force intensity, $l_r = l_i\hat{r}_i$. The subscript *ret* denotes the retarded time and the integration is performed over the actual blade surface $f$ = 0. For a moving surface the retarded time $\tau$ is determined by the expression

$$|\mathbf{x} - \mathbf{y}(\eta,\tau)| = a_0(t - \tau),$$

where $t$ and $\tau$ are the observer and the source time, and $\eta$ is the Lagrangian variable of a moving surface point.

Following the Farassat 1A formulation, the expression (1) can be rewritten in the form [2]:

$$p'(\mathbf{x}, t) = p'_T(\mathbf{x}; t) + p'_L(\mathbf{x}; t). \quad (2)$$

Here

$$p'_T(\mathbf{x}; t) = \frac{1}{4\pi}\int_{f=0}\left[\frac{\rho_0 v_n}{r|1-M_r|^2} + \frac{\rho_0 v_n \dot{M}_r}{r|1-M_r|^3}\right]_{ret} dS \quad (3)$$

is the thickness noise, and

$$p'_L(\mathbf{x}; t) = \frac{1}{4\pi a_0}\int_{f=0}\left[\frac{p\cos(\theta)}{r|1-M_r|^2} + \frac{p\cos(\theta)\dot{M}_r}{r|1-M_r|^3}\right]_{ret} dS \quad (4)$$

is the load noise. In these equations $\cos(\theta) = \boldsymbol{n}\cdot\hat{\boldsymbol{r}}$ ($\boldsymbol{n}$ is a normal to emission surface), a dot over a variable implies the source time derivative.

As it was mentioned above the considered mathematical model of expressions (2) - (4), corresponds to the classical impermeable formulation of FW–H equations: the source surface $f$ = 0 in this case corresponds to the blade surface. This mathematical model for the high blade tip Mach numbers usually gives under predicted values of the noise level. Nevertheless the classical FW–H formulation was used in the resent publications for relatively low blade tip Mach numbers values (see reference [10], for example). Then the acoustic pressure (based on the Formulation 1A) can be modified with the empirical adjustment, based on the radiation Mach number.

As the rotor blade behavior is characterized by complex motion in the computer program it is assumed that the blade is the rotating not deformed rigid body.

The employed algorithm is similar to the algorithm, described, for example, in [2]. The first stage of FWH solution algorithm is envisaged dividing the rotor blade surface into a number of panels. Integration over each panel is approximated using the value at the center.

Rotation matrices and an angular velocity vector are used to describe the blade rotation and the effects of pitch, cone and rotor disc tilt angles. Positions and velocities of points are determined in a blade-fixed stationary frame by combining the matrices/vectors with rigid body surface relations for rotating reference frames.

The summed effect of all the panels at the observer location gives the acoustic pressure time history. Described algorithm was implemented to the in-house HRS (Helicopter Rotor Sound) solver.

Figure 5 presents the results of numerical simulation for an observer located at the rotor disc plane on the distance of 3$R$ from rotor. From the comparison the total noise signatures predicted by the Fluent and HRS codes for an observer located in the rotor plane a good agreement is seen.

The results of numerical simulation were compared also to the experimental data. Time observation shown in Figure 6 is 0.002 sec for the rotation period of 0.0241 sec.

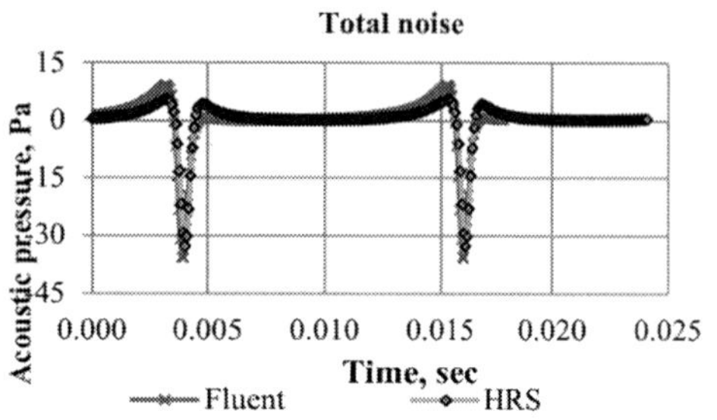


**Fig. 5.** Total noise predicted in the Fluent and HRS codes

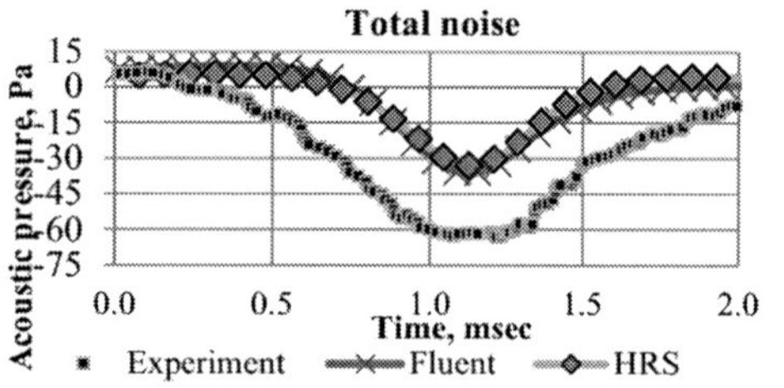


**Fig. 6.** Total noise predicted in the Fluent and HRS codes vs experimental data ([7])

Presented comparison of the Sound Pressure Level (SPL) predicted in the HRS solver corresponds to the SPL predicted in the Fluent code. It can be noted also a discrepancy between the experimental data and predicted by both (Fluent and HRS) solvers. This discrepancy is typical (see, for example, reference [2]) for accepted in the present work formulation of aeroacoustic task without taking into account the quadrupole term in FW–H equation.

Figures 7, 8 present the total noise predicted in Fluent and HRS codes for the observers located out of the rotor plane. Reference [7] presents for hover mode results for the in-plane observer only. For arbitrary observation angle the HRS solver results are compared to the data obtained with the FW–H method in Fluent.

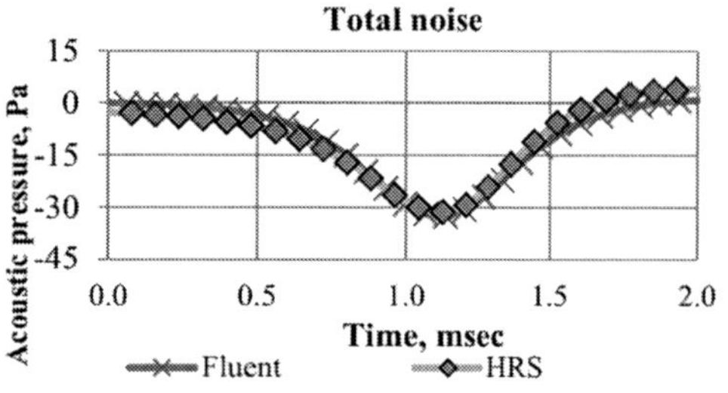


**Fig. 7.** Total noise predicted in the Fluent and HRS codes for the observation angle of 13 degrees

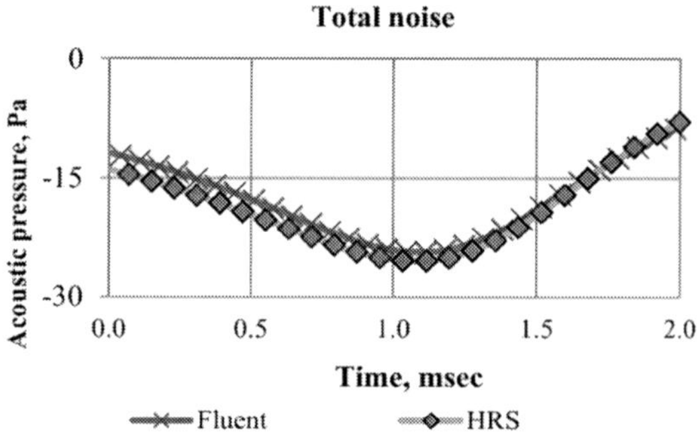


**Fig. 8.** Total noise predicted in Fluent and HRS codes for the observation angle of 26 degrees

It can be noted here that unlike the observation angle variation of the rotor blade collective angle for the values $\theta_{0.75}$ =0; 3.3 and 6 degrees did not lead to strong change of the pick SPL values.

One can in general conclude that the results of numerical simulation of the UH–1H rotor in hover with HRS (HMB) codes are in good agreement to results of numerical simulation with the ANSYS Fluent code. Some discrepancy between the results of both solvers can be explained, in particular, by the presence of the discrepancy between predicted aerodynamical loads on the rotor blades.

## 4 Conclusions and future work

Validation of the in-house HRS code for the FW–H (Ffowcs Williams – Hawkings) numerical solution was considered. The total rotor noise was considered as the sum of the thickness and load noise based on linear formulation of aeroacoustics. To determine the nearfield flow parameters for the UH–1H rotor in hover mode, the HMB and Fluent CFD codes are used. The aeroacoustic FW–H option of the Fluent code was used also to determine the SPL (Sound Pressure Level) for comparison to HRS results at arbitrary observation angles. The results of aeroacoustic numerical simulation of the UH–1H rotor in hover with HRS (HMB) codes are in good agreement to the Fluent results. Discrepancy to the experimental data for the both solvers is explained by the linear formulation of the considered task (without the quadrupole term for the high speed noise).

In the future, the in-house HRS program for the rotor aeroacoustics will be revised for high blade tip Mach numbers.

Acknowledgement

This work is supported by the "Project part of state task in the field of the scientific activity" Grant (No 9.1694.2014/K).

## References

1. B. Edwards, Ch. Cox, NASA/CR–2002–211650 (May 2002).

2. K. S. Brentner, F. Farassat, *Progress in Aerospace Sciences*, **39** (2003).

3. J. E. Ffowcs Williams, D. L. Hawkings, *Philosophical Transactions of the Royal Society*, **264**, no. 1151 (1969).

4. K. S. Brentner, NASA-TM-111251 Report, NASA Langley Research Center, Hampton, VA (1996).

5. A. S. Lyrintzis, *Aeroacoustics*, **2**, no. 2 (2003).

6. A. Kusyumov, S. Mikhailov, L. Garipova, A. Batrakov and G. Barakos, International conference EFM–2015, Czech Republic (November 18–21, 2014).

7. F.H. Shmitz and Y.H., Yu, NASA Technical memorandum 84390 (1983).

8. N. A. Mohd, G. N. Barakos, RAeS Aerodynamics Conference, University of Bristol, UK (July 27–28, 2010).

9. J. Seddon, Basic helicopter aerodynamics, BSP professional books, 137 pp. (1990).

10. K. Sharma, K. S. Brentner, AHS 72nd Annual Forum, West Palm Beach, Florida, (May 17–19, 2016).

# Modification of the Riemann problem and the application for the boundary conditions in computational fluid dynamics

*Martin* Kyncl[1,a] and *Jaroslav* Pelant[1,b]

[1] Aerospace Research and Test Establishment (VZLÚ),
Beranových 130, 199 05 Praha - Letňany, Czech Republic

**Abstract.** We work with the system of partial differential equations describing the non-stationary compressible turbulent fluid flow. It is a characteristic feature of the hyperbolic equations, that there is a possible raise of discontinuities in solutions, even in the case when the initial conditions are smooth. The fundamental problem in this area is the solution of the so-called Riemann problem for the split Euler equations. It is the elementary problem of the one-dimensional conservation laws with the given initial conditions (LIC - left-hand side, and RIC - right-hand side). The solution of this problem is required in many numerical methods dealing with the 2D/3D fluid flow. The exact (entropy weak) solution of this hyperbolical problem cannot be expressed in a closed form, and has to be computed by an iterative process (to given accuracy), therefore various approximations of this solution are being used. The complicated Riemann problem has to be further modified at the close vicinity of boundary, where the LIC is given, while the RIC is not known. Usually, this boundary problem is being linearized, or roughly approximated. The inaccuracies implied by these simplifications may be small, but these have a huge impact on the solution in the whole studied area, especially for the non-stationary flow. Using the thorough analysis of the Riemann problem we show, that the RIC for the local problem can be partially replaced by the suitable complementary conditions. We suggest such complementary conditions accordingly to the desired preference. This way it is possible to construct the boundary conditions by the preference of total values, by preference of pressure, velocity, mass flow, temperature. Further, using the suitable complementary conditions, it is possible to simulate the flow in the vicinity of the diffusible barrier. On the contrary to the initial-value Riemann problem, the solution of such modified problems can be written in the closed form for some cases. Moreover, using such construction, the local conservation laws are not violated. Algorithms for the solution of the modified Riemann problems were coded and used within our own developed code for the solution of the compressible gas flow (the Euler, the Navier-Stokes, and the RANS equations). Numerical examples show superior behaviour of the suggested boundary conditions. Constructed boundary conditions are robust and accelerate the convergence of the method. The original result of our work is the analysis of various modifications of the Riemann problem and its applications.

## 1 Introduction

Undoubtedly, the boundary conditions play important role in the computational fluid dynamics (CFD). We work with the system of equations describing non-stationary compressible turbulent fluid flow, i.e. the Reynolds-Averaged Navier-Stokes (RANS) equations in 2D and 3D. We focus on the numerical solution of these equations, and on the boundary conditions. We suggest the original approach to the boundary conditions. The aim is to satisfy the conservation laws in the close vicinity of the boundary. Usually, the boundary problem is being linearized, or roughly aproximated. The inaccuracies implied by these simplifications may be small, but it has a huge impact on the solution in the whole studied area, especially for the non-stationary flow. In our approach we try to be as exact as possible. Therefore we use the analysis of the so-called Riemann problem for the 2D/3D split Euler equations in order to construct the boundary values (for the density, velocity, pressure). We solve the boundary modifications of this initial-value problem. The crucial problem of many numerical methods for the solution of the compressible flow lies in the evaluation of the so-called fluxes through the edges/faces $\Gamma_{ij}$ of the particular elements. The state values in the vicinity of the edge $\Gamma_{ij}$ are known at time instant $t_k$, and these values form the initial conditions (LIC - left-hand side, and RIC - right-hand side) for the so-called Riemann problem for the 2D/3D split Euler equations. The exact (entropy weak) solution of this problem cannot be expressed in a closed form, and has to be computed by an iterative process (to given accuracy). Therefore various approximations of this solution are usually analyzed. At the boundary faces we deal with the local modified Riemann problem, where the LIC is given, while the RIC is not known. In some cases (far field boundary) it is wise to choose the RIC here as the solution of the local Riemann problem with given far field values, which gives better results than the solution of the linearized Riemann problem, see [3]. Another boundary condition based on the exact Riemann problem solution, simulating the impermeable wall on move, was shown in

[a] e-mail: kyncl@vzlu.cz
[b] e-mail: pelant@vzlu.cz

[13, pages 221-225], where the RIC is constructed in a special way, in order to obtain the desired solution. Using the analysis of the Riemann problem we show, that the RIC for the local problem can be partially replaced by the suitable complementary condition. Some of the suggested boundary conditions were shown in [8] (by preference of pressure, temperature, velocity,...). We complement the boundary problem suitably, and we show the analysis of the resulting uniquely-solvable modified Riemann problem. We construct own algorithm for the solution of this boundary problem. It can be used within various methods in CFD. The algorithm was coded and used within our own developed code for the solution of the compressible gas flow (the Euler, NS, and RANS equations).

## 2 The Riemann Problem for the Split Euler Equations

For many numerical methods dealing with the two or three dimensional equations, describing the compressible flow, it is useful to solve the Riemann problem for the *3D split* Euler equations. We search the solution of the system of partial differential equations in time $t$ and space $(x, y, z)$

$$\begin{aligned} \frac{\partial\varrho}{\partial t}+\frac{\partial\varrho u}{\partial x}&=0\\ \frac{\partial\varrho u}{\partial t}+\frac{\partial(p+\varrho u^2)}{\partial x}&=0\\ \frac{\partial\varrho v}{\partial t}+\frac{\partial\varrho uv}{\partial x}&=0\\ \frac{\partial\varrho w}{\partial t}+\frac{\partial\varrho uw}{\partial x}&=0\\ \frac{\partial E}{\partial t}+\frac{\partial u(E+p)}{\partial x}&=0, \end{aligned} \tag{1}$$

equipped with the initial conditions

$$\varrho(x,t)=\varrho_L, \boldsymbol{v}(x,t)=\boldsymbol{v}_L, p(x,t)=p_L, \quad x<0, \tag{2}$$

$$\varrho(x,t)=\varrho_R, \boldsymbol{v}(x,t)=\boldsymbol{v}_R, p(x,t)=p_R, \quad x>0. \tag{3}$$

Vector $\boldsymbol{v}=(u,v,w)$ denotes the velocity, $\varrho$ density, $p$ pressure, $E=\varrho e+\varrho|\boldsymbol{v}|^2$ is the total energy, with $e$ denoting the specific internal energy. We assume the equation of state for the calorically ideal gas

$$e=\frac{p}{\varrho(\gamma-1)}.$$

'Split' means here that we still have 5 equations in 3D, but the dependence on $y, z$ (space coordinates) is neglected, and we deal with the system for one space variable $x$. The system (1) is considered in the set $Q_\infty=(-\infty,\infty)\times(0,+\infty)$.

The solution of this problem is fundamentally the same as the solution of the Riemann problem for the 1D Euler equations, see [9, page 138]. In fact, the solution for the pressure, the first component of the velocity, and the density is exactly the same as in one-dimensional case. It is a characteristic feature of the hyperbolic equations, that there is a possible raise of discontinuities in solutions, even in the case when the initial conditions are smooth, see [12, page 390]. Here the concept of the classical solution is too restrictive, and therefore we seek a weak solution of this problem. To distinguish physically admissible solutions from nonphysical ones, entropy condition must be introduced, see [12, page 396]. By the solution of the problem (1),(2),(3) we mean the *weak entropy solution* of this problem in $Q_\infty$. The analysis to the solution of this problem can be found in many books, i.e. [9], [12], [13]. The general theorem on the solvability of the Riemann problem can be found in [9, page 88].

The solution is piecewise **smooth** and its general form can be seen in Fig. 1, where the system of half lines is drawn.

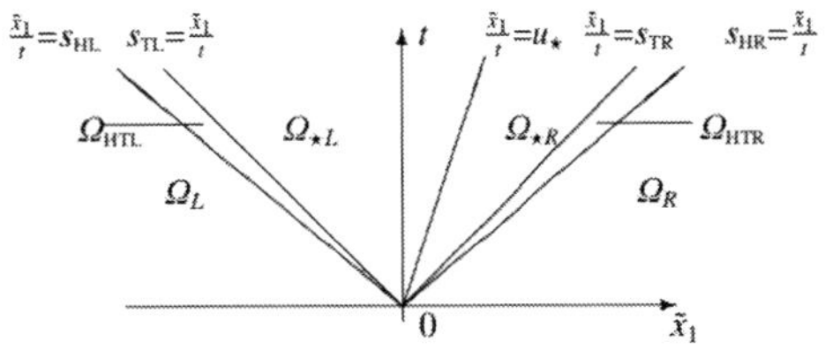


**Fig. 1.** Structure of the solution of the Riemann problem (1),(2),(3)

These half lines define regions, where solution is either constant or given by a smooth function. Let us define the open sets called **wedges** (see Fig. 1):

$$\begin{aligned} \Omega_L&=\text{wedge}(L_{-\infty}, s_{\text{HL}}), & \Omega_{\star R}&=\text{wedge}(u_\star, s_{\text{TR}}),\\ \Omega_{\text{HTL}}&=\text{wedge}(s_{\text{HL}}, s_{\text{TL}}), & \Omega_{\text{HTR}}&=\text{wedge}(s_{\text{TR}}, s_{\text{HR}}),\\ \Omega_{\star L}&=\text{wedge}(s_{\text{TL}}, u_*), & \Omega_R&=\text{wedge}(s_{\text{HR}}, L_{+\infty}). \end{aligned}$$

Using the theory in [9], [12], [13], we can write the solution for the primitive variables as

$$\begin{aligned} (\varrho,u,v,w,p)|_{\Omega_L}&=(\varrho_L,u_L,v_L,w_L,p_L),\\ (\varrho,u,v,w,p)|_{\Omega_{\star L}}&=(\varrho_{\star L},u_\star,v_L,w_L,p_\star),\\ (\varrho,u,v,w,p)|_{\Omega_{\star R}}&=(\varrho_{\star R},u_\star,v_R,w_R,p_\star),\\ (\varrho,u,v,w,p)|_{\Omega_R}&=(\varrho_R,u_R,v_R,w_R,p_R). \end{aligned}$$

The folowing relations for these variables hold:

$$u_\star=u_L+\begin{cases} -(p_\star-p_L)\left(\dfrac{\frac{2}{(\gamma+1)\varrho_L}}{p_\star+\frac{\gamma-1}{\gamma+1}p_L}\right)^{\frac{1}{2}}, p_\star>p_L\\ \frac{2}{\gamma-1}a_L\left[1-\left(\frac{p_\star}{p_L}\right)^{(\gamma-1)/2\gamma}\right], p_\star\le p_L \end{cases} \tag{4}$$

$$\varrho_{\star L}=\begin{cases} \varrho_L\dfrac{\frac{\gamma-1}{\gamma+1}\frac{p_L}{p_\star}+1}{\frac{p_L}{p_\star}+\frac{\gamma-1}{\gamma+1}}, p_\star>p_L\\ \varrho_L\left(\frac{p_\star}{p_L}\right)^{\frac{1}{\gamma}}, \quad p_\star\le p_L \end{cases} \tag{5}$$

$$s^1_{TL}=\begin{cases} u_L-a_L\sqrt{\frac{\gamma+1}{2\gamma}\frac{p_\star}{p_L}+\frac{\gamma-1}{2\gamma}}, p_\star>p_L\\ u_\star-a_L\left(\frac{p_\star}{p_L}\right)^{(\gamma-1)/2\gamma}, \quad p_\star\le p_L \end{cases} \tag{6}$$

$$u_\star=u_R+\begin{cases} (p_\star-p_R)\left(\dfrac{\frac{2}{(\gamma+1)\varrho_R}}{p_\star+\frac{\gamma-1}{\gamma+1}p_R}\right)^{\frac{1}{2}}, p_\star>p_R,\\ -\frac{2}{\gamma-1}a_R\left[1-\left(\frac{p_\star}{p_R}\right)^{(\gamma-1)/2\gamma}\right], p_\star\le p_R \end{cases} \tag{7}$$

$$\varrho_{\star R}=\begin{cases} \varrho_R\dfrac{\frac{p_\star}{p_R}+\frac{\gamma-1}{\gamma+1}}{\frac{\gamma-1}{\gamma+1}\frac{p_\star}{p_R}+1}, p_\star>p_R\\ \varrho_R\left(\frac{p_\star}{p_R}\right)^{\frac{1}{\gamma}}, \quad p_\star\le p_R \end{cases} \tag{8}$$

$$s^3_{TR}=\begin{cases} u_R+a_R\sqrt{\frac{\gamma+1}{2\gamma}\frac{p_\star}{p_R}+\frac{\gamma-1}{2\gamma}}, p_\star>p_R\\ u_\star+a_R\left(\frac{p_\star}{p_R}\right)^{(\gamma-1)/2\gamma}, \quad p_\star\le p_R \end{cases} \tag{9}$$

Here $a_L = \sqrt{\gamma p_L/\varrho_L}$, $a_R = \sqrt{\gamma p_R/\varrho_R}$, and $\gamma$ denotes the adiabatic constant. Further $s^1_{TL}$ denotes "unknown left wave speed", $s^3_{TR}$ "unknown right wave speed". Note, that the system (4) - (9) is the system of 6 equations for 6 unknowns $p_\star, u_\star, \varrho_{\star L}\ \varrho_{\star R}, s^1_{TL}, s^3_{TR}$.

### Solution for the Pressure $p_\star$

The combination of the equations (4), (7) gives the implicit equation for the unknown pressure $p_\star$

$$u_L + F_{1L}(p_\star) = u_R + F_{3R}(p_\star),$$

with

$$F_{1L}(p) = \begin{cases} -(p - p_L)\left(\frac{\frac{2}{(\gamma+1)\varrho_L}}{p + \frac{\gamma-1}{\gamma+1}p_L}\right)^{\frac{1}{2}}, & p > p_L \\ \frac{2}{\gamma-1}a_L\left[1 - \left(\frac{p}{p_L}\right)^{(\gamma-1)/2\gamma}\right], & p \leq p_L. \end{cases},$$

$$F_{3R}(p) = \begin{cases} (p - p_R)\left(\frac{\frac{2}{(\gamma+1)\varrho_R}}{p + \frac{\gamma-1}{\gamma+1}p_R}\right)^{\frac{1}{2}}, & p > p_R, \\ -\frac{2}{\gamma-1}a_R\left[1 - \left(\frac{p}{p_R}\right)^{(\gamma-1)/2\gamma}\right], & p \leq p_R. \end{cases}$$

This is a nonlinear algebraic equation, and one cannot express the analytical solution of this problem in a closed form. The solution $p_\star$ can be found as the root of the function $F(p)$ defined as

$$F(p) = F_{3R}(p) - F_{1L}(p) + u_R - u_L. \tag{10}$$

The analysis of this function can be found in [9]. Here we state, that $F(p)$ is monotone and concave. Also the analytic expression for its derivative is available. For a positive solution for the pressure $F(0) < 0$ is required. This gives the *pressure positivity condition*

$$u_R - u_L < \frac{2}{\gamma - 1}(a_L + a_R). \tag{11}$$

The Newton method can be used to find the root of $F(p) = 0$. With the pressure $p_\star$ known, we use the equations (4)-(9) to obtain the whole solution.

### Remarks

- Once the pressure $p_\star$ is known, the solution on the left-hand side of the contact discontinuity depends only on the left-hand side initial condition (2). And similarily, with $p_\star$ known, only the right-hand side initial condition (3) is used to compute the solution on the right-hand side of the contact discontinuity.
- The solution in $\Omega_L \cup \Omega_{HTL} \cup \Omega_{\star L}$ (across 1 wave) (solvability in general case)
  The solution components in $\Omega_L \cup \Omega_{HTL} \cup \Omega_{\star L}$ region must satisfy the system of equations (4)-(6). It is a system of three equations for four unknowns ($\varrho_{\star L}$, $p_\star$, $u_\star$, $s^1_{TL}$). We have to add another equation in order to get the uniquely solvable system in $\Omega_L \cup \Omega_{HTL} \cup \Omega_{\star L}$. This is the key problem for the outlet boundary condition.
- The equation (4) can be written as the *equation for pressure*, see [8]

$$p_\star = P(u_\star), \tag{12}$$

$$P(u) = \begin{cases} P_S(u), & u < u_L, \\ P_R(u), & u_L \leq u < u_L + \frac{2}{\gamma-1}a_L, \end{cases}$$

with

$$P_S(u) = \frac{2p_L + \frac{\gamma+1}{2}\varrho_L(u_L - u)^2 + (u_L - u)\sqrt{4\varrho_L\gamma p_L + \varrho_L^2(\frac{\gamma+1}{2})^2(u_L - u)^2}}{2},$$

$$P_R(u) = p_L\left(\frac{-u + u_L + \frac{2}{\gamma-1}a_L}{\frac{2}{\gamma-1}a_L}\right)^{\frac{2\gamma}{\gamma-1}}.$$

In our work we use the analysis of this problem also for the solution of the initial-boundary value at the boundary faces.

## 3 Boundary Condition by Preference of Pressure

Using this boundary condition, we try to prescribe given static pressure at the face. This boundary condition corresponds to the real-world problem, when we deal with the experimentally obtained pressure distribution at the boundary. The conservation laws must be satisfied in close vicinity to the boundary face. We use the analysis of the incomplete Riemann problem to construct the values for the density and the pressure at each boundary face. The following notation is used (some values are used only for an INLET case)

- $\varrho_B$ static density at the boundary (unknown)
- $p_B$ static pressure at the boundary (unknown)
- $\boldsymbol{v}_B$ velocity vector at the boundary (in global coordinates) (unknown)
- $u_B$ normal component of velocity at the boundary (local coordinates) (unknown)
- $\boldsymbol{n}$ unit outer normal to face $\boldsymbol{n} = (n_1, n_2, n_3)$
- $\varrho_L$ static density near the wall, time 0
- $p_L$ static pressure near the wall, time 0
- $\boldsymbol{v}_L$ velocity vector, state near the wall at time 0
- $u_L$ normal component of velocity, state near the wall at time 0, $u_L = \boldsymbol{v}_L \cdot \boldsymbol{n}$
- $p_{\text{GIVEN}}$ given value for the pressure (=desired pressure at the face)
- $\varrho_{\text{GIVEN}}$ given value for the density (=desired INLET density at the face)
- $\boldsymbol{d}$ given INLET velocity direction $\boldsymbol{d} = (d_1, d_2, d_3)$

At the vicinity of the boundary we solve the modified Riemann problem for the split Euler equations (1), introduced in Section 2, with the left-hand side initial condition (2) and the **complementary conditions** (prescribed pressure).

$$p_\star = p_{\text{GIVEN}}, \tag{13}$$

$$p_R = p_\star, \tag{14}$$

Here $p_\star$ denotes the pressure in $\Omega_{\star L} \cup \Omega_{\star R}$ part of the solution of the Riemann problem for the split Euler equations, shown in Section 2. This way we have prescribed the desired pressure value whenever it is possible (see the general form of the solution in Section 2). Now it is possible to use (4),(5),(6) Further we must discuss all the possibilities of the left (shock, rarefaction) and the middle (discontinuity) wave in the solution, which is shown in figure 2. In the case of **outlet** ($u_\star \geq 0$), the transformation of the velocity into the global coordinates is

$$\boldsymbol{v}_B = \boldsymbol{n}(u_B - u_L) + \boldsymbol{v}_L.$$

Here $u_B$ denotes the computed velocity in the normal direction. In the case of an **inlet** ($u_\star < 0$) we must **prescribe another conditions** in order to obtain a uniquely solvable system. Here we may choose (for example) arbitrary density $\varrho_{\text{GIVEN}}$ and the direction of the velocity $\boldsymbol{d}$.

$$\varrho_{\star R} = \varrho_{\text{GIVEN}}, \ \boldsymbol{v}_B = \boldsymbol{d}|\boldsymbol{v}|, \text{ with } \boldsymbol{v}_B \cdot \boldsymbol{n} = u_\star \tag{15}$$

Then it is

$$p_B = p_{\text{GIVEN}}, \ \varrho_B = \varrho_{\text{GIVEN}}, \ \boldsymbol{v}_B = \frac{\boldsymbol{d}u_\star}{\boldsymbol{d} \cdot \boldsymbol{n}}.$$

The analysis of this problem was shown also in [8].

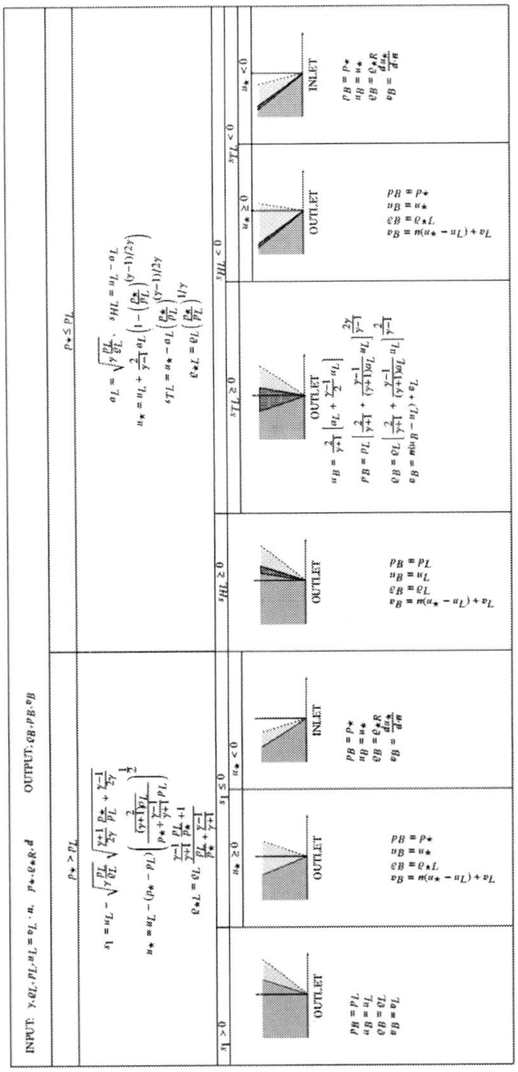


**Fig. 2.** Algorithm for the solution of the problem (1),(2),(13),(14),(15) at the half line $S_B = \{(0,t);\ t > 0\}$. The value of the pressure $p_\star$ is prescribed. Possible situations are illustrated by the pictures showing $\Omega_L \cup \Omega_{HTL} \cup \Omega_{\star L}$ region. The sought boundary state located at the time axis, marked red.

| | input data | | solution | | |
|---|---|---|---|---|---|
| Test | $u_L$ | $p_\star$ | $\varrho_B$ | $u_B$ | $p_B$ |
| 1 | 1.0 | 4.57991 | 1 | -0.50000 | 4.57991 |
| 2 | -1.0 | 0.12913 | 0.23175 | 0.50000 | 0.12913 |
| 3 | 2.0 | 0.5 | 1.0 | 2.0 | 1.0 |
| 4 | 2.0 | 3.0 | 1.0 | 2.0 | 1.0 |
| 5 | 2.0 | 6.77045 | 3.2593 | 0.00002 | 6.77045 |
| 6 | 2.0 | 8.0 | 1 | -0.23607 | 8.0 |

**Table 1.** The initial data $\varrho_L = 1.00$, $u_L$, $p_L = 1.00$, $p_\star$ and the solution $\varrho_B$, $u_B$, $p_B$ for the test problems.

## 4 Boundary Condition by Preference of Velocity

Here we show the boundary condition prefering the prescribed given velocity at the face. The conservation laws must be satisfied in close vicnity to the boundary face. Therefore we use the analysis of the incomplete Riemann problem to construct the values for the density and the velocity at each boundary face. The following notation is used (some values are used only for an INLET case)

- $\varrho_B$ static density at the boundary (unknown)
- $p_B$ static pressure at the boundary (unknown)
- $\boldsymbol{v}_B$ velocity vector at the boundary (in global coordinates) (unknown)
- $u_B$ normal component of velocity at the boundary (local coordinates) (unknown)
- $\boldsymbol{n}$ unit outer normal to face $\boldsymbol{n} = (n_1, n_2, n_3)$
- $\varrho_L$ static density near the wall, time 0
- $p_L$ static pressure near the wall, time 0
- $\boldsymbol{v}_L$ velocity vector, state near the wall at time 0
- $u_L$ normal component of velocity, state near the wall at time 0, $u_L = \boldsymbol{v}_L \cdot \boldsymbol{n}$
- $\boldsymbol{v}_{\text{GIVEN}}$ given velocity vector (=desired value at the face)
- $u_{\text{GIVEN}}$ given value for the normal vcomponent of velocity (=desired normal velocity at the face) $u_{\text{GIVEN}} = \boldsymbol{v}_{\text{GIVEN}} \cdot \boldsymbol{n}$
- $\varrho_{\text{GIVEN}}$ given value for the density (=desired INLET density at the face)

At the vicinity of the boundary we solve the modified Riemann problem for the split Euler equations (1), introduced in Section 2, with the left-hand side initial condition (2) and the **complementary conditions** (prescribed normal component of velocity).

$$u_\star = u_{\text{GIVEN}}, \tag{16}$$

$$u_R = u_\star, \tag{17}$$

Here $u_\star$ denotes the normal component of velocity in $\Omega_{\star L} \cup \Omega_{\star R}$ part of the solution of the Riemann problem for the split Euler equations, shown in Section 2. This way we have prescribed the desired normal component of velocity whenever it is possible (see the general form of the solution in Section 2). Now it is possible to use (12),(5),(6). Further we must discuss all the possibilities of the left (shock, rarefaction) and the middle (discontinuity) wave in the solution, which is shown in figure 3. The tangential velocities are conserved (same as "left" values) in the case of **outlet** ($u_\star \geq 0$), and the velocity vector in the global coordinates is

$$\boldsymbol{v}_B = \boldsymbol{n}(u_B - u_L) + \boldsymbol{v}_L.$$

Here $u_B$ denotes the computed velocity in the normal direction. In the case of an **inlet** ($u_\star < 0$) we must **prescribe another conditions** in order to obtain a uniquely solvable system. Here we may choose (for example) arbitrary density $\varrho_{\text{GIVEN}}$ and the whole vector of velocity

$$\varrho_{\star R} = \varrho_{\text{GIVEN}}, \quad \boldsymbol{v}_R = \boldsymbol{v}_{\text{GIVEN}}. \tag{18}$$

For example, it is possible to choose the density $\varrho_{\star R}$ using the given entropy index $\frac{p_o}{\varrho_o^\gamma}$ ($p_o, \varrho_o$ are some reference values for the pressure and the density)

$$\varrho_{\star R} = p_\star^{1/\gamma} / \left(\frac{p_o}{\varrho_o^\gamma}\right)^{1/\gamma} = \varrho_o \left(\frac{p_\star}{p_o}\right)^{1/\gamma}$$

Another possibility is to conserve (choose) the total temperature $\theta_0$ for the inlet case. Then, using thermodynamic relations (energy equation for a perfect gas), it is

$$\varrho_{\star R} = \frac{p_\star}{R\theta_0 \left(1 - \frac{\gamma-1}{2} \frac{|\boldsymbol{v}_{\text{GIVEN}}|^2}{\gamma R \theta_0}\right)}. \tag{19}$$

Let $\theta_0 = \theta_{0R} = p_R / (R\varrho_R \left(1 + \frac{\gamma-1}{2} \frac{|\boldsymbol{v}_R|^2}{\gamma p_R/\varrho_R}\right))$ and $\boldsymbol{v}_{\text{GIVEN}} = \boldsymbol{v}_R$. Then, using (19), it is $\varrho_{\star R} = \varrho_R \frac{p_\star}{p_R}$.

The algorithm for the construction of the primitive variables $\varrho_B, \boldsymbol{v}_B, p_B$ at the boundary face is shown in Fig. 3. The complete analysis of this problem is shown in [8].

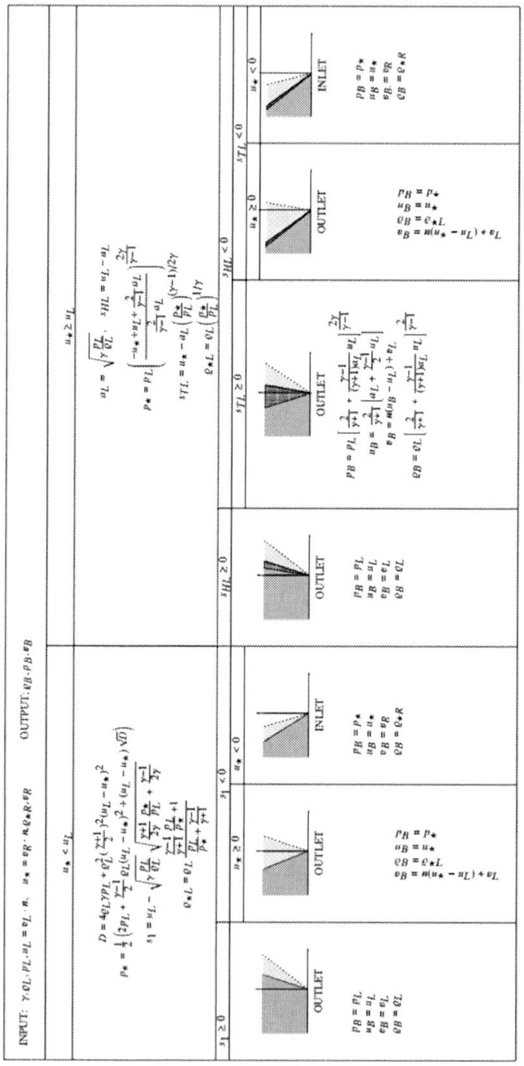


**Fig. 3.** Algorithm for the solution of the problem (1),(2),(16),(17),(18) at the half line $S_B = \{(0,t);\ t > 0\}$. Possible situations are illustrated by the pictures showing the region $\Omega_L \cup \Omega_{HTL} \cup \Omega_{\star L}$ with the sought boundary state located at the time axis.

## 5 Boundary Condition by Preference of Temperature

Using this boundary condition, we try to prescribe given static temperature at the face, comlete analysis was shown in [8]. The conservation laws must be satisfied in close vicnity to the boundary face. We use the analysis of the incomplete Riemann problem to construct the values for the density, pressure and velocity. The following notation is used (some values are used only for an INLET case)

- $\varrho_B$ static density at the boundary (unknown)
- $p_B$ static pressure at the boundary (unknown)
- $\boldsymbol{v}_B$ velocity vector at the boundary (in global coordinates) (unknown)
- $u_B$ normal component of velocity at the boundary (local coordinates) (unknown)
- $\boldsymbol{n}$ unit outer normal to face $\boldsymbol{n} = (n_1, n_2, n_3)$
- $\varrho_L$ static density near the wall, time 0
- $p_L$ static pressure near the wall, time 0
- $\boldsymbol{v}_L$ velocity vector, state near the wall at time 0
- $u_L$ normal component of velocity, state near the wall at time 0, $u_L = \boldsymbol{v}_L \cdot \boldsymbol{n}$
- $\theta_{\text{GIVEN}}$ given value for the pressure (=desired temperature at the face)
- $\varrho_{\text{GIVEN}}$ given value for the density (=desired INLET density at the face)
- $\boldsymbol{d}$ given INLET velocity direction $\boldsymbol{d} = (d_1, d_2, d_3)$

At the vicinity of the boundary we solve the modified Riemann problem for the split Euler equations (1), introduced in Section 2, with the left-hand side initial condition (2) and the **complementary conditions** (prescribed pressure).

$$\theta_{\star L} = \theta_{\text{GIVEN}}, \tag{20}$$

$$\theta_R = \theta_{\star L}, \tag{21}$$

Here $\theta_{\star L}$ denotes the static temperature in $\Omega_{\star L}$ part of the solution of the Riemann problem for the split Euler equations, shown in Section 2. This way we have prescribed the desired temperature value whenever it is possible (see the general form of the solution in Section 2). Now it is possible to use (4),(5),(6) together with the equation of state $\varrho_{\star L} = \frac{p_\star}{R\theta_{\star L}}$. We must discuss all the possibilities of the left (shock, rarefaction) and the middle (discontinuity) wave in the solution, which is shown in figure 4. In the case of **outlet** ($u_\star \geq 0$), the transformation of the velocity into the global coordinates is

$$\boldsymbol{v}_B = \boldsymbol{n}(u_B - u_L) + \boldsymbol{v}_L.$$

Here $u_B$ denotes the computed velocity in the normal direction. In the case of an **inlet** ($u_\star < 0$) we use the condition (21), which ensures that $p_R = p_\star, u_R = u_\star, \varrho_R = \varrho_{\star R} = \varrho_{\star L}$. Further it is possible to prescribe the direction of the velocity $\boldsymbol{d}$

$$\boldsymbol{v}_B = \boldsymbol{d}|\boldsymbol{v}|, \text{ with } \boldsymbol{v}_B \cdot \boldsymbol{n} = u_\star. \tag{22}$$

Then it is

$$\boldsymbol{v}_B = \frac{\boldsymbol{d} u_\star}{\boldsymbol{d} \cdot \boldsymbol{n}}.$$

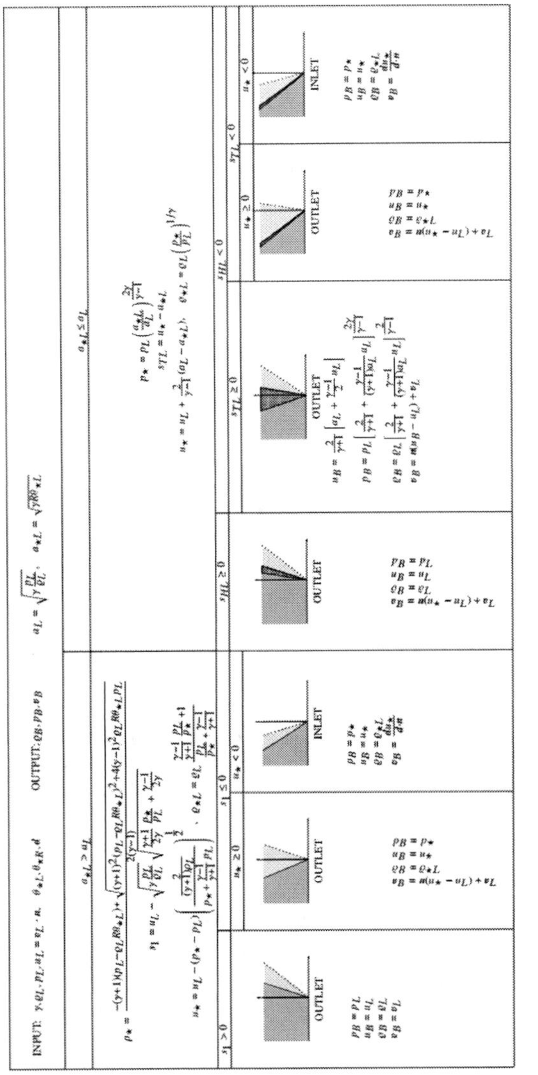

**Fig. 4.** Algorithm for the solution of the problem (1),(2),(20),(21),(22) at the half line $S_B = \{(0,t);\ t > 0\}$. The value of the pressure $p_\star$ is prescribed. Possible situations are illustrated by the pictures showing $\Omega_L \cup \Omega_{HTL} \cup \Omega_{\star L}$ region. The sought boundary state located at the time axis, marked red.

| input data | | solution (rounded) | | | |
|---|---|---|---|---|---|
| $u_L$ | $\theta_{\star L}$ | $\varrho_B$ | $u_B$ | $p_B$ | $\theta_B$ |
| 0.0 | 4500.0 | 7.0492 | -2434.5 | 9105332.5 | 4500.0 |
| 0.0 | 300.00 | 1.5000 | -62.370 | 129171.2 | 300.00 |
| 0.0 | 270.00 | 1.1547 | 26.345 | 89486.3 | 270.00 |
| 0.0 | 200.00 | 0.54528 | 255.83 | 31303.2 | 200.00 |
| 0.0 | 100.00 | 0.50235 | 278.89 | 27908.2 | 193.55 |
| 200.0 | 240.00 | 0.88342 | 312.22 | 61512.3 | 242.58 |
| 200.0 | 260.00 | 1.0507 | 257.13 | 78413.4 | 260.00 |

**Table 2.** The initial data $u_L$, $\theta_{\star L}$ and the solution $\varrho_B$, $u_B$, $p_B$ for test problems with $\varrho_L = 1.25$, $p_L = 100000$.

## 6 Boundary Condition by Preference of Total Quantities and Direction of Velocity

Using this boundary condition, we try to prescribe given total quantities at the face, comlete analysis was shown in [8], [5]. The conservation laws must be satisfied in close vicnity to the boundary face. We use the analysis of the incomplete Riemann problem to construct the values for the density, pressure and velocity. The following notation is used (some values are used only for an INLET case)

- $\varrho_B$ static density at the boundary (unknown)
- $p_B$ static pressure at the boundary (unknown)
- $\boldsymbol{v}_B$ velocity vector at the boundary (in global coordinates) (unknown)
- $u_B$ normal component of velocity at the boundary (local coordinates) (unknown)
- $\boldsymbol{n}$ unit outer normal to face $\boldsymbol{n} = (n_1, n_2, n_3)$
- $\varrho_L$ static density near the wall, time 0
- $p_L$ static pressure near the wall, time 0
- $\boldsymbol{v}_L$ velocity vector, state near the wall at time 0
- $u_L$ normal component of velocity, state near the wall at time 0, $u_L = \boldsymbol{v}_L \cdot \boldsymbol{n}$
- $p_o$ given value for the total pressure (=desired total pressure at the face)
- $\theta_o$ given value for the total temperature (=desired total temperature at the face)
- $\boldsymbol{d}$ given INLET velocity direction $\boldsymbol{d} = (d_1, d_2, d_3)$

At the vicinity of the boundary we solve the modified Riemann problem for the split Euler equations (1), introduced in Section 2, with the left-hand side initial condition (2) and the **complementary conditions** (prescribed direction of velocity and given total quantities).

$$\boldsymbol{v}_B = \boldsymbol{d}|\boldsymbol{v}_B|, \text{ with } \boldsymbol{v}_B \cdot \boldsymbol{n} = u_\star. \tag{23}$$

$$\theta_{\star R} = \theta_o\left(1 - \frac{\gamma - 1}{2a_o^2}u_\star^2\right), \text{ with } a_0^2 = \gamma R\theta_0(\boldsymbol{d} \cdot \boldsymbol{n})^2, \tag{24}$$

$$p_\star = p_o\left(\frac{\theta_{\star R}}{\theta_0}\right)^{\gamma/(\gamma-1)} \tag{25}$$

$$u_\star < 0. \tag{26}$$

Here $\theta_o > 0, p_o > 0$ are given constants, $R$ denotes the characteristic gas constant, and $\gamma$ is the adiabatic constant. The equations (24),(25) are considered for the velocity $u_\star \in \left(-\sqrt{\frac{2a_o^2}{(\gamma-1)}}, \sqrt{\frac{2a_o^2}{(\gamma-1)}}\right)$. The equations (24),(25) come from the idea, that the *total pressure* $p_o$ and the *total temperature* $\theta_o$ are known (prescribed) in the region $\Omega_{\star R}$.

According to analysis shown in [5], there are two possibillities for the 1-wave, either there is a **1-shock wave** and the sought velocity $u_\star$ can be found as a root of the function

$$F(u) = p_o\left(1 - \frac{\gamma - 1}{2a_o^2}u^2\right)^{\gamma/(\gamma-1)} - P_S(u), \tag{27}$$

considered for $u < u_L, -\sqrt{\frac{2a_o^2}{(\gamma-1)}} < u < 0$, and with $P_S(u)$ defined in (12).

Or there is a **1-rarefaction wave**, and the velocity $u_\star$ can be computed combining (24),(25),(12) as

$$u_\star = \frac{2\left(u_L + \frac{2}{\gamma-1}a_L\right) - \sqrt{DIS}}{2\left[1 + \frac{2a_L^2}{(\gamma-1)a_o^2}\left(\frac{p_o}{p_L}\right)^{(\gamma-1)/\gamma}\right]}, \tag{28}$$

$$DIS = 4\left(\frac{p_o}{p_L}\right)^{(\gamma-1)/\gamma}\frac{2a_L^2}{(\gamma-1)a_o^2}\left[\frac{2a_o^2}{(\gamma-1)} + \frac{4a_L^2}{(\gamma-1)^2}\left(\frac{p_o}{p_L}\right)^{(\gamma-1)/\gamma} - \left(u_L + \frac{2a_L}{\gamma-1}\right)^2\right].$$

Summarizing the above possibilities of the 1-shock and the 1-rarefaction wave we can construct the algorithm in figure 5. for the computation of the velocity $u_\star$ in the *star region*.

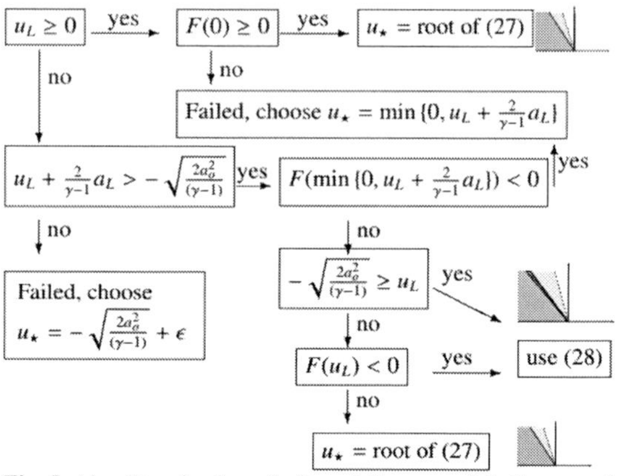


**Fig. 5.** Algorithm for the velocity solution $u_\star < 0$.In the case of failure, the problem doesn't have a solution, and the values are chosen.

Knowing the velocity $u_\star$ we compute the pressure $p_\star$ using (25), and the density $\varrho_{\star R} = \frac{p_\star}{R\theta_{\star R}}$. At last we set the boundary values as

$$p_B = p_\star, \quad \varrho_B = \varrho_{\star R}, \quad \boldsymbol{v}_B = \frac{\boldsymbol{d}u_\star}{\boldsymbol{d} \cdot \boldsymbol{n}}.$$

The problem (1),(2), (25), (24) has a unique solution for the initial data satisfying $u_L + \frac{2}{\gamma-1}a_L > -\sqrt{\frac{2a_o^2}{(\gamma-1)}}$, and $F(\min\{0, u_L + \frac{2}{\gamma-1}a_L\}) > 0$. In the case of $u_L + \frac{2}{\gamma-1}a_L \leq -\sqrt{\frac{2a_o^2}{(\gamma-1)}}$ there is no solution. In this case we prescribe the velocity $u_\star = -\sqrt{\frac{2a_o^2}{(\gamma-1)}} + \epsilon$, with $\epsilon > 0$ being a small positive constant. If $F(\min\{0, u_L + \frac{2}{\gamma-1}a_L\}) \leq 0$ then the problem does not have a negative solution, and for the practical applications we choose the velocity $u_\star = \min\{0, u_L + \frac{2}{\gamma-1}a_L\}$, or we use the boundary condition preferring the static pressure $p_\star = p_o$, see [2,8], in this case.

Let $\tilde{p}_0, \tilde{\theta}_0$ be the given total quantities together with the given **tangential velocity components** : $v_t$ and $w_t$ are the velocity components in the direction of $\boldsymbol{o}$ and $\boldsymbol{p}$. Here $\boldsymbol{o} \cdot \boldsymbol{n} = 0$, $|\boldsymbol{o}| = 1$, and $\boldsymbol{p} = \boldsymbol{n} \times \boldsymbol{o}$. Then we set

$$\boldsymbol{d} = -\boldsymbol{n}, \quad \theta_0 = \tilde{\theta}_o - \frac{\gamma - 1}{2\gamma R}(v_t^2 + w_t^2), \quad p_0 = \tilde{p}_0\left(\frac{\theta_0}{\tilde{\theta}_0}\right)^{\frac{\gamma}{\gamma-1}},$$

and we use the same algorithm (figure 5.) for the computation of the velocity $u_\star$. Then it is $p_B = p_\star$, $\varrho_B = \varrho_{\star R}$, and

$$\boldsymbol{v}_B = (n_1 u_\star + o_1 v_t + p_1 w_t, n_2 u_\star + o_2 v_t + p_2 w_t, n_3 u_\star + o_3 v_t + p_3 w_t, ).$$

## 7 Inlet Boundary Condition by Preference of Massflow

Using this boundary condition, we try to prescribe given total quantities at the face, comlete analysis was shown in [7]. The conservation laws must be satisfied in close vicnity to the boundary face. We use the analysis of the incomplete Riemann problem to construct the values for the density, pressure and velocity. The following notation is used (some values are used only for an INLET case)

- $\varrho_B$ static density at the boundary (unknown)
- $p_B$ static pressure at the boundary (unknown)
- $\boldsymbol{v}_B$ velocity vector at the boundary (in global coordinates) (unknown)
- $u_B$ normal component of velocity at the boundary (local coordinates) (unknown)
- $\boldsymbol{n}$ unit outer normal to face $\boldsymbol{n} = (n_1, n_2, n_3)$
- $\varrho_L$ static density near the wall, time 0
- $p_L$ static pressure near the wall, time 0
- $\boldsymbol{v}_L$ velocity vector, state near the wall at time 0
- $u_L$ normal component of velocity, state near the wall at time 0, $u_L = \boldsymbol{v}_L \cdot \boldsymbol{n}$
- $G_\star$ given inlet massflow (=desired massflow at the face)
- $\theta_0$ given value for the total temperature (=desired total temperature at the face)
- $\boldsymbol{d}$ given INLET velocity direction $\boldsymbol{d} = (d_1, d_2, d_3)$

At the vicinity of the boundary we solve the modified Riemann problem for the split Euler equations (1), introduced in Section 2, with the left-hand side initial condition (2) and the **complementary conditions** prescribing the mass flow, **direction of velocity**, and total temperature in $\Omega_{\star R}$ given as

$$u_\star \varrho_{\star R} = G_\star, \tag{29}$$

$$\boldsymbol{v}_B = \boldsymbol{d}|\boldsymbol{v}_B|, \text{ with } \boldsymbol{v}_B \cdot \boldsymbol{n} = u_\star. \tag{30}$$

$$\theta_{\star R} = \theta_o\left(1 - \frac{\gamma - 1}{2a_o^2}u_\star^2\right), \text{ with } a_0^2 = \gamma R\theta_0(\boldsymbol{d}\cdot\boldsymbol{n})^2, \tag{31}$$

$$u_\star < 0, \tag{32}$$

where $G_\star \le 0$ is given constant ($G_\star = \frac{mass flow}{face\ area}$). Here $\gamma$ and $R$ are gas constants. The equation (31) is considered for the velocity $u_\star \in \left(-\sqrt{\frac{2a_o^2}{(\gamma-1)}}, 0\right)$. In general, there are two possibilities of the wave pattern, which may interest us. We seek the unknown velocity $u_\star$ as a root of the function

$$F(u) = G(u) - G_\star, \quad \text{where } G(u) = u\frac{P(u)}{R\theta(u)}, \tag{33}$$

with $P(u)$ defined in (12), and

$$\theta(u) = \theta_0\left(1 - \frac{\gamma - 1}{2}\frac{u^2}{a_0^2}\right), \quad u \in \left(-\sqrt{\frac{2a_0^2}{\gamma - 1}}, 0\right).$$

The figure 6. shows example of such function for the chosen values. The function $F(u)$ is monotone, and it has a unique root if $u_L + \frac{2}{\gamma-1}a_L > -\sqrt{\frac{2a_0^2}{\gamma-1}}$.

Once the velocity $u_\star < 0$ is known, it is possible to compute the remaining values

$$\varrho_B = G_\star/u_\star, \; p_B = p(u_\star), \boldsymbol{v}_B = \frac{du_\star}{\boldsymbol{d}\cdot\boldsymbol{n}}.$$

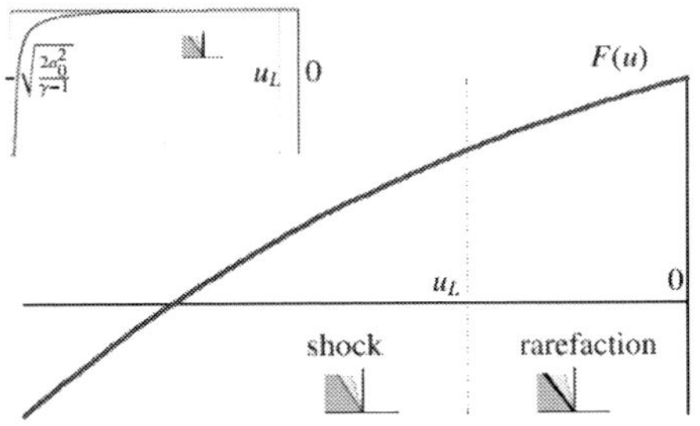


**Fig. 6.** Graph of the function $F(u)$ (red line) for given values $\theta_0 = 273.15, \gamma = 1.4, \varrho_L = 1.25, p_L = 100000, u_L = -50, G_\star = -200$.

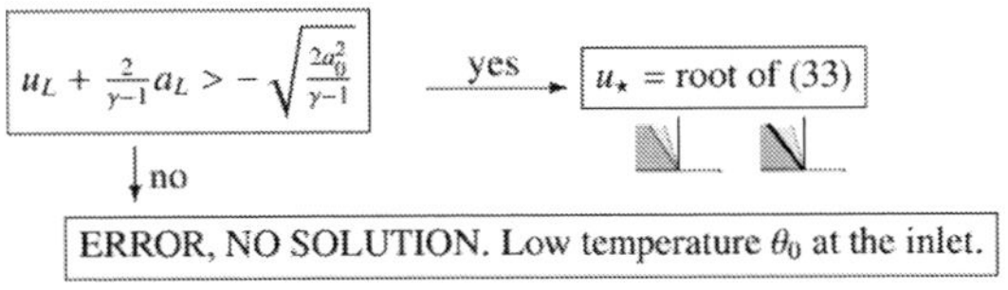


**Fig. 7.** Algorithm for the velocity solution $u_\star < 0$.In the case of failure, the problem doesn't have a solution, and the values are chosen.

Let $\tilde{\theta_0}$ be the given total temperature together with the given **tangential velocity components** : $v_t$ and $w_t$ are the velocity components in the direction of $\boldsymbol{o}$ and $\boldsymbol{p}$. Here $\boldsymbol{o}\cdot\boldsymbol{n} = 0$, $|\boldsymbol{o}| = 1$, and $\boldsymbol{p} = \boldsymbol{n}\times\boldsymbol{o}$. Then we set

$$\boldsymbol{d} = -\boldsymbol{n}, \quad \theta_0 = \tilde{\theta_o} - \frac{\gamma - 1}{2\gamma R}(v_t^2 + w_t^2),$$

and we use the same algorithm (figure 7.) for the computation of the velocity $u_\star$. Then it is $p_B = p_\star$, $\varrho_B = \varrho_{\star R}$, and

$$\varrho_B = G_\star/u_\star, \; p_B = p(u_\star),$$

$$\boldsymbol{v}_B = (n_1u_\star + o_1v_t + p_1w_t, n_2u_\star + o_2v_t + p_2w_t, n_3u_\star + o_3v_t + p_3w_t, ).$$

| input data | | | solution | | |
|---|---|---|---|---|---|
| $u_L$ | $p_L$ | $G_\star$ | $\varrho_B$ | $u_\star$ | $p_\star$ |
| 40 | 70000 | -44.3198 | 1.2853 | -34.481 | 100558.8 |
| -100 | 70000 | -171.582 | 1.1693 | -146.74 | 88080.1 |
| -200 | 100000 | -185.547 | 1.1427 | -162.38 | 85285.7 |
| -300 | 70000 | -244.851 | 0.89857 | -272.49 | 60920.9 |
| -600 | 70000 | -207.065 | 0.49508 | -418.24 | 26444.9 |

**Table 3.** The initial data $u_L$, $p_L$, $G_\star$ and the solution $\varrho_B$, $u_\star$, $p_\star$ for the test problems with $\varrho_L = 1.25$, $\theta_0 = 273.15$, $\boldsymbol{d}\cdot\boldsymbol{n} = 1$.

## 8 Outlet Boundary Condition by Preference of Massflow

Using this boundary condition, we try to prescribe given total quantities at the face, comlete analysis was shown in [6, 7]. The conservation laws must be satisfied in close vicnity to the boundary face. We use the analysis of the incomplete Riemann problem to construct the values for the density, pressure and velocity. The following notation is used (some values are used only for an INLET case)

- $\varrho_B$ static density at the boundary (unknown)
- $p_B$ static pressure at the boundary (unknown)
- $\boldsymbol{v}_B$ velocity vector at the boundary (in global coordinates) (unknown)
- $u_B$ normal component of velocity at the boundary (local coordinates) (unknown)
- $\boldsymbol{n}$ unit outer normal to face $\boldsymbol{n} = (n_1, n_2, n_3)$
- $\varrho_L$ static density near the wall, time 0
- $p_L$ static pressure near the wall, time 0
- $\boldsymbol{v}_L$ velocity vector, state near the wall at time 0
- $u_L$ normal component of velocity, state near the wall at time 0, $u_L = \boldsymbol{v}_L \cdot \boldsymbol{n}$
- $G_\star$ given inlet massflow (=desired massflow at the face)

At the vicinity of the boundary we solve the modified Riemann problem for the split Euler equations (1), introduced in Section 2, with the left-hand side initial condition (2) and the **complementary conditions** prescribing the mass flow

$$u_\star \varrho_{\star R} = G_\star, \tag{34}$$

where $G_\star \geq 0$ is given constant ($G_\star = \frac{massflow}{face\ area}$). We are interested in the solution with $u_\star > 0$, and $s_{L?} < 0$, which guarantees the possibility of the values to be prescribed at the boundary. In general, there are two possibilities of the wave pattern, which may interest us.

- **1-shock wave**
  Here, we are interested in the solution with $s_1 < 0, u_\star > 0$. Let us construct the function $F_S(u)$, using the relations (4),(5)

$$F_S(u) = u\varrho_{\star L}(u) - G_\star, \tag{35}$$

  where

$$\varrho_{\star L}(u) = \varrho_L \frac{(\gamma - 1)p_L + (\gamma + 1)P_S(u)}{(\gamma + 1)p_L + (\gamma - 1)P_S(u)},$$

  and $P_S(u)$ defined in (12). The sought velocity $u_\star$ is the root of this function $F_S(u)$. It is $s_1 < 0$ for $u_\star < u_X = u_L \frac{2+(\gamma-1)M_L^2}{(\gamma+1)M_L^2}$, with $M_L^2 = u_L^2 \varrho / \gamma p$, see [8]. The first derivative is $F_S' > 0$ for $u < min\{u_L, u_X\}$. The problem has a solution with the 1-shock wave if $F_S(min\{u_L, u_X\}) > 0$.
- **1-rarefaction wave**
  Let us assume the solution with the 1-rarefaction wave. Using (4),(5),(34), the velocity $u_\star$ is the root of the function

$$F_R(u) = u\varrho_L \left[1 - \frac{\gamma - 1}{2a_L}(u - u_L)\right]^{\frac{2}{\gamma-1}} - G_\star, \tag{36}$$

defined in the interval $u_L < u < u_L + \frac{2a_L}{\gamma-1}$. It is $F_R(u_L) = \varrho_L u_L - G_\star$. The first derivative is zero at the points $u_1 = u_L + \frac{2a_L}{\gamma-1}$, $u_0 = \frac{\gamma-1}{\gamma+1}\left(u_L + \frac{2a_L}{\gamma-1}\right)$. Derivative $F_R'(u) > 0$ in the interval $(0, u_0)$ and $F'(u) < 0$ for $u \in (u_0, u_1)$. The maximum of the function $F_R$ is at the point $u_0$. We are interested in the solution with $s_{TL} < 0$. This is satisfied only fo $u_\star < u_0$.

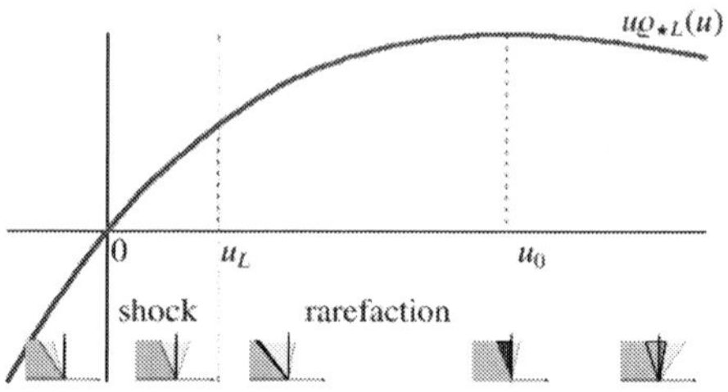


**Fig. 8.** Example of the function $u\varrho_{\star L}(u)$, here $u_L = 20$, $\gamma = 1.4$, $\varrho_L = 1.25$, $p_L = 100000$.

**Algorithm**
Here we present the possible algorithm for the solution of the sought values at the boundary.

```
F(u_L) = u_L ϱ_L − G_⋆
if F(u_L) > 0 then
  u_⋆ = root of F_S(u), u ∈ (0, min{u_X, u_L}), (35).
else
  if u_L > u_0 then
    u_⋆ = u_L
  else
   if(u_1 < u_0)
      u_⋆ = u_1 − ε
   else
     if F(u_0) < 0 then
      u_⋆ = u_0
     else
      u_⋆ = root of F_R(u), u ∈ (u_L, u_0), (36).
     end
   endif
  end
end
Compute p_B = p_⋆, ϱ_B = ϱ_⋆L using (12),(5).
```

The resulting boundary velocity in the global coordinates is $\boldsymbol{v}_B = \boldsymbol{n}(u_B - u_L) + \boldsymbol{v}_L$.

| input data | | solution | | | |
|---|---|---|---|---|---|
| $u_L$ | $G_\star$ | $G_B$ | $\varrho_B$ | $u_B$ | $p_B$ |
| -20 | 1 | 1.00 | 1.174034 | 0.8517646 | 91596.49 |
| -20 | 50 | 50.00 | 1.011280 | 49.44231 | 74326.85 |
| -20 | 150 | 130.35 | 0.4730351 | 275.5533 | 25655.28 |
| 20 | 1 | 1.00 | 1.323527 | 0.7555568 | 108333.1 |
| 20 | 100 | 100.00 | 0.9684777 | 103.2548 | 69960.27 |
| 20 | 200 | 150.45 | 0.5330942 | 282.2200 | 30328.54 |
| 400 | 10.0 | 10.00 | 3.218839 | 3.106711 | 421919.0 |
| 400 | 100 | 100.00 | 3.059145 | 32.68887 | 385170.7 |
| 400 | 600 | 500 | 1.25 | 400 | 100000 |

**Table 4.** Outlet boundary condition by preference of massflow. The input data $u_L$, $G_\star$ and the solution (rounded) $\varrho_B$, $u_B$, $p_B$, $G_B = \varrho_B u_B$ for the test problems with $\varrho_L = 1.25$, $p_L = 100000$.

# 9 EXAMPLES

## 9.1 The Flow Around the Obstacle

Here we present a computational result of the 2D non-stationary inviscid channel flow at Mach number $M = 0.67$. A body immersed in the flowing fluid establishes a certain wave pattern which evolves in time and eventually exits the channel. At figure 1 we show, that the fixed (values are fixed at the boundary) and linearized (as described in [9]) boundary conditions do not give the expected result in time. The inlet is located left, outlet right, other boundaries are considered as wall. The fixed boundary conditions give incorrect results near boundaries. The linearized boundary condition reflects the waves into the domain, leading to the oscillations in the solution. The new suggested boundary conditions do not suffer from these drawbacks. The residual behavior (shown right) demonstrates this result.

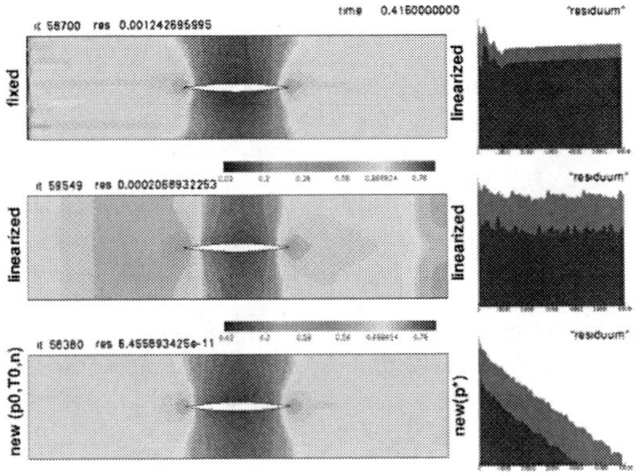


**Fig. 9.** Incompressible flow, body in the channel. Comparison of the various boundary conditions.

## 9.2 The Shortened Domain Example

The following numerical example shows superior behavior of the suggested boundary condition. This is the test case of the the inviscid flow through the Double Circular Arc (DCA) blade cascade DCA08. The blades of this cascade are composed of two circular arcs with the relative thickness 8%. At the inlet we use the boundary condition conserving the total pressure $p_o = 101325$ Pa, the total temperature $\theta_o = 273.15$ K, and the direction of the velocity $\alpha_{IN} = 5.2°$. At the outlet, the outlet boundary condition with the averaging technique described in [8], preferring the pressure $p_\star = 45722.351$ Pa in average. The part of the inlet flow is supersonic. It can be seen in figure 10 (left), that the used inlet boundary condition does not reflect the shock waves into the computational area. The right picture shows that this boundary condition can be used on the shortened computational domain with the similar result. Constructed boundary condition is robust and accelerates the convergence of the method, it can be also used on the shortened domains. This is the original result of our work.

**Fig. 10.** The compressible gas flow, transonic regime. Two computations with the same boundary data. The bottom picture shows the computation with the shortened inlet part. Inlet located left. Pressure isolines, density and Mach number isolines, highlighted Mach number 1.

## 9.3 The Outlet Mass Flow Example

The developed software with presented boundary conditions was used for the simulation of the compressible turbulent flow in the 3D axis-symmetrical channel. Axis $x$ is the axis of symmetry. Example is shown in figure 11. At the inlet we used the boundary condition shown in section 6, conserving the total pressure, total temperature, and zero tangential velocity, with $\theta_o = 273.15$ K , $p_o = 101325$ Pa. The boundary condition shown in section 8 was used at the outlet, with the given massflow $G = 4.0$ kg·s$^{-1}$·m$^{-2}$. At each face, the value $G_\star$ was computed (in each iteration) in order to match the massflow across the whole boundary. The boundary condition preffering the zero normal velocity was used in the case of the inviscid flow. For the viscous turbulent flow, this condition was modified by the zero velocity at the wall, and wall temperature $\theta_{WALL} = 273.15$ K was set. Further $k_{WALL} = 0$ m$^2$·s$^{-2}$ and $\omega_{WALL} = c_\omega \frac{6\mu}{\beta \varrho y_s^2}$, $c_\omega \frac{6}{\beta} = 120$. Here by $y_s$ we mean the distance between the face and the center of the neighbouring element.

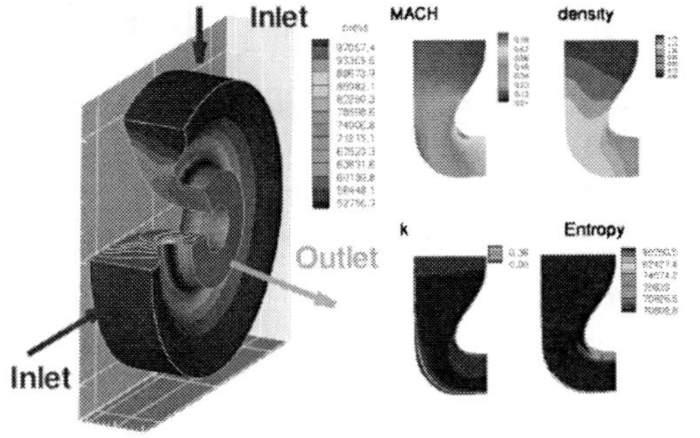


**Fig. 11.** 3D axis-symmetrical turbulent flow, 3D geometry shape, results at chosen 2D crosscuts.

Another example demonstrates the capabilities of the new boundary condition in the use with the unstationary gas flow. Here $x$ is the axis of symmetry. At the geometry

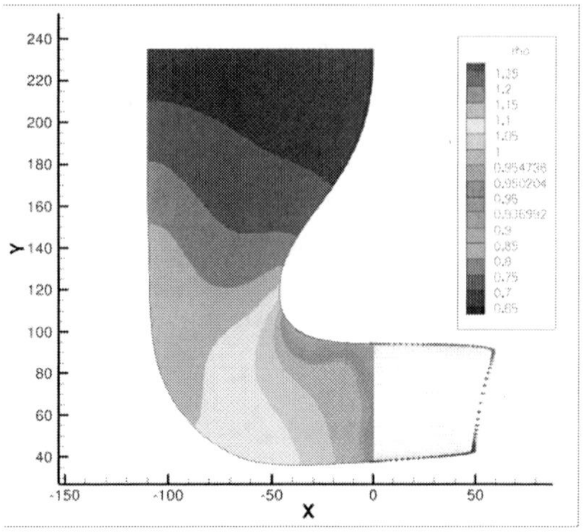


**Fig. 12.** 3D axis-symmetrical turbulent flow, density isolines and velocity profile at the outlet, 2D crosscut.

crosscut shown at figure 13, the inlet is located at the left part of the boundary, the outlet points away from axis $x$. Inlet boundary condition with the preference of total quantities was used at the inlet,$\theta_o$ = 875.00 K, $p_o$ = 101905 Pa. The mass flow $G_\star = 0.75 \text{kg}\cdot\text{s}^{-1}\cdot\text{m}^{-2}$ was preferred at the outlet (in average). Wall temperature was set to $\theta_{WALL}$ = 875.00 K. The computational mesh in 2D crosscut consisted of 89x87 quadrilaterals.

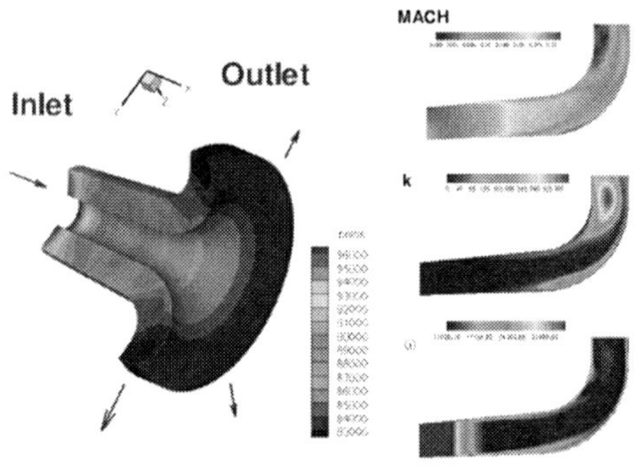


**Fig. 13.** 3D axis-symmetrical turbulent flow, 3D geometry shape, results at chosen 2D crosscuts.

## Conclusion

This paper is focused on the boundary conditions for 3D compressible flow, and summarizes the work presented in [2], [8], [5], [6], [7], and other publications. The algorithms for the construction of 3D boundary conditions are presented. The work is based on the analysis of the Riemann Problem for the split Euler equations and the modifications of this problem. Here the left hand side initial condition is replaced by various complementary conditions. The solution of such problems is shown.

**Acknowledgment**
The results originated with the support of Ministry of the Interior of the Czech Republic, project SCENT, Grant MSM 0001066902 of the Ministry of Education of the Czech Republic, and Ministry of Industry and Trade of the Czech Republic for the long-term strategic developmentof the research organization. The authors acknowledge this support.

## References

1. C. J. Kok. AIAA Journal, Vol. **38.**, No. 7., (2000).
2. M. Kyncl and J. Pelant. *Applications of the Navier-Stokes equations for 3d viscous laminar flow for symmetric inlet and outlet parts of turbine engines with the use of various boundary conditions.* Technical report R3998, VZLÚ, Beranových 130, Prague, (2006).
3. V. Dolejší *Discontinuous Galerkin method for the numerical simulation of unsteady compressible flow.* WSEAS Transactions on Systems, 5(5):1083-1090, (2006)
4. M. Kyncl and J. Pelant. *Numerical Simulation of the Flow Around Diffusible Barriers.* Experimental fluid mechanics 2012, EFM 2012, Hradec Králové (Czech Republic) (2012).
5. M. Kyncl and J. Pelant. *The Initial-Boundary Riemann Problem for the Solution of the Compressible Gas Flow.* ECCOMAS 2014, ECFD VI, 20.-25.7.2014, Barcelona, Spain (2014).
6. M. Kyncl and J. Pelant. *The Initial-Boundary Riemann Problem for the Solution of the Compressible Gas Flow.* Applied Mechanics and Materials, Vol. 821, pp. 70-78, 2016.
7. M. Kyncl and J. Pelant. *Analysis of the Boundary Problem with the Preference of Mass Flow.* ECCOMAS 2016, 5.-10.6.2016, Crete Island, Greece (2016).
8. M. Kyncl. *Numerical solution of the three-dimensional compressible flow.* Master's thesis, Prague, (2011). Doctoral Thesis.
9. M. Feistauer, J. Felcman, and I. Straškraba. *Mathematical and Computational Methods for Compressible Flow.* Oxford University Press, Oxford, (2003).
10. M. Kyncl and J. Pelant. *Implicit method for the 3d Euler equations.* Technical report R5375, VZLÚ, Beranových 130, Prague, (2012).
11. M. Kyncl and J. Pelant. *Implicit method for the 3d RANS equations with the k-w (Kok) Turbulent Model.* Technical report R5453, VZLÚ, Beranových 130, Prague, (2012).
12. M. Feistauer. *Mathematical Methods in Fluid Dynamics.* Longman Scientific & Technical, Harlow, (1993).
13. E. F. Toro. *Riemann Solvers and Numerical Methods for Fluid Dynamics.* Springer, Berlin, (1997).

# Evaluation of the turbulent kinetic dissipation rate in an agitated vessel

*Bohus* Kysela[1a], *Jiri* Konfrst[1], *Zdenek* Chara[1], *Radek* Sulc[2], and *Darina* Jasikova[3]

[1] Institute of Hydrodynamics, AS CR, v.v.i.
Pod Patankou 30/5, 166 12 Prague 6, Czech Republic
[2] Department of Process Engineering, Faculty of Mechanical Enginnering, Czech Technical University in Prague, Technicka 4, 166 07 Prague 6, Czech Republic
[3] Institute for Nanomaterials, Advanced Technology and Innovation, Technical University of Liberec, Studentska 1402/2, 461 17 Liberec, Czech Republic

**Abstract.** The design of agitated tanks depends on operating conditions and processes for that are used for. An important parameter for the scale-up modelling is the dissipation rate of the turbulent kinetic energy. The dissipation rate is commonly assumed to be a function of the impeller power input. But this approach gives no information about distribution of the dissipation rate inside the agitated volume. In this paper the distributions of the dissipation rate inside the agitated vessels are estimated by evaluations of the CFD (Computational Fluid Dynamics). The results obtained from RANS (Reynolds Averaged Navier-Stokes equations) k-$\varepsilon$ turbulent model and LES (Large Eddy Simulations) with Smagorinsky SGS (Sub Grid Scale) model are compared. The agitated vessels with standard geometry equipped with four baffles and stirred by either a standard Rushton turbine or a high shear impeller were investigated. The results are compared with mean dissipation rate estimated from the total impeller power input.

## 1 Introduction

The main aim of this paper is to map distribution of dissipation rate inside an agitated vessel and compared commonly used k-$\varepsilon$ RANS model with LES results. The discrepancy between results treated by averaging processes and results of transient simulation is indisputable. Such discrepancies of a simulation of velocity distributions (breakup of trailing vortices) were indicated in [1]. The improvement of distribution of dissipation rate inside the vessels is also useful for future development in particular processes and on aggregates or bubbles breakup modelling [2]. A lot of papers have been devoted to a turbulence modelling in an agitated vessel [3–6] and this process is still open. The proposed simulations were performed on a geometry shown in Fig. 1. The simulations were performed with two types of impeller: Rushton turbine (RT) - Fig. 2 (left) and high shear impeller (HSI) - Fig. 2 (right).

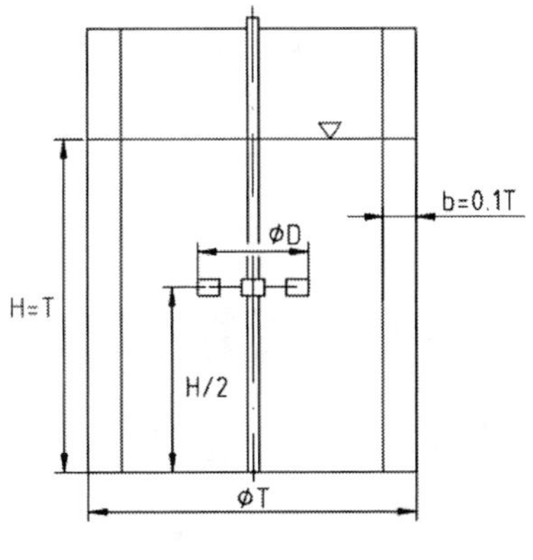


**Fig. 1.** Geometry of mixing tank (T = 300 mm, H/T = 1, T/D = 3, b/T = 1/10, four baffles). Two impellers: Rushton turbine (RT) and high shear impeller (HSI).

## 2 Simulations

The pressure-based solver pimpleDyMFoam from Open-FOAM 4.0 package (for incompressible flow) was used for calculations in the agitated vessel. The impeller movement was simulated by SM (Sliding Mesh) method. The diameter of the sliding region is twice as large as the diameter of the impeller. Turbulence was modelled by either LES approach, with Smagorinsky SGS model or standard k-$\varepsilon$ model. The final non-structured (mainly hexagonal) meshes consisted of 3.3 mil. cells for RT and 1.5 mil. cells for HSI. Only the part of the impeller shaft outside the sliding region was defined as a moving wall with impeller speed velocity, the others walls were set to the no slip condition. The rotation speed was 300 rpm (for both

**Fig. 2.** Impellers: Rushton turbine (left) and high shear (right).

[a] e-mail: kysela@ih.cas.cz

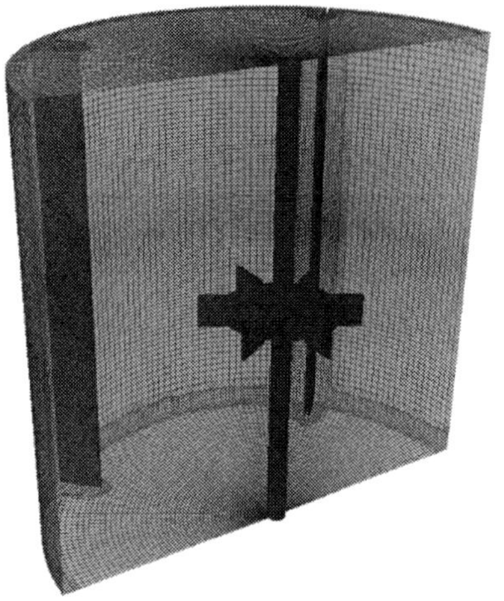

**Fig. 3.** Tank mesh.

**Fig. 4.** Rushton turbine impeller mesh.

**Fig. 5.** High shear impeller mesh.

impellers). The time step was 0.0001 s that corresponds to 0.18° of the impeller movement.

## 3 Methods and evaluation

The advantage of batch systems represented by the agitated vessels is the closed domain where the input energy is delivered through the impeller and dissipated by viscose forces onto heat. Hence, it is possible to determine the total power and compare it with dissipation of turbulent kinetic energy.

**Table 1.** Power number for turb. regime ($Re_M > 10^4$):

| Impeller | $P_o$ |
|---|---|
| Rushton turbine | 5.0 |
| high shear impeller | 0.6 |

### 3.1 Impeller power input

The power consumption of impeller operating in the agitated vessel is given by a lot of parameters. In that case it was simplified and so-called high speed impellers with commonly used vessel geometry were employed. The fully turbulent regime ($Re_M > 10^4$) was proposed. The mixing Reynolds number is defined:

$$Re_M = \frac{nD^2}{\nu}, \tag{1}$$

where $D$ is impeller diameter, $n$ impeller speed and $\nu$ is kinematic viscosity. The power input should be calculated with

$$P = P_o \rho n^3 D^5, \tag{2}$$

where $P_o$ is power number dependent on impeller type (see tab. 1) and $\rho$ is density of agitated fluid. The mean dissipation rate inside agitated vessel is determined by

$$\bar{\varepsilon} = \frac{P}{\rho V_T}, \tag{3}$$

where $V_T$ is tank volume.

### 3.2 Evaluation of the energy dissipation

Assuming a local equilibrium between production and dissipation of turbulent kinetic energy at the cut-off scale the dissipation rate of TKE from LES results was estimated as:

$$\varepsilon_{LES} = \frac{\nu + \nu_t}{2}\left(\frac{\partial u_i}{\partial x_j} + \frac{\partial u_j}{\partial x_i}\right)^2. \tag{4}$$

Evaluation of power was performed by the calculations in each cell volume $V_i$ with knowledge of dissipation rate of turbulent kinetic energy $\varepsilon_i$ and the mean dissipation rate should be obtained by the summation of values inside all cells:

$$\bar{\varepsilon}_c = \frac{1}{V_c}\sum \varepsilon_i V_i, \tag{5}$$

where $V_c$ is total volume $\sum V_i$ .
The distribution of dissipated energy was calculated as power input transformed by dissipation in each cell:

$$P_i = \varepsilon_i V_i \rho, \tag{6}$$

where the total dissipated power in vessel is $P_c = \sum P_i$. Consecutively the data were sorted by the values of the dissipation rate and the distribution was evaluated in the

DOI: 10.1051/epjconf/201714302062

form of the dependency of the cumulative curve on the tank volume:

$$Power\,input/Total\,power\,input = \frac{\sum P_j}{P_c}, \qquad (7)$$

that is expressed as sum of cell volumes per total volume:

$$Volume/Total\,volume = \frac{\sum V_j}{V_c}. \qquad (8)$$

## 4 Results

Four basic simulations were performed: k-$\varepsilon$ and LES combined with two impellers (RT and HSI). The results of velocity contours on the impeller disk are depicted in Fig. 6. It is obvious, that the distributions of velocities are different, moreover, the velocity distribution represented by averaged values never occurs in real cases. The deviation of the results is also observable in Fig. 7 where the iso-volumes of regions with values of $\varepsilon$ greater than 1 $m^2 \cdot s^{-3}$ are depicted and in Fig. 8, where the iso-volumes of regions with values of $\varepsilon$ greater than 0.1 $m^2 \cdot s^{-3}$ are shown.

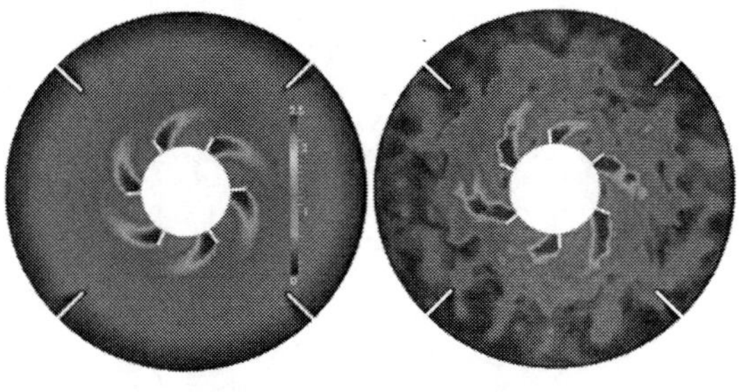

**Fig. 6.** Velocity results obtained in disk plane by LES for RT: Angle averaged results - similar as from RANS (left) and instantaneous (right).

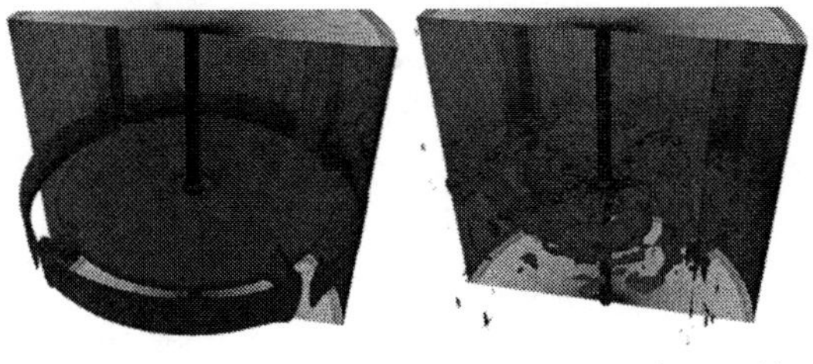

**Fig. 7.** Comparison of iso-volumes obtained for Rushton turbine impeller with conditions: $\varepsilon > 1\ m^2 \cdot s^{-3}$ for k-$\varepsilon$ (left) and LES (right).

Hence, the results were processed according to the evaluation of distributions approach. The first method is to evaluated energy dissipated in vessel in relation with volume. The cumulative distribution curves were calculated and the results are shown in Fig. 9. The curves are normalized by the total power input and the total volume. The distribution curves of k-$\varepsilon$ model show the lower values in small volumes (30% of energy is dissipated in 1/100 of volume)

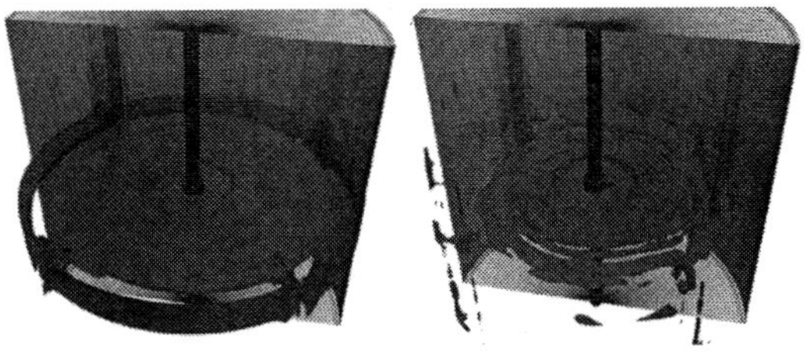

**Fig. 8.** Comparison of iso-volumes obtained for high shear impeller with conditions: $\varepsilon > 0.1\ m^2 \cdot s^{-3}$ for k-$\varepsilon$ (left) and LES (right).

**Table 2.** Comparison of mean dissipation rates $\overline{\varepsilon}$ ($m^2 \cdot s^{-3}$) obtained from LES or k-$\varepsilon$ calculation and estimated from power input (eq. 3).

| Impeller | $\overline{\varepsilon_{LES}}$ | $\overline{\varepsilon_{k-\varepsilon}}$ | $\overline{\varepsilon}$ |
|---|---|---|---|
| RT | 0.292 | 0.287 | 0.295 |
| HSI | 0.022 | 0.020 | 0.035 |

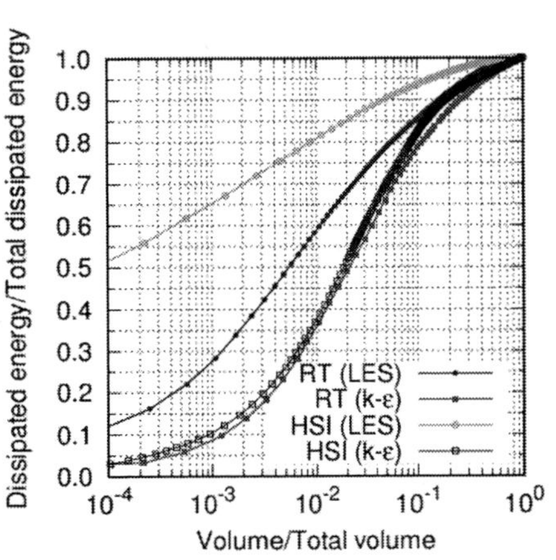


**Fig. 9.** Cumulative distributions obtained from simulations based on k-$\varepsilon$ and LES.

and similar shape for both impellers. In contrary the results of LES calculations indicate that more energy has been dissipated in very small volumes (60% of energy is dissipated in 1/100 of volume for RT and 80% for HSI), moreover, the distribution of both impeller are quite different. This corresponds with conception, that for HSI the energy is dissipated only near the disk surface, while for RT the dissipation regions are also inducted by trailing vortices. The similar results were obtained by the method based on iso-volumes evaluation (Figs. 7 and 8). The volumes were calculated from selected data of iso-volumes of $\varepsilon$ ranging from 0.001 to 5, see Fig. 10. The predicted values of the dissipation rate of the TKE are higher for k-$\varepsilon$ in a middle section of the volume distribution, while close to the lateral sections the k-$\varepsilon$ results approach the LES results. As can be expected the dissipation rates of the RT impeller are about ten times higher compared to the HSI data. Nevertheless, for each type of the impeller the total powers evaluated from both simulations are almost the same (see Tab. 2). The differences are only between the energy cal-

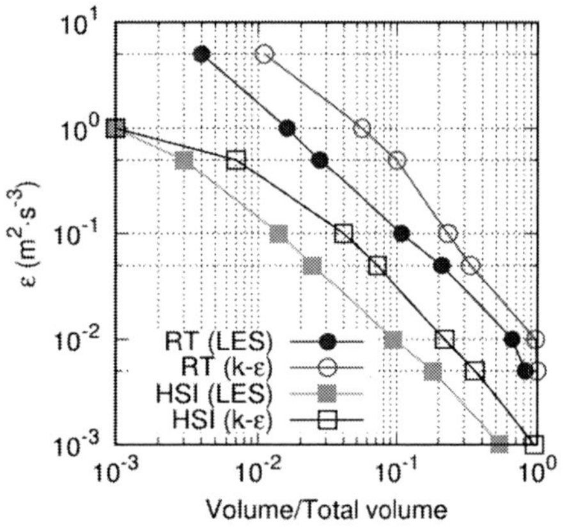


**Fig. 10.** Values of iso volume size for values of $\varepsilon$ grater than depicted ones obtained from simulations based on k-$\varepsilon$ and LES.

culated from the power input and from the dissipation rate. This could be caused by the using of more coarse mesh near the surface.

## 5 Conclusions

The distribution of dissipation rate of turbulent kinetic energy was estimated for Rushton turbine impeller and high shear impeller. Resulted distributions from k-$\varepsilon$ turbulence model seem to be more narrow but with generally higher shifted values in comparison with LES ones. Both simulation methods give satisfactory results of total dissipated energy in agitated vessel, but the distribution is rather affected by the method of turbulence modelling.

## Acknowledgments

This research has been subsidized by the research project No. GA ČR 16-20175S and RVO: 67985874.

## References

1. Z. Chara, B. Kysela, J. Konfrst, I. Fort, Applied Mathematics and Computation **272, Part 3**, 614 (2016)
2. M. Soos, R. Kaufmann, R. Winteler, M. Kroupa, B. Lüthi, AIChE Journal **59**, 3642 (2013)
3. A. Bakker, L. Oshinowo, Chemical Engineering Research and Design **82**, 1169 (2004), in Honour of Professor Alvin W. Nienow
4. M. Jahoda, M. Mostek, A. Kukukova, V. Machon, Chemical Engineering Research and Design **85**, 616 (2007)
5. J. Gimbun, C.D. Rielly, Z.K. Nagy, J.J. Derksen, AIChE Journal **58**, 3224 (2012)
6. J.J.J. Gillissen, H.E.A. Van den Akker, AIChE Journal **58**, 3878 (2012)

# Simulation of collagen solution flow in rectangular capillary

*Bohus* Kysela[1a], *Jan* Skocilas[2], Rudolf Zitny[2], Jaromir Stancl[2], *Milan* Houska[3], and *Ales* Landfeld[3]

[1] Institute of Hydrodynamics, AS CR, v.v.i., Pod Patankou 30/5, 166 12 Prague 6, Czech Republic
[2] Department of Process Engineering, Faculty of Mechanical Enginnering, Czech Technical University in Prague, Technicka 4, 166 07 Prague 6, Czech Republic
[3] Food Research Institute Prague, Radilova 7, 102 00 Prague 10, Czech Republic

**Abstract** The viscoelastic properties of foods and polymers can be evaluated from flow of the material in capillary with specified dimension and shape. The extrusion rheometer equipped by capillary with rectangular cross-section was used for determination of the rheological behaviour of water collagen solution. The measurements of the axial profiles in longitudinal direction of the total stresses at capillary wall were performed for various shear rates. The linear viscoelastic model of Oldroyd B type: White-Metzner model was used for simulation of fluid flow in OpenFOAM software package. The simulations describe the effect of relaxation time on wall total stress in convergent-divergent capillary.

## 1 Introduction

The aim of this study is to solve the flow of viscoelastic fluid (water collagen solution) using the *OpenFOAM - extended* with application of "viscoelasticFoam" solver and its modifications. This solver was developed and implemented by Favero at al. [1]. The main goal is to obtain dependency of total normal stress $\sigma$ on bottom wall of the rectangular capillary:

$$\sigma = \tau_{yy} - p \tag{1}$$

where $\tau_{yy}$ is normal extrastress component and $p$ is pressure inducted by the flow of viscoelastic fluid. Rheological properties are modelled by the constitutive model White-Metzner [2] combined with power law fluid model.
In this study the capillary with convergent-divergent part was chosen for investigation of relaxation time effect on wall total stress. This simple geometry consist of stabilizing part 200 *mm*, convergent part 200 *mm*, and divergent part 200 *mm* (see fig. 1). On base of the previous exper-

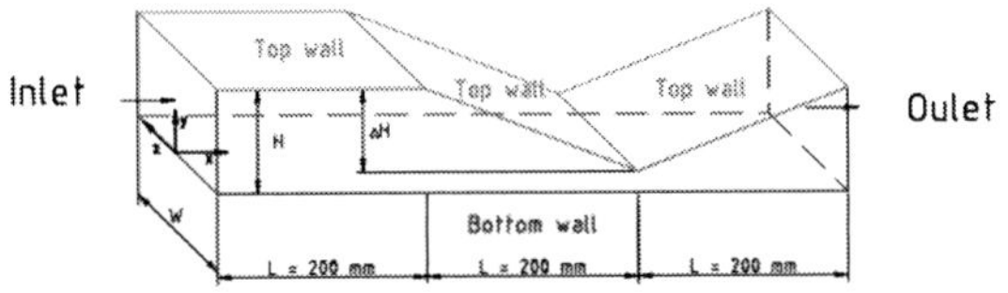


**Figure 1.** Sketch of geometry which was used for creation of computational domain. ($H$ = 5 *mm*, $\Delta H$ = 4 *mm*, $L$ = 200 *mm*, $W$ = 8 *mm*).

imental investigations [3,4] material constants were proposed for used water collagen solution, that are necessary as input to computational model. The proposed material properties for that case were: density $\rho$ = 1000 $kg \cdot m^{-3}$; Newtonian solvent viscosity $\eta_s$ = 7 $Pa \cdot s$; coefficient of consistency $k$ = 1353 $Pa \cdot s^n$; flow index $n$ = 0.25; the relaxation time varied from $\lambda$ = 0.003 to 0.1 $s$ (for water $\lambda \approx 10^{-12}$ $s$).

## 2 Theory of solution method

The method was originally implemented on finite volume method based on solving following system of equations:
Continuity equation:

$$\nabla \cdot (U) = 0 \tag{2}$$

Momentum equation:

$$\frac{\partial (\rho U)}{\partial t} + \nabla \cdot (\rho U U) = -\nabla p + \nabla \cdot \tau_s + \nabla \cdot \tau_p, \tag{3}$$

where $\tau_s$ is shear tensor acting from Newtonian solution and $\tau_p$ is extra elastic shear tensor.

$$\tau_s = 2\eta_s D, \tag{4}$$

where $\eta_s$ is viscosity of Newtonian solution and $D$ is strain rate tensor:

$$D = \frac{1}{2}\left(\nabla U + [\nabla U]^T\right). \tag{5}$$

Constitutive model White-Metzner:

$$\tau_{p_K} + \lambda_K (II_D) \overset{\nabla}{\tau}_{p_K} = 2\eta_{p_K} (II_D) D, \tag{6}$$

where $\overset{\nabla}{\tau}_{p_K}$ represents ***upper convected derivatives*** and second invariant of strain rate tensor is

$$(II_D) = \dot{\gamma} = \sqrt{2D : D}. \tag{7}$$

For power law model it was defined:

$$\eta_{p_K} (II_D) = k (II_D)^{n-1}, \tag{8}$$

and relaxation time was defined as constant value:

$$\lambda_K (II_D) = \lambda = const. \tag{9}$$

[a] e-mail: `kysela@ih.cas.cz`

**Table 1.** Boundary conditions.

| Boundary | Velocity | Pressure | Tau |
|---|---|---|---|
| Inlet | (0.1 0 0) | zeroGradient | (0 0 0 ...) |
| Outlet | zeroGradient | 0.0 | zeroGradient |
| Walls | (0 0 0) | zeroGradient | zeroGradient |

**Table 2.** Material model constants of water.

| Name | Variable | Value |
|---|---|---|
| Density | $\rho\ [kg \cdot m^{-3}]$ | 1000 |
| Newtonian solvent viscosity | $\eta_s\ [Pa \cdot s]$ | 0.001 |
| Coefficient of consistency | $k\ [Pa \cdot s^n]$ | 0.001 |
| Flow index | $n$ [-] | 1.0 |
| Relaxation time | $\lambda\ [s]$ | $\approx 10^{-12}$ |

**Table 3.** Material model constants of water collagen solution [4].

| Name | Variable | Value |
|---|---|---|
| Density | $\rho\ [kg \cdot m^{-3}]$ | 1000 |
| Newt. solv. viscosity | $\eta_s\ [Pa \cdot s]$ | 7.0 |
| Coef. of consistency | $k\ [Pa \cdot s^n]$ | 1353 |
| Flow index | $n$ [-] | 0.25 |
| Relaxation time | $\lambda\ [s]$ | $10^{-12}$; 0.003; 0.01; 0.1 |

## 3 Simulation set-up and results

The OpenFOAM-extended package v3.2 was used. The mesh 1000 × 30 × 20 was generated and optimized by *blockMesh* utility. The boundary conditions are given in table 1. The inlet is defined by constant velocity $U_x = 0.1\ m \cdot s^{-1}$ (where $x$ coordinate corresponds with capillary axis, while $y$ coordinate corresponds with the capillary height) and with zero shear. The coordinate system origin (x, y, z) = (0, 0, 0) is set in the middle of the channel width at the bottom wall at velocity inlet. The task was divided into two parts, where the first one was focused on **the validation of the solver results**, and the second one on **the flow simulations of water collagen solution** with variable $\lambda$ values from $10^{-12}$ to 0.1 $s$ (see table 3). The validation was performed with material constants of water (see tab. 2).

### 3.1 Flow simulation with water properties

The initial calculations were performed for basic verification of proposed model. Newtonian fluid (water) was used as basic fluid, where the relaxation time $\lambda \approx 10^{-12}\ s$ (see tab. 2). The velocity profiles and distribution of total shear stress evaluated perpendicularly to the axis of rectangular capillary in distance 0.2 m of length (in the end of the linear section) and in distance 0.4 m of length (in contraction) are depicted in figs. 2 and 3. The velocity profile obtained in the developing section (the first linear section) is parabolic in shape and it is in agreement with presumptions as well as the evaluated total shear stress values are in agreement.

Then the aim was to obtain the distribution of the total normal stress $\sigma$ along the capillary (eq. 1). The distribution of pressure and shear stress component $\tau_{yy}$ acting on bottom wall (see fig. 1) is depicted in fig. 4. The behaviour of pressure distribution is also in agreement with presumptions, where the pressure linearly decreased with length

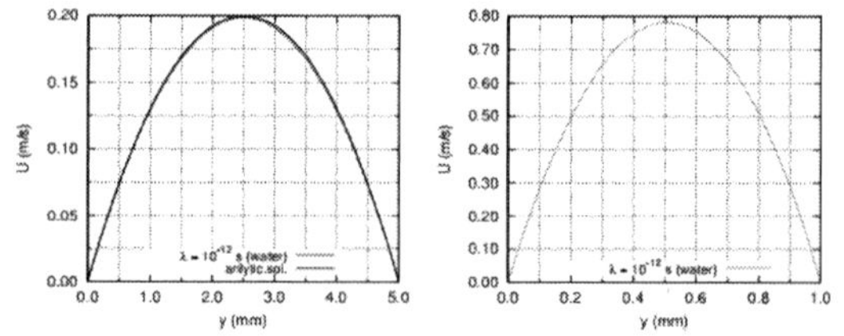

**Figure 2.** Velocity profiles in two distances L = 0.2 (left); 0.4 m (right) for water properties. The first profile is compared with analytical solution [5].

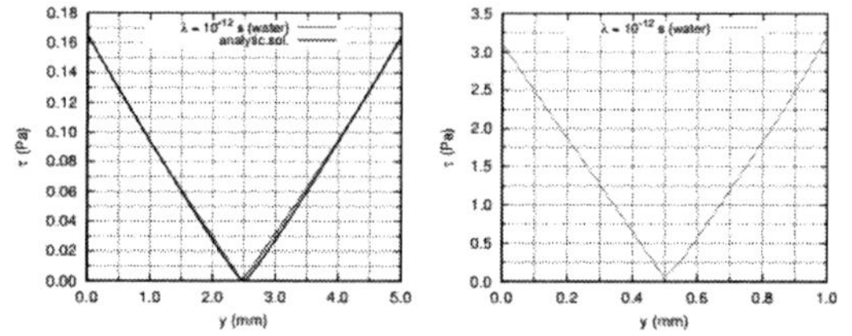

**Figure 3.** Profiles of shear stress for two distances L = 0.2 (left); 0.4 m (right) for water properties. The first profile is compared with analytical solution [5].

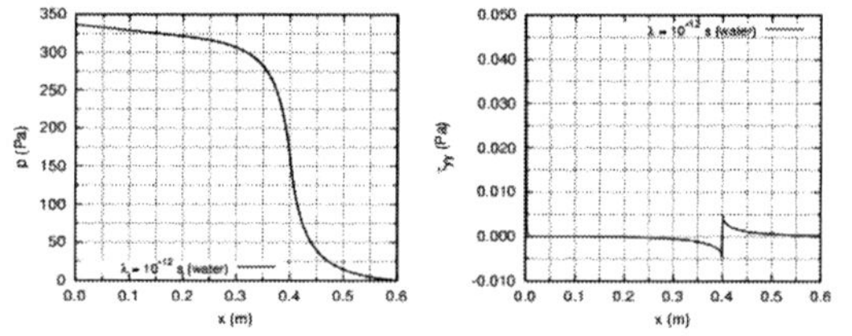

**Figure 4.** Evaluated pressure (left) and shear stress component $\tau_{yy}$ (right) acting on wall for "water" flow through capillary.

(in the first linear region). Then the convergent and divergent regions are represented by symmetrical pressure drop curves with inflection point in contraction point. Moreover, shear stress component $\tau_{yy}$ calculated from water properties seems to be negligible in comparison with pressure values.

### 3.2 Results of collagen solution flow simulation

The following calculations were performed with properties corresponding to experimental investigations [4]. The proposed parameters are given in tab. 3. The example of obtained distribution of pressure $p$ and shear stress component $\tau_{yy}$ acting on bottom wall (see fig. 1) is depicted in fig. 5 for relaxation time was $\lambda = 0.003\ s$. The velocity profiles and profiles of total normal stress for all simulated relaxation times in distance 0.2 m of length (in the end of the linear section) and in 0.4 m of length (in contraction) are depicted in figs. 6 - 7 and 8-9.

The shapes of velocity profile in linear developing section are very similar and it seems that the relaxation time has no effect on its shape, while the profile in contraction shows the deformation. The evaluation of total shear stress indicates the increase of shear stress near walls with increasing relaxation time in both profiles.

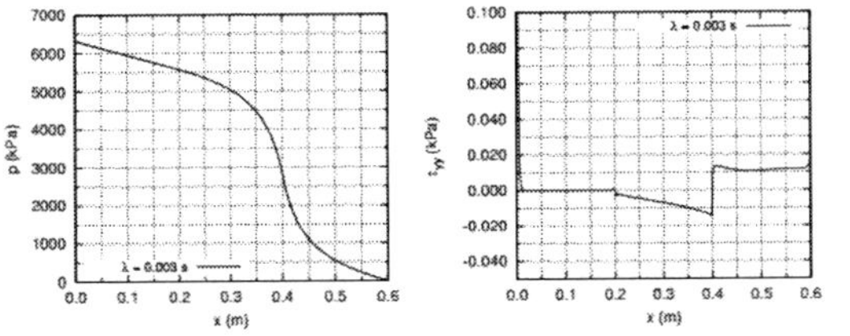


**Figure 5.** Evaluated pressure (left) and shear stress component $\tau_{yy}$ (right) on wall for flow of collagen solution through capillary: $\lambda = 0.003\ s$.

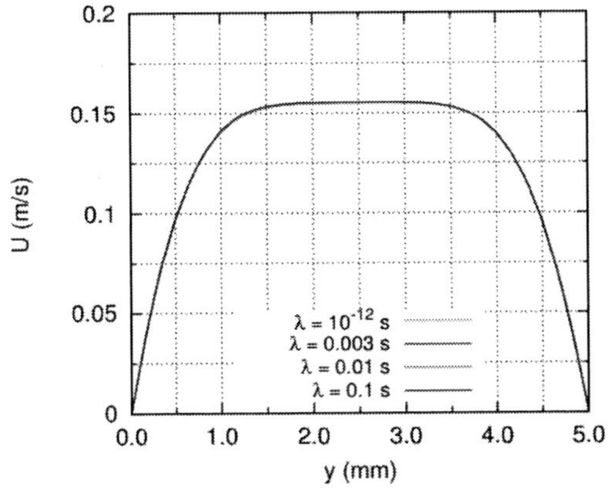


**Figure 6.** Velocity profile of collagen solution flow simulation ($\lambda = 10^{-12}$; 0.003; 0.01; 0.1 $s$) in distance L = 0.2 m.

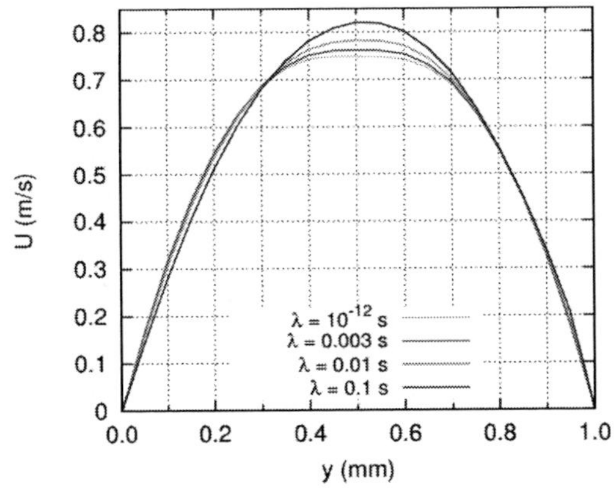


**Figure 7.** Velocity profile of collagen solution flow simulation ($\lambda = 10^{-12}$; 0.003; 0.01; 0.1 $s$) in distance L = 0.4 m.

### 3.3 Comparison of pressure and $\tau_{yy}$ dependencies acting on wall

The distinct disproportion between values of pressure and shear stress component $\tau_{yy}$ was also found for collagen solution. This concludes that the shear stress component $\tau_{yy}$ is negligible in comparison with pressure values. The dependency on relaxation time is depicted in fig. 10.

The shear stress $\sigma$ acting on capillary wall is proportional to the pressure. The distributions of pressure are not quite identical consequently the values were normalised by the

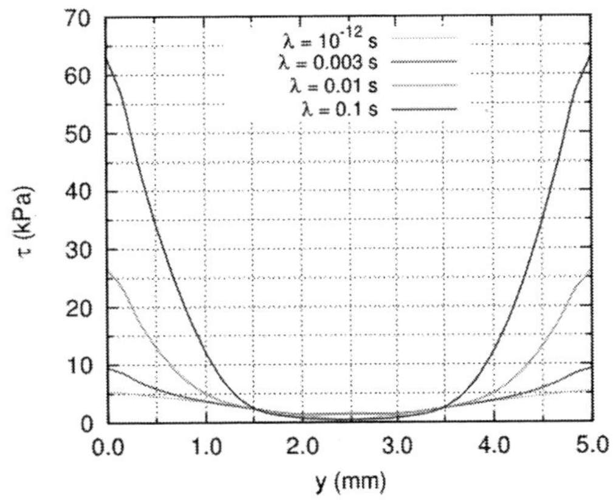


**Figure 8.** Profile of shear stress of collagen solution flow simulation ($\lambda = 10^{-12}$; 0.003; 0.01; 0.1 $s$) in distance L = 0.2 m.

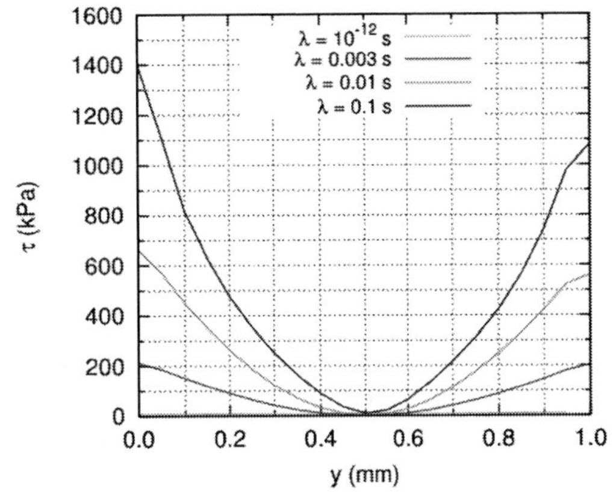


**Figure 9.** Profile of shear stress of collagen solution flow simulation ($\lambda = 10^{-12}$; 0.003; 0.01; 0.1 $s$) in distance L = 0.4 m.

value $p_{0.2m}$ where the developing linear section ends (see fig. 1). The normalized pressures acting on wall are shown in fig. 11. The discrepancy for water (the slope in the developing linear section) is given by parameters of fluid.

The effect of relaxation time and the deformation of the pressure curves in convergent and divergent sections is presented in fig. 12, where the detailed view onto contraction point (L = 0.4 m) shows the discrepancy and irregularity of pressure curves due to shift from point $p/p_{0.2m} = 0.5$ caused by the change of the relaxation time.

## 4 Conclusions

The proposed model of viscoelastic fluid was implemented and the calculation was validated for Newtonian fluid (data for water) in rectangular capillary. It could be assumed that the basic calculations of the flow are valid.

The obtained results indicate that the shear stress component $\tau_{yy}$ is negligible in comparison with pressure values for calculations of total stress $\sigma$ measured on wall of the capillary.

The effect of relaxation time is shown on comparison of

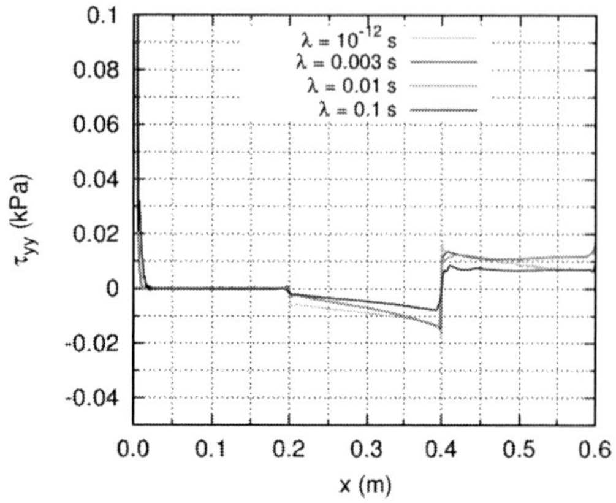


**Figure 10.** Shear stress component distributions $\tau_{yy}$ acting on wall ($\lambda = 10^{-12}$; 0.003; 0.01; 0.1 $s$).

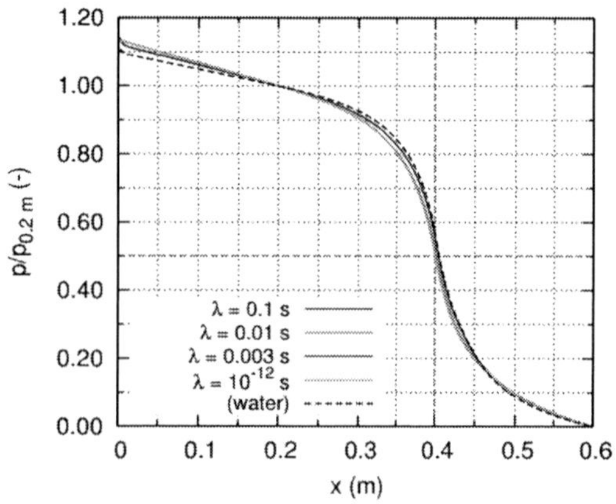


**Figure 11.** Comparison of the pressure distributions normalized by value $p_{0.2m}$ i.e. pressure at the end of linear section where L = 0.2 m (see fig. 1).

the pressure dependency (fig. 11 and 12), where its shift in inflection point of pressure curve is proportional to the relaxation time. It indicates that the deformation of pressure curves and its irregularity for convergent and divergent channel section is caused by the relaxation time effect.

## Acknowledgments

This research has been subsidized by the research project No. GA ČR 14-23482S and RVO: 67985874.

## References

1. J. Favero, A. Secchi, N. Cardozo, H. Jasak, Journal of Non-Newtonian Fluid Mechanics **165**, 1625 (2010)
2. J.L. White, A.B. Metzner, Journal of Applied Polymer Science **7**, 1867 (1963)
3. R. Zitny, A. Landfeld, J. Skocilas, J. Stancl, V. Flegl, M. Houska, Acta Polytechnica **55**, 352 (2015)

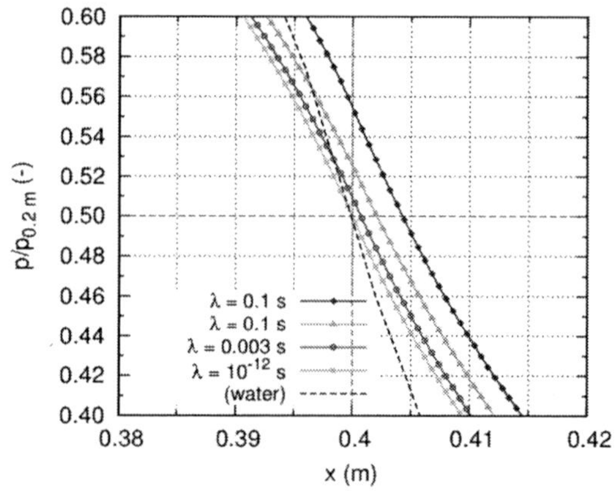


**Figure 12.** Comparison of the pressure distributions normalized by value $p_{0.2m}$ i.e. pressure at the end of linear section where L = 0.2 m (see fig. 1). Detail in vicinity of inflexion point (L = 0.4 m; $p/p_{0.2m} = 0.5$).

4. J. Skočilas, R. Žitný, J. Štancl, M. Dostál, A. Landfeld, M. Houška, Journal of Texture Studies (2016), (in print)
5. R. Shah, A. London (Academic Press, 1978), ISBN 978-0-12-020051-1

# Proportional valve with a piezoelectric actuator

*Pawel Andrzej* Laski [1,*]

[1] Kielce University of Technology, Faculty of Mechatronics and Mechanical Engineering, Department of Automation and Robotics, Aleja Tysiaclecia Panstwa Polskiego 7, 25-314 Kielce, Poland

**Abstract.** The article concerns a slotted proportional valve for use in pneumatic and hydraulic systems. There is a growing demand for both hydraulic and pneumatic ultrafast proportional valves. The conducted analysis of literature confirms the lack of such solutions for proportional valves. The currently used pneumatic systems for selection and segregation of parts and objects require ultrafast valves. The presented solution for the proportional valve can significantly improve and accelerate this type of technological processes. Furthermore, fast proportional valves can be successfully used for positional control of pneumatic and hydraulic drives. The article presents the design of a slotted divide valve and sets the maximum mass flow rate for service roads.

## 1 Introduction

Currently, the series and large series production requires the removal of defective or damaged parts and objects from the production line without turning them off. One of the methods applied to achieve such an effect is blowing out an object moving in front of the nozzle arrangement by a strong air flow [1,2]. The inclusion and exclusion of air flow should be done in the shortest possible time and the value of flow rate should be tailored to the size and mass of the object [3,4].

The nozzle arrangement is most frequently controlled by spool valves. Their task is to separate the air stream between flow windows of the valve and directing the flow of air in the direction of the object. Separation of the air stream takes place through suitably shaped edges of the valve spool and its body. Such a solution has several advantages, but also one drawback associated with the mass of the valve spool that causes a relatively long switching time. A new solution was proposed in order to shorten the response time and improve the parameters of classic valves. The valve design uses an ultrafast piezoelectric actuator with mechanical amplifier which guarantees a proportional shift of elements that separate the stream and thus the controlled flow of air. This resulted in a design of the proportional valve type 2/2 (two-way, two-position). The design constructed with the use of CFD software required the determination and presentation of a vector velocity field and pressure distribution in the cross-sections of flow windows [5,6]. The maximum flow values for the valve spool were determined on the basis of simulation studies. Based on simulation studies, it can be concluded that it is possible to build a proportional pneumatic valve with high values of the flow rate.

## 2 Actuators used in valves.

There are two types of piezoelectric actuators used in industrial and prototypical valves to control the fluid systems [7]. One of the solutions is a piezoelectric actuator in the form of a plate with thin thickness restrained to the body by one end, while the second end of the piezoelectric actuator opens or closes the flow window. In this solution the piezoelectric actuator moves at a high frequency with significant displacement of a few millimetres but the resulting force does not exceed a few Newtons [8]. The valves of this structure are used in preliminary stages to control the pressure relief valve and divide valve. A similar structure solution is used in a valve wherein the control element has a discoidal shape with a small thickness in relation to the diameter. As a result of deformation within the piezoelectric actuator, the central portion of the disc moves thus obtaining an opened flow window of the valve. Another solution that uses piezoelectric actuators involves structures in which the piezoelectric actuator has the form of a multi-layered stack. It uses structures in which piezoelectric actuators in the form of a stack open and close the flow windows, but in that case the movement of the propulsion system does not exceed 100μm [9]. A slight movement hinders the structure of valves in a significant way. This solution was used in injectors and unit pump injectors of diesel engines. Currently, researchers are trying to develop valve designs with multi-layered piezoelectric stacks, where slight movement of the piezoelectric actuator in the form of a stack is multiplied by mechanical levers and cams. This results in a significant movement of the whole drive, but decreases the force and operating frequency. The drive and control component in the designed structure of the valve is a multi-layered piezoelectric actuator with a mechanical amplifier in the form of an elliptical bracket. The displacement of drive

* Corresponding author: pawell@tu.kielce.pl

amounts to 500μm, the maximum locking force amounts to 570N and the operating frequency is up to 460Hz. The multi-layered actuator of the valve in the form of a piezoelectric stack allows for the ultrafast movement of separation inserts. That is how the valve is opened and closed.

## 2.1 Operation of a slotted divide valve type 2/2

The designed flow valve and distribution of the air stream is carried out by a suitably shaped assemblies that form flow windows. Such a solution lacks the typical separation slider. The operation of valves was shown in figure 1 and figure 2.

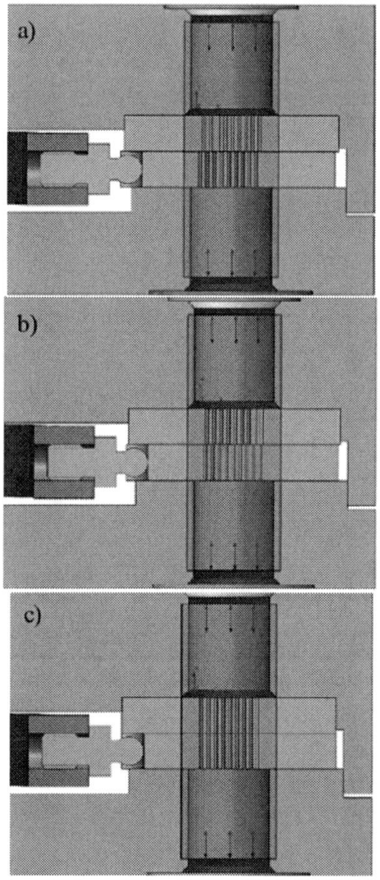


**Fig. 1.** A fragment of the cross section of the 5/3 valve. a) neutral position – closed valve, b) 50% displacement of the actuator, c) the maximum flow – open position.

The present valve contains a singular slot in the form of a rectangle with one side equal to the displacement of a piezoelectric actuator. Whereas the size of the sides of a flow window in the form of a rectangular is adjusted to the designed flow rate and the total amount of slots. The discharge window of the valve accumulates the flows from individual slots and the obtained nominal flow of the valve. The number of slots corresponds to the surface of the target flow window for the path of 2/2 valve. The valve comprises a flow path corresponding to the supply port 1 of the flow 2, in the indicated body (6,7) which are embedded in the slotted core plates (5) and corresponding groups of slots of the lower insert. The lower slotted insert is treated separately from one of the arms (4) of the amplifier. The second arm of the amplifier is attached to the body (6) of the valve by a screw. In a neutral position (figure 1a) the flow paths of the valve are completely closed. Figure 1b shows the valve at 50% of the flow corresponding to the displacement of the insert with slots of 250 μm. Figure 1c shows the end position of the slotted core corresponding to the valve in the open position. The piezoelectric actuator enables the movement of inserts with slots in a proportionate manner, the lower plate insert can occupy intermediate position relative to the upper plate insert and the flow through flow paths will depend on the degree of slots opening.

The total area of the slots, creating a flow window, is equal to the surface of the valve connection G1/4" (~34mm2). An increase of the flow window surface does not constitute a technical problem for the designed structure. The flow window may be increased by adding more slots while maintaining the displacement of piezoelectric actuators equal to 500 μm. Figure 2 shows a three-dimensional model of the valve with a partial semi-transparent view.

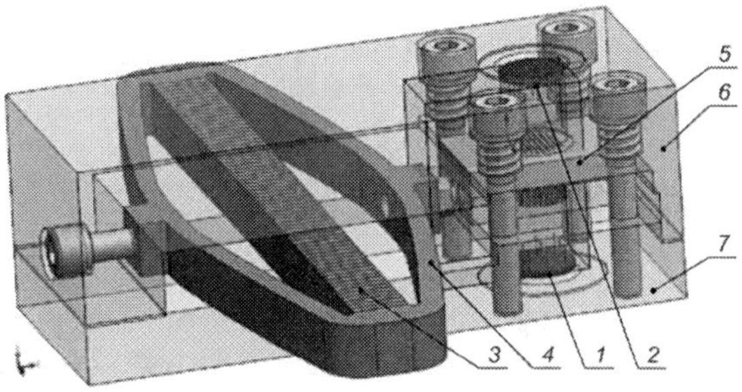


**Fig. 2.** Solid model of the valve. 1,2 – connections, 3 – piezoelectric actuator, 4 – mechanical amplifier of the displacement, 5 – slotted separation inserts, 6,7 – valve body.

# 3 Flow simulation (CFD)

Figure 4 shows the results of simulation studies in the CFD ANSYS software. A spatial computing grid with irregular structure was set in order to conduct the analysis. The crucial area of the set grid was a group of slots where the minimum mesh size was 1 μm. It was assumed that the valve was used to control gases, including compressed filtered air with a maximum solid particle size not exceeding 40 μm. For the purpose of the study we assumed working pressure of the valve ranging from 0.25 to 0.7 MPa and operating temperature ranging from $-10 \div +60$ $^{\circ}C$. The CFD ANSYS software used in the modelling process uses a mathematical model describing the physics of flow for the turbulent motion of viscous fluid described by Reynolds equations and equations of stream continuity. The equations form the basis of an interrelationship, allowing the determination of pressure distribution and velocity field of the flow. The software was used to develop a computing grid, where the number of nodes of the analysed volume of fluid (air) was 283280 which corresponds to 1318183 components. Figure 3 shows the grid and the enlarged fragment within slots of the flow window. The following physical quantities were assumed for calculations: the ambient temperature $T = 20\,^{o}\mathrm{C}$, density $\rho = 1.185\ \mathrm{kg/m^3}$, supply pressure 0.6 MPa, reference pressure $p = 101325\ \mathrm{Pa}$. Fig. 4, 5 shows the results of simulation research in the CFD software.

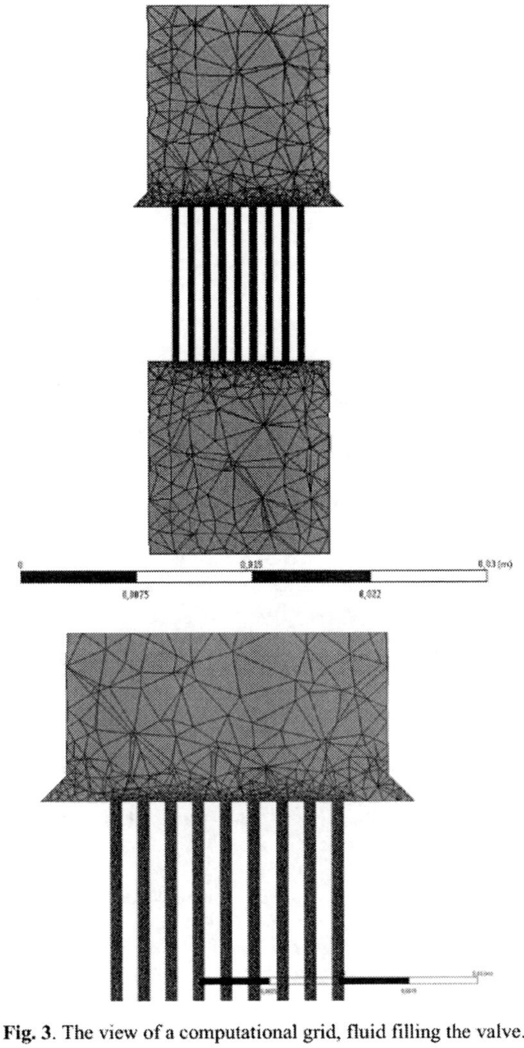

**Fig. 3**. The view of a computational grid, fluid filling the valve.

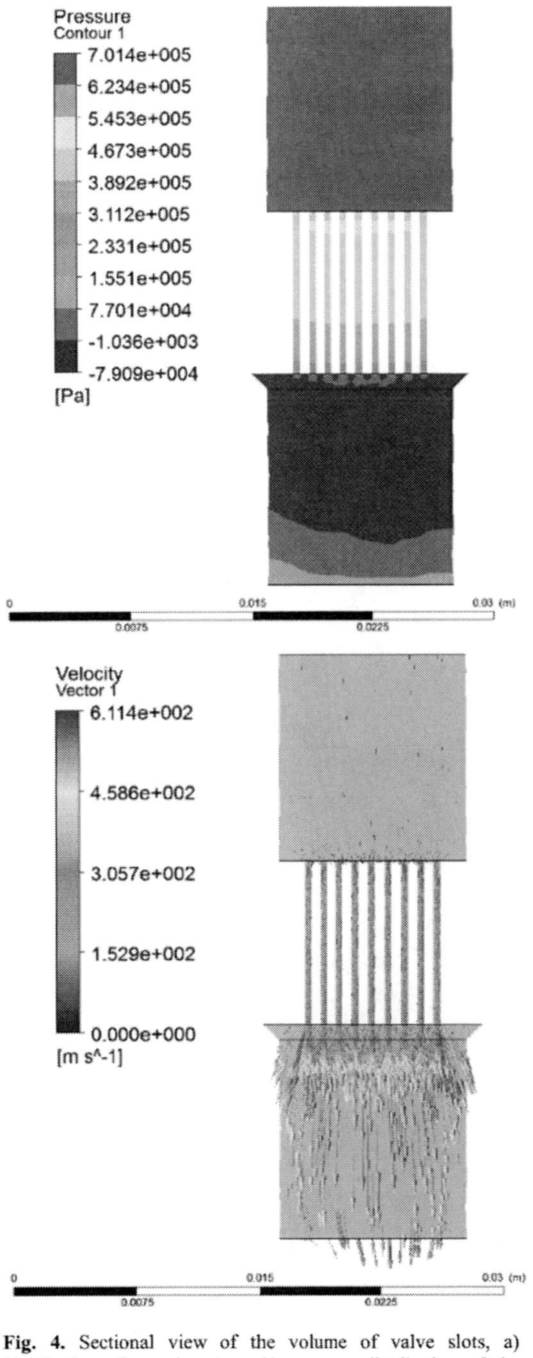


**Fig. 4.** Sectional view of the volume of valve slots, a) distribution of speed rectors, b) pressure distribution of the fluid.

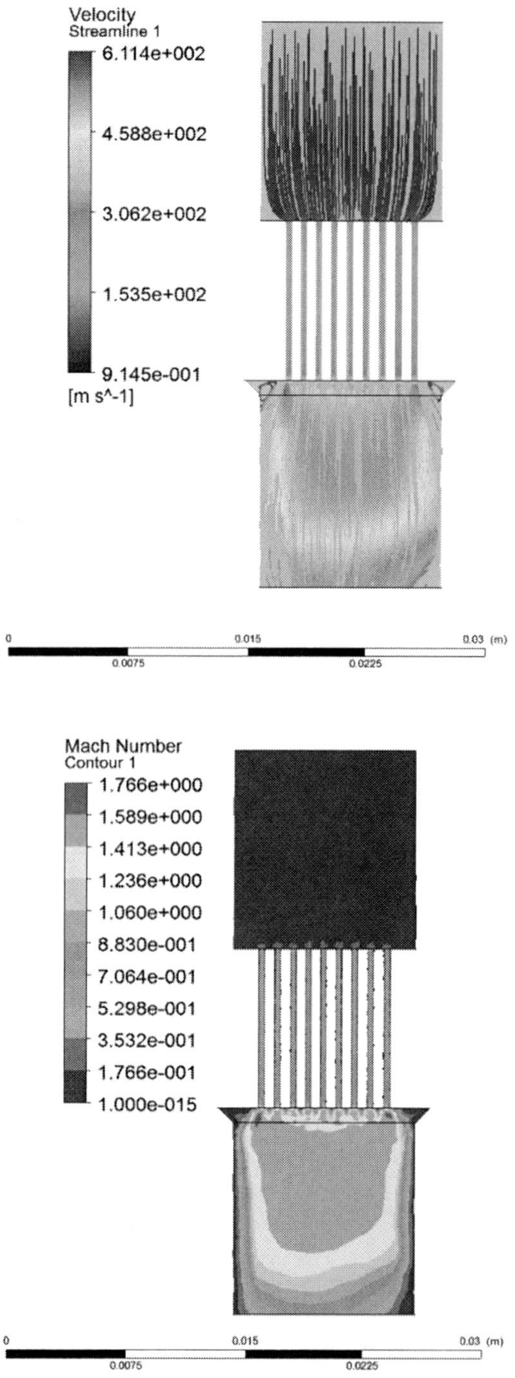


**Fig. 5.** Sectional view of the volume of valve slots, a) distribution of streamline, b) distribution of Mach number.

The modelling studies have set the maximum mass flow rate on the road 1 – 2 amounting to 0.0343 kg/s.

## 4 Conclusions

The numerical methods of fluid mechanics used in modelling of the valve allowed to determine the flow parameters of the valve prototype in the initial phase of the design concerning the slots of a mobile insert. The use of numerical methods CFD can confirm the constructive assumptions and avoid time-consuming and expensive bench studies. The presented prototype of the valve has a number of features that distinguish it from the produced proportional valves. The analysis of the market of pneumatics and patent solutions shows the lack of a solution that would enable to obtain flows through the flow windows of the valve above 0.0343 kg/s and the response time above 120 Hz. The slotted structure of separation plates of the valve allows to accumulate the flow from individual windows and thus to construct the valve with so far unattainable flows of thousands litres per minute. In addition, the innovative structure of flow windows makes it possible to design valves of any flows adjusted to technological processes.

This solution is protected by patent PL 213836 B1.

This research was supported in part by PLGrid Infrastructure.

## References

1. J.E. Takosoglu, (Acad Sci Czech Republic, Inst Thermomechanics, Dolejskova 5, Prague 8, 182 00, Czech Republic, 2016).
2. J.E. Takosoglu, P.A. Laski, S. Blasiak, G. Bracha, D. Pietrala, Measurement {&} Control. **49** (2) (2016), 62–71.
3. D.S. Pietrala, (Acad Sci Czech Republic, Inst Thermomechanics, Dolejskova 5, Prague 8, 182 00, Czech Republic, 2016).
4. J. Zwierzchowski, (Acad Sci Czech Republic, Inst Thermomechanics, Dolejskova 5, Prague 8, 182 00, Czech Republic, 2016).
5. S. Blasiak, C. Kundera, Procedia Engineering. **39** (2012), 315–326.
6. S. Blasiak, A. Pawinska, International Journal of Heat and Mass Transfer. **90** (2015), 710–718.
7. G.F. Bracha, (Acad Sci Czech Republic, Inst Thermomechanics, Dolejskova 5, Prague 8, 182 00, Czech Republic, 2016).
8. M. Blasiak, R. Kotowski, Przeglad Elektrotechniczny. **85** (12) (2009), 40–43.
9. D. Janecki, J. Zwierzchowski, (Imeko, Po Box 457, H-1371 5 Budapest, Hungary, 2009).

 DOI: 10.1051/epjconf/201714302065

# Study of the effect of temperature on the positioning accuracy of the pneumatic muscles

*Pawel Andrzej* Laski [1], *Slawomir* Blasiak [1], *Jakub Emanuel* Takosoglu [1], *Dawid Sebastian* Pietrala [1], *Gabriel Filip* Bracha[1], *Jaroslaw* Zwierzchowski [1], *Lukasz* Nowakowski [1] *Krzysztof* Borkowski [1], *Malgorzata* Blasiak [1]

[1] Kielce University of Technology, Faculty of Mechatronics and Mechanical Engineering, Aleja Tysiaclecia Panstwa Polskiego 7, 25-314 Kielce, Poland

**Abstract.** The article concerns experimental studies of the effect of temperature on the positioning accuracy of pneumatic muscles. It presents results of experimental studies in the form of thermal images from thermal imaging camera. Pneumatic artificial muscles have unique operational characteristics and because of that they are used in industrial production processes, where classic drives do not work. During operation of muscles with large frequencies above 60 Hz, one can observe a significant increase in temperatures on the bladder surface. The article concerns a study aimed at the determination of the maximum temperature which can be achieved and whether it affects the accuracy of their positioning.

## 1 Introduction

Pneumatic drive is widely used in all branches of industry and is the most environmentally friendly energy carrier [1–11]. It uses a vacuum or compressed air, which after the operation is returned directly to the atmosphere without the need for using filtering devices, as it does not contain any impurities [12–20]. Artificial muscle drives have been known since 1929. Dimitri Sensaud de Lavaud was the first to invent the pneumatic artificial muscle and then the muscle invented by Garasiev could be powered either by compressed air or other liquids. It was built from a stretchable membrane surrounded by a non-stretchable braid on the outside.

Because of the unique functional characteristics of artificial muscle drives when compared to conventional pneumatic actuators, they are used as drive elements in anthropomorphic robots, bionic and humanoid bioprosthetics, manipulators used for rehabilitation and physiotherapy, and more often in the automation of production processes [21–31]. Pneumatic artificial muscles PAM can be compared with single-acting actuators due to the way they function [32].

Pneumatic artificial muscles which, in comparison with the known and widely used systems, are distinguished primarily by soft start and soft stop. In addition, pneumatic artificial muscles have a high resistance to overload, they are characterized by high dynamics, the lack of indexing working movement (the phenomenon of stick–slip) and, most importantly, very low consumption of compressed air, which places them in the group of energy efficient drives. The pneumatic artificial muscle is a highly energy efficient drive and has ten times the initial force in comparison with the power actuator of the same diameter.

Currently, there are two most commonly used types of pneumatic muscles, in which the tight and elastic bladder is braided with small mesh (net or sleeve) of small geometric dimensions and a certain angle of weave and suitable elastic properties. In the latter case, the braid is vulcanized with an elastic bladder. The braid (fine mesh) is mounted at the ends of the muscle, creating a certain kind of artificial tendon. The value of generated force depends on the pressure of gas, its original length, the degree of shortening and material properties.

## 2 Experiential stand

An experiential stand, whose diagram was shown in Figure 1 and Figure 2, was designed in order to conduct experimental studies of the effect of temperature on the positioning accuracy of pneumatic muscles. Pneumatic muscles on the experimental stand are attached to the fixed base on one side and mounted in the moving beam connected to the base by a rotary joint on the other side. The measurement of muscle shortening is done indirectly as a result of the rotation angle of the beam relative to the base and depending on the geometrical size connecting the beam and the muscle. A precise servo valve from Festo MPYE-5-1/8-HF-010-B with 1/8'' connector was used to control the pneumatic muscles. Despite the fact that muscle drives are elements controlled by pressure, they required the use of a throughflow valve, as the valve used in the central switching zone has a linear characteristic of pressure. In addition, the use of such a valve enables the simultaneous control of two muscle drives. At the time of switching the valve to one of the extreme positions, one of the muscles is being filled and the opposite

* Corresponding author: pawell@tu.kielce.pl

DOI: 10.1051/epjconf/201714302065

muscle is being drained. After switching the valve to the second extreme position, the process is reversed and the drained muscle is being filled while the filled muscle is being drained. The experiential stand was designed to imitate the most common way of using the muscles in driving mechanisms [30,33]. Since the pneumatic artificial muscle has only the ability of shortening, in order to achieve the return movement one uses the opposite muscle acting antagonistically as it does in living organisms. Piezoresistive pressure transmitters with analog output of 0–10 V were used to measure the pressure inside of the pneumatic muscle. The reading of pressure inside of the pneumatic muscle and the value of its shortening is performed using real-time system, built on the basis of the Matlab/Simulink package and measurement cards. Measurement cards (AD and DA) [34–36] are installed on the PC Target from speedgoat [7,8,37–42]. Experimental studies were carried out for the pneumatic muscle from Festo Fluidic Muscle DMSP-10-300. The nominal length of the muscle before shortening amounted to 300 mm and the diameter amounted to 10mm. Figure 3 shows a cross section of the pneumatic muscle from Festo.

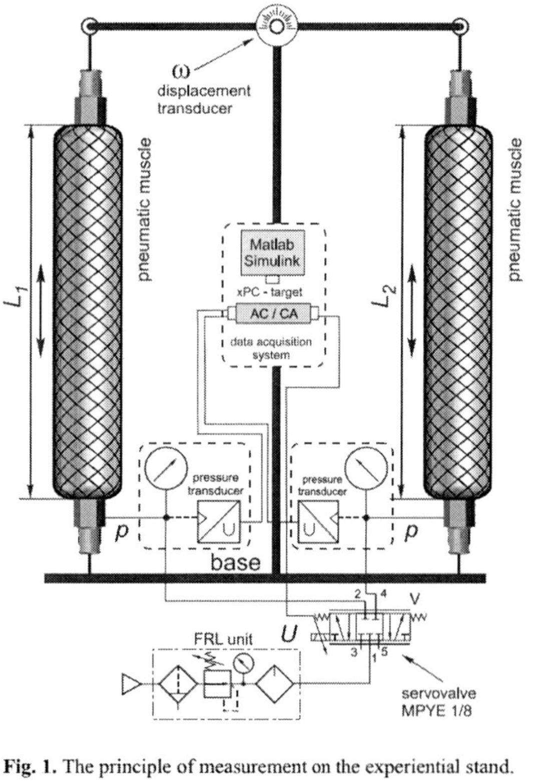


**Fig. 1.** The principle of measurement on the experiential stand.

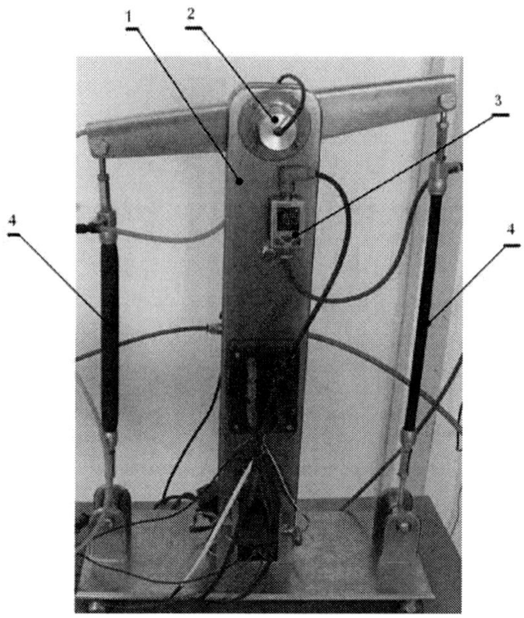


**Fig. 2.** General view of the experiential stand used to study the effect of temperature on the positioning accuracy of pneumatic muscles: 1 – bearing structure, 2 – position sensor, 3 – pressure sensor, 4 – tested muscles.

## 2.1 Operation of the pneumatic muscle

The operation of a muscle drive (shortening) is caused by the increase in air pressure within the elastic bladder. The increase in pressure causes the simultaneous increase of the circuit at the expense of reducing the length of a muscle. As a consequence, it increases the shrinkage of a muscle and forms an axial tensile force corresponding to stresses in the braid (mesh). The greatest force of the muscle drive is achieved in the initial phase of the movement, which decreases to zero at the maximum of its shrinkage, but only in the presence of constant pressure.

Through the control of pressure changes, one can control the force generated by the muscle, and the degree of its shortening.

The use of pneumatic muscle drives allows for efficient and economical management of the consumed compressed air in circuits that control the device. The consequence of this is the reduction of emitted carbon dioxide and other pollutants generated during the production of electricity. However, as a result of the operation of muscles with high frequencies, a fraction of the energy is dissipated and transformed into heat. One can observe a visible temperature increase of the working muscles.

The tested muscle was made in a specific manner and consists of layers of elastic rubber forming a tight bladder and braid (mesh) embedded and vulcanised inside the bladder (Fig. 3). This construction allows only to shrink the muscle but not to stretch it in the initial phase of the movement, since it can cause permanent damage to its structure.

**Fig. 3.** Cross section of the DMSP muscle from Festo.

## 3 Experimental study of the DMSP muscle drive

The studies of the positioning accuracy of muscles drives involved an alternant shrinkage of muscles working antagonistically. The inducement signal had the form of a sine wave that ensures a gentle way of filling and draining the muscles. The muscles were working alternately at a frequency of 20 Hz (they reached full shortening). They were unilaterally powered by compressed air. The studies were carried out to the point of reaching temperatures in the range of 31 ÷ 51°C with a step every 5°C.

Figures 4-8 show thermographic images [43–45] for temperatures in the range of 31 ÷ 51°C, obtained using the VIGO cam V50 thermal imaging camera.

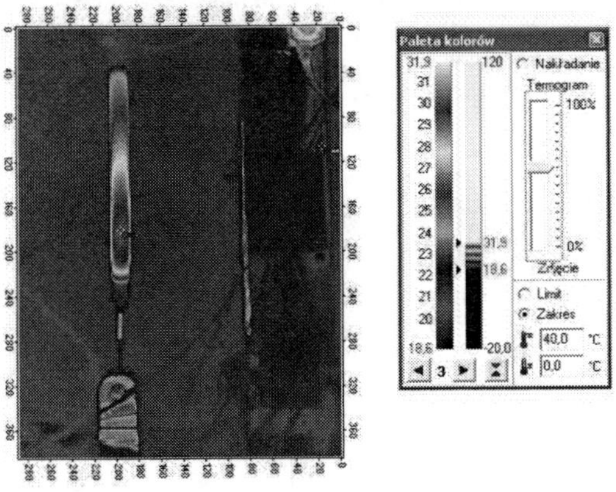


**Fig. 4.** An image of the DMSP muscle at the temperature of ~31 °C obtained using thermal imaging camera.

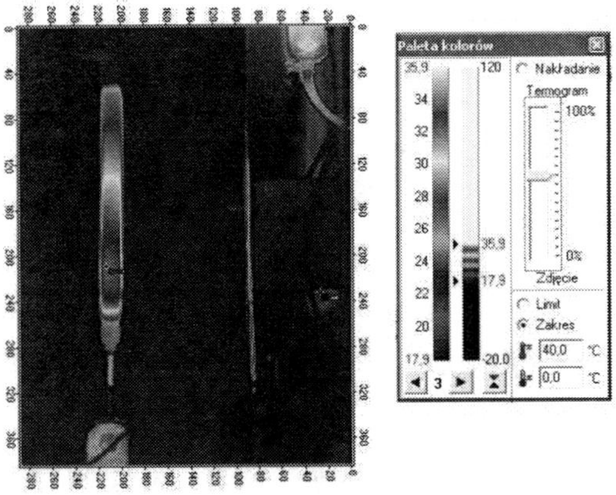


**Fig. 5.** An image of the DMSP muscle at the temperature of ~36 °C obtained using thermal imaging camera.

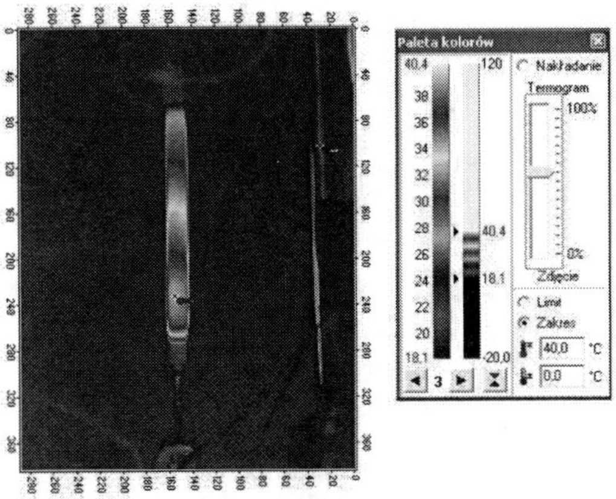


**Fig. 6.** An image of the DMSP muscle at the temperature of ~41 °C obtained using thermal imaging camera.

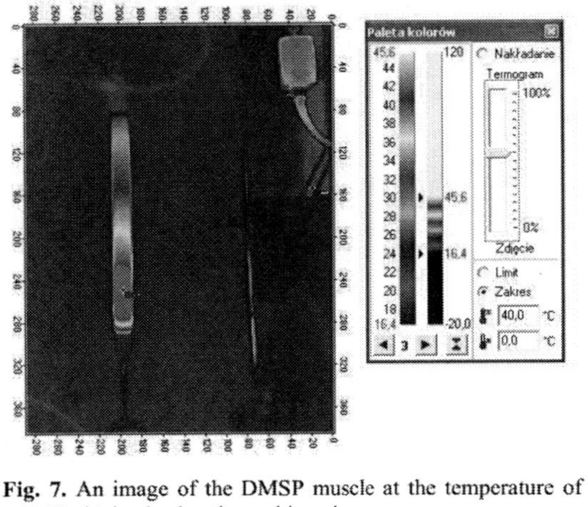


**Fig. 7.** An image of the DMSP muscle at the temperature of ~46 °C obtained using thermal imaging camera.

DOI: 10.1051/epjconf/201714302065

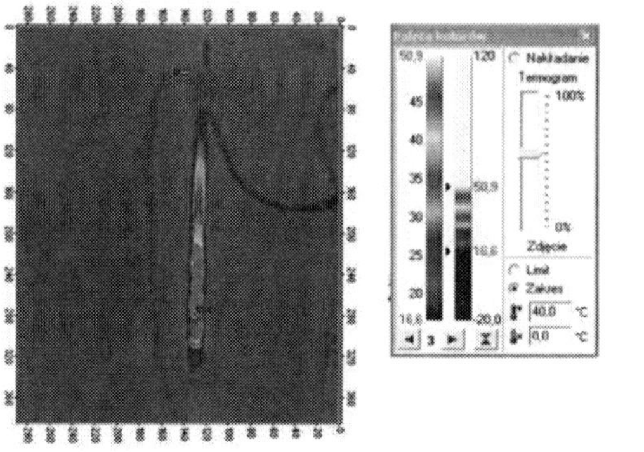

**Fig. 8.** An image of the DMSP muscle at the temperature of ~51 °C obtained using thermal imaging camera.

Figure 9 shows an image of the experiential stand obtained using thermal imaging camera [9,46], with visible muscles where the highest temperature is observed at fittings of the muscle.

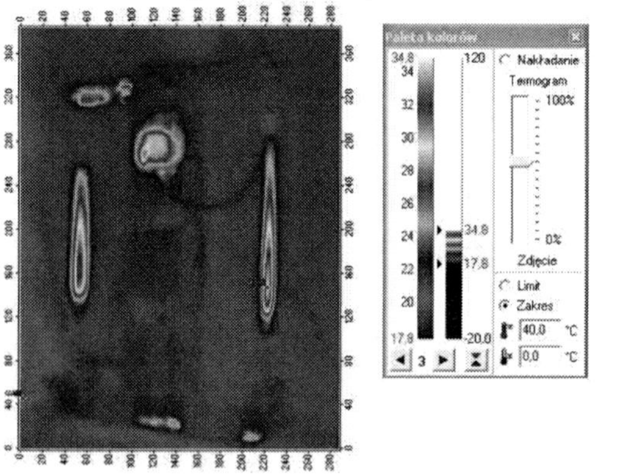

**Fig. 9.** View of the experiential stand in a thermographic image.

Dynamic characteristics were presented in figure 10 and figure 11 in order to compare the studies conducted for temperatures reached by muscles and their influence on the positioning accuracy.

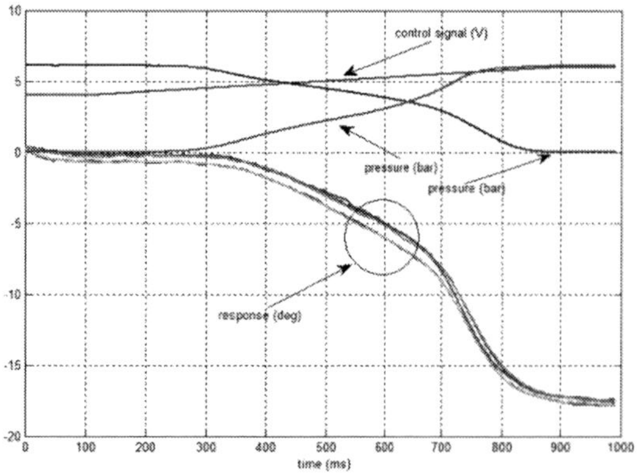


**Fig. 10.** Dynamic characteristics of the DMSP muscle for temperatures in the range of 31°C, 36°C, 41°C, 46°C, 51°C

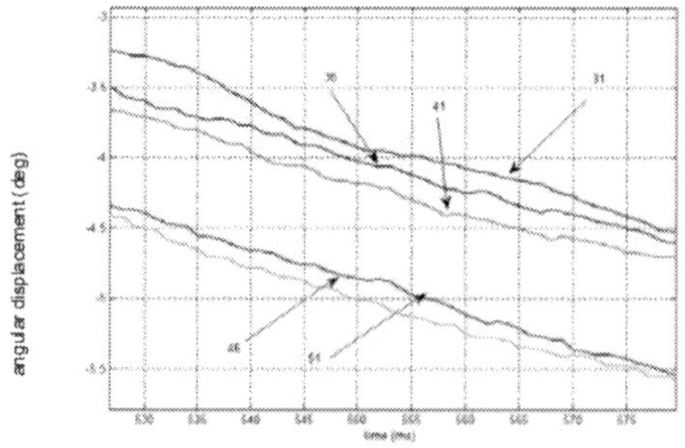


**Fig. 11.** Zoom in on part of dynamic characteristics of the DMSP muscle for temperatures in the range of 31°C, 36°C, 41°C, 46°C, 51°C.

## 4 Conclusions

Because of their unique characteristics, muscle drives find new applications in medicine and industry. It has been shown that the increase in temperatures of muscles as a result of dynamic operation have little effect on the achieved positioning accuracy of muscle drives (figure 10 and figure 11). The increase in muscle temperature results from friction of the braid (mesh) on the elastic bladder. In addition, the muscle was warming up to a greater extent at the bottom connection. This results directly from the heat exchange processes. The upper part of the muscle was filled with new "cold" volume of air 20 times within one second. This resulted in a greater dissipation of thermal energy. It was also observed that attempts to increase the frequency of switching the muscles to achieve higher surface temperature did not result in its increase. Thermal equilibrium was achieved spontaneously. This knowledge allows for safe use of pneumatic muscles with considerable frequencies of operation without concern of damage resulting from the increase in temperature of the drive.

The experiential stand used to determine the dynamic characteristics of muscle drives is protected by patent PL 223042 B1.

## References

1. J.E. Takosoglu, (Acad Sci Czech Republic, Inst Thermomechanics, Dolejskova 5, Prague 8, 182 00, Czech Republic, 2016).
2. J.E. Takosoglu, P.A. Laski, S. Blasiak, (Acad Sci Czech Republic, Inst Thermomechanics, Dolejskova 5, Prague 8, 182 00, Czech Republic, 2014).
3. J.E. Takosoglu, R.F. Dindorf, P.A. Laski, International Journal of Advanced Manufacturing Technology. **40** (3–4) (2009), 349–361.
4. J.E. Takosoglu, P.A. Laski, S. Blasiak, Proceedings of The Institution of Mechanical Engineers Part I-Journal of Systems And Control Engineering. **226** (I10) (2012), 1335–1343.
5. J.E. Takosoglu, R.F. Dindorf, P.A. Laski, International Journal Of Advanced Manufacturing Technology. **40** (3–4) (2009), 349–361.

6. J.E. Takosoglu, P.A. Laski, S. Blasiak, G. Bracha, D. Pietrala, Measurement & Control. **49** (2) (2016), 62–71.
7. R. Trochimczuk, in: Gosiewski, Z and Kulesza, Z (Ed.), Mechatronic Systems And Materials Iv, Trans Tech Publications Ltd, Laublsrutistr 24, Ch-8717 Stafa-Zurich, Switzerland, 2013: pp. 3–8.
8. R. Trochimczuk, M. Gawrysiak, in: Gosiewski, Z and Kulesza, Z (Ed.), Mechatronic Systems And Materials III, Trans Tech Publications Ltd, Laublsrutistr 24, Ch-8717 Stafa-Zurich, Switzerland, 2009: pp. 107–112.
9. S. Wawrzyniak, K. Peszynski, (Acad Sci Czech Republic, Inst Thermomechanics, Dolejskova 5, Prague 8, 182 00, Czech Republic, 2014).
10. J. Zwierzchowski, (Acad Sci Czech Republic, Inst Thermomechanics, Dolejskova 5, Prague 8, 182 00, Czech Republic, 2016).
11. D.S. Pietrala, (Acad Sci Czech Republic, Inst Thermomechanics, Dolejskova 5, Prague 8, 182 00, Czech Republic, 2016).
12. S. Blasiak, J.E. Takosoglu, P.A. Laski, Journal of Thermal Science and Technology. **9** (2) (2014), JTST0011--JTST0011.
13. S. Blasiak, A. Pawinska, International Journal of Heat and Mass Transfer. **90** (2015), 710–718.
14. S. Blasiak, P.A. Laski, J.E. Takosoglu, International Journal of Heat and Mass Transfer. **57** (1) (2013), 22–31.
15. S. Blasiak, C. Kundera, J. Bochnia, Procedia Engineering. **39** (2012), 366–378.
16. S. Blasiak, C. Kundera, Procedia Engineering. **39** (2012), 315–326.
17. S. Blasiak, Journal of Thermal Science and Technology. **10** (1) (2015), JTST0016--JTST0016.
18. S. Blasiak, *International Journal of Heat and Mass Transfer*. **81** (2015), 90–102.
19. S. Blasiak, A. V Zahorulko, Tribology International. **94** (2016), 126–137.
20. S. Blasiak, J.E. Takosoglu, P.A. Laski, (Acad Sci Czech Republic, Inst Thermomechanics, Dolejskova 5, Prague 8, 182 00, Czech Republic, Svratka, Czech Republic, 2014).
21. P. Suranek, M. Mahdal, J. Tuma, (IEEE, 345 E 47TH ST, New York, NY 10017 USA, 2013).
22. L. Smutny, M. Mahdal, J. Skuta, (World Scientific And Engineering Acad And Soc, Ag Loannou Theologou 17-23, 15773 Zographou, Athens, Greece, 2009).
23. L. Nowakowski, M. Wijas, (Acad Sci Czech Republic, Inst Thermomechanics, Dolejskova 5, Prague 8, 182 00, Czech Republic, 2016).
24. L. Nowakowski, E. Miko, M. Skrzyniarz, (Acad Sci Czech Republic, Inst Thermomechanics, Dolejskova 5, Prague 8, 182 00, Czech Republic, 2016).
25. I. Krzysztofik, (Acad Sci Czech Republic, Inst Thermomechanics, Dolejskova 5, Prague 8, 182 00, Czech Republic, 2016).
26. D. Janecki, J. Zwierzchowski, (Imeko, Po Box 457, H-1371 5 Budapest, Hungary, 2009).
27. D. Janecki, J. Zwierzchowski, Measurement Science & Technology. **26** (8) (2015),.
28. D. Janecki, L. Cedro, J. Zwierzchowski, Metrology And Measurement Systems. **22** (2) (2015), 289–302.
29. D. Janecki, J. Zwierzchowski, L. Cedro, Bulletin of The Polish Academy of Sciences-Technical Sciences. **63** (3) (2015), 771–779.
30. T. Huscio, W. Kolodziejczyk, in: Gosiewski, Z and Kulesza, Z (Ed.), Mechatronic Systems And Materials IV, Trans Tech Publications Ltd, Laublsrutistr 24, Ch-8717 Stafa-Zurich, Switzerland, 2013: pp. 313–317.
31. G.F. Bracha, (Acad Sci Czech Republic, Inst Thermomechanics, Dolejskova 5, Prague 8, 182 00, Czech Republic, 2016).
32. Łukasz Terelak, Badanie wpływu temperatury na dokładność pozycjonowania napędów mięśniowych z użyciem kamery termograficznej VIGO, Politechnika Świętokrzyska, 2011.
33. T. Huscio, F. Siemieniako, in: Gosiewski, Z and Kulesza, Z (Ed.), Mechatronic Systems And Materials IV, Trans Tech Publications Ltd, Laublsrutistr 24, Ch-8717 Stafa-Zurich, Switzerland, 2013: pp. 372–377.
34. J. Zavadil, J. Tuma, J. Valicek, M. Mahdal, J. Los, (IEEE, 345 E 47TH ST, NEW YORK, NY 10017 USA, 2013).
35. P. Suranek, M. Mahdal, J. Tuma, (IEEE, 345 E 47TH ST, New York, NY 10017 USA, 2014).
36. P. Suranek, M. Mahdal, J. Tuma, (IEEE, 345 E 47TH ST, New York, NY 10017 USA, 2015).
37. A. Laski Pawel, E. Takosoglu Jakub, S. Blasiak, Robotics And Autonomous Systems. **72** (2015), 59–70.
38. P.A. Laski, (Acad Sci Czech Republic, Inst Thermomechanics, Dolejskova 5, Prague 8, 182 00, Czech Republic, 2016).
39. P.A. Laski, J.E. Takosoglu, S. Blasiak, (Acad Sci Czech Republic, Inst Thermomechanics, Dolejskova 5, Prague 8, 182 00, Czech Republic, 2014).
40. P.A. Laski, J.E. Takosoglu, S. Blasiak, Robotics and Autonomous Systems. (2014),.
41. P.A. Laski, J.E. Takosoglu, S. Blasiak, Robotics and Autonomous Systems. **72** (2015), 59–70.
42. I. Krzysztofik, Z. Koruba, Journal of Automation and Information Sciences. **44** (5) (2012), 38–47.
43. P. Kolber, D. Perczynski, B. Landowski, S. Wawrzyniak, (Acad Sci Czech Republic, Inst Thermomechanics, Dolejskova 5, Prague 8, 182 00, Czech Republic, 2016).
44. K. Migawa, L. Knopik, S. Wawrzyniak, (Acad Sci Czech Republic, Inst Thermomechanics, Dolejskova 5, Prague 8, 182 00, Czech Republic, 2016).
45. K. Peszynski, W. Szmyt, S. Wawrzyniak, D. Perczynski, (Acad Sci Czech Republic, Inst Thermomechanics, Dolejskova 5, Prague 8, 182 00, Czech Republic, 2016).
46. Z. Travnicek, K. Peszynski, J. Hosek, S. Wawrzyniak, International Journal Of Heat And Mass Transfer. **46** (7) (2003), 1265–1278.

# Mathematical modelling of non-isothermal flow in buildings

*Richard* Lenhard[1] and *Tomáš* Puchor[1]

[1]University of Žilina, Faculty of Mechanical Engineering, Department of Power Engineering, Univerzitná 1, 010 26 Žilina, Slovakia

**Abstract.** Numerical simulations are used in many areas of science and technology. In last time are numerical methods used not only in small scale applications, in fluid mechanics, but are also used in civil engineering for simulation of air flow, for simulation of heat transfer. This paper deals with simulation of air flow in building for municipal use. The core of the paper is the calculation of the input parameters for the creation of numerical model and description of the model itself. The conclusion provides an analysis and comparison of simulation results with calculated values for the building.

## 1 Introduction

The numerical model is defined on the basis of various physical phenomena in complex differential equations. In this case, we will observe the flow of air in a rather complex geometry. In view of the fact that this is the 3D model of flow which can be described as a system of partial differential equations that can be solved by numerical methods. The calculation of the following equations are typically designed for a number of cells.

For the solution of ventilation can be used Fluent program, part of the software ANSYS package. When creating a numerical model it must be determined in the applicable geometry created with mesh computing. It is necessary to know all the input data and set them as boundary conditions that are necessary for correct calculation. After performing the calculation, it can be a graphical displayed of results.

In the numerical solution we choose a general model of the building. Another criterion was the choice of the ventilation system, the selected building should be ventilated with air to circulate. The layout of the proposed construction building is in Fig. 1. Numerical simulation is performed in order to determine if it is possible to achieve the natural air flow and the natural exchange of different combinations of open and closed windows and doors [1, 13] in the building.

## 2 Modeling and mesh computing

3D model of the building was made up from drawings for construction of the building. The model has been adapted and simplified to the need to perform CFD simulations.

Network is a systemic breakdown of computational domain into sub interlinked cells in 2D space or in 3D space [2].

For numerical simulation have been chosen, these mesh computing:
- hexagonal,
- tetragonal,
- polyhedral.

**Table 1.** Parameters of the building.

| **Building dimensions** | |
|---|---|
| Width | 19 m |
| Length | 22.6 m |
| Ground clearance | 9.3 m |
| Area | 343 $m^2$ |
| Volume | 3,310 $m^3$ |

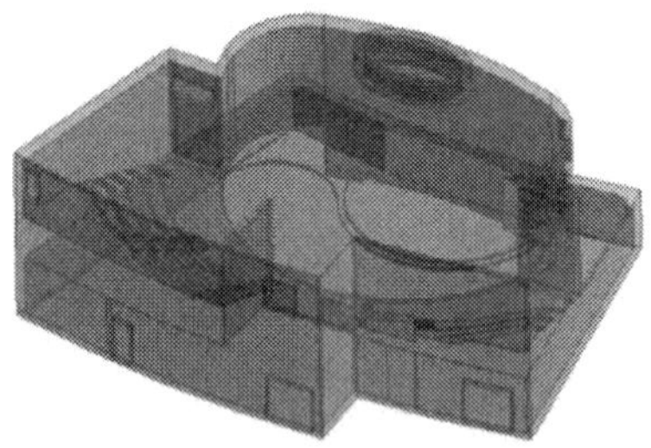

**Fig. 1** Building model geometry before treatment

Because it is a very complex model geometry the mesh must be created quickly. Of course computational mesh had to be of sufficient quality to get the best results from CFD simulations. [13]

### 2.1 Hexagonal mesh

The mesh is made up of mostly by hexagonal elements Fig. 2. Model building has a complex geometry, the mesh is made up of a quadrilateral and by triangonal elements. Given the complexity of the

* Corresponding author: richard.lenhard@fstroj.uniza.sk

DOI: 10.1051/epjconf/201714302066

geometry was aggravation elements toward the center of the volume of the model and to create a network composed of the elements of hexagonal and tetragonal.

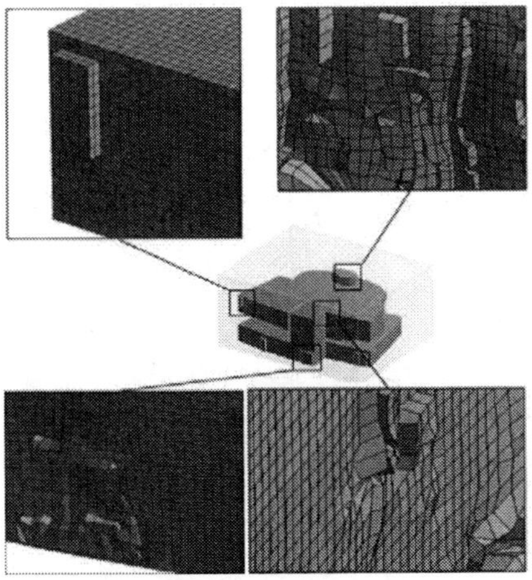

**Fig. 2** Cuts hexagonal mesh

**Table 2.** Parameters of a hexagonal mesh and quality

| Mesh | Number of elements | Orthogonal Quality | Ortho Skew |
|---|---|---|---|
| **Hexagonal mesh** | 1,536,370 | $1.005.10^{-12}$ | 0.999 |

## 2.2 Tetragonal mesh

Fig. 3 shows a network element tetragonal and was allowed automatically in the whole volume.

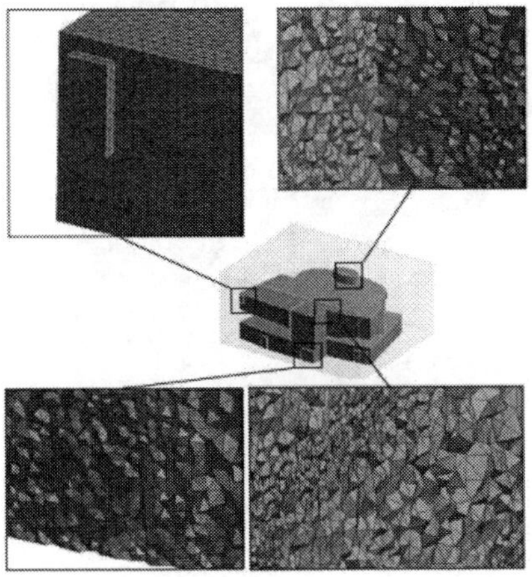

**Fig. 3** Cuts tetragonal mesh

**Table. 3.** Parameters of a tetragonal mesh and quality

| Mesh | Number of elements | Orthogonal Quality | Ortho Skew |
|---|---|---|---|
| **Tetragonal mesh** | 5,484,930 | $3.125.10^{-2}$ | 0.955 |

## 2.3 Polyhedral mesh

Polyhedral mesh was made up of a mesh consisting of elements of tetragonal, as a command converting polyhedral fig. 4.

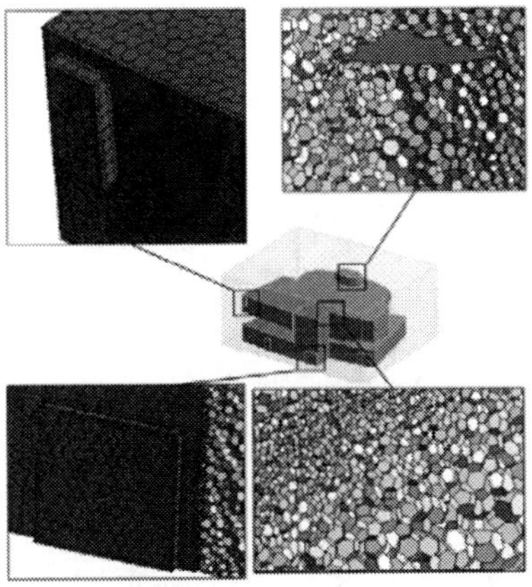

**Fig. 4** Cut polyhedral mesh

**Table 4.** Parameters of a polyhedral mesh and quality

| Mesh | Number of elements | Orthogonal Quality | Ortho Skew |
|---|---|---|---|
| **Polyhedral mesh** | 1,137,173 | $1.059.10^{-1}$ | 0.894 |

Hexagonal mesh showed the worst quality mesh. Tetragonal mesh did not meet the quality or number of elements, but after converting to mesh polyhedral significantly decreased the number of elements and improve the quality of the mesh. For simulation was chosen polyhedral mesh.

# 3. Boundary conditions

Building has heated bench with a constant temperature at 36 °C and heated floor temperature at 25 °C. Outdoor design temperature is set to 15 °C. Design temperature for the walls is 5 °C. In the building will be simulated airflow with closed windows and doors at first. Based on simulation results, will be decided on whether the natural flow will be sufficient for air exchange in the building or installation of ventilation elements will be required [1].

In the programming environment of FLUENT have been set initial and boundary conditions. Among the software simulations, Ansys Fluent software allows, the simulation of physical and thermophysical effects to the greatest interface. Modeling airflow simulations were transferred by employing three models:

- Laminar,
- k - ε model,
- k - ω model.

The procedure for entering the initial and boundary conditions were as follows:

1) input of material parameters,
2) enter the input and output values of the flowing medium,
3) set points,
4) adjusting the calculation,
5) initialization calculation,
6) start of calculation.

### 3.1 CFD simulation results

CFD simulation result is achieved by making the natural flow (Tab. 5) of air in the building. The resulting rates were compared with closed and open windows and doors using three calculation models.

**Table 5.** Scores air speed of CFD simulations

| **Velocity [m.s$^{-1}$]** | **Simulation** | |
|---|---|---|
| Model | open | closed |
| laminar | 1.5253 | 3.5131 |
| k-ε | 0.1547 | 0.2282 |
| k-ω | 0.1403 | 0.4157 |

### 3.1.1 The results of numerical simulation model Laminar (open windows and doors)

Fig. 5 shows the temperature fields in laminar model. After completion of iterations can see the method and direction of mixing cold outside air with warmer air inside. Air distribution through building can be seen by streamlines and vectors of the air in the premises through the low-lying inlet openings (doors and windows) and exit through the skylight.

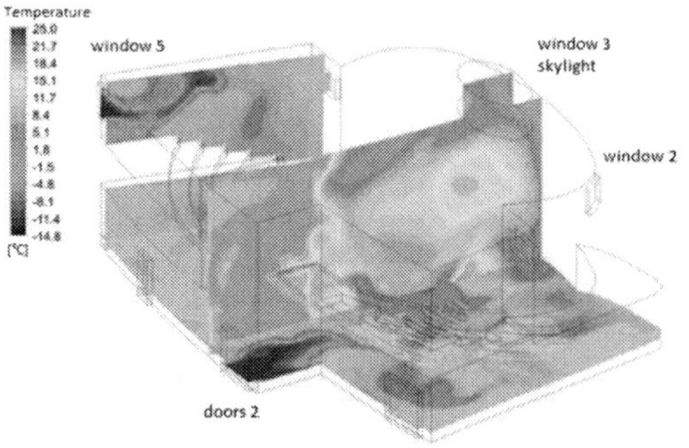


**Fig. 5** Temperature ratio in laminar building model

Fig. 4 shows the different temperatures that range between – 14.8 °C to 25 °C in a church building section. Low visible mainly at the door and at the window 2, 5. Air temperature 25 °C is accumulated in the heated benches and rises to the ceiling.

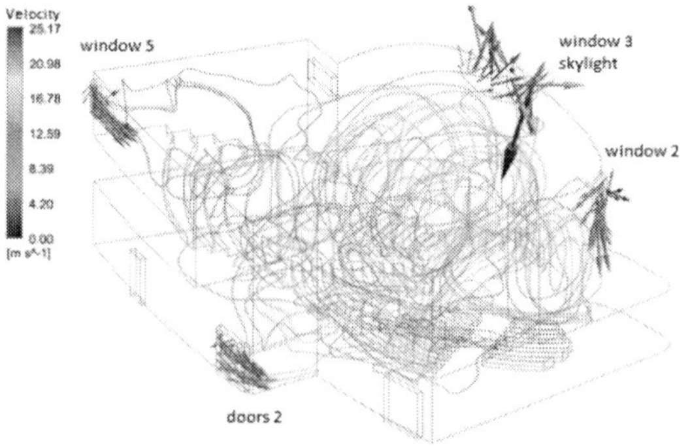


**Fig. 6** Shows streamlines and the vectors in laminar model

Fig. 6 are shown the ventilation velocity laminar model; max. air velocity reaches 25.17 m.s$^{-1}$. In the middle of the building can be seen mixing of air at lower speeds than the speed of the air inlet and outlet openings in each.

### 3.1.2 The results of numerical simulation model k- ε

Fig. 7 are shown the temperature fields in the model k-ε. In this model, the expiration of iterations was cold air entering through the lower area of the door and set down in the upper part of the door and the window and the window 1, 5. Also, you can see the temperature conditions through the headroom space. The movement of the nozzle and direction vectors is shown in Figure 8.

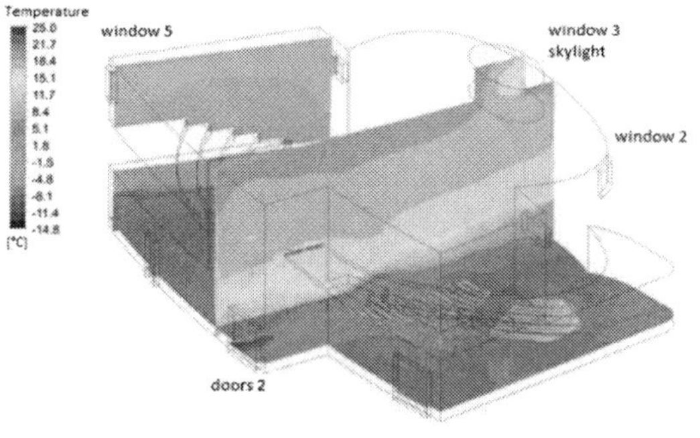


**Fig. 7** Temperature ratio for model k-ε

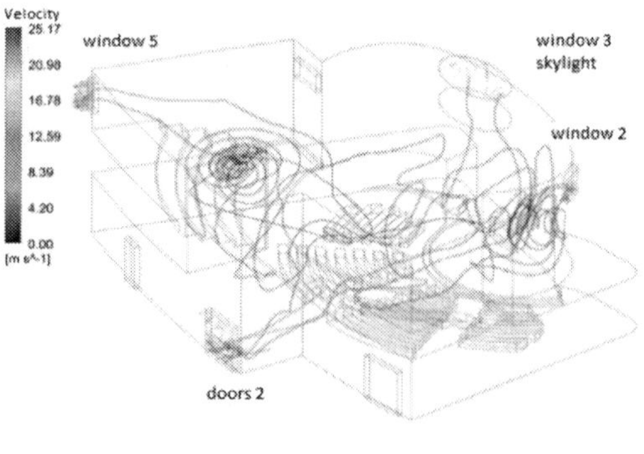


**Fig. 8** Velocity in k-ε

DOI: 10.1051/epjconf/201714302066

### 3.1.3 The results of the numerical simulation model k – ω

The model k – ω was undesirable chimney effect-facing the flow of air inside the building, to a greater extent taking place thereto, and the skylight window of 3, and to a lesser extent through the open doors 2.

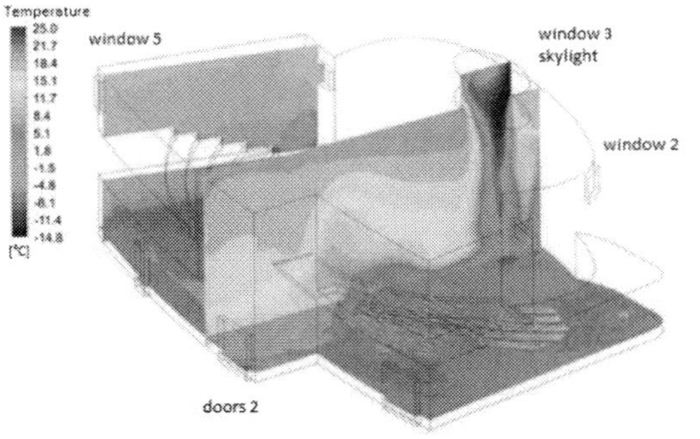


**Fig. 9** Temperature ratio in the k – ω model

## 4 The results of numerical simulation model Laminar (open windows and doors)

The numerical simulation with closed windows and doors have been selected sections to display the contours of temperature and air velocity.

### 4.1 The results of numerical simulation model Laminar

Fig. 10 shows the internal temperature status with doors and windows closed as there is for heating the surrounding air from the benches on the ground floor and first floor and the air cooling of the walls.

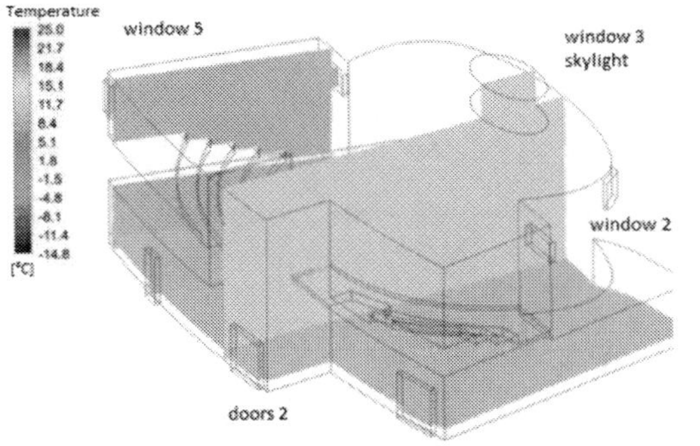


**Fig. 10** Temperature conditions in the model Laminar

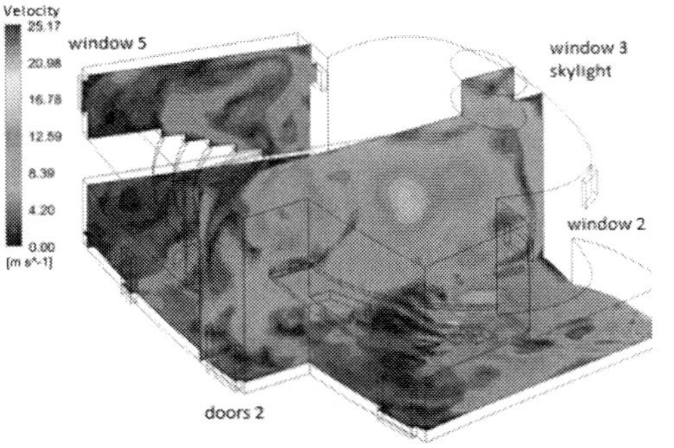


**Fig. 11** Velocity ratio for laminar model

### 4.2 Výsledky numerickej simulácie modelu k-ε a k–ω

Graphic documentation of the result of applying a numerical model to a ε-ω, in Fig. 12 and 14 are detected by the temperature fields in the building sections. Fig. 13 and 15 are the results of the speed range of flow shown by contours.

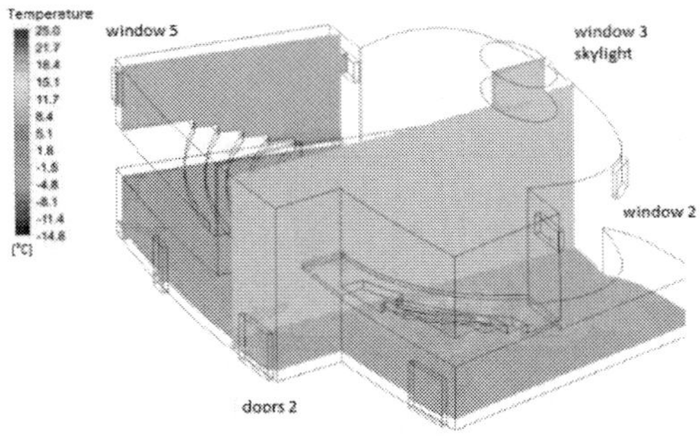


**Fig. 12** Temperature ratio for model k-ε

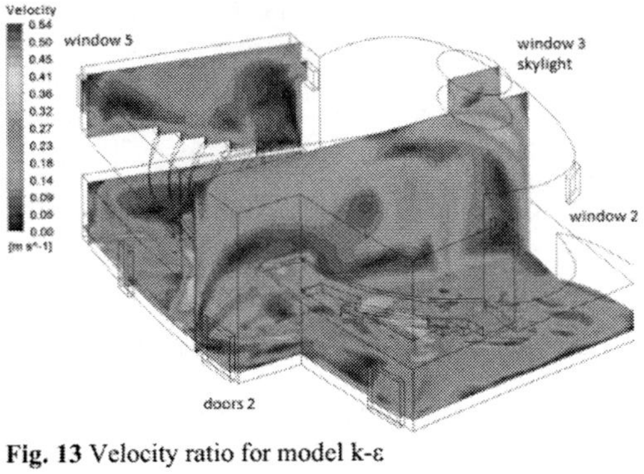


**Fig. 13** Velocity ratio for model k-ε

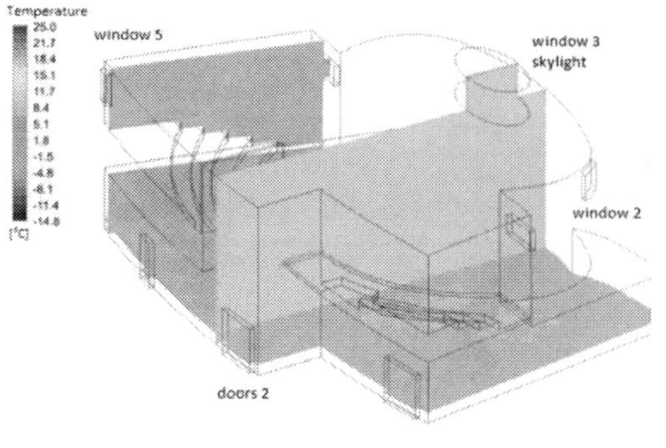


**Fig. 14** Temperature ratio for the k-ω model

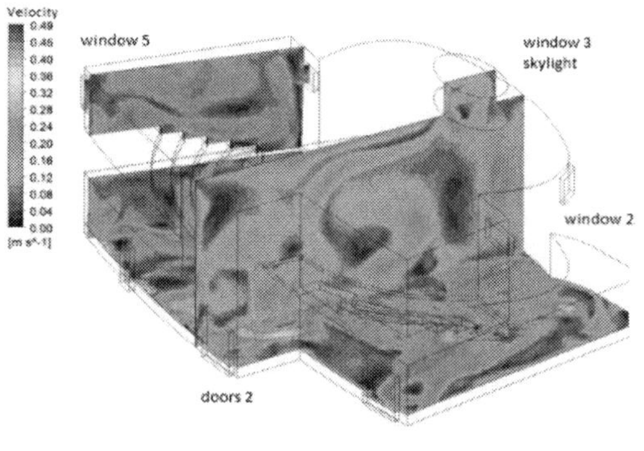


**Fig. 15** Velocity ratio for the k- ω model

## 5 Conclusion

Three types of meshes was created and compared after simulations.

The results of the comparison:

- A hexagonal network, increase magnification cell towards its center (the worst network). When reducing the number of components quality of network decreased.

- Tetragonal network was due to poor quality converted to grid "polyhedral", thereby reducing the number of components and quality of the network increased.

The result of the comparison network was chosen for simulation hexagonal grid and three models of turbulence "laminar", "k-ε", "realizable" of a wall with an extended feature "enhanced wall treatments" a model "for ε-standard". The simulations were carried out with closed and then opened selected windows and doors.

The main goal was to handle numerical simulation of flow in the building and pointing out that it is possible to perform such analyzes even before the position of the object itself.

Laminar model was best to simulate natural air flow through building.

## Acknowledgement

This contribution is the result of the project: "Limits of radiative and convective cooling through the phase changes of working fluid loop thermosiphon"; APV 15 - 0778

## References

1. Daniels, K.: *Technika budov - Príručka pre architektov a projektantov*. Bratislava: Jaga grup, s. r. o., (2003).
2. ANSYS Help. 2015 *Ansys Help Viewer 16.2*. SAS IP, Inc. (2015).
3. Kaduchová, K., Lenhard, R., Gavlas, S., Jandačka, J.: The structural design of the experimental equipment for unconventional heating water using heat transfer surfaces located in the heat source, EPJ Web of Conf. **45**, (2013).
4. Kaduchová, K., Lenhard, R., Gavlas, S., Jandačka, J.: Numerical simulation of indirect heating of water by heat transfer surface area located at the heat source, EPJ Web of Conf. **45**, (2013).
5. Gatlin, E. *Natural ventilation tricks to cool off your summer*. [online].
6. Hehhálek, J. *Spôsoby vetrania bytov a rodinných domov*. [online].
7. Ďurčanský, P., Lenhard, R., Jandačka, J. Comparison of mathematical models for heat exchangers of unconventional CHP units. *Acta Polytechnica: journal of advanced engineering*, (2015).
8. Kasanický, M., Lenhard, R., Malcho, M. Optimization of atypical heat exchanger with using CFD simulation. Žilina, Transcom 2015, (2015).
9. Gavlas, S., Ďurčanský, P., Lenhard, R., Jandačka, J. Mathematical simulation of heat exchanger working conditions. EPJ Web of Conference, (2015).
10. Kapjor, A., Hužvár, J., Ftorek, B., Smatanová, H. Analysis of the heat transfer from horizontal pipes at natural convection. AIP Conf. Proc. **1608**, (2014).
11. Skočilasová, B., Skočilas, J. Determination of pressure drop coefficient by CFD simulation. AIP Conf. Proc. **1608**, (2014).
12. Siažik, J., Malcho, M, Rezničák, R.: Proposal of bypass in heat recovery system with sucking air, AIP Conf. Proc. **1745**, (2016).
13. Lenhard, R., Ďurčanský, P.: Use of numerical simulation in building ventilation, XX. AEaNMiFMaE 2016, (2016)
14. Lenhard, R., Malcho. M.: The second device for measuring of the thickness of the falling condensate in the gravity assisted heat pipe. AIP Conf. Proc. **1745**, (2016).
15. Kozubková, M., Šťáva, P., Drábková, S.: Matematické modely nestlačitelného a stlačitelného proudění: metoda konečných objemů. Ostrava, (1999).
16. Novomestský, M., Smatanová, H., Kapjor, A.: Natural convective heat transfer from square cylinder, AIP Conf. Proc. **1745**, (2016).

# Prediction of localized aerosol deposition in a realistic replica of human airways using experimental data and numerical simulation

*Frantisek* Lizal[1,*], *Jakub* Elcner[1], *Miloslav* Belka[1], *Jan* Jedelsky[1], and *Miroslav* Jicha[1]

[1]Brno University of Technology, Faculty of Mechanical Engineering, Energy Institute, Technicka 2896/2, Brno 616 69, Czech Republic

**Abstract.** The presence of aerosol deposition hot-spots in human airways presumably contributes to development of various diseases. The overall aerosol deposition in human lungs can be predicted with sufficient accuracy nowadays. However, the prediction of localized aerosol deposition poses arduous challenge, namely in diseased lungs. Numerical simulation is considered to be a promising tool for the successful prediction. Yet, the validation of such simulations is difficult to perform, as not enough experimental data acquired using realistic airway replicas is available. This paper presents a first comparison of localized deposition measurement and simulation performed on the identical realistic geometry. The analysis indicates that both approaches yield similar results for low Reynolds number flows.

## 1 Introduction

Although the inhalation therapy is widely used among patients with asthma, cystic fibrosis or chronic obstructive pulmonary disease nowadays, its potential is far from being completely utilized. It could be used for delivery of pharmaceutical aerosols suitable for treatment of systemic diseases such as diabetes, or for targeted cancer treatment. However, the problem of current inhalation devices is that they cannot be easily tailored for each individual patient. Yet, especially human upper airways vary widely among people, which leads to strong variability in the delivered dose in the case that the same type of inhaler is used by all patients.

Moreover, despite the fact that the overall aerosol deposition in human lungs can be predicted with sufficient accuracy, the prediction of localized aerosol deposition poses arduous challenge, namely in diseased lungs. It should be emphasized, that the localized deposition prediction is essential not only for pharmaceutical aerosols, but, as several authors demonstrated [1,2], the presence of aerosol deposition hot-spots in human airways presumably contributes to development of various diseases, such as bronchial carcinoma.

It is obvious that the future of inhalation therapy lies in our ability to predict the localized deposition of particles in patient-specific lung geometries. One approach is to use stochastic lung deposition models. This type of simulation has been applied recently by Farkas et al. [3] to calculate the deposition distribution of two marketed aerosol drugs. Another approach utilizes the fast development of computational fluid and particle dynamics (CFPD). It seems achievable to calculate the precise local deposition in resolution of few $mm^2$ in the near future. The advances in CFPD are tightly linked with the innovation of medical imaging techniques, as the acquisition of high-quality geometry of airways and the precise knowledge of boundary conditions remain challenging tasks even nowadays.

Many studies documenting the development of numerical simulations of flow and particle deposition in human lungs have been published in last two decades (see e.g. the reviews of Hofmann [2] or Rostami [4]). However, only rarely have they been performed on realistic lung geometries and validated by reliable experimental data on the identical geometries. Yet, the validation represents extremely important step and only validated simulations can be used for realistic predictions in pharmaceutical or toxicological applications.

The difficulty of acquiring the experimental data grounds in complicated and hardly accessible geometry of lungs. The most realistic way of validation would be the *in vivo* measurement of the patient whose lung geometry was used for the simulation. However, several problems complicate this option. First, the medical imaging methods often require the application of radioactive substances, while the higher dose of radiation applied the better resolution of the data. Regrettably, the radiation dose is limited due to its negative health effect. The methods that do not use radioactivity are usually even less precise in measurement of deposited particles location. Second, it is extremely difficult to obtain the high quality images, as the human body moves during the scanning and also the lung geometry and the location of deposited particles inevitably change in time due to the physiological processes in the body. Third, the whole process is also complicated by the administrative requirements (informed consent, ethical committee approval, etc.) which are necessary to prevent the public health.

* Corresponding author: lizal@fme.vutbr.cz

The above mentioned obstacles lead to the application of *in vitro* approach. It means that the real human lung geometry is used for fabrication of physical cast, that is used for experiments. After that, the simulations and experiments can be performed on the identical geometry. This method was used by our team and will be described later in detail.

Among the recent authors who validated numerical simulations by in vitro data we can mention Kerekes et al. [5], who focused on the flow field in a simplified model of tracheobronchial airways and validated their results by laser Doppler velocimetry (LDV). However, their geometry was simplified to facilitate the application of the LDV technique. Nonetheless, the curvature and tortuosity of airways is important to be able to simulate the flow division in the tracheobronchial tree correctly, as was documented e.g. by Bauer and Brucker [6]. Validated simulation of lung flow performed on the curved realistic geometry was published e.g. by Elcner et al. [7].

There are very few papers dealing with the CFPD simulation of localized particle deposition so far. We should mention the work of Holbrook et al. [8], who focused on the quality of the wall surface and its influence on the results.

In this paper we compare the first results of our numerical simulation and measured deposition of particles in a realistic model of human airways consisting of the oral cavity and following airways down to the seventh generation of branching.

## 2 Methods

The method for measurement of localized aerosol deposition presented in this paper divides the airways into segments of sizes between 2 to 5 cm. It is generally possible to go down to the resolution of approximately 5×5×5 mm, which corresponds to the maximal resolution of the positron emission tomography (PET) scanner. The numerical simulation is able to go even to the μm-scale and indeed such a good resolution is needed for real understanding of the pathological processes connected with the deposition hot-spots in human lungs. However, even the cm-scale is a good step towards the validated maximal resolution of CFPD calculations.

### 2.1 Experimental setup

The experimental setup has been described in detail in [9], therefore only a brief description will be given here. The measurement was performed on a realistic replica of human airways, whose dimensions and development have been published in [10]. The geometry of airways used for fabrication of our model comes from a computed tomography (CT) of the tracheobronchial tree excised from a cadaver (for detail please see Schmidt et al. [11]), which was combined with a 3D scan of the wax cast of the so called Lovelace Respiratory Research Institute (LRRI) "A" model [12]. The combined airway geometry was then processed in a computer aided design (CAD) software Rhinoceros to form a segmented physical model of human lungs (Fig. 1). The model was originally divided into segments to facilitate other experimental methods, such as optical counting of deposited particles. The deposited particles were consequently evaluated for each segment also in the case of the presented PET method, although this method could generally go down to smaller portions of the model.

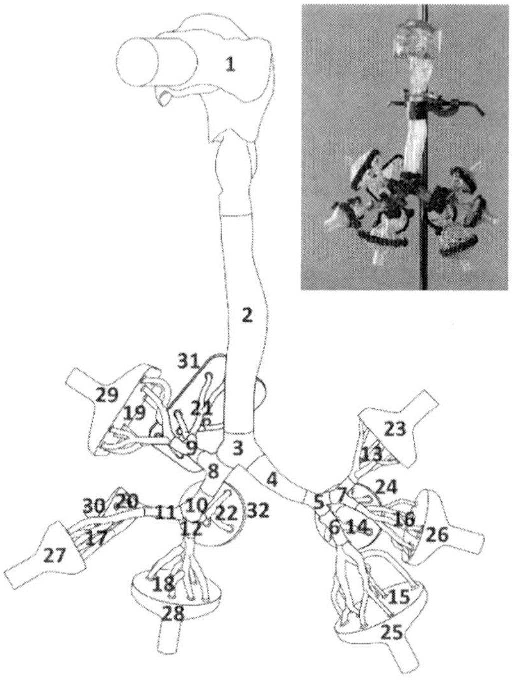


**Fig. 1.** A scheme with the designation of the segments and the photograph of a segmented realistic model of human lungs.

Both the experiment and the simulation were performed for steady inhalation flow rates of 15 and 30 liter/min. The whole experimental setup is schematically depicted in Fig. 2.

The realistic physical model was fixed in a wooden box and all ten outputs were connected to filters which were collecting the non-depositing particles. The aerosol particles in sizes of 2.5 and 4.3 μm were generated using condensation monodisperse aerosol generator (CMAG) TSI 3475 and tagged by positron emitting isotope $^{18}$F.

The exposed model was promptly transported to the hybrid PET/CT scanner Siemens Biograph Truepoint 64 and immediately scanned. Both the PET and CT images were used for evaluation in the software Carimas (Turku PET centre, Finland). Each segment of the model was enclosed in a separate volume of interest (VOI) in the Carimas software and its volume radioactivity in Bq/ml was calculated. Consequently, the deposition fraction as a ratio of radioactivity in a given segment to the total radioactivity entering the model was determined.

DOI: 10.1051/epjconf/201714302067

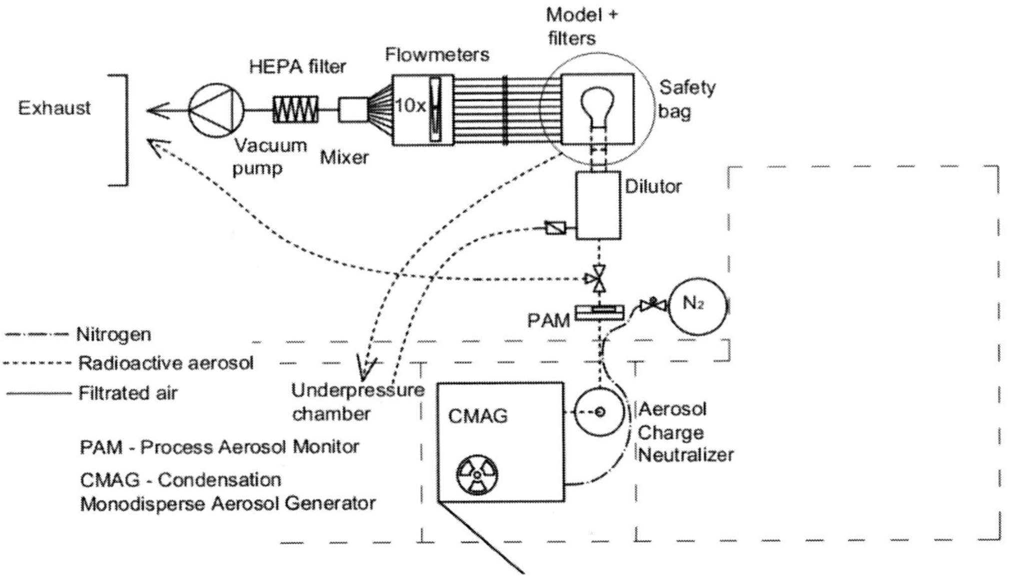


**Fig. 2.** A scheme of the setup for radioactive aerosol deposition measurement

## 2.2 Numerical simulation

The identical geometry of the realistic human airways was used also for the numerical simulations. The air flow was calculated using steady Reynolds-averaged Navier-Stokes (RANS) method with shear stress transport (SST) Menter's k-ω model of turbulence with treated boundary layer, which satisfies the conditions for low Reynolds number.

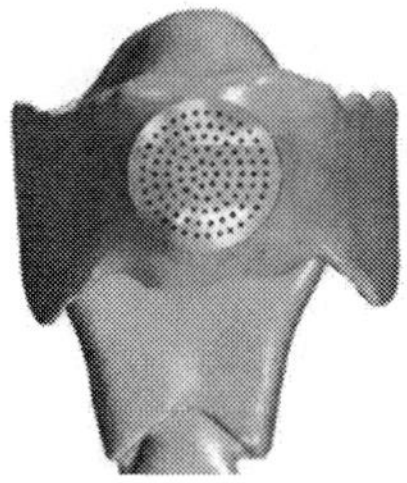

**Fig. 3.** The injection points of parcels representing particles in the numerical simulation.

The simulations were performed in a commercial solver Star-CCM+. The model had 6 million cells in polyhedral mesh. Particle deposition was simulated using embedded Lagrangian multiphase model with 10,000 parcels with particles of sizes 2.5, 4.3, 6.0, and 8.0 μm. The particles were distributed to the model from a plane at the beginning of the inlet cylinder. The plane contains 100 equally distributed injection points (see Fig. 3). 100 parcels were emitted from each of the points. Initial velocity magnitude of the parcel is identical with the mean cross-sectional inlet velocity and direction is determined by a vector parallel with the axis of the inlet cylinder.

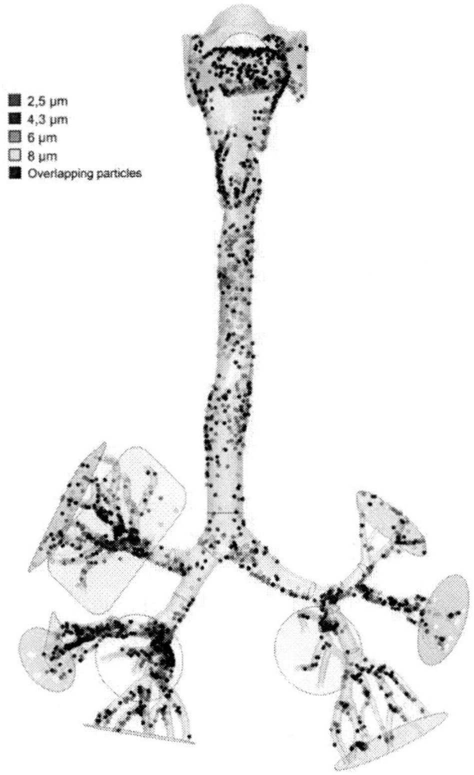


**Fig. 4.** The numerically simulated deposition sites of various sizes of particles for steady inhalation flow rate 30 liter/min.

## 3 Results and discussion

As the numerical simulation enables more detailed tracing of particles compared to the experimental method, we plotted a figure with deposition sites of particles (see Fig. 4). In this way the possible deposition hot-spots could be identified. Particles of different sizes are represented by dots of a different colours. The deposition sites with overlaps of two or more different sizes of particles are represented by the black colour. It is obvious that the oral cavity presents a significant obstacle for the inhaled particles of all sizes. Then many particles also deposited in the glottal area. Interestingly, particles were separated based on their size downstream of the glottis. The biggest 8 μm particles deposited in the upper part of the trachea, whereas the smaller 6 μm particles deposited in the middle part. It is a consequence of the laryngeal jet, which arises in the glottis. As the biggest particles have the highest inertia, they cannot avoid the collision with the wall after being accelerated by the jet.

Surprisingly, only minor deposition occurred at the carina of the first bifurcation, where major deposition hot-spot was expected. The particles apparently managed to follow the streamlines closely at this location for the flow rate of 30 liter/min and continued to the lower airways.

On the other hand, unexpectedly high local deposition was found in the entrance regions of segments 18, 20 and 19. As all sizes of particles deposited there, it seems that not only the impaction but also turbulent dispersion influences the transport of particles.

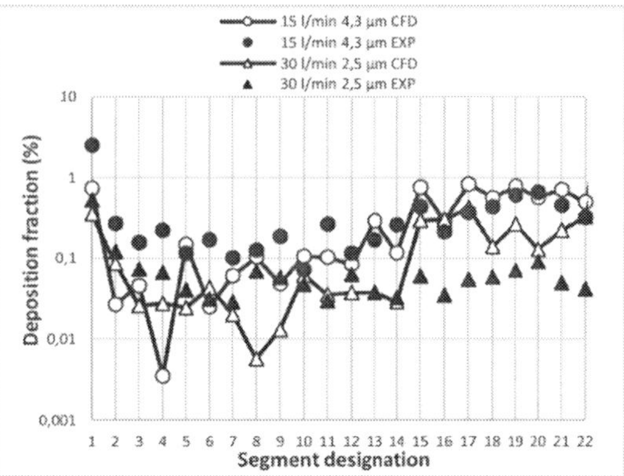


**Fig. 5.** A comparison of experimentally measured and numerically simulated deposition fractions in all segments of the realistic model of lungs.

However, the numerically simulated data does not agree well with the experimentally obtained data. As can be seen in Fig. 5, the simulation underestimates the deposition fraction in the upper part of the model (segments 1 to 11) for the lower flowrate (15 liter/min), whereas the segments with multiple branching (segments 12 to 22) were simulated correctly. On the contrary, for the higher flowrate (30 liter/min) the upper part (except the segment no. 8) was simulated with sufficient precision, but the lower segments were significantly overestimated by the simulation.

The above mentioned differences do not necessarily mean that the simulation is inaccurate. It should be noted that the Langrangian deposition model contains a stochastic element and as this simulation was performed only once, it must be seen as one of statistically random results. It is necessary to perform the simulation repeatedly and only after then calculate the average deposition fraction and compare the simulations with the experiments. This procedure will be therefore accomplished as a next step.

## 4 Conclusions

The comparison of experimental and numerically simulated deposition results demonstrated that the computational model was able to simulate general trends, although notable differences were found after the first run of simulation. Therefore, further repeated simulations are needed to receive statistically relevant data. Only after the successful validation can the numerical simulation serve for the prediction of localized aerosol deposition in human lungs.

Acknowledgement: This work was supported by the Czech Science Foundation under the grant GA16-23675S and by the project FSI-S-14-2355.

## References

1 A. Churg and S. Vedal, Occup Environ Med **53** (8), 553 (1996).
2 W. Hofmann, Journal of Aerosol Science **42** (10), 693 (2011).
3 A. Farkas, A. Jokay, P. Furi, I. Balashazy, V. Muller, B. Odler, and A. Horvath, Aerosol and Air Quality Research **15** (6), 2466 (2015).
4 A. A. Rostami, Inhalation Toxicology **21** (4), 262 (2009).
5 A. Kerekes, A. Nagy, M. Veres, I. Rigo, A. Farkas, and A. Czitrovszky, Measurement **90**, 144 (2016).
6 K. Bauer and C. Brucker, Journal of Biomechanical Engineering-Transactions of the Asme **137** (8) (2015).
7 J. Elcner, M. Forman, J. Jedelský, F. Lízal, and M. Jícha, presented at the International conference Experimental Fluid Mechanics 2010, Liberec, 2010 (unpublished).
8 L. T. Holbrook and P. W. Longest, Journal of Aerosol Science **59**, 6 (2013).
9 F. Lizal, M. Belka, J. Adam, J. Jedelsky, and M. Jicha, Proceedings of the Institution of Mechanical Engineers Part H-Journal of Engineering in Medicine **229** (10), 750 (2015).
10 F. Lizal, J. Elcner, P. K. Hopke, J. Jedelsky, and M. Jicha, Proceedings of the Institution of Mechanical Engineers Part H-Journal of Engineering in Medicine **226** (H3), 197 (2012).
11 A. Schmidt, S. Zidowitz, A. Kriete, T. Denhard, S. Krass, and H. O. Peitgen, Computerized Medical Imaging and Graphics **28** (4), 203 (2004).
12 Y. Zhou and Y. S. Cheng, Aerosol Science and Technology **39** (6), 492 (2005).

# Graphene field-effect transistor application for flow sensing

*Maciej* Łuszczek[1,*], *Dariusz* Świsulski[1], *Robert* Hanus[2], *Marcin* Zych[3], and *Leszek* Petryka[4]

[1]Gdańsk University of Technology, Faculty of Electrical and Control Engineering, 80-233 Gdańsk, Poland
[2]Rzeszów University of Technology, Faculty of Electrical and Computer Engineering, 35-959 Rzeszów, Poland
[3]AGH University of Science and Technology, Faculty of Geology, Geophysics and Environmental Protection, 30-059 Kraków, Poland
[4]AGH University of Science and Technology, Faculty of Physics and Applied Computer Science, 30-059 Kraków, Poland

**Abstract.** Microflow sensors offer great potential for applications in microfluidics and lab-on-a-chip systems. However, thermal-based sensors, which are commonly used in modern flow sensing technology, are mainly made of materials with positive temperature coefficients (PTC) and suffer from a self-heating effect and slow response time. Therefore, the design of novel devices and careful selection of materials are required to improve the overall flow sensor performance. In this work we propose graphene field-effect transistor (GFET) to be used as microflow sensor. Temperature distribution in graphene channel was simulated and the analysis of heat convection was performed to establish the relation between the fluidic flow velocity and the temperature gradient. It was shown that the negative temperature coefficient (NTC) of graphene could enable the self-protection of the device and should minimize sensing error from current-induced heating. It was also argued that the planar design of the GFET sensor makes it suitable for the real application due to supposed mechanical stability of such a construction.

## 1 Introduction

There is a number of potential applications for micro sensors, e.g. for monitoring of gas or fluid flow including flow cytometry, cleanroom environmental monitoring, wind, gas chromatography, viscosity measurements and so on (see e.g. Ref. [1] for the review). These sensors complement technologies such as microfluidic channels, valves, pumps, and heaters that are assembled together in the so-called lab on a chip devices or micro total analysis systems.

Thermal-based flow sensors usually utilize the hot-wire configuration. The sensing principle of such a device is based on the dependence of the sensor heat transfer on the fluid velocity, temperature and composition. The conventional hot-wire sensor consists of a thin metal wire or a film made of a material with a high temperature coefficient of resistance such as tungsten, platinum, platinum-rhodium and platinum-iridium. At the first approximation, when the wire is heated by an electrical current I, the thermal energy generated by the wire equals the energy loss due to convective heat transfer at the equilibrium state, $I^2R_w = A_w (T_w - T_f)h$, where $A_w$ denotes the surface of the wire, $T_w$ and $T_f$ are the wire and fluid temperatures, respectively, h is the heat transfer coefficient and $I^2R_w$ represents Joule heat. The resistance of the hot wire ($R_w$), naturally, depends on the wire temperature ($T_w$). At the steady state, this resistance can be expressed using a linear approximation of the wire temperature, $R_w = R_0 [(1+ \alpha (T_w - T_f)]$, where $R_0$ is the baseline resistance of the wire and $\alpha$ is the TCR. The heat transfer coefficient ($h$) can be expressed as the function of fluid velocity ($V_f$) according to King's law, $h = a + bV_f c$, where $a$, $b$ and $c$ are the constants obtained from experimental calibrations. Finally, the fluid velocity $V_f$ can be evaluated from the above-mentioned equations. It is worth to note at this place, that the sensitivity of the device to the flow velocity is provided by having the wire or film temperature $T_w$ substantially higher than the fluid temperature $T_f$. Additionally, the measurement of voltage drop across the wire allow also to determine local fluid velocity, if only the physical properties of the gauge material and the mechanism of heat transfer are well known. Generally, the hot-wire devices operates in two regimes, that is, (1) a constant-current regime ($I_w$ = const) when the gauge voltage pulsations are attributable to temperature variation and, hence, the wire resistance, and (2) the regime of constant temperature ($T_w$ = const), maintained by the feedback system with a variable current heating the sensor. One or the other regime is used depending on the specific character of the measurements. However, the hot-wire sensors, which are commonly used in modern flow sensing technology, exhibit some essential limitations, because they are mainly made of materials with positive temperature coefficients (PTC) and suffer from a self-heating effect and slow response time. Therefore, the design of novel devices and careful selection of materials are required to improve the overall flow sensor performance.

Graphene is considered as an interesting new material in this context. It is a two-dimensional honeycomb crystal made of carbon atoms, exhibiting electrical mobility and thermal conductivity over an

* Corresponding author: maciej.luszczek@pg.gda,pl

order of magnitude greater than silicon. The electronic properties of graphene layer can be easily tuned with the use of an external gate. The variation of the gate (G) voltage with respect to source (S) or drain (D) terminals influence electron and hole densities and as a result we obtain an ambipolar graphene field effect transistor (GFET). The principle of its operation is the same as in the case of the conventional metal-oxide semiconductor field-effect transistor (MOSFET), that is, it is based on the so-called field effect which enables to control the conductivity in the graphene channel region by an electric field generated by the applied external voltage. More precisely, the shape of the channel and the flow of the charge carriers from the source to the drain is influenced by the voltage applied across the insulated gate and the source. Electronic structure of graphene indicates that it is a zero-gap semiconductor with characteristic linear dispersion near the Fermi level. The type of carriers and the carrier density in the channel depend on the potential difference between the channel and the gate. As a consequence, the two branches of the transfer characteristics, separated by the charge neutrality point (so-called Dirac point), are visible. At the Dirac point the minimum conductivity and maximum resistance are observed [2]. This material has a large surface-to-volume ratio, low thermal inertia and high sensitivity to temperature variation. What is more, the negative temperature coefficient (NTC) of graphene should enable the self-protection of the sensors and the minimization of the sensing errors caused by the current-induced heating. Recently, it has been demonstrated that the graphene-based "hot wire" device could be used as a nanoscale flow and temperature sensor [3].

In this work we propose graphene field-effect transistor (GFET) to be used as microflow sensor. Theoretical prediction of behavior of the newly designed graphene sensor is an essential task because the realization of the prototype seems to be costly and difficult due to many technological problems. It is why the simulation was performed at the first stage of the investigation. Here, we present the results of the preliminary study of the GFET device for the flow sensing application.

## 2 Computational method

We performed the calculations with the use of the GFET Tool program [4] which enables simulation of the electrical characteristics of a top-gated graphene field-effect transistor. The drift-diffusion approach was employed to calculate self-consistently *I-V* characteristics at different initial temperatures. Also the carrier density in the channel could be evaluated in this program. The velocity saturation, "puddle" charge density distribution, caused by different impurities and thermally excited carriers, Seebeck effect and device breakdown due to thermal self-heating are taken into account in the code, too. It is possible to examine the GFET device with various channel geometry $W \times L$ ($W$ - width, $L$ - length) different thicknesses of the top-gate dielectric and $SiO_2$ layers and various values of mobility of the both type of carriers (electrons and holes) could be set up. Also Dirac voltage can be declared at the beginning of the calculation.

In our study we simulated the GFET devices with $W$ and $L$ parameters from the range of 0.5 - 10 μm, the top-gate dielectric thickness of 10 nm and the 300 nm $SiO_2$ layer. The initial temperatures were from 300K to 400K. The mobility of carriers was assumed to be 3000 $cm^2$/Vs. Generally, we simulated the performance of the devices at the Dirac voltage of 0 V.

## 3 Results and discussion

Output characteristics (not presented here), $I_D(U_{DS})$, of all the modelled GFET with different graphene channel geometry, are similar to those of typical FETs. Either linear or saturation regions are noticeable at various initial temperatures, especially for gate-source voltages, $U_{GS}$, much higher than the Dirac voltage, $U_{Dirac}$. The drain saturation current depends directly on $U_{GS}$, what is typical for all types of FETs. There are some differences, however, in transfer characteristic, $I_D(U_{GS})$, when compared with the typical MOSFET. The threshold voltage is not observed, so the GFET is in a conduction state for all $U_{GS}$ [5]. The drain current, $I_D$, reaches the minimum for $U_{GS}$ close to the Dirac point voltage. One can also see a shift of the point of the minimal conductance towards positive values with increasing drain voltage $U_{DS}$. This effect can be explained by the influence of $U_{DS}$ on the channel potential [6]. Maximum of channel resistance, $R_m$, corresponding to the minimum of the conductivity, is visible for $U_{GS}$ close to Dirac voltage and it is in order of kΩ. The values of drain-to-source resistance, $R$, depend on $U_{DS}$ voltage and the smallest values of the resistance are in the linear region of the output characteristics. All the obtained characteristics are in agreement with the previous simulations [2].

In Fig. 1 channel resistance vs. graphene dimensions with $L \times W$ up to 10 μm x 10 μm is shown. As is easy to notice, the resistance increase with $L$ and decrease with $W$.

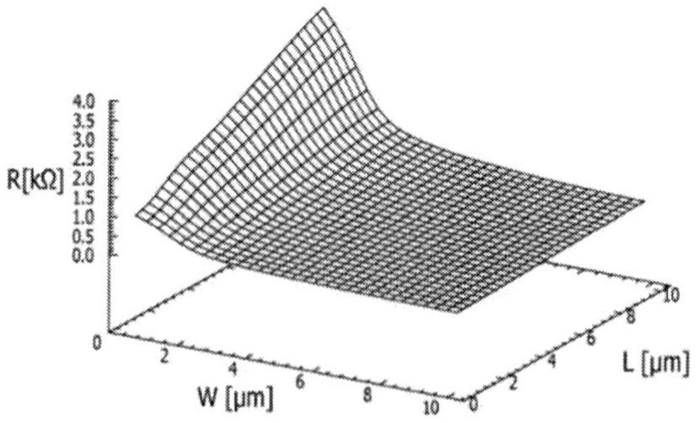


**Fig. 1.** Resistance vs. channel dimensions at room temperature ($U_{DS}$ = 0.1 V, $U_{GS}$ = 5 V, $V_{Dirac}$ = 0 V).

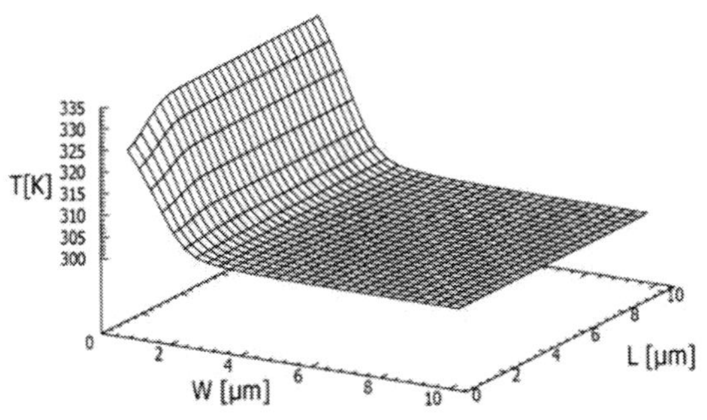

**Fig. 2.** Channel temperature vs. device dimensions ($T_{ini}$ = 300K, $U_{GS}$ = 5 V, $U_{Dirac}$ = 0 V, $I_{Dmax}$ = 0.5 mA).

Obviously, an electrical drain current is responsible for heating of the device and this effect depends strongly on the geometry of graphene sheet as it is presented in Fig. 2 for the case of GFET device at initial temperature of $T_{ini}$ = 300K. The temperature of narrow (small $W$) and long (large $L$) graphen stripe rises significantly even for relatively low currents. To avoid self-heating one should design GFET with large W and small $L$, instead. On the other hand, this effect could be utylized when the graphene channel works as a "hot-wire" element for microflow sensing. The general idea of the device operation is as follows. The GFET element device should be heated by the drain current, $I_D$, and subjected to fluid flow. Electrical resistance, $R$, is supposed to change according to the channel temperature $T$. Thus, heat transfer rate can be transduced into a signal that depends on the fluid flow rate. Indeed, our simulations indicate that the resistivity of GFET device is temperature-dependent (Fig. 3). At the first approximation, the $R$-$T$ dependence could be treated as a linear one for temperatures from the range of 300-400K. Our calculations also confirm the negative temperature coefficient (NTC) of graphene with the estimated value of $\alpha = -8.7 \cdot 10^{-4}$ K$^{-1}$ for $L$ x $W$ = 3 µm x 10 µm. In the next stage, the as-prepared GFET-based device should be characterized by imposing a known fluid flow with stable temperature to calibrate the sensor by measuring the resulting resistance or voltage change. The fluids used to characterize the sensor should be the same as fluids which are planned to be measured since the thermal conductive properties of the fluid are integral to the transduction mechanism.

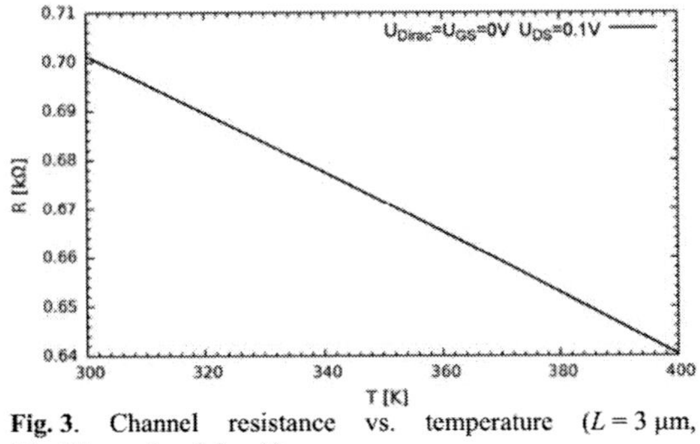

**Fig. 3.** Channel resistance vs. temperature ($L$ = 3 µm, $W$ = 10 µm, $I_D$ = 0.5 mA).

Reported in the literature temperature maps of graphene field-effect transistors with distinct "hot spots" that appear along the channel and vary with the applied voltage imply that the primary heating mechanism is due to energy loss of carriers within the graphene channel and not due to contact resistance [7]. The GFET device goes from hole to electron-hole and electron conduction when the gate voltage changes, what is also reflected in the temperature profiles along the channel. Our calculations confirm that temperature hot spots are always located at the position of minimum carrier density (which is the "crossing point" of electron and hole concentrations – not presented here) along the graphene channel. The temperature profiles for different drain currents are collected in Fig. 4. As can be seen, the temperature and its distribution could be easily controlled by the drain current, $I_D$. (Note that GFET is heated to temperatures above 450K by 7 mA and 9 mA drain currents, what probably should lead to the breakdown of the real element.) Additionally, the location of the maximum could be changed by the gate voltage, $U_{GS}$.

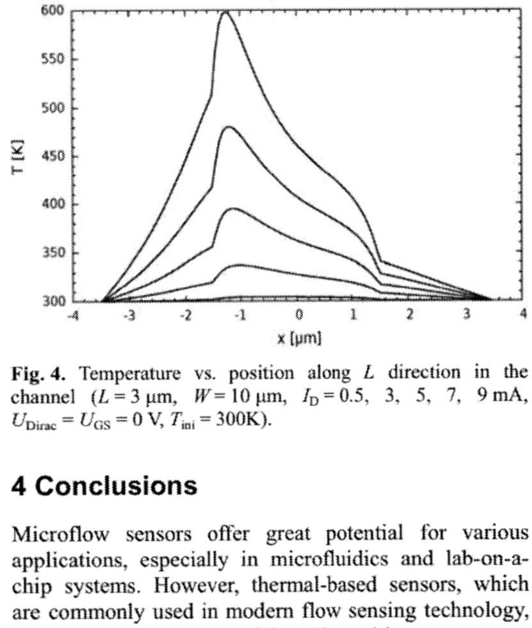

**Fig. 4.** Temperature vs. position along $L$ direction in the channel ($L$ = 3 µm, $W$ = 10 µm, $I_D$ = 0.5, 3, 5, 7, 9 mA, $U_{Dirac}$ = $U_{GS}$ = 0 V, $T_{ini}$ = 300K).

## 4 Conclusions

Microflow sensors offer great potential for various applications, especially in microfluidics and lab-on-a-chip systems. However, thermal-based sensors, which are commonly used in modern flow sensing technology, are mainly made of materials with positive temperature coefficients and suffer from a self-heating effect and slow response time. Therefore, the design of novel devices and careful selection of materials are required to improve the overall flow sensor performance. Here, we propose graphene field-effect transistor (GFET) to be used as microflow sensor. Before concluding, it is relevant to summarize our findings. In this study the results of the simulation of a top-gated graphene field-effect transistor (GFET) were presented. We succeeded in reproducing typical GFET characteristics. Temperature distribution in graphene channel was simulated and it was demonstrated that the temperature of graphen channel, which could be treated as a "hot wire" for microflow sensing, can be controlled with the choice of voltages applied on the three terminals. The negative temperature coefficient (NTC) of graphene was confirmed in our simulations, too, what could enable the self-protection of the device and should minimize sensing error from current-induced heating.

To evaluate flow sensitivities, the prototype of graphene "hot wire" GFET sensor should be placed in a kind of package with capillaries as upstream and downstream flow interconnections. During the measurement, the temperature of the analyzed fluid should be stabilized (e.g. at room temperature). The flow sensing responses of the GFET devices could be studied by monitoring the current and resistance changes of the graphene channel (for instance in the Wheatstone bridge configuration). In our opinion, GFET devices could be successfully employed for the flow sensing application due to their simplicity of design, ease of mass production in high-density array and potential mechanical stability of the planar construction. In the future the measurements of the real GFET-based sensors are planned what will allow for the direct comparison between the simulation and the experimental data.

## References

1. J.T.W. Kuo, L. Yu, E. Meng, Micromachines, **3** (3), 550-573 (2012)
2. M. Łuszczek, M. Turzyński, D. Świsulski, Przegląd Elektrotechniczny, **91** (10), 170-172 (2015)
3. H. Al-Mumen, F. Rao, L. Dong, W. Li, Micro and Nano Letters, **8**, 681–685 (2013)
4. E. Pop, F. Lian, GFET Tool, DOI:10.4231/D3QF8 JK5T, https://nanohub.org/resources/gfettool, (2014)
5. F. Schwierz, Nature Nanotechnology, **5**, 487–496 (2010)
6. S. Rodriguez, S. Vaziri, M. Ostling, A. Rusu, E. Alarcon, M.C. Lemme, ECS Solid State Lett., **5** (1), Q39–Q41 (2012)
7. M.-H. Bae, Z.-Y. Ong, D. Estrada, E. Pop, Nano Lett., **10**, 4787–4793 (2010)

# A numerical solution to an inverse unsteady-state heat transfer problem involving the Trefftz functions

*Beata* Maciejewska[1,*], *Magdalena* Piasecka[2]

[1]Faculty of Management and Computer Modelling
[2]Faculty of Mechatronics and Mechanical Engineering
Kielce University of Technology, Al. 1000-lecia P.P. 7, 25-314 Kielce, Poland

**Abstract.** This paper shows the results concerning flow boiling heat transfer in an asymmetrically heated vertical minichannel. The heated element for FC-72 Fluorinert flowing in that minichannel was a thin foil. The foil surface temperature was monitored continuously at 18 points by K-type thermocouples from the outer foil surface. Fluid temperature and pressure in the minichannel inlet and outlet, current supplied to the foil and voltage drop were also monitored. Measurements were carried out at 1 s intervals. The objective was to determine the heat transfer coefficient on the heated foil–fluid contact surface in the minichannel. It was obtained from the Robin boundary condition. The foil temperature was the result of solving the nonstationary two-dimensional inverse boundary problem in the heated foil. Using the FEM combined with Trefftz functions as basis functions solved the problem. The unknown temperature values at nodes were calculated by minimising the adequate functional. The values of local heat transfer coefficients were consistent with the results obtained by the authors in their previous studies when steady-state conditions were analysed. This time, however, these values were analysed as time dependent, which facilitated observation of coefficient changes that were impossible to observe under the steady-state conditions.

## 1 Introduction

Confronted with the recent broadening of the application range for narrow flow boiling heat transfer systems, researchers are devoting extensive experimental and theoretical studies to better understand heat transfer processes in mini gaps. The review of relevant literature is provided in [1-4]. In their previous works, the authors focused on steady-state flow boiling heat transfer. Their current research projects are designed to evaluate flow boiling heat transfer in minichannels under unsteady state conditions. A growing body of research is using advanced numerical analysis in simulations of heat and mass transport processes in channels. Experiments concerning time-dependent flow parameters are shown and discussed in, among others, [5-9].

Heat transfer coefficient identification belongs to the family of inverse heat conduction problems [10-12] solved with the use of many numerical methods, including the Trefftz method [13]. The method involves approximating unknown solution to a differential equation by means of a linear combination of functions strictly satisfying the governing differential equation. Details related to the Trefftz method are provided in [14-24]. This method is suitable for solving direct [13-15, 22], inverse [17-24], stationary [17, 21, 24] and non-stationary [18, 19, 20, 22] problems. This paper generalizes the results showed in [18].

Based on Trefftz functions, the authors 2D time-and-space basis functions in FEM was constructed. The experimental time-dependent measurements were used to develop a method for determining the heat transfer coefficient.

## 2 Experimental research

The essential part of the experimental stand is the measurement module with two parallel minichannels. The schematic diagram of the measurement module, is shown in figure 1. Each minichannel was 1.7 mm deep, 16 mm wide, and 180 mm long. The heated element for FC-72 Fluorinert flowing in the minichannels was a thin foil (0.1 mm thick) made of Haynes-230 alloy. In the tested minichannel, the temperature of the outer surface of the foil was measured by means of thermoelements. The other side of the minichannel was observed through a glass pane to identify the two-phase flow patterns. Two thermocouples and two pressure converters were installed at the inlet and outlet of the minichannels. The infrared temperature measurements realized in the other minichannel was not the case in this work but it was used for rough comparison. The experimental stand was discussed in [1-4].

* Corresponding author:beatam@tu.kielce.pl

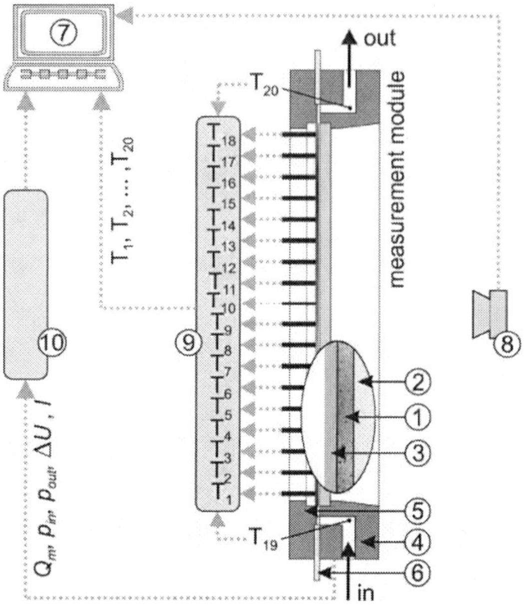


**Fig. 1.** The schematic diagram of the measurement module: 1-minichannel, 2-glass pane, 3-heated foil, 4-channel body, 5-front cover, 6- copper electrode, 7-DaqLab 2005 data acquisition station, 8-high speed camera, 9 -pc computer, 10- MCC data acquisition device, experimental parameters: $Q_m$ - mass flow rate , $p_{in/out}$ - pressure at the inlet/outlet to the minichannel, $\Delta U$ - voltage drop, $I$ - current.

In the experimental series, the current supplied to the heated foil was gradually increased within 240 s. The temperature of the foil surface were recorded continuously at 18 points by K-type thermocouples placed at the outer surface along the central, axially symmetric axis of the channel. The signals from these thermocouples and thermocouples controlling fluid temperature at the inlet and outlet to the minichannel were recorded using a DaqLab 2005 data acquisition station. The thermocouples recording foil temperature at the selected points were marked in figure 1 as $T_1$ .. $T_{18}$, two thermocouples installed at the inlet and outlet of the minichannel - as $T_{19}$ and $T_{20}$. The current supplied to the foil, the voltage drop across the foil, the pressure of the fluid at the minichannel inlet and outlet and mass flow rate were recorded using an MCC data acquisition device with analogue signal conditioning. All measurements were performed at 1-second intervals.

## 3 Analysis and modelling

Two-dimensional nonstationary heat flow in the minichannel was assumed, neglecting temperature variation along the width of the minichannel. The dimensions taken into account included dimension $x$ along the flow direction and dimension $y$, perpendicular to the direction of flow, related to the heated foil thickness. Only the subcooled boiling region was analysed. Local heat transfer coefficients were calculated from the third type boundary condition:

$$\alpha(x,t) = \frac{-\lambda \dfrac{\partial T}{\partial y}(x,\delta,t)}{T(x,\delta,t) - T_f(x,t)} \quad (1)$$

where: $\delta$ is the thickness of the heated foil, $\lambda$ is the thermal conductivity of the heated foil, $T$ is the foil temperature, $T_f$ is the fluid temperature calculated from the assumption of the linear distribution of the fluid temperature along the minichannel.

To find the foil temperature, $T$, the inverse heat conduction problem in the heated foil was solved. Foil temperature variation over time is described by the non-homogeneous heat equation

$$\frac{\partial^2 T}{\partial x^2} + \frac{\partial^2 T}{\partial y^2} - \frac{\rho c_p}{\lambda}\frac{\partial T}{\partial t} = -\frac{q_V(t)}{\lambda}, (x,y) \in \Omega, t > 0 \quad (2)$$

and the initial and the boundary conditions:

$$T(x,y,0) = T_0(x,y) \quad (3)$$

$$\frac{\partial T}{\partial x}(0,y,t) = 0 \quad (4)$$

$$\frac{\partial T}{\partial x}(L,y,t) = 0 \quad (5)$$

$$\lambda \frac{\partial T}{\partial y}(x,0,t) = q_{loss}(t) \quad (6)$$

$$T(x_p,0,t_m) = T_p(t_m) \text{ for } p = 1,2,\ldots P; m = 1,2,\ldots M \quad (7)$$

where: $\Omega = \{(x,y) \in R^2 : 0 < x < L, \quad 0 < y < \delta\}$, $L$- heated foil length, $\rho$ - heated foil density, $c_p$ - heated foil specific heat, $q_{loss}$ - heat losses to the surroundings, estimated as in [2], $T_p$ - thermocouple-measured foil temperature, P- number of measurements, $M$ - number of time intervals, $\lambda, \delta$ - defined as for Eq. (1).

The volumetric heat flux $q_V$ was calculated from

$$q_V(t) = \frac{I(t)\Delta U(t)}{A\,\delta} \quad (8)$$

where $I$ - current supplied to the heated foil, $\Delta U$ - drop in voltage along the length of the heated foil, $A$ - surface area of the heated foil in contact with the fluid, $\delta$ - defined as for Eq. (1).

The inverse problem, Eqs. (2-7), was solved using the FEMT (the FEM with Trefftz functions as basis functions) in time-space elements. In order to find an approximate solution of this problem, the time interval $<0,t>$ was divided into $M$ subintervals $<(m-1)\,\Delta t, m\,\Delta t>$, for $m = 1,2,\ldots,M$, where $\Delta t$ denote the length of the time interval. The time-space region $\overline{\Omega} = \Omega \times <(m-1)\,\Delta t, m\,\Delta t>$, for a fixed $m$, was divided into $L1 \cdot L2$ cuboidal elements with 8 nodes in each of them, where $L1$ was the number of elements in the $x$-axis direction and $L2$ was the number of elements in the $y$-axis direction. In each element $\Omega_j$, $j = 1, 2, \ldots, L1 \cdot L2$, the set of basis functions were characterized in nodes $(x_i, y_i, t_i)$ by the following property

$$f_{jk}(x_i, y_i, t_i) = \delta_{ki}, \quad i = 1,2,\ldots,N \qquad (9)$$

where: $\delta_{ki}$ denotes the Kronecker delta, $j$ is the element number, $k$ is the basis function number in $j$-th element, $N$ is the number of nodes in the element.

The temperature of the heated foil in each element $\Omega_j$ for a fixed $m$ was approximated with a formula containing a linear combination of the basis functions $f_{jk}(x, y, t)$:

$$T_m^j(x, y, t) = \tilde{u}(x, y, t) + \sum_{k=1}^{N}\left(\hat{T}^n - \tilde{u}(x_n, y_n, t_n)\right) f_{jk}(x, y, t) \quad (10)$$

where $n$ denotes the node number in the entire domain $\overline{\Omega}$, $\tilde{u}(x_n, y_n, t_n)$ is the particular solution in the $n$-th node of domain $\overline{\Omega}$, $\hat{T}^n$ is the unknown temperature in the $n$-th node of domain $\overline{\Omega}$.

From the basis function properties it follows that coefficients $\hat{T}^n$ are the values of temperature in the nodes of $\overline{\Omega}$.

All unknown coefficients $\hat{T}^n$ of linear combination in Eq. (10) were calculated by minimizing the functional that describes the mean square error of the approximate temperature and its derivatives on the boundary in the initial time and along common edges of the adjacent elements.

## 4 Results and discussion

The study was devoted to the heat transfer coefficient identification in the regions of subcooled and saturated boiling. The coefficient was found by solving the inverse heat conduction problem with used the FEM based on Trefftz-type basis functions in time-space subdomains. The calculations were performed using the time-dependent temperature measurements.

The relationship between the temperature of the heat transfer coefficient and the distance from the minichannel inlet, were determined for three times: $t_1$ = 10 s, $t_{50}$ = 50 s and $t_{80}$ = 80 s is shown in figure 2. These data were selected to present the results obtained for the subcooled boiling region. The values of the heat transfer coefficient in the saturated boiling region, calculated for three other times, i.e., $t_{125}$ =125 s, $t_{185}$=185 s and $t_{245}$=245 s, as a function of the distance from the minichannel inlet, are presented in figure 3.

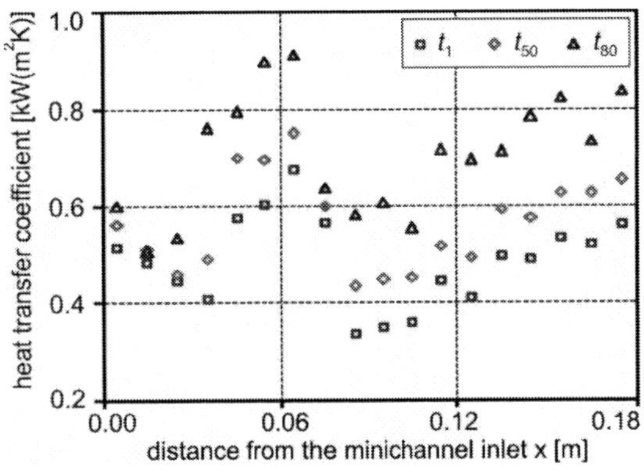


**Fig. 2.** Heat transfer coefficient vs. the distance from the minichannel inlet for $t_1$ = 10 s, $t_{50}$ = 50 s, $t_{80}$ = 80 s, the subcooled boiling region, experimental parameters:
$q_V$ =1.1·10$^5$ kW/m$^3$, $p_{in}$ =113.8 kPa for $t_{10}$ = 10 s,
$q_V$ =1.4·10$^5$ kW/m$^3$, $p_{in}$ =114.4 kPa for $t_{50}$ = 50 s,
$q_V$ =1.8·10$^5$ kW/m$^3$, $p_{in}$ =115.4 kPa for $t_{80}$ = 80 s;
average $Q_m$= 0.012 kg/s.

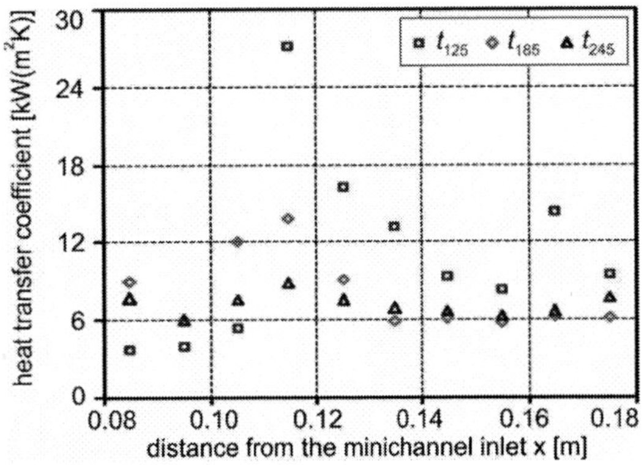


**Fig. 3.** Heat transfer coefficient vs. the distance from the minichannel inlet for $t_{125}$ =125 s, $t_{185}$=185 s and $t_{245}$=245 s, the saturated boiling region, experimental parameters:
$q_V$ =2.79·10$^5$ kW/m$^3$, $p_{in}$ =118.5 kPa for $t_{125}$ = 125 s,
$q_V$ =3.93·10$^5$ kW/m$^3$, $p_{in}$ =125.1 kPa for $t_{185}$ = 185 s,
$q_V$ =3.85·10$^5$ kW/m$^3$, $p_{in}$ =12.99 kPa for $t_{245}$ = 245 s;
average $Q_m$= 0.012 kg/s.

These results confirm those obtained by the authors in their earlier studies when steady state conditions were analysed [1-4]. Under the subcooled boiling, local heat transfer coefficients had relatively low values up to 1000 W/(m$^2$K), figure 2. A scatter of the heat transfer coefficient values in the range 300 to 900 W/(m$^2$K) can be observed in this boiling region, in particular in the beginning minichannel length up the distance of 0.8 m from the minichannel inlet. Further in the channel, the heat transfer coefficient developed uniformly showing increasing values for the three times under analysis, $t_1$, $t_{50}$ and $t_{80}$. Under the saturated boiling, local heat transfer coefficients achieved high values, figure 3. The heat transfer coefficient was the highest at the beginning of the nucleate saturated boiling up to 30 kW/(m$^2$K). Results obtained for $t_{125}$ and $t_{185}$ indicate sinusoid fluctuation as a function of the distance from the minichannel inlet. At the final stage of the experiment, the results obtained for $t_{245}$ confirmed that in the fully developed boiling region for higher vapour phase fractions, the heat transfer coefficient decreased to 5-7 kW/(m$^2$K) at the minichannel outlet. This means that the coefficients in the subcooled boiling region were several times as big as those in the fully developed boiling region.

Figures 4 and 5 show the heat transfer coefficient vs. distance with the increasing heat flux being supplied to the foil, separately for the subcooled and saturated boiling regions, respectively.

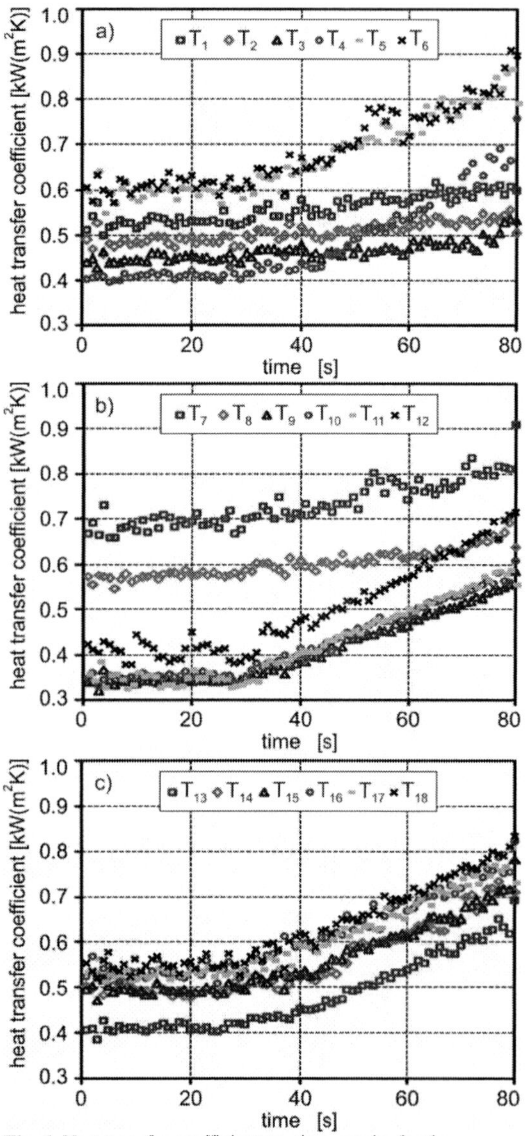


**Fig. 4.** Heat transfer coefficient vs. time, results for the subcooled boiling region calculated on the basis of temperature measurements from 18 thermocouples: a) thermocouples $T_1$ to $T_6$; b) thermocouples $T_7$ to $T_{12}$; a) thermocouples $T_{13}$ to $T_{18}$;; experimental parameters: $q_V$ from $1.12 \cdot 10^5$ kW/m$^3$ to $1.8 \cdot 10^5$ kW/m$^3$, $p_{in}$ from 113.7 kPa to 115.4 kPa, average $Q_m$= 0.012 kg/s.

In detail, the results obtained for:

- the subcooled boiling region are shown in figure 4 as local values of the heat transfer coefficient calculated on the basis of the foil temperature measured at all selected points along the minichannel (18 points),
- the saturated boiling region are shown in figure 5 as local heat transfer coefficients calculated on the basis of the foil temperature measured at 10 points near the minichannel outlet (thermocouples from $T_9$ to $T_{18}$, see figure 1).

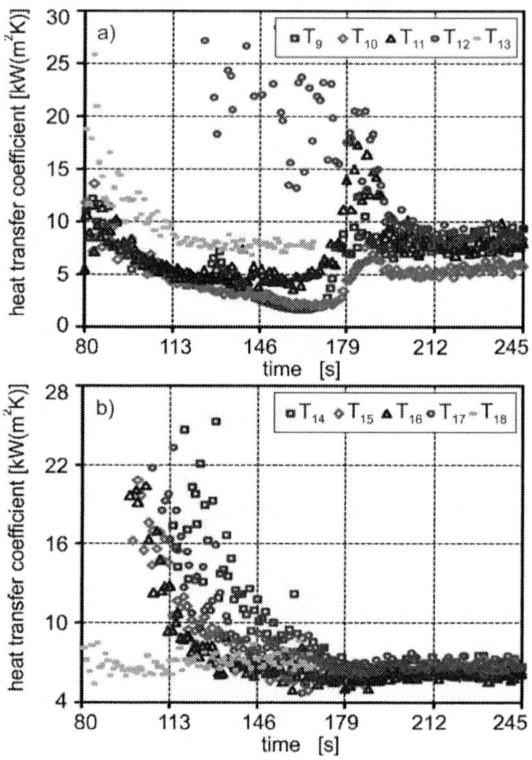


**Fig. 5.** Heat transfer coefficient vs. time, results for the saturated boiling region calculated on the basis of temperature measurements from thermocouples $T_9$ to $T_{18}$ ,experimental parameters: $q_V$ from $1.8 \cdot 10^5$ kW/m$^3$ to $3.86 \cdot 10^5$ kW/m$^3$, $p_{in}$ from 115.4 kPa to 128.9 kPa, average $Q_m$= 0.012 kg/s.

During the analysis of time-dependent values of the coefficient, figures 4 and 5, the surface temperature recorded by thermocouples $T_9$, $T_{10}$ and $T_{11}$ indicates coefficient value increase following its drop. Probably the fully developed boiling region was not reached within 150-170 seconds at the channel mid-length. Thermocouple $T_{12}$ suggests an incremental increase in the coefficient value at that time and channel length. The results from thermocouples $T_{13}$ to $T_{18}$ indicate a substantial reduction in the value of the coefficient, which at the end stage of the experiment levelled off at a low level just at the channel outlet.

## 5 Conclusions

The authors of this paper investigated flow boiling heat transfer in a vertical minichannel in subcooled and saturated boiling regions. A thin foil was the heated element for FC-72 Fluorinert flowing in the minichannel. Mounted at the outer surface of the foil, K-type thermocouples observed the surface temperature at 18 points. Data acquisition stations monitored the experimental parameters. The measurements were performed at 1 s intervals.

The numerical calculations aimed at determining the value of heat transfer coefficient at the foil-fluid interface in the minichannel. A two-dimensional unsteady-state inverse boundary problem was solved in time-dependent subdomains to determine the foil temperature. The method used to solve the inverse problem was FEMT. Unlike the classical FEM, in which shape functions depend only on space variables, the FEMT method applied in the paper used two-dimensional space-time basis functions. Whereas the FEM requires formulating boundary conditions on all boundaries, the proposed method, FEMT, can be used to solve both direct and inverse problems. Although the approximate foil temperature satisfies differential equation strictly, it meets the boundary conditions only approximately. The calculation results show that to identify the time-dependent heat transfer coefficient, the approximate solution can be constructed with the use of time-dependent Trefftz functions.

The values of local heat transfer coefficient confirm the results obtained by the authors in their earlier studies when steady-state conditions were analyzed. Under the subcooled boiling, local heat transfer coefficients had relatively low values. Under the saturated boiling local heat transfer coefficients reached high values. At the final phase of the experiment, the result obtained confirmed that heat transfer coefficient values decreased in the fully developed boiling region for higher fractions of the vapour phase. The results, in the form of the time-dependent local heat transfer coefficient values, showed the local coefficient heat flux increase-related changes that were impossible to observe under steady-state conditions. The authors are planning to verify the results in their further works.

**ACKNOWLEDGMENTS**

The research was supported in part by the Polish National Science Centre (No. DEC-2013/09/B/ST8/02825).

## References

1. M. Piasecka, Int. J. Refrig. **56**, 198-212 (2015)
2. M. Piasecka, K. Strąk, B. Maciejewska, Heat Transf. Eng. **38** (3), 332-346 (2017)
3. M. Piasecka, B. Maciejewska, Exp. Therm. Fluid Sci. **68**, 459-467 (2015)
4. M. Piasecka, Int. J. Heat Mass Transf. **81**, 114-121 (2015)
5. A. Majumder, B. Mehta, S. Khandekar, Int. J. Therm. Sci, **66**, 8-18 (2013)
6. B. Mehta, S. Khandekar, Int. J. Therm. Sci, **91**, 157-166 (2015)
7. Z. Zhou, X. Xu, X. Liang, Exp. Therm. Fluid Sci., **34**, 1409-1414 (2010)
8. J. Taler, B. Wójtowicz, Arch. Thermod., **28**, 3-14 (2007)
9. P. Ocłoń, S. Łopata, M. Nowak, Heat Mass Transf., **51**, 553-566 (2015)
10. A. Tikhonov, V. Arsenin, *Solution of ill-posed problems* (Wiley, New York, 1977)
11. J. V. Beck, B. Blackwell, C. R. St. Clair, *Inverse heat conduction. Ill-posed Problems* (Wiley - Interscience Publ., New York, 1985)
12. M. N. Özisjk., H. R. B. Orlande, *Inverse heat transfer: fundamentals and applications. Fundamentals and Applications* (Taylor & Francis, New York, 2000)
13. E. Trefftz, Proc. Int. Kongress für Technische Mechanik, Zürich, 131-137 (1926)
14. I. Herrera, Meth. Partial Diff. Eqs., **16**, 561-580 (2000)
15. V. Kompis, F. Konkol, M. Vasko, Comp. Assis. Mech. Eng. Sci., **8**, 385-395 (2001)
16. Z.-C. Li, T.-T. Lu, H.-T. Huang, H.-D Cheng Alexander, Num. Meth. Part. Diff. Eqs., **23**, 1-52 (2006)
17. K. Grysa, A. Maciąg, A. Pawińska, Int. J. Heat Mass Transf., **55**, 7336–7340 (2012)
18. M. J. Ciałkowski, J. Therm. Sci., **11**, 163-171 (2002)
19. M. J. Ciałkowski, K. Grysa, J. Inv. Ill-Posed Prob., **18**, 595-616 (2010)
20. K. Grysa, B. Maciejewska, J. Theor. Appl. Mech., **51**, 251-264 (2013)
21. S. Hożejowska, R. Kaniowski, M.E. Poniewski, Exp. Therm. Fluid Sci., **78**, 18–29 (2016)
22. A. Maciąg, Int. J. Num. Meth. Biomed. Eng., **27**, 1107-1125 (2011)
23. S. Blasiak, A. Pawinska, Int. J. Heat Mass Transfer, **90**, 710-718 (2015)
24. B. Maciejewska, M. Piasecka, Heat Mass Transfer, DOI: 10.1007/s00231-016-1895-1 (2016)

# Comparison of FEM calculated heat transfer coefficient in a minichannel using two approaches: Trefftz base functions and ADINA software

*Beata* Maciejewska[1,*], *Paweł* Łabędzki[1], *Artur* Piasecki[2], *Magdalena* Piasecka[3]

[1]Faculty of Management and Computer Modelling
[2] contractor of the project from the Polish National Science Centre (No. DEC-2013/09/B/ST8/02825)
[3]Faculty of Mechatronics and Mechanical Engineering
Kielce University of Technology, Al. 1000-lecia P.P. 7, 25-314 Kielce, Poland

**Abstract.** The paper presents the methods of heat transfer coefficient determination for boiling research during FC-72 flow in a minichannel. The boundary condition in the form of distributions of temperature on the outer side of the minichannel heated wall was obtained using infrared thermography. It was assumed two-dimensional steady-state heat flow. The local values of the heat transfer coefficients on the surface between the heated foil and boiling liquid, were determined from the Robin boundary condition. Data necessary for the heat transfer coefficient evaluation were obtained from numerical computations using two approaches: calculation procedure based on the Trefftz functions and FEM simulations by ADINA software. The shape functions were linear combinations of the Trefftz functions. Combinations of the Trefftz functions exactly satisfy the differential equation. Coefficients of the linear combination of the shape function in the approximate solution were chosen to minimize residuals on domain boundary and along common edges of adjacent elements. Temperature measurement points were located in boundary nodes. During FEM simulations 4-node FCBI elements were used, fluid flow was assumed to be laminar, incompressible and material constants of the fluid and of the foil were independent on temperature. The results of the comparative analysis were presented and discussed.

## 1 Introduction

Recently, there is a demand of using the heat exchangers characterized by small size. At the same time, heat transfer with the change of phase as boiling and condensation processes leads to intensification of resulting in increased efficiency of operation. The problem of heat transfer in heat exchangers with mini spaces was analyzed by a number of researchers. Review of the relevant literature is provided in [1-4] in which the authors presented their investigation results for flow boiling heat transfer in minichannels.

## 2 Main aim

The main aim of this work is to propose a method of heat transfer coefficient determination for FC-72 flow boiling in a minichannel asymmetrically heated. The boundary condition in the form of distributions of temperature on the outer side of the heated minichannel wall and main thermal and flow parameters are obtained from the experiment. It is assumed two-dimensional steady-state heat flow. The local values of heat transfer coefficients on the surface between the heated foil and boiling liquid were determined from the Robin boundary condition. Data necessary for the heat transfer coefficient evaluation were obtained from numerical computations in which two approaches were used: a calculation procedure based on the Trefftz functions [5-16] and FEM simulations by ADINA software

## 3 Experimental data

The method of heat transfer coefficient determination is presented for boiling research during FC-72 flow in a rectangular minichannel asymmetrically heated. This minichannel is 1.7 mm deep, 16 mm wide and 180 mm long and is vertically oriented. The measurement module with minichannels is the essential part of the research set up. The schematic diagram of the module is shown in figure 1. The working fluid flowing in the minichannel is FC-72 Flourinert. The heated element is a thin foil (0.1 mm in thickness) made of Haynes-230, which comes into contact with fluid in the minichannel. The temperature of the outer surface of the heated foil was measured with the use of FLIR E60 infrared camera. Thermal accuracy of the camera is ± 1 °C or ± 1% within the temperature range of 0 - 120°C. Example IRT thermogram is shown in the top of the figure 1. Before experiment, the foil was coated with a black paint to

* Corresponding author: beatam@tu.kielce.pl

achieve known emissivity (of 0.83) [2]. It was possible to observe the other side of the foil through a glass pane to visualise the two-phase flow patterns. K-type thermocouples and pressure converters are placed at the inlet and outlet of the measurement module. The experimental data was collected using a data acquisition station and a computer with appropriate software. The research stand was presented in detail in [2].

During experimental series, there is a laminar flow of FC-72 Fluorinert along a minichannel. It is accompanied by a gradual increase in the electric power supplied to the foil followed by an increase in the heat flux transferred to the fluid. Data for the selected setting of the heat flux supplied to the foil during subcooled boiling is analysed. In this boiling region, the liquid flowing in the minichannel was superheated at the interface with the foil and subcooled at the core of the flow. The experimental thermal and flow parameters, physical properties of the heated minichannel wall (made of Haynes-230) and working fluid (FC-72) properties for this setting are presented in Table 1. FC-72 properties were estimated for the average temperature of the fluid (22.5 C) flowing in the minichannel. Thermal conductivity and the other properties of Haynes-230 were taken from the manufacturer data sheet (Haynes Int.) for the average temperature of the heated wall (60 C).

**Table 1.** Main experimental thermal and flow parameters, physical properties of Haynes-230 and FC-72 properties.

| **Main thermal and flow experimental parameters** | |
|---|---|
| Heat flux (density) | 6 369 kW m$^{-2}$ |
| Flow velocity | 0.15 m s$^{-1}$ |
| Mass flux | 255.3 kg m$^{-2}$ s$^{-1}$ |
| Reynolds number | 1 230 (laminar flow) |
| Fluid temperature at the inlet | 17.8 C |
| Fluid temperature at the outlet | 27.2 C |
| Pressure at the inlet | 112 738 Pa |
| Pressure at the outlet | 102 096 Pa |
| Inlet liquid subcooling | 42.2 C |
| **Physical properties of the heated wall (made of Haynes-230)** | |
| Density | 8 970 kg m$^{-3}$ |
| Thermal conductivity | 9.7 W m$^{-1}$ K$^{-1}$ |
| Specific Heat | 397 J kg$^{-1}$ K$^{-1}$ |
| **Physical properties of the working fluid (FC-72)** | |
| Liquid density | 1685 kg m$^{-3}$ |
| Dynamic viscosity | 0.0006978 Pa s |
| Liquid Specific Heat | 1044 J kg$^{-1}$ K$^{-1}$ |
| Liquid Thermal Conductivity | 0.0561 W m$^{-1}$ K$^{-1}$ |

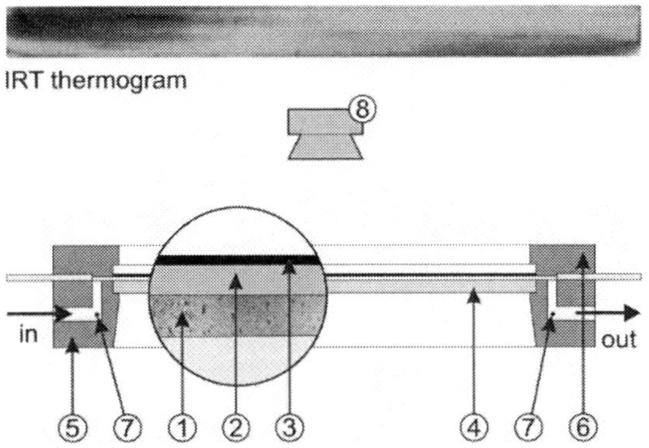


**Fig. 1.** The schematic diagram of the measurement module: 1-minichannel, 2-heated foil, 3-black paint, 4-glass pane, 5-channel body, 6-front cover, 7-thermocouple, 8-infrared camera.

## 4 The methods to determine the heat transfer coefficient (FEM)

### 4.1 The mathematical formulation

The heat flow in the minichannel is assumed to be a two-dimensional steady-state flow. Temperature variability along the minichannel width is not taken into account. The $x$ dimension in the flow direction and the $y$ dimension perpendicular to the flow direction (referring to the thickness of the heated foil) are considered. The local values of heat transfer coefficients are determined from the Robin boundary condition:

$$\alpha(x) = \frac{-\lambda_F \dfrac{\partial T_F(x,\delta_F)}{\partial y}}{T_F(x,\delta_F) - T_f(x,\delta_F + y_i)}, \quad i = 1, 2 \qquad (1)$$

where $\lambda_F$ – the thermal conductivity coefficient of the foil, $\delta_F$ – the foil thickness, $T_F$ – the foil temperature, $T_f(x, y_i)$ – the fluid temperature at a selected distance from the minichannel depth, calculated in two variants: at 1/4 distance of the minichannel depth from the inner foil surface, $y_1 = d/4$ and at 1/2 distance of the minichannel depth from the foil surface, $y_2 = d/2$, $d$ – the depth of the minichannel (see figure 2).

Those coefficients, on the surface between the heated foil and boiling liquid, are calculated numerically by FEM, with the aid of ADINA software and the calculation procedure based on Trefftz functions.

The following equations are used to describe the boundary-value problem in two adjacent regions:
- Poisson's equation in the heated foil [12]:

$$\nabla^2 T_F = -\frac{I \cdot \Delta U}{A \cdot \delta_F \cdot \lambda_F} \quad \text{for } (x,y) \in \Omega_F \qquad (2)$$

- the Fourier–Kirchhoff equation in fluid [16]:

$$\kappa \nabla^2 T_f = w_x(y)\frac{\partial T_f}{\partial x} \text{ for } (x,y)\in \Omega_M \quad (3)$$

The boundary conditions have the form:

$$\frac{\partial T_F}{\partial x}(0,y)=0 \quad (4)$$

$$\frac{\partial T_F}{\partial x}(L,y)=0 \quad (5)$$

$$\lambda\frac{\partial T_F}{\partial y}(0,y)=q_{loss} \quad (6)$$

$$T_F(x_p,0)=T_p \text{ for } p=1,2,\ldots,P \quad (7)$$

$$T_f(x,\delta_F)=T_F(x,\delta_F) \quad (8)$$

$$T_f(0,y)=T_f^{in} \quad (9)$$

$$T_f(L,y)=T_f^{out} \quad (10)$$

$$\lambda_f\frac{\partial T_f}{\partial y}(x,\delta_F+d)=0 \quad (11)$$

where $\Omega_F=\{(x,y)\in R^2: 0<x<L,\ \ 0<y<\delta_F\}$, $\Omega_M=\{(x,y)\in R^2: 0<x<L,\ \ \delta_F<y<\delta_F+d\}$, $L$ – the minichannel length, $I$ – current, $\Delta U$ – the voltage drop, $A$ – the surface area of the heated foil determined for IRT, $P$ – the number of measurements obtained from infrared thermography on the outer surface of the heated foil, $T_p$ – the foil temperature measured by infrared thermography at the boundary $y=0$, $q_{loss}$ – the heat loss to the surroundings calculated as in [2], $w_x(y)$ – component of vector fluid velocity, parallel to the minichannel heated surface [16], $\lambda_F$, $\delta_F$, $d$ – defined as for Eq. (1).

### 4.2 Using the Trefftz functions

To solve the system of equations (2-11), in each element $\Omega_j$, $j=1,2,\ldots,L1\cdot L2$ of the domain $\Omega=\Omega_F\cup\Omega_M$, the foil approximate temperature was defined by the formula:

$$T_F^j(x,y)=u(x,y)+\sum_{k=1}^{N}\left(\widehat{T}_F^n-u(x_n,y_n)\right)f_{jk}(x,y) \quad (12)$$

and the fluid approximate temperature was described by the expression:

$$T_f^j(x,y)=\sum_{k=1}^{N}\widehat{T}_f^n g_{jk}(x,y) \quad (13)$$

where: $u(x,y)$ – the particular solution to Eq. (2), $f_{jk}(x,y,)$ – the basis functions determined as linear combinations of the Trefftz functions according to Laplace's equation [9], $g_{jk}(x,y,)$ – the basis functions referred to the Fourier–Kirchhoff equation, $\widehat{T}_F^n$ – the unknown temperature in the $n$-th node of domain $\Omega_F$, $\widehat{T}_f^n$ – the unknown temperature in the $n$-th node of domain $\Omega_M$, $j$ – element number, $k$ – basis function number in $j$-th element, $n$ – node number in the entire domain $\Omega$, $N$ – the number of nodes in the element, $L1$ – the number of elements in the $x$-axis direction and $L2$ – the number of elements in the $y$-axis direction.

The basis functions $g_{jk}(x,y,)$ were constructed with the use of the Trefftz functions (that satisfy the Fourier–Kirchhoff equation) given in [16]. The unknown coefficients $\widehat{T}_F^n$ and $\widehat{T}_f^n$ were calculated by minimizing the appropriate functional [2].

### 4.3 Using the ADINA software

ADINA software in version 9.0 was used for FEM simulations. Two dimensional model of fluid flow was considered. The same boundary conditions as in calculations based on Trefftz functions were used (see equations (4)-(11)). Laminar, incompressible fluid flow was assumed and that material constants of fluid and of the foil were independent on temperature (see Table 1). Under these assumptions FEM model was the same as one represented by equations (2) and (3) [17, section 2.6]. FE mesh is presented on figure 2.

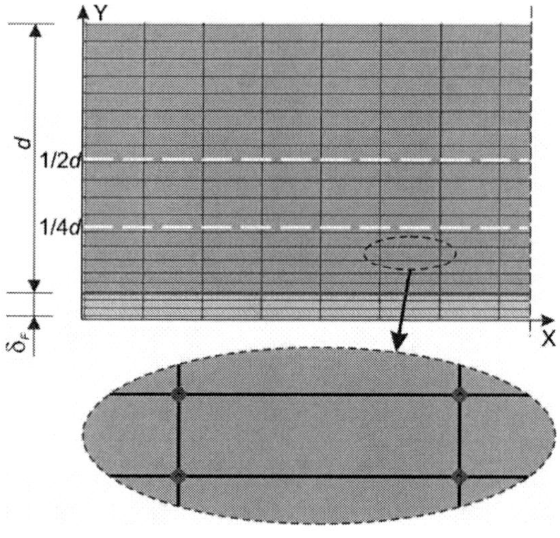


**Fig. 2** Scheme of FE mesh fragment (mesh in fluid is coloured in blue, mesh in foil - in yellow).

During simulations 4-node planar FCBI (flow condition-based interpolation) elements were used [17]. Mesh density was established in such a way to decrease mesh quality indicator for heat flux in y direction, which measure jumps of the flux values on adjacent elements, to 10% level (final mesh had 10 000 elements and 10 521 nodes). The temperature was calculated in the nodes, while heat fluxes at the Gauss integration points. Smoothing techniques provided by ADINA were used to obtain flux values in the nodes. As the mesh along that line is sufficiently dense, accuracy of the results remains unaffected.

Temperatures and heat fluxes for the heat transfer coefficient calculation according to Eq. (1) were taken from the FEM calculation, from the lines shown in figure 2 corresponding to: $T_F$ (x, $\delta_F$) - red line (at the foil-fluid boundary), $T_f$ $(x, y)$ - white lines, for the minichannel depths $y=d/4$ or $d/2$.

## 5 Results and discussion

The results are presented graphically in figure 4 as relationship between the heat transfer coefficient and the distance from the minichannel inlet to compare the values calculated using the Trefftz functions (red points) from with that obtained using the ADINA software (green points). Separate plots were generated for two different distances of the minichannel depth: at $d/4$ distance (figure 3a) and at $d/2$ distance on the minichannel depth from the foil-fluid boundary (figure 3b), areas of the listed distances are shown in figure 2. The results of the heated surface temperature measurement at the inlet and outlet of a minichannel region (about 10% from each side) could be ignored due to the occurrence of the largest measurement errors when using infrared thermography, as reported in [18]. Additionally, 2D temperature distribution of the heated foil and fluid in the minichannel obtained using the ADINA software was shown in figure 4.

In the subcooled boiling region, the heat transfer coefficient was relatively low. The results obtained using the ADINA software indicate that the local values of the heat transfer coefficient decreased slightly with the distance from the minichannel inlet (figure 4a,b). The highest values of the heat transfer coefficient (above 600 W $m^{-2}$ $K^{-1}$) are observed very close to the minichannel inlet at the distance 0÷0.0015 m from the channel inlet. The analysis of the data calculated using the Trefftz functions, for channel axis, presented in figure 3b indicates that the heat transfer coefficient in the subcooled boiling region decreased slightly with the distance from the minichannel inlet. The results calculated at the $d/4$ distance on the minichannel depth from the foil-fluid boundary are higher than those calculated for the $d/2$ distance and minor data value fluctuations in the range 230÷280 W $m^{-2}$ $K^{-1}$ are observed at the outlet from the minichannel.

Summarizing, in the analyzed boiling region, the heat transfer coefficient calculated using the Trefftz functions did not differ significantly from that obtained using the ADINA software omitting results obtained very close to the minichannel inlet. Too small a number of mesh nodes introduced into the Trefftz functions-based calculation may lead to discrepancies in relation to the results obtained from ADINA. The highest discrepancy, observed at the minichannel inlet, results from the discontinuity of the heated foil temperature field and the mean temperature at the boundary in the fluid in the ADINA model. The boundary conditions used in the calculations should be corrected in both approaches, which would require separating a thermal layer for the calculations and setting in the region the corrected values of the fluid temperature at the channel inlet.

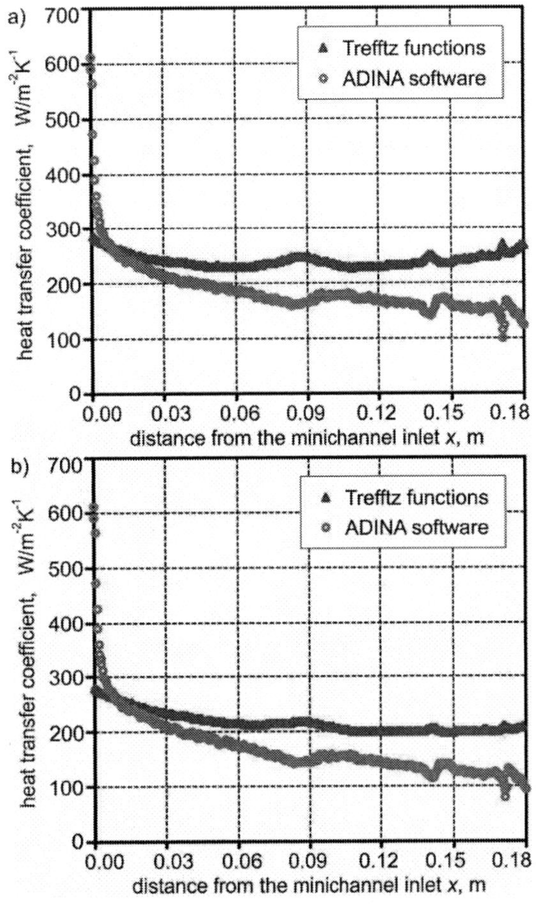


**Fig. 3.** Heat transfer coefficient vs. the minichannel length, calculated with the aid of Trefftz functions (red points) and obtained by ADINA software (green points), generated at: a) $d/4$ distance and b) $d/2$ distance, on the minichannel depth from the foil-fluid boundary.

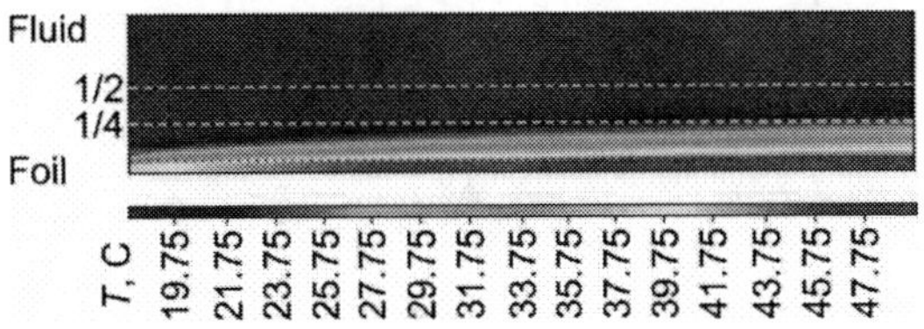


**Fig. 4.** 2D temperature distributions of the heated foil and fluid in the minichannel obtained using the ADINA software.

## 6 Conclusions

The results presented in this paper confirm that different FEM methods are suitable for calculations, as verified through experiment. Two methods of heat transfer coefficient determination for the flow boiling of FC-72 in a minichannel were proposed. Data from the experiment were used in the FEM to calculate the heat transfer coefficient in the minichannel using two approaches: Trefftz base functions and ADINA software. The boundary condition in the form of temperature distributions on the outer side of the minichannel heated wall was obtained using infrared thermography. Two-dimensional steady-state heat flow was assumed. The local values of heat transfer coefficients on the surface between the heated foil and boiling liquid were determined from the Robin boundary condition.

The results of the comparative analysis were presented and discussed. In the analysed subcooled boiling region, the heat transfer coefficient calculated using the Trefftz functions did not differ significantly from that obtained using the ADINA software omitting results obtained very close to the minichannel inlet. Boundary conditions near the heated surface at the channel inlet should be adjusted appropriately to improve compatibility of the proposed calculation methods.

ACKNOWLEDGMENTS

The research reported herein was supported in part by a grant from the Polish National Science Centre (No. DEC-2013/09/B/ST8/02825).

## References

1. M. Piasecka, B. Maciejewska, Exp. Therm. Fluid Sci. **68**, 459-467 (2015)
2. M. Piasecka, K. Strąk, B. Maciejewska, Heat Transf. Eng. **38** (3), 332-346 (2017)
3. M. Piasecka, Int. J. Refrig. **56**, 198-212 (2015)
4. M. Piasecka, Int. J. Heat Mass Transf. **81**, 114-121 (2015)
5. E. Trefftz, Proc. Int. Kongress für Technische Mechanik, Zürich, 131-137 (1926)
6. I. Herrera, Meth. Partial Diff. Eqs., **16**, 561-580 (2000)
7. K. Grysa, A. Maciąg, A. Pawińska, Int. J. Heat Mass Transf., **55**, 7336–7340 (2012)
8. M. J. Ciałkowski, A. Frąckowiak, J. Therm. Sci., **11**, 148-162 (2002)
9. M. J. Ciałkowski, J. Therm. Sci., **11**, 163-171 (2002)
10. M. J. Ciałkowski, K. Grysa, J. Inv. Ill-Posed Prob., **18**, 595-616 (2010)
11. K. Grysa, B. Maciejewska, J. Theor. Appl. Mech., **51**, 251-264 (2013)
12. S. Hozejowska, R. Kaniowski, M. E. Poniewski, Int. J. Num. Methods Heat Fluid Flow, **24**, 811–824 (2014)
13. A. Maciąg, Int. J. Num. Meth. Biomed. Eng., **27**, 1107-1125 (2011)
14. S. Blasiak, A. Pawinska, Int. J. Heat Mass Transfer, **90**, 710-718 (2015)
15. B. Maciejewska, M. Piasecka, Heat Mass Transfer, DOI 10.1007/s00231-016-1895-1, (2016)
16. S. Hożejowska, J.Theor. Appl. Mech, 53, 969-980 (2015)
17. ADINA CFD & FSI, Raport ARD 13-10, December 2013, ADINA R & D, Inc.
18. M. Piasecka, D. Michalski, K. Strąk, The European Physical Journal Web of Conf. **114**, paper No. 02094 (2015)

EPJ Web of Conferences **143**, 02071 (2017) DOI: 10.1051/epjconf/201714302071
*EFM 2016*

# Determination of the temperature distribution in a minichannel using ANSYS CFX and a procedure based on the Trefftz functions

*Beata* Maciejewska[1,*], *Sławomir* Błasiak[2], *Magdalena* Piasecka[2]

[1]Faculty of Management and Computer Modelling
[2]Faculty of Mechatronics and Mechanical Engineering
Kielce University of Technology, Al. 1000-lecia P.P. 7, 25-314 Kielce, Poland

**Abstract.** This work discusses the mathematical model for laminar-flow heat transfer in a minichannel. The boundary conditions in the form of temperature distributions on the outer sides of the channel walls were determined from experimental data. The data were collected from the experimental stand the essential part of which is a vertical minichannel 1.7 mm deep, 16 mm wide and 180 mm long, asymmetrically heated by a Haynes-230 alloy plate. Infrared thermography allowed determining temperature changes on the outer side of the minichannel walls. The problem was analysed numerically through either ANSYS CFX software or special calculation procedures based on the Finite Element Method and Trefftz functions in the thermal boundary layer. The Trefftz functions were used to construct the basis functions. Solutions to the governing differential equations were approximated with a linear combination of Trefftz-type basis functions. Unknown coefficients of the linear combination were calculated by minimising the functional. The results of the comparative analysis were represented in a graphical form and discussed.

## 1 Introduction

Recent advances in technology have produced devices with mini spaces. The trend toward miniaturization of components is the driving force behind better cooling technologies designed to prevent going over the maximum allowable operating temperatures. Phase-change processes, such as flow boiling, are very important for their ability to provide highest possible heat fluxes at a low temperature difference between a heated surface and a boiling liquid over a small heat transfer area. Extensive efforts to recognize boiling phenomena in small channels include theoretical analyses based on experimental measurements. The results can be applied to the construction of devices with minichannels and in heating, cooling, thermostabilization and thermoregulation applications.

## 2 Main aim

This work discusses a mathematical model of the heat transfer occurring between the flowing coolant, minichannel walls and surroundings. The boundary conditions in the form of temperature distributions on the outer sides of the minichannel walls were determined from the experimental measurement data The problem was analysed numerically using either the ANSYS CFX program or special calculation procedures based on the finite element method and Trefftz functions [5-16] in the thermal boundary layer.

## 3 Experimental data

The essential part of the experimental rig is a measurement module with a vertical rectangular minichannel, 1.7 mm deep, 16 mm wide and 180 mm long. The working fluid flowing in the minichannel is Flourinert FC-72. The schematic diagram of the measurement module is shown in figure 1. The heated wall of the minichannel is a metal plate made of Haynes-230 alloy, about 0.45 mm thick. The other wall of the minichannel is a 6 mm thick stainless steel (unheated) plate. Temperatures on the minichannel walls outer surfaces were measured simultaneously using infrared thermography. The temperature of the heated plate was monitored with the aid of the FLIR E60 infrared camera and the unheated plate - with the FLIR SC640 infrared camera. Before each experimental series, the metal plates were coated with a black paint to achieve an emissivity of 0.83 [1]. K-type thermocouples and pressure converters were installed at the inlet and outlet of the minichannel.

The flow loop comprises (besides the measurement module): a gear pump, a compensating tank, a tube-type heat exchanger, a filter, a Coriolis mass flowmeter and a deaerator. A data acquisition station and a computer with appropriate software were used for experimental data

* Corresponding author: beatam@tu.kielce.pl

collection. The supply and control system consists of an inverter welder, shunt, ammeter and a voltmeter. The experimental rig is presented in [1-4].

During experiment, there is a FC-72 Fluorinert laminar flow in a minichannel. Data for the selected setting of the heat flux supplied to the heated plate during subcooled boiling is analysed. In the subcooled boiling region, the liquid flowing in the minichannel was superheated at the interface with the heated plate and subcooled at the core of the flow. The experimental thermal and flow parameters, physical properties of the heated minichannel wall (made of Haynes-230) and the properties of the working fluid (Fluorinert FC-72) for this setting are taken to be the same in the two calculation approaches and shown in Table 1.

**Table 1.** Main experimental thermal and flow parameters, physical properties of Haynes-230 and FC-72 properties.

| **Main thermal and flow experimental parameters** | |
|---|---|
| Volumetric heat flux | $1.22 \cdot 10^5$ kW m$^{-3}$ |
| Flow velocity | 0.2 m s$^{-1}$ |
| Mass flux | 344 kg m$^{-2}$ s$^{-1}$ |
| Reynolds number | 1 846 (laminar flow) |
| Fluid temperature at the inlet | 14.3 C |
| Fluid temperature at the outlet | 25.2 C |
| Pressure at the inlet | 89.9 kPa |
| Inlet liquid subcooling | 39 C |
| **Physical properties of the heated plate (made of Haynes-230)** | |
| Density | 8 970 kg m$^{-3}$ |
| Thermal conductivity | 9.7 W m$^{-1}$ K$^{-1}$ |
| Specific Heat | 397 J kg$^{-1}$ K$^{-1}$ |
| **Physical properties of the unheated plate (made of stainless steel)** | |
| Density | 7 750 kg m$^{-3}$ |
| Thermal conductivity | 15.1 W m$^{-1}$ K$^{-1}$ |
| Specific Heat | 480 J kg$^{-1}$ K$^{-1}$ |
| **Physical properties of the working fluid (FC-72)** | |
| Liquid density | 1 737 kg m$^{-3}$ |
| Dynamic viscosity | 0.0006918 Pa s |
| Liquid Specific Heat | 1044 J kg$^{-1}$ K$^{-1}$ |
| Liquid Thermal Conductivity | 0.0561 W m$^{-1}$ K$^{-1}$ |

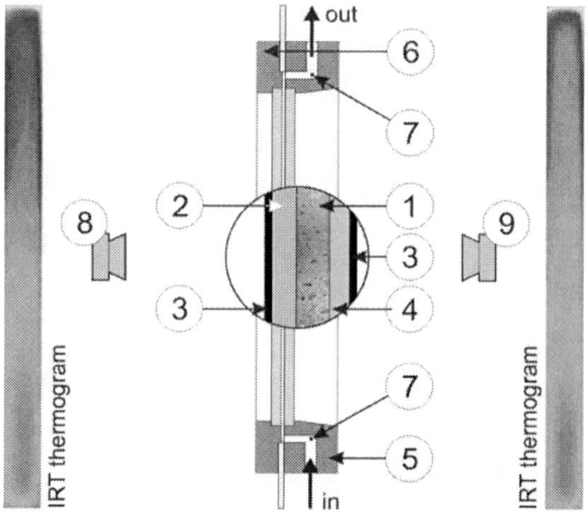


**Fig. 1.** The schematic diagram of the measurement module: 1-minichannel, 2-heated plate, 3- black paint with known emissivity, 4-unheated plate, 5-channel body, 6-front cover, 7-thermocouple, 8,9 - infrared cameras

## 4 Numerical methodology

### 4.1 The mathematical formulation

Two-dimensional steady-state heat flow in the minichannel is assumed. Temperature variation along the minichannel width is negligible for this study. The $x$ dimension along the flow direction and the $y$ dimension - perpendicular to the flow direction (related to the thickness of the heated plate) are considered.

The boundary-value problem is described in two adjacent regions by:

- Poisson's equation in the heated plate [12]:

$$\nabla^2 T_{p\_h} = -\frac{I \cdot \Delta U}{A \cdot \delta_{p\_h} \cdot \lambda_{p\_h}} \quad \text{for } (x, y) \in \Omega_{p\_h} \tag{1}$$

- Fourier–Kirchhoff's equation in the thermal boundary layer of the fluid [16]:

$$\frac{\lambda_f}{c_p \rho} \nabla^2 T_f = w_x(y) \frac{\partial T_f}{\partial x} \quad \text{for } (x, y) \in \Omega_T \tag{2}$$

The boundary conditions have the form:

$$T_{p\_h}(0, y) = T^1_{p\_h} \tag{3}$$

$$T_{p\_h}(L, y) = T^{P_h}_{p\_h} \tag{4}$$

$$\lambda_{p\_h} \frac{\partial T_{p\_h}}{\partial y}(0, y) = q_{loss} \tag{5}$$

$$T_{p\_h}(x_m, 0) = T^m_{p\_h} \quad \text{for } m = 1, 2, \ldots, P_h \tag{6}$$

$$T_f(x, \delta_{p\_h}) = T_{p\_h}(x, \delta_{p\_h}) \tag{7}$$

$$T_f(0,y) = T_f^{in} \text{ for } \delta_{p\_h} < y < \delta_{p\_h} + \delta_T \quad (8)$$

$$T_f(L,y) = T_f^{out} \text{ for } \delta_{p\_h} < y < \delta_{p\_h} + \delta_T \quad (9)$$

where $\Omega_{p\_h} = \{(x,y) \in R^2 : 0 < x < L, \quad 0 < y < \delta_{p\_h}\}$, $\Omega_T = \{(x,y) \in R^2 : 0 < x < L, \quad \delta_{p\_h} < y < \delta_{p\_h} + \delta_T\}$, $T_{p\_h}$ – the heated plate temperature, $T_f$ – the fluid temperature, $L$ – length of the minichannel, $I$ – current, $\Delta U$ – voltage drop, $A$ – surface area of the heated plate, $P_h$ – the number of measurements obtained using infrared thermography (IRT) on the outer surface of the heated plate, $T_{p\_h}^m$ – the heated plate temperature measured by infrared thermography at the boundary $y = 0$, $q_{loss}$ – heat loss to the surroundings calculated as in [1], $\delta_{p\_h}$ – heated plate thickness, $\delta_T$ – the thickness of the thermal boundary layer [16], $\lambda_{p\_h}$ – coefficient of thermal conductivity of the heated plate, $\lambda_f$ – coefficient of thermal conductivity of the fluid, $c_p$ – specific heat of the fluid, $\rho$ – density of the fluid $w_x(y)$ – component of vector liquid velocity, parallel to the minichannel heated surface [16].

## 4.2 The use of Trefftz functions

In each element $\Omega_j$, $j = 1,2,\ldots,L1 \cdot L2$ of the domain $\Omega = \Omega_{p\_h} \cup \Omega_T$, the system of equations (1-9) was solved. The approximate temperature $T_{p\_h}$ was given by:

$$T_{p\_h}^j(x,y) = u(x,y) + \sum_{k=1}^{N} \left(\widetilde{T}^n - u(x_n, y_n)\right) f_{jk}(x,y) \quad (10)$$

the fluid approximate temperature was described by the expression:

$$T_f^j(x,y) = \sum_{k=1}^{N} \widetilde{T}^n g_{jk}(x,y) \quad (11)$$

where: $u(x,y)$ – the particular solution of Eq. (1), $f_{jk}(x,y,)$ – the basis functions determined as linear combinations of the Trefftz functions according to Laplace's equation [9], $g_{jk}(x,y)$ – the basis functions referred to the Fourier–Kirchhoff equation (the Trefftz functions that satisfy the Fourier–Kirchhoff equation are given in [16]), $\widetilde{T}^n$ – the unknown temperature in the $n$-th node of domain $\Omega$, $j$ – element number, $k$ – basis function number in $j$-th element, $n$ – node number in the entire domain $\Omega$, $N$ – the number of nodes in the element, $L1$ – the number of elements in the $x$-axis direction and $L2$ – the number of elements in the $y$-axis direction.

To find unknown coefficients $\widetilde{T}^n$, the appropriate functional was minimized [1].

## 4.3 The use of the ANSYS CFX program

The conditions (3)-(9) and equality of heat fluxes on the contact surface between the heated plate and the fluid were taken into account in the computational module by using the Conservative Interface Flux option for fluid-solid interfaces. The calculation results from solving the presented model on boiling heat transfer during FC-72 flow in a rectangular minichannel with the ANSYS CFX are shown to compare with results obtained from a special calculation procedure based on the FEM and Trefftz functions (presented in the previous chapter).

Numerical calculations were performed in the ANSYS CFX environment. The shape of the minichannel with the flowing fluid and its structural elements, i.e., the 0.45 mm thick heated plate and the 6 mm thick unheated plate restraining the channel space, were modelled with the aid of the Geometry-Design Modeler, as shown in figure 2.

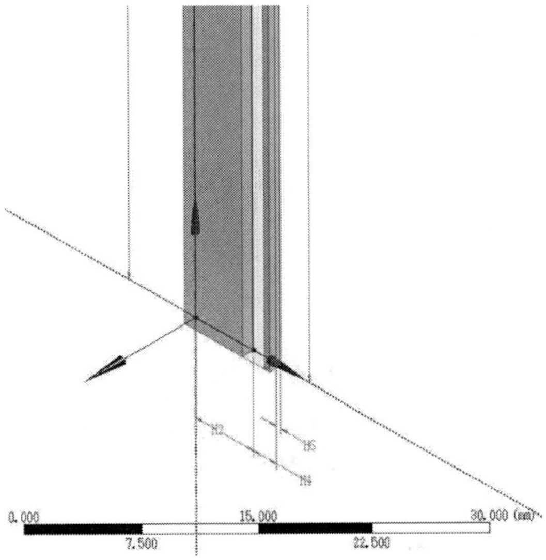


**Fig. 2.** The schematic diagram of the measurement module in Geometry-Design Modeler.

The next stage involved generating a mesh in all elements of the measurement module model. The Advanced Size function was used with the criteria based on Proximity and Curvature, with parameters as in figure 3.

The mesh consisted of 589 138 elements and 716 940 nodes. For the sake of the character of numerical calculations and the need to combine the heat transfer phenomena in solids (the heated plate and the unheated plate), the computational module according to the project schematic was used, as shown in figure 4.

DOI: 10.1051/epjconf/201714302071

| | |
|---|---|
| Use Advanced Size Function | On: Proximity and Curvature |
| Relevance Center | Fine |
| Initial Size Seed | Full Assembly |
| Smoothing | High |
| Transition | Fast |
| Span Angle Center | Fine |
| Curvature Normal Angle | Default (12.0 °) |
| Num Cells Across Gap | 6 |
| Proximity Size Function ... | Faces and Edges |
| Min Size | 5.e-005 m |
| Proximity Min Size | Default (1.8017e-005 m) |
| Max Face Size | 5.e-004 m |
| Max Size | 5.e-003 m |
| Growth Rate | Default (1.50 ) |
| Minimum Edge Length | 4.5e-004 m |

**Fig. 3.** The details of the mesh.

Another stage of preparing numerical calculations in ANSYS Workbanch involved determining boundary conditions, which were adopted in compliance with the relationships described in Chapter 4, formulas (3)-(9). The boundary conditions being the temperature distributions along the outer wall of the heated plate and the unheated plate were taken from the experimental data and used in the Fluid Flow (CFX) module in the form as shown in figure 5.

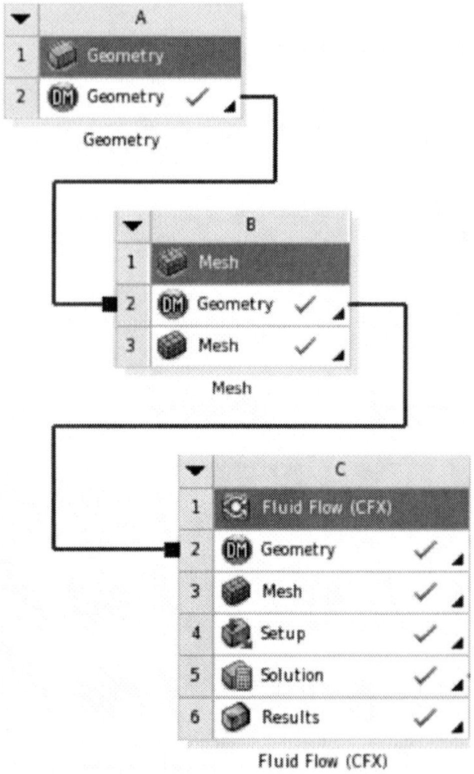


**Fig. 4.** The project schematic.

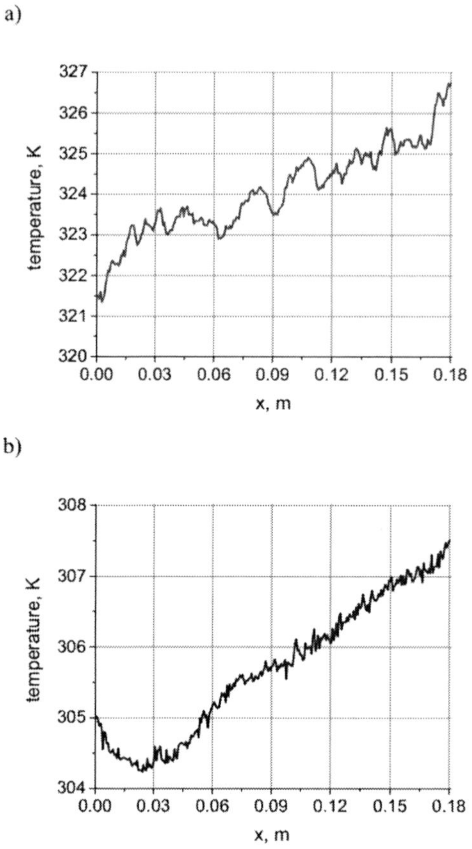


**Fig. 5.** Experimental measurement data: a) temperature of the heated plate, b) temperature of the unheated plate.

## 5 Results and discussion

The results are presented graphically in figure 6 as relationship between temperature and the distance from the minichannel inlet to compare the values calculated using the Trefftz functions (red points) from with that obtained using the ANSYS software (blue points), at $\delta_M/40$ distance, where $\delta_M$ denotes the minichannel depth. Results obtained near the heated plate-fluid boundary, which is comprised of a thermal boundary layer, indicates that the temperature distribution calculated using the Trefftz functions and obtained using the ANSYS software are very similar. The temperatures increased slightly with the distance from the minichannel inlet and their values are high, similarly to the heated plate temperature measured from infrared thermography.

Minor differences in the results may be due to:

- simplifications used in the mathematical model (1)-(9) relative to the ANSYS CFX model (a laminar model governed by the Navier–Stokes equations which describe the processes of momentum, heat and mass transfer);
- the use of various approximate calculation methods in ASNYS CFX and the FEM based on Trefftz functions.

It should be added that the subcooled boiling region took place during experiments. In this region temperatures are relatively low, as the heat transfer coefficient [1,3,4].

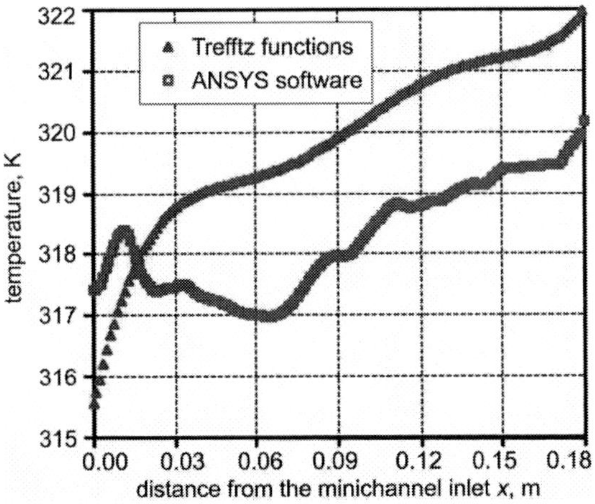


**Fig. 6.** Temperature vs. the minichannel length, obtained using the Trefftz functions (red points) and the ANSYS software (blue points), generated at $\delta_M$ /40 distance on the minichannel depth from the foil-fluid boundary.

## 6 Conclusions

This work discusses the mathematical model for laminar-flow heat transfer in a minichannel. The boundary conditions in the form of temperature distributions on the outer sides of the channel walls were determined from experimental data. The data were collected from the experimental stand the essential part of which is a vertical minichannel 1.7 mm deep, 16 mm wide and 180 mm long, asymmetrically heated by a Haynes-230 alloy plate. Infrared thermography allowed determining temperature changes on the outer side of the minichannel walls. The problem was analysed numerically through either ANSYS CFX software or special calculation procedures based on the Finite Element Method and Trefftz functions. Results were presented graphically as temperature vs. the minichannel length, near the heated plate-fluid boundary, which is comprised of a thermal boundary layer. The temperature distributions calculated using the Trefftz functions and obtained using the ANSYS software were very similar.

ACKNOWLEDGMENTS

The research reported herein was supported in part by a grant from the Polish National Science Centre (No. DEC-2013/09/B/ST8/02825).

## References

1. M. Piasecka, K. Strąk, B. Maciejewska, Heat Transf. Eng. **38** (3), 332-346 (2017)
2. M. Piasecka, Int. J. Refrig. **56**, 198-212 (2015)
3. M. Piasecka, B. Maciejewska, Exp. Therm. Fluid Sci. **68**, 459-467 (2015)
4. M. Piasecka, Int. J. Heat Mass Transf. **81**, 114-121 (2015)
5. E. Trefftz, Proc. Int. Kongress für Technische Mechanik, Zürich, 131-137 (1926)
6. I. Herrera, Meth. Partial Diff. Eqs., **16**, 561-580 (2000)
7. K. Grysa, A. Maciąg, A. Pawińska, Int. J. Heat Mass Transf., **55**, 7336–7340 (2012)
8. M. J. Ciałkowski, A. Frąckowiak, J. Therm. Sci., **11**, 148-162 (2002)
9. M. J. Ciałkowski, J. Therm. Sci., **11**, 163-171 (2002)
10. M. J. Ciałkowski, K. Grysa, J. Inv. Ill-Posed Prob., **18**, 595-616 (2010)
11. K. Grysa, B. Maciejewska, J. Theor. Appl. Mech., **51**, 251-264 (2013)
12. S. Hozejowska, R. Kaniowski, M. E. Poniewski, Int. J. Num. Methods Heat Fluid Flow, **24**, 811–824 (2014)
13. A. Maciąg, Int. J. Num. Meth. Biomed. Eng., **27**, 1107-1125 (2011)
14. S. Blasiak, A. Pawinska, Int. J. Heat Mass Transf., **90**, 710-718 (2015)
15. B. Maciejewska, M. Piasecka, Heat Mass Transf., DOI: 10.1007/s00231-016-1895-1 (2016)
16. S. Hożejowska, J.Theor. Appl. Mech, 53, 969-980 (2015)
17. M. Piasecka, D. Michalski, K. Strąk, The European Physical J. Web Conf. **114**, paper No. 02094 (2015)

# Influence of operating conditions and atomizer design on circumferential liquid distribution from small pressure-swirl atomizer

*Milan* Malý[1,*], *Lada* Janáčková, *Jan* Jedelský and *Miroslav* Jícha [1]

[1] Brno University of Technology, Faculty of Mechanical Engineering, Energy Institute, Technická 2896/2 Brno 616 69, Czech Republic

**Abstract.** The spray symmetry is an important aspect in most practical applications. However, it is often an overlooked parameter. A measurement of circumferential distribution was carried out by a circular-sectored vessel on several pressure-swirl atomizers with spill-line over a wide range of injection pressure. The obtained results show that the spray uniformity improves markedly with the injection pressure. The increase in a number of tangential entry ports has only a minor effect on the spray uniformity. Even a small mechanical corruption of the atomizer internal parts negatively affects the spray patternation.

## 1 Introduction

The symmetry of spray pattern produced by the atomizer is an important parameter in many applications. E.g.: non-uniformities in circumferential fuel distribution in gas turbine combustors can lead to a rise of local pockets of the low fuel-air mixture in which burning rates are low, thereby produce high concentrations of carbon monoxide and unburned hydrocarbons. Moreover, it may lead to non-uniform flame distribution and eventually to local overheating of the combustion chamber. The reasons for spray a non-uniformity are all different: asymmetrical design of the atomizer, uneven size of the inlet ports, poor surface finish. Geometrical dimensions of various parts and operating regimes also affects the spray symmetry.

The number of publications related to the spray patternation is surprisingly low; probably the most complex study was carried out in 1993 by Chen and Lefebvre [1]. They investigated several simplex atomizers with various swirl chamber geometries. Khavkin [2] summarized possible reasons of the spray asymmetry of simplex atomizers.

This study investigates several simplex atomizers differing in the swirl chamber shape and the number of the tangential inlet ports, and more over two spill-return atomizers are also represented in tests. The spill-return atomizers differ due to the presence of a passage in the back wall of the swirl chamber. Liquid, injected via tangential ports into the swirl chamber, is divided into two streams, one of them is discharged outside and atomized while the second is spilled back to a reservoir. This feature greatly expands a turn-down ratio of these atomizers compared to the simplex atomizers, because the liquid is always supplied to the swirl chamber at high pressure guaranteeing a good atomization at various injection flow rates. No relevant literature dealing with the patternation of the spill-return atomizers was found.

## 2 Experimental setup

Experiments were performed at a specially designed facility for cold spray testing in Spray laboratory at the Brno University of Technology.

### 2.1 Tested atomizers

Six simplex and two spill-return atomizers conformed to the schematic drawing in Figure 1 were used in this study, all had the same total area of the inlet ports, $A_p = 0.72$ mm$^2$, and the diameter of the exit orifice, $d_o = 0.42$ mm. Four different swirl chambers differed in the shape of the convergent part – see Figure 2. The shapes under study were as follows: A - hemispheric, B - conical; C - curved convex and D - flat conical. In addition, simplex atomizers with two (S2), three (S3) and four (S4) inlet ports were used. Two spill-return atomizers with spill-line in a different position were used as well: SA – axially-positioned, SB – off-axis spill orifices.

### 2.2 Test bench and mechanical patternator

The tested liquid was supplied to the atomizer by a specially designed fuel supply track (schematic layout in Figure 3). The liquid, stored in the fuel tank (1), was pumped via a filter (2) by a gear pump (3) to the atomizer (8). The liquid mass flow rate was regulated by driving the pump speed. The amount of pumped fuel was measured in the inlet line by a Coriolis mass flow meter (4). Liquid overpressure was measured by a piezo-resistive sensor (7). The inlet line was also equipped

* Corresponding author: milan.maly@vutbr.cz

DOI: 10.1051/epjconf/201714302072

with a temperature reading. The spill-line was equipped with a static pressure meter (9), regulating needle valve (11) and the positive displacement flow meter (10).

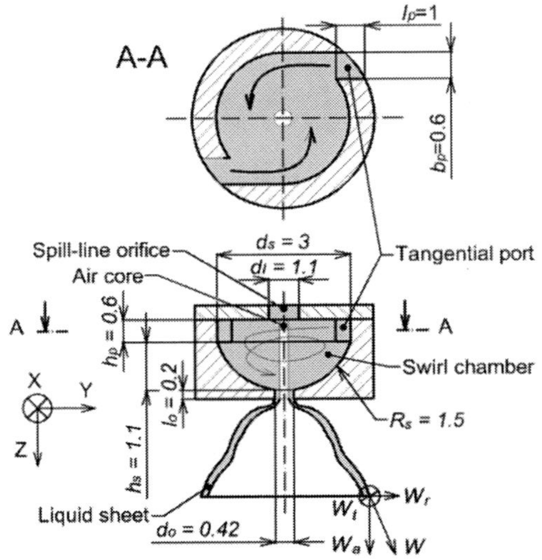


**Fig. 1.** Schematic layout of tested atomizer

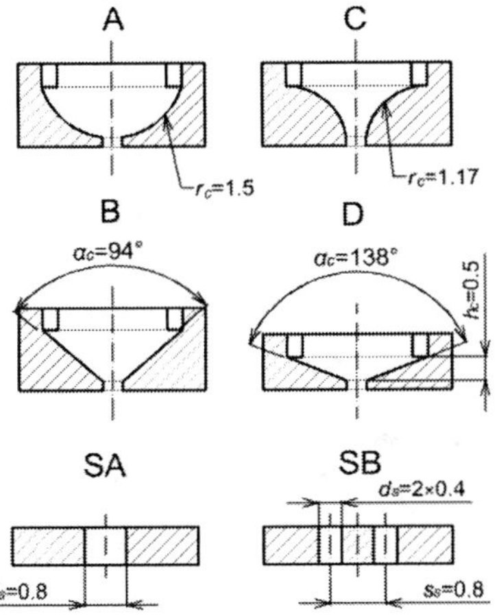


**Fig. 2.** Schematic drawings of tested variants

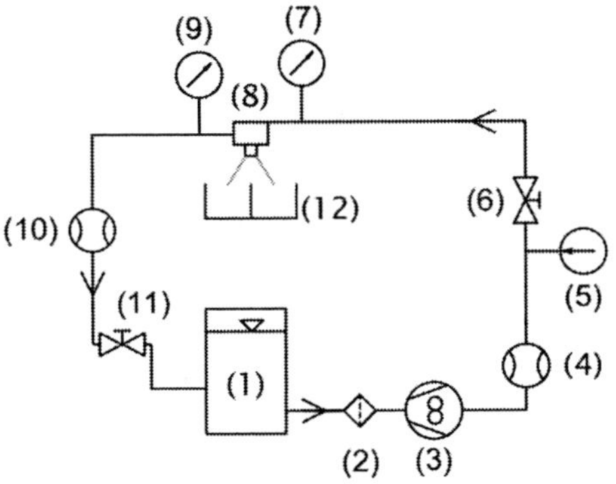


**Fig. 3.** Schematic layout of the test bench

The spray circumferential distribution was rated by a simple circular-sectored vessel (12), often called a mechanical patternator, with 16 pie-shaped sectors, as shown in Figure 4. The patternator was placed 50 mm downstream from the exit orifice of the measured atomizer. The volume of each sector was 60 ml. The duration of each test was determined by the time required for one of the sectors to become nearly full. Usually, it was between two and four minutes. The height of the liquid from the upper edge in each sector was measured with a caliper, and the liquid height was calculated. These values were consequently averaged to get the mean height, and the levels of the sectors were normalized against the mean height. The standard deviation of normalized values was calculated as a single parameter describing the non-uniformities:

$$\sigma = \sqrt{\frac{\sum_{i=1}^{n} (y_i - \bar{y})^2}{n}} \quad (1)$$

where $n$ is the number of sectors (16), $y_i$ is the normalized height of the liquid in sector i and $\bar{y}$ is the liquid average height.

Jet-A1 fuel was used as the tested liquid. Physical properties of the Jet-A1 at room temperature (20 °C) were: $\sigma = 0.029$ kg/s$^2$, $\mu_l = 0.0016$ kg/(m·s), $\rho_l = 795$ kg/m$^3$.

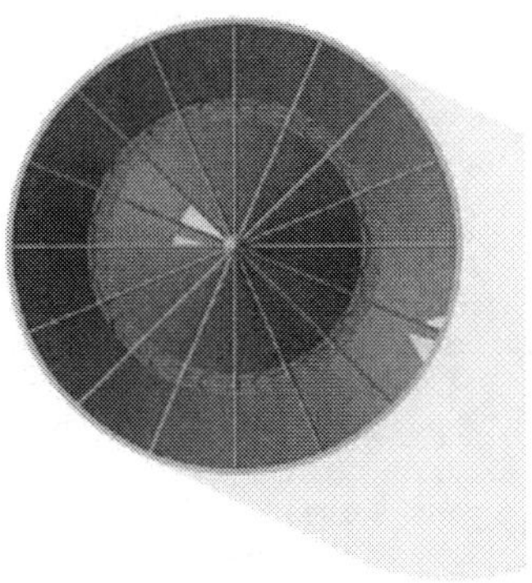

**Fig. 4.** Mechanical patternator

## 3 Results and discussion

Each atomizer was tested at three pressure regimes, the atomizers with the spill-line were also tested at three spill regimes at constant inlet pressure. Each regime was measured several times, and data consequently averaged.

### 3.1 Number of tangential ports

The effect of the number of the tangential inlet ports is shown in dependence on the circumferential position in Figure 5 and its standard deviation along the three pressure regimes is plotted in Figure 6. The data presented in Figure 5 are typical of the results obtained.

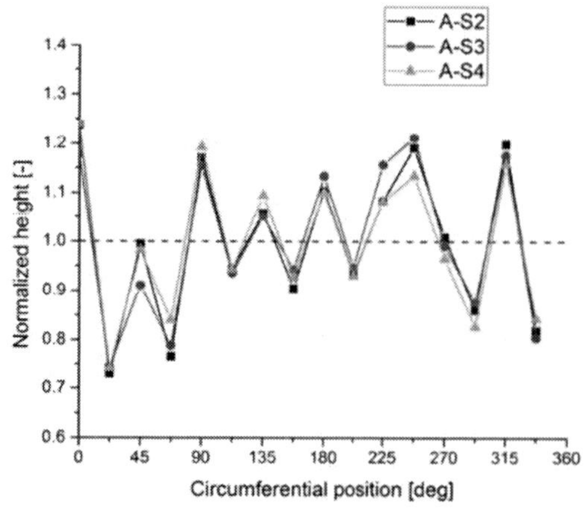


**Fig. 5.** The liquid distribution from the simplex atomizers with various tangential ports number, inlet pressure 1 MPa.

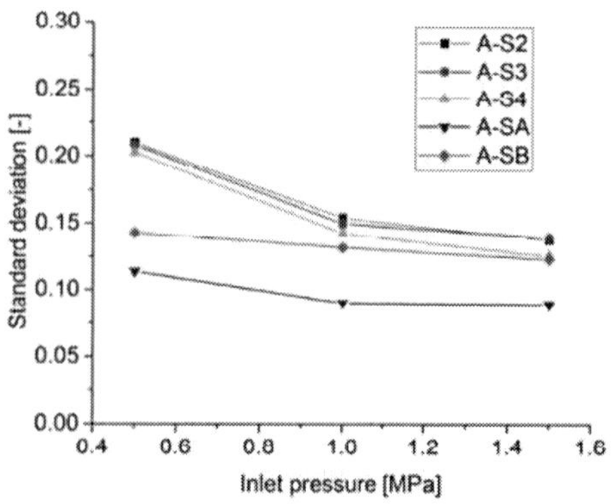


**Fig. 6.** The influence of tangential ports number and presence of spill-line orifice. Spill-to-Feed ratio (SFR) 0.

However, in this particular example, the atomizers have almost identical behaviour, even if the number of tangential ports is changed. The change in the standard deviation is almost negligible, an increase in a number of the ports from two to four enhances the standard deviation by less than 10 %. Similar results were obtained by Chen and Lefebvre [1]. They found the difference in spray patternation between the two and three-port atomizers as insignificant; however, the one-port atomizer under their test had a noticeably rougher liquid distribution. As it is obvious from Figure 6, the increasing liquid pressure has a beneficial effect for all atomizers in terms of liquid distribution.

Different results were obtained with the spill-return atomizers (SA and SB). They are the two-port types but their liquid distribution shows better results compared to the simplex S2. SB with the off-axis spill-orifices has the standard deviation converging to the simplex type at higher pressures. However, SA shows the most uniform spray pattern. This atomizer, the only one in this study, has a visually unstable spray in the same way as it was described in [3]. Because of the long-time measurement compared to the spray fluctuating frequency (which was about 5 Hz), data from the patternator are time-averaged and the spray may be not so symmetrical in the short time observation. However, when the spill is applied, the unstable atomizer turns into the stable one at approx. spill to feed ratio (SFR) of 0.1. For SFR higher than 0.1, the SA atomizer becomes stable and results are not distorted by the spray fluctuations. The influence of SFR on patternation at constant inlet pressure of 1 MPa is shown in Figure 7. Increasing SFR has a positive effect on the droplet size [4] but circumferential distribution rapidly worsens. This behaviour was observed for both spill-return atomizers; their standard deviation reaches almost an identical value at SFR 0.4. Patternation rapidly deteriorates with further increases in SFR over 0.4. This phenomenon may be related to the decreasing discharge velocity with increasing SFR. The droplets are migrating from one sector to another with lower activity, and thus the atomizer is more sensitive to any imprecisions.

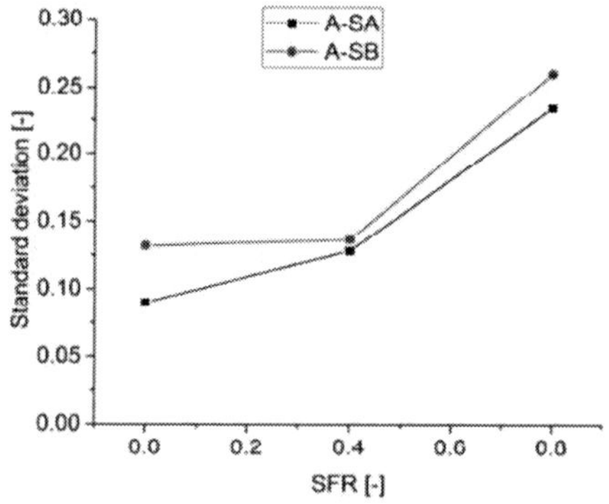


**Fig. 7.** The influence of the spill regime on patternation at inlet pressure of 1 MPa

### 3.2 Influence of swirl chamber shape

Four swirl chambers were tested in the three-ports simplex atomizer. Results, presented in Figure 7, show a considerable influence of the swirl chamber shape on the spray patternation. The most uniform distribution was accomplished by the high conical chamber B. The chamber with the lower convergent part (D) has about 25 % higher standard deviation which is almost undependable on the inlet pressure. A similar characteristic was observed in the curved convex chamber – C. Unlike to the C and D chambers, patternation of the hemispheric chamber A strongly depends on the inlet pressure.

### 3.3 Mechanical corruption of the atomizer

Apart from the atomizer design itself, geometrical inaccuracies, surface quality, and mechanical corruptions have an impact on the spray uniformity [1, 2, 5]. These problems were discussed in the appointed literature. However, impact of manufacturing quality was described only theoretically.

DOI: 10.1051/epjconf/201714302072

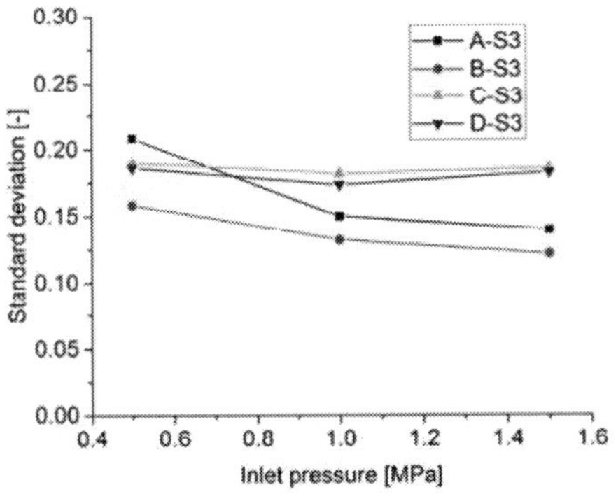


**Fig. 8.** The influence of the swirl chamber shape

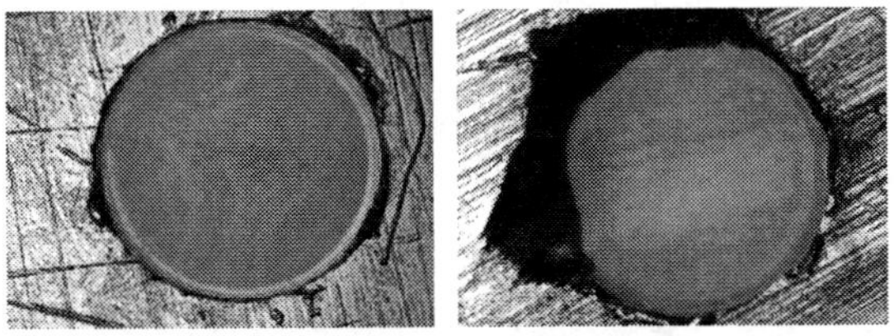

**Fig. 9.** Left: fine exit orifice, right: damaged exit orifice

Due to a difficult assessment of the surface quality and geometrical accuracy of the atomizer itself, this part focuses primarily on the mechanical corruption of the atomizer parts.

In this paper, two atomizers damaged by mishandled installation are investigated, the first one has a minor mechanical corruption of the exit orifice while the second one has a corrupted tangential port. They are compared to the identical types of regular shape. A geometrical difference between the corrupted and original exit orifice is shown in Figure 9; the original part is on the left, the corrupted one on the right. The images were taken using an optical microscope Nikon Eclipse E200. The depth of the scratch is approximately 0.1 mm. The tangential port was damaged at its exit cross-section and the liquid flow may be deflected from original direction.

The scratch on the outer edge of the exit orifice caused an increase in the standard deviation by 86 %. The atomizer with corrupted tangential port has almost 3 times higher standard deviation compared to the fine one.

## 4 Conclusions

The experimental investigation was undertaken to study the effect of several parameters on the spray uniformity. Besides simplex atomizers, spill-return types were also presented in this study.

- The circumferential uniformity of spray pattern improves with increasing liquid pressure for the most of the tested atomizers.
- An increase in the number of the tangential ports from two to three or four has a negligible effect.
- The shape of the swirl chamber convex part contributes greatly to the spray uniformity.
- The unstable atomizer has a more uniform spray in the long-time measurement than the stable one.
- Poor finish quality or mechanical damage of the exit orifice or the inlet tangential ports deteriorate the spray symmetry.

## Acknowledgement

This work has been supported by the project No. GA15-09040S funded by the Czech Science Foundation, the project LO1202 NETME CENTRE PLUS with the financial support from the Ministry of Education, Youth and Sports of the Czech Republic under the "National Sustainability Programme I" and project Reg. No. FSI-S-14-2355 funded by the Brno University of Technology.

## References

1. S. Chen, A. Lefebvre, J. Rollbuhler, J. of eng. for gas turbines and power, **115**, pp. 447-452. (1993)
2. Y. Khavkin, *The Theory and Practice of Swirl Atomizers* (2004)
3. M. Maly, L. Janackova, J. Jedelsky,M. Jicha, *EPJ Web of Conferences* (2016)
4. J. Jedelsky, M. Maly, L. Janackova, M. Jicha, ILASS p. 12 (2016)
5. A. H. Lefebvre, *Atomization and sprays* (1989)

EPJ Web of Conferences **143**, 02073 (2017) DOI: 10.1051/epjconf/201714302073
*EFM 2016*

# Numerical and experimental modelling of the centrifugal compressor stage – setting the model of impellers with 2D blades

*Richard* Matas[1,*], *Tomáš* Syka[1], and *Ondřej* Luňáček[2]

[1]University of West Bohemia in Pilsen, New Technologies Research Centre, Univerzitní 8, 306 14 Plzeň, Czech Republic
[2]Howden ČKD Compressors s.r.o., Klečákova 347, 190 00 Praha 9, Czech Republic

**Abstract.** The article deals with a description of results from research and development of a radial compressor stage. The experimental compressor and used numerical models are briefly described. In the first part, the comparisons of characteristics obtained experimentally and by numerical simulations for stage with vaneless diffuser are described. In the second part, the results for stage with vanned diffuser are presented. The results are relevant for next studies in research and development process.

## 1 Introduction

Centrifugal compressors are widely used in the industry. A traditional producer of multi-stage centrifugal compressors is company Howden ČKD Compressors s.r.o. (HCKD). These compressors are commonly used in many applications across the oil and gas sector, storage and processing, in the metallurgy, mining and also in chemical industry.

The description of processes, operations and performance of the centrifugal compressors are described in many publications, for example in [1]. Performances of the centrifugal compressors stages are analysed namely on account of two main properties: pressure ratio and efficiency. These two parameters vary with volume flow and with Mach number.

A new family of compressor stages is being implemented nowadays. The development project is realized in cooperation with PCA Engineers Limited and University of West Bohemia in Pilsen (UWB). The main goal of the project is to achieve higher stage efficiency and better operational properties. The verification of the new stages properties requires detailed measurement of performance characteristics to obtain reliable data for compressor design. Measured data are also important for validation of CFD numerical models. Validated numerical models are useable for reliable prediction of the stage performance. The validated numerical models can give also the complex information about the flow in the stage and help to design the profile of the channel.

The new stages family consists of stages with 3D impellers and 2D impellers depending on required flow coefficient. The typical stage with 3D impeller (RTK01) and obtained results are described in [2]. This contribution is focused on stages with 2D impellers. For the study was selected two stages, the first one with vaneless diffuser (RTK04) and the second one with vanned diffuser (RTK06).

## 2 Experimental facilities

### 2.1 Experimental rig

The current development of the centrifugal compressor stages follows the systematic research and development program in the former ČKD company in the past; see for example the paper [3]. The experimental compressor stages verification is being realized in the development test facility HCKD built in Pilsen in 2013, where the single-stage test compressor DARINA IV with the exchangeable flow part is placed. The compressor consists of an inlet cone, an impeller, a diffuser, a return bend, a de-swirl channel and an exit part.

This quite unique compressor is placed on the shared frame with the coupling integrated torquemeter and the gearbox. This machine set is driven by the electromotor with maximum input wattage 1.2 MW, which is driven by the frequency converter for the continuous revolution change. The test rig is shown in the upper part of Fig. 1.

The test rig is designed for impellers with diameter of 440 mm. It is possible to change shaft revolutions from 4 000 to 17 500 rpm. A set of throttle-valves for the aerodynamic resistance adjusting is placed in the discharge piping.

### 2.2 Experimental equipment

The parameters of the measured stage are obtained by complex measurement and data acquiring system. Mass-flow rates of the air through the suction and discharge piping are measured by orifices (max. 42 000 kg/hour).

* Corresponding author: mata@ntc.zcu.cz

 DOI: 10.1051/epjconf/201714302073

**Fig. 1.** Centrifugal compressor stage test rig with evaluation planes.

The flow parameters are measured in the evaluation planes as shown in the Fig. 1 down. The description of the planes in the stage is long-term used, see [4]. There is a number of static and total pressure and total temperature probes (multiple sensors are placed in main evaluation planes 0 and 7). Two three-hole pressure probes combined with the positioning system for measuring the return vanes inlet angles are placed in the return channel. The torquemeter gives the value of torque and rotating speed of the compressor rotor.

Physical models of the new designed stages were prescribed in HCKD. The impeller of the stage RTK04 and the vanned diffuser for the stage RTK06 are depicted in the Fig. 2.

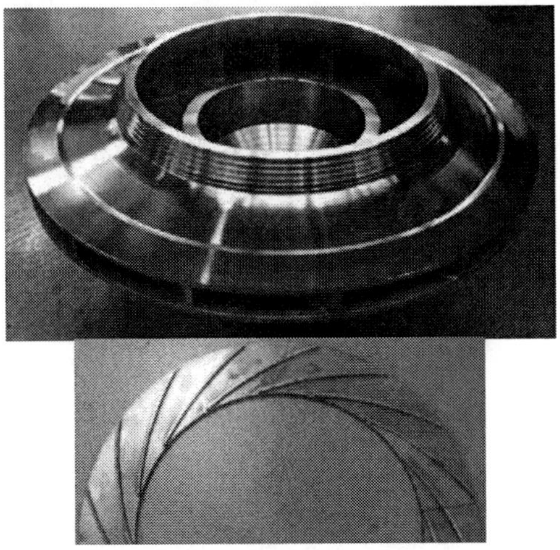

**Fig. 2.** 2D impeller of the stage RTK04 for experimental compressor and the diffuser vanes for the stage RTK06.

## 3 Numerical simulations

### 3.1. Software

CFD simulations are now an essential tool for the design of all types of devices in turbomachinery. They enable to gain much time between the concept and the test. The verification of simulations is carried out virtually in all development turbomachinery groups and widely published; for example in [5] and many others. The development of the stages has been supported by the CFD simulations in ANSYS/CFX (CFX) and NUMECA FINE/Turbo (NFT) systems.

### 3.2 Simulations of stages with 2D impellers

The methodology of simulation and properties of model has been also evaluated and improved [6]. The mentioned stages were modelled without and with leakage. The geometry of numerical models and computations domains of the stages with leakage can be seen in Fig. 3.

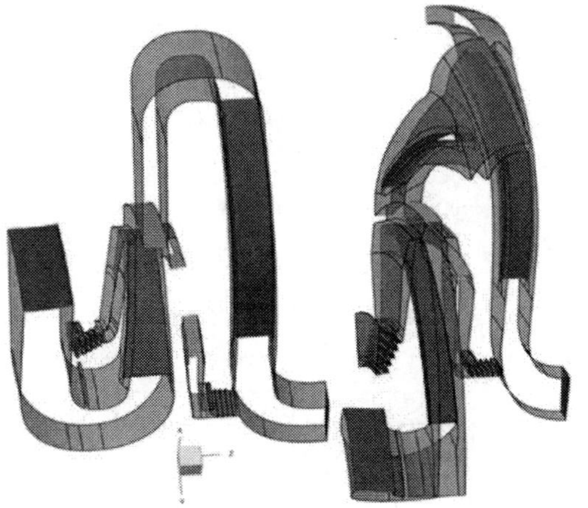

**Fig. 3.** Stages geometries of the computation domains.

SST k-ω turbulence model was used for most of the described simulations. For selected simulations of the stage RTK04 was used the EARSM turbulence model in NUMECA FINE/Turbo system.

### 3.3 Parameters of stages

The evaluation of various compressor stages parameters was done according to formulas below. Values of evaluated parameters were averaged by mass flow in postprocessing. The parameters are: specific total work (1), specific isentropic work (2), pressure ratio (3), isentropic efficiency (4) and flow coefficient (5). The indices represent: *t* – total, *s* – static, *is* – isentropic, *p* – pressure, *0* – plane 0 (stage inlet), *7* – plane 7 (stage outlet), *2* – output from the impeller.

$$\Delta h_t = c_p \cdot (T_{t7} - T_{t0}) \quad (1)$$

$$\Delta h_{is} = z \cdot r \cdot T_{t0} \cdot \frac{\kappa}{\kappa - 1}\left[\left(\frac{p_{t7}}{p_{t0}}\right)^{\frac{\kappa-1}{\kappa}} - 1\right] \quad (2)$$

$$PI = \frac{p_{t7}}{p_{t0}} \quad (3)$$

$$ETA = \frac{\Delta h_{is}}{\Delta h_t} \quad (4)$$

$$FI0 = \frac{4 \cdot Q_0}{\pi \cdot D_2^2 \cdot u_2} \quad (5)$$

The obtained characteristics include the confidental data of HCKD and only relative values are presented in the next chapters consequently. The values $PI_r$, $ETA_r$ and $m_r$ are related to the design point in the measured characteristic.

## 4 Comparison of results - stage RTK04

The results from CFD stages simulations were compared to the measured data converted to the standard inlet conditions. The Fig. 4 shows the computational mesh for NFT simulations with leakage. The numbers of cells for CFD models are shown in the Table 1.

**Table 1.** Numbers of cells for CFD models of RTK04.

| CFD model | Number of cells |
|---|---|
| CFX | 1,653,523 |
| CFX with leakage | 2,695,255 |
| NFT | 2,996,000 |
| NFT with leakage | 6,056,000 |

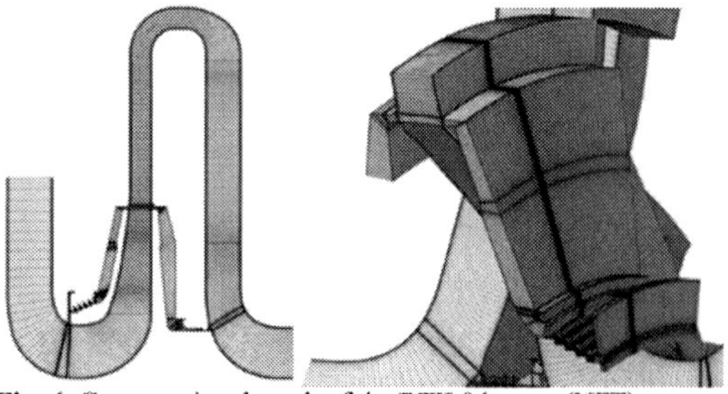

**Fig. 4.** Computational mesh of the RTK 04 stage (NFT)

The Fig. 5 shows the averaged pressure distribution in the meridional view. In the Fig. 6 you can see the performance characteristics. The Fig. 7 depicts the inlet flow angle into return channel vanes. Both figures are for 12 666 rpm. It is possible to compare the results of described CFD simulations with measured data.

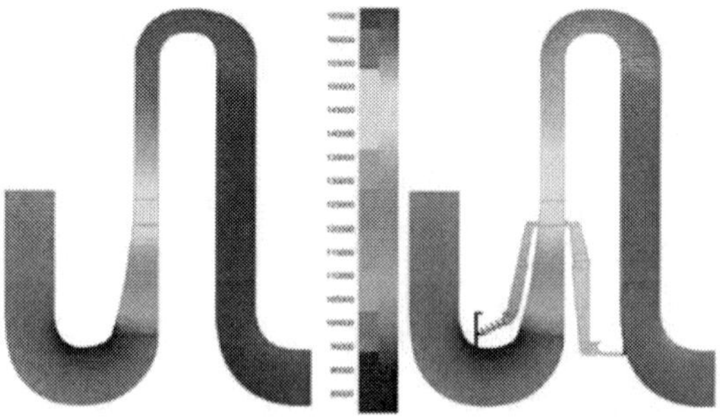

**Fig. 5.** Pressure distribution [Pa] in the stage RTK04 with and without leakage for the design point – meridional averaging

The results of both CFD systems are similar. The results of the NFT cases with the EARSM turbulent model are very close to the measured values of *PI* and *ETA*. This configuration of the CFD model can be used for reliable simulations of radial compressor stage characteristics of similar geometric and flow parameters. The evaluated values in models without leakage are overpredicted.

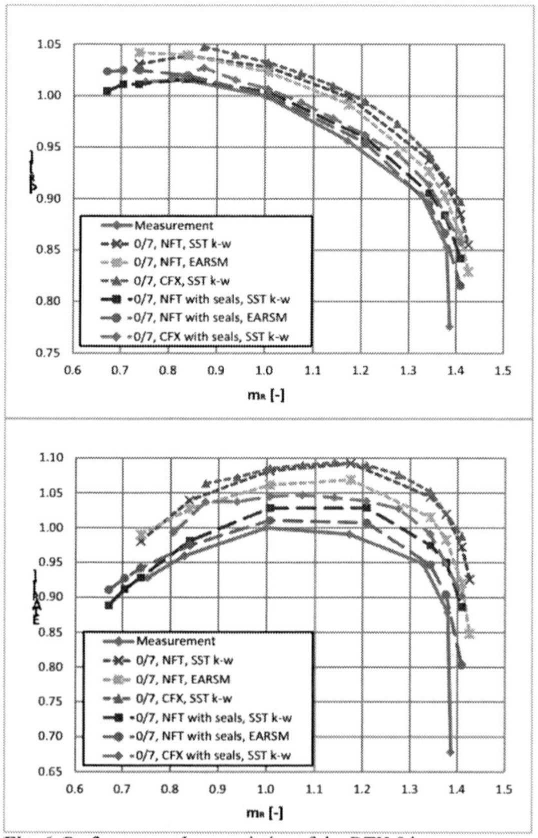


**Fig. 6.** Performance characteristics of the RTK 04 stage

The inlet flow angles into return channel vanes are very similar for all computed models. The values for higher mass flows are satisfying for simulations in comparison to measured values. The comparison for lower values of mass flow shows that the values in experiments rise, whereas in simulations sink. The reason can be the unsteady phenomena in the flow field or asymmetric flow in various channels which can't be detect in one channel model of the stage.

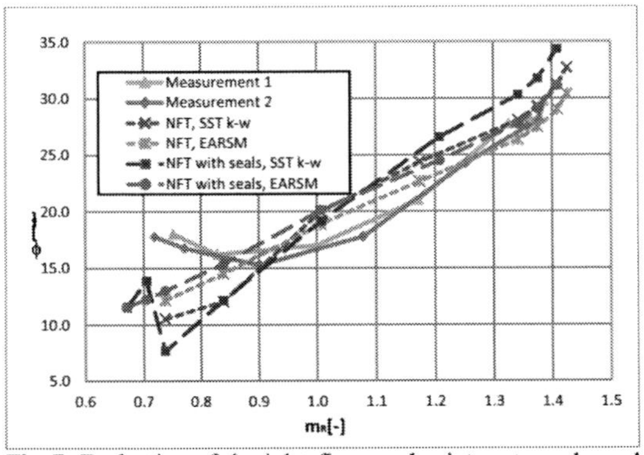


**Fig. 7.** Evaluation of the inlet flow angles into return channel vanes of the RTK 04 stage

## 5 Comparison of results - stage RTK06

The same process as for the stage RTK04 was implemented for the stage RTK06. The numerical model is more sophisticated due to vanned diffuser, because the model of the diffuser vane has to be added.

The Fig. 8 shows the computational mesh for NFT computations with leakage. The numbers of cells for CFD models are shown in the Table 2.

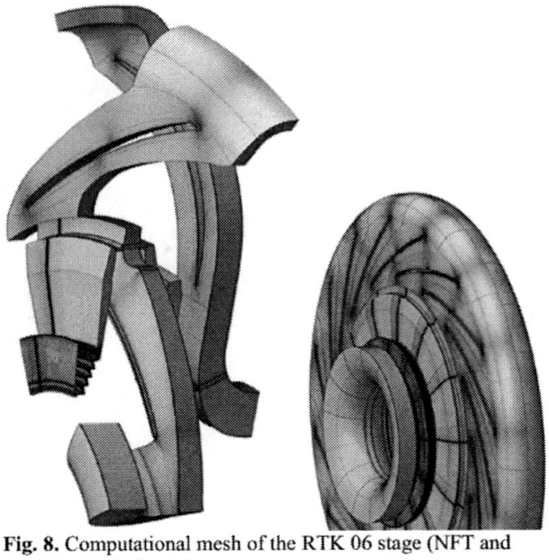

**Fig. 8.** Computational mesh of the RTK 06 stage (NFT and CFX II with leakage)

**Table 2.** Number of cells for numerical models of RTK06.

| CFD model | Number of cells |
|---|---|
| CFX I | 885,660 |
| CFX I with leakage | 1,132,269 |
| NFT and CFX II | 6,120,000 |
| NFT and CFX II with leakage | 11,070,000 |

The Fig. 9 shows the flow field in the meridional view. In the Fig. 10 are presented the performance characteristics. The results of numerical simulations can be compared sufficiently to the measured values.

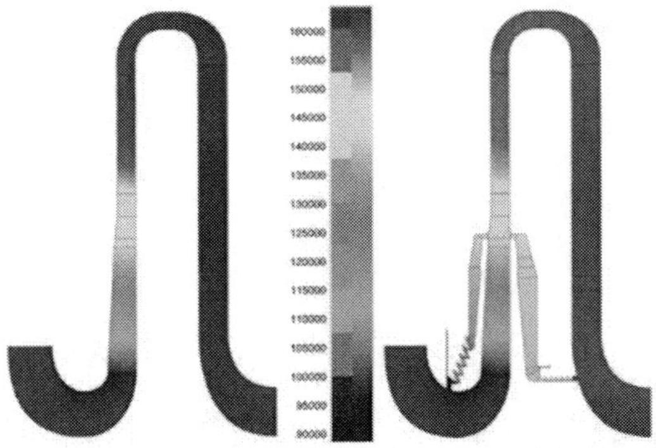

**Fig. 9.** Pressure distribution [Pa] in the stage RTK06 with and without leakage for the design point – meridional averaging

The performance characteristics in models without leakage are overpredicted again. The results evaluated from coarse mesh simulations show relatively high deviation from measured values. The results of both CFD systems with leakage and fine meshes are similar. The results of these cases are very close to the measured values of *PI* and *ETA*. These configurations of the CFD models can be used for the reliable simulations of characteristics for radial compressor stages with vanned diffusers of similar geometric and flow parameters. The disadvantage is relatively complicated mesh generation and relatively high computational costs.

## 6 Conclusions

The verification of the compressor stages with the relatively wide range of parameters is possible to perform on the HCKD DARINA IV compressor with good repeatability. The test facility is able to evaluate complete integral and some local quantities in order to prepare appropriate data for thermodynamic design. It is possible to verify CFD models and their applicability by mutual comparison of experimental data and CFD simulations results.

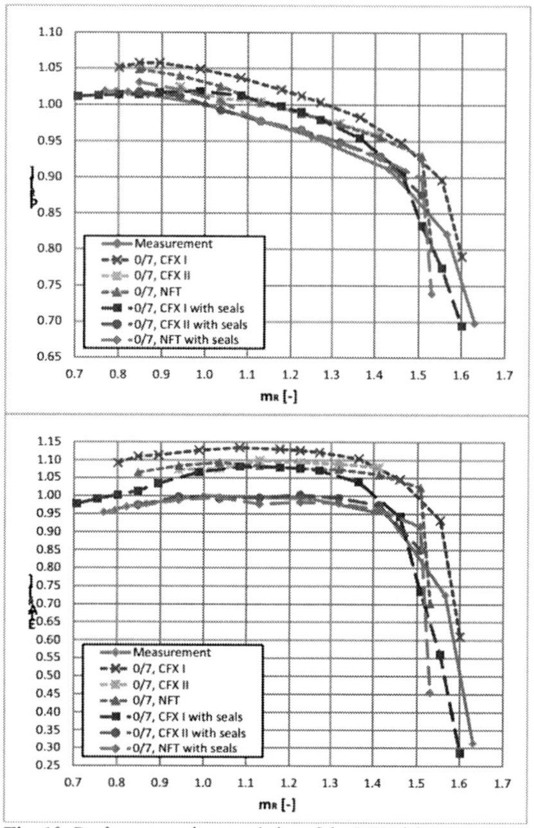


**Fig. 10.** Performance characteristics of the RTK 06 stage

The methodology for the centrifugal compressor stages CFD simulations was verified in the NUMECA FINE/Turbo and ANSYS/CFX CFD software. Based on the verified numerical simulations, it is possible to additionally enhance experimental measurement and also improve the quality of the development process, because CFD results cover complete quantities distribution in the stage. Beside this, it also enables to simulate various modifications of stage quickly and to assess them without the need of excessive testing.

The values from models without leakage are overpredicted. For 2D impellers are necessary to use correct coefficients for using the results in the development process. The benefit of these models is relatively rapid generating of the computational meshes and low computational costs.

The presented analyses of the 2D stages and the verifications of the numerical models are the first steps for further works. In the case of the stage RTK04 it is the study, which has been focused on the influence of the technological holes for welding in various positions on the rotor blades to the stage characteristics. The stage RTK06 served to the experimental and numerical modelling of the leakage clearances and influences of their values on the stage characteristics. Results of both mentioned studies will be published subsequently.

The results were developed within the CENTEM PLUS (LO1402) by financial means from the Ministry of Education, Youth and Sports under the "National Sustainability Programme I." and in the framework of the FR-TI3/421 project (Ministry of Industry and Trade of the Czech Republic) and specific research. The authors would like to thank the Howden ČKD Compressors Ltd. for permission to publish this paper and all colleagues who participated on this work.

## References

1. M. P. Boyce, "Principles of Operation and Performance Estimation of Centrifugal Compressors," Proceedings of the twenty-second turbomachinery symposium, Dallas, TX. (1993)
2. Numerical and Experimental Modelling of the Radial Compressor Stage Syka, Tomáš and Matas, Richard and Luňáček, Ondřej, AIP Conference Proceedings, **1745, 020059** (2016).
3. J. Paroubek and J. Kynčl, "The Influence of Impeller Flow Channel on Aerodynamic Performance of a Centrifugal Compressor Stage," ASME Paper **98-GT-40** (1998).
4. J. Paroubek and V. Cyrus and J. Kynčl, „The Influence of Outlet Stator Part Surface Roughness on the Performance of a Very Low Flow Coefficient Centrifugal Compressor Stage", ASME Paper **96-OT-179** (1996)
5. P. Le Sausse and P. Fabrie and D. Arnou and F. Clunet, "CFD comparison with centrifugal compressor measurements on a wide operating range," EPJ Web of Conferences, Volume **45** (2013)
6. T. Syka, "Geometry structure influence on the radial compressor stage efficiency," Ph.D. thesis, in Czech, ZČU in Pilsen (2015)

EPJ Web of Conferences **143**, 02074 (2017) DOI: 10.1051/epjconf/201714302074
*EFM 2016*

# Wind tunnel measurement of flow and dispersion over urban area

*Petr* Michálek[1,*] and *David* Zacho[1]

[1]Department of Aerodynamics, Aerospace Research and Test Establishment (VZLU), Beranových 130, Praha 9, Czech Republic

**Abstract.** Measurements of flow and dispersion over urban area model were conducted in the boundary layer wind tunnel in VZLU. Scale model of an industrial area with its residential neighborhood was equipped with ground-level emission source and the flow and dispersion of emissions was measured. Flow field was studied using hot wire anemometer and concentration field was studied using flame ionization detector. These measurements were used for the verification of a new computational model of emission flow and dispersion, which was developed in VZLU.

## 1 Introduction

There are many industrial areas near historical centers of cities or residential areas in the Czech Republic. This fact presents a possible threat to the lives and properties of people living in these areas by escape of dangerous gases.

A new computational model for flow and gas dispersion in the atmospheric boundary layer was developed in VZLU. The computational model is based on numerical simulation of compressible gas flow in the gravitational field and finite volume method of Reynolds-averaged Navier-Stokes equations. The software uses pre-calculated results of flow and dispersion in chosen areas near chemical plants in Czech Republic together with actual meteorological data to estimate the most dangerous area. The estimate will be used by rescue services and firemen to make the rescue action more effective or by the local authorities for preparation of emergency evacuation plans.

This field of research was studied e.g. by Hall et al. [1], Lawton and Robins [2], Yassin et al. [3]. Previous dispersion studies in VZLU dealt with simple dispersion cases, i.e. change in roughness [4], rectangular building [5], terrain wave and slope [6]. Large terrain model in scale 1:1500 was used in [7]. Details about the computational model can be found in [8, 9].

## 2 The experiment

The experimental work was done in the boundary layer wind tunnel (BLWT) in VZLU, a. s. The BLWT is an open-circuit wind tunnel with cross section 1.8 m wide and 1.5 m high with adjustable ceiling. The test section for boundary layer development is 13.6 m long. The tunnel is powered by 55 kW fan. Maximal velocity above the boundary layer is ca. 24 m $s^{-1}$. Reference velocity above the boundary layer is measured with Pitot-static probe and hot-wire probe. The model space is equipped with 1.75 m dia. turntable and traversing device for moving the probe in all three directions x, y, z. The simulation means can create agricultural, suburban or urban boundary layer according to Eurocode 1 [10]. The boundary layer developed over the entire test section with plastic sheet with 7 mm high truncated cones and a rectangular barrier 140 mm high, which served as turbulence generator. This boundary layer corresponds with terrain type III (suburban).

### 2.1 The urban area model

The model of the area of interest was manufactured from lightweight boards on the basis of a digital terrain model of the Liberec city. The model was made in scale 1:300 and presents an area of ca. 870 m × 500 m in full scale. The building models were simplified into blocks and cut out of synthetic wood. The model is shown in Figure 1, where wind direction is from upper left to bottom right corner. The modeled area followed the valley of Lužická Nisa river in southeast-northwest direction. The emission source is a small chamber 25 × 25 × 10 mm covered with filtration fabric on its top and built in the model. The source presents a ground level point or small area source of gas emissions. Two possible locations of the source were studied, i.e. behind the chimney and behind the gas tanks. This paper presents the results of the case where the emission source was located near the gas tanks. The map of the modeled area with marked position of the sources is presented in Figure 2.

* Corresponding author: michalek@vzlu.cz

**Fig. 1**. The model in wind tunnel

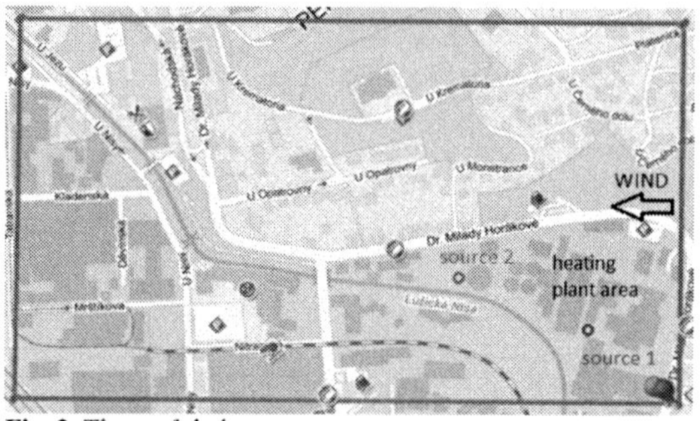


**Fig. 2**. The modeled area

## 2.2 Flow and dispersion measurements

The flow field of the incoming boundary layer was measured with dual-sensor hot-wire probe and the concentration field was measured with flame ionization detector. The vertical mean velocity profile is shown in Figure 3, where U (m $s^{-1}$) is longitudinal component of velocity vector and V (m $s^{-1}$) is perpendicular velocity vector component, $I_u$ (%) is turbulence intensity of the component U, $I_v$ (%) is turbulence intensity of the component V and z (mm) is vertical height above the ground. $U_{red}$ is reduced part of U, which was used for the logarithmic fit of the mean velocity profile.

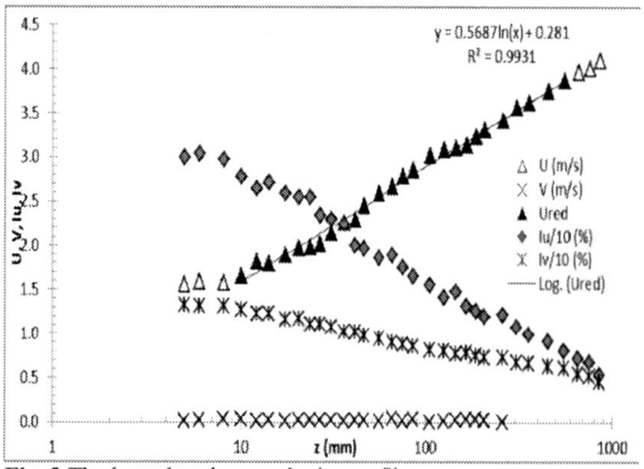


**Fig. 3** The boundary layer velocity profile

The wind tunnel velocity was set to ca. 4.3 m $s^{-1}$ above the boundary layer and this reference velocity was used in the dispersion experiments as well. For this size of the modeled area, the so-called small-scale similarity criteria applies, which means that the minimal Reynolds roughness number Re* has to be satisfied. This ensures invariance of the flow pattern around the structure on the main flow velocity. The Reynolds roughness number is defined as

$$Re^* = z_0 \times u^* / \nu = 9.3 \geq 1 \quad (1)$$

where ν is kinematic viscosity of air ($m^2$ $s^{-1}$), u* is frictional velocity (m $s^{-1}$) and $z_0$ is aerodynamic roughness (m). Values of u* = 0.23 m $s^{-1}$ and $z_0$ = 0.61 mm were derived from the logarithmic fit of the mean velocity profile.

The emission source was connected to the gas tank with ethane as tracer gas and to the compressed air source via flow controllers. For neutral (non-buoyant) emission modeling, a mixture of compressed air with volume flow 5.1 l $min^{-1}$ and ethane with flow 0.25 l $min^{-1}$ was used. Concentration field was measured using suction tube mounted on a traversing device, necessary tubing, peristaltic pump and a flame ionization detector (FID). FID calibration was made with calibration gas mixture with precise concentration 100 ppm (parts per million) of ethane in air. Sampling time was set to 60 s for one point, and then the probe moved to next point and waited 30 s to clean the tubing from previous sample. The response of this system is up to 10 Hz that is why only mean concentration values were evaluated. FID signals were recorded in the computer using analog to digital input card and LabView software.

Regarding the terrain configuration, which includes high slope in its northern part, measurements were made in available area in the nearest neighborhood of the heating plant and the river valley. The local flow velocity was studied in horizontal and vertical profiles. Concentration fields were studied in a network of points at heights 10, 25, 50 … 200 mm above the source level. The practical realization of the experiment was quite difficult, because all model buildings and terrain had to be avoided; therefore some areas of the model were not measured.

# 3 Results and discussion

The resulting image from measurements within a network of points processed as contour graph at the lowest height 10 mm (3 m in full scale) above the source level is shown in Figure 4. The coordinates xFS (m) and yFS (m) are in full scale values. The source was at the right tip of the measured area and is marked with black square. Crosses indicate measuring points and the main flow direction is from right to left. This height was the lowest height above the model surface, where measuring was possible in practice. The plume extension behind the source can be observed. Figure 5 shows concentration field at height 25 mm (7.5 m in full scale). The maximal concentration is about half of the one in Figure 4. Figure 6 presents the concentration field at height 50 mm (15 m in full scale) and maximal concentration value almost one third of the maximal value in Figure 5. One can see that the concentration is rapidly decreasing with height. The concentration field from Figure 4 (lowest level) was placed over a satellite image, so that the extent of the plume can be evaluated in real environment, as shown in Figure 7.

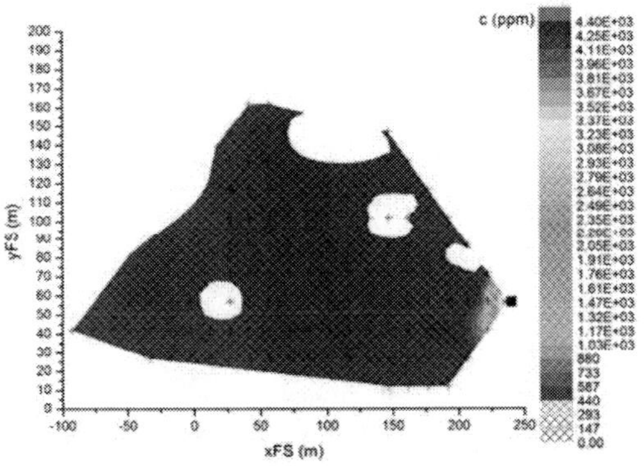


**Fig. 4.** Concentration field at z = 10 mm

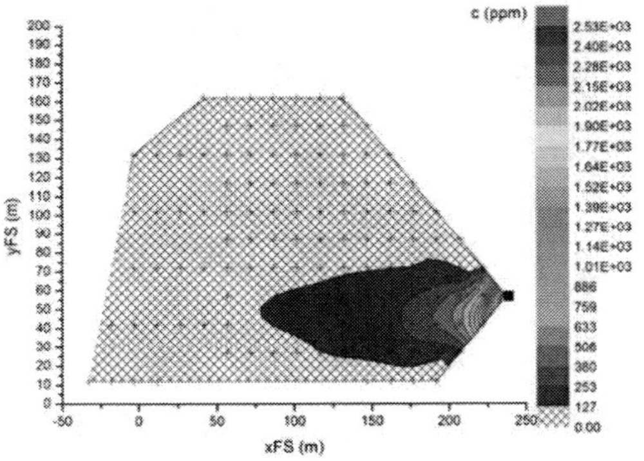


**Fig. 5.** Concentration field at z = 25 mm

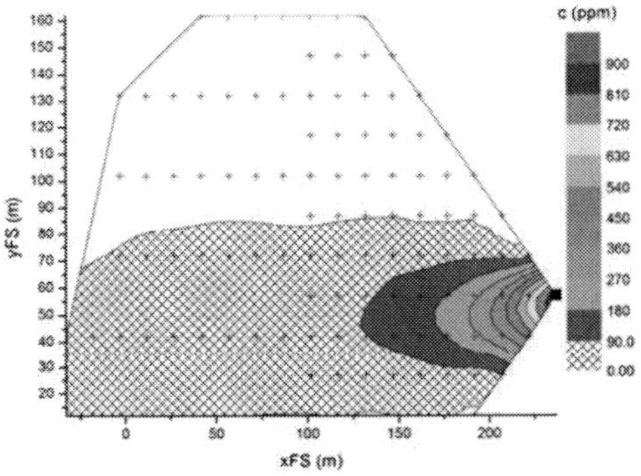


**Fig. 6.** Concentration field at z = 50 mm

## 4 Conclusion

Concentration fields over a detailed "city scale" model of urban area in Liberec were measured in the BLWT in VZLU. Mean concentration and velocity profiles and fields were measured with the use of tracer gas emission, flame ionization detectors and hot-wire probes. The extent of the plume from simulated gas leakage from a heating plant in Liberec was determined horizontally and vertically. The mean concentration fields were measured in order to use them for verification and validation of the numerical dispersion model.

Authors thank the support from the Ministry of the Interior of Czech Republic, project VG20122015098.

## References

1. D.J. Hall, V. Kukadia, S. Walker, P. Tilz, G.W. Marsland, *Proceedings of Physmod 2007* (2007)
2. T. Lawton, A. Robins, *Proceedings of Physmod 2007* (2007)
3. M.F. Yassin, M. Ohba, H. Tanaka, *Proceedings of 14th joint conference on application of air pollution meteorology with the air and waste management* Assoc, Atlanta, Georgia, USA, 1-9, (2006)
4. P. Michálek, D. Zacho, *Air Pollution XX* (2012)
5. P. Michálek and D. Zacho, *Int. J. Sust. Dev. Plan.* **10**, p. 203-206 (2015)
6. P. Michálek, D. Zacho, *Experimental Fluid Mechanics 2014*, p. 366 (2014)
7. P. Michálek, D. Zacho, *Engineering Mechanics 2015, p.85* (2015)
8. M. Kyncl, Numerical solution of the three-dimensional compressible flow. *Ph.D. thesis*, MFF UK, Prague (2011)
9. M. Kyncl, J. Pelant, *Technical report R-5453*, VZLU, Prague (2012)
10. Eurocode 1: Actions on structures – Part 1-4: General actions – Wind loads (EN 1991-1-4), *European Committee for Standardisation*, Brussels (2005)

DOI: 10.1051/epjconf/201714302074

**Fig. 7.** Concentration field compared with photomap

DOI: 10.1051/epjconf/201714302075

# Comparison of two surface temperature measurement using thermocouples and infrared camera

*Dariusz* Michalski*, *Kinga* Strąk, *Magdalena* Piasecka

Faculty of Mechatronics and Mechanical Engineering
Kielce University of Technology, Al. 1000-lecia P.P. 7, 25-314 Kielce, Poland

**Abstract.** This paper compares two methods applied to measure surface temperatures at an experimental setup designed to analyse flow boiling heat transfer. The temperature measurements were performed in two parallel rectangular minichannels, both 1.7 mm deep, 16 mm wide and 180 mm long. The heating element for the fluid flowing in each minichannel was a thin foil made of Haynes-230. The two measurement methods employed to determine the surface temperature of the foil were: the contact method, which involved mounting thermocouples at several points in one minichannel, and the contactless method to study the other minichannel, where the results were provided with an infrared camera. Calculations were necessary to compare the temperature results. Two sets of measurement data obtained for different values of the heat flux were analysed using the basic statistical methods, the method error and the method accuracy. The experimental error and the method accuracy were taken into account. The comparative analysis showed that although the values and distributions of the surface temperatures obtained with the two methods were similar but both methods had certain limitations.

## 1 Introduction

Instruments used to measure temperature can be divided into contact and contactless, according to the way the heat is transferred from the body or medium to be measured to the thermometer.

This paper compares the results of temperature measurement realized by contact method based on thermocouples and obtained from infrared thermography, which is an example of a contactless method. In their previous research, the authors performed a comparative analysis for two temperature contactless methods: infrared thermography and liquid crystal thermography [1]. Liquid crystal thermography was used in a study of the FC-72 flow boiling heat transfer in rectangular minichannels with enhanced heated surfaces [2, 3].

The thermocouple-based contact method is commonly used for measuring the surface temperature of liquids and gases. In 1821, T. J. Seebeck discovered that it was possible to create an electrical voltage by soldering two different metals (the Seebeck effect). When two metallic elements having different Seebeck coefficients are joined together at one end, a voltage called electromotive force (EMF) can be detected using a voltmeter at the other end [4]. Calibration of thermocouples is very important. The temperature calibration uncertainty can be evaluated, for instance, using the conventional method described in the Guide to the expression of Uncertainty in Measurement (GUM) or the numerical Monte Carlo method [5]. The numerical method is generally more flexible than the conventional method. Characteristic temperatures of type E thermocouples and a simple approximation of the relationship between thermal electromotive force (EMF) and temperature $T$ were discussed in [6]. The characteristic temperatures of the thermocouples were approximated using quadratic polynomials assuming an accuracy of ±0.5 K over the temperature range from 0 to 1000 °C. Genix et al. [4] describe results obtained with an intrinsic thermocouple used to perform ultra-local temperature measurements. The intrinsic thermocouple, which includes a conducting tip placed in contact with a conducting sample, is designed to measure rapid changes in the surface temperature. A fast-response thermocouple is used to measure, for example, the surface temperature of internal combustion engines, as described by Marr et al. [7]. Heichal et al. [8] discuss the use of thin film sensors to detect a 15 ns laser pulse.

In the paper by Kus et al. [9] the tool-chip interface temperature was estimated by simultaneously employing a type K thermocouple and an infrared radiation pyrometer. The finite element method was used to compare the temperature results; the experimental results showed good agreement.

Infrared thermography is a powerful technique that can be used to quantify the thermal field on a measured surface. The characteristic features of the IR camera include quick response time and high sensitivity. The fully two-dimensional IR camera offers colour images and is able to identify surface temperature fields. Errors are easy to detect and correct because of tangential

* Corresponding author: *michalski@tu.kielce.pl*

DOI: 10.1051/epjconf/201714302075

distortions. The IR camera is well-suited to study fluid flows that are complex in nature. It can be applied to analyse convective heat fluxes on the heating element or changes in the flow field [10]. Infrared thermography has many applications. Mikielewicz et al. [11] discuss its use to analyse the increase in the dryout heat flux for four fluids flowing through two small-diameter silver tubes. The method has also been found suitable to study the thermo-hydrodynamics of the initial phase of the hydrodynamically fully developed but thermally developing internal laminar flow of water in a square cross-sectional minichannel. Moreover, infrared thermography can be used for high-resolution measurement of the local temperature on the heated wall for one-dimensional conduction in the transverse direction as well as for the determination of the heat transfer coefficient [12].

## 2 Experimental stand

The major parts of the experimental setup presented in figure 1 are: a main loop, a supply and control system and a data and image acquisition system. The flow loop comprised a measurement module (1), an infrared camera (2), a DaqLab 2005 data acquisition station (3), a MCC data acquisition station (SC-1608G) (4) and a PC computer (5). The measurement module with two parallel vertically oriented minichannels (IRT minichannel and TC minichannel), each 1.7 mm deep, 16 mm wide and 180 mm long, was the most important element of the experimental setup.

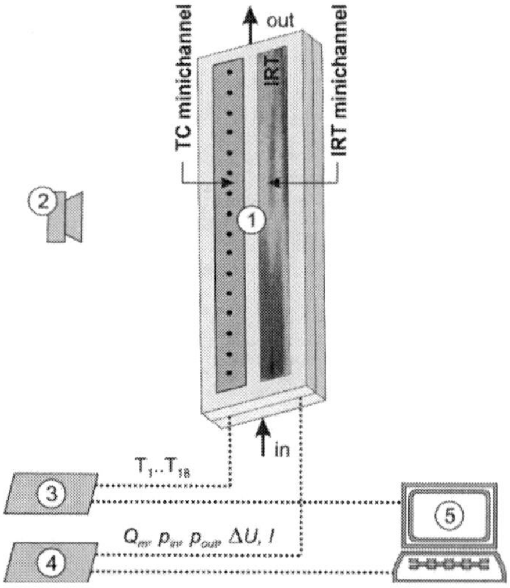


**Fig. 1.** Schematic diagram of the data acquisition system: 1 - measurement module, 2 - infrared camera, 3 - DaqLab 2005 data acquisition station, 4 - MCC data acquisition station (SC-1608G), 5 - PC computer.

In both minichannels, the heating element was a 0.1 mm thick foil made of rolled acid-proof Haynes-230 alloy with the surface roughness $Ra$ = 0.121 µm. The working fluid flowing in minichannels was distilled water. In one minichannel (IRT minichannel), the temperature on the outer side of the foil was measured using infrared thermography. The measurements were taken in the central axially symmetric part of the minichannel (4 mm x 180 mm). The surface was coated with a black paint with an emissivity of 0.83 [2]. Narrow micanite panels were placed on both sides of the foil to prevent its deformation. In the other minichannel (TC minichannel), the temperature was monitored by means of thermocouples attached at 16 points on the outer surface of the foil along the central, axially symmetric axis of the minichannel. Two thermocouples were mounted in the minichannel: one at the inlet and the other at the outlet. The signals from the thermocouples were recorded by the data acquisition station.

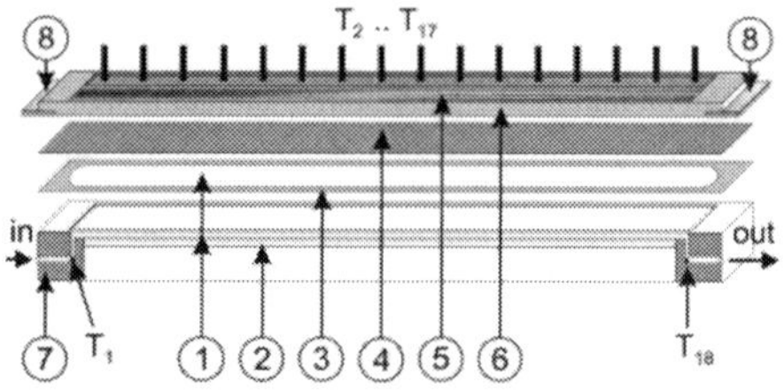


**Fig. 2.** Schematic diagrams of a measurement module; 1 - minichannel, 2 - glass pane, 3 - PTFE spacer, 4 - heated foil, 5 - IRT thermogram, 6 - front cover, 7 - channel body, 8 - copper electrode.

## 3 Two methods of surface temperature measurement

### 3.1 Contact method using thermocouples

Thermocouples are temperature sensors that respond to a change in the temperature with a change in the thermoelectric force. The thermocouple consists of two conductors differing in material joined together at one end. A thermoelectric effect is used to measure temperature. It consists in creating a thermoelectric force, which is a result of the difference in temperatures between two junctions: the measurement junction (the connected ends of the conductors) and the reference junction (the free ends). The reference temperature is usually fixed at 0 °C. There are different types of thermocouples, each with unique characteristics. The most common are T, E, J, K, N, S, R and B. The types used for high temperatures are G, D and C. Each type has a specific temperature measurement capability so that the results can be represented linearly in the desired measurement range [13].

The thermocouples used in the experiments were manufactured by Czaki Thermo-Product, Poland (type TP-221) with $T_{max}$ = +600 °C. They have a NiCr-NiAl type K sheathed thermocouple sensor with an outer diameter of 0.5 mm. The measurement junction is galvanically isolated from the sheath (type B).

The NiCr-NiAl type K thermocouple is oxidation-proof resistant to higher temperatures, a reducing atmosphere and sulphur compounds. The type K thermocouple can operate at very high temperatures (up to 1000 °C or even 1100 °C); it is more resistant to high temperatures than other thermocouples made of non-noble metals [13].

### 3.2 Contactless method by Infrared Thermography

Infrared thermography is a method that involves the detection of infrared radiation emitted by an object with a temperature higher than absolute zero. Thermal radiation occurs in the region between visible light and microwaves [14].

One of the important parameters in the IR measurement is the emissivity of the object measured (0÷1). Emissivity can be defined as the ratio of the radiation emitted by a physical body to the radiation emitted by a black body at the same temperature [15, 16].

Infrared cameras produce thermograms, which provide qualitative and quantitative data about the thermal conditions and thermodynamic processes taking place in the observation area. An IR camera can be used to record static conditions or monitor dynamic processes taking place under laboratory conditions. Correct interpretation of thermograms requires knowledge of the properties of the material analysed and the effect of the surroundings. A good knowledge of the measurement technique is also useful as in practice it is often necessary to identify and eliminate the sources of interference [16, 17].

## 4 Results

The main aim of this study was to compare the results of the temperature measurements performed simultaneously in two parallel minichannels using two different methods: the contact method (thermocouples - TC) and the contactless method (infrared thermography - IRT).

Piasecka et al. [1] describe a similar analysis, but it was made for two contactless surface temperature measurement methods: infrared thermography and liquid crystal thermography. It was found that the values and distributions of the temperatures obtained with both methods were similar. However, the IR camera used in that study was different from the one employed in this research.

The experiments described in this paper were conducted under constant laboratory conditions with an ambient temperature of 300 K, a relative humidity of 40% and an atmospheric pressure of 0.095 MPa. In the IR measurements, the distance between the IR camera and the foil surface was 0.4 m and the foil was coated with a black paint with an emissivity of 0.83.

### 4.1 Uncertainty and limitations of thermocouples and IRT

The type K thermocouples applied to measure the temperature of the foil surface in one minichannel were 0.5 mm in diameter. They were capable of measuring temperatures in the range from -40 °C to 600 °C. In this study, the measurement accuracy assumed for the type K thermocouples was about 1.5 °C in the range from -40 °C to 375 °C [13, 18].

The thermocouple-based method has some limitations. One of the major problems is attaching a thermocouple to the surface to be measured. Incorrect bonding may cause the device to show a temperature lower than the actual one. When a thermocouple is used to measure the temperature of an element in a high frequency radio circuit, it may change the electrical characteristics of that circuit.

A thermocouple fixed to a small electronic element may be responsible for the removal of heat from the element. As a result, the temperature displayed is lower than the actual temperature, i.e. one measured before the thermocouple was attached. Decalibration may lead to an error in measurement; it may change the electrical characteristics, which will make the reading inaccurate. Finally, an open thermocouple junction may also cause errors in temperature measurement [19].

The images taken with a FLIR E60 infrared camera system operating in the spectral range from 7.5 μm to 13 μm were recorded in a digital system with an image frequency of 60 Hz [20]. As specified by the producer, the FLIR E60 camera has a thermal accuracy of ± 2°C or ± 2%. In an additional certificate, however, we read that a higher accuracy of ± 1 °C or ± 1% can be achieved in the temperature range from 0 to 120°C [21].

In the IRT measurements, the effect of the external factors on the measurement accuracy was substantial. It was found that the measurement uncertainty was dependent on several parameters. The most important were the emissivity and moisture of the object, its distance from the camera, the angle of observation, the air supply and, finally, the effect of the sources of interference such as the heat produced by the heating elements or by the lamps lighting the surface.

## 5 Comparative statistical analyses of temperature data

The temperature data obtained for the contact (TC) method and the contactless (IRT) method were compared statistically. Two settings of the increasing heat flux supplied to the foil and three distances from the minichannel inlet were considered. The analysis was performed according to the following procedure:

- graphical comparison of the measurement results concerning the foil temperature (figures 3 and 4);
- determination of the experimental error for the temperature measurement method according to the principles of statistical inference (taking account of the foil temperature) followed by the estimation of the mean value of the experimental error, the estimation of the mean deviations of the experimental error [22, 23] and the estimation of the confidence interval for a single experimental error and the measurement accuracy [24].

### 5.1 Graphical Comparison

The values of the heated foil temperature measured along the central line in the axially symmetric part of the TC and IRT minichannels were compared graphically. The data obtained by means of thermocouples and infrared thermography for two settings of the increasing heat flux supplied to the foil (denoted as #1 and #2) and three distances from the minichannel inlet are shown in figure 3. Figure 4 illustrates differences in the surface temperature of the heated foil $\Delta T$ between the TC and IRT measurements against the time for the selected distances from the minichannels inlet.

### 5.2 Basic statistical parameters

Table 1 provides all the basic statistical parameters used in the comparative statistical analysis of the temperature results obtained from the contact TC and the contactless IRT methods.

**Table 1.** Basic statistical parameters

| Basic parameters for #1 | Distances from the minichannels inlet | | |
|---|---|---|---|
| | **0.04 m** | **0.09 m** | **0.14 m** |
| *MSE* | 18.45 | 15.01 | 7.97 |
| *MRSE* | 0.53 | 0.52 | 0.34 |
| *MRAE* | 0.63 | 0.58 | 0.42 |
| **Basic parameters for #2** | **Distances from the minichannels inlet** | | |
| | **0.04 m** | **0.09 m** | **0.14 m** |
| *MSE* | 11.96 | 19.25 | 8.87 |
| *MRSE* | 0.42 | 0.59 | 0.37 |
| *MRAE* | 0.55 | 0.73 | 0.45 |

where: *MSE* –Mean Squared Error, *MRSE* – Mean Relative Squared Error, *MRAE* – Mean Relative Absolute Error.

Two settings of the heat flux supplied to the foil (#1, #2) and three distances from the minichannel inlet were considered. The statistical parameters under analysis were: the mean squared error *(MSE)*, the mean relative squared error *(MRSE)*, and the mean relative absolute error *(MRAE)* [25].

### 5.3 Relative experimental error

The investigations included determining the measurement accuracy of both methods. As IRT was assumed to be the reference method, the experimental results obtained with the thermocouples were compared to those recorded with the infrared camera [23, 24].

The relative experimental error *(EME)* was calculated using the following relationship [22]:

$$EME = \frac{T_{TC} - T_{IRT}}{T_{IRT}}, \qquad (1)$$

where: $T_{TC}$ - foil temperature obtained with the thermocouples (the test method), $T_{IRT}$ - foil temperature measured by IRT (the reference method).

The IRT method was selected to be the reference method for the measurement of the foil temperature because the measurement error obtained for IRT was lower [16] than that reported for TC [13, 18]. The relative error of the TC (test) method is shown below.

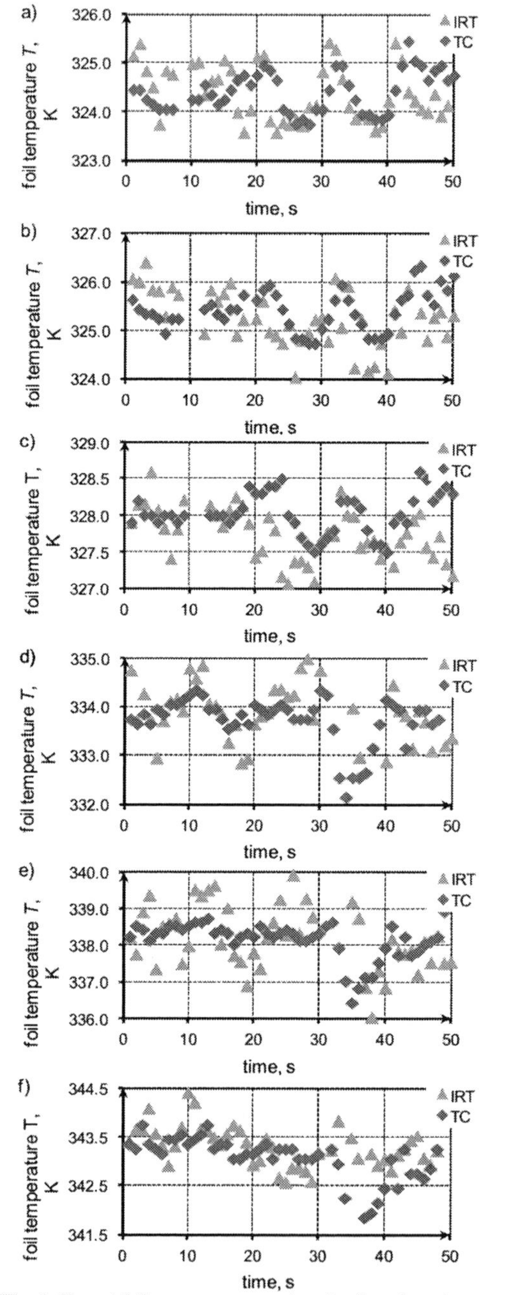


**Fig. 3.** Heated foil temperature vs. time for the selected distances from the minichannels inlet: a, d) 0.04 m, b, e) 0.09 m and c, f) 0.14 m; data obtained from IRT (green points) and TC (blue points); settings of the heat flux: a-c) #1, d-f) #2.

 DOI: 10.1051/epjconf/201714302075

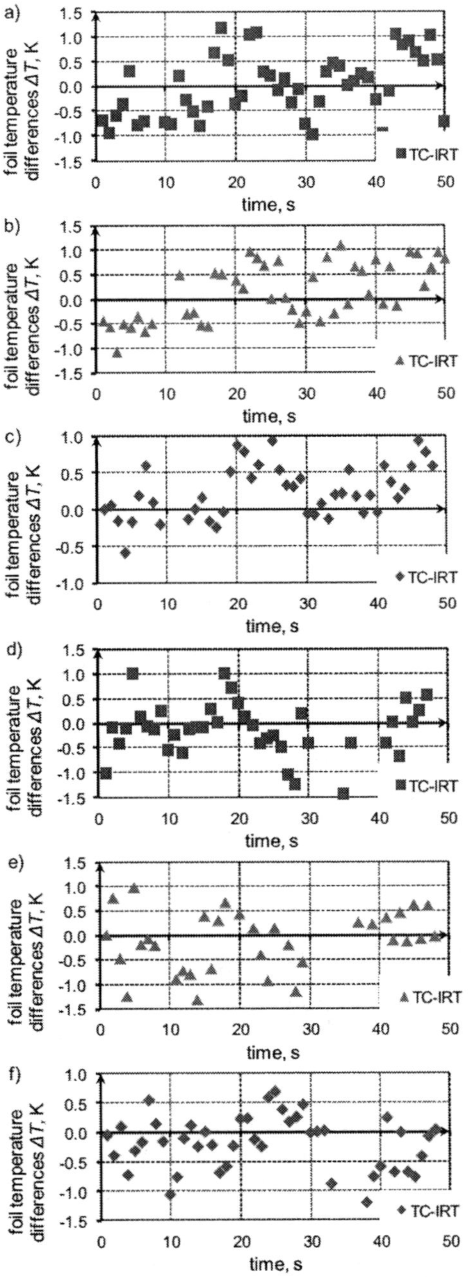


**Fig. 4.** Differences in the surface temperature of the heated foil between the TC and IRT measurements vs. time for the selected distances from the minichannels inlet: a, d) 0.04 m (red points), b, e) 0.09 m (green points) and c, f) 0.14 m (blue points); data obtained from IR (green points) and TC (blue points); settings of the heat flux: a-c) #1, d-f) #2.

The mean value of the experimental error ($\overline{EME}$) and the confidence interval *(CI)* for the mean value of the experimental error were provided in [1]. In the analysed population, $\overline{EME}$ was estimated using the following procedure [24]: determine the experimental errors for the TC and IRT methods, calculate the mean values of the experimental errors and determine the intervals of confidence for the mean values of the experimental errors with normal distribution.

### 5.4. Accuracy of the TC and IRT methods

The values of the experimental errors were used to estimate the accuracy of the methods. The calculations were based on the following relationship:

$$MA = \left|\overline{EME} \pm u_p \cdot s\right|_{\max} \cdot 100\%, \qquad (2)$$

where: $\overline{EME}$ – mean value of the relative experimental error, $u_p$ – the expansion coefficient for the level of significance $\alpha = 0.05$, $u_p = 2$ [26], and $s$ –mean deviation of the experimental error.

The method accuracy (*MA)* determined with Eq. (2) [22] was used to qualitatively assess the measurement accuracy of the Czaki Thermo-Product TP-221 type K thermocouples employed in the tests.

Table 2 shows results of the statistical analysis including the relative experimental error obtained by comparing the TC and IRT methods.

**Table 2.**Method accuracy and basic statistical parameters of the experimental error

| **Basic parameters for #1** | **Distances from the minichannels inlet** | | |
|---|---|---|---|
| | **0.04 m** | **0.09 m** | **0.14 m** |
| Sample size *n* | 48 | 45 | 47 |
| *Min EME* | -0.0030 | -0.0033 | -0.0018 |
| *Max EME* | 0.0036 | 0.0034 | 0.0041 |
| *Median* | -0.00011 | 0.00028 | 0.00058 |
| $\overline{EME}$ | -0.000020 | 0.00047 | 0.00089 |
| *s* | 0.0019 | 0.0018 | 0.0013 |
| *CI* | 0.00002 ± 0.0038 | 0.00047 ± 0.0036 | 0.00089 ± 0.0026 |
| *MA* | 0.39% | 0.41% | 0.34% |
| **Basic parameters for #2** | **Distances from the minichannels inlet** | | |
| | **0.04 m** | **0.09 m** | **0.14 m** |
| Sample size *n* | 40 | 38 | 43 |
| *Min EME* | -0.0043 | -0.0057 | -0.0035 |
| *Max EME* | 0.0030 | 0.0029 | 0.0020 |
| $\overline{EME}$ | -0.00048 | -0.00071 | -0.00057 |
| Median | -0.00031 | -0.00033 | -0.00042 |
| *s* | 0.0017 | 0.0022 | 0.0013 |
| *CI* | -0.00048 ± 0.0034 | -0.00071 ± 0.0044 | -0.00057 ±0.0026 |
| *MA* | 0.38% | 0.50% | 0.33% |

where*: EME* – experimental error, $\overline{EME}$ – mean values of the relative experimental error, *s* – mean deviation, *CI* – confidence interval, *MA* – method accuracy.

From the method accuracy *(MA)* it is evident that the two methods used for the measurement of the foil temperature are similar. In certain ranges, they can be used interchangeably. Under laboratory conditions, the maximum accuracy of the measurement system can reach 15% [22, 23]. In the case considered, the maximum value of *MA* was 0.5%, which is very low when compared with the literature data [22, 23].

### 5.5 Results and discussion

Comparing the time-temperature curves in figures 3 a-c obtained for heat flux #1, we can see that the foil temperature measured with thermocouples is similar to that obtained with IRT. However, the curve created for the TC method is smoother. Since IRT is more sensitive to changes in the foil temperature, the scatter of results was greater. The curves obtained for TC and IRT are similar in shape.

Comparing the time-temperature curves in figures 3 d-f, obtained for heat flux #2, we can notice that the foil temperature measured with the TC method is similar to that obtained with IRT. As thermocouples are less sensitive to changes in the foil temperature, the curves are smoother. Moreover, there are no local or global extremes as is the case of curves created on the basis of the IRT results. The differences between the TC and IRT curves are more visible for heat flux #1. The maximum differences in the surface temperature of the heated foil (figure 4) between heat fluxes #1 and #2 are less than 3 K The differences are acceptable for the experimental error obtained for the thermocouples [13, 18] and the infrared camera [21].

As can be seen from Table 1, the values of the mean squared error *(MSE)* are relatively high. However, the values of the mean relative squared error *(MRSE)* and the mean relative absolute error *(MRAE)* are rather low. The values of the median provided in this table are lower than the average in all the cases considered. The values of the standard deviation are high, which may indicate difficult experimental conditions – presence of a large number of sources of disturbance affecting the measurement results. The tests were continued after some structural modifications had been made to the experimental setup, which enabled elimination of most sources of disturbance.

## 6 Conclusions

The paper has compared two methods for measuring the surface temperature of the heated foil. The experiments were conducted using a special test stand designed to analyse fluid boiling heat transfer in parallel minichannels. Two methods were employed: contact and contactless. The surface temperature of the heated foil in one minichannel was measured with thermocouples (the contact method). The foil temperature in the other was determined using infrared thermography (the contactless method). Two sets of measurement data obtained for different values of the heat flux were analysed by applying the basic statistical method, the experimental error and the method accuracy. The comparative analysis indicated that the values of the foil temperature obtained with the two methods were similar.

## References

1. M. Piasecka, D. Michalski, K. Strąk, EPJ Web of Conf. **114**, paper 02094 (2016)
2. M. Piasecka, K. Strąk, B., Heat Transf. Eng. **38** (2017) (to be published)
3. M. Piasecka, Int. J. Refrig. **56**, 198–212. (2015)
4. M. Genix, P. Vairac, B. Cretin, Int. J. Therm. Sci. **48**, 1679–1682 (2009)
5. J. V. Pearce, P.M. Harris, J.C. Greenwood, **31**, 1517–1526 (2010)
6. V.A. Drebushchak, Thermochim. Acta. **603**, 218–226 (2015)
7. M.A. Marr, J.S. Wallace, S. Chandra, L. Pershin, J. Mostaghimi, Exp. Therm. Fluid Sci. **34**, 183–189 (2010)
8. Y. Heichal, S. Chandra, E. Bordatchev, Exp. Therm. Fluid Sci. **30**, 153–159 (2005)
9. A. Kus, Y. Isik, M. Cemal Cakir, S. Coşkun, K. Özdemir, Sensors (Switzerland) **15**, 1274–1291 (2015)
10. G.M. Carlomagno, G. Cardone, Exp. Fluids. **49**, 1187–1218 (2010)
11. D. Mikielewicz, J. Wajs, M. Gliński, A.B.R.S. Zrooga, Exp. Therm. Fluid Sci. **44**, 556–564 (2013).
12. B. Mehta, S. Khandekar, Exp. Therm. Fluid Sci. **42**, 219–229 (2012)
13. https://www.czaki.pl/en/catalogue
14. A. Rogalski, K. Chrzanowski, Optoelectron. Rev. **10**, 653–691 (2002)
15. T. Karwat, Insulations **5**, 33–36 (2008) (in Polish)
16. K. Ziętala, *Application of infrared camera in civil engineering*, Proc. 11th Eur. Conf. Young Res. Sci. Work. Transcom 2015, Žilina, 347–352 (2015)
17. M. Pešek, O. Pech, EPJ Web Conf. 67, paper 02091 (2014)
18. PN-EN 60584-1:2014-04 - *Termoelementy Część 1: Specyfikacje i tolerancje EMF*
19. B.R.J. Rogers, I.T.C. Senior, T. Course, Inframation **3**, 10–12 (2002)
20. http://www.flir-direct.com
21. Calibration certificate No. K1501035, *Calibration laboratory No. 2372, accredited by Czech accreditation Institute under ČSN EN ISO/IEC 17025:2005* for: Calibration of non-contact temperature measuring instruments
22. S. Adamczak, D. Michalski, Metrol. Meas. Syst. **11**, 221–234 (2004)
23. S. Adamczak, *Measurement of surface texture. Form profiles, waviness and roughness* (in Polish), (WNT, Warsaw, Poland, 2008)
24. M.L. Tiku, Aust. J. Stat. **22**, 260–275 (1980)
25. G. Bohm, *Introduction to Statistics and Data Analysis for Physicists* (Verlag Deutsches Elektronen-Synchrotrofn, 2010)
26. ISO 10012:2003 *Measurement management systems -Requirements for measurement processes and measuring equipment*

EPJ Web of Conferences **143**, 02076 (2017)
*EFM 2016*
DOI: 10.1051/epjconf/201714302076

# Application of a particle swarm optimization for shape optimization in hydraulic machinery

*Prokop* Moravec[1,a], *Pavel* Rudolf[1]

[1]Viktor Kaplan Dept. of Fluid Engineering, Faculty of Mechanical Engineering, Brno University of Technology, Czech Republic

**Abstract.** A study of shape optimization has become increasingly popular in academia and industry. A typical problem is to find an optimal shape, which minimizes (or maximizes) a certain cost function and satisfies given constraints. Particle Swarm Optimization (PSO) has received a lot of attention in past years and is inspired by social behaviour of some animals such as flocking behaviour of birds. This paper focuses on a possibility of a diffuser shape optimization using particle swarm optimization (PSO), which is coupled with CFD simulation. Influence of main parameters of PSO-algorithm and later diffuser shapes obtained with this method are discussed and advantages/disadvantages summarized.

## 1 Introduction

A study of size and shape optimization is very popular in industry and academia. There are a variety of methods to get the desired size/shape. These methods include nowadays well-liked particle swarm optimization (PSO) method.

This paper builds upon knowledge of an article [1] and extends the possibilities of optimizing a hydraulic diffuser, which is located behind a runner of a swirl turbine and where a recovery of kinetic energy into pressure energy takes place (Figure 1).

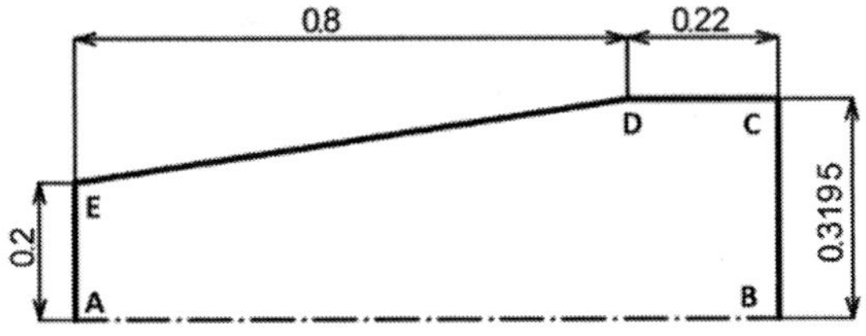


**Fig. 1.** Hydraulic turbine diffuser (dim. - metres) [1, 13]

Main goal in [1] and also in this paper is to modify the shape of the diffuser for the purpose of maximizing a coefficient of pressure recovery $c_p$ (1):

$$c_p = \frac{p_2 - p_1}{\frac{1}{2}\rho v_1^2}, \qquad (1)$$

where $p_1$ (Pa) is static pressure at inlet of the computational domain, $p_2$ (Pa) is static pressure at outlet of the domain; $\rho$ (kg/m$^3$) is density of water, $v_1$ (m/s) is bulk velocity at inlet of the domain.

Note: The value of velocity at inlet is constant and does not match with the real velocity profile behind the runner of the swirl turbine [1].

## 2 Particle swarm optimization

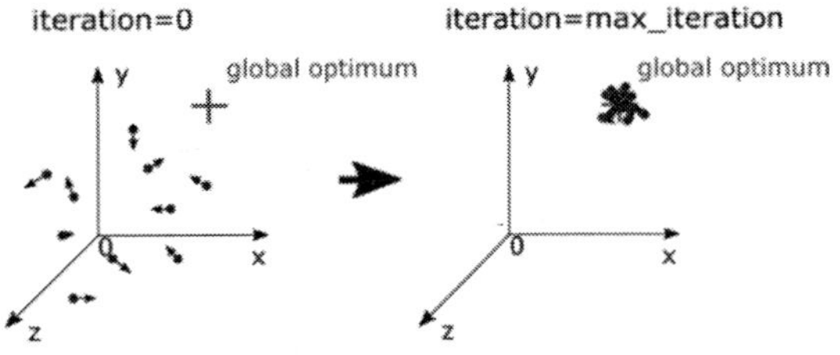


**Fig. 2.** Particle swarm optimization – concept [14]

The PSO algorithm (PSOA) [2] was introduced in 1995 by J. Kennedy and R. Eberhart and was discovered through a simulation of a simplified social model. This algorithm is based on a bird flocking, fish schooling and swarming theory in particular.

The PSO algorithm uses a swarm of particles (subjects/individuals) randomly distributed in computational area. Individual particles exchange information about their velocity (= step size), position and fitness (fitness = solution value of the examined function).

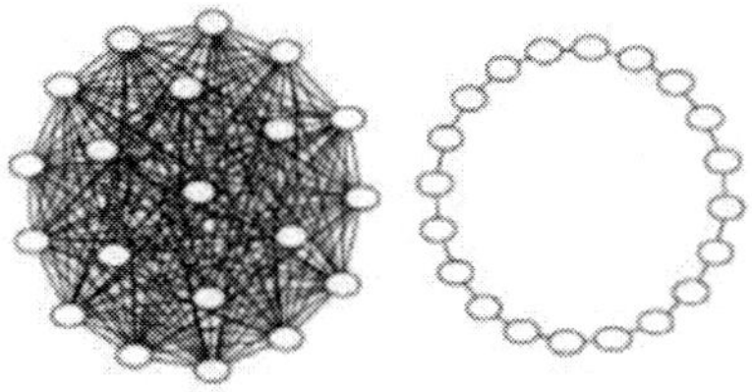

**Fig. 3.** GBEST model (left) and LBEST model [7]

Each particle in PSO algorithm is treated as a point in $n$-dimensional space. The $i$th particle is represented as

[a] Corresponding author: Prokop.Moravec@seznam.cz

$X_I = (x_{i1}, x_{i2}, \dots, x_{in})$. The best previous position of any particle is recorded and represented as $P_I = (p_{i1}, p_{i2}, \dots, p_{in})$. The index of the best particle in the swarm is represented by index *g*, and velocity for *i*th particle is $V_I = (v_{i1}, v_{i2}, \dots, v_{in})$ [3, 4].

There are two basic PSO concepts [3]: Global Best PSO (GBEST model) and Local Best PSO (LBEST model). They differ in swarm topology (Figure 3) [7].

In GBEST model (Figure 3) all particles are directly connected to the best solution (of the examined function) in the swarm, but on the other hand in LBEST model (Figure 3) each particle is influenced by a smaller number of neighbouring particles of the swarm array. Typically LBEST neighbourhoods consist of two neighbours, one on each side - ring lattice [7].

### 2.1 Global Best PSO

The particles of the swarm are moving according to the following equation [3, 4]:

$$v_{in} = v_{in} + c_1 * rand() * (p_{in} - x_{in}) + c_2 * Rand() * (p_{gn} - x_{in}), \quad (2)$$

where $c_1$ and $c_2$ are positive constants; *rand()* and *Rand()* are two random vectors from range *(0, 1)*; $p_{gn}$ is the best value of the examined function obtained by the whole swarm (GBEST).

The positions of particles are afterwards computed as follows [3, 4]:

$$x_{in} = x_{in} + v_{in}. \quad (3)$$

### 2.2 Local Best PSO

In this concept, velocity is computed similar as in Global Best PSO [3, 4]:

$$v_{in} = v_{in} + c_1 * rand() * (p_{in} - x_{in}) + c_2 * Rand() * (p_{ln} - x_{in}), \quad (4)$$

only difference is $p_{ln}$, which is the best value of the examined function in the particle neighbourhood (LBEST).

Note: Random vectors *rand()* and *Rand()* were later in this study (in MATLAB code) created by function *rand*.

## 3 Improvements and modifications of Global Best PSO algorithm (PSOA)

Some references prefer Global Best PSO (pros: fast convergence, for simple problems; cons: liable to being trapped in local minimum than Local Best PSO) [3, 9, 12]. On the other hand, some references favour Local Best PSO (pros: for complex problems; cons: slower algorithm) [3, 9, 12]. In the middle stands [9] – both of these concepts show similar performance therefore GBEST or LBEST cannot be preferred. This all leads to conclusion that the advantage of the chosen PSO mainly depends on experience of user and on the type of examined problem.

Global Best PSO algorithm is used in this paper. It can be modified in several ways for example [4, 5, 6, 7, 8]. In this study were considered only four alternations, which are mentioned in the following text.

### 3.1 Maximal velocity $v_{max}$

In just few iterations of PSO algorithm particle velocity can increase its value to the large magnitudes. This problem can be solved with velocity constraint – the maximal velocity $v_{max}$ [8]:

$$v_{in} = \begin{cases} -v_{max}, & v_{in} < -v_{max} \\ v_{max}, & v_{in} > v_{max} \end{cases} \quad (5)$$

### 3.2 Inertia weight *w*

In [4] was introduced "A modified Particle Swarm Optimizer" and the inertia weight *w* was implemented into the equations (2, 3). Equation (2) changed to:

$$v_{in} = w * v_{in} + c_1 * rand() * (p_{in} - x_{in}) + c_2 * Rand() * (p_{gn} - x_{in}). \quad (6)$$

The inertia weight *w* is used to balance the global and local search abilities in the swarm. Parameter *w* could be treated in two possible ways: constant or adaptive.

#### *3.2.1 Constant inertia weight w*

In [4] authors suggested to choose inertia weight from interval $0.9<w<1.2$ to ensure, that PSOA will have the best chance to find the global optimum (maximum of coefficient $c_p$) in moderate number of iterations.

Authors in [8] claimed that inertia weight $w=0.8$ is good "starting" choice for the algorithm, but it depends mainly on the magnitude of maximal velocity $v_{max}$.

#### *3.2.2 Adaptive inertia weight w*

The inertia weight can be also computed adaptively. Many researchers have recommended that *w* should be large in the exploration state (the beginning of the algorithm) and small in the exploitation state (the end of the algorithm). In [5] authors proposed that *w* decreases linearly:

$$w = w_{max} - (w_{max} - w_{min}) * \frac{iteration}{\max\_iteration}, \quad (7)$$

where *iteration* represents current iteration of PSOA; *max_iteration* represents maximal number of iterations of PSOA; $w_{max}$ and $w_{min}$ are maximal/minimal inertia weights and they are usually set to 0.9 for $w_{max}$ and 0.4 for $w_{min}$ [4, 5].

DOI: 10.1051/epjconf/201714302076

### 3.3 Parameters $c_1$, $c_2$

Parameter $c_1$ represents "self-cognition". It can maintain diversity of the swarm (helps particle to explore computational area) [2].

Parameter $c_2$ represents "social influence". This parameter forces the swarm to converge to the current global best solution of the examined function. It can also assist to the faster convergence [2].

#### 3.3.1 Constant $c_1$ and $c_2$

In [2] authors recommended constant values of the parameters $c_1$ and $c_2$ equal to integer 2. This setting ensures that PSOA will have the best chance to find global optimum (maximum of the coefficient $c_p$ in our case).

#### 3.3.2 Adaptive $c_1$ and $c_2$

Parameters $c_1$ and $c_2$ ($c_{1,2}$ shortly) can be treated in the same ways as inertia weight in – linearly [10]:

$$c_1 = (c_{1f} - c_{1i}) * \frac{iteration}{\max\_iteration} + c_{1i}, \quad (8)$$

and

$$c_2 = (c_{2f} - c_{2i}) * \frac{iteration}{\max\_iteration} + c_{2i}, \quad (9)$$

where *iteration* represents current iteration of PSOA; *max_iteration* represents maximal number of iterations of PSOA; $c_{1i}$ and $c_{1f}$ are maximal/minimal values of $c_1$ parameter and are usually set to 2.5 and 0.5 respectively; $c_{2i}$ and $c_{2f}$ are minimal/maximal values of $c_2$ parameter and are usually set to 0.5 and 2.5 respectively [10].

### 3.4 Specific computational area

In this study it was necessary to define computational area (the swarm territory) for the swarm due to possibility of creation of degenerated and inadmissible designs (geometries) – red rhomboid "FGHI" in Figure 4. If any particle emerged outside this area during the algorithm, it was immediately assigned to the relevant area border.

In this paper the computational area was built in the sufficient distance around the best solution (control point position) found in [13] – Figure 4.

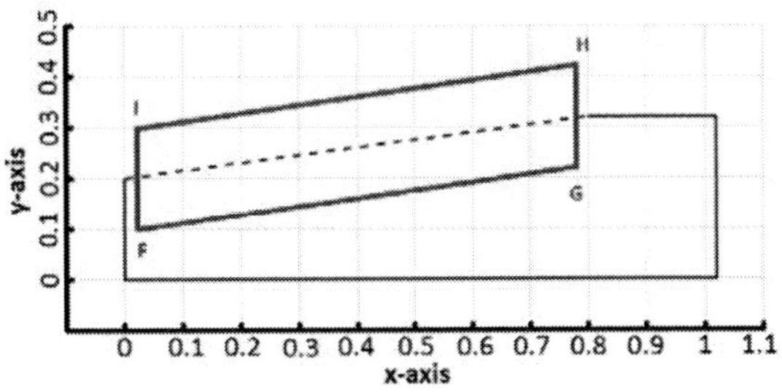


**Fig. 4.** Specific computational area

The positions of points (x-position; y-position), which characterize computational area for the swarm, are:

F = (0.02; 0.1); G = (0.78; 0.2195); H = (0.78; 0.3195); I = (0.02; 0.3).

## 4 PSOA - Overview

In MATLAB software a master code was created, which calls and handles particle swarm optimization algorithm (PSOA – also programmed in MATLAB), mesh creation in Ansys ICEM CFD and CFD simulation in Ansys FLUENT.

### 4.1 Particle swarm optimization algorithm (PSOA)

Shape optimization using PSOA consists of several main steps (Figure 5).

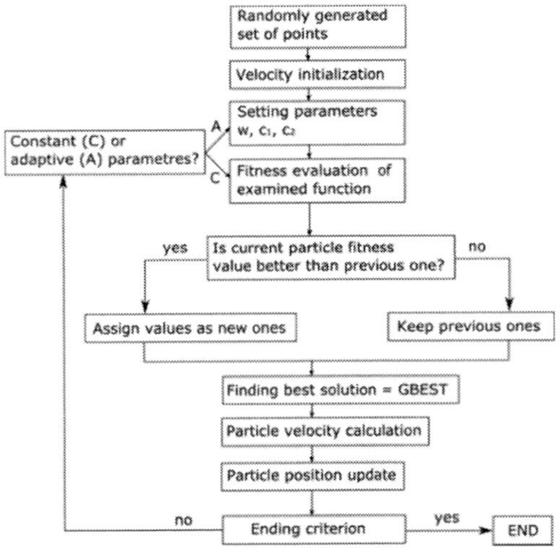


**Fig. 5.** PSOA

For this study, the ending criterion of the algorithm was set to the maximal number of iterations (max_iteration). Initial velocities (velocity initialization – Figure 5) were set as zeros [12]. Every particle of the swarm contains information about locations of control points and value of $c_p$ and looks like this: ($x_1$; $y_1$; $x_2$; $y_2$; $c_p$).

### 4.2 Computational mesh for CFD

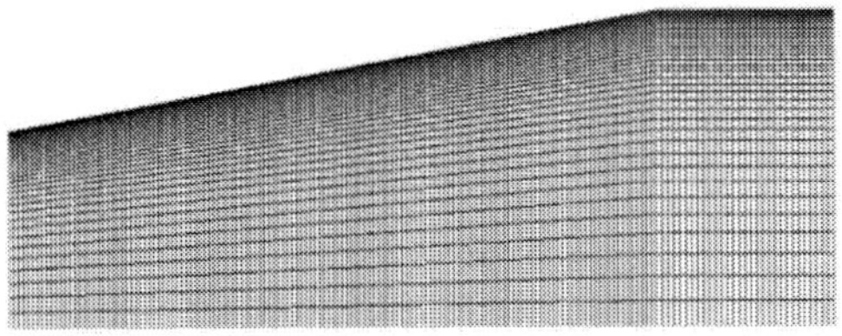

**Fig. 6.** Initial 2D domain mesh [1]

During the shape optimization the computational 2D mesh was created in Ansys ICEM CFD. The mesh creation was handled with the text script, which ICEM can easily read and be controlled with. Mesh always met necessary quality requirements - quad type of mesh, skewness, and higher mesh resolution in near wall region (Figure 6).

## 4.3 CFD

Ansys FLUENT was also controlled via text script – journal. Total number of iterations was set to 1500 to ensure the adequate computational convergence. After each simulation, FLUENT saves essential variables ($p_1$, $p_2$ - to the transcript file) for coefficient of pressure recovery $c_p$ - equation (1).

Chosen discretization schemes and turbulence model were set as follows [1]:

| | |
|---|---|
| Pressure: | Standard |
| Momentum: | Quick |
| Turbulent diss. rate: | Second order upwind |
| Turbulent kinetic energy: | Second order upwind |
| Turbulence model: | Realizable k- ε |

Table 1 displays types of boundary conditions, their locations (see Figure 1), magnitudes and turbulence intensity values.

**Table 1.** Boundary conditions [1]

| | Type | Magnitude | Turbulent intensity |
|---|---|---|---|
| A-E | Velocity inlet | 2 m/s (constant) | 5 % |
| B-C | Pressure outlet | 0 Pa | 7 % |
| A-B | Axis | - | - |
| E-D-C | Wall | - | - |

Note: The value of velocity at inlet is constant and does not match with the real velocity profile behind the runner of the swirl turbine [1].

# 5 Results

## 5.1 Evaluation procedure

Shape change was confined to section in between points **E** and **D** (Figure 1). This line could be parameterized in many different ways. This study focuses only on parameterization with Bézier curves, which have two free control points (i.e. four optimization parameters).

Several test cases of chosen parameters were made (Figure 7). Every test case was repeated 5 times to capture the basic behaviour of the swarm. The ending criterion was set to the maximal number of iterations *max_iteration=50.*

Two main attributes of the swarm were examined: the magnitudes of $c_p$ of the GBEST particle (Figure 8, 10,..., 30) and the positions of the control points of the GBEST particle (Figures 9, 11,...,31).

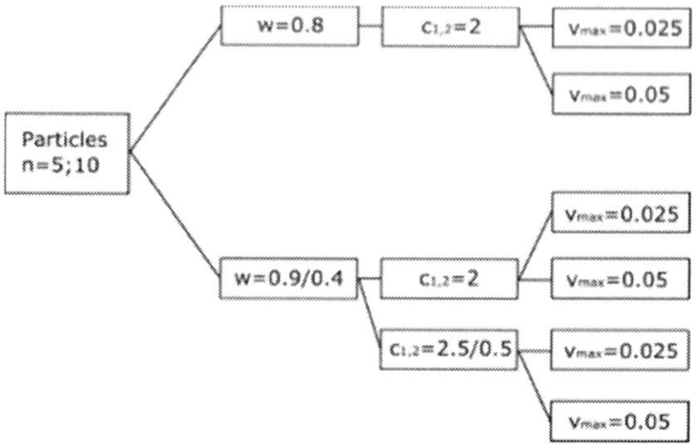


**Figure 7.** Selections of the main parameters

## 5.2 Test cases

Initial shape

The value of coefficient of pressure recovery $c_p$ for initial shape (Figure 1) was **0.736824** [13].

### *5.2.1 Swarm of 5 particles*

Five particles in the swarm were chosen as the smallest reasonable number for the first six test cases. Average computational time for this choice was 28107 *(s)*.

Free control points of Bézier curve were randomly generated in computational area at every beginning of PSOA (Figure 5). Every particle consists of two control points - one control point is represented as "**x**" and the other one as "+" (Figure 8).

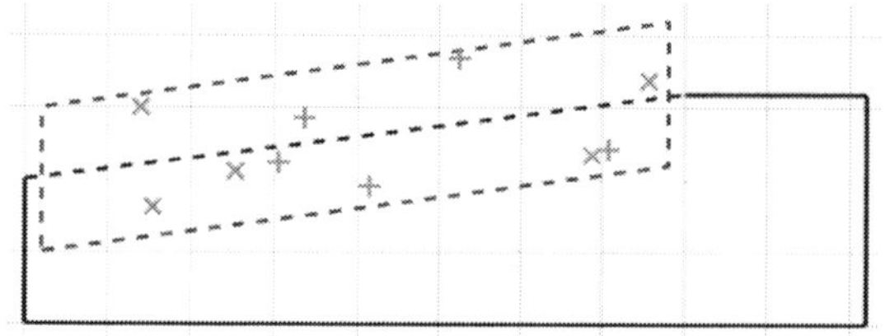

**Fig. 8.** Random sets of control points – 5 particles

Note: Randomness of point generation was ensured in MATLAB by function *rand. Rand* returns random number from interval *(0, 1)*.

Constant parameters $w$, $c_{1,2}$

Figures 9 - 12 capture the examined attributes of the swarm of five particles with constant parameters $w$ and $c_{1,2}$. The only difference between these two test cases was in maximal velocity $v_{max}$=0.025 ($v_{max}$=0.05).

*a) Maximal velocity $v_{max}$=0.025*

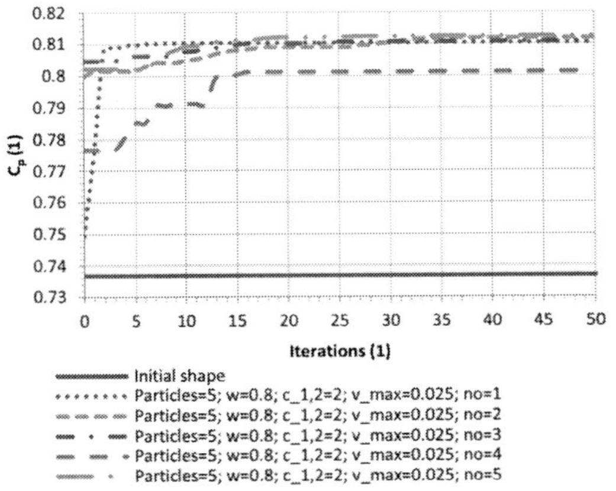


**Fig. 9.** Magnitudes of $c_p$: Particles=5; w=0.8, $c_{1,2}$=2; $v_{max}$=0.025

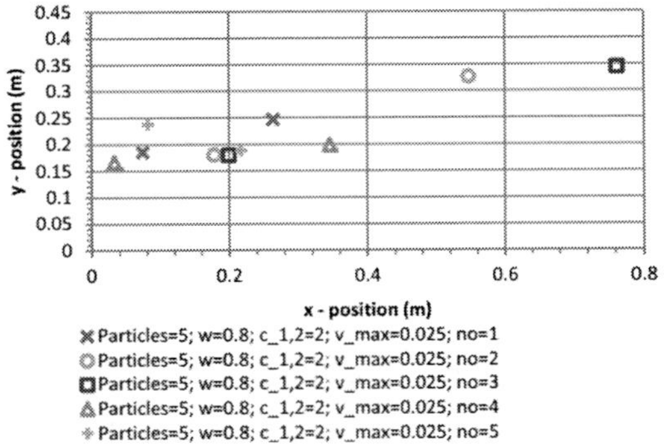


**Fig. 10.** Control points positions: Particles=5; w=0.8, $c_{1,2}$=2; $v_{max}$=0.025

$C_p$ value of the GBEST particle in the swarm limited by $\mathbf{v_{max}=0.025}$ reached up to **0.812454**.

*b) Maximal velocity $v_{max}$=0.05*

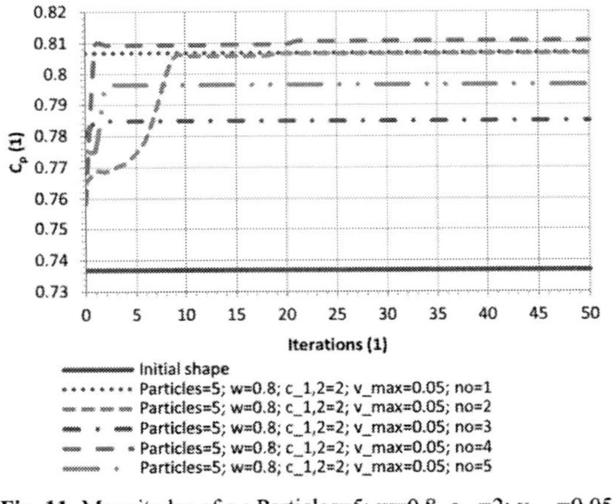


**Fig. 11.** Magnitudes of $c_p$: Particles=5; w=0.8, $c_{1,2}$=2; $v_{max}$=0.05

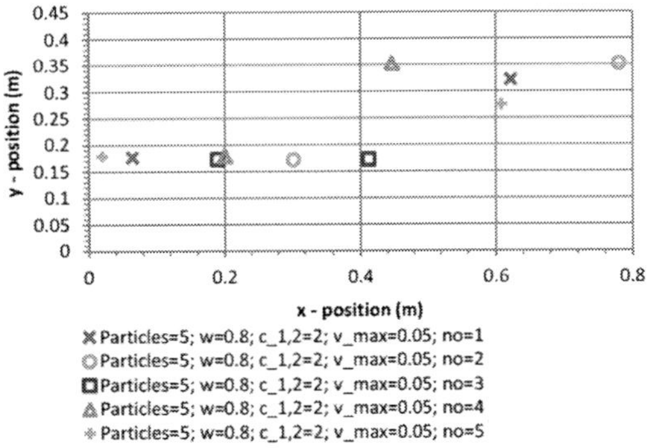


**Fig. 12.** Control points positions: Particles=5; w=0.8, $c_{1,2}$=2; $v_{max}$=0.05

This selection of parameters shows the widest spread of final values of the coefficient of pressure recovery (Figure 11). Despite this disadvantage examined function has improved compared to the initial shape. $C_p$ value of the GBEST particle in the swarm limited by $\mathbf{v_{max}=0.05}$ reached up to **0.810670**.

Adaptive *w*, constant $c_{1,2}$

In these test cases only inertia weight *w* is adaptive – linearly decreases its value through the optimization process – equation (7). This fact should in theory help the swarm more precisely locate global maximum of $c_p$ at the end of the PSO algorithm than the swarm with constant *w*.

Figures 13 - 16 capture the examined attributes of the swarm of five particles with adaptive parameter *w* and constant parameters $c_{1,2}$. The only difference between these two test cases was again in maximal velocities $v_{max}$=0.025 and $v_{max}$=0.05.

Note: Only one combination of $w_{max}$ and $w_{min}$ was tested and they were set to 0.9 and 0.4 respectively. This combination was based on [4].

*a) Maximal velocity $v_{max}$=0.025*

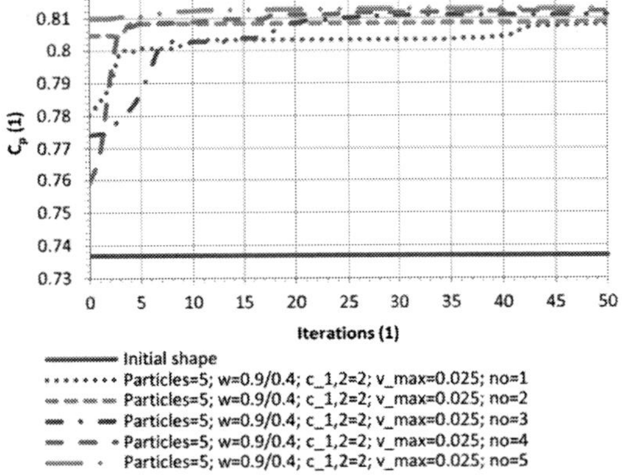


**Fig. 13.** Magnitudes of $c_p$: Particles=5; w=0.9/0.4, $c_{1,2}$=2; $v_{max}$=0.025

 DOI: 10.1051/epjconf/201714302076

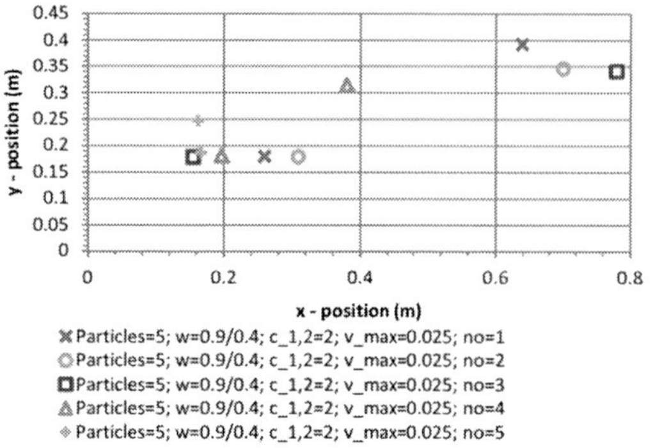


**Fig. 14.** Control points positions: Particles=5; w=0.9/0.4, $c_{1,2}$=2; $v_{max}$=0.025

$C_p$ value of the GBEST particle in the swarm limited by $\mathbf{v_{max}}$**=0.025** reached up to **0.812733**.

*b) Maximal velocity $v_{max}$=0.05*

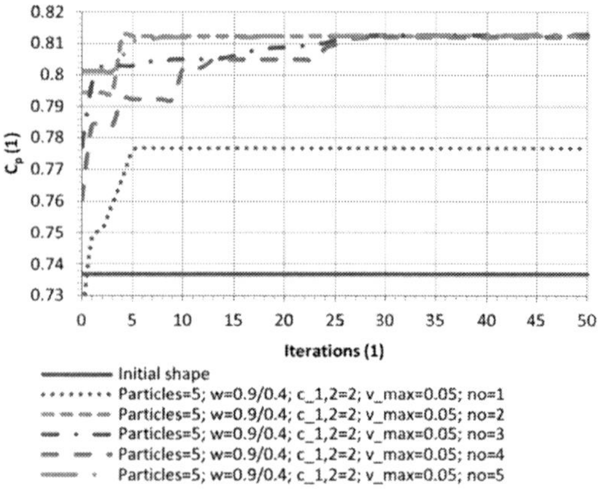


**Fig. 15.** Magnitudes of $c_p$: Particles=5; w=0.9/0.4, $c_{1,2}$=2; $v_{max}$=0.05

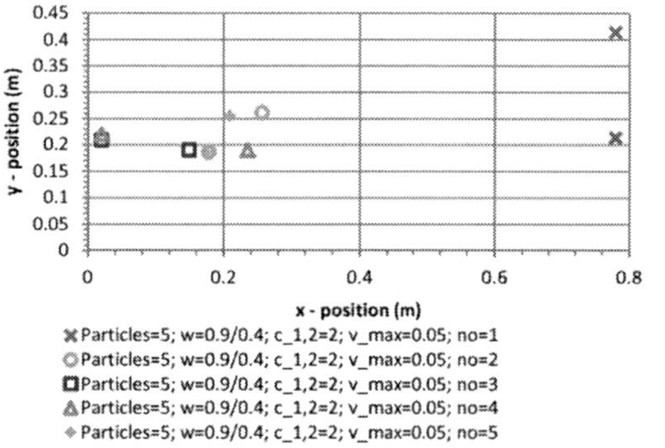


**Fig. 16.** Control points positions: Particles=5; w=0.9/0.4, $c_{1,2}$=2; $v_{max}$=0.05

$C_p$ value of the GBEST particle in the swarm limited by $\mathbf{v_{max}}$**=0.05** reached up to **0.812915**.

Adaptive parameters $w$, $c_{1,2}$

The last two test cases contain swarms with adaptive parameters $w$ and $c_{1,2}$. The modification of these parameters was realized via linear decrease – equations (7, 8, 9). This combination of parameters forces the swarm to quick convergence to the global maximum of $c_p$, but one danger could emerge – convergence to some local minimum without proper "search" of the computational area.

Figures 17 - 20 capture the examined attributes of the swarm of five particles with adaptive parameter $w$ and adaptive parameters $c_{1,2}$. The only difference between these two test cases was once again in maximal velocities $v_{max}$=0.025 and $v_{max}$=0.05.

Note: Only one combination of $w_{max}$, $w_{min}$ [4] and one combination of maximal/minimal values of $c_1/c_2$ [10] was tested.

*a) Maximal velocity $v_{max}$=0.025*

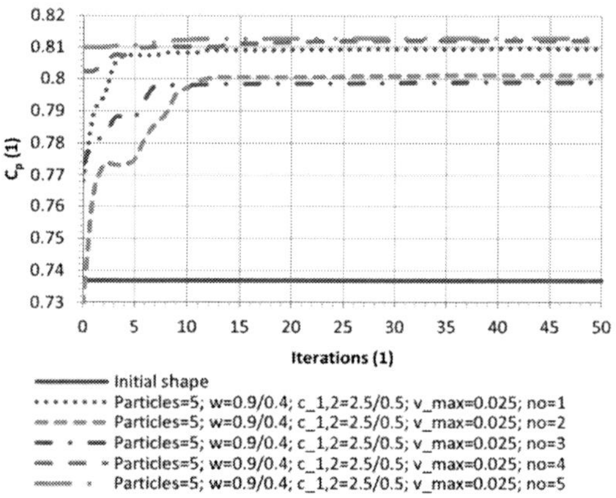


**Fig. 17.** Magnitudes of $c_p$: Particles=5; w=0.9/0.4, $c_{1,2}$=2.5/0.5; $v_{max}$=0.025

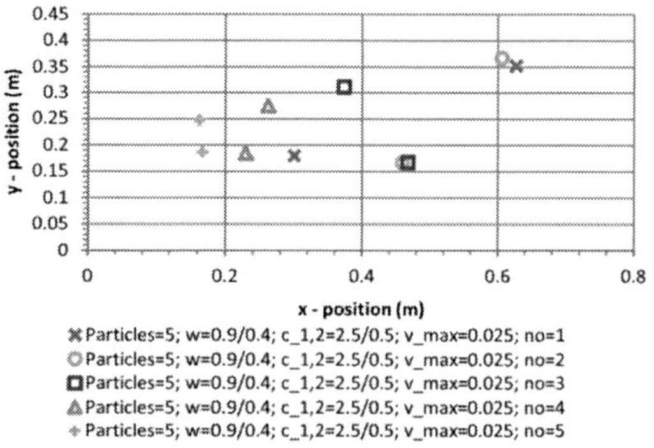


**Fig. 18.** Control points positions: Particles=5; w=0.9/0.4, $c_{1,2}$=2; $v_{max}$=0.025

$C_p$ value of the GBEST particle in the swarm limited by $\mathbf{v_{max}}$**=0.025** reached up to **0.812733**.

*b) Maximal velocity $v_{max}$=0.05*

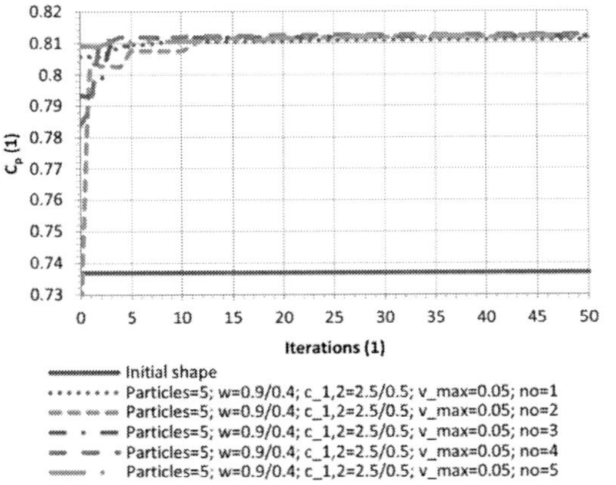


**Fig. 19.** Magnitudes of $c_p$: Particles=5; w=0.9/0.4, $c_{1,2}$=2.5/0.5; $v_{max}$=0.05

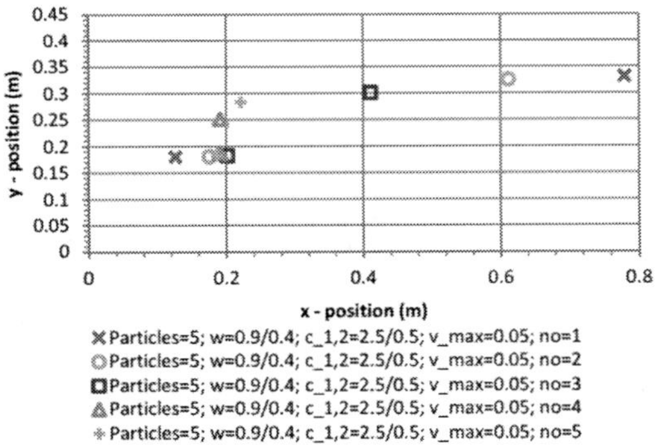


**Fig. 20.** Control points positions: Particles=5; w=0.9/0.4, $c_{1,2}$=2; $v_{max}$=0.025

$C_p$ value of the GBEST particle in the swarm limited by $\mathbf{v_{max}=0.05}$ reached up to **0.812543**.

**Partial results**

Every test case improved the value of the coefficient $c_p$ compared to the initial shape – Table 2. The best diffuser design is shown in Figure 21.

For constant parameters $w$ and $c_{1,2}$ is better choice smaller maximal velocity $v_{max}$=0.025. This fact strongly depends on the size of computational area for the swarm. On the other hand - when at least one of the parameter is adaptive, the bigger maximal velocity ($v_{max}$=0.05) seems to be a good choice to achieve better results.

Test case with adaptive $w$ and $c_{1,2}$ shows the most stable values of the coefficient $c_p$ (Figure 19), on the other hand the most unstable test case was with parameters $w$=0.8; $c_{1,2}$=2; $v_{max}$=0.05 (Figure 10).

The disadvantage of each test case is in final positions of the control points of Bézier curves – in every single test they were spread out across the whole computational area and they never converged to a specific location.

**Table 2.** Partial results of 5-particle swarm

| 5 particles | |
|---|---|
| Parameter selection | Best $c_p$ (1) |
| w=0.8; $c_{1,2}$=2; $v_{max}$=0.025 (no. 5) | **0.812454** |
| w=0.8; $c_{1,2}$=2; $v_{max}$=0.05 (no. 4) | **0.810670** |
| w=0.9/0.4; $c_{1,2}$=2; $v_{max}$=0.025 (no. 5) | **0.812733** |
| w=0.9/0.4; $c_{1,2}$=2; $v_{max}$=0.05 (no. 3) – Figure 21. | **0.812915** |
| w=0.9/0.4; $c_{1,2}$=2.5/0.5; $v_{max}$=0.025 (no. 5) | **0.812733** |
| w=0.9/0.4; $c_{1,2}$=2.5/0.5; $v_{max}$=0.05 (no.4) | **0.812543** |

Note: The coefficient of pressure recovery $c_p$ increases as a result of weaker velocity gradients and shear layers (it was confirmed by theoretical analysis in [11]).

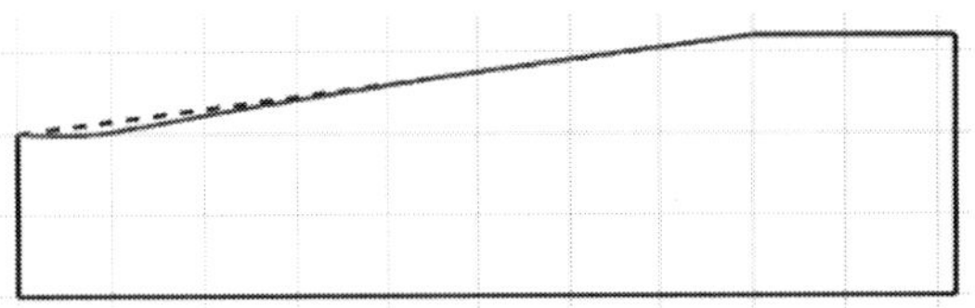

**Fig. 21.** The best diffuser design achieved by swarm of five

### *5.2.2 Swarm of 10 particles*

The last six test cases contain swarms of ten particles. Average computational time of this choice was 56925 *(s)*.

Once again: Free control points of Bézier curve were randomly generated in computational area at every beginning of PSOA (Figure 5). Every particle consists of two control points - one control point is represented as "x" and the other one as "+" (Figure 22).

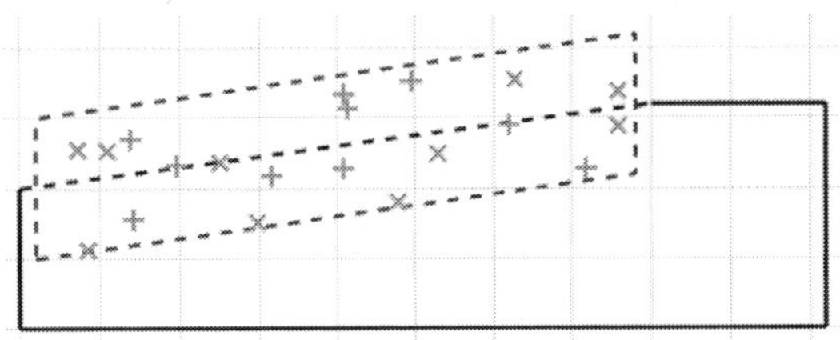

**Fig. 22.** Random sets of control points in computational area – 10 particles

Note: Swarms with larger number of particles were not tested due to enormous size of computational time.

Test procedure was the same as in case of swarm of five – evaluation technique was executed according to the diagram in Figure 7.

Constant parameters $w$, $c_{1,2}$

*a) Maximal velocity $v_{max}$=0.025*

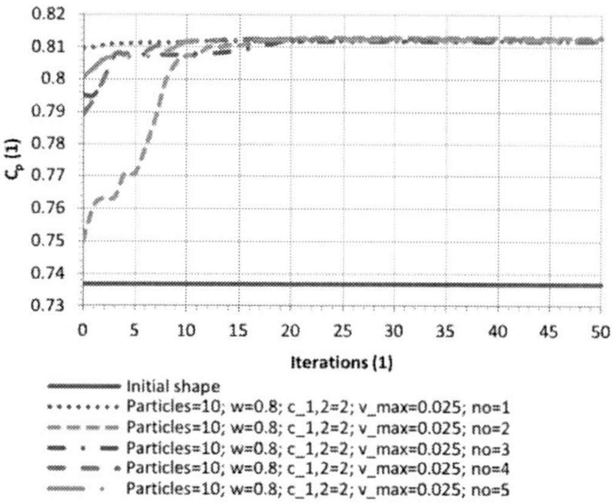


**Fig. 23.** Magnitudes of $c_p$: Particles=10; w=0.8, $c_{1,2}$=2; $v_{max}$=0.025

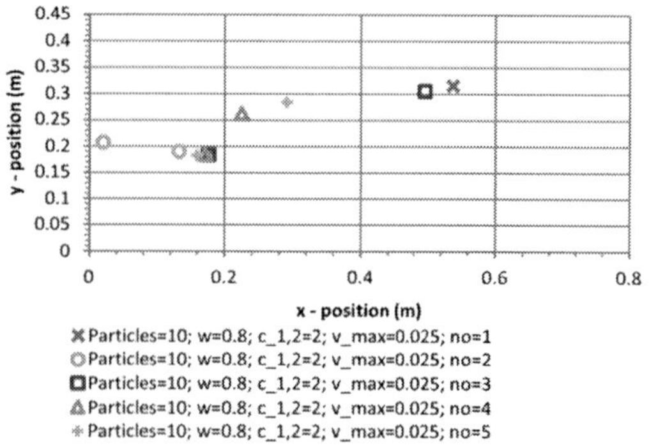


**Fig. 24.** Control points positions: Particles=10; w=0.8, $c_{1,2}$=2; $v_{max}$=0.025

$C_p$ value of the GBEST particle in the swarm limited by $\mathbf{v_{max}=0.025}$ reached up to **0.812857**.

*b) Maximal velocity $v_{max}$=0.05*

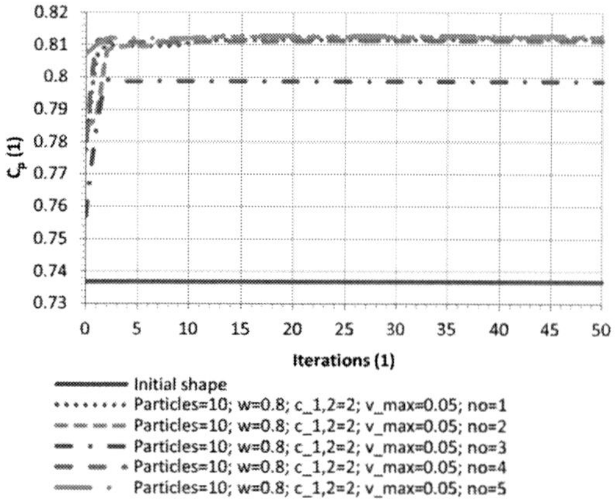


**Fig. 25.** Control points positions: Particles=10; w=0.8, $c_{1,2}$=2; $v_{max}$=0.05

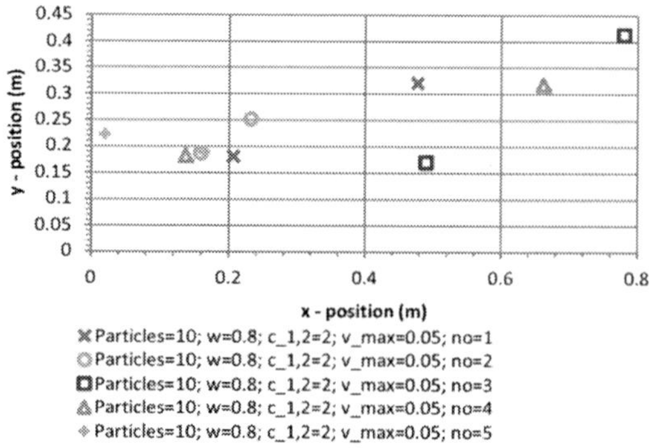


**Fig. 26.** Control points positions: Particles=10; w=0.8, $c_{1,2}$=2; $v_{max}$=0.05

$C_p$ value of the GBEST particle in the swarm limited by $\mathbf{v_{max}=0.05}$ reached up to **0.812707**.

Adaptive $w$, constant $c_{1,2}$

*a) Maximal velocity $v_{max}$=0.025*

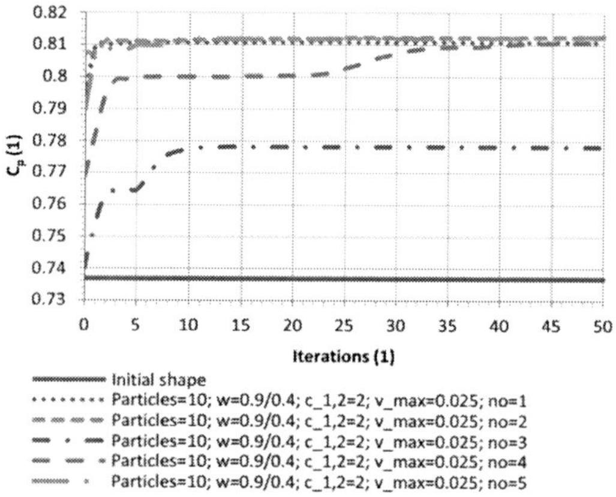


**Fig. 27.** Magnitudes of $c_p$: Particles=10; w=0.9/0.4, $c_{1,2}$=2; $v_{max}$=0.025

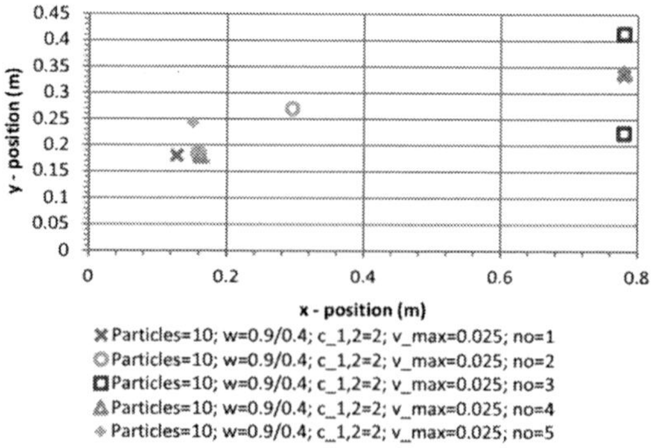


**Fig. 28.** Control points positions: Particles=10; w=0.9/0.4, $c_{1,2}$=2; $v_{max}$=0.025

$C_p$ value of the GBEST particle in the swarm limited by $\mathbf{v_{max}=0.025}$ reached up to **0.812698**.

*b) Maximal velocity* $v_{max}=0.05$

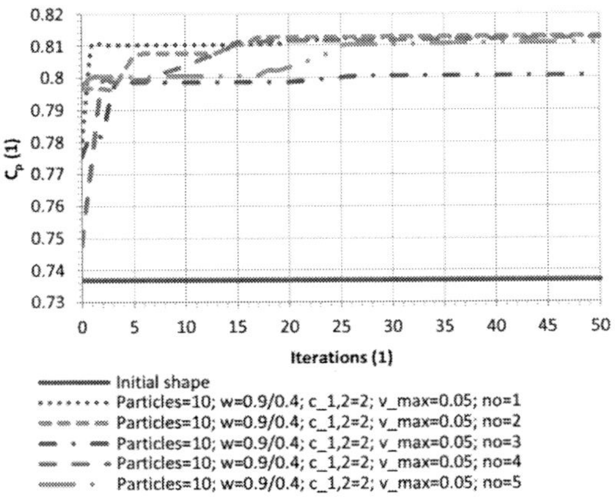


**Fig. 29.** Magnitudes of $c_p$: Particles=10; w=0.9/0.4, $c_{1,2}$=2; $v_{max}$=0.05

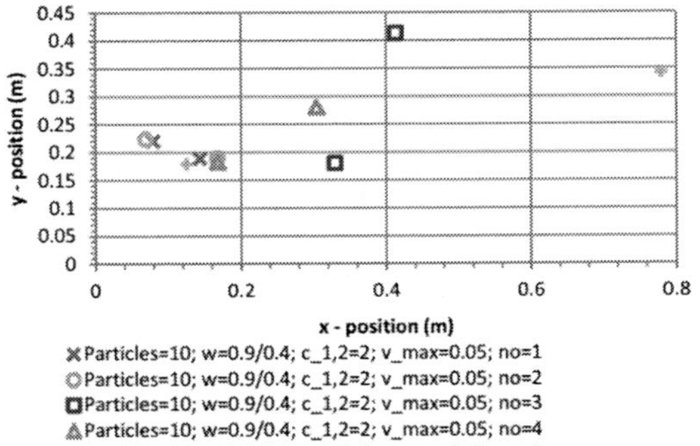


**Fig. 30.** Control points positions: Particles=10; w=0.9/0.4, $c_{1,2}$=2; $v_{max}$=0.05

$C_p$ value of the GBEST particle in the swarm limited by $\mathbf{v_{max}=0.05}$ reached up to **0.812831**.

Adaptive parameters *w*, $c_{1,2}$

*a) Maximal velocity* $v_{max}=0.025$

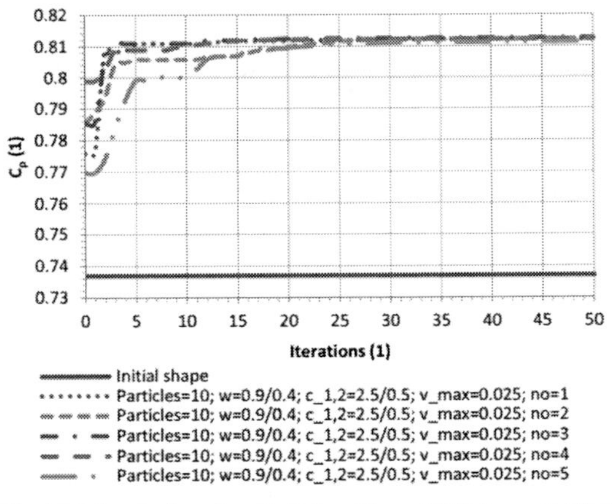


**Fig. 31.** Magnitudes of $c_p$: Particles=10; w=0.9/0.4, $c_{1,2}$=2.5/0.5; $v_{max}$=0.025

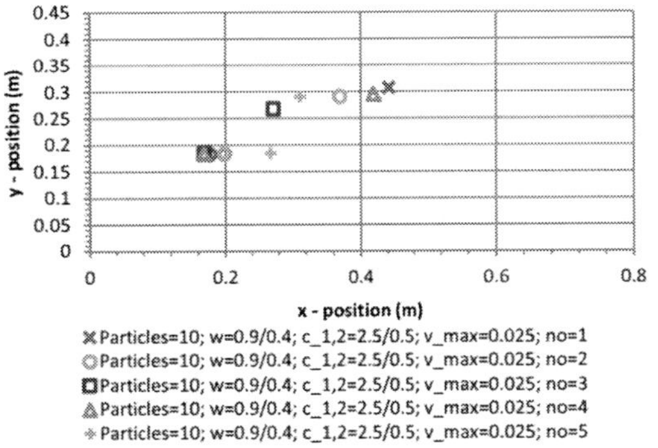


**Fig. 32.** Control points positions: Particles=10; w=0.9/0.4, $c_{1,2}$=2; $v_{max}$=0.025

*a) Maximal velocity* $v_{max}=0.05$

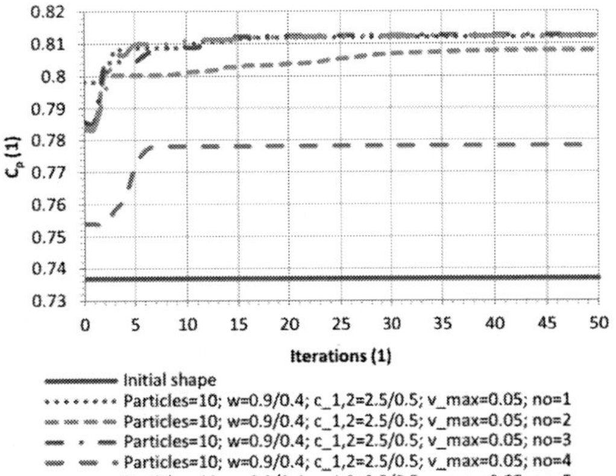


**Fig. 33.** Magnitudes of $c_p$: Particles=10; w=0.9/0.4, $c_{1,2}$=2.5/0.5; $v_{max}$=0.05

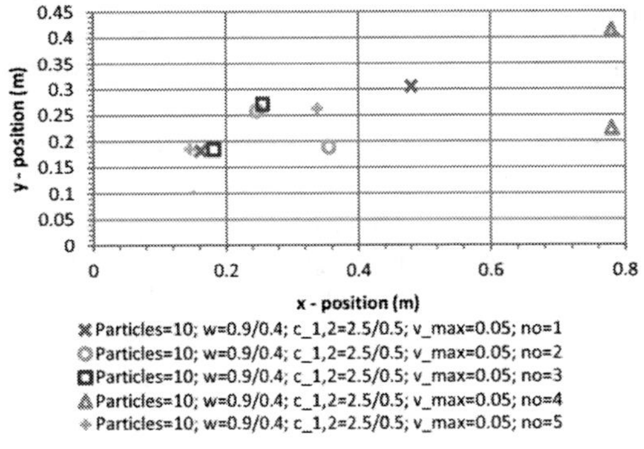


**Fig. 34.** Control points positions: Particles=10; w=0.9/0.4, $c_{1,2}$=2; $v_{max}$=0.05

**Partial results**

Every test case improved the value of $c_p$ coefficient compared to the initial shape – Table 3. The design of the best attempt is shown in Figure 35.

Trend of smaller maximal velocity for constant parameters continues also for the swarms of ten particles – Figure 23.

Ten particles should generally search the computational area thoroughly (compared to five particles), but the highest value of $c_p$ falls behind the swarms of five (for comparison Tables 2 and 3).

The disadvantage of almost every test case is in final positions of the control points of Bézier curves, but compared to the swarms of 5 particles, there is a promising combination of parameters $w$=0.9/0.4; $c_{1,2}$=2.5/0.5; $v_{max}$=0.025 – see Figures 31 and 32. This combination shows stable $c_p$ values and stable positions of control points in the end of the test; on the other hand this test case converged to some local maximum and did not bring the highest value of the examined function (coefficient of $c_p$).

**Table 3.** Partial results of 10-particle swarm

| 10 particles | |
|---|---|
| Parameter selection | Best $c_p$ (1) |
| w=0.8; $c_{1,2}$=2; $v_{max}$=0.025 (no. 2) – Figure 35. | 0.812857 |
| w=0.8; $c_{1,2}$=2; $v_{max}$=0.05 (no. 5) | **0.812707** |
| w=0.9/0.4; $c_{1,2}$=2; $v_{max}$=0.025 (no. 5) | **0.812698** |
| w=0.9/0.4; $c_{1,2}$=2; $v_{max}$=0.05 (no. 2). | **0.812831** |
| w=0.9/0.4; $c_{1,2}$=2.5/0.5; $v_{max}$=0.025 (no. 3) | **0.812530** |
| w=0.9/0.4; $c_{1,2}$=2.5/0.5; $v_{max}$=0.05 (no.3) | **0.812563** |

Note: The coefficient of pressure recovery $c_p$ increases as a result of weaker velocity gradients and shear layers (it was confirmed by theoretical analysis in [11]).

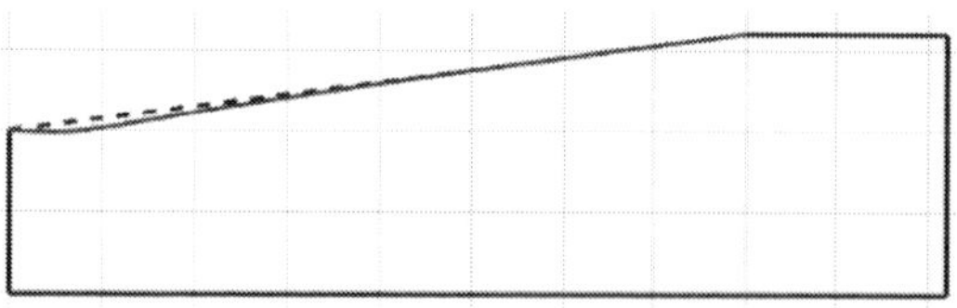

**Fig. 35.** Best diffuser design achieved by swarm of ten

## 6 Conclusions

Particle Swarm Optimization (PSO) is quite interesting and productive non-gradient type of optimization, which is based on social behaviour of various animals and uses swarm of particles to find global optimum.

Application of this optimization method strongly depends on examined problem. Present paper builds upon knowledge of [1] and focuses its energy on optimization of the hydraulic turbine diffuser. The shape optimization of the turbine diffuser became a basic model prototype of various methods [1, 13].

Two types of swarms were constructed – with five and ten particles and with different combinations of parameters (Figure 7). Every time PSO managed to improve value of $c_p$ compared to the initial shape (Figure 1) - $c_p$ increases as a result of weaker velocity gradients and shear layers. Final diffuser designs (Figure 21 and 35) had similar shape and achieved similar $c_p$ - **0.812915** for swarm of five and **0.812857** for the swarm of ten particles.

The main advantage of PSO: high value of coefficient of pressure recovery, because particles can "search" the computational area quite effectively. But on the other hand there are also some disadvantages: a need of creation of the computational area (Figure 4) to avert of production of degenerated designs and also a computational time was an issue – every iteration a great number of CFD simulation had to be performed.

Overall the PSO shows promising results – Table 4.

**Table 4.** Comparison of optimization methods

| Method | $c_p$ (1) | |
|---|---|---|
| Adjoint solver- 4. order polynomial [1] | **0.812354** | dotted line |
| Nelder-Mead – Bézier curve with 2 free control points [13] | **0.812146** | dashed line (with "o") |
| Particle Swarm Optimization - Bézier curve with 2 free control points | **0.812915** | solid line (with "x") |

It shows that the PSO algorithm produced slightly better values of $c_p$ than both Nelder-Mead algorithm and Adjoint solver.

Control points positions do not match, when PSO and Nelder-Mead are compared – Figure 36. But in general – all three methods show similar foundation of the final design – the tapered shape at the inlet of the diffuser and thus the crucial difference lies in the rest of the parametrized curve (Figure 36).

**Fig. 36.** Design comparison - detail

The comparison of computational time for all mentioned methods (Table 4) has not been done due to use of different computers (clusters).

There are some ways for improvement - future work will focus on some possible modifications of PSO e.g.:

1. Particle reduction during the algorithm – exploit a benefit of a large number of particles at the beginning of the algorithm and then if a certain criterion is met, algorithm will start to reduce the size of the swarm. This alternation brings potential savings in the computational time.
2. Combination of two optimization methods – at the beginning of the optimization process usage of PSO for its robustness and later in the end of the process an application of a straight forward method as Nelder-Mead to ensure good convergence. Also this modification brings potential savings in computational time.

## References

1. P. Moravec, J. Hliník, P. Rudolf: Optimization of Hydraulic Turbine Diffuser. EFM 2015, Prague, 2015, pp. 1-7.
2. J. Kennedy, R. Eberhart: Particle swarm optimization. The IEEE International Conference on Neural Networks, Perth, 1995, vol. 4, pp. 1942-1948.
3. J. Kennedy, R. Eberhart: A new Optimizer Using Particles Swarm Theory. The Sixth International Symposium on Micro Machine and Human Science, Nagoya, 1995, pp. 39-43.
4. Y. H. Shi, R. Eberhart: A Modified Particle Swarm Optimizer. The Int. Conf. on Evolutionary Computation, 1998, pp. 69-73.
5. Y. H. Shi, R. Eberhart: Empirical Study of Particle Swarm Optimization. IEEE 1999, pp. 1945-1950.
6. Z.-H. Zhan, J. Zhang, Y. Li, H. S.-H. Chung: Adaptive Particle Swarm Optimization. The IEEE Transactions on Systems, Man, and Cybernetics, 2009, vol. 39, pp. 1362-1381.
7. J. Kennedy, R. Mendes: Population Structure and Particle Swarm Performance. Proceedings of the Evolutionary Computation 2002, vol. 2, pp. 1671-1676.
8. Y. H. Shi, R. Eberhart: Parameter Selection in Particle Swarm Optimization. Evolutionary Programming VII: Proceedings of the Seventh Annual Conference on Evolutionary programming, San Diego, 1998, pp. 591-600.
9. A.P. Engelbrecht: Particle swarm optimization: Global or Local Best? BRICKS Congress on Computational Intelligence and 11$^{th}$ Brazilian Congress on Computational Intelligence 2013, pp. 124-135.
10. A. Ratnaweera, S. K. Halgamuge, H.C. Watson: Self-organizing hierarchical particle swarm optimizer with time-varying acceleration coefficients. IEEE Transactions on Evolutionary Computation 2004, pp. 240-255.
11. P. Rudolf: Study of the shear layers for swirl draft tube optimization. PhD thesis. In Czech
12. A.P. Engelbrecht: Particle Swarm Optimization: Velocity initialization. IEEE World Congress on Computational Intelligence, Brisbane, 2012.
13. J. Hliník: Shape optimization of hydraulic turbine diffuser. Bachelor thesis. In Slovak. 2015, Brno.
14. Conceptual diagram of PSO in action with a single global minimum. Figure: PSO_concept.png [online] [cit 2016-09-01]. Available online: http://wirelesstechthoughts.blogspot.cz/2013/06/an-introduction-to-particle-swarm.html

## Acknowledgement

This work is an output of cooperation between Centre of competence of Technology Agency of the Czech Republic TE02000232 "Rotary machines" and NETME Centre, regional R&D centre built with the financial support from the Operational Programme Research and Development for Innovations within the project NETME Centre (New Technologies for Mechanical Engineering), Reg. No. CZ.1.05/2.1.00/01.0002 and, in the follow-up sustainability stage, supported through NETME CENTRE PLUS (LO1202) by financial means from the Ministry of Education, Youth and Sports under the National Sustainability Programme I.

# A study of the flow boiling heat transfer in an annular heat exchanger with a mini gap

Tomasz Musiał[1,a], Magdalena Piasecka[1] and Sylwia Hożejowska[2]

[1] *Kielce University of Technology, Faculty of Mechatronics and Mechanical Engineering, Al. 1000-lecia PP 7, 25-314 Kielce, Poland*
[2] *Kielce University of Technology, Faculty of Management and Computer Modelling, Al. 1000-lecia PP 7, 25-314 Kielce, Poland*

**Abstract.** In this paper the research on flow boiling heat transfer in an annular mini gap was discussed. A one-dimensional mathematical approach was proposed to describe stationary heat transfer in the gap. The mini gap 1 mm wide was created between a metal pipe with enhanced exterior surface and an external tempered glass pipe positioned along the same axis. The experimental test stand consists of several systems: the test loop in which distilled water circulates, the data and image acquisition system and the supply and control system. Known temperature distributions of the metal pipe with enhanced surface and of the working fluid helped to determine, from the Robin boundary condition, the local heat transfer coefficients at the fluid - heated surface contact. In the proposed mathematical model it is assumed that the cylindrical wall is a planar multilayer wall. The numerical results are presented on a chart as function of the heat transfer coefficient along the length of the mini gap.

## 1 Introduction

In recent years, many works have been published to address the issues of the heat transfer process in flow boiling in mini gaps. The major advantage of the heat exchangers with mini gap technology is improving the effectiveness of the heat transfer processes. Effective heat transfer is very important for the cooling processes in miniature equipment, as it prevents overheating and permanent damage to components. The cooling technology that uses flow boiling processes in mini channels will contribute to further development of current trends regarding the miniaturisation of products and devices.

In Ref. [1] authors showed the results of experimental investigation into heat transfer of helium-xenon mixtures in cylindrical channels. In Ref. [2], in order to determine the heat transfer coefficient, the researches performed experimental tests during nanofluid (based on water and copper oxide particles) flow in a cylindrical channel. Ref. [3] described determination of the heat transfer coefficient in a mini channel evaporator with R-134a as a refrigerant. In Ref. [4], authors presented the study of R134a flow boiling heat transfer and evaluation of existing correlations was reported. Flow boiling heat transfer of R134a in a multiport minichannel heat exchangers was presented in Ref. [5]. In Ref. [6], authors proposed an improved semi-empirical method for determining heat transfer coefficient in flow boiling in conventional and small diameter tubes. Papers [7] and [8],. presented investigations into heat transfer bubbly boiling process, harnessing environmentally friendly refrigerating media in conventional channels. Research on heat transfer coefficients was presented in [9], where pressure drop and 'dryout' for flow boiling of water in an oil heated minichannel was discussed.

In authors' earlier research [10-13], flow boiling heat transfer in rectangular minichannels were studied while Fluorinert FC-72 was heated by an enhanced surface. Liquid crystal thermography and/or infrared thermography were applied to measure the temperature of the heated minichannels wall. Local values of the heat transfer coefficients were calculated numerically using one- and two-dimensional methods.

In the present paper, one-dimensional method is proposed to determine the heat transfer coefficient in flow boiling of distilled water in an annular mini gap.

## 2 Experimental setup and data

### 2.1 Experimental stand and procedure

The experimental stand is shown in Fig. 1. It is composed of several systems: the test loop with an annular mini gap in which the working fluid (distilled water) circulates, the data and image acquisition system, the supply and control system, and the lighting system. The test loop consist of: a heat exchanger, mass flow meter, a compensating tank, a gear pump, a deaerator and a filter. The data and image acquisition system comprises: a data acquisition station, a high speed camera, the lighting and a computer with special software.

[a] Corresponding author: tmusial@tu.kielce.pl

After deareation, there is a gradual increase in the electric power supplied to cartridge heater. Data for three settings of increasing heat flux, during laminar flow of distilled water along a mini gap, are analysed. The data used were obtained from the subcooled boiling region, in which the working fluid was superheated only at the interface with the heater and subcooled at the core of the flow.

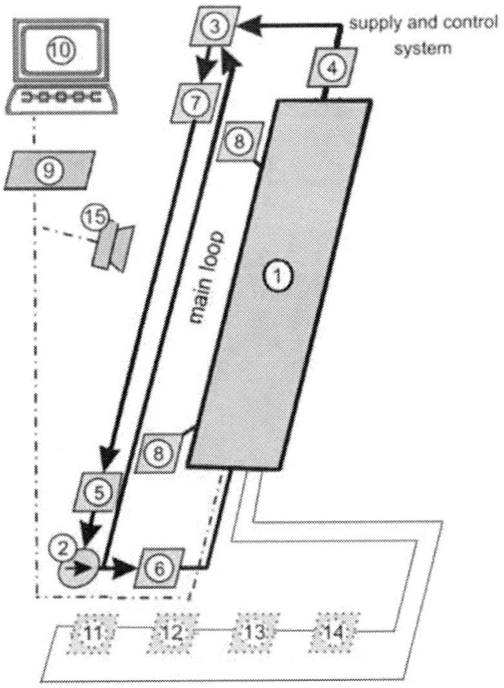


**Fig. 1.** The schematic diagram of the main loops at the experimental setup, 1-measurement module with a an annular mini gap; 2- gear pump; 3-compensating tank/pressure regulator; 4-tube-type heat exchanger, 5-filter, 6-mass flow meter; 7-deaerator; 8-pressure transducer; 9-data acquisition station; 10-pc computer; 11-inverter welder; 12-voltmeter; 13-ammeter; 14-shunt;15- high-speed camera.

## 2.2 The testing module with an annular mini gap

Figure 2 presents the testing module with an annular mini gap. The mini gap 1 mm wide was created between the metal pipe (2) and the glass pipe (1) positioned along the same axis. Inside the metal pipe (2), a cartridge heater (4) was located axially and symmetrically, powered by autotransformer with adjusted current intensity. Longitudinal slots was made in the inside surface of the pipe to fix the thermoelement sensors with a spacing of approximately 1 cm in the flow line. Thermoelement wires run in the gap between the heater and the metal pipe in the thermal conductive filler layer. The module header had a sealing (5) and the sensors of successive thermoelements and pressure transducers to measure the coolant temperature and positive pressure in the inlet and outlet to/from the mini gap.

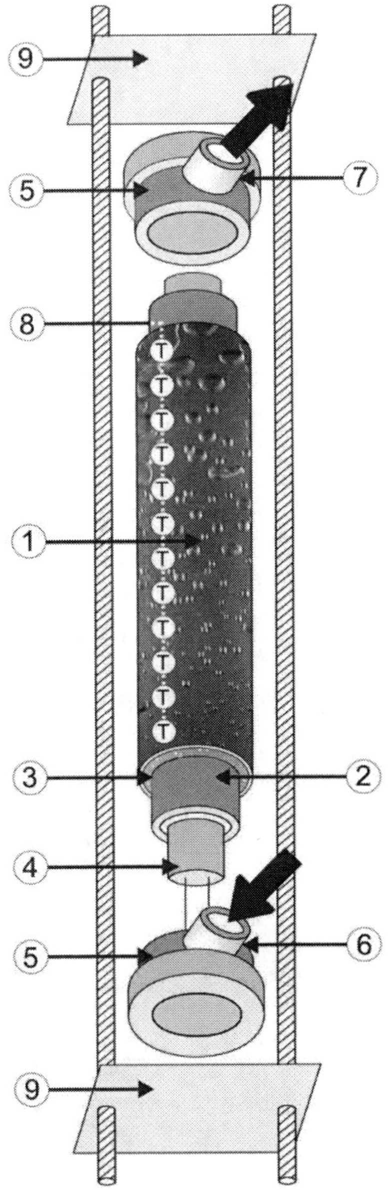


**Fig. 2.** The testing module with an annular mini gap layout: 1-glass pipe for viewing flow patterns, 2-metal pipe with enhanced surface, 3-annular mini gap, 4-cartridge heater, 5-module header with sealing, 6-inlet port for the medium, 7-outlet port for the medium, 8-thermoelements between the heater (4) and the metal pipe (2) in the thermal conductive filler layer, 9-stirrup bolts.

# 3 Heat transfer coefficient determination

It is assumed that the steady state heat transfer process in the testing module is stationary. Particular elements of the module create a system of planar layers with different thickness and thermal conductivity. Diagram illustrating the assumptions adopted for the planar multilayer wall is presented in Fig. 3.

The cartridge heater (with radius $r_1$) is the heat source with power output $q_V$.

For $r = r_1$ at each point $z_1, z_2, \ldots, z_M$, the condition below is satisfied

$$\tfrac{1}{2} r_1 q_V = \lambda_h \frac{\partial T_h(r_1, z_i)}{\partial r} = \lambda_s \frac{\partial T_s(r_1, z_i)}{\partial r} \tag{1}$$

Since difference $r_2 - r_1$ is very small, it is possible to replace the partial derivative with a finite difference:

$$\frac{\partial T_s(r_1, z_i)}{\partial r} \approx \frac{T_{s,i} - T_h(r_1, z_i)}{r_2 - r_1} \tag{2}$$

where $T_{s,i} = T_s(r_2, z_i)$ are the temperatures measured using the thermocouples at points $(r_2, z_i)$, Fig. 3a.

Combining equations (1) and (2) we obtained formula for the temperature $T_h(r_1, z_i)$:

$$T_h(r_1, z_i) = T_{s,i} + q_V \frac{(r_2 - r_1) r_1}{2\lambda_s} \tag{3}$$

The relationship between the cartridge heater temperature and the metal surface temperature for the planar multilayer wall Fig. 3a can be written as:

$$T_h(r_1, z_i) - T_F(r_3, z_i) = \frac{q_V r_1}{2} \left( \frac{(r_2 - r_1)}{\lambda_s} + \frac{(r_3 - r_2)}{\lambda_F} \right) \tag{4}$$

From equations (3) and (4) we get

$$T_F(r_3, z_i) = T_{s,i} - \frac{q_V r_1}{2} \frac{(r_3 - r_2)}{\lambda_F} \tag{5}$$

From equation (5) and from the Robin boundary condition, we got the formula for the local heat transfer coefficients at points $z_1, z_2, \ldots, z_M$, at the interface between the heated surface and the fluid

$$\alpha(z_i) = \alpha_{1,i} = \frac{0.5 r_1 q_V}{T_{s,i} - 0.5 r_1 q_V \frac{(r_3 - r_2)}{\lambda_F} - T_f} \tag{6}$$

In equation (6) it was assumed that the fluid temperature $T_f$ changes linearly in the mini gap, from the inlet temperature $T_{f,in}$ to the outlet temperature $T_{f,out}$. Both temperatures are known from the measurements.

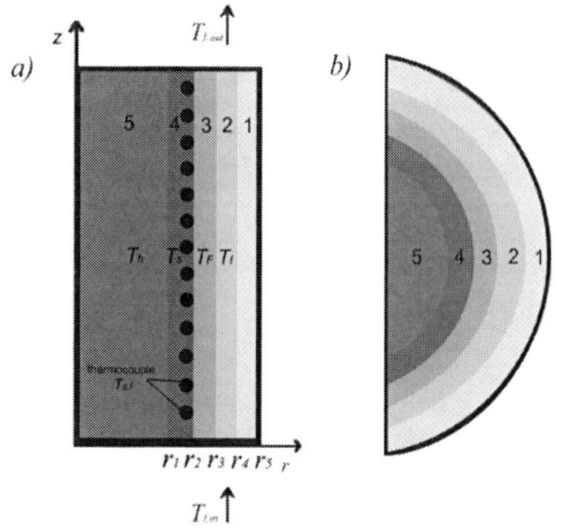


**Fig. 3.** Diagram illustrating the assumptions adopted for the planar multilayer wall: a) horizontal cross-section, b) vertical cross-section, 1- glass pipe, 2 - mini gap, 3- metal pipe, 4- thermal paste, 5- cartridge heater.

## 4 Results

Calculations were conducted for the subcooled boiling region at three settings of the volumetric heat flux, with all parameters of the experiment given in Fig. 4. The local values of heat transfer coefficient were found to decrease with the mini gap length. The increasing heat flux supplied to the heated surface caused heat transfer coefficients to decrease from above 4 kWm$^{-2}$ K$^{-1}$ to below 3 kWm$^{-2}$ K$^{-1}$ (for higher values of the volumetric heat flux). At the volumetric heat flux of 57.5 [MW/m$^3$], the heat transfer coefficient shows a more inclined plot along the mini gap length, takes the highest value at the beginning sections of the mini gap, with the lowest value of all three cases at the end section.

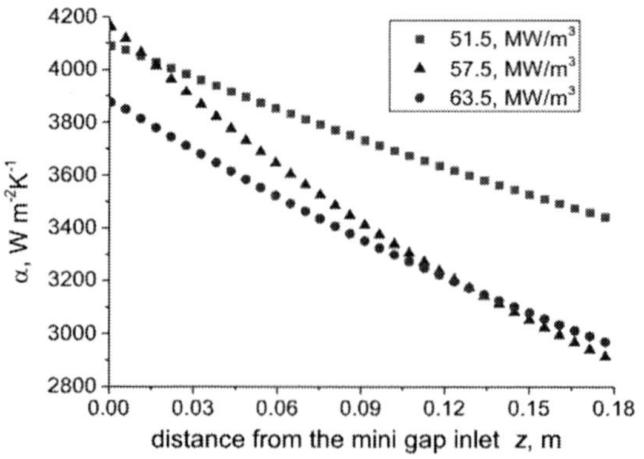


**Fig. 4.** Heat transfer coefficient vs. the mini gap length, experimental parameters (average): mass flow rate of 3.4 kg·s$^{-1}$, inlet pressure of 130 kPa, inlet liquid subcooling of 26 K, three settings of the volumetric heat flux are shown in figure.

## 5 Conclusions

The present paper discusses the flow boiling heat transfer research conducted in an annular w mm wide mini gap with distilled water as the working fluid. The studies were carried out for three settings of the increasing heat flux supplied to the heated surface, in the subcooled boiling region. The one-dimensional mathematical method was proposed to describe stationary heat transfer in the gap. Known temperature distributions in the heated surface and in the working fluid helped to determine, from the Robin boundary condition, the local heat transfer coefficients at the fluid - heated surface contact. In the proposed mathematical approach, the cylindrical wall is assumed to be a planar multilayer wall. The results were presented on the chart as a function of the heat transfer coefficient along the mini gap length. Observations indicated that the heat transfer coefficient decreased with the distance from the gap inlet.

In their further research, the authors intended to extend the results obtained over the saturated boiling region and focus on numerical modelling of heat transfer with the aid of the CFD module.

#### Acknowledgments

The research has been financially supported by the National Scientific Center from funds granted by virtue of decision No. DEC-2013/09/B/ST8/02825.

## References

1. S. L. Elistratov, O. V.Vitovskii, E. Y. Slesareva, *J. Eng. Thermophys.* **24,** 33–35 (2015).
2. A. V. Minakov, V. Y. Rudyak, D. V.Guzei, & A. S. Lobasov, *High Temp.* **53,** 246–253 (2015).
3. E. B. Dollera, E. P. Villanueva, *IOP Conf. Ser. Mater. Sci. Eng.* **88,** 12027 (2015).
4. Y. Xu, X. Fang, G. Li, D. Li, Y. Yuan, *Int. J. Heat Mass Transf.* **92,** 1143–1157 (2016).
5. J. Kaew-On, K. Sakamatapan, S. Wongwises, *Exp. Therm. Fluid Sci.* **35,** 364–374 (2011).
6. D. Mikielewicz, J. Mikielewicz, J. Tesmar, *Int. J. Heat Mass Transf.* **50,** 3949–3956 (2007).
7. T. Bohdal, *Annu. Set Environ. Prot.* **15,** 107–126 (2013).
8. T. Bohdal, *Int. J. Heat Fluid Flow* **21,** 449–455 (2000).
9. L. A. Campbell, S. G. Kandlikar, *Proc. 2nd Int. Conf. Microchannels Minichannels* (2004).
10. M. Piasecka, *Int. J. Refrig.* **56,** 198–212 (2015).
11. M. Piasecka, *Int. J. Heat Mass Transf.* **81,** 114–121 (2015).
12. S. Hożejowska, M. Piasecka, *Heat Mass Transf.* **50,** 1053–1063 (2014).
13. M. Piasecka, K. Strąk, B. Maciejewska, *Heat Transf. Eng.* **38** (3)**,** DOI 10.1007/s00231-016-1895-1 (2017).

# Experimental measurement, calculation and thermal visualization condenser temperature of cooling device with a heat pipe technology

*Patrik* Nemec[1,*], *Milan* Malcho[1], *Katarina* Kaduchová[1]

[1]University of Zilina, Faculty of mechanical engineering, department of Power Engineering, Univerzitna 8215/1, 01026 Zilina, Slovakia

**Abstract.** This work deal with evaluation of condenser temperature by experimental measurement, calculation and thermal visualization of cooling device working with a heat pipe technology. The referred device in the article is cooling device capable transfer high heat fluxes from electric elements to the surrounding. One from many things influenced the heat flux amount transferred from electronic elements through the cooling device to the surrounding is condenser construction, its capacity and option of heat removal. The work contain description, working principle and construction of cooling device. Experimental part describe the measuring method and mathematical calculation to condenser temperature evaluation of cooling device depending on the loaded heat of electronic components in range from 250 to 750 W. The mathematical calculation is based on physical phenomena of boiling, condensation and natural convection heat transfer. The results of experimental measurement and mathematical calculation are verified by thermal imagining of device condenser by IR camera.

## 1 Introduction

The cooling device with heat pipe technology may be visualised as a long hollow pipe, bent and the ends joined to form a continuous loop, filled with working fluid and orientated in a vertical plane. If the one side of the loop is heated and the other side cooled, the average density of the fluid in the heated side is less than in the cooled side. An essentially hydrostatic pressure difference, as a result of the thermally induced temperature gradient between the hot and the cold sides, rives the fluid flows around the loop. The 'buoyancy' force, as it is often termed, driving the fluid is in turn counteracted by an opposing frictional force that tends to retard the flow [1]. Cooling device can transfer heat from the interior of a microelectronic system to a central location where space limitations are less stringent. The advantages that a cooling device with heat pipe technology enjoys over a conventional refrigeration system include: (1) absence of moving parts leading to a more reliable system operation, (2) increased choices for selecting a working fluid compatible with microelectronics chips since it does not have to go through a refrigeration cycle, (3) reducing the decomposition rate of the working fluid as the higher temperatures at the compressor discharge in a vapour compression refrigeration system are not encountered, (4) clean operation as no oil is circulated through the system. In comparison to pool boiling systems employing vapour space condensation, a heat pipe technology offers more flexibility in terms of providing a centralized condenser with different feed lines to individual evaporator stations. Further, with the addition of a liquid circulating pump in a cooling device, higher heat transfer coefficients associated with flow boiling systems could be realized [2].

## 2 Cooling device construction

The main parts of cooling device are:

- evaporator,
- condenser,
- pipeline,
- filling and closing valve,
- working fluid

The model of cooling device is shown on Fig. 1.The evaporator enables on the base of a phase change (boiling) of the working fluid an intensive heat removal from its surface. It has to be constructed so that it will prevent the leakage of the working fluid, maintain pressure differences in all the walls and enable heat transfer from the electronic component into the working fluid as well as a suitable distribution of liquid and vapour phases of the working fluid. When choosing the material suitable for the construction of evaporator, it is necessary to pay attention to its thermokinetic characteristics. To provide a minimal temperature drop between a heat source and evaporator, the evaporator material must feature high thermal conductivity. To prevent escape of vapour, it should not be porous. The material should have high strength but, at the same time, it should be easily machineable and compatible with the working fluid [3]. The evaporator body is a plate with

* Corresponding author: patrik.nemec@fstroj.uniza.sk

DOI: 10.1051/epjconf/201714302078

dimensions 116 x 206 x 30 mm. To provide the working fluid circulation there are two 12 mm openings drilled horizontally on the plate and connected with twenty one 6 mm vertical connecting channels. They provide the transport of heated fluid vapour from the bottom to the top section of the evaporator. On the outer contact surface of the evaporator and electronic component there are three grooves with a mounted temperature sensor.

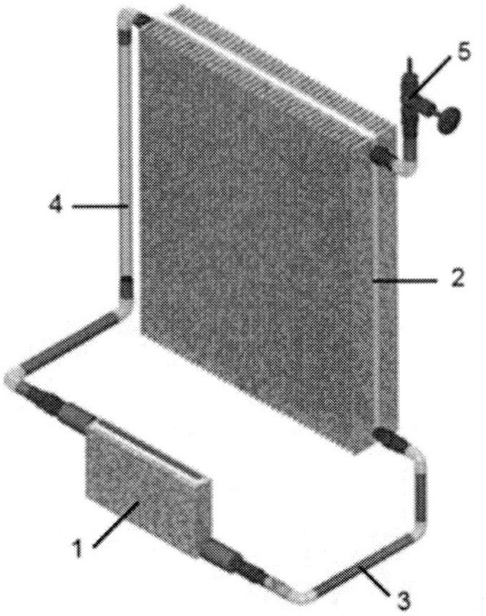


**Fig. 1.** Model of loop thermosiphon 1 – evaporator, 2 – condenser, 3 – vapour pipeline, 4 – liquid pipeline, 5 – filing and closing valve.

The condenser is proposed so that it was possible removed heat in to the surrounding by natural convection. It is created from two alumina fin coolers with dimensions 400 x 480 mm and fin height 45 mm. The each cooler has on the back side along the length ninety 7.5 mm deep and wide vertical grooves, and on the top and bottom two 10 mm deep and wide horizontal grooves. The cooler are soldered together on the back side and through the grooves can the working fluid flow down to the evaporator. This robust condenser construction allow remove high heat fluxes loaded to the evaporator [4].

The vapour pipeline and liquid pipeline connect evaporator with condenser and are transportation parts through the working fluid circulate in closed loop. The pipelines are made from copper material, due to its good installation properties and compatibility with almost all mediums used in heat pipe [5].

The working fluid in cooling device is Fluorinert FC 72, due to its compatibility with most metals, low boiling temperature (56 °C) and solidification (-90 °C) and, first of all, due to its excellent dielectric characteristics [6].

## 3 Mathematic calculation

The condenser temperature calculation is based on heat transfer calculation of the film condensation and laminar flow down condensate on the condenser wall and heat transfer by conduction to the condenser fin and heat transfer at free convection.

Assumption of the mathematical calculation condenser temperature is:

- Film condensation and vapour friction neglecting,
- substance properties of the condensate are not influenced by temperature,
- vapour density is lower than condensate density ($\rho_p << \rho_c$)

Basic of the mathematical calculation are heat transfer equations and criterial equations. The condensate flow down character is define by Reynolds criterion.

$$Re = \omega_c \, \delta_c / \upsilon_l \qquad (1)$$

Where $\omega_c = (g/\upsilon_l)\,(\delta_c^2/3)$ is condensate flow down velocity of film condensation, $\delta_c = (m_c \upsilon_l 3 / g \rho_l c_c n_k)^{1/3}$ is condensate film width, $c_c$ is channel circumference, $n_k$ is number of channels and $m_c$ is heat flux defined by $m_c = Q/l_v$.

Conductive heat transfer through the condensate film at the laminar flow down is define by Fourier equation

$$q_c = (\lambda_l/\delta_c)(t_v - t_{w1}) \qquad (2)$$

Where $q_c$ is je heat flux density on condenser inner surface given by $q_c = Q/A_{c1}$, $t_v$ is condensing vapour temperature equal boiling temperature and $t_{w1}$ is condenser inner surface temperature.

Condensation heat flux is given by Newton equation

$$Q_c = \alpha_c(t_v - t_{w1})A_{c1} \qquad (3)$$

Where $A_{c1}$ is condenser inner area, $\alpha_c$ is condensation heat transfer coefficient.

Condensation heat transfer coefficient is defined by Nusselt criterions.

$$Nu = 1.14(GaPr_cKu)^{1/4}, \text{ where } Re_{crit} < 400 \qquad (4)$$

$$Nu = H_c\alpha_c/\lambda_l \qquad (5)$$

Where $Ga = gH_c^3/\upsilon_l^2$ is Galileo criterion, $Pr_c = \upsilon_l/a_l$ is Prandtl criterion, $Ku = l_v/c_p(t_v - t_{w1})$ is Kutaladze criterion.

It can be expressed by relation.

$$\alpha_c = 1.14(\lambda_l^3 \rho_l g l_v / H_c(t_v - t_{w1})\nu_l)^{1/4} \qquad (6)$$

Where $H_c$ is characteristic parameter in direction of condensate flow.

Heat transfer at a free convection express Newton formula.

$$Q_{fc} = \alpha_{fc}\,(t_{w2} - t_s)A_{c2} \qquad (7)$$

Where $Q_{fc}$ is heat flux at free convection, $A_{c2}$ is condenser outer area, $t_{w2}$ is condenser outer surface temperature, $t_s$ is surrounding temperature, $\alpha_{fc}$ is a heat transfer at free convection.

Condenser outer surface temperature $t_{w2}$ is determined by Fourier formula.

$$t_{w2} = t_{w1} - (\lambda_c/\delta_{wc})q_c \qquad (8)$$

Where $\delta_{wc}$ is condenser wall thickness.

Heat transfer coefficient at free convection is defined by Nusselt criterions.

$$Nu_{fc} = 0.55(Gr_{fc}Pr)^{1/4}(Pr/Pr_{fc})^{1/4} \quad \text{at } Gr_{fc}Pr < 10^9 \qquad (9)$$

$$Nu_{fc} = H_{wc}\alpha_{fc}/\lambda_t \quad (10)$$

Where $Gr_{fc} = \gamma\theta Ga$ is Grashoff criterion, $Pr_{fc} = \nu_t/\alpha_t$ is Prandtl criterion at free convection, $Ga = gH_{wc}^3/\nu_t^2$ is Galileo criterion, $H_{wc}$ is condenser wall high at direction of acceleration due to gravity, $\theta$ is temperature difference of wall surface and surrounding, $\gamma$ thermal volume expansion [7, 8].

## 4 Experiment

Fig. 2 show scheme to measure cooling efficiency of cooling device. Measurement consist from heat load of evaporator part and temperature detection on the evaporator and condenser part. On the evaporator part of the cooling device are fixed electric components connected to the DC power supply source and load heat flux to cooling device. Between surface of electronic element and evaporator surface are inserted three thermocouples (on the left side, on the right side and in the centre) to scanning electronic component temperature because the highest admissible temperature on the contact surface is 100 °C. The temperature of the condenser is measured by two pair's temperature sensors. One thermocouple pairs is placed in the centre of condenser and second thermocouple pairs is placed on the top of condenser. On the condenser is connected pressure sensor to measure pressure in cooling device. All sensors are connected on the measuring unit where measured parameters convert and transmit signal to the PC. The measurement of the cooling efficiency of the cooling device starts at the input electric power 250 W and continuously increased by step of 100 W till the 750 W. Value 750 W was limited by the maximal electric power capacity of the DC power supply source. The measurement take place in the laboratory conditions at surrounding temperature around 25 °C [9].

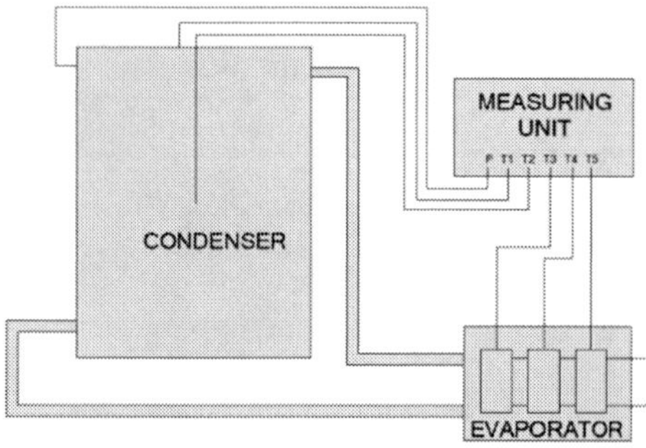


**Fig. 2.** Measuring scheme of loop thermosiphon cooling efficiency.

During the cooling efficiency of the cooling device measurement was the condenser of the cooling device scanned by IR camera to monitoring condenser wall temperature. Goal of the temperature monitoring is comparison measured temperature values and if the temperature distribution in the condenser is uniform or if there are some places with higher and lower temperature. To results guarantee scanned temperatures by IR camera was condenser painting by black colour with emissivity value 0.96 [10, 11].

## 5 Results

In the figure 3 is shown measured and calculated condenser temperature of cooling device depending on heat load. In the figure is seen that the calculated and measured results are very similar. This mean that the formulas chosen for condenser temperature calculation in this case was correct and can be applied in other similar cases of heat transfer calculations.

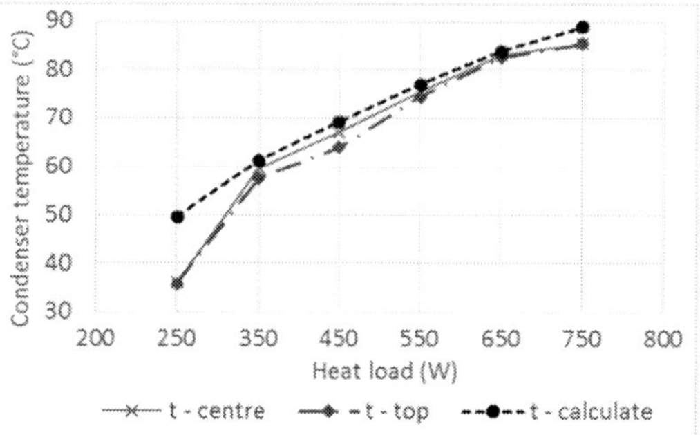


**Fig. 3.** Measured and calculated condenser temperature of cooling device depending on heat load

In the figure 4 and 5 are shown results of condenser temperature scanned by IR camera. The image in the figure 4 was made at heat load 500 W and image in the figure 5 was made at 750 W. In the figure 4 is seen non-uniform temperature distribution in the condenser, there are place on the right side with much more high temperature than in the centre or left side condenser. In the figure 5 is seen more uniform temperature distribution in the condenser. This result confirm measured and calculated results and is seen that the condenser of the cooling device is able remove much more heat flux to the surrounding.

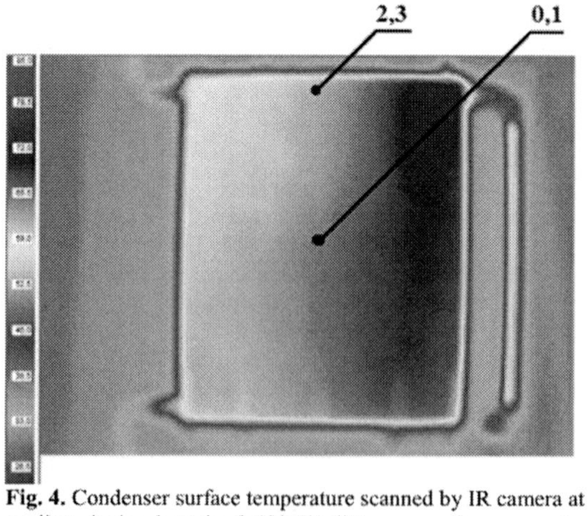


**Fig. 4.** Condenser surface temperature scanned by IR camera at cooling device heat load 500 W (0,1 – area thermocouples location, 2,3 – area thermocouples location).

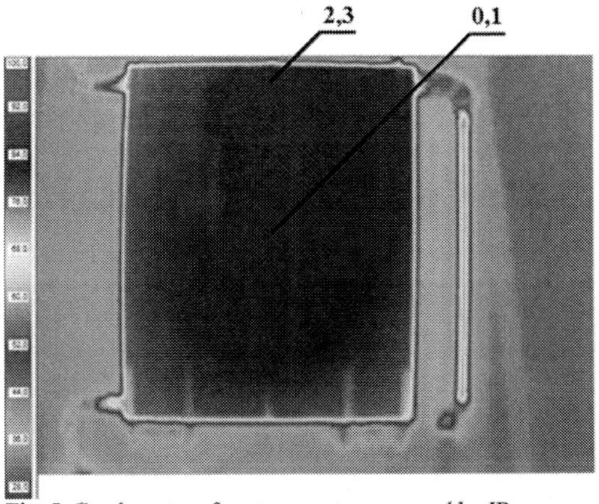


**Fig. 5.** Condenser surface temperature scanned by IR camera at cooling device heat load 750 W (0,1 – area thermocouples location, 2,3 – area thermocouples location).

## 6 Conclusion

Experiment was aim at evaluation of condenser temperature by experimental measurement, calculation and thermal visualization of cooling device working with a heat pipe technology. Heat flux transferred by cooling device to the surrounding is limiting with construction condenser. This was observed at the thermal images. Thermal visualization of cooling device condenser show that the condenser construction is capable transfer high heat fluxes around 750 W from electric elements to the surrounding and much more. Conclusion of results comparison obtained by calculation and measurement is that in the calculation were chosen correct formulas and they can be applied in other similar cases of heat transfer calculations.

This work is supported by the financial assistance of the project APVV – 15-0778 "Limits of Radiative and Convective Cooling through the Phase Changes of Working Fluid in Loop Thermosyphon".

## References

1. C. Ruppersberg, R.T. Dobson, J. of Energy in South Africa, **18**, 3 (2007)
2. J. Hartenstine, R. Boner III, J. Montgomery, T. Semenic, Proc. of ASME Inter PAC conf., (2007)
3. P. Pilat, M. Patsch, S. Papucik, M. Vantuch, AIP Conf. Proc., **1608** (2014)
4. L. J. Orman, R. Chatys, Transport Means - Proceedings of the International Conference (2011)
5. R. Lenhard, J. Jandačka, AIP Conf. Proc., **1558** (2013)
6. M. Holubčik, P. Ďurčanský, J. Jandačka, R. Nosek, Scient. World J., **2014** (2014)
7. J. Cernecky, J. Koniar, Z. Brodnianska, J. of Heat Transfere, **136**, 4 (2014)
8. M. Mahdavi, S. Qiu, S. Tiari, Apl. Therm. Eng., **81** (2015)
9. Ľ. Šooš, P. Križan, M. Matúš, J. Beniak, Int. J. of Mechanics, **9** (2015)
10. T. Orzechowski, L. J. Orman, Polution Engineering, **38**, 7 (2006)
11. T. Kruczek, Energy, **91** (2015)

# Visualization of various working fluids flow regimes in gravity heat pipe

*Patrik* Nemec[1,*]

[1]University of Zilina, Faculty of Mechanical Engineering, Department of Power Engineering, Univerzitna 8215/1, 01026 Zilina, Slovakia

**Abstract.** Heat pipe is device working with phase changes of working fluid inside hermetically closed pipe at specific pressure. The phase changes of working fluid from fluid to vapour and vice versa help heat pipe to transport high heat flux. Amount of heat flux transferred by heat pipe, of course depends on kind of working fluid. The article deal about visualization of various working fluids flow regimes in glass gravity heat pipe by high speed camera and processes casing inside during heat pipe operation. Experiment working fluid flow visualization is performed with two glass heat pipes with different inner diameter (13 mm and 22 mm) filled with water, ethanol and fluorinert FC 72. The working fluid flow visualization explains the phenomena as a working fluid boiling, nucleation of bubbles, and vapour condensation on the wall, vapour and condensate flow interaction, flow down condensate film thickness on the wall occurred during the heat pipe operation.

## 1 Introduction

The thermosiphon (gravity heat pipe) is shown in figure 1. The bottom is a heating (evaporation) section, the middle an adiabatic section, and the upper part a condensation section.

A small quantity of liquid is placed in a tube from which the air is then evacuated and the tube sealed.

The lower end of the pipe is heated, this causing that the vapour flows from the bottom to the condensation section where it is cooled and condensed into liquid. Then the liquid flows downwards along the wall as a very thin film. However, liquid plugs may form at high heat input. This entire cycle repeats in heat pipe again and again.

The Thermosiphon works on both conduction and convection mechanism. The heat from the heat source is transferred to the thermosiphons through conduction and inside thermosiphon heat convection occurs [1].

Since the latent heat of evaporation is large, considerable quantities of heat can be transported with a very small temperature difference from end to end. Thus, the structure will also have a high effective thermal conductance.

One limitation of the basic thermosiphon is that in order for the condensate to be returned to the evaporator region by gravitational force, the evaporation section must be situated at the lowest point [2].

The amount of heat that can be transported by these systems is normally several orders of magnitude greater than pure conduction through a solid metal [3].

A thermosiphon needs only a temperature difference to transfer large amount of heat, and are widely used in various areas of industry, such as chemical engineering, thermal engineering or thermal management systems with limited space applications and other heat recovery systems.

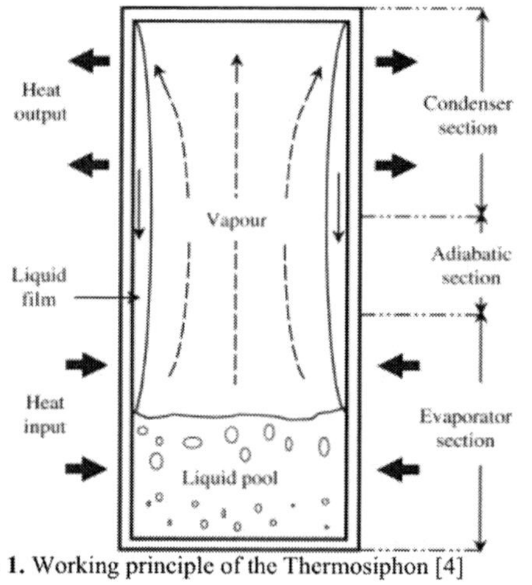


**Fig. 1.** Working principle of the Thermosiphon [4]

## 2 Boiling and condensation

The theory of boiling is complex and not yet fully understood. The boiling process depends on factors such as mass flow, vapour content and the temperature difference between the working fluid and the heating surface. The available research is based on empirical experiments on smooth tubes. However, the processes of evaporation could be explained by a combination of the available research on boiling and flow regimes [5, 6].

* Corresponding author: patrik.nemec@fstroj.uniza.sk

### 2.1 Boiling

The heat transfer coefficient for pool boiling has a characteristic curve shown in figure 2 that displays heat flux versus the temperature difference between the evaporating working fluid and heating surface. As the temperature difference increases, the energy flow per heat transfer area (kW/m2) will increase until an unstable maximum value for heat transfer is reached, the critical heat flux. At this point, vapour will form to such an extent that it hinders contact between the liquid of working fluid and the heat transfer area. As the temperature difference increases, a continuous gas film will form and the heat flux will reach a minimum. In the figure 2 are shown different boiling regimes.

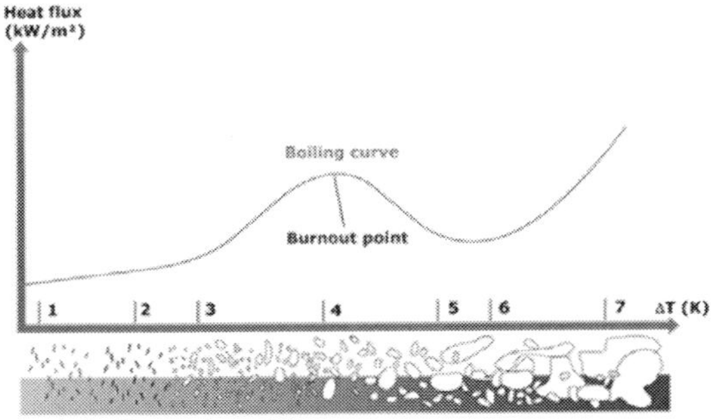


**Fig. 2.** Boiling regimes [7]

*Free convection evaporation:* Heat is transferred between the wall and the working fluid without bubble formation. The liquid close to the surface wall becomes slightly superheated and evaporates in the interface between liquid and gas.

*Sub-cooled nucleate boiling region*: Heat transfer between the wall and the working fluid is sufficiently large to create bubbles, but they collapse in contact with the liquid bulk.

*Nucleate boiling region*: This is the most important boiling region for technical applications. The temperature difference required to enter this region is approximately 3K. Superheated liquid overcomes the surface tension, forming unstable vapour bubbles that collapse in contact with the sub-cooled liquid of working fluid. The additional turbulence caused by forming and collapsing bubbles increases the heat transfer. The heat transfer increases as the temperature difference between the working fluid and the heating surface wall increases to reach a maximum critical heat flux at the burnout point.

*Partial film boiling:* Increasing the temperature difference further, beyond the burnout point, will cause the working fluid to evaporate too quickly to allow new liquid of working fluid access to the heat transfer surface. The partial vapour blanket acts as insulation, causing the heat transfer coefficient to drop greatly and reducing the overall heat flux. This is an unstable region that should be avoided, because performance is uncertain and may fluctuate substantially.

*Complete film boiling*: With a very high temperature difference, a stable film of working fluid vapour will form on the heat transfer area. The film will act as an effective insulation layer hindering any direct contact between the heating surface and the liquid working fluid. Although the heat transfer coefficient reaches a minimum here, this evaporation region is preferable over the unstable partial film-boiling region, because predictions of the heat transfer coefficient are more reliable.

### 2.2 Condensation

Condensation is the phase transformation from vapour to liquid state. It may be carried in a space or on a solid surface and is provided with a heat and mass transfer. The heat pipe efficiency is supported by high rates of heat transfer during the condensation. The condensation on a solid surface can occur in the two different ways shown in figure 3 as a film condensation or dropwise condensation.

When the film condensation occurs, the condensate is form in a continuous liquid film that covers and wets the solid wall cold surface. The liquid film flows down the surface under gravity or other forces (e.g. capillarity). Heat transfer to the solid surface extends through the liquid film [8].

When the dropwise condensation occur, the working fluid vapour reach the wall cold surface, reduces its energy and causing the working fluid liquefaction and droplets formation on the cold surface that grow with direction of the vapour condensation. The small neighbour's drops are collecting to the bigger drops until they are tearing down from the surface by the gravity or other forces. As drops are flow down the wall, they leaving clean surface where condensation can proceed [9].

The heat transfer coefficient of the dropwise condensation is 10 times larger than the film condensation, but its duration is short and from this reason is difficult usable in practice. If the dropwise condensation has been continuous or if the thickness of the liquid film has been reduced can be achieved increasing heat transfer coefficient [10].

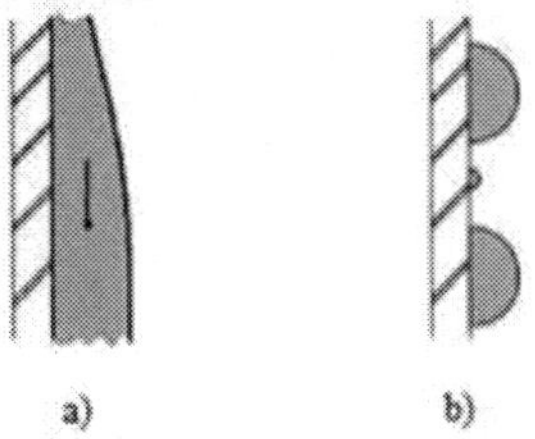


**Fig. 3.** Condensation regimes: a) Film condensation, b) Dropwise condensation [11]

## 3 Experiments

Experiments deal about flow visualization of working fluids (water, ethanol and fluorinert FC 72) in glass gravity heat pipes with inner diameter 13 and 22 mm and total length 500 mm. In the figure 4 is shown positioning of the experimental apparatus. The working fluid flow inside the glass heat pipe placed on the heat source with heat output regulator was scanned by high speed video camera. The visualization of the working fluid flow in heat pipes can be helpful in solution or CFD simulation of the flow dynamics or heat and mass transfer of the heat pipes [12, 13].

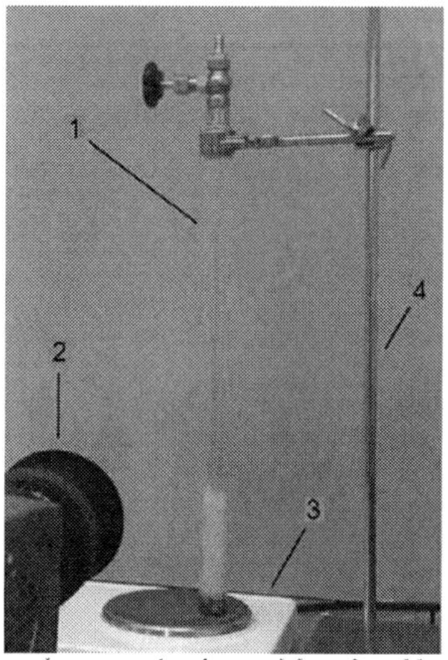


**Fig. 4.** Measuring setup: 1 – thermosiphon, 2 – video camera, 3 – heat source, 4 – stand

### 3.1 Heat pipe construction

In the figure 5 are containers of gravity heat pipes manufactured for experiment of working fluid visualization. The body of heat pipes is created form borosilicate glass tube. The borosilicate glass is suitable for laboratory use because is more resistant to thermal shock than others glass. It has a smoother transition between temperatures, and can even withstand different temperatures at the same time. On the bottom is Cu cap and on the top is filling and closing valve.

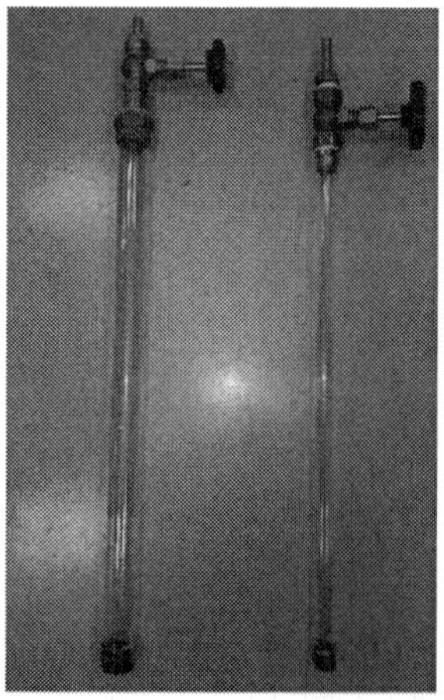

a) b)

**Fig. 5.** Glass containers of gravity heat pipes: a) Inner diameter 13 mm b) Inner diameter 22 mm

The heat pipes manufacturing was performed by method of working fluid evaporation. This method covers filling the total volume of heat pipe with working fluid and then evaporation of working fluid to 20 % of heat pipe. This method guaranty that inside heat pipe is only pure working fluid and no air is inside.

### 3.2 Working fluid flow visualization

In the figures 6, 7, 8, 9, 10 and 11 are shown two-phase flow regimes of working fluid (water, ethanol and fluorinert FC 72) in gravity heat pipe at evaporator temperature 105 °C scanned by high speed video camera.

In the figure 6 are shown evaporation and condensation sections of glass gravity heat pipe with inner diameter 13 mm and water working fluid. In the figure 6a is shown film boiling and creating big bubbles of the working fluid on bottom heat pipe evaporator section. On the top is seen as a vapour pushes the liquid to the adiabatic section. In the figure 6b is shown dropwise condensation of the working fluid on the condensation section surface. The small droplets are collected to the big drops and they flow down to the evaporator section due to gravitation action. Thus is created free space for a new dropwise condensation on the clear surface.

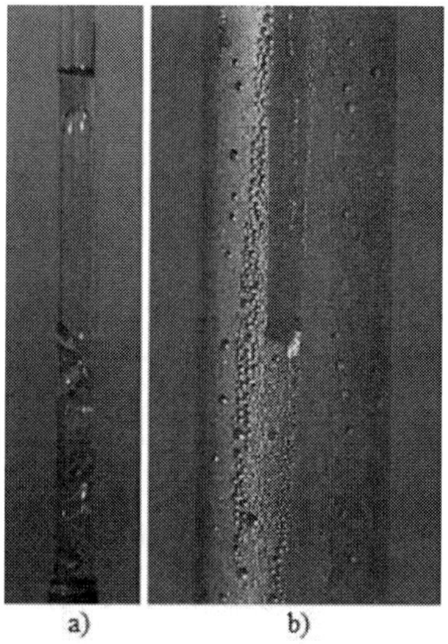

a) b)

**Fig. 6.** Glass gravity heat pipe - inner diameter 13 mm, working fluid water: a) Evaporation section b) Condensation section.

In the figure 7 are shown evaporation and condensation sections of glass gravity heat pipe with inner diameter 22 mm and water working fluid. In the figure 7a is shown film boiling and creating big bubbles of the working fluid on bottom heat pipe evaporator section. It is similar action as in the heat pipe with inner diameter 13 mm, but in this case liquid is not pushes by vapour to the adiabatic section so high. This may cause bigger diameter of heat pipe. In the figure 7b is shown dropwise condensation of the working fluid on the condensation section surface. In this case this action is same as at heat pipe with inner diameter 13 mm.

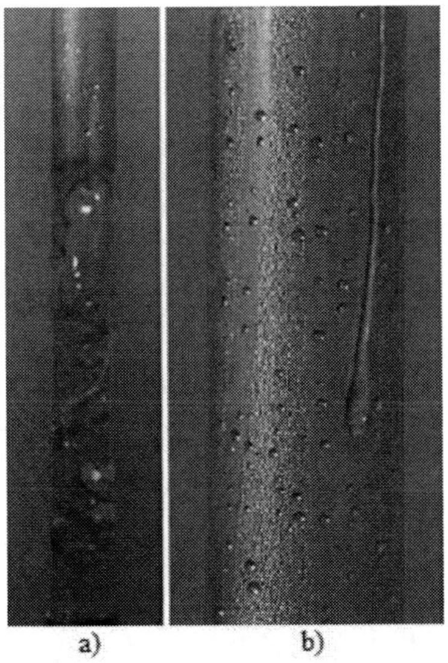

**Fig. 7.** Glass gravity heat pipe - inner diameter 22 mm, working fluid water: a) Evaporation section b) Condensation section.

In the figure 8 are shown evaporation and condensation sections of glass gravity heat pipe with inner diameter 13 mm and ethanol working fluid. In the figure 8a is shown pool boiling of the working fluid in whole evaporator section. In the figure 8b is shown condensation section. The heat from evaporator to condenser was transferred well but there was observed even not dropwise condensation either film condensation of the working fluid in this kind of heat pipe.

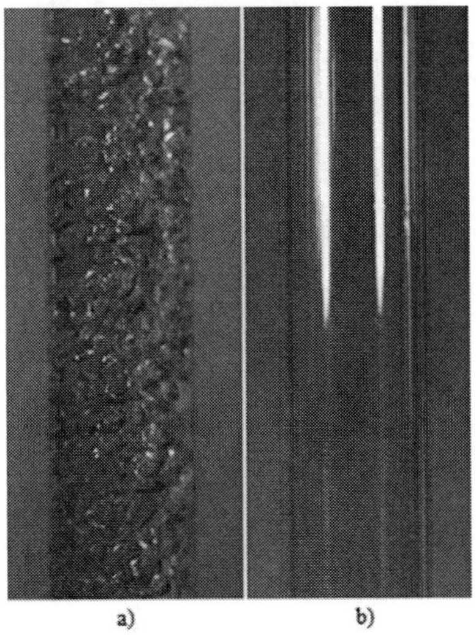

**Fig. 8.** Glass gravity heat pipe inner diameter - 13 mm, working fluid ethanol: a) Evaporation section b) Condensation section.

In the figure 9 are shown evaporation and condensation sections of glass gravity heat pipe with inner diameter 22 mm and ethanol working fluid. In the figure 9a is shown mixed pool and film boiling, where the bubble boiling is creating in whole evaporation section and film boiling creation on the bottom and then rise up to the top of the evaporation section. In the figure 9b is not so good shown the film condensation but the film condensation of the working fluid was observed only in the right hand side heat pipe wall kind.

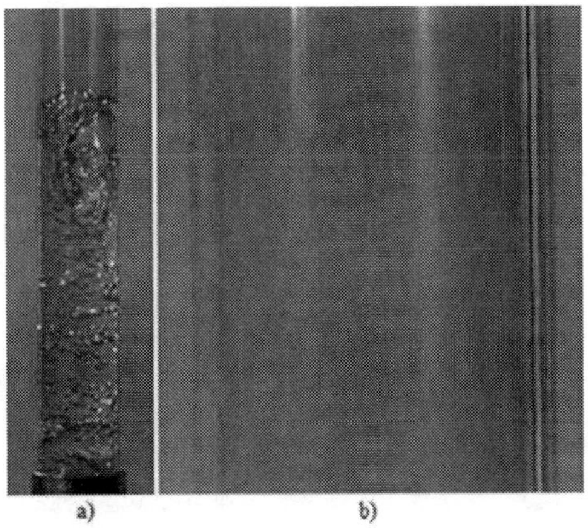

**Fig. 9.** Glass gravity heat pipe - inner diameter 22 mm, working fluid ethanol: a) Evaporation section b) Condensation section.

In the figure 10 are shown evaporation and condensation sections of glass gravity heat pipe with inner diameter 13 mm and fluorinert FC 72 working fluid. In the figure 10a is shown pool boiling of the working fluid in whole evaporator section. In the figure 10b is shown film condensation of the working fluid. Condense of working fluid flown down throughout entire inner surface heat pipe wall.

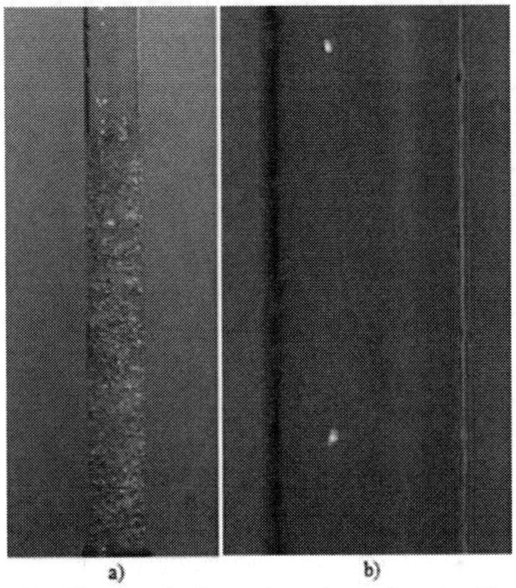

**Fig. 10.** Glass gravity heat pipe - inner diameter 13 mm, working fluid fluorinert FC 72: a) Evaporation section b) Condensation section.

In the figure 11 are shown evaporation and condensation sections of glass gravity heat pipe with inner diameter 22 mm and fluorinert FC 72 working fluid. In the figure 11 a) is shown pool boiling of the working fluid in whole evaporator section. In the figure 11b is shown the same film condensation of the working fluid as in the figure 10b. Condense of working fluid flown down throughout entire inner surface heat pipe wall.

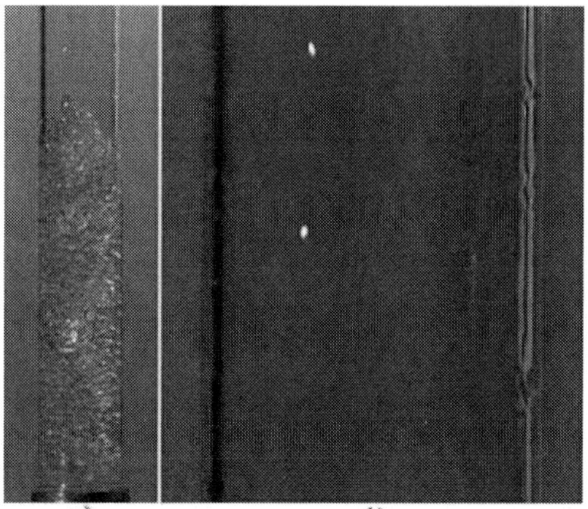
a) b)

**Fig. 11.** Glass gravity heat pipe - inner diameter 22 mm, working fluid fluorinert FC 72: a) Evaporation section b) Condensation section.

## 4 Conclusions

Aim of this experimental work was flow visualization of various working fluid kinds in the gravity heat pipe and working fluid flow comparison considering to different inner diameter of heat pipe.

The experiment with various working fluid show that the investigated working fluids in heat pipe have various boiling and condensation regimes. At water heat pipe was observed film boiling and dropwise condensation regimes, at ethanol heat pipe was observed pool boiling and film condensation but the film condensation wetting only one side inner surface of the heat pipe wall, at the fluorinert FC 72 heat pipe was observed pool boiling and film condensation fully wetting entire inner surface of the heat pipe wall.

The experiment with different diameter dimensions show that the heat pipe dimeter dimension have not influence on the condensation regime but have influence on the richness of the flowing down condensate. Experiment show that the with the changing heat pipe diameter from small to big the boiling regime can change from pool boiling to mix of pool boiling and partially film boiling. Other findings of the experiment is that the boiling regime at heat pipe with the bigger diameter occur only in the evaporator section and at the heat pipe with smaller diameter occur in evaporator and partially at the adiabatic section.

This work is supported by the financial assistance of the project APVV – 15-0778 "Limits of Radiative and Convective Cooling through the Phase Changes of Working Fluid in Loop Thermosyphon".

## References

1. A. S. Sundaram, A. Bhaskaran, J. of El. Cool. and Therm. Cont., **1**, 15-21, (2011)
2. D. Ray, P. Kew, *Heat pipes, fifth eddition* (Elseviere, Oxford, 2006)
3. P. Dunn, D. Ray, *Heat Pipes, 4th Edition*, (Pergamon Press, Oxford, 1994)
4. L. J. Orman, AIP Conf. Proc., **1608** (2014)
5. L. J. Orman, EPJ Web of Conf., **67** (2014)
6. S. Liu, J. Li, Q. Chen, Flow Measur. and Instrum., **18**, 216 – 222, (2007)
7. http://handbooks.swep.net/RefrigerantHandbook/chapter_06/Pages/6.2-Boiling-Regimes.aspx
8. R. Nosek, M. Holubčík, Š. Papučík, Sci. World J., **2014**, 487549, (2014)
9. M. Čarnogurská, M. Příhoda, J. Molínek, J. of Mech. Sci. and Tech., **25 (11)**, pp. 2935-2941, (2011)
10. T. Brestovič, N. Jasminská, R. Pyszko, M. Lázár, M. Puškár, Measurement: J. of the Inter. Measur. Confed., **72**, 3380, pp. 52-60, (2015)
11. http://ltcm.epfl.ch/files/content/sites/ltcm/files/shared/import/migration/COURSES/TwoPhaseFlowsAndHeatTransfer/lectures/Chapter_7.pdf
12. R. Lenhard, J. Jandačka, AIP Conf. Proc., **1558**, 2138-2141, (2013)
13. R. Lenhard, K. Kaduchová, Š. Papučík, J. Jandačka, EPJ Web of Conf., **67**, 02067, (2014)

# Evaluating the moisture transfer of membranes for total heat exchanger

Nguyen Van Vu [a] and *Václav* Dvořák

Department of Power Engineering Equipment, Faculty of Mechanical engineering, Technical University of Liberec, Czech Republic

**Abstract.** This paper presents an experimental evaluation of moisture transfer of a novel membrane which is used in enthalpy heat exchanger. The moister transfer through a membrane was investigated in a model of counter flow heat exchanger and the effectiveness of moister transfer was evaluated for various flow velocities. It was found that the measuring error, i.e. difference between moister absorbed and released is quite high. Therefore, measuring uncertainty analyses were performed. The resulting uncertainty of effectiveness of moister transfer was derived using partial derivatives of all influencing quantities and considering both random and systematic uncertainties. It was found that measuring of mass flow rate has the biggest influence on resulting uncertainty of moister transfer effectiveness. Therefore, accuracy of measured air flow rates measuring must be improved to reach lower resulting uncertainty.

## 1 Introduction

Air ventilation is important to maintain healthy condition for residential building, especially in cold climate areas. However, ventilation causes a huge energy lost, which has to be compensated by heating. Therefore, heat exchangers are used to get back part of heat energy from exhaust air and convey it to the fresh air. In the last decades, researchers focused on conventional heat exchangers which can recover only sensible heat. On the demand of increasing indoor air quality and decreasing energy consumed by the ventilation system, total heat exchanger (or enthalpy heat exchanger) has been investigated because of the following advantages: Total heat exchanger can recover not only sensible but also latent heat from exhaust air which means higher potential use. Moreover, by using enthalpy heat exchanger, we can avoid or reduce condensation and frost inside a heat exchanger, which is one of the most important disadvantages of aluminium or plastic based conventional heat exchangers when working at temperatures lower than dew point or freezing point temperature, respectively [1], [2]. Among several types of total heat exchangers, membrane-based enthalpy heat exchangers have great potential thanks to compactness and lightweight.

Due to the complex coupling between heat and moisture transfer, research and development of membranes for membrane-based heat exchanger is still in an early state and need more effort from researchers. Huge number of mathematical and numerical models were conducted to estimate effectiveness of heat and moisture transfer, pressure drop and so on, such as in works [2]–[5]. Many researches also focus on experimental method to evaluate prototype models or novel membranes, such as [1], [6]–[8]. Many investigations have been done, but most of researchers focused on their own configurations, there were no general model due to complex combinations of different factors.

As continuation of preliminary work [6], this paper presents evaluation of moisture transfer of novel membranes and a decent analysis of uncertainty in calculating moisture transfer effectiveness.

## 2 Moisture transfer

Structure of membranes prevents transfer of dust or polluted air while enable molecules of water to pass through. There are several driving forces, which cause the transfer of moisture though membrane, such as the difference of density and pressure between both sides of the membrane.

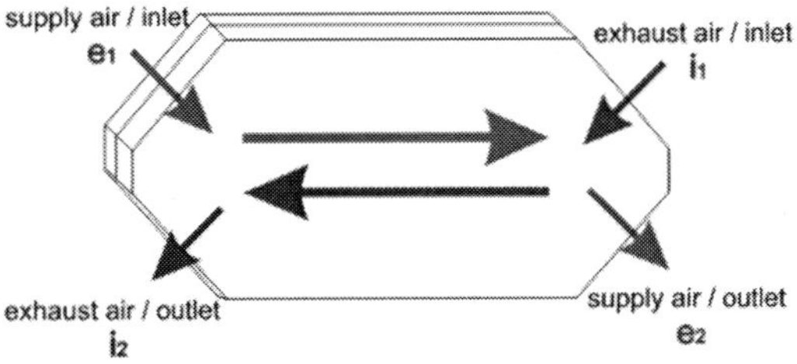


**Figure 1**: Counter flow plate heat exchanger.

[a] Corresponding author: nguyen.van.vu@tul.cz

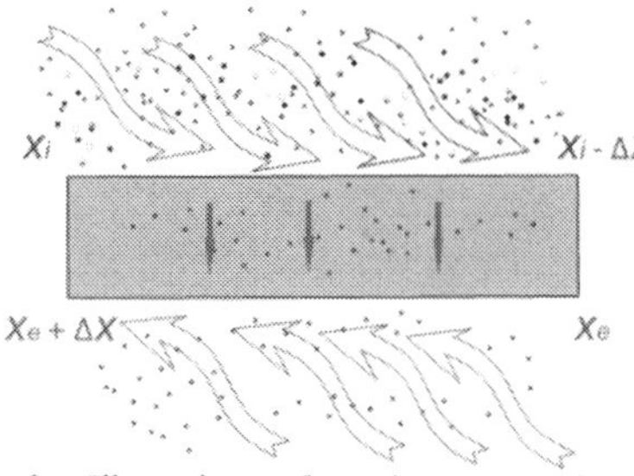


**Figure 2**: Illustration of moisture transfer through membrane.

Membrane separates the exhaust and supplied flow, prevents dust, bacterium and poisonous substances like CO, $CO_2$, $CH_4$ and so on, meanwhile it allows moisture from exhaust air pass through to supplied air.

A specific experimental enthalpy exchanger for testing moisture transfer through membrane was build. Due to the real-life working condition of the enthalpy exchanger, we consider the driving force comes only from the density's difference. Therefore, we tried to ensure temperature, pressure of both flows (dry and moist airflow) are equal to each other during the whole measurement.

Psychometric diagram (figure 3) shows amount of specific humidity $\Delta x$ resp. specific enthalpy of the supply $\Delta h$ and the exhaust air, which are transferred to supply air flow. Their relation is given by equation

$$\Delta h_w = \Delta x(l_0 + c_w\, t), \quad (1)$$

where $l_0$ is specific latent heat of vaporization

$l_0 = 2260\ kJ/kg$ while $c_w = 4180\ J \cdot kg^{-1} \cdot K^{-1}$ is the specific heat capacity of water. By comparing terms in equation (1), it can be rewritten as

$$\Delta h_w = \Delta x\ l_0. \quad (2)$$

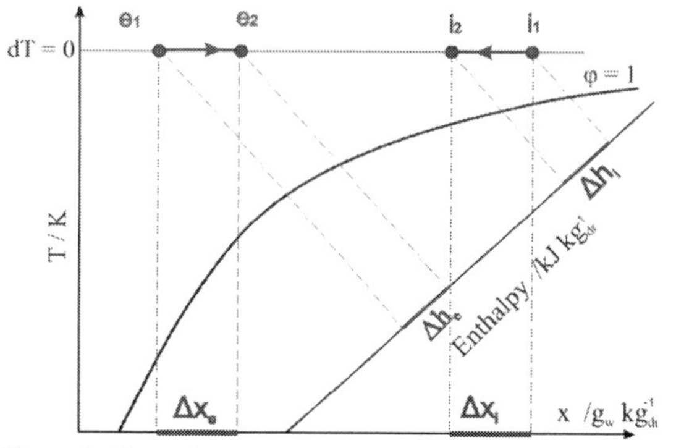


**Figure 3**: Illustration of recovered latent heat in psychometric diagram.

## Experimental arrangement

The figure 4 describes experimental arrangement. Dry compressed air flows to T-junction (4) where it is divided into two ways. One of them is humidified by flowing through a vaporizer (7). By doing that, two levels of humidified air were created: high humidity of average around 75% and low humidity of averaged around 60%. Volumetric flows of humidified and dry airs are controlled by valves (6) and then flow into experimental enthalpy exchanger, where they flow in counter flow arrangement but separated from each other by novel membrane (11). Volumetric flows were set from 200 litres per hour to 2000 litres per hour which correspond to typical flow velocity of (0.26 ÷ 2.58) m/s. We used high precision FHAD36R sensors to measure relative

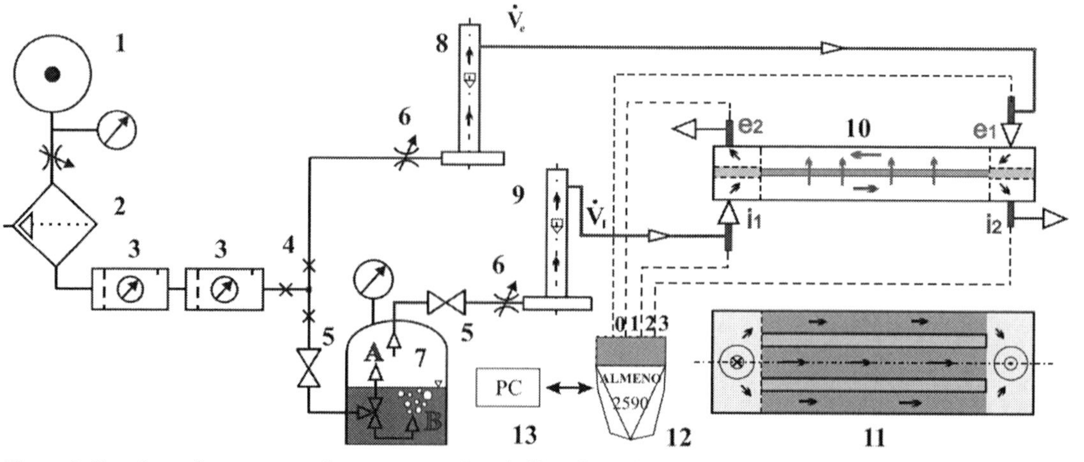


**Figure 4**: Experimental arrangement: 1 - compressor, 2 – air filter, 3 – valves, 4 – T-conjunction, 5 - flap, 6 – smooth valves, 7 - vaporizer, 8 & 9 - rotameters, 10 - experimental enthalpy exchanger, 11 – novel membrane, 12 - Almeno 2590, 12.0 - sensor for dry airflow inlet , 12.1 - sensor for dry airflow outlet , 12.2 - sensor for humidified-inlet airflow, 12.3 - sensor for humidified-outlet airflow, 13 - PC

humidity, temperature, barometric pressure of dry and humidified airflows before and after moisture transfer through novel membrane. Device Almeno 2590 was used to log these values to PC for every second during measurement.

### Effectiveness of moisture transfer

From the equation (2), effectiveness of latent heat transfer could be defined from [9] as

$$\eta_x = \frac{x_{e2} - x_{e1}}{x_{i1} - x_{e1}}, \tag{3}$$

where $x$ is specific humidity defined as

$$x = \frac{m_{moisture}}{m_{dry\ air}}. \tag{4}$$

For calculation specific humidity from relative humidity, equation below is usually used [9]

$$x = 0.622\,\frac{\varphi\, p''_p}{p_b + \varphi\, p''_p}, \tag{5}$$

where $p''_p$ is the partial pressure of saturated vapor calculated by

$$p''_p = e^{23.58 - \frac{4044.2}{235.6+t}}, \tag{6}$$

in which $t$ is temperature of air flow.

Process of moisture transfer through membrane is a sequence of two "sub-process". Firstly, absorption moisture of membrane from exhaust air with more humidity and secondly, moisture release from membrane to supply, dryer air. Relative measuring error of effectiveness of moisture transfer is given by ratio of difference of transferred humidity from both sides of membrane, i.e. difference between absorption and release, and average transferred humidity. For equal mass flow rates of dry air, the relative measuring error of effectiveness of moisture transfer is defined by ratio

$$\varepsilon_x = 2\frac{|\Delta x_e + \Delta x_i|}{|\Delta x_e| + |\Delta x_i|}. \tag{7}$$

## 3 Uncertainty

The uncertainties of experimental measurements are connected to the accuracy, reliability of measuring instruments and the way they are manipulated. Uncertainty of every measuring device is defined as function of two values: random uncertainty and systematic uncertainty.

Standard deviation is defined as

$$s = \sqrt{\frac{\sum(x-\bar{x})^2}{(n-1)}}, \tag{8}$$

where $\bar{x}$ is the mean value and n is the number of measurements

Random uncertainty $\delta_r$ (also known as uncertainty type A) is caused by unknown reasons and can be calculated from standard deviation. It means that random uncertainty follows from measured data as

$$\delta_r = \pm\sqrt{\frac{\sum(x-\bar{x})^2}{(n-1)n}} = \pm\frac{s}{\sqrt{n}}. \tag{9}$$

The systematic uncertainty $\delta_S$ (also known as uncertainty type B) relates to several factors, first is given by the manufacturer and secondly by calibrations. The systematic uncertainty is specified by the manufacturer of the measuring device.

The total uncertainty then is

$$\delta_T = \pm\sqrt{\delta_r^{\,2} + \delta_S^{\,2}}. \tag{10}$$

If quantity $A$ is not measured directly, but it is a function of some other quantities $f(x_i)$ , then uncertainty of quantity $A$ is according work [10]

$$\delta_A = \sqrt{\sum\left(\frac{\partial f(x_i)}{\partial x_i}\delta x_i\right)^2}, \tag{11}$$

where $\frac{\partial f(x_i)}{\partial x_i}$ is partial derivation of quantity $x_i$ and $\delta x_i$ is the uncertainty of $x_i$. In following text, we used these formulas to specify uncertainty of effectiveness of moisture transfer.

Effectiveness of moisture transfer is defined as a ratio of actually transferred (released) humidity into dryer air and maximum theoretical humidity, which can be absorbed from moister air

$$\eta = \frac{\dot{m}_e(x_{e2} - x_{e1})}{\dot{m}_i(x_{i1} - x_{e1})}. \tag{12}$$

The uncertainty of effectiveness is calculated

$$\delta_\eta = \sqrt{\left(\frac{\partial \eta}{\partial \dot{m}_e}\delta_{\dot{m}_e}\right)^2 + \left(\frac{\partial \eta}{\partial \dot{m}_i}\delta_{\dot{m}_i}\right)^2 + \left(\frac{\partial \eta}{\partial x_{i1}}\delta_{x_{i1}}\right)^2 + \left(\frac{\partial \eta}{\partial x_{e1}}\delta_{x_{e1}}\right)^2 + \left(\frac{\partial \eta}{\partial x_{e2}}\delta_{x_{e2}}\right)^2} \quad (13)$$

where

$$\frac{\partial \eta}{\partial \dot{m}_e} = \frac{(x_{e2} - x_{e1})}{\dot{m}_i(x_{i1} - x_{e1})}, \frac{\partial \eta}{\partial \dot{m}_i} = -\frac{\dot{m}_e(x_{e2} - x_{e1})}{\dot{m}_i^2 (x_{i1} - x_{e1})},$$

$$\frac{\partial \eta}{\partial x_{e2}} = \frac{m_e}{\dot{m}_i(x_{i1} - x_{e1})}, \frac{\partial \eta}{\partial x_{e1}} = \frac{\dot{m}_e(x_{e2} - x_{i1})}{\dot{m}_i(x_{i1} - x_{e1})^2}$$

and $\frac{\partial \eta}{\partial x_{i1}} = -\frac{\dot{m}_e(x_{e2} - x_{e1})}{\dot{m}_i(x_{i1} - x_{e1})^2}$ are partial derivations of equation (11) and $\delta_{\dot{m}_e}$, $\delta_{\dot{m}_i}$, $\delta_{x_{i1}}$, $\delta_{x_{e1}}$, $\delta_{x_{e2}}$ are total uncertainties of separated quantities.

From here, we can see that in order to get the uncertainty of moister transfer we have to calculate uncertainties of mass flows and specific humidity. Before that, we need to know other related uncertainties. Calculations of these uncertainties will be shown in following text.

## Temperatures, relative humidity, atmospheric pressure

Device Almeno and sensors FHAD36R were used to monitor and log relative humidity, temperature of airflows and barometric pressure. Sensor FHAD36R has accuracy ± 1% of relative humidity and long-term stability is lower than 1% of relative humidity [11].

Temperatures, relative humidity, atmospheric pressure are three quantities that we got directly from raw data. Their uncertainties then can be derived using equations (8) and (9).

## Saturation pressure of water vapour of moist air

Quality $p''_p$ can be calculated using equation specified in [6], and its uncertainty is function of temperature of the air. We can obtain this value by using partial derivation with the variable of temperature

$$\frac{\partial p''_p}{\partial t} = -\frac{4044.2}{(235.6 + t)^2} e^{23.58 - \frac{4044.2}{235.6+t}}, \quad (14)$$

Then uncertainty of $p''_p$ is then

$$\delta_{p''_p} = \frac{\partial p''_p}{\partial t}\delta t \quad (15)$$

**Specific humidity** $x$ is given by equation

$$x = 0.622 \frac{\varphi\, p''_p}{p_b + \varphi\, p''_p}, \quad (16)$$

where $\varphi$ is relative humidity. The uncertainty of specific humidity is derived as

$$\delta_x = \sqrt{\left(\frac{\partial x}{\partial \varphi} \cdot \delta_\varphi\right)^2 + \left(\frac{\partial x}{p_b} \cdot \delta_{p_b}\right)^2 + \left(\frac{\partial x}{p''_p} \cdot \delta_{p''_p}\right)^2}, \quad (17)$$

## Volumetric flow

It is important to maintain flow rates in two flows at given values during measurement. We use two variable area meters (rotameters) to monitor mass flows; their factory's uncertainty is ± 3%. We also verified their accuracy by using another high precise volume meter. We absorbed that, these rotameters work well at flow rates from 400÷1600 l/h but the position of rotating float became unstable. In addition, pressure of airflow before and after passing the float differed. The actual volumetric flow is given by read nominal volumetric flow and by both, nominal and actual densities by relation

$$\dot{V}_m = \dot{V}_n \cdot \sqrt{\frac{\rho_n}{\rho_m}}, \quad (18)$$

where $\dot{V}_n$ is the nominal value of flow, which is read on the scale of variable area meter, $\rho_n$ is nominal density and $\rho_m$ is actual density of flowing air.

The uncertainty of volumetric flow is

$$\delta_{\dot{V}_m} = \sqrt{\left(\frac{\partial \dot{V}_m}{\partial \dot{V}_n} \cdot \delta_{\dot{V}_n}\right)^2 + \left(\frac{\partial \dot{V}_m}{\partial \rho_n} \cdot \delta_{\rho_n}\right)^2 + \left(\frac{\partial \dot{V}_m}{\partial \rho_m} \cdot \delta_{\rho_m}\right)^2}, \quad (19)$$

### Mass flows

Relation between mass flow rate of dry air $\dot{m}_{dry\,air}$ and measured volumetric flow rate $\dot{V}_m$ is

$$\dot{m}_{dry\,air} = \frac{\rho\,\dot{V}_m}{1+x}\,. \tag{20}$$

The uncertainty of mass flow rate is

$$\delta_{\dot{m}} = \sqrt{\left(\frac{\partial \dot{m}}{\partial \dot{V}}\cdot\delta_{\dot{V}}\right)^2 + \left(\frac{\partial \dot{m}}{\partial \rho}\cdot\delta_{\rho}\right)^2 + \left(\frac{\partial \dot{m}}{\partial x}\cdot\delta_{x}\right)^2}\,, \tag{21}$$

where partial derivatives are

$$\frac{\partial \dot{m}}{\partial \dot{V}} = \frac{\rho}{1+x}, \tag{22}$$

$$\frac{\partial \dot{m}}{\partial \rho} = \frac{\dot{V}}{1+x}, \tag{23}$$

and

$$\frac{\partial \dot{m}}{\partial x} = -\frac{\rho\,\dot{V}}{1+x^2}, \tag{24}$$

### Density of moist air

Density of moist air according to [9] is calculated from equation of state of ideal gas as a mixture of dry air and water vapour

$$\rho_{moist} = \frac{p_{dry\,air}}{r_{dry\,air}\,T_0} + \frac{p_{moist}}{r_{moist}\,T_0}\,. \tag{25}$$

With room condition we can use bellow equation [9]

$$\rho_{moist} = \frac{1.316\cdot 10^{-3}}{T_0}(2.56\cdot p + \varphi\cdot p''_m)\,. \tag{26}$$

In the same way, we calculated the density's uncertainty of moist air.

## 4 Results

### Uncertainties' source analysis

First, we evaluate the uncertainty of the moisture transfer. Effectiveness and its uncertainty at the flow rate 200 l/h is $(63.89 \pm 1.53)\%$. Contributions of uncertainty of individual quantities to resulting uncertainty of effectiveness are in table 1. Measurement of mass flow rates contributes to the resulting relative uncertainty by 2.4% each. The rest uncertainty is caused by measuring values of specific humidity and the highest contributions of 51.8% and 41.8% are given by measurement of $x_{i1}$ and $x_{e2}$ respectively, while influence of $x_{e1}$ is negligible and of 2%.

That means the uncertainties of specific humidity have the dominant impact on uncertainty of measuring moisture transfer. Analysis components of uncertainty of measuring specific humidity unveils the dominant causes (table 2.). Which are the uncertainty of relative humidity, contributes 28.6% and the uncertainty of partial saturated moist air, contributes 71.4%. In other hand, partial saturated moist air is the function of only temperature (equation14.). Therefore, reducing the uncertainty of measuring flows' temperature and relative humidity will help reducing uncertainty of efficiency measurement.

### Moisture transfer's efficiency of membranes

Figure 5 shows the curve of effectiveness of presented membrane: the lower the velocity of airflow is, the greater is the effectiveness. It is clear that the velocity of airflows has great impact on the value of effectiveness. Along with that, graph also presents the uncertainty of measurement. The greater efficiency is, the bigger uncertainties values are.

### Error and uncertainty of measurement

Error in this paper presents the difference between moisture absorbed and released by membrane. The measurement was carried out when the moisture absorption and release process of membrane are stable. However, it is difficult to prove that these processes are parallel, especially due to accumulation of water inside membranes.

While uncertainty of measurement is relevant, the error values is considerable (table 4). This can be expressed in term of "high precision but less accuracy", means that measured values are close to each other but far from the "true" or acceptable values, read [12]. For this particulate sample, the error values are fairly small but for some others the errors are greater. Empirically, we noticed that the error values has strong connection to the ability of moisture transfer of membrane - if a membrane has a high moisture transfer effectiveness, then low measuring error is obtained and vice versa (with other conditions remain unchanged).

**Table 1**: Individual contributions in the uncertainty budget of effectiveness measuring (resulting of equation 12)

| **Function** | $\left(\frac{\partial \eta_m}{\partial m_i}\delta_{m_i}\right)^2$ | $\left(\frac{\partial \eta_m}{\partial m_e}\delta_{m_e}\right)^2$ | $\left(\frac{\partial \eta_m}{\partial x_{i1}}\delta_{x_{i1}}\right)^2$ | $\left(\frac{\partial \eta_m}{\partial x_{e1}}\delta_{x_{e1}}\right)^2$ | $\left(\frac{\partial \eta_m}{\partial x_{e2}}\delta_{x_{e2}}\right)^2$ |
|---|---|---|---|---|---|
| **Value** | 0.0004 | 0.0004 | **0.0079** | **0.0003** | **0.0064** |
| **Contribution** | 2.4% | 2.4% | **51.4%** | 2% | **41.8%** |

**Table 2**: Contributions to uncertainty of specific humidity measuring

| **Function** | $\left(\frac{\partial x}{p_b}\cdot\delta_{p_b}\right)^2$ | $\left(\frac{\partial x}{p''_p}\cdot\delta_{p''_p}\right)^2$ | $\left(\frac{\partial x}{\partial \varphi}\cdot\delta_{\varphi}\right)^2$ |
|---|---|---|---|
| **Value** | $1.5\cdot10^{-13}$ | $3.7\cdot10^{-9}$ | $1.5\cdot10^{-9}$ |
| **Contribution** | $2.8\cdot10^{-3}$% | **71.4%** | **28.6%** |

**Table 3**: Correlation between absolute uncertainty and flow's velocity

| **Flow's velocity *m/s*** | 0.26 | 0.52 | 0.78 | 1.03 | 1.29 | 1.55 | 1.81 | 2.07 | 2.33 | 2.58 |
|---|---|---|---|---|---|---|---|---|---|---|
| **Efficiency** % | 63.89 | 45.93 | 39.38 | 33.72 | 30.07 | 26.88 | 24.75 | 23.62 | 22.62 | 21.64 |
| **Uncertainty %** | 1.53 | 1.39 | 1.15 | 0.98 | 0.82 | 0.73 | 0.63 | 0.55 | 0.48 | 0.43 |
| **Error %** | 4.29 | 7.33 | 6.44 | 5.21 | 5.85 | 5.37 | 5.35 | 4.99 | 4.86 | 3.4 |

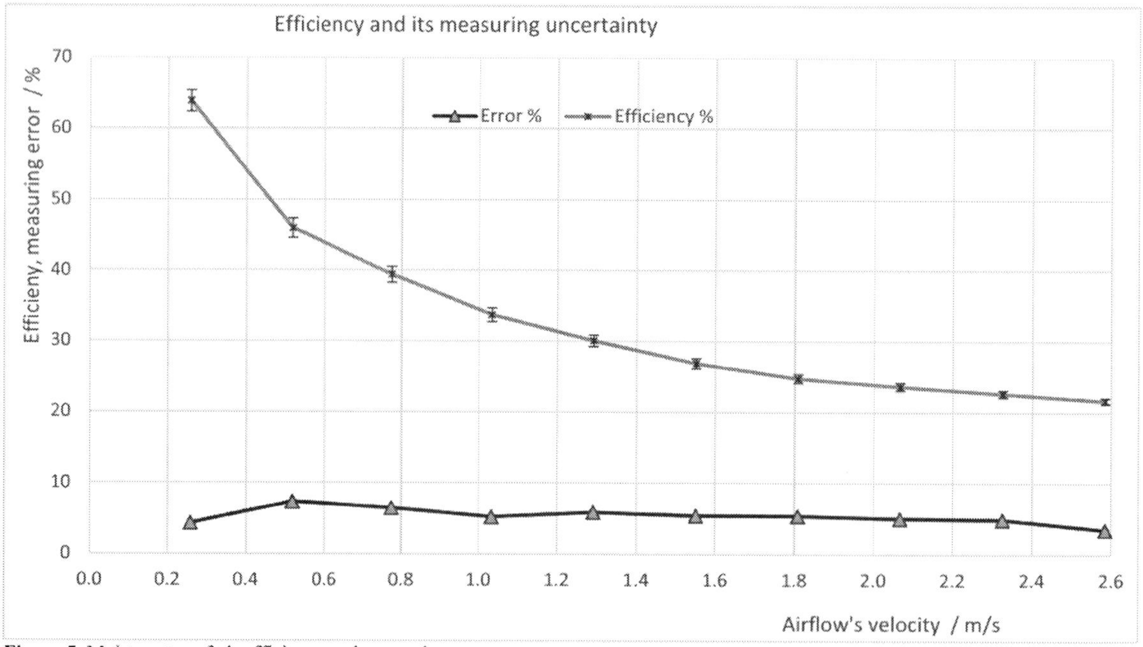


**Figure 5**: Moisture transfer's efficiency and measuring error

## 5 Discussion and conclusion

By analysis measuring uncertainties we unveiled that the uncertainty of sensors FHAD36R cause the imprecision. Actually, according to manufacturer, those sensors are very accuracy and precision. We may need to perform a calibration on them, or try other devices to compare results.

Although the measuring error was acceptable, there are still possible ways to improve measuring procedure. Based on experiences during this experimental investigation, we would like to consider improving our testing device. First, we need to log mass flow rates continuously into PC during the entire testing process by precise digital mass flow meters. By doing this, we can reduce uncertainty of flow rates.

The change of inlet relative humidity due to change in airflow velocity is carried out in table 4. Upper row is the velocity of airflow (m/s) and the lower row is the corresponding relative humidity (%). This phenomenon makes the evaluation of effectiveness less impact. Ideally, humidity of airflows should be stable and constant across for all flow velocities and all tested membranes.

**Table 4**: Variation of inlet relative humidity relevant to airflow's velocity

| Velocity of flow (m/s) | 0. 26 | 0. 52 | 0. 78 | 1. 03 | 1. 29 | 1. 55 | 1. 81 | 2. 07 | 2. 33 |
|---|---|---|---|---|---|---|---|---|---|
| RH (%) | 63 .0 | 64 .6 | 65 .9 | 67 .7 | 69 .4 | 71 .0 | 73 .9 | 76 .0 | 76 .3 |

Second, we will add a soft-seat valve right after the flap (5) to control more precisely the mass flow rate of air entering the vaporizer or design and use more convenient vaporizer.

Third it would be beneficial to use different approach to specify partial pressure of saturated vapour than equation (6).

Ambient temperature should be maintained unchanged during measurement (sometime lab's temperature changed significantly during a single measurement).

## Acknowledgment

We gratefully acknowledge the support of the Student Grant Agency of the Technical University of Liberec (project no. 21124).

## References

[1] S. M. Aarnes, "Membrane Based Heat Exchanger.," 2012.

[2] L. Z. Zhang and J. L. Niu, "Effectiveness Correlations for Heat and Moisture Transfer Processes in an Enthalpy Exchanger With Membrane Cores," *J. Heat Transfer*, vol. 124, no. 5, p. 922, 2002.

[3] W. Yaïci, M. Ghorab, and E. Entchev, "Numerical analysis of heat and energy recovery ventilators performance based on CFD for detailed design," *Appl. Therm. Eng.*, vol. 51, no. 1–2, pp. 770–780, 2013.

[4] Z.-X. Li, T.-S. Zhong, J.-L. Niu, F. Xiao, and L.-Z. Zhang, "Conjugate heat and mass transfer in a total heat exchanger with cross-corrugated triangular ducts and one-step made asymmetric membranes," *Int. J. Heat Mass Transf.*, vol. 84, pp. 390–400, May 2015.

[5] L. Z. Zhang and J. L. Niu, "Effectiveness Correlations for Heat and Moisture Transfer Processes in an Enthalpy Exchanger With Membrane Cores," *J. Heat Transfer*, vol. 124, no. 5, p. 922, 2002.

[6] V. Dvořák. P. Novotý, N. Van Vu, "Measurement of Moisture Transport in the Membrane–Based Enthalpy Exchanger," *EFM*, 2013.

[7] L. Z. Zhang, "Heat and mass transfer in a quasi-counter flow membrane-based total heat exchanger," *Int. J. Heat Mass Transf.*, vol. 53, no. 23–24, pp. 5499–5508, 2010.

[8] N. Van Vu, "Evaluation properties of membranes for enthalpy heat exchanger," Technical univesity of Liberec, 2015.

[9] J. Chyský, K. Hemzal a kol., *Větrání a klimatizace, Technický průvodce*. Praha: SNTL, 1993.

[10] J. Sulc, V. Dvorak, "Technická měření - návod k cvicení," Liberec: Vysokoškolský podnik, 2005, p. 60.

[11] Ahlborn, "Sensor fhad36r," 2012. .

[12] J. R. Taylor, *An Introduction to Error Analysis: The Study of Uncertainties in Physical Measurements , 2d Edition*. University Science Books, 1997.

EPJ Web of Conferences **143**, 02081 (2017) DOI: 10.1051/epjconf/201714302081
*EFM 2016*

# Design of MSR primary circuit with minimum pressure losses

*Tomáš* Noga[1,*], *Pavel* Žitek[1], and *Václav* Valenta[1]

[1]University of West Bohemia, Faculty of Mechanical Engineering, Department of Power System Engeneering, Univerzitni 8, 306 14 Plzen, Czech Republic, {zitek, noga, valenta}@kke.zcu.cz

**Abstract.** This article describes a design of a MSR primary circuit with minimum pressure losses. It includes a brief description of this type of a reactor and its integral layout, properties, purpose, etc. The objective of this paper is to define problems of pressure losses calculation and to design a proper device for a primary circuit of MSR reactor, including its basic dimensions. Thanks to this, it can become an initial project for a construction of a real piece of work. This is the main contribution of the carried out study. Of course, this article is not a detailed solution, but it points out facts and problems, which future designers may have to face. The further step of our work will be a reconstruction of the current experiment for a two-stage flowing.

## 1 IV Generation Molten Salt Reactor

The development policy for these reactors has been planned so that at least some of them are prepared to be used in 2020 to 2030, when a service life of currently used reactors will have expired. They are reactors able to provide us with sufficient energy safely and economically. Molten salts will be used as coolants; substances enabling an operation under much higher temperature and consequently efficiency as well.

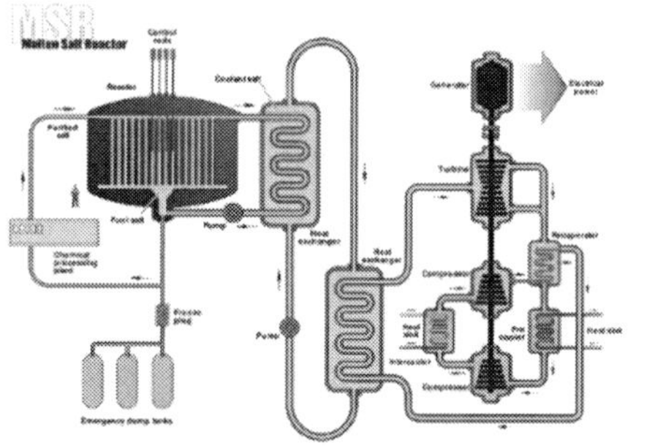

**Fig. 1.** MSR reactor

From a long-term point of view of nuclear power use, it is necessary to assure sufficient fuel for reactors. Therefore it is necessary to introduce fast reactors, which can use plutonium Pu as well as uranium 238U. They are capable to provide enough fuel for milleniums in a configuration of a fast breeder reactor [1,2].

## 2 MSR Primary circuit

Nuclear fuel is dissolved directly in a fluoride molten salt of a primary circuit and it flows around a graphite moderator along with dissolved fission products. This type of a reactor has several benefits. It allows for broader possibilities of processing nuclear fuel ($Th^{-233}$ U cycle of denatured $Th^{-233}$), including the use of a fast breeder reactor regime for $Th^{-233}$U cycle. It provides a better "on-line" control of the whole reactor (fuel is liquid, homogeneous; so-called campaigns related to the fuel replacement in reactors with solid propellants are dropped) [5].

**Table 1.** Reference parameters of MSR system

| Reactor parameters | Reference value |
|---|---|
| Assumed power | 1000 MWe |
| Coolant inlet/outlet temperature | 565/700 °C at pressure of approx. 0,5 MPa |
| Specific volume power | 22 MWt/m$^3$ |
| Net efficiency | 44 to 50 % |

On the other hand, this type of a reactor is very sensitive to corrosion and material compactibility (continuous operation, increased corrosion activity of molten salts with dissolved nuclear fuel and dissolved fission products to construction material). This reactor concept offers probably the largest complex benefit potentially out of all the IV generation reactor types (particularly in its fast breeder version), but also the biggest technological challenge, despite successful testing. These advanced reactors should include all so far known positive experience; they represent the needed safety standard, high operational reliability, long-term life cycle and economical competitivness with other sources. Fuel consisting of molten fluorides of uranium

* noga@kke.zcu.cz

$UF_4$ or plutonium $PuF_3$ dissolved in a mix of molten fluorides LiF + $BF_2$ + NaF circulates in the MSR reactor. The fuel will be in a form of molten fluorides of uranium, sodium and zirconium, a moderator is solid graphite. The salts used have melting points at 425 to 510 °C. They are perfectly liquid at working temperatures of approx. 550 to 750 °C. The salts don't actively react with air nor water, which significantly reduces construction problems. Steam tension is very low and thus the system can be non-pressure operated. The salts have very good corrosion characteristics and some materials conforming these systems have already been tested, e.g. MoNiCr (Škoda JS) [6].

### 2.1 Determination and composition of material properties of fluoride salts for a primary circuit

The melting temperature was required to be ≤ 500 °C for the selection of a salt mixture. Minimum operational temperature is considered $T_{min} = T_{tav} + 100$ °C and its determination was based on experience from projects of fast sodium reactors as well as a course of solubility of $P_uF_3$. In the MSBR (Molten Salt Breeder Reactor) project, working in a thorium cycle, the temperature of 66 °C was used for a difference of a minimum operational temperature and the melting temperature. It can influence favourably a heat exchanger design but a reduction of a minimum operational temperature will also decrease solubility of tri – fluorides. Maximum temperature of a long-term run is $T_{max} \leq 710$ °C and it was determined on the basis of Hastelloy N type material properties [8].

**Table 2.** Values usable for thermo-physical analysis

| Characteristics and temperature | Equation considering M [g mol$^{-1}$] |
|---|---|
| **Density ρ [g/cm$^3$]** | Relative error ± 0,9 % |
| 800 K = 527 °C | $\rho = 1{,}9801 + 0{,}00421$ M |
| 1000 K = 727 °C | $\rho = 1{,}8106 + 0{,}00578$ M |
| **Viscosity η [cP]** | Relative error ± (4 až 6)% |
| 880 K = 527 °C | $\eta = -\ 24{,}606 + 0{,}85068$ M |
| 900 K = 627 °C | $\eta = -\ 9{,}041 + 0{,}37085$ M |
| 1000 K = 727 °C | $\eta = -\ 4{,}949 + 0{,}22002$ M |
| **Thermal capacity $c_p$ [kJ kg$^{-1}$K$^{-1}$]** | Relative error ± (10 až 20)% |
| 800 – 1000 K (527 – 724 °C) | $c_p = 3{,}73 - 0{,}037$ M |
| **Thermal conductivity λ [W m$^{-1}$ K$^{-1}$]** | Relative error ± 15% (total) |
| 800 K = 527 °C | $\lambda = 1{,}58 - 0{,}01796$ M |
| 900 K = 627 °C | $\lambda = 1{,}63 - 0{,}01796$ M |

We will determine the basic properties of fluoride salts according to the previous table, where *M* is molar weight. The purity of salts is important for their use and a failure to keep it may lead to increased corrosion on construction materials. Purity of inert gas is required as well. When using helium, the content of impurities (O, N, C, $CO_2$, water vapour) should be lower than 0,01% mol, as apparent from experience of operating helium-cooled reactors [8].

#### *2.1.1 Advantages and disadvantages of fluoride salts*

A large number of salt mixtures of various concentrations of particular elements is possible to use to cool down the reactor core. Single component salts are not used due to their higher melting points compared to two and multi component salts. Advantages of molten salts, in comparison with most of used coolants, can be summarized generally in four points:

- high specific thermal capacity
- high boiling point
- low pressures of saturated vapours under high temperatures
- low melting point

The following points rank among the main disadvantages of fluoride salts:

- high demands on liquid salts purity
- need of a perfect moisture removal from liquid salts – when in a contact with $H_2O$, hydrofluoric acid (HF) is created, which has strong corrosive effects
- relatively high solidification temperature of liquid salts
- high demands on construction materials – temperature, corrosive activity of liquid salts
- need of an embedded secondary circuit for radiation safety [8]

### 2.2 Design of an integral layout of a MSR primary circuit

An integral layout design of a homogeneous MSR with a gas lift is necessary when preparing a project. We have to provisionally design a pressure vessel considering the size of a core. We will apply experience from the MSBR project to select the wall thickness t = 60 mm. We are limited in the height since a compressive chimney is recommended to be H < 10 m. The tank diameter is chosen Dn = 6 m, so that we can place 6 plate exchangers in the upper part and the reflector zone.

Let's not forget that the gas lift efficiency is influenced by the height, in which the exchanger and the active zone are located. The problem is a volume compensator. We will use the analogy with a gas lift model where we obtain a space for the released gas (He). We suggest channels, between which there is enough space for regulating rods and He supply into the gas lift, to lead primary salts to particular heat exchangers. Delivery of secondary salts into the pressure vessel requires sensors on a reactor pressure vessel. There is a freezing valve located in the bottom part of the pressure vessel design for draining primary salts. The preliminary design enables to solve another tasks, such as: :

- He management
- solution of passively cooled subcritical vessels with a freezing valve providing emergency shutdown.
- solution of a reactor regulation. Solution of solid fissile products filtration.
- computations of a pressure vessel wall shielding.

The gas lift efficiency goes up along with minimizing pressure dissipations, i.e. the pipeline design, 90 degrees arcs, bevelling, narrowing, extension, etc.

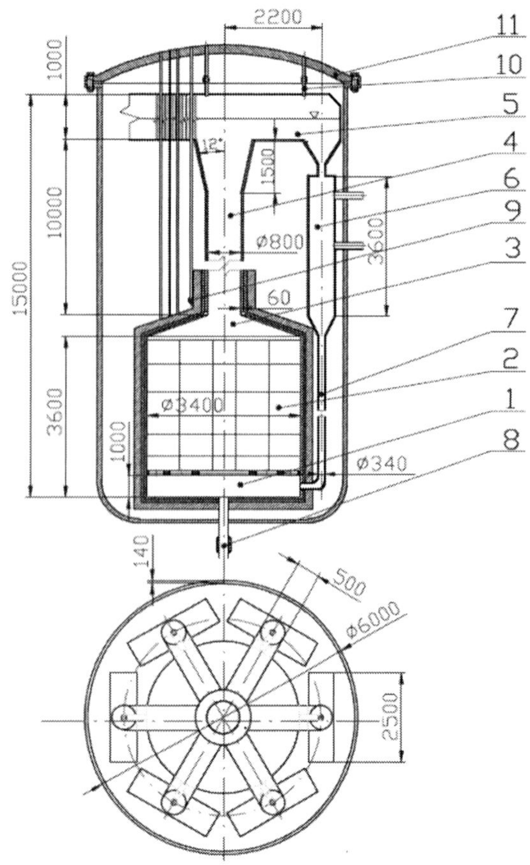


**Fig. 2.** Design of component layout of a MSR primary circuit with minimum pressure dissipations

We will place a ring in the bottom part of the chimney, into which we will deliver inert gas (helium) and inject it in the chimney through holes. As a result, density of fluoride salts reduces and a circulation improves. The gas will exhaust, being recycled and purified from gaseous fissile products. Consequently, it will be again used in the circuit.

**Table 3.** Description of a primary circuit components

| Position | Description | Position | Description |
|---|---|---|---|
| 1 | Mixing chamber | 7 | Cold leg |
| 2 | Core | 8 | Outlet valve |
| 3 | Transition to chimney | 9 | Ring - helium inlet |
| 4 | Chimney | 10 | Helium outlet |
| 5 | Volume compensator | 11 | Top header |
| 6 | Counter-flow exchangers | | |

The use of a gas – lift in a primary MSR circuit offers a lot of benefits. The removal of a mechanical device, which needs a power source for its operation, significantly contributes to the increase of a reliability and safety of a nuclear power device.

All mechanical devices are outside the core, therefore it is possible to carry out repairs and adjustments while the reactor is running by switching to a back-up device. Furthermore, another advantage is a lower demand on energy and the fact that it is possible to be used even in reactors, which work with a medium of a temperature higher than 1000 °C, for which a pump doesn´t exist yet. Current industrial pumps can function up to 600 °C and new generations of pumps capable of working up to 700 °C are beginning to be introduced [4].

## 3 Calculation of pressure losses of a MSR primary circuit

Thermal and hydraulic calculation of a nuclear power reactor along with physical, strength and technical-economic calculations of all components and the primary circuit itself serve to proceed the primary circuit project aiming at optimization and assuring nuclear safety.

Thermal and hydraulic calculation determine, along with a physical calculation, thermal power released by a nuclear reaction along the cross-section and height of the core. Further calculations determine the heat medium flow through the core, temperatures on the mixture surface and inside it, pressures in a circulation circuit and other parameters of the medium flowing through the primary circuit. Therefore it is necessary to know or estimate thermohydraulic characteristics of the primary circuit and its regime parameters, along with construction material, thermal and hydraulic characteristics of particular reactor components. [3]

The calculation of a designed primary circuit proceeds as follows:

### 3.1. Process of a calculation of hydraulic losses of a primary circuit

First, we will define basic dimensions of a primary circuit (*Fig. 2*). We will consider an integral setup in a reactor vessel of a homogeneous MSR reactor. The main construction material for a high-temperature reactor primary circuit is graphite and its composites.

Before proceeding, it is worth preparing a list of local losses and giving some thoughts to construction modifications leading to their minimizations. Dimensions, height data, temperature, subcriticality check outside the core, that all should be listed for individual parts of a primary circuit.

We have to specify the main parameters of the core based on requirements – e.g. we can determine the thermal power $P_t$ from the requirement on electric power production with the help of efficiency $\eta$.

$$P_t = \frac{P_{el}}{\eta} \tag{1}$$

To determine the thermal power, we will choose the mean specific power $q$, it is for the gas - lift:

$$q = 25 \div 40 \left[\frac{MW_t}{m^3}\right] \tag{2}$$

Based on the above mentioned data we can define the core volume $V$, furthermore a suitable diameter $D$ and height $H$. The diameter of a chimney $d$ can be determined from a subcriticality condition ( $k_{ef} \leq 0{,}95$).

We will proceed the calculation of the theoretical mass flow $W_{\infty T}$ from the free flow according to the primary circuit design:

$$W_{\infty T} = \frac{P_t}{\Delta T . c_p} \tag{3}$$

We know how to determine the mass flow rate from the above stated formula. We will choose a methodology to calculate the gas – lift. The most suitable model for nuclear reactors is a disperse bubble flow model, as it provides maximum intensification of natural flow and ensures an impact-free operation.

We can determine velocities in individual parts by the following relation:

$$v_p = \frac{W_{\infty T}}{\rho_p . A_p} \tag{4}$$

Friction losses must be considered in each part in order to calculate the total losses

$$\Delta p_{p\,fr} = \lambda . \rho_p . \frac{L_p . v_p^2}{D_p . 2} \tag{5}$$

and local losses

$$\Delta p_{p\,l} = \xi_p . \rho_p . \frac{v_p^2}{2} \tag{6}$$

The total friction losses in a $p$ part will be:

$$\chi_{tf} = \Delta p_{p\,fr} + \sum_{n=1}^{N} \Delta p_{p\,l} \tag{7}$$

After we have determined the total pressure losses, we are able to determine the real stabilized mass flow $W_{\infty R}$ for a natural flow in a primary circuit by using the gas - lift [3,4,7].

$$W_{\infty R} = \left[\frac{2 . \rho^2 . g . \beta . P_t}{\chi_{tf} . c_p} . (z_e - z_{pc})\right]^{\frac{1}{3}} \tag{8}$$

where : $W_{\infty R}$ real mass flow rate in a balanced state
$\rho$ mean density of a coolant in a circuit
$g$ gravity acceleration
$\beta$ thermal expansion coefficient
$z_e$ , $z_{pc}$ mean heights of locations of a heat exchanger and core in a primary circuit
$c_p$ thermal capacity of the coolant
$\chi_{tf}$ total hydraulic losses in a primary circuit

The relation (8) shows the dependence of the zone $P$, flow-carried, on the cube of the mass flow rate $W_{\infty R}$. When using the "gas-lift", the following applies for the chimney with a cross-section

$$A_{ch} = \frac{\pi}{4} . D_{ch}^2 \, [m^2] \tag{9}$$

according to the law of conservation of mass

$$W_{\infty}^{gas-lift} = (W_{\infty R} + W_g) \left[\frac{kg}{s}\right] \tag{10}$$

For the use of $W_{\infty}^{gas-lift}$, it is generally necessary to use $c_p$ and ρ for a double-phase mixture, as well as to determine new losses at the same time, as $W_{\infty}^{gas-lift}$ enlarges the flow velocity in the whole primary circuit. The volume of gas supplied by $m$ holes into the chimney of the gas – lift $W_g \left[\frac{kg}{s}\right]$ can be determined by the following expressions:

This relation was derived for $W_g$

$$W_g = C_d . \frac{\pi}{4} . d^2 . m . p_1 . \sqrt{\frac{1}{r . T_g} - \left(\frac{2 . \kappa}{\kappa - 1}\right)} . \sqrt{\left(\frac{p_2}{p_1}\right)^{\frac{2}{\kappa}} - \left(\frac{p_2}{p_1}\right)^{\frac{\kappa-1}{\kappa}}} \tag{11}$$

Where: $m$ is a number of holes of a diameter $d$
$p_1$ gas pressure in front of the throttle hole
$p_2$ gas pressure behind the throttle hole

According to the anglo-saxon literature, the dissipation coefficient is determined by the relation:

$$C_d = \mu . \sqrt{\left[1 - \frac{D^2}{d^2}\right]^{-1}} \tag{12}$$

Where: μ outlet coefficient

$D$ channel diameter

The following relation applies for the critical coefficient

$$K_{cr} = \left(\frac{p_2}{p_1}\right)_{crit} = \left(\frac{2}{\kappa+1}\right)^{\frac{\kappa}{\kappa-1}} \tag{13}$$

### 3.3. Evaluation and conclusion

The meaning of the calculation is to optimize basic circuit dimensions and summarize concrete local and friction losses, see Fig. 2, and particular local and friction losses are summarized, whereas they were calculated onto the whole primary circuit of the reactor, see the results stated in Tab. 3. Dissipation coefficients $\xi_p$ , $\lambda$ for each part of the primary circuit were calculated as well. Let's point out that losses are summed up when counting presure losses in a series primary circuit, however, they are considered only once in case of a parallel system. It means that when having six loops, pressure loss from six or more loops will be the same since it is a parallel system.

The following table summarizes pressure losses in particular components. As apparent, the biggest local pressure loss occurs at the inlet to a mixing chamber, this loss could be reduced by a gradual expansion of an inlet neck; it is a detailed issue which requires more attention and a more detailed and precise solution in the future.

**Table 3.** Concrete pressure losses in particular places

| Component | Name of the component | Pressure loss |
|---|---|---|
| 1. | Core | 3,3 Pa |
| 2. | Inlet into chimney (cone restriction) | 27,7 Pa |
| 3. | Chimney | 14550,7 Pa |
| 4. | Outlet from chimney (cone expansion) | 721,6 Pa |
| 5. | Channel | 360,7 Pa |
| 6. | Inlet into heat exchanger (cone restriction) | 512,2 Pa |
| 7. | Heat exchanger | 22800 Pa |
| 8. | Outlet from heat exchanger (cone expansion) | 256,5 Pa |
| 9. | Cold leg | 24943,3 Pa |
| 10. | Cold leg arc | 10477,1 Pa |
| 11. | Inlet into mixing chamber | 62451,7 |
| 12. | Mixing chamber screen | 55000 Pa |
| 13. | Total pressure losses of primary circuit | 0,196 MPa |

This calculation aimed at more concrete examples of local and friction dissipations as well as relations defining them and proving which parameters or dimensions influence the given dissipations the most. As my hydraulic calculation must correspond with physical and thermal calculations I certainly acknowledge individual minimal changes in parameters and dimensions. The main objective was to design a circuit for the further project, which will provide a more detailed solution of this matter.

This work has been supported by the project LQ1603 (Research for SUSEN) and student project SGS-2016-045 (Improving the efficiency, reliability and service life of power machines and equipment 4).

## References

1. A. Mlčúch, *Jaderné reaktory 4. Generace,* (Diplomová práce, Brno, 2012)
2. M. Katzer, *Jaderné reaktory IV. Generace,* (Bakalárska práce, Brno, 2011)
3. V. Valenta, *Postupy výpočtu pro řešeni „gas-liftu" pro MSR,* zpráva ZČU-FST-KKE, (červen 2012)
4. V. Valenta, *Dvoufázové proudení v problémech jaderné energetiky a techniky, ZČU-FST,* (leden 2012)
5. S. Goldberg, R. Rosner, *Nuclear Reactors : Generation to Generation,* (2012-05-21)
6. J. Uhlíř, P. Souček, *MSR Technologie,* Ústav jaderného výzkumu Řez a.s., (2013)
7. J. Linhart, *Termomechanika – stručné učební texty.* Plzeň-ZČU-FST
8. V. Valenta, *Návrh vhodné směsi fluoridových solí pro transmutor $PuF_3$ a vyšších aktinidu,* Praha : Ústav jaderné fyziky, (2005)

EPJ Web of Conferences 143, 02082 (2017) DOI: 10.1051/epjconf/201714302082
*EFM 2016*

# Influence of the cutting parameters on the workpiece temperature during face milling

*Lukasz* Nowakowski[1,*], *Michal* Skrzyniarz[1], *Edward* Miko[1], *Jakub* Takosoglu[1], *Slawomir* Blasiak[1], *Pawel* Laski[2], *Gabriel* Bracha[2], *Dawid* Pietrala[2], *Jaroslaw* Zwierzchowski[2] and *Malgorzata* Blasiak[3]

[1] Chair of Mechanical Engineering and Metrology, Kielce University of Technology; Aleja Tysiąclecia Państwa Polskiego 7; 25-314 Kielce; Poland
[2] Faculty of Mechatronics and Mechanical Engineering, Department of Automation and Robotics, Kielce University of Technology, Aleja Tysiaclecia Panstwa Polskiego 7, 25-314 Kielce, Poland
[3] Department of Mechanics, Kielce University of Technology; Aleja Tysiąclecia Państwa Polskiego 7; 25-314 Kielce, Poland

**Abstract.** This thesis presents the outcome of experimental research of the impact of changes in cutting speed and volume of material processed during a face milling process on the temperature of the processed object made of copper of M1Ez4 class. Measurement of the temperature of the processed object was conducted in six points with K-type thermocouples. The theoretical amount of released heat per unit of time for particular parameters of machining was also calculated.

## 1 Introduction

During the mechanical processing of the machining process, over 99.5% of machining work is transformed into heat [1-3]. The amount of heat is released per unit of time, so the stream of heat corresponds to the approximate machining power [3-6]. The source of heat in the machining process includes: work of plastic deformations and division of material (formation of chips), friction on the contact surface and rake face [2-11]. Heat generated as a result of the machining process is dissipated through chips, acquired by the tool and the processed material, and raised up to the ambient atmosphere or acquired by the cooling and lubricating liquid [4-6]. In the case of ISO N materials – so non-ferrous metals such as aluminum, copper, brass, etc. the percentage distribution of heat streams may amount to about 70% for the processed object, 20% for chips, and 2% for the cutting edge [12-14]. The uncontrolled temperature increase of the processed object may lead to a performance decrease of parts of machines, machine tools [15-19], and devices i.e. guides, proximity face seals, pneumatic valves [20-27], and precision devices used to control flying objects [28-30]. Increased temperature of the processed object during processing may lead to changes in the structure of the surface layer [31-33], decrease of processing accuracy [15-19], [34, 35] related to thermal expansion of the processed material, as well as its deformation [36, 37] resulting from uneven heating. Mistakes in performance of precision parts caused by uncontrolled heating of the processed object during face milling may result in improper operation of the device e.g. by faster wear of surfaces that work together, which may cause vibration and friction resistances [38 - 40].

## 2 Methods

The subject of studies was to determine temperature changes of the processed object during face milling of copper. The research was performed on samples made of M1Ez4 copper (99.95% Cu) with six predrilled holes for thermocouples. The size and placement of holes has been presented in Fig. 1.

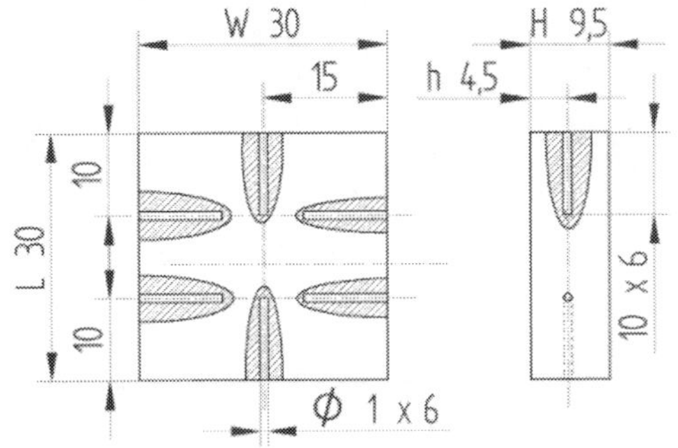


**Fig. 1.** View of workpiece

The process of face milling of 30 x 30 mm surface of copper samples has been conducted on a vertical machining center AVIA VMC800 (fig. 2) with the usage of the cutting head R245-080Q27-12M manufactured by Sandvik Coromant with machining diameter of $D_c$ = 80 mm, facilitated with 245-12T3M-PL4230 plates. Machining attempts have been conducted at two machining velocities $v_c$= 250 and 375 m/min, at a fixed feed rate on the edge of $f_z$=0.05 mm/tooth and fixed depth of cut of $a_p$=0.5 mm for which the distance from thermocouples "h" and volume of the processed material

* Corresponding author: lukasn@tu.kielce.pl

were variable (table 1). Detailed parameters of machining have been presented in table 2.

**Table 1.** Workpiece dimensions

| | Length L, mm | Width W, mm | Height H, mm | Distance h, mm | Workpiece vol., mm$^3$ |
|---|---|---|---|---|---|
| 1 | 30 | 30 | 9.5 | 4.5 | 8550 |
| 2 | | | 9 | 4 | 8100 |
| 3 | | | 8.5 | 3.5 | 7650 |
| 4 | | | 8 | 3 | 7200 |
| 5 | | | 7.5 | 2.5 | 6750 |
| 6 | | | 7 | 2 | 6300 |

Measurement of the temperature of the processed object was conducted through six K-type thermocouples with diameter of 1 mm placed in six holes of 10 mm depth distributed as in fig. 1. Fig. 3 presents the measurement station and the sequence of placing thermocouples in the sample.

**Table 2.** Parameters of machining

| | | |
|---|---|---|
| **Cutting speed** | $v_c$ =250 m/min | $v_c$, 375 m/min |
| **Spindle rotation** | n = 855 rpm | n = 855 rpm |
| **Feed per tooth** | $f_z$ =0.05 mm/tooth | |
| **Feed rate** | $v_f$=513 mm/min | |
| **Depth of cut** | $a_p$ =0.5 mm | |
| **Width of cut** | $a_e$= 30 mm | |

**Fig. 2.** View of vertical machining center AVIA VMC 800

**Fig. 3.** Measurement station and the sequence of placing thermocouples in the sample.

## 3 Calculations

Total heat in machining released per unit of time (heat stream) corresponds to the approximate machining power and may be determined by the following formula:

$$Q = F_c \cdot v_c \;\; W \tag{1}$$

where: $F_c$ - machining force N, $v_c$ - cutting speed m/s.

Determination of the value of machining force $F_c$ may be conducted on the basis of a formula based on knowing the specific machining resistance $k_c$.

$$F_c = A \cdot k_c \;\; N \tag{2}$$

where: A - cross sectional area of the machined layer mm$^2$ N, $k_c$ - specific machining resistance N/mm$^2$.

Specific machining resistance $k_c$ must be calculated depending on the average thickness of the machined layer $h_m$.

$$k_c = \frac{k_{c1.1}}{h_m{}^m} \;\; N/mm^2 \tag{3}$$

where: $k_{c1.1}$ - the value of specific resistance for the unit of thickness of the machined layer N/mm$^2$ , $h_m$ - the average thickness of the machined layer in mm, m - exponent.

The average thickness of chips ($h_m$) for a simple cutting edge will be determined depending on:

$$h_m = \frac{180 \cdot \sin \kappa_r \cdot a_e \cdot f_z}{\pi \cdot D_{cap} \cdot arcsin\left(\frac{a_e}{D_{cap}}\right)} \;\; mm \tag{4}$$

where: $\kappa_r$ - setting angle °, $D_{cap}$ - the effective machining diameter to the actual machining depth in mm.
The efficient machining diameter for the actual depth of machining and the specific setting angle is calculated depending on:

$$D_{cap} = D_c \cdot \frac{2 \cdot a_p}{\tan \kappa_r} \;\; mm \tag{5}$$

Input data and calculation results of total heat released per unit of time during face milling for specific conditions of machining have been presented in table 3.

**Table 3.** Input data and calculation results

| Data | Results |
|---|---|
| Dc = 80 mm | $D_{cap}$ = 81 mm |
| $a_p$ = 0.5 mm | |
| $\kappa_r$ = 45 ° | |
| $f_z$ =0.05 mm/tooth | $h_m$ = 0.034 mm |
| $a_e$= 30 mm | |
| $k_{c1.1}$= 780 N/mm$^2$ | $k_c$ = 1695.6 N/mm$^2$ |
| m = 0.23 | |
| A = 0.025 mm$^2$ | $F_c$ = 42.39 N |
| $v_c$ = 4.16 m/s | **Q = 176.34 W** |
| $v_c$ = 6.25 m/s | **Q = 264.94 W** |

As a result of conducted theoretical calculations, it has been determined that during face milling with a cutting head R245-080Q27-12M made of M1Ez4 copper, the amount of released heat per unit of time for machining velocity of $v_c$ =250 m/min is Q = 176.34 W, while for $v_c$ =375 m/min - Q = 264.94 W

## 4 Results of experimental studies

Figure 4 presents graphs for the change of temperature within the period recorded by 6 thermocouples during processing of the sample made of M1Ez4 copper with machining velocity of $v_c$ =250 m/min for the distance of the tool from thermocouples at h = 4 mm. By analyzing results presented in figure 4, it has been observed that the maximum temperature of the sample during the process of face milling was obtained for thermocouple no. 5 - it amounted 66.1 °C. In the case of thermocouple no. 1, the maximum temperature was 58.6 °C, which was the lowest maximum value measured for all thermocouples. Table 4 presents a comparison of results for maximum temperatures recorded by individual thermocouples for various "h" values of the tool distance from thermocouples.

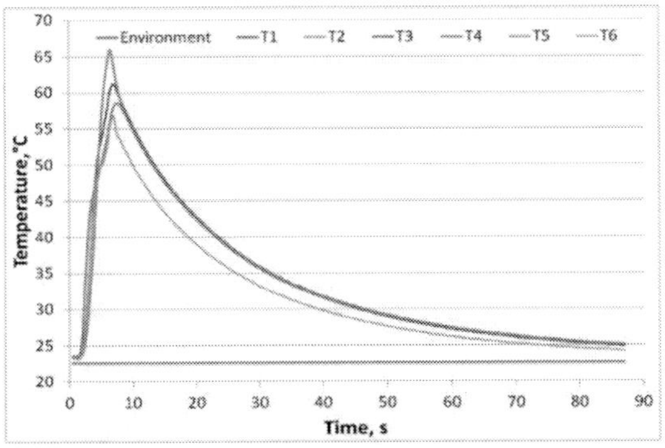


**Fig. 4.** Graphs of temperature during processing of the sample with cutting speed of $v_c$ =250 m/min for the distance of the tool from thermocouples at h = 4 mm

Results presented in table 4 indicate uneven heating of the processed object during the process of face milling. For all samples, the highest temperatures were recorded in thermocouple 5 that was located in the left upper corner of the milled sample, so the cutting plates had to travel the longest distance in order to remove the material in that zone. The lowest values of the maximum temperature were recorded by thermocouple no. 1 that was located in the upper right corner. The only deviation was recorded during the 1st attempt of machining, during which the maximum temperature of 47.1 °C was recorded for thermocouple no. 6, which was caused by improper fixing of the thermocouple in the measurement hole.

Figure 5 presents charts of temperature changes of the copper sample that was face milled with machining velocity of $v_c$ =375 m/min for the tool distance from thermocouples of h = 4 mm. The calculated amount of released heat per unit of time for machining velocity $v_c$ =375 m/min was Q = 264.94 W. When analyzing results presented in figure 5, it has been observed that the maximum sample temperature at 73.5 °C was obtained for the thermocouple no. 4. On the other hand, the lowest maximum value was recorded for thermocouple no. 2, for which the maximum temperature was 60.4 °C. Change of machining velocity from 250 to 375 m/min caused an increase of 88.6 W the amount of released heat per unit of time, which led to a change of heating zones of the processed sample.

**Table 4.** Maximum temperatures recorded by individual thermocouples during processing of the sample with cutting speed of $v_c$ =250 m/min

| | Temperature, °C | | | | | |
|---|---|---|---|---|---|---|
| **h, mm** | **T1** | **T2** | **T3** | **T4** | **T5** | **T6** |
| **4.5** | 56.7 | 56.8 | 59.3 | 63.3 | 63.6 | 47.1 |
| **4** | 58.6 | 58.6 | 61.3 | 65.9 | 66.1 | 56.8 |
| **3.5** | 60.2 | 60.3 | 63.0 | 67.8 | 68.1 | 62.1 |
| **3** | 61.6 | 61.7 | 64.7 | 69.9 | 70.3 | 62.9 |
| **2.5** | 63.4 | 63.6 | 66.8 | 72.3 | 72.6 | 66.3 |
| **2** | 64.6 | 64.8 | 68.3 | 73.8 | 74.2 | 63.2 |

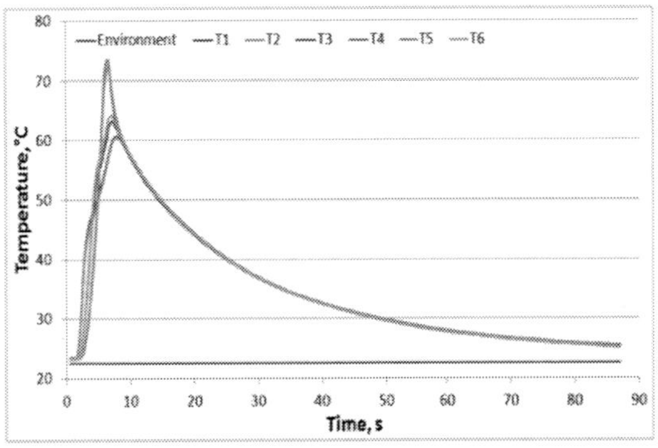


**Fig. 5.** Graphs of temperature during processing of the sample with cutting speed of $v_c$ =375 m/min for the distance of the tool from thermocouples at h = 4 mm

Analogously to table 4, table 5 presents a comparison of results for maximum temperatures recorded by individual thermocouples for various "h" values of the tool distance from thermocouples.

**Table 5.** Maximum temperatures recorded by individual thermocouples during processing of the sample with cutting speed of $v_c$ =375 m/min

| | Temperature, °C | | | | | |
|---|---|---|---|---|---|---|
| **h, mm** | **T1** | **T2** | **T3** | **T4** | **T5** | **T6** |
| **4.5** | 58.90 | 58.60 | 61.20 | 71.40 | 70.20 | 62.40 |
| **4** | 60.60 | 60.40 | 63.20 | 73.50 | 72.50 | 64.20 |
| **3.5** | 62.30 | 62.00 | 64.90 | 75.80 | 74.80 | 66.00 |
| **3** | 63.40 | 63.10 | 66.50 | 79.90 | 77.40 | 67.80 |
| **2.5** | 65.10 | 64.80 | 68.40 | 81.60 | 79.80 | 69.70 |
| **2** | 66.60 | 66.30 | 70.30 | 84.30 | 82.50 | 71.70 |

In the case of all samples. the highest temperature was recorded for thermocouple no. 4 that was located in the

DOI: 10.1051/epjconf/201714302082

left lower corner of the milled sample. while the lowest values of the maximum temperature were recorded by thermocouple no. 2 that was located in the right lower corner. When comparing data presented in tables 4 and 5, it has been determined that the increase of machining velocity from 250 to 375 m/min resulted in an increase of the maximum temperature of the machined object by 10.1 °C.

Figure 6 presents the charts of temperatures recorded during the process of face milling on copper samples with two machining velocities for h=4.5 mm. The continuous line stands for temperatures recorded for machining velocity of 250 m/min. while dotted line - for machining velocity equal 375 m/min.

When analyzing the chart presented in figure 6, we may notice that the increase of machining velocity results in an increase of the flow of released heat. which heats the processed object faster and to higher temperatures.

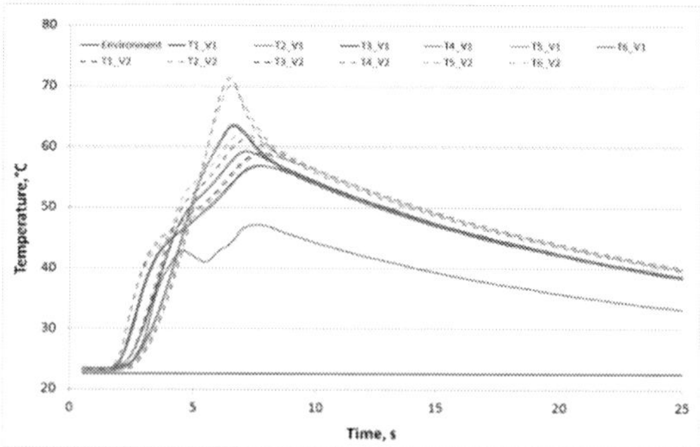


**Fig. 6.** Graph of maximum temperatures for different cutting speed

Table 6 presents example results of the impact of the change in machining velocity on the temperature of the processed object for h=4.5 mm. Change of machining velocity from 250 to 375 m/min caused an increase of 88.6 W the amount of released heat per unit of time. which in the case of the distance of the tool from thermocouples h=4.5 resulted in an increase of the maximum temperature of the processed object by $\Delta T$ = 8.1 °C in comparison to processing at a lower machining velocity.

**Table 6.** Comparison of maximum temperature for different cutting speed

| | Temperature, °C | | | | | |
|---|---|---|---|---|---|---|
| $v_c$, m/min | T1 | T2 | T3 | T4 | T5 | T6 |
| 250 | 56.7 | 56.8 | 59.3 | 63.3 | 63.6 | 47.1 |
| 375 | 58.9 | 58.6 | 61.2 | 71.4 | 70.2 | 62.4 |
| ΔT | 2.2 | 1.8 | 1.9 | 8.1 | 6.6 | 15.3 |

Figure 7 presents the comparison of lists of temperatures (maximum. average. and minimum) of milled samples depending on machining velocity and the distance of the tool from thermocouples. When analyzing data from figure 7, it has been observed that together with the decrease of the "h" value of the tool distance from thermocouples. the maximum values of recorded temperatures were increasing. which also resulted from the decrease of volume of processed samples after each cycle of the tool, which has been presented in table 1. For machining velocity equal 250 m /min, loss of material of 450 mm$^3$ caused an increase of the maximum average temperature of the sample by 2.12 °C, while for $v_c$=375 m/min by 2.58 °C.

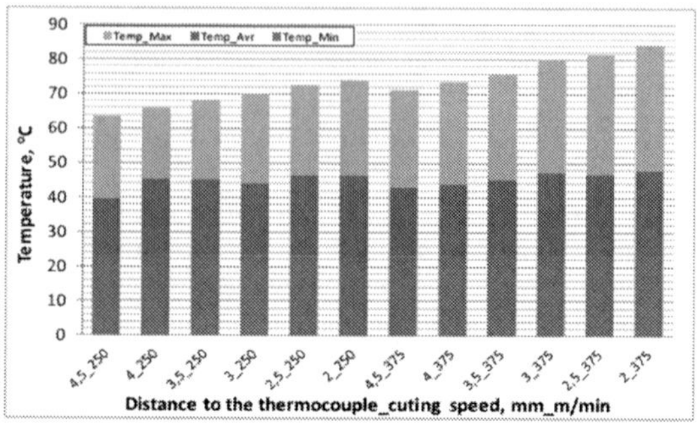


**Fig.7.** Comparison of temperatures (maximum, average, and minimum) of milled samples

## 5 Summary

Conducted research allowed to analyze the impact of changes in cutting speed and volume of material processed during a face milling process on the temperature of the processed object made of copper of M1Ez4 class.

Conducted research proves that:

1. Increased machining velocity during the process of face milling causes an increase of performance of the process itself, but also results in an increased amount of released heat per unit of time, which may lead to uncontrolled heating of the processed object to high temperatures.
2. The process of face milling with $v_c$=250 m/min, $f_z$=0.05 mm/tooth, $a_p$=0.5 mm, $a_e$=30 mm led to heating of the processed object to a maximum temperature of 74.2°C.
3. Increase of the machining velocity from 250 to 375 m/min resulted in an increase of the maximum temperature of the processed object from 10.1 °C to 84.3°C.
4. During the process of face milling, the processed object was heated unevenly.
5. Along with the decrease of the volume of processed samples and, consequently, of the "h" value of tool distances from thermocouples, the maximum values of recorded temperatures increased.
6. For machining velocity of 250 m/min. loss in material of 450 mm$^3$ caused an increase of the average maximum temperature of the sample by 2.12 °C. while for $v_c$=375 m/min - 2.58 °C.

## References

1. M. Bartoszuk, W. Grzesik, Edited by: Outeiro, Modelling Of Machining Operations Advan. Mat. Res. **223**, 231 (2011)
2. W. Grzesik, M. Bartoszuk, P. Nieslony, Machining Science And Technology **9**, (4), 529 (2005)
3. W. Grzesik, P. Nieslony, M. Bartoszuk,Trans. of the North Ameri. Manuf. Res. Inst. of Sme **XXIX**, 343 (2001)
4. S. Hożejowska, R. Kaniowski, M. E. Poniewski, Exper. Thermal and Fluid Science **78**, 18 (2016)
5. S. Hożejowska, R. Kaniowski, M. E. Poniewski, Inter. J. of Num. Meth. for Heat & Fluid Flow **24**, (4), 811 (2014)
6. R. Pastuszko, R. Kaniowski, EPJ Web of Conferences (*Proc. Int. Conf. Experimental Fluid Mechanics 2010*) **25**, 02019 (2012)
7. S. Blasiak, J. E. Takosoglu, P. A. Laski, J. Therm. Sci. Techn. **9**, 2 (2014)
8. S. Blasiak, P. A. Laski, J. E. Takosoglu, Int. J. Heat Mass Tran. **57**, 1 (2013)
9. S. Błasiak, C. Kundera, Procedia Eng. **39**, 315 (2012)
10. S. Blasiak, A. Pawinska, Int. J. Heat Mass Transf. **90,** 710 (2015)
11. S. Blasiak, C. Kundera, J. Bochnia, Procedia Eng. **39**. 366 (2012)
12. W. Depczyński, R. Kazała, K. Ludwinek, K. Jedynak, Mat. **9** (7), 567, 1 (2016)
13. W. Depczyński *METAL 2014 23rd Inter. Conf. on Metal. and Mat.* 1219 (2014)
14. S. Spadło, P. Młynarczyk, W. Depczyński, *Proceedings of 24th Inter. Conf. on Metall. and Mat. METAL 2015* 863 (2015)
15. L. Nowakowski, E. Miko, M. Skrzyniarz, Edited by Zolotarev I., Radolf V. *Proceedings of 22nd International Conference on Engineering Mechanics 2016* (Academy of Sciences of the Czech Republic. 2016)
16. L. Nowakowski, M. Wijas, Edited by Zolotarev I., Radolf V. *Proceedings of 22nd International Conference on Engineering Mechanics 2016* (Academy of Sciences of the Czech Republic. 2016)
17. E. Miko, L. Nowakowski, Procedia Eng. **39** (2012a)
18. E. Miko, L. Nowakowski, Procedia Eng. **39** (2012b)
19. S.Adamczak, E. Miko, F. Cus, Strojniski Vestnik-Jour. of Mech. Eng. **55**, (1), 45 (2009)
20. J.E. Takosoglu, Edited by Zolotarev I., Radolf V. *Proceedings of 22nd International Conference on Engineering Mechanics 2016* (Academy of Sciences of the Czech Republic. 2016)
21. J.E. Takosoglu, P.A. Laski, S. Blasiak, G. Bracha, D. Pietrala, Meas. Cont. **49**, 2 2016
22. P.A. Laski, J.E. Takosoglu, S. Blasiak, Rob Auton Syst.**72**, 59 (2015)
23. G.F. Bracha. Edited by Zolotarev I., Radolf V. *Proceedings of 22nd International Conference on Engineering Mechanics 2016* (Academy of Sciences of the Czech Republic. 2016)
24. J.E. Takosoglu, P.A. Laski, S. Blasiak, Edited by Fuis V., *Proceedings of 20th International Conference on Engineering Mechanics 2014* (Brno University of Technology. Czech Republic. 2014)
25. J.E. Takosoglu, P.A. Laski, S. Blasiak, P. I. Mech. Eng. I-J. Sys. **226**, 10 (2012)
26. P.A. Laski, Edited by Zolotarev I.. Radolf V. *Proceedings of 22nd International Conference on Engineering Mechanics 2016* (Academy of Sciences of the Czech Republic. 2016)
27. D.S. Pietrala, Edited by Zolotarev I.. Radolf V. *Proceedings of 22nd International Conference on Engineering Mechanics 2016* (Academy of Sciences of the Czech Republic. 2016)
28. I. Krzysztofik, Z. Koruba, J. of App. Math., Vol. **2014**, 11 (2014).
29. I. Krzysztofik, Edited by Zolotarev I., Radolf V. *Proceedings of 22nd International Conference on Engineering Mechanics 2016* (Academy of Sciences of the Czech Republic. 2016)
30. Z. Koruba, I. Krzysztofik, P. I. Mech. Eng. K-J. Multi-Body Dyn. **227**, K1 (2013)
31. J. Zwierzchowski, Edited by Zolotarev I.. Radolf V. *Proceedings of 22nd International Conference on Engineering Mechanics* 2016 (Academy of Sciences of the Czech Republic. 2016)
32. D. Janecki, L. Cedro, J. Zwierzchowski, Metrology Meas. Sys. **22**, 2 2015
33. D. Janecki, J. Zwierzchowski, L. Cedro. B. Pol. Acad. Sci-Tech. Sci. **63**, 3 2015
34. P. Michalik. J. Zajac. M. Hatala. D. Mital. V. Fecova. Measur. **58**, 416 (2014)
35. P. Michalik. J, Zajac, Manuf. Eng. **9** (4), 63, (2010)
36. M. Graba. J, of Theor. And Appl. Mech. **50,** (1), 23,(2012)
37. M. Graba., J. of Theor. And Appl. Mech. **51,** (2), 349 (2013)
38. L. Nowakowski, M. Miesikowska, M. Blasiak, Edited by Zolotarev I., Radolf V. *Proceedings of 22nd International Conference on Engineering Mechanics 2016* (Academy of Sciences of the Czech Republic. 2016)
39. M. Blasiak, Edited by Zolotarev I., Radolf V. *Proceedings of 22nd International Conference on Engineering Mechanics 2016* (Academy of Sciences of the Czech Republic. 2016)
40. M. Blasiak, R. Kotowski, Przeglad Elektrotechniczny. **85**, 12 (2009)

# The impact of cooling methods on the maximum temperature of the processed object during side milling

*Lukasz* Nowakowski[1,*], *Michal* Skrzyniarz[1], and *Edward* Miko[1]

[1]Kielce University of Technology, Department of Manufacturing Engineering and Metrology, Aleja Tysiąclecia Państwa Polskiego 7; 25-314 Kielce; Poland

**Abstract.** This article presents the changes in temperature distribution concerning the processed object during side milling. The analysis included the influence of selected cooling methods (cold air from vortex tube, cooling with compressed air, and without cooling) on the maximum temperature of the processed object during the process of milling. Measurement of the maximum temperature was conducted with the use of thermocouples and a thermal camera.

## 1 Introduction

During the process of machining, heat is generated as a result of elastic and plastic deformation of the material and of friction occurring between the contact surface and the processed material[1-8]. An additional source of providing heat to the area of machining it the work of friction of chips against the surface of the tool. The temperature present within the area of machining influences the shape and size accuracy of created objects [9-14]. It has a significant meaning during performance of precision parts of machines with high reliability[15-23], as well as prototypes [24] i.e. contactless face seals [4-8] of gyroscopes etc. [24-26]. It may, additionally, cause deformation of the processed object and the tools [27-35]. Excessive amount of heat delivered to the tool results in a loss of cutting properties of the tools and, consequently, in damaging the tool, which has a significant impact on production costs and efficiency of performed processing. Monitoring of temperature is very important due to the need for selection of proper conditions to perform the process of machining. Heat is dissipated from the machining area through the tool, processed object, created chips, as well as other sources, e.g. by lifting or usage of liquids: cooling liquids and cooling and lubricating liquids. The largest amount of heat is dissipated from the machining area through chips. That is why its shape, cross-section, and method of tearing off is one of the essential factors that influence distribution of temperatures within the area of machining. Conducting the process of machining with the use of cutting fluid cools the tool and the processed object. Additionally, it decreases friction that occurs within the area of machining, which leads to a lower amount of released heat and lower demand for power required to conduct processing [36-38].

The measurement of average temperatures within the machining area has been mainly based on contact methods by using thermocouples, as well as contactless methods that are based on the phenomenon of heat radiation [39].

## 2 Methods

Tests were performed on a vertical machining center AVIA VMC 800. The tool used for the experiment was a tow-edge solid end mill produced by Sandvik Coromant - R216.32-12025-AK38A H10F with diameter of 12 mm. Table no. 1 presents the basic geometrical dimensions of the tool used for processing.

**Table 1.** Geometrical dimensions of the tool.

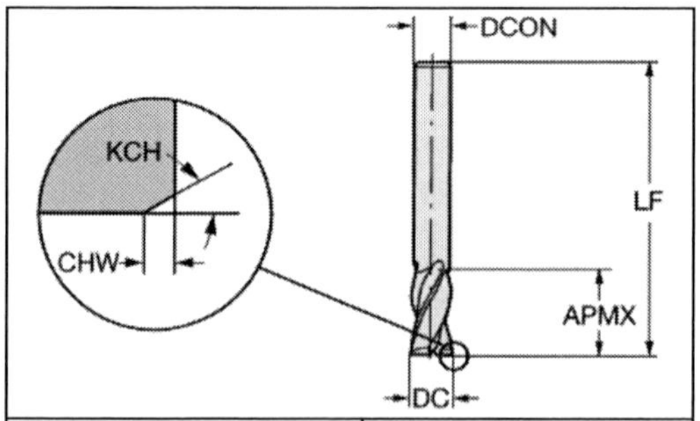


| | |
|---|---|
| DC = 12 mm | DOC = 12 mm |
| LF = 100 mm | KCH = 45° |
| APMX = 38 mm | CHW = 0.1 mm |

The process of side milling was performed at depth of $a_p$ = 30 mm, side feed $a_e$ = 0.5 mm and feed rate of f = 300 mm/min. Turning speed of the spindle was 10,000 rpm. The tested material was a 30x40x10 mm rectangular sample made of copper. Figure no. 1 presents geometrical dimensions of the sample along with the arrangement of thermocouples. K-type thermocouples of

* Corresponding author: skrzyniarzmichal@gmail.com

1 mm in diameter were used to measure the temperature. They were placed 10 mm inside the processed object.

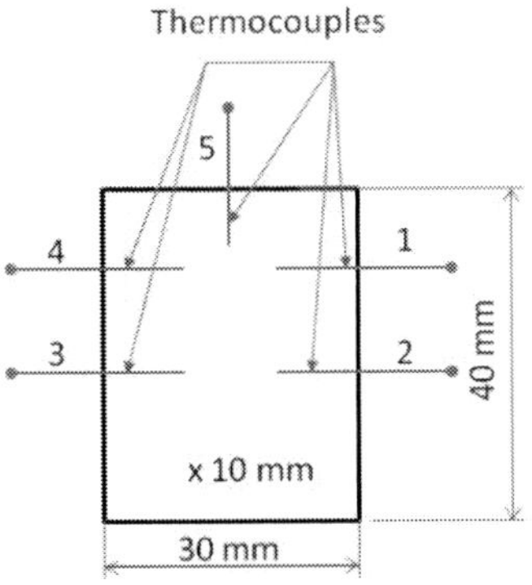


**Fig. 1.** Geometrical dimensions of the sample.

Apart from contact measurement conducted with the use of thermocouples, the experiment station also included a thermal camera Flir i7. Specification of that camera has been presented in table no. 2.

**Table 2.** Specification Flir i7 camera

| **Resolution** | **140x140 pixels** |
|---|---|
| Thermal Sensitivity | 0.1 °C |
| Temperature Range | - 20 ÷ 250°C |
| Accuracy | ± 2°C |
| Spectrum | 7.5 ÷13 μm |
| Frame Rate | 9 kHz |
| Emissivity | 0.1 ÷ 1.0 |

**Fig. 2.** Experiment station used for measurement.

The experiment was to conduct machining attempts and measure the average and maximum temperatures inside the processed object and on its front wall. The machining processed was conducted several times on one sample in order to determine the distribution of temperatures together with changes in the distance of the measurement point from the source of heat. Additionally, it was decided to conduct an experiment without the use of cooling, with the use of cooling for the machining area by using compressed air, and by using a vortex tube. That device allows division of the stream of compressed air into a stream of hot and cold air that is directed at the machining area. Figure no. 2 presents the experiment station used for measurements.

## 3 Results

Table no. 3 presents the average and maximum temperature values obtained during processing without cooling. It was observed that the highest values of average temperatures were noted at the distance of 2mm of thermocouples from the machining area. The farther the source of heat from thermocouples, the lower the temperature. The analogous dependency has been noted for measurement made by the thermal camera.

**Table 3.** Temperature values obtained Turing process without cooling.

| **Temp, °C** | **Distance of thermocouples from the machining area, mm** | | | | | |
|---|---|---|---|---|---|---|
| | **4.5** | **4** | **3.5** | **3** | **2.5** | **2** |
| Thermo. 1 | 101 | 108 | 108 | 128 | 134 | 140 |
| Thermo. 2 | 90 | 98 | 108 | 114 | 119 | 123 |
| Thermo. 3 | 75 | 80 | 88 | 93 | 95 | 98 |
| Therm. 4 | 82 | 87 | 96 | 103 | 105 | 110 |
| Thermo. 5 | 90 | 97 | 108 | 115 | 119 | 124 |
| Camera | 97 | 104 | 115 | 124 | 131 | 142 |

Figure 3 presents charts that describe the change in temperatures during the processing without cooling at the distance of thermocouple distribution of 2.5 mm from the source of heat in form of the operating tool. It was noticed that the maximum temperature of 134°C was obtained for thermocouple no. 1 that was the farthest from the source of heat. Accumulation within that area is caused by the fast and constant flow of heat from the beginning of the processing. Average results of temperature measurement for thermocouples 1 and 2 differ from one another although they have been placed on the same wall. It is caused by the distance of thermocouples from the upper and lower edge where some part of the heating energy directed to thermocouple no. 2 is obtained by the grip, as well as the lower unprocessed part of the sample. Analogous situations occur in the case of other measurements.

Table no. 4 contains results of measurement of the average and maximum temperatures for processing with the same technological parameters. Additionally in that case, cooling of the machining area with a vortex tube was applied. During measurement, the noted temperature of air at the outlet from the tube was about – 29 °C. After analyzing the results, it was noticed that cooling of the machining area with cold air does not significantly affect the results of the average temperatures noted by the vortex tube and the maximum temperature recorded with the thermal camera.

DOI: 10.1051/epjconf/201714302083

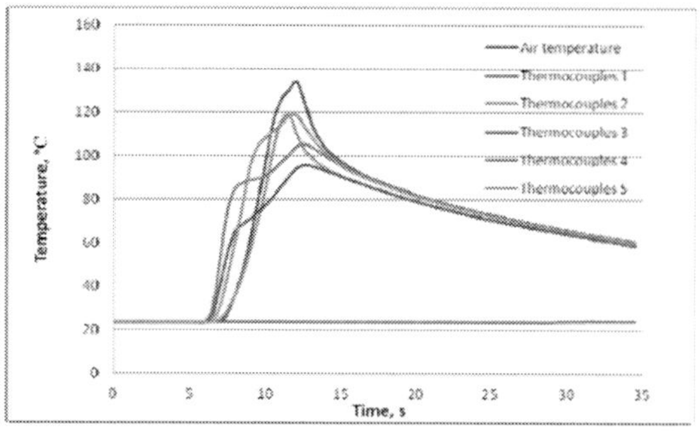


**Fig. 3.** Change the temperatures during the processing without cooling at the distance of thermocouple distribution of 2.5 mm.

**Table 4.** Temperature values obtained Turing process with cooling by Vortex tube.

| Temp, °C | Distance of thermocouples from the machining area, mm | | | | | |
|---|---|---|---|---|---|---|
| | **4.5** | **4** | **3.5** | **3** | **2.5** | **2** |
| Thermo. 1 | 109 | 117 | 126 | 128 | 147 | 154 |
| Thermo. 2 | 100 | 104 | 112 | 114 | 120 | 126 |
| Thermo. 3 | 83 | 84 | 90 | 91 | 95 | 98 |
| Thermo. 4 | 89 | 93 | 99 | 100 | 105 | 110 |
| Thermo. 5 | 98 | 105 | 110 | 112 | 118 | 124 |
| Camera | 99 | 115 | 120 | 120 | 126 | 142 |

The maximum average temperature for processing with cooling is by 14 °C higher than in the case of processing without cooling. It may be caused by lack of dissipation of heat from machining area through the tool that heated to a high temperature due to conducting the previous processing.

**Table 5.** Temperature values obtained Turing process with cooling by compressed air.

| Temp, °C | Distance of thermocouples from the machining area, mm | | | | | |
|---|---|---|---|---|---|---|
| | **4.5** | **4** | **3.5** | **3** | **2.5** | **2** |
| Thermo. 1 | 108 | 113 | 125 | 131 | 137 | 146 |
| Thermo. 2 | 96 | 99 | 109 | 114 | 119 | 126 |
| Thermo. 3 | 81 | 84 | 89 | 86 | 97 | 101 |
| Thermo. 4 | 88 | 91 | 100 | 104 | 108 | 113 |
| Thermo. 5 | 97 | 100 | 110 | 115 | 120 | 127 |
| Camera | 97 | 101 | 118 | 120 | 125 | 138 |

In the case of processing with cooling in the form of compressed air, values of temperatures were recorded in table no. 5. The recorded air temperature at the nozzle outlet was 24°C. The average and maximum recorded temperatures in that case of processing do not significantly deviate from the values recorded for previous processing attempts with cooling by the use of compressed air from a vortex tube or without cooling.

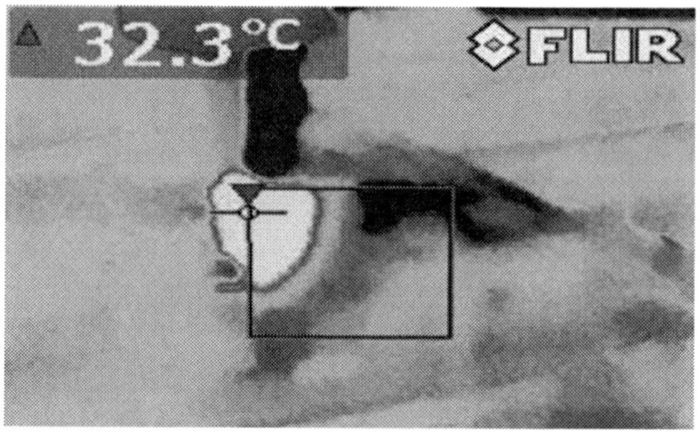


**Fig. 4.** Example image obtained with the thermal camera during measurement.

Figure no. 4 presents the example image obtained with the thermal camera during measurement. The noted temperature in that case was 32.3 °C and was observed during the start of machining. In the picture, one may clearly notice the direction of release of heat to the remaining part of the sample, while the temporary maximum temperature was in places where the material was divided.

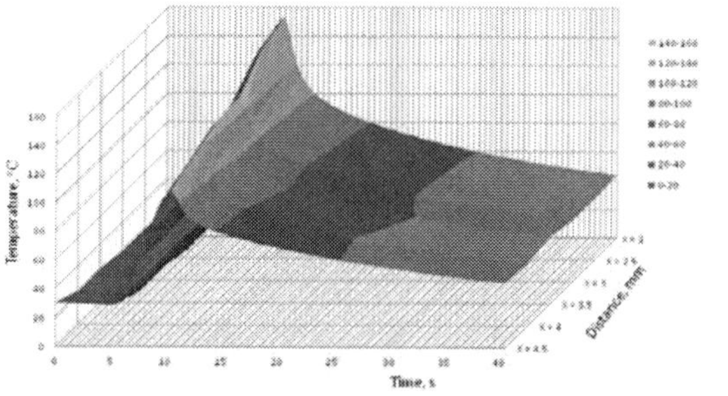


**Fig. 5.** Changes of the temperature during the processing with cooling by the use of a vortex tube.

Figure 5 presents the chart describing changes of temperature during the processing with cooling by the use of a vortex tube. The highest temperature was noted for that type of machining and amounted 154 °C, while the cooling process of the processed object was the fastest and amounted 51 seconds to reach temperature of the processed object below 30 °C. Application of cooling in the form of compressed air required 77 seconds after finishing the process to decrease temperature of the processed object below 30 °C, while in the case of processing without cooling - 104 seconds.

## 4 Conclusions

The analysis of the experiment allowed formulation of the following conclusions:

1. Processing by using the cooling method with compressed air or cold air from a vortex tube does not

decrease the maximum average value of temperatures of the processed object.

2. The highest temperature was noted for thermocouple no. 1, which results from accumulation of heat and no dissipation of it from the area by contact with the machining grip.
3. Analysis of temperatures inside the sample performed with thermocouples is similar to values of temperatures recorded by the thermal camera.
4. The cooling method of the machining process significantly affects dissipation of heat from the processed object after the machining process is finished.
5. Maximum temperatures recorded during processing with cold compressed air result from limited heat capacity of the tool arising from accumulation of heat stored from the previous processing.

## References


1. S. Hożejowska, R.Kaniowski, M. E. Poniewski, Exper. Thermal and Fluid Scienc 78, 18 (2016)
2. S. Hożejowska, R.Kaniowski, M. E. Poniewski, Inter. J. of Num. Meth. for Heat & Fluid Flow 24, (4), 811 (2014)
3. R. Pastuszko, R. Kaniowski, EPJ Web of Conferences (*Proc. Int. Conf. Experimental Fluid Mechanics 2010*) 25, 02019 (2012)
4. S. Blasiak, J. E. Takosoglu, P. A. Laski, J. Therm. Sci. Techn. 9, 2 (2014)
5. S. Blasiak, P. A. Laski, J. E. Takosoglu, Int. J. Heat Mass Tran. 57, 1 (2013)
6. S. Blasiak, C. Kundera, Procedia Eng. 39, 315 (2012)
7. S. Blasiak, A. Pawinska. Int J Heat Mass Transf. 90, 710 (2015)
8. S. Blasiak, C. Kundera, Bochnia J. Procedia Eng. 39, 366 (2012)
9. L. Nowakowski, E. Miko, M. Skrzyniarz, Edited by Zolotarev I., Radolf V. *Proceedings of 22nd International Conference on Engineering Mechanics 2016* (Academy of Sciences of the Czech Republic, 2016)
10. L. Nowakowski, M. Wijas, Edited by Zolotarev I., Radolf V. *Proceedings of 22nd International Conference on Engineering Mechanics 2016* (Academy of Sciences of the Czech Republic, 2016)
11. L. Nowakowski, M. Miesikowska, M. Blasiak, Edited by Zolotarev I., Radolf V. *Proceedings of 22nd International Conference on Engineering Mechanics 2016* (Academy of Sciences of the Czech Republic, 2016)
12. E. Miko, L. Nowakowski, Procedia Eng. 39 (2012a)
13. E. Miko, L. Nowakowski, Procedia Eng. 39 (2012b)
14. S. Adamczak, E. Miko, F. Cus, Strojniski Vestnik-Jour. of Mech. Eng. 55,(1),45 (2009)
15. J.E. Takosoglu, Edited by Zolotarev I., Radolf V. *Proceedings of 22nd International Conference on Engineering Mechanics 2016* (Academy of Sciences of the Czech Republic, 2016)
16. J.E. Takosoglu, P.A. Laski, S. Blasiak, G. Bracha, D. Pietrala. Meas. Cont. 49, 2, 2016
17. P.A. Laski, J.E. Takosoglu, S. Blasiak, Rob Auton Syst.72, 59 (2015)
18. G.F. Bracha. Edited by Zolotarev I., Radolf V. *Proceedings of 22nd International Conference on Engineering Mechanics 2016* (Academy of Sciences of the Czech Republic, 2016)
19. J.E. Takosoglu, P.A. Laski, S. Blasiak, Edited by Fuis V. *Proceedings of 20th International Conference on Engineering Mechanics 2014* (Brno University of Technology, Czech Republic, 2014)
20. J.E. Takosoglu, P.A. Laski, S. Blasiak. P. I. Mech. Eng. I-J. Sys. 226, 10 (2012)
21. P.A. Laski, Edited by Zolotarev I., Radolf V. *Proceedings of 22nd International Conference on Engineering Mechanics 2016* (Academy of Sciences of the Czech Republic, 2016)
22. D.S. Pietrala, Edited by Zolotarev I., Radolf V. *Proceedings of 22nd International Conference on Engineering Mechanics 2016* (Academy of Sciences of the Czech Republic, 2016)
23. S. Adamczak, B. Kaczmarska, J. Bochnia, Metrology and Measurement Systems, 22 (2015)
24. I. Krzysztofik, Z. Koruba J. of App. Math., Vol. 2014, 11 (2014),
25. I. Krzysztofik, Edited by Zolotarev I., Radolf V. *Proceedings of 22nd International Conference on Engineering Mechanics 2016* (Academy of Sciences of the Czech Republic, 2016)
26. Z. Koruba, I. Krzysztofik, P. I. Mech. Eng. K-J. Multi-Body Dyn. 227, K1, 2013
27. J. Kumaszewski. I Zagórski, Management and Production Engineering Review, 4 (2013)
28. M. Graba, J. of Theor. And Appl. Mech. 50 (1) 23 (2012)
29. M. Graba, J. of Theor. And Appl. Mech. 51 (2) 349 (2013)
30. W. Depczyński, R. Kazała, K. Ludwinek, K. Jedynak, Mat. 9 (7), 567, 1 (2016)
31. W. Depczyński *METAL 2014 23rd Inter. Conf. on Metal. and Mat.* 1219 (2014)
32. S. Spadło, P. Młynarczyk, W. Depczyński *Proceedings of 24th Inter. Conf. on Metall. and Mat. METAL 2015* 863 (2015)
33. J. Zwierzchowski, Edited by Zolotarev I., Radolf V. *Proceedings of 22nd International Conference on Engineering Mechanics* 2016 (Academy of Sciences of the Czech Republic, 2016)
34. D. Janecki, L. Cedro, J. Zwierzchowski, Metrology Meas. Sys. 22, 2, 2015
35. D. Janecki, J. Zwierzchowski, L. Cedro, B. Pol. Acad. Sci-Tech. Sci. 63, 3, 2015
36. M. Bartoszuk, W. Grzesik, Edited by: Outeiro, Modelling Of Machining Operations Advan. Mat. Res. 223, 231 (2011)
37. W. Grzesik,; M. Bartoszuk, P. Nieslony, Machining Science And Technology , 9, (4), 529 (2005)
38. W. Grzesik,; P. Nieslony, M. Bartoszuk, Trans. of the North Ameri. Manuf. Res. Inst. of Sme, XXIX, 343 (2001)
39. P. Niesłony, W. Habrat, P. Laskowski, Mechanik, 8/9, 71-81 (2015)

# Computational investigation of heat transfer and pressure drop in a typical louver fin-and-tube heat exchanger for various louver angles and fin pitches

*Abdulkerim* Okbaz[1*], *Ali Bahad r* Olcay[2], *Mehmet Salih* Cellek[1] and *Ali* Pınarbaşı[1]

[1]Yildiz Technical University , Mechanical Engineering Department , Besiktas, 34349, Istanbul, Turkey
[2]Yeditepe University, Mechanical Engineering Department , Atasehir, 34755, Istanbul, Turkey

**Abstract.** In this study 3-D numerical simulations on heat transfer and pressure drop characteristics for a typical louver fin-and- double-row tube heat exchanger were carried out. The heat transfer improvement and the corresponding pressure drop amounts were investigated depending on louver angles, fin pitch and Reynolds number, and reported in terms of Colburn *j*-factor and Fanning friction factor *f*. The heat transfer improvement and the corresponding pressure drop amounts were investigated depending on louver angles between 20° ≤ϴ≤ 30°, louver pitch of $L_p$=3.8 mm and frontal velocities of U between 1.22 m/s - 3 m/s. In addition, flow visualization of detailed flow features results, such as velocity vectors, streamlines and temperature counters have been shown to understand heat transfer enhancement mechanism. The present results indicated that louver angle and fin pitch noticeably affected the thermal and hydraulic performance of heat exchanger. It has been seen that increasing louver angle, increases thermal performance while decreasing hydraulic performance associated to pressure drop for fin pitches of 3.2 mm and 2.5 mm. Fin pitch determines the flow behaviour that for fin pitch of 2 mm, increasing louver angle decreased heat transfer and pressure drop. Velocity vectors and streamlines give considerable information about the flow whether it is duct directed or louver directed. For all conditions the flow is louver directed.

## 1 Introduction

As it well known, the air side of heat exchangers have much more thermal resistance. Decreasing attempts of air side thermal resistance play an important role to improve the performance of the heat exchangers. Louver fins provide enhanced and interrupted surfaces to increase convective heat transfer. The louvers on the surface of the fins provide restarting of boundary layer and the air flow to travel across the neighbour fins and tubes. There forms a thinner thermal boundary layer and air can flow much more distance in the heat exchanger. Thus, the thermal performance of the heat exchanger increases. On the other hand, interrupted surfaces and travelling of air in longer distance increases pressure drops. Therefore, a suitable design of louvered fins is needed to enhance thermal performance of heat exchanger without prohibitive increase in the pressure drop. Researchers and engineers attempt to design more efficient louver fin heat exchangers studying various geometrical parameters experimentally and numerically.

In their experimental study, Aoki et al. [1] concluded that heat transfer coefficients at low air velocity conditions decreased with increasing fin pitch while heat transfer coefficient reached a maximum value at angle of 28°-30°. They also found that at wider angles of louvers heat transfer coefficient started to decrease. Webb and Trauger [2] defined a flow efficiency as the ratio of actual transverse distance to ideal transverse distance. Wang et al. [3] conducted experiments using commercially available louver fin-and-tube heat exchangers. They studied heat transfer and pressure drop characteristics depending on number of tube rows, fin pitch, and tube size. Kim and Bullard [4] stated that flow depth was one of the most important parameter for pressure drop characteristics of louver fin heat exchangers in their experimental study. Dejong and Jacobi [5] found that when the flow was far from bounding walls of heat exchanger, it showed spatially periodic behaviour and high heat transfer and low pressure drop occurred. In the region close to the bounding walls, large flow separation zones occurred and heat transfer decreased about 50%. Dong et al. [6] conducted an experimental study testing air side heat transfer and pressure drop characteristics of 20 types of multi-louvered fin and flat tube heat exchangers. Vaisi et al. [7] experimentally showed that the configuration of the louvered fins had the dominant influence on the heat transfer and pressure drop. They found that a symmetrical arrangement of louvered fins provides more advantage than asymmetrical arrangement in terms of heat transfer and pressure drop. Dogan et al. [8] compared heat transfer and pressure drop characteristics of double-row and triple-row multi-louvered fin heat

* Corresponding author: aokbaz@yildiz.edu.tr

exchangers. They found that the overall performance of double-row fin heat exchanger was more effective. Rugh et al. [9] experimentally studied heat transfer and friction coefficients for high-density louvered fin and flat tube heat exchangers used for passenger compartment heating in automobiles. Suga and Aoki [10] concluded that smaller louver angles were found to provide an overall better performance due to behaviour of thermal wakes. Springer and Thole [11] concluded that to do an experiment with louver fins, enough number of fin rows is needed to simulate flow behaviour of real size heat exchanger. Zhang and Tafti [12] classified thermal wakes interference of multi-louvered fins as inter-fin interference which occurs between adjacent rows of louvers for louver directed flows and intra-fin interference which occurs on subsequent louvers of the same row or fin for duct directed flows. Perrotin and Clodic [13] compared numerical results of 2D and 3D louver fin heat exchanger models with wind tunnel tests. The results of 3D models were closer to the experimental results and 2D models overestimated significantly the heat transfer coefficient. Hsieh and Jang [14] reported that successively variable louver angles effectively enhance the heat transfer performance. Okbaz et al. [15] studied to determine minimum number of rows which provide periodic flow conditions with adequate flow field to conduct experimental measurements of their test facility. They discussed the flow structure depending on the number of rows

According to literature survey there are several studies which investigate heat transfer and pressure drop characteristics on basic geometrical structures. In this study, thermal and hydraulic performance of a louvered fin heat exchanger that is to be designed for an particular industrial application have been investigated numerically. Results are given in terms of Colburn ***j*** factor and Fanning factor ***f***, temperature counters and stream-lines results.

## 2 Numerical method

The three dimensional flow field and heat transfer were solved using commercial code ANSYS Fluent. Several turbulence models such as: Standard k-ϵ turbulence model, RNG k-ϵ turbulence model, Realizable k-ϵ turbulence model, SST-k- ω turbulence model and RSM were tested. For this louver fin heat exchanger model, experimental study have not been done yet. But, for other louver fin models which are the products of FRİTERM Company and similar to current model, There are experimental results to validate numerical model. These experiments have been conducted in the test lab of FRITERM Company. When the heat transfer results of numerical and experimental studies were compared, it was seen that Standard k-ϵ turbulence model over predicted convective heat transfer coefficient. RNG k-ϵ turbulence model and Realizable k-ϵ turbulence model predicts convective heat transfer coefficient closer to the experimental results. Reynolds Stress Model (RSM) results are also close to experimental results. However, more CPU time and memory were required to solve via RSM. Hsieh and Jang [16], used extended k-ϵ turbulence model of Chen and Kim [17] in their numerical study. In this study, we considered separated flow effects of tubes when solving three dimensional flow fields. Due to realizable model provides the best performance of all the k-ϵ turbulence models of separated flows (ANSYS Fluent Theory Guide [18]), its low cost of running time and its well-known good performance on solving heat transfer problems, Realizable k-ϵ turbulence model was used to solve three-dimensional flow field and heat transfer. To solve more accurately wall shear stress and wall heat transfer Enhanced Wall Treatment option for Realizable k-ϵ turbulence model was applied. Enhanced Wall Treatment is a near wall modelling method to resolve viscous sublayer and fully turbulent region. A pressure based predictor/multi-corrector solution procedure was employed to enhance velocity–pressure coupling and continuity-satisfied flow filed. The Couple algorithm was applied for solving flow field. The discretization of the convective terms in the governing equations was done via a second order upwind scheme, order central differencing scheme was applied for the diffusive terms. The pressure gradient in the momentum equations was treated via a second order discretization scheme.

The dimensionless equations for continuity, momentum (Reynolds- averaged Navier–Stokes equations) and energy may be expressed in tensor form as:

$$\frac{\partial U_i}{\partial X_i}=0 \tag{1}$$

$$\frac{\partial}{\partial X_j}(U_iU_j)=-\frac{\partial P}{\partial X_i}+\frac{1}{\mathrm{Re}}[\text{"}^{2}\,U_i]-\frac{\partial}{\partial X_j}(\overline{u_iu_j}) \tag{2}$$

$$\frac{\partial}{\partial X_j}(\theta U_j)=-\frac{\partial P}{\partial X_i}+\frac{1}{\mathrm{Re\,Pr}}[\text{"}^{2}\,\theta]-\frac{\partial}{\partial X_j}(\overline{u_j\theta}) \tag{3}$$

The modeled transport equations for k and ϵ in the realizable k-ϵ model are (ANSYS Fluent Theory Guide[18]):

$$\frac{\partial}{\partial t}(\rho k)+\frac{\partial}{\partial x_j}(\rho k u_j)=\frac{\partial}{\partial x_j}\left[(\mu+\frac{\mu_t}{\sigma_k})\frac{\partial k}{\partial x_j}\right]+G_k+G_b-\rho\varepsilon-Y_M+S_k \tag{4}$$

$$\frac{\partial}{\partial x}(\rho\varepsilon)+\frac{\partial}{\partial x_j}(\rho\varepsilon u_j)=\frac{\partial}{\partial x}\left[\left(\mu+\frac{\mu_t}{\sigma_\varepsilon}\right)\frac{\partial\varepsilon}{\partial x_j}\right]$$
$$+\rho C_1 S\varepsilon-\rho C_2\frac{\varepsilon^2}{k+\sqrt{v\varepsilon}}+C_{1\varepsilon}\frac{\varepsilon}{k}C_{3\varepsilon}G_b+S_\varepsilon \tag{5}$$

$$C_1 = \max\left[0.43, \frac{\eta}{\eta+5}\right], \eta = S\frac{k}{\varepsilon}, S = \sqrt{2S_{ij}} \quad (6)$$

Here, $G_k$ represents the generation of turbulence kinetic energy due to the mean velocity gradients. $G_b$ is the generation of turbulence kinetic energy due to buoyancy, $Y_m$ represents the contribution of the fluctuating dilatation in compressible turbulence to the overall dissipation rate. $C_{1\varepsilon}$, $C_{2\varepsilon}$ and $C_{3\varepsilon}$ are constants. $S_k$ and $S_\varepsilon$ are user defined source terms (ANSYS Fluent Theory Guide [18]).

The quality of the mesh was carefully assessed during the meshing. Meshes with five layers of inflation region were established in the close region of the fin wall and tube wall while unstructured tetrahedron meshes were used outer region. The flow field was separated several subfields to satisfy good mesh quality and sensitivity. 12 million elements were used to solve flow field. The near-wall mesh with y+ $\leq 1$ was established that fine enough to resolve viscous sublayer. No slip boundary conditions were applied on tube wall surfaces. The walls of louver fins were at constant temperature of $T_w$=313.15 K. The inlet velocity of the flow were 1.22 m/s, 1.52 m/s, 2.28 m/s and 3 m/s and the temperature of the inlet fluid was set at 30°C. The turbulence intensity at the inlet is set at 10%. At the exit, pressure outlet boundary condition at operational pressure of 0 Pa was applied. The distance of the inlet boundary to geometry of interest and the distance of the outlet boundary to geometry of interest is 33 mm and 99 mm, respectively. Before starting all analysis effects of this distance length were tested. Periodic boundary conditions were applied left- right sides and top-bottom sides of solution domain.

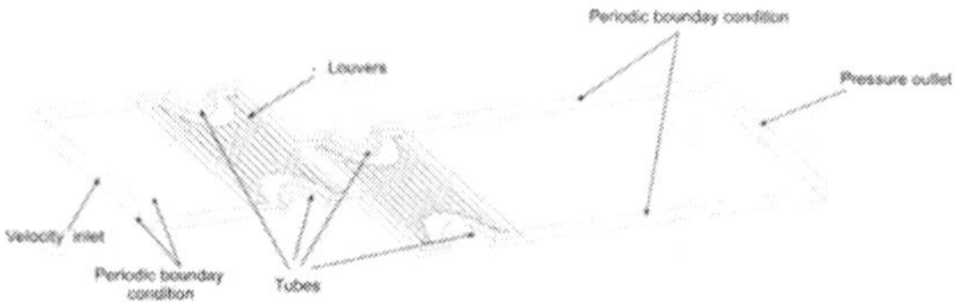

**Fig. 1** 3D computational domain and boundary conditions.

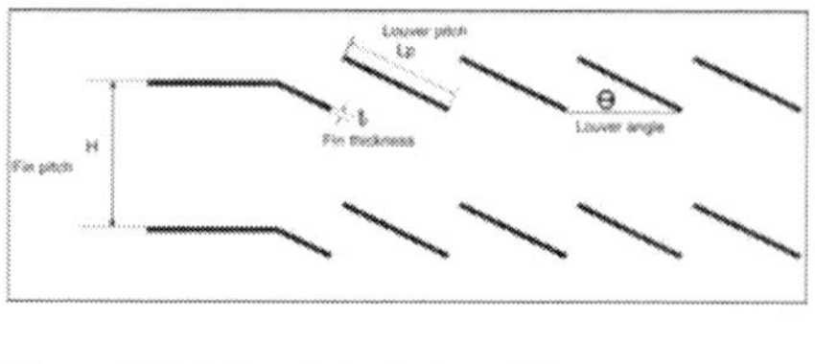

**Figure 2** Details of physical model

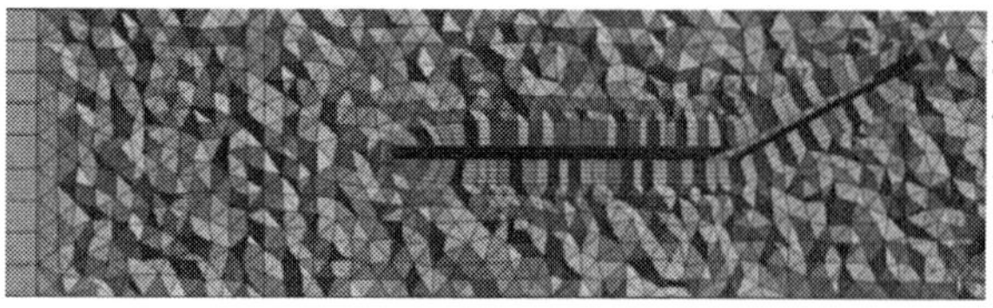

**Fig. 3** Detailed view of computational grid system

The Reynolds number is defined as:

$$\mathrm{Re} = \frac{\rho U H}{\mu} \quad (7)$$

Where U is the frontal velocity and H is the fin pitch.

The averaged heat transfer coefficient h is defined as:

$$T_b = \frac{T_{in} + T_{out}}{2} \quad (8)$$

$$q'' = h.(T_w - T_b) \quad (9)$$

Here, q″ is the total heat flux and $T_b$ is the bulk mean temperature of the fluid. The averaged heat transfer coefficient can be expressed in dimensionless form by Nusselt number Nu.

The Fanning friction factor $f$ and Colburn factor $j$ are defined as:

$$j = \frac{Nu}{\mathrm{Re}.\mathrm{Pr}^{2/3}} = \frac{h}{\rho.U.C_p}.\mathrm{Pr}^{2/3} \quad (10)$$

$$f = \frac{\Delta P}{\frac{1}{2}.\rho.U^2}.\frac{H}{4.L} \quad (11)$$

**Table 1.** Geometric and operation parameters

| | | |
|---|---|---|
| $L_p/F_p$ (Ratio of louver pitch to fin pitch) | 1.9, 1.52, 1.187 | |
| Louver tilt angle (Θ) | 22°≤ Θ ≤30° | |
| Fin thickness ($t_f$): | 0.12 | mm |
| Flow length (L): | 66 | mm |
| Inlet air | | |
| Temperature :($T_{in}$) | 303.15 | K |
| Frontal velocity (U) | 1.22≤U≤3 | m/s |
| Wall temperature ($T_w$): | 313.15 | K |

## 3 Results and discussion

The hydraulic and thermal performances of a louver fin heat exchanger depend on geometrical details and flow conditions. In Figure 4 and Figure 5 velocity vectors and related streamline patterns coloured with velocity magnitudes are shown, respectively. These visualization results were obtained from top plane view at the periodic boundary region. There occurs a wake region as a result of flow separation around tubes with weaker velocity magnitudes. As seen in Figure 6, the temperature of the cold fluid increases faster in the louvered region of the fin.

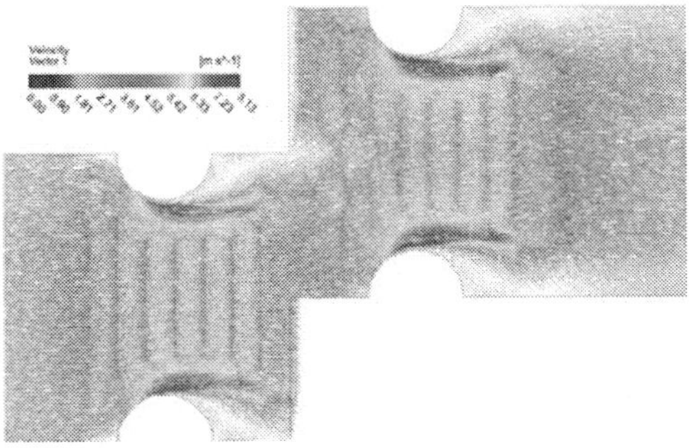

**Fig. 4** Velocity vectors for heat exchanger with louver angle of $\Theta$=30°, $L_p/F_p$ = 1.1875 and frontal air velocity of U= 3 m/s. (Obtained from top periodic plane)

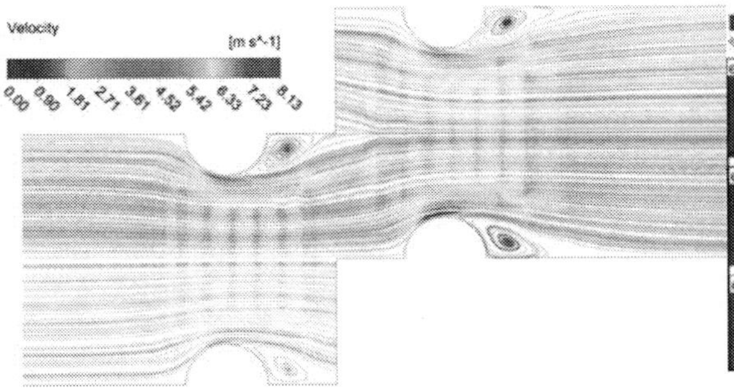


**Fig. 5** Streamline patterns for heat exchanger with louver angle of $\Theta$=30°, $L_p/F_p$ = 1.1875 and frontal air velocity of U= 3 m/s. (Obtained from top periodic plane)

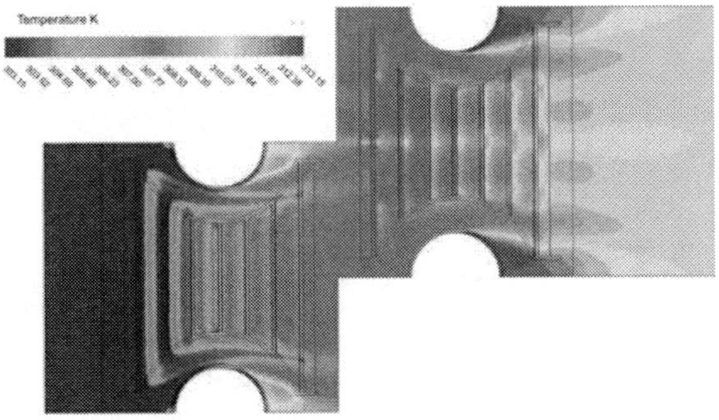

**Fig. 6** Temperature counters for heat exchanger with louver angle of $\Theta$=30°, $L_p/F_p$ = 1.1875 and frontal air velocity of U= 3 m/s. (Obtained from top periodic plane)

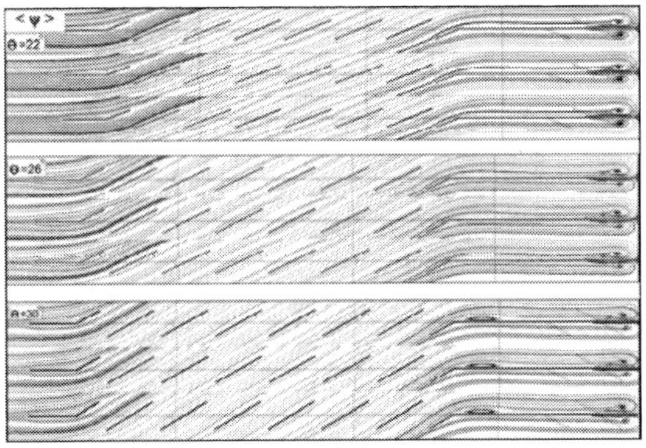

**Fig. 7** Streamlines for heat exchanger with louver angle of $\Theta$ =22°, 26° and 30°, $L_p/F_p$ = 1.1875 and frontal air velocity of U= 3 m/s (Obtained from middle plane of first louver row.)

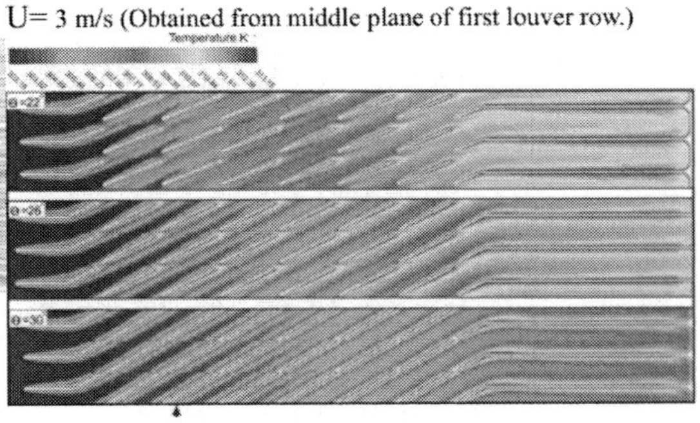

**Fig. 8** Temperature counters for heat exchanger with louver angle of $\Theta$ =22°, 26° and 30°, $L_p/F_p$ = 1.9 and frontal air velocity of U= 3 m/s (Obtained from middle plane of first louver row.)

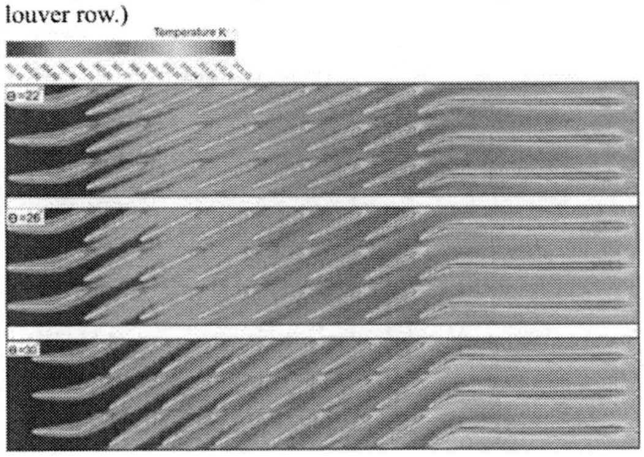

**Fig. 9** Temperature counters for heat exchanger with louver angle of $\Theta$ =22°, 26° and 30°, $L_p/F_p$ = 1.52 and frontal air velocity of U= 3 m/s (Obtained from middle plane of first louver row.)

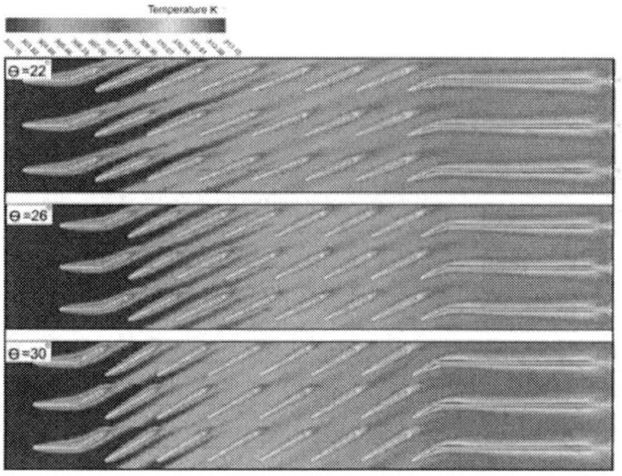


**Fig. 10** Temperature counters for heat exchanger with louver angle of Θ =22°, 26° and 30°, $L_p/F_p$= 1.1875 and frontal air velocity of U= 3 m/s (Obtained from middle plane of first louver row.)

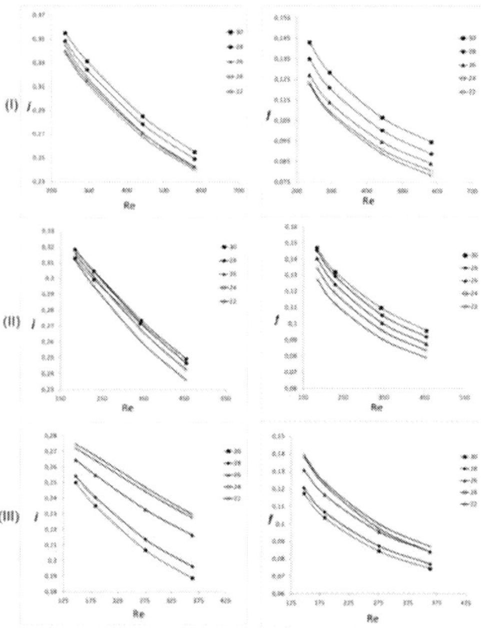


**Fig. 11** Colburn factor ***j*** and Fanning friction factor ***f*** graphics for $L_p/F_p$= 1.1875 (row I), $L_p/F_p$= 1.52 (row II) and $L_p/F_p$= 1.9 (row III).

In Figure 7 streamline patterns of inlet louvers are given for $L_p/F_p$ = 1.1875 and louver angles of 22°, 26° and 30° and frontal air velocity of 3 m/s. These streamline patterns were determined from side plane in the middle of first row of louvers (inlet louver region). The flow is louver directed for all louver angles. As seen from the streamline patterns, when the louver angle increases, the distance of the flow that air travel across the fins increases. At the end of the inlet louver region and upstream of the tubes a large recirculation region patterns of horseshoe vortex was formed. Sahin et al. [19] experimentally showed the formation of vortexes which are called as horseshoe vortexes for plate-fin and tube heat exchangers via particle image velocimetry technique. They reported that well-known horseshoe vortexes magnified the entrainment process which occurred between the main flow and the wake-flow regions and enhanced the heat transfer rate in the case of fin-tube heat exchangers. For the louver angle of 30°, a flow separation region on downstream direction of the last louver occurred due to high louver angle. This formation increases pressure drop penalty.

The temperature distribution along the louvers is given in Figure 8, 9 and 10 for $L_p/F_p$= 1.9, 1.52, 1.1875 respectively at frontal air velocity of 3 m/s. These temperature counters were also taken from side plane in the middle of first row of louvers. These results give considerable information about heat transfer characteristics. The heating process of cold fluid along the louvers and mixing of cold and hot fluids can be easily seen. As seen from the figures louver angle affects heat transfer differently depending on fin pitches.

In Figure 7, the temperature counters for louver angles of 22°, 26° and 30° are given for $L_p/F_p$= 1.9 at frontal air velocity of 3 m/s. The exit temperature for the case the louver angle of 22° is higher due to mixing of cold and hot air is magnified. Increasing the louver angle to 30° made the distance closer the leading edge and trailing edge of upper and lower fins' louvers. Thus, boundary layer was formed thicker when the fluid flowed along the louvers across the fins. When $L_p/F_p$ was decreased to 1.52 , in the case for the louver angle of 22°, heat transfer was higher. In Figure 9 the results of temperature counters are given for $L_p/F_p$ is 1.1875. At this distance of lower and upper fins, the trailing edge and leading edge of louvers are far enough to restart of boundary layer. Thus, increasing louver angle increased heat transfer due to increasing of flow length.

To evaluate quantitatively heat transfer and hydraulic performance for all louver angles, frontal velocities of air and fin pitches, Colburn factor *j*, Fanning friction factor *f* are given in Figure 11. In the first row of Figure 10, Colburn factor *j* and Fanning friction factor *f* is given $L_p/F_p$=1.1875. Colburn factor *j* increased with increasing louver angles. Increased louver angles provide the flow in longer distance by travelling between the neighbour fins. Fanning friction factor *f* also increased while the louver angle increased as a result of longer flow length. For smaller fin pitch ($L_p/F_p$=1.52) Colburn factor *j* decreased compared to case of $L_p/F_p$=1.1875. The highest heat transfer occurred for louver angle of 28° and it was too close the value of 30°. However, the pressure drop is highest for the case the louver angle of 30°. When $L_p/F_p$=1.9, the heat transfer and pressure drop characteristics completely changed depending on louver angle. The highest heat transfer value in terms of Colburn factor *j* was for the case of louver angle of 22° while the smallest value of it was for the case of louver angle of 30°. Moreover, the lowest pressure drop in terms of Fanning friction factor *f* was for louver angle of 24° and it was very close to value of 22° while the highest value was for the case of louver angle of 30°. For louver pitch to fin pitch $L_p/F_p$ is equal to 1.9 heat transfer and pressure drop characteristics reversed. The reason of this situation was explained looking temperature distribution results.

## 4 Conclusion

In this study air-side heat transfer and pressure drop characteristics of multi-louvered fin heat exchangers were studied numerically. 3-D numerical simulations was done via ANSYS Fluent software. The heat transfer and pressure drop characteristics were reported in terms of Colburn *j*-factor and Fanning friction factor *f*. The louver angles were changed in the range of $20° \leq \Theta \leq 30°$, and frontal velocities of U were between 1.22 m/s and 3 m/s. To understand the effect of flow structure on heat transfer and pressure drop characteristics streamline patterns, velocity vectors and temperature counters were given. The following conclusions have been achieved:

- For all louver angles, ratio of louver pitch to fin pitch and operation conditions of frontal velocities, the flow was full louver directed.
- For ratios of louver pitch to fin pitch 1.1875 and 1.52, heat transfer capacity and pressure drop penalties increased with increasing louver angle.
- The effect of louver angle on heat transfer and pressure drop was decreased for $L_p/F_p$ =1.52 compared to case of $L_p/F_p$ =1.1875
- For the ratio of louver pitch to fin pitch $L_p/F_p$=1.9, Colburn *j*-factor and and Fanning friction factor *f* decreased with increasing louver angle

### ACKNOWLEDGMENTS

The authors would like to acknowledge the funding of Republic of Turkey Ministry of Science, Industry and Technology and FRITERM Company under project contract number of 0649.STZ.2014. This study has been prepared from ongoing Ph.D. thesis studies of Abdulkerim OKBAZ.

## References

1. H. Aoki, T. Shinagawa, K.K. Suga, Exp. Thermal and Fluid Sci. **2**, 293–300 (1989)
2. R.L. Webb, P. Trauger, Exp. Thermal and Fluid Sci. **4**, 205–17 (1991)
3. C.C. Wang, C.J. Lee, C.T. Chang, S.P. Lin, International Journal of Heat and Mass Transfer **42**, 1945-1956 (1999)
4. M.H. Kim, C.W. Bullard, International Journal of Refrigeration **25**, 390–400 (2002)
5. N.C. DeJong, A.M. Jacobi, Experimental Thermal and Fluid Science **27**, 237–250 (2003)
6. J. Dong, J. Chen, Z. Chen, W. Zhang, Y. Zhou, Energy Conversion and Management **48**, 1506–1515 (2007)
7. A. Vaisi, M. Esmaeilpour, H. Taherian, Applied Thermal Engineering **31**, 3337-3346 (2011)
8. B. Dogan, O. Altun, U. Ugurlubilek, M. Tosun, T. Sarıçay, L. B. Erbay, Applied Thermal Engineering **91**, 270-278 (2015)
9. J.P. Rugh, J.T. Pearson, S.A. Ramadhyani, *ASME Symp Ser HTD* **201**, 15-24 (1992)
10. K. Suga, H. Aoki, *Proc. of ASME/JSME Thermal Eng. Joint Conf.* **4**, 361–368 (1991)
11. M.E. Springer, K.A. Thole, Experimental Thermal and Fluid Sci. **18**, 258–69 (1998)
12. X. Zhang, D.K. Tafti, International Journal of Heat and Mass Transfer **44**, 2461-2473 (2001)
13. T. Perrotin, D. Clodic, International Journal of Refrigeration **27**, 422-432 (2004)
14. C.T. Hsieh, J.Y. Jang, Applied Thermal Engineering **26**, 1629-1639 (2006)
15. A. Okbaz, A.B. Olcay, A. Pinarbasi, *Experimental Fluid Mechanics 2014* 393-399 (2014)
16. C.T. Hsieh, J.Y. Jang, Applied Thermal Engineering **26**, 1629-1639 (2006)
17. Y.S. Chen, S.W. Kim, *NASA CR* 179204, (1987)
18. *ANSYS Fluent Theory Guide*, ANSYS Inc. Canonsburg, 52-53 (2013)
19. B. Sahin, N.A. Ozturk, C. Gurlek, International Journal of Heat and Fluid Flow **29**, 340–351 (2008)

# Boiling on fins with wire screen of variable effective conductivity

*Tadeusz* Orzechowski[1]

[1]Kielce University of Technology, IŚGiE Department, Al. Tysiąclecia PP7, Poland

**Abstract.** The high scale of integration of modern equipment used for medical, military and other purposes puts heavy demands as regards the removal of great heat fluxes. This can be achieved only in exchangers that apply the phase change phenomena. Among many methods to improve boiling heat transfer, the wire mesh covering demonstrates some advantages due to the possibilities of designing the desired microstructure parameters, availability on the market, and low cost. The wire mesh microstucture with specified geometrical parameters produces anisotropy in conductivity. The different arrangement of the mesh layers relative to the direction of the heat flux is a cause of the change of temperature distribution within the layer. The consequence is a respective change in the discharge conditions of the gas phase and liquid feed. The experiments were conducted on fins covered with a single layer of copper mesh with lumen of 38 % and boiling FC-72 at ambient pressure. Compared with the smooth surface, the wire mesh structures yield an increase in the heat transfer rate at boiling. It is also shown that nucleate boiling is initiated at lower wall superheat. Formulas for longitudinal and perpendicular thermal conductivity are given for different mesh structure arrangements.

## 1 Introduction

These Modern heat exchangers used in electronics, cooling and chemical industry need to have compact structure and remove large amounts of heat. The machines generate so much heat that simple surface enhancement with ribs stopped to provide a sufficient solution long ago. It was found that phase change processes offer new possibilities as regards the construction of high-efficiency heat exchangers. An additional, yet very desirable feature they have is the fact that they maintain almost constant temperature. That prevents overheating that is unwanted, and also allows setting the optimum technological parameters. Therefore, exchangers of that type have gained increasing popularity [1-4].

In the mid 1960s, in studies by Haley and Westwoter [5], it was shown that boiling heat transfer coefficient is related to the fin temperature along its height, and can be described by the boiling curve. However, when fins are packed in the exchanger, different behaviour of those can be observed, which depends on packing density and fin geometry [6].

To dissipate increasingly high heat fluxes, higher wall superheats are required, which at too large values at the base, leads to the quick development of film boiling. That poses a threat of the surface burn-out [7]. One of the ways to prevent such phenomena is to deposit an additional insulation layer with low heat conductivity, or to make fins that have variable cross-section along the height [8]. Both options offer a limited possibility of increasing the amount of dissipated heat. It becomes necessary to deposit additional layers that enhance heat transfer surface. Here, the following techniques can be distinguished:

- various structures obtained by pressing or bending ribs [9], [10];
- modifying of the heating surface by increasing its roughness mechanically or technologically [11], [12];
- imposing additional laminar-composite layers with preset parameters [13], [14].

The application of many proposed solutions is limited because of technology requirements and the costs. The other factor that has to be taken into account is the repeatability due to the heat transfer conditions. Mesh structures constitute a specific group. Their main advantage is that they are relatively readily available on the market, and can be made from different materials, and have different geometrical parameters to suit the purposes of heat transfer. This surface enhancement can be provided with desirable geometry to meet the anticipated operating conditions. Such structures allow constructing layers with selected parameters, such as pore diameters and the coating thickness. Wire mesh packing increases efficiency of heat exchangers, as well as the heat flux and heat transfer coefficient, while simultaneously lowering the temperature of the heating surface [15]. Heat dissipation from such systems is a result of complex processes that occur inside the layer. The construction factors that mostly affect the process include porosity, mesh size and the wire thickness, which was discussed in, e.g. [16], [17] and many other studies.

The design of mesh layers that have desirable parameters is relatively simple. That involves the deposition of successive mesh layers onto one another.

* Corresponding author: todek@tu.kielce.pl

Depending on the needs, those are pressed against the base [19], soldered or sintered. The conductivity of such structures depends on their arrangement, and can vary in value in different directions [18]. That also refers to spatially extended structures [19]. The magnitude of the thermal conductivity coefficient is also affected by the technology used to deposit layers onto the base. In the process, contacts of different quality, between the structure skeleton elements are formed. They are decisive for the internal transport of the heat. In study [20], the example of square-shaped and diamond-shaped wire screens was used to present the methodology for calculating volumetric porosity, specific surface area, and in-plane thermal conductivity. The dependences given in the study concern the wire mesh screens that are cold compressed. In study [21], the existing models of computing the effective conductivity of mesh screens were classified into two groups. The first category includes those dependences, on the basis of which relatively high values are obtained, which results from strong contact between wires. The other group comprises those formulas which produced low effective conductivity due to weak contacts. It was demonstrated that, under the assumption of isotropy, divergence in calculations amounts to two orders of magnitude. A new theoretical model was proposed to determine effective thermal conductivity taking into account contact conditions between wires. The model, however, does not account for the lengths of wire-to-wire or wire-to-base contacts that formed in the sintering process. The contacts depend on the process parameters, whereas their surface characteristics is different from that resulting exclusively from the layer geometry [22], [23].

Although the formulas for calculating mesh conductivity differ much in various studies, they some of them indicate variable conductivity in the direction longitudinal and perpendicular to the surface. If mesh layers are deposited on the fin, a part of heat is also transported parallel to the surface. Consequently, when calculating heat transfer in such systems, it is not possible to disregard the anisotropic properties of mesh screens. The aim of this study is to analyze the effect of the geometry of the mesh structures on the heat transfer process.

## 2 Experimental facility

Investigations into heat transfer on the fin with mesh covering were conducted using the facility shown, in a form of diagram, in Fig. 1. The facility was installed in a large volume vessel in such a way so that one side with the mesh cover could be in contact with the boiling liquid, whereas the other side is exposed to ambient atmosphere. Such fixing of the sample is necessary because of the adopted measurement procedure that is based on the analysis of the thermal field measured with the thermal imaging camera.

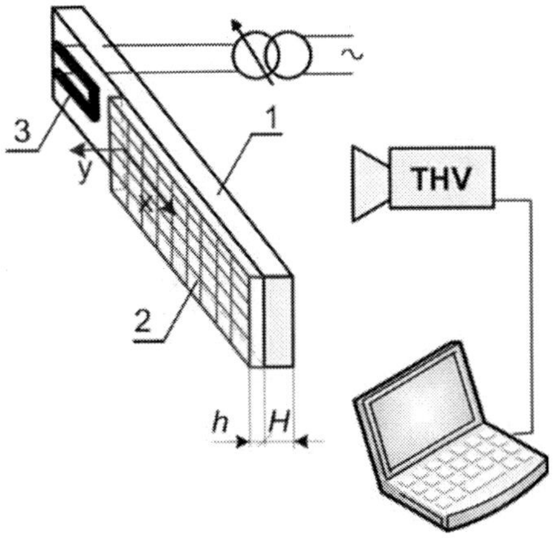


**Fig. 1.** Diagram of the test facility: 1 - examined element, 2 - mesh structure, 3 - main heater.

A majority of liquids used in boiling exchangers is almost entirely opaque to infrared radiation of thermal imaging cameras. Consequently, it is not possible to directly observe the surface on which heat transfer occurs. Therefore, the only possible measurement to be taken is the one on the side in contact with atmosphere. In this study, the investigations concerned a long and thin copper fin that had the thickness of $H$, selected in such a way so that, under measurement conditions, the number $Bi<0.1$. In such conditions, it is possible to assume one-dimensional model of heat transfer in the fin [24]. Otherwise, a substantial temperature drop can occur in the mesh layer [25], depending on its thickness $h$ (see Fig.1). The phenomenon is also dependent on the type of contact with the base and the geometry of the arrangement.

## 3 Thermal conductivity of the mesh structure - physical model

The market offers various meshes, which depending on the needs, are made from different materials and have different weave. In heat transfer applications, plain-weave copper meshes are most frequently used. They are deposited on the surface, one onto the other, and compressed or sintered, depending on the technology. Due to the process repeatability, it is assumed that meshes are symmetrically arranged relative to heat flow direction. The geometric arrangement of either squares or diamonds is found, as shown in Fig. 2. A repeatable unit cell in the orthogonal projection is also marked in the figure.

 DOI: 10.1051/epjconf/201714302085

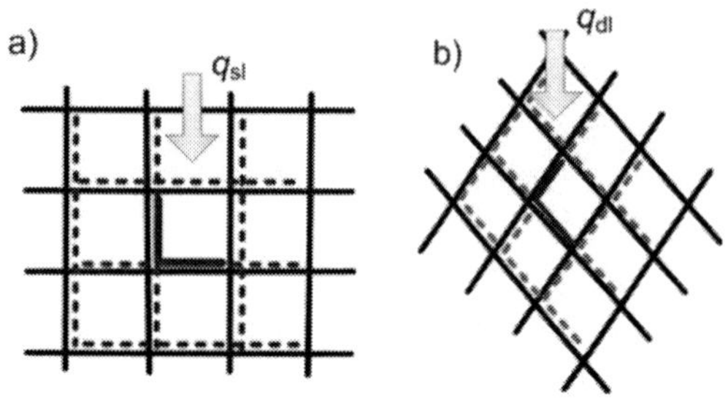


**Fig. 2.** Structure arrangement relative to the longitudinally flowing heat flux: a) square-shaped screen relative to $q_{sl}$, b) diamond-shaped screen relative to $q_{dl}$.

The symmetric structure of diamond-shaped wire screen with a constant opening width $w$ and wire diameter $d$ can be characterised by the weave angle $\beta$ (see Fig. 3). Dimensions $a$ and $b$, which determine the size of the repeatable unit cell of the mesh, result from the geometric quantities given above.

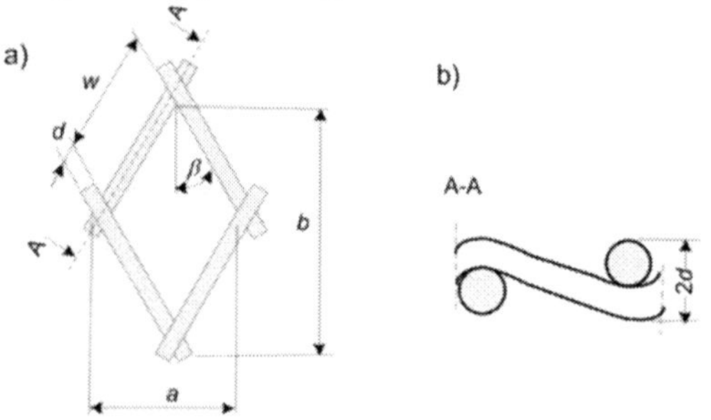


**Fig. 3.** Geometry of a unit cell: plain view and A-A section.

For diamond-shaped plain-weave wire mesh shown in Figs. 2b and 3, respectively, the values of the thermal conductivity coefficients can be computed from the following dependences:

- a ratio of the mean value of the thermal conductivity coefficient in the perpendicular direction $\lambda_{dp}$ to the mesh material conductivity coefficient $\lambda_n$:

$$\frac{\lambda_{dp}}{\lambda_n} = \frac{\delta(\delta + d)}{(w + d)^2 \sin 2\beta} \quad (1)$$

- in the longitudinal direction:

$$\frac{\lambda_{dl}}{\lambda_n} = \frac{\delta^2}{d + 2w\sin\beta} \cdot \left( \frac{1}{d + 2w\cos\beta} + \frac{d + 2w\cos\beta}{2d(w + d)} \right) \quad (2)$$

where $\delta \approx 0.9d$.

For square-shaped wire mesh, Fig 2a, the formula for the mean value of the thermal conductivity coefficient in the longitudinal direction $\lambda_{sl}$ is as follows:

$$\frac{\lambda_{sl}}{\lambda_n} = \frac{\delta^2}{d(d + w)} \quad (3)$$

The mean value of the thermal conductivity coefficient of the structure in the perpendicular direction $\lambda_{sv}$ is calculated in the same way in both cases, namely from formula (1), which for square-shaped mesh takes on the following form:

$$\frac{\lambda_{sv}}{\lambda_n} = \frac{\delta(\delta + d)}{(w + d)^2} \quad (4)$$

All the dependences above were derived while disregarding the thermal conductivity of vapour that fills the structure, and of thin liquid film that feeds structure. Both quantities are negligibly small compared with the mesh material conductivity.

# 4 Results and discussion

Figure 4 shows an exemplary result of investigations into boiling heat transfer on a long fin, with Fluorinert FC-72 as a working medium.

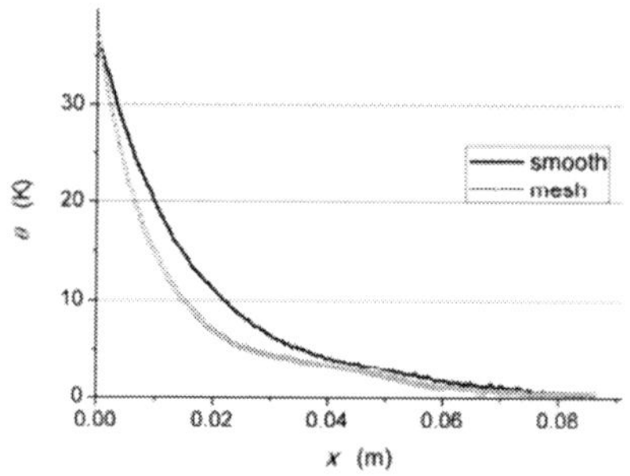


**Fig. 4.** Wall superheat distribution for smooth and single-layered copper specimens.

The local boiling curve for the copper fin, 3 mm in thickness, was determined in accordance with the procedure described in study [24]. The technique utilises fin vertical surfaces with wire mesh structures. Mesh with lumen of 38 %, made of 99.9% copper wire, 0.2 mm in diameter, was applied. The experiment was conducted for refrigerant boiling at ambient pressure. The measurement results yield the temperature distribution along the fin length. Each specimen was measured at least three times, at different values of electric power supplied to the main heater. In each case, the other parameters remained unchanged and constant. Local values of heat flux were determined on the basis of the results of measurements carried out on specimens having a smooth surface and with a single layer of superimposed mesh structure. The values are correspondingly shown in Fig. 5.

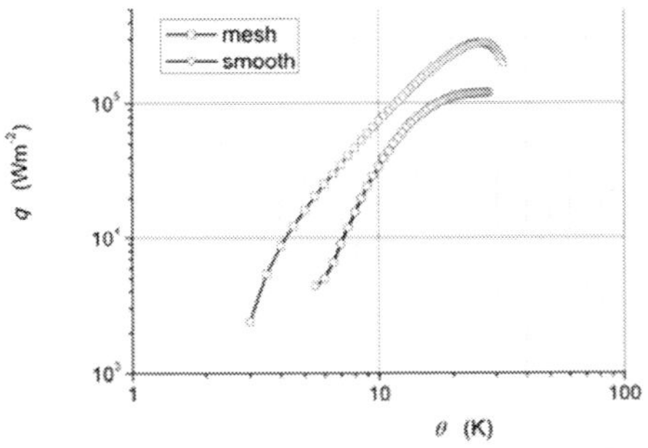


**Fig. 5.** Boiling curve defined locally for smooth and single-layered copper specimens.

Compared with the smooth surface, the wire mesh structure produces an increase in the heat transfer rate at boiling. From Fig. 5, it is clear that nucleate boiling is initiated at lower superheat. The results of measurements show that the imposition of mesh onto the surface reduces, by ~2÷3 K, the initiating point of nucleate boiling as compared with the smooth one.

Mesh structures are characterised by conductivity anisotropy, which was discussed in Section 3. Figure 6 shows a change in the mean value of the thermal conductivity coefficient in the longitudinal direction (the value decreases) and the perpendicular one (the graph line is symmetrical with respect to the angle $\beta$=45°) for diamond-shaped mesh and different weave angles. With a decrease in $\beta$, the structure becomes more close-packed, and the change in the dimensions of the balanced unit cell leads to an increase in the heat transfer coefficient.

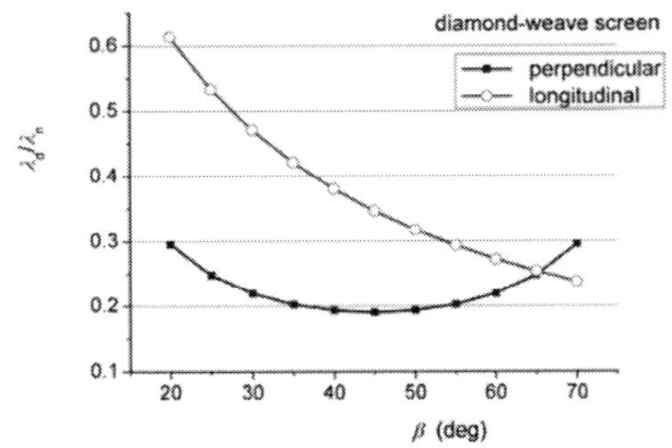


**Fig. 6.** Conductivity coefficient in the longitudinal and perpendicular directions in diamond-weave screens.

Substantial changes in the amount of heat dissipated by the surface with a single sintered layer compared with the smooth surface results from complex process taking place inside the structure. The conductivity model is one of many computational tools available for boiling heat transfer on enhance surfaces. For this model, it is necessary to precisely give the effective conductivity, which is very difficult. The literature provides a lot of dependences, however, the results of calculations obtained with those formulas often differ by as much as a few orders of magnitude [21].

For the structure arranged as shown in Fig. 2a, thermal conductivity in the longitudinal direction, calculated acc. formula (3) is 72.2 W/mK. The value calculated acc. (4), in the direction perpendicular to the fin surface is 102.6 W/mK. Figure 7 presents the distribution of superheat inside a single layer of copper mesh, which was obtained on the basis of the values quoted above. Computations were conducted numerically acc. to the procedure discussed in [25]. For the sake of comparison, superheat distributions are shown for the same distances from the fin base, under the assumption of isotropy, and the effective conductivity of the mesh equal to 10.9 W/mK.

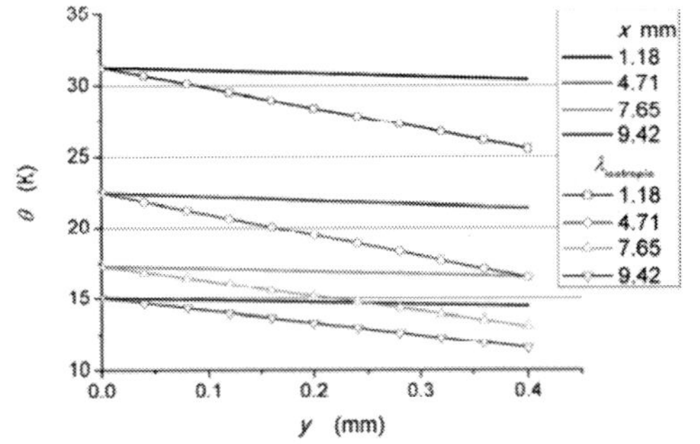


**Fig. 7.** Superheat distribution in the mesh layer in selected cross-sections.

## 5 Conclusions

The result of the investigations into FC-72 boiling at ambient pressure comes in the form of the boiling curve, i.e. heat flux change with wall superheat $\theta$, as presented in Fig. 5. The experiments were conducted on fins covered with a single layer of copper mesh with lumen of 38 %. Compared with the smooth surface, the wire mesh structures yield an increase in the heat transfer rate at boiling. It is also shown that nucleate boiling is initiated at lower wall superheat.

Mesh structures are widely available, which makes them an option of choice when the surfaces of boiling heat exchangers are enhanced. Unlike coverings of other types, e.g. metal – porous ones, they are characterised by different thermal conductivity. That feature must be considered on individual basis, while taking into account the geometry of the finned surfaces. If the calculations are made for incomplete data on the structure properties, the results produced will considerably differ from the real ones. As shown in Fig. 7, neglecting the mesh anisotropic properties results in up to 5 K difference in the fin external surface superheat, which directly affects the amount of dissipated heat.

## References

1. H.N. Chaudhry, B.R. Hughes, Appl. Energ. **136**, 32 (2014)

2. S. Hożejowska, M. Piasecka, Heat Mass Transfer **50**, 1053 (2014)
3. P. Naphon, Energ. Convers. Manage. **51**, 2362 (2010)
4. A. Tyburczyk, EPJ Web of Conferences **114**, (2016) 02127
5. K.W. Haley, J.W. Westwater, *Proc. III Int. Heat Transfer Conf. Chicago* **13**, 245 (1966)
6. G.J. Klein, J.W. Westwater, AICH Journal **17**, 1050 (1971)
7. Z.W. Liu, W.W. Lin, D.J Lee, J.P. Hsu, J. Therm. Sci. **10**, 79 (2001)
8. C. Pais, P.E. Webb, ASME J. Heat Trans. **97**, 79 (1991)
9. N.I. Pecherkin, A.N. Pavlenko, O.A. Volodin, Int. J. Heat Mass Tran. **90**, 149 (2015)
10. B. Niezgoda-Żelasko, J. Żelasko, Int. J. Refrig. **54**, 151 (2015)
11. Y.-H. Kim, K.-J. Lee, D. Han, Heat Mass Transfer **45**, 55 (2008)
12. J. Kim, S. Jun, R. Laksnarain, S.M. You, Int. J. Heat Mass Tran. **101**, 992 (2016)
13. S.H. Kim, G.C. Lee, J.Y. Kang, K. Moriyama, H.S. Park, M.H. Kim. Int. J. Heat Mass Tran. **102**, 756 (2016)
14. L. Dong, X. Quan, P. Cheng, Int. J. Heat Mass Tran. **71**, 182 (2014)
15. M. Dyga, R. Płaczek, Int. J. Heat Mass Tran. **53**, 5499 (2010)
16. Z.G. Xu, C.Y. Zhao, Int. J. Heat Mass Tran. **85**, 824 (2015)
17. T. Orzechowski, A. Tyburczyk, *8th World Conf. Exp. Heat Transfer, Fluid Mech., Thermodynamics*, Lisbon **143** (2013)
18. J. Xu, R.A. Wirtz, ASME J. Electron Packaging **127**, 353 (2005)
19. Z. Zhao, Y. Peles, M.K. Jensen, Int. J. Heat Mass Tran. **57**, 690 (2013)
20. J. Xu, R.A. Wirtz, *19-th IEEE SEMI-THERM*, **IEEE 03CH37437**, 53 (2004)
21. C. Li, G.P. Peterson, Int. J. Heat Mass Tran. **49**, 4095 (2006)
22. R. Chatys, T. Orzechowski, J. Mech. Compos. Mater. **40**, 159 (2004)
23. W. Depczyński, R. Kazała, K. Ludwinek, K. Jedynak, Materials **9**, 1 (2016)
24. T. Orzechowski, Exp. Therm. Fluid Sci. **31**, 947 (2007)
25. T. Orzechowski, A. Tyburczyk, EPJ Web of Conferences **67**, (2014) 02088

# Leidenfrost boiling of water droplet

*Tadeusz* Orzechowski[1]
[1] Kielce University of Technology, IŚGiE Department, Al. Tysiąclecia PP7, Poland

**Abstract.** The investigations concerned a large water droplet at the heating surface temperature above the Leidenfrost point. The heating cylinder was the main component of experimental stand on which investigations were performed. The measurement system was placed on the high-sensitivity scales. Data transmission was performed through RS232 interface. The author-designed program, with extended functions to control the system, was applied. The present paper examines the behaviour of a large single drop levitating over a hot surface, unsteady mass of the drop, and heat transfer. In computations, the dependence, available in the literature, for the orthogonal droplet projection on the heating surface as a function of time was employed. It was confirmed that the local value of the heat transfer coefficient is a power function of the area of the droplet surface projection. Also, a linear relationship between the flux of mass evaporated from the droplet and the droplet orthogonal projection was observed.

## 1 Introduction

Evaporation of the droplets is studied in many areas of technology [1], [2], [3]. Various investigations conducted into the subject focus on the physics of the phenomenon, which depends strongly on the liquid and its components, thermal and flow conditions, interaction with the surface, etc. Although the problem has been thoroughly analysed, the mechanisms of evaporation are still not fully clear.

Physical processes involved in droplet evaporation are closely related to the surface. At low temperatures, below the saturation temperature, diffusion is a dominant phenomenon in evaporation. The density of the vapour generated at the liquid surface is different from that of the ambient atmosphere. Under the conditions of ideal gas, vapour density is proportional to its molecular mass. If it is water evaporation in the air, rising lighter vapours enhance the convection processes. When the molecular mass of the liquid is greater than ~30 kg/kmol (roughly for air), the liquid vapours linger around the surface, high saturation state prevails, and mass exchange is almost entirely diffusive. Such a mechanism was experimentally demonstrated, on the example of water and hexane droplets, in study [4]. At wall superheat of a few degrees, single vapour bubbles can be seen, which initiate nucleate boiling. The droplet lifetime diminishes abruptly, and latent heat of vaporisation allows dissipating large heat fluxes. As regards cooling efficiency, this range is the most advantageous one. Investigations into the dynamics of nucleate boiling in droplets deposited on the rough copper surface and the polished surface with thermal sprayed gold coating were reported in [5]. Heat transfer on the polished surface was found to be considerably lower than on the rough surface. Around the Leidenfrost point, the concentration of vapour inside the droplet is much lower than for pool boiling of water under the same conditions.

Clear differences are observed between nucleate pool boiling and boiling in the droplet. Intense nucleate boiling in the droplet is accompanied by continuous changes in its shape and the area of contact with the substrate. Those phenomena are caused by vapour bubbles that flow inside and depart. The rate and frequency of bubble generation is random in character, which makes evaporation from the droplet unstable and spatially unsteady in this range. Observations of droplets in this range do not reveal structures that are typical for nucleate boiling, namely, depending on the wall superheat, bubble columns, vapour mushrooms and vapour clouds [6], [7]. At wall superheat values below 7 K, vapour bubbles grow relatively slowly and depart not causing significant changes in the droplet shape. Only at the wall superheat above 10 K, bubbles visibly change the droplet circumference. At the bubble growth stage, the resultant force of its buoyancy can be resolved into two components, namely vertical and horizontal ones. The vertical component force causes a change in height, the other – in the boundary of droplet contact with the substrate. The state of the heating surface is decisive for the number of active nucleation centres, which initiate individual bubbles. As the density of nucleation centres on the surface grows with superheat, the nucleate pool boiling becomes more intense. In the limited droplet volume, that becomes a source of instability. When the ratio of the distance between bubbles S to the maximum diameter of the departing bubble $d_b$ is in the range $1 < S/d_b < 3$, this instability is substantial. At higher values of this ratio, bubbles do not affect one another. However, when the S value is lower than unity, bubbles start to coalesce [8], [9]. The maximum heat flux removed from the surface is found at the critical point temperature. Bubbles are formed in densely spaced nucleation centres

* Corresponding author: todek@tu.kielce.pl

in such a way that, at further increase in superheat, bubble columns start coalescing. As a result, the surface is locally dried up, and the density of the removed heat is considerably reduced. Surface temperature grows and the inflowing liquid evaporates abruptly, which makes the whole process unstable. This range is termed a transition boiling regime [10]. The final stage of this regime is the formation of a stable vapour film, which supports the droplet above the heating surface [11], [12]. The first observation of that effect was reported by Johann Gottlob Leidenfrost in *A Tract About Some Qualities of Common Water* in 1756 and it is named after him to this day. In many technological processes, the element that needs to be cooled has the initial temperature that is higher than Leidenfrost temperature. At the same time, the cooling rate is the parameter that is decisive for the final properties of the top layer and its structure [13], [14]. In technical systems, appropriate control contributes to the optimal conditions of machine operation [15], [16]. To solve the task thus posed, it is necessary to have possibly complete information on the physical processes that occur in droplet evaporation, which despite numerous investigations have not been fully explained.

The aim of this study is to provide the analysis of heat transfer between a large water droplet at atmospheric pressure, and a hot surface at a temperature higher than Leidenfrost temperature.

## 2 Experimental Set-up

Investigations into droplet evaporation at a temperature above the Leidenfrost point were conducted at the stand, the main component of which is a cylindrical copper heater. (see Fig. 1).

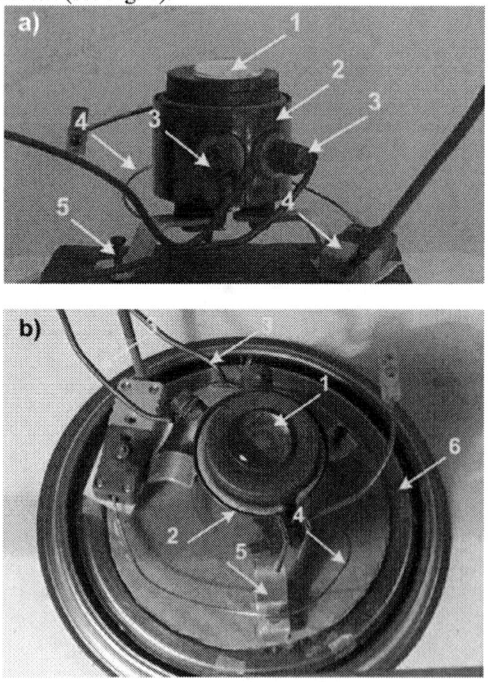


**Fig. 1.** Photograph of the experimental set-up: a) side view, b) top view 1- droplet, 2 - heater, 3 - electrical power supply, 4 - thermocouple, support 5- support pin, 6 – scales-pan.

The cylinder upper surface is shaped like a bowl having a very large radius, which allows centric, stable positioning of the liquid droplet under investigation. Directly under the surface, in the symmetry axis, k-type sheathed thermocouple is placed. It is connected, via suspended conductors, to analogue-digital converter, from which the measured signal is recorded in the computer memory. The heating system consists of a band heater with a resistor placed in the thermal insulation that is capable of continuous operation at the temperatures up to 500 $^{o}$C. The heater was installed on the cylinder outer surface. The heater is supplied with electric power from the autotransformer by suspended conductors. The output power of the heater is set by voltage, adjusted according to the voltmeter readings. Three support pins were fixed to the heating cylinder. Those support the heating cylinder with the heater, upholding it at approx. 2 cm above the surface, which prevents excessive heating of the substrate. Additionally, two screws were installed on pin ends. The tightening and loosening of the screws allows precise levelling of the measurement system so that the central position of droplet could be maintained, as shown in Fig.1b.

The complete measurement set was placed on the electronic weighing scales, with the maximum capacity of 500 g. The scales readability is 0.02 g, and sensitivity is 0.001 g. The magneto-electric transducer converts the load to electric signal. The factory setting of the transducer is responsible for precise conversion of time-variable load. Actual readings of the scales are displayed on the liquid crystal panel and the digital signal is available through the RS232 interface, with which the scales is equipped. The interface is used for direct communication with the computer. To that end, the author-designed program was developed to perform the control. The program has, among others, functions for the scales switching on/off, tarring, setting of the sampling frequency and storing the scales readings. The program also makes it possible to temporarily stop the registration and re-start, e.g. when a successive measurement is recorded, after previously tarring the scales, if necessary.

The supply conductors and the compensation one of the thermocouple are suspended from coupled supports that make a rigid and stable construction. The resilience of conductors produces time-variable load, the value of which also depends on temperature. The impact of this phenomenon on the scales readings was effectively eliminated due to several-hour conditioning of the measurement system with electric power supplied. An additional and desirable effect produced in this way was thermal stabilisation. Prior to the start of measurements, a few series of the scales readings were registered to check the stability of readings and repeatability of results. The detailed description of the measurement procedure, together with the analysis of uncertainties were presented in [11] and [17].

## 3 Results and discussion

The investigations outcome is a continuous measurement of the mass of evaporating droplet deposited on the upper surface of the heating cylinder (see Fig.1b.), the temperature of which is higher than Leidenfrost temperature. An exemplary set of measurement data is shown in Fig. 2.

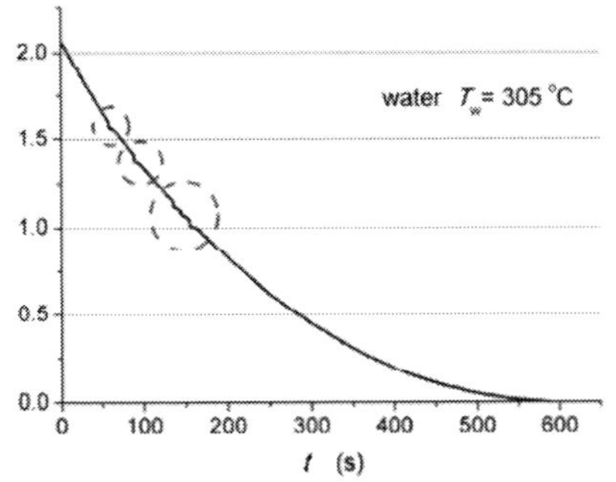


**Fig. 2.** Evaporation of the water droplet on the wall at the temperature $T_w$=305 °C with marked regions of departing bubbles.

The change in the droplet mass over time is shown for the pre-set temperature of the heating cylinder. The mean value of the latter was $T_w$=305 °C. The analysis of the measurement always started after a while, at the instant when the droplet mass was ~2 g. In this way, it was possible to eliminate the period of the initial stabilization of the system, related to the drop deposition. The frequency of the scales readings was always the same and equalled 1 Hz. Because of a growing relative error related to the accuracy of the scales used in the experiment, the measurements ended when the droplet was a few times greater than the scales sensitivity [11].

For large mass droplets, different types of instabilities are observed, which result from complex convective motions inside the droplet [18], [19]. Among others, bubbles appear, which grow and escape. An exemplary instant of a departure of a few bubbles is indicated in Fig. 2. The phenomenon fades away with time as the droplet mass decreases.

Evaporation of the droplet levitating above a hot surface is an extremely complex process, in which the phenomena of heat and mass transfer coincide. However, the greatest portion of heat transfer proceeds between the heating surface and the droplet bottom. If it is assumed that evaporation from the droplet upper surface is small compared with the amount of heat removed from the bottom surface, and overall heat transfer coefficient contains convective and radiative components, i.e. $\alpha=\alpha_c+\alpha_r$, heat balance can be written, which implies change of droplet mass $m$ over time $t$:

$$\alpha(T_w - T_d)A = -\left(c_p(T_s - T_d) + h_{fg}\right)\frac{dm}{dt} \quad (1)$$

where $T_w$, $T_d$ and $T_s$ are surface, droplet and saturation temperatures, respectively, $A$ – the droplet orthogonal projection onto the heating surface, $c_p$ – specific heat, and $h_{fg}$ – phase change enthalpy at the saturation temperature, here, at atmospheric pressure.

Using the measured change in the droplet mass over time, on the basis of heat balance (1), it is possible to determine local values of the heat transfer coefficient acc. to the procedure reported in [17]. To this end, by employing smoothing and differentiation procedures available in many computational packages, a change in mass flux over time is determined. The results obtained for the droplet of concern are presented in Fig. 3. Distinct disturbances can also be identified in the figure for the first 170 seconds. They occur for the droplet mass above ~1.3 g. Below that value, the disturbances tend to terminate.

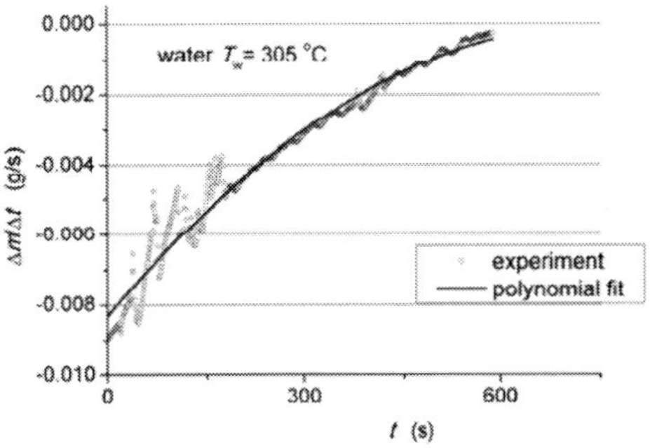


**Fig. 3.** Changes in the mass flux during evaporation from the droplet.

Different formulas for expressing a change in the droplet orthogonal projection onto the heating surface can be found in literature. Usually, those are time-dependent power functions. One of the formulas has the following form [20]:

$$\frac{A(t)}{A_0} = C_B\left(a^t - C_A\right) \quad (2)$$

where $A_0$ is the area at the initial instant, whereas $C_A$, $C_A$>0 and a<1 are constants under pre-set and steady state conditions of the experiment.

Figure 3 shows almost exponential increase, in time, of the flux of the mass evaporated from the droplet. Simultaneously, an exponential decrease in the area of the droplet orthogonal projection onto the heating surface (see Eq. 2) is confirmed by the experiment. As can be seen in Fig. 4, an almost linear relationship holds between those two quantities. The figure also shows the region of departing bubbles, the activity of which in the presented measurement series subsides when the droplet mass is below 1.2 g.

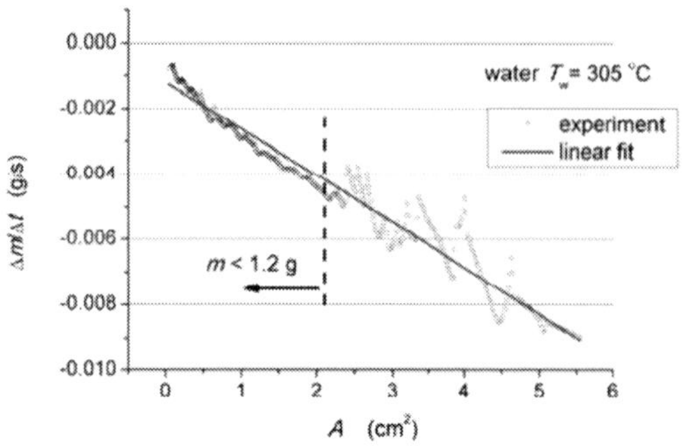


**Fig. 4.** Mass flux as a function of the droplet orthogonal projection onto the heating surface.

The dependence (2) was used to determine, from equation (1), the local value of the heat transfer coefficient, which as shown in Fig. 5, varies exponentially with the droplet area $A(t)$.

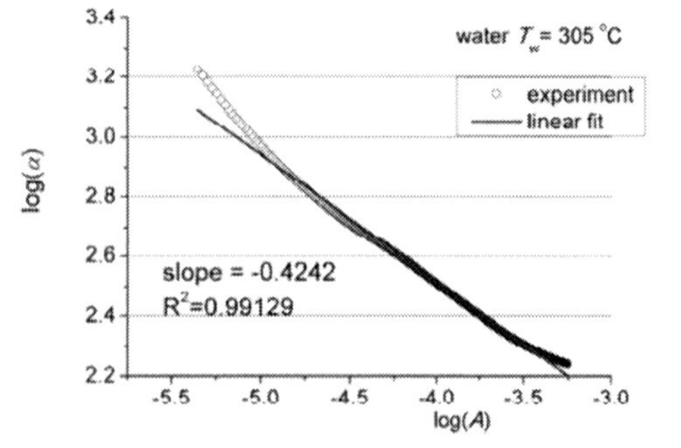


**Fig. 5.** Local value of the heat transfer coefficient as a function of the area of droplet projection onto the heating surface.

## 3 Conclusions

The investigations concerned a large water droplet at the heating surface temperature above the Leidenfrost point. The heating cylinder was the main component of experimental stand on which investigations were performed. The measurement system was placed on the scales with the sensitivity of 0.001 g and capacity of up to 500 g (see Fig. 1). After a few hours of the system conditioning, measurements of the droplet mass in time were taken (see Fig. 2). Data transmission was performed using RS232 interface and the author-designed software with extended functions for the system control. The analysis of results was carried out employing the droplet heat balance, under the assumption that heat transfer results only from droplet interaction with the heating surface. There, intense evaporation occurs and the remaining components (e.g. evaporation from the droplet upper surface) are negligibly small (see Eq. 1). To make calculations, the dependence, available in the literature, for the droplet orthogonal projection as the function of time was applied (see Eq. 2). It was confirmed that the local value of heat transfer coefficient is a power function of the area of the droplet surface projection (see Fig. 5). The power exponent obtained is approx. -0.42, which is a slightly lower value than that reported for droplets up to 1 g in weight [17]. The discrepancy may result from the formula for droplet area dependence on time, that was different from the adopted one. The droplet orthogonal projection was calculated acc. to study [20]. The dependence was given for water droplet deposited over a copper substrate heated to 400 °C, which differs from the quoted studies. As a result, inaccuracies, which are difficult to estimate, may occur in the calculations of temporary values of the heat transfer coefficient (see Fig. 5). Additionally, for large water droplets, above 1g, local generation of bubbles inside droplets is observed. When departing, bubbles disturb the droplet stability (see Figs. 3 and 4). Additionally, a linear relationship between the flux of mass evaporated from the droplet and the droplet orthogonal projection was observed (see Fig. 4).

## References

1. I. Khamis, K.C. Kavvadias, Nucl. Eng. Des. **248**, 48 (2012)
2. K.H.M. Abdalrahman, Sabariman, E. Specht, Int. J. Heat Mass Tran. **78**, 76 (2014)
3. Y. Wang, N. Zhou, Z. Yang, Y. Jiang, Appl. Therm. Eng. **103**, 510 (2016)
4. N. Shahidzadeh-Bonn, S. Rafa, A. Azouni1, D. Bonn, J. Fluid Mech. **549**, 307 (2006)
5. S.Ya. Misyura, Int. J. Heat Mass Tran. **71**, 197 (2014)
6. S. Nishio, T. Goto, N. Nagaib, Int. J. Heat Mass Tran. **41**, 3191 (1998)
7. T.M. Wójcik, EPJ Web of Conferences **25**, (2012) 01100
8. M. Shoji, Int. J. Heat Mass Tran. **47**, 1105 (2004)
9. V.V. Chekanov, Teplofiz. Vys. Temper. **15**, 121 (1977)
10. Y. Mitsutake , S. Illias, K. Tsubaki, M. N. Hasan, M. Monde, Procedia Eng. **105**, 5 ( 2015 )
11. T. Orzechowski, S. Wciślik, Energ. Convers. Manage. **76**, 918 (2013)
12. T. Orzechowski, S. Wciślik, Exp. Therm. Fluid Sci. **59**, 230 (2014)
13. R. Wendelstorf, K.-H. Spitzer, J. Wendelstorf, Int. J. Heat Mass Tran. **51**, 4892 (2008)
14. M. Chabičíovský, M. Hnízdil, A.A. Tseng, M. Raudenský, Int. J. Heat Mass Tran. **88**, 236 (2015)
15. M.R. Islam, B. Shabani, G. Rosengarten, Appl. Energ. **178**, 660 (2016)
16. D. Liu, F.-Y. Zhao, H.-X. Yang, G.-F. Tang, Energy **83**, 29 (2015)
17. T. Orzechowski, S. Wciślik, Int. J. Heat Mass Tran. **73**, 110 (2014)
18. D. Brutin, B. Sobac, F. Rigollet, C. Le Niliot, Exp. Therm. Fluid Sci. **35**, 521 (2011)
19. V.E. Nakoryakov, S.Ya. Misyura, S.L. Elistratov, Int. J. Heat Mass Tran. **55**, 6609 (2012)
20. G. Paul, P. K. Das, I. Manna, Exp. Therm. Fluid Sci. **60**, 346 (2015)

# Vortex shedding control of circular cylinder by perforated shroud in deep water

*Gokturk M.* Ozkan[1], *Tahir* Durhasan[1*], *Engin* Pinar[2], *Muhammed M.* Aksoy[3], *Huseyin* Akilli[1] and *Beşir* Sahin[1]

[1]Cukuruova University, Mechanical Engineering Department, 01330 Adana, Turkey
[2]Cukuruova University, Ceyhan Mechanical Engineering Department, 01950 Adana, Turkey
[3]Osmaniye Korkut Ata University, Energy Systems Department, 80000 Osmaniye, Turkey

**Abstract.** The aim of the present study is to control the vortex shedding downstream of a circular cylinder (inner cylinder) by the existence of outer perforated cylinder concentrically located around the inner cylinder in deep water. The flow characteristics downstream of concentrically placed coupled cylinders were investigated quantitatively by the Particle Image Velocimetry (PIV) technique. Diameter of the outer perforated cylinder and inner cylinder were kept constant as $D_o$=100 mm and $D_i$=50 mm. The depth-averaged free-stream velocity was also kept constant as U=100 mm/s which corresponded to the Reynolds number of $Re_{Do}$=10,000 based on the outer cylinder diameter. Experiments were conducted for six porosities (β = 0.3, 0.4, 0.5, 0.6, 0.7 and 0.8) in order to show the effect of these parameters on the flow control. Maximum values of both Reynolds shear stress, <uv> and turbulence kinetic energy, <TKE> significantly decreased with the existence of outer perforated cylinder and also, the location of peak magnitudes of turbulence statistics occurred at locations further downstream compared to the bare cylinder case. The most effective control was revealed for the porosity of β=0.7.

## 1 Introduction

It is well-known phenomena that flow over a circular cylinder which causes acoustic noise, vibration, fatigue and shortens the life of cylinder. To end this, the flow control around a cylinder has been an attractive topic for more than a century due to its engineering significance and in part due to its tempting simplicity in setting up arrangements in experimental or numerical studies.

Ikeda and Takaishi [1] showed that the stable wake shear layers of the perforated cylinder and suppression of Aeolian tone are achieved due to the jets emitted from the holes at regular intervals. Boorsma et. al. [2] performed experimental study to investigate and optimize the effect of different perforated fairing on the noise control at different location. The fairing self-noise is reduced remarkably by breakdown of the vortex shedding process, resulting in a reduction of associated broadband noise level. Zho and Cheng [3] demonstrated that lift reduction of circular cylinder can be achieved by properly choosing the porous material. Kleissl and Georgakis [4] performed experimental study in order to show the effect of modifying bridge cable shape and surface on suppression of the vortex induced vibration. They revealed that shroud cylinder significantly reduced the vortex-induced oscillating lift forces. Yu et al. [5] studied numerically on the steady flow around a porous circular cylinder. They found that contrary to that of the solid cylinder, the recirculating wake develops downstream of or within the porous cylinder, but not from the surface of it.Ozkan et al. [6] expressed that the change in flow characteristics downstream of circular cylinder surrounded by a permeable cylinder in shallow water using particle image velocimetry technique. They observed that the peak magnitude of turbulent kinetic energy and Reynolds stress decrease remarkably due to the presence of the outer cylinder compare with natural cylinder case. Gozmen and Akilli [7] studied on the control of flow around cylinder surrounded by permeable cylinder in deep water. They noted that parameters, which are the porosity and diameter ratio, play important role on the suppression of vortex shedding. Pinar et al. [8] experimentally investigated flow structure of perforated cylinder in shallow water. The results of their study show that porosity had a significant effect on the control of large-scale vortical structures downstream of cylinder. The effect of perforated fairing on the vortex street suppression of circular cylinder was studied by Durhasan et al. [9]. Their results indicated that the wake region of cylinder was elongated remarkably along the main flow direction and vortex shedding frequency reduced substantially by the presence of perforated fairing for porosity in the range of β=0.3-0.6.

An experimental study was carried out for purpose of controlling the vortex shedding of the circular cylinder using perforated shroud in deep water. The velocity field of downstream of the circular cylinder were determined by using PIV measurement technique.

* Corresponding author: tdurhasan@cu.edu.tr

DOI: 10.1051/epjconf/201714302087

Flow structures and turbulent statistics were analyzed and also compared to bare cylinder in detail.

## 2 Experimental setup

Experiments were carried out in a low turbulence closed-loop, opened surface water channel test section having the following dimensions: a length of 8000 mm, width of 1000 mm and depth of 750mm. This facility was built in the Fluid Mechanics Laboratory at Cukurova University.

In the present study, perforated outer cylinders constructed from a chrome-nickel steel sheet having 1 mm thickness. The diameter of outer perforated cylinder and inner cylinder were kept constant as $D_o$=100 mm and $D_i$=50 mm, respectively. Six different outer cylinder ($\beta$=0.3, 0.4, 0.5, 0.6, 0.7 and 0.8) were used in order to investigate effect of porosity of the perforated shroud. The porosity ratio, $\beta$ was defined as the ratio of the total open area on cylinder to whole cylinder surface area. During the experiments perforated shrouds were located concentrically around the inner cylinder in shown figure 1. Free stream velocity was U=100 mm/s, corresponding Reynolds number Re=10000 based on the outer cylinder diameter.

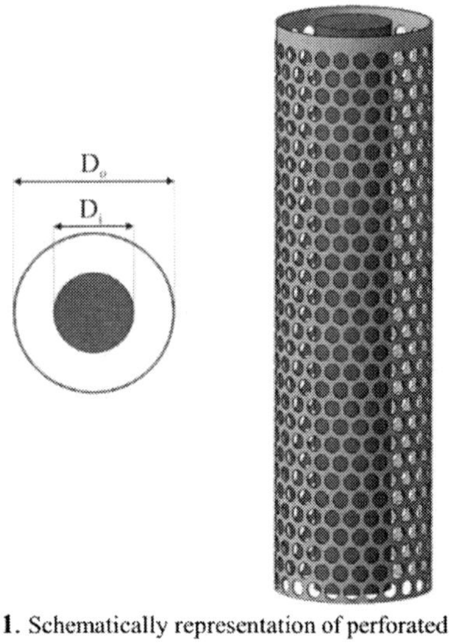


**Fig. 1**. Schematically representation of perforated shroud

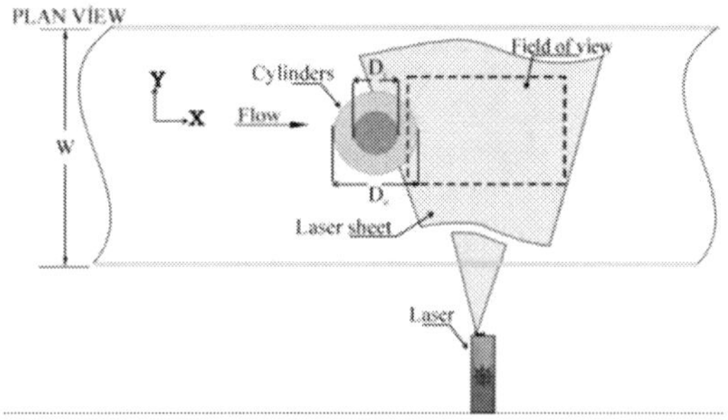


**Fig. 2**. Schematic view of test section

All measurement were carried out on the plain view for water height of 400 mm and laser sheet was located parallel to the bottom water channel at height of 200 mm. A schematic view of test section is shown in figure 2.

The experiments were performed and the measured data were processed using Dantec Dynamics PIV system and FLOW MANAGER Software installed on a computer. The illumination of measurement field was supplied with a thin and an intense laser light sheet by using a pair of double-pulsed Nd:YAG (yttrium aluminum garnet) laser source energy level of 120 mJ at 532 nm wavelength. The time interval between pulses was 1.75 ms for all experiments and the thickness of the laser sheet was approximately 2 mm. The time interval and the laser sheet thickness were selected such that the maximum amount of particles in the interrogation window was obtained. The image capturing was performed by an 8 bit cross-correlation charge-coupled device (CCD) camera having a resolution of 1016×1016 pixels, equipped with a Nikon AF Micro 60 *f* / 2.8 *D* lenses. In the image processing, 32×32 integration window pixels were used and an overlap of 50% was employed. A total of 3844 (62×62) velocity vectors were obtained for an instantaneous velocity field at a rate of 15 frames / s. The field of view of measuring planes covers the downstream of the cylinder with the area of 200mm x 200mm for all configurations and the image magnification was 0.0709 mm / pixel. The water was seeded with 12 μm diameter hollow glass sphere particles. The number of particles in an interrogation area was in between 20 and 25. The uncertainty in velocity relative to depth averaged velocity is about 2% in the present experiments. The instantaneous images were captured, recorded, and stored in order to obtain averaged-velocity vectors and other statistical properties of the flow field. Spurious velocity vectors (less than 3%) were removed using the local median-filter technique and replaced by using a bilinear least squares fit technique between surrounding vectors.

## 3 Results

First of all, the wake region of bare (inner) cylinder is present in fig. 3 in order to comparison with the controlled situations. Minimum and incremental values of time averaged vorticity $<\omega>$ were taken as ±2 and 2, respectively. The solid and dashed lines present positive (counter-clockwise) and negative (clockwise) contours, respectively. Both positive and negative contours of the time averaged vortices were emanated from the lower and upper shoulders of the cylinder and get close to each other at near wake region for bare cylinders. This results present a well-known situation which was called von Karman vortex street. On the other hand, the minimum and incremental values of turbulent kinetic energy, <TKE> contours were taken as 0.01. Turbulent kinetic energy, <TKE> contours of bare cylinder clearly indicate that flow around the cylinder causes turbulence in the near wake region of bare (inner) cylinder.

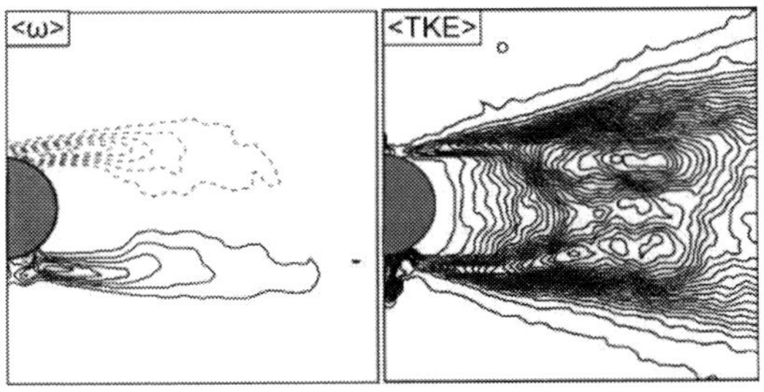


**Fig. 3.** Distribution of time averaged vorticity, <ω> and <TKE> counters for bare cylinder.

Fig. 4. show that distribution of time averaged vorticity counters for different porosity of perforated shroud. Almost symmetrical shear layer structure can be observed for all porosity ratios. It is observed that presence of perforated shroud affects the formation of shear layers on the both sides of cylinder which are gradually elongated with the increasing porosity of the shroud, so that interaction of the shear layer is delayed in the downstream direction. Also, strength of shear layer is decreased with the increasing porosity compared with bare cylinder case. Effect of the perforated shrouds are more dominant on the flow structure downstream of the cylinder in the range of $0.5 \leq \beta \leq 07$. On the other hand, vorticity clusters occur in the near wake region for porosity value of $\beta=0.8$ and this indicated that the perforated have insignificant effect on the flow control for porosity value of $\beta=0.8$.

Time averaged turbulent kinetic energy, <TKE> contours are presented in fig. 5 for different porosity of perforated shroud. Porosity, β, has a significant effect on the flow structure compared to bare cylinder. Especially, porosity of $\beta \leq 0.6$, TKE intensity along the shear layers decrease with the increasing porosity, β, and interact with each other in the downstream region at the end of second field of view and peak magnitudes approximately located $7D_i$ of the cylinder away from the inner cylinder base. The flow emanating through the holes on the perforated shroud causing a jet flow prevents the vorticity on each side of the coupled cylinder by intercepting the interaction of vortices and thus momentum transfer increases into the wake region. As a result, TKE values in the wake region decrease in comparison to the bare cylinder case. The reason of this kinetic energy intensification in shear layers is the result of prevention of Karman vortices; however, the Kelvin-Helmholtz vorticity becomes more dominant on the shear layers between the free-stream region and the wake region. Therefore, the most effective flow control is achieved for $\beta=0.7$. Moreover, the peak magnitude of TKE value of $D_i=50$ mm bare cylinder is found to be 0.2886 and attenuation rate of the maximum value of TKE is approximately %68 at most, compared to bare cylinder.

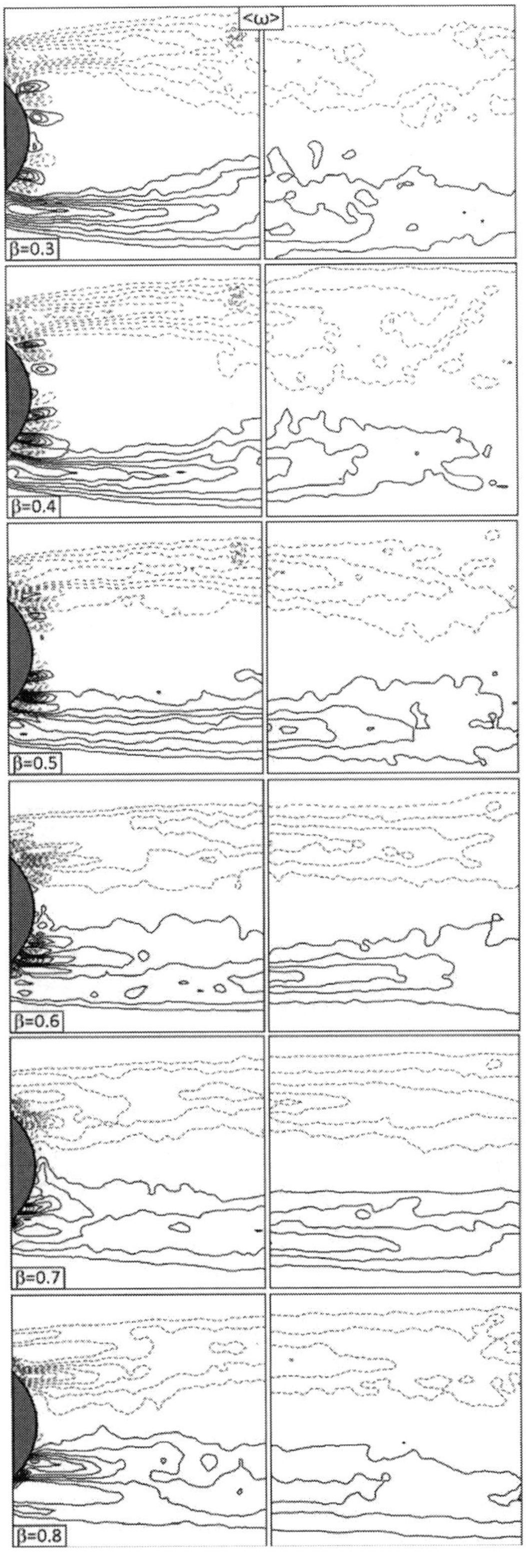


Fig. 4. Distribution of time averaged vorticity, <ω> counters for different porosity of perforated shroud.

 DOI: 10.1051/epjconf/201714302087

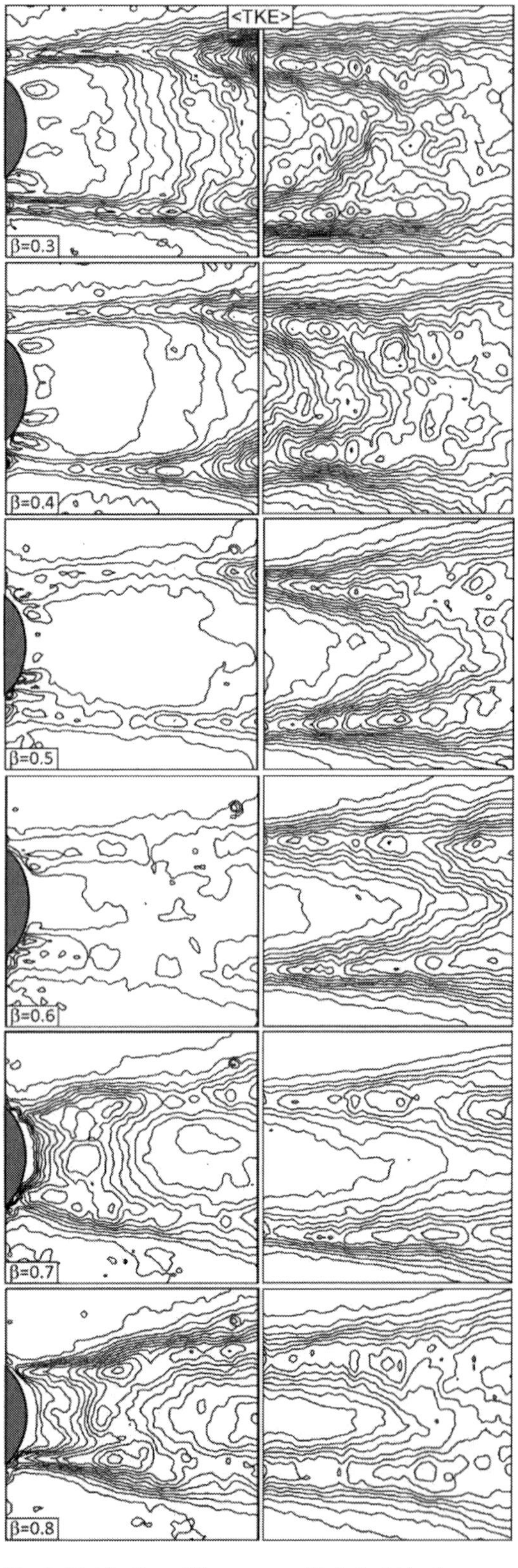


Fig. 5. Distribution of time averaged turbulent kinetic energy, <TKE> counters for different porosity of perforated shroud.

The variation of peak magnitude of Reynolds shear stress, <uv> with respect to porosity, β are presented in fig.6. It is observed that the peak magnitude of Reynolds shear stress, <uv> gradually decreased until the porosity value of β=0.7. Maximum reduction of the peak magnitude of Reynolds shear stress, <uv> is obtained for porosity value of β=0.7 and it is reduced by % 68 for in compared with bare (inner) cylinder. On the other hand, the peak magnitude of Reynolds shear stress, <uv> for porosity of β=0.8 is increased to compare with porosity of β=0.7, but it has still lower value of the peak magnitude of Reynolds shear stress, <uv> than value of bare cylinder case.

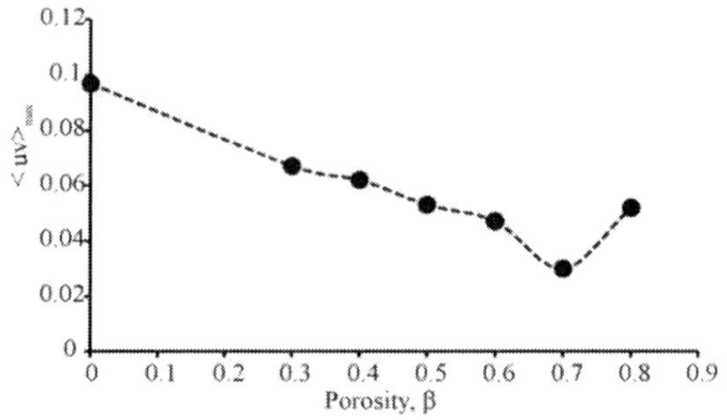


Fig.6. The variation of Reynolds shear stress, <uv> with porosity

## 4 Conclusion

In this study a passive flow control method was applied in order to suppression of vortex shedding downstream of circular cylinder by the existence of perforated shroud concentrically located around the inner cylinder in deep water. To reveal the effect of perforated shroud on the vortex shedding control six different porosity, β were used as β = 0.3, 0.4, 0.5, 0.6, 0.7 and 0.8.

It was observed that perforated shroud has a significant effect on the flow structure of circular cylinder. Especially porosity value of β<0.7, the shear layers of cylinder are elongated gradually due to jet-alike flow through the holes on surface of perforated shroud. Also, turbulence statics are reduced by % 68 for porosity of β=0.7 in compared with bare cylinder.

## Acknowledgements

This study supported by the funding of the Scientific and Technological Research Council of Turkey under contract no. 109R001.

## References

1. M. Ikeda, T. Takaishi, Perforated pantograph horn aeolian tone suppression mechanism, Q. Report of RTRI (Jpn), **45,** 169-174 (2004)
2. K. Boorsma, X. Zhang, N. Molin, L. C. Chow, Bluff body noise control using perforated fairings, AIAA J., **47** 33-43 (2009)

3. M. Zhao, L. Cheng, Finite element analysis of flow control using porous media, Ocean Eng., **37** 1357-1366 (2010)
4. K. Kleissl, C. T. Georgakis, Aerodynamic control of bridge cables through shape modification: A preliminary study, J. Fluids Struc., **27** 1006-1020 (2011)
5. P. Yu, Y. Zeng, T. S. Lee, X. B. Chen, H. T. Low, Steady flow around and through a permeable circular cylinder, Comp. Fluids, **42** 1-12 (2011)
6. G. M. Ozkan, V. Oruc, H. Akilli, B. Sahin, Flow around a cylinder surrounded by a permeable cylinder in shallow water, Exp. Fluids, **53** 1751-1763 (2012)
7. B. Gozmen, H. Akilli, Flow control downstream of a circular cylinder by a permeable cylinder in deep water, Wind Struc. Int. J., **19** 389-404 (2014)
8. E. Pinar, G. M. Ozkan, T. Durhasan, M. M. Aksoy, H. Akilli, B. Sahin, Flow structure around perforated cylinders in shallow water, J. Fluids Struc., **55** 52-63 (2015)
9. T. Durhasan, M. M. Aksoy, E. Pinar, G. M. Ozkan, H. Akilli, B. Sahin, Vortex street suppression of a circular cylinder using perforated semi-circular fairing in shallow water, Exp. Term. Fluid Sci, **79** 101-110 (2016)

EPJ Web of Conferences **143**, 02088 (2017) DOI: 10.1051/epjconf/201714302088
*EFM 2016*

# Control of the flow in the annular region of a shrouded cylinder with splitter plate

*Gokturk Memduh* Ozkan[1,*], *Tahir* Durhasan[1], *Engin* Pinar[2], *Arda* Yenicun[1], *Huseyin* Akilli[1] *and Besir* Sahin[1]

[1] Faculty of Engineering and Architecture, Department of Mechanical Engineering, Çukurova University, 01330, Adana, Turkey
[2] Faculty of Ceyhan Engineering, Department of Mechanical Engineering, Çukurova University, Ceyhan, Adana, Turkey

**Abstract.** In the present study, the flow control with a splitter plate was studied considering the annular region of a shrouded cylinder. The effect of splitter plate angle, $\alpha$ which was defined according to the cylinder centreline is investigated experimentally in deep water using Particle image Velocimetry (PIV) technique and flow visualization by dye injection method. The range of splitter plate angle was selected within $60^o \leq \alpha \leq 180^o$ with an increment of $30^o$. The porosity of the shroud which is a perforated cylinder was selected as $\beta$=0.7 in order to have larger fluid entrainment through the cylinder. The results were compared with the no-plate case and showed that the splitter plate located in the annular region of shrouded cylinders is effective on reducing the turbulence levels just behind the cylinder base, as well as the near wake of the perforated shroud.

## 1 Introduction

The flow around bluff bodies has been extensively studied in order to analyse the effects of unsteady flow on the integrity of even large engineering structures. The cylinder as a bluff body has been preferred by many researchers where the vortex shedding from the cylinder causes fluctuations in pressure forces. According to this condition, vibration, noise or resonance are produced on the cylinder which should be controlled. The aims of controlling the flow around cylinder are to reduce drag, magnitude of the fluctuating force and time-varying flow-induced forces acting on bluff bodies. Flow control is normally classified into active and passive control in which the active methods require energy input whereas the passive methods modify the flow structure of interest and no energy input required. Therefore the passive control techniques are more preferred in practical applications. There is an abundance of studies in the literature on the flow past a circular cylinder starting with Karman [1] and Strouhal [2]. Roshko [3] used splitter plate placed behind the cylinder to control the flow structure. Price [4] used different shaped shrouds in order to control vortex induced vibration in his thesis which is about the effect of different geometries on flow control. Every et al. [5] reviewed the practical aspects of vortex excited structural vibrations in sub-sea water applications with particular emphasis on methods of their suppression. In a part of their study, they focused on the effect of shrouds on the flow control, reported a comprehensive literature and stated that shrouds with the optimum dimension gave a 50% reduction in oscillations of a plain study on suppressing flow induced vibrations.

A similar logic is being used in recent years called porous media which includes the idea of generating a permeable medium around a bluff body. An emerging approach of flow control is to attach a porous layer to the external surface of a structure. Bruneau and Mortazavi [6, 7] investigated the flow control by using porous layer around a cylinder numerically. They used penalization method and found that adding a porous ring around a riser pipe can damp the vortex induced vibrations. The study of Ozkan et al. [9] presents an example of flow control by shrouded cylinder in shallow water. They investigated the flow around a cylinder (inner cylinder) by a permeable outer cylinder having different porosities, $\beta$ and diameter ratios, D/d. They stated that the permeable outer cylinder suppresses the organized vortex street by reducing the velocity fluctuations in the near wake of the cylinder. The flow structure around perforated circular cylinders was studied experimentally by Pinar et al. [10]. Their study has shown that the porosity, $\beta$ has a significant impact on the unsteady flow structure downstream of the cylinder. The jet-alike flow through the holes on the surface of the perforated cylinders effectively prevents the formation of a well-organized Karman vortex street. With increasing porosity, the vortices formed on the upper and lower shear layers are elongated along the flow direction by losing their magnitudes compared to the plain cylinder case. Recently, Durhasan et al. [11] controlled the flow around a circular cylinder using semi-circular cylinders which are concentrically located. They performed their experiments considering various porosity values and arc angles and found that the perforated semi-circular fairings are effective on reducing turbulent statistics. Molin [8] performed a numerical study about perforated

* Corresponding author: gmozkan@cu.edu.tr

shrouds used to reduce vortex-induced vibrations of cylinders. He concluded that the proposed theory cannot be said to predict reliable values of the drag of shrouded cylinders since it fails to reproduce correctly the flow in the downstream part of the annular region, in-between the shroud and cylinder.

Within the light of Molin's [8] conclusion and the results of Ozkan et al. [12] the focus on the annular region of shrouded cylinder is taken into consideration for the current study. A splitter plate is located in the annular region of perforated shroud with varying angle to prevent the interaction of vortices generated by the inner cylinder.

## 2 Experimental Method

Both PIV and dye visualization experiments were performed in a large-scale closed-loop recirculating water channel, located in the Mechanical Engineering Department of Çukurova University having a test section of 1m width x 0.75m height and 8m length made of a 15mm thick transparent Plexiglas sheet. Free stream velocity was controlled by a 15kW centrifugal pump with speed control unit which was used to adjust pump frequency. Honeycomb arrangements were located inside the channel in order to minimize the free-stream turbulence, which is expected to be less than 1%. The free stream velocity was taken as 100mm/s which corresponds to a Reynolds number of $Re_d$=5000 according to the cylinder diameter. All experiments were performed above a platform, having a length of 2.3m. Perforated shroud is placed at the distance of 1.8m from the leading edge of the platform to provide fully-developed boundary layer flow. During all experiments, the total depth of the water in the channel was kept constant as 0.6m. A schematic view of the experimental setup is shown Fig.1.

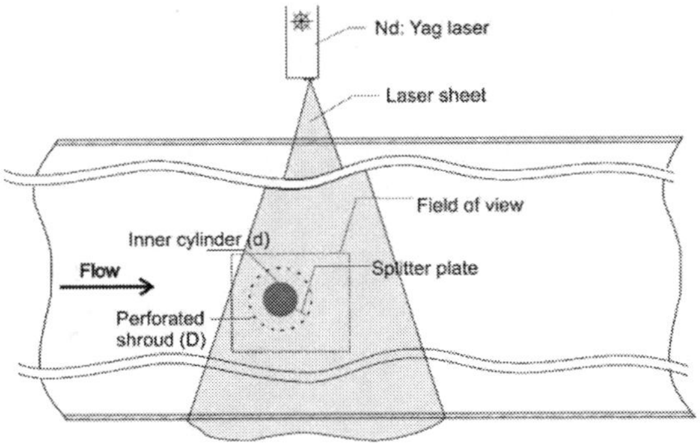


**Fig. 1.** Plan view of the experimental set-up

The geometric blockage of the perforated shroud was 10% with respect to the width of test section. Chrome–nickel metals sheets with 1mm thicknesses were drilled using laser cutting machine to have d=10mm holes on the cylinder surface, then they were rolled to manufacture the perforated shroud. The number of the holes on the shroud is expected to have significant effects on the flow structure. Therefore, the porosity ratio, β defined as the ratio of the total open area on the cylinder to the whole cylinder surface area was chosen as a dimensionless parameter. Fig.2 shows the schematic representation of the model and calculation of the porosity, β. Here the inner cylinder, splitter plate and perforated shroud dimensions are d=0.05m, L=0.05m and D=0.1m, respectively. The porosity, β value was taken constant as β=0.7 through the experiments.

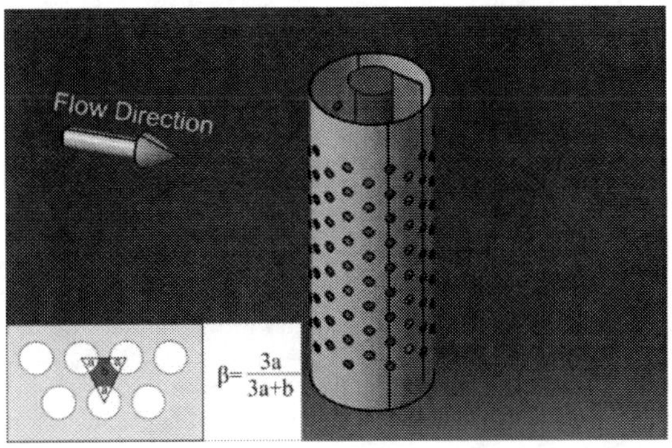


**Fig. 2.** Schematic representation for the model and the definition of porosity, β

The laser sheet was adjusted at the half of the water level and parallel to the bottom wall of the channel. Experiments were conducted in two steps: In the first step, flow visualization experiments were performed using Rhodamine type dye that shines under the continuous laser beam in the desired flow field. Throughout the experiments, dye was supplied at a location where the flow was not affected by the injection. Visualization of the experimental results was recorded with a high speed digital video recorder (SONY 80X handycam) to examine the flow characteristics in detail. Next, the Particle Image Velocimetry (PIV) technique was used in order to measure the velocity vector fields. Experiments were focused on the region between the cylinder and perforated shroud (which will be called as annular region). The measurement field was illuminated by a thin and intense laser light sheet using double-pulsed Nd:YAG laser units. Each pulse has a nominal energy output of 120mJ at 532nm wave length. The time interval between pulses was 1.75ms for all experiments and the thickness of the laser sheet illuminating the measurement plane was approximately 2mm. The time interval and the laser sheet thickness were selected such that the maximum amount of particle displacement in the interrogation window was obtained. The water was seeded with 10 μm-diameter seeding particles and the particle density was about 1100kg/m$^3$. The number of particles in an interrogation area was in between 15 and 20. The experimental measurements were performed and the data were processed using Dantec Dynamics PIV system and Dynamics Studio Software installed on a computer. In each experiment, 1000 instantaneous images were captured at a rate of 15Hz, recorded and stored in order to obtain time-averaged velocity vectors. Assuming isotropic flow, the values of turbulent kinetic energy (TKE) were evaluated using the estimation of Sheng et al. [13] which states that the third component has been supposed to be equal to the half of $(<u'u'>+<v'v'>)$ and TKE can be calculated using $<TKE>=3/4(<u'u'>+<v'v'>)$. The image capturing was performed by an 8-bit cross-correlation CCD camera having a resolution of 1600×1200 pixels, equipped with a Nikon AF Micro 60 f/2.8D lens. In the image processing, 32×32 rectangular interrogation pixels were used and an overlap of 50% was employed. Shadows, reflections or laser sheet distortions etc. cause spurious velocity vectors in the flow field, therefore a local median-filter technique was employed to erase spurious vectors (less than 3%) in which they were replaced using bilinear interpolation between neighbouring velocity vectors. The velocity vector field was also smoothed to avoid dramatic changes in the velocity field using the Gaussian smoothing technique. To minimize distortion of the velocity field, a smoothing parameter of 1.3 was chosen. The details of the effects of these factors can be found in the studies of Fouras and Soria [14], Adrian [15] and Keane and Adrian [16]. Finally, the instantaneous and mean vorticity maps (vorticity value at each grid point was calculated from the circulation around eight neighbouring point) and turbulence statistics were determined as a result of post-processing operation.

## 3 Results and Discussion

The pictures evaluated from dye visualization experiments are presented in Fig. 3 where the flow behaviour is clarified by the red lines showing the vortical flow structures. Here the dashed outer line presents the perforated shroud located concentrically around the inner cylinder. In order to interpret the difference, the results of shrouded cylinder without any splitter plate (no-plate case) are also presented on the first picture. An incoherent vortex pair is observed within the annular region downstream of the inner cylinder which might increase the turbulence. As a result, even the vortex shedding is controlled downstream of the shrouded cylinder, there would be vortex shedding in the near wake of inner cylinder. To get rid of this undesirable effect, a splitter plate is located in between inner cylinder and the shroud and presented by letter pictures in Fig. 3 for various plate angles.

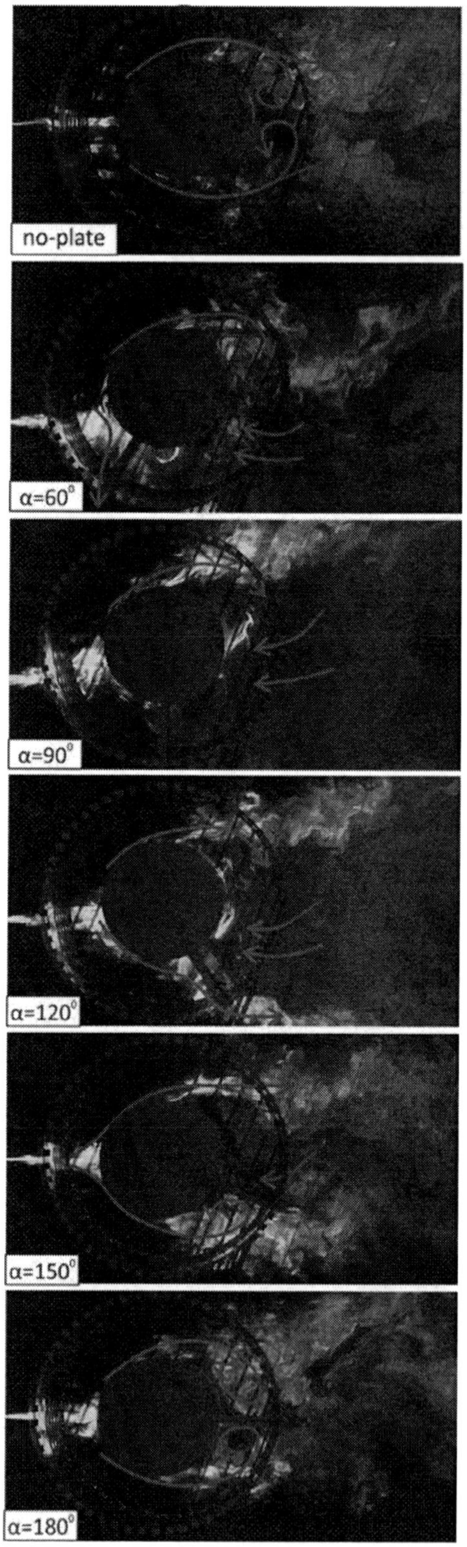


**Fig. 3.** Instantaneous pictures of the dye visualization experiments

In Fig. 2, for the splitter plate angles of $60^o \leq \alpha \leq 120^o$, the incoming flow through perforated shroud is blocked by the plate where some amount of fluid leaves from the perforated shroud along the lower shear layer. Besides, a huge amount of fluid, hence momentum is transferred to the other side, i.e. upper shear layer of the inner cylinder. Since the porosity of the perforated shroud has a high value, most of the fluid emanate through the holes on the upper side and then recirculates in the wake of perforated cylinder, finally re-enters into the annular region as can be seen by $60^o \leq \alpha \leq 120^o$. Increasing the splitter plate angle to $\alpha=150^o$ brings out slight changes in flow structure, i.e. a recirculating flow exists in the lower side of the annular region which breaks through the holes. The angle value of $\alpha=180^o$ presents the well-known splitter plate method within the literature and might be the best way of preventing the interaction of vortices formed by the inner cylinder within the annular region. Here the vortices are more distinct compared with the no-plate case and their interaction is eliminated. Furthermore, the generated vortices are decomposed by the holes on the perforated shroud and cannot be seen in the near wake of the shroud.

The qualitative dye visualization results were analysed to understand the general behaviour of the flow which actually cannot be determinant for the accuracy of the method. Therefore it is compulsory to analyse the quantitative PIV results. Fig. 4 includes the time-averaged streamlines superimposed with the vorticity contours for all cases. In consistent with the flow visualization pictures, a clear recirculation occurs within the annular region of shrouded cylinder for no-plate case. The focal points generated here are designated by $F_1$ and $F_2$ within annular region whereas another focal point, $F_3$ is generated in the downstream of perforated shroud. According to this result one can conclude that there exist vortex shedding within the annular region of the shrouded cylinder, as well as downstream of the perforated shroud. Therefore, use of a shroud with high porosity cannot be an accurate flow control technique. Because in this case both the inner cylinder and the cylinder-shroud configuration is under the effect of vortex shedding. To prevent this, locating a splitter plate may bring out a solution for the problem. Use of the splitter plate having a plate angle of $\alpha=60^o$ makes the two focal points away from each other, i.e. one is formed on the upper side close to the perforated shroud and the other is formed on the lower side close to the splitter plate. The results of $\alpha=90^o$ presents the best case where no recirculation exists for this plate angle, however the vorticity magnitude on the lower side of perforated shroud contacting with the leading edge of the plate dramatically increases due to the jet flow through the holes in that region. This is also valid for the case of $\alpha=60^o$ with lower vorticity magnitudes. The results of $\alpha=120^o$ similar with the lower plate angles except the formation of high magnitude vorticity layer on the leading edge of the plate. It should be pointed out that for $60^o \leq \alpha \leq 120^o$ the wake of the perforated shroud is highly effected by the use of splitter plate in the annular region, i.e. the width and length of the recirculation region is increased compared with the no-plate case.

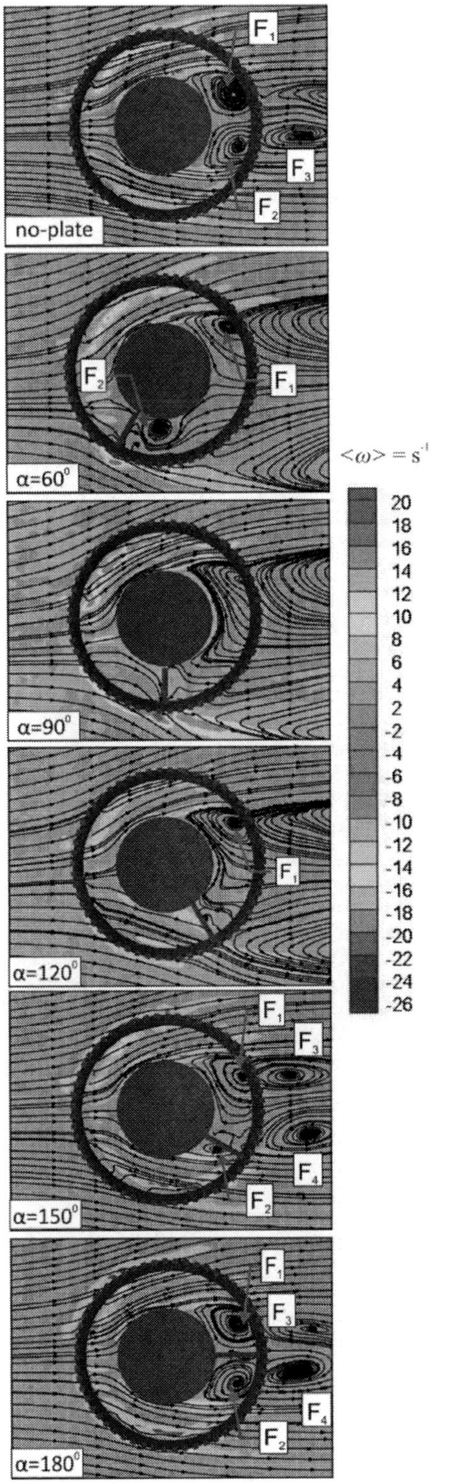


**Fig. 4.** Time-averaged streamlines superimposed with the vorticity contours for all cases

For α>120°, similar flow structures with that of the no-plate case are obtained where two focal points are generated within the annular region and another pair of focal points are generated in the wake of perforated shroud. The use of a splitter plate with α=180° completely prevents the interaction of vortex pairs generated by the cylinder, however there still exists the vortex pairs in the wake of perforated shroud which should also be eliminated.

In order to interpret the effectiveness of splitter plate on flow control within annular region, the focus on quantitative results passed to turbulence statistics by Fig. 5. Here the time-averaged dimensionless turbulent kinetic energy, <TKE> concentrations are presented for the whole flow field where the minimum and incremental values for the turbulent kinetic energy were both taken as 0.005. The aim of the study was to suppress the vortex shedding downstream of the inner cylinder within the annular region. One can easily see that the <TKE> values at upstream of the perforated shroud presenting the effect of incoming are highly concentrated for all cases. For the no plate case, the <TKE> concentrations are formed on upper and lower shear layer of the inner cylinder within the annular region. Furthermore, there exists intensive concentrations in the wake of perforated cylinder which also need to be diminished. When looking at the controlled cases with splitter plate, it can be seen that the lowest <TKE> values are achieved for the plate angle of α=120°, for both the annular region and the near wake of the perforated shroud. For larger plate angles, the TKE concentrations within the annular region are located in the very near wake of inner cylinder, however the upper and lower shear layer concentrations are separated from each other by the existence of splitter plate. The unfavourable condition for α=150° and α=180° is that the turbulence levels in the wake of the perforated shroud is still high as evaluated by the no-plate case. Therefore even the use of a splitter plate is preferable within the annular region, the flow generated in the wake of perforated shroud should also be controlled. It can be concluded by the TKE results that the best plate angle to reduce turbulence in the annular region downstream of the inner cylinder is α=120° where the turbulence level is decreased remarkably compared with the no-plate case.

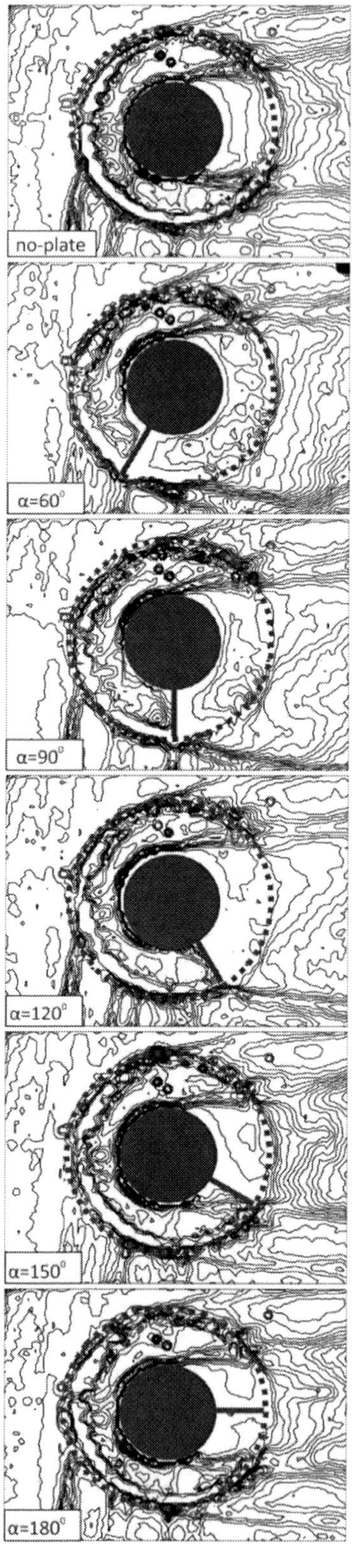


**Fig. 5.** Time averaged turbulent kinetic energy, <TKE> distributions for all cases

## 4 Conclusions

The flow within the annular region of a shrouded cylinder was studied and the control of the unsteady vortical flow generated by the inner cylinder was analysed within this work. A splitter plate was located in the annular region and the angle of the plate was varied with respect to the cylinder centreline. The effect of the splitter plate angle on the flow structure within annular region, as well as the near wake of the perforated shroud were analysed using flow visualization and PIV techniques.

The results revealed that the use of a splitter plate with the plate angles of $60^o \leq \alpha \leq 120^o$ increases the wake width in the near wake of the perforated shroud by decreasing the turbulence levels in the annular region. The best reduction in the turbulence was evaluated for the plate angle of $\alpha=120^o$ where the turbulence concentrations are lower in the near wake of the cylinder, as well as the perforated shroud. Therefore the splitter plate at an angle of $\alpha=120^o$ could be advised for the current method.

Furthermore, the cases of $\alpha=150^o$ and $\alpha=180^o$ also be advised within the aid of a complete elimination of upper and lower vorticity layer interactions in the annular region. However for these cases, the turbulence levels were found to be still high in the near wake of the perforated shroud which remains as a problem to be solved or controlled by the future works.

## Acknowledgements

The funding of the Scientific and Technological Research Council of Turkey under contract number 114R087 is greatly appreciated.

## References

1. T. Karman, Matematisch-Physikalische Klasse, Göttingen Nachrichten, *Über der Mechanismus des Wiederstandes, den eine bewegter Körper in einer Flüssigkeit erfahrt.* 547-556 (1912)
2. V. Strouhal. Annelen für Physik und Chemie, *Über eine besondere Art der Tonerregung, (On an unusual sort of sound excitation)*, **241**, 216-251 (1878).
3. A. Roshko, Journal of Wind Engineering and Industrial Aerodynamics, *Perspectives on bluff body aerodynamics.* **49**, 79–100 (1993).
4. P. Price, J. Eng. Mech. Div., Am. Soc. Civ. Eng, Suppression of the fluid-induced vibration of circular cylinders, **82**, 1030 (1956).
5. M. J. Every, R. King, D. S. Weaver, Ocean Engineering, *Vortex-Excited vibration of cylinder and cables and their suppression,* **9**, 135-157 (1982).
6. C-H Bruneau, I. Mortazavi, International Journal for Numerical Methods in Fluids, *Passive control of the flow around a square cylinder using porous media,* **45**, 415–433 (2004).
7. C-H Bruneau, I. Mortazavi, Computers and Fluids, *Numerical modelling and passive flow control using porous media*, **37**, 488–498 (2008).
8. B. Molin, Journal of Fluids and Structures, *A potential flow model for the drag of shrouded cylinders*, **7**, 29-38 (1993).
9. G.M. Ozkan, V. Oruc, H.Akilli, B. Sahin, Experiments in Fluids, *Flow around a cylinder surrounded by a permeable cylinder in shallow water,* **53,** 1751-1763 (2012).
10. E. Pinar, G.M. Ozkan, T. Durhasan, M.M. Aksoy, H. Akilli, B. Sahin, Journal of Fluids and Structure, *Flow structure around perforated cylinders in shallow water*, **55,** 52-63 (2015).
11. T. Durhasan, M.M. Aksoy, E. Pinar, G.M. Ozkan, H. Akilli, B. Sahin, *Vortex street suppression of a circular cylinder using perforated semi-circular fairing in shallow water* Experimental Thermal and Fluid Science, **79**, 101-110 (2016).
12. G.M. Ozkan, H.Akilli, B. Sahin, EPJ Web of Conferences, *Effect of High Porosity Screen on the Near Wake of a Circular Cylinder*, **45**, 01071 (2013).
13. J.Sheng, H. Meng, R.O. Fox, Chemical Engineering Science, *A large eddy PIV method for turbulence dissipation rate estimation,* **55**, 4423-4434 (2000)
14. A. Fouras, J. Soria, Exp. Fluids, *Accuracy of out-of-plane vorticity measurements derived from in-plane velocity field data*, **25**, 409–430 (1998).
15. R.J. Adrian, Annul. Rev. Fluid Mech., *Particle-imaging techniques for experimental fluid mechanics* **23**, 261–304 (1991).
16. R.D. Keane, R.J. Adrian Opt. Methods Flow Particle Diagnostics, *Optimization of particle image velocimeters*, **68**, 139–159 (1989).

# A numerical study of the influence of solidity on the performance of vertical axis turbine

*Teresa* Parra-Santos[1,a], *Diego J.* Palomar-Trullen[1], *Mario* Fernandez-Gomez[1], *Armando* Gallegos-Muñoz[2], *Cristobal N.* Uzarraga[3], *Francisco* Castro-Ruiz[1]

[1]University of Valladolid, Department of Energy and Fluid Mechanics, Valladolid, Spain
[2]University of Guanajuato, Salamanca, Mexico
[3]Technologic Institute of Durango, Mexico

**Abstract.** The paper is developed in the framework of CFD to study the performance of Vertical Axis Turbines (VAT). One direct application is to establish design trends regarding the solidity. The model is validated with benchmarks from the literature for fluvial turbine. As for turbulence models, transitional SST version of k-w is used. The model includes two additional conservation equations for intermittency and critical Reynolds that establishes the transition from laminar to turbulent. Flow pattern are analyzed at intermediate positions along the revolution. A reduction in solidity increases the operation conditions.

## 1 Introduction

Vertical Axis Turbines (VAT) can operate at low velocities, hence they are suitable to extract energy from both: wind and water flows. The fluid mechanic design in both cases is quite similar, the only difference is the operating condition. While wind is characterised by velocities around 7 m/s for these applications, the water velocities are in the order of 1 m/s, however, there is a slight change on the range of variation of Reynolds number when operating with air or water. In both cases, Reynolds number has an intermediate value in the transition regime between laminar and turbulent.

This collaboration of three teams from universities of Valladolid (Spain), Guanajuato and Durango (Mexico) started in 2011. Previous milestones involved the characterization of vertical axis wind turbines (VAWT) using 2D models in a non-arm configuration, some examples are:

- The influence of the twist angle variation at the chord ends in symmetric airfoils NACA0015, [1].
- The analysis of the camber effect in non-symmetric airfoils NACA2425 and NACA7425, [2-3].
- The study of the solidity of the rotor NACA 0025, [3-4].
- Influence of fixed pitch angle in NACA0015 [5-6].

The turbulence models that have a better performance in 2D simulations of VAT are the RNG k-ε used in [2-4] and the transitional SST k-ω applied with success in [5-6] and recommended by other references of the literature [7-8].

This paper is devoted to the validation of the 2D model of a three straight blades turbine type H- Darrieus, see figure 1, and the analysis of the influence of solidity on the performance of the rotor.

The manuscript is structured as follows: section 2 summarizes the main details of the numerical model (mesh, operating conditions, and turbulence model), section 3 shows the validation of the model, in section 4 the flow patterns are analysed and section 5 shows the influence of the solidity in the power coefficient curves.

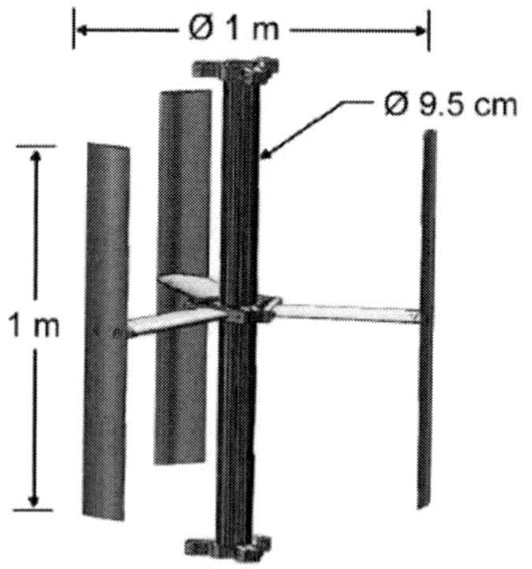


**Fig. 1.** Scheme of the dimensions of the UNH-RVAT [9-10].

## 2 Numerical model

Navier Stokes equations for two-dimensional, transient, incompressible and turbulent flow are to be solved using

[a] Corresponding author: mariateresa.parra@uva.es

moving reference mesh in the rotor and steady mesh in its surroundings.

Equations were solved with a second order scheme. Pressure and velocity coupled resolution uses the method SIMPLE.

Temporal resolution corresponds to 600 time steps per revolution; hence it is recalculated when the rotation speed is modified.

$$\Delta T = \frac{2\cdot\pi}{\omega\cdot 600} \quad (1)$$

## 2.1 Boundary conditions

The geometrical details of the turbine are summarized in table 1, and the operating conditions are indicated in table 2.

**Table 1.** Summary of geometrical details for VAT.

| Airfoil | NACA 0020 |
|---|---|
| **Shaft Diameter** | 0.095 m |
| **Rotor Radius (R)** | 0.5 m |
| **Length (b)** | 1 m |
| **Chord (c)** | 0.136 m |
| **Number of Blades (Z)** | 3 |
| **Solidity(σ)** | 0.4 |

**Table 2.** Summary of operating conditions for VAT.

| Airfoil | NACA 0020 |
|---|---|
| **Density (kg/m$^3$)** | 1000 |
| **Viscosity (kg/(ms))** | 1.1 $10^{-3}$ |
| **Water Velocity (m/s)** | 1 |
| **Turbulence Intensity (%)** | 10 |
| **Averaged Reynolds no. (-)** | 9.07 $10^5$ |
| **Depth of the turbine (m)** | 1 |
| **Reference Pressure (pa)** | 111125 |

## 2.2 Turbulence model

Transitional SST version of the k-ω model performs adequately when the flow regime is transient between laminar and turbulent regimes [7, 8]. It solves two additional conservation equations: one for intermittency, ϒ that is the ratio of time for turbulent flow and another for the transient Reynolds number.

$$\frac{\partial}{\partial t}(\rho\Upsilon) + \mathrm{div}(\rho\vec{v}\Upsilon) = \mathrm{div}\left(\left(\mu + \frac{\mu_T}{\sigma_\Upsilon}\right)\mathrm{grad}(\gamma)\right) + S \quad (2)$$

$$S = P_{\Upsilon 1} - D_{\Upsilon 1} + P_{\Upsilon 2} - D_{\Upsilon 2} \quad (3)$$

The source term has different contributions of production, P, and destruction, D, of the laminar regime. Subindex 1 is related to the strain tensor whereas subindex 2 is function of control functions to establish the laminar or turbulent regime are based on 2 definitions of Reynolds number indicated in equations (4) and (5).

$$Re_v = \frac{\rho y^2 S}{\mu} \quad (4)$$

$$R_T = \frac{\rho K}{\mu\omega} \quad (5)$$

Where S is the strain tensor, ω is the turbulent frequency scale and k is the turbulent kinetic energy.

## 2.3 Non dimensional parameters

The performance of the turbine is expressed in terms of non dimensional parameters. The power coefficient provides information of the efficiency of the turbine, being its maximum value 59% based on the Betz theory.

Power coefficient depends on the torque T, rotation speed ω and fluid velocity $V_\infty$, see equation (6),

$$C_p = \frac{T\cdot\omega}{\frac{1}{2}\cdot\rho\cdot v_\infty^3\cdot A} \quad (6)$$

TSR stands for Tip Speed Ratio and is defined in equation (7).

$$TSR = \frac{\omega\cdot R}{v_\infty} \quad (7)$$

Solidity was assumed to be the equation (8).

$$\sigma = \frac{Z\cdot c\cdot b}{2\cdot R\cdot b} \quad (8)$$

## 2.4 Sensibility to the mesh

Three different meshes are used to test that the results are independent from the spatial resolution. The main difference is the spatial resolution near the foils. Table 3 shows the number of cells of the three models as well as evidences of the quality of the mesh.

One parameter related to the mesh quality is the wall $y^+$ defined as a function of the friction velocity $u_\tau$, the distance of the cell center to the wall and the kinematic viscosity. The transitional STT k-ω turbulence model requests $y^+$ values around the unit in order to solve the laminar boundary sub-layer.

**Table 3.** Details of the mesh near the blades.

| Mesh | No. cells | Y$^+$ | Detail of the mesh |
|---|---|---|---|
| **Initial** | 139 kc | (7,185) | 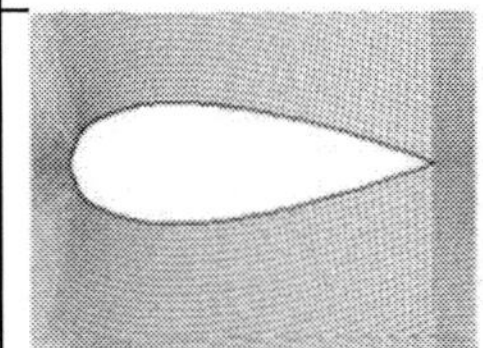 |
| **R-1** | 157 kc | (3,100) | 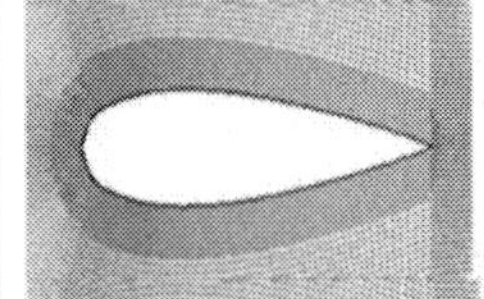 |
| **R-2** | 183 kc | (1,60) | 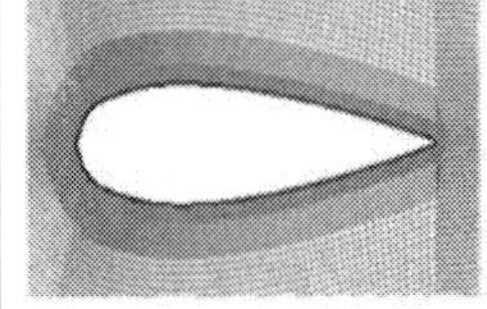 |

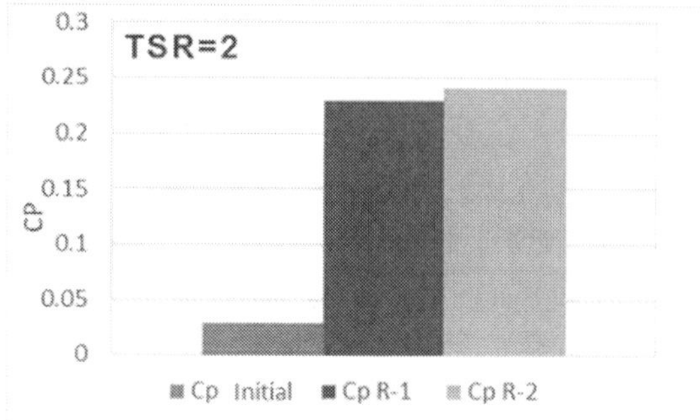


**Fig. 2.** Sensibility to the mesh of the peak power coefficient.

Figure 2 depicts the power coefficients for the three meshes at design operating conditions, i.e. with the TSR that provides the maximum power coefficient. Looking for a balance between precision and computational cost, any of the R-1 and R-2 meshes are suitable because of their low variance.

## 3 Validation of the model

The validation was carried out using the benchmark UNH-RVAT (University of New Hampshire- reference vertical axis turbine). This test case has a rotor with three blades NACA0020 and it is properly referenced by Bachant and Wosnik [9-10].

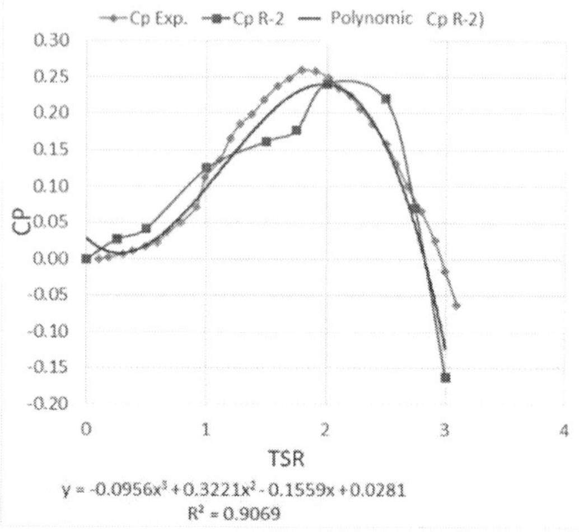


**Fig. 3.** Validation of the numerical power coefficient with experimental results from [9-10].

Each simulation must converge in a periodic behaviour with three periods in every revolution of the rotor. Once full periodical performance is achieved, the averaged of the instantaneous power coefficient provides one point of the characteristic curve. When modifying the rotation speed of the rotor, the curve is built up. Figure 3 shows the experimental results for the model UNH-RVAT, ref [9-10], in green diamonds as well as numerical results in blue squares.

A similar type of turbines composed by two counter-rotating rotors is the model DOE RM2 (Department of Energy - Reference Model 2) characterised at the Sandia Laboratories [11-12]. This has blades NACA 0021 and the application field is energy production from fluvial streams.

## 4 Flow pattern at design conditions

This section is devoted to the analysis of the flow pattern. The rotor spins counter clockwise.

**Table 4.** Identification of pressure and suction side of the airfoils.

| Position from 12 hours in counter wise clock | High pressure side | Low pressure side |
|---|---|---|
| 30° | External | Inner |
| 45° | External | Inner |
| 60° | External | Inner |
| 75° | External | Inner |
| 90° | External | Inner |
| 105° | External | Inner |
| 120° | External | Inner |
| 195° | Inner | External |
| 210° | Inner | External |
| 225° | Inner | External |

Figure 4 depicts the contours of pressure from the initial position with blades in locations 0°, 120° and 240° every 15°. The instantaneous flow patterns shows that the suction side (low pressure side) of the blades changes from the inner side to the outer side as indicated in table 4. As a result, the instantaneous power coefficient changes from a minimum value in location with blades at 0°, 120° and 24°, to a maximum value when blades are at 60°, 180° and 300°. The averaged power coefficient is the one represented in the characteristic curve of figure 3.

## 5 Influence of the solidity

When modifying the rotor radius, it is possible to modify the solidity, reducing it as the rotor radius becomes larger. Literature shows evidences of an increase of the TSR range were the turbine operates as power generator as the solidity is decreased. Also, the point in which the maximum power conditions are achieved, the design conditions, tends to appear at higher TSR. This means the turbine operates in more efficient conditions for energy generation.

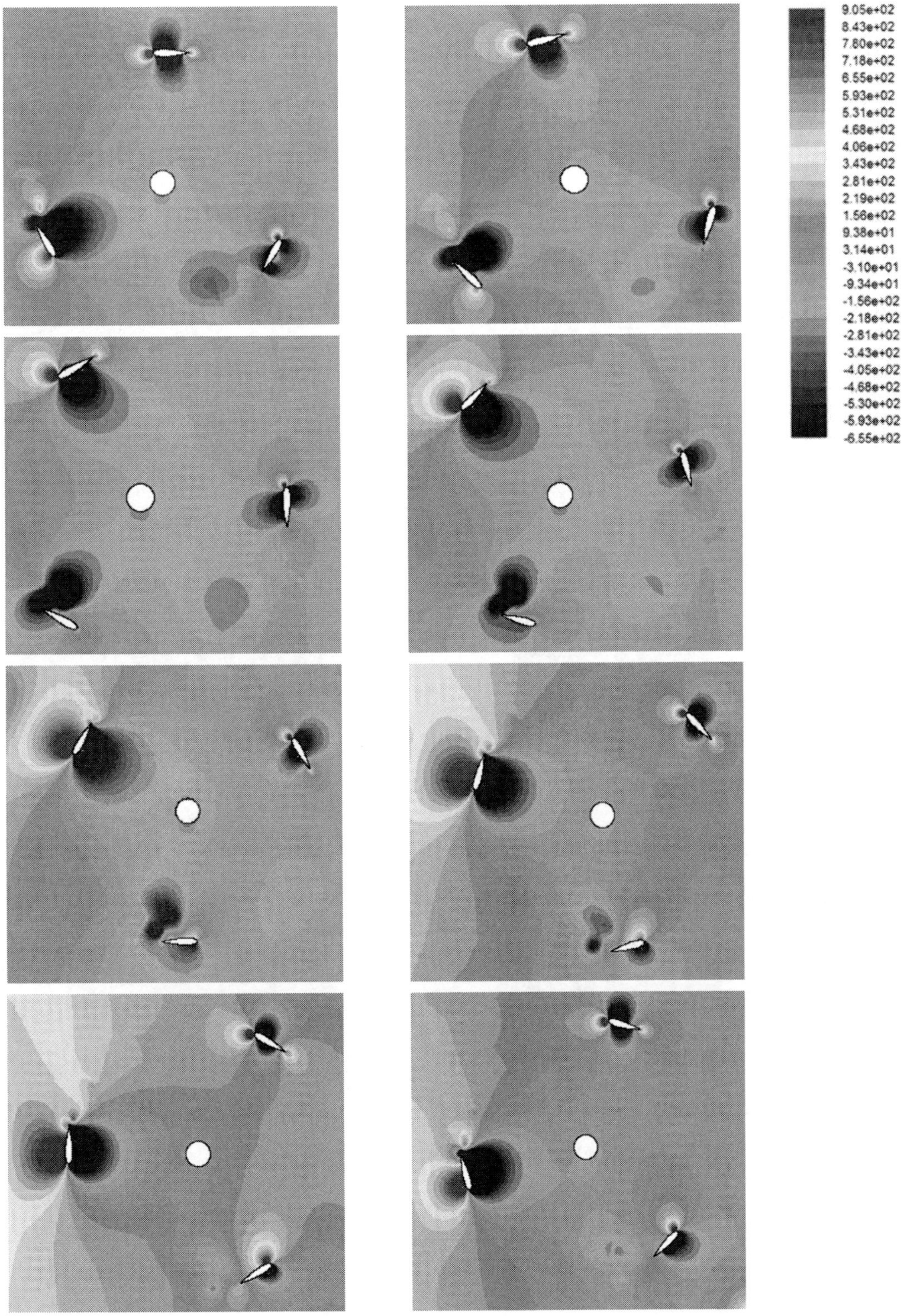


**Fig. 4.** Instantaneous Pressure Contours TSR = 2, every 15°.

Table 5 shows the conditions of the NACA 0020 operating as well as an estimation of the power of energy generated when operating with a flow of water of 1 m/s.

**Table 5.** Details design conditions at different solidities.

| **Solidity** | **σ = 0.5** | **σ = 0.3** |
|---|---|---|
| TSR range | 1.05 – 2.7 | 1 – 3.5 |
| Cp of design | 0.22 | 0.23 |
| TSR of design | 1.4 | 2.37 |
| Power (w) | 110 | 192 |

## Conclusions

This work is the collaboration of three research groups: University of Guanajuato, Durango and Valladolid. The background was based on H-Darrieus operating at low wind velocities.

This work is the first one with cross turbines suitable to operate with marine currents or fluvial flows. The model is validated using the benchmark UNH-RVAT from the University of New Hampshire. It is a relatively new test case with available experimental results that allows validation of numerical models.

A simple 2D model using transitional SST k-omega model is suitable to validate with the experimental results from other sources.

A brief study of the influence of the solidity on the efficiency of the turbine has been developed, showing how in the case studied, decreasing the rotor solidity increases the operating range and the TSR of design.

Future works consider the design optimization based on different geometric parameters as solidity, pitch angle, torsion, and number of blades among others.

## References

1. N.C. Uzarraga, A. Gallegos, M. T. Parra, J. M. Beltran. ASME ESFuelCell2012-91113 (2012)
2. T. Parra, C. Uzarraga, A. Gallegos, M.A. Rodriguez, F. Castro. FEDSM2014-21356 (2014)
3. T. Parra, C. Vega, A. Gallegos; et al, EPJ Web of Conferences, **92** (2015) 02058
4. T. Parra, M. Esteban, F. Gonzalez, C. Uzarraga, A. Gallegos, F. Castro. International Conference of Advanced Technologies and Sciences (2014) pp. 770-772
5. T. Parra-Santos, E. Chamoso, V. Gomez, A. Gallegos, C.N. Uzarraga and M. A. Rodriguez. EPJ Web of Conferences, **114** (2016) 2088
6. T. Parra-Santos, D. J. Palomar Trullen, A. Gallegos, C. N. Uzarraga, M. Regidor-Sanchez, F. Castro. FEDSM2016-7503 (2016)
7. R. Lanzafame, S. Mauro, M. Messina Energy Procedia **45**, pp 131 – 140 (2014)
8. R. Lanzafame, S. Mauro, M. Messina. Renewable Energy **52**, pp. 31-39 (2013).
9. P. Bachant and M. Wosnik "Effects of Reynolds Number on the Energy Conversion and Near-Wake Dynamics of a High Solidity Vertical-Axis Cross-Flow Turbine" Energies (2016)
10. P. Bachant and M. Wosnik "Reynolds Number Dependence of Cross-Flow Turbine Performance And Near-Wake Characteristics" Proceedings of the 2nd Marine Energy Technology Symposium METS-(2014)
11. C. Hill, V. Neary, B. Gunawan, M. Guala, F. Sotiropoulos. "Reference Model Program RM2: Experimental Results" Report June 24, (2014)
12. V. S. Neary, A. A. Fontaine, P. Bachant, B. Gunawan, M. Wosnik, C. Michelen, R.J. Meyer. W. A. Stranka. Sandia National Laboratories (2014).

EPJ Web of Conferences **143**, 02090 (2017) DOI: 10.1051/epjconf/201714302090
*EFM 2016*

# Reconstruction of 3D PIV data in complicated experimental arrangements

*David* Pavlík[1], *Václav* Uruba[2], *Václav* Kopecký[1]

[1] *Institute of New Technologies, Technical Univesity of Liberec, Studentská 2, Liberec, Czech Republic*
[2] *Institute of Termomechanics of the Academy of Sciences of the Czech Republic, Dolejšova 5, Praha 8, Czech Republic*

**Abstract.** In this paper a three-dimensional reconstruction of flow field behind flat plate representing a wing is presented. The reconstruction is always performed for pair of 2D vector maps obtained by 3D PIV with two cameras which record measurement area from different locations. Three-dimensional reconstruction can be obtained in various ways. This paper summarizes two: the reconstruction based on the known correspondences and the reconstruction based on the knowledge of intrinsic and extrinsic parameters of cameras. The methods can be used in the cases when it is impossible to use a calibration pattern or when reconstruction by commercial software fails.

## 1 Introduction

The 3D PIV is a contactless measurement method which is advisable to use for the comprehensive description actions behind flat plate with rounded edges. The method is based on the records of displacements of seeding particles carried by the fluid stream. Satiating particles are very small and their ideal size or other properties are determined by many experiments. During a measurement pulsed lasers are used in order to illuminate particles in predetermined measuring area. For recording of the particle images two cameras are mostly used. The optimal angle between axies of the cameras lens is 90° [1]. Two sets of images are used for computation of two-dimensional vector maps of flow field corresponding to each camera. The first set is captured at the time $t$ while second set of images is recorded several microseconds later $(t+\Delta t)$. The calculation is based on correlation methods. Subsequently, there are reconstructed 3D vector maps from the two-dimensional vector maps, which characterize fluid flow in the measurement area. The reconstruction can be done in several ways.

A calibration procedure of stereo-view has to precede each measurement. The calibration can be done based on the knowledge of geometry layout and parameters of cameras or by recording of a calibration pattern. The calibration pattern may be a plate with black/white dots on white/black background and with defined dimensions, see Fig. 4. This type of pattern is firstly recorded in the place of the laser cut and furthermore in several positions in front of the cut and behind the cut, see in paragraph 2.1. Another possibility is the calibration pattern in the form of a chessboard, which is positioned and recorded in place of the laser cut, as in the previous case, and is turned into random positions around the laser cut. The calibration pattern called multilevel target is another frequently used form of pattern. On the surface of the multilevel target, as the name suggests, dots are placed at least at two levels (eg. 0 mm and 1 mm). The multilevel target does not require multiple recording during the reconstruction procedure. The recording at laser cut position is sufficient for the complete calibration [2].

The vast majority of laboratories, which deals with measurement method 3D PIV, use a commercial software, like Dynamic Studio, for the evaluation of measured data. Despite their many advantages, in some cases the three-dimensional reconstruction algorithms are not sufficiently robust in order to correctly evaluate the measurements. This is mainly in cases where the calibration target is recorded through components which significantly distort the resulting image of the pattern or particles, see paragraph 2.2. A typical and practical example can be measurement of flow within a cylindrical pipe. Another problem arises when it isn't possible to place the calibration pattern into the measurement area (eg. in a long pipe, tunnel) or its movement isn't possible. The reconstruction by standard commercial programs fails in these cases, see paragraph 3.2.

Two approaches for extraction of 3D information from 2D PIV data have been developed and they are introduced in this paper. The approaches have been used and tested in experimental research of flow behind flat plate in cooperation with Institute of Termomechanics, Academy of sciences of the Czech Republic, see paragraph 1.1. The measurement was done in distortion free environment in order to have relevant data for comparison with results obtained by commercial available Dynamic Studio software. The applicability of approaches was verified in the other measurement when the distortion in the observation direction was presented (the experimental research of flow inside cylindrical pipe).

[a] Ing. David Pavlík: david.pavlik@tul.cz

### 1.1 Experimental research of flow behind flat plate

The motivation of the presented research is supporting new ideas about principle of flight by Hoffman and Johnson from KTH Stockholm, see [3,4]. The new hypothesis of physical mechanism of flight relies on existence of streamwise vortical structures on the suction side of the airfoil and within its wake. The vortices origin is supposed to be the instability of the boundary layer subjected to adverse pressure gradient on the airfoil pressure side (i.e. upper) [5].

The simple experiment is made with the simplest airfoil possible represented by a flat plate in uniform flow and moderate angle of attack. However the experimental results are of a qualitative nature only, they reveals existence of streamwise vortices above the plate without any doubt.

The measuring plane has been located perpendicular to the main flow, behind the plane trailing edge. The situation is shown in Fig. 1. Some other information and conclusions from the research are shown in [6].

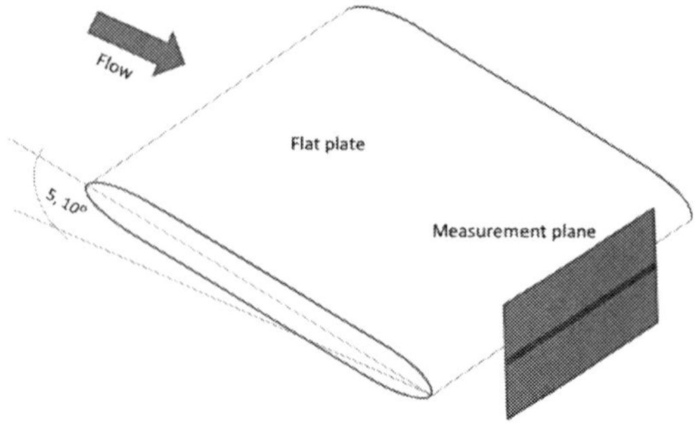


**Figure 1.** Experimental research behind flat plate

## 2 Reconstruction based on known correspondences

For the three-dimensional reconstruction of vector maps a method based on the knowledge of the point positions in object coordinate system *(X,Y,Z)* and the positions of their projections in the image planes *(u,v)* was used [7]. The object coordinate system is also called a scene. For that were used calibration patterns and their recordings. This approach can be used in cases where records of measured area are subject to great optical distortion. A preparation of the experiment is described in following paragraph.

### 2.1 Experimental setup

The setup is schematically shown in Fig.2. The cameras used in measuring are Nano HISENSE 3 (Dantec) with parameters: CCD, maximal frame rate 600 Hz at resolution 1280x1024 px, objectives Micro Nikor 60mm. Cameras were mounted on a traverse system, which allowed precise shift in three axes. Angle between axes of their objectives were 78 °.

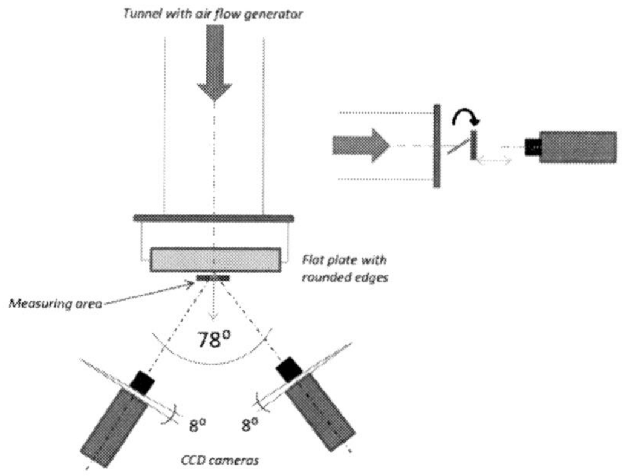


**Figure 2.** Experimental setup

The cameras have a very small depth of field. If we want to focus the whole field of view corresponding to the measuring plane (object) on the image plane (CCD detector), the Scheimpflug condition must be met, see Fig. 3. The condition is met if the image plane, object plane and camera lens plane cross each other at a straight line in space [8]. A special camera mount, which allows rotation of camera body towards camera lens, was used for this purpose.

A laser beam of double cavity pulse laser New wave Pegasus PIV was shaped into a so-called laser sheet by cylindrical lens. The laser sheet was a vertical plane with thickness of 1 mm. Parameters of the laser: 10mJ at 1kHz, maximal 10kHz, wavelength 527nm. The laser sheet defines a measuring plane (area) in the flow field. The measurement was carried out in defined locations behind the flat plate. Dantec Dynamic components were used to synchronize cameras with the laser.

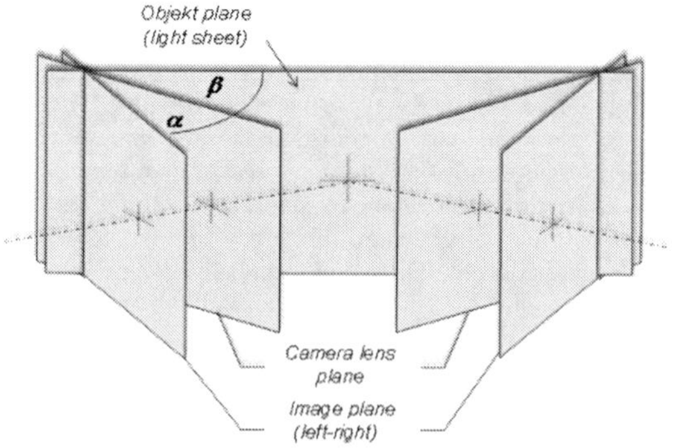


**Figure 3.** Scheimpflug condition

The first step of measurement was to place calibration pattern with small traverse system into the measuring area. Black dots form a pattern on a white background. Pattern was illuminated by a very powerful source of light. This was done because we need uniform distribution of light intensity along a pattern. The calibration pattern was recorded in five positions: in front of and behind the laser sheet. The pattern was moving along axis perpendicular to the plane of the measurement area, see Fig. 4.

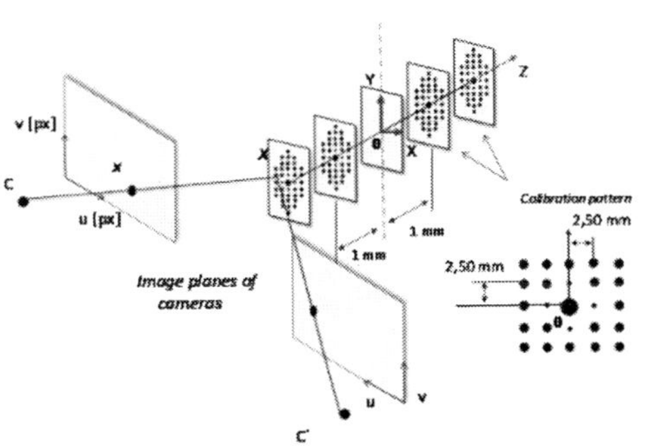


**Figure 4.** Recording of calibration pattern

The calibration pattern was removed after recording and the tunnel was saturated by particles (Safex). The measurement was performed at air flow speed: 5m/s, 10m/s. The raw images were processed in Dantec software, where vector maps of flow velocity were calculated and validated.

### 2.2 Reconstruction

For the reconstruction the relation between positions of points in the scene *(X,Y,Z)* and positions of their projections in the image planes *(u,v)* was used, as mentioned above. The calibration patterns and their recordings were used for this purpose. Positions of calibration points in the scene were described in the object coordinate system. The origin of the coordinate system was identical to the intersection of the central axis of laser sheet and line from zero points (calibration images centers placed in defined positions inside tunnel (0,0,Z)). The orientation of the object coordinate system and the image coordinate system is shown in Fig.4. Further, the dimensions of the calibration pattern, as well as an example of the projection of one point into image planes are shown in the same figure.

The relation between positions of points in scene *(X,Y,Z)* and positions of their projections in the image plane *(u,v)* is expressed by polynomial model n-th order [7], which is able to describe nonlinearities caused by the image distortion:

$$F(u,v) = \sum_{i=0,j=0,k=0}^{i=n,j=n,k=n} A_{i,j,k} x^i y^j z^k \tag{1}$$

The polynomial model of the third order (in X (i= 0; 1; 2; 3),Y (j= 0; 1; 2; 3), in Z second order (k= 0; 1; 2)), which is sufficient for most experiments, was used:

$$\begin{aligned}\begin{bmatrix} u \\ v \end{bmatrix} = A_{000} &+ A_{100}X + A_{010}Y + A_{001}Z \\ &+ A_{110}XY + A_{101}XZ + A_{011}YZ \\ &+ A_{200}X^2 + A_{020}Y^2 + A_{002}Z^2 \\ &+ A_{300}X^3 + A_{210}X^2Y + A_{201}X^2Z \\ &+ A_{030}Y^3 + A_{120}XY^2 + A_{021}Y^2Z \\ &+ A_{102}XZ^2 + A_{012}YZ^2 + A_{111}XYZ\end{aligned} \tag{2}$$

In Eq. (2) coefficients A are two-dimensional vectors. Nineteen unknown vectors for each camera were necessary to be calculated.

First step of reconstruction was to find positions of the calibration points in the targets images *(u,v)* and assign them to corresponding points in the object coordinate system *(X,Y,Z)*. The coefficients A were calculated by last squares method for each camera.

Next step was the reconstruction of a three-dimensional flow model from the vector maps. The reconstruction was resolved by triangulation. For this it was necessary to identify a common field of view of the cameras. A vector from two-dimensional velocity vector map, computed from the particles images recorded by left camera, was assigned to the corresponding vector from the two-dimensional velocity vector map captured by the right camera. It resulted in a set of four nonlinear equations with three unknowns, see Eq. (2). Solution of the equations was a three-dimensional velocity vector in a certain interrogation area. Same procedure was used to all correspondences and finally we calculated the three-dimensional velocity vector map.

## 3 Reconstruction based on knowledge of cameras intrinsic and extrinsic parameters

Reconstruction based on knowledge of cameras intrinsic and extrinsic parameters enables to reconstruct the three-dimensional flow velocity vector maps in cases where the calibration target cannot be put into the measuring area or where it isn't possible to move with the calibration target inside tunnel or pipe. However the approach cannot be used in the presence of the optical distortions. This reconstruction is based on a linear transformation unlike to the polynomial model.

The experimental setup was the same as in previous case (Fig.2), except for a placement of the calibration target into tunnel. Before starting the measurement it was necessary to get parameters which are listed below.

### 3.1 Theory of pinhole camera model

A pinhole camera model (also perspective camera) approximates very well properties of real camera and in this approximation is sufficient in this case. The camera transforms points *X(X,Y,Z)* from the object coordinate system to points *x(u,v)* in image plane, such that:

$$x = PX, \tag{3}$$

In Eq. (3) P is a camera projection matrix [9], with dimensions $3x4$ and rank 3.
The projection matrix of cameras can also be expressed by equations:

$$P = \begin{bmatrix} f\alpha & 0 & p_x & 0 \\ 0 & f\beta & p_y & 0 \\ 0 & 0 & 1 & 0 \end{bmatrix} = \begin{bmatrix} f\alpha & 0 & p_x \\ 0 & f\beta & p_y \\ 0 & 0 & 1 \end{bmatrix} [\,\mathrm{I}\,|\,0\,] = $$
$$= K\,[\,\mathrm{I}\,|\,0\,] \tag{4}$$

In (4) $p_x$ and $p_y$ are coordinates of principal point in the image plane, $\alpha$ is the width and height ratio of pixels, $\beta$ is degree bevel of pixels, $f$ is focal length. These parameters are collectively called Intrinsic camera parameters and are expressed by calibration matrix $K$ [9]. Situation, when during projection the optical center is not located at the origin of the object coordinate system and main axis of camera is not identical with the axis $Z$, is described by the extrinsic parameters (5). The extrinsic parameters accordingly define orientation and position of camera coordinate system in the scene. The orientation represents matrix $R$, with dimensions $3x3$ (rotation of the camera around all three axes of object coordinate system) and translation vector $t$ which specifies position of camera optical center (focus of lens).

$$P = KR\,[\,\mathrm{I}\,|\,\text{-}\boldsymbol{t}\,] \tag{5}$$

### 3.2 Reconstruction

The aim of this approach was to find the projection matrices of both cameras. It was necessary to determine the matrices $K,R$ and vector $t$. The matrices $K_1, K_2$ (for each camera) were calculated from several records of chessboard with defined grid, see Fig.5. Camera calibration toolbox for Matlab was used for calculation [10]. The recording was done before the measurement, when cameras were located outside traverse system. Next step was to determine the extrinsic parameters. The orientation of cameras coordinate systems represents matrices $R_1, R_2$. The translation vectors $t_1, t_2$ specify the position of the optical centers of cameras (focus of lens), as noted above. The center of the coordinate system was identical with center of measurement area, see Fig. 4.

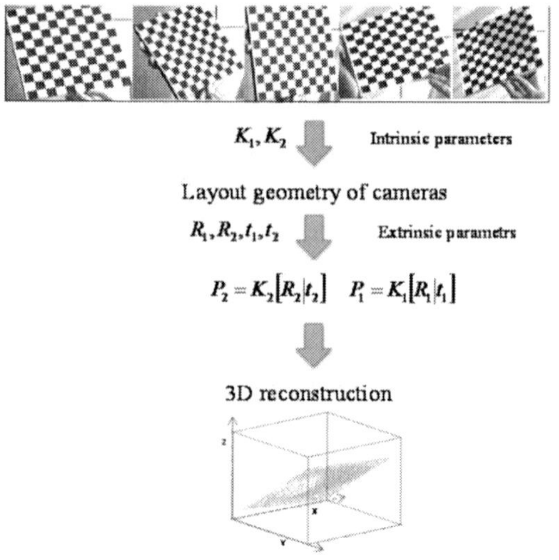


**Figure 5.** The reconstruction process

The vector of the two-dimensional velocity vector map, computed from the particles images recorded by left camera, was assigned to corresponding vector from the right camera two-dimensional velocity vector map, same as in paragraph 2.2. The three-dimensional velocity flow vector in certain interrogation area was calculated according to equation (3) and subsequent triangulation [10]. Same approach was used for all correspondences and it was possible to reconstruct the three-dimensional vector map. The reconstruction is not invariant to rotation, translation and evinces certain scale to reality. The scale is possible to determine from other measures, eg. from 2D PIV method or Pitot tube point measurement.

## 4 Results of reconstruction

The results of the reconstructions are presented by 2D vector maps of flow velocity with scalar maps (Z component). The measuring plane has been located perpendicular to the main flow, 40 mm behind the plane trailing edge, 10 angle of attack and flow velocity was 5m/s. The flow was captured with the sampling period 1,25ms. The upper part of vector maps corresponds to the location behind the plane trailing edge.

The accuracy of the reconstruction method based on the known correspondences was approximately +/- 1% (for all components) compared with the results from Dynamic Studio.

The accuracy of the reconstruction method based on the knowledge of intrinsic and extrinsic parameters of cameras was approximately +/- 2% in Z component, and +/- 30% in X, Y component compared with the results from Dynamic Studio. The error in the results is mainly caused by high sensitivity to inaccuracy of the measured parameters *(R,t,K)*.

 DOI: 10.1051/epjconf/201714302090

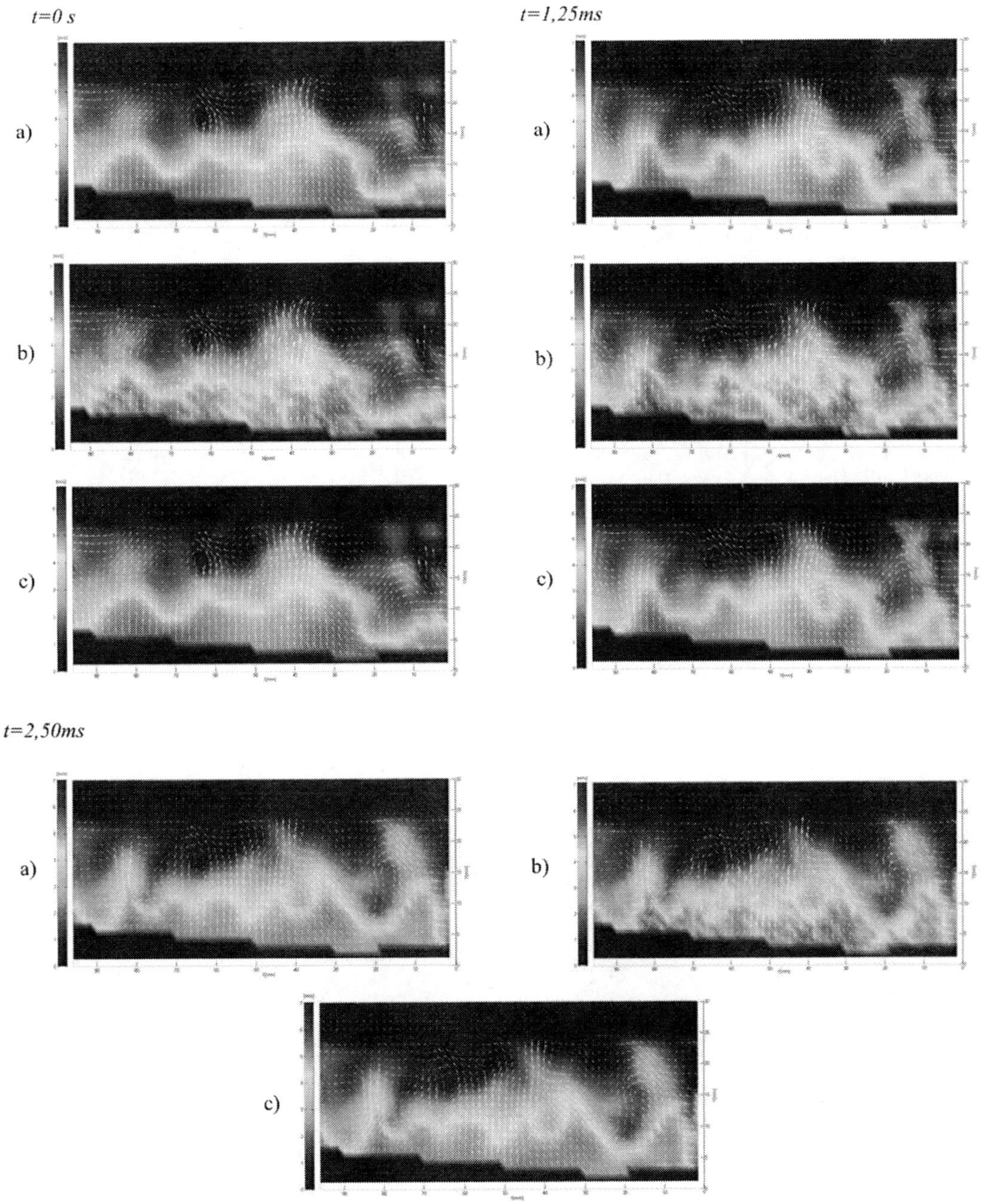


**Figure 6.** 2D vector maps of flow velocity with scalar map (Z component) in time *t*: (a) the reconstruction in paragraph 2., (b) the reconstruction in paragraph 3., (c) Dynamic Studio

## 4 Conclusion

The paper describes two approaches for evaluation of data obtained by 3D PIV method that can be used in situations where commercial available softwares can fail. The reconstruction based on known correspondence (paragraph 2.) is applicable in cases where the measurement area recorded through objects (eg. cylindrical, convex, concave glass) which significantly distort the images of calibration pattern and satiating particles. The reconstruction by commercial software (eg. Dynamic Studio) often fails in these cases. It was shown that the third order of the polynomial fitting model (paragraph 2.2) is sufficient in most cases. Otherwise it is possible to choose a higher order.

The reconstruction based on intrinsic and extrinsic parameters of cameras (paragraph 3.) is applicable in case where the calibration target can´t be put into the measuring area or where is not possible to move with the calibration target inside the area (eg. long pipe, tunnel). The reconstruction cannot be performed by conventional commercial softwares in such cases. The error in the results is mainly caused by high sensitivity to inaccuracy of the measured parameters $(R,t,K)$.

## Acknowledgment

This work was supported by the Ministry of Education, Youth and Sport of the Czech Republic within the SGS project no. 21066/115 on the Technical University of Liberec.

## References

1. C. TROPEA, A. L. YARIN, J. F. FOSS: *Handbook of experimental fluid mechanics, Volume 1*, Springer, 1556 pgs, 2007.
2. L. M. SUÁSTEGUI: *Overview on stereoscopic Particle Image Velocimetry*, Instituto Politécnico Nacional, Santa Catarina, Mexico, 16 pgs, 2012.
3. J. HOFFMAN, C. JOHNSON, *The Mathematical Secret of Flight*, Normat 57, 4, pp.1–25, (2009)
4. J. HOFFMAN, C. JOHNSON, *Resolution of d'Alembert's paradox*, Journal of Mathematical Fluid Mechanics, Vol.12 (3), pp.321-334, (2010)
5. P. MANNEVILLE, *Instabilities, Chaos and Turbulence.* Imperial College Press, 2004
6. V. URUBA, *On 3D Flow-Structures behind an Inclined Plate*, EFM 2016, *Experimental fluid mechanics*. in Mariánské Lázně, 15.11-18.11., Czech Republic, 6p. (2016)
7. M. BSIBSI: *Advances in stereoscopic Particle Image Velocimetry and Application to Supersonic Base Flow Exhaust Plume Interaction,* TUDelft, 129 pgs, 2004.
8. V. KOPECKÝ: *Laser anemometri in fluid Mechanics,* Ed. TRIBUN EU s.r.o., 204 pgs, Brno 2008, in Czech Republic.
9. B. TUROŇOVÁ. *3D Recnstruction of Scene from Stereo Images,* 48 pgs, Charles Univesity in Prague, 2008.
10. J. Y. BOUGUET, *Computational Vision* [online] Camera Calibration Toolbox for Matlab, 2005, URL: http://www.vision.caltech.edu/bouguetj/calib_doc/ .

DOI: 10.1051/epjconf/201714302091

# Heat transfer in a confined impinging jet with swirling velocity component

Karel Petera[a], Martin Dostál

Czech Technical University in Prague, Department of Process Engineering, Technická 4, Praha 6, Czech Republic

**Abstract.** Heat transfer measurements based on an infrared experimental method (TOIRT) are compared with CFD simulations of a confined impinging jet with tangential velocity component. The tangential velocity component added to a pure impinging jet introduces into the flow field and heat transfer some similarities with real industrial processes like agitated vessels with axial-flow impellers. The tangential velocity component significantly influences the velocity field and heat transfer intensity in the stagnant region when compared to the classic impinging jet characteristics. Several turbulence models were used in numerical simulations of an agitated vessel with axial-flow impeller in a draft tube. Heat transfer coefficients at the vessel bottom were evaluated using the TOIRT method and compared with numerical results. The lateral heat conduction in the impinged wall was analysed with the conclusion that it has relatively small impact on the measured heat transfer coefficients. Quite good agreement of experimental data and simulation results was achieved concerning the size and position of the heat transfer maximum at the vessel bottom.

## 1 Introduction

High heat transfer intensities at relatively small surfaces can be achieved with impinging jets putting them in the focus of industrial interest. Various experimental, numerical and analytical data concerning the hydrodynamics and heat or mass transfer in impinging jets can be found in the literature, mainly in a configuration where the fluid (liquid) jet impinges a smooth plane surface [1–4]. In practice, many impinging jets are generated by devices with rotary parts therefore some tangential (swirling) velocity component is always present at the jet nozzle which has a significant impact on the velocity profile. Figure 2 illustrates the influence of the tangential velocity component on the axial velocity profile. A tangential velocity component linearly increasing from zero at the center to the maximum value at the jet radius was superposed on a developed axial velocity profile in a pipe at Reynolds number Re = 23000. It was then used as the inlet boundary condition of the jet nozzle. Due to centrifugal forces induced by the swirling velocity component, the parabolic-shaped velocity profile in the axial direction was transformed to a triangle-shaped velocity profile at the nozzle outlet (length $L = 2d$), see Figure 2. It is clearly visible there that the axial velocity component approaches zero values near the center.

A similar velocity profile can be observed at the outlet of the draft tube in an agitated vessel with an axial-flow impeller, see Figure 3. Similarities of velocity profiles between an unconfined impinging jet with a swirling velocity component and a real industrial equipment like the agitated vessel can be clearly identified here. The question is how much they influence the temperature field and if some similarities can be identified in the heat transfer as well.

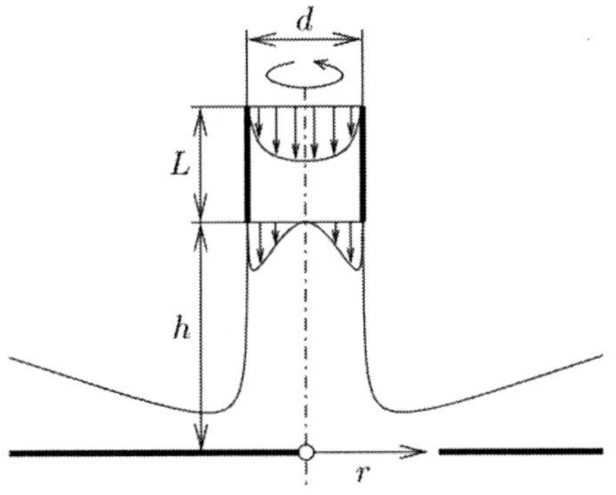


Figure 1: A scheme of a jet with nozzle length $L$, diameter $d$, and distance $h$ from the impinged plate.

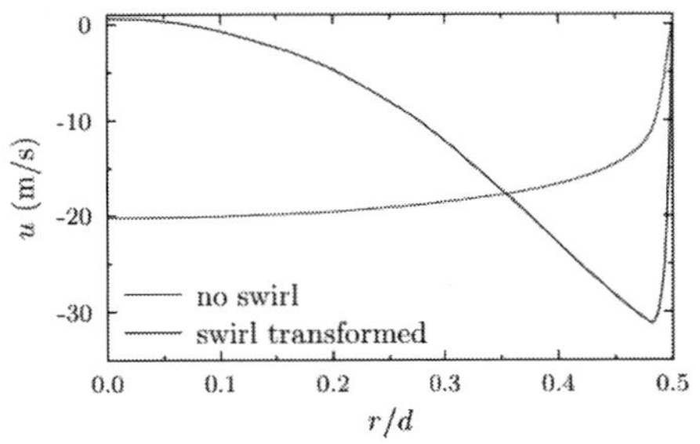


Figure 2: The axial velocity profile transformation due to centrifugal forces induced by the tangential (swirling) velocity component. $h/d = 2$, Re = 23000, Swirl number $S = 1.02$ (Eq. 3).

[a] Corresponding author: karel.petera@fs.cvut.cz

DOI: 10.1051/epjconf/201714302091

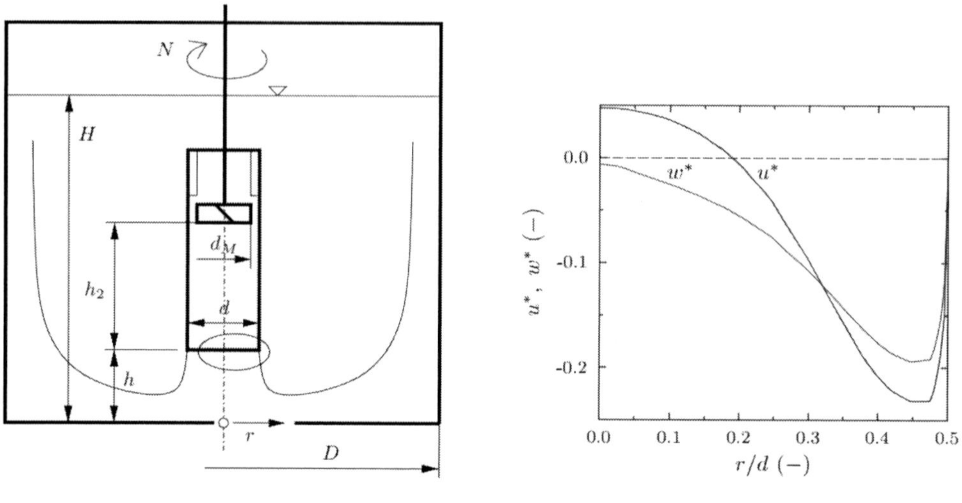


Figure 3: Geometrical configuration of a confined impinging jet generated by an axial-flow impeller in a draft tube, with an example of typical axial and tangential velocity profiles at the outlet of the draft tube [5].

As demonstrated on Figure 2, the axial velocity profile of a swirling impinging jet differs significantly from a non-swirling case which should have a large impact on heat transfer, especially in the stagnation region where the axial velocities are quite small or even negative. Figure 4 illustrates the difference between the swirling and non-swirling jet for frequently published case with Reynolds number Re = 23000 and distance to nozzle diameter ratio $h/d = 2$. A large drop in Nusselt number values is visible at the center of the stagnation region which is a consequence of small values of the axial velocity components there. It is in accord with some of the literature data [6], even though as pointed out by [7], the flow and heat transfer characteristics show drastically different results depending on the type of swirl generators. In our case, we have used an axial-flow impeller with six blades, 45° pitch angle, and diameter $d_M = 61$ mm. It was placed in a draft tube with inner diameter 70 mm. The distance of the draft tube from the vessel bottom was 70 mm, giving the ratio of height and jet diameter as $h/d = 1$. The vessel diameter was $D = 392$ mm, and liquid height $H = 430$ mm. The position of the impeller in the draft tube was $h_2 = 215$ mm above the tube outlet. The total tube length was 300 mm, therefore the top part of the tube was 60 mm below the liquid surface. In the upper part of the draft tube, small 60 mm long baffles were installed to prevent a whirl formation.

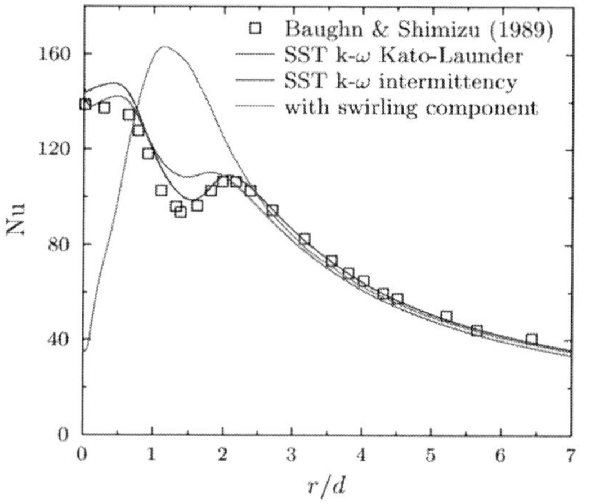


Figure 4: Comparison of the Nusselt number values for air impinging jets with and without tangential/swirling velocity component. Re = 23000, $h/d = 2$, Swirl number at the outlet of the draft tube $S = 1.02$ (1.27).

The axial-flow impeller produced a jet with tangential (swirling) velocity component at the outlet of the draft tube, which, in addition, belongs to the category of confined impinging jets because it is placed in a cylindrical vessel. In this paper, we compared CFD simulation results with experimental data measured by the TOIRT (infra-red temperature-oscillation) method [8].

## 2 Numerical simulation of agitated vessel

We have used ANSYS CFD with Fluent solver version 16.0 to perform our numerical simulations and get comparison with experimental data. ANSYS Meshing component was used to create a mesh on the geometry depicted on Figure 3. We have made a mesh independence study with several mesh sizes and we finally used a mesh with 2.2 million of cells. Inflation layers were used near the vessel and draft tube walls, as well as the impeller itself. As we were interested in the heat transfer at the vessel bottom, the inflation layers were created there so that $Y^+$ quantity was close to the recommended value 1 and the near-wall region was fully resolved with no approximations by wall functions. In our case, the maximum value of $Y^+$ was 1.8 which we found sufficient.

We used RANS (Reynolds averaged Navier-Stokes) turbulence models [9] which are less computationally expensive than LES-based or even DNS approaches. With RANS models, the Reynolds stress tensor in the time-averaged

Navier-Stokes equation must be somehow defined, i.e. the system of equations must be closed. A simpler definition is based on the Boussinesq eddy-viscosity approximation [10] which assumes that the Reynolds (turbulent) stress tensor is proportional to mean strain-rate tensor $S_{ij}$ and the proportional factor is eddy (turbulent) viscosity $\mu_t$.

$$-\varrho\overline{u_i' u_j'} = 2\mu_t S_{ij} - \frac{2}{3}\varrho k \delta_{ij} \tag{1}$$

In this work, we mainly used Realizable $k-\varepsilon$ and SST $k-\omega$ models which uses another set of (transport) equations for quantities like turbulence kinetic energy $k$ and energy dissipation $\varepsilon$ or specific dissipation rate $\omega$ to express turbulent viscosity $\mu_t$ in the Equation (1), for example

$$\mu_t = \alpha^* \frac{\varrho k}{\omega} \tag{2}$$

According to our previous work [11], we put more preferences to SST $k-\omega$ model with Kato-Launder limiter [12] as it resolves the excessive production of turbulence kinetic energy caused by a very high level of shear stress rate in the stagnation regions [13]. Without this limiter, the Nusselt number values were usually overpredicted near the stagnation point in our numerical simulations of an unconfined impinging jet with no swirling component [11]. Kato-Launder limiter is enabled by default in the ANSYS Fluent implementation of the SST $k-\omega$ Intermittency Transition model which gave us best agreement with experimental data for free impinging jets [11]. With SST $k-\omega$ model, Kato-Launder limiter must be turned on manually in the Fluent solver.

We used the MRF (Moving/Multiple Reference Frame) [13] approach to get the velocity field in the agitated vessel with the draft tube. First, we started with a steady-state simulation, and then we switched to the transient case when the vessel bottom was heat by a constant heat flux. After an initial transient period (2 seconds) we gathered the temperature field (wall) statistics over a larger time period (10 – 20 seconds), and we evaluated the heat transfer coefficients (Nusselt numbers) at the vessel bottom. Figure 5 presents a comparison of the Nusselt number values at the bottom of the agitated vessel for several turbulence models. A large drop in the heat transfer intensity near the stagnation point is in accordance with our simulations of the unconfined impinging jet (see Figure 4). Of course, there are huge differences from the unconfined jet near the vertical wall where the stream of fluid decelerates due to the presence of the wall and changes its direction up along the vertical wall of the vessel. A recirculation zone is visible in the flow field here, see Figure 6, which has a direct consequence in the jumps of the heat transfer intensities near the vertical wall.

The results of the numerical simulations presented in this section are then used in comparison with experimental data in the subsequent text as well as in the analysis of the TOIRT method. The impeller speed 400 rpm corresponded to the swirl intensity represented by Swirl number $S = 1.1$ in the case of the following definition [14]

$$S = \frac{2}{d}\frac{G_\varphi}{G_x} \tag{3}$$

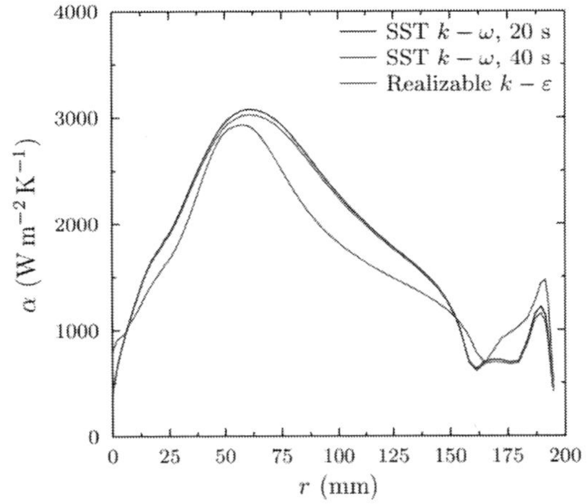


Figure 5: Comparison of the Nusselt number values at the bottom of the agitated vessel for impeller speed 400 rpm (mixing Reynolds number 25 000), constant heat flux boundary condition and several turbulence models. A comparison of two time intervals (20s and 40s) used in transient statistics collection for SST $k-\omega$ model with Kato-Launder limiter [12] is presented here.

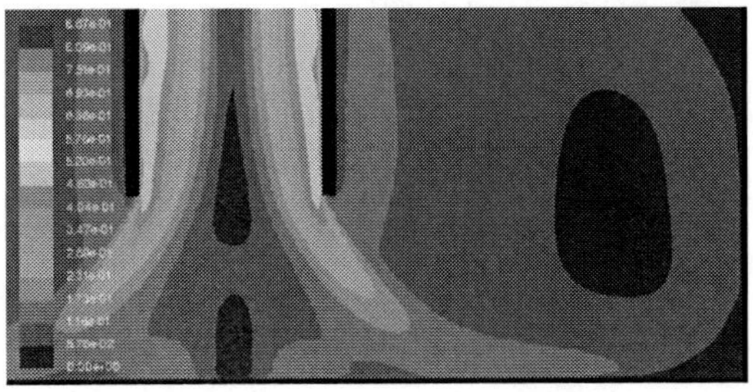

Figure 6: Contours of velocity magnitude below the draft tube and near the vessel bottom, 400 rpm, Swirl number $S = 1.1$ (0.7), SST $k-\omega$ turbulence model.

where $G_\varphi$ and $G_x$ represent the flux of angular and axial momentum's, respectively.

$$G_\varphi = \int_A \varrho r u w \,\mathrm{d}A\,, \quad G_x = \int \varrho u^2 \,\mathrm{d}A \tag{4}$$

In the literature, a different and simpler definition of the Swirl number is also used [7]

$$S = \frac{w_b}{u_b} \tag{5}$$

where $w_b$ and $u_b$ represent the bulk tangential (azimuthal) and axial velocities, respectively.

$$u_b = \frac{1}{A}\int_A u \,\mathrm{d}A\,, \quad w_b = \frac{1}{A}\int_A w \,\mathrm{d}A \tag{6}$$

This simpler definition (6) of the Swirl number gives value 0.7 for our case of impeller speed 400 rpm.

## 3 TOIRT experimental method

The temperature oscillation method accompanied by the contactless temperature measurements using infra-red ther-

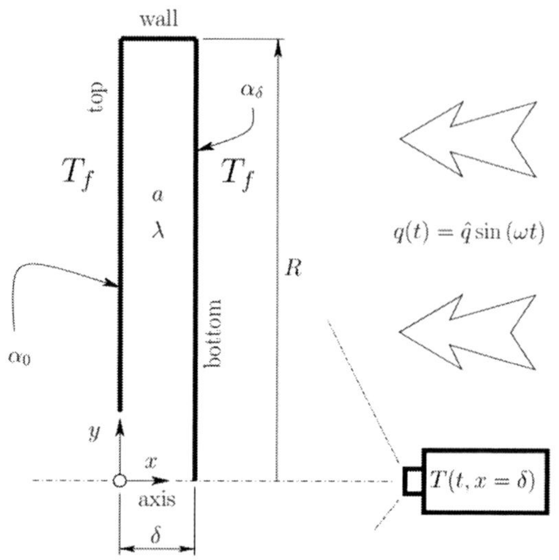


Figure 7: The principle of the TOIRT method by Wandelt and Roetzel [8]. The left side of the wall (denoted as "top") is impinged by a stream of fluid, the right side ("bottom") is heated by an oscillating heat flux. The temperature response $T(t, x = \delta)$ is then measured by an infrared camera.

mography (Temperature Oscillation Infra-Red Thermography, TOIRT) was used to measure and verify the heat transfer coefficient along the impinged heat transfer surface. Freund [15] measured the temperature response of a plate induced by an oscillating heat flux impinging the plate which was an extension of work by [8]. See [16–18] for more details and applications of this method.

The principle of the TOIRT method is depicted on Figure 7. It is based on measuring the time dependency of $T|_{x=\delta}$ (or other corresponding characteristic) at the surface plate of width $\delta$ induced by oscillating heat flux $q|_{x=\delta}$. The temperature response depends on thermophysical properties of the plate and heat transfer coefficients on both sides of the plate, i.e. $\alpha_0$ and $\alpha_\delta$. Transient temperature field $T$ in a homogeneous plate with thermal diffusivity $a$ is described by the Fourier equation

$$\frac{\partial T}{\partial t} = a\left(\frac{\partial^2 T}{\partial x^2} + \frac{\partial^2 T}{\partial y^2} + \frac{\partial^2 T}{\partial z^2}\right). \tag{7}$$

Neglecting the lateral heat conduction described by the last two terms on the right-hand side and adding boundary conditions on both sides of the plate, it is possible to find an analytical solution for the case of a periodically oscillating impinging heat flux. The boundary conditions of the third kind can be applied on both sides, with a periodic oscillation of the heat flux on one side. Assuming zero temperatures $T_f$ on both sides, they can be described as

$$\lambda \left.\frac{\partial T}{\partial x}\right|_{x=0} = \alpha_0 \, T|_{x=0} \,, \tag{8}$$

$$\lambda \left.\frac{\partial T}{\partial x}\right|_{x=\delta} = \hat{q}\sin \omega t - \alpha_\delta \, T|_{x=\delta} \,, \tag{9}$$

where $\lambda$ is thermal conductivity of the plate, $\omega$ is frequency of the oscillating heat flux with amplitude $\hat{q}$, and $\alpha_0$, $\alpha_\delta$ are the already mentioned heat transfer coefficients on both sides of the plate. Using the Laplace transform, a temperature solution on both sides of the plate for amplitude $A$ and phase shift $\varphi$ can be found [8]

$$T(x,t) = A(x)\sin(\omega t - \varphi(x)) \,. \tag{10}$$

Then, the phase shift at the surface with induced temperature oscillations ($x = \delta$) can be described [8] as

$$\tan\varphi|_{x=\delta} = \frac{c_1 + 2\xi\psi_0 c_2 + 2\xi^2\psi_0^2 c_3}{2\xi\psi_0(1+r)c_0 + 2\xi^2\psi_0^2(1+2r)c_1 + 4\xi^3\psi_0^3 r c_2 + c_3} \tag{11}$$

where the following dimensionless variables and parameters are used

$$r = \alpha_\delta/\alpha_0 \,, \quad \psi_0 = \frac{\alpha_0}{\delta\varrho c_p\omega} = \frac{\alpha_0 a}{\delta\lambda\omega} \,, \quad \xi = \delta\sqrt{\frac{\omega}{2a}} \tag{12}$$

$$\begin{aligned} c_0 &= \cosh^2\xi\cos^2\xi + \sinh^2\xi\sin^2\xi \,, \\ c_1 &= \cosh\xi\sinh\xi + \cos\xi\sin\xi \,, \\ c_3 &= \cosh^2\xi\sin^2\xi + \sinh^2\xi\cos^2\xi \,, \\ c_4 &= \cosh\xi\sinh\xi - \cos\xi\sin\xi \,. \end{aligned} \tag{13}$$

It is maybe not clear from the complicated solution (11), but the phase shift of the temperature response depends on the size of unknown heat transfer coefficient $\alpha_0$, and, on the contrary, it does not depend on the amplitude of the heat flux. Furthermore, it does not depend on the accuracy of the oscillating temperature field measurements. In this case with the neglected lateral heat conduction, it is not necessary to measure the surface temperature precisely, only the exact time dependency between the impinging heat flux and the measured temperature response is required.

## 4 Experimental

As mentioned in the previous section denoted to the numerical simulations of the agitated vessel, the heat transfer coefficients at the surface impinged by a stream of fluid (liquid) with a superposed tangential (swirling) velocity component differ significantly from the case with no tangential component. In this section, the aim is to verify the numerical results for the case of the confined swirling impinging jet generated by the axial-flow impeller in a cylindrical vessel with the draft tube.

The heat transfer coefficient in an impinging jet was measured on a circular plate (bottom of the vessel) with diameter 392 mm and thickness 1 mm. The plate was made of steel with thermal conductivity 14.6 $\mathrm{W\,m^{-1}\,K^{-1}}$, density 7800 $\mathrm{kg\,m^{-3}}$ and specific heat capacity 501 $\mathrm{J\,kg^{-1}\,K^{-1}}$. Water was used as the fluid and its temperature was within the range 25°C to 29°C during all measurements. The temperature change of water was up to 2°C during one measurement.

The oscillating heat flux impinging the other (bottom) side of the plate was generated by 3×500 W halogen lamp. The light intensity modulation, that is the size of the impinging heat flux, was maintained using the voltage variation within range 20% – 90% of the maximum possible

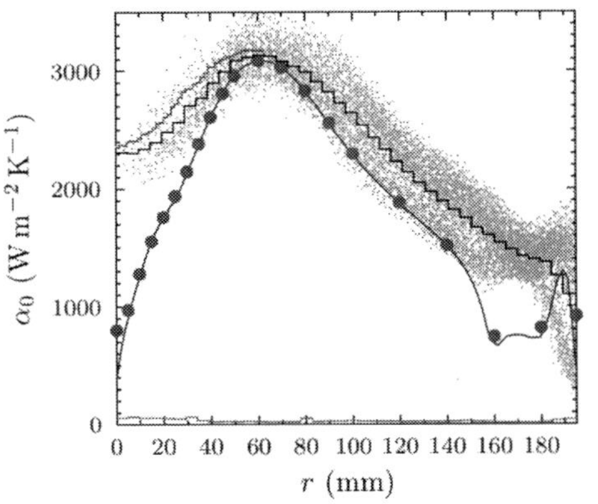


Figure 9: Experimental values of heat transfer coefficients on the surface (vessel bottom) impinged by a jet with tangential velocity component (black stepwise line) compared with the results of numerical simulations in section 2 (blue line). The shorter stepwise green line represents heat transfer coefficients measured independently with a different focus length of lens. The magenta line in the figure describes errors of the averaged experimental points (gray) in a specific location. Circles represent theoretical values which should resulted from the TOIRT measurement taking into account the lateral heat conduction.

value. The modulation proceeded with frequency 0.1 Hz and it was controlled by a precise sine-impulse generator. The temperature response of this plate side was measured by the infrared camera thermoIMAGER TIM 160 with sampling frequency 10 Hz (camera resolution 160 x 120 points, the real distance between the measured points corresponds to approximately 3.41 mm and 1.04 mm, depending on the focus length of the lens).

The measured results for two focus lengths of the lens are depicted on Figure 9. The bunch of points (10 000 approx.) represents the experimental data transformed from the Cartesian (infrared camera measurement, see Figure 8) to the cylindrical (polar) coordinate system describing the dependency of the heat transfer coefficient on the radius.

The black stepwise line in the Figure 9 represents average values in a corresponding range of the radial coordinate. In this case, the whole range of the radial coordinate has been divided into 40 equidistant sections. The blue line in the figure represents the heat transfer coefficients coming from the numerical simulation of the agitated vessel described in section 2. The measured data show very good agreement with the numerical data as for the peak position and its magnitude. Larger discrepancies are visible on the left as well as on the right of the maximum. In both cases, the experimental values are larger than the simulated ones, especially near the stagnation point where we wanted to verify our numerical results and the literature data.

Figure 9 shows clearly that the differences between the experimental values of the heat transfer coefficients are larger in regions with smaller heat transfer intensities and in regions with larger gradients of the heat transfer coefficients which is closely related to the radial gradients of the wall temperature. In such case, the heat conduction in the heat transfer wall appears not only in the normal direction, but also in the radial direction. And the heat conduction in the radial direction (lateral conduction) is neglected in the TOIRT method by Wandelt and Roetzel [8]. To evaluate the errors yielding from the neglect of the lateral heat conduction, we performed numerical simulations of this case in ANSYS CFD which are described in the next section.

## 5 Numerical simulation of the Wandelt and Roetzel method

We used ANSYS CFD with Fluent solver to simulate the heat conduction in the heat transfer surface represented by circular plate of diameter 392 mm and thickness 1 mm (the bottom of the agitated vessel). Figure 7 depicts the basic geometry of this task. It was modeled as an axisymmetric case. A perfect insulation, that is zero heat flux was considered as the boundary condition on the upper wall edge of the plate. The boundary condition of the third kind based on liquid temperature 300 K and the heat transfer coefficients gained in the numerical simulation of the agitated vessel was specified at the edge denoted as "top". A mixed boundary condition was applied to the edge "bottom". It combined the third-kind boundary condition and an oscillating heat flux. The surrounding temperature $T_f$ was again 300 K (this temperature was also used as the initial condition of the transient solution), and the heat transfer coefficient was set to 3 W m$^{-2}$ K$^{-1}$ (natural convection). The heat flux impinging the "bottom" had a sine time dependency with 10 Hz frequency and 10 kW m$^{-2}$ amplitude. The following UDF function was used to define this boundary condition in the Fluent solver.

```
DEFINE_PROFILE(heat_flux_alpha, thread, position)
{
  real r[ND_ND];
  real y;
  face_t f;

  real flow_time = RP_Get_Real("flow-time");
  real wTemp, q;

  real freq = 0.1;
  real alpha_delta = 3.0;
  real q_amplitude = 10000;

  begin_f_loop(f, thread)
  {
    F_CENTROID(r,f,thread);
    y = r[1];
    wTemp = F_T(f,thread);
    q = q_amplitude*sin(2.0*PI*freq*flow_time) +
                    alpha_delta*(300.0-wTemp);
    F_PROFILE(f, thread, position) = q;
  }
  end_f_loop(f, thread)
}
```

The problem was solved as a transient case with time step 0.01 s (first, time step 0.1 s was used but the resulting values of heat transfer coefficients showed 10% difference from the theoretical solution by Wandelt and Roetzel [8] even in cases when the lateral conduction can be neglected). A 2-D mesh with 40 thousand elements was created on the surface representing the axisymmetric geometry of the wall (Figure 7). During the transient solution of the temperature field in the wall, the temperature time dependency on the "bottom" edge was monitored at several discrete points just as it is monitored by the infrared camera. The black oscillating line on Figure 10 represents an example of such time dependency. The time dependency of the oscillating heat flux is represented by the blue line on the same figure (with a scaled amplitude to fit the vertical axis range which has no influence on the phase shift).

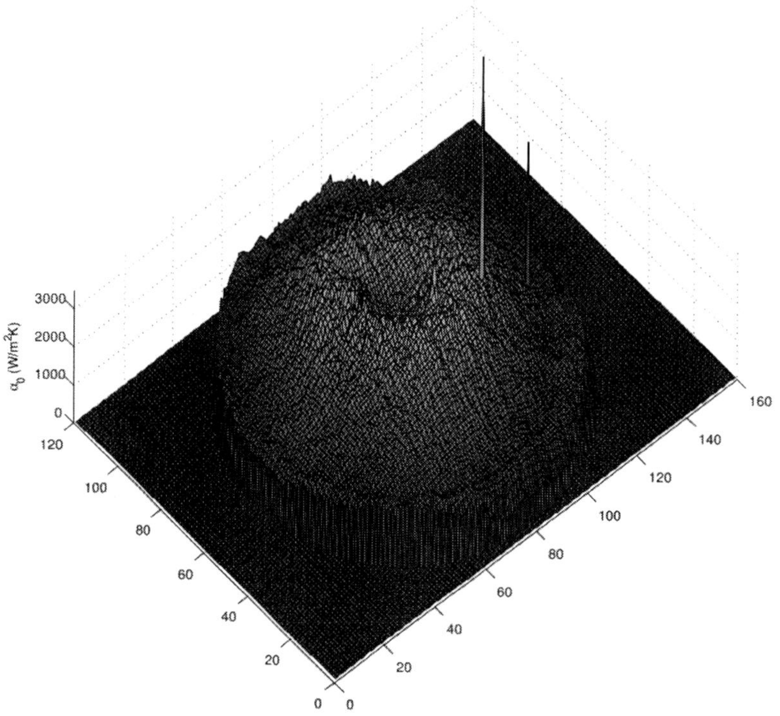


Figure 8: An illustration of the measured heat transfer coefficients at the heat transfer surface in individual points taken by the infrared camera (the size of the surface is expressed in pixels given by the camera resolution).

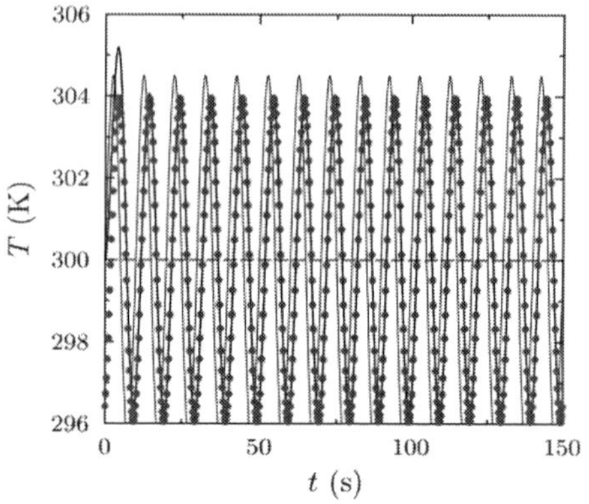


Figure 10: ANSYS CFD (Fluent) simulation results of Wandelt and Roetzel method [8] for heat transfer coefficients 1000 W m$^{-2}$ K$^{-1}$ at the impinged "top" surface and 3 W m$^{-2}$ K$^{-1}$ at the illuminated "bottom" surface. The blue line represents the oscillating heat flux from halogen lamps. The black line corresponds to the calculated temperature profile at radius 100 mm of this "bottom" surface. The red points represent the sine dependency of this response (for larger time values). The phase shift between the actuating (source) signal and the response is obvious here and it can be used in evaluation of the heat transfer coefficient at the impinged "top" surface.

Applying Wandelt and Roetzel method [8] on these time dependencies of the wall surface temperatures, the phase shift between the oscillating source signal and oscillating response temperature can be evaluated using Eq. (11). The red points in Figure 10 follow the measured temperature response and represent the theoretical temperature response described by Eq. (10). In this case, it can be evaluated as

$$\begin{aligned} T(t) &= 300 + 3.9492 \sin\left[2\pi\, 0.1\,(t - 1.8007)\right] = \\ &= 300 + 3.9492 \sin\left(2\pi\, 0.1\, t - 1.1314\right). \end{aligned} \tag{14}$$

The phase shift is 1.1314 radians in this case which corresponds to heat transfer coefficient 1000 W m$^{-2}$ K$^{-1}$ for the given geometrical and thermophysical parameters.

Processing the temperature time dependencies in all monitored points at the "bottom" edge will give us the radial profile of the heat transfer coefficients at the "top" edge, see Figure 11. The results presented here show that the values based on Wandelt and Roetzel method [8] have quite good agreement with correct heat transfer coefficients in regions with values greater than 1000 W m$^{-2}$ K$^{-1}$, that is in the prevailing part of the measured profile. Let us remind that this experimental method neglects the lateral conduction. Figure 11 demonstrates that the lateral conduction can be neglected along the whole surface except small regions near the stagnation point and the vessel wall. In these regions, the values of heat transfer coefficients fall below 1000 W m$^{-2}$ K$^{-1}$ and the lateral conduction due to the radial temperature gradients affects the measured and

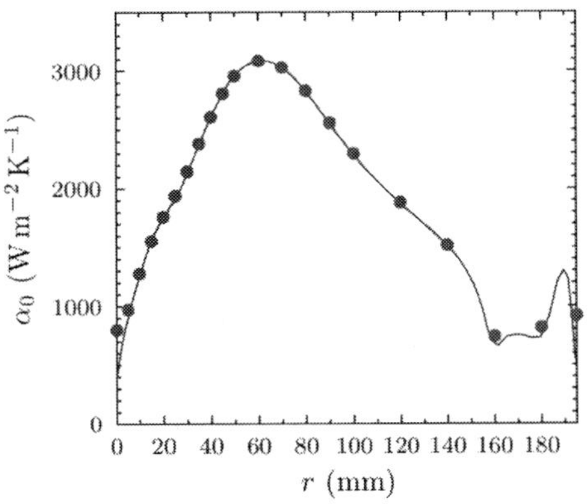


Figure 11: Numerical results including the heat conduction in the wall illuminated by an oscillating heat flux and consequent evaluation based on Wandelt and Roetzel method [8]. The continuous line represents the applied heat transfer coefficients at the "top" edge, and the discrete points were evaluated from the simulated temperature dependencies according to the Wandelt and Roetzel method [8].

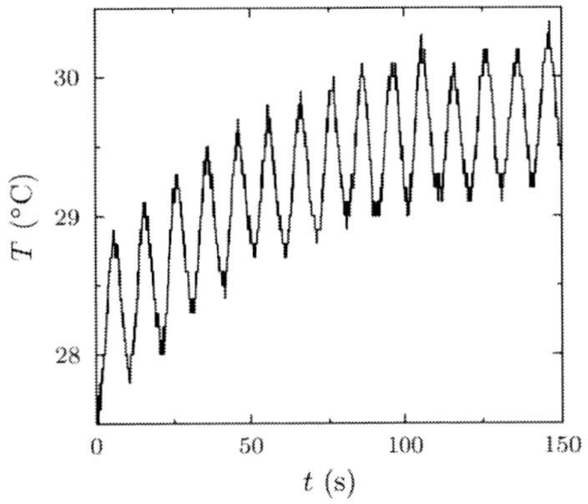


Figure 12: Time dependency of temperature taken by the infrared camera using the TOIRT method in one point on the wall. The oscillating temperature response with period 10s (frequency 0.1 Hz) superposed on an "exponentially" increasing the mean temperature of the liquid due to the heat accumulation in the system. This dependency has to be transformed to a temperature response oscillating around a constant value.

evaluated values which are larger than real ones (they are almost twice larger near the stagnation point).

## 6 Conclusions

The values of heat transfer coefficients measured by the TOIRT method and evaluated according to Wandelt and Roetzel method [8] which neglects the lateral heat conduction are presented on Figure 9 (black line). The values of heat transfer coefficients including the influence of the lateral conduction (distinct points in the figure) do not improve the agreement of the simulation results with experimental data in the given system. Hence, the lateral heat conduction is probably not the reason of relatively large experimental values near the stagnation point.

A possible source of the disagreement between experimental and numerical results could lie in relatively small amplitudes of temperature oscillations, see Figure 12. This could be resolved by a more intensive heat source.

Quite good agreement was achieved in the size and position of the heat transfer maximum. The measurements have been performed several times and all experimental data were within ±8% interval, indicating a relatively good repeatability of the measuring procedure.

## Acknowledgment

This work has been supported by research project of the Grant Agency of Czech Republic No. 14-18955S.

## Nomenclature

| | |
|---|---|
| $a$ | thermal diffusivity, $a = \lambda/\varrho c_P$ (m$^2$ s$^{-1}$) |
| $A$ | surface, area of impinged plate or vessel bottom (m$^2$) |
| $c_{0,1,...,4}$ | constants (parameters) used in Equation (11), see Equation (13) for their definitions (–) |
| $c_P$ | specific heat capacity (J kg$^{-1}$ K$^{-1}$) |
| $d$ | diameter of jet nozzle or draft tube (m) |
| $d_M$ | diameter of impeller (m) |
| $D$ | outer diameter of impinged plate or vessel bottom (m) |
| $f$ | frequency (Hz, s$^{-1}$) |
| $G_x$ | flux of axial momentum in the definition of Swirl number, see Eq. (4) (kg m s$^{-2}$) |
| $G_\varphi$ | flux of angular momentum in the definition of Swirl number, see Eq. (4) (kg m$^2$s$^{-2}$) |
| $h$ | jet distance from the plate (m) |
| $h_2$ | distance of the impeller from the outlet of the draft tube (m) |
| $H$ | height of liquid in the vessel (m) |
| $k$ | turbulent kinetic energy (m$^2$s$^{-2}$, J kg$^{-1}$) |
| $N$ | impeller speed (s$^{-1}$, rpm) |
| Nu | Nusselt number, Nu = $\alpha d/\lambda$, where heat transfer coefficient $\alpha$ is evaluated as $\alpha = q/(T_w - T_b)$ (–) |
| $q$ | heat flux (W m$^{-2}$) |
| $\hat{q}$ | amplitude of oscillating heat flux (W m$^{-2}$) |
| $r$ | radial coordinate (m) |
| $r$ | dimensionless parameter in Equation (11), see Equation (12) for its definition (–) |
| $R$ | outer radius of impinged plate or vessel bottom (m) |
| Re | Reynolds number, $u_b D/\nu$ (–) |
| $S_{ij}$ | strain-rate tensor (s$^{-1}$) |
| $t$ | time (s) |
| $T$ | temperature (°C, K) |
| $T_b$ | bulk temperature (average) at the jet nozzle (°C, K) |
| $T_f$ | fluid (liquid) temperature (°C, K) |
| $T_w$ | wall (boundary) temperature (°C, K) |
| $u$ | velocity, axial velocity (m s$^{-1}$) |
| $u_b$ | bulk (average) velocity, see Eq. (6) (m s$^{-1}$) |
| $u^*$ | dimensionless axial velocity related to the tip velocity of the impeller, $u^* = u/(\pi d_M N)$ (m s$^{-1}$) |
| $w$ | tangential (swirl) velocity (m s$^{-1}$) |
| $w_b$ | bulk tangential velocity, see Eq. (6) (m s$^{-1}$) |
| $u'$ | velocity fluctuation (m s$^{-1}$) |

$w^*$ dimensionless tangential velocity related to the tip velocity of the impeller, $w^* = w/(\pi d_M N)$ $(\mathrm{m\,s^{-1}})$
$S$ Swirl number, see Eqs. (3), (5) (–)
$x, y, z$ spatial coordinates of the Cartesian coordinate system (m)

$\alpha$ heat transfer coefficient $(\mathrm{W\,m^{-2}K^{-1}})$
$\gamma$ intermittency (–)
$\Gamma$ diffusivity in the user-defined scalar equation $(\mathrm{kg\,m^{-1}s^{-1}})$
$\delta$ width of impinged plate (m)
$\delta_{ij}$ Kronecker unity tensor (–)
$\varepsilon$ dissipation rate of turbulent kinetic energy $(\mathrm{m^2\,s^{-3}})$
$\lambda$ thermal conductivity $(\mathrm{W\,m^{-1}\,K^{-1}})$
$\mu$ dynamic viscosity (Pa s, $\mathrm{kg\,m^{-1}s^{-1}}$)
$\mu_t$ turbulent (eddy) dynamic viscosity (Pa s, $\mathrm{kg\,m^{-1}s^{-1}}$)
$\nu$ kinematic viscosity, $\nu = \mu/\varrho$ $(\mathrm{m^2\,s^{-1}})$
$\xi$ dimensionless width, $\xi = \delta\sqrt{\omega/2a}$ (–)
$\varrho$ density $(\mathrm{kg\,m^{-3}})$
$\phi$ general scalar variable in the user-defined scalar equation (–)
$\varphi$ phase shift (–)
$\psi_0$ dimensionless parameter used in Equation (11), see Equation (12) for its definition (–)
$\omega$ specific dissipation rate of turbulent kinetic energy, $\omega = \varepsilon/k$ $(\mathrm{s^{-1}})$
$\omega$ angular frequency of oscillating heat flux, $\omega = 2\pi f$ $(\mathrm{s^{-1}})$

## References


1. V. Katti, S. Prabhu, Int. J. Heat Mass Transfer **51**, 4480 (2008)
2. M. Behnia, S. Parneix, Y. Shabany, P.A. Durbin, Int. J. Heat Fluid Flow **20**, 1 (1999)
3. D. Lytle, B. Webb, Int. J. Heat Mass Transfer **37**, 1687 (1994)
4. J. Baughn, S. Shimizu, ASME Journal of Heat Transfer **111**, 1096 (1989)
5. P. Vlček, T. Jirout, *CFD simulation of flow in mixing equipment with draft tube*, in *CHISA conference* (Seč, Czech Republic, 2015)
6. V.I. Terekhov, Y.M. Mshvidobadze, Thermal Science **20**, suppl. 1, 35 (2016)
7. Z.U. Ahmed, Y.M. Al-Abdeli, M.T. Matthews, Computers & Fluids **118**, 255 (2015)
8. M. Wandelt, W. Roetzel, *Lockin thermography as a measurement technique in heat transfer*, in *QIRT 96 - Eurotherm Series 50* (1997), pp. 189–194
9. D. Wilcox, *Turbulence Modeling for CFD* (DCW Industries, Inc., 2006)
10. J.O. Hinze, *Turbulence* (McGraw-Hill Publishing Co., New York, 1975)
11. Petera, K., EPJ Web of Conferences **114**, 02091 (2016)
12. M. Kato, B. Launder, *The modelling of turbulent flow around stationary and vibrating square cylinders*, in *Ninth Symposium on "Turbulent Shear Flows"* (1993)
13. ANSYS Fluent, *ANSYS Fluent Theory Guide*, ANSYS, Inc., Release 16.0 (2015)
14. M. Mahmood, *Heat transfer from swirling impinging jets*, Ph.D. thesis, Cranfield Institute of Technology (1980)
15. S. Freund, *Local Heat Transfer Coefficients Measured with Temperature Oscillation IR Thermography*, Ph.D. thesis, Hamburg (2007)
16. S. Freund, S. Kabelac, *Measurement of local convective heat transfer coefficients with temperature oscilation IR thermography and radiant heating*, in *Proceedings of HT2005* (2005), pp. 1–7
17. S. Freund, A.G. Pautsch, T.A. Shedd, S. Kabelac, Int. J. Heat Mass Transfer **50**, 1953 (2007)
18. S. Freund, S. Kabelac, Int. J. Heat Mass Transfer **53**, 3764 (2010)

# Improvement of radiotracers experiments in mass transfer processes

*Leszek* Petryka[1,a], *Volodymyr* Mosorov[2], *Marcin* Zych[1], *Robert* Hanus[3], *Jerzy* Sobota[4], *Marek* Jaszczur[1], and *Dariusz* Świsulski[5]

[1]AGH University of Science and Technology, Kraków 30-059, Poland
[2] Łodz University of Technology, Łodz 90-924, Poland
[3] Rzeszów University of Technology, Rzeszów 35-959, Poland
[4]Wrocław University of Environmental and Life Sciences, Wrocław 50-375, Poland
[5]Gdańsk University of Technology, 80-233 Gdańsk, Poland

**Abstract.** The greatest part of industrial mass transfer is based on continuous transportation of different substances throughout especially constructed installations. In that kind of practice, a flow of such mass in each part determines the process unit and can be described by the residence time distribution (RTD) of the substance inside this unit. This parameter may be evaluated by the properly arranged tracer experiment. It is worthwhile to notice that flow components are labeling for a sufficiently long time for offering a lot of advanced techniques fitted to the particular circumstances.

## 1 Introduction

An analysis of a selected particle or a component movement in any flow is possible using tracers. In this approach to investigation, especially convenient are radioactive elements induced or placed in the examined substance and consequently detected by a probe located outside of the process [1, 2]. That opportunity ensures radioisotopes emitting gamma radiation, when energy of photons exceeds a hundred kVe. The above mentioned simple idea has been applied for almost a hundred years and presently may offer many additional improvements [3].

The paper focused on a case of the hydraulic transportation of a solid phase in pipes and in such processes as a mineral industry where descriptions of the mass transfer by a residence time or velocity distribution are applied [4].

## 2 Velocity Measurements

A popular example of radiotracer application is solid particles transportation by liquid through a pipeline, presented in Fig. 1. When we imagine that a single labelled particle is introduce to the cross section $S_0$ with the axial momentum equal to zero. Consequently during transportation through initial section $L_1$, provided the same axial velocity of that grain, as the most in the vicinity. In practice, the $L_1$ distance may vary from several centimeters to meters, depending on accuracy demanded and turbulence of the flow [5, 6].

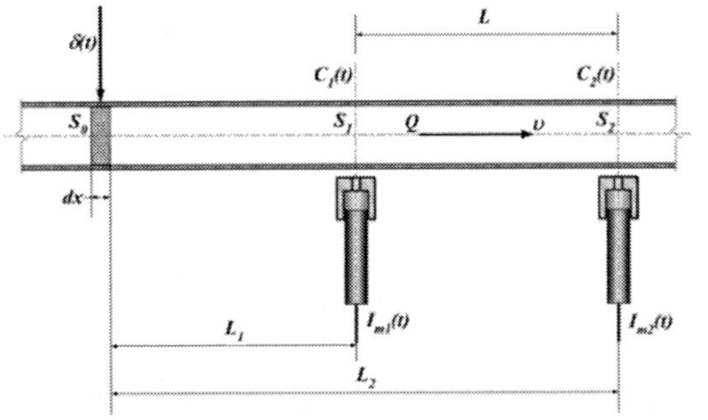


**Fig. 1.** Radiotracer measurement in a pipe: $S_0$ – tracer injection section, $S_1$ – beginning of measuring section, $S_2$ – end of measuring section, $C(t)$ – tracer concentration, $Q$ – volumetric flow rate, $I_m$ – measured count rate.

After fulfilling that condition, the residence time $t$ in the pipe section between $S_1$ and $S_2$ allows determining the local axial velocity $v_L$ of the marked grain representing the solid phase, according to:

$$v_L = L / t \qquad (1)$$

In case of the one particle experiment there is no problem with determination of time in which a tracer is passing by both the $S_1$ and $S_2$ cross sections of the pipe, so the residence time:

$$t = t_2 - t_1 \qquad (2)$$

The experiment described above would be repeated several times, for the mean velocity $v_m$ of entire labelled phase determination. That method is comparatively

[a] Corresponding author: Leszek.Petryka@fis.agh.edu.pl

simple, but long lasting and during all the time the same stable condition of the flow is demanded

$$v_m = \frac{1}{n}\sum_{i=1}^{n} v_i \tag{3}$$

Huge reduction of the time spent for the velocity measurement may be achieved by collecting significant number of marked particles in the single portion injected to $S_0$ section in a manner resembling Dirac's $\delta(t)$ distribution. Nevertheless, those facilities are covered by more difficult measurement and smaller accuracy of results [7, 8].

In such cases it should be assumed that:

$$C_{in}(t) = \delta(t) \tag{4}$$

is uniformly distributed in a total cross-section of a stream.

Additionally during this kind of velocity measurement, the activity of tracers in the injected portion should be reduced to a level accepted by the applied probes and safe for the staff involved in the investigation.

Moreover, the injected sample should be statistically significant but still below the level devastating the process stability.

Furthermore, in this method, the recorded data processing is more complicated by the background determinations and residence time evaluation procedures. In consequence a few groups in the world are sufficiently experienced for the proper arrangement of such measurements.

## 3 Single particle's velocity measurement

Fig. 2 shows the measurement part of the hydraulic installation build up in the Central Mining Institute in Katowice, Poland.

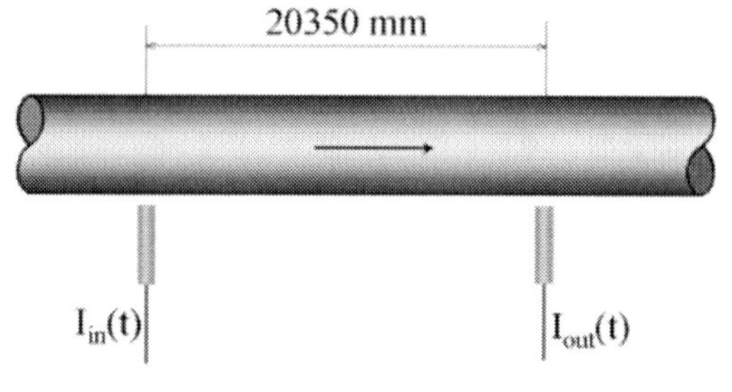


**Fig. 2.** The idea of the transit time measurement.

During experiments testing solid particles' velocity determination, the sandstone grains, about 10 mm diameter, were transported by water through a steel pipe over 40 m long, and of 185 mm inner diameter. The installation used for the testing is presented in Fig. 3, where the rotation pump forced the flow containing a mixture of almost equal sandstone grains and water with 1.2 Mg/m$^3$ density. In that case, the solid phase was represented by a single grain containing radioactive isotopes of $^{24}$Na and $^{46}$Sc. Due to that, probes 1 and 2 provided "yes" or "no" information only.

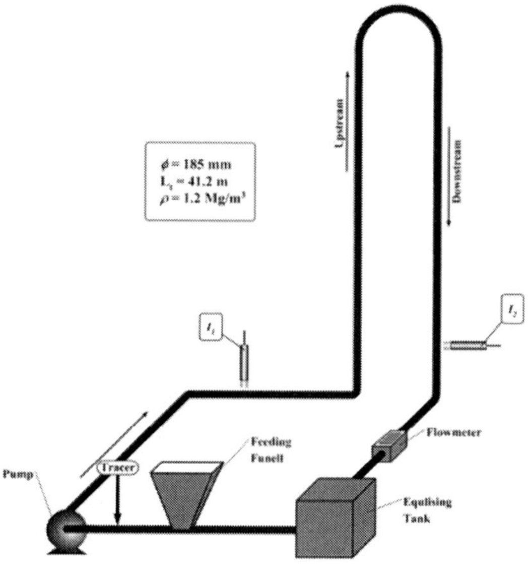


**Fig. 3.**The hydraulic installation used for the velocity measurement's testing.

This means that the data acquisition set is responding "no" almost all of the time, except of the situations where the labeled particle is just in the particular $S_i$ cross-section. Such binary information is practically not influenced by the background radiation. Moreover, due to the mixture recirculation in the closed loop, the recorded data, allowed at least 30 times to observe the labelled grain passing through the measuring section. Due to that, the set of fluctuated residence times, shown in Fig. 4 was recorded. Their histogram was very similar to the residence time distribution $E_{12}(t)$ presented in Fig. 5.

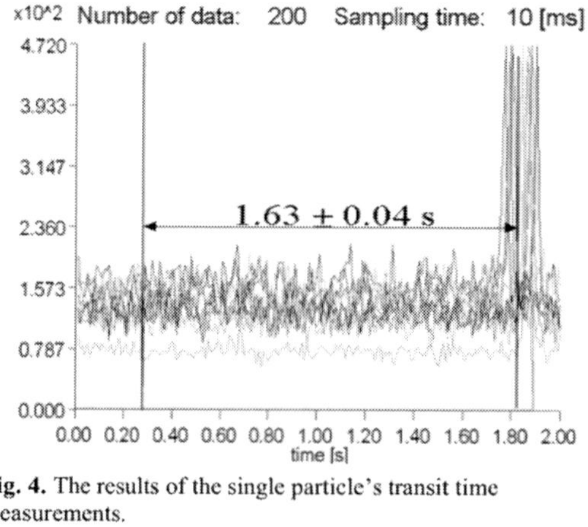


**Fig. 4.** The results of the single particle's transit time measurements.

## 4 Traditional solid particle's velocity measurement

The mostly applied radiotracer's velocity measurement based on the join injection of the sample, which in our case consisted of 30 grains in roughly 10 mm, labelled by the radioactive isotopes. Moreover, in that case, also the probes 1 and 2 may record recirculation of total radioactive sample.

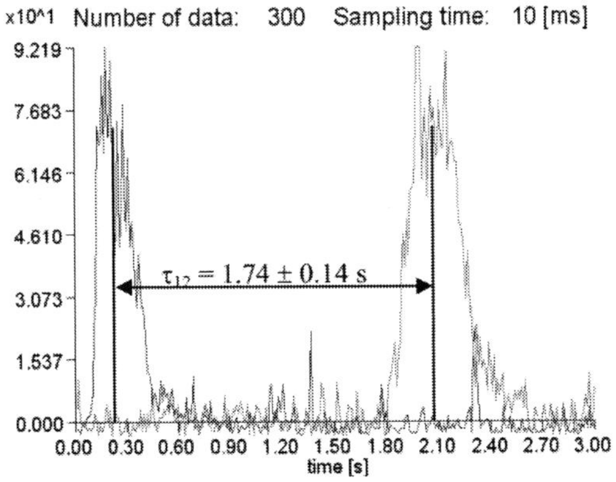


**Fig. 5.** The first peaks recorded by probes 1 and 2 in the measurement LA4003.

It is worth to notice that the ideal injection of the tracer in the examined flow should have a form of $\delta(t)$ (Dirac function of time, but uniformly distributed in total cross-section of a stream). In practice, this condition of injection is difficult to fulfill, nevertheless it is readily applied due to limited amount of the tracer activity and quick measurement. Often the assumption of impulse injection is satisfied when the tracer is introduced in short time in relation to mean axial velocity and much shorter than its residence time distribution $E_i(t)$ in the mixing distance $L_1$.

Though as yet, in practice it is hard to guarantee uniform distribution of the tracer in the cross-section of the stream, but the same effect may be achieved by assuring sufficiently long $L_i$ distance which should allow even distribution of the tracer and give the same velocity as for the surrounding particles.

In the presented investigation, the above conditions were fulfilled with the use of radiotracers' injection into the inlet of the pump. Subsequently in the result of proper selection of tracer and measurement's geometry, the linear relation between tracer concentration in the $i$ cross-section $C_i(t)$ and count rate originated from the tracer $I_i(t)$ may be obtained:

$$C_i(t) = DI_i(t) \tag{5}$$

and for the $i$ probe

$$I_i(t) = I_{mi}(t) - I_{Bi} \tag{6}$$

where:
$D_i$ – calibration coefficient,
$I_i(t)$ – effective counts rate originated by the tracer,
$I_{Bi}$ – count rate representing the background radiation.

The classical tracer velocity measurement mostly applies the moment's method. The mean axial velocity in this way may be calculated according to Danckwerts by the residence time distributions of tracer derived from the following equations [9]:

$$E_i(t) = \frac{C_i(t)}{\int_0^\infty C_i(t)dt} = \frac{I_i(t)}{\int_0^\infty I_i(t)dt} \tag{7}$$

and the ordinary $j$ moment of the above distribution

$$M_{ij} = \int_0^T t^j E_i(t)dt \tag{8}$$

where $T$ – time of observation.

In practice, the $T$ should last until significant part of the tracer will leave the process, so when in consequence the radiation emitted by the tracer may be neglected.

For these calculations, the most important is the first moment, representing mean residence time of the tracer in the mixing section of the stream $L_1$:

$$\tau_i = \int_0^T tE_i(t)dt \tag{9}$$

Similarly $\tau_2$ may be calculated for the section $L_2$ and then mean transportation delay:

$$\tau_{12} = \tau_2 - \tau_1 \tag{10}$$

according to symbols presented in Fig. 1,

$$v = L/\tau_{12} \tag{11}$$

and the central $j$ moment of $E_i(t)$:

$$\mu_{ji} = \int_0^T (t-\tau_i)^j E_i(t)dt \tag{12}$$

The velocities uncertainty $u(v_i)$ may be approximated by $\sigma_i$ derived from the second central moment:

$$\sigma_i^2 = \mu_{2i} = \int_0^T (t-\tau_i)^2 E_i(t)dt \tag{13}$$

In consequence, the uncertainty of the transportation delay, according to Petryka and Keblowski [10]:

$$u(\tau_{12}) = \sqrt{(\Delta\tau_1)^2 + (\Delta\tau_2)^2} \tag{14}$$

for the data delivered by the first probe, when low indexes denote to the moments order:

$$\Delta\tau_1 = \sqrt{\left(\frac{\Delta M_1}{M_0}\right)^2 + \left(\frac{\Delta M_0}{M_0^2}M_1\right)^2} \tag{15}$$

where

$$\Delta M_1 = \sqrt{M_2 + \left[\frac{1}{2}\Delta t n(n+1)\sigma_B\right]^2} \qquad (16)$$

and

$$\Delta M_0 = \sqrt{M_0 + (n\sigma_B)^2} \qquad (17)$$

$$\sigma_B = \sqrt{\frac{\sum_{i=1}^{n}(I_{Bi} - I_B)^2}{n(n-1)}} = \sqrt{\frac{\sum_{i=1}^{n} I_{Bi}^2 - nI_B^2}{n(n-1)}} \qquad (18)$$

$$I_B = M_{B0}/n \qquad (19)$$

$n$ – total number of the background data,
$M_{BO}$ – zero moment of the background.

In the same way $u(\tau_2)$ may be calculated, based on data delivered by the second probe.

The example illustrating the above calculations was selected from the measurements conducted in the Main Mining Institute Laboratory in Laziska, Poland. The schema of this installation is presented in Fig. 3. During applied measurement the flow consisted of sandstones transported by water. As the tracer was selected a sample consisting of 30 sandstone grains in roughly 10 mm, activated in the Nuclear Reactor, so radiation of this marker was emitted mostly by $^{24}$Na and $^{46}$Sc. The established distance between probes $L$ = 20.350 m, provided transportation delay $\tau_{12}$ = 1.74 ± 0.14 s, and mean velocity $v_{12}$ = 11.6 ± 0.9 m/s. The data recorded during that experiment is shown in Fig. 5, and deliver results illustrating the described method according to Petryka and Keblowski [ibid.]. An example of $I_i$(t) distributions obtained in radiotracer measurement in Laziska Laboratory, is shown in Fig. 6 [ibid.].

## 5 Deconvolution calculations

In the deconvolution method, the radiation recorded by the first probe $I_{m1}$(t) may by assumed as the input tracer to the pipes' measuring section $L$.

Actual radiotracer investigations provided signals $I_{m1}$(t) and $I_{m2}$(t) originated by the tracer but also containing noise. Some part of that noise is caused by measuring equipment and electronics. Completed measurements showed that the noise level in recorded signals may be significantly reduced by an improvement of the measuring geometry and proper sampling time selection, Petryka [11].

Moreover, the deconvolution method is sensitive mostly to the high frequency disturbances, so its influence may be reduced by cutoff frequencies introduced to both $I_{mi}$(t) signals. These filtrations give $I_{m1}$'(t) and $I_{m2}$'(t) signals, according to Proakis and Manolakis [12]. Result of this procedure application to the La4003 measurement is presented in Fig. 7 and Table 1.

According to that procedure, the residence time distribution $E_{12}$(t) may be derived from the Duhamel integral:

$$I_{m2}'(t) = \frac{1}{A_f}\int_0^t I_{m1}'(t-\xi)E_{12}(\xi)\,d\xi = I_{m1}'(t) * E_{12}(t) \qquad (20)$$

where:
$A_f$ – normalizing coefficient,
$I_{m1}$'(t) and $I_{m2}$'(t) – signals representing tracer, prepared for convolution calculations.

In practice, the above integral might be solved with help of the Fourier transform.

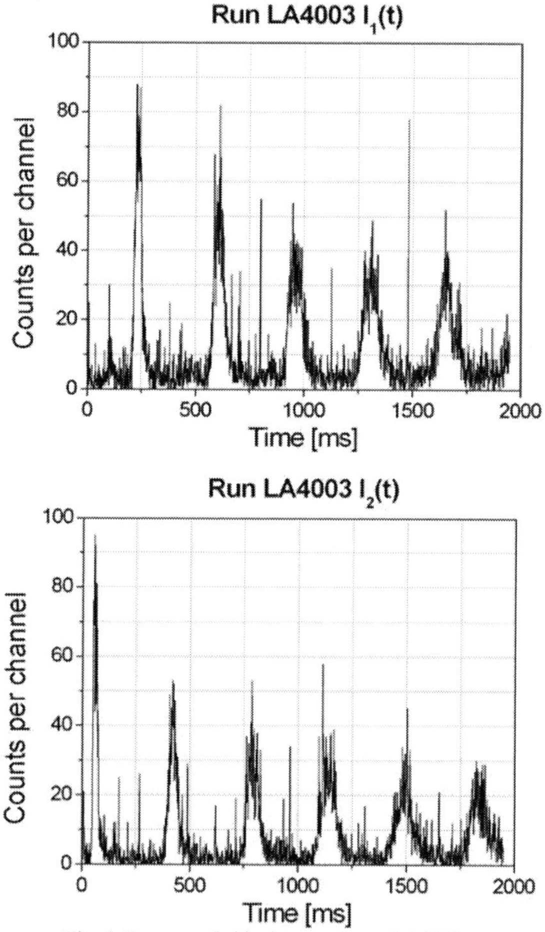

**Fig. 6.** Data recorded in the experiment LA4003.

For signal from the first probe:

$$I_{m1}'(\omega) = \int_0^{\infty} I_{m1}'(t)e^{-i\omega t}\,dt \qquad (21)$$

Similarly, the signal from the second probe may be transformed and the residence time distribution in the frequency domain may be obtained from the following equation:

$$E_{12}(\omega) = A_f \frac{I_{m2}'(\omega)}{I_{m1}'(\omega)} \tag{22}$$

and at last:

$$E_{12}(t) = \frac{1}{2\pi i} \int_{-\infty}^{\infty} E_{12}(\omega) e^{i\omega t} d\omega \tag{23}$$

Distribution obtained in this way may be directly applied to (9) for the mean residence time $\tau_{12}$ evaluation, spent for transportation of tracer through the distance $L$.
As a result, the mean velocity may be obtained from (11) and interpreted in the same way as one described in the following chapter.

## 6 Cross-correlation analysis

The Cross-correlation method recalls directly to the "single particle experiment" when the same, small statistical fluctuations of the tracer concentration $C_1(t)$, may be found in the $C_2(t)$ distribution after a delay necessary for the tracer transportation. The properly established measurement provides $I_{m1}(t)$ and $I_{m2}(t)$ signals related to both concentrations by (5) and (6). After removing low frequency part and high frequency disturbances, it is possible to receive $I_{m1}''(t)$ and $I_{m2}''(t)$ signals which would be ergodic and still related to the small statistical fluctuations of the tracer concentration, as it was proposed by Petryka and Oszajec [13].

For sufficiently long observations $T$, the time $\tau$ spent for transportation may be observed by the following cross-correlation distribution:

$$R_{12}(\tau) = \frac{1}{T}\int_0^T I_{m1}''(t) I_{m2}''(t-\tau)dt = \tag{22}$$
$$= I_{m1}''(t) \otimes I_{m2}''(t)$$

The above interpretation of the cross-correlation distribution suggests similarities between both $R_{12}(t)$ and $E_{12}(t)$ distributions, as it was proposed by Petryka [14].
Due to that, evaluation of the mean residence time $\tau_{12}$ in this method may be calculated from an equation similar to (7) and consequently the mean velocity may be obtained from (9) and compared to the previous methods. Described above procedures were tested in a numerous radiotracer experiments and one of them is accessible in the following chapter.

## 7 Example of velocity measurement

Distributions $I_{m1}(t)$ and $I_{m2}(t)$ recorded during experiment LA4003 are given in Fig. 6. In both parts of that figure it is easy to recognize the recirculation of the marked sand sample, and find that the maximal count rate reaches almost 10000 cps where sampling time was established to 10 ms.

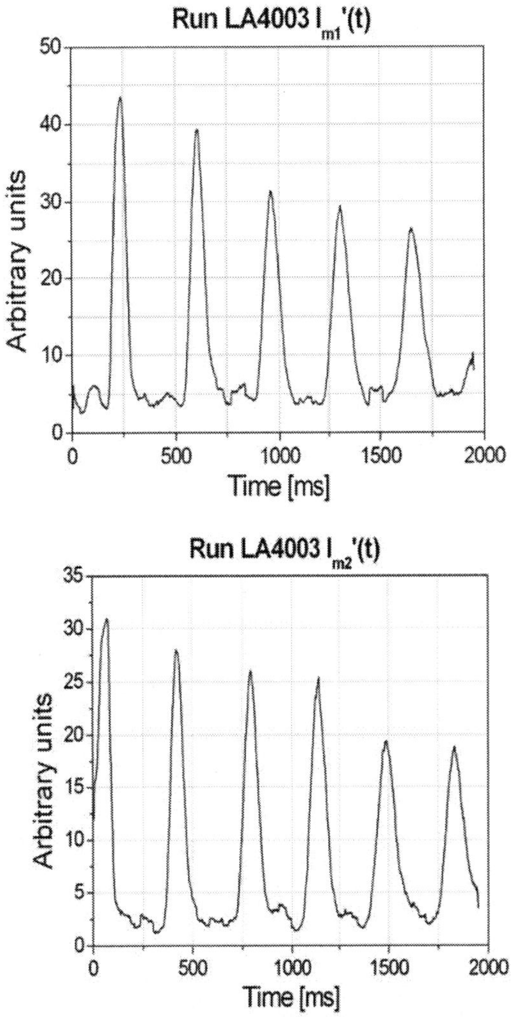


**Fig. 7.** Signals prepared for deconvolution calculations in the experiment LA4003.

In result of the Fourier transform according to (19) and after elimination of high frequencies above 65 internal units, both distributions $I_{m1}'(t)$ and $I_{m2}'(t)$, shown in Fig. 7 were obtained. Consequently, by deconvolution calculations of these data according to (18) up to (21), the residence time distribution $E_{12}(t)$ was derived.
Similarly, signals for correlation analysis were prepared. In this case, both low frequencies below 9 units and high one above 81 units were removed, producing signals $I_{m1}''(t)$ and $I_{m2}''(t)$ shown in Fig. 8.

Results of both approaches give very similar distributions shown in Fig. 9. Their processing by (9) and (11) provided results collected in Table 1. The uncertainty of transportation delay in case of the last two methods was evaluated as half of the selected sampling time. Consequently the relative uncertainty was obtained from the following equation:

$$\delta = \frac{u(\tau_{12})}{\tau_{12}} \cdot 100\% \tag{23}$$

 DOI: 10.1051/epjconf/201714302092

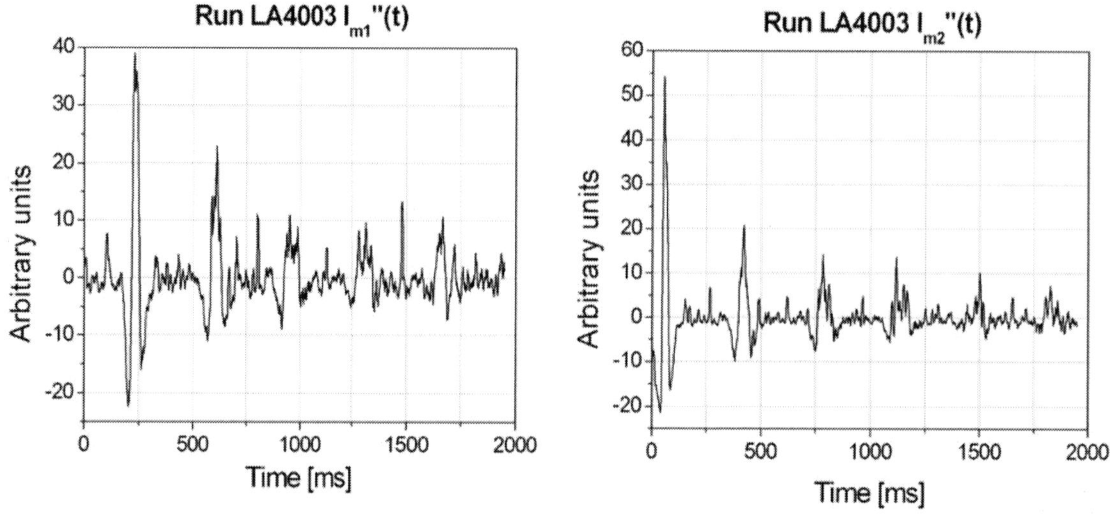


**Fig. 8.** Signals prepared for correlation calculations in the experiment LA4003.

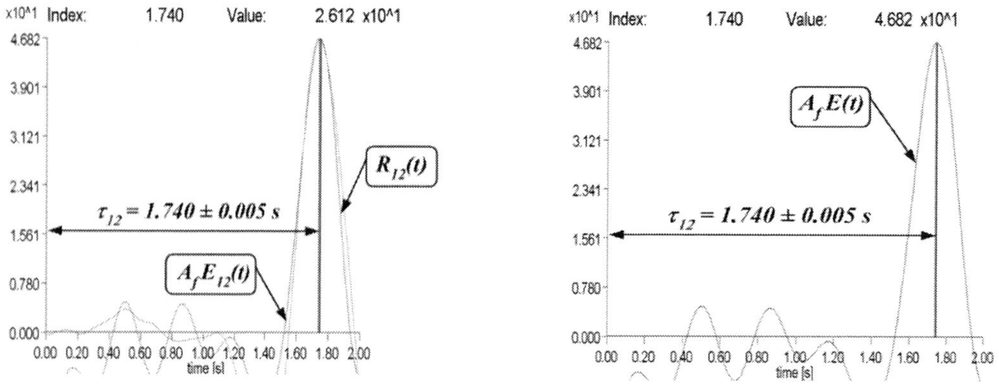


**Fig. 9.** Comparison of the cross correlation $R_{12}(t)$ with Residence Time Distribution $E_{12}(t)$.

Table 1. Comparison of the experiment's LA4003 results

| Method | $\tau_{12}$ [s] | $v$ [m/s] | $\sigma$ [%] |
|---|---|---|---|
| Moments | 1.74 ± 0.14 | 11.6 ± 0.9 | 8 |
| Deconvolution | 1.740± 0.005 | 11.69 ± 0.03 | 0.3 |
| X-correlation | 1.740± 0.005 | 11.69 ± 0.03 | 0.3 |

The presented experiment illustrates application of the proposed procedures in comparison to the traditional one. It is easy to recognize how in this case, the propose processing of the same experiment's data, improves 30 times accuracy of the velocity measurement. Moreover, it allows better interpretation of tracer's experiments and extending range of possible flow investigations.

## 8 Conclusions

The presented methods and described experiment, as well as attached literature, allow to formulate the following proposals:

- radioisotopes may profitably supplement traditional method of multiphase flow analysis [15-17],
- reduction of the necessary assumption demanded for radiotracer experiments allows applying easier measuring procedures and extending their range of application [18-25],
- huge development in radiotracer methods, especially in flow velocity measurements, allows both reduction of the radioactivity and increase of the measurements' accuracy [20, 26-29],

- in many cases, application of radioisotopes may bring smaller hazard due to mass of radioisotopes reduction [30-33],
- in some measurements radiotracers may be replaced by sealed radioactive sources [10, 16, 26]
- equipment based on a pair of sealed radioactive sources appeared to be convenient for two-phase flow control [34-36],
- simultaneous application of tracers and sealed sources gives possibility of more detailed analysis of complex flows [37-40].

## Acknowledgements

All presented investigations were possible due to support under the project P105/10/1574 of the Grant Agency of the Czech Republic, and RVO: 67985874 of the ASCR, as well as the Project No N N523 755340 of the National Science Centre of Poland, which are gratefully acknowledged.

## References

1. C.M. Allen, E.A. Taylor, Trans. Am. Soc. Mech. Eng. **45**, 285-341 (1923)
2. J.S. Charlton et al. *Guidebook on Radioisotope Tracers in Industry* (IAEA Tech. Rep. Ser. 316, Vienna, 1990)
3. C.G. Clayton, J. Brit. Nucl. Eng. Soc. 252-68 (1964)
4. C.W. Sheppard, *Principle of the Tracer Method* (John Wiley and Sons, NY, 1962)
5. G. Falcone, G.F. Hewitt, C. Alimonti, *Multiphase flow metering: principles and applications* (Elsevier, Amsterdam, 2009)
6. G. Johansen, P. Jackson, *Radioisotope gauges for industrial process measurements* (John Wiley and Sons, NY, 2004)
7. V. Mosorov, Appl. Rad. Isotop. **103**, 65 (2015)
8. V. Mosorov, M. Zych, R. Hanus, L. Petryka, Appl. Rad. Isotop. **112** 136–140 (2016)
9. P.V. Danckwerts, Chem. Eng. Sci. **2**, 1 (1953)
10. L. Petryka, R. Keblowski, Precise velocity measurements in two-phase flow by means of sealed radioactive sources. Report of Contract 9326 to IAEA 1-28 (1999)
11. L. Petryka, *Proc. of Int. Conf. on Transport and Sedimentation of Solid Particles*, Prague Jan. 24-26, E5-1 ÷ E5-6 (1995)
12. J.G. Proakis, D.G. Manolakis, *Digital Signal Processing* (Macmillan Publishing Company, N. Y. 1992)
13. L. Petryka, J. Oszajec, Nucl. Geoph. **7**, 323-333 (1993)
14. L. Petryka, *IAEA Consultant's Meeting on "Radiometric cross-correlation techniques for flow rate measurement"*, Krakow, Poland, 1 – 31, (2001)
15. B.K. Arvoh, R. Hoffmann, M. Halstensen, Flow. Meas. Instrum. **23**, 56-65 (2012)
16. L. Petryka, et al, *EPJ WoC* **92,** 02065 (2015)
17. M. Kutz (ed.), *Handbook of Measurement in Science and Engineering* (John Wiley and Sons, NY, 2013)
18. J. Sobota, S. Boczarski, L. Petryka, M. Zych, *Proc. of OMS-2005,* Changsha, Hunan, China, Oct. 9-13, (2005)
19. J.S. Bendat, A.G. Piersol, *Random data - analysis and measurement procedures* (John Wiley and Sons, NY, 2010)
20. L. Petryka, M. Zych, R. Hanus, J. Sobota, P. Vlasak, *Proc. of 10th ISOPE OMS-2013*, Szczecin, Poland, Sept. 22-26, 230-233 (2013)
21. L. Petryka, M. Zych, R. Murzyn, Nukleonika **50**, 43–46 (2005)
22. R. Hanus, L. Petryka, M. Zych, Flow Meas. Instrum. **40**, 58–63 (2014)
23. R. Hanus, M. Zych, L. Petryka, V. Mosorov, P. Hanus, *EPJ WoC* **92,** 02020 (2015)
24. http://www.oilsandsmagazine.com/
25. P. Vlasak, Z. Chara, J. Krupicka, J. Konfrst, J. Hydrol. Hydrom. **62**, 241-247 (2014)
26. L. Petryka, *Keynote Lecture. Proc. 12th Int. Conf. on T&S,* Prague, Czech Republic, Sept. 20-24, Vol. 1, 59-74 (Ed. AR, Wrocław, 2004)
27. O. Dubé, D. Dubé, J. Chaouki, F. Bertrand, Appl. Radiat. Isot. **89**, 109 (2014)
28. J.S. Charlton, *Radioisotope Techniques for Problem Solving in Industrial Process Plants* (Leonard Hill, Glasgow and London, 1986)
29. S.-H. Jung, J.-S. Kim, J.-B. Kim, T.-Y. Kwon, App. Appl. Rad. Isotop. **67** 1254–1258 (2009)
30. L. Furman, L. Petryka, Z. Stęgowski, A. Wierzbicki, Nucl. Instr. & Meth. in Phys. Res. B. **211** (3) 436-442 (2003)
31. J. Max, Methodes et techniques du traitement du signal et application aux mesures physiques, Vol 1, (Paris 1972)
32. F. Mesch, R. Fritsche, H. Kipphan, J. Dyn. Meas. Control. ASME Trans. **96** 414-20 (1974)
33. L. Petryka, M. Zych, R. Murzyn, *Tracers and tracing methods,* Ciechocinek, Poland, June 22-24, 34-39 (2004)
34. K. Przewłocki, L. Furman, L. Petryka, Z. Stęgowski, J. Kamiński, Hydrol. Hydromech. **44** (1) 3-11(1996)
35. G.J. Roach, T.S. Whitaker, M.J. Millen, Duet Multiphase Flow Meter. The APEA Journal **40** Part 1, 492-498 (2000)
36. H. Santos-Cottin, These, Universite Paul Sabatier, Toulouse (1976)
37. J. Sobota, S. Boczarski, L. Petryka, *Proc.of 4th ISOPE* IOM, Szczecin Poland, 127-131 (2001)
38. J. Thyn, R. Zitny, J. Kluson, T. Cechak, *Analysis and diagnostics of industrial processes by radiotracers and radioisotope sealed sources I* (CTU Prague, 2000)
39. J. Thyn, R. Zitny, *Analysis and diagnostics of industrial processes by radiotracers and radioisotope sealed sources II* (CTU Prague, 2002)
40. R. Zitny, *Tracers and tracing methods,* Ciechocinek, Poland, June 22-24, 52-61 (2004)

EPJ Web of Conferences **143**, 02093 (2017) DOI: 10.1051/epjconf/201714302093
*EFM 2016*

# The characteristics of a pneumatic muscle

*Dawid* Pietrala

Department of Automation and Robotics, Kielce University of Technology, 25-314 Kielce, Poland

**Abstract.** The article presents static and dynamic characteristics of pneumatic muscles. It presents the structure of the laboratory stand used to test pneumatic muscles. It discusses the methodology for determination of static and dynamic characteristics. The paper also illustrates characteristics showing the relationship of pneumatic muscles length and operating pressure, at a constant loading force (isotonic characteristics). It presents characteristics showing the relationship of pneumatic muscles shortening and values of loading forces, at a constant operational pressure (isobaric characteristics). It also shows the dependence of force generated by the muscle on the operating pressure, at a constant value of pneumatic muscles shortening (isometric characteristics). The paper also presents dynamic characteristics of a pneumatic muscle showing the response of an object to a gradual change in the operating pressure, at a constant loading force acting on the pneumatic muscle.

## 1 Introduction

The pneumatic artificial muscle is a type of single acting pneumatic actuator. It is made by placing a stretchable bladder inside of the elastic braid. Both elements are connected by permanent ends. One of them is used to supply pressurized air into the muscle. The increase in internal pressure of the muscle causes an increase in its diameter, and thus its shortening. The change in length of the muscle can be used as an operating movement in order to drive the machinery. The return of the muscle to the starting position must be achieved by the application of external forces (spring return, the force of gravity or the second pneumatic artificial muscle) [1,2].

## 2 The methodology for determining the characteristics of pneumatic muscles

The work of a pneumatic muscle is characterized by three basic parameters: internal pressure $P$, loading force $F$ and relative shortening $\varepsilon$. Relative shortening, or the degree of muscle shortening, is expressed as the percentage dependence of the change in the muscle length with respect to its nominal length [1,3,4].

$$\varepsilon = \frac{L - L_n}{L_n} \cdot 100\% \qquad (1)$$

where:
$\varepsilon$ – relative shortening of the muscle,
$L$ – actual length of the muscle,
$L_n$ – nominal length of the muscle.

The static isotonic characteristic shows the dependence of muscle shortening $\varepsilon$ on changes in the internal pressure $p$ at a constant loading force $F = const$. This characteristic was determined in the following manner: one end of the muscle was securely fixed to the stationary base. The other end of the muscle was affected by a known constant loading force, acting in the muscle axis. The return of that force was directed in a way that stretched the muscle. Then the internal pressure of the muscle was increased and the value of shortening degree was read. The pressure was reduces after reaching the maximum pressure and the value of shortening degree was read again. This procedure was repeated for successive values of loading forces.

The static isobaric characteristic shows the dependence of muscle shortening $\varepsilon$ on changes in the loading force $F$, at a constant internal pressure $p = \text{const}$. When determining this characteristic one end of the muscle was also permanently immobilized. The muscle was subjected to a known constant internal pressure. Then the loading force of the muscle was gradually increased and the shortening degree was read. The force was gradually reduced after reaching the maximum loading force and the value of shortening degree was read again. This procedure was repeated for successive values of internal pressure of the muscle [1,5,6].

The static isometric characteristic shows the dependence of loading force $F$ on changes in the internal pressure $p$, at a constant shortening of the muscle $\varepsilon = const$. This characteristic was determined in the following manner. One end of the muscle was permanently immobilized. The other end of the muscle was immobilized in a way that made it possible to read the actual length of the muscle, and thus to read the actual shortening degree of the muscle. Then the internal pressure of the muscle was gradually increased and the actual force generated by the muscle was read. The pressure was gradually reduced after reaching the

* Corresponding author: dpietrala@tu.kielce.pl

maximum value and the actual force generated by the muscle was read again. This procedure was repeated for successive shortening degrees of the muscle.

The dynamic characteristics of pneumatic muscles illustrate shortening of the muscle $\varepsilon$ in a time function $t$, as a response of the object to a gradual change of internal pressure of the muscle $p$, at a constant loading force $F = const$ [7–10].

## 3 The laboratory stand used to test pneumatic muscles

The characteristics presented in the article were determined on a laboratory stand for testing pneumatic muscles. This stand was designed and made by the author of this article. It is adapted to conduct studies on an individual muscle or pairs of muscles operating antagonistically. The stand makes it possible to determine static and dynamic characteristics of pneumatic muscles. The scheme and actual view of the stand was shown in fig. 1. The stand includes the supporting structure, linear guides that adjust the stand to the length of the tested muscle and bearing block with gear belt drive, turning the plane motion of the muscle on the rotary motion. An important element of the stand is the AC MT-C06401C2NT3D servomotor from Servotronix with the CDHD-0032AAP1 servo controller and a precise planetary gear [11–14]. It is used to load the axis of a gear belt drive with a known torque, and thereby to load the muscle with the force of known value. The chosen servomotor has the capability of producing a continuous torque with the value of 26 Nm, and an instantaneous torque with the value of up to 78 Nm. This translates into a continuous loading force of the muscle with the value of up to 812 N, or an instantaneous torque with the value of up to 2436 N. In addition, the stand was equipped with a strain gauge with a measuring range of ±1000 N with an analogue voltage output. The change in internal pressure is achieved by means of a piezoelectric proportional pressure valve Hoerbiger Tecno Plus with an analogue voltage input. It allows to change the pressure in range of 0–10 bars. The measurement of the muscle length is implemented as a measurement of the angle of rotation of the gear belt drive by means of an optical absolute encoder Posital/Fraba OCD-S101G-1416-S060-CAW with a resolution of 16bit with the SSI digital output. The control of stand components and data acquisition is performed by means of an electronic circuit with the STM32F407VGT6 microcontroller equipped with circuits of 16 analogue inputs and 16 analogue outputs with a resolution of 16 bit each, and an interface supporting up to 6 sensors with the SSI output [15–18].

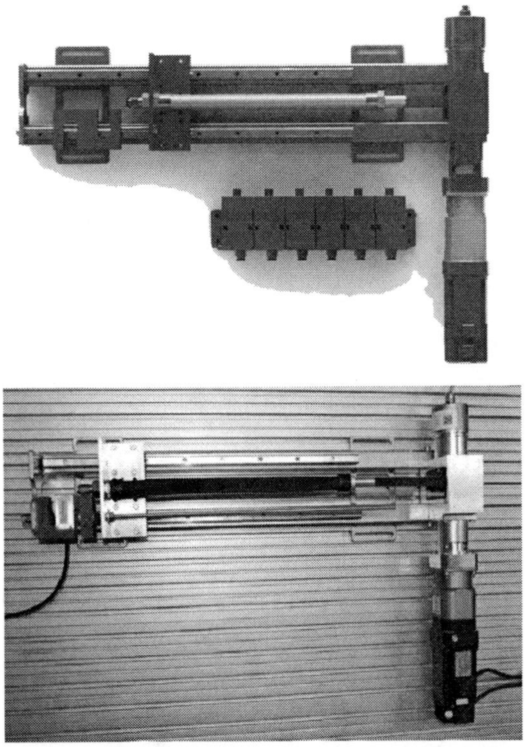

**Fig. 1** The stand used to test pneumatic muscles

## 4 Static and dynamic characteristics of pneumatic muscles

The paper presents static characteristics: isotonic, isobaric and isometric as well as dynamic characteristics of pneumatic muscles. These dependencies were designated on the laboratory stand described above, and the tested object was a pneumatic artificial muscle made according to the developer's project. The muscle contains a latex bladder placed on an elastic braid and modelled on pneumatic muscles from Shadow Air Muscle, which means that the braid is connected to the bladder only at the ends of the muscle. The nominal length of the muscle is 350mm.

Fig. 2 and fig. 3 present a number of static and isotonic characteristics of the tested muscle. These dependencies were designated for loading forces $F = const$ in the range from 50 N to 550 N. Each characteristic was designated by gradual increase of the internal pressure $P$ in the muscle from 0bar to 3bars in 0.1 bar increments [3,4,19].

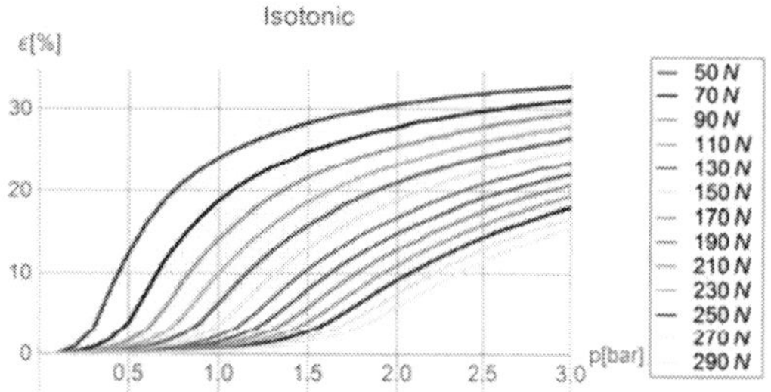


**Fig. 2** Isotonic characteristics for constant loading forces from 50 N to 290 N at a pressure increase.

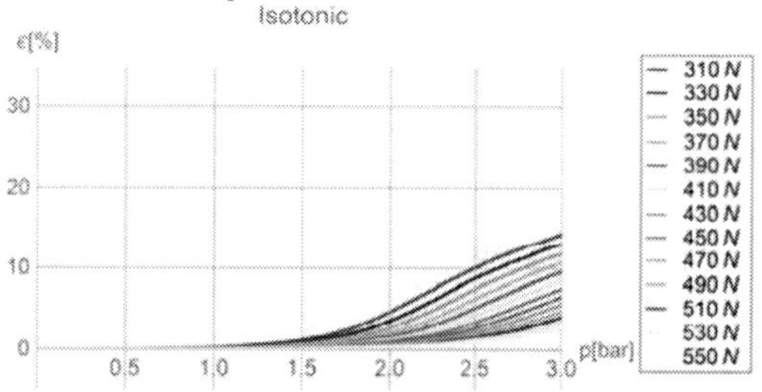


**Fig. 3** Isotonic characteristics for constant loading forces from 310 N to 550 N at a pressure increase.

Fig. 4 and fig. 5 present a number of static isotonic characteristics of the tested muscle for the same loading forces $F = const$. Each characteristic was designated by gradual decrease of the internal pressure $p$ in the muscle from 3bars to 0bar in 0.1 bar increments.

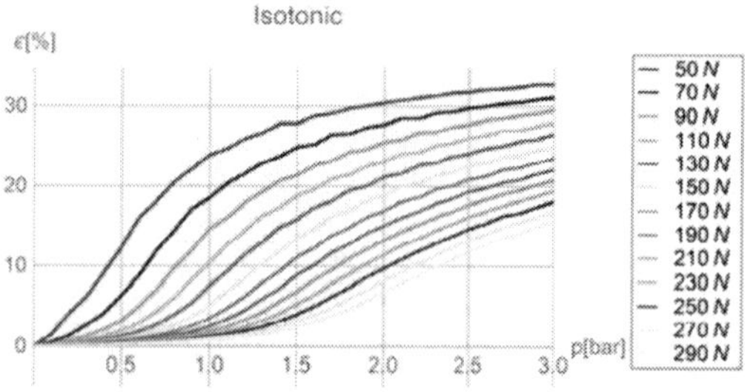


**Fig. 4** Isotonic characteristics for constant loading forces from 50 N to 290 N at a pressure decrease.

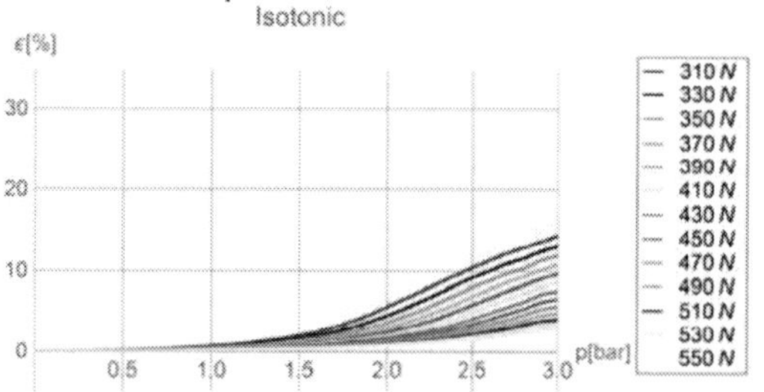


**Fig. 5** Isotonic characteristics for constant loading forces from 310 N to 550 N at a pressure decrease.

When comparing two figures one can see the difference between the corresponding characteristics which is the hysteresis of a working muscle. This means that when the muscle shortens, it moves in a different way than when it increases its length. The hysteresis for the two isotonic characteristics can be seen in fig. 6 and fig. 7.

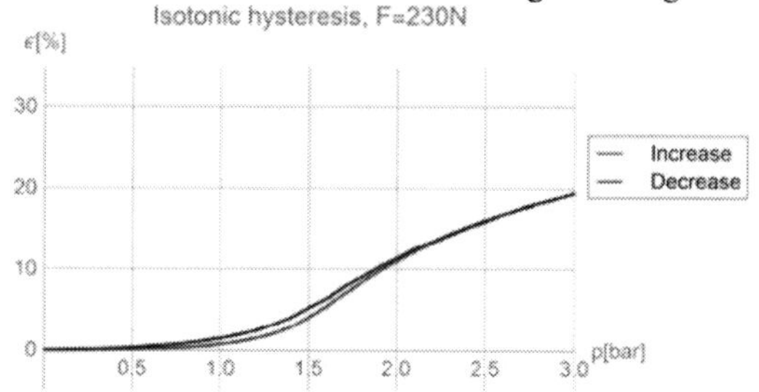


**Fig. 6** Hysteresis for the selected isotonic characteristic.

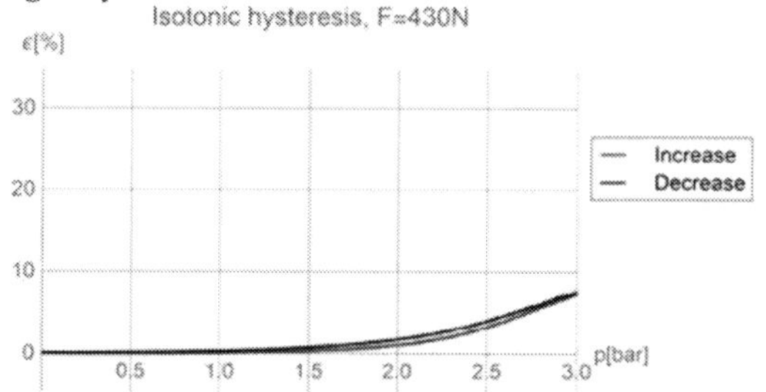


**Fig. 7** Hysteresis for the selected isotonic characteristic.

Fig. 8 and fig. 9 present a series of static isobaric characteristics. These dependencies were designated for constant internal pressures of the muscle $p = const$ in the range from 0bar to 3bars. Each characteristic was designated by gradual increase of the loading force $F$ of the muscle from 50 N to 550 N in 20 N increments.

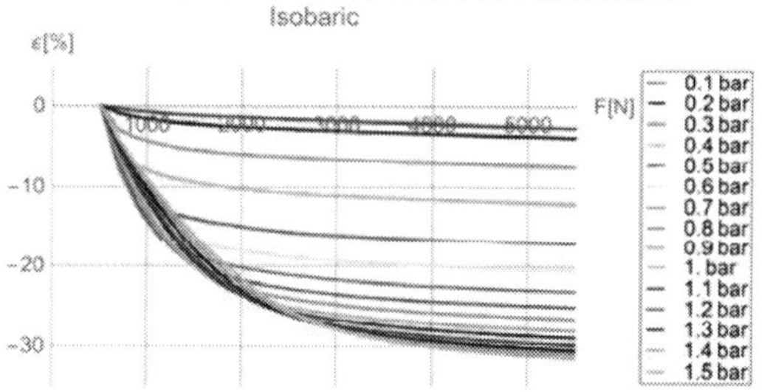


**Fig. 8** Isobaric characteristics for constant internal pressures from 0.1 bar to 1.5 bar at a force increase.

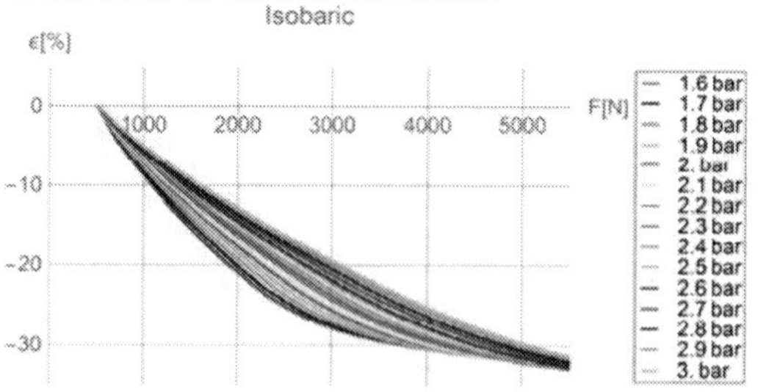


**Fig. 9** Isobaric characteristics for constant internal pressures from 1.6 bar to 3 bars at a force increase.

Fig. 10 and fig. 11 present a number of static isobaric characteristics of for the same constant internal pressures $p = const$. Each characteristic was designated by gradual decrease of loading forces $F$ in the range from 550 N to 50 N in 20 N increments.

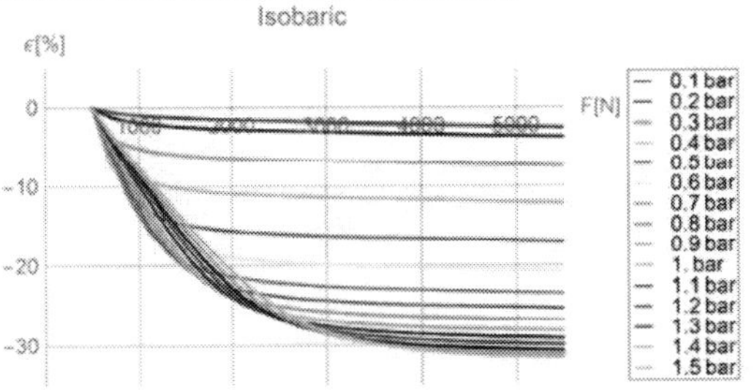


**Fig. 10** Isobaric characteristics for constant internal pressures from 0.1 bar to 1.5 bar at a force decrease.

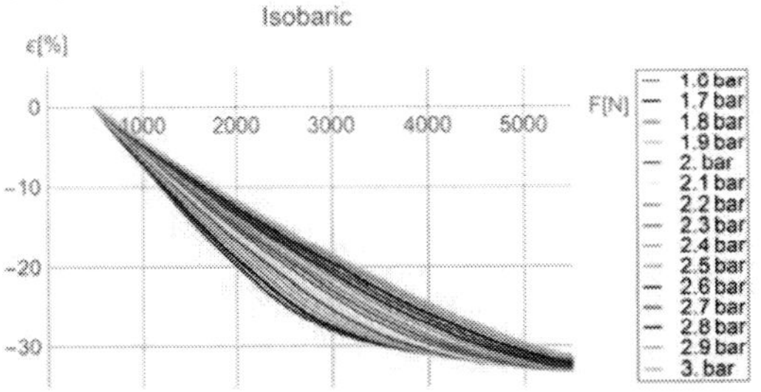


**Fig. 11** Isobaric characteristics for constant internal pressures from 1.6 bar to 3 bars at a force decrease.

As in the case of isotonic characteristics, the isobaric characteristics also show hysteresis between the corresponding dependencies. Hysteresis for two exemplary isobaric characteristics was shown in fig. 12 and fig. 13.

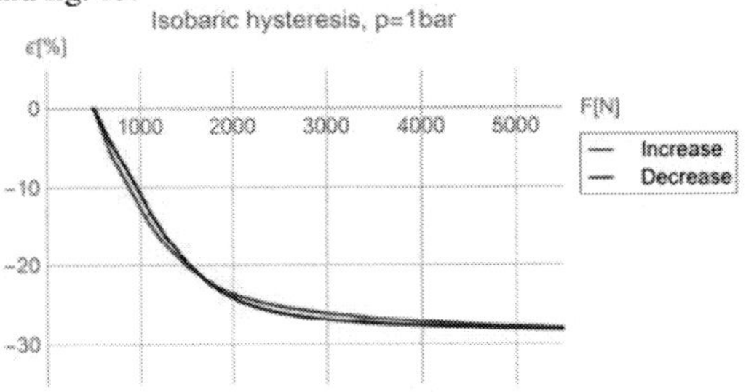


**Fig. 12** Hysteresis for the selected isobaric characteristic.

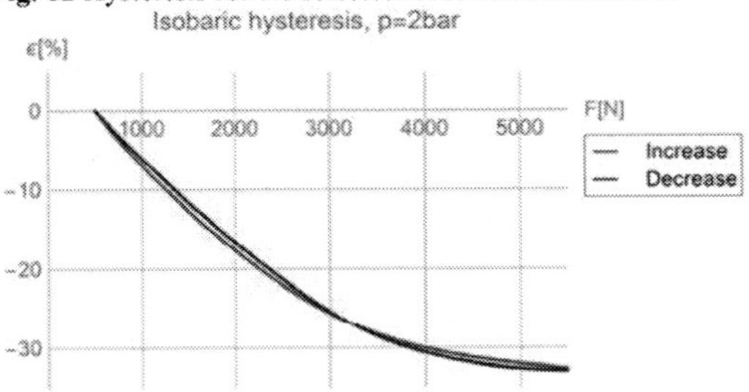


**Fig. 13** Hysteresis for the selected isobaric characteristic.

The paper also presents isometric characteristics showing the dependence of force $F$ produced by the muscle in a function of internal pressure $p$ at a constant shortening degree $\varepsilon = const$. Fig. 14 and fig. 15 present a series of isometric characteristics. These dependencies were designated for constant values of the muscle shortening. The muscle length varied from a nominal value of 350 mm to 250 mm in 5 mm increments, which provided the degree of muscle shortening in the range from 0% to 35.71%. Each characteristic was designated by gradual increase of the internal pressure $p$ in the muscle from 0 bar to 3 bars, in 0.1 bar increments.

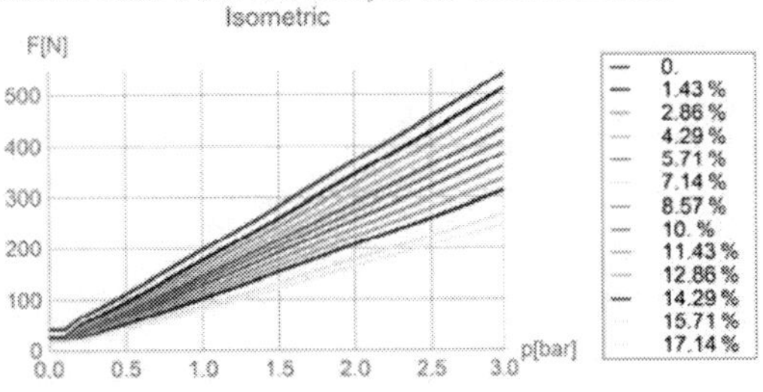


**Fig. 14** Isometric characteristics for constant shortening of the muscle from 0% to 17.14% at a pressure increase.

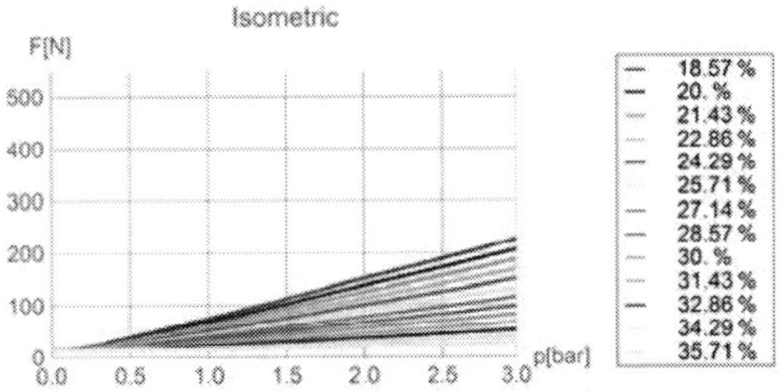


**Fig. 15** Isometric characteristics for constant shortening of the muscle from 18.57% to 35.71% at a pressure increase.

Fig. 16 and fig. 17 present a number of static isometric characteristics of the tested muscle for the same values of the shortening degree $e = const$. Each characteristic was designated by gradual decrease of the internal pressure $p$ in the muscle from 3 bars to 0 bar in 0.1 bar increments.

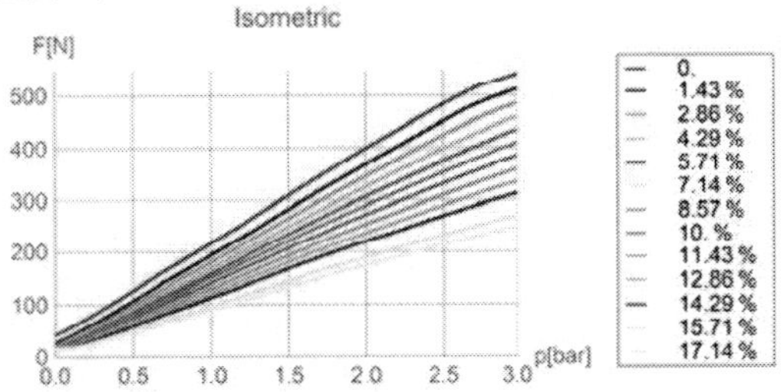


**Fig. 16** Isometric characteristics for constant shortening of the muscle from 0% to 17.14% at a pressure decrease.

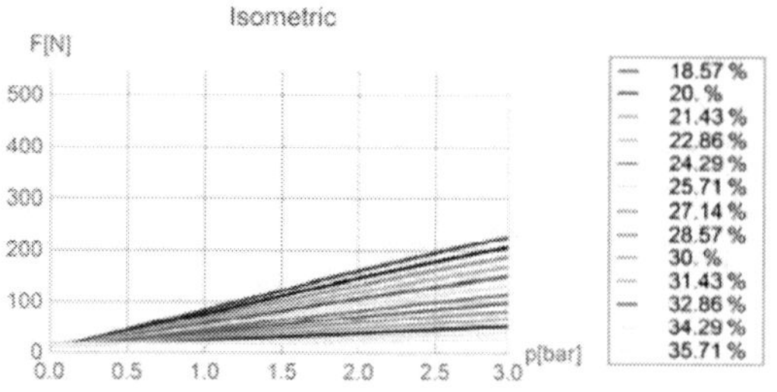


**Fig. 17** Isometric characteristics for constant shortening of the muscle from 18.57% to 35.71% at a pressure decrease.

In the case of isometric characteristics one can also observe the difference between the corresponding characteristics which is the hysteresis of a working muscle. Hysteresis for two selected isometric characteristics can be seen in fig. 18 and fig. 19.

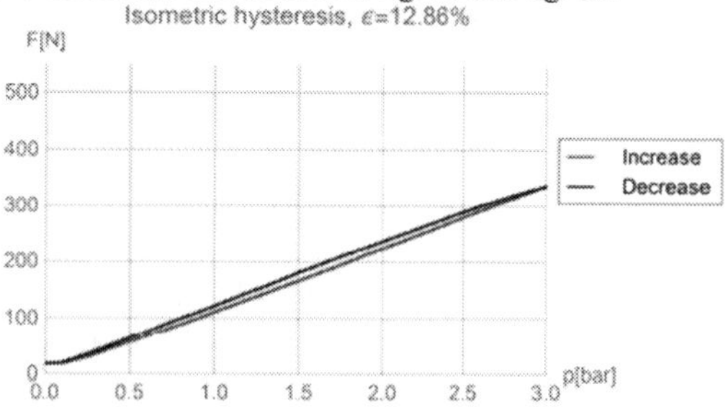


**Fig. 18** Hysteresis for the selected isometric characteristic.

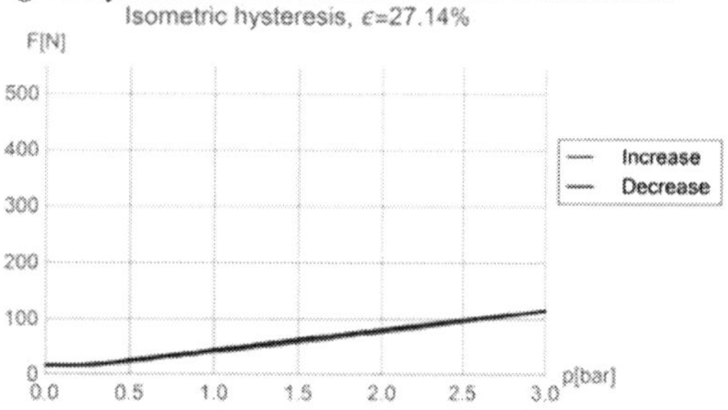


**Fig. 19** Hysteresis for the selected isometric characteristic.

The paper also presents dynamic characteristics of pneumatic muscles. These dependencies show the muscle shortening $\varepsilon$ as the response of the object to a gradual change in the internal pressure $p$ from 0 bar to the set pressure value, at constant values of loading forces $F = const$. Sampling period when determining dynamic characteristics was 5 ms. fig. 20–fig. 23 present the selected dynamic characteristics of pneumatic muscles [3,20].

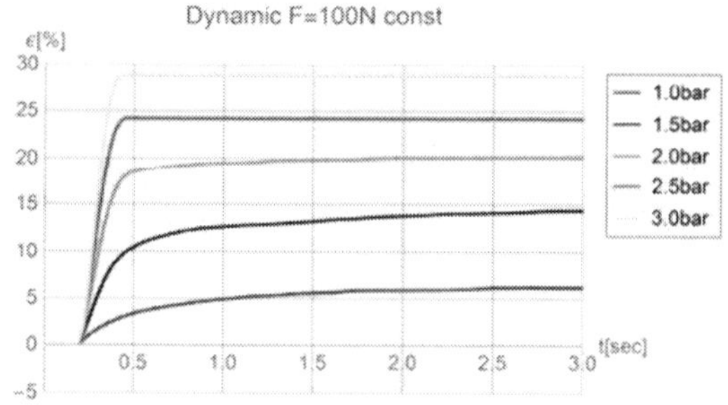


**Fig. 20** The muscle response to gradual changes in pressure from 0bar to the indicated value at a constant loading force F=100 N

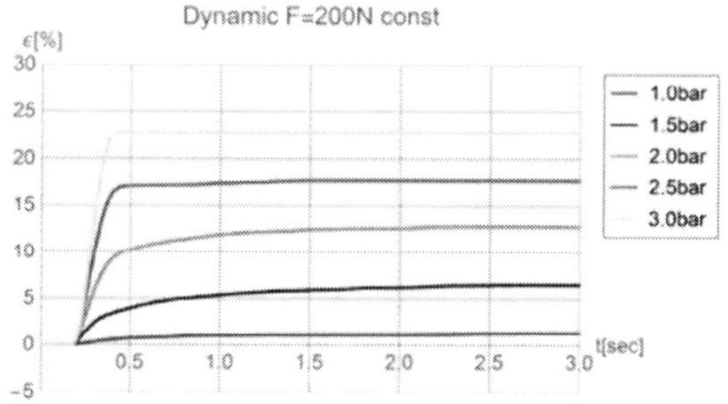


**Fig. 21** The muscle response to gradual changes in pressure from 0bar to the indicated value at a constant loading force F=200 N

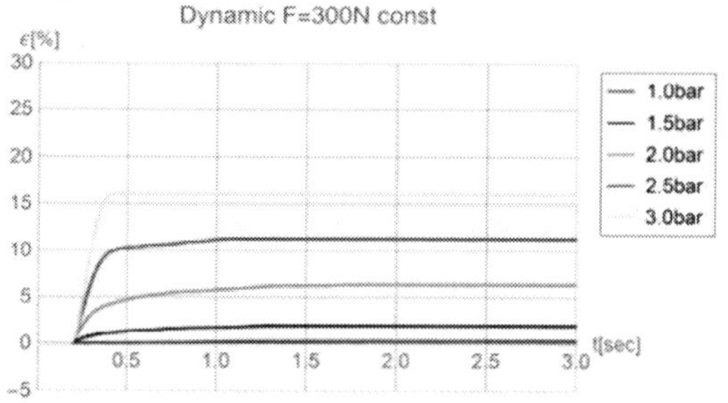


**Fig. 22** The muscle response to gradual changes in pressure from 0bar to the indicated value at a constant loading force F=300 N

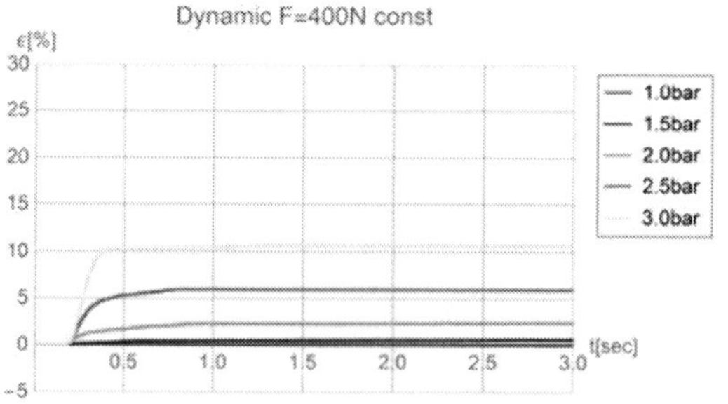


**Fig. 23** The muscle response to gradual changes in pressure from 0bar to the indicated value at a constant loading force F=400 N

## 5 Conclusions

Pneumatic muscles seen as drive elements have a number of features that distinguish them from conventional drives. The total tightness, lack of stick-slip phenomenon and generation of smooth movements have a positive impact on the ability to use the muscles as drive elements. Non-linear nature of the muscles hinders the precise control of such drives. For this purpose, it is extremely important to know all the dependencies that bind basic parameters of pneumatic muscles: the loading force, internal pressure and shortening degree.

The article presents basic concepts concerning pneumatic muscles. It presents methodologies for determining static and dynamic characteristics of muscles. It describes the structure of laboratory stand used to test pneumatic muscles. The paper also presents static and dynamic characteristics of an exemplary pneumatic muscle.

## References

1. P.A. Laski, J.E. Takosoglu, S. Blasiak, (ACAD SCI CZECH REPUBLIC, INST THERMOMECHANICS, DOLEJSKOVA 5, PRAGUE 8, 182 00, CZECH REPUBLIC, 2014).
2. D.S. Pietrala, (ACAD SCI CZECH REPUBLIC, INST THERMOMECHANICS, DOLEJSKOVA 5, PRAGUE 8, 182 00, CZECH REPUBLIC, 2016).
3. J.E. Takosoglu, P.A. Laski, S. Blasiak, G. Bracha, D. Pietrala, *MEASUREMENT & CONTROL*. **49** (2) (2016), 62–71.
4. P.A. Laski, J.E. Takosoglu, S. Blasiak, *Robotics and Autonomous Systems*. (2014),.
5. P.A. Laski, (ACAD SCI CZECH REPUBLIC, INST THERMOMECHANICS, DOLEJSKOVA 5, PRAGUE 8, 182 00, CZECH REPUBLIC, 2016).
6. J.E. Takosoglu, P.A. Laski, S. Blasiak, (ACAD SCI CZECH REPUBLIC, INST THERMOMECHANICS, DOLEJSKOVA 5, PRAGUE 8, 182 00, CZECH REPUBLIC, 2014).
7. S. Blasiak, *Journal of Thermal Science and Technology*. **10** (1) (2015), JTST0016–JTST0016.
8. S. Blasiak, C. Kundera, *Procedia Engineering*. **39** (2012), 315–326.
9. S. Blasiak, C. Kundera, J. Bochnia, *Procedia Engineering*. **39** (2012), 366–378.
10. S. Blasiak, A. Pawinska, *International Journal of Heat and Mass Transfer*. **90** (2015), 710–718.
11. D. Janecki, J. Zwierzchowski, *MEASUREMENT SCIENCE {&} TECHNOLOGY*. **26** (8) (2015),.
12. J. Zwierzchowski, (ACAD SCI CZECH REPUBLIC, INST THERMOMECHANICS, DOLEJSKOVA 5, PRAGUE 8, 182 00, CZECH REPUBLIC, 2016).
13. D. Janecki, J. Zwierzchowski, L. Cedro, *BULLETIN OF THE POLISH ACADEMY OF SCIENCES-TECHNICAL SCIENCES*. **63** (3) (2015), 771–779.
14. E. Miko, L. Nowakowski, (ELSEVIER SCIENCE BV, SARA BURGERHARTSTRAAT 25, PO BOX 211, 1000 AE AMSTERDAM, NETHERLANDS, 2012).
15. D. Janecki, J. Zwierzchowski, (IMEKO, PO BOX 457, H-1371 5 BUDAPEST, HUNGARY, 2009).
16. D. Janecki, L. Cedro, J. Zwierzchowski, *METROLOGY AND MEASUREMENT SYSTEMS*. **22** (2) (2015), 289–302.
17. B. Nowakowski, A.L. Kawczynski, A. Kolbus, A. Lemarchand, *EUROPEAN PHYSICAL JOURNAL B*. **84** (1) (2011), 137–145.
18. A. Nowakowski, M. Cybulski, A. Sliwczynski, A. Chil, Z. Teter, P. Seroczynski, et al., *BMC CANCER*. **15** (2015),.
19. J.E. Takosoglu, R.F. Dindorf, P.A. Laski, *INTERNATIONAL JOURNAL OF ADVANCED MANUFACTURING TECHNOLOGY*. **40** (3-4) (2009), 349–361.
20. G.F. Bracha, (ACAD SCI CZECH REPUBLIC, INST THERMOMECHANICS, DOLEJSKOVA 5, PRAGUE 8, 182 00, CZECH REPUBLIC, 2016).

# The effects of perforated cylinders on the vortex shedding on the suppression of a circular cylinder

*Engin* Pinar[1,*], *Tahir* Durhasan[2], *Göktürk M.* Ozkan[2], *Muhammed M.* Aksoy[2], *Huseyin Akilli*[2], *Besir* Sahin[2]

[1] Department of Mechanical Engineering, Faculty of Ceyhan Engineering, Cukurova University, 01950, Adana/ Turkey
[2] Department of Mechanical Engineering, Faculty of Engineering and Architecture, Cukurova University, 01330, Adana/ Turkey

**Abstract.** The aim of this study is the control of unsteady vortical flow occurred downstream of a circular cylinder located in shallow water flow using concentrically located outer perforated cylinder. The porosities, β have been changed between 0.1 and 0.8 in the present study. The increments of porosity β were taken as 0.05 in the range of 0.1 and 0.8 with a hole diameter of d=10 mm. The ratio of inner cylinder diameter to outer cylinder diameter, $D_i/D_o$ was selected as 0.25, 0.3, 0.4, 0.5 and 0.6 the inner cylinder diameter is $D_i$=50mm where the outer cylinder diameter is $D_d$=100mm. Experiments were performed at a constant depth of the water level as h=50mm (half of the outer cylinder diameter). Free stream velocity was taken as $U_\infty$=100 mm/s corresponding to a Reynolds number of Re $D_o$=10000 based on the outer cylinder diameter. It has been observed that the inner circular cylinder was highly affected by the existence of surrounding outer perforated cylinders. It is observed that the intensity of Reynolds shear stress correlating, $<u'v'/U^2>$ is completely attenuated in the region both downstream of concentric cylinder and between the concentric cylinders. It is determined from the experiments that porosity, β=0.55 is the most effective parameter for control of flow structure that is occurred from the inner cylinder.

## 1 Introduction

A bluff body such as a cylinder which is placed in a flow alters the general characteristics of the flow structure.. This alteration causes more complex and unsteady flow characteristics like resonance and high-amplitude of fluctuating forces and have important implications for flow-induced vibration and noise generation. The source of vibrations and fluctuating forces is the flow separation followed by vortex shedding after a critical value of Reynolds number. The situation is of practical interest in many engineering applications as chimney stacks, long- span bridges, marine structures, tall buildings and towers etc. and has been steadily investigated by the researchers Bearman, [1, 2, 3, 4, 5]; Matsumato [6]; Williamson [7]; Williamson and Govardhan, [8, 9]. In order to reduce or eliminate the adverse effects of vortex-induced vibrations (VIV), it is inevitably important to control the flow structure around the bluff body and/or alter the mechanical properties of the material used in terms of damping and natural frequency. Since the material enhancement might be high-priced, various techniques about the flow control have been studied extensively in recent years. Every et al. [10] stated in their review that the flow can be controlled by means of an alteration of the structural profile in order to affect the separation of the boundary layer to prevent the occurrence of vortex shedding along the length of the structure or alter the formation of the wake vortices and their interactions. Unsteady flow formed downstream of cylinder causes serious problems in many engineering applications. In order to solve the problems mentioned, vortices should be suppressed using various flow control techniques. Price [11] investigated various porous shrouds (a higher diameter cylinder placed coincidently around an inner bare cylinder to understand their effects on suppression of the vortex-excited vibrations. He concluded that the vibration suppression can be achieved by shrouded cylinders. Perforated cylinder shrouds on the flow control has been started to be investigated by other researchers. Every et al. [10] performed a review to point out the practical aspects of vortex excited structural vibrations in a marine environment with particular emphasis on methods of their suppression. Molin [12] worked on a numerical study about perforated shrouds used to reduce vortex- induced vibrations of cylinders. The drag reduction of the circular cylinder was investigated experimentally by Wang et al. [13] in terms of the pressure measurement at a Reynolds number of 82,000 based on the cylinder diameter. A rod was placed upstream of the cylinder to control the flow around the cylinder. The interaction of the rod with the flow around the cylinder was examined to realize the drag reduction mechanism with an upstream rod. Barranow et al. [14] performed a numerical study for controlling of wake flow around a circular cylinder encased in a thin cylindrical perforated sheath. Their studies were carried out unsteady two-dimensional laminar flow conditions. They demonstrated that diverting a part of the flow from

* Corresponding author: epinar@cu.edu.tr

the front stagnation region to the perforation orifices, generating there a flow stabilising the flow field. Sahin [15] carried out an experimental study using perforated plate as a flow control device. He concluded that non-uniformity in the velocity is increased as the porosity is decreased. Turhal and Cuhadaroglu [16] performed experiments to investigate some aerodynamic parameters of the flow around perforated-surface square, horizontal, and diagonal cylinders. The surface pressures of the perforated surface square cylinders were measured for the cases of injections through the top, rear, top-rear, and all surfaces separately at Re values of 10,000, 16,000, and 24,000. The $C_D$ and St were determined based on pressure coefficient measurements and vortex shedding frequency measurements, respectively. Their experimental results showed that the injection through the various surfaces of a square cylinder has some effects on the aerodynamic characteristics of the square cylinder, depending on the position of the perforated surface. They concluded that $C_D$ of a horizontal square cylinder can be reduced by injection through all the surfaces, and for a diagonal square cylinder; it also can be reduced by injection through the rear surfaces. They also revealed that the vortex shedding frequency of the flow around a horizontal square cylinder can also be decreased by high-momentum injection through the rear surface. Gozmen et.al. [17] conducted experimental work on the control of vortical flow of circular cylinder using a splitter plate having different lengths and heights: how a submerged attachment changes the flow structure downstream of bluff bodies in shallow water flow and the possibility of contamination dispersion in transverse direction in shallow flows, due to Karman vortices, be prevented by suppressing the flow oscillations downstream of a bluff body. They found that as the ratio of $h_p/h_w$ increases, the effect of the splitter plate on the suppression of the vortex shedding increases. Flow characteristics and examination of spectra indicate that Karman vortex shedding is attenuated pronouncedly for the cases of $L/D \geq 1$ and $hp/hw \geq 0.75$. The transverse Reynolds normal stress is more effective on the attenuation of turbulent kinetic energy than the stream wise Reynolds normal stress, $<u'u'/U^2>$. The value of peak transverse Reynolds normal stress, $<u'u'/U^2>$ is reduced to 90% of that of the bare cylinder at most. Oruc et. al. [18] were investigated suppression of asymmetric flow behavior downstream of two side-by-side circular cylinders with a splitter plate in shallow water. The results revealed that splitter plate had little effect on the flow structure for $L/D < 3$ such that the wake structure downstream of each of the cylinders was narrow and wide. However, the symmetrical flow pattern was noticed downstream of the cylinders for $L/D = 3$. In these cases, the gap-flow remained unbiased and it was parallel to the flow axis. The peak concentration of Reynolds shear stress, $<u'v'/U^2>$ also diminished substantially. Increasing splitter plate length also caused the location of peak magnitude of $\langle u'v' \rangle$ to move slightly in downstream direction as well as elongation of vorticity contours in stream wise direction. The deflection of the wake and thereby the instability of the wake was considerably prevented with the presence of the splitter plate for which resulted in two well symmetric, stable wakes having approximately the same order of magnitudes of vortex shedding frequencies around the cylinders. Pinar et al. [19] stated in their review that the experimental investigations were carried out in order to have detailed information on the flow structure around perforated cylinders using high-image density Particle Image Velocimetry technique in shallow water flow. In order to analyze the effect of porosity, β on the flow structure, the porosities in the range of $0.1 \leq \beta \leq 0.8$ with an increment of 0.1 were used and the results were compared with the bare cylinder case by means of velocity and vorticity contours, turbulent kinetic energy, Reynolds shear stress and streamline topologies. It was concluded that the porosity, β had a substantial effect on the control of large-scale vortical structures downstream of the cylinder in which the shear layers were elongated, fluctuations were significantly attenuated and formation of Karman Vortex Street was successfully prevented by the use of perforated cylinders.

## 2 Material and method

All experiments were carried out in the test section of water channel which has 8000mm length x 1000mm wide x 750mm deep was constructed from a 15 mm thick transparent plexiglass sheet with upstream and downstream fiberglass reservoirs. Experiments have been studied in a uniform flow by means of both flow visualization with laser illumination of Rhodamine dye technique and digital particle image velocimetry technique (PIV). Visualization of the experimental results were recorded with a high speed SONY 9.2 mega pixels HDR-PJ670 handycam type digital video recorder to examine the flow behaviors in detail. The video camera is used to capture the instantaneous video images of the vortex flow structures. Each experiment is recorded approximately three minutes. The images were analyzed with frame grabber. Digitized images were enhanced for analysis using Adobe Photoshop software. Particle Image Velocimetry technique was used to measure the velocity vector fields. The PIV technique is one of the most reliable methods for flow velocity measurement in the modern fluid mechanics and aerodynamics. By this measurement technique, two-dimensional instantaneous velocity profiles downstream of the cylinder were measured and flow properties were defined by quantitative images such as average vorticity, streamline topologies, and turbulence statistics of the flow. Flow characteristics were examined by the PIV technique in the annular region between the inner and outer cylinders at different values of porosity (β=0.25, 0.4, 0.55, 0.65 and 0.8) and diameter ratios ($D_i/D_o$ =0.25, 0.3, 0.4, 0.5 and 0.6). Figure 1 shows schematic representation for inner cylinder, perforated outer cylinder.

DOI: 10.1051/epjconf/201714302094

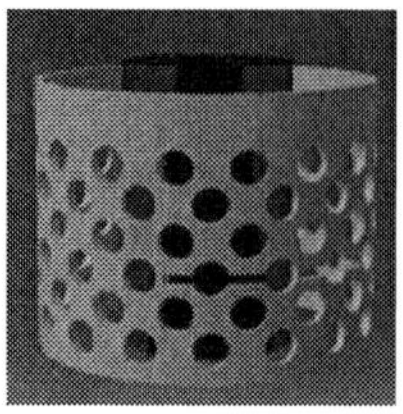

**Fig 1.** Schematic representation for inner cylinder, perforated outer cylinder.

The porosity ratio, β, is defined as the ratio of the gap area on the body to the whole body surface area. Perforated outer cylinder is manufactured from metal plate of 1mm thickness. For 2mm thick laser sheets' illumination of the region between perforated outer cylinder and inner cylinder, 180 degree 1mm thick groove is slotted behind the perforated outer cylinder according to flow direction. Diameter ratio, $D_i/D_o$, is the ratio of inner cylinder diameter to the outer cylinder diameter.

Figure 2 shows representation of water channel. The total depth of the water in the channel was adjusted to a 600mm height. All experiments were carried out above a platform, having a length of 2300mm to obtain fully developed shallow flow conditions. The water height between the base of the platform and the free surface was adjusted as 50mm which corresponds to half of the outer cylinder diameter ($D_o$=100mm). Free surface velocity is U=100mm/s, which corresponds to the Reynolds number of Re $D_o$= 10000 based on the outer cylinder diameter. The measurements were performed and the data were processed using Dantec Dynamics PIV system and Flow Manager Software installed on a computer. The measurement field was illuminated by a thin and an intense laser light sheet by using a pair of double-pulsed Nd: YAG laser units each having a maximum energy output of 120mJ at 532nm wavelength. The laser sheet was oriented parallel to the bottom surface of the water channel and the experiments were carried out at the midsection of the water height. The image capturing was performed by an 8-bit cross-correlation CCD camera having a resolution of 1600 x 1200 pixels, equipped with a Nikon AF Micro 60 f/2.8D lens. In the image processing, 32x32 rectangular interrogation pixels were used and an overlap of 50% was employed. A total of 7128 (99x72) velocity vectors were obtained for an instantaneous velocity field at a rate of 15 frames per second. The time interval between pulses was 1.5 ms for all experiments and the thickness of the laser sheet illuminating the measurement plane was approximately 2 mm. The time interval and the laser sheet thickness were selected such that the maximum amount of particle displacement in the interrogation window was obtained.

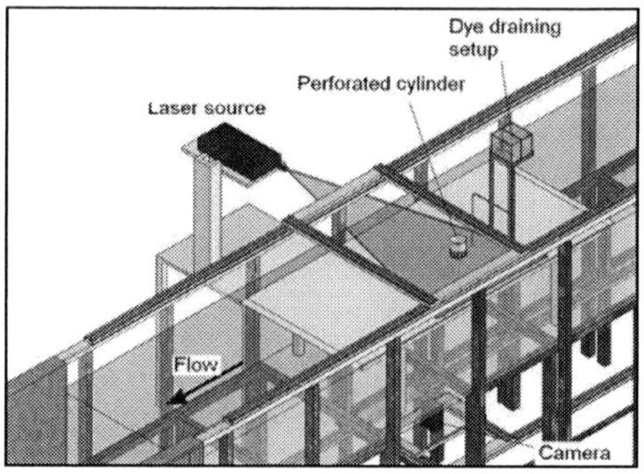


**Fig 2.** Schematic representation of water channel

## 3 Result and Discussion

First of all, flow characteristics for the bare cylinders are investigated in order to have a reference for the controlled cases. Instantaneous dye images for the bare cylinders case obtained using the flow visualization technique for the inner cylinders are shown in Figure 3. The separated and recirculating flow in the near wake region with the help of visualizing is clearly seen in the case of bare cylinders. It is seen from the dye picture that alternating vortex shedding and the well-known Karman Vortex Street is evident downstream of the cylinders. It is shown that the Von Karman vortex street formed along the downstream of the cylinders have a certain frequency which decreases by increasing diameters of the bare cylinders. The wake is highly unsteady and large eddies or vortices are shed from the downstream region of the cylinders. The large eddies are formed at a regular frequency cause pressure disturbances in the flow. In order to control the near wake region of inner cylinder (bare cylinder), perforated cylinders having various porosity, β were used as a control parameters.

Figure 4 presents the result of dye visualization experiments for the porosity, β=0.55. of the perforated cylinder. Von Karman vortices are controlled easily for low diameter ratios ($D_i/D_o$) due to the existence of outer perforated cylinder. Kelvin-Helmholtz vortices appear along the shear layer of the concentric cylinders. It can be seen clearly from dye experiments that diameter ratio is an important parameter on the control of Karman vortices formed by the inner cylinder or outer perforated cylinder. It is determined from the dye experiments that porosity, β and diameter ratio of inner and outer cylinders, $D_i/D_o$ are the two significant parameters on the control of flow structure generated by the inner cylinder. The effect of these parameters on the flow structure formed downstream of the inner cylinder is determined in detail with the Particle Image Velocimetry Technique (PIV).

Figure 5a 5b show the variation of maximum Reynolds shear stress, $<u'v'/U^2>$ values that occurs downstream of the different bare cylinders diameter and concentric cylinder with different diameter ratios, respectively. As it can be seen from figure 5b, maximum values of Reynolds shear stress, $<u'v'/U^2>$ decrease

DOI: 10.1051/epjconf/201714302094

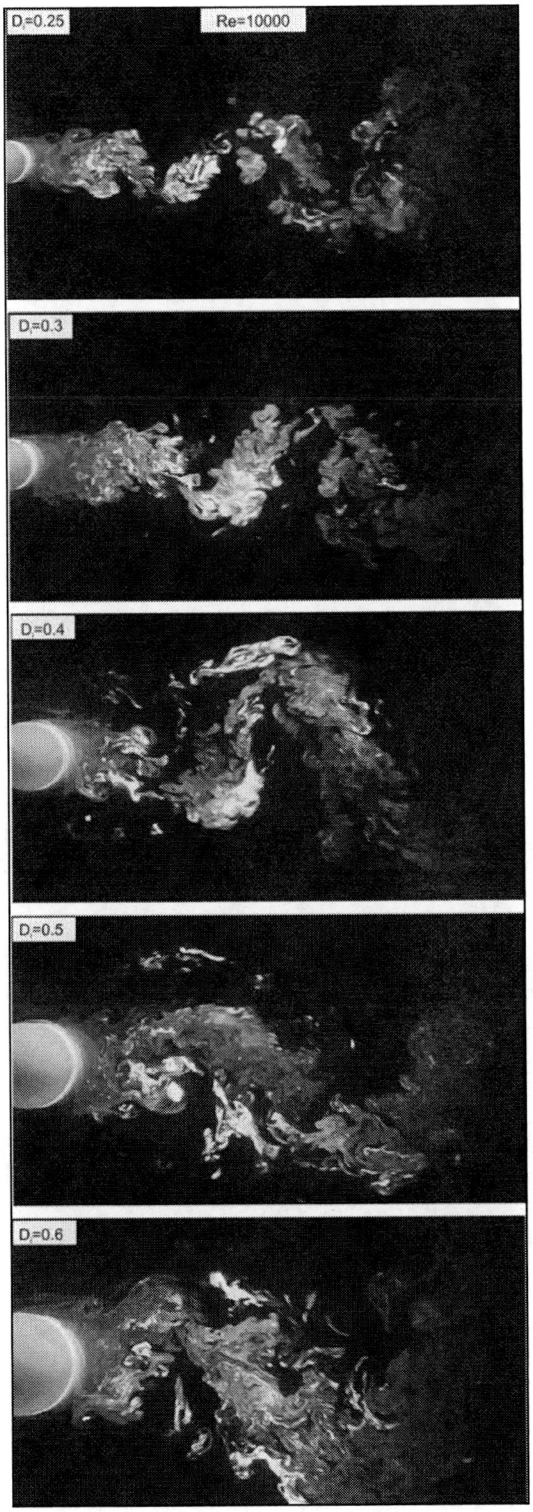


**Fig 3.** Dye visualization of flow structure around inner (bare) cylinders at Re=10000

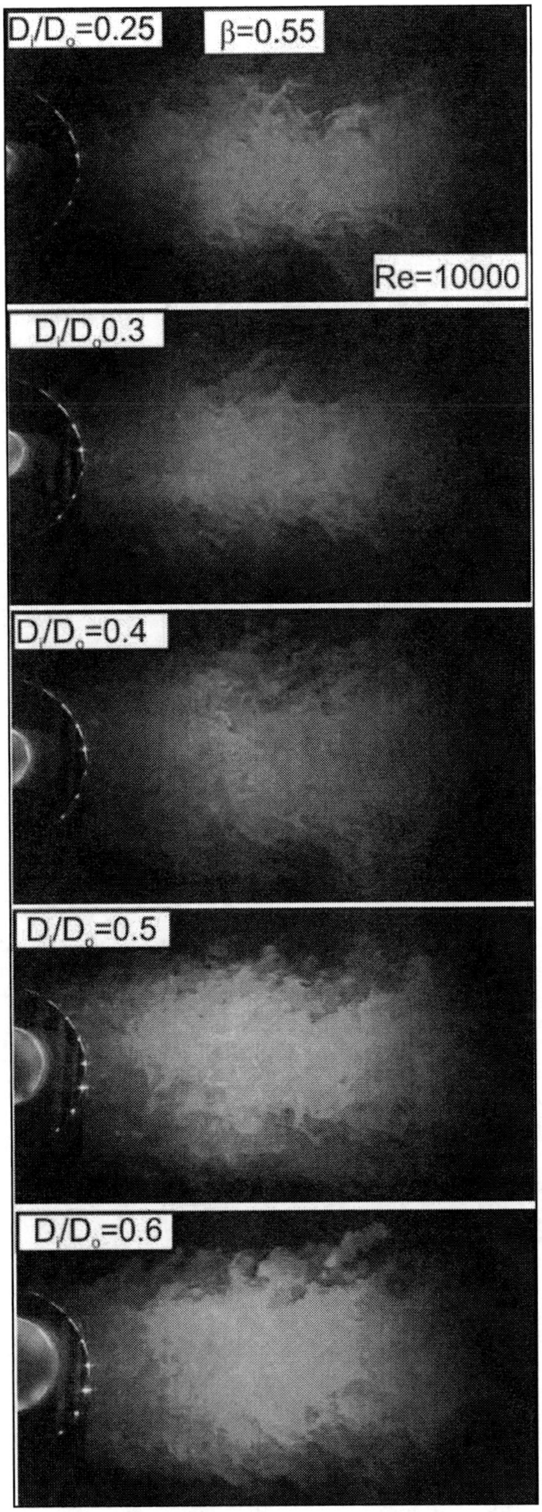


**Fig 4.** Dye visualization of flow structure downstream of concentric cylinders for various diameter ratios at Re=10000 and porosity, β=0.55

with increasing porosity, β up to porosity, β=0.55 for all diameter ratio. There are sudden increases in the maximum values of Reynolds shear stress, $<u'v'/U^2>$ for β≥0.6 cases for the diameter ratios of 0.5 and 0.6. Experiments reveal that outer perforated cylinders are significantly effective to reduce values of maximum Reynolds shear stress, $<u'v'/U^2>$ compared to the bare cylinders case in Figure 5a and 5b. Formations of the Karman vortices are suppressed and so that a vortical flow structure downstream of the cylinder with a certain frequency does not exist anymore. The situation indicates that the flow structure changes dramatically.

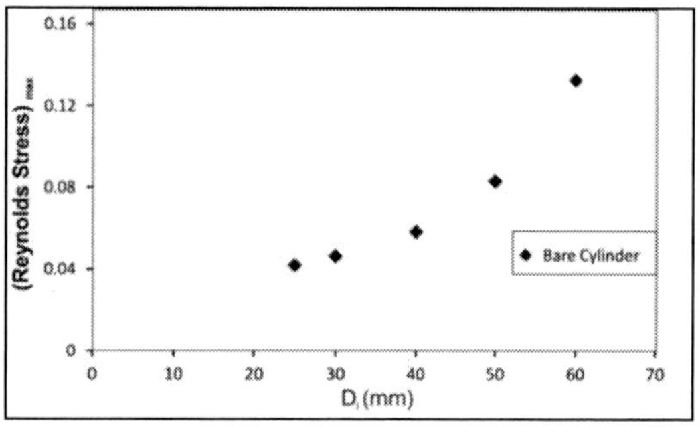


**Fig 5a.** Variation of Reynolds Shear stress$_{max}$, $<u'v'/U^2>$ with the downstream of bare cylinder diameter.

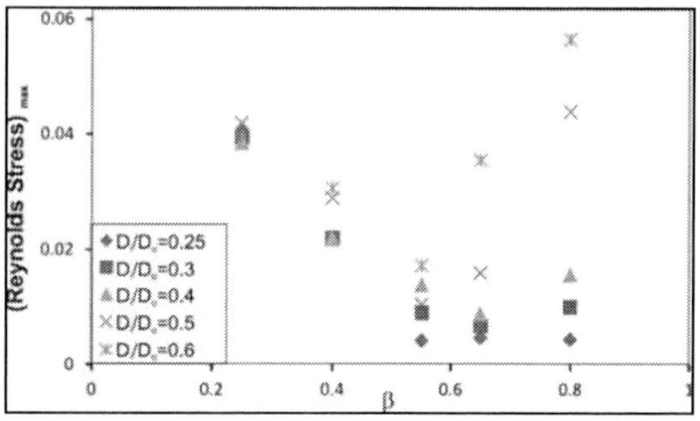


**Fig 5b.** Variation of Reynolds Shear stress$_{max}$, $<u'v'/U^2>$ with porosity, β, downstream of concentric cylinder.

After this part of the study, all attention was focused on the flow structures between inner and outer cylinder. Especially in smaller diameter ratios, the flow structure in the region between the inner and outer cylinders was investigated in terms of time-averaged Reynolds shear stress, $<u'v'/U^2>$ and streamlines, <Ψ>, superposed with the vorticity contours <ω> in order to understand whether unsteady Karman vortices are formed or not. The time-averaged streamline topologies, <Ψ>, superposed with the vorticity contours between inner and outer perforated cylinders for various porosities, β at $D_i/D_o$=0.25 are given together in Figure 6 The minimum and incremental values of time-averaged vorticity contours, <ω> were taken as ±2 and 2. For low porosity, β values, it is seen that the separated flow region in the downstream of the inner cylinder expands in lateral direction which is perpendicular to the main flow direction. Width of the separated flow region decreases with the increasing porosity β. It is clearly seen from the patterns of streamlines, <Ψ> that two foci ($F_1$ and $F_2$) are formed in the downstream of inner cylinder. These flow structures are also observed for the bare cylinder case. Locations of the critical points (foci, F) move toward the outer perforated cylinder from the inner cylinder with increasing porosity, β. Figure 7 shows the vorticity, <ω> contours and streamlines, <Ψ> obtained for $D_i/D_o$ =0.3. The results obtained from the previous diameter ratio, $D_i/D_o$= 0.25 are also valid for this case. However, the results obtained for $D_i/D_o$=0.6 differ from the result of low diameter ratios. The time-averaged streamline topologies, <Ψ>, superposed with the vorticity contours between inner and outer perforated cylinders are shown in Figure 8 for $D_i/D_o$ =0.6 case. When the vorticity, <ω> contours are investigated, the peak magnitude of vorticity, <ω> formed due to inner cylinder increases with increasing the porosity, β. When the streamlines, <Ψ>, are investigated, it is seen that critical points named as focus, F does not occur in between the inner and outer perforated cylinders due to the small gap ratio. Flow structure between inner and outer perforated cylinders has changed significantly.

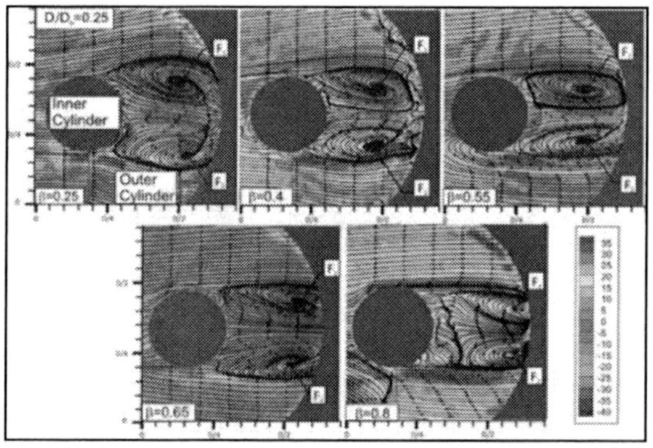


**Fig 6.** The time-averaged streamline topologies, <Ψ>superposed with the vorticity contours <Δω>=2 between inner and outer perforated cylinders for various porosities, β at $D_i/D_o$=0.25

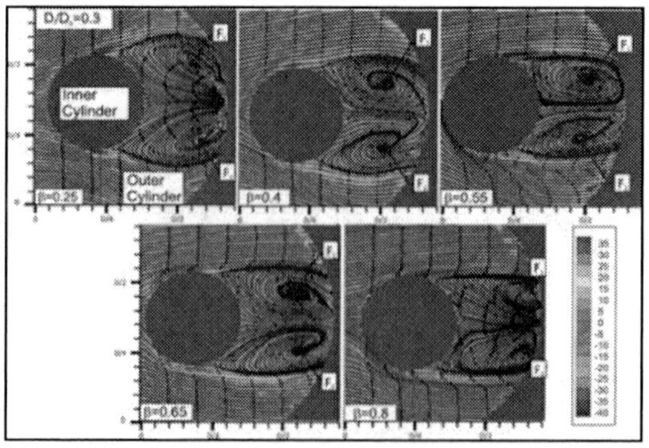


**Fig 7.** The time-averaged streamline topologies, <Ψ> superposed with the vorticity contours <Δω>=2 between inner and outer perforated cylinders for various porosities, β at $D_i/D_o$=0.3

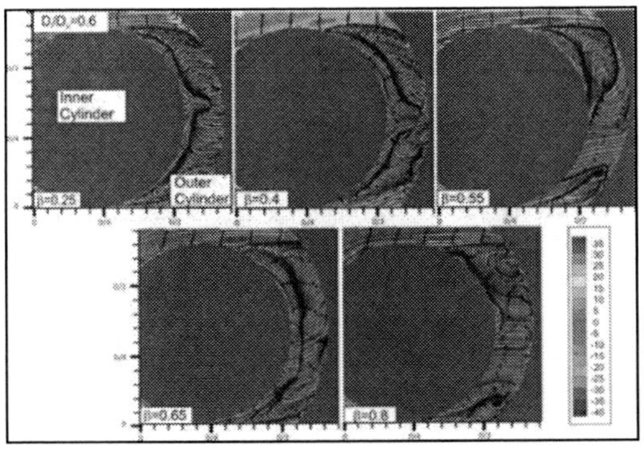


**Fig 8.** The time-averaged streamline topologies, <Ψ> superposed with the vorticity contours <Δω>=2 between inner and outer perforated cylinders for various porosities, β at $D_i/D_o$=0.6.

The variation of the maximum values of Reynolds shear stresses, <u′v′/U$^2$> in the region between inner and outer cylinders with the porosity, β is shown in Figure 9. There is a significant increase in maximum value of Reynolds shear stress, <u′v′/U$^2$>for β=0.8. The maximum values of Reynolds shear stress, <u′v′/U$^2$> were found in the other geometric conditions to be close to each other and independent of the diameter ratio, $D_i/D_o$.

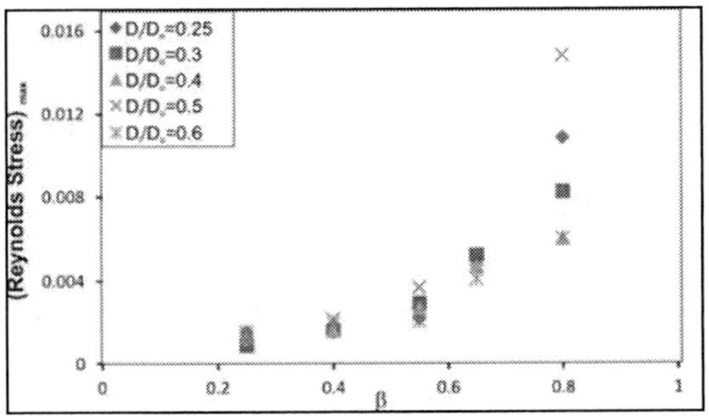


**Fig 9**. Variation of Reynolds Shear stressmax, <u′v′/U$^2$> with the porosity, β, at the region between inner bare cylinder and outer perforated cylinder.

When Figures 5a, 5b and 9 are compared with each other it can be seen that the existence of the outer cylinder is determined to be highly effective on the reduction of the maximum value of Reynolds shear stress, <u′v′/U$^2$> intensities in the region between the concentric cylinders. In the case of bare cylinder, the increase of cylinder diameter magnitudes the maximum values of Reynolds shear stress, <u′v′/U$^2$> significantly, but maximum Reynolds shear stress, <u′v′/U$^2$> formed in the region between the inner and outer perforated cylinder is not affected by the diameter of inner cylinder (except for β=0.8). Due to excessive number of the hole on the perforated cylinder on the cause of this condition is the loss of effect on the flow structure of the perforated cylinder. For small diameter ratios $D_i/D_o$= 0.25 and 0.3 and low porosity, β= 0.25, the max Reynolds shear stress, <u′v′/U$^2$> value in the downstream of the concentric cylinders has substantially the same value obtained from inner cylinders.

## 4 Conclusion

The effects of porosity, β and diameter ratio, $D_i/D_o$ were investigated and results were discussed to find out the optimum cases for effective flow control. When the perforated cylinder takes place around the bare cylinder, experiments reveal that the flow structure in the region both downstream of concentric cylinder and between the concentric cylinders is significantly affected compared to the bare cylinder case. We understood from qualitative and quantitative experiments that diameter ratios ($D_i/D_o$) does not have a significant influence on the flow structure for region between the inner and outer perforated cylinder unlike downstream of concentric cylinder.

In conclusion, maximum values of Reynolds shear stress, <u′v′/U$^2$> formed between the inner and outer cylinders have a lower value than the one occurs downstream of bare cylinder or outside of the concentric cylinder which indicates that the volume flow rate entering fluid through the perforated cylinder becomes lower due to pressure increase annular region.

It is determined from the experiments that porosity, β=0.55 suggestible parameter for reduction of magnitude of Reynolds Shear Stress correlations, <u′v′/U$^2$> and control of flow structure generated by the inner cylinder both in the region of downstream of concentric cylinder and between inner and outer cylinders.

## Acknowledgments

The authors would like to acknowledge the financial support of this work by the Scientific and Technology Research Council of Turkey under contract no 109R001.

## References

1. G.R.S. Assi, P.W. Bearman, N. Kitney, Journal of Fluids and Structures, *Low drag solutions for suppressing vortex-induced vibration of circular cylinders*, **25**, 666–675 (2009).
2. G.R.S. Assi, P.W. Bearman, N. Kitney, M.A. Tognarelli, Journal of Fluids and Structures, *Suppression of wake-induced vibration of tandem cylinders with free-to-rotate control plates*, **26**, 1045–1057 (2010).
3. P.W. Bearman, Journal of Fluids and Structures, *Circular cylinder wakes and vortex-induced vibrations*. **27**, 648–658 (2011).
4. P.W. Bearman, Journal of Fluid Mechanics, *Understanding and predicting vortex-induced vibrations*. **634**, 1–4 (2009).
5. P.W. Bearman, Annual Review of Fluid Mechanics *Vortex shedding from oscillating bluff bodies*. **16**, 195–222(1984).

6. M. Matsumato, Journal of Fluids and Structures, Vortex *shedding of bluff bodies,* **13**,791–811 (1999).
7. C.H.K. Williamson, Annual Review of Fluid Mechanics, *Vortex dynamics in the cylinder wake.* **28**, 477–539 (1996).
8. C.H.K. Williamson, R. Govardhan. Annual Review of Fluid Mechanics, *Vortex-induced vibrations*, **36**, 413-455 (2004).
9. C.H.K. Williamson, R. Govardhan, Journal of Wind Engineering and Industrial Aerodynamics, *A brief review of recent results in vortex-induced vibrations.* **96**, 713–735 (2008).
10. M.J. Every, R. King, D.S. Weaver., Ocean Engineering. *Vortex-excited vibrations of cylinders and cables and their suppression*, **9**, 135-157 (1982).
11. P. Price, J. Eng. Mech. Div., Am. Soc. Civ. Eng., Suppression of the fluid-induced vibration of circular cylinders, **82**, 1030 (1956).
12. B. Molin, Journal of Fluids and Structures, *A potential flow model for the drag of shrouded cylinders.* **7**, 29–38 (1993).
13. J.J. Wang, P.F. Zhang, S.F. Lu, K. Wu, Flow Turbul Combust, *Drag reduction of a circular cylinder using an upstream rod. 76, 83–101 (2006).*
14. *P. A. Baranov, V. L. Zhdanov, S. A. Isaev, V. B. Kharchenko, A. E. Usachov in 2003 published in Fluid Dynamics, Vol. 28, p203.*
15. B. Sahin, International Journal of Mechanical Science, *Pressure losses in an isolated perforated plate and jets emerging from the perforated plate,* **31**, 51-61 (1989).
16. A.O. Turhal, B. Cuhadaroglu, Experimental Thermal and Fluid Science, *The effects of surface injection through a perforated square cylinder on some aerodynamic parameters*, **34**, 725–735 (2010).
17. Gozmen, B., Akilli, H., B. Sahin, Passive control of circular cylinder wake in shallow flow. Measurement, **46**, 1125-1136 (2013).
18. G. M. Ozkan, V. Oruc, H. Akilli, B. Sahin, Experiments in Fluids, *Flow around a cylinder by a permeable outer cylinder in shallow water.* **53**, 1751–1793 (2012).
19. E. Pinar, G.M. Ozkan, T. Durhasan, M.M. Aksoy, H. Akilli, B. Sahin, Journal of Fluids and Structure, *Flow structure around perforated cylinders in shallow water*, **55**, 52-63 (2015).

# On the recirculation zone suppression behind HUMP profile using the DBD actuator

*Pavel* Procházka[1,a], *Václav* Uruba[1]

[1]Institute of Thermomechanics, Czech Academy of Sciences, Czech Republic

**Abstract.** Previously, the DBD (dielectric barrier discharge) plasma actuator was used in rectangular channel to modify the properties of the boundary layer in spanwise and in streamwise orientation. The actuator was redesigned for using on the surface of the Glauert-Goldschmied body in different position to influence the point of the separation and the reattachment point as well as the total extent of the separation bubble. The most intensive effect occurs when the actuator takes effect in the point of separation. Further downstream, the ionic wind produced by DBD causes complex coherent structures in the wake for spanwise orientation in both direction. The effect of streamwise orientation is studied also. Actuator is operated in steady regime to produces continuous ionic wind as well as in unsteady regime when the vortex street is generated. The properties of vortex street is given by modulation parameters (frequency and duty cycle). The effect of these parameters on the wake is evaluated. This experiment is realized in the perspex channel connected to the blow-down wind tunnel. The HUMP profile is flush-mounted to the bottom side. The wire electrode of the actuator is situated in x/L = 0.63, 0.66, 0.69 and 0.72 of the chord length. The time-resolved PIV (Particle Image Velocimetry) is used as a main measurement technique. The flow field behind the profile is captured in longitudinal plane as well as in cross-section planes using 3D PIV. The results based on statistical quantities will be presented in this paper. More, next part will be devoted to the decomposition analysis of the flow dynamics (BOD, OPD).

## 1 Introduction

This article is dealing with the possibility to control the form drag of the streamlined profile using plasma actuators. The plasma actuator belongs to the group of active flow control devices, such synthetic jets or moving surface devices. They have many advantages in comparison with other devices as low volume resulting in simple assembly, no mechanical breakable moving parts, low need for power and very high response time needed for sophisticated active flow control. On the other hand, they have quite low efficiency of energy conversion. However, their implementation into real application is relatively slow, since they have been developing for no longer than two decades. So that is why, the physical principle of the plasma discharge and its effect on the flow is not fully clarified. There are several numerical studies [1] of their influence on the boundary layer but the main lack of that approach is certain underestimation of the ionic wind thrust. From this perspective, the experimental studies are very important to supplement required pieces of knowledge. Nowadays, the most used type is called DBD (dielectric barrier discharge) actuator, which is powered by alternating current of very high voltage value.

There was tested a wire type of DBD actuator in steady air as well as in zero-pressure gradient boundary layer developed in rectangular channel previously [2]. The plasma actuator was driven in steady regime as is typical (e. g. [3]) and also in unsteady regime [4]. This regime seems to be very convenient for the purpose of active flow control utilization and it will be presented in the text further. The plasma actuator can generate necessary thrust in spanwise orientation (the same or opposite sense as channel flow) or in streamwise orientation, when the thrust is generated in perpendicular direction [5]. It has been shown that plasma actuation can stabilize or destabilize the boundary layer. The biggest effect was observed for channel velocities between 5 and 10 $m \cdot s^{-1}$ when the velocity gain was about 10% in normalized velocity profile.

The plasma actuator can work in two distinct regimes. Plasma actuator generates continuous ionic wind during steady regime. The ionic wind has character of wall-jet-like flow. The unsteady regime, which occurs after that rectangular modulation of voltage waveform is applied, is characterized by the existence of vortical structures inside the ionic wind which is called as vortex street. The properties of this phenomenon are determined by the modulation parameters [2]. The plasma actuator unsteady regime is more efficient and do not need so much power (lower energy consumption) and should be in the center of interest to implement it for active flow control of the wake behind streamlined profile.

[a] Corresponding author: prochap@it.cas.cz

DOI: 10.1051/epjconf/201714302095

The HUMP profile, also GG profile (Glauert-Goldschied), was chosen as a test case for plasma actuation, since this profile - established by Seifert [6] – was used many times (e.g. [7]) in numerical studies as well as in experimental ones to control the flow around it. The results described in available literature will be helpful for comparison with other authors. The flow naturally separates in the rear section of the profile and the size of the recirculation zone is dependent on Reynolds number.

The approach of this study is fully experimental. The data treatment will be set in dependency on statistical or dynamical point of view. To get statistical indicators, the data acquisition time will be longer with lower acquisition frequency. The dynamics of this phenomenon will be studied from time-resolved data. Since the unsteady regime produces pseudo-periodical structures, the phase-locked measurement will be employed to suppress the random phenomena. The energy decomposition will be utilized and the results also from stereo PIV measurements will be shown.

## 2 Experiment description

### 2.1 Actuator description

The DBD actuator has to consist of at least two electrodes and a dielectric layer between them. The wire type of DBD actuator used for this study was developed and tested during long research along flat plate [5]. This type of actuator has proved that it can generate sufficient strong ionic wind to influence boundary layer and its plasma discharge is homogenous and constant in time. However, some changes in actuator design had to be made to enable the assembly on the hump surface.

The plasma actuator is flush-mounted onto the surface so it does not affect the flow field at all. Since the curvature of the Hump profile is significant, the use of the silica glass dielectric was not possible any more. Now, the dielectric is made from several layers of kapton foils (total thickness is 0,48 mm) which are curved perfectly to respect the original shape. The upper electrode is still thin wire (30μm) and is power by high-voltage high-frequency waveform. The lower electrode is made from copper foil and is grounded and encapsulated to set the discharge only at the top side. The electrode gap is set to approximately 3 mm. Several anchor points were utilized to lead the wire in desirable position. The wire was preloaded using spiral torsion spring which ensures the straight shape and constant distance from the lower electrode. The actuator was designed to enable fast change between four distinct positions on the hump surface. The actuator was applied in spanwise orientation to produce the ionic wind in the same direction as the channel flow and also in opposite direction.

### 2.2 Power source

The power source was designed especially for the power supply of this actuator type. It is conceived as a half-bridge circuit excitating output high voltage ferrite transformer. This power source can give constant voltage value independently of the load. The voltage value can be set also independently and its maximal value is 12 kV in peaks ($U_{eff}$ = 8,5 kV). The frequency is also adjustable in some range but for all time was set to value of 16 kHz. This source disposes of so-called shut-down function. This is used to modulate the voltage waveform. The rectangular modulation brings another two parameters – modulation frequency and duty cycle DC. These are used to control the vortical structures generated during unsteady regime.

The plasma actuator produces continuous ionic wind in steady regime where the power consumption is at maximal level. The maximal mean velocity in the ionic wind is up to 2 m·s-1. The dependency between applied voltage value and maximal velocity of the ionic wind is linear. After that rectangular modulation is applied, the vortices inside ionic wind can be observed (in steady air) which is the main property of the unsteady regime. Modulation frequency determines the length of one period. Modulation frequency determines also the vortex shedding frequency and their diameter. The rate between the time, when the plasma discharge is on and off, is determined by the duty cycle. Duty cycle determines the time during which the vortex is supplied by the energy of the ionic wind. Note, that DC=50% means that the power consumption of the electricity is half of the original value. Duty cycle of 30% means that plasma discharge is active for 70% of one period.

### 2.3 Experimental setup

The hump profile was mounted inside perspex rectangular channel which was connected to the blow-down wind tunnel. The total length of the channel was 3000 mm and cross-section dimensions were 100 x 250 mm. The channel had rectangular corners and smooth walls made from transparent plate to enable optical access for laser and cameras. The turbulence intensity was 0.1% behind the contraction [8]. The profile was placed at the bottom side so that the distance between leading edge and the channel inlet was x = 1800 mm. The channel velocities were 5, 10 and 20 $m \cdot s^{-1}$. The boundary layer in front of the profile was fully developed turbulent. The boundary layer thickness at the leading edge and related Reynolds number are given at table 1.

**Table 1.** Boundary layer thickness and Reynolds number

| | $\delta_{0.99}$ | $Re_x$ | $Re_\delta$ |
|---|---|---|---|
| 5 $m \cdot s^{-1}$ | 31.2 mm | $6 \cdot 10^5$ | $10 \cdot 10^3$ |
| 10 $m \cdot s^{-1}$ | 28 mm | $12 \cdot 10^5$ | $18 \cdot 10^3$ |
| 20 $m \cdot s^{-1}$ | 24.2 mm | $24 \cdot 10^5$ | $32 \cdot 10^3$ |

The hump size was designed so that the height is 25 mm which is 10% of the channel height. The congestion level is then satisfied and the profile is big enough to assembly the actuator. The resulting chord length is approximately L = 194 mm (figure 1). Previously, the flow field behind hump was studied experimentally without actuation. The separation point still remains at the same point for all flow regimes. Many authors have found that the actuation should take effect as close as possible to the place where flow separates. However, it is not easy task to determine what is exactly the right spot of the ionic wind activity.

Four distinct actuator locations were tested. The wire position was chosen as the reference point. The first position located the most upstream was at x/L = 0.63 (63% of the chord length). This is 122 mm in horizontal sense from the leading edge. This position can be considered as slightly upstream of the separation point. Next three positions were at x/L = 0.66, 0.69 and 0.72. The second position is located slightly downstream of the natural separation point. And the last two positions are even further. These were established especially to test unsteady modulation as well as ionic wind oriented in opposite sense.

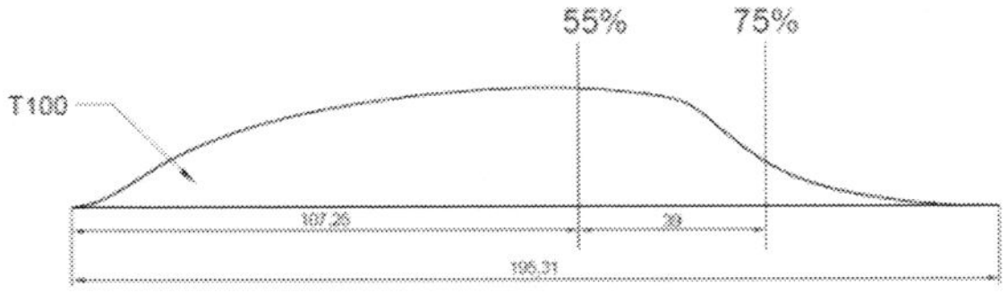


**Figure 1.** The hump profile, dimensions, zone of the actuation

### 2.4 Measurement technique

Time-resolved Particle Image Velocimetry (TR-PIV) was used as measurement method. The most measurement was acquired by standard 2D PIV in longitudinal plane perpendicular to the channel bottom side. To illuminate the tracing particles (safex generator of oil particles, diameter approx. 1 μm), the laser New Wave Pegasus Nd:YLF was utilized. The maximal repetition rate is 10 kHz. The energy of one pulse is 10 mJ for 1 kHz (10 W per head). This laser has double-head, cylindrical optics to make laser sheet with thickness of 1 mm and it emits the coherent light with wavelength of 527 nm. The CCD camera Phantom V711 was used to acquire the images in the plane of view. The total resolution is 1280x 800 and corresponding maximal acquisition frequency is 6 kHz in single frame mode. The camera memory is 8 GB. This camera was shifted along the field of view by a traverser to get sufficient space resolution.

The stereo PIV in cross-section plane was suggested to supplement the previous mentioned measurement. The goal was to reveal potential asymmetry of the flow. Two CMOS camera NanoSense MKIII with maximal resolution 1280 x 1024 pixels and corresponding acquisition frequency 512 double-images per second were utilized. The camera memory was 4 GB. These factors were limiting to get sufficient dynamical image of the flow. The Scheinpflug criterion was fulfilled. The precision of that method is within 1-2 %, if specific requirements are fulfiled [9]. The both cameras and laser sheet was shifted simultaneously by traverser system to measure in four distinct cross-section planes (x/L = 0.69, 0.87, 1.00 and 1.50).

Three regimes were investigated. The base case is without plasma actuation. There are results in the table 2. The steady and especially unsteady regimes are used to suppress the size of the separation bubble. The plasma discharge is driven by maximal voltage value (12 kV for unsteady case, 8 kV for steady case). The effect of modulation frequency and duty cycle was studied. The frequency was measured for range from 10 to 500 Hz and duty cycle varied from 10 to 90%.

The mean flow field characteristics were evaluated from data acquired by 100 Hz and at least 6 seconds. The POD (Proper Orthogonal Decomposition) analysis was performed from data acquired by 2 kHz. Also phase-locked measurement was applied to highlight the periodical process by unsteady regime. There were acquired 20 phases per each period and then the averaging over at least 80 periods was performed.

## 3 Results

The table 2 shows the results of the different separation bubble size for three channel velocities. The separation point is still at the same place but the reattachement point is moving upstream with increasing Reynolds number. The main goal of the plasma actuation is to minimize the dimension of the recirculation zone – either by reattachement point shifting or by reduction of the zone thickness.

**Table 2.** Separation and reattachement point location without plasma actuation

| | **separation** | **reattachement** |
|---|---|---|
| 5 $m \cdot s^{-1}$ | x/L = 0.65 | x/L = 1.31 |
| 10 $m \cdot s^{-1}$ | x/L = 0.65 | x/L = 1.12 |
| 20 $m \cdot s^{-1}$ | x/L = 0.65 | x/L = 0.97 |

The figure 2 reveals the small section of the profile with the actuator placed at x/L = 0.63. The mean flow field is coming only from thrust generated by plasma actuator, channel flow is not present. Notice, that maximal velocity is close to the hump surface approx. 4 mm downstream. This implies that the position of the wire electrode is not crucial parameter by the making decision what is the right place to take effect. On the other hand, the plasma actuation evokes sufficiently strong suction effect to generate positive velocities of about half original magnitude in the vicinity of the wire electrode. The ionic wind is closely attached to the curved surface due to Coanda effect.

The mean flow field behind hump profile which is developed under natural condition can be seen from the figures 3,4 and 5. Figure 3 is plotted for u-component of velocity. The white area denotes the zone of back-flow (negative velocity). The flow is made visible by streamline in the figures 4 and 5. The flow reattaches the surface at x/L = 1.31 and x/L = 1.12 for 5 $m \cdot s^{-1}$ and 10 $m \cdot s^{-1}$, respectively. More, the thickness of the separation bubble was also decreasing with channel velocity. The flow field evoked under 20 $m \cdot s^{-1}$ is not plotted here. The recirculation zone is present but very tiny and hardly evaluated from PIV data.

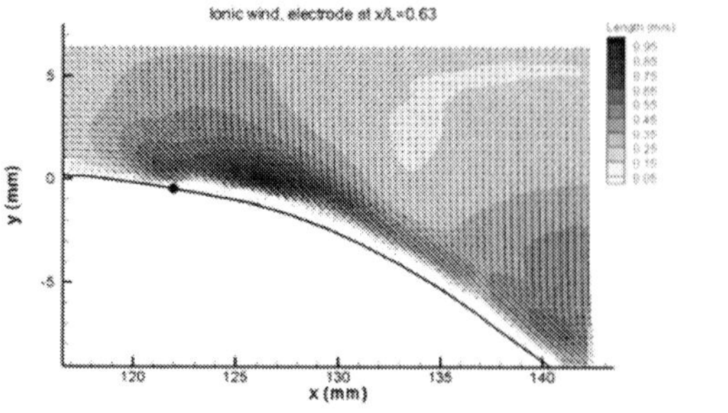


**Figure 2.** Ionic wind without channel flow

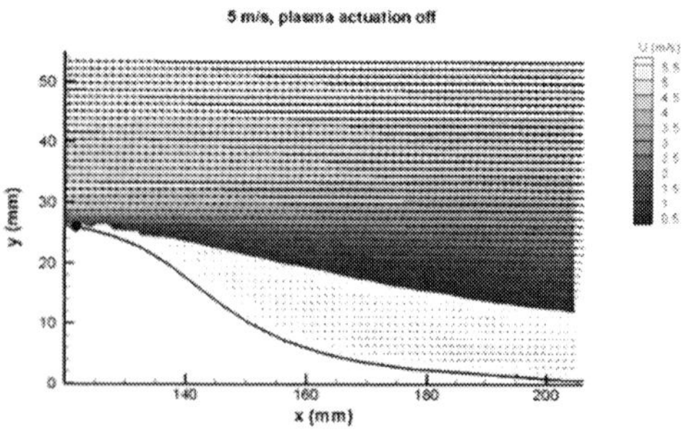


**Figure 3.** 5 m·s$^{-1}$, plasma actuation off

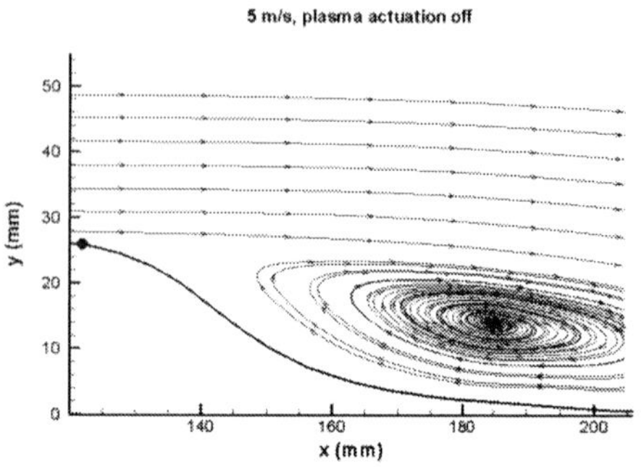


**Figure 4.** 5 m·s$^{-1}$, plasma actuation off

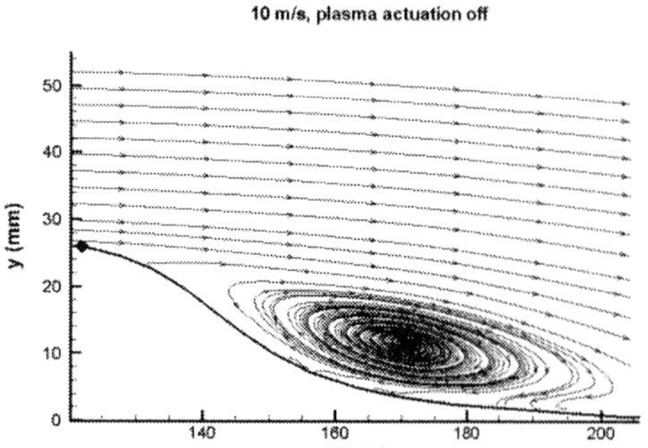


**Figure 5.** 10 m·s$^{-1}$, plasma actuation off

The plasma actuation applied in the first and second position has led to strong reduction of the separation (figure 6). The actuator in 63% of the chord length was able to fully suppress the separation. There is only small separation up to 0.77 for second position of the actuator. The situation is different, when the actuator is placed further downstream. In the third position, the ionic wind in not able to decrease this zone, there is only uncertain change of the separation bubble shape. There is the actuator placed in the last position (0.72) in the figure 7. The actuator is not able to attach the flow any more because even the suction effects are far away from the point of separation. The size of the bubble is approximately the same. However, the topology of the wake is not the same.

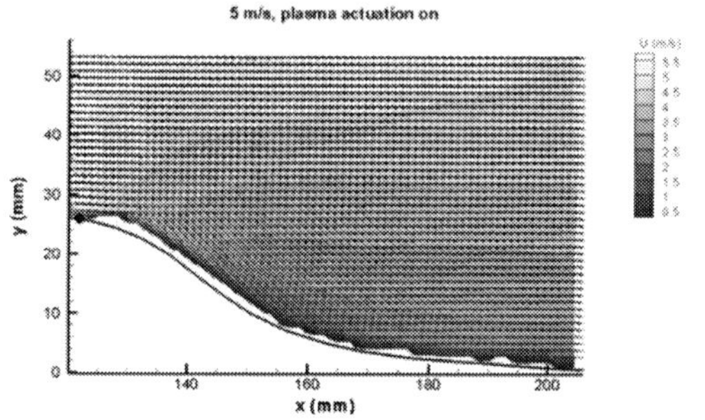


**Figure 6.** 5 m·s$^{-1}$, plasma actuation on at 0.63

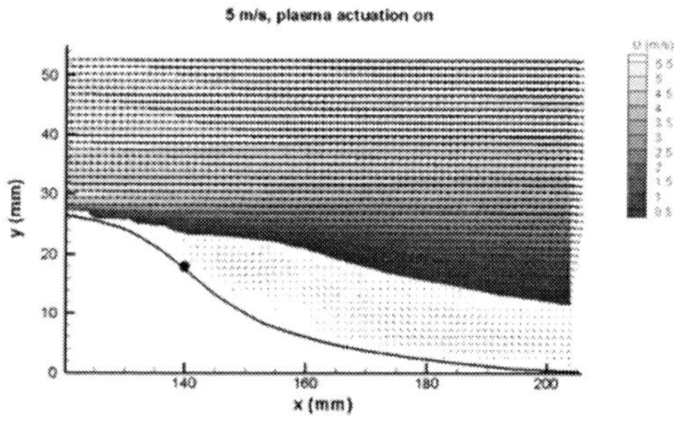


**Figure 7.** 5 m·s$^{-1}$, plasma actuation on at 0.72

The secondary pseudo-steady structures can be seen very well at the figure 8. This image is acquired from phase-averaged measurement and it represents the 4th phase during one period. The secondary vortical structure is visible about 15 mm downstream from the wire electrode. The origin and the existence of that vortex is closely related to the modulation parameters. It seems that the frequency does not have any effect on the shape and size of this structure. On the other hand, the frequency determines the periodicity of this phenomenon. Duty cycle is a crucial parameter which determines the time after which the vortex is fed by the energy. The unsteady modulation with higher duty cycle value (shorter time of plasma action) has led to almost no secondary structure in mean flow field. The structure is also not constant in space but it is moving very slightly away from the electrode. This convective velocity is in order of tenths of m·s$^{-1}$. The tendency is weakly growing with applied modulation frequency.

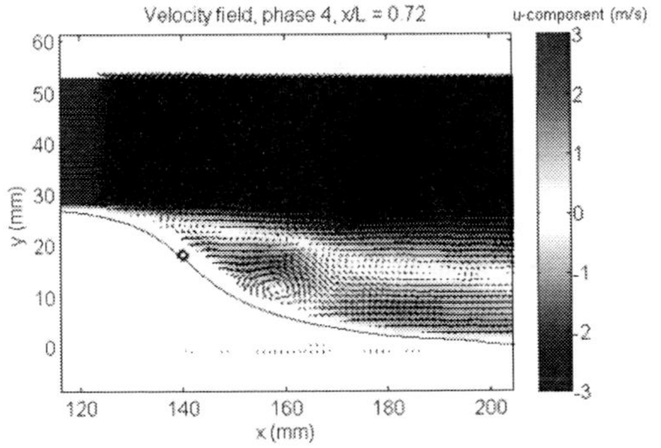


**Figure 8.** 5 m·s$^{-1}$, actuation at 0.72, unsteady (10Hz/30%)

Also the effect of the ionic wind generated in opposite direction was tested. Here, the results obtained from the second position of the actuator will be presented. The separation bubble became greater in size of both dimensions for all three channel velocities. The separation point under 5 m·s$^{-1}$ was shifted to x/L = 0.57 (instead of 0.65) and the reattachement point was shifted to x/L = 1.37 (instead of 1.31). The reattachement point of the double channel velocity was x/L = 1.26. The effect of opposite ionic wind is more dominant for the actuator positions located more upstream.

The effect of duty cycle is also crucial at the first position also. Duty cycle of 90% does not evoke the ionic wind for desirable time and flow separates at the identical point as in the base case and it creates quite huge recirculation zone (figure 10). Nevertheless, the size of separation bubble is approximately twice smaller and the reattachement point was shifted upstream to x/L = 1.08. Of course, this regime is not so competitive in comparison with steady actuation or actuation under DC = 30% but its energy consumption is only a fraction of the original value, more precisely one tenth.

The effect of modulation frequency value is demonstrated at the figures 11 and 12. It is obvious (especially from data acquired by phase-averaged measurement) that if the higher value of modulation frequency is used, the mean flow field is more attached to the surface. The flow is periodically attached to the surface and again separated from the surface according to modulation parameters. There is one switching reattachement/separation per period. With the higher modulation frequency applied (observed from 50 Hz), the flow is not able to follow this switching due to inertia forces and stays more or less attached to the surface for all time. Notice, that the frequency value is not responsible for power consumption, unlike the duty cycle.

The last figure represents extracted second POD mode of the wake flow influenced by the actuator placed at the last position and driven by 10 Hz and 30% DC, channel velocity 5 m·s$^{-1}$. The extracted POD mode contains probably the secondary vortical structure mentioned

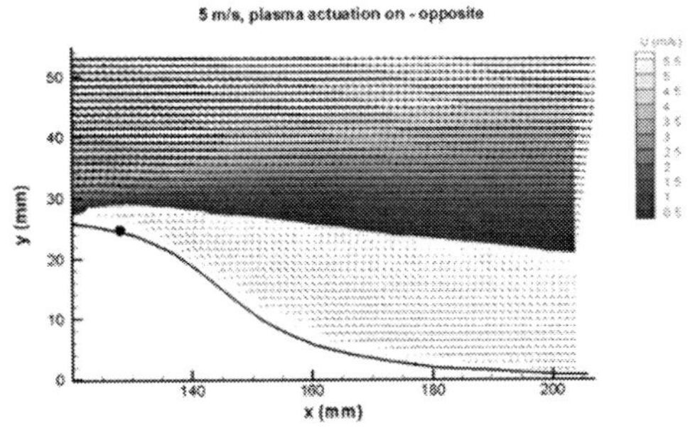


**Figure 9.** 5 m·s$^{-1}$, opposite plasma at 0.66

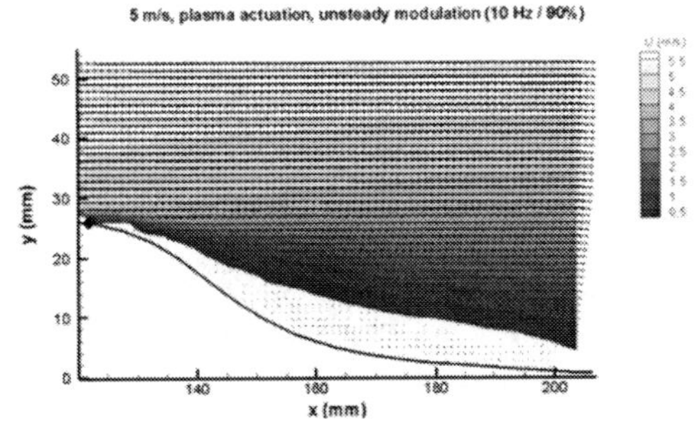


**Figure 10.** 5 m·s$^{-1}$, actuation at 0.63, unsteady (10Hz/90%)

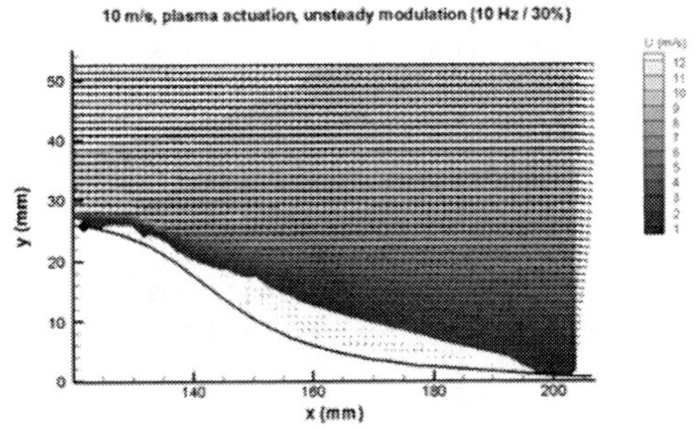


**Figure 11.** 5 m·s$^{-1}$, actuation at 0.63, unsteady (10Hz/30%)

above. The first mode is actually mean flow field so the second mode is the most important concerning kinetic energy content (see table 3). The energy of this mode is caused by the plasma ionic wind. The table 3 shows another four examples of POD mode extraction. There are just five first modes ordered according to the energy content. The difference between 30% and 70% of DC should be analysed more in detail. Probably the regime with DC = 70% means that the ionic wind importance is not so significant in comparison with channel flow. The decomposition of naturally developed wake has showed that the first two modes were more dominant than the rest of them. That is why the situation seems to be similar for unsteady regime driven by higher duty cycle value.

 DOI: 10.1051/epjconf/201714302095

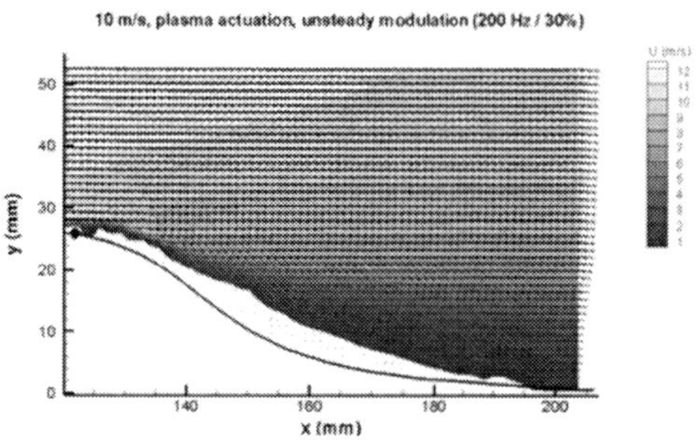


**Figure 12.** 5 m·s$^{-1}$, actuation at 0.63, unsteady (200Hz/30%)

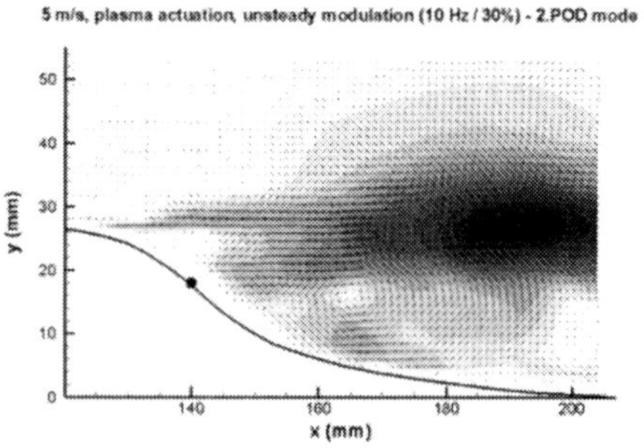


**Figure 13.** Second POD mode

**Table 3.** Five most important modes sorted by kin. energy

| | 1 | 2 | 3 | 4 | 5 |
|---|---|---|---|---|---|
| x/L = 0.72, 10 Hz | 18.6% | 9.1% | 6.6% | 4.8% | 3.9% |
| x/L = 0.72, 20 Hz | 26.3% | 10.3% | 6.5% | 4.0% | 2.5% |
| x/L = 0.72, 50 Hz | 15.3% | 9.7% | 6.5% | 5.9% | 4.6% |
| x/L = 0.66, 30% | 26.6% | 16.2% | 5.5% | 4.0% | 3.5% |
| x/L = 0.66, 70% | 34.5% | 17.7% | 5.5% | 3.8% | 2.6% |

## 4 Conclusion

The flow field behind hump profile was investigated experimentally using time-resolved Particle Image velocimetry. To control the separation point and the size of the recirculation bubble, the DBD wire actuator was used in steady and unsteady regime. The flow field was investigated from statistical and also from dynamical point of view. POD analysis was performed. Phase-locked measurements were conducted to detect secondary vortical structures present in the wake which are incorporated into the flow by the energy of plasma actuation.

The ionic wind is strength enough to fully suppress the area of negative velocity for lower channel velocities about 5 m·s$^{-1}$. It was observed that the optimal actuator position was at x/L = 0.63 or 0.66 which is in a good agreement with other authors. Nevertheless, the thrust coming from plasma actuation placed further downstream can significantly change the properties of the wake due to existence of secondary pseudo-steady vortical structure. The dimension of the separation bubble can be easily magnified by the ionic wind affecting the flow in the opposite sense.

The role of duty cycle is crucial. Higher value of duty cycle does not lead to considerable improvement but on the other hand, the energy consumption of these regimes is tiny. The effect of modulation frequency is also important. Higher frequency is more appropriate, since the flow is not able to follow the rapid changes and stay more or less attached to the surface.

## Acknowledgement

The authors gratefully acknowledge financial support of the Grant Agency of the Czech Republic, No. GP14-25354P.

## References

1. P. C. Dörr, M. J. Kloker, Procedia UITAM 14, 469-478 (2015)
2. P. Procházka, Dissertation thesis, ČVUT (2013)
3. Y. Li, X. Zhang, X. Huang, Exp. Fluids 49, 367-377 (2010)
4. N. Bénard, N. Balcon, G. Touchard, E. Moreau, Exp. Fluids 45, 333-355 (2008)
5. P. Procházka, V. Uruba, EPJ Web of Conferences 114, 02099 (2016)
6. A. Seifert, L. G. Pack, AIAA Journal 40, No. 7, 1363-1372 (2002)
7. C. He, T. C. Corke, M. P. Patel, AIAA Journal, 0935 (2007)
8. P. Jonáš, O. Mazur, V. Uruba, IV. Watermanagement conference, Prague (2004)
9. M. Raffel, C. Willert, S. Wereley, J. Kompenhans, Particle image velocimetry, Springer-Verlag (2007)

DOI: 10.1051/epjconf/201714302096

# Research on high-temperature heat receiver in concentrated solar radiation system

*Przenzak* Estera[1,*], *Basista* Grzegorz[1], *Bargiel* Paulina[1], *Filipowicz* Mariusz[1]

[1]AGH University of Science and Technology, Faculty of Energy and Fuels, Department of Sustainable Energy Development, A. Mickiewicza Av. 30, 30-059 Kraków, Poland

**Abstract.** The article presents the results of experimental and computer simulations studies of the high temperature heat receiver working in the concentrated solar radiation system. In order to study the radiation absorption process and heat exchange, the two types of computer simulations were carried out. The first one was used to find the best location for absorber in the concentrating installation. Ray Tracing Monte Carlo (RTMC) method in Trace Pro software was used to perform the optical simulations. The results of these simulations were presented in the form of the solar radiation distribution map and chart. The data obtained in RTMC simulations were used as a second type boundary conditions for Computational Fluid Dynamics (CFD) simulations. These studies were used to optimize the internal geometry of the receiver and also to select the most effective flow parameters of the working medium. In order to validate the computer simulations, high temperature heat receiver was tested in experimental conditions. The article presents the results of experimental measurements in the form of temperature, radiation intensity and power graphs. The tests were performed for varied flow rate and receiver location. The experimental and computer simulation studies presented in this article allowed to optimize the configuration of concentrating and heat receiving system.

## 1 Introduction

Nowadays, great emphasis in each of the activities is put on sustainable development. One of its most important aims is to ensure affordable and clean energy. There are many different ways to gain energy from resources that meet high demands from both ecological and economical sides.

Among all renewable resources more and more attention is paid to solar energy, which huge potential still remains untapped. Energy gained from the Sun is widely accessible and may be easily converted into for e.g. heat or electricity. Moreover, it is commonly use in small and medium-small buildings as a source of cheap and safe energy. There are two main application used to obtain energy from solar radiation. Solar thermal collectors receive heat for domestic applications and photovoltaic (PV) technologies produce electricity. It is also possible to connect this both systems into one called hybrid photovoltaic-thermal system, what improve their performance. However, the power density of energy obtained from the solar radiation is relatively low compared to other energy sources. Due to this disadvantage it is important to develop various kinds of concentrating systems, which provide amplification of the intensity of solar radiation incident on the surface of the receiver.

Preparing a properly operating installation requires performing a series of simulations. For solar systems, it is necessary to analyse the calculation of energy flux distribution on the receiver. One of the widely used technique is a ray tracing simulation. This universal tool can be exploit to survey reflection, refraction, diffraction and scatter of the light beams. In [1] ray tracing method allows to perform the simulation of the optical behaviour of three cavity receivers of various geometry complexity (spherical, cylindrical and conical). Ray tracing was also used to simulate a multi-faceted point focus concentrator composed of 18 spherical mirrors [2]. Due to this model it was available to achieve a homogenous flux distribution on the surface of the receiver. Moreover, in [3] has been presented a project of a solar concentrator, wherein the heat is transferred by bundled optical fibre instead of a heat receiver [3]. The simulation was performed using modified ray tracing method so-called Monte Carlo, in which the rays are randomly defined according to a probability density.

Computational Fluid Mechanics (CFD) plays a significant role in analysis of complex energetic processes. In case of solar heat receivers, due to CFS simulations some of the crucial operational parameters can be obtained, for e.g. heat transfer or distribution of the temperature on the absorbing surface. This method is widely used and very popular because of achieving good simulation results in compare to taken measurements. In [4] solar volumetric – receiver for high temperatures was designed using CFD calculations. The results obtained on the basis of tests carried out on the experimental

* Corresponding author: esterap@agh.edu.pl

 DOI: 10.1051/epjconf/201714302096

prototype and CFD model proved to be very similar, the temperature differences between the measurements and CFD simulations was from 3.2 to 8.4%, which is an acceptable result. Solar cavity receiver used to transfer the energy absorbed from solar radiation to the working medium in solar tower system [5] was simulated using CFD method. Simulation results allowed to perform optimisation of surface heat flux distribution inside the receiver, especially its uniformity. Influence of porosity on the thermal analysis and design of the solar received was studied in [6]. Temperature and velocity distribution were simulated using CFD software in terms of various configurations of the absorber. Due to porous structure (constant and gradual porosity) used material feature highly resistant to thermal stresses, but the thermal efficiency depends strongly on the distribution and depth of pores.

## 2 Description of the solar system

At the Department of Sustainable Energy of the AGH University of Science and Technology, research on the hybrid solar energy cooling system is conducted. The system consists of solar radiation concentrator combined with a classic solar collector system. The aim of the research is verification of the possibility of cold generation during high demand periods. Main components of the system installed on the roof of the D9 building (the Department of Sustainable Energy of the AGH University in Kraków) are: a parabolic solar radiation concentrator, which reheats the working medium (thermal oil) flowing between the solar collectors and the tank [7], two flat solar collectors connected to a 150l buffer tank.

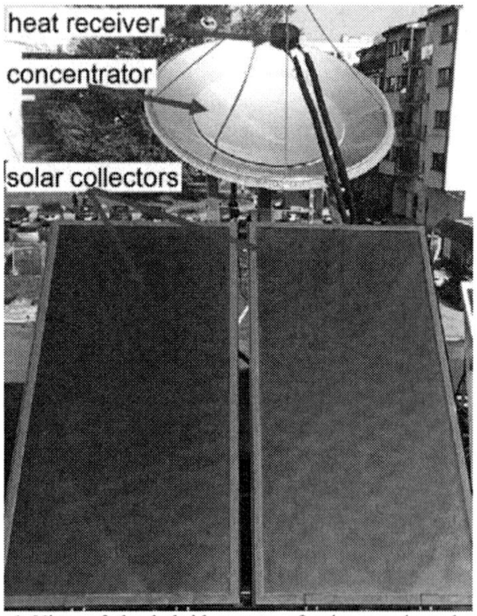


**Fig. 1** View of the hybrid system of solar panels and solar radiation concentrator mounted on the roof of Department of Sustainable Energy Development, AGH University in Krakow.

Due to safety reasons, the system has been equipped with a dry-cooler protecting again overheating of the collectors and the utility water tank. It allows for conducting long-term research. Ultimately, the entire system shall be equipped with an absorption chiller. Curretly, efficiency of high temperature heat generation in term of absorption chiller driven is in progress.

### 2.1 Solar radiation concentrator

The solar radiation concentrator used in the research has been composed of two concentrating elements: A dish covered with a two-layer reflective foil (1.8 m in diameter) and a mirror mounted on the dish with a diameter of 1.2 m. The system in concern is characterized by two focal points of rays (in the distance of 70 cm and 85 cm from the mirror's surface), which results from the different geometries of the focusing elements and the fact that the mirror has been installed at a distance of 2 cm from the dish surface. The concentrator's system follows the location of the sun using a differential sensor and two linear actuators. This solution enables the changing of the azimuth and elevation angle in automatic mode.

### 2.2 High-temperature heat receiver

An important element of the installation in terms of generation of high-temperature heat is a receiver, which is placed in a focal point of concentrated solar radiation. The main impact on radiation absorption efficiency has the location and shape of the receiver. In turn, the efficiency of high-temperature heat removal by working medium depends on the inner geometry of the receiver.
The study has been conducted on a flat heat receiver with a diameter of 18 cm, which absorbing surface was covered by high-temperature black paint. Back and side walls of the receiver were insulated using mineral wool with a thickness of 2 cm. The receiver comprised of a aluminium block with milled internal canal, a cover, a seal and stub-pipes.

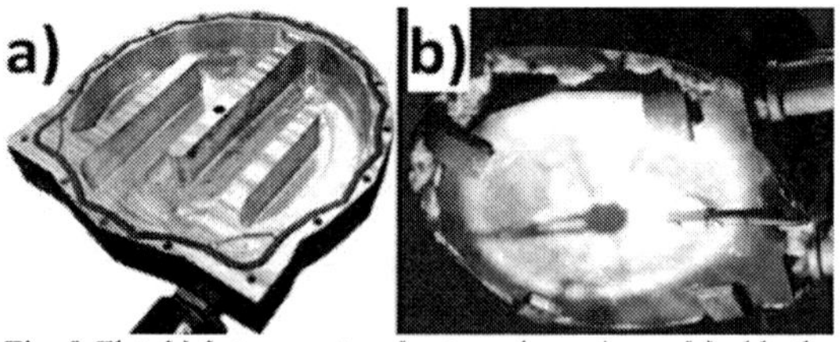


**Fig. 2** Flat, high-temperature heat receiver: a) canal inside the receiver, b) receiver lighted by concentrated radiation.

In figure 2a the interior of the receiver is shown. The working medium canal, a seal, screw used to fasten the cover and one of the stub-pipe are visible. Figure 2b presents the receiver operating in concentrated solar radiation. A focused rays falling on the absorbing surface and also a mineral wool insulating against heat loss are visible.

Currently, receivers with concave and convex surfaces are tested. However, this article presents the results related only to the flat receiver, which features the simplicity of the construction and flawlessly exploitation.

### 2.3 Measurement and control system

The installation was equipped with the automatic control and measurements system based on the programmable logic controller (PLC). A number of measurement elements, like sensors of temperature, pressure, solar radiation intensity, wind and a flow meter are connected to the controller. The 370W oil pump, controlled by the inverter, is used as a circulation pump. Moreover, the automatic system allows to control the performance of the cooler fan used as a protection against overheating of the installation.

The installation is equipped with 10 temperature sensors installed in important points of the system together with 8 solar radiation intensity sensors. Detailed description of the system is presented in paper [8].

Experimental studies of the hybrid installation were conducted during the summer period. The cloudless, characterized by low intensity of the scattered radiation were selected. A series of measurements of the energy parameters was conducted for 10 different locations of the receiver in the geometrical system of the concentrator. The distance of the receiver from the surface of the mirror was changed every 2cm from 72 to 90cm. For each selected distance, the power of circulation pump was varied. The relative pump capacity was changed every 5% in the range of 80% - 60% (55% if the flow was still too high). In this way the temperature variation was examined for medium flow from approx. 1,4 to 4,7 l/min. Each single measurement (the given distance, pump capacity) took approx. 20min necessary for the system stabilisation.

## 3 Experimental research

All raw measured points are collected in general figure (fig.3) in temperature difference – flow coordinates. In further part of the paper detailed analyses including also data form solar radiation sensors is presented.

As shown in figure 3, the highest temperature difference was achieved for two distances of the receiver, i.e. for 76cm and 78cm. For low medium flows of approx. 1.4-3 l/min the temperature difference was achieved at the level of 12-20 . This value is sufficient to increase the temperature of the working medium behind solar collectors to the values necessary for the effective operation of an absorption cooler.

The change of the distances to 72, 74, 80 and 82 cm causes a decrease the temperature differences to the level of approx. 8-15 . However, these values can also be sufficient for powering the absorption chiller, but with lower efficiency.

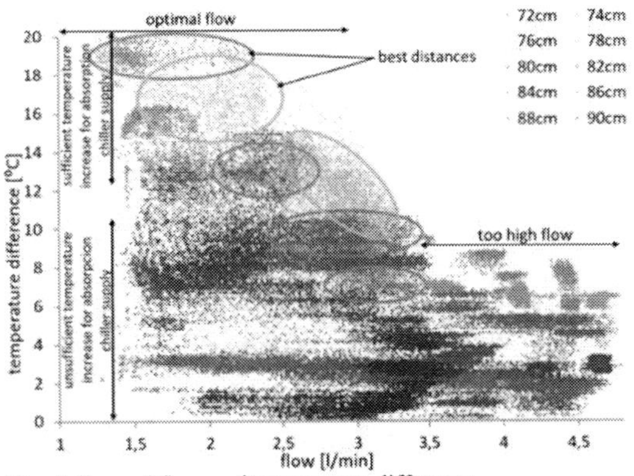


**Fig. 3** General figure of temperature difference

The worst results (less than 8 ) were achieved for distances 84-90cm. It can be seen that the change of the flow has a very slight effect on the temperature increase. At the same time, we can see that the flows over 3.5l/min for each location of the receiver are too high for the effective heat generation.

### 3.1 Temperature variation

The temperatures measured at selected parts of the installation depend on the circulation pump capacity, as is shown in fig 4.

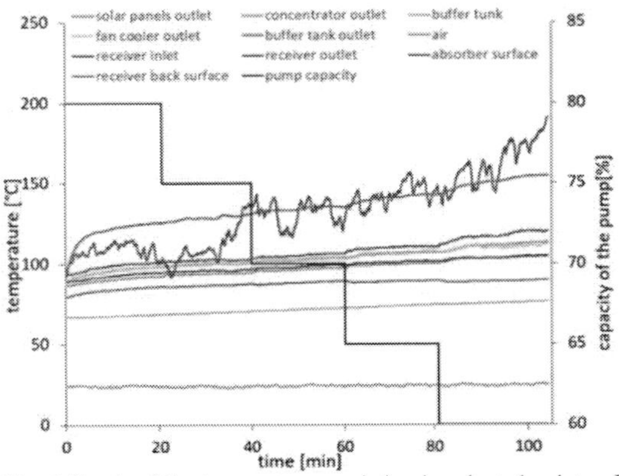


**Fig. 4** Graph of the temperatures variation in selected points of the installation.

The increase of the working medium temperature is observed with the decrease of the flow. It is also seen, that the temperature on the absorber surface changes quite rapidly. It is caused by the inaccuracy of the solar tracking system. The focal point of concentrated radiation escapes from the central part of the receiver. This is one of the problem associated with the use of concentrated solar systems. It should be included that a sensor positioned on the absorbing surface (seen on fig.2b) is additionally heated by the concentrated solar radiation.

### 3.2 Solar radiation intensity

In fig.5 the solar radiation variation registered by five sensors corresponding to the temperature measurements presented in fig.4 is shown. In approximately 35 minute a rapid change of the direct radiation intensity, caused by the inaccuracy of the solar tracking system is visible. It can be seen that the measurement was carried out in a sunny, cloudless day – that is a cause of low values of scatter radiation and the lack of dynamic changes of the radiation intensity.

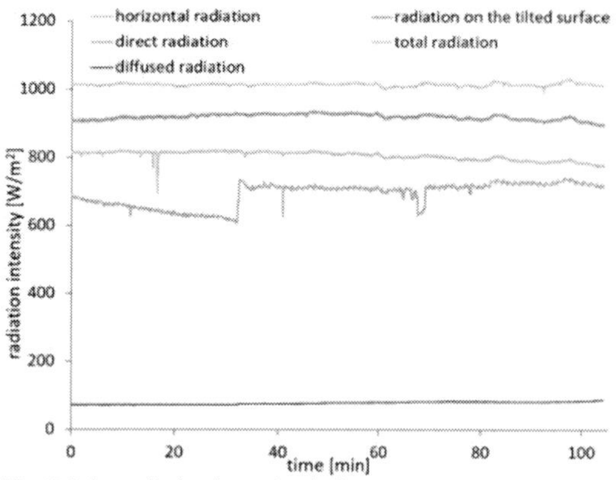


**Fig. 5** Solar radiation intensity during measurements.

### 3.3 Power of the receiver

Based of the registered data, the power of the receiver (considered as a heat source) was calculated. The results are shown in fig. 6.

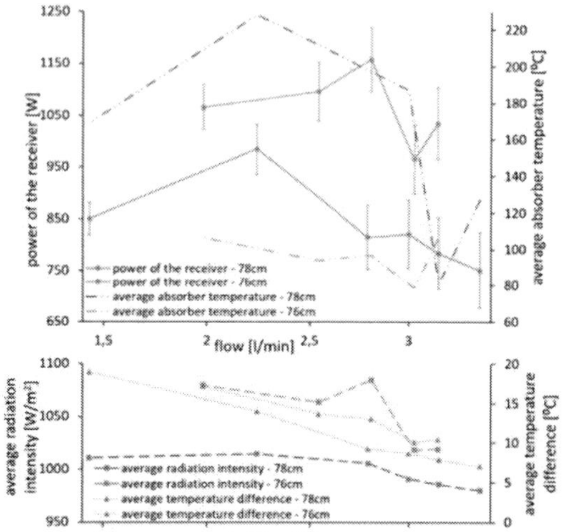


**Fig. 6** Fluctuation of the receiver power

It can be seen that in the case of the receiver located at a distance of 78cm from the mirror surface, the maximum power was achieved for the flow with a value of approx. 2.3l/min. Lower flow (1.5l/min) causes the loos of the power even despite achieving a large temperature difference (approx. 20 ).

The maximum power for the distance of 76cm was achieved for the flow approx. 2.8l/min. However, it should be noted, that during the flows higher than 2.5l/min, and therefore for the pump efficiency in the range of 65-60%, a significant dynamic changes of the solar radiation intensity happened (fig.7).

In the graph fig.7 one of the basic disadvantages of the systems using the concentrated solar radiation can be observe. Namely, it can work effectively only on sunny cloudless days.

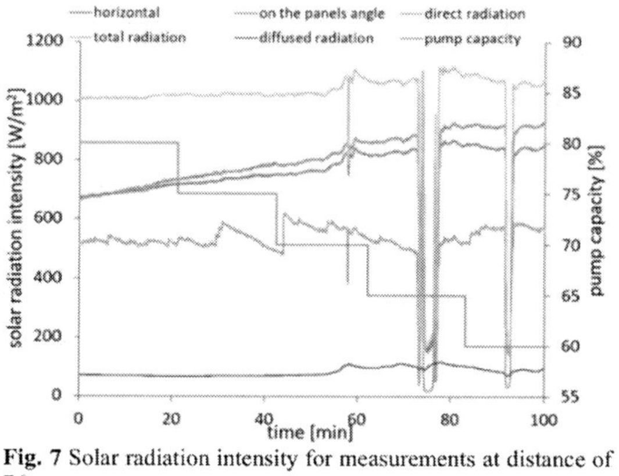


**Fig. 7** Solar radiation intensity for measurements at distance of 76cm

A visible decrease in solar radiation intensity has a significant impact on the achieved value of the receiver's power. Comparing the two distances, it can be noted that the receiver located closer to the mirror (76cm) achieves an output almost 200W higher compared to the receiver moved 2cm back. In some part it is effect of the higher average radiation intensity comparing intensity for the distance 78cm. This effect will be included in computer simulations presented in the further part of the paper.

## 4 Computer simulations

To explain observed data, the computer simulations were carried out. For this purpose, two simulation methods were used. In the first one, the Ray Tracing Monte Carlo (RTMC) method was applied to trace solar rays in the optical system. The second method – Computational Fluid Dynamics – was used for creating the model of the working medium flow inside the high-temperature receiver.

### 4.1 Ray Tracing

The Ray Tracing method was used to determine energy of solar radiation delivered to the surface of the receiver in function of the distance from the parabolic mirror. [9]. On this basis the optimal distance was selected – 76cm. Computer simulations were made using the software of the Lambda Research company [10].

For the purposes of conducting the simulation, 4 geometric blocks were designed: a canopy and a concentrator's mirror, flat receiver of the high-temperature heat and the Sun. For the bowl covered with reflective foil, the reflectance ratio was set at the level of 70%, while for the mirror surface this rate was set to 95%. The absorbency rate was adopted for the absorption surface of the receiver at the level of 90%. An important parameter is also the spectrum of solar radiation emitted by the block of "the Sun". The spectrum in conformity with the standard ASTM E-490-00 was used [11].

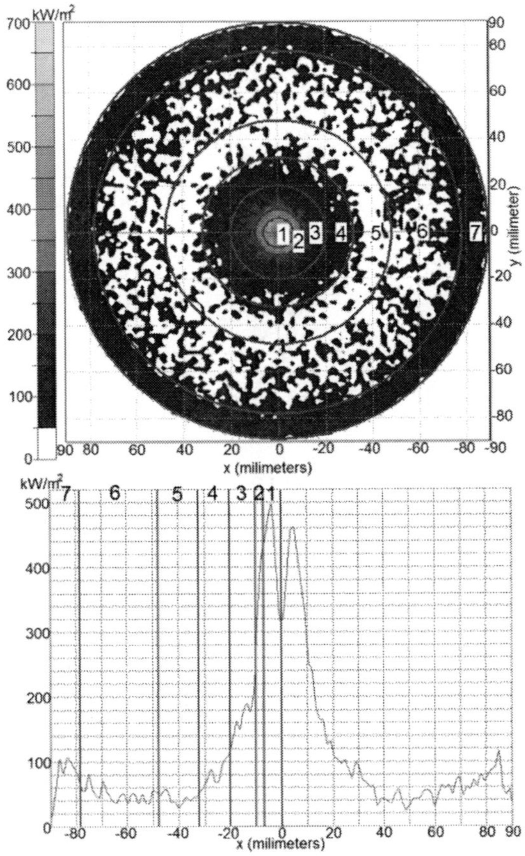


**Fig. 8** Map and graph of radiation intensity on the absorbing Surface of high temperature heat receiver.

The values of the solar radiation intensity on the absorption surface were generated in the form of maps [9]. Fig.8 presented a sample map for a distance of 76cm, where the intensity of radiation emitted from the surface of the Sun was $1025 W/m^2$.

In fig.8 it can be seen that the intensity of solar radiation on the entire absorption surface of the receiver is not uniform. The highest intensity was noted on a small area (less than $1 cm^2$) in the central part of the receiver – approx. $600\text{-}700 kW/m^2$. The largest part of the receiver absorbs the radiation with the intensity of approx. $50\text{-}100 kW/m^2$.

The data obtained through the simulation using the Ray Tracing Monte Carlo method was used for creating a numerical model of the working medium flow inside the receiver using the Computational Fluid Dynamics method. For this purpose, the surface of the receiver was divided with circles presented in fig. H. The average values of the radiation intensity in each marked ring were used as the value constituting the boundary condition of the second type in the CFD modelling.

**Tab. 1**. Boundary conditions for RTMC model

| Experiment | | | | RTMC model |
|---|---|---|---|---|
| flow [l/min] | receiver inlet temp. | receiver outlet temp. | average radiation intensity $[W/m^2]$ | total flux on the absorber [W] |
| 3.1 | 55.6 | 66.2 | 1020 | 1733 |
| 2.9 | 60.2 | 70.3 | 1025 | 1743 |
| 2.7 | 64.9 | 77.5 | 1081 | 1838 |
| 2.5 | 56.9 | 67.4 | 1020 | 1733 |
| 2.2 | 74.2 | 92.6 | 1077 | 1831 |

For the distance indicated on the path of the computer simulations of the distances of 76cm five experimental measurements were selected with a different value of the working medium flow. The intensity of the solar radiation in the simulation was set as a value read from the measurement transducers during the experimental research. As a result of the conducted analyses, based on the results of the simulations, the total flux absorbed by the receiver's surface was calculated. The results are presented in table 1.

### 4.2. Thermal analysis

The experimental studies have allowed the measurement of the temperature difference of the working medium between the outlet and inlet to the device. The spatial numerical model using the computational fluid dynamics model (CFD) was developed for a more detailed way of analysing the operation of the device and location of the potential imperfections. The model describes the phenomena related with the flow of the working medium, and with the heat exchange between the receiver and the environment. The model in the steady state was made using the commercial software ANSYS Fluent [12].

In the first stage of works, the spatial geometry was prepared, representing the real receiver and the domain of liquid filling it. Then, the computing grid was prepared, for the purposes of the numerical model. After the final adjustment, it consisted of about 3,1 million elements, mostly hexagonal in nature. The boundary layer consisted of 8 layers with the growth factor equal 1.2, and height of the first element equal 0.25 mm. The cross-section of the computational grid of the liquid domain with the visible boundary layer was presented in fig.9.

The material of the fluid domain was defined using data of the thermal oil manufacturer. The properties of the oil, such as specific heat, density and viscosity were defined as the temperature functions, using higher order polynomials. The domain material of the receiver was aluminium, which properties were defined based on the database of the program.

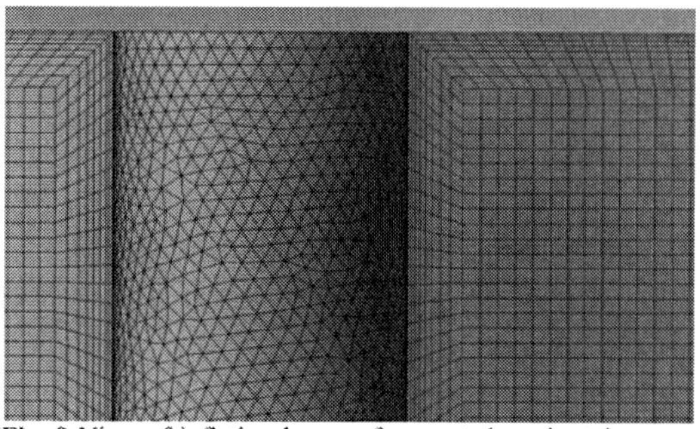

**Fig. 9** View of inflation layers of generated mesh at the cross section of fluid domain

2-equation turbulence model SST k-omega was used for the calculations. The contact surfaces between the domains were described using the two-directional interfaces.

The constant boundary conditions of the model were the relative pressure at the outlet of the domain, equal 0 Pa, and parameters of the heat exchange on the rear and side walls of the receiver. The heat transfer coefficient was defined as 10 W/($m^2$K), while the coefficient of emissivity of the surface was equal 0.85. These conditions were fixed for each of the calculation cases. However, the real conditions of the environment during individual measurement series of the experimental studies were variable. This may have a slight impact on differences between the results of the model and the experiment.

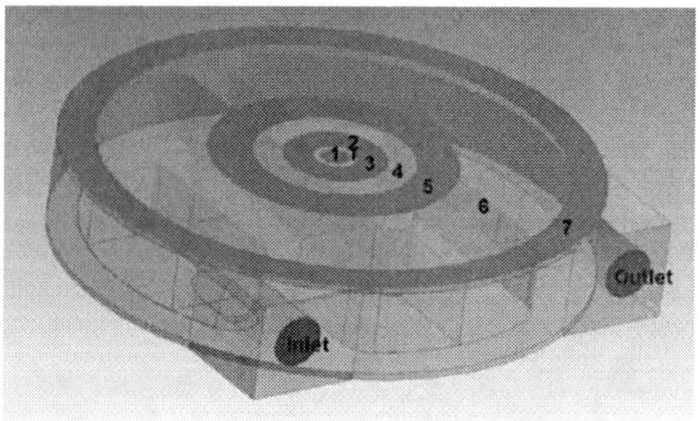


**Fig. 10.** Surfaces for boundary conditions

The remaining boundary conditions were defined using the parameters in order to perform the variant analysis. Six cases (DP0 – DP5) with different boundary conditions defined based on the experimental and RTMC data, have been developed, to prepare devices characteristic. The full set of boundary conditions for all design points was presented in the table 2. The areas, on which the boundary conditions were defined for each case, on the other hand were presented in fig. 10. The areas 1-7 on the absorber's surface correspond the boundary conditions "In1-In7", while "Inlet" and "Outlet" are the inlet and outlet boundary conditions for the fluid domain. For each case also the pressure in the system was defined, based on the results of the experimental research.

**Tab. 2**. Boundary conditions for design points DP0-DP5

| DP0 | DP1 | DP2 | DP3 | DP4 | DP5 |
|---|---|---|---|---|---|
| Inlet Velocity [m/s] | | | | | |
| 0.130 | 0.144 | 0.164 | 0.177 | 0.190 | 0.200 |
| Inlet Temperature [K] | | | | | |
| 371.0 | 347.3 | 331.6 | 338.0 | 330.0 | 328.7 |
| Operating Pressure ($\cdot 10^{-5}$) [Pa] | | | | | |
| 1.76 | 1.76 | 1.75 | 1.76 | 1.80 | 1.78 |
| In1 [kW/$m^2$] | | | | | |
| 294 | 308 | 301 | 301 | 294 | 287 |
| In2 [kW/$m^2$] | | | | | |
| 224 | 259 | 266 | 273 | 259 | 252 |
| In3 [kW/$m^2$] | | | | | |
| 112 | 126 | 126 | 126 | 126 | 119 |
| In4 [kW/$m^2$] | | | | | |
| 49 | 56 | 56 | 56 | 56 | 56 |
| In5 [kW/$m^2$] | | | | | |
| 21 | 28 | 35 | 35 | 28 | 28 |
| In6 [kW/$m^2$] | | | | | |
| 35 | 42 | 35 | 35 | 42 | 35 |
| In7 [kW/$m^2$] | | | | | |
| 56 | 63 | 56 | 56 | 63 | 56 |

Calculations were conducted for six cases, selected on the basis of the mass flow values at the inlet (DP0-DP5, table 2). The values of the average velocity and the fluid temperature were defined as the inlet boundary condition on the "Inlet" surface. These values were selected based on the results of experimental studies. The same way was used for defining the pressure in the system, for each point. On the basis of the radiation schemes, obtained using the model used in accordance with the RTMC method, also the boundary conditions were defined on the absorber's surface, divided into 7 rings. The heat flux derived from the radiation reflected by the concentrator incident on the absorber's surface was lowered by 30% of its value in order to obtain the complete heat flux passing through the surface.

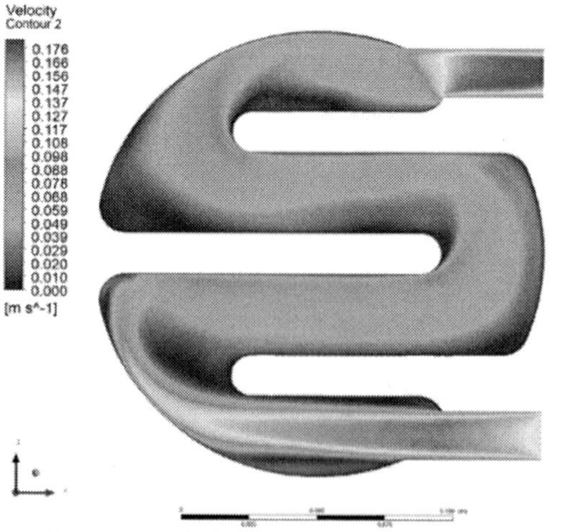


**Fig. 11** Average velocity value at symmetry cross-section plane of fluid domain for DP0

DOI: 10.1051/epjconf/201714302096

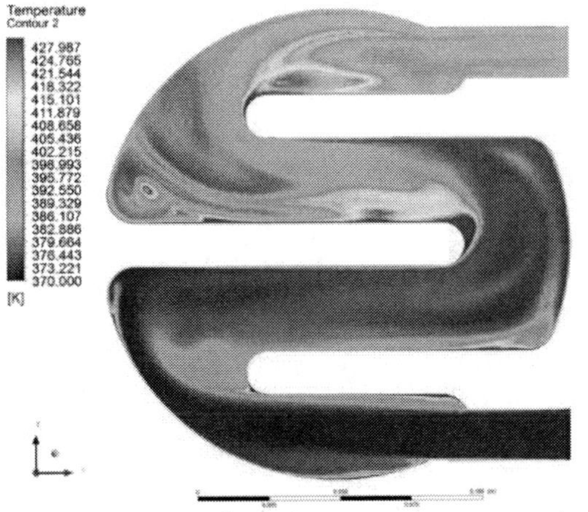


**Fig. 12** Temperature distribution at symmetry cross-section plane of fluid domain for DP0

Fig. 11 shows the average value of the velocity on the symmetry surface of the fluid domain. Some zones, in which the oil flow is inefficient, and the average speed is near 0 can be observed on the bends of the channel. These zones are the largest in the area of the first and the third return of the fluid. Figure 12 in turn shows the temperature distribution on the same symmetry plane. It can be observed how these two parameters relate to each other. Due to the insufficient heat transfer in the areas, where the flow velocity is close to zero, the temperature value is clearly higher. In certain areas (in the middle part of the receiver) these values are even almost 100K higher than the average temperature value of the fluid. This phenomenon not only affects the reduction of the efficiency of the heat exchange, but it can cause the local sources of overheating of the fluid. The results of numerical modelling therefore suggest that some modifications of geometry, which aim to align the flow, could have a positive impact on the increase of the efficiency of the receiver.

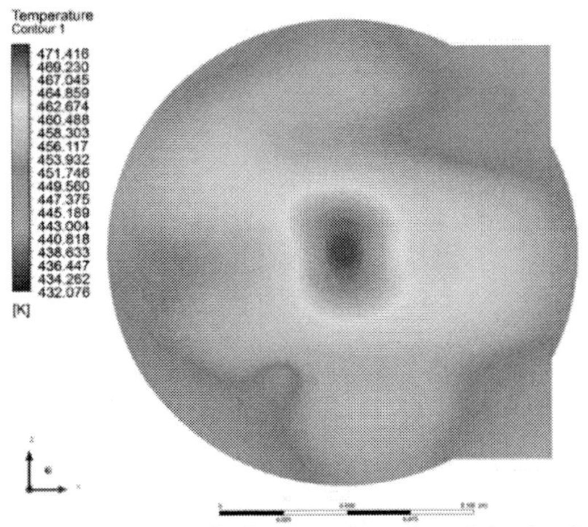


**Fig. 13** Temperature distribution at the surface of absorber for DP0

Figure 13 presents the temperature distribution on the absorber's surface. These values are very difficult to measure in the experimental manner, due to the large impact on the operation of measuring instruments falling on the surface of solar radiation. The average temperature on the surface of the absorber was 460K. The highest value obtained in the vicinity of the focal point of the concentrator and exceeded 470K.

## 5 Discussion

As part of the research the experimental measurements were performed using the installations of the solar collectors connected with the solar radiation collector, and calculations of the system using the numerical methods, RTMC and CFD.

As a result of experimental studies, the potentially most beneficial location of the receiver was determined in the geometrical system of the concentrator. For two selected distances the energy parameters of the system were measured, based on which the receiver's capacity was calculated.

The results obtained in the way of the experiment were then used as the input data for conducting computer simulations with the Ray Tracing Monte Carlo method. This way the values of the intensity of the radiation absorbed by the surface of the high-temperature heat receiver was determined. These values were then used as boundary conditions for creating the numerical model with the Computational Fluid Dynamics method.

The completed research works allowed the performance of the validation of the results of the computer simulations and the ones obtained through the experimental research. The results were presented in a graph in fig. 14.

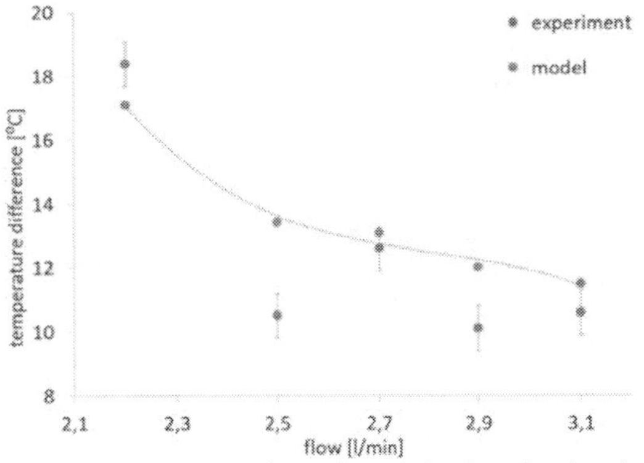


**Fig. 14** Temperature at the surface of absorber placed at the distance of 76cm from the mirror surface

The graph in fig. 14 presents the values of temperature difference between the outlet and inlet obtained for different values of the working medium flow through the receiver, obtained in way of the experiment and numerical modelling. The highest values of reheating the medium obtained in the case of the volumetric stream equal approx. 2.2 l/min. Because of the varied atmospheric conditions, the power at this point was lower than the power obtained for the flow contained in the range of 2.5 – 3 l/min (fig. 6). However, the studies showed that when using the system of the concentrator it is possible to increase the temperature of the factor by nearly 20°C, what is sufficient to effectively power the absorption chiller.

The studies conducted in diverse approaches (experimental measurements and computer simulations) made it possible to optimise the hybrid system of solar collectors with the solar radiation concentrator. On the path of the experimental studies and simulations of the Ray Tracing type it was indicated that the best location of the receiver in the geometrical system of the concentrator is the distance of 76cm from the mirror's surface. The comparison of the experimental results with the numerical model implemented with the CFD method allowed to point out that the best value of the working medium flow for the studied system is approx. 2l/min.

Further researches will be related to cool production in sorption chiller powered by high temperature heat generated in concentrated solar radiation system.

## Acknowledgement

The work has been completed as part of the statutory activities of the Faculty of Energy and Fuels at the AGH University "Studies concerning the conditions of sustainable energy development".

## References

[1] A. M. Daabo, S. Mahmouda, R. K. Al-Dadah, Appl. Energ. **179**, 1081–1096, (2016)

[2] R. Perez-Enciso, A. Gallo, D. Riveros-Rosas, E. Fuentealba-Vidal, C. Perez-Rábago, Renew. Energ. **93**, 115-124, (2016)

[3] M. Rahou, A. Mojiri, G. Rosengarten, J. Andrews, Sol. Energy, **124**, 15–25, (2016)

[4] M.I. Roldán a, E. Zarza, J.L. Casas, Renew. Energ. **76**, 608-618, (2015)

[5] N. Tu, J. Wei, J. Fang, Sol. Energy **112**, 128–143 (2015)

[6] M.I. Roldán, O. Smirnova, T. Fend, J.L. Casas, E. Zarza, Renew. Energ., **62**, 116-128, (2014)

[7] E. Przenzak, M. Szubel, M. Filipowicz, Energ. Convers. Manage., **125**, 97-106, (2016)

[8] E. Przenzak, M. Filipowicz, EPJ – Web of Conferences, **114**, 02100 (2016), DOI: http://dx.doi.org/10.1051/epjconf/201611402100

[9] E. Bożek, M. Filipowicz, J Civil Eng. Env. Arch., **32,** vol.62 no2, 19-28, (2015)

[10] TracePro software, Lambda Research Corporation, http://www.lambdares.com/, last access 06.10.2016

[11] Allen C. W., Astrophysical Quantities, The Athlone Press, London, 1963

[12] ANSYS software, Inc, http://www.ansys.com/, last access 06.10.2016

# Validation of digital holographic tomography in flow measurement

*Pavel* Psota[1*], *Roman* Doleček[1], *Vít* Lédl[1], *Tomáš* Vít[2], *Pavel* Mokrý[1], and *Petra* Dančová[2]

[1]Regional Centre for Special Optics and Optoelectronic Systems (TOPTEC), Institute of Plasma Physics, Academy of Sciences of the Czech Republic, Za Slovankou 1782/3, 182 00 Prague 8, Czech Republic
[2]Institute of Energetic Devices (KEZ), Faculty of Mechanical Engineering, Technical University of Liberec, Studentská 2, 461 17 Liberec, Czech Republic

**Abstract.** This paper reports on measurements by digital holographic tomography in flow measurement. It is demonstrated capability of the method by measurement of general temperature field consisting of system of three orifices. Results obtained by digital holographic tomography are compared to results obtained by hot wire anemometry CTA.

## 1 Introduction

Digital holographic interferometry (DHI) for measurement and visualization of temperature fields is well established tool in the field of the fluid mechanics [1-4]. DHI is very sensitive to faint temperature variations that are measured by a laser beam propagating through the inspected area and thus the method is noninvasive. However, due to integral nature of interferometric techniques, for general temperature fields the conventional DHI must be replaced by more complex method called digital holographic tomography (DHT) [1,5]. DHT provides information of the temperature variations within the whole volume with great lateral and temperature resolution. Since DHT is quite new approach in the fluid engineering the authors want to lay the DHT on the firm foundation by comparison of DHT to generally accepted technique like Constant Temperature Anemometry (CTA) [6]. This paper briefly introduces basic principles of DHT, shows some possible application in fluidics and the results are verified by CTA.

## 2 Digital holographic tomography in fluid mechanics

### 2.1 Digital Holography

Digital holography comprises recording of the digital hologram and its reconstruction. For recording, CCD or CMOS sensors are used to capture microinterference pattern (digital hologram) in which the information about the phase of the incoming waves is encoded. The hologram $H$ is formed by a superposition of a reference wave $U_r$ and an object wave $U_o$ :

$$H \approx |U_o + U_r|^2 = |U_o|^2 + |U_r|^2 + U_o U_r^* + U_r U_o^* . \qquad (1)$$

Digitally recorded hologram is transferred to a computer as an array of numbers that represent the wave field at the hologram plane. Reconstruction of the digital hologram is done numerically by a computer. The propagation of optical fields is completely described by diffraction theory, which allows numerical reconstruction of the image as an array of complex numbers representing the amplitude and phase of the optical field. The free space propagation can be solved by the convolution theorem [1]

$$U(x,y) = \\ = \Im^{-1}\left\{\Im(H(\xi,\eta) r^*(\xi,\eta))\Im(g(x,y))\right\} \qquad (2)$$

with the impulse response:

$$g(x,y) = \frac{1}{j\lambda}\frac{\exp(jk\sqrt{d^2 + x^2 + y^2})}{\sqrt{d^2 + x^2 + y^2}} . \qquad (3)$$

The Fourier transform of *g(x,y)* can be calculated and expressed analytically as well. This saves one Fourier transform for reconstruction.

### 2.2 Temperature fields measurement by DHT

It is well know that a temperature variation in fluids is accompanied by a variation of refractive index of the fluid. This feature can be exploited for measurement of the temperature through the knowledge of the refractive index change. When two waves propagate through a medium with certain refractive indexes their phase difference $\Delta\varphi$ varies with respect to the change of the

* Corresponding author: psota@ipp.cas.cz

refractive indexes $\Delta n$. This is mathematically expressed by formula

$$\Delta\varphi = \frac{2\pi}{\lambda}\oint_L \Delta n dl, \quad (4)$$

where $dl$ denotes the differential distance along the line $L$ and $\lambda$ is the wavelength of the wave. Therefore, to retrieve information about the refractive index change one has to solve (4).

The solution is quite straightforward for two dimensional temperature fields (integral is simplified to multiplication) and for symmetrical temperature fields (leads to Abel transform), however, for general distribution of temperature fields the solution of (4) without any additional information is impossible. Therefore it is required to use tomographical approach.

The refractive index change in (4) can be written in polar coordinates $(t,\Phi)$ as

$$\Delta n(y,z) = \Delta n(t\cos\Phi, t\sin\Phi). \quad (5)$$

The determination of general fields requires the analysis of a large number of phase projections by methods of tomography. In this case the line integral transform (1) is mathematically equivalent to Radon Transform:

$$\Delta\varphi(s,\theta) = \iint \frac{2\pi}{\lambda}\Delta n(y,z)\delta(y\cos\theta + z\sin\theta - s)dydz, \quad (6)$$

where $\Delta n(y,z)$ is the refractive index distribution in Cartesian coordinates, δ is the Dirac pulse, θ is the angle of projection and s is the shift distance from the origin. The reconstruction of the refractive index change is based on back-projection mathematically described by Inverse Radon Transform [1]:

$$\Delta n(y,z) = \frac{\lambda}{2\pi}\int_0^{\pi} g(y\cos\theta + z\sin\theta, \theta)d\theta. \quad (7)$$

The back-projection operation propagates the measured phase information back into the image space along the projection paths in order to get 3D volume data of the refractive index change.

The phase at one projection can be measured by digital holographic interferometry where two digital holograms $H_1, H_2$ are captured by a digital sensor. The first hologram corresponds to an initial (reference) state of a temperature field and the second hologram is captured when the temperature field has been varied. Two wave fields are reconstructed from the digital holograms as complex fields represented by formula:

$$\begin{aligned} U_1(n\Delta x, m\Delta y) &= |U_1(n\Delta x, m\Delta y)| \times \\ &\times \exp[\varphi_1(n\Delta x, m\Delta y)], \\ U_2(n\Delta x, m\Delta y) &= |U_2(n\Delta x, m\Delta y)| \times \\ &\times \exp[\varphi_2(n\Delta x, m\Delta y)], \end{aligned} \quad (8)$$

respectively. When the wave fields are divided one by another $U_1/U_2$, their phase difference $\Delta\varphi$ is subtracted, and it can be shown that for the phase difference $\Delta\varphi$ holds an expression:

$$\Delta\varphi = \arg\left(\frac{\mathrm{Im}\{U_1\}\mathrm{Re}\{U_2\} - \mathrm{Im}\{U_2\}\mathrm{Re}\{U_1\}}{\mathrm{Im}\{U_1\}\mathrm{Im}\{U_2\} + \mathrm{Re}\{U_1\}\mathrm{Re}\{U_2\}}\right). \quad (9)$$

## 3 Experimental arrangements

The DHT measurement was realized by Twymann-Green interferometer, see Fig. 1. The spatially filtered (SF) and collimated (CO) laser beam is split by polarizing beam splitter BS1. The reference beam (RW) goes through mirrors M1, M2 and beam splitter BS2 to the digital camera (CAM). The object beam (OW) goes twice (before and after reflection at M3) through the measured area (SAMPLE) and after the beam splitter BS2 impinges the CAM where the both beams are superposed creating the digital hologram.

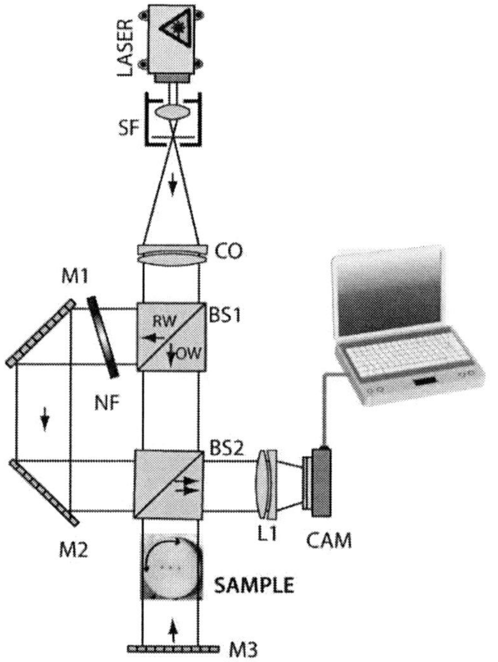


**Fig. 1.** Experimental holographic arrangements schematics of interferometer with double pass for tomographic measurement. BS1-polarizing beam splitter, SF-spatial filter, CO-collimation objective, BS2-nonpolarizing beam splitter, L1-objective lens, M-mirror, NF-neutral filter.

This setup forces the beam to pass through the measured object twice which results in higher sensitivity [3,5].

To demonstrate the reliability of the method we measured a non-symmetrical temperature field of heated air flowing out from a three orifices system. The outgoing air was heated inside the cavity under the orifice by a heating cartridge in order to increase its temperature difference to surrounded fluid. The

DOI: 10.1051/epjconf/201714302097

temperature of heating cartridge was set to 100°C and regulated by PID regulator.

The measured object was placed on the 360° rotation stage and rotated step by step at angle of 10° from 0° to 170° to get projections from different viewing angles. At every step a sequence of digital holograms is captured and reconstructed. The reconstructed phase fields at the same projection angle were averaged out in order to suppress the random phenomena. This measurement procedure was repeated for every angle step. Before the measurement we captured a sequence of reference digital holograms when the heated system was off.

The representation of the reconstructed phase data as a function of the angle is known as a sinogram. The 3-D temperature distribution can be reconstructed from the sinograms by a filtered backprojection algorithm, see (7). For this purpose, the standard inverse Radon transform (iradon command) in MATLAB was used with Ram-Lak filter and implemented slice-by-slice along the rotation axis. Reconstructed temperature distribution in 3D space is shown in Fig. 2.

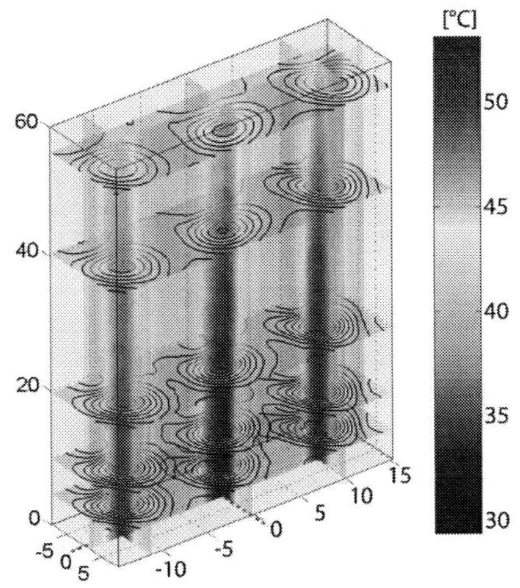


**Fig. 2.** 3D temperature distribution measured by DHT. Axes units are mm

## 4 Results and Discussions

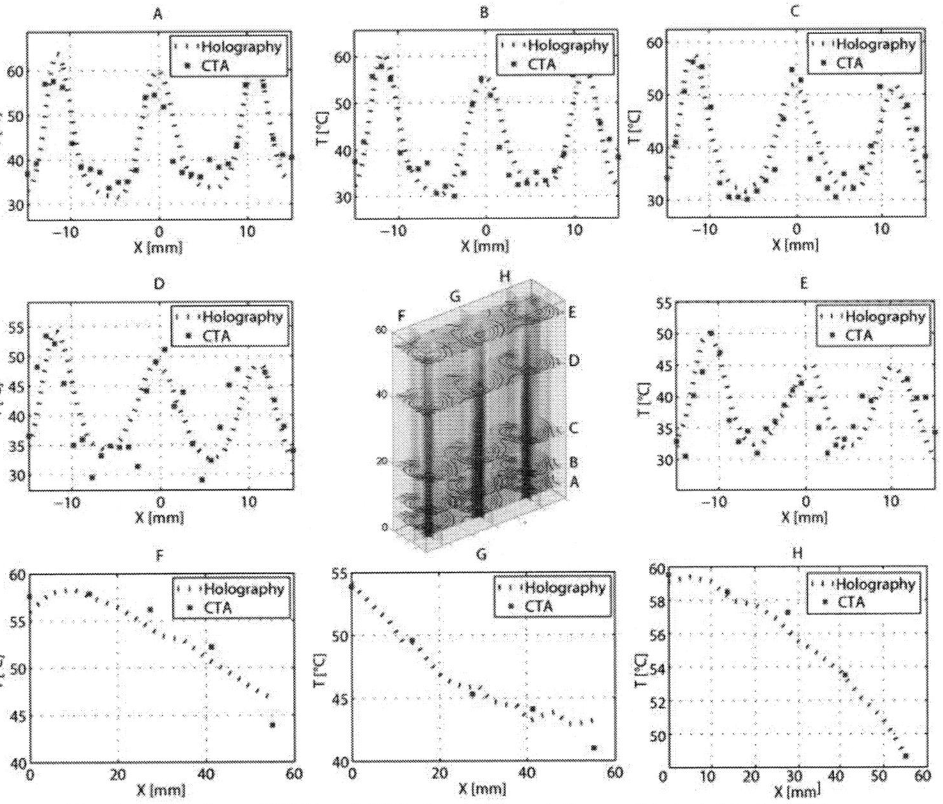


**Fig. 3.** Comparison of DHT and CTA along different slices as illustrated in the 3D image.

To verify the reliability of digital holographic tomography we compared results of DHT with measurement results of well-established CTA (Constatnt Temperature Anemometry). CTA is single point method and therefore we chose some representative points

within the measured area. The CTA measurements were carried out along horizontal lines at different heights. The results at different heights (5 mm, 10 mm, 20 mm, 40 mm and 55 mm) are denoted by the letters A-E in Fig. 3. The distance of measuring step by CTA in horizontal direction was 1 mm. DHT is a full-field method and therefore we had to pick voxels of the measured temperature distribution corresponding to those values measured by CTA.
Results are presented in Fig. 3. 1D plots denoted A-E present results obtained by DHT and CTA along horizontal lines with the same label. The graphs denoted by letters F-H shows how the temperature varies along vertical line (axis of the orrifices).

To quantitatively determine the difference of the method we computed RMS difference $\Delta T_{RMS}$ at all measured points $p$:

$$\Delta T_{RMS} = \sqrt{\frac{\sum_{p=1}^{N}\left(T_{DHT}(p)-T_{CTA}(p)\right)^2}{N}} = 3.8°C \quad (10)$$

The RMS (root-mean-square) difference is 3.8 °C that can be expressed as an agreement in the range of cca 12%.

It is worthwile mentioning the main sources of the disrepancies in results between DHT and CTA. The major contributor is fact that the both measurements had to be performed separately at different time due to invasive feature of CTA. The enviromental conditions as well as behaviour of the phenomenon itself had been probably slightly varying between both meausurement procedures.

The uncertainty of measurement by DHT is also affected by the number of projection used in inverse Radon transform. The angle step between succesive projections was set to be 10°. This generates an error of cca 0.4 °C for temperature values typical for our system, see Fig. 4.

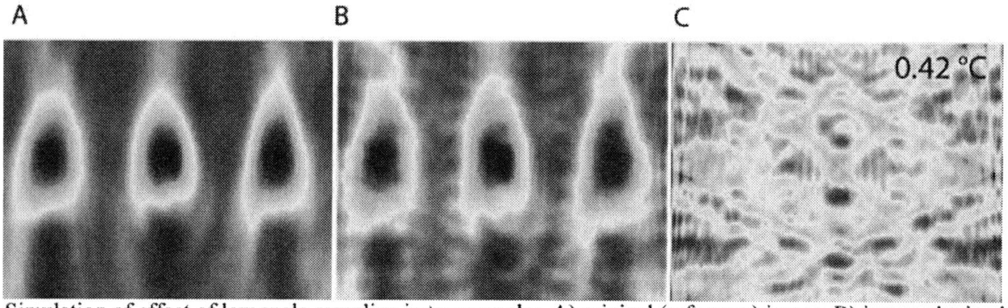


**Fig. 4.** Simulation of effect of low angle sampling in tomography. A) original (reference) image. B) image obtained by inverse Radon transform with step of 10°. C) RMS difference (error) between A and B. The theoretical error introduced by low angle sampling in our experiments is 0.42°C .

We gratefully acknowledge the support of the Grant Agency of the Czech Republic - Czech Science Foundation (projects no. 14-08888S and 14-32228S).

## References

1. T. Kreis, *Handbook of holographic interferometry: optical and digital methods. 2005.*
2. C. Shakher and A. K. Nirala, *Applied optics*, **33**, 11, (1994).
3. V. Lédl, T. Vít, R. Doleček, and P. Psota, *EPJ Web of Conferences*, **25,** (2012)
4. D. Wang and T. Zhuang, *Optics and lasers in engineering*, **36**, 3, (2001).
5. Roman Dolecek, Pavel Psota, Vít Lédl, Tomáš Vít, Jan Václavík, and Václav Kopecký, *Applied Optics*, **52**, 1, (2013).
6. J. Smits, A. E. Perry, and P. H. Hoffmann, *Journal of Physics E: Scientific Instruments*, **11**, 9, (2001)

EPJ Web of Conferences **143**, 02098 (2017) DOI: 10.1051/epjconf/201714302098
*EFM 2016*

# Experimental investigation of hydrodynamic cavitation through orifices of different geometries

*Pavel* Rudolf[1,*], *Dávid* Kubina[1], *Martin* Hudec[1], *Jiří* Kozák[1], *Blahoslav* Maršálek[2], *Eliška* Maršálková[2] and *František* Pochylý[1]

[1]Brno University of Technology, Faculty of Mechanical Engineering, V. Kaplan Department of Fluid Engineering, Technická 2896/2, 61669 Brno, Czech Republic
[2]Institute of Botany – Academy of Sciences of the Czech Republic, Lidická 25/27, 60200 Brno, Czech Republic

**Abstract.** Hydrodynamic cavitation in single and multihole orifices was experimentally investigated to assess their hydraulic characteristics: loss coefficients, inception cavitation number, cavitation number for transition to supercavitation. Significant difference for singlehole and multihole orifices was observed in terms of the measured loss coefficient. It is significantly more effective to use multihole orifices, where energy dissipation is much lower.It was found that using scaling factor given by ratio of orifice thickness suggests linear behaviour of both loss coefficient and inception cavitation number. Orifices seem to be convenient choice as flow constriction devices inducing cavitation due to their simplicity.

## 1 Introduction

Cavitation is process of liquid evaporation due to sudden pressure drop down to the the level of saturated vapor pressure. When the cavitation bubbles are subjected to higher pressure, so called cavitation collapse occurs, which is very fast condensation [1]. Cavitation is very undesirable in hydraulic machines, where severe erosion of the surfaces, vibrations and pulsations take place. On the other hand positive utilization of cavitation has been reported in cleaning processes, sonochemistry or food processing industry [2].

Cavitation can be induced by fast moving liquid stream, ultrasonic waves, focused laser beam or electrical spark. The latter methods are not very practical, because they only produce single bubbles. Ultrasonic generation of cavitation is very effective, but only in small volumes of liquid. Therefore the only way of liquid treatment by cavitation in large bulks of liquid is hydrodynamic cavitation.

Hydrodynamic cavitation has been successfully used for cell disruption [3, 4] and disinfection (i.e. elimination of microorganisms). The effects of cavitation are both physical and chemical. Physical effects are due to pressure waves emitted during cavitation bubble collapse and microjet formation. Chemical effects are caused by generation of free radicals [3].

In our previous research Venturi tube has been extensively used to induce cavitation [6-8]. Thorough description of cavitation regimes was provided based on high-speed camera records and records of pressure pulsations. However in research of [3] orifice is used as a simple flow constriction device, which induces local pressure drop. In [5] also multihole variants of orifice are investigated.

In our present research we focus on using multihole orifices for cyanobacteria mitigation. In the first stage several types of single and multihole orifices were experimentally investigated to assess their hydraulic parameters, namely loss coefficients, inception cavitation number and regime maps.

## 2 Experimental setup

A closed hydraulic circuit in laboratories of V. Kaplan Department of Fluid Engineering was used for all experimental tests (see Fig.1). Circuit is equipped with frequency controlled pump, pressure vessel and vacuum pump. Common pressure transducers based on strain gauge principle (DMP331, BD Sensors, accuracy ±0.25% of the range) were used. Flow rate was measured by induction flowmeter (MQI99, ELA, accuracy ±0.5% of the range). Temperature data were acquired by resistance thermometer (HSO-502 1A2L, HIT, accuracy ±0.1% of the range). Electric signals from all transducers and sensors are collected via data acquisition system NI9184 (National Instruments). Measurements were conducted in LabVIEW environment.

Several pressure sensors were mounted (p3, ..., p6) to assess influence of sensor position on the measured pressures. Circuit was operated at least for half an hour prior the measurement, in regime of maximum underpressure achievable by vacuum pump to ensure elimination of air from water.
Test section is made from plexiglass to provide visualization of the cavitation pattern.

* Corresponding author: rudolf@fme.vutbr.cz

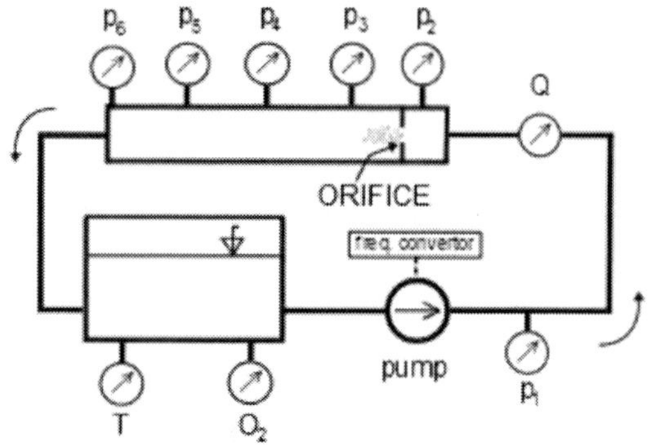


**Fig. 1.** Hydraulic circuit.

Rest of the circuit was made from plastic pipelines with inner diameter 53 mm. The orifice was manufactured by milling and drilling from stainless steel of 11 mm thickness with outer diameter 53 mm. All orifices have the same total flow area 452 mm$^2$, only number of the holes changes from 1 up to 36 and obviously their diameters change too (from 24 mm for singlehole to 4 mm for 36-hole orifice). Reynolds numbers corresponding to different regimes vary in range 50000 – 200000.

Required cavitation numbers were set mostly using flow rate control, for very low cavitation numbers also with help of the vacuum pump.

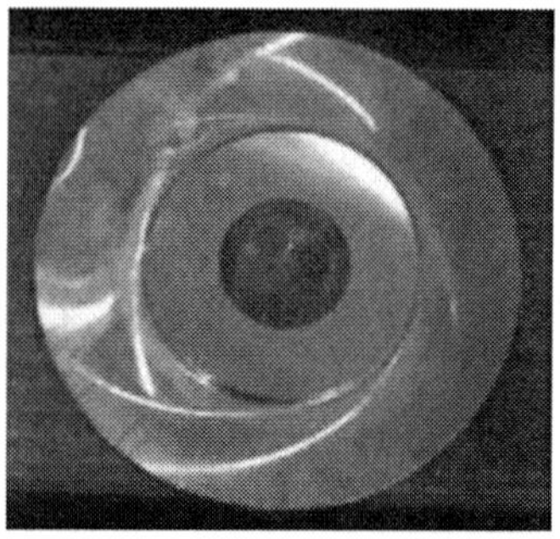

**Fig. 2.** Single hole orifice ∅ 24 mm.

**Fig. 3.** Multihole orifice 9 x ∅ 9 mm (circular arrangement).

**Fig. 4.** Multihole orifice 9 x ∅ 9 mm (square arrangement).

**Fig. 5.** Multihole orifice 16 x ∅ 6 mm.

**Fig. 6.** Multihole orifice 36 x ∅ 4 mm.

## 3 Results

Two loss coefficients were computed. The first one from pressure difference from sensors $p_2$ and $p_3$ and scaled by bulk velocity through the flow area (i.e. area of all holes):

$$\zeta_A = \frac{p_2 - p_3}{\rho \frac{\bar{v}^2}{2}} \quad (1)$$

Sensor $p_2$ is placed 50 mm upstream of the orifice plate, whereas sensor $p_3$ is 50 mm downstream of the orifice.

It means that pressure sensor $p_3$ is in the position where the flow constriction occurs and for some regimes is in region of saturated vapor.

Loss coefficient $\zeta_B$ is evaluated using pressure from sensor $p_6$ (950 mm downstream of the orifice), which is placed in region of stream regeneration.

$$\zeta_B = \frac{p_2 - p_6}{\rho \frac{\bar{v}^2}{2}} \qquad (1)$$

Cavitation number is defined from pressure $p_3$ downstream of the orifice and vapor pressure:

$$\sigma = \frac{p_3 - p_{vapor}}{\rho \frac{\bar{v}^2}{2}} \qquad (2)$$

Vapor pressure is evaluated from temperature according to equation (3):

$$p_{vapor} = 100.\, e^{53{,}6796 - \frac{6743.69}{T} - 4{,}8451 lnT} \qquad (3)$$

Figures 7, 8 present plots of loss coefficients $\zeta_A$ and $\zeta_B$ for all types of orifices.

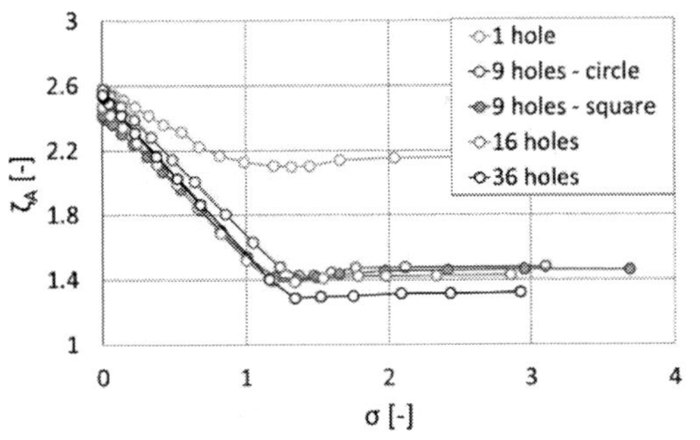


**Fig. 7.** Loss coefficient $\zeta_A$ for .all orifice variants.

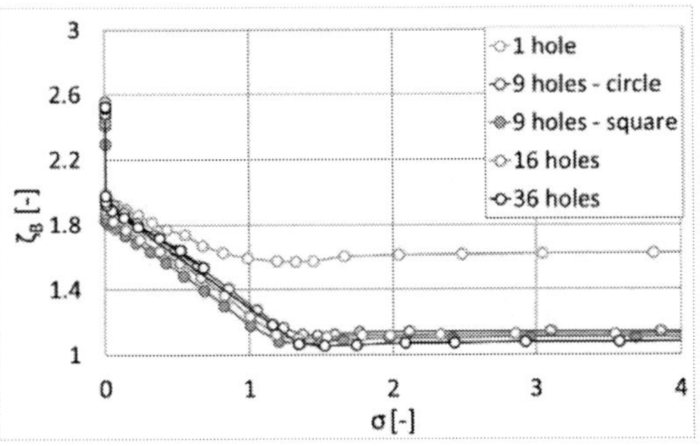


**Fig. 8.** Loss coefficient $\zeta_B$ for .all orifice variants.

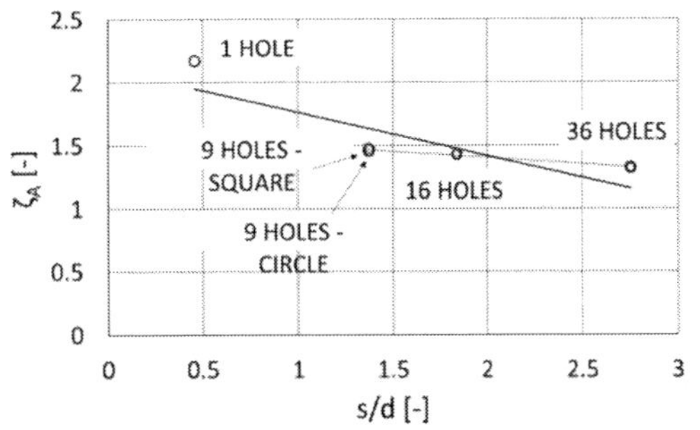


**Fig. 9.** Loss coefficient $\zeta_A$.vs. scaling factor s/d.

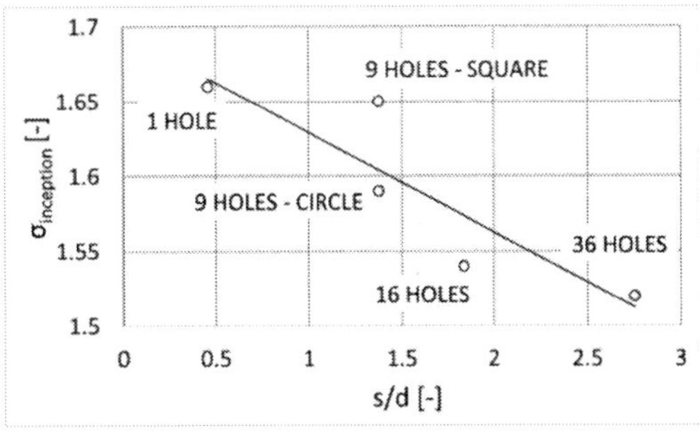


**Fig. 10.** Inception cavitation number vs. scaling factor s/d..

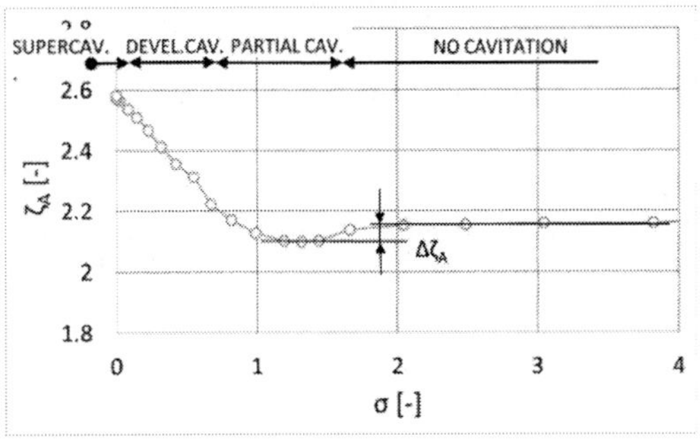


**Fig. 11.** Singlehole orifice – flow regimes.

For both loss coefficients it is apparent that regardless whether it is non-cavitating or cavitating regime there is significant reduction of hydraulic losses for multihole orifices compared with the singlehole one.

Singlehole orifice was selected as a representative to show all the regimes, which are encountered within the operation of the orifice over whole range of flow rates, see Fig.11. Before substantial pressure drop is obtained, the orifice is operated in non-cavitating regime. Then cavitation inception occurs accompanied by slight decrease of loss coefficient. Further increase of flow rate leads to steep increase of dissipated energy within regimes of partial and developed cavitation, where the clouds of bubbles fill whole pipe cross-section. Even higher flow rates, i.e. very low cavitation numbers, are associated with onset of supercavitation. By supercavitation we understand situation, when a liquid jet appears, which is surrounded by saturated vapor.

To compare the different orifices a ratio of orifice thickness (s) and hole diameter (d) was selected as a nondimensional parameter. Figure 9 shows that loss coefficient $\zeta_A$ linearly changes with s/d ratio. The same scaling parameter is also suitable for description of the cavitation inception. Cavitation number corresponding to inception observed by naked eye as appearance of first cavitation bubbles scales linearly with s/d ratio for different variant of multihole orifices.

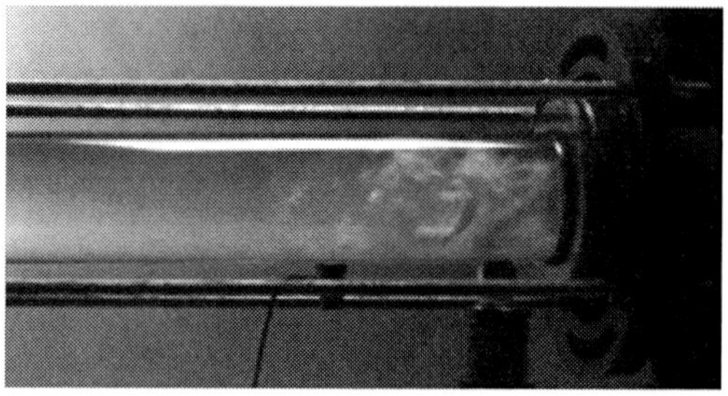

**Fig. 12.** Singlehole orifice – developed cavitation, $\sigma$ = 0.42.

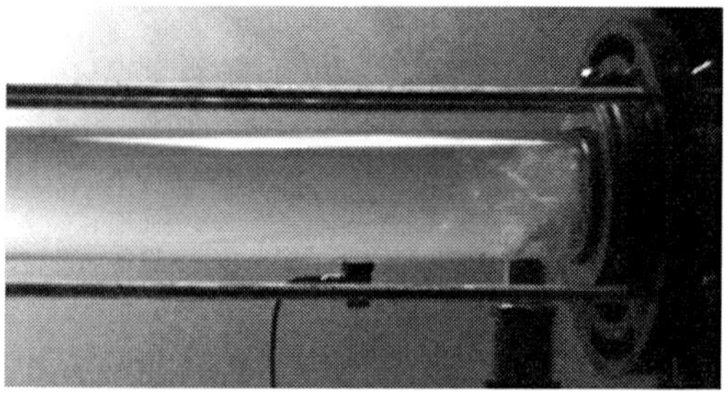

**Fig. 13.** Multihole orifice (9 holes – circular arrangement) – developed cavitation, $\sigma$ = 0.42.

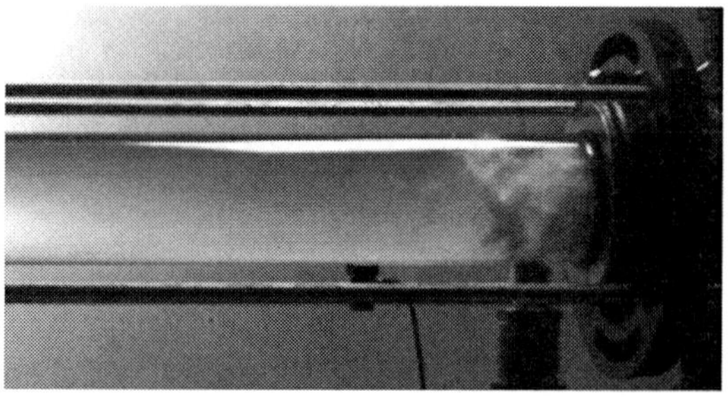

**Fig. 14.** Multihole orifice (9 holes – circular arrangement) – developed cavitation, $\sigma$ = 0.13.

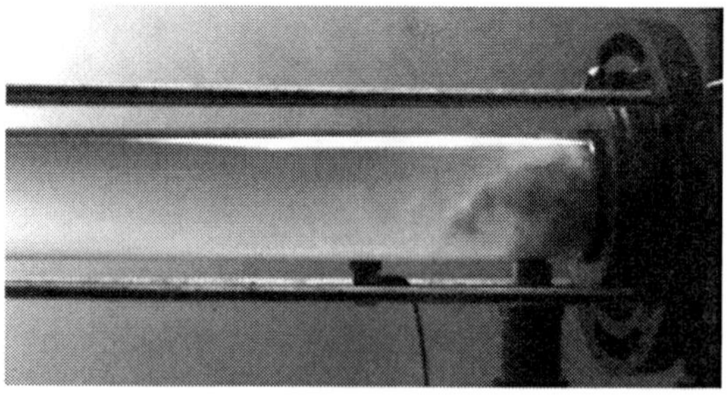

**Fig. 15.** Multihole orifice (36 holes) – developed cavitation, $\sigma$ = 0,13.

Another interesting feature is drop of the loss coefficient just after the cavitation inception (see Fig.11). Usually this phenomenon is explained by wall lubrication due to the tiny cavitation bubbles. However Fig. 7 suggests that the loss coefficient drop is more significant for singlehole orifice than for the multihole ones although the wetted area of the holes is larger for the multihole variants. It should also be noted that the loss coefficient drop is also observed on $\zeta_B$ coefficient computed from the pressure $p_6$ far downstream of the orifice. Absolute and relative loss coefficient drops are summarized in Table 1. Using scaling by the value of loss coefficient in the noncavitating regime it is apparent that the drop is roughly by 3.5% of the value in noncavitating case for all types of orifices.

**Table 1.** Loss coefficient drop in initial stage of partial cavitation

| | $\Delta\zeta_A$ | $\Delta\zeta_A/\zeta_{A,NONCAV}$ |
|---|---|---|
| 1 hole | 0.073 | 0.034 |
| 9 holes - circle | 0.056 | 0.038 |
| 9 holes - square | 0.050 | 0.034 |
| 16holes | 0.048 | 0.034 |
| 36 holes | 0.042 | 0.032 |

## 4 Conclusion

An extensive experimental research was carried with five types of orifices. It was proved that multihole orifices bring significant drop of loss coefficient (down to about 60% compared to singlehole orifice) both in cavitating and non-cavitating regimes. With decreasing cavitation number this difference becomes less pronounced.

Experiments also showed that the initial stage of partial cavitation is characterized by decreasing loss coefficient. This difference makes up about 3.5% compared to the noncavitating case.

Further research will be focused on evaluation of spectral characteristics of the flow, i.e. identification of the significant frequencies associated with the unsteady nature of the cavitating flow behaviour.

Aside the hydraulic characteristics, the efficiency in cyanobacteria elimination will be evaluated for different types of orifices and selected operating regimes.

## Acknowledgement

Czech Science Foundation is gratefully acknowledged for support of this research under project No 16-18316S „Principles and mechanisms causing microorganism elimination by hydrodynamic cavitation".

## References

1. J. P. Franc, *Fundamentals of cavitation*, (2006)
2. P. Gogate, A. Kabadi, Biochem. Eng. Jour., **44**, 1, (2009)
3. B. Balasundaram, S.T.L. Harrison, Biotech. and Bioeng., **94**, 2, (2006)
4. P. Gogate, Chem. Eng. and Proc., **47**, (2008)
5. P. Testud, P. Moussou, A. Hirschberg, Y. Auregan, Proc.PVP2005, (2005)
6. D. Jančula, P. Mikula, B. Maršálek, P. Rudolf, F. Pochylý, Aquacult. Int., **22**, 2, (2014)
7. P. Rudolf, M. Hudec, P. Zubík, D. Štefan, EPJ Web of Conf., 25, 01081, (2012)
8. P. Rudolf, M. Hudec, M. Gríger, D. Štefan, EPJ Web of Conf., 67, 0,12002, (2015)

EPJ Web of Conferences **143**, 02099 (2017)
*EFM 2016*
DOI: 10.1051/epjconf/201714302099

# Flow structures in end-view plane of slender delta wing

*Besir* Sahin[1]*, *Mehmet* Oguz Tasci[1], *Ilyas* Karasu[2] and *Huseyin* Akilli[1]

[1]Department of Mechanical Engineering, Faculty of Engineering and Architecture, Cukurova University, 01330, Adana/ Turkey
[2]Departments of Aerospace Engineering, Faculty of Aeronautics and Aerospace, Gaziantep University, 27310, Gaziantep/Turkey

**Abstract.** Present investigation focuses on unsteady flow structures in end-view planes at the trailing edge of delta wing, X/C=1.0, where consequences of vortex bursting and stall phenomena vary according to angles of attack over the range of $25° \leq \alpha \leq 35°$ and yaw angles, $\beta$ over the range of $0° \leq \beta \leq 20°$. Basic features of counter rotating vortices in end-view planes of delta win with 70° sweep angle, $\Lambda$ are examined both qualitatively and quantitatively using Rhodamine dye and the PIV system. In the light of present experiments it is seen that with increasing yaw angle, $\beta$ symmetrical flow structure is disrupted continuously. Dispersed wind-ward side leading edge vortices cover a large part of flow domain, on the other hand, lee-ward side leading edge vortices cover only a small portion of flow domain.

## 1 Introduction

There are several factors that influence delta wing aerodynamics, particularly formations of vortical flow over delta wing. For example, primary factors are angle of attack $\alpha$, yaw angle $\beta$, sweep angle $\Lambda$, roll angle, $\theta$ wing thickness, t leading edge geometry and conditions of free-stream as well as Reynolds Number, Re. More precisely the geometry of delta wing has close relations with the formations of leading edge vortices and their bursting incidences. Lambourne and Bryer (1961) and Sarpkaya (1971) visualized a couple of vortex bursting events over a delta wing qualitatively which were called spiral and bubble type vortex bursting as shown in Figure 1.

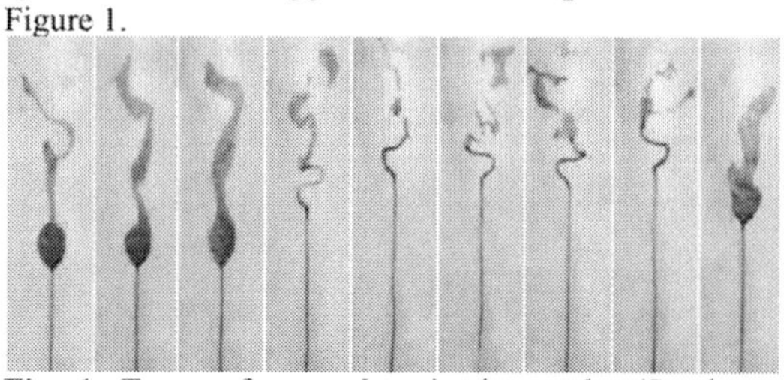

**Fig. 1.** Types of vortex bursting in a tube (Sarpkaya, 1971)

Vortex bursting phenomena is key source of striking delta wing which triggers large mechanical vibrations leading to heavy fatigue destruction and causes the loss of load-bearing ability of a material under periodic load application.

Delta wings are differed from the other type of wings because a couple of leading edge vortices, separations, vortex breakdowns and chaotic vortical flows are generally available. Representation of flow field transformations over a delta wing at a certain angle of attack, $\alpha$ is given in Figure 2 (Anderson, 2001).

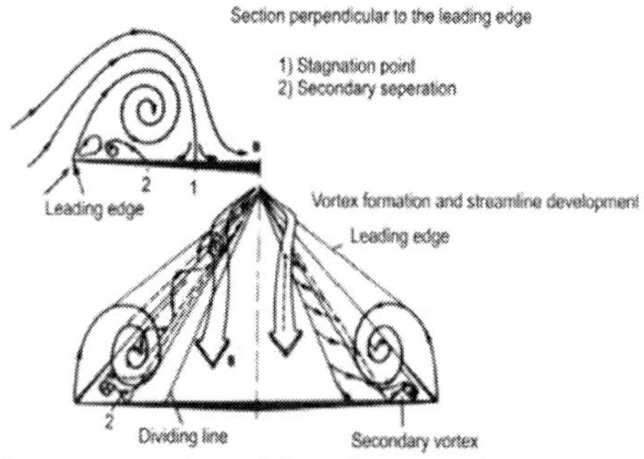


**Fig. 2.** Representation of flow field transformations over a delta wing under angle of attack, $\alpha$ (Anderson, 2001)

It is identified that aerodynamics of high angle of attack, $\alpha$ is one of the most important elements in aircraft design for the view of lift force $F_L$. A delta wing would also ensure high lift coefficient, $C_L$ at a larger angle of attack, $\alpha$ comparing to the wing of passenger aircrafts. Leading edge vortices at high angles of attack, $\alpha$ create most portion of lift forces, $F_L$. But, vortex bursting close to the surface of delta wing deteriorates maneuverability of air vehicles and causes material fatigues. Comprehension of these types of flows is very important for better aircraft maneuverability. A delta wing furnishes the Unmanned Combat Air Vehicles (UCAVs) with capability of sharp maneuvers and tactical advantages. Maneuverability of all these geometrical styles are restricted by the occurrence of vortex bursting and stall influencing both overall forces and moments of aircrafts.

Ozgoren et al. (2002) reported that there are five distinct layers of vorticity in the absence of vortex breakdown at low angle of attack, $\alpha$. Their delta wing

* Corresponding author: bsahin@cu.edu.tr

model was highly swept which had $\Lambda=75°$ and they monitored flow structure at attack angles of attack of $\alpha=24°$, 30°, 32° and 35°. Co-existing distributions of azimuthal vorticity were classified.

Akilli et al. (2001) conducted an experimental examination to search vortex bursting sensitiveness over a delta wing at high angles of attack $\alpha$, via placing very thin wire orthogonally to the center of leading-edge vortex. It is shown that a thin wire could move the onset of vortex bursting by as much as fifteen vortex diameters in the upstream direction.

Cross-flow studies in end-view planes of delta wings have not been conducted in detail. There are only a few studies available. For example, Yaniktepe and Rockwell (2005) performed experimental investigations on non-slender diamond and lambda type wings to analyze flow structures at the trailing edge zones. In both wings, vortical flow structures in the cross-flow planes of trailing edge change rapidly with angles of attack, $\alpha$ when delta wings have high sweep angle, $\Lambda$ more than 40 degree.

Although substantial scientific and technological information was obtained about instantaneous and time-averaged flow structures of delta wings with relatively large sweep angles, $\Lambda$ in recent years, but influences of yaw angle, $\beta$ were not studied at all. Recent demands on UCAVs have encouraged researchers to pay more attentions in characterizing flow structures of delta wings. Present investigation focuses on the unsteady flow structure which occurs downstream of onset of vortex breakdown and services as a source of buffeting and stall that reduces the lift force, $F_L$.

Emphasizes in current investigation is given on the cross-flow in end-view planes at the trailing edge of delta wing, X/C=1.0 where consequences of vortex bursting and stall phenomena vary according to angles of attack, $\alpha$ and yaw angle, $\beta$. In the past, a great deal of effort was spent on the studies of delta wing aerodynamics with a high sweep angle $\Lambda$. It is possible to say that flow structure over this kind of wings is understood very well. But, research work in the case of low and moderate sweep angles, $\Lambda$ of delta wing are quite limited comparing to the delta wings with high sweep angles, $\Lambda$. On the other hand, it is worth to emphasize that the effect of yaw angles, $\beta$ on a delta wing aerodynamics have not been studied in detail yet. Particularly, there is no published work on the aerodynamics of slender delta wings as a function of yaw angle, $\beta$. For this reason, as a motivation of present work, further studies are required in order to understand flow structures and aerodynamics of slender delta wings in detail. A low and moderate sweep angled, $\Lambda$ delta wings, which are generic plan-forms of MAVs and UCAVs, have serious control and instability problems due to not having conventional aerodynamic control surfaces. For these wings; coactions between the leading edge vortex and boundary layer, the leading edge vortex bursting, a localized surface flow separation, effects of these incidents on a wing surface vibration and buffeting are among fundamental research topics. It is possible to discover a method to control instability problems by making further investigations on these topics and to understand flow phenomena better. Locations of vortex bursting over delta wings is not stable and displays fluctuation along the vortex axis.

Overall scene of vortex bursting close to the delta wing surface is presented in Figure 3 (Karasu et al., 2015). As seen in the figure, yaw angle, $\beta$ influences onset of vortex location in dramatically.

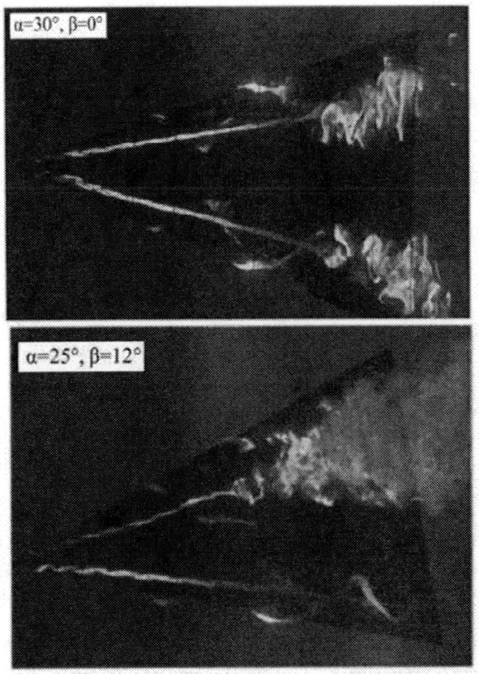


**Fig. 3.** Representation of symmetrical vortex bursting over a high sweep angle of delta wing, $\Lambda$ (Karasu et al., 2015)

Flow structures over the surface of non-slender diamond wing were studied by Yayla et al. (2009) using dye visualization and the Stereo Particle Image Velocimetry (PIV) System. In their study flow structures, turbulence statistics and vortex bursting transformation were examined by changing the data of yaw angle, $\beta$ within the range of $0° \leq \beta \leq 15°$ for attack angle of $\alpha=7°$ at $Re=1x10^4$. They stated that when yaw angle, $\beta$ increases, the vortex bursting location on one side approaches to the wing apex, on the opposite side moves towards the trailing edge. Canpolat et al. (2009) monitored changes of flow structures occurred over a delta wing with respect to the angle of attack, $\alpha$ and yaw angle, $\beta$ having sweep angle, $\Lambda$ of delta wing as 40°. Flow characteristics over a delta wing in plan-view plane was reported for angles of attack, $\alpha=7°$, 10°, 13°, and 17°, and yaw angles, $\beta=0°$, 6°, 8°, and 15°. Dye visualizations of flow in cross flow planes were also reported at locations of X/C=0.6, 0.8, and 1. 0.

Sohn et al. (2004) performed an experimental study on a yawed delta wing with inclusion of LEX employing dye visualizations. Interactions between Leading Edge Extension (LEX) and delta wing were investigated at some attack angles, $\alpha$ and yaw angles, $\beta$. They concluded

 DOI: 10.1051/epjconf/201714302099

that wing vortex and LEX vortex are coated around each other to have combined vortex which are comparably strengthened with a new identity when yaw angle, β is taken as 0°. Visualization of vorticity consecrations over a double delta wing with and without yaw angle, β was also reported by Sohn et. al, 2004. Experimental work on structures of flow over surface of a non-slender delta wing with 40° sweep angle, Λ was conducted by Canpolat et al. (2009). They stated that when a delta wing has certain value of a yaw angle β, symmetrical flow structures disappear and vortex bursting goes on earlier at the windward side with respect to the leeward side. They also discovered that main vortices in cross flow planes occurred inside of vortices close to central axis of delta wing. Yayla et al. (2010) performed experimental studies on an aerodynamics of a non-slender diamond wing which has 40° sweep angle, Λ. They pointed out that yaw angle β has substantial effects on the vortex bursting using the dye visualization technique. It was concluded that up to 4° yaw angle, β there are no clear changes in the vortex bursting locations, but at a higher yaw angle, β after 4°, the point of vortex bursting moves towards the leading edge on the windward side, while this location moves further downstream on the leeward side. Locations of asymmetrical vortex bursting are seen over a delta wing in plain-view plane. Sohn and Chang (2010) observed the effect of central body on a yawed double delta wing using flow visualizations over a surface and wing-surface pressure measurements. They confirmed that up to 24° attack angle, α availability of central body has a small effectiveness on the pressure distribution of suction side on the upper surface of wing, even at a large yaw angle of β=20°.

## 2 Material and Method

Experiments were made in a close -loop open water channel in Fluid Mechanics Laboratory at Çukurova University. Dimensions of water channel are set as 8000 mm length, 1000 mm width, 750 mm height and test section of water tunnel is made by a transparent Plexiglas which has thickness of 15 mm. The free-stream velocity is 120 mm/s which corresponds to the Reynolds number of $Re=2.10^4$ based on the cord length, C which is 120 mm. The Delta wing was kept stationary in the test section by a special mechanism. The Delta wing was hold in horizontal position by a slender support arm that stretched from mid-chord of delta wing vertically. A few dye visualization experiments were conducted in order to have an idea, for example, how flow patterns behave in end-view planes of the delta wing at a location, x/c=1.0. Fluorescent rhodamine dyes which reflect laser light were used. Rhodamine dye was conveyed in a thin slot located along the surface of delta wing. SONY HD-SR1 model video camera was employed. Schematic view of delta wing, measuring cross-section and experimental setup are shown in Fig. 4.

A 2D Particle Image Velocimetry (PIV) was used for capturing images of velocity fields. The PIV technique captures a general view of instantaneous flow data in a specified flow field. This feature lets user to examine even existence of small eddies with their effects and to provide opportunity for computing vorticity fields quantitatively. Particle Image Velocimetry (PIV) records the distance that particles move in time between two pulses of laser illuminations.

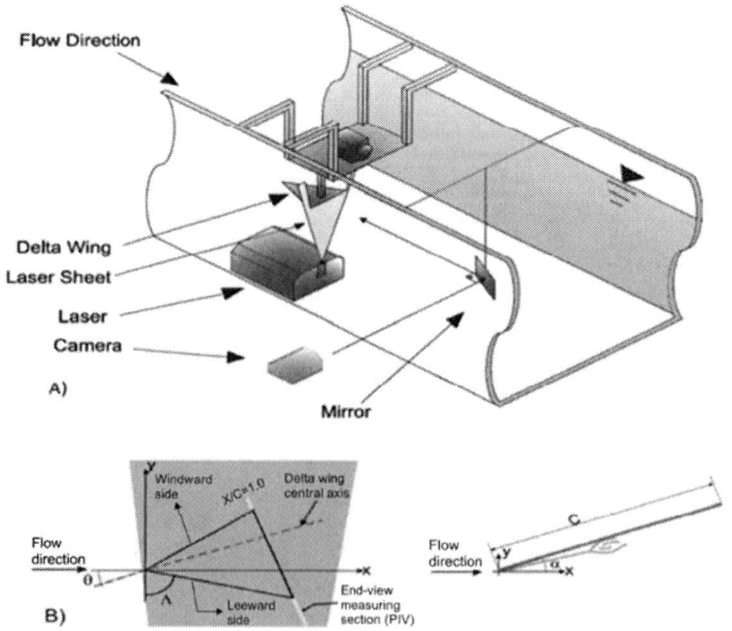


**Fig. 4.** A) Experimental Set-up of the PIV system, B) Schematic representation of the delta wing

In general, the user can obtain image acquisition and data processing from the PIV system combined with the software and computer. Seeding particles moving in water flow were flashed by pulsating laser beams. A successive light sheet pulses provided the opportunity for the camera to construct an image map. Images were received by the CCD camera which has resolution of 1600×1200 pixels at a rate of 15 frames per second. Time delay between frames was taken as 170 μs. To eliminate bad vectors the CLEANVEC software was used (Meinhart and Soloff (1999)) in order to remove unwanted vectors. After that to fill the places of deleted vectors bilinear interpolation with least squares fit is used. During each test, a total 1050 images were captured. The laser light sheet was set normal to the flow direction. Mirror which had 600mm-700 mm surface area located further downstream of the delta wing and was turned by 45° from the free-stream flow direction in order to deliver the vision of delta wing and cross-flow domain to the camera.

## 3 Result and Discussion

The dye visualization experiments were firstly conducted in order to study vortical flow structures further downstream of the vortex bursting at a cross-flow plane, X/C=1.0. As soon as the leading edge vortices break down, a complex flow structure is developed. A pair of main counter rotating vortices take place in end-view planes at X/C=1.0. Figs. 5, 6 and 7 demonstrate that as soon as the location of vortex breakdown comes closer to the present measuring plane, X/C=1.0 the magnitude of interactions increases violently.

A small size vorticity concentrations occurred in the counter rotating main vorticities are also well-defined. Under high angles of attach, α the vortex break down occurs at an earlier stage of X/C. In this case, magnitude of large and small scales of vortices attenuated at X/C=1.0. As reported in the review work of the present study, Yaw angle, β is very effective parameter in deteriorating the symmetrical flow structures leading to asymmetrical flow characteristics. Namely, on the leeward side ( Left hand side) of the delta wind, the locations of vortex break down moves further downstream while the oppositely oriented leading edge vortices breakdowns earlier. Either the leeward side or the windward side (Right hand side) leading edge vortex which ever breakdown close to end-view planes of the delta wing the magnitude of newly forming vortices become stronger. Here, free-stream flow direction is taken as a reference line for yaw angle, β and locations of End-view plane or cross flow measuring plane, X/C=1 is fixed according to zero yaw angle, β. A quantitative visualization of vortical flow provides detailed physics of flow structures. In order to see the effect of yaw angle, β on the flow structures in end-view plane of the delta wing time-averaged velocity vectors, <V>, streamline <Ψ> and vorticity, <ω> are defined using instantaneous velocity data. As seen in figures, laser sheet is passed through the cross-section at X/C=1.0 because dye visualization tests were conducted on the same cross-section.

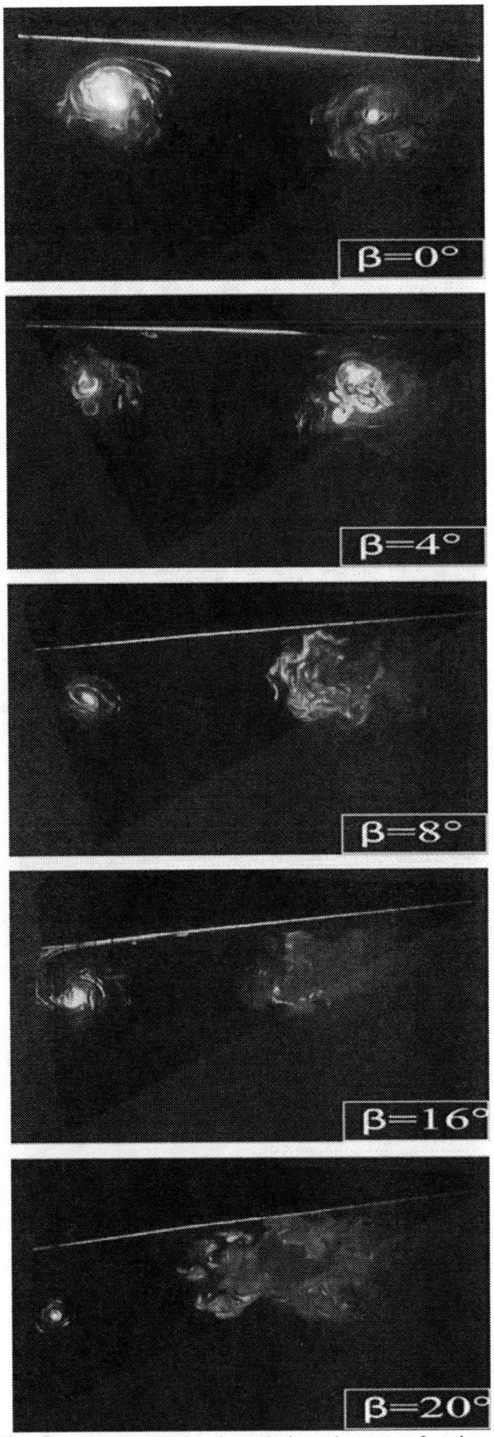


**Fig. 5**. Vortex formation in end-view plane as a function of yaw angle, β at angle of attack, α=25°. The laser sheet is located at x/c=1.0

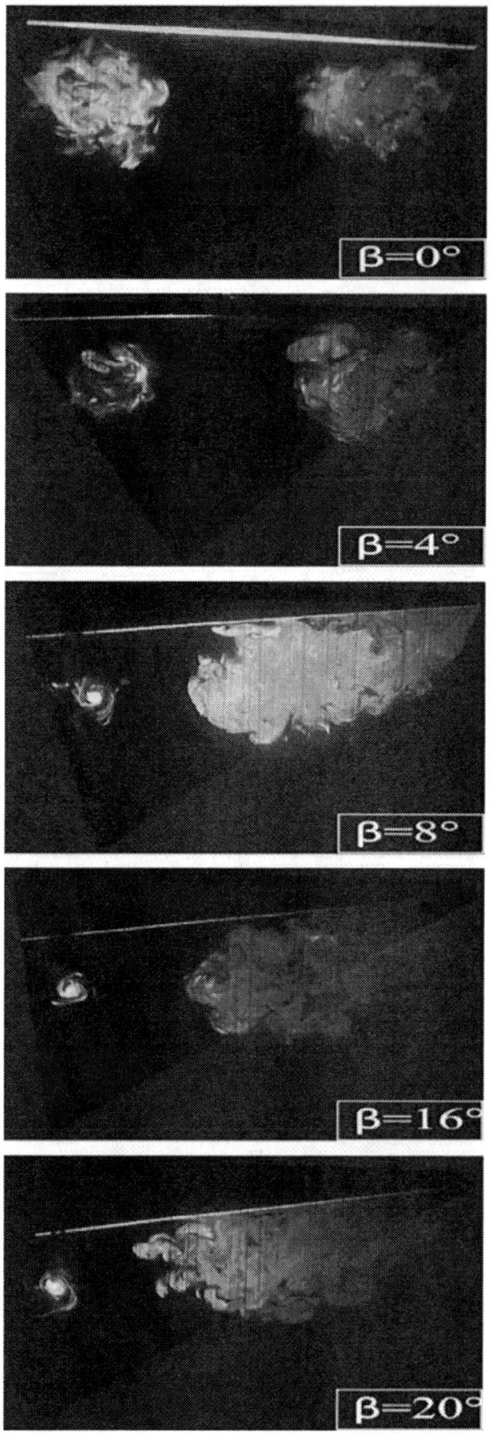


**Fig. 6.** Vortex formation in end-view plane as a function of yaw angle, β at angle of attack, α=30°. The laser sheet is located at x/c=1.0

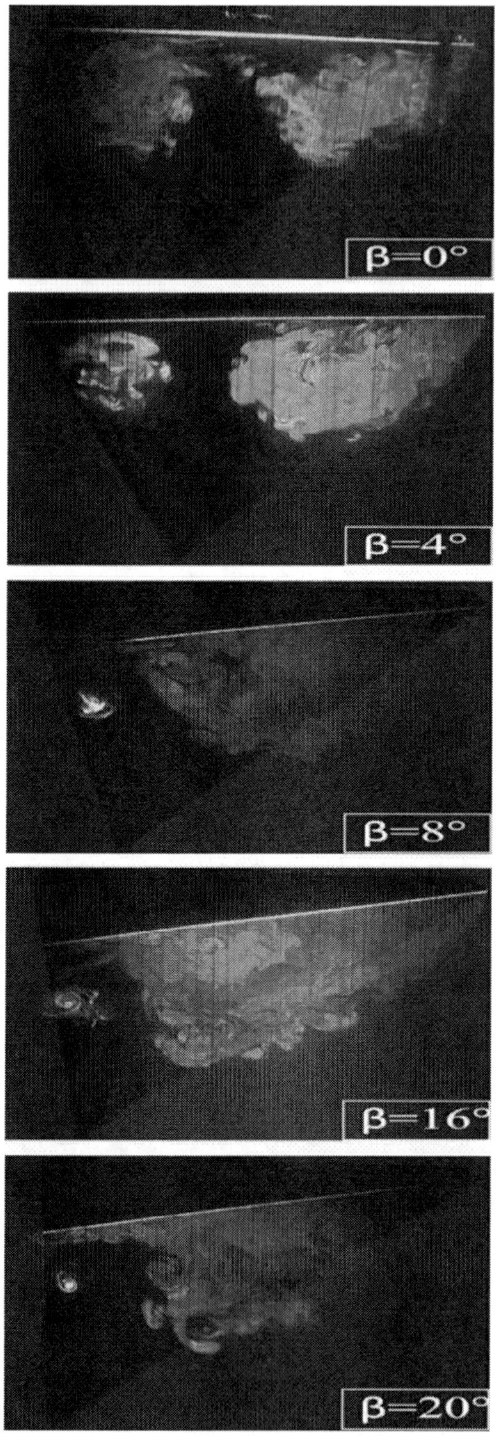


**Fig. 7.** Vortex formation in end-view plane as a function of yaw angle, β at angle of attack, α=35°. The laser sheet is located at x/c=1.0

It is also reported in the review of this text, for example, taking the dimensionless cord length as X/C=1.0 and yaw angle, β=0 there is a symmetrical flow structures in the observed vertical plane crossing the free-stream flow direction as seen in Fig. 8. These flow patterns are demonstrated by dye observations, animations of instantaneous flow data as well as time-averaged flow data.

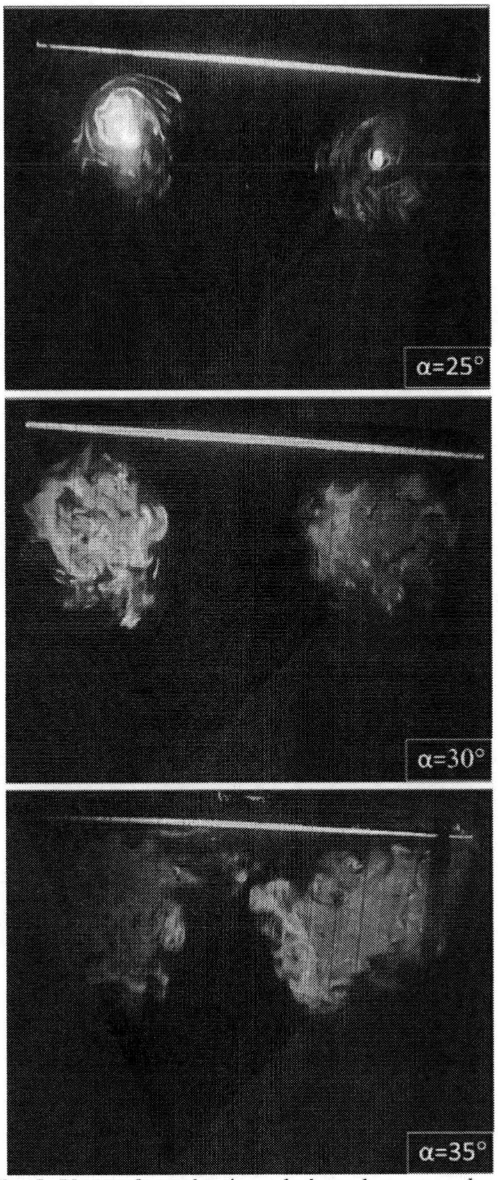


**Fig. 8.** Vortex formation in end-view plane at angle of attack α=25°, 30° and 35° and yaw angle, β=0. The laser sheet is located at x/c=1.0

Patterns of time-averaged streamline, <Ψ>, for angles of attack of $\alpha$=25° 30° and 35°and varying yaw angles, β over the range of 0° ≤ β ≤ 20° are present in Figs. 9-11. The laser sheet is located at a location of x/c=1.0. Inspection of these images in figures shows that center of well-formed foci, $F_1$ and $F_2$, saddle points; $S_1$ and $S_2$ distinctly designate symmetrical flow structures and the domain of main rotating vorticity concentrations. Central points of swirl patterns of streamlines, $F_1$ and $F_2$ move to the left hand side of wing because of rising yaw angle, β from 0° to 20°. It is observed that the swirling patterns of streamlines, <Ψ> indicating flow circulations regions are not symmetrical. Actually, saddle points $S_1$ and $S_2$ take place on a boundary. This borderline identifies the border between weak flow region and free-stream flow region. Flow structures is extremely susceptible about alteration of yaw angle, β and angles of attack, α. Well-defined foci, $F_1$ and $F_2$, are presented by contour of streamlines, <Ψ>. Below foci, $F_1$ and $F_2$, two saddle points, $S_1$ and $S_2$ are developed.

Patterns of velocity vectors, <V> clearly show locations of swirling of flow which are coincident with the foci of streamline patterns, $F_1$ and $F_2$. As seen from Figs. 12-14, distribution of velocity vectors, <V> reveals that as the yaw angle β, increases, one of the leading edge vortices moves to the left hand side of delta wing and the other leading edge vortex also moves to the left hand side as well close to the center line of delta wing. To study patterns of time-averaged flow data, for angles of attack of α= 25°, 30°, 35° and five different values of yaw angle β, a laser sheet is located at X/C=1.0 throughout the experiment for PIV measurements. The distribution of time-averaged velocity vectors, <V> exhibits a localized swirl flow patterns with a high velocity magnitude around the centers of foci, F1 and F2 developing two identifiable recirculation cells. When angles of attack, α is set as α=25°, 30° and 35° keeping yaw angle, β at 0° value, it is seen in all cases that the domain and the magnitude of time-averaged velocity vectors, <V> are symmetrical as seen in the first images of Figs. 12-14. But, the magnitude of time-averaged velocity vectors is gradually attenuated as angle of attack, α is increased. As also seen in the dye visualization experiment, when yaw angle is set to β=0°, there are symmetrical flow structures on both sides of central cord axis of the delta wing. But, the magnitude of velocity vectors, <V> are higher and size of recirculation flow field is larger in the leeward side for higher angles of attack, α and higher yaw angles, β at X/C=1.0. Both central points of foci, F1 and F2 move towards the leeward side of the wing.

DOI: 10.1051/epjconf/201714302099

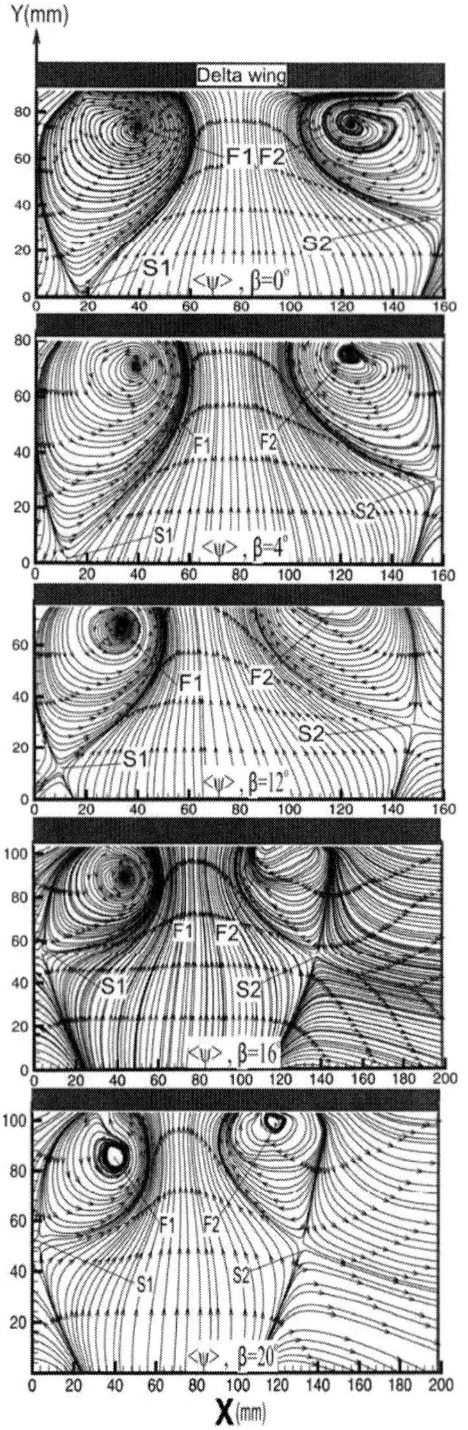


**Fig. 9.** Patterns of time-averaged streamlines, <Ψ>, with variation of yaw angle, β for angle of attack, $\alpha$=25°. The laser sheet is located at x/c=1.0

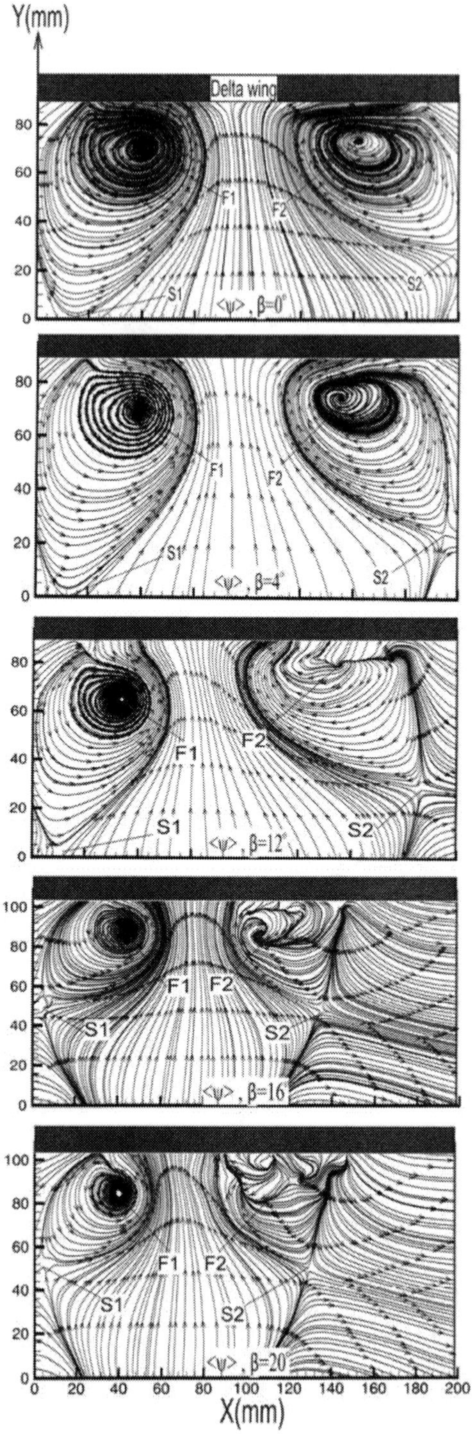


**Fig. 10.** Patterns of time-averaged streamline, <Ψ>, with variation of yaw angle, β for angle of attack, $\alpha$=30°. The laser sheet is located at x/c=1.0

 DOI: 10.1051/epjconf/201714302099

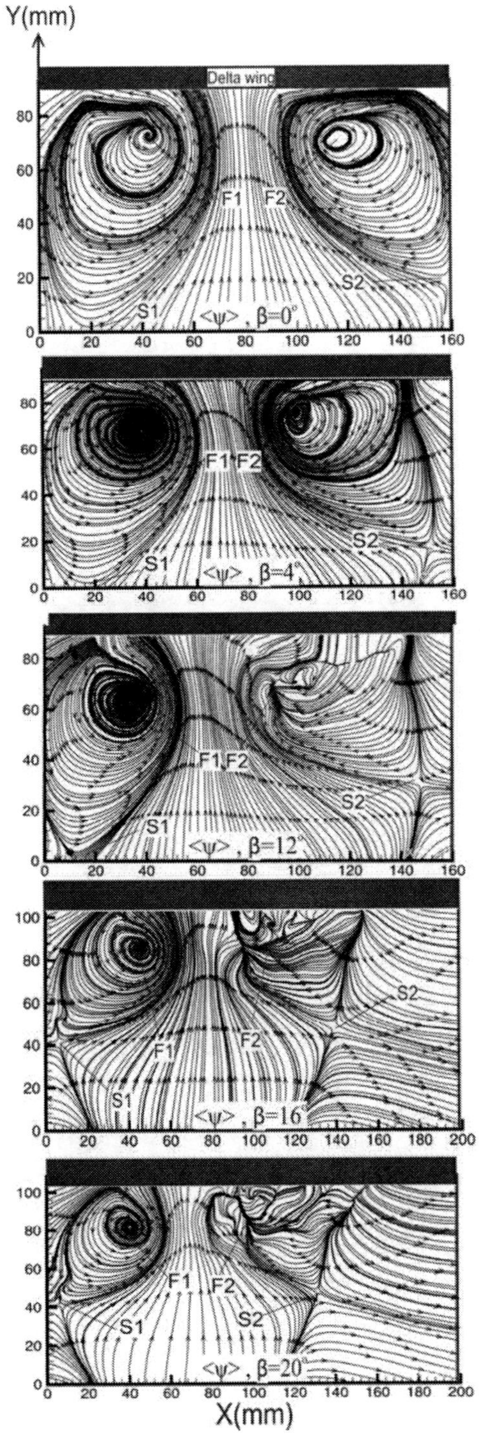


**Fig. 11.** Patterns of time-averaged streamline, <Ψ>, with variation of yaw angle, β for angle of attack $\alpha$=35°. The laser sheet is located at x/c=1.0

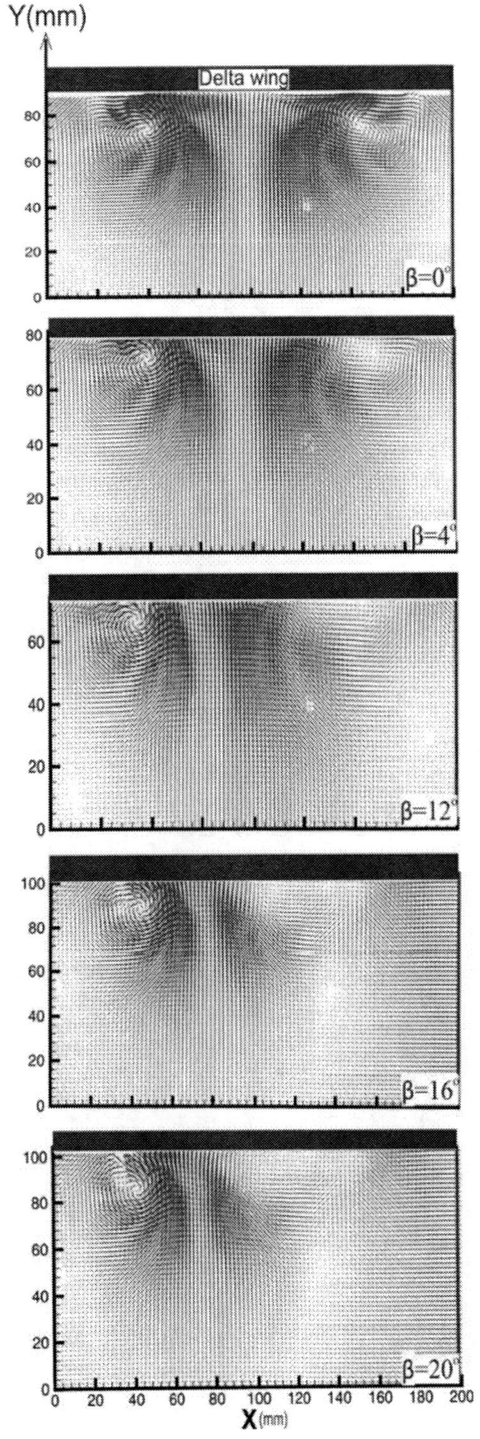


**Fig. 12**. Patterns of time-averaged distribution of velocity vectors, <V> with variation of yaw angle, β for angle of attack, $\alpha$=25°. The laser sheet is located at x/c=1.0.

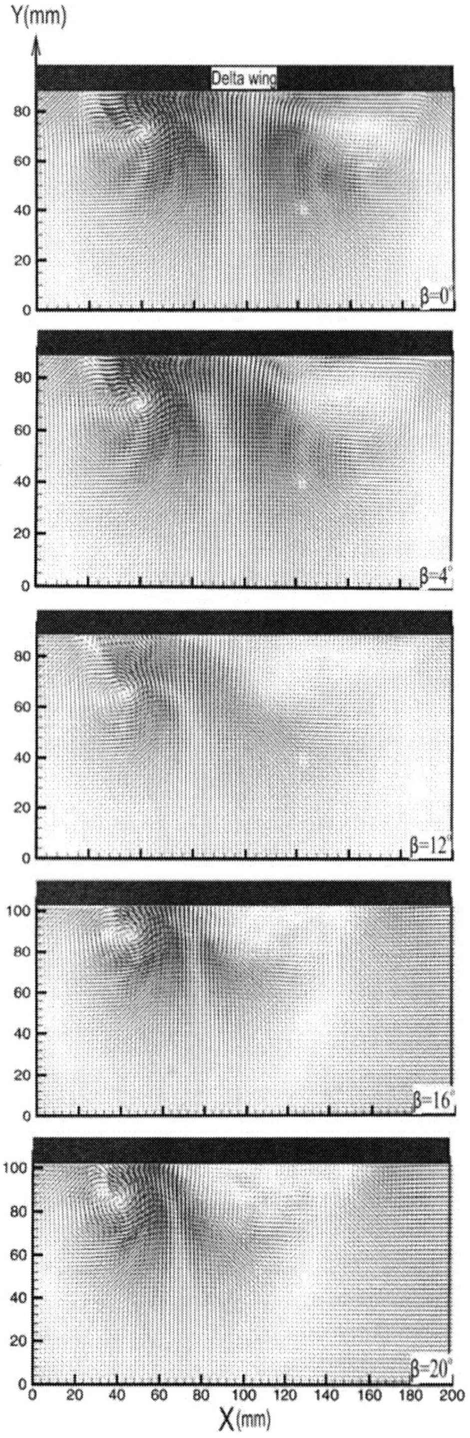


**Fig. 13.** Patterns of time-averaged distribution of velocity vectors, <V> with variation of yaw angle, β for angle of attack, $\alpha$=30°.The laser sheet is located at x/c=1.0

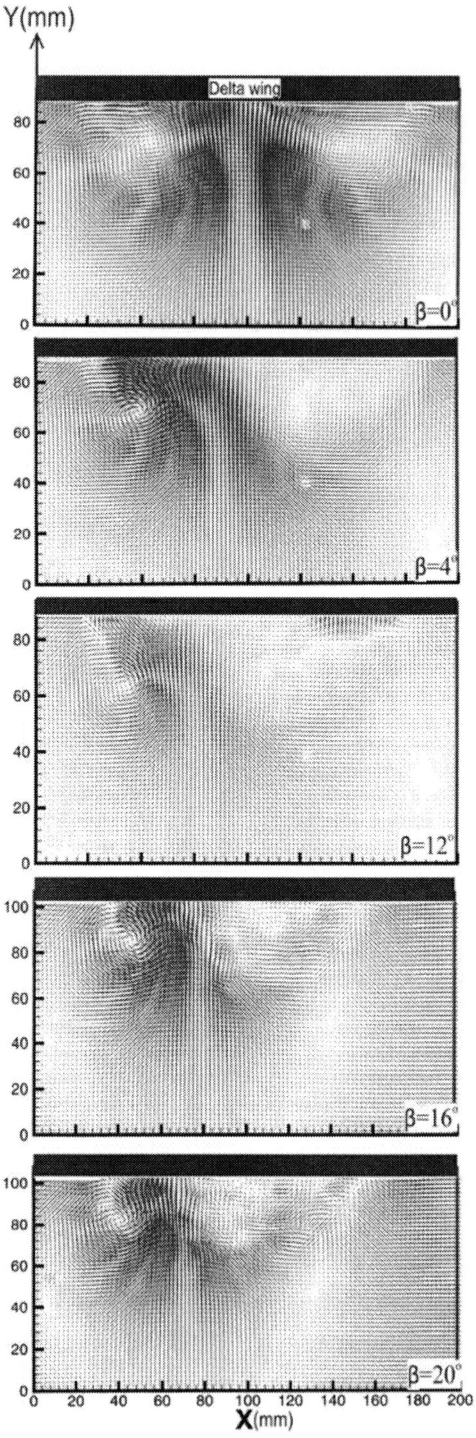


**Fig. 14.** Patterns of time-averaged distribution of velocity vectors, <V> with variation of yaw angle, β for angle of attack, $\alpha$=35°. The laser sheet is located at x/c=1.0

## 4 Conclusion

In this work, basic features of counter rotating leading edge vortices in end-view planes of the delta wing with 70° sweep angle, Λ were studied both qualitatively and quantitatively using Rhodamine dye and the PIV system. Experiments were conducted by altering angles of attack over the range of $25° \leq \alpha \leq 35°$ and yaw angles, β over the range of $0° \leq \beta \leq 20°$for the present study. Rhodamine dye visualizations were performed in end-view planes passing through rear end of delta wing at X/C=1.0 in order to observe alterations of vortical flow structures under the effect of yaw angles, β.

In end-view plane, symmetrical flow structures are developed in the case of zero yaw angles, β for all case of angles of attack, α. In the light of present experiments it is seen that with increasing yaw angle, β symmetrical flow structure is disrupted continuously.

The streamline topology <Ψ> shows that there are two saddle points, $S_1$ and $S_2$, and foci, $F_1$ and $F_2$. Saddle points, $S_1$ and $S_2$, are clearly seen and located below the foci, $F_1$ and $F_2$. The streamline patterns, <Ψ> show a well-defined swirl pattern. The saddle points, $S_1$ and $S_2$, and center of foci, $F_1$ and $F_2$, gradually move downward and get closer to each other as angle of attack, α is increased because of stall flow regions expends in size as a function of both angle of attack, α and yaw angle, β. Patterns of streamlines, <Ψ> of counter rotating flow recirculation are dissimilar in terms of size and magnitude. Differences between size and severity of vortices are easily seen in PIV results that are quantitatively visualized in terms of streamlines, <Ψ> and velocity vectors, <V>.

Time-averaged velocity vectors, <V> indicates a well-defined a pair of identical swirling flow cells for the case of zero yaw angle, β for all cases of angles of attack, that are considered in the present work. But, a well-defined recirculating flow region on the windward side gradually attenuates and diminishes and finally a single flow circulating loop is only seen in the image of velocity vector distributions <V> on the leeward side in the case of higher yaw angle, $\beta \geq 12^0$. Increasing yaw angle, β from β=0° to β=4° flow structures are subjected to small changes. Symmetrical flow structures are dramatically altered at a yaw angle of $\beta=20^0$ compared with the case of yaw angles from $\beta = 0^0$ to $\beta=4^0$.The distribution of time-averaged velocity vectors, <V> exhibits several small scales circulations presented by localized swirl patterns of streamlines.

## Acknowledgments

This study was supported by The Scientific and Technological Research Council of Turkey (TÜBİTAK) under contract number of 114M497 and Çukurova University Scientific Research Unit under contract number of FLY-2016-5845.

## References

1. Akilli, H., Sahin, B., and Rockwell, D., (2001), "Control of Vortex Breakdown by a Transversely-Oriented Wire", Physics of Fluids, 13(2):452-463
2. Anderson, J.D., (201), "Fundamentals of aerodynamics", McGraw-Hill Higher Education, ISBN 0-07-118146-6
3. Canpolat C., Yayla S., and Sahin B., (2012), "Observation of the Vortical Flow over a Yawed Delta Wing, Journal of Aerospace Engineering",25(4):613-626
4. Canpolat, C., Yayla, S., Sahin, B., and Akilli, H., (2009), "Dye Visualization of the Flow Structure over a Yawed Nonslender Delta Wing", The Journal of Aircraft, 46(5), pp. 1818-1822
5. Karasu I., Sahin B., Akilli H., and Canpolat C., (2015) "Dye Visualization of a Yawed Slender Delta Wing", Journal of Thermal Engineering, Yildiz Technical University Press, Istanbul, Turkey, 1(2-7): 646-654
6. Lambourne N. C., and Bryer D. W., (1961), "The Bursting of Leading Edge Vortices--Some Observations and Discussion of the Phenomenon", Aeronautical Research Council Reports And Memoranda, Reports and Memoranda No. 3282
7. Meinhart and Soloff, (1999), "Cleanvec software", Turbulence Laboratory, University of Illinois at Urbana-Champaign
7. Ozgoren, M., Sahin, B., and Rockwell, D., (2002), "Vortex Structure on a Delta Wing at High Angle-of-Attack", AIAA Journal, 40(2):285- 292
9. Sarpkaya, T., (1971), "On stationary and travelling vortex breakdowns", Journal of Fluid Mechanics 45(03):545-559
10. Sohn, M. H., Lee, K. Y., and Chang, J. W., (2004), "Vortex Flow Visualization of a Yawed Delta Wing with Leading-Edge Extension", Journal of Aircraft, 41(2)
11. Sohn, M. H., and Chang J. W., (2010), "Effect of a centerbody on the vortex flow of a double-delta wing with leading edge extension", Aerospace Science and Technology 14(1):11-18
12. Yaniktepe, B. and Rockwell D., (2005), "Flow Structure on Diamond and Lambda Planforms: Trailing-Edge Region", AIAA Journal, 43(7), pp. 1490-1500
13. Yayla, S., Canpolat, C. Sahin, B., and Akilli, H., 82010), "Yaw angle effect on flow structure over the nonslender diamond wing", AIAA journal, 48(10):2457-2461
14. Yayla, S., Canpolat, C., Sahin, B., and Akilli, H., (2010), "Effect of Yaw Angle on the Formation of Vortex Breakdown over the Diamond Wing", Journal of Thermal Science and Technology, 30(1): 79-89

 DOI: 10.1051/epjconf/201714302100

# Laboratory modelling of the transfer processes between the ocean and atmosphere in the boundary layers

*Daniil* Sergeev[1,2*], *Alexander* Kandaurov[1,2], *Yuliya* Troitskaya[1,2] and *Maxim* Vdovin[1,2]

[1]Institute of applied physics RAS, Geophysical Research Division, 603950 Ulyanova st. 46, Nizhny Novgorod, Russia
[2]Lobachevsky state university of Nizhny Novgorod, 603950 Prospect Gagarina 23, Nizhny Novgorod, Russia

**Abstract.** The processes of momentum and heat transfer between ocean and atmosphere in the boundary layer were investigated within laboratory modeling for a wide range of wind speed and surface wave including hurricane conditions. Experiments were carried out on the Wind-Wave Flume of the Large Thermostratified Tank of IAP RAS. A special net located under the surface at different depths allows to vary parameters of surface waves independently on wind parameters. Theory of self-similarity of air flow parameters in the flume was used to calculate values aerodynamic and heat transfer coefficients from the measured velocity and temperature profiles by Pito and hotfilm gauges respectively. Simultaneous measurements of surface elevation with system wire allow to obtain spectra and integral parameters of waves. It was demonstrated that in contrast to the drag coefficient, heat transfer coefficient is virtually independent of wind speed and wave parameters to the moment of the beginning of spray generation and then increases rapidly.

## 1 Introduction

The main quantitative characteristics of atmosphere and ocean turbulent exchange are fluxes of momentum, heat and moisture. These fluxes play an important role in many aspects of meteorological and oceanographic research, including climate modeling, weather forecasting, modeling of boundary-layer processes etc. Turbulent exchange of energy and momentum between the ocean surfaces and the atmosphere to a large extent controls the energy and water cycle and general circulation of the ocean and the atmosphere. At the boundary-layer scale, the fluxes in question control generation of waves and development of the upper ocean mixed layer. In numerical weather prediction and climate modeling, the air–sea fluxes are parameterized through mean-flow meteorological parameters. Turbulent momentum $\langle u'w'\rangle = u_*^2$, heat $\langle T'w'\rangle$ and moisture $\langle q'w'\rangle$ fluxes are expressed via bulk formulas through the following meteorological parameters measured at a reference level (usually $h_{10}$=10 m above the water surface): the 10 m wind speed $U_{10}$, the difference between the 10 m air and water temperature $\Delta T_{10}$ and the difference of relative moisture at the water level and 10 m above the water surface $\Delta q_{10}$:

$$\begin{aligned} \langle u'w'\rangle &= C_D U_{10}^2 \\ \langle T'w'\rangle &= C_T U_{10} \Delta T_{10} \\ \langle q'w'\rangle &= C_E U_{10} \Delta q_{10} \end{aligned} \quad (1)$$

The dimensionless exchange coefficients; drag coefficient $C_D$, the Stanton number $C_T$ and Dalton number $C_E$ are dependent both on wind speed and density stratification of the marine atmospheric boundary layer (MABL) and on the effective roughness of the water surface. The heat and moisture transfer coefficients are usually considered equal (see [1]), then here only the Stanton number and grad coefficient will be investigated.

According to Monin-Obukhov similarity theory [2] the vertical profiles of velocity and temperature in a stratified MABL are:

$$u(z) = \frac{u_*}{\kappa} \ln \frac{z}{z_0} + \Phi_m\left(\frac{z}{L}\right); \quad (2)$$

$$T(z) - T(0) = \frac{\langle T'w'\rangle}{\kappa u_*} \Pr_t \ln \frac{z}{z_{0T}} + \Phi_T\left(\frac{z}{L}\right). \quad (3)$$

Here

$$L = \frac{u_*^3}{\kappa \frac{g}{T} \langle Tw'\rangle} \quad (4)$$

Monin-Obukhov length scale (see [2]), $Pr_t$ – turbulent Prandtl number, $z_0$ и $z_{0T}$ are roughness momentum and heat lengths. When $z << L$ functions $\Phi_m(z/L) = \Phi_T(z/L)$ cease, then following from (2), (3) the velocity and temperature profiles $u(z)$ and $T(z)$ are logarithmic.

* Corresponding author: daniil@hydro.appl-sci.nnov.ru

It is easy to see that, comparing definitions for the exchange coefficients (1) with expressions for the velocity profile (2) and temperature profiles (3) gives that the $C_D$ and $C_T$ depend on the characteristics of the stratification of the MABL, defined by the Obukhov scale, and on the properties of the underlying surface parameterized by roughness height. A convenient expression for the latter are the coefficients of turbulent exchange at neutral stratification:

$$C_{D10N} = \frac{\kappa^2}{\left(\ln\left(h_{10}/z_0\right)\right)^2} \tag{5}$$

$$C_{T10N} = \frac{\kappa^2}{\Pr_t \ln\left(h_{10}/z_0\right)\ln\left(h_{10}/z_{0T}\right)} = \frac{\kappa\sqrt{C_{D10N}}}{\Pr_t \left|\ln\left(\dfrac{h_{10}}{z_{0T}}\right)\right|} \tag{6}$$

where $h_{10}$=10 m.

Obviously, the equations (5) and (6) can be regarded as alternative definitions of roughness parameters $z_0$ and $z_{0T}$

Parameterization of roughness parameters or the exchange coefficients used in the models of atmospheric circulation, are reduced to their dependencies on wind speed [3], and in the advanced models [4] on the wave age. The problem of dependencies of the exchange coefficients in MABL on wind speed is closely connected with the question of the influence of surface waves and spray, i.e. the consequence of wind forcing, on the exchange of the momentum and mass between the atmosphere and the ocean. The effect of waves on the momentum transfer has been studied by many authors (see, for example, [5-8] and references therein). It is shown that the momentum transfer by wave disturbances, or the form drag, increases with the increasing wind speed accompanied by the widening of the wave spectra. The decrease of the turbulent momentum flux near the water surface leads to the decrease of the wind speed at the standard height and increase of the drag coefficient and the surface roughness parameter.

The influence of wave disturbances on the heat transfer is weaker than on the momentum transfer, because even at moderate wind speeds the heat flux controlled by wave-induced disturbances is a fraction of percentage of the turbulent heat flux (see, discussion of this issue, for example, in [9]). In connection with it, the theoretical model [10, 11] does not regard the wave heat flux, considering that the heat transfer is determined only by turbulent transport. In this case, the calculated dependence $C_{T10N}$ on the wind speed is significantly affected by the model of turbulent exchange. At the same time, all of the above mentioned models predict the increasing dependence of $C_{T10N}$ ($U_{10}$).

The experimental data confirm weaker dependence of the heat transfer coefficient $C_{T10N}$ on wind speed compared to $C_{D10N}$ ($U_{10}$) (see, for example, the field [1] and laboratory [12] data), but the character of the dependence is different in different models. A number of studies (e.g. the algorithm COARE 3.0 [1], field data [13] and laboratory data [12]) indicate weak growth of $C_{T10N}$ with increasing wind speed $U_{10}$. Also, a weak growing dependence of the enthalpy transfer coefficient on the wind speed was found in the recent laboratory experiment [14], which studied heat transfer at a strong wind. At the same time, in [15] the dependence of $C_{T10N}$ on the wind speed was not found.

Increasing wind speed causes an exponential increase in the amount of spray in the atmospheric boundary surface layer (see, for example, [16] and references therein). The effect of the impact of spray on the drag coefficient is ambiguous. A number of studies argue that spray cause reducing of the aerodynamic resistance of the water surface at hurricane conditions. For example, according [17, 18] the mechanism of reducing the aerodynamic drag caused by the fact that spray create in the atmosphere surface layer efficient stable stratification in density, reducing turbulent exchange. In [19] another mechanism of reducing the resistance, associated with the peculiarities of the spray momentum exchange with the air flow, is analyzed. On the contrary, as is estimated in [20], the exchange of spray momentum with turbulent flow leads to an increase in surface drag. The effect of spray on the heat transfer in the atmospheric surface layer is discussed in the theoretical work [11], where in the case of stable stratification is shown a strong growth of heat transfer coefficient at a wind speed of 15-20 m/s.

The purpose of this study is to investigate the influence of wind speed and the water surface disturbances (waves and spray) on the exchange of the momentum and heat in MABL. The difficulty of such a study is connected with the fact that the waves are excited by the wind, and it is impossible to distinguish the contribution to the momentum and heat transfer caused by wave disturbances, from the influence of the peculiarities of turbulent transport, determined by the speed of the wind. The main difference of this study from the previous experimental studies and field observations is in the use of artificial, independent of wind speed, control of parameters of waves at a certain wind speed. This helped to obtain dependencies of the transfer coefficients on the wind speeds, parameters of waves and spray. It should be mentioned that flux transfer coefficient and heat transfer coefficient were measured simultaneously. It allowed to define scale of the temperature roughness $z_{0T}$ characterizing heat transfer and independent on the flux transfer, as well as scale of surface wave roughness $z_0$. So it follows from (6) that the increase of $C_{T10N}$ is possible due to the growth of both $C_{D10N}$ and the temperature roughness $z_{0T}$, which is determined by the properties of the underlying surface, quantifying the characteristics of waves and spray. Otherwise simultaneous measurment of $C_{D10N}$ and $C_{T10N}$ allowed to estimate its ratio, which an important parameter of hurricane development is the ratio of heat and momentum exchange coefficients, which determines the condition for maintaining a hurricane formulated by [21] as $C_T / C_D > 0.75$.

According to the experiments [15], the ratio is close to 0.75, but uncertainties of the field experiments are too large for definite conclusions.

The paper has following structure. In the Section 2 experimental setup and a scheme of the adjustment of the surface roughness independently on the wind force are described. Section 3 is devoted to description of the measuring of the transfer coefficients by profiling method. The method of retrieving 3D spectra of the waves is described in the section 4. In the sections 5 and 6 dependencies of the transfer coefficients on the parameters of wind flow, waves and influence of the sprays are discussed. Conclusion is in the section 7.

## 2 The description of the experimental setup for the study of heat and momentum exchange at strong winds

The experiments were carried out on the Wind-Wave Flume of the Large Thermostratified Tank of IAP RAS (the overview see on Fig.1). The airflow channel with 10 m length has the section of 0,4 × 0,4 m above the water surface. The airflow is created by a centrifugal fan, the frequency or rotation $f$ ranges from 7.5 to 50 Hz. The airflow velocity at the axis of the channel is proportional to $f$ and varies from 4 m/s to 20 m/s (see Fig 3), which corresponds to the equivalent speed of 7-42 m/s. The detailed description of the experimental facility, the principles of creating and controlling the airflow is given in [22]. The general scheme of the experiments is shown in Fig. 2.

To create the temperature stratification of the surface layer of the wind, the air entering the channel, was heated to 35-40 C° (depending on the air flow speed). The temperature on the water surface for all the experiments was maintained constant of about 15 C°.

**Fig. 1.** Wind-Wave Flume of the Large Thermostratified Tank of IAP RAS (air channel in the center).

A special feature of this experiment is the ability to control surface waves irrespectively of the speed of the wind flow in the channel. For this purpose, the plastic net (mesh) 0.25 mm thick with a cell of 1.6 x 1.6mm has been stretched along the entire channel (first time was used in [23]). The net did not affect the heat exchange, but the characteristics of surface waves varied depending on its depth: the waves were absent when the net was located at the level of the undisturbed surface of the water, but at maximum depth (33cm), it had practically no effect on the parameters of the surface disturbances for all wind speeds implemented on the facility.

Velocity and temperature profiles in the working section of the flume (at a distance of 6.5 m from the entrance channel) were measured simultaneously with the help of respectively the Pitot tube and hot film gauge, mounted on the vertical scanner. The L-shaped Pitot tube with the differential pressure transducer Baratron MKS 226 A provided the accuracy of velocity measurement 3 cm $s^{-1}$. The accuracy of temperature measurements with hot film is 0.1 °C.

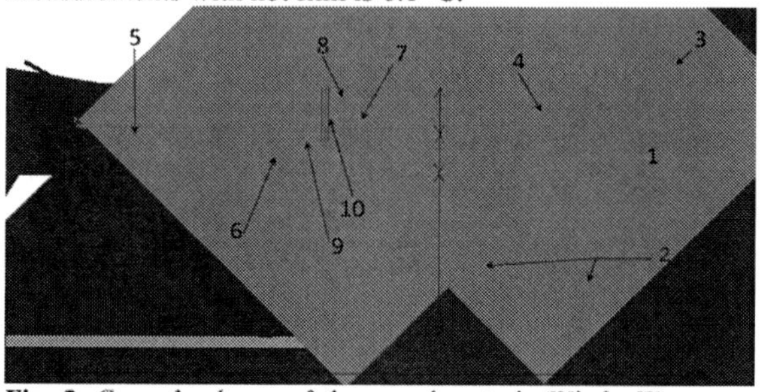


**Fig. 2.** General scheme of the experiments in Wind –Wave Flume. The sizes noted in cm. (1)Wind–wave channel body, (2) vertical bearings ,(3) convergent – diffusion section with a honeycomb, (4) hot film anemometer at the entrance, (5) wave absorber, (6) a net along the channel installed on the different depths (7) Pitot tube on a scanning system, (8) hot film thermometer on the same scanning, (9) a gauge for water temperature measurements. (10) three channel wire wave-gauge.

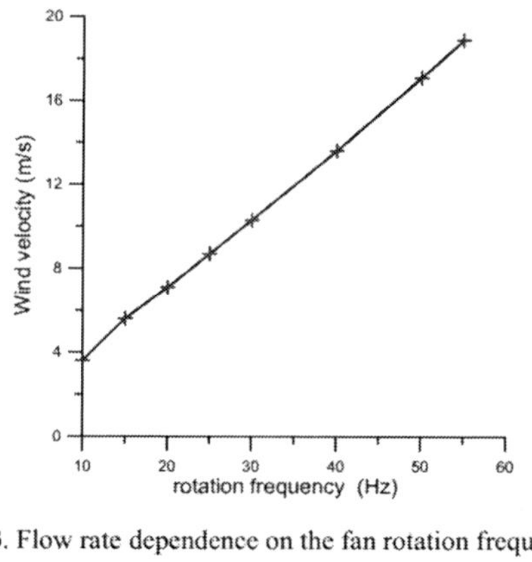


**Fig. 3.** Flow rate dependence on the fan rotation frequency

The scanning method with the consecutive height increment of 3 – 5 mm and accruing time of 1 minutes at each point was used. For each fixed wind parameters 2 profiles of velocity and temperature were measured for subsequent averaging. The lower level of scanning was located at a distance of 1 cm from the crests of the waves and depended on the wind speed, while the upper layer was 38 cm (in 2 cm below the upper lid of the channel). The temperature and wind speed at the inlet of the flume were controlled with the additional hot film gauge. Also, the temperature gauge was placed under water in the working section to measure the temperature of the surface layer of water.

## 3 The determination of heat and momentum flows on the velocity profiles of the air flow and temperature measured in the channel

To determine the parameters of the atmospheric boundary layer the algorithm suggested in [22] was used, which generalizes the model of the boundary layer near a flat surface [24]. It is well known that the boundary layer on a smooth flat plate is composed of the viscous sub-layer having a thickness of about $\delta_{vis} = \nu_a / u_*$ (less than 1 mm), positioned above the layer of constant fluxes having a thickness of about 0,15$\delta$, where $\delta$ is the boundary layer displacement thickness; and "the wake" part of the turbulent boundary layer, located higher up to the border of the displacement layer. In the turbulent boundary layer over a waved surface, an additional layer is generated, in which there is a transfer of momentum from the airflow to the wave perturbations on the water surface and the sum of turbulent and wave momentum fluxes is constant:

$$\tau_{turb}(\eta) + \tau_{wave}(\eta) = u_*^2 . \quad (7)$$

and here $\tau_{wave}(\eta)$ decreases with increasing distance from the boundary on the scale $h_{wave}$, which, according to [25] can be determined from the following relationship

$$kh_{wave}\left(\frac{\kappa V_p}{u_*} + \ln kz_0 - \ln h_{wave}k\right) = \kappa^2 \quad (8)$$

Here $V_p$ and $k$ are the phase velocity and the wave number corresponding to the frequency of the peak of elevation surface spectrum. The numerical values of the parameters of wind and waves in the flume obtained below, give an estimate of $kh_{wave}$ ~0.01. Here the condition $h_{wave} < 0,15\delta$ is observed, and within the layer of the constant momentum flux there is a layer of constant turbulent momentum flux $\tau_{turb}(\eta) = u_*^2$, in which the velocity depends on the vertical coordinate logarithmically.

As a rule, in wind tunnels and wind and wave channels, the thickness of the layer of constant fluxes is very small, for example, in the flume it was about 3 cm. The measurement of the wind speed and temperature at such a small distance from the surface of waved water surface especially at strong winds is a difficult problem, mainly due to the effect of spray torn off the wave crests. However, the parameters of the layer of constant fluxes can be obtained from the measurements in the 'wake" part of the turbulent boundary layer, as the profiles of velocity defect in the near-wall turbulent flows are self-similar:

$$\frac{U_{max} - U(z)}{u_*} = F\left(\frac{z}{\delta}\right), \quad (9)$$

where $U_{max}$ is the maximum speed in a turbulent boundary layer. According to [24] for non-gradient turbulent boundary layer at a flat plate or in the tube, the following approximation of the self-similar velocity profile

$$U_{max} - U(z) = \begin{cases} u_*\left(-\frac{1}{\kappa}\ln(z/\delta) + \alpha\right); z/\delta < 0.15, \\ \beta u_*(1 - z/\delta)^2 ; z/\delta > 0.15. \end{cases} \quad (10)$$

where $\kappa$=0.4 is the van Karman constant.

The constants $\alpha$ and $\beta$ can be obtained from the best fitting of the experimental data. In [24], the values of the constants ($\alpha$=1, $\beta$=9.6) for no-gradient turbulent boundary layer and ($\alpha$=1, $\beta$=7.1) for the turbulent Poiseuille flow in a pipe. The experiments in the wind - wave flume show that the profile of the velocity defect in the air flow above the waved water surface is also self-similar, and the velocity profile can be approximated by the expression (10). Determining the constants $\alpha$ and $\beta$ included the following data processing. The profiles of the flow velocity defect measured at a certain frequency of rotation of the fan and different positions of the net controlling waves, were expressed in terms of the self-similar coordinate $y=z/\delta$ and normalized by the curvature of the velocity profile $\beta u_*$. Thus obtained dimensionless velocity defect profiles are shown in Fig. 4 for a number of fan rotation frequencies. It can be seen

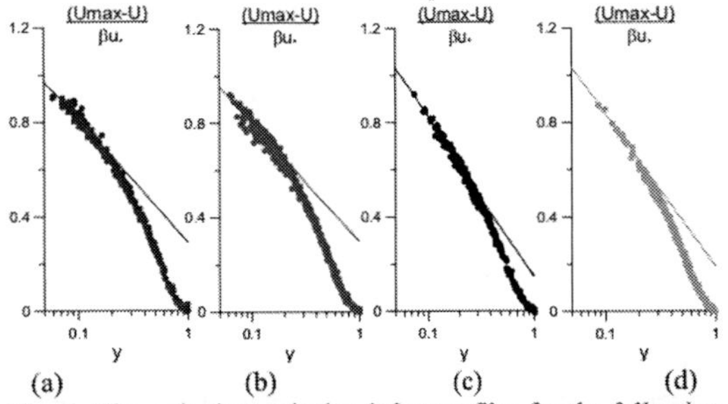


**Fig. 4.** Dimensionless velocity defect profiles for the following values of the fan speed: (a) - 20 Hz, (b) - 30, (c)- 35 Hz, (d) - 40 Hz.

that the experimental points collapse on certain curves, confirming the self-similarity of the profile of the flow velocity defect in the channel above the water surface. With this normalization the logarithmic part of the self-similar profile of the velocity defect has the form

$$\frac{U_{max} - U(z)}{\beta u_*} = \frac{1}{\beta}\left(-\frac{1}{\kappa}\ln(y) + \alpha\right) \quad (11)$$

The parameters of the logarithmic boundary layer can be obtained from the measurements in the wake part of the turbulent boundary layer, first, retrieving the parameters of turbulent boundary layer ($U_{max}$ and $\delta$) from best fit of the experimental data by equation (10) for $z/\delta$>0.15 and then calculating the parameters of the logarithmic boundary layer by the following expressions:

$$U(z) = \frac{u_*}{\kappa}\ln(z/z_0) \quad (12)$$

where

$$z_0 = \delta \exp\left(-\kappa U_{max} / u_* + \alpha\kappa\right) \quad (13)$$

Using (12, 13) it is possible to determine the wind speed at the standard height $H_{10}$ = 10m and calculate the drag coefficient

$$C_{D10N} = \frac{\kappa^2}{\left(\kappa U_{max} / u_* - \alpha\kappa + \ln\left(H_{10} / \delta\right)\right)^2} \quad (14)$$

Similar processing carried out with the temperature profiles allows determining the Stanton number and temperature roughness from the temperature profiles in the channel. We used in this case the self-similarity of the profile of "the temperature defect", which was defined similarly to the expression (11) for the velocity defect:

$$\frac{T_{max} - T(z)}{T_*} = G\left(\frac{z}{\delta_T}\right), \quad (15)$$

where $T_* = \dfrac{\langle T'w' \rangle}{u_*}$.

For the approximation of the self-similar dependence for the profile of "the temperature defect" we used the expression similar to (10):

$$T_{max} - T(z) = \begin{cases} T_*\left(-\frac{1}{\kappa}\Pr_t \ln\left(z/\delta_T\right) + \alpha_T\right) \\ z/\delta_T < 0.15, \\ \beta_T T_* \left(1 - z/\delta_T\right)^2 \\ z/\delta_T > 0.15. \end{cases} \quad (16)$$

For implementation the profiling method for measuring turbulent heat flux in MABL we need the turbulent Prandtl number $Pr_t$. According to [3] $Pr_t$ was assumed to be 0.85. This value was also confirmed by direct numerical simulation of turbulent boundary layer above waved water surface [26]. To determine the constants $\alpha_T$ and $\beta_T$ data processing was carried out as described above for the velocity profile. Similarly, the profiles of "the temperature defect" of the air flow were expressed in terms of the self-similar coordinate $y=z/\delta_T$ and normalized by the curvature of the temperature profile $\beta_T T_*$. Thus obtained dimensionless profiles of "the temperature defect" are shown in Fig. 5 for various fan rotation frequencies.

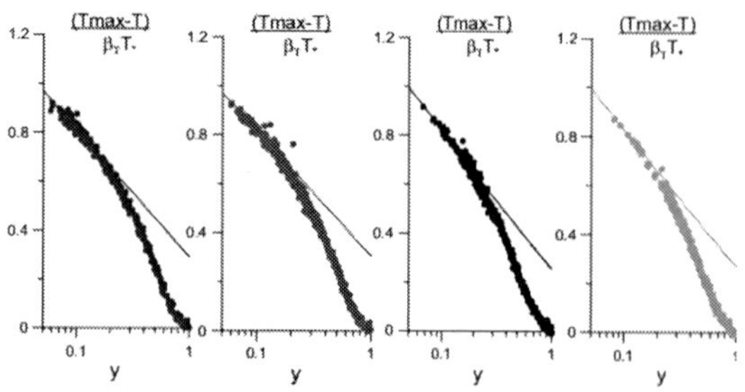


**Fig. 5.** Dimensionless profiles of "the temperature defect" for the following values of the fan speed: (a) - 20 Hz, (b) - 30, (c)- 35 Hz, (d) - 40 Hz.

The turbulent heat flux and temperature roughness were calculated by the measurements of the temperature in the "wake" part similar to the parameters of the logarithmic boundary layer (friction velocity and roughness parameter). The profile of the temperature defect was fitted by a polynomial of order 2, which determines 3 parameters $\beta_T T_*$, $T_{max}$ and $\delta_T$.

Knowing the constant $\beta_T$, taking into account the values of the van Karman constant and the Prandtl turbulent number, it is possible to find $T_*$, and taking the friction velocity found above , to determine the heat flux $\langle T'w' \rangle = T_* u_*$. Then the parameter of temperature roughness $z_{0T} = \delta_T \exp\left(-\frac{\kappa}{\Pr_t}\frac{\left(T_{max} - T_w\right)}{T_*} + \alpha_T\kappa\right)$ follows from (16), where $T_w$ - water surface temperature and the temperature profile in the layer of constant fluxes takes the form:

$$T(z) = T_w + \frac{T_*}{\kappa}\Pr_t \ln\left(\frac{z}{z_{0T}}\right) \quad (17)$$

Using equation (17) for calculating the temperature difference between the surface water and air at a standard height $H_{10}$=10м and equation (12) for $U_{10}$ gives then the Stanton number:

$$C_{T10N} = \kappa^2 \Bigg/ \left\{\left(\frac{\kappa U_{max}}{u_*} - \alpha\kappa + \ln\left(\frac{H_{10}}{\delta}\right)\right) \times \right. \\ \left. \times\left(\kappa\frac{\left(T_{max} - T_w\right)}{\Pr_t T_*} - \alpha_T\kappa + \ln\left(\frac{H_{10}}{\delta}\right)\right)\right\} \quad (18)$$

It follows directly from (17), that:

$$\frac{T_*}{T(H_{10}) - T_w} = \frac{\kappa}{\Pr_t \ln\left(\frac{H_{10}}{z_{0T}}\right)} \quad (19)$$

depends only on temperature roughness. We will use the dimensionless parameter (19) instead of the dimensional temperature roughness for convenience of comparing with the available experimental data.

Comparing (14), (18) and (19) easy gives

$$C_{T10N} = \frac{T_*}{T(H_{10}) - T_w}\sqrt{C_{D10N}} \quad (20)$$

The expression (20) is useful for the analysis of dependences of the Stanton number, the parameter of temperature roughness and the coefficient of resistance of the wind and waves.

## 4 Measuring parameters of surface waves

The wind wave field parameters in the flume were measured by three wire gauges positioned in the corners of an equal-side triangle with 2.5 cm side, the data sampling rate was 100 Hz. Three-dimensional space-time spectra were obtained from the measured data by the algorithm FDM (Fourier Directional Method) [22], similar to Wavelet Directional Method (WDM), suggested in [27].

Then, the three-dimensional space-time spectrum $S(\omega,k,\theta)$ can be obtained similar to [26] by using the conditional averaging of the spectral amplitudes by $k$ and $\theta$.

Integrating the spectra obtained $S(\omega,k,\theta)$ by the wave number or frequency, gives temporary $S(\omega,\theta)$ or spatial $S(k,\theta)$ angular spectra, respectively. Integrating by $\theta$ gives the omnidirectional frequency and wave number spectra. The upper limit of the wave number spectrum is prescribed by the distance between the wave staffs $d$, $k_u=\pi/d$, in the used configuration $k_u$=1,25 cm$^{-1}$.

The advantage of FDM against WDM is due to its higher spectral resolution (see [22]). Judging by the balance of the frequency resolution and signal accumulation time later was selected the window size of 512 samples (5,12 seconds).

To quantify the wave field we used below the dimensionless mean square slope (m.s.s.) defined as the integral of the omnidirectional slope spectrum over the range of measured wave numbers, as it was determined in the [28]:

$$m.s.s. = \int_{k_0}^{k_u} k^2 S(k,\theta)\, k dk d\theta \qquad (21)$$

where $k_0$ is a wave number below the spectral peak.

The dependences of the wind speed, significant wave height and spray marker on the control parameters of the experiment (the fan rotation frequency and control net depth) are shown in Fig. 6. The figures show that the fan rotation frequency determines mainly the wind speed (Fig. 6a), the latter relatively weakly depends on the depth of the net (Fig. 6c), which on contrary strongly affects the wave height (see Fig. 6b and d). Low dependency of wind speed on the net position can be explained by the low water surface drag under any wave conditions.

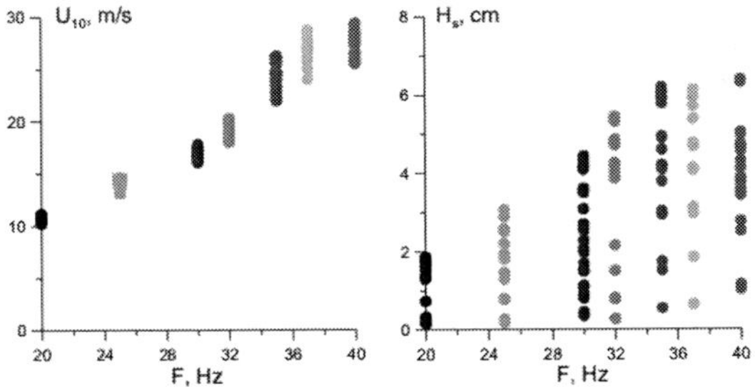


(a) (b)

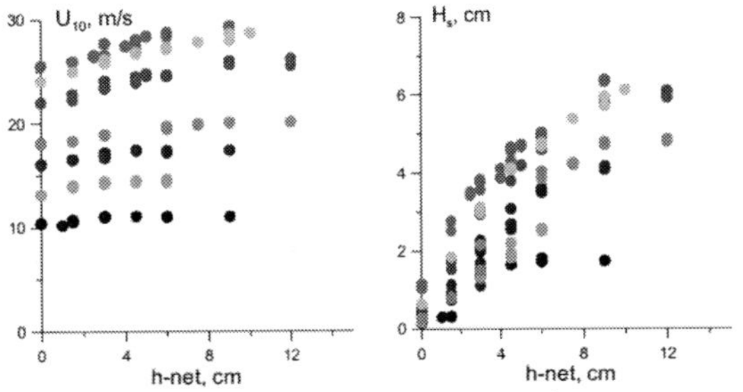


(c) (d)

**Fig.6.** (a) equivalent wind speed $U_{10}$ on the $F$ (fan rotation speed), (b) significant wave height $H_s$ on the $F$, (c) equivalent wind speed $U_{10}$ on the $h_{net}$ (depth of the net), (d) significant wave height $H_s$ on the $h_{net}$. Black symbols corresponds fan speed 20 Hz, green - 25Hz, blue - 30 Hz, orange - 32 Hz, red - 35 Hz, sky - 37 Hz, violet - 40 Hz.

## 5. Experimental dependencies of the turbulent exchange coefficients on wind velocity

The obtained dependencies of exchange coefficients on parameters of wind are discussed in this section and compared with available field [29, 31-35] and laboratory [14, 30] experiments. Fig. 7 shows the dependency of drag coefficient on 10 m wind speed. It is obvious that, alternatively to the previous data, the noticeable spreading of the values of $C_{D10N}$ is observed at a certain frequency of fan rotation, which mainly prescribes the air-flow velocity (see Fig. 7a).

A similar strong spreading of the experimental points is seen in the dependency of the Stanton number on wind speed in Fig. 7b. The comparison with Fig. 7c shows, that the experimental spreading of the temperature roughness is significantly lower than for the Stanton number. It indicates the fact that the Stanton number, which is the product of $\sqrt{C_{D10N}}$ and the logarithm of the temperature roughness (see eq. 20), depends on the wave regime mostly through the dependence on surface roughness of $C_{D10N}$.

It is interesting to note that at $U_{10}$>25 m/s the Stanton number and the temperature roughness increases sharply, as can be seen below,due to the presence of spray. As a result, the ratio $C_{T10N} / C_{D10N}$ exceeds the marginal value 0.75 for all wind speeds (see Fig. 7d) confirming the Emanuel criterion of the hurricane maintaining.

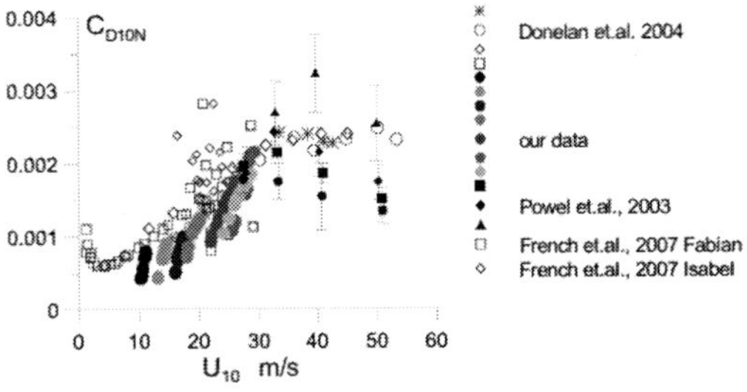

 DOI: 10.1051/epjconf/201714302100

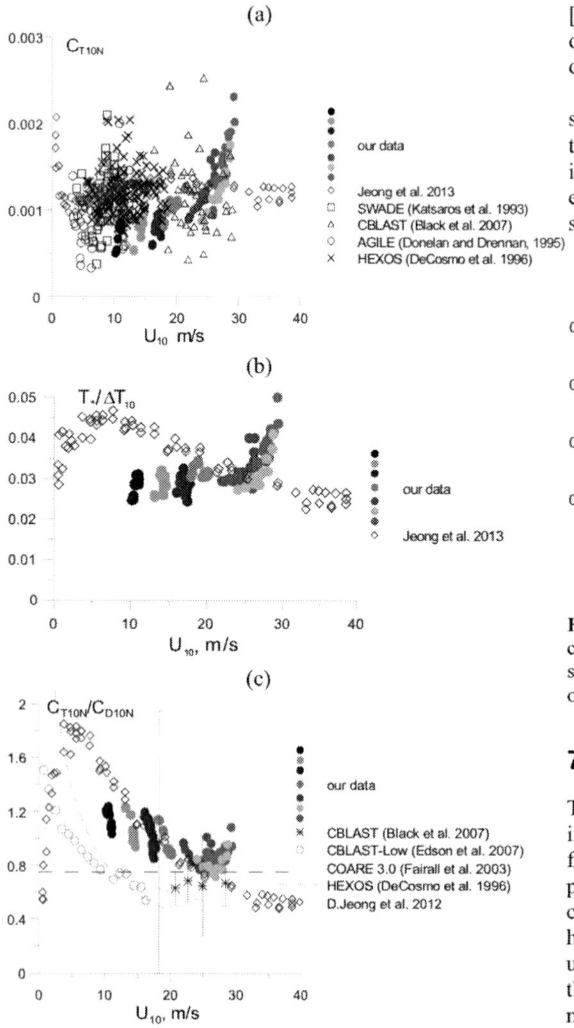


(d)

**Fig.7.** Dependencies: (a) of drag coefficient $C_{D10N}$ on $U_{10}$, (b) heat transfer coefficient $C_{T10N}$ on $U_{10}$, (c) parameter of the temperature roughness on $U_{10}$. (d) the ratio $C_{T10N} / C_{D10N}$ on $U_{10}$. Red dashed line $C_{T10N} / C_{D10N}$=0.75. Colors closed circles corresponds results of this study (the same fan speed as on Fig. 6).

## 6. The dependency of exchange coefficients on the parameters waves and spray presence

In this section we discuss dependence of the coefficients on the effective surface roughness, which is controlled by the waves and sprays. For the estimation of the wave parameters mean square slope (m.s.s.) is used (see (21)). Fig. 8 shows the dependency of drag coefficient and the Stanton number on m.s.s. As it was mentioned before sprays play a key role in the transfer processes. Presence of sprays in the air could be controlled by the detecting spikes in the signals of hot film close to the way used in [36]. Thus the closed symbols on the plot show the data obtained in no-spray conditions, and the open symbols denote the data obtained in the presence of spray.

It can be seen from the figures that in the absence of spray the drag coefficient increase with the steepness of the wave, the heat transfer coefficient is almost independent of m.s.s.. The same figure shows that both exchange coefficient strongly increase in the presence of spray (open symbols).

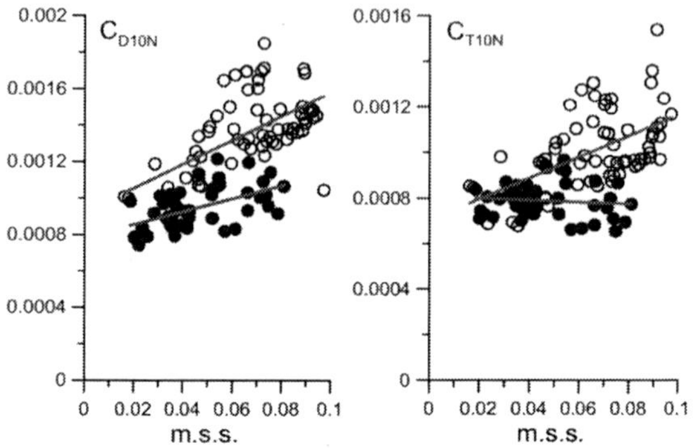


**Fig. 8** Dependencies of the drag coefficient (a), heat transfer coefficient (b) on m.s.s Red line - linear approximation. Closed symbols corresponds absence of the sprays in experiments, open symbols corresponds presence of spray.

## 7. Conclusion

The most important characteristics that determine the interaction between the atmosphere and the ocean are flows of momentum, heat and moisture. For their parameterization the dimensionless exchange coefficients (the surface drag coefficient $C_{D10N}$ and the heat transfer coefficient or the Stanton number $C_{T10N}$) are used. Numerous field and laboratory experiments show that $C_{D10N}$ increases with increasing wind speed at moderate and strong wind, and, as was recently shown, $C_{D10N}$ decreases at hurricane wind speed. Waves are known to increase the sea surface resistance due to enhanced form drag, the sea spray is considered a possible mechanism of the "drag reduction" at hurricane conditions. The dependence of the heat transfer coefficient $C_{T10N}$ on the wind speed is not so certain, and the role of the mechanism associated with the wave disturbances in the mass transfer is not completely understood. The observations and laboratory data show that this dependence is weaker than for the $C_{D10N}$, and there are differences in the character of the dependence in different data sets.

A series of experiments to study the processes of turbulent exchange of momentum and heat in a stably stratified temperature turbulent boundary layer air flow over the waved water surface was carried out on the Wind-Wave Flume of the Large Thermostratified Tank of IAP RAS. The peculiarity of this experiment was the option to change the surface wave parameters regardless of the speed of the wind flow in the channel. For this purpose a plastic net has been stretched along the channel. The net does not affect the heat exchange, but the characteristics of surface waves depended on the position of the net. To create a stable temperature stratification of the wind, the air entering the flume was heated but the surface water temperature was maintained constant. Experiments were performed for a wide range of wind speed up to hurricane conditions.The turbulent flows of heat and momentum and roughness parameters were retrieved from the measured velocity and temperature profiles and subsequent data processing exploiting the self-similarity of the temperature and velocity profiles. As a result the surface drag and heat exchange coefficients as well as roughness parameters were obtained. Wind wave spectra and integral parameters (significant wave height, mean square slope) were retrieved from measurements by a 3-channel array wave gauge by coherent spectral data processing. The dependences of the exchange coefficients on the winds speed and wave parameters were obtained. It is shown that the exchange coefficients increase with the wind speed and wave height. It was found, that a sharp increase of heat exchange coefficient at wind speeds exceeding 25m/s was accompanied by the emergence of the spray in the air flow over water.

This work was supported by the Russian Foundation of Basic Research (No. 15-35-20953 17-05-00958, 14-05-91767 AF_a, 16-55-52022), providing experiments was partial supported by Russian Science Foundation (Agreement No. 14-17-00667) and processing of the data was partially supported by Russian Science Foundation (Agreement No. 15-17-20009).

## References

1. C.W. Fairall, E.F. Bradley, J.E. Hare, A.A. Grachev, J.B. Edson, J. Climate. **16** (2003)
2. Monin, A.S., Yaglom A.M., (Statistical hydrodynamics, NAUKA. 1965).
3. Zilitinkevich, S. S., Izvestiya AN SSSR. FAO, **23**, 593 (1976)
4. V.N. Lykossov, Wind stress over the ocean, (Cambridge University Press, 2002)
5. Janssen, P. A. E. M.,J. Phys. Oceanogr., **19**, 745 (1989)
6. Makin, V. K., Kudryavtsev, V. N., and Mastenbroek, C., Boundary- Layer Meteorol., **73**, 159 (1995)
7. Makin, V. K. and Kudryavtsev, V. N. J. Geophys. Res., **104(C4)**, 7613(1999)
8. Yu. I. Troitskaya, G. V. Rybushkina, Izvestiya, Atmospheric and Oceanic Physics, **44(5)**, 621 (2008)
9. Yu.I., Ezhova E.V., Zilitinkevich S.S. Nonlin. Processes Geophys. **20, (**2013)
10. Makin V. K. and Mastenbroek C., Boundary-Layer Meteorol., **79**, 279 (1996)
11. Makin, V. K., J. Geophys. Res., **103(C1)**, 1137 (1998)
12. Ocampo-Torres F. J., Donelan M. A., Merzi N., and Jia F, Tellus, Ser. B, **46**, 16 (1994).
13. Brut, A., Butet, A., Durand, P., Caniaux, G., and Planton, S., Q.J.R. Meteorol. Soc., **131**, 2497 (2005)
14. Dahai Jeong, Brian K. Haus, and Mark A. Donelan,Spray. J. Atmos. Sci., **69**, 2733 (2012).
15. Drennan, W. M., Zhang, J., French, J. R., McCormick, C., and Black, P. G., J. Atmos. Sci., **64**, 1103 (2007)
16. Andreas E. L., J. Phys. Oceanogr., **28**, 2175 (1998)
17. Makin V. K. Boundary Layer Meteorol., **115 (1)**, 169 (2005)
18. Kudryavtsev V. N. J. Geophys. Res., **111**, C07020 (2006)
19. Bortkovskii etl.. (Processi perenosa tepla vblizi poverhnosti ocean-atmosphera, Hydrometizdat, 1974)
20. Kudryavtsev V. N., and V. K. Makin, Boundary Layer Meteorol., **140(3)**, 383 (2011)
21. Emanuel, K. A., J. Atmos. Sci., **52**, 3969 (1995)
22. Yu. I. Troitskaya, D.A. Sergeev, A.A. Kandaurov, G.A Baidakov, M.A. Vdovin, V.I. Kazakov J. Geophys. Res, **117**, C00J21 (2012)
23. Yu.I.Troitskaya, D.A.Sergeev, A.A.Kandaurov, M.I. Vdovin, A.A. Kandaurov, E.V.Ezhova, S.S.Zilitinkevich Nonlinear. Geoph. Processes, **20**, (2013)
24. Hinze J. O. (Turbulence: An Introduction to its Mechanism and Theory, McGraw-Hill. 1959)
25. Belcher, S.E., Newley, T.M.J. and Hunt J.C.R., J. Fluid Mech., **249**, 557 (1993)
26. O. A. Druzhinin Y. I. Troitskaya and S. S. Zilitinkevich Q. J. R. Meteorol. Soc. DOI:10.1002/qj.2677 (2015)
27. Donelan, M.A., W.M. Drennan and A.K. Magnusson, J. Phys. Oceanogr., **26**, 1901 (1996)
28. Kudryavtsev V. N., and V. K. Makin Boundary-Layer Meteorology, **98**, 155 (2001)
29. DeCosmo, J., K. B. Katsaros, S. D. Smith, R. J. Anderson, W. A. Oost, K. Bumke, and H. Chadwick, J. Geophys. Res., **101**, 12,001 (1996)
30. Donelan, M. A., B. K. Haus, N. Reul, W. J. Plant, M. Stiassne, H. C. Graber, O. B. Brown, and E. S. Saltzman (2004), Geophys. Res. Lett., **31**, L18306 (2004)
31. Black, P. G. et al., Bull. Am. Meteorol. Soc., **88**, 357 (2007)
32. French, J. R., W. M. Drennan, J. A Zhang, and P. G. Black J. Atmos. Sci., **64**, 1089 (2007)
33. Powell, M. D., P. J. Vickery, and T. A. Reinhold Nature, **422**, 279 (2003)
34. Donelan M.A., Drennan W. M. (Proc. Symp. on Air-Water Gas Transfer, Heidelberg University, 677, 1995)
35. Edson etl Bulletin of the American Meteorological Society. 88 (3), 341 (2007)
36. Lai, R. J., and O. H. Shemdin, J. Geophys. Res., 79, 3055 (1974).

# Laboratory modelling of the wind-wave interaction with modified PIV-method

*Daniil* Sergeev[1,2*], *Alexander* Kandaurov[1,2], *Yuliya* Troitskaya[1,2], *Guillemette* Caulliez[3], *Maximilian* Bopp[4], *Bernd* Jaehne[4]

[1]Institute of applied physics RAS, Geophysical Research Division, 603950 Ulyanova st. 46, Nizhny Novgorod, Russia
[2]Lobachevsky state university of Nizhny Novgorod, 603950 Prospect Gagarina 23, Nizhny Novgorod, Russia
[3]Mediterranean Institute of Oceanography, Avenue de Luminy 163, 13009, Marseille, France
[4]Insititut fur Umweltphysik Im Neuenheimer Feld 229 69120, Heidelbeg, Germany

**Abstract.** Laboratory experiments on studying the structure of the turbulent air boundary layer over waves were carried out at the Wind-Wave Flume of the Large Thermostratified Tank of the Institute of Applied Physics, Russian Academy of Sciences (IAP RAS), in conditions modeling the near water boundary layer of the atmosphere under strong and hurricane winds and the equivalent wind velocities from 10 to 48 m/s at the standard height of 10 m. A modified technique of Particle Image Velocimetry (PIV) was used to obtain turbulent pulsation averaged velocity fields of the air flow over the water surface curved by a wave and average profiles of the wind velocity. The main modifications are: 1) the use of high-speed video recording (1000-10000 frames/sec) with continuous laser illumination helps to obtain ensemble of the velocity fields in all phases of the wavy surface for subsequent statistical processing; 2) the development and application of special algorithms for obtaining form of the curvilinear wavy surface of the images for the conditions of parasitic images of the particles and the droplets in the air side close to the surface; 3) adaptive cross-correlation image processing to finding the velocity fields on a curved grid, caused by wave boarder; 4) using Hilbert transform to detect the phase of the wave in which the measured velocity field for subsequent appropriate binning within procedure obtaining the average characteristics.

## 1 Introduction

The interaction between wind flow and surface waves is one of central problems of the study and parameterization of exchange processes in boundary layers of the atmosphere and ocean [1]. Of special interest is the case of steep and breaking waves forming at a strong wind. But carrying out field measurements of air flows, wave parameters for such conditions is very difficult problem (see. [2]). That is why laboratory modeling is used [3–5] to simulate conditions up to severe weather. The main difficulties in the experimental study of a turbulent air flow over a wavy water surface in laboratory conditions are connected with measuring wind characteristics near the water surface, especially in wave troughs, where one can expect the appearance of the most interesting features of this flow, such as shielding and separation of the flow. The technique Particle Image Velocimetry PIV is best fit for measuring the air flow in wave troughs and water flows close to the surface. In [6–8], the experience of using the PIV technique for measuring the air flow velocity over a wavy surface was presented. In [8], the structure of average velocity fields in the air flow and their perturbations initiated by waves, as well as the structure of turbulent stresses, were successfully studied. However, those measurements were carried out at low wind velocities.

The subject of this work is studying the characteristics of air flows, waves, under conditions of breaking of waves with the formation of sprays near a wavy surface, in particular, in wave troughs due to high wind speeds.

Earlier, measurements under conditions of strong winds with an equivalent wind velocity $U_{10} > 25$ m/s in the laboratory modeling of extreme meteorological conditions were carried out only using contact gages (Pitot tubes and hotwires/hotfilms) at a consider able distance from surface wave crests [3–5, 9]. The measurement region was positioned above the layer of constant fluxes, where the velocity profile is logarithmic and its parameters (the friction velocity and roughness height $z_0$) could be determined directly using the profile method. Therefore, and $z_0$ have to be determined using the self-similarity property of the velocity profile of the flow in channels, as it was done in [4]. However, to refine the self-similar form of the velocity profile, it is necessary to measure the flow velocity as close to the water surface as possible [10].

The measurement can be implemented only using noncontact methods basing on visualisation, e.g., PIV. This work presents measurement results for the air flow velocity over waves in laboratory conditions modeling

* Corresponding author: daniil@hydro.appl-sci.nnov.ru

DOI: 10.1051/epjconf/201714302101

wide range of the winds (up to strong and hurricane conditions) with use of a specially modified PIV technique.

## 2 Experimental setup and measuring technique

Experiments were carried out on the Wind-Wave Flume of the Large Thermostratified Tank of IAP RAS (overview on Fig. 1 a). The length of the air flow channel with a cross section of 0.4 × 0.4 m is 10 m over the water surface. A detailed description of this setup and principles of design and control for the air flow in it were presented in [4, 5]. The general scheme of the experiments is presented in Fig. 1 b. In addition to PIV methods, which were the main tool in the presented investigations, earlier approbated contact ways of measurements were used. In the working section of the channel, at a distance of 7 m from the entrance, average profiles of the air flow velocity were measured using a Pitot tube. The air flow in the channel was visualized using polyamide particles with an average diameter of 10 μm and density of 1.02 g/m3. The inertial time was $2 \times 10^{-4}$ s. The seeding device was similar to that used in [8] and positioned at the entrance of the channel at a distance of 6 m from the detection area. The test experiments demonstrated that the system brings no distortions to the wind flow in the region of measurements. To apply the PIV method (principal scheme shown on the Fig. 1 c), the motion of particles in the air in the region of measurements was illuminated by a laser sheet along the channel axis 8 m from the inlet. The laser sheet is formed by a cylindrical lens from a vertical laser beam (a continuous Nd YAG laser, 532 nm, 4 W). The width of the illuminated area was varied choosing the radius of the cylindrical lenses and their mutual arrangement. The motion of particles introduced into the air flow and water surface which were illuminated by the laser sheet was taken from the side by VideoSprint high speed camera positioned horizontally in a sealed box (see Fig. 1b). The camera was positioned horizontally and the level of its optic axis was above the level of the water surface by 8 cm. The focal plane was at a distance of 77 cm from the laser knife. The size of the taken area was from (66.4 – 16 mm) × 256 mm (the horizontal size of a frame decrease with an increase in filming speed). A droplet removal system in the form of a metal tube blown with supplied compressed air was mounted from the internal side of the channel side wall through which the filming was performed. According to test experiments, the system introduced no distortions to the wind flow in the region of measurements.

The experiments were carried out at four values of air flow rates in the channel: 1.1, 1.6, 2.2, and 2.7 $m^3/s$; as it is shown below, this corresponds to equivalent wind velocities at heights of 10 m, $U_{10}$, 11, 20, 37, and 48 m/s, respectively. In two last cases, a strong breaking of waves with the formation of white caps and intense generation of sprays was observed (see photographs in Fig. 2). For each wind velocity, 30 implementations were taken, 3000–4500 frames in each implementation. The filming speed was 1500, 3000, 5000, and 6000 fps; the exposure time was from 50 to 14 μs.

(a)

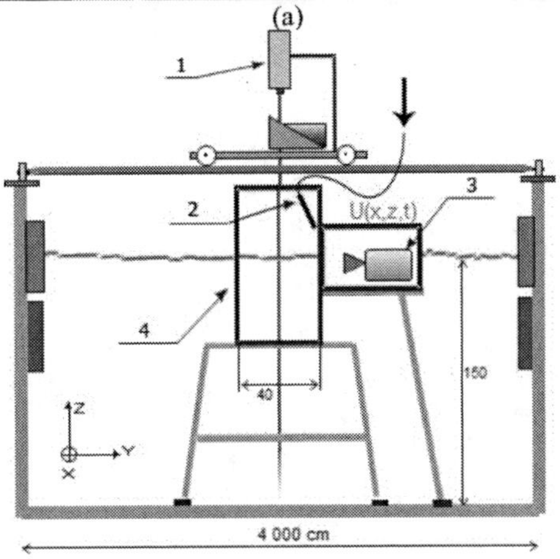


(b)

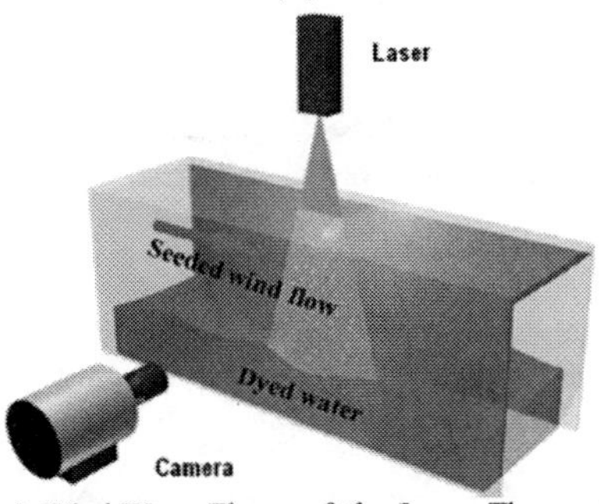


**Fig. 1.** a) Wind-Wave Flume of the Large Thermostratified Tank of IAP RAS (air channel in the center) b) principal scheme of experimental setup in the working cross-section of the wind channel 1- laser, 2 – droplets blow-off system, 3 – high-speed camera, 4 – wind-wave channel. c) scheme of PIV-method using.

## 3 Experimental data processing

Determining the shape of wave surface for each frame is necessary for finding the velocity field by cross-correlation processing with the adaptive PIV algorithm over a curvilinear grid in the immediate vicinity from the water surface. Earlier, to determine the surface shape by images from a highspeed camera, a stepwise algorithm [11] was developed based on the Canny method [12]. That method operated well under conditions of weakly breaking waves. However, with an increase in the air flow velocity, a transition to a rather intense breaking with the formation of sprays and whitecaps was observed. In this order, a combined method of measuring the elevation of the water surface was used. In this method, optical measurements were completed by contact measurements using a wire wave recorder mounted on the channel axis near the edge of the laser sheet. The records of the level elevation and high speed camera were synchronized. The final form is a combination of data obtained by contact and noncontact ways; with an increase in wind velocity, the role of contact measurements increased up to a full substitution of optical measurements for the case of flow rate of 2.7 $m^3/s$ (see Fig. 3).

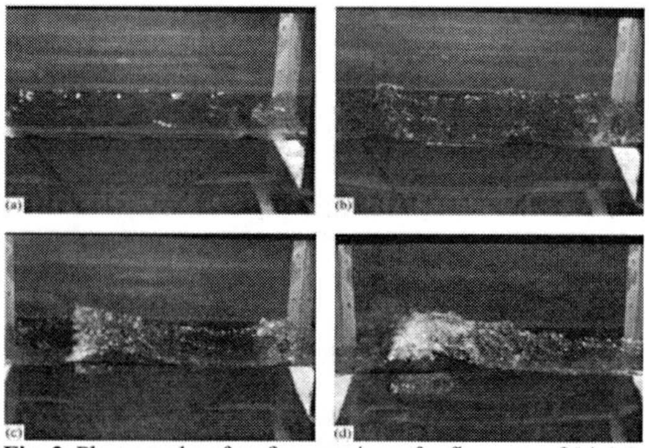

**Fig. 2.** Photographs of surface waviness for flow rate of (a) 1.1, (b) 1.6, (c) 2.2, and (d) 2.7 $m^3/s$.

After finding the surface shape, velocity fields were calculated by the cross-correlation method over a curvilinear grid taking into account the current shape of the surface [8]. We used a modified PIV processing method based on adaptive finding for the shift of the cross-correlation function (from now on, briefly CCF) peak for rectangular elements of the image for two sequential frames. The space on the grid of the velocity field calculation was 3.2 mm. To increase the algorithm accuracy due to a decrease in the search window size, the processing was performed in two stages. At the first stage, using a window with dimensions of 128 × 64 px, the profile of the average horizontal shift of particles for each wind velocity was found.

At the second stage, the comparison window was shifted to the magnitude of the shift found at the first stage. The size of the search window could be less than the final shift of particles, which increased the spatial resolution of the method, accelerated the processing, and made it possible to obtain more data. The CCF maximum was searched by the adaptive method during two iterations by analogy with [8]. At the first passage, the shift over the window with a larger size (32 × 32 px or 6.4 × 6.4 mm) was approximately determined; then, an after search was performed with allowance for the calculated shift for a window with a lesser size (8 × 8 px or 1.6 × 1.6 mm). To increase the accuracy at the last step of cross correlation, an algorithm of subpixel approximation of the peak by a Gaussian two-dimensional profile (see [8]) was applied; this allows one to take into account CCF values at points near the maximum.

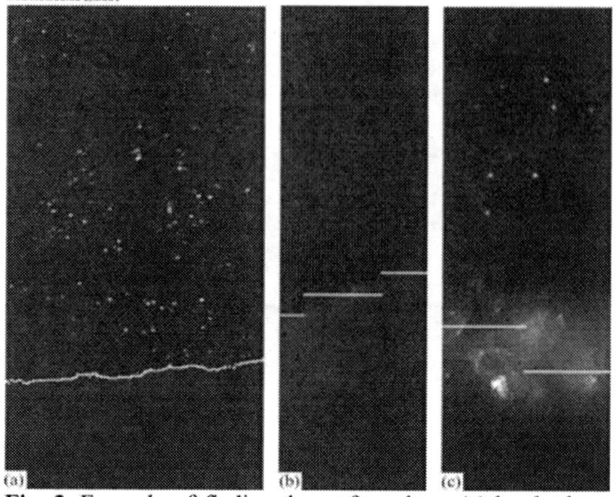

**Fig. 3.** Example of finding the surface shape (a) by the laser optical method and by data from a wave recorder in the case of (b) a foam line and (c) the rear front of the wave.

Data that do not satisfy the quality criteria were eliminated from further processing. In particular, windows with an insufficient quantity of particles were eliminated from the processing. For each image, points corresponding to high values of the brightness gradient were found using the Sobel operator. These points correspond to edges of particle images in the frames. For each window, the number of similar boundary pixels was calculated. The threshold value of the brightness gradient in the process of searching the edges was matched so that no boundary pixels fell in regions without particles. The cross correlation was performed only with windows in which at least one boundary pixel was present. In this process, if the number of boundary pixels considerably exceeded the average number, flares, spindrifts, or a foam line were most often present in part of the image corresponding to this region; as a consequence, the shift calculated for it was not taken into account in the process of further averaging. Regions for which the value of the CCF maximum considerably differs from the typical value in the implementation were also not taken into account, because such regions most often corresponded to the case of a particle escaping from the laser sheet during the interval between frames.

At strong winds, the inhomogeneity of seeding increased, which resulted in considerable discontinuities (the absence of data) in time dependences of velocities at fixed horizontal coordinates. Despite measures for the removal of droplets, optical distortions increased with an increase in the wind velocity. This resulted in decreasing of accuracy.

To correctly calculate average velocity fields and profiles, a method based on phasing the measured wind velocities at different levels from a wavy surface was used. The phase of the wave over which the point of velocity measurement is positioned was determined using the Hilbert transform over time realizations of water surface elevations presented in several equidistant vertical cross sections for each frame. There were from three to seven such cross sections in a frame (depending on the flow rate).The position of the surface for each vertical column of the velocity field was calculated using linear interpolation. Thus, each column has its own dependence of the surface position on time.

Wave number spectra of waving have a pronounced peak, and deviation of the water surface from the horizontal is determined mainly by the fundamental wave harmonic. For this reason, averaging over the phase of the fundamental wave harmonic was performed. Frequency filtering of time realizations obtained in the aforementioned way was performed over a band with a width of 2 Hz near the peak frequency determined for each wind velocity by time realizations from the wave recorder. Such filtered realizations with a removed constant component fit well for determining the wave phase by use of the Hilbert transform. Using the economical algorithm of the fast Fourier transform, the Hilbert transform permits one to determine the amplitude and phase of surface elevation as a function of time.

To obtain velocity fields averaged over turbulent fluctuations, conventional averaging at a fixed phase was performed. To reduce errors connected with the insufficient number of measurements, the data were binned in phase intervals of 18°, which yields 20 different binns of the phase.

The averaging was performed in two ways:

(i) Averaging over fixed horizons (Fig. 4a). The velocity fields were divided into horizons with a step of 32 px (6.4 mm of the real height). For each horizon and each phase, velocities were accumulated from the whole realization and then averaged.

(ii) Averaging over a fixed distance from the surface (Fig. 4b). For each frame in each column of the grid, a cell with the same number was chosen; in the curvilinear coordinate system used, this meant a fixed distance of the point from the surface. The data from corresponding similar cells over the height and phase were accumulated from the whole realization and then averaged.

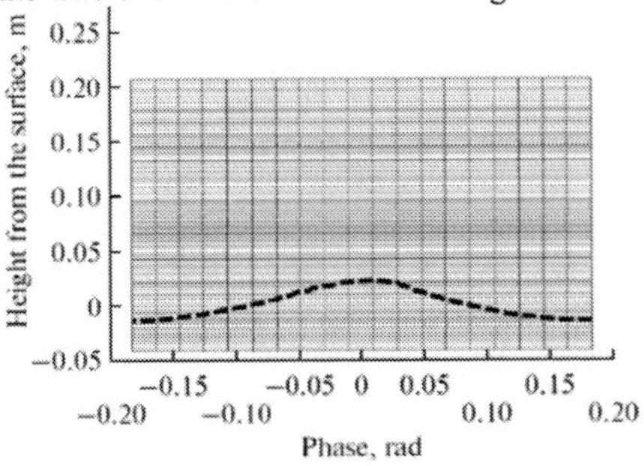


(a)

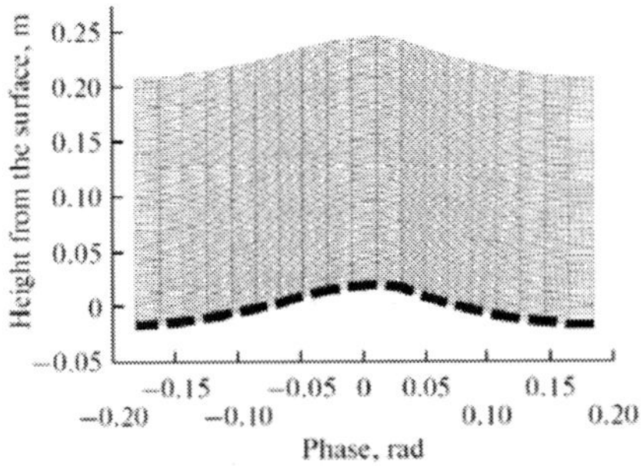


(b)

**Fig. 4.** Examples of a grid over which the velocity was averaged: (a) averaging over fixed horizons and (b) averaging over the distance from the surface.

## 4 Results

Pictures of turbulent-pulsation averaged velocity fields of the air flow were obtained for both ways of averaging. In curvilinear coordinates, by analogy with [8], the vertical coordinate is counted from the position of the water surface at each instant of time. The position of the average surface level was taken as zero in rectangular coordinates. The horizontal coordinate is the wave phase for a given point $\varphi$ recalculated using the value of the wavelength $\lambda$ determined by the dispersion relation for waves on deep water for a frequency corresponding to the peak in the center of the spectrum for each value of the wind velocity:

$$x = -\frac{\varphi}{2\pi}\lambda$$

Examples of flow fields for both averaging are presented in Fig. 5.

The number of points satisfying the criteria of data quality is different for the same distance from the surface in different phases. This difference is especially seen near the surface, where the decrease in the quantity of data satisfying the quality criterion from the leeward side of the wave crest (negative values of the phase in Fig. 6) caused by decrease in the number of tracing particles in this region at the instant of shooting due to the shielding of the wind flow by the wave crest. Note that, for the cases of two high wind velocities ($U_{10}$ of 37 and 48 m/s), an inverse picture is observed: the number of points satisfying the criteria of data quality becomes larger from the leeward side of the crest. This can be connected with the appearance of sprays, which are intensely generated at velocities $U_{10}$ > 25 m/s. The droplets concentration from the windward side of the wave crest is considerably higher than from the leeward side, because the droplets are generated mainly near the top of the wave and whirled away by the wind. The cross-correlation algorithm for cells in this region shows a shift of droplets, not tracing particles, which are relatively not numerous there.

DOI: 10.1051/epjconf/201714302101

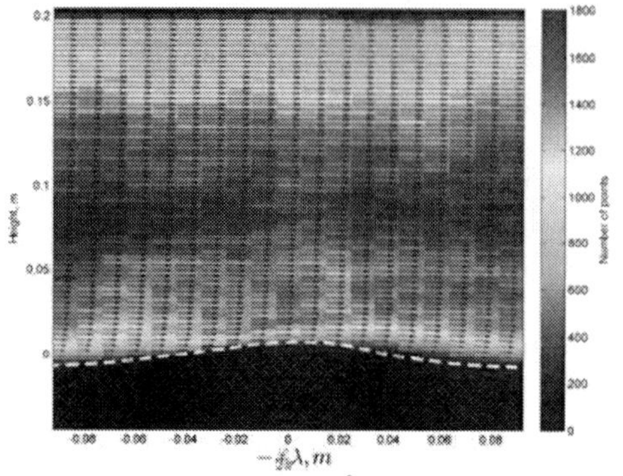
(a) 1.1 $m^3/s$

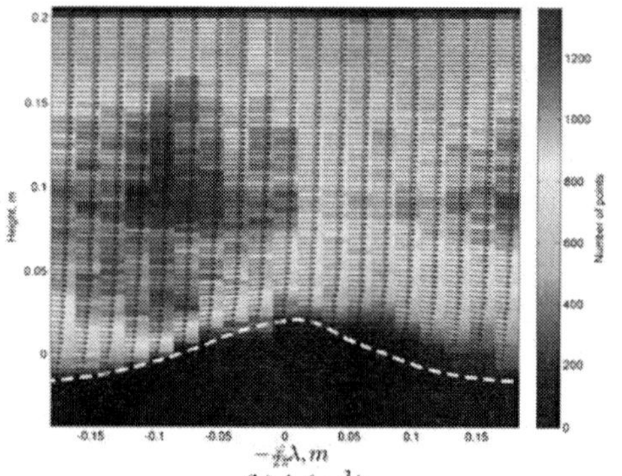
(b) 1.6 $m^3/s$

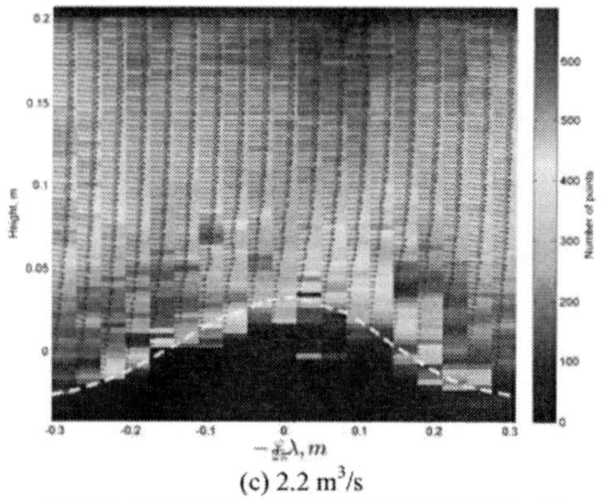
(c) 2.2 $m^3/s$

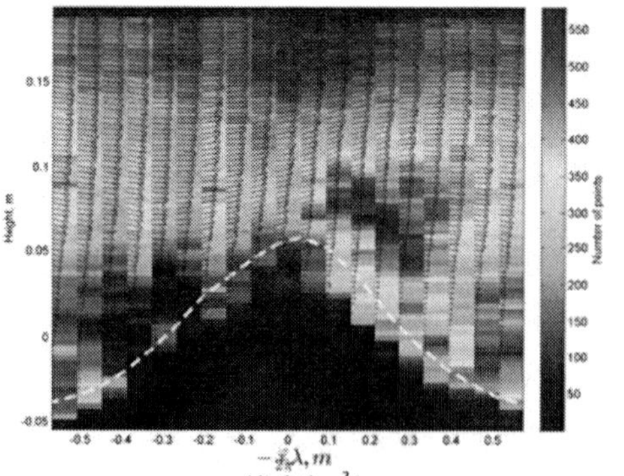
(d) 2.6 $m^3/s$

**Fig. 5** (a) –(d)Velocity fields averaged over ensemble, for fixed phase along the horizons in Cartesian coord. Color indicates number of points in ensemble of averaging for each cell.

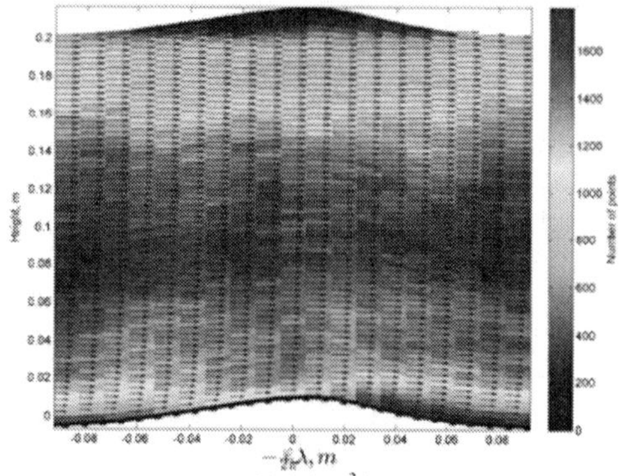
(e) 1.1 $m^3/s$

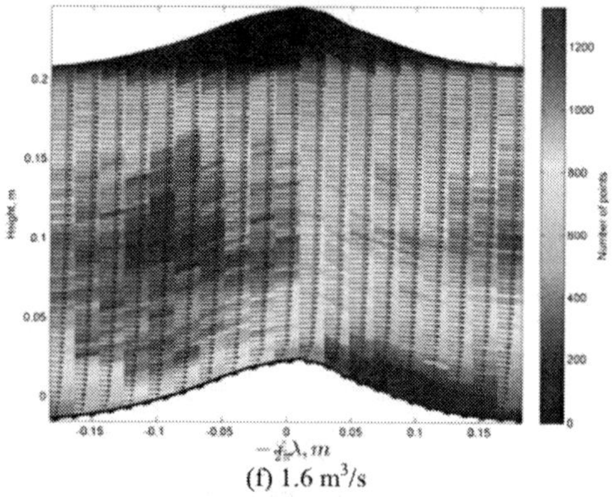
(f) 1.6 $m^3/s$

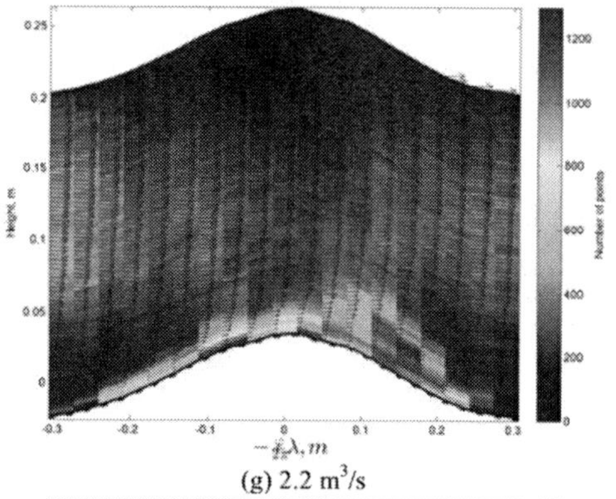
(g) 2.2 $m^3/s$

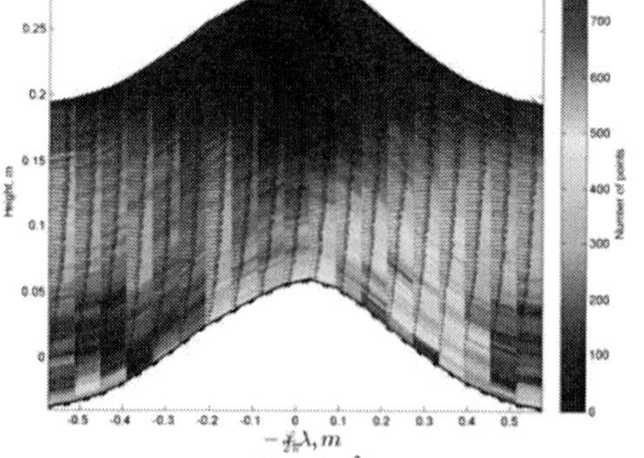
(h) 2.6 $m^3/s$

**Fig. 5** (e) – (h) Velocity flow fields averaged over ensemble, for fixed phase in curvilinear coordinate. Color indicates number of points in ensemble of averaging for each cell.

Since the droplet velocity is lower than the wind velocity, this leads to an erroneous underestimation of the air flow velocity near the surface. The different number of particles in different phases makes it necessary to apply conventional averaging over the phase to obtain correct results.

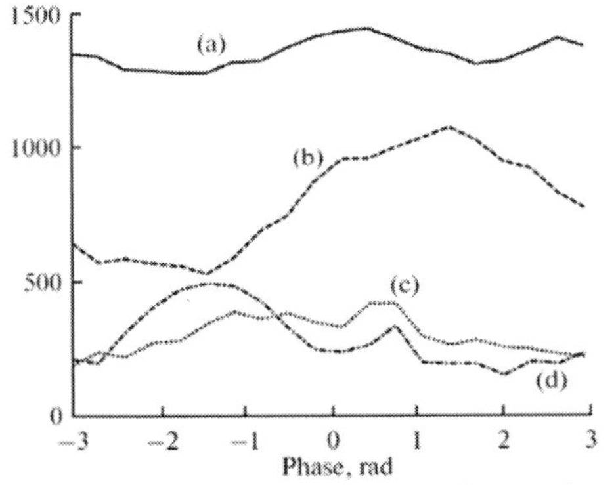


**Fig. 6.** Number of cells taken into account in the averaging as a function of the wave phase for different values of air flow rate: (a) 1.1, (b) 1.6, (c) 2.2, and (d) 2.7 $m^3/s$. The function is averaged over height inside the nearwater layer with a thickness of 5 cm.

Velocity values obtained using the Pitot tube always turned to be lower than those obtained in PIV measurements. Note that the PIV technique is a direct method of determining the air flow velocity; in contrast to it, the Pitot tube is an indirect method of estimating the flow velocity by pressure and the underestimation of the flow velocity by the Pitot tube probably indicates the presence of a systematic error.

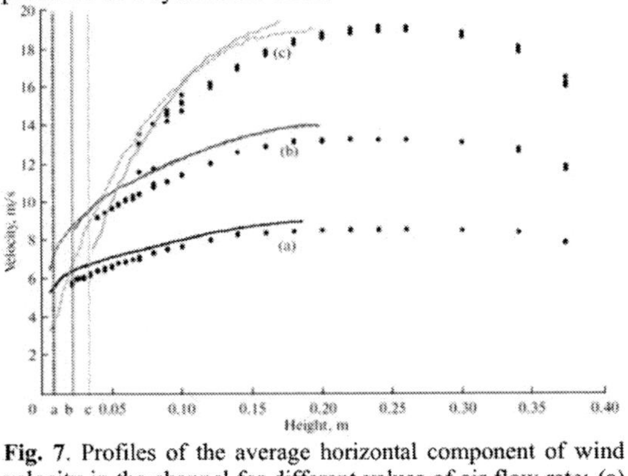


**Fig. 7.** Profiles of the average horizontal component of wind velocity in the channel for different values of air flow rate: (a) 1.1, (b) 1.6, (c) 2.2 $m^3/s$. The profiles were obtained by averaging PIV data over curvilinear coordinates (dashed lines), Cartesian coordinates (solid lines), and based on data from the Pitot tube (symbols). The dotted vertical lines shows average levels of wave crests for each case. For the flow rate 2.7 $m^3/s$.

Fig 7 shows also the air flow velocity profiles obtained by averaging measurement data using the PIV method. Two ways of averaging the velocity field were used: over horizons (the solid line) and curvilinear coordinates (the dashed line).

Both the velocity profiles have a similar shape and close values; they agree with each other well in the region above wave crests. The main difference between them is that averaging over curvilinear coordinates can yield air flow velocities in wave troughs, while averaging over horizons can yield the vertical profile of the average wind velocity only above wave crests. In connection with this, the aerodynamic drag factor of the water surface was determined by dependences obtained by averaging over curvilinear coordinates.

## 5. Conclusion

Laboratory experiments on studying the structure of the turbulent air boundary layer over waves were carried out at the Wind-Wave Channel, IAP RAS, under conditions modeling the near water boundary layer of the atmosphere under strong and hurricane winds. Air flow in the channel with a square cross section of 0.16 m2 was created by fan. The air flow rate took values of 1.1, 1.6, 2.2, and 2.7 $m^3/s$, which corresponded to estimate values of equivalent wind velocities from 10 to 48 m/s at the standard height of 10 m.

A modified technique of Particle Image Velocimetry (PIV) was used to obtain turbulent pulsation averaged velocity fields of the air flow over the water surface curved by a wave. We emphasize that such remote methods permit one to obtain the air flow velocity field also below wave crests in wave troughs. Using wave phase averaging at fixed distances from the water surface, average profiles of the air flow velocity were obtained in curvilinear coordinates following the wave. Note that measurements could be reliably performed only using the PIV technique. Using contact measurements (Pitot tube), profiles of the air flow velocity over wave crests were measured in Cartesian coordinates. To compare measurement results obtained by two techniques, measurement data obtained by the PIV technique were also expressed in Cartesian coordinates. Both experimental methods yielded close results at distances of more than 10 mm from wave crests. But introducing the Pitot tube with a diameter of about 10 mm into the flow could resulted in considerable distortions of the flow close to the water surface. Parameters of the air flow (friction velocity and drag coefficient of the surface) could be further determined by extrapolating the logarithmic part of the velocity profile, and their dependencies on the wind speed could be obtained.

This work was supported by the Russian Foundation of Basic Research (No. 15-35-20953 17-05-00958, 14-05-91767 AF_a, 16-55-52022), providing experiments was partial supported by Russian Science Foundation (Agreement No. 14-17-00667) and processing of the data was partially supported by Russian Science Foundation (Agreement No. 15-17-20009). Also work is supported by international project Air-Sea Interaction under Stormy and Hurricane Conditions: Physical Models and Applications to Remote Sensing (ASIST) ASIST within the FP7 program and of the Federal Target Program "Research and development in priority areas of Russian scientific and technological complex for 2014-2020" (grant agreement 14.616.21.0059, a unique identifier RFMEFI61615X0059 project)

## References

1. C.W. Fairall, E.F. Bradley, J.E. Hare, A.A. Grachev, J.B. Edson, Climate., **16**(4). 571 (2003)
2. M.D. Powell, P.J. Vickery, T.A., Nature. **422**., 279(2003).
3. M.A. Donelan, B.K. Haus, N. Reul, W.J. Plant, M. Stiassnie, H.C. Graber, O.B. Brown, E. S. Saltzman, Geophys. Res. Lett., **31**, L18306 (2004)
4. Yu. I. Troitskaya, D.A. Sergeev, A.A. Kandaurov, G.A Baidakov, M.A. Vdovin, V.I. Kazakov JOURNAL OF GEOPHYSICAL RESEARCH, **117**(C00J21), 13(2012)
5. Yu. Troitskaya, D. Sergeev, A. Kandaurov and V. Kazakov., (Recent Hurricane Research - Climate, Dynamics, and Societal Impacts, InTech, 2011)
6. Reul, N., H. Branger, and J.-P. Giovanangeli, Bound.-Layer Meteor., **126**, 477 (2008)
7. Veron, F., G. Saxena, and S. K. Misra, Geophys. Res. Lett., **34**, L19603(2007).
8. Yu. Troitskaya, D. Sergeev, O. Ermakova, G. Balandina, J. Phys. Oceanogr., **41**, 1421(2011)
9. 9. C.T. Hsu, E.Y. Hsu, R.L. Street, J. Fluid **105**, 87(1981)
10. Hinze J. O. (Turbulence: An Introduction to its Mechanism and Theory, McGraw-Hill. 1959)
11. D. Sergeev, A. Kandaurov Yu. Troitskaya, Proc. Opt. Meth. Flows Invest. 90 (2011).
12. Canny, J. A., IEEE Trans. Pattern Analysis and Machine Intelligence. **8(6),** 679(1986)

# Feasibility and accuracy assessment of light field (plenoptic) PIV flow-measurement technique

*Chandra* Shekhar[1,*], *Syo* Ogawa [2], and *Tatsuya* Kawaguchi [2]

[1]Heat and Fluid Dynamics Laboratory, IHI Corporation, 1 Shin-Nakahara-cho, Yokohama-2358501, Japan
[2]Department of Mechanical Engineering, Tokyo Institute of Technology, 2-12-1 Ookayama, Tokyo-1528550, Japan

**Abstract.** A light field camera can enable measurement of all the three velocity components of a flow field inside a three-dimensional volume when implemented in a PIV measurement. Due to the usage of only one camera, the measurement procedure gets greatly simplified, as well as measurement of the flows with limited visual access also becomes possible. Due to these advantages, light field cameras and their usage in PIV measurements are actively studied. The overall procedure of obtaining an instantaneous flow field consists of imaging a seeded flow at two closely separated time instants, reconstructing the two volumetric distributions of the particles using algorithms such as MART, followed by obtaining the flow velocity through cross-correlations. In this study, we examined effects of various configuration parameters of a light field camera on the in-plane and the depth resolutions, obtained near-optimal parameters in a given case, and then used it to simulate a PIV measurement scenario in order to assess the reconstruction accuracy.

## 1 Introduction

Light field cameras allow recording three dimensional features of a given object by recording information about the angle of the incoming rays on the image sensors. It was initially developed by Ives [1] and Lippman [2] about a century ago, and the underlying science was explored by Gershun [3] when he defined the concept of light field. In modern light field cameras, recording of the angular information is achieved by placing a micro-lens array between the main lens and the image sensor. The raw image that forms on the sensor happens to be utterly complex, which are mathematically processed to obtain a human-perceivable image. The historical foundations and working principle of the camera are well described by Levoy [4].

Since a light field camera can record three dimensional features through angular information, it can be used to identify a seeded three-dimensional fluid flow as well. Thus, it can be used to conceptualize a three-dimensional PIV measurement technique to obtain flow velocity. Despite this possibility, its usage in flow measurement has started only recently [5, 6]. Fundamentally, there are two ways of using the camera in PIV flow measurements. The first way is to refocus the camera at a given depth plane and obtain a two-dimensional particle image there. Thereafter, cross-correlation between two such focused particle images coming from two snapshots of a seeded flow, respectively, would yield the two-dimensional velocity field in the focused plane. After refocusing the camera in multiple planes, multiple such two-dimensional velocity fields can be obtained, effectively yielding a 3D-2C flow field. Kawaguchi et al [7] used this concept to obtain velocity fields in two parallel planes, after performing a dual-plane PIV measurement. Although conceptually easy, the results suffer from severe inaccuracies, which stem from appearance of blurred out-of-the-plane particles in the refocused plane. The second way of using a light field camera is to reconstruct the three-dimensional instantaneous particle distribution of the seeded flow by using a tomographic algorithm, followed by performing a three-dimensional cross-correlation in order to obtain all the three components of the velocity in the reconstructed volume. Nowadays, this latter method is being mostly studied due to its superior properties.

Lynch [5] and Lynch and Thurow [6] described usage of light field cameras in PIV measurements in detail. They simulated PIV-like scenarios, performed a three-dimensional reconstruction of the particle intensity field, discretized two such intensity fields in interrogation volumes, and then performed three-dimensional cross-correlation to obtain velocity vectors. In subsequent studies, Fahringer and Thurow [8, 9] reconstructed similar intensity fields using MART algorithm [10], which resulted in more realistic particles in the reconstructed volume. Despite that, locations of the reconstructed particles in the study by Fahringer and Thurow [8] matched with those of the original particles only approximately. Fahringer and Thurow [9] estimated positional accuracy of the reconstructed particles through simulations, after considering 40 spherical particles lying on the optical axis of the camera system. The results showed that difference between the reconstructed particle positions

* Corresponding author: chandra_shekhar@ihi.co.jp

and those of the original ones are unacceptably large in the depth direction. Hadfield and Nobes [11] also observed large inaccuracies in the position of the seeded particles in a particle tracking velocimetry (PTV) measurement. Kawaguchi et al [7] and Ogawa et al [12] discussed the MART algorithm and used it to reconstruct a three-dimensional synthetic intensity field, from where they demonstrated trade-off between the thickness of the measured volume and the spatial resolution in the depth direction.

Many of the above studies yielded velocity vector fields also, but their accuracy is highly questionable. The severity became apparent from a study by Thurow et al [13], who compared results obtained by a light field PIV measurement with the results obtained by a stereo PIV measurement. The flow measurements were performed above a wing. The comparison showed that the two results differ greatly, sometimes even in a qualitative sense. There are various sources of such large error, e.g. large elongation of the reconstructed particles in the depth direction, which leads to accordingly large uncertainty in the location of their centroid (which represents their positions), imprecise knowledge of the distances between the main lens and the microlens array, optical distortions present in the main lens and lenses in the microlens array, imperfect alignment of the microlens array and the image-sensor plane, to name some. In the literature, studies trying to remove some of these inaccuracies are reported [8, 14, 15]. Foy and Vlochos [16] simulated scenarios where multiple light field cameras are used, and reported that high-fidelity measurements can be performed even with only two cameras.

It is evident from the literature survey that further accuracy assessment of usage of light field cameras in PIV measurements is required. In the present study, we have assessed effects of some camera construction parameters on the in-plane and the depth resolutions, followed by assessing positional accuracy of the reconstructed particles in a three-dimensional volume. We performed this study using computer simulations.

## 2 Fundamentals, theory, and simulation details

Three typical modern configurations of light field cameras are schematically shown in Fig.1. These are called Plenoptic-1, Plenoptic-2 (Galilean), and Plenoptic-2 (Keplerian) configurations, respectively. As shown in the figure, a light field camera consists of a main lens, a microlens array, and an image sensor. In the Plenoptic-1 arrangement, the main lens is focused on the microlens array. In Plenoptic-2 arrangements, the microlenses are focused on the image plane. In the Galilean type arrangement, the image plane of the main lens lies behind the image sensor, whereas it lies before the sensor in the Keplerian type of the arrangement. In the present study, the attention is paid on the Keplerian arrangement, because it offers more flexibility when configured physically.

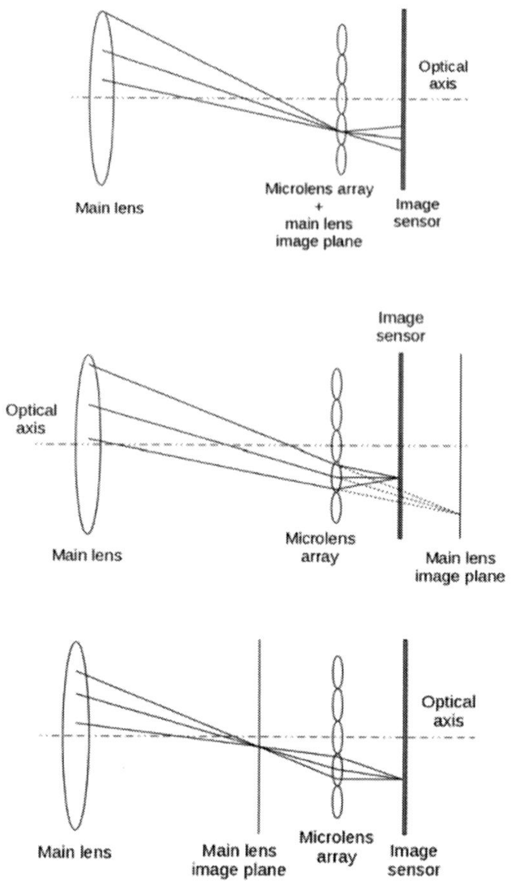


**Fig.1.** Schematic ray diagrams of plenoptic-1, plenoptic-2 (Galilean), and plenoptoc-2 (Keplerian) configurations of the light field cameras, from top to bottom, respectively.

A detailed diagram of the studied camera system is shown in Fig.2. As shown, there are eight configuration parameters, namely, $A$, $B$, $a$, $b$, $F$, $f$, $D$, and $d$. Simultaneously, there are three optical constrains. Two of these constrains are the lens equations $1/A + 1/B = 1/F$ and $1/a + 1/b = 1/f$, and the third one is $F/D \approx f/d$, which stems from the fact that, for optimal imaging by microlenses (i.e. maximum but non-overlapping micro images), the F-number (indicated by F#) of the main lens and the microlenses should be almost equal [7]. Furthermore, $d$ is kept constant in this study at 0.25 *mm*. This leaves us with four parameters that can be chosen freely.

In the next section, we will examine the effects of the four freely chosen parameters $B/A$ ($\equiv M$), $b/a$ ($\equiv m$), $F$, and $f$ on the in-plane and the depth resolutions of the camera system. The two resolutions are represented by $R_{xy}$ and $R_z$, respectively. In order to obtain quantitative values of $R_{xy}$ and $R_z$, we consider a spherical particle of diameter 15 $\mu m$ present at the location where the optical axis of the camera intersects the object plane (see Fig.2).

Light rays emerging from this particle pass through the main lens and the microlens array, and then recorded on the image sensor. The pixel size of the image sensor, $\delta \times \delta$, is set to 5.15×5.15 $\mu m^2$, which is approximately same as that typically found in commercially available cameras. We ensured that the spot size yielded by the particle is smaller than the pixel size, so that the calculated $R_{xy}$ and $R_z$ are independent of the size of the particle and represent true resolutions (if the particle size is significantly larger and its spot size covers multiple pixels, the calculated $R_{xy}$ and $R_z$ would depend on the particle size). In general, the light rays are recorded on multiple pixels, qualitatively similar to that shown in Fig.2. The recorded light intensity is averaged over the pixel area, followed by reconstructing the particle using MART algorithm [10]. The reconstructed particle is elongated in the depth direction. The intensity of the reconstructed particle is maximum on its centre, and it decreases in a Gaussian fashion along any radial line that extends from the centre. The cut-off intensity value is set to 3. The cut-off results into a well-defined elongated particle with clearly recognized boundary. Its width in the $Z = 0$ plane would be same both along the $X$ and the $Y$ axes, which is taken as the quantitative value of $R_{xy}$. Similarly, its length along the $Z$ axis is taken as the value of $R_z$.

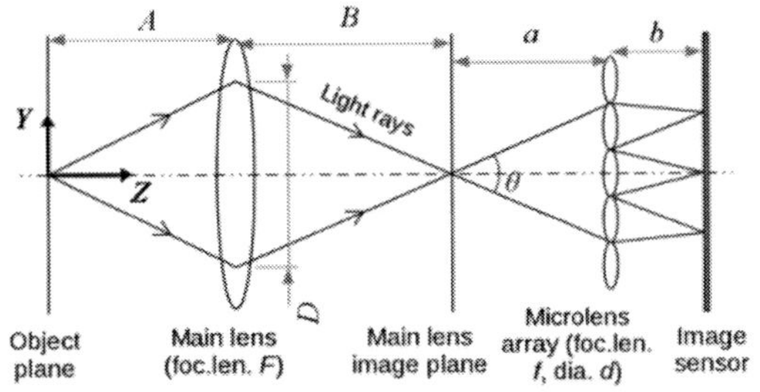


**Fig.2.** Schematic ray diagram of the light field camera that is used in the present study. Various distances and the used right-handed Cartesian coordinate system are also shown.

## 3 Effect of the freely-varying camera parameters on the resolutions

Each of the freely varying parameters, namely $M$, $F$, $m$, and $f$, is varied one by one, while keeping the remaining three constant. $M$ and $m$ are chosen around the values which are typically used in short-distance flow measurements, and $F$ and $f$ are chosen around the values which are commercially available, so that these can be used to assemble a prototype light field camera. For a given $M$, $F$, $m$, $f$, and $d$ (where $d$ = 0.25 $mm$, as noted earlier), the three optical constrains can be used to write mathematical relationships for $A$, $B$, $a$, $b$, $D$, and F#, as follows:

$$A = (1 + M)\, F/M \quad (1)$$

$$B = (1 + M)\, F \quad (2)$$

$$a = (1 + m)\, f/m \quad (3)$$

$$b = (1 + m)\, f \quad (4)$$

$$D \approx Fd/f \quad (5)$$

$$\mathrm{F\#} = f/d \quad (6)$$

where F# is the common F-number of the main lens and the micro lenses, and $D$ represents the effective diameter of the main lens (see Fig.2).

In order to understand the way the resolutions depend on the configuration parameters, we analysed the ray traces. We calculated equations of the paths that a ray emerging from the location $(X, Y, Z) = (0, h, 0)$ at angle $\varphi$ from the $Z$ axis follows (i) between the main lens and the microlens array and (ii) between the microlens array and the image sensor. The two equations, respectively, can be written as follows:

$$Y = S_1\, Z + C_1 \quad (7)$$

$$Y = S_2\, Z + C_2 \quad (8)$$

where:

$S_1 = (1 - A/F) \tan \varphi - h/F,$

$S_2 = -(A/f) \tan \varphi - \{(a + B)/f - 1\}S_1 - (h - Y_C)/f,$

$C_1 = h + (\tan \varphi - S_1)\, A,$

$C_2 = h + (\tan \varphi - S_2)\, A + (S_1 - S_2)(a + B),$

$Y_C$ = $Y$ coordinate of the centre of the microlens on which the ray incidents, and

$-\tan^{-1}\{(D + 2h)/2A\} \leq \varphi \leq +\tan^{-1}\{(D - 2h)/2A\}$.

The constrain on $\varphi$ ($-\varphi_1 \leq \varphi \leq \varphi_2$) stems from the fact that the rays emerging from the location $(0, h, 0)$ must fall inside the effective region of the main lens (see Fig.2).

For the rays emerging from the centre of the 15 $\mu m$ spherical particle, $h = 0$. First, we used equations (7) and (8), and calculated the ray traces for a reference configuration where $(M, m, F, f) = (1, 0.1, 30\ mm, 1\ mm)$, which yields $A = B = 60\ mm$, $a = 11\ mm$, $b = 1.1\ mm$, $D = 7.5\ mm$, and F# = 4. The ray traces near the microlens array and the image sensor are plotted in Fig.3. As evident, the rays emerging from different microlenses fall at different locations of the image sensor.

In order to qualitatively examine the way $M$ affects the resolutions, we plotted the ray traces for $M = 5$ in Fig.4, while keeping $m$, $F$, and $f$ same as those used in the reference configuration. It yields $A = 36\ mm$, $B = 180\ mm$, $a = 11\ mm$, $b = 1.1\ mm$, $D = 7.5\ mm$, and F# = 4. As evident, the ray traces are different from that observed in Fig.3. After examining values of $R_{xy}$ and $R_z$ for multiple values of $M$, we found that $R_{xy} = (0.055\ mm)/M$ and $R_z = (1.214\ mm)/M$. These values imply that the in-plane resolution is more than 20 times better than the depth resolution. Note that $R_{xy} \approx \delta/(M{\cdot}m)$. The inversely proportional dependence of $R_z$ on $M$ is due to the fact that, as $M$ increases, the angle that the rays make onto the centre of the spherical particle (i.e. $\varphi_1 + \varphi_2$) also increases (see Fig.2).

We examined dependence of the resolutions on $F$ by plotting ray traces for $F = 150\ mm$, while keeping $M$, $m$, and $f$ same as that in the reference configuration. It

 DOI: 10.1051/epjconf/201714302102

yields $A = 300$ *mm*, $B = 300$ *mm*, $a = 11$ *mm*, $b = 1.1$ *mm*, $D = 37.5$ *mm*, and F# = 4. The ray traces are shown in Fig.5. After comparing this figure with Fig.3, it is evident that the two sets of the ray traces are identical, except that the ray traces in Fig.5 are physically shifted in $Z$ direction. Also, note that the effect of the increased $F$ has been to increase $A$, $B$, and $D$ linearly by a same proportion; thus, $\varphi_1+\varphi_2$ remains unaltered. Consequently, the resolutions are also expected to remain independent of $F$. Indeed, after examining multiple values of $F$, we confirmed $R_{xy} = 0.055$ *mm* and $R_z = 1.214$ *mm*.

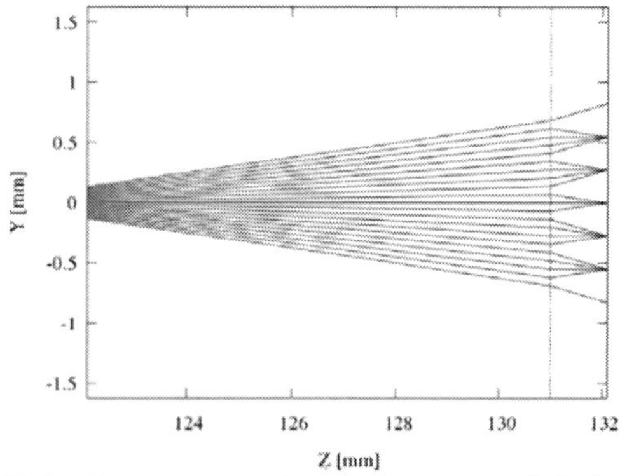


**Fig.3.** Ray traces near the microlens array and the image sensor, for $M = 1$, $m = 0.1$, $F = 30$ *mm*, and $f = 1$ *mm* (the reference configuration). Location of the microlens array is indicated by the dashed line. The image sensor lies at the right-hand side boundary.

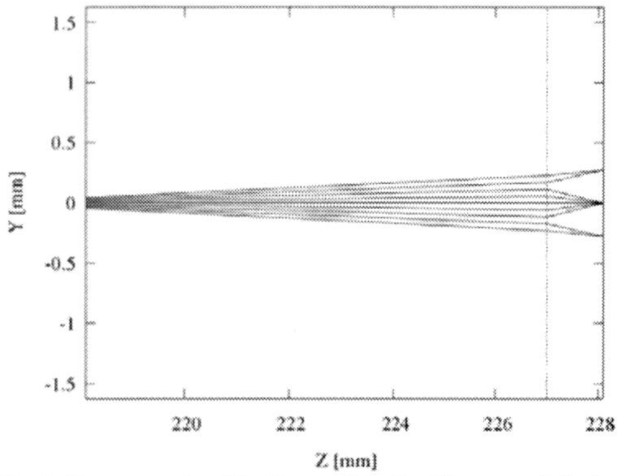


**Fig.4.** Ray traces for $M = 5$, $m = 0.1$, $F = 30$ *mm*, and $f = 1$ *mm*. Location of the microlens array is indicated by the dashed line. The image sensor lies at the right-hand side boundary.

Next, we examined effect $m$ on the resolutions. The ray traces for $m = 0.5$ is plotted in Fig.6. The other three parameters are kept same as that in the reference configuration. For $m = 0.5$, we obtain $A = 60$ *mm*, $B = 60$ *mm*, $a = 3$ *mm*, $b = 1.5$ *mm*, $D = 7.5$ *mm*, and F# = 4. After examining resolutions for multiple values of $m$, we obtained $R_{xy} = (0.0055\ mm)/m$ and $R_z \approx (0.0037/m^3 - 0.0743/m^2 + 0.5411/m - 0.4838)$ *mm*. Here also, $R_{xy} \approx \delta/(M{\cdot}m)$. $R_z$, in general, is expected not to depend on $m$, as $\varphi_1+\varphi_2$ also does not. In the present case, however, the effect appears perhaps due to some optical complexity arising from the small value of $a$.

We can observe the effect of $f$ by comparing the ray traces for $f = 3$ *mm* with those in the reference configuration (see Fig.7 and Fig.3). For $f = 3$ *mm*, we obtain $A = 60$ *mm*, $B = 60$ *mm*, $a = 33$ *mm*, $b = 3.3$ *mm*, $D = 2.5$ *mm*, and F# = 12. It is evident from the comparison that the ray traces in Fig.7 are much flatter than those in Fig.3. After quantitatively examining the resolutions for multiple values of $f$, we obtained $R_{xy} = 0.055$ *mm* and $R_z = 1.214f$. The directly proportional dependence of $R_z$ on $f$ results from smaller $\varphi_1+\varphi_2$, which is caused by the larger F#.

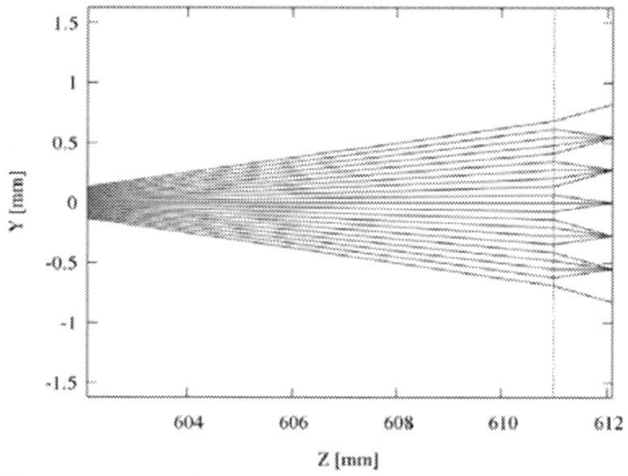


**Fig.5.** Ray traces for $M = 1$, $m = 0.1$, $F = 150$ *mm*, and $f = 1$ *mm*. Location of the microlens array is indicated by the dashed line. The image sensor lies at the right-hand side boundary.

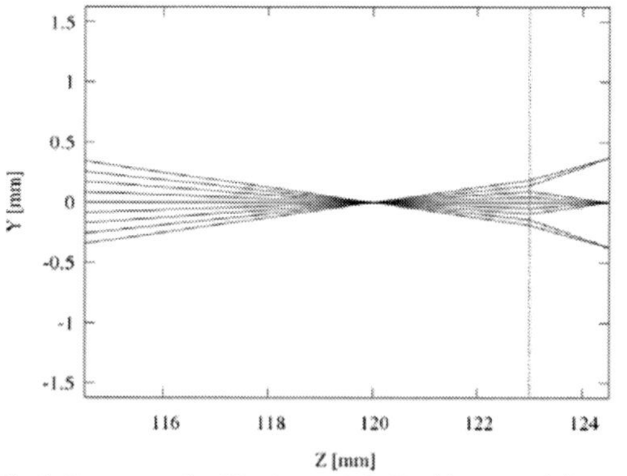


**Fig.6.** Ray traces for $M = 1$, $m = 0.5$, $F = 30$ *mm*, and $f = 1$ *mm*. Location of the microlens array is indicated by the dashed line. The image sensor lies at the right-hand side boundary.

The above dependences of the resolutions show that (i) $R_{xy}$ decreases when $M$ increases, (ii) $R_{xy}$ remains constant when $F$ varies, (iii) $R_{xy}$ decreases when $m$ increases, and (iv) $R_{xy}$ remains constant when $f$ varies. $R_z$

also exhibits the same qualitative dependences on *M*, *F*, and *m*, but increases when *f* increases. Although not presented, we observed similar qualitative dependences also when the camera is arranged in the Galilean configuration. Quantitatively, however, $R_{xy}$ and $R_z$ are little smaller than those in the Keplerian configuration.

In the next section, based on the qualitative insights obtained on the way $R_{xy}$ and $R_z$ depend on *M*, *F*, *m*, and *f*, we will obtain near-optimal set of the configuration parameters so that the values of $R_{xy}$ and $R_z$ are close to minimum.

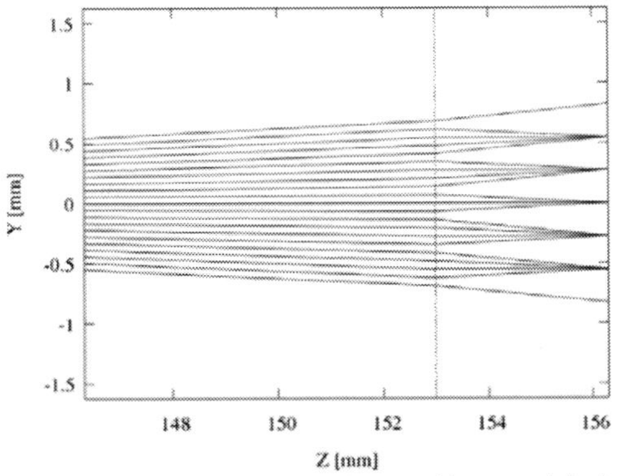


**Fig.7.** Ray traces for $M = 1$, $m = 0.1$, $F = 30$ *mm*, and $f = 3$ *mm*. Location of the microlens array is indicated by the dashed line. The image sensor lies at the right-hand side boundary.

It is also important to note that a larger divergence angle at the image plane of the main lens results in a smaller $R_z$. Therefore, we analysed the divergence angle $\theta$ (see Fig.2) for three types of commercially available lenses, whose (*F*, F#) = (35 *mm*, 1.4), (50 *mm*, 1.4), and (85 *mm*, 1.4). We also used two types of commercially available microlense arrays with (*f*, *d*) = (2.08 *mm*, 0.25 *mm*) and (0.6 *mm*, 0.20 *mm*). The results showed that $\theta$ depends weakly on the main lens, but strongly on the microlenses. For the three types of the main lenses used, $\theta$ varies in the range 4.5°~6° when (*f*, *d*) = (2.08 *mm*, 0.25 *mm*) and in the range 12°~16° when (*f*, *d*) = (0.6 *mm*, 0.20 *mm*). For the Raytrix R5 camera, we found that $\theta$ is about 10°.

## 4 Reconstruction accuracy

Finally, we set $F = 43.39$ *mm*, $D = 14.46$ *mm*, $A = 144.68$ *mm*, $B = 61.98$ *mm*, $a = 2.45$ *mm*, and $b = 0.79$ *mm* as the configuration parameters, which is close to optimum when (*d*, *f*) = (0.2 *mm*, 0.6 *mm*). These values of *d* and *f* are chosen based on commercially available microlenses of smallest-size category. The corresponding values of *M*, *m*, and F# are about 0.42, 0.32, and 3, respectively. The divergence angle, $\theta$, in this case is more than 15°. With these parameters, we examine accuracy of the positions of the reconstructed particles. For this purpose, the size of a pixel in the image sensor is set to $5.15 \times 5.15$ $\mu m^2$ and the number of the pixels is set to $1280 \times 1024$. We image distribution of a set of spherical particles whose diameter is 50 $\mu m$ and which are arranged in the $Z = 0$ plane. The particles are arranged on 1 *mm* spaced rectangular grids, as shown in Fig.8.

We perform MART reconstruction from the imaged particle distribution and obtain particle distribution in the nominal image plane of the main lens, followed by back-projecting it across the main lens, in the nominal object plane. It allows comparison between the positions of the original particles and the position of the reconstructed particles, respectively. Details of the used MART algorithm are described elsewhere [7, 12]. Next, we shift the plane of the particle distribution to $Z = +1$ *mm* and then repeat the imaging, the MART reconstruction, and the back-projection process. In total, the particles plane is shifted to 10 locations in the +Z directions and 10 location in the −Z direction, allowing us to examine the position error in the three-dimensional space $-10\ mm \leq X \leq 10\ mm$, $-8\ mm \leq Y \leq 8\ mm$, $-10\ mm \leq Z \leq 10\ mm$. The combined size of the reconstructed volume becomes about $21.0 \times 16.8 \times 16.8$ $mm^3$, which we divided into $412 \times 330 \times 330$ *voxels*. Consequently, the size of a voxel becomes about $51 \times 51 \times 51$ $\mu m^3$.

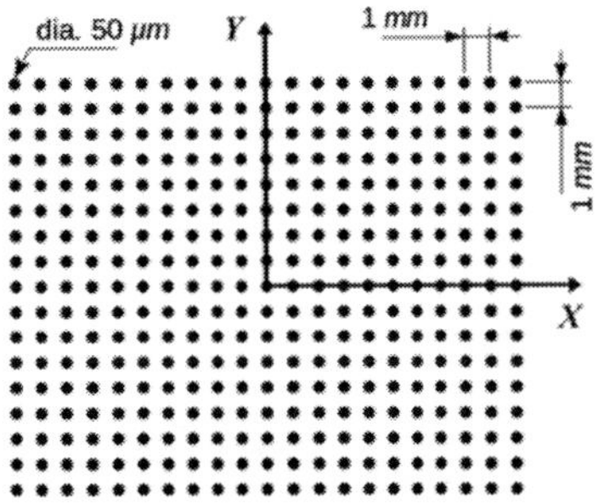


**Fig.8.** The particle distribution that is imaged by the camera.

The *X* and the *Y* positions of the centres of the particles are plotted in Fig.9 (see the top plot). The horizontal and the vertical axes show the positions of the original particles and the positions of the reconstructed particles, respectively. Ideally, the plotted data should lie on the dashed diagonal line. However, as evident, the data deviate from the line. The *Z* positions are also plotted in the same figure (see the bottom plot). The RMS deviation in the *X* and the *Y* directions is about 0.14 *mm*, and about 0.35 *mm* in the *Z* direction. Despite the significant deviation in the *Z* direction, our preliminary calculation of the displacement vectors based on two such reconstructed particle distributions show that the errors introduced in the vectors are rather small, implying that accuracy of PIV obtained velocity vectors perhaps depend only weakly on the position errors, making PIV flow measurement using such a light field camera possible. However, in order to assess the vector errors properly, a thorough investigation is required.

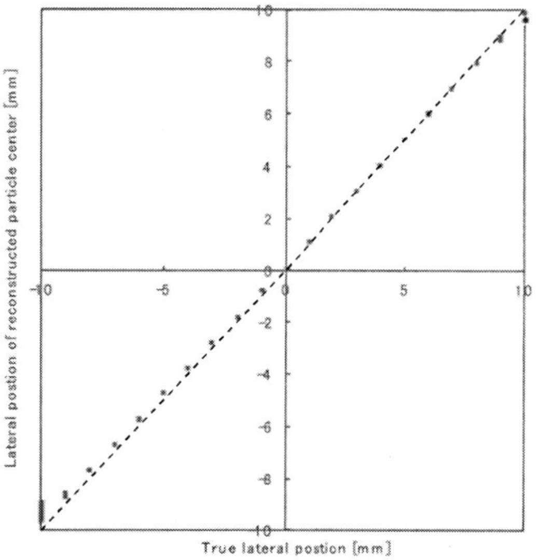


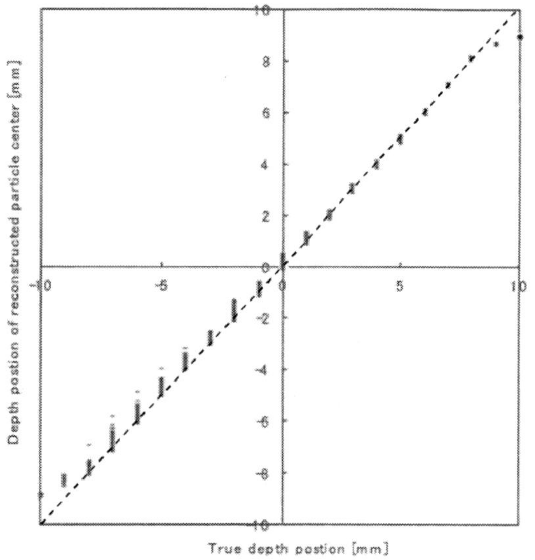


**Fig.9.** Positions of original particles (horizontal axis) versus positions of the reconstructed particles (vertical axis). The top plot shows the $X$ and the $Y$ positions, and the bottom plot shows the $Z$ positions, respectively.

## 5 Conclusions

Dependence of $R_{xy}$ and $R_z$ on $M$, $F$, $m$, and $f$ are examined. The results show that both $R_{xy}$ and $R_z$ are inversely proportional to $M$ and $m$, but independent of $F$. As for dependence on $f$, $R_{xy}$ remains constant, whereas $R_z$ is directly proportional to $f$. The dependences on $m$ and $f$ are stronger than that on $M$. The near-optimal configuration parameters for a commercially available microlens array with $f = 0.6$ $mm$ and $d = 0.2$ $mm$ can be given as $F = 43.39$ $mm$, $D = 14.46$ $mm$, $A = 144.68$ $mm$, $B = 61.98$ $mm$, $a = 2.45$ $mm$, and $b = 0.79$ $mm$. After simulating a light field PIV measurement scenario by using a light field camera configured with the optimal parameters, the difference between the positions of the MART reconstructed particles and the original particles came out to be small in the $X$ and the $Y$ directions, but significant in the $Z$ direction. A preliminary calculation of the displacement vectors, however, did not exhibit large error. A more detailed investigation is required to thoroughly assess this error.

## References

1. F. E. Ives, *U.S. Patent 725567* (1903).
2. G. Lippmann, Comp. Rendus de l'Academie des Sciences **146**, 446 (1908).
3. A. Gershun, *The light field* (Moscow, 1936), translated by P. Moon and G. Timoshenko, J. Math. Phys., **18**, 51 (1939).
4. M. Levoy, Computer **39**, 46 (2006).
5. K. P. Lynch, *Master's Thesis* (Auburn University, USA, 2011).
6. K. P. Lynch, B. S. Thurow, *41st AIAA Fluid Mechanics Conference and Exhibit* (USA, 2011).
7. T. Kawaguchi, T. Saito, I. Satoh, *17th International Symposium on Applications of Laser Techniques to Fluid Mechanics* (Portugal, 2014).
8. T. W. Fahringer, B. S. Thurow, *42nd AIAA Fluid Dynamics Conference and Exhibit* (USA, 2012).
9. T. W. Fahringer, B. S. Thurow, *52nd Aerospace Sciences Meeting* (USA, 2014).
10. G. T. Herman, A. Lent, Comp. Bio. Med. **6**, 273 (1976).
11. J. Hadfield, D. S. Nobes, *18th International Symposium on Application of Laser and Imaging Techniques for Fluid Mechanics* (Portugal, 2016).
12. S. Ogawa, T. Kawaguchi, I. Satoh, T. Saito, *16th International Symposium on Flow Visualization* (Japan, 2014).
13. B. S. Thurow, T. Fahringer, K. Johnson, S. Hellman, *17th International Symposium on Applications of Laser Techniques to Fluid Mechanics* (Portugal, 2014).
14. O. Johannsen, C. Heinze, B. Goldluecke, C. Perwass, *Time-of-Flight and Depth Imaging. Sensors, Algorithms, and Applications*, 302.
15. C. M. Thomason, T. F. Fahringer, B. S. Thurow, *52nd Aerospace Sciences Meeting* (USA, 2014).
16. R. R. L. Foy, P. Vlachos, *10th International Symposium on Particle Image Velocimetry* (The Netherlands, 2013).
17. R. Perwass and C. Perwass, *US Patent US20120050562 A1* (2012).

# Laboratory modeling of flow regimes in a draft tube of Francis hydro-turbine

*Sergey* Shtork[1,2*], *Ivan* Litvinov[1,2], *Andrey* Mitryakov[1,2], and *Kemal* Hanjalic[2,3]

[1]Institute of Thermophysics SB RAS, 630090 Novosibirsk, Russia
[2]Novosibirsk State University, 630090 Novosibirsk, Russia
[3]Delft University of Technology, Delft, Netherlands

**Abstract.** The paper reports on some results of the experimental study of flow and pressure pulsations in a laboratory model of the draft tube (DT) of Francis-99 hydro-turbine over a broad range of operating regimes. Velocity distributions at the model inlet varied within 866 modes of the turbine load conditions, including those with maximum coherent pressure pulsations on the model walls. The contact and non-contact methods were used to measure pressure pulsations on the model walls using acoustic sensors and to measure the averaged velocity distribution with a laser Doppler anemometer "LAD-06i". Analysis of the results have showed that in the model cone there are flow modes with forming of precessing vortex cores, accompanied by a sharp increase in the amplitude of coherent pressure pulsations on the wall and the velocity field rearrangement. It is shown that the vortex core starts forming with an increase in the integral parameter of swirl up to S>0.5. A novelty of the work is the combination of the rapid prototyping (3-d printing) of the inflow swirl generators and computerized measurement techniques that makes it possible to acquire rapidly a large amount of experimental data for a variety of designs and operating conditions. The results can provide insight into the effect of various design and operating parameters on the flow physics, as well as serve for verification of the numerical simulations.

## 1 Introduction

The growing energy consumption nowadays imposes high requirements to the hydroelectric power plants. This is because electricity production must be adjusted, considering ups and downs of the load of electric mains. Varying the generator load results in short-term changes in rotation frequency of the turbine impeller, thereby the flow behind it becomes swirling. This leads to the formation of unsteady flow regimes under which the vortex instability of the swirling flow behind the impeller is formed. It is known as precessing vortex core (PVC) [1-3]. This effect causes strong periodic pressure pulsations that propagate throughout the water column due to the presence of an elbow in the HPP drainage tract (draft tube) [4-6]. The vortex rotation frequency is sufficiently low and lies in the same range with fundamental frequencies of the hydraulic unit structure, which can lead to uncontrolled resonance and vibrations. In addition to the risk of equipment damage and reduced performance, this effect may decrease the turbine efficiency. These facts make the PVC effect an important and urgent problem of hydropower engineering and hydrodynamics.

To date, computer simulation is the main tool for modeling complex three-dimensional unsteady flows in the flow path of the turbine [7,8]. However, numerical methods need to be validated by experimental data. Since the experiments on full-scale plants are difficult or impossible, the verification of numerical methods is usually realized on a smaller laboratory model of the turbine flow path. For simplifying the model design the experiments may be performed with an air as the working medium instead of water [9,10]. Besides, to specify the velocity distribution realized behind a real turbine it is not necessary to model the entire hydraulic path of the hydraulic turbine. Instead, a combination of two swirlers with a special profile of vanes may be used. For example, the work [11,12] presents a method that using a combination of fixed and rotating vane swirlers allows reproducing velocity distributions at the inlet of the draft tube model, which are very close to those behind the full-scale Francis turbine.

To date, the mechanisms of the PVC formation have not been revealed in full detail. A clear relationship between the velocity distribution behind the impeller and the level of pressure pulsations caused by the PVC are not fully understood either. It is in this context, the present work investigates the flow regimes with strong pressure pulsations in the model draft tube of hydraulic turbine, the vortex nature of these pulsations and relation to the PVC effect. Analysis of the velocity distributions allowed identifying indicators of the PVC formation by varying the input parameters of the test stand.

*Corresponding author: shtork@itp.nsc.ru

DOI: 10.1051/epjconf/201714302103

## 2 Experimental techniques

As already mentioned, using a combination of the two vane swirlers it is possible to obtain the velocity distribution close to that, implemented behind a real impeller, while avoiding the need in reproducing the entire turbine path. This approach not only greatly simplifies and cheapens the experimental rig construction, but also provides additional opportunities to expand the range of input conditions in the draft tube. Among other things, it enables obtaining and investigating the modes, which cannot yet be implemented on existing designs of turbines, but they are promising in terms of reducing level of pulsations and improving energy efficiency.

In these experiments, we used a pair of swirlers, whose geometry is presented in [13-14]. The blades were designed for optimum performance of Francis hydraulic turbines and calculated for volumetric flow rate $Q_c$ = 48.5 l/s and rotation frequency of the runner $n_c$ = 40.5 Hz. A set of swirlers was made by rapid prototyping using a 3D-printer Cube X Duo. As the draft tube design for Francis turbines we chose the model geometry Francis-99, which is widely used to validate numerical calculations [15,16].

The experiments were carried out on an open aerodynamic setup with the test section, produced by the method of rapid prototyping (Fig. 1). The air flow was generated by a blower with a rated flow rate and pressure $Q_{max}$ = 550 m$^3$/h, $P$ = 0.4 bar, respectively. The airflow rate in the wind tunnel circuit was regulated using a frequency converter and measured by an ultrasonic flowmeter. Air flow was supplied to the flow path of the working area over six axisymmetric nozzles and passed through a couple of honeycombs; then the flow was further leveled using a profiled nozzle. The initial swirl was imparted to the flow at a stationary swirler (guide vane); then the rotating swirler (impeller) redistributed the swirl, thereby simulating the desired velocity distribution at the inlet to the draft tube. The servomotor drove the impeller into rotation with a frequency in the range from 0 to 50 Hz. Operating parameters, flow rate $Q$ and rotation frequency $n$ were set with the accuracy of 1.5 and 0.5%, respectively, using original software, providing the feedback control.

To register pressure pulsations we used acoustic sensors based on microphones Behringer ECM 8000, whose signals were digitized by the ADC L-Card E-440. Profiles of the averaged velocity and RMS velocity fluctuations of the flow were measured using the laser Doppler anemometer "LAD-06i". As tracers we used particles of paraffin oil aerosol, produced by Laskin atomizer, which allows obtaining droplets with a characteristic size of 1-3 μm [17].

## 3 Results

The initial stage of the work was determining the modes with maximum pressure pulsations that appear in a draft tube (DT) cone at deviating from the design mode ($Q_c$ and $n_c$) due to the generation of the precessing vortex core. The calculated parameters corresponding to the optimal mode of the turbine operation were taken from [13-14]. In the experiments, two acoustic sensors were placed in the middle of the cone from opposite sides flush with the cone walls (Fig. 1). Operating parameters of the setup ranged for the flow rate from 0.39$Q_c$ to 1.23 $Q_c$, and for the frequency of the impeller rotation from 0.2$n_c$ to 1.23$n_c$. The signal from the sensors was recorded for 10 seconds with a sampling frequency of 2 kHz.

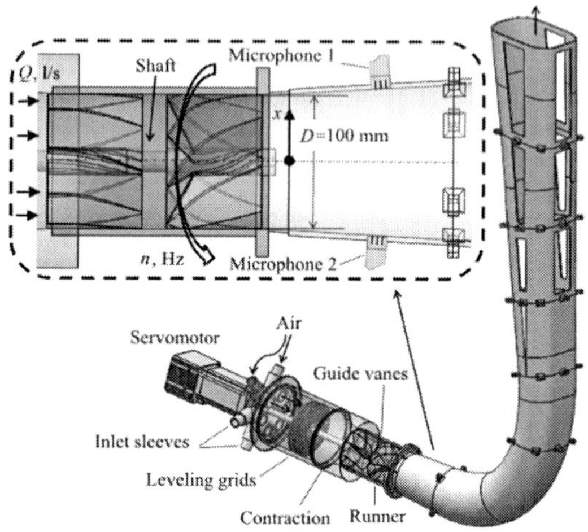


**Fig. 1.** Scheme of experimental setup (the insert shows location of the measuring microphones and the measuring section *I*, spaced from the rear edge of the central body to 4 mm; *D* is the inlet diameter of the conical part of the draft tube).

Fig. 2 presents a diagram based on the measuring results for the 866 operation modes of the unit corresponding to nonoptimal conditions. The diagram shows maximum amplitudes in the spectra of the differential signal of two sensors depending on the flow rate and rotation frequency of the impeller. The differential signal was used to enhance the contribution from the precessing vortex and suppress the in-phase component of pulsations, whose source was noises of the aerodynamic contour. It may be seen that in the upper left part of the diagram there is the region corresponding to high levels of pressure pulsations.

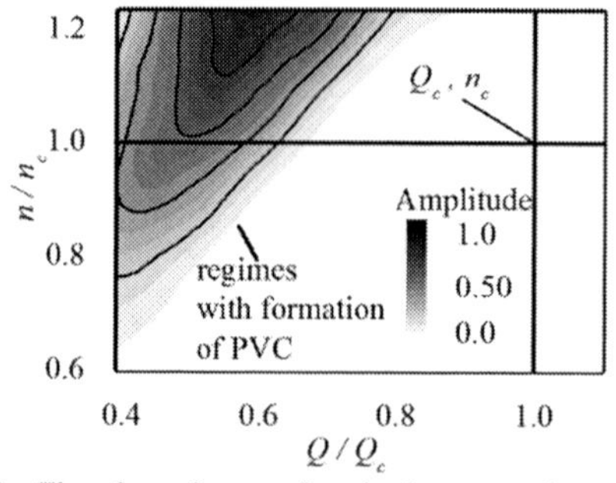


**Fig. 2.** The dependence of pulsations on the operating parameters *Q* and *n*. The scale is given in arbitrary units.

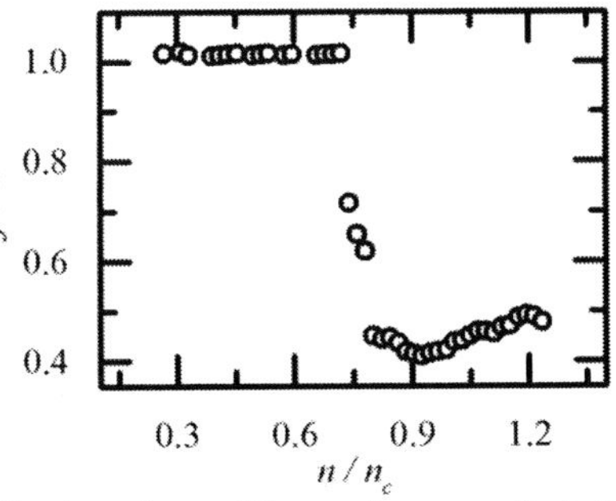


**Fig. 3.** The dependence of the maximum amplitude frequency in the spectrum of the difference signal of two sensors on rotation frequency of the swirler.

These pulsations are most likely caused by the precessing vortex core, which is formed in the cone of the draft tube at nonoptimal operation modes. Fig. 3 presents the dependence of the maximum amplitude frequency in the spectrum of the difference signal of the microphones upon the impeller frequency at a fixed flow rate $0.59Q_c$. According to this figure, up to a frequency of $0.75n_c$ in the spectra there is a dominating peak corresponding to the frequency of the swirler rotation $n$. Then, starting from the rotation frequency $0.8n_c$ in the spectrum begins to dominate the frequency of about 0.42 $n$, associated with the development of vortex instability in the form of PVC. With further increase in the rotation frequency, the dominant frequency slowly grows.

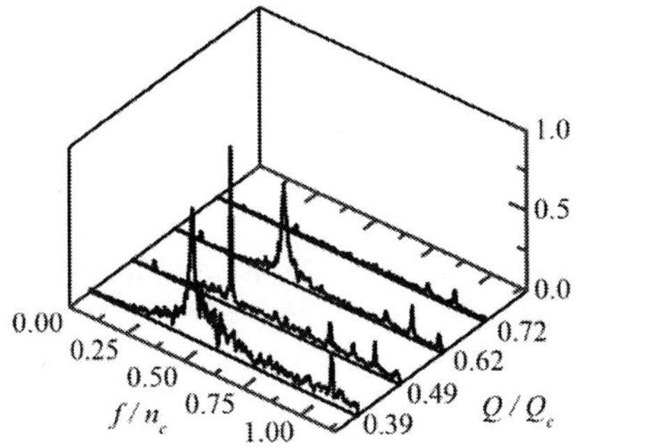


**Fig. 4.** Waterfall diagram of the spectra of the difference signal of two acoustic sensors. The vertical axis is given in relative units.

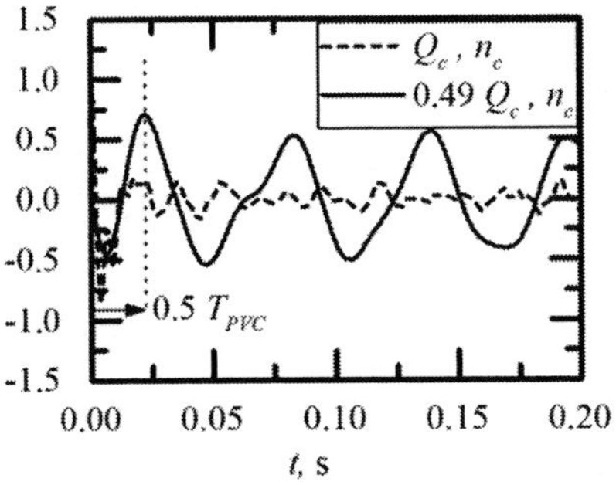


**Fig. 5.** Cross-correlation functions for different modes of $Q$ (relative units are plotted on y-axis).

Fig. 4 shows evolution of the spectra of the differential signal for different flow rates at the fixed rotation frequency $n_c$. It is seen that the dominant peak in the spectra, which is associated with the rotation frequency of the precessing vortex, first increases in its level, and starting from the flow rate $0.39Q_c$ reaches maximum at the flow rate of $0.49Q_c$, and then at $0.79Q_c$ disappears completely from the spectrum.

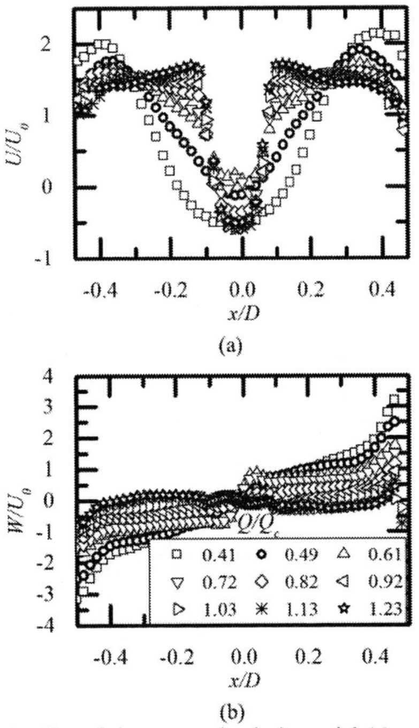


**Fig. 6.** Profiles of the averaged velocity: axial (a), tangential component (b).

To confirm that the precessing vortex motion gives the largest contribution to pressure pulsations at non-optimal modes Fig. 5 shows the cross-correlation functions, built for two regimes (optimal and non-optimal, with maximum pulsations) with flow rates $Q_c$ and $0.49Q_c$, respectively, at a constant rotation frequency of the swirler equal to $n_c$. It is seen that for the optimal mode the maximum cross-correlation function caused by the noise is only present in zero, and no periodicity of the cross-correlation function is observed. For a signal corresponding to the non-optimal regime, the maximum of the function falls on ½ of the PVC period. This is because the sensors detect the pressure disturbance in the opposite phase. I.e. one sensor registers the pressure pulse caused by the vortex passage, and then, in a half period of the PVC precession, the pulse is registered by the second sensor. Thus, in the cone the vortex loses its axial position and executes a precessional movement, coming alternately to one and then to another sensor.

To analyze the velocity profiles, of the whole set of investigated operating conditions we selected the flow rates corresponding to 0.41, 0.49, 0.61, 0.72, 0.82, 0.92, 1.03, 1.13 and $1.23Q_c$, at a fixed rotation frequency of the swirler equal to $n_c$. The averaged distributions of axial and tangential velocity components were measured in the control section, located at a distance of 4 mm behind the bluff body. Fig. 6a presents profiles of averaged axial velocity. For profiles, corresponding to 0.41 and $49Q_c$, a wide area of the reverse flow and the parabolic shape of the profile may be seen. They serve as a good indicator of the PVC formation [18-19]. With further increase in the flow rate the reverse flow area is reduced to the size of the bluff body – $0.2D$ ($D$ is the diameter of the inlet orifice of the DT cone). Distribution of tangential velocity is characterized by a decrease in the level of maximum velocity (Fig. 6b). For the modes with 1.13 and $1.23Q_c$ the regions with the opposite sign of swirl are observed.

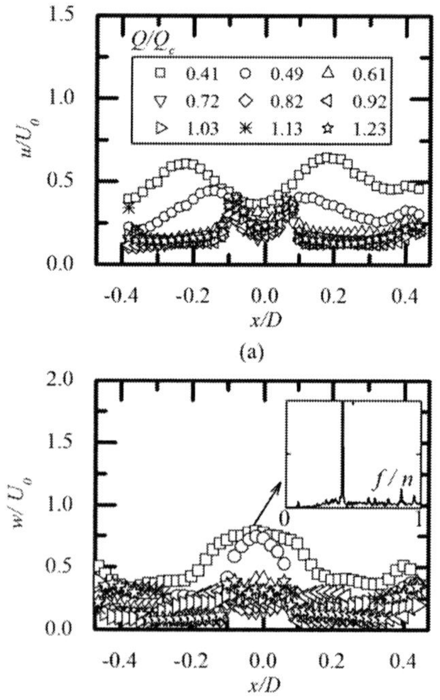


(b)

**Fig. 7.** Profiles of the RMS averaged velocity: axial (a), tangential components (in the inset there is Lomb spectrum for the mode with 0.49 $Q_c$, $n_c$) (b).

Characteristic profiles of mean square deviations (RMS) of axial and tangential velocity components are presented in Fig. 7 for the flow rate $0.82Q_c$ and the rotation frequency n = $n_c$. Distributions of axial velocity pulsations show that at low flow rate the level of pulsations reaches $0.5U_0$, and at the flow rate over 0.61 $Q_c$ it falls to $0.2U_0$. Moreover, the positions of the maxima of the axial velocity RMS change from $0.2D$ to $0.1D$, which is associated with the flow rearrangement and the PVC effect suppression due to reduction in the radius of the vortex precession. RMS of tangential velocity shows that the pulsations change from $0.63U_0$ to $0.25U_0$ at increasing flow rate, and the position of the maximum tangential velocity is shifted to the axis of flow rotation.

To prove the vortex nature of the pulsations we present the spectrum of pulsations of the tangential velocity component in the vicinity of the cone center. To build the spectrum of non-equidistant velocity signal use the Lomb algorithm [20]. Insert in Fig. 7b shows the relative amplitude of LS – the Lomb spectrum of pulsations of tangential velocity component for the flow rate of $0.49Q_c$ and rotation frequency $n_c$. The presence of a narrow peak at the PVC frequency correlates with the corresponding spectrum of the differential signal, which shows the vortex nature of the pulsations.

Non-optimal flow regimes behind the hydro-turbine, distinguished by a strong residual swirl, can be characterized by the swirl parameter:

$$S = \frac{\int_{\Sigma} WUr d\Sigma}{(D/2)\int_{\Sigma} U^2 d\Sigma}, \qquad (1)$$

where $U$ is the axial velocity component, $W$ is the tangential velocity component, $d\Sigma$ is the element of area, and $D$ is the inlet diameter of the draft tube cone.

Fig. 8 shows the dependence of the integral parameter of the swirl, calculated from the averaged velocity profiles (Fig. 6).

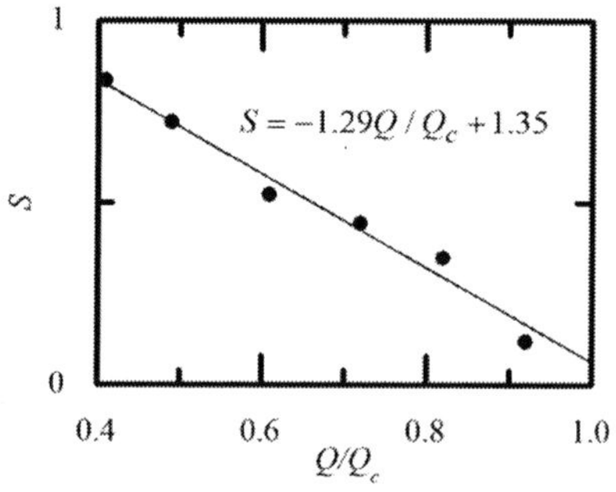


**Fig. 8.** The dependence of the integral parameter of the swirl on the flow rate at a fixed rotation frequency $n_c$.

The dependence with good accuracy represents a linear function $S=1.29Q/Q_c+1.35$. It becomes zero at the flow rate $1.04Q_c$, which is very close to optimal conditions, and the maximum value $S=0.83$ is obtained for $Q=0.41Q_c$. The formation of a vortex core is observed for $S>0.5$, which corresponds to the critical value known from the literature, after which the vortex breakdown, the formation of the reverse flow zone and the PVC generation begin [1].

Thus, the paper illustrates the usefulness of the rapid prototyping of physical models for experimental study of flow unsteadiness and emergence of the non-stationary vortex structures for a variety of swirler designs and a range of operating conditions. Such results can provide insight into the effect of various design and operating parameters on the flow physics, as well as serve for verification of the numerical simulations. Despite the neglect of cavitation and possible scaling effects, the approach can be regarded as useful for fast gathering of information at much reduced expenses compared to large-scale experiments or massive numerical simulations.

## 4 Conclusions

The work has shown that the deviation from the optimal calculated regime to the area, corresponding to the modes of real turbine underload, results in generating the vortex instability in the form of the PVC. The presence of precession motion of the vortex and its relationship with the regime of maximum pressure pulsations in the draft tube cone have been demonstrated through the evolution of spectra and analysis of cross-correlation functions of the pressure pulsations.

The analysis of the measured profiles of the averaged axial velocity component has revealed a wide area of the reverse flow of the parabolic shape for the modes with PVC formation. When approaching the optimal regime (increasing flow rate) the central valley on the axial velocity profiles is narrowed to the transverse dimension of the bluff body. The tangential velocity component in the vicinity of the optimal regime tends to zero value. Pulsation spectra of the tangential velocity confirm the PVC formation with precession frequency of 0.4-0.42$n_c$ for low flow rates 0.41 and 0.49$Q_c$, corresponding to non-optimal regimes. The swirl parameter calculated on velocity profiles for different flow rates is a linear dependence on the flow rate. At that the increased level of pulsations, caused by the PVC, and the central area of the reverse flow appear at S>0.5, which corresponds to the critical value known from the literature. If this level is exceeded in a swirling flow the breakdown of the vortex and the generation of the PVC take place.

The study was performed with financial support of the Russian Science Foundation (project No. 14-29-00203).

## References

1. A. Gupta, D. Lilley, and N. Syred, *Swirl flows* (Abacus Press, London, 1984)
2. E. Fernandes, M. Heitor, and S. Shtork, Exp. Fluids, **40** 2006
3. I. Litvinov, D. Sharaborin, S. Shtork, Thermophys. Aeromech., **22** (2015)
4. P. Deriaz, 1st Symposium IAHR, (1960)
5. O. Kirschner, A. Ruprecht, and S. Riedelbauch, 3rd Symposium IAHR, (2009)
6. B. Mulu, P. Jonsson, and M. Cervantes, Appl. Energy, **93** (2012)
7. A. Minakov, D. Platonov, A. Dekterev, A. Sentyabov, and A. Zakharov, Int. J. Heat Fluid Flow, **53** (2015)
8. Z. Zuo, S. Liu, D. Liu, D. Qin, and Y. Wu, J. Hydr. B, **27** (2015)
9. M. Nishi, K. Yoshida, M. Yano, M. Okamoru, K. Miyagawa, and S. Liu, 2nd Symposium IAHR, (2007)
10. N. Povh, *Aerodynamic experiment in machinery* (Leningrad, 1974, in Russian)
11. R. Susan-Resiga, S. Muntean, and A. Bosioc, The 4th International Symposium on Fluid Machinery and Fluid Engineering, (2008)
12. S. Skripkin, M. Tsoy, S. Shtork, and K. Hanjalić, J. Hydraul. Res., **54** (2016)
13. P. Kuibin, I. Litvinov, V. Sonin, A. Ustimenko, and S. Shtork, Sc. Bull. Novosibirsk State University, **11** (2016), (in Russian)
14. I. Litvinov, A. Mitryakov, and S. Shtork, Power Tech. Eng., **50** (2016)
15. M. Cervantes, C. Trivedi, O. Dahlhaug, and T. Nielsen, J. Phys., **579** (2015)
16. A. Gavrilov, A. Sentyabov, A. Dekterev, adn K.Hanjalić, Int. J. Heat Fluid Fl., (2016), (available online).
17. C. Tropea, A. Yarin, and J. Foss, *Springer Handbook of Experimental Fluid Mechanics* (Springer-Verlag, Berlin Heideberg, 2007)
18. P. Yazbadabi, Griffiths A., and N. Syred, Exp. Fluids, **17** (1994)
19. I. Litvinov, S. Shtork, P. Kuibin, S. Alekseenko, and K. Hanjalić, Int. J. Heat Fluid Fl., **42** (2013)
20. V. Vityazev, *Analysis of irregular time series* (S. Petersburg university, 2001, in Russian)

DOI: 10.1051/epjconf/201714302104

# The design of axial shaftless pump

*Michal* Schmirler[1,*], *Hana* Netrebska[1]

[1]CTU in Prague, Faculty of Mechanical Engineering, Department of Fluid Mechanics and Thermodynamics

**Abstract.** The axial shaftless pump with a rotary casing has been proposed. The pump is unique in its small space requirements and the ability to draw a liquid with a high content of impurities and fibers. Modern motor with an external commutation was used to propel the pump rotor. The pump can be used for both pumping of liquids and marine propulsion.

## 1 Introduction

The blade devices belong to the group of hydrodynamic machines and generally can be divided into machines and engines. The group of blade machines includes pumps, fans, blowers and compressors. The group of blade engines includes turbines (water, steam and gas turbines) and marine propulsions - thrusters. All blade machines can be further subdivided according to the direction of the flow of the pumped fluid to radial, axial and diagonal ones. Generally, the axial blade machines are able to operate with lower pressure drop and higher flow rates than centrifugal machines [1]. Therefore, because of this advantage, the axial blade machines are used as pumps for pumping large volume flows, turbines for the processing of high flow rates at low level differences (Kaplan turbine), and as marine engines.
This paper focuses on a structural design of an axial shaftless pump with a rotated casing. This innovative design has the following advantages, compared with conventional shaft machines:

a) Simpler design, reliability
b) Compact design, small size
c) The possibility of pumping liquids with long fibers, cables - resistant to the entanglement of fibers on the shaft
d) Easy modification of the performance parameters.

## 2 Design requirements

The rotor of an axial pump consists of a hub and a small amount of spatially curved runner blades. The blades do not create a closed channels, their profile is analogous to a propeller of the aircraft, or the wing profile. Therefore, they are also called propeller pumps. The arrangement of the conventional axial pump is shown in Fig.1. Usually the hub of the pump has smooth teardrop shape. The blades of the impeller may be a solid part of the hub or tilting by axis perpendicular to the rotation axis of the hub. The tilting can be performed during an operation or at standstill of the machine - sometimes even by installing a set of new blades. The axis of tilting blades may be firmly connected with a mechanism inside the hub, which is mechanically connected to a control device external to the pump. The disadvantages of such solutions are the dimensions and complexity.

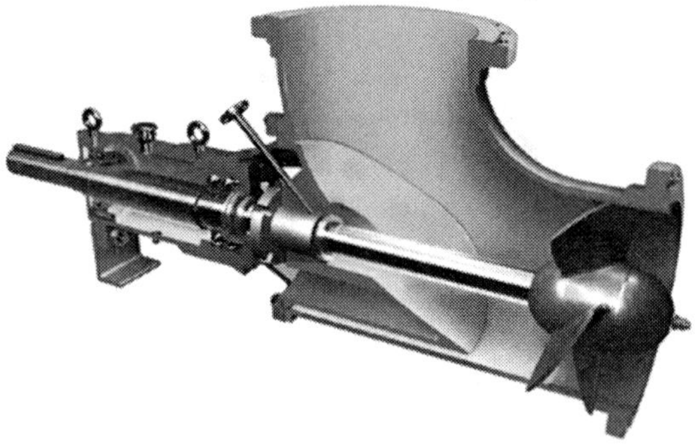

**Fig. 1.** A conventional configuration of axial pump, here the AFH900 pump (www.flowserve.com)

The requirements for the design proposal of the new pump resulted from the generally known design principles and from the benefits listed above. They are as follows:

a) Simple design - the simplicity of design allows for lower production and service costs. This point was achieved by using mostly cylindrical and round parts, which are based on commercially available blanks and can be produced by conventional machining methods. The need for sealing the output shaft is also eliminated by the use of the shaftless design.
b) Compact design - a very important element in terms of the possible use of the machine. Small dimensions were achieved especially by applying advanced motors with an external electronic commutation.
c) The pumping of liquids with fibers - this requirement is met by using a system with a rotary casing and with the absence of the drive shaft.

* Corresponding author: michal.schmirler@fs.cvut.cz

d) Modification of the performance parameters - the possibility to adjust the parameters according to the customer requirements. The change of performance parameters can be performed simply by replacing the blades of the rotor. One pump casing covers a wide range of operating points.

## 3 Design proposal

So far, this design was used as a marine propulsion and as a propeller of the Micro autonomous underwater vehicles (AUVs). The AUVs need small-scale, powerful and safe propulsion systems especially when they are performing missions in pipes and other confined environments. However, the most conventional propulsion systems do not satisfy all of these three requirements: small, powerful and safe. A micro propulsion systems meeting those requirements are developed based on the RIM propeller concept.

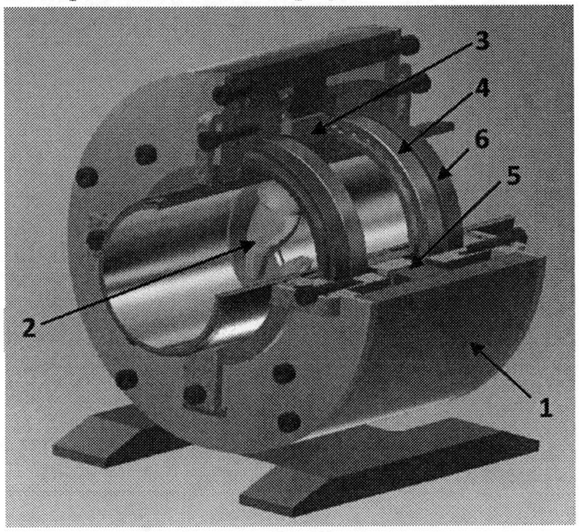


**Fig.2.** The shaftless pump proposal – schematic view; 1 – housing, 2 – impeller, 3 – rotor, 4- bearing, 5 - motor, 6 – rotor seals

The proposed pump (Fig.2.) comes from the same principle as those propellers. The core element of the pump body is the outer shell. In this dimension, it has not been possible to find a suitable tubular duralumin blank, therefore, a thick-walled tube of stainless steel was used. The side flanges are bolted to the shell. For practical reasons, the decision to separate the bearing housings from the bushing flange (lower machining costs) was made. The amount of removed material was also reduced by separating the bearing housings. Both side flanges are identical, made from aluminum rods having an outer diameter of 190 mm. The suction and discharge pipes are inserted outside the flanges. Due to a too small outer diameter of the tubes, it was not possible to mount components screwed directly. The pipes are secured by specially curved washers. The bearing housings are inserted from the inner side of the flanges. They are made of aluminum tubes as well as the flanges. A special edge was used in one of the bearing housing. This edge catches the axial forces in the bearing (Fig.3.). The houses are fitted in the front flange bolts. Half of the screws was chosen longer and extends the curved washers of suction and discharge tubes.

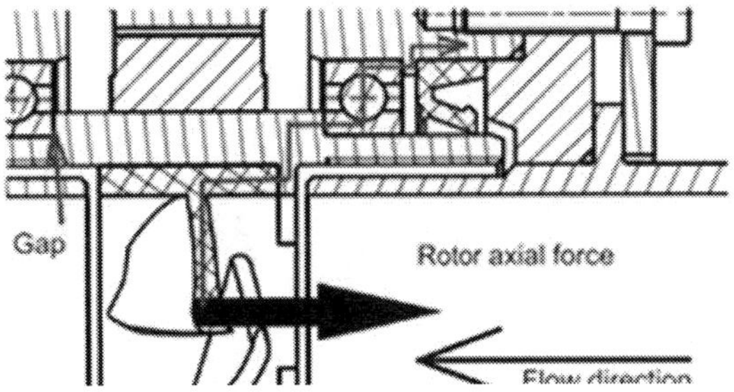


**Fig. 3.** The axial force reactions in the pump.

Therefore, they can be removed without causing the undesirable release of housings. The sealing of the flange, the suction, the pressure pipe and the bearing housing has been realized by O-rings. The ring with the winding of the electric motor is inserted into the housing and suitably secured, for example, by an adhesive. The wreath with permanent magnets is mounted on the rotor, and again, secured against movement.

The electric motor arranged in a narrow ring will be a part of the machine body. The transmission device and the drive shaft will be omitted. The issue of the pump and a possible utilization of the pump is appropriately described in [2]. This arrangement would reduce the weight and the size of the machine. The ring electric motor, which is a direct part of the machine, is already used in the propulsion water craft. The performance characteristics of this constructed electric motor are fully satisfactory. Relatively to the bigger diameter, a fairly large number of poles is mounted to the circumference of the ring. Therefore, the electric motor is aptly low-speed and achieves high torque values. The rotor of the electric motor comprises an inner ring, which carries several impeller blades on its inner side. They are oriented perpendicularly to the axis of the rotation. The outer ring (stator) allows the engines to turn along their vertical axis. The turning of the entire drive improves the maneuverability of the vessel. The simplicity of this system has the advantage of a peaceful and quiet operation. It is not necessary to insulate the rotor, only the functional part of the engine has to be hermetically sealed.

## 4 Rotor design

The rotor is composed of a support portion and the impeller. The support portion is mounted in bearings and sealed with shaft seals. Drawing diagram of the entire device is shown in Fig.4

 DOI: 10.1051/epjconf/201714302104

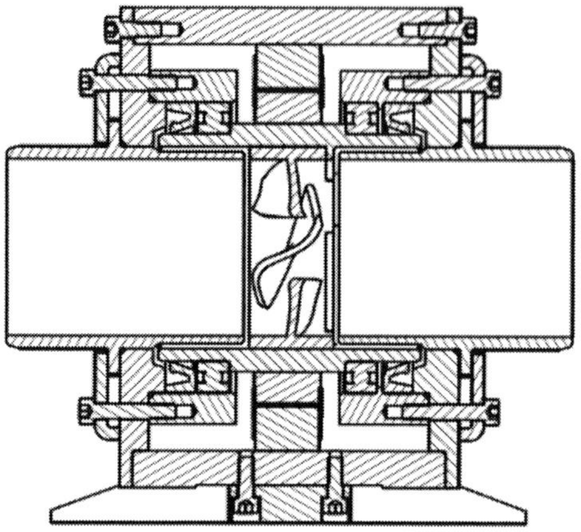

**Fig. 4.** The drawing of new designed pump with rotated casing.

The pump impeller is designed as a separate part that can be replaced if needed (Fig.5.). The impeller is made by using a 3D printer.

**Fig. 5.** The model view to the impeller ring.

## 4.1 Derivation of the blade geometry

In this case, the finding of a suitable template to derive the blades from, is not easy. The geometric similarity is not valid, mainly due to the changes in the form (attachment to the outer ring and not to the shaft). In the method described in the preceding paragraph, it is assumed that the experimentally determined parameters of the model (which, in addition, must have a minimal size) are used to convert the parameter to a larger piece, or the entire range of machines. Thus, the conversion may not work in the case of the reduction of the model. The blade thickness is another problem. After the reduction to the desired size, the wall of the blade may be too thin. which could be a complication during manufacturing, in addition to the strength problems that could occur. In the 80s, an axial pump for suction dredger was developed [3]. The requirement was issued by the Czech Prague shipyard. It was a modification of the pump AA-69-9.9 Sigma Lutin [4], which, at the time, was the top model of the axial pump. Since the medium for a suction pump is a suspension of water with sand and other materials, the main adjustment in the moving blades was their gain to a constant thickness. This changed the blade profile and decreased efficiency, but greatly increased durability. The reinforced profile of the blade is now suitable for a correction to shaftless version and a subsequent reduction. There are a lot of experimental data, including the drawing of the blade [5] for this pump. Original blade geometry is Fig. 6:

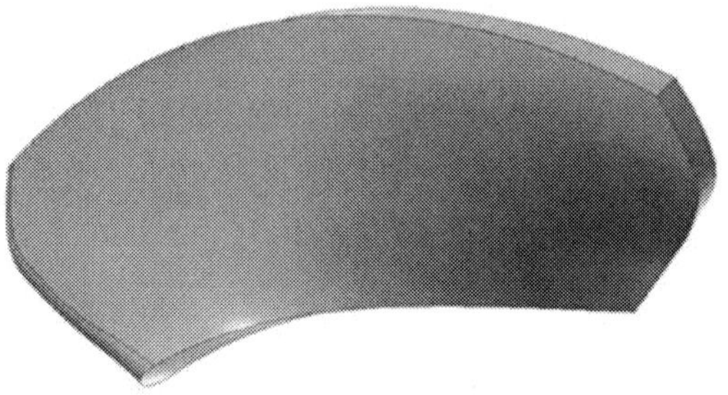

**Fig. 6.** The original blade from the axial pump AA-69-9.9 Sigma Lutin.

This blade geometry belongs to the wheel with a central shaft. For purposes of the proposed shaftless machine, it was necessary to modify the geometry of the original blade. The new geometry of the blade is shown in Fig. 7.

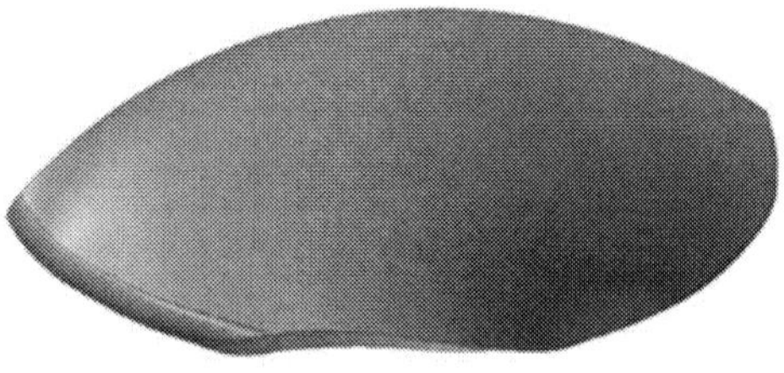

**Fig. 7.** The modified blade geometry used for new shaftless axial pump.

## 4.2 Operating parameters

The calculation of the pump parameters usually follows the completion of the design of the pump. Those parameters are directly related to the type of used movable blades. The calculation and modeling of a completely new shape of the blades is not appropriate in this case. The solution in the form of geometric similarity of existing blades is far more convenient. Analysis of the problems is described in detail in [6]. The mechanics of material bodies is based on three basic physical quantities. They are length $l$ , time $t$ and mass $m$ . All the other physical quantities, which are essential for hydraulic machinery, are derived from the basic ones. If the original is labeled as a model (M) and the new machine as the real part (D), then the dimensionless basic modules of basic physical quantities can be defined as follows:

$$\lambda = \frac{l_D}{l_M}, \quad (1)$$

$$\tau = \frac{t_D}{t_M}, \quad (2)$$

$$\mu = \frac{m_D}{m_M}. \quad (3)$$

The formulas for recalculation of the model parameters to the real part can be derived using the formulas (1,2,3), if the model parameters are known. For the conversion of specific energy $Y\left(J \cdot kg^{-1}\right)$, flow rate $Q\left(m^3 \cdot s^{-1}\right)$ and power $P\left(W\right)$, the following formulas are deduced:

$$Q_D = \left(\frac{D_D}{D_M}\right)^3 \cdot \frac{n_D}{n_M} \cdot Q_M \cdot \frac{\eta_{VD}}{\eta_{VM}}, \quad (4)$$

$$Y_D = \left(\frac{D_D}{D_M}\right)^2 \cdot \left(\frac{n_D}{n_M}\right)^2 \cdot Y_M \cdot \frac{\eta_{hD}}{\eta_{hM}}, \quad (5)$$

$$P_D = \frac{\rho_D}{\rho_M} \cdot \left(\frac{D_D}{D_M}\right)^5 \cdot \left(\frac{n_D}{n_M}\right)^3 \cdot P_M \cdot \frac{\eta_{hD}}{\eta_{hM}} \cdot \frac{\eta_{VD}}{\eta_{VM}}. \quad (6)$$

The quantities are indicated by indices - M denotes the model and D denotes the real part. $D$ is the diameter of the impeller, $\rho$ is the liquid density. The revolutions of the machine are labeled $n$. An equally important parameter is the efficiency η. We distinguish three types of efficiency - volumetric, hydraulic and mechanical. The hydraulic efficiency is experienced primarily at the hydrodynamic machine. Mechanical efficiency is, for simplicity, identical for the model and for the work. According to [7], the volumetric efficiency can also be neglected, because it is not readily determinable and the resultant error is still negligible. Thanks to these simplifying, it is possible to omit the last fraction from

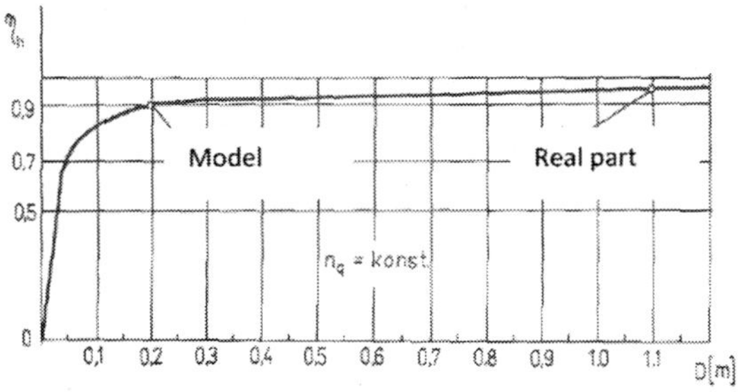


**Fig. 8.** The dependence of the hydraulic efficiency on the rotor diameter.

experimentally) is plotted in Fig. 8. It is recommended to keep the minimum impeller diameter $D_{MIN} = 300mm$, when modeling the test pump [7]. Unfortunately, the new designed impeller has a diameter of only 80 mm. Therefore, the values of the performance parameters obtained by a model similitude were outside the scope. Actual performance values of the new pump will be verified experimentally in the next step of the research.

## 5 Conclusions

A unique shaftless pump with a rotary casing and with an estimated output of 1000W has been designed within the project. The main advantages of engineered construction include a small dimensions, simple construction and easy manufacturing. Currently, the documentation is ready and the manufacturing of the pump have started. Upon completion of the production, the performance measurements will be carried out and necessary modifications will be suggested.

## Acknowledgement

This research project has been conducted as part of the industrial research and development project "Advanced Technologies for Production of Heat and Electricity - TE01020036" (programme Competence centres), implemented in cooperation between 13 participants and the Czech Technical University in Prague, Faculty of Mechanical Engineering as leading organizational unit, in 2012-2019.

## References

1. J. Blaha, K. Brada, Hydraulické stroje. Praha: SNTL - Nakladatelství technické literatury, 1992. ISBN 80-03-00665-1.
2. P.M. Tuohy, Development of Canned Line-start Rim-driven Electric Machines. Manchester: The University of Manchester, Faculty of Engineering and Physical Sciences, 2011. PhD thesis
3. J. Melichar, Návrh lopatkování a zkoušky axiálního bagrovacího čerpadla. Bratislava: STU, 1999. Strojné inžinierstvo 99 – Conference proceeding. Pp. 313-318. ISBN 80-227-1259-0.
4. J. Spunda, Hydraulický model čerpadla AA-68-9,9. Research report VZ-083/67-Vz: VÚ SIGMA Olomouc, 1968.
5. J. Melichar, Výkonové charakteristiky axiálního modelového čerpadla s upraveným lopatkováním oběžného kola. Prague: CTU in Prague, 1984. Report České loděnice Praha.
6. J. Blaha, K. Brada, Metoda experimentů a modelování. Prague: CTU in Prague 1995. ISBN 80-01-01236-0.

# Substitute fluid examinations for liquid manure

*Kevin* Schrader[1,*], *Marco* Riedel[1] and *Helmut* Eichert[1]

[1]University of Applied Sciences Zwickau, Department of Automotive Engineering, Dr.-Friedrichs-Ring 2A, 08056 Zwickau (Germany)

**Abstract.** For the farming industry it is essential to use liquid manure as natural fertilizer. Through new agricultural regulation 2015 in Germany the industry must develop new liquid manure spreader systems because the ammonia and methane emission are limited. In a research project the University of Applied Sciences Zwickau and some other industry partners will develop such a new innovative liquid manure spreader. The new liquid manure spreader should use pulsating air to distribute the liquid manure exactly. The pulsating air, which flows through the pipelines, should be analysed at a test station. For examinations at this test station it is important to find another substitute fluid because liquid manure smells strong, is not transparent and is also not homogeneous enough for scientific investigations. Furthermore it is important to ensure that the substitute fluid is, like liquid manure, a non-Newtonian fluid. The substitute fluid must be a shear-thinning substance - this means the viscosity decrease at higher shear rate. Many different samples like soap-water-farragoes, jelly-water-farragoes, agar-water-farragoes, soap-ethanol-farragoes and more are, for the project, examined in regard of their physical properties to find the best substitute fluid. The samples are examined at the rotational viscometer for viscosity at various shear rates and then compared with the viscosity values of liquid manure.

## 1 Introduction

New regulations from 2015 forced the agricultural industry to optimize the old liquid manure spreader systems. In a research project the University of Applied Sciences Zwickau together with industry partners optimized the liquid manure spreader and examined the impact of pulsating air to distribute the slurry exactly. Therefore it was important to find a substitute fluid for slurry because it smells strong and is not transparent. The substitute fluid has to be transparent because of examinations with the laser Doppler anemometer (short form: LDA). Furthermore the medium and the boundary of the pipe have to be diaphanous [1]. The following article explains the measurement principle of the "cup and bob" rotational viscometer. After that explanation the examination steps for the substitute fluid should be explained.

One kind of the rotational viscometer is the "cub and bob" viscometer, which works with an exact defined volume sample. A "cup and bob" rotational viscometer consists of three important parts. The first part is the cup (1) and the second is the bob or rotor (2). The third important component is the measuring volume (3), which contains the measuring fluid. The measuring substance is filled in the cup. After that the cup is fixed in the rotational viscometer. The component bob immersed in the fluid. The measuring fluid is now located between the cup and the rotor. In figure 1 the setup of a "cup and bob" rotational viscometer is shown.

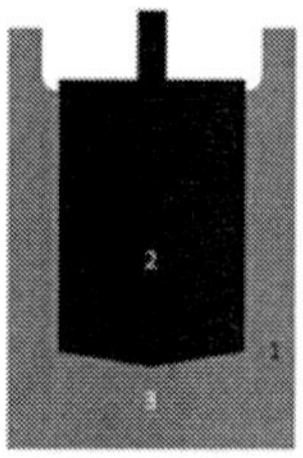


**Fig. 1**- Setup of a "cup and bob" rotational viscometer, (1) cup, (2) bob/rotor, (3) fluid

There are two different kinds of "cup and bob" rotational viscometers. The first kind is called Couette system. At a Couette system the cup moves and the bob stands still. The other type is named Searle system. The Searle system is characterized by a moving rotor and a cup which is standing still.

The space between the cup and the rotor is called annular gap. This gap forms the measurement volume. The fluid adheres to the boundary of the cup and the rotor. Through the movement of the rotor, the fluid will be moved too. The examining fluid is build up from serval layers. This layers slides on top of each other (laminar flow). Between these layers there is a velocity difference, which is called shear gradient or shear rate.

At the "cup and bob" rotational viscometer it is usually to adjust the angular speed and to measure the torque ($M_d$) at this speed. With the geometric dimensions (surface size - A/ bob radius - r) and the physical sizes of the angular gap (y), force (F) and velocity (v) the dynamic viscosity can be determined. For the

* Corresponding author: kevin.schrader.1d4@fh-zwickau.de

equation (3) it is important to calculate the shear rate and the shear stress. With the force and the dimensions (A and r), the shear stress can be calculated (equation (1)). Furthermore with the velocity and dimensions of the annular gap it is possible to calculate the shear rate (equation (2)).

Shear Stress:

$$\tau = \frac{M_d}{A \cdot r} = \frac{F}{A} \quad (1)$$

Shear Rate:

$$\dot{\gamma} = \frac{\omega}{y \cdot r} = \frac{v}{y} \quad (2)$$

The following equation is used to calculate the viscosity ($\mu$) from shear rate ($\dot{\gamma}$) and shear stress ($\tau$) :

$$\mu = \frac{\tau}{\dot{\gamma}} \quad (3)$$

## 2 Examinations of cattle and pig manure

The first task to characterize the mechanical properties of the liquid manure was to analyse a sample in the viscometer. For this study a cattle manure sample was used and examined in a rotational viscometer that was produced by the company Physika. Unfortunately, the measurements led to no plausible values because the fluid contained too much large particle like sand and straw. Therefore to specify slurry publications and reports ([2, 3, 4]) from the past years were used. It was important to find articles which contain information about the viscosity in dependent of the shear rate. The next figure shows the values of shear rate in dependent of viscosity for pig manure (shear stress and shear rate were used form the publication [2] – viscosity were calculate of these physical sizes). According to this publication the measurement environment, and so the liquid manure too, had a temperature of 20 *°C*. The dry matter content (DM) varied from 8 - 22 *%-DM*.

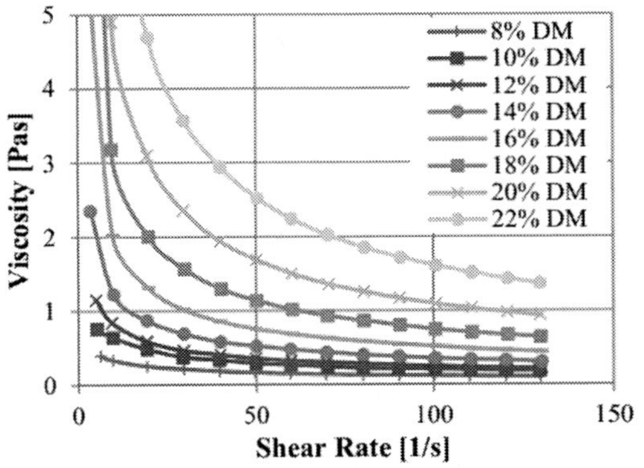


**Fig. 2** - Description of the dynamic viscosity in dependent of shear rate from pig manure with several dry matter contents (calculate from source [2])

As in figure 2 shown, the viscosity decrease with increasing shear rate. Fluids which have such a mechanical property are referred as shear-thinning fluids. The shear-thinning liquids and gases belong to the non-Newtonian fluids. This means that the shear rate is not direct proportional to the shear stress. The following substances belong to the representatives of shear-thinning fluids:

- ketchup,
- toothpaste,
- paints,
- buttermilk with flakes,
- shampoo.

Also it is apparent from figure 2 that with increasing dry matter content the dynamic viscosity of liquid pig manure increases. The poorer flow behaviour can be attributed to the more viscous phase of the slurry. The dynamic viscosity varies from about 0.50 to 7.00 *Pa s* at low shear rates of about 0…10 *1/s*. At higher shear rates (from about 75 to 130 *1/s*) the values of the dynamic viscosity are in the range of 0.25 to 1.50 *Pa s*.

In addition to the value of the dry matter content, the temperature can also vary. In figure 3 [4] the viscosity and the shear rate at different temperatures is shown. The slurry temperature was raised from 30 to 60 *°C*. The dry matter content was at a constant value of 9.1 *%-DM*. Instead of pig manure cattle manure was used for the investigation.

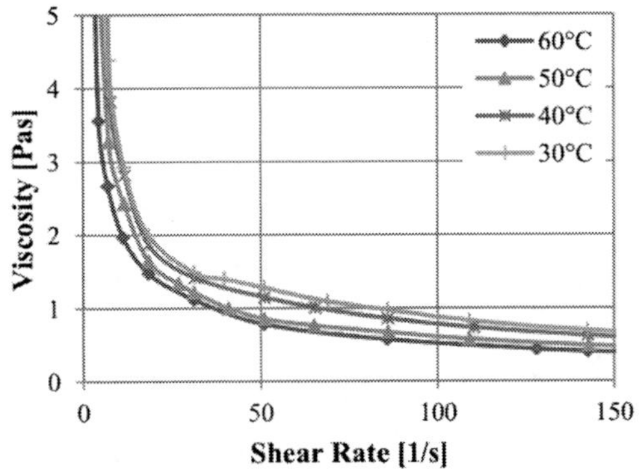


**Fig. 3** - Description of the dynamic viscosity in dependent of shear rate from cattle manure with several temperatures (source [4])

In figure 3 it is obvious that cattle manure is like pig manure a shear-thinning substance. The variation of the temperature led to an alteration of the dynamic viscosity. With increasing temperature the dyn. viscosity decreases, see figure 3. The value for the dynamic viscosity is about 3.0 *Pa s* at a shear rate from 1 to 10 *1/s*. For shear rates above 120 *1/s* the viscosity decrease to about 0.5 *Pa s*.

In the following step, cattle and pig manure have to be compared to know how different the viscosity values of these two kinds of slurry are. In order to allow a qualitative comparison of the two manure types, the dry matter content and the temperature of the slurries have to be approximately equal. The dry matter content of the two liquid manures was 11.6 *%-DM*. There was no information about the slurry temperature but both measurements were realized at the same measurement series, so constant boundary conditions are assumed.

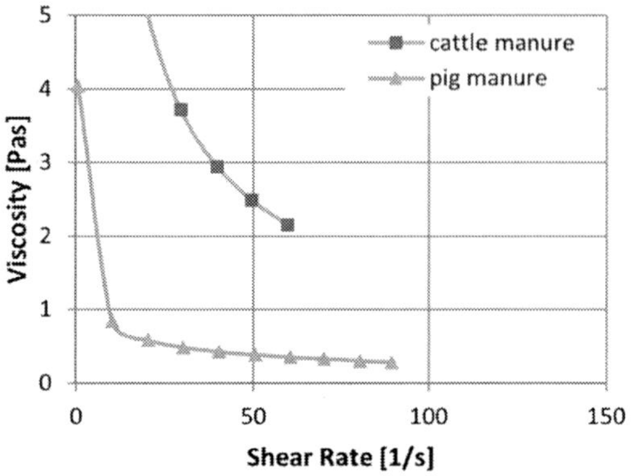


**Fig. 4** - Description of the comparison between cattle and pig manure at the same dry matter content (source [3])

It is obvious that the dynamic viscosity of pig manure is lower compared to cattle manure, which is shown in figure 4. Partially the deviation of the viscosity is about 80 %.

*Conclusion:*
The substitute fluid must be a shear-thinning substance. Furthermore the viscosity values should be about 3.0 *Pa s* at lower shear rates from about 0 to 10 *1/s*. Higher shear rates (about 200 *1/s*) require a dynamic viscosity from approximately 0.5 *Pa s*. On the top the fluid must be transparent because a laser Doppler anemometer measurement needs a diaphanous substance.

## 3 Examinations of a substitute fluid for cattle and pig manure

After the mechanical properties are known, the next task is to identify a substitute fluid. The shear-thinning substances which are mentioned in the pre-text can't be used because for a variety of reasons. For example ketchup can't be used because it is not transparent and so it's not possible to use it for a laser Doppler anemometer measurement. Or buttermilk with flakes of butter is a shear-thinning medium which has the same mechanical properties (viscosity) [5] like pig or cattle liquid manure. But because of the colour and the perishability buttermilk with flakes of butter can't be used as a substitute medium. In search for another substitute fluid many shear-thinning substances were examined. Not only their rheological properties but also some subjective characteristic were important to evaluate the substitute fluids. The subjective criterion includes transparency, the foaming behaviour and the settling of liquids. The following fluid variations were examined for the experiment:

- soap or shampoo
- soap-ethanol-mixture
- water-glue-mixture
- water-soap-mixture
- water-gelatine-mixture
- water-agar-mixture

Through subjective examinations it became apparent that most of the above mentioned fluids are not relevant for the rheological analyses. It was also apparent that the mixture ratios are of great importance. The above mentioned mixture (regardless which mixture ratio the composition has) soap, soap-ethanol and water-soap proved to be unsuitable. The water-gelatine-mixture convinced at the subjective examination. After a long standing time, it led however to a strong odour. That is the reason why it wasn't analysed in rheological studies. The water-glue-mixture convinced with very good subjective properties. But after a long period of disuse the fluid hardened and so this mixture is unusable as substitute medium. All subjective evaluation criteria have been complied with a water-agar-mixture. The convinced mixture ratio is 98.8% water and 1.2% agar. In the next step this mixture is examined in a rotational viscometer (the results will follow in the article).

### 3.1 Substitute fluid – shampoo/soap

The literature research indicated that shampoo has a similar rheological behaviour such as liquid manure. Therefore a sample of this fluid was examined in a rotational viscometer at the University of Applied Sciences Zwickau. The figure 5 below shows the values of this measurement in comparison to slurry.

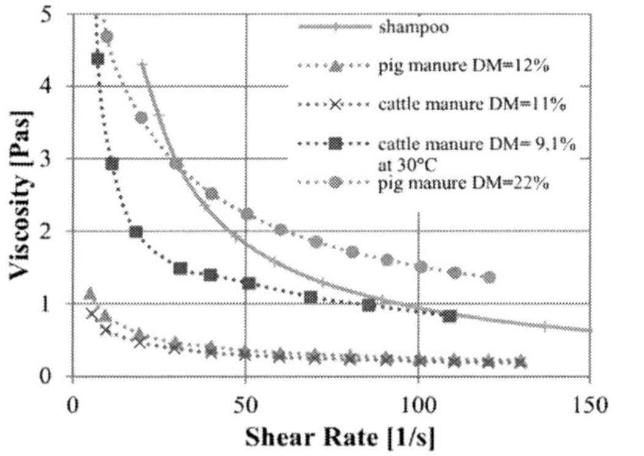


**Fig. 5** - Description of the dynamic viscosity form shampoo and some different slurries

As it is apparent in figure 5, the dynamic viscosity values of shampoo is in the range of pig and cattle manure. The examined medium was evaluated subjectively after the rheological analysis. Unfortunately, the subjective investigation led to the conclusion that the pure shampoo or soap can't be used as substitute fluid because it has a bad foaming behaviour. Also a mixture of soap and ethanol can't be used for further investigations because it foams strong too. Furthermore, soap led to strong corrosion on the test rig parts. During a pumping test the soap changed the colour from colourless to yellow resulting from corrosion of the pump. The reason for that is the basic character of the soap. The usage of soap as substitute medium is not recommended.

### 3.2 Substitute fluid – water-agar-mixture

The subjective examination of a water-agar-mixture led to good results, so this mixture was analysed with the rotational viscometer. Figure 6 includes the graph of viscosity at different shear rates for the substitute fluid water-agar-mixture.

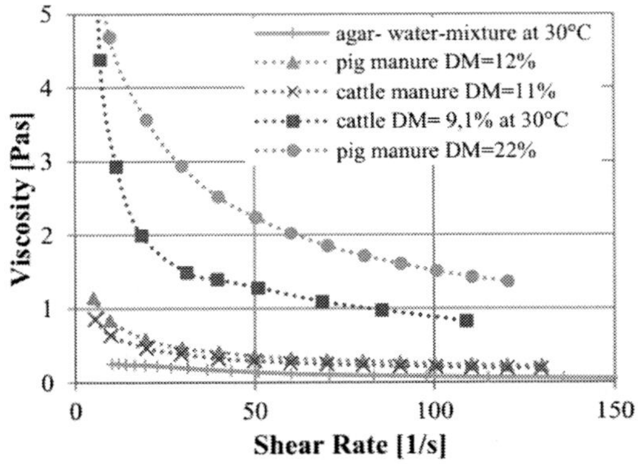


**Fig. 6** - Description of the viscosity-course from a water-agar-mixture

As it is apparent in figure 6, the water-agar-mixture has a slightly shear-thinning behaviour. At higher shear rates the viscosities of the mixture and pig manure approach each other. To get similarity between the flow of cattle and pig manure with high dry matter content in the distribution, the Reynolds number must be maintained. So the following equation must be complied:

$$Re_{manure} = Re_{substitute\,fluid} \quad (4)$$

A closer look at the Reynolds number equation shows that by changing the velocity and dynamic viscosity at the same liquid manure spreader the dimensionless quantity could be the same.

$$Re = \frac{v \cdot d \cdot \rho}{\eta} \quad (5)$$

Through adjusting the velocity at the liquid manure spreader, the mixture of water-agar can represent the characteristics of cattle or pig manure at high shear rates.

## 4 Conclusions

There are some different substitute fluids for pig or cattle manure. The problem decides which substitute fluid is to use. Shampoo and soap has the best mechanical properties compared to slurry. But these two fluids have bad optical properties and they lead to corrosion on metallic components because of their chemical composition. Furthermore shampoo and soap have a bad foaming behaviour. Therefore other substitute fluids were subjective examined. The result is that especially the subjective properties of a water-agar-mixture were advantageous in the case of the problem. Therefore the mixture was analysed in a rotational viscometer. Under the condition of a constant Reynolds number (by changing the flow velocity in the distributor) the properties of the water-agar-mixture and liquid manure is comparable.

## Nomenclature

| | |
|---|---|
| Re | Reynolds number |
| v | Velocity |
| d | Characteristic length |
| ρ | Density |
| η | Dynamic viscosity |
| dyn. | Dynamic |
| DM | Dry matter |
| Pas | Pascal-Seconds |
| s | Second's |
| LDA | Laser-Doppler-Anemometer |

## References

1. M. Breuer: Laser-Doppler-Anemometrie. *Vertiefungspraktikum Strömungsmechanik,* 2 (2014)
2. M. Türk.: Das Fließverhalten von trockensubstanzreicher Schweinegülle. *agrartechnik*, **36**, 33 (1986)
3. M. Türk.: Zum Fließverhalten von Güllemischungen. *agrartechnik*, **37**, 64 (1987)
4. H.M. El-Mashad, W.K.P. van Loon, G. Zeeman, G.P.A. Bot: Rheological properties of dairy cattle manure. *Bioresource Technology* (2004)
5. M. Breitwieser: Fließverhalten von Buttermilch – mit Butterflocken. *IMETER Methode Nr. 5*, (2013)

EPJ Web of Conferences **143**, 02106 (2017) DOI: 10.1051/epjconf/201714302106
*EFM 2016*

# Proposal of experimental device for the continuous accumulation of primary energy in natural gas hydrates

*Ján* Siažik[1,*], *Milan* Malcho[1] and *Richard* Lenhard[1]

[1] University of Žilina, Faculty of Mechanical Engineering, Department of Power Engeneering, Univerzitna 1, 010 26 Žilina, Slovakia

**Abstract.** Hydrates of the natural gas in the lithosphere are a very important potential source of energy that will be probably used in the coming decades. It seems as promising accumulation of the standard gas to form hydrates synthetically, stored, and disengage him when is peak demand. Storage of natural gas or biomethane in hydrates is advantageous not only in terms of storage capacity, but also from the aspect of safety storage hydrates. The gas stored in such form may occurs at relatively high temperatures and low pressures in comparison to other Technologies of gas- storage. In one cubic meter of hydrate can be stored up to 150 $m^3$ of natural gas, depending on the conditions of thermobaric hydrate generation. This article discusses the design of the facility for the continuous generation of hydrates of natural gas measurement methodology and optimal conditions for their generation.

## 1 Introduction

Hydrates have been for decades treated as a potential energy source especially for countries with limited access to conventional hydrocarbon fuels or for countries with strategic interest for the creation of alternative supplies of natural gas. Hydrate is generally a substance containing water and hydrocarbons (or other gas). The storage of natural gas hydrates methane in particular, is preferable particularly in view of storage capacity, but also from the aspect of safety storage of gas, due to higher temperatures and lower pressures in comparison to other storage technologies. Gas hydrate is a solid state, which compose from about 85 % water and 15 % gas clathrate bound in the lattice. The hydrates also involve some uncertainty due to the ongoing research and their use as potential sources of revenue. Gas hydrate can be a great source of energy that will be included in the consideration as supply of natural gas for next decades. It is estimated that 99 % of the world reserves of gas hydrates are found in marine sediments at appropriate temperature and pressure conditions prevailing in the upper meters at sea bed, and in the sediments at depths of up to 300 m and 4000 m. Hydrates can also occur on the mainland, but only in areas of permafrost, thus the long-term frozen surface areas with year-round temperatures below 0 °C. Saving energy in suitable form for safe gas storage and its subsequent release consists of many challenges. At the moment, we have emphasized the overall efficiency of these processes, the light use efficiency, increasing the efficiency and safety of the energy storage. The aim of the use of methane hydrates in practice could be, for example ensuring coverage of peaks in the cogeneration of electricity and heat with gas released from hydrates.

In our prototype equipment for the hydrate generation the gas hydrate will be prepared from water and gas as a solid hydrate. The work will be using natural gas as gas hydrates forming gas and energy requirements of hydrates. It is further proposed a high-pressure circuit diagram of an experimental apparatus for production of gas hydrates with a design pressure of 25 MPa, which would maintain the necessary state quantities to determine suitable parameters for the storage of gas hydrate structures. The device, which will be implemented in the continuation of this work should also serve to identify specific energy consumption for the formation of hydrates.

## 2 Natural gas as gas forming hydrates

A number of different gases can produce clathrates of binding with a molecule of water. In the article the focus will be on creating the most widely used gas hydrate energy carriers - natural gas.

### 2.1 Natural gas

Natural gas from the physical point of view is a combination of gas, the main component is methane. Natural gas is often found together with oil or coal. Natural gas is a colourless, in itself odourless, flammable gas. Belongs to a group of heating gas and is used for heating, cooking, water heating, also in heating plants, cogeneration units and transport (such as motor vehicles). It can be used also in other forms. One form of a CNG (Compressed Natural Gas), that is, compressed natural gas at a pressure of 20 MPa. The second form of the LNG (Liquefied Natural Gas), liquefied natural gas

* Corresponding author: jan.siazik@fstroj.uniza.sk

at -162 °C. It consists primarily of methane and higher hydrocarbons with a small admixture of inert gases. Natural gas is not toxic, rarefied and lighter than air. [1]

## 2.2 Hydrates of natural gas

Gas hydrates are ice-like crystalline solids that form from mixtures of water and light natural gases such as methane, carbon dioxide, ethane, propane and butane. Methane was the dominant component among other hydrocarbon gases in the sediments. Gas hydrates, potentially one of the most important energy resources for the future. Methane gas hydrates are increasingly considered a potential energy resource. In the fig. 1 is shown a burning of methane hydrate sample.

**Fig. 1.** Burning of methane hydrate sample

Enormous reserves of hydrates can be found under continental shelves and on land under permafrost. Gas hydrate or clathrate consists of three general structure types. Depending on the size of the guest molecule, natural gas hydrates can consist of any combination of three crystal structures: (1) Structure I or sI, (2) Structure II or sII and (3) Structure H or sH. When pure liquid water freezes it crystallizes with hexagonal symmetry, but when it freezes as a hydrocarbon hydrate it does so with cubic symmetry for sI and sII, reverting to hexagonal symmetry for sH. Methane hydrates are widespread in sea sediments in depths of hundreds meters below the sea floor along the outer continental margins and are also found in Arctic permafrost. Some deposits are close to the ocean floor and at water depths as shallow as 150 m, although at low latitudes they are generally only found below 500 m. The deposits can be 300–600 m thick and cover large horizontal areas. Hydrates may affect climate because when warmed or depressurized, they decompose and dissociate into water and methane gas, which is one of the greenhouse gases that contributes to global warming. Methane hydrates hold the danger of natural hazards associated with sea floor stability, release of methane to ocean and atmosphere and gas hydrates disturbed during drilling pose are a safety problem. [2]

## 2.3 Formation of natural gas hydrates

Hydrates gas forms water ice containing a large amount of methane clathrate bound in the form - in the cavities of the crystal lattice. This is a solid ice-like compound whose existence on earth has been known since the 50s of the 20th century. Cubic meter of solid hydrate is containing an average of about 164 $m^3$ of methane in the gaseous form. Hydrate density varies around 900 kg. $m^{-3}$ at room temperature, the pressure is unstable. Their formation can be partially attributed to the thermal decomposition of organic matter, mostly oil. [3] The stability of natural gas hydrates is relate to the structure these substances and interactions, that underway of certain physical conditions. Conditions of hydrates are characterized by:

- the presence of water and the concentration that is necessary for the formation of the hydrate lattice, the appropriateness of using the geometry of the existence of the water molecules and the hydrogen bonds to form a grid,
- presence, kind and concentration of the guest molecules,
- the pressure and temperature of the medium in which hydrates will occur. [1]

Hydrates are thus crystalline solid generated by the contact of liquid water with the hygroscopic small gas molecules such as methane, ethane, hydrogen sulfide or carbon dioxide under specific pressure and temperature. The initial conditions for hydrate formation is the presence of water and the existence of ties necessary for creating a grid that fills the guest molecule. This creates basic cell. The sequence of the gas hydrate is in fig. 2.

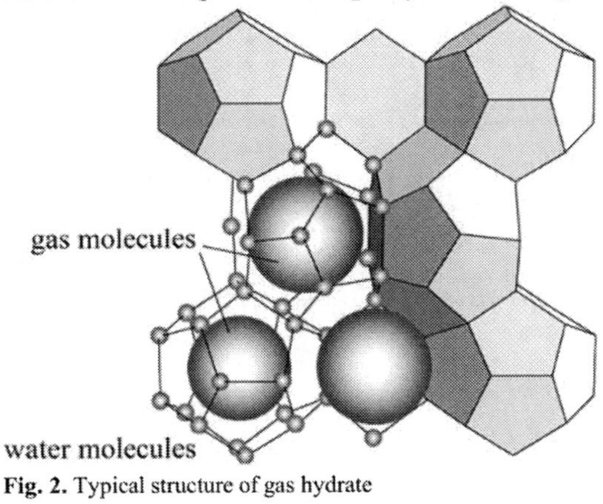


**Fig. 2.** Typical structure of gas hydrate

In relation to the conditions of formation and growth is known that hydrate is produced at high pressure and low temperature. Completion of the grid requires the presence of molecules of gas ($CH_4$, $C_2H_6$ ...) for creating a water grid. [4]

The creation of hydrates is affected further by:
- turbulence - at high flow rates is the formation of

hydrates is supported in places where a high flow rate is present (gas flow control at points). Increasing the velocity of the gas is due to the reduced flow through the valve space. On reducing the temperature due to the Joule-Thomson effect, - at stirring gas - gas hydrates can be formed due to mixing of gases in pipes or warehousing

• site creation - it´s a volume where it is supported by a phase transition that occurs in the formation of the solid phase (from the liquid phase). Formation of hydrates may occur with errors in pipe welds, bends, valves, etc. Also, where are deposits, contaminants, such as mud and sand. [5]

The hydrates may form hydrogen bonds along the edges of the grids accordingly, to the impact energy of hydrogen bonds as in the ice. Water molecules in clathrates have fewer configuration degrees of freedom as they have in the case of ice. Grid ice is stable and does not contain a guest molecule. In the case of an empty grid unstable hydrates to the structure have to be stabilized by van der Walls interactions to the presence of the molecule within the grid. [6]

Depending on the type of gas composition - by the host in the hydrate- the density of the hydrate is from 800 - 1200 kg. $m^{-3}$.

Hydrates features:

• 1 $m^3$ of water may be bound 207 $m^3$ of methane in the form of a solid hydrate with volume of 1.26 $m^3$, whereas without the gas comprises about 1 $m^3$ of water freezing ice into 1.09 cubic meters,

• the unit volume of methane hydrate at a pressure of 26 bar and a temperature of 0 °C is containing 164 units of gas,

• the hydrates is by 80 % of the volume filled with water and 20 % gas. [7]

Gas hydrates form three crystal structures (shown in fig. 3) cubic structure I, structure II, cubic and hexagonal structure H. The structure of I is the simplest and is composed of 46 water molecules. It consists of two small spherical cavities with twelve pentagonal faces ($5^{12}$), followed by six large cavities with two hexagonal faces and twelve pentagonal faces ($5^{12}6^2$). It is occupied by methane, $CO_2$ and hydrogen sulfide in small cavities. For large cavities it is occupied by ethane. The structure of II is composed of 136 water molecules. It consists of sixteen small cavities with twelve pentagonal faces ($5^{12}$), followed by eight large cavities with twelve pentagonal and hexagonal four walls ($5^{12}6^4$). All are located in a spherical model. The structure is occupied by nitrogen which includes large and small cavities, propane and isobutane. These molecules make up only a cavity. The structure H is formed by 34 water molecules. It consists of small chambers, in some cases, large chamber. The structure H hydrates produces a 2-methylbutane, 2,2-dimethylbutane, methylcyclohexane.

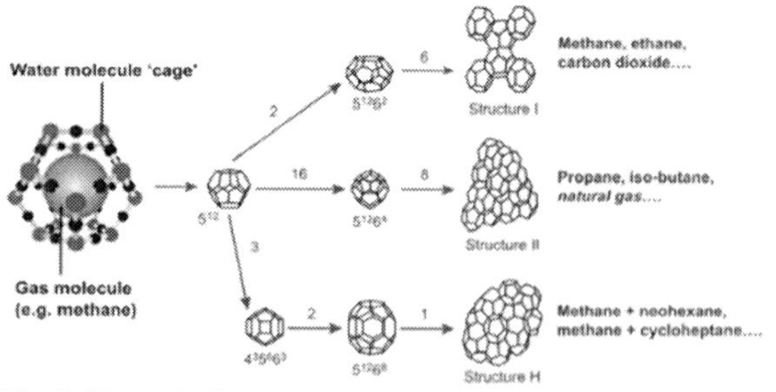


**Fig. 3.** The gas hydrate structures

## 2.4 The occurrence of natural gas hydrates

Gas hydrates are formed in a medium at high pressure and low temperature in the presence of sufficient water, and gas. On the following figure 4 is a phase diagram of the formation of methane hydrate.

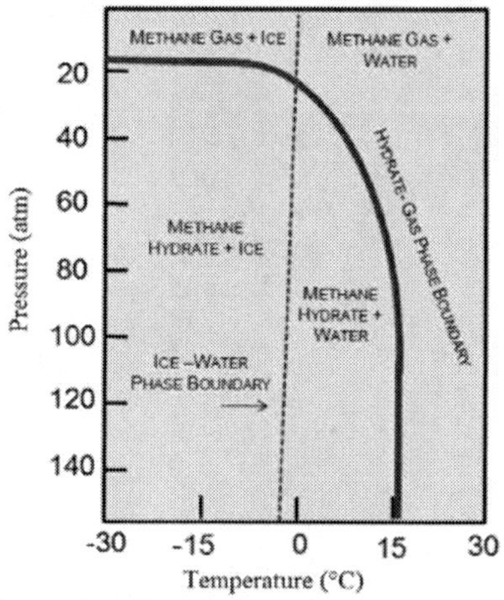


**Fig. 4.** Phase diagram for methane hydrate formation.

The requirements of hydrate formation restrict the occurrence of natural gas hydrates to two types of geologic locations:

- under permafrost in the polar continental shelves
- in sediment beneath the ocean floor.

# 3 Proposal experimental facility

Phenomenon of typical power consumption is for example uneven consumption of natural gas and biomethane in cogeneration units at peak load. Just energy-efficient accumulate of natural gas hydrates in a controlled manner to release the gas could be as a source of primary energy and thus could cover the peak load. High-pressure experimental apparatus for production of gas hydrates is proposed, which would maintain the necessary state variables for the determination of appropriate parameters of the accumulation of gas in the hydrate structure with a design pressure of 250 bar. Under the scheme of fig. 6 is made a device with which it will be possible to determine the specific energy consumption for the formation of hydrates. Identifying the parts of the scheme is in Tab. 9.

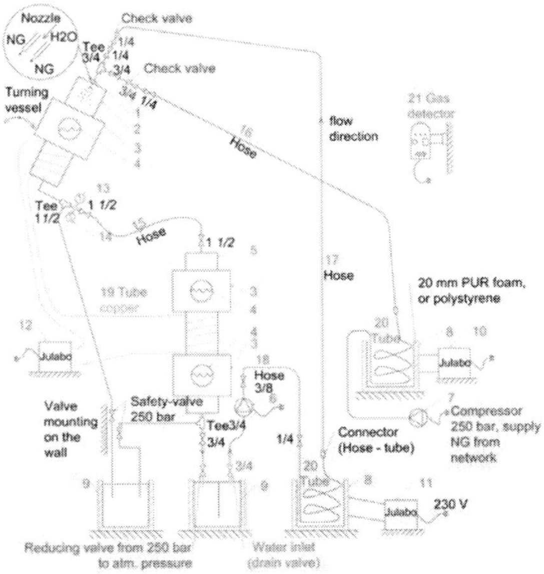


**Fig. 6.** Schematic diagram of experimental equipment

Experimental facility is under development and procurement of components and it's possible to start measurements with pressure under 100 bar. Pump 6 will suck distilled water which is compressed to a pressure of 100 bar, and which will pass through stainless steel tube 19. The container 8 will be connected to 11 Julabo - circulation thermostat with pum which will ensure cooling water in the tube by pressing the set temperature. The water pressure of 100 bar will continue to the hose 17 toward the nozzle. Second, the gas compressor 7, squeeze the gas to a pressure of 100 bar, which will flow into the hose 20. The stainless steel tubing 20 is wound into a coil and positioned in a container 8 which is connected to Julabo 10. This is conducted to provide the cooling of the gas after its compression to the desired temperature. The compressed gas is mixed with the compressed water and together they enter the nozzle 1. The nozzle is designed to atomize the mixture of water and compressed gas into small particles, which will make possible a better binding to the natural gas-hydrate form of molecular bonds. The mixture of water and gas should flow from the first container 2 through a hose 15 into a high pressure vessel 5 after its fulfilment. After filling of the container with a mixture of high-pressure gas and water visible area can be observed through the sapphire viewing windows 4, with sealed shut-off valve ¾ (located under high pressure vessel 5). From that moment will run hydrate formation germs, and will be measured in time for hydrate formation. After nucleated hydrates, the entire device can be depressurized and remove them through the flange and further investigate under the microscope. The electronic manometer can be connected to the measurement panel. The entire facility are 4 sapphire visors (2 pieces per container 2 and 2 per container 5), the other holes are covered, with blind flanges. Other undescribed valves in schematic diagram are shut off valves with the right pressure to 250 bar. Description of the proposed valve is in Tab. 10. In the laboratory will be located gas detector 21, to monitor the environment. Accidental increase of the gas concentration in the air will warn the audio response sensor.

**Table 1.** Position on the diagram of experimental equipment

| Poz. | Title | Model/type or performance |
|---|---|---|
| 1 | Nozzle | Spray angle 51 ° and 155 °, external thread ¼′′ |
| 2 | High pressure vessel | $\phi$175/146,9 mm, m=42 kg, 5.7l G1½′′, l=600 mm, p=250 bar |
| 3 | Flange on the container | Steel flange 255x255 mm |
| 4 | Sapphire sight glass | $\phi$ 50 mm +0/ -0.1 mm thickness 20.0 mm ± 0.1 mm |
| 5 | High pressure vessel | $\phi$175/146.9 mm, m=73 kg, 15l G1½′′, l=1159 mm, p=250 bar |
| 6 | Plunger pumps | $p_{max}$ = 450 bar, $T_{max}$ = 70 °C, Q = 5,1 l.min$^{-1}$ |
| 7 | Natural gas compressor | CNG, 3 kW, 400 V, 50 Hz |
| 8 | Container refrigeration | Material plastic |
| 9 | Container discharge | Material plastic |
| 10 | JULABO FL 2503 | $p_{max}$ = 27 bar, refrigerant R 404a, 230 V, 50 Hz |
| 11 | JULABO F 32, JULABO MA | $p_{max}$ = 27 bar, refrigerant R 134a, 230 V, 50 Hz |
| 12 | JULABO FP 40, JULABO HE | $p_{max}$ = 25 bar, refrigerant R 404a, 230 V, 50 Hz |
| 13 | Pressure gauge pointer | PTL GC0015, STN EN 837 40 MPa |
| 14 | Electronic manometer | 0-300 bar, 4-20 mA |
| 15 | Hose for natural gas | WP 385 bar, SAE 100, length 2m, DN ¾′′, DIN EN 856 4SP |
| 16 | Hose for natural gas | WP 400 bar, 1Q16, length 3m, EN 853 2SN6 |
| 17, 18 | Hose | WP 400 bar, DN 1/8′′, EN-853 2SN 17 (l=3 m), 18 (l=4 m) |
| 19 | Seamless tube | cupper D/d/l $\phi$10/$\phi$8 mm/50 m, STN EN 1057 |
| 20 | Seamless tube | Stainless steel D/d/l $\phi$10/$\phi$7mm/30 m; do 30 MPa, DIN 17 458 |
| 21 | Gas detector | Detection LPG, NG, coal-gas, 100-240 V, 50/60 Hz, volume >85dB, range -10 °C to 50 °C |

### 3.1 Energy intensity of hydrate creation

Energy consumption can be generally expressed through coefficient EROEI which reflects energy returned on energy invested. If the ratio is equal to or less than 1, source of energy becomes as energy sink and may not be already used as the primary power source. The following are the equations needed to estimate energy intensity of creation of methane hydrate. The following are mentioned nomenclature and method of calculation. Volume flow of natural gas is $\dot{V}_{NG} = 1.388 \times 10^{-3}\ m^3.s^{-1}$, volume flow water has been established $\dot{V}_W = 8.325 \times 10^{-6}\ m^3.s^{-1}$, operating water pressure $\Delta p_W = 4.9 \times 10^6$ Pa, pump efficiency $\eta_P = 0.6$, primary energy factor $f_e = 2.57$. Physical input power consumption of pump $N_P$ [W] is calculated by using eq. (1) and substituting into eq. (2) is calculated primary energy of pump $E_{PP}$ [W].

$$N_P = \frac{\Delta p_W.\dot{V}_W}{\eta_P} \tag{1}$$

$$E_{PP} = N_P.f_e \tag{2}$$

It is also known inlet pressure $p_1 = 1 \times 10^5$ Pa, outlet pressure $p_2 = 1 \times 10^7$ Pa, polytropic exponent $n = 1.2$, compressor efficiency $\eta_C = 0.6$. Compression power of compressor $A_{TC}$ [W] is given by eq. (3) and substituting into the eq. (4) is obtain input power consumption of compressor $P_C$ [W]. Primary energy of compression of natural gas $E_{PNG}$ [W] is given by eq. (5).

$$A_{TC} = \frac{n}{n-1}.p_1.\dot{V}_1.\left[\left(\frac{p_2}{p_1}\right)^{\frac{n-1}{n}} - 1\right] \tag{3}$$

$$P_C = \frac{A_{TC}}{\eta_C} \tag{4}$$

$$E_{PNG} = P_C.f_e \tag{5}$$

Another well-known quantities are density of natural gas $\rho_{NG} = 0.72$ kg.m$^{-3}$, water density $\rho_W = 1000$ kg.m$^{-3}$, specific heat capacity of natural gas $c_{NG} = 1609$ J.m$^{-3}$.K$^{-1}$, specific heat capacity of water $c_W = 4187$ J. kg$^{-1}$.K$^{-1}$, cooling factor $f_R = 3.5$, temperature difference $\Delta t = 30$ °C, it includes the ambient temperature 10 °C. Energy for cooling the natural gas $\dot{Q}_{NG}$ [W] is given by the equation (6), input power consumption of cooling natural gas $P_{NG}$ [W] is given by eq. (7). Energy for cooling the water $\dot{Q}_W$ [W] is shown by eq. (8) and input power consumption of cooling water $P_W$ [W] is calculated by eq. (9).

$$\dot{Q}_{NG} = \rho_{NG}.\dot{V}_{NG}.c_{NG}.\Delta t \tag{6}$$

$$P_{NG} = \frac{\dot{Q}_{NG}}{f_R}.f_e \tag{7}$$

$$\dot{Q}_W = \rho_W.\dot{V}_W.c_W.\Delta t \tag{8}$$

$$P_W = \frac{\dot{Q}_W}{f_R}.f_e \tag{9}$$

Calorific value of natural gas is $Q_{CV} = 34 \times 10^6$ J.$m^{-3}$, primary usable energy $E_{PU}$ [W] is given by eq. (10). Relationship for calculating the coefficient EROEI generally is shown eq. (11). Equations (2), (5), (7), (9), (10) were put into adjusted equation for calculating EROEI (12) under our case for methane hydrate formation.

$$E_{PU} = \dot{V}_{NG}.Q_{CV} \tag{10}$$

$$EROEI = \frac{\text{Energy returned}}{\text{Energy invested}} \tag{11}$$

$$EROEI = \frac{E_{PU}}{E_{PP} + E_{PNG} + P_{NG} + P_W} \tag{12}$$

Estimated value coefficient given by eq. (12) has a coefficient value of EROEI = 10.6 after rounding number. This value is favorable however, actual values of coefficient will be streamed after evaluaiton of the measurement. The calculation was framework with limited data and energy inputs are related to primary energy. The following are mentioned studies, which can be seen as moving EROEI value depending on the case. Study by [8] where was created model the heating of submarine methane hydrate deposits located at depths between 1000 and 1500 meters, when the heating is supplied by hot water flowing through a horizontal pipe located inside the reservoir. EROEI factors begin at 28 as the water starts flowing, and reduce to a value of 2.8 after fifty years of production, but it is quite a different case. Another study by [9] says that for crude oil, a very wide range of EROEI values apply, depending inter alia on the depth of the well and on whether the well is on- or offshore. In these early years of the 21st century a range of 3-10 is often expressed. Next study by [10] says about fuel sources and regions oil and gas production for Canada EROEI value of 15 in year 2010.

## 4 Conclusion

Interesting way how to accumulate energy of natural gas is using hydrates of natural gas. According to previous studies, it appears to be advantageous to store natural gas into hydrates and in terms of operational or relatively safe conditions compared to CNG – where is required higher pressures or LNG - where is required lower temperatures. Both of these options are energy intensive. Estimated factor EROEI for the formation of methane hydrate under laboratory conditions was calculated to value of 10.6. If the EROEI that we determined on this paper had been less than number 1, production methane hydrate would be useless.

Work on article has been financially supported by the project VEGA-1/0718 / 15th Accumulation of high potential energy through the process of generating hydrates of natural gas and biomethane.

## References

1. V. Foltin, *Natural Gas Hydrates – Challenge and Opportunity* (2013)
2. A. Demirbas, *Energy Conversion and Management*, **51**, 7, (2010)
3. S. Depowski, *Surowce energetyczne,* **10** (1998)
4. A. Kvenvolden, *Gas hydrates as a potential energy* resource, **1570** (1993)
5. J. Carroll, *Natural gas hydrates: A guide for engineer*, Burlington (2003)
6. F. Franks, *Water a Matrix of Life,* 2nd edition, (2000)
7. F. Makogon, *Natural Gas Hydrates – A promising Source of Energy,* **2**,1 (2010)
8. B. Sunden, C. A. Brebbia, D. Poljak, Advanced Computational Methods and Experiments in Heat Transfer XII, (2012)
9. C. J. Jones, IJMEE, **36**, 6 (2008)
10. Ch. A. S. Hall, *EP*, **64**, 12 (2014)
11. M. Holubčík, J. Jandačka, Š. Papučík, P. Pilát, *Manufacturing Technology*, **15** (2015)
12. J. Jandačka et al., *Emisné zaťaženie životného prostredia*, 1 (2014)
13. Š. Rezničák, M. Malcho, J. Siažik, *Time of formation of methane hydrates*, AeaNMiFMaE 2016 (2016)
14. M. Novomestský, Š. Papučík, J. Siažik, *Heat performance of heat pipe thermosyphon*, AeaNMiFMaE 2016 (2016)
15. J. Siažik, M. Malcho, *Energy transformation in industry* 2015 (2015)
16. P. Ďurčanský, Š. Papučík, J. Jandačka, M. Holubčík, R. Nosek, *Scientific World Journal*, **55**, 138254 (2014)

# Turbulence intensity measurement in the wind tunnel used for airfoil flutter investigation

*Petr* Šidlof[1,a] , *Pavel* Antoš[2], *David* Šimurda[2], *Martin* Štěpán[1]

[1]Technical University of Liberec, NTI FM, Studentská 2, 461 17 Liberec 1, Czech Republic
[2]Institute of Thermomechanics, Academy of Sciences of the Czech Republic, Dolejškova 5, 182 00 Prague 8, Czech Republic

**Abstract.** The paper reports on hot wire turbulence intensity measurements performed in the entry of a suction-type wind tunnel, used for investigation of flow-induced vibration of airfoils and slender structures. The airfoil is elastically supported with two degrees of freedom (pitch and plunge) in the test section of the wind tunnel with lateral optical access for interferometric measurements, and free to oscillate. The turbulence intensity was measured for velocities up to M = 0.3 i) with the airfoil blocked, ii) with the airfoil self-oscillating. Measurements were performed for a free inlet and further with two different turbulence grids generating increased turbulence intensity levels. For the free inlet and static airfoil, the turbulence intensity lies below 0.4%. The turbulence grids G1 and G2 increase the turbulence level up to 1.8% and 2.6%, respectively. When the airfoil is free to oscillate due to fluid-structure interaction, its motion disturbs the surrounding flow field and increases the measured turbulence intensity levels up to 5%.

## 1 Introduction

The aircraft lifting surfaces (wings, flaps and ailerions, stabilators, elevons and rudders), helicopter blades, propellers, turbine or compressor blades are slender structures subjected to unsteady aerodynamic forces and moments. As such, they are potentially susceptible to aeroelastic instability such as coupled-mode flutter, stall flutter or dynamic stall. A comprehensive review on flutter and dynamic stall is given in [1], a practical guide for technical designers can be found in [2]. Fluid-structure interaction can lead to catastrophic consequences, and thus the aircraft components must be carefully designed and tested in order to ensure that the aeroelastic instability never occurs within the operating conditions.

During last years, the fundamental properties of a self-oscillating NACA0015 airfoil with two degrees of freedom have been studied in a high-speed suction-type wind tunnel of the Institute of Thermomechanics in Nový Knín [3]. For low inlet air speeds, damped oscillations were observed. When the elastically supported airfoil is subjected to flow of higher velocity, the system becomes aeroelastically unstable and any motion of the airfoil induced by an initial impulse or disturbance in the flow exponentially increases. Due to nonlinearity of the system, the oscillations finally stabilize in limit cycle oscillations (LCO) of certain frequency and amplitude, strongly dependent on the mechanical properties of the system (stiffness of the springs, mass distribution, mechanical damping, eigenfrequencies of the pitching and plunging modes). The experimental setup allows dynamic measurement of the flow-induced oscillations and time-resolved measurements of the static pressure at four pressure ports on the surface of the airfoil. Synchronously, the flow fields around the airfoil can be monitored either using time-resolved interferometric measurements [3,4] or Schlieren visualizations [5].

The test section is located shortly downstream of the free entry of the wind tunnel. In this configuration, the turbulence intensity of the flow should be quite low. The purpose of the current study is to measure the turbulence intensity levels when different turbulence grids are fixed upstream of the test section, and compare them to the case when no turbulence grid is present.

## 2 Experimental setup

The model design builds on previous experience obtained in the Institute of Thermomechanics in the years 2006-2012, when self-oscillating models of the NACA0015 and DCA18% airfoils with two degrees of freedom were designed and the methodology of the measurements developed [6,7,3]. The airfoil, supported by a linear and torsional springs, is fixed in a suction-type wind tunnel with optical access through the side walls for Schlieren visualizations and interferometric measurements of the flow field.

### 2.1 Mechanical setup

The NACA0015 airfoil is supported by vertical guides, freely moving in the vertical direction, and rotating around the elastic axis in 1/3 of the chord length (see Fig. 1 and 2).

* Corresponding author: petr.sidlof@tul.cz

**Fig. 1.** Model of the pitch-plunge airfoil in the wind tunnel (side view).

The test section has a cross-section of 80 x 210 mm, with the free opening of the wind tunnel 310 mm upstream of the airfoil leading edge.

**Fig. 2.** Airfoil and hot-wire probe (view from inlet part of wind tunnel).

The airfoil has a chord length of 59.5 mm, considerable smaller than comparable self-oscillating airfoil designs, e.g. [8-13]. The stiffness of the flat springs is 16 383 N/m, resulting in the still-air natural frequency of the plunging mode $f_{plunge}$ = 14.6 Hz. In current configuration, the eigenfrequency of the torsional mode is $f_{pitch}$ = 17.8 Hz.

## 2.2 Motion and pressure sensors

Due to the size of the airfoil (chord 59.5 mm, maximum thickness 9 mm), there is a very limited space for installation of the sensors inside the model. The airfoil is equipped with a miniature rotary magnetic encoder (RLS - Renishaw RM08) measuring the pitch angle, and four dynamic piezoresistive pressure sensors with built-in preamplifiers (Freescale Semiconductor MPXH6115) mounted flush with the airfoil surface. The pressure sensors are denoted $p_{1U}$, $p_{2U}$ (upper surface) and $p_{1L}$, $p_{2L}$ (lower surface) and are located 10.8 mm and 29.5 mm from the leading edge, respectively (see Fig. 3).

Outside of the test section, the vertical deflection of the airfoil is measured by a laser triangulation sensor (MicroEpsilon optoNCDT ILD-2200). An additional strain gauge bridge is mounted on the flat spring supporting the airfoil frame and calibrated to provide a redundant signal of the deflection.

The signals from the airfoil pressure transducers, strain gauge bridge and the triangulation sensor are digitized, monitored online and stored by a Dewetron data acquisition system. The software includes a special module for triggering of the high-speed camera, which enables perfect synchronization of the recorded interferograms with the signals from all the sensors.

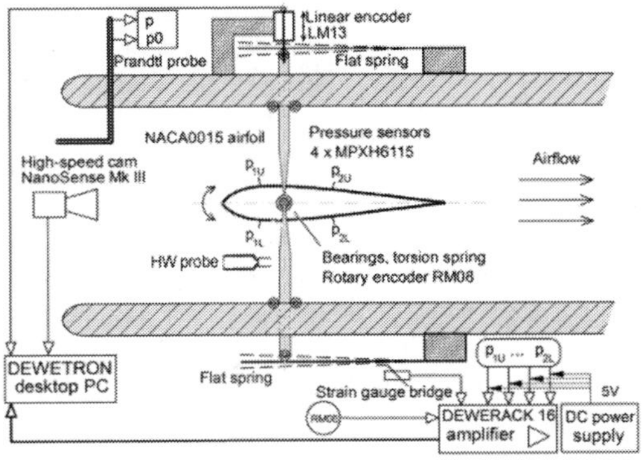


**Fig. 3.** Mechanical sensors, HW probe and electrical connections

## 2.3 Optical measurements

The test section of the wind tunnel and its special lateral optical glasses are designed for measurements of the flow field using the Mach-Zehnder interferometer and for Schlieren visualizations. The former method, used within current study, is sensitive to the air density which has local variations due to compressibility effects, the latter to the density gradient [14]. More details regarding the optical configuration can be found in [5].

## 2.4 Grid generators

Two grid turbulence generators denoted as G1 and G2 were tested during the measurements (see Fig. 4). The grids are manufactured from an interwoven steel wire, and are fixed 140mm downstream of the wind tunnel entry. The parameters of the grids are summarized in Tab. 1.

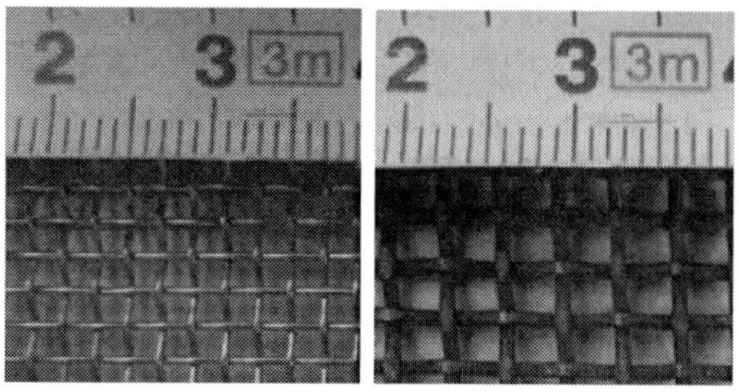

**Figure 4.** Grids G1 (left) and G2 (right)

**Table 1.** Wire diameter, grid spacing and mesh size.

| | diam [mm] | spacing [mm] | mesh size [mm] |
|---|---|---|---|
| *G1* | 0.5 | 1.6 | 2.1 |
| *G2* | 0.95 | 2.0 (vertically)<br>2.4 (horizontally) | 2.95 (vertically)<br>3.4 (horizontally) |

## 2.5 Hot-wire measurements

The hot-wire probe was placed 260 mm downstream of the wind tunnel entry. It was in the plane of the leading edge of the airfoil, i.e. 120 mm downstream of the turbulence grid.

A single-sensor probe DANTEC 55P01 was used for the measurement in flow. The sensor is a tungsten wire with gold-plated ends of the diameter $d_w$=5 μm and the active length $l_w$=1.25 mm. The probe was calibrated in the rig with variable flow velocity in the range of velocities $U$=(2-140) m/s. Sensor was operated at CTA mode with wire temperature $T_w$=513 K. Details about principle of HWA measurement and its evaluation can be found in [15].

The CTA system DANTEC Streamline was used. The output signals are then digitalized using the A/D transducer (National Instruments data acquisition system 16 bit) with sampling frequency 250 kHz.

# 3 Results

The turbulence intensity measurements were performed for a set of inlet flow velocities ranging from Ma = 0.1 to Ma = 0.3. Three cases were tested: without turbulence grid, and with grids G1 and G2 mounted. In most cases, the airfoil motion was blocked to measure the turbulence intensity in the flow field undisturbed by airfoil oscillations, but three cases were also tested when the airfoil was free and under flow-induced vibration with quite large amplitudes (up to 8.3 mm in vertical motion and 27.6 deg in rotation).

The results of mechanical and optical measurements are reported and summarized in Tab. 2.

**Table 2.** Measurement by mechanical sensors. Used grid generator of turbulence, inlet flow Mach number, status of the airfoil support. Frequency of flow-induced vibration, amplitude of the plunging and pitching mode.

| No. | Ident. | grid | Ma [-] | airfoil | $f_{airfoil}$ [Hz] | $A_y$ [mm] | $A_\varphi$ [deg] |
|---|---|---|---|---|---|---|---|
| 5 | 2958-5 | - | 0.1 | blocked | - | - | - |
| 6 | 2958-6 | - | 0.15 | blocked | - | - | - |
| 7 | 2958-7 | - | 0.15 | free | - | - | - |
| 8 | 2958-8 | - | 0.2 | blocked | - | - | - |
| 9 | 2958-9 | - | 0.2 | free | 17.5 | 2.5 | 29.5 |
| 10 | 2958-10 | - | 0.25 | blocked | - | - | - |
| 11 | 2958-11 | - | 0.25 | free | 17.5 | 4.1 | 28.9 |
| 12 | 2958-12 | - | 0.3 | blocked | - | - | - |
| 13 | 2958-13 | - | 0.3 | free | 16.7 | 8.3 | 27.6 |
| 14 | 2958-14 | G1 | 0.1 | blocked | - | - | - |
| 15 | 2958-15 | G1 | 0.15 | blocked | - | - | - |
| 16 | 2958-16 | G1 | 0.2 | blocked | - | - | - |
| 17 | 2958-17 | G1 | 0.25 | blocked | - | - | - |
| 18 | 2958-18 | G1 | 0.3 | blocked | - | - | - |
| 19 | 2958-19 | G2 | 0.1 | blocked | - | - | - |
| 20 | 2958-20 | G2 | 0.15 | blocked | - | - | - |
| 21 | 2958-21 | G2 | 0.2 | blocked | - | - | - |
| 22 | 2958-22 | G2 | 0.25 | blocked | - | - | - |
| 23 | 2958-23 | G2 | 0.3 | blocked | - | - | - |

The graph of velocity U and turbulence intensity $I_u$ during measurements No. 5, 6, 8, 10, 12 is given in Figure 5. This is the case of flow without grid generator with blocked airfoil.

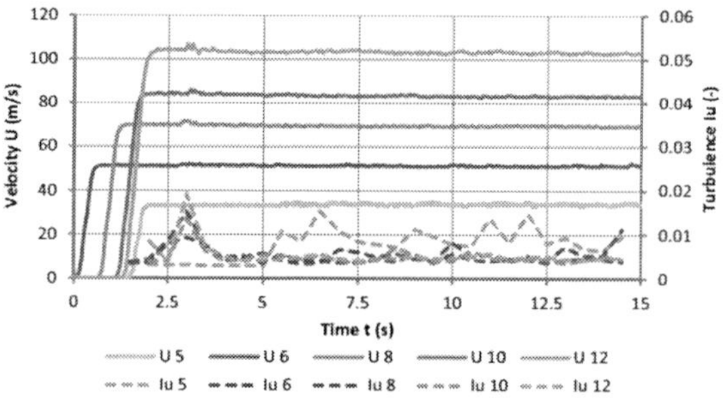

**Fig. 5.** Measurements No. 5, 6, 8, 10, 12 with blocked airfoil and no grid generator.

The plot in Fig. 6 shows measurements No. 7, 9, 11, 13 with free airfoil. The oscillations in flow field induced by the airfoil motion are clearly visible. For computation of turbulence intensity, the hot-wire signal was processed by high-pass filter at 500 Hz to extract amplitudes from low frequencies.

DOI: 10.1051/epjconf/201714302107

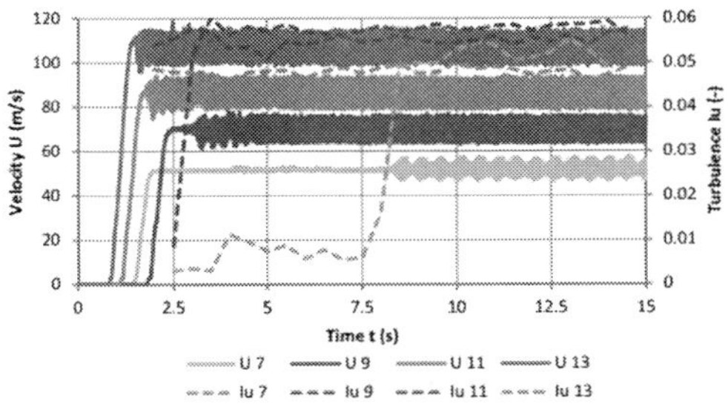


**Fi. 6.** Measurements No. 7, 9, 11, 13 with free airfoil and no grid generator.

Next two graphs in Fig. 7 and 8 show measurements No. 14-22 with blocked airfoil and grid generators of turbulence G1 and G2.

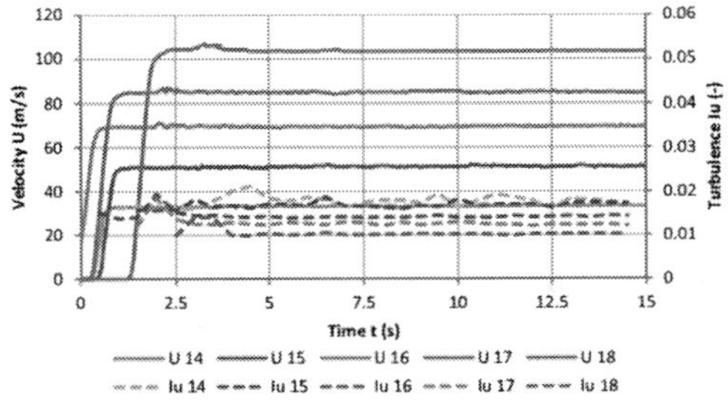


**Fig. 7.** Measurements No.14-18 with blocked airfoil and grid generator G1.

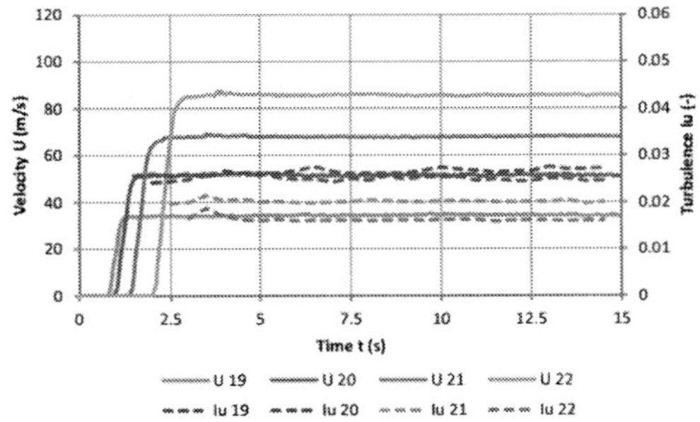


**Fig. 8.** Measurements No. 19-22 with blocked airfoil and grid generator G2.

Spectra of flow velocity fluctuations from the HW signal are shown below. The FFT spectrum in case of blocked airfoil and grid G1 is shown in Fig. 9.

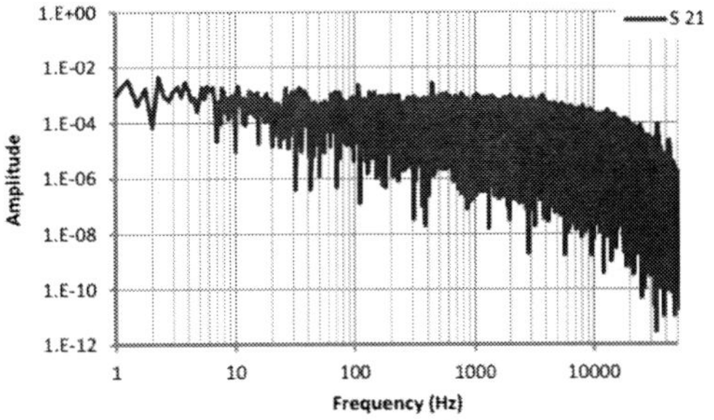


**Fi. 9.** Spectrum of the HW signal in measurement No. 21. Blocked airfoil, grid G2.

Spectra of flow with free airfoils are plotted for measurements No. 7, 9, 11 and 13. The spectrum of measurements No. 7 in Fig. 10 shows significant peaks at frequencies 15.750 Hz (100%A); 17.746 Hz (38%A); 13.783 Hz (9%A); 19.711 Hz (2%A). In this case, the airfoil is free, but the flow velocity is not sufficient to trigger the aeroelastic instability. In all other measurements (with higher velocities) a single dominant frequency is seen, matching perfectly the frequency of airfoil oscillation obtained from mechanical sensor measurements.

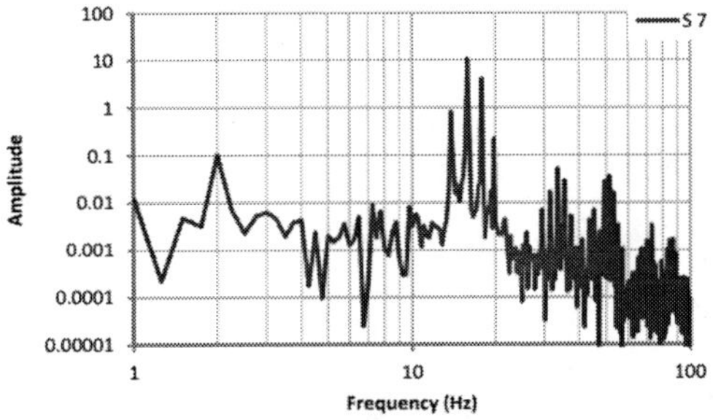


**Fig. 10.** Spectrum from measurement No. 7. Free airfoil, no airfoil oscillation.

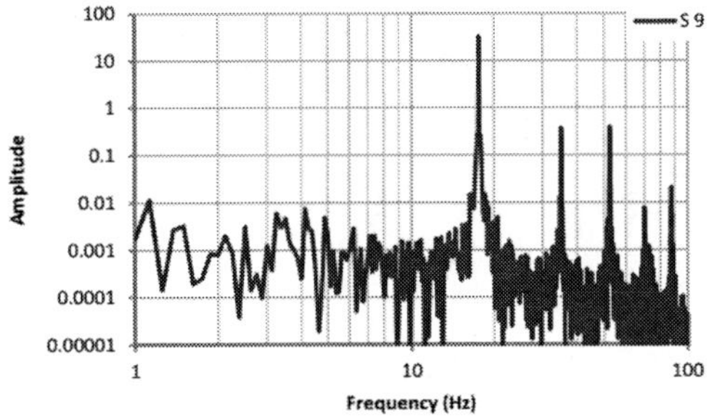


**Fig. 11.** Spectrum from measurement No. 9. Free airfoil, limit cycle oscillations at 17.49 Hz.

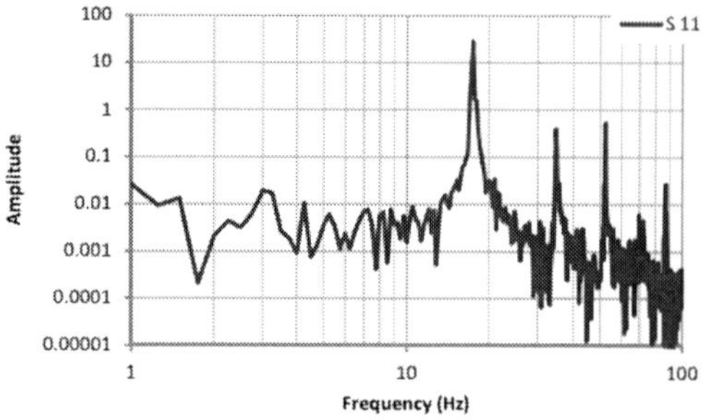


**Fig. 12.** Spectrum from measurement No. 11. Free airfoil, limit cycle oscillations at 17.47 Hz.

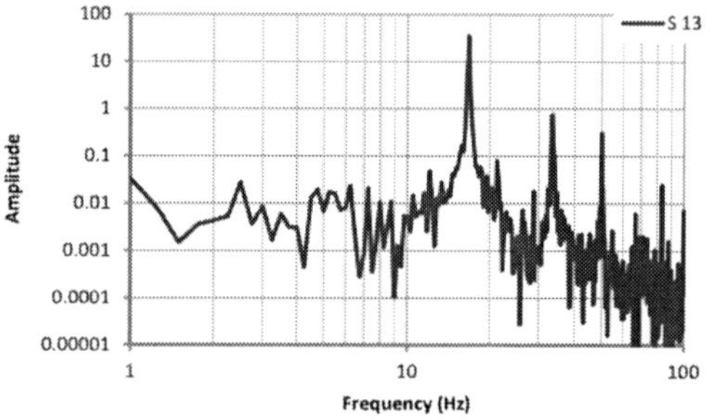


**Fig. 13.** Spectrum from measurement No. 13. Free airfoil, limit cycle oscillations at 16.73 Hz.

The results of the turbulence intensity measurements are summarized in Table 3.

**Table 3.** HWA measured data. Mean velocity, turbulence intensity level and frequency of peak in spectrum of the flow around the airfoil.

| No. | grid | Ma [-] | airfoil | $I_u$ [%] | $f_{flow}$ [Hz] |
|---|---|---|---|---|---|
| 5 | - | 0.1 | blocked | 0.28* | - |
| 6 | - | 0.15 | blocked | 0.33* | - |
| 7 | - | 0.15 | free | 5.2 | 15.750 |
| 8 | - | 0.2 | blocked | 0.35* | - |
| 9 | - | 0.2 | free | 5.78 | 17.499 |
| 10 | - | 0.25 | blocked | 0.35* | - |
| 11 | - | 0.25 | free | 5.5 | 17.471 |
| 12 | - | 0.3 | blocked | 0.37* | - |
| 13 | - | 0.3 | free | 4.78 | 16.736 |
| 14 | G1 | 0.1 | blocked | 1.79 | - |
| 15 | G1 | 0.15 | blocked | 1.68 | - |
| 16 | G1 | 0.2 | blocked | 1.41 | - |
| 17 | G1 | 0.25 | blocked | 1.25 | - |
| 18 | G1 | 0.3 | blocked | 1.01 | - |
| 19 | G2 | 0.1 | blocked | 2.65 | - |
| 20 | G2 | 0.15 | blocked | 2.51 | - |
| 21 | G2 | 0.2 | blocked | 2.02 | - |
| 22 | G2 | 0.25 | blocked | 1.62 | - |
| 23 | G2 | 0.3 | blocked | - | - |

* Minimal value of Iu

## 4 Discussion and conclusions

The turbulence intensity levels in the entry of the wind tunnel test section used for investigation of the aeroelastic instability of an airfoil with two degrees of freedom were investigated. In case of blocked airfoil, when the inlet part of the wind tunnel is free, the turbulence level at the airfoil leading edge ranges from 0.28% to 0.37%. When the fine turbulence grid G1 is mounted, the turbulence intensity is 1.79% for the lowest flow speed Ma = 0.1 and 1.01% for the highest possible flow speed Ma = 0.3. The coarse turbulence grid G2 induces turbulence levels ranging from 2.65% to 1.62%, decreasing with increasing flow velocity.

In the case of self-oscillating airfoil, the motion heavily disturbs the surrounding flow field and the measured intensity of turbulence were of levels of 4.78% - 5.78%.

In spectra of flow obtained from HW measurements were found significant peaks at frequencies of 15.750 Hz, 17.499 Hz, 17.471 Hz, 16.736 Hz, which match the frequencies of airfoil oscillation from mechanical measurements (N/A, 17.5 Hz, 17.5 Hz, 16.7 Hz).

## Acknowledgements

The research has been supported by the Czech Science Foundation, project 13-10527S "Subsonic flutter analysis of elastically supported airfoils using interferometry and CFD". The experimental data were acquired during measurements in the wind tunnel of the Institute of Thermomechanics, Academy of Sciences of the Czech Republic in Nový Knín, in cooperation with Martin Luxa, Jaromír Horáček and Václav Vlček.

## References

1. W. J. McCroskey, Annual Review Of Fluid Mechanics **14**, 1 (1982)
2. L. Ericsson, J. Reding, Journal of Fluids and Structures **2,** 1 (1988)
3. P. Šidlof, V. Vlček, M. Štěpán, J. Horáček, M. Luxa, D. Šimurda, J. Kozánek, *Proceedings of the ASME PVP 2014 conference* (ASME, 2014)
4. P. Šidlof, Š. Riss, V. Vlček V., *Topical Problems of Fluid Mechanics* (2016),
5. P. Šidlof, V. Vlček, M. Štěpán, Journal of Fluids and Structures (2016, in press)
6. V. Vlček, J. Kozánek, Acta Technica **56** (2011)
7. J. Kozánek, V. Vlček, I. Zolotarev I., International Journal of Structural Stability and Dynamics **13,** 7 (2013)
8. T. O'Neil, T. W. Strganac T. W., Journal of Aircraft **35,** 4 (1998)
9. C. Marsden, S. Price S, Journal of Fluids and Structures **21,** 3 (2005)
10. E. Tinar, O. Cetiner, Experiments in Fluids **41,** 2 (2006)
11. D. Poirel, Y. Harris, A. Benaissa, Journal of Fluids and Structures **24,** (2008)

DOI: 10.1051/epjconf/201714302107

12. M. Schwartz. S. Manzoor, P. Hémon, E. de Langre., Journal of Fluids and Structures **25**, 8 (2009)
13. N. A. Razak, T. Andrianne, G. Dimitriadis, AIAA Journal **49**, 10 (2011)
14. A. H. Shapiro, *The Dynamics and Thermodynamics of Compressible Fluid Flow* (The Ronald Press Company, 1954)
15. P. Antoš, EPJ Web of Conferences **25**, 01004 (2012)

DOI: 10.1051/epjconf/201714302108

# Design of a test facility for probe calibration

Jan Šimák[1,a]

[1] *Aerospace Research and Test Establishment, Department of Aerodynamics, Beranových 130, 199 05 Prague - Letňany, Czech Republic*

**Abstract.** A possibility to easily calibrate probes for flow field measurements is always welcome. From this reason, a design of a test facility for probe calibration was made. The probes will be calibrated in a free jet of known properties, which is created by an exchangeable nozzle to cover a wide range of Mach numbers up to Mach 2. The most important is to create a homogeneous flow across the test section. This is accomplished by a precise design of the nozzles carried out by numerical tools. The convergent nozzle part is common for all subsonic flow regimes while the divergent part (forming a de Laval nozzle) is suited for a specific supersonic Mach number. These parts are designed using the method of characteristics. Numerical simulations performed by a CFD code show a feasibility and quality of the proposed test facility.

## 1 Introduction

Evaluation of experiments from internal aerodynamics is still dependent on measurements by pressure probes. However, the probes cannot measure the flow velocity and angles directly. Every probe can be slightly different due to manufacturing inaccuracy or a wear level and thus a calibration needs to be done before using. To calibrate a probe we need a flow of well known properties within a wide range of Mach and Reynolds number. Having such a device can significantly shorten the time needed to perform an experiment.

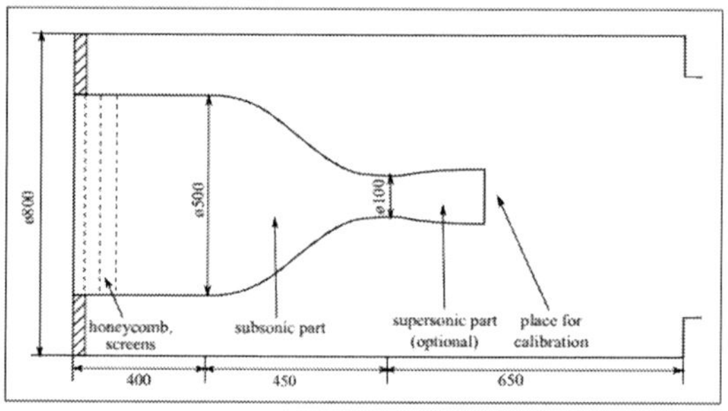


Figure 1: Scheme of the calibrating device with the nozzle.

## 2 Design of the device

The calibrating device is intended to be a part of a closed circuit wind tunnel. A round nozzle creating a stream of air with required parameters will be installed inside a testing chamber. For subsonic regimes, the nozzle has a convergent shape. For supersonic regimes, a divergent extension is attached to form a de Laval type nozzle. The subsonic (convergent) part remains the same for the whole range of assumed velocities. The supersonic extension has an individual shape for each specific outgoing Mach number.

On the inlet of the compressed air into the nozzle, there are a honeycomb and screens to reduce the inlet turbulence and to homogenize the flow. The probe will be inserted into the free jet behind the nozzle in the place of the expected homogeneous flow (figure 1).

### 2.1 Subsonic nozzle

The inlet diameter of the convergent nozzle is $d_{in}$ = 500 mm and the exit diameter is $d_{out}$ = 100 mm. Thus the compression ratio is CR = 25. The shape of the subsonic nozzle is described with a polynomial of the 5th degree [? ],

$$r(x) = r_{in} - (r_{in} - r_{out})\left[6(x/L)^5 - 15(x/L)^4 + 10(x/L)^3\right], \tag{1}$$

where $r$ is the radius at the distance $x$ from the beginning, $r_{in}$ and $r_{out}$ are the inlet and exit radius and $L$ is the nozzle length.

The exit Mach number is determined by a ratio of the total pressure before the nozzle $p_{in,c}$ and the static pressure $p_{out}$ behind the nozzle exit. In the case of a subsonic flow, this pressure is equal to the chamber static pressure $p_b$. The Mach number can be expressed for an isentropic flow as

$$M_{out} = \sqrt{\frac{2}{\kappa - 1}\left[\left(\frac{p_{in,c}}{p_{out}}\right)^{(\kappa-1)/\kappa} - 1\right]}. \tag{2}$$

In the numerical simulations, the pressure $p_{out}$ is set to 102.315 kPa and the total temperature $T_{in,c}$ = 300 K and

[a] Corresponding author: simak@vzlu.cz

DOI: 10.1051/epjconf/201714302108

Table 1: Quality of laminar flow for various nozzle length, $M_{out} = 0.5$ (boundary layer properties, Reynolds number based on momentum thickness, uniformity).

| $L/d_{in}$ | $\delta$ [mm] | $\delta^*$ [mm] | $\theta$ [mm] | $H$ | $Re_\theta$ | uniformity [%] |
|---|---|---|---|---|---|---|
| 0.4 | 0.615 | 0.365 | 4.322e-2 | 8.439 | 494 | 3.020 |
| 0.8 | 0.444 | 0.153 | 5.326e-2 | 2.875 | 610 | 0.650 |
| 0.9 | 0.459 | 0.156 | 5.543e-2 | 2.809 | 635 | 0.484 |
| 1.4 | 0.544 | 0.178 | 6.599e-2 | 2.690 | 756 | 0.161 |

Table 2: Pressure ratio, mass and volumetric flow rates.

| $M_{out}$ | 0.2 | 0.5 | 0.9 | 1.1 | 1.5 | 2.0 |
|---|---|---|---|---|---|---|
| $p_{in,c}/p_{out}$ | 1.028 | 1.186 | 1.691 | 2.135 | 3.671 | 7.824 |
| $Q_m$ [kg/s] | 0.641 | 1.644 | 3.127 | 3.990 | 6.865 | 14.644 |
| $Q_V$ [m$^3$/s] | 0.525 | 1.166 | 1.556 | 1.573 | 1.574 | 1.575 |

no heat flow through walls is assumed. For simplicity, the results presented below are obtained for an outflow into a free space and the influence of the restricted space of the chamber is discussed later.

We expect a low inlet velocity up to 8 m/s and the incoming turbulent structures are reduced with the set of a honeycomb and screens. The high CR and the resulting pressure gradient increase the chance of a relaminarization of a turbulent boundary layer inside the convergent nozzle. From this reasons, it is justifiable to solve the flow in the subsonic nozzle part as a laminar problem.

The length of the nozzle was selected according to preliminary results as $L = 450$ mm (ratio $L/d_{in} = 0.9$). This length is a compromise between a growth of a boundary layer (and increased losses) and a decrease of a uniformity of the exit stream. The uniformity is evaluated as a standard deviation of the Mach number outside the boundary layer. The results for $M_{out} = 0.5$ indicate, that the shortest variant causes separation of the flow in the throat. The longer variants do not show this undesirable condition. The other critical region for a possible flow separation is the beginning of the contraction. The flow separates there (and becomes attached later) for $L/d_{in} < 0.8$ at $M_{out} = 0.5$. With lower $M_{out}$, the sufficient length rises. However, the influence of this separation is not so strong due to the favourable pressure gradient.

## 2.2 Numerical method - method of characteristics

The supersonic nozzle is determined by the method of characteristics (MOC), which is a very popular tool for solving problems of supersonic flows. It is based on favourable properties of equations which describe a supersonic flow field. We assume a compressible, inviscid and stationary flow of a perfect gas with a negligible vorticity and heat transfers. That means an isentropic flow. The problem can be considered as axisymmetric and so a formulation in cylindrical coordinates is natural. Further we assume that the domain where the flow is solved (i.e. inside of the supersonic nozzle) has suitable properties, namely is bounded and simply connected, and the state variables $(u, v_r, v_\omega)$, $p$ and $\varrho$ have continuous first derivatives. Hence there exists a velocity potential and the flow can be described by a quasilinear partial differential equation of a second order, which is hyperbolic for supersonic velocities. From this equation, we derive a system of equations for the so called characteristics (see [? ],[? ] chap. 13). A characteristic is a line (or surface in three dimensions), along which the first derivatives of state variables are indeterminate although the state variables itself are continuous. Along a characteristic we can transform the governing partial differential equation into an ordinary differential equation.

Right characteristic $C^+$:

$$\frac{d\varphi}{dx} + F(M)\frac{dM}{dx} - Q^+(M,\varphi)\frac{1}{r} = 0, \tag{3}$$

$$\frac{dr}{dx} = \tan(\varphi - \mu). \tag{4}$$

Left characteristic $C^-$:

$$\frac{d\varphi}{dx} - F(M)\frac{dM}{dx} + Q^-(M,\varphi)\frac{1}{r} = 0, \tag{5}$$

$$\frac{dr}{dx} = \tan(\varphi + \mu), \tag{6}$$

where $M$ is the Mach number, $\varphi$ the direction of the flow, $\mu = \arcsin(1/M)$ the Mach angle, $\gamma$ the adiabatic constant,

$$F(M) = \frac{\sqrt{M^2-1}}{M\left(1+\frac{\gamma-1}{2}M^2\right)}, \tag{7}$$

$$Q^+(M,\varphi) = \frac{\tan\varphi}{\sqrt{M^2-1}+\tan\varphi}, \tag{8}$$

$$Q^-(M,\varphi) = \frac{\tan\varphi}{\sqrt{M^2-1}-\tan\varphi}. \tag{9}$$

That means, the direction of characteristics is the same as the direction of the Mach waves propagation.

Every point of the computational domain can be considered as an intersection of the corresponding left and right characteristic. Using the notation from the figure 2

DOI: 10.1051/epjconf/201714302108

we can evaluate unknown variables at the point 3 from the known values at points 1 a 2:

$$\varphi_3 - \varphi_2 + F(M_{23})(M_3 - M_2) - Q^+(M_{23}, \varphi_{23}) \frac{x_3 - x_2}{r_{23}} = 0, \quad (10)$$

$$\varphi_3 - \varphi_1 - F(M_{13})(M_3 - M_1) + Q^-(M_{13}, \varphi_{13}) \frac{x_3 - x_1}{r_{13}} = 0, \quad (11)$$

$$r_3 - r_2 - (x_3 - x_2)\tan(\varphi_{23} - \mu_{23}) = 0, \quad (12)$$

$$r_3 - r_1 - (x_3 - x_1)\tan(\varphi_{13} + \mu_{13}) = 0, \quad (13)$$

where the subscript $ij$ denotes an average of values at points $i$ and $j$. The characteristics are thus approximated by piecewise linear curves. This system is perfectly valid inside the domain and, using the symmetry, near the axis as well.

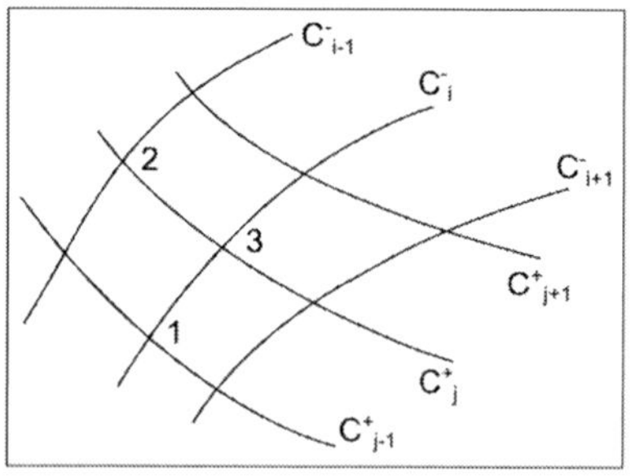


Figure 2: Grid of characteristics inside the computational domain.

The nonlinear system can be rewritten into a vector form

$$S(q) = 0, \; q = (x_3, r_3, M_3, \varphi_3)^T, \quad (14)$$

and solved by a damped Newton method

$$\frac{DS(q_k)}{Dq}(q_{k+1} - q_k) = -\alpha S(q_k), \quad (15)$$

with $0 < \alpha \leq 1$ a damping factor and $q_0$ a suitable initial condition. In this case we chose $q_0 = (x_{12}, r_{12}, M_{12}, \varphi_{12})^T$. If the grid is sufficiently dense, the initial condition is close to the solution and the method rapidly converges. The Jacobian can be expressed analytically (see [? ]).

The situation is slightly different when the point 3 is on a wall (figure 3). Assume some point from the wall, then the flow direction follows from the geometry (i.e. $x_3$, $r_3$ and $\varphi_3$ are known). The only unknown is the Mach number $M_3$. Suppose we know some previous characteristic $C^+_{j-1}$. We construct a characteristic $C^-_p$ going through the point 3. The intersection of $C^+_{j-1}$ and $C^-_p$, denoted as 1, is a linear combination of points $\bar{1}$ and $\bar{2}$ with a parameter $t \in \mathbb{R}$.

The equations (11) and (13) can be therefore rewritten as

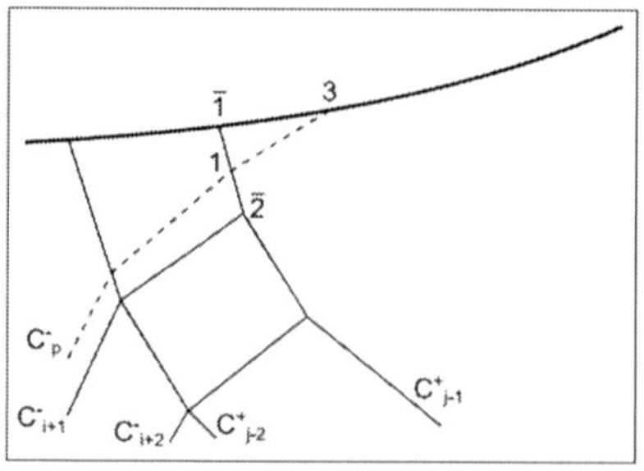


Figure 3: Characteristic on a wall.

$$\varphi_3 - \varphi_1(t) - F(M_{13}(t))(M_3 - M_1(t)) + Q^-(M_{13}(t), \varphi_{13}(t)) \frac{x_3 - x_1(t)}{r_{13}(t)} = 0, \quad (16)$$

$$r_3 - r_1(t) - (x_3 - x_1(t))\tan(\varphi_{13}(t) + \mu_{13}(t)) = 0 \quad (17)$$

for two unknowns $M_3$ and $t$.

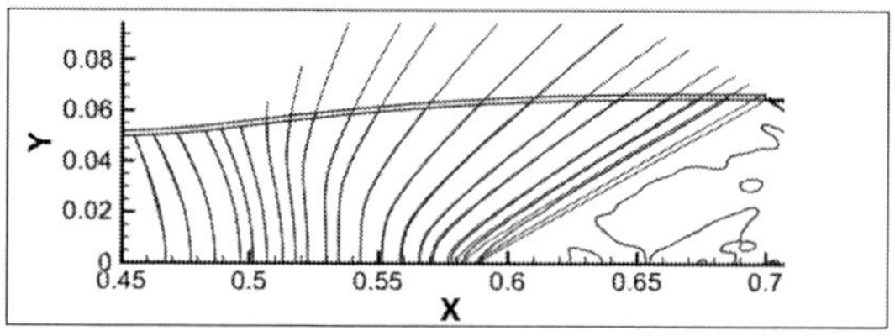


Figure 4: Comparison of Mach number distribution obtained by MOC (red) and inviscid CFD (black). Nozzle geometry for $M_{out} = 2$.

The comparison of flow fields obtained by MOC and CFD shows good agreement (figure 4).

### 2.3 Getting the supersonic nozzle shape

The supersonic exit Mach number is determined by a ratio of the throat and exit area $A_t/A_{out}$,

$$\frac{A_t}{A_{out}} = M_{out}\left[\frac{2}{\gamma+1}\left(1 + \frac{\gamma-1}{2}M_{out}^2\right)\right]^{-\frac{\gamma+1}{2(\gamma-1)}}. \quad (18)$$

With the knowledge of the throat radius $r_t = 50$ mm and the target $M_{out}$ it is possible to determine the exit radius $r_{out,id}$ for an isentropic flow. However, boundary layers have an impact on the optimal nozzle shape in a real flow. To preserve the exit area $A_{out}$, the real exit radius $r_{out,vis} > r_{out,id}$ must be found. This correction will be discussed later.

Four main regions can be identified inside the nozzle (figure 5). It is a subsonic region before the throat, a region of an initial expansion, the straightening of the flow and finally a region of homogeneous flow. Some authors divide the nozzle into more parts, which may ensure more flexibility of the design (e.g. [? ? ]) but we consider the mentioned partitioning sufficient. The origin of the coordinate system is located in the centre of the geometrical throat.

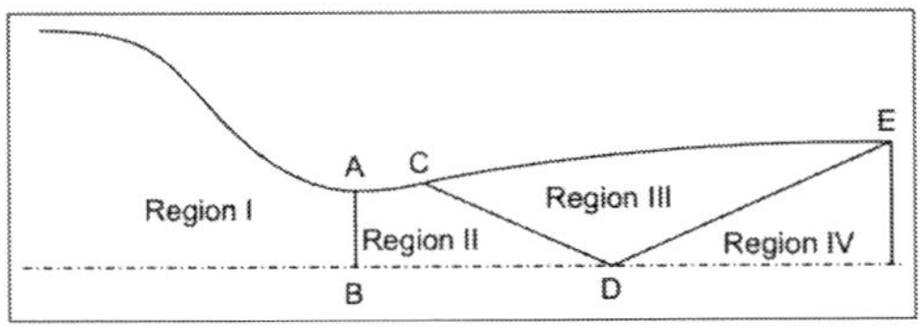


Figure 5: Principal regions of a supersonic nozzle.

The region I is responsible for an acceleration of the flow to a sonic speed and its shape is not so important for the quality of the exit flow unless it causes a flow separation near walls. The convergent geometry (1) used as the subsonic nozzle seems to be well designed to be also used as a part of the supersonic nozzle.

**Region II**

The region II is an area between the sonic line AB and a right characteristic through the point D, which is a point on the axis where the demanded $M_{out}$ is reached. The position of the point D is unknown beforehand and must be obtained from the calculations.

We assume for simplicity that the sonic line is straight, i.e. the line segment AB, and the flow on that line is parallel to the axis. This can be achieved by a nozzle with zero curvature of the throat. Otherwise, the sonic line is curved and the flow structure down the throat is more complex. However, if the curvature, compared to the throat radius, is not high as in our case, it is possible to make this assumption without an excessive loss of precision.

We need to set appropriate boundary conditions to be able to start with the creation of the characteristics grid. Usually, these include values on the sonic line, which follows from our assumption, and values either on the nozzle wall or on the axis. The presented method uses the first option, a prescribed shape of the initial expansion. The choice of the geometry is not determined uniquely and it influences the resulting shape of the region III contour.

The gridpoints are evaluated step by step starting from the point 1 (see Figure 6). We know values on the sonic line and the coordinates of points on the nozzle wall together with the tangent. We will construct the right characteristics $C_j^+$, $j = 1, \dots, n$. The process always begins on the wall, using the equations (16) and (17) to determine the value of the Mach number. After that, intersections with the left characteristics $C_i^-$, $i = 1, \dots, j$ are found by solving the system (14). In this way, we get a characteristics grid in the area of the region II. During the grid construction it may happen that some of the left characteristics intersect the nozzle wall. This indicates a reflection of expansion waves, now weaker, back into the nozzle.

The starting points of right characteristics are chosen such that the grid sufficiently covers the given area, with

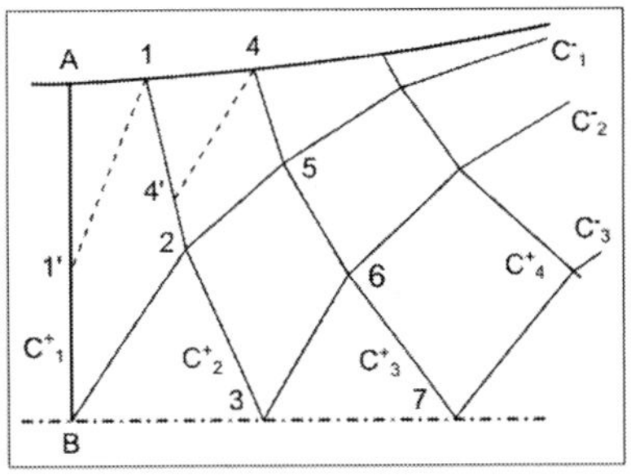


Figure 6: Process of grid construction in region II.

an emphasis on a higher grid density near the throat. This is achieved by a demand to a constant increment $\Delta\mu$ of the Mach angle between two wall points. The distribution of the Mach number along the wall is unknown, but it can be approximated by the assumption $\nu \approx \varphi$, where $\nu$ is the Prandtl-Meyer expansion angle defined as

$$\nu(M) = \sqrt{\frac{\gamma+1}{\gamma-1}} \arctan \sqrt{\frac{\gamma-1}{\gamma+1}(M^2-1)} - \arctan \sqrt{M^2-1}. \tag{19}$$

The shape of the nozzle in the region II is described by a function

$$r(x) = r_t + \frac{x^2}{x_{\bar{C}}}\left(1 - \frac{x}{3x_{\bar{C}}}\right)\tan\varphi_{\bar{C}}, \quad x \in \langle 0, x_C \rangle. \tag{20}$$

The length $x_{\bar{C}}$ and the angle $\varphi_{\bar{C}}$ must be large enough to ensure that $M_{out}$ can be reached. It holds true, that

$$r(0) = r_t, \quad r'(0) = 0, \quad r'(x_{\bar{C}}) = \varphi_{\bar{C}} \quad \text{a} \quad r''(x_{\bar{C}}) = 0 \tag{21}$$

and also $x_C \le x_{\bar{C}}$, $\varphi_C \le \varphi_{\bar{C}}$ which is in no contradiction to the fact, that the point C is not necessarily an inflection point in axisymmetric flow.

If we choose the distance $x_{\bar{C}} = x_C = 0$ we get a so-called minimum length nozzle. The initial expansion is then reduced directly into the throat and the flow is sharply bent by an angle of $\varphi_C$. As follows from the theory for a 2D flow this angle for a planar nozzle is uniquely defined by $\varphi_C = \nu(M_{out})/2$. Unfortunately, for an axially symmetric flow no such analytical expression exists. The value $\varphi_C$ needs to be sought iteratively to achieve the equality $M_D = M_{out}$. According to [? ] it holds $\varphi_C < \nu(M_{out})/4$ for $x_C = 0$. Another difference to a planar flow is that the point C is not an inflection point in an axisymmetric flow.

The problems are similar in the case $x_C > 0$. The dividing characteristic CD is determined iteratively such that $M_D = M_{out}$.

The length $x_C$ and the resulting overall length of the nozzle must be chosen with regard to practical requirements. In practice, the length of the nozzle is often limited by the available space or weight limits. On the other hand, the minimum length nozzle is not completely perfect in terms of the flow quality. The throat curvature is high and areas of compression can occur downstream of the curved sonic line. The sharp deflection of the flow in the throat leads to a production of a Prandtl-Meyer expansion fan between the first and last characteristic (i.e. Mach line). The waves are more condensed and their elimination is more geometry sensitive. In a real flow a boundary layer forms on the wall which also influences the direction of Mach lines. Any deviation from the evaluated grid causes imperfect elimination of expansion and compression waves and the disruption of the free stream. In contrast, longer nozzles, where initial expansion is more gradual, damp expansion waves by reflections from the walls and thus the sensitivity to the exact shape is reduced.

**Region III**

The region III is an area restricted by the right-running characteristic CD and the left-running characteristic DE. This part is responsible for a gradual straightening of the flow into the axial direction and formation of the homogeneous flow. The values on the left boundary are obtained as a result from the region II. The right boundary is next to the region IV where we expect a uniform, axial flow with the prescribed $M_{out}$. Therefore $\varphi = 0$ and $M = M_{out}$ on DE and the slope is $\tan\mu_{out}$. The grid points are constructed from the point D, as shown in figure 7. The grid should be constructed beyond the point E, which position is determined by the equation (18). The resulting grid can have, for example, a pattern shown in Figure 8 (the reflected characteristics in the region II are removed from the picture).

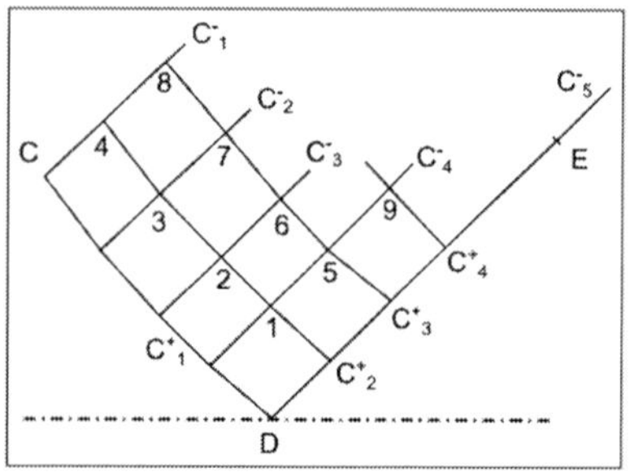


Figure 7: Creation of grid in the region III.

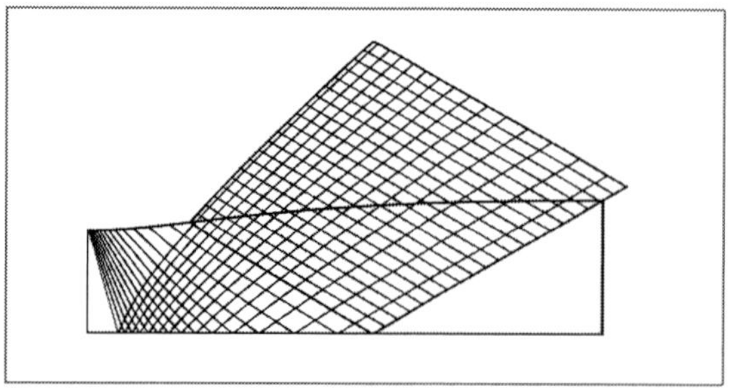

Figure 8: Complete characteristic grid in the nozzle.

After the characteristic grid is constructed and the flow values are evaluated in all nodes, the contour of the nozzle remains to be determined. The contour forming the nozzle is a streamline going through the point C. The streamline is approximated by a piecewise linear curve that intersects the left and right characteristics. The direction of each line segment is determined by the value of the angle $\varphi$ in the intersection with a characteristic. The values in intersections are approximated by a linear combination of values in nodal points.

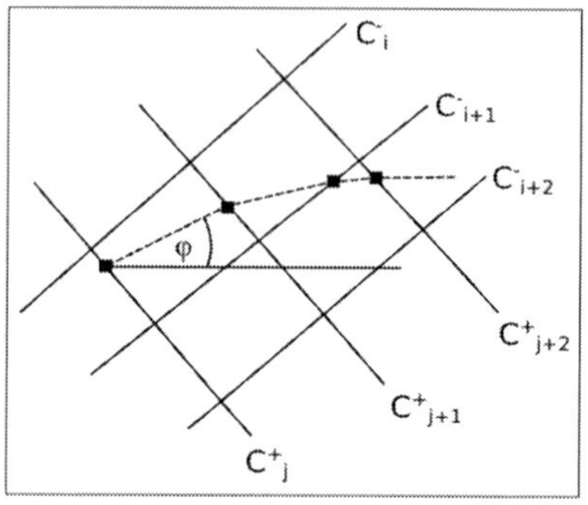


Figure 9: Construction of the streamline in region III.

The resulting contour is a piecewise linear curve which can contain some unwanted roughness, that unnecessarily disrupt the uniformity of flow. From that reason the curve is smoothed using a cubic B-spline.

## 2.4 Viscous correction

The above presented method for design of a nozzle assumes an inviscid flow. However, as was already mentioned, the real situation must deal with boundary layers which influence the effective cross-section. From this reason, the exit Mach number would be lower for such a nozzle. To overcome this problem, a correction is introduced. The correction is based on an estimate of the displacement thickness $\delta^*$ which broadens the nozzle geometry. This estimate is obtained with a modified Head method for a turbulent, compressible, axisymmetric flow [? ].

The basis is the von Karman momentum equation in a form for compressible axisymmetric flow,

$$\frac{d\theta}{dx} = \frac{C_f}{2} - (H + 2 - M_e^2)\frac{\theta}{u_e}\frac{du_e}{dx} - \frac{\theta}{r}\frac{dr}{dx}, \quad (22)$$

with an equation for the shape parameter $H_1 = \Delta/\theta$,

$$\theta\frac{dH_1}{dx} = C_E - H_1\left(\frac{C_f}{2} - (H + 1)\frac{\theta}{u_e}\frac{du_e}{dx}\right). \quad (23)$$

These equations relate the momentum thickness $\theta$, mass flow thickness $\Delta = \int_0^\delta \frac{\varrho u}{\varrho_e u_e} dy$, shape parameter $H$, further the coefficients of skin friction and entrainment, $C_f$ and $C_E$, the Mach number $M_e$ and tangential velocity $u_e$ on the boundary layer edge.

The unknown required variables $C_f$, $C_E$ and $H$ are evaluated from empirical relations derived from results on a flat plate with a zero pressure gradient and corrected to include the compressibility effects.

## 3 Numerical verification

**Subsonic flow**

The stream leaving the nozzle undergoes a homogenization over a short distance from the nozzle exit ($\sim r_{out}$) and creates a uniform core. This core goes thinner with the increasing distance, however, at the distance 450 mm from the nozzle the core thickness is 50 mm (for $M_{out} = 0.5$).

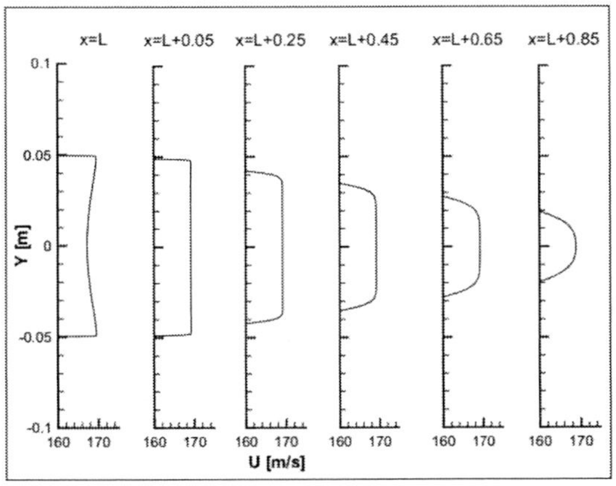


Figure 10: Velocity profiles across the stream behind the nozzle ($M_{out} = 0.5$).

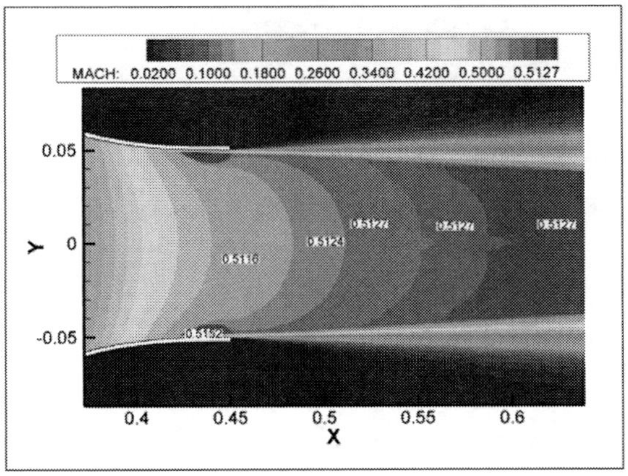


Figure 11: Mach number in the jet, higher pressure ratio due to the restricted space ($M_{out} = 0.5$).

**Supersonic flow**

The results confirm the expected problems for low supersonic speeds. The difference between $r_t$ and $r_{out}$ is very small for low supersonic Mach numbers, namely 0.2 mm for $M_{out} = 1.1$ and 1.6 mm for $M_{out} = 1.3$. The boundary layer thickness is 1.5 mm and the displacement thickness $\delta^* = 0.3$ mm for $M_{out} = 1.3$. As a result, the Mach number is heavily dependent on the boundary layer and every imperfection of nozzle walls.

An area of a uniform flow exists behind the nozzle exit, restricted by an expansion or shock wave, which can be identified in figures 13 – 15. The direction of this wave is given by the Mach angle $\mu = \arcsin(1/M_{out})$. In the case $p_{out} < p_b$ an oblique shock wave arises from the nozzle outlet and the flow is called over-expanded. If $p_{out} > p_b$ an expansion wave arises and the flow is called under-expanded. In the case of equality of $p_{out}$ and $p_b$, the design conditions are meet and a uniform flow is behind the nozzle. The figures 13 – 15 show the Mach number distribution behind the nozzle in the case of an outflow into a free space.

### 3.1 Influence of the testing chamber on the nozzle outflow

The above-mentioned results were obtained as a simulation of a flow into an open space. However, the real concept includes a restricted space of the test chamber. Depending on the jet velocity, a venturi pump effect occurs which decreases the static and total pressure $p_b$ and $p_{b,c}$ inside the chamber. In the case of a subsonic jet, the nozzle exit velocity is higher then expected according to (2). Otherwise for supersonic regimes, the flow is under-expanded and the homogeneity of the jet is corrupted. The values of the static pressure $p_b$ evaluated on the outer side of the nozzle exit wall are shown in the figure 12. In reality it would be appropriate to operate with the real and actual value of $p_b$ and include a correction in the evaluation of the pressure ratio $p_{in,c}/p_b$. The effect of the chamber is illustrated in the figure 16.

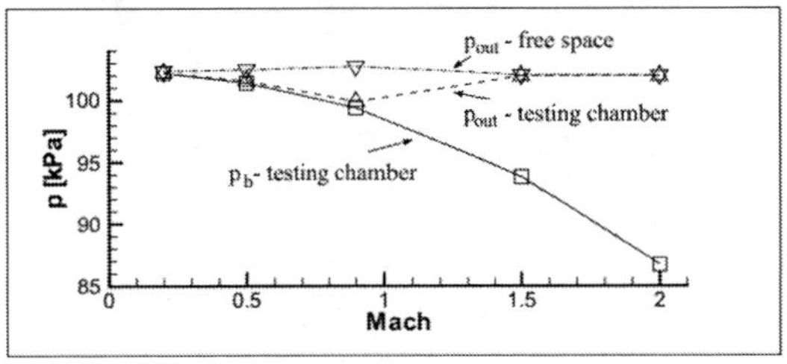


Figure 12: Effect of the restricted area behind the nozzle on the static pressure.

## 4 Conclusion

This paper presents a design of a probe calibration devise, namely the most crucial part, which is the nozzle and the testing chamber. The CFD evaluations show a suitability of the proposed solution, the resulting jet is homogeneous enough and of required parameters. The only problems

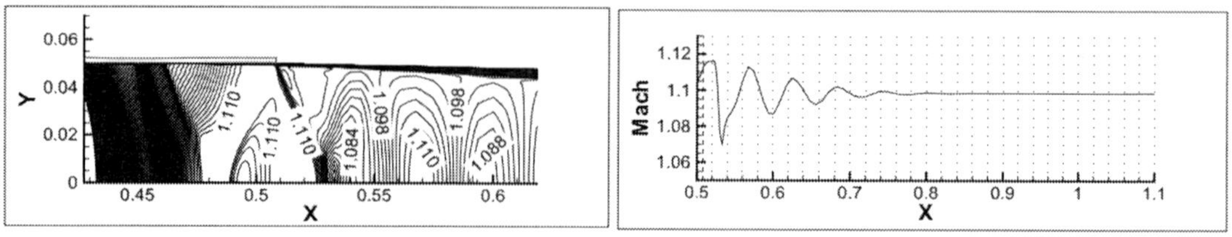


Figure 13: Mach number distribution behind a nozzle and on the axis at Mach 1.1.

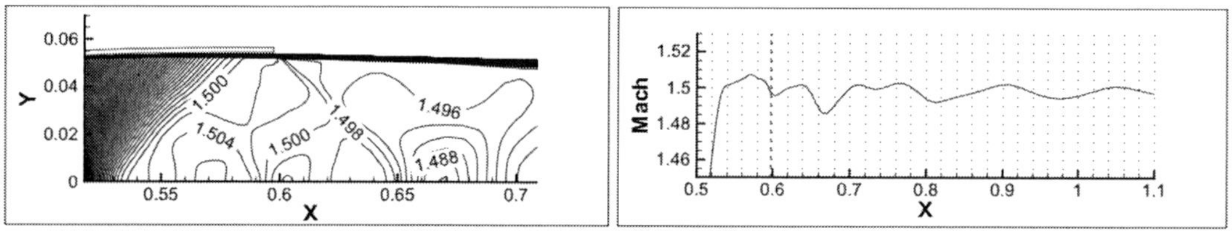


Figure 14: Mach number distribution behind a nozzle and on the axis at Mach 1.5.

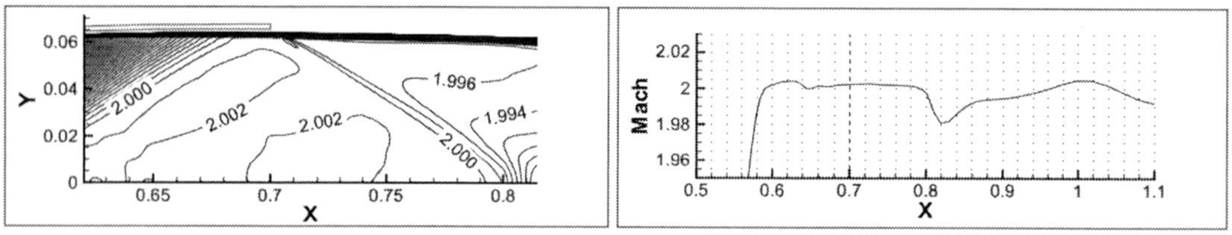


Figure 15: Mach number distribution behind a nozzle and on the axis at Mach 2.0.

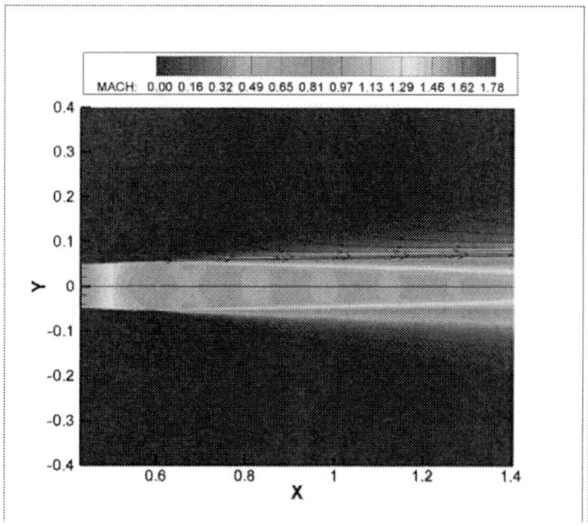


(a) free space

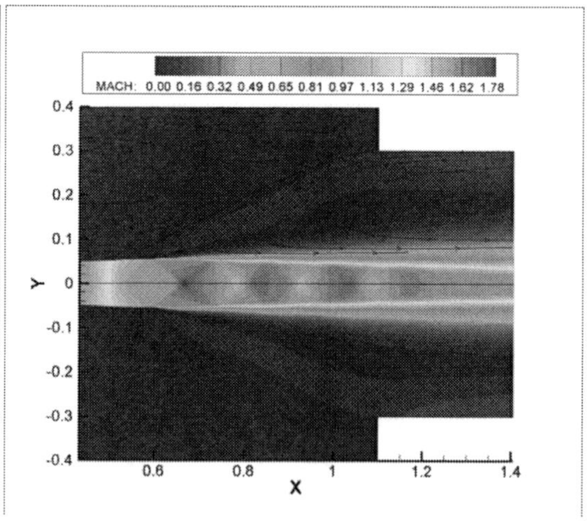


(b) test chamber

Figure 16: Effect of the test chamber on the jet ($M_{out}$ = 1.5).

DOI: 10.1051/epjconf/201714302108

could arise for low supersonic speeds when the sensitivity to the nozzle geometry can influence the outflow conditions. The jet is wide enough to accommodate various probes.

The obtained results will serve as a basis for manufacturing of the device. In the first stage, it will deal with subsonic regimes. After that, the experimental data will be compared to the numerical results.

**Acknowledgement**
Presented results were obtained with the support of the Ministry of Industry and Trade of the Czech Republic, aimed at a long-term development of a research organization.

EPJ Web of Conferences **143**, 02109 (2017) DOI: 10.1051/epjconf/201714302109
*EFM 2016*

# Numerical calculations of 2D transonic flow in GAMM channel and over the profile

*Martin* Slouka[1,*], *Karel* Kozel[1,2]

[1]Dept. of Technical Mathematics, Faculty of Mechanical Eng., CTU in Prague, 121 35 Prague 1, Czech Rep.
[2]Institute of Thermomechanics, Academy of Science CR, 182 00 Prague 8, Czech Rep.

**Abstract.** The aim of this work is to make a 2D numerical model of the solution of the transonic inviscid and viscous compressible flow around the profile. In a case of viscous flow several turbulent models are used. For the verification of the calculation Baldwin-Lomax model is compared with Wilcox k-omega model and SST turbulent model. Calculations are done in GAMM channel computational domain with 10% DCA profile and in the turbine cascade computational domain with 8% DCA profile. Numerical methods are based on a finite volume solution. Comparisons are done with the experimental data for the 8% DCA profile.

## 1 Introduction

Mathematical models used in this work are based on the solution of the two-dimensional transonic inviscid and viscous compressible flow. In a case of inviscid flow system of Euler equations is equipped. In a case of viscous flow, the simulation is based on the solution of the system of RANS equations. The system is completed by the algebraic Baldwin-Lomax model of turbulence, Wilcox k-ω model and the SST k-ω model.

Computational domain represents GAMM channel with 10% DCA profile and blade cascade with 8% DCA profile. The results show differences between the uses of each numerical model and experimental results.

## 2 Governing equations

The two-dimensional viscous turbulent compressible flow is prescribed by the system of RANS in the following conservation form

$$\mathbf{W}_t + \mathbf{F}_x + \mathbf{G}_y = \mathbf{R}_x + \mathbf{S}_x, \tag{1}$$

where

$$\mathbf{W} = (\rho, \rho u, \rho v, e)^T, \tag{2}$$

$$\mathbf{F} = \left(\rho u, \rho u^2 + p, \rho uv, (e+p)u\right)^T, \tag{3}$$

$$\mathbf{G} = \left(\rho v, \rho uv, \rho v^2 + p, (e+p)v\right)^T, \tag{4}$$

$$\mathbf{R} = \left(0, \tau_{xx}, \tau_{xy}, u\tau_{xx} + v\tau_{xy} - q_x\right)^T, \tag{5}$$

$$\mathbf{S} = \left(0, \tau_{xy}, \tau_{yy}, u\tau_{xy} + v\tau_{yy} - q_y\right)^T, \tag{6}$$

and

$$\tau_{ij} = \mu\left(\frac{\partial u_i}{\partial x_j} + \frac{\partial u_j}{\partial x_i}\right) - \frac{1}{3}\delta_{ij}\frac{\partial u_k}{\partial x_k}, \tag{7}$$

$$q_i = -\frac{\kappa}{\kappa - 1}\frac{\mu}{Pr}\frac{\partial}{\partial x_i}\left(\frac{p}{\rho}\right). \tag{8}$$

This system is closed by following equation

$$p = (\kappa - 1)\left[e - \frac{1}{2}\rho\left(u^2 + v^2\right)\right]. \tag{9}$$

In the equations above ρ denotes density, $(u,v)$ are components of local velocity in $x$ and $y$ direction respectively, $e$ denotes total energy per unit volume, $p$ is pressure given by previous equation of state, $\tau_{ij}$ is shear stress, $q_i$ represents heat flux and $\mu$ represents dynamical viscosity.

Following constants were used in previous equations: $\kappa$ is isentropic exponent equaled to 1.4, $Pr$ is Prandtl number equaled to the constant value 0.7. Inviscid flow represented by the system of Euler equations is reached after neglecting of viscosity in the viscous fluxes on the right hand side in the system (1).

## 3 Turbulence modeling

Viscous compressible turbulent flow represents the system of Reynolds averaged Navier-Stokes equations (RANS) which is formally the same as (1) with the model of turbulence enclosed for the new variable of the turbulent viscosity. The equations from (7) and (8) are now modified as

* Corresponding author: m.slouka@seznam.cz

$$\tau_{ij} = (\mu + \mu_t)\left(\frac{\partial u_i}{\partial x_j} + \frac{\partial u_j}{\partial x_i}\right) - \frac{1}{3}\delta_{ij}\frac{\partial u_k}{\partial x_k}, \quad (10)$$

$$q_i = -\frac{\kappa}{\kappa - 1}\left(\frac{\mu}{Pr} + \frac{\mu_t}{Pr_t}\right)\frac{\partial}{\partial x_i}\left(\frac{p}{\rho}\right), \quad (11)$$

where $\mu_t$ represents turbulent viscosity and $Pr_t$ is turbulent Prandtl number which is the constant value 0.9.

### 3.1 Baldwin-Lomax model

Baldwin-Lomax model [1] is the modification of basic Cebeci and Smith [2] algebraic turbulent model. The boundary layer is divided into two regions, inner region (close to the wall) and outer region. Turbulent viscosity in the inner part is given by

$$\mu_{ti} = \rho F_d^2 \kappa^2 y^2 |\Omega| \quad (12)$$

where vorticity is in 2D case defined as

$$\Omega = \frac{\partial u}{\partial y} - \frac{\partial v}{\partial x}, \quad (13)$$

van Driest function given as follow

$$F_d = 1 - \exp\left(-\frac{1}{A}\frac{u_\tau y}{v_w}\right) \quad (14)$$

and friction velocity by formula

$$u_\tau = \sqrt{\frac{\tau_w}{\rho}} = \left(v\frac{\partial u}{\partial y}\right)_w^{1/2}. \quad (15)$$

Turbulent viscosity in outer region is given by

$$\mu_{to} = \alpha C_{cp} F_w F_k. \quad (16)$$

Function $F_w$ is determined by the relation

$$F_w = y_{max} F_{max} \quad (17)$$

for the solid wall and by the relation

$$F_w = C_{wk} y_{max} \frac{\Delta U^2}{F_{max}} \quad (18)$$

for the wake after the profile. $F_{max}$ is the maximum of the function

$$F = y|\Omega| F_d \quad (19)$$

and $y_{max}$ is the distance from the wall in which $F(y_{max}) = F_{max}$ holds and

$$F_k = \left[1 + 5.5\left(C_{KL}\frac{y}{y_{max}}\right)^6\right]^{-1}. \quad (20)$$

In Baldwin-Lomax turbulent model following constants are used:

$\kappa = 0.4$, $A = 26$, $\alpha = 0.0168$,
$C_{cp} = 1.6$, $C_{KL} = 0.3$ and $C_{wk} = 0.3$.

### 3.2 Wilcox k-ω model (2006)

The two-equation model is given by transport equations for two characteristic scales of turbulent motion, in this case turbulent energy $k$ and specific dissipation rate $\omega$. Generally, the turbulent viscosity is defined as follow

$$\mu_t = \rho\frac{k}{\omega}. \quad (21)$$

The Wilcox k-ω model [3], [4] is given by the equations

$$\frac{\partial(\rho k)}{\partial t} + \frac{\partial(\rho u_j k)}{\partial x_j} = P_k + \frac{\partial}{\partial x_j}\left[\left(\mu + \sigma^* \mu_t\right)\frac{\partial k}{\partial x_j}\right] - \beta^* \rho k\omega, \quad (22)$$

$$\frac{\partial(\rho\omega)}{\partial t} + \frac{\partial(\rho u_j \omega)}{\partial x_j} = \gamma\frac{\omega}{k}P_k + \frac{\partial}{\partial x_j}\left[(\mu + \sigma\mu_t)\frac{\partial\omega}{\partial x_j}\right] - \beta\rho k\omega^2 + C_D, \quad (23)$$

where

$$C_D = \sigma_d \frac{\rho}{\omega} max\left(\frac{\partial k}{\partial x_j}\frac{\partial\omega}{\partial x_j}, 0\right), \quad (24)$$

$$P_k = \tau_{ij}\frac{\partial u_i}{\partial x_j}, \quad (25)$$

$$\tau_{ij} = \mu_t\left(2S_{ij} - \frac{2}{3}\frac{\partial u_k}{\partial x_k}\delta_{ij}\right) - \frac{2}{3}\rho k\delta_{ij}, \quad (26)$$

$$S_{ij} = \frac{1}{2}\left(\frac{\partial u_i}{\partial x_j} + \frac{\partial u_j}{\partial x_i}\right) \quad (27)$$

and turbulent viscosity as given by

$$\mu_t = \rho\frac{k}{\hat{\omega}}, \quad (28)$$

where

$$\hat{\omega} = \max\left(\omega;\ C_{lim}\sqrt{\frac{2\hat{S}_{ij}\hat{S}_{ij}}{\beta^*}}\right), \quad (29)$$

$$\hat{S}_{ij} = S_{ij} - \frac{1}{3}\frac{\partial u_k}{\partial x_k}\delta_{ij}. \quad (30)$$

The constants are:

$\gamma = 13/25$, $\beta_0 = 0.0708$, $\beta^* = 9/100$,
$\sigma^* = 3/5$, $\sigma = 1/2$ and $C_{lim} = 7/8$.

The auxiliary functions are:

$$\beta = \beta_0 f_\beta, \quad (31)$$

$$f_\beta = \frac{1 + 85\chi_\omega}{1 + 100\chi_\omega}, \quad (32)$$

$$\chi_\omega = \left|\frac{\Omega_{ij}\Omega_{jk}\hat{S}_{ki}}{(\beta^*\omega)^3}\right|, \quad (33)$$

$$\hat{S}_{ki} = S_{ki} - \frac{1}{2}\frac{\partial u_m}{\partial x_m}\delta_{ki}, \quad (34)$$

$$\Omega_{ij} = \frac{1}{2}\left(\frac{\partial u_i}{\partial x_j} - \frac{\partial u_j}{\partial x_i}\right). \quad (35)$$

### 3.3 SST k-ω model

SST model [5] is combination of k-ω and k-ε model in the following modification

$$\frac{\partial(\rho k)}{\partial t}+\frac{\partial(\rho u_j k)}{\partial x_j}= P_k+\frac{\partial}{\partial x_j}\left[\left(\mu+\sigma^*\mu_t\right)\frac{\partial k}{\partial x_j}\right]-\beta^*\rho k\omega, \quad (36)$$

$$\frac{\partial(\rho\omega)}{\partial t}+\frac{\partial(\rho u_j\omega)}{\partial x_j}= \gamma\frac{\omega}{k}P_k+\frac{\partial}{\partial x_j}\left[(\mu+\sigma\mu_t)\frac{\partial\omega}{\partial x_j}\right]-\beta\rho k\omega^2+ 2\rho(1-F_1)\sigma_2\frac{1}{\omega}\frac{\partial k}{\partial x_j}\frac{\partial\omega}{\partial x_j}. \quad (37)$$

Turbulent viscosity is then given by

$$\mu_t=\min\left(\rho\frac{k}{\omega},\rho\frac{a_1 k}{F_2\Omega}\right), \quad (37)$$

where $\Omega_{ij}$ is given by (35) and functions $F_1$ and $F_2$ are defined as follow

$$F_1=\tanh\left(\Gamma_1^4\right), \quad (38)$$

$$\Gamma_1=\min\left[\max\left(\frac{\sqrt{k}}{\beta^*\omega y};\frac{500\nu}{\omega y^2}\right);\frac{4\rho\sigma_2 k}{D_\omega y^2}\right], \quad (39)$$

$$D_\omega=\max\left(2\rho\sigma_2\frac{1}{\omega}\frac{\partial k}{\partial x_j}\frac{\partial\omega}{\partial x_j};10^{-20}\right), \quad (40)$$

$$F_2=tanh\left(\Gamma_2^2\right), \quad (41)$$

$$\Gamma_2=\max\left(2\frac{\sqrt{k}}{0{,}09\omega y};\frac{500\nu}{\omega y^2}\right). \quad (42)$$

Model constant are calculated according the following general formula

$$\Pi=F_1\Pi_1+(1-F_1)\Pi_2. \quad (43)$$

Constants of k-ω model are:

$\beta_1=0.075$, $\sigma_1=0.5$ and

$$\gamma_1=\beta_1/\beta^*-\sigma_1\kappa^2/\sqrt{\beta^*}. \quad (44)$$

Constant of k-ε model are:

$\beta_2=0.0828$, $\sigma_2=0.85$ and

$$\gamma_2=\beta_2/\beta^*-\sigma_2\kappa^2/\sqrt{\beta^*}. \quad (45)$$

General constants for both are:

$\beta^*=0.09$, $\sigma^*=0.85$, $\kappa=0.41$ and $a_1=0.31$.

## 4 Numerical method

### 4.1 Numerical scheme

A Finite volume method was applied on the cell centered non-orthogonal structured grid with quadrilateral cells.

For the numerical solutions Lax-Wendroff scheme in McCormack modification [6] was used with predictor step

$$\mathbf{W}_{i,j}^{n+1/2}=\mathbf{W}_{i,j}^{n}-\Delta t\mathrm{Res}\mathbf{W}_{i,j}^{n} \quad (46)$$

and corrector step as follow

$$\mathbf{W}_{i,j}^{n+1}=\frac{1}{2}\left(\mathbf{W}_{i,j}^{n}+\mathbf{W}_{i,j}^{n+1/2}-\Delta t\mathrm{Res}\mathbf{W}_{i,j}^{n+1/2}\right). \quad (47)$$

The Jameson's artificial dissipation model was applied to damp the oscillations.

$$\tilde{\mathbf{W}}_{i,j}^{n+1}=\mathbf{W}_{i,j}^{n+1}+AD\left(\mathbf{W}_{i,j}^{n}\right), \quad (48)$$

$$AD\left(\mathbf{W}_{i,j}^{n}\right)=C_1\psi_1\left(\mathbf{W}_{i+1,j}^{n}-2\mathbf{W}_{i,j}^{n}+\mathbf{W}_{i-1,j}^{n}\right)+ C_2\psi_2\left(\mathbf{W}_{i,j-1}^{n}-2\mathbf{W}_{i,j}^{n}+\mathbf{W}_{i,j-1}^{n}\right) \quad (49)$$

where the pressure correction functions are

$$\psi_1=\frac{\left|p_{i+1,j}^{n}-2p_{i,j}^{n}+p_{i-1,j}^{n}\right|}{\left|p_{i+1,j}^{n}\right|+\left|p_{i,j}^{n}\right|+\left|p_{i-1,j}^{n}\right|}, \quad (50)$$

$$\psi_2=\frac{\left|p_{i,j+1}^{n}-2p_{i,j}^{n}+p_{i,j-1}^{n}\right|}{\left|p_{i,j+1}^{n}\right|+\left|p_{i,j}^{n}\right|+\left|p_{i,j-1}^{n}\right|}. \quad (51)$$

### 4.2 Boundary conditions

Inlet boundary conditions were realized for as follow: inlet velocity $Ma_\infty$ together with angle of attack $\alpha$, density $\rho_\infty$ and total energy per volume $e_\infty$ were set; pressure $p_\infty$ was extrapolated from the flow field for the inlet Mach number $Ma_\infty<1$, for the inlet Mach number $Ma_\infty\geq 1$ pressure $p_\infty$ was set.

Outlet boundary conditions were done as follow: in a case of Mach number $Ma_\infty<1$ outlet pressure was set and other variables were extrapolated from the flow field; in a case of Mach number $Ma_\infty\geq 1$ all the variables were extrapolated from the flow field.

Solid wall condition was realized by the adding of virtual cells. In a case of inviscid flow velocity components were prescribed so that sum of velocity vectors equals zero in its tangential component. In a case of viscous flow, no-slip condition for the velocity was prescribed so that sum of velocity vectors equals zero on the wall.

Following parameters were chosen for the setting of turbulence models at the wall

$$k_w=0, \quad (52)$$

$$\omega_w=\frac{60\mu}{\rho\beta y^2}, \quad (53)$$

and for the inlet flow

$$k_\infty=\frac{3}{2}(Int_T U_\infty)^2, \quad (54)$$

$$\omega_\infty=C_\mu^{-1/4}\frac{\sqrt{k}}{L}, \quad (55)$$

where the intensity of turbulence was prescribed as $Int_T$=0.05, constant $C_\mu$=0.09 and dimension $L$ is a length of the used geometry.

## 5 Numerical results

The isolines of the Mach number are shown in all pictures. Flow parameters at the inlet are prescribed - $Ma_\infty$ corresponds to the inlet Mach number and $\alpha$ is an angle of the flow attack. Numerical results were chosen from a set of different setting of the inlet Mach number and angle of attack. Also the impact of the artificial dissipation model was checked to damp the oscillations and in a case of viscous model to eliminate the influence on the physical character.

First six figures correspond with the results of the calculations of transonic inviscid and viscous turbulent flow in the GAMM channel. This configuration was set as a reference for the testing of the behavior of both models. The calculation of the inviscid flow shows good result, the maximal value of Mach number $Ma_{max}$ = 1.326 is in comply with the expectation. Mach number along the profile is shown and have a good correlation with the result presented by Fürst [7]. The results of the viscous flow calculations have the good agreement to the each other, k-ω models shows the shock wave slightly higher than the algebraic model, maximal value of Mach number is around $Ma_{max}$ = 1.166 for the k-ω models.

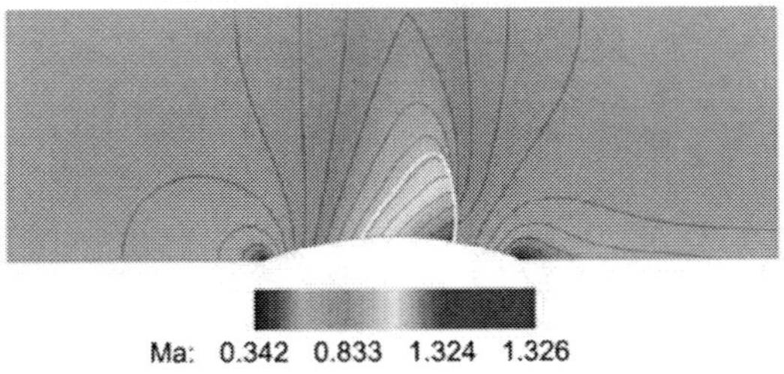


**Fig. 1.** *GAMM channel, inviscid flow, $Ma_\infty$ = 0.675, α = 0.0°*

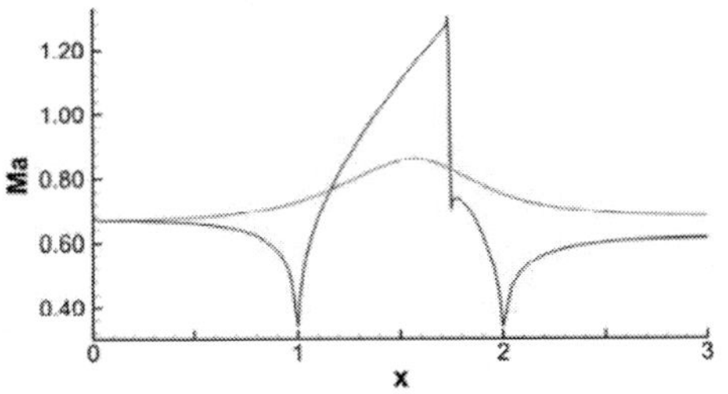


**Fig. 2.** *GAMM channel, inviscid flow, $Ma_\infty$ = 0.675, α = 0.0°, distribution of Mach number along the profile*

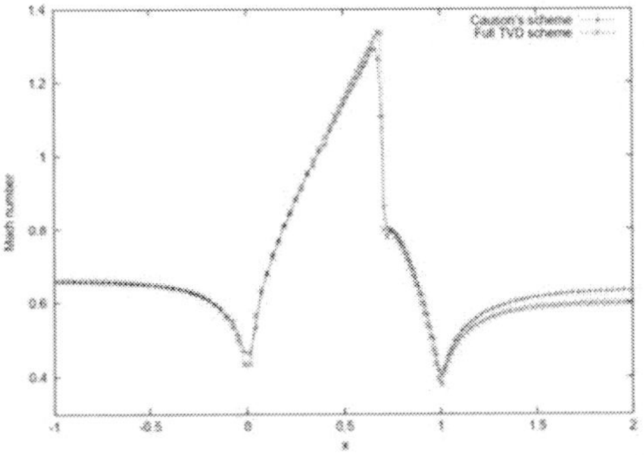


**Fig. 3.** *GAMM channel, inviscid flow, $Ma_\infty$ = 0.675, α = 0.0°, distribution of Mach number along the profile, Fürst [7]*

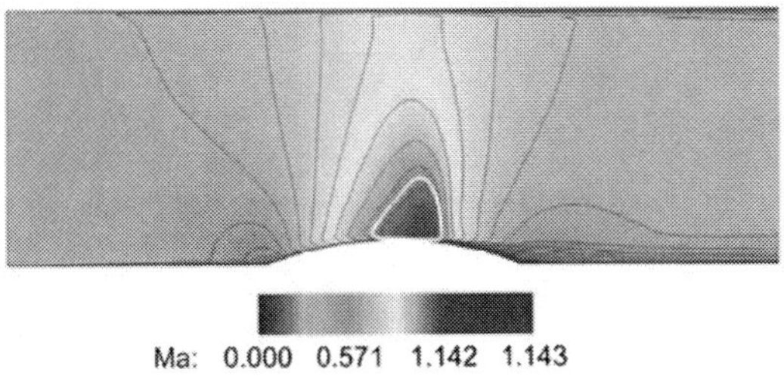


**Fig. 4.** *GAMM channel, viscous turbulent flow, $Ma_\infty$ = 0.675, α = 0.0°, Re = $10^6$, Baldwin-Lomax model*

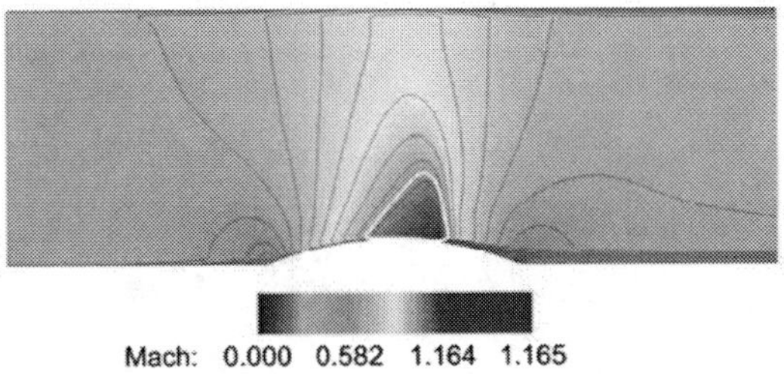


**Fig. 5.** *GAMM channel, viscous turbulent flow, $Ma_\infty$ = 0.675, α = 0.0°, Re = $10^6$, Wilcox k-ω model*

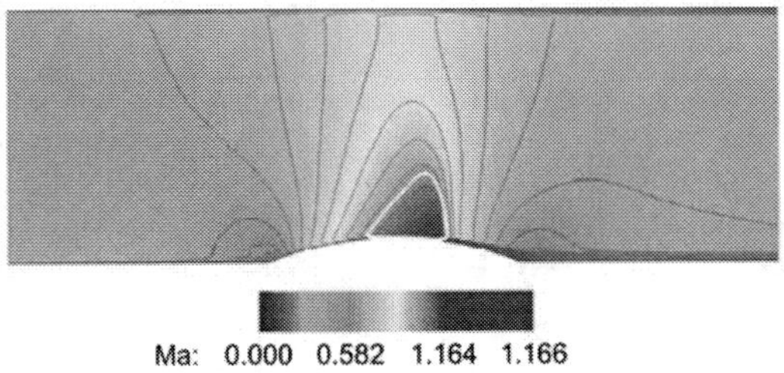


**Fig. 6.** *GAMM channel, viscous turbulent flow, $Ma_\infty$ = 0.675, α = 0.0°, Re = $10^6$, SST k-ω model*

The next set of figures shows the comparison between the used inviscid and viscous turbulent flow models and with experimental data of the measurements of the transonic flow through DCA cascade done by Institute of Thermomechanics AS CR in Prague and published in [8]. In this work there are shown the numerical results for the three different test cases and the comparison with the experimental data.

The inlet Mach number in the calculations was set slightly higher compare to the experiment. By the small change of the stream flow angle in the interval from -3° to 3° the proper shape of the shock wave can be reached. These facts were considered in all calculations.

Figures 7 and 8 are corresponding with the experimental data shown on the figure 9. In a case of both numerical calculations the sonic line is visible on the upper profile and also on the lower profile although the sonic line in the experiment is only on the upper profile.

 DOI: 10.1051/epjconf/201714302109

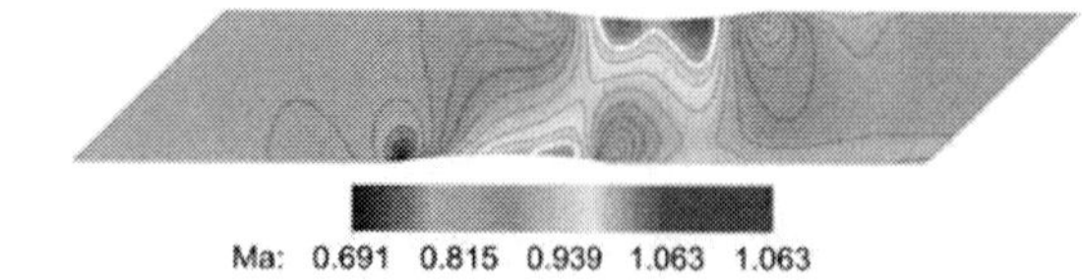


**Fig. 7.** *DCA blade cascade, inviscid flow, $Ma_\infty$ = 0.830, $\alpha$ = 0.5°*

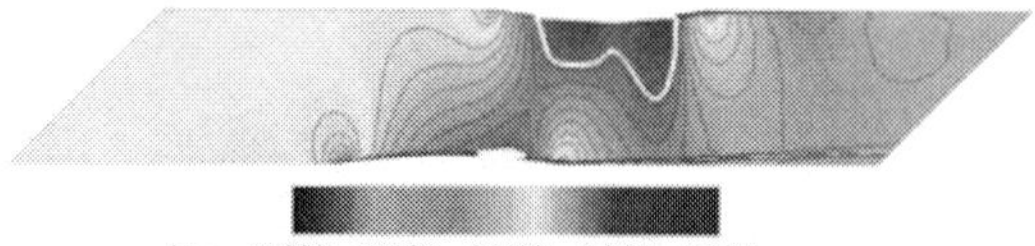


**Fig. 8.** *DCA blade cascade, viscous turbulent flow, $Ma_\infty$ = 0.835, $\alpha$ = 0.5°, Re = $10^6$, SST k-ω model*

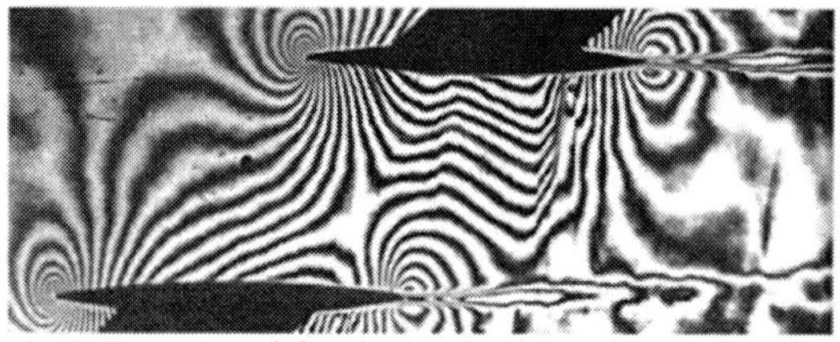

**Fig. 9.** *Experimental data, $Ma_\infty$ = 0.813, $\alpha$ = 0.0°*

Figures 10, 11 and 12 are corresponding with the experimental data shown on the figure 13 with the inlet Mach number $Ma_\infty$ = 0.946. The position of the sonic line is similar for the all the numerical results in the upstream region and shows the good agreement with the upstream part of the experiment.

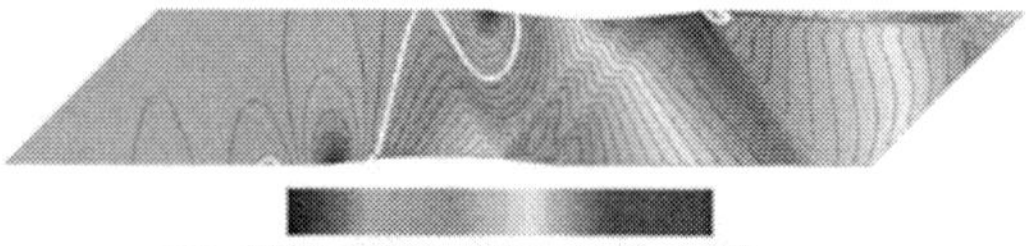


**Fig. 10.** *DCA blade cascade, inviscid flow, $Ma_\infty$ = 0.980, $\alpha$ = 2.0°*

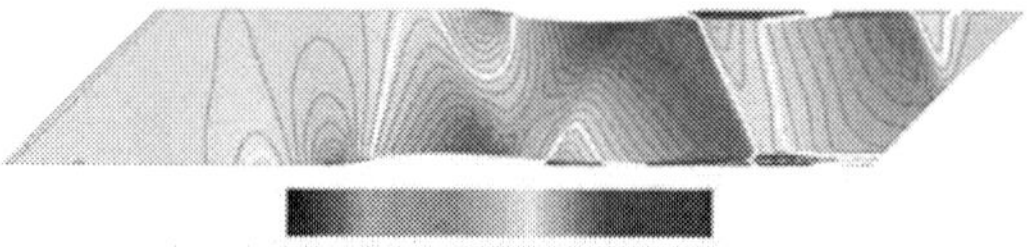


**Fig. 11.** *DCA blade cascade, viscous turbulent flow, $Ma_\infty$ = 1.02, $\alpha$ = 0.0°, Re = $10^6$, Wilcox k-ω model*

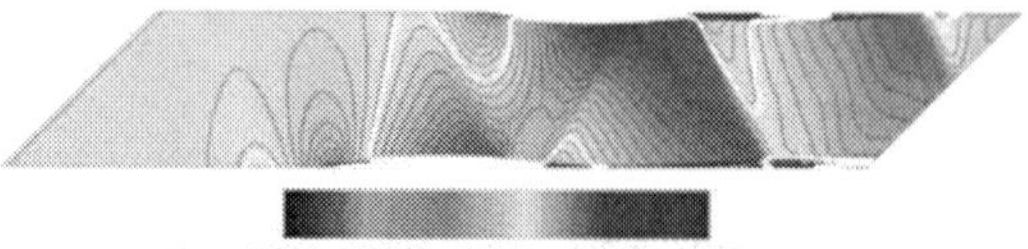


**Fig. 12.** *DCA blade cascade, viscous turbulent flow, $Ma_\infty$ = 1.02, $\alpha$ = 0.0°, Re = $10^6$, SST k-ω model*

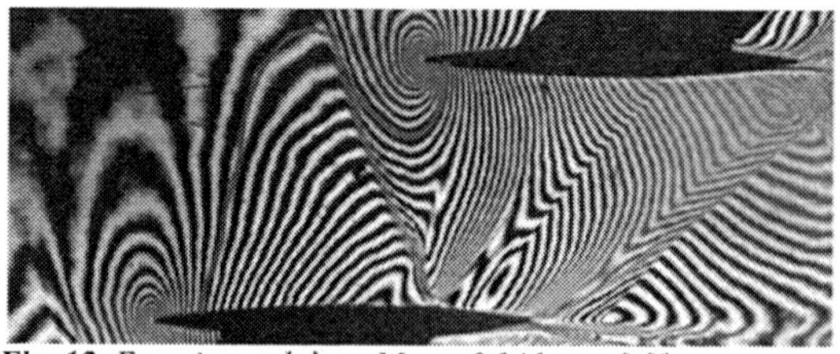

**Fig. 13.** *Experimental data, $Ma_\infty$ = 0.946, $\alpha$ = 0.0°*

The last set of the figures illustrates the comparison between the numerical results of the viscous flow calculations and the experimental data for the inlet Mach number $Ma_\infty$ = 1.073. The sonic line is again presented in upstream and downstream part of the profiles and also the discontinuity in flow is visible. The quite good agreement with the experimental data is clear.

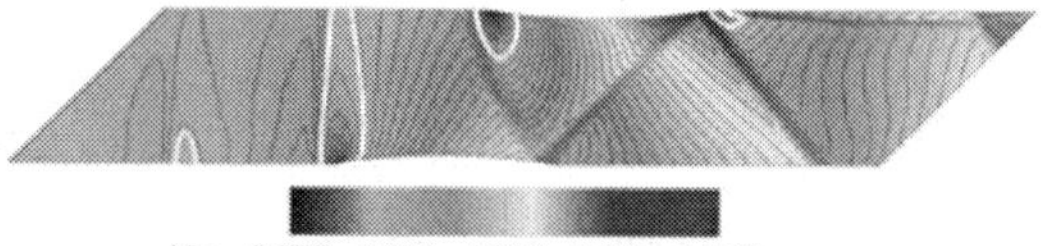


**Fig. 14.** *DCA blade cascade, inviscid flow, $Ma_\infty$ = 1.130, $\alpha$ = 0.0°*

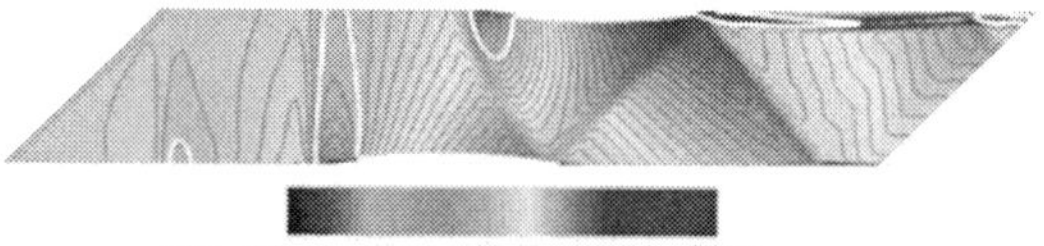


**Fig. 15.** *DCA blade cascade, viscous turbulent flow, $Ma_\infty$ = 1.135, $\alpha$ = 0.0°, Re = $10^6$, Wilcox k-ω model*

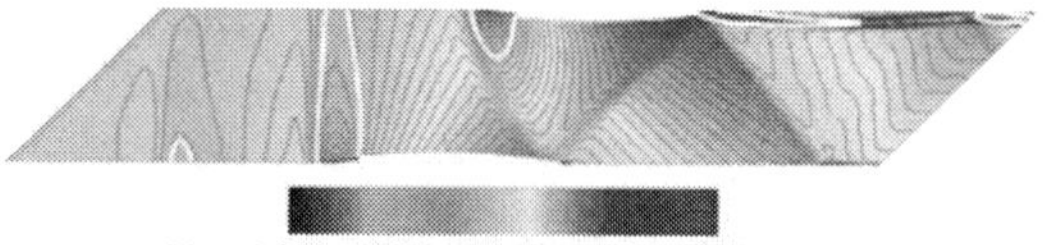


**Fig. 16.** *DCA blade cascade, viscous turbulent flow, $Ma_\infty$ = 1.135, $\alpha$ = 0.0°, Re = $10^6$, SST k-ω model*

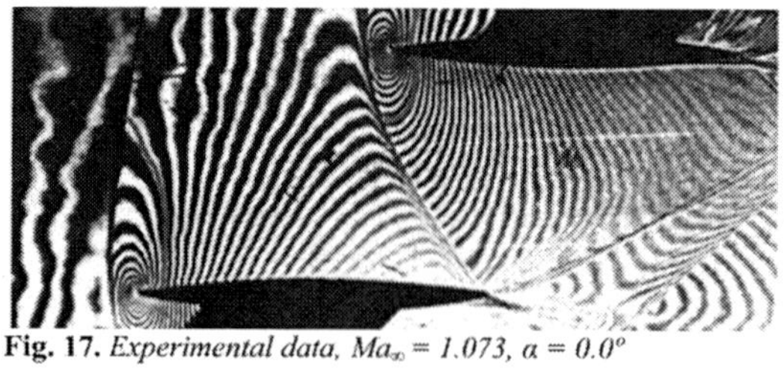

**Fig. 17.** *Experimental data, $Ma_\infty$ = 1.073, $\alpha$ = 0.0°*

## 6 Conclusion

The developed software was successfully tested for the calculations of inviscid and viscous turbulent flow in the GAMM channel and in the DCA blade cascade with the small region of the separation. The possibilities of different numerical approaches were described. The good correlation with the experimental data was shown. Achieved results are comparable with the other authors, e.g. [9], [10].

The research has been financed in part by grants P 101/10/1329 and P 101/12/1271 and by the Education for Competitiveness Operation Program.

## References

1. B. Baldwin, H. Lomax, *Thin layer approximation and algebraic model for separated turbulent flows*, AIAA P., **78**, 257 (1978)
2. T. Cebeci, A. M. O. Smith, *Analysis of turbulent boundary layers*, Acad. Press, New York (1974)
3. D. C. Wilcox, *Turbulence modeling for CFD*, DCW Ind., La Canada, Cal., 3rd edition (2006)
4. D. C. Wilcox, *Formulation of the k-ω turbulence model revisited*, AIAA J., **46**, 2823-2838 (2008)
5. F. R. Menter, *Two-equation eddy-viscosity turbulence models for engineering applications*, AIAA J., **32**, 1598-1605 (1994)
6. R. Dvorak, K. Kozel, *Mathematical modelling in aerodynamics*, CTU in Prague (1996)
7. J. Fürst, K. Kozel, *Numerical solution of inviscid and viscous flows using modern schemes and quadrilateral or triangualar mesh*, Math. Bohem., **2**, 379-393 (2001)
8. R. Dvorak, *Transonic flows*, Academia, Prague, (1986)
9. J. Trefilik, K. Kozel, *Proceedings: Recent advances in fluid mechanics and heat & mass transfer*, WSEAS, Florence, 193-196 (2011)
10. J. Trefilik, J. Huml, K. Kozel, J. Prihoda, *Proceedings: Czech - Japanese seminar in applied mathematics 2010*, CTU in Prague, 94-107 (2010)

EPJ Web of Conferences **143**, 02110 (2017) DOI: 10.1051/epjconf/201714302110
*EFM 2016*

# Influence of blockage effect on measurement by vane anemometers

*Jan* Sluse[1,*]

[1]Czech Metrology Institute, Okruzni 31/772, 638 00 Brno, Czech Republic

**Abstract.** The article deals with influence of blockage effect caused by vane anemometer in the wind tunnel by measurement via this anemometer. The influences will be represented by correction coefficient. The first part of this article is focused on the design of the impeller of vane anemometers. The impellers are printed on 3D printer with variable parameters. The anemometer is fixed in an open section of the wind tunnel with closed loop and the velocity profile is measured by Laser Doppler velocimetry (LDV) in front and behind it for all impellers. The experimental data are compared with the numerical simulation in OpenFOAM. The results are correction coefficients.

## 1 Introduction

Automatization in companies increases the claims on measurement, either operating measurement or testing measurement. This fact puts emphasis on accuracy of measurement and low uncertainty of measurement. The gauges of common units are on higher technical level than in old days. Now it is necessary to focus on the way how the gauges are used and on the measurement as a systematic process with many conditions. For example vane anemometers have significant differences between measurement inside the pipe with small diameter or in the open space. In the pipe a phenomena called blockage effect is created and the value on the gauge is not correct.

This article is focused on blockage effect created in wind tunnel during the calibration of anemometer. In this case testing gauge is putted in the measuring space and indicated value is compared with the value of primary standard. In fact the blockage effect is created in all cases but influence on the result is not significant in all cases. It depends on many conditions (diameter, blades, velocity, method of measurement etc.). For reducing the effect it is necessary to carry out the correction. The negligence of blockage effect may have significant influence on the result of calibration.

The aim of this article is to carry out the comparison of numerical simulation with experimental data. Validated numerical simulation will be used for calculation of correction factor for another design of anemometers (blades, diameter etc.).

## 2 The principle of calibration

The principle of calibration the anemometers is based on comparing the velocity of flow measured by testing gauge with velocity measured by primary standard. The calibration in CMI (Czech Metrology Institute) is carried out in close loop aerodynamics tunnel with open measuring space (Fig. 1). This tunnel was developed by Germany company Westenberg Engineering.

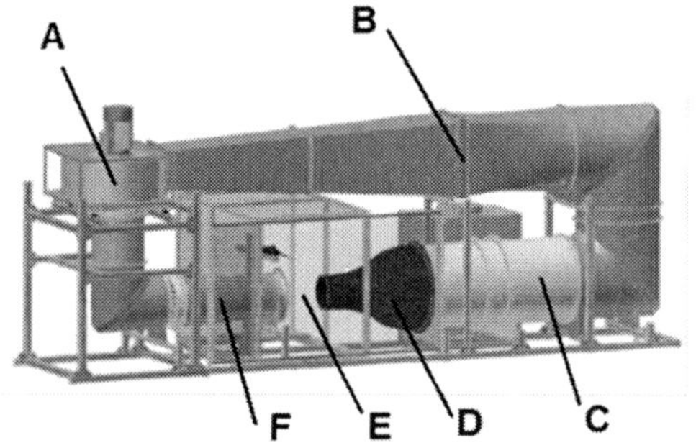


**Fig. 1.** Design of tunnel in CMI.

The legend of the elements (see figure 4) is the following: A – fan, B – heat exchanger, C – honeycomb, D – convergent nozzle, E – measuring space, F – diffuser.

The length of measuring space is 63 cm and the diameter of the nozzle is 45 cm. The area contraction of the nozzle is 6 times. The velocity profile on the outlet of nozzle is flat (like piston profile). The intensity of turbulence is maximal 3 % for the lowest speed and maximal 0.3 % for rest speed. The velocity of air can be set up in velocity range of (0.3 – 50) m/s. The expanded uncertainty type B of measurement in velocity range (0.3 – 5) m/s is (0.01 m/s + 0.3% of measured value) and in velocity range of (5 – 50) m/s is 0.5 % of measured value. The set-up of the velocity can be carried out in two ways. First way is manual mode (with fix rotation speed). The second way is automatic mode with

* Corresponding author: jsluse@cmi.cz

feedback. The feedback is based on measuring the velocity by Prandtl tube which is also located in measuring space. The rotation speed is set up based on the comparison of measured value by Prandtl tube and velocity value in program for automatic mode.

The primary standard in this tunnel is LDA (Laser Doppler Anemometer). It is one direction Nd:YAG laser with wavelengths 532 nm and the laser power 75 mW. The focal length is 500 mm. As the particles for measurement DEHS (di-ethylhexyl-sebacate) were used.

For measurement of velocity profiles can be used traverser motion system in whole three axes.

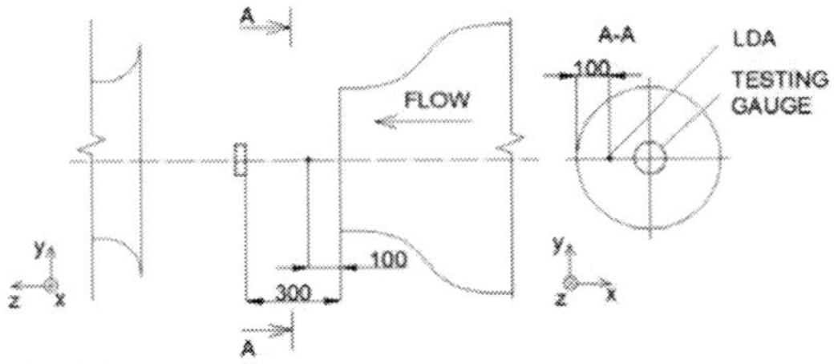


**Fig. 2.** The measuring space.

The testing gauge is fixed in the middle of the measuring space (Fig. 2) and the value of primary standard is measured 10 cm behind the nozzle through. This composition is necessary to minimalize the blockage effect. For this case it is important to know the correction factor for position (difference between the middle of the measuring space and place where the primary standard measures). Another necessary correction is the correction of influence of blockage effect (aim of this article).

Thermal probes, Prandtl tubes and vane anemometers are calibrated very often. The rest of the article is focused only on the vane anemometers which are suitable for studying of blockage effect due to their shape and proceeding.

## 3 Design and construction of the testing anemometer

The tested anemometer wasn't chosen from serial manufactured anemometers but it was made bespoke. For the construction a 3D printer was chosen which created product of good quality for testing blockage effect for good price. The advantage of the design of the anemometer is the possibility to change the shape of the impeller (shape of blades, diameter of the impeller, angle of pitch of the blades, width of impeller etc.).

The shape of main body was derived from the conception of the renowned manufacturers of anemometers. It consists of half ellipsoid with ball-bearing fasten inside. There are five arms coming out from the ellipsoid into the ring which creates cover of the impeller. The main body is divided into two halves which enables changing of the impeller. Both halves fit into each other due to indentations or mirror indentations. The main body is created by resistant ABS plastics. Bottom of the main body contains screw-thread for fixing of the monopod and hole for placement of optical sensor of rotations. The impeller is made of black PLA plastic which is more suitable for printing of difficult shaped parts. The impeller includes 6 blades. The impeller was made in three different variants with different angle of the blades of 20°, 30° a 40° (Fig. 3).

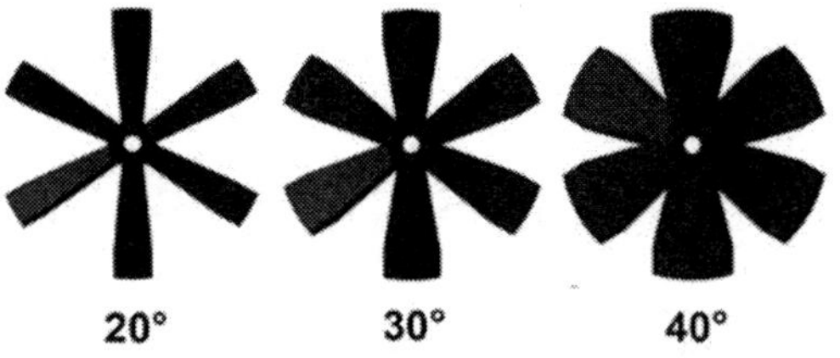


**Fig. 3.** The impeller variants.

Other versions of impellers were created with different diameters of 50 mm, 100 mm and 150 mm. Shaft was made of carbon rod which is tough enough and easy to machine. Detecting of the rotation of the impeller was carried out with optical part which was located in the main body of the anemometer. The optical part evaluated average value of frequency of rotation for 1 s, 5 s and 30 s. This was carried out with the help of the Arduino desk and the program which is based on the knowledge of the number of blades. Fig. 4 shows the assembly of the anemometer.

**Fig. 4.** The assembly of anemometer.

## 4 Experiment

### 4.1. Setup

The measurement was necessary to carry out with manual mode (set up fix rotation speed) of fan to avoid changes of rotation speed caused by blockage effect. In case of automatic mode this effect would change rotation speed because in space where the Prandtl tube is fixed is another velocity via blockage effect. This fact might devaluate the whole experiment.

The velocity was measured on grid points. The grid includes 10 points on x axis and 10 points on y axis. The distance between points was 50 mm (Fig. 5). This grid was measured in 5 surfaces beginning 100 mm from through of nozzle and distance between the surfaces was 100 mm.

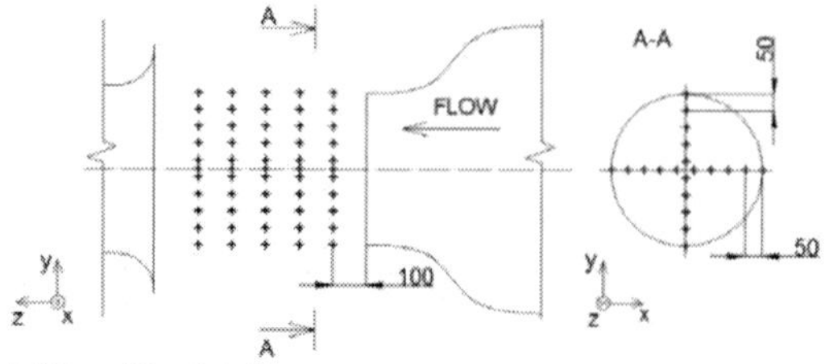


**Fig. 5.** The grid points for measurement.

The measurement began without anemometer to measure initial conditions (velocities of air). The measurement was repeated 10 times to determine uncertainty. The measurement continued with fixed anemometer. The result was difference between average velocities measured with and without anemometer.

### 4.2. The results

By the measurement in boundary points of grid high value of turbulence was indicated. This fact was caused by friction between stream of air from nozzle and claim air inside measuring box. These points were removed from results.

The difference of velocity profiles in each surface is shown on Fig. 6. In this case anemometer with diameter 50 mm and angle of blades 20° was used; velocity was 5 m/s. On the graph only left side of the velocity profiles is shown because monopod for fix anemometer was on the opposite side and velocity was affected.

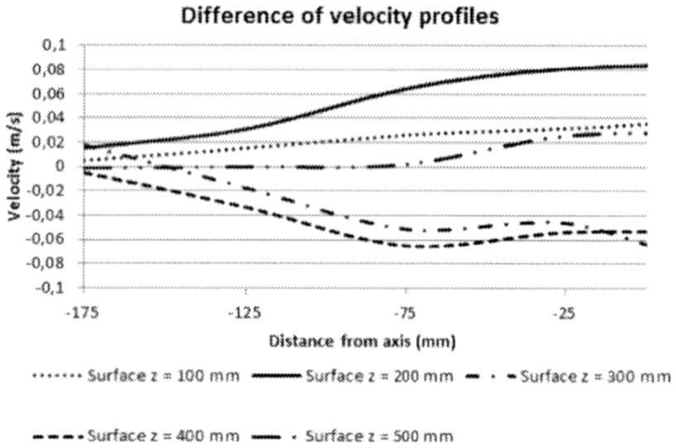


**Fig. 6.** Difference of velocity profiles in each surface.

Next graph shows the comparison between variable angle of blades for velocity of air 5 m/s and for diameter of anemometer 50 mm. The influence of angle of blades on the result is shown on the graph (Fig. 7). The angle of blades influences the magnitude of blockage effect and the size of space affected by blockage effect. This result was expected because the angle of blades influences permeability of air flow through the anemometer. Velocity difference between impeller 20° and impeller 40° in tracking point was 0.0035 m/s. It is necessary to take into account the influence of angle of blades on magnitude of correction of blockage effect or increase the uncertainty of measurement.

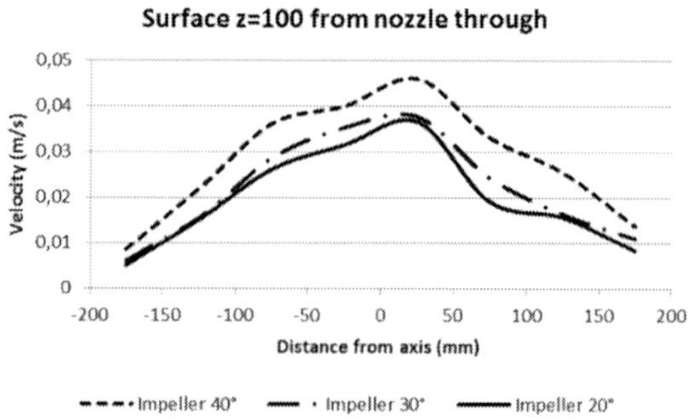


**Fig. 7.** Difference of velocity for all variables of angle of blades.

The correction and uncertainty is shown in tables bellow. The results are for anemometer with diameter 50 mm (Table 1), 100 mm (Table 2) and 150 mm (Table 3) and for variable velocities.

**Table 1.** Diameter of impeller 50 mm.

| Velocity (m/s) | Correction (m/s) | Uncertainty (m/s) |
|---|---|---|
| 1 | 0,002 | 0,002 |
| 3 | 0,000 | 0,002 |
| 5 | 0,001 | 0,002 |
| 10 | 0,001 | 0,003 |
| 20 | 0,001 | 0,006 |
| 30 | 0,005 | 0,009 |
| 40 | 0,002 | 0,012 |
| 50 | 0,000 | 0,015 |

**Table 2.** Diameter of impeller 100 mm.

| Velocity (m/s) | Correction (m/s) | Uncertainty (m/s) |
|---|---|---|
| 1 | 0,002 | 0,002 |
| 3 | 0,000 | 0,002 |
| 5 | 0,001 | 0,002 |
| 10 | 0,002 | 0,003 |
| 20 | 0,002 | 0,006 |
| 30 | 0,006 | 0,010 |
| 40 | 0,002 | 0,012 |
| 50 | 0,000 | 0,015 |

**Table 3.** Diameter of impeller 150 mm.

| Velocity (m/s) | Correction (m/s) | Uncertainty (m/s) |
|---|---|---|
| 1 | 0,002 | 0,002 |
| 3 | 0,000 | 0,002 |
| 5 | 0,001 | 0,002 |
| 10 | 0,002 | 0,003 |
| 20 | 0,003 | 0,007 |
| 30 | 0,007 | 0,011 |
| 40 | 0,003 | 0,013 |
| 50 | 0,000 | 0,017 |

# 5 CFD simulation

## 5.1. Introduction and geometry

The aim of this part was to simulate the flow around the anemometer. The results were used for comparison between experimental data and simulation and will be used for the creation of the correction table for more variables of anemometers. The geometry is shown on Fig. 8. The air flows from left side to right side. On the inlet the velocity is pre-described as a constant value.

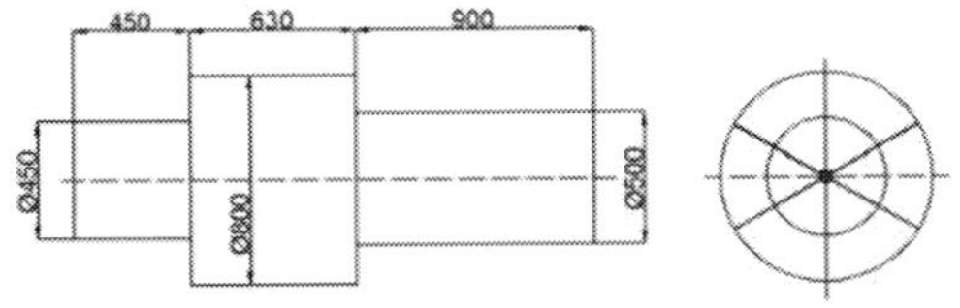


**Fig. 8.** Geometry of simulation.

## 5.1. The mesh, boundary condition, solver

Meshes have been realized by using a capability of OpenFOAM called blockMesh. The setup is more difficult than in commercial programs but the mesh is fully in the user's hands. The refinement of the mesh was carried out near the anemometer and inside the anemometer. Table 4 shows the details about the mesh.

**Table 4.** The mesh properties.

| Mesh number | Cells |
|---|---|
| 1 / 3D | 2 431 904 |
| 2 / 3D | 4 376 704 |
| 3 / 3D | 8 041 728 |

The numerical problem was solved as a steady, viscous, turbulent, incompressible flow by a simpleFoam solver with k-ε and k-ω models of turbulence. Converged results were those results which had residuals lower than $10^{-10}$ for velocity in z axis and $10^{-8}$ for velocity in y and x axis.

## 5.2. Mesh convergence and turbulence model

As a final mesh the mesh number 3 was chosen because next refinement the mesh does not influence the result. Numerical simulation with two different turbulence models was carried out for this mesh. The result from surface 100 mm from nozzle through is shown on Fig. 9.

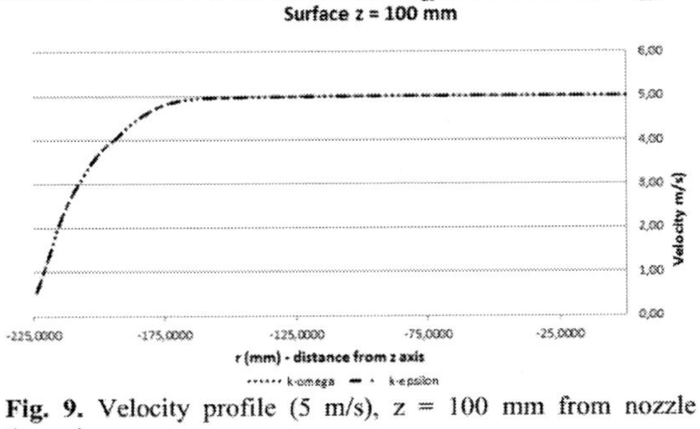


**Fig. 9.** Velocity profile (5 m/s), z = 100 mm from nozzle through.

The result is very similar for all surfaces.

## 5.3. The results

The simulations were carried out only for velocity 5 m/s. The boundary conditions and initial data were being changed in this case to get conformity between numerical model and experimental data. The conformity is received if the results from numerical simulation lay in area of uncertainty of measurement. The difference between numerical simulation and experimental data was 30 % (Table 5). In the table the difference between flow with and without the anemometer is stated.

**Table 5.** Comparison numerical simulation and experimental data.

| Velocity (mm/s) | Numerical simulation (mm/s) | Experimental data (mm/s) |
|---|---|---|
| 5 000 | 0,7 | 1,0 |

# 6 Conclusion

Phenomena called blockage effect and problems connected with this effect were explained in first part of the article. Integral part contains principle of calibration and testing facility for experiments.

In the next part the design of testing anemometer was shown. The design of anemometers was derived from the conception of the renowned manufacturers of anemometers but it is adjusted so that it enables usage of another impellers. It also contains part for detecting of the rotation. The values from the sensor will be used as boundary conditions for simulations.

Experimental part of this article shows the extent of area around anemometer influenced by blockage effect. The results from comparison of different types of impellers show that blockage effect depends on anemometer's permeability of air.

Final part of this chapter shows the correction table for one type of anemometer, including uncertainty which must be added to the measurement. The part dealing with the simulation is using only steady solver. This solver is faster for calculation but it does not reflect the processes inside of the case. The difference between the experiment and numerical simulation is 30 %. It is necessary to use unsteady solver with AMI interface in the future and add the rotation of impeller. This solver is time demanding because it needs dynamics mesh which is recalculated in each time step.

DOI: 10.1051/epjconf/201714302111

# Interference in axisymmetric synthetic jet actuator

*Emil* Smyk*

UTP University of Science and Technology, Faculty of Mechanical Engineering, Control Group, Poland

**Abstract.** The paper presents research of phenomena of mechanical wave interference during generate synthetic jets in actuator with two executive elements (speakers). Two speakers work with the same or other phase using the phenomenon of mechanical wave interference (sound wave). In effect the volume of ingestion and expulsion flow is a sum of two speakers' work. In this paper is presented analysis of operation so constructed axisymetric actuator and influence of phase shift - between work of the speakers - on actuator's operation.

## 1 Introductions

Synthetic jets are formed by periodically suction and blowing of fluid to some chamber by nozzle. Device to produce synthetic jets is named synthetic jet actuator. It is realized by replacement one of actuator's wall by movable element. Depending on the method of accomplishing the movement of wall, can distinguish piezoelectric, electromagnetic, acoustic and mechanical synthetic jet actuators. Example of synthetic jet actuator and idea of synthetic jets forming are shown in fig.1.

Name of "synthetic jet" was firstly used in [1]. It is to suggest that the flow is synthesised by creation of the succeeding vortex rings. It is worth of noted that because time- averaged flow by a nozzle of synthetic jet actuator is zero, the same volume of flow is ingestion and expulsion, synthetic jet are named zero-net mass-flux (ZNMF) jet. Synthetic jet actuator doesn't need additional fluid supply to correct functioning, what distinguishes synthetic jet from continuous jets.

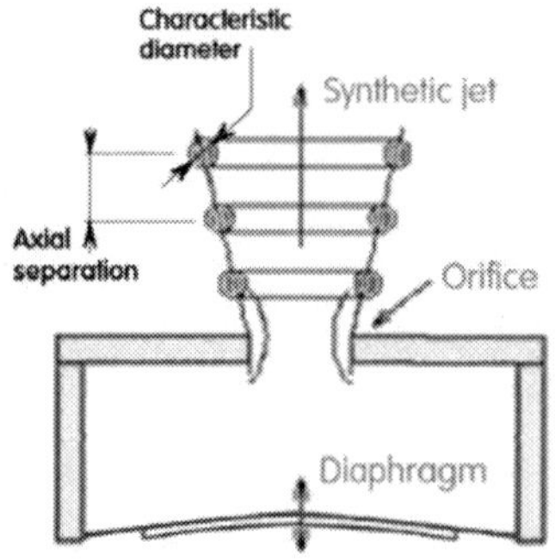


**Fig. 1.** Example of the idea of synthetic jets forming – from [2]

Property of synthetic jet are the reason of widespread interesting theme of their application. They are used in cooling applications, chemical engineering, active flow control etc. Particularly active flow control is a synthetic jet's applications field which is dynamically developed, demonstrated by a large number of papers appearing in recent years [2-6].

Interest of SJ widened not only knowledge that issue [7-10], but conducive to creating new synthetic jet's applications [11]. Interesting method of generation synthetic jet was presented in [12, 13, 14]. In all of paper actuators was construct by replacement two of opposed walls by a diaphragm of speakers (fig. 2). Such construction of actuators should increase velocity of synthetic jet.

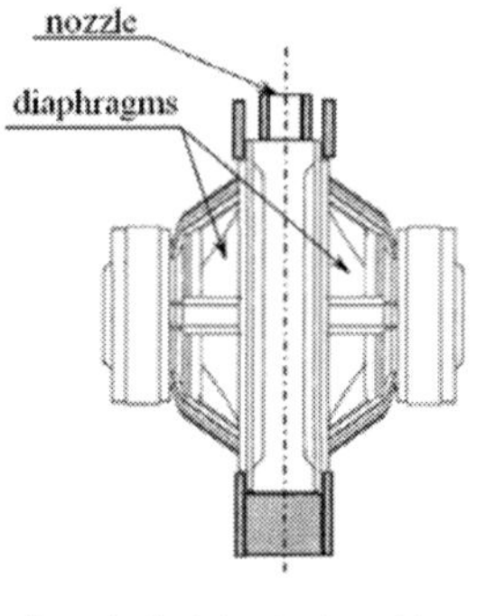


**Fig. 2.** Example of synthetic jet actuator with two speakers

Two speakers operate with the same phase using the phenomenon of mechanical wave interference (sound wave). In effect the volume of ingestion and expulsion flow is a sum of two speakers' operation. It must be noted that in [12] speakers works in this same way, what is provided by actuators construction. However, in [13] can be seen that work of one of speakers is suppressed. This is caused by putting the speaker in closed chamber. During his work preassure behind speaker periodically decreases (when fluid is expulsion from actuator by movement speaker diaphgram 'forward') and increses (when fluid is ingestion to actuator by movement speaker diaphgram 'backward').

Although in [12, 13, 14] was used this specifical type of actuators, they do not include information about

* Corresponding author: emil.smyk@utp.edu.pl

DOI: 10.1051/epjconf/201714302111

interferenc of wave, and how it influence at parameters of synthetic jet.

The aim of this work is investigate how phenomene of interferenc influeces on synthetic jets.

## 2.Parameters

The synthetic jets are produced by periodically expulsion and ingestion of fluid. Therefore, their velocity can't be determined in the same way as continuous jets velocity. The time-mean velocity $U_0$ define from the orifice centerline at the axis:

$$U_0 = \frac{1}{T}\int_0^{T_E} u_0 dt \tag{1}$$

Where $T$ is time period ($T = 1/f$, where $f$ is frequency), $T_E$ is time of flow expulsion ($T_E = T/2$), and $u_0$ is velocity of flow.

Another parameters used in case of synthetic jet actuator is nondimensional stroke ratio $L/d$ and Reynolds number $Re$. They have been identified as key actuator operational parameters that influence synthetic jet [1]. Stroke ratio $L/d$ is a proportion between length and diameter of expulsion fluid volume (fig. 3). So it is ratio of distance over which the fluid is blown and the size of nozzle. Stroke ratio $L/d$ can be calculated from:

$$\alpha\frac{\pi D^2}{4}\Delta = \frac{\pi d^2}{4} \tag{2}$$

Where $\alpha$ is the fraction of the volume displaced by an imaginary piston undergoing a peak to peak deflection of $\Delta$. $D$ is inner diameter of actuator and $d$ is diameter of nozzle.

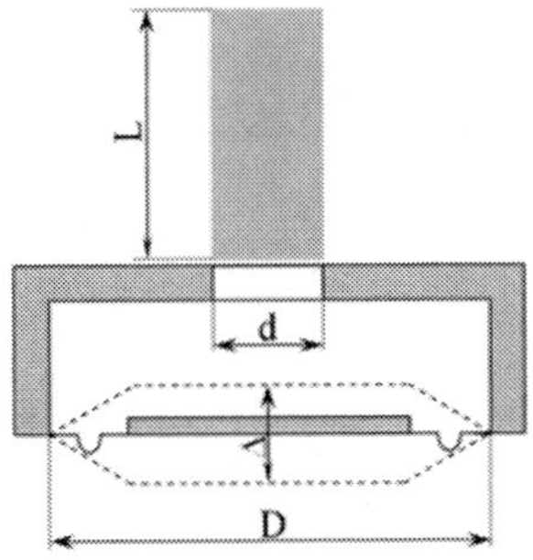


**Fig. 3.** Schema of actuator, where volume of fluid displaced by the diaphragm is ejected through the orifice in the form of slug

The definitions of the nondimensional parameters Reynolds numbers $Re$ can be introduced as follows:

$$Re = \frac{U_0 d}{\nu} \tag{3}$$

Where $\nu$ is kinematic viscosity of fluid.

It must be noted that these formulas can be used for incompressible fluids. In case of compressible fluids (example air) this formulas can be used for Mach number less than 0.3. In such a case the error should be less than 3%. Mach number can be calculated from:

$$Ma = \frac{u}{c} \tag{4}$$

Where $u$ is local flow velocity fluid (gas), and $c$ is speed of sound in the medium.

## 3. Research subject

Aim of the study is examination of the application of interference phenomena to synthetic jet. Based on the work [12, 13] was made synthetic jet actuator presented in fig. 4.

Actuator consist of two opposite speakers Monacor SP-6/8SQ mounted on a metal ring. On the circuit therefore (ring) was done 24 holes in diameter $\phi 3$. On ring are attached two sleeves so as to create between them a gap of 0.5mm. One of sleeve creates from the back of the speaker closed chamber. Its task is suppression of speaker operation. By this actuator structure, speakers generates other wave, through the same power supply.

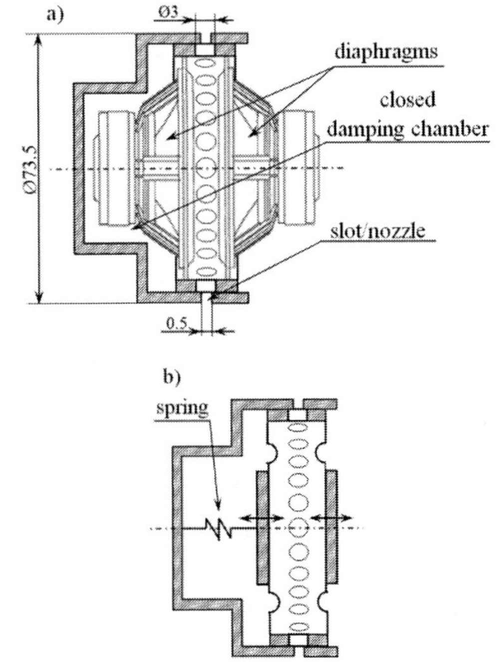


**Fig. 4.** Schema of synthetic jet actuator a) drawing, b) physical model

In fig. 4b was showed physical model of synthetic jet actuator. In this model closed chamber on the left is replaced by spring. The restoring force of spring results from the chamber volume changes (the mass of air in chamber is constant). Bellow was presented analysis of spring properties. The analysis is based on fig. 5.

To present the closed chamber as spring is needed the comparison of forces acting on the actuator diaphragm in the initial and final state. On the fig. 5 is presented model used to this analysis. The constant volume of air closed in chamber behind the speaker presses the diaphragm with forces $F_0$. It is due to air pressure. If the chamber is presented as a cylinder with fixed base area $A$ (area of speaker diaphragm) and a variable height $l$ ($l_0$ is a initial height) then force $F_0$ can be calculated from ideal gas law:

$$p_0 l_0 A = RT \quad (5)$$

where $R$ is a universal gas constant ($R$=8,314 J/(mol·K)), $T$ is a temperature in Kelvin. Then converted equation (5):

$$p_0 A = \frac{RT}{l_0} \quad (6)$$

$$F_0 = \frac{RT}{l_0} \quad (7)$$

However, if the membrane is ejected from a distance $\Delta l$ force will decrease and will equal:

$$F = \frac{RT}{l_0 + \Delta l} \quad (7)$$

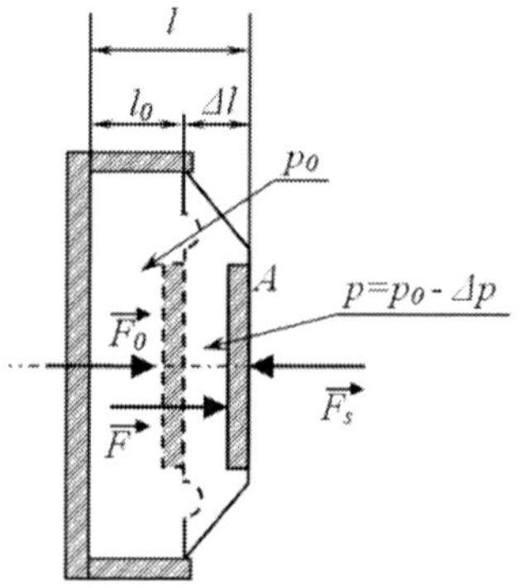


**Fig. 5.** Simplified model of the speaker with a closed chamber

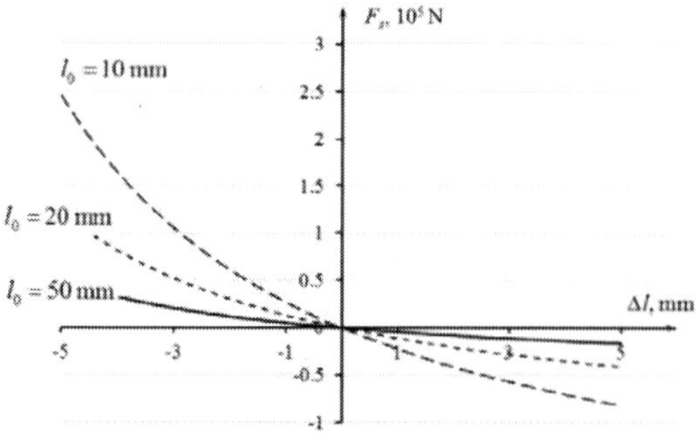


**Fig. 6.** Force characteristic for different chamber length $l_0$

The force acting on the diaphragm decreases when the chamber volume increases (or increases if the volume of chamber decreases). The change of this force can by presented by force $F_s$ which resist movement of the diaphragm.

$$F_s = F - F_0 \quad (8)$$

$$F = \frac{RT}{l_0 + \Delta l} - \frac{RT}{l_0} \quad (9)$$

$$F = -RT\frac{\Delta l}{l_0\left(l_0 + \Delta l\right)} \quad (10)$$

Analyzing the equation (10) can be seen that the closed chamber behind the speaker can be treated as a non-linear spring, which modulus of elasticity is dependent on the volume of the chamber (the bigger $l$ the lower the modulus of elasticit). Sample plots the resilient force of the spring shown in fig. 6.

The phenomena of mechanical wave interference occurs in the effect of the imposition a two wave with sundry amplitude or phase. Therefore test actuator was powered by arbitral waveform generator Rigol DG4162, with possibility of changing phase of current supply, for each from two supply channel. Electrical power was identified by measuring the voltage and current supply, and set by stereo amplifier SeoUm Pa-940/2. Electric scheme of actuator connection was showed in fig. 7.

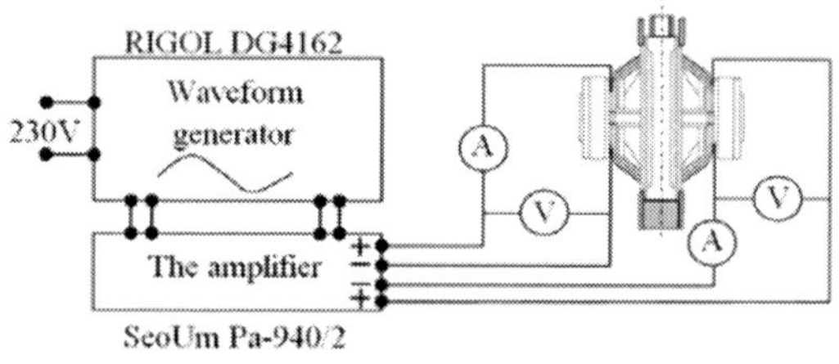


**Fig. 7.** Electric scheme of the actuator connection

All measurements of synthetic jet velocity was done by using hot-wire anemometer. Anemometer contained probe 55P16 connected through two modules MiniCTA and NI9215 with personal computer. Hot-wire probe was placed two millimeter from nozzle actuator in the nozzle axis. Measurements was carried out by temperature $T = 22°$ and pressure $p = 1010\text{hPa}$.

## 4.Results

Before the main tests, there were carried out preliminary tests of the actuator. During primary testing, it was determined characteristic frequency of actuator. Then it was examined the operation of the synthetic jet actuator when power is one of the speakers. Furthermore examined how phase shift $\Delta\varphi$ of speakers work influence on synthetic jet velocity. In main tests, it was determined operating characteristic of actuator by different phase shift $\Delta\varphi$ of speakers operation.

It must be noted that as phase shift $\Delta\varphi = 0°$ understands the situation when both speakers operate in the same phase (example both speakers perform a movement forward and they push fluid from the actuator).

 DOI: 10.1051/epjconf/201714302111

### 4.1 Primary tests

The first step of tests was designation of characteristic frequency of actuator. It was identified as $f$ = 147 Hz. In the time of all further studies, actuator was powered by electric current with such frequency.

Then it was test whether phase shift of speakers operation has a influence on synthetic jet actuator. For this purpose was measured synthetic jet velocity for different phase shift $\Delta\varphi$ by a power supply $P$ = 3W. The experiment data and sinusoidal approximation was showed in fig. 8.

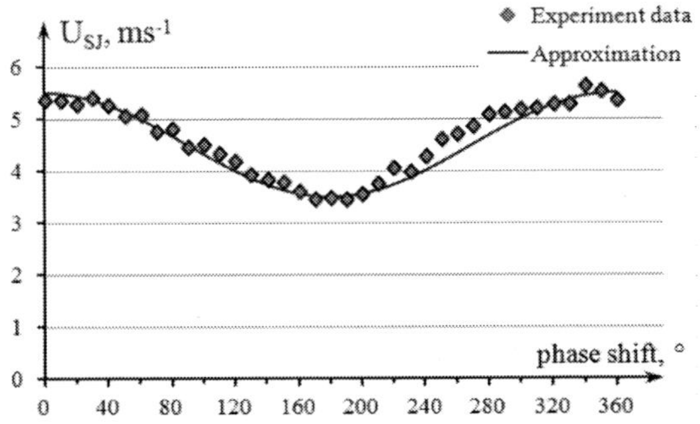


**Fig. 8.** Influence of phase shift on synthetic jet velocity for power supply $P$ = 3W

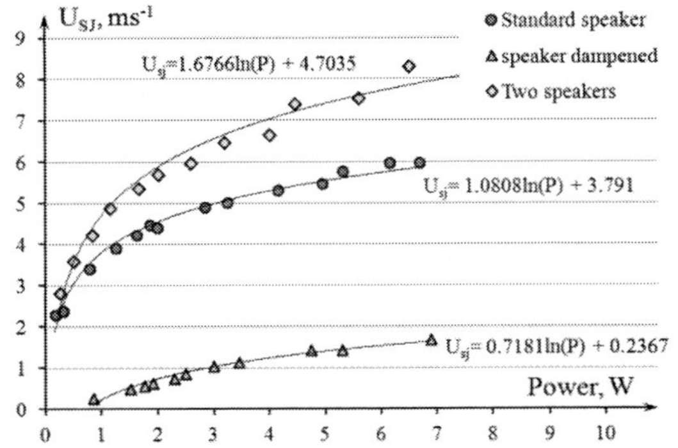


**Fig. 9.** Operating characteristic of actuator for speakers working separately and together

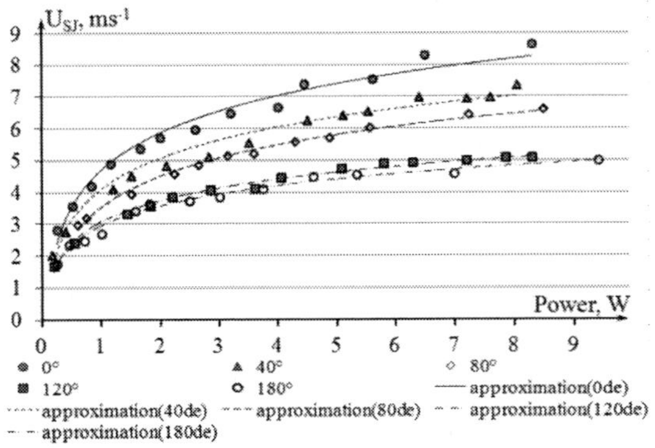


**Fig. 10.** Actuator characteristic for selected phase shift of speakers operation.

The next step was defined the operating characteristic for actuator by supply speakers separately and together. The purpose of these measurements is also to determine whether the closed damping chamber has effect on the loudspeaker operation in the actuator. Characteristic of actuator is presented in fig. 9.

### 4.2 Main tests

The main part of the test is designate characteristic of actuator for different phase shift $\Delta\varphi$ of speakers operation. Fig. 10 presents actuator characteristic for selected phase shift $\Delta\varphi$ and logarithmic approximation of experimental data. It must be noted that $U_{SJ}$ is a selected resultant of velocity and can't be used (in this case) to determine whether interference in actuator can be treated in a standard way.

In order to check how interference occurs in actuator, on the basis of measured velocity has been designated course of velocity in the actuator slot (fig. 4). The velocity change in the slot is a sine wave at a given frequency ($f$ = 147 Hz) but unknown amplitude $U_{MAX}$ . This amplitude was calculated as average of maximum velocity from the measurements.

Then compare the amplitude waveforms (depends of electrical power $P$) calculated for real data and theoretical interference waves generated by speakers. Comparison of theoretical and experimental data is presented in fig. 11.

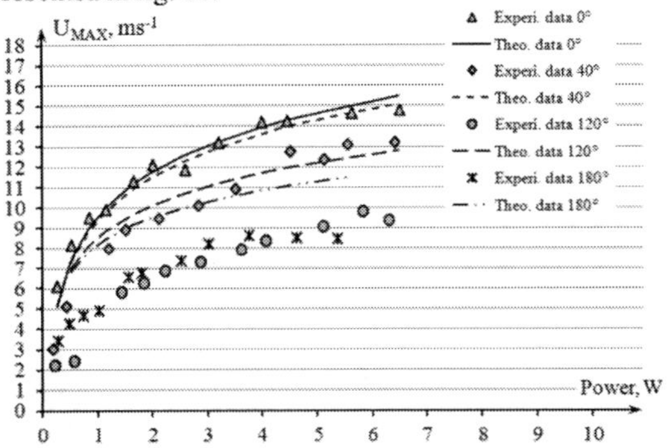


**Fig. 11.** Experimental and theoretical amplitude of air velocity in actuator slot

## 5. Discussion of results

Fig. 8 presents change of synthetic jet velocity depending of phase shift $\Delta\varphi$ of speakers. Trend of this change is as expected and can be approximated by sine wave. It exhibits some asymmetric however, it is negligibly small. This allowed for reduction of measurements (phase shift range between $\Delta\varphi$ = 0° to $\Delta\varphi$ = 180°).

The next step of primal measurements was determination of operating characteristics of the actuator. It is presented in fig. 9. The damping chamber used in actuator has significant influence on operation of speaker. It is worth remembering that characteristic 'standard speaker' and 'dampened speaker'(fig. 9) is operating characteristic of the same speaker(the same model), but working in different conditions. The speaker with damping chamber generates synthetic jet only the power $P = 1$ W, while the speaker without suppression produces synthetic jet from $P = 0.2$ W.

As it was excepted the velocity, when powered two speakers, was greater then, when powered one speaker. It must be noted that, when low power, the synthetic jets are generated mainly by speaker without suppression. However, the influence of interference, occurring in actuator, is visible even for power less than $P = 1$ W.

The hitherto measurements shows that the application of actuators with two executive elements is preferred because of increased of increased velocity of synthetic jet. In case of works [12,13] applications this actuator improved properties of test object without magnification of size.

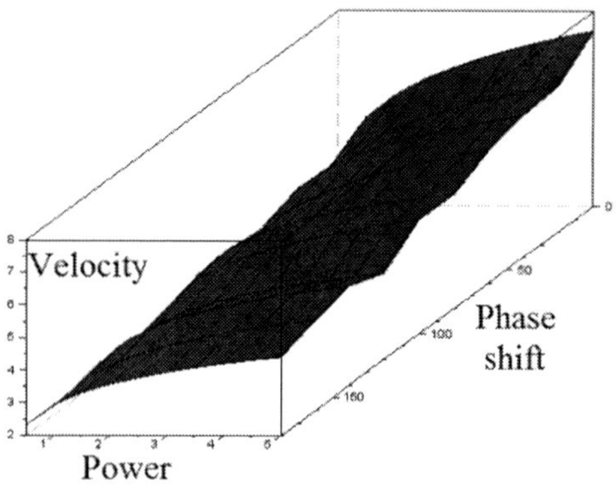


Fig. 12. Solution family of actuator operating characteristic

The next measurements involved in determining of operating characteristic of actuator with other phase shift $\Delta\varphi$. This solution helped designate the plane, which presents dependence between synthetic jet's velocity $U_{SJ}$ and parameters of supply (power $P$ and phase shift $\Delta\varphi$ of speakers operation ). This dependence is showed in fig. 12. Examples operating characteristic are presented in fig. 10.

Fig. 12. presents some dependence between velocity and phase shift and power ($U_{SJ}$ ($P$, $\Delta\varphi$ )). As can be seen function $U_{SJ}$ ($P$, $\Delta\varphi$ ) have a maximum with $\Delta\varphi = 0°$ (for specific power). This results is compatible with theory.

The above fact and data don't give sufficient information about the phenomena of interference in actuator. For this purpose was appointed the velocity amplitude $U_{MAX}$ in slot of the actuator (fig. 4). $U_{MAX}$ was designated as the average of maximum velocity of six next expulsion. It was calculated in case when speakers operated separately and together, whit other phase shift $\Delta\varphi$ and power $P$. Theoretical value of amplitude $U_{MAX}$ was calculated through addition two sine wave, first with velocity amplitude $U_{MAX}$ of damped speaker, and second of unsuppressed speaker, and with specified phase shift $\Delta\varphi$. The results are presented in fig. 11.

It can be seen that theoretical and experimental velocity amplitudes $U_{MAX}$, with phase shift $\Delta\varphi = 0°$, are the same. But the greater phase shift $\Delta\varphi$ between speakers the greater different between experimental and theoretical data. Even for phase shift $\Delta\varphi = 40°$ difference between data is big. It is connected with uncorrected operation of actuator.

Speakers, which operate in different phase have tend to interfere with their mutual actions. Measurements shows increasing of turbulent of generated synthetic jet with simultaneous drop of velocity. It is undesirable effect. Moreover noted that the actuator operated with phase shift $\Delta\varphi$ is very sensitive on time of switch on.

Although, the speakers was powered by current with steady shape shift $\Delta\varphi$, the actuator generated synthetic jet on different velocity if the speakers wasn't switch on in the same time. It means that executive, elements in synthetic jet actuator which used phenomena of interference, have strong damping properties to each other if they don't operate in the same phase. It needs to be heighted that all measurements was carried out for speakers switched on in the same time. In case speaker switching at different times it happened that actuator produce synthetic jet on low-amplitude ($U_{MAX} = 2$ m·s$^{-1}$) even for high power ($P = 6$ W). It mean that acoustic actuator using the interference phenomena is very sensitive on the desynchronization of speakers. Moreover, it should be noted that sensitivity is the greater the higher is the phase shift.

## 66Conclusion

Presented results confirmed the reasonableness of the use of the actuator with two executive element. Moreover, the phenomena of interference, in produce of synthetic jet, was confirmed and discuss.

Additionally, in experiment was used closed damping chamber behind one of speakers. The chamber has significant and adverse influence on work of speaker and so on synthetic jet generated by him.

## Acknowledgements

Scientific work financed from the budget for science in the years2015/2018 as a research project within the program under the name "Diamond Grant ".

## References

1. B.L. Smith, A. Gezler, *The formtion and evoluation of synthetic jets*, Phisic of Fluids **31**, p. 2281 (1998)
2. H. Tang, S. Zhong, *Incompressible flow model on synthetic jet actuator*, AIAA J. **44**, p. 908 (2006)
3. D. Tamburello, A. Michael, *Active control of a free jet using a synthetic jet*, Int. J. of Heat and Fluid Flow, **29(4)**, p. 967-984 (2008)
4. D. Tamburello, M. Amitay, *Three-dimensional interactions of a free jet with a perpendicular synthetic jet*, J. of Turbul. **8(35)**, pp.1-22 (2007)
5. L.G. Pack, A. Seifert, *Periodic excitation for jet vectoring and enhanced spreading*, J. of Aircraft, **38(3)**, pp. 486–495 (2001)

6. S.A. Davis, A. Glezer, *The manipulation of large- and small-scales in coaxial jets using synthetic jet actuators*, AIAA Paper 2000-0403 (2000)
7. P. Krejčí, Z. Trávníček, V. Tesař, J. Kordík, M. Pavelka, *Frequency Characteristics of an Annular Synthetic Jet Actuator of Large Inner Diameter*, Proceedings of Conf. 'Experimental Fluid Mechanics 2007' (2007)
8. Z. Travnicek, Z .Broucková, J. Kordík, *Formation criterion for axisymmetric synthetic jets at high Stokes numbers*, AIAA J. **50(9)**, pp. 2012 (2012)
9. V. Tesař, K. Peszynski, *Experiments with synthetic jet for detecting potential terrorists*, J. of Pol. CIMAC (2010)
10. P. Gil, P. Strzelczyk, *Performance and efficiency of loudspeaker driven synthetic jet actuator*, Exp. Therm. and Fluid Science **78**, pp.163-174 (2016)
11. V. Tesař, K. Peszynski, *Synthetic jet applied to detect potential terrorists*, J. of Pol. CIMAC (2010)
12. V. Tesař, M. Pavelka, E. Smyk, K. Peszyński, *Control of flow separation by vestigial" synthetic jet*, Colloq. Fluid Dyn. (2014)
13. J. Kordik, T. Zdeněk M. Pavelka, *Energetic efficiencies of synthetic and hybrid synthetic jet actuators driven by electrodynamic transducers*, Exp. Therm. and Fluid Science **69**, pp. 119-126 (2015)
14. V. Tesař, Z. Broučková, J. Kordik, Z. Trávníček, K. Peszyński, *Valves with flow control by synthetic jets*, EPJ Web of Conf. **25**, EDP Sciences (2012)

# Numerical simulation of axisymmetric valve operation for different outer cone angle

*Emil* Smyk*

UTP University of Science and Technology, Faculty of Mechanical Engineering, Control Group, Poland

**Abstract.** One of the method of flow separation control is application of axisymmetric valve. It is composed of nozzle with core. Normally the main flow is attached to inner cone and flow by preferential collector to primary flow pipe. If through control nozzle starts flow jet (control jet) the main flow is switched to annular secondary collector. In both situation the main flow is deflected to inner or outer cone (placed at the outlet of the valve's nozzle) by Coanda effect. The paper deals with the numerical simulation of this axisymetric annular nozzle with integrated synthetic jet actuator. The aim of the work is influence examination of outer cone angle on deflection on main stream.

## 1 Introductions

One of the method of flow separation control is application of axisymmetric valve. The valve of this type was described first time in 2010 [1] and its operation is described precisely in [2, 3]. Exemplary application of axisymmetric valve is presented in fig.1.

Axisymmetric valve is composed from nozzle with core. Normally the main flow is attached to inner cone (fig. 2) and flows by preferential collector to primary flow pipe. If through control nozzle flows jet (control jet), the main flow is switched to annular secondary collector. In both situation the main flow is deflected to inner or outer cone (placed at the outlet of the nozzle, fig.2) through Coanda effect. Generally as control jet is used synthetic jet generated by actuator embedded in the core of nozzle [1, 2, 3].

This type of valve uses two phenomena – Coanda effect and synthetic jet. Both of phenomena have influence on main flow switching and they cooperate in this process. Synthetic jet (control jet) detaches the main flow from inner cone and the main flow additionally is deflected by outer cone (Coanda effect). Only the occurrence of these two phenomena redirects the main flow.

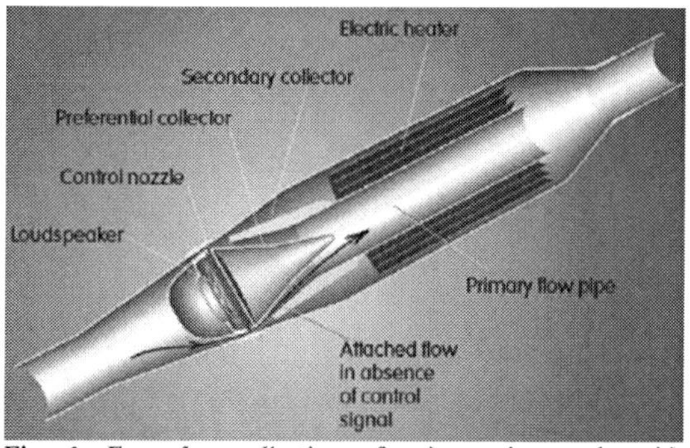


**Fig. 1.** Example application of axisymetric nozzle with synthetic jet actuator – from [2]

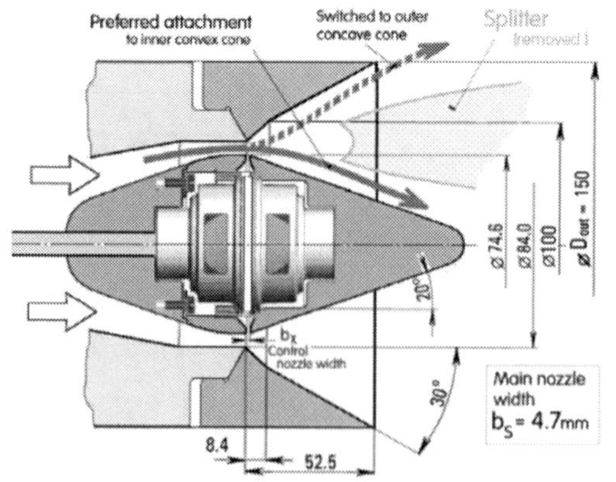


**Fig. 2.** Drawing of axisymmetric valve taking account of the inner and outer cone (outer cone angle α=30°) – from [3]

### 1.1 Synthetic jet

Synthetic jets are generated by periodically ingestion and expulsion of fluid at the exit of a nozzle or across orifice [4]. Because is no net flux of fluid during one full cycle of operation, synthetic jet is named zero– net mass– flux (ZNMF) jet. At some distance from exit of nozzle are created edge vortices. The idea of synthetic jet generation is presented in fig. 3.

* Corresponding author: emil.smyk@utp.edu.pl

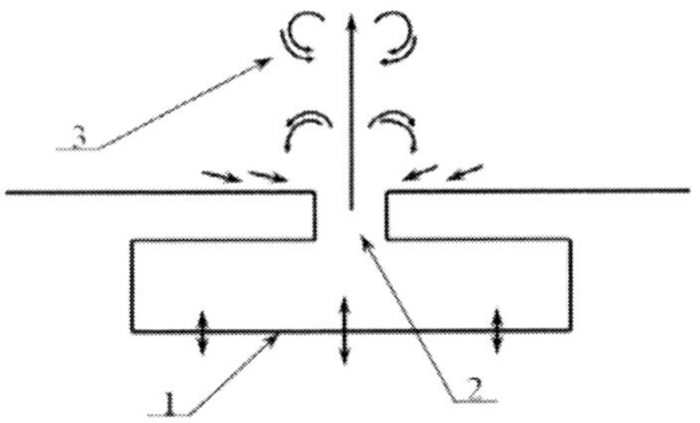


**Fig. 3.** Idea of the synthetic jet formation, 1– moving diaphragm, 2– nozzle, 3– edge vortices

Synthetic jet has many applications: in aviation, aeronautics, refrigeration systems and active flow control etc. One of synthetic jet applications examples is axisymmetric nozzle presented in this paper.

### 1.2 Coanda effect

Jets have tendency to adhesion and to flow nearby solid boundaries. This phenomena is known as Coanda effect. Flows deflected by a curved surface are used in many applications [5, 6]. Coanda effect is still not exactly known, as evidenced by the extensive investigation [7]. The reason for this state of affairs is a high level of complexity this phenomena. The Coanda effect depends on parameters such as: roughness and temperature of the solid surface, temperature, viscosity, density of fluid end other. Exact influence of these parameters on Coanda effect is still under investigation [8].

Fluids deflected by the surface adhesion force. If this force is predominant then fluid follows along the surface. This phenomena and Coanda effect is used to active flow control in axisymmetric value [1, 2, 3] (fig.1).

The Canada effect is very important for the proper operation of the axisymmetric valve. It affects the behaviour of the stream and it can determine on the correct valve operation. One of the element which has important influence on Coanda effect is outer cone angle (fig.2) and this issue is the topic of this paper.

This investigation presents numerical simulation of axisymetric nozzle (used to valve construction) and the influence of the outer cone angle on behaviour of main flow.

## 2.Numerical investigation

### 2.1. Simulation of synthetic jet

Main problem of numerical investigation of axisymmetric valve is simulation of synthetic jet. Recently, numerical research of synthetic jet was conducted in many publications [9-14]. Deserves special attention [10], in which was presented four different methods of synthetic jet simulation. Fig. 4 shows variants of synthetic jet models.

Fig. 4a shows typical method of synthetic jet simulation. In this solution, moving diaphragm of synthetic jet actuator is simulated by sinusoidal velocity profile. The velocity (normal to the wall) is defined as:

$$v(x, y=0, t) = v_0 f(x) \sin(\omega \cdot t) \qquad (1)$$

Where $t$ is time, $v_0$ is amplitude of velocity, ω is angular frequency. Three different spatial distribution function *f(x)* were considered:

$$f(x) = \begin{cases} 1 \\ \sin(\pi \cdot x \cdot d^{-1}) \\ \sin^2(\pi \cdot x \cdot d^{-1}) \end{cases} \qquad (2)$$

Such a simple boundary conditions representation eliminates the need to simulate of the diaphragm.

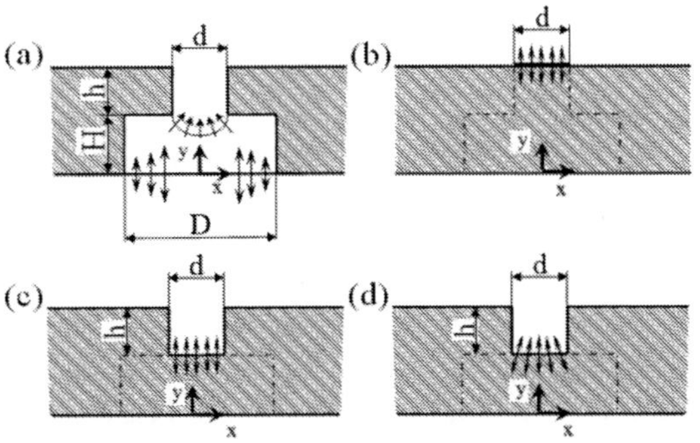


**Fig. 4.** Schema of different method on synthetic jet simulation (a)typical solution – SJ slot for a full cavity configuration, (b) the modified boundary conditions, (c) and (d) slot-only models

The easiest model of synthetic jet simulation is shown in fig. 4b. In this case the chamber and nozzle of actuator are omitted and only nozzle exit is taken into account. It is inaccurate but uncomplicated model. The intermediate solution is simulation of all nozzle (fig. 4c, d), so called slot-only actuator model [10]. The sinusoidal velocity can be defined as parallel to the axis of the nozzle (fig. 4c) or depend on distance from the axis (fig. 4d). The detailed discussion of presented models and their comparison was conducted in [10]

Due to the complexity of the analyzed object (fig. 2) was decided to use slot-only model with velocity normal to boundary. The velocity was defined as:

$$v(t) = v_0 \sin(\omega \cdot t) \qquad (3)$$

### 2.2 Model geometry and mesh

As mentioned this investigation shows how the outer cone angle affect on behaviour of main flow. Therefore, it was prepared different models of axisymmetric valve with different cone angle. Example model with basics dimensions is presented in fig. 5. The value of investigated outer cone angle was presented in table 1.

The model based on the geometry of axisymmetric nozzle presented in fig. 2. In each model presents only half of the nozzle, because the symmetry boundary conditions was used in order to simplify the simulations.

As mentioned the synthetic jet actuator was replaced by slot–only model (see fig. 4b).

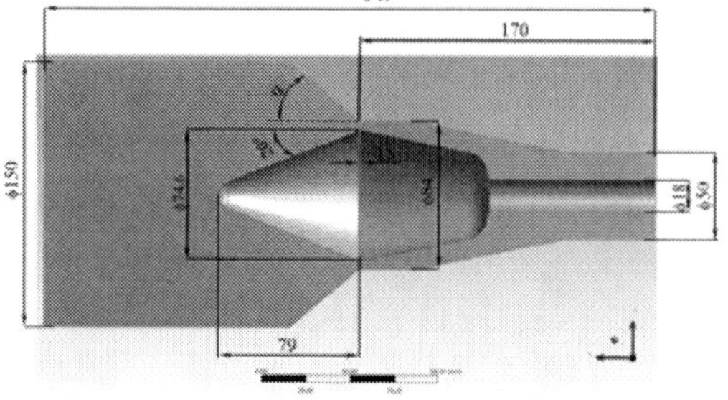

**Fig. 5.** Model of axisymmetric valve – front view, for α = 40°

Computational meshes for each model was generated in ANSYS Meshing. For each model was defined the same face sizing for the same walls. The difference of nodes and elements numbers is caused by different outer cone angle. The element quantity was very similar for each case. Example element quantity was shown in fig. 6. The models exploit tetrahedron meshes (Tet4).

**Table 1.** Models parameters

| Outer cone angle α | Number of nodes | Number of elements |
|---|---|---|
| 90° | 580 605 | 3 023 419 |
| 80° | 388 676 | 1 993 636 |
| 70° | 385 616 | 1 977 427 |
| 60° | 381 609 | 1 956 745 |
| 50° | 378 915 | 1 942 058 |
| 40° | 373 193 | 1 911 981 |
| 30° | 366 726 | 1 877 306 |
| 20° | 356 014 | 1 819 285 |
| 10° | 329 108 | 1 682 436 |

## 2.3. Boundary conditions

Boundary conditions are shown in fig. 6. It must be noted, that all unlabelled surfaces belong to the Walls group, and they was omitted to preserve legibility of figure. On this surfaces, the boundary condition "wall" was applied. Velocity in nozzle of synthetic jet actuators was modelled by UDF (user defined function), which is defined as dependency of velocity on time. All boundary conditions are described in table 2.

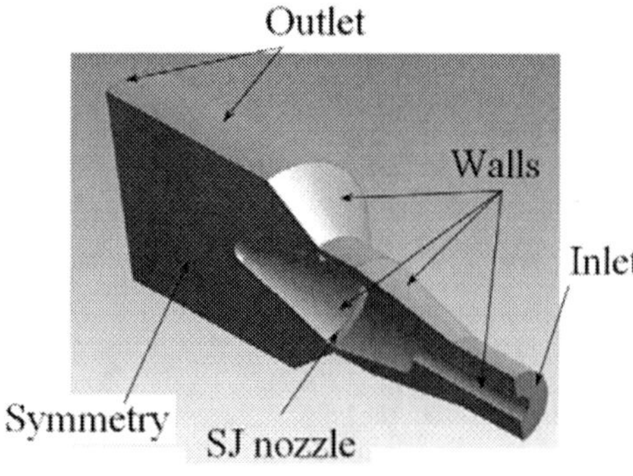


**Fig. 6.** Boundary conditions used in numerical modelling

**Table 2.** Boundary conditions

| Label | Boundary condition type | Parameters, description |
|---|---|---|
| Inlet | mass – flow – inlet | Inlet mass flow: $5{,}78 \cdot 10^{-3}$ kg/s, or $2{,}835 \cdot 10^{-3}$ kg/s |
| Outlet | pressure – outlet | Determined model boundary, outlet |
| SJ nozzle | velocity – inlet | Velocity profile defined by UDF |
| Symmetry | symmetry | Symmetry plane |
| Walls | Wall | Walls of nozzle and outer cone |

It must be noted that inlet was defined as "mass–flow– inlet" with two values. To verify the accuracy of the simulation were compared the simulation results with results presented in [3]. Accordingly, the inlet parameters was used such as in [3] ($2{,}835 \cdot 10^{-3}$ kg/s, it is mass flow rate for half of valve). In case of main simulation was used twice as much mass flow rate ($5{,}78 \cdot 10^{-3}$ kg/s).

## 2.4. Simulation parameters

Numerical simulation was done in ANSYS Fluent 16. It was used simulation without gravity with standard k-ε viscous model with enhanced wall treatment, and with automatic model constants. It was used SIMPLE solution method and standard solution initialization. The need of synthetic jet simulation in simulation resulted in the use of transient model with time step equal 1/20 of velocity $v(t)$ period ($T = 2\pi/\omega$, see formula (3)).

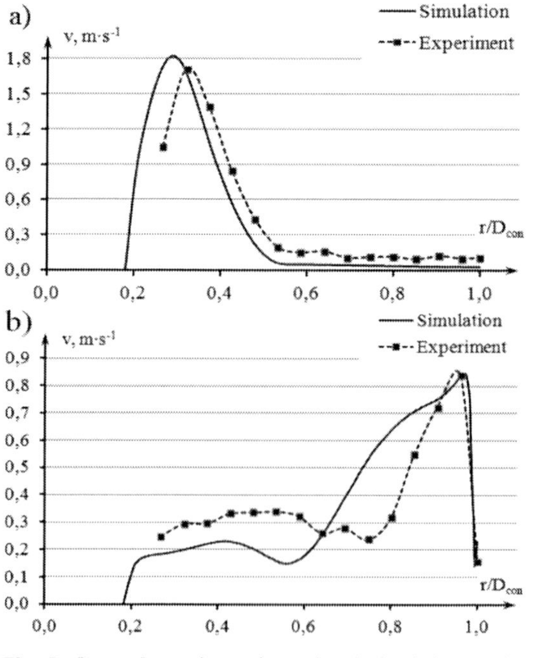


**Fig. 7.** Comparison of experimental and simulation results. The velocity profile of main stream in the outlet plane from the outer cone with angle α=30°, a) main stream velocity profile without synthetic jet, b) main stream velocity profile with activated synthetic jet.

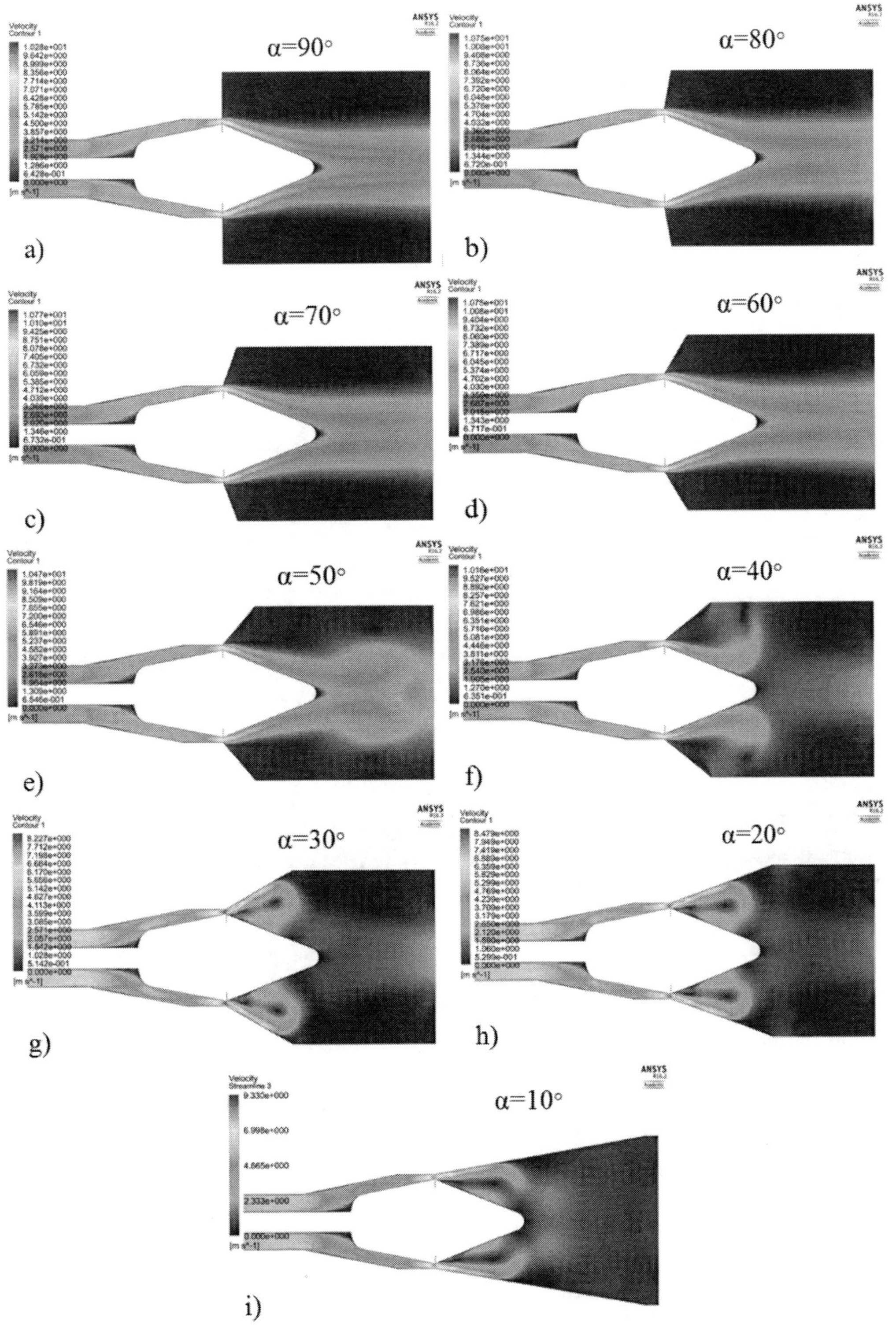


**Fig. 8.** Conturs of velocity on symmetry surfaces for different outer cone angle

## 3. Simulation

### 3.1 Simulation model verification

The first step of test was comparison of velocity profile measured on the real object with results of simulation. The purpose of this comparison is verification of simulation model. The verification was carried out for outer cone angle $\alpha = 30°$, and the velocity profile was measured in the plane of the stream outlet from the outer cone (slot between inner cone and the outer cone ending). The experimental and simulation results are presented in fig. 7.

In fig. 7a was shows comparison of experimental and simulation velocity profile received for a model/test stand without synthetic jet. The main stream deflects to the inner cone by a Coanda effect. Therefore, the velocity near the inner cone is noticeably higher. The difference between simulation and experimental data is slight and velocity profile shift can be due by probe positioning errors. Experiment and simulation are compatible.

Fig. 7b shows velocity profile comparison with activated synthetic jet. The main stream is pushed back toward the outer cone and deflects to it and thus the velocity is higher near the outer cone. In experiment stream is more condensed (near the outer cone) than in simulation. However, this discrepancy is probably case not an accurate the synthetic jet representation in simulation (inaccurate definition of UDF).

Because, the difference between simulation and experimental data was inconsiderable, it was decided that selected simulation model was chosen correctly.

### 3.2. Influence of outer cone angle on stream

In fig. 8 was presented contours of velocity. These pictures shows calculated flow field on symmetry. The detachment of the stream (to inner cone) appears with outer cone angle equal 30° or less. It should pay attention in fig. 8f ($\alpha = 40°$). In this case stream returns in the direction of outer cone. This effect is probably caused by the formation of a vacuum between the stream and outer cone.

The stream attachment to the outer cone takes place only for $\alpha = 30°$ or less. In these cases the stream returns in the direction of inner cone, but initially it is concentrated close to the outer cone wall. The reason for this is a vortex formed between the flowing stream and the inner cone (fig. 9). The vortex causes the stream return to the inner cone. However, it probably helps to leads of stream to outer cone, near the annular gap. It must be noted that the existence of vortex has two consequence. Initially it detaches the stream from the inner cone and then it deflects the main stream towards the inner cone.

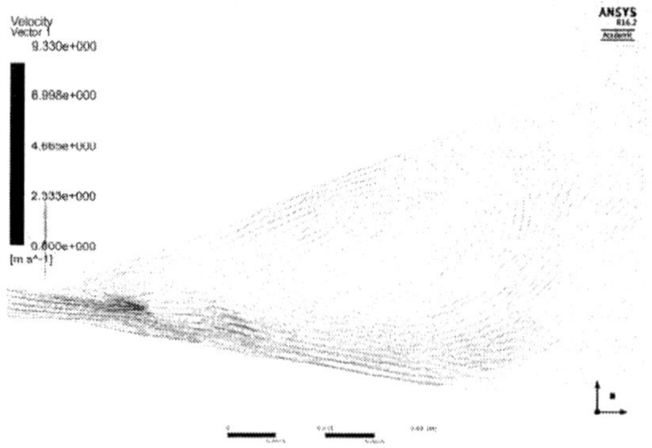


**Fig. 9.** Velocity vectors between inner and outer cone on symmetry plane for $\alpha = 10°$

## 4. Conclusion

In paper was done the simulation of axisymetric annular nozzle for different outer cone angle $\alpha$. It has been demonstrated that for the selected simulation model the numerical data are compared with experimental data. The diminution of the outer cone angle $\alpha$ has a positive effect on the stream redirection. Additionally has been shown the existence of vortices, which has considerable influence on the main stream.

The next step of research should be a investigate the influence of partition (splitter), between inner and outer cone, on the stream. More attention must be devoted to velocity vectors, and SJ boundary conditions.

## Acknowledgements

Scientific work financed from the budget for science in the years 2015/2018 as a research project within the program under the name "Diamond Grant ".

## References

1. V. Tesař, Z. Trávníček, *Způsob dvourežimového řízení průtoku tekútlny a zařízenik prováděni tohoto způsobu* (Ways and means for two-regime control of fluid flow; in Czech), Czech Republic Patent Document Nr. 303280, DOI: 10.13140/ 2.1.4160.8963 filed 30th Jul 2010
2. T. Václav, Z. Broučková, J. Kordik, Z. Trávníček, K. Peszyński, *Valves with flow control by synthetic jets*, EPJ Web of Conf. **25**, EDP Sciences (2012)
3. V. Tesař, M. Pavelka, E. Smyk, K. Peszyński, *Control of flow separation by vestigial" synthetic jet*, Colloq. Fluid Dyn. (2014)
4. B.L. Smith, A. Gezler, *The formtion and evaluation of synthetic jets*, Physic of Fluids **31**, p. 2281 (1998)
5. F. Lalli, A. Bruschi, R. Lama, L. Liberti, S. Mandrone, V. Pesarino, *Coanda effect in coastal flows*, Coastal Engineering **57(3)**, pp.278-289 (2010)
6. N. K. Singh, K. Ramamurthi, *Formation of Coanda jet from sharp-edged swirl nozzle with base plate,*

DOI: 10.1051/epjconf/201714302112

Exp. Therm. and Fluid Science **33.4**, pp.675-682 (2009)

7. A. Dumitrache, F. Frunzulica, T. C. Ionescu, *Mathematical modelling and numerical investigations on the Coanda effect,* INTECH Open Access Publisher (2012)
8. A. Dumas, M. Subhash, M. Trancossi, J.P. Marques, *The influence of surface temperature on Coanda effect*, Energy Procedia **45**, 626-634 (2014)
9. T. Hyhlík, Z. Travnicek, F. Marsik, *Numerical simulation of a synthetic jet into quiescent air*, ISTP-16 (2005)
10. R. Raju, E. Aram, R. Mittal, L. Cattafesta, *Simple models of zero-net mass-flux jets for flow control simulation*, International Journal of Flow Control **1(3)** (2009)
11. V. Dvořák, P. Dančová, P. Švarc, *Experimental investigation into flow in an ejector with four synthetic jets*, EPJ Web of Conf. **25**, EDP Sciences (2012)
12. Z. Trávníček, P. Dančová, J.H. Lam, V. Timchenko, J. Reizes, *Numerical and experimental studies of a channel flow with multiple circular synthetic jet*, EPJ Web of Conf. **25**, EDP Sciences (2012)
13. A.C. Macovei, F. Frunzulica, *Numerical simulations of synthetic jets in aerodynamic applications*, INCAS Bulletin **6(1)** (2014)
14. P. Dančová, J. Novosád, T. Vít, Z. Trávníček, *Experimental and numerical research of synthetic jet array*, EPJ Web of Conf. **114**, EDP Sciences (2016)

EPJ Web of Conferences **143**, 02113 (2017) DOI: 10.1051/epjconf/201714302113
*EFM 2016*

# Heat transfer measurements with TOIRT method

S. Solnař, K. Petera, M. Dostál, T. Jirout

Czech technical university in Prague, Faculty of Mechanical Engineering, Department of Process Engineering

Received: date / Revised version: date

**Abstract.** Temperature Oscillation Infra-Red Thermography (TOIRT) method was used to measure heat transfer coefficients between a flat surface and a confined impinging jet generated by an impeller in a difusor and baffled vessel. The TOIRT method is based on measuring a phase-lag between the oscillating heat flux applied to the heat transfer surface and the surface temperature response using a contactless infra-red camera. The phase lag is in a direct relationship with the heat transfer coefficient.

## 1 Introduction

Within the technical progress we try among the other things to maximize heat transfer in the apparatuses in which heat transfer happens. The intensity of convective heat transfer is characterized by the size of the heat transfer coefficient $\alpha$ which can be found in Newton's law

$$\dot{Q} = \alpha \, S \, \Delta T, \tag{1}$$

where $\dot{Q}$ is convective heat flux, $S$ heat transfer surface and $\Delta T$ temperature difference between surface and surroundings.

There are two main groups of heat transfer measuring methods: time dependent (where the temperature time response is measured) and time independent (where the system is in thermodynamic equilibrium and the heat flux is measured). TOIRT method stands somewhere between these two groups.

TOIRT method was used by S. Freund [1] to measure the temperature response of wall which was heated by an oscillating heat flux. In this work author also measured heat transfer coefficients of various geometries like impinging jet falling to flat surface or water flow in tube. Another usage of TOIRT method [2] measured the heat transfer coefficients of falling water to water level.

Experimental data given by this method show very good agreement with data measured by conventional methods.

## 2 Temperature oscillation method

The principle of Wandelt and Roetzel's TOIRT method [3] is shown on Fig. 1. The method is based on measuring the temperature field on the wall with IR camera. Temperature field depends on two main factors: modulated heat flux and heat transfer coefficients on both sides of the wall with the thickness $\delta$. The oscillating heat flux is modulated by the sine function and the temperature response is also a sine function. An information that we are trying to find is phase lag between these two sine functions.

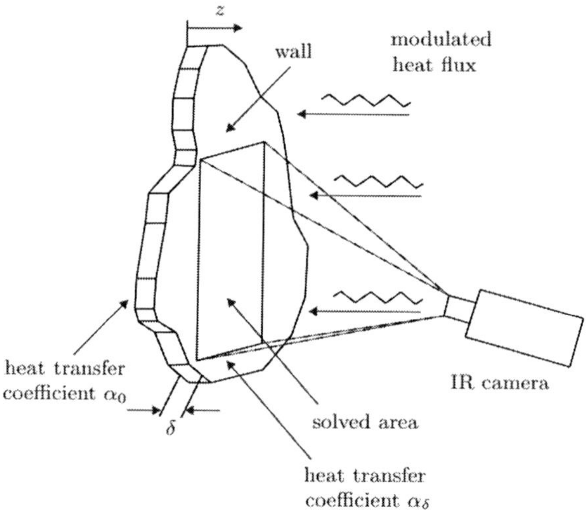


Fig. 1: The principle of Wandelt and Roetzel's method [3]

Time change of a temperature of a homogeneous wall (defined by temperature conductivity $a$) is described by Fourier's equation

$$\frac{\partial T}{\partial t} = a \left( \frac{\partial^2 T}{\partial x^2} + \frac{\partial^2 T}{\partial y^2} + \frac{\partial^2 T}{\partial z^2} \right). \tag{2}$$

While neglecting the lateral heat conduction in the wall (first two partial derivations on right side) it is possible to find an analytical solution in the case of periodical oscillation of heat flux. The boundary conditions are of the third kind (Newton's) and can be written as

$$\lambda \left.\frac{\partial T}{\partial z}\right|_{z=0} = \alpha_0 \, T|_{z=0} \; , \tag{3}$$

$$\lambda \left.\frac{\partial T}{\partial z}\right|_{z=d} = \hat{q} \sin \omega t - \alpha_\delta \, T|_{z=\delta} , \tag{4}$$

where $\alpha_0$ is heat transfer coefficient we are looking for, $\alpha_\delta$ heat transfer coefficient on the other side, $\omega$ angular speed of oscillating heat flux and $\hat{q}$ amplitude of heat flux. Authors used a Laplace transformation for solving this system of equations and found solution

$$T(z,t) = A(z) \sin(\omega t - \varphi(z)). \tag{5}$$

The phase-lag $\varphi(z)$ on the surface with periodic oscillating heat flux can be written as

$$\tan \varphi|_{z=\delta} = \frac{c_1 + 2\xi\psi_0 c_2 + 2\xi^2\psi_0^2 c_3}{2\xi\psi_0(1+r)c_0 + 2\xi^2\psi_0^2(1+2r)c_1 + 4\xi^3\psi_0^3 r c_2 + c_3} \tag{6}$$

where dimensionless parameters $r$, $\psi_0$, $\xi$ and $c_{0,1,2,3}$ are

$$r = \frac{\alpha_\delta}{\alpha_0}, \quad \psi_0 = \frac{\alpha_0 a}{\delta \lambda \omega}, \quad \xi = \delta \sqrt{\frac{\omega}{2a}} \tag{7}$$

$$\begin{aligned} c_0 &= \cosh^2\xi \cos^2\xi + \sinh^2\xi \sin^2\xi \\ c_1 &= \cosh\xi \sinh\xi + \cos\xi \sin\xi \\ c_2 &= \cosh^2\xi \sin^2\xi + \sinh^2\xi \cos^2\xi \\ c_3 &= \cosh\xi \sinh\xi - \cos\xi \sin\xi \end{aligned} \tag{8}$$

These equations show no dependency on the amplitude of heat flux. On the other hand, experimental results are affected by the precise reading of the time lag between the heat flux and surface temperature.

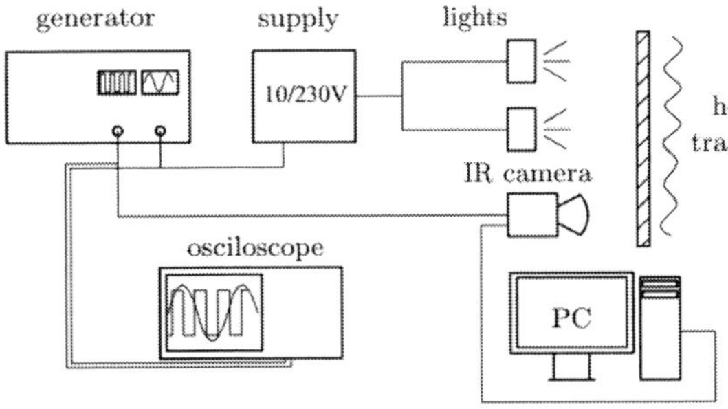


Fig. 2: Wiring schematics

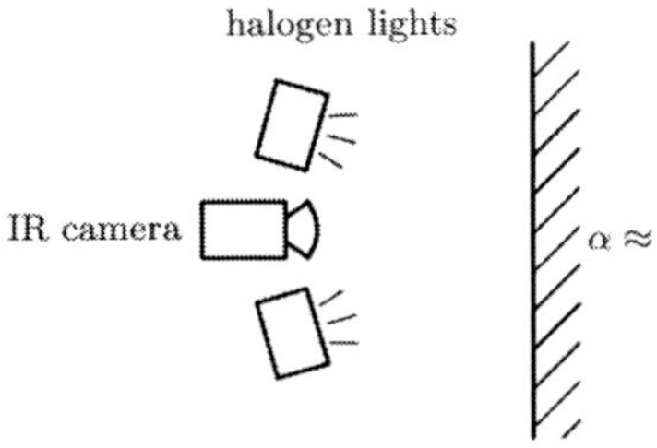


Fig. 3: Time synchronisation

Table 1: Time synchronisation results

| | phase-lag (°) | time delay (s) |
|---|---|---|
| zero heat transfer method | 8.271 | 0.229 |
| recording method | 8.377 | 0.232 |

## 3 Application of method

For TOIRT method, it is needed to generate sinusoidally modulated heat flux and to monitor temperature fields in precisely defined time intervals. The sine function is modulated in the first channel of function generator BK Precision 4052. The voltage output 1 to 10 V is connected with stabilized power supply and controls halogen lights. The halogen lights are used for generating heat flux, because they have relatively small efficiency, so most of the electric power is transformed to heat. We use four 500W halogen evenly distributed lights so the lateral conduction can be neglected.

Freund [1] wrote that if dimensionless thickness of the wall $\xi < 0.5$, the lateral conduction is negligible compared to the heat transfer rate through the wall and the spatial resolution is then expected to be limited by the resolution of the camera. In our case, the $\xi = 0.114$ therefore we were in accordance with this condition.

Second channel of BK Precision 4052 is used for generating impulses that are used for triggering the IR camera thermo IMAGER TIM 160 with resolution 160×120 points.

The sine heat flux with frequency 0,1 Hz was used in our experiments. Sample frequency of IR camera was 10 Hz.

### 3.1 Time synchronisation

This method is based on precise time measurements so it is necessary to take into account the time delay of measuring system itself. The biggest amount of time delay lies in the halogen lights and their thermal capacity, which has to be overcome when the signal changes.

The first method for time synchronisation measures the same case of experiment but on the semi-infinite body. The schema of time synchonization is on Fig. 3. The semi-infinite body leads to almost zero heat transfer and the measured phase-lag is taken as a sync phase-lag.

The second method records halogen lights at a step change of control voltage with speed camera. The results of time synchronisation are summarized in Tab. 1.

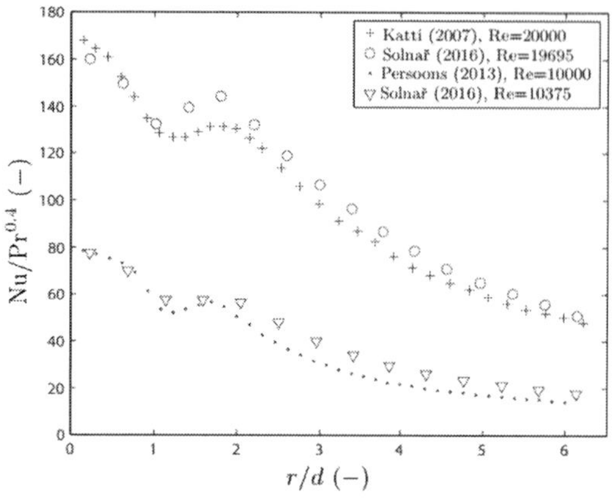


Fig. 4: Results comparsion of impinging jet, $H/D = 1$

## 4 Validation

The method was validated by results measured on a jet impinging perpendicularly the plate. This geometry was measured many times, see Katti and Prabhu [4] or Persoons et al. [5], for example.

Katti and Prabhu [4] measured the jet impinging perpendicularly the target plate (80×160 mm) made of stainless steel foil with thickness 0.06 mm. Air was supplied by an compressor through a calibrated orifice flow meter and was cooled to the required temperature. The target plate was heated by ohmic heating and thanks to the low thickness the lateral conduction was neglected. The temperature field was measured with IR camera positioned on the other side of target plate. The heat transfer coefficients were established from the temperature time response.

Persoons et al. [5] also measured impinging jet with air, but for lower Reynolds numbers. They used a heat flux sensor which measured the temperature difference across a well-defined thermal barrier. The heat transfer coefficients were determined from measured heat flux.

Our impinging jet experiment was made with water. The jet tube is made of aluminium with inner diameter $d = 22$ mm and corresponding length to $45\,d$. The water circuit is opened above the bottom of unbaffled vessel (diameter $D = 392$ mm). The circuit is powered by a centrifugal water pump with speed control via electric frequency inverter. The pump inlet is connected with the water in the vessel so the height of water level is constant. The pump outlet is connected with the induction flow meter (Krohne OptiFlux 5300). The volumetric flow is used for calculating the mean velocity in the tube and Reynolds numbers according to equation

$$\mathrm{Re} = \frac{u\,d\,\rho}{\mu} \tag{9}$$

where $u$ is mean velocity in tube, $\rho$ water density and $\mu$ dynamic viscosity. Temperature of water was measured with Pt1000 platinum ohmic temperature sensor and was $20.5 \pm 1.5$°C through all the measurings. Corresponding water properties at this temperature are: density $\rho = 998.21$ kg m$^{-3}$, dynamic viscosity $\mu = 1{,}002$ mPa s and Prandtl number 6.98.

## 5 Experiment setup

Scheme of our experiments is depicted on Fig. 5 and Fig 6. For both experiments an plastic vessel with diameter $D = 392$ mm is used. The bottom of the vessel is made of stainless steel with thickness 0.99 mm and it is coated by a black matt color on the outer side of the vessel bottom.

Thermo-physical properties of the vessel bottom are: heat conductivity 14.6 W m$^{-1}$ K$^{-1}$, density 7800 kg m$^{-3}$ and specific heat capacity 501 J kg$^{-1}$ K$^{-1}$. The vessel is filled with water to height $H$. Height $h$ of difusor or impeller above botoom is adjusted with accuracy of 1 mm.

The IR camera was set to monitor temperature fields on the entire bottom of the vessel. The temperature fields were measured with accuracy 0.1°C and spatial resolution about 3.3 mm/pixel.

Temperature of water in the vessel has been measured with Pt1000 platinum ohmic temperature sensor.

### 5.1 Impeller in difusor

In the first case (Fig. 5 left), an impeller (six-blade, 45°, diameter $d = 61$ mm) in difusor (diameter $d_v = 70$ mm) was measured. Experiments were performed for ratio $h/d_v = 1$. The water level was maintained at $H = D = 392$ mm. Reynolds numbers were calculated according to equation

$$\mathrm{Re} = \frac{n\,d^2\,\rho}{\mu} \tag{10}$$

where $n$ is rotation speed of impeller and $d$ is diameter of the impeller. The water temperature during measurements was $25.2 \pm 1.1$°C. With corresponding thermophysical properties of water, the Reynolds numbers were from 6 000 to 26 000.

### 5.2 Impeller in baffled vessel

In the second case (Fig. 5 right), an impeller (six-blade, 45°, diameter $d = 131$ mm) in a baffled (4 baffles $B = 40$ mm) vessel was measured. Experiments were performed for ratio $h/d = 1$. The height of water level was maintained at $H = D = 392$ mm. Reynolds numbers were calculated according to equation (10) and they were within range 17 000 − 95 000. The water temperature during measurements was $24.6 \pm 1.8$°C.

DOI: 10.1051/epjconf/201714302113

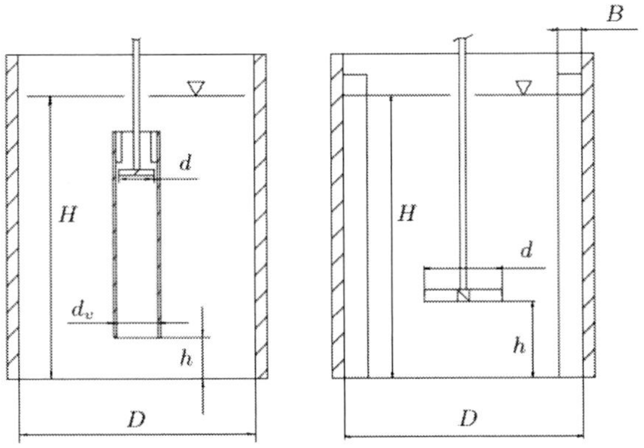


Fig. 5: Drawing of experiments, Left: impeller in difusor, Right: impeller in baffled vessel

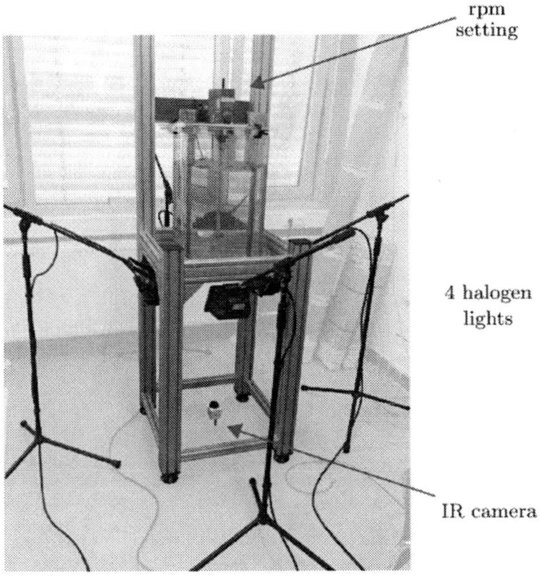


Fig. 6: Experimental setup photo

## 6 Data reduction

The temperature fields obtained from IR camera are transformed into a 3D matrix (160px × 120px × number of shots) in MATLAB. Time dependence of temperature in points corresponds with sinusoid and exponential function (water in vessel is slowly heated). Wandelt and Roetzel [3] invented the method for steady-state temperature oscillations so the exponential part of time dependence of temperature has to be removed (see Fig. 7).

The phase lags $\varphi(x, y)$ are evaluated by using a nonlinear regression procedure on the revised data. After this procedure, the phase lag from time synchronisation is subtracted and the phase lags $\varphi(x, y)$ are transformed to heat

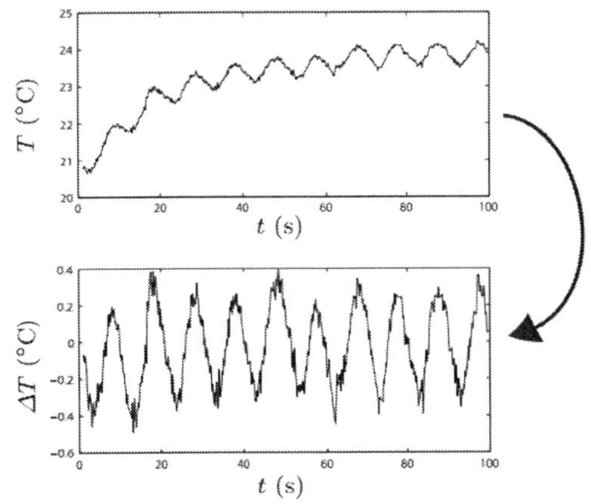


Fig. 7: Example of removing the exponential dependency of measured data

transfer coefficients $\alpha(x, y)$. Nusselt numbers are calculated by

$$\mathrm{Nu} = \frac{\alpha\, d_{\mathrm{char}}}{\lambda} \tag{11}$$

where $\lambda$ is heat conductivity coefficient and $d_{\mathrm{char}}$ is characteristic length ($d$ for impeller in baffled vessel, $d_v$ for difusor). The Nusselt number is then expressed in the form of correlation relationship

$$\frac{\mathrm{Nu}}{\mathrm{Pr}^{0.4}} = C\,\mathrm{Re}^m \tag{12}$$

where $C$ is geometric parameter. The Prandtl number has not been investigated in these experiments.

## 7 Results

Nusselt number dependencies on dimensionless coordinate $r/d$ are on Fig. 8 and Fig. 10. Evaluated parameters of equation (12) are summarized in Tab. 2.

### 7.1 Impeller in baffled vessel

The distributions of Nusselt number in the case of impeller in a baffled vessel show a visible increase around dimensionless coordinate $r/d \approx 0.5$, but this increase is not very significant. This coordinate coresponds with the end of the impeller blades.

Fig. 9 illustrates a surface graph of heat transfer coefficients. As we can see there are areas with higher intensities of heat transfer due to swirling around baffles. These swirling effects are taken into count in averaged Nusselt number $\overline{\mathrm{Nu}}$.

DOI: 10.1051/epjconf/201714302113

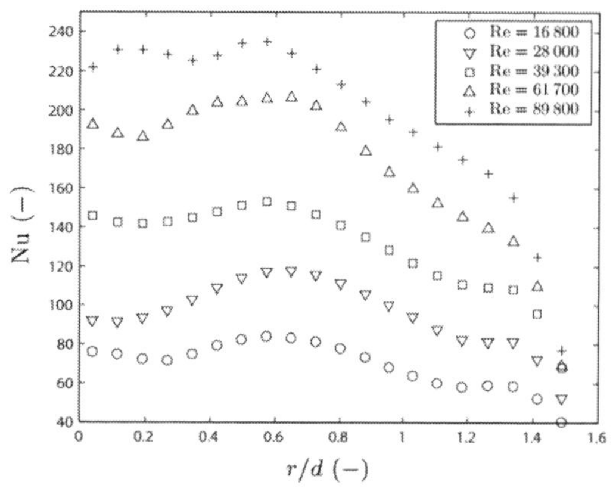


Fig. 8: Nusselt number distribution, impeller in baffled vessel, $h/d = 1$, $\mathrm{Pr} = 6.98$ (water)

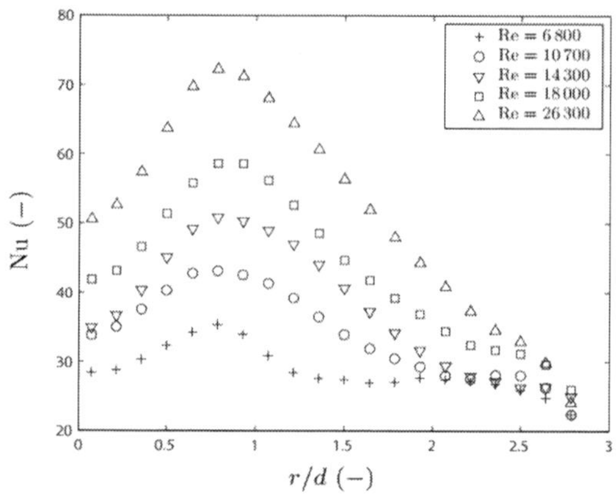


Fig. 10: Nusselt number distribution, difusor, $h/d = 1$, $\mathrm{Pr} = 6.98$ (water)

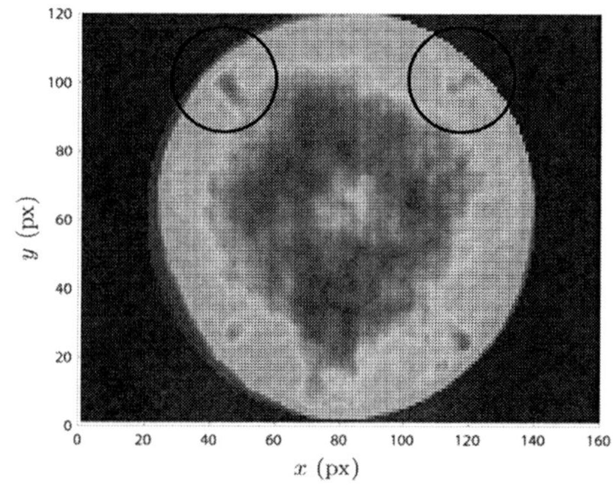


Fig. 9: Intensity of convective heat transfer with highlighted swirling in baffle areas

## 7.2 Impeller in difusor

The distributions of Nusselt number in the case of difusor show significant peaks arround $r/d \approx 0.8$ where it is about 50% higher than the Nusselt number at dimensionless coordinate $r/d = 0$. In this case, we can conclude some similarities between the impinging jet and impeller in difusor. The peak of convective heat transfer in the case of impinging jet is located at dimensionless coordinate about $r/d = 2$. In the case of impeller in difusor, this peak is located at coordinate about $r/d = 1$.

# 8 Measurement accuracy

This measurement technique has no dependency on the amplitude of heat flux so the minimum of the heat flux is given by a sensitivity of IR camera.

On the other hand, experimental results are affected in time scale. One degree error in the phase-lag (in our case that means: $1\,\mathrm{deg} = 0.02\overline{7}\,\mathrm{s}$ of time-lag) can cause an error in percent of heat transfer coefficient.

As we can see in (7) results are in a relationship with thermo-physical properties of the wall and with the thickness of the wall. Thermo-physical properties have no significant importance for the results, but the thickness of the wall has. Half a milimeter error in measuring the wall can cause an error in tens of percents of heat transfer coefficient.

Table 2: Experimental results in the form of equation (12)

| | Impeller in baffled vessel | | Impeller in difusor | |
|---|---|---|---|---|
| | $C$ | $m$ | $C$ | $m$ |
| $\mathrm{Nu}_0$ | 0.0907 | 0.6811 | 0.5261 | 0.4487 |
| $\overline{\mathrm{Nu}}$ | 0.1016 | 0.5822 | 0.6466 | 0.4188 |

# 9 Conclusion

The TOIRT method was used for heat transfer measuring and results of impinging jet experiments were compared with the literature data to validate the method. Results are in good agreement with both here cited references [4], [5].

Heat transfer coeficients on a flat bottom of a vessel were measured in wide range of Reynolds numbers for two geometries: impeller in baffled vessel and impeller in difusor. In the case of the impeller in baffled vessel, the effect of swirling in baffle areas was found. These swirling effects were taken into count in averaged Nusselt number $\overline{\mathrm{Nu}}$.

In the case of impeller in difusor, similarities with impinging jet falling to flat surface were found. The peak of convective heat transfer in case of impinging jet is located at dimensionless coordinate about $r/d = 2$. In the case of impeller in difusor this peak is located at coordinate about $r/d = 1$.

## Acknowledgment

This work is supported by research project of the Grant Agency of Czech Republic No. 14-18955S.

## Nomenclature

| | |
|---|---|
| $a$ | thermal diffusivity, $a = \lambda/\rho\, c_\mathrm{p}$ ($\mathrm{m^2\, s^{-1}}$) |
| $A$ | amplitude (-) |
| $B$ | baffle dimension (m) |
| $C$ | geometric constant (-) |
| $c_{0,1,2,3}$ | dimensionless parameters |
| $d$, $D$ | diameter (m) |
| $d_\mathrm{char}$ | characteristic dimension (m) |
| $d_v$ | cylinder diameter (m) |
| $h$, $H$ | height (m) |
| $m$ | exponent (-) |
| $n$ | revolutions ($\mathrm{s^{-1}}$) |
| Nu | Nusselt number, Nu$= \alpha\, d_\mathrm{char}/\lambda$, (-) |
| $\mathrm{Nu_0}$ | Nusselt number at $r/d = 0$ |
| $\overline{\mathrm{Nu}}$ | averaged Nusselt number |
| Pr | Prandtl number, Pr$= \rho\, c_\mathrm{p}\, \nu/\lambda$ (-) |
| $\dot{Q}$ | heat flux ($\mathrm{W\, m^{-2}}$) |
| $\hat{q}$ | amplitude of heat flux ($\mathrm{W\, m^{-2}}$) |
| Re | Reynolds number, Re$= n\, d^2\, \rho/\mu = u\, d\, \rho/\mu$ (-) |
| $S$ | surface ($\mathrm{m^2}$) |
| $t$ | time (s) |
| $T$ | temperature (°C) |
| $u$ | velocity ($\mathrm{m\, s^{-1}}$) |
| $\alpha$ | heat transfer coefficient ($\mathrm{W\, m^{-2}\, K^{-1}}$) |
| $\delta$ | dimension (m) |
| $\lambda$ | heat conductivity ($\mathrm{W\, m^{-1}}$) |
| $\mu$ | dynamic viscosity (Pa s) |
| $\nu$ | kinematic viscosity ($\mathrm{m^2\, s^{-1}}$) |
| $\rho$ | density ($\mathrm{kg\, m^{-3}}$) |
| $\varphi$ | phase lag (°) |
| $\omega$ | annular speed ($\mathrm{s^{-1}}$) |
| $r$, $\psi_0$, $\xi$ | dimensionless parameters (-) |

## References

1. S. Freund, *Local Heat Transfer Coefficients Measured with Temperature Oscillation IR Thermography* (Hamburg, 2008)
2. S. Freund, A. G. Pautsch, T. A. Shedd, S. Kabelac, International Journal of Heat and Mass Transfer **50**, (2007), p. 1953–1962
3. M. Wandelt and W. Roetzel, Quantitaive Infrared Thermography 96, (1997) p. 189–194
4. V. Katti and S. Prabhu, International Journal of Heat and Mass Transfer **51**, (2008), p. 4480–4495
5. T. Persoons, K. Balgazin, K. Brown, D. B. Murray, Journal of Heat Transfer **135**, (2013)
6. S. Freund, S. Kabelac, International Journal of Heat and Mass Transfer **53**, (2010), p. 3764–3781
7. N. Zuckerman, N. Lior, Advances in Heat Transfer **39**, (2006), p.565–631
8. K. Rehme, *Handbook of single-phase convective heat transfer* (A Wiley-Interscience Publication, 1987)
9. K. W. Treuren, Z. Wang, P. T. Ireland, T. V. Jones, The American society of mechanical engineers **47**, (1993)
10. D. Lytle, B. W. Webb, International Journal of Heat and Mass Transfer **37**, (1994), p. 1687–1697
11. M. Hogler, Advances in Heat Transfer **13**, (1977), p. 1–59
12. R. Viskanta, Experimental Thermal and Fluid Science, (1993), p. 111–134
13. V. V. Katti, S. N. Yasaswy, S. V. Prabhu, Heat and Mass Transfer, (2011), p. 237–244
14. A. Herchang, J. J. Yuh, Y. Jer-Nah, International Journal of Heat and Mass Transfer **45**, (2002), p. 4096–4078

# Low cycle fatigue numerical estimation of a high pressure turbine disc for the AL-31F jet engine

*Miroslav* Spodniak[1,*], *Marek* Klimko[2], *Marián* Hocko[1], and *Pavel* Žitek[2]

[1]Technical University of Kosice, Department of Aviation Engineering, 7 Rampová, 04221 Košice, Slovakia
[2]University of West Bohemia, Department of Power System Engineering, 22 Univerzitní, 30614 Pilsen, Czech Republic

**Abstract.** This article deals with the description of an approximate numerical estimation approach of a low cycle fatigue of a high pressure turbine disc for the AL-31F turbofan jet engine. The numerical estimation is based on the finite element method carried out in the SolidWorks software. The low cycle fatigue assessment of a high pressure turbine disc was carried out on the basis of dimensional, shape and material disc characteristics, which are available for the particular high pressure engine turbine. The method described here enables relatively fast setting of economically feasible low cycle fatigue of the assessed high pressure turbine disc using a commercially available software. The numerical estimation of accuracy of a low cycle fatigue depends on the accuracy of required input data for the particular investigated object.

## 1 Introduction

A low cycle fatigue (LCF) is a very common task not only in engineering industry but in aviation industry as well. The first LCF's occurrence in aviation industry along with its association with new technologies and new materials has demanded the requirement for a special attention dedicated to this theme. As it is well known, the LCF of jet engines is linked with more factors and effects.

A jet engine is a heat machine in which thermal energy is delivered from the combustion of fuel with air and consequently converted to useful work. [1] A jet engine and its constructional parts are subject to very high loads and moments, which depend on the engine regime. Considering the angle of changing load in jet engine parts in dependence on time, it is possible to distinguish two kinds of loads; quasi static and dynamic load. In calculation, it is necessary to examine each part individually because of a variety of the load. Jet engine parts and construction assemblies are subject to load and aerodynamic forces, which are due to streaming of air and hot burning gases through the engine sections. The actual pattern of aerodynamic force depends on the difference of pressures and a variation of momentum. This force has two compounds, axial and tangential. Other load is caused by centrifugal load and momentum, which arise by the movement of mass and fly maneuvers. [2] On rotating parts of jet engine, centrifugal forces emerge, which have influence on the LCF. Centrifugal forces are subject to the rotor and its parts, such as a shaft, a disc, a compressor and turbine blades etc. Spinning of these parts sets off centrifugal forces, which have to be included in numerical counts. Another meaningful factor for counting a life estimation is temperature. A mixture of fuel and compressed air is burned in a combustion chamber under very high temperature. From the combustion chamber, hot gases flow across high pressure and low pressure turbine to exhaust a nozzle. Besides transition, these hot gases create thermal stress, which has the effect on cooling the pressure turbine disc and cooling the turbine rotor blades. [3] The condition of jet engine parts influence several factors, either directly or indirectly. It also includes the atmospheric conditions, in which a jet engine is operated, then the speed of flight, flight level etc.

Jet engine parts load, as it is known, is related to high temperatures. The fatigue of a pressure turbine disc is caused by cyclic changes of load, which are induced by the temperature gradient, centrifugal forces of spinning parts of the disc and rotor blades as well. Cyclic temperature relief has a significant impact during transitional regimes of a jet engine. These varieties of the temperature cause thermal fatigue of material, the beginning of a disc cracks and their gradual increasing.

The most unfavourable effect of cyclic fatigue results in the shutdown of a jet engine in high altitudes with a subsequent autorotation. This process is associated with severe cooling down of hot jet engine parts. As a result, this process creates very a high strain inside and outside the material. Standardly, the highest thermal gradients are reached in time from five to ten minutes, from the beginning of a transitional process. The cycle fatigue is linked and accompanied also with jags, holes, disc rims, in which stress is concentrated.

* Corresponding author: miroslavspodniak1@gmail.com

## 2 Constructional characteristic of the pressure turbine disc AL-31F jet engine

A pressure turbine of a bypass turbofan jet engines like AL-31F, that is used in Su-27 aircraft, is axial, two shafts. The pressure turbine consists of a one stage high pressure turbine and a one stage low pressure turbine. Both of them have rotor blades and stator vanes mounted, which are cooled by air. On lower regimes it is for the purpose to increase the engine thrust used partly for disconnection of cooling both of pressure turbines. [4]

A high pressure turbine consists of a disc, rotor blades, gudgeon and a high pressure turbine shaft.

A high pressure turbine disc is made by ironwork with consequent chip machining. The disc inner rim rabbets are made in order to ensure attaching rotor blades. There are 90 rotor blades with rabbets to locate elastic blade locks given to ensure axial movement and slant holes for cooling the air input. On the right side of the disc, there is located a labyrinth seal rim and a rim, which is used to remove disc. In the disc flange section there are roll holes drilled to fit screws that are connecting the shaft, the disc and the high pressure turbine rotor pivot. The high pressure turbine AL-31F prime parameters:

- The overall ratio of expansion: $\pi_{Tc} = 2.9$.
- Top efficiency: $\eta_{Tc} = 0.87$.
- Circumferential velocity: $u = 510$ m.s$^{-1}$.
- RPM: $n = 13\ 300$ min.$^{-1}$.
- The temperature of the inlet gas: $T_{3c} = 1665$ K.
- Weight: $G = 70$ kg.
- Speed ratio: $u/c_1 = 0.46$.
- High pressure turbine gas flow rate: $Q_{pl.VTT} = 70$ kg.s$^{-1}$.
- Low pressure turbine gas flow rate: $Q_{pl.NTT} = 72$ kg.s$^{-1}$.

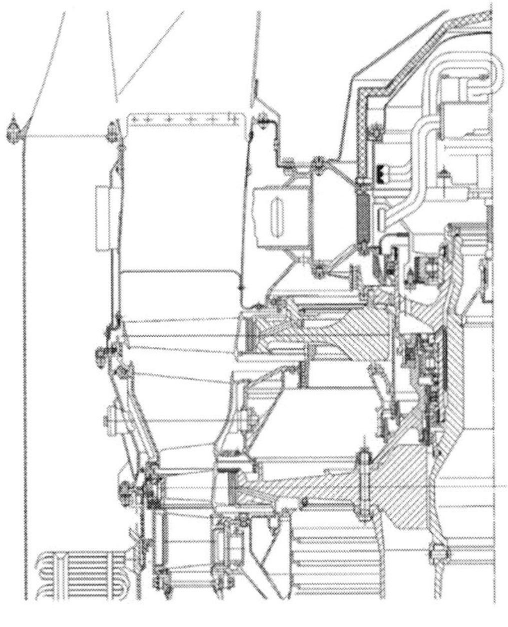

**Fig. 1.** Two stage pressure turbine configuration

## 3 Geometry and finite element model of the AL-31F high pressure turbine disc

The first step in the life evaluation of the jet engine AL-31F high pressure turbine disc was a creation of a 3D model in the CAD program. Modeling was carried out by Solid modeling in 3D program SolidWorks. The program SolidWorks allows Solid modeling and to execute needed analysis to use for a life fatigue counting. A life cycle fatigue analysis of AL-31F jet engine high pressure turbine disc was carried out by the finite element method in SolidWorks. The information from the available resources and technical documentation were used to create an identical 3D high pressure turbine model of the AL-31F jet engine. In order to simplify the calculation, the segment from 3D modeling was formed. The segment is shown on the Fig. 2; the meshing of the segment was made in SolidWorks program, too. In order to be able to perform static analyses in SolidWorks simulation, it is necessary to apply boundary conditions (Fig. 3). [5]

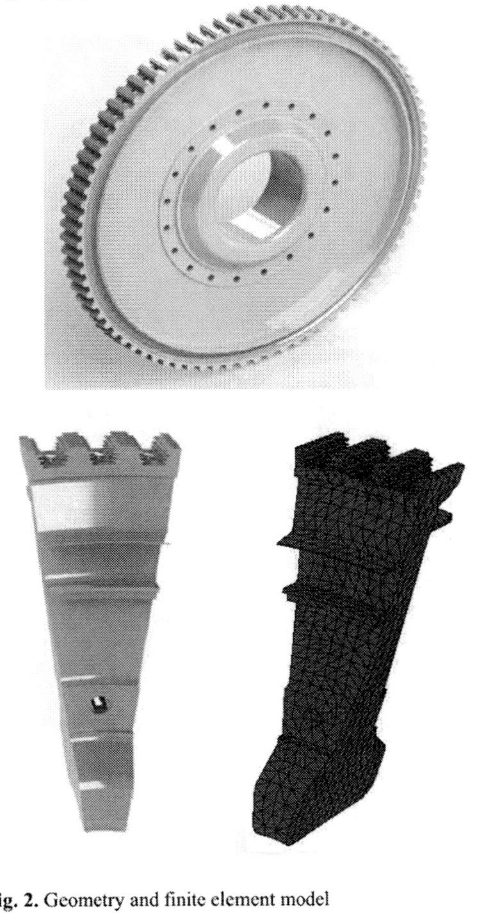

**Fig. 2.** Geometry and finite element model

The life fatigue calculation of the jet engine AL-31F high pressure turbine has been simplified in terms of a commercial engine mission. Two fundamental regimes of a jet engine have been used, idle run and maximal regime. As well as engine regimes, material properties of the disc are necessary. Other simplification lay in material properties dependent on temperature.

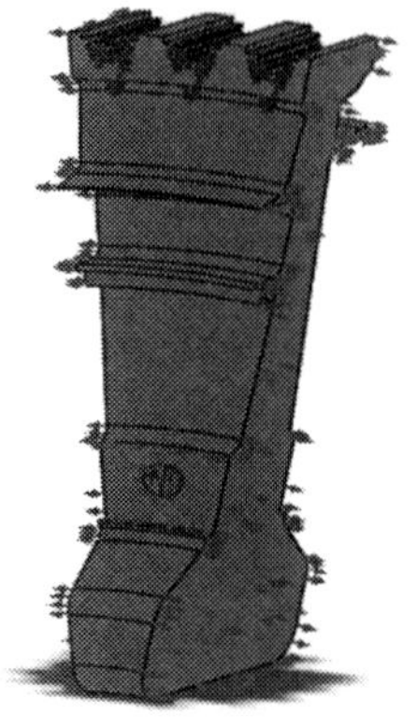

**Fig. 3.** Boundary conditions of the disc element

The static analyses have been carried out in the finite element method; these analyses dictate the stress of the disc that is operated on the idle run and maximal regime. Before we execute any analyses, it is necessary to find out the forces system of the disc, as described in the first chapter. The forces system of the jet engine AL-31F high pressure disc is shown in the Fig. 4. For the purpose of this numerical estimation, there were two thermal analyses proceeded – "thermal 1" and "thermal 2". The thermal analysis marked as "thermal 1" represents the heat load of the disc running on the idle run. The thermal analysis "thermal 2" defines the heat load on the maximal regime. In Fig. 4, we can see forces system of the AL-31F disc, where its resulting load is produced by the resulting centrifugal load caused by blades $F_o$, centrifugal load of the disc, centrifugal load of screws $F$, then axial forces on screws $F_A$, $F_t$ friction force, $M_t$ torque flange, $P_s$ overlap pressure of screws. [3], [6]

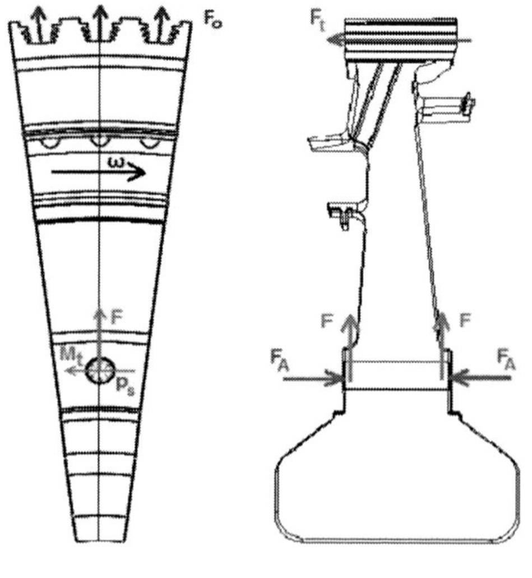


**Fig. 4.** Forces system AL-31F disc element

Fatigue strength properties are necessary to evaluate the total life, according to the mathematical formula:

$$\frac{\Delta\varepsilon}{2} = \underbrace{\frac{\sigma'}{E}(2N_f)^b}_{\text{ELASTIC}} + \underbrace{\varepsilon_f'(2N_f)^c}_{\text{PLASTIC}} \quad (1)$$

The fatigue failure under LCF is based on a characteristic plastic deformation:

$$\frac{\Delta\varepsilon}{2} = \varepsilon_f'(2N_f)^c \quad (2)$$

where:
- $\Delta\varepsilon$: local strain range
- $\sigma'$: cyclic stress amplitude
- $E$: Young's modulus
- $c$: fatigue ductility exponent
- $N_f$: number of cycle to failure
- $\varepsilon$: fatigue ductility coefficient [5]

## 4 Thermal analysis of the AL-31F high pressure turbine disc

As discussed in the previous section, temperature has a major effect on LCF. The disc is subject to high thermal gradients, which are given by the difference of high temperature. Another influence brings about the fact, that the disc is cooled by air from the compressor. The temperature of the disc AL-31F jet engine achieves values within 370°C till 590°C on the maximal regime. The biggest thermal difference occurs between the outer rim, respectively in a part of the blade mounting and the inner rim. Shutting down the jet engine in high altitude has a particularly dangerous impact, when the engine parts with high temperature are sharply cooled, and vice versa, after repeated shutting down, they are swiftly heated, which results in the induction of an extremely high load. In order the disc to be totally overheated for the working temperature, the required period is 5 to 10 minutes, after finishing the transition process.

The temperature distribution at the radius of the disc and its temporal changes is dependent on constructional parameters rather than on the method of cooling, on the perfection of the jet engine regulatory system and flight cycles of the jet engine. During the starting process, the thermal gradients grow at the radius of the disc and are associated with growing of compressive stress and thermal stress in the disc rim, which is heated faster. In the central part of the disc, which has a higher thermal inertia, tensile stress arises on a start-up.

The thermal analyses for the purposes of this article have been carried out twice for two different regimes; first it is the idle run and, second, the maximal regime. In this task, the numerical estimation has been simplified with the temperature of the sectors at the disc surface. The result of the thermal analysis is shown in the picture 5, in terms of the temperature dependence and the diameter of the disc. [7]

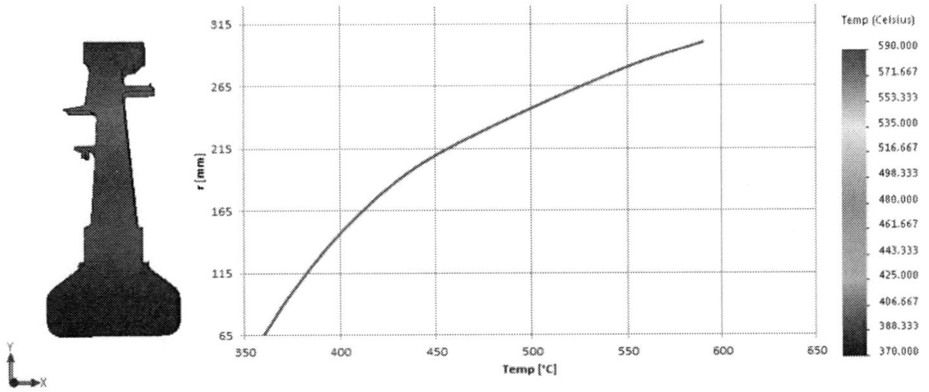


**Fig. 5.** Temperature profile

## 5 Finite element method analyses

The finite element method analysis (FEA) for life fatigue determination has been carried out. In this analysis, information acquired in thermal and structural studies has been used, as discussed in the previous sections. These studies are in accordance with the idle run and the maximum regime. Thermal load combined with a centrifugal load of a disc, the centrifugal load of blades, screws and other loads were counted in two strength analyses the for idle run and maximal regime as well as for thermal analyses.

The most critical values of the reduced stress have been occurring in both static analyses, where the most critical section was concentrated on the edge of the disc flange in the inner rim and in holes for mounting screws. The outcome of this study is represented by Fig. 6.

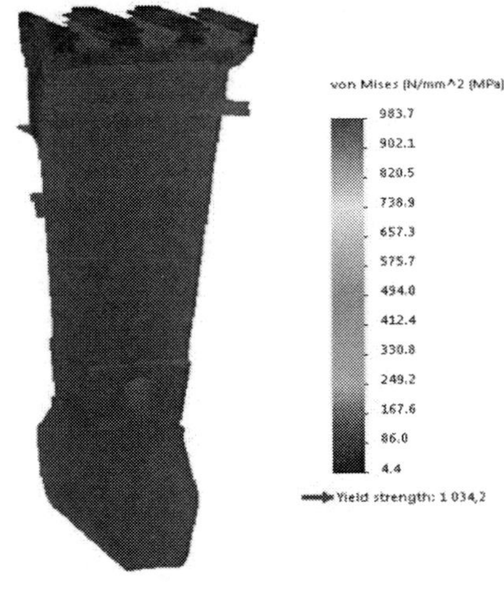


**Fig. 6.** Stress of von Mises

## 6 Fatigue life assessment of the AL-31F jet engine high pressure turbine disc

The strain and stress obtained from static analyses are used for LCF life analyses. In order to carry out the life analysis, there is another information required. Apart from linear static analyses, there is material information, which is necessary for life analyses, such as Yield Strength, Tensile Strength, Young Modulus, Poisson ratio and SN curve. For the purposes of this study, the data was generated by 3D CAD software. The cyclic load is based on a periodic load between the two static analyses, the idle run and the maximal regime.

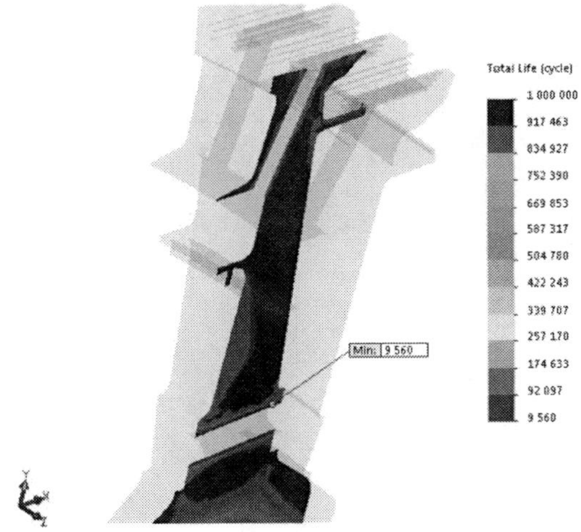


**Fig. 7.** Total life

The linear static analyses correspond with started regimes in CAD software, which provides information about the most stressed part of the disc, thus the highest concentration of the reduced strain in the disc. This information is necessary for a life estimation along with other material information, as discussed previously.

The result of the fatigue analysis has indicated a number of cycles, which overrun, and would theoretically mean the beginning of the fatigue crack propagation. In the section view, there are shown the most critical parts of the element disk, which acquire the minimum number of cycles. In Figure 7, there is clearly shown the most critical part of AL 31f engine disc. Thus the hole for a screw is the most critical part, which reaches minimum Total Life to crash. The most critical load is concentrated in the section of the disc flange.

## 7 Conclusion

The exact determination of the LCF jet engine high pressure disc, in particular the type of the jet engine, largely affects the economic efficiency of the jet engine.

The accuracy of results obtained from the static analyses and from the life fatigue study of the AL-31F jet engine high pressure disc is slightly influenced in terms of some simplifications, which have been discussed in the previous sections. For an accurate evaluation, essential exact data are required, most of which it is possible to acquire just from a manufacturer. As for temperature data, this is the most important fact for the LCF disc evaluation. To by exact in the life estimation, it is necessary to have data about temperature of the sectors at the disc surface. Despite these facts, this article presents a procedure of LCF estimation for a concrete jet engine and states critical sections of a high pressure disc.

LCF determination is a demanding and lengthy process, numerical counts not always correspond with real situations. The accuracy of the LCF estimation is possible to verify by practical tests, in which there are simulated real conditions of high pressure turbine operating. Such tests are compared with the numerical method of LCF determination more demanding in terms of time as well as economic and ecological viewpoint, but despite this fact in practice, these tests have to be executed.

This work has been supported by the project SGS-2016-045 (Increasing the efficiency, reliability and service life of power system machines and devices 4) and the Ministry of Education, Youth and Sport Czech Republic Project LQ1603 (Research for SUSEN). This work has been realized within the SUSEN Project (established in the framework of the European Regional Development Fund (ERDF) in project CZ.1.05/2.1.00/03.0108).

## References

1. F.C. Rogers, Y.R. Mayehew, *Gas turbine theory - 4th edtion* (London, 1996), ISBN 0-582-23632-0
2. J. Kamenický, Z. Linhart, *Konstrukce leteckých motoru II.* (Brno, 1989)
3. J.L. Kerrebrock, *Aircraft engines and gas turbines II* (Massachusetts Institute of Techonology, 1992)
4. S.V. Falaleev, *Construction AL-31F turbofans* (Samara)
5. M. Vable, *Mechanics of materials II* (Michigan Technological University, 2012)
6. P.P. Walsh, P. Fletcher, *Gas turbine performance* (2004), ISBN 0-632-06434-X
7. K. Semrád, *Pevnostná analýza disku plynovej turbiny motora DV-2* (Košice, 2001)

DOI: 10.1051/epjconf/201714302115

# Numerical simulation of flood barriers

Pavel Srb[1,*], Michal Petrů[1], and Petr Kulhavý[1]

[1] Institute for Nanomaterials, Advanced Technologies and Innovation, Technical University of Liberec, Studentská 2, 461 17, Liberec 1, Czech Republic

**Abstract.** This paper deals with testing and numerical simulating of flood barriers. The Czech Republic has been hit by several very devastating floods in past years. These floods caused several dozens of causalities and property damage reached billions of Euros. The development of flood measures is very important, especially for the reduction the number of casualties and the amount of property damage. The aim of flood control measures is the detention of water outside populated areas and drainage of water from populated areas as soon as possible. For new flood barrier design it is very important to know its behaviour in case of a real flood. During the development of the barrier several standardized tests have to be carried out. Based on the results from these tests numerical simulation was compiled using Abaqus software and some analyses were carried out. Based on these numerical simulations it will be possible to predict the behaviour of barriers and thus improve their design.

## 1 Introduction

Floods in terms of their occurrence are inseparable, spatially and temporally irregular part of the water cycle. Its erosion, sedimentation, transports, mechanical and other effects are an important element of landscape development process [1].

According to the European directive [2], the term "flood" means the temporary covering by water of land not normally covered by water. This shall include floods from rivers, mountain torrents, Mediterranean ephemeral water courses, and floods from the sea in coastal areas, and may exclude floods from sewerage systems.

The term "flood risk" means the combination of the probability of a flood event and of the potential adverse consequences for human health, the environment, cultural heritage and economic activity associated with a flood event.

In the Czech Republic there are floods caused especially by torrential rains, prolonged rains, melting snow and moving of ice floe. In 1997 the Czech Republic was hit by flood, whose size and damaging impact was surprising [3]. The flood of a comparable size had not affected the Czech Republic perhaps in several centuries. Nobody could have expected that there would be a flood of similar magnitude over a very short period of several years. Since 1997 the Czech Republic has been hit by several other very devastating floods. Both loss of life and material damage are shown in table 1.

These recent floods disproved the myth that from natural point of view the next large flood can be expected in the longer period of time. Further flood can be expected at a distance of few years. Documented historical floods in the territory of the Czech Republic from eg. 1890, 1897 and 1903 demonstrate that the grouping of flood in a shorter period of time is not so unusual.

**Table 1.** Loss of life and material damage [4]

| Year | Loss of lives | Damage [mil. CZK] |
|---|---|---|
| 1997 | 60 | 62,600 |
| 1998 | 10 | 1,800 |
| 2000 | 2 | 3,800 |
| 2001 | 0 | 1,000 |
| 2002 | 19 | 75,100 |
| 2006 | 9 | 6,200 |
| 2009 | 15 | 8,500 |
| 2010 | 8 | 15,200 |
| 2013 | 15 | 15,400 |

Floods in the Czech Republic has always occurred, and therefore we must reckon with them in the future also. Each flood is different, formation mechanisms of same type of flood may be similar, but the course and effects on the environment are unique. Follows from the foregoing it is necessary to learn and expand our knowledge and experience from previous floods, for reduction of life lost and damage caused by future floods [5,6].

* Corresponding author: pavel.srb@tul.cz

## 2 Flood barriers

There are many requirements on flood barriers and their components [7]. All materials used in flood abatement equipment shall be suitable for the intended application. Flood abatement equipment have to be constructed using corrosion resistant materials. The materials used for the seals must be resistant to biological degradation. Manually operated caps and valves, and mechanisms, must have a securing, locking, and supervising component. The design of automatic opening barriers shall be such that major debris that could impact the barrier's functionality is prevented from entering the system. Other debris shall be able to be flushed from the system during periodic maintenance. All components of the flood abatement equipment products need to be regularly tested. The components required for the flood abatement equipment shall be designed for proper system functionality, and maximum long term reliability. Flood abatement equipment designs should take into account the possibility of component failure and the potential of that failure to impair the effectiveness of the system. Such impairments shall be minimized through fail-safe, redundant component, over-design, de-rating, or other means demonstrating equivalent reliability.

Figure 1. shows one typical design of flood barrier with ground structure. This barrier contains 2 portable pillars, several stop logs and tightening devices. The portable pillars are aluminium H-shaped poles which stop logs are inserted into. The stop logs are hollow extruded profiles with sealing in the bottom.

**Fig. 1.** Flood barrier design.

## 3 Barrier testing

Flood barriers must be subjected to many tests like hydrostatic strength test, system leakage, vibration resistance, impact and wear resistance, salt spray corrosion and many others. Performance testing on components may be conducted for an individual component, component assembly and also for the entire barrier system. This paper focuses especially on the debris impact test. A flood barrier should withstanding multiple impacts from floating debris. The leakage rate shall not exceed 3.1 litres per minute per meter length. The barrier's length is measured along centre point of the barrier's seal to the ground. In addition, the permanent deflection of the barrier shall be less than or equal to 15 cm. Limit water level should be 2/3 of maximum water depth. Two tests shall be conducted with two different size logs of diameter 30 and 43 cm, weight 277 and 358 kg.

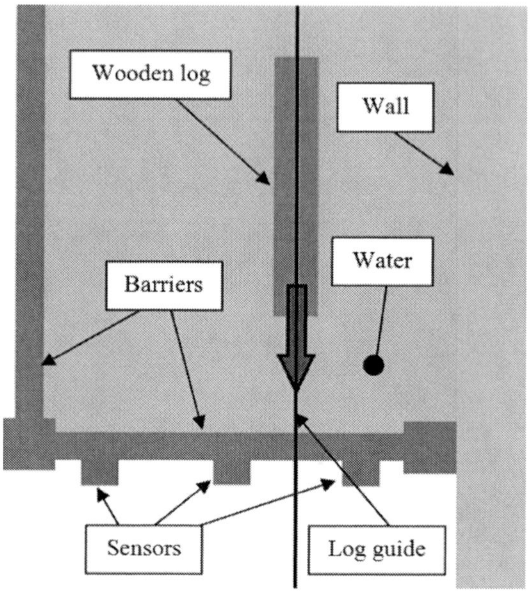


**Fig. 2.** Layout of experiment

The experiment [8] (Fig. 2 and 3) was assembled in hydrodynamic laboratory at a company called JAP-Jacina in the best possible accordance to the standard [7]. Impact test was conducted by impactor (wooden log) several times at speed rating from 0.2 to 2 $m.s^{-1}$.

**Fig. 3.** Real assembly of experiment.

The whole testing process was measured by system DEWETRON DEWE 500. The measured values provided information of barrier displacements in three positions, deformation of the barrier by tensometer and movement of the wooden log by wire displacement sensor. Acquired data including the synchronized with a high speed video camera were recorded to measuring device. The maximum deflection value of flood barrier at the moment of impact was determined as the difference of values in the centre sensor and corner sensors. This consideration is based on the assumption that the corner sensors measure the deformation of the rubber seal in the anchoring pillars. This approach eliminates the deformation of the seal, and shows the actual deflection of the barrier.

## 4 Numerical simulations

A problem involving fluid flow interaction with structures is called fluid-structure problem (FSI), for example, the flow around a structure interacting with the structure and vice versa. The flow usually undergoes large deformation, but the structure does not have to be fixed and rigid but can be moveable and deformable [9]. Examples are sloshing of tanks in launchers, limit cycle oscillations of wings or instabilities in the wind interaction with cable stayed bridges.

### 4.1 CEL method

Coupled Eulerian Lagrange Analysis technique combines two mesh approaches Lagrangian and Eulerian in one analysis [9, 10]. This method was developed initially for the simulation of the inflation of an airbag, and then followed problems like tank sloshing due to impact loads, collisions between a bird and aircraft and aquaplaning.

**Fig. 4.** The water-filled bottle drop test.

The example of CEL simulation is shown in Fig 4. there is the result of the water filled bottle drop test [11]. The bottle hit the floor at a skew angle, with one of the bottom corners experiencing the initial impact. A simulation for the bottle accounts both the exterior forces on the bottle from the floor impact, as well as the interior forces of the water pushing against the bottle.

This method consists of mesh treatment, surface reconstruction and solving fluid and structure equations. The purpose of this technique is to avoid mesh problems when performing simulations that involve extreme deformations. The most common mesh approach is the Lagrangian one, but that is not suitable for excessive distortion cases. For those cases Eulerian meshing approach is more suitable. CEL technique captures the strengths of both Lagrangian and Eulerian methods.

The CEL method enables the user to selectively mesh the analysis components. The large deformed bodies (fluid, gas, etc.) can be meshed using Eulerian technique and remaining moving structure using the conventional Lagrangian technique. The interaction behavior between them is given by contact definition

In Lagrangian simulations each element of the mesh is used to represent a volume of material. The material move is possible only with mesh deformation. The Lagrangian approach is suitable for the majority of structural problems, but in simulations where the material is under large deformations, elements will become distorted and the simulation cannot continue. Figure 5 shows the Lagrangian simulation, on the left undeformed object, on the right deformed object.

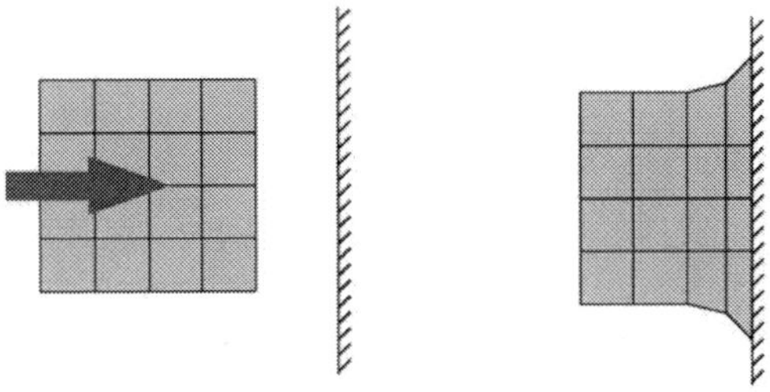

**Fig. 5.** Lagrangian simulation

In Eulerian simulations, the grid remains stationary during the simulation and material flows through elements that do not deform. The Eulerian mesh is typically a simple rectangular grid of elements constructed to extend well beyond the Eulerian material boundaries, giving the material space to move and deform. Figure 6 shows Eulerian simulation, the initial shape on the left and the deformed one on the right.

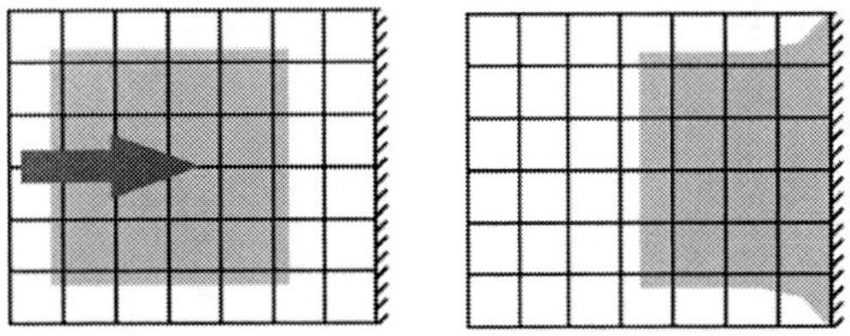

**Fig. 6.** Eulerian simulation

### 4.2 Assigning materials

For assigning material in Eulerian analysis it is necessary to use an assignment predefined field. Material assignment predefined fields rely on the concept of material volume fractions. During an Eulerian analysis there is a tracked material presentation in each element in terms of a volume fraction. The volume fraction represents the percentage of the element's volume that is occupied by a given material. For elements that are partially filled with material the exact geometric composition of the material within the element is not known. Abaqus interpolates the material volume fractions from adjacent elements to estimate the material boundaries within the element. Volume fractions are expressed as a number between zero and one, this number means the percentage of filling, e.g. 0.32 means that 32 % of element is filled by assigned material. Figure 7 shows an example of a volume fraction.

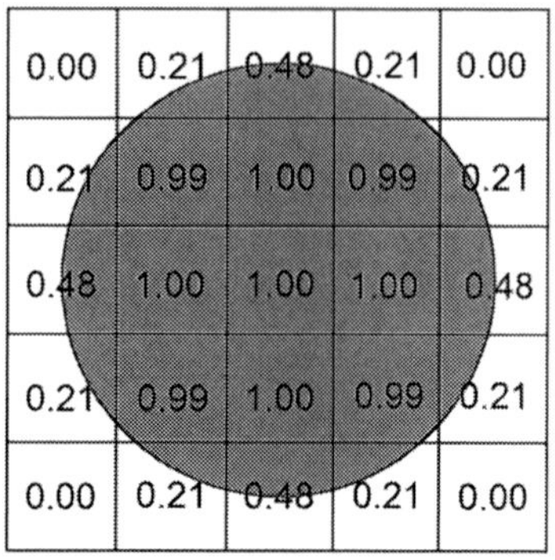


**Fig. 7.** Volume fraction example.

The elements where volume fractions are not defined were assigned as a void. Similarly, for the elements where volume fraction is less than one, a void is assigned for remainder of the volume fraction. The void elements do not have material properties, but other material can flow into and through void elements during the analysis.

## 4.3 Material model of water

Formulation of Equation of state Mie Grüneisen $U_S - U_P$ form [9, 10] was used as a material model for water in the Abaqus simulation. The most common form of Mie-Grüneisen is:

$$p - p_H = \Gamma \rho (E_m - E_H) \tag{1}$$

where $p_H$ and $E_H$ are the Hugoniot pressure and specific energy (per unit mass) and they are functions of density only. $\Gamma$ is the Grüneisen ratio, which describes the alteration in a crystal lattice's vibration frequency, based on the lattice's increase or decrease in volume as a result of temperature change. It is used in quantum mechanics and it is defined as:

$$\Gamma = \Gamma_0 \frac{\rho_0}{\rho} \tag{2}$$

Where $\Gamma_0$ is the initial Grüneisen ratio (material constant) and $\rho_0$ is the reference density. The Hugoniot energy $E_H$ is related to the Hugoniot pressure $p_H$ by:

$$E_H = \frac{p_H \eta}{2 \rho_0} \tag{3}$$

Where $\eta = 1 - \rho_0/\rho$ is the nominal volumetric compressive strain. Elimination of $\Gamma$ and $E_H$ from the above equations yields:

$$p = p_H \left(1 - \frac{\Gamma_0 \eta}{2}\right) + \Gamma_0 \rho_0 E_m \tag{4}$$

The state and the energy equations represent coupled equations for pressure and internal energy. Abaqus solves these equations simultaneously at each material point.

A common form of the Hugoniot data is given by

$$p_H = \frac{\rho_0 c_0^2 \eta}{(1 - s\eta)^2} \tag{5}$$

Where $c_0$ and s define the linear relationship between the shock velocity $U_S$ and the particle velocity $U_P$ as follows:

$$U_S = c_0 + sU_p \tag{6}$$

With the above assumptions the linear $U_S - U_P$ Hugoniot form is written as

$$p = \frac{\rho_0 c_0^2 \eta}{(1 - s\,\eta)^2} \left(1 - \frac{\Gamma_0 \eta}{2}\right) + \Gamma_0 \rho_0 E_m \tag{7}$$

Where

$$\rho_0 c_0^2 > 0 \tag{8}$$

is equivalent to the elastic bulk modulus at small nominal strains. There is a limiting compression given by the denominator of this form of the equation of state

$$\eta_{lim} = \frac{1}{s} \tag{9}$$

or

$$\rho_{lim} = \frac{s\,\rho_0}{s - 1} \tag{10}$$

In case of modeling water according to [10] the parametres given by table 2 were used.

**Table 2.** Parameters used for material model

| Parameter | Value |
|---|---|
| Density (ρ) | 1,000 kg/m$^3$ |
| Viscosity (η) | 0.001 Pa s |
| $C_0$ | 1483 m/s |
| s | 0 |
| $\Gamma_0$ | 0 |

## 4.4 Model definitions

A simplified CAD model of flood barriers was used to define the numerical simulation. The real flood barrier contains two portable pillars, a certain number of stop logs depending of on the height of the whole barrier and tightening device. For the simulation the entire barrier was replaced by only two horizontal stop logs formed by the rectangular hollow profile with a length of 1.2 m and wall thickness of 3 mm. The other stop logs have minimal effect on the experiment therefore can be neglected. The effect of fixed pillars was replaced with fixed constraints to both ends of used stop logs. Eulerian volume (where water can move) was created as a block of 5x1x0.5 m. The initial volume of water formed by the block 4x1x0.35 m was placed to Eulerian volume. a log with a diameter of 0.2 m and 3 m length with a density of 900 kgm$^{-3}$ (weight 85 kg), which corresponds approximately to the log used in real experiment, was placed into the water area. The assembly of the simulation is shown in Figure 8.

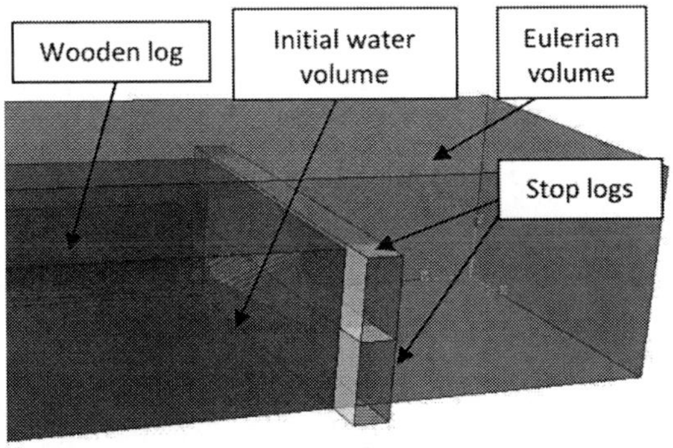


**Fig. 8.** Assembly of simulation.

The mesh used in the simulation was quite coarse. The Eulerian volume was meshed by 32,500 brick elements with edge size 40 mm, the wooden log contains 1,200 brick elements with the shortest edge 28.4 mm and the stop log contains 240 shell quad elements with edge lenght 40 mm.

The gravitational acceleration of 9.81 ms$^{-2}$ was applied on the whole assembly. The log movement was controlled in order to reach desired speed (from 1 to 2ms$^{-1}$) during 1 s, after that the forced move was deactivated and the log was moved its own momentum. The entire simulated action lasted about 2 seconds, depending on the speed of the log.

Explicit method and CEL technique is intended primarily for fast processes lasting usually in the order of milliseconds. In this case where the length is about 2 s problem with contact stability between water and structure may occur.

Volume fraction information on the nodes is determined by the arithmetic mean of the volume fraction values of the cell surrounding the considered node. When volume fraction value on the node is under 0.5, the contact of fluid and structure is not applied on this node. The liquid may penetrate into the Lagrangian body. Leaking problem is self-recovering. Once it is leaking into nearby node with low volume fraction thus leading to increased mean of volume fraction in the surrounding cells. Once the value of the volume fraction rises to 0.5 on node, contact is applied again. This recapture of contact may result in rapid displacement of liquid which cause unrealistic acceleration of liquid.

## 5 Results

The simulation was carried out in the first place to see whether it is even possible to simulate water in such spatial and time range. From a visual comparison of animation of simulation results and the record of high speed camera it can be concluded that both waveforms are very similar. Figures 9 and 10 show the behavior of water just before impact of the log in the simulation and real experiment.

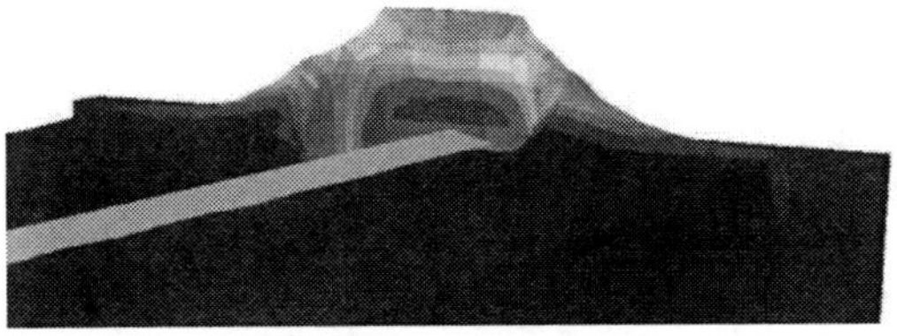

**Fig. 9.** Shape of water splash - simulation.

**Fig. 10.** Shape of water splash - experiment.

It can be said that the shape of the splashed water in the simulation is relatively in good agreement with the real experiment. The next question is about precision of simulation results. In Figure 11 shows the results of deformation distribution of the stop logs affected by the wooden log.

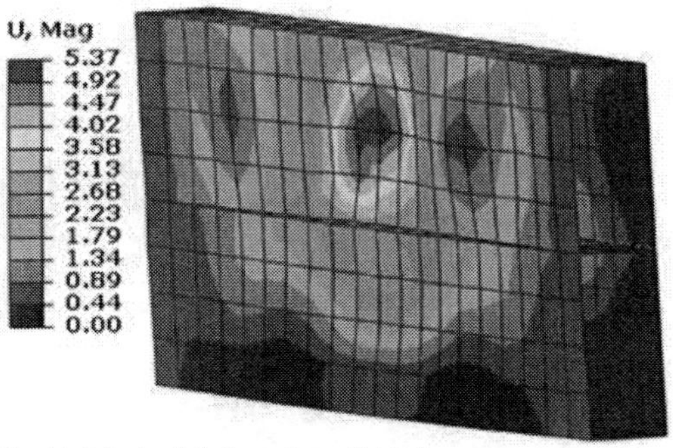


**Fig. 11.** Maximal deformation of stop log.

The maximal value of deformation in the simulation was about 5.4 mm, in the experiment under the same conditions it was 4.6 mm. Both values are in a quite good agreement considering the simplifications of the simulation.

## 6 Conclusion

In the case of a flood, it is necessary to ensure the protection of people and property as soon as possible. One of possibilities is using mobile flood barriers. Mobile barriers must fulfil many requirements, such as low weight, easy installation, durability, low price etc. All these requirements must considered for a development of a new design of flood barrier. During a development, it is therefore necessary to carry out many tests, which are often very expensive. Some of these tests can be replaced by numerical simulations. Numerical simulations were performed using Abaqus software by CEL (Coupled Eulerina Lagrangian) technique. The results of the simulations of simplified models show a relatively good agreement with experiments. Thus proving that numerical simulations are a good auxiliary tool in examining the behaviour of water interacting with the flood barrier. The numerical simulations can be used to predict the behaviour of the barrier during a flood, which is very important but hardly feasible by experiment.

In future work, similar simulations will be used in the development and optimization of new types of flood barriers without ground structure.

### Acknowledgements

The results of this project VI20152018005 were obtained with co-funding from the Ministry of the Interior of the Czech republic as part of targeted support from the project "Vývoj protipovodňových systémů pro zvýšení ochrany obyvatelstva a infrastruktury".

## References

1. J. Hladný, *Sborník příspěvků závěrečného semináře Hodnocení vlivu změn přírodního prostředí na vznik a vývoj povodní, Fakta a mýty o povodních.* (2007)
2. Official Journal of the European Union, *Directive 2007/60/ec,* **288**, 27, (2007)
3. J. Hladný, et al, *Vyhodnocení povodňové situace v červenci 1997, Souhrnná zpráva*, (1998)
4. P. Puncochar, Flood protection in the czech republic and international, Anti flood defenses today's problems, (2012)
5. J. Hladný, M. Kratka, L. Kašpárek, *August 2002 catastrophic flood in the Czech Republic* (2004)
6. L. Čamrová, J. Jílková, *Povodňové škody a nástroje k jejich snížení.* (2006)
7. Approval Standard for Flood Abatement Equipment, Class Number 2510, FM Approvals LLC. (2013)
8. M. Petrů, L. Ševčík, A. Lufinka, T. Martinec, EAN, **54**, (2016)
9. A. Sillem, *Feasibility study of a tire hydroplaning simulation in a finite element code using a coupled Eulerian-Lagrangian method,* (2008)
10. Simulia, Abaqus version 6.13-EF manual, USA, (2013)
11. A. Skzat, Scientific Letters of Rzeszow University of Technology, 25-35, **84** (2012),

# Boiling heat transfer during flow of distilled water in an asymmetrically heated rectangular minichannel

*Kinga* Strąk*, *Magdalena* Piasecka

Faculty of Mechatronics and Mechanical Engineering
Kielce University of Technology, Al. 1000-lecia P.P. 7, 25-314 Kielce, Poland

**Abstract.** This paper discusses test results concerning flow boiling heat transfer in a minichannel 1.7 mm in depth, 16 mm in width and 180 mm in length. The essential part of the experimental stand was a vertically oriented rectangular minichannel, which was heated asymmetrically with a plate made of Haynes-230 alloy. Distilled water was used as the cooling fluid. Changes in the temperature on the outer side of the heated plate in the central, axially symmetric part of the channel were measured using infrared thermography. Simultaneously, the other side of the heated plate in contact with the fluid was observed through a glass pane to identify the two-phase flow patterns. The one-dimensional model used for the heat transfer analysis took into account the heat flow direction, which was perpendicular to the direction of the fluid flow in the minichannel. The study involved determining local values of the heat transfer coefficient and generating boiling curves. The data for water were compared with the findings reported for the FC-72 fluid.

## 1 Introduction

Flow boiling heat transfer of water in minichannels and microchannels has been described by various researchers (see, for example, [1-2] and [3-5], respectively). Reference [1] investigates the effect of pressure (ranging from 2 to 16 bar) on the flow boiling of water. Reference [2] analyses the influence of surface wettability on the two-phase pressure drop during flow boiling of water at atmospheric pressure in a single rectangular channel 0.5 mm high, 5 mm wide and 180 mm long. References [3-4] deal with the flow boiling of deionised and degassed water for different mass fluxes in a vertical narrow rectangular microchannel with a hydraulic diameter of $d_h$ = 0.48 mm [3] and $d_h$ = 1.2 mm [4]. A study of the flow boiling of deionized water in silicon microchannels in a single annular flow boiling regime is presented in [5].

One of the key parameters affecting flow boiling heat transfer in a mini space is the spatial orientation of the test module. Reference [6] discusses complex mechanisms of the critical heat flux (CHF) in flow boiling for different orientations of a rectangular channel (with inclination angles ranging from 0° to 270°) under standard gravity conditions with heat supplied to one side or two sides of the heating foil. The influence of gravity on the CHF in flow boiling was negated in Ref. [7]. The negating criteria for two-phase inlet conditions were: to overcome the influence of gravity perpendicular to the heating wall, to overcome the influence of gravity parallel to the heating wall and to ensure that the heated wall is sufficiently long to ensure liquid contact. Reference [8] focuses on subcooled flow boiling. The measurements of the CHF were conducted over a range of flow orientations: 0°, 45°, 90°, 135°, 180°, 225°, 270° and 315°. The analysis of the measurement results reveals that the CHF increases with increasing velocity and that at the highest velocity there is a considerable consolidation of data for different orientations.

The influence of minichannel position on heat transfer coefficient and the occurrence of flow structures were investigated in the research conducted at the Kielce University of Technology [9-11]. Enhanced surfaces were used in the experiments. The use of microstructured surfaces produced in thermal processes (as described in [12-15]) is known to intensify the heat transfer process. Studies have been conducted on pool boiling [16-20], flow boiling [9-11, 21-23] heat transfer and confined spaces as face seals [24-25].

## 2 Experimental stand and methodology

### 2.1 Experimental set-up

The major parts of the experimental setup presented in Fig. 1 were: a main loop, a supply and control system and a data and image acquisition system. The flow loop comprised a test section (1), a gear pump (2), a compensating tank (3), a heat exchanger (4), a filter (5), a mass flow meter (6), a deaerator (7) and pressure converters (8). The data and image acquisition system was made up of an infrared camera (17), a quick shot camera (9), a lighting system (10), a data acquisition

* Corresponding author: *kzietala@tu.kielce.pl*

DOI: 10.1051/epjconf/201714302116

station (11) and a computer with specialized software (12). The supply and control system consisted of an inverter welder (13), a shunt (14), an ammeter (15) and a voltmeter (16).

The test module (Fig. 2a) with a rectangular minichannel, 1.7 mm in depth, 16 mm in width and 180 mm in length, was the most important element of the experimental setup. As shown in Fig. 2b, different orientations of the minichannel at different inclination angles were analysed.

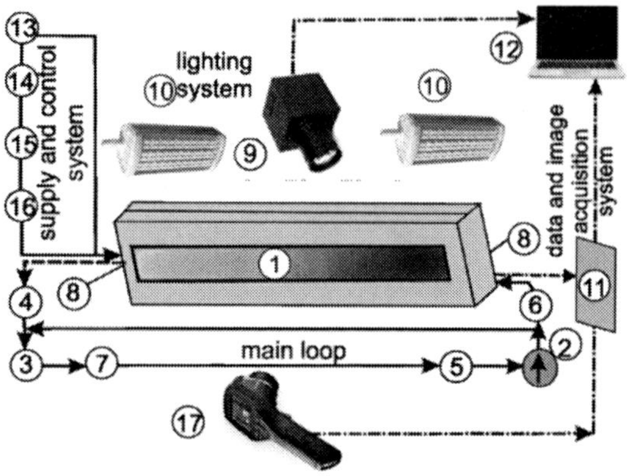


**Fig. 1.** The main systems of the experimental setup, 1-test section with a minichannel; 2- gear pump, 3-compensating tank/pressure regulator, 4-tube-type heat exchanger, 5-filter, 6-mass flowmeter, 7-deaerator, 8-pressure converter, 9-quick shot camera, 10-high power LEDs (400 W), 11-data acquisition station, 12-personal computer (PC), 13-inverter welder, 14-shunt, 15-ammeter, 16-voltmeter, 17-infrared camera.

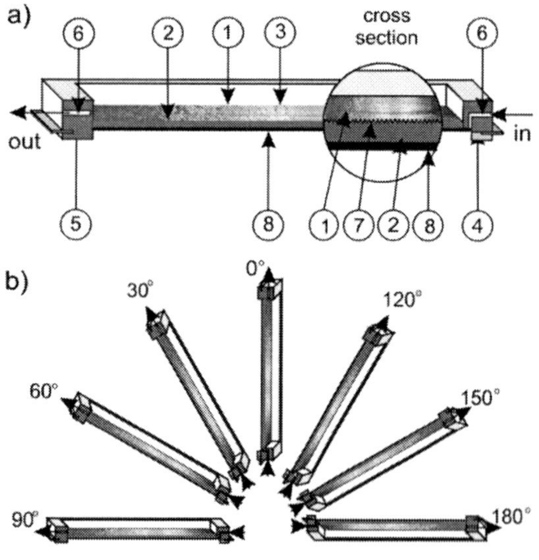


**Fig. 2.** a) The test module with the minichannel: 1-minichannel, 2-heated plate, 3-glass panel, 4-channel body, 5-front cover, 6-thermocouple, 7-enhanced surface of the plate, 8-black paint layer, b) orientations of the test module.

The surface of the 0.45 mm thick heated plate made of Haynes-230 alloy was modified on one side. The cooling fluid flowing in the minichannel was distilled water. Infrared thermography was employed to measure changes in the temperature on the outer smooth side of the plate. It was necessary to coat the surface with a black paint to obtain an emissivity of 0.83 [21].

The temperature measurements were limited to the central axially symmetric part of the minichannel (4 mm x 180 mm). On both sides of the minichannel there were narrow micanite reinforcement panels to prevent the heated plate from deforming.

The other side of the plate, i.e. the modified surface in contact with the fluid was observed through a glass pane. The surface was illuminated by high power LEDs. The two-phase flow patterns were recorded using a quick shot camera. The surface enhanced by vibration-assisted laser surface texturing had a roughness parameter $Ra = 1.46$ μm.

K-type thermocouples and pressure converters were installed at the inlet and outlet of the minichannel. The E60 FLIR infrared camera had an accuracy of ± 1 °C or ± 1% in the temperature range 0 ÷ 120 °C.

### 2.2 Methodology

The tests involved deaerating the distilled water before it flowed laminarly along the minichannel. The system operated at a predetermined pressure and a predetermined flow rate. As the electric power supplied to the heated plate was increased gradually there was a gradual increase in the heat flux transferred to the fluid in the minichannel. That resulted in the onset of nucleate boiling and enhancement of heat transfer. The distribution of temperature on the smooth surface of the heated plate was measured using infrared thermography. Simultaneously, the flow pattern was observed on the other side of the minichannel (enhanced surface of the heated plate in contact with the fluid).

## 3 Determination of the heat transfer coefficient

Local values of the heat transfer coefficient $\alpha(x)$ were established using a one-dimensional approach. Two boiling regions, subcooled and saturated, were considered. The heat flow direction, which was perpendicular to the direction of the fluid flow in the minichannel, was taken into account and it was determined using the following equation [21]:

$$\alpha(x) = q_w / (T_p(x,\delta_p) - T_l(x) - q_w \cdot \frac{\delta_p}{\lambda_p}) \quad (1)$$

where $x$ – the distance from the minichannel inlet along the direction of the flow, $q_w$ – the heat transferred to the fluid in the minichannel (heat flux density), $T_p(x)$ – the plate temperature measured by infrared thermography, $T_l(x) = T_f(x)$ in the subcooled boiling region, $T_f(x)$ – the fluid temperature determined on the basis of the linear distribution of the fluid temperature along the length of the minichannel from the inlet to the outlet, $T_f(x) = T_{sat}(x)$ in the saturated boiling region, $T_{sat}(x)$ – the saturation temperature determined on the basis of the linear distribution of pressure along the length of the minichannel from the inlet to the outlet, $\lambda_p$ – the coefficient of thermal conductivity of the plate, $\delta_p$ – the plate thickness.

As assumed in [21], the heat transferred to the fluid in the minichannel $q_w$ is equal to the difference between the heat generated by the heated plate and the heat lost to the surroundings.

The temperature measurements by infrared thermography were performed for the central part of the heated plate, where there was no insulation and, consequently, the loss of heat to the surroundings was the greatest. The heat loss was reported to constitute 1.3% of the heat flux density [21].

## 4 Results and analysis

The measurements were recorded for the assumed heat flux supplied to the heated plate, under similar stable thermal and flow conditions. The experimental results pertain to the steady state (stationary condition). The analysis includes the data from the experiments performed for seven various angular positions of the test section with the minichannel and mass flow rate $Q_m$ = $3{\cdot}10^{-3}$ $kg{\cdot}s^{-1}$. The minichannel was inclined at 0, 30, 60, 90, 120, 150 and 180 degrees, see Fig. 2b. The results are presented graphically as the heat transfer coefficient versus the distance from the minichannel inlet obtained for two boiling regions: (the subcooled boiling region, Fig. 3 and the saturated boiling region, Fig. 4) and as boiling curves (Figs. 5-6).

Local heat transfer coefficients were determined using the one-dimensional method described in [21].

The boiling curves shown in Figs. 5-6 were constructed at seven inclination angles from ranges 0°÷90° (Figs. 5a,c, 6a,c) and 120° ÷ 180° (Figs. 5b,d, 6b,d). The curves were plotted as the heat flux density against differences $T_p$-$T_f$ (Figs. 5a-b, 6a-b) and $T_p$-$T_{sat}$ (Figs. 5c-d, 6c-d) for two selected distances: 1/3 from the inlet - 0.06 m (Figs.5a-d) and 1/3 from the outlet - 0.12 m (Figs. 6a-d).

During the experiments, boiling started and developed at higher heat fluxes when distilled water was used as the working fluid in the minichannel. This is related to different properties of the fluids and their boiling points, 56 °C for Fluorinert FC-72 and 100 °C for water at atmospheric pressure.

Experimental parameters of the tested experimental series are presented in Tables 1 and 2 as representative data for the subcooled and saturated boiling regions.

In the subcooled boiling region (Fig. 3), heat transfer coefficients were relatively high compared with the results obtained with FC-72 as the working fluid [9,10] but at substantially higher heat fluxes. The local heat transfer coefficient values increased with the distance from the minichannel inlet (0.12 m) to the outlet (0.18 m) and achieved maximum of about 10 $kW{\cdot}(m^{-2}K^{-1})$ at the highest heat flux. The highest local heat transfer coefficients were achieved for orientation 0° and 30° for all heat fluxes analysed.

In the saturated boiling region (Fig. 4), local heat transfer coefficients decreased with the distance from the minichannel inlet. The local heat transfer coefficient values were comparison to values in the subcooled boiling region. The coefficients and their dependence on the distance from the minichannel inlet varied from those obtained in the experiments using FC-72 as the working fluid. The highest coefficients were recorded in the saturated boiling region (FC-72). With FC-72, local heat transfer coefficients were several times higher at the saturated boiling region in comparison to the values obtained in the subcooled boiling region. In the saturated boiling region the coefficient values were typically observed to decrease with the growing distance from the minichannel inlet [24-25]. The differences between the values obtained with distilled water and FC-72 Fluorinert are the result of different physical properties of the two fluids, such as wettability, viscosity and surface tension. The two-phase flow patterns observed during the experiments with water also varied considerably from those identified earlier for FC-72. Developed heated surfaces require proper selection of geometrical parameters for specific liquids. The study on pool boiling [17] indicates that smaller dimensions of cavities are suitable for liquids with lower surface tensions (as FC-72) at heat fluxes above 25 $kW{\cdot}m^{-2}$.

**Table 1** Experimental parameters for the subcooled boiling region

| Set of heat flux (No.) | Average mass flux $G$ ($kg\ m^{-2}s^{-1}$) | Average inlet pressure $p_{in}$ (kPa) | Average inlet liquid subcooling $\Delta T_{sub}$ (K) | Heat flux density $q_w$ ($kWm^{-2}$) |
|---|---|---|---|---|
| 1 | 98 | 126 | 15 | 125 |
| 2 | | | | 135 |
| 3 | | | | 145 |
| 4 | | | | 155 |

**Table 2** Experimental parameters for the saturated boiling region

| Set of heat flux (No.) | Average mass flux $G$ ($kg\ m^{-2}s^{-1}$) | Average inlet pressure $p_{in}$ (kPa) | Average inlet liquid subcooling $\Delta T_{sub}$ (K) | Heat flux density $q_w$ ($kWm^{-2}$) |
|---|---|---|---|---|
| 2 | 98 | 128 | 14 | 135 |
| 3 | | | | 145 |
| 4 | | | | 155 |
| 5 | | | | 165 |

 DOI: 10.1051/epjconf/201714302116

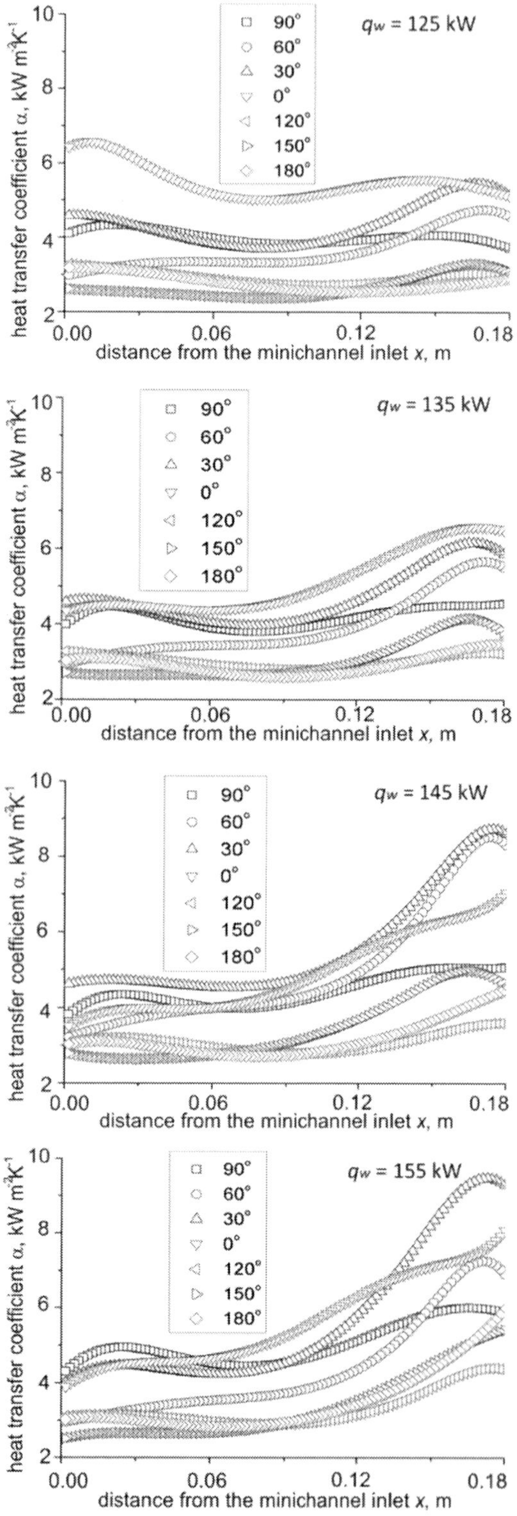


**Fig. 3.** Heat transfer coefficient vs. distance from the minichannel length obtained for the subcooled boiling region for all tested minichannel orientations and four heat fluxes.

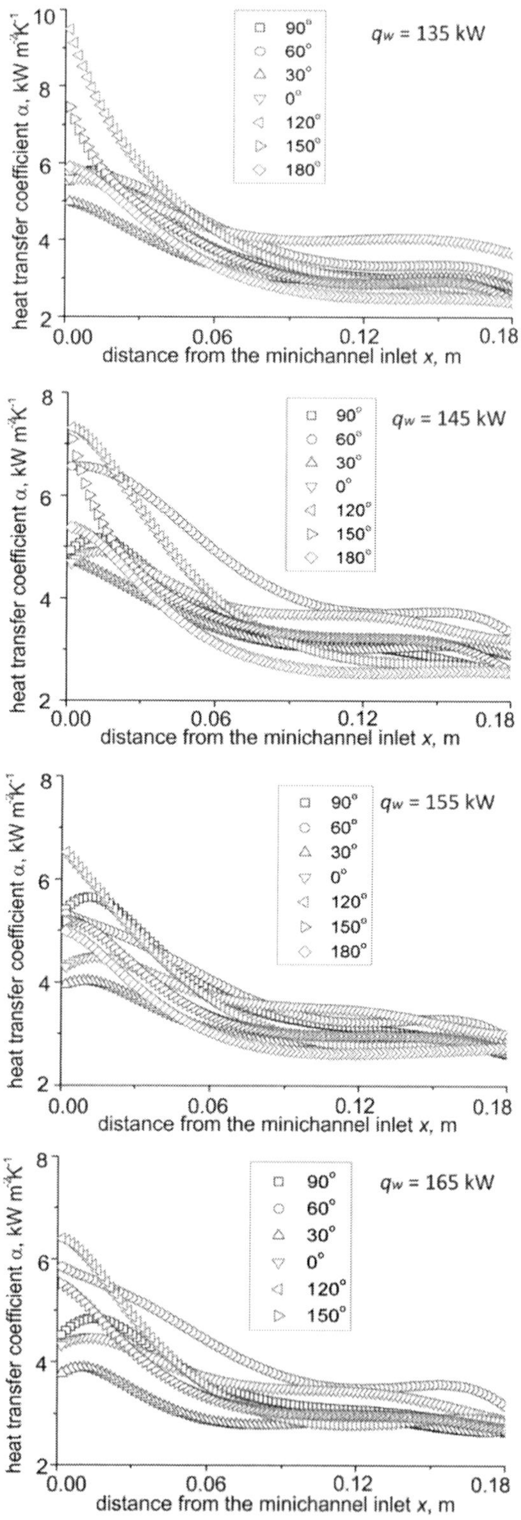


**Fig. 4.** Heat transfer coefficient vs. distance from the minichannel length obtained for the saturated boiling region for all tested minichannel orientations and four heat fluxes.

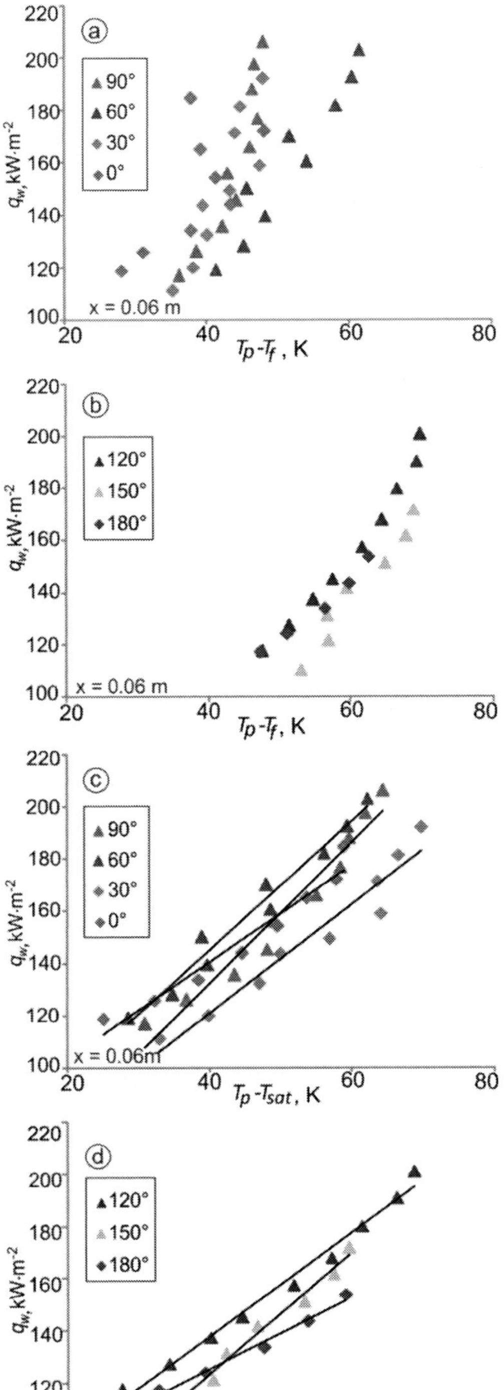


**Fig. 5.** Boiling curves for the 0.06 m distance from the minichannel inlet, orientation: a, c) 0°-90°, b, d) 120°-180°, $T_p$ - heated plate temperature, $T_f$ - bulk fluid temperature, $T_{sat}$ - fluid saturation temperature.

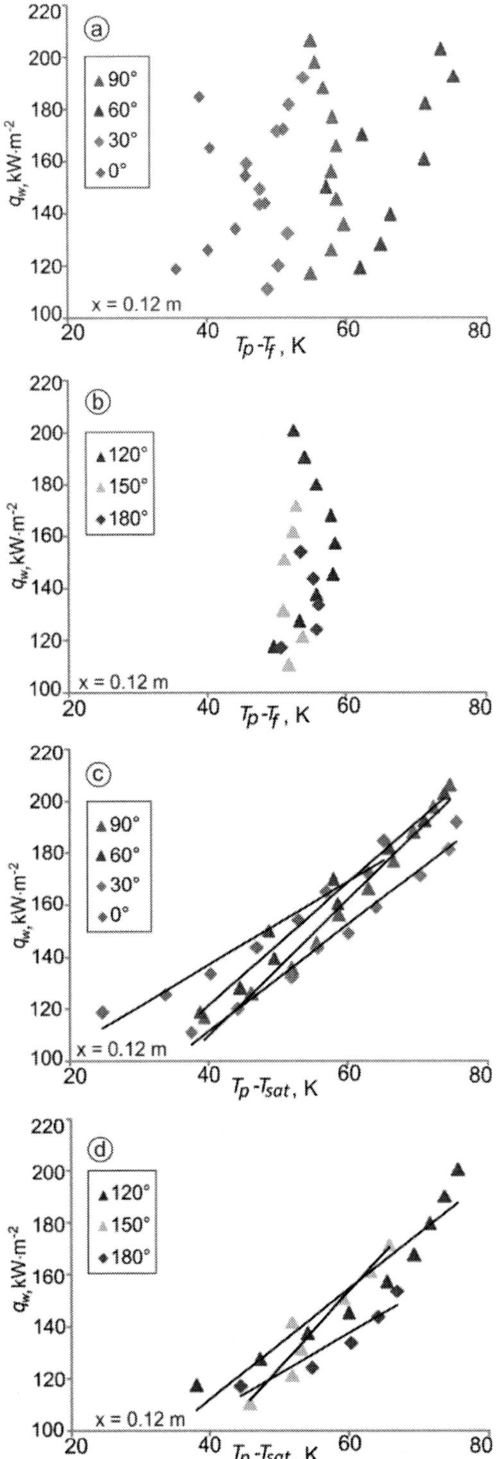


**Fig. 6.** Boiling curves for the 0.12 m distance from the minichannel inlet, orientation: a, c) 0°-90°, b, d) 120°-180°, $T_p$ - heated plate temperature, $T_f$ - bulk fluid temperature, $T_{sat}$ - fluid saturation temperature.

As for the influence of the angular position of the minichannel on the value of heat transfer coefficient in the saturated boiling region, the results indicate that:

- the maximum values were achieved at orientation 120° (Figs. 4 a-d), at the inlet of the minichannel.
- the lowest were obtained at orientation 30° (Figs. 4a-d), at the inlet of the minichannel and at orientation 180° (Figs. 4a-c), at about 1/3 of the distance from the minichannel inlet.
- high values of the heat transfer coefficient were also recorded at orientation 60° (Fig. 4b,d) and orientation 0° at about inlet to 2/3 of the distance from the minichannel inlet (Fig. 4a).

The arrangement of points on the boiling curves plotted vs. $T_p$-$T_f$ (Fig. 5a and Fig. 6a) shows the highest scatter and variability of the data obtained for four orientations: 0°, 30°, 60° and 90°. The boiling curve at 60° orientation was constructed for the highest $T_p$-$T_f$ difference. The curves for the remaining three orientations, i.e., 120°, 150° and 180°, were constructed for the highest difference $T_p$-$T_f$ and were similar. At orientations 0°÷90°, $T_p$-$T_f$ is mainly in the range from 20÷60 K (Fig. 5a) or 20÷80 K (Fig. 6a), whereas at 120°÷180° in the range from 40÷80 K (Fig. 5b) or 40÷60 K (Fig. 6b).

Analysis of the boiling curves plotted vs. $T_p$-$T_{sat}$ (Figs. 5c-d and Figs. 6c-d) has a course close to linear and parallel for various inclination angles (the highest for orientation 120° - Figs. 5d, 6d and orientation 60° - Figs. 5c, 6c) and the heated surface superheat differences are substantial, with $T_p$-$T_{sat}$ of 77 K for the highest heat flux.

## 5 Conclusions

This paper discusses test results concerning flow boiling heat transfer in a minichannel with a surface modified by vibration-assisted laser texturing. Distilled water was used as the cooling fluid. Local values of the heat transfer coefficient were calculated using a one-dimensional method. The analysis includes the data from the experiments performed for seven various angular positions of the test section with the minichannel. The minichannel was inclined at 0, 30, 60, 90, 120, 150 and 180°. The results are presented graphically as the heat transfer coefficient vs. the distance from the minichannel inlet obtained for two boiling regions (subcooled and saturated), and as boiling curves.

The highest local heat transfer coefficients were achieved for orientations 0° and 30° in the subcooled boiling region and orientations 60° and 120° in the saturated boiling region.

The boiling curves plotted vs. $T_p$-$T_{sat}$ has a course close to linear and parallel for various inclination angles and vs. $T_p$-$T_f$ shows the highest scatter and variability of the data obtained for four orientations: 0°, 30°, 60° and 90°.

The data for water were compared with the findings reported for FC-72 to identify the influence of working fluid types on flow boiling heat transfer in minichannels with enhanced heated surface. The differences between the values of the heat transfer coefficient obtained in the experiments with distilled water and FC-72 Fluorinert resulted from different physical properties of these two fluids. The heated surface of the minichannel used in the experiments was enhanced, which led the initiation and development of the flow boiling process. The developed heated surfaces require proper selection of geometrical parameters for specific liquids to intensify heat transfer processes with a phase change.

## Acknowledgements

The research reported herein was partly supported by a grant from the Polish National Science Centre (No. DEC-2013/09/B/ST8/02825).

## References

1. K. H. Bang, K. K. Kim, S. K. Lee, B. W Lee, Int. J. Therm. Sci. **50** (3), 280-286 (2011)
2. H. T. Phan, N. Caney, P. Marty. S. Colasson, J. Gavillet, Appl. Therm. Eng. **31** (11-12), 1894-1905 (2011)
3. E. Sobierska, R. Kulenovic, R. Mertz, Int. J. Therm. Sci. **46**, 1172-1181 (2007)
4. E. Sobierska, R. Kulenovic, R. Mertz, M. Groll, Exp. Therm. Fluid Sci. **31**, 111-119 (2006)
5. F. Yang, X. Dai, Y. Peles, P. Cheng, J. Khan, Ch. Li, Int. J. Heat Mass Transf. **68**, 703-715 (2014)
6. C. Konishi, Int. J. Heat Mass Transf. **92**, 467–482 (2016)
7. C. Konishi, I. Mudawara, M.M. Hasan, Int. J. Heat Mass Transf. **65,** 203–218 (2013)
8. J.S. Brower, J.F Klausner, *Proc. HT-FED2004 Westin Charlotte & Convection Center Charlotte*, 1–6 (2004)
9. M. Piasecka, B. Maciejewska, Exp. Therm. Fluid Sci. **68**, 459–467 (2015)
10. M. Piasecka, Int. J. Heat Mass Transf. **81**, 114–121 (2015)
11. M. Piasecka, Int. J. Refrig. 56, 198-212 (2015)
12. W. Depczyński, S. Spadło, P. Młynarczyk, E. Ziach, P. Hepner, *Proc. 24th Int. Conf. Metall. Mater. Met. 2015*, Poland (2015)
13. B. Grabas, Exp. Therm. Fluid Sci. **74**, 100–109 (2016)
14. B. Grabas, Exp. Therm. Fluid Sci. **68**, 499–508 (2015)
15. N. Radek, K. Bartkowiak, Phys. Procedia, 499–505 (2011)
16. R. Pastuszko, Exp. Therm. Fluid Sci. **38**, 149–164 (2012)
17. R. Pastuszko, R Kaniowski, EPJ Web Conf. **25**, paper No. 02019 (2012)
18. R. Pastuszko, M. Piasecka, J. Physics Conf. Ser. **395**, paper No. 012137 (2012)
19. T. Orzechowski, Exp. Therm. Fluid Sci. **31** (8), 947–955 (2007)
20. T. Orzechowski, A. Tyburczyk, ISI Master Journal List **87** (7), 48–51 (2011)
21. M. Piasecka, K. Strąk, B. Maciejewska, Heat Transf. Eng. **38** (3), 332-346 (2017)

 DOI: 10.1051/epjconf/201714302116

22. S. Hozejowska, R. Kaniowski, M.E. Poniewski, Int. J. Numer. Methods Heat Fluid Flow, **24**, 811–824 (2014)
23. R. Kaniowski, M.E. Poniewski, Arch. Thermodyn. **34** (2), 3–22 (2012)
24. S. Blasiak, A. V. Zahorulko, Tribol. Int. **94**, 126–137 (2016)
25. S. Blasiak, Int. J. Heat Mass Transf. **81**, 90–102 (2015)

 DOI: 10.1051/epjconf/201714302117

# Investigation of flow in axial stage of experimental turbine

*Petr* Straka[1,*], *Jaroslav* Pelant[1], *Martin* Němec[1], *Tomáš* Jelínek[1], and *Petr* Milčák[2]

[1]Aerospace Research and Test Establishment, Plc., Beranových 130, 19905 Prague – Letňany, Czech Republic
[2]Dossan Škoda Power Co., Ltd., Tylova 1/57, 30128 Plzeň, Czech Republic

**Abstract.** The contribution deals with investigation of flow in the axial stage of the experimental turbine with reaction blading. The main intention is to evaluate an effect of the secondary flows on the efficiency. Most important source of the secondary flows is an outflow from the seal above the rotor shroud. Other sources of the secondary flows are axial gaps between the rotor disc and the side walls of the stator blade channel and the outlet diffuser. Results of the numerical simulation are compared with the experimental data. In this work there is also studied possibility of transfer the outlet dissipation structures at the inlet of the stage as an approach to modelling of the multistage configuration.

## 1 Introduction

Flow in an axial turbine stage of an experimental turbine is investigated under the project of applied research TA04020129 [1], which deals with research of flow in multistage configuration of the axial turbine stages.

In case of multistage configuration of the axial turbine stage the inlet flow field to the stage is influenced by the flow through previous stages. Affected are primarily peripheral parts of the span.

In initial stage of the project TA04020129 [1] it was studied a simple possibility of modification of the measuring stand [2] for axial turbine stages in single stage configuration. The influence of previous stages was simulated by increasing of the boundary layer thickness on the side walls of the inlet diffuser and by adding the turbulizing grid inside the inlet diffuser. Some results of the numerical simulation and the experimental investigation can be found in [3,4].

The influence of previous stages is in this work simulated using transfer of the radial distribution of the outlet flow field parameters to the inlet boundary of the same stage. This approach is, of course, feasible only using numerical simulation, but not experimentally. Figure 1 shows a scheme of transfer of the outlet flow field parameters to the inlet boundary of the computational domain. In this work it is used the approach where the radial distributions of the normalized dynamic pressure $p_{dyn.,norm.}$, the turbulence intensity $Tu$, the velocity vector angle in the tangential plane $\alpha$ and the velocity vector angle in the meridian plane $\gamma$ are transferred from chosen section in the outlet part of the axial turbine stage to the inlet boundary. The normalized dynamic pressure is the ratio of the difference between

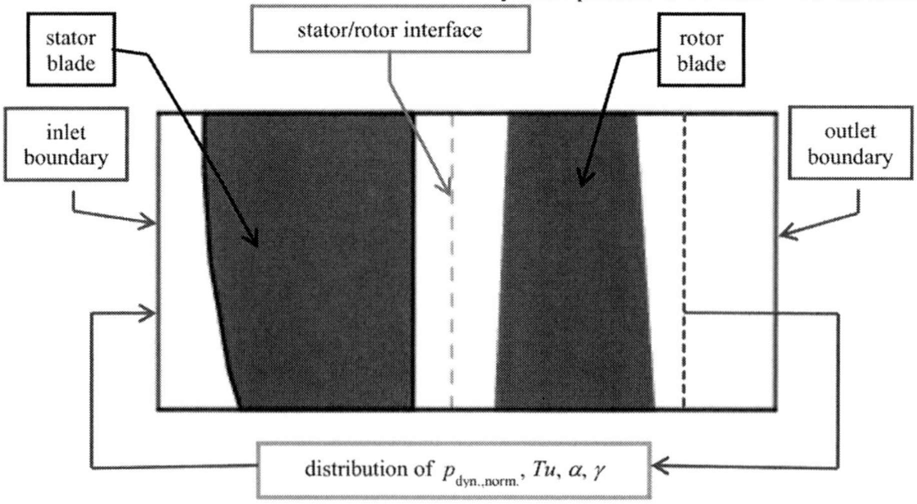


**Fig. 1.** Scheme of transfer of the outlet flow field parameters to the inlet boundary.

* Corresponding author: straka@vzlu.cz

the total pressure and the static pressure to the maximum value at given section. The fact, that the inlet and the outlet sections may have different dimensions, is respected using a scaling method. The calculation of flow in the axial turbine stage is performed iteratively. In the first step it is performed calculation without influence of previous stages, as the stage would be the first in a row. Then the radial distributions of $p_{dyn.,norm.}$, $Tu$, $\alpha$, $\gamma$ are evaluated in chosen outlet section. In the next step there is performed calculation of flow with prescribed radial distributions of $p_{dyn.,norm.}$, $Tu$, $\alpha$, $\gamma$ at the inlet boundary of the computational domain, as the stage would be the second in a row. The cycle can be repeated several times to simulate the order of the axial turbine stage in multistage configuration. Results of calculations of flow through the axial turbine stage called "REAC_v2-2", which is one of several configurations of reaction turbine stages designed in Doosan Škoda Power company and which are investigated under the project TA04020129 [1], are presented in this work.

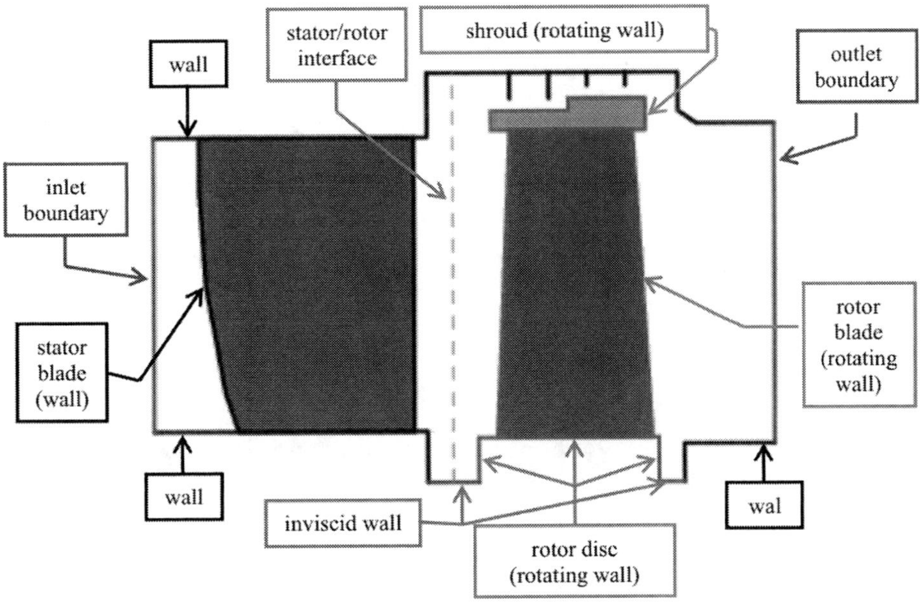


**Fig. 2.** Scheme of the computational domain.

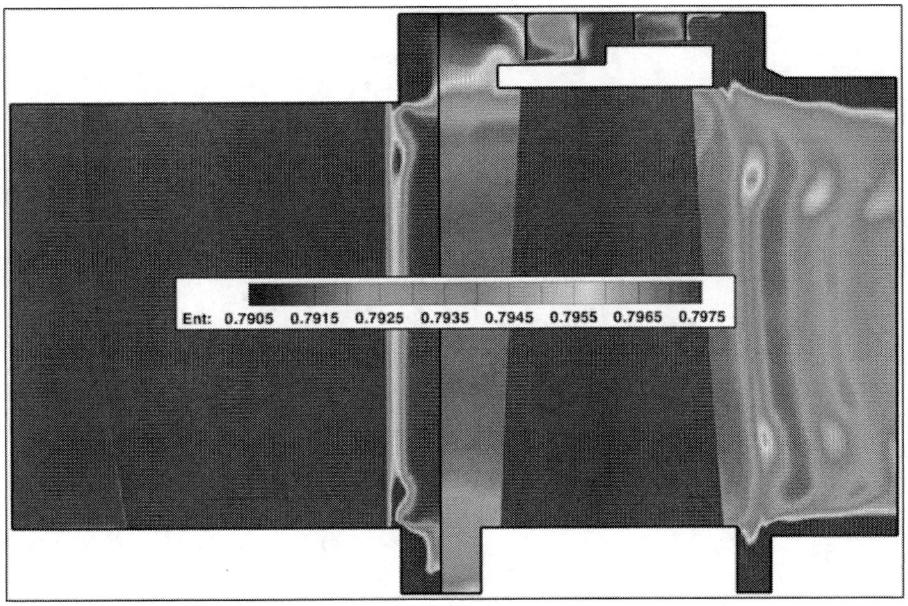


**Fig. 3.** Distribution of the normalized entropy index (Ent $= p / \rho^{\kappa}$) in meridian section in case without simulation of influence of the previous stage.

Calculations of flow were performed using the in-house numerical software, which is based on solution of RANS equations closed with two-equations turbulence model (for more details see [5,6,7]). The nonlinear explicit algebraic model of the Reynolds stress is used in this work [8]. The flowing medium is diatomic perfect gas. In figure 2 there is shown a scheme of the computational domain with highlighting the types of the boundary condition on various parts of the boundary. We can see that the rotor blade is equipped with the shroud. In the area above the shroud we can see the strips of the above-shroud seal. The rotor blades are carried on the rotating disc. Between the hub-wall of the stator and the rotor disc as well as between the rotor disc and the hub-wall of the outlet diffuser there are an axial gaps. These axial gaps are in this work replaced by cavities with a depth of about 15% blade height. On the bottom side of these cavities there is prescribed the slip boundary conditions (so called "inviscid wall"). The inlet boundary of the computational domain is placed in distance about 20% of the axial chord of the stator blade in front the stator leading edge (on the tip-diameter). The outlet boundary of the computational domain is placed in distance about 80% of the axial chord of the rotor blade behind the rotor trailing edge (on the hub-diameter). The axial distance between the stator trailing edge and the rotor leading edge (on the hub-diameter) is about 25% of the blade height. This dimension corresponds to the experimental configuration, where the flow field behind the stator wheel was traversed by the pneumatic probes. The radial gap between the rotor shroud and the strips of the above-shroud seal is about 1.06% of the blade height. The plane of evaluation of the radial distribution of the outlet flow field parameters was chosen in distance above 30% of the axial chord of the rotor blade behind the rotor trailing edge (on the hub-diameter). The radial

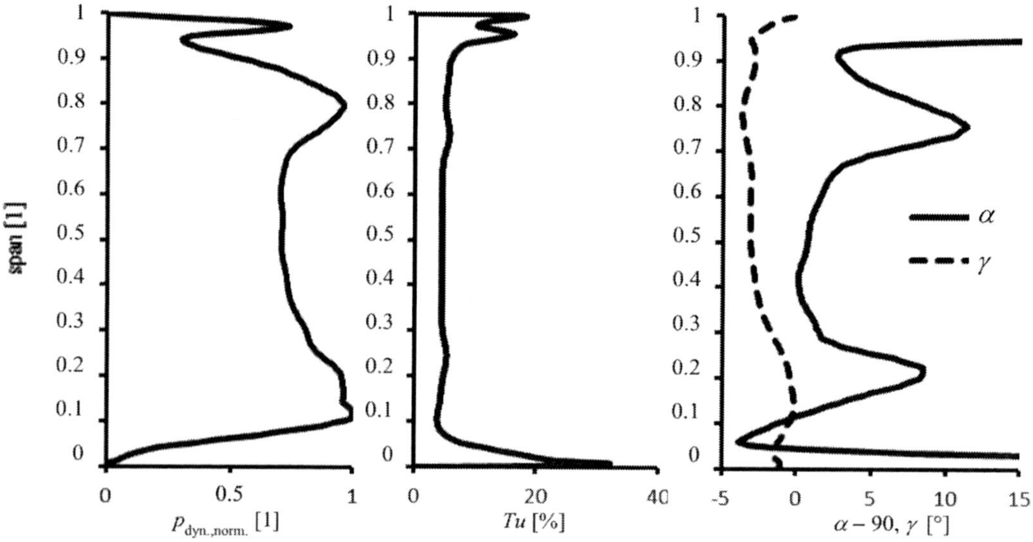


**Fig. 4.** Radial distribution of $p_{\text{dyn.,norm.}}$, $Tu$, $\alpha$, $\gamma$.

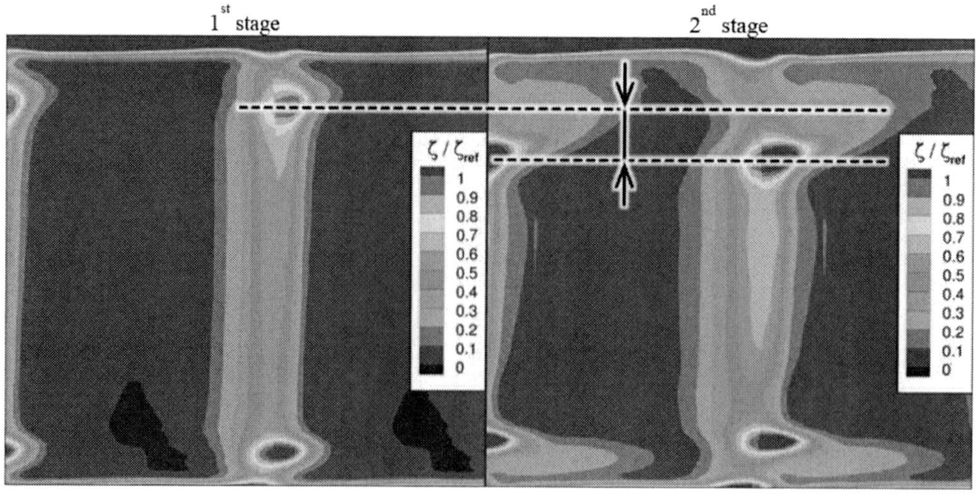


**Fig. 5.** Field of normalized kinetic energy losses in section behind the stator blades.

distribution of the outlet flow field parameters in this plane is influenced by the flow across the axial gap between the rotor disc and the hub-wall of the outlet diffuser and in tip-region it is influenced by the stream from the above-shroud seal.

## 2 Results

Flow in the axial turbine stage "REAC_v2-2" was calculated for the pressure ratio $p_{T0} / p_{S2} = 1.065$ ($p_{T0}$ is the inlet total pressure, $p_{S2}$ is the static outlet pressure), the velocity ratio $u / c_{2is} = 0.61$ ($u$ is the circumferential speed, $c_{2is}$ is the outlet isentropic velocity), the outlet isentropic Mach number $M_{2is} = 0.299$, revolution speed 2159.7 RPM, the total inlet pressure $p_{T0} = 100.931$ kPa, the total inlet temperature $T_{T0} = 297.5$ K. For calculation without the simulation of influence of previous stages the inlet turbulence intensity $Tu_0 = 5\%$ and the axial direction of the velocity vector were chosen.

In figure 3 there is shown field of the non-dimensional entropy index ($p / \rho^{\kappa}$, where $p$ is the pressure, $\rho$ is the density and $\kappa$ is the adiabatic exponent) in the meridian section in case of flow without the simulation of influence of previous stages. Note here that all flow fields in this work are shown in normalized cylindrical coordinates ($x / x_{\text{ref}}$, $\varphi / \varphi_{\text{ref}}$, $r / r_{\text{ref}}$) instead of the Cartesian coordinates ($x$, $y$, $z$). The entropy index shows the dissipative structures in the flow field. It is clear that the major dissipation occurs due to flow through the above-shroud seal, due to mixing the stream from the above-shroud seal with the main stream in blade channel and due to flow across the axial gap before and behind the rotor disc. In Figure 3 there is also possible to detect projections of wakes to the meridian plane.

In figure 4 there are shown radial distributions of the normalized dynamic pressure $p_{\text{dyn.,norm.}}$, the turbulence intensity $Tu$ and the velocity vector angles $\alpha$ and $\gamma$ in section placed in distance above 30% of the axial chord of the rotor blade behind the rotor trailing edge (on the

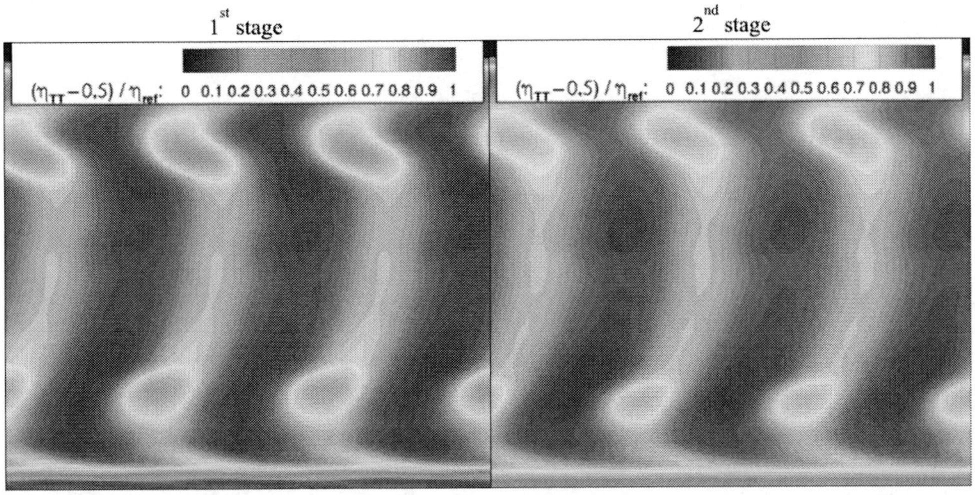


**Fig. 6.** Field of normalized total-total efficiency in section behind the stator blades.

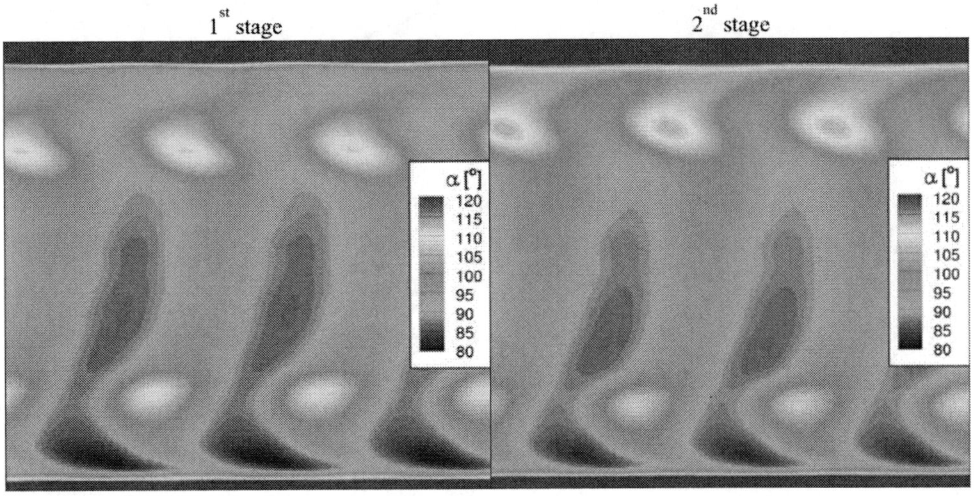


**Fig. 7.** Field of velocity vector angle $\alpha$ in section behind the stator blades.

hub-diameter). These distributions are evaluated from calculation without the simulation of influence of previous stages and are used for the inlet boundary condition in calculation of flow in stage which would be the second in a row in multistage configuration.

Way, how to prescribe the radial distribution of the turbulence intensity $Tu$ and velocity angles $\alpha$, $\gamma$, is straightforward. The radial distribution of the normalized dynamic pressure is used to define the local value of the total pressure on the inlet boundary: $p_T(r,\varphi) = p_S(r,\varphi) + p_{\mathrm{dyn.,norm.}}(r)\ [p_{T0} - p_S(r,\varphi)]$. Note that $(r, \varphi)$ are the cylindrical coordinates, $p_{T0}$ is prescribed value of the inlet total pressure and $p_S$ is local value of the static pressure which is extrapolated from the computational domain to the inlet boundary.

The influence of previous stage on the flow field in the second stage in multistage configuration is shown in figure 5 where is shown field of normalised kinetic energy losses in the axial gap between the stator and the rotor in plane placed in distance of 2 mm behind the stator trailing edge. From figure 5 it is clear that the previous stage causes increasing of the kinetic energy losses in peripheral parts of the span and it also causes shift of location of maximum value of the kinetic energy losses in wake in tip area. This demonstrates that there is an effect on development of the secondary vortices inside the inter-blade channel in tip area of the stator blades. The influence of the previous stage on the flow field at the outlet from the second stage is documented in figures 6 to 9. Figures 6, 7 and 8 show fields of the normalized total-total efficiency and two angles of the velocity vector (in tangential and meridian plane) in plane located in distance of 8 mm behind the rotor blade trailing edge (on the hub-diameter). The total-total

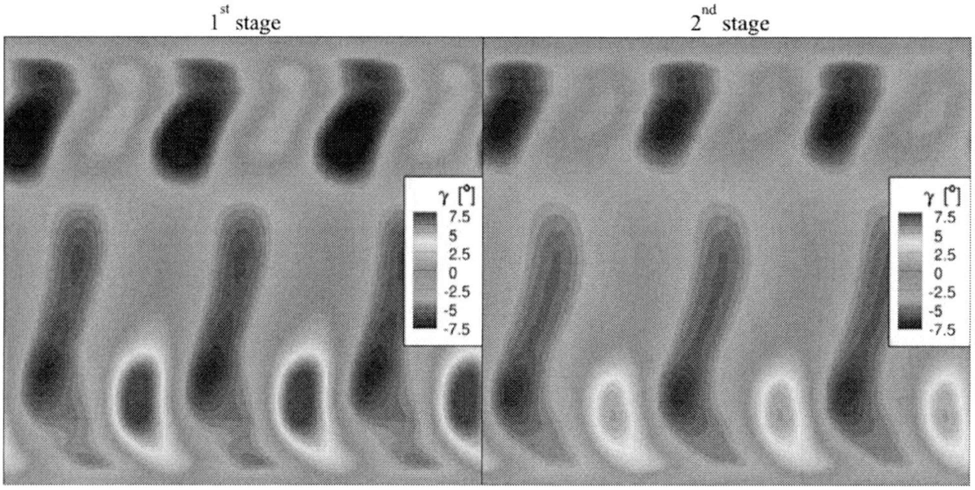


**Fig. 8.** Field of velocity vector angle $\gamma$ in section behind the stator blades.

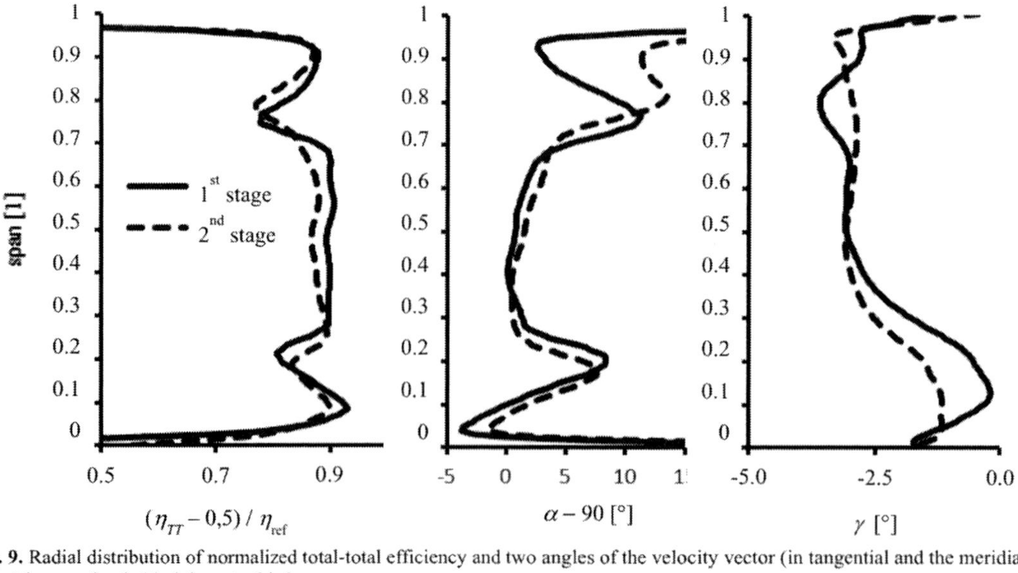


**Fig. 9.** Radial distribution of normalized total-total efficiency and two angles of the velocity vector (in tangential and the meridian plane) in a section behind the rotor blades.

efficiency is defined as $\eta_{TT} = (T_{T0} - T_T) / ( T_{T0} - T_{Tis})$ where $T_{T0}$ is the inlet total temperature, $T_T$ is local total temperature and $T_{Tis}$ is the total isentropic temperature.

On fields of the normalized total-total efficiency there are not apparent any significant differences between the first and the second stage in a row. I case of the angle $\alpha$ there are some differences in upper part of the span, whilst in case of the angle $\gamma$ there are some differences in lower part of the span. This shows also figure 9 where are shown radial distributions of the circumferentially averaged values.

Figures 10 and 11 illustrate the influence of previous stages on flow field in axial turbine stage, which would be the third in multistage configuration. In figure 10 there is shown field of normalised kinetic energy losses in the axial gap between the stator and the rotor in plane placed in distance of 2 mm behind the stator trailing edge. From figure 10 it is evident that once again there is a change in the position of maximum of the kinetic energy losses in wake in the tip area. However, the difference between the second and third stage is not nearly as great as the difference between the first and second stage. This is also documented in figure 11 where are shown radial distributions of circumferentially averaged normalized total-total efficiency and two angles of the velocity vector in plane located 8 mm behind the trailing edge of the rotor blade (on the hub-diameter). Figure 11 demonstrates that the outlet flow field from the second stage in a row is almost identical with the outlet flow field from the third stage in a row in multistage configuration. Therefore, it can be assumed that the simulation of the fourth stage in a row should not already yield any new information.

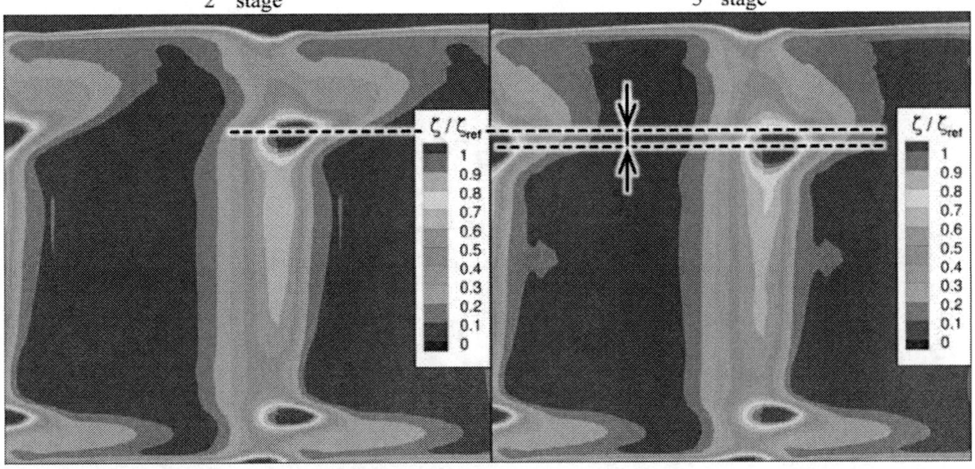


**Fig. 10.** Field of normalized kinetic energy losses in section behind the stator blades.

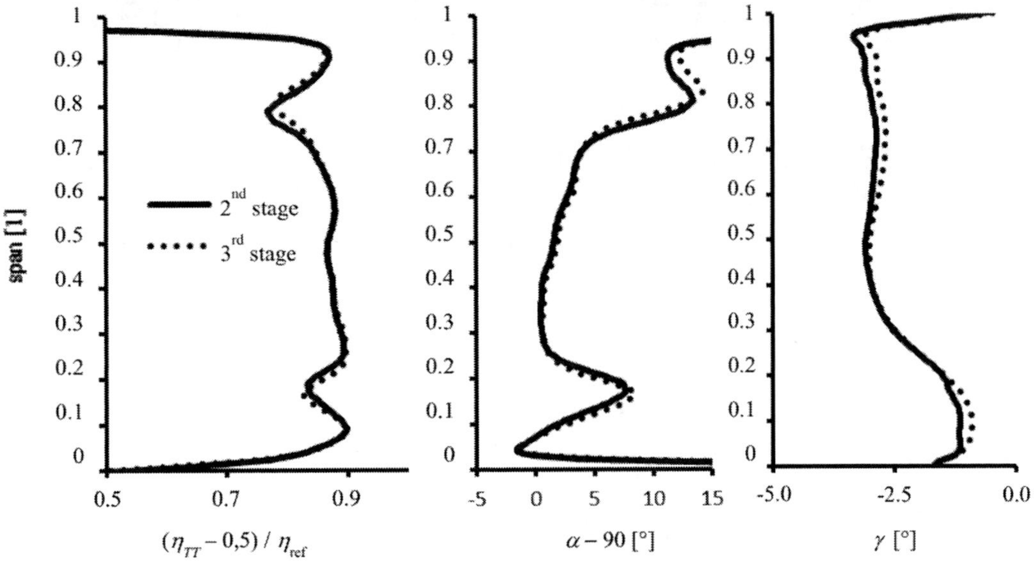


**Fig. 11.** Radial distribution of normalized total-total efficiency and two angles of the velocity vector (in tangential and the meridian plane) in a section behind the rotor blades.

## 3 Comparison of experimental and numerical results

As it was mentioned above, the axial distance between the stator blade trailing edge and the rotor blade leading edge corresponds with the configuration in which the flow field behind the stator blade was traversed with pneumatic probes in experimental research. It is therefore possible to compare results of the numerical simulation with the experimental data.

The experimental measurement was carried out on the test stand which is a part of the high-speed close-loop wind tunnel. This facility allows setting of the Reynolds and Mach number independently each other. A concept of the rotatable stator is used in consequence of linear probe manipulator. More details can be found in [2] and in [9-11].

Figures 12 to 15 show numerically calculated and measured flow fields in plane located in distance of 2 mm behind the trailing edge of the stator blade.

In figure 12 there are compared fields of the normalized kinetic energy losses, figure 13 shows comparison of the normalized Mach number fields and in figures 14 and 15 there are compared fields of two angles $\alpha$ and $\gamma$ of the velocity vector.

One can see, that the numerical results fairly good correspond with the experimental data. There is a little difference in slope of the wake behind the stator blade as shown in figures 12 and 14. Figure 15 shows little more significant difference in distribution of the angle $\gamma$. To explain this fact will be paid attention in the follow-up work.

## 4 Conclusions

Calculations of flow in axial stage "REAC_v2-2" of the experimental reaction turbine were performed. The order of the axial turbine stage in multistage configuration was simulated through transferring of the radial distributions of the outlet flow field parameters on the inlet boundary

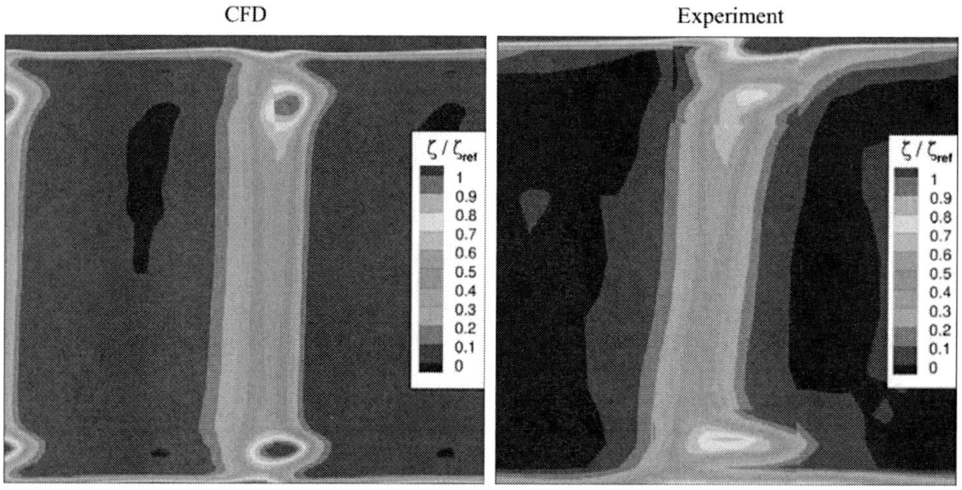


**Fig. 12.** Field of normalized kinetic energy losses in section behind the stator blades.

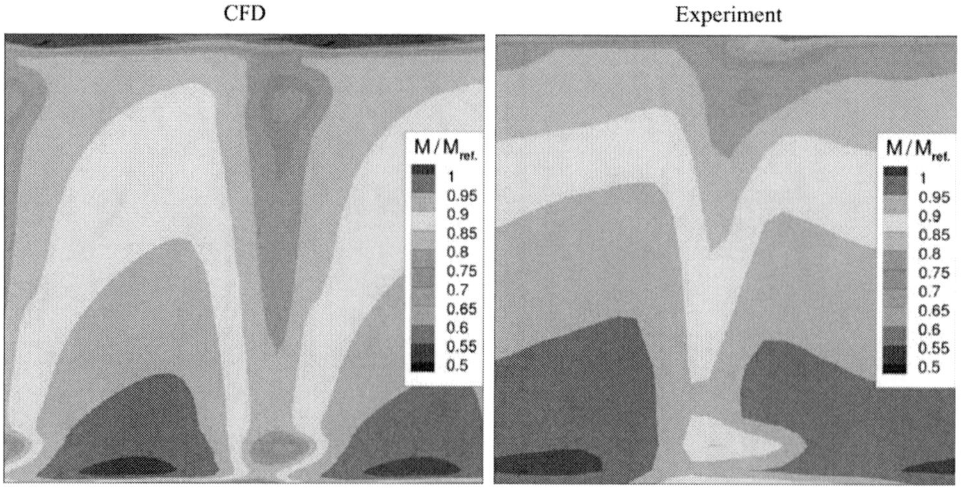


**Fig. 13.** Field of Mach number in section behind the stator blades.

DOI: 10.1051/epjconf/201714302117

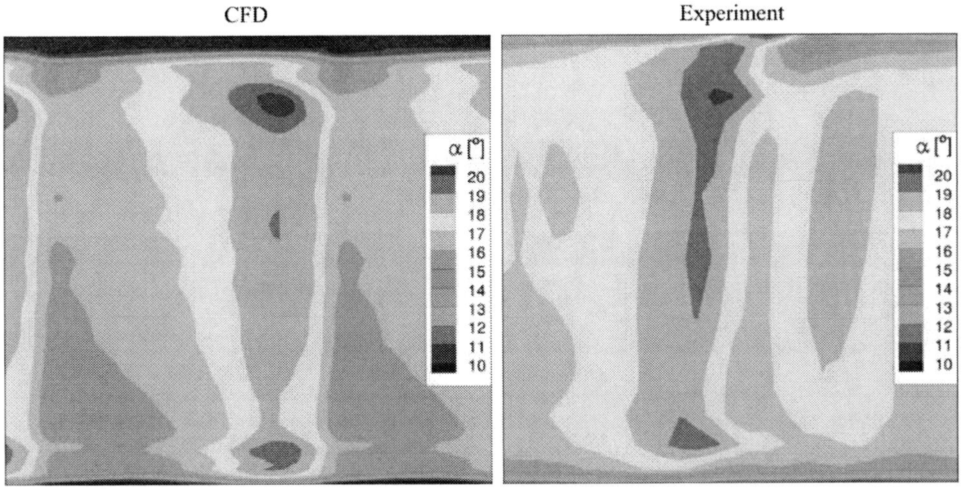


**Fig. 14.** Field of the angle $\alpha$ in tangential plane in section behind the stator blades.

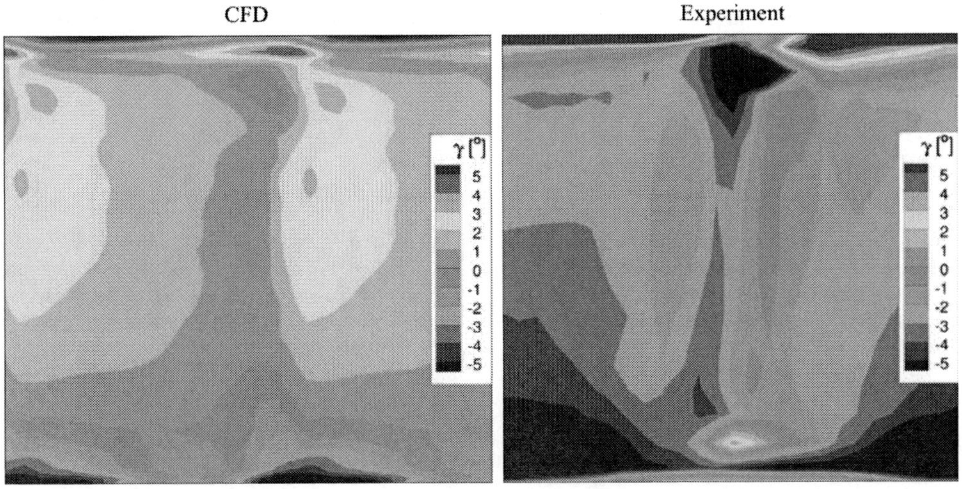


**Fig. 15.** Field of the angle γ in meridian plane in section behind the stator blades.

of the computational domain. It was found that the influence of the previous stage has the most effect on the flow field of the second stage. There were found some differences in the flow field of the third stage in area behind the stator blades, however, there are only minimal differences between outlet flow fields of the second and the third stage.

This work was supported by the Long-term Framework Advancement Plan provided by the Ministry of Industry and Trade of the Czech Republic and by the project TA04020129 of the Technology Agency of the Czech Republic.

## References

1. https://www.rvvi.cz/cep?s=rozsirene-vyhledavani&ss=detail&n=0&h=TA04020129
2. http://www.vzlu.cz/cs/tunel-pro-turbomachinery-c74.html
3. T. Jelínek, P. Straka, V. Uruba, Proc. Conf. AENMFME 2016, pp. 77-80 (2016)
4. T. Jelínek, M. Němec, V. Uruba, Proc. Conf Engineering Mechanics 2016, pp. 254-257 (2016)
5. P. Straka, EPJ Web of conferences, **25**, 01090 (2012)
6. P. Straka, Report VZLÚ R-5904, Prague (2013), (in Czech)
7. P. Straka, Report VZLÚ R-4910, Prague (2010), (in Czech)
8. C.L. Rumsay, T.B. Gatski, Jour. Of Aircraft, **83**, pp. 904-910 (2001)
9. P. Straka, M. Němec, T. Jelínek, EPJ Web of conferences, **92**, 02088 (2015)
10. T. Jelínek, M. Němec, Turbomachinery 2010, Plzeň (2010) (in Czech)
11. M. Němec, T. Jelínek, J. Benetka, Report VZLÚ R-4939, Prague (2010) (in Czech)

# Simulation of transitional flows through a turbine blade cascade with heat transfer for various flow conditions

*Petr* Straka[1,a], *Jaromír* Příhoda[2], *Martin* Kožíšek[3], *Jiří* Fürst[3]

[1] Aerospace Research and Test Establishment, Plc, Prague, Czech Republic
[2] Institute of Thermomechanics AS CR, v.v.i., Prague, Czech Republic
[3] Faculty of Mechanical Engineering, Czech Technical University in Prague, Czech Republic

**Abstract.** The contribution deals with the simulation of the transitional flows with heat transfer by means the EARSM turbulence model of Hellsten [1] completed by the algebraic transition model of Straka and Příhoda [2] and by the three-equation model of Walters and Cokjlat [3]. The both mathematical models were tested for the flat plate flow on a heated wall measured by Sohn and Reshotko [16] and then applied to the simulation of compressible flow through the VKI turbine blade cascade according to measurements of Arts et al. [4]. The simulations were carried out for subsonic and transonic regimes at various free-stream turbulence levels. The best agreement of numerical results with experimental data was achieved by the URANS approach applied for the EARSM model with the algebraic transition model giving good results for both subsonic and transonic regimes as well.

## 1 Introduction

The numerical simulation of compressible flows in turbomachinery by means of the RANS approach should be based not only on the advanced turbulence model but also on adequate models of the laminar/turbulent transition and the turbulent heat transfer. The bypass transition models come out mostly from the intermittency concept using the algebraic and/or transport equation for the intermittency coefficient, see e.g. Thermann and Niehuis [5], Straka and Příhoda [2] and/or Langtry and Menter [6], Kubacki et al. [7], and Menter et al. [8]. Besides, the three-equation *k-kL-ω* model by Walters and Cokljat [3] with the equation for the energy of non-turbulent fluctuations can be used for transition modelling as well. Nevertheless both types of transition models need empirical relations describing the onset and the length of the transition region. The turbulent heat transfer is mainly modelled by the simplest assumption of the constant turbulent Prandtl number.

The contribution deals with the numerical simulation of the compressible flow through a turbine blade cascade with heat transfer by means two different turbulence models. The EARSM turbulence model of Hellsten [1] was completed by the algebraic bypass transition model of Straka and Příhoda [2] and by the model of turbulent heat transfer based on the generalized gradient hypothesis of Launder [9]. Futher the *k-kL-ω* model of Walters and Cokjlat [3] where the relation for the turbulent thermal diffusivity corresponds to the variable turbulent Prandtl number was used. Both models were tested by means of the flow over a heated flat plate and applied for simulation of flow through a turbine blade cascade.

[a] Corresponding author: straka@vzlu.cz

## 2 Mathematical model

The governing equations for compressible flow are given by the Favre-averaged Navier-Stokes equations and by constitutive relations. The system of equations was closed partly by the EARSM model with the algebraic transition model implemented into the in-house numerical code and partly by the three-equation *k-kL-ω* model implemented into the open-source code OpenFOAM.

The EARSM model is used in the form corresponding to turbulent-viscosity models with the turbulent stress given by the relation

$$\tau_{ij} = \mu_t\left(2S_{ij} - \frac{2}{3}\delta_{ij}\frac{\partial U_k}{\partial x_k}\right) - \frac{2}{3}\delta_{ij}\rho k - a_{ij}^{(ex)}\rho k \quad (1)$$

where the extra-anisotropy tensor $a_{ij}^{(ex)}$ is dependent on strain-rate and vorticity tensors and their invariants. Transport equations for the turbulent energy $k$ and the specific dissipation rate $\omega$ are given by the SST model according to Menter [11] in the form

$$\frac{\partial(\rho k)}{\partial t} + \frac{\partial(\rho U_j k)}{\partial x_j} = P_k + \frac{\partial}{\partial x_j}\left[(\mu + \sigma_k\mu_t)\frac{\partial k}{\partial x_j}\right] - \beta^*\rho\omega k \quad (2)$$

$$\frac{\partial(\rho\omega)}{\partial t} + \frac{\partial(\rho U_j\omega)}{\partial x_j} = \frac{\omega}{k}P_k - \beta\rho\omega^2 + \frac{\partial}{\partial x_j}\left[(\mu + \sigma_\omega\mu_t)\frac{\partial\omega}{\partial x_j}\right] + C_D \quad (3)$$

Turbulent viscosity is given by the relation

$$\mu_t = C_\mu\rho k\tau_t \quad (4)$$

with the turbulent time scale

$$\tau_t = \max\left(\frac{1}{\beta^* \omega}; C_\tau \sqrt{\frac{\nu}{\beta^* \omega k}}\right) \tag{5}$$

where the Kolmogorov viscous time scale is used near the wall and $\beta^* = 0.09$ and $C_\tau = 6$. The coefficient $C_\mu$ is obtained from the equation

$$C_\mu = -\left(\beta_1 + II_\Omega \beta_6\right)/2 \tag{6}$$

where coefficients $\beta_1$ and $\beta_6$ depend on strain-rate and vorticity tensors and their invariants $II_S$ and $II_\Omega$. The detailed description of the EARSM model is given by Hellsten [1].

The production term in the turbulent energy equation was modified for the reduction of the undesirable over-production of the turbulent energy in the stagnation region.

The turbulent heat transfer is modelled according to Launder [9] by the generalized gradient hypothesis in the form

$$q_i = -C_t \frac{\tau_{ij}}{\omega} \frac{\partial T}{\partial x_j} \tag{7}$$

with the constant $C_t = 0.3$. For the prediction of the transitional flows, the production and destruction terms in the $k$-equation are multiplied by the intermittency coefficient $\gamma$. Similarly, the effective viscosity is given by $\mu_{ef} = \mu + \gamma\mu_t$ in the transition region. The transition model is based on the concept of different values of the intermittency coefficient in the boundary layer $\gamma_i$ and in the free stream $\gamma_e$. The intermittency coefficient in the boundary layer $\gamma_i$ is expressed by the relation

$$\gamma_i = 1 - \exp\left[-\hat{n}\sigma\left(Re_x - Re_{xt}\right)^2\right] \tag{8}$$

The transition onset is given by the empirical correlation for the momentum Reynolds number $Re_{\theta t} = f\ (Tu, \lambda_t)$ where $Tu$ (%) is the free-stream turbulence level and $\lambda_t$ is the pressure-gradient parameter. The length of the transition region is expressed using the parameter $N = \hat{n}\sigma\, Re_{\theta t}^3$ where $\hat{n}$ is the spot generation rate and $\sigma$ is the spot propagation rate introduced by Narasimha [12]. The effect of the free-stream turbulence and the pressure gradient on the parameter $N$ is correlated by an empirical relation proposed for the attached flow by Solomon et al. [13]. The onset of transition in separated flow is given by the correlation proposed by Mayle [14] in the form $Re_{xt} = f\,(Re_{\theta s}, Re_{xs})$ where $Re_{\theta s}$ is the momentum Reynolds number at the separation and $Re_{xs}$ is the Reynolds number related to the distance of the separation from the leading edge.

The application of local variables is necessary for prediction of transitional flows in complex geometries. Therefore the momentum Reynolds number is replaced by the maximum of the vorticity Reynolds number according to Langtry and Menter [6]. The vorticity Reynolds number is given by the relation

$$Re_\Omega = y^2 |\Omega| / \nu \tag{9}$$

where $y$ is the distance from the wall and $\Omega$ is the absolute value of the vorticity tensor. The link between both Reynolds numbers is expressed by the relation $Re_\theta = Re_{\Omega max}/C$ where the parameter $C$ depends on the pressure gradient expressed by the parameter

$$L = Re_\Omega^2 \frac{\nu}{U_e^2} \frac{dU_e}{dx} \tag{10}$$

The three-equation model of Walters and Cokljat [2] is based on the assumption that the transition process can be expressed by the transfer from the energy of non-turbulent velocity fluctuations $k_L$ to the turbulent energy $k_T$ of three-dimensional turbulent velocity fluctuations.

The transport equations for the turbulent energy $k_T$, laminar energy $k_L$ and the specific dissipation rate $\omega$ are given by equations

$$\frac{D(\rho k_T)}{Dt} = \rho\left(P_{kT} + R_{BP} + R_{NAT} - \omega k_T - D_T\right) + \frac{\partial}{\partial x_j}\left[\left(\mu + \frac{\rho\alpha_T}{\sigma_k}\right)\frac{\partial k_T}{\partial x_j}\right] \tag{11}$$

$$\frac{D(\rho k_L)}{Dt} = \rho\left(P_{kL} - R_{BP} - R_{NAT} - D_L\right) + \frac{\partial}{\partial x_j}\left(\mu \frac{\partial k_L}{\partial x_j}\right) \tag{12}$$

$$\frac{D(\rho\omega)}{Dt} = C_{\omega 1}\frac{\rho\omega}{k_T}P_{kT} + \left(\frac{C_{\omega R}}{f_w} - 1\right)\frac{\rho\omega}{k_T}\left(R_{BP} + R_{NAT}\right) - C_{\omega 2}\rho\omega^2 + C_{\omega 3} f_\omega \alpha_T f_w^2 \rho \frac{\sqrt{k_T}}{y^3} + \frac{\partial}{\partial x_j}\left[\left(\mu + \frac{\rho\alpha_T}{\sigma_\omega}\right)\frac{\partial \omega}{\partial x_j}\right] \tag{13}$$

The terms $P_{kT}$ and $P_{kL}$ express the production of turbulent and non-turbulent velocity fluctuations respectively. Similarly, terms $D_T$ and $D_L$ express the diffusion. The transfer of energy between non-turbulent and turbulent vortices is modelled by terms $R_{BP}$ and $R_{NAT}$ expressing the effect of the decay of laminar fluctuations during natural and bypass transitions. The turbulent heat transfer in the energy equation is expressed in the form

$$q_i = -\rho\alpha_\theta \frac{\partial h}{\partial x_i} \tag{14}$$

where $h$ is the enthalpy and $\alpha_\theta$ is the turbulent thermal diffusivity used in the form corresponding to the variable turbulent Prandtl number. The both mathematical models are described in detail by Fürst et al. [15].

## 3 Results

The mathematical models were tested using the transitional boundary layer on a heated flat plate according to Sohn and Reshotko [16]. Measurements were carried out in a low-speed closed-circuit wind tunnel. The heated flat plate was formed by the lower wall of the test section with the suction of boundary layer upstream of the leading edge. Measurements were accomplished at the mean velocity $U_e \approx 31$ m/s and the constant wall heat flux $q_w \approx 410$ W/m$^2$ for several free-stream turbulence levels increased by grids. The initial unheated length of the flat plate was $x_o = 0.035$ m.

For simulations were used experimental data for the grid 2 with nominal free-stream turbulence $Tu = 2.4\%$. The computational domain corresponds to the experimental arrangement. The rectangular computational domain was 1.5 m long and 0.15 m high. The inlet was in the distance $x = -0.05$ m upstream of the leading edge. Standard inlet and outlet boundary conditions were applied. The symmetry boundary condition was used on the upper boundary of the domain. The constant temperature gradient was prescribed at the wall according to the Fourier law $q_w = \lambda(\partial T/\partial y)_w$ where $\lambda$ is the thermal conductivity.

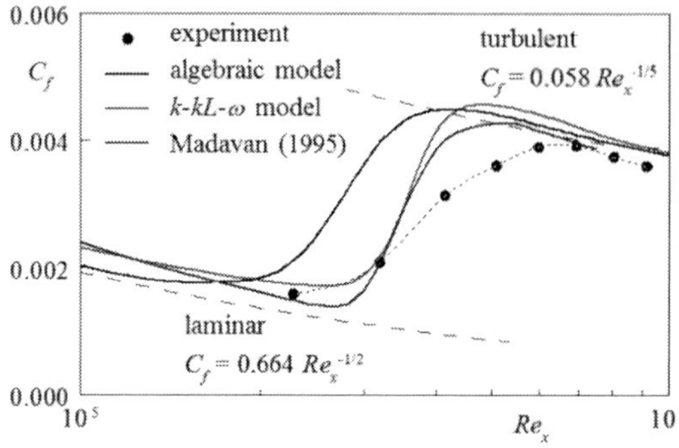


**Fig. 1** Skin friction distribution on the heated flat plate

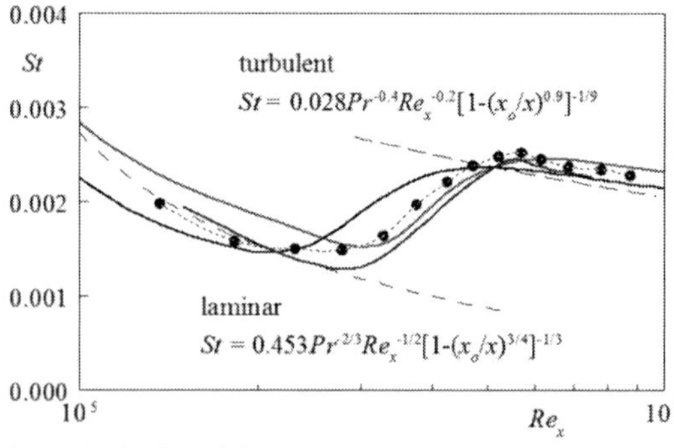


**Fig. 2** Distribution of the Stanton number on the heated flat plate

The skin friction distribution on the heated flat plate is compared in Fig. 1 with experimental data of Sohn and Reshotko [16] and with results of numerical simulation by means of the DNS method completed by Madavan [17]. For comparison, well-known skin friction laws for the flat-plate laminar and turbulent boundary layers are shown as well.

It is surprising that the algebraic transition model gives the skin friction distribution with a rather early onset of the transition. The *k-kL-ω* model and the DNS method agree well with experimental data, but the predicted length of the transition region is rather shorter.

The heat transfer at the wall is described by the Stanton number determined by the relation

$$St = \frac{q_w}{\rho c_p U_e (T_w - T_\infty)} \tag{15}$$

On the basis of the Reynolds analogy, the Stanton number for the constant wall heat transfer $q_w$ = const. is given according to Kays and Crawford [5] for the laminar flat-plate boundary layer by the relation

$$St_L = 0.453\,Pr^{-0.67}\,Re_x^{-0.5}\left[1-\left(\frac{x_o}{x}\right)^{0.75}\right]^{-0.333} \tag{16}$$

and for the turbulent flat-plate boundary layer

$$St_T = 0.028\,Pr^{-0.4}\,Re_x^{-0.2}\left[1-\left(\frac{x_o}{x}\right)^{0.9}\right]^{-0.111} \tag{17}$$

where $x_o$ is the entrance unheated length and the Prandtl number $Pr$ = 0.708. The predicted distribution of the Stanton number on the heated flat plate is compared with experimental data of Sohn and Reshotko [16] in Fig. 2. The transition onset predicted by the algebraic model by means of the Stanton number distribution is again slightly upstream similarly as for the skin friction distribution. The *k-kL-ω* model gives a good agreement with experimental data comparable with DNS results.

The both models were further applied for the numerical simulation of the compressible flow through the plane VKI high pressure turbine blade cascade with the relative spacing $t/c$ = 0.85 and the zero inlet flow angle according to measurements of Arts et al. [5]. Experiments were carried out in the test facility enabling an independent selection of both Mach and Reynolds numbers. Besides measurements of the blade velocity distribution, the loss coefficient and exit flow angle, experiments were focused on measurements of the blade convective heat transfer.

Predictions were carried out for two subsonic (MUR 132 and MUR 218) and two transonic regimes (MUR 210 and MUR 213) at the isentropic outlet Reynolds number $Re_{2is} = 1\times10^6$, free-stream turbulence $Tu$ = 1 and 4 %, and the wall temperature $T_w$ = 300 K. Other boundary conditions were prescribed according to individual test cases, i.e. total inlet pressure and total inlet temperature, and the outlet static pressure. The overview of free-stream conditions for selected VKI test cases is given in Table 1.

**Table 1** Free-stream conditions for VKI test cases

| Case | $Re_{2is}$ | $M_{2is}$ | $Tu$ | $\mu_t/\mu$ |
|---|---|---|---|---|
| MUR132 | $0.97\times10^6$ | 0.68 | 0.8 | 10 |
| MUR210 | $1.10\times10^6$ | 1.076 | 1 | 10 |
| MUR218 | $1.01\times10^6$ | 0.76 | 4 | 50 |
| MUR213 | $1.09\times10^6$ | 1.068 | 4 | 50 |

In the case of the *k-kL-ω* model, the computational grid containing about 62000 cells is formed by rectangular cells in the boundary layer and triangles and rectangles in the free stream. The nearest node was in the distance from the wall $y^+ \approx 0.2$. The onset of the grid is in the distance 0.9*c* upstream of the leading edge. The detail of the computational grid near the trailing edge is shown in Fig. 3.

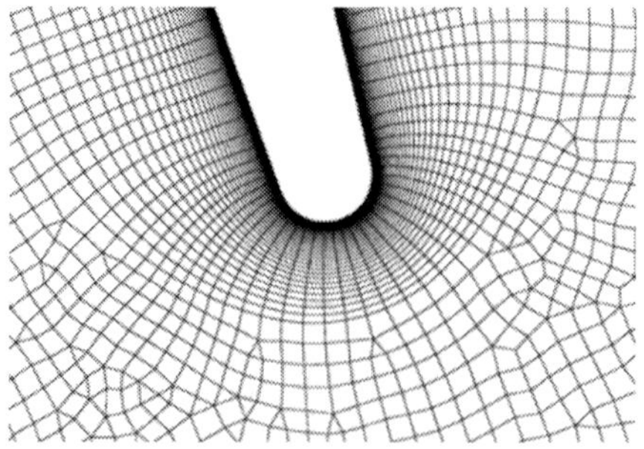

**Fig. 3** Detail of the grid near the trailing edge

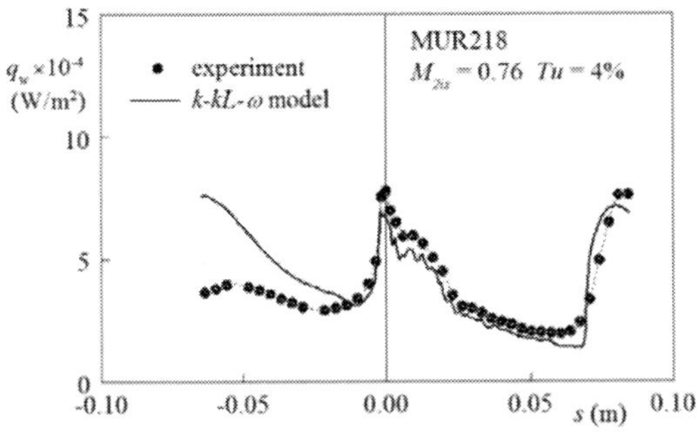


a) MUR 218

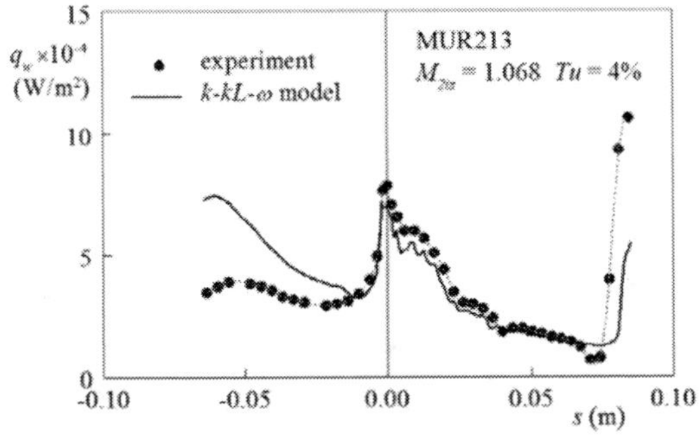


b) MUR 213

**Fig. 4** Distribution of the heat flux (*k-kL-ω* model)

The distribution of the wall heat flux along the blade obtained by the *k-kL-ω* model for the free-stream turbulence about $Tu = 4\%$ is shown in Fig. 4. Numerical results are very similar for subsonic and transonic regimes. The agreement with experimental data is quite good on the suction side ($s > 0$) where the transition onset is near upstream of the trailing edge. However, the *k-kL-ω* model overestimates the heat flux on the pressure side at higher free-stream turbulence.

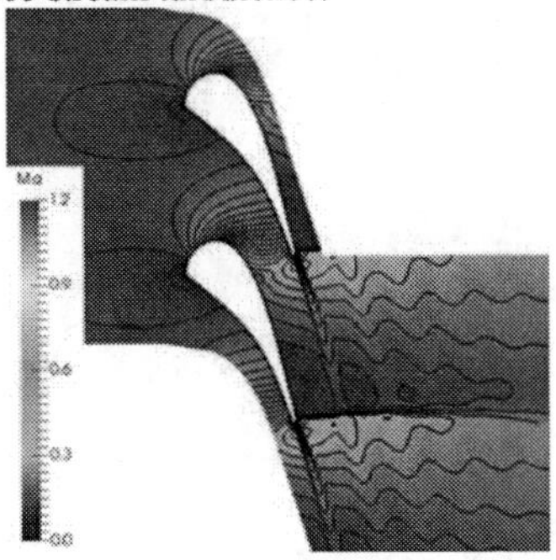

**Fig. 5** Isolines of the Mach number (MUR 213)

The Mach number isolines for the case MUR 213 obtained by the *k-kL-ω* model are shown in Fig. 5 where the shock wave is apparent at the suction side near upstream of the trailing edge.

The distribution of the wall heat flux obtained by the *k-kL-ω* model for the free-stream turbulence about $Tu \approx 1\%$ is shown in Fig. 6. For the lower level of free-stream turbulence, the agreement of numerical results with experiment is quite good on the pressure and suction side, but the transition onset on the suction side in the case MUR 132 is predicted as early as at $s \approx 0.06$ m. The difference between the prediction and experimental data can be caused by the bypass-transition threshold function which is not dependent on the pressure gradient.

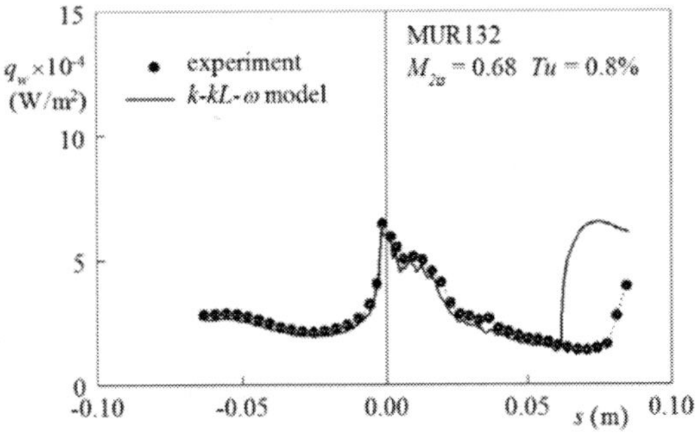


a) MUR 132

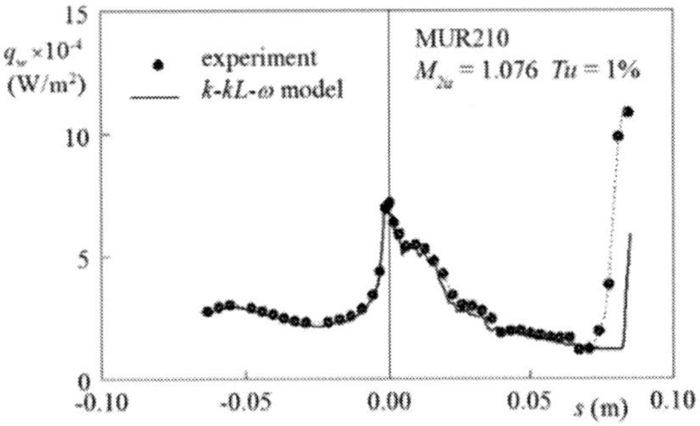


b) MUR 210

**Fig. 6** Distribution of the heat flux (*k-kL-ω* model)

Numerical simulations of compressible flow through the VKI turbine blade cascade by the EARSM model completed by the algebraic transition model were carried out using the computational structured mesh with a block overlapping containing about 55000 quadrilateral cells consists of the O-type block around the blade profile and the H-type block which covers one pitch of blade cascade. The detail of the grid near the leading edge is shown in Fig. 7.

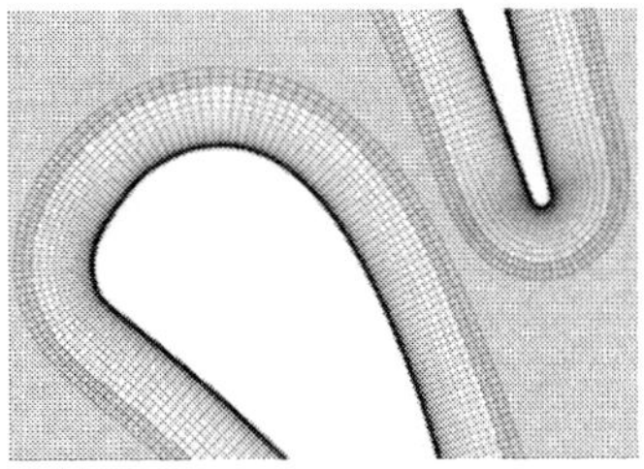

**Fig. 7** Detail of the multi-block structured grid

Due to finite non-negligible thickness of the blade trailing edge, simulations were carried out both by the RANS and by the URANS approaches. Predictions of the compressible flow with heat transfer were accomplished for test cases given in Tab. 1.

The distribution of the wall heat flux obtained by the EARSM model with the algebraic transition model for the free-stream turbulence $Tu = 4\%$ and $1\%$ is shown in

Fig. 8 and 9. The distribution of the wall heat flux is given for the averaged URANS solution. A very good agreement with experimental data was obtained on the pressure and suction sides for both subsonic regimes where the RANS and URANS approaches give the practically same results.

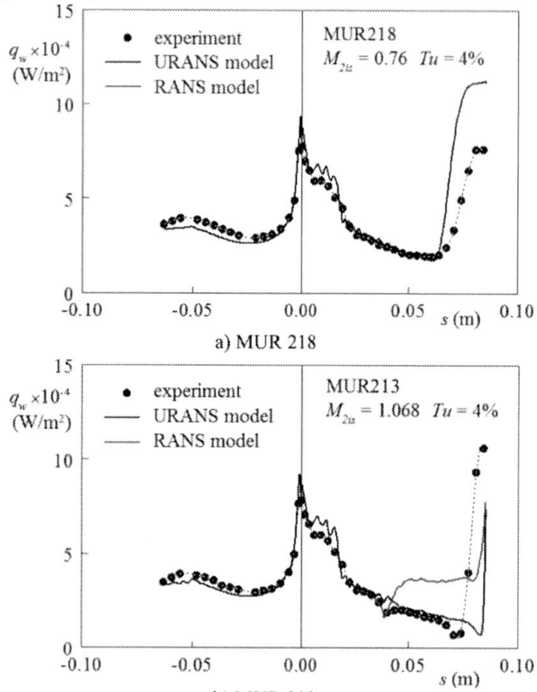


**Fig. 8** Distribution of the heat flux (algebraic model)

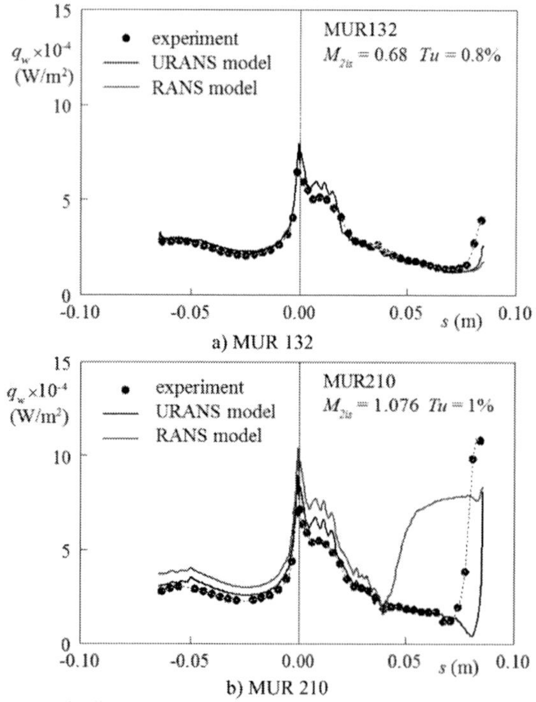


**Fig. 9** Distribution of the heat flux (algebraic model)

The flow field in the blade cascade is demonstrated by isolines of the Mach number shown in Fig.10 for the RANS solution (a), the time-averaged URANS solution (b) and the instantaneous URANS solution (c).

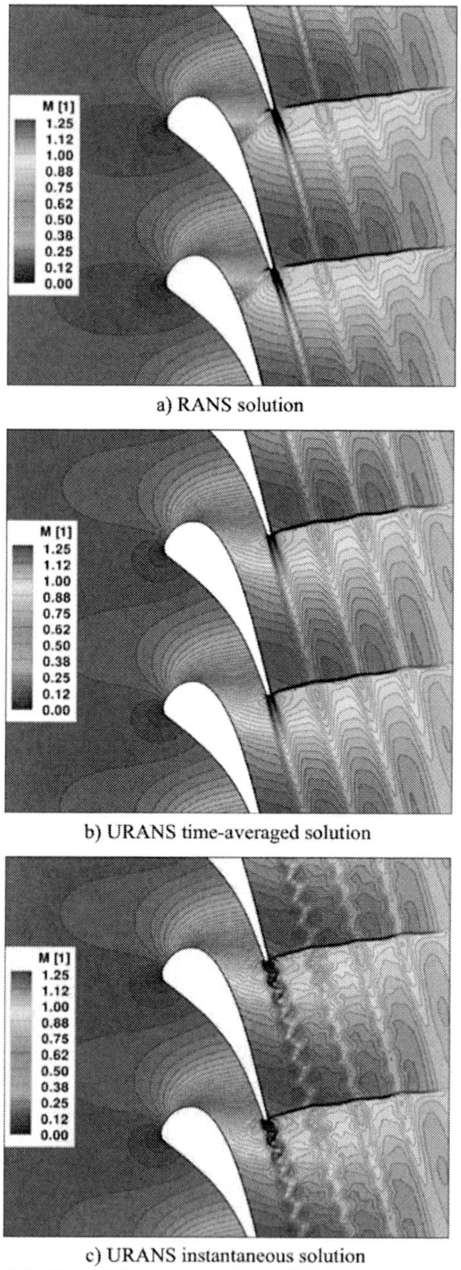


**Fig. 10** Isolines of the Mach number (MUR 213)

The RANS approach gives the transition onset on the suction side at the distance about $s \approx 0.04$ m for the both transonic regimes MUR 213 and MUR 210 similarly as the $\gamma$-$Re$ model of Langtry and Menter [6] for another VKI test case. This transition onset can be caused by the interaction of the boundary layer with the inner branch of the exit shock wave visible for the RANS solution (see

Fig. 10a) while the time-averaged URANS solution does not show this shock wave.

The URANS approach gives a better agreement with experimental data even though the transition onset on the suction side at $s \approx 0.083$ m is probably caused by the interaction with the exit shock wave. Besides, the instantaneous field of the Mach number isolines obtained by the URANS method shows the noticeable vortex line behind the trailing edge of the blade (see Fig. 10c). Similar flow structures including vortex shedding were found out by the RANS and LES simulations of flow with heat transfer in the VKI blade cascade accomplished by Gourdain et al. [18].

## Conclusions

The transitional flow with heat transfer was simulated by the *k-kL-ω* model of Walters and Cokljat [3] and by the EARSM model of Hellsten [1] completed by the algebraic transition model of Straka and Příhoda [2] and the turbulent heat transfer model based on the generalized gradient hypothesis. The both mathematical models were tested by experimental data of Sohn and Reshotko [16] for the boundary-layer flow on a heated plat plate and then applied to the simulation of compressible flow through the VKI turbine blade cascade according to measurements of Arts et al. [4].

The simulations were carried out for subsonic and transonic regimes at various values of free-stream turbulence. Due to finite thickness of the blade trailing edge, the RANS and URANS approach was applied for the EARSM model with the algebraic transition model. A good agreement was obtained by both approaches for subsonic regimes but only the URANS approach gives adequate results of numerical simulations for transonic regimes where the RANS approach gives a pseudo shock wave whose interaction with the boundary layer leads to the premature transition on the suction side.

## Acknowledgement

The work was supported by the institutional support RVO 61388998 and by grants TA CR TA03020277 and GA CR P101/12/1271.

## References

1. A. Hellsten, *New two-equation turbulence model for aerodynamics applications*, PhD. Thesis, Helsinki University of Technology (2004)
2. P. Straka, J. Příhoda, Proc. Conf. Experimental Fluid Mechanics, Liberec, 636-641 (2010)
3. D. K. Walters, D. Cokljat, Jour. Fluids Engineering., **130**, 121401-121414 (2008)
4. T. Arts, M. Lambert de Rouvroit, A. W. Rutheford, *Aero-thermal investigation of a highly loaded transonic linear turbine guide vane cascade*, Technical Note 174, von Karman Institute for Fluid Dynamics, Rhode Saint Genèse (1990)
5. H. Thermann, R. Niehuis, Jour. Turbomachinery, **128**, 474-483 (2006)
6. R. Langtry, F. R. Menter, AIAA Jour., **47**, 2894-2906 (2009)
7. S. Kubacki, K. Lodefier, R. Zarzycki, W. Elsner, E. Dick, Flow, Turbulence and Combustion, **83**, 539-568 (2009)
8. F. R. Menter, P. E. Smirnov, T. Liu, R. Avancha, Flow, Turbulence Combustion, **95**, 583-619 (2015)
9. B. E. Launder, Jour. Heat Transfer, **110**, 1112-1128, (1988)
10. W. M. Kays, M. B. Crawford, *Convective heat and mass transfer*, McGraw Hill, New York (1993)
11. F. R. Menter, AIAA Jour., **32**, 1598-1605 (1994)
12. R. Narasimha, Prog. Aerospace Science, **22**, 29-80, (1985)
13. W. J. Solomon, G. J. Walker, J. P. Gostelow, Jour. Turbomachinery, **118**, 744-751 (1996)
14. R. E. Mayle, Jour. Turbomachinery, **113**, 509-537 (1993)
15. J. Fürst, J. Příhoda, P. Straka, Computing, **95**, S163-S182 (2013)
16. K.-H. Sohn, E. Reshotko, *Experimental study of boundary layer transition with elevated free-stream turbulence on a heated flat plate*, NASA CR-187068 (1991)
17. N. K. Madavan, *Direct numerical simulation of transitional and turbulent flow over a heated flat plate using finite-difference schemes*, NASA CR-200759 (1995)
18. N. Gourdain, L. Y. M. Gicquel, E. Collado Morata, Jour. Propulsion and Power, **2**, 847-862 (2011)

EPJ Web of Conferences **143**, 02119 (2017) DOI: 10.1051/epjconf/201714302119
*EFM 2016*

# The improvement of the surface hardness of stainless steel and aluminium alloy by ultrasonic cavitation peening

*Styková* Janka[1], *Müller* Miloš[1,*], and *Hujer* Jan[1]

[1]Technical University of Liberec, Faculty of Mechanical Engineering, Studentská 2, 461 17 Liberec 1, Czech Republic

**Abstract.** This article presents first results of the experimental investigation of the influence of the cavitation shot less peening process on the properties of stainless steel and aluminium alloy specimens. The cavitation field was generated by an ultrasonic horn submerged in water and operated by an ultrasonic generator. The temperature of the water was controlled by thermometer and adjusted by separate water cooling system. The mass loss, the mass loss rate and the modification of the surface hardness are evaluated for different cavitation exposure intervals. The mass loss was measured by micro weighing scale and the surface hardness by the micro-hardness meter. The presented results indicates the significant improvement in the surface hardness for both tested materials.

## 1 Introduction

In liquid systems the term cavitation usually refers to the formation and collapse of cavities in the free liquid or in the proximity of submerged surfaces. Due to undesirable effects connected to the cavitation erosion the term cavitation is mainly connected to the field of the hydraulic machinery [1]. However the cavitation phenomena can also be used in many industrial applications. One of these applications is the cavitation shotless peening. The cavitation shotless peening is a method used to improve the fatigue strength and to introduce compressive residual stresses to the metallic materials.

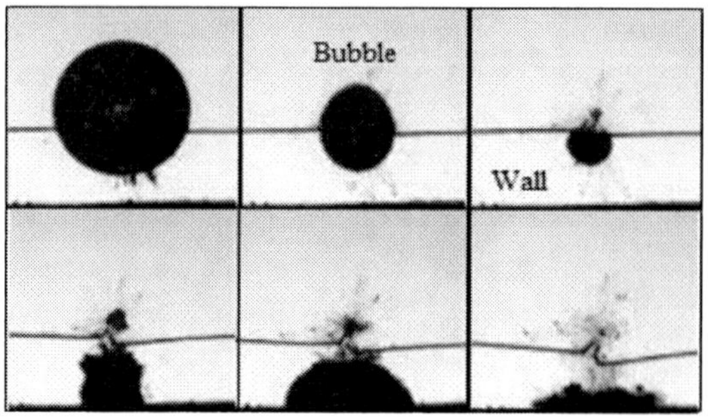


**Fig. 1.** Cavitation bubble collapse close to a solid surface.

The method is named "shotless peening", as no shots are required and the cavitation bubbles collapses are used instead of shooting particles [2], [3]. For instance, during the micro shot peening the particles of size 0.03 to 0.5 mm made of high speed tool steel, cemented carbide or ceramics are bombarding the treated surface [6] causing plastic deformation of the material.

The plastic deformation induces compressive stresses in the peened surface and tensile stresses within the material interior. The mechanical effects responsible for the cavitation erosion are produced mainly during the cavitation bubble collapse.

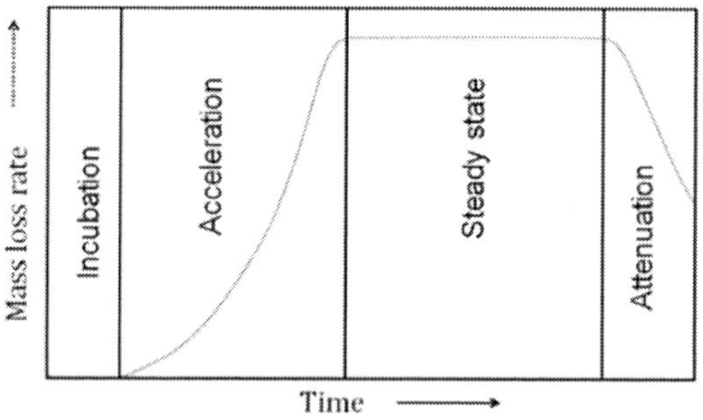


**Fig. 2.** Characteristic curves of the mass loss rate during the cavitation erosion.

Cavitation bubble produces during the collapse close to a solid surface a high speed liquid jet and a high pressure shock wave against the surface [1], [7]. Both these effects act as the mechanical micro hammers on treated surfaces and cause the plastic deformation similarly to the micro shot peening. The typical consequence of the cavitation bubble collapse is shown in Figure 1. Current methods used for the cavitation shot peening are classified according to the way how the cavitation is generated. One method uses cavitating jet, which is produced by the injection of a high speed water jet into the water [2]. Second method uses an ultrasonic generator to produce an ultrasonic cavitation field close

* Corresponding author: milos.muller@tul.cz

to the specimen surface [3]. It was shown, that in the comparison with the classical shot peening methods the cavitation shot less peening exhibits considerably higher improvement in the fatigue strength [4]. The cavitation shot less peening also introduces the compressive residual stresses without considerable inverse in the surface roughness unlike the shot less peening [5]. This effect is caused by the smoother profile of pits produced by collapsing bubble unlike the sharp pit profile caused by the impacting particles.

The time evaluation of the cavitation erosion rate is shown in Figure 2. The cavitation erosion process can be divided into four periods [7]. In the incubation period no measurable mass loss is indicated. It this period, plastic deformations can occur in the impacted region and the surface roughness increases. In the acceleration period, the erosion rate increases to its maximum level. The mass loss of the material can be detected in this period. In the steady state period, the cavitation erosion rate is almost constant. In the attenuation period the erosion rate decreases. This is caused by the decrease in the cavitation collapses pressure in the vicinity of the rough surface.

It can be seen that any processes for the improvement of the material characteristic using the cavitation shot less peening should by realized during the incubation period, where no mass loss takes place. The duration of the incubation period differs based on the material properties and the preceding surface treatment.

## 2 Experiment and measurement

### 2.1 Tested specimens

For the measurement two materials stainless steel 1.4301 EN 10088-3 and aluminium alloy EN AW-2030 T3 were selected. Basic mechanical properties of selected materials are given in Table 1. The specimen geometry is shown in Figure 3.

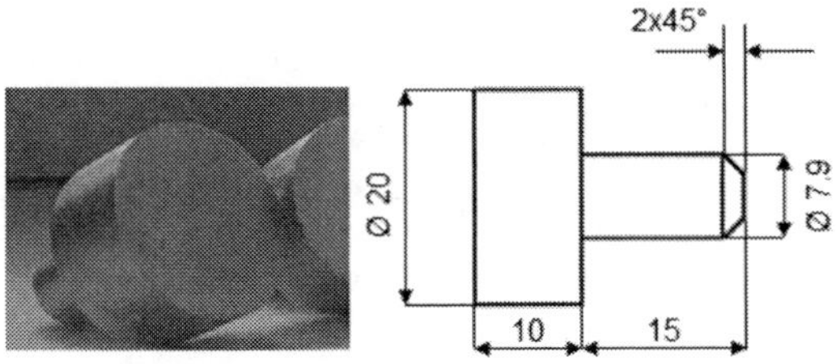


**Fig. 3.** Specimen geometry.

Each specimen was finished to reach the surface roughness lower than 0.8 μm. Number of specimens for aluminium alloy was 8 and number of specimens for stainless steel was 9. The fixation of the specimens was provided to eliminate a possible mass loss during the specimen replacement.

### 2.2 Experimental setup

The scheme of the experimental setup used for the measurement is shown in Figure 4. For the generation of the ultrasonic field of frequency 20 kHz ±1kHz, the ultrasonic generator UIP1000hd (Hielscher) was used. The amplitude of the horn vibrations was set for each measurement to value of 150 μm.

**Table 1.** Mechanical properties of tested materials.

| Material | Minimum Yield Strength [MPa] | Ultimate Tensile Strength [MPa] | Ductility [%] |
|---|---|---|---|
| EN 10088-3 | 350 | 700-850 | 20 |
| EN AW-2030 T3 | 250 | 400 | 10 |

The specimen was placed 20 mm under the free liquid level. The working distance between the horn tip and the exposed specimen surface was set to 0.5 mm.

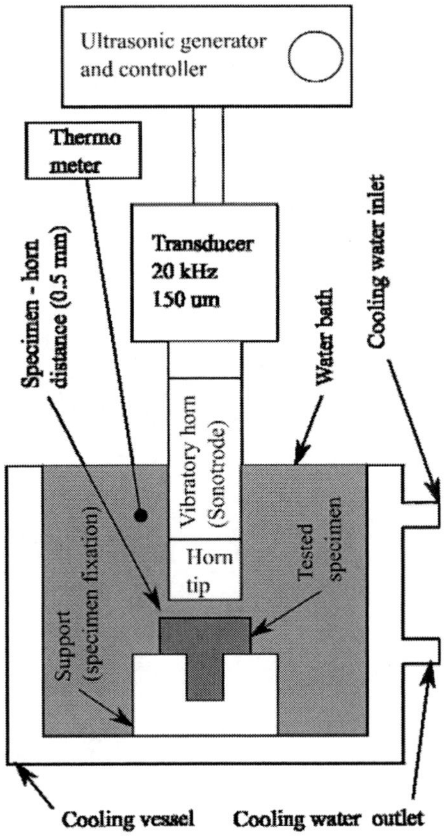


**Fig. 4.** Experimental setup.

Due to the intensive heat production during the testing process, water cooling system was used to stabilize the water temperature. The cooling water was introduced directly into the channels build in the vessel wall. The water temperature was controlled by temperature detector. Tap water of temperature 14°C was used for the cooling. The flow rate of the cooling liquid was set by test to reach the water temperature in range 20±2°C. The mass of specimens was measured by micro weighing scale (Kern). The surface hardness measurement was realized by the test according to Vickers using hardness meter Zwick 3212.

### 2.3 Measurement

Each specimen for the measurement was exposed by cavitation field for a different time period. Then the mass loss and the change in the surface hardness after the exposure period were evaluated. The key problem was to set the time periods for the specimen exposure as the increase in the surface hardness is connected only to the incubation phase. Based on measurements realised by Li [8], the exposure time for the incubation period of stainless steel was estimated to be 75 minutes. The first specimen was exposed by the cavitation field for 45 minutes and for each other specimen the time period was extended by 15 minutes against the previous specimen period. The exposure time for aluminium alloy was set shorter due to lower mechanical properties. The specimens were exposed for 2, 3, 4, 5, 6, 10 and 15 minutes. For the mass loss evaluation each specimen was weighted before and after the exposure period. The micro-hardness of the surface was measured using a diamond indenter of pyramid geometry with the top angle of 136°. The duration of the measurement was 45-60s for stainless steel specimens and 10-15s for aluminum alloy specimens. The surface hardness was evaluated from the length of diagonals of the imprinted rectangular shape of the indenter according to the formula

$$HV = 0,189\,F/d^2\,, \qquad (1)$$

where $d$ is an averege value of lenghts of diagonals and $F$ is the force load.

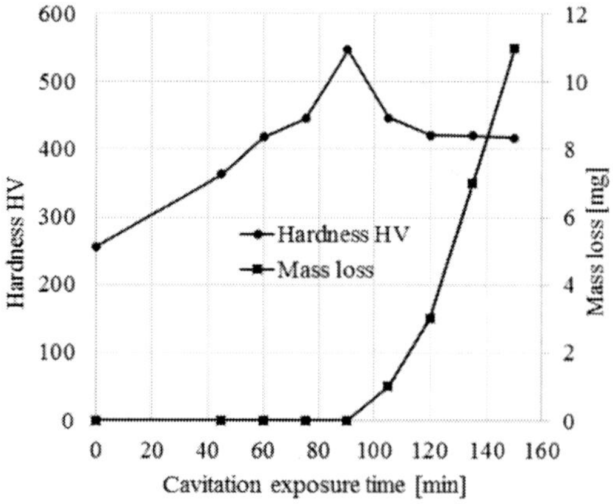


**Fig. 5.** Evaluation of the cavitation erosion for the stainless steel specimen.

The lenght of diagonals was measured by an optical microscope. The force load for the measurement was set to 0,2 N to ensure the diagonals visibility on the specimen after the cavitation exposure period. For each specimen the micro hardness was measured at 15 different points of the specimen surface.

## 3 Results evaluation

For each specimen the mass loss after the cavitation exposure period and the modification of the surface hardness were evaluated. Figure 5. shows the mass loss and the surface hardness change for the stainless steel specimen.

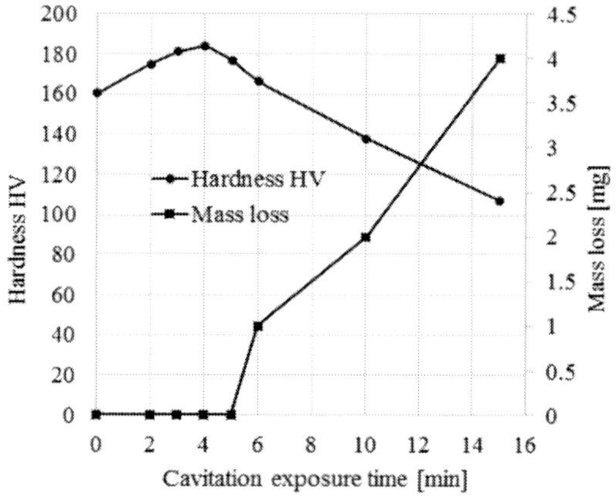


**Fig. 6.** Evaluation of the cavitation erosion for the aluminium alloy specimen.

The incubation period for the tested specimens was about 90 minutes. In this interval no mass loss was detected. The mass loss rate for this material is shown in Table 2. The mass loss rate increases after the incubation period and reaches the maximum after 135 minute of the exposure time and then is constant. The specimen hardness increases during the incubation period and reaches the maximum HV 547 at the end of the incubation period. After this point the surface hardness decreases and then remains almost constant.

Figure 6. shows the mass loss and the surface hardness change for the aluminium alloy specimens. The incubation period for aluminium alloy specimens was about 5 minutes. After this time the mass loss rate increases. The surface hardness increases approximately till the end of the incubation period to value of HV 184 and then decreases.

It can be seen that the incubation period for the aluminium alloy is significantly shorter than the incubation period for the stainless steel. The mass loss rate at the beginning of the cavitation erosion process is slower for the stainless steel, which corresponds to the significantly better mechanical properties of this material. The increase in the surface hardness is more than two times higher for the stainless steel than for the aluminium alloy.

**Table 2.** The mass loss rate evaluation for different periods of the cavitation exposure.

| Aluminum alloy | | Stainless steel | |
|---|---|---|---|
| Cavitation exposure time | Mass loss rate | Cavitation exposure time | Mass loss rate |
| [min] | [mg/h] | [min] | [mg/h] |
| 2 | x | 45 | x |
| 3 | x | 60 | x |
| 4 | x | 75 | x |
| 5 | x | 90 | x |
| 6 | 60 | 105 | 4 |
| 10 | 15 | 120 | 8 |
| 15 | 24 | 135 | 16 |
| x | x | 150 | 16 |

It can be also seen that for the stainless steel specimens the surface hardness increases more than two times from its initial value, whilst the surface harness for the aluminium alloy specimens increases only about 14 %.

## 4 Conclusion

Presented article shows the first results of the investigation of the possible improvement of the surface hardness by the ultrasonic cavitation peening. The results shows the possibility of using the ultrasonic cavitation to improve the surface hardness of the tested material.

In the measurement the mass loss and the modification of the surface hardness were evaluated on different specimens exposed to the cavitation field for different time periods. This method however can introduce some errors connected to the specimen replacement and different structure of each specimen. On the other hand this measurement procedure improves the quality of the micro hardness measurement. It is also clear that the surface hardness is only one of many parameters, which can be influenced by the peening process.

The future work will be connected to the testing of other methods for the evaluation of the surface hardness and to the evaluation of the other material properties which can be influenced by the cavitation peening process.

## Acknowledgement

This publication was written at the Technical University of Liberec as part of the project "Experimental and numerical investigation in applied fluid mechanics and energy devices" with the support of the Specific University Research Grant, as provided by the Ministry of Education, Youth and Sports of the Czech Republic in the year 2016, internal no. 21124.

## References

1. Ch. E. Brennen, *Cavitation and Bubble Dynamics*, Oxford University Press (1995)
2. D. Odhiambo, H. Soyama, Int. J. Fatigue, **25** (2006)
3. Y. Gao, B. Wu, Z. Liu, Y. Zhou, N. Shen, H. Ding, J. Manuf. Sci. E-T ASME, 136 (2014)
4. H. Soyama, Y. Sekine, Int. J. Sustainable Eng., **3** (2010)
5. H. Soyama, D. O. Macodyio, S. Mall, Tribol. Lett., **17**, 3 (2004)
6. Y. Gao, M. Yao, J. Li, Metall. Trans., **33**, 6 (2002)
7. J. P. Franc, J. M. Michel, Fundamentals of cavitation, *Kluwer academic publisher* (2005)
8. Z. Li, J. Han, J. Lu, J. Zhou, J. Chen, Wear, **321** (2014)

# Local velocity scaling in T400 vessel agitated by Rushton turbine in a fully turbulent region

*Radek* Šulc[1,*], *Pavel* Ditl[1], *Ivan* Fořt[1], *Darina* Jašíkova[2], *Michal* Kotek[2], *Václav* Kopecký[2], and *Bohuš* Kysela[3]

[1]Czech Technical University in Prague, Faculty of Mechanical Engineering, Department of Process Engineering, Technická 4, 166 07 Prague, Czech Republic
[2]Technical University of Liberec, Institute for Nanomaterials, Advanced Technology and Innovation, Studentská 1402/2, 461 17 Liberec, Czech Republic
[3]Czech Academy of Sciences, Institute of Hydrodynamics v.v.i, Pod Paťankou 30/5, 166 12 Prague, Czech Republic

**Abstract.** The hydrodynamics and flow field were measured in an agitated vessel using 2-D Time Resolved Particle Image Velocimetry (2-D TR PIV). The experiments were carried out in a fully baffled cylindrical flat bottom vessel 400 mm in inner diameter agitated by a Rushton turbine 133 mm in diameter. The velocity fields were measured in the zone in upward flow to the impeller for impeller rotation speeds from 300 rpm to 850 rpm and three liquids of different viscosities (i.e. (i) distilled water, ii) a 28% vol. aqueous solution of glycol, and iii) a 43% vol. aqueous solution of glycol), corresponding to the impeller Reynolds number in the range 50 000 < Re < 189 000. This Re range secures the fully-developed turbulent flow of agitated liquid. In accordance with the theory of mixing, the dimensionless mean and fluctuation velocities in the measured directions were found to be constant and independent of the impeller Reynolds number. On the basis of the test results the spatial distributions of dimensionless velocities were calculated. The axial turbulence intensity was found to be in the majority in the range from 0.388 to 0.540, which corresponds to the high level of turbulence intensity.

## 1 Introduction

It is important to know the flow and the flow pattern in an agitated vessel in order to determine many impeller and turbulence characteristics, e.g. impeller pumping capacity, intensity of turbulence, turbulent kinetic energy, convective velocity, and the turbulent energy dissipation rate. The information and data that are obtained can also be used for CFD verification.

In PIV studies, the spatial distributions of various properties are often presented. However, the inspection analysis shows that the validity of the spatial distribution of any dimensionless property for arbitrary process conditions in geometrically similar agitated vessels requires independence of a given dimensionless property from the impeller Reynolds number. The spatial distribution that is obtained is in general valid only when this independence is both theoretically predicted and experimentally verified. Unfortunately, in many studies the results are presented only for one impeller speed and, in addition, when a Rushton turbine is used as an impeller, the experiments are often carried out in the transient flow regime.

Secondly, the flow discharging from an impeller has been investigated mainly, while the regions outside the impeller region have not been treated with the same level of interest ([1]). In our previous work [2] the local velocity profiles in the zone in upward flow to the impeller for three impeller rotation speeds in a vessel 300 mm in the inner diameter filled by a water and agitated by a Rushton turbine were presented.

The aim of this work is to study scaling of the velocity field outside the impeller region in a vessel 400 mm in inner diameter mechanically agitated by a Rushton turbine in a fully turbulent region for three liquids of different viscosities. The independency of dimensionless spatial velocity distribution on the impeller Reynolds number was tested statistically. The hydrodynamics and the flow field were measured in an agitated vessel using Time Resolved Particle Image Velocimetry (TR PIV).

## 2 Theoretical background

### 2.1 Inspection analysis of flow in an agitated vessel

The flow of a Newtonian fluid in an agitated vessel has been described by the Navier – Stokes equation:

$$\rho \cdot \left( \frac{\partial \vec{U}}{\partial t} + \vec{U} \bullet \nabla \vec{U} \right) = -\nabla p + \mu \cdot \nabla^2 \vec{U} + \rho \cdot \vec{g} \qquad (1)$$

* Corresponding author: radek.sulc@fs.cvut.cz

This equation can be rewritten into dimensionless form, as follows (e.g. [3]):

$$\left(\frac{\partial \vec{U}^*}{\partial t^*} + \vec{U}^* \bullet \nabla^* \vec{U}^*\right) = -\nabla^* p^* + \frac{1}{\mathrm{Re}} \cdot \nabla^{*2} \vec{U}^* + \frac{1}{Fr} \cdot \vec{n}^* \quad (2)$$

where the dimensionless properties are defined as follows:

- dimensionless instantaneous velocity:

$$\vec{U}^* = \frac{\vec{U}}{N \cdot D}, \quad (3a)$$

- dimensionless instantaneous pressure:

$$p^* = \frac{p}{\rho \cdot N^2 \cdot D^2}, \quad (3b)$$

- dimensionless space gradient: $\nabla^* = \frac{\nabla}{D}$, (3c)

- impeller Reynolds number $\mathrm{Re} = \frac{N \cdot D^2}{\nu}$, (3d)

- impeller Froude number: $Fr = \frac{N^2 \cdot D}{g}$, (3e)

and where $\vec{n}$ is a unity vector.

Similarly, the equation of continuity for stationary flow of a non-compressible fluid is given

$$\nabla \bullet \vec{U} = 0 \quad (4)$$

and one can be rewritten into the dimensionless form as follows:

$$\nabla^* \bullet \vec{U}^* = 0 \quad (5)$$

The following relations can be obtained for the dimensionless velocity components and the dimensionless pressure, respectively, by inspection analysis of Eqs. (2) and (5), as follows:

$$\vec{U}^* = f_1(\vec{x}^*, t^*, \mathrm{Re}, Fr), \quad (6)$$

$$p^* = f_2(\vec{x}^*, t^*, \mathrm{Re}, Fr), \quad (7)$$

where $\vec{x}^*$ is a dimensionless location vector.

For stationary flow with a periodic character the velocity time dependence can be eliminated by substituting the velocity and pressure by time-averaged properties. For highly turbulent flow in a baffled vessel, the viscous and gravitational forces can be neglected and finally the time-averaged dimensionless velocity components and the pressure are independent of the impeller Reynolds and the Froude numbers, and depend on location only:

$$\overline{U_i^*} = f_1(\vec{x}^*), \quad (8)$$

$$p^* = f_2(\vec{x}^*). \quad (9)$$

Reynolds decomposition of instantaneous velocity components has been applied for the velocity profiles studied in this work.

### 2.2 Mean and fluctuation velocity

Using PIV, the instantaneous velocity data set $U_i\ (t_j)$ in the i[th] direction for $j = 1, 2, \ldots N_R$ at observation times $t_j$ with equidistant time step $\Delta t_S$ (i.e. $\Delta t_S = t_{j+1} - t_j$) was obtained in a given location. Assuming the ergodic hypothesis the time-averaged mean velocity $\overline{U_i}$ was determined as the average value of velocity data set $U_i\ (t_j)$:

$$\overline{U_i} = \frac{1}{N_R} \cdot \sum_j U(t_j) \quad \text{for } j = 1, 2, \ldots, N_R, \quad (10)$$

where $\overline{U_i}$ is mean velocity in the i[th] direction, $U_i\ (t_j)$ is instantaneous velocity in the i[th] direction at the observation time $t_j$, and $N_R$ is the number of data items in the velocity data set.

Consequently, the fluctuation velocity in the i[th] direction $u_i\ (t_j)$ at observation time $t_j$ is obtained by decomposition of the instantaneous velocity:

$$u_i(t_j) = U_i(t_j) - \overline{U_i} \quad \text{for } j = 1, 2, \ldots, N_R, \quad (11)$$

where $u_i\ (t_j)$ is the fluctuation velocity in the i[th] direction at observation time $t_i$, $\overline{U_i}$ is mean velocity in the i[th] direction, $U_i\ (t_i)$ is instantaneous velocity in the i[th] direction at the observation time $t_i$.

The root mean squared fluctuation velocity is determined as follows:

$$\overline{u_i} = \left(\frac{1}{N_R} \cdot \sum_i u_i^2(t_j)\right)^{1/2} \quad \text{for } j = 1, 2, \ldots, N_R, \quad (12)$$

where $\overline{u_i}$ is the root mean squared fluctuation velocity, and $u\ (t_i)$ is the fluctuation velocity at observation time $t_i$.

## 3 Experimental

The hydrodynamics and the flow field were investigated in an agitated vessel using Time Resolved Particle Image Velocimetry (TR PIV). The experiments were carried out in a fully baffled cylindrical flat bottom vessel 400 mm in the inner diameter. The tank was agitated by a Rushton turbine 133 mm in diameter, i.e. the dimensionless impeller diameter $D/T$ was 1/3. The dimensionless impeller clearance $C/D$ taken from the lower impeller edge was 0.75. The tank was filled with degassed liquid, and the liquid height was 400 mm, i.e. the dimensionless liquid height $H/T$ was 1.

**Table 1.** Operational parameters.

| Run # | liquid | N (rpm) | ν ($m^2/s$) | Re (-) | Record time $T_R$ (s) |
|---|---|---|---|---|---|
| w-N300 | water | 300 | $9.35 \times 10^{-7}$ | 94 641 | 24.8 |
| w-N600 | water | 600 | $9.35 \times 10^{-7}$ | 189 282 | 24.8 |
| g28-N600 | 28 % glycol | 600 | $2 \times 10^{-6}$ | 88 445 | 24.8 |
| g28-N750 | 28 % glycol | 750 | $2 \times 10^{-6}$ | 110 556 | 24.8 |
| g43-N600 | 43 % glycol | 600 | $3 \times 10^{-6}$ | 58 963 | 24.8 |
| g43-N750 | 43 % glycol | 750 | $3 \times 10^{-6}$ | 73 704 | 24.8 |
| g43-N850 | 43 % glycol | 850 | $3 \times 10^{-6}$ | 83 531 | 24.8 |

The dimensionless baffle width *B/T* was 1/10. To prevent air suction the vessel was covered by a lid.

Three liquids of different viscosities were used as the agitated liquid: i) distilled water ($\nu = 9.35\times10^{-7}$ $m^2/s$), ii) a 28% vol. aqueous solution of glycol ($\nu = 2\times10^{-6}$ $m^2/s$), and iii) a 43% vol. aqueous solution of glycol ($\nu = 3\times10^{-6}$ $m^2/s$). The velocity fields were measured at an impeller rotation speed in the range from 300 rpm to 850 rpm, which covers the impeller Reynolds number range from 50 000 to 189 000. This Re range secures the fully-developed turbulent flow. The operational conditions are summarized in Table 1.

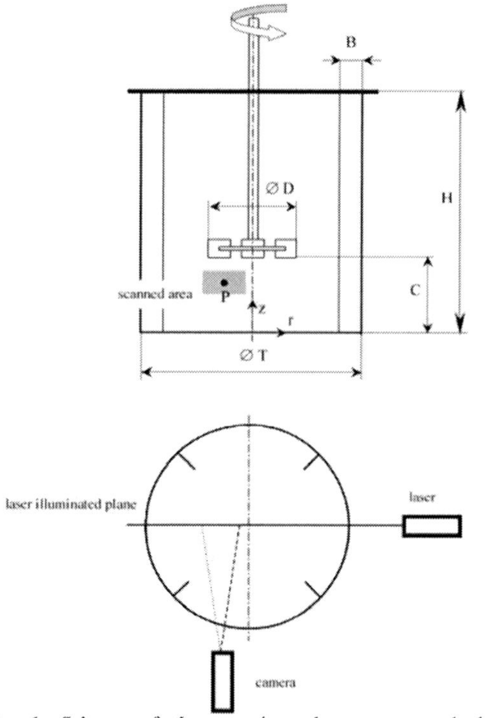


**Fig. 1.** Scheme of the experimental apparatus and the investigated area.

The time-resolved LITRON LDY 302 2D-PIV system (Dantec Dynamics (Denmark)) consists of a Neodyme-YLF laser (light wave length 532 nm, impulse energy 2x15 mJ), a SpeedSence 611 high speed PIV-regime camera (working on frequency 1kHz with resolution 1280 x 800 pixels) with a Nikon Macro 200 optical lens system equipped with an optical filter of wave length 570 nm. Rhodamine B fluorescent particles 11.95 ± 0.25 μm in mean diameter were used as seeding particles. The fluorescent particles lit by 532 nm light emit 570 nm light. In this way, non-seeding particles such as impurities and bubbles are separated, and are not recorded. The operating frame rate was 1 kHz (1000 vector fields per second), i.e. the sampling time $\Delta t_s$ was 1 ms. The measured vertical plane was located in the center of the vessel and in the middle of the baffles. The plane was illuminated by a laser sheet. The investigated area was 43 x 27 mm, and the PIV spatial resolution $\Delta$ = 0.74 mm was maintained the same during the experiment.

The position of middle point P of the investigated area corresponds to dimensionless radius $2r/T$ = 0.224 and dimensionless height $z/T$ = 0.138. The scheme of the experimental apparatus and the investigated area is depicted in Fig. 1. The camera was positioned orthogonally to the laser sheet. The investigated area corresponds to region F according to the classification given by Fořt et al. [4]. The method of ensemble averaged velocity field can be used due to the axisymmetric character of the flow in this region.

## 4 Experimental data evaluation

According to the inspection analysis, the dimensionless velocities normalized by the product of impeller speed *N* and impeller diameter *D* should be independent of the impeller Reynolds number.

The effect of impeller Reynolds number on dimensionless velocities was tested by hypothesis testing ([5]). The statistical method of hypothesis testing can estimate whether the differences between the predicted parameter values (e.g. according to some proposed theory) and then the parameter values evaluated from the measured data are negligible. In this case, we assumed dependence of the tested parameter on the impeller Reynolds number, described by the simple power law correlation *parameter* = $B.Re^{\beta}$, and the difference between predicted exponent $\beta_{pred}$ and evaluated exponent $\beta_{calc}$ was tested. The hypothesis test characteristics are given as $t = (\beta_{calc} - \beta_{pred})/\ s_{\beta}$ where $s_{\beta}$ is the standard error of parameter $\beta_{calc}$.

**Table 2.** Dimensionless velocities – effect of the impeller Reynolds number

| | Hypothesis: parameter = B.(Re)$^0$ percentage / t-characteristics \|t\| | |
|---|---|---|
| Parameter | acceptable | not acceptable |
| $\overline{U}_r/(ND)$ (-) | 97.9 % / 0.62 | 2.1 % / 3.1 |
| $\overline{U}_{ax}/(ND)$ (-) | 87.2 % / 1.3 | 12.8 % / 3.4 |
| $\overline{u}_r/(ND)$ (-) | 92 % / 0.62 | 8 % / 3.4 |
| $\overline{u}_{ax}/(ND)$ | 98.7 % / 1.0 | 1.3 % / 2.8 |
| T.I.$_{ax}$ (-) | 86.1 % / 1.3 | 13.9 % / 3.1 |

If the calculated $|t|$ value is less than the critical value of the t-distribution for ($m$-2) degrees of freedom and significance level $\alpha$, the difference between $\beta_{calc}$ and $\beta_{pred}$ is statistically negligible (statisticians state: "the hypothesis cannot be rejected"). In our case, the independence of dimensionless velocities of the impeller Reynolds number was tested as the hypothesis, i.e. *parameter* = $B.(Re)^0$ = const., i.e. $\beta_{pred} = 0$. The t-distribution coefficient $t_{(m-2),\ \alpha}$ for seven impeller Reynolds numbers and significance level $\alpha = 0.05$ is 2.5706. The hypothesis testing was done for each point in the investigated area. The hypothesis test results are presented for investigated area in Table 2 by the percentage of points in which the above-formulated hypothesis *parameter = const.* is satisfied and by the percentage of points in which the hypothesis *parameter = const.* can not be accepted. For illustration, the average values of calculated $|t|$ value are presented here also.

The dimensionless radial mean velocities were found to be close to zero, in the range from 0.008 to 0.074. These findings correspond to characteristics of the given zone, according to Fořt et al. [4]. This region contains predominantly ascending flow along the vessel axis towards the impeller. The dimensionless axial mean velocity was found to be in the range from 0.317 to 0.511. These values are higher than the dimensionless values for radial mean velocity as expected for this region. The tested hypothesis can be accepted in the majority of profile points of the investigated area.

The dimensionless radial r.m.s. fluctuation velocities were found to be in the range from 0.189 to 0.323. The tested hypothesis can be accepted in the majority of profile points, as is signalized by the very low calculated $|t|$ values. The dimensionless axial r.m.s. fluctuation velocities were found to be in the range from 0.153 to 0.295. The tested hypothesis can be accepted in the majority of profile points as is again signalized by the low calculated $|t|$ values.

On the basis of the results of this hypothesis test, we assume that all dimensionless velocities can be statistically taken as constant and independent of the impeller Reynolds number. The spatial distributions of dimensionless velocities for seven operational conditions were averaged and are presented in Figures 2 – 5. For illustration, the radial profiles of velocity components are presented in Figures 6 - 9 for selected horizontal positions $z^*$ = 0.120, 0.135 and 0.150. The profiles obtained are in accordance with velocity profiles typical for region F as reported by Fořt et al. [4].

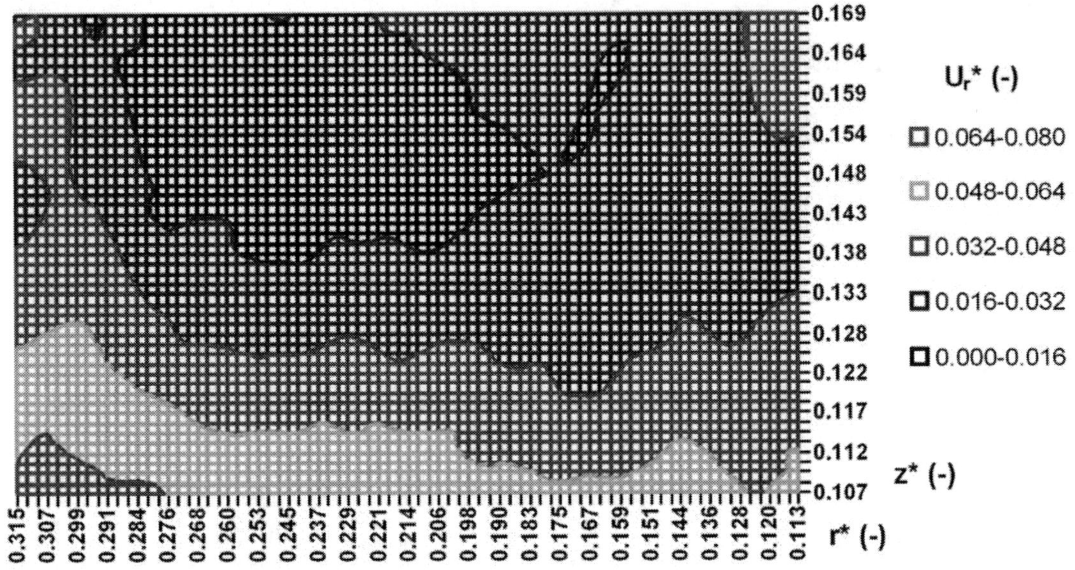


**Fig. 2.** Dimensionless radial mean velocity distribution $U_r^* = \overline{U}_r/(ND) = f(r^*, z^*)$.

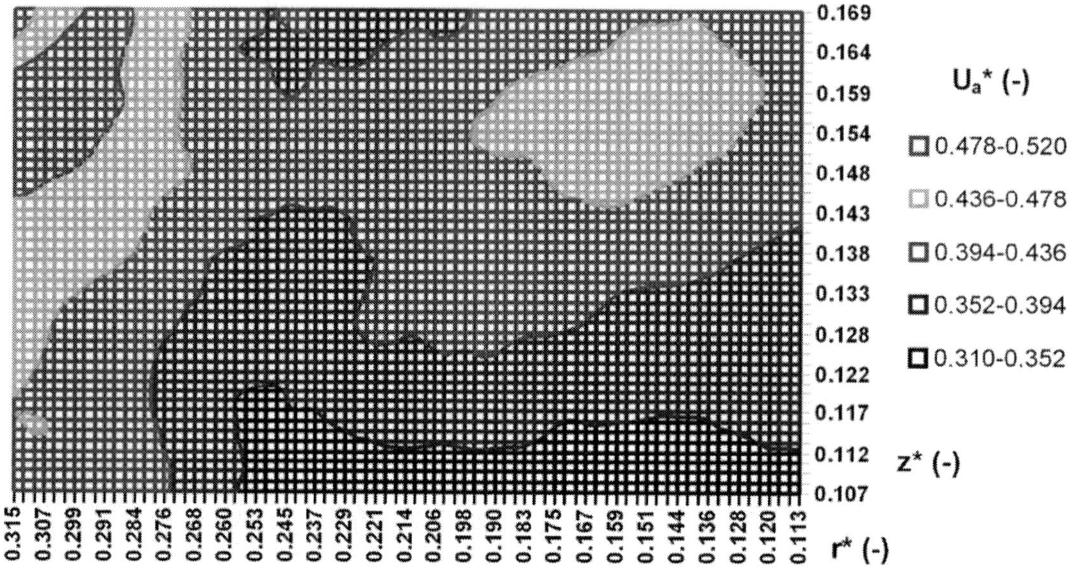


**Fig. 3.** Dimensionless axial mean velocity distribution $U_{ax}^* = \overline{U}_{ax}/(ND) = f(r^*, z^*)$.

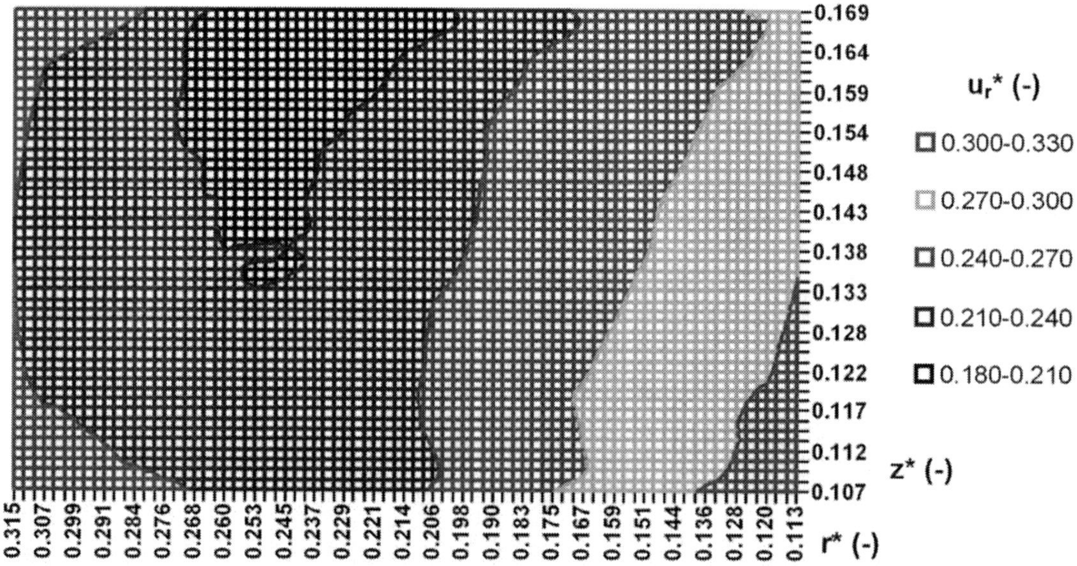


**Fig. 4.** Dimensionless radial fluctuation velocity distribution $u_r^* = \overline{u}_r/(ND) = f(r^*, z^*)$.

DOI: 10.1051/epjconf/201714302120

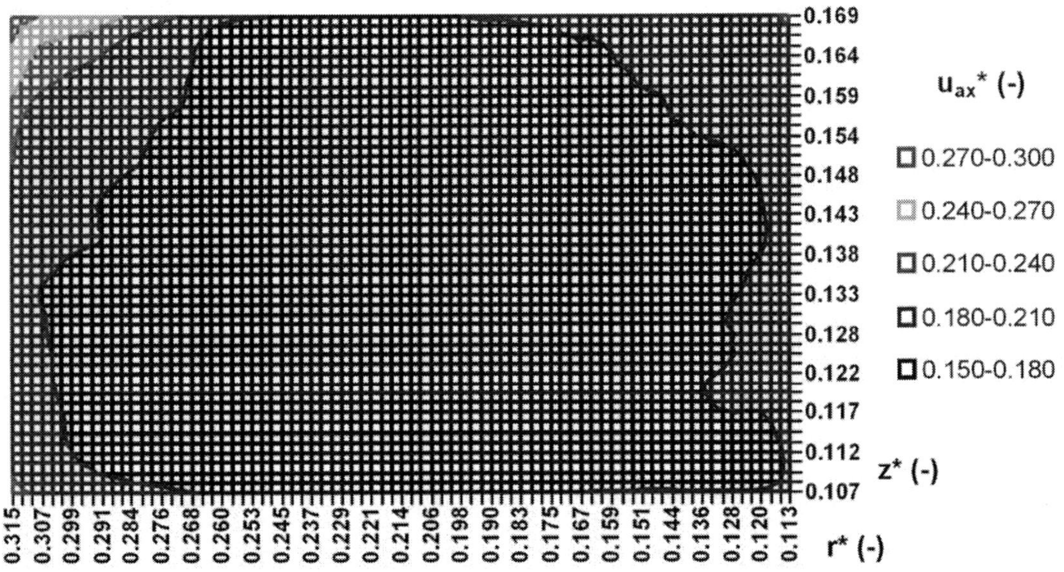


**Fig. 5.** Dimensionless axial fluctuation velocity distribution $u_{ax}^* = \overline{u}_{ax}/(ND) = \mathrm{f}(r^*, z^*)$.

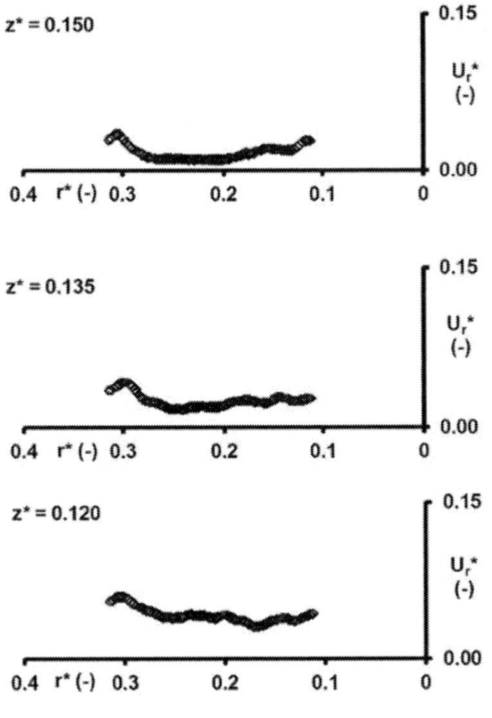


**Fig. 6.** Dimensionless radial mean velocity profiles $\overline{U}_r/(ND) = \mathrm{f}(r^*)$.

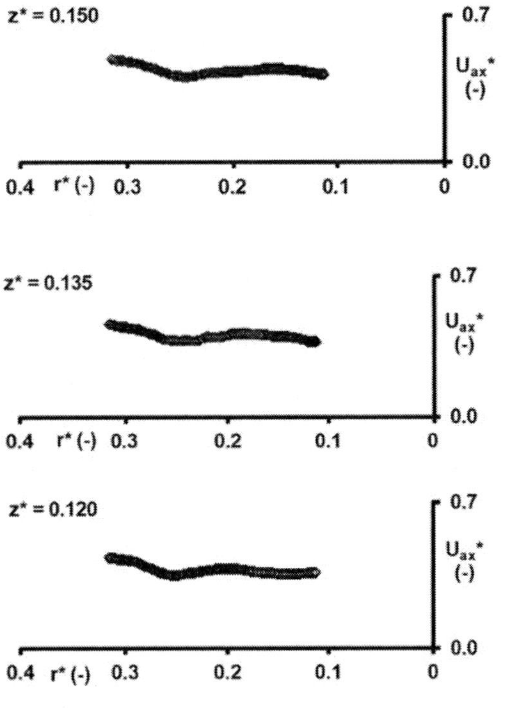


**Fig. 7.** Dimensionless axial mean velocity profile $\overline{U}_{ax}/(ND) = \mathrm{f}(r^*)$.

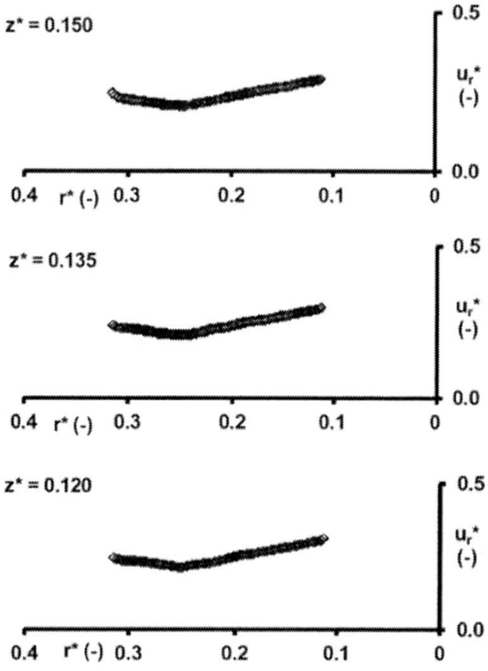


**Fig. 8.** Dimensionless radial fluctuation velocity profile $\overline{u}_r/(ND) = f(r^*)$.

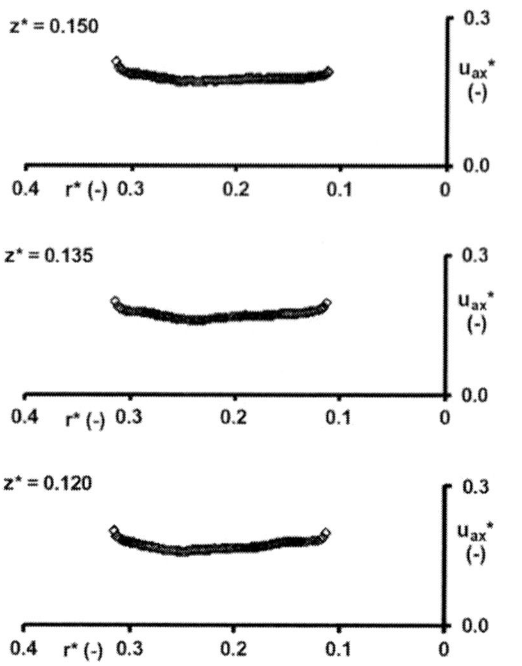


**Fig. 9.** Dimensionless axial fluctuation velocity profile $\overline{u}_{ax}/(ND) = f(r^*)$.

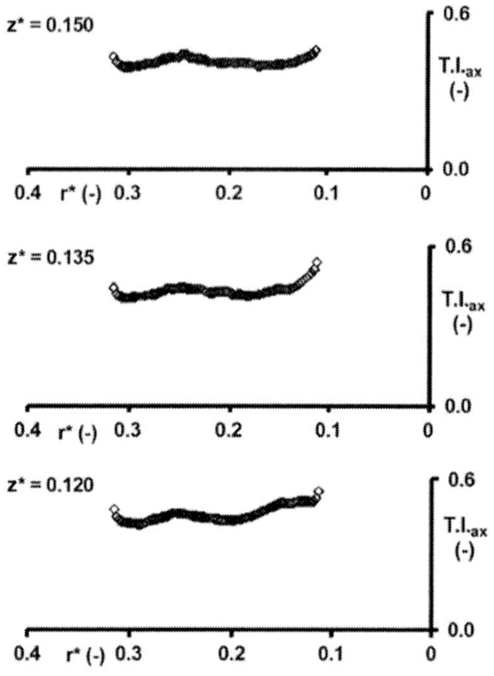


**Fig. 10.** Axial turbulence intensity profile $T.I._{ax} = f(r^*)$.

**Intensity of turbulence**

The axial turbulence intensity was calculated for each point in the investigated area as follows:

$$T.I._{ax} = \frac{\overline{u}_{ax}}{\overline{U}_{ax}} \tag{13}$$

where $\overline{u}_{ax}$ is axial r.m.s. fluctuation velocity, $\overline{U}_{ax}$ is axial mean velocity. For dimensionless velocities independent of the impeller Reynolds number, the turbulence intensity should also be independent of this quantity. The independence of dimensionless velocities from the impeller Reynolds number was again tested by hypothesis testing. The t-distribution coefficient $t_{(m-2),\ \alpha}$ for seven impeller Reynolds numbers and significance level $\alpha = 0.05$ is 2.5706. Hypothesis testing was done for each point in the investigated area. The hypothesis test results are presented in Table 2, as well.

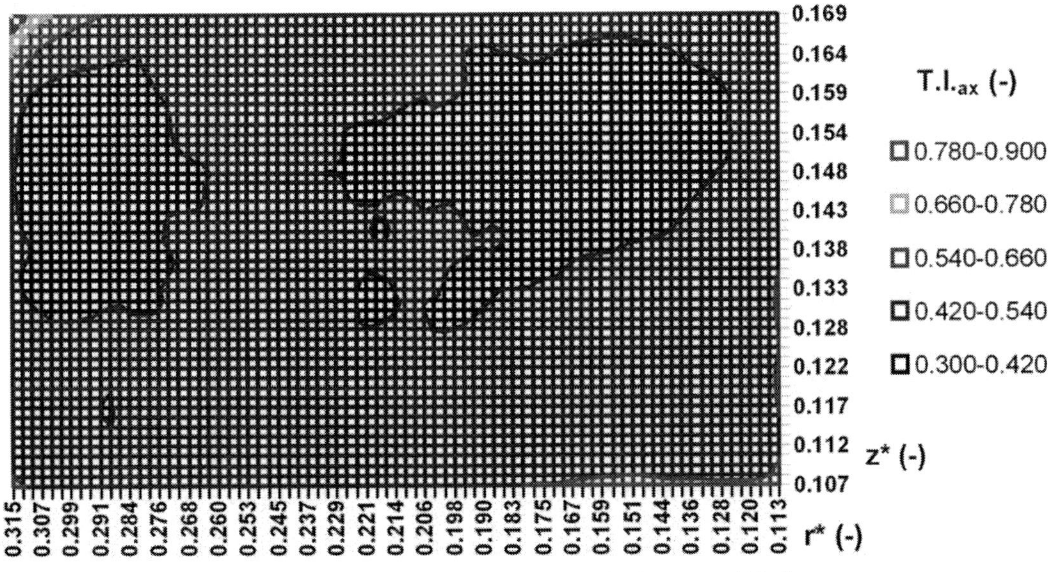


**Fig. 11.** Axial turbulence intensity distribution $T.I._{ax} = f(r^*, z^*)$.

The tested hypothesis can be accepted in the majority of profile points On the basis of these hypothesis test results, we assume that the axial turbulence intensity can be statistically taken as constant and independent of the impeller Reynolds number, as expected. As shown, the calculated values are in the majority of points in the range from 0.388 to 0.540. These values correspond to high turbulence intensity. As expected, the highest turbulence intensity was found close to the impeller axis in the ascending flow core. The radial profiles of axial turbulence intensity are presented in Figure 10 for selected horizontal positions $z^*$ = 0.120, 0.135 and 0.150. The spatial distribution of quantity $T.I._{ax}$ is shown in Figure 11.

## 5 Experimental data evaluation

The following results have been obtained in this study:

The hydrodynamics and flow field were measured in a vessel 400 mm in the inner diameter agitated by a Rushton turbine using the 2-D Time Resolved Particle Image Velocimetry (2-D TR PIV). The velocity fields were measured in the zone in upward flow to the impeller for impeller rotation speeds from 300 rpm to 850 rpm and three liquids of different viscosities, corresponding to the impeller Reynolds number in the range 50 000 < Re < 189 000.

The dimensionless radial mean velocities were found to be close to zero. These findings correspond to the characteristics of the given zone according to Fořt et al. [4]. This region contains predominantly ascending flow along the vessel axis towards the impeller.

In accordance with the theory of mixing, the dimensionless mean and fluctuation velocities in measured directions were found to be constant and independent of the impeller Reynolds number. On the basis of the test results the spatial distributions of dimensionless velocities were calculated.

The axial turbulence intensity was calculated and was found to be in the majority of points in the range from 0.388 to 0.540, which corresponds to high level of turbulence intensity. As expected, the highest turbulence intensity was found close to the impeller axis in the ascending flow core. It was found that the axial turbulence intensity can be statistically taken as constant and independent of the impeller Reynolds number.

This research has been supported by Grant Agency of the Czech Republic project No. 16-20175S and by Ministry of Education, Youth and Sports of the Czech Republic project No. LO1201 (National Programme for Sustainability I) and RVO:67985874.

## Symbols

B baffle width, m
C impeller clearance, m
D impeller diameter, m
H liquid height, m
m number of experimental points, -
N impeller rotational speed, 1/s
$N_R$ number of data in velocity data set, -
r radius (radial coordinate), m
$r^*$ dimensionless radius; $r^* = 2r/T$, -
Re impeller Reynolds number; $Re = N.D^2/\nu$, -
t time, s
$t_i$ observation time, s
$t_R$ record time, s
$t_{(m-2), a=0.05}$ t-distribution for (m-2) degrees of freedom and significance level $\alpha$, -
T tank diameter, m
T.I. turbulence intensity, -
$u_i$ fluctuation velocity in $i^{th}$ direction, m/s

$\overline{u}_i$ root mean squared fluctuation velocity in $i^{th}$ direction, m/s
$U_i$ instantaneous velocity in $i^{th}$ direction, m/s
$\overline{U}_i$ mean velocity in $i^{th}$ direction, m/s
z actual height (axial coordinate), m
$z^*$ dimensionless height; $z^* = z/T$, -

## References

1. B. Kysela, J. Konfršt, I. Fořt, M. Kotek, Z. Chára, Chem. and process Eng., **35**, 137-147 (2014)
2. R. Šulc, V. Pešava, P. Ditl, Acta Polytechnica, **54**, 430-438 (2014)
3. V. Novák, F. Rieger, K. Vavro, *Hydraulické pochody v chemickém a potravinářském průmyslu* (SNTL, Prague, 1989)
4. I. Fořt, A. Obeid, V. Březina, Coll. Czechoslov. Chem. Comm., **47**, 226-239 (1982)
5. B.L. Bowerman, R.T.O'Connell, *Applied statistics: improving business processes* (Richard D. Irwin, USA, 1997)

# The minimum record time for PIV measurement in a vessel agitated by a Rushton turbine

*Radek* Šulc[1,*], *Pavel* Ditl[1], *Ivan* Fořt[1], *Darina* Jašíkova[2], *Michal* Kotek[2], *Václav* Kopecký[2], and *Bohuš* Kysela[3]

[1]Czech Technical University in Prague, Faculty of Mechanical Engineering, Department of Process Engineering, Technická 4, 166 07 Prague, Czech Republic
[2]Technical University of Liberec, Institute for Nanomaterials, Advanced Technology and Innovation, Studentská 1402/2, 461 17 Liberec, Czech Republic
[3]Czech Academy of Sciences, Institute of Hydrodynamics v.v.i., Pod Paťankou 30/5, 166 12 Prague, Czech Republic

**Abstract.** In PIV studies published in the literature focusing on the investigation of the flow field in an agitated vessel the record time is ranging from the tenths and the units of seconds. The aim of this work was to determine minimum record time for PIV measurement in a vessel agitated by a Rushton turbine that is necessary to obtain relevant results of velocity field. The velocity fields were measured in a fully baffled cylindrical flat bottom vessel 400 mm in inner diameter agitated by a Rushton turbine 133 mm in diameter using 2-D Time Resolved Particle Image Velocimetry in the impeller Reynolds number range from 50 000 to 189 000. This Re range secures the fully-developed turbulent flow of agitated liquid. Three liquids of different viscosities were used as the agitated liquid. On the basis of the analysis of the radial and axial components of the mean- and fluctuation velocities measured outside the impeller region it was found that dimensionless minimum record time is independent of impeller Reynolds number and is equalled $N.t_{Rmin}$ = 103 ± 19.

## 1 Introduction

The knowledge of the flow and the flow pattern in an agitated vessel is important for determination of many impeller and turbulence characteristics, e.g. impeller pumping capacity, intensity of turbulence, turbulent kinetic energy, convective velocity, and the turbulent energy dissipation rate. The information and data that are obtained can also be used for CFD verification.

The hydrodynamics and flow field are investigated in an agitated vessel usually using Particle Image Velocimetry (PIV) or Laser-Dopler Anemometry (LDA). In PIV studies published in the literature the record time is ranging from the tenths and the units of seconds. Kysela et al. [1] investigated the hydrodynamics in a fully baffled cylindrical flat bottom vessel 300 mm in inner diameter filled by water agitated by a Rushton turbine with impeller speed 300 rpm. They analysed time record of radial mean- and fluctuation velocities measured in the impeller discharge flow. They reported that the analysed velocities were relatively stable values after 5 s of the record. From this study, it follows that the record time is one of the important factors affecting the relevance and accuracy of measured data. However, it should be noted that this result was observed only for one liquid and one impeller speed.

The aim of this work is to determine minimum record time for PIV measurement in an agitated vessel that is necessary to obtain relevant results of velocity field. The hydrodynamics and flow field were measured in an agitated vessel using 2-D Time Resolved Particle Image Velocimetry (2-D TR PIV). Three liquids of different viscosities were used as the agitated liquid: i) distilled water ($\nu$ = 9.35x10$^{-7}$ m$^2$/s), ii) a 28% vol. aqueous solution of glycol ($\nu$ = 2x10$^{-6}$ m$^2$/s), and iii) a 43% vol. aqueous solution of glycol ($\nu$ = 3x10$^{-6}$ m$^2$/s). The velocity fields were measured at an impeller rotation speed in the range from 300 rpm to 850 rpm, which covers the impeller Reynolds number range from 50 000 to 189 000. This Re range secures the fully-developed turbulent flow of agitated liquid.

## 2 Theoretical background

For highly turbulent flow of Newtonian fluid in a baffled vessel, the viscous and gravitational forces can be neglected and finally the time-averaged dimensionless velocity components are independent of the Reynolds number and the Froude number, and depend on the location only.

Using PIV, the instantaneous velocity data set $U_i$ ($t_j$) in the i$^{th}$ direction for $j$ = 1, 2, … $N_R$ at observation times $t_j$ with equidistant time step $\Delta t_E$ (i.e. $\Delta t_E = t_{j+1} - t_j$) was obtained in a given location. Using obtained data set $U_i$ ($t_j$) in the i$^{th}$ direction the time-averaged mean velocity $\overline{U}_i$ and root mean squared fluctuation velocity $\overline{u}_i$ were calculated in given direction for averaging interval of various lengths $t_P$. The time-averaged mean velocity

* Corresponding author: radek.sulc@fs.cvut.cz

$\overline{U}_i(t_p)$ for averaging interval of time length $t_P$ was determined using the ergodic hypothesis as the average value of velocity data items $U_i$ ($t_j$) in the averaging interval of time length $t_P$:

$$\overline{U}_i(t_P) = \frac{1}{N_P} \cdot \sum_j U(t_j) \quad \text{for } j = 1, 2, \ldots, N_P, \quad (1)$$

where $\overline{U}_i(t_P)$ is mean velocity in the i$^{th}$ direction for averaging interval $t_P$, $U_i$ ($t_j$) is instantaneous velocity in the i$^{th}$ direction at the observation time $t_j$, and $N_P$ is the number of data items in velocity data set for averaging interval of time length $t_P$.

Consequently, the fluctuation velocity in the i$^{th}$ direction $u_i$ ($t_j$) at observation time $t_j$ for averaging interval of time length $t_P$ is obtained by decomposition of the instantaneous velocity:

$$u_i(t_j) = U_i(t_j) - \overline{U}_i(t_P) \quad \text{for } j = 1, 2, \ldots, N_P, \quad (2)$$

where $u_i$ ($t_j$) is the fluctuation velocity in the i$^{th}$ direction at observation time $t_i$, $\overline{U}_i(t_P)$ is mean velocity in the i$^{th}$ direction for averaging interval $t_P$ and $U_i$ ($t_i$) is instantaneous velocity in the i$^{th}$ direction at the observation time $t_i$.

The root mean squared fluctuation velocity for averaging interval of time length $t_P$ is determined as follows:

$$\overline{u}_i(t_P) = \left( \frac{1}{N_P} \cdot \sum_i u_i^2(t_j) \right)^{1/2} \quad \text{for } j = 1, 2, \ldots, N_P, \quad (3)$$

where $\overline{u}_i(t_P)$ is the root mean squared fluctuation velocity in the i$^{th}$ direction for averaging interval $t_P$, and $u$ ($t_i$) is the fluctuation velocity at observation time $t_i$ for time interval $t_P$.

## 3 Experimental

The hydrodynamics and the flow field were measured in an agitated vessel using Time Resolved Particle Image Velocimetry (TR PIV). The experiments were carried out in a fully baffled cylindrical flat bottom vessel 400 mm in the inner diameter. The tank was provided with a Rushton turbine 133 mm in diameter, i.e. the dimensionless impeller diameter $D/T$ was 1/3. The dimensionless impeller clearance $C/D$ taken from the lower impeller edge was 0.75. The liquid height was 400 mm, i.e. the dimensionless liquid height $H/T$ was 1. The dimensionless baffle width $B/T$ was 1/10. To prevent air suction, the vessel was covered by a lid.

Three liquids of different viscosities were used as the agitated liquid: i) distilled water ($\nu = 9.35 \times 10^{-7}$ m$^2$/s), ii) a 28% vol. aqueous solution of glycol ($\nu = 2 \times 10^{-6}$ m$^2$/s), and iii) a 43% vol. aqueous solution of glycol ($\nu = 3 \times 10^{-6}$ m$^2$/s). The velocity fields were measured at an impeller rotation speed in the range from 300 rpm to 850 rpm, which covers the impeller Reynolds number range from 50 000 to 189 000. This means that fully-developed turbulent flow of agitated liquid was reached. The operational conditions are summarized in Table 1.

The time-resolved LITRON LDY 302 2D-PIV system (Dantec Dynamics (Denmark)) consists of a Neodyme-YLF laser (light wave length 532 nm, impulse energy 2x15 mJ), a SpeedSence 611 high speed PIV-regime camera (working on frequency 1kHz with resolution 1280 x 800 pixels) with a Nikon Macro 200 optical lens system equipped with an optical filter of wave length 570 nm. Rhodamine B fluorescent particles 11.95 ± 0.25 μm in the mean diameter were used as seeding particles. The fluorescent particles lit by 532 nm light emit 570 nm light. In this way, non-seeding particles such as impurities and bubbles are separated, and are not recorded. The operating frame rate was 1 kHz (1000 vector fields per second), i.e. the sampling time $\Delta t_s$ was 1 ms. The measured vertical plane was located in the center of the vessel and in the middle of the baffles. The plane was illuminated by a laser sheet. The investigated area was 43 x 27 mm, and the PIV spatial resolution $\Delta = 0.74$ mm was maintained the same during the experiment.

The scheme of the experimental apparatus and the investigated area is depicted in Fig. 1. The camera was positioned orthogonally to the laser sheet. The investigated area corresponds to region F according to the classification given by Fořt et al. [2]. The method of an ensemble averaged velocity field can be used due to axisymmetric character of the flow in this region. The selected data obtained in position P corresponding to dimensionless radius $2r/T$ = 0.224 and dimensionless height $z/T$ = 0.138 are presented in this paper.

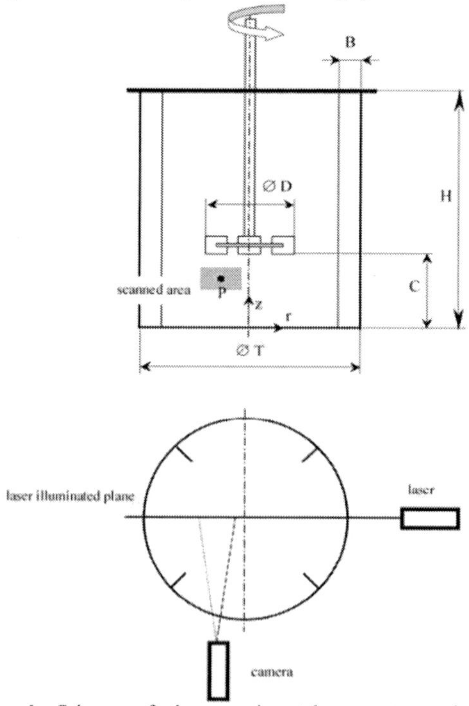


**Fig. 1.** Scheme of the experimental apparatus and the investigated area.

**Table 1.** Operational parameters.

| Run # | liquid | N (rpm) | ν ($m^2$/s) | Re (-) | Record time $T_R$ (s) |
|---|---|---|---|---|---|
| w-N300 | water | 300 | $9.35 \times 10^{-7}$ | 94 641 | 24.8 |
| w-N600 | water | 600 | $9.35 \times 10^{-7}$ | 189 282 | 24.8 |
| g28-N600 | 28 % glycol | 600 | $2 \times 10^{-6}$ | 88 445 | 24.8 |
| g28-N750 | 28 % glycol | 750 | $2 \times 10^{-6}$ | 110 556 | 24.8 |
| g43-N600 | 43 % glycol | 600 | $3 \times 10^{-6}$ | 58 963 | 24.8 |
| g43-N750 | 43 % glycol | 750 | $3 \times 10^{-6}$ | 73 704 | 24.8 |
| g43-N850 | 43 % glycol | 850 | $3 \times 10^{-6}$ | 83 531 | 24.8 |

## 4 Experimental data evaluation

The time courses of mean- and fluctuation velocities in radial and axial direction calculated for averaging intervals of various time lengths are presented in Figures 2 - 8 for each experimental run. The time courses of radial mean velocity are close to zero and have irregular shape. These findings correspond to characteristics of the given zone, according to Fořt et al. [2]. This region contains predominantly ascending flow along the vessel axis towards the impeller.

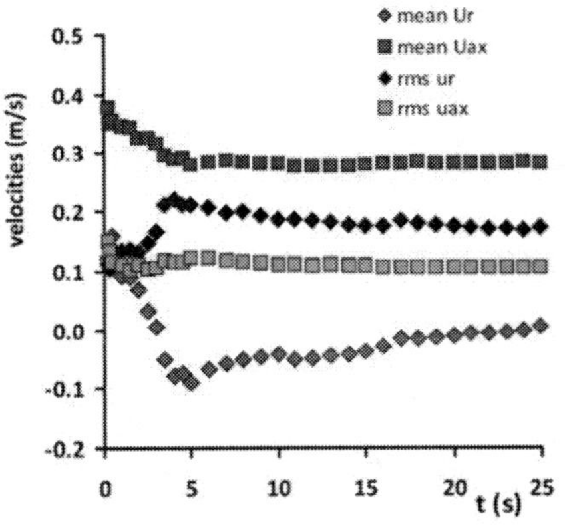


**Fig. 2.** Velocities vs. time length of averaging interval – w-N300; Re = 95 102.

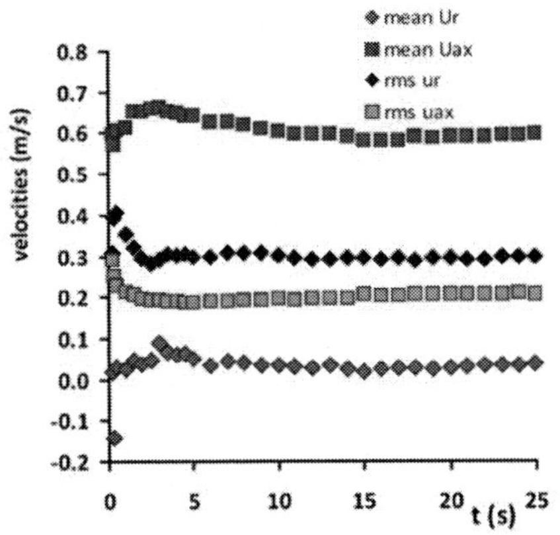


**Fig. 3.** Velocities vs. time length of averaging interval – w-N600; Re = 189 292.

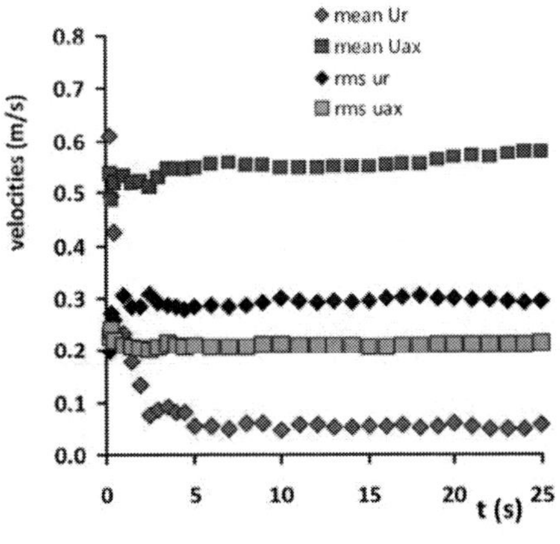


**Fig. 4.** Velocities vs. time length of averaging interval – g28-N600; Re = 88 445.

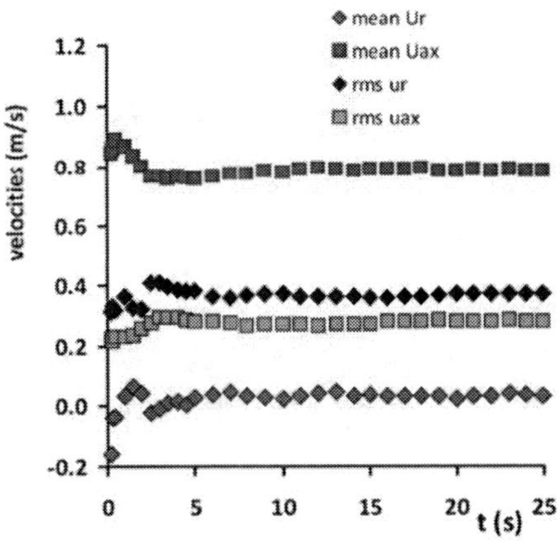


**Fig. 5.** Velocities vs. time length of averaging interval – g28-N750; Re = 110 556.

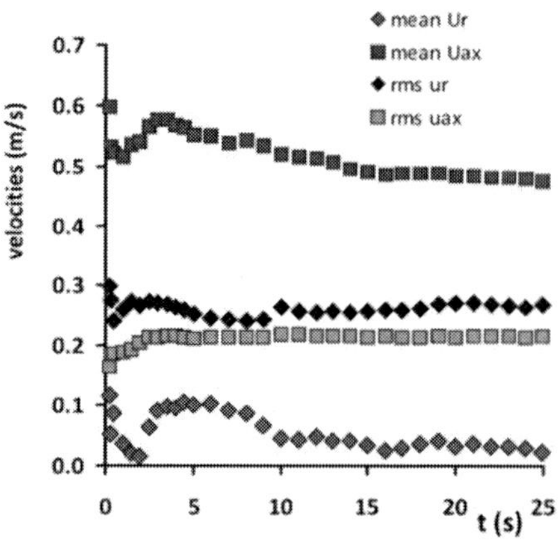


**Fig. 6.** Velocities vs. time length of averaging interval – g43-N600; Re = 58 963.

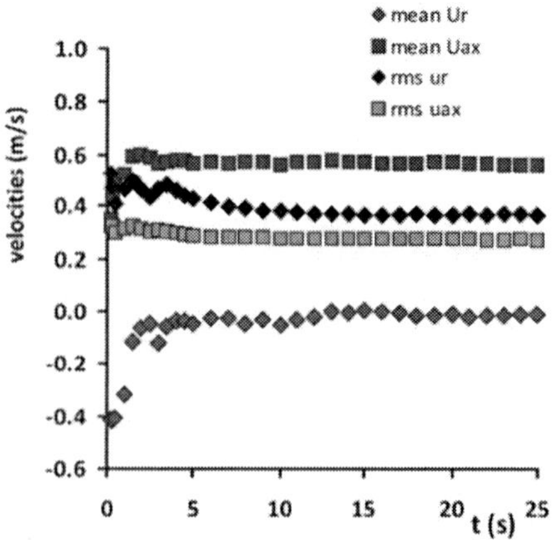


**Fig. 7.** Velocities vs. time length of averaging interval – g43-N750; Re = 73 704.

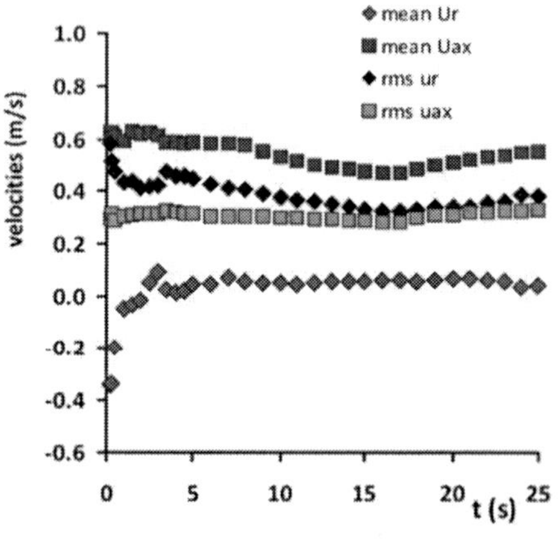


**Fig. 8.** Velocities vs. time length of averaging interval – g43-N850; Re = 85 531.

The time length of averaging interval that is necessary to stabilize the velocity line (i.e. to reach the stable velocity value) is called the stabilization time. Thus principally, the minimum record time must the same or longer than this stabilization time.

For determination of stabilization time the same method that is used for determination of the blending time [3] was used. Using this method the stabilization time was found at the moment when the volatility of time dependence of time-averaged velocity is less than defined percentage of stable value. Usually, the ± 2% range is applied. The principle of evaluation of stabilization time is demonstrated in Fig. 9 for illustration.

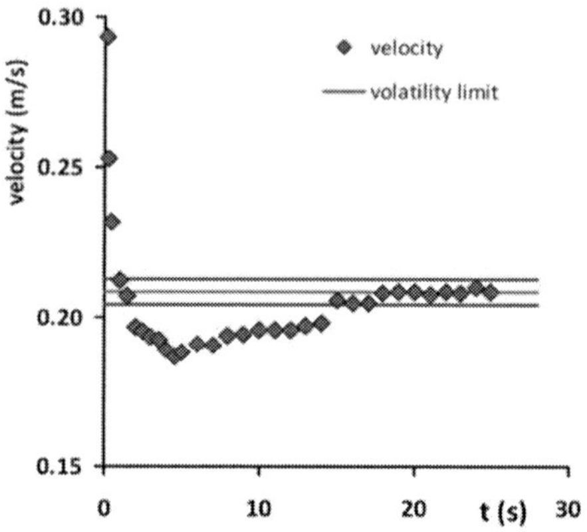


**Fig. 9.** Example of evaluation of stabilization time.

The stabilization time was determined for each velocity (i.e. for the mean- and fluctuation velocities in radial and axial direction) separately. In some cases (radial mean velocity for w-N300 and g43-N600) the velocity time course was not stabilized within whole measured record time. In these cases the stabilization time was roughly estimated however is not taken into account during following treatment. In some other cases (e.g. axial mean velocity and fluctuation velocities for g43-N850) the time courses exhibit quasi-stable wave character instead of stable line. In these cases the stabilization time could not be determined.

The stabilization times determined are presented in Table 2 in dimensionless form as a product of time and impeller speed for each velocity and experimental run. The dimensionless time defined as a product of impeller rotational speed and time corresponds to total number of rounds of the impeller during given time.

The effect of time length of averaging interval on calculated time-averaged velocity can be spectacularly demonstrated plotting data in dimensionless form. The dependences of dimensionless velocity on dimensionless time length of averaging velocity are shown for illustration in Figures 10 and 11 for the radial mean velocity and axial fluctuation velocity, respectively.

As follows from Table 2 the values of dimensionless stabilization time oscillate in the range from 25 in minimum value to 188 in maximum value depending on impeller rotational speed and viscosity, i.e. depending on the impeller Reynolds number.

**Table 2.** Dimensionless stabilization time $N.t_S$ for ± 2% volatility

| Run # | Re (-) | N.t$_S$ (-) | | | |
|---|---|---|---|---|---|
| | | $\overline{U}_r$ | $\overline{U}_{ax}$ | $\overline{u}_r$ | $\overline{u}_{ax}$ |
| w-N300 | 94 641 | *110* | 25 | 100 | 80 |
| w-N600 | 189 282 | 100 | 100 | 110 | 150 |
| g28-N600 | 88 445 | 50 | 100 | 100 | 50 |
| g28-N750 | 110 556 | 100 | 100 | 125 | 188 |
| g43-N600 | 58 963 | *150* | 160 | 100 | 30 |
| g43-N750 | 73 704 | 163 | 38 | 150 | 100 |
| g43-N850 | 83 531 | 142 | n.a. | n.a. | n.a. |

Note: Values printed by italic letters were excluded from data treatment.

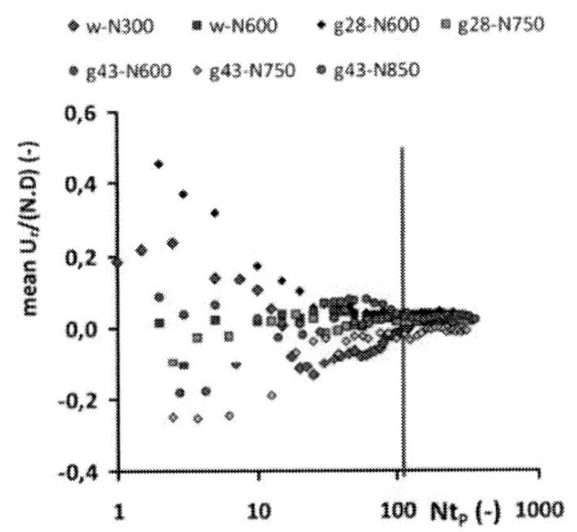


Fig. 10. Dimensionless velocity as a function of dimensionless time length of averaging interval - radial mean velocity

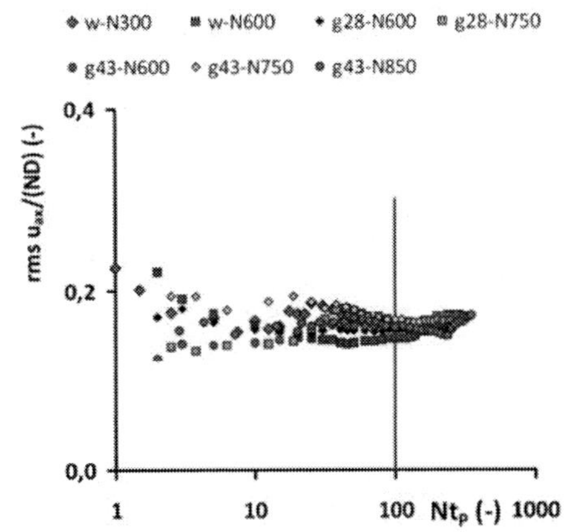


**Fig. 11.** Dimensionless velocity as a function of dimensionless time length of averaging interval - axial fluctuation velocity

**Table 3.** Dimensionless stabilization time for ± 2% volatility – effect of the impeller Reynolds number

| Velocity | m (-) | t-distribution $t_{(m-2),\ \alpha=0.05}$ | Relation $N.t_S = B_R.(Re)^{\beta}$ $\beta_{calc}$ (-) | t-characteristics \|t\| Hypothesis $N.t_S = B_H.(Re)^0$ | Average value |
|---|---|---|---|---|---|
| $\overline{U}_r$ | 5 | 3.1825 | -0.231 | 0.3 (acceptable) | 111 ± 54 |
| $\overline{U}_{ax}$ | 6 | 2.7764 | 0.047 | 0.1 (acceptable) | 87 ± 52 |
| $\overline{u}_r$ | 6 | 2.7764 | -0.010 | 0.1 (acceptable) | 114 ± 21 |
| $\overline{u}_{ax}$ | 6 | 2.7764 | 1.258 | 2.2 (acceptable) | 100 ± 63 |
| Overall data | 23 | 2.0796 | 0.319 | 1.01 (acceptable) | 103 ± 19 |

Note: The t-distribution for (m-2) degrees of freedom and significance level $\alpha = 0.05$.

The effect of impeller Reynolds number on dimensionless stabilization time was tested by hypothesis testing ([4]) for each velocity separately. The statistical method of hypothesis testing can estimate whether the differences between the predicted parameter values (e.g. according to some proposed theory) and whether the parameter values evaluated from the measured data are negligible. In this case, we assumed dependence of the tested property on the impeller rotational speed, described by the simple power law correlation $property = B.Re^{\beta}$, and then the difference between predicted exponent $\beta_{pred}$ and evaluated exponent $\beta_{calc}$ was tested. The hypothesis test characteristics are given as $t = (\beta_{calc} - \beta_{pred})/\ s_{\beta}$ where $s_{\beta}$ is the standard error of parameter $\beta_{calc}$. If the calculated $|t|$ value is less than the critical value of the t-distribution for ($m$-2) degrees of freedom and significance level $\alpha$, the difference between $\beta_{calc}$ and $\beta_{pred}$ is statistically negligible (statisticians state: "the hypothesis cannot be rejected"). In our case, the independence of dimensionless stabilization time from the impeller Reynolds number was tested as the hypothesis, i.e. $N.t_S = B.(Re)^0$ = const., i.e. $\beta_{pred} = 0$. The hypothesis test results are presented in Table 3. For illustration, the values of calculated $|t|$ value are presented here also.

On the basis of the results of this hypothesis test, we assume that dimensionless stabilization times determined for each velocity separately can be statistically taken as constant and independent of the impeller Reynolds number. The average values were calculated and are presented in Table 3.

Since the values of dimensionless stabilization time $N.t_S$ are comparable regardless of the velocity type and velocity direction, we can assume that the stabilization time should be the same for both the mean- and fluctuation velocity in both radial and axial direction. Assuming the independency of dimensionless stabilization time on velocity type we evaluated all data together. The results are presented in Table 3.

The independency on impeller Reynolds number was confirmed and averaged value of $N.t_S = 103 \pm 19$ was calculated. It means that the minimum record time must cover at least 103 rounds of the impeller to obtain relevant results of velocity field.

For comparison, by the same procedure the dimensionless stabilization time $N.t_S = 81 \pm 16$ can be determined for ± 5%

## 5 Conclusions

The following results have been obtained:

The hydrodynamics and flow field were investigated in a vessel 400 mm in the inner diameter agitated by a Rushton turbine using the 2-D Time Resolved Particle Image Velocimetry (2-D TR PIV). The velocity fields were measured in the zone in upward flow to the impeller for impeller rotation speeds from 300 rpm to 850 rpm and three liquids of different viscosities, corresponding to the impeller Reynolds number in the range 50 000 < Re < 189 000.

The effect of time length of the averaging interval on calculated time-averaged velocity was investigated. The time courses of mean- and fluctuation velocities in radial and axial direction calculated for averaging intervals of various time lengths were calculated in selected position in agitated system for each experimental run.

Further, the stabilization time (the time length of averaging interval that is necessary to stabilize the velocity line) was determined for the mean- and fluctuation velocities in radial and axial direction and for each experimental run. Principally, the minimum record time must be the same or longer than this stabilization time.

On the basis of the statistical analysis it was found that minimum record time $t_{Rmin}$ must satisfied the following condition $N.t_{Rmin} \propto N.t_S = 103 \pm 19$ for the investigated area and ± 2% volatility.

This research has been supported by Grant Agency of the Czech Republic project No. 16-20175S and by Ministry of Education, Youth and Sports of the Czech Republic project No. LO1201 (National Programme for Sustainability I) and RVO:67985874.

## Symbols

B baffle width, m
C impeller clearance, m
D impeller diameter, m
H liquid height, m
m number of experimental points, -
N impeller rotational speed, 1/s
$N_P$ number of data items in velocity data set for averaging interval, -
$N_R$ number of data in velocity data set, -
r radius (radial coordinate), m
$r^*$ dimensionless radius; $r^* = 2r/T$, -
Re impeller Reynolds number; $Re = N.D^2/\nu$, -
t time, s
$t_i$ observation time, s
$t_P$ time length of averaging interval, s
$t_R$ record time, s
$t_S$ stabilization time, s
$t_{(m-2),\, a=0.05}$ t-distribution for (m-2) degrees of freedom and significance level $\alpha$, -
T tank diameter, m
$u_i$ fluctuation velocity in i$^{th}$ direction, m/s
$\overline{u}_i$ root mean squared fluctuation velocity in i$^{th}$ direction, m/s
$U_i$ instantaneous velocity in i$^{th}$ direction, m/s
$\overline{U}_i$ mean velocity in i$^{th}$ direction, m/s
z actual height (axial coordinate), m
$z^*$ dimensionless height; $z^* = z/T$, -

## References

1. B. Kysela, J. Konfršt, Z. Chára, M. Kotek, EPJ Web of Conferences **67**, 02065 (2014)
2. I. Fořt, A. Obeid, V. Březina, Coll. Czechoslov. Chem. Comm., **47**, 226-239 (1982)
3. I. Fořt, T. Jirout, F. Rieger, R. Allner, R. Sperling, Acta Polytechnica, **41**, 7-13 (2001)
4. B.L. Bowerman, R.T.O'Connell, *Applied statistics: improving business processes* (Richard D. Irwin, USA, 1997)

EPJ Web of Conferences **143**, 02122 (2017) DOI: 10.1051/epjconf/201714302122
*EFM 2016*

# On numerical approximation of incompresible fluid flow with free surface influenced by surface tension

Petr Sváček[a]

Czech Technical University in Prague, Faculty of Mechanical Engineering, Dep. of Technical Mathematics, Karlovo nám. 13, Praha 2, Czech Republic

**Abstract.** This paper deals with the numerical approximation of two immiscible incompressible fluids flow. The two-dimensional problem is mathematically described with the aid of the incompressible Navier-Stokes equations. The motion of the two phases is treated with the aid of level set method. The surface tension effects are taken into account as well as the contact angle influence. The problem is formulated in a weak sense and discretized with the aid of the finite element method. The details of the numerical methods are given and several possibilities of the realization of the surface tension term are discussed. Numerical results are shown.

## 1 Introduction

The numerical approximation of two-phase flows with free surface is very important in various scientific as well as in technical applications or industrial processes, see e.g. [1], [2], [3]. The kinematics and dynamics of the fluid interface plays an important role in many of these processes. The approximation of free surface flow influenced by the high surface tension and possibly contact angles is very difficult problem, see e.g. [4], [5], [6], [7], [8].

In order to approximate free surface flows there are several fast, robust and accurate numerical methods available dealing with fixed computing grids. Particularly, these methods are the marker and cell method [9], volume of fluid method [10] and level set method [11], [12], [13]. There are many papers describing the application of the mentioned methods, but its reliability is usually not tested.

In this paper, the two-dimensional flow of two immiscible fluids is considered, mathematically described. In order to take into account the surface tension effects the variational formulation is introduced and the discretization with the the finite element method is applied. The motion of the interface between these two fluids (free surface) is realized using the level set method, cf. [11]. A modification of the standard finite element method is used in order to properly address the arising spurious currents nearby the free surface caused by the discontinuous pressure, see [7] or [5]. The implemented method is verified by solution of a benchmark problem, cf. [4].

## 2 Mathematical model

In this section two immiscible fluids are assumed to occupy a domain $\Omega \subset \mathbb{R}^2$. The domain $\Omega$ is a bounded computational domain with the Lipschitz continuous boundary decomposed into three mutually disjoint parts $\Gamma_W$ (representing fixed wall) , $\Gamma_S$ (representing part of boundary where the slip boundary condition is prescribed) and $\Gamma_O$ (outlet part of the boundary). Each fluid is assumed to occupy a time dependent subdomain ($\Omega^1_{(t)}$ or $\Omega^2_{(t)}$) with the moving interface $\hat{\Gamma}_t$ (free surface), i.e. the interface is the common boundary of both subdomains $\hat{\Gamma}_t = \partial\Omega^1_{(t)} \cap \partial\Omega^2_{(t)}$. Both incompressible fluids are characterized by the density ($\rho_1$ and $\rho_2$) and its viscosity ($\mu_1$ and $\mu_2$).

In order to describe the mathematical model in a single equation, the Heaviside function $H(x,t)$ is introduced

$$H(x,t) = \begin{cases} 1 & \text{for } x \in \Omega^1_{(t)}, \\ \frac{1}{2} & \text{for } x \in \hat{\Gamma}_t, \\ 0 & \text{for } x \in \Omega^2_{(t)}, \end{cases} \tag{1}$$

Further, the density $\rho(x,t)$ and viscosity $\mu(x,t)$ functions are defined using the Heaviside functions as functions giving either the fluid local density or viscosity at the point $x \in \Omega$ at time $t \in (0,T)$. Thus the density function reads

$$\rho(x,t) = \rho_1 H(x,t) + (1 - H(x,t))\rho_2, \tag{2}$$

and the viscosity function is given by formula

$$\mu(x,t) = \mu_1 H(x,t) + (1 - H(x,t))\mu_2. \tag{3}$$

Similarly, the pressure and the flow velocity functions defined on the domain $\Omega$ are denoted by $p(x,t)$ and $\boldsymbol{u} = \boldsymbol{u}(x,t)$, respectively. The flow motion in $\Omega^i_{(t)}$, $i = 1,2$ is then described by the incompressible Navier-Stokes equations

$$\rho\frac{\partial \boldsymbol{u}}{\partial t} + \rho(\boldsymbol{u}\cdot\nabla)\boldsymbol{u} - \nabla\cdot\boldsymbol{\sigma} = \rho\boldsymbol{f}. \tag{4}$$

[a] Corresponding author: petr.svacek@fs.cvut.cz

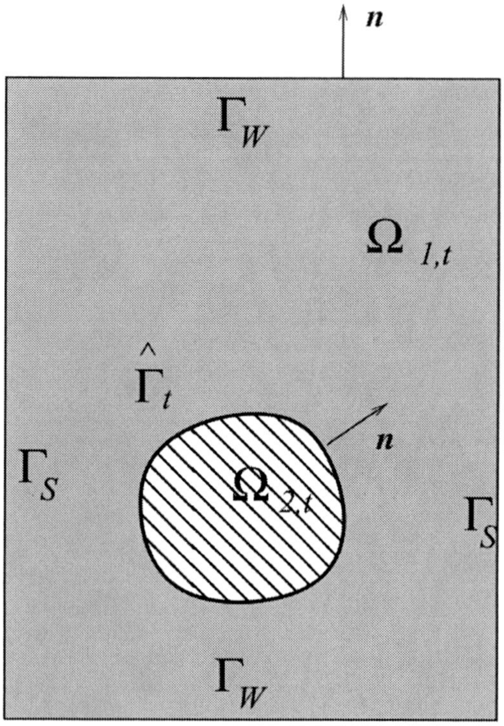


**Fig. 1.** Sketch of the computational domain.

On the interface $\hat{\Gamma}_t$ the following boundary conditions including the surface tension effects needs to be specified

$$\begin{aligned} a) \quad & \boldsymbol{u}|_{\Omega^1_{(t)}} = \boldsymbol{u}|_{\Omega^2_{(t)}} \\ b) \quad & [\boldsymbol{\sigma}\cdot\boldsymbol{n}] = \gamma\kappa\mathbf{n}, \end{aligned} \tag{5}$$

where $\gamma$ is the surface tension coefficient, $\kappa$ denotes the curvature of the interface $\Gamma_I$ and $\boldsymbol{n}$ here denotes the normal to the $\Gamma_I$ pointing into $\Omega^2_{(t)}$.

The Navier-Stokes equations (4) together with the interface condition (5) can be written in $\Omega = \Omega^1_{(t)} \cup \hat{\Gamma}_t \cup \Omega^2_{(t)}$ as

$$\rho\frac{\partial \boldsymbol{u}}{\partial t} + \rho(\boldsymbol{u}\cdot\nabla)\boldsymbol{u} - \nabla\cdot\boldsymbol{\sigma} = \rho\boldsymbol{f} + \gamma\kappa\boldsymbol{n}\delta_{\hat{\Gamma}_t}, \tag{6}$$

where $\delta_{\hat{\Gamma}_t}$ is the delta function of the interface $\hat{\Gamma}_t$ and $\boldsymbol{n}$ is the unit normal vector to the interface $\hat{\Gamma}_t$ (oriented into $\Omega^2_{(t)}$). The meaning of Eq. (6) is clear in the weak formulation: we seek $\boldsymbol{u}$ such that for any $\boldsymbol{v} \in \boldsymbol{H}^1(\Omega)$ holds

$$\int_\Omega \rho\left(\frac{\partial \boldsymbol{u}}{\partial t} + (\boldsymbol{u}\cdot\nabla)\boldsymbol{u}\right)\cdot\boldsymbol{v} + \boldsymbol{\sigma}\cdot(\nabla\boldsymbol{v})\,\mathrm{d}x = \int_{\hat{\Gamma}_t}\gamma\kappa\boldsymbol{n}\cdot\boldsymbol{v}\,\mathrm{d}S + \int_\Omega \rho\boldsymbol{f}\cdot\boldsymbol{v}\,\mathrm{d}x. \tag{7}$$

Furthermore, system (6) is equipped with an initial condition $\boldsymbol{u} = 0$ and with suitable boundary conditions.

Further, on the boundary $\partial\Omega$ the following boundary conditions are prescribed

$$\begin{aligned} &a)\ \boldsymbol{u} = 0 && && \text{on } \Gamma_W, \\ &b)\ \boldsymbol{u}\cdot\boldsymbol{n} = 0, && \boldsymbol{t}\cdot(\boldsymbol{\sigma}\cdot\boldsymbol{n}) = 0 && \text{on } \Gamma_S, \\ &c)\ \boldsymbol{\sigma}\cdot\boldsymbol{n} = 0 && && \text{on } \Gamma_O. \end{aligned} \tag{8}$$

**Level set method.** The free surface motion is numerically treated with the aid of the *level set* method, see e.g. [11]. The level set function generally can be considered as the signed distance function being zero on the interface $\hat{\Gamma}_t$ at any time instant $t$, positive in $\Omega^1_{(t)}$ and negative in $\Omega^2_{(t)}$. In practical realization the level set function is used only to implicitly define the interface $\hat{\Gamma}_t$ as the set of points satisfying the condition $\phi(x,t) = 0$. With the aid of a level set function $\phi(x,t)$ the Heaviside function $H(x,t)$ satisfy

$$H(x,t) = \begin{cases} 1 & \text{for } \phi(x,t) > 0, \\ \frac{1}{2} & \text{for } \phi(x,t) = 0, \\ 0 & \text{for } \phi(x,t) < 0. \end{cases}$$

The motion of the interface $\hat{\Gamma}_t$ is then realized by solving the transport equation for the function $\phi(x,t)$, i.e.

$$\frac{\partial \phi}{\partial t} + \boldsymbol{u}\cdot\nabla\phi = 0. \tag{9}$$

Equation (9) means that the interface $\hat{\Gamma}_t$ moves in time with the flow velocity $\boldsymbol{u}$.

**Surface tension.** The surface tension realized with the aid of the weak reformulation, see [4], [14]. To this end the tangent derivative $\nabla_\Gamma$ is defined as

$$\nabla_\Gamma g = \nabla g - (\boldsymbol{n}\cdot\nabla g)\boldsymbol{n}$$

and similarly the second order tangent derivative, i.e. the Laplace-Beltrami operator, (see [14]) as

$$\triangle_\Gamma = \nabla_\Gamma\cdot\nabla_\Gamma.$$

Now, using the relation

$$\kappa\boldsymbol{n} = \triangle_\Gamma x,$$

and applying the integration by parts on $\hat{\Gamma}_t$ we get

$$\int_{\hat{\Gamma}_t}\gamma\kappa\boldsymbol{n}\cdot\boldsymbol{v}\,\mathrm{d}S = \int_{\partial\hat{\Gamma}_t}\gamma(\nabla_\Gamma x)\cdot\boldsymbol{v} - \int_{\hat{\Gamma}_t}\gamma(\nabla_\Gamma x)\cdot(\nabla_\Gamma\boldsymbol{v})\,\mathrm{d}S, \tag{10}$$

where on the right hand side the integral over $\partial\hat{\Gamma}_t$ denotes the integral over the contact lines (in 3D). For the two-dimensional problem the integral over $\partial\hat{\Gamma}_t$ is the sum over the contact points, which - in the case of $\hat{\Gamma}_t$ being a closed curve - is the empty set and equals zero. In the other case this reformulation naturally includes the contact angles into the weak formulation of the problem.

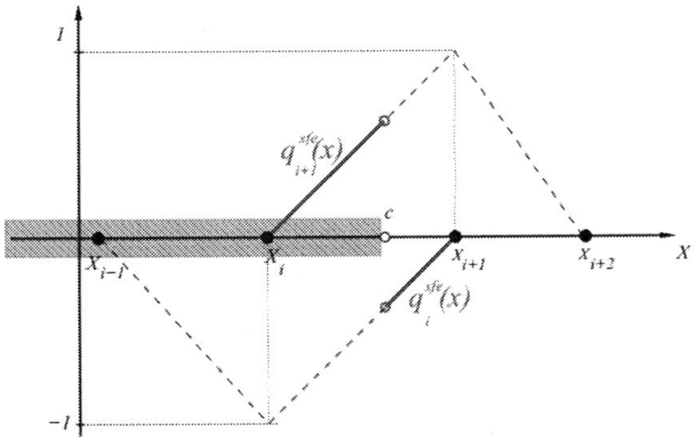


**Fig. 2.** The base functions of the extended finite element space shown for 1D case on the grid given by points $x_{i-2}$, $x_{i-1}$, $x_i$ and $x_{i+1}$. The interface at time $t$ is located between the nodes $x_i$ and $x_{i+1}$ at $x = c$ and the interval where the levelset function is positive ($\phi(x,t) > 0$) is marked by the dashed part of the x-axis.

## 3 Numerical approximation

**Flow step.** In order to discritize the problem in time an equidistant division $t_n = n\Delta t$ of the time interval $[0,T)$ is considered with a constant time step $\Delta t = T/N$. Furthermore, the approximations of the flow velocity $\boldsymbol{u}$, the presure $p$, the density $\rho$ and the viscosity $\mu$ at the time instant $t_n$ are denoted by $\boldsymbol{u}^{(n)}$, $p^{(n)}$, $\phi^{(n)}$, $\rho^n$ and $\mu^n$, respectively. The time derivatives are then approximated by the second order backward difference formula as

$$\frac{\partial \boldsymbol{u}}{\partial t}|_{t=t_{n+1}} \approx \frac{3\boldsymbol{u}^{(n+1)} - 4\boldsymbol{u}^{(n)} + \boldsymbol{u}^{(n-1)}}{\Delta t},$$

$$\frac{\partial \phi}{\partial t}|_{t=t_{n+1}} \approx \frac{3\phi^{(n+1)} - 4\phi^{(n)} + \phi^{(n-1)}}{\Delta t}.$$

The time discretized weak formulation of Eq. (6) then reads: Find $\boldsymbol{u} = \boldsymbol{u}^{n+1} \in \boldsymbol{V}$ and $p = p^{n+1} \in Q$ such that

$$\int_\Omega \rho^{n+1}(x)\left(\frac{\boldsymbol{u}-\boldsymbol{u}^n}{\Delta t} + (\boldsymbol{u}\cdot\nabla)\boldsymbol{u}\right)\cdot\boldsymbol{v}\,\mathrm{d}x \quad (11)$$

$$+\int_\Omega \mu^{n+1}(x)\nabla\boldsymbol{u}\cdot\nabla\boldsymbol{v} - p(\nabla\cdot\boldsymbol{v})\,\mathrm{d}x + \int_\Omega(\nabla\cdot\boldsymbol{u})q\,\mathrm{d}x$$

$$= -\int_{\Gamma^{n+1}}\gamma(\nabla_\Gamma x)\cdot(\nabla_\Gamma\boldsymbol{v})\,\mathrm{d}S + \int_\Omega \rho^{n+1}(x)\boldsymbol{f}\cdot\boldsymbol{v}\,\mathrm{d}x$$

holds for all $\boldsymbol{v} \in \boldsymbol{V}$ and $q \in Q$.

In the practical computations the polygonal computational domain $\Omega$ is used and the spaces $\boldsymbol{V}$ and $Q$ are approximated by the FE subspaces $\boldsymbol{V}_h$ and $Q_h$, respectively, defined over an admissible triangulation $\mathcal{T}_h$ of the domain $\Omega$. The Taylor-Hood finite elements are used for approximation of velocity and pressure, i.e. the velocity components are sought in the space $\boldsymbol{V}_h = [H_h]^2 \subset \boldsymbol{V}$, where the space $H_h$ is defined by

$$H_h = \{\phi \in C(\overline{\Omega}); \phi|_K \in P_2(K) \text{ for each } K \in \mathcal{T}_h\}, \quad (12)$$

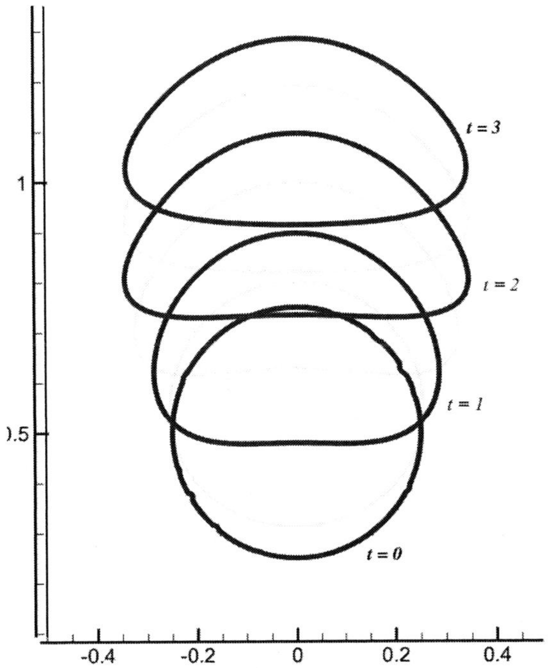


**Fig. 3.** The result of the rising bubble test for the first test case. The shape of the interface is shown at the time instants $t \in \{0, 0.5, 1, 1.5, 2, 2.5, 3\}$. The effect of much higher surface tension leads to a development of a stable bubble shape.

and $P_k(K)$ denotes the space of all polynomials of degree less or equal to $k$ on $K$. Further, the pressure (as well as the level set function) is approximated in the space

$$Q_h = \left\{\phi \in C(\overline{\Omega}) : \phi|_K \in P_1(K) \text{ for each } K \in \mathcal{T}_h\right\}. \quad (13)$$

The discrete flow problem then reads: Find $\boldsymbol{u}_h = \boldsymbol{u}_h^{n+1} \in \boldsymbol{V}_h$ and $p_h = p_h^{n+1}$ such that Eq. (11) holds for any test function $\boldsymbol{v} := \boldsymbol{v}_h \in \boldsymbol{V}_h$ and $q := q_h \in Q_h$. In order to treat the discontinuity of the pressure due to the presence of the surface tension the extended finite element method (XFEM) is applied, see e.g. [7].

**Extended finite element method.** The XFEM method is used to treat the possible numerical instability caused by the surface tension: the pressure is discontinuous along the interface due to the applied surface tension term. The numerical approximation with the aid of Taylor-Hood finite element means that the presure is continous. In order to capture the pressure discontinuity the pressure finite element space $Q_h$ is enriched using the localization of a enrichment function, in this case the enrichment of the Heaviside function $H_\Gamma(x)$ given as the Heaviside function $H_\Gamma(x) = H(x, t_{n+1})$ at the time instant $t_{n+1}$.

In order to enrich the finite element space, the original base functions of $Q_h$ are used, i.e. we denote the index

DOI: 10.1051/epjconf/201714302122

set $\mathcal{J} = \{1, \dots, n\}$, $n = \dim Q_h$ and the mesh nodes by $x_j$, $j \in \mathcal{J}$. The nodal base functions are then denoted by $q_i \in Q_h$, $i \in \mathcal{J}$ and satisfy $q_i(x_j) = \delta_{ij}$. The $\mathcal{J}'$ is the subset of all the neighbours of the interface $\hat{\Gamma}_t$, i.e.

$$\mathcal{J}' = \{j \in \mathcal{J} : \operatorname{supp} q_j \cap \hat{\Gamma}_t \neq \emptyset\}.$$

Finally, the enrichment of the space $Q_h$ is made using the discontinuous base functions $q_j^{xfem}$ defined by

$$q_j^{xfem}(x) = q_j(x)\left(H_\Gamma(x) - H_\Gamma(x_j)\right), \tag{14}$$

where the term $-q_j(x)H_\Gamma(x_j)$ makes the support of the function $q_j^{xfem}(x)$ localized to the single element containing the interface, see Fig. 2.

Here, $H_\Gamma(x_j)$ can be left out from the right hand side as this only adds a constant multiple of the continuous base function $q_j(x)$. On the other hand, this term makes the function $q_j^{xfem}(x)$ being zero at every node $x_i$, $i \in \mathcal{J}$ and also makes the support of $q_j^{xfem}(x)$ localized only to the elements containing the interface $\hat{\Gamma}_t$, which simplifies the practical discretization of the problem. The FE space $Q_h$ is then replaced by the extended FE space

$$Q_h^{xfem} = Q_h \oplus \operatorname{span}\{q_j^{xfem} : j \in \mathcal{J}'\}.$$

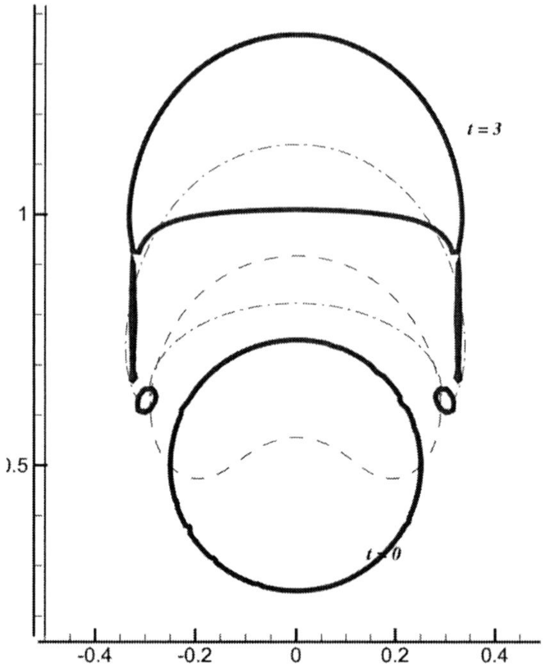


**Fig. 4.** The result of the rising bubble test for the second test case. The shape of the interface is shown at the time instants $t \in \{0, 1, 2, 3\}$.

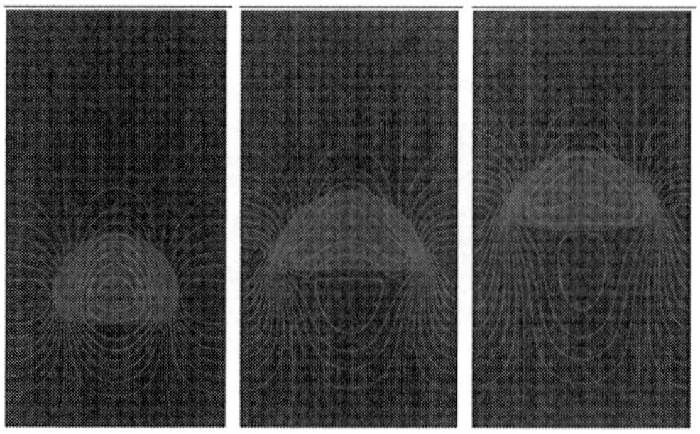

**Fig. 5.** The result of the rising bubble case I: The shape of the interface and the isolines of the vertical component of the flow velocity at time instants $t \in \{1, 2, 3\}$ computed at the grid with $h = 1/80$.

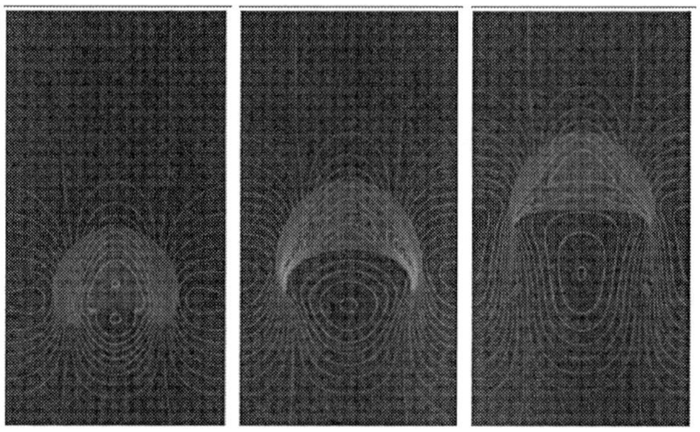

**Fig. 6.** The result of the rising bubble case II: The shape of the interface and the isolines of the vertical component of the flow velocity at time instants $t \in \{1, 2, 3\}$ at the finest grid with the spatial step $h = 1/80$.

**Level set step and coupled problem.** The level set equation (9) is numerically approximated with the aid of finite element method. In order to get the numerically stability of the scheme, the algebraic flux corrections are applied, see [15]. Moreover, in order to keep the level set function as the signed distance function, a re-initialization step is applied, see also [4]. The solution of the coupled problem is then performed by a de-coupled algorithm with strong coupling. This means, that both problems are solved several times in order to get a convergence criteria satisfied. In practical computations the level set step is solved first followed by solution of equation (11) for approximation of the flow velocity $\boldsymbol{u}^{n+1}$ and the pressure $p^{n+1}$.

## 4 Numerical results

In order to verify the described numerical method, the two benchmarks for the case of a rising bubble considered in [4] were numerically approximated. The first test case (case I) used the following values of the fluid's densities

$\rho_1 = 1000\,\mathrm{kg\,m^{-3}}$, $\rho_2 = 100\,\mathrm{kg\,m^{-3}}$,the fluid's viscosities $\mu_1 = 10\,\mathrm{Pa\,s}$, $\mu_2 = 1\,\mathrm{Pa\,s}$, the volume force represented by the gravitational acceleration $\boldsymbol{f} = (0, -0.98)\,\mathrm{m\,s^{-2}}$ and the surface tension $\gamma = 24.5\,\mathrm{N/m}$. The height of the considered computational domain was chosen as $H = 2\,\mathrm{m}$ and the width $W$ was equal to $1\,\mathrm{m}$. The fluid 2 was originally located inside of the circle with the diameter of $0.5\,\mathrm{m}$ with its center located $0.5\,\mathrm{m}$ up from the bottom of the domain, see 1 with the boundary $\Gamma_W$ being the bottom and top of the domain, and the side (left and right) boundaries were chosen as $\Gamma_S$ (i.e. $\Gamma_O = \emptyset$). Due to the gravitational force , the second fluid with the lower density starts to rise and deform its shape, see Fig. 3 However, due to high surface tension a stable shape is developed after a short time period. Since this time the fluid keeps rising in an undeformed shape, see Fig. 5.

Following [4] also the second benchmark test was approximated (test case II). In this case the fluid's densities were $\rho_1 = 1000\,\mathrm{kg\,m^{-3}}$ and $\rho_2 = 1\,\mathrm{kg\,m^{-3}}$, fluid's viscosities were $\mu_1 = 10\,\mathrm{Pa\,s}$ and $\mu_2 = 0.1\,\mathrm{Pa\,s}$. The same gravitational acceleration was used $\boldsymbol{f} = (0, -0.98)\,\mathrm{m\,s^{-2}}$ but lower surface tension $\gamma = 1.96\,\mathrm{N/m}$ was considered. The development of the free surface is shown in Fig. 4. The development of the shape as well as the vertical component of the flow velocity is shown in Fig. 6.

The computations were performed on triangular meshes with the equidistant partition. The spatial steps $h = 1/20, 1/40$ or $1/80$ were used. The step $h = 1/40$ was the coarsest mesh used in [4]. The time step used in the computation was $\Delta t = 0.002$. The motion of the domain $\Omega^2_{(t)}$ with the area $\mathcal{A}(t)$ was tracked in terms of the $y-$coordinate of the center of mass

$$T_y(t) = \int_{\Omega^2_{(t)}} x_2 \,\mathrm{d}x / \mathcal{A}(t),$$

the circularity defined by

$$C(t) = 2\sqrt{\pi \mathcal{A}(t)} / \int_{\partial\Omega^2_{(t)}} 1 \,\mathrm{d}S$$

and the rise velocity

$$V = \int_{\Omega^2_{(t)}} u_2 \,\mathrm{d}x / \mathcal{A}(t).$$

In order to verify the presented numerical method the values of $T_y$, $C$ and $V$ were computed at every time instant for both the test cases. The graphs of $T_y$, $C$ and $V$ in dependence on time shown in Fig. 7 for the test case I agrees well with the results in [4]. The computations were performed on three grids and the differences were not significant. Similarly, the graphs of $T_y$, $C$ and $V$ in dependence on time shown in Fig. 8 for the test case II is similar to the results presented in [4]. Nevertheless for the benchmark II the results are much more sensitive (particularly for the circularity) to the mesh refinement. See also the quantitative comparison for both cases in Tables 1, 2.

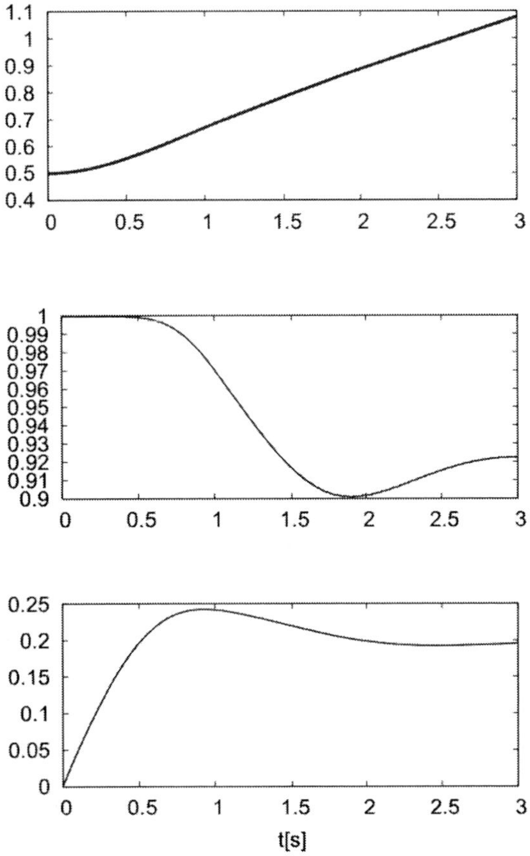


**Fig. 7.** The quantitative results for the rising bubble case I: The graphs of the center of mass $T_y$ (top), the circularity $C$ (middle) and the rise velocity $V$ (bottom) in dependence on time.

**Table 1.** The quantitative results for the rising bubble case I: the comparison of the computed and the reference quantities.

| | ref. [4], case I | present study, case I |
|---|---|---|
| $T_y(3)$ | 1.0813 | 1.0801 |
| $C_{min}$ | 0.9013 | 0.9025 |
| $t_{C_{min}}$ | 1.9041 | 1.898 |
| $V_{max}$ | 0.2417 | 0.2421 |
| $t_{V_{max}}$ | 0.9213 | 0.92 |

## 5 Conclusion

The motion of two immiscible fluid was mathematically modelled with the aid of a variational formulation which includes the surface tension and possibly also the contact angles effects. The time discretization of the problem was performed with the aid of the second order backward difference formula. The spatial discretization was done with the aid of the finite element method, where the XFEM was applied in order to capture the disconti-

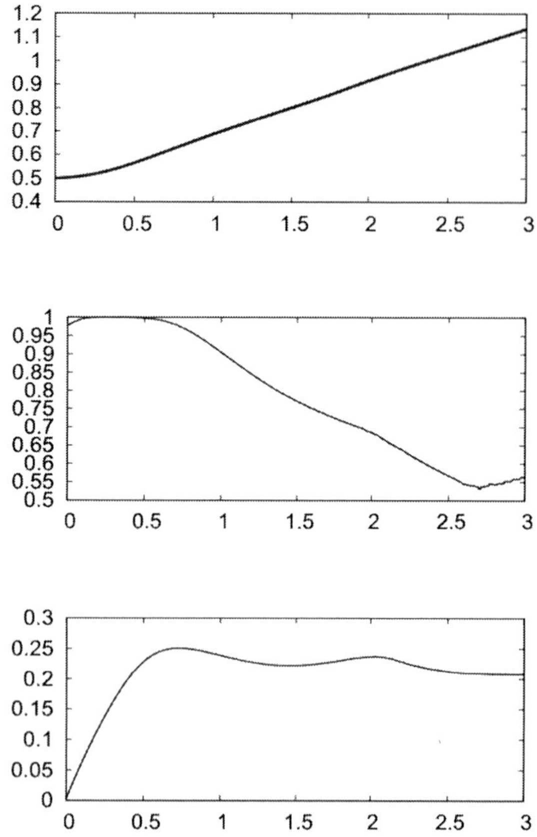


**Fig. 8.** The quantitative results for the rising bubble case II: The graphs of the center of mass $T_y$ (top), the circularity $C$ (middle) and the rise velocity $V$ (bottom) in dependence on time.

**Table 2.** The quantitative results for the rising bubble case II: the comparison of the computed and the reference quantities.

| | ref. [4], case II | present study, case II |
|---|---|---|
| $T_y(3)$ | 1.1380 | 1.13618 |
| $C_{min}$ | 0.5869 | 0.53 |
| $t_{C_{min}}$ | 2.4 | 2.7 |
| $V_{max}$ | 0.2524 | 0.2507 |
| $t_{V_{max}}$ | 0.7332 | 0.7298 |
| $V_{max,2}$ | 0.2434 | 0.237 |
| $t_{V_{max,2}}$ | 2.0705 | 2.03 |

nuity of the pressure along the free surface. The motion of the free surface was treated with the aid of the level set method. The presented numerical method was tested on two benchmark problems from [4] and the obtained numerical results were compared to the reference values.

**Acknowledgment** This work was supported by the *Czech Science Foundation* under the Grant No. 13-00522S

## References

1. A. Maiwald, P. Scheller, C. Brücker, R. Schwarze, *Flow-induced emulsification of mold powder slag into liquid steel*, in *STEELSIM 2009 3rd International Conference Simulation and Modelling of Metallurgical Processes in Steelmaking* (Leoben, 2009), pp. 162 – 167
2. A. Bansal, P. Chapelle, Y. Delannoy, E. Waz, P. Le Brun, J.P. Bellot, Metallurgical and Materials Transactions B **46**, 2096 (2015)
3. A.V. Minakov, M.V. Pervukhin, D.V. Platonov, M.Y. Khatsayuk, Computational Mathematics and Mathematical Physics **55**, 2066 (2015)
4. S. Hysing, S. Turek, D. Kuzmin, N. Parolini, E. Burman, S. Ganesan, L. Tobiska, International Journal for Numerical Methods in Fluids **60**, 1259 (2009)
5. J.W. Barrett, H. Garcke, R. Nürnberg, Computer Methods in Applied Mechanics and Engineering **267**, 511 (2013)
6. S.R. Ransau, Tech. Rep. 3/2002, Norwegian University of Science and Technology, Trondheim (2002)
7. H. Sauerland, T.P. Fries, Computers & Fluids **87**, 41 (2013)
8. E. Bänsch, S. Weller, *A comparison of several time discretization methods for free surface flows*, In *ALGORITMY 2012, 19th Conference on Scientific Computing*, edited by A. Handlovicova, Z. Minarechova, D. Sevcovic (Vysoke Tatry - Podbanske, Slovakia, 2012), pp. 331 – 341
9. F. Harlow, J. Welch, Physics of Fluids **8**, 2182 (1965)
10. J. Klostermann, K. Schaake, R. Schwarze, International Journal for Numerical Methods in Fluids **71**, 960 (2013)
11. J.A. Sethian, *Level Set Methods and Fast Marching Methods*, Cambridge Monograph on Applied and Computational Mathematics, $2^{nd}$ edn. (Cambridge University Press, Cambridge, U.K., 1999)
12. S. Basting, M. Weismann, Journal of Computational Physics **255**, 228 (2013)
13. M. Raessi, H. Pitsch, Computers & Fluids **63**, 70 (2012)
14. E. Bänsch, Numer. Math. **88**, 203 (2001)
15. D. Kuzmin, Journal of Computational Physics **219**, 513 (2006)

# Numerical approximations of flow induced vibrations of vocal folds

Petr Sváček [a]

Czech Technical University in Prague, Faculty of Mechanical Engineering, Dep. of Technical Mathematics, Karlovo nám. 13, Praha 2, Czech Republic

**Abstract.** The paper focus on mathematical modelling of incompressible fluid flow interacting with vibrations of an elastic vocal fold. The flow in moving domain is modelled by the incompressible Navier-Stokes equations written in the Arbitrary Lagrangian-Eulerian (ALE) form. The channel geometry is an approximation of the human glottal region. The flow model is coupled with a simplified structure model. The problem is mathematically described and the resulting fluid-structure interaction problem is discretized by a stabilized finite element method. A strong coupling algorithm is applied for solution of the coupled fluid-structure problem. The choice of boundary conditions is discussed, particularly the choice of different artificial inlet/outlet boundary conditions is described in details. The numerical results are shown.

## 1 Introduction

The mathematical modelling of fluid-structure interactions is a complex difficult task. Nowadays it has become to be important also in biomechanics of human voice, see [1]. Naturally in medicine it is important to understand the basic principles of voice production, see e.g. [2]. However, the voice production is a complex mechanism which includes complex airflow, vocal fold vibrations, acoustic resonances, etc. The voice sound is created in the glottis - the narrow part of the channel between the two vocal folds. The vocal folds vibrations leads to periodical (almost or complete) closure of the channel. The collision of the vocal folds creates the sound, but mathematical and numerical modelling of such a complex phenomenon encounters many difficulties as e.g. the correct choice of boundary conditions at the artificial inlet/outlet parts of boundary or the treatment of the mesh deformation.

One possibility how to study the complex problem of voice creation is the use of experimental methods. The experiments partly focus on measurement of the externally driven models, e.g. [3], [4], [5]), or on self-oscillating models, e.g. [6], [7], [8]. However, there are many difficulties encountered in both in vivo and in vitro measurements.

The mathematical models on the other hand can be used to provide important information about the voice production mechanisms. In this case several simplifications are usually made for the purpose of the computational models, for example see the two–mass model of the vocal folds, see [9] or several works using the same concept as [10]. These simple mass-spring models coupled with a quasi–1D airflow proved to be very useful, see e.g. [10]), and they can be used to estimate the frequencies or vocal folds loading by impact stress, see [11]. The main advantage of these models is the ability to perform nearly real–time simulations. However, these mathematical models usually do not address the complex problem in all details and uses several simplifications.

Particularly, the simplified flow models do not describe the glottal airflow details. That is why recently the numerical solution of the 2D Navier–Stokes equations has been used. For instance see [12], where the finite volume approximations of the Navier–Stokes equations were coupled with the two–mass dynamic model. See also [13], where the 2D finite element approximations of the Navier-Stokes equations coupled with a two degrees of model with the vibrating vocal fold described by a mechanically equivalent two degrees of freedom system. The fluid–structure interaction using the coupled finite element (FE) models of the vocal folds and airflow was published in [6], [14]. Nevertheless also in these high-fidelity models several simplifications are used, e.g. no contact problem is considered.

In this paper we shall focus on addressing the proper inlet/outlet boundary conditions applicable in the case of closing of the channel due to the vocal folds vibrations. First, the mathematical model addressing the fluid structure interactions is described. Further, the numerical approximation of the described mathematical model by the finite element method is given. The numerical results are presented and the choice of boundary conditions and its implementation is discussed.

## 2 Mathematical model

The mathematical description of the interaction of incompressible flow and vocal folds consists of equations of

[a] Corresponding author: petr.svacek@fs.cvut.cz

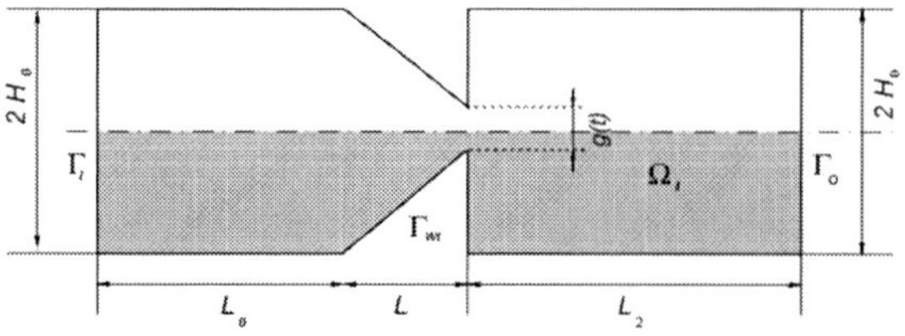


**Fig. 1.** The computational domain for fluid flow for male case geometry. The symmetric channel with vibrating vocal folds is considered. The inlet, outlet and vibrating parts of the boundary are shown. The two configurations of the channel for two different vocal folds models are shown.

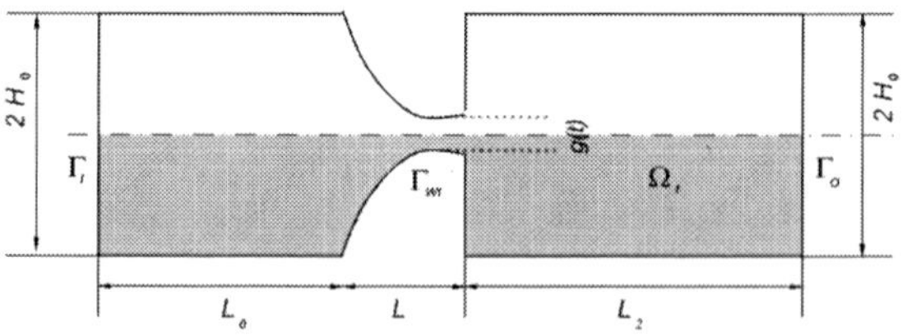


**Fig. 2.** The computational domain for fluid flow for the female case geometry. The symmetric channel with vibrating vocal folds is considered. The inlet, outlet and vibrating parts of the boundary are shown. The two configurations of the channel for two different vocal folds models are shown.

motion for the vocal folds coupled with the incompressible Navier-Stokes equations via interface conditions, see Figs. 1, 2. First, the fluid flow in the computational domain surrounding the vibrating vocal fold is described. To this end the Arbitrary Lagrangian Eulerian method is applied in order to treat the motion of the computational domain. Further the motion of the vocal folds is described as linear 2D elastic continuum on the reference domain, i.e. Lagrangian concept is applied. In order to simplify the model a two degrees of freedom structure is used. The interface conditions are presented.

### 2.1 Arbitrary Lagrangian Eulerian method

First, the time dependence of the computational fluid flow domain $\Omega_t^f$ needs to be addressed. Here, the so-called *arbitrary Lagrangian-Eulerian* (ALE) is used, see e.g. [15]. In order to apply this method a regular smooth one-to-one ALE mapping $\mathcal{A}$ of the reference configuration $\Omega_0^f$ onto the current configuration $\Omega_t^f$, i.e it maps the point $\boldsymbol{X}$ from the reference configuration $\boldsymbol{X} \in \Omega_0^f$ on the point $\boldsymbol{x}$ of the current configuration $\boldsymbol{x} \in \Omega_t^f$ at any time instant $t$. The point $\boldsymbol{x}$ with a fixed reference $\boldsymbol{X}$ moves in time $\boldsymbol{x} = \mathcal{A}(\boldsymbol{X}, t)$ with the domain velocity defined by

$$\boldsymbol{w}_D(\boldsymbol{x}, t) = \frac{\partial \mathcal{A}}{\partial t}(\boldsymbol{X}, t), \qquad \boldsymbol{x} = \mathcal{A}(\boldsymbol{X}, t). \tag{1}$$

Further, the time derivative with respect to the reference frame is introduced as ALE derivative. The ALE derivate is then given as

$$\frac{D^{\mathcal{A}} f}{Dt} f(\boldsymbol{x}, t) = \frac{\partial \tilde{f}}{\partial t}(\boldsymbol{X}, t), \tag{2}$$

where the function $\tilde{f}$ is the transformation of the function $f$ onto the reference frame, i.e. $\tilde{f}(\boldsymbol{X}, t) = f(\mathcal{A}(\boldsymbol{X}, t), t)$ for any $\boldsymbol{X} \in \Omega_0^f$. The ALE derivative then satisfy (see [15])

$$\frac{D^{\mathcal{A}} f}{Dt}(x, t) = \frac{\partial f}{\partial t}(x, t) + \boldsymbol{w}_D(x, t) \cdot \nabla f(x, t). \tag{3}$$

### 2.2 Incompressible Navier-Stokes equations

The flow of an incompressible viscous fluid in the domain $\Omega_t^f$ is described by the system of the incompressible Navier-Stokes equations (see, e.g., [16]) written in the ALE form

$$\rho \frac{D^{\mathcal{A}} \boldsymbol{v}}{Dt} + \rho((\boldsymbol{v} - \boldsymbol{w}_D) \cdot \nabla)\boldsymbol{v} = \operatorname{div} \boldsymbol{\tau}^f, \qquad \nabla \cdot \boldsymbol{v} = 0. \tag{4}$$

Here $\boldsymbol{v} = (v_1, v_2)$ is the fluid velocity vector, $\rho$ is the constant fluid density, and $\boldsymbol{\tau}^f = (\tau_{ij}^f)$ is the fluid stress tensor given by

$$\boldsymbol{\tau}^f = -p\mathbb{I} + 2\mu\mathbb{D}(\boldsymbol{v}), \tag{5}$$

where $p$ is the pressure, $\mu > 0$ is the constant fluid viscosity, $\delta_{ij}$ denotes the Kronecker symbol and $\mathbb{D}$ is the symmetric part of the velocity gradient given by

$$\mathbb{D}(\boldsymbol{v}) = \frac{1}{2}(\nabla \boldsymbol{v} + (\nabla \boldsymbol{v})^T). \tag{6}$$

For the system (4) the initial condition $\boldsymbol{v}(\boldsymbol{x}, 0) = \boldsymbol{v}^0(\boldsymbol{x})$ for $\boldsymbol{x} \in \Omega_0^f$ is specified. Further the boundary conditions needs to be prescribed on the boundary $\partial\Omega_t^f$ of the computational domain. The boundary $\partial\Omega_t^f$ is assumed to be formed by mutually disjoint parts $\partial\Omega_t^f = \Gamma_I \cup \Gamma_O \cup \Gamma_{Wt}$, where $\Gamma_I$ denotes the inlet part of the boundary, $\Gamma_O$ is the outlet part of the boundary and $\Gamma_{Wt}$ denotes the (fixed and moving) wall. The following boundary conditions are prescribed

$$\begin{array}{ll} \text{a)} & \boldsymbol{v} = \boldsymbol{v}_I \ \text{ on } \Gamma_I, \\ \text{b)} & \boldsymbol{v} = \boldsymbol{w}_D \ \text{ on } \Gamma_{Wt}, \\ \text{c)} & -\boldsymbol{n} \cdot \boldsymbol{\tau}^f + \frac{1}{2}\rho(\boldsymbol{v} \cdot \boldsymbol{n})^- \boldsymbol{v} = p_{ref}\boldsymbol{n}, \ \text{on } \Gamma_O, \end{array} \tag{7}$$

where $p_{ref}$ denotes a reference mean pressure value at the outlet (we set $p_{ref} = 0$) and $\boldsymbol{n}$ is the unit outward normal vector to $\partial\Omega_t$ and $\boldsymbol{v}_I$ is the inlet velocity. For the analysis of the boundary condition 7c) see e.g. [17], [18]. Particularly this type of boundary condition treats well the cases of the (backward) inflow of the fluid at the outlet part of the boundary $\Gamma_O$. For this reason this boundary condition can be also used at the inlet boundary. This means that the condition 7, a) can be replaced by the condition

$$-\boldsymbol{n}\cdot\boldsymbol{\tau}^f + \frac{1}{2}\rho(\boldsymbol{v}\cdot\boldsymbol{n})^-\boldsymbol{v} = p_I\boldsymbol{n}. \tag{8}$$

The boundary condition (8) means that on the inlet instead of the inlet flow velocity $\boldsymbol{v}_I$ the the inlet pressure $p_I$ or, as $p_{ref} = 0$, the pressure drop $\Delta p = p_I - p_{ref}$ between the inlet and the outlet part of the boudnary. Let us also mention here, that the practical realization of particulary the Dirichlet boundary condition (inlet velocity) can be realized either as the Dirichlet lift function, i.e. $\boldsymbol{v} = \boldsymbol{v}_I + \boldsymbol{v}_0$ with $\boldsymbol{v}_0$ approximations being zero on the inlet part of the boundary, or using a penalization. In this paper both approaches were tested.

## 2.3 Structure model

**Elastic structure.** The deformation of vocal folds is described as deformation of elastic structure. Let us consider a bounded domain $\Omega^s \subset \mathbb{R}^2$ representing the vocal fold. The displacement of the structure is denoted by $\boldsymbol{u}$, which is a function $\boldsymbol{u} = \boldsymbol{u}(\boldsymbol{X}, t)$ defined at a time instant $t \in (0, T)$ and a point $\boldsymbol{X} = (X_1, X_2) \in \Omega^s$.

The deformation of the elastic body is modelled by the dynamic elasticity. The equations of motion of an elastic body have the form

$$\rho^s \frac{\partial^2 \boldsymbol{u}}{\partial t^2} - \operatorname{div}\boldsymbol{\tau}^s = \boldsymbol{f}, \tag{9}$$

where $\boldsymbol{f} = (f_1, f_2)$ is the density of the volume force, $\rho^s$ denotes the density of the structure and the stress tensor $\boldsymbol{\tau}_s$ is given by the generalized Hooke law for isotropic material in the form

$$\boldsymbol{\tau}^s = \lambda^s e_{kk}(\boldsymbol{u})\,\mathbb{I} + 2\mu^s \boldsymbol{e}(\boldsymbol{u})x \tag{10}$$

where $\lambda^s$ and $\mu^s$ are the so-called Lamé coefficients. The small strain tensor $\boldsymbol{e} = (e_{ij})$ then is given by

$$\boldsymbol{e}(\boldsymbol{u}) = \frac{1}{2}(\nabla\boldsymbol{u} + (\nabla\boldsymbol{u})^T), \tag{11}$$

where $\nabla$ denotes the gradient operator with respect to variables $X_1, X_2$.

Similarly as for fluid, the initial and boundary conditions are prescribed for system (9, i.e. the initial conditions read $\boldsymbol{u}(\cdot, 0) = 0$ and $\frac{\partial \boldsymbol{u}}{\partial t}(\cdot, 0) = 0$. The boundary $\partial\Omega_s$ is then decomposed in two disjoint parts, the moving interface $\Gamma^s_W = \Gamma_{W_0}$ and the other (fixed) part of boundary $\Gamma^s_D$. The boundary conditions then read

$$\text{a) } \boldsymbol{n}\cdot\boldsymbol{\tau}^s = \boldsymbol{n}\cdot\boldsymbol{\tau}^f \text{ on } \Gamma^s_W \qquad \text{b) } \boldsymbol{u} = 0 \text{ on } \Gamma^s_D. \tag{12}$$

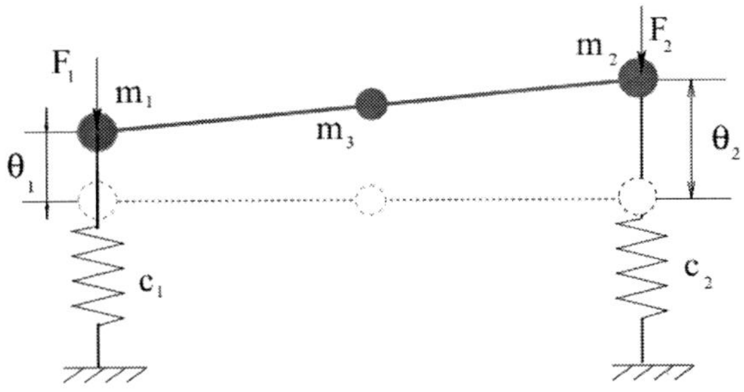


**Fig. 3.** Aeroelastic two degrees of freedom model (with masses $m_1$, $m_2$, $m_3$) in displaced position (displacements $\theta_1$ and $\theta_2$) and resulting aerodynamic forces $F_1$ and $F_2$.

**Reduced order model.** The vibrations of the vocal folds usually exhibit two fundamantal frequencies. Such vibrations can be described with an simplified aeroelastic two degrees of freedom model. The motion of $\Gamma_{Wt}$ is governed by the displacements $\theta_1(t)$ and $\theta_2(t)$ (upward positive) of the two masses $m_1$ and $m_2$, respectively (see Fig. 3). The displacement vector $\boldsymbol{\theta} = (\theta_1, \theta_2)^T$ is obtained by the solution of the following equations (see [10] for details)

$$\mathbb{M}\ddot{\boldsymbol{\theta}} + \mathbb{B}\dot{\boldsymbol{\theta}} + \mathbb{K}\boldsymbol{\theta} = -\boldsymbol{F}, \tag{13}$$

where $\mathbb{M}$ is the mass matrix, $\mathbb{K}$ is the diagonal stifness matrix with spring constants $c_1, c_2$ on its diagonal and $\mathbb{B}$ is the matrix of the proportional structural damping. The mass matrix is given by

$$\mathbb{M} = \begin{pmatrix} m_1 + \frac{m_3}{4} & \frac{m_3}{4} \\ \frac{m_3}{4} & m_2 + \frac{m_3}{4} \end{pmatrix} \tag{14}$$

where $m_1, m_2, m_3$ are masses. The components of $\boldsymbol{F} = (F_1, F_2)^T$ are the aerodynamical forces (downward positive). The proportional damping matrix is chosen as

$$\mathbb{B} = \varepsilon_1\mathbb{M} + \varepsilon_2\mathbb{K}.$$

The aerodynamic forces $F_1$ $F_2$ in Eq. (13) depends on flow velocity $\boldsymbol{v}$ and pressure $p$.

$$F_1(t) = \frac{h}{L}\int_0^L (L - x)p_f(x,t)n_2(x,t)\,dx,$$

and

$$F_2(t) = \frac{h}{L}\int_0^L x p_f(x,t)n_2(x,t)\,dx,$$

where $L$ is the length of the vocal fold, $h$ is the depth of the structure (i.e. third dimension in $z$ axis), value $p_f(x,t)$ denotes the value of the pressure at time instant $t$ on the surface of the vocal fold positioned at $x$ and $n_2(x,t)$ denotes the second component of the unit outward normal to the surface of the vocal fold at position $x$ and time $t$.

**Table 1.** Structural parameters considered for the aeroelastic system in numerical examples; $f_1, f_2$ are the corresponding natural frequencies of the structure in vacuo.

| | Model F | Model M |
|---|---|---|
| $m$ [kg] | $3.274 \times 10^{-4}$ | $4.812 \times 10^{-4}$ |
| $I$ [kg/m$^2$] | $1.341 \times 10^{-9}$ | $2.351 \times 10^{-9}$ |
| $e$ [m] | $1.133 \times 10^{-3}$ | $0.771 \times 10^{-3}$ |
| $c_1$ [N/m] | 44.8 | 56 |
| $c_2$ [N/m] | 84.6 | 174.3 |

## 3 Numerical approximation

The presented fluid-structure problem is first decoupled. The motion equations are time integrated with the aid of Newmark method. The fluid flow is approximated by the stabilized finite element method. The coupled fluid-structure model is solved with the aid of strongly coupled scheme.

The main attention is paid to the approximation of the fluid flow model. However, the application of the finite element method for incompressible Navier-Stokes equations requires to carefully treat several issues. The first one is the dominating advection in the momentum equations. The second is the compatibility of the velocity/pressure spaces. Moreover, in order to obtain physically admissible solution it is usually necessary to apply suitable mesh refinement (e.g. anisotropically refined mesh, see [19]) combined with a stabilization technique, see [20], [21], [22]. The applied stabilization scheme needs to be modified in the case of the application on moving domains (see [23]).

First the problem is time discretized, i.e the time interval is divided $0 = t_0 < t_1 < \cdots < T$, $t_k = k\Delta t$ with a constant time step $\Delta t > 0$. The velocity $\boldsymbol{v}$ and the pressure $p$ at time $t_n$ are then approximated by $\boldsymbol{v}^n$ and $p^n$, respectively. Furthermore, the ALE velocity $\boldsymbol{w}_D(t_{n+1})$ is approximated by $\boldsymbol{w}_D^{n+1}$ computed at a point $\boldsymbol{x} = \mathcal{A}(\boldsymbol{X}, t_{n+1})$ by

$$\boldsymbol{w}_D^{n+1}(\boldsymbol{x}) \approx \frac{3\mathcal{A}(\boldsymbol{X}, t_{n+1}) - 4\mathcal{A}(X, t_n) + \mathcal{A}(X, t_{n-1})}{2\Delta t}. \tag{15}$$

The ALE derivative of the velocity $\boldsymbol{v}$ at time $t_{n+1}$ is approximated using the second-order backward difference formula, i.e.

$$\frac{D^A \boldsymbol{v}}{Dt} \approx \frac{3\boldsymbol{v}^{n+1} - 4\hat{\boldsymbol{v}}^n + \hat{\boldsymbol{v}}^{n-1}}{2\Delta t}, \tag{16}$$

where at a given time instant $t_{n+1}$ we denote $\hat{\boldsymbol{v}}^i(\boldsymbol{x}) = \boldsymbol{v}^i(\mathcal{A}(\boldsymbol{X}, t_i))$ with $\boldsymbol{x} = \mathcal{A}(\boldsymbol{X}, t_{n+1})$.

The time discretized problem is then formulated weakly with the aid of the velocity space $W = \boldsymbol{H}^1(\Omega)$, the pressure space $Q = L^2(\Omega)$ and the space of test functions

$$X = \{\boldsymbol{v} \in W; \boldsymbol{v}|_{\Gamma_I \cup \Gamma_{W_t}} = 0\}. \tag{17}$$

Now, the weak formulation of the time discretized problem (4) is obtained with the multiplication by the test functions $\boldsymbol{\varphi} \in X, q \in Q$, sumation, integration over $\Omega_t^f$ and transformation with the aid of Green's theorem. The weak solution $U = (\boldsymbol{v}, p)$ then satisfy equation

$$a(U, V) = L(V), \qquad \text{for all } V = (\boldsymbol{\varphi}, q) \in X \times Q, \tag{18}$$

where the form $a(U, V)$ is defined by

$$\begin{aligned} a(U,V) &= \frac{3\rho}{2\Delta t}(\boldsymbol{v}, \boldsymbol{\varphi})_\Omega + \left(\boldsymbol{\tau}^f, \mathbb{D}(\boldsymbol{\varphi})\right)_\Omega + c(\boldsymbol{v}, \boldsymbol{v}, \boldsymbol{\varphi}) \\ &+ \int_{\Gamma_O} \frac{1}{2}\rho(\boldsymbol{v}\cdot\boldsymbol{n})^+ \boldsymbol{v}\cdot\boldsymbol{\varphi}\,\mathrm{d}S. + (\nabla\cdot\boldsymbol{v}, q)_\Omega, \qquad (19) \\ L(V) &= \frac{\rho}{2\Delta t}\left(4\hat{\boldsymbol{v}}^n - \hat{\boldsymbol{v}}^{n-1}, \boldsymbol{\varphi}\right)_\Omega - \int_{\Gamma_O} p_{\text{ref}}\boldsymbol{\varphi}\cdot\boldsymbol{n}\,\mathrm{d}S, \end{aligned}$$

and the convective form $c(\boldsymbol{v}^*, \boldsymbol{v}, \boldsymbol{\varphi})$ reads

$$c(\boldsymbol{v}^*, \boldsymbol{v}, \boldsymbol{\varphi}) = \int_\Omega \frac{1}{2}\rho(\boldsymbol{b}\cdot\nabla\boldsymbol{v})\boldsymbol{\varphi} - \frac{1}{2}\rho(\boldsymbol{v}^*\cdot\nabla\boldsymbol{\varphi})\boldsymbol{v}\mathrm{d}x, \tag{20}$$

where $\boldsymbol{b} = \boldsymbol{v}^* - 2\boldsymbol{w}_D^{n+1}$. The form $c(\boldsymbol{v}^*, \boldsymbol{v}, \boldsymbol{\varphi})$ is obtained from the convective term $(\boldsymbol{v}^*\cdot\nabla\boldsymbol{v}, \boldsymbol{\varphi})_\Omega$ by integration by parts and using the boundary conditions.

In order to apply the Galerkin finite element method (FEM) to the discretization of the problem 18, we approximate the spaces $W$, $X$, $Q$ by their finite dimensional subspaces $W_h$, $X_h$, $Q_h$, i.e.

$$X_h = \{\boldsymbol{v}_h \in W_h; \boldsymbol{v}_h|_{\Gamma_I \cap \Gamma_{W_t}} = 0\}. \tag{21}$$

The couple $(X_h, Q_h)$ of the finite element spaces is chosen based on a regular triangulation of the computational domain such that it satisfy the Babuška-Brezzi (BB) inf-sup condition (see, e.g., [24], [25]). The spaces $W_h$, $X_h$ and $Q_h$ are formed by the well-known Taylor-Hood $P_2/P_1$ conforming finite elements used for the velocity/pressure approximation. This means that $p_h$ is a linear function and $\boldsymbol{v}_h$ is a quadratic vector-valued function on each element $K \in \mathcal{T}_h$. The fully stabilized scheme is used (see, e.g., [26], [21]).

The *stabilized discrete problem* reads: Find $U_h = (\boldsymbol{v}_h, p_h) \in W_h \times Q_h$ such that $\boldsymbol{v}_h$ satisfies approximately conditions (7), a), b), c)(i) and

$$\begin{aligned} a(U_h, V_h) + \mathcal{L}_h(U_h, V_h) + \mathcal{P}_h(U_h, V_h) &= L(V_h) + \mathcal{F}_h(V_h) \\ \text{for all } V_h = (\boldsymbol{\varphi}_h, q_h) &\in X_h \times Q_h. \qquad (22) \end{aligned}$$

The stabilization terms are defined using the modified test function defined on each element $K$ as

$$\psi(V) = \delta_K(\boldsymbol{b}\cdot\nabla)\boldsymbol{\varphi} + \nabla q \tag{23}$$

with $\boldsymbol{b} = (\boldsymbol{v} - \boldsymbol{w}_D^{n+1})$. The stabilization terms then reads as

$$\begin{aligned} \mathcal{L}(U,V) &= \sum_{K\in\mathcal{T}_h}\left(\frac{3\rho v}{2\Delta t}\boldsymbol{v} - \operatorname{div}\boldsymbol{\tau}^f + \rho(\boldsymbol{b}\cdot\nabla)\boldsymbol{v}, \psi(V)\right) \\ \mathcal{F}(V) &= \sum_{K\in\mathcal{T}_h}\left(\frac{\rho}{2\Delta t}(4\hat{\boldsymbol{v}}^n - \hat{\boldsymbol{v}}^{n-1}), \psi(V)\right) \qquad (24) \end{aligned}$$

and the div-div stabilization reads

$$\mathcal{P}_h(U,V) = \sum_{K\in\mathcal{T}_h} \tau_K(\nabla\cdot\boldsymbol{v},\nabla\cdot\boldsymbol{\varphi})_K \qquad (25)$$

Here, $\delta_K \geq 0$ and $\tau_K \geq 0$ are suitable chosen parameters. The choice of the parameters $\delta_K$ and $\tau_K$ is carried out according to [26] and [22]. The parameter $\delta_K$ is defined on the basis of the local transport velocity, local element size and local viscozity. In the case of the Taylor-Hood finite elements and a bounded convection $\boldsymbol{v} \approx 1$ the following choice of parameters appears suitable:

$$\tau_K = \tau_*, \qquad \delta_K = \delta^* h_K^2, \qquad (26)$$

where $\tau^* > 0$ and $\delta^* > 0$ are fixed constants.

## 4 Numerical results

For the computations the fluid density $\rho = 1.2$ kg m$^{-3}$, and fluid kinematic viscosity $\nu = 1.58 \times 10^{-5}\,\mathrm{m}^2/\mathrm{s}$ were chosen. The distance of the masses from the center was $l = L/2$, the lengths of sub- and supra- glottal regions were $L_0 = 2L$ and $L_2 = 2L$, respectively. The height of the channel was $2H_0$, where $H_0 = \frac{1}{2}g_0 + \max_{x\in\langle 0,L\rangle} a(x)$ and $g_0$ is the initial gap $g(0) = g_0$. The initial gap $g_0$ was chosen 0.4 mm and 0.6 mm, respectively.

Using the given shape and dimension of the vocal fold and using the structural density $\rho_h$ the total mass $m$, the moment of inertia $I$ and the excentricity $e$ is computed. The system is then replaced by the equivalent three mass system where $m_{1,2} = \frac{1}{2l^2}(I + m\,e^2 \pm m\,e\,l)$ and $m = m_1 + m_2 + m_3$. The structural parameters were chosen according [27], see Tab. 1. The presented results shows the computed case with the prescribed inlet velocity. The prescribed inlet pressure did not lead to an instability in the considered range of velocities/pressures.

The solution of the aeroelastic system was performed for two cases represented by Tab. 1. The solution of the flow model (4), structure model (13) and interface conditions was numerically approximated in a simplified geometry shown in Figs. 1, 2. The aeroelastic response for the is shown in Fig. 4 for *Model F* and in Fig. 5 for *Model M* in time domain in terms of displacements $\theta_1(t)$ and $\theta_2(t)$. The vibrations of the structure die in time after a time period due to both structural and strong aerodynamic damping for lower flow velocities. With increasing flow velocity the flutter type of instability can be observed. The critical velocity is in a good agreement with results presented in [27]. The physical meaning of the instability is so-called phonation onset which is an important voice production characteristic in humans. The flow velocity patterns are shown in Fig. 6, where the typical pressure distribution and flow velocity magnitude pattern can be observed.

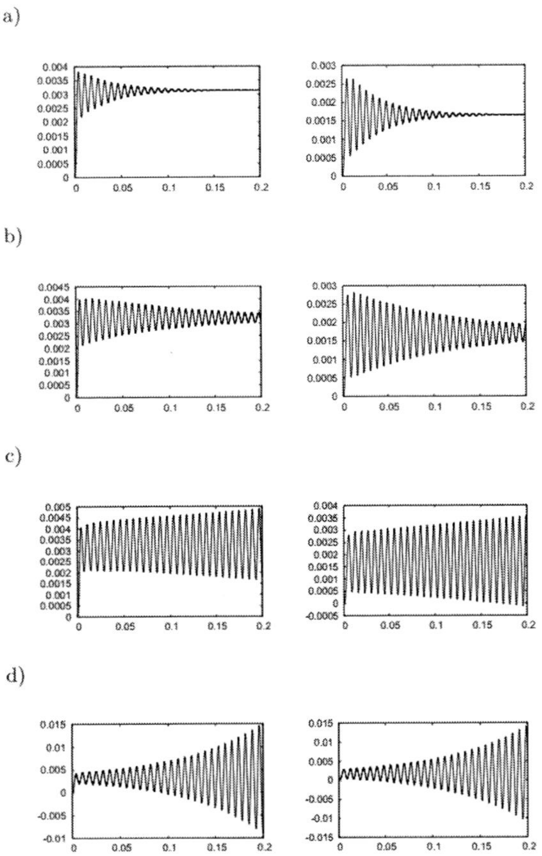


**Fig. 4.** The aeroelastic response of the system $\theta_1$ (left), $\theta_2$ (right) for the model F, gap $g_0 = 0.6$ mm and the inlet velocity a) $V_0 = 0.55m/s$, b) $V_0 = 0.58m/s$, c) $V_0 = 0.6m/s$ and d) $V_0 = 0.62m/s$.

## 5 Conclusion

In this paper a fluid structure interaction problem of flow induced vibrations of vocal folds was described. The numerical approximations based on the finite method were presented. The choice of the inlet and the outlet boundary condition was discussed. On the considered model problem only the prescribed inlet velocity formulations were successful. The prescribed pressure difference lead to an additional non-physical damping of the resulting aeroelastic system. This is particularly caused by the fact, that the prescribed inlet pressure boundary condition is imposed only weakly, and during the computation the pressure is not constant in time, but depends on the vocal fold vibrations.

**Acknowledgment** This work was supported by the *Czech Science Foundation* under the *Grant No. 16 - 01246S*.

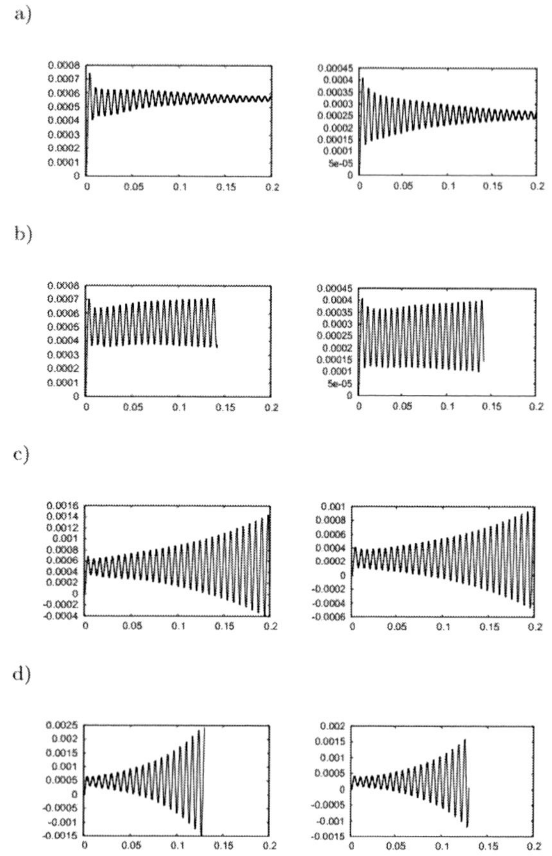


**Fig. 5.** The aeroelastic response of the system $\theta_1$(left),$\theta_2$(right) for the model M, gap $g_0 = 0.4\,\mathrm{mm}$ and the inlet velocity a) $V_0 = 0.60 m/s$, b) $V_0 = 0.63 m/s$, c) $V_0 = 0.65 m/s$ and d) $V_0 = 0.68 m/s$

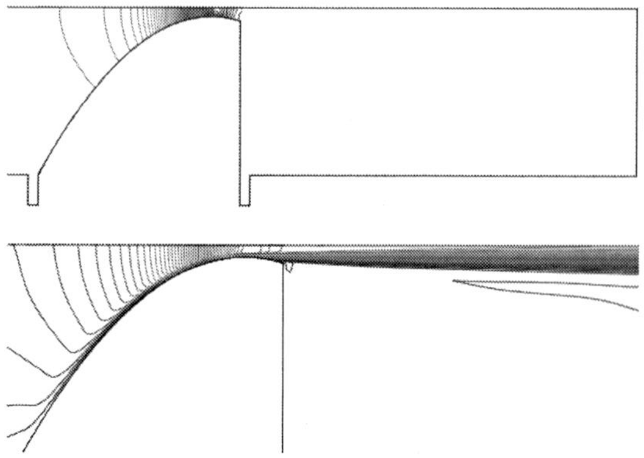

**Fig. 6.** The pressure and magnitude of velocity distributions nearby the modelled vocal fold for the gap $g_0 = 0.60\,\mathrm{mm}$, the inlet flow velocity $U_\infty = 1.2\mathrm{m}\ /\ \mathrm{s}$, the channel lengths $L_0 = L_2 = 2L$ at the time instant $t = 0\,\mathrm{s}$.

## References

1. I.R. Titze, *The Myoelastic Aerodynamic Theory of Phonation* (National Center for Voice and Speech, U.S.A., 2006)
2. S. Thomson, J. Tack, G. Verkerke, Journal of Biomechanics **40**, 3598 (2007)
3. M. Triep, C. Brücker, W. Schröder, Exp Fluids **39**, 232 (2005)
4. B.R. Kucinschi, R.C. Scherer, K.J. DeWitt, T.T.M. Ng, The Journal of the Acoustical Society of America **119**, 3011 (2006)
5. B. Erath, M. Plesniak, Exp Fluids **49**, 131 (2010)
6. S. Thomson, L. Mongeau, S. Frankel, J Acoust Soc Am **118**, 1689 (2005)
7. P. Šidlof, O. Doaré, O. Cadot, A. Chaigne, Exp Fluids **51**, 123 (2011)
8. J. Horáček, V. Uruba, V. Radolf, J. Veselý, V. Bula, Appl Comput Mech **5**, 21 (2011)
9. K. Ishizaka, J.L. Flanagan, The Bell System Technical Journal **51**, 1233 (1972)
10. J. Horáček, P. Šidlof, J.G. Švec, Journal of Fluids and Structures **20**, 853 (2005)
11. J. Horáček, P. Sváček, M. Růžička, M. Feistauer, Applied and Computational Mechanics **1**, 43 (2007)
12. M.P. de Vries, H.K. Shutte, A.E.P. Veldman, G. Verkerke, Journal of Acoust. Soc. Am. **111**, 1847 (2002)
13. P. Sváček, J. Horáček, Communications in Computational Physics **12**, 789 (2012)
14. G. Link, M. Kaltenbacher, M. Breuer, M. Döllinger, Computation Methods in Applied Mechanical Engineering **198**, 3321 (2009)
15. T. Nomura, T.J.R. Hughes, Computer Methods in Applied Mechanics and Engineering **95**, 115 (1992)
16. M. Feistauer, *Mathematical Methods in Fluid Dynamics* (Longman Scientific & Technical, Harlow, 1993)
17. C.H. Bruneau, P. Fabrie, International Journal for Numerical Methods in Fluids **19**, 693 (1994)
18. J.G. Heywood, R. Rannacher, S. Turek, Int. J. Numer. Math. Fluids **22**, 325 (1992)
19. V. Dolejší, East-West Journal of Numerical Mathematics **9**, 1 (2001)
20. R. Codina, Computational Method in Applied Mechanical Engineering **190**, 1579 (2000)
21. T. Gelhard, G. Lube, M.A. Olshanskii, J.H. Starcke, Journal of Computational and Applied Mathematics **177**, 243 (2005)
22. P. Sváček, M. Feistauer, *Application of a Stabilized FEM to Problems of Aeroelasticity*, in *Numerical Mathematics and Advanced Application* (Springer, Berlin, 2004), pp. 796–805, ISBN 3-540-21460-7
23. P. Sváček, M. Feistauer, J. Horáček, Journal of Fluids and Structures **23**, 391 (2007)
24. V. Girault, P.A. Raviart, *Finite Element Methods for the Navier-Stokes Equations* (Springer, Berlin, 1986)
25. P.M. Gresho, R.L. Sani, *Incompressible Flow and the Finite Element Method* (John Wiley & Sons, Chichester, 1998)
26. G. Lube, Numerical Analysis and Mathematical Modelling **29**, 85 (1994)
27. J. Horáček, J.G. Švec, Journal of Fluids and Structures **16**, 931 (2002)

# Methods of measurement signal acquisition from the rotational flow meter for frequency analysis

*Dariusz* Świsulski[1,*], *Robert* Hanus[2], *Marcin* Zych[3], and *Leszek* Petryka[4]

[1]Gdańsk University of Technology, Faculty of Electrical and Control Engineering, 80-233 Gdańsk, Poland
[2]Rzeszów University of Technology, Faculty of Electrical and Computer Engineering, 35-959 Rzeszów, Poland
[3]AGH University of Science and Technology, Faculty of Geology, Geophysics and Environmental Protection, 30-059 Kraków, Poland
[4]AGH University of Science and Technology, Faculty of Physics and Applied Computer Science, 30-059 Kraków, Poland

**Abstract.** One of the simplest and commonly used instruments for measuring the flow of homogeneous substances is the rotational flow meter. The main part of such a device is a rotor (vane or screw) rotating at a speed which is the function of the fluid or gas flow rate. A pulse signal with a frequency proportional to the speed of the rotor is obtained at the sensor output. For measurements in dynamic conditions, a variable interval between pulses prohibits the analysis of the measuring signal. Therefore, the authors of the article developed a method involving the determination of measured values on the basis of the last inter-pulse interval preceding the moment designated by the timing generator. For larger changes of the measured value at a predetermined time, the value can be determined by means of extrapolation of the two adjacent inter-pulse ranges, assuming a linear change in the flow. The proposed methods allow analysis which requires constant spacing between measurements, allowing for an analysis of the dynamics of changes in the test flow, eg. using a Fourier transform. To present the advantages of these methods simulations of flow measurement were carried out with a DRH-1140 rotor flow meter from the company Kobold.

## 1 Principles of methods

One of the simplest and commonly used instruments for measuring the flow of homogeneous substances is the rotational flow meter.

An essential part of rotary flow metres is the rotor rotating at a speed proportional to the fluid flow rate. The rotor may be vane or screw type.

A vane rotor (Fig. 1) has a symmetrical blade. A flow meter is used for measuring low flow rates in pipelines with a diameter of 40 mm.

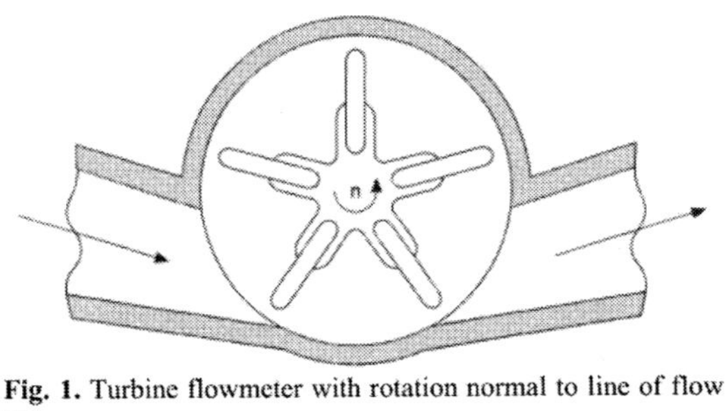


**Fig. 1.** Turbine flowmeter with rotation normal to line of flow [1].

Screw flow meters (Fig. 2) have a rotor in the shape of a multiple turn screw through which fluid flows axially. Such flow meters may be used for higher flows and diameters.

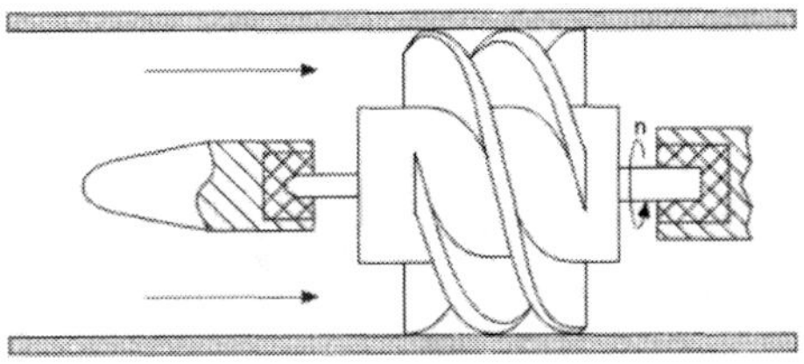


**Fig. 2.** Turbine flowmeter with rotation in line of flow [1].

The rotational speed of the rotating element is measured for example by an induction sensor. It is proportional to the mean fluid flow rate and its volume. At the output of the inductive sensor we receive a pulse string.

The impulse at the output of the incremental encoder appears at time $t_i$, where the average speed at the time of the previous pulse $t_{i-1}$ is equal to the ratio of the angle of rotation of the rotating referred to the inverse of the resolution (sensitivity) $C_e$ (number of pulses per revolution) to time $t_i$-$t_{i-1}$, wherein the disk is rotated by this angle, that is, when the condition [2]:

$$\frac{1}{t_i - t_{i-1}} \int_{t_{i-1}}^{t_i} \omega(t)dt = \frac{\frac{1}{C_e}}{t_i - t_{i-1}} \qquad (1)$$

where $\omega(t)$ – the value of instantaneous speed.

* Corresponding author: dariusz.swisulski@pg.gda.pl

The result is a pulse signal $w(t)$ of the distance between the pulse proportional to the volume of the fluid flow (Fig. 3) [3, 4].

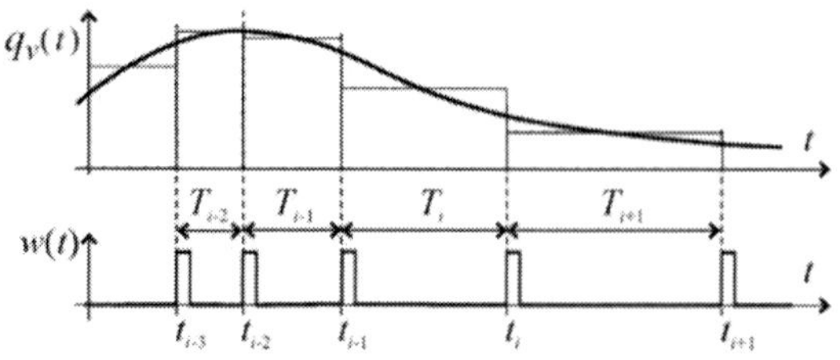


**Fig. 3.** The pulse signal $w(t)$ of frequency proportional to the flow rate of the fluid flow $q_v$.

By measuring the pulse frequency we can calculate the volume flow of the flowing fluid.

## 2 The measurements in dynamic conditions

Traditional methods of digital recording of pulse signals may be divided into two groups: measuring the number of periods of the pulse signal in a predetermined reference time, and measuring the number of pulses of the reference generator during one period or an integer multiple of the pulse signal. For the methods of the second group, the number of signal periods during the measurement of the pulse can be constant or dependent upon their length.

The choice of the appropriate methods is usually dependent on the maximum quantization error. Also, the time taken for the measurement method is dependent of the assumed maximum quantization error. This approach is acceptable when signal frequency is constant in time. The problem arises when during the measurement, the frequency of the signal pulse changes. When measuring flow, when its value changes during the measurement process, this would change the distance between successive pulses. Therefore, the measurements should be used other methods described below.

For measurements in dynamic conditions, a variable interval between pulses does not allow direct frequency analysis of the measurement signal [5].

The methods presented further allow analysis requiring constant spacing between measurements, allowing for an analysis of the dynamics of changes in the test flow, e.g. using a Fourier transform.

Another feature of the measurement systems using such methods is the ability to integrate the channels of voltage, in which the voltage signal is converted to digital form in the analog-digital channels with a frequency signal [6]. Measurements in both types of channels can be sampled using a common timing generator which makes it possible to obtain information about the values measured at the same points in time. This allows e.g. to present the dependence of the measured flow of other physical quantities.

Pulse signal processing methods, depending on the time for which the measurement result is determined can be divided into two groups. In on-line measurements the measured value is determined based on the position of the pulse preceding the moment of measurement. When measuring off-line, first you write to memory the location of all pulses in the time of registration, and after the completion of the measurement determines a value measured in the moments of measurement adopted. Therefore, in this case the measured value is determined not only by the position of the pulse preceding the moment of measurement, but also occurring after that moment.

## 3 On-line acquisition

The easiest way to register a pulse signal $w$ lies in the fact that the value of the signal at any time $t_k$ (defined by the pulse from the timing generator, common for analog and frequency channels) is calculated based on the length of the last interval inter-pulse $T_i$ preceding the moment $t_k$ (Fig. 4).

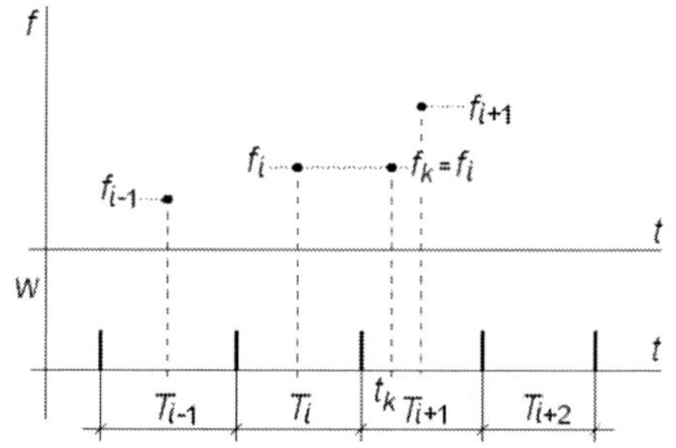


**Fig. 4.** Measuring the frequency of the last interval between pulses.

If the value of the measured signal changes during the measurement (whereby also the interval between pulses changes) and the longer period of the pulse signal comes to pass (and consequently the longer the time between the end of the period $T_i$ and the time $t_k$) value obtained from the measurement may deviate from that which was at the time of sampling.

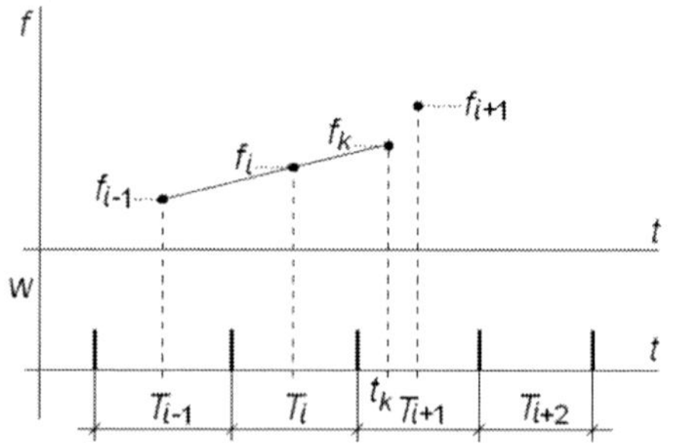


**Fig. 5.** Frequency measurement with extrapolation of the last two interval periods $T_{i-1}$ and $T_i$.

Therefore, in case of change of the measured value, the value at time $t_k$ can be calculated from the formula (2) using extrapolation of two adjacent inter-pulse compartments preceding the moment $t_k$. A linear change of frequency is assumed and that frequency value $f_i$

obtained from the measurement interval $T_i$ is equal to the instantaneous frequency of the midpoint of this range (Fig. 5).

$$f_k = f_{i-1} + \frac{(f_i - f_{i-1})\left(t_k - \left(\sum_{j=1}^{i-2} T_j + \frac{1}{2}T_{i-1}\right)\right)}{\left(\sum_{j=1}^{i-1} T_j + \frac{1}{2}T_i\right) - \left(\sum_{j=1}^{i-2} T_j + \frac{1}{2}T_{i-1}\right)} = \frac{1}{T_{i-1}} + \frac{2\left(\frac{1}{T_i} - \frac{1}{T_{i-1}}\right)\left(t_k - \sum_{j=1}^{i-2} T_j - \frac{1}{2}T_{i-1}\right)}{T_{i-1} + T_i} \tag{2}$$

However, such a solution may lead to an increase in measurement error when the frequency does not vary in a manner similar to a linear, but random manner. Frequency determined on the basis of the last two periods may differ significantly from its current value.

## 4 Off-line acquisition

In off-line mode, one must first register the location of pulses, and only after the registration one can determine the measured values in times of sampling. Therefore, to determine the measured value of the pulse signal at any time $t_k$ one can use not only the position of the pulse preceding the moment $t_k$, but also after they occur. With this solution, instead of using extrapolation, interpolation can also be used.

The simple method implementation in off-line mode is to determine the value measured on the basis of the inter-pulse interval, in which the moment $t_k$, i.e. the pulse position before this moment and after that moment. The value of the time $t_k$ is calculated from the average value of the frequency between pulses. In the case of linear change in the measured value, the mean value of the time interval $T_{i+1}$ is equal to the instantaneous value in the middle of this range, namely: $f_k = f_{i+1}$ (Fig. 6). The method gives satisfactory results when the measured value is of low dynamics.

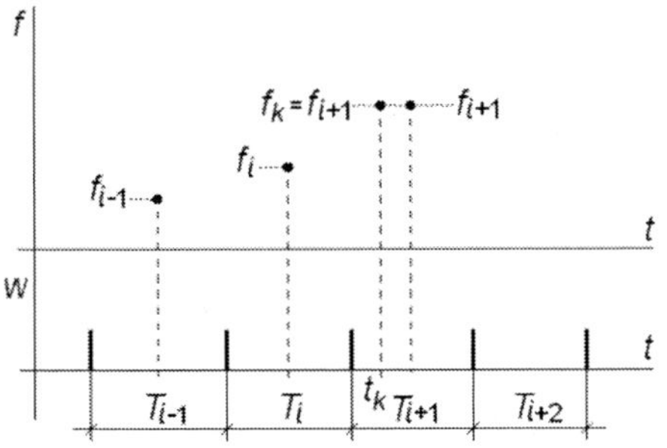


**Fig. 6.** Measuring the frequency of the current interval between pulses.

If the change of the measured value during the measurement is bigger, the better solution is to determine the frequency of two adjacent inter-pulse periods. One should assume a linear variation of frequencies and that the value of the frequency $f_i$ derived from the measurement interval $T_i$ is equal to the instantaneous frequency of the midpoint of this range.

Depending on the position of the time $t_k$, intervals which determine the frequency $f_k$ are $T_i$ and $T_{i+1}$ (Fig. 7) or $T_{i+1}$ and $T_{i+2}$ (Fig. 8).

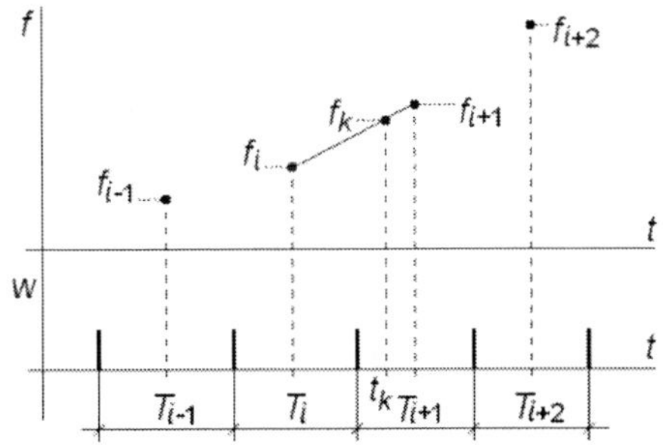


**Fig. 7.** Frequency measurement based on the length of intervals between two pulses $T_i$ and $T_{i+1}$.

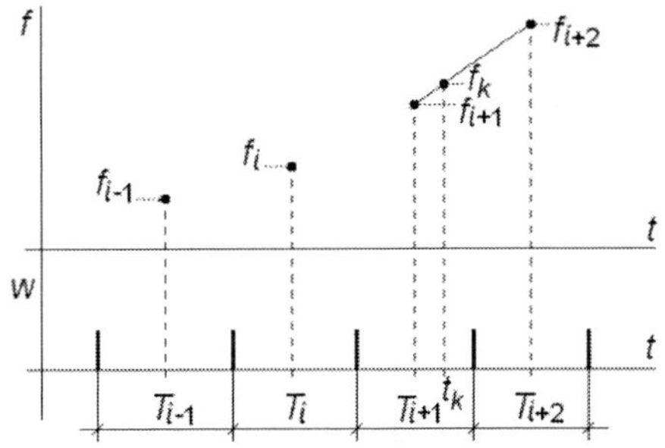


**Fig. 8.** Frequency measurement based on the length of intervals between two pulses $T_{i+1}$ and $T_{i+2}$.

For the moment $t_k$ occurring in the first half of the inter-pulse interval (Fig. 7) the frequency $f_k$ is calculated from the formula (3), and for the case where time $t_k$ is present in the second half of the interval $T_{i+1}$ (Fig. 8) from equation (4).

$$f_k = f_i + \frac{(f_{i+1} - f_i)\left(t_k - \left(\sum_{j=1}^{i-1} T_j + \frac{1}{2}T_i\right)\right)}{\left(\sum_{j=1}^{i} T_j + \frac{1}{2}T_{i+1}\right) - \left(\sum_{j=1}^{i-1} T_j + \frac{1}{2}T_i\right)} = f_i + \frac{2(f_{i+1} - f_i)\left(t_k - \sum_{j=1}^{i-1} T_j - \frac{1}{2}T_i\right)}{T_i + T_{i+1}} \tag{3}$$

$$f_k = f_{i+1} + \frac{(f_{i+2} - f_{i+1})\left(t_k - \left(\sum_{j=1}^{i} T_j + \frac{1}{2}T_{i+1}\right)\right)}{\left(\sum_{j=1}^{i+1} T_j + \frac{1}{2}T_{i+2}\right) - \left(\sum_{j=1}^{i} T_j + \frac{1}{2}T_{i+1}\right)} =$$
$$= f_{i+1} + \frac{2(f_{i+2} - f_{i+1})\left(t_k - \sum_{j=1}^{i} T_j - \frac{1}{2}T_{i+1}\right)}{T_{i+1} + T_{i+2}} \quad (4)$$

The sampling rate is determined by the period of time between sampling times $t_k$ and $t_{k+1}$.

Simultaneously with the measurement of the frequency at the time $t_k$ the voltage channels may be sampled. Such a need occurs when apart from a converter with an output frequency one uses other transducers with voltage output and requires information on the values measured in the same moments of time.

## 5 Simulation of recording the signal from the flow meter

To present the advantages of these methods, simulations of the flow measurements were performed with the DRH-1140 rotor flow meter (Fig. 9) manufactured by Kobold [7].

**Fig. 9.** Rotating Vane Flow Meter DRH-1140 manufactured by Kobold [7].

The range of this flow meter is 26 l/min, and the frequency of the output frequency signal at a maximum flow rate of 240 Hz. The flow meter uses a fixed geometry impeller blades. Inside the rotor are magnets hermetically isolated from the flowing medium. The Hall effect sensor is located in the housing by the magnetic field magnet, and generates a frequency signal with a frequency proportional to the flow [7].

It was assumed that the flow rate varies sinusoidally in the range of (20 ± 4) l/min, which corresponds to a frequency of the pulse signal at the output of the transducer in the range of 147.7 Hz to 221.5 Hz.

Figure 10 shows the length of the intervals between successive pulses as a function of their number. Inter-pulse times vary in the range of 0.00451 s to 0.00677 seconds.

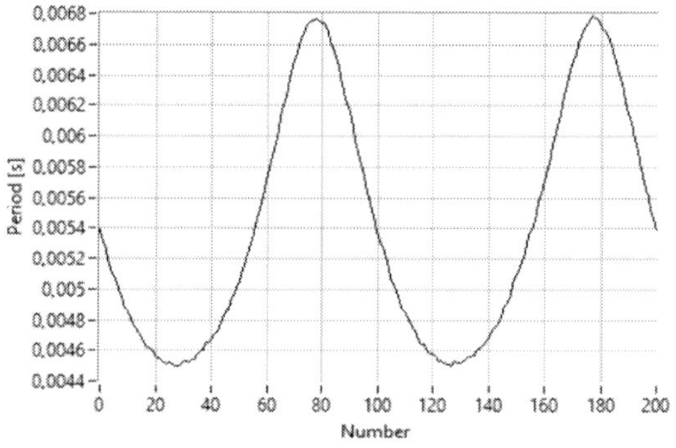


**Fig. 10.** The length of intervals between pulses as a function of their sequence number.

Due to the varying distances between the moments for which information on the value of the successive inter-pulse intervals was obtained obvious there is fact, that from the string of such obtained values one cannot directly determine the spectrum amplitude. This would result in the appearance of components in the spectrum which do not occur in reality (Fig. 11).

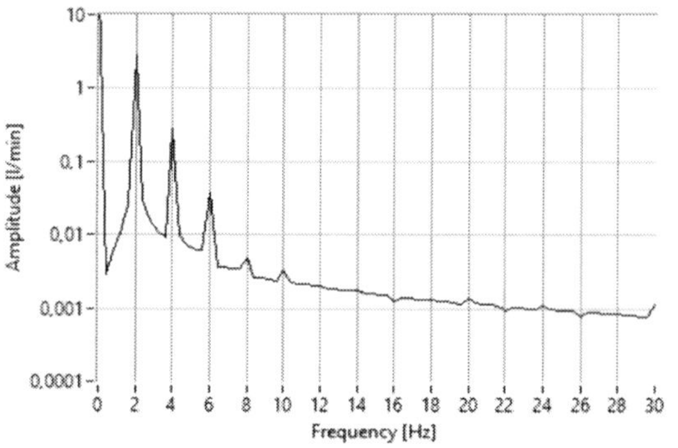


**Fig. 11.** The spectrum of changes of the measured value determined by the length of intervals between successive pulses.

In order to obtain the spectrum, one can use the previously mentioned methods of acquisition of pulse signal modulated by frequency.

The presented results will be obtained by on-line acquisition of the last inter-pulse interval (as shown in Fig. 4) and the extrapolation of the last two inter-pulse periods (as shown in Fig. 5 and equation (2)).

It was assumed, that the sampling rate is fixed at 0.005 s. The results of the simulation are shown in figures 12 and 13.

On the basis of these graphs one can reach two conclusions. First, results obtained using the last inter-pulse interval they are delayed in time relative to the actual changes of the measured value. This follows from the values obtained from the subsequent sampling moments based on the length of intervals preceding the moment of inter-pulse sampling. Avoiding this disadvantage is made possible by the use of the inter-pulse interval, but it is possible for the off-line method (as shown in Fig. 6). This delay does not occur for a method wherein the measured values are determined from extrapolation of the last two inter-pulse periods.

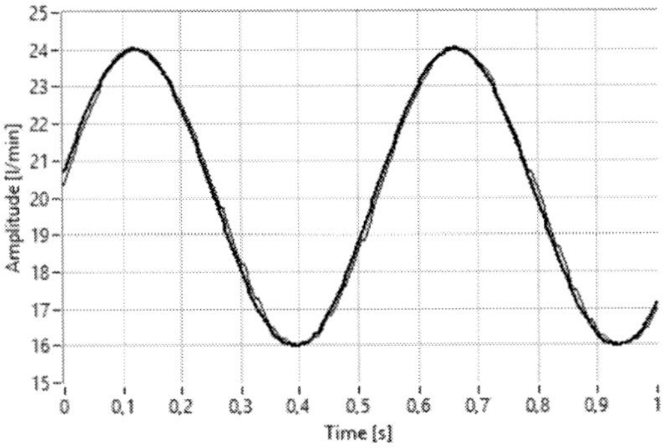


**Fig. 12.** The course of the measured value as a function of the time appointed for the last inter-pulse interval (thin line) and the extrapolation of the last two inter-pulse periods (bold line).

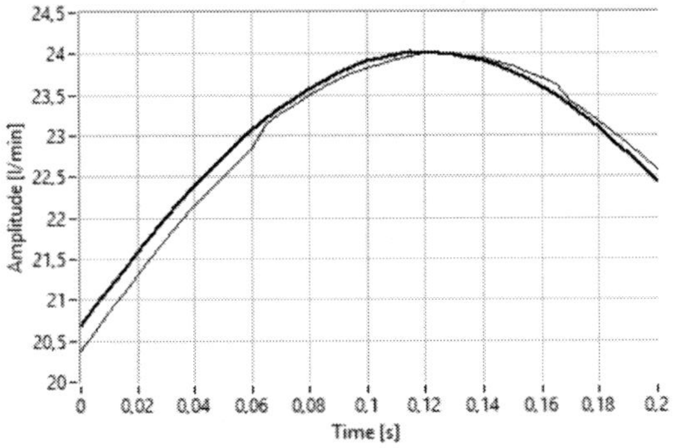


**Fig. 13.** The enlarged portion of the graph in figure 12.

The second conclusion that can be drawn from the resuting waveform is a discontinuity in the course of the measured value obtained from the last inter-pulse interval. A discontinuity does not arise in the method in which the mesured values are determined from extrapolation of the last two inter-pulse periods. This is even more evident when one shortens the sampling time. Figure 14 shows the waveforms for the same conditions, as shown in figure 13, but with a reduced sampling time of 0.005 seconds to 0.0005 seconds.

A stepped waveform obtained is due to the fact that the sampling rate is several times smaller than the time of the inter-pulse intervals. Due to this, several successive sampling moments for a given inter-pulse interval give the same value of the measured value determined from the previous inter-pulse interval.

Designated courses of changes of the measured value of the presented methods, thanks to the constant sampling frequency, can be used to determine the amplitude spectrum. This spectrum for both methods is shown in figure 15.

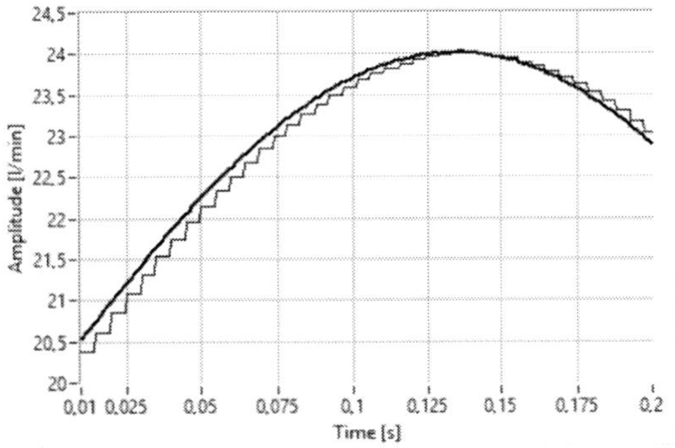


**Fig. 14.** The course of the measured value as a function of the time appointed for the last inter-pulse interval (thin line) and the extrapolation of the last two inter-pulse periods (bold line) for the sampling period of 0.0005 seconds.

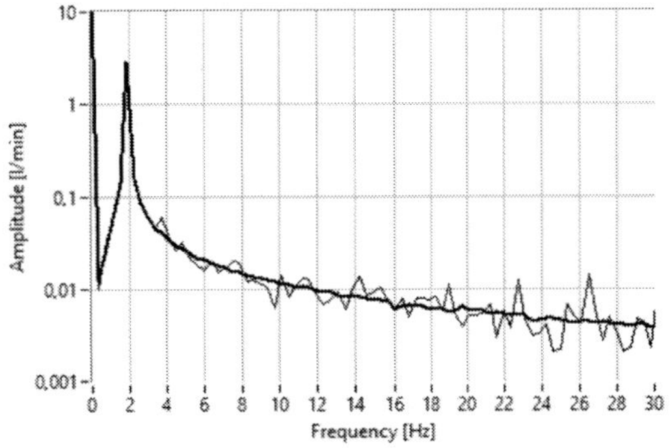


**Fig. 15.** The amplitude spectrum of the mesured value determined on the basis of the last inter-pulse interval (thin line) and the extrapolation of the last two inter-pulse periods (bold line).

From the graph it can be concluded that the presented methods are well suited for the frequency analysis, although due to the discontinuity of the method discussed earlier, in which one uses single inter-pulse interval, they give inferior results.

The measurement technique should still take into account the interference of the measured value and measurement errors (in this case flow meter errors – the measurement accuracy is 2.5% of the measuring range given by the manufacturer, and the processing errors of inter-pulse time interval to a digital value). To illustrate the impact of errors and distortions simulated by adding them to the inter-pulse times the Gaussian White Noise with standard deviation of the Gaussian probability density function is equal to $1 \cdot 10^{-5}$. The resulting time-domain waveforms are shown in figures 16 and 17, while in the frequency domain in figure 18.

Comparing the obtained results with the previous ones one can conclude that the errors of measurement and interference have a large impact on final results for the method, in which the measured values are determined from the extrapolation of the last two inter-pulse periods. This influence is so significant that the results can be even worse than in the case of the method using a single inter-pulse period.

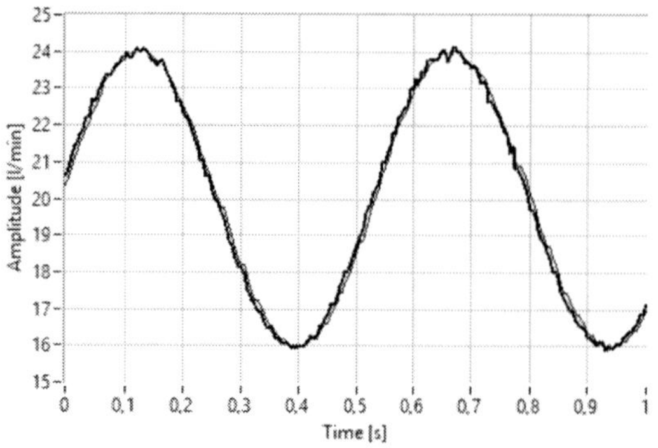


**Fig. 16.** The course of the measured value as a function of the time appointed for the last inter-pulse interval (thin line) and the extrapolation of the last two inter-pulse periods (bold line) with Gaussian White Noise.

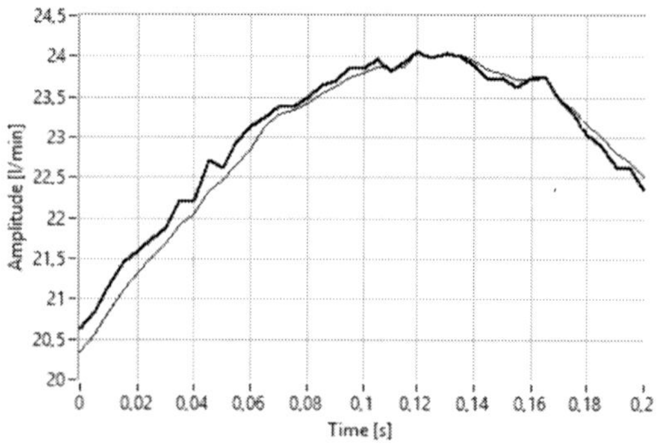


**Fig. 17.** The enlarged part of the graph in figure 16.

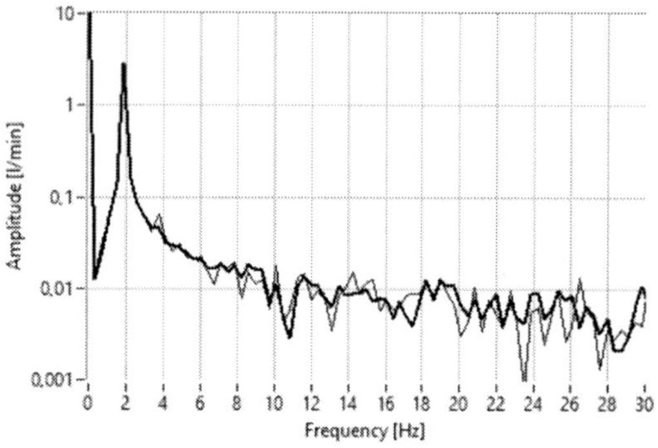


**Fig. 18.** The amplitude spectrum of the meaured value determined on the basis of the last inter-pulse interval (thin line) and the extrapolation of the last two inter-pulse periods (bold line) with Gaussian White Noise.

## 6 Conclusions

In rotary flow meters the flowrate of the fluid flow is calculated by measuring the frequency of the pulses generated by the incremental encoder.

The presented review developed by the authors of the pulse signal processing methods in certain moments of sampling shows that the choice of method depends on the measurement mode (on-line or off-line) and the nature of the change of the measured value. The simplest solution is to determine the value measured on the basis of a single inter-pulse interval, more complex methods use more periods.

The methods allow to obtain information about the flow rate measured at the same times for the voltage and frequency of the channel, and the samples that were obtained from the frequency channels can be processed by methods requiring a constant sampling frequency (e.g. Fourier Transform).

The results of the flow measurement simulation done with the use of the DRH-1140 rotary flow meter produced by Kobold confirm the advantages of the presented methods.

## References

1. Yu.L. Shekhter, DOI: 10.1615/AtoZ.f.flow_metering, http://www.thermopedia.com/content/770/
2. D. Świsulski, Zeszyty Naukowe Wydziału Elektrotechniki i Automatyki Politechniki Gdańskiej **34**, 67-70 (2013) (in Polish)
3. N.V. Kirianaki, S.Y. Yurish, N.O. Shpak, V.P. Denega, *Data Acquisition and Signal Processing for Smart Sensors* (John Wiley & Sons, New York, 2002)
4. D. Świsulski, E. Pawłowski, Wiadomości Elektrotechniczne **54** (7), 3-9 (2016) (in Polish)
5. E. Pawłowski, Pomiary Automatyka Kontrola **60** (8), 607-609 (2014)
6. D. Świsulski, Przegląd Elektrotechniczny **88** (10b), 29-31 (2012) (in Polish)
7. KOBOLD, *Rotating Vane Flow Meter for Liquids*, http://fluida-bg.com/assets/files/7817/s4gb_drh.pdf

EPJ Web of Conferences **143**, 02125 (2017) DOI: 10.1051/epjconf/201714302125
*EFM 2016*

# Homogenous and heterogeneous combustion in the secondary chamber of a straw-fired batch boiler

*Mateusz* Szubel[1,*], *Wojciech* Adamczyk[2], *Grzegorz* Basista[1], *Mariusz* Filipowicz[1]

[1]AGH University of Science and Technology, Faculty of Energy and Fuels, al. Mickiewicza 30, 30-059 Krakow, Poland
[2]Silesian University of Technology, Institute of Thermal Technology, Konarskiego 22, 44-100 Gliwice

**Abstract.** Currently, the attention of the producers of biomass batch boilers is mostly focused on the problem of the total efficiency of energy conversion, CO emissions as well as particulate matter emissions. Due to the regulations of the European Union, the emissions referred to above have to be kept at certain levels because of health considerations, but also because of the necessity to increase the efficiency of the devices. The paper presents the process of analysis of a straw-fired small-scale boiler. In this study, the early stage CFD model presented in a previous paper [1] has been improved and evaluated. Based on [2], an additional set of specimens participating in homogeneous gas reactions was assumed to describe the combustion process sufficiently. Associated Arrhenius parameters have been applied for the description of these reactions. ANSYS Fluent 16 has been used to perform the analysis and the analysis was focused on the CO emissions level as well as on the impact of the modelling approach on the result of the computing. Moreover, losses related to incomplete combustion have been calculated for each of the considered cases.

## 1 Biomass combustion – selected aspects of theory and Computational Fluid Dynamics

Biomass combustion involves many advanced physical and chemical processes. The characteristics of the process depend on the properties of the applied fuel as well as on the technology of combustion [3].

From the point of view of fuel properties, special consideration should be given to the high humidity content and to the volatiles. Due to this fact, four general combustion stages can be listed [4, 5, 6]: warming and drying, thermal decomposition of the fuel components, combustion of the products of the thermal decomposition and combustion of charcoal.

Biomass combustion involves many advanced physical and chemical processes. Depending on the gasifying agent, the composition of the mixture that is the product of gasification varies. However, the same processes have to be taken into account in each case: $H_2$ - $O_2$ system, oxidation of carbon monoxide as well as conversion of methane. Oxidation of higher hydrocarbons is of lower significance. Moreover, reactions resulting in the creation of nitrogen oxides are also important.

### 1.1 Selected mechanisms of combustion in the gaseous phase

The $H_2$-$O_2$ system is important in itself, but also as a part of some schematics or stage of the oxidation of hydrocarbons and in some specific cases also CO. A detailed description of hydrogen oxidation is presented in [7, 8]. The mechanism of $H_2$ oxidation differs depending on temperature, pressure and degree of reaction. When it comes to the modelling of the $H_2$-$O_2$ system, as much as 40 reactions can be taken into account, including eight compounds, such as $H_2$, $O_2$, $H_2O$, OH, O, H and $H_2O_2$ [8].

The oxidation of carbon monoxide is important as a stage of the combustion of the volatile products of the gasification. It is also an extremely important element of the oxidation of hydrocarbons. A general, simplified description of the oxidation of CO should include the following stages: the decomposition of the fuel with the release of CO and the final oxidation of CO to CO2 [3].

The process of CO oxidation is really slow [7, 8], except for reactions in which some compounds of hydrogen are present. Small amounts of $H_2O$ and $H_2$ have a huge impact on the oxidation rate. The reason for that is the participation of hydroxyl radicals in the oxidation of CO [9]. When the compound including hydrogen is $H_2O$, this process can be described in a simplified way using the equations below [8, 10]:

$$CO + O_2 \rightarrow CO_2 + O \quad (1)$$

$$O + H_2O \rightarrow OH + OH \quad (2)$$

$$CO + OH \rightarrow CO_2 + H \quad (3)$$

$$H + O_2 \rightarrow OH + O \quad (4)$$

The reaction described by equation 1 occurs slowly and it is insignificant in the $CO_2$ formation process. Reaction 2 results in the formation of hydroxyl radicals, which have to be provided as a substrate in the next step. Moreover, this stage causes other processes due to the production of hydrogen atoms, which are able to react

* Corresponding author: mszubel@agh.edu.pl

with $O_2$ (production of OH groups). Oxygen radicals are an additional product of the described process. Both hydroxyl and oxygen radicals are present in the third stage (described by equation 3), which is a key step in the CO oxidation [8].

The situation is a little different when hydrogen instead of water is the oxidizing factor of CO. The following reactions take place [8]:

$$O + H_2 \rightarrow OH + H \quad (5)$$

$$OH + H_2 \rightarrow H_2O + H \quad (6)$$

However, it is necessary to take into account the whole H2-O2 system to describe the full process of oxidation of CO in the presence of hydrogen.

Methane ($CH_4$) is unique in the group of alkanes due to the special tetrahedral molecular structure with C-H chemical bonds, which are characterized by high energy. This fact results in some uncommon features, such as relatively high ignition temperature and low flame speed. The kinetics of $CH_4$ combustion are well known and the directions of the process are described by a number of reactions. For example, in [8] the author used 177 reactions to describe the complex mechanism of combustion of CH4 in a C-H-O system as well as 102 reactions with N2.

Processes presented above show that even the oxidation of the simple chemical compounds that are taken into consideration is a complicated process. The description of hydrocarbons is encumbered with a lot of problems and limitations. However, it is possible to implement some empirical global models, which include series of simplifications and estimations. Examples of single step and multistep global models as well as values of important kinetic parameters can be found in the literature [11, 12].

### 1.2 Particulate emissions from biomass combustion in small size biomass boilers

The environmental impact resulting from the conversion of solid biomass into energy in the combustion process in small-scale installations causes particulate matter (PM) emissions, which becomes a critical aspect if compared to the considerably lower PM emissions from methane boilers [15].

PM can consist of mineral substances as well as organic material. Due to the fact that the content of different components in the biomass varies greatly, the particulate emissions also vary depending on the device technology and the applied fuel.

A few significant factors influencing the PM emissions have to be listed. Apart from the number of problems connected to the construction of the combustion chamber, it is the direct relationship between the PM emissions and the value of the excess air, as well as the improper content of moisture in the fuel. High moisture content has a negative impact on PM emissions. However, straw that is too dry (<10%) is not recommended for the same reason.

Considering the heat balance of the boiler, transportation of the solid particles with the exhaust is undesirable due to the content of combustible organic components. The amount of energy related to the material leaving the combustion chamber with the exhaust is called the loss of fly ash. However, while the organic part of the PM causes problems, the inorganic ones are also problematic due to the possible presence of potassium, sulphur, sodium and zinc, which could lead to increased emissions. The greatest amount of PM emissions from biomass combustion consists in particles with an aerodynamic diameter smaller than 2.5 [15].

### 1.3 Simplified approach to modelling of combustion and reduced mechanisms of chemical reactions

The computational cost of chemical kinetics evaluation and the associated species transport equations is something that has to be taken into account when designing a combustion model. Moreover, the costs grow rapidly with the number of applied reactions as well as the level of complexity [8, 13, 14].

To allow for more efficient performance of simulations of advanced combustion problems, simplified processes called reduced mechanisms have been developed. A reduced mechanism of a chemical reaction includes a relatively low number of equations, which is enough to achieve a satisfying agreement of the theoretical description of the reaction and the actual process. A more detailed description of specific reduced mechanisms of combustion can be found in [13].

The simplified approach to the combustion process used in CFD numerical modelling is useful in relatively simple cases, like homogenous premixed combustion, as well as in complicated problems, such as simulations of operation of combustors – both conventional fuel-fired and biomass-fired. Moreover, it is possible to extend the functionality of commercial CFD-solvers in terms of solid phase descriptions and solid-gas interactions by defining key components of the solid phase (so-called pseudocomponents) and developing appropriate equations [2]. Of course, all stages preceding the direct combustion, such as drying and warming, devolatilization and interactions of the products with the solid phase have to be taken into account in the model.

### 1.4 Computational Fluid Dynamics as a tool in studies of biomass thermo–chemical conversion

In the paper by Xue, Heindler and Fox [16], the CFD numerical model has been developed to describe the process of fast biomass pyrolysis in a reactor with a fluidized bed. The authors took into account the many stages of the process by properly defining the kinetics of the reactions. This analysis was focused on the dynamics of heating the fuel, the reactions in the solid phase (including char formation) as well as the spatial distribution of tar and gases in the reactor. Similar studies have been performed by Blondeau and Jeanmart in [17], where an advanced model describing the pyrolysis of fuel particles in a pulverized coal-fired boiler has been proposed.

Results of numerical simulations as well as data obtained in experimental studies are often used in more advanced numerical analyses, including numerical studies of combustors with a fixed bed. Studies described in [18] are devoted to a subgrid-scale model, which has been developed to analyse the intraparticle gradients, the interactions with other particles and the gas phase. The Euler–Euler CFD framework has been applied. According to the remark in subsection 3.2, to simulate the entire system, the authors of the paper have combined the main model with other submodels describing the gas phase, the bed reaction and the interactions between them.

A slightly different approach has been presented by Buczyński, Weber and Szlęk in [18], where a CFD-based model has been developed for predicting the performance of a central heating boiler fired with solid fuels. A unique feature of this model is that it does not require a priori distinction between the fixed-bed region and the free-board region. The model was used to optimize both the operation algorithm and the design of the heating unit.

The development of the numerical model describing the heterogeneous combustion of biomass based on experimental data and validation of computations allows to perform variant analysis, which can be focused i.a. on parameters such as the impact of the excess air, air staging or the number and direction of the air nozzles on the process of combustion and the heat balance of the boiler. Such works are described in many papers, i.e. [1, 2, 19, 20].

Based on information contained in the literature, it is possible to prepare numerical simulations of low power straw-fired batch boilers. Currently, the application of the CFD codes to study this type of boilers is still under development, but some elaborations describing adequate approaches are already available [2]. The further part of the paper is devoted to the description of an extension of the existing CFD model of a 180 kW straw-fired batch boiler.

## 2 Experimental studies of the post-combustion in the secondary combustion chamber of a 180 kW biomass boiler

### 2.1 Characteristics of the examined boiler

The analyses described in this paper were devoted to a solid biomass batch boiler (Ekopal RM40 model). The EKOPAL RM40 boiler (figure 1) is a 180 kW unit classified in the group of gasification boilers. The dimensions of the primary combustion chamber are: 2405/3075/1880 mm (height/width/depth). The device is designed to burn straw in the form of cubes or bales.

The combustion of fuel proceeds in two stages: in the primary combustion chamber and the secondary combustion chamber. In the first stage, gasification of the fuel in the primary chamber takes place under controlled air deficiency. Combustible gaseous products of the first stage flow into the secondary chamber, where the post-combustion occurs. The next element behind the secondary combustion chamber and the ash separator is a heat exchanger, consisting of tubes arranged in three rows. The separation of the combustion stages as well as the application of the exhaust heat exchanger causes an increase in the total efficiency of energy conversion (for the nominal conditions of operation approximately 84 %).

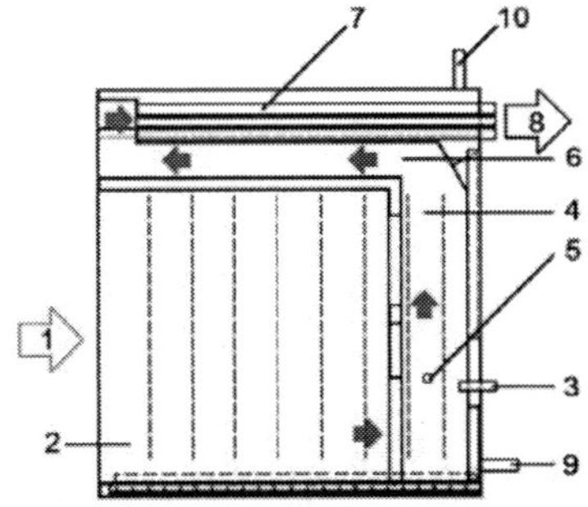


**Figure 1.** Cross section of the boiler: 1) direction of the fuel loading (door), 2) primary combustion chamber, 3) primary air nozzles connected with the fan through the air manifold, 4) secondary combustion chamber, 5) secondary air nozzle on the side wall of the secondary combustion chamber, 6) dust separator, 7) heat exchanger, 8) outlet of the exhaust, 9) inlet of the cold water, 10) outlet of the hot water.

It is important to provide a sufficient distribution of air in the areas of the two combustion chambers. The boiler system is equipped with an air supply system, which uses a centrifugal fan and an air manifold in the form of pipes arranged in series, connected with the main pipe responsible for the transport of the air provided by the fan to each duct terminated by the nozzle. The air distributor of the examined boiler consist of seven pipes, welded on the front of the main duct, connected with the area of the primary combustion chamber. Additional two ducts are responsible for the feeding of air to the secondary combustion chamber. The combustion chambers are surrounded by a water jacket filled with water. More detailed information concerning the boiler can be found in [1].

To perform measurements of the temperature inside the combustion chambers, a PTTK thermocouple has been applied. The sensor that was to control the temperature during the experiment was situated on the rear wall of the boiler, just behind the inlet to the secondary combustion chamber. A MRU ECO 3000 exhaust analyzer has been applied during the experiments to control oxygen, carbon monoxide and carbon dioxide contents in the exhaust. The device is adapted to analyze all popular solid fuels, including biomass.

A PLC set is the heart of the control and measurement system. It includes a controller connected to a PC. The controller consist of modules which receive (from the sensors) and send (to the PC and other devices in the system) an analogue and digital signal.

### 2.2 Description of the experiment

Experimental tests of the combustion in case of the current state of the heating device have been carried out as the first stage of the study in order to analyze the dynamics of the process in respect of temperature changes at the inlet to the secondary combustion chamber, as well as the concentration of carbon monoxide in the exhaust. Before the measurement, warming of the device has been carried out to provide standard operation conditions during the experiment. Based on the procedure developed in previous works, it was decided to perform three series of experiments, which was deemed enough to confirm the repeatability of the process.

Straw in the form of cubes (700x500x500 mm) has been used. Estimated fuel weight for the individual test was 40 kg, which is about ~ 1/3$^{th}$ of the nominal load (full primary combustion chamber). Moisture content in each straw cube has been measured by drying the sample of the biomass to a constant weight (using a dedicated stove). Water content was expressed as the percentage, based on the weight, of the sample before drying and weight of dry sample. The average value has been calculated based on results from three experimental series and amounted to 8.44%.

Ignition has been initiated through a special orifice located at the rear bottom corner of the boiler side wall. After the ignition, the fan was run and its efficiency has been increased gradually (in ~ 2 minutes) to its maximum value (250 dm3/s). Each experiment has been carried out till the moment when the temperature in the considered measurement point was lower than 50°C (45 minutes).

### 2.3 Results of the experiment

Based on the Arrhenius equation it can be said that, apart from the presence and type of oxidizer, temperature is the key factor determining the dynamics of the gasification process. Moreover, it has to be mentioned that carbon monoxide is the most important product of the gasification. Due to this fact, studies of relations between the concentration of carbon monoxide and the process temperature provide valuable information about the efficiency of energy conversion.

Figure 2 shows the dynamics of changes of the two aforementioned variables during the combustion process (averaged curves created based on data from three experiments). It is visible that the period of high CO concentration occurs before the phase of nominal combustion. It is a widespread phenomenon in the operation of batch boilers. Furthermore, the presented results are in agreement with the data presented in the previous paper, devoted to the analysis of the same heating unit [1].

This phenomenon is connected with the necessity of the stabilisation of the conditions in the combustion chamber, as well as with the highest intensity of the biomass devolatilisation during the first stage of combustion – directly after ignition. After the first 6~7 minutes of the experiment, the concentration of CO decreases, although the high temperature persists due to the fact that most volatiles have been already released and the balance of the process is moved to the combustion of charcoal.

Due to the direct impact of the presence of an oxidizer and the concentration of the combustion process, it was necessary to calculate the excess air ratio. The value of this parameter in the period of the highest CO generation has been determined indirectly, based on equation 7

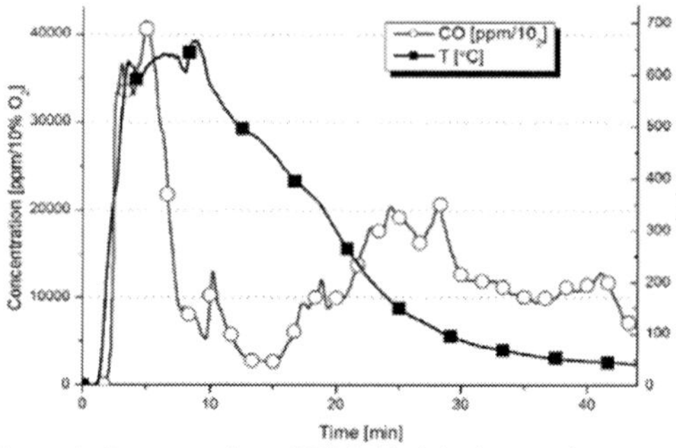


**Figure 2.** Concentration of CO behind the heat exchanger (outlet from the unit) and dynamics of temperature changes on the inlet to the secondary combustion chamber of the examined boiler.

$$\lambda = \frac{CO_{2MAX}}{CO_2}, [-] \tag{7}$$

Where:
$CO_{2MAX}$ – maximum theoretical content of $CO_2$ in the exhaust [%]
$CO_2$ – current concentration of $CO_2$ in the exhaust [%]

$CO_{2max}$ has been calculated using the equation presented below:

$$CO_{2MAX} = \frac{21C^r}{C^r+2{,}37(H^r-\frac{O^r-S^r}{8})}, [\%] \tag{8}$$

Where:
C – carbon content in the biomass [%]
H – hydrogen content in the biomass [%]
O – oxygen content in the fuel [%]
S – sulphur content in the biomass [%]

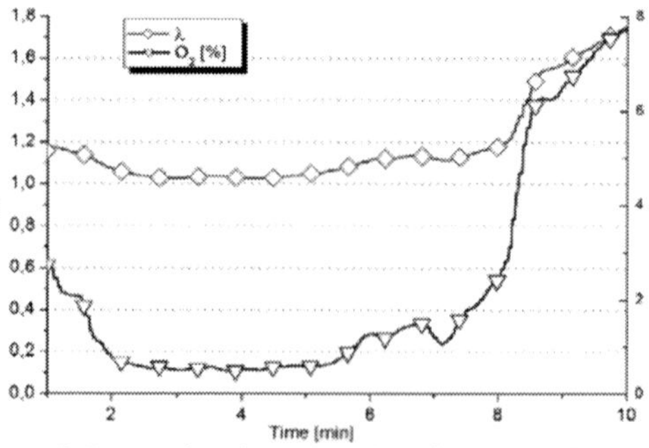


**Figure 3.** Excess air and concentration of oxygen in the exhaust in the first 10 minutes of the combustion process.

The results of the calculation in comparison with the concentration of oxygen in the exhaust are presented in Figure 3, wherein the first minute is considered as the ignition stage and skipped. Despite reaching the full efficiency of the fan, excess air in the whole period of high CO emission is relatively low – between a little above 1 and 1.2. A low level of excess air is positive when it comes to stack losses. Moreover, it has to be taken into account that providing combustion air without preheating may in some cases lead to undesirable temperature reduction in the post-combustion region.

On the other hand, an excess air level that is too low can seriously impede the oxidation of the combustible gases in the secondary combustion chamber. It can result in problems with meeting the regulations concerning the range of pollutions emissions and in the increase of losses resulting from imperfect combustion.

It is important that in the considered period of the experiment (the first ten minutes), the combustion process occurs mainly in the secondary combustion chamber, because the primary chamber is still filled by the fuel and direct access to this part of the device is still closed.

Based on the aforementioned results of the experimental tests it was concluded that it is required to perform an evaluation of the impact of the excess air in the secondary combustion chamber on the key parameters of the operation, which determine the efficiency of the heating unit.

In case of technical limitations (such as parameters of the fan installed on the experimental stand), CFD provides the possibility to carry out simplified variant analysis of conditions of CO post-combustion depending on the excess air. The results of numerical modelling described in the paper were useful in deciding on the process of increasing the efficiency of the boiler.

## 3 Numerical studies of the post-combustion in the secondary combustion chamber of a 180 kW biomass boiler

### 3.1 Development of the numerical model for the homogenous combustion

A numerical CFD model has been developed to analyse the phenomena occurring at the beginning of the combustion in the area of the secondary combustion chamber, especially in respect of the impact of the excess air on the selected operational parameters. A fluid dynamics model applied at the previous stage of the research [1] has been implemented. The kinetics of the chemical reactions were based on "Finite Rate Eddy Dissipation". Furthermore, the previous simple combustion model was developed based on the set of equations describing the chemical reactions [2]:

$$CO + \frac{1}{2}O_2 \xrightarrow{E_{a1}, A_1} CO_2 \tag{9}$$

$$CO_2 \xrightarrow{E_{a2}, A_2} CO + \frac{1}{2}O_2 \tag{10}$$

$$H_2 + \frac{1}{2}O_2 \xrightarrow{E_{a3}, A_3} H_2O \tag{11}$$

$$CH_4 + 2O_2 \xrightarrow{E_{a4}, A_4} 2H_2O + CO_2 \tag{12}$$

$$NH_3 + O_2 \xrightarrow{E_{a5}, A_5} NO + H_2O + \frac{1}{2}H_2 \tag{13}$$

$$NH_3 + NO \xrightarrow{E_{a6}, A_6} N_2 + H_2O + \frac{1}{2}O_2 \tag{14}$$

$$HCN + O_2 \xrightarrow{E_{a7}, A_7} NO + CO + \frac{1}{2}H_2 \tag{15}$$

$$HCN + NO \xrightarrow{E_{a8}, A_8} N_2 + CO + \frac{1}{2}H_2 \tag{16}$$

The applied values of the activation energy as well as the pre- exponential factors corresponding to the respective chemical reactions are collected in Table 1.

**Table 1.** Activation energies ($E_a$) and pre-exponential factors (A) applied in reactions described by equations 3–10.

| **No.** | **1** | **2** | **3** | **4** |
|---|---|---|---|---|
| $E_a$ $[\frac{J}{kg \cdot mol}]$ | $1.7 \cdot 10^8$ | $1,7 \cdot 10^8$ | $3.1 \cdot 10^7$ | $1.27 \cdot 10^8$ |
| A | $2.239 \cdot 10^{12}$ | $5 \cdot 10^8$ | $9.87 \cdot 10^8$ | $1.0729 \cdot 10^{10}$ |
| **No.** | **5** | **6** | **7** | **8** |
| $E_a$ $[\frac{J}{kg \cdot mol}]$ | $6.65 \cdot 10^7$ | $6.65 \cdot 10^7$ | $2.8 \cdot 10^8$ | $2.5 \cdot 10^8$ |
| A | $12.1 \cdot 10^3$ | $8.73 \cdot 10^{14}$ | $1 \cdot 10^{10}$ | $3 \cdot 10^{12}$ |

Spatial geometry has been performed in Autodesk Inventor 2015 software. Subsequently, the geometry has been exported to the ANSYS Workbench, first to the ANSYS DesignModeler module, which allows to perform additional edition of the geometric model. Then a discretisation of the continuous domain has been carried out using ANSYS Meshing. The process of mesh generation has been performed in a way that guarantees the predominance of hexagonal grid elements. The discretisation has been carried out using the "Multizone", the "Hex dominant" and the automatic method. The total number of grid elements was $1.5 \cdot 10^6$. K – epsilon "realizable" model of turbulence has been applied.

Thermal properties of the wall materials have been taken into account by a function which is based on the definition of the material (steel), the wall thickness (8 mm) and the temperature of the operating medium on the other side of the wall. A "Discrete Ordinate" model of radiation, which is based on angular discretisation, has been applied to take into account the spectral properties (absorption coefficient) of molecules consisting of three atoms.

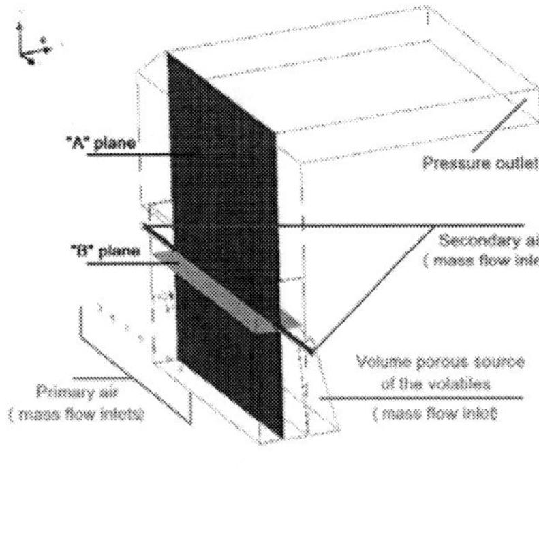

**Figure 4.** Sketch of the computational domain considered in the numerical analysis with a simplified description of the basic boundary conditions and the position of the cross sectional planes ("A" and "B") used in the analysis of the results.

Figure 4 is a sketch of the computational domain and simplified description of boundary conditions. The whole domain has been defined as a continuous fluid, except for the inlet region, which has been defined as a porous body and set as a volume source of the volatiles. This part of the domain represents piece of the straw bale. The gas at the inlet has been defined as a mixture of compounds which are reagents in the previously mentioned reactions.

Nine air inlets have been defined – seven on the rear wall (R1–R7) of the secondary combustion chamber (primary air) and two (D1, D2) on the side walls (secondary air). The air mass flow for each duct has been set based on the results of experimental tests and numerical analysis of the air manifold carried out by the author, which are described in [21].

To include the impact of the chimney draft on the process, negative pressure on the outlet from the computational domain (-20 Pa) has been set. Ten variants of the excess air (1.25 – 3,5) have been analyzed to determine the basic relations between this coefficient and the selected operational parameters of the heating device.

A steady state analysis has been carried out. The residues for the governing equations have been set on the level of $10^{-5}$. A "First order upwind" advection scheme has been applied in the first approach. Then it has been changed to "Second order upwind" in case of the momentum and energy equations in the final computation.

### 3.2 Results of the numerical simulation of the homogenous combustion

Results obtained in the process of computations for the first case of excess air (lambda = 1.1) are fully comparable with the experimental data when it comes to CO concentration (Fig. 5). Due to the fact that the real combustion in the considered stage (stabilization of the combustion just after ignition, in the first few minutes) is strongly unstable, it is impossible to reproduce this process faithfully. The agreement of the numerical and experimental data, however, is fully satisfying.

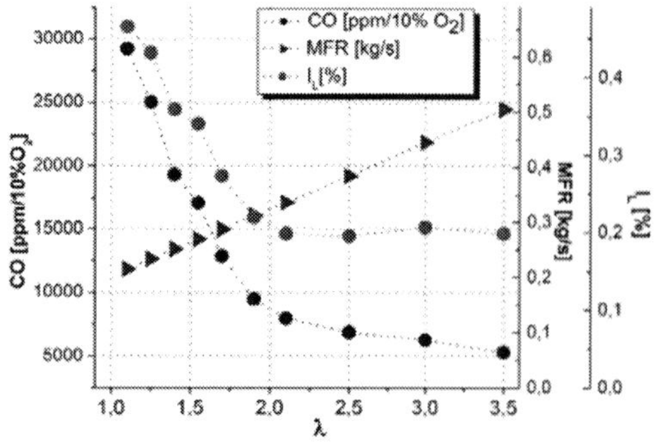


**Figure 5.** CO concentration, mass flow rate of the exhaust (MFR) as well as the loss caused by imperfect combustion as a function of the excess air.

The results of the variant analysis show that it is possible to control the efficiency of the combustion by the modulation of air streams. From the second variant on, the concentration of CO in the exhaust increases. The increase of excess air results in a better access to the oxidizer as well as a higher mixing rate, which is a key issue in case of CO combustion. After the critical point ($\lambda$ = 1.9), the dynamics of decrease of the CO content become significantly lower. It has to be underlined that the concentration of CO in each chart in the paper has been recalculated to 10% of the air concentration in the exhaust.

Apart from CO concentration, Figure 5 presents also the changes of relative amount of energy connected to the losses resulting from imperfect combustion, which is the result of the presence of combustible gases in the exhaust. The value of this parameter behaves analogously to the concentration of CO, due to the fact that this compound has been taken into account as the most significant combustible component of the mixture in the secondary combustion chamber.

The loss resulting from imperfect combustion has been calculated based on the equation below:

$$I_L = \frac{V_{SS} \cdot Q_{CO} \cdot CO}{Q_w}, [\%] \tag{17}$$

Where:

$V_{ss}$ – exhaust volume flow rate $[\frac{m^3}{s}]$

$Q_{CO}$ – calorific value of the CO $[\frac{MJ}{kg}]$

CO – CO content in the exhaust [%]

$Q_w$ – calorific value of the biomass $[\frac{MJ}{kg}]$

The calorific value of the biomass fuel has been calculated using the equation below:

$$Q_w = 339 \cdot C + 1214{,}2\left(H_2 - \frac{O}{8}\right) + 104{,}7S - 25{,}1W, [\tfrac{MJ}{kg}] \tag{18}$$

Where:

C, $H_2$, O, S – content of the compound in the biomass [%]

W – content of moisture in the biomass [%]

The specifics of the changes of the imperfect combustion loss can be explained based on the results presented in Figure 6, which displays the distribution of the CO2 mass fraction on the plane constituting the "B" cross section (see Fig. 4) of the computational domain.

In the reference case (lambda = 1.1), the area of the combustion, which is characterised by a relatively high concentration of CO2, is located in the centre of the chamber. Due to a relatively low mass air flow rate, the velocity of the mixture in the aforementioned section is low and the reactions are occurring between the secondary air nozzles.

Along with the increase of the excess air, the combustion area is moving to the corners of the combustion chamber. In case of $\lambda$ = 1.9 the intensity of the process is the greatest. The sections characterised by the highest $CO_2$ concentration are the widest, despite a much higher oxygen content than in the example "a". In case of excess air amounting to 3.5, the conditions in the chamber undergo deterioration.

The increase of a relatively cold air stream results in the cooling down of the combustion chamber. It has to be said that the concentration of $CO_2$ is directly tied to the temperature at the outlet from the computational domain, which can be noted based on Figure 6 and 7.

In case of the three first variants, the temperature decreases slightly due to the increasing amount of air in the combustion area. A reversal is noted when the excess air reaches 1.55. Between lambda = 1.55 and 1.9 the range of the best conditions is observed. It has to be noted that the excess air values which cause the highest temperatures are tied to the most dynamic reduction of CO.

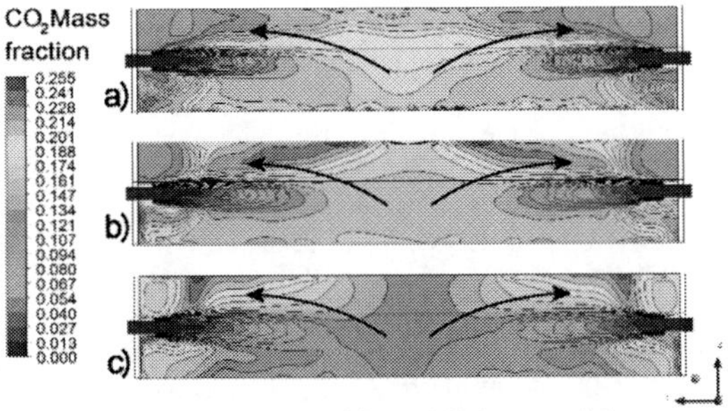


**Figure 6.** CO2 mass fraction (plane "B") in case of the excess air amounting to 1.1 (a), 1.9 (b) and 3.5 (c). The black arrows show the direction of the "movement" of the region of high combustion intensity.

Moreover, relatively low temperatures calculated for excess air greater than 2.1 explain the stabilisation of the level of CO concentration.

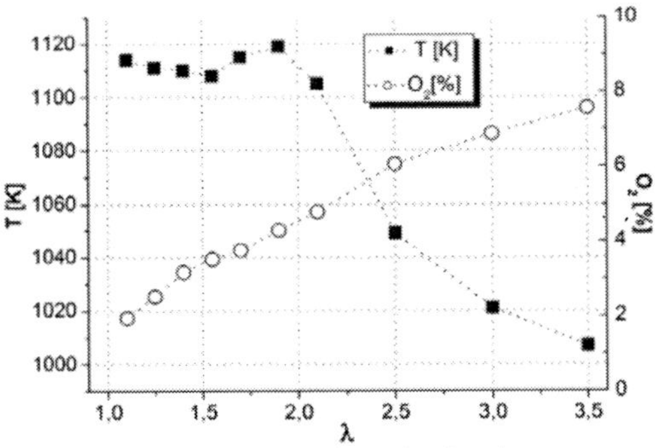


**Figure 7.** Temperature of the exhaust leaving the computational domain and concentration of oxygen as a function of the excess air.

The increase of the velocity of the air streams results in a significant change of the temperature distribution in the secondary combustion chamber. Figure 8 shows the change of location of the regions of high temperature, which corresponds to the increase of the excess air.

Two extreme cases (1.1 and 3.5) as well as the optimal one (1.9) have been presented. Due to the more efficient penetration of the domain by the cold air, the process of combustion in the centre of the chamber is disturbed. The areas of high temperature are moving to the side walls of the combustion chamber.

Comparison of Figure 6c and 9c allows to note that in case of the considered plane, the regions which are characterised by the highest kinetic energy of turbulence are also the coldest areas. Due to this fact, efficient post-combustion of carbon monoxide is very difficult. Based on the discussed results it is possible to conclude that in case of the studied process the decisive factor is temperature.

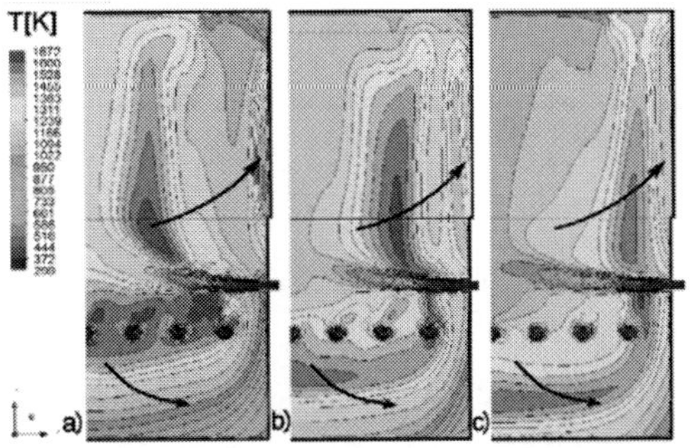


**Figure 8.** Distribution of temperature ("A" plane) in case of the excess air amounting to 1.1 (a), 1.9 (b) and 3.5 (c). The black arrows show the direction of "movement" of the region of high combustion intensity.

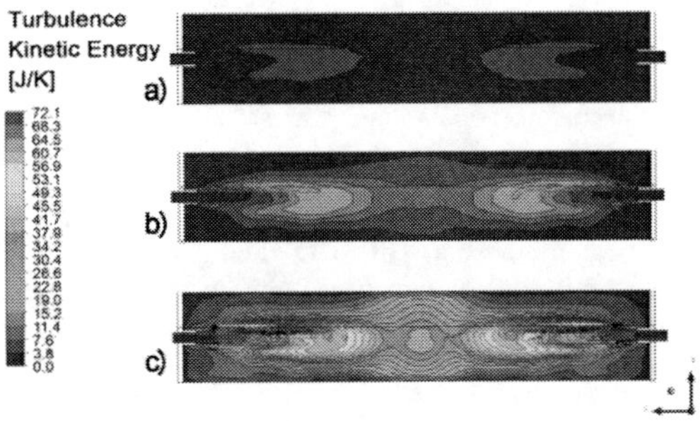


**Figure 9.** Kinetic energy of turbulence ("B" plane) in case of the excess air amounting to 1.1 (a), 1.9 (b) and 3.5 (c).

The level of turbulences has a positive impact on the combustion efficiency but it is necessary to avoid excessive temperature reduction. Due to this fact, it is recommended to provide efficient mixing by appropriate design of the nozzles instead increasing of the inlet air velocity.

However, the higher the temperature of the gases leaving the combustion chamber, the greater is the amount of energy that may be lost due to stack loss. Based on the results of the numerical simulation it was possible to estimate the amount of energy which is carried out of the secondary combustion chamber by the exhaust stream (Fig. 9). It was calculated using the equation below.

$$E_L = \sigma \cdot \frac{(T_1-T_2)+0{,}59\cdot CO}{CO_2+CO}, [\%] \qquad (19)$$

Where:

σ – Siegert constant (calculated based on the content of moisture in the fuel and the average $CO_2$ emissions during the combustion)
$T_1$ – exhaust temperature [K]
$T_2$ – ambient temperature [K]
CO, $CO_2$ – content of the compound in the exhaust [%]

As it can be seen in Figure 10, the exhaust loss increases with the increase of the excess air and in last case it reaches ~50%. Two factors determine the permanent trend of the exhaust loss curve. Up to λ = 1.9 the increase of the loss is the result of the increasing difference between gas temperature and ambient temperature. Based on equation (19) it can be noted that in case of greater values of the excess air, $CO_2$ concentration is the key factor determining the value of energy loss.

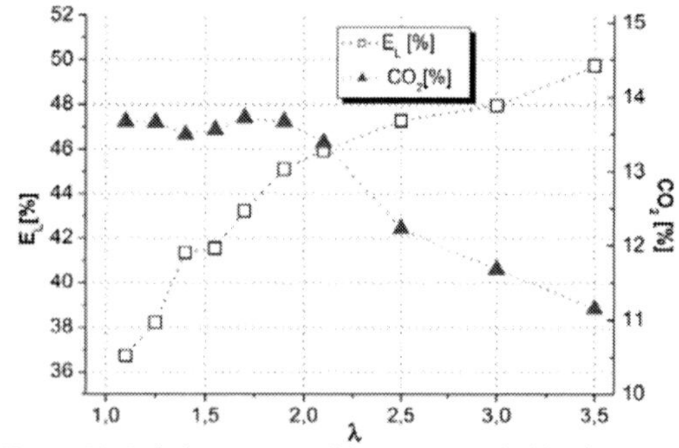


**Figure 10.** Relative amount of the energy carried by the exhaust and concentration of $CO_2$ in the exhaust as a function of the excess air.

The concentration of $CO_2$ calculated at the outlet of the secondary combustion chamber decreases along with the post-combustion of CO and consequently the production of $CO_2$ increases with the increase of the excess air. It is connected to a dynamic increase of $O_2$ concentration in the region of the highest intensity of the chemical reactions.

### 3.3 Development of the numerical model for the heterogeneous combustion

The second step of the research described in this paper was inclusion of the transport of the solid phase in the previous numerical model of the homogenous combustion. Moreover, a simplified model of particle matter combustion has been developed.

A "Discrete Phase Model" (DPM) available in the ANSYS Fluent Solver has been applied to describe the transport of the particles lifted by the gases in the secondary combustion chamber from the fuel bed. The first approach assumed the definition of twenty point sources (so called "injectors") of discrete phase particles. The injectors have been located in three parallel rows characterized by different distances from the floor of the combustion chamber. Sources were situated in the area of the porous domain, which represents a piece of the straw bale.

The discrete phase was set as interacting with the continuous phase. The physical properties of the individual particles have been defined based on the values characteristic for the straw. A "kinetics/diffiuson-limited" model of combustion has been implemented. Due to the simplified approach, a uniform diameter of the particle amounting to 1um has been assumed.

The previous set of reactions has been expanded. Based on tools available in ANSYS Fluent as well as the description of the elementary reactions, an additional equation describing the combustion of the volatiles released from the solid particles has been developed:

$$volatiles + 0.622O_2 \xrightarrow{E_{a9},A_9} 0.924CO + 1.316H_2O + 0.0085N_2 + 0.001SO_2 \quad (20)$$

Where:
$E_{a9} = 2.027 \cdot 10^9 \; [\frac{J}{kg \cdot mol}]$
$A_9 = 2.119 \cdot 10^{11}$

It has to be noted that the thermochemical conversion of the solid particles has been considered as wet combustion. Due to this fact, the liquid fraction (0.04169) has been defined in case of each injector. However, some mechanisms, such as turbulent dispersion, have been skipped at the current stage of the model development.

Two cases of discrete phase mass flow rate have been considered. Mass flow rates were calculated based on the assumption that in the first case, the chemical energy of the total discrete phase stream constitutes 1% of the chemical energy of the gasification products ($E_{DPM}$ = 1%), while in the second case it is 10% ($E_{DPM}$ = 10%).

The chemical energy of the discrete phase particles has been calculated based on equation 18. The chemical energy of the gasification products ($Q_g$) has been determined using the following equation:

$$Q_g = \dot{m}_f \cdot (\textstyle\sum_{i=1}^{n} x_i \cdot Q_i) \quad (21)$$

Where:
$\dot{m}_f$ – mass flow rate of the gasification products $[\frac{kg}{s}]$
$x_i$ – mass fraction of the combustible component
$Q_i$ – calorific value of the combustible component $[\frac{MJ}{kg \cdot K}]$

The mass flow rate of the discrete phase, which corresponds to 1% and 10% of the gas mixture energy, has been calculated using appropriate proportions and it amounted to $2{,}86 \cdot 10^{-4}$ kg/s and $2{,}86 \cdot 10^{-3}$ kg/s, respectively. Based on the results of studies of homogeneous combustion it was decided that the reference case with lambda = 1.1 as well as three promising cases of the excess air: 1.55, 1.7 and 1.9 should be considered in the further simulations.

### 3.4 Results of the numerical simulation of the heterogeneous combustion

As the first step of the investigation of the impact of the presence of the discrete phase on the combustion in the secondary combustion chamber, the $CO_2$ concentration has been analysed. Figures 11 and 12 show the distribution of the component depending on the percentage of the chemical energy of the discrete phase (EDPM) and excess air.

Based on comparison with the areas of high temperature in case of homogenous combustion which were presented in Figure 8 it can be noted that the gasification of the particulate matter occurs in a higher section of the chamber. The effect of the expansion of the areas of high concentration of $CO_2$ with the increase of the mass flow rate of the discrete phase is directly connected to the combustion of carbon monoxide released from the solid particles.

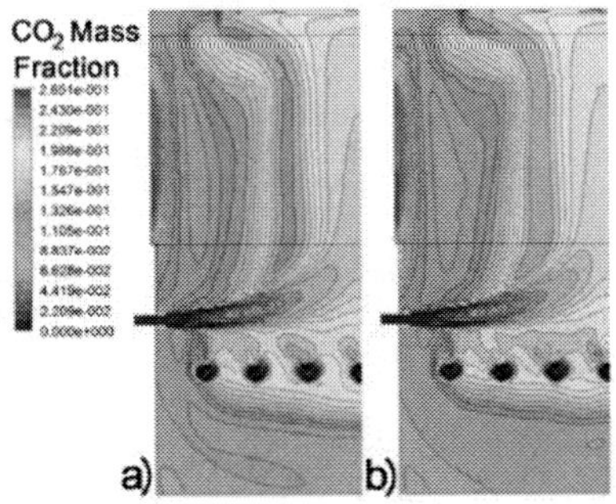


**Figure 11.** Distribution of $CO_2$ ("A" plane) in the secondary combustion chamber: a) $\lambda = 1.1$, $E_{DPM} = 1\%$, b) $\lambda = 1.1$, $E_{DPM} = 10\%$.

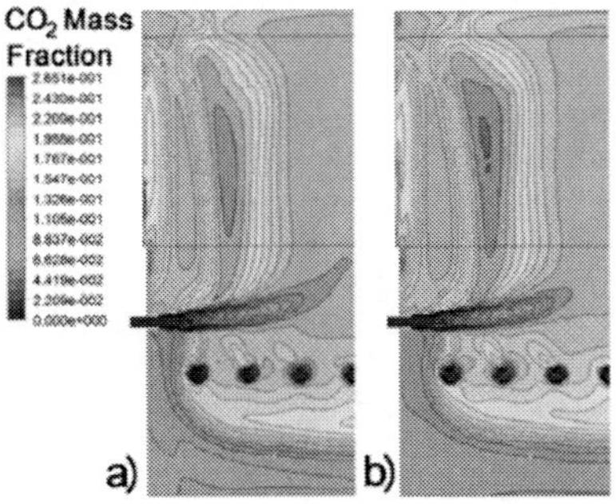


**Figure 12.** Distribution of $CO_2$ ("A" plane) in the secondary combustion chamber: a) $\lambda = 1.9$, $E_{DPM} = 1\%$, b) $\lambda = 1.9$, $E_{DPM} = 10\%$.

The inclusion of the gasification process in the area of the secondary combustion chamber requires the consumption of a certain amount of energy, so it has to influence the heat balance of the reaction. However, the exhaust temperature calculated for the two considered variants of the discrete phase content (Figure 13) is fully comparable with the results obtained in the analysis of the homogenous combustion. It can be concluded that the thermal effect of the gasification is compensated by the effect of combustion of the released CO.

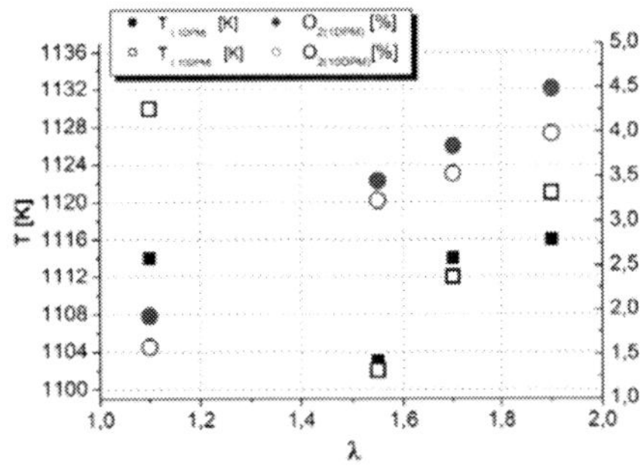


**Figure 13.** Temperature of the exhaust leaving the computational domain and concentration of oxygen in case of $E_{DPM} = 1\%$ and $E_{DPM} = 10\%$ as a function of the excess air.

The exhaust temperature was again directly related to the excess air. In the whole range of the analysed values of the excess air, the mole fraction of oxygen is lower for variants including higher content of the discrete phase in the chamber. The combustion of the released CO becomes effective from the point where the excess air amounts to 1.55, which results in an increased temperature at the outlet of the computational domain.

Figure 14 presents the impact of the particulate matter content and excess air value on the phenomena occurring at the height of the secondary air nozzles. The effect of the expansion of the intensive combustion regions is visible here as well. However, it is significant in case of lambda = 1.1 due to the influence of the relatively cold air on the temperature at the centre of the considered section of the chamber. In each variant, the gas phase reactions are critical, which results in the influence of the excess air on the process analogous to the one taking place in case of homogeneous combustion.

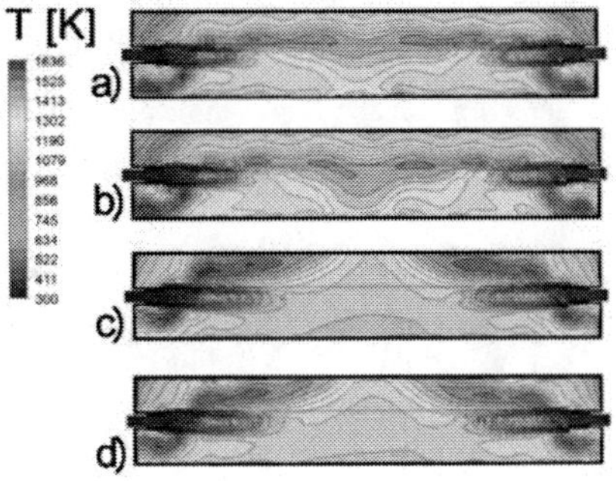


**Figure 14.** Distribution of temperature ("B" plane): a) $\lambda = 1.1$, $E_{DPM} = 1\%$, b) $\lambda = 1.1$, $E_{DPM} = 10\%$, c) $\lambda = 1.9$, $E_{DPM} = 1\%$, d) $\lambda = 1.9$, $E_{DPM} = 10\%$.

The values of exhaust loss (Figure 15) determined both for $E_{DPM} = 1\%$ and $E_{DPM} = 10\%$ are slightly lower than for the homogeneous combustion, except for lambda = 1.1. The values calculated for the incomplete combustion loss in case of $E_{DPM} = 10\%$ are the highest from among all the cases considered in the described research. It is the result of the release of a relatively big amount of CO in the higher sections of the combustion chamber.

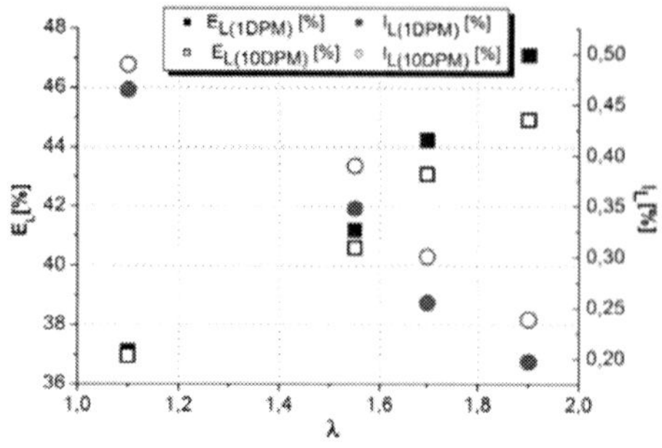


**Figure 15.** Relative amount of the energy carried by the exhaust ($E_L$) and loss resulting from imperfect combustion ($I_L$) as a function of the excess air in case of $E_{DPM} = 1\%$ and $E_{DPM} = 10\%$.

Due to the location of the secondary air nozzles on the side walls of the chamber, during the flow to the outlet solid particles are pushed from the centre of the domain to the walls. This is one of the significant reasons for changes of the range of the combustion region in these parts of the computational domain.

However, the movement of the discrete phase elements is relatively undisturbed outside the region of the contact of streams blown from the secondary air nozzles. Due to this fact, it is recommended to implement modified nozzles which would allow for the prolongation of the path of particles to the outlet by means of intensifying the turbulences [1].

Moreover, it has to be taken into account that the implementation of the distribution of the diameter of the particulate matter influences the flow parameters of the particles and consequently results in specific tracks of the discrete phase elements. However, the research presented in this paper was focused primarily on the impact of the presence of discrete phase on the changes of conditions in the secondary combustion section of the boiler.

## 4 Summary and conclusions

Based on the numerical modelling which has been carried out using the ANSYS Fluent solver it was discovered that an increase of the excess air in case of combustion occurring in the secondary chamber of the straw-fired batch boiler leads to a significant reduction of CO emissions. However, an increase of the exhaust stream resulting from the higher excess air causes higher levels of losses. The results of the conducted computations allow to conclude that in the examined boiler type it is recommended to apply excess air on the level between 1.55 and 1.9, which is in general agreement with the recommendations of the producer (λ = 1.25 - 1.75). The implementation of a discrete phase model results in changes in temperature distribution as well as different local concentrations of important products of chemical reactions. However, in case of a relatively low amount of particulate matter applied in the model, no significant changes in heat balance are observed. The inclusion of a discrete phase model in the study of the operation of biomass-fired batch boilers is necessary when studying issues related to the reduction of the particulate matter emissions. Furthermore, the gasification of the solid particles lifted up from the fuel bed may be considered an important mechanism of CO production. Due to the combustion dynamics in case of batch heating units it is recommended to apply a dynamic control of the air feed, based on measurements of the exhaust temperature, excess air and a selected compounds such as CO or $O_2$.

This study was carried out under the "BioORC: Construction of cogeneration system with small to medium size biomass boilers" project.

## References

1. M. Szubel, EFM 2015: Proceedings of the International Conference, Prague, Czech Republic
2. M. Miltner, A. Makaruk, M. Harasek, A. Friedl, Cl. Tech. and Env. Pol., **5** (2008)
3. B. Sørensen, *Renewable Energy. Its physics, engineering, environmental impacts, economics & planning* (Academic Press, Boston, 2000)
4. L. Rosendahl, *Biomass combustion science, technology and engineering* (Woodhead Publishing, Oxford 2013)
5. T. B. Johansson, H. Kelly, A. K. N. Reddy, R. H. Williams, *Renewable Energy Sources for Fuels and Electricity* (Island Press, Washington 1993)
6. S. van Loo, J. Coppejan, *The handbook of biomass combustion and co-firing* (EARTHSCAN, London 2008)
7. P. Basu, *Biomass gasification and pyrolysis practical design* (Academic Press, Boston, 2010)
8. S. R. Turns, *An introduction to combustion. Concepts and applications*, (McGraw-Hill, Boston 2000)
9. I. Glassman, *Combustion* (Academic Press, Orlando 1987)
10. R. A. Yetter, F. L. Dryer, H. Rabitz, *A comprehensive reaction mechanism for carbon monoxide/hydrogen/oxygen kinetics*, Comb. Sc. and Tech., **79** (1981)
11. D. J. Hautman, F. L. Dryer, K. P. Schug, I. A. Glassman, *A multiple step overall kinetic mechanism for the oxidation of hydrocarbons*, Comb. Sc. And Tech. **25** (1981)
12. C. K. Westbrook, F. L. Dryer, *Simplified reaction mechanisms for the oxidation of hydrocarbon fuels in flames*, Comb. Sc. And Tech., **27** (1981)
13. H. K. Versteeg, W. Malalasekera, *An Introduction to Computational Fluid Dynamics. The Finite Volume Method* (Pearson Education Limited, Harlow, 2007)
14. L. Chaoqun, T. Jiyuan, Y. Guan-Heng, *Computational Fluid Dynamics. A Practical Approach* (Butterworth-Heinemann, Oxford 2013)

15. A. Bianchini, F. Cento, L. Golfera, M. Pellegrini, C. Saccani, *Performance analysis of different scrubber systems for removal of particulate emission from a small size biomass boiler*, Biom. and Bioen. **92** (2016)
16. Q. Xue, T. J. Heindel, R. O. Fox, *A CFD model for biomass fast pyrolysis in fluidized-bed reactors*, Chem. Eng. Sc., **66** (2011)
17. J. Blondeau, H. Jeanmart, *Biomass pyrolysis in pulverized-fuel boilers: Derivation of apparent kinetic parameters for inclusion in CFD codes*, Proc. of Com. Inst., **33** (2011)
18. R. Baczyński, R. Weber, A. Szlęk, *Innovative design solution for small-scale domestic boilers: Combustion improvements using a CFD-based mathematical model*, Journ. of En. Inst., **88** (2015)
19. J. Chaney, H. Liu, L. Li, *An overview of CFD modelling of small-scale fixed-bed biomass pellet boilers with preliminary results from simplified approach*, En. Conv. and Man., **63** (2012)
20. J. Collazo, J. Porteiro, J.L. Miguez, E. Granada, M.A. Gómez, *Numerical Simulation of a small-scale biomass boiler*, En. Conv. and Man., **64** (2012)
21. M. Szubel, T. Siwek, O*ptimization of the operation of energy devices on the example of an air distribution system inside the biomass boiler*, Tech. Trans. **19** (2014)

# Determination of flow-rate characteristics and parameters of piezo pilot valves

*Jakub* Takosoglu[1,*], *Paweł* Laski[2], *Slawomir* Blasiak[1], *Gabriel* Bracha[2], *Dawid* Pietrala[2], *Jaroslaw* Zwierzchowski[2], *Lukas* Nowakowski[1]

[1]Faculty of Mechatronics and Mechanical Engineering, Department of Manufacturing Engineering and Metrology, Kielce University of Technology, Aleja Tysiaclecia Panstwa Polskiego 7, 25-314 Kielce, Poland
[2]Faculty of Mechatronics and Mechanical Engineering, Department of Automation and Robotics, Kielce University of Technology, Aleja Tysiaclecia Panstwa Polskiego 7, 25-314 Kielce, Poland

**Abstract.** Pneumatic directional valves are used in most industrial pneumatic systems. Most of them are two-stage valves controlled by a pilot valve. Pilot valves are often chosen randomly. Experimental studies in order to determine the flow-rate characteristics and parameters of pilot valves were not conducted. The paper presents experimental research of two piezo pilot valves.

## 1 Introduction

Pneumatically driven devices are used almost in every industry (medical, transport, aviation [1-4], automotive, light and heavy industries, mining, metallurgy, CNC [5-9], robotics [10-16]) as well as in our everyday lives [17-20]. In order to control pneumatic drives, mainly pneumatic directional valves are used [21, 22], which can be: mechanically, manually, pneumatically and electro-magnetically operated. Due to the dynamic development of electronics [23] and programmable controlling [24, 25], production processes mainly use PLCs [26, 27]. Hence, pneumatic directional valves electromagnetically operated are widely applied in the pneumatic devices. Pneumatic directional valves are built of two primary components:

- pneumatic distributor, which is responsible for the subsequent stage (action) of the valve,
- pilot valve(s), which is responsible for the initial stage (action) of the valve [28].

The pilot valves of the directional valves are usually selected randomly. Catalogues of companies of the pneumatic industry do not include the parameters of pilot valves as such, only parameters of a complete set - the distributor and the pilot valve constituting the directional valve. Two piezo pilots [29-31] by Hoerbiger-Origa: P20 381RF-NG-CPN (Fig. 1) and P8 385RF-NG-SPN61-B (Fig. 2) are the subject of the research. The parameters thereof are shown in Table 1 and Table 2 respectively. Those elements were selected for research due to the possibility to compare the data provided by the producer in the catalogue with the results obtained.

**Table 1.** Parameters of P20 381RF-NG-CPN pilot valve.

| Parameters | Value |
|---|---|
| Nominal size | 2 mm |
| Valve function | 3/2-way, normally closed |
| Nominal pressure | 0.6 MPa |
| Operating pressure range | 0.12-0.8 MPa |
| Critical pressure ratio $b$ | 0.33 |
| Sonic conductance $C$ | $4.42 \cdot 10^{-9}$ $m^3/(s \cdot Pa)$ |
| Nominal flow rate | $18.33 \cdot 10^{-4}$ $m^3/s$ |

a)    b)

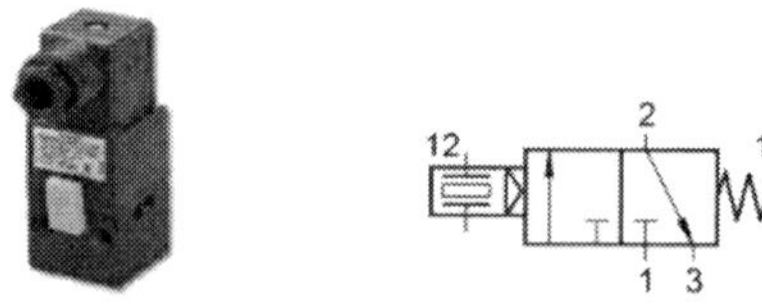


**Fig. 1.** P20 381RF-NG-CPN: a) general view, b) symbol

**Table 2.** Parameters of P8 385RF-NG-SPN61-B pilot valve.

| Parameters | Value |
|---|---|
| Nominal size | 1.6 mm |
| Valve function | 3/2-way, normally closed |
| Nominal pressure | 0.6 MPa |
| Operating pressure range | 0.12-0.8 MPa |
| Critical pressure ratio $b$ | 0.34 |
| Sonic conductance $C$ | $2.08 \cdot 10^{-9}$ $m^3/(s \cdot Pa)$ |
| Nominal flow rate | $8.33 \cdot 10^{-4}$ $m^3/s$ |

a)    b)

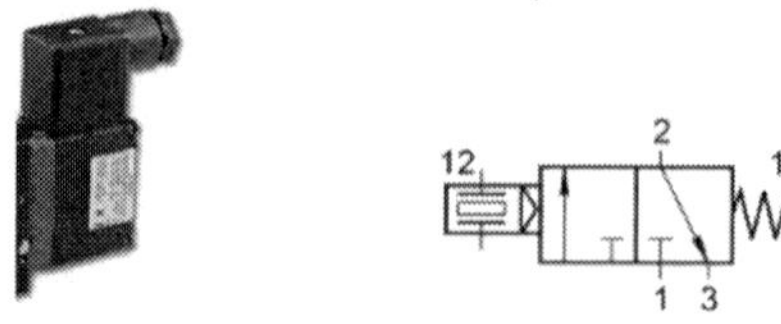


**Fig. 2.** P8 385RF-NG-SPN61-B: a) general view, b) symbol

* Corresponding author: qba@tu.kielce.pl

## 2 Flow parameters

Pneumatic systems via working medium, most often compressed air, host the exchange of energy and control signals [32-35]. Pneumatic devices are characterized by resistance [36-38], which influences the flow rate of the compressed air, and hence, its flow intensity. Those characteristics are described by flow parameters, which can be determined on the basis of experiments [39-41]. The method of studying pneumatic components that host the exchange of compressible gases is described by ISO 6358 [42].

### 2.1 Choked flow

Critical flow is a state, in which the gas flow rate in a certain area of the component equals the local speed of sound. It occurs when inlet pressure $p_1$ is high enough in relation to outlet pressure $p_2$. Massive gas flow intensity is then proportional to inlet pressure $p_1$ and inversely proportional to the square root of the temperature at inlet $T_1$, and does not depend on outlet pressure $p_2$:

$$q_m^* = C \cdot p_1 \cdot \rho_0 \sqrt{\frac{T_0}{T_1}} \; for \; \frac{p_2}{p_1} \leq b \qquad (1)$$

$$q_m = C \cdot p_1 \cdot \rho_0 \sqrt{\frac{T_0}{T_1}} \sqrt{1 - \left(\frac{\frac{p_2}{p_1} - b}{1-b}\right)^2} \; for \; \frac{p_2}{p_1} > b \qquad (2)$$

where:
$q_m^*$ – mass flow intensity under choked flow conditions,
$q_m$ – mass flow intensity under subsonic flow,
$p_1$ – inlet gas pressure,
$p_2$ – outlet gas pressure,
$\rho_0$ – gas density,
$C$ – sonic conductance,
$b$ – critical pressure ratio,
$T_0$ – normalized reference temperature,
$T_1$ – inlet temperature.

### 2.2 Critical pressure ratio

Critical pressure ratio $b$ is the highest value of $p_2/p_1$ pressure ratio, which triggers choked flow in the tested component [42]. If the inlet temperature is constant during research, then value $b$ for each obtained value $q_m$ in terms of subsonic flow [43, 44] is calculated using the formula presented below:

$$b = 1 - \frac{\frac{\Delta p}{p_1}}{1 - \sqrt{1 - \left(\frac{q_m}{q_m^*}\right)^2}} \qquad (3)$$

where:
$\Delta p$ – pressure drop.
If during the research, a change in $p_1$ and $T_1$, occurs, then for each value $q_m$ received in terms of subcritical flow, $b$ is calculated on the basis of the following formula:

$$b = 1 - \frac{\frac{\Delta p}{p_1}}{1 - \sqrt{1 - \left(\frac{q_m}{C \rho_0 p_1} \sqrt{\frac{T_1}{T_0}}\right)^2}} \qquad (4)$$

### 2.3 Sonic conductance

Sonic conductance $C$ is the ratio of mass gas flow intensity $q_m^*$ through the component, to the product of inlet pressure $p_1^*$ and the density of that gas $\rho_0$ in normalized reference atmosphere conditions during choked flow:

$$C = \frac{q_m^*}{p_1^* \cdot \rho_0} \sqrt{\frac{T_1^*}{T_0}} \qquad (5)$$

where:
$p_1^*$ – inlet gas pressure under choked flow conditions,
$T_1^*$ – inlet temperature under choked flow conditions.

The symbols and units shall be applied in accordance with Table 3. The numerals used in subscripts or superscripts or marked with an asterisk (*) with regard to the symbols listed in Table 2 shall be used as specified in Table 4.

**Table 3.** Symbols and units.

| Description | Symbol | SI unit |
|---|---|---|
| Critical pressure ratio | $b$ | - |
| Sonic conductance | $C$ | $m^3/(s{\cdot}Pa)$ |
| Absolute stagnation pressure | $p$ | Pa |
| Mass flow rate | $q_m$ | kg/s |
| Volume flow rate at standard reference atmosphere | $q_v$ | $m^3/s$ (ANR) |
| Gas constant (for a perfect gas) | $R$ | J/(kg·K) |
| Absolute stagnation temperature | $T$ | K |
| Mass density | $\rho$ | $kg/m^3$ |

**Table 4.** Subscripts and superscripts.

| Superscript | Subscript | Meaning |
|---|---|---|
| | 0 | Conditions of standard reference atmosphere defined in ISO 8778, i.e.: $T_0$=293.15 K $p_0$=0.1 MPa $\rho_0$=1.185 kg/m$^3$ 65% relative humidity RH |
| | 1 | Upstream conditions |
| | 2 | Downstream conditions |
| * | | Conditions during choked flow tests |

If a volumetric flow meter was used for the measurements, and the volumetric flow intensity is to be used, then the formulae (1), (2), (3), (4), (5) will take the following form:

$$q_m = \rho \cdot q_v \qquad (6)$$

$$b = 1 - \frac{\frac{\Delta p}{p_1}}{1 - \sqrt{1 - \left(\frac{q_v}{q_v^*}\right)^2}} \qquad (7)$$

$$b = 1 - \frac{\frac{\Delta p}{p_1}}{1 - \sqrt{1 - \left(\frac{q_v}{C p_1} \sqrt{\frac{T_1}{T_0}}\right)^2}} \qquad (8)$$

$$C = \frac{q_v^*}{p_1^*}\sqrt{\frac{T_1^*}{T_0}} \tag{9}$$

## 3 Test stand

The test stand used to determine the flow rate characteristics of piezo pilot valves is shown in Fig. 3, whereas the overview of the stand in Fig.4.

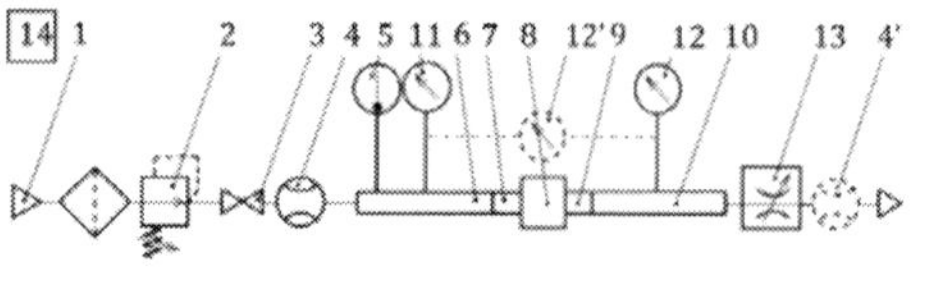


**Fig. 3.** The diagram of the test stand applied to determine the flow rate characteristics: 1 – compressed gas source and filter, 2 – adjustable pressure regulator, 3 – shut-off valve, 4 – flow-rate measuring device, 5 – temperature-measuring instrument, 6 – upstream pressure-measuring tube, 7 – upstream transition connector, 8 – component under test, 9 – downstream transition connection, 10 – downstream pressure-measuring tube, 11 – upstream pressure gauge or transducer, 12 – downstream pressure gauge or transducer, 13 – flow control valve, 14 – barometer.

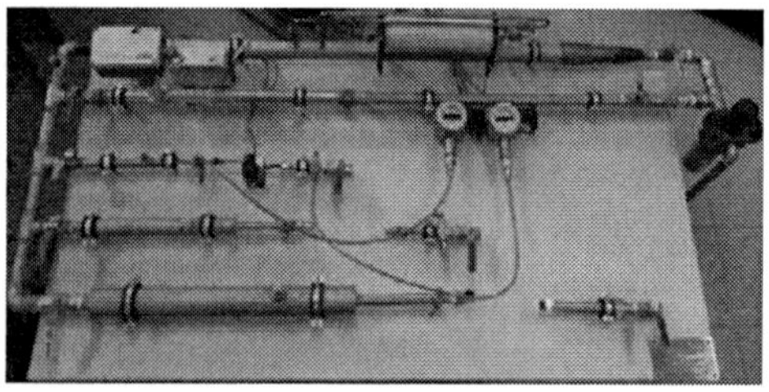

**Fig. 4.** General view of the test stand

Flow characteristic $p_2=f(q)$ of the component is the dependency of the overpressure behind component $p_2$, in the flow function of volume or weight $q$, under constant overpressure in front of component $p_1$. Company catalogues present it in a graphical form, primarily with $p_1$=0.6MPa or $p_1$=0.63MPa. Such characteristic can be obtained by maintaining constant overpressure in front of the component with the use of a reducing valve and changing the flow behind the component with the use of a ball valve. The individual characteristic is presented also as the dependency of pressure drop $\Delta p=p_1-p_2$ in component $\Delta p=f(q)$. Other characteristics are also common, such as $p_2/p_1=f(q)$ and $q=f(p_2/p_1)$. The family of the ordinary flow-rate characteristics is comprised of characteristics $p_2=f(q,p_1)$ obtained under different constant values of overpressure $p_1$ in front of the component. These values are usually taken from a standardized chain: 0.16; 0.25; 0.4; 0.63; 0.8; 1.0; 1.2 MPa. The family of the characteristics is also presented in $\Delta p\%=f(q,p_1)$, where $\Delta p\%=((p_1-p_2)/p_1)\cdot 100\%$. Similarly to the individual characteristic, the families of the ordinary characteristics are shown graphically as well.

## 4 Experimental research results

The experimental research was conducted under constant inlet pressure $p_1$=const and constant temperature $T_0=T_1$= const. Fig. 5 and 6 show a graphical interpretation [45] of the results of the experimental research on piezo pilot valves concerning the determination of critical pressure ratio $b$ and sonic conductance $C$. The charts in Fig.7-10 show different flow rate characteristics of the pilots. The charts in Fig. 11-12 present characteristics $C=f(p_2/p_1)$ of sonic conductance $C$ in the function of pressure ratio $p_2/p_1$.

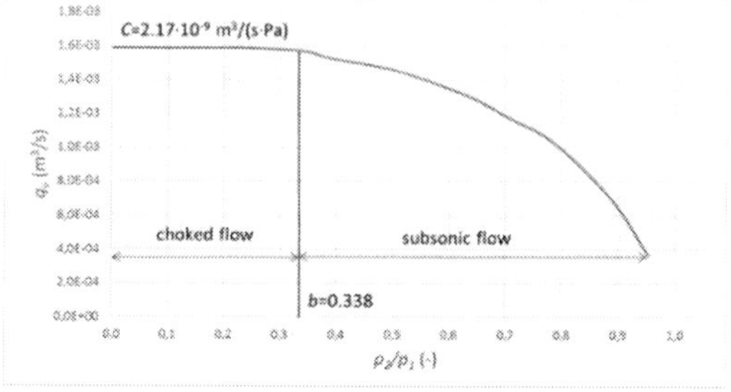


**Fig. 5.** Graphical interpretation of critical pressure ratio $b$ and sonic conductance $C$ of P8 385RF-NG-SPN61-B pilot.

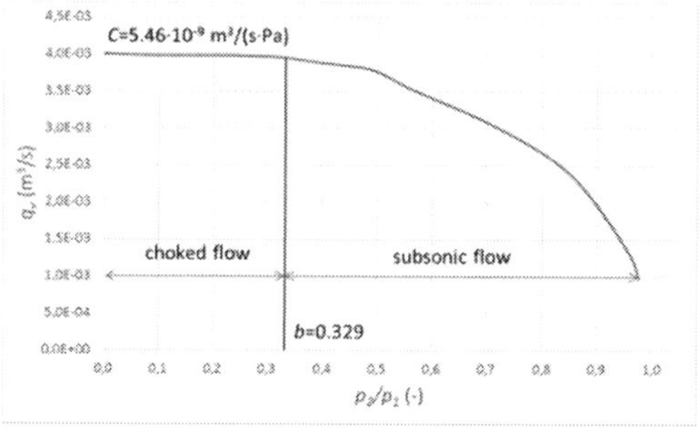


**Fig. 6.** Graphical interpretation of critical pressure ratio $b$ and sonic conductance $C$ of P20 381RF-NG-CPN pilot.

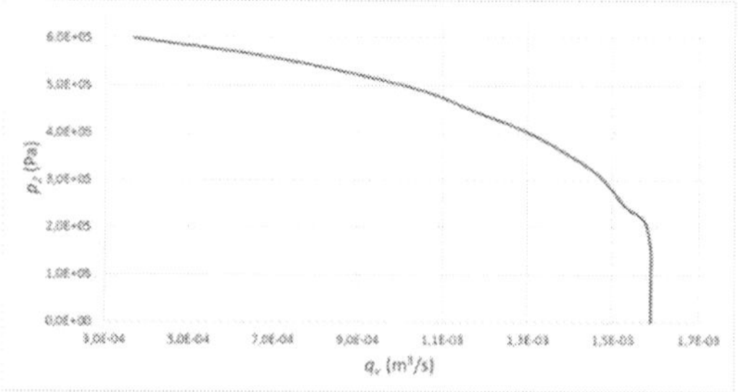


**Fig. 7.** Flow-rate characteristic $p_2=f(q_v)$ of P8 385RF-NG-SPN61-B pilot.

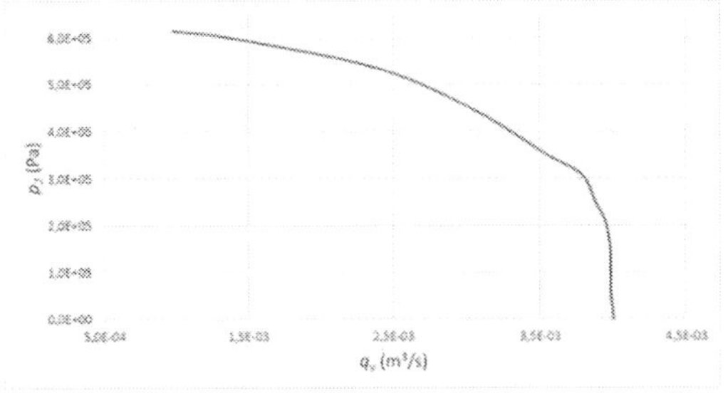

**Fig. 8.** Flow-rate characteristic $p_2=f(q_v)$ of P20 381RF-NG-CPN pilot.

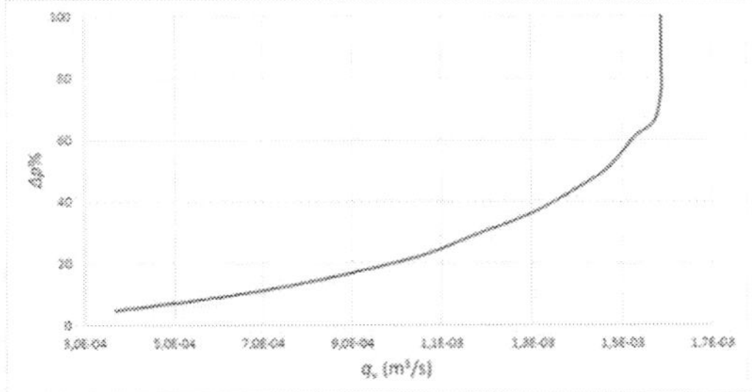

**Fig. 9.** Flow-rate characteristic $\Delta p\%=f(q_v)$ of P8 385RF-NG-SPN61-B pilot.

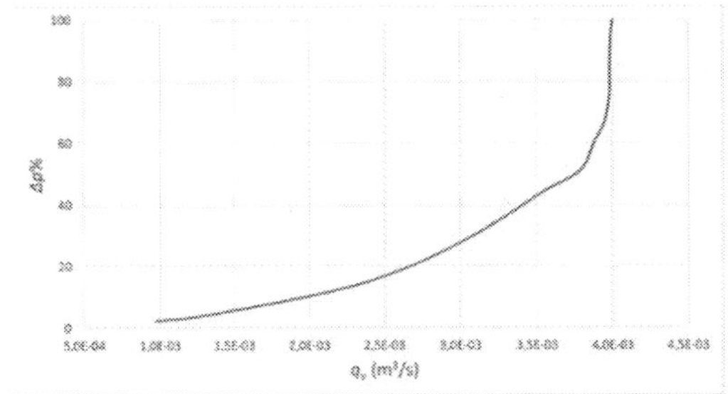

**Fig. 10.** Flow-rate characteristic $\Delta p\%=f(q_v)$ of P20 381RF-NG-CPN pilot.

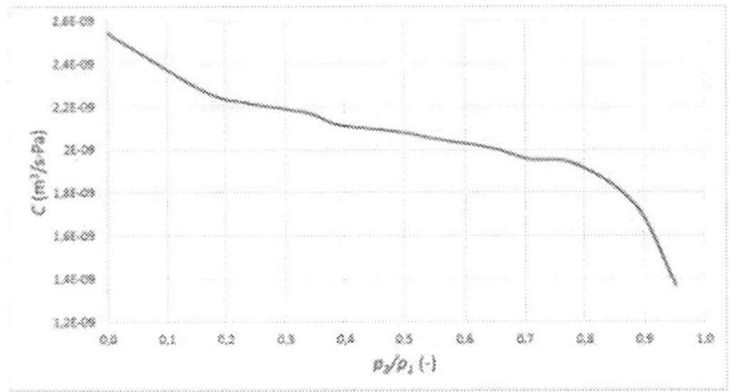

**Fig. 11.** Characteristic $C=f(p_2/p_1)$ of P8 385RF-NG-SPN61-B pilot.

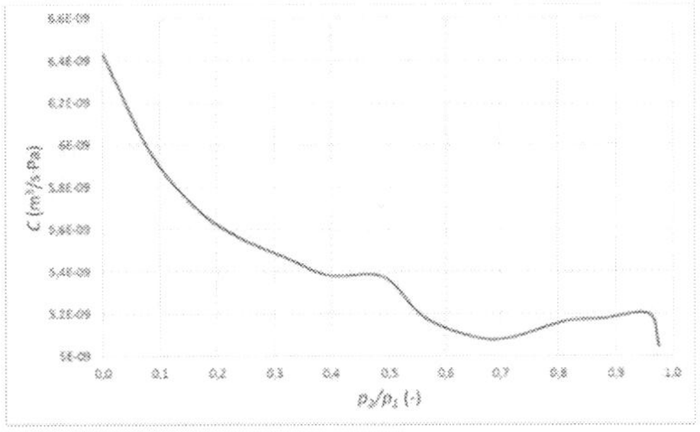

**Fig. 12.** Characteristic $C=f(p_2/p_1)$ of P20 381RF-NG-CPN pilot.

On the basis of formulae (7) and (8), the ratio of critical pressure $b$ under subcritical flow conditions as well as sonic conductance $C$ under choked flow conditions were calculated. The results were average values out of four measurements conducted in accordance with ISO 6358. Table 5 shows the comparison of the experimental research results made on the basis of two methods, and the catalogue data provided by the producer (Table 1 and 2).

**Table 5.** Comparison of the results.

| | I method | | II method | | producer's data | |
|---|---|---|---|---|---|---|
| | $b$ | $C\cdot10^{-9}$ | $b$ | $C\cdot10^{-9}$ | $b$ | $C\cdot10^{-9}$ |
| P8 pilot | 0.338 | 2.17 | 0.283 | 1.91 | 0.34 | 2.08 |
| P20 pilot | 0.329 | 5.46 | 0.342 | 5.19 | 0.33 | 4.42 |

## 5 Summary

The article presents the methods and the test stand to determine the flow-rate characteristics and parameters of piezo pilot valves. Flow-rate parameters $b$ and $C$ can be calculated on the basis of experimental research with the use of two methods. The selection of the method depends on the possibility to achieve the critical flow rate in the studied component. On the basis of the experimental research conducted, both methods and the data provided by the producer (Table 5) were compared. As it can be seen, the results achieved are quite similar to each other and to the data provided by the producer of the piezo pilots.

## References

1. D. Gapinski, Z. Koruba, I. Krzysztofik, Mech. Syst. Signal. Pr. **45,** 2 (2014)
2. Z. Koruba, Z. Dziopa, I. Krzysztofik, J. Theor. Appl. Mec. **48,** 1 (2010)
3. I. Krzysztofik, Solid State Phenom. **180** (2012)
4. I. Krzysztofik, Z. Koruba, J. Automat. Inform. Sciences **44**, 5 (2012)
5. E. Miko, L. Nowakowski, L. Procedia Eng. **39** (2012a)
6. E. Miko, L. Nowakowski, L. Procedia Eng. **39** (2012b)

7. L. Nowakowski, M. Miesikowska, M. Blasiak, Edited by Zolotarev I., Radolf V. *Proceedings of 22nd International Conference on Engineering Mechanics 2016* (Academy of Sciences of the Czech Republic, 2016)
8. L. Nowakowski, E. Miko, M. Skrzyniarz, Edited by Zolotarev I., Radolf V. *Proceedings of 22nd International Conference on Engineering Mechanics 2016* (Academy of Sciences of the Czech Republic, 2016)
9. L. Nowakowski, M. Wijas, Edited by Zolotarev I., Radolf V. *Proceedings of 22nd International Conference on Engineering Mechanics 2016* (Academy of Sciences of the Czech Republic, 2016)
10. R. Trochimczuk, M. Gawrysiak, Solid State Phenom. **147-149** (2009)
11. R. Trochimczuk, Solid State Phenom. (2013)
12. R. Trochimczuk, Int. J. Appl. Mec. Eng. **19**, 9 (2014a)
13. R. Trochimczuk, T. Kuźmierowski, Int. J. Appl. Mec. Eng. **19**, 9 (2014b)
14. D.S. Pietrala, Edited by Zolotarev I., Radolf V. *Proceedings of 22nd International Conference on Engineering Mechanics 2016* (Academy of Sciences of the Czech Republic, 2016)
15. J. Zwierzchowski, Edited by Zolotarev I., Radolf V. *Proceedings of 22nd International Conference on Engineering Mechanics 2016* (Academy of Sciences of the Czech Republic, 2016)
16. P.A. Laski, J.E. Takosoglu, S. Blasiak, Edited by Fuis V. *Proceedings of 20th International Conference on Engineering Mechanics 2014* (Brno University of Technology, Czech Republic, 2014)
17. J.E. Takosoglu, R.F. Dindorf, P.A. Laski, Int. J. Adv. Manuf Tech. **40**, 3-4, 2009
18. J.E. Takosoglu, Edited by Zolotarev I., Radolf V. *Proceedings of 22nd International Conference on Engineering Mechanics 2016* (Academy of Sciences of the Czech Republic, 2016)
19. J.E. Takosoglu, P.A. Laski, S. Blasiak, G. Bracha, D. Pietrala. Meas. Cont. **49**, 2, 2016
20. J.E. Takosoglu, P.A. Laski, S. Blasiak. P. I. Mech. Eng. I-J. Sys. **226**, 10, 2012
21. J.E. Takosoglu, P.A. Laski, S. Blasiak, Edited by Fuis V. *Proceedings of 20th International Conference on Engineering Mechanics 2014* (Brno University of Technology, Czech Republic, 2014)
22. P.A. Laski, J.E. Takosoglu, S. Blasiak, Robot. Auton. Sys. **72**, 2015
23. R. Farana, B. Walek, M. Janosek, J. Zacek, *Proceedings of the 15th International Carpathian Control Conference (ICCC)*, 2014
24. R. Nadolski, K. Ludwinek, Przeglad Elektrotechniczny, **4**, 2012
25. H. Wisniewski, L. Plonecki, *Proceedings of the 16th International Carpathian Control Conference (ICCC)*, 2015
26. P. Wos, R. Dindorf, Int. J. Dynam. Control, 2014
27. P. Wos, R. Dindorf, Int. J. Appl. Mec. Eng. **17**, 3, 2012
28. S. Blasiak, J.E. Takosoglu, P.A. Laski, Edited by Fuis V. *Proceedings of 20th International Conference on Engineering Mechanics 2014* (Brno University of Technology, Czech Republic, 2014)
29. P.A. Laski, Edited by Zolotarev I., Radolf V. *Proceedings of 22nd International Conference on Engineering Mechanics 2016* (Academy of Sciences of the Czech Republic, 2016)
30. M. Blasiak, R. Kotowski, Przeglad Elektrotechniczny, **85**, 12, 2009
31. M. Blasiak, Edited by Zolotarev I., Radolf V. *Proceedings of 22nd International Conference on Engineering Mechanics 2016* (Academy of Sciences of the Czech Republic, 2016)
32. D. Janecki, L. Cedro, J. Zwierzchowski, Metrology Meas. Sys. **22**, 2, 2015
33. D. Janecki, J. Zwierzchowski, L. Cedro, B. Pol. Acad. Sci-Tech. Sci. **63**, 3, 2015
34. I. Krzysztofik, Edited by Zolotarev I., Radolf V. *Proceedings of 22nd International Conference on Engineering Mechanics 2016* (Academy of Sciences of the Czech Republic, 2016)
35. Z. Koruba, I. Krzysztofik, P. I. Mech. Eng. K-J. Multi-Body Dyn. **227**, K1, 2013
36. G.F. Bracha. Edited by Zolotarev I., Radolf V. *Proceedings of 22nd International Conference on Engineering Mechanics 2016* (Academy of Sciences of the Czech Republic, 2016)
37. S. Blasiak, Int. J. of Heat Mass Tran. **81**, 2015
38. S. Blasiak, A. Pawinska, Int. J. of Heat Mass Tran, **90**, 2015
39. P. Wos, R. Dindorf, Asian J. Contr. **15**, 4, 2013
40. P. Wos, R. Dindorf, Mechatronics: Ideas for Industrial Applications, **317**, 2015
41. P. Wos, R. Dindorf, Dynamical Systems - Applications, 2013
42. ISO 6358. Pneumatic fluid power - Determination of flow-rate characteristics of components using compressible fluids
43. S. Blasiak, J. E. Takosoglu, P. A. Laski, J. Therm. Sci. Techn. **9**, 2, 2014
44. S. Blasiak, P. A. Laski, J. E. Takosoglu. Int. J. Heat Mass Tran. **57**, 1, 2013
45. I. Krzysztofik, J. Takosoglu, Z. Koruba. Annual Reviews in Control. http://dx.doi.org/10.1016/j.arcontrol.2016.10.003

DOI: 10.1051/epjconf/201714302127

# Experimental research of flow servo-valve

*Jakub* Takosoglu[1,*]

[1]Faculty of Mechatronics and Mechanical Engineering, Department of Manufacturing Engineering and Metrology, Kielce University of Technology, Aleja Tysiaclecia Panstwa Polskiego 7, 25-314 Kielce, Poland

**Abstract.** Positional control of pneumatic drives is particularly important in pneumatic systems. Some methods of positioning pneumatic cylinders for changeover and tracking control are known. Choking method is the most development-oriented and has the greatest potential. An optimal and effective method, particularly when applied to pneumatic drives, has been searched for a long time. Sophisticated control systems with algorithms utilizing artificial intelligence methods are designed therefor. In order to design the control algorithm, knowledge about real parameters of servo-valves used in control systems of electro-pneumatic servo-drives is required. The paper presents the experimental research of flow servo-valve.

## 1 Introduction

Pneumatic proportional valves (servo-valves) are the components characterised by great enhancement of the input signal [1-3]. They aim to strengthen and transform incoming signals into strong outgoing signals [4-7] continuously - strong in terms of their flow intensity (flow valves) or pressure (pressure valves) [8, 9]. The greatest enhancements can be gained using electric signals which serve as incoming signals. Hence, proportional valves electrically controlled are applied the most commonly. Such valve consists of two basic systems:

- electro-mechanic transducer, which processes voltage or current signal into linear (angular) shift or force (momentum) [10-12],
- one-, two-, or three-stage amplifier.

Proportional valves are primarily used in control systems [13-17] of electric quick-change production processes [18-22]. The design of electro-pneumatic servo-valves is not substantially different from electro-hydraulic ones [23, 24]. These are most often directional valves with four or two control ends. The slider-sleeve system of pneumatic servo-valves has no elastic seals for lowering frictional drag and consequential hysteresis [25-27]. Lack of sealing results in significant losses of compressed air [28, 29], which can be prevented by using additional flow canal cover in the slider-sleeve system, which, however, creates 'a callousness zone' in the servo-valve around the central slider. Table 1 shows the parameters of examined MPYE-5-1/8-HF-010-B proportional valve manufactured by Festo.

**Table 1.** Proportional valve parameters

| Parameters | Values |
|---|---|
| Nominal size | 6 mm |
| Valve function | 5/3-way, normally closed |
| Operating pressure range | 0-1 MPa |
| Critical pressure ratio *b* | 0.26 |
| Sonic conductance *C* | $3.1 \cdot 10^{-8}$ $m^3/(s \cdot Pa)$ |
| Nominal flow rate $q_v$ | $11.67 \cdot 10^{-3}$ $m^3/s$ |
| Critical frequency *f* | 100 Hz |
| Max. histeresis | 0.4 % |
| Power supply | 17-30 V |
| Setpoint value | 0-10 V |

The overview of the proportional valve is shown in Fig. 1a, whereas the schematic diagram - in Fig. 1b.

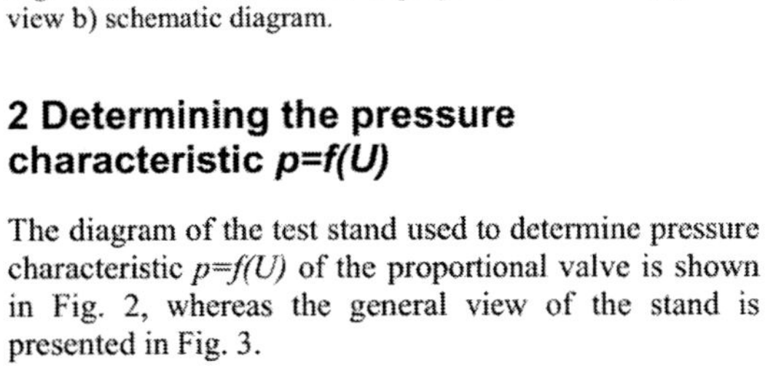


**Fig. 1.** MPYE-5-1/8-HF-010-B proportional valve: a) general view b) schematic diagram.

## 2 Determining the pressure characteristic *p=f(U)*

The diagram of the test stand used to determine pressure characteristic *p=f(U)* of the proportional valve is shown in Fig. 2, whereas the general view of the stand is presented in Fig. 3.

* Corresponding author: qba@tu.kielce.pl

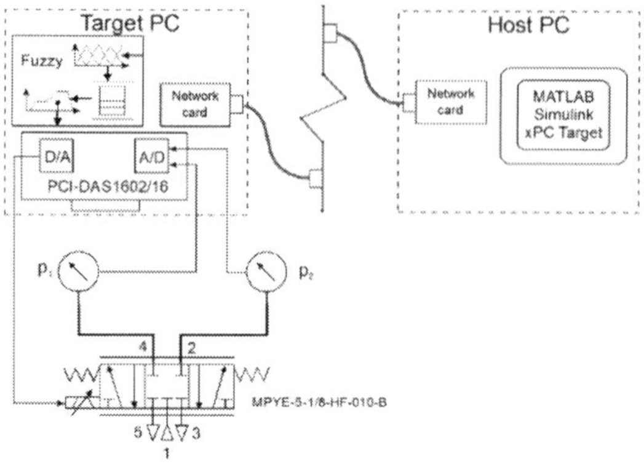


**Fig. 2.** Diagram of the test stand applied to determine pressure characteristic $p=f(U)$.

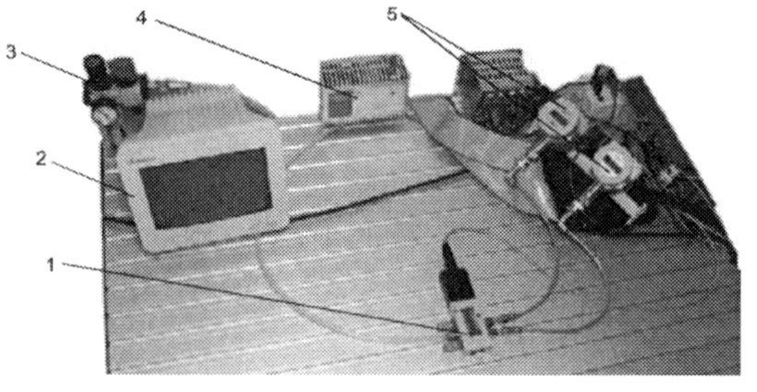


**Fig. 3.** General view of the test stand: 1 – proportional valve, 2 – xPC Target monitor, 3 – air service unit, 4 – power supply, 5 – pressure transducer

For pressure measurement at 1-2 and 1-4 flow ways of the proportional valve, two digital pressure sensors of PXW series (Peltron) with one voltage-out slot were used. Calibration curve $p=f(Up)$ (pressure $p$ in output voltage function $Up$) of PXW pressure sensor is shown in Fig. 4.

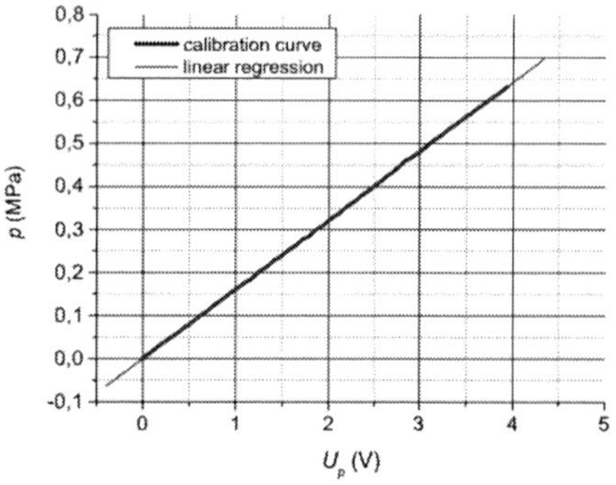


**Fig. 4.** Calibration curve $p=f(U_p)$ of the pressure sensor.

The linear regression of the calibration curve of PXW pressure sensor was determined with the following formula:

$$p=(p_p\pm\delta p_p)+(K_p\pm\delta K_p)U_p \qquad (1)$$

where:
$p$ – pressure,
$U_p$ – pressure sensor output voltage,
$K_p$ – directional coefficient of the calibration curve,
$\delta K_p$ – error of the calibration curve directional coefficient,
$p_p$ – free term indicating the intersection of calibration curve with the $p$ axis,
$\delta p_p$ – error of the calibration curve free term.

Constituents ($p_p\pm\ \delta p_p$) and ($K_p\pm\ \delta K_p$) of the linear regression of the calibration curve were determined using OriginPro software:

$$p_p+\delta p_p=1.9\cdot10^{-4}\pm1.8\cdot10^{-4} \qquad (2)$$

$$K_p+\delta K_p=\frac{p_p}{U_p}=0.16056\pm7\cdot10^{-5} \qquad (3)$$

After taking the results (2) and (3) into account, the linear regression of the calibration curve of PXW pressure converter is expressed in the following way:

$$p=1.9\cdot10^{-4}\pm1.8\cdot10^{-4}+(0.16056\pm7\cdot10^{-5})\,U_p \qquad (4)$$

The measurement data was analysed using Matlab- - Simulink package [30-33] of xPC Target software with a 16-bit AD/DA PCI-DAS 1602/16 card manufactured by Measurement Computing and fitted with 16 input slots and two output slots [34-38]. During the measurements, the voltage signal was introduced in a linear and incremental manner onto the solenoid of the proportional valve at 0 up to 10 V. By taking advantage of the calibration curve (Fig. 4), pressure characteristic $p=f(U)$ of MPYE-5-1/8-HF-010-B proportional valve was determined as shown in Fig.5.

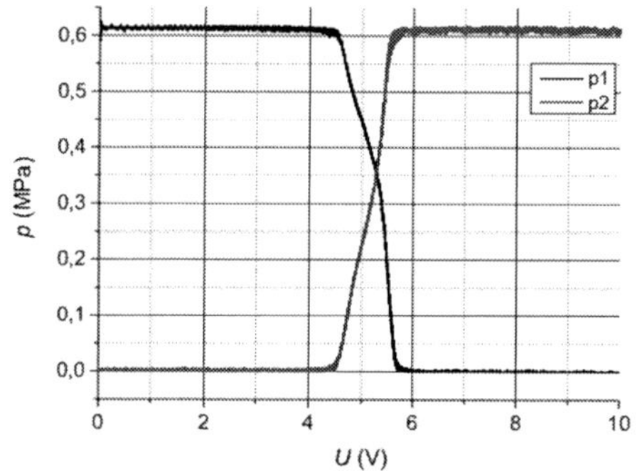


**Fig. 5.** Pressure characteristic $p=f(U)$ of MPYE-5-1/8-HF-010-B.

## 3 Determining the flow characteristic $q_v=f(U)$

The diagram of the test stand which is used to determine flow characteristic $q_v=f(U)$ of the proportional valve is shown in Fig. 6, whereas the general view of the stand in Fig. 7.

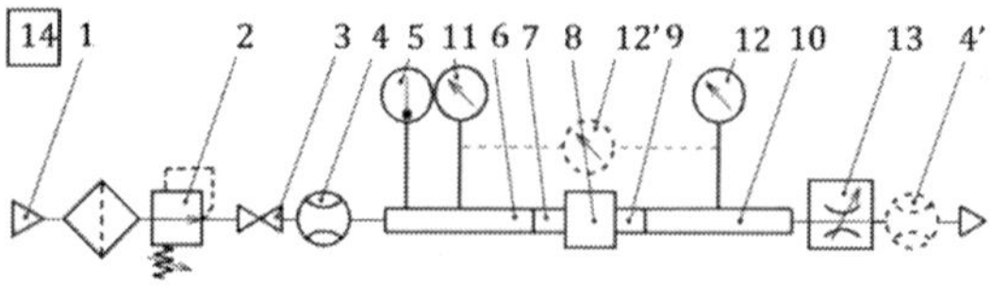

**Fig. 6.** Diagram of the test stand applied to determine flow characteristic $q_v=f(U)$: 1 – compressed gas source and filter, 2 – adjustable pressure regulator, 3 – shut-off valve, 4 – flow-rate measuring device, 5 – temperature-measuring instrument, 6 – upstream pressure-measuring tube, 7 – upstream transition connector, 8 – component under test, 9 – downstream transition connection, 10 – downstream pressure-measuring tube, 11 – upstream pressure gauge or transducer, 12 – downstream pressure gauge or transducer, 13 – flow control valve, 14 – barometer.

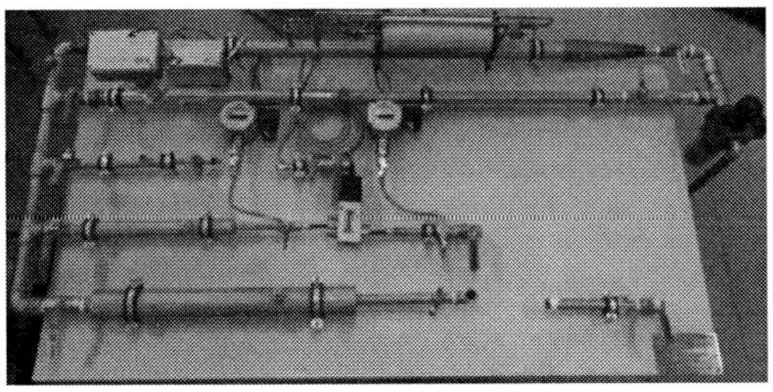

**Fig. 7.** General view of the test stand.

In order to select the diameters and lengths of measuring tubes and laminar flow tubes, the dependencies from ISO 6358 were used. The calculations were made on the basis of the diagram shown in Fig. 8.

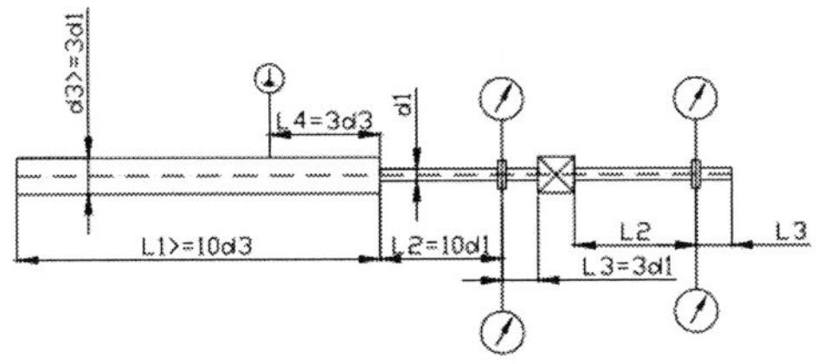


**Fig. 8.** Dependencies of the selection of pipe lengths and diameters (according to ISO 6358 [39])

The measurement data was analysed with the use of Sensyflow D software of ABB that comes with a flowmeter, which allows measurement of mass flow and volumetric air flow. The flow-rate characteristic of the component is the dependency of air overpressure behind component $p_2$ in volume flow function $q_v$ under constant overpressure $p_1$ in front of component $p_1$. In the case of proportional valves, the flow-rate characteristics are most often shown by drive voltage function. Fig. 9 presents flow-rate characteristic $q_v=f(U)$ under constant inlet pressure $p_1$=0.6 MPa.

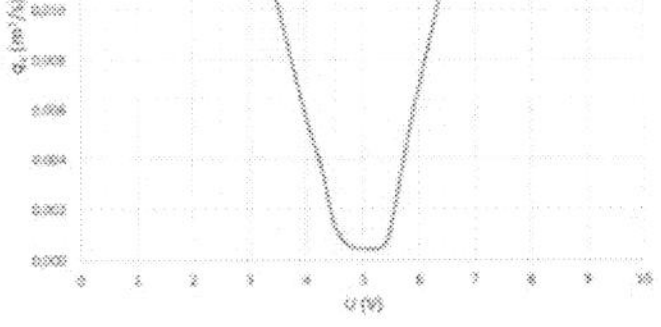

**Fig. 9.** Flow-rate characteristic $q_v=f(U)$ of MPYE-5-1/8-HF-010-B proportional valve.

## 4 Determining the response time and the idle time

The response time and the idle time of MPYE-5-1/8-HF-010-B proportional valve were measured at ITE-made PT-2 test stand, the diagram of which is shown in Fig. 10, while Fig. 11 presents its overall view.

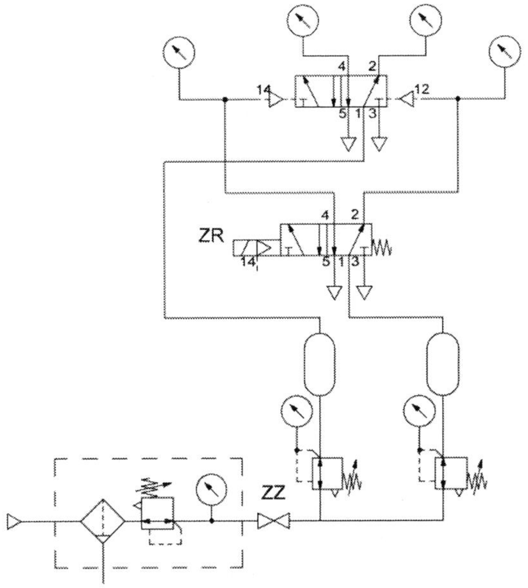


**Fig. 10.** Diagram of the test stand applied to determine the response time and the idle time.

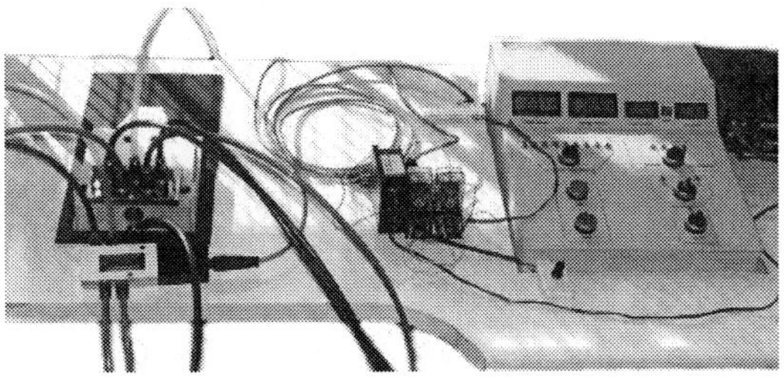

**Fig. 11.** General view of the PT-2 test stand.

PT-2 test stand is dedicated to study 3/2, 5/2 and 5/3 directional valves which are pneumatically controlled on both sides, pneumatically controlled on one side with a spring actuated return, electromagnetically controlled on both sides and magnetically controlled on one side with a spring actuated return. Because of the fact that the stand was not adapted to study the proportional valves, it had been appropriately modified. A custom-made voltage adjuster of the proportional valve was replaced with the pilots of the solenoid valves. The study at PT-2 test stand consisted in determining the valve response time and the idle time during valve adjustment. The determination of the response time of valves controlled electrically consisted in measuring a time gap between opening or closing the electrical circuit feeding the electromagnet controlling the valve, and reaching an appropriate pressure level at the outlet. The determination of the idle time of valves controlled electrically consisted in measuring a time gap between opening or closing the electrical circuit feeding the electromagnet [40, 41], which controls the valve, with output pressure values amounting to zero (the flow route is then closed). At PT-2 test stand, the pressure change (behind the valve) in a time function, as well as the electrical current in the solenoid of the electromagnet controlling the valve were measured. To measure quick pressure changes over time, Minitron MPXG20 with a measuring range between 0 and 1 MPa and accuracy ± 1% was used. A series of 20 measurements was conducted during shifting the proportional valve from extreme positions which corresponded to voltage of 0 V and 10 V, and back – 10 V to 0 V. Table 1 shows the average results of the experimental measurements of the response and idle times for the proportional valve type MPY5-1/8-HF-010-B. The measurements performed at PT-2 stand are consistent with RP 82 P method "Response time characteristic of pneumatic control valves" described in the guidelines of the European Fluid Power Committee (CETOP).

**Table 2.** Measurement results

| valve slider position | response time | idle time |
|---|---|---|
| 10V→0V | 14.4 ms | 2.0 ms |
| 0V→10V | 15.1 ms | 1.3 ms |

## 4 Summary

Pneumatic proportional valves are mainly used in control systems of quick-change production processes that use advanced control systems, e.g. fuzzy logic, artificial neural networks or state space variables [42-45]. When designing such control systems, true parameters of the proportional valves applied are especially crucial. This article presents the methods and test stands dedicated to determine static characteristics: pressure $p=f(U)$ and flow $q_v=f(U)$, as well as the response time and the idle time of MPYE-5-1/8-HF-010-B proportional valve. These methods can successfully be used in further research on control valves.

## References

1. J.E. Takosoglu, R.F. Dindorf, P.A. Laski, Int. J. Adv. Manuf Tech. **40**, 3-4, 2009
2. P.A. Laski, J.E. Takosoglu, S. Blasiak, Edited by Fuis V. *Proceedings of 20th International Conference on Engineering Mechanics 2014* (Brno University of Technology, Czech Republic, 2014)
3. P.A. Laski, J.E. Takosoglu, S. Blasiak, Robot. Auton. Sys. **72**, 2015
4. R. Trochimczuk, M. Gawrysiak, Solid State Phenom. **147-149** (2009)
5. R. Trochimczuk, Solid State Phenom. (2013)
6. R. Trochimczuk, Int. J. Appl. Mec. Eng. **19**, 9 (2014a)
7. R. Trochimczuk, T. Kuźmierowski, Int. J. Appl. Mec. Eng. **19**, 9 (2014b)
8. P.A. Laski, Edited by Zolotarev I., Radolf V. *Proceedings of 22nd International Conference on Engineering Mechanics 2016* (Academy of Sciences of the Czech Republic, 2016)
9. D.S. Pietrala, Edited by Zolotarev I., Radolf V. *Proceedings of 22nd International Conference on Engineering Mechanics 2016* (Academy of Sciences of the Czech Republic, 2016)
10. P. Wos, R. Dindorf, Mechatronics: Ideas for Industrial Applications, **317**, 2015
11. M. Blasiak, R. Kotowski, Przeglad Elektrotechniczny, **85**, 12, 2009
12. M. Blasiak, Edited by Zolotarev I., Radolf V. *Proceedings of 22nd International Conference on Engineering Mechanics 2016* (Academy of Sciences of the Czech Republic, 2016)
13. J.E. Takosoglu, Edited by Zolotarev I., Radolf V. *Proceedings of 22nd International Conference on Engineering Mechanics 2016* (Academy of Sciences of the Czech Republic, 2016)
14. J.E. Takosoglu, P.A. Laski, S. Blasiak, G. Bracha, D. Pietrala. Meas. Cont. **49**, 2, 2016
15. L. Nowakowski, M. Miesikowska, M. Blasiak, Edited by Zolotarev I., Radolf V. *Proceedings of 22nd International Conference on Engineering Mechanics 2016* (Academy of Sciences of the Czech Republic, 2016)
16. L. Nowakowski, E. Miko, M. Skrzyniarz, Edited by Zolotarev I., Radolf V. *Proceedings of 22nd International Conference on Engineering Mechanics 2016* (Academy of Sciences of the Czech Republic, 2016)
17. L. Nowakowski, M. Wijas, Edited by Zolotarev I., Radolf V. *Proceedings of 22nd International Conference on Engineering Mechanics 2016* (Academy of Sciences of the Czech Republic, 2016)
18. R. Farana, B. Walek, M. Janosek, J. Zacek, *Proceedings of the 15th International Carpathian Control Conference (ICCC)*, 2014
19. R. Nadolski, K. Ludwinek, Przeglad Elektrotechniczny, **4**, 2012
20. H. Wisniewski, L. Plonecki, *Proceedings of the 16th International Carpathian Control Conference (ICCC)*, 2015
21. E. Miko, L. Nowakowski, L. Procedia Eng. **39** (2012a)
22. E. Miko, L. Nowakowski, L. Procedia Eng. **39** (2012b)
23. P. Wos, R. Dindorf, Int. J. Dynam. Control, 2014
24. P. Wos, R. Dindorf, Int. J. Appl. Mec. Eng. **17**, 3, 2012
25. G.F. Bracha. Edited by Zolotarev I., Radolf V. *Proceedings of 22nd International Conference on Engineering Mechanics 2016* (Academy of Sciences of the Czech Republic, 2016)

26. S. Blasiak, J. E. Takosoglu, P. A. Laski, J. Therm. Sci. Techn. **9**, 2, 2014
27. S. Blasiak, P. A. Laski, J. E. Takosoglu. Int. J. Heat Mass Tran. **57**, 1, 2013
28. S. Blasiak, Int. J. of Heat Mass Tran. **81**, 2015
29. S. Blasiak, A. Pawinska, Int. J. of Heat Mass Tran, **90**, 2015
30. D. Gapinski, Z. Koruba, I. Krzysztofik, Mech. Syst. Signal. Pr. **45,** 2 (2014)
31. Z. Koruba, Z. Dziopa, I. Krzysztofik, J. Theor. Appl. Mec. **48,** 1 (2010)
32. I. Krzysztofik, Solid State Phenom. **180** (2012)
33. I. Krzysztofik, Z. Koruba, J. Automat. Inform. Sciences **44,** 5 (2012)
34. P. Wos, R. Dindorf, Asian J. Contr. **15**, 4, 2013
35. P. Wos, R. Dindorf, Dynamical Systems - Applications, 2013
36. J. Zwierzchowski, Edited by Zolotarev I., Radolf V. *Proceedings of 22nd International Conference on Engineering Mechanics 2016* (Academy of Sciences of the Czech Republic, 2016)
37. D. Janecki, L. Cedro, J. Zwierzchowski, Metrology Meas. Sys. **22**, 2, 2015
38. D. Janecki, J. Zwierzchowski, L. Cedro, B. Pol. Acad. Sci-Tech. Sci. **63**, 3, 2015
39. ISO 6358. Pneumatic fluid power - Determination of flow-rate characteristics of components using compressible fluids
40. J.E. Takosoglu, P.A. Laski, S. Blasiak, Edited by Fuis V. *Proceedings of 20th International Conference on Engineering Mechanics 2014* (Brno University of Technology, Czech Republic, 2014)
41. S. Blasiak, J.E. Takosoglu, P.A. Laski, Edited by Fuis V. *Proceedings of 20th International Conference on Engineering Mechanics 2014* (Brno University of Technology, Czech Republic, 2014)
42. I. Krzysztofik, Edited by Zolotarev I., Radolf V. *Proceedings of 22nd International Conference on Engineering Mechanics 2016* (Academy of Sciences of the Czech Republic, 2016)
43. Z. Koruba, I. Krzysztofik, P. I. Mech. Eng. K-J. Multi-Body Dyn. **227**, K1, 2013
44. J.E. Takosoglu, P.A. Laski, S. Blasiak. P. I. Mech. Eng. I-J. Sys. **226**, 10, 2012
45. I. Krzysztofik, J. Takosoglu, Z. Koruba. Annual Reviews in Control. http://dx.doi.org/10.1016/j.arcontrol.2016.10.003

# Taxonomic trees of fluidic oscillators

Václav Tesař [1,a]

[1] *Institute of Thermomechanics v.v.i., Czech Academy of Sciences*
*Dolejškova 5, 182 00 Praha-Kobylisy, Czech Republic*

**Abstract.** Fluidic no-moving-part oscillators generating pulsation in fluid flow became recently a very popular subject of investigations. The reason is their capability to increase efficiency of various chemical and physico/chemical processes. Also control of flows past bodies is more effective with the pulsation. Advantages of fluidic oscillators used for the task are low cost, robustness, long life, no maintenance, and other similar factors associated with absence of mechanical components. New oscillator principles - as well as old, nearly forgotten and now re-discovered - are currently developed. Sheer number of possible alternatives makes them difficult to survey. This paper attempts at clarifying the situation.

## 1 Introduction

Dictionary definition says:

Taxonomy = *classification into groups based on similarities of structure or relationships.*

Fluidic oscillators are devices supplied by a steady fluid flow and converting it into a self-excited periodic flow. Although some of their working principles were known (or at least their existence was recognised) already in 19$^{th}$ century [2], the configurations as we know them today were designed, tested, and patented later, around the middle of the 20$^{th}$ century. The inventors, Warren [8], Zalmanzon [9], and others were then guided by knowledge then already acquired with electronic oscillators — consisting of an electronic amplifier and electrically conducting loop forming a negative feedback signal path. In fluidics, the configurations were analogous: a fluidic oscillator consists of at least one fluidic amplifier together with channels for feedback loop fluid flows. From their originally expected applications, control and signal processing systems, fluidics was displaced by electronics and, became practically forgotten. A new lease of life they obtained recently when they were found useful for generating pulsation in various chemical and physico/chemical process engineering operations, the efficiency of which was demonstrated to be significantly improved by the pulsation superposed on the flow. Also in control of fluid flows past objects, e.g. by suppressing turbulence or flow separation on aircraft wings, the acting control jets are substantially more efficient if they pulsate and this may be achieved with fluidics. Compared with a mechanical counterparts the fluidic oscillators exhibit lower cost, easy manufacturing, robustness, long life, no maintenance, and other beneficial factors stemming from the absence of mechanical, moved or deformed valve components.

[a] Corresponding author: tesar@it.cas.cz

The growing demand has led to inventing various new oscillator working principles - as well as studying agsin the old ones. Number of existing and possible alternatives has now reached a stage making orientation in this field difficult. Considering possible combinations and permutations makes them and the relations between them quite confusing. The present survey paper is an attempt to clarify the situation.

## 2. Standard: amplifier and loop(s)

Most present-day fluidic oscillators consist from two separately identifiable components: a no-moving-part fluidic amplifier – mostly of the bistable diverting type, as shown in Fig. 1 - and one or two feedback loop channels for fluid flow. The bistable two-loop configuration (Fig.1) may be seen as two monostable devices of the block diagram in Fig. 2, with common supply nozzle.

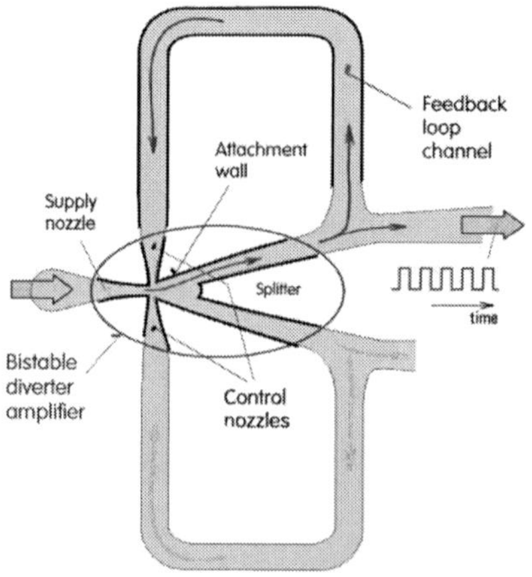


**Figure 1** Oscillators with bistable jet-deflection amplifier and two feedback loops as shown here are now practically standard.

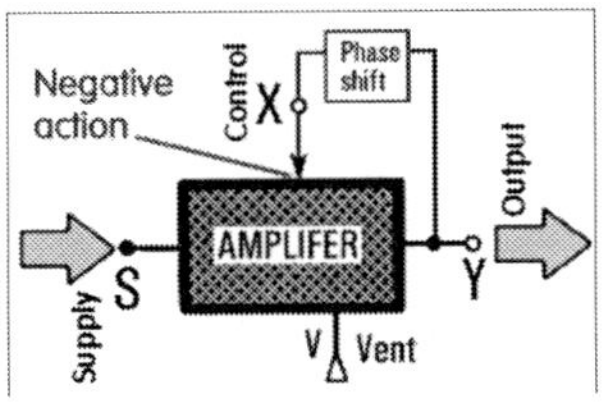


**Figure 2** Basic principle of self-excited oscillation with amplifier and negative feedback loop. The design in Fig. 1 may be viewed upon as an anti-parallel configuration with common supply nozzle of two integrated amplifiers & feedbacks.

The bistability is capability of the jet (generated in supply nozzle) to stay deflected in one of two different regimes without any external deflecting action. In the jet-type amplifier it is the consequence of the Coanda attachment effect: the jet clings to a nearby suitably placed solid wall. The negative feedback flow tends to separate the jet from its wall, so that it cannot remain steady and oscillates between the two regimes. Its use in the oscillator means the generated output flow has the character of a train of more or less rectangular pulses (Fig. 1). It is possible to use instead the proportional, continuously operating amplifier – generating a harmonic outp[ut flow - but the versions with jet attachment are generally more reliable and easier to set up.

There are two alternative possibilities, shown in Fig. 2, how to achieve the jet separation from the wall to which it is attached. For the explanation the picture uses a simpler amplifier version, with asymmetric geometry and only one attachment wall (so that it is monostable). In addition to the mono- and bistable versions, Fig. 3 shows yet another factor that multiplies the number of potential amplifier alternatives. There are two cases, A and B. The first A is presented in the top part of the illustration. At it is shown with the jet attached while at right the jet is displaced by the flow through the control nozzle oriented perpendicularly. This control action is more common. In the bottom part of Fig. 3 is presented the other alternative B. The separation of the jet from the wall is there achieved by making it more difficult for the main flow to get into the exit terminal. This is similar to the effect

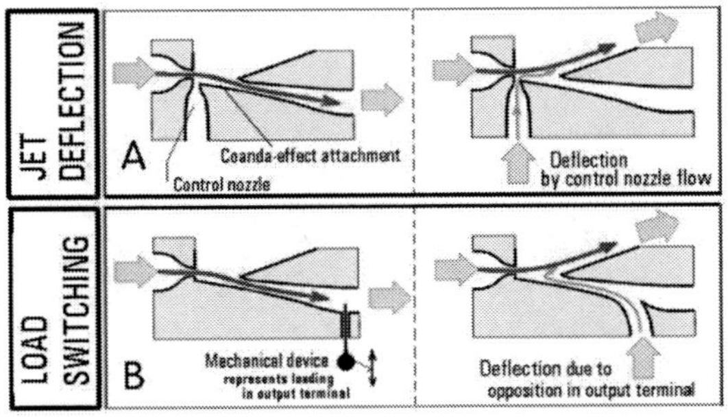


**Figure 3** Explained here on the example of simple one-sided monostable amplifier are the two alternative modes, A and B, of controlling the Coanda attachment. Apart from the more usual A, jet separation from the wall by a perpendicular control nozzle flow, the jet may be also forced away by blocking, B, the flow in output terminal.

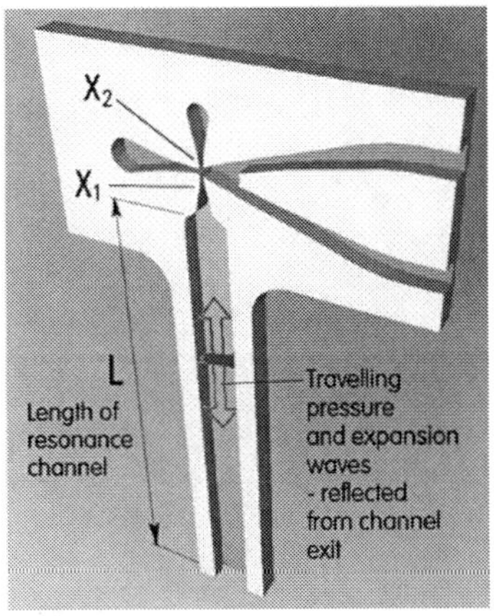


**Figure 4** Recently introduced [1] oscillator principle using travelling waves in a resonator. It is an attractive solution for high oscillation frequencies

achieved by insertion of a mechanical flow-blocking device as is schematically represented in the left-hand drawing B in Fig. 3.

Adding to the number of possible oscillator alternatives is another type of feedback presented in Fig. 4. There is no channel forming a loop. The negative feedback action is generated in a resonator channel (or tube – or any resonant cavity) attached to thet control nozzle $X_1$. Instead of running a flow around the loop the feedback action here results from the return of compression and expansion waves reaching the ends of the resonator. The pressure wave at the end of the resonator generates an rarefaction wave moving by the

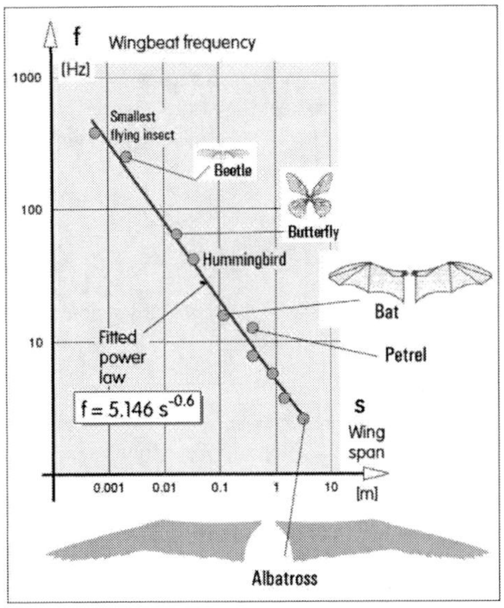


**Figure 5** Universal law of nature: higher frequency oscillation is generated in smaller sized oscillator. It is valid not only for flying organisms as shown here, but applies also to fluidic oscillators.

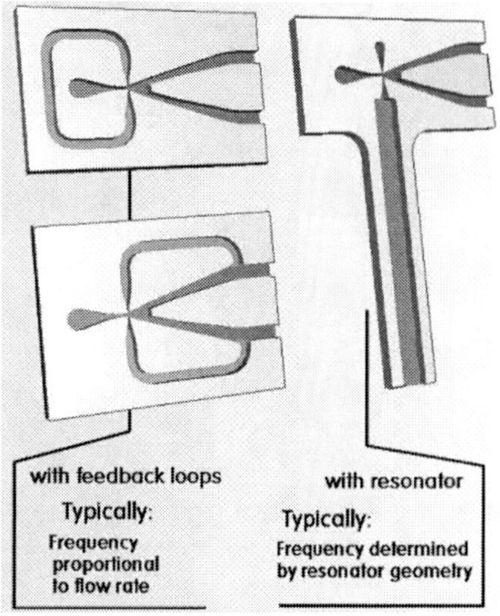


**Figure 6** Two basic behaviour modes of fluidic oscillators with bistable diverter amplifiers: either constant Strouhal number [2] (at left) or constant frequency (right).

speed of back to the amplifier. There its arrival causes jet separation from its attachment wall. Of course, this travelling wave character of the feedback, may be applied not only with the bistable amplifier but also in other configurations (such as the monostable amplifier or load-switching from Fig. ). As a result, the number of possible combinations increases.

The reason for use of the acoustic feedback is the capability to achieve high oscillation frequency, which is desirable in some applications. Sound waves propagate rather fast and the resonating channel may be quite short, both factors contributing to the high frequency. In the feedback loop versions typical top frequency in a palm-top size oscillator [1] is somewhere near to f = 300 – 400 Hz. Higher frequency would require high flow velocities and very small size, both inconvenient factors for some oscillator uses. The problem with the small characteristic size scale s is explained by the interesting general law presented in Fig. 6,

$$f \sim s^{-0.6} \qquad \ldots (1)$$

valid in nature universally for oscillating objects – and in principle also valid for the fluidic oscillators with feedback loops governed by the Strouhal number constancy (Fig. 6).

There is yet another factor complicating any survey of oscillators. It is the alternative configuration with only a single feedback loop channel connecting both control nozzles (top left in Fig. 6). This not immediately apparent configuration used to be attributed to Spyropoulos [9] but it is in fact another invention of Warren [10]. While requiring less space and simpler control circuit it is less reliable and more difficult to get operating. Increasing the number of possible permutations in general surveys of oscillators, at adds to the complexity of the taxonomic tree discussed here.

## 3. Less known feedbacks

Apart from the most popular designs presented in Fig. 6, there are many other self-excited fluid flow oscillation phenomena that may be also used in fluidics. The reason for their slower acceptance was absence of obvious feedback mechanisms and general difficulties associated with their identification – compared with the immediately apparent feedback loop principle analogy to the electronic oscillators.

A similar direct analogy exists also between the negative-slope oscillator in Fig. 7 and the electronic Esaki diode oscillator [16]. The fluidic version is rarely used and in fact little known.

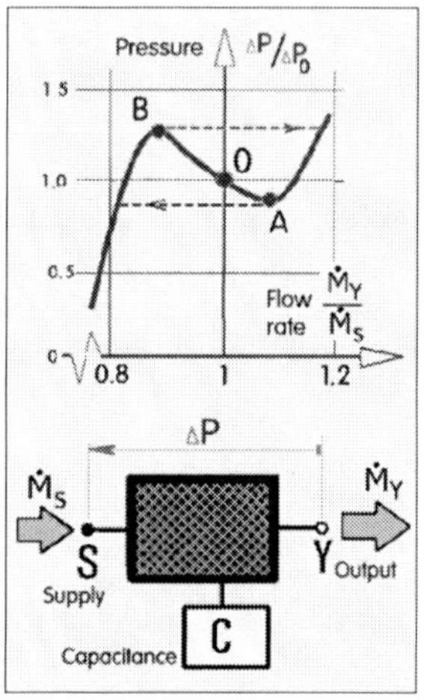


**Figure 7** Hydrodynamic instability with consequent self-excited oscillation exists in fluidic devices exhibiting negative slope on a part of their characteristic.

Similar characteristics with negative slope sections – but having no electronic counterpart — are encountered in the colliding jets, Fig. 8. The tendency towards oscillation is there so strong that investigating experimentally the part of the curve between A and B is rather difficult and calls for efficient flow regulators. This oscillator is not used in the typical oscillator role,

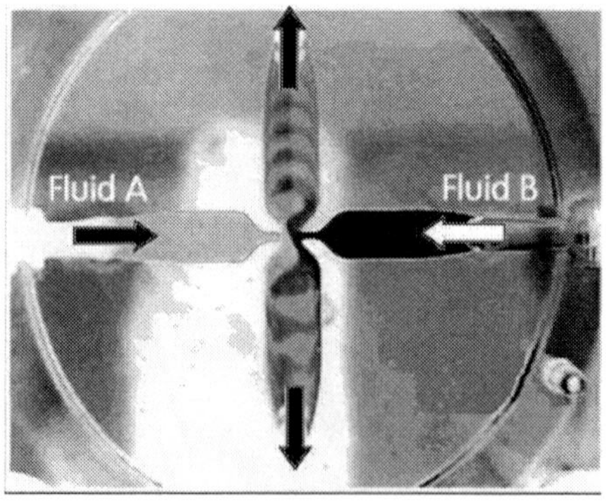


**Figure 8** Hydrodynamic instability (and oscillation) is encountered also in the flowfield of colliding jets [6].

details of the processes taking place in the interaction space were instead investigated with the intention of using the device in small-scale, microfluidic mixers [6]. Visualisations of the flowfield (with interesting procedure of posterisation of images – drastic reduction of the colour scale) have actually led to identification of the fact that the jets turn back, following what is in effect a feedback loop path.

A typical example of naturally occurring fluidic oscillator, visualisation of the flow in which is presented in Fig. 9, is the generation of periodically vortices shed downstream from a blunt body (in Fig. 9 the body is a circular cylinder). As long ago as in the 19[th] century (ref. [2]) the periodicity in this wake flow was recognised and measured. However, understanding where exactly the feedback action (which obviously must be present) takes place is still rather difficult.

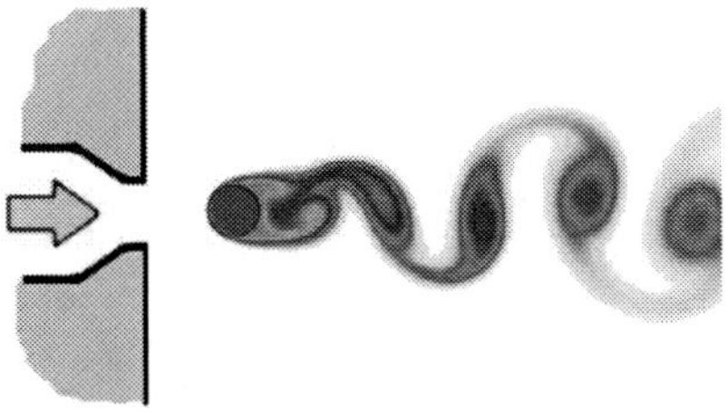

**Figure 9** An example of a fluidic oscillator existing in nature – a blunt body (cylinder) with trailing von Kármán vortex street [2]. The internal feedback effect is not simply identified.

The inflow as well as outflow from the device pictured in Fig. 10 is constant – the unsteady periodic character exists only in the vicinity of the target body. This device was developed not for generating a fluidic output signal, but for use in a digital flowmeter with electronic output — electric signal generated by a periodically cooled thermistor placed at the target. The condition for success of this idea is achieving in the cavities in Fig. 10 an almost perfect constancy of Strouhal number

$$\mathrm{Sh} = \frac{f\,b}{w} \quad \ldots (2)$$

- where b and w are the nozzle exit width and velocity.

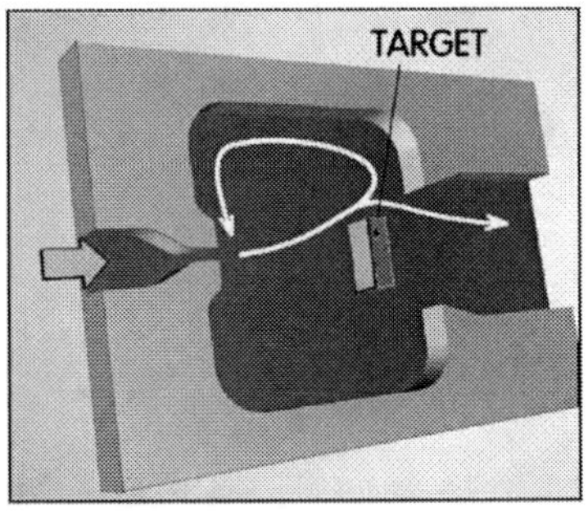


**Figure 10** Oscillator with the blunt-body target, essentially similar to Fig. 9 but with oscillation enhanced by vortical motions in the cavities provided for the purpose on both sides of the jet. Device geometry was developed for Sh = const.

The flow rate $\dot{M}$, simply proportional to w, is evaluated by counting the oscillation pulses per unit of time.

Deviations from desirable perfect Strouhal number constancy at very low velocities in the flow shown in Fig. 9 were identified to stem from the Reynolds-number

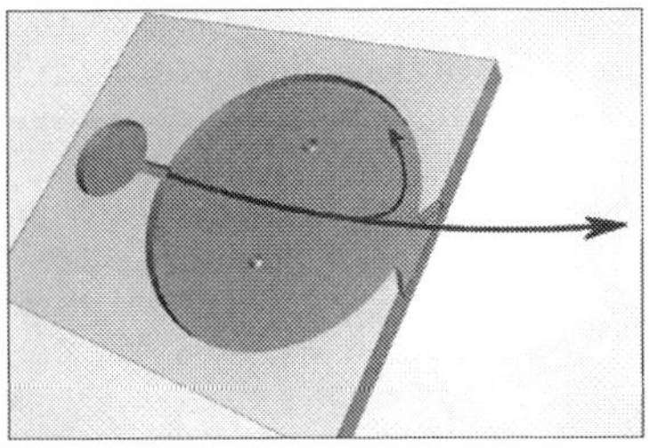

**Figure 11** Nakayama oscillator [5] with internal loops driven by jet-edge separating scraper .

dependent variations of the location at which the cylinder boundary layer separates from the solid surface. This is the reason for replacing in Fig. 10 the cylinder by the target with sharp edges that fix the separation location. Stabilisation of Sh value in the flowmeters is also helped by providing the cavities in which form of alternating vortices.

At low Reynolds numbers, typical for fluidics, the vortices forming on both sides of a submerged jet can play a significant role in the oscillator behaviour. In the Nakayama oscillator [5], Fig. 11, they are peeled off by the edges of the scraper and form a vortical pattern inside the cavity, causing alternating bending of the jet.

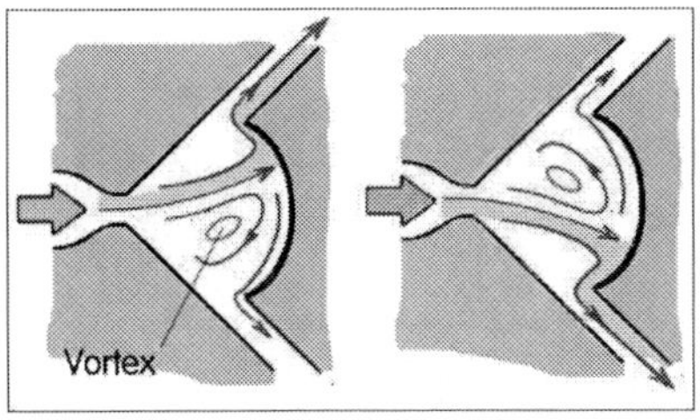


**Figure 12** Naturally arising self-excited oscillatory phenomena are also based on instability of jet impingement, especially on a concave surface [3].

Another phenomena with instability and resultant periodic motions are found in impinging jets. Impingement on a flat wall is characterised by only small sideways "dancing" of the stagnation point. More powerful are the pulsations of impingement in a cavity – or the concave wall as shown in Fig. 12. Pressure differences between both sides of the jet are increased if there is a suitable cavity for accommodating there the vortical motions. The vortices deflect the jet and the amplitudes of the "dance" may become quite large. Obvious disadvantage of the configuration from Fig. 12 in practical applications is poor energetic efficiency, unavoidable in flows with long submerged jet paths and sharp changes of flow directions

 DOI: 10.1051/epjconf/201714302128

# 4. Relaxation oscillators

There is a family of "feedback-less" oscillator principles little known nowadays – although one of the family members was actually the first ever patented [9] no-moving-part fluidic oscillator. The basic idea is gradually increasing level of some parameter (pressure, flow rate, or hydraulic resistance) until it reaches a limit. At thta moment the flowfield can no more keep the initial regime conditions and the flow regime collapses. The character of the flow in the collapse is such that it makes possible a return of the whole fluidic system to the starting condition. Thereafter the process of parameters increase can start again. Typically, such a collapse of the flowfield takes place in the load-switched diverter (case B in Fig. 3) if its load in the active flowpath reaches the level at which the Coanda effect no more suffices in keeping the jet at its attachment wall and the jet is switched away.

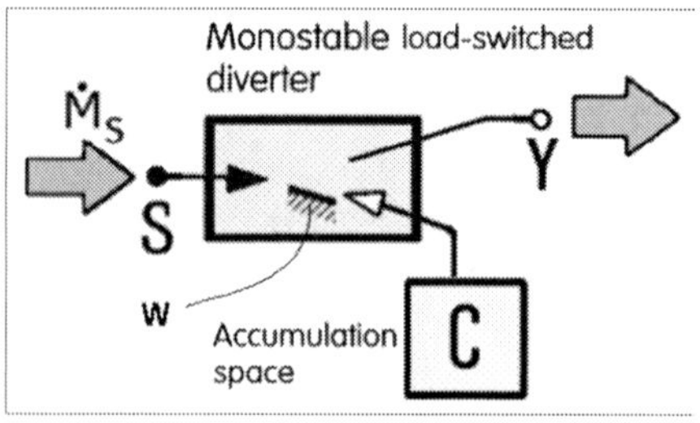


**Figure 13** Schematic representation of the earliest ever patented fluidic oscillator [9] based on relaxation phenomena in monostable jet attachment.

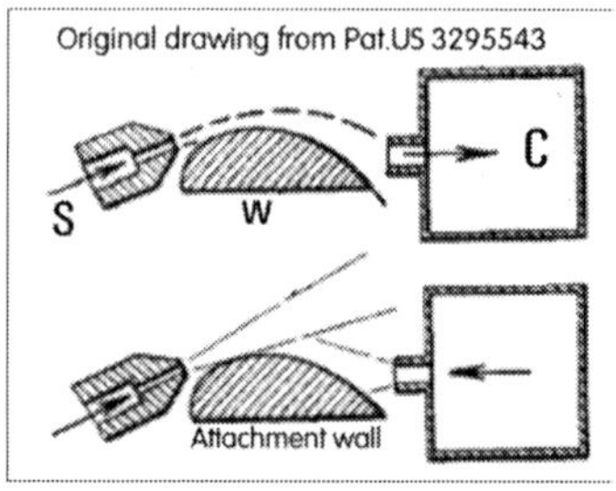


**Figure 14** The tested configuration of the relaxation oscillator [9] according to Fig. 13.

Still the best example is the Zalmanzon's oscillator [9], the schematic circuit diagram of which is presented in Fig. 13. It corresponds to the drawing in Fig. 14 taken from the Patent [9], showing its origin in stalling an aeroplane wing. The monostable diverter has its single attachment wall W positioned so that it guides the attached jet (generated in the supply nozzle S) into the accumulation chamber C. The fluid is air (or gas). Being compressible, it is gradually accumulated in the chamber and its pressure there increases. Of course, the accumulation is limited. Finally, the pressure in chamber C reaches the level at which it represents untenable loading for the amplifier output terminal. The jet switches away from W and is directed into the other terminal Y – where it remains deflected by the return flow issuing from the chamber C. Only when the pressure in the emptied chamber reaches the starting condition, the jet returns to W and the chamber filling starts anew.

As a matter of fact, even in the principles described as feedback-less, analysis of the processes makes possible discovering some action that plays the feedback role.

# 5. Multivibrator with two amplifiers

## 5.1 Mutual blocking of anti-parallel amplifiers

Another little known fluidic oscillator configuration is represented schematically in Fig. 15. It corresponds to an analogous circuit in electronics called "multivibrator", [7] invented at the time of WW I. From there the idea was taken over into fluidics. It may be possible to see the bistable fluidic amplifier presented in Fig. 1, as an integral configuration of two monostable amplifiers (with common supply nozzle and also common interaction cavity, sharing also the outlet terminals).

In Fig. 15 there are the two amplifiers, in the direct analogy with electronics both of the turn-down type. They are connected in anti-parallel, so that the flow through one of them corresponds to no-flow regime in the other, opposite one. There are two main flowpaths, one from the supply terminal S through the hydraulic resistor $R_1$ and the Amplifier 1. The other main flowpath passes through the hydraulic resistor $R_2$ and the Amplifier 2. If an amplifier, say the Amplifier 1, is in the OPEN regime, almost all pressure supplied from S drops on the resistor $R_1$. The node $u_1$ is almost at the low, atmospheric pressure. If, however, the Amplifier 1 is in the CLOSED regime, the flow rate through $R_1$ is small so that the pressure drop on it is small. Almost all pressure supplied from S drops on the Amplifier 1 and the node $u_1$ is almost at the supply pressure level. The high pressure from the node $u_1$ is led into the second control terminal $X_2$ of the Amplifier 2. It tends to switch it from the OPEN regime into the CLOSED one, but this effect is delayed by the phase-shift element. Behaviour of the Amplifier 2 is

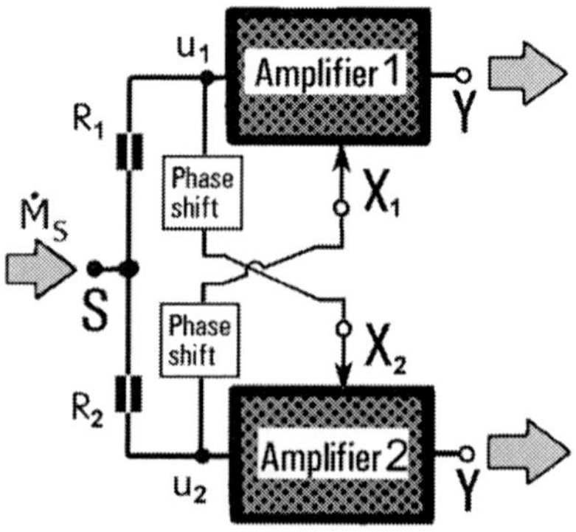


**Figure 15** Fluidic oscillators may be assembled from two amplifier devices blocking each other's flow from the supply S to the exit Y. This principle is a direct parallel to the electronic flip-flop multivibrator invented by Abraham and Bloch [7].

thus the very opposite to what happens in Amplifier 1, only that it is delayed. The amplifiers simply block one another. Depending on the adjustment of the phase shift, the circuit either behaves as a flip-flop binary memory – or oscillates, with the pressure and flow in the output terminals Y pulsating between the two regimes.

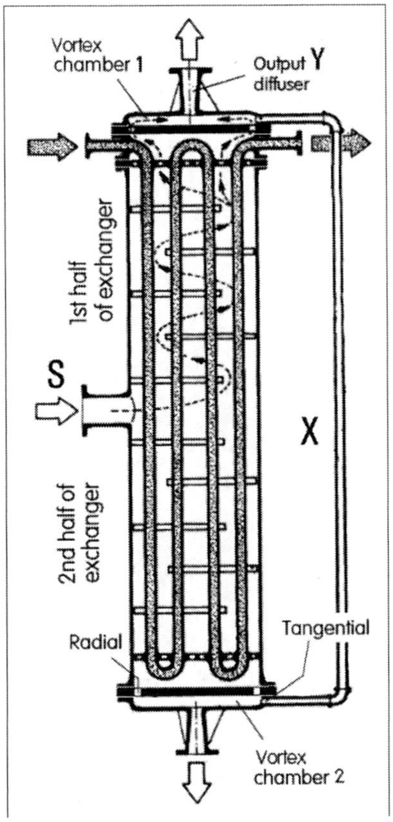


**Figure 16** The heat exchanger with the flow oscillation on gas side as schematically represented in Fig. 17.

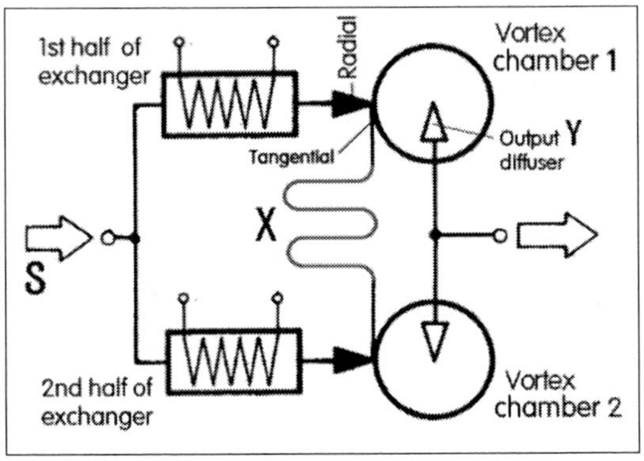


**Figure 17** Heat transfer intensification by oscillation generated in a mutually blocked vortex amplifiers – actually end-spaces of exchanger body [12]. This is the earliest known fluidic oscillator based on the multivibrator principle Fig. 15.

The earliest known realisation of this oscillator principle with anti-parallel amplifiers was described in [12]. The object was a pulsating heat exchanger – the pulsation on the gas side destroying the thermal boundary layer and therefore increasing the exchanger efficiency. Upper and lower halves of the exchanger took over the role of the resistors $R_1$ and $R_2$ in Fig. 15. Spaces at ends of the echanger's cylindrically shaped body served as vortex chambers of the active devices - the vortex amplifiers.

Recently experimentally investigated oscillator presented in Fig. 18 is essentially analogous. Again, it was the turn-down capability in the amplifiers that was obtained by applying the idea of vortex amplifier. the amplifiers discussed in [15] are the CSVA version (Coanda-switched vortex amplifier) – in fact an integral combination of a bistable diverter and vortex amplifier. Upstream from the vortex chamber there are ante-chambers with a pair of mutually inclined attachment walls to which attaches the jet, formed by issuing from the supply nozzle.

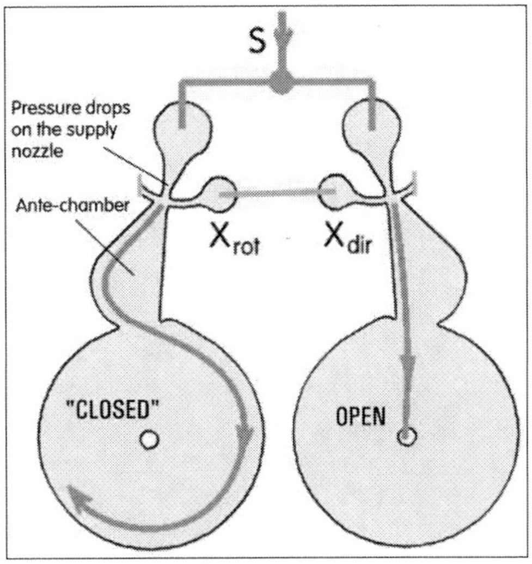


**Figure 18** One version of the oscillator described in [15], with Coanda-switched turn-down valves.

Also worth noting in Fig. 18 is the fact that the large area contractions in the supply nozzles generate the pressure drops corresponding to those in Fig. 15 that were obtained by the resistors $R_1$ and $R_2$.

### 5.2 Combining the ideas of relaxation and two anti-parallel devices:

The latest designs of this type of oscillators have as their starting point the version from Fig. 18, but were developed towards simplicity. The idea is presented schematically in Fig. 19. Note that the bottom drawing there, with the monostable diverter, corresponds directly to the Zalmanzon' s design presented in Fig. 13 – only replacing the time delay by gradual compression of air in the accumulation chamber by the gradual spin-up of the rotation in the vortex chamber. General experience with monostable and bistable diverters prefers the former. This is why the actually built and tested was the design shown in the top part of Fig. 19. As shown in Fig. 20, the resultant fluidic oscillator was capable of operating at very low frequencies – without the inconvenient need of stowing the very extremely long feedback loop tubing, that would be necessary with the standard feedback loops of Fig. 1.

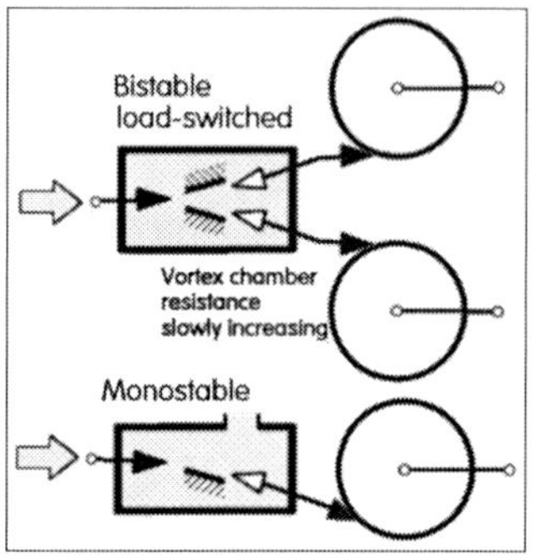


**Figure 19** The basic principle of the recent relaxation oscillator [13, 14]. The monostable version is similar to Fig. 13, accumulation delay by compressibility replaced by the spinning-up of rotation in vortex chamber.

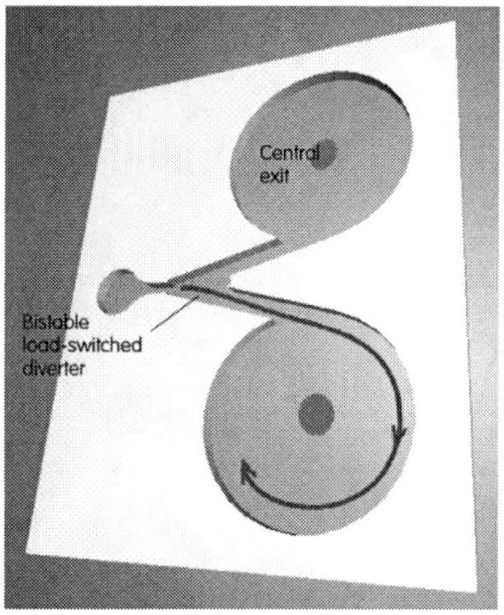


**Figure 20** The main plate of fluidic oscillator according to [14, 13]. In experimental evaluation it was capable of oscillating at extremely low frequencies. Symmetric configuration with bistable switching (top of Fig. 19).

The spin-up time delays in vortex chamber is a quite slow process and yet the oscillator is quite compact. Instead of the two parallel flowpaths of Fig. 15, with the division at the supply inlet into the circuit, there is here upstream from the vortex chambers the common jet-type diverter. It is load-switched when the rotation in the active vortex chamber attains the speed at which the Coanda effect cannot overcome the centrifugal force.

A demand for the low-frequency oscillators arose recently in bioreactors with supply of $CO_2$ to cultivating algae in bioreactors [18] developed for the potential production of liquid fuels. The low frequency makes possible a return of generated microbubbles into the passage inside the aerator during each cycle. This prevents the microbubbles from their conjunction and keeps them small – with advantageous surface-area to volume ratio.

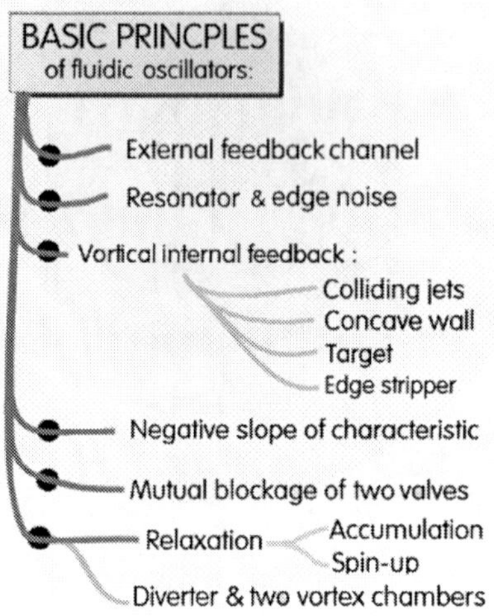


**Figure 21** Taxonomic tree of oscillators arranged on the basis of operation mechanism principles.

## 6. Classifications criteria for taxonomic trees

As is certainly apparent already from the brief survey above that mere enumeration of the known oscillator principles produces a quite long list. It would be useful to have for general orientation in such lists a tree of mutual relations. As is the general problem in taxonomy, the key difficulty lies in the choice of the applied classification criterion. The resultant trees are different when formed on different bases.

Somewhat resembling the classification of organisms in biology, in a response to requests the present author has found three alternative approaches to the classification:

I) Feedback principle criterion
II) Criterion based on the working mode
III) Number of amplifiers

### 6.1 Classification according to the used feedback principle

The simple taxonomic tree resulting from the consideration of working principles discussed in the previous Sections of this paper is presented in the diagram Fig. 21.

No particular ordering of the items was there applied and this may result in certain shortcomings. For example, the final part of the previous Section 5 discussed an oscillator related to the relaxation type as well as to the oscillator with mutual blockage of two amplifiers. Such cross-linkage, however, was difficult to include the the tree graph – and the solution in Fig. 21 is mere addition of yet another function principle, related to the relaxation oscillators.

### 6.2 Classification by the number of active devices (amplifiers)

In another attempt at the classification as the basis for setting up the tree, presented in Fig. 22, the basic criterion applied was the number of active elements — i.e. the number of fluidic amplifiers as devices that could be separately investigated. There is no such device (at least not immediately apparent) in the tree branches from S1 to S3 in Fig. 22. Also the load-switched diverter in the branch R would meet difficulties if it were to be used separately as an amplifier – despite the fact that some amplification effects are present.

### 6.3 Working mode classification

In another, earlier attempt at setting up a graphical survey of oscillators, the relations between the tree members were taken more into focus — as is apparent in its reproduction here in Fig. 23. More attention is there given to the relaxation switching, which seemed to offer a number of advantages for the algae growing bioreactors [18]. The permutations of bistability and monostability on one hand — and switching by a signal transported through the feedback channel vs. the load-switching — has led to the undesirable crossing of the relevant branches.

The tree as presented in Fig. 23 was used as a basis for detailed discussions in which the coloured number indicating the classification was helpful. An example of such local detailed development of the tree is presented in the following Fig. 24. Individual configurations are there drawn schematically. As in the schematic representations above, the black (filled) triangles in this representation represent the nozzles (converting pressure energy into increased flow velocity) while the white (empty) triangles are represent diffusers (flow slowed down while pressure rises). Because of the flow direction changes of some of the represented flows, these representations have to be considered with care. Also schematically represented are the Coanda-effect attachment walls.

The devices shown in the top row of Fig. 24 oscillate being switched by flow from the perpendicularly oriented control nozzles. These nozzles are connected to the feedback loops (i.e. this genera in the tree are equal to the genus A as shown in Fig. 3). In the middle horizontal row are relaxation type oscillators, with load-switched amplifiers and accumulation chambers C in which air (or a compressible fluid in general) is accumulated for a part of the cycle to be released in the subsequent part.. The bottom row cases are the present author's novel configurations, with load-switched diverters and vortex amplifiers — in some cases reduced to mere passively behaving vortex chambers. As for the vertical columns, in the left-hand column are monostable relaxation oscillators, in the middle column are symmetrically switched configurations with bistable diverter and, finally, at right are cases of anti-parallel oscillators with mutually blocking two fluidic amplifiers. They are currently investigated for microbubble generation applications — which were made possible (and in some cases economically feasible) by oscillation generated by fluidics.

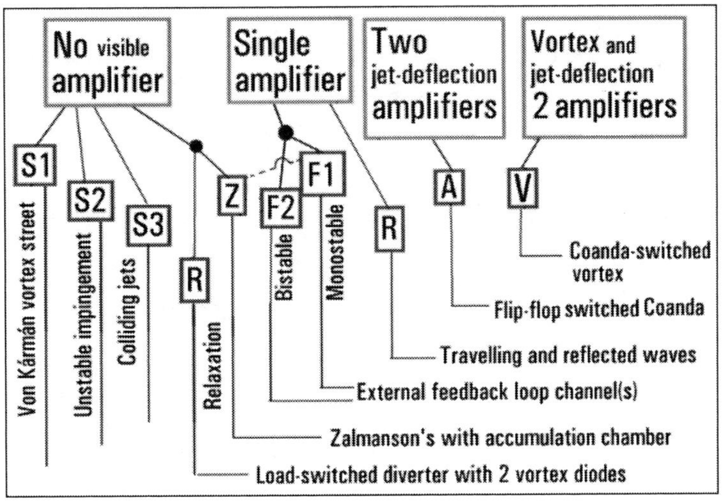


**Figure 22** Taxonomic tree of fluidic oscillators with classification based on the number of active (i.e. amplifying) devices.

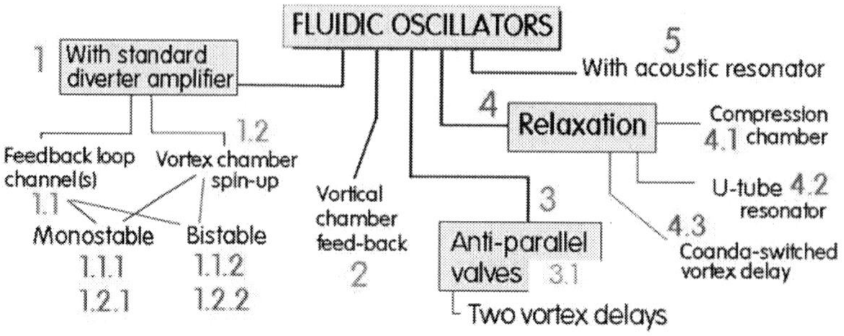


**Figure 23** Alternative taxonomic tree of fluidic oscillators with classification numbering.

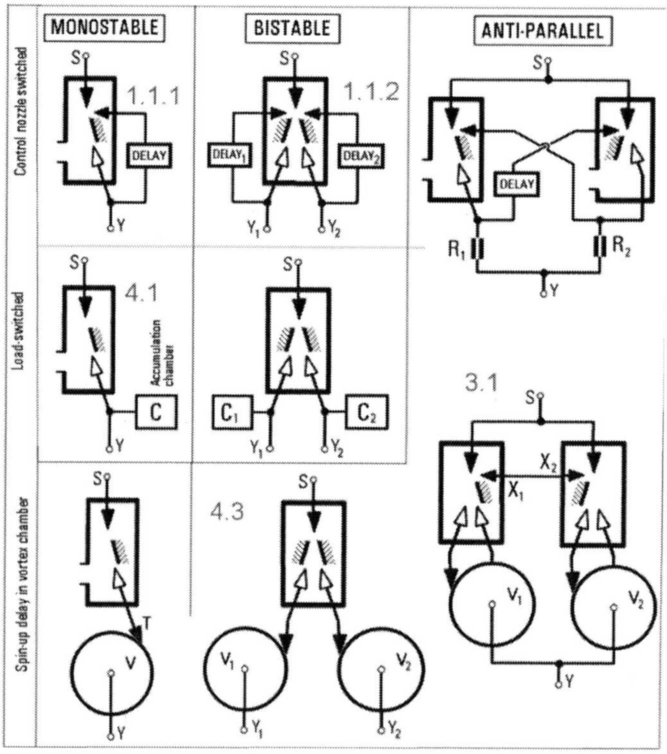


**Figure 24** A part of taxonomic tree: table of genetically related oscillator configurations with single-jet diverter bistable and monostable amplifiers.

## Conclusions

The starting point in scientific investigations is listing of investigated objects. The next one is to classify them and put them into order, using it in subsequent detailed studies. Fluidic oscillators are objects that recently became of importance. The number of applicable principles of flow pulsation is large and the mutual relations between them not clearly discernible. Following the analogous approach used in biological sciences, present author provided a list of working principles coming into consideration and set up an analogy to taxonomic trees.

## Acknowledgements

Author obtained support from GAČR - Grant Agency of the Czech Republic, by research grant Nr. 13-23046S. There was also institutional support RVO:61388998.

## References

1. V. Tesař, S. Zhong, F. Rasheed, *AIAA Journ.* **51** 397 (2013)
2. V. Strouhal. Annalen der Physik und Chemie **5** 216251 (1878)

3. C.-L. Sun, C.-Y. Sun C.-Y., *Microsystem Technologies* **17** 911 (2011)
4. O. Uzol, C. Camci, Proc. *of 4th Internat. Symp. on Particle Image Velocimetry*, paper 1029, Göttingen (2001)
5. Nakayama Y. et al., *Bulletin of JSME* **29** 3313 (1986)
6. V. Tesař, *Chemical Engineering Journal* **155** 789 (2009)
7. H. Abraham, E. Bloch, *Annales de Physique* **9** 237 (1919)
8. R. W. Warren, *US Patent 3,016,066* (1960)
9. L. A. Zalmanzon, *US Patent 3,295,543* (1960)
10. C. E. Spyropoulos, *Proc. of the Fluid Amplification Symp., Harry Diamond Laboratories*, Washington D.C. (1964)
11. R. W. Warren, US Patent 3,016,066, 1960
12. V. Tesař, *Czechoslovak Patent Nr. 262367* (1986)
13. V. Tesař, K. Peszynski, Smyk E., *EPJ Web of Conferences* **114** 02121 (2016)
14. V. Tesař, *Czechoslovak Patent Nr. 306064* (2014)
15. V. Tesař, EPJ Web of Conferences 114 02123 (2016)
16. R. Esaki, Y. Kurose Y., US Patent 3033714 (1957)
17. V.Tesař, C.-H. Hung, W. Zimmerman, *Sensors and Actuators A* **125** 159 (2006)
18. W.B. Zimmerman, B.N. Hewakandamby, V. Tesař, H.C.H. Bandulasena, O.A. Omotowa *International Sugar Journal* **112** 90 (2010)

EPJ Web of Conferences 143, 02129 (2017) DOI: 10.1051/epjconf/201714302129
*EFM 2016*

# What can be done with microbubbles generated by a fluidic oscillator? (survey)

Václav Tesař [1,a]

[1] *Institute of Thermomechanics v.v.i., Czech Academy of Sciences Dolejškova 5, 182 00 Praha-Kobylisy, Czech Republic*

**Abstract.** Recently discovered possibility of generating sub-millimetre bubbles by means of pulsating the inlet gas flow by a fluidic oscillator has opened a way to many interesting and advantageous engineering applications. Some of them have been known before, but were too inefficient and nearly forgotten. Now they deserve discussing in the new perspective. Other, applications, new ones, appear now in literature at a rapid pace. This paper, by providing a survey of already existing possibilities and advantages, aims at providing the readers an inspiration for their own developments.

## 1 Introduction

Two-phase operations between gas and liquid in chemical and process engineering are often performed in the form of the gas blown into the liquid as bubbles. It has been always recognised that many (if not all) of these processes would be more efficient if the given gas volume is distributed into a large number of smaller bubbles - because this means an increase of the total collective area of the mutual contact surface between the phases. Another potential advantage of the smaller size would be their slow rise in the liquid, making available a longer contact time. There are other advantages, nowadays discovered and reported increasingly often. Until recently it was, however, practically impossible to use them. Despite all the attempts, there has been almost always a failure due to too large bubble size. The desirable sub-millimetre microbubbles, Fig. 1, could not be produced economically.

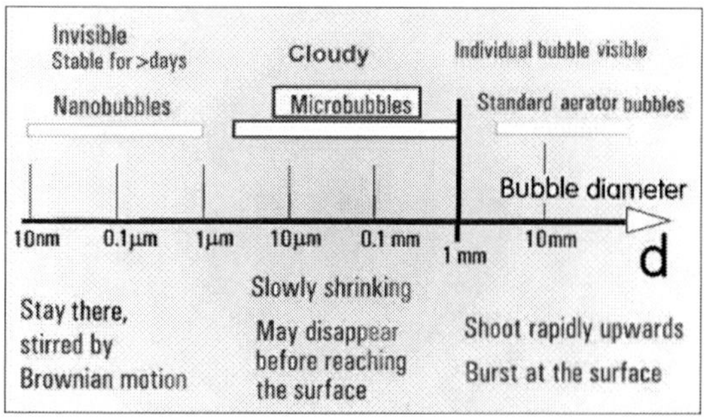


**Figure 1** Three size classes of gas bubbles in a liquid. Until recently, it was possible to generate efficiently only the large, millimetre-sized bubbles. Little is known at the small size end of the spectrum about the rather enigmatic nano-bubbles [1], less than 1 nm. Between them are **microbubbles,** discussed in the present article. Generating them became possible with fluidic oscillators.

**Figure 2** A typical example of the microbubble paradox: small air flow pores – here of equivalent exit diameter only 0.12 mm – do not produce correspondingly small bubbles: average bubble size in this case was 5.7 mm.

Existing bubble generation techniques are presented in recent review [2]. The most popular and actually simplest among them is the use of an aerator body with a large number of parallel small passages.- an example of which shows the photograph in Fig. 2. What seemed to be an obvious way how to make the microbubbles is to use an aerator with sub-millimetre passage exits. This approach, however, proved to be particularly elusive. For reasons that were not always obvious the generated bubbles are large then the passage exits — usually by as much as one or two decimal orders of magnitude. Other methods than the percolation through an aerator, like e.g. applying ultrasound or de-compressing the liquid containing dissolved gas were also found either not

[a] Corresponding author: tesar@it.cas.cz

working as expected or were ineffective and hence not acceptable economically.

The situation has changed after a chance discovery of the present author, then staying at the University of Sheffield, U. K.. He could generate surprisingly small bubbles with a standard aerator by placing into the gas inlet lines a simple and inexpensive fluidic oscillator. This device [3] was actually designed for a wholly different purpose (separation control of flow past a wind-tunnel model). In a series of simple experiments the pulsation generated in the airflow under some conditions (gas flow rates and oscillation frequencies were to be closely adjusted) the generated bubbles were indeed in the desirable sub-millimetre microbubble range.

## 2 Basic idea of fluidic oscillators

This solution of the microbubble problem, for which was later obtained a patent [4], was soon followed by similar positive experience of other researchers. It should be said that the basic idea of oscillation as the way to smaller bubbles is not totally new and was tried earlier. It was the advantages offered by fluidic oscillators, Fig. 3, that opened the way to the new possibilities.

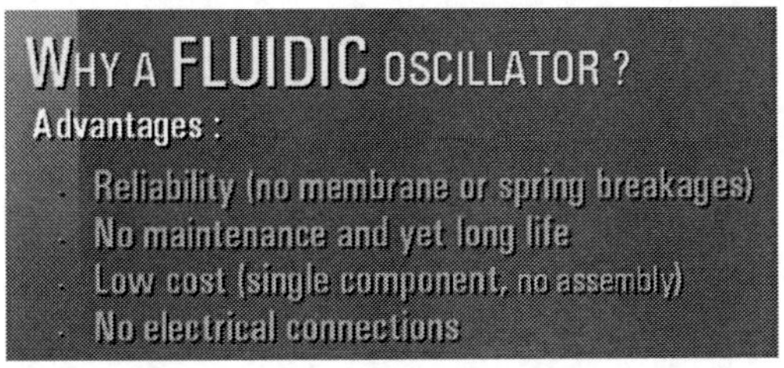


**Figure 3** What has now really opened the road towards the discussed applications of microbubbles are the advantages, here listed, that result from the absence of mechanical (moving or deformed) components in fluidics.

There are two essential components of which the fluidic oscillators consist. One of them is fluid flowfield with amplification properties. This is obtained by shaping the cavities for the flow so that it possesses a sensitive spot. The second component is introduction of a negative feedback.

The amplification means that a quite weak input flow causes a substantial overall change of the whole flowfield. Once the importance of such sensitive spots in fluid flow became recognised, they were discovered present in many flow cases. A typical example of such sensitivity may be a laminar flow with Reynolds number adjusted to critical transition conditions. In this case the input flow causes a transition into turbulence – with character quite different than the original laminar flow. This particular turbulisation sensitivity is today rarely used in fluidic amplifiers (since it is too sensitive even to various mechanical or acceleration disturbances and hence difficult to adjust). The best experience has been so far obtained with another sensitive flow case: the deflection of a jet generated by flow from a nozzle by input flow directed to it perpendicularly.

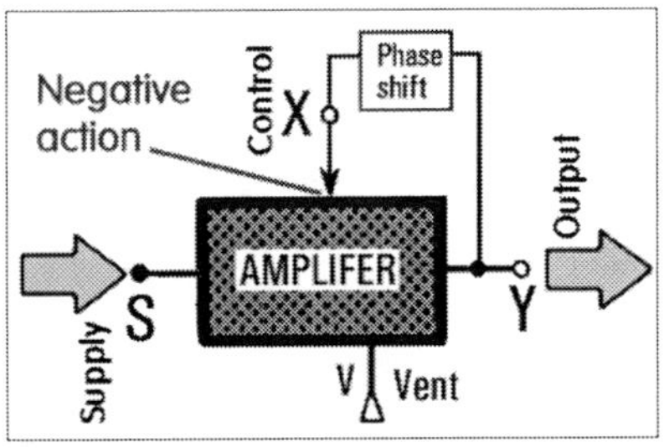


**Figure 4** Basic configuration principle of fluidic oscillators. The key component is a fluidic amplifier, responding to a small input by substantial change in the output flow. Oscillation is then generated by addition of feedback loop. It destabilises the conditions by returning back a part of output fluid flow to act negatively on the flow in the control terminal X.

This jet deflection is currently the common working principle in devices specifically made to serve as fluidic amplifiers. Basic principles of their configuration are in Fig. 4. There are several terminals for the fluid flow into and out the amplifier. At left in Fig. 4 is supply terminal S for bringing into the device a more or less constant fluid flow. Then there is the input terminal X. Into it enters the weak signal that is to be amplified and generate the change – decrease – in the output flow in the terminal Y. Very often (though not absolutely necessarily) there is also a vent terminal V through which leaves the fluid that is directed away from the output terminal Y.

Without any control flow in X in the jet-deflection type amplifier the whole flow coming through S is allowed to continue into Y. There are amplifiers with proportional character of the response: the higher is the magnitude of the control flow admitted into X, the more is the supplied flow deflected into the vent V. Many amplifiers used in fluidic oscillators respond differently: by switching the flow between Y and V. This is achieved by using the Coanda effect of the main flow attaching to a nearby attachment wall.

The other necessary effect, which turns the amplifier into the oscillator, is feedback action. It is present in such self-excited oscillation processes like the *"friction tones"* in trailing vortex streets past bluff bodies [5], but there it is not immediately apparent. Its typical form in fluidic oscillators is a flow channel connecting the output terminal Y with the control inlet X, as schematically presented in Fig. 4. The feedback enables a part of the output flow from Y returning back into X. There are two aspects this feedback must have if it is to be of use in an oscillator. First, the control action on the main flow coming from S must be such that an increasing control flow into X causes a decrease (and not increase) of the output flow in Y. The other aspect - which may be also fulfilled quite automatically - is the presence of some phase delay acting in the loop. This may be simply due to the time it takes to the fluid flow to get from Y to X.

To provide an example how the fluidic oscillators are arranged, the next Fig. 5 presents a simple example. The particular case shown there, with a monostable amplifier is not, as a matter of fact, usual. It has, nevertheless, two advantages for the present discussion.

DOI: 10.1051/epjconf/201714302129

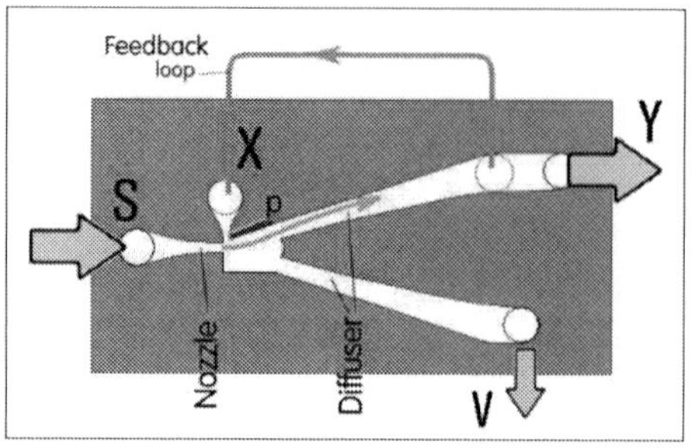


**Figure 5** An example of the simplest fluidic oscillator. It consists of cavities and channels made – as now usual — by photoetching in a thin flat plate. The jet, generated in the supply nozzle, attaches by the Coanda effect to the preferred attachment wall p – from which it is switched away into V by the feedback flow coming through the loop.

First, it is simple so that the character of fluid flows may be in Fig. 5 easily imagined. Second, it was exactly this configuration which in the schematic representation - Fig. 6 - was originally accompanying the description of invention in the Patent [4].

Characteristically for present-day fluidics, the amplifier of the oscillator in Fig. 5 is made by etching the channels for fluid flow in a small rectangular plane plate. For closing the channels (and thus preventing the fluid in them from escaping), the cavities are covered on top and bottom by similarly shaped rectangular flat cover plates. The terminals S, X, Y, and V were in this case arranged as holes drilled in one cover plate so that the fluid enters them perpendicularly.

Let us follow in Fig. 5 the air flow entering through the supply terminal S. By issuing from the supply nozzle the air forms a jet, the flow of which is shown in Fig. 5 as the curved blue line. By the Coanda attachment

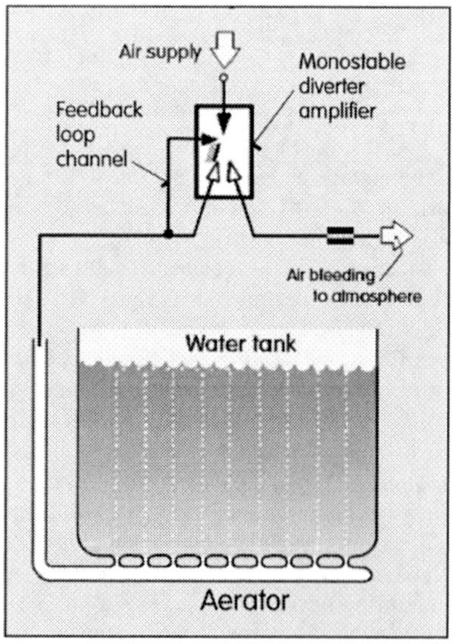


**Figure 6** Microbubble generator principle. This drawing was originally prepared for the Patent Application [4].

mechanism, this jet attaches to the attachment wall p and is led into the output terminal Y. This channel leading into Y is shaped as a diffuser — i.e. its cross-section area gradually increases in the flow direction. Note that this is an opposite effect to what happens in the nozzle connected to S, where the cross-section area decreases in the direction of flow. It input pressure energy is in the nozzle gradually converted into kinetic energy. Then, in the diffuser, the air flow is slowed down, its pressure increases. Thus in the interaction region near the wall p there is a pressure minimum, with pressure lower than in Y. It is this pressure difference by which the air flow is driven through the feedback loop channel.

Upon arrival into X the feedback flow impinges upon the main jet at a right angle. This forces the jet to stop its attaching to the attachment wall p and separate. All the flow, from S as well as from X is thus diverted into the vent V. This, of course, means there is no flow through the output terminal Y — and hence also no flow through the feedback channel. The deflecting action of the feedback flow (after a short phase delay) stops acting. The jet from S returns to the attachment wall p. It attaches there and is led to the output terminal Y. The whole circle of event can begin anew in another oscillation cycle. The flow intoY thus oscillates. The use of this oscillation for microbubble formation is schematically represented in Fig. 6.

There are two obvious disadvantages of this simple oscillator with the monostable amplifier of Figs. 4 and 5, i.e. with the single attachment wall p. One of the shortcomings is less reliable switching. The effect of the flow separation from the attachment wall is easily influenced by various disturbances. As a result, the oscillation pulses tend to be of unequal duration. More important is the second shortcoming, immediately visible in Fig. 6. A large amount of the compressed driving air is lost by escaping through the vent V into the atmosphere. This problem is solved in oscillators which are now in use — but they are somewhat more complex, unsuitable for this introductory description.

For some time after the discovery of the microbubble generation the actual mechanism of processes taking place at the exits from the aerator remained not clearly understood. Significant progress in the understanding brought study of images obtained using high-speed video-recorder, as they are discussed in both parts of ref. [6]. It was discovered that the essential influence on the size of generated bubbles has the conjunction phenomenon. Small microbubbles rise in the liquid so slowly, that a number of them while they are still at or near to the aerator exits. They get into contact and coalesce. Moreover, after each coalescence the rising motion is practically stopped. This means more bubble contacts and more conjunctions — the one leading the air into the vent V (Part I in [6]). A way from this effect of unwelcome bubble size growth was found by returning the already formed microbubble back into the aerator channel. There they remain for a part of the generation cycle. This protects them from immediate contact. Also, this return movement causes bringing into the aerator channel some liquid. The inertia of this liquid column means that in the subsequent expulsion phase of the microbubble it is accelerated so far from the aerator passage exit, that the undesirable conjunction becomes impossible.

Obviously, this returning movement of the microbubble and its neighbour liquid column in the passage is possible only if for the some part of the cycle the pressure in the amplifier cavity inverts its sign – i.e. generates suction. This is made possible by another diffuser in the oscillator — the one leading the air into the vent V. This importance of the other diffuser was initially not recognised. It was its discovery (in Part II of ref. [6]) that made the adjustment of oscillator properties a fully reliable procedure.

## 3 Qualitatively different properties of microbubbles

Microbubbles are generally believed to be essentially the same as the larger bubbles, differing only by their smaller size. Also the limit size by which they are defined, the diameter d = 1 mm, is by many researchers considered just a simple and convenient number, without any physical importance.

Perhaps rather surprisingly, detailed studies of some processes involving microbubbles have revealed a behaviour in some aspects qualitatively different from that of larger bubbles. First of all, the larger bubbles do not posses a strictly spherical shape. Even if they are stationary, their geometry is deformed by the hydrostatic pressure difference between the bubble top and bottom. In microbubbles the internal pressure is so much higher that a deviation from sphericity is negligible.

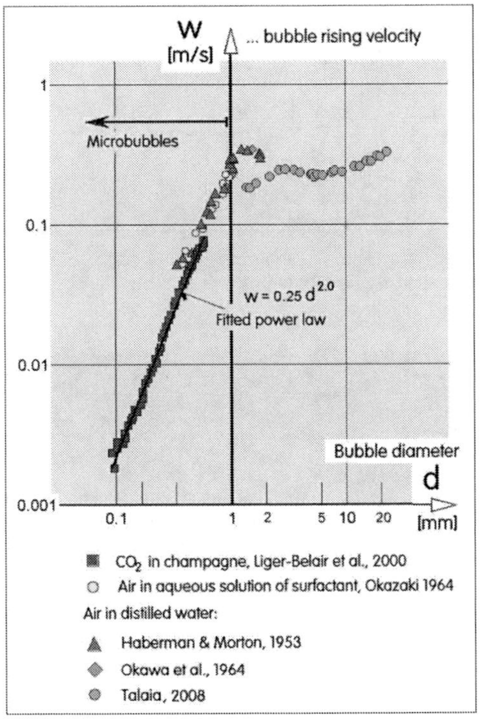


Figure 7 Collection of experimental data about rising velocity of bubbles and microbubbles in the liquid. Quite important fact is the change in the behaviour at the microbubbles' upper size limit d = 1 mm - though it was originally chosen only for convenience and simplicity.

Even more shape differences between the practically always spherical microbubbles and larger bubbles are apparent in studies of the microbubble ascent in the liquid towards the surface above. Large bubbles tend to be deformed by hydrodynamic forces. Also the character of the ascent velocity w [m/s] dependence on the bubble diameter d is different in the two cases. It is remarkable visible in Fig. 7, which presents collected experimental data of various researchers. Microbubbles rise at a velocity practically proportional to the square of their diameter. Ascent velocity of larger bubbles varies in a complex way, with the velocity being roughly the same

Other differences in behaviour are associated with the propagation character of various processes, like the heat transfer schematically represented in Fig. 8. In larger bubbles the propagation inwards takes place in the form of boundary layer with progressively growing thickness – while the microbubbles, being very small, became affected practically simultaneously throughout their full internal space.

Based on a similar factor, demonstration by Al-Yaqoobi, Hogg, and Zimmerman [7] have found another remarkable difference. They have shown potentially very important qualitative properties of microbubbles in separating ethanol-water mixture — or even azeotropic mixtures not separable by traditional distillation process. In the industrially important ethanol-water case, the traditional distillation approach requires large amounts of heating energy to get the liquid to its boiling point. The concept of isothermal microbubble distillation is to heat the gas phase rather than the liquid phase to achieve separation. The advantage of the microbubble distillation was in achieving vapour concentration higher than that what can be obtained under the equilibrium conditions.

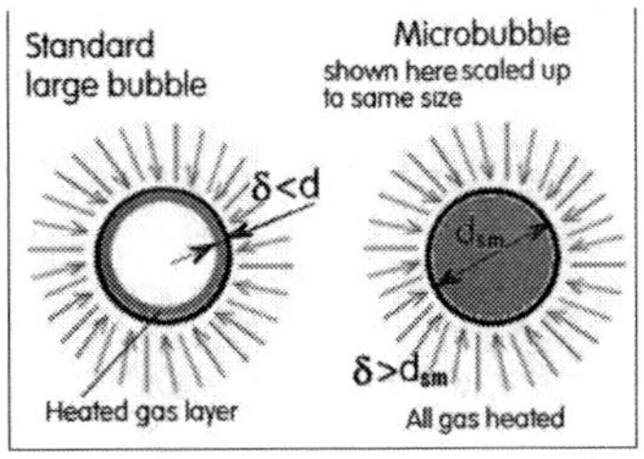


**Figure 8** What appears to be outwardly another qualitatively different property of microbubbles is the simple consequence of their diameter $d_{sm}$ smaller than the distance $\delta$ of diffusive transport in the gas.

## 4 Applications review

The first impression from consideration of already existing applications of microbubbles – as well as the new ones currently suggested in various literature sources as potentially attractive — is their number is surprisingly large. Some idea about the numbers is available in the review illustration Fig. 9. The application cases are there grouped into families with their names in red ellipses. Individual uses belonging to a family are named inside blue ellipses.

Admittedly, some of the applications listed there are just first ideas, not having yet progressed to an industrial scale. As a matter of fact, some of the listed cases (typically among them is the idea of changing ocean albedo) would require so huge financial investments that they are not likely to be pursued in foreseeable future. Fig. 9 takes this into account by arranging the applications in the degree of their realisation possibilities. Those already used and developed, like the case A (note the increased shade of the background), are drawn near the left top corner – while those at present less likely to employ industrially are at right in the bottom pf the image.

What should be also noted in this illustration are the connecting lines drawn between some application families. They put together those cases that, although belonging to different use areas, are based on physically related physical principles. A typical example are the microbubbles use to separate components from a mixture. There are such separation processes in which the obtained substances are discarded as unwanted – yet similar separation processes, on the other hand, are cases where the extracted substance is the very object of the whole application effort.

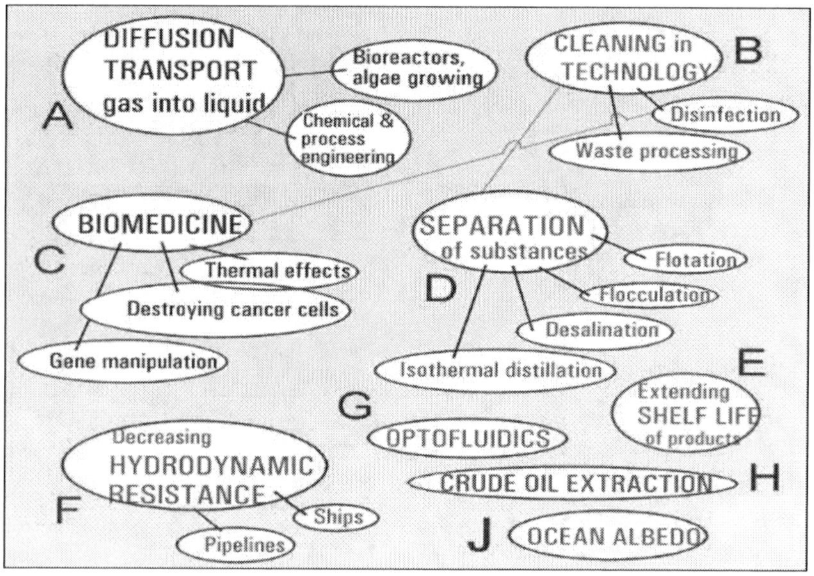


**Figure 9** Survey of currently investigated microbubble applications that were made possible (and in some cases economically feasible) by oscillation generated by fluidics. Currently the development is particularly rapid in bioreactors and in medicine.

## 5 Enhanced transport phenomena

The family **A** in Fig. 9 was the initial reason behind the renewed (or increased) interest in microbubbles. The idea of increasing the total contact surface by division of the given total volume into many small-size objects is obvious and thus the first generating of microbubbles for this purpose, Figs. 10 and11, was immediately accepted without a need for deeper investigation. Recent interest is primarily in quantifying the effect microbubbles have on the overall gas transport efficiency in the liquid. A paper on this subject – for the particular case of $O_2$ transport to water in which it dissolves – is e.g. ref. [8]. The dissolved oxygen in water is, in particular, needed to support existence of aerobic bacteria, presence of which is necessary in various biochemical processes. Of recent interest are also the interphase transport processes in microbubble-sparged airlift reactors, discussed in [9]. Fluidic oscillators can generate in the airlift loop a microbubble cloud. Mass transfer through such clouds

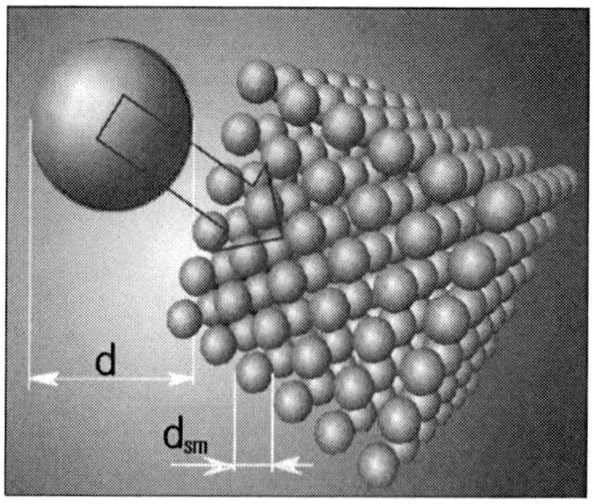


**Figure 10** Increase of total collective microbubble surfaces as a function of their number of bubbles. What is assumed is isometry, i.e. the same size of all small bubbles obtained by the division of initial air volume.

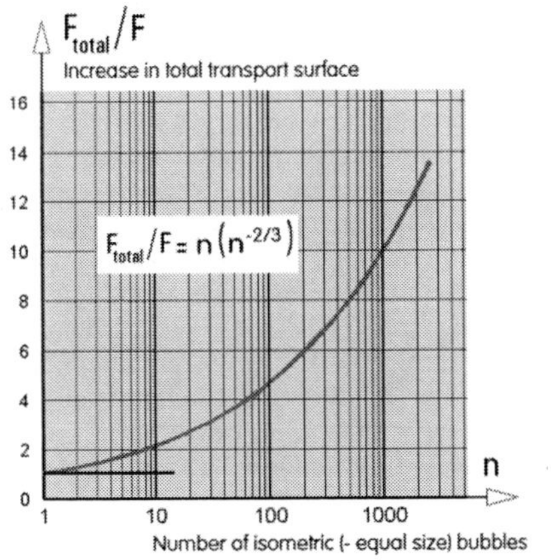


**Figure 11** Dependence on the number n of the small bubbles from Fig. 10 of the transport surface F increase (assuming the same size and perfectly spherical shapes).

in the case of carbon dioxide diffused into water was investigated, e.g., in [10]. The efficient delivery of $CO_2$ to simple, unicellular plants living in water is a subject of extremely high contemporary importance. Green plants, including the simple algae, possess the extraordinary capability of generate by photosynthesis various hydrocarbon compounds needing only water, light, and $CO_2$ from water. It is no problem for contemporary chemistry to convert the resultant product into the liquid fuel for powering e.g. automobiles. The advantage of simple organisms used for this purpose is their extremely fast reproduction cycle – providing they obtain suitable conditions. This idea is not at all surprising. The photosynthesis is, after all, the same process by which were generated simple plankton organisms, from which the crude oil was made ~ 300 million years ago. Imitating the same procedure at an industrial scale can solve the grave problems of geopolitics – the unbalance between the countries possessing the crude oil sources and those that are the largest consumers. A recent comprehensive review of the tasks and problems aiming at the particular case of bioethanol is available in [11] and more general recent progress in developing oil from algae is discussed in [12]. Pilot scale studies and specifically a design of an airlift bioreactor for the algae growing purpose with fluidic oscillator induced microbubbles are described in detail in. [13].

## 6 Microbubbles used for cleaning

Important increase in effectiveness over earlier techniques is offered by microbubbles also in the applications summarised in Fig. 9 as the case **B**. It is the cleaning and waste removal as well as processing. Typical wastewater processing plant [14] actually entrusts the cleaning process to aerobic bacteria (anaerobic ones can perform the procedure as well, but slowly and with roughly 10-times lower efficiency). The basic problem of the aerobic organisms is general absence in household wastes of oxygen they need – and without which they die. This has been recognised long time ago and practically all present-day plants include some form of aeration, which consumes some 70 % of the typical plant running costs. Unfortunately, present day facilities usually employ for aeration some forms of paddlewheels, with efficiency which is very poor An early development of microbubble aerator for higher-efficiency waste water treatment using aerobic activated sludge was described by Terasaka et al. in [15].

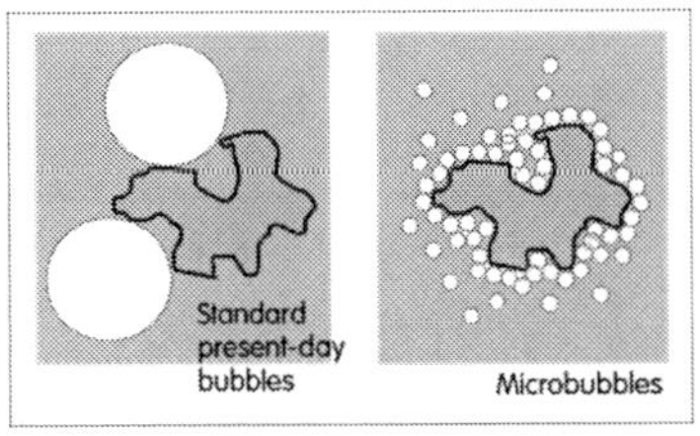


**Figure 12** The removal of dirt - or, generally, different components - involves complicated action of ionts on the bubble surface. Nevertheless at least some of the intensive removal effect by microbubbles is explained by their small size and geometrically easier attachment to the complex surfaces.

Another process which microbubbles take part in cleaning tasks is using their strong clinging to contaminants like oil or grease, cf. Fig. 12. The fact of substantial importance is they make possible an removal of dirt from processed water environmentally friendly. The process does not work with any chemical detergents. This was discussed in [16]. As an example, in [17] is quantitatively evaluated the use of microbubbles to de-contamination of manufactured silicon wafers. Their yield losses over 50% are so far caused by micro-contamination, difficult to remove by present methods. It is demonstrated in [17] to be easily and environment friendly removable by microbubbles. In fact, microbubbles were demonstrated by Tsuge et al., [18] even to perform disinfection by removal of pathogens. The mechanism presented in Fig. 12 is nowadays offered on a commercial basis to cleaning skin and hairs of home pets, according to Fig. 13. Human use of this disinfection is now still in a stage of administrative approval tests – suggesting, nevertheless, the potential use in biomedical

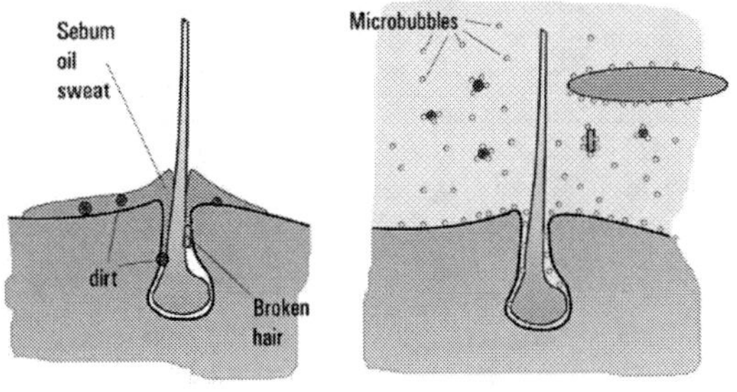


**Figure 13** Cleaning skin and hairs by microbubbles is based on their capability to dislodge dirt particles and secreted sebum. Advantage, apart from more intensive effect, is the total absence of any chemical cleaners.

applications. Schematically represented in Fig. 14 is another essentially cleaning type application. It is a use in a technologically quite distant area: extracting crude oil (or natural gas) from what has been so far considered to be „aged" and no more useful wells — especially those in which the extracted fluid is firmly held in tiny pores and fissures in the rocks. Microbubbles generated in the downhole oscillator can enter the crevices and displace the exploitable fluids that would otherwise remain there as not mineable. In addition, the existence of tiny microbubbles, only extremely slowly moving relative to the surrounding crude oil, form together with it a mixture

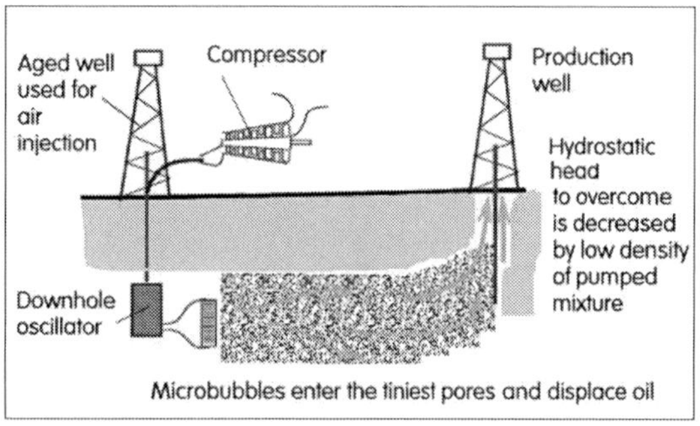


**Figure 14** Although belonging to a very much different application family – the proposed crude oil and gas extraction should use, as underlying principle, the separation and cleaning effect by creeping into tiny cleavages and displacing from there the extracted fluid.

exhibiting very small overall density. This is reflected in the correspondingly very low hydrostatic pressure that has to be overcome in the extraction process in the production well.

## 7 Biomedical uses

Microbubbles found application **C** in medicine very early – and in a wide spectrum of uses. Again, some of uses belong to the diffusion and/or reaction controlled transport phenomena – for example ref. [19] investigated transport and reaction-controlled dissolution of oxygen microbubbles in flowing blood.

The most successful early application of microbubbles present in human body fluids was due to their ability to scatter and reflect ultrasound – enhancing substantially this diagnostic method [20]. Besides the pre-clinical ultrasound imaging, There are clinical application, like the assessment of myocardial perfusion reported in [21]. Microbubbles can also convert energy of the ultrasonic vibration into a local thermal therapeutic effect [22]. Streaming effect on microbubbles in ultrasonic field can destroy cells [23] and can cause permeabilisation of membranes for drugs [24] in particular the anti-cancer ones [25, 26]. Microbubbles of suitable composition possess ability to target specific cellular markers of disease and deliver localised chemotherapy. Recently demonstrated was capability to switch on mechanisms of gene therapy and enhance lesion ablation through cavitation. Microbubbles as drug carriers are traced inside the body using a low-intensity ultrasound and burst at the target site by applied high-intensity ultrasound. As a particular success is considered the recently demonstrated drug delivery to the brain by focused ultrasound and microbubble-mediated disruption of blood-brain barrier [27, 28]. Of large importance are the recent uses for tumor ablation, sonothrombolysis, and stroke therapy. Microbubbles were also demonstrated in use for measuring absolute blood pressure [29]. Systems with microbubbles have been developed for screening of peptides specific to colon cancer cells and also for capturing circulating tumor cells to provide early cancer diagnosis [30].

## 8 Separation techniques

This family **D** of microbubble applications includes mainly the techniques already mentioned above as belonging to the respective application families, as shown in Figs. 12, 13, and 14. There are, however, uses in areas that are seemingly quite unrelated. One of them is mineral engineering [31]. The clinging of microbubbles to small particles produced by grinding is used for separation by flotation technique. This enhanced separation by flotation, together with advances in grinding, allow now economic exploitation of deposits that a decade ago would be described as too low-grade and therefore uninteresting.

Flotation on one hand – and its opposite, gravity sedimentation — are currently an object of intensive research associated with the farming of algae discussed above in sect. 5. as of immense interest for generating renewable petrol. While harvesting of some heavier algae is more efficient by use of sedimentation, other algae taxonomic group have their settling velocity in water so low that microbubble-facilitated flotation is a better alternative. Ref. [32] presents an example of microflotation through microbubble clouds.

Another separation technique that may benefit from use of algae is flocculation. Also in this case the interest concentrates on identification of the least expensive harvesting of algae. In [33] is discussed separation from urban landscape water using a high density microbubble layer enhanced by micro-flocculation.

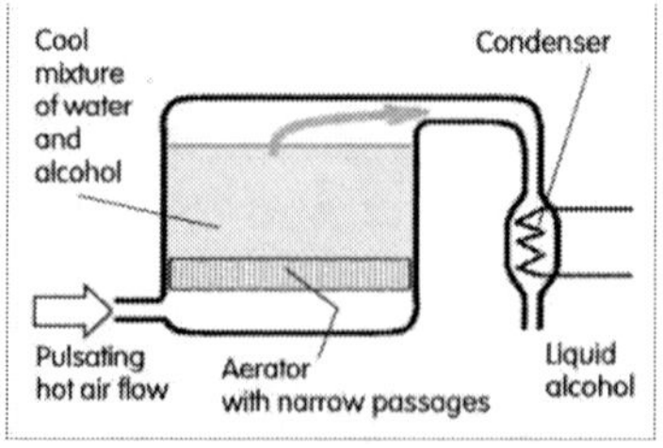


**Figure 15** Schematic principle of cold-liquid distillation. Microbubbles generated in aerator by pulsating hot gas flows contain evaporated surrounding mixture of water and ethyl alcohol and saturate with them. As they gradually cool, the water condenses and leaves the bubble.

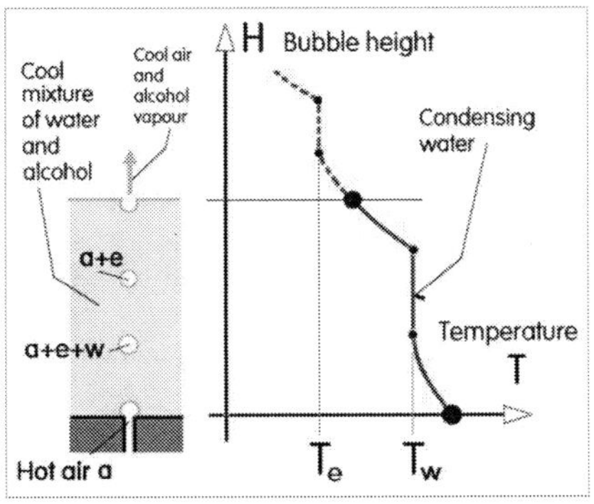


**Figure 16** Schematic temperature dependence on height of rising initially hot air microbubble in the distillation process. Water condenses and leaves the bubble before reaching the surface. The advantage is no heating of most of the liquid, saving the energy.

Energetically very important distillation by microbubbles may be explained by means of Figs. 15 and 16. The fluidic pulsation of the hot air fed into the aerator generates hot microbubbles. They become soon saturated by vapours of water (boiling at $T_w$ = 100 °C) and ethyl alcohol (boiling at $T_e$ =78.36 °C) from the surrounding liquid mixture. Soon the microbubbles, exchanging heat with the liquid, cool to $T_w$ and the condensing water leaves the bubble. Remaining air and alcohol vapour microbubble reaches the liquid surface before its temperature decreases below $T_e$. It bursts there and frees the alcohol, which is then converted into the liquid in the condenser, Fig. 15.

Finally, eat another separation technique of high and increasing importance is producing potable water from sea water by desalination. The highest perspective is seen for reverse osmosis desalination by semi-permeable membranes [34] with membrane efficiency improvement by microbubbles.

## 9 Food production

Microbubbles also find applications in producing food, **E**. They can bring two advantages over the current state of art. One of them is significant improvement of sensory qualities of dietary products, especially the smoothness that is demonstrably higher. US Patent [35] also covers potential of microbubbles as fat replacer The low-fat food contains aqueous phase, oil droplets, and protein coated air microbubbles. The other way towards improvements in food industry is extending the shelf-life of products of foamy character – such as whipped cream, salad dressings, sorbets, and mousses. Some of these products, typically ice-creams, have to be made by a specialised machinery on the spot to retain their immediately recognisable freshness – which is rather expensive. It is a considerable advantage that a foam of microbubbles was demonstrated to remain stable for up to a year. Gas-to-liquid mass transfer being the rate-limiting step in aerobic fermentations, improving the efficiency by using microbubbles is obvious. Also of interest for food industry is water evaporation ability of microbubbles [36] without the heat reaching the dried product.

## 10 Reducing the friction drag

Decreasing the resistance to motion of a body in water by blowing air bubbles on the water contact surfaces **F** – such as under the ship or boat bottom - has been a quite old idea, Fig. 17. It was intuitively felt a way to significant improvement of transportation efficiency – although it should not be forgotten that it can decrease only the friction resistance and not the other source of the total drag – the generation of waves on water surface. Tests of the idea, actually already with microbubbles, were performed and described in [37] as long time ago as in 1973, i.e. before the era microbubbles generated fluidically. A quite large reduction in the frictional resistance component under laboratory conditions (no wave drag) was demonstrated.

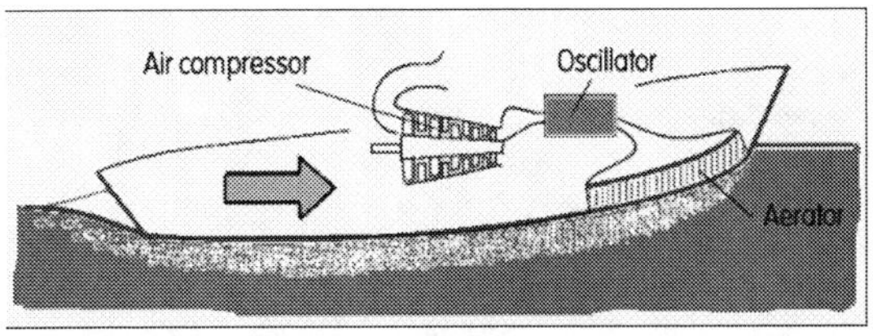


**Figure 17** Schematic principle of increasing the speed of a fast boat by injecting microbubbles into water at the bow.

Discontinuation of the progress was due to the inefficient method used to microbubbles generation. The gas was actually not air but hydrogen, generated by water electrolysis. The generation unit required and spent more energy than was saved on the propulsion side. The same problem of inefficient microbubble generation at the quite large scale, required to obtain reasonable Reynolds numbers, plagued also other later attempts, including those that were able to show remarkable friction drag reductions up to 80% — i.e. decrease to mere one fifth of original value [38, 39] – of course with large gas or air flow rates. Ship model tests with microbubbles have been already made at size scale near to that of practical use — there are reports of practical tests of Indonesian Navy fast patrol boat [40].

For process engineers may be important the corresponding decrease by microbubbles of fluid flow frictions in pipes and tubes [41, 42].

## 11 Applications at microfluidic scale

Many of the currently developed techniques of using microbubbles aim at a very small, microfluidics scale applications. A use was found in gene manipulation [46], biosensors [47], mixing of reactants for chemical microreactors [48] or sorting small object transported in a fluid [49].

Of these more or less potential applications, real importance is expected in the new field of optofluidics,

**G**, where ordered microbubble configurations, e.g. microbubble linear arrays, can create tuneable optical components [50, 51] for control of laser light.

## 12 Extracting crude oil

One particular aspect of this application **H** was already discussed in Sect. 6 in association with Fig. 14. It is the capability of microbubbles to displace the mined fluids. Another aspect — also in association with the so called "aged" oil wells — should be also mentioned. The large depths from which oil is obtained necessarily generate correspondingly large hydrostatic pressure that must be overcome. As the oil is removed, the naturally present pressure inevitably decreases until at a certain stage the extraction ceases to be useful — even though down there still remains a considerable amount of the oil. The remedy may be bringing to the bottom of the hole compressed air. The mixture of oil and air has substantially lower density and the oil flow may continue — as long, however, as the air remains in the mixture. On the very long travel upwards standard air bubbles coalesce. This finally leads to the situation shown schematically at the left-hand side of Fig. 18. The air separated from the mixture rises through the hole centre while at the walls forms an annular layer of the oil flowing down and not reaching the required upwards location.

Replacing the standard large bubbles with microbubbles, is a reasonable solution. With their slower motion in the oil - and also the very long vertical path before their coalescence reaches any dangerous level - makes extraction of the oil (at right in Fig. 18) again profitable.

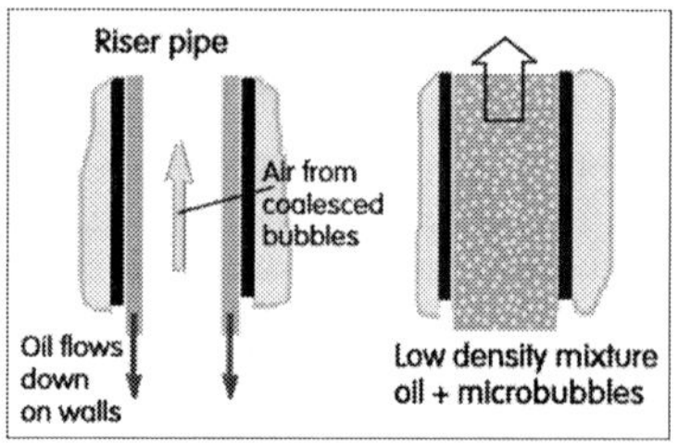


**Figure 18** Blowing air downhole does not secure getting the oil to the surface. Standard large bubble coalesce (left) and the result is blowing up the released air alone while the oil flows down on riser pipe walls. With microbubbles (right) is blown up a mixture causing smaller hydrostatic pressure.

## 13 Increasing albedo of oceans

Albedo is the term used to characterise the "whiteness" of planetary surface exposed to sun light. The higher is the albedo value, the larger is the amount of thermal power that is reflected from the surface (and therefore the smaller is the remaining amount that is absorbed).

Albedo of ice - especially when covered, as usual, with fresh snow, Fig. 19 - is very much near the top of the scale from 0.0 to 1.0. On the contrary, albedo of deep ocean is, near the very bottom of this scale. In the glaciers and ice melting under the present condition of global warming there is an interesting negative feedback phenomenon. The more ice is melt in the Arctic (and the lower is the decreasing albedo of the water surface there), the more thermal power is absorbed and the faster is the ice removal.

In principle at least, there is a possibility discussed in [51] how to prevent this feedback from closing itself: generating microbubbles on the ocean surface can replace the missing ice albedo.

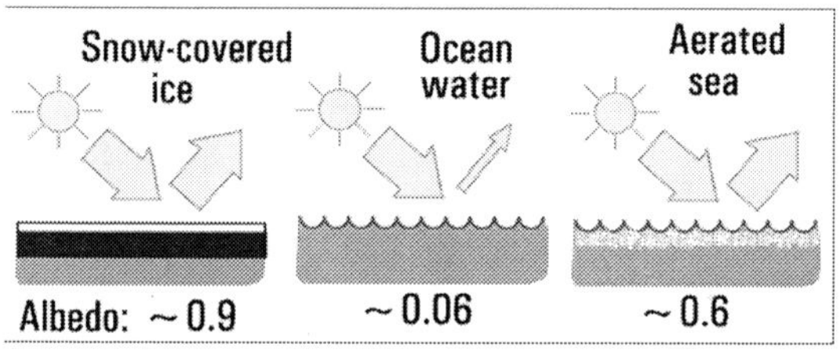


**Figure 19** The global warming may be suppressed by sunlight reflecting microbubbles blown into an ocean surface layer to increase their albedo.

## Conclusions

Fluidics, the technique of generating and handling fluid flows – preferably without an action of mechanical components – came recently with a so far not widely known possibility how to generate very small, sub-millimetre sized gas bubbles in liquids. components. Perhaps even less that the microbubble generation principle has been so far known the reason why it should be useful and profitable to generate them. Present paper contains a survey and classification of the microbubble applications that may provide an inspiration for further research in this area.

## Acknowledgements

Author obtained support from GAČR - Grant Agency of the Czech Republic, by research grant Nr. 13-23046S. There was also institutional support RVO:61388998.

## References

1. Prevenslik T., Proc. of Conf. *'Topical Problems of Fluid Mechnics'*, p. 113, Prague 2014
2. J. Rodríguez-Rodríguez et al., *Annual Review of Fluid Mechanics*, 47, 405,(2015)
3. V. Tesař V., C.-H. Hung, W.B.J. Zimmerman, *Sensors and Actuators A*, 125, 159, (2006)
4.. W. B. J. Zimmerman, V. Tesař, *European Patent.* EP2081666, filed Oct. 2007

5. Z. Trávníček, *Engineering Mechanics* **13,** 67, (2006)
6. V. Tesař, *Chemical Engineering Science*, 116, 843, (2014)
7. A. Al-Yaqoobi A., D. Hogg, W. B. Zimmerman, *International Journal of Chemical Engineering*, ID 5210865, 2016
8. T. Andinet, I. Kim, J.-Y. Lee, *Desalination and Water Treatment*, in Press (2016)
9. A.M. Al Taweel, A.O. Idhbeaa, A. Ghanem A., *Chemical Engineering Science* **100** 474 (2013)
10. M.K.H. Al-Mashhadani, H.C.H. Bandalusena, W.B. Zimmerman, *Industrial and Engineering Chemistry Research* **51** 1864 (2012)
11. A. R. Sirajunnisa, D. Surendhiran, *Renewable and Suistainable Energy Reviwes* **66** 248 (2016)
12. K. Kumar et al., *Renewable and Sustainable Energy Reviews* **65** 235 (2016)
13. W. B. Zimmerman et al., *Appl. Energy* **88** 335 (2011)
14 Zhang L. et al., *Water Science and Technology* **74** 138 (2016)
15 K. Terasaka et al., *Chemical Engineering Science* **66** 3172 (2011)
16. K. Wataneabe et al.,: *Proc. of FLUCOME 2013, 12th Intern. Conf.*, Nara, Japan, November 2013
17. Xi, X., *PhD Thesis, Mechanical Engineering Department*, Imperial College London, 2012
18 Tsuge H., et al., *Kagaku Kogaku Ronbunshu* **35** 548 (2009)
19. M. Fischer, I. Zinovik, D. Poulikakos, *International Journal of Heat and Mass Transfer* **52** 5013 (2009)
20. P. Kogan, R.C. Gessner, P.A. Dayton, *Bubble Sci. Eng. Technol*, 2 3 }2010)
21. L. Rotaru, T. Nanea, *Journ. Med. Life* **8** 471(2015)
22. T. Kanagawa, *Proc. of FLUCOME 2013, 12th Intern. Conf.*, Nara, Japan
23. K. Kooiman et al.: *Journal of Controlled Release* **154**, 35, (2011)
24. J. S. Oh J. S. et al., *Computers in Biology and Medicine* **44** 37 (2014)
25 Y. Watanabe et.al., Cancer Science **99** 2525 (2008)
26. S.-T. Ren et al., *Physics of Fluids*, **27** 356 (1984)
27. W.M. Pardridge, *NeuroRx*. **2** 3 (2005)
28. T. Q. Nan, PhD Thesis, Univ of Toronto, Canada (2005)
29. C. Tremblay-Darveau et al., *Ultrasound in Medicine and Biology* **40** 775 (2014)
30. Y.-H. Hsiao, *Scientific Reports*, **6** 32454 (2016)
31. E. Waters, K. Hadler, J.J. Cilliers, *Minerals Engineering* **21** 918 (2008)
32 J. Hanotu, H.C.H. Bandulasena, W.B. Zimmerman, *Biotechnology and Bioengineering*, DOI **10**.1002/bit.24449
33. S. Chen et al., *Water Sci. Technol.*, **70** 811 (2014)
34. H. Hasegawa, S. Otsu, *Transactions of the Japan Society of Mechanic Engineers, Ser. B* **77** 2049 (2011)
35. I.T. Norton, P.W.Cox, Tchuenbou-Magaia F. L., *US Pat. US8647696 B2*, filed 2008
36. W. B. Zimmerman, M. K. H. Al-Mashhadani, H.C.H Bandulasena, *Chemical Engineering Science* **101** 865 (2013)
37. M. E. McCormick, R. Bhattacharyya, *Naval Engineeris Journal* **85** 11 (1973)
38. V.G. Bogdevich et al., *Proc of2nd Second Internat. Conf. on Drag Reduction* BHRA, Cambridge, (1977)
39. N. K. Madavan et al., *Phys. of Fluids* **27** 356 (1984).
40. Yanuar et al., Mar. Sci. Technol **17** 301 (2012)
41 O. Watababe, A. Masuko, Y. Shirose, *J. Soc. Naval Archit.* **183** 53 (1998)
42 Y. Moriguchi, H. Kato, *Journal of Marine Science and Technology* **7** 79 2002
43 M. J. Pang, J. J. Wei, B. Yu, *Ocean Engineering* **81** 58 (2014)
44 M.M. Shams, M. Dong, N. Mahinpey, *Chemical Engineering Science* **112** 72 (2014)
45 R.R. Sun et al., *Journal of Controlled Release* **182** 111 (2014)
46 L. A. Kuznetsova, W. T. Coakley, *Biosensors and Bioelectronics*, **22** 1567 (2007)
47 J. H. Lee et al., *Sensors and Actuators A* 182 153 2012
48 C. Wang, *Biomicrofluidics* **6** 012801 (2012)
49 M. Hashimoto et al., *Small* **2** (2006)
50 A. Allouch, *Microfluidics and Nanofluidics* **1** (2004)
51 N. Mengis, D.P. Keller, A. Oschlies, *Climate Engineering Conference*, Berlin, Germany (2014)

EPJ Web of Conferences **143**, 02130 (2017) DOI: 10.1051/epjconf/201714302130
*EFM 2016*

# Verification of ANSYS Fluent and OpenFOAM CFD platforms for prediction of impact flow

*Petra* Tisovská[1], *Pavel* Peukert[1], *Jan* Kolář[1]

[1]VÚTS Liberec, a. s., 460001 Liberec XI, Czech Republic

**Abstract.** The main goal of the article is a verification of the heat transfer coefficient numerically predicted by two CDF platforms – ANSYS-Fluent and OpenFOAM on the problem of impact flows oncoming from 2D nozzle. Various mesh parameters and solver settings were tested under several boundary conditions and compared to known experimental results. The best solver setting, suitable for further optimization of more complex geometry is evaluated.

## 1 Introduction

An impact flow is widely used for a very intensive drying effect of wet surfaces. Although it's good performance is generally well known, there are still many geometric parameters to be exploit regarding their influence on drying. CFD simulation of nozzle designs may be accepted as a technically easiest approach to find the most efficient geometry of the nozzle. Unfortunately the fluid mechanics of the impact flow is very complex and demanding to be numerically simulated. Particularly, the prediction of near-wall turbulence parameters of the flow of various regimes along the wall for, is the most obvious problem.

For the examination of the impact flow were used two highly sophisticated CFD packages. The commercial software ANSYS-Fluent 17.1 and open source package OpenFOAM 4.0. Both solvers computed the same flow problems of impact flow oncoming to a normal wall and results are later compared with experimental data.

## 2 Investigated task

For investigation purposes the geometry based on the experimental work of Korger and Křížek [1] was used. They determined averaged local mass transfer coefficients of an impact flow from a slit nozzle. The impact flow is used commonly to increase the heat and mass transfer from the wall (Polat[2], Martin[3]). Because the heat and mass transfer physics in the boundary layer are similar one to the other, like mentioned Ghiaasiaan [4] or Zoot [5], it is possible to find a correlation between them.

The slit nozzle scheme of Korger and Křížek's [1] experiment is shown in the Fig.1. Where b is the nozzle orifice width, alpha the nozzle opening angle, L the height between the nozzle orifice and upper wall and s the height between the nozzle orifice and affected wall. The dimension x represents the horizontal coordinate from the centre of the nozzle. The problem is fully two-dimensional.

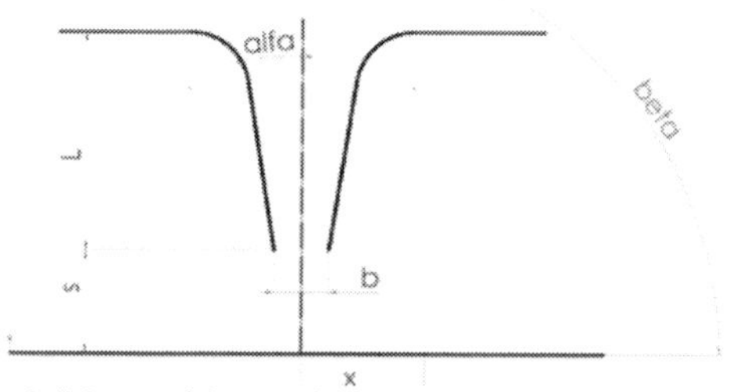


**Fig. 1.** Scheme of the tested nozzle

The resulting mass transfer coefficients for various parameters from their experiment [1] are shown in the Fig.2. Gaining the maximal total surface heat and mass transfer for a limited minimal wall to nozzle orifice distance the case with the s/b ratio of 4 was chosen. Interesting for that case is the presence of two local minimums and two smaller local maximums and a higher total mass transfer coefficient. Which is caused by increased turbulence a resulting mixing at the wall region.

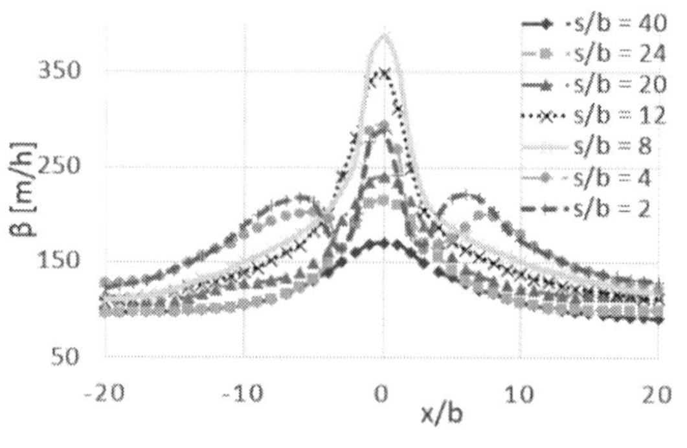


* Corresponding author: pavel.peukert@vuts.cz

**Fig. 2.** Local mass transfer coefficient from Korger and Křižek's experiment [1].

For the computations was used a 4 mm nozzle width with an open angle of 7°. The computational grid (Fig.3) was created in Ansys-meshing and was used for all computations presented. The mesh in near-wall and impact regions is refined to ensure proper resolution for the simulation. Because the computations were done within a parametric study only a 2D computational grid with 150 thousand elements was used.

## 3 Computation conditions

For the Fluent and OpenFOAM the same settings were set where possible. To get the similar conditions to the experiment the pressure at the inlet was set to an overpressure of 480 Pa which causes an average flow velocity of 30 m/s in the nozzle orifice and a temperature of 300 K. The pressure outlet was set to 0 Pa overpressure and a backflow temperature of 300 K. The wall temperature was set to 320 K to ensure a reasonable range to use constant air properties for normal pressure of 1 atmosphere and 300 K.

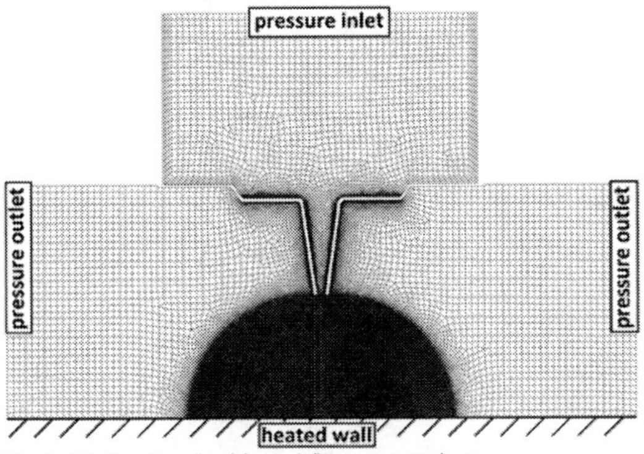

**Fig. 3.** Unstructured grid used for computations

Considering that the flow had significantly turbulent regions, few k-omega and k-epsilon turbulence models were tested in both program environments, because of their relative unobtrusive computing requirements for engineering computations [7].

### 3.1 Used turbulence models

The models of turbulence (eddy viscosity) had different approach how to handle the turbulence and turbulent energy dissipation. Due the different computation platforms some differences in the setting are present. Still the settings and schemes were try to by similar as possible.

In the both environments the RANS equations were fulfilled with the k-omega standard and SST formulation. Further k-epsilon standard, RNG and Realizable were included. Moreover, the difference of the standard and enhanced wall treatment for resolving the near-wall flow parameters was tested for the k-epsilon models in ANSYS-Fluent.

## 4 Heat and mass transfer relations

The boundary-layer conservation equations for momentum, thermal energy and mass species are mathematically similar so an analogy between these three parts was proposed and described in various literature [4], [8]. Due the time the analogy proposes developed from the relative simple Reynolds analogy over few stages to the most accurate Chilton-Colburn Analogy which defines dimensionless j parameters for heat ($j_{th}$) and mass transfer ($j_{ma}$) which were assumed to be near equal.

$$j_{th} = \frac{\mathrm{Nu_l}}{\mathrm{Re_l} Pr_l^{1/3}} \tag{1}$$

$$j_{ma} = \frac{\mathrm{Sh_l}}{\mathrm{Re_l Sc}^{1/3}} \tag{2}$$

That assumption provides a tool to estimate the heat flux or mass transfer values if the main flow parameters and one of the mentioned values are known.

The experimental value of the mass transport coefficient was recomputed to dimensionless heat flux $\dot{q}^*$ which is defined by equation

$$\dot{q}^* = \dot{q}/\dot{q}_{wam}, \tag{3}$$

where $\dot{q}$ is the local heat transfer and $\dot{q}_{wam}$ the weighted averaged mean of the investigated case. The choice to observe heat transfer was done with regard to following research investigations.

## 5 Results from ANSYS-Fluent

From various computations the local heat transfer rates which were converted to dimensionless heat flux using equation 3 were obtained. They were compared to the experiment in Fig.4. The chart shows similar progress with the local minimum in a similar position for the k-omega standard and SST. While the k-epsilon model tendency is to smoothing the curves. The total heat transfer comparison for the k-epsilon RNG with enhanced wall functions had the best match, while other k-epsilon models overestimate the heat flux.

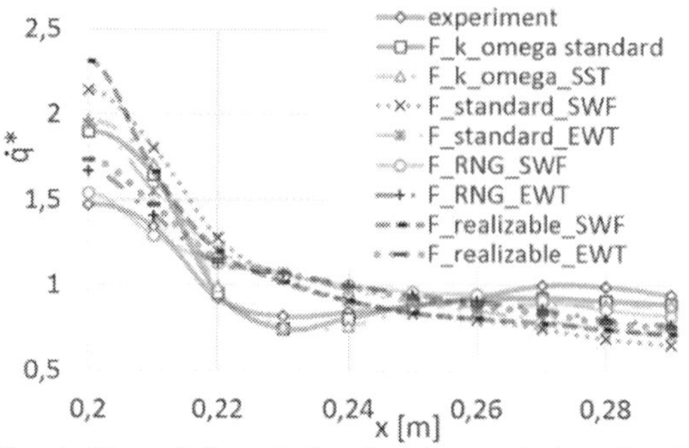


**Fig. 4.** Chart of dimensionless heat flux on horizontal wall distance from the experiment and Ansys-Fluent simulation

In Fig.5 the pressure field from ANSYS-Fluent for the k-epsilon model is shown, which has a relative similar shape like the result in Fig.8 from OpenFOAM with small value deflections. The graphic outlay is a little different, because the pictures for the Fluent solutions were post-processed in ANSYS CFD-post and the solution from OpenFOAM were processed in Paraview.

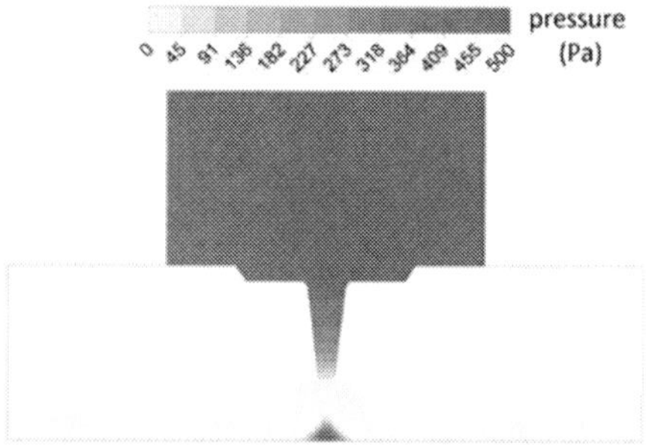


**Fig. 5**. Pressure field from Ansys-Fluent, k-esilon standard

Fig. 6 shows the velocity field from Ansys-Fluent and Fig. 9 from OpenFOAM. With small value deflections.

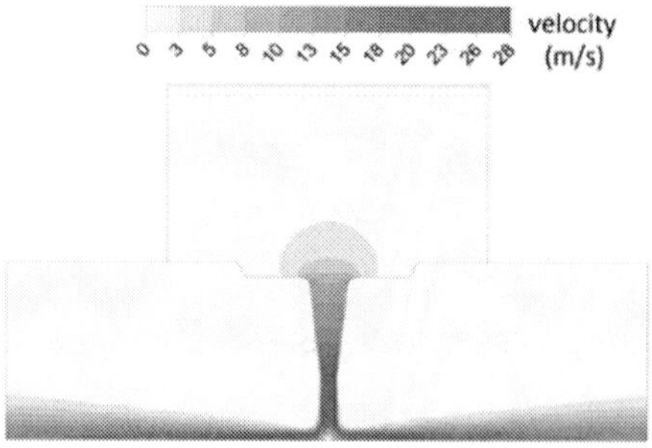


**Fig. 6.** Velocity magnitude from Ansys-Fluent, k-epsion standard

## 5 Results from OpenFOAM

Local dimensionless heat flux for the OpenFOAM results is shown in Fig.7. The progress curves of the dimensionless heat flux is for all cases similar with a little different tilt. None of them matches the experimental results.

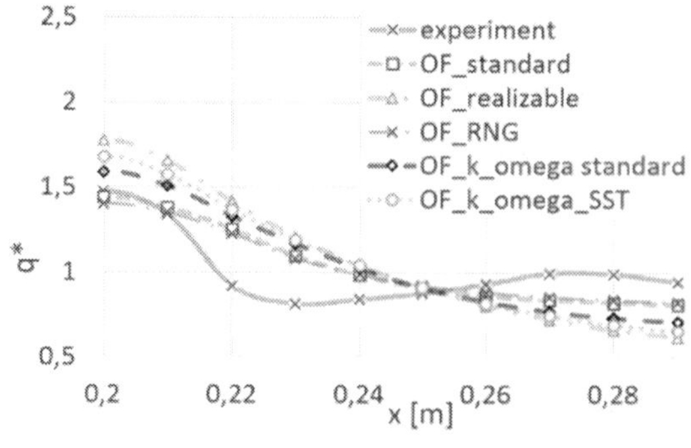


**Fig. 7.** Chart of dimensionless heat flux on horizontal wall distance from the experiment and OpenFOAM simulation

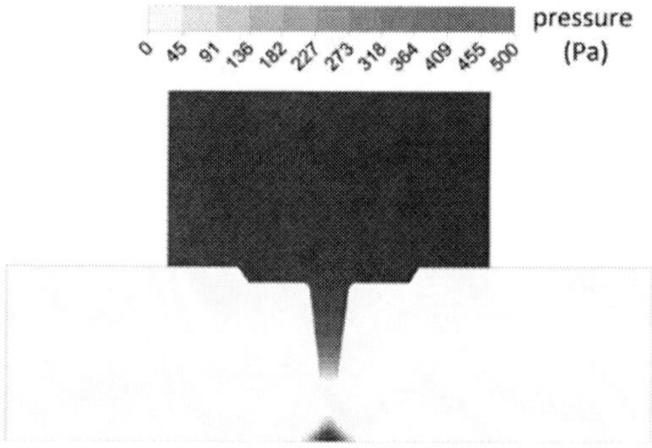


**Fig. 8.** Pressure field from OpenFOAM, k-epsilon standard

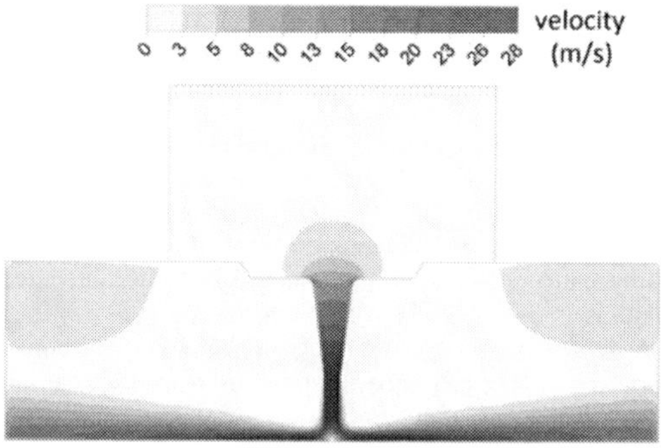


**Fig. 9.** Velocity field from OpenFOAM, k-epsilon standard

## 6 Facing the results from experiment ANSYS-Fluent and OpenFOAM

The k-omega standard (Fig.10) model in ANSYS-Fluent gives tendency compared to the experiment with an overestimation in the central area and underestimation at local minimums, but the positions of these are relative correct. From OpenFOAM a smooth curve with only one global maximum is the result of solution. The SST modifications (Fig.11) gives similar result like the standard model. A small change at the far-sides where the heat flux holds a constant value shows in ANSYS-Fluent results only.

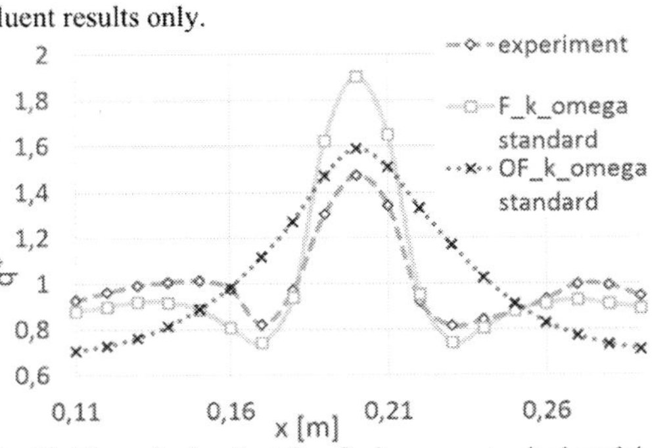


**Fig. 10.** Dimensionless heat flux for k-omega standard model

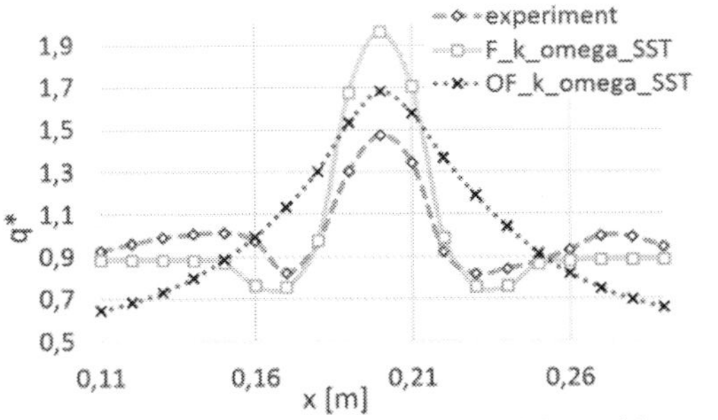


**Fig. 11.** Dimensionless heat flux for k-omega SST model

All k-epsilon turbulent models smoothed out the local minimums. And even when they had a similar total heat flux of the area like the experiment they were not able to resolve properly the problem. The standard k-epsilon (Fig.12) from Ansys-Fluent and OpenFOAM varies especially in the area of the global maximum.

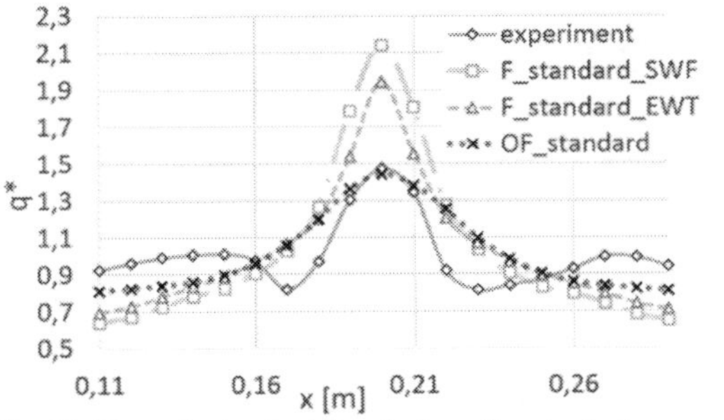


**Fig. 12.** Dimensionless heat flux for k-epsilon standard

The modified formulations of RNG (Fig.13) and realizable (Fig.14) match between both platforms relatively well. Which allows a possible direct comparison between cases computed with these models on both platforms.

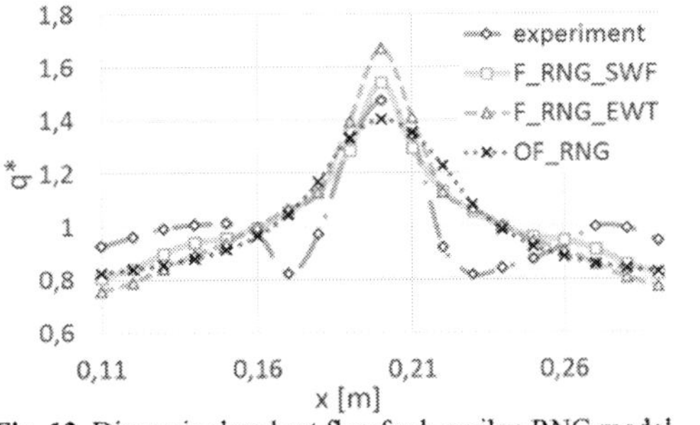


**Fig. 13.** Dimensionless heat flux for k-epsilon RNG model

## 7 Summary

The considered numerical solutions never get the exactly same shape of the heat transfer coefficient like the one obtained from the conversion from the experiment. That is given partially due to some state of simplification in the analogy between mass and heat transfer rates and the simplifications in the turbulence models (2D modelling, RANS approach, boundary layer regimes). It is not possible to do a relative exact simulation within a reasonable time. But for engineering purposes these models were often sufficient.

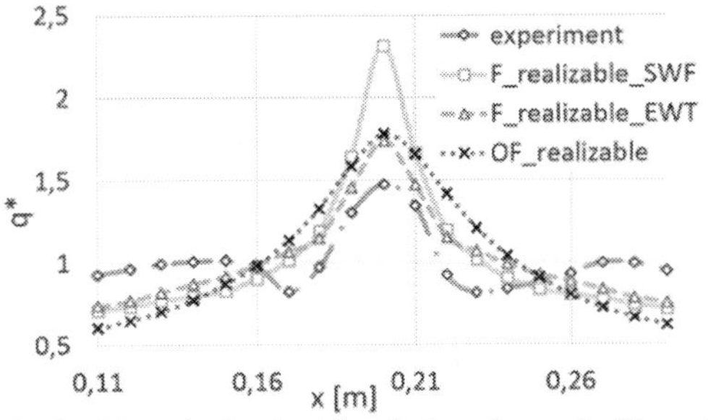


**Fig. 14.** Dimensionless heat flux for k-epsilon realizable model

The k-epsilon models generally smoothed the progress and overestimated heat flux. The dimensionless heat flux curves indicate the best match for the standard k-omega model in Ansys-Fluent. In OpenFOAM for all cases were the local minima smoothed out.

Regarding the total heat flux over the resolved space, the most accurate results is provided by the RNG modification of k-epsilon model with the enhanced wall treatment. But it is necessary to mention that a lot of information about the flow was smoothed out (the local minima) what could affect a similar based computation and distort significantly the results.

The computation times to solve the tasks are often very different for both platforms. This is partially given due to worse solution stability in OpenFOAM for more skewed meshes or more complex flow patterns for some solvers. As an example, it could be mentioned, that to obtain a similar solution for the same case OpenFOAM take 75 minutes while Fluent needed only 20 minutes. This is partially given by the time domain of both approaches. Fluent solves only statistically steady state (steady solution), while the OpenFOAM fully resolves all the time steps leading to statistically the steady state.

## References

1. M. Korger, F. Kořínek. *Součinitele přenosu hmoty při impaktním proudění ze štěrbinových trysek.* Strojírenství (17), 536-541. [retrieved 2016-06-28]
2. S. Polat, *Heat and Mass Transfer in Impingement Drying*, (Drying Technology, **11**, 1993)
3. H. Martin, B. Spalding, S. Whitaker, H. H. Winter, *Advances in Heat Transfer*, (v**13**, 1977)
4. S. M. Ghiaasiaan, *Convective Heat and Mass Transfer*, Georgia Institute of Technology, (2011)
5. M. Zogg, *Wärme- und Stofftransportprozesse*, (Otto Salle Verlag, Frnkfurt am Main,1983)
6. Kolář, Adámek, *Evaporation and impact flows in CFD*
7. J. Příhoda, P. Louda, *Matematické modelování turbulentního proudění.* (ČVUT, Praha, 2007)
8. Ch. J. Geankoplis. *Transport Processes and Separation Process Principles.* (Prentice hall, 2003)
9. ANSYS Fluent Theory Guide r.17.0, (2016)

# Direct simulation Monte Carlo method for gas flows in micro-channels with bends with added curvature

*Tomáš* Tisovský[a,1], *Tomáš* Vít[b,1]

[1]Department of Power Engineering Equipment, Faculty of Mechanical Engineering, Technical University of Liberec, Czech Republic

**Abstract.** Gas flows in micro-channels are simulated using an open source Direct Simulation Monte Carlo (DSMC) code dsmcFOAM for general application to rarefied gas flow written within the framework of the open source C++ toolbox called OpenFOAM. Aim of this paper is to investigate the flow in micro-channel with bend with added curvature. Results are compared with flows in channel without added curvature and equivalent straight channel. Effects of micro-channel bend was already thoroughly investigated by White et al. Geometry proposed by White is also used here for refference.

## 1 Introduction

Considerable progress in field of miniaturization over recent years, most notably the advancements in development of MEMS (micro-electro-mechanical systems) have led to the need of investigation of gaseous flow in micro-scale geometries.

Gas flow can be modeled either at macroscopic level (Navier-Stokes equations) or at microscopic level.
The Direct Simulation Monte Carlo method, as described by Bird [1], is the dominant numerical method for solving engineering problem involving rarefied gas flow.

Degree of rarefaction of a gas is classified throughout the dimensionless Knudsen number

$$Kn = \frac{\lambda}{L}, \tag{1}$$

where λ is mean free path of the gas molecules and L is the characteristic dimension of the flow.

As can be seen from equation (1), high degree of rarefaction can be achieved by either lowering the density of the gas or decreasing characteristic dimension of the fluid region.

The continuum fluid assumption is valid roughly for $Kn < 0.001$ [1]. In present study, flows with Knudsen number around 0.05 are considered.

For sufficiently high Knudsen numbers, the continuum hypothesis breaks down and solution of the Navies-Stokes equations yields physically incorrect results. Micro-scale flows exhibit many features that are not to be observed in similar cases of macro-scale flows. Most notable difference is that while the continuum solution gives zero velocity at walls (the so called no slip condition), rarefied gas does not behave in this manner and there is a considerable velocity slip at walls. Similarly, there is also temperature jump at walls. Other considerable difference are the non-linear pressure gradients that have been observed in micro-channel flows [3].

Based on $Kn$, gas flow can be classified into four regimes [4].

**Table 1.** Flow classification.

| Regime | *Kn* |
|---|---|
| Continuum flow | $Kn < 0.001$ |
| Slip flow | $0.001 < Kn < 0.1$ |
| Transition flow | $0.1 < Kn < 10$ |
| Free molecule flow | $Kn > 10$ |

The slip flow can be well modeled with the Navier-Stokes equations with specific velocity slip and temperature jump boundary conditions. For transition flow, no such approach can be used and methods based on molecular gas dynamics are the only alternative.

Since DSMC is the dominant method for solving rarefied gas flows, there has been many DSMC studies dealing with straight microchannel gas flows [2, 3, 7, 8, 9]. However, there is significant lack of data concerning microchannels of more complex geometry. Nevertheless, White et al. [2] used DSMC and carried out series of numerical simulations for flow in microchannels with

[a] Corresponding author: tomas.tisovsky@tul.cz
[b] Corresponding author: tomas.vit@tul.cz

bends. No work on flow in microchannels with bend with added curvature has been found in available literature.

In this paper, DSMC investigation of flow in micro-channel with rounded-off bends is carried out. This research is largely based on work of White et al. Apart from their results regarding micro-channel with bends, they also conducted many simulations in order to validate the dsmcFOAM solver [5] against previously published results.

## 2 Direct simulation Monte Carlo

The only chemical elements which form stable monoatomic molecules at standard conditions are the noble gases. Therefore, to simplify the simulation, working gas used in this paper is argon. As a result, no internal modes (rotational or vibrational) are considered.

### 2.1 Physics of micro-scale flow

Macroscopic properties of a flows at low Knudsen numbers are defined as averages of appropriate molecular quantities.

Important quantity in DSMC is the number density n which gives the number of molecules per unit volume. This can be calculated from the ideal gas law if pressure and temperature are known. Mass of individual molecules is denoted $m$.

Results will be viewed in terms of pressure, mean (macroscopic) velocity and wall shear stress. These quantities are calculated at each time step by following relations and then averaged over large number of samples. In present study, about one million samples were used.

#### *2.1.1 Calculating flow properties*

Each molecule in gas has velocity $\boldsymbol{c}$. By averaging over large number of molecules, mean molecular velocity $\bar{\boldsymbol{c}}$ can be obtained. Mean molecular velocity is also the macroscopic stream velocity $\boldsymbol{c_0}$. From this, the random thermal velocity component of each molecule can be obtained as

$$\boldsymbol{c}' = \boldsymbol{c} - \boldsymbol{c_0}. \quad (2)$$

Scalar pressure tensor can be defined as

$$p = \frac{1}{3}\rho\overline{c'^2}. \quad (3)$$

Finally, the viscous stress tensor is defined as

$$\tau_{ij} = \rho\overline{c'_i c'_j} - \delta_{ij}p. \quad (4)$$

### 2.2 Modeling physical effects in DSMC

To model rarefied gas flow, it is necessary to compute typical intermolecular collisions and molecule-surface interactions.

In order to perform Direct Simulation Monte Carlo, computation geometry of the flow domain needs to be defined which is then divided into cells. Since the mesh cells are used mainly to determine potential collision partner for given cell and otherwise have no effect to the computation, requirements for cell quality (measured most notably by skewness, non-orthogonality and aspect ratio) are, to some extent, not of concern. Typical cell size of the mesh should be a fraction of the mean free path. Similarly, the time step should be a fraction of the mean collision time of the gas molecules. [2] It is the generally low value of the time step that represents the greatest drawback for DSMC with respect to the computational effort needed to arrive to the converged solution.

Basic quantities associated with molecular model are the number of molecules per unit volume and mass, size (diameter) of the molecules, velocity and internal state of each molecule in the system. Those parameters are related to mean free path and collision frequency so that distance and time scales of the effects due to collisions can be evaluated.

### 2.3 Simulation procedure

The DSMC method is a particle method in which each simulated particle represents large number of real gas molecules. The physics of the gas is captured by uncoupling the motion of the particles from their collisions. Collisions between particles are evaluated stochastically after all particles have been moved. Collision then cause discrete changes in individual particles velocities and energies. The algorithm of DSMC is presented in figure 1.

Particles are moved through computational domain during each time step according to their individual velocities and the time step. In the absence of external force field, particles move in straight lines. If 2D flow is to be simulated, individual particle displacement in homogeneous direction is simply ignored.

Particles which cross boundary of the flow are either removed from the simulation or reflected back into the domain.

Upcoming collision step requires that the particles are grouped into the cells of the computational grid. Particles need to be newly indexed after each move.

Collisions are calculated statistically between particles residing in the same computational cell. Particles are paired in each cell to form a list of possible collision pairs. An accept-reject method is used to determine which pairs will actually collide. Probability of a collision occurring is dependent on the collision model and selection scheme being employed, and is the function of the relative velocity, collision cross section and gas density in that cell. Probability is also proportional to the time step and particle weight. The collision between two particles is accepted if the probability is greater than a random number.

Number of pairs tested and the collision probabilities are such that the average collision rate is equal to that given by the kinetic theory. In OpenFOAM, the No Time Counter method is used.

Once a pair of particles has been selected for collision, momentum and energy are exchanged (they are, of course, conserved in each collision). Actual trajectories and dynamics of the particles are not considered when determining mechanics of the collision. Post collision properties are generated statistically. Variable hard sphere (VHS) model is used to model individual molecules.

Momentum and translational energy transfer is considered in the center of mass frame of reference. Velocity of the center of mass is held constant to conserve momentum while the magnitude of relative velocity may be changed as a result of internal energy transfer. For VHS collision model, scattering angle is isotropic.

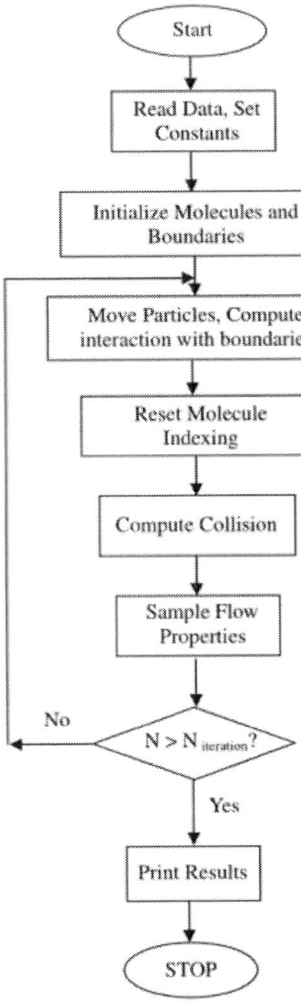


**Fig. 1.** DSMC algorithm

Macroscopic properties of a gas are local averages of the properties of molecules of the gas. Since DSMC directly simulates the distribution function, flow field results can be determined by sampling particle properties and calculating the moments of sampled data.

Computational grid is used to group particles for the purpose of sampling their properties. Flow field properties are calculated at each cell.

Calculation of smooth flow properties requires a large sample size to accurately model the distribution function. Magnitude of statistical fluctuations varies according to the inverse square root of sample size. Real gas is also subject to these fluctuations, but the magnitudes are much smaller due to the sample size which is much larger than the one available in simulations. It is generally necessary to sample over a number of iterations to increase the sample size and therefore reduce fluctuations.

Flows in micro-channels occur generally roughly at normal conditions (i.e. at pressures around atmospheric and at temperatures around 300 *K*). Individual particles have therefore thermal velocities orders of magnitude larger than the overall average velocity (i.e. stream velocity). This makes it difficult to compute macroscopic properties of the flow accurately. Generally, many samples are needed and the DSMC method is expensive.

### 2.3.1 Pressure boundary conditions

The problem with DSMC is, as for virtually any other method for simulating real flow of some fluid, in boundary conditions. The velocity profiles at the inlet and outlet are not known due to current experimental limitations at the micro-scale. Usually, pressure and temperature at inlet and pressure at outlet are the only known thermo-mechanical properties at the boundaries.

Employed pressure inlet boundary condition was the one suggested by Wang [7]. In this approach, income macroscopic velocity for new particles is computed from statistical macroscopic velocity obtained from particles in cells near the inlet boundary. For the inlet face perpendicular to *x* axis, this gives

$$(c_{0x_{in}})_j = (c_{0x})_j + \frac{p_{in}-p_j}{\rho a}. \quad (5)$$

Here, $c_{0x}$ is the x component of macroscopic stream velocity and a is the speed of sound.
Velocity component perpendicular to inlet face is

$$(c_{0y_{in}})_j = (c_{0y})_j. \quad (6)$$

To determine inlet number flux density, number density corresponding to inlet pressure is also needed. This is given by ideal gas equation of state

$$n_{in} = \frac{p_{in}}{kT_{in}}. \quad (7)$$

Outlet boundary condition is of the form suggested by Liou and Fang [8, 9]. Density is

$$(\rho_{out})_j = \rho_j + \frac{p_{out}-p_j}{a^2}. \quad (8)$$

Since the density is just number density multiplied by mass of one molecule, equation (8) is the equation for number density. Considering again the boundary face perpendicular to direction of flow, velocity components are

$$(c_{0x_{out}})_j = (c_{0x})_j + \frac{p_j - p_{out}}{\rho a}, \tag{9}$$

$$(c_{0y_{out}})_j = (c_{0y})_j. \tag{10}$$

And finally, temperature is given by equation of state of ideal gas

$$(T_{out})_j = \frac{p_{out}}{r(\rho_{out})_j}. \tag{11}$$

Modification of above equations to boundary faces perpendicular to y axis is straight forward.

## 3 Micro-channels with rounded bends

Adding curvature to the bends enables investigation of the effects of competing concave and convex curvature on the flow.

Curved surface of a channel is regarded convex if all the line segments between each pair of points belonging to the surface intersect the interior of the channel. If this condition is not satisfied, the curved surface is concave. Concave surfaces cause the reflected molecules to spread (i.e. diverge) while convex surfaces cause them to meet (i.e. converge).

Diverging character of a reflection in collision between molecule and concave channel surface, as well as in intermolecular collision between two molecules, is the cause of chaos in hard sphere system. In such reflection, small changes in initial trajectories, caused by finite precision of our computers, are magnified at each collision. This justifies the collisions to be calculated statistically between particles residing in the same computational cell.

### 3.1 Setting up the simulation

As in work of White et al. [2], the straight channel length L is *15 μm* and its width B is 1 μm. Concequently the bends and round offs are added in a way that the channels volume and surface area do not change.

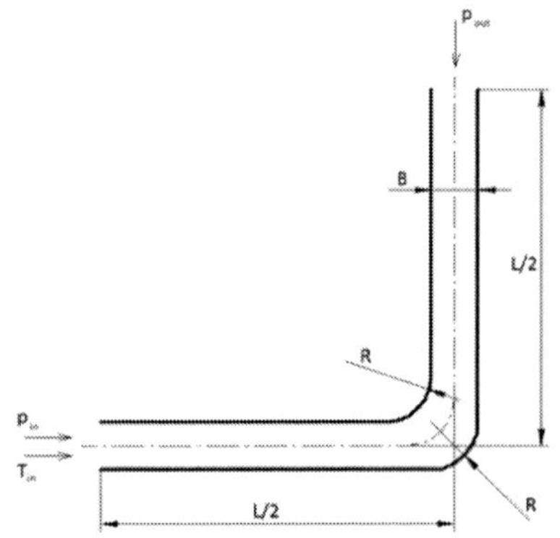


**Fig. 2.** Micro-channel with rounded-off bend

Geometry of channel with rounded-off bend is shown in figure 2. For future reference and according to figure 2, wall with bend with convex curvature will be reffered to as lower wall, similarly, wall with bend with concave curvature will be reffered to as upper wall.

The working gas is argon and the inlet gas temperature and wall temperature are both 300 *K*. The walls of the channels are modeled as fully diffuse boundaries.

Computational domain was also meshed in accordance with the work of White. It is recommended that the cell size in DSMC is less than mean free path of the gas molecules. However, it is possible to use elongated cell sizes in the streamwise direction of micro-channel flows [2], where macroscopic gradients are relatively small. In channels with bend, the corner region is meshed with cells of sizes smaller than mean free path in both directions.

Value of mean free path is needed to determine flow regime. For VHS molecules, mean free path is [1]

$$\lambda = \frac{2(5-2\omega)(7-2\omega)}{15}\sqrt{\frac{\rho}{2\pi nkT}}\frac{\mu}{\rho}, \tag{12}$$

where $\omega$ is the viscosity idex of the gas.

From this expression and from the characteristic dimension of the problem, which is here the width of the channel, inlet and outlet Knudsen numbers can be determined.

Two cases were modeled, one partially (at inlet) in slip regime and partially (at outlet) at transition regime. Second one is fully in transition regime.

**Table 2.** Simulated cases

| Case | $p_{in}$ [Pa] | $p_{out}$ [Pa] | $Kn_{in}$ [1] | $Kn_{out}$ [1] |
|---|---|---|---|---|
| Case 1 | 100,000 | 33,333 | 0.055 | 0.158 |
| Case 2 | 18,500 | 6,166 | 0.300 | 0.862 |

Different computational meshes need to be used for Case 1 and Case 2.

### 3.2 Results

Averaged flow fields were obtained for four geometries, each with the boundary conditions corresponding to case 1 and case 2. As a result, 8 numerical simulations were performed and results were sampled from approximately one milion samples. It should be noted that computational meshes need to be adjusted in accordance to given set of boundary conditions. Consequently, 8 computational meshes were created.

#### *3.2.1 Mach number*

Mach number was calculated based on averaged channel centre-line velocity and on speed of sound of perfect gas. It is found that local Mach number decreases in bend region and the decrease is more significant as the curvature of the bend gets smaller. Maximum decrease is found for sharp bend.

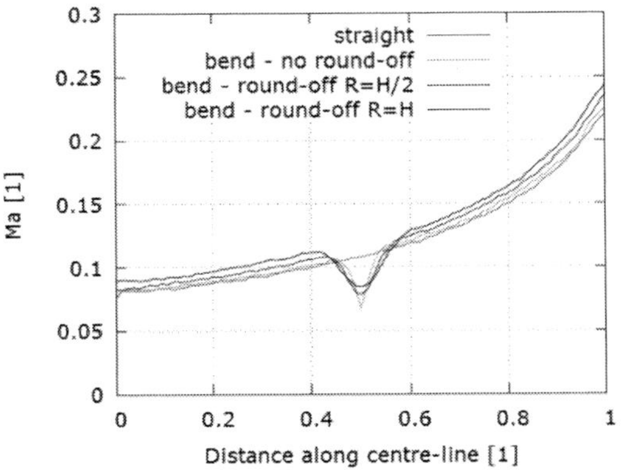


a) Case 1 ($Kn_{in} = 0.055$)

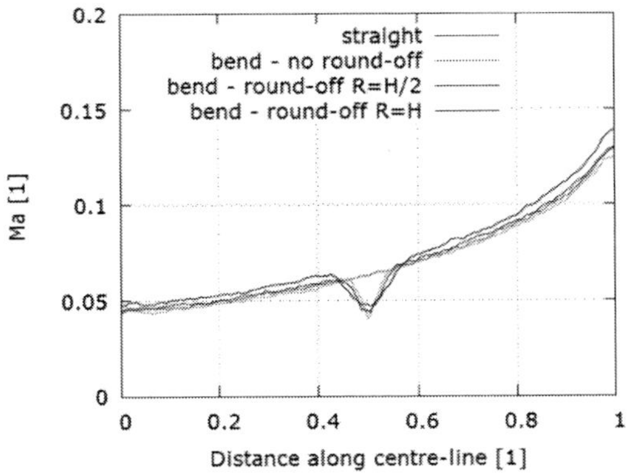


b) Case 2 ($Kn_{in} = 0.300$)

**Fig. 3.** Mach number centre-line profiles

### *3.2.1 Pressure*

Pressure profiles along channel centre-line closely resemble linear profile. The deviation from linear distribution is plotted in figure 4 and it is greater for larger Knudsen number flows as the rarefaction effects start to dominate over the compressibility effects. At the middle point of the centre-line, the pressure deviations for channels with bends exhibit significant changes. Those are the results of changes in velocity due to the channel geometry (as shown in figure 3) and are therefore mostly compressibility effects. The other effect which influences the pressure jump is the flow separation, which can occur in some cases and which lowers the effective cross-section of the flow in corner region. However, this effect is not significant and does not occur for all considered cases (as is demonstrated further).

The pressure jumps for channels with rounded-off bends are greater than in cases of sharp bends and are smoothed of to larger region of space.

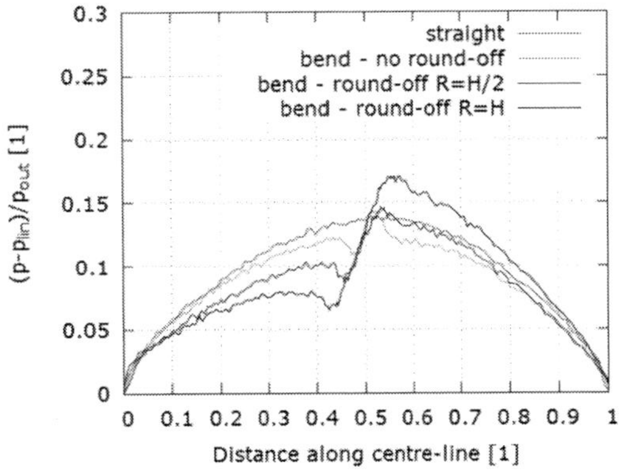


a) Case 1 ($Kn_{in} = 0.055$)

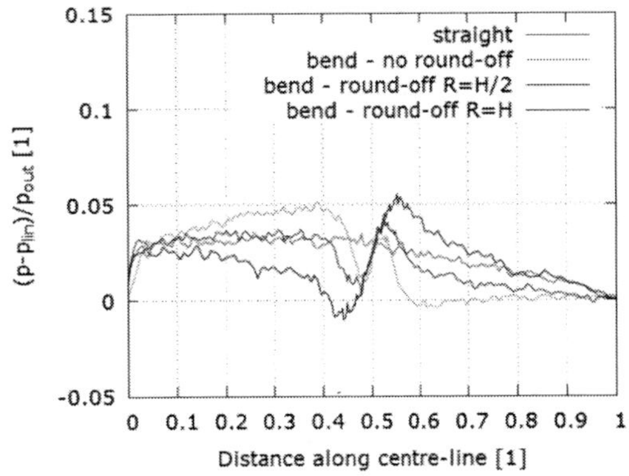


b) Case 2 ($Kn_{in} = 0.300$)

**Fig. 4.** Deviation of pressure distribution from linear profile

### *3.2.3 Wall shear stress*

Averaged shear stresses in channels with bends are compared with the averaged shear stress in corresponding straight channel. For both cases in consideration, channel with sharp bend exhibited lower shear stress compared to straight channel. With added curvature, the shear stress increased above the value for straight channel.

Care must be taken for the averaging to be taken as weighted average to acount for different width of computational cells in region of bend as compared to rest of the geometry.

Actual shear stresses profiles along both walls for both cases are shown in figures 5 and 6 with 5a and 6a corresponding to lower walls and 5b and 6b corresponding to upper walls.

**Table 3.** Normalized average shear stress $\frac{\tau}{\tau_{straight}}$

| $Kn_{in}$ | $R = 0$ | $R = H/2$ | $R = H$ |
|---|---|---|---|
| 0.055 | 0.979 | 1.011 | 1.041 |
| 0.300 | 0.946 | 0.980 | 1.019 |

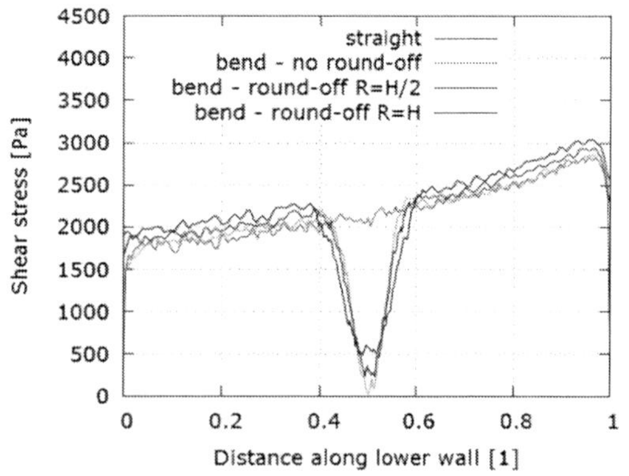


a) wall with convex curvature (lower wall)

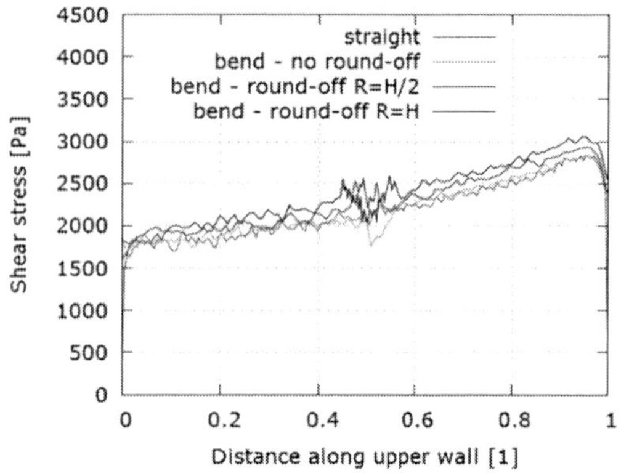


b) wall with concave curvature (upper wall)

**Fig. 5.** Wall shear stress for Case 1 ($Kn_{in} = 0.055$)

Case of slip to transition regime exhibits significant drops in wall shear stress at lower wall in region of the bend. Consequently, if there is a reduction in averaged wall shear stress, most of this reduction is achieved there. As curvature is added to the bend, the decrease in wall shear stress can become insufficient to compensate for the modest increase in wall shear stress which is evident for channels with bends along straight parts of its walls and which becomes more significant as the radius of curvature increases. Apart from this increase, there is also little bit more significant increase in regions close to the edges of curvature at upper wall. Interestingly, this increase becomes significant only for large enough curvatures. For sharp bend, there is a decrease in wall shear stress in this region.

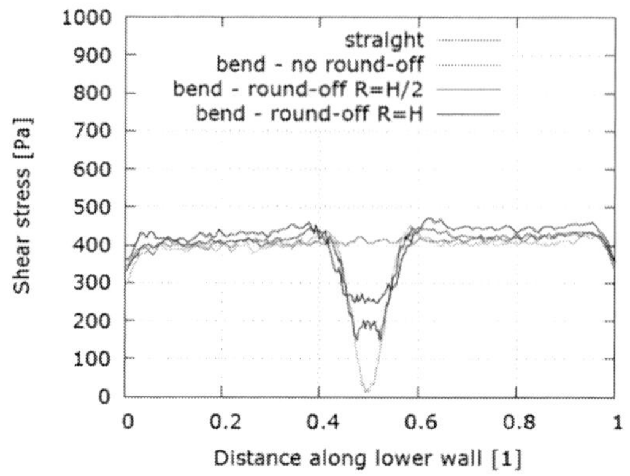


a) wall with convex curvature (lower wall)

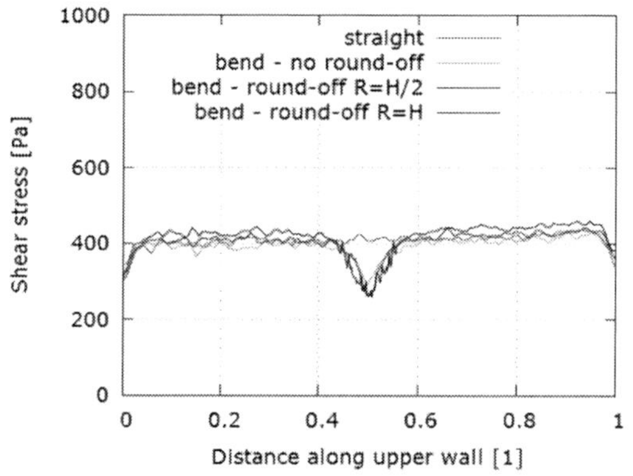


b) wall with concave curvature (upper wall)

**Fig. 6.** Wall shear stress for Case 2 ($Kn_{in} = 0.300$)

On the other hand, case of fully transition regime exhibits significant wall shear stress drops in regions of the bend for both upper and lower wall. As in case of slip to transition regime, reduction of wall shear stress can only be achieved if the decrease in bend region compensates for the increase along the straight parts of walls of channel with bend. Since this increase once again becomes more significant with increasing radius of curvature of the bend, there is once again a threshold of radius of curvature above which is the averaged wall shear stress greater than in corresponding case of straight channel. For fully transition regime, this radius of curvature is larger than in case of slip to transition regime.

Since wall shear stress at upper wall for case of slip to transition regime for sharp bend has the same character as wall shear stress at upper wall for case of fully transition regime for bends with added curvature, it can be predicted that, for sufficiently large radius of curvature, wall shear stress profile similar to that of slip to transition regime with curved bend may be achieved in fully transition regime for sufficiently large radius of curvature.

### *3.2.4 Slip velocity*

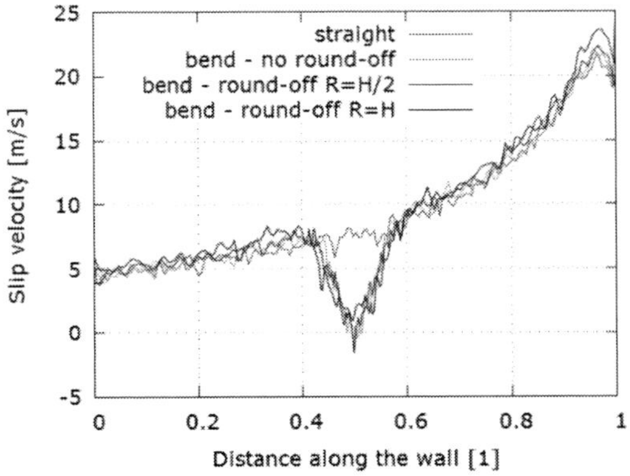


a) wall with convex curvature (lower wall)

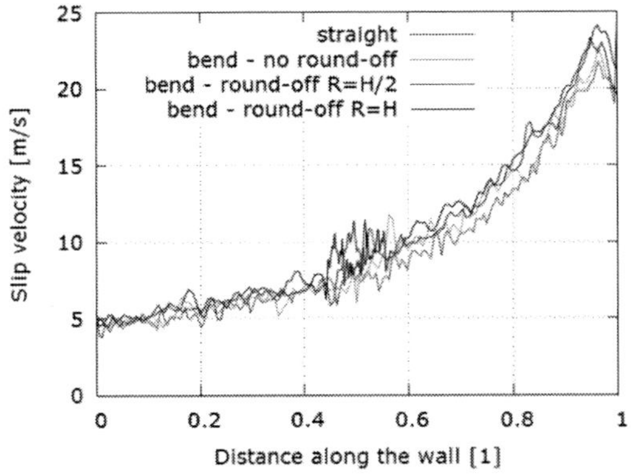


b) wall with concave curvature (upper wall)

**Fig. 7.** Slip velocity for Case 1 ($Kn_{in} = 0.055$)

Figures 7 and 8 show slip velocity along both the channel walls for case 1 and case 2. Possitive values of slip velocity are those in direction from inlet to outlet (i.e. in the negative direction of the pressure gradient along the channel). Case of $Kn_{in} = 0.055$ for sharp bend and for $R = H/2$ and case of $Kn_{in} = 0.300$ for sharp bend both exhibit slip velocity drop below zero for the wall with convex curvature in region of the bend. This suggests existence of a small region of separated flow in a form of vortex.

Slip velocity along the wall with concave curvature shows modest increase in the region of the bend for the case of more pressurized flow ($Kn_{in} = 0.055$) and modest decrease in the same region in case of less pressurized flow ($Kn_{in} = 0.300$).

Furthermore, there is strong correlation between slip velocity and corresponding wall shear stress.

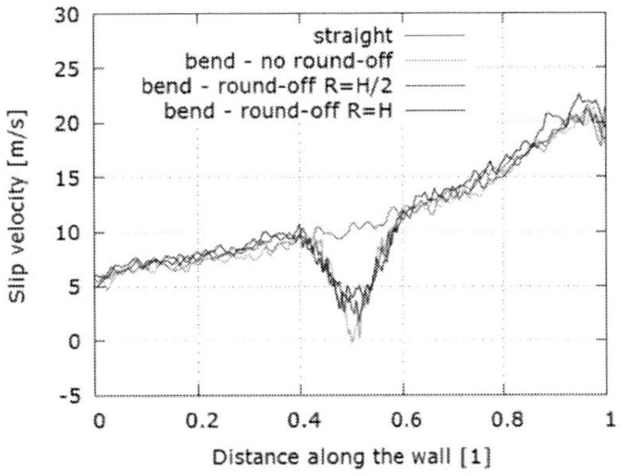


c) wall with convex curvature (lower wall)

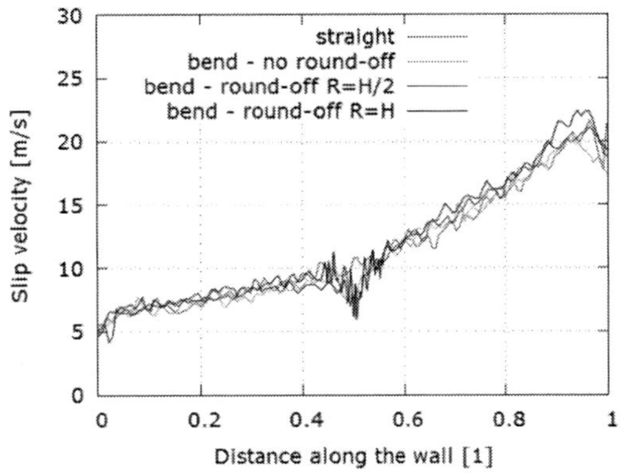


d) wall with concave curvature (upper wall)

**Fig. 8.** Slip velocity for Case 1 ($Kn_{in} = 0.300$)

### *3.2.5 Mass flow rate*

To investigate the effect of rounded-off bends to the mass-carrying capacity of the microchannel, mass flow rates of particular microchannels are normalized with respect to the mass flow rate through corresponding straight microchannel.

It was shown by White [2] that based on boundary conditions, adding bend to microchannel can lead to either increase or decrease of mass flow rate through the microchannel. In this paper, curvature is added to both the above cases and it is found that for investigated cases, this always results in increase of mass flow rate as compared to straight channel. This holds also for case where sharp bend results in decrease in mass flow rate.

Furthermore, it was shown that microchannels with bend in fully slip regime (which were not modeled here) can also exhibit mass flow rate enhancement. This result was obtained separately with DSMC [2], with lattice Boltzmann method [10] and also with continuum fluid dynamic simulation with specialized boundary conditions [11].

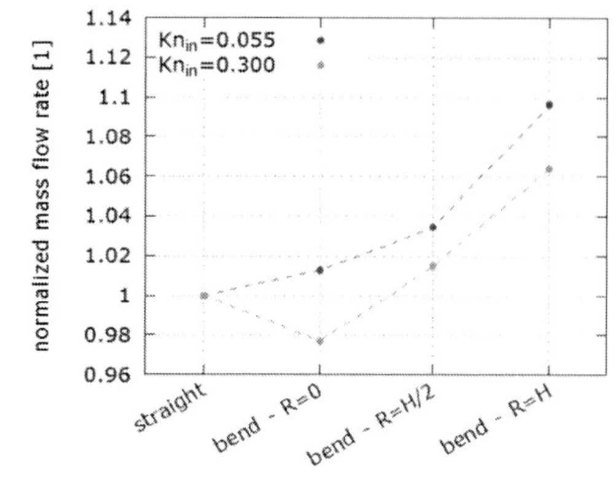


**Fig. 9.** Mass flow rate through microchannels

Comparing table 3 with figure 9 shows that while the averaged wall shear stress of microchannel with sharp bend is below the corresponding value for straight microchannel, the mass flow rate through the microchannel can be either greater or smaller as compared to that of straight microchannel.

## 4 Conclusions

The investigation of pressure driven flows in microchannels with bends with added curvature was performed for the case of flow in fully transition regime and for the case of flow in slip to transition regime. Each of the cases investigated consisted of simulations of straight channel, channel with sharp bend and two channels with curved bends. In total, 8 DSMC simulations were performed and special care was taken to model the bend region with cells sufficiently small to capture possible regions of separated flow. Such region was subsequently confirmed even for one of the cases with curved bend.

Most notably, it was found that adding curvature to the bend can eventually lead to increase in mass flow rate through the channel even in the case where, for the same boundary conditions, adding sharp bend would result in mass flow rate decrease. This is an important result which can prove valuable in practical applications such as design of MEMS devices.

For future work, more throughout investigation of gas flow in microchannels should be conducted. Especially in fully slip regime which present paper does not deal with. This regime offers the possibility to model with both the molecular dynamics methods and with continuum fluid dynamic methods with specialized boundary conditions. In this case DSMC method is more reliable but also more computationally demanding than simulation of more rarefied regimes. Apart from this, greater curvatures should also be added to the bend to obtain better picture of the character of the change in mass flow rate.

## Acknowledgement

We gratefully acknowledge the support of the Grant Agency of the Czech Republic (Project No. 16-16596).

## References

[1] Bird G. A. Molecular gas dynamics and the direct simulation of gas flows. New York: Oxford Science Publications, Oxford University Press Inc., 1994.

[2] White C., Borg M. K., Scanlon T. J., Reese J. M. A DSMC Investigation of Gas Flows in Micro-channels with Bends. Journal of Computers & Fluids 2012.

[3] Arkilic E. B. Measurement of the Mass Flow and Tangential Momentum Accommodation Coecient in Silicon Micromachined Channels. Ph.D. Thesis. Massachusetts Institute of Technology, Cambridge 1997.

[4] Tsien H. Superaerodynamics, The Mechanics of Rarefied Gas. Journal of the Aeronautical Sciences 13:653–664, 1946.

[5] Micro & Nano Flows for Engineering, dsmcFOAM [Computer code]. Available at http://www.micronanoflows.ac.uk/

[6] OpenFOAM Foundation, 2016. Available at http://www.openfoam.org/

[7] Wang M, Li Z. Simulations for gas flows in microgeometries using the direct simulation Monte Carlo method. International Journal of Heat and Fluid Flow 2004.

[8] Liou WW, Fang Y. Heat transfer in microchannel devices using DSMC. Journal of Microelectromechanical Systems 2001.

[9] Fang Y, Liou WW. Computations of the flow and heat transfer in microdevices using DSMC with implicit boundary conditions. Journal of Heat Transfer 2002.

[10] Agrawal A, Djenidi L, Agrawal A. Simulation of gas flow in microchannels with a single ninety-degree bend. Computational Fluids 2009.

[11] Rovenskaya O, Croce G. Numerical analysis of rarefaction and compressibility effects in bent microchannels. ASME Conference Proceeding 2010.

# Heat transfer enhancement of a single row of tube

*Takayuki* Tsutsui[1,*]

[1] Department of Mechanical Engineering, The National Defense Academy, 1-10-20 Hashirimizu, Yokosuka , Kanagawa 238-8686 Japan

**Abstract.** A rod was positioned upstream of a circular cylinder to enhance its heat transfer and reduce its drag in air stream. The results are increasing the overall heat transfer by 40% over and reducing the drag by 30% the value for a single circular cylinder. In the present, this technique applied to a single row of tubes. Heat transfer enhancement and drag reduction are important factor for multi-tube heat exchanger. The present study investigated the heat transfer and surface pressure characteristics of each tube for single row of tubes under various numbers of tubes, pitch and Reynolds number. The tube diameter, D, was 20 mm, and the rod diameter was 5 mm. The distance between the center axes of the rod and the tube was 30mm. The Reynolds number based on D ranged from $5.3\times10^3$ to $2.1\times10^4$. The result is that this technique is effective for heat transfer enhancement and drag reduction of row of tube.

## 1 Introduction

Heat transfer enhancement and drag reduction are important in the design of multi-tube heat exchangers. Ribs [for example 1], rib roughness [for example 2] and winglets [for example 3] are well known as good manners. But the machine production on surface of tubes and fin is necessary for them. It has been previously shown that heat transfer and drag can be respectively enhanced and reduced simply by positioning a rod upstream from a circular cylinder. [4, 5] Under the optimum conditions, for a single circular cylinder with a rod positioned upstream, the overall heat transfer is increased by 40%, while the drag of the cylinder is reduced by 30%.

Figure 1 (a) shows the flow visualization of a circular cylinder with a rod under the optimum conditions of L / D = 1.5 and d / D = 0.25, where D, d, and L are the cylinder diameter, rod diameter, and distance between the center axes of the rod and the cylinder, respectively. The separated shear layer from the rod reattaches at the front face of the cylinder, and a quasi-static vortex is formed between the rod and the cylinder. For comparison, Figure 1 (b) shows the flow visualization of a circular cylinder without a rod.

Figure 2 (a) shows the Nusselt number (Nu) distribution. Nu is considerably higher at approximately $\phi = \pm 40° \sim \pm 50°$, which corresponds to the separated shear layer reattachment points. However, Nu in the frontal stagnation area ($\phi = \pm 20$ °) is lower than for a single cylinder due to the quasi-static vortex. The overall value of Nu at a Reynold's number of $Re = 2.1 \times 10^4$ is over 40% higher than the corresponding value for a single cylinder [1]. Figure 2 (b) shows the pressure coefficient (Cp) distribution on the front region ($\phi = \pm 25$ °). Here Cp is negative because a quasi-static vortex has formed between the rod and the cylinder. Cp reaches a maximum at the reattachment region ($\phi = \pm 35$ °) of the shear layer that separated from the rod.

There are many applications of this technique to cylinder clusters in the fields of mechanical, civil and electrical engineering, e.g., in heat exchangers, chimney stacks, and electric transmission lines. One example of such a heat exchanger incorporates several tubes. Heat transfer enhancement and drag reduction are important in multi-tube heat exchangers. Several arrangements of cylinder clusters and multi-tube arrays were reviewed by Zdravkovich [6], and the most effective arrangement for applications of this technique was reported to be a single row of tubes.

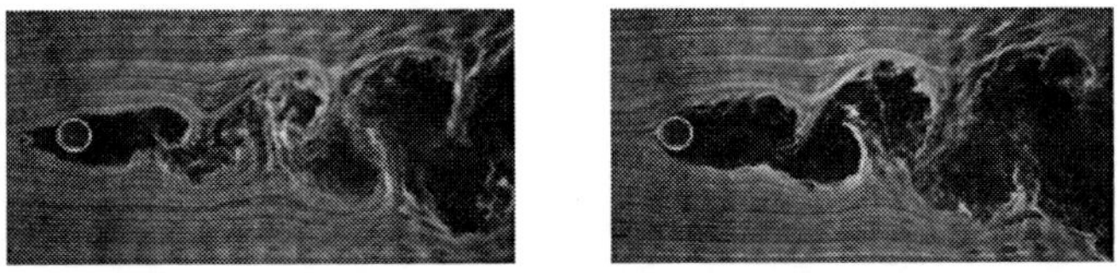

(a) (b)

**Fig. 1.** Flow visualization of a single cylinder ($D$ = 20 mm, Re = $1.1 \times 10^4$) (a) with and (b) without a rod

* tsutsui@nda.ac.jp

The goal of the present study is to confirm the effectiveness of this technique for multi-tube arrangements. In the present study, we investigate the heat transfer and surface pressure characteristics of each tube in a single row of tubes for various numbers of tubes, pitches, and Reynolds numbers.

## 2 Experimental apparatus and methods

The configuration of the row of tubes and the experimental setup are shown in Figure 3. Experiments were performed in a low-speed wind tunnel with a 400-mm-high, 300-mm-wide, and 400-mm-long working section. The tube diameter, D, was 20 mm, and the rod diameter, d, was 5 mm. The longitudinal distance L between the center axes of the rod and the tube was 30 mm. The ratios d / D and L / D were 0.25 and 1.5, respectively, which correspond to the optimal conditions for heat transfer enhancement and drag reduction [4, 5]. The pitch, S, between adjacent tubes was varied from 21 to 60 mm. The freestream velocity U was varied between 4 and 16 m / s, and the turbulent intensity was approximately 0.4 % in this range. The value of Reynolds number, Re, based on D ranged from $5.3 \times 10^3$ to $2.1 \times 10^4$.

Figures 4 (a) and 4 (b) show the constant heat flux model and measurement system respectively, used in the present study. The models were used for local heat transfer measurements and were fabricated from 3-mm-thick acrylic resin pipe, which was covered with two 8-μm-thick stainless steel sheets. The sheets were electrically connected in series and were heated by applying a direct current to produce a constant heat flux condition. Copper-constantan (C-C) thermocouples with diameters of 0.1 mm were attached to the underside of Figures 4 (a) and 4 (b) show the constant heat flux model and measurement system respectively, used in the present study. The models were used for local heat transfer measurements and were fabricated from 3-mm-thick acrylic resin pipe, which was covered with two 8-μm-thick stainless steel sheets. The sheets were electrically connected in series and were heated by applying a direct current to produce a constant heat flux condition. Copper-constantan (C-C) thermocouples with the stainless steel sheets in a spiral arrangement at ten-degree intervals. The local temperatures on the surfaces of the tubes were measured simultaneously with respect to the free-stream velocity and gap parameters. The surface pressure on the tubes was measured by a multi-tube pressure measurement system connected to 0.6-

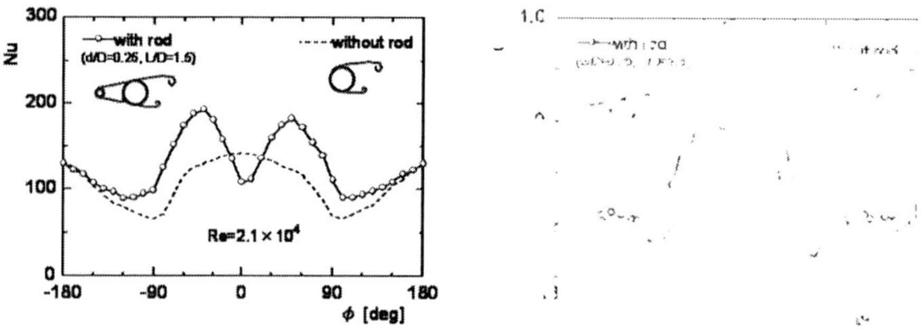


(a) (b)

**Fig. 2.** Effect of the rod ($D = 20$ mm, Re = $2.1 \times 10^4$) (a) Nusselt number distribution and (b) Pressure coefficient distribution

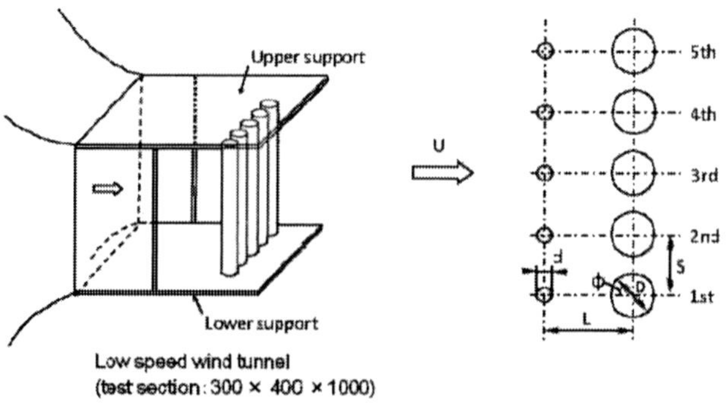


**Fig. 3.** Coordinate system and experimental setup

DOI: 10.1051/epjconf/201714302132

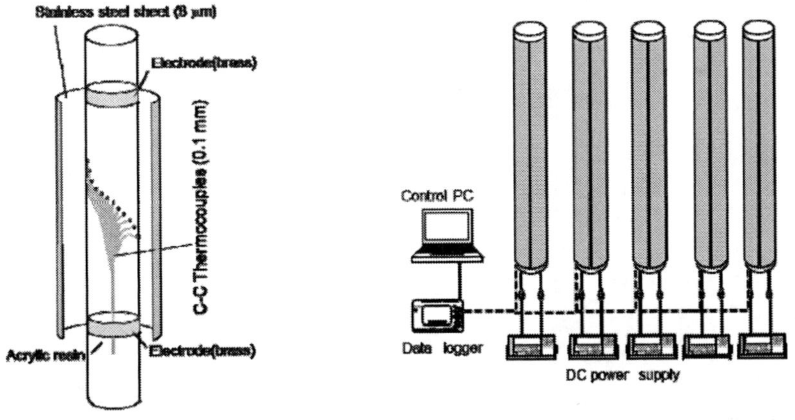


(a) (b)

**Fig. 4.** Schematic illustrations of (a) Constant heat flux model and (b) measurement system.

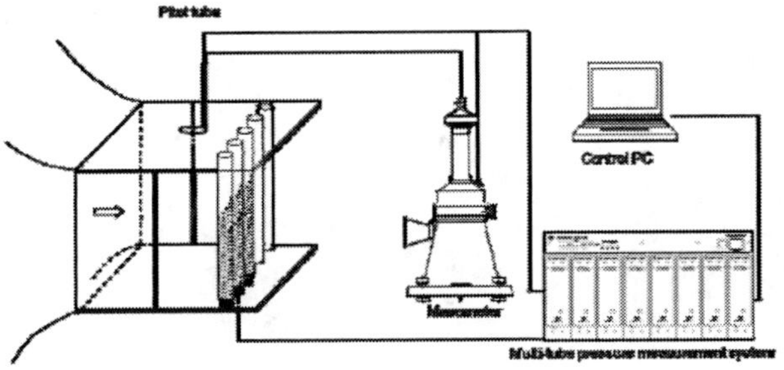


**Fig. 5.** Schematic illustration of pressure measurement system.

mm-diameter pressure taps in a spiral arrangement at ten-degree intervals on the surfaces of the tubes.

Figure 5 shows the pressure measurement system. We prepared five identical models and measured the surface pressures for all of the tubes simultaneously.

## 3 Results and discussion

### 3.1. Flow visualization and wake

Figures 6 shows photographs of the flow visualizations, velocity, u / U, and turbulence intensity, Δu / U, profiles in the wakes of rows of two, three, four, and five tubes, where u and Δu are the velocity and root mean square (RMS) value of fluctuating velocity, respectively. The flow visualizations were obtained using the smoke wire method, and the photographs show the flows with and without rods. The profiles show the velocity and turbulence intensity in the upper and lower parts of the figure. The u / U and Δu / U profiles reveal the differences in the gap flow and the wake for the 'with rods' and 'without rods' conditions.

As shown in Figure 6 (a), for the two-tube case without rods, the flow around the tube is a typical biased flow [7]. The gap flow between the tube switches intermittently from one side to the other side of the tubes. For the 'with rods' conditions, the wakes from the rods are biased toward the upper and lower sides (outer sides), and these wakes reattach on the upper and lower sides of the front face of the tube. The gap flow between the tubes becomes biased toward the upper tube.

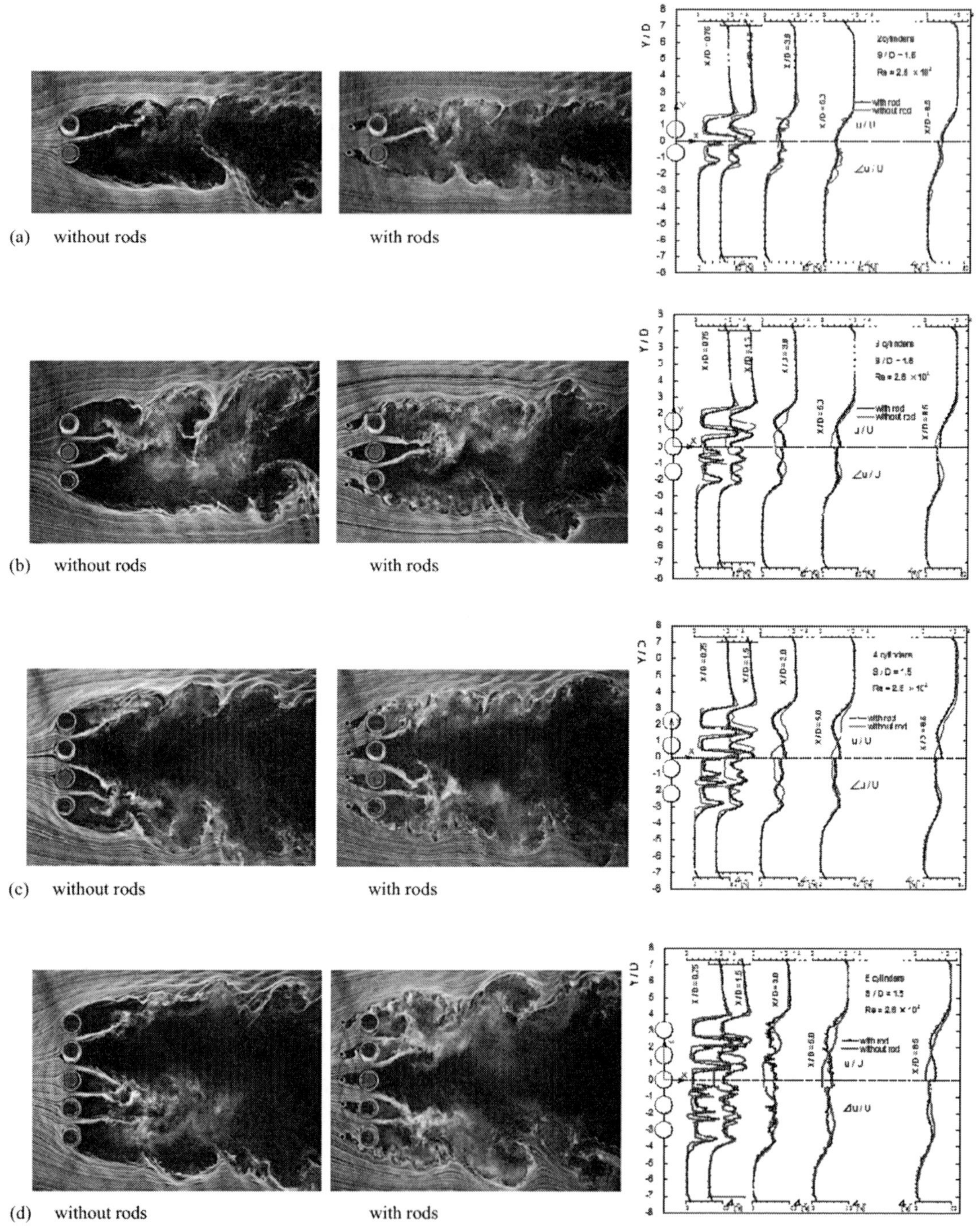


**Fig. 6**. Flow visualization and wake (D = 20 mm, Re = $2.1 \times 10^4$, S/D = 1.5) for (a) two tubes, (b) three tubes, (c) four tubes, and (d) five tubes.

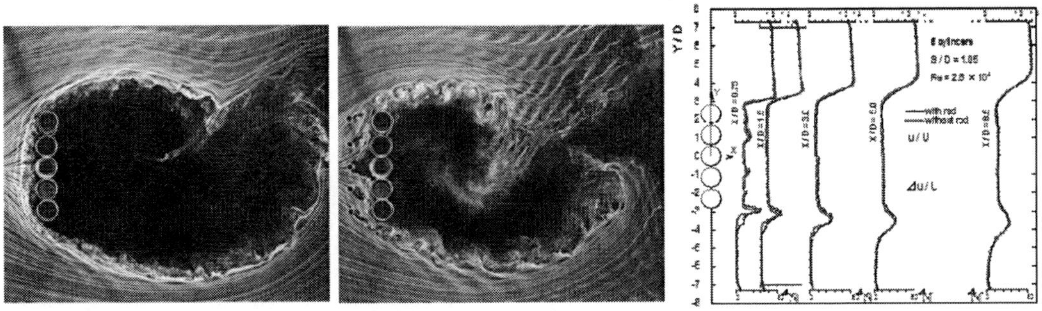

(a) without rods with rods

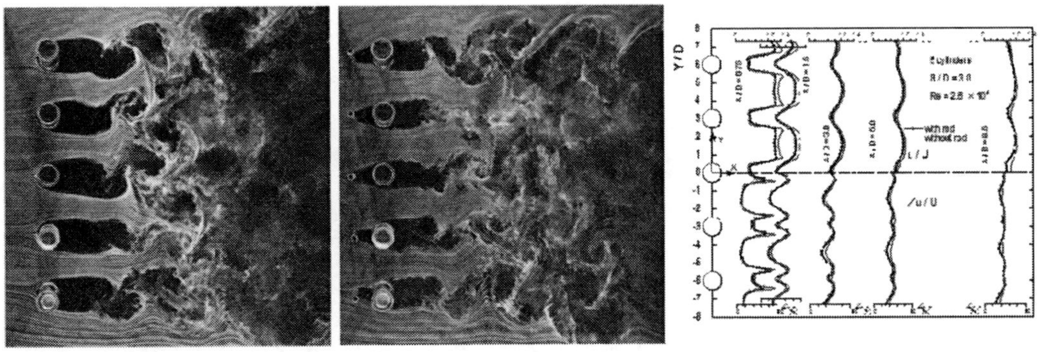

(b) without rods with rods

**Fig. 7**. Flow visualization and wake (effect of pitch in the five-tube system, D = 20 mm, Re = 2.1 × $10^4$) for (a) S/D = 1.05 and (b) S/D = 3.0.

For rows of three- to five-tube without rods, as biased in every case. For the 'with rods' conditions, the separated shear layer from the center rod or from a more centrally located rod reattaches on the front face of the tube positioned directly behind the rod. The wakes of the uppermost and lowermost rods are biased toward the upper and lower sides, respectively.

Based on the flow visualizations and the u / U and Δu / U profiles, as a general tendency, for the 'with rods' conditions, the gap flow is biased toward the center, and the wake is narrower than that for the 'without rods' conditions

Figure 7 shows the flows for two different pitch conditions for the rows of rods and tubes for the case of five tubes. For S / D = 1.05, a single vortex street is formed behind the tubes. There is no gap flow between the tubes and the differences in the profiles of u / U and Δu / U between the cases with and without rods are slight. For S / D = 3.0, a vortex street is formed by each tube, and the flow around each tube is independent. The values of u / U and Δu / U for the tubes under the 'with rods' condition are lower than those for the 'without rods' condition.

### 3.2 Local and overall heat transfer

The time averaged local Nu distributions of the first, second, and third tubes in the five-tube arrangements with and without rods for various gap ratios are shown in Figure 8. The fourth and fifth tubes are omitted, because the first and fifth and second and fourth tubes are symmetric for third tube.

For S / D = 1.05, the distribution of the third tube is symmetric for $\phi = 0$ °, because it is positioned in the center of the row. The distribution has peaks at approximately $\phi = \pm 80$ ° because of the reattachment of the separated shear layer from the rod. The increase in Nu is small for the first and second tubes, and the heat transfer of primarily the third tube (center tube) is enhanced.

For S / D = 1.5, Nu increases for the entire surface of the second and third tubes and for the rear face of the first tube because the separated shear layer from the rod reattaches on the front face of the tube and the gap flow between the tubes reattaches on the rear face of the tube. The heat transfer enhancement is increased in order of the third, second, and first tubes.

DOI: 10.1051/epjconf/201714302132

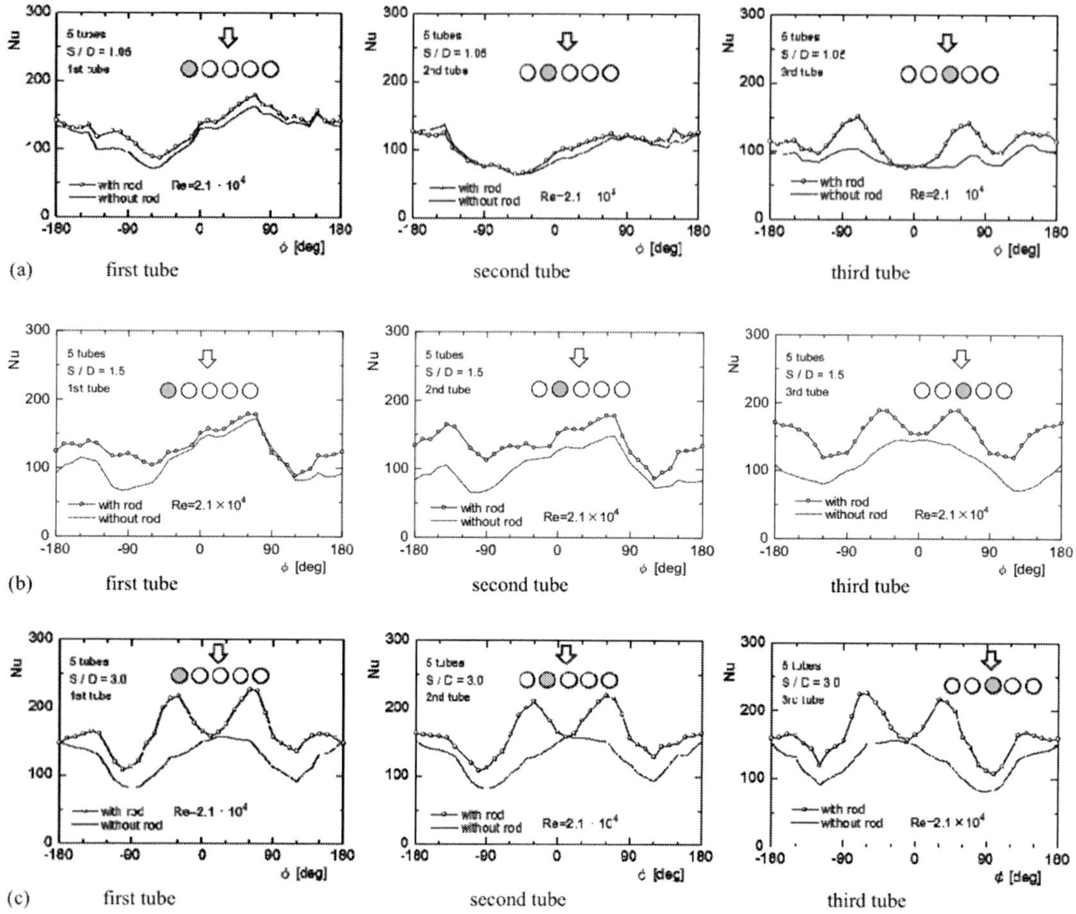


**Fig. 8**. Nusselt number distribution (effect of pitch in the five-tube system, D = 20 mm, Re = 2.1 × $10^4$) for (a) S/D = 1.05, (b) S/D = 1.5, and (c) S/D = 3.0.

For S / D = 3.0, the distributions of the first, second, and third tubes are the same as that of a single tube. For the 'with rods' condition, all of the tubes in a row have enhanced heat transfer.

### 3.3 Pressure distribution

Figure 9 shows the Cp distributions under the same conditions as those of Figure 8.

For S / D = 1.05, Cp is high in the region of $\phi = \pm 90°$ for the second and third tubes, because the incoming flow was stagnated by the row. There is little difference between the Cp distributions of the first, second, and third tubes for the 'with rods' and 'without rods' conditions.

The values of Cp on the side and rear faces of the tubes for S / D = 1.5 become lower than those for S / D = 1.05, because of the gap flow through the narrow gap between the tubes.

For S / D = 3.0, the Cp distributions of the first, second, and third tubes are the same as that of a single tube.

### 3.4 Heat transfer enhancement and drag reduction

We also evaluated the effectiveness of applying this technique to multiple tubes arranged in a single row.

The average Nu ratio, $Nu_R$ / $Nu_O$, of the row of tubes is shown in Figure 10. Subscripts R and O indicate 'with rods' and 'without rods', respectively. The value of $Nu_R/Nu_O$ increases as the number of tubes and the size of the gap increase.

Figure 11 shows the drag coefficient ratio, $C_{DR}$ / $C_{DO}$, of the row of tubes. Based on the drag and momentum balance principle, the drag coefficient was obtained by integrating the velocity distribution of the wake. For the narrow gap (S / D = 1.05), the effect on

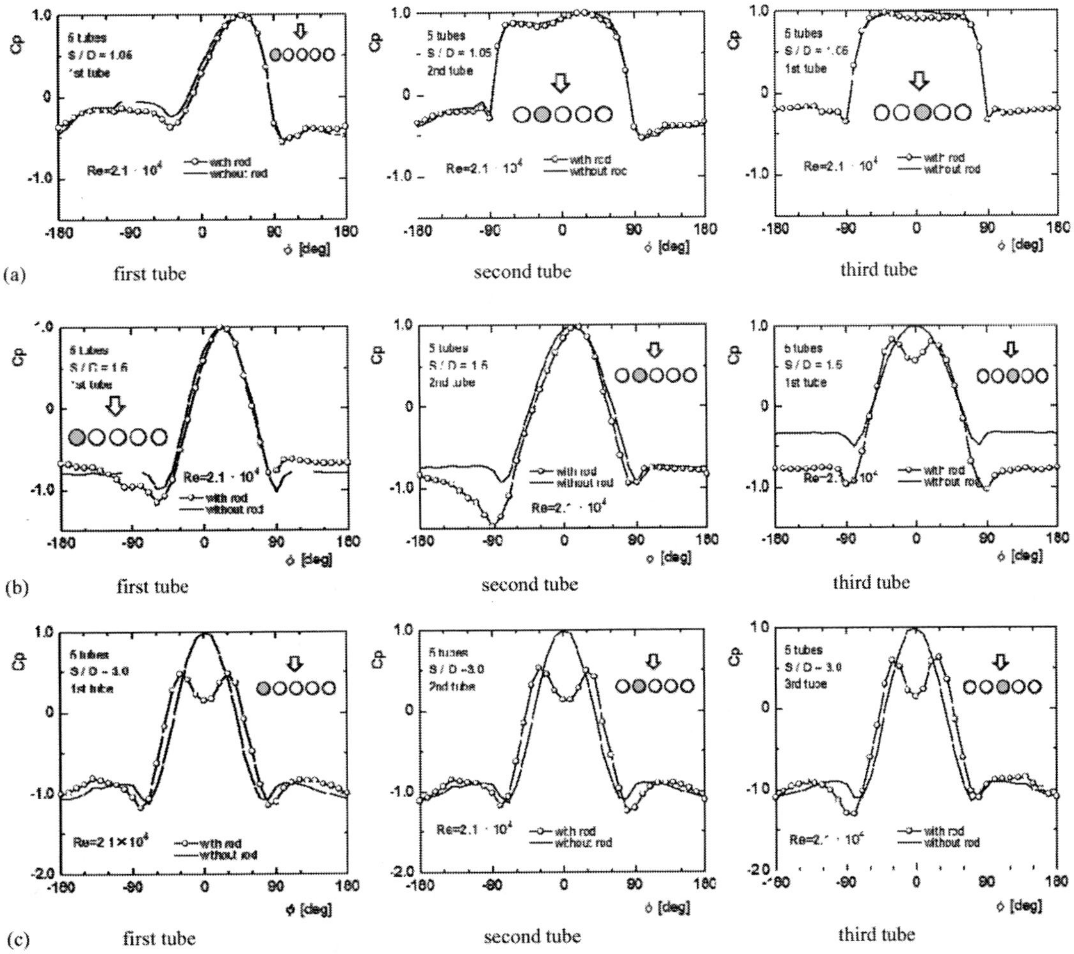


**Fig. 9.** Pressure coefficient distribution (effect of pitch in the five-tube system, D = 20 mm, Re = $2.1 \times 10^4$) for (a) S/D = 1.05, (b) S/D = 1.5, and (c) S/D = 3.0.

drag reduction is slight. The value of $C_{DR}$ / $C_{Do}$ decreases as the number of tubes and the size of the gap increase.

The results for the overall evaluation of heat transfer enhancement and drag reduction ($Nu_R$ / $Nu_o$) / ($C_{DR}$ / $C_{Do}$) are shown in Figure 12. A higher value of ($Nu_R$ / $Nu_o$) / ($C_{DR}$ / $C_{Do}$) indicates higher efficiency. Increasing the number of tubes and the size of the gap results in higher efficiency.

## 4 Conclusions

A row of small rods was placed upstream from a row of circular tubes in order to reduce drag and enhance heat transfer.

Experiments were performed to confirm the effectiveness of this technique. The experimental results revealed the effectiveness of this technique for enhancing the local heat transfer of a row of circular tubes. Moreover, increasing the number of tubes and the gap between the rows of rods and tubes provided greater heat transfer enhancement and increased drag reduction.

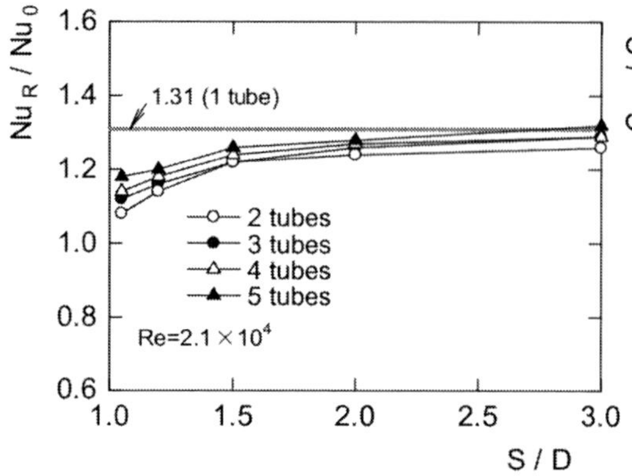


**Fig. 10**. Average Nusselt number of the row of tubes

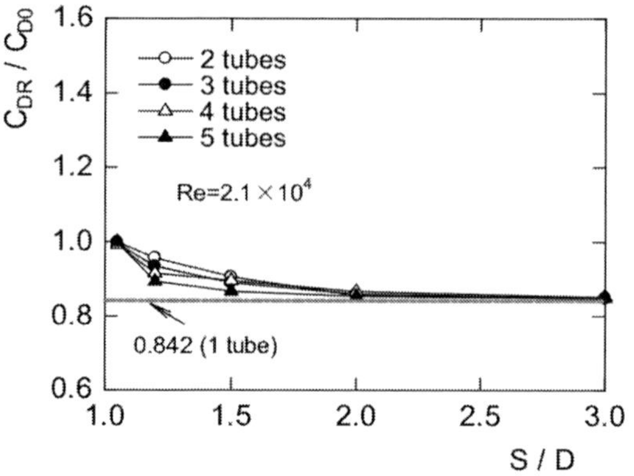


**Fig. 11**. Drag coefficient of the row of tubes

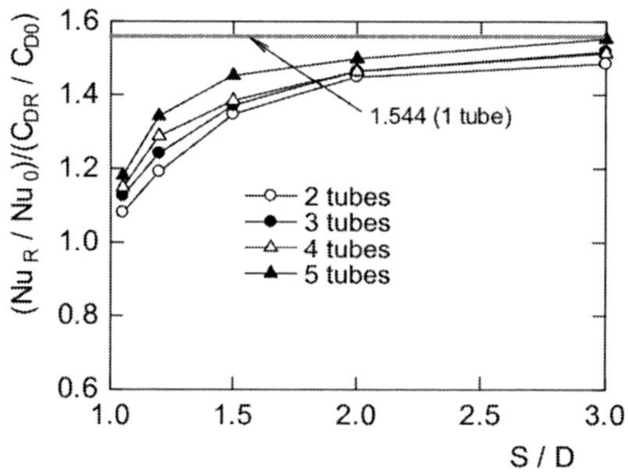


**Fig. 12**. Heat transfer enhancement and drag reduction

## References

1. Wen-Chieh Huang, Cheng-An Chen, Chi Shen, Jung-Yang San, Effects of characteristic parameters on heat transfer enhancement of repeated ring-type ribs in circular tubes, Exp. Thermal and Fluid Science. 68, pp. 371-380 (2015)
2. Subhankar Saha, Sujoy Kumar Saha, Enhancement of heat transferof laminar flow through a circular tube having integral helical rib roughness and fitted with wavy strip inserts, Exp. Thermal and Fluid Science. 50, pp. 107-113 (2013)
3. K. M. Kwak, K. Torii, K. Nishino, Simultaneous heat transfer enhancement and pressure loss T reduction for finned-tube bundles with the fiest or two transverse rows of built-in winglets, Exp. Thermal and Fluid Science. 29, pp. 625-632 (2005)
4. T. Tsutsui, T. Igarashi, Heat Transfer Enhancement of a Circular cylinder, Trans. ASME J. Heat Transfer. 128, pp. 226-233 (2010)
5. T. Tsutsui, T. Igarashi, Drag Reduction of a Circular cylinder in an Air-stream, J. Wind Engineering and Industrial Aerodynamics, 90, pp. 527-541 (2002)
6. M. M. Zdravkovich, Flow Around Circular Cylinders Vol.2: Applications (Oxford University Press, 2005)
7. Tsutsui, An Experimental Study on Heat Transfer around Two Side-by-side Closely Arranged Circular cylinders, Trans. ASME J. Heat Transfer. 132, pp. 111704-1-111704-8 (2010)

EPJ Web of Conferences **143**, 02133 (2017) DOI: 10.1051/epjconf/201714302133
*EFM 2016*

# Experimental validation of mathematical model for small air compressor

*Ján* Tuhovčák[1,*], *Jiří* Hejčík[1], *Miroslav* Jícha[1], and *Ladislav* Šnajdárek[1]

[1]Brno University of Technology, Faculty of Mechanical Engineering, Energy Institute, Technicka 2896/2, Brno 616 69, Czech Rep.

**Abstract.** Development process of reciprocating compressors can be simplified by using simulation tools. Modelling of a compressor requires a trade-off between computational effort and accuracy of desired results. This paper presents experimental validation of the simulation tool, which can be used to predict compressor behaviour under different working conditions. The mathematical model provides fast results with very good accuracy, however the model must be calibrated for a certain type of compressor. Small air compressor was used to validate an in-house simulation tool, which is based on mass and energy conservation in a control volume. The simulation tool calculates pressure and temperature history inside the cylinder, valve characteristics, mass flow and heat losses during the cycle of the compressor. A test bench for the compressor consisted of pressure sensors on both discharge and suction side, temperature sensor on discharge side and flow meter with calorimetric principle sensor.

## 1 Introduction

Reciprocating compressors are used in wide range of application and in every application they are used in broad spectrum of operating conditions. Refrigeration compressors are produced in big series, however they have to fulfil various requirements of cooling demand. Large scale industrial compressors are mostly different piece by piece as the application changes from chemical processing, gas transportation and storage to compressed air technologies. Keeping the efficiency of a compressor at highest level requires using sophisticated technologies and smart regulation, which must be designed and tested during development process. Simulation tools play a significant role in this process.

There are several approaches how to simulate the cycle of a compressor. First category of models is based on polytropic or isentropic principle inside the cylinder. In the work of Li [1] and Posch [2] there is a deep review of these models and their experimental validation. The models are used to predict mass flow rate, electrical power and discharge temperature, however Posch [2] concluded that no model became apparent to fulfil accuracy demands.

Another models for compressors are based on the energy balance over a control volume. Numerical solution is not complicated, although the results are correct and the models are easy to use. They are often used in combination with different tools, e.g. one dimensional model for suction/discharge line [3], thermal model [4] or electrical model of motor [5]. One of the problem is that these models must be experimentally calibrated for specific type of machine, which reduces the universality for all compressors.

Most advanced tools use complex numerical simulations to predict fluid flow inside the cylinder of a reciprocating compressor. Posch [6] analysed the whole compressor using CFD. Aigner [7] developed his own three dimensional numerical code for compression chamber and suction and discharge line. These models offer very detailed results, however their demand for computational power and time cost is much higher compared to previous models.

## 2 Simulation model

The model used in this work is based on the principle of energy (1) and mass conservation (2).

$$dW + dQ + \sum_i dm_i h_i = dU \qquad (1)$$

$$\frac{dm_{cyl}}{dt} + \sum_i m_i = 0 \qquad (2)$$

$Q$ stands for heat transfer from/to the cylinder, $dm_i h_i$ represents inflow or outflow through the valves, $dU$ is inner energy, $dW$ is the work of the piston and the change of inner energy is $dU$. Movement of valves is described by a single degree of freedom equation (3) [8],

$$m\ddot{x} + d\dot{x} + kx = \sum_i F_i \qquad (3)$$

where $m$ is valve mass, $d$ is damping constant and $k$ is spring stiffness. $F$ is the force acting on the valve and $x$ is its position. Flow through the valves is considered as steady, one dimensional and isentropic, solved by following equation.

* Corresponding author: jan.tuhovcak@vutbr.cz

$$\dot{m} = \phi_{eff} \cdot \rho_1 \cdot \left(\frac{p_2}{p_1}\right)^{\frac{1}{\kappa}} \sqrt{\frac{2\kappa}{\kappa-1}\frac{p_1}{\rho_1}\left(1-\left(\frac{p_2}{p_1}\right)^{\frac{\kappa-1}{\kappa}}\right)} \quad (4)$$

Effective flow area $\varphi_{eff}$ is determined using results from numerical simulations of particular valves. Heat transfer inside the cylinder is calculated by improved Annand approach [9], eq. (5)

$$Nu = 0.7\,\mathrm{Re}^{0.7}\,\mathrm{Pr}^{0.7} \quad (5)$$

Thermodynamic properties of the working fluid are gathered from CoolProp library [10]. More information on the simulation model could be found in [11].

## 3 Experimental setup

The compressor used in this work is one-cylinder air compressor. Parameters are shown in Table 1.

**Table 1.** Compressor parameters.

| | Value | Unit |
|---|---|---|
| Theoretical volumetric flow | 7.4 | $m^3$/hr |
| Actual volumetric flow | 4.3 | $m^3$/hr |
| Maximum discharge pressure | 10 | bar |
| Weight | 20 | kg |
| Dimensions | 460x300x380 | mm |

Compressor was mounted with rubber silent blocks to a test bench in order to reduce vibration and noise level, however it still produced a lot of vibrations. Therefore the suction and discharge line were connected to sensors through rubber hoses. Figure 1 illustrates a schematic of a test bench composed of pressure sensors, mass flow meter, temperature sensors and wattmeter.

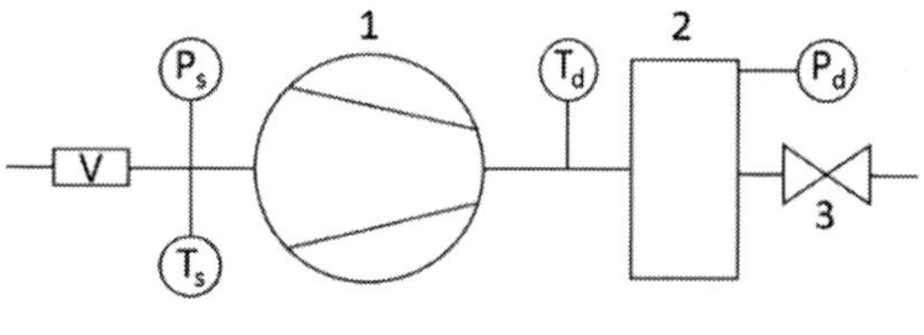


**Fig. 1.** Schematic of test bench.

**Table 2.** Test bench parts.

| | | | |
|---|---|---|---|
| 1 | Air compressor | $T_s$ | Suction temperature |
| 2 | Pressure vessel | $T_d$ | Discharge temperature |
| 3 | Control valve | P | Power consumption |
| m | Flow meter | $P_d$ | Discharge pressure |
| $P_s$ | Suction pressure | | |

The compressor is sucking fresh air from surroundings through mass flow meter. Temperature sensor is part of the mass flow meter and it measures the inlet temperature, yet it does not change significantly from room temperature (22°C). Absolute pressure is measured directly in front of suction port of the compressor. Inside the cylinder, the gas is compressed and discharged through the discharge valve into the pipeline, where temperature sensor PT100 is positioned. Following safety valve is used to protect compressor from higher pressure loads than 10 bar. As the reciprocating compressor is not pumping compressed air continuously, but rather in specific intervals, it is necessary to reduce pressure waves before the discharge pressure is measured. Therefore a 20 litres pressure vessel was inserted in the test rig and the pressure is sampled there.

### 3.1 Methodology

The measurements were performed in two different ways: steady state measurements and continuous measurements. First option means, that the discharge pressure was set to a specific value and measurement lasted long enough to stabilize the discharge pressure and temperature. Stabilization took around 20-30 minutes, especially due to the temperature sensor, which has large thermal inertia and therefore needs more time to reach actual discharge temperature. Second option means that the compressor was started from "no-load" condition and it was run until the discharge pressure reached 10 bars, when the safety valve opened. Temperature was not evaluated in this case.

All measurements started with control valve fully opened. This settings are considered as "no-load" case. In steady state measurements the control valve was manually moved towards closed position to increase the pressure in the vessel. Several points were measured between opened and closed position. Both types of measurements were performed three times, however it was not possible to reach the same discharge pressures in steady state measurements, as the control valve was operated manually. Nevertheless the trends of all measured variables were almost identical.

Suction pressure and discharge pressure obtained from the experiments were used as a boundary conditions for the simulation tool. Validation of the model was based on comparing the mass flow, discharge temperature and power consumption (efficiency) of the compressor.

## 4 Results and discussion

The air compressor is supplying compressed gas at various pressure levels, depending on particular application. Therefore it is essential to examine the compressor behaviour at different discharge pressures.

Mathematical model of the compressor provides results of cylinder pressure and temperature through the cycle of crankshaft. Ideal validation would be a pressure indication, however the experimental equipment is much more complicated and rather expensive. Comparing "global" parameters of the compressor could still give an answer to model accuracy.

### 4.1 Volumetric flow

Main parameters that are influencing mass flow are suction and discharge pressure, suction temperature and valve properties. Suction pressure and suction temperature were kept constant during the experiments. Discharge pressure was the parameter that was changed. Valve properties were evaluated by series of CFD simulations and valve spring stiffness measurements.

The flowmeter is measuring volumetric flow under normalized conditions, meaning pressure equal to 1013 hPa and temperature equal to 273.15 K. Density of air at this condition is 1.293 kg/m$^3$. Mass flow obtained from mathematical model was recalculated for normalized conditions as well and the results are displayed in Figure 2.

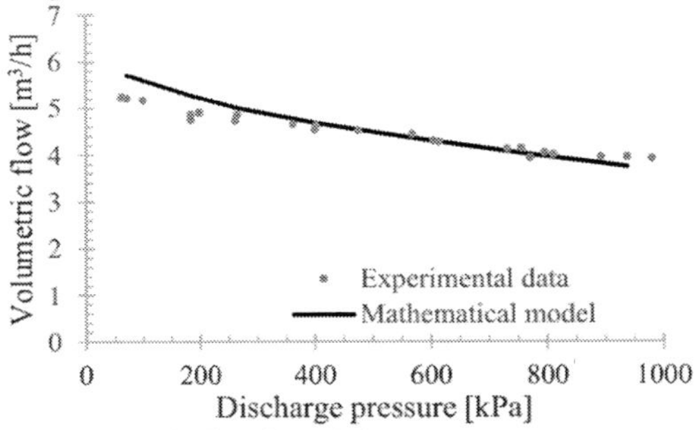


**Fig. 2** Volumetric flow through the compressor.

The results from mathematical model show good agreement with experimental data, especially for higher discharge pressures. The difference is below two percent for discharge pressures above 300 kPa. Lower discharge pressures show bigger difference, reaching up to twelve percent for 70.4 kPa discharge pressure. Decreasing volumetric flow is also causing decrease in volumetric efficiency of the compressor. Theoretical volumetric flow of the compressor is 7.4 m$^3$/hr. This value is dependant just on dimensions of the compressor and its speed. It does not include any losses caused by valves, death volume or superheating of working gas. Following Figure 3 shows decrease in efficiency caused by increasing discharge pressure.

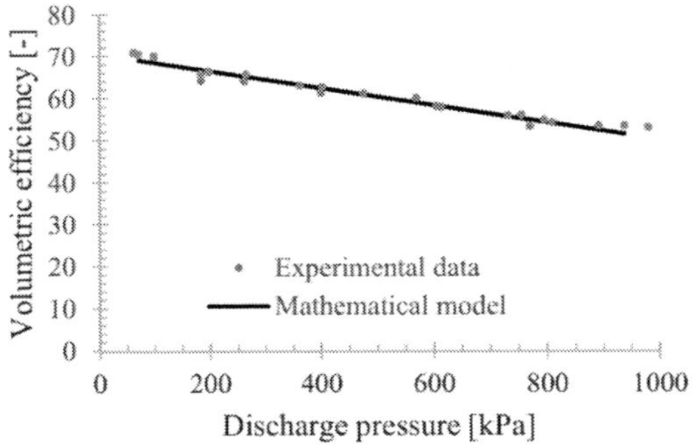


**Fig. 3** Volumetric efficiency of the compressor.

### 4.2 Compressor efficiency

The performance of the compressor is formed by three main factors: electrical efficiency of motor, mechanical efficiency of drive shaft and thermodynamic efficiency of the processes inside the cylinder. However, the character of measurement did not allow us to evaluate every efficiency individually. Overall compressor efficiency was evaluated, comparing power consumption and work done by the compressor, eq. (6).

$$\eta = \frac{W_c}{W_{in}} \tag{6}$$

Actual work of the compressor $W_c$ is calculated using 1$^{st}$ Law of Thermodynamics, equation (7).

$$W = Q - m \cdot \Delta h \tag{7}$$

$W$ is compressor work, $Q$ is heat transferred between the cylinder wall and working gas, $m$ is the mass inside the cylinder and $h$ represents enthalpy.

Results in Figure 4 show experimentally measured power consumption multiplied by estimated efficiency and the work done by the compressor. The efficiency was estimated from series of measurement and was set to 50%. As it can be seen in the Figure 4, the estimated efficiency seems to fit quite adequately with results from mathematical model above 300 kPa discharge pressure. In the region under 300 kPa discharge pressure the work done by compressor decreases significantly, but the power consumption stays under the same trend through the whole measured interval. This means that the overall efficiency is lower when the discharge pressure is decreasing, Figure 5.

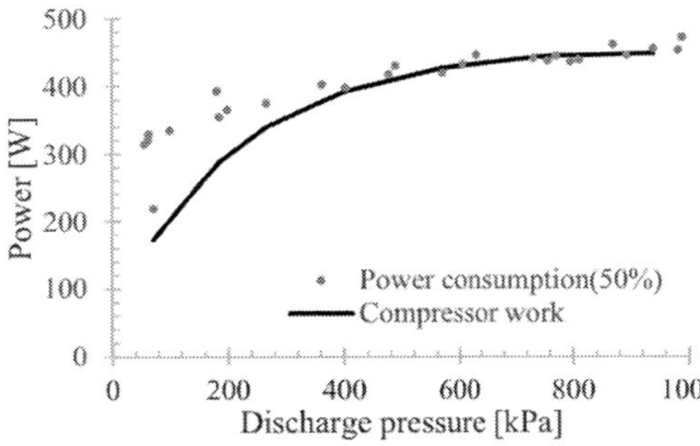


**Fig. 4** Power consumption and compressor work.

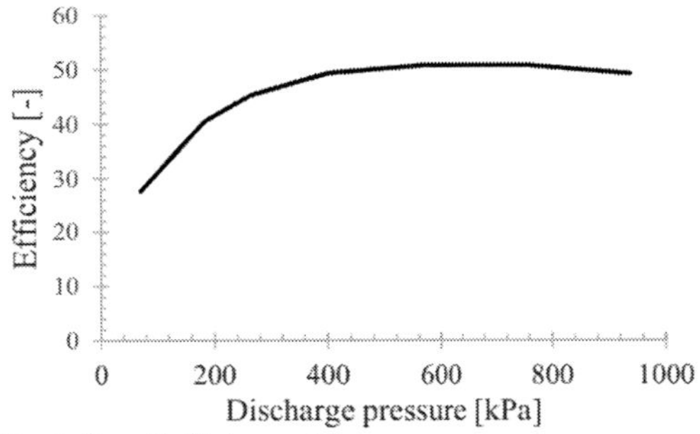


**Fig. 5.** Overall efficiency of the compressor.

The reason for that could be the electrical efficiency of the motor, which is decreasing as the load on the motor is dropping down [12]. Burt [12] stated, that the efficiency of the motor changes by approximately 10%, when the relative load decreases from 60% to 20%. This is also shown in Figure 5, where the decreasing efficiency of actual compressor is shown with a respect to the discharge pressure. The electric motor should be examined more deeply to prove this decrease in overall efficiency of the compressor. Heat transfer inside the compressor might also cause inaccuracy in efficiency calculation. Lower load on motor and higher mass flow through the compressor cause lower thermal load on walls, which are surrounding working gas. The effect of gas superheating is lower and the work done by compressor will be affected. The same reason could be behind the discrepancies in volumetric flow rate at lower pressures, see Figure 2. The mathematical model uses constant suction temperature, but during the experiments, the gas in the suction line is heated up by the surrounding wall, e.g. valve plate or cylinder head. Measuring the temperature right in front of suction valve could help to predict volumetric flow and work done by compressor more accurately.

### 4.3 Discharge temperature

The highest gas temperature occurs when the discharge valve is opened. Mathematical model calculates temperature in the cylinder, however this temperature is difficult to measure as it changes very fast. Therefore the probe was positioned in the discharge line. Omitting the effect of re-expansion of compressed gas in discharge chamber and heat transfer through the walls of discharge line caused rather big discrepancies between mathematical and experimental results in Figure 6. A CFD simulation of fluid flow in discharge line proved that the temperature in measuring point will not significantly overcome 100 °C, even for the highest discharge pressure. As there is a significant temperature drop in discharge line a thermal model must be developed and included in the presented mathematical model. Also the experimental setting has to be changed and the thermal probe must be positioned right behind the discharge valve.

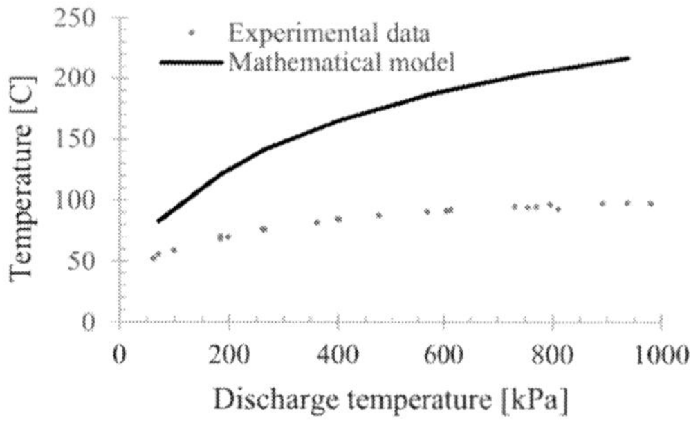


**Fig. 6** Discharge temperature.

## Conclusion

In this work a validation of developed mathematical model for reciprocating compressor is presented. Comparison of calculated and measured volumetric flow shows good agreement in whole range of working conditions, which is indicating that the mathematical model is giving reasonable results. In case of power-work comparison or discharge temperature the results were not completely satisfying. The efficiency of the compressor was estimated, but it was found out that it is not a constant value and at lower loads it decreases. Therefore it is necessary to perform more experiments focused on this problem. Another issue was found out with discharge temperature. The mathematical model calculates the temperature inside the cylinder, however the in the discharge line of the compressor. The influence of heat transfer in discharge line and gas expansion must be examined more deeply in order to calculate temperature distribution in discharge line accurately.

Another thing which is influencing the calculation is the wall temperature of the cylinder. In the mathematical model the temperature was set to 80 °C, which may not be the actual temperature. Moreover, it may change with varying discharge temperature. Due to this fact, the wall temperature will be measured exactly with thermocouples in future work.

More experimental data are necessary, especially the temperature profile of the compressor in different points. Performing experiments and development of new models for reciprocating compressor will be the goal of future works.

## Acknowledgement

The presented research was supported by the Brno University of Technology project FSI-S-14-2355. The authors gratefully acknowledge funding from the Ministry of Education, Youth and Sports of the Czech Republic under the National Sustainability Programme I (Project LO1202 NETME CENTRE PLUS)

## References

1. W. Li, in *Int. J. Refrig.* (2012), pp. 1722–1733.
2. S. Posch, E. Berger, M. Heimel, R. Almbauer, and A. Stupnik, (2014).
3. W. Lang, R. Almbauer, A. Burgstaller, and D. Nagy, Eng. Conf. 1 (2008).
4. S. K. Lohn, M. C. Diniz, and C. J. Deschamps, IOP Conf. Ser. Mater. Sci. Eng. **90**, 12068 (2015).
5. T. Dutra and C. J. Deschamps, Int. J. Refrig. **59**, 168 (2015).
6. S. Posch, J. Hopfgartner, M. Heimel, E. Berger, R. Almbauer, and S. Stangl, Int. Compress. Eng. Refrig. Air Cond. High Perform. Build. Conf. 1 (2016).
7. R. Aigner, Internal Flow and Valve Dynamics in a Reciprocating Compressor, TU Vienna, 2007.
8. M. Costagliola, J. Appl. Mech. **17**, 415 (1950).
9. W. J. D. Annand, Proc. Instn. Mech. Engnrs. **117**, 973 (1963).
10. I. H. Bell, J. Wronski, S. Quoilin, and V. Lemort, Ind. Eng. Chem. Res. **53**, 2498 (2014).
11. J. Tuhovcak, J. Hejcik, and M. Jicha, Appl. Therm. Eng. **103**, 607 (2016).
12. C. M. Burt, X. Piao, F. Gaudi, B. Busch, and N. F. Taufik, J. Irrig. Drain. Eng. **134**, 129 (2008).

# Experimental verification of methane local concertations formation in a confined space

*Aleš* Tulach[1,a], *Miroslav* Mynarz [1] and *Milada* Kozubková[2]

[1]VSB – Technical University of Ostrava, Faculty of Safety Engineering, Department of Fire Protection, Lumirova 630/13, Ostrava - Vyskovice, 700 30, The Czech Republic
[2]VSB – Technical University of Ostrava, Faculty of Mechanical Engineering, Department of Hydromechanics and Hydraulic Equipment, 17.listopadu 15/2172, Ostrava -Poruba, 708 33, The Czech Republic

**Abstract.** The paper describes the experimental measurements concerning the formation of critical concentrations of methane in a closed space. Measurements were carried out in order to monitor the formation, propagation and distribution of methane-air mixture in a confined space. Experimental data are also used for verification of numerical simulations of methane distribution and formation of its local critical concentrations in a closed space. Within experiments, the spaces with low velocities of flow of flammable gas were primarily watching and the measurements provided information about local changes of methane concentration in time. During experiments more points could be monitored at the same time so compact description of flammable gas propagation through given space was obtained. Gained results will be used for choice of suitable mathematical model for numerical simulations.

## 1 Introduction

Leakages of hazardous gases or vapours of hazardous liquids occur increasingly today. In case of determination of an ignition source, which may cause inflammation of the formed gas mixture, it is necessary to perform more detailed analyses. One of the possibilities consists in the use of CFD numerical simulations [1][2] describing the propagation of the flammable gas through the investigated space and the formation of local critical concentrations depending on time.

However, mathematical models used in numerical simulations must be after their completion filled with sufficiently precise input data (determined often experimentally), and simulation results must be verified by experimental measurements. Verification of calculations and experiments helps to choose the appropriate mathematical model, which can accurately simulate the actual cases of leakages and propagation of flammable gases.

## 2 Experimental part – Modification of physical model

For verification of numerical simulations, physical measurements were taken on complex geometry of the building. Measurements were performed with use of a measuring set, which imitated the circulating area of common block of flats, in which a leakage of gas from the main gas valve located in the basement can occur. The measuring set consisted of a staircase area and a part of the basement. Special paper [8] concerned with detailed description of the measuring set. Due to certain problems and inaccuracies observing at verification of mathematical model, physical measurements should also be put more exactly. Previous measurements involved several insufficiencies, mainly due to specificity of the experiment. Therefore some significant modifications were made before following measurements. The first change compared to previous measurements consisted in replacement of the source of leaking gas. Instead of Transit natural gas (98.39 vol. % of methane [3][4]), only pure gas was used (concentration 99.95 vol. % of methane). Pressure tank was used as a source of methane (natural gas) which provided methane flow under constant pressure of 75 kPa using defined reducing valve. For previous measurement, public gas distribution system was used; however pressure oscillated during the day due to various consumption of natural gas. Intensity of leaking methane was controlled by regulation of flow rating which was setting in constant value Q = 0,871 $dm^3/s$. Value of flow was chosen in relation to dimensions of the model and used gas detectors.

The second and the most significant improvement of physical measurement was to specify the detection of local methane concentrations in investigated confined space.

[a] Corresponding author: ales.tulach@vsb.cz

 DOI: 10.1051/epjconf/201714302134

## 2.1 Measurement of local methane concentrations

In previous measurements five two-stage detectors GC20-CH4 and five three-stage detectors GI30-CH4 calibrated to methane $CH_4$ (Figure 1) were inserted into external walls of the measuring kit. Detectors signalled occurrence of 0.5, 1 and 2.5 % of the methane volume fraction at the monitored points [7].

**Figure 1** Detector GC20-CH4 (left) and the detector GI30-CH4 (right)

Detectors were equipped with non-selective hot-semiconductor sensors, which react to the presence of oxidizing or reducing gases. It concerns surface absorption detection where exchange of electrons proceed between conduction and valence band.

New detectors were developed and calibrated to improve the quality of experimental measurements of dynamic behaviour of local methane concentrations. Semiconductor sensors were used no longer after literature search (study of advantages and disadvantages of various sensors) and attention was paid to detectors with sensors on principle of catalytic combustion of flammable gases (detection proceeds through comparison of electric resistances of measuring spiral). Sensor MI-02 (Figure 2) by Hankook Gas kiki comp. was chosen from available sensors.

Newly developed detecting system enabled to obtain more detailed course of formation of local concentrations and immediate measured concentration was recorded in seconds intervals. The aim was to develop the detector for recognition of gas concentration within the range of lower than 1 vol. % to 5 vol. % (lower explosive limit of methane).

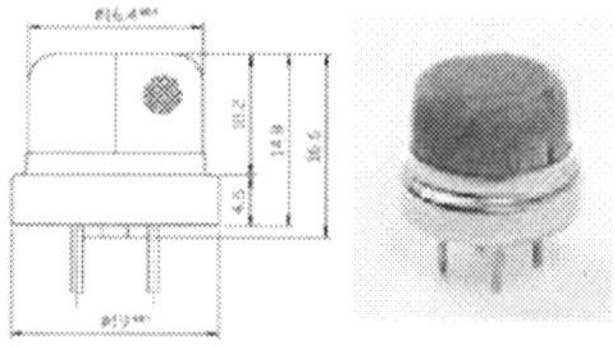

**Figure 2** Catalytic sensor MI-02

Parameters of sensor MI-02:
Power supply 2.3 VDC ± 10 %; current lower than 210 mA ± 10 %; output loss 0,46 W ± 10 %; time of primary stabilization 30 s ± 10 %; speed of response 10 s ± 10 %; temperature at catalytic reaction 425 °C ± 10 %.

Newly developed detectors were constructed with the help of printed circuit (Figure 3). Total of a ten measuring circuits were made.

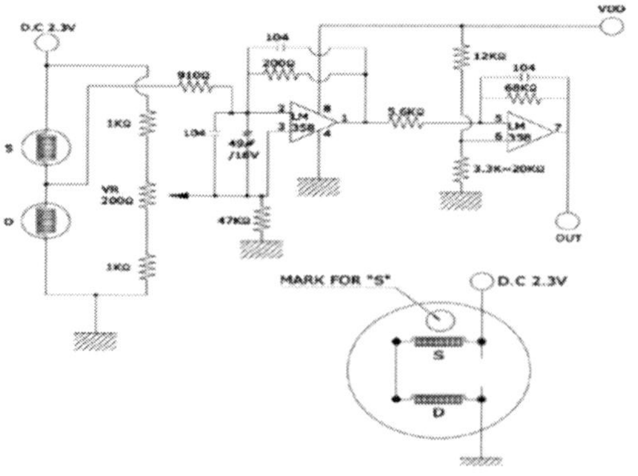


**Figure 3** Circuit diagram

Testing of the prototype led to findings that temperature of a cooler does affect constructed measuring circuit and thus resulting measured voltage (measured concentration). Therefore more coolers were used and they were purposely placed farther from measuring circuits as it is seen in Figure 4. These changes (compares to the prototype) caused decrease of temperature and minimalization of influence on measuring circuits by ambient temperature.

**Figure 4** Five newly developed detectors

All used detectors were connected to the data loggers OMEGA RD8800 which afterwards recorded in second intervals changing voltage (Figure 5).

**Figure 5** Data loggers of detectors

## 2.2 Calibration of detectors

Before measurement, newly made detectors were calibrated. The same gas (99.95 vol. % methane) was used both for calibration and physical model. The most available and exact method of calibration was to determine required methane concentration in the air using partial pressures in the explosive autoclave VA20 (Figure 6) in which homogeneous methane-air mixture was formed. Pressure sensor MSD100MRE was used for measurement of partial pressures. Measurements uncertainty of this sensor is ± 0,2 %.

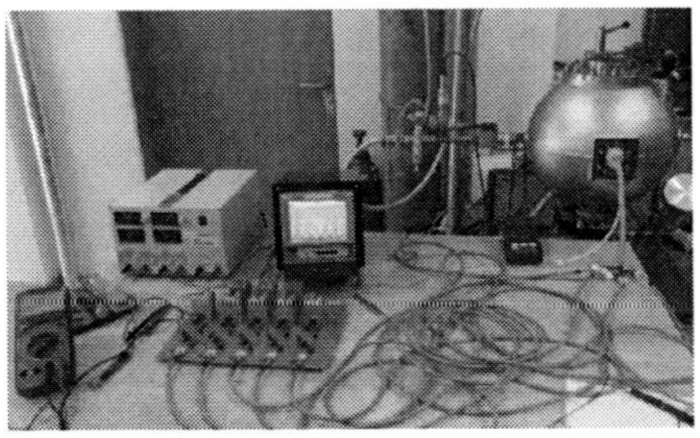

**Figure 6** Calibration of detectors

For calibration, mixtures were prepared one after another with 1 to 5 vol. % of methane with air. Measured voltages were inscribed in the graph (Figure 7).

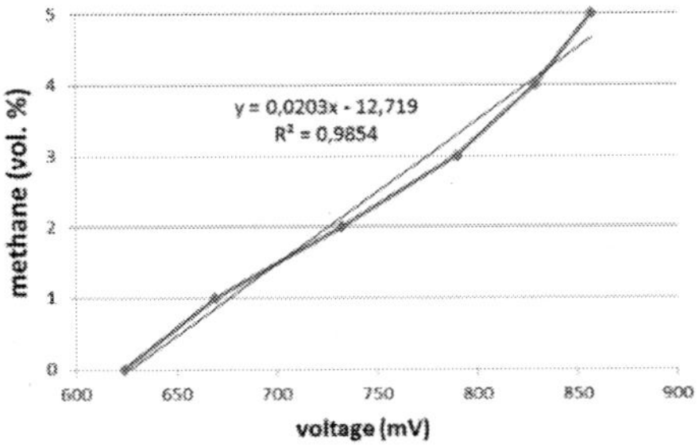


**Figure 7** Calibration function for monitoring point 5

Measured points were interleaved with straight line using statistical methods. On the basis of this straight line, the function for conversion between measured voltage and methane concetration was generated. Graph of calibration function shows approximately linear course of voltage increase with concentration increase.

### 2.3 Process of measurement of local concentrations

Ten measurements of local concentration of methane were taken in confined space (Figure 9).

**Figure 8** Location of new sensors

Current detectors were located at the same places as in previous measurements [8]. New detectors were added next to current detectors approximately to the same places (Figure 8).

Result of these measurements was to determine time between beginning of gas leakage and moment of occurrence of 5 % vol. methane concentration in all monitored points. The latest time to reach 5 % vol. methane concentration was in monitored points No. 2 and 5 located in the lowest and highest levels of the staircase area (Figure 10).

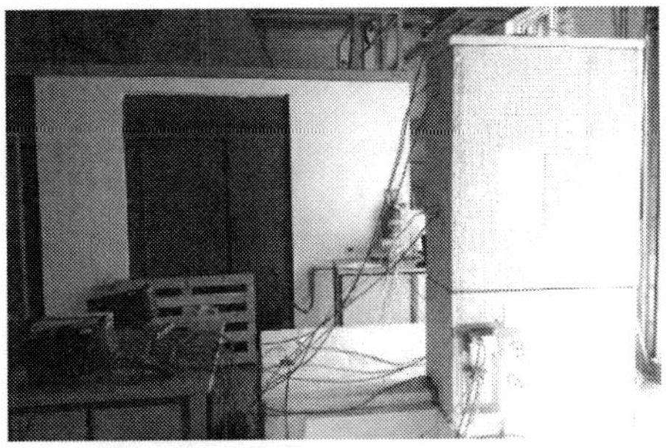

**Figure 9** Measuring the formation of local concentrations of methane

## 3 Theoretical part – numerical model

The program ANSYS Fluent [1] was used for numerical modelling of physical measurements described above, as it is suitable especially for solution of the flows in the given space. The selected program uses for mathematical description of the ongoing physical processes following laws:

Navier-Stokes equations [5] express the conservation law of momentum and describe the flow as follows.

$$\frac{\partial \rho u}{\partial t}+\frac{\partial \rho uu}{\partial x}+\frac{\partial \rho uv}{\partial y}+\frac{\partial \rho uw}{\partial z}=\rho a_x-\frac{\partial p}{\partial x}+\nu\left(\frac{\partial^2 u}{\partial x^2}+\frac{\partial^2 u}{\partial y^2}+\frac{\partial^2 u}{\partial z^2}\right)+S$$
$$\frac{\partial \rho v}{\partial t}+\frac{\partial \rho vu}{\partial x}+\frac{\partial \rho vv}{\partial y}+\frac{\partial \rho vw}{\partial z}=\rho a_y-\frac{\partial p}{\partial y}+\nu\left(\frac{\partial^2 v}{\partial x^2}+\frac{\partial^2 v}{\partial y^2}+\frac{\partial^2 v}{\partial z^2}\right)+S_y \quad (1)$$
$$\frac{\partial \rho w}{\partial t}+\frac{\partial \rho wu}{\partial x}+\frac{\partial \rho wv}{\partial y}+\frac{\partial \rho ww}{\partial z}=\rho a_z-\frac{\partial p}{\partial z}+\nu\left(\frac{\partial^2 w}{\partial x^2}+\frac{\partial^2 w}{\partial y^2}+\frac{\partial^2 w}{\partial z^2}\right).$$

The continuity equation [5] is the fourth equation, which deals with conservation of mass flow.

$$\frac{\partial(\rho)}{\partial t}+\frac{\partial(\rho u_x)}{\partial x}+\frac{\partial(\rho u_y)}{\partial y}+\frac{\partial(\rho u_z)}{\partial z}=S_z \quad (2)$$

Transfer of admixtures (mass fraction) is solved by the balance equation [5], which in a changing time computes with the values of mass fractions of admixture "$Y_i$" and with components of flow velocity of present gases "$u_i$". It takes into account the diffusion flux of the $i^{th}$ component of " $J_{j,i}$ " mixture, the rate of production of admixtures resulting from chemical reaction " $R_i$ " and the rate of formation of increments from the distributed admixtures "$S_i$". Distribution of admixtures varies in dependence on the diffusion flux [5].

$$\frac{\partial}{\partial t}(\rho Y_i)+\frac{\partial}{\partial x_j}(\rho u_j Y_i)=-\frac{\partial}{\partial x_j}J_{i,j}+R_i+S_i \quad (3)$$

Particular description of geometry and mesh is dealt in detail in different contribution [8]. The mesh was created in the program ANSYS Meshing and it consisted of **1 869 187 elements.** Parameter for **determination of quality of 3D element** (size of its deformation) was **0,899**.

Current monitoring points remained in ANSYS Fluent without changes. Then ten monitoring points were created in the model with use of the program ANSYS Fluent (Figure 10).

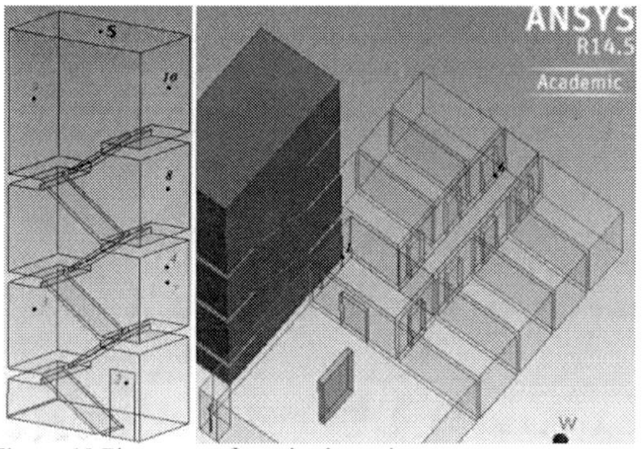


**Figure 10** Placement of monitoring points

Location of monitoring points was identical with the placement of new detectors on the outer walls of the experimental model enclosing the space.

On the basis of these points the program ANSYS Fluent then made an evaluation of the dependence between the duration of leakage and the methane concentration in the air mixture.

## 4 Results

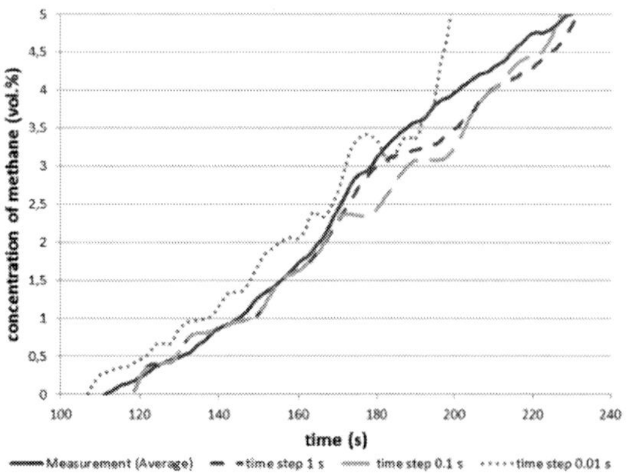


**Figure 11** Comparison measurements with numerical simulation (model k-ε RNG) at the point 5

Comparison of measured and calculated values is presented in the graph (Figure 11). It describes describe the change of volume concentration of methane in monitoring point No. 5 in time.

The graph shows average values from ten measurements and values calculated in the program ANSYS Fluent (mathematical model k-ε RNG) at various time steps (1000; 100 and 10 ms).

## 5 Conclusions

Gradually five turbulence models (k-ε Standard, k-ε RNG, k-ε Realizable, k-ω Standard, k-ω SST) were tested. The best match with the results of experimental measurements was achieved by numerical simulations with mathematical models k-ε RNG.

In general, flammable gas first filled the spaces under the flight of stairs and stairs landings and only then the borders of the monitored concentrations decreased in the lower part of the staircase space, where they were detected by the sensor located in the monitoring point No. 2. This resulted in weak flow with low Reynolds number. After the last improvement of the quality of physical measurement using newly developed detectors for concentration, good agreement was achieved between results from experimental measurements and numerical simulations.

### Acknowledgements

This paper was financially supported by the project of grant ministry of interior the Czech Republic under the Id. No. VI20152019047, entitled "Development of the rescue destructive bombs for the disposal of statically damaged buildings".

### References

[1] Ansys, Inc. *ANSYS FLUENT 14.5 - Theory Guide.* (2013)
[2] M. Bojko, *3D Flow – ANSYS Fluent: textbook (in Czech).* 1. issue. Ostrava: Edit centre VŠB – TUO, (2012)
[3] J. Fík, *Natural gas: tables, diagrams, equations, calculations.* Praha : Agentura ČSTZ, s.r.o., 355, (2006)
[4] V. Koza, L. Čapla, Determination of leak gas amount from damaged gas pipes. Gas: professional periodical for gas manufacture. Vol. XC, **2**, 38-42, (2010)
[5] M. Kozubková, *Modelling of liquids flow, FLUENT, CFX.* 1. issue. Ostrava: Edit centre VŠB – TUO, (2008)
[6] I. Skiteva, V. Seleznev, Numerical analysis of methane-air pollution. Russia, Computation Mechanics Technology Center (2006)
[7] Technical conditions and service manual for detectors GC20N and GC20K, 8, (2002)
[8] A. Tulach, M. Mynarz, M. Kozubková, Contribution to formation of explosive mixture in a closed dissected space as a result of natural gas leak. In EPJ Web of Conferences. EDP Sciences. (2015)

DOI: 10.1051/epjconf/201714302135

# Analysis of pressure losses in the diffuser of a control valve

*Petr* Turecký[1], *Lukáš* Mrózek[2*], *Ladislav* Tajč[2], and *Michal* Kolovratník[3]

[1]ENVIROS, s.r.o., Dykova 53/10, 101 00 Praha 10-Vinohrady, Czech Republic
[2]Univerzity of West Bohemia, Department of Power System Engineering, Univerzitní 22 Pilsen 306 14, Czech Republic
[3]Czech Technical University in Prague, Department of Energy Engineering, Technická 4, Prague 6, Czech Republic

**Abstract.** The pressure loss in the diffuser of a control valve is evaluated by using CFD computations. Pressure ratios and lifts of a cone for the recommended flow characteristics of an experimental turbine are considered. The pressure loss in a valve is compared with the pressure loss in a nozzle, i.e. the embodiment of the valve without a cone. Computations are carried out for the same mass flow. Velocity profiles are evaluated in both versions of computations. Comparison of computed pressure losses, with the loss evaluated by using relations for diffusers with the ideal velocity conditions in the input cross-section, is carried out.

## 1 Introduction

Control valves of steam turbines are basically nozzles with a variable cross-section. A confuser as well as a diffuser channel is applied here. The change of the mass flow of the steam is enabled by adjustment of the flow area and pressure ratio across the valve. The flow area depends on the lift of a cone and the pressure ratio is determined by the operating parameters of a turbine. A sudden change of the cross-section arises in the valve under the cone. A wake vortex arises under the cone, which distorts the velocity field in the diffuser. At a certain stage of the operation control, i.e. when the turbine approaches the power, a pressure loss across the valve is required. In this case the operational reliability is most important. Emphasis is placed on the control of pressure pulsations and thus extreme vibrations are controlled too. However, during nominal conditions of the turbine it is required the minimum pressure loss in the entire flow system before the turbine. In general, the pressure loss is related to friction and shape modifications of the valve. The decisive proportion of the pressure loss is caused by a sudden change of the cross-section. An outlet diffuser is used in the valve, therefore it is impossible to specify clearly for what change of the flow area this loss must be considered. Behind the cone of the valve the wake vortex arises with the flow separation. In the diffuser at small lifts of a cone the backflow arises, which may fill up its substantial part. There is a significant deformation of velocity field in the throat as well as in the entire diffuser. Available literature data for computation of the pressure loss in diffusers are based on the existence of the balanced velocity field at the inlet into the diffuser. The aim of this computation study is to evaluate the impact of the wake vortex on the increase of the pressure loss in the diffuser. Geometry of a valve with a cone and without a cone is modelled and the pressure losses are computed for the same mass flow.

## 2 Operational parameters of a control valve

Model version of a control valve is shown in figure 1.

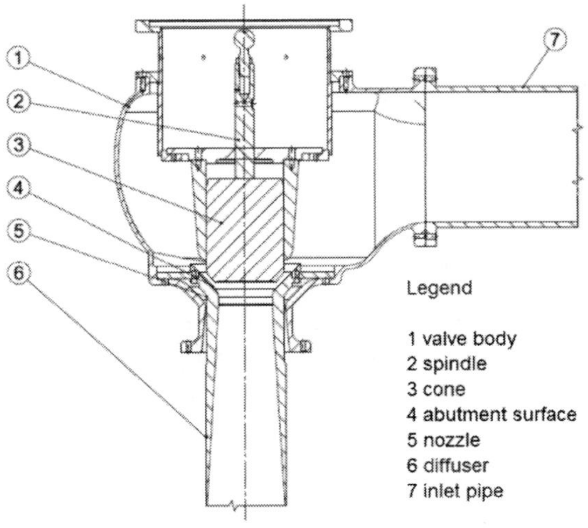


**Fig. 1.** Scheme of control valve

With the change of the cone lift of a valve $h$ the flow area of the valve changes too. Mass flow through a valve is determined by the flow area, input parameters and pressure ratio across that valve. Figure 2 shows the general flow characteristics of the typical embodiment of a valve in dimensionless parameters [5]. The real mass flow $G$ is related to the critical flow $G_*$, which is considered for the area of the throat of the diffuser $D_h$.

* Corresponding author: lukas.mrozek@doosan.com

DOI: 10.1051/epjconf/201714302135

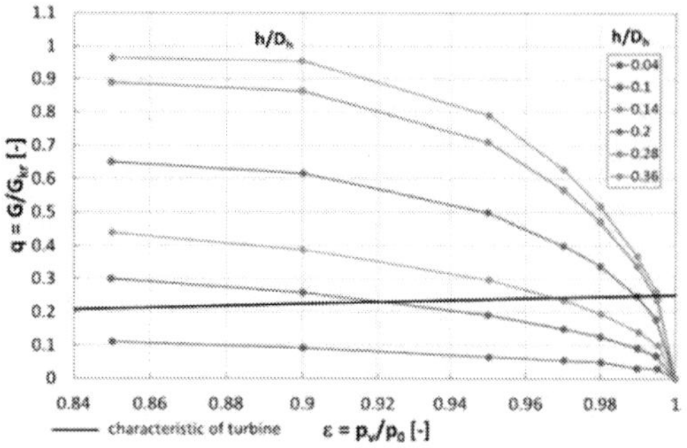


**Fig. 2.** Flow characteristics of valves

There is also presented expected operational characteristics of the experimental turbine. Relevant pressures and temperatures are shown in table 1. The computation study is carried out for these recommended pressure ratios and cone lifts. In table 2 are shown other characteristic data of the tested valve, respectively of a computational model of the diffuser.

**Table 1.** Specified parameters for the simulation

| $h$ [mm] | $(p_v)_v$ [Pa] | $p_{0c}$ [Pa] | $T_0$ [K] |
|---|---|---|---|
| 4 | 40968 | | |
| 6 | 51051 | | |
| 8 | 67717 | | |
| 10 | 79989 | | |
| 14 | 91086 | 100000 | 294 |
| 20 | 96067 | | |
| 24 | 97114 | | |
| 28 | 97655 | | |
| 36 | 97792 | | |

**Table 2.** Initial values

| | | | |
|---|---|---|---|
| Inlet pressure | $p_{0c}$ | 100 | kPa |
| Inlet temperature | $T_0$ | 294 | K |
| Heat capacity ratio | $\kappa$ | 1.4 | - |
| Specific gas constant | $r$ | 287 | J/kgK |
| Length | $l$ | 400 | mm |
| Diameter of diffuser throat | $D_h$ | 100 | mm |
| Angle of diffuser opening | $\gamma$ | 7 | ° |
| Cross section of diffuser throat | $S_h$ | 7854 | mm$^2$ |
| Density | $\rho_0$ | 1.185 | kg/m$^3$ |
| Area ratio | $n$ | 2.218 | - |

## 3 Computational models

A computational model of a valve is presented in figure 3. It is a simplified version of the normal performance of a valve. Instead of the real 3D flow in the valve chamber and in the diffuser, only the rotationally symmetrical arrangement is considered, i.e. only the variant with the 2D flow.

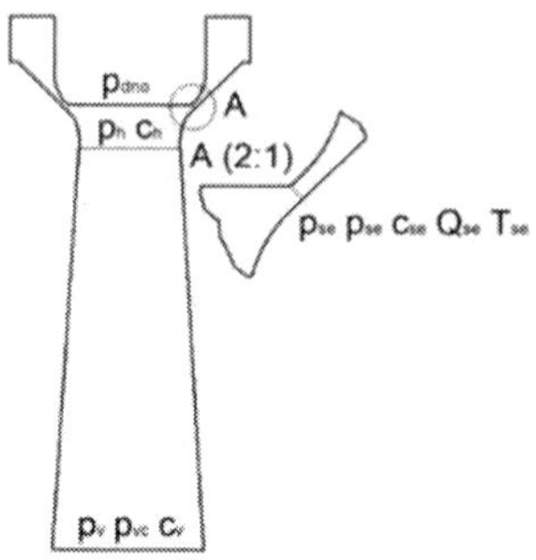


**Fig. 3.** A computational model of a valve

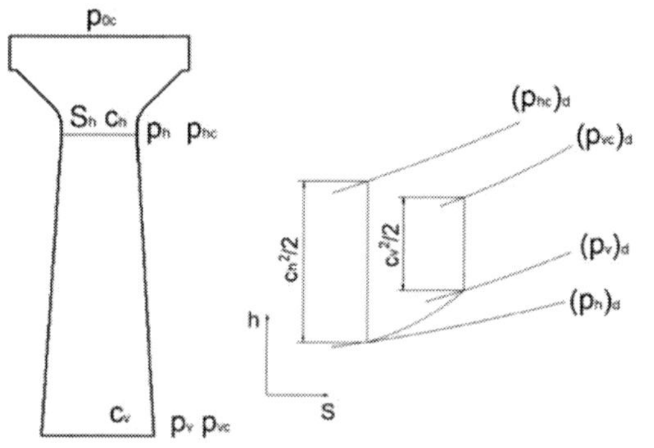


**Fig. 4.** A computational model of a nozzle

Extension of the flow channel arises already from the area of the saddle under the cone. In the variant of a nozzle as shown in figure 4, only the diffuser itself is considered with the steady area of the neck $S_h$. In order to achieve the same mass flow in a nozzle as in a valve, an appropriate back pressure $p_v$ behind the diffuser must be present. As a result of the wake vortex formation and a sudden change of the cross-section, the pressure loss in the variant of a valve $\Delta p_z = p_{0c} - p_{vc}$ is higher than in a nozzle. CFD simulation of the 2D model of a control valve and a nozzle is carried out by using the ANSYS Fluent software. For individual sections of the flow channel the different density of the mesh was chosen. As a result, there are 614000 cells in a valve model and 436000 cells in a nozzle.

## 4 Results

Nine CFD simulations for the stated cone lifts and recommended pressure ratios were carried out in the variant of a valve. Nine computations of flowing in a nozzle with the same mass flow as in the variant with the

DOI: 10.1051/epjconf/201714302135

valve were added to it.

The demonstration of the velocity distribution in the variant of a valve and a nozzle is shown in figure 5. These are variants corresponding to the cone lift 6 mm ($h/D_h = 0.06$).

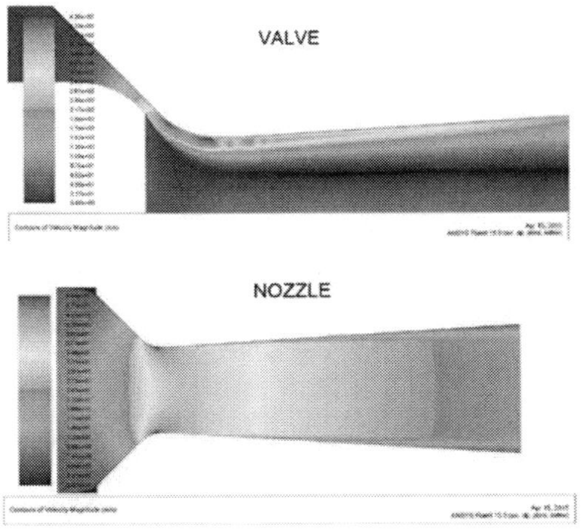


**Fig. 5.** The velocity field in a variant of a valve and a nozzle at the same mass flow

From the detailed distribution of the velocity vectors in figure 6 the backflow and significant changes of the velocity in a narrow strip by the diffuser wall is clearly seen. The backflow does not occur in a nozzle. Computation also confirms the pressure fluctuation near by the diffuser wall at relative cone lift $h/D_h < 0.08$.

**Fig. 6.** The vector velocity distribution

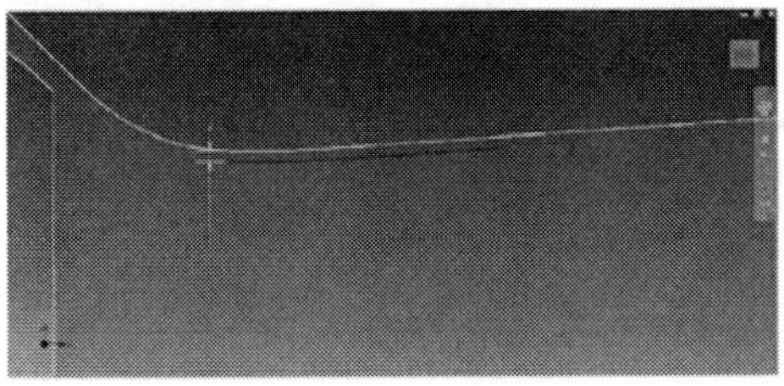

**Fig. 7.** Section of velocity changes near the wall

Analysis of pressure pulsations and velocity changes for the section of the diffuser behind its throat in the zone 2 mm from the wall is carried out. This is shown in figure 7. Fluctuation of the static pressure and velocity is presented in the graphs in figures 8 and 9. Occurrence of pressure pulsations is connected to local occurrence of supersonic velocities.

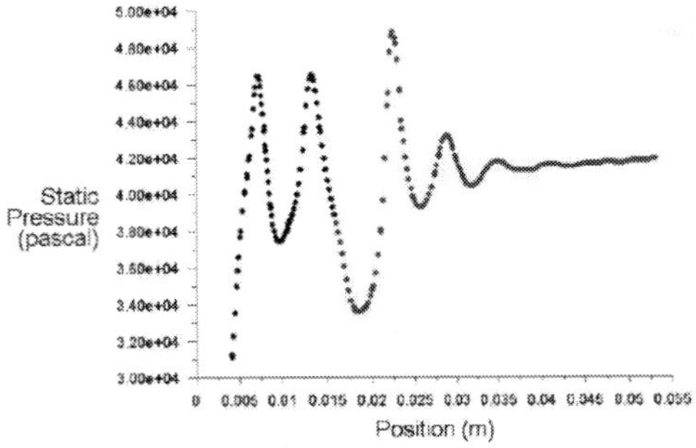


**Fig. 8.** Fluctuation of the static pressure

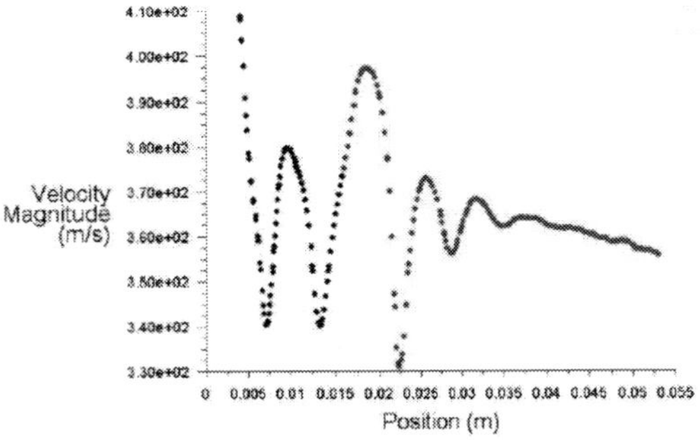


**Fig. 9.** Fluctuation of velocities by the wall

After the transition to higher cone lifts when Mach number in the diffuser throat is already $Ma < 1$, the pressure pulsation failed to verify by using computations. However, as the result of flushing the vortexes behind the cone the pressure changes undoubtedly occur.

Due to the wake vortex behind a cone the velocity field in the diffuser throat is significantly distorted. Maximum velocity occurs near the diffuser wall while in its middle section the velocity may have even a negative value. Velocity ratios for a valve with proportional cone lift $h/D_h = 0.04$ and for a nozzle with the same mass flow are shown in figure 10. Considerably uneven velocity field also occurs at the outlet from the diffuser. Even at the cone lift $h/D_h = 0.2$ in the diffuser throat a backflow occurs. Velocity ratios of this variant of computation are presented in figure 11. If the nozzle embodiment (a valve without a cone) is considered, then the velocity field is very balanced in individual cross-sections.

Values of computed averaged quantities in the saddle, in the throat and at the outlet from the valve are presented in table 3. The data for a nozzle are shown in table 4.

DOI: 10.1051/epjconf/201714302135

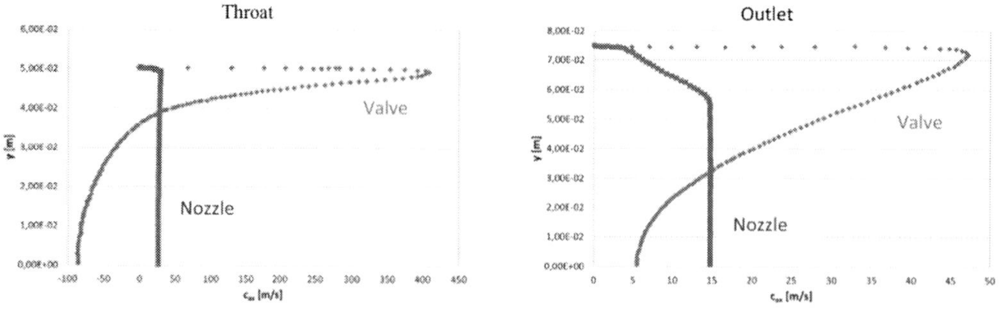


**Fig. 10.** Velocity profile in the cross section of the throat and outlet from the diffuser at the lift $h/D_h = 0.04$

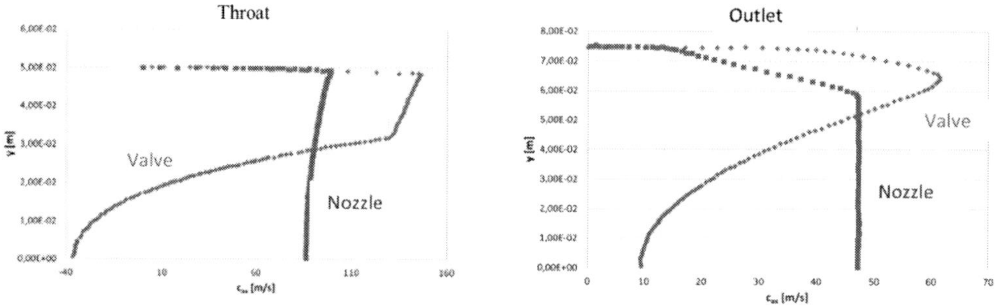


**Fig. 11.** Velocity profile in the cross section of the throat and from the diffuser outlet at the lift $h/D_h = 0.2$

**Table 3.** Table of averaged quantities in the saddle, in the throat and at the outlet from the valve

| $h$ | $S_{se}$ | $G$ | $c_{se}$ | $c_v$ | $(p_{se})_v$ | $(p_{dno})_v$ | $(Ma_{se})_v$ | $\rho_{se}$ | $T_{se}$ | $(p_h)_v$ | $(p_{sec})_v$ | $(p_{vc})_v$ | $(p_v)_v$ |
|---|---|---|---|---|---|---|---|---|---|---|---|---|---|
| [mm] | [mm$^2$] | [kg/s] | [m/s$^2$] | [m/s$^2$] | [Pa] | [Pa] | [-] | [kg/m$^3$] | [K] | [Pa] | [Pa] | [Pa] | [Pa] |
| 4 | 1094 | 0.252 | 313.45 | 35.53 | 52201 | 26963 | 1.000 | 0.742 | 245 | 28000 | 98816 | 41303 | 40968 |
| 6 | 1654 | 0.383 | 305.98 | 44.20 | 54104 | 37642 | 0.971 | 0.762 | 247 | 35417 | 99113 | 52119 | 51051 |
| 8 | 2223 | 0.508 | 276.3 | 44.92 | 61094 | 56487 | 0.862 | 0.831 | 256 | 51208 | 99321 | 68614 | 67717 |
| 10 | 2801 | 0.589 | 224.11 | 44.86 | 72870 | 70341 | 0.682 | 0.944 | 269 | 66284 | 99549 | 81049 | 79989 |
| 14 | 3984 | 0.677 | 161.83 | 45.66 | 85105 | 83539 | 0.482 | 1.055 | 281 | 80841 | 99784 | 92338 | 91086 |
| 20 | 5824 | 0.828 | 129.76 | 47.70 | 90245 | 88846 | 0.383 | 1.101 | 286 | 87582 | 99880 | 97468 | 96067 |
| 24 | 7096 | 0.909 | 114.97 | 50.00 | 92288 | 90902 | 0.338 | 1.119 | 287 | 89591 | 99912 | 98677 | 97114 |
| 28 | 8403 | 0.929 | 97.67 | 50.37 | 94390 | 93228 | 0.287 | 1.137 | 289 | 90795 | 99938 | 99233 | 97655 |
| 36 | 11123 | 0.996 | 77.94 | 53.14 | 96401 | 95484 | 0.228 | 1.154 | 291 | 90463 | 99962 | 99523 | 97792 |

**Table 4.** Quantities in the throat and at the outlet from nozzle

| h [mm] | $(p_{hc})_d$ [Pa] | $(p_{vc})_d$ [Pa] | $(p_v)_d$ [Pa] | $(p_h)_d$ [Pa] |
|---|---|---|---|---|
| 4 | 99995 | 99981 | 99871 | 99551 |
| 6 | 99989 | 99960 | 99711 | 98961 |
| 8 | 99981 | 99933 | 99498 | 98157 |
| 10 | 99975 | 99912 | 99329 | 97507 |
| 14 | 99968 | 99885 | 99118 | 96684 |
| 20 | 99952 | 99830 | 98679 | 94932 |
| 24 | 99943 | 99796 | 98411 | 93838 |
| 28 | 99940 | 99787 | 98338 | 93537 |
| 36 | 99931 | 99755 | 98085 | 92481 |

From the computations it is possible to specify, by using different methods, the value of the loss coefficient for the computation of occurred pressure loss $(\Delta p_z)_v = p_{sec} - p_{vc}$. We can consider the dynamic pressure in the valve saddle, in the diffuser throat, at the diffuser outlet or the pressure on the bottom of the cone.

$$(\Delta p_z)_v = \zeta_{se}(p_{sec} - p_{se})_v \quad (1)$$

$$(\Delta p_z)_v = \zeta_{dno}(p_{sec} - p_{dno})_v \quad (2)$$

$$(\Delta p_z)_v = \zeta_h(p_{sec} - p_h)_v \quad (3)$$

$$(\Delta p_z)_v = \zeta_v(p_{sec} - p_v)_v \quad (4)$$

Their course depending on the relative lift of a cone is shown in figure 12. The pressure on the bottom of the cone is similar to the pressure in the diffuser throat. The course of loss coefficients corresponds to it.

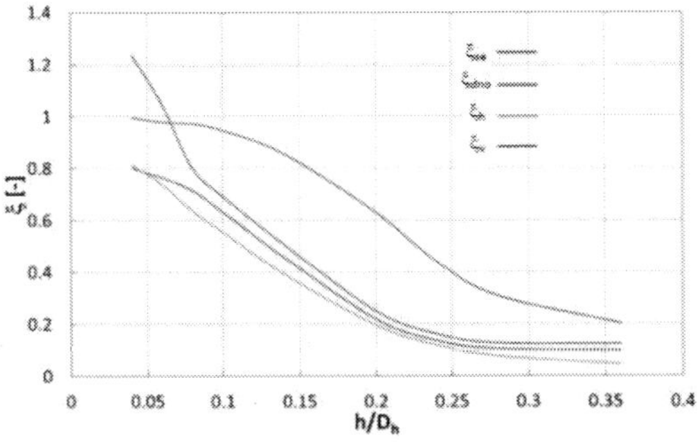


**Fig. 12.** The course of loss coefficients

With the gradual opening of the valve the pressure loss decreases and therefore the value of all pressure coefficients decreases too. At the maximum lift of a cone the pressure loss reaches up to 20% of the total pressure differential across the diffuser. At small lifts of a cone the critical pressure ratio is set in the saddle of the valve and the increase of loss coefficient $\zeta_{se}$ corresponds to it.

In embodiment of a nozzle it is on the contrary. The highest pressure loss arises at the maximum flow of the media through a nozzle. The pressure loss at the nozzle is $(\Delta p_z)_d = (p_{hc} - p_{vc})_d$. From comparison of pressure losses for a valve and for a nozzle we can evaluate the relevant loss coefficient $\zeta_d$, where $(\Delta p_z)_v = \zeta_d(\Delta p_z)_d$. It is the flow for considered relative lifts $h/D_h$ .It is presented in figure 13. The loss coefficient $\zeta_d$ appears to change from infinity to zero.

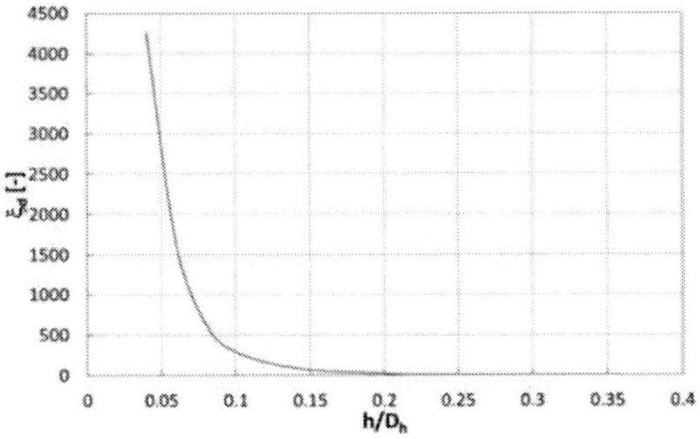


**Fig. 13.** Dependence of the loss coefficient of a nozzle proportional lift

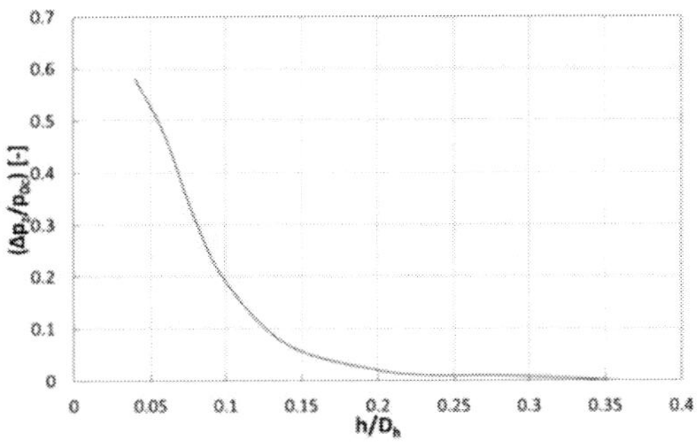


**Fig. 14.** Dependence of the proportional pressure loss on the proportional lift

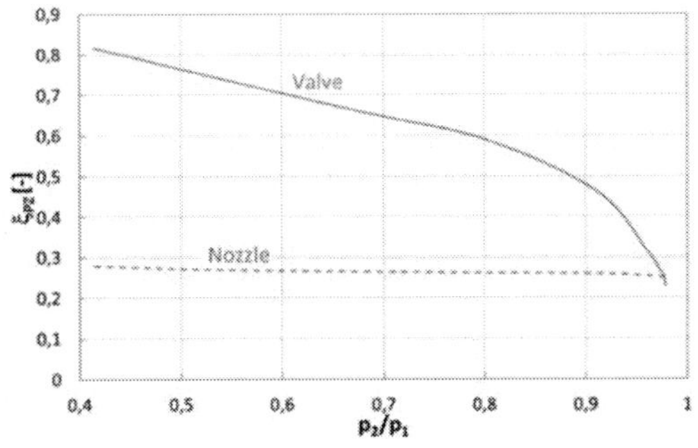


**Fig. 15.** Dependence of the loss pressure coefficient on the proportional pressure at the outlet

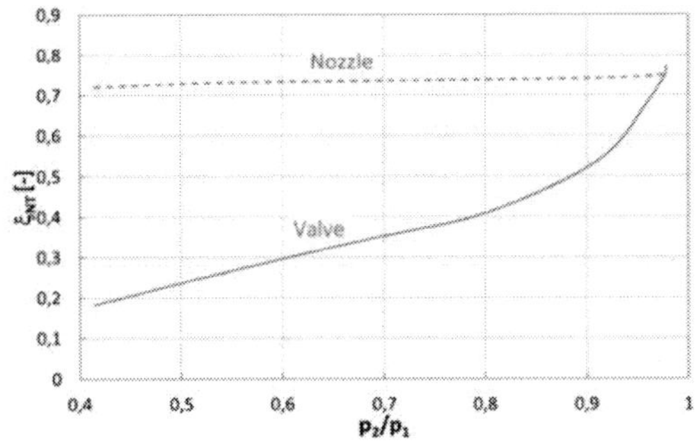


**Fig. 16.** Dependence of the coefficient of pressure increase on the proportional pressure at the outlet

Expression of the pressure loss in the valve using the pressure loss in the nozzle in the stated form is not for practical purposes suitable. In turbine practice the pressure loss is evaluated as the proportional pressure loss of the input pressure. In this form the pressure loss in the diffuser valve is presented in figure 14. The pressure loss at the lift $h \rightarrow 0$ depends on the return pressure. For condensing turbines this pressure is close to zero, therefore throttling the steam in the valve obstructs the entire pressure gradient.

The quality of processes in diffusers may also be evaluated by using the coefficients of the pressure loss or coefficient of pressure increase. The pressure loss is presented in the graphs in figures 15 and 16. They show the differences in embodiment of a valve and a nozzle.

- Valve

$$\zeta_{PZ_v} = \frac{1 - \frac{p_v}{p_{sec}}}{1 - \frac{p_h}{p_{sec}}}$$

$$\zeta_{NT_v} = \frac{\frac{p_v}{p_{sec}} - \frac{p_h}{p_{se}}}{1 - \frac{p_h}{p_{sec}}}$$

- Nozzle

$$\zeta_{PZ_d} = \frac{1 - \frac{p_v}{p_{hc}}}{1 - \frac{p_h}{p_{hc}}}$$

$$\zeta_{NT_d} = \frac{\frac{p_v}{p_{hc}} - \frac{p_h}{p_{hc}}}{1 - \frac{p_h}{p_{hc}}}$$

The pressure loss at the valve decreases gradually, whereas the nozzle is constant. The coefficient of the pressure increase shows a similar course. When a valve is fully open the data for a valve and a nozzle are consistent.

In literature [1,2] it is possible to find experimentally evaluated materials for computation of pressure losses in diffusers with ideal velocity ratios at the inlet into the diffuser. Thus for the variant of a nozzle it is possible to compare computed pressure losses with the prediction of the pressure loss according to the literature sources. The loss coefficient according to [2] is considered,

$$\zeta_P = \left[\frac{\lambda}{8\sin(\gamma/2)}\frac{n+1}{n-1} + 3{,}2(\mathrm{tg}(\gamma/2))^{1{,}25}\right]\left(1-\frac{1}{n}\right)^2$$

where $\gamma$ is the angle of the diffuser opening, $n$ is ratio of the outlet area to the area of the throat and $\lambda$ is the coefficient of friction.

The pressure loss is evaluated as the proportion of inlet dynamic pressure.

$$\Delta p_z = \zeta_P \frac{1}{2}\rho c^2$$

Good conformity with the prediction of the pressure loss is confirmed. Comparison of prediction of the pressure loss with the computed loss is presented in figue 17.

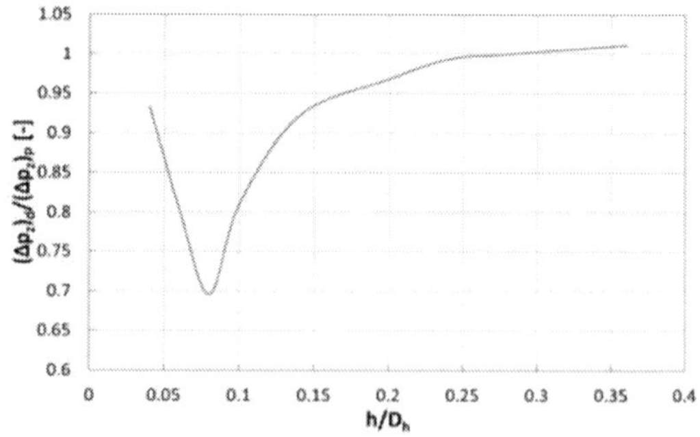


**Fig. 17.** Comparison of the pressure loss according to literature and from the CFD simulation

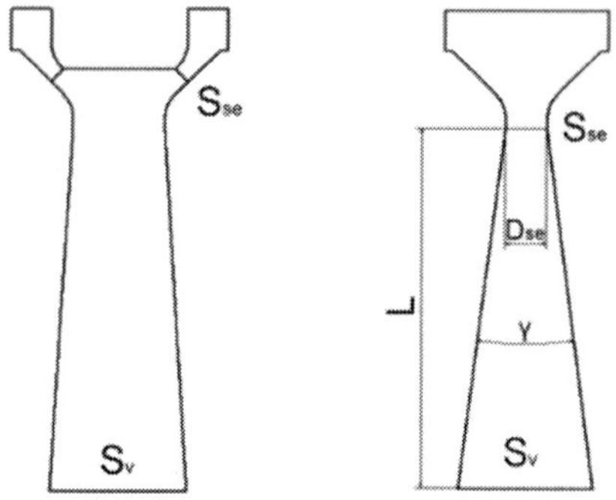


**Fig. 18.** Adjustment of diffuser geometry

In order to estimate the pressure loss in the diffuser of the valve, besides the loss caused by friction and by extension of the channel, the pressure loss due to the wake vortex must be evaluated too. To achieve this, the geometry of the valve must be adjusted and must be close to embodiment of a nozzle, which means the diffuser with the ideal inlet velocity field. Adjustment of valve geometry is shown in figure 18. The area of the throat changes. the nozzle adapts to this adjustment so that it could correspond to the area of the saddle under the cone of the valve and the angle of the diffuser expansion γ.

The entire pressure loss has three parts

$$\Delta p_z = \Delta p_T + \Delta p_R + \Delta p_U$$

The pressure loss caused by friction

$$\Delta p_T = \zeta_T \frac{1}{2}\rho c_{se}^2 = \zeta_T(p_{sec} - p_{se})$$

$$\zeta_T = \varphi_T\left(1-\frac{1}{n}\right)^2 = \left(\frac{\lambda}{8\sin(\gamma/2)}\frac{n+1}{n-1}\right)\left(1-\frac{1}{n}\right)^2$$

The pressure loss and loss coefficients by channel expansion $\Delta p_R$ and $\zeta_R$ is evaluated similarly.

$$\zeta_R = \varphi_R\left(1-\frac{1}{n}\right)^2 = 3{,}2(\mathrm{tg}(\gamma/2))^{1{,}25}\left(1-\frac{1}{n}\right)^2$$

The pressure loss caused by wake vortex $\Delta p_U$ can be reached when from computed pressure loss in the valve is the loss $\Delta p_T + \Delta p_R$ deducted.

$$\Delta p_U = (\Delta p_Z)_V - (\Delta p_T + \Delta p_R) = \varphi_U \left(1-\frac{1}{n}\right)^2 \frac{1}{2}\rho c_{se}^2$$

$$\zeta_U = \varphi_U \left(1-\frac{1}{n}\right)^2$$

In figure 19 the friction loss coefficient $\zeta_T$ is presented. With a gradual increase of the flow area through the valve, and with the increase of the mass flow, its value increases. However, compared with other factors impacting the final pressure loss it has the least impact.

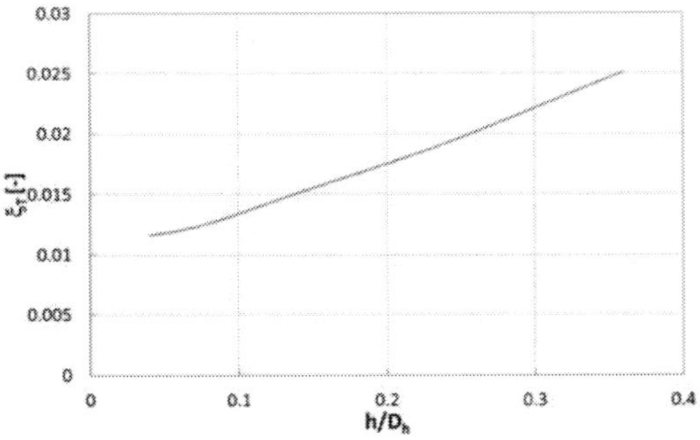


**Fig. 19.** Dependence of the loss coefficient by friction on the proportional lift

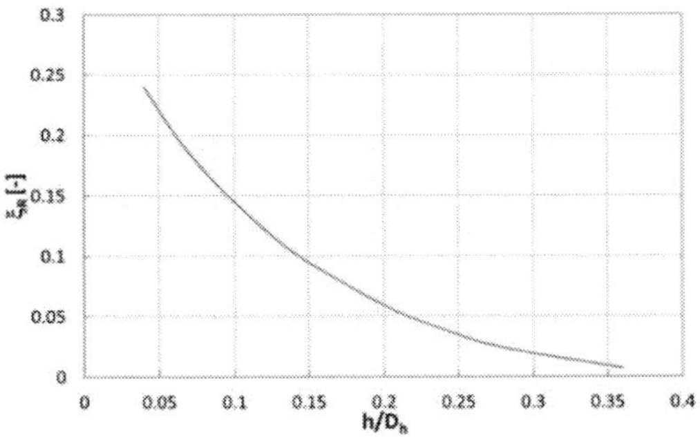


**Fig. 20.** Dependence of the loss coefficient by expansion on the proportional lift

The loss coefficient of diffuser expansion and the loss coefficient of wake vortex with the cone lift decreases. Their course is presented in figure 20 and figure 21. The impact of wake vortex on the pressure loss is the most significant. The pressure loss depends on the velocity of steam flowing through a valve. The smaller the dimensions of a valve, the higher are the velocities at a given certain quantity of the steam in the flow section of a valve, which causes the increase of the final pressure loss. The choice of dimension of the flow may impact the pressure loss in the valve under the certain operational conditions of the turbine. Figure 22 shows the range of choice of relative flow rate $q$. If a valve with lower value $q_{max}$ is used, $\varepsilon_{max}$ is higher too and thus the pressure loss on the valve is lower.

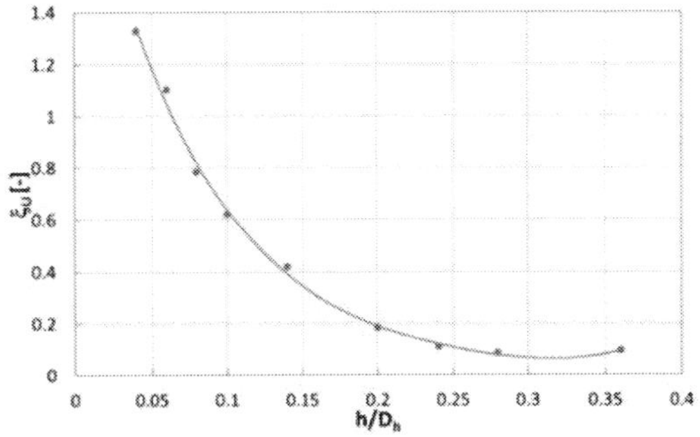


**Fig. 21.** Dependence of the loss coefficient by wake vortex on the proportional lift

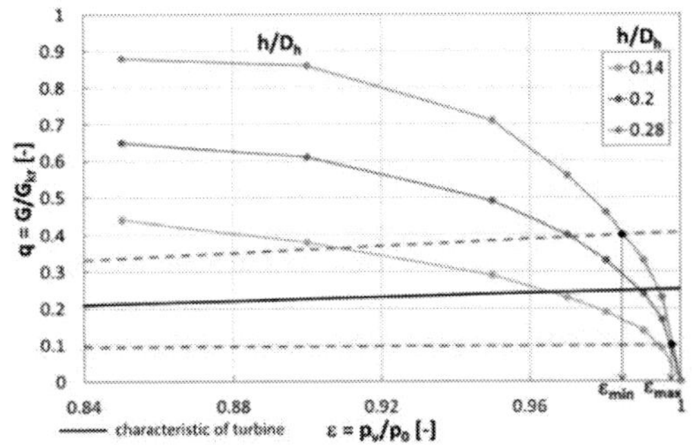


**Fig. 22.** Range of operating pressure ratios

## 5 Conclusion

- Due to wake vortex under the cone in the diffuser of a control valve the velocity field gets deformed. Backflow occurs in the diffuser. Flow separation from the diffuser wall may occur.
- The highest velocities occur by the diffuser wall close to its throat. This is the place with the occurrence of pressure pulsations at the supersonic steam flow.
- Uneven velocity field also occurs at the outlet from the diffuser.
- The final pressure loss can be evaluated using loss coefficients for a friction, expansion of the diffuser and for impact of the wake vortex. The wake vortex has the highest impact on the pressure loss in the valve.
- Choice of dimensions of the valve can influence the pressure loss in the valve under the certain operational conditions of a turbine.

## Nomenclature

| | |
|---|---|
| $c$ | velocity |
| $h$ | cone lift |
| $n$ | area ratio |
| $p$ | pressure |
| $q$ | proportional flow |

| | |
|---|---|
| $D$ | diameter |
| $G$ | mass flow |
| $Ma$ | Mach number |
| $S$ | flow area |
| $T$ | temperature |
| $Re$ | Reynolds number |
| $\gamma$ | angle of diffuser opening |
| $\varepsilon$ | pressure ratio |
| $\lambda$ | coefficient of friction loss |
| $\rho$ | specific weight |
| $\zeta$ | loss coefficient |

## Subscripts

| | |
|---|---|
| $c$ | total |
| $h$ | throats |
| $o$ | inlet |
| $v$ | outlet |
| $se$ | saddle |
| $dno$ | bottom of the cone |
| $d$ | nozzle |
| $ax$ | axial |
| $z$ | loss |
| $T$ | friction |
| $U$ | wake vortex |
| $R$ | dimensional |
| $PZ$ | pressure loss |
| $NT$ | pressure increase |
| $P$ | prediction |

## References

1. M. Dejč, *Technická dynamika plynů,* Praha: SNTL, 659s. (1967)
2. И. Л. Повх, *Аэродинамический эксперимент в машиностроении, Ленинград*, Машиностроеные Ленинградский отделение, (1974)
3. J. Nožička, *Základy termomechaniky*, ČVUT Praha, 187s. ISBN 9788001040225, (2008)
4. R. Dvořák, K. Kozel, *Matematické metody v aerodynamice*, ČVUT Praha, 153s. ISBN 8001008517
5. J. Šesták, *Přenos hybnosti, tepla a hmoty*, ČVUT Praha, 299s. ISBN 800101715x, (1998)
6. J. Ambrož, Parní turbíny a kondenzace, ČVUT Praha, 281s., (1984)

DOI: 10.1051/epjconf/201714302136

# Coanda effect in valves

*Václav* Uruba[1,2,*], *Pavel* Procházka[1], *Vladislav* Skála[1]

[1]Institute of Thermomechanics, AS CR, v. v. i., Dolejškova 5, Praha 8, Czechia
[2]University of West Bohemia, Faculty of Mechanical Engineering, Universitní 22, Plzeň, Czechia

**Abstract.** Coanda effect takes place in flow within valves diffuser for certain conditions. The valve plug in half-closed position forms wall-jet, which could be stable or instable, depending on geometry and other conditions. This phenomenon was subject of experimental study using time-resolved PIV technique. For the acquired data analysis the special spatio-temporal methods have been used.

## 1 Introduction

Recently the Coanda principle has become increasingly used in a wide variety of applications in many fields of applied dynamics of fluids – see e.g. [2]. In addition, this effect could be observed in the natural world. Devices containing a jet flow employing this principle usually offer substantial flow deflection, and enhanced turbulence levels and entrainment compared with a free jet flow. However, some disadvantages are connected with this phenomenon as jet flow detachment, considerable local increase of turbulence level, and appearance of quasi-periodical structures in the flow-field resulting in generation of pressure fluctuations associated with increased noise level. Physical mechanisms connected with those phenomena has not been explored in details yet.

The presented paper is contribution to study of the Coanda effect in some special cases which could occur in valves.

## 2 Coanda effect

The Coanda effect is a phenomenon that was first observed in 1910 by Romanian mathematician and engineer Henri Coanda. He discovered that when air was ejected from a rectangular nozzle, it would attach itself to an inclined flat plate connected to the nozzle exit. Emphasizing the need for a sharp angle between the nozzle and the flat plate, Coanda then applied the principle to a series of deflecting surfaces, each at a sharp angle to the previous one, and succeeded in turning flows through angles as large as 180 degrees. He stated that "when a jet of fluid is passed over a curved surface, it bends to follow the surface, entraining large amounts of air as it does so", and this phenomenon has become known as the "Coanda Effect".

Ability of the jet flow to follow the curved surface is determined by relevant conditions. The most important among them is the jet flow structure, and, of course, the surface shape which is to be followed. As a result, the jet flow follows the curved surface (or not) and the topology and dynamics of the emerging flow-field defined by the wall-jet flow. The same condition determine the structure of the flow-field and possible presence of instabilities.

The Coanda effect produces the attached boundary layer on a curved surface, however its ability has limits. For certain conditions the Coanda effect stops working and the boundary layer detaches. This process is of stability nature and could occur suddenly and random or quasi-periodical in time.

## 3 Experimental setup

The valve in question is depicted schematically in Fig. 1. It is a typical valve with side inlet used e.g. in a steam turbines. All parts of the valve are of circular cross-section. The diffuser angle is 7 degrees.

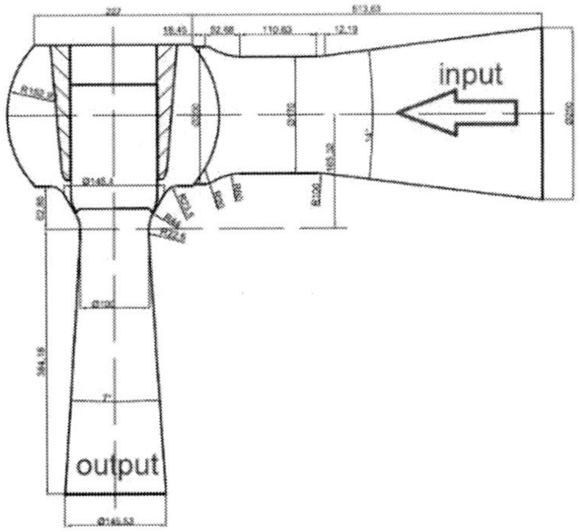


**Fig. 1.** The valve schema.

* Corresponding author: uruba@it.cas.cz

For the purpose of experimental research, the simplified model of rectangular cross-section and only right-hand valve half was designed – see Fig. 2

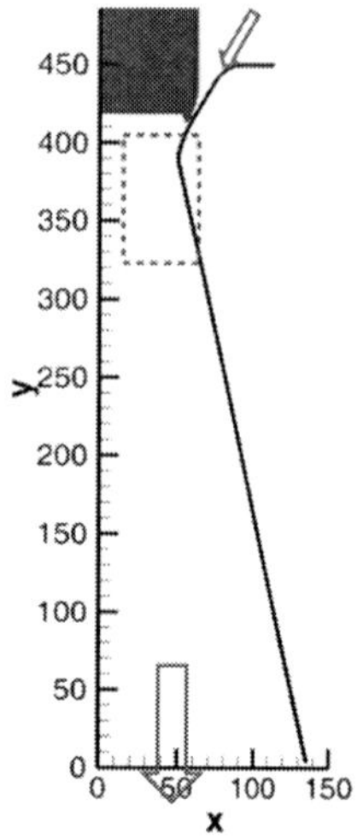


**Fig. 2.** The valve model.

In Fig. 2 the valve plug is shown in the two extremal positions, fully closed (in red) fully open (in blue). Position of the plug is described by the shift value $h$ in y direction with origin in fully closed position. So, the fully closed position corresponds to $h = 0$ mm and fully open $h = 30$ mm. The shift $h$ determines the jet width, in fact. The arrows indicate flow directions and dashed red rectangle the measuring area. The angle of the half-diffuser of the model was chosen 12.5 degrees.

The existing blow-down facility of Institute of Thermomechanics, Prague, has been used for experiments. The model was located on the output from the contraction nozzle of a cross-section of 250 x 100 mm$^2$. Velocity of air approx. 10 m/s was used for most experiments.

## 4 Methods and Procedures

Time-resolved PIV measuring technique has been applied. The single camera method provided instantaneous in-plane velocity components in the plane of symmetry of the valve parallel to the mean flow. Frequency of evaluated velocity fields was 2 kHz (i.e. 2 000 measurements per second). The records consist of 4 000 measurements, covering 2 s of time evolution. For seeding particles, a standard SAFEX fog generator has been used.

The instantaneous velocity fields have been evaluated from double-snaps using Dantec Dynamics A/S PIV software DynamicStudio version 3.40. The standard 3-step adaptive correlation method with consequent reduction of interrogation area has been used. The final size of interrogation area was 32 x 32 pixels with 50 % overlap. This method considerably reduces random error of resulting velocities, it can be estimated within 1.5 % of maximum velocity (see [1]). However, all methods used for the data analysis are of statistical nature, therefore random errors of evaluated modes are negligible. Possible systematic error is connected with used tracing particles with typical size of 1 μm. For the questioned case, the systematic error is within 1 % of the maximum velocity.

The time-mean statistics have been evaluated. For the dynamical behaviour the Oscillation Pattern Decomposition (OPD) method has been adopted. The OPD method allows for extraction of quasi-periodical modes. The OPD method is described in details in [3].

Each OPD mode is characterised by complex topology and frequency $f$, e-folding time $\tau$ and periodicity $p$. Frequency of quasi-periodical mode evolution, e-folding time is mean time to the mode amplitude reduction by factor e, it could be understood as reciprocal value of the mode damping. The periodicity evaluates the e-folding time in multiples of the process periods. If the $p$ is greater than 0.343 the mode is considered to be oscillating, otherwise it is non-oscillating or decaying in nature (see [5]).

## 5 Results

The results are to be shown evaluated within the rectangular area – see Fig. 2. The plug shift $h$ was changed in broad range: 2.5, 5, 7.5, 10, 12.5, 15, 17.5, 20 and 30 mm respectively. For presentation only selected results have been chosen. However only the case $h = 30$ mm shows failure of the Coanda effect with fully detached boundary layer.

### 5.1 Time mean results

For the presentation of time-mean characteristics the two typical cases have been chosen: the case of $h = 10$ mm, when the flow is attached to the curved wall and the Coanda effect is working and $h = 30$ mm, detached flow and Coanda effect failure. The first case will be referred hereinafter as (a), the latter as (b).

In Fig. 3 there are time-mean vector-fields with vector-lines added in arbitrary manner (blue colour) for the cases (a) and (b).

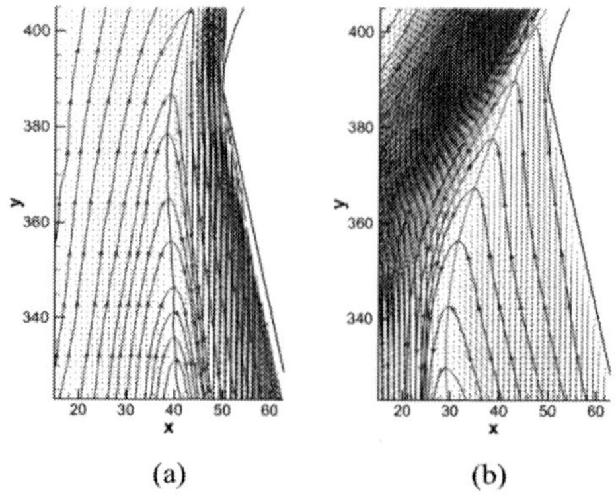

**Fig. 3.** Vector-lines of time-mean velocity vector field, (a) $h = 10$ mm, (b) $h = 30$ mm.

The brown and green lines represent boundary of area with streamwise (down) and back (up) flow, corresponding to the vanishing streamwise velocity component. For the case (a) the back-flow is located in the left-hand part, while for the case (b) is back-flow in the right. Please note that even if the back-flow region is relatively big in area extend, the backward oriented velocity moduli are relatively very small and thus the backward oriented flow-rate is also small (compare with Fig. 5, where moduli distributions are shown).

The back-flow boundary position is function of the valve plug shift, this is demonstrated in Fig. 4, were boundaries for all examined shifts are shown. The $h$ value is put close to relevant curve.

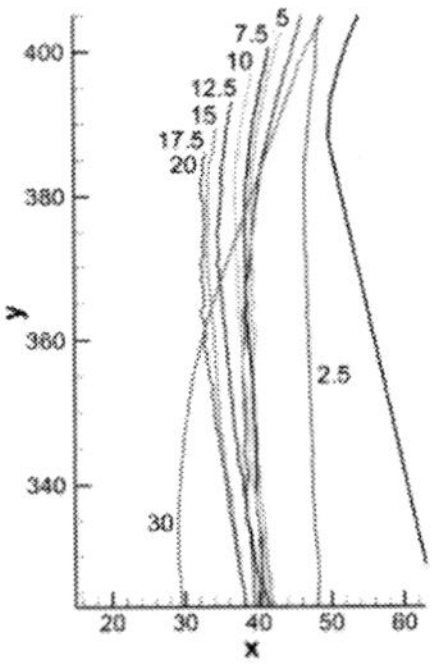


**Fig. 4.** Zero streamwise velocity component lines, $h$ is variable.

The position of the zero streamwise velocity component depends systematically on the plug shift (i.e. width of the jet). This is evident in the upper part ($y = 390$ mm): the bigger shift $h$ and width of the jet, the thicker the area of attached flow in the left. However more-or-less the same attached area thickness is established very quickly, at the lower area border ($y = 330$ mm). Only the case of $h = 2.5$ mm shows considerably thinner attached area. This effect could be attributed to viscose effects, when the gap between plug and valve wall is very thin.

The case $h = 30$ mm (green line) represents an exception, of course, as the flow is detached.

In Fig. 5 there are mean velocity moduli distributions for (a) and (b) cases respectively. Much higher velocities are indicated in streamwise direction flow regions than in the backward flow.

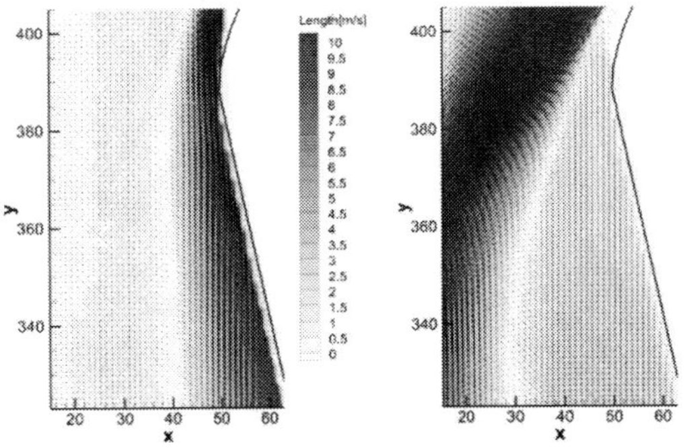


(a) (b)

**Fig. 5.** Time-mean velocity vector fields and moduli distributions, (a) $h = 10$ mm, (b) $h = 30$ mm

To indicate the intensity of dynamical activity, the variances of velocity components are evaluated. The variances of both evaluated velocity components in a given point were added, distributions of sum of variances are shown in Fig. 6.

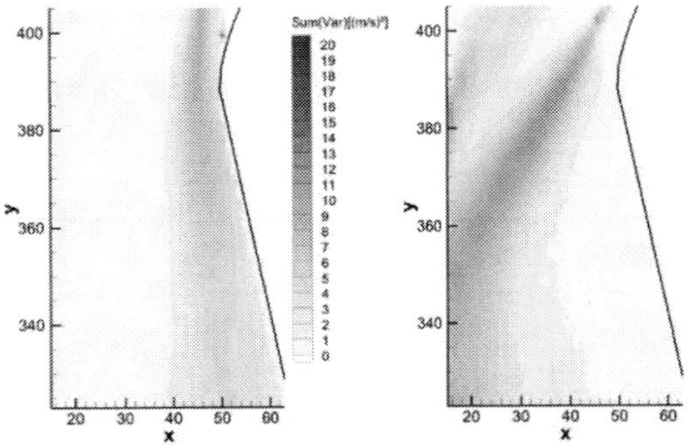


(a) (b)

**Fig. 6.** Distributions of sum of velocity components variances, (a) $h = 10$ mm, (b) $h = 30$ mm

The distributions of velocity variance indicate high fluctuating activity within whole volume of the wall-jet in the case (a), while for the case (b) maximum activity is located to the lower border of the free jet, less intensive local maximum within the upper jet border. The same conclusion could be made from the correlation coefficient distributions in Fig. 7. Correlation coefficient indicates locations with high production of turbulence.

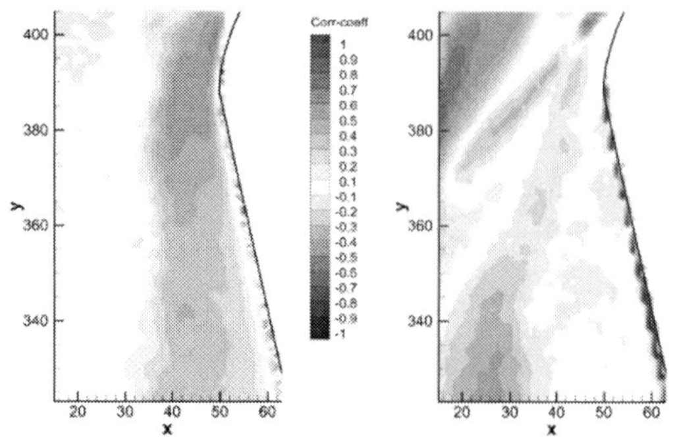


(a) (b)

**Fig. 7.** Correlation coefficient distributions, (a) $h = 10$ mm, (b) $h = 30$ mm

The information connected with flow dynamics is however very limited. More detailed information will be provided in the following paragraph.

### 5.2 Dynamics

The detailed study of dynamics of flow-field has been performed using the OPD method. The mode with strong oscillating capacity are to be presented here, see the Tab. 1.

**Table 1.** Results of OPD analysis.

| $h$ [mm] | OPD No. | $f$ [Hz] | $\tau$ [ms] | $p$ [1] |
|---|---|---|---|---|
| 5 | 2 | 34.8 | 17.30 | 0.601 |
| 5 | 4 | 82.3 | 7.94 | 0.653 |
| 5 | 7 | 187.9 | 3.174 | 0.596 |
| 10 | 4 | 67.5 | 7.94 | 0.535 |
| 10 | 5 | 146.9 | 5.94 | 0.872 |
| 10 | 11 | 129.4 | 2.48 | 0.533 |
| 15 | 8 | 123.1 | 5.14 | 0.633 |
| 15 | 10 | 198.5 | 3.74 | 0.742 |
| 15 | 12 | 231.9 | 3.40 | 0.788 |
| 20 | 3 | 101.7 | 7.03 | 0.714 |
| 20 | 7 | 211.1 | 5.41 | 1.143 |
| 20 | 8 | 152.8 | 4.55 | 0.695 |
| 30 | 4 | 221.1 | 7.57 | 1.674 |
| 30 | 6 | 126.2 | 6.28 | 0.793 |
| 30 | 9 | 331.6 | 5.64 | 1.870 |

In Table 1 all relevant parameters of selected OPD modes are shown. For each $h$ position 5, 10, 15, 20 and 30 mm respectively, the mode frequency $f$, e-folding time $\tau$ and periodicity $p$ is given for the 3 selected OPD modes.

The OPD topology will be presented in form of real and imaginary parts, velocity vector field and corresponding vorticity field (red – positive, blue – negative). As an example the OPD mode 2 for $h = 5$ mm is depicted in Fig. 8.

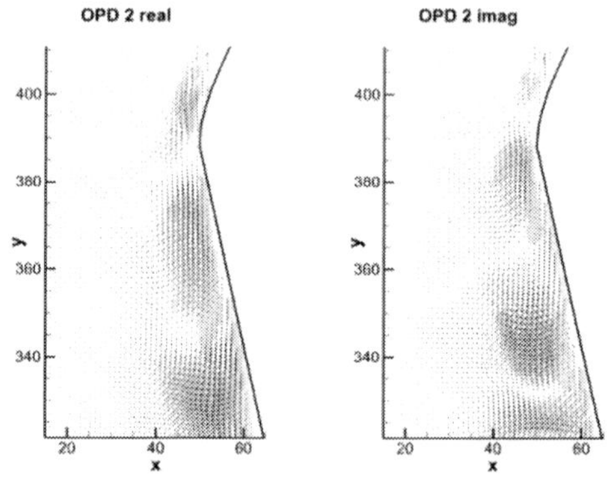


**Fig. 8.** OPD mode 2 for $h = 5$ mm, real and imaginary parts.

The oscillating mode in Fig. 8 is a typical traveling mode, the vortical structures travel in space with a small modification within the oscillation period. The real and imaginary parts of the mode shows very similar topology, the main difference is in positions of vortical structures.

All other modes are of very similar nature, so we will present only real parts of OPD modes, the imaginary parts are similar, only shifted in space.

In Fig. 9 there are OPD modes 4 and 7 for $h = 5$ mm.

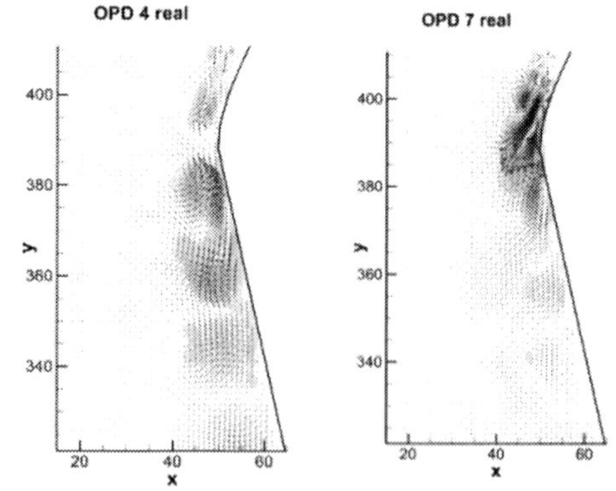


**Fig. 9.** OPD modes 4 and 7 for $h = 5$ mm, real parts.

The traveling vortices could be recognized within the wall jet (mode OPD 4), in the case of OPD 7 very intense vortical structures are located just in the region of the valve seat.

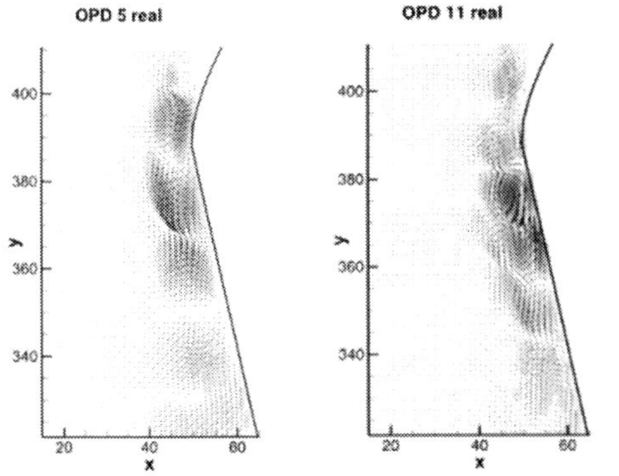


**Fig. 10.** OPD modes 5 and 11 for $h$ = 10 mm, real parts.

The OPD modes for $h$ = 10 mm in Fig. 10 show the vortical structures shifted downstream in the streamwise direction. Further downstream shift could be visible for even bigger shifts $h$ = 15 mm and 20 mm in Figs. 11 and 12. The dynamical structures became weaker.

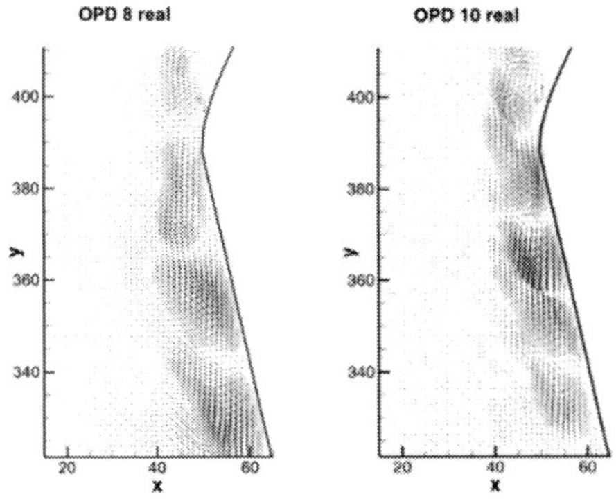


**Fig. 11.** OPD modes 8 and 10 for $h$ = 15 mm, real parts.

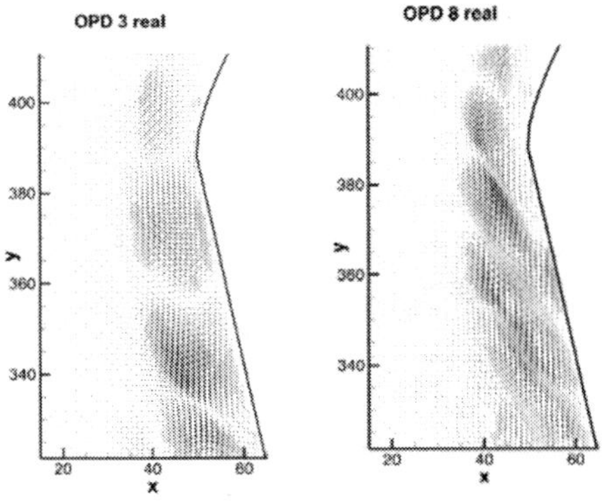


**Fig. 12.** OPD modes 3 and 8 for $h$ = 20 mm, real parts.

The last plug position $h$ = 30 mm exhibits failure of Coanda effect and detachment of the jet. The dynamics is completely different. The travelling dynamical structures are moving together with the jet, staying attached to it.

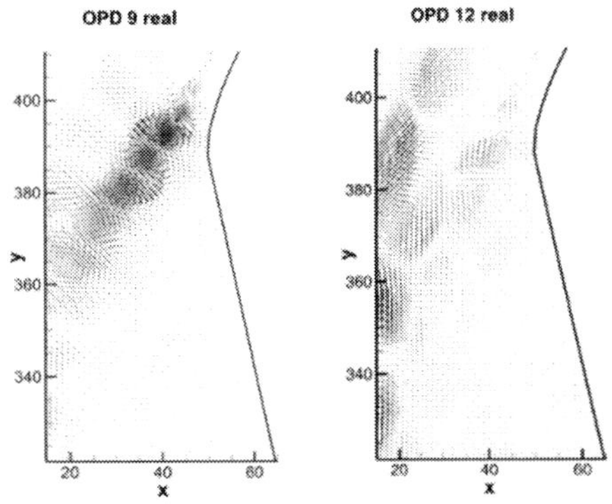


**Fig. 13.** OPD modes 9 and 12 for $h$ = 30 mm, real parts.

In Fig. 13 the OPD mode 9 shows very distinct and well defined vortex train in the region of the free jet lower border. The mode OPD 12 represents dynamics of whole free jet.

## 6 Conclusions

The mean flow characteristics and dynamical modes of the flow-field in the valve diffuser were shown. Coanda effect is active for narrow wall jets, i.e. small valve plug shifts. The flow-field dynamics is typically limited to the wall-jet region. With growing wall-jet width the dynamical vortical structures shift further downstream and became weaker. For big shifts of the valve plug the Coanda effect fails and the jet detaches from the wall. The resulting free jet exhibits very intense dynamics. On the lower border of the jet-flow the distinct vortex street with alternative orientation was detected.

## Acknowledgement

The authors gratefully acknowledge the support of the Technology Agency of the Czech Republic by means of projects TA04020129 and TA04011437.

## References

1. DynamicStudio User's Guide, 2015
2. C. Lubert, On Some recent applications of the Coanda effect, *The Journal of the Acoustical Society of America* **128** (4) :2286, October 2010
3. V. Uruba, Decomposition methods in turbulent research, EFM11, *EPJ Web of Conferences* **25** 01095 (2012)
4. V. Uruba, O. Hladík, P. Jonáš, Dynamics of secondary vortices in turbulent channel flow, ETC13, *Journal of Physics Conference Series* **318** 062021 (2011)
5. V. Uruba, Near wake dynamics around a vibrating airfoil by means of PIV and Oscillation Pattern Decomposition at Reynolds number of 65000, *Journal of Fluids and Structures* **55** pp. 372–383 (2015)

EPJ Web of Conferences **143**, 02137 (2017) DOI: 10.1051/epjconf/201714302137
*EFM 2016*

# On 3D flow-structures behind an inclined plate

*Václav* Uruba[1,3,*], *David* Pavlík[2], *Pavel* Procházka[1], *Vladislav* Skála[1] and *Václav* Kopecký[2]

[1]Institute of Thermomechanics, AS CR, v.v.i., Dolejškova 5, Praha 8, Czechia
[2]Institute of New Technologies, Technical University of Liberec, Studentská 2, Liberec, Czechia
[3]University of West Bohemia, Faculty of Mechanical Engineering, Universitní 22, Plzeň, Czechia

**Abstract.** Stereo PIV measurements has been performed behind the inclined plate, angle of attack 5 and 10 deg. Occurrence and dynamics of streamwise structures behind the plate trailing edge have been studied in details using POD method. The streamwise structures are represented by vortices and low- and high-velocity regions, probably streaks. The obtained results support the hypothesis of an airfoil-flow force interaction by Hoffman and Johnson [1,2].

## 1 Introduction

The motivation of the presented study is supporting new ideas about principle of flight by Hoffman and Johnson from KTH Stockholm, see [1,2]. The new hypothesis of physical mechanism of flight relies on existence of streamwise vortical structures on the suction side of the airfoil and within its wake. The vortices origin is supposed to be the instability of the boundary layer subjected to adverse pressure gradient on the airfoil suction side (i.e. upper) – see [3]. Some consequences of the hypothesis are shown in [4,5,6].

The simple experiment is made with the simplest airfoil possible represented by a flat plate in uniform flow and moderate angle of attack. However the experimental results are of a qualitative nature only, they reveals existence of streamwise vortices above the plate without any doubt.

## 2 Experimental setup

Flat plate inclined with angle of attack 5 and 10 degs has been placed in a uniform low turbulence stream. The blow-down facility produces a jet with uniform velocity distribution, mean velocity about 5 m/s, intensity of turbulence less than 0.2 %. The plate of thickness 2 mm had rounded edges, chord 100 mm and span 300 mm.

As for the measuring technique, the time-resolved stereo PIV method was used for the experiments. The measuring system DANTEC consists of laser with cylindrical optics and two CMOS cameras. The software Dynamics Studio 3.4 was used for velocity-fields evaluation. Laser New Wave Pegasus Nd:YLF, double head, wavelength 527 nm, maximal frequency 10 kHz, a shot energy is 10 mJ for 1 kHz (corresponding power 10 W per head). Two cameras NanoSense MkIII, resolution 1280 x 1024 pixels and frequency 500 and 800 double-snaps per second, 1000 double-snaps were acquired in sequence corresponding to 2 s of the record time. The stereo-PIV method has been used for evaluation all 3 velocity components in the measuring plane.

The measuring plane has been located perpendicular to the main flow, 35 mm behind the plane trailing edge covering the span 60 mm in the flow-field middle position. The situation is shown in Fig. 1.

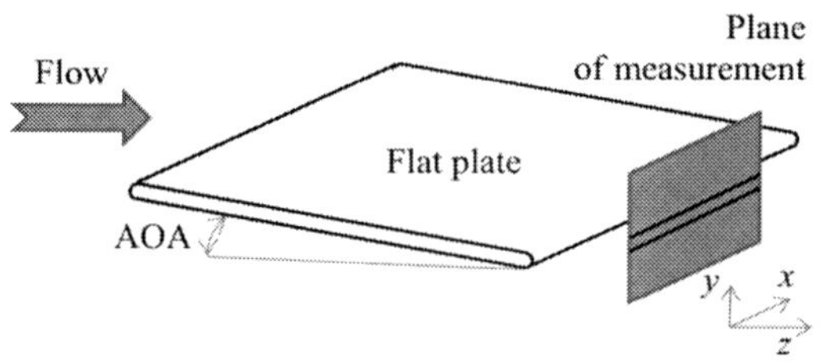


**Fig. 1.** Schematics of the experiment.

In Fig. 1 the Cartesian coordinate system is introduced, plane $xy$ is the measuring plane, mean flow is in the $z$ direction. AOA stands for angle of attack.

## 3 Results

Two variants are considered characterized by the angle of attack 5 and 10 degrees.

In Fig. 2 three instantaneous velocity fields arbitrarily chosen are shown, black vectors represent the projection of velocity vectors to the $xy$ measuring plane, colour denotes the third velocity component $w$ in the $z$ direction. The two horizontal white lines (for $y = 0$ and -2 mm) represent projection of the plate trailing edge. To visualize the velocity field structure, the vector-lines are added arbitrarily in white.

* Corresponding author: uruba@it.cas.cz

The 3 consecutive snapshots are presented in Fig. 2 with double acquisition period.

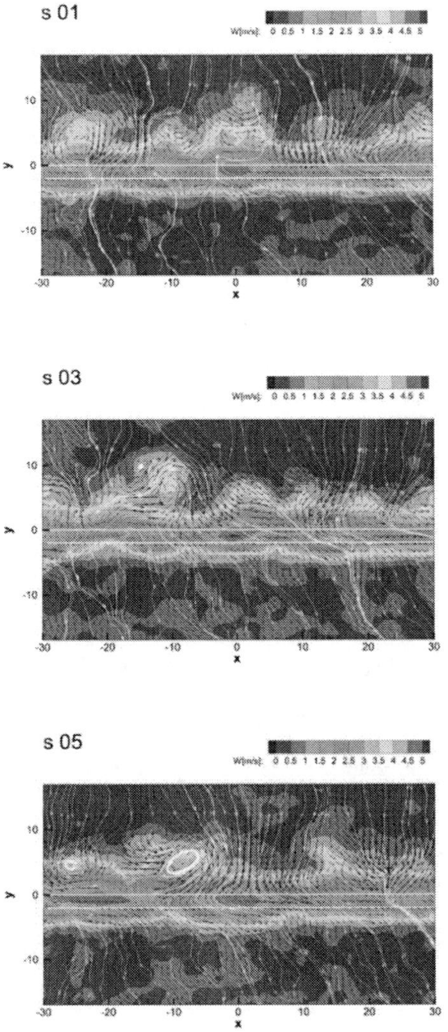


**Fig. 2.** 3 snapshots, angle of attack 5 deg.

The flow seems to be perturbed by presence of the plate, behind the plate trailing edge the low-velocity wake is visible. The flow disturbances in the upper half (i.e. above the plate) are of character vortices, while flow below the plate is more regular.

Similar conclusions could be made from the instantaneous velocity fields generated by the plate with angle of attack 10 deg in Fig. 3. Wake is in this case much bigger, the location of minimum velocity is a little bellow the trailing edge position due to the distance of the measuring plane. In this case the flow above the plate is much more perturbed, the flow bellow is regular, but deflected downwards due to the plate downwash effect.

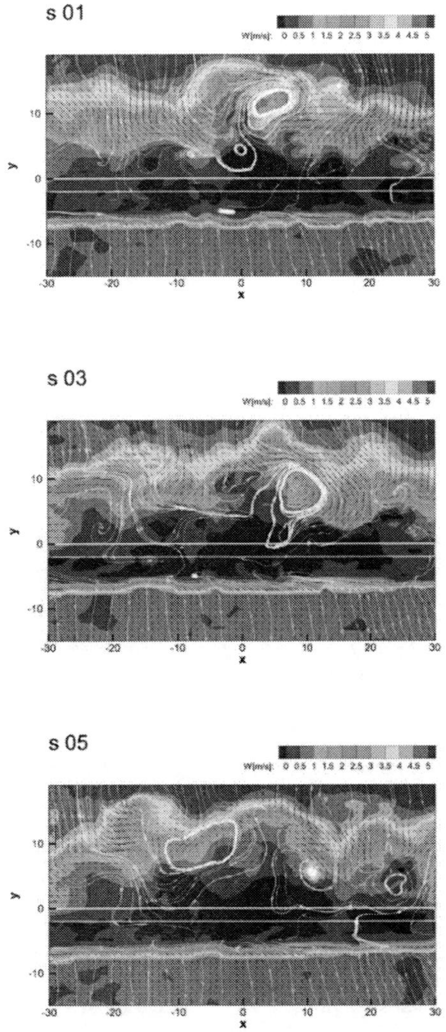


**Fig. 3.** 3 snapshots, angle of attack 10 deg.

The high-frequency dynamical behaviour of the flow above the plate is visible, as the flow patterns differ considerable from each other.

### 3.1. Mean flow picture

First, the statistical characteristics of the flow have been evaluated. In Fig. 4a and 4b the time-mean velocity field is shown for angle of attack 5 deg and 10 deg respectively. The low-velocity wake is located behind the trailing edge, little bellow it due to downwash effect. The wake is much more pronounced for the AOA = 10 deg, it is well extended above the plate.

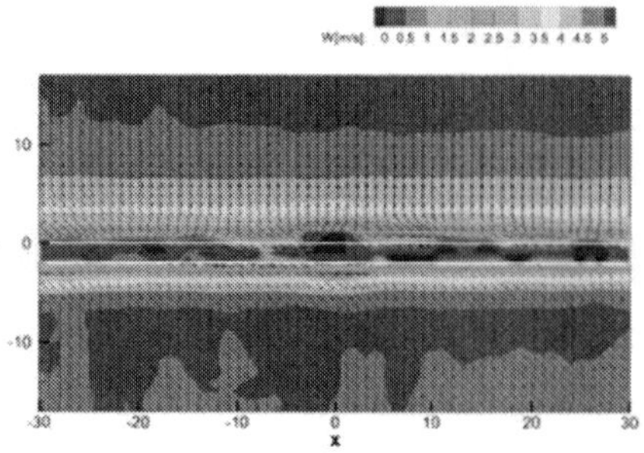


(a)

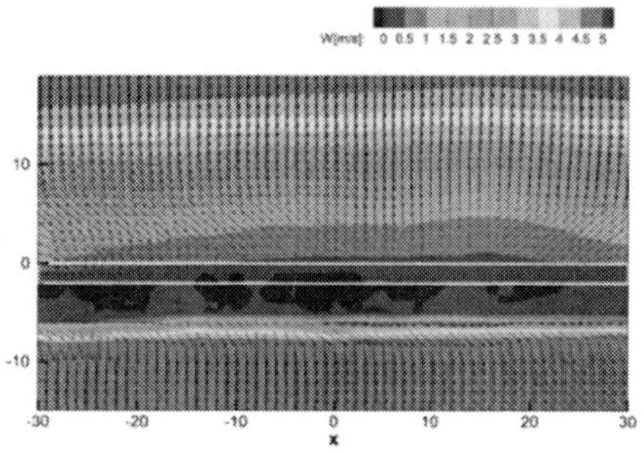


(b)

**Fig. 4.** Time-mean velocity distributions for angle of attack 5 deg (a) and 10 deg (b)

Please note that no vortical structures are detectable in the time-mean velocity field topology.

The dynamical activity could be characterized by velocity variance. In Fig. 5a and 5b there are distributions of sums of all 3 velocity components variances. The scale for both cases is the same, it is clear that the maximal value of the sum of velocity components variance is for the angle of attack 10 deg more than one order of magnitude higher than for 5 deg.

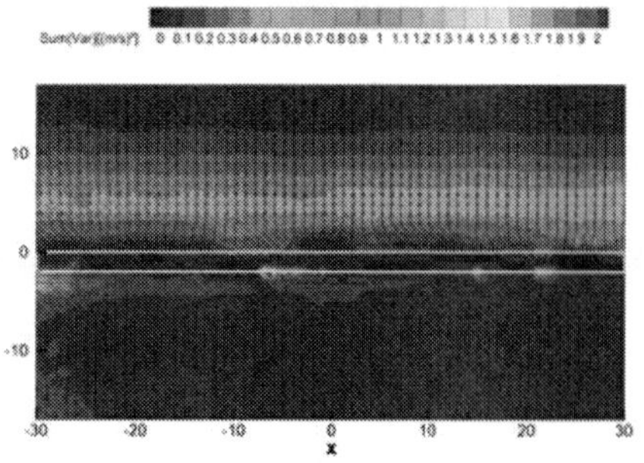


(a)

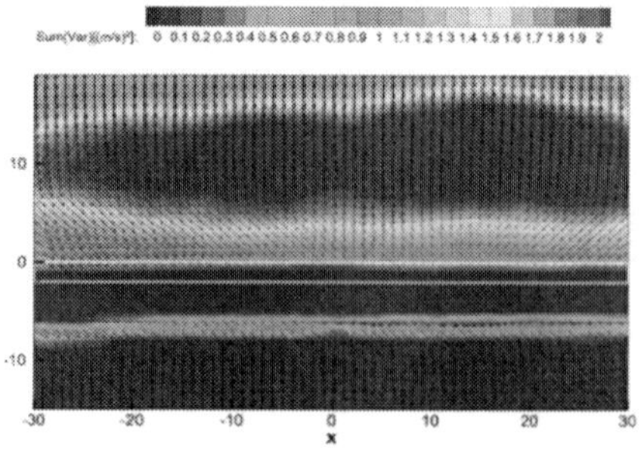


(b)

**Fig. 5.** Sum of velocity components variance distributions for angle of attack 5 deg (a) and 10 deg (b)

Maximum of dynamical activity is located about 5 mm above the trailing edge in the case of AOA = 5 deg, while for AOA = 10 deg the maximum is located 10 mm above the plate trailing edge.

## 3.2 Flow dynamics

The flow-field dynamics could be quantified from velocity variances distribution in Fig. 5. Here we could distinguish locations with low dynamical activity (deep blue) and higher dynamical activity (green, yellow) and highest (red).

To study the topology dynamics in details, the Proper Orthogonal Decomposition method (hereinafter POD) has been applied to the vector-field snaps. The POD method indicates the topological orthogonal modes containing maximum fluctuation energy (i.e. velocity fluctuation variance). More details on the POD method see e.g. [7,8].

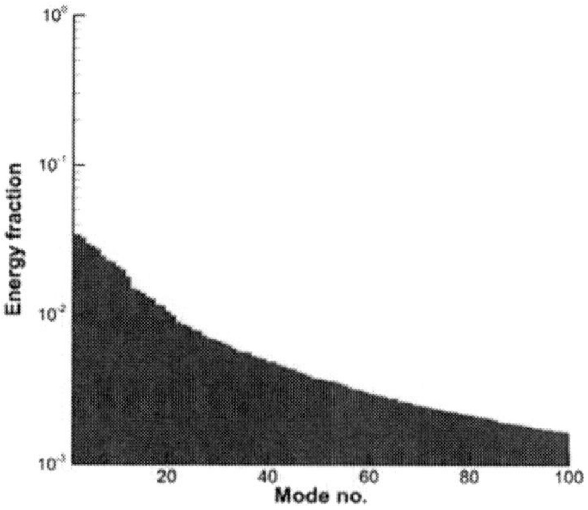


**Fig. 6.** Energy fraction of the POD modes, angle of attack 5 deg.

First the case of angle of attack 5 deg has been analysed, the POD modes have been computed. The POD modes are ordered according to theirs energy beginning from POD 1 with the highest energy. In Fig. 6 there is energy fraction of first 100 POD modes, the most energetic mode POD 1 contains about 3.5 % of the total energy of velocity fluctuations only.

The topology of POD modes will be presented in the form of vector-distribution patterns, only in-plane velocity components are to be shown. Colour denotes the streamwise velocity component $w$ in $z$ direction, red colour means positive velocity fluctuation (i.e. high velocity region), while blue colour means negative velocity fluctuation (i.e. low velocity region). As a vortex is typical element of the velocity field, the swirling strength has been used for identification. Swirling strength is defined as the imaginary part of the complex eigenvalue of the velocity gradient tensor. Local minima of negative-valued swirling strength indicate a vortex core position, while positive values indicate areas of the flow, where shear may be present, but no swirling motion. We will show isolines of negative-valued swirling strength in green, typically forming closed curves around a vortex core. Centre of a vortex could be considered as the centre of the structure, number of isolines indicates the vortex strength (more isolines, higher vortex strength).

In Fig. 7 there are a few selected POD modes with typical topologies. The position of the plate trailing edge is indicated by black lines.

POD 1

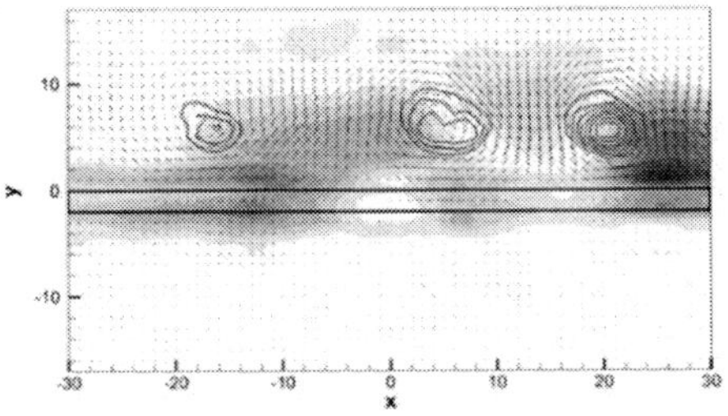

POD 2

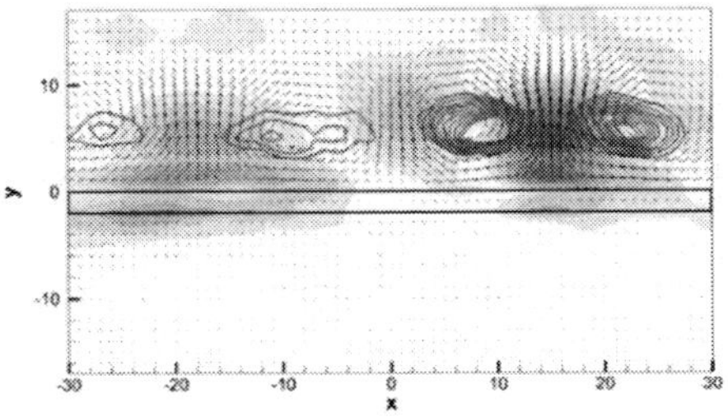

POD 4

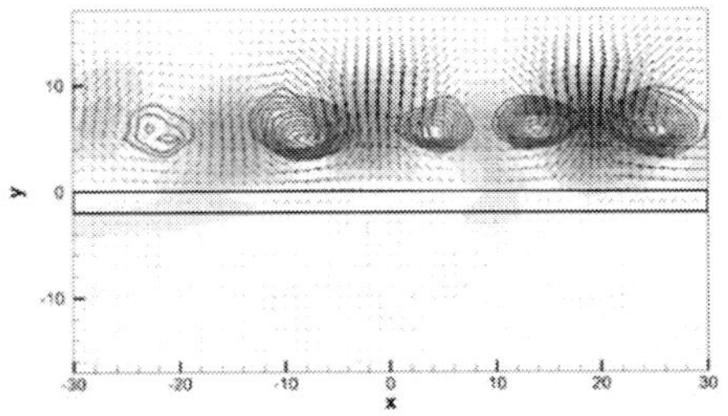

POD 11

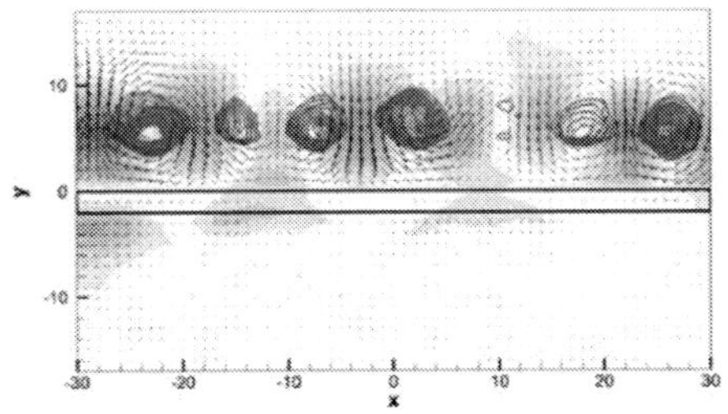

POD 21

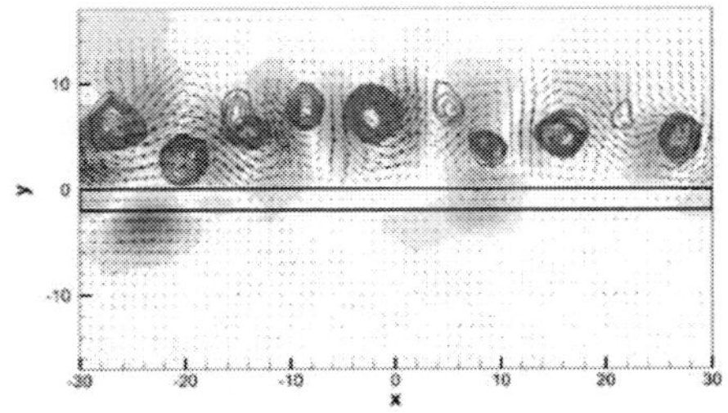

POD 58

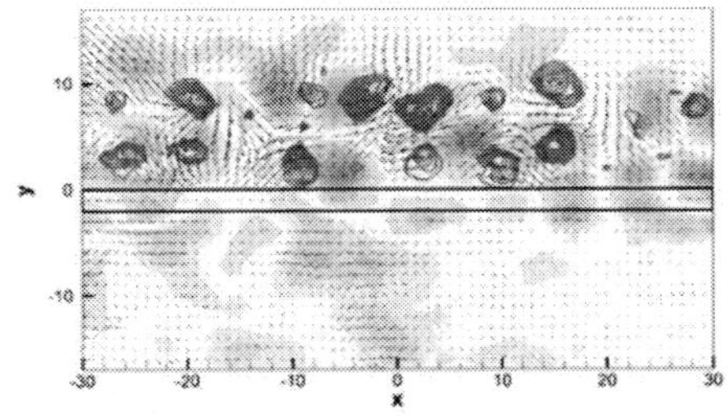

**Fig. 7.** Selected POD modes topologies, angle of attack 5 deg.

DOI: 10.1051/epjconf/201714302137

The POD modes in Fig. 7 show generally system of vortices located along the trailing edge, above it. Typically the adjacent vortices are of opposite orientation, the space between the neighbouring vortices is filed by fluid of either higher or lower streamwise velocity regions. This could indicate low- and high-velocity streaks, however the analysis is provided only in one plane perpendicular to the flow, we have no information about topology in $z$ direction. The dynamical structures could be modulated also in the streamwise direction, or not. Anyway, the streamwise velocity component variations $w$ are determined by distribution of mean velocity component $v$ in vertical ($y$) direction, see Fig. 4. Generally, if the fluid element is transported by the vortex motion from the region of higher mean velocity (on the top) to low velocity region (wake), the streamwise velocity fluctuation is positive and vice versa. It seems to be a rule for the fluctuation velocity components within the upper half of the pictures ($y > 0$): for the positive spanwise fluctuation velocity component $v$ (upward vectors orientation), the fluctuation of streamwise velocity component $w$ is negative and for negative $v$ fluctuation velocity component (downward vectors orientation) the of streamwise velocity component $w$ is positive.

The high energy modes carry long streamwise structures, both vortices and streaks, while higher order (low energy) modes are more structured also in streamwise direction, very probably. The details should be subject of further research.

Anyway, it is clear that the high energy modes (low order POD modes) contain bigger structures than the low energy modes.

The same analysis has been carried out for the case of angle of attack 10 deg. In Fig. 8 there is energy fraction of first 100 POD modes, energy of the mode 1 is about 12.3 % of the total energy of velocity fluctuations.

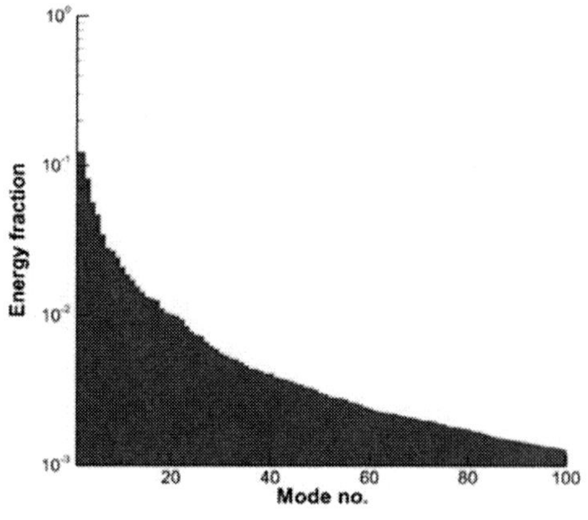


**Fig. 8.** Energy fraction of the POD modes, angle of attack 10 deg.

Selected POD modes topologies with typical vortical structures and streaks are in Fig. 9.

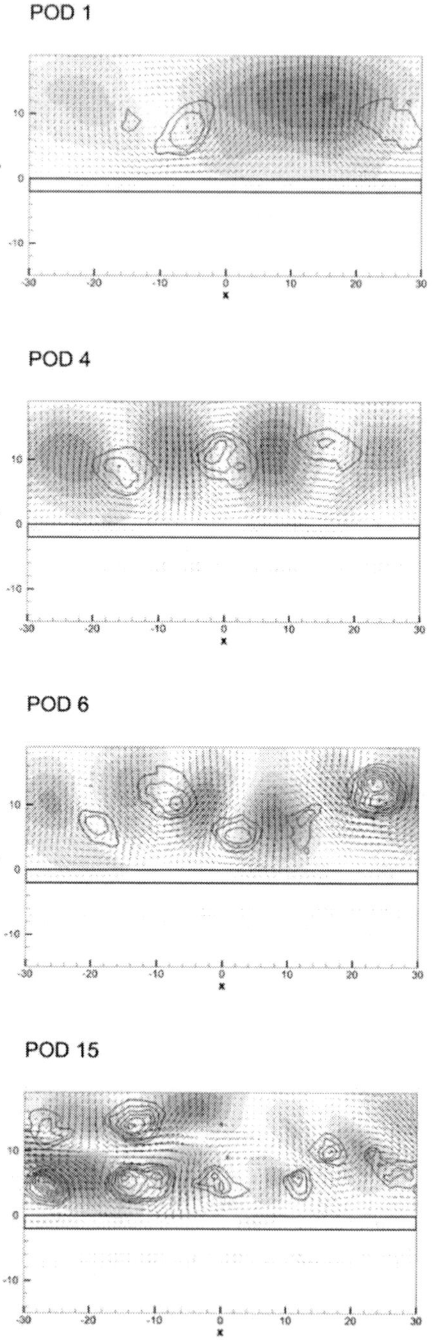

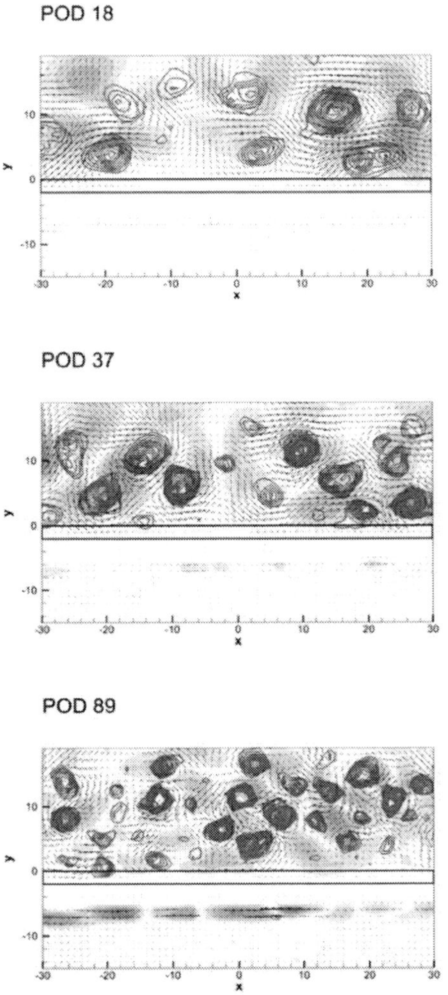


**Fig. 9.** Selected POD modes topologies, angle of attack 10 deg.

The observation of the POD modes topology is similar with the case of angle of attack 5 deg, however the touched area is of much bigger extend, especially in the upper part above the plate. The structures, both vortical and streaks, are distributed in more rows above the plate, for the higher order modes the space is completely filled by the structures (see e.g. POD mode 89). Lower order modes are characterized by a few number of very big structures in space (e.g. POD 1,4,6). For the lower order modes the vortices strength seems to be relatively week, however it should be noticed, that for the purpose of presentation the vector field is normalized and does not reflect the mode energy. Thus, the topology figures could provide qualitative information only.

## 4 Conclusions

The experimental results reveals existence of dynamical structures within the boundary layer on the plate suction side. Two types of structures are identified interacting with each other: streamwise vortices and low- and high-velocity regions, probably streaks. However experiments have been carried out in a single measuring plane behind the plate trailing edge, evolution of the dynamical structures should be studied in future.

The time-mean topology is vortex free.

The presented results support the theory of lift force generation by Hoffman and Johnson published in [1,2], however they could not be considered as the final and decisive confirmation.

## Acknowledgement

The authors acknowledge the support of the Technology Agency of the Czech Republic by means of projects TA04020129 and TA04011437.

## References

1. J. Hoffman, C. Johnson, The Mathematical Secret of Flight, *Normat 57*, 4, pp.1–25, (2009)
2. J. Hoffman, C. Johnson, Resolution of d'Alembert's Paradox, J. Math. Fluid Mech., Vol.12 (3), pp.321-334, (2010)
3. P. Manneville, *Instabilities, Chaos and Turbulence.* (Imperial College Press, 2004)
4. V. Uruba, Force Interaction of an Airfoil in Fluid-Flow, *15th conference on Power System Engineering, Thermodynamics & Fluid Flow - ES 2016*, June 09 - 10, 2016, Pilsen, Czech Republic, 7375, 7p., (2016)
5. V. Uruba, On Aerodynamic Forces Physical Mechanism, *SKMTaT 2016*, 20-23 June, 2016, Slovak University of Technology in Bratislava (R. Olšiak, K. Prikkel), pp. 258-262 (2016)
6. V. Uruba, On 3D Instability of Wake behind a Cylinder, *AEaNMiFMaE 2016*, 27-29 April 2016, University of Žilina, Department of Power Engineering (R. Lenhard, K. Kaduchová) pp. 253-256 (2016)
7. V. Uruba, Near Wake Dynamics around a Vibrating Airfoil by Means of PIV and Oscillation Pattern Decomposition at Reynolds Number of 65 000, *Journal of Fluids and Structures*, **55**, pp. 372-383 (2015)
8. V. Uruba, Decomposition Methods in Turbulent Research, EFM11, *EPJ Web of Conferences*, **25** 01095 (2012)

# Possibility of derivation of energetic state variables from simplified BWR thermal state equation for water and water steam

*Magda* Vestfálová[1,*]

[1]KEZ, TU Liberec, Studentská 1402/2, 461 17 Liberec 1, Czech Republic

**Abstract.** Thermal equation of state, along with the basic heat capacities are the most important documents needed to description the behavior of substances. Any other thermodynamic properties of a substance can already be deduced from these relations. The aim of this paper is to derive the dependence of energetic properties of water steam on the temperature and density from simple thermal equation of state based on original BWR thermal equation of state which describes the behavior of $H_2O$ in the gaseous phase in region of usual pressures and temperatures with sufficient accuracy.

## 1 Introduction

Water and water vapor are commonly found in nature. Moreover, they are substances which are frequently used in various technical devices. Therefore, it is necessary to know the properties and behavior of the substance. Properties of water and steam are thoroughly investigated, measured and tabulated in great detail [1]. Research of properties of water and water steam began to organize on an international level since 1929 to provide reliable and consistent data on their properties. In 1972 there was founded the International Association for the Properties of Water and Steam - IAPWS which deals with verification and interpolation of experimental data and with expressing of the thermodynamic and transport properties of water and steam. In 1995, there was accepted Pruß - Wagner equation as the international formulation of thermodynamic properties of water and steam for scientific purposes IAPWS-95 [2]. The equation is valid for the liquid and gas in a wide range of temperatures and pressures. It contains a large number of constants. International Standards properties of water and steam for example [3] are based on Scientific formulation IAPWS-95.

In many routine technical cases, there is more appropriate to use some "simple" equation for calculation of thermodynamic properties of water steam. "Simple" form of thermal equation of state will be applicable only on a limited range of temperatures and pressures, and its accuracy will be less.

In 1940 Manson Benedict, G. B. Webb and L. C. Rubin constructed "simple" so called original Benedict - Webb - Rubin (hereinafter BWR) thermal equation of state [4].

$$\begin{aligned} p = & \, r T \rho + r T \rho^2 B_0 + r T \rho^3 b - \\ & - A_0 \rho^2 - a \rho^3 + a \alpha \rho^6 - \frac{\rho^2 C_0}{T^2} + \\ & + \frac{c \rho^3}{T^2} \left(1 + \gamma \rho^2\right) e^{-\gamma \rho^2}. \end{aligned} \tag{1}$$

There is some physical basics of BWR equation in the state behavior of gases. Equation contains an addition to the specific gas constant $r$ still eight parameters: $A_0$, $B_0$, $C_0$, $a$, $b$, $c, \alpha$ and $\gamma$, that characterize the individual properties of the gas. They must be determined from experimental dates by regression. So, as well as other multi-parameter equations, BWR equation can be only used to description of properties of substances for which there is sufficient amount of experimental data in wide range of pressures and temperatures. This requirement is fulfilled for water and water steam undoubtedly.

The original BWR equation was suggested for calculation of behavior of hydrocarbons. For these substances values obtained from BWR equation are very close to the experimental date [5]. For these substances BWR equation describes the behavior of materials in single phase areas (it is in superheated steam and in the super cooled liquid) very well. In heterogeneous areas (below the limit curve), as expected, there is a local maximum and a local minimum and the results can not be explained by equilibrium thermodynamics. Parameters of BWR equation for hydrocarbons are presented for example in work [6] too.

In [7] there was shown the using of BWR thermal equation of state for $H_2O$. There was described derivation of the isochoric specific heat capacity and enthalpy from the thermal equation BWR. It has been shown that in the area where the error of the $p$-$v$-$T$ values obtained from BWR thermal state equations is small, the values of enthalpy, calculated from the relation derived

* Corresponding author: magda.vestfalova@tul.cz

using equation BWR are sufficiently precise too. The disadvantage is that the application area (the area with small errors) is very small and the equations for isochoric heat capacity and enthalpy is still quiet difficult.

In [8] there was found that in BWR thermal equation of state there can by neglected some members and for suitable values of constants of the new, more simple thermal equation of state we can obtain the *p-v-T* values with reasonable accuracy, it means the deviation of the values obtained from equation and tabulated data according to formulation for scientific purposes IAPWS-95 is less than *1 %*.

The aim of this paper is to derive the equations for calculation enthalpy and entropy from simple thermal state equation derived in [8] and to control the accuracy of these values in comparison with the tabulated data according to formulation for scientific purposes IAPWS-95 [2].

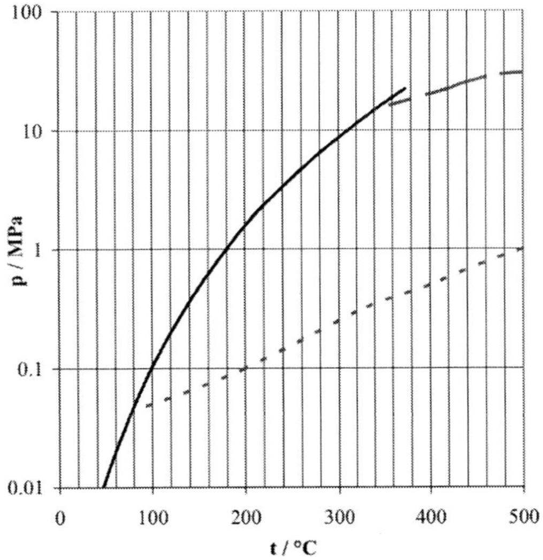


**Fig. 1.** The area of applicability of the different models for calculation of state properties of $H_2O$ with the accuracy to *1 %*: thick line - the limit curve; red dotted line - limit of applicability of ideal gas model; green dashed line - limit of applicability of simplified BWR equation [11].

## 2 Equation of state

In [8] there was derived for water vapor a simplified BWR equation in the form

$$p = rT\rho + rT\rho^2 B_0 + rT\rho^3 b - \\ - A_0\rho^2 - a\rho^3 - \frac{\rho^2 C_0}{T^2}. \quad (4)$$

This equation gives for routine engineering calculations sufficiently accurate results. When we use simplified BWR equation with coefficients derived for water steam, we can calculate the properties of water vapor to a state of beginning of condensation until the pressure about *15 MPa* (temperature about *340°C*). For high temperatures, equation can be used for even higher pressures (see Figure 1.), but not more than probably *30 MPa*.

The fundamental problem while using BWR equation for water and steam is that the critical point is not captured correctly: critical point under BWR equation is in the *p-v* diagram shifted significantly "to the right", i.e. to the higher values of specific volumes (less densities) and the "up", i.e. to the higher pressures, and is located on the higher isotherm than are the corresponding real values. The limit curve constructed using BWR equation

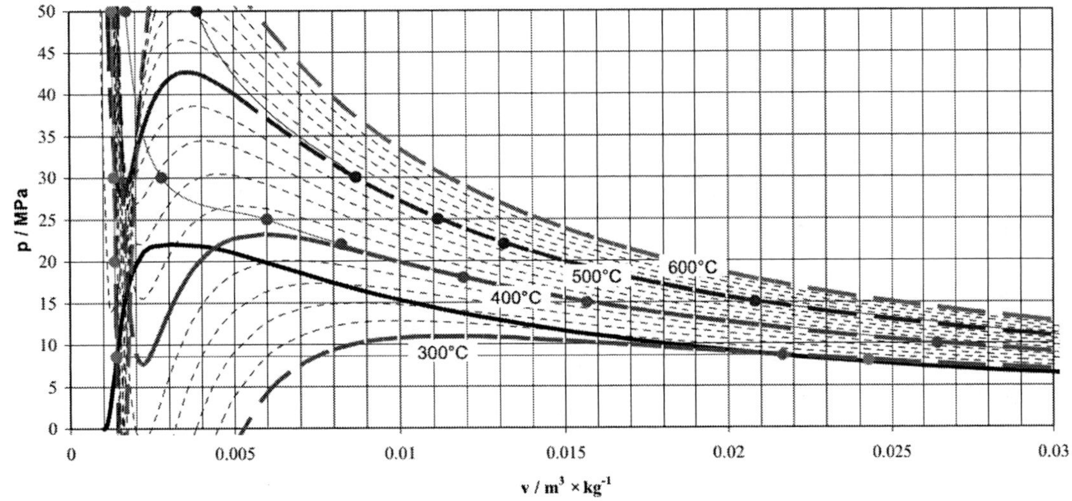


**Fig. 2.** Isotherms of simplified BWR equation [8] for $H_2O$: *300°C, 400°C, 500°C* a *600°C* (thick dashed lines) and theirs comparison with real isotherms according to IAPWS´95 data (points).

would be in the phase *p-T* diagram extended to a higher values of pressures and temperatures, the curve would continue beyond the real critical condition even further. In the *p-v* diagram would be the limit curve constructed from BWR equation shifted "to the right and up" from the real limit curve. Therefore we can use the BWR equation for water and water steam in the single-phase field - in the gas phase (the behavior of liquid phase does not coincide with the results obtained from BWR equation) and at greater distances from the critical point. The results obtained from the new simplified equation (without correction terms with coefficients "*c*" and "γ") have for a lower temperature the similar variation of limit curve as BWR equation, moreover, for temperatures greater than about *500°C* and high densities the results are totally unrealistic. For very high pressures equation can not be used either for high temperatures (see Figure 2.).

## 3 The dependence of the specific heat capacity on the volume of a substance described by simplified BWR equation

The dependence of isochoric specific heat capacity on the volume can be obtained from the thermal equation of state [7] by the following procedure:

$$\left(\frac{\partial c_V}{\partial v}\right)_T = \left\{\frac{\partial}{\partial T}\left[T\left(\frac{\partial p}{\partial T}\right)_V - p\right]\right\}_V = T\left(\frac{\partial^2 p}{\partial T^2}\right)_V . \tag{5}$$

Partial derivatives $\left(\frac{\partial p}{\partial T}\right)_V$ and $\left(\frac{\partial^2 p}{\partial T^2}\right)_V$ we obtain from the simplified BWR equation (4) as

$$\left(\frac{\partial p}{\partial T}\right)_V = \rho r + \rho^2 r B_0 + \rho^3 r b + \frac{2\rho^2 C_0}{T^3} \tag{6}$$

and

$$\left(\frac{\partial^2 p}{\partial T^2}\right)_V = \left[-\frac{6\rho^2 C_0}{T^4}\right] \tag{7}$$

and so

$$\left(\frac{\partial c_V}{\partial v}\right)_T = T\left(\frac{\partial^2 p}{\partial T^2}\right)_V = -\frac{6\rho^2 C_0}{T^3}. \tag{8}$$

The dependence of isochoric specific heat capacity on the temperature and volume we can obtain by integration the relation (8) over the volume between "known" and the current state. The "known" state: the state at $v \to \infty$, i.e. $p \to 0$, it is the idealized state, for which the specific heat capacity does not depend on volume, but only on the temperature [9]

$$\begin{aligned} c_V(T,v) &= c_V^{id}(T) + \int_\infty^v \left(\frac{\partial c_V}{\partial v}\right)_T dv = \\ &= c_V^{id}(T) + \int_\infty^v T\left(\frac{\partial^2 p}{\partial T^2}\right)_V dv = \\ &= c_V^{id}(T) + \int_\infty^v \left[-\frac{6\rho^2 C_0}{T^3}\right] dv = \\ &= c_p^{id}(T) - r + \frac{6\rho C_0}{T^3}. \end{aligned} \tag{9}$$

To illustrate the applicability of the above equation, we show (Figure 3) calculated courses of the isochoric specific heat capacities on temperature for $H_2O$ for pressures (from top to bottom): *20MPa, 10MPa, 1MPa, 0.1MPa* and *611.657 Pa*. As the values of isobaric specific heat capacity in the "idealized state" $c_p^{id}(T)$ we used the values at a pressure *611.657 Pa* according to [3]. As in the previous case, the experimentally

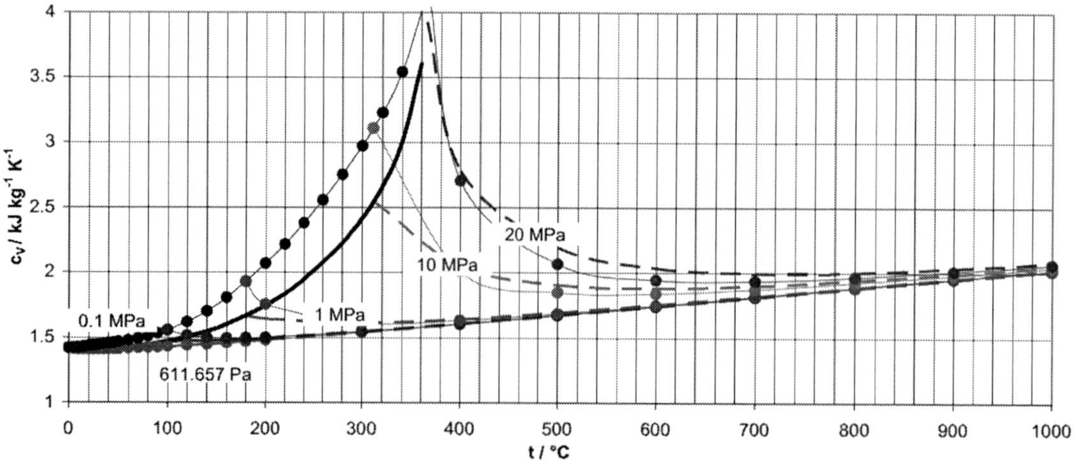


**Fig. 3.** Dependence of the isochoric heat capacity of $H_2O$ on the temperature for pressures: *611.657 Pa* (red), *0.1MPa* (blue), *1MPa* (green), *10MPa* (purple) and *20MPa*(brown) for real substance - comparison of results according simplified BWR equation (4) (thick dashed lines) and according IAPWS´95 data (points on the thin lines).

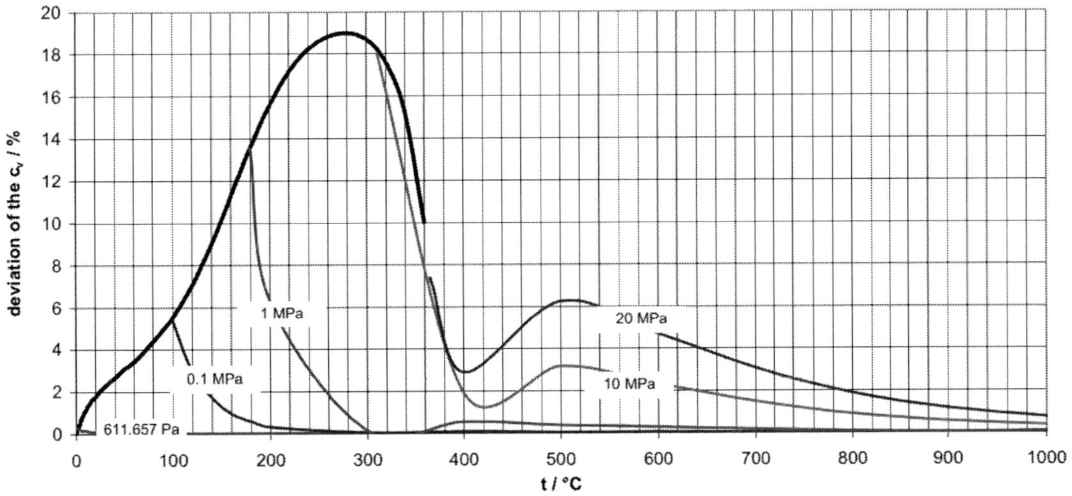


**Fig. 4.** Deviation of the calculated values of isochoric heat capacity of $H_2O$ from IAPWS'95 data in percentage for pressures: *611.657 Pa, 0.1MPa, 1MPa, 10MPa, 20MPa.*

determined course [3] is also plotted (circle points on the thin lines) and the limit curve is indicated in the diagram, too. Again, it is a nice match in locations distant from the limit boundary curve. But near the boundary curve the calculated values are far from IAPWS'95 data even in the pressures and temperatures in which *p-v-T* dates obtained from the simplified BWR equation show a good agreement with IAPWS'95 data.

Figure 4 shows the deviation of the calculated values of isochoric specific heat capacity from IAPWS'95 data in percentage.

The deviations of isochoric specific heat capacities calculated from equation derived from a simplified BWR state equation are too large even for low values of pressures.

## 4 Energy values

An important quantity in the calculation of the processes taking place in technical equipment is the enthalpy. Because we are working with state equation which was derived from BWR equation, it means with the state equation of the physical type, we will work again with the enthalpy depending on the temperature and volume. We will use the relation [7]

$$dh(T,v)=c_V\,dT+v\left(\frac{\partial p}{\partial T}\right)_V dT+ + v\left(\frac{\partial p}{\partial v}\right)_T dv+T\left(\frac{\partial p}{\partial T}\right)_V dv. \tag{10}$$

Partial derivatives $\left(\frac{\partial p}{\partial v}\right)_T$ and $\left(\frac{\partial p}{\partial T}\right)_V$ it is possible to obtain from the simplified BWR equation (4):

$$\left(\frac{\partial p}{\partial v}\right)_T = -T\rho^2 r - 2T\rho^3 r B_0 - -3T\rho^4 r b + 2A_0\rho^3 + 3a\rho^4 + \frac{2\rho^3 C_0}{T^2}. \tag{11}$$

The derivation $\left(\frac{\partial p}{\partial T}\right)_V$ was derived previously (6).

The dependence of enthalpy on the temperature and volume is obtained by integration the relation (10) between "known" and the current state over the volume and temperature. As in the case of specific heat capacity, it is chosen as the "known" state the idealized state ($v \to \infty, p \to 0$ , so the enthalpy does not depend on volume, but only on the temperature [12]) and the temperature at which we know the value of the enthalpy ($T_0$)

$$h(T,v)=h_{T_0}^{\infty}+\int_{T_0}^{T}\left[c_V^{id}(T)+v^{id}\left(\frac{\partial p}{\partial T}\right)_V^{id}\right]dT+ +\int_{\infty}^{v}\left[v\left(\frac{\partial p}{\partial v}\right)_T+T\left(\frac{\partial p}{\partial T}\right)_V\right]dv. \tag{12}$$

Due to the large errors in the calculation of isochoric specific heat capacity from the relation, which was derive from our state equation, we will integrate the equation (23) only over the volume and we will use as the "known" state the dependence of enthalpy on temperature for " $v \to \infty, p \to 0$", practically the values of enthalpy at a pressure of *611.657 Pa* according IAPWS'95 date:

$$h(T,v)=h^{\infty}(T)+ +\int_{\infty}^{v}\left[v\left(\frac{\partial p}{\partial v}\right)_T+T\left(\frac{\partial p}{\partial T}\right)_V\right]dv \tag{13}$$

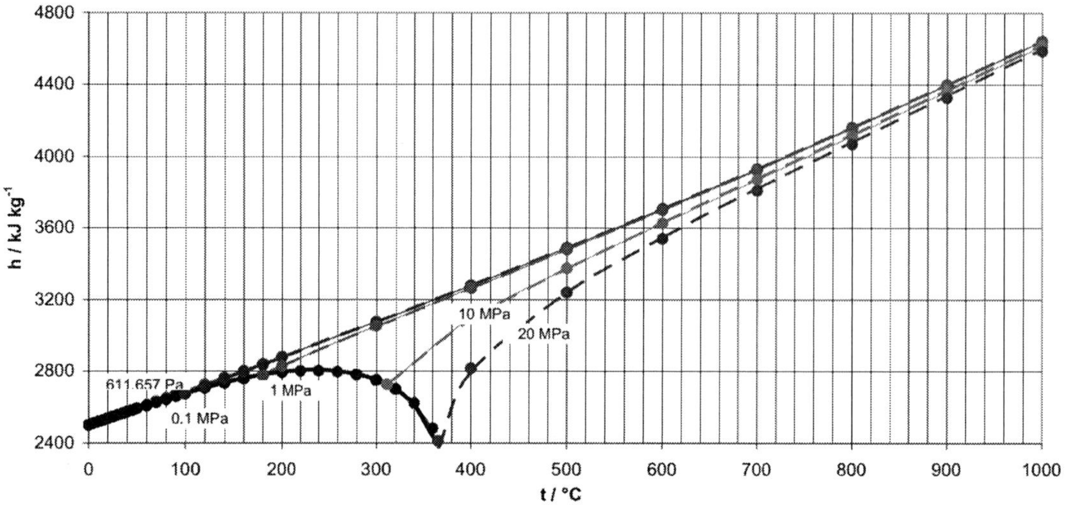


**Fig. 5.** Dependence of the enthalpies of $H_2O$ on the temperature for pressures: *611.657 Pa, 0.1MPa, 1MPa, 10MPa, 20MPa* - comparison of results according simplified BWR equation (26) (thick dashed lines) and according IAPWS´95 data (points on thin lines).

Substituting the partial derivatives (13) and (25) [12]

$$h(T,v)=h^{\infty}(T)+\rho\left[rT B_0-2A_0-\frac{4C_0}{T^2}\right]+ \\ +\frac{\rho^2}{2}\left[2rTb-3a\right] \tag{14}$$

To illustrate the applicability of the above equation, see on Figure 5 calculated courses of the enthalpies on temperature for $H_2O$ for pressures (from the button to the top): *20MPa, 10MPa, 1MPa, 0.1MPa* and *611.657 Pa.* As the values of enthalpy "idealized state" $h^{\infty}(T)$ we took values at a pressure *611.657 Pa* according to [3].

As in the previous case, the experimentally determined course [3] is also plotted (circle points on the thin lines) and the limit curve is indicated in the diagram, too. In case of enthalpy, there is a very nice match of the calculated values with IAPWS´95 data even on the boundary curve.

Figure 6 shows the deviation of the calculated values of enthalpies from IAPWS´95 data in percentage.

The deviations of enthalpies calculated from equation derived from a simplified BWR state equation are very small even for the values on the boundary curve. The significant error occurs only near the critical point.

Now, let's derive the relationship for the absolute value of entropy.

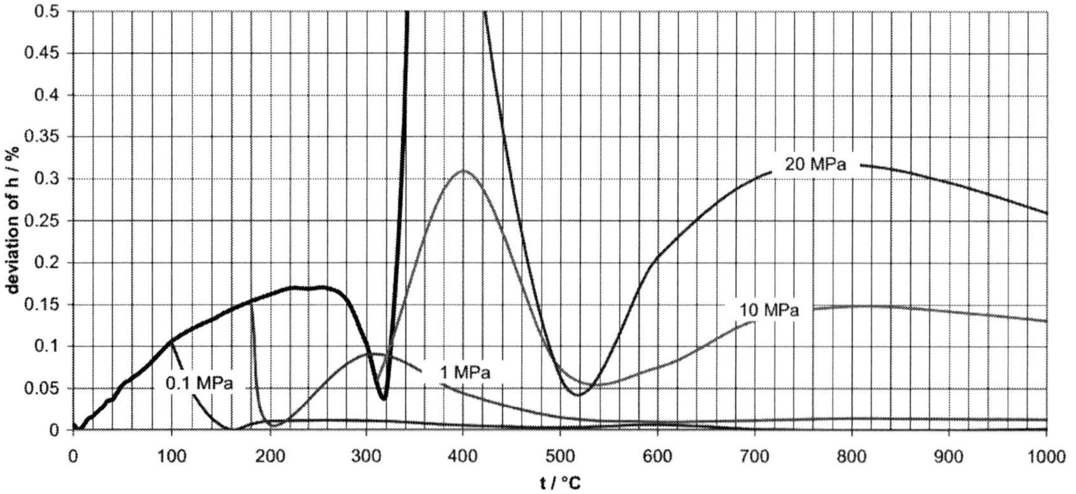


**Fig. 6.** Deviation of the calculated values of enthalpies of $H_2O$ from IAPWS´95 data in percentage for pressures: *611.657 Pa, 0.1MPa, 1MPa, 10MPa, 20MPa.*

The full differential of entropy we can write the usual way as

$$ds(T,v)=\left(\frac{\partial s}{\partial T}\right)_v dT+\left(\frac{\partial s}{\partial v}\right)_T dv. \qquad (15)$$

The first member on the right hand side we can express by comparison of the Gibbs equation

$$du=T\,ds+p\,dv \qquad (16)$$

with the total differential of the internal energy as the function of characteristic independent variables

$$du(T,v)=\left(\frac{\partial u}{\partial s}\right)_v ds+\left(\frac{\partial u}{\partial v}\right)_s dv. \qquad (17)$$

The first partial derivation of internal energy with respect to entropy at constant volume is

$$\left(\frac{\partial u}{\partial s}\right)_v=T, \qquad (18)$$

its second order partial derivation

$$\left(\frac{\partial^2 u}{\partial s^2}\right)_v=\left(\frac{\partial T}{\partial s}\right)_v=\frac{T}{c_v}. \qquad (19)$$

The second member on the right hand side we can express using Maxwell equation

$$\left(\frac{\partial s}{\partial v}\right)_T=\left(\frac{\partial p}{\partial T}\right)_v. \qquad (20)$$

So, the total differential of entropy it is possible to write as

$$ds(T,v)=\left(\frac{c_V}{T}\right)dT+\left(\frac{\partial p}{\partial T}\right)_v dv. \qquad (21)$$

The dependence of entropy on the temperature and volume is obtained by integration the relation (21) between "known" and the current state over the volume and temperature. As in the case of specific heat capacity and enthalpy, it is chosen as the "known" state the idealized state: $v \to \infty$, $p \to 0$, so the entropy does not depend on volume, but only on the temperature [9] (practically the state with "very low" pressure: $p_0 = 611.657\ Pa$, $v = v_0$) and at temperature at which we know the value of the entropy ($T_0$):

$$s(T,v)=s^0_{T_0}+\int_{T_0}^{T}\left[\frac{c_V^{id}(T)}{T}\right]dT+ \\ +\int_{v_0}^{v}\left[\left(\frac{\partial p}{\partial T}\right)_V^{id}\right]dv+\int_{v}^{\infty}\left[\left(\frac{\partial p}{\partial T}\right)_V^{id}\right]dv+ \qquad (22) \\ +\int_{\infty}^{v}\left[\left(\frac{\partial p}{\partial T}\right)_V^{BWR}\right]dv.$$

The first term on the right hand side is the known value of entropy at the temperature $T_0$ and volume $v_0$. The second member expresses the deviation of entropy in an idealized state of gas when the temperature changes from $T_0$ to general temperature $T$. We can use the Mayer equation to its expression. The third member is a deviation of entropy along the ideal isotherm between the state of the volume $v_0$ and the state of the general

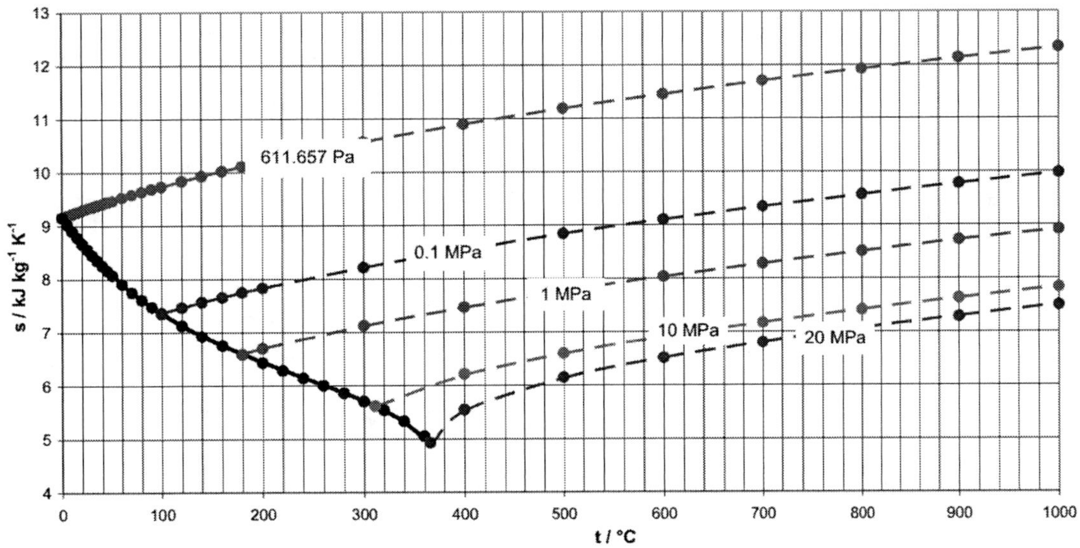


**Fig. 7.** Dependence of the entropies of $H_2O$ on the temperature for pressures: *611.657 Pa, 0.1MPa, 1MPa, 10MPa, 20MPa* - comparison of results according simplified BWR equation (29) (thick dashed lines) and according IAPWS'95 data (points on thin lines).

volume $v$. To its expression we use the ideal gas state equation. The fourth member is a deviation of the entropy along the ideal isotherm between the state of the general volume $v$ and the state where $v \to \infty$, it is a state where the ideal isotherm coincides with the real isotherm. To its expression we use again the ideal gas state equation. The last member is the deviation of entropy between the state, where $v \to \infty$, and the actual real state:

$$s(T,v)=s^0_{T_0}+\int_{T_0}^{T}\left[\frac{c_p^{id}(T)}{T}\right]dT - r\ln\frac{T}{T_0}+r\ln\frac{v}{v_0}+\int_{\infty}^{v}\left[-\frac{r}{v}+\left(\frac{\partial p}{\partial T}\right)_V^{BWR}\right]dv. \quad (23)$$

Due to the large errors in the calculation of isochoric specific heat capacity from the relation, which was derive from our state equation, we will integrate the equation (26) only over the volume and we will use as the "known" state the dependence of entropy on temperature for " $v \to \infty$, $p \to 0$", practically the values of entropy at a pressure of *611.657 Pa* according IAPWS´95 date:

$$s(T,v)=s^0(T)+r\ln\frac{v}{v_0}+\int_{\infty}^{v}\left[-\frac{r}{v}+\left(\frac{\partial p}{\partial T}\right)_V^{BWR}\right]dv. \quad (24)$$

After substituting the partial derivative of pressure with temperature at constant volume for BWR equation (13), integration and adjustments we obtain [9]

$$s(T,v)=s^0(T)+r\ln\frac{p_0}{\rho r T}-\rho\left(rB_0+\frac{2C_0}{T^3}\right)-\frac{br\rho^2}{2}. \quad (25)$$

To illustrate the applicability of the above equation, we show (Figure 7) calculated courses of the entropies on temperature for $H_2O$ for pressures (from the button to the top): *20MPa, 10MPa, 1MPa, 0.1MPa* and *611.657 Pa.*

As in the previous cases, the experimentally determined course [3] is also plotted (circle points on the thin lines) and the limit curve is indicated in the diagram, too. In case of entropy, there is a very nice match of the calculated values with IAPWS´95 data even on the boundary curve. The significant error occurs only near the critical point, too.

Figure 8 shows the deviation of the calculated values of entropies from IAPWS´95 data in percentage.

## 6 Conclusions

When calculating isochoric specific heat capacity of $H_2O$ in the gaseous state according to the equation derived from a simplified BWR equation [8] the errors are large in the low-pressure area too. So, it is unusable approach for practical calculations of the properties of water vapor. Conversely, a simplified BWR equation of state can be used with sufficient accuracy to the derivation and calculation of energy values, resp. theirs dependence on the volume. If we start from the real value of the enthalpy and entropy of $H_2O$ at low pressure (in a state

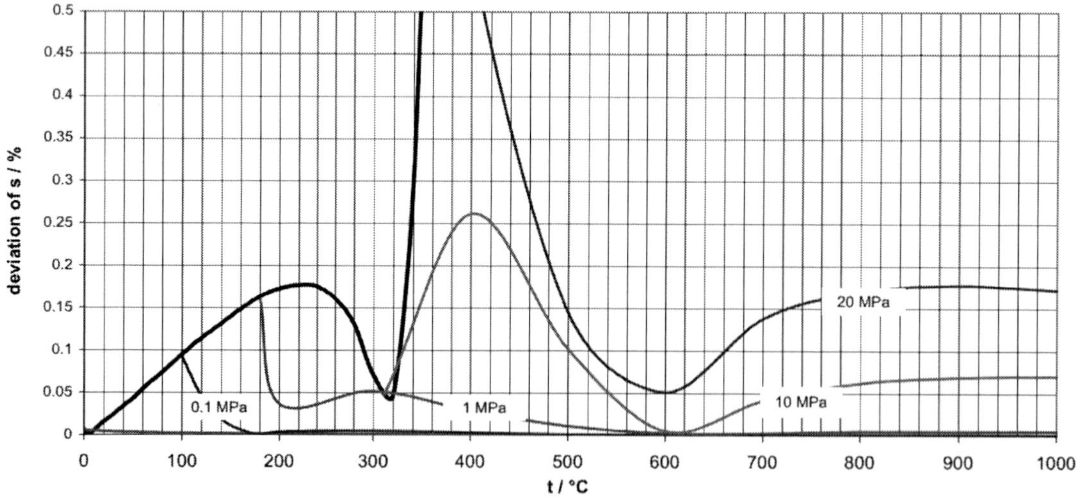


**Fig, 8.** Deviation of the calculated values of entropies of $H_2O$ from IAPWS´95 data in percentage for pressures: *611.657 Pa, 0.1MPa, 1MPa, 10MPa, 20MPa.*

approaching to "idealized state") we can use the simple formulas (26) and (29) derived from the simplified BWR equation of state (4) for the practical calculation their values at higher pressures successfully. The calculated values are sufficiently precise to a state of beginning of condensation until the pressure about *15 MPa* (temperature about *340°C*).

## Acknowledgment

This publication was written at the Technical University of Liberec, Faculty of Mechanical Engineering with the support of the Institutional Endowment for the Long Term Conceptual Development of Research Institutes, as provided by the Ministry of Education, Youth and Sports of the Czech Republic in the year 2016.

## References

1. M.Vestfálová, P.Šafařík, Experimental Fluid Mechanics (2015)
2. W.Wagner, A.Pruß, *The IAPWS Formulation 1995 for the Thermodynamic Properties of Ordinary Water Substance for General and Scientific Use*, Journal of Physical and Chemical Reference Data, 31: 387-535 (2002)
3. O.Šifner, J.Klomfar, *Mezinárodní standardy termofyzikálních vlastností vody a vodní páry*, (1996)
4. M. Benedict, G. Webb, L. C. Rubin, Journal of Chemical Physics 10 (12): 747-758 (1942)
5. M. Benedict, G. Webb, L. C. Rubin, Chemical Engineering Progress (CEP*)* 47 (8): 419–422 (1951)
6. J. Novák, *Termodynamické vlastnosti plynů*, (2007)
7. M. Vestfálová, Experimental Fluid Mehcanics (2014)
8. M.Vestfálová, M.Petříková, Experimental Fluid Mechanics (2015)
9. I. Středa, *Stavová rovnice BWR a termodynamické vlastnosti relných plynů a jejich směsí*, (1987)

# Tools for determination of humid air state for experimental and numerical investigation of equilibrium physical processes

*Magda* Vestfálová[1,*] and *Pavel* Šafařík[2]

[1]KEZ, TU Liberec, Studentská 1402/2, 461 17 Liberec 1, Czech Republic
[2]CTU in Prague, FME, Department of Fluid Dynamics and Thermodynamics, Technická 4, 166 07 Praha 6, Czech Republic

**Abstract.** The presented article deals with the solution of processes in humid air involving the particular elimination of moisture contained in the air. In the case that all of the moisture in the humid air is in the gaseous state, it is a homogeneous mixture. In the case that a part of the moisture in the humid air is in the state of liquid or solid, it is a heterogeneous mixture. In the presented paper processes, where the partial elimination of moisture from humid air (either as a liquid or as a solid phase) can appear, are studied. They are the processes, where the moist air mixture becomes heterogeneous. Based on the theory of ideal mixtures the computational methods enabling determination of the thermodynamic state of the beginning of the elimination of moisture are developed. The thermodynamic properties of dry air and water are applied. Subsequently there are designed diagrams that can be used as an aid to identify interfaces of homogeneous and heterogeneous humid air.

## 1 Introduction

Moist air is a mixture of dry air and moisture. To simplify the physical description of humid air which occurs in nature or participates in technical processes, it is possible to consider only those of its constituents. The presence of other substances in the air (such as various gaseous impurities, a liquid component of substances from different technologies, solid dust) will not be considered.

If all of the moisture in the humid air is in the state of gas, it is homogeneous mixture. Moist air as the heterogeneous mixture contains moisture
- partially in the vapour state and partially in the liquid state, or
- partially in the vapour state and partially in the solid state, or
- partially as a vapour state, partially in the liquid state and partially in the solid state.

Moisture in the atmospheric air is present in relatively small amounts, but it is of great importance.

A very common problem is the determination of the parameters of the state when moist air as a homogeneous mixture becomes heterogeneous mixtures, i.e. a state in which a moist air begins to secrete moisture. This condition is usually at isobaric conditions, referred to as dew point. This paper focuses on the theoretical tools to determine the equilibrium conditions on the border of homogeneous and heterogeneous mixtures.

## 2 Components of moist air

Dry air can be considered as an ideal gas in the wider area around the atmospheric conditions. For the needs of aviation, the composition of dry air was adopted according to Table 1 by the international convention [1]. It is the standard used as a reference, especially for instrument calibration.

**Table 1.** Composition of dry air.

| Gas | chemical symbol | volume fraction [%] |
|---|---|---|
| Nitrogen | $N_2$ | *78* |
| Oxygen | $O_2$ | *21* |
| other gases | | *1* |

Basic physical and thermodynamic properties of dry air as the ideal gas:

- average molecular weight: $M = 28.9583 \frac{kg}{kmol}$
- specific gas constant: $r = 287.12 \frac{J}{kg \cdot K}$
- isobaric specific heat capacity: $c_p = 1005.9 \frac{J}{kg \cdot K}$
- isochoric specific heat capacity: $c_v = 718.8 \frac{J}{kg \cdot K}$

* Corresponding author: magda.vestfalova@tul.cz

- Poisson ratio: $\kappa = 1.399$

A theoretical model of an ideal gas for dry air near atmospheric conditions is sufficient for working calculations. For the purposes of exact solutions, especially for scientific applications, it is appropriate to apply the theoretical model of semi-ideal or real gas for dry air. For cases of high pressure dry air (pressure above *1 MPa*) and dry air at a higher temperature (the temperature above *250 °C*) it is necessary to use the model of a real gas, for example NIST (National Institute for Standards and Technology, USA) [2].

Water is thermodynamically relatively complex matter. It is most widely used and best available medium for energy transfer. Therefore, extensive research concentrates and focused at her. For work of professionals are gradually approved documents by the International Association for the Properties of Steam (IAPWS) as a reference for determining the values of thermodynamic and thermophysical properties of water and steam. Important documents are:

- 1995 (Revision 2014) - Formulation of thermodynamic properties of ordinary water substance for general and scientific use (abbreviated IAPWS-95)
- 1997 - New international formulation of the thermodynamic properties of water and steam for industrial purposes (abbreviated IAPWS IF-97).

The equation of state for water and steam under those two documents are substantially distant ideal gas.

Basic physical properties of water and water vapour, according to [3]:

- molar weight: $M = 18.015 \frac{kg}{kmol}$
- specific gas constant: $r = 461.26 \frac{J}{kg \cdot K}$
- triple point:

  the temperature of the triple point:
  $t_{tr} = 0.01°C$, $T_{tr} = 273.16 K$
  the pressure at the triple point:
  $p_{tr} = 611.657 Pa$

- critical point:

  the temperature of the critical point:
  $t_{cr} = 373.946°C$, $T_{cr} = 647.096 K$
  the pressure at the critical point:
  $p_{cr} = 22.064 MPa$

Thermodynamic behaviour of moisture occurring in a homogeneous moist air (it is water vapour) may be at normal atmospheric conditions described by the model of an ideal gas.

Basic thermodynamic properties of steam as an ideal gas:

- isobaric specific heat capacity: $c_p = 1898.56 \frac{J}{kg \cdot K}$
- isochoric specific heat capacity: $c_V = 1398.56 \frac{J}{kg \cdot K}$
- Poisson ratio: $\kappa = 1.33$

In field of conditions, when the moist air becomes a heterogeneous mixture, it is, when it begins to eliminate of moisture from the moist air in the form of liquid or solid phase, it is necessary to take into account the real properties (behaviour) of $H_2O$ when describing the properties of moisture.

For the thermodynamics of moist air it is very important determination of thermodynamic properties of water vapour at saturation limits.

The dependence of the pressure of saturated steam at a temperature function on the saturation limit vapour-liquid from the triple point to the critical point determines the scientific formulation (IAPWS 95) [4]

$$p'' = p_{cr} \cdot \exp\left[\frac{1}{1-\tau} \left(a_1 \cdot \tau + a_2 \cdot \tau^{1.5} + a_3 \cdot \tau^3 + a_4 \cdot \tau^{3.5} + a_5 \cdot \tau^4 + a_6 \cdot \tau^{7.5}\right)\right], \tag{1}$$

where:

$$\tau = 1 - \frac{T}{T_{cr}} \tag{2}$$

is the temperature function,

| | |
|---|---|
| $T$ | is the temperature *[K]*, |
| $T_{cr} = 647{,}096 K$ | is the critical temperature of water vapour, |
| $p_{cr} = 22{,}064 MPa$ | is the critical pressure of water vapour, |

the constants in equation (1) are listed in Table 2.

**Table 2.** The constants in equation (1).

| $a_1$ | $a_2$ | $a_3$ |
|---|---|---|
| −7.59517783 | 1.84408259 | −11.7866497 |
| $a_4$ | $a_5$ | $a_6$ |
| 22.6807411 | −15.9618719 | 1.80122502 |

Pressure depending on the temperature from *50 K* to the triple point on the saturation limit between the solid and gas phase $H_2O$, it is for sublimation and desublimation, is solved according to equation [5]

$$p'' = p_{tr} \cdot \exp\left(\theta^{-1} \cdot \sum_{i=1}^{3} a_i \cdot \theta^{b_i}\right), \tag{3}$$

where:

$$\theta = \frac{T}{T_{tr}} \tag{4}$$

is the temperature function

| | |
|---|---|
| $T$ | is the temperature *[K]*, |
| $T_{tr} = 273.16 K$ | is the temperature of triple point of water, |
| $p_{tr} = 611.657 Pa$ | is the pressure of triple point of water, |

the constants in equation (3) are shown in Table 3.

**Table 3.** The constants in equation (3).

| | |
|---|---|
| $a_1$ | $-21.2144006$ |
| $a_2$ | 27.3203819 |
| $a_3$ | $-6.1059813$ |
| $b_1$ | 0.00333333333 |
| $b_2$ | 1.20666667 |
| $b_3$ | 1.70333333 |

## 3 Moist air

Moist air as a thermodynamic system is a mixture of dry air and water substance. Moisture (water) contained in the moist air is always in the form of steam, but also (in the case of a heterogeneous mixture) may be part of the moisture in the moist air also in a state of liquid or solid.

The object of this paper is to find a boundary at which ends the existence of moisture in the air only in the form of steam.

The temperature of the moist air $t_{MA}\left[{}^\circ C\right]$ is equal to the temperatures of its components

$$t_{MA}=t_{DA}=t_V=t\,. \tag{5}$$

The pressure of the moist air $p_{MA}\left[Pa\right]$ according to Dalton's law is equal to the sum of the partial pressures of dry air and water vapour

$$p_{MA}=p_{DA}+p_V\,. \tag{6}$$

The weight of the moist air $m_{MA}\left[kg\right]$ is

$$m_{MA}=m_{DA}+m_{H_2O}\,. \tag{7}$$

The density of the moist air $\rho_{MA}\left[\frac{kg}{m^3}\right]$ is a ratio of weight of moist air to its volume, it is assuming that the moisture in humid air only as vapour

$$\rho_{MA}=\rho_{DA}+\rho_V\,. \tag{8}$$

The pressure of saturated steam $p_V''\left[Pa\right]$ is for moist air a significant magnitude determining the limit among of moisture in the air for the existence of a homogeneous mixture:

- if $p_V<p_V''$, the humid air is unsaturated and the mixture is homogenous,
- if $p_V=p_V''$, the humid air is saturated and the mixture is homogenous,
- if $p_V>p_V''$, the humid air is supersaturated and the mixture is heterogeneous.

The pressure of saturated steam is an unequivocal function of temperature

$$p_V''=p_V''(t), \tag{9}$$

(see equations (1), (3))

Dew point temperature $t_{DP}\left[{}^\circ C\right]$ is the temperature of moist air, in which the humid air reaches the saturation state after isobaric cooling

$$p_V(t)=p_V''(t_{DP}), \tag{10}$$

it is when all water as vapour, which is contained therein, is in a state of saturated steam.

Absolute humidity of moist air $a\left[\frac{kg}{m^3}\right]$ is the mass of vapour contained in a unit volume of air. It is the density of vapour $\rho_V$ at a pressure of vapour $p_V$ and at the temperature $t$. Absolute humidity is defined only for a homogeneous mixture, and can be at a given temperature at the range $\left\langle 0;\rho_V''(t)\right\rangle$.

Relative humidity of moist air $\varphi\left[-\right]$ is the ratio of the absolute humidity to the absolute humidity of saturated air at a given temperature

$$\varphi=\frac{a}{a''}\,. \tag{11}$$

Relative humidity is defined only for a homogeneous mixture and can at the range $\left\langle 0;1\right\rangle$. Assuming that the water vapour containing in the humid air can be described by the model of an ideal gas

$$\varphi=\frac{p_V}{p_V''}\,. \tag{12}$$

Specific humidity $x\left[\frac{kg_V}{kg_{DA}}\right]$ is the ratio of the mass of water to the weight amount of dry air

$$x=\frac{m_{H_2O}}{m_{DA}}\,. \tag{13}$$

In the case of a homogeneous mixture the contained moisture is in vapour form and

$$x=\frac{m_V}{m_{DA}}=\frac{\rho_V}{\rho_{DA}}\,. \tag{14}$$

In the supersaturated air, it is heterogeneous mixture, a part of moisture is as saturated steam and the remaining moisture will be excluded. For $t>0$ °C will be portion of the moisture

$$x'=x-x'' \tag{15}$$

excluded as a liquid. For $t<0$ °C, a portion of the moisture

$$x'''=x-x'' \tag{16}$$

will be excluded as solid phase. For $t=0$ °C may be part of the moisture excluded as a liquid phase and a part as a solid phase

$$x'''+x'=x-x''\,. \tag{17}$$

If the moist air forms a homogeneous mixture and its components (dry air and steam) can be described by the model of an ideal gas, it is possible to derive the relationship between the specific and the relative humidity

$$x=\frac{\rho_V}{\rho_{DA}}=0{,}622\cdot\frac{\varphi\,p_V''}{p_{MA}-\varphi\,p_V''}\,. \tag{18}$$

For specific humidity of saturated homogeneous mixture $(\varphi=1)$

$$x''=0.622\cdot\frac{p_V''(t)}{p_{MA}-p_V''(t)}=x''(t)\,. \tag{19}$$

# 4 Equilibrium *p - t* and *t - x* diagrams for moist air

For a given value of specific humidity of moist air it is possible to solve the dependence of the pressure of saturated moist air on the temperature from equations (19) and (1) respectively (3), it means the curve at steady $p$ - $t$ diagram (with parameter specific humidity $x$) at which there occur the elimination of moisture from humid air, see Figure 1. In the diagram there is marked the triple point for the dry air - $tr_{DA}$ , the critical point for the dry air - $cr_{DA}$ , the saturation line of the vapour with the triple point for the vapour - $tr_V$ and the critical point for the vapour - $cr_V$ and curves of pressure of saturated moist air on the temperature for selected values of specific humidity $x$. Figure 2 is a detail of $p$ - $t$ diagram which is thus a tool for displaying the state of moist air and the equilibrium processes in moist air, and for distinguishing whether the moist air is a homogeneous or heterogeneous mixture.

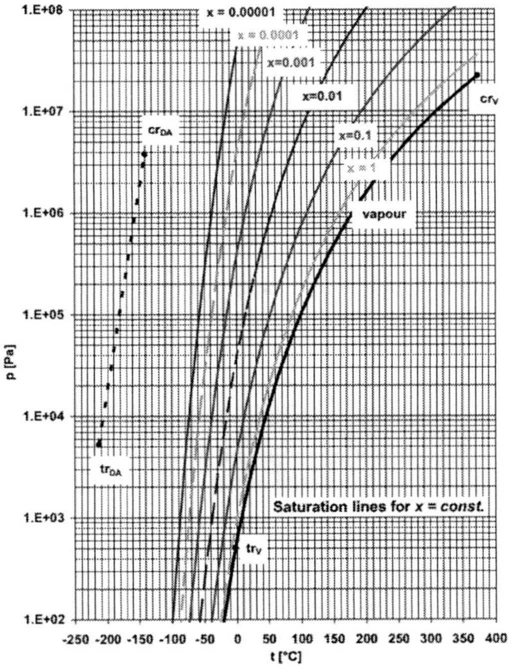


**Fig 1.** Curves of saturated moist air for different values of specific humidity $x$ *[kg$_V$/kg$_{DA}$]* in $p - t$ diagram.

For a given value of the pressure of moist air it is possible to solve the dependence of the temperature of saturated moist air on specific humidity from equation (19), it means the curve in the $t$ - $x$ diagram (with parameter total pressure $p_{MA}$) at which there occur the elimination of moisture from humid air, see Figure 3. Such a diagram is a tool for displaying the state of moist air and the equilibrium processes in humid air, and for distinguishing whether the humid air is a homogeneous or heterogeneous mixture.

It should be noted that for the pressures and temperatures greatly differing from normal atmospheric conditions is no longer possible to describe the components of the moist air by a model of ideal gas and the shape of the curves of saturation would deviated in $p$ - $t$ and $t$ - $x$ diagram from the ideal shapes counted on the basis of these simplifying assumptions in paragraph 2.

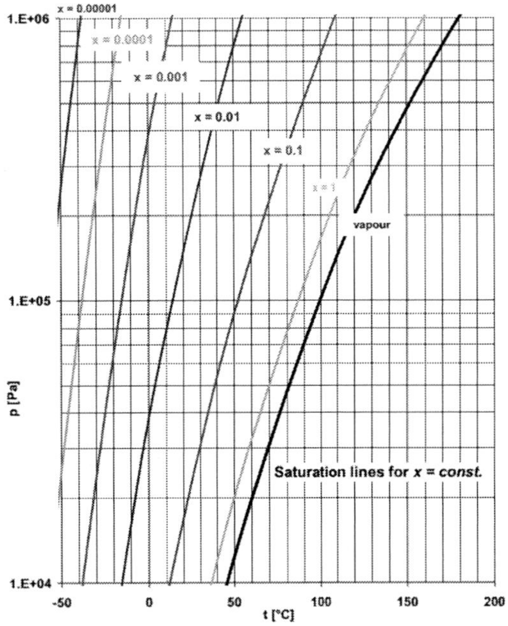


**Fig. 2.** Curves of saturated moist air for different values of specific humidity $x$ *[kg$_V$/kg$_{DA}$]* in $p - t$ diagram (detail).

# 5 Basic reversible processes in unsaturated moist air of unchanging composition

In moist air different thermodynamic processes can run. Under certain conditions the originally unsaturated - homogeneous - moist air can get on the border between homogeneous and heterogeneous mixtures during thermodynamic process, it is to the state of saturated moist air. The above-mentioned diagrams can be used for finding those conditions advantageously.

To the equilibrium $p$ - $t$ diagram can be drawn isochoric network, it is lines of constant specific volume of moist air (dashed curves in Figure 4), respectively isentropic network, it is lines of constant entropy moist air (dotted curves in Figure 4).

If the isobaric process carried in the unsaturated moist air, it can be displayed in the $p$ - $t$ diagrams in Figure 1, Figure 2 and Figure 4, as a horizontal line ($p$ = *const.*). If it is an isobaric heating, the process runs in diagrams "to the right", it is to the higher temperature values, conversely, isobaric cooling runs in diagrams "to the left". In $t$ - $x$ diagram on Figure 3, each action occurring in the moist air at constant composition can be

DOI: 10.1051/epjconf/201714302139

displayed as vertical line, whose position is determined by the value of its specific humidity. When isobaric heating (temperature increase), the amount of moisture which may contain a homogenous moist air, increases, the saturation therefore can not occur. Conversely, when cooling, the quantity of moisture which may contain homogeneous moist air decreases, i.e. when a certain temperature value is achieved, the moist air becomes saturated. The value of this temperature can be easily detected from any of the diagrams.

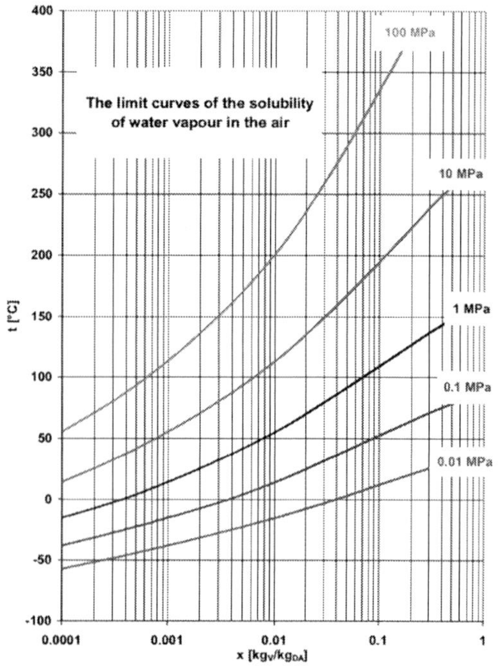


**Fig. 3.** The curves of saturated moist air for different values of the total pressure $p_{MA}$ in the $t$ - $x$ diagram.

If the isothermal process carried in the unsaturated moist air, it can be displayed in the $p$ - $t$ diagrams in Figure 1, Figure 2 and Figure 4, as a vertical line (*t = const.*). If it is an isothermal expansion, the process runs in diagrams toward "down", it is to the lower pressure values, conversely, isothermal compression toward "up". In $t$ - $x$ diagram on Figure 3 is each isothermal process displayed as a single point, given by the temperature and composition. When isothermal expansion (pressure reduction), the amount of moisture which may contains homogeneous moist air, increases, saturation can not occur. Conversely, when isothermal compression (pressure increase), the quantity of moisture which may contain homogeneous moist air decreases, it is when a certain pressure value is achieved, the moist air becomes saturated. The value of this pressure can be easily detected from any of the diagrams.

Isochoric processes occurring in the unsaturated moist air are shown in Figure 4 as a set of dashed lines - isochors (*v = const.*). If it is an isochoric heating, the process runs in diagrams "to the right and up", it is to the higher temperature values and higher pressure values, conversely, isochoric cooling runs in diagrams "to the left and down". When isochoric heating (temperature increase), the amount of moisture which may contain a homogenous moist air, increases, the saturation therefore can not occur. Conversely, when cooling, the quantity of moisture which may contain homogeneous moist air decreases, it is when a certain temperature value is achieved, and the moist air becomes saturated. The value of this temperature can be easily detected from diagrams in Figure 4.

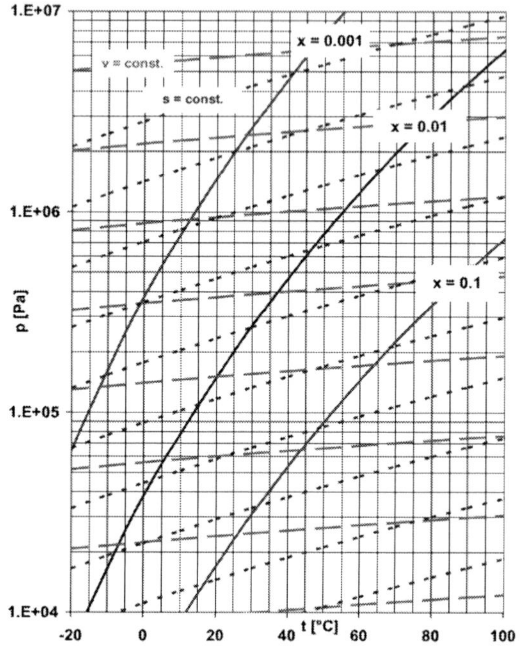


**Fig. 4.** The curves of saturated moist air for different values of specific humidity $x$ *[kgV/kgDA]* in the $p$ - $t$ diagram supplemented with network of isentropic lines (dotted curves) and isochoric lines (dashed curves).

Isentropic processes occurring the unsaturated moist air are shown in Figure 4 shows a set of dotted lines - isentropic lines (*s = const.*). If it is an isentropic compression, the process runs in diagrams toward "up and right", it is to the higher pressure values and higher temperature values, conversely, isentropic expansion toward "down and left". When isentropic compression (pressure increase), the amount of moisture which may contains homogeneous moist air, increases, saturation can not occur. Conversely, when isentropic expansion (pressure reduction), the quantity of moisture which may contain homogeneous moist air decreases, it is when a certain pressure value is achieved, the moist air becomes saturated. The value of this pressure can be easily detected from diagrams in Figure 4.

For illustration, the particular processes are compared in $p$ - $t$ diagram in Figure 5: It is assumed unsaturated moist air at *50 °C, 100 kPa* and a specific humidity *0.01* $kg_V/kg_{DA}$. When the isobaric cooling the saturation is reached at about *14 °C*. When isochoric cooling, the saturation is achieved at about *12 °C*, the pressure is about *90 kPa*. If the isentropic process proceeds in this moist air, the state of saturation is reached by air expansion to a pressure of *55 kPa* and a temperature of *5 °C*. If the isothermal process proceeds in this moist air, the state of saturation is reached by air compression to a pressure of *600 kPa*.

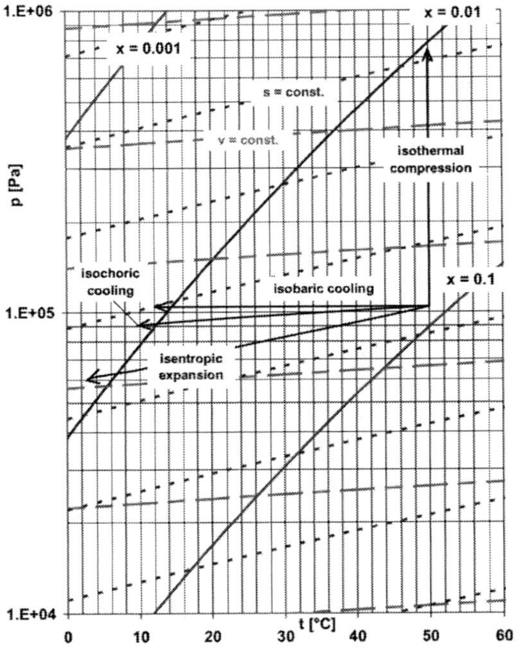


**Fig. 5.** The curves of saturated moist air for different values of the total pressure $p_{MA}$ in the $p$ - $t$ diagram supplemented with network of isentropic lines (dotted curves) and isochoric lines (dashed curves). (Detail with illustrative examples) (Specific humidity $x$ $[kg_V/kg_{DA}]$).

## Conclusions

Assuming ideal behaviour of the gaseous components of moist air (dry air and water vapour) there were constructed diagrams (equilibrium $p$ - $t$ diagram and $t$ - $x$ diagram) allowing for experimental and numerical investigation of the processes in the humid air easily to find a boundary at which the existence of moisture air only as vapour ends. Upon reaching this limit (it is upon reaching the states lying on curves of saturation in the diagrams), the moisture begins exclude from the moist air, it is moist air becomes the heterogeneous mixture, which is illustrated in the cases of the isobaric cooling, isothermal compression, isochoric cooling and isentropic expansion for $x = const.$ in unsaturated moist air.

## Acknowledgment

This publication was written at the Technical University of Liberec, Faculty of Mechanical Engineering with the support of the Institutional Endowment for the Long Term Conceptual Development of Research Institutes, as provided by the Ministry of Education, Youth and Sports of the Czech Republic in the year 2016.

The second author, P.Šafařík, expresses thanks for support from the Technology Agency of the Czech Republic in the frame of the Competence Centre Advanced Technology of Heat and Electricity Output, No.TE01020036.

## References

1. International Standard Atmosphere, ISO 2533:1975, International Organization for Standardization, (1975)
2. Lemmon, E.W., Jacobsen, R.T., Penoncello, S.G., Friend, D.G., *Thermodynamic Properties of Air and Mixtures of Nitrogen, Argon, and Oxygen from 60 to 2000 K at Pressures to 2000 MPa*, Journal of Physical and Chemical Reference Data, Vol. 29, No. 3: 331-385 (2000)
3. O.Šifner, J.Klomfar, *International Standards of Thermophysical Properties of Water and Steam* (in Czech) (1996)
4. IAPWS: Revised Release on the IAPWS Formulation 1995 for the Thermodynamic Properties of Ordinary Water Substance for General and Scientific Use (2014)
5. IAPWS: Revised Release on the Pressure along the Melting and Sublimation Curves of Ordinary Water Substance (2011)

# Investigation of droplet nucleation in CCS relevant systems – Design and testing of a $CO_2$ branch of the mixture preparation device

Václav Vinš[1,*], Miroslav Čenský[1,*], Jiří Hykl[1], and Jan Hrubý[1]

[1]Institute of Thermomechanics of the CAS, Dolejškova 1402/5, Prague 8, 182 00, Czech Republic

**Abstract.** A unique in-house designed experimental apparatus for investigation of nucleation of droplets in CCS relevant systems is being developed by our team. The nucleation is measured with the help of a rapid pressure drop within an expansion chamber. In this study, a $CO_2$ branch representing an important part of the mixture preparation device (MPD) was designed, assembled and tested. MPD is intended for preparation of $CO_2$-rich mixtures in a sense of accurate setting of flow rates of individual gaseous components through the experimental section with the expansion chamber. In the $CO_2$ branch, the saturated liquid $CO_2$ coming from a cylinder with a dip tube is pressurized by a supercritical pump with a constant flow rate to the pressures of up to 340 bar. The required mass flow of the supercritical $CO_2$ of up to 180 g/h at pressures between 74 bar and 150 bar is accurately tuned with a help of a set of stainless steel capillary tubes with inner diameter of 0.1 mm. Some of the tubes are submerged in a thermostatic bath set to a constant temperature of 70 °C in order to assure the supercritical flow conditions of $CO_2$. Needed lengths of the capillary tubes were determined with a help of a one-dimensional numerical model allowing prediction of both isothermal and adiabatic flow conditions.

## 1 Introduction

The team of the Laboratory of Phase Transition Kinetics at the Institute of Thermomechanics of the CAS is developing a unique experimental apparatus for investing the formation of droplets in $CO_2$-rich systems. Motivation for this research is to obtain new valuable data for nucleation of droplets (the initial stage of droplet formation) in binary and multicomponent mixtures important in the design of new carbon capture and storage (CCS) technologies [1]. At the current state, the research of our team is focused mainly on binary systems of carbon dioxide with nitrogen and argon. Later, binary systems with water and ternary systems will be studied.

A unique mixture preparation device (MPD) for preparation of $CO_2$-rich mixtures with up to three components was designed by our team. Its function is based on precise setting of flow rates of individual components within the measuring system. In this study, the $CO_2$-branch representing the most important part of MPD was designed, assembled and successfully tested.

## 2 New apparatus for investigation of nucleation in $CO_2$-rich systems

In the new experimental apparatus, the nucleation of droplets in $CO_2$-rich systems is measured with the help of a rapid pressure drop inside a stainless steel expansion chamber [2-4]. The apparatus allows measurement under wide operating conditions at temperatures from –20 °C to +150 °C and pressures up to 130 bar. Consequently, the nucleation of droplets can be investigated both in a low pressure vapor and in a high pressure supercritical fluid. We note that the critical point temperature and pressure of $CO_2$ are 30.98 °C and 73.8 bar, respectively. The apparatus allows simulating various scenarios relevant in CCS technologies including sudden rupture of the pressurized pipeline with $CO_2$-rich fluid.

The expansion within the chamber, filled with a $CO_2$-rich mixture in the form of the superheated vapor or supercritical fluid, can be considered as isentropic due to the fast pressure drop. The heat transfer from the walls of chamber and the ambient into the fluid can be neglected in this case. As the decreasing pressure crosses the equilibrium two-phase envelope, the fluid becomes thermodynamically metastable, i.e. at supersaturated state. At certain conditions given by the initial state and the kinetics of the pressure drop, the supersaturation of the original phase is large enough such that the phase transition is initiated. Liquid droplets start to nucleate and subsequently grow.

Conditions inside the expansion chamber are measured with a fast-response pressure transducer and accurate temperature probes. The nucleation of droplets and their subsequent growth is detected optically by using an in-house designed interferometer.

* Corresponding authors: vins@it.cas.cz and censky@it.cas.cz

## 3 Mixture preparation device

Gaseous mixtures with $CO_2$ are prepared within MPD using the principle of an accurate adjustment of flow rates of individual components. The flow rates are tuned with a supercritical fluid pump (SCP) for $CO_2$, a set of flow controllers and stainless steel capillary tubes used as elements with defined pressure drop at a given flow rate. The temperatures and pressures of flowing gases are varied and measured with the help of a set of thermostatic baths, backpressure regulators, pressure transducers and precise temperature probes.

Preparation of mixtures by fine tuning of flow rates has an important advantage for the investigation of nucleation, which is highly sensitive to a presence of impurities. The gaseous mixtures can flow through MPD and the expansion chamber for longer time (even several hours if needed). This helps to equilibrate the adsorption of gas molecules on the walls of the MPD system and the expansion chamber and to flush away possible impurities. We note that the system is carefully evacuated to a low vacuum before each test. Most of the MPD tubing is made from stainless steel tubes with outer diameter of ¼ inch and wall thickness of 0.036 inch and 0.049 inch for maximum pressures of 350 and 515 bar, respectively.

At the current state, the MPD system consists of three main parts: the $CO_2$ branch, the gaseous branch with carrier gas (nitrogen or argon), and the final section with mixed gases containing the expansion chamber. In this study, the design and testing measurements of the $CO_2$ branch are described in detail.

### 3.1 Gaseous branch with nitrogen or argon

The carrier gas, either nitrogen or argon, is delivered from the 50 liter gas cylinder with a maximum load pressure of 200 bar. The output pressure is varied with the pressure regulating valve and measured with 300 bar pressure transducer of 33 series from the Keller[a] manufacturer. The flow rates are set by two thermal mass flow meters of F-23xM series designed for high pressure operation up to 400 bar from the Bronkhorst[b] manufacturer. The two flow meters have different ranges from 0.01 to 0.50 $l_n$/min and from 0.04 to 2.0 $l_n$/min. Unit $l_n$/min denotes flow rate of gas calculated at the normal conditions, i.e. at the density of given gas at temperature of 0 °C and atmospheric pressure of 101 325 Pa. Nominal flow of 1.0 $l_n$/min corresponds to mass flow of 75 g/h and 107 g/h in case of nitrogen and argon, respectively. The gaseous branch is designed to operate with outlet pressures from 38 up to 160 bar.

The carrier gas is mixed with carbon dioxide coming from the $CO_2$ branch in ¼ inch tubes placed inside a thermostatic bath set to temperatures between 5 and 150 °C depending on the conditions for nucleation of droplets investigated within the expansion chamber.

[a] www.keller-druck.com

[b] www.bronkhorst.com

### 3.2 $CO_2$ branch

The $CO_2$ branch allows fine adjustment of the $CO_2$ mass flow up to 180 g/h at the outlet pressures of the supercritical fluid from 74 to 150 bar. A simplified scheme of the setup is provided in Fig. 1. The saturated liquid $CO_2$ coming from cylinder (1) equipped with a dip tube enters the system at an approximate pressure of 63 bar corresponding to the vapor pressure of $CO_2$ at room temperature 24 °C. The inlet pressure is detected with pressure gauge (2). The saturated liquid $CO_2$ is compressed by supercritical fluid pump (SCP)[c] (3) up to pressure of 340 bar representing the allowed maximum set on the safety valve after the pump. Typical operating pressure of SCP is around 270 bar. The two-piston supercritical pump operates at a constant flow rate of up to 24.0 ml/min of the saturated liquid $CO_2$. In order to eliminate the pressure fluctuations caused by the operation of the supercritical pump, the system is equipped with a high-pressure buffer tank (accumulator – AC) with inner volume of 150 ml (item 4 in Fig. 1). Part of the supercritical $CO_2$ from the buffer tank returns back to the supercritical pump inlet through a backflow capillary tube (CAP 2) with inner diameter of 0.10 mm and length of 0.85 m. Within the backflow capillary tube, the supercritical $CO_2$ is adiabatically throttled to the saturated liquid state. At the SCP operating pressure of 270 bar, the mass flow through the capillary tube is around 340 g/h. The backflow capillary tube allows setting SCP approximately to the middle of its range, i.e. to the flow rate around 12.0 ml/min of the saturated liquid $CO_2$ corresponding to the mass flow of 511 g/h. The mass flow going to the rest of MPD and the expansion chamber varies typically between 50 g/h and 180 g/h. This part of the $CO_2$ flow is expanded to the required outlet pressure in two stainless steel capillary tubes (CAP 3 and CAP 4) placed in a thermostatic bath (5). The thermostatic bath with water is set to the temperature around 70 °C, i.e. far above the critical temperature of $CO_2$. The capillary tubes CAP 3 and CAP 4 have equal inner diameters of 0.1 mm and lengths of 3.5 m and 7.5 m, respectively. The capillary tubes CAP 3 and CAP 4 deliver mass flow of 112 g/h and 73 g/h at the SCP operating pressure of 270 bar and the $CO_2$ branch outlet pressure of 100 bar. Each of the two capillary tubes can be open / closed by a needle valve. Temperature along the system is measured with three small Pt100 resistance thermometers (15–17) attached to the outer surface of the ¼ inch tubing.

In the final design of MPD, the $CO_2$ branch will end at the location of the temperature sensor No. 17 where it will be connected to the final section with mixed gases. For the individual testing of the $CO_2$ branch, the setup was additionally equipped with two pressure transducers Keller[a] of the 33 type with the range from 1 to 300 bar and accuracy better than 0.45 bar (items 7 and 8 in Fig. 1). The transducers measured pressures after the buffer tank AC (7) and after the isothermal capillary tubes CAP 3 and CAP 4 (8). The mass flow through the system was measured with the Coriolis mass flow meter

[c] http://ssihplc.com/hplc-pumps/supercritical-24-pump-cf/

mini CORI-FLOW[b] (18) with the range from 2 to 200 g/h and accuracy better than 0.2% of reading. The flow meter operates at pressures from 3 up to 140 bar and can measure both the gases and the liquids. The pressure drop on the flow meter is between 1.7 and 24 bar depending on the flow rate and the fluid properties, i.e. mostly compressible/incompressible fluid. The testing $CO_2$ branch is terminated by an exhaust capillary tube CAP 5 where the supercritical $CO_2$ is adiabatically throttled to the atmospheric pressure. The capillary tube has inner diameter of 0.1 mm and length of 0.77 m.

The supercritical fluid pump requires approximately 30 minutes of flushing with pure $CO_2$ before its operation in order to remove air humidity and other impurities from its internal volumes. The $CO_2$ branch is therefore equipped with a flushing capillary tube CAP 1 with inner diameter 0.1 mm and a length of 0.33 cm. In CAP 1, the saturated liquid $CO_2$ is throttled from the vapor pressure of 63 bar to the atmospheric pressure by the maximum proposed flow rate below 150 g/h. Fig. 2 shows photo of the main part of the $CO_2$ branch in the laboratory.

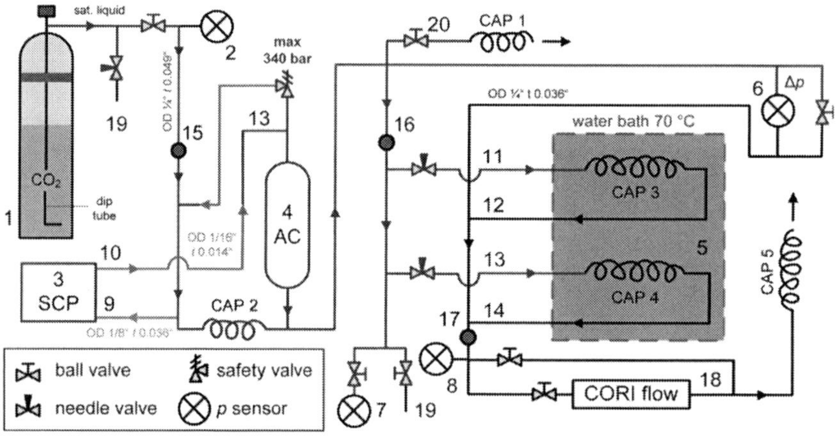


**Fig. 1.** Scheme of the $CO_2$ branch of the MPD system. Description: 1 – liquid $CO_2$ cylinder, 2 – pressure gauge, 3 – supercritical fluid pump (SCP), 4 – accumulator (AC), 5 – thermostatic bath, 6 – differential pressure transducer, 7 – absolute pressure transducer – high *p*, 8 – absolute pressure transducer – low *p*, 9 – SCP inlet, 10 – SCP outlet, 11 – CAP 3 inlet, 12 – CAP 3 outlet, 13 – CAP 4 inlet, 14 – CAP 4 outlet, 15 – *T* sensor (from the $CO_2$ cylinder), 16 – *T* sensor (after AC), 17 – *T* sensor (after bath), 18 – mass flow meter, 19 – connection for vacuum pump, 20 – flushing valve.

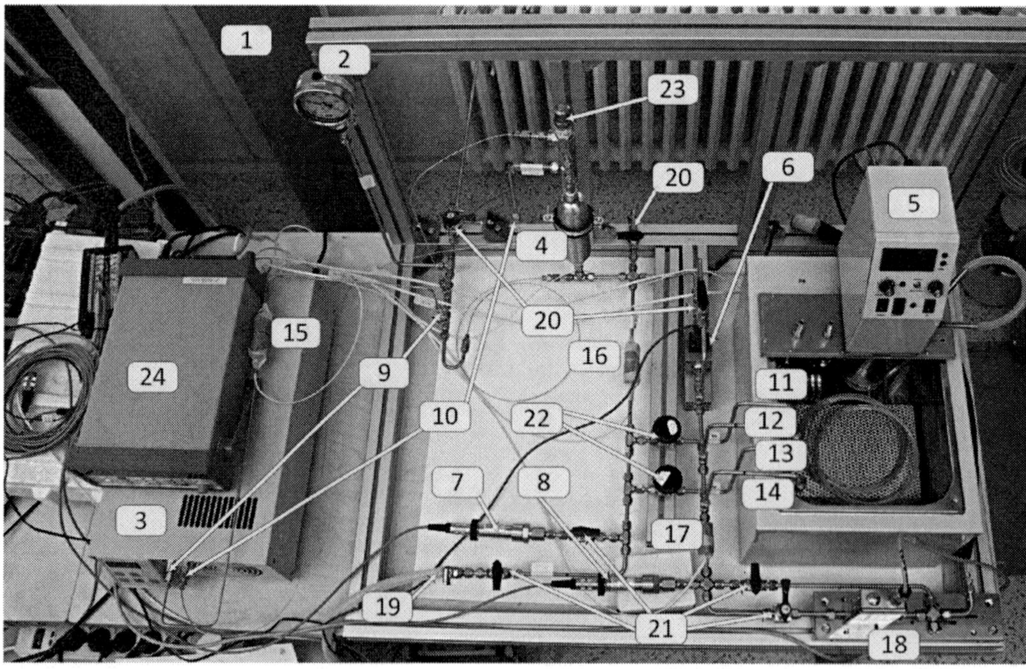


**Fig. 2.** Photo of the $CO_2$ branch of the MPD system. Description: points from 1 to 20 are same as in Fig. 1; 21 – ball valves, 22 – needle valves, 23 – safety valve, 24 – DAQ system.

### *3.2.1 Capillary tubes on $CO_2$ branch*

A set of various capillary tubes CAP 1 to CAP 5 operating at different flow regimes is used on the $CO_2$ branch of MPD; see Fig. 1. Table 1 summarizes parameters of all capillary tubes and their designed flow regimes. Capillary tubes CAP 1, CAP 2 and CAP 5 operate at adiabatic flow conditions, while the flow in CAP 3 and CAP 4 placed inside the thermostatic bath can be considered as isothermal. In the exhaust capillary tubes CAP 1 and CAP 5, the two-phase flow of vapor-liquid mixture occurs as $CO_2$ is throttled from pressures above 60 bar down to the atmospheric pressure. In case of remaining three capillary tubes, the $CO_2$ fluid stays under the supercritical conditions along the entire length of the capillary tube.

**Table 1.** List of capillary tubes (CAP) employed on the $CO_2$ branch and their operating flow conditions.

| CAP | 1 | 2 | 3 | 4 | 5 |
|---|---|---|---|---|---|
| $d$ [mm] | 0.1 | 0.1 | 0.1 | 0.1 | 0.1 |
| $L$ [cm] | 33 | 85 | 350 | 750 | 75 |
| $T_{in}$ [°C] | 25 | 25 | 40 | 40 | 38 |
| $p_{in}$ [bar] | 64 | 220 | 220 | 220 | 100 |
| $p_{out}$ [bar] | 1 | 64 | 100 | 100 | 1 |
| flow condition[1] | A | A | iT | iT | A |
| $CO_2$ state[2] | TP | S | S | S | S + TP |

[1] A – adiabatic, iT – isothermal
[2] TP – two-phase vapor + liquid, S – supercritical

The flow performance of the isothermal capillary tubes CAP 3 and CAP 4 was predicted with a simple one-dimensional numerical model. The defined pressure drop over the capillary tube from the inlet pressure $p_{in}$ to the outlet pressure $p_{out}$ was divided in $N$ segments, i.e. $\Delta p_{i-1} = p_{i-1} - p_i$ with $i = 1 \div N$. Length of the $i$-th segment of the capillary tube $\Delta z_i$ could be obtained from the following equation

$$\Delta z_i = \frac{p_{i-1} - p_i - \frac{\dot{m}^2}{A^2}\left(v_i - v_{i-1}\right)}{\frac{4 f_{i-1} \dot{m}^2 \left(v_i + v_{i-1}\right)}{\pi^2 d^5}}, \qquad (1)$$

where $\dot{m}$ is the mass flow in kg/s, $d$ is the capillary tube inner diameter, $A$ is the cross section area $A = \pi d^2/4$, $v$ is the specific volume in m³/kg and $f$ is the friction factor calculated, e.g., from the Colebrook equation [5]

$$\frac{1}{\sqrt{f}} = -2\log\left(\frac{\varepsilon}{3.7d} + \frac{2.51}{Re\sqrt{f}}\right). \qquad (2)$$

In equation (2), $Re$ is the Reynolds number and $\varepsilon$ is the inner wall roughness of the capillary tube. In all our calculations, a relative wall roughness $\varepsilon/d = 0.003$ was considered. Thermophysical properties of $CO_2$, i.e. the density $\rho = 1/v$ and viscosity, were taken as functions of pressure $p_i$ at a constant temperature. The properties at each computational grid point $i$ were calculated using the property packages REFPROP [6] and TREND [7].

The two-phase flow of vapor-liquid mixture within an adiabatic capillary tube was predicted by the numerical model similar to that employed in our previous studies [8-10]. More details about the design of the capillary tubes and the flow behavior of $CO_2$ at various inlet and outlet conditions will be provided in other article.

## 4 Preliminary tests

Functionality of the $CO_2$ branch was successfully tested in this study. Signals from the pressure transducers, the mass flow meter, and the temperature sensors were recorded by the automatic data acquisition system (DAQ) Keysight[d] 34970A (item 24 in Fig. 2).

Fig. 3 shows temperatures along the system recorded during the flushing of SCP. The saturated liquid $CO_2$ coming from the cylinder flows through SCP and leaves the $CO_2$ branch via the exhaust capillary tube CAP 1. Three temperature sensors T CAP 1 to 3 were additionally placed on the capillary tube in the distances 140, 60, and 4 mm from the capillary tube outlet. As can be seen in Fig. 3, the temperature drops along the capillary tube from 24 °C to 7 °C due to the evaporation of liquid $CO_2$ during the adiabatic throttling. We note that the outlet temperature is significantly below the minimum measured temperature of 7 °C. The recorded temperatures along the capillary tube have only informative character as their uncertainties are quite high due to the relatively low flow rates, very small inner diameter of the capillary tube of 0.1 mm compared to its wall thickness around 0.75 mm and poor thermal contact between the temperature probes and the capillary tube with outer diameter of only 1.6 mm. Nevertheless, the measured temperatures show physically correct trend.

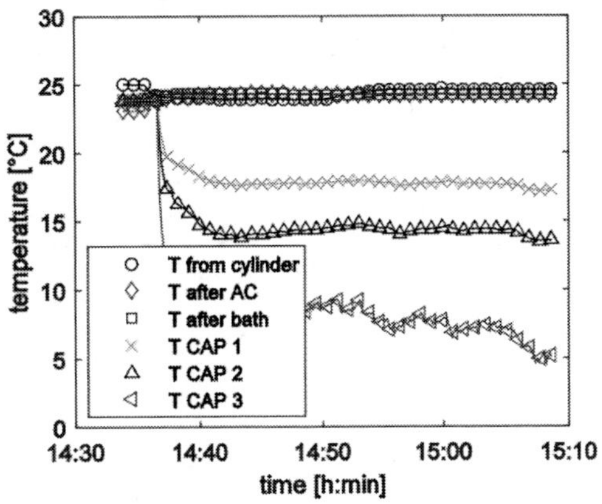


**Fig. 3.** Temperatures along the system and the exhaust capillary tube CAP 1 during the flushing of SCP with the saturated liquid $CO_2$.

Fig. 4 shows typical conditions during the startup of SCP. The flow rate through the pump was set to

[d] http://www.keysight.com

 DOI: 10.1051/epjconf/201714302140

7.0 ml/min of the saturated liquid $CO_2$ in this case. As can be seen, the pressure at the pump outlet, i.e. after the buffer tank AC, is gradually increasing until it stabilizes at a final value of 180 bar. The pressure after the isothermal capillary tubes CAP 3 and CAP 4 together with the mass flow are accordingly increasing to the equilibrated conditions of 77 bar and 105 g/h, respectively. The system got stabilized after approximately 30 minutes from the pump startup due to the relatively low flow rate of 7.0 ml/min set on the pump compared to the proposed operating conditions at 12.0 ml/min.

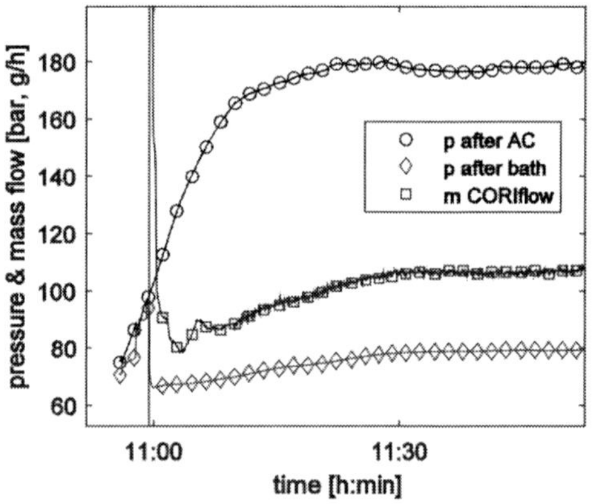


**Fig. 4.** Pressure in bar after the buffer tank (AC) and after the thermostatic bath with capillary tubes CAP 3 and CAP 4 and the mass flow in g/h measured by mini CORI-FLOW after the startup of SCP with the set point of 7.0 ml/min.

Fig. 5 shows temperatures measured along the $CO_2$ branch and additionally at the outer surface of the exhaust capillary tube CAP 5 during the startup of the supercritical pump. The flow conditions correspond to the case depicted in Fig. 4. Temperature in the thermostatic bath with capillary tubes CAP 3 and CAP 4 was set to a constant value of 55 °C. It is important to note that the tubing of the $CO_2$ branch was not insulated during the preliminary tests. Consequently, the temperature after the thermostatic bath varied relatively slowly due to the heat losses into the ambient. The temperature sensors T CAP 1 to 3 were placed on the outer surface of capillary tube CAP 5 in a distance of 229, 74, and 3 mm from its outlet, respectively. Similarly as in case shown in Fig. 3, the temperatures measured along the capillary tube have only informative character. Fluctuations of temperature T CAP 1 are caused by a presence of water ice formed from the condensed air humidity close to the capillary tube outlet.

Performance of the $CO_2$ branch was investigated at several stabilized flow conditions. Some of these tests are summarized in Table 2. First two cases were measured with SCP turned off. The maximum pressure in the system was given by the vapor pressure of $CO_2$ at the room temperature. First case represents ordinary flushing of the $CO_2$ branch before the startup of SCP and corresponds to the results shown in Fig. 3. In the second case and the third case, the exhaust capillary tube for flushing CAP 1 was placed after the mass flow meter instead of the exhaust capillary tube CAP 5 shown in Fig. 1. The third case was measured with the flow rate of the saturated liquid $CO_2$ through SCP of 5.0 ml/min by both isothermal capillary tubes CAP 3 and CAP 4 open. The thermostatic bath (5) was set to 40 °C in this case. The mass flow through the system and the pressure after SCP stabilized at 87.6 g/h and 137 bar, respectively.

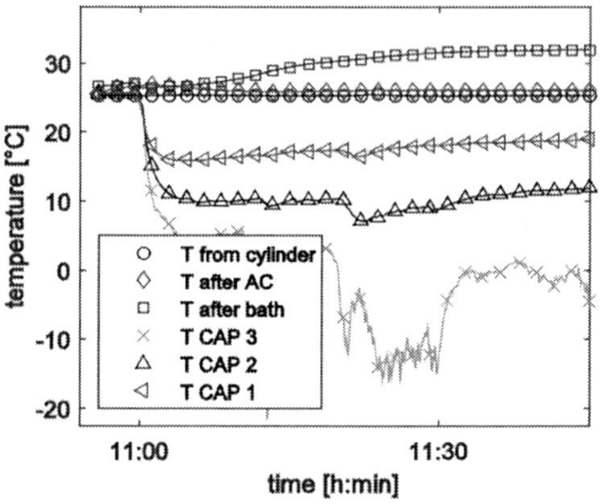


**Fig. 5.** Temperatures in the system measured after the startup of SCP with the flow rate of the saturated liquid $CO_2$ set to 7.0 ml/min. Temperatures T CAP 1 to 3 were measured at the outer surface of the exhaust throttling capillary tube CAP 5.

Last two cases in Table 2 were measured at the proposed operating conditions, i.e. at the flow rate set to 12.0 ml/min on SCP and the thermostatic bath set to 70 °C. The capillary tube CAP 5 was employed as the exhaust tube after the mass flow meter in both cases. In the first case, the isothermal capillary tube CAP 3 with the length of 3.5 m was open while in the second case the capillary tube CAP 4 with the length of 7.5 m was active. Results in Table 2 show an interesting phenomenon. The pressure drop over the shorter capillary tube CAP 3 equal to 184 $bar_d$ was higher than in case of the longer capillary tube CAP 4 equal to 158 $bar_d$. Consequently, the absolute pressure before the mass flow meter was 10 bars higher in case of open capillary tube CAP 4 than in case of open capillary tube CAP 3. As a result, the mass flow through the system was higher by 13.7 g/h in case of open capillary tube CAP 4. The result is in contradiction with the numerical model described in section 3.1.1. At the flow conditions provided in Table 2, the numerical model predicts maximum mass flow of 112 g/h and 74 g/h for capillary tube CAP 3 and CAP 4, respectively. Possible explanation for this behavior can be that the capillary tubes do not have a constant inner diameter of 0.10 mm. An increase of the inner diameter to 0.11 mm results in the mass flow predicted by the model of 144 g/h for CAP 3 and 95 g/h for CAP 4. The other explanation could be that the flow performance of the $CO_2$ branch is strongly affected by the exhaust capillary tube CAP 5, i.e. the last element of the $CO_2$ system where the pressure drops from approximately 100 bar to the atmospheric pressure. The exhaust capillary tube behaves

**Table 2.** Description of stabilized flow conditions within the $CO_2$ branch at different set points on SCP and different configurations of the exhaust tube to atmosphere

| Pump set point | iso-*T* CAPS | *p* after AC | *p* after bath | $\dot{m}$ CORI-FLOW | *T* from cylinder | *T* after AC | *T* set on bath | *T* after bath | *T* CAP 1 | *T* CAP 2 | *T* CAP 3 | Exhaust tube |
|---|---|---|---|---|---|---|---|---|---|---|---|---|
| [ml/min] | | [bar] | [bar] | [g/h] | [°C] | [°C] | [°C] | [°C] | [°C] | [°C] | [°C] | |
| off[1] | both | 63.21 | 63.15 | 0.00 | 24.6 | 24.2 | – | 24.2 | 17.7 | 14.3 | 7.3 | |
| off[2] | both | 63.39 | 38.21 | 25.05 | 25.0 | 24.2 | – | 24.5 | 17.7 | 16.4 | 10.5 | CAP 1 |
| 5 | both | 137.19 | 70.42 | 87.55 | 25.6 | 25.9 | 40 | 28.7 | – | – | – | |
| 12 | CAP 3 | 283.45 | 99.15 | 135.43 | 24.7 | 26.2 | 70 | 35.7 | 23.5 | 16.2 | 0.7 | CAP 5 |
| 12 | CAP 4 | 268.36 | 110.64 | 149.12 | 24.8 | 26.0 | 70 | 36.3 | 23.7 | 18.2 | 1.3 | |

[1] Flushing of the system – exhaust capillary tube CAP 1 placed after the flushing valve (item 20 in Fig. 1)
[2] Exhaust capillary tube CAP 1 placed after the mini CORI-FLOW mass flow meter instead of CAP 5

similarly as a chocked nozzle at these conditions. The detected performance of the isothermal capillary tubes CAP 3 and CAP 4 combined with the exhaust capillary tube CAP 5 require further investigation. However, in the final design of MPD, the mass flow meter and the exhaust capillary tube will not be present. A new test is therefore planned in the close future. The exhaust capillary tube CAP 5 will be removed from the system. The outlet of the mass flow meter will be connected back to the $CO_2$ cylinder outlet (item 1 in Fig. 1) after pressure gauge (2). In this case, the pressure after the mass flow meter will be equal to the vapor pressure of $CO_2$, i.e. approximately 63 bar, which is relatively close to the designed conditions when the outlet pressure from the $CO_2$ branch shall vary between 74 and 150 bar. Another option is to replace the current capillary tubes CAP 3 and CAP 4 with another capillary tubes having the same dimensions and to verify the results measured in this study. The inner diameter can significantly vary along the capillary tube length [9,10].

## 5 Conclusions

A $CO_2$ branch representing the main part of a unique mixture preparation device (MPD) was designed, assembled and successfully tested in this study. The MPD system is intended for preparation of $CO_2$-rich mixtures in a sense of accurate setting of flow rates of individual components of the gaseous mixture. MPD is part of the larger in-house designed experimental apparatus for investigation of nucleation of droplets in CCS relevant systems.

The $CO_2$ branch comprises of several high-tech components such as supercritical fluid pump, accurate pressure transducers and Coriolis mass flow meter. A set of various stainless steel capillary tubes are used in the $CO_2$ branch for fine tuning of the output mass flow of the supercritical $CO_2$. Parameters of the capillary tubes were determined with the help of a one-dimensional numerical model for isothermal and adiabatic capillary tubes. Preliminary measurements proved functionality of the $CO_2$ branch. However, some further tests are required, especially, in order to clarify performance of the isothermal capillary tubes placed inside a thermostatic bath.

## Acknowledgement

The research leading to these results has received funding from the Norwegian Financial Mechanism 2009–2014 under Project Contract no. 7F14466 and from the institutional support RVO:61388998.

## References

1. IPCC Special Report: Carbon Dioxide Capture and Storage – Summary for Policymakers. Intergovernmental Panel on Climate Change, Eighth Session of IPCC Working Group III, Montreal, (Canada) 22-24 September 2005
2. C.R.T. Wilson: Condensation of Water Vapour in the Presence of Dust-free Air and other Gases. Phil. Trans. R. Soc. A., London 189 (1897) 265-307.
3. J. Wölk, R. Strey, J. Phys. Chem. B **105**, 11683-11701 (2001)
4. B.M. Cwilong, M.H. Edwards, Phys. Rev. **85**, 380-381 (1952)
5. C.F. Colebrook, J. Institution of Civil Engineers **11**, 133-156 (1939)
6. E.W. Lemmon, M.L. Huber, M.O. McLinden, REFPROP, NIST Standard Reference Database 23 Version 9.1 (2013)
7. R. Span, T. Eckermann, S. Herrig, S. Hielscher, A. Jäger, M. Thol, TREND. Thermodynamic Reference and Engineering Data 2.0, Lehrstuhl Fuer Thermodynamik, Ruhr-Universitaet Bochum, Bochum, Germany, 2015.
8. V. Vacek, V. Vinš, Int. J. Thermophys. **28**, 1490-1508 (2007)
9. V. Vinš, V. Vacek, J. Chem. Eng. Data **54**, 2395-2403 (2009)
10. V. Vacek, V. Vinš, Int. J. Thermophys. **30**, 1179-1196 (2009)

# Temperature and pressure correlation for volume of gas hydrates with crystal structures sI and sII

Václav Vinš[1,*], Andreas Jäger[2], Sebastian Hielscher[3], Roland Span[3], Jan Hrubý[1], and Cornelia Breitkopf[2]

[1]Institute of Thermomechanics of the CAS, Dolejškova 1402/5, Prague 8, 182 00, Czech Republic
[2]Faculty of Mechanical Science and Engineering, Institute of Power Engineering, Technische Universität Dresden, Helmholtzstr. 14, 01069 Dresden, Germany
[3]Thermodynamics, Ruhr-Universität Bochum, Universitätsstr. 150, 44780 Bochum, Germany

**Abstract.** The temperature and pressure correlations for the volume of gas hydrates forming crystal structures sI and sII developed in previous study [Fluid Phase Equilib. 427 (2016) 268-281], focused on the modeling of pure gas hydrates relevant in CCS (carbon capture and storage), were revised and modified for the modeling of mixed hydrates in this study. A universal reference state at temperature of 273.15 K and pressure of 1 Pa is used in the new correlation. Coefficients for the thermal expansion together with the reference lattice parameter were simultaneously correlated to both the temperature data and the pressure data for the lattice parameter. A two-stage Levenberg Marquardt algorithm was employed for the parameter optimization. The pressure dependence described in terms of the bulk modulus remained unchanged compared to the original study. A constant value for the bulk modulus $B_0 = 10$ GPa was employed for all selected hydrate formers. The new correlation is in good agreement with the experimental data over wide temperature and pressure ranges from 0 K to 293 K and from 0 to 2000 MPa, respectively. Compared to the original correlation used for the modeling of pure gas hydrates the new correlation provides significantly better agreement with the experimental data for sI hydrates. The results of the new correlation are comparable to the results of the old correlation in case of sII hydrates. In addition, the new correlation is suitable for modeling of mixed hydrates.

## 1 Introduction

Our team is developing advanced models for thermophysical properties and phase behavior of fluid and solid systems relevant especially in carbon capture and storage (CCS) technologies. Most of the research is related to the modeling of gas hydrates, i.e. non-stoichiometric solid mixtures of water and various gases. Gas hydrates form so called clathrate structures, in which the hydrogen bonded water molecules build a crystal lattice with cavities containing the gas molecules [1]. Nowadays, gas hydrates are being investigated for several reasons, e.g., in the natural gas industry where they can cause pipeline blockage and other operational failures, as potential energy resource in the form of methane hydrates at the sea bottom or in relation to CCS technologies [2].

A new thermodynamic model for eight gas hydrate formers relevant mostly in CCS has recently been developed by our team [3-5]. The model is based on the van der Waals and Platteeuw approach [6] combined with highly accurate multiparameter equations of state for other phases than the hydrate [7,8]. Pure gas hydrates (water with a single gas species) of carbon dioxide, argon, oxygen, nitrogen, methane, carbon monoxide, ethane and propane can be predicted. An important advantage of the new model is that it is capable to predict a large variety of phase equilibria including those with pure solid phases, i.e. water ice and solid $CO_2$ (dry ice), over wide temperature and pressure ranges with very good accuracy. Current activities are focused on a further extension of the pure hydrate model to hydrate mixtures. New complex phase equilibria algorithms, e.g., for three-phase and four-phase equilibria including both fluid and solid phases, are being developed for this purpose [9]. The pure hydrate model also needs to be modified such that it allows consistent modeling of gas hydrates and phase equilibria for various overall compositions of multicomponent aqueous mixtures. Among other, these modifications include a revision of the current correlation for the volume of gas hydrates.

In this study, the volume of gas hydrates forming cubic crystal structures of type sI and sII was investigated. The correlation for the lattice parameter employed in the pure hydrate model [3] was revised and modified such that it can be used for modeling of mixed hydrates. The reference conditions, i.e. the reference temperature $T_0$ and the reference pressure $p_0$, of the new volume correlation were set constant for both hydrate structures. Their values $T_0 = 273.15$ K and $p_0 = 1$ Pa are the same as the reference state for the Gibbs energy and enthalpy of hydrate employed in the model for mixed

* Corresponding author: vins.vaclav@seznam.cz

hydrates [9]. Parameters of the new correlation for 11 sI hydrates and 18 sII hydrates were correlated to the experimental data for both the temperature and the pressure dependence of the volume of gas hydrates using a two-stage Levenberg–Marquardt algorithm [10,11].

## 2 Volume of gas hydrates

In case of sI and sII hydrates, the volume can be determined from the lattice parameter $a$ of the unit cell as both structures form a cubic crystal system. In the hydrate models based on the van der Waals and Platteeuw approach [6], a hypothetical crystal structure called empty $\beta$-lattice is considered. This structure is formed by water molecules connected through hydrogen bonds and does not contain any gas (guest) molecules in the cavities. Knowing the lattice parameter and the number of water molecules $N_w$ forming the unit cell, which is constant for each hydrate structure, the molar volume of the empty $\beta$-lattice can be determined as follows

$$v^{\beta}\left[\mathrm{m^3\,mol^{-1}}\right]=\left(a\left[\mathrm{\AA}\right]10^{-10}\right)^3\frac{N_{\mathrm{AV}}\left[\mathrm{mol^{-1}}\right]}{N_{\mathrm{w}}}, \quad (1)$$

where $N_{AV}$ stands for the Avogadro constant. The molar volume of hydrate can be calculated from the molar volume of the empty $\beta$-lattice in the following manner

$$v=v^{\beta}\left(1+\sum_i \upsilon_i \sum_J \theta_{i,J}\right)^{-1}. \quad (2)$$

In eq. (2), $\upsilon_i$ is the number of cavities of type $i$ divided by the number of water molecules in a unit cell $N_w$ and $\theta_{i,J}$ is the fractional occupancy of cavity $i$ by guest molecule of type $J$. The cage occupancy $\theta_{i,J}$ needs to be determined from a convenient hydrate model, e.g., ref. [3-5,12,13] or experimentally by diffraction methods [14-16].

**Table 1.** Summary of experimental data for the lattice parameter of sI hydrates

| Guest | *T* – data | | *p* – data | |
|---|---|---|---|---|
| | *N* | Ref. | *N* | Ref. |
| EtO | 28 | [20,21] | - | |
| $CO_2$ | 70 | [16,17, 22-26] | - | |
| $C_2H_6$ | 11 | [17] | 10 | [27] |
| $0.85CH_4 + 0.15\ CO_2$ | 4 | [17] | - | |
| $0.47\ CH_4 + 0.53\ C_2H_6$ | 2 | [17] | - | |
| TMO | 8 | [28] | - | |
| CO | 1 | [29] | - | |
| $H_2S$ | 1 | [30,31] | - | |
| $CH_4$ | 53 | [16,31-36] | 52 | [37,38] |
| Xe | 26 | [16,23,29] | 16 | [27,39] |
| $N_2$ (sI) | - | | 6 | [14,15] |

Tables 1 and 2 summarize experimental data points available for the lattice parameter of sI hydrates and sII hydrates, respectively. The annotation "$T$ – data" denotes data for the temperature dependence of the lattice parameter measured at low pressures below 1 MPa; typically at atmospheric pressure. "$p$ – data" marks data measured at higher pressures up to 3 GPa. Tables 1 and 2 contain also data for several binary mixed hydrates measured mostly by Hester et al. [17].

**Table 2.** Summary of experimental data for the lattice parameter of sII hydrates

| Guest | *T* – data | | *p* – data | |
|---|---|---|---|---|
| | *N* | Ref. | *N* | Ref. |
| Air | 5 | [29,40] | 4 | [15] |
| Ar | 1 | [29] | - | |
| Kr | 3 | [30] | | |
| $O_2$ | 1 | [29] | 9 | [15] |
| $N_2$ | 18 | [16,29] | 18 | [14,15] |
| TMO | 9 | [28] | - | |
| $C_3H_8$ | 21 | [17,33,41] | - | |
| $0.60\ CH_4 + 0.4\ C_3H_8$ | 3 | [17] | - | |
| $0.30\ C_2H_6 + 0.7\ C_3H_8$ | 5 | [17] | | |
| $0.182\ CO_2 + 0.818\ C_3H_8$ | 5 | [17] | | |
| $0.876\ CH_4 + 0.124$ i-$C_4H_{10}$ | 7 | [17] | | |
| Natural gas | 3 | [17] | - | |
| $0.95\ CH_4 + 0.05\ C_5H_{10}O$ | 10 | [17] | | |
| THF | 18 | [21,42] | | |
| $0.82\ CH_4 + 0.18\ C_2H_6$ | 12 | [43] | | |
| THF + $H_2S$ | 1 | [44] | | |
| THF + Xe | - | | 13 | [27] |
| $CH_4$ (sII) | - | | 13 | [45] |

### 2.1 Correlations for the temperature dependence

Most of the hydrate models, e.g., by Klauda and Sandler [46] or Martín and Peters [47], use correlations for the thermal expansion of lattice parameter determined by Tse et al. on the basis of their experimental measurements with ethylene oxide (EtO) hydrates for structure sI [20,21]

$$a(T)=11.835+2.2173\cdot10^{-5}T+2.2415\cdot10^{-6}T^2, \quad (3)$$

and with tetrahydrofuran (THF) hydrates [21] for structure sII

$$a(T)=17.130+2.249\cdot10^{-4}T+2.013\cdot10^{-6}T^2-1.009\cdot10^{-9}T^3. \quad (4)$$

Both correlations provide reasonable results over a wide temperature range from 20 K up to 260 K. However, application of eqs. (3) and (4) to the modeling of various hydrates different from ethylene oxide and tetrahydrofuran represents a rather rough simplification. As it is shown later in Tables 5 and 6 in section 3.1, the lattice parameter at a given temperature and pressure may significantly vary for various hydrate formers. The reason is that the interactions between the guest molecules and the water cavities and subsequently the volume of the unit cell depend on the guest type.

A universal and accurate correlation for thermal expansion of sI and sII hydrates was developed by Hester et al. [17]. Based on the literature data and their own experimental measurements for a large variety of hydrates including binary mixed hydrates, Hester et al. found out that the thermal expansion can be considered dependent only on the hydrate structure. Consequently, the temperature dependence of the relative lattice parameter $a/a_0$ can be universally described by the following function for all hydrates forming the same crystal structure.

$$\frac{a(T)}{a_0} = \exp\begin{bmatrix} \alpha_1 (T - T_0) + \\ \alpha_2 (T - T_0)^2 + \alpha_3 (T - T_0)^3 \end{bmatrix} \quad (5)$$

In eq. (5), $a_0$ denotes the lattice parameter at the reference temperature $a(T_0)$. Introduction of the reference lattice parameter in eq. (5) made the correlation by Hester et al. universal for all sI and sII hydrates unlike the correlations by Tse et al. [20,21], eqs. (3) and (4), which are based solely on the data for two specific hydrates.

Hansen et al. [16] have recently measured new accurate data for the thermal expansion of hydrogenated and deuterated hydrates of $CO_2$, $CH_4$, Xe and $N_2$ over wide temperature ranges from 10 K up to the stability limit around 280 K. A high-pressure cell was used for the measurements at the temperatures above 200 K. The data were correlated by polynomial fits of up to the forth order in case of $CO_2$, $CH_4$, and $N_2$ hydrates. The data for Xe hydrates and for $CO_2$ hydrates (for comparison) were correlated by Hansen et al. [16] by a modified Einstein model for thermal expansion.

$$a(T) = a_0 + (e_0 + e_1 T)/[\exp(\theta_E / T) - 1] \quad (6)$$

In eq. (6), $a_0$ is the lattice parameter at 0 K, $e_0$ and $e_1$ are coefficients of the internal energy of the crystal, and $\theta_E$ is a characteristic Einstein temperature. Correlation (6) provided better results only for Xe hydrates compared to the polynomial fits. Based on their new data, Hansen et al. [16] stated that the thermal expansion depends on the hydrate former and that a universal model for all hydrates forming the same crystal structure proposed by Hester et al. [17] can hardly be justified. On the other hand, major differences between the new data of Hansen et al. and the universal correlation (5) occur mostly in the high temperature region above approximately 220 K. The correlation by Hester et al. [17] was based mostly on the low temperature data as the data above 230 K were rather limited at that time. As a consequence, results of correlation (5) by Hester et al. [17] are more or less pure extrapolation to temperatures above 230 K. In this work, we employed the universal approach of Hester et al. [17], because the new temperature and pressure correlation for the volume of gas hydrates is intended for the modeling of various gas hydrates with large differences in the amount and quality of experimental data for the lattice parameter.

## 2.2 Correlations for the pressure dependence

The data situation for the pressure dependence of the lattice parameter corresponding to the isothermal compressibility of hydrates is significantly worse compared to the thermal expansion data; see the amount of data points and the number of references indicated with "$p$ – data" in Tables 1 and 2.

Most of the hydrate models use various quadratic or cubic functions of pressure in the correlations of hydrate volume, e.g., the correlation employed by Klauda and Sandler [46] based on the data for methane hydrates

$$a(T,p) = \sqrt[3]{a(T)^3 - 8.006 \cdot 10^{-9} p + 5.448 \cdot 10^{-12} p^2}, \quad (7)$$

where $a(T)$ can be obtained from eqs. (3) or (4) and $p$ denotes the pressure in MPa. The main weakness of similar correlations is that they are valid over limited pressure ranges and their extrapolating behavior is rather questionable, e.g., the compressibility of hydrates can increase with increasing pressure or become negative at a certain pressure.

In the accurate model for natural gas hydrates by Ballard and Sloan [1,12], a constant compressibility $\kappa$ was considered.

$$a(T,p) = a(T)\exp[-\kappa(p - p_0)] \quad (8)$$

However, as it is discussed in our previous work [3], Ballard and Sloan correlated the compressibility mostly to the hydrate three-phase equilibrium data and not to the crystallographic data for the lattice parameter. Consequently, in the model of Ballard and Sloan [1,12] the compressibility was employed as a sort of correcting parameter improving the predictive ability of the hydrate model at high pressures. Values of $\kappa$ differ by several orders of magnitude for various hydrate formers in this case.

In our model for pure $CO_2$ hydrates [13], a non-constant compressibility was introduced.

$$a(T,p) = a(T)\left[1 - \frac{\kappa_1 (p - p_0)}{1 + \kappa_2 (p - p_0)}\right]. \quad (9)$$

The compressibility given by eq. (9) is monotonically decreasing to zero at an infinite pressure, which represents physically reasonable behavior. Nevertheless, fitting two parameters $\kappa_1$ and $\kappa_2$ for the compressibility of various hydrate formers appeared as rather difficult task according to unfavorable situation with the experimental data.

In our previous work [3-5] focused on the development of model for eight pure gas hydrates relevant in CCS, it was shown that the pressure dependence of the volume of sI and sII gas hydrates can be well described with the Murnaghan [18] equation of state for solids. The Murnaghan equation is a simplified version of the Birch-Murnaghan [19] equation, which is commonly used for the representation of experimental lattice parameter data. The Murnaghan equation considers a bulk modulus, i.e. the inverse value of the isothermal compressibility, that is linearly increasing with pressure.

$$B(p) = B_0 + B_0' p = \frac{1}{\kappa(p)} \quad (10)$$

In eq. (10), $B_0$ is the bulk modulus in Pa at zero pressure and $B_0'$ is a constant bulk modulus derivative.

The Murnaghan equation of state results in the following correlation for the pressure dependence of the lattice parameter

$$a(T,p)=a(T,p_0)\left(\frac{B_0+B_0'p}{B_0+B_0'p_0}\right)^{\frac{-1}{3B_0'}}, \qquad (11)$$

where the reference pressure $p_0$ is arbitrarily set to 1 Pa in this study. Based on the comparison of the Murnaghan equation and the Birch-Murnaghan equation, the bulk modulus derivative can usually be taken constant $B_0'=4$. As it was shown in ref. [3], the pressure data available for hydrates forming structures sI and sII can be well correlated with a constant bulk modulus $B_0$ = 10 GPa. This universal value is also used here.

## 3 New correlation for the lattice parameter of sI and sII gas hydrates

The new correlation for the lattice parameter intended for modeling mixed hydrates has the same mathematical form as the correlation for pure gas hydrates developed in the previous work [3]; see eq. (12). However, the values of the parameters $T_0$, $p_0$, $a_0$ and coefficients $\alpha_1$ to $\alpha_3$ are different for the pure hydrate model and the mixed hydrate model.

$$a(T,p)=a_0\left(\frac{B_0+4p}{B_0+4}\right)^{\frac{-1}{12}} \qquad (12)$$
$$\exp\left[\alpha_1(T-T_0)+\alpha_2(T-T_0)^2+\alpha_3(T-T_0)^3\right]$$

The volume correlation employed in the pure hydrate model has several disadvantages making it inconvenient for the modeling of mixed hydrates. The correlation uses different values for the reference temperature $T_0$ for each hydrate structure. Moreover, the reference temperatures and the reference pressure differ from the reference state conditions of the water empty $\beta$-lattice at which the reference Gibbs energy and the reference enthalpy are given, i.e. $T_0^\beta=273.15$ K and $p_0^\beta=1$ Pa ; for more detail see ref. [3-5,9]. In the model of pure gas hydrates, the thermal expansion coefficients $\alpha_1$ to $\alpha_3$ were evaluated only from the lattice parameter data for $CO_2$ hydrates in case of sI structure [13] and taken from Hester et al. [17] in case of sII structure. The $CO_2$ correlation [13] and the correlation by Hester et al. [17] were evaluated solely from the temperature data. Main advantage of the new correlation (12) is that it can simultaneously be fitted to both the temperature and the pressure data. Consequently, all available data summarized in Tables 1 and 2 can be considered in the evaluation of the reference parameter $a_0$ and the thermal expansion coefficients $\alpha_1$ to $\alpha_3$ in the new correlation (12). As a result, the new correlation shall provide a more reliable and universal prediction for the lattice parameter of a large variety of sI and sII hydrates.

For the purpose of the new mixed hydrate model, a universal reference temperature $T_0$ = 273.15 K and a universal reference pressure $p_0$ = 1 Pa are proposed in eq. (12). The bulk modulus has a constant value of $B_0=10\cdot10^9$ Pa for all the selected hydrates. Parameter $a_0$ and coefficients $\alpha_1$ to $\alpha_3$ were determined with the help of a two-stage Levenberg-Marquardt optimization algorithm [10,11]. The main algorithm optimizes values for the thermal expansion coefficients $\alpha_1$ to $\alpha_3$, which are universal for a given hydrate structure. For each iteration of the main algorithm, nested Levenberg-Marquardt algorithm optimizes the reference lattice parameter $a_0$ for each hydrate former. The nested algorithm is called $N$-times, where $N$ is the number of hydrate formers equal to 11 in case of sI hydrates and 18 in case of sII hydrates; see data summarized in Tables 1 and 2. The Levenberg-Marquardt algorithm optimizes following objective function

$$\sum_{i=1}^{N} w_i\left(a_{\exp,i}-a_{\mathrm{cor},i}\right)^2=\min, \qquad (13)$$

where $a_{\exp}$ is the experimental value and $a_{\mathrm{cor}}$ is the correlated value of the lattice parameter. $w_i$ stands for the weight of the given experimental data point. For most data points, the weight was set to unity. The weight was reduced to 0.25 for the data measured with deuterated hydrates, i.e. with heavy water ($D_2O$), and for high pressure data above 700 MPa. The data for deuterated hydrates by Hansen et al. [16] were not considered in the parameter optimization. The weight for the data taken from the molecular simulations for methane hydrates by Docherty et al. [36] was reduced down to 0.0001.

Table 3 shows a comparison of the average deviation AD defined by eq. (14) and the absolute average deviation AAD given by eq. (15) for the correlation employed in the pure hydrate model [3] and for the new correlation. In case of sI hydrates, the new correlation provides significantly lower AD and AAD. Main reason is that the new correlation was fitted to all available data, while the pure hydrate correlation was correlated only to the temperature dependence for $CO_2$ hydrates. Results of both correlations are comparable in case of sII hydrates. The new correlation is slightly better.

$$\mathrm{AD}=\frac{1}{N}\sum_{i=1}^{N}a_{\exp,i}-a_{\mathrm{cor},i} \qquad (14)$$

$$\mathrm{AAD}=\frac{1}{N}\sum_{i=1}^{N}\left|a_{\exp,i}-a_{\mathrm{cor},i}\right| \qquad (15)$$

**Table 3.** Number of experimental data, absolute deviation AD and average absolute deviation AAD for the correlation used in the model for pure gas hydrates [3] and the new correlation eq. (12)

| | Vinš et al. [3] | this study |
|---|---|---|
| **Structure sI** | | |
| *N* | 288 | |
| AD / Å | -1.122 | -0.550 |
| AAD / Å | 4.316 | 3.388 |
| **Structure sII** | | |
| *N* | 179 | |
| AD / Å | -0.449 | -0.155 |
| AAD / Å | 2.128 | 2.029 |

### 3.1 Final parameters

Table 4 contains the final values for thermal expansion coefficients $\alpha_1$ to $\alpha_3$ for hydrate structures sI and sII. The reference temperature $T_0$ in eq. (12) equals 273.15 K.

The optimized values for the reference lattice parameter $a_0$ are given in Table 5 for sI hydrates and in Table 6 for sII hydrates. The values are being used in the model for mixed hydrates [9].

**Table 4.** Thermal expansion coefficients for the lattice parameter of sI and sII gas hydrates correlated by eq. (12)

| Struct. | $\alpha_1$ / K$^{-1}$ | $\alpha_2$ / K$^{-2}$ | $\alpha_3$ / K$^{-3}$ |
|---|---|---|---|
| sI | $8.416128 \cdot 10^{-5}$ | $7.141596 \cdot 10^{-8}$ | $-2.106472 \cdot 10^{-10}$ |
| sII | $7.988807 \cdot 10^{-5}$ | $1.686500 \cdot 10^{-7}$ | $1.363783 \cdot 10^{-10}$ |

**Table 5.** Reference lattice parameter $a_0$ at 273.15 K and 1 Pa in eq. (12) for sI hydrates

| Guest | $a_0$ / Å |
|---|---|
| EtO | 11.997604 |
| $CO_2$ | 11.979675 |
| $C_2H_6$ | 12.067015 |
| 0.85$CH_4$ + 0.15 $CO_2$ | 11.978807 |
| 0.47 $CH_4$ + 0.53 $C_2H_6$ | 12.042290 |
| TMO | 12.138742 |
| CO | 12.017111 |
| $H_2S$ | 12.082851 |
| $CH_4$ | 11.970005 |
| Xe | 11.987297 |
| $N_2$ (sI) | 11.969756 |

**Table 6.** Reference lattice parameter $a_0$ at 273.15 K and 1 Pa in eq. (12) for sII hydrates

| Guest | $a_0$ / Å |
|---|---|
| Air | 17.283598 |
| Ar | 17.232725 |
| Kr | 17.237294 |
| $O_2$ | 17.262769 |
| $N_2$ | 17.275606 |
| TMO | 17.226661 |
| $C_3H_8$ | 17.288930 |
| 0.60 $CH_4$ + 0.4 $C_3H_8$ | 17.313328 |
| 0.30 $C_2H_6$ + 0.7 $C_3H_8$ | 17.276450 |
| 0.182 $CO_2$ + 0.818 $C_3H_8$ | 17.327134 |
| 0.876 $CH_4$ + 0.124 i-$C_4H_{10}$ | 17.374086 |
| Natural gas | 17.313652 |
| 0.95 $CH_4$ + 0.05 $C_5H_{10}O$ | 17.435520 |
| THF | 17.334988 |
| 0.82 $CH_4$ + 0.18 $C_2H_6$ | 17.284555 |
| THF + $H_2S$ | 17.336587 |
| THF + Xe | 17.424193 |
| $CH_4$ (sII) | 17.277118 |

### 3.2 Results – temperature dependence

Fig. 1 shows a comparison of the experimental data for the temperature dependence of the lattice parameter of sI hydrates with the correlation by Jäger et al. [13] used in the pure hydrate model [3-5] and the new correlation (12). The deviation of both correlations from the experimental data is provided in Fig. 2.

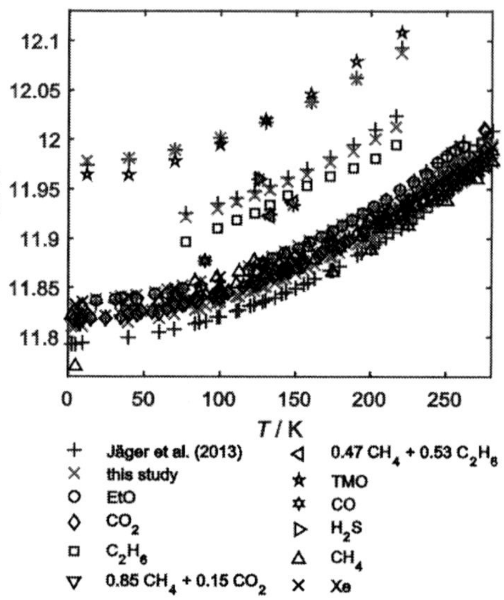


**Fig. 1.** Experimental data for the temperature dependence of the lattice parameter of sI gas hydrates at low pressures (typically at atmospheric pressure) compared to the correlation by Jäger et al. [1] and the new correlation.

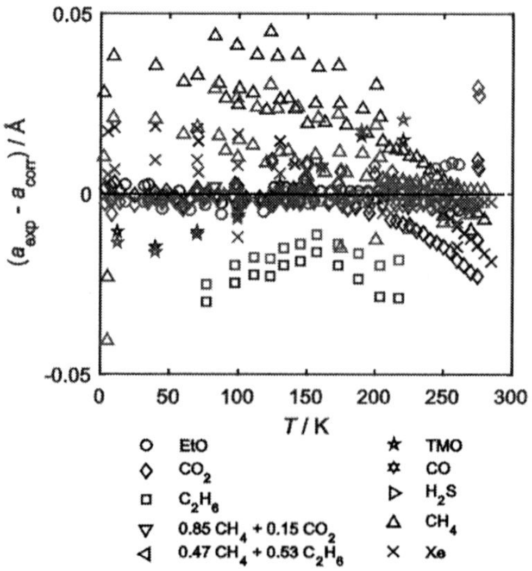


**Fig. 2.** Deviation of the correlation by Jäger et al. [13] (blue symbols) and the new correlation (red symbols) from the experimental data for the temperature dependence of the lattice parameter of sI gas hydrates at low pressures (typically at atmospheric pressure).

DOI: 10.1051/epjconf/201714302141

In the low temperature region below 200 K, both correlations provide relatively good reproduction of the experimental data. Larger deviations can be seen only by some data sets for methane and ethane hydrates caused by slight discrepancies between the temperature data sets and the pressure data sets of these gas hydrates. The new correlation provides significantly better reproduction of the high temperature data for $CO_2$, $CH_4$ and Xe hydrates by Hansen et al. [16] from 2016. As already shown in Table 3, the new correlation is in general in better agreement with the experimental data; see Fig. 2.

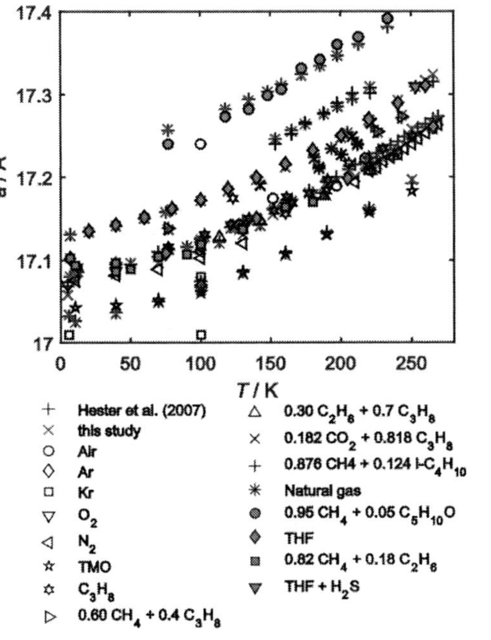


**Fig. 3.** Experimental data for the temperature dependence of the lattice parameter of sII gas hydrates at low pressures (typically at atmospheric pressure) compared to the correlation by Hester et al. [17] and the new correlation.

The experimental data for the temperature dependence of the lattice parameter for sII hydrates are compared to the correlation by Hester et al. [17] and the new correlation (12) in Fig. 3. The correlations provide very good reproduction of the experimental data over a wide temperature range. The differences between both correlations are almost negligible in this case, except for the new temperature data for nitrogen hydrates by Hansen et al. [16]. The new correlation is in much better agreement with these data compared to the correlation by Hester et al. [17]. The deviations of both correlations from the experimental data are shown in Fig. 4.

## 3.3 Results – pressure dependence

The pressure dependence of the lattice parameter for sI hydrates is shown in Fig. 5. The experimental data for four hydrate formers are compared with the correlation employed in the pure hydrate model [3] and with the new correlation (12).

At extremely high pressures above 500 MPa, the results of the new correlation are better for xenon hydrates while slightly worse for ethane hydrates compared to the original correlation [3]. As can be seen in Fig. 6 showing the deviations of both correlations from the experimental data, a noticeable improvement of the new correlation was achieved at pressures below 150 MPa. The new correlation agrees better with the data for methane hydrates and for xenon hydrates in this region.

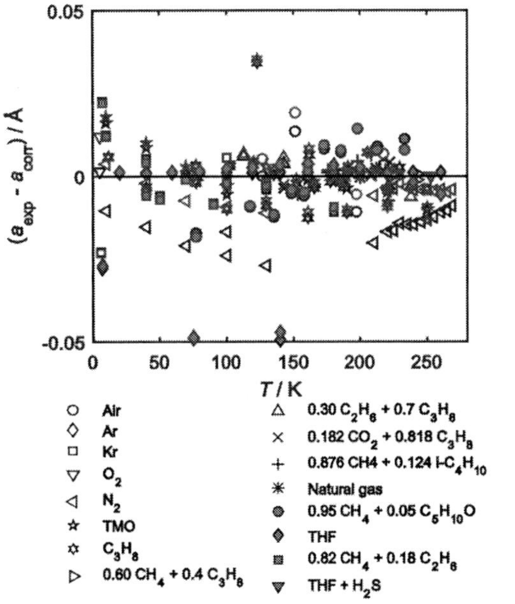


**Fig. 4.** Deviation of the correlation by Hester et al. [17] (blue symbols) and the new correlation (red symbols) from the experimental data for the temperature dependence of the lattice parameter of sII gas hydrates at low pressures (typically at atmospheric pressure).

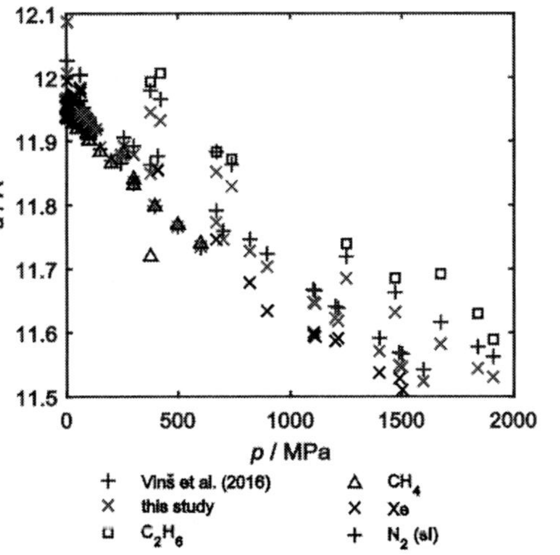


**Fig. 5.** Experimental data for the pressure dependence of the lattice parameter of sI gas hydrates compared to the correlation by Vinš et al. [3] and the new correlation.

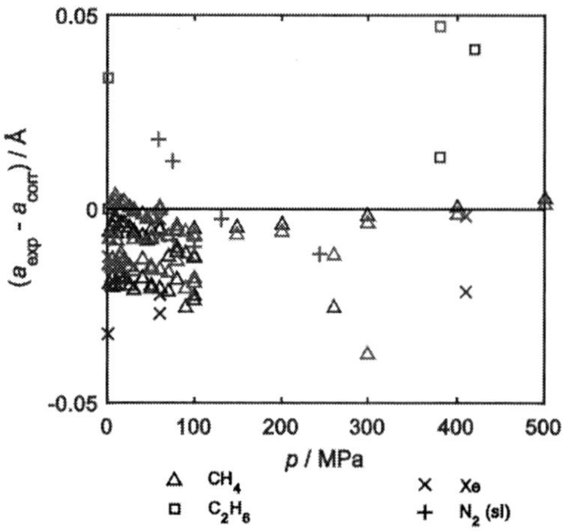


**Fig. 6.** Deviation of the correlation by Vinš et al. [3] (blue symbols) and the new correlation (red symbols) from the experimental data for the pressure dependence of the lattice parameter of sI gas hydrates.

Fig. 7 shows the pressure dependent data for the lattice parameter of sII hydrates. The two correlations provide comparable results, which are in good agreement with the experimental data in both cases. We note that the data for THF + Xe mixed hydrates up to 3 GPa were considered in this study.

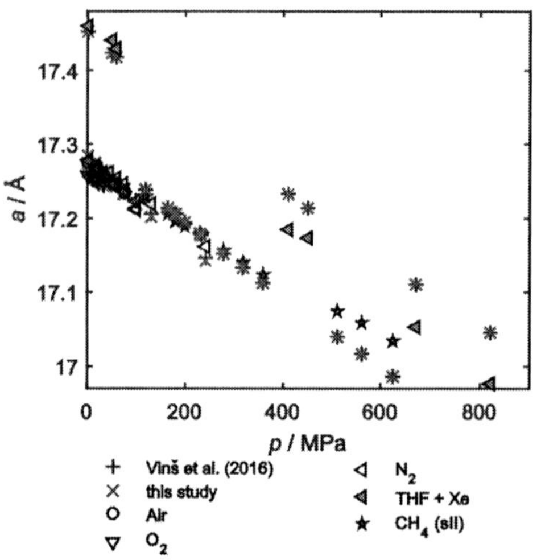


**Fig. 7.** Experimental data for the pressure dependence of the lattice parameter of sII gas hydrates compared to the correlation by Vinš et al. [3] and the new correlation.

## 4 Conclusions

The correlation for the temperature and pressure dependences of the lattice parameter of sI and sII gas hydrates developed in the previous study [3] was revised and modified such that it can be used for modeling of mixed hydrates. The new correlation was simultaneously fitted to the experimental data for both the temperature dependence and the pressure dependence of the hydrate volume using a two-stage Levenberg-Marquardt algorithm. A set of values for the thermal expansion coefficients $\alpha_1$ to $\alpha_3$ and for the reference lattice parameter $a_0$ was evaluated at the reference temperature of 273.15 K and the reference pressure of 1 Pa. The bulk modulus was considered constant and equal to 10 GPa (with $B_0' = 4$) for all selected hydrates.

The new model for the volume of sI and sII gas hydrates is in good agreement with the experimental data, including the recent data of Hansen et al. [16], over wide temperature and pressure ranges. Therefore, the universal modeling approach for sI and sII formers for the temperature dependence of the lattice parameter as proposed by Hester et al. [17] and for the pressure dependence as proposed by Vinš et al. [3] with the new parameters proposed in this work will be used for modeling mixed hydrates.

## List of abbreviations

| | |
|---|---|
| AAD | absolute average deviation |
| AD | average deviation |
| Ar | argon |
| CCS | carbon capture and storage |
| $CH_4$ | methane |
| $C_2H_6$ | ethane |
| $C_3H_8$ | propane |
| $C_5H_{10}O$ | tetrahydropyran |
| CO | carbon monoxide |
| $CO_2$ | carbon dioxide |
| EtO | ethylene oxide |
| $H_2S$ | hydrogen sulphide |
| i-$C_4H_{10}$ | isobutane |
| Kr | krypton |
| $N_2$ | nitrogen |
| $O_2$ | oxygen |
| THF | tetrahydrofuran |
| TMO | trimethylene oxide |
| Xe | xenon |

## Acknowledgement

The study was conducted with the financial help of the Czech Science Foundation grant GJ15-07129Y and the institutional support RVO:61388998.

## References

1. E.D. Sloan, C.A. Koh, Clathrate Hydrates of Natural Gases, third ed. CRC Press, Taylor & Francis Group, New York, USA, 2008
2. H. Li, J.P. Jakobsen, Ø. Wilhelmsen, J. Yan, Appl. Energy **88** 3567–3579 (2011)
3. V. Vinš, A. Jäger, R. Span, J. Hrubý, Fluid Phase Equilib. **427**, 268-281 (2016)

4. V. Vinš, A. Jäger, J. Hrubý, R. Span, Fluid Phase Equilib. (2016) (submitted for publication)
5. A. Jäger, V. Vinš, R. Span, J. Hrubý, Fluid Phase Equilib. **429**, 55-66 (2016)
6. J.H. van der Waals, J.C. Platteeuw, Clathrate solutions, Adv. Chem. Phys. **2** (1959) 1-57.
7. J. Gernert, R. Span, EOS-CG, J. Chem. Thermodyn. **93**, 274-293 (2016)
8. O. Kunz, W. Wagner, J. Chem. Eng. Data **57**, 3032-3091 (2012)
9. S. Hielscher, A. Jäger, V. Vinš, C. Breitkopf, J. Hrubý, R. Span, to be presented at Thermodynamik Kolloquium 2016, Technische Universität Kaiserslautern, Germany, October 5-7 (2016)
10. K. Levenberg, Q. Appl. Math. **2**, 164 (1944)
11. D.W. Marquardt, J. Soc. Ind. Appl. Math. **11**, 431 (1963)
12. A.L. Ballard, E.D. Sloan Jr., Fluid Phase Equilib. **194**, 371-383 (2002)
13. A. Jäger, V. Vinš, J. Gernert, R. Span, J. Hrubý, Fluid Phase Equilib. **338**, 100-113 (2013)
14. W.F. Kuhs, B. Chazallon, P.G. Radaelli, F. Pauer, J. Inclusion Phenom. **29** 65 (1997)
15. B. Chazallon, W.F. Kuhs, J. Chem. Phys. **117**, 308-320 (2002)
16. T.C. Hansen, A. Falenty, W.F. Kuhs, J. Chem. Phys. **144**, 054301 (2016)
17. K.C. Hester, Z. Huo, A.L. Ballard, C.A. Koh, K.T. Miller, E.D. Sloan, J. Phys. Chem. B **111**, 8830-8835 (2007)
18. F.D. Murnaghan, Proc. Natl. Acad. Sci. U. S. A. **30** 244-247 (1944)
19. F. Birch, Phys. Rev. **71**, 809-824 (1947)
20. J.S. Tse, W.R. McKinnon, M. Marchi, J. Phys. Chem. **91**, 4188-4193 (1987)
21. J.S. Tse, J. Phys. **48**, 543–548 (1987)
22. T. Ikeda, O. Yamamuro, T. Matsuo, K. Mori, S. Torri, T. Kamiyama, F. Izumi, S. Ikeda, S. Mae, J. Phys. Chem. Solids **60**, 1527-1529 (1999)
23. T. Ikeda, S. Mae, O. Yamamuro, T. Matsuo, S. Ikeda, R.M. Ibberson, J. Phys. Chem. A **104**, 10623 (2000)
24. Z. Huo, M. Eaton, K.T. Miller, E.D. Sloan, Int. J. Thermophys. **26**, 107-114 (2005)
25. K.A. Udachin, C.I. Ratcliffe, J.A. Ripmeester, J. Phys. Chem. B **105**, 4200-4204 (2001)
26. S. Circone, L. A. Stern, S. H. Kirby, W. B. Durham, B. C. Chakoumakos, C. J. Rawn, A. J. Rondinone, Y. Ishii: J. Phys. Chem. B **107**, 5529-5539 (2003)
27. A.Y. Manakov, A.Y. Likhacheva, V.A. Potemkin, A.G. Ogienko, A.V. Kurnosov, A.I. Ancharov, ChemPhysChem **12**, 2476-2484 (2011)
28. A.J. Rondinone, B.C. Chakoumakos, C.J. Rawn, Y. Ishii, J. Phys. Chem. B **107**, 6046-6050 (2003)
29. D.W. Davidson, M.A. Desando S.R. Gough, Y.P. Handa, C.I. Ratcliffe, J.A. Ripmeester, J.S. Tse, Nature **328**, 418-419 (1987)
30. D. W. Davidson, Y. P. Handa, C. I. Ratcliffe, J. S. Tse, B. M. Powell, Nature **311**, 142-143 (1984)
31. D.W. Davidson, S.K. Garg, S.R. Gough, Y.P. Handa, C.I. Ratcliffe, J.S. Tse and J.A. Ripmeester, J. Incl. Phenom. **2**, 231-238 (1984)
32. C. Gutt, B. Asmussen, W. Press, M.R. Johnson, Y.P. Handa, J.S. Tse, J. Chem. Phys. **113**, 4713-4721 (2000)
33. M.T. Kirchner, R. Boese, W.E. Billups, L.R. Norman, J. Am. Chem. Soc. **126**, 9407-9412 (2004)
34. A.G. Ogienko, A.V. Kurnosov, A.Y. Manakov, E.G. Larinov, A.I. Ancharov, M.A. Sheromov, A.N. Nesterov, J. Phys. Chem. B **110**, 2840-2846 (2006)
35. S. Takeya, M. Kida, H. Minami, H. Sakagami, A. Hachikubo, N. Takahashi, H. Shoji, V. Soloviev, K. Wallmann, N. Biebow, A. Obzhirov, A. Salomatin, J. Poort, Chem. Eng. Sci. **61**, 2670-2674 (2006)
36. H. Docherty, A. Galindo, C. Vega, E. Sanz, J. Chem. Phys. **125**, 074510 (2006)
37. A. Klapproth, E. Goreshnik, D. Staykova, H. Klein, W.F. Kuhs, Can. J. Phys. **81**, 503-518 (2003)
38. H. Hirai, T. Kondo, M. Hasegawa, T. Yagi, Y. Yamamoto, T. Komai, K. Nagashima, M. Sakashita, H. Fujihisa, K. Aoki, J. Phys. Chem. B **104**, 1429-1433 (2000)
39. C. Sanloup, H.-K. Mao, R. Hemley, J. Proc. Natl. Acad. Sci. U.S.A. **99**, 25-28 (2002)
40. S. Takeya, H. Nagaya, T. Matsuyama, T. Hondoh, V.Y. Lipenkov, J. Phys. Chem. B **104**, 668-670 (2000)
41. C.J. Rawn, A.J. Rondinone, B.C. Chakoumakos, S. Circone, L.A. Stern, S.H. Kirby, and Y. Ishii, Can. J. Phys. **81**, 431-438 (2003)
42. C.Y. Jones, S.L. Marshall, B.C. Chakoumakos, C.J. Rawn, Y. Ishii, J. Phys. Chem. B **107**, 6026-6031 (2003)
43. C.J. Rawn, A.J. Rondinone, B.C. Chakoumakos, S.L. Marshall, L.A. Stern, S. Circone, S.H. Kirby, C.Y. Jones, B.H. Toby, Y. Ishii, Proc. 4th Int. Conf. Gas Hydrates, Yokohama, Japan, May 19-23, 595-598 (2002)
44. T.C.W. Mak, R.K. McMullan, J. Chem. Phys. **42**, 2732-2737 (1965)
45. J. Shu, X. Chen, I.M. Chou, W. Yang, J. Hu, R.J. Hemley, H. Mao, Geosci. Front. **2**, 93-100 (2011)
46. J.B. Klauda, S.I. Sandler, Ind. Eng. Chem. Res. **39**, 3377-3386 (2000)
47. A. Martín, C.J. Peters, J. Phys. Chem. C **113**, 422-430 (2009)

# Influence of distance between water distribution and grid fill on water drop size

*Pavol* Vitkovic[1,*]

[1]Czech Technical University in Prague, Department of Fluid Dynamics and Thermodynamics, 16607 Prague, Czech Republic

**Abstract.** The article deals with the experimental study of water drop size measurement below the grid fill. During experiment was changing the distance between the water distribution and grid fill. It's changing the length of impact trajectory on the grid fill. Measurement of the water drop size was located below the first layer of the grid fill.

## 1 Introduction

The splash type fill or pack is designed to break the mass of water falling through the cooling tower into a large number of drops [1]. The film fill is more efficient in heat transfer in a smaller volume than the splash fill. But the splash fill has a small surface area exposed to water, it is non-fouling splash fill. It is used in refineries, sugar and steel mills and food processing plants.
The mathematical model of a splash fill is described in [2]. The mathematical model of a rain zone is described in [3, 4]. All of those models needed a measure a diameter of the water droplets. The small spherical drops can be measured by the IPI method [5]. The splash – grid fill generates non-spherical water droplets. In this case is better to use the laser-induced fluorescence [6].

## 2 Experimental facility and setup

The experimental facility is consisting of a basin, a test section, an extending part, a water pump, a flow meter and a droplet generator. The test section is made from a plexiglass. The dimension of the test section is 0.6 x 0.6 m and a total height of this section is 1 m. The extending part is consisting of several modules with a height of 1 m and 0.5 m. The modules are made from a galvanized metal sheet. And it can be arbitrarily combined with each other. The water droplets generator is supplied by the water feed pump connected to the water basin below the test section. The water flow rate is measured by the electromagnetic flow meter. The droplets generator was made from plexiglass sheets [6]. It is consisting of a three layers (Fig. 1). The last layer has holes with 1 mm diameter arranged into the grid 0.01 x 0.01 m (Fig 2). The grid of the holes is creating an active zone in the test section. It dimension is 0.2 x 0.2 m (Fig. 2). The droplets generator was placed on the top of the experimental facility (Fig. 3). Droplet generator drop size distribution based on volume is on the Fig. 4. It is measured in [6].

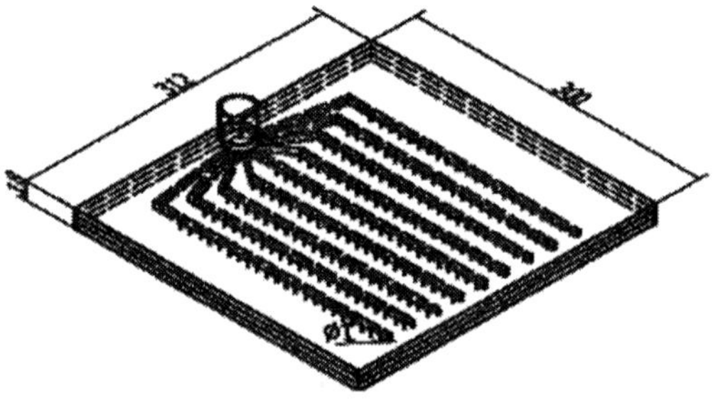

**Fig. 1.** Droplet generator with the 1 mm holes.

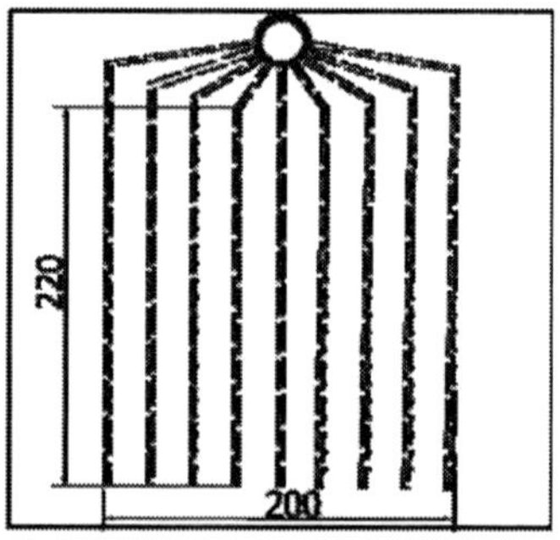


**Fig. 2.** The active zone of the droplet generator 200 x 200 mm.

**Fig. 3.** Droplet generator made from the plexiglass.

* Corresponding author: pavol.vitkovic@fs.cvut.cz

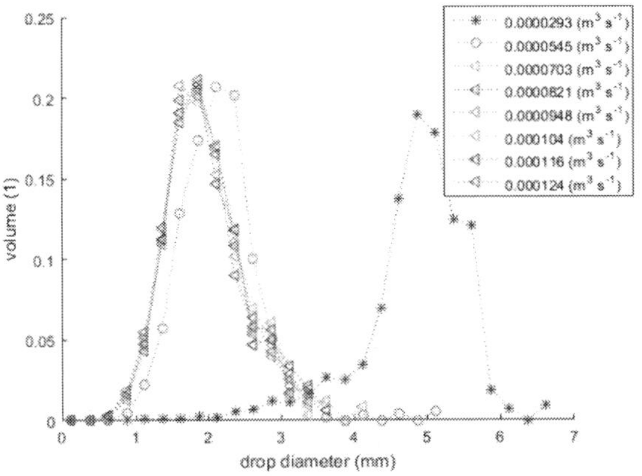


**Fig. 4.** Droplet generator - drops size distribution based on volume [6].

For all experimental points described below is used the splash (grid) fill with the rhombus shape bars (Fig. 5). The side length of the rhombus is 50 mm and the thickness of the bar is 5 mm. The layer of grid fill (0.6 x 0.6 m) is placed on the top of the test section (Fig. 6). Above the fill was extending piece. The extending piece has defined the distance between the water drops generator and the grid fill layer. This distance was chosen 0.5 m, 1.0 m, and 1.5 m.

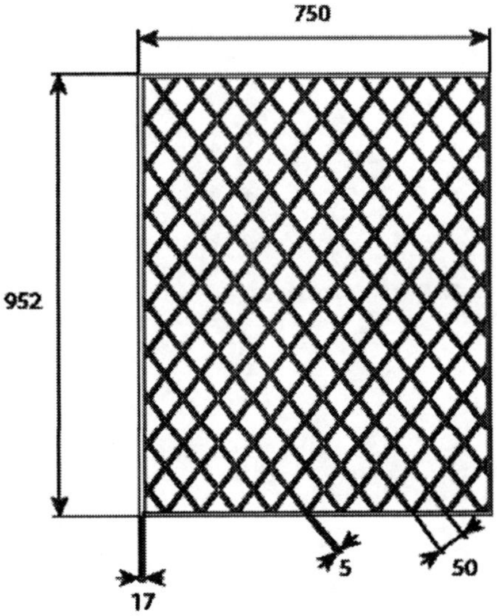


**Fig. 5.** The splash (grid) fill with the rhombus shape bars.

The test section is located 0.178 m below the splash (grid) fill. The laser beam is expanded into a sheet using laser optics in the test section (Fig. 7). The Rhodamine G as tracer dye after absorbing the laser light with a wavelength of 523 mm emitted the light of wavelength between 500 and 700 nm. Excitation light was filtered by a red filter. The images were captured by the camera SONY XCG/H280E with 85 mm lens. The exposure time was set to 1/100 sec. The shooting speed was 33 fps. The aperture was set at the f/4. The Coherent Verdi G Series semiconductor laser was used. The laser power was set at 1 W.

The water flow rate was chosen from 0.00009 $m^3/s$ until 0.00047 $m^3/s$. Totally was 6 measurement points for 3 different distance between the water droplet generator and the splash (grid) fill (Table 1). The first 2 points are covered with an experiment described in [6]. The points from 3 until 6 represent the water flow rate commonly used in the real cooling tower systems.

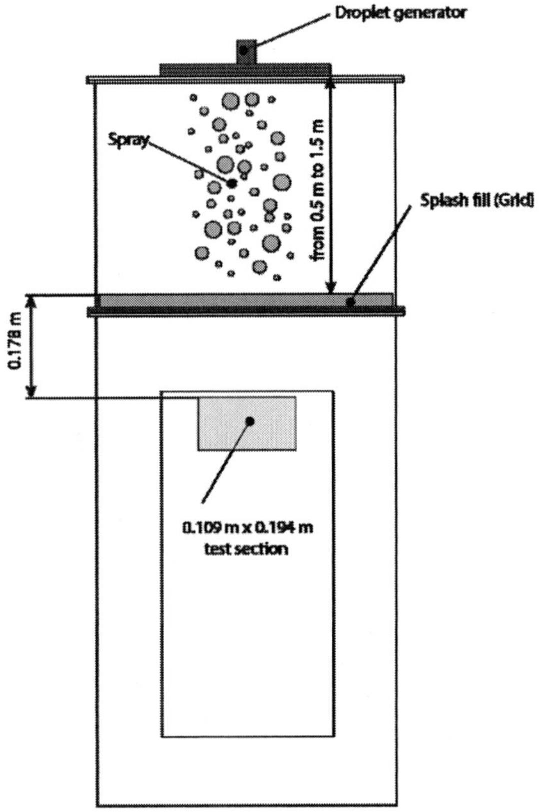


**Fig. 6.** Experiment setup.

**Fig. 7.** The laser sheet with the water droplets saturated by Rhodamine G.

 DOI: 10.1051/epjconf/201714302142

**Table 1.** Water flow rate for each measurement points and the distance between the drops generator and the splash fill.

| No. | H = 0.5 m | H = 1.0 m | H = 1.5 m |
|---|---|---|---|
| 1 | 0.0000926 $m^3/s$ | 0.0000912 $m^3/s$ | 0.0000891 $m^3/s$ |
| 2 | 0.0001243 $m^3/s$ | 0.0001224 $m^3/s$ | 0.0001251 $m^3/s$ |
| 3 | 0.0001500 $m^3/s$ | 0.0001524 $m^3/s$ | 0.0001529 $m^3/s$ |
| 4 | 0.0002810 $m^3/s$ | 0.0002808 $m^3/s$ | 0.0002819 $m^3/s$ |
| 5 | 0.0003879 $m^3/s$ | 0.0003885 $m^3/s$ | 0.0003875 $m^3/s$ |
| 6 | 0.0004766 $m^3/s$ | 0.0004730 $m^3/s$ | 0.0004708 $m^3/s$ |

The water flow rate was set up approximately at the same value (Fig. 8) for each distance between the water droplet generator and the fill layer.

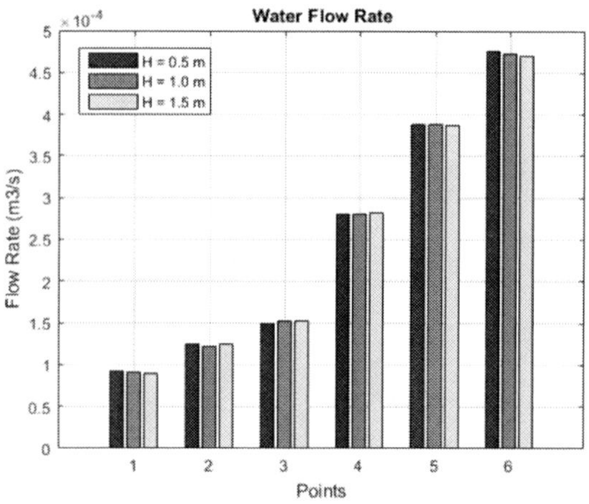


**Fig. 8.** Water flow rate during the experiment

## 3 Experimental results

The signal from the fluorescent light is converted to the white blobs on a black background. From the blob area is calculated an equivalent spherical drop diameter. The water drop size distribution based on volume are at following figures (Fig. 9 – Fig. 14) for different water flow rates and different between the droplet generator and grid fill. Each histogram represents the amount of water for individual diameters of the water drops.

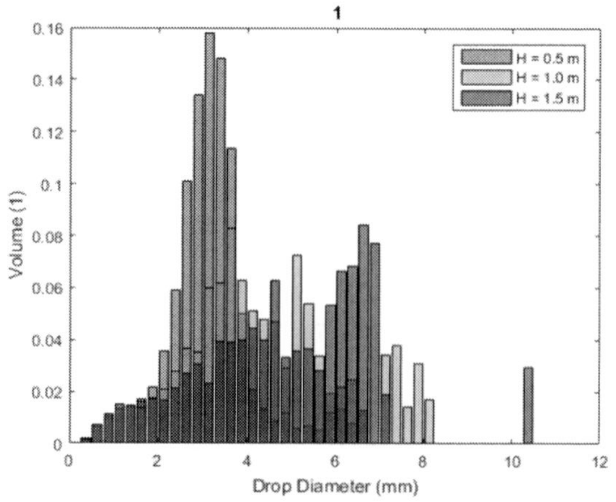


**Fig. 9.** Drop size histogram based on volume for the mean water flow rate 0.0000910 $m^3/s$

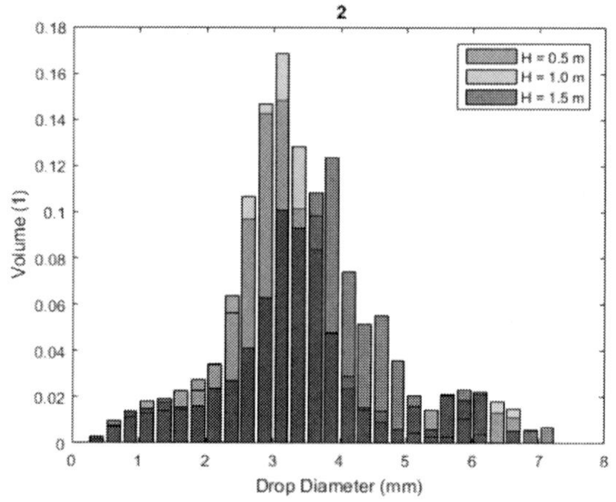


**Fig. 10.** Drop size histogram based on volume for the mean water flow rate 0.0001239 $m^3/s$

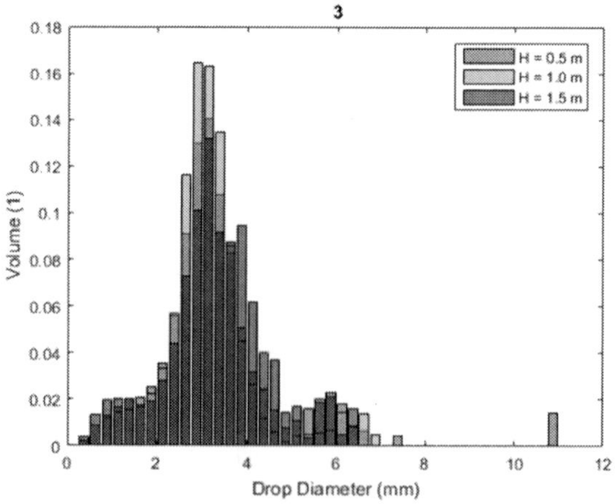


**Fig. 11.** Drop size histogram based on volume for the mean water flow rate 0.0001517 $m^3/s$

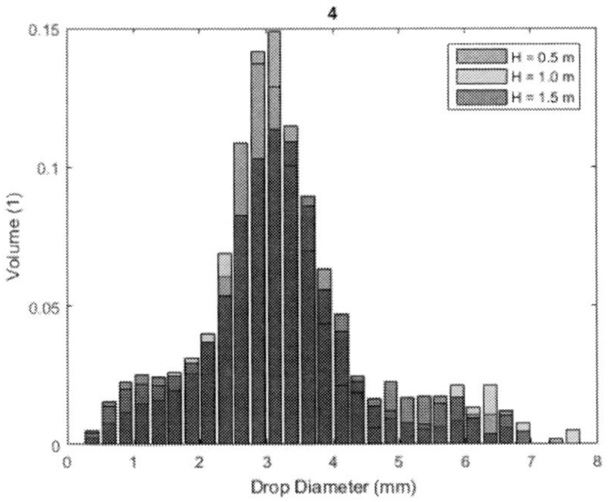


**Fig. 12.** Drop size histogram based on volume for the mean water flow rate 0.0002812 $m^3/s$

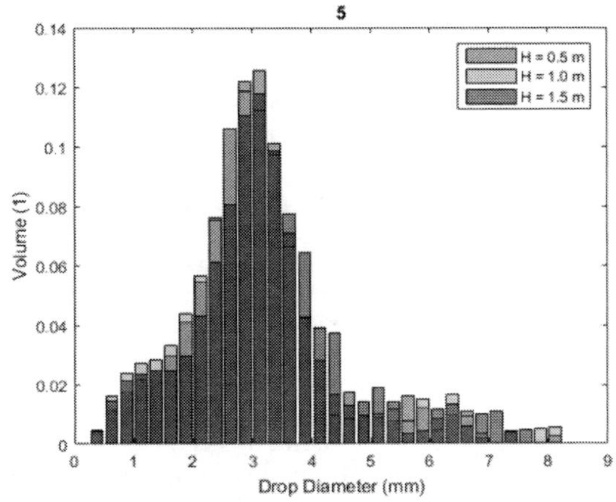


**Fig. 13.** Drop size histogram based on volume for the mean water flow rate 0.0003880 $m^3/s$

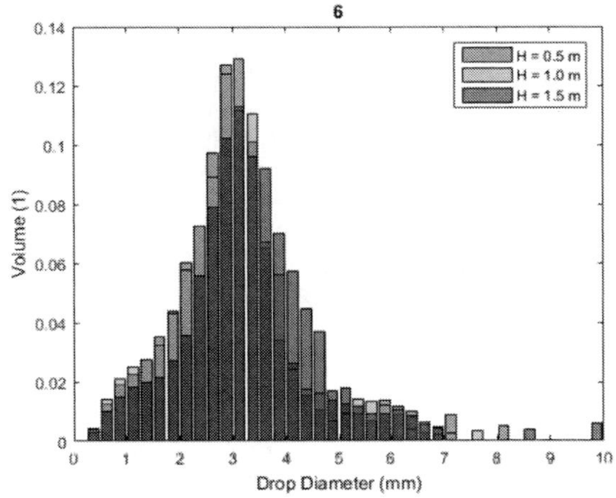


**Fig. 14.** Drop size histogram based on volume for the mean water flow rate 0.0004735 $m^3/s$

## 4 Conclusion

The experimental data was obtained for splash (grid) fill with rhombus shaped bars. They have chosen 3 distances between the water droplet generator and the fill layer. For study was used 1 layer of the splash fill. The splash (grid) fill with the rhombus shape bar generate the biggest water droplet like they come at the fill. It is confirming the experimental results in [6]. The water droplet generator is creating a mean droplet with diameter 2 mm. The splash fill is creating a mean droplet with a diameter of 3 mm and more. The study shows that the influence of the distance between the spray device (in this case the water droplet generator with 1mm holes) is very weak. The histograms on Fig. 11 till Fig. 14 are covered for each distance between the droplet generator and the fill layer. It means that no matter the chosen distance. However, the influence is discovered in Fig. 9 and significantly less on Fig. 10 at the lowest water flow rates. This water flow rates are not commonly used in the real cooling tower.

## Acknowledgments

The author acknowledges the support received from Advanced technologies for heat and electricity production TE01020036, Centre for research of multiphase flow and thermodynamics processes in renewable sources and energetics – NEW ENERGY supported by the European Union (CZ.2.16/3.1.00/22130) and Centre of 3D volumetric velocimetry – COLA supported by the European Union (CZ.2.16/3.1.00/21569).

## References

1. D.G Kröger, *Air-Cooled Heat Exchangers and Cooling Tower*, (2004)
2. A. A. Dreyer, P. J. Erens, J. Heat Mass Transfer, ***39***, 109-123, (1996)
3. T. Hyhlík, Engineering Mechanics, **21**, 240-243, (2014)
4. T. Hyhlík, ICCT 2014 International Conference, 219-226, (2014)
5. J. Stodůlka, R. Vitkovicova, EPJ Web of Conferences, **114**, 02111, (2016)
6. P. Vitkovic, EPJ Web of Conferences, **114**, 02137, (2016)

EPJ Web of Conferences **143**, 02143 (2017) DOI: 10.1051/epjconf/201714302143
*EFM 2016*

# Influence of detergents on water drift in cooling towers

*Rut* Vitkovicova[1,*]

[1] CTU in Prague, FME, Department of Fluid Dynamics and Thermodynamics, Technická 4, 166 07 Prague, Czech Republic

**Abstract.** An influence of detergents on the water drift from the cooling tower was experimentally investigated. For this experimental measurements was used a model cooling tower, especially an experimental aerodynamic line, which is specially designed for the measurement and monitoring of processes taking place around the eliminators of the liquid phase. The effect of different concentrations of detergent in the cooling water on the drift of water droplets from a commonly used type eliminator was observed with visualization methods.

## 1 Introduction

The research and development of wet cooling towers could be divided into three parts: Part One, which deals with the most efficient heat transfer between air and water and tries to optimize especially the function of the cooling fills and spraying nozzles as described in [1] where the author closely examines for the calculations and descriptions of processes from the viewpoint Thermodynamics [2], where the authors analyse experimental measurements of the thermal characteristics of the grid fills in the tower. The second part is the environmental issue. Cooling tower as a heat exchanger may more or less affect the environment and their activities have a negative effect on it. The formation of the plume is one of the problems and therefore it pays attention to the predictions using numerical models, as stated in [3], [4]. Great attention is also paid to drift of cooling water droplets. Eliminators of a liquid phase drift from cooling tower together with fills, nozzles and water distribution systems are essential elements of wet cooling towers. Eliminators are placed above spray nozzles on the upper floors of the tower. Their main task, as their name suggests, is to reduce the water drift of cooling water from the system. Reasons for seeking to minimize drift are both reducing the loss of cooling water from the system and also ecological reasons. Drift water may cause-consequence of adverse weather conditions a build up of ice and fog near the tower, can chemicals contaminate neighborhood which is contained in the cooling water, and ultimately may contain bacteria which live in refrigeration systems. Therefore, many countries are given the value of the maximum drift of the cooling water (approximately 0.005% of the cooling water or less).

Sufficient pure water is becoming recently an increasingly common problem in the world. Therefore it is encountered increasingly in cooling using with poor quality water. Deterioration of water quality has a significant influence on its properties and behavior. The water with reduced surface tension has a particularly negative effect on the eliminators. This occurs when the cooling water contains detergents.

## 2 Principles of drift eliminators function

Effective reduction of unpolluted water mass from air flow is mainly due to the curved channels of the eliminator. The larger is the droplet size, the larger inertia forces actuate on it in a curved channel, which leads to the fact that the droplet is no longer able to follow the flow field and adheres to the wall of the eliminator and streams down by gravity into the space of the cooling fills. The maximum droplet size, which passes through the eliminator, thus largely depends on the shape of the channel, which can be designed by CFD simulation [5]. However, if the water is contaminated with detergents, its behavior significantly changes. Due to detergents the surface tension of water decreases, which has two significant negative consequences for the effectiveness of eliminators: 1) reduces the diameter of the droplets, 2) increases the adhesion to the surface of the eliminator. As a result of these phenomena, the eliminators efficiency decreases significantly. Given that this is a new problem in a wet cooling branch, it is at first necessary for the successful solution to experimentally analyze, what processes and phenomena occur in the eliminators depending on the concentration of detergent contained in the water.

## 3 Measurement

### 3.1. Experimental facilities and measurement techniques

Visualization method was used for a monitoring of the phenomena, which occurs in the eliminators during the

* Corresponding author: rut.vitkovicova@fs.cvut.cz

passing of water with a deteriorated quality. This experiment was conducted in an experimental aerodynamic line specially adapted for measuring of eliminators characteristics. In Fig. 1 is a scheme of the experimental line. Below the inlet nozzle has placed a reservoir with water, which is supplied by a pump to the nozzle in the area below the test section.

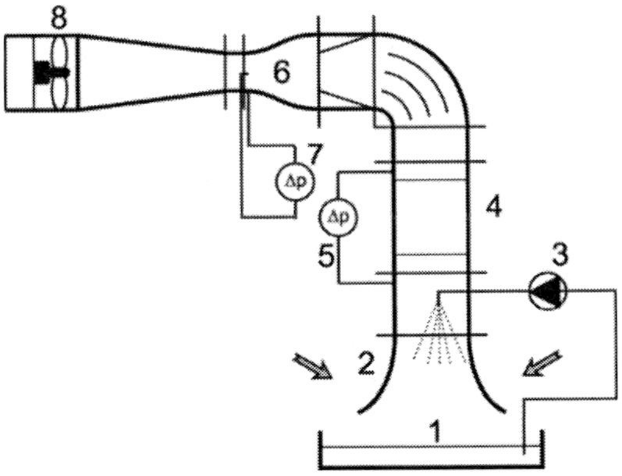


**Fig. 1.** The scheme of the experimental line: 1 – water reservoir, 2 - air inlet contraction, 3 – pump, 4 - replaceable test section, 5 - pressure drop measurement, 6 - contraction for the mass flow rate measurement, 7 - dynamic pressure measurement, 8 - driving fan

For this measurement, it was used as the spray nozzle a commonly used nozzle in cooling towers from BETE Fog Nozzle, Inc. The test section has a cross section 400x400 mm and a height 600 mm and is made from transparent plexiglass, order to be able to carry out optical measurements and visualization. The indisputable advantage of this experimental line is a relatively small reservoir, which allows easy replacement of water sprayed.

A visualization based on the principle specified in [6] was chosen for a basic analysis of the behaviour of water with detergents in the eliminators. A continuous laser was used as the lighting in this case. The images were obtained by a camera Nikon D 80 with a lens with a focal length of 85 mm. A surface tension of water with various concentrations of surfactant was determined by the method described in [7].

### 3.2 Experimental set – up

Test measurements were carried out for the blade type eliminators, which were placed in a special box (Fig.2). This box allows the optical access to the channel between the eliminators. Order to be able to monitor this area, eliminator was specially adapted, when the middle part was made from a transparent polycarbonate. This allowed to laser sheet an access to the channel at an angle of about 45 °.

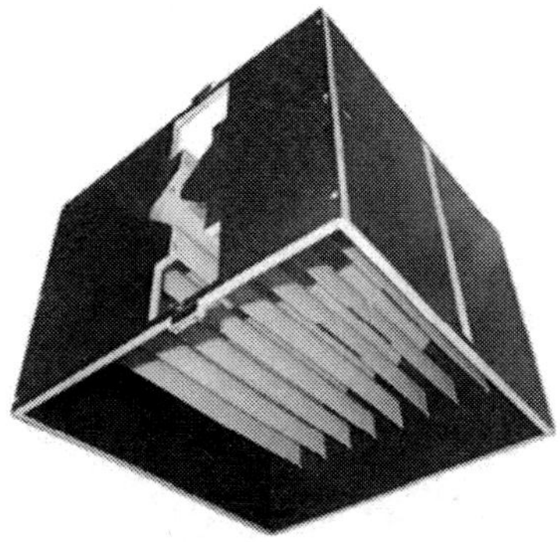

**Fig. 2.** The box with eliminators

It was observed 9 regimes as indicated in Table 1. For the three air velocity was watching the behavior of tap water droplets, water droplets with a detergent in a ratio of 1: 1000 and water droplets with a detergent in a ratio of 1:500.

**Table 1.** The regimes of measurement

| Regime | Air Velocity [m/s] | Water | Surface tension σ [ $10^{-3}$ N.m$^{-1}$] |
|---|---|---|---|
| 1 | 3 | Tap water | 72.7 |
| 2 | 4.2 | Tap water | 72.7 |
| 3 | 4.8 | Tap water | 72.7 |
| 4 | 3 | Detergent 1:1000 | 61.7 |
| 5 | 4.2 | Detergent 1:1000 | 61.7 |
| 6 | 4.8 | Detergent 1:1000 | 61.7 |
| 7 | 3 | Detergent 1:500 | 53.9 |
| 8 | 4.2 | Detergent 1:500 | 53.9 |
| 9 | 4.8 | Detergent 1:500 | 53.9 |

## 4 Results

The first test measurements were the measurement with clean tap water, the surface tension values reached 72.68. $10^{-3}$ N.m $^{-1}$. In this case, pure water catches on the sides of the eliminators, where it forms a water film which extends up to half of the height of the channel between the eliminators (Fig.3). From the obtained images, it is seen that the situation may arise whereby smaller droplet detached from a water film, but even that is captured in the channel of eliminator (see Fig. 4 at the end of this article). The used method of visualization enables to capture only the larger droplets trace, however, the presence of smaller droplets can detect by changes of brightness in the channel and in places just above it. It has been found from the measurement by IPI method that droplets which may pass through this eliminator, have a size on the order of microns to about 60 microns. [8].

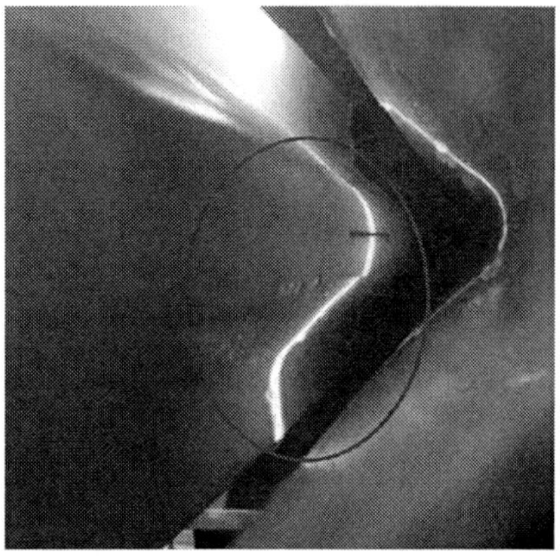

**Fig. 3.** The eliminators after the measuring with pure water (in the red ellipse it is shown the wet part of eliminators wall)

When increasing the velocity slightly increases the number of droplets entering into the eliminators channel, and thus slightly decreasing the efficiency of elimination (Fig. 4a – 4c). However, a water film did not form even at the highest measured velocity at the top of the channel.

In a second test measurement water contained detergent in a ratio of 1: 1000 and its surface tension dropped to 61.7. $10^{-3}$ N.m $^{-1}$. In this case, it was already a considerable difference in the behavior of droplets. In Figures 4d – 4f are images for different inlet airflow velocities. It is apparent that the number of entering droplets increased and so has increased the number of droplets from the eliminator. It was observed on the eliminators walls the water film even in areas that were not wetted in the case of zero concentration of detergent in the water and at the same air flow velocity. This fact has one of the major impacts on the ability of the drift eliminator to capture water. When reaching a certain velocity, called the break throw velocity, it happens that the droplets will begin to detach from a water film which is then entrained in air stream out of the tower. It wasn't observed detaching of droplets from water film during measurements with clean water for a given velocity, however at the time of the second set of measurements with water with reduced surface tension, it was observed detaching of the droplets from water film at the same velocities.

The third test measuring was carried out for the concentration of detergent in water in the ratio of 1: 500, and the surface tension of this solution decreased to 53.9. $10^{-3}$ N.m $^{-1}$. In this measurement, it hasn't been a substantial difference by this visualization method in passing droplets through the channel compared to the previous case. The water film is formed over the whole surface of the eliminator. Images from this measurement are shown in Figures 4g – 4i. The more obvious difference was observed only above the eliminators, where it could see more flooding. Due to the lower surface tension, the droplet diameter decreased, therefore, a larger amount may go through eliminator.

## 5 Conclusion

This work deals with an introduction to the issue of cooling systems with using of water with deteriorated quality. There was studied the influence of changes in surface tension of water on the ability eliminators capture water droplets. The assumptions confirm on basic experimental measurements, that water with a slightly reduced surface tension is significantly worse captured using conventional eliminators. It was verified from visualizations and observations that more drops of water passing through the channel of eliminator already when the surface tension was reduced by about the 10. $10^{-3}$ N.m$^{-1}$, thus to the value 61.7. $10^{-3}$ N.m$^{-1}$. These were either captured on the wall, wherein it formed the water film or, if their size was small enough, flew out of the eliminator. Just increased the number of smaller droplets and higher wettability is due to the smaller surface tension. It was observed higher drift of water over eliminators with increasing air velocity and decreasing surface tension. To obtain detailed information, such as droplet size distribution above and below the eliminator and the overall efficiency of the eliminator relative to the water flow, it is necessary to conduct further experiments.

## Acknowledgements

The author acknowledges the support received from Advanced technologies for heat and electricity production TE01020036 and the Centre of 3D volumetric anemometry - COLA, reg. č. CZ.2.16/3.1.00/21569 - Operating Program Prague competitiveness.

## References

1. D.G. Kröger, *Air-Cooled Heat Exchangers and Cooling Tower*, (2004)
2. P. Vitkovic, L. Dvorak, HTFF, **127**, (2016)
3. A. Kouchi, R. Ohba, K. Okabayashi, T. Nonaka, Mitsubishi heavy industries Ltd. technical review, Vol. 36, **3**, (1999)
4. J. Cizek, J. Nozicka, AIP Conf. Proc. 1745, 020004 (2016)
5. J. Stodůlka, R. Vitkovičová, EPJ Web of Conferences, **114**, 02111 (2016)
6. Pavol Vitkovic, EPJ Web of Conferences, **92** 02109 (2015)
7. S. V. Chichkanov, V. E. Proskurina, V. A. Myagchenkov, Butlerov Communications. **3** (9): 33–35, (2002)
8. J. Cizek, L. Novakova, J. Nozicka, *Reducing Liquid Phase Drift from Cooling Towers*, (Gradient, Prague 2009)

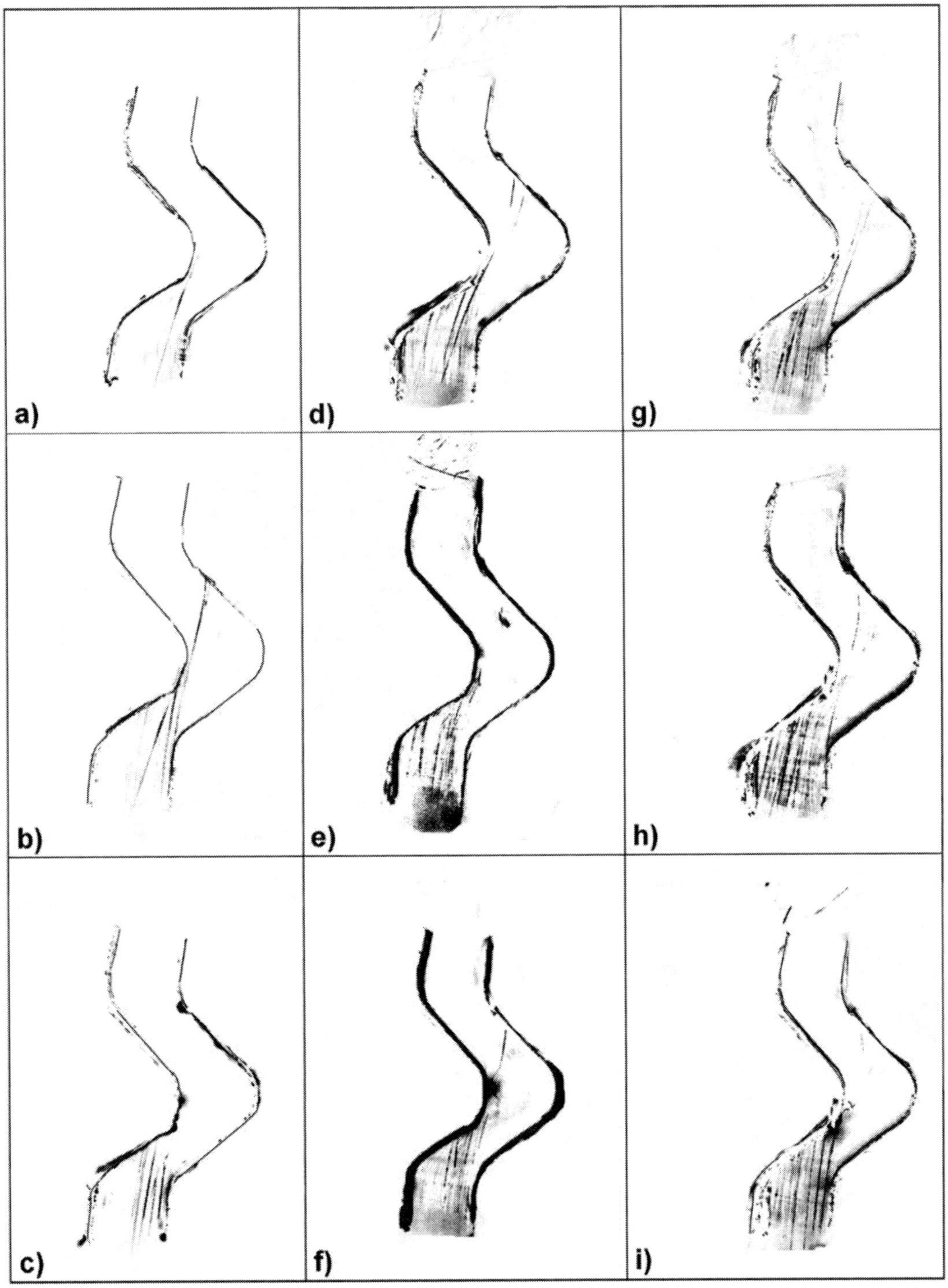


**Fig. 4.** Visualisations of the dependences of the surface tension on inlet velocity: a) tap water, $\sigma = 72.68.\ 10^{-3}$ N.m $^{-1}$, v = 3 m/s, b) tap water, $\sigma = 72.68.\ 10^{-3}$ N.m $^{-1}$, v = 4.2 m/s, c) tap water, $\sigma = 72.68.\ 10^{-3}$ N.m $^{-1}$, v = 4.8 m/s, d) detergent 1: 1000, $\sigma = 61.7.\ 10^{-3}$ N.m $^{-1}$, v = 3 m/s, e) detergent 1: 1000, $\sigma = 61.7.\ 10^{-3}$ N.m $^{-1}$, v = 4.2 m/s, f) detergent 1: 1000, $\sigma = 61.7.\ 10^{-3}$ N.m $^{-1}$, v = 4.8 m/s, g) detergent 1: 500, $\sigma = 53.9.\ 10^{-3}$ N.m $^{-1}$, v = 3 m/s, h) detergent 1: 500, $\sigma = 53.9.\ 10^{-3}$ N.m $^{-1}$, v = 4.2 m/s, i) detergent 1: 500, $\sigma = 53.9.\ 10^{-3}$ N.m $^{-1}$, v = 4.8 m/s

# Energy efficiency and economic analysis of the thermomodernization of forest lodges in the Świętokrzyski National Park

*Sylwia* Wciślik[1,*]

[1]Kielce University of Technology, Department of Piped Utility Systems, Aleja Tysiąclecia Państwa Polskiego 7, 25-314 Kielce, Poland

**Abstract.** This paper analyses energy efficiency of thermomodernization project on the example of three forest lodges located in the Świętokrzyski National Park. Currently, one of the basic requirements posed for the buildings subjected to modernization is to reduce carbon dioxide emissions even above 80% in comparison with the original values. In order to fulfil such criteria, it is necessary to apply alternative solutions based on renewable energy sources. Due to limited budget, low cubic capacity and location of the buildings, solar collectors with storage tanks and biomass boilers provide a rational option. For such a case, the emissions of basic pollutants such as $CO_2$, $SO_x$, $NO_x$ or particulates is obtained. The study also gives the results of calculations of payback time (SPBT) for the investment for exemplary forest lodge.

## 1 Introduction

Emissions of harmful substances into the air, associated with various activities and manufacturing processes, are currently under tight state control all over the world [1]. According to [2], anthropogenic activities are responsible for 78% greenhouse gases emissions in the EU countries. Construction industry, broadly understood, is one of the economy sectors which generates considerable pollution [3]. Therefore, different tasks are undertaken to increase energy efficiency by reducing energy consumption for heating, ventilation, and domestic hot water production. Also, it is aimed to reduce energy losses in local heating plants and local energy sources that power them. Conventional energy sources could be replaced, completely or partially, with the renewable ones. Some activities are intended to control lighting intensity and maintain it at the required level, and to make improvements in the equipment efficiency and in control of the power supply [4].

Better efficiency of end-use energy can be obtained by the development of new innovative technologies [5]. As regards existing buildings, to achieve that outcome, it is necessary to perform energy audits, and also collect inventory data on heat and cooling sources, and sanitary installations with respect to technical and construction aspects [6].

It is therefore time-consuming to perform an energy audit, and the auditors need to be professionals with expertise and experience in the theme. In [7], a simplified methodology was developed to determine the energy efficiency in buildings with small cubic capacity (single-family houses), and to analyse the return on thermal upgrading investment.

The tool employed to perform engineering diagnostics and heat loss balance is a thermal imaging camera [8, 9]. Study [10] shows how to correctly assess and interpret the results of thermography measurements for facilities located in open areas. To that end, the authors developed a method that involves the determination of the equivalent ambient temperature. That allows computing the substitute radiative ambient temperature. This value is further used to determine local heat losses through radiation in accordance with recommendations in the European standard [11].

Contactless measurement of the temperature field and qualitative thermographic analysis combined with visual examination also provides an effective tool to reveal moisture anomalies in the building envelope elements [12].

Study [13] presented a review of the existing thermography methods for diagnostics of construction faults, which contribute to increased thermal energy consumption.

In addition to reduction in the end-use energy consumption, one of major requirements for buildings intended for thermal upgrading is a reduction in carbon dioxide emissions and particulate matter emissions into the atmosphere [3]. Depending on the source of funding, it is required that after thermal upgrading the quantities mentioned above should decrease even by as much as 80% compared with the existing data [14]. It is possible to meet that requirement when alternative solutions are used, including renewable energy sources. Some solutions of that kind, however, require heavy investments, which represents a considerable limitation, especially for institutions financed from the state budget.

* Corresponding author: sylwiazw@tu.kielce.pl

DOI: 10.1051/epjconf/201714302144

This paper presents an analysis of energy efficiency and environmental effect of thermal upgrading. The project involved three forest lodges located in the Świętokrzyski National Park (central Poland). The paper also discusses economic issues related to the selection of the optimal improvements.

## 2 Characteristics of the buildings

This study provides the analysis of the results of the energy and environmental audits carried out for three settlements located in the Świętokrzyskie Mountains (legal status as of 1 Dec. 2010). The settlements mostly provide residential facilities for foresters employed by the Świętokrzyski National Park. Some of the area of the buildings is used as office space.

Table 1 shows basic data on the buildings of concern. Two of them were constructed in the 1950s. Lodge 1 is a timber building, whereas two other buildings, namely Lodges 2 and 3, are traditional masonry constructions.

**Table 1.** General data on the buildings.

| Description | Forest Lodge 1 | Forest Lodge 2 | Forest Lodge 3 |
|---|---|---|---|
| Structure | timber | masonry | masonry |
| Year of construction | 1926 | 1958 | 1960 |
| Cubic capacity, $m^3$ | 459.7 | 830.3 | 571.7 |
| Heated volume of the building, $m^3$ | 459.7 | 830.3 | 571.7 |
| Floor space, $m^2$ | 179.9 | 261.8 | 194.5 |
| Usable floor area of the residential part, $m^2$ | 129.1 | 216.0 | 148.7 |
| Form factor, $m^{-1}$ | 0.98 | 0.43 | 0.57 |
| Building users, number | 4.00 | 4.00 | 3.00 |
| Ventilation | natural | | |

It was decided that the settlements needed major refurbishment because of the age of the buildings, the condition of the building envelope elements and that of window and door joinery. The refurbishment was extended to include thermal upgrading. Table 2 lists heat transfer coefficients for the building envelope elements before thermomodernisation and after thermal upgrading.

Because of the buildings location in the vicinity of deciduous trees, the area layout needed to be altered in such a way as to prevent further biological degradation of the walls that is observable in Figure 1. Biological corrosion was found on the surfaces of interior timber ceilings.

**Table 2.** Heat transfer coefficients of the building envelope elements with thermal upgrading, $U$ W/m$^2$K

| | | Before thermomodernization | After thermomodernization |
|---|---|---|---|
| No. | **Forest Lodge 1** | | |
| 1 | Exterior walls | 0.77, 0.80 | 0.24, 0.25 |
| 2 | Windows | 2.20, 2.20 | 1.10, 2.20 |
| 3 | Doors/traffic doors | 2.60, 2.60 | 2.00, 2.00 |
| 4 | Floors on the ground | 1.28 | 0.28 |
| 5 | Ceilings over the passage | 1.32 | 0.29 |
| | **Forest Lodge 2** | | |
| 1 | Exterior walls | 0.73, 0.51 | 0.24, 0.21 |
| 2 | Roof/flat roof | 1.35, 2.36 | 0.19, 0.47 |
| 3 | Doors/traffic doors | 2.60 | 2.20 |
| 4 | Interior ceilings | 0.61 | 0.22 |
| | **Forest Lodge 3** | | |
| 1 | Exterior walls | 0.73, 0.81, 0.51 | 0.24, 0.25, 0.21 |
| 2 | Roof/flat roof | 0.60 | 0.21 |
| 3 | Windows | 2.20 | 1.10 |
| 4 | Doors/traffic doors | 2.60 | 2.20 |
| 5 | Interior ceilings | 0.49 | 0.22 |

**Fig. 1.** A view of the western façade of Forest Lodge 2 after partial thermal upgrading (roof replacement), the photograph by Piotr Szafraniec, Świętokrzyski National Park.

Additionally, computations concerning thermal and moisture conditions, carried out acc. [15], demonstrated that the exterior wall in Forest Lodge 2 ($U$ = 0.733 W/m$^2$K, $R_c$ = 1.365 W/(m$^2$K)) was incorrectly designed with regard to avoiding mould growth. The effective value of the temperature coefficient of the building envelope element, $f_{Rsi}$ = 0.712, was determined on the basis of the heat transfer coefficient $U$ and the total thermal resistance of the element $R_c$. Steam condensation is observed in the months of the space heating season from November till March. February turned out to be a critical month as the temperature coefficient was $f_{Rsi.max}$ = 0.732.

To avoid the accumulation of moisture, the building envelope element should be designed in such a way so that the condition $f_{Rsi} > f_{Rsi.max}$ could be satisfied. Figure 2 shows pressure distribution in the element of concern. It can be seen that in the second and third layers of the wall, the saturation pressure $P_{n.sat}$ takes on the values lower than steam partial pressure $P_n$, which results in precipitation.

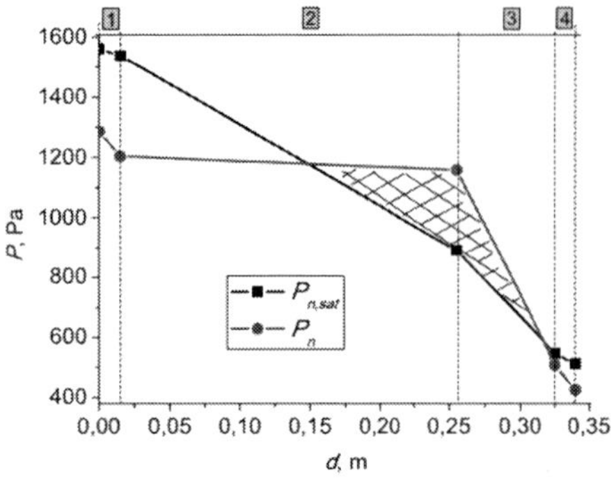


**Fig. 2.** Graph of moisture condensation in the critical month in the exterior wall shown in Figure 1; 1-4 – the envelope components in succession: cement-lime plaster, hollow masonry unit, cement-bonded particle board, cement-lime plaster.

In all three buildings, the heat source for central heating installations is a ~25 kW boiler fired by solid fuel, mainly wood. Because of good condition of the boilers, replacements are not planned. In all buildings, domestic hot water, is delivered due to gravitation to the 100 $dm^3$ tank via the boiler. In the audit, it was planned to modernise DHW installation and to set up solar collectors.

## 3 Analysis of the optimum thermal upgrading option

To have a part of the loan, taken out for thermal upgrading, written off, it was necessary to show that the planned works would result in at least 25% reduction in thermal energy demand, and also lower emissions of greenhouse gases. Having a limited budget for thermomodernization, to meet the conditions above, it was decided to upgrade the building envelope elements with the highest heat transfer coefficient $U$ and to modernise the domestic hot water system. In all settlements, flat solar collectors, relying on renewable energy sources, with the area of $6m^2$, 300 $dm^3$ DHW tank, 18 $dm^3$ diaphragm expansion vessel and other minor pieces of hydraulic equipment, together with thermal insulation. were installed. It was assumed that solar collectors should provide approx. 70% of DHW demand. The remaining portion should be obtained by biomass combustion in the existing boiler.

Figure 3 shows Forest Lodge 2 after thermal upgrading with solar collectors installed on the south side of the building. The building envelope elements were appropriately designed to avoid mould growth.

Table 3 lists selected improvements included in the optimum thermal upgrading option, ranked in the ascending order of the value of return on the investment expressed as Simple Pay-Back Time (SPBT) for an exemplary settlement, namely Forest Lodge 2.

**Fig. 3.** A view of the eastern and southern façades of Forest Lodge 2 after thermal upgrading (solar collectors installed), the photograph by Piotr Szafraniec, Świętokrzyski National Park.

**Table 3.** Improvements included in the optimum thermal upgrading option and Simple Pay-Back Time (SPBT) for Forest Lodge 2.

| Item | Improvement/ modernisation | Planned costs of works, EUR | SPBT, years |
|---|---|---|---|
| 1 | Ceiling above the cellar | 2832.15 | 10.93 |
| 2 | Domestic hot water installation | 3698.22 | 15.05 |
| 3 | Roof | 8216.79 | 22.01 |
| 4 | Exterior wall 1 | 5465.39 | 44.72 |
| 5 | Exterior doors 1 | 507.23 | 52.13 |
| 6 | Exterior doors 2 | 507.23 | 52.13 |
| 7 | Exterior wall 2 | 1296.45 | 73.67 |
| 8 | Foundation wall | 881.72 | 148.02 |

High SPBT values should be given some attention. It turned out that thermal upgrading of old deteriorated buildings could not be performed without a major retrofit. That generates high unit costs that can significantly affect the pay-back time of the investment outlays. If the audit were conducted strictly in accordance with the legal regulations [16], the works not directly related to improvements in thermal conditions, or those not considered economically viable (e.g. modernization of the foundation wall in Forest Lodge 2), should not be classified as thermal upgrading. In reality, it is difficult to clearly distinguish between thermal upgrading and retrofit.

Additionally, thermal energy demand was established by means of computations [17]. For central heating system, especially in old buildings, it is difficult to unambiguously determine the demand value without earlier monitoring of the internal environment parameters [18]. The rooms are most often underheated, the design internal temperature values are not reached during the heating period, and moisture accumulations are found on the envelope elements (see Figs. 1 and 2).

It should be added that the optimum thermal upgrading option that was selected was the one that could generate the greatest energy costs savings. The costs were calculated to meet the annual energy demand for heating and hot domestic water production. The savings, expressed as percentage values for individual settlements, are presented in Table 4.

As a result of thermal upgrading work, energy performance of the buildings will be also considerably altered. Major computational parameters are presented in Table 5.

**Table 4.** Savings that result from the choice of the optimum thermal upgrading option.

| Settlement | Annual savings in thermal energy demand, % | Annual savings in energy costs, % |
|---|---|---|
| Forest Lodge 1 | 35.23 | 33.18 |
| Forest Lodge 2 | 32.19 | 31.70 |
| Forest Lodge 3 | 37.66 | 36.31 |

**Table 5.** Energy performance of the buildings.

| | | before thermomodernization | | | after thermomodernization | | |
|---|---|---|---|---|---|---|---|
| No. | **Forest lodge** | 1 | 2 | 3 | 1 | 2 | 3 |
| 1 | Design thermal output of the heating system, kW | 26.88 | 23.41 | 23.17 | 18.02 | 16.66 | 17.20 |
| 2 | Design thermal power for DHW production, kW | 2.03 | 1.63 | 2.03 | 2.03 | 2.03 | 2.03 |
| 3 | Annual heat demand for the building space heating (disregarding the heating system efficiency and space heating seasonal variation), GJ/year | 363.83 | 290.25 | 202.56 | 262.83 | 206.31 | 158.88 |
| 4 | Computation of energy consumption for DHW production, GJ/rok | 65.69 | 72.26 | 65.69 | 16.80 | 12.60 | 14.40 |
| 5 | Design non-renewable primary energy demand (PE), kWh/m$^2$a | | | | 61.7 (145.07)* | 64.6 (146.98)* | 46.7 (185.63)* |

* PE for a reference building after upgrading acc. [19]

## 4 Environmental effect

The measurable effect of the thermal upgrading of concern is a reduction in emissions of harmful substances into the atmosphere. Table 6 lists specific emission factors for different pollutants, pertinent to different fuel types [20].

The computations of yearly amounts of harmful substances emitted into the atmosphere, including the most important ones, i.e. carbon emissions, together with the environmental effect for exemplary Forest Lodge 2 are shown in Figure 4.

**Table 6.** Specific emission factors for different pollutants.

| Fuel type | Unit | SO2 | NOX | CO | CO2 | Particulates | Soot | Benzo(a)pyrene (BaP) |
|---|---|---|---|---|---|---|---|---|
| Solar collectors | kg/Mg | 0 | 0 | 0 | 0 | 0 | 0 | 0 |
| Electric energy | kg/kWh | 0.91 | 0.0023 | 0.00069 | 1 | 0.0015 | $27 \cdot 10^{-7}$ | $54 \cdot 10^{-9}$ |
| Biomass | kg/Mg | 0.69 | 19.97 | 1.17 | 0 | 0.69 | 0 | 0 |

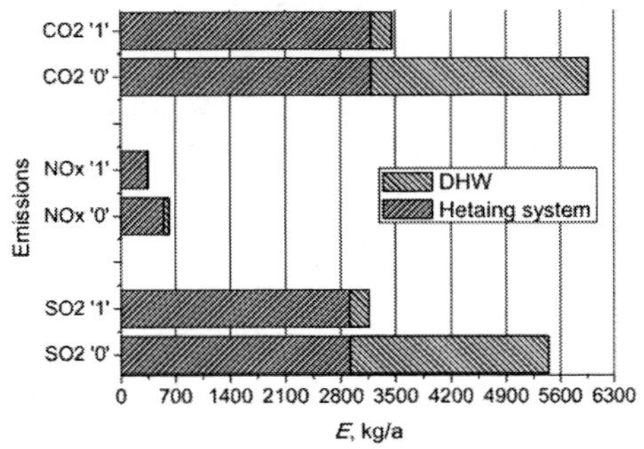


**Fig. 4.** Environmental effect of thermal upgrading for exemplary Forest Lodge 2; '0' – before, '1' – after thermomodernization.

Multiplying the computed emissivities by the so-called toxicity index $K$, equivalent emissions for individual substances before ($E_{r0}$) and after ($E_{r1}$) thermal upgrading are calculated. The environmental effect for equivalent emissions was determined from the dependence: $E = E_{r0} - E_{r1}$. The results are presented in Table 7.

**Table 7.** Equivalent emissions for the optimal thermal upgrading option in Forest Lodge 2.

| Pollutants emitted | Toxicity index K | Emissions | | Equivalent emissions | |
|---|---|---|---|---|---|
| | | Before | After | Before, $E_{r0}$ | After, $E_{r1}$ |
| | | upgrading, kg/year | | upgrading, kg/year | |
| $SO_2$ | 1.00 | 5448.57 | 3158.91 | 5448.57 | 3158.91 |
| $NO_X$ | 0.75 | 618.55 | 353.71 | 463.91 | 265.28 |
| Particulates | 0.75 | 29.84 | 17.13 | 22.38 | 12.85 |
| Soot | 3.75 | 0.02 | 0.01 | 0.06 | 0.04 |
| BaP | 30000.0 | 0.00 | 0.00 | 9.66 | 5.60 |
| Total equivalent emissions | | | | 5944.58 | 3442.68 |
| Environmental effect for equivalent emissions, kg/year | | | | 2501.90 | |
| Environmental effect for equivalent emissions, % | | | | 42.09 | |

## 5 Conclusions

The study presents the analysis of the energy efficiency of a thermal upgrading project that covered three forest lodges located in the Świętokrzyski National Park (Poland), and constructed in the years 1958, 1960 and 1926. In the study, attention was paid to high values of SPBT of the investment, which was shown in Table 3. For instance, thermal upgrading of the exterior walls in one of the lodges will recoup in as long as 45 years. High SPBT values are related to large outlays on repairs and ancillary works. Additionally, static indicators of economic efficiency are usually overestimated and they do not account for the whole period of the project operation. They are intended to provide initial assessment of the investment profitability.

It was shown that in all cases thermal upgrading results in the reduction in the annual thermal energy demand by at least 30%. Moreover, annual energy costs will be higher than also 30%.

Additionally, computational demand for non-renewable primary energy in lodges 1, 2 and 3 is 47, 62 and 65 kWh/m$^2$a, respectively.

The energy audit demonstrated that the environmental effect produced by thermal upgrading is a reduction in greenhouse gasses emissions. The total equivalent emissions are 2501 kg/year, which amounts to 42.09 %.

## References

1. United Nations framework convention on climate change
2. Directive 2006/32/EC of the European Parliament and of the Council of 5 April 2006 on energy end-use efficiency and energy services and repealing Council Directive 93/76/EEC.
3. H.P. Ebert, EPJ Web of Conferences 98. 08001 (2015)
4. M. Ajmone-Marsan, D. Arrowsmith, W. Breymann, O. Fritz, M. Masera, A. Mengolini, A. Carbone, Eur. Phys. J. Special Topics 214. 547 (2012)
5. M. Iglesiasla, J. Rodriguez, D. Franco, EPJ Web of Conferences 33, 05004 (2012)
6. J. Gorzynski, Basics of energy analysis of buildings, OWPW Publishing 2012 (in Polish)
7. K.H. Kim, J. S. Haberl, Build. Simulation 8(5), 515 (2015)
8. T. Kruczek, Energy 91, 35 (2015)
9. S. Zwierzchowska (Wciślik), T. Orzechowski, A. Tyburczyk, Instal. Theory and practice in installations 12, 32 (2010) (in Polish)
10. T. Kruczek. Arch. of Thermodynamics 36. 67 (2015)
11. PN-EN ISO 12241 Thermal insulation for building equipment and industrial installations – Calculation rules. 2001
12. J.R. Kominsky, J.S. Luckino, T.F. Martin, Passive infrared thermography—a qualitative method for detecting moisture anomalies in building envelopes. Tedford and Pond 1 (2007).
13. M. Fox, D. Coley, S. Goodhew, P. Wilde, Thermography methodologies for detecting energy related building defects, Renew. Sust. Energ. Rev. 40, 296 (2014)
14. http://www.eog.gov.pl
15. PN-EN ISO 13788 Hygrothermal performance of building components and building elements. Internal surface temperature to avoid critical surface humidity and interstitial condensation. Calculation methods, 2012
16. Regulation of the Minister of Infrastructure of 17 March 2009 on the detailed scope and forms part of the energy audit and the audit of repair, design audits cards, as well as the algorithm assessing the profitability of the thermomodernization project, as amended (in Polish).
17. PN-EN 12831 Heating systems in buildings, 2006
18. M. Bagona, M. Lopusniak, D. Katunsky, Tech. Trans. Civil Eng. 2-B/2010, issue 4, Cracow University of Technology Publishing (2010)
19. Regulation of the Minister of Infrastructure of 12 April 2002 on the technical conditions to be fulfilled

DOI: 10.1051/epjconf/201714302144

by buildings and their location (Journal of Laws No 75. item. 690, as amended) (in Polish)
20. Information and instruction materials of MOŚZNiL 1/96 'The emission of pollutants into the air from the processes of fuels combustion' Journal of Laws 2004 No. 281, item. 2784

EPJ Web of Conferences **143**, 02145 (2017) DOI: 10.1051/epjconf/201714302145
*EFM 2016*

# Laminar-turbulent transition regimes in the conical Taylor-Couette flow system

*Fatma* Yahi[1, 2,*], *Ahcene* Bouabdallah[1], *Yasmina* Hamnoune[1], *Valéry* Botton[3], *Daniel* Henry[3], *Séverine* Millet[3], *François* Rousset[4] and Takahiro Adachi[5]

[1]*Thermodynamics and Energetic Systems Laboratory*, Faculty of Physics, University of Sciences and Technology Houari Boumediene, B.P. 32 El Alia 16111 Bab Ezzouar, Algiers, Algeria.
[2]*Genie Physical of Hydrocarbons Laboratory*, Faculty of hydrocarbons and chemistry, University M'Hamed Bougara, 35000 Boumerdes, Algeria.
[3]*Laboratoire de Mécanique des Fluides et d'Acoustique*, CNRS/ Université de Lyon, Ecole Centrale de Lyon/Université Lyon 1/INSA de Lyon, ECL, 36 Avenue Guy de Collongue, 69134 Ecully Cedex, France.
[4]*CETHIL, UMR CNRS 5008*, Université de Lyon, INSA de Lyon/Université Lyon 1, INSA, Bâtiment Sadi Carnot, 9 rue de la Physique, 69621 Villeurbanne Cedex, France.
[5]*Department of Mechanical Engineering*, Akita University, Tegata-Gakuen 1-1, Akita 010-8502, Japan.

**Abstract.** The present work is intended to experimentally study the Taylor-Couette flow between coaxial cones. The inner cone is rotated and the outer cone is maintained fixed. Both cones have the same apex angle $\Phi$ =12°, giving a constant annular gap $\delta$ =0.12. The height of the fluid column is H=155 mm. The working fluid is assumed as Newtonian and has constant properties (as density and viscosity) within the range of the required experimental conditions. By means of visualization techniques, the critical thresholds related to the onset of various instabilities have been obtained and the corresponding flow modes have been identified. Using images processing, spatio-temporal diagrams have also been calculated, showing the characteristics (wavelength, drift velocity) of the downward helical motion. The results obtained for these transition regimes are compared to those of Wimmer et al. [1-3].

## 1 Introduction

The study of instability phenomena in rotating systems remains very crucial for many industrial applications, as those concerned by mixing. Indeed, a better control of the processes could be obtained by a better understanding of the flow mechanisms involved and the way they vary with the governing parameters and the geometrical factors.

The investigation of the hydrodynamic instabilities in such systems has both objectives of predicting the critical parameters responsible for their onset and characterizing the flow which are thus triggered.

This kind of study in rotating systems can be helpful for the design of rotating machinery such as multiple concentric drives or turbine rotor, but also for the optimization of chemical equipment, such as compact rotating heat exchangers and mixers.

The viscous flow between concentric rotating cylinders has been widely studied. Concerning the conical geometry, Wimmer [1, 2, and 3] intensively investigated the Taylor-Couette flow in such geometry, showing experimentally that the flow is sensitive to the initial and boundary conditions, the gap size, the rotation velocity and the apex angle. Wimmer [1] demonstrated the coexistence of steady and unsteady flow regimes, which is not observed in the cylindrical geometry. Noui-Mehidi et al. [4, 5, 6, 7] found, numerically, that when the apex angle $\Phi$ tends to 0°, the flow structure is identical to the classical Taylor-Couette flow and when the apex angle is increased towards $\Phi$= 180°, the vortex pair structure starts to break down and thereafter disappears. It is also observed that the flow develops similar characteristics as the flow between disks for an apex angle $\Phi > 60°$ [5].

Noui-Mehidi et al. [8] examined experimentally the acceleration effect of the inner cone rotation on the transition properties in a conical system, characterized by an apex angle equal to $\Phi$ = 16.38°. They noted that the acceleration rate defined as dRe/dt has a significant effect on the final flow: if the acceleration rate is less than 6.08, a downward helical motion was observed, while for accelerations greater than 6.08, an upward helical motion took place. The acceleration rate can also affect the vortices number. It is also observed that, for dRe/dt = 0.06, six steady pairs of vortices develop at Re = 340, whereas when the acceleration is dRe/dt = 1.3, seven steady pairs of vortices appear at Re = 520 and eight pairs at Re = 730. It is mentioned that the vortices number is very sensitive to the apex angle [5-8]. More recently, Xu et al. [11] showed in their numerical simulations that the maximum velocity and the pressure magnitudes decrease with increasing the cone apex angle. Xu et al. [12] observed that the behavior of the

* Corresponding author: yahi.fatma@gmail.com

conical flow is dominated by a competition between the meridional flow and the radial flow. Li et al. [12] studied the transition to Taylor vortices between rotating conical cylinders in the case of rigid boundaries, where both end plates are stationary.

The main objective of this paper is to highlight the critical thresholds corresponding to the onset of the different instabilities occurring in the flow between two vertical coaxial cones. By means of visualization techniques, we were able to show the different transition modes with their characteristics. Using images processing, we were able to plot the spatio-temporal diagrams corresponding to these instabilities of the laminar-turbulent transition regime. The results are compared to those of Wimmer et al. [1-3].

## 2 Experimental setup

The experimental device consists of two coaxial cones made of insulating and transparent material (Plexiglas) in order to allow a good visualization of flow regime. Both cones have the same apex angle $\Phi$=12° giving a constant annular gap $\delta = d/R_{1max}$ where d= (4.85± 0.2) mm. The inner cone can rotate and the outer cone is maintained at rest [14].

Our system is characterized by an outer cone with largest radius $R_{2max}$ = (45 ± 0.2) mm and lowest radius $R_{2min}$ = (12± 0.2) mm. The largest radius of the inner cone is $R_{1max}$= (40.15 ± 0.2) mm, while the lowest radius is $R_{1min}$= (7.15 ± 0.2) mm. The length of the fluid column is fixed at H = (155 ± 0.2) mm. The inner cone is driven by a DC motor connected to the rotating axis by a flexible in order to avoid the adverse effects of vibration (Fig. 1).

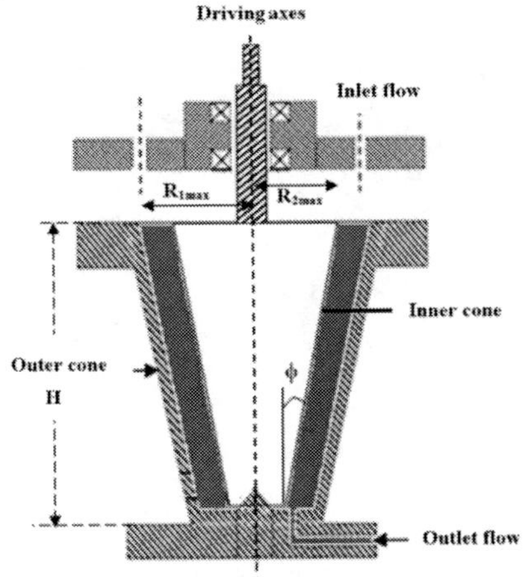


**Fig. 1:** Conical Taylor-Couette flow system.

The working fluid is a solution of 20% of Vaseline oil CHALLALA, favoring a better suspension of the particles in the fluid visualization, which is added to 80% of a petroleum product SIMILI to reduce the viscosity of the oil, with a concentration of 2 g/l of aluminum flakes. Such a mixture constitutes a Newtonian fluid characterized by its kinematic viscosity $\nu$=4.8 $10^{-6}$ m²/s and its density $\rho$ = 777.23 kg/m³ with an accuracy of 1%.

In order to characterize the onset of the hydrodynamic instabilities, it is necessary to introduce dimensionless numbers involving viscous forces, which play a stabilizing role, and centrifugal forces, which have a destabilizing effect. These dimensionless numbers, which serve as control parameters of the flow, are the Reynolds number Re, the Taylor number Ta and the Froude number Fr defined in Table 1.

**Table 1.** Control parameters.

| | |
|---|---|
| Reynolds number | $Re = \frac{R_{1max}\Omega d}{\nu}$ |
| Taylor number | $Ta = Re\delta^{1/2}$ |
| Froude number | $Fr = \frac{R_{1max}\Omega}{\sqrt{gd\cos\phi}}$ |

The manifestation of a given wave form or instability was identified with three visualization techniques [14], which are now presented.

### 2.1 Reflection of natural light

This method is based on the light beam reflection on the seeding particles. The light is supplied by an external source located in front of the experimental device. The intensity of the reflected light will depend on the orientation of the aluminium flakes, which are aligned with the local velocity vector. As a consequence, if the velocity vector is axial, the flakes will reflect light strongly and bright zones will appear on the images. On the contrary, if the velocity has a significant radial component, the flakes will be oriented parallel to the light rays and will let the light pass without reflection, giving dark zones on the images. However, as shown by Wimmer [3], two successive vortices have different orientations (Fig. 2), so that only one dark zone out of two is clearly visible on the images. In the interval between these well visible dark zones, two counter-rotating vortices are then present.

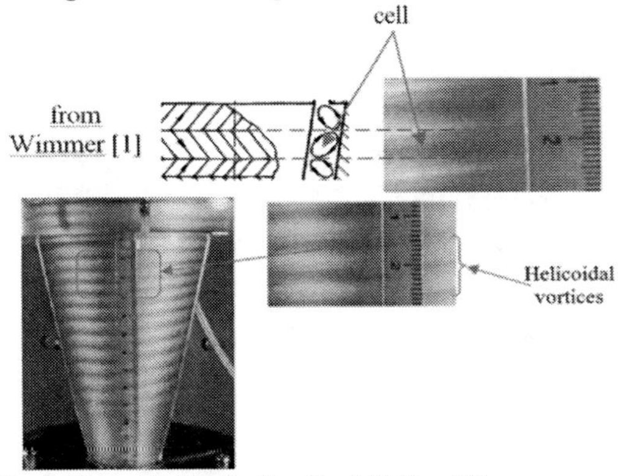


**Fig. 2:** Natural light reflection ($\delta$ = 0.12, Ta = 79).

DOI: 10.1051/epjconf/201714302145

## 2.2 Transmission of natural light

This method of visualization is based on the optical transmission of the light source, which is placed behind the experimental device. The light rays pass through the flow and provide in-depth structure of the flow. In contrast with the previous technique, dark zones correspond to axial velocity vectors and bright zones to radial orientations, and all the dark and bright zones are well visible (Fig. 3).

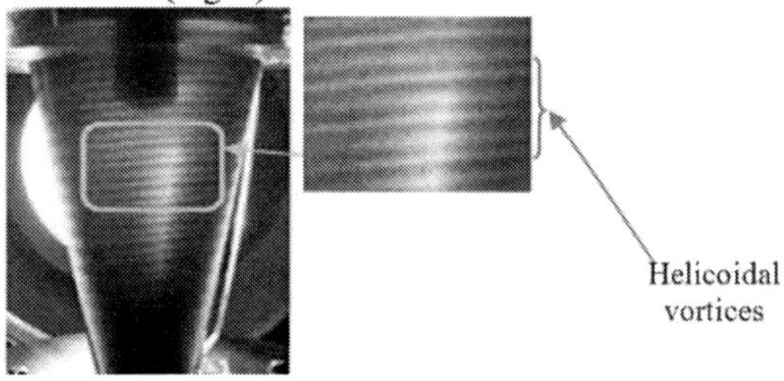


**Fig. 3:** Natural light transmission ($\delta = 0.12$, Ta = 72).

## 2.3 Laser transmission

Finally, in order to examine the local structure of the flow and better see the vortices, we have used a coherent light (laser He, 1mW and with wavelength $\lambda$=623.8nm), which is expanded as a laser sheet thanks to a cylindrical lens. This laser sheet illuminates an azimuthal plane in order to see the global vortex structure in this plane or to isolate certain vortices. Adequate settings of the optics are necessary in order to obtain good visualizations with a rigorous stigmatism of the images (Fig. 4a-4b).

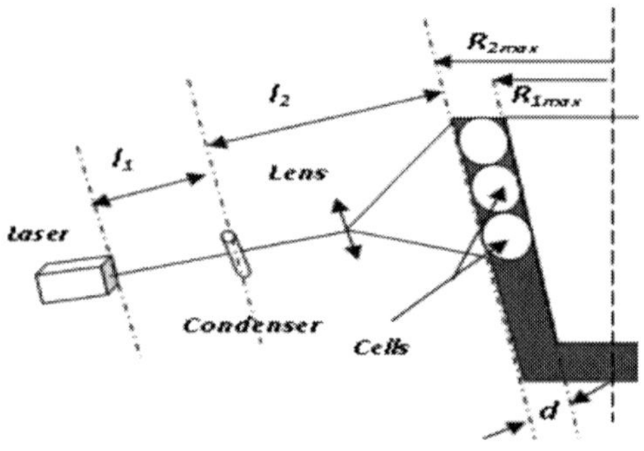


a)

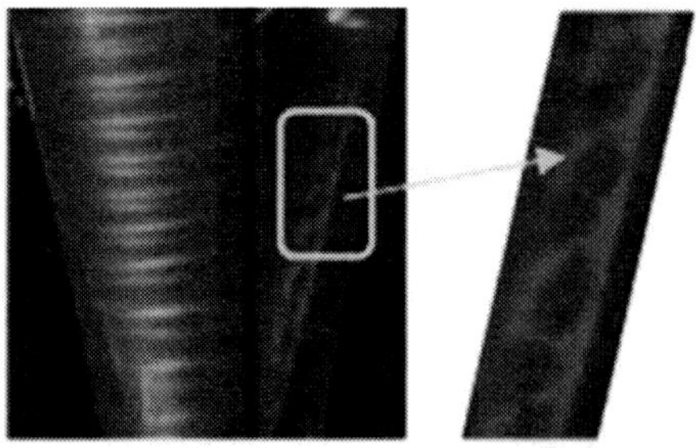

b)

**Fig. 4: a)** Laser transmission visualization layout; **b)** Laser transmission visualization ($\delta = 0.12$, Ta = 88.5).

# 3 Results and discussions

For the small gap configuration, $\delta$ = 0.12, and for different angular velocity, we investigate the basic flow and the laminar-turbulent transition regime for Taylor numbers up to Ta=318.1.

## 3.1 Basic flow

The basic flow, in the absence of any disturbances, is laminar, three-dimensional ($V_\Theta$, $V_r$ and $V_z$ velocity components) and axi-symmetric (Fig. 5). The three-dimensional character of the flow is the result of the imbalance of the centrifugal forces in the annulus: indeed, the linear variation of the radius versus the conical axial position z induces a linear variation of these centrifugal forces. The flow observed in any azimuthal plane corresponds to a large loop going up along the inner cone and down along the outer cone and turning at the end plates [1]. These end plates being fixed with no-slip conditions, the centrifugal forces decrease in their vicinity. This effect will promote end recirculations, in the direction of the main loop at the bottom and in the opposite direction at the top. The recirculation at the bottom will only reinforce the main loop. In contrast, the recirculation at the top will progressively generate a vortex in counter-rotation with the main loop. Due to the axi-symmetry of the flow, this vortex (known as Ekman vortex) is in fact a toroidal cell. In the small gap situations considered here, it has not been possible to provide evidence of this Ekman vortex.

A reason could be the difficulty to visualize the upper zone of the annulus, which is behind a large thickness Plexiglas zone (Fig. 1).

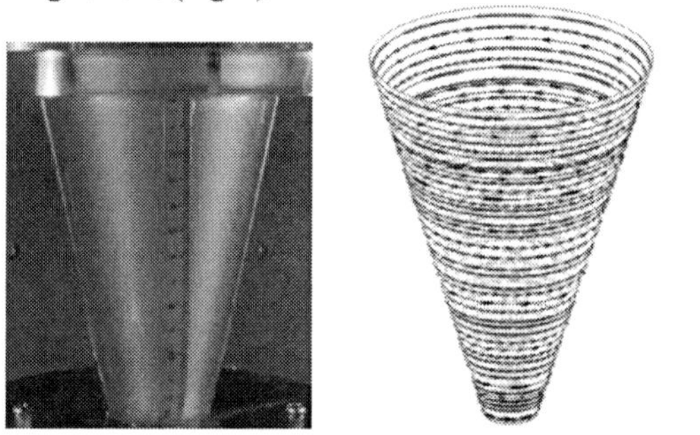

**Fig. 5:** Basic flow ($\delta = 0.12$, Ta < $Ta_{c1}$).

## 3.2 Transition regimes

By progressively increasing the angular velocity of the inner cone (corresponding to an increase of the Taylor number), we observe the development of disturbances in the upper part of the basic flow for Ta>$Ta_{c1}$=42.3. These disturbances correspond to the well-known Taylor vortices (TV), which generally appear as steady regular closed vortex cells. In our case, the first Taylor cell appears in the vicinity of the upper edge, where the centrifugal forces are greater. This toroidal cell is stationary and its size is of the order of the annular gap. It forms a pair of cells with the rest of the flow, which is unperturbed and remains a large loop (UF) (Fig. 6). Only

DOI: 10.1051/epjconf/201714302145

the first Taylor cell will be visible, before the development of the next instability.

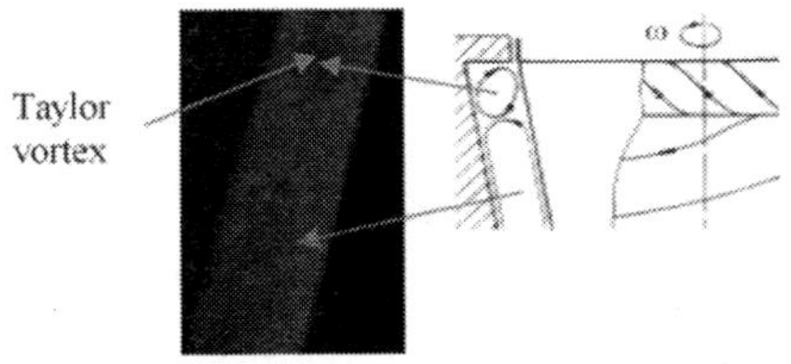


**Fig. 6:** Laser transmission visualization of the Taylor vortex (TV) (Ta = $Ta_{c1}$=42.3).

By still increasing the inner cone velocity gradually, we observe the appearance of new cells, which propagate downwards as a helical motion (DHM) (Fig. 7). This phenomenon occurs at Ta = $Ta_c^{DHM}$ = 47.2. The downward movement of the vortices seems to be related to the corresponding decrease of the radius in our geometry, as experiments performed in the up side down configuration have given an upward movement of the vortices. Note that, as for the Taylor vortices, there remain an unperturbed zone (UF), corresponding to a large loop, at the bottom part of the annulus. This unperturbed zone will only disappear for far larger Taylor numbers, as for Ta=192 (see Fig. 9). For Ta above $Ta_c^{DHM}$ and before the next instability, three different flow structures can be observed in the annulus: from top to bottom, the steady Taylor vortex (TV), the downward helical vortices (DHM) and the unperturbed flow loop (UF).

**Fig. 7:** Visualization of the downward helical motion (DHM) at Ta = 61.1.

A wavy mode (WM) corresponding to an azimuthal wave appears in the very upper part of the annulus with a very low amplitude at Ta = $Ta_{c2}$ = 68.8. The other parts of the annulus keep the same flow properties as those described above. The wavy mode first affects the Taylor vortex, but for increasing Ta, it will also affect the downward helical motion (see Fig. 9f). Finally, for Ta = $Ta_c^{UHM}$ = 72, we observe the appearance of upward helical motions of cells. At this value of Ta, we observe the co-existence of the Taylor vortex with wavy mode, the downward helical motion, the upward helical motion and the unperturbed flow at the bottom (Fig. 8). At larger Ta, we will note the possible superposition of three flow instabilities in some zones of the annulus, namely, the downward helical motion (DHM), the upward helical motion (UHM) and the wavy mode (WM).

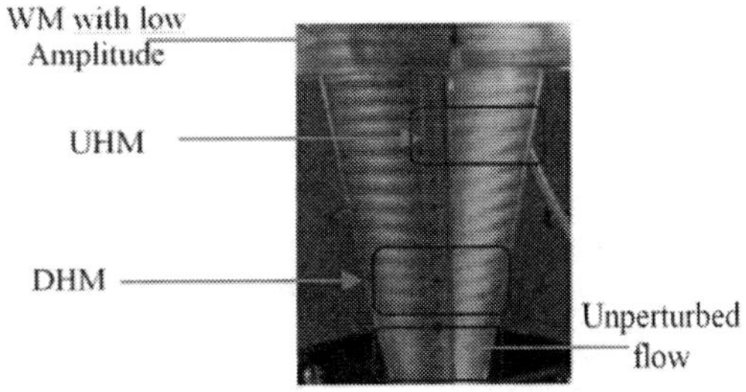


**Fig. 8.** Co-existence of four flow regimes: wavy mode (WM) affecting the Taylor vortex, the upward helical motion (UHM), the downward helical motion (DHM), and the unperturbed flow (UF) (Ta = 80.6).

By observation of Table 2, we can see that the results obtained with our experimental device for δ = 0.12 are in good agreement with those of Wimmer [1] obtained for δ = 0.1. Wimmer [1] observed the Taylor vortex, the downward helical motion and the wavy mode for values of Ta close to ours.

**Table 2:** Comparison of the critical Taylor number values for different gap configurations (completely filled flow system).

| **Critical Taylor number** | $Ta_{c1}$ | $Tac^{DHM}$ | $Ta_{c2}$ | $Ta_c^{UHM}$ |
|---|---|---|---|---|
| **Flow regimes** | **TV** | **DHM** | **WM** | **UHM** |
| Wimmer [1] δ = 0.1 | 41.6 | 50 | 74.9 | |
| Present work δ = 0.12 | 42.3 | 47.2 | 68.8 | 72 |

Some flow states showing the instability modes in a larger range of Ta values are presented in Fig. 9. We can note the larger amplitude of the wavy mode for the larger values of Ta.

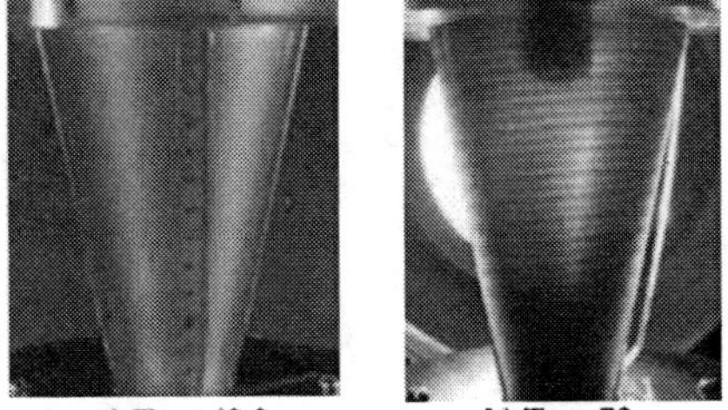

**a)** Ta < 42.3
basic laminar flow

**b)** Ta = 72
DHM

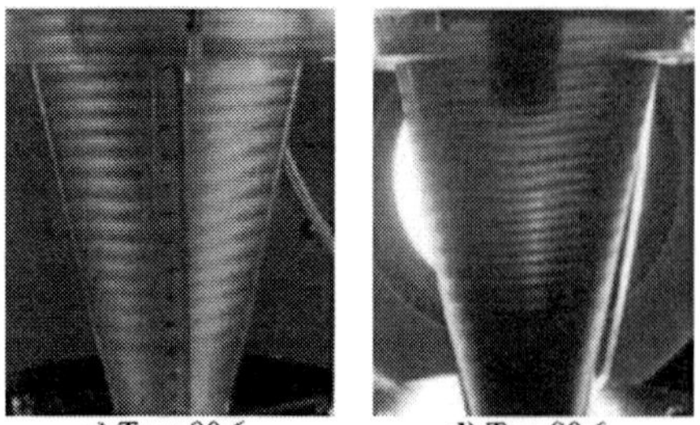

**c)** Ta = 80.6 **DHM+WM+UHM** **d)** Ta= 89.6 **DHM+WM+UHM**

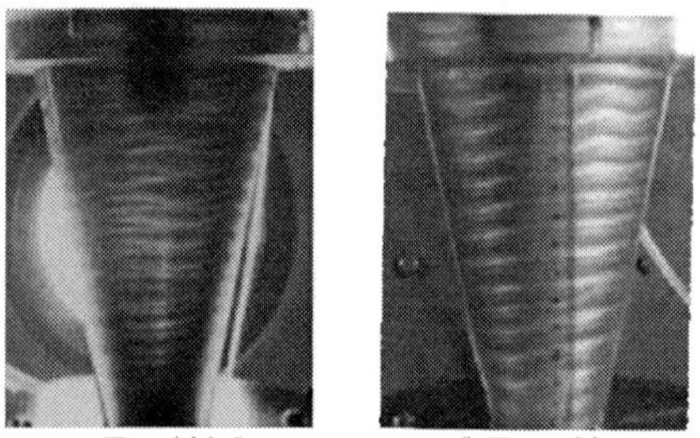

**e)** Ta =130.5 **DHM+WM+UHM** **f)** Ta =192 **DHM+WM+UHM**

**Fig. 9:** Flow state visualizations for different Ta showing the different flow regimes: **a)**, **c)** and **f)** Natural light reflection; **b)**, **d)** and **e)** Natural light transmission ($\delta = 0.12$).

Finally, we can remark that the helical structure appears on previously obtained axi-symmetric regimes (i.e. the basic flow and Taylor vortex regimes) and breaks this axi-symmetry. This indicates that the corresponding instability must be an azimuthal mode with a non-zero azimuthal wavenumber m. In fact, by accurately analyzing the images of downward helical motion (as in Fig. 10), we have found that a pair of counter-rotating vortices is shifted down by a height equal to its wavelength after a -azimuthal rotation (which is what we can see on the front view of the cone on the images). By extrapolation (to take into account the shift occurring on the back face of the cone), we deduce that the pair of vortices is shifted by two wavelengths for a turn around the cone, which indicates that the helical instability corresponds to an azimuthal mode m=2. Comparisons, however, were not possible as we did not find any indication of azimuthal wavenumbers in the previous publications on rotating cones. Numerical stability analyses could be performed to clarify this point.

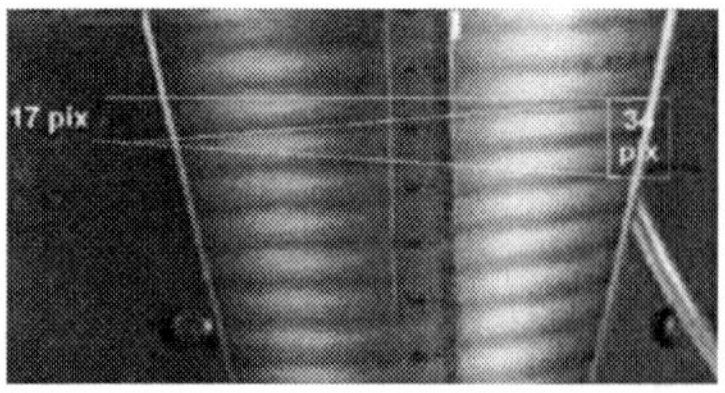


**Fig. 10:** Analysis of the downward helical motion to determine the azimuthal wavenumber of the governing instability.

### 3.3 Evolution of cells number and wavelength for different flow regimes

The analysis of the results allowed us to determine the evolution of the cells number n according to the Taylor number Ta and the flow regime. The curve showing n versus Ta is shown in Fig. 11. The vortices number n starts at one (corresponding to the Taylor vortex) in the vicinity of the first instability at Ta=42.3. It then increases quite quickly to reach its maximum value n=36 at Ta=78.3. This steep increase corresponds to the progressive downward extension of the helical structure as Ta is increased. The vortices number n keeps the same value n=36 on a large Ta gap, up to Ta=161, indicating that, for all these states, the helical structure has reached the bottom of the cone. At Ta=192, m slightly decreases to m=34, with the disappearance of a pair of counter-rotating cells. Something similar is observed at Ta=276.7, and the new structure with m=32 is obtained up to Ta=318.1, the limit Ta value in our study.

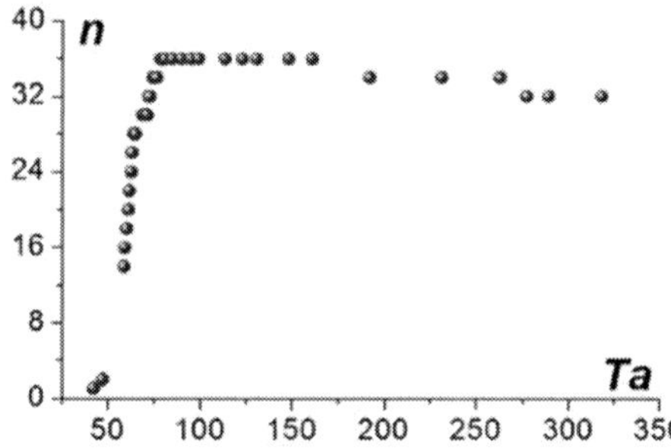


**Fig. 11:** Variation of the cell number n versus the Taylor number Ta.

By using image processing with Matlab, we were able to extract spatio-temporal diagrams from the videos available for some of the flow states (sequences of 100 to 200 images are chosen, covering time intervals of 9 to 13 seconds). For each image of the videos, we have registered the grey levels along a given generatrix of the cone (280 pixels along a generatrix). If the generatrix determined by the vertical plane perpendicular to the image is taken as a reference (this generatrix corresponds to the vertical line in the middle of the image), the generatrix chosen for the spatio-temporal diagrams presented in Fig. 12 is the generatrix at - /6 (on the left part of the image) and for the diagram in Fig. 13, it is the generatrix at /8 (on the right part of the image). These diagrams are presented as color levels in a domain with the time (in seconds) in abscissa and the position along the generatrix (in mm and with the origin at the bottom of the cone) in ordinate. They will allow us to determine the wavelength and the drift velocity of the vortices.

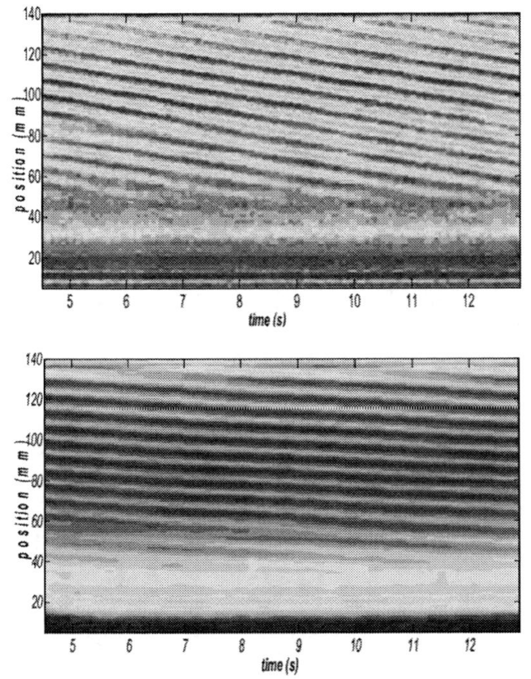


**Fig. 12:** Spatio-temporal diagrams showing the downward helical motion for Ta=64 (first view) and 74 (second view).

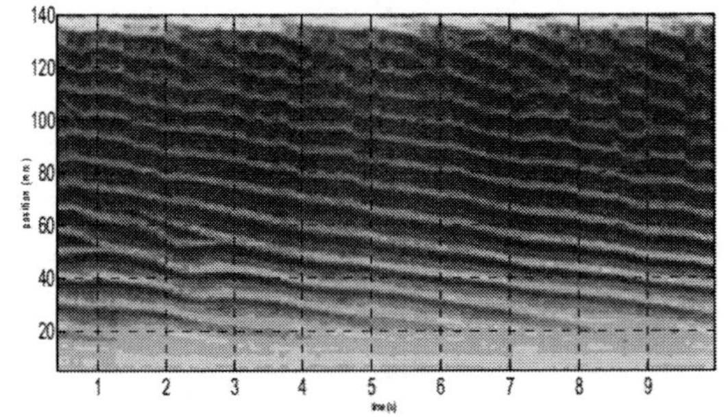


**Fig. 12:** Spatio-temporal diagram for Ta = 89.6.

In figure 12, the downward helical motion is clearly put in evidence for Ta=64 and 74. In both cases, there also remains an unperturbed flow zone at the bottom of the cone. Comparisons between the two cases (which are presented in domains with the same scale) show that the drift velocity decreases when Ta is changed from 64 to 74. At Ta=74, a case above the different instability thresholds, the wavy mode is hardly visible (except, may be, close to the top) and the upward helical motion not clearly present.

In contrast, for Ta=89.6, we clearly observe the co-existence of the downward helical motion, the wavy mode and the upward helical motion. The wavy mode affects almost all the cells: it has, however, a large amplitude for the cells in the upper part of the cone and this amplitude decreases as the cells reach lower zones.

The upward helical motion can be seen through the sudden disappearance of a pair of counter-rotating vortices, which appears in the diagram as a dislocation. From the time interval between two successive dislocations ($T \approx 1.64\ s$) and the fact that it generally affects the next but one pair of cells (two wavelengths apart, i.e. L=2 $\lambda \approx 3.87$ mm), we can estimate a propagation velocity for this upward helical motion at 2.36 mm/s.

The analysis of these spatio-temporal diagrams allowed us to determine the wavenumber k of the cell structure by Fourier transform and to deduce the wavelength λ. For example, for Ta=61 (a case with a downward helical motion of the cells), the wavenumber is k=0.1067 giving a wavelength λ=9.366 mm, which is close to 2d = 9.7 mm. The results obtained for $\lambda^* = \lambda/(2d)$ in a Ta range from 60 to 90 are given in Fig. 14. The increase of the Taylor number induces a decrease of the wavelength $\lambda^*$, which becomes almost constant for Ta above approximately 70 at a value close to 0.77 ($\lambda \approx 7.5$ mm).

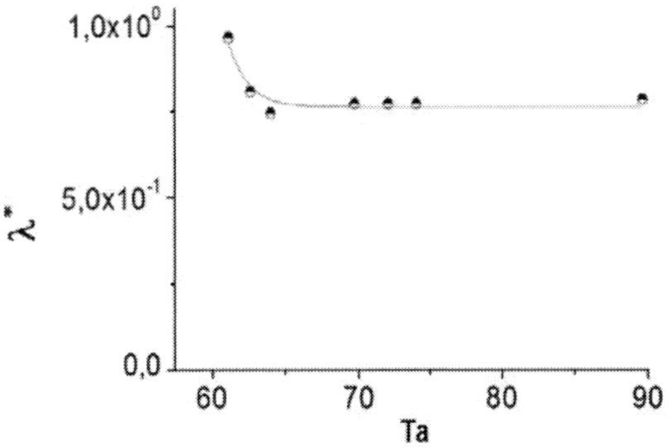


**Fig. 14:** Evolution of the dimensionless wavelength $\lambda$ versus the Taylor number.

The dimensionless drift velocity is given by $V_d{}^* = V_d/(R_{1max}\Omega)$, where $V_d = \lambda/T$, λ is the wavelength and T is the period of the flow signal. The variation of this drift velocity with the Taylor number is presented in Fig. 15. $V_d$ is found to decrease when the Taylor number is increased. For Ta in the range 60 to 80, this decrease is close to a linear decrease, in good agreement with what was observed by Wimmer [1].

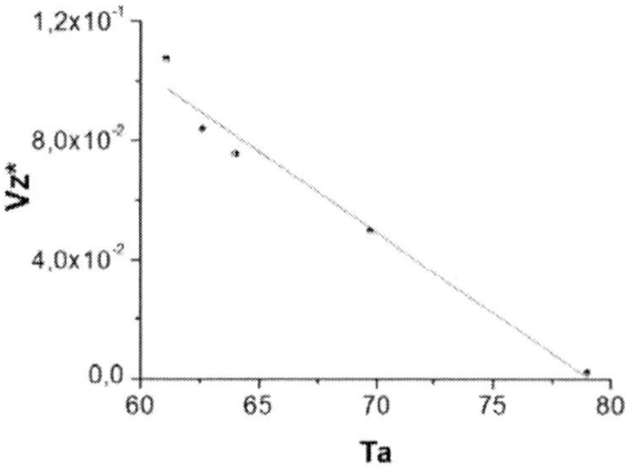


**Fig. 15:** Evolution of the dimensionless drift velocity versus the Taylor number.

## Conclusion

The study of the laminar-turbulent transition in the conical Taylor-Couette flow system was motivated by the interest for different industrial applications and by the hope to discover a specific laminar-turbulent transition. The present experiments, based on optical visualization techniques, indicate that the chaotic regime occurs after a small number of transitions to distinct modes: the Taylor vortex mode, the downward helical motion, the wavy mode and the upward helical motion. The results are in good agreement with those obtained by Wimmer [1-3] in the case $\delta = 0.11$. Spatio-temporal diagrams were also obtained by image processing. The spectral analysis performed on such diagrams enabled us to determine the wavelength and the corresponding drift velocity of the downward moving cells in the interval $60 < Ta < 90$. It was noted that, when Ta is increased, the wavelength decreases towards a constant value, whereas the drift velocity follows a linear decrease.

F.H. acknowledges the PROFAS B+ fellowship allowing this international collaboration work.

## References

1. M. Wimmer, J. Fluid Mech. **292**, 205-227 (1995).
2. M. Wimmer, J. Zierep, Acta Mech. **140**, 17-30 (2000).
3. M. Wimmer, Prog. Aerospace Sci. **25**, 43-103 (1988).
4. M.N. Noui-Mehidi, M. Wimmer, Acta Mech. **135**, 13-25 (1999).
5. M.N. Noui-Mehidi, N. Ohmura, K. Kataoka, Int. J. Fluid Dyn. **5**, Article 1 (2001).
6. M.N. Noui-Mehidi, N. Ohmura, K. Nishiyama, T. Takigawa, Commun. Nonlinear Sci. Num. Sim. **14**, 613-621 (2009).
7. M.N. Noui-Mehidi, N. Ohmura, K. Kataoka, 12th Int. Couette-Taylor Workshop, September 6-8 (2001).
8. M.N. Noui-Mehidi, N. Ohmura, K. Kataoka, J. Fluids Struct. **16**, 247–262 (2002).
9. M.N. Noui-Mehidi, N. Ohmura, K. Kataoka, J. Fluids Struct. **20**, 331-334 (2005).
10. N. Ohmura, M.N. Noui-Mehidi, K. Sasaki, K. Kitajima, K. Kataoka, J. Chem. Eng. Japan **37**, 546-550 (2004).
11. X.F. Xu, L.X. Xu, Commun. Nonlinear Sci. Num. Sim. **14**, 2670-2676 (2009).
12. X.F. Xu, P. Wen, L.X. Xu, D.P. Cao, Commun. Nonlinear Sci. Num. Sim. **15**, 1228–1239 (2010).
13. Q.S. Li, P. Wen, L.X. Xu, J. Hydro. **22**, 241-245 (2010).
14. F. Yahi, Y. Hamnoune, A. Bouabdallah, S. Lecheheb, F. Mokhtari, J. Applied Fluid Mech. 9, N°6 (to be published, 2016).

# Experimental investigation of the mutual interference flow of two circular cylinders by flow visualization

*Yoshifumi* Yokoi[1,*], and *Rut* Vitkovičová[2]

[1]National Defense Academy of Japan, Department of Mechanical Engineering, 1-10-20 Hashirimizu, Yokosuka 239-8686, Japan
[2]CTU in Prague, FME, Department of Fluid Dynamics and Thermodynamics, Technická 4, 166 07 Prague, Czech Republic

**Abstract.** In order to understand the aspect of the mutual interference flow from two circular cylinders, the visual observation experiment was performed by use a water flow apparatus. The purpose of this study is accumulation of the basic image data for comparing with numerical computation or previous experimental results. In this report, the intervals of two circular cylinders were varied, the visualization experiment was performed, and the vortex shedding characteristics and the flow pattern in each case were investigated. The cylinder setting conditions were seven kinds (the position of the rear-side circular cylinder is changed). The cylinder diameter ratios were four kinds ($D/d$=1.0, 1.67, 2.5 and 5.0). The variation of Reynolds number was three kinds ($Re$=548.7, 1200 and 2500). The dye oozing streak method was used in this visualization experiment. Although the previous PIV experimental result and present result obtained the same flow feature, the aspect of an interference flow became clear by changing the color of tracer ink.

## 1 Introduction

Flow visualization is one of the useful study tools which can overlook the behaviour and phenomenon of fluid visually. The behaviour of fluid is shown by a stream line, a particle pass line, or a streak line, and the aspect of the flow is grasped by them. If the flow is a steady flow which does not change in time, those characteristic lines are in agreement. However, if the flow is an unsteady flow which changes in time, they are not in agreement and show a respectively different aspect. The case where the Karman vortex street produced behind a circular cylinder is expressed with the stream line, the particle pass line, and the streak line as an example of the unsteady flow known well is mentioned. Since the aspect of the flow which changes with approach methods of visualization in the unsteady flow is observed, when solving one flow phenomenon, it becomes important to change the visualization method and an observation viewpoint and to observe the phenomenon. However, there are few examples which compared the result obtained by the flow visualization by a different technique about the same unsteady flow phenomenon. In the aspect of the interference flow from two bodies, it is one of the most interesting matters to get to know the aspect of the flow by the difference in the visualization technique. As the study relevant to the interference problem of two bodies, Zdravkovich [1] introduces the studies that solve the drag and lift in the tandem arrangement case and parallel arrangement case and influence of two identical cylinders on the frequency of vortex shedding. Strykowski and Sreenivasan [2] deal experimentally and numerically the influence of the small cylinder (rear cylinder) and the formation of the vortex street behind the main cylinder at low Reynolds numbers. In their work indicate that there are positions of secondary small cylinder (rear cylinder) in which it can control the vortex street. Choi et al. [3] classify methods of flow control for the bluff cylinder and present results of comparing regions of attenuating instabilities using a small secondary cylinder. Mittal and Raghuvanshi [4] solve the influence of a small secondary cylinder on the main cylinder (primary cylinder) using numerical simulations. They compare with their results and previous experimental results and focus on the causes of the influence of the control cylinders on the main cylinder. Yokoi and Hirao [5-9] is reporting the result of the study shown by the streak line about the vortex shedding from the circular cylinder which is oscillating in the direction of flow using the flow visualization technique based on a pouring streak method. Rut et al. [10, 11] is reporting the result of study which showed the interference flow from two circular cylinders arranged staggered using the flow visualization technique based on a suspension method by the particle pass line. Then, the visualization experiment by the streak line method was performed on the same conditions as a previous experiment for the database creation in comparison with the previous experimental result [10].

In this study, the position and diameter ratio of two circular cylinders which are arranged staggered were varied using the closed circuit water channel apparatus, the tracer ink of different colour was made to ooze from each circular cylinder, and flow visualization by the streak method was performed. As a result, the aspect of

* Corresponding author: yokoi@nda.ac.jp

DOI: 10.1051/epjconf/201714302146

**Fig. 1**. Aspect of experimental apparatus

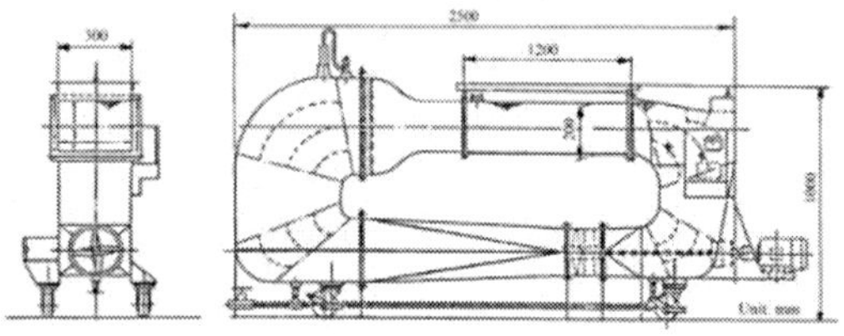

**Fig. 2.** Schematic diagram of closed circuit water channel

vortex shedding from each circular cylinder and the aspect of mutual interference flow were shown vividly.

## 2 Experimental apparatus and method

### 2.1 Outline of experimental apparatus

The experimental apparatus consists of a closed circuit water channel experimental apparatus, a visualization equipment, and an image record equipment. The aspect of experimental apparatus is shown in Fig. 1. The closed circuit water channel is a vertical circulation type of 2.5m in length, 0.3m in width, and 1m in height, and the volume of water is 0.4 tons. The water channel consisted of water flow generation equipment (1 axial flow type pump), rectification device, test section and 4 corner parts with guide vanes. The test section is 1.2m in length, 0.3m in width, 0.2m in depth, and the flume structure with the surface of the water. The test section is a product made from transparent acrylics. The maximum flow velocity in the test section was 1.3m/s. The schematic diagram of the closed circuit water channel is shown in Fig. 2.

The visualization apparatus consists of a tracer ink feed unit and a lighting installation. The ink tank is installed above the water surface, and a set of the clamps and opening-and-closing clips for flow regulation are formed in the vinyl tube which connects the circular cylinder to the ink tank.

As for the lighting installation, one halogen light of 500W and one stroboscope were used. Here, the halogen light was used as a main light source, and the stroboscope was used as a sub light source. And the stroboscope was used for measurement of vortex shedding frequency.

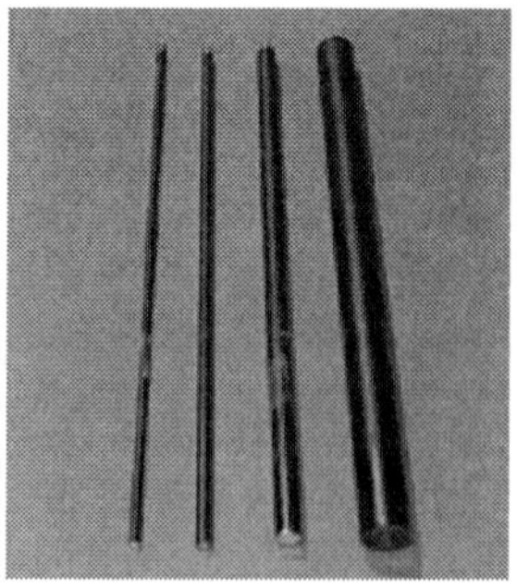

**Fig. 3**. Test circular cylinders, 4mm, 6mm, 10mm, 20mm

The video recording apparatus consists of a small size 3CCD video camera, a video recorder, and a monitor TV. Photography of an image installed a mirror aslant under the test section, and recorded the aspect of the flow projected on the mirror. Since the photography from the test section bottom side is not influenced by water surface wave for an observation picture, it is convenient.

### 2.2 Test cylinders

The length of all test circular cylinders is 300mm. As for the outer diameter, 4mm, 6mm, 10mm, and 20mm are prepared. The screw for circular cylinder fixation (M4) is prepared in the circular cylinder end, and the tracer ink introduction way is provided in the inside of the circular cylinder from the section of the screw to the tracer oozing port on the surface of circular cylinder. Two tracer ports for visualization were provided in the position of $\pm 60$ degrees from the cylinder stagnation point. They are located near 100mm from the circular cylinder other end. When installing the circular cylinder in the water tank test section, the tracer oozing position is 100mm under the water surface. The arrangement of circular cylinders is shown in Table 1. For convenience, front side and rear side cylinders are called the "1st cylinder" and "2nd cylinder", respectively.

In order to reduce mixture by tracer oozing, the depth positions of the oozing port differ in the 1st cylinder and the 2nd cylinder. The oozing port of 1st cylinder is provided in the water surface side, and the oozing port of 2nd cylinder is provided in the bottom side. The gap between both is 2mm. The photograph of these circular cylinders is shown in Fig 3.

### 2.3 Flow visualization method

The flow visualization technique used in this experiment is a pouring streak method. Tracer ink is led to the inside of the circular cylinder using a vinyl tube from a tracer tank, and it carries out by adjusting the valve travel of the clamp which provided that the tracer ink of the grade which does not destroy a boundary layer from the oozing port on the circular cylinder surface oozed in the middle of the vinyl tube. The used tracer ink is the poster color thinned moderately. "Pink" and "turquoise blue" were

Table 1 Setting positions of 2nd cylinder

| *D/d* | 1.0, 1.67, 2.5 | 5.0 |
|---|---|---|
| | (*x/D*, *y/D*) | (*x/D*, *y/D*) |
| Posi1 | (4.30, -1.20) | (4.05, 1.25) |
| Posi2 | (3.20, -0.80) | (3.20, -0.80) |
| Posi3 | (2.40, -0.80) | (2.40, -0.80) |
| Posi4 | (1.95, -0.80) | (1.95, -0.80) |
| Posi5 | (2.00, -0.60) | (2.00, -0.60) |
| Posi6 | (1.50, -0.35) | (1.50, -0.35) |
| Posi7 | (1.50, 0.00) | (1.50, 0.00) |

used with the 1st cylinder and the 2nd cylinder, respectively. In the case of the single circular cylinder, the white poster color was used.

### 2.4 Experimental parameters

The main experiment parameters are Reynolds number *Re*, the diameter ratio of two circular cylinders *D/d*, and the arrangement position of two circular cylinders. Reynolds number *Re* is three kinds ($Re$=548.7, 1200 and 2500), and is obtained from the diameter of the 1st circular cylinder *D*, and main flow velocity *U*. The diameter ratios of two circular cylinders is four kinds ($D/d$=1.0, 1.67, 2.5 and 5.0). The combinations of the diameter of two circular cylinders are 10mm with 10mm, 10mm with 6mm, 10mm with 4mm, and 20mm with 4mm, respectively. Arrangement of two circular cylinders is seven kinds. The 1st cylinder position was put on origin (x, y) = (0, 0), and the position of the 2nd cylinder was changed as shown in Table 1.

### 2.5 Experimental procedure

As preparation, water is filled in the small closed circuit water channel, and condition is prepared so that experiment operation can be performed. The circular cylinder diameter ratio is determined. The 1st cylinder is installed in an origin, and the position of the 2nd cylinder is determined and attached in a circular cylinder attachment plate from the arrangement position of two circular cylinders. The plate equipped with circular cylinders is installed in the predetermined position of the test section. Test Reynolds number is set up by operating the axial flow pump of the closed circuit water channel. Quantity of the tracer ink which oozes from each circular cylinder is adjusted, and video recording of the flow pattern image of the visualization experiment is performed about 1 minute. If recording is completed, washing of the test section and exchange of water will be performed, and it will be set as the following test Reynolds number. If the variation of Reynolds number is completed, the arrangement position of two circular cylinders will be changed and the visualization experiment will be repeated in the same procedure. If the variation of the arrangement position of two circular cylinders is completed, the circular cylinder diameter ratio will be re-determined, and the visualization experiment will be gone on in the same procedure.

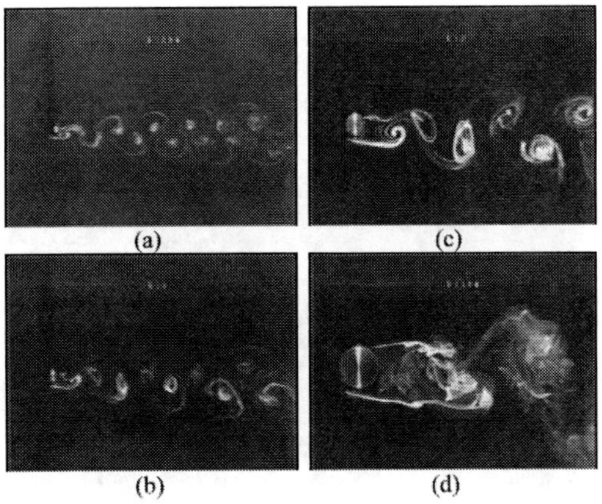


Fig. 4. Flow features of vortex shedding, single cylinder case, (a) $D$=4mm, (b) $D$=6mm, (c) $D$=10mm, (d) $D$=20mm

In video recording, photography of the plate which wrote the experimental condition before photography of the flow pattern was performed. The image data taken into the video recorder is copied to DVD, it reproduces with a personal computer, and a still picture is created by capturing operation.

## 3 Experimental results and discussion

### 3.1 Single circular cylinder case

Visual observation of the vortex shedding from a single circular cylinder was performed for operating evaluation of the experiment equipment. The photographs of the instantaneously flow pattern from the circular cylinder of each diameter are shown in Fig. 4. Here, the main flow velocity *U* is in the same state ($U$= 0.0556m/s). In each photograph, it is shown that vortices are alternatery shed from the circular cylinder and Karman vortex street is formed in the wake. It seems on photograph that the number of the vortexes of the Karman vortex formed with increase of the circular cylinder diameter *D* is decreasing. In the experiment of each circular cylinder, the luminescence period of a stroboscope was synchronized with the vortex shedding period, the Karman vortex shedding frequency *f* was measured, and it determined for a Strouhal number *St*. The Strouhal number is defined by $St=fD/U$ based on the outer diameter of cylinder *D* and vortex shedding frequency *f*. As a result, the Strouhal number of each circular cylinder was about 0.21. Generally it is well known that the Strouhal number of circular cylinder is about 0.2. Since the value by experiment equivalent to the general known value was obtained, it turns out that the apparatus and the measurement technique of this experiment are appropriate.

### 3.2 Visualized flow pattern of Two cylinders

It is one of the most interesting things of this study to conduct visualization investigation of an aspect that install another circular cylinder behind cylinder and the

DOI: 10.1051/epjconf/201714302146

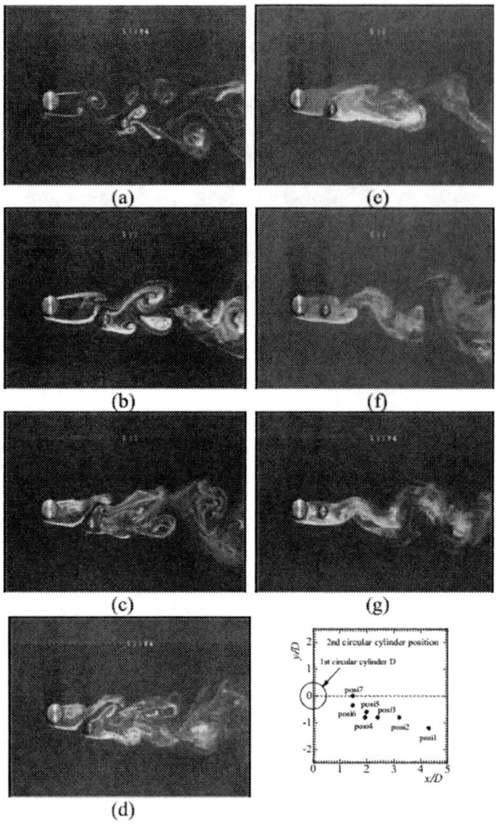


**Fig. 5.** Aspect of the flow when varying the position of 2nd cylinder, *Re*=548.7, *D/d*=1.67, (a) Posi1, (b) Posi2, (c) Posi3, (d) Posi4, (e) Posi5, (f) Posi6, (g) Posi7

vortex flow is controlled. It was observed that the aspect of the flow of the 2nd cylinder completely differs from the aspect of the flow when the 2nd cylinder is placed independently through all experimental conditions. In staggered arrangement like this study, flow pattern can divide roughly into the following three types by the behaviour of the separating shear layer of the 1st cylinder by the side of the 2nd cylinder. (1) The state completely rolled by the separating shear layer before the 2nd cylinder. (2) The state of colliding with the 2nd cylinder before having been rolled by the separating shear layer since the 2nd cylinder is located in the outside of the separating shear layer. (3) The state where the flow from the 2nd cylinder is taken into the re-circulation flow of the 1st cylinder since the 2nd cylinder is located inside the separating shear layer. Figure 5 shows the aspect of the flow when varying the position of the 2nd cylinder in circular cylinder diameter ratio *D/d*=1.67 and Reynolds number *Re*=548.7, as an example. In "Posi1" case, the vortex is alternately discharged from the 1st cylinder, and the vortex by the side of the 2nd cylinder has collided with the 2nd cylinder. It has united with the vortex street which could draw near to the 1st cylinder side the vortex shedding from the 2nd cylinder under the influence, and was discharged from the 1st cylinder. In "Posi2" case, before having been rolled by the separating shear layer discharged from the 1st cylinder, it has collided with the 2nd cylinder. Therefore, the Karman vortex was not produced from the 1st cylinder, but the vortex shedding from the 1st cylinder has assimilated to the vortex shedding of the 2nd cylinder. In "Posi3" case, the separating shear layer from the 1st cylinder is flowing into the re-circulation region behind the 1st cylinder. In "Posi4" case and "Posi5" case which the position of the 2nd cylinder approaches, the phenomenon seems to be more remarkable. In "Posi6" case or "Posi7" case, the 2nd cylinder is placed inside roll up of the separating shear layer discharged from 1st cylinder. Therefore, the separating shear layer from 2nd cylinder is carried to the re-circulation flow of the 1st cylinder, and forms the stagnation region. The vortex shedding of two circular cylinders forms the Karman vortex just like the vortex shedding from one compound body.

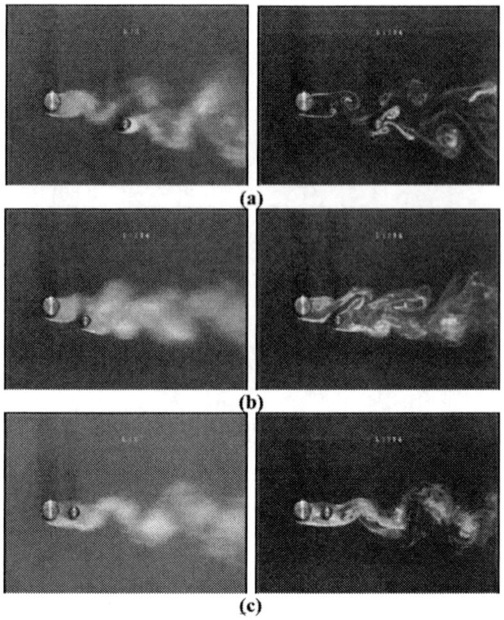


**Fig. 6.** Aspect of the flow when varying the flow velocity, (a) Posi1, *Re*=2500, 548.7, (b) Posi4, *Re*=2500, 548.7, (c) Posi7, *Re*=2500, 548.7,

In the range of the Reynolds number as for which the value of the Strouhal number becomes almost constant, roll up of the separating shear layer occurs generally in the same position. So, even when varying the Reynolds number in the experiment range, it is expected that the same flow pattern is obtained. Figure 6 shows the flow pattern of circular cylinder arrangement in which the flow pattern of three types mentioned above is shown. Here, it is circular cylinder diameter ratio *D/d*=1.67 and Reynolds number *Re*=2500. It turns out that the aspect of an interference flow does not change even if it changes Reynolds number compared with Fig. 5.

If its attention is paid to the behaviour of the separating shear layer of 1st cylinder, it will be expected that the aspect of an interference flow does not change

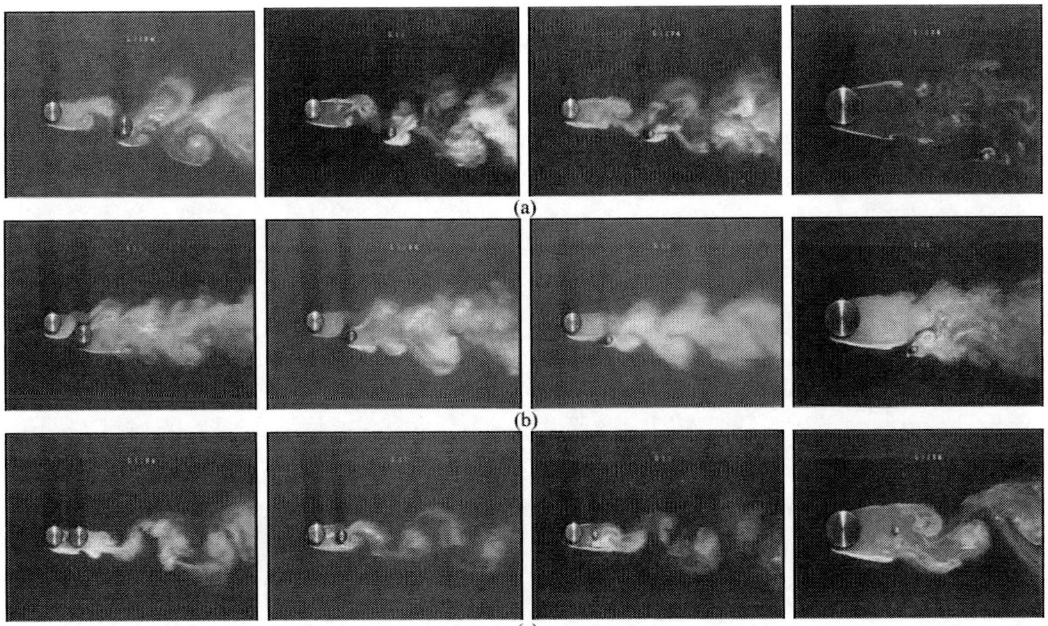

(c)

**Fig. 7**. Aspect of the flow when varying the cylinder diameter ratio *D/d* in *Re*=1200, (a) Posi1, *D/d*=1.0, 1.67, 2.5, 5.0, (b) Posi4, *D/d*=1.0, 1.67, 2.5, 5.0, (c) Posi7, *D/d*=1.0, 1.67, 2.5, 5.0,

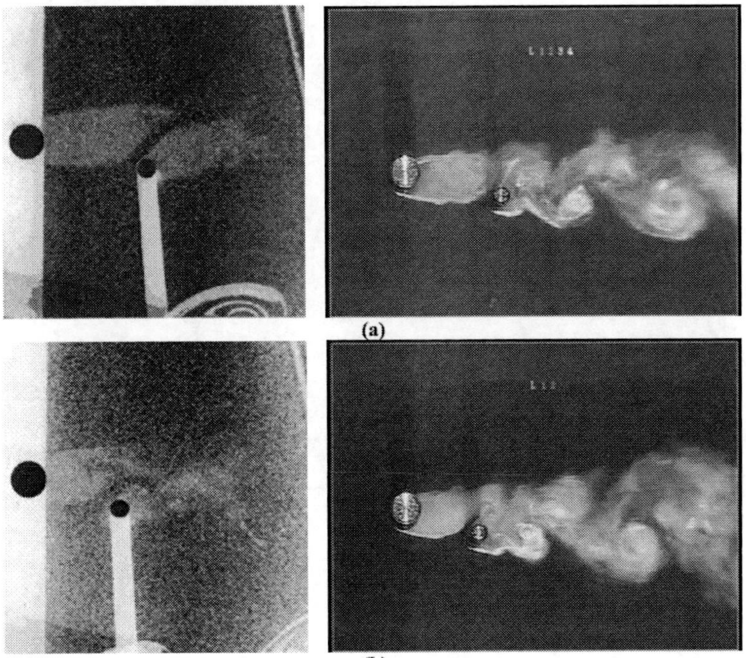

(b)

**Fig. 8**. Comparison of the visualization technique in the particle pass method and the streak method, left-hand side visualization photograph is in the particle pass method in reference 6, (a) *Re*=1394, *D/d*=2, Posi2 and *Re*=1200, *D/d*=1.67, Posi2, (b) *Re*=1394, *D/d*=2, Posi3 and *Re*=1200, *D/d*=1.67, Posi3

even if it changes the diameter ratio *D/d*. Figure 7 shows the result of having changed the circular cylinder diameter ratio *D/d* in the situation which constitutes the typical flow pattern, respectively. When the position which the 2nd cylinder places is the same, it turns out that the aspect of an interference flow does not change even if it changes the diameter ratio *D/d*.

### 3.3 Comparison of aspect of flow by different visualization technique

It is also one of the most interesting things of this study to compare the result of having visualized the same phenomenon by different technique. Figure 8 shows a previous visualization result [10] and the visualization result obtained in this experiment. Here, although both have some difference at experimental conditions, it is a comparatively near case. The object of the comparison is in the state where the circulation region existed behind the 1st cylinder, and the direction of the vortex shedding from the 2nd cylinder has turned to the 1st cylinder side. The previous visualization technique [10] is based on the PIV technique by particle mixing, and shows movement of fluid by the particle pass line. On the other hand, present visualization technique is a pouring streak method by tracer ink oozing, and shows movement of fluid by the streak line. The particle pass line is the route which the same particles follow, and the streak line is the aspect of different particles which follow the same route. So, at the unsteady flow field, visible aspects differ in both. However, in the aspect of the flow of the 1st cylinder behind, and the aspect of the flow of the 2nd cylinder behind, even if it uses the different visualization technique, it is visible to the same aspect. This matter means that there is little fluctuation like the steady flow in the flow of this region.

## 4 Conclusions

Visualization observation of the flow around two circular cylinders which arranged staggered was carried out using the streak line method. The following conclusions were obtained.

(1) The aspect of interference flow changes with the position of 2nd cylinder.

(2) Even if Reynolds number changes, the aspect of an interference flow does not change.

(3) When the position of the 2nd cylinder is the same, even if the diameter ratio changes, the aspect of the interference flow does not change.

(4) The flow of the circulation region around two circular cylinders has little fluctuation so that it can be regarded as the steady flow.

## References

1. M. M. Zdravkovich, J. Fluids Eng. **99**(4). 618-633(1977)
2. P. J. Strykowski, K. R. Sreenivasan, J. Fluid Mech. **218**, 71-107(1990)
3. H. Choi, I. Lee, W. P. Jeon, Annu. Rev. Fluid Mech. **40**, 113-139(2008)
4. S. Mittal, A. Raghuvanshi, Int. J. Numer. Math. Fluids **35**, 421-447(2001)
5. Y. Yokoi, K. Hirao, J. Fluid Science and Technology **4**-2, 401-414(2009)
6. Y. Yokoi, K. Hirao, EPJ Web of Conf. **45**, 01098(2013)
7. Y. Yokoi, K. Hirao, EPJ Web of Conf. **67**, 02131(2014)
8. Y. Yokoi, K. Hirao, J. Fluid Science and Technology **9**-3, 0057(2014)
9. Y. Yokoi, K. Hirao, EPJ Web of Conf. **92**, 02116(2015)
10. R. Vitkovicova, J. Nozicka, J. Cizek, V. Skala, EPJ Web of Conf. **92**, 02110(2015)
11. R. Vitkovicova, Vladislav Skala, EPJ Web of Conf. **114**, 02138(2016)

EPJ Web of Conferences **143**, 02147 (2017) DOI: 10.1051/epjconf/201714302147
*EFM 2016*

# Power of dissipation in a rotating machine

*Pavel* Žitek[1,*], *Marek* Klimko[1], and *Tomáš* Noga[1]

[1]University of West Bohemia in Pilsen,Univerzitní 8, 306 14 Pilsen, Czech Republic

**Abstract.** This article deals with a possibility of how to determine a power dissipation in a rotating machine. It divides losses in three parts according to their origin – losses arising around the rotating disc, in bearings, and by a friction of a shaft surface due to viscosity. A precise determination of the power dissipation is important when evaluating the efficiency of the whole device. The procedure is graphically illustrated on the example of a single-stage experimental turbine.

## 1 Introduction

It is important to know dissipations emerging in a rotor to properly determine a rotating system power. We will demonstrate a procedure of determination of this power on the rotor of the air turbine VT-400.

The core is to detect the power of the turbine stage to determine its efficiency. (1). The power can be determined from the dynamometer connected to the turbine rotor. However, the dynamometer power is lower by the dissipation power, compared to the stage power, therefore it has to be included in the calculation of the efficiency

$$\eta_{obv} = \frac{\frac{P_{st}}{m_v}}{h_0 - \Delta h_{c2}} \quad [1] \tag{1}$$

$$\eta_{obv} = \frac{\frac{P_D + P_f}{m_v}}{h_0 - \Delta h_{c2}} \quad [1] \tag{2}$$

where

$\eta_{obv}$ stage efficiency
$P_{st}$ stage power
$m_v$ air mass flow rate
$h_0$ isoentropic drop on the stage
$\Delta h_{c2}$ kinetic energy in the air leaving the stage
$P_D$ power obtained from the dynamometer
$P_f$ dissipation power

Furthermore, we will consider dissipations caused by a friction of the disc, bandage, shaft and bearings.

## 2 Disc friction dissipation power

Literature mentions various mathematical relations to calculate the friction dissipation of rotating discs. Basically, they are a result of a dual approach to this problem. One group of equations originates from results of experiments, whereas the second one applies the basic laws of fluid mechanics on a rotating disc friction in the environment.

### 2.1 Experimental research

The mutual sign of equations obtained experimentally is almost the same functional dependency of dissipations on the peripheral velocity (revolutions) and the disc diameter. Yet, the constants in the equations differ, due to the fact that experiments have been carried out with discs of a different thickness; distances of limiting walls were different as well, both in the axial and radial directions. Another difference lies is a different value of Reynolds number during measurements. Higher values of constants correspond to small Re values and vice versa.

Generally, all relations (except one) can be written in the following form:

$$P_{fd} = k \cdot \left(\frac{u}{10}\right)^3 \cdot d^2 \cdot \rho \quad [W] \tag{3}$$

where:

$d$ disc diameter
$u$ peripheral velocity
$k$ coefficient resulting from a particular experimental research

We have focused on four authors out of all studied approaches.

The first is prof. Stodola [1]. He determined the coefficient $k = 1$ (for superheated steam) and $k = 1.25$ for saturated steam. Based on experiments carried out later on, prof. Stodola defined the coefficient $k = 2.7$.

The next author is Zalf [2]. He determined the coefficient oscillating within the range of $k \in \langle 0.865, 1.73 \rangle$, depending on the distance of limiting surfaces. The bottom boundary of the coefficient applies for small distances from limiting surfaces, upon increasing the distance the coefficient value grows. The coefficient values have come out of precise

* Corresponding author: zitek@kke.zcu.cz

measurements on real turbine stages within the area of $Re > 5 \cdot 10^6$, whereas Reynolds number for the disc is defined:

$$Re = \frac{d \cdot u}{\nu} \quad [1] \tag{4}$$

where $\nu$ is kinematic viscosity of the environment surrounding the rotating disc.

The third reviewed author is Scegliaev [3]. The equation, unlike the others in this group, includes also a friction dissipation of a bandage connected to moving blades by rivets with smaller heads. He defines the coefficient as $k = 44$.

The last research we have studied was led by prof. Miskovsky [4]. He presents the equation in a slightly different form:

$$P_{fd} = k \cdot \left(\frac{u}{10}\right)^{2.8} \cdot d^{1.4} \cdot \rho \quad [W] \tag{5}$$

The coefficient value varies within the range of $k \in \langle 0.08, 0.1 \rangle$.

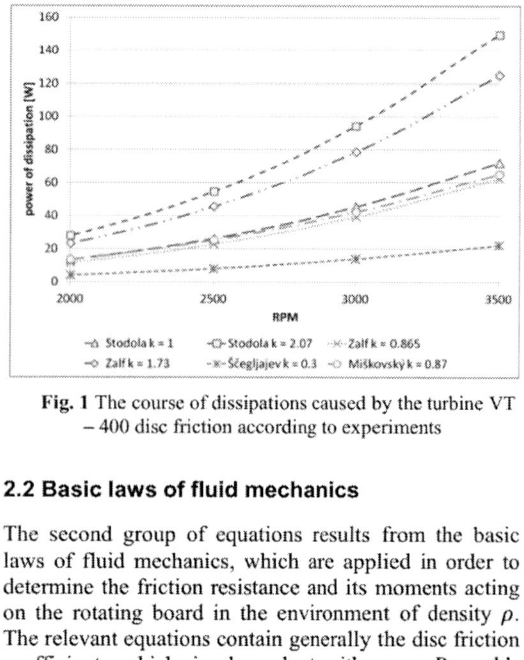


**Fig. 1** The course of dissipations caused by the turbine VT – 400 disc friction according to experiments

## 2.2 Basic laws of fluid mechanics

The second group of equations results from the basic laws of fluid mechanics, which are applied in order to determine the friction resistance and its moments acting on the rotating board in the environment of density $\rho$. The relevant equations contain generally the disc friction coefficient, which is dependent either on Reynolds number only or also on the proportional distance of limiting walls from the disc. The results can be summarized in one general relation:

$$P_{fd} = k \cdot c_t \cdot u^3 \cdot r^2 \cdot \rho \ [W] \tag{6}$$

where:

$r$ disc diameter
$c_t$ friction coefficient
$k$ coefficient defined by particular authors

According to Traupel [5], $c_t$ is defined as follows (for $Re > 2 \cdot 10^5$; whereas the Re number definition is according to the equation (4)):

$$c_t = \frac{0.003}{Re^{0.2}} \quad [1] \tag{7}$$

According to Scegliaev [3], the friction coefficient for turbulent flow within the disc area is

$$c_t = 2.5 \cdot 10^{-2} \cdot \left(\frac{s}{r}\right)^{0.1} \cdot Re^{-0.2} \quad [1] \tag{8}$$

where $s$ is the distance of the limiting wall from the disc.

According to Tuliszky [6], the friction coefficient is again (beside others) a function of Reynolds number. However, it is a function of the radius instead of the diameter in this case.

$$c_t = \frac{C_R}{Re^{0.2}} = \frac{C_R}{\left(\frac{u \cdot r}{\nu}\right)^{0.2}} \quad [1] \tag{9}$$

where $C_R$ is so-called reduced disc friction coefficient depending on the proportional distance $s/r$ (fig. 2).

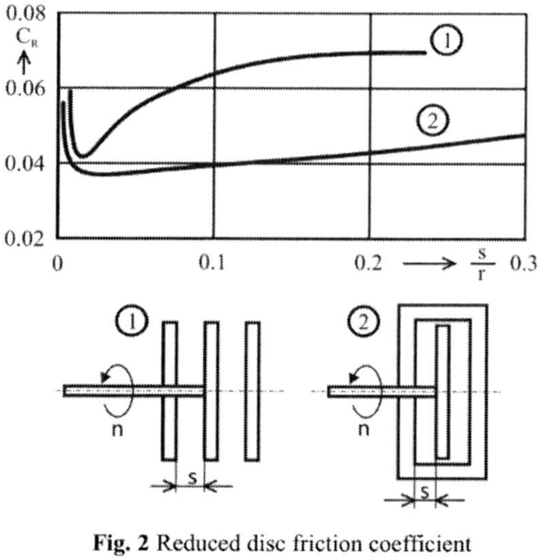


**Fig. 2** Reduced disc friction coefficient

Linnecken [7] has provided a very detailed study. He has deduced the following relation for the friction coefficient:

$$c_t = \frac{0.5 \cdot \pi}{\frac{s}{r} \cdot Re} + \frac{0.04}{Re^{0.2}} \cdot \frac{1 + \frac{s}{r}}{2 + \frac{s}{r}} \quad \left| Re = \frac{u \cdot r}{\nu} \right. \tag{10}$$

The dissipation power is a function of a peripheral velocity, a disc diameter, the friction coefficient and the coefficient $k$ in all four cases. However, that is the only compliance. As demonstrated above, the difference is not only in the coefficient $k$, but also in the disc friction coefficient $c_t$. Traupel, for instance, doesn't take the limiting surfaces influence into consideration. Nevertheless, the ratio $s/r$ is of a not inconsiderable influence, according to the remaining three authors.

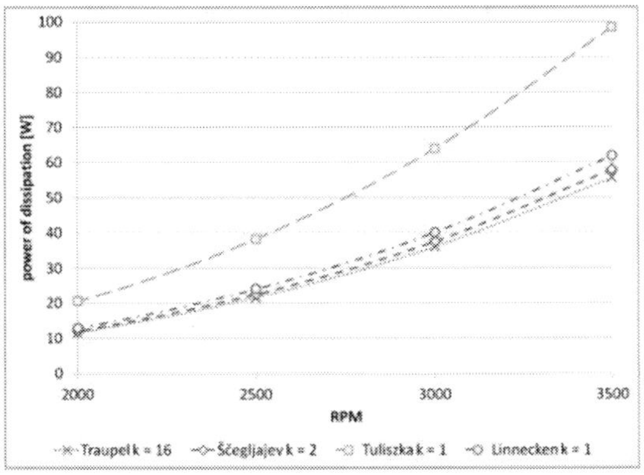


**Fig. 3** The course of dissipations caused by the turbine VT – 400 disc friction according to fluid mechanics

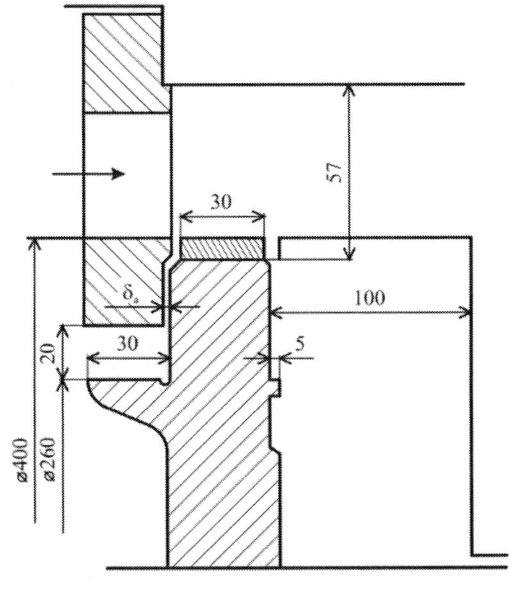


**Fig. 4** Scheme of the stage for a friction dissipation calculation

The equations obtained by both approaches have naturally been applied on VT-400 (fig. 1, fig. 3). We have chosen the ratio $s/r = 1$ given the geometry of the flow part of the turbine (fig. 4).

### 2.3 Summary

Taking into account the conditions used by individual authors, limiting walls influence, the flow part geometry and other circumstances, we have chosen the relation deduced by Linnecken (6), (10). The course of the dissipation power is presented in fig. 5.

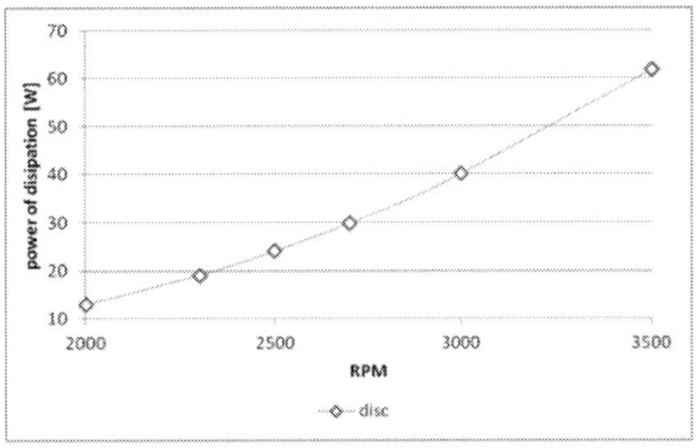


**Fig. 5** Disc dissipation power

## 3 Dissipation power caused by the bandage friction and projections on the disc

### 3.1 Bandage

The most detailed calculation of the bandage friction dissipation is mentioned by Linnecken again:

$$P_{fb} = c_t \cdot u_b^3 \cdot r_b \cdot b \cdot \rho \;\; [W] \tag{11}$$

where:

$u_b$ bandage peripheral velocity
$r_b$ bandage diameter
$b$ disc width

$$c_t = \frac{2\cdot\pi}{\frac{t}{r_b}\cdot Re_b} + \frac{0.15}{{Re_b}^{0.2}} \cdot \frac{1+\frac{t}{r_b}}{2+\frac{t}{r_b}} \; \Big| Re = \frac{u_b \cdot r_b}{\nu} \tag{12}$$

where:

$t$ distance of the limiting wall from the bandage

As evident (12), $c_t$ is a function of the flow part geometry and Reynolds number. As a matter of interest, we are adding the geometry dependency of $c_t$ for different values of Reynolds number, see fig. 6.

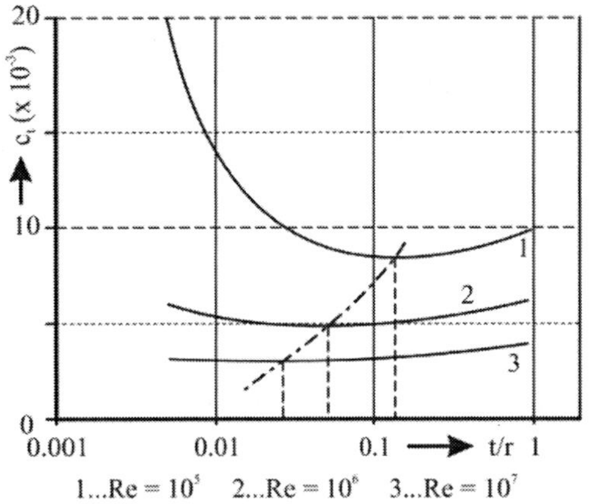


**Fig. 6** Dependence of $c_t$ on the bandage geometry

### 3.2 Projections on the disc

Another type of dissipation taken into account are these caused by the friction of projections in the surrounding environment. The calculation is entirely the same as for bandage dissipation power, yet the difference is in substituted values, which is apparent in fig. 4. The resulting value of the dissipation power is lower with respect to smaller diameter of the projection, leading to a lower peripheral velocity, and also given a larger distance from limiting surfaces (see fig. 7)

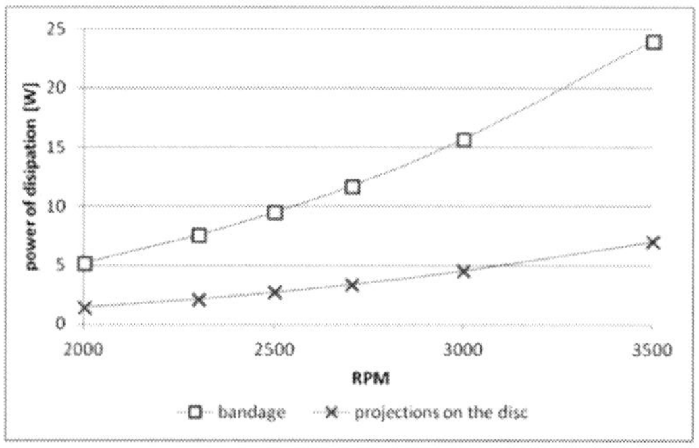


**Fig. 7** Dissipation power of the bandage and projections on the disc

## 4 Dissipation power in bearings

This chapter will deal with the second part of the total friction losses, particularly dissipations in bearings. These losses can be determined by a calculation and are given by the following relation:

$$P_L = \frac{\pi \cdot d \cdot n \cdot F_e \cdot f_t}{60} \quad [W] \tag{13}$$

where:

$d$ inner diameter of a bearing

$f_t$ rolling friction coefficient (single row ball bearings $f_t = 0.0015$, tilting double row bearings $f_t = 0.002$)

$F_e$ equivalent force in a particular bearing

### 4.1 Axial force

Axial force acting on bearings is defined by the difference of pressures in front of and behind a blade wheel. It means that it is a function of the stage reaction $\rho$, chosen $u/c$ (since $c$ is defined by the pressure drop) and also the speed $n$ (it is necessary to change $\Delta p$ along with the change of speed to maintain optimal $u/c$).

Therefore, $F_A = f(\rho, u/c, n)$.

As the determination of axial dissipations depends on several variables, their determination can be proceeded only for concrete conditions set by experimental measurements.

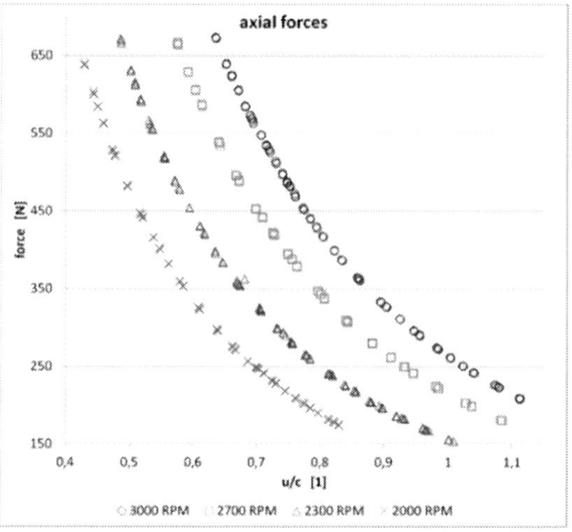


**Fig. 8** Influence of u/c and speed on axial force

### 4.2 Radial forces

We will use the scheme in fig. 9 to count up reaction forces in bearings.

We will calculate radial forces in bearings, i.e. reactions $R_A$ and $R_B Z$, from the equilibrium of forces and moments. We will use the detected radial and axial forces to calculate so-called equivalent force acting on the bearing.

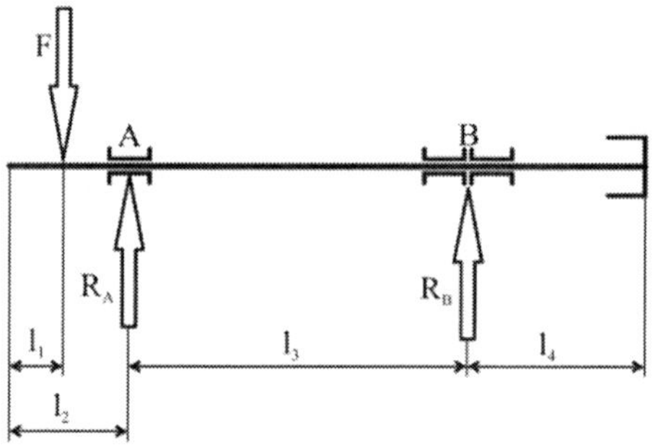


**Fig. 9** Scheme of the shaft

$$\sum F_{ix} = 0 \tag{14}$$

$$\sum F_{iy} = 0 : R_A - F - q \cdot l + R_B = 0 \tag{15}$$

$$\sum F_i\,(B) = 0 : R_A \cdot l_3 - -F \cdot [l_3 + (l_2 - l_1)] - -q \cdot \frac{(l_2 + l_3)^2}{2} + q \cdot \frac{{l_4}^2}{2} \tag{16}$$

$$R_A = \frac{F\,[l_3 + (l_2 - l_1)] + q\frac{(l_{2+l_3})^2}{2} - q\frac{l_4^2}{2}}{l_3} \tag{17}$$

$$R_B = F - R_A + q.l \tag{18}$$

DOI: 10.1051/epjconf/201714302147

### 4.3 Equivalent force

Equivalent force can be determined several ways. Three of them are listed as follows:

Niemann determines the equivalent force

$$F_e = F_R + 2.75\ F_A \qquad (19)$$

Fronius considers

$$F_e = \sqrt{F_R^2 + F_A^2} \qquad (20)$$

The equivalent force according to Swen

$$F_e = X_o \cdot F_R + Y_o \cdot F_A \qquad (21)$$

Apart from $F_A$ and $F_R$, there are also coefficients $X_0$ and $Z_0$ appearing in the relation (21). We assume the front tilting double row bearing (fig. 10), with an angle $\alpha = 10°$ and relevant coefficients $X_0 = 1$, $Y_0 = 2.6$, to capture the whole axial force. An example can demonstrate that dissipations in bearings are progressively 54.21 (19) - 41.41 (20) – 53.5 (21) watts instead of 41.09, not considering the axial force, for revolutions of 4000 min$^{-1}$, according to the above mentioned authors. As apparent, the axial force has basically no influence on the equivalent force in Fronius's relation. We have use Niemann's relation for our computation.

**Fig. 10** Tilting double row bearings [8]

### 4.4 Summary of dissipations in bearings

As already presented, axial force changes depending on several variables, whereas the radial force is determined by the rotor geometry, or its weight. Should we consider one version of blading (i.e. steady reaction stage) and measure using the same $u/c$, we will obtain different axial forces for different revolutions (see table 1).

**Table 1** Axial and radial forces

| rpm | 2 000 | 2 300 | 2 700 | 3 000 |
|---|---|---|---|---|
| $F_A$ $[N]$ | 260 | 346 | 479 | 580 |
| $R_A$ $[N]$ | 272 | | | |
| $R_B$ $[N]$ | 169 | | | |

The dissipation power due to the friction in bearings can be viewed in fig. 11.

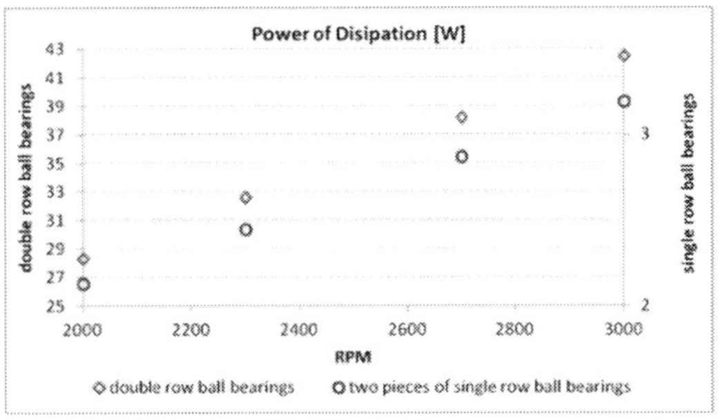


**Fig. 11** Dissipation power in bearings

## 5 Shaft friction dissipations

We have derived the relations to calculate the shaft dissipation power from the basic laws of mechanics and fluid mechanics. We will come out of Fig. 12 .

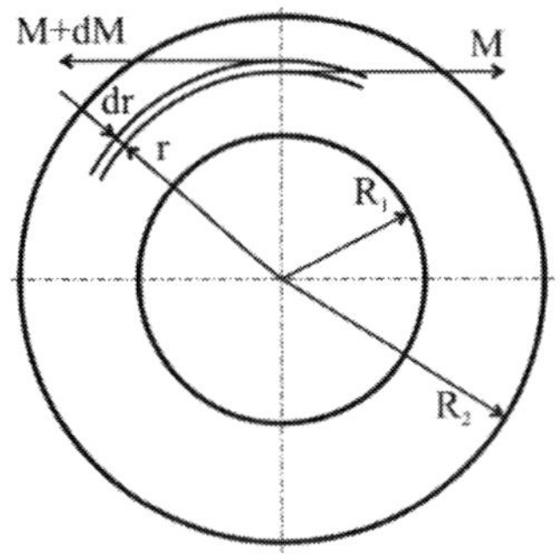


**Fig. 12** Scheme for computation

$$M + dM - M = 0 \qquad (22)$$

$$\int dM = \int 0 \;\Rightarrow\; M = C_1 \qquad (23)$$

$$M = F_f \cdot r = \tau \cdot S \cdot r = \tau \cdot 2 \cdot \pi \cdot r \cdot L \cdot r \qquad (24)$$

$$\tau = \eta \cdot \frac{dw}{dr} \qquad (25)$$

$$\eta \cdot \frac{dw}{dr} \cdot 2 \cdot \pi \cdot L \cdot r^2 = C_1 \qquad (26)$$

$$\int_{w_2}^{w} dw = \int_{R_2}^{r} \frac{C_1}{2 \cdot \eta \cdot \pi \cdot L} \cdot \frac{dr}{r^2} \quad |w_2 = 0 \qquad (27)$$

$$w = \frac{C_1}{2 \cdot \pi \cdot \eta \cdot L} \cdot \left(\frac{1}{R_2} - \frac{1}{r}\right) \qquad (28)$$

$$C_1:\ r = R_1; w = w_o \qquad (29)$$

$$w_o = \frac{C_1}{2 \cdot \pi \cdot \eta \cdot L} \cdot \left(\frac{1}{R_2} - \frac{1}{R_1}\right) \qquad (30)$$

$$P_{fs} = M \cdot \omega = w_o \cdot 2 \cdot \pi \cdot \eta \cdot L \cdot \frac{R_1 \cdot R_2}{R_1 - R_2} \cdot \omega \qquad (31)$$

The resulting relation shall be:

$$P_{fs} = M \cdot \omega = \frac{8 \cdot \upsilon \cdot \rho \cdot \pi^3 \cdot L \cdot R_1^2 \cdot R_2 \cdot n^2}{R_1 - R_2} \quad (32)$$

where:

$M$ torsion moment
$F_f$ friction force
$\tau$ shear stress
$\eta$ dynamic viscosity
$S$ shaft surface
$v$ kinematic viscosity (of air)

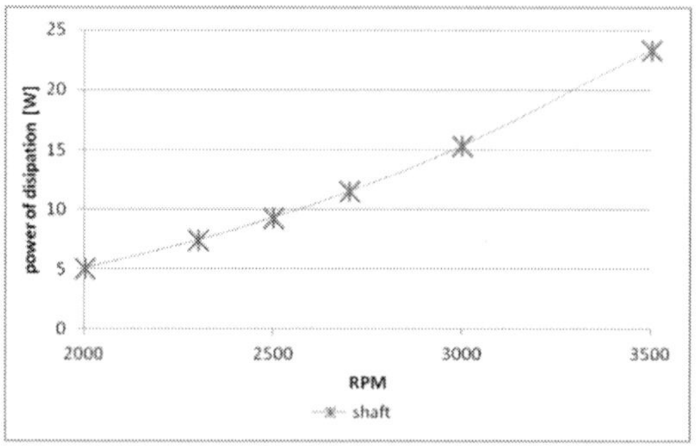


**Fig. 13** Shaft dissipation power

## Conclusion

As stated in the introduction, it is necessary to determine the power of rotor dissipation correctly in order to determine the efficiency of the rotating system. This article presents a procedure of dissipation determination on the example of the power of dissipation of a particular rotor (the rotor of the experimental VT-400 turbine). We will have to include also dissipations in the dynamometer into this calculation. However, it is not so simple, therefore there is an option to use a moment flange, due to which the dynamometer losses can be omitted.

The dissipation analysis followed by the power of dissipation determination clearly shows the influence of particular sources of dissipations. (see fig. 13). The highest proportion – almost a half – belongs to losses occured in bearings (at lower speed), mainly in the axial force absorbing bearing (fig. 11), and losses caused by the disc friction in the surrounding environment.

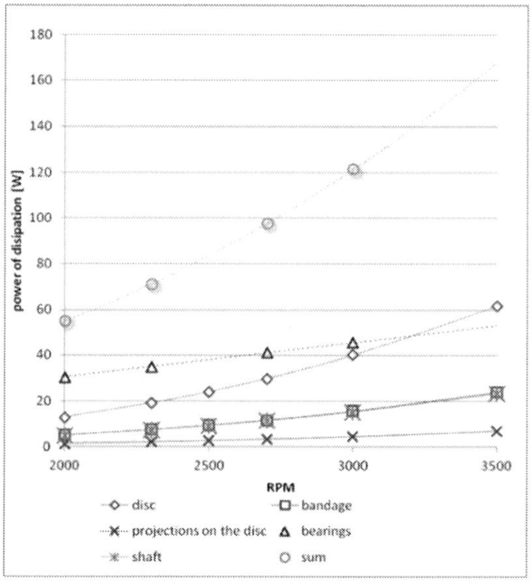


**Fig. 14** Rotor dissipation power

As manifested in this article, it is an analytical solution of this matter only and will be followed by an experimental verification of results in the near future. The objective is to find and verify such a methodology which we could use for various types of blade wheels without performing new complex experimental verifications of the dissipation power for every change of operational parameters or blading replacement.

The presented work was financially supported by the Ministry of Education, Youth and Sport Czech Republic Project LQ1603 (Research for SUSEN). This work has been realized within the SUSEN Project (established in the framework of the European Regional Development Fund (ERDF) in project CZ.1.05/2.1.00/03.0108) and student project SGS-2016-045 (Improving the efficiency, reliability and service life of power machines and equipment 4).

## References

1. A. Stodola, *Dampf u. Gasturbinen*, 1924
2. G.A. Zalf, V.V. Zvjagincev *Těplovoj rasčet parovych turbin*. Mašgiz Moskva, 1961
3. A.V. Ščegljajev, *Parovyje turbiny*. Energija Moskva, 1967
4. L. Miškovský, *Parní a plynové turbíny*. Praha, 1949
5. W. Traupel, *Termische Turbomaschinen*. Springer, 1958
6. E. Tuliszka, *Turbiny cieplne*. Wydawnictwa Naukovo – Techniczne Warszawa, 1973
7. H. Linnecken, *Der Radreibungsverlust insbesondere bei Turbomaschinen*. AEG Mitteilungen, 1957
8. [online]. 15.8.2016 [cit. 2016-08-15]. Available on: *http://www.elotechnik.cz/lozisko-2217-k-zvl.html#zalozka-2*

 DOI: 10.1051/epjconf/201714302148

# On the effect of segregation on intense bimodal bed load

*Štěpán* Zrostlík[a] and *Václav* Matoušek

Department of Hydraulics and Hydrology, Civil Engineering, Czech Technical University in Prague, 166 29 Prague 6, Czech Republic

**Abstract.** Open-channel two-phase flow above a granular mobile bed is studied experimentally and theoretically. In the two-phase flow, water serves as a carrying liquid for plastic grains transported as collisional contact load in the upper-stage plane bed regime. The investigation evaluates friction- and transport characteristics of the flow under the condition of intense collisional transport of grains and links them with the internal structure of the two-phase flow. The paper focusses on the effect of bimodal solids (mixed two fractions of grains of similar density and different size and shape) on the flow characteristics and internal structure. Hence, experimental results obtained for the bimodal mixture are compared with results for individual grain fractions. The experiments show that the bimodal character of the transported solids affects the layered internal structure of the flow as a result of fraction segregation due primarily to gravity (kinetic) sieving during transport. The segregation also affects the friction- and transport characteristics of intense bed load. In the paper, the effects are described and quantified.

## 1 Introduction

In natural hazard management of rivers and streams, sediment transport is one of the key elements to assess. It has direct consequences for morphology and stability of river channels and is responsible for infrastructural damages caused by extreme flood events [1].

Natural sediment forming a river bed and being transported during flood events is composed of grains of different sizes. The composition of bed material is described by the grain size distribution (GSD). Mobility of grains of different sizes is different in flood flow and the interaction of different grains with each other and with the flowing liquid is a complex process difficult to describe physically [2]. As a first step in investigating the complex process of size-affected granular interactions, some authors study transport of bimodal sediments (sediments composed of two grain fractions of two different sizes) [1,3]. Vertical sorting of the grain fractions is typical for transport of bimodal sediments.

In river engineering, the vertical sorting is usually associated with armouring of a channel bed. The mechanisms governing erosion of finer grains from the bed and resulting armouring of the bed surface have been described by numerous authors, e.g. [2,4], and the phenomena are modelled in contemporary river morphology software tools, for instance HEC-RAS. The erosion/armouring phenomena preferably occur at bed shear conditions suitable for a development of bed forms (particularly dunes).

However, little attention has been paid to vertical sorting in the upper plane bed (UPB) regime typical for extreme flood events. In the UPB regime, there are no bed forms and the stationary armor layer is replaced by a mobile transport layer dominated by interparticle collisions. The vertical sorting in the collision-dominated granular flows can be modelled using the kinetic theory of granular flows [5] but the modelling and its experimental verification has been done for dry granular flows only.

Our previous work on aqueous monodisperse flows showed that the UPB flows stratified into layers, which affected velocity distribution, flow friction and sediment transport. In this work, we discuss how a bimodal character of sediment affects the flow stratification and the associated flow properties. The discussion is based on own experiments in a tilting flume.

## 2 Methods

The bimodal sediment experiments were carried out in the same tilting flume as our previous tests with monodisperse sediments, i.e. in the 0.2-m flume in the Water Engineering Laboratory of the Czech Technical University (CTU) in Prague. A detailed description of the setup, measuring equipment and experimental procedure is given elsewhere [6-9]. For the experiments with bimodal mixtures, the procedure was extended with sieving analyses of samples of transported sediments collected at the outlet of the flume during each test run.

[a] Corresponding author: stepan.zrostlik@fsv.cvut.cz

## 2.1 Experimental work

Tests were carried out for different installed conditions of steady-state uniform flow in the upper plane bed regime. First, experiments were carried out with individual monodisperse sediment fractions separately and then bimodal sediment was tested composed of the previously tested sediment fractions. Both fractions were plastic grains of similar density and different size and shape. They also differed in colour to make visual observations of the individual fractions in the bimodal mixture more recognisable. Properties of the fractions and the bimodal sediment are summarized in Table 1. Plastic grains of the finer fraction (Hostaform, HSF3) are white, its shape is ellipsoidal, the equivalent sphere diameter $d_e = 3.18$ mm and the density $\rho_s = 1359$ kg.m$^{-3}$. Plastic grains of the coarser fraction (Tiulit, TLT50) are black and of a cylindrical shape with the equivalent diameter $d_e = 5.41$ mm and the density $\rho_s = 1307$ kg.m$^{-3}$.
Bimodal sediment (HT = TLT50 + HSF30) was mixed of the same volumes of the individual fractions and so loaded to the flume recirculating system. In order to manage as uniform as possible distribution of the individual fractions along the system, the mixture of water and bimodal sediment was circulated at high velocity through the system before each test run. After grains of both fractions were sufficiently suspended and mixed, the discharge of mixture, $Q$, was set to a required value for the test. Uniform flow with flow depth, $h$, above deposit developed in the flume. Grains of both fractions were well mixed and proportionally represented in the deposit. Segregation of grains started to be visible at the top of the deposit in a few minutes.

**Table 1.** Properties of sediments

| Properties | HSF3 | TLT50 | HT |
|---|---|---|---|
| $d_e$ (mm) | 3.18 | 5.41 | |
| $v_t$ (m.s$^{-1}$) | 0.131 | 0.149 | 0.140 |
| $\rho_s$ (kg . m$^{-3}$) | 1359 | 1307 | 1333 |
| *Percentage of HSF3 in sample* | 100 | 0 | approx 32 |

**Table 2.** Range of values of integral parameters of flow

| Properties | | HSF3 | TLT50 | HT |
|---|---|---|---|---|
| $h$ (mm) | min | 29 | 47 | 38 |
| | max | 115 | 101 | 91 |
| $Q$ (l.s$^{-1}$) | min | 4.6 | 7.8 | 5.7 |
| | max | 18,9 | 16.2 | 14.1 |
| $I_e$(-) | min | 0.002 | 0.003 | 0.008 |
| | max | 0.058 | 0.055 | 0.054 |
| $C_{vd}$(-) | min | 0.01 | 0.01 | 0.01 |
| | max | 0.251 | 0.294 | 0.0260 |

Test runs were carried out within a broad range of flow conditions (see Table 2). As a proportion of the individual fractions in the bimodal sediment could vary for different flow conditions it was necessary to determine the proportion for each test by sampling the sieving the transported bimodal sediment. The proportion influenced calculated representative grain size, density, and settling velocity of the bimodal sediment. Values of the density $\rho_s$ and the terminal settling velocity $v_t$ were calculated as the normal arithmetic mean. The characteristic grain size was determined as the effective mean diameter, $d_m$, from the GSD of sediment fraction, $d_m = \sum d_i \Delta p_i / \sum \Delta p_i$ where $d_i$ is the diameter of sieve opening and $\Delta p_i$ denotes the weight percentage of particles passed.

## 2.2 Measuring instruments

Standard measuring instruments were used to measure integral parameters of the flow (depth, discharge, slope, delivered concentration) and they are described elsewhere [6-7]. We used three different measurement techniques to measure a vertical velocity profile of the longitudinal component of the velocity in the axis of the flume. The Prandtl probe (PT) was used primarily to validate velocity profiles obtained from the other methods. A sensitivity of PT is limited with a minimal value near 1 cm.s$^{-1}$. Two others methods were acoustic: the ultrasonic velocity profiler (UVP) with a probe directed opposite to the flow and Acoustic Doppler Velocity Profiler (ADVP) with a probe oriented downwards. A disadvantage of the UVP method in this particular application is that particles are in sporadic contact with the measuring membrane of the probe. The great advantage of ADVP is that it is non-invasive and does not disturb flow in the transport layer. Also velocity measuring instruments are described in more details elsewhere [6-7, 9].
An estimation of measuring errors of individual instruments is given in [6]. Processing of measured data is described in [6-7, 9].

## 2.3 Observation of development of flow structure

We investigate a development of a layered structure of flow with bimodal contact-load sediment in the UPB regime. Our previous work on aqueous monodisperse flows showed that the UPB flows stratified into three layers [6-9]. The uppermost layer contained clear water and occasionally saltating particles. The bottom layer was formed by stationary bed deposit. The transport layer developed between the water layer and the deposit layer. A vast majority of sediment was transported in a collisional mode through the transport layer. The layer of smaller particles was observed at interface between deposit and transport layers during experiment with mixtures.
Figure 1 shows schematically the three layers and also an additional interfacial layer which we observed to develop when the bimodal sediment was transported in the UPB regime. Figure 2 is a photograph of typical flow schematically depicted in Figure 1. The development of the interfacial layer composed of the finer particles (the white layer in Figures 1 and 2) separating the deposit and the transport layer was a result of vertical sorting in the

lower part of the transport layer. The interfacial layer itself could be split into two sublayers. The upper sublayer contained moving particles while the lower sublayer did not move. The interface between the deposit and the transport layer was always within the white layer of finer particles.

The observed pattern seems to support the picture of the process of two-size grain segregation as sensitive to the grain size ratio [1]. If the ratio is high, finer grains are smaller than voids among coarse particles and the fines naturally settle no matter what the nature of interaction is among the coarser grains. Otherwise, the finer grains are too big to settle naturally through the voids and they percolate down only through spaces which open and close due to collisions among the coarser grains in sheet flow. This results in vertical sorting in which the finer grains move down and the coarser grains migrate deeper into the transport layer as they are squeezed up by the finer ones. It seems that this mechanism is responsible for a development of the interfacial layer observed in our bimodal sediment flows.

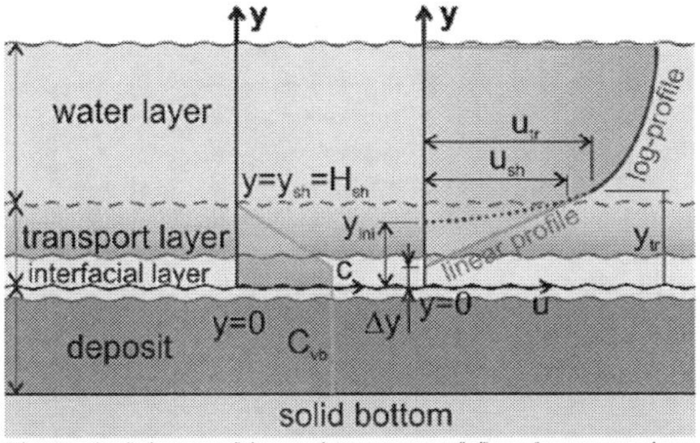


**Figure 1.** Scheme of layered structure of flow in upper plane bed regime

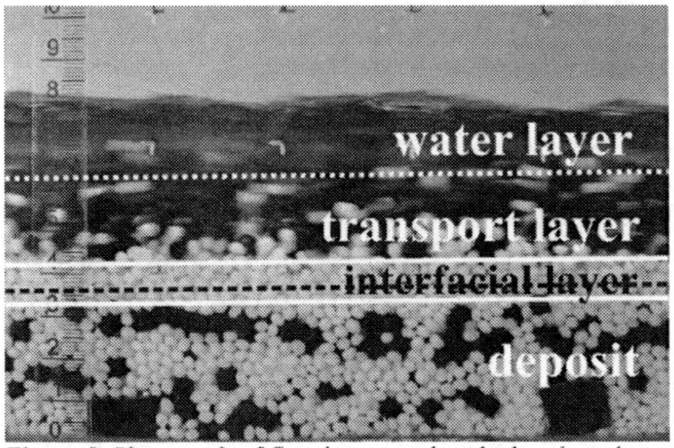


**Figure 2.** Photograph of flow in upper plane bed regime shows distribution of grains of bimodal sediment (black – coarser grains, white – finer grains). Legend: white solid lines – boundaries of interfacial layer, black dashed line – top of stationary deposit, white dashed line – top of transport layer.

### 2.4 Description of velocity distribution

Our earlier work [6-9] also shows that a vertical distribution of local longitudinal velocities is influenced by the flow stratification. Basically, a velocity profile can be considered as composed from a linear part in the transport layer and a logarithmic part in the layer with clear water (Figure 1).

The logarithmic part of a velocity profile can be described using the logarithmic law of the wall [6],

$$\frac{u}{u_b^*} = 2.5\ln\left(B\frac{y-y_{ini}}{k_s}\right) \quad (1)$$

where $u$ is longitudinal velocity at vertical position $y$, $u_b$* is the bottom shear velocity, $y_{ini}$ is position of origin of logarithmic profile, $k_s$ is the boundary roughness and $B$ is an integration constant for open channel flow ($B = 30$).

The velocity distribution is linear in the transport layer and it can be described by following equation [6],

$$\frac{u}{u_{tr}} = \left(\frac{y-\Delta y}{y_{tr}-\Delta y}\right)^n \quad (2)$$

where $y_{tr}$ is vertical position, at which a profile smoothly transforms its shape from linear to logarithmic, $u_{tr}$ is velocity at position $y_{tr}$, $\Delta y$ is initial position of linear profile and $n$ is an exponent (n=1).

Taking the above assumptions into account, an optimization procedure is carried out on measured velocity profiles with as a result values for five velocity-profile-associated parameters ($u_b$*, $y_{ini}$, $k_s$, $y_{tr}$, $\Delta y$). The key results for the bimodal sediments are discussed in chapter 4.

## 3 Evaluation of integral data

Integral quantities of sediment laden flows are mutually compared for flows of individual fractions (HSF3 and TLT50) and for the mixture of the two fractions (HT). In plots, they are related to the dimensionless form of the bed shear velocity $u$* defined as

$$u^* = \sqrt{g \cdot R_b \cdot I_e} \quad (3)$$

in which $R_b$ is hydraulic radius associated with bed, $g$ is gravitation acceleration and $I_e$ is slope of energy grade line of flow. The dimensionless form is called Shields parameter, $\theta$. For the form of the parameter we use in our evaluations, see [6-9]. Fore individual grain fractions, the equivalent diameter of grain, $d_e$, is used to calculate the Shields parameter. For bimodal sediments, the effective mean diameter, $d_m$, is used.

### 3.1 Quantification of distribution of fractions in bimodal mixture

The thickness of the interfacial layer is sensitive to the Shields parameter (the bed shear velocity) as Figure 3 shows. It tends to increase with increasing bed shear indicating that bed shear promotes vertical sorting at the interface between the deposit and the transport layer. Also, Figure 3 shows that the sliding upper part of the interfacial layer tends to be thicker than the stationary lower part, particularly at high bed shear, leaving just one or two layers of the finer grains at the top of the deposit. It is interesting to find out which effect the presence of the interfacial layer and its splitting into the stationary part and the sliding part has on bed friction and transport of sediment. Before we discuss this, it is important to determine how proportionally the two fractions are distributed in the flowing bimodal mixture.

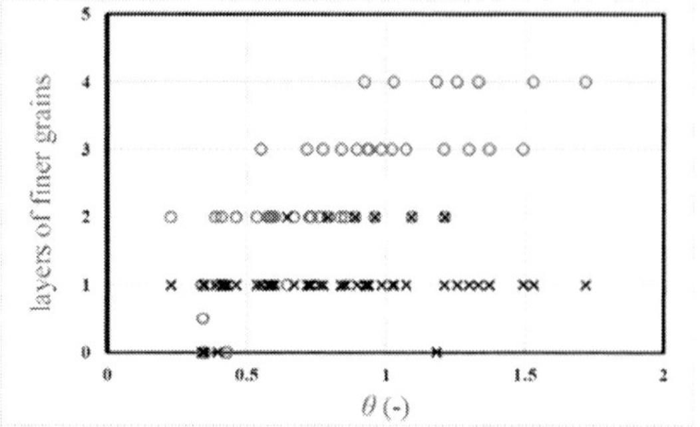


**Figure 3.** Number of layers of finer grains at interface between deposit and transport layer. Legend: green circles – sliding layers, black crosses – stationary layers.

Figure 4 shows the results of mixture sampling at the flume outlet. At the highest Shields parameters, the proportion of the finer fraction (HSF3) is approximately constant (about 32 per cent by weight, meaning that 68 per cent of the total weight of the sampled sediment is due to the coarser grains). At the lowest Shields parameters, virtually no fines are transported and transport of the coarser fraction dominates.

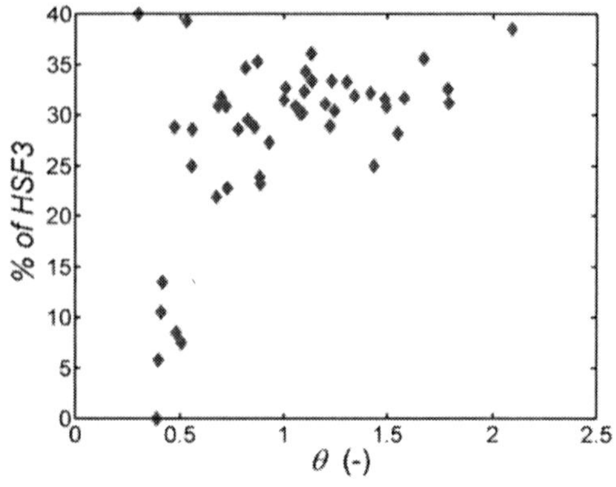


**Figure 4.** Weight percentage of finer fraction (HSF3) in sample of transported bimodal sediment

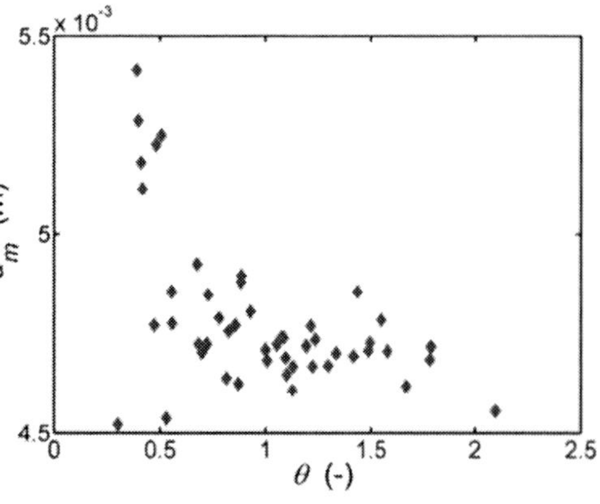


**Figure 5.** Mean diameter of sample of transported bimodal sediment

In Figure 5, the same effect is expressed using the variation of the mean diameter of the transported bimodal sediment with the Shields number. The plot confirms that the proportion of the fractions in the mixture remains constant for Shields parameters bigger than say one.

### 3.2 Discharges of sediment and mixture

In our previous work, we used a modified Meyer-Peter and Müller (MPM) formula to compare measured discharges of sediment with predictions. The MPM formula relates the dimensionless discharge of sediment (called Einstein parameter) to the Shields parameter. In Figure 6, plots of the Einstein parameter, $\Phi$, for individual fractions and the bimodal sediment exhibit very similar values except for very high Shields parameters. At very high bed shear, where the transport is intense, the coarse TLT5 is more mobile than the finer HSF3. The bimodal sediment exhibits the same mobility as the coarse fraction alone.

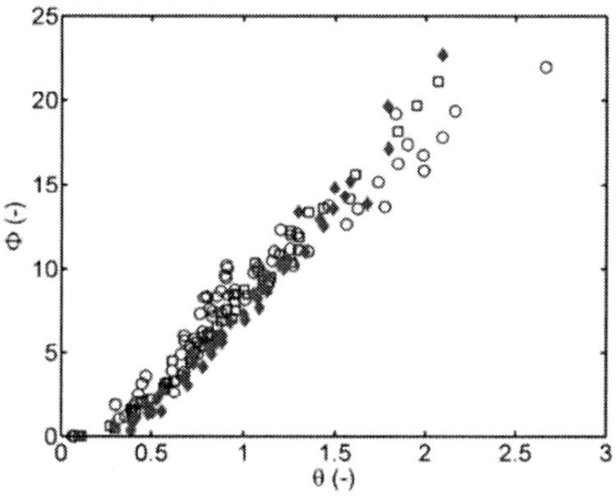


**Figure 6.** Dimensionless discharge of sediment versus Shields parameter. Legend: blue cylinders - HSF3, black squares - TLT50, red full diamonds – HT.

In order to relate the sediment discharge with the discharge of mixture (water+sediment), we employ the volumetric delivered concentration, $C_{vd}$, which is defined as a ratio of the two discharges. Figure 7 relates $C_{vd}$ with the specific discharge of mixture (defined as the volumetric discharge of mixture divided by the flow width). The plot is instrumental in delimiting the range of discharges which could be reached in the UPB regime in our experimental setup. At low discharges, the boundary is limited by the setup itself (a danger of settling of grains in connecting pipes of the setup). A transition from the UPB regime to the wavy regime [6-7] is the upper boundary at high discharges. The plot shows that higher delivered concentrations and thus sediment discharges can be reached for the coarser sediment flow than for the finer sediment flow. Bimodal sediment flows can reach similar concentrations as the coarser fraction flows. Apparently, a presence of coarse grains enables to extend the range of the UPB regime to higher discharges of both sediment and mixture. We hypothesize that the coarse grains in the transport layer actually stabilize the layer and prevent the plane bed from undulation (transition to wavy regime) which occurs earlier if the layer contains only finer grains.

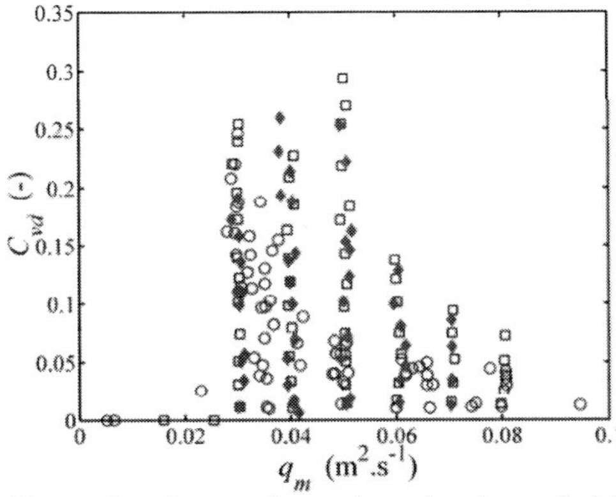


**Figure 7.** Range of experimental values of delivered concentration and specific discharge of mixture in upper plane bed regime. Legend: blue cylinders - HSF3, black squares - TLT50, red full diamonds – HT.

Also associated with the discharge of mixture is Froude number, $Fr_b$, defined as

$$Fr_b = \frac{U}{\sqrt{g \cdot R_b}} \tag{4}$$

in which $U$ = mean velocity of mixture. In [6-7], we discussed its relationship with the Shields parameter which exhibit a special shape with a maximum value of $Fr_b$ reached at the Shields parameter value of about one and with a drop in $Fr_b$ values at higher Shields parameters. We identified the left-hand side of the plot below the threshold $\theta \approx 1$ as the region of weak transport and the right-hand side above the threshold $\theta$ as the region of intense transport. Figure 8 shows this shape for flows of both individual fractions and reveals that for bimodal sediment the maximum reached but no drop is observed in the region of intense transport.

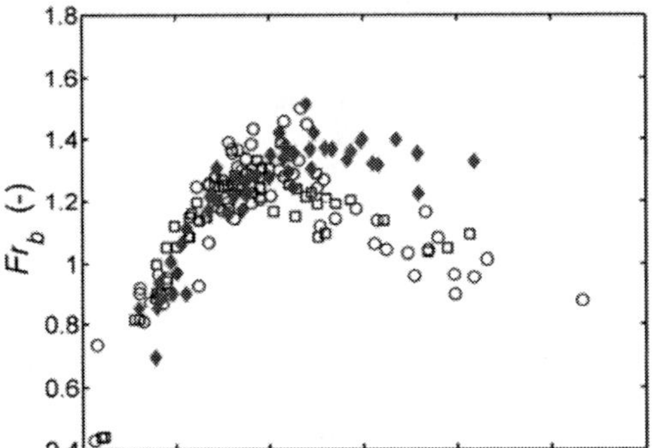


**Figure 8.** Froude number versus to Shields parameter. Legend: blue cylinders - HSF3, black squares - TLT50, red full diamonds – HT.

The absence of the $Fr_b$ drop with the increasing $\theta$ at intense transport of bimodal sediment may indicate lower bed resistance in bimodal flow than in flows of individual fractions. The drop is due primarily to the increase in the flow depth caused by increased resistance at more intense transport of sediment. If the drop is not there, than the increase of the flow depth is not there because the bed resistance is not high enough.

### 3.3 Bed resistance

The overall parameter expressing the bed resistance is the friction coefficient of the bed, $\lambda$. It is defined as

$$\lambda = \frac{8 \cdot g \cdot R_b \cdot I_e}{U^2} \tag{5}$$

and can be determined from our measured integral parameters of flow. Plotted in Figure 9 the coefficient is indeed smaller for bimodal flow than for flows of individual fractions at the condition of very intense transport associated with values of Shields parameter higher than the threshold value.

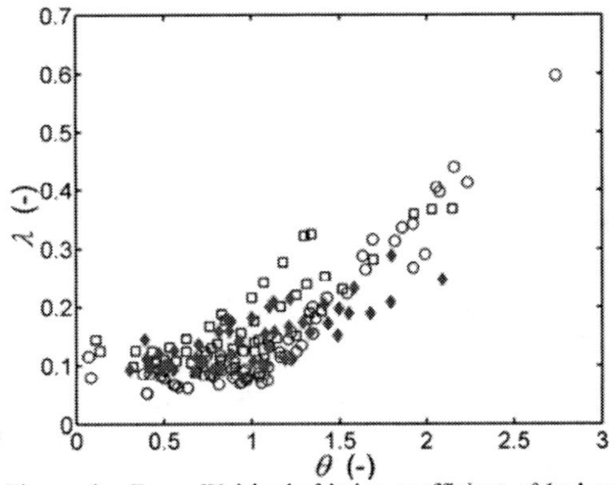


**Figure 9.** Darcy-Weisbach friction coefficient of bed versus Shields parameter. Legend: blue cylinders - HSF3, black squares - TLT50, red full diamonds – HT.

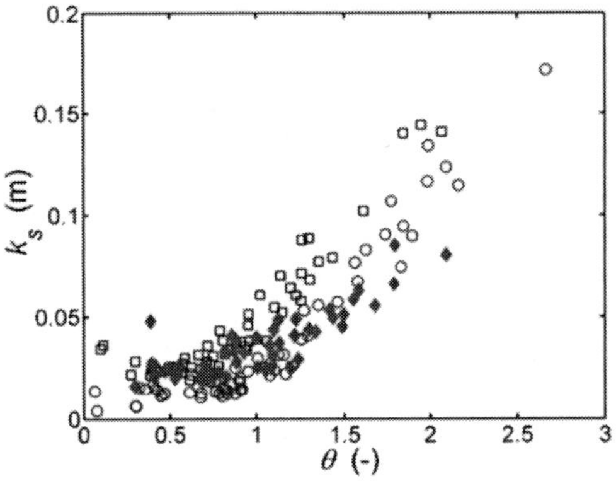


**Figure 10.** Bed roughness for log law of the wall versus Shields parameter. Legend: blue cylinders - HSF3, black squares - TLT50, red full diamonds – HT.

Figure 10 shows the same experimental results expressed as the bed roughness $k_s$ for the logarithmic law of the wall. It demonstrates that the bed roughness considerably increases with the Shields parameter if $\theta$ exceeds the threshold value of one due to the influence of the intense transport of collisional contact load on the bed resistance. For some reasons, the increase is the modest for the bimodal flow.

We assume a link between the observed smaller resistance of the bed and the presence of the interfacial layer in bimodal flows. Camera records and measurements of velocity profiles must be employed to identify possible reasons for the smaller resistance, e.g. for smoothening of the bed surface in bimodal flows.

## 4 Evaluation of video records and velocity distribution measurements

### 4.1 Video records

Each test run was recorded by a standard low-speed video camera located near a glass side wall of the flume. The record took about 10 second and contained frames capturing the entire depth and the width of about 15 centimeter of the flow (see Figure 2, which shows one of the recorded frames). Analyses of the records enable to identify positions of interfaces in the layered structure of the flow and observe a flow behavior in regions of particular interest in the flow.

Visual observations focused on the flow behavior at the interface between the transport layer and the bed deposit reveal some important differences between the bimodal flow and flows with individual fractions. One would think that the presence of the interfacial layer composed of the finer grains (HSF3) in the bimodal flow bears a close resemblance to the flow with the HSF3 sediment alone as this flow also exhibits an interfacial layer composed of HSF3 grains. Nevertheless, the observations suggest that the interfacial layer behaves quite differently in HSF3 flow and in bimodal flow. While in the bimodal flow the interfacial layer remains relatively thin and stable even at very high bed shear, it tends to be thicker and quite unstable in the HSF3 flow at high bed shear. The interfacial layer changes its thickness in time and exhibits temporal sliding interruptions causing intermittent motion of grains the layer and a variation in the position of the top of deposit.

### 4.2 Velocity distribution

Measured velocity profiles follow the hybrid linear-logarithmic shape quite well for both mono-disperse flows and bimodal flows. In our analysis of flow conditions at the interface with the deposit we focus on the linear part of a velocity profile.

The initial position of the linear profile, $\Delta y$, above the visually observed position of the top of deposit is an appropriate indicator of conditions at the interface. Basically, a high value of $\Delta y$ indicates a thick interfacial sliding layer with intermittent motion of grains. Figure 11 shows that $\Delta y$ increases with Shields parameter and tends to become much thicker for HSF3 flows than for bimodal flows at high bed shear. This confirms our conclusion based on the visual observations that the interfacial layer is thicker, and hence more unstable, in the HSF3 flows than in the bimodal flows.

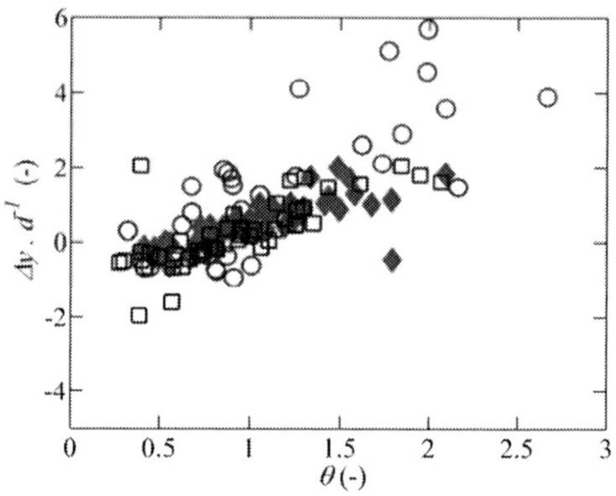


**Figure 11.** Relative initial position of linear profile versus Shields parameter. Legend: blue cylinders - HSF3, black squares - TLT50, red full diamonds – HT.

Another important characteristic of a linear profile is its gradient,

$$\frac{du}{dy} = \frac{u_{tr}}{y_{tr} - \Delta y} \tag{6}$$

derived from Eq. (2). It gives the slope of the profile across a dominant part of the transport layer and provides information about a measure of bed friction and collisional interactions as quantified in the kinetic theory. A steep velocity gradient indicates low bed friction and intense collisional interactions. Figure 12 shows that the velocity gradient increases with Shields parameter and reaches a certain limiting value at $\theta$ exceeding the threshold value for intense transport. The limiting value is considerably smaller for the coarser TLT50 flow than for the finer HSF3 flow and for the bimodal flow.

The general trends in the development of the velocity gradient correspond with the trends obtained from a comparison of the bed friction coefficient and the equivalent roughness of the bed. The roughness is small for HSF3 and HT compared to TLT50. The gradients are high for HSF3 and HT compared to TLT50, showing that lower roughness of the bed is associated with a more effective velocity development above the bed and thus the higher velocity gradient. The fact that the bimodal flow does not exhibit a higher value of the limiting maximum gradient than the HSF3 flow, although the bed friction is lower, may be attributed to the higher delivered concentration of the bimodal flow than of the HSF3 flow at the same value of Shields parameter.

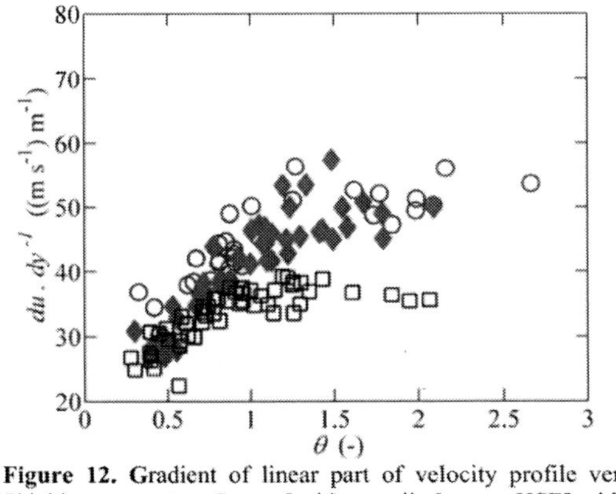


**Figure 12.** Gradient of linear part of velocity profile versus Shields parameter. Legend: blue cylinders - HSF3, black squares - TLT50, red full diamonds – HT.

The finding that the gradient of the linear velocity profile increases with the Shields parameter offers a hypothesis explaining the thickening of the interfacial layer in bimodal flow at higher bed shear (Figure 3). In the transport layer, the bigger velocity gradient leads to more intense collisions among coarser grains (as the kinetic theory of granular flows suggests) and larger voids among the colliding grains through which the finer grains can fall more easily and finally produce a thicker interfacial layer. A measurement of local concentrations in the lower part of the transport layer would reveal whether more intense collisions lead to lower local concentrations (and hence larger voids). A development of an appropriate measuring technique is work currently in progress. However recent local concentration measurements in a transport layer above deposit in a pressurized pipe [10] support the above assumption.

## 5 Conclusions

A proportion of individual fractions in transported bimodal sediment is independent of Shields parameter except for Shields values near the lower boundary of the upper plane bed regime.

The interfacial layer develops at the top of deposit below bimodal flow due to vertical sorting, it thickens at more intense bed shear.

Bed friction is smaller in bimodal flow than in flows of the bimodal sediment fractions alone at high bed shear (value of Shields parameter above one). This effect is attributed to the development of the interfacial layer which seems to smoothen the bed in the bimodal flow.

The gradient of a linear velocity profile increases with more intense bed shear. It offers a hypothesis explaining a thickening of the interfacial layer under more intense bed shear. The hypothesis is based on principles of the kinetic theory of granular flows and requires a further verification.

## Acknowledgments

The research has been supported by the Czech Science Foundation through the grant project No. 16-21421S and by Faculty of Civil Engineering of the Czech Technical University in Prague through the student grant project No. SGS14/179/OHK1/3T/11.

## References

1. C. Bel, K. Rorsman,P. Frey, *THESIS, (Symposium on Two-phase Modelling for Sediment dynamics in Geophysical Flows), Chatou (France)* (2013)
2. G. Parker, ASCE Manual 54 Chapter 3 Transport of gravel and sediment mixture
3. R. A. Kuhnle, J.K Horton, S. J. Bennett, *Sedimentology*, **53**, 631-654, (2006)
4. A. Bloom, J. S. Ribberink, ,G. Parker, J. Goephys. Res. , **113**, (2008)
5. M. Larcher, J. T. Jenkins, *J. Fluid Mech.*, **782**, 405-429, (2015)
6. V. Matoušek, V. Bareš. J. Krupička, T. Picek, Š. Zrostlík, *J. Hydrol. Hydromech.*, **63**(4), (2015)
7. V. Matoušek, V. Bareš. J. Krupička, T. Picek, Š. Zrostlík, *Can. J. Chem. Eng.*, **94**(6), (2016)
8. Š. Zrostlík, V. Bareš, J. Krupička, T. Picek, V. Matoušek, *EPJ Web of Conferences 92*, 02120 (2015)
9. V. Bareš, Š. Zrostlík, T. Picek, J. Krupička, V. Matoušek , *Flow Meas. Instrum.*, **51**, 68-78, (2016)
10. V. Matoušek, *J. Hydraul. Eng.*-ASCE, **135**(12), 1101-1106, (2009)

# Design type air engine Di Pietro

*Jarosław* Zwierzchowski[1,*]

[1]Faculty of Mechatronics and Mechanical Engineering, Kielce University of Technology, Al. 1000-PP 7; 25 – 314 Kielce; PL

**Abstract.** The article presents a pneumatic engine constructed by Angelo Di Pietro. 3D solid models of pneumatic engine components were presented therein. A directional valve is a key element of the control system. The valve functions as a camshaft distributing air to particular engine chambers. The construction designed by Angelo Di Pietro is modern and innovative. A pneumatic engine requires low pressure to start rotary movement. With the use of CFD software, the fields of velocity vectors' distribution were determined. Moreover, the author determined the distribution of pressure values in engine inlet and outlet channels. CFD model studies on engine operation were conducted for chosen stages of operating cycles. On the basis of simulation tests that were conducted, the values of flow rates for the engine were determined. The distribution of pressure values made it possible to evaluate the torque value on the rotating shaft.

## 1 Introduction

**Fig. 1.** Overview of the pneumatic engine designed by Di Pietro [13, 15, 16].

The air engine is used in industry and is a competition in relation to the electric motors. It is used mostly in dangerous explosive work atmospheres or in the vicinity of strong magnetic eclectic fields.

An important factor determining the profitability of its application is suitable efficiency and productivity. This article describes an innovative pneumatic engine designed by Angelo Di Pietro. According to his assumptions, this engine has 94.5% efficiency, and constant high torque. Additionally, this engine does not generate vibrations and has a very low friction, allowing the engine to operate at a low supply pressure of only 1 PSI (approx. 0.07 bar) [14].

Described engine has the following structure and work rules. Rotational motion is generated on a shaft positioned eccentrically in the cylindrical engine body. To the shaft there are attached specially arranged piston bearings, which exerts a force to interior cylinder, causing one sinusoidal rotation. The rotational movement of the cylinder forces expanding air in sequentially inflating air chambers. The tightness of the chambers provides cams with flat springs attached to the engine body. The cams closely adjacent to the cylinder. Below in Figure 1 is a plan view of the air engine designed by Angelo di Pietro. Figures show internal structure of the engine described above and also assembled structure ready to use [1-6].

In the literature there are not reported accurate control methods dedicated to this type of engines. Therefore, in the next chapter we will present the 3D project of pneumatic engine with air flow simulation and its model with numbers of actuators, valves, air distributors. This engine would be improved by using new numerical analyses described in papers [7-12]. This papers are also good base of theoretical and industrial use.

* Corresponding author: j.zwierzchowski@tu.kielce.pl

## 2 Design of the engine air control system

In the paper there will be described prototype design of the air engine having 9 working chambers. We will also show control system and do the simulation of the air flow in the engine.

### 2.1 3D model of the pneumatic engine

Below on figure 2 is a general view of designed air motor. There is chosen a structure having 9 working chambers.

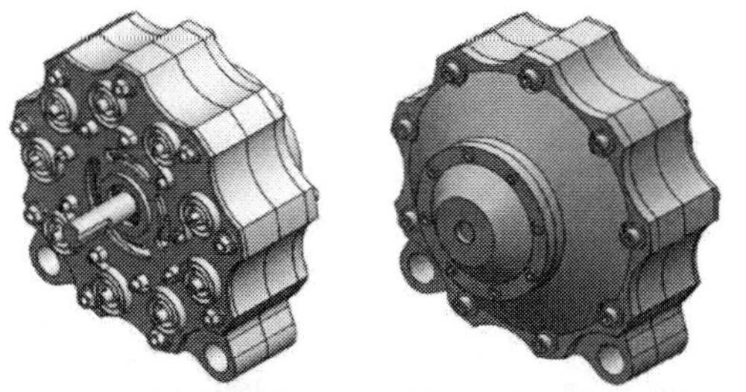

**Fig. 2.** 3D model of the air motor: a) front view, b) rear view.

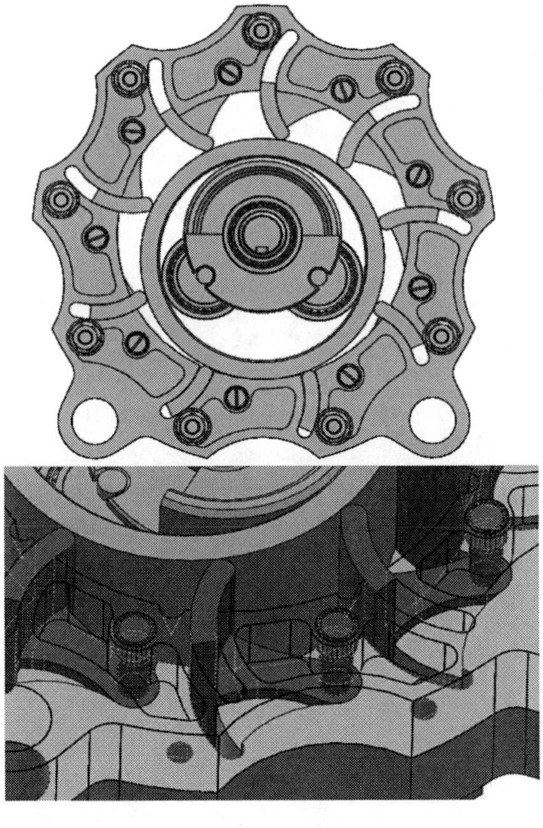

**Fig. 3.** Mounting cams in the engine.

We can see that by watching the front part of the engine (Fig. 2a), where are the springs attached in order to assist in internal piston oscillating movement. Figure 2b shows the back cover of the air engine. Figure 3 shows a method of mounting cams. It is important that the springs pressing the cams to the mentioned previously cylinder, should produce large enough force, that the cams do not lose contact with the cylinder at high speeds. Wrong selected force of the springs would cause the air flow between chambers and reduce engine efficiency. Cams clamp should be selected experimentally.
In order to minimize the adverse impact of friction between the internal components of the engine, there was care taken, that contact of the piston outer surface was not directly associated with the surface of the cylinder body. Occurring space helps keeping friction at the appropriate level. The rotary movement of the whole inner mechanism are provided by ball bearings. Designed structure ensures proper engine works. The next movement stages of the engine are shown below in Figure 4. There is showed how the piston and engine cams work, depending on rotation of the engine shaft.

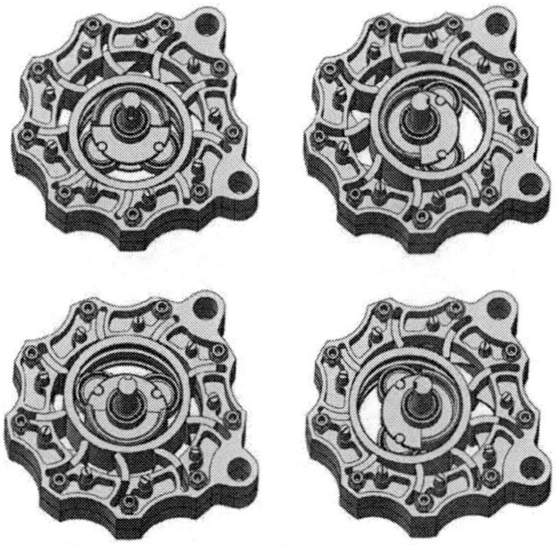

**Fig. 4.** The subsequent stages of work pneumatic engine.

### 2.2 The control system project of the pneumatic engine

The control system is designed in that manner, so at any given time only one chamber of the engine is supplied, two neighbouring chambers are closed and the other ones are connected with the atmosphere.
On the top of rear engine cover is located air distributor. This cover has air supply threaded opening toward to the engine.
This section will show how to control 9-chamber air motor invented by Di Pietro. In this describing design there is no electronics control in favour of a rotary distributor and a fixed rear cover, what is shown in Figure 5.

In the described engine air distributor has a conical shape (see Figure 6a). This element during rotation transmits compressed air to the respective consecutive chambers, and also receives air from the chambers already did not use.

**Fig. 5.** Engine control system: 1 - air channels cover, 2 - distributor, 3 - rear cover.

On Figure 6c shows a cross-sectional of the distributor, there can be distinguished inlet, outlet channels and shaft hole. The width of the inlet channel is expressed in degrees and is equal 30 degrees. This channel directs air from the connection of the engine body to the air channels.

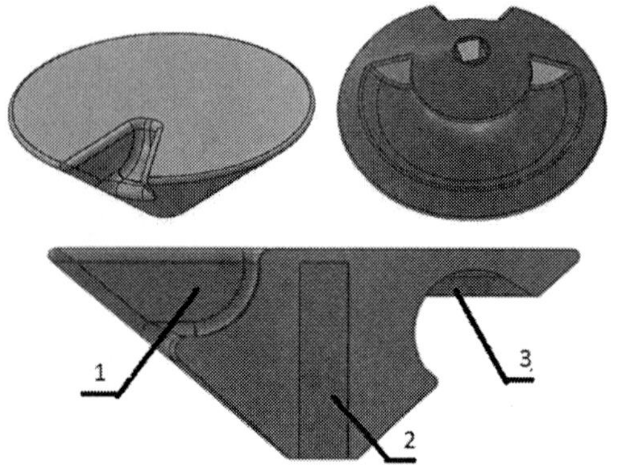


**Fig. 6.** Distributor a) view from the top, b) view from bottom c) cross-section of the distributor; 1 - inlet channel, 2 - outlet channel, 3 - shaft hole.

The outlet channel is formed so as to minimize energy loss. This is possible because in the passive chambers there is no changes in pressure. The channel is located on the underside of the distributor and connects the supply channels in back cover with vent channels. The width of this channel is equal 220 degrees. Figure 6b shows a front view of the distributor.

In the rear part of the engine, centrally, are located notches of air channels, having its origin at the element base and end in engine working chambers. Those channels are used for filling and emptying of the working chambers. These channels are arranged symmetrically in a plane of circle and positioned at angle of 23.3 degrees, respect to the mentioned base. Their size is 8 mm in diameter. Figure 7, below, shows a cross-section of the bonnet, where we can distinguish: 1 air channel, 2 air inlets, 3 insert, 4 vent channels.

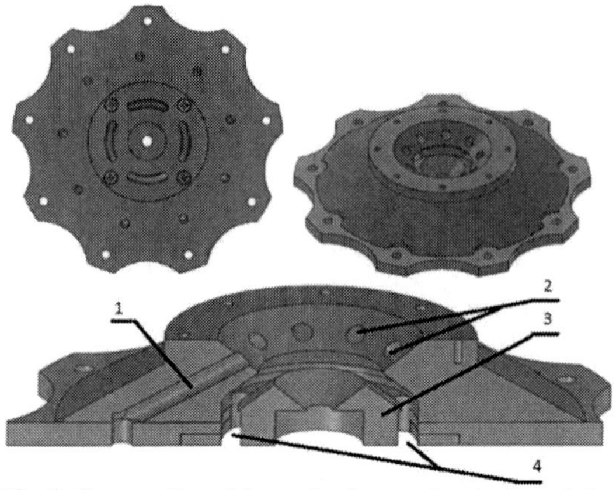


**Fig. 7.** Cross section of the engine bonnet: 1 - air channel, 2 - air inlet channels, 3 - insert, 4 outlet channels.

An important parameter is the width of the distributor channel, expressed in degrees. It effects on cycle time length of the working chamber, which is translated into the air consumption, power, and torque of the engine. In describing design this value is set to 30° and was selected to assure some overlapping in air supply in cycles between adjacent chambers.

The angular distance between the inner edges of the air channel holes in the lid is equal 25.18°. This is so called dead zone. For instance, when the designer would make a mistake, and the width of the inlet channel would be smaller than the width of the zone, there would be a risk of engine stop in a position in which the individual start-up would be impossible.

Furthermore, in the work cycloramas there would be visible moments in which the engine would not produce torque, because none of the chambers would be air powered at that time. Such a break will result torque and work culture decrease of the engine.

The overlap between two air channels on the supply side is equal (see fig. 8):

$$x = 30° - 25{,}18° = 4{,}82°\ 26 \qquad (1)$$

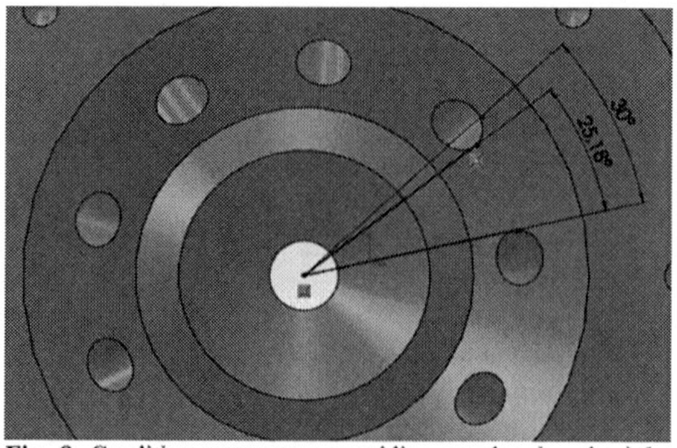


**Fig. 8.** Condition parameters providing overlapping the inlet channel.

This means that two adjacent chambers will be powered simultaneously by a time of 4.82° shaft rotation angle.

## 2.3 Air flow simulation

In order to illustrate air flow in the control components, we used CFD Flow Simulation software, which is extra added to basic SolidWorks software.
The first simulation shows a moment of the engine operating cycle. It is state, when one of the working chambers is air supply. The initial condition of engine work is to have total pressure equal 0.3MPa in inlet of the distributor body.
The resulting flow trajectories is shown in Figure 9. There is also illustrate the air speed, which increases dramatically in inlet channel to achieve maximum value in the working chamber.

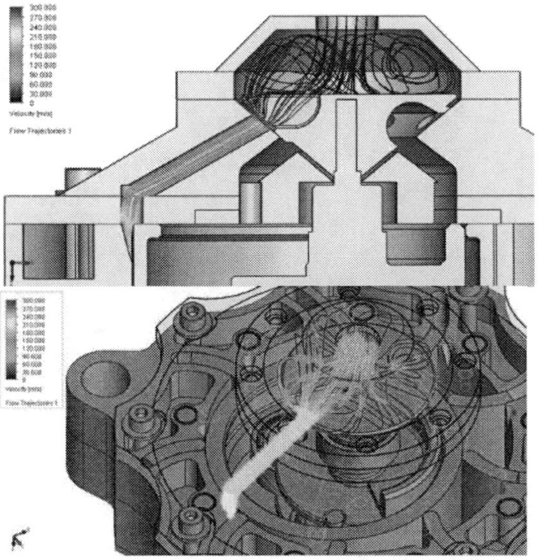

**Fig. 9.** View of the trajectories of inlet air streams in cross-section through the air channel.

Flow Simulation CFD also allows us to estimate parameters such as mass and volume flow, which are crucial in engine work.

**Table 1.** Supply simulation parameters of working chamber.

| parameter | Values |
|---|---|
| Pressure [MPa] | **0,3** |
| Speed [m / s] | **13,5** |
| Mass flow [kg / s] | **0,018** |
| The volumetric flow rate [l / s] | **4,97** |

The obtained values are shown in Table 1 and describes the conditions of the inlet channels from distributor.
The second case shows the flow of outlet air. The figures 10 below show the simulation cycle, emptying the working chamber. The initial condition is the total pressure prevailing in the interior of engine, where the value is equal to 0.15MPa.

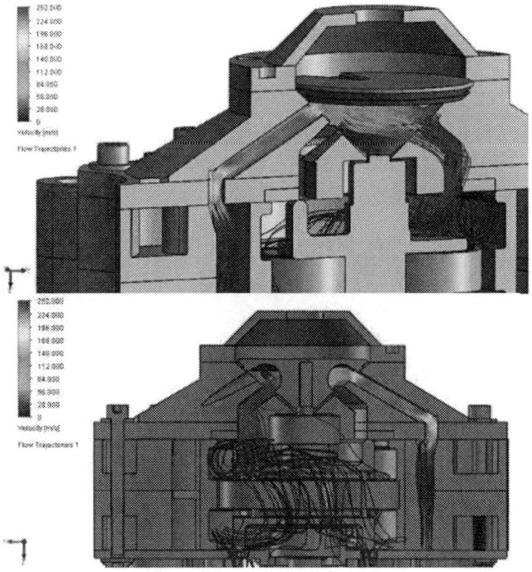

**Fig. 10.** View of the inlet air trajectory.

Fig. 11 shows, how the air way out from inside the engine through the vent channels, witch are located in the front cover. The main vent channel is the interior of the cylindrical piston.

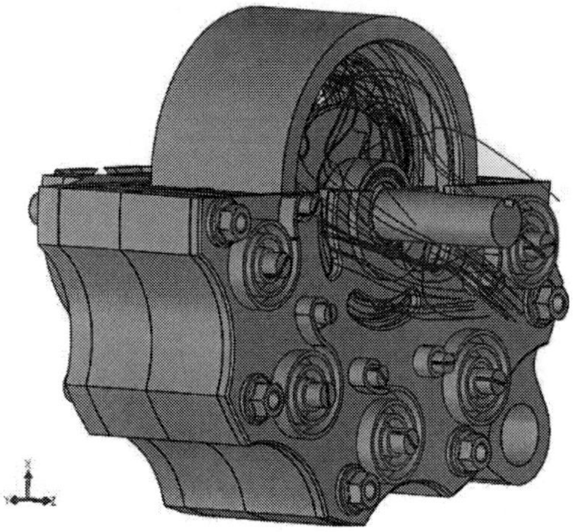

**Fig. 11.** View of air flow through the vent channels.

## 3 Model research of pneumatic engine control

In theory, presented control system would be able to manage air engine.

We discussed about that in chapter two. In case of building a model of the engine distributor, we try replacing its key part using numbers of pneumatic elements, lists below.

This example shows how a huge role does the distributor play and how many elements are needed to carry out its functions. It also allows us to find out the manner and results of its actions. In order to present the model of the control system, there was used FluidSIM program.The control system consists of following elements:

- single acting actuator with return spring (9 units),
- sequential switcher "Quickstepper" (1 unit),
- delay valve (9 units),

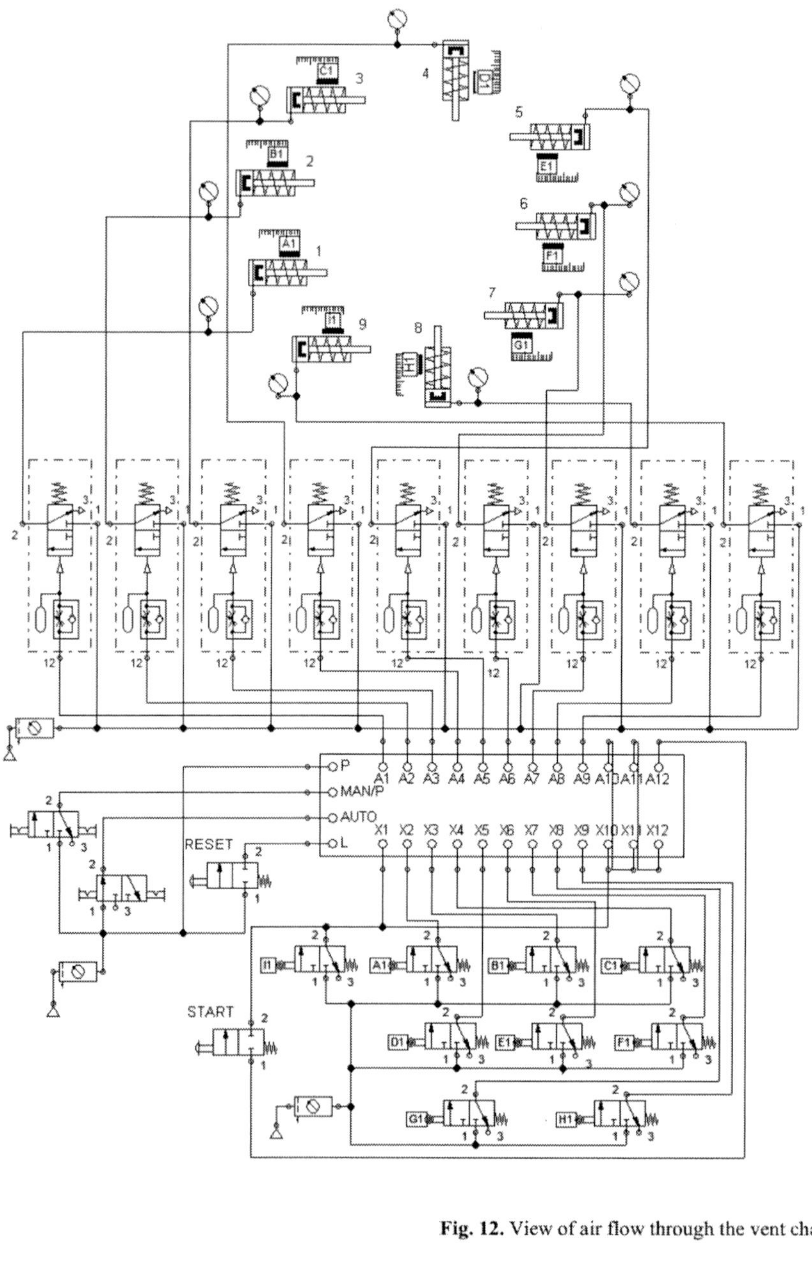


**Fig. 12.** View of air flow through the vent channels.

- 3/2 limit valve controlled by roll back spring (9units.),
- 2/2 valve controlled by button with spring return (2 units),
- 2/2 bistable valve controlled by buttons (2 units),
- manometer (9 units),
- a source of compressed air with the FRL (3 units).

Working nine cylinders simulate the work chambers of the engine. Sequential switcher controlled from end valve signals acts as a distributor of compressed air. To get the effect of power supply cycles overlapping of subsequent chambers we used set of delay valves. Valves changes are followed by a signal from the sequential switch (see fig. 12).
Fig. 13 shows cyclorama operation of the above system. We generated it in the "state diagram" block.
It is used to visualize the successive power cycles of chambers in the engine. The last line shows the angular position of the drive shaft and is added manually.

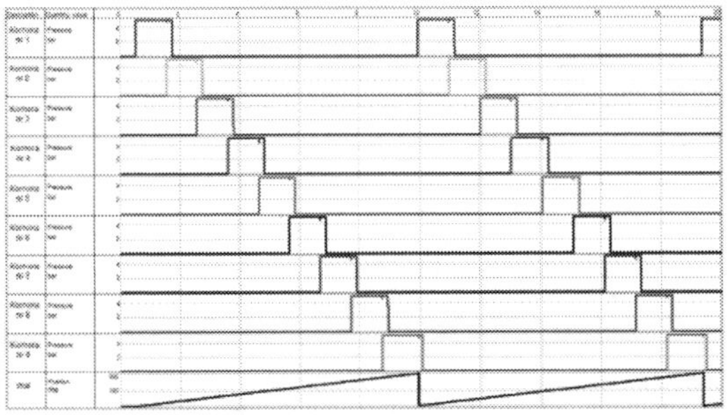

**Fig. 13.** Cyclorama of control system.

## 4 Summary and conclusions

Aim of the work, which was control system design of the air engine has been realized through the development of a 3D model. It was not an easy task, due to very small information about the dilates of the system, provided by the inventor of the engine. The proposal described in this paper, was preceded by a number of more or less successful prototype models. The second important aspect of this work was bunch of information about air flow, which take place in engine. We have shown pictures telling about air behaviour in inlet and outlet channels. In paper there are also figures describing interior concept of the pneumatic engine. We have written about cams and springs attached to them.
Finally, there have been showing model of directional valve building with numbers of pneumatic elements. We have focused on it, because in assumption it plays a crucial role in increasing efficiency and other desired parameters.
Good work of this model has been proved via cyclorama. All prototype designed, which we have presented here need to be checked and it will be next step in our research map.

## References

1. P.A. Laski, J.E. Takosoglu, S. Blasiak, Robot. Auton. Sys. **72**, 2015
2. P.A. Laski, Edited by Zolotarev I., Radolf V. *Proceedings of 22nd International Conference on Engineering Mechanics 2016* (Academy of Sciences of the Czech Republic, 2016)
3. D.S. Pietrala, Edited by Zolotarev I., Radolf V. *Proceedings of 22nd International Conference on Engineering Mechanics 2016* (Academy of Sciences of the Czech Republic, 2016)
4. M. Blasiak, Edited by Zolotarev I., Radolf V. *Proceedings of 22nd International Conference on Engineering Mechanics 2016* (Academy of Sciences of the Czech Republic, 2016)
5. J.E. Takosoglu, Edited by Zolotarev I., Radolf V. *Proceedings of 22nd International Conference on Engineering Mechanics 2016* (Academy of Sciences of the Czech Republic, 2016)
6. J.E. Takosoglu, P.A. Laski, S. Blasiak, G. Bracha, D. Pietrala. Meas. Cont. **49**, 2, 2016
7. L. Nowakowski, E. Miko, M. Skrzyniarz, Edited by Zolotarev I., Radolf V. *Proceedings of 22nd International Conference on Engineering Mechanics 2016* (Academy of Sciences of the Czech Republic, 2016)
8. H. Wisniewski, L. Plonecki, *Proceedings of the 16th International Carpathian Control Conference (ICCC)*, 2015
9. S. Blasiak, C. Kundera. Procedia Eng. **39**:315-326, 2012
10. E. Miko, L. Nowakowski, L. Procedia Eng. **39** (2012a)
11. G.F. Bracha. Edited by Zolotarev I., Radolf V. *Proceedings of 22nd International Conference on Engineering Mechanics 2016* (Academy of Sciences of the Czech Republic, 2016)
12. Z. Koruba, I. Krzysztofik, P. I. Mech. Eng. K-J. Multi-Body Dyn. **227**, K1, 2013
13. Significant new rotary engine design runs on compressed air - Image 5 of 8. [13.05.2016]; Available from: http://www.gizmag.com/go/3185/picture/6137/
14. Engine Air. [20.06.2016]; Available from: http://www.engineair.com.au/
15. WWF 2 Degrees Project. [13.05.2016]; Available from: http://2degreesproject.com.au/.
16. Green Speed Motorcycle nouveau concept de moto à air comprimé (Moto-Net du 26/11/2009). [12.05.2015]; Available from: http://www.moto-net.com/p_article_print.php?RefArticle=3442

EPJ Web of Conferences **143**, 02150 (2017) DOI: 10.1051/epjconf/201714302150
*EFM 2016*

# Model tests of wind turbine with a vertical axis of rotation type Lenz 2

*Jaroslaw* Zwierzchowski[1,*], *Pawel Andrzej* Laski[1], *Slawomir* Blasiak[1], *Jakub Emanuel* Takosoglu[1], *Dawid Sebastian* Pietrala[1], *Gabriel Filip* Bracha[1], and *Lukasz* Nowakowski[1]

[1] Faculty of Mechatronics and Mechanical Engineering, Kielce University of Technology, Al. 1000-PP 7; 25 – 314 Kielce; PL

**Abstract.** A building design of vertical axis wind turbines (VAWT) was presented in the article. The construction and operating principle of a wind turbine were described therein. Two VAWT turbine models were compared, i.a. Darrieus and Lenz2, taking their strengths and weaknesses into consideration. 3D solid models of turbine components were presented with the use of SolidWorks software. Using CFD methods, the air flow on two aerodynamic fins, symmetrical and asymmetrical, at different angles of attack were tested. On the basis of flow simulation conducted in FlowSimulation, an asymmetrical fin was chosen as the one showing greater load bearing capacities. Due to the uncertainty of trouble-free operation of Darrieus turbine on construction elements creating the basis thereof, a 3D model of Lenz2 turbine was constructed, which is more reliable and makes turbine self-start possible. On the basis of the research, components were designed and technical docu mentation was compiled.

## 1 Introduction

Turbines convert kinetic energy into electrical energy using a rotor mounted on a shaft connected to the generator. Rotating turbine rotor decreases wind speed from $v_1$ before turbine to $v_2$ behind turbine. Difference between those two wind speeds is a processed measure value of kinetic energy of moving rotor connected with the generator of electricity. Generally, the maximum theoretical wind power can be calculated with assuming, that wind speed behind the turbine is equal to zero:

$$P_{max,theoretical} = \frac{1}{2}\rho A U_0^3, \qquad (1)$$

where: $\rho$ – air density, $U$ – wind speed, A – surface area reduction of wind speed.
In practice, it is impossible to reduce the wind speed behind the turbine to zero, therefore $C_p$ power factor is added into theoretical model. Power factor represents the ratio between the actual and maximum power. Theoretical $C_p$ maximum is labelled to Betz's limit and is equal to 16/27 (or 0.593) [2]. In real condition power factor does not reach even 0.5. This fact is due to aerodynamic losses. $C_p$ factor reaches the maximum value when wind speed $v_2$ behind the turbine is equal to 1/3 of $v_1$ before one.

### 1.1 Aerodynamic properties

When an object is exposed to the wind, it experiences impact of aerodynamic force. Occurrence of this force is caused by wind pressure and viscous friction between air and outer surface of the object. Aerodynamic force in the direction of the wind is called the drag. When streamline shape object, such as an aircraft wing panel, is exposed to the wind, the aerodynamic force is distributed on the two components:
• drag – component parallel to the air flow direction,
• lift – component perpendicular to the air flow direction.
This is shown in Figure 1, where are moment and forces acting on wing in two-dimensional space.

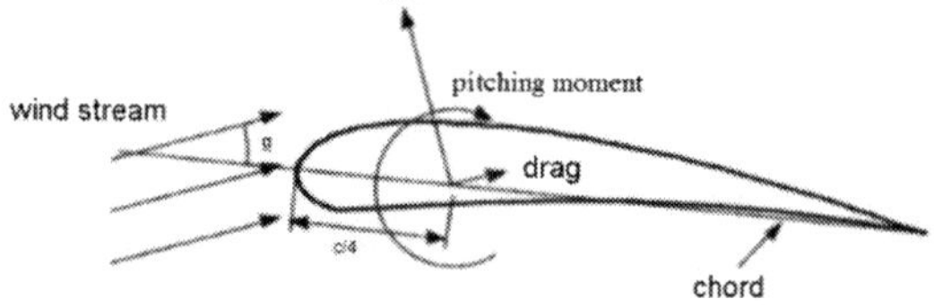


**Fig. 1.** Torque and forces acting on wing.

The flow velocity at the upper convex part of the panel is increased, it creates a low pressure area over the air panel in accordance with Bernoulli's principle. This will result in lift force that is perpendicular to the direction of the incoming air. The application point, where forces act, is considered on the chord, within 1/4 of the chord length from the leading edge. To describe the force as a whole there is added pitching moment which is generally located at the same reference point as components of forces [2].
Another characterizing parameter of the aerodynamic properties is tip-speed ratio, also known as specific speed ratio, defined as follows:

* Corresponding author: j.zwierzchowski@tu.kielce.pl

$$\lambda = u_t / U, \qquad (2)$$

where: $u_t$ – the tangential component of the rotor speed, $U$, – wind speed.
Flight speed factor is related to the $C_p$ power factor, whose optimal level is obtained for the specific TSR (tip-speed ratio). For devices operating on the basis of the drag, TSR will never exceed the value of 1, but for aircraft wings or helicopter panels based on the strength of lift reaches higher values [1].

### 1.2 VAWT turbine

The first practical wind machines were windmills with vertical axis marketing invented in eastern Persia in the ninth century. Those turbines had a rod construction in which, were mounted sails made of sugar cane bundles as shown in Fig. 2.

**Fig. 2.** Windmill on the vertical axis of the ninth century [6].

In modern times there have been developed several designs of vertical axis wind turbines (VAWT). Their main advantage is that the mechanical and electrical parts such the gear or generator can be installed close to the ground. This ensures greater economic efficiency in the tower construction and facilitates maintenance. Additionally, VAWT are non-directional, which means that they take the wind from all sides and do not need control mechanism deviations. Their drawback is the fact that most them work on lift strength must have automatic start-up. There is necessary external source power to carry the turbine in motion.

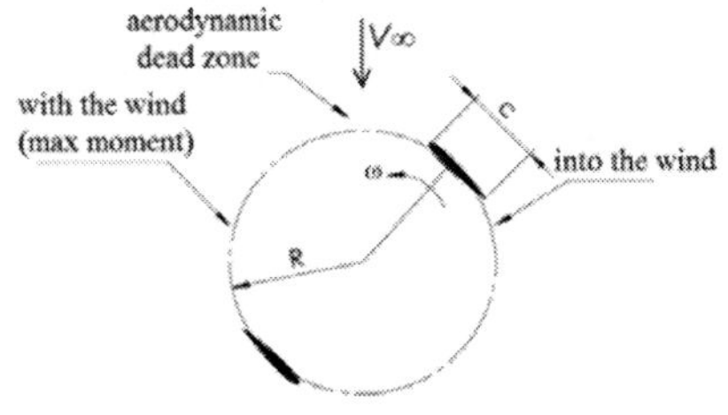


**Fig. 3.** VAWT - view from the top.

Performance of wind turbines with vertical axis is generally less compred to their horizontal counterparts because rotor panels during rotation pass through aerodynamic dead zone shown in Fig. 3.

#### *1.2.1 Type Darrieus*

H-Darrieus turbine (Type H-rotor) can be characterized by straight panels which, together create the letter H. The model is shown in Fig. 4. It has improved efficiency, simpler construction and relatively high power unit. The rotor aerofoils are spaced up from the center of the turbine, so that the rotation continuously changes the attack angle relative to the wind. However, when one of the blades is operated with the wind, the second opposed resists. This in consequence reduces the efficiency of the design.

**Fig. 4.** Turbine H – rotor [3].

#### *1.2.2 Type Lenz*

All wind turbines type VAWT were inspired by two basic types: Savonius'a and Darrieus'a. In 2007, was published an article of a wind turbine with greater efficiency of approx. 40% compare to ones above. This turbine was invented by Ed Lenz. The project introduced a lot of innovation, for example he combined Darrieus turbine working on the aerodynamic lift force and Savonius turbine relying on the drag (see Fig 5).

**Fig. 5.** VAWT turbine Lenz2 type [5].

Tip speed ratio of the turbine described in the article is between 0.8 and 1.2 at wind speeds of 6.7 m/s. This means that the turbine uses the lift force. The rotational speed of the turbine should be approx. 130 r / min.

## 2 Simulation studies and model construction

CFD analysis of two turbine aerofoils Darrieus NACA0012 and NACA4415 (Fig. 6) was carried out using Flow Simulation module, which is an integral part of the SolidWorks. When comparing the two aerodynamic panels $C_l$ lift coefficient was obtained for different wind wings attack angles ranging from 0° to 18°. It allows to select optimum wing attack angle. It was also a pressure factor distribution and velocity distribution calculated around the wing. Those two wings were inspired by a model from the windgenkits.com website with modifications using SolidWorks 3D.

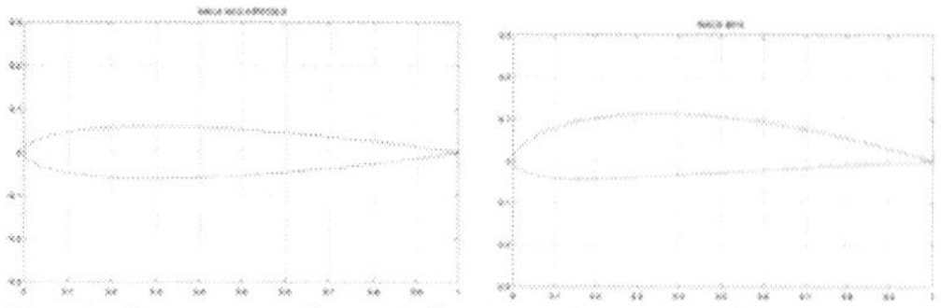

**Fig. 6.** Geometry of aerofoil: a) NACA 4415, b) NACA 0012 [4].

The main parameters that characterize those aerofoils are a chord length $c$ (chord) and the thickness, which is a function of the length. For aerofoil NACA0012 its thickness is 12% $c$. The second aerofoil is asymmetrical profile NACA4415. It is often used for testing and simulation. Its thickness is 15% $c$. In the next part of the paper is compared pressure distribution in upper and lower sections of both panels using different attack angles.
The rotational speed ω is represented by a factor of flight λ (TSR). This parameter is expressed as the ratio of airfoil speed to the of $V_{in}$ air stream speed:

$$\lambda = \frac{\omega \cdot R}{V_{in}}. \quad (3)$$

Power factor $C_p$ is determined experimentally by measuring the power produced by the turbine and dividing it by the theoretical turbine power, which is calculated from formula (2):

$$C_p = \frac{P}{\frac{1}{2}\rho A U^3}. \quad (4)$$

The wind speed is set to 5 m/s. Pressure, temperature and humidity are average values. The analysis was performed for several angles of attack to determine the appropriate position of the aerofoil relative to the inflowing air stream. It was done in order to compare the properties and behaviour of both wings. Simulation parameters are stored in Table 1.

**Table 1.** Simulation parameters.

| | |
|---|---|
| angle of attack | 0°, 2°, 4°, 6°, 8°, 10°, 12°, 14°, 16°, 18° |
| wind speed | 5 m/s |
| type analysis | Outside |
| flow model | turbulent and laminar; model: k-ε |
| Pressure | 101325 Pa |
| Temperature | 293.2 K |
| Humidity | 50,00% |

## 2.1 Simulation parameters – Darrieus turbine

Due to contours of pressure distribution and velocity vectors the point of stagnation is shown. Fig. 5a shows equal pressure distribution on both sides of the airfoil due to its symmetry. The values in the upper left corner of drawings is a dimensionless coefficient of pressure distribution calculated from the formula

$$C_p = \frac{p_i - p}{\frac{1}{2}\rho U^2}. \quad (5)$$

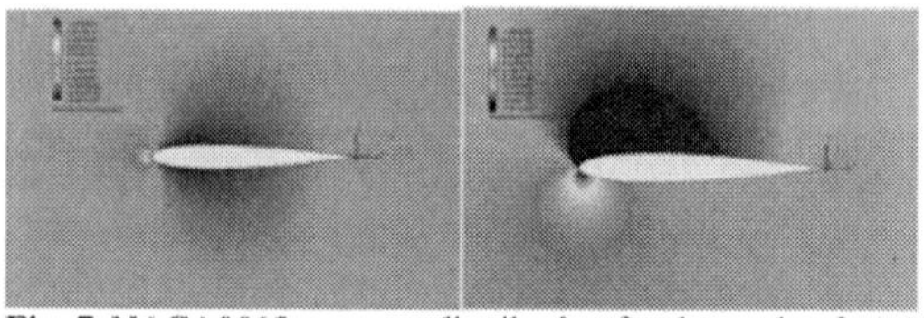

**Fig. 7.** NACA0012 pressure distribution for the angle of: a) 0°, b) 10°.

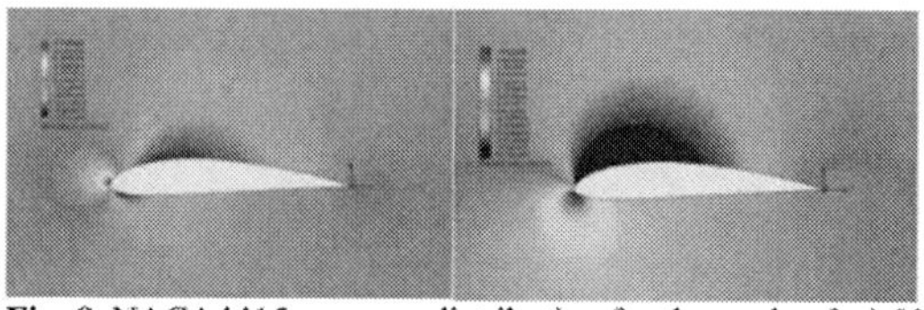

**Fig. 8.** NACA4415 pressure distribution for the angle of: a) 0°, b) 10°.

Fig. 7b shows a pressure distribution for the attack angle of 10°, There is in upper side of the airfoil elevated pressure zone, the stagnation point is moved down. If attack angle increases airfoil lump becomes less streamlined. The same can be seen with panel, with the difference that the pressure difference for asymmetric airfoil is higher than for symmetrical one (see Fig. 8).
The most important parameter obtained through simulation is lift $L$, actually, lift coefficient $C_l$. For example, from Fig. 7a lift $L = 9{,}047798e - 006\ [N]$. The $L$ value can be read from Table 2, there are surface parameters exported from FlowSimulation. Lift coefficient is from formula

$$C_l = \frac{L}{\frac{1}{2}\rho U^2 bc} = 0.00050102. \quad (6)$$

**Table 2.** Simulation parameters – surface parameters.

| Parameter | Minimum | Maximum | Average |
|---|---|---|---|
| Pressure [Pa] | 101319,46 | 101344,85 | 101322,60 |
| Density (Fluid) [kg/m^3] | 1,2004570 | 1,2007300 | 1,2004885 |
| Shear Stress [Pa] | 0 | 0,6173043 | 0,1277972 |
| Temperature (Fluid) [K] | 293,19976 | 293,21306 | 293,2069 |
| Relative Pressure [Pa] | -5,531395 | 19,859177 | -2,3966333 |

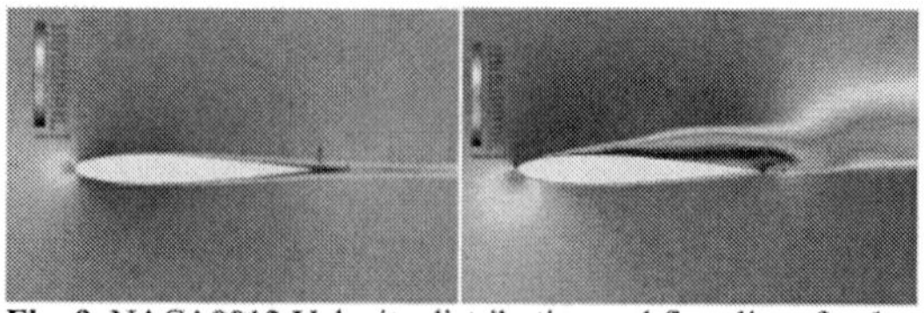

**Fig. 9.** NACA0012 Velocity distribution and flow lines for the angle of: a) 0°, b) 10°.

In Fig. 9 despite velocity distribution there is also added flowing lines illustrating air behavior at descent of the airfoil. We can see jet, which is "glued" to convex part of the wing. On steam down edge is creating elevated pressure zone, as shown in Fig. 9a and 10a. This phenomenon is desirable.

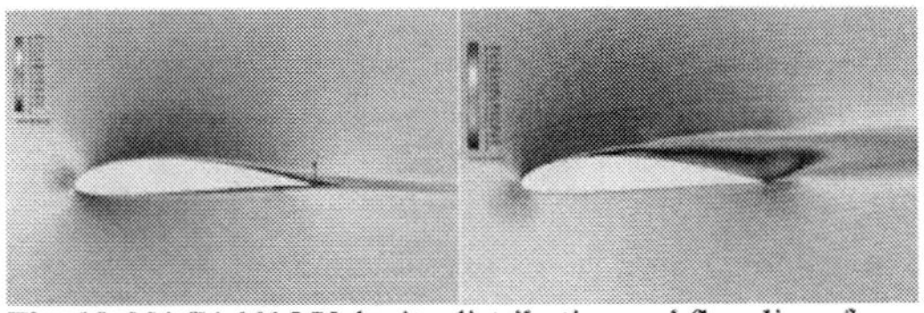

**Fig. 10.** NACA4415 Velocity distribution and flow lines for the angle of: a) 0°, b) 10°.

Fig. 9b again shows the velocity distribution around the airfoil, assisted flow lines to illustrate the behavior of the jet. Clearly stall zone. The angle for the lift reaches the highest value is called the critical angle. In Fig. 10b is shown the velocity distribution of an angle of 10 °, the in the separation area speed falls to zero, while in the convex part of airfoil maximum value of the velocity is higher than in the case of a symmetric panel.

## 2.2 Results comparison from Flow Simulation and Xfoil

The results of the lift coefficient $C_l$ are shown in Tables 3 and 4 and compared with the values taken from the website airfoiltools.com, the site is a good source of information on the airfoils.

**Table 3.** Simulation - NACA0012 surface parameters.

NACA0012 Flow Simulation Xfoil

| Atack Angle | turbulent flow | | | laminar & turbulent flow | | | | | | |
|---|---|---|---|---|---|---|---|---|---|---|
| | $C_l$ | Error $C_l$ | $C_l/C_d$ | $C_l$ | Błąd $C_l$ | $C_l/C_d$ | AoA | $C_l$ | $C_d$ | $C_l/C_d$ |
| 0 | 0,001 | | 0,037 | 0,001 | | 0,000 | 0,000 | 0,000 | 0,017 | 0,000 |
| 2 | 0,217 | 19,04% | 13,759 | 0,189 | 36,72% | 11,980 | 2,000 | 0,268 | 0,016 | 16,379 |
| 4 | 0,448 | 18,29% | 30,816 | 0,378 | 1,33% | 26,228 | 4,000 | 0,374 | 0,014 | 25,888 |
| 6 | 0,667 | 19,42% | 43,812 | 0,558 | 3,85% | 36,788 | 6,000 | 0,537 | 0,015 | 35,298 |
| 8 | 0,879 | 21,83% | 44,893 | 0,728 | 5,51% | 37,098 | 8,000 | 0,686 | 0,020 | 35,046 |
| 10 | 1,07% | 21,20% | 37,470 | 0,87% | 3,22% | 30,511 | 10,000 | 0,847 | 0,029 | 29,526 |
| 12 | 1,247 | 21,97% | 21,165 | 0,960 | 2,52% | 16,298 | 11,000 | 0,905 | 0,059 | 16,798 |
| 14 | 1,092 | 23,29% | 14,520 | 0,978 | 14,30% | 12,997 | 11,500 | 0,838 | 0,075 | 11,130 |
| 16 | 0,920 | 14,92% | 10,007 | 0,878 | 10,02% | 10,006 | 11,750 | 0,790 | 0,087 | 9,075 |

**Table 4.** Simulation – NACA4415 surface parameters.

NACA4415 Flow Simulation Xfoil

| Atack Angle | turbulent flow | | | laminar & turbulent flow | | | | | |
|---|---|---|---|---|---|---|---|---|---|
| | $C_l$ | Error $C_l$ | $C_l/C_d$ | $C_l$ | Błąd $C_l$ | $C_l/C_d$ | $C_l$ | $C_d$ | $C_l/C_d$ |
| 0 | 0,169 | 199,02% | 8,339 | 0,231 | 119,26% | 11,373 | 0,507 | 0,020 | 24,936 |
| 2 | 0,484 | 46,48% | 24,174 | 0,598 | 18,86% | 29,837 | 0,710 | 0,020 | 35,404 |
| 4 | 0,662 | 36,55% | 31,026 | 0,826 | 9,47% | 38,701 | 0,904 | 0,021 | 42,366 |
| 6 | 0,842 | 26,57% | 35,581 | 1,062 | 0,35% | 44,890 | 1,055 | 0,024 | 45,049 |
| 8 | 0,998 | 22,76% | 38,745 | 1,239 | 1,13% | 48,105 | 1,225 | 0,026 | 47,562 |
| 10 | 1,154 | 15,85% | 40,516 | 1,345 | 0,58% | 47,214 | 1,337 | 0,028 | 48,938 |
| 12 | 1,214 | 13,19% | 35,268 | 1,430 | 3,90% | 41,542 | 1,374 | 0,034 | 39,922 |
| 14 | 1,397 | 0,10% | 30,781 | 1,470 | 4,88% | 32,380 | 1,399 | 0,045 | 30,811 |
| 16 | 1,471 | 5,85% | 23,314 | 1,510 | 8,25% | 23,924 | 1,386 | 0,063 | 21,951 |
| 18 | 1,524 | 11,62% | 17,069 | 1,501 | 10,31% | 16,819 | 1,347 | 0,089 | 15,085 |

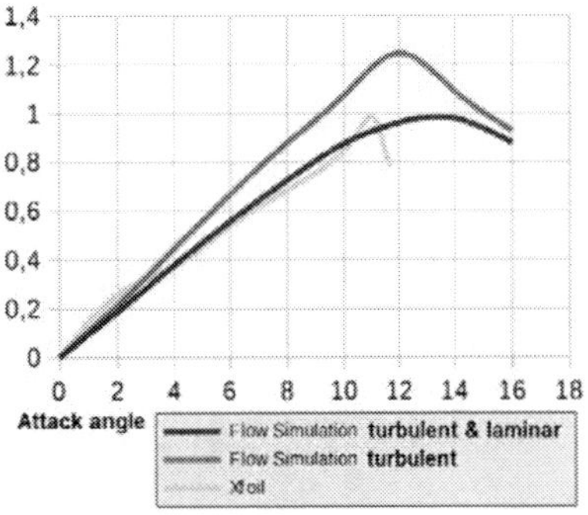


**Fig. 11.** NACA0012 $C_l$ vs. attack angle.

All currently available charts on this site have been made with the application Xfoil, written by researchers at the Massachusetts Institute of Technology, the application is a professional tool for design and analysis of wing.

Figure 11 compares the results of the Flow Simulation for laminar turbulent and only turbulent flow of NACA0012 symmetrical panel with the values from Xfoil program. Laminar and turbulent flow to angle of 10° gives results consistent with the program, lift coefficient is linear. 13° angle is a critical angle of attack for this flow.

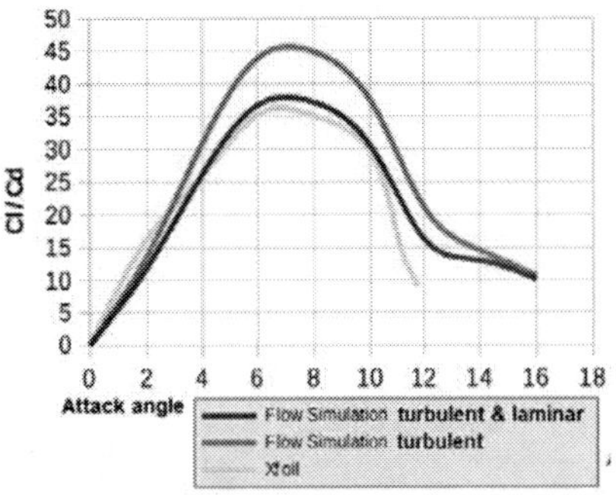


**Fig. 12.** NACA0012 $C_l/C_D$ vs. attack angle.

Figure 12 shows the ratio of $C_l$ to $C_d$ with respect to the attack angle for the airfoil NACA 0012. The highest value read from the graph determines the optimum angle which should be set for the airfoil to maximize aerodynamic force, in this case 7°. The value of the $C_d$ was taken from www.airfoiltools.com. We can see the convergence of the results of flow laminar and turbulent analysis results of Xfoil.

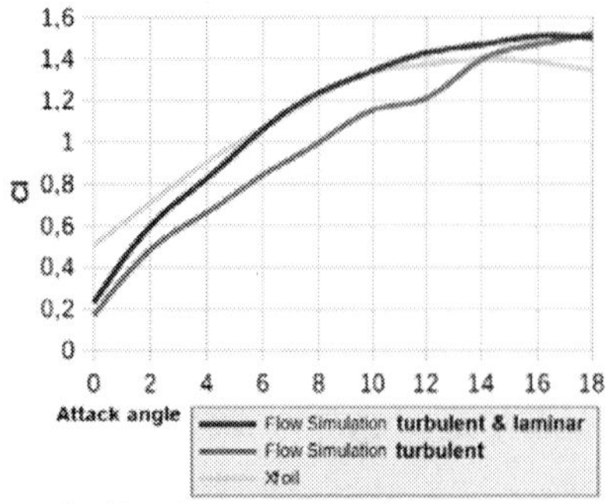


**Fig. 13.** NACA4415 $C_l$ vs. attack angle.

Figure 13 shows the results for the $C_l$ coefficient NACA 4415 airfoil. Comparing the values from Tables 3 and 5 showing that predicted lift coefficient is higher for this unsymmetrical airfoil. For angle equal to 8-9° the turbulent-laminar flow chart is almost linear, with increasing angle of attack values of gain coefficient $C_l$ decreases, while increasing the drag coefficient $C_d$. The consequence is to break the jet illustrated in Figure 10b. From the Fig. 14 can be read the optimum angle of attack equal to 9 degrees. The angle value coincides with the simulation results from Xfoil program. Again, we can see that the analysis of turbulent flow does not give satisfactory results, although for higher value of the angle results begin to cover what may well be a matter of chance.

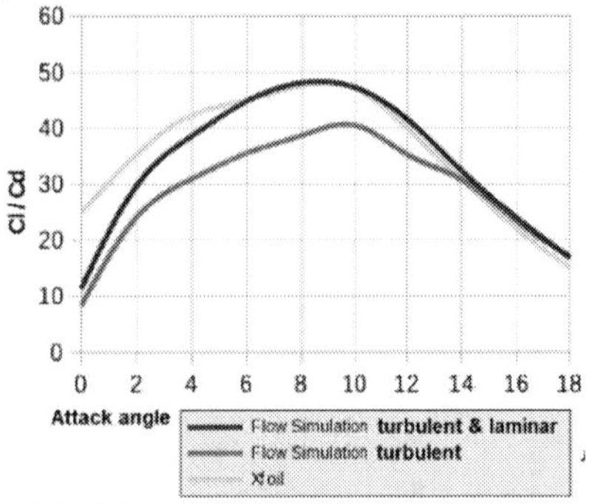


**Fig. 14.** NACA4415 $C_l / C_D$ vs. attack angle.

## 2.3 Calculation of airfoil aerodynamic force of Darrieus turbine

The optimum angle of attack for the airfoil NACA0012 is 7 degrees, the value of the resultant force is 0,012 [N] for the NACA 0045 at an angle of 9 degrees East is 0.023 [N]. Figure 15 is a graphical representation of the results obtained as a ratio of aerodynamic force to attack angle. For optimum attack angles of simulated individual airfoils data give consistent results analysis Xfoil.

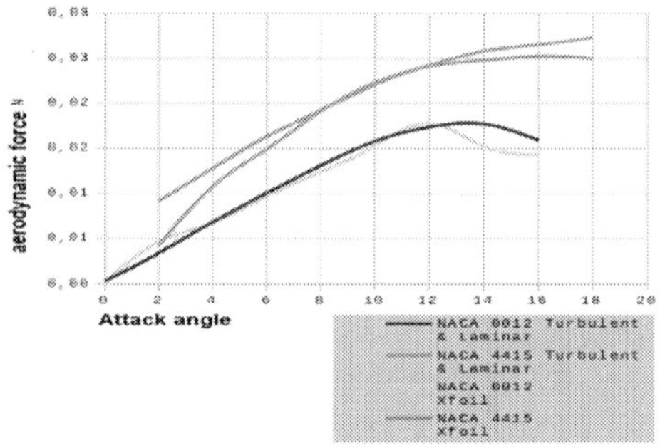


**Fig. 15.** Aerodynamic force – summary results.

On the basis of the results, decided to use the NACA4415 airfoil. For the project Darrieus vertical wind turbine key parameter is generated lift, the only one option to be considered is the use of asymmetric wing of attack angle 9°.

## 2.4 Lenz2 turbine

Leaving the core design: shaft, hub bearings, stator and the rotors, developed a second model type Lenz2. Darrieus turbine using lift is very efficient compared to the turbine Savonius's, but has problems with self-start. While Savonius's turbine trouble-free start it does not reach higher speeds, so Lenz2 model seems to be the ideal solution, combining the two concepts.
Figure 16 illustrates the pressure distribution of airfoil Lenz2 at angle of attack 9°. According to the turbine author Ed Lenz, after many trials and tests, he was able to determine the angle of attack for the turbine achieves the highest performance, quite accidentally 9° angle coincides with the angle unsymmetrical airfoil. It is showing a low pressure area at the front, upper part of panel.

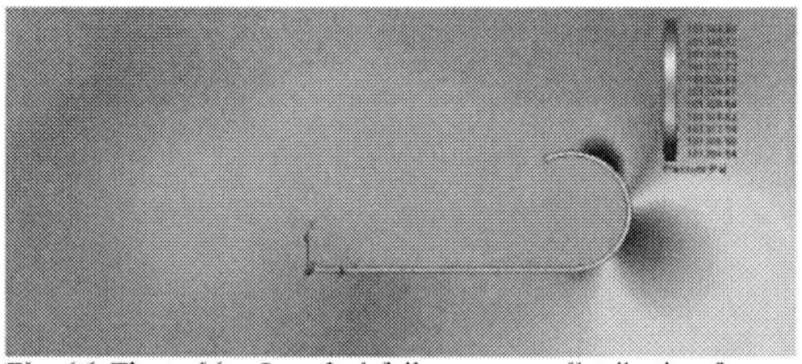

**Fig. 16.** The turbine Lenz2 airfoil - pressure distribution for angle of attack 9°.

Figure 17 shows the velocity distribution on flow lines. After wind left from streamlined side of airfoil swirls are created accordance with turbulent flow characteristics. In place of the lower pressure showing that the speed reaches the maximum value.

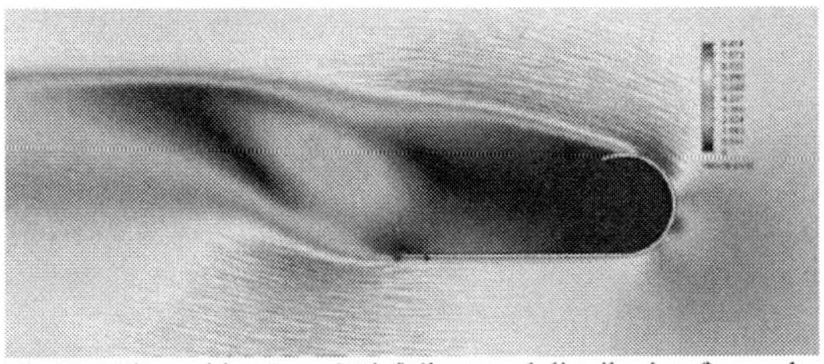

**Fig. 17.** The turbine Lenz2 airfoil - speed distribution for angle of attack 9°.

Based on flow simulation it can be seen that the airfoil exhibits aerodynamics.

Figure 18 illustrates the behaviour of the air mass entering the turbine and around it. The arrows in the figure are the velocity vectors of blowing air showing the direction and value. The rotational speed of the turbine is $\omega = 100\, rev/min$, the wind speed is $v = 5\, m/s$. The greater the wind speed decrease in the turbine, the turbine use it better. On the front parts of the aerofoil can be seen that the velocity reaches a maximum value.

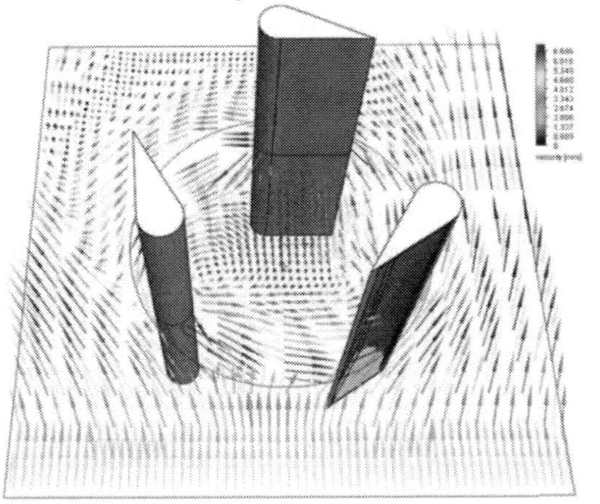

**Fig. 18.** The turbine Lenz2 airfoil - distribution of wind speed.

Strength analysis of bent sheet is taken incumbent on the plating of aluminum stiffeners and attachments weighing a total of 4.5 kg. SolidWorks automatically created shell element based on the characteristics of the model (see Fig. 19).

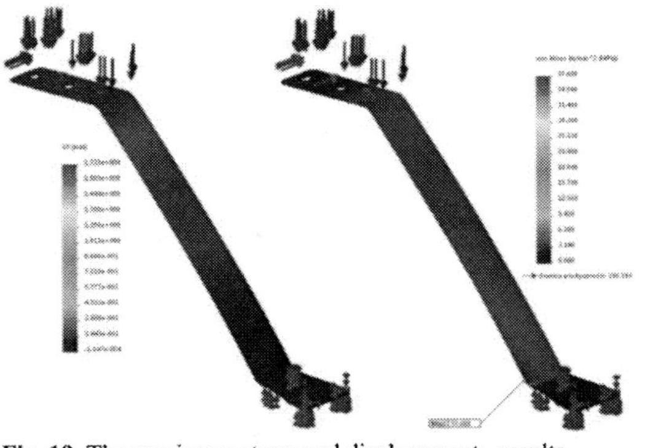

**Fig. 19.** The maximum stress and displacement - results.

The figure 20 shows the results of analysis for adverse winds, the results are satisfactory, although tension rose nearly doubled to 61.8 [MPa] but still are far below the yield stress, which is below 26%, good supply. Vertical movement also increase, nearly doubled, now is 3 mm, which still does not constitute grounds for concern. When the structure is assembled gently, and when working in difficult conditions, lowering the wing of 2-3 mm does not matter.

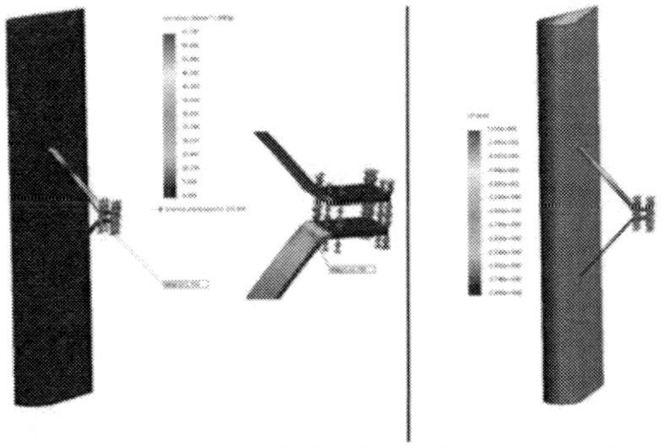

**Fig. 20.** Results of the analysis - the maximum stress and displacement under wind pressure.

## 3 Summary and conclusions

A rapidly growing research area of using wind power turbines provides great opportunities for research and development of existing ones and force to create new solutions in this field. Darrieus turbine project had been initial idea, but after building 3D model and simulating aerodynamic, as the information exploration on it, we changed the model of the turbine to Lenz2. This was because, H-rotor Darrieus model did not give self-start assurance at low wind speed and the self-excitation was not reach at all. Turbine Lenz's combines the use of drag force and aerodynamic force. This turbine is also self-started at low wind speed and its wings can reach greater speed than the speed of the wind.

HAWT (Horizontal Axis Wind Turbine) turbines are used in the production of electricity on an industrial scale, but if we want to have a backyard system, which would help save on water heating or garden lighting, Lenz's turbine model is a good option. We can use for more research papers [7-11].

## References

1. E. Hau, Fundamentals, Technologies, Application and Economics. Springer, (2000)
2. M. O. L. Hansen, Aerodynamics of Wind Turbines. Earthscan (2009).
3. M. Utkin, Młody technik **02** p. 26-29 (2005),
4. Website http://airfoiltools.com/airfoil/
5. Website http://aquaponicfun.com/step-by-step/wind-turbine/
6. Research Institute for Sustainable Energy, RISE Information Portal-Windmill http://www.rise.org.au /info/Tech/wind .html

7. H. Wisniewski, L. Plonecki, Proceedings of the 16th International Carpathian Control Conference (ICCC), 2015
8. J. Zwierzchowski, Edited by Zolotarev I., Radolf V. Proceedings of 22nd International Conference on Engineering Mechanics 2016 (Academy of Sciences of the Czech Republic, 2016)
9. M. Blasiak, R. Kotowski, Przeglad Elektrotechniczny, **85**, 12, 2009
10. M. Blasiak, Edited by Zolotarev I., Radolf V. Proceedings of 22nd International Conference on Engineering Mechanics 2016 (Academy of Sciences of the Czech Republic, 2016)
11. L. Nowakowski, E. Miko, M. Skrzyniarz, Edited by Zolotarev I., Radolf V. Proceedings of 22nd International Conference on Engineering Mechanics 2016 (Academy of Sciences of the Czech Republic, 2016)
12. E. Miko, L. Nowakowski, L. Procedia Eng. **39** (2012b)
13. Z. Koruba, I. Krzysztofik, P. I. Mech. Eng. K-J. Multi-Body Dyn. **227**, K1, 2013
14. J.E. Takosoglu, P.A. Laski, S. Blasiak, G. Bracha, D. Pietrala. Meas. Cont. **49**, 2, 2016

# AUTHOR INDEX

# AUTHOR INDEX

# AUTHOR INDEX